MODERN ENGINEERING PHYSICS

For the Students of B.E./B.Tech. of All Technical Universities of India

A. S. VASUDEVA

Prof. & Head, Deptt. of Physics
B.B.S.B. Engg. College
Fateh Garh Saheb
Punjab
Former Head, Department of Physics and Applied Sciences
Guru Nanak Dev Engineering College
Ludhiana (Punjab)

S. CHAND
PUBLISHING

S. Chand And Company Limited
(ISO 9001 Certified Company)

S. Chand And Company Limited
(ISO 9001 Certified Company)
Head Office: D-92, Sector–2, Noida – 201301, U.P. (India), Ph. 91-120-4682700
Registered Office: A-27, 2nd Floor, Mohan Co-operative Industrial Estate, New Delhi – 110 044, Phone: 011-49731800
www.**schandpublishing.com;** e-mail: **info@schandpublishing.com**

Marketing Offices:

Chennai	:	Ph: 23632120; chennai@schandpublishing.com
Guwahati	:	Ph: 2738811, 2735640; guwahati@schandpublishing.com
Hyderabad	:	Ph: 40186018; hyderabad@schandpublishing.com
Jalandhar	:	Ph: 4645630; jalandhar@schandpublishing.com
Kolkata	:	Ph: 23357458, 23353914; kolkata@schandpublishing.com
Lucknow	:	Ph: 4003633; lucknow@schandpublishing.com
Mumbai	:	Ph: 25000297; mumbai@schandpublishing.com
Patna	:	Ph: 2260011; patna@schandpublishing.com

First Edition 1992
Subsequent Editions and Reprints 2001, 2003, 2004, 2006, 2007, 2008 (Twice), 2010 (Twice)
Sixth Revised Edition 2013
Reprints 2016, 2017, 2020, 2021, 2022, 2023

Reprint 2024

ISBN: 978-81-219-1757-5 **Product Code:** H3EPH68PHYS10ENAF13O

PRINTED IN INDIA

By Vikas Publishing House Private Limited, Plot 20/4, Site-IV, Industrial Area Sahibabad, Ghaziabad – 201 010 and Published by S. Chand And Company Limited, A-27, 2nd Floor, Mohan Co-operative Industrial Estate, New Delhi – 110 044.

Dedicated to the
memory of
my father

PREFACE TO THE SIXTH EDITION

The author feels obliged to his esteemed readers of the country and abroad for the confidence shown in the book. The book has been prescribed as standard textbook for various universities, N.I.T's and even referred as a standard textbook by some I.I.T's. It has found its place in various libraries of the country and abroad. The author feels honoured by the tremendous confidence shown in the book. The author feels more humbled and gets added responsibility to update the book. The revised edition has been thoroughly checked and new chapter on **dielectrics** has been added as it is prescribed in most of the universities of the country. All other chapters have also been thoroughly checked.

However, inspite of all effects, some discrepancies/omissions may still exist. The author will personally thank all the worthy readers who bring to his notice such omissions so that the book becomes more suitable and useful for the students and teachers alike.

A.S. VASUDEVA

PREFACE TO THE FOURTH EDITION

It is a matter of great satisfaction that during the last three to four years the book is receiving over-whelming response from the students and teachers alike. The book has been found quite useful and has been prescribed as standard textbook/reference book in the various universities of the country. The author feels humbled by the confidence shown by the readers and intends to put more efforts to make the book more useful for its readers.

This edition has been thoroughly checked and misprints, omissions and errors have been rectified to a large extent. A new chapter on *Particle Accelerators* has been added and many more examples and questions have been incorporated so that book finds more acceptance by its readers. Many numericals and additional topics have been added at relevant chapters to make the matter more comprehensive and understandable. Almost all the chapters have been checked and wherever needed, recast of the matter and its improvement has been undertaken with lots of care. Few more topics have been added, so that it caters to the needs of the students of **Osmania University, Rajasthan Technical University** and many other universities of the country.

In spite of all the efforts, still some shortcomings may exist and the author will be personally thankful to all those who give their suggestions to make it a better and more useful book.

Any suggestion for the improvement of the book will be gratefully accepted and acknowledged in the future editions of the book.

A.S. VASUDEVA

PREFACE TO THE FIRST EDITION

The book in its present form is due to my interaction with the students for quite a long time. It had been my long-cherished desire to write a book covering most of the topics that form the syllabii of the Engineering and Science students at the degree level. I tried my best to introduce the various topics from elementary level, so that those students who have not got sufficient knowledge of the subject may not exert too much to grasp the various topics.

Many students, although able to understand the various topics of the book, may not be able to put their knowledge to use. For this purpose a number of questions and problems are given at the end of each chapter.

I am grateful to Mr. Ravindra Kumar Gupta, Director, who after having healthy discussion encouraged me to take up this project. I am also thankful to Shri V.S. Johri and his team of Staff for taking personal interest so that the book is published well in time.

I am also thankful to my family members, friends and colleagues, who constantly inspired me to work more fruitfully. I am really obliged to all those who helped me in collecting some useful study material for preparing this book.

Though much care has been taken so that book is error free, some mistakes or errors might have crept in due to oversight. I will be personally grateful to all those who bring to my knowledge any shortcoming of the book. All the healthy suggestions will be gratefully accepted and all efforts will be made to make the book more useful for all levels of reading.

A. S. VASUDEVA

SYLLABUS

ENGINEERING PHYSICS — I

Unit – I

Relativistics Mechanics : Inertial & non-inertial frames, Galilean transformation equations, Michelson-Morley experiment, Einstein's postulates, Lorentz transformation equations, Length contraction & time dilation, Addition of velocities, Variation of mass with velocity, Mass energy equivalence.

07 Hrs

Unit – II

Electromagnetics: Recapitulation of vector product, Gradient, Divergence & Curl, Statement and explanation of Gauss divergence & Stokes theorems, iseful vector identities.

Maxwell's equations (Integral and differential forms), Equation of continuity, Transverse nature of EM waves, EM – wave equation and its propagation characteristics in free-space, Poynting vector.

07 Hrs

Unit – III

Interference: Spatial and temporal coherence, Interference in thin films of uniform thickness and in wedge-shaped film (qualitative), Newton's rings, anti reflection and high reflection coatings (qualitative), interference filters (qualitative).

Diffraction: Single and N-slit diffraction, Grating spectra, Rayleigh's criterion of resolution, Resolving power of grating. 08 Hrs

Unit – IV

Polarization: Polarization by reflection and refraction, Double refraction, Nicol prism, Sheet polarizer, Production and analysis of plane, circularly and elliptically polarized lights, Basic concepts of optical activity, Polarimeter (Half shade).

Laser: Spontaneous and stimulated emission of radiation, Einstein's coefficients, Construction and working of Ruby, He-Ne and semiconductor lasers, Important laser applications (qualitative).

07 Hrs

Unit – V

Fiber Optics and Holography :

Fundamental ideas about optical fibers, Types of fibers, Acceptance angle and cone, Numerical aperture, Propagation mechanism and communication in optical fiber, Attenuation, Signal loss in optical fiber and dispersion.

Basic principles of holography, Construction of hologram and wave reconstruction, Applications of holography (qualitative). 06 Hrs

Reference books:

(*i*) J.W. Jewett Jr., R.A. Serway, Physics for scientists & engineers with modern physics (Cengage learning 2nd Indian reprint 2011).

(*ii*) A. Beiser, Concepts of Modern Physics, (Mc-Graw Hill).

(*iii*) Robert Resnick, Introduction to special theory of relativity (Wiley).

(*iv*) Ajoy Ghatak, Optics (Tata Mc-Graw Hill).

(*v*) Resnick, Haliidey and Walker Fundamental of Physics (Wiley).

(*vi*) David J. Griffith, Introduction to Electrodynamics (Prentice-Hall India).

(*vii*) S.D. Jain and G. Sahasrabudhe, Engineering Physics (Universities Press).

(*viii*) K. Rajagopal, Engineering Physics (Prentice-Hall India).

(*ix*) G. Aruldhas, Engineering Physics (Prentice-Hall India).

FUNDAMENTAL CONSTANTS

Physical Quantity	Symbol	Value
Atomic mass unit	(u)	$1.660\ 540\ 2 \times 10^{-27}$ kg $931.494\ 32$ MeV/c^2
Avogadro's number	N_A	$6.022\ 136\ 7 \times 10^{23}$ particles/mol
Bohr magneton	$\mu_B = \frac{eh}{2m_e}$	$9.274\ 015\ 4 \times 10^{-24}$ J/T
Bohr radius	$a_0 = \frac{h^2}{m_e e^2 k_e}$	$5.291\ 772\ 49 \times 10^{-11}$ m
Boltzmann's constant	$k_B = R / N_A$	$1.380\ 658 \times 10^{-23}$ JK^{-1}
Compton wavelength	$\lambda_C = \frac{h}{m_e c}$	$2.426\ 310\ 58 \times 10^{-12}$ m
Coulomb constant	$k_e = \frac{1}{4\pi\varepsilon_0}$	$8.987\ 551\ 787 \times 10^9$ N . m^2/C^2
Deuteron mass	m_d	$3.343\ 586\ 0 \times 10^{-27}$ kg $2.013\ 553\ 214$ u
Electron mass	m_e	$9.109\ 389\ 7 \times 10^{-31}$ kg $5.485\ 799\ 03 \times 10^{-31}$ kg $0.510\ 999\ 06$ MeV/c^2
Electron-volt	eV	$1.602\ 177\ 33 \times 10^{-19}$ J
Elementary charge	e	$1.602\ 177\ 33 \times 10^{-19}$ C
Gas constant	R	8.313 510 J/K mol
Gravitational constant	G	$6.672\ 59 \times 19^{-11}$ N . m^2/kg^2
Hydrogen ground state engery	$E_1 = \frac{e^2 k_e}{2a_0}$	– 13.605 698 eV
Josephson frequency-voltage ratio	$2e/h$	$4.835\ 976\ 7 \times 10^{14}$ Hz/V
Magnetic flux quantum	$\Theta_0 = \frac{h}{2e}$	$2.067\ 834\ 61 \times 10^{-15}$ T . m^2
Neutron mass	m_n	$1.674\ 928\ 6 \times 10^{-27}$ kg $1.008\ 664\ 994$ u 939.565 MeV/c^2

Nuclear magneton	$\mu_n = \dfrac{eh}{2m_p}$	5.050 786 6 × 10–27 J/T
Permeability of free space	μ_0	$4\pi \times 10^{-7}$ T . m/A
Permittivity of free space	$\varepsilon_0 = 1/\mu_0 c^2$	$8.854\ 187\ 817 \times 10^{-12}$ $C^2/N\ .\ m^2$
Planck's constant	h	$6.626\ 075 \times 10^{-34}$ J.s
	$\hbar = h / 2\pi$	$1.054\ 572\ 66 \times 10^{-34}$ J.s
Proton mass	m_p	$1.672\ 623 \times 10^{-27}$ kg 1.007 276 470 u 938.272 3 (28) MeV/c^2
Rydberg constant	R_h	$1.097\ 373\ 153\ 4 \times 10^7\ m^{-1}$
Spped of light in vacuum	c	$2.997\ 924\ 58 \times 10^8$ m/s

CONTENTS

PART I. PROPERTIES OF MATTER

PART II. THERMAL PHYSICS

PART III. OPTICS

PART IV. SOUND

PART VII. MODERN PHYSICS

PART I
PROPERTIES OF MATTER

Sir Isaac Newton (1642-1727)

Issac Newton, considered by many the greatest scientist who ever lived, was born in 1642 at Woolsthorpe, Lincolnshire, England. He was a delicate child who lived a somewhat solitary life, preferring to construct a variety of mechanical models and devices rather than play games with his schoolmates.

In 1687 Newton published his *Principia*. This was probably the greatest single event in the history of science. The *Principia* combined Newton's three laws of motion and his law of universal gravitation into a system capable of explaining a great variety of phenomena both on earth and in the heavens. Here was the great synthesis of mechanics that scientists had hoped for.

In 1703 he was elected president of the Royal Society of London, and held that post until his death in 1727. His best scientific days were now behind him, but in 1704 he published the *Opticks*, an account of some of his ingenious work on the physics of the spectrum, interference, color vision, and the rainbow. Increasingly, however, his thoughts were directed to biblical and theological studies and to historical chronology rather than to natural science.

Newton never married, and at one time protested bitterly to John Locke, the famous philosopher, about Locke's efforts to "embroil him with women". Newton was absent-minded and inconsistent in his behaviour, sometimes open and generous at other times irritable and spiteful.

Newton died in 1727 at the age of 84, and was buried with great pomp and ceremony in Westminster Abbey, London.

Nature to him (Newton) was an open book, whose letters he could read without effort. In one person he combined the experimenter, the theorist, the mechanic and, not least, the artist in expression.

[Albert Einstein (1879-1955)]

The inner logic of Galilean Mechanics was so strong that Newton was able to take the great step of applying it to the motion of the stars.

[Max. Born (1882-1970)]

From time to time in the kingdom of mankind a man arises who is of universal significance, whose work changes the current of human thought or of human experience, so that all that comes after him bears evidence of his spirit. Such a man was Shakespeare. Such a man was Beethoven, such a man was Newton, and, of the three, his kingdom is most widespread.

[E.N. Dac. Andrude]

CHAPTER 1

UNITS AND MEASUREMENT

1.1. INTRODUCTION

In Physics, much importance is given to the measurements and it is due to this reason that physics is called **Science of Measurement**. In general, Physics is that science which deals with nature and natural phenomenon.

1.2. UNITS FOR MEASUREMENT

Measurement of a physical quantity involves its comparison with a chosen standard of the same kind as the physical quantity.

The chosen standard of the same kind which is selected as reference to measure a physical quantity is called the unit of that quantity.

The measurement of a physical quantity means to find out the number of times its unit is contained in that physical quantity. The process of measurement of a physical quantity involves:

(*a*) selection of a unit;

(*b*) to find the number of times that unit is contained in the physical quantity.

1.3. FUNDAMENTAL AND DERIVED UNITS

The quantities mass, length and time are called fundamental physical quantities and their units are called the *fundamental units*.

We define fundamental units as those units, which can neither be derived from one another, nor can they be further resolved into other units.

For measurement of mass length and time, there are independent units such as *kilogram, metre and second*. To express other physical quantities we express them in terms of fundamental units of mass, length and time.

Units of all physical quantities **which can be expressed in terms of fundamental units of mass, length and time are called derived units.**

1.4 SYSTEMS OF UNITS

There are three major systems of units based on fundamental units:

(*a*) **C.G.S. System:** In this system, unit of length is cm, unit of mass is gram and unit of time is second.

(*b*) **F.P.S. System:** In which *foot, pound, second* are the units of length, mass and time respectively. This system is no longer in use.

(*c*) **M.K.S. System:** In which *metre, kilogram* and *second* are the respective units of the measurement of length, mass and time.

A coherent system *of units is a system based on a certain set of basic units' from which all derived units are obtained by multiplication or division, without introducing numerical factors. M.K.S. system is a coherent system of units.*

1.5. S.I. SYSTEM OF UNITS

The name S.I. system of the units is an abbreviation of "System International". It is an internationally accepted system of units which was introduced by the General Conference of Weights and Measures held in October, 1960. The system is, infact, a modification over M.K.S. system and is sometimes also known as Rationalized M.K.S. system. The various units of this system are rational and coherent in nature. In S.I. system, three additional dimensional quantities are used so as to cover all the fields of Physics.

The S.I. system of units thus consist of seven fundamental units and two supplementary units.

(a) Fundamental Units

1. Unit of length ... Metre (m)
2. Unit of mass ... Kilogram (kg)
3. Unit of time ... Second (s)
4. Unit of current ... Ampere (A)
5. Unit of thermodynamical temperature ... Kelvin (K)
6. Unit of luminous intensity ... Candela (cd)
7. Unit of quantity ... Mole

We may define these units as follows:

***(i)* Metre (m):** It is defined as the length equal to 1,650,763.73 wavelengths in vacuum of a certain orange red light emitted by Krypton isotope of mass number 86.

***(ii)* Kilogram (kg):** It is defined as the mass of a standard block of Plantinum-Iridium alloy (90% Pt + 10% Ir) kept in the International Bureau of Weights and Measures at Sevres near Paris.

***(iii)* Second (s):** It is defined as the duration of 9,192,631,770 periods of radiation corresponding to transition between two hyperfine levels of Cs^{133}.

***(iv)* Ampere (A):** It is defined as that constant current which when flowing through two parallel, straight, infinite conductors of negligible cross-section held 1 meter apart in vacuum, produces a force = 2×10^{-7} Newton/metre between them.

***(v)* Kelvin (K):** It is defined as 1/273.16 of thermodynamic temperature of the triple point of water.

***(vi)* Candela (cd):** It is defined as the luminous intensity in a $\perp$ direction of $\frac{1}{600,000}$ sq. metre area of a black body kept at temperature of freezing platinum (1773°C) under a pressure of 101,325 Newton/sq. metre.

***(vii)* Mole:** It is defined as the amount of substance containing same number of elementary units as there are atoms in 0.012 kg of Carbon-12.

So a mole contains 6.023×10^{23} specified elementary units such as atoms, molecules, ions etc.

(b) Supplementary Units

1. Unit of plane angle ... Radian (rad)
2. Unit of solid angle ... Steradian (Sr)

These are defined as follows:

***(i)* Radian (rad):** It is defined as is the angle subtended at the centre of a circle by an arc equal in length to the radius of the circle.

***(ii)* Steradian (Sr):** It is defined as the solid angle subtended at the centre of a sphere by its surface whose area is equal to the square of the radius of the sphere.

(c) Derived Units

Some of the derived units on S.I. System of units are given below:

1. Area ... Square Metre (m^2)
2. Volume ... Cubic Metre (m^3)
3. Density ... Kilogram/cubic metre (kgm^{-3})
4. Velocity ... Metre/second (ms^{-1})
5. Acceleration ... Metre/second/second (ms^{-2})
6. Angular velocity ... Radian/second (rad s^{-1})
7. Angular acceleration ... Radian/second/second (rad s^{-2})
8. Force ... Newton (N)

9. Work, Energy ... Joule (J)
10. Power ... Watt (W)
11. Pressure ... Newton/Square Metre (Nm^{-2})
12. Frequency ... Hertz (Hz)
13. Specific heat ... Joule/Kilogram/Kelvin ($Jkg^{-1}\ K^{-1}$)
14. Luminous flux ... Lumen (l m)
15. Intensity of illumination ... Lux (lx) or l $m.m^{-2}$
16. Electric charge ... Coulomb (C)
17. Potential, Potential difference ... Volt (V)
18. Capacitance ... Farad (F)
19. Resistance ... Ohm (Ω)
20. Strength of electric field ... Volt/metre (Vm^{-1})
21. Magnetic flux ... Weber (Wb)
22. Strength of magnetic-field ... Tesla or Weber/metre^2 ($Wb\ m^{-2}$).

1.6. MERITS OF S.I. SYSTEM OF UNITS

(*i*) Every unit in use can be derived from the seven base units and two supplementary units which makes S.I. units coherent.

(*ii*) Since the number of units is quite small (seven base and two supplementry units) so the system is rational.

(*iii*) The units being internationally selected, have the advantage that the same set of units are used everywhere.

(*iv*) It is a decimal system *i.e.*, the units of a given quantity are related to one another in power of 10. So conversions are easy.

(*v*) 'Joule' is the unit of all kinds of energy.

(*vi*) 'Watt' is the unit of all kinds of power.

1.7. IMPORTANT CONVERSIONS

Some of the important conversions of the metric units of length are:

1 kilometre = 10^3 metres
1 metre = 10 decimetres
1 metre = 100 centimetres
1 metre = 1000 millimetres

Some of the important conversions of the units of mass are:

1 metric tonne = 10 quintals
1 quintal = 100 kilograms
1 kilogram = 1000 grams
1 gram = 10 decigrams
1 gram = 100 centigrams
1 gram = 1000 milligrams

Some commonly used metric pre-fixes are:

Giga (G) .... $\times 10^9$
Mega (M) .... $\times 10^6$
Kilo (K) .... $\times 10^3$
Deci (d) .... $\times 10^{-1}$
Centi (c) .... $\times 10^{-2}$
Milli (m) .... $\times 10^{-3}$
Micro (μ) .... $\times 10^{-6}$
Nano (n) .... $\times 10^{-9}$
Pico (p) .... $\times 10^{-12}$
Femto (f) .... $\times 10^{-15}$
Atto (a) .... $\times 10^{-18}$

1.8. CONCEPT OF LENGTH

It is the extent of the separation i.e., distance between two points in space.

While measuring length we come across objects widely differing in sizes. Some of them are as large as a mountain whereas others may be as small as grain of dust. The former objects constitute what is called *Macrocosm* ("large world") and the latter are called *Microcosm* ("small world"). The earth, the sun, the stars and other heavenly bodies belong to first category whereas bacteria, electrons, protons etc., belong to the second category.

***(i)* Macrocosm measurement:** Different units used to measure large distances are:

***(a)* Astronomical Unit (A.U.):** *It is the average distance from earth to the sun.*

$$\therefore \qquad \text{One A.U.} = 1.496 \times 10^{11} \text{ m}$$

***(b)* Light year:** *It is the distance travelled by light in one year.*

$$1 \text{ light year} = 946 \times 10^{15} \text{ m}$$

***(c)* Parsec:** *It is defined as the distance at which a star would have a parallex of 1 second of arc.* It stands for **Paralactic-second.**

Parallax θ of a star is the angle subtanded by semi-major axis of the earth's orbit ⊥ to the direction of the star as show in Fig. 1.1.

$$1 \text{ Parsec} = 3.26 \text{ light years}$$

or $$1 \text{ Parsec} = 3.08 \times 10^{16} \text{ m}$$

Parsec is used to express the distance of stars.

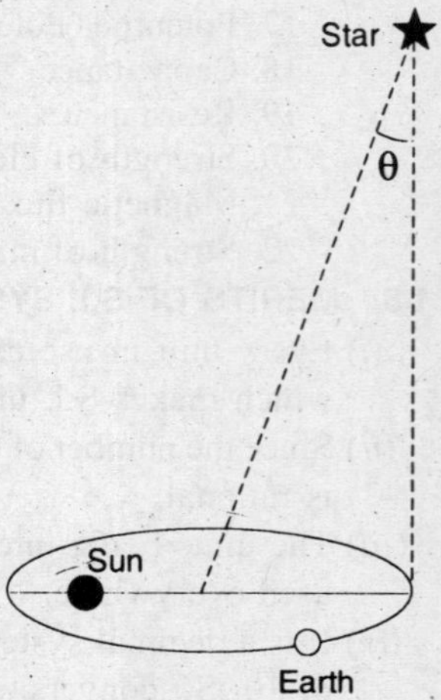

Fig. 1.1

***(ii)* Microcosm measurement:** On the other hand on atomic and subatomic scales, the distances are very-very small. For example, the diameter of the nucleus is about 10^{-14} m, electron has a diameter of about 10^{-15} m and is about 10^{-10} m from the nucleus and the effective radius of proton is 1.2×10^{-15} m.

Different units used to measure small distances are:

(a) One micron $= 10^{-6}$ m
(b) One Angstrom (Å) $= 10^{-10}$ m
(c) One Fermi $= 10^{-15}$ m.

1.9. DIMENSIONAL EQUATION

Dimensions of a derived unit are the powers to which the fundamental units of mass, length and time are raised to represent that particular unit.

Dimensional equation corresponds to the relation between the physical quantity and its dimensional formula.

Uses of Dimensional Equation

With the help of dimensional equation we can:

(i) convert one system of unit to another,
(ii) check the correctness of the physical equation,
(iii) derive important relations,
(iv) select some experiments to acquire important information.

Limitations of Dimensional Equation

(i) With Dimensional Equation, we can not derive an exact form of a physical relation which depends on more than three quantities.
(ii) Relations involving trigonometric and exponential functions cannot be derived from these equations.
(iii) These equations do not give any information about dimensionless constants and pure numbers, which have to be determined indirectly with separate experiments.

1.10 PRINCIPAL OF HOMOGENEITY

Principle of homogeneity states that the *powers of fundamental units on the one side of the equation must be equal to their respective powers on the other side of the equation.*

In order to apply the principle of homogeneity, let us take up the case of a simple pendulum.

Time period T of a simple pendulum depends on (*i*) mass of the pendulum (*ii*) length of pendulum (*iii*) acceleration due to gravity g (*iv*) angle of swing of the pendulum.

We may write the dimensional equation for time period as

$$t \propto m^a . l^b . g^c . \theta^d$$

where a, b, c and d are constants, may be zero, positive or negative.

$$\therefore \quad t = km^a . l^b . g^c . \theta^d \qquad ...(1)$$

where k is dimensionless constant and θ has no dimension, $\therefore$ $d = 0$.

So Equation (1) may be written as $M^0 L^0 T^1 = M^a L^b [L.T^{-2}]^c = M^a L^{b+c} T^{-2c}$

Applying principle of Homogeneity, we get

$$a = 0,\ b + c = 0,\ -2c = 1 \qquad \therefore c = -1/2,\ b = 1/2,\ a = 0$$

So Equation (1) becomes $t = = k\sqrt{l/g}$ experimentally, the value of $k = 2\pi$, so the time period of a simple pendulum is $t = 2\pi\sqrt{l/g}$.

1.11. DIMENSIONS OF SOME OF THE IMPORTANT QUANTITIES ARE GIVEN AS UNDER:

1. Angular momentum = Moment of Inertia × Angular velocity $= M^1L^2 \times T^{-1} = M^1L^2T^{-1}$

2. Specific heat $= \dfrac{\text{Heat Energy}}{\text{Mass} \times \text{Temperature rise}} = \dfrac{M^1L^2T^{-2}}{M^1} = M^0L^2T^{-2}\theta^{-1}$

3. Latent heat $= \dfrac{\text{Heat Energy}}{\text{Mass}} = \dfrac{M^1L^2T^{-2}}{M^1} = M^0L^2T^{-2}$

4. Coeff. of thermal conductivity $= \dfrac{\text{Heat Energy} \times \text{Distance}}{\text{Area} \times \text{Temp diff} \times \text{Time}} = \dfrac{M^1L^2T^{-2}L^1}{L^2 \times \theta \times T^1} = M^1L^1T^{-3}\theta^{-1}$

5. Gas constant $= \dfrac{\text{Pressure} \times \text{Volume}}{\text{Temperature}} = \dfrac{M^1L^{-1}T^{-2} \times L^3}{\theta^1} = M^1L^2T^{-2}\theta^{-1}$

6. Boltzmann's constant $= \dfrac{\text{Energy}}{\text{Temperature}} = \dfrac{M^1L^2T^{-2}}{\theta} = M^1L^2T^{-2}\theta^{-1}$

7. Planck's constant $= \dfrac{\text{Energy}}{\text{Frequency}} = \dfrac{M^1L^2T^{-2}}{T^{-1}} = M^1L^2T^{-1}$

8. Coeff. of viscosity $= \dfrac{\text{Force}}{\text{Area} \times \text{Velocity gradient}} = \dfrac{M^1L^1T^{-2}}{L^2 \times \dfrac{L^1T^{-1}}{L^1}} = M^1L^{-1}T^{-1}$

9. Thermal Resistance $= \dfrac{\text{Temp. difference at two ends}}{\text{Rate of flow of heat through it}}$

$$= \frac{\text{Length or thickness of material}}{\text{Thermal conductivity} \times \text{area}}$$

$$= \frac{L^1}{(M^1L^1T^{-3}\theta^{-1})\,L^2} = M^{-1}L^2T^3\theta^1$$

10. Entropy $= \dfrac{\text{Heat Energy}}{\text{Temperature}} = \dfrac{M^1L^2T^{-2}}{\theta} = M^1L^2T^{-2}\theta^{-1}$

11. Electric charge = Current × Time $= A^1T^1$

12. Electric potential $= \dfrac{\text{Work}}{\text{Charge}} = \dfrac{M^1 L^2 T^{-2}}{AT} = M^1 L^2 T^{-3} A^{-1}$

13. Capacitance $= \dfrac{\text{Charge}}{\text{Potential}} = \dfrac{AT}{M^1 L^2 T^{-3} A^{-1}} = M^{-1} L^{-2} T^4 A^2$

14. Resistance $= \dfrac{\text{Potential}}{\text{Current}} = \dfrac{M^1 L^2 T^{-3} A^{-1}}{A} = M^1 L^2 T^{-3} A^{-2}$

15. Inductance $= \dfrac{\text{Induced e.m.f.}}{\text{Change in current/Unit time}} = \dfrac{M^1 L^2 T^{-3} A^{-1}.T}{A} = M^1 L^2 T^{-2} A^{-2}$

16. R.C. = Resistance × Capacitance $= M^1 L^2 T^{-3} A^{-2} \times M^{-1} L^{-2} T^4 A^2 = T$

17. $\dfrac{L}{R} = \dfrac{\text{Inductance}}{\text{Resistance}} = \dfrac{M^1 L^2 T^{-2} A^{-2}}{M^1 L^2 T^{-3} A^{-2}} = T$

(Dimensions of RC and L/R are same and equal to time T).

18. $\dfrac{1}{\sqrt{LC}} = \dfrac{1}{(M^1 L^2 T^{-2} A^{-2} \times M^{-1} L^{-2} T^4 A^2)^{1/2}} = \dfrac{1}{T} = \text{Frequency}$

19. Electric permittivity $= \dfrac{\text{Charge} \times \text{Charge}}{\text{Force} \times \text{Distance}^2} = \dfrac{AT \times AT}{M^1 L^1 T^{-2} \times 2} = M^{-1} L^{-3} T^4 A^2$

20. Electric field intensity $= \dfrac{\text{Force}}{\text{Charge}} = \dfrac{\text{Electric Potential}}{\text{Distance}} = \dfrac{M^1 L^2 T^{-3} A^{-1}}{L^1} = M^1 L^1 T^{-3} A^{-1}$

21. Magnetic field = Magnetic Induction = Flux density

$$= \frac{\text{Force}}{\text{Charge} \times \text{Velocity}} = \frac{\text{Force}}{\text{Current} \times \text{Distance}} = \frac{M^1 L^1 T^{-2}}{AL^1} = M^1 T^{-2} A^{-1}$$

22. Magnetic permeability $= \dfrac{\text{Magnetic Induction} \times \text{Distance}}{\text{Current}}$

$$= M^1 T^{-2} A^{-1} \times \frac{L^1}{A^1} = M^1 L^1 T^{-2} A^{-2} = M^1 L^1 Q^{-2}$$

23. Magnetic flux = Magnetic Induction × Area × cos θ

$$= M^1 T^{-2} A^{-1} \times L^2 = M^1 L^2 T^{-2} A^{-1} = M^1 L^2 T^{-1} Q^{-1}$$

24. Magnetic Field Intensity $= \dfrac{\text{Magnetic Induction}}{\text{Magnetic permeability}} = \dfrac{M^1 T^{-2} A^{-1}}{M^1 L^1 T^{-2} A^{-2}} = \dfrac{A}{L}$

25. Product of permeability and electric permittivity $(\mu_0 \varepsilon_0)$

$$= M^1 L^1 T^{-2} A^{-2} \times M^{-1} L^{-3} T^4 A^2 = \frac{T^2}{L^2} = \frac{1}{(\text{Velocity})^2}$$

$$\therefore \quad \text{Velocity} = \frac{1}{\sqrt{\mu_0 \varepsilon_0}}$$

So $\dfrac{1}{\sqrt{\mu_0 \varepsilon_0}}$ correspond to velocity of light in free space.

Example 1. *If the units of length, mass and force are chosen as fundamental unit, what will* *be the dimensions of time in terms of dimensions of these units?*

Solution: Let the dimensions of length, mass, force and time be L, M, F and T respectively. So we have

$$\text{Force} = \text{Mass} \times \text{Acceleration}$$

$$= \text{Unit of mass} \times \frac{\text{Unit of velocity}}{\text{Unit of time}}$$

$$= \text{Unit of mass} \times \frac{\text{Unit of distance}}{\text{Unit of time} \times \text{Unit of time}}$$

$$\therefore \quad F = \frac{M^1 L^1}{T^2}$$

$$T^2 = \frac{M^1 L^1}{F} = M^1 L^1 F^{-1}$$

$$\therefore \quad T = \mathbf{M^{1/2}\ L^{1/2}\ F^{-1/2}}.$$

Example 2. *A gas bubble from an explosion under water oscillates with a period and proportional to $P^a d^b E^c$ where P is hydrostatic pressure, d is the density of water and E is the energy of explosion. Find the values of a, b and c.*

Solution: Let $t = k\, P^a d^b E^c$, where k is dimensionless constant.

Writing in dimensional form

$$M^0 L^0 T^1 = [M^1 L^{-1} T^{-2}]^a\, [M^1 L^{-3}]^b\, [M^1 L^2 T^{-2}]^c$$

$$= M^{a+b+c}\, L^{-a+3b+2c}\, T^{-2a-2c}$$

Equating powers on two sides of the equation

$$a + b + c = 0,\ -a - 3b + c = 0,\ -2a - 2c = 1$$

Solving for a, b and c, we get

$$a = -\frac{\mathbf{5}}{\mathbf{6}},\ b = \frac{\mathbf{1}}{\mathbf{2}},\ c = \frac{\mathbf{1}}{\mathbf{2}}.$$

Example 3. *Wavelength λ associated with a moving particle depends on its mass m, velocity of motion v and Planck's constant h. Show dimensionally, that $\lambda \propto \frac{h}{mv}$.*

Solution. Let $\lambda \propto m^a v^b h^c$, where a, b, c are constants.

$$\therefore \quad \lambda = k\, m^a\, v^b\, h^c \qquad ...(1)$$

where k is dimensionless constant. Writing Equation (1) in dimensional form, we get,

$$L^1 = M^a\, [L^1\, T^{-1}]^b\, [M^1\, L^2\, T^{-1}]^c \text{ or } M^0\, L^1\, T^0 = M^{a+c}\, L^{b+2c}\, T^{-b-c}$$

Form principle of homogeneity

$$a + c = 0, \quad b + 2c = 1, \quad -b - c = 0$$

$$\therefore \quad a = -c, \quad b = 1 - 2c, \quad b = -c$$

Solving for a, b, c, we get $a = -1$, $b = -1$, $c = +1$

So Equation (1) may be written as

$$\lambda = k.\, m^{-1}\, v^{-1}\, h^1 = k \cdot \frac{h}{mv} \quad \text{or} \quad \lambda \propto \frac{\mathbf{h}}{\mathbf{mv}}.$$

Example 4. *Obtain by the method of dimensional analysis an expression for the surface tension of a liquid rising in a capillary tube. Assume that the surface tension depends on mass, pressure and radius of the tube. The constant $k = 1/2$.*

Solution: We have the relation

$$\text{Surface Tension} = k\, m^a\, p^b\, r^c \qquad ...(1)$$

Writing in dimensional form

$$\frac{M^1 L^1 T^{-2}}{L^1} = [M^1]^a \left[\frac{M^1 L^1 T^{-2}}{L^2}\right]^b [L^1]^c$$

$\therefore \quad [M^1\ T^{-2}] = [M^{a+b}]\ [L^{-b+c}]\ [T^{-2b}]$

From principle of homogeneity, we have

$$(a + b) = 1, -b + c = 0, -2b = -2 \quad \therefore\ b = 1,\ c = 1,\ a = 0$$

$\therefore$ Equation (1) becomes

Surface tension $= k\,.\,p\,.\,r.$

Also $k = 1/2$ and pressure $p = hdg$

$$\therefore \text{ Surface tension } = \frac{h.d.g.r}{2}.$$

Example 5. *A planet moves round the sun in circular orbit. The time period of revolution t of the planet depends on (i) Radius of the orbit R (ii) Mass of the sun M (iii) Gravitational constant G. Show dimensionally that $t^2 \propto R^3$.*

Solution: We are given $\quad t = kR^a\ M^b\ G^c \quad$ ...(1)

Writing in dimensional form

$$[M^0\ L^0\ T^1] = [L]^a\ [M]^b\ [M^{-1}\ L^3\ T^{-2}]^c = [M^{b-c}]\ [L^{a+3c}]\ [T^{-2c}]$$

From principle of homogeneity, we have

$$b - c = 0, \quad a + 3c = 0, \quad -2c = 1$$

$$\therefore \quad c = -\frac{1}{2}, \quad b = -\frac{1}{2}, \quad a = \frac{3}{2}$$

From Equation (1) $\quad t = k\ R^{3/2}\ M^{-1/2}\ G^{-1/2} \quad \therefore\ \boldsymbol{t^2 \propto R^3}$

Example 6. *Find the number of watts in 1 h.p. (Given 1 inch = 2.54 cm and 1 dB = 453.6 gm)*

Solution: 1 horse power is given by

$$1 \text{ HP} = 550 \times 32.2 \text{ ft. poundals/sec.}$$

$$\text{Dimension of power} = \frac{\text{Work}}{\text{Time}} = M^1 L^2 T^{-3}$$

Now in F.P.S. system unit of M, L and T are pounds, ft and second while in c.g.s. system they are gm, cm and sec.

$$\therefore \quad 500 \times 32.2\ [M_1]\ [L_1]^2\ [T_1]^{-3} = k\ [M_2]\ [L_2]^2\ [T_2]^{-3}$$

Now

$$\frac{M_1}{M_2} = \frac{\text{pound}}{\text{gm}} = 453.6 \text{ (given)}$$

$$\frac{L_1}{L_2} = \frac{\text{foot}}{\text{cm}} = 12 \times 2.54 \text{ (given)}$$

$$\frac{T_1}{T_2} = \frac{\text{second}}{\text{second}} = 1$$

$$\therefore \quad k = 550 \times 32.2 \times 453.6 \times [12 \times 2.54]^2 \times 1 = 746 \times 10^7$$

$$\therefore \quad 1 \text{ HP} = 746 \times 10^7 \text{ cm} \times \text{dynes/sec}$$

$$= 746 \times 10^7 \text{ ergs/sec} = 746 \text{ J/sec} = \mathbf{746\ Watt.}$$

***Example** 7. Density of mercury is 13.6 gm/cm^3. If mass is measured in kg and length in meter, then what will be the density of mercury in new system of units?*

Solution: Dimensions of density = $M^1\ L^{-3}$

$$\therefore \quad 13.6 M_1 L_1^{-3} = k\ M_2 L_2^{-3}$$

$$k = 13.6 \times \left(\frac{M_1}{M_2}\right)\left(\frac{L_1}{L_2}\right)^{-3}$$

$$\frac{M_1}{M_2} = \frac{\text{gm}}{\text{kg}} = \frac{1}{1000}$$

$$\frac{L_1}{L_2} = \frac{\text{cm}}{\text{meter}} = \frac{1}{1000}$$

$$\therefore \qquad k = 13.6 \times \left(\frac{1}{1000}\right)\left(\frac{1}{1000}\right)^{-3} = 13.6 \times 10^3 \text{ kg/m}^3.$$

QUESTIONS AND PROBLEMS

1. What are fundamental and derived units? Give some examples.

2. What is principle of homogeneity? Explain your answer with the example of time period of a simple pendulum.

3. What are the various systems of units? How are they used?

4. What are S.I. system of units? Give their importance.

5. What are dimensional equations? Give their merits and limitations.

6. In Poiseuille's equation $V = \frac{\pi}{8} \frac{pr^4}{\eta l}$ find the dimensional of η. **[Ans.** $m^1 L^{-1} T^{-1}$]

7. Check the correctness of the relaton $\rho = \frac{3g}{4\pi r G}$ where letters have usual meanings. [**Ans.** Correct]

8. Deduce the dimensional formula for

(*i*) Surface tension
(*ii*) Gas constant
(*iii*) Gravitational constant
(*iv*) Permittivity
(*v*) Permeability
(*vi*) Magnetic flux
(*vii*) Planck's constant
(*viii*) Magnetic intensity
(*ix*) Conductance
(*x*) Inductance

CHAPTER

2

GRAVITATION

2.1. NEWTON'S LAW OF GRAVITATION

According to Newton's law of Gravitation "every body in the universe attracts every other body with a force which is directly proportional to the product of their masses and inversely proportional to the square of the distance between them." **The line of action of the force being the line joining the bodies.**

If m_1 and m_2 are the masses of two bodies and r is the distance between them, then the force of attraction between them is given by

$$F \propto \frac{m_1 m_2}{r^2}$$

or $$F = G\frac{m_1 m_2}{r^2} \quad ...(1)$$

where G is the **Universal constant** having the same value for all pairs of particles. G is also known as *Gravitational constant.* The gravitational forces between two bodies are an action and reaction pair. The first body exerts a force on the second body that is directed towards the first body along the line joining them. Similarly, the second body exerts a force on the first body that is directed towards the second body along the line joining them. These forces are equal in magnitude but are oppositely directed.

In Equation (1) if $m_1 = m_2 = 1$ and $r = 1$, then $F = G$. So gravitational constant G is defined as *numerically equal to the force* with which two particles each of unit mass and placed a unit distance apart attract each other. Its value is 6.67×10^{-11} Nm^2 kg^{-2}.

Since G is very small, so its effect cannot be experienced in daily life. However in the case of **celestial bodies** which have very large masses, the forces of attraction are very large. In fact, in their cases this force provides the necessary centripetal force to keep them in specific orbits.

Dimension of G are $$[G] = \frac{M^1 L^1 T^{-2} \times L^2}{M^2} = L^3 M^{-1} T^{-2}$$

So units of G may be expressed as m^3 kg^{-1} s^{-2}.

2.2. VERIFICATION OF GRAVITATION LAW FROM MOTION OF THE MOON

Newton gave the proof of gravitation law from the motion of moon. Moon revolves around the earth once in 27.3 days. Its path is nearly a circle of radius 3.80×10^8 m (which corresponds to the distance of moon from the earth). Its centripetal acceleration a which is directed toward the centre of earth is given by

$$a = r\omega^2$$

where r is the distance of moon from the centre of the earth and ω its angular velocity.

Also $$\omega = \frac{2\pi}{T}$$

where T is is the period, so $$a = r \cdot \left(\frac{2\pi}{T}\right)^2$$

Putting the values of r and T, we get

$$a = \frac{4 \times (3.142)^2 \times 3.80 \times 10^8}{27.3 \times 24 \times 60 \times 60} \text{ ms}^{-2} = 0.0027 \text{ ms}^{-2}$$

This is the acceleration of moon towards the centre of earth. The ratio of this acceleration a to the acceleration g (= 9.8 ms^{-2}) for a free falling body at earth's surface is given as

$$\frac{a}{g} = \frac{0.0027}{9.8} = \frac{1}{3600} = \frac{1}{(60)^2}$$

Also moon is 60 times away from the centre of earth as compared to a body on the earth's surface (As the earth's radius is 6.37×10^6 m and distance of moon from earth's centre is 3.8×10^8 m). So, if the Newton's law is correct, the moon's gravitational acceleration should be $\left(\frac{1}{60}\right)^2$ times the acceleration acting on a body near the earth's surface. This is what we got above and so the **Newton's law is verified.**

2.3. KEPLER'S LAWS OF PLANETARY MOTION

Kepler gave **three** laws of planetary motion which are stated as under:

***(i)* Law of Orbits:** All planets move in elliptical orbits with the Sun as focus. *This is called law of orbits.*

In the Fig. 2.1, the points F_1 and F_2 are far apart. However, in case of our planet earth, F_1 and F_2 are very close together and so orbit is practically a circular one.

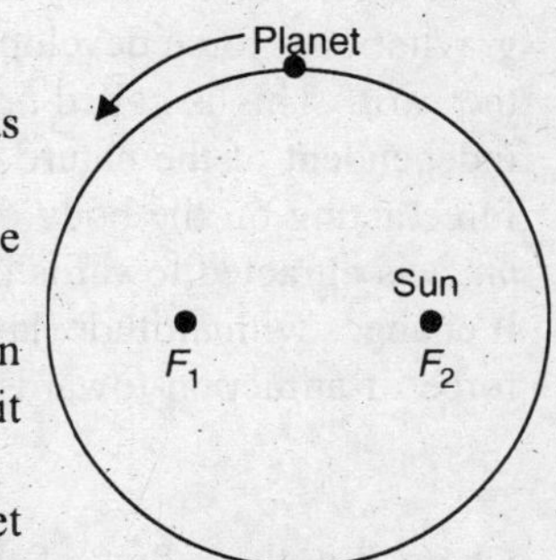

Fig. 2.1

***(ii)* Law of Areas:** According to this law a line joining any planet to the Sun sweeps equal areas in equal times. As shown in Fig. 2.2, let P_1, P_2, P_3 and P_4 represent the positions of a planet at different times in its orbit and let S be the position of the Sun.

Kepler's second law states that if time interval between P_1 and P_2 equals the time interval between P_3 and P_4 then area A_1 and A_2 must be equal. Also planet moves faster from P_1 to P_2 than from P_3 to P_4.

***(iii)* Law of Periods:** The square of the period of any planet about the Sun is proportional to the cube of planet's mean distance from the sun. This is called *law of periods.*

Let T_1 and T_2 be the periods of two planets around the Sun and let R_1 and R_2 be their respective mean distances from the sun. Then we have from law of periods

$$T_1^2 \propto R_1^3$$

and

$$T_2^2 \propto R_2^3$$

Combining these two, we get

$$\frac{T_1^2}{T_2^2} = \frac{R_1^3}{R_2^3}$$

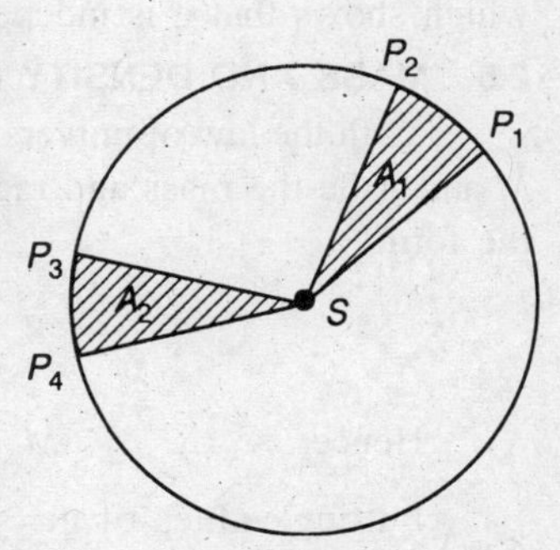

Fig. 2.2

If we consider the earth's orbit as circular, the centripetal force acting on the earth and directed towards the centre (Sun) is given by

$$F = \frac{mv^2}{R} \qquad ...(1)$$

where m is the mass of earth v its linear as speed and R = radius of circular orbit.

If T be the period of revolution of the earth, then

$$v = R\omega = R \cdot \frac{2\pi}{T}$$

Putting in Equation (1)

$$F = m \cdot \frac{4\pi^2 R}{T^2}$$

According to Kepler's third law,

$$T^2 = kR^3, \text{ where } k \text{ is constant.}$$

$$F = \frac{m.4\pi^2.R}{KR^3} \cdot = \frac{4\pi^2}{K} \cdot \left(\frac{m}{R^2}\right)$$

or

$$F \propto \frac{m}{R^2}$$

This formed the basis for Newton to develop his *laws of Gravitation.*

2.4. RELATION OF *g* AND *G*

The property due to which earth attracts a body towards its centre is called **gravity** and the force with which a body is attracted towards the centre of earth is called *gravitational force.* The gravitational force develops an *acceleration* in the body which is falling freely near the surface of the earth. This is called *acceleration due to gravity* and is represented by g. This acceleration is independent of the nature of body, its shape and size etc. So if the mass of a body is m, then the force acting on the body due to gravity is mg. So if $m = 1$ then g is the *force with which a unit mass* is attracted towards the earth's centre. g is a *vector* quantity and is not a *universal constant.* It changes with altitude, latitude etc., and have the units of LT^{-2}. If m is the mass of the earth, then force of attraction towards the centre on a body of mass m place on the surface of earth will be

$$F = \frac{GMm}{R^2}$$

where R is the radius of the earth.

Also

$$F = mg$$

$\therefore$

$$mg = \frac{GMm}{R^2}$$

or

$$g = \frac{GM}{R^2}$$

which shows that g is independent of shape, size and nature of the body.

2.5. MASS AND DENSITY OF EARTH

With the law of universal gravitation, we can calculate the mass and density of the earth. Let M and R be the mass and radius of earth respectively. Then the equation of g may be written in the form

$$g = \frac{GM}{R^2}$$

Hence,

$$M = \frac{gR^2}{G}$$

Putting values of $g = 9.80$ m/s^2, $R = 6.37 \times 10^6$ metres and $G = 6.67 \times 10^{-11}$ Newton m^2/kg^2, we get

$$M = \frac{9.80 \times (6.37 \times 10^6)^2}{6.6 \times 10^{-11}} = 6.0 \times 10^{24} \text{ kg approximately.}$$

Now, if ρ be the average density of earth then

$$\rho = \frac{\text{Mass}}{\text{Volume}} = \frac{M}{\frac{4}{3}\pi R^3}$$

$\therefore$

$$M = \frac{4}{3}\pi R^3 \rho$$

or $$\rho = \frac{6.0 \times 10^{24}}{\frac{4}{3}\pi \times (6.37 \times 10^6)^3} = 5.5 \times 10^3 \text{ kg/m}^3.$$

2.6. INERTIAL AND GRAVITATIONAL MASS

Let two bodies be of gravitational masses m_1 and m_2 respectively and let F_1 and F_2 be their respective forces of attraction towards earth then, we have

$$\frac{F_1}{F_2} = \frac{m_1}{m_2}$$

Newton's second law of motion gives a relation between the gravitational and inertial masses. If m_1 and m_2 be the inertial masses of the body whose gravitational masses are m_1 and m_2 respectively, then

$$M_1 = \frac{F_1}{g} \text{ and } M_2 = \frac{F_2}{g} \quad \therefore \quad \frac{F_1}{F_2} = \frac{M_1}{M_2}$$

which means $$\frac{m_1}{m_2} = \frac{M_1}{M_2}$$

i.e., $$m \propto M$$

or gravitational mass is proportional to the inertial mass of the body.

2.7. VARIATION OF ACCELERATION DUE TO GRAVITY

As stated above the value of g varies with altitude, depth, latitude etc. We will consider this variation as under.

(*a*) Variation with altitude: We know $g = \frac{GM}{R^2}$ at the surface of the earth so at altitude h above earth's surface g would be

$$g_h = \frac{GM}{(R+h)^2}$$

where $m \to$ Mass of the earth

and $R \to$ Radius of the earth

$$\therefore \quad \frac{g_h}{g} = \frac{R^2}{(R+h)^2}$$

or $$g_h = g \cdot \frac{R^2}{(R+h)^2} = g \cdot \frac{1}{\left(1+\frac{h}{R}\right)^2} = g\left(1 - \frac{2h}{R}\right)$$

So the value of g decreases as we go above the earth's surface.

(*b*) Variation with depth. We have the relation

$$g = \frac{GM}{R^2}$$

If earth is considered a homogeneous sphere of uniform density ρ, then

$$M = \frac{4}{3}\pi r^3 \cdot \rho$$

$$\therefore \quad g = \frac{G}{R^2} \cdot \frac{4}{3}\pi R^3 \rho = \frac{4}{3}\pi G \rho R$$

Now at a depth d below the earth's surface, the earth's mass becomes

$$M' = \frac{4}{3}\pi (R-d)^3 \cdot \rho$$

So new value of g will be

$$g' = G \cdot \frac{4}{3}\pi \frac{(R-d)^3 . \rho}{(R-d)^2} = \frac{4}{3}\pi G (R-d)\rho$$

$$\therefore \quad \frac{g'}{g} = \frac{\frac{4}{3}\pi G (R-d)\rho}{\frac{4}{3}\pi G R\rho} = \frac{R-h}{R} = 1 - \frac{h}{R}.$$

$$\therefore \quad g' = g\left(1 - \frac{h}{R}\right)$$

which shows that value of g goes on decreasing with depth and becomes *zero at the centre of the earth.*

(*c*) Variation with latitude: The value of g goes on changing from place to place on the earth's surface. It is maximum at the poles and decreases towards the equator. It has minimum value at the equator.

There are two reasons for this *viz.* (*i*) shape of the earth, and (*ii*) Rotation of the Earth.

(*i*) Due to shape of the Earth: Earth is not a perfect sphere. It is more nearly a spheroid having its *polar radius* smaller than the *equatorial radius.* So from Newton's law of Gravitation, a body at the pole would attracted more strongly than the body on the equator so the value of g is more at poles and least at the equator.

(*ii*) Due to Rotation of Earth: As earth is revolving around its own axis so every body placed on the Earth's surface is describing a circle. If the Earth is revolving with angular velocity ω, then a body P on the surface of earth is also moving with angular velocity ω and has centripetal accel. $r\omega^2$ directed towards the centre of the circle. This accel. is provided by the acceleration due to gravity. So the effective value of accel, due to gravity is decreased to $g \,.\, r\omega^2$

i.e., $\quad g' = g \,.\, r\omega^2$.

Now at the equator r is maximum (equal to the radius R of the earth) and at pole $r = 0$, so g is minimum at equator and maximum at the pole. As shown in Fig. 2.3

$$PQ = r = PO \cos \lambda = R \cos \lambda$$

Centripetal force at P due to rotational motion is

$$F' = mr\omega^2 = mR \cos \lambda\omega^2$$

This force is directed towards the centre Q of the circular path. Let it be represented in magnitude and direction by $\overrightarrow{PA}$.

Similarly force due to gravity acting on body would be mg and directed towards O, the centre of earth. Let it be represented in magnitude and direction by $\overrightarrow{PC}$.

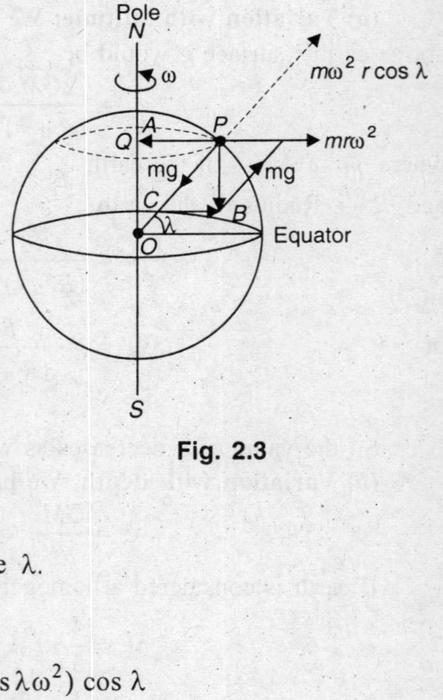

Fig. 2.3

A part of this is used in giving centripetal force F'

From figure $\quad mg = F' + mg'$

where g' is the acceleration due to gravity at the latitude λ.

From ΔPBC, we have

$$PB^2 = PC^2 + BC^2 - 2PC \,.\, BC \cos\ (\angle PCB)$$

$$(mg')^2 = (mg)^2 + (mR \cos\lambda\omega^2)^2 - 2\,(mg)\,(mR \cos\lambda\omega^2) \cos\lambda$$

or $\quad g'^2 = g^2 + R^2 \cos^2 \lambda\omega^4 - 2gR \cos^2 \lambda\omega^2$

Since ω is very small, so $R^2 \cos^2 \omega^4$ may he neglected as compared to other terms.

$$\therefore \quad g^{12} = g^2 - 2gR \cos^2 \lambda\omega^2$$

$$= g^2\left(1 - \frac{2R}{g}\cos^2 \lambda\omega^2\right)$$

$$\therefore \quad g' = g\left(1 - \frac{2R}{g}\cos^2 \lambda\omega^2\right)^{1/2}$$

$$= g\left(1 - \frac{R}{g}\cos^2 \lambda\omega^2\right)$$

(neglecting higher terms)

$$= g - R\omega^2 \cos^2 \lambda$$

which shows that g' is less than g.

At equator $\lambda = 0°$; $\cos = 1$

$\therefore$ $g' = g - R\omega^2$ (minimum)

at poles $\lambda = 90°$; $\cos \lambda = 0$

$\therefore$ $g' = g$ (maximum)

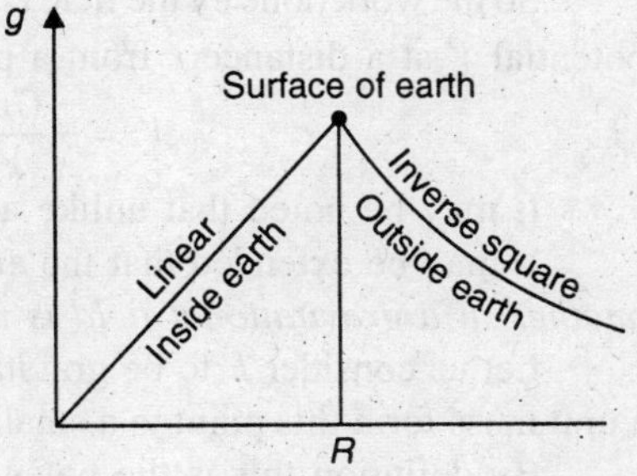

Fig. 2.4

The above results can be represented as shown in Fig. 2.4.

As g changes from place to place and also going above or below the earth's surface, its standard value is taken at the sea level at a latitude of 45° and the standard value is 9.8 ms^{-2}.

2.8. GRAVITATIONAL FIELD AND GRAVITATIONAL ATTRACTION

Gravitational field: *A particle of matter attracts every other particle with a force given by Newton's law of gravitation. The space surrounding the attracting particle within which its gravitational force of attraction is perceptible is called its gravitational field.*

Gravitational Attraction: *The gravitational attraction at a point in a gravitational field is the force experienced by a unit mass placed at that point, it being assumed that the unit mass does not bring about any change in the field.* Thus the attraction F at a point due to a particle of mass m at a distance r from the point is by Newton's law given by

$$F = G\frac{m \times 1}{r^2} = \frac{GM}{r^2}$$

and is a vector quantity.

2.9. GRAVITATIONAL POTENTIAL

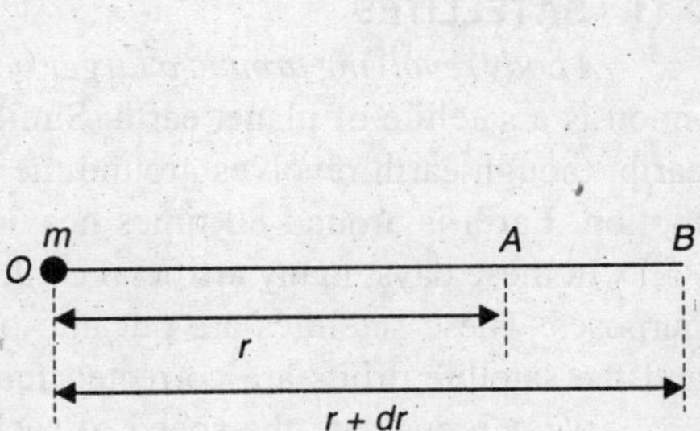

Fig. 2.5

In order to move a mass in a gravitational field work must be done. *If moved against the gravitational field, work must be done to overcome the gravitational attraction.* If moved in the direction of the field work is done by the field itself.

So gravitational potential at a point in a gravitational field is defined as the *work done in moving a unit mass from infinity to that point.* We consider infinity to be that point where gravitational field is assumed to be zero.

Let a point mass m be placed at O and let a unit mass be placed at A at a distance r from O, then we have the gravitational attraction given by

$$\frac{GM}{r^2} \text{ acting towards } O$$

If the unit mass is moved from A to B against the force of attraction, the work done against attraction is $= \frac{GM}{r^2}\, dr.$

So the total work done in moving mass from A to ∞ is

$$\int_r^{\infty} \frac{Gm}{r^2}\,dr = Gm\left[-\frac{1}{r}\right]_r^{\infty} = \frac{Gm}{r}$$

So the work done by the field (*i.e.*, attraction) in moving from ∞ to A is $-\frac{Gm}{r}$. So gravitational potential V at a distance r from a point mass is given by the relation

$$V = -\frac{Gm}{r}$$

It may be noted that unlike attraction, *potential is a scalar quantity.*

It may be extended that the amount of *work done in moving a unit mass from one point to another in a gravitational field is equal to the difference of potential at those points.*

Let us consider F to be gravitational attraction at a point, work done by the field in moving a unit mass form this point to a small distance dr to another point in the direction of field is $-F\,.\,dr$.

By defintion this is the potential difference dV at the two points.

$$\therefore \qquad dV = -F\,.\,dr \quad \text{or} \quad F = -\frac{dV}{dr}$$

So, we have the *force of attraction at any point in a gravitational field is equal to the negative rate of change of potential with distance (also called potential gradient) at that point.*

2.10. GRAVITATIONAL POTENTIAL ENERGY

The work *evolved* in bringing a body form ∞ to a point in a gravitational field (or other body) is called *gravitational potential energy* of the body at that point. So by definition, we have the work done in bringing a body of mass m form ∞ to a point at a distance r (in this case r is the radius of earth) given by

$$U = -\frac{GmM}{r}$$

where M is the mass of planet earth and negative sign indicates that work has been got from the gravitational force. So the potential at infinity is zero and also the gravitational work done is independent of the actual path connecting the points.

2.11. SATELLITES

A body revolving around a larger body under certain influence is called a satellite. For example moon is a satellite of planet earth. Similarly in some other planets satellites exist. In the case of earth, though earth revolves around the sun, it may be considered stationary to study the moon's motion. Earth is around 80 times heavier than moon, so moon is termed as *celestial satellite.*

In these days, many artificial earth satellites are used for various scientific and technological purposes. These satellites are put into orbits with the help of rockets, which fall down in stages and the satellite orbits are corrected for various scientific purposes.

We can evaluate the speed of such a satellite in the following way.

2.12. ORBITAL VELOCITY

Orbital velocity of a satellite is the velocity required to put the satellite into its orbit.

Let a satellite of mass m be put into a circular orbit around earth at a height x above its surface. Let the earth be of mass M and radius R. Radius of orbit of the satellite will be $(R + x)$ and let the velocity with which the satellite move in an orbit be V.

Gravitational force exerted by earth on satellite (body) will be

$$F = \frac{GMm}{r^2}$$

and centripetal force is $\quad F = \dfrac{mv_0^2}{r}$

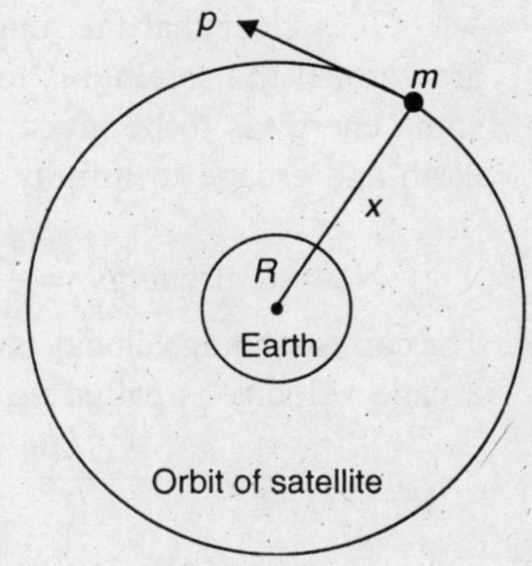

Fig. 2.6

$$\therefore \quad \frac{mv_0^2}{r} = \frac{GMm}{r^2} \quad \text{or} \quad v_0^2 = \frac{GM}{r}$$

In this case $r = R + r$

$$\therefore \quad v_0 = \sqrt{\frac{GM}{R+x}}$$

also $g = \dfrac{GM}{R^2}$

$$\therefore \quad v_0 = \sqrt{\frac{R^2 G}{R+x}} = R\sqrt{\frac{g}{R+x}}$$

if $h << R$ *i.e.*, satellite is at a few hundred kilometer above earth's surface in that case

$$v_0 = \sqrt{gR}$$

Time period of satellite is given by

$$T = \frac{2\pi r}{v_0} = \frac{2\pi r}{R}\sqrt{\frac{R+x}{g}}$$

$$= 2\pi\sqrt{\frac{(R+x)}{gR^2}}$$

For satellite with $x << R$

$$T = 2\pi\sqrt{\frac{R}{g}}$$

As is clear from above equation, $v_0 = R\sqrt{\dfrac{g}{(R+x)}}$ represents the minimum velocity v_m to keep the satellite in a circular orbit of radius r. Hence, if $v_0 < v_m$, the satellite will fall back on earth. If $v_0 > v_m$, but less than critical velocity (called escape velocity), the satellite executes an elliptical orbit. If $v_0 >> v_m$ escape velocity, the satellite will fall itself from gravitational pull of the earth and moves away along a parabola ($v_0 >$ escape velocity) or hyperbola ($v_0 =$ escape velocity).

2.13. ESCAPE VELOCITY

A satellite revolving around Earth will possess both kinetic energy and potential energy. Kinetic energy will be due to its motion and potential energy will be due to the gravitational field of the Earth.

P.E. will be given as

$$\text{P.E.} = -\frac{GMm}{r}$$

and K.E. is given by K.E. $= 1/2mv^2$

$\therefore$ Total Energy of the satellite is

$$E = \text{P.E.} + \text{K.E.}$$

$$= \frac{GMm}{r} + \frac{1}{2}mv^2$$

also $v^2 = \dfrac{Gm}{r}$

$$\therefore \quad E = -\frac{GMm}{r} + \frac{1}{2}m \cdot \frac{GM}{2}$$

$$= -\frac{GMm}{2r}$$

It is clear that the total energy possessed by a satellite is *netative.* So at ∞, P.E. and K.E. are zero. If the satellite is to be sent to infinity, *i.e.*, it is made to escape the orbit of its motion, some energy is to be given to it. The energy required for a satellite to leave its orbit around the earth and escape to infinity is called **Binding Energy.**

Now total energy $= \frac{1}{2} mv^2 - \frac{GMm}{r}$ should be zero if the satellite leaves the earth's orbit. The corresponding velocity is called *escape velocity* of the satellite and the energy required for acquiring escape velocity is called *escape energy* of satellite. We have for escape velocity v_e, the relation

$$\frac{1}{2} mv_e^2 - \frac{GMm}{r} = 0$$

$$\therefore \quad v_e^2 = \frac{2GM}{r}$$

or $$v_e = \sqrt{\frac{2GM}{r}}$$

In this case $r = R + x$ if $x << R$, then

$$v_e = \sqrt{\frac{2GM}{R}} = \sqrt{\frac{2gR^2}{R}} = \sqrt{2gR} \qquad \left[\because g = \frac{GM}{R^2}\right]$$

It is clear from above that the escape velocity is independent of mass of the body. However, force required to accelerate a body till it acquires *escape velocity depends on its mass.* Also orbital velocity is given by

$$v_0 = \sqrt{\frac{GM}{R}} = \sqrt{gR} \qquad [\because GM = gR^2]$$

$$\therefore \quad \frac{v_e}{v_0} = \sqrt{2} \text{ or } v_e = \sqrt{2}v_0$$

So if the orbital velocity of a satellite increases to $\sqrt{2}v_0$, it escapes form the earth's field. If ρ is the density of earth, then

$$M = \frac{4}{3}\pi R^3 \rho$$

So we have $$v_e = \sqrt{\frac{2G}{R} \times \frac{4}{3}\pi R^3 \rho}$$

$$= R \cdot \left(\frac{8}{3}\pi G \rho R^2\right)^{\frac{1}{2}}$$

For earth $R = 6.38 \times 10^6$ m

$g = 9.81$ ms^{-2}

$$\therefore \quad v_e = \sqrt{2gR}$$

$$= \sqrt{2 \times 9.8 \times 6.38 \times 10^8}$$

$$= 11.2 \times 10^3 \text{ ms}^{-1}$$

$$= 11.2 \text{ km s}^{-1}$$

In actual parctice the escape velocity is much higher than this value due to resistance offered by the earth's atmosphere.

It may be interesting to note that escape velocity of moon is around 2.37×10^3 ms^{-1} while the rms value of the velocity of molecules at moon temperature is greater than this value, *it is due to this reason that moon has no atmosphere.*

At earth, the escape velocity is 11.2 km s^{-1} while the average thermal velocity of hydrogen molecules at 500 K would be 2.5 km s^{-1} and so the molecules of hydrogen and other gases do not leave the earth's atmosphere. **So the planet Earth has atmosphere.**

Example 1. (*a*) *The gravitational force acting on a rocket at a height from the earth surface is 1/3rd of the force acting on a body at sea level. What is the relation between h and R_e (radius of the earth)?*

(*b*) *The distance of pluto from the sun is 40 times the distance of earth. If the mass of earth and pluto be equal, then compare the gravitational forces of sun on these planets.*

Solution: (*a*) At sea level, gravitational force acting on a body is given by

$$F_1 = \frac{GM_e m}{R_e^2}$$

where m is mass of body. At height h the gravitational force between Rocket and earth will be

$$F_2 = \frac{GM_e m'}{(R_e + h)^2}$$

Now, $$F_2 = \frac{1}{3} F_1$$

$\therefore$ $$\frac{GM_e m'}{(R_e + h)^2} = \frac{1}{3} \frac{GM_e m}{R_e^2}$$

which gives $$m' . 3R_e^2 = m\,(R_e^2 + h^2 + 2R_e h)$$

If $m \approx m'$, then, we have

$$3R_e^2 = R_e^2 + h^2 + 2R_e h$$

or $$h^2 + 2R_e h - 2R_e^2 = 0$$

or $$h = \frac{-2R_e \pm \sqrt{(4R_e^2 + 8R_e^2)}}{2}$$

$$= \frac{-2R_e \pm 3.464R_e}{2}$$

$\therefore$ $$h = \mathbf{0.732}\boldsymbol{R_e}$$

(*b*) The gravitational force between sun and earth = F_1

$$F_1 = \frac{GM_s M_e}{R_{es}^2}$$

The gravitational force between sun and pluto = F_2

$$F_2 = \frac{GM_s M_p}{R_{ps}^2}$$

Now $R_{ps} = 40R_{es}$ and $M_e = M_p$, so we have

$$\frac{F_1}{F_2} = \frac{1600}{1} \quad \text{or} \quad F_1 = \mathbf{1600}\boldsymbol{F_2}.$$

Example 2. *Caculate the orbital velocity of satellite revolving around the earth near its surface. The radius of earth is 6.4×10^6 meter and acceleration due to gravity is 10 m/sec^2. What will be the orbital velocity if the satellite be at a height of 2000 km from the earth's surface.*

Solution: Orbital velocity of satellite is given by the relation

$$v_0 = \sqrt{\frac{GM_e}{R_e + x}} \quad \text{In this case } x = 0$$

so, $$v = \sqrt{\frac{GM_e}{R_e}}$$

Since $$g = \frac{GM}{R_e^2}$$

so, $$v = \sqrt{R_e g}$$

$$= \sqrt{6.4 \times 10^6 \times 10}$$

$$= 8 \times 10^3 \text{ m/s} \approx 8 \text{ km/s}$$

At a height of 2000 km from the earth surface the speed is

$$v = \sqrt{\frac{GM_e}{6.4 \times 10^6 + 2 \times 10^6}} = \sqrt{\frac{GM_e}{8.4 \times 10^6}}$$

also $$GM_e = R_e^2\, g$$

$\therefore$ $$v = \sqrt{\frac{(6.4 \times 10^6)^2 \times 10}{8.4 \times 10^6}} = \sqrt{\frac{64 \times 6.4}{8.4}}$$

$$\approx \mathbf{6.98 \times 10^3 \approx 7 \times 10^3\ ms^{-1}}.$$

Example 3. (*a*) *The mean distance of sun from a planet is four times the distance of the sun from the earth. In how many years will that planet complete one revolution around the sun?*

(*b*) *On the planet venus a body has* -7.5×10^6 *Joules of gravitational potential energy. Work out the energy required for the body to escape from the planet.*

Solution: (*a*) Period of revolution of a planet round the sun is given by

$$T = 2\pi \sqrt{\frac{r^3}{GS}}$$

where r is the distance of the planet from the centre of the sun and S is the mass of sun.

So we have for the earth

$$T_e = 2\pi \sqrt{\frac{r_{es}^3}{GS}}$$

where r_{es} distance of earth from sun for other planet

$$T_p = 2\pi \sqrt{\frac{r_{ps}^3}{GS}} \quad \text{where } r_{ps} =$$

distance of planet from sun,

$$r_{ps} = 4R_e, \text{ so we have}$$

$$\frac{T_p}{T_e} = \sqrt{\frac{(4r_e)^3}{(r_e)^3}}$$

$$= \sqrt{64} = 8$$

or, $T_p = 8\,T_e$, since $T_e = 1$ year, so, which means the period of revolution of planet round the sun is **8 years.**

(*b*) Escape energy = – gravitational PE

So at the planet venus the escape energy would be

$-(-7.5 \times 10^6 \text{ Joule}) = \mathbf{7.5 \times 10^6}$ **Joules.**

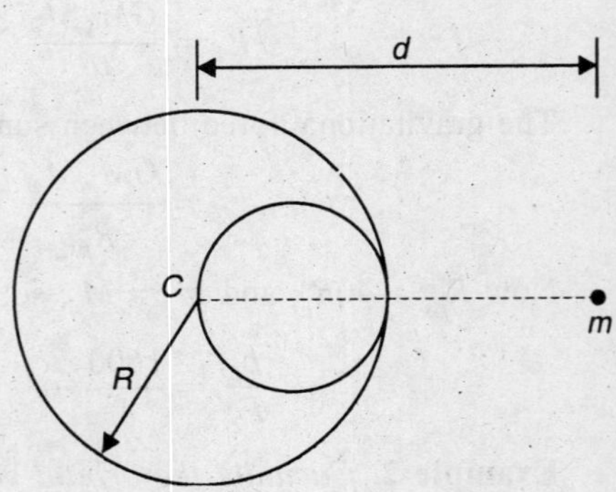

Fig. 2.7

Example 4. *A spherical hollow is made in a lead sphere of radius R such that its surface touches the outside surface of the lead sphere and passes through its centre. The mass of the lead sphere before hollowing was M. What is the force of attraction that this sphere would exert on a particle of mass m which lies at a distance d from the centre of the lead sphere on the straight line joining the centres of the sphere and the hollow?*

Solution: Volume of lead sphere $= \frac{4}{3}\pi R^3$

Volume of hollow sphere $= \frac{4}{3} p \left(\frac{R}{2}\right)^3$

$$= \frac{1}{8}\left(\frac{4}{3}\pi R^3\right)$$

To calculate the force of attraction on the point mass m, we must calculate the force due to the solid sphere and subtract from this force that the mass of the hollowed sphere would have exerted on m. The distance from the centre of the hollow sphere to mass m is $(d - R/2)$.

So, the net force on m is given by

$$F = G\frac{mM}{d^2} - \frac{Gm\,(M/8)}{(d - R/2)^2} = \boldsymbol{G\frac{mM}{d^2}\left[1 - \frac{1}{8\,(1 - R/2d)^2}\right]}.$$

Example 5. *If a planet were suddenly stopped in its orbit, supposed to be circular, show that it would fall into the sun in a time $\frac{\sqrt{2}}{8}$ times the period of planet's revolution.*

Solution: Let r be the radius of the planets' orbit. Then, we have

$$\frac{GMm}{r^2} = m\omega^2 r \text{ or } \omega = \sqrt{\frac{GM}{r^3}}$$

Hence, time period $T = \frac{2\pi}{\omega} = 2\pi\sqrt{\frac{r^3}{GM}}$

If the planet is at a distance x from the sun and is moving towards the sun, then by the conservation of energy, we may write

$$\frac{1}{2}\left(\frac{dx}{dt}\right)^2 - \frac{GM}{x} = -\frac{GM}{r}$$

or
$$\left(\frac{dx}{dt}\right)^2 = \frac{2GM}{r}\left(\frac{r}{x} - 1\right)$$

$$= \frac{2GM}{r}\left(\frac{r-x}{x}\right)$$

$$\therefore \quad \frac{dx}{dt} = \sqrt{\frac{2GM}{r}}\sqrt{\frac{r-x}{x}}$$

or
$$\sqrt{\left(\frac{2GM}{r}\right)}\cdot dt = \sqrt{\frac{x}{r-x}}\,dx$$

Integrating, we get

$$\sqrt{\frac{2GM}{r}}\int dt = \int_0^r \sqrt{\frac{x}{r-x}}\,dx$$

or
$$\sqrt{\frac{2GM}{r}}\cdot t = \frac{\pi r}{2}$$

or $$t = \frac{\pi r}{2}\sqrt{\frac{r}{2GM}} = \frac{\sqrt{2}}{8}\,T.$$

Example 6. *Three particles each of mass m are situated at the vertices of an equilateral triangle of side length a. The only forces acting on the particles are their gravitational forces. It is desired that each particle moves in a circle while maintaining the original mutual separation a. Find the initial velocity that should be given to each article and also the time period of the circular motion.*

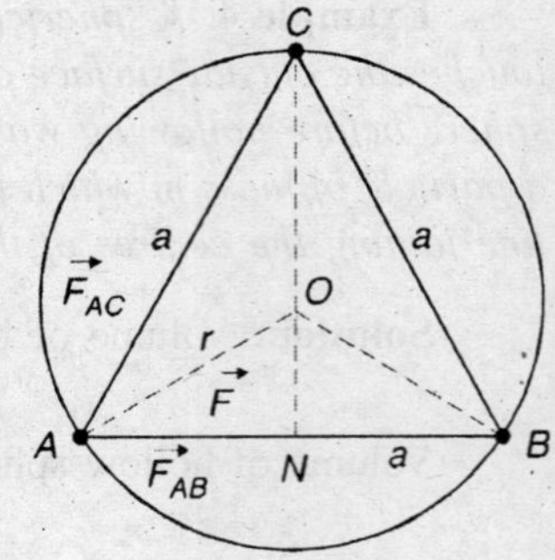

Fig. 2.8

Solution: The geometry of problem is as that shown in Fig. 2.8. Here the radius of the circular path is given as

$$r = OA = OB = OC = (AN/\cos 30^{\circ})$$

$$= \frac{a/2}{\sqrt{(3/2)}} = \frac{a}{\sqrt{3}}$$

The gravitational force at A, due to mass m at C is

$$F_{AC} = \frac{Gmm}{a^2} \text{ along } AC$$

Similarly, the gravitational force at A due to mass at B is given by

$$F_{AB} = \frac{Gmm}{a^2} \text{ along } AB$$

The resultant F of these forces would be along AO. As the angle between the two forces is 60°, the resultant force F would be given by the relation

$$F = \sqrt{F_{AC}^2 + F_{AB}^2 + 2F_{AC}\,F_{AB}\cos 60^{\circ}}$$

$$= \sqrt{\left(\frac{Gmm}{a^2}\right)^2 + \left(\frac{Gmm}{a^2}\right)^2 + 2\left(\frac{Gmm}{a^2}\right)\left(\frac{Gmm}{a^2}\right)\times\frac{1}{2}}$$

$$= \frac{Gmm}{a^2}\sqrt{3}$$

This force provides the necessary centripetal force, which means.

$$\frac{mv^2}{a} = \frac{Gmm}{a^2}\sqrt{3}.$$

$$v = \left(\frac{Gm}{a}\right)^{3/2}$$

So, time period $$T = \frac{2\pi r}{v} = \frac{2\pi\,(a/\sqrt{3})}{(Gm/a)^{1/2}} = \frac{2\pi a^{3/2}}{(3Gm)^{3/2}}.$$

Example 7. (*a*) *What will be the escape velocity on the surface of the moon if it is 11.2 km/sec on the surface of the earth? Assume that the value of g at the surface of the moon is 1/5th of that at the earth and the radius of moon is 1/4th the radius of the earth.*

(*b*) *The speed of revolution of an artificial satellite revolving very near the earth is 8 km/s. What will be the escape velocity for a body on earth?*

Solution: (*a*) Escape velocity $= \sqrt{2gR_e}$

$$11.2 \times 10^3 = \sqrt{2gR_e}$$

or $$2gR_e = (11.2)^2 \times 10^6$$

At moon, $$v_e = \sqrt{\frac{g}{5}\times\frac{R_e}{4}\times 2}$$

or $$v_e^2 = \frac{gR_e}{10}$$

or $$gR_e = 10v_e^2$$

Hence, $$20v_e^2 = (11.2)^2 \times 10^6$$

Hence, $$v_e = \frac{11.2 \times 10^3}{\sqrt{20}} \text{ km/sec}$$

$$\approx \textbf{2.5 km/sec.}$$

(*b*) Orbital velocity $= \dfrac{\text{Escape velocity}}{\sqrt{2}}$

Hence, escape velocity $= \sqrt{2} \times 8$ km/sec $\approx$ **11.3 km/sec.**

Example 8. *(a) A particle of mass m is revolving in a horizontal circle of radius r under a centripetal force K/r² where K is a constant. What is the total energy of the particle?*

(b) The astronauts in a satellite orbiting the earth feel weightlessness. Does the weightlessness depend upon the distance of the satellite from the earth? If so how? Explain your answer.

Solution: (*a*) Centripetal force is given by

$$= \frac{K}{r^2} = \frac{mv_s^2}{r} = \frac{Gm_e m}{r^2}$$

Now $$\text{P.E.} = -\frac{GM_e m}{r}$$

Also $$\frac{GM_e m}{r} = +\frac{K}{r}$$

Hence, $$\text{P.E.} = -\frac{K}{r}$$

$$\text{K.E.} = \frac{1}{2}mv_s^2 = \frac{K}{2r}$$

Hence, Total energy $= \dfrac{K}{2r} - \dfrac{K}{r} = -\dfrac{\boldsymbol{K}}{\boldsymbol{2r}}.$

(*b*) The reaction force felt by the astronaut is zero and so there is weightlessness. This is not dependent on the distance of satellite from the earth.

Example 9. (*a*) *Taking moon's period of revolution about the earth as 30 days (and neglecting the effect of sun and other planets on its motion), calculate its distance r from the earth (G = 6.67 × 10⁻¹¹ Nm² kg⁻², mass of earth M_e = 6 × 10²⁴.*

(*b*) *How much energy would be needed for a 100 kg body to escape from the earth? (g = 10 m/s² and radius of the earth R_e = 6.4 × 10⁶ meter).*

Solution: We have the relation

(*a*) $$T = 2\pi\sqrt{\frac{r^3}{GM_e}}$$

where T is period of revolution of moon, r is distance from earth, M_e = Mass of earth.

In this case T = 30 days

= 30 × 24 × 60 × 60 seconds

= 2.592 × 10⁶ seconds.

So, $$(2.592 \times 10^6)^2 = \frac{4\pi^2 \times r^3}{6.67 \times 10^{-14} \times 6 \times 10^{24}}$$

or $$r^3 = \frac{(2.592 \times 10^6)^2 \times 6.67 \times 6 \times 10^{13}}{4\pi^2}$$

$$= 68.1 \times 10^{24}$$

$$\therefore \quad r = 4.05 \times 10^{8} \text{ m}$$

$$= \mathbf{4.08 \times 10^{5} \text{ km}}$$

(*b*) Escape energy $= \dfrac{GM_e m}{R_e} = mgR_e$ $\qquad (\because GM_e = gR_e^2)$

$$\therefore \quad E = 100 \times 10 \times 6.4 \times 10^{6} \text{ J}$$

$$= \mathbf{6.4 \times 10^{9} \text{ J.}}$$

Example 10. (*a*) *If the diameter of the earth becomes twice its present value but the mass remains unchanged, then how would be the weight of an object on the earth's surface affected?*

(*b*) *The mass and diameter of a planet are twice those of the earth. What will be the time-period of that pendulum on this planet, which is a second's pendulum on the earth?*

Solution: (*a*) Value of g initially is

$$g = \frac{GM_e}{R_e^2} \qquad ...(1)$$

Now M_e remains same. Initial diameter was $2R_e$ and now it becomes $4R_e$ and so the radius is now $2R_e$. So new value of g is

$$g' = \frac{GM_e}{(2R_e)^2} \qquad ...(2)$$

From (1) and (2) $\dfrac{g'}{g} = \dfrac{\mathbf{1}}{\mathbf{4}}$

Hence, new weight is reduced to 1/4 of its original value.

(*b*) Since, $\quad g = GM_e/R_e^2$

If M and R of a planet are $2M_e$ and $2R_e$ respectively then g at the planet will be half to the value at earth. So the time period of seconds pendulum of the earth will be $\mathbf{2\sqrt{2}}$ **sec** on the planet.

Example 11. *Get an expression for the period of revolution of a planet around the Sun. Hence show how the mass of the Sun can be calculated?*

Solution: Let the planet of mass m move around the Sun of mass M in circular orbit of radius r. The gravitational force of attraction exerted by Sun on planet will be

$$F = G \cdot \frac{Mm}{r^2}$$

Centripetal force on the planet and directed towards sun is $\dfrac{mv^2}{r}$ where v is the linear velocity of planet.

So we have $\quad \dfrac{GMm}{r^2} = \dfrac{mv^2}{r}$

Since, $v = \omega r$ where ω is angular velocity of planet, so we have

$$\frac{GMm}{r^2} = \frac{m\omega^2 r^2}{r} \quad \text{or} \quad \omega^2 = \frac{GM}{r^3}$$

Since, $\quad \omega = \dfrac{2\pi}{T}$ where T is period of revolution of planet,

$$\frac{4\pi^2}{T^2} = \frac{GM}{r^3}$$

or $$T^2 = \frac{4\pi^2 r^3}{GM}$$

or $$T = 2\pi\sqrt{\frac{r^3}{GM}}$$

Mass of sun may be evaluated from above equation, and is given by

$$M = \frac{4\pi^2 r^3}{GT^2}.$$

Example 12. *Two satellites S_1 and S_2 revolve round a planet in coplaner circular orbits in the same sense. Their periods of revolution are 1 hour and 8 hours respectively. The radius of orbit of S_1 is 10^4 km.*

(*a*) *When S_2 is closest to S_1 find the speed of S_2 w.r.t. S_1*

(*b*) *Also find the angular speed of S_2 as actually observed by an asronaut in S_1.*

Solution: (*a*) We have the relation

$$\frac{mv^2}{R} = \frac{GMm}{R^2} \quad \therefore \quad v^2 = \frac{GM}{R} \qquad ...(1)$$

Also
$$v = \frac{2\pi R}{T} \quad \therefore \quad v^2 = \frac{4\pi^2 R^2}{T^2} \qquad ...(2)$$

From (1) and (2)

$$\frac{4\pi^2 R^2}{T^2} = \frac{GM}{R} \quad \therefore \quad T^2 = \frac{4\pi^2}{GM} \cdot R^3$$

If T_1 and T_2 are the time periods for satellites S_1 and S_2 respectively, then

$$\frac{T_1^2}{T_2^2} = \frac{R_1^3}{R_2^3} \quad \therefore \quad R^2 = R_1 \cdot \left(\frac{T_2}{T_1}\right)^{2/3}$$

Now T_1 = 1 hr., T_2 = 8 hr and $R_1 = 10^4$ km.

$$\therefore \quad R^2 = 10^4 \times \left(\frac{8}{1}\right)^{2/3} = 4 \times 10^4 \text{ km}.$$

$$v_1 = \frac{2\pi R_1}{T_1} = \frac{2\pi \times 10^4}{1} = 2\pi \times 10^4 \text{ km hr}^{-1}$$

$$v_2 = \frac{2\pi R_2}{T_2} = \frac{2\pi \times 4 \times 10^4}{8} = \pi \times 10^4 \text{ km hr}^{-1}$$

Relative velocity of S_2 w.r.t. S_1, is

$$v = v_2 - v_1 = (\pi \times 10^4 - 2\pi \times 10^4) \text{ km hr}^{-1}$$

$$\therefore \quad |v| = \pi \times 10^4 \textbf{ km/hr}$$

(*b*)
$$\omega = \frac{v}{R_2 - R_1} = \frac{\pi \times 10^4}{4 \times 10^4 - 1 \times 10^4} = \frac{\pi}{3} \textbf{ rad/hr.}$$

Example 13. *A tunnel is dug from one side of earth's surface to the other end along a chord. If a particle is dropped from one end, it oscillates within the tunnel. Will it attain the same maximum speed along the chord as it attains along the diameters?*

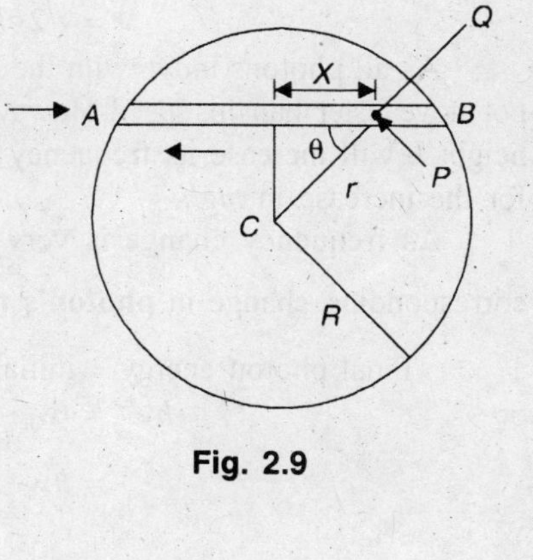

Fig. 2.9

Solution: Let the particle of mass m be at a position Q in the tunnel at any instant.

Force on particle at that position

$$F = -\frac{GMmr}{R^3}$$

where r is the distance of particle from centre of earth and R is the radius of earth.

This force is directed towards the centre C of earth along QC.

Force experienced by the particle along chord = $F \cos \theta$.

$$= -\frac{GMm.r}{R^3} \cdot \frac{x}{r} = -\frac{GMm}{R^3} \cdot x$$

which has the form $F = -kx$

So the motion is Simple Harmonic and force constant is $\frac{GMm}{R^3}$.

$$\therefore \quad \text{Time period } T = 2\pi \sqrt{\frac{m}{k}} = 2\pi \sqrt{\frac{R^3}{GM}}$$

Which shows that the time period is independent of the mass of the particle that is dropped into the tunnel.

Speed gained by the particle will not be the same if it is dropped in the tunnel along the diameter and the chord. As the difference of P.E. at point P and C will be different than that at B and C. So the K.E. gained by a particle will be more when it moves along diameter than that along the chord.

Putting the values of R, M and G, we get the expression for time period

$$T = 2\pi \sqrt{\frac{(6.4 \times 10^6)^3}{6.6 \times 10^{-11} \times 6 \times 10^{24}}} = \mathbf{84.2 \ min.}$$

2.14. PHOTONS AND GRAVITY

It is observed that light is affected by gravity. We can understand the gravitational behaviour of light from the fact that although a photon has zero rest mass, it **interacts with electrons** as though it has inertial mass. Photon mass may be given as

$$m = \frac{p}{v} = \frac{h\nu}{c^2} \quad \text{...(1)}$$

as incase of photon, $p = = \frac{h\nu}{c}$ and $\nu = c$.

From the principal of equivalence, gravitational mass is equal to the inertial mass. It means that a photon of frequency ν will be gravitationally acting like a particle of mass $h\nu/c^2$.

We can demonstrate the gravitational behaviour of a photon in laboratory. We know that when a body of mass m falls from height h near earth's surface, the body gets accelerated due to earth's gravitational pull and gains energy mgh on reaching the earth's surface. Final energy gained by the body is $\frac{1}{2} mv^2$ and from the relation

$$mgh = \frac{1}{2} mv^2$$

We get the final speed v

$$v = \sqrt{2gh}$$

As all photons move with the speed of light and so can not move faster than this speed. However a photon falling through height h will increase its frequency from ν to ν' to account for the increase in mgh.

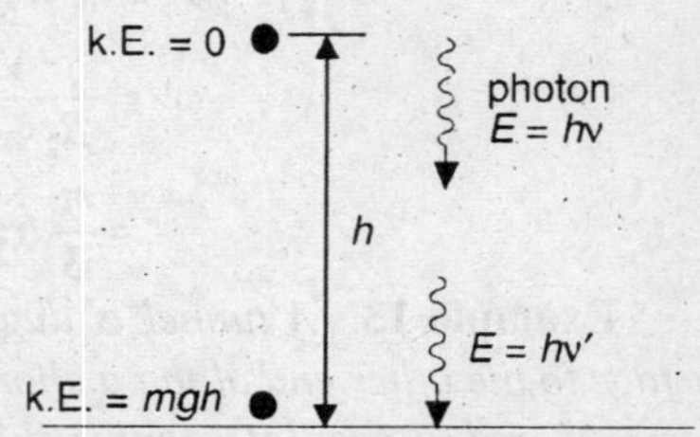

Fig. 2.10

As frequency change is very small in the laboratory experiment, so we can neglect the corresponding change in **photon's mass** $\frac{h\nu}{c^2}$. So we have

Final photon energy = initial photon energy + increase in potential energy

or
$$h\nu' = h\nu + mgh$$

$$= h\nu + \frac{h\nu}{c^2} \cdot gh$$

$$= h\nu\left[1+\frac{gh}{c^2}\right] \quad ...(2)$$

which is the energy acquired by a photon through a height h.

2.15. GRAVITATIONAL RED SHIFT

We can account for an interesting astronomical effect due to the gravitational behaviour of light *i.e., if the frequency associated with a photon moving towards earth increases then the frequency of a photon moving away from earth should decrease.*

We know that the earth's gravitational field is not as strong, as that of many stars. Let a star of mass m and radius R emits a photon of initial frequency ν.

P.E. of a mass m on the surface of star is given by

$$\text{P.E.} = -\frac{GMm}{R}$$

Negative sign is because of the attractive force between M and m.

So the P.E. of a photon of mass $\frac{h\nu}{c^2}$ on the surface of the star will be

$$\text{P.E.} = -\frac{GM}{R}\cdot\frac{h\nu}{c^2}$$

Total energy of the photon is the sum of its Quantum energy and P.E.

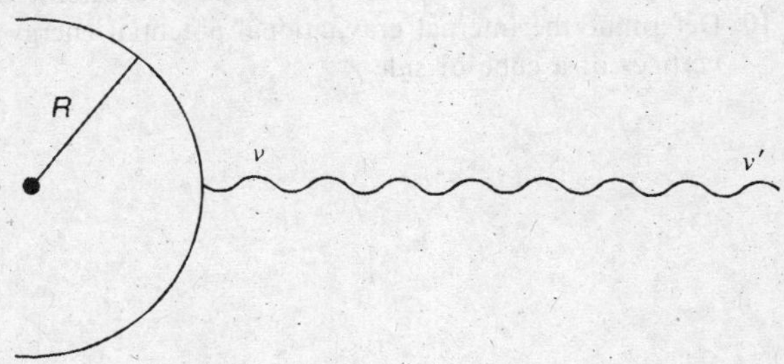

Fig. 2.11

$$\therefore \quad E = h\nu - \frac{GM}{R}\cdot\frac{h\nu}{c^2}$$

$$= h\nu\left[1-\frac{GM}{R.c^2}\right] \quad ...(1)$$

At a very far off distance from the star, the photon has total energy as electromagnetic energy given by

$$E = h\nu' \quad ...(2)$$

Comparing (1) and (2) as the P.E. of the photon is negligible on earth's surface as compared to that on star's surface.

$$\therefore \quad h\nu' = h\nu\left[1-\frac{GM}{Rc^2}\right]$$

$$\therefore \quad \frac{\nu'}{\nu} = 1-\frac{GM}{Rc^2}$$

or

$$1-\frac{\nu'}{\nu} = \frac{GM}{Rc^2}$$

i.e.,

$$\frac{\Delta\nu}{\nu} = \frac{\nu-\nu}{\nu} = \frac{GM}{Rc^2} \quad ...(3)$$

Which means that the photon has lower frequency at earth as it looses some energy when it leaves the star. It means that a photon in the visible range of spectrum shifts towards red end (longer wavelength side).

This phenomenon is called the *gravitational red shift.* It is different than the *Doppler's red shift* which is due to *expanding universe theory.*

The validity of Equation (3) can be checked by comparing the frequencies of stellar spectra with those observed in laboratory. In most of the cases, since M/R is too small to observe the red shift. However, in case of *white dwarfts*, this shift is observable. (A white dwarf is an old star whose interior consists of atoms whose electronic structures have collapsed and so it is very small). A typical *white dwarf* has the *mass of the sun and size of the Earth.*

QUESTIONS AND PROBLEMS

1. State the laws of gravitation? How will you give evidence to support these laws?
2. What is acceleration due to gravity? Derive an expression in terms of mass of earth and gravitational constant.
3. How value of g is affected by altitude and depth?
4. Give Kepler's laws of planetary motion. How Newton's law of gravitation can be derived from planetary laws?
5. How g is affected due to this rotation of earth around its own axis?
6. What is gravitational potential? Derive an expression for gravitational potential at a point.
7. What is orbital velocity? Derive an expression for the same.
8. What is a satellite? Derive an expression for the escape velocity.
9. Calculate the velocity and time period of a satellite assuming it to move in a circular path.
10. Determine the internal gravitational potential energy of 8 particles, each of mass m located at the vertices of a cube of side r. **[Ans.** 22.8 Gm^2/r]

CHAPTER

3

CENTRE OF MASS AND COLLISIONS

3.1. CENTRE OF MASS

Newton's laws of motion can be understood and applied in a better way if the knowledge of *centre of mass* of a body is grasped clearly. It is due to the fact that even when a body vibrates or rotates as it is moving, there is one point on the body, the *centre of mass* that behaves in the same way as a single particle subjected to an external force would move. For example if an irregular body (say a bat) is projected in the air so that it moves and vibrates simultaneously to cause a parabolic motion, the *centre of mass* will behaves as simple parabolic motion. As the bat rotates, the centre of mass performs a simple motion.

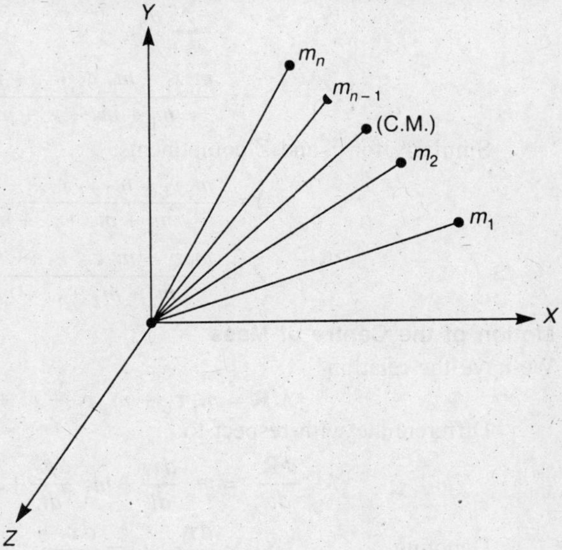

Fig. 3.1

We may, also define the centre of mass of a body as the point at which an *applied force produces acceleration but no rotation.* So the centre of mass may be considered as a point where the *whole mass of the body may be supposed to be concentrated.* In fact it can be proved that *centre of mass of a system of particles moves as things all the mass of the system were concentrated at the centre of mass and all the external forces were applied at that point.*

The centre of mass of a system of N particles (as shown in Fig. 3.1) is defined as a point whose position vector is given by

$$\mathbf{R} = \frac{m_1 \mathbf{r}_1 + m_2 \mathbf{r}_2 + \ldots + m_n \mathbf{r}_n}{m_1 + m_2 + \ldots + m_n}$$

Let M be the total mass of particles,

$$\therefore \qquad M = \Sigma m = m_1 + m_2 + \ldots + m_n$$

and $$M\mathbf{R} = m_1 \mathbf{r}_1 + m_2 \mathbf{r}_2 + \ldots + m_n \mathbf{r}_n$$

or $$\mathbf{R} = \frac{\sum_{i=1}^{n} m_i \mathbf{r}_i}{\sum_{i=1}^{n} m_i}$$

In case of two particles system, we have

$$\mathbf{R} = \frac{m_1 \mathbf{r}_1 + m_2 \mathbf{r}_2}{m_1 + m_2}$$

If (x, y, z) be the coordinates of centre of mass then

$$\mathbf{R} = \mathbf{i}x + \mathbf{j}y + \mathbf{k}z$$

where **i, j** and **k** are unit vectors along x, y and z directions respectively. X component of position vector of centre of mass is given by

$$X = \frac{\sum_{i=1}^{n} m_i x_i}{\sum_{i=1}^{n} m_i}$$

$$= \frac{m_1 x_1 + m_2 x_2 + \ldots + m_n x_n}{m_1 + m_2 + \ldots + m_n} \quad \ldots(1)$$

Similarly for Y and Z components

$$Y = \frac{m_1 y_1 + m_2 y_2 + \ldots + m_n y_n}{m_1 + m_2 + \ldots + m_n} \quad \ldots(2)$$

$$Z = \frac{m_1 z_1 + m_2 z_2 + \ldots + m_n z_n}{m_1 + m_2 + \ldots + m_n} \quad \ldots(3)$$

Motion of the Centre of Mass

We have the relation

$$M\mathbf{R} = m_1 \mathbf{r}_1 + m_2 \mathbf{r}_2 + \ldots + m_n \mathbf{r}_n$$

Differentiate with respect to t

$$M \frac{d\mathbf{R}}{dt} = m_1 \frac{d\mathbf{r}_1}{dt} + m_2 \frac{d\mathbf{r}_2}{dt} + \ldots$$

Denoting $\mathbf{v}_1 = \frac{d\mathbf{r}_1}{dt} \quad \mathbf{v}_2 = \frac{d\mathbf{r}_2}{dt}$

where v_1 and v_2 are velocities of particles and $\mathbf{V} = (d\mathbf{R}/dt)$ is the velocity of centre of mass, then,

$$M\mathbf{V} = m_1 \mathbf{v}_1 + m_2 \mathbf{v}_2 + \ldots + m_n \mathbf{v}_n$$

$$= \sum_{i=1}^{n} m_i \mathbf{v}_i \quad \ldots(4)$$

As, the right hand side of above equation is constant, so in the absence of external forces the velocity of centre of mass **V** is constant.

Differentiating Equation (4) w.r.t. t, we have

$$\frac{Md\mathbf{V}}{dt} = \sum_{i=1}^{n} \frac{d}{dt}(m_i \mathbf{v}_i)$$

$$= \sum_{i=1}^{n} F_i$$

where $F_i = m_i \frac{d\mathbf{v}_i}{dt}$.

From Newton's third law all the internal forces will occur in equal and opposite pairs so, they will contribute nothing to the total force.

Hence $\sum_{i=1}^{n} F_i$ represents the sum of only the external forces acting on all the particles.

Which may be written as

$$M\,(d\mathbf{V}/dt) = \mathbf{F}_{ext}$$

or

$$M\mathbf{a}_{cm} = \mathbf{F}_{ext} \quad ...(5)$$

where $\mathbf{a}_{cm}$ = acceleration of centre of mass.

Equation (5) clearly proves that the *centre of mass of a system of particles moves as though all the mass of the system were concentrated at the centre of mass and all the external forces were applied at the point.*

Again the linear momentum of a system of particles may also be written as

$$\begin{aligned} \mathbf{p} &= \mathbf{p}_1 + \mathbf{p}_2 + \dots \mathbf{p}_n \\ &= m_1\mathbf{v}_1 + m_2\mathbf{v}_2 + \dots + m_n\mathbf{v}_n \\ &= M\mathbf{V}. \end{aligned} \quad ...(6)$$

So, the total momentum of a system of particles is equal to the product of the total mass of the system and the velocity of its centre of mass.

Now differentiating Equation (6) with respect to t we get

$$(d\mathbf{P}/dt) = M\,(d\mathbf{V}/dt) = \mathbf{F}_{ext}$$

Which actually is Newton's second law for a system of particles and may be written as

$$\mathbf{F}_{ext} = (d\mathbf{P}/dt) \quad ...(7)$$

which is a generalization of single particle equation.

$\mathbf{F} = d\mathbf{P}/dt$ to a system of many particles and no mass is entering or leaving the system. It should he noted that the principle of conservation of momentum can be shown to be valid for above system of particles.

Example 1. *The centre of mass of three particles of masses 10, 20 and 30 gm is located at the point (1, 1, 1) in the three dimensional coordinate system. Where should a fourth particle of mass 40 gm be placed so that the centre of mass may shift to the origin (0,0,0)?*

Solution: Let the positions of particles of masses 10 gm, 20 gm and 30 gm be defined by the coordinates (x_1, y_1, z_1), (x_2, y_2, z_2) and (x_3, y_3, z_3) respectively.

$$\bar{x} = \frac{m_1 x_1 + m_2 x_2 + m_3 x_3}{m_1 + m_2 + m_3}$$

$$1 = \frac{10x_1 + 20x_2 + 30x_3}{10 + 20 + 30}$$

$$\therefore \quad x_1 + 2x_2 + 3x_3 = 6 \quad ...(1)$$

Let the fourth mass of 40 gm be placed at (x_4, y_4, z_4) so that the centre of mass becomes at (0, 0, 0), then

$$0 = \frac{10x_1 + 20x_2 + 30x_3 + 40x_4}{10 + 20 + 30 + 40}$$

or $\quad x_1 + 2x_2 + 3x_3 + 4x_4 = 0$

Subtracting (1) from (2), we get

$$4x_4 = -6 \quad \text{or} \quad x_4 = -3/2$$

Similarly, $\quad y_4 = -\frac{3}{2}, \quad z_4 = = -\frac{3}{2}$

So 40 gm mass should be placed at $\left(-\frac{3}{2}, -\frac{3}{2}, -\frac{3}{2}\right)$.

Example 2. *A system consists of particles of masses 5, 2 and 3 gms located at (1, 0, –1), (1, 2, 1) and (1, 1, –3) respectively. Find the centre of position of mass.*

Solution: Position vector **R** is of centre of mass is given by

$$\mathbf{R} = \frac{m_1\mathbf{r}_1 + m_2\mathbf{r}_2 + m_3\mathbf{r}_3}{m_1 + m_2 + m_3}$$

$$= \frac{5\,(\mathbf{i} - \mathbf{k}) + 2\,(\mathbf{i} + 2\mathbf{j} + \mathbf{k}) + 3\,(\mathbf{i} + \mathbf{j} - 3\mathbf{k})}{5 + 2 + 3}$$

$$= \frac{10\mathbf{i} + 7\mathbf{j} - 12\mathbf{k}}{10}$$

which gives the position of centre of mass as **[1, (7/10), (–12/10)]**.

Example 3. *Three particles of mases 1, 2 and 3 gm have its centre of mass at the point (2, 2, 2), what must be the position of fourth particle of mass 4 gm, so that the combined center of mass may now be at the point (0, 0, 0).*

Solution: We have the position vector of the centre of mass of system of three particles as $(2\mathbf{i} + 2\mathbf{j} + 2\mathbf{k})$. Let $\mathbf{r}_1$, $\mathbf{r}_2$ and $\mathbf{r}_3$ be the position vectors of masses 1 gm, 2 gm and 3 gm respectively. So we have

$$2\mathbf{i} + 2\mathbf{j} + 2\mathbf{k} = \frac{1 \times \mathbf{r}_1 + 2 \times \mathbf{r}_2 + 3 \times \mathbf{r}_3}{1 + 2 + 3}$$

or $$\mathbf{r}_1 + 2\mathbf{r}_2 + 3\mathbf{r}_3 = 6\,(2\mathbf{i} + 2\mathbf{j} + 2\mathbf{k})$$

Now, on inclusion of 4th particle the centre of mass of the system becomes zero, *i.e.*, we have

$$0 = \frac{\mathbf{r}_1 + 2\mathbf{r}_2 + 3\mathbf{r}_3 + 4\mathbf{r}_4}{1 + 2 + 3 + 4}$$

or $\mathbf{r}_1 + 2\mathbf{r}_2 + 3\mathbf{r}_3 + 4\mathbf{r}_4 = 0$

or $6\,(2\mathbf{i} + 2\mathbf{j} + 2\mathbf{k}) + 4\mathbf{r}_4 = 0$

Hence, $\mathbf{r}_4 = -\,3\mathbf{i} - 3\mathbf{j} - 3\mathbf{k}$

So the fourth particle must have been located at **(–3, –3, –3)**.

Example 4. *Two particles of mass 6 gm and 3 gm having position vectors (2i + 2j + 2k) and (– 4i – 4j – 4k) are moving with velocities (–i – j) and (4i + 4j) respectively.*

Find the linear momentum of mass of the system of these particles. Also deduce the position of centre of mass.

Solution: Position of centre of mass is given by

$$\mathbf{R} = \frac{m_1\mathbf{r}_1 + m_2\mathbf{r}_2}{m_1 + m_2}$$

or $$\mathbf{R} = \frac{6\,(2\mathbf{i} + 2\mathbf{j} + 2\mathbf{k}) + 3\,(-4\mathbf{i} - 4\mathbf{j} - 4\mathbf{k})}{m_1 + m_2} = 0$$

which shows that centre of mass of the system is at the origin.

Now, the velocity of centre of mass **V** is given by

$$= \frac{m_1\mathbf{v}_1 + m_2\mathbf{v}_2}{m_1 + m_2}$$

or $$\mathbf{V} = \frac{6\,(-\mathbf{i} - \mathbf{j}) + 3\,(4\mathbf{i} + 4\mathbf{j})}{6 + 3} = \frac{6\mathbf{i} + 6\mathbf{j}}{9} = \frac{2}{3}\,\mathbf{i} + \frac{2}{3}\,\mathbf{j}$$

So linear momentum of centre of mass of the system will be

$$\mathbf{P} = M\mathbf{V} = (m_1 + m_2)\,\mathbf{V}$$

$$= \frac{(6 + 3)\,(6\mathbf{i} + 6\mathbf{j})}{9}$$

$$= (6\mathbf{i} + 6\mathbf{j}).$$

Example 5. *Two particles of masses 100 gm and 300 gm have at a given time, position 2i + 5j + 13k and –6i + 4j – 2k cm respectively and velocities 10i – 7j – 3k and 7i – 9j + 6k cm/sec. respectively. Deduce the instantaneous position of the centre of mass, the velocity of second particle in a frame of reference travelling with the centre of mass.*

Solution: (*i*) Centre of mass is given by

$$\mathbf{R} = \frac{m_1\mathbf{r}_1 + m_2\mathbf{r}_2}{m_1 + m_2}$$

$$= \frac{100\,(2\mathbf{i} + 5\mathbf{j} + 13\mathbf{k}) + 300\,(-6\mathbf{i} - 4\mathbf{j} - 2\mathbf{k})}{100 + 300}$$

$$= \frac{-16\mathbf{i} + 17\mathbf{j} + 7\mathbf{k}}{4} \text{ cm}$$

(*ii*) Velocity of centre of mass is given by the relation

$$\mathbf{V} = \frac{m_1 \mathbf{v}_1 + m_2 \mathbf{v}_2}{m_1 + m_2}$$

$$= \frac{100\,(10\mathbf{i} - 7\mathbf{j} - 3\mathbf{k}) + 300\,(7\mathbf{i} - 9\mathbf{j} + 6\mathbf{k})}{100 + 300}$$

$$= \frac{31\mathbf{i} - 34\mathbf{j} + 15\mathbf{k}}{4} \text{ cm/sec.}$$

We have the velocity of the particle in centre of mass frame as

$$= \mathbf{V}_2 - \mathbf{V} = 7\mathbf{i} - 9\mathbf{j} + 6\mathbf{k} - [31\mathbf{i} - 34\mathbf{j} + 15\mathbf{k})/4]$$

$$= \mathbf{(-3i - 2j + 9k)/4 \text{ cm/sec.}}$$

Example 6. *Show that total linear momentum of a system consisting of N number of particles is zero about the centre of mass.*

Solution: Consider *i*th particle of the system whose position vector be $\mathbf{r}_i$ Fig. 3.2. Now, the position vector of the centre of mass of the system is given by the relations

$$\mathbf{R} = \frac{m_1\mathbf{r}_1 + m_2\mathbf{r}_2 + \ldots + m_n\mathbf{r}_n}{m_1 + m_2 + \ldots + m_n}$$

$$= \frac{\sum_{i=1}^{n} m_i\mathbf{r}_i}{\sum_{i=1}^{n} m_i} = \frac{\sum_{i=1}^{n} m_i\mathbf{r}_i}{M}$$

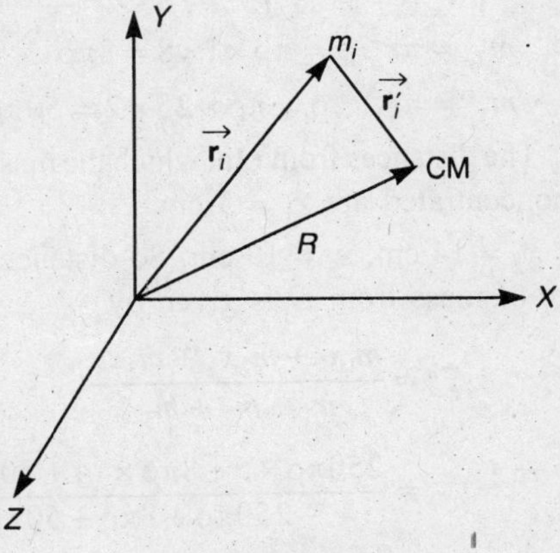

Fig. 3.2

where M = total mass of system. If r_1' be the position vector of the *i*th particle with respect to the centre of mass then from Fig. 3.2.

$$r_i = \mathbf{R} + \mathbf{r}_i'$$

So
$$\mathbf{R} = \frac{\sum_{i=1}^{n} m_i\,(\mathbf{R} + \mathbf{r}_i')}{M}$$

or
$$\mathbf{R} = \frac{\sum_{i=1}^{n} m_i\,\mathbf{R}}{M} + \frac{\sum_{i=1}^{n} m_i\,\mathbf{r}_i'}{M}$$

Also
$$\sum_{i=1}^{n} m_i = M$$

$\therefore$
$$\mathbf{R} = \frac{\sum_{i=1}^{n} m\mathbf{R}}{M} + \frac{\sum_{i=1}^{n} m_i\,\mathbf{r}_i'}{M}$$

or
$$\mathbf{R} = \mathbf{R} + \frac{\sum_{i=1}^{n} m_i\,\mathbf{r}_i'}{M}$$

or
$$\sum_{i=1}^{n} m_i\,\mathbf{r}_i' = 0$$

On differentiating with respect to t, we get

$$\frac{\sum_{i=1}^{n} m_i \, d\mathbf{r}_i'}{dt} = 0$$

or
$$\sum_{i=1}^{n} m_i \, \mathbf{v}_i' = 0$$

where $\mathbf{v}_i'$ is the velocity of ith particle in the centre of mass frame, which proves the relation.

Example 7. *A machine part is shown in the Fig. 3.3. It consists of a 10 cm long and 10 cm diameter solid cyliner attached at one end of a 8 cm long, 2 cm diameter cylindrical rod. At other end of the rod, a 10 cm diameter and 2 cm wide disc is attached. Find the centre of mass of the machine part assuming it to be made of the homogeneous material.*

Solution: We measure the distances from free end of the cylinder. Let ρ be the density (homogeneous) of the system, then the masses of three components are given by

$$m_1 = \pi r^2 l\rho = \pi\rho \times 25 \times 10 = 250\pi\rho$$

$$m_2 = \pi r^2 l\rho = \pi\rho \times 1 \times 8 = 8\pi\rho$$

$$m_3 = \pi r^2 \, t \, . \, \rho = \pi\rho \times 25 \times 2 = 50\pi\rho$$

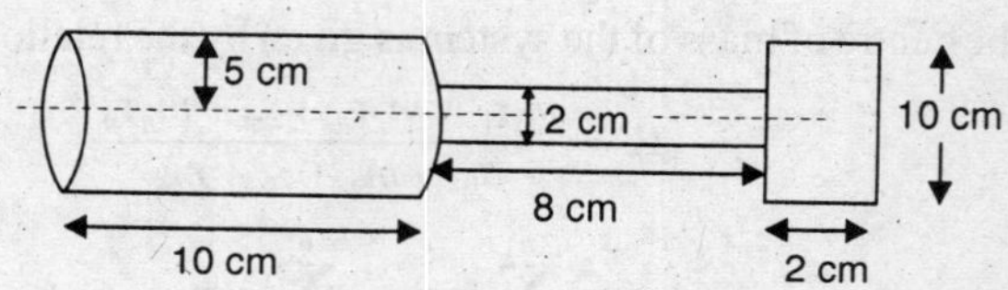

Fig. 3.3

The distances from O at which the masses are concentrated are $x_1 = 5$ cm.

$x_2 = 14$ cm, $x_3 = 19$ cm. So distance of centre of mass from O is given by

$$\bar{x} = \frac{m_1x_1 + m_2x_2 + m_3x_3}{m_1 + m_2 + m_3}$$

$$= \frac{250\pi\rho \times 5 + 8\pi\rho \times 14 + 50\pi\rho \times 19}{250\pi\rho + 8\pi\rho + 50\pi\rho}$$

$= \mathbf{7.5 \ cm.}$

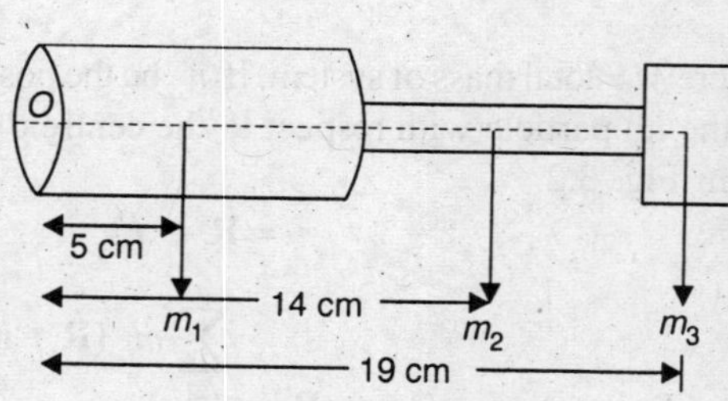

Fig. 3.4

Example 8. *A thin homogeneous lamina is in the form of a circular disc of radius R. From it a circular hole is cut off having exactly half the radius of the lamina and touching the lamina's circumference (see Fig. 3.5). Find the location of centre of mass of the disc.*

Solution: Take the centre of disc O as origin (0, 0), the coordinates of the centre O' of the circular hole are $\left(\frac{R}{2}, 0\right)$. The mass of the disc

$$m \propto R^2$$

So that mass of material removed from the circular hole $m' \left(\frac{R}{2}\right)^2$, $m' = \frac{m}{4}$.

The centre of mass of the remaining portion will be on the line OO' on the other side of O'. The mass m' of the

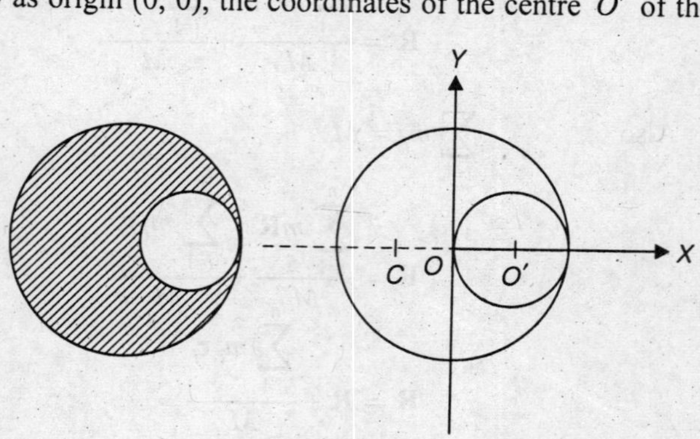

Fig. 3.5

removed portion will be considered negative. So distance of the centre of mass from O is given by

$$OC = \frac{m \times 0 - \frac{m}{4} \times \frac{R}{2}}{m - \frac{m}{4}} = -\frac{\mathbf{R}}{\mathbf{6}}$$

The negative sign shows that C is on the side opposite to that of O'.

Example 9. *Locate the centre of mass of a system of particles of masses m_1 = 1 kg, m_2 = 2 kg and m_3 = 3 kg, situated at the corners of an equilateral triangle of side 1.0 m.*

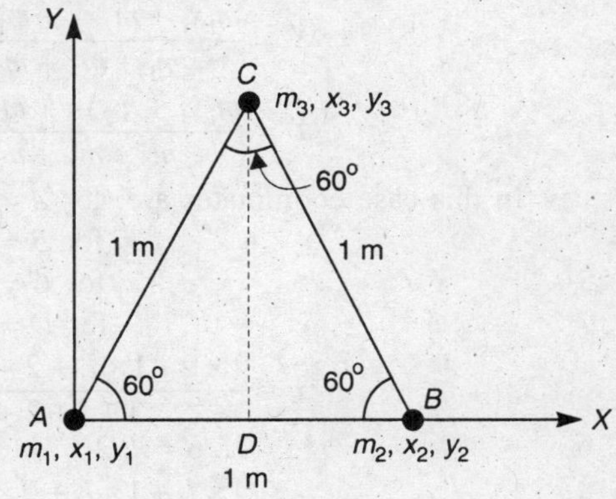

Fig. 3.6

Solution: We have from Fig. 3.6.

The centre of mass is given by

$$x_{cm} = \frac{m_1x_1 + m_2x_2 + m_3x_3}{m_1 + m_2 + m_3}$$

$$y_{cm} = \frac{m_1y_1 + m_2y_2 + m_3y_3}{m_1 + m_2 + m_3}$$

In this case m_1 = 1 kg, m_2 = 2 kg, m_3 = 3 kg

$x_1 = 0,\ y_1 = 0,\ x_2 = 1,\ y_2 = 0;$

$x_3 = 0.5,\ y_3 = \left(\frac{\sqrt{3}}{2}\right)$

So, we have $$x_{cm} = \frac{1 \times 0 + 2 \times 1 + 3 \times 0.5}{1 + 2 + 3} = \frac{3.5}{6} \text{ m}$$

$$y_{cm} = \frac{1 \times 0 + 2 \times 0 + 3\ \sqrt{3}/2}{1 + 2 + 3} = \frac{\sqrt{3}}{4} \text{ m}$$

which gives the coordinates of centre of mass as **(35/6, $\sqrt{3}$/4).**

Example 10. *A circular plate of uniform thickness has a diameter of 56 cm. A circular portion of diameter 42 cm is removed from one edge of the plate as shown in Fig. 3.7. Find out the position of centre of mass of the remaining portion.*

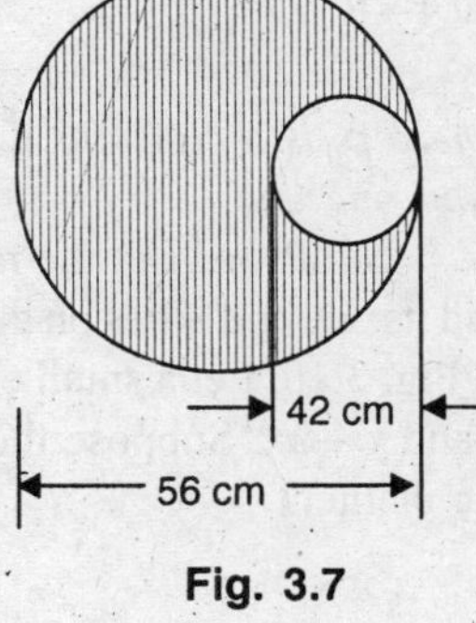

Fig. 3.7

Solution: Redrawing the above figure in Fig. 3.8 let the plate be uniform and O be the centre of the mass of whole plate. Let G_1 be the centre of mass of the cutout circular portion and the centre of mass of the rest out plate lie on the line G_1 O. Let G_2 be the centre of mass of the remaining portion. Since the plates are uniform, weights would be proportional to their areas.

So, Area of whole plate $= \pi\ (28)^2\ cm^2$

and Area of the cutout portion $= \pi\ (21)^2\ cm^2$

Area of the remaining portion $= \pi\ [28)^2 - (21)^2]$

$= \pi\ [(28 + 21)\ (28 - 20)]$

$= 343\ cm^2$

Also $$\frac{\text{Weight of the cutout portion}}{\text{Weight of the remaining portion}} = \frac{\pi \times (21)^2}{\pi \times 343} = \frac{9}{7}$$

Taking moments about O, we have

Weight of the cutout portion × OG_1

= weight of the remaining portion × OG_2

or OG_2 = (9/7) × 7 = **9 cm**

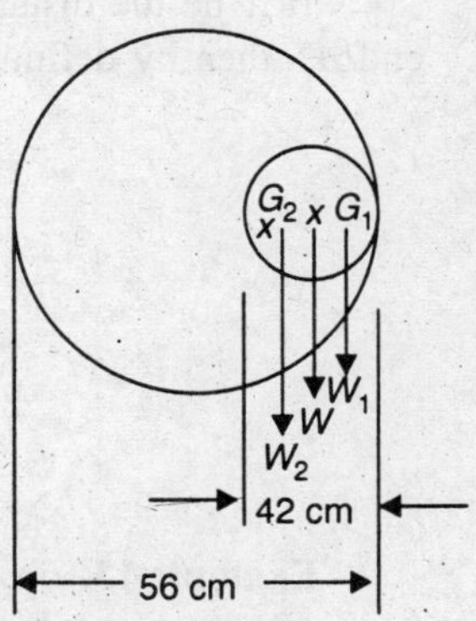

Fig. 3.8

The centre of mass of the remaining portion of the circular plate will be at a distance of 9 cm from the centre of the plate.

Example 11. *Particles of masses 1 kg, 1 kg, 2 kg and 2 kg are placed at the corners A, B, C, D of a square of side L as shown in Fig. 3.9. Find the centre of mass of the system.*

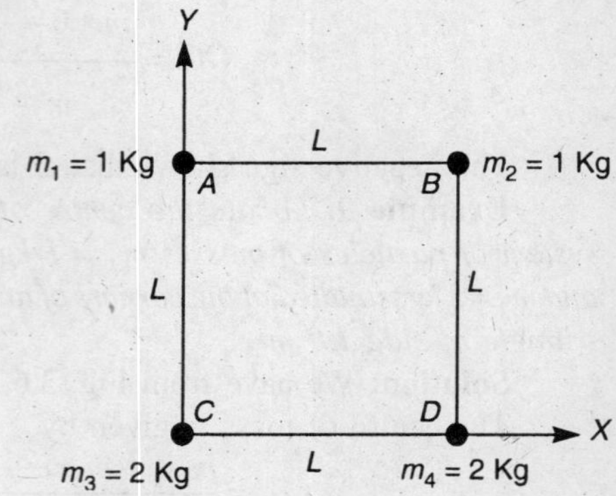

Fig. 3.9

Solution: Centre of mass of the system is given by

$$X = \frac{m_1x_1 + m_2x_2 + m_3x_3 + m_4x_4}{m_1 + m_2 + m_3 + m_4}$$

$$Y = \frac{m_1y_1 + m_2y_2 + m_3y_3 + m_4y_4}{m_1 + m_2 + m_3 + m_4}$$

In this case coordinates are for $A = O, L$
for $B = L, L$
for $C = O, O$
for $D = L, O$

$$\therefore \quad x = \frac{1 \times 0 + 1 \times L + 2 \times 0 + 2 \times L}{1 + 1 + 2 + 2}$$

$$= 3L/6 = L/2$$

$$y = \frac{1 \times L + 1 \times L + 2 \times 0 + 2 \times 0}{1 + 1 + 2 + 2}$$

$$= 2L/6 = L/3$$

So the coordinates of centre of mass are given by
$= L/2, L/3$.

Example 12. *Calculate the centre of mass of a non-uniform rod whose mass per unit length varies as:*

$$\rho = \rho_0 \ (x^2/L)$$

where ρ_0 is a constant, L is the length of rod and x is the distance of any point on rod measured from one end.

Solution: Let the rod be placed along the x-axis and its one end A lies on the origin of the x-axis as shown in Fig. 3.10. Let a small element CD of rod lie between x and $x + dx$. Suppose the dm be the centre of mass of the element.

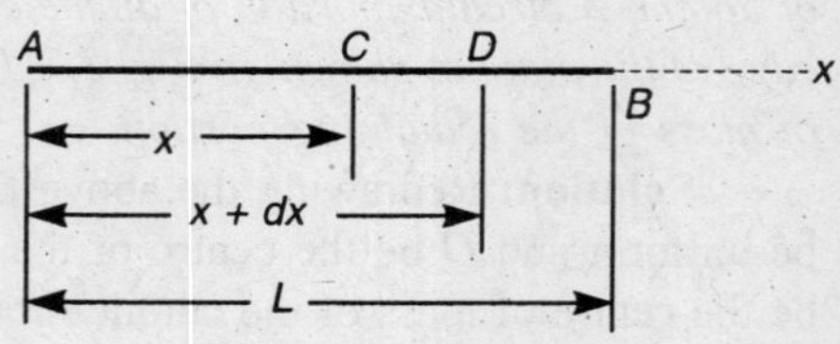

Fig. 3.10

$$\therefore \quad dm = \rho\, dx$$

$$= (\rho_0/L)\, x^2\, dx$$

If X be the distance of the centre of mass from end A, then by definition, we have

$$X = \frac{\int_0^L x\, dm}{\int_0^L dm} = \frac{(\rho_0/L) \int_0^L x^3\, dx}{(\rho_0/L) \int_0^L x^2\, dx}$$

$$X = \frac{L^4/4}{L^3/3} = \mathbf{(3/4)\ L}.$$

Example 13. *Two particles A and B initially at rest, move towards each other under a mutual force of attraction. At the instant when the speed of A is v and the speed of B is 2v, find out the speed of the centre of mass of the system, what happens, if mass of particle A is twice that of particle B.*

Solution: The speed of the centre of mass is given by the relation

$$\mathbf{v}_{cm} = \frac{m_1\mathbf{v}_1 + m_2\mathbf{v}_2}{m_1 + m_2}$$

Here the velocity of m_1 is v and that of m_2 is 2**v**, so, we have

$$\mathbf{v}_{cm} = \frac{m_1\mathbf{v} - m_2.2\mathbf{v}}{m_1 + m_2}$$

$$= \frac{\mathbf{v}\,(m_1 - 2m_2)}{m_1 + m_2}$$

As $m_1 > m_2$, and in case $m_1 \cong 2m_2$

$\therefore$ $\mathbf{v}_{cm} = \mathbf{0}$

i.e., centre of mass is at rest.

Example 14. *The mass of the moon is about 0.013 times the mass of earth and the distance from the centre of the moon to the centre of the earth is about 60 times the radius of the earth. How far is the centre of mass of the earth-moon system from the centre of the earth? Take the earth's radius to be 6400 km.*

Solution: Let the distance of the centre of mass of earth-moon system be r_E from the centre of earth and r_H from the centre of moon. If r be the distance between earth and moon then

$$r_E = \frac{m_M r}{m_E + m_M} = \frac{0.013 m_E r}{m_E + 0.013 m_E}$$

$$= (0.013/1.013)\ (60 \times 6400 \text{ km})$$

$$= \mathbf{4928\ km}.$$

Example 15. (*a*) *Two blocks of masses 5 kg and 2 kg are placed on a frictionless surface and connected by a spring. An external kick gives a velocity 14 m/s to the heavier block in the direction of lights one. Deduce (a) the velocity gained by the centre of mass and (b) the separate velocities of the two blocks in the centre of mass coordinates just after the kick.*

Solution: (*a*) From Fig. 3.11, we have

The velocity of centre of mass $\mathbf{v}_{cm}$ is given by

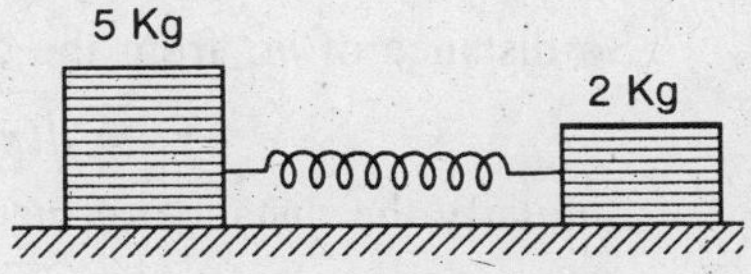

Fig. 3.11

$$\mathbf{v}_{cm} = \frac{m_1\mathbf{v}_1 + m_2\mathbf{v}_2}{m_1 + m_2}$$

In this case $m_1 = 5$ kg, $m_2 = 2$ kg

$\mathbf{v}_1 = 14$ m/s, $\mathbf{v}_2 = 0$

$$\therefore \quad \mathbf{v}_{cm} = \frac{5 \times 14 + 2 \times 0}{5 + 2} = 10 \text{ m/s}.$$

(*b*) It is given that the centre of mass coordinate reference frame is one in which centre of mass is at rest. So the velocity of heavier block in the frame just after the kick is given by

$$\mathbf{v}_1 = \mathbf{v}_2 - \mathbf{v}_{cm} = 14 - 10 = 4 \text{ m/s}$$

and that of lighter block is given by

$$\mathbf{v}_2 = \mathbf{v}_1 - \mathbf{v}_{cm} = 0 - 10 = \mathbf{-10\ m/s}.$$

Example 16. *Two blocks of masses* m_1 *and* m_2 *are connected by a weightless spring of stiffness k rest on a smooth horizontal plane. Block 2 is shifted a small distance x to the left and then released. Find the velocity of the centre of mass of the system after Block 1 breaks the wall.*

Solution: Geometry of problem is shown in Fig. 3.12.

Since *PE* of compression = (1/2) kx^2

When the block m_1 breaks off from the wall, the spring has its unstratched length and the *KE* of block m_2 is given by

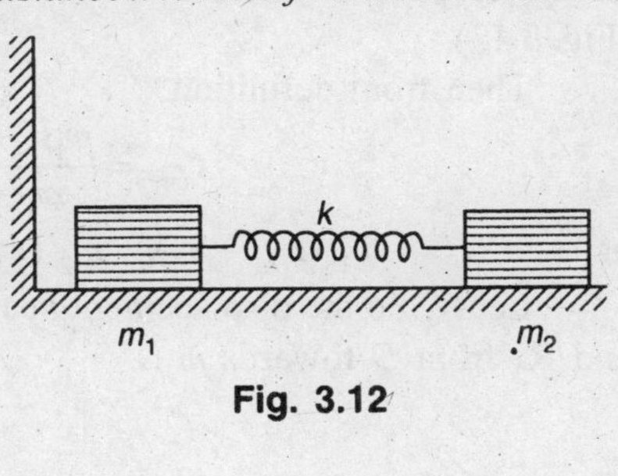

Fig. 3.12

$$(1/2)\ m_2\ v_2^2 = (1/2)\ kx^2$$

Hence, $\mathbf{v}_2^2 = (kx^2/m_2) \quad \therefore \quad \mathbf{v}_2 = x\sqrt{k}/m_2$

for centre of mass $$x_{cm} = \frac{m_1 x_1 + m_2 x_2}{m_1 + m_2}$$

The distance x_1 and x_2 are measured from the wall, so we have

$$\frac{dx_{cm}}{dt} = \frac{m_1}{m_1 + m_2}\frac{dx_1}{dt} + \frac{m_2}{m_1 + m_2}\frac{dx_2}{dt}$$

At start, we have $$\frac{dx_{cm}}{dt} = \frac{m_2 \mathbf{v}_2}{m_1 + m_2} = \left(\frac{m_2 x}{m_1 + m_2}\right)\sqrt{\left(\frac{k}{m_2}\right)}$$

$$= \frac{x\,(\sqrt{km_2})}{m_1 + m_2}.$$

Example 17. (*a*) *Show that the ratio of the distance of two particles from their centre of mass is the inverse ratio of their masses.*

(*b*) *Show that the centre of mass of two particles is on the line joining them at a point whose distance from each particle is inversely proportional to the mass of that particle.*

Solution: (*a*) Centre of mass is given by as

$$x_{cm} = \frac{m_1 x_1 + m_2 x_2}{m_1 + m_2}$$

$$y_{cm} = \frac{m_1 y_1 + m_2 y_2}{m_1 + m_2}$$

$$z_{cm} = \frac{m_1 z_1 + m_2 z_2}{m_1 + m_2}$$

Let the centre of mass lie at the origin.

Then $$x_{cm} = y_{cm} = z_{cm} = 0$$

$$\therefore \quad x_2 = -(m_1 x_1/m_2),\ y_2 = -(m_1 y_1/m_2),\ z_2 = -(m_1 z_1/m_2)$$

So distance of m_1 from the centre of mass (situated at origin) is given by

$$\mathbf{r}_1 = \sqrt{(x_1^2 + y_1^2 + z_1^2)}$$

Similarly the distance of m_2 from the centre of mass (situated at origin) would be

$$\mathbf{r}_2 = \sqrt{(x_2^2 + y_2^2 + z_2^2)}$$

$$= \sqrt{(m_1^2/m_2^2)\,(x_1^2 + y_1^2 + z_1^2)}$$

$$= (m_1/m_2)\sqrt{(x_1^2 + y_1^2 + z_1^2)}$$

$$= (m_1/m_2)\,.\,\mathbf{r}_1$$

Hence, the ratio of distance of m_1 and m_2 from the centre of mass will be given as

$$(\mathbf{r}_1/\mathbf{r}_2) = (m_2/m_1).\ \mathbf{r}_1$$

Hence the result proved.

(*b*) Let the position vectors of masses m_1 and m_2 be $\mathbf{r}_1$ and $\mathbf{r}_2$ respectively and that of centre of mass D be $\mathbf{r}_{cm}$ (Fig. 3.13).

Then from definition

$$\mathbf{r}_{cm} = \frac{m_1 \mathbf{r}_1 + m_2 \mathbf{r}_2}{m_1 + m_2}$$

or, $$m_1(\mathbf{r}_{cm} - \mathbf{r}_1) = m_2(\mathbf{r}_2 - \mathbf{r}_{cm}) \quad ...(1)$$

Let $\mathbf{R}_1$ be the displacement vector from D towards m_1 and $\mathbf{R}_2$ from D towards m_2.

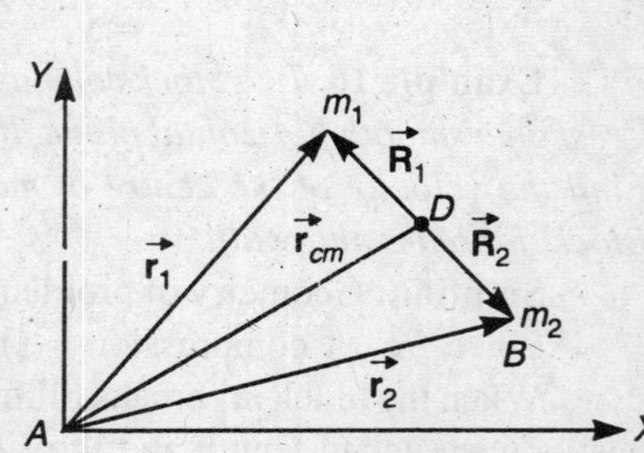

Fig. 3.13

$$\therefore \qquad \mathbf{R}_1 = \mathbf{r}_1 - \mathbf{r}_{cm} \qquad ...(2)$$

and
$$\mathbf{R} = \mathbf{r}_2 - \mathbf{r}_{cm} \qquad ...(3)$$

Then using Equations (1), (2) and (3), we get

$$-m_1 \mathbf{R}_1 = m_2 \mathbf{R}_2$$

which means $\mathbf{R}_1$ and $\mathbf{R}_2$ are oppositely directed. In other words the centre of mass lies on the line joining m_1 and m_2.

Also we have form above equations

$$m_1 \mathbf{R}_1 = m_2 \mathbf{R}_2 = \text{constant}$$

or
$$\mathbf{R}_1 \propto (1/m_1), \text{ and } \mathbf{R}_2 \propto (1/m_2)$$

i.e., the distance of the centre of mass from each particle is inversely proportional to the mass of the particle.

Example 18. *Locate the centre of mass of the machine part, consisting of a disc of 4 cm in diameter and 2 cm long, and a cylindrical rod 2 cm in diameter and 12 cm long coaxial with the disc, both made of a homogeneous material.*

Solution: By symmetry the centre of mass lies on the axis (Fig. 3.14) and the centre of mass of each part is midway between its end.

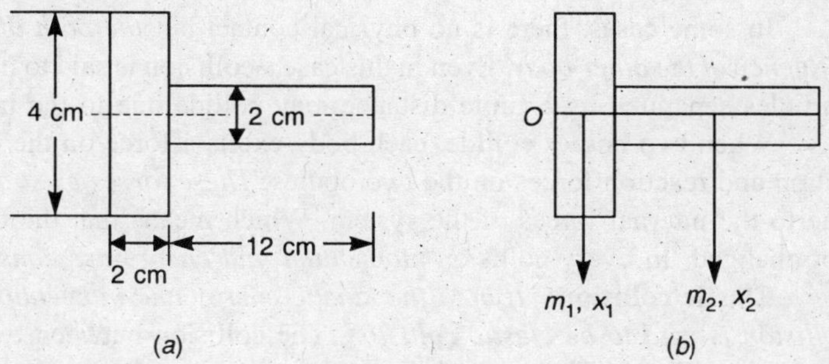

Fig. 3.14

The volume of the disc is 8π cm^3 and that of rod is 12π cm^3. Since the material is homogeneous, the masses of the two parts are proportional to their volumes, *i.e.*, we have

$$\therefore \qquad (m_1/m_2) = (8\pi/12\pi) = (2/3)$$

$$\therefore \qquad m_2 = (3m_1/2)$$

If the origin is taken at the left face of the disc on the axis then $x_1 = 1$ cm and $x_2 = 8$ cm (Fig. 3.14), then centre of mass is given by

$$x_{cm} = \frac{m_1 x_1 + m_2 x_2}{m_1 + m_2} = \frac{m_1 \text{ (1 cm)} + (3/2)\, m_1 \text{ (8 cm)}}{m_1 + (3/2)\, m_2} = \mathbf{5.2 \text{ cm.}}$$

Example 19. *Find the centre of mass of a homogeneous semicircular plate. Let a be the radius of the circle.*

Solution: Geometry of problem is that shown in Fig. 3.15.

The centre of mass must lie on the axis of symmetry *i.e.*, on the line AB perpendicular to the flat base of the semi-circle and passing through the centre of base. Select the base along the y-axis so that the line AB is along x-axis, and let A be the origin.

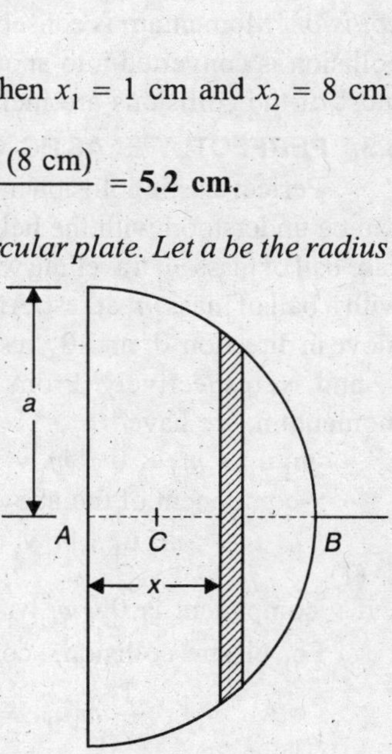

Fig. 3.15

For continuous mass distribution, we have

$$x_{cm} = (1/M) \int x\, dm$$

Let σ be the surface density, *i.e.*, mass/unit area, so that

$$M = (\pi a^2 \sigma)/2$$

Consider a rectangular strip of width dx at distance x from the origin and parallel to y-axis (as given by shaded part)

Area of strip $= 2\sqrt{(a^2 - x^2)}\, dx$

$$dm = 2\sigma \sqrt{(a^2 - x^2)}\, dx$$

Also, $\int x\,dm = 2\sigma \int_0^a x\sqrt{(a^2 - x^2)}\,dx$

Let $x = a\sin\theta,\ dx = a\cos\theta\,d\theta$

$\therefore\quad \int x\,dm = 2\sigma a^2 \int_0^{\pi/2} \cos^2\theta\sin\theta\,d\theta$

$= (2\sigma a^3/3)$

Hence, $x_{cm} = -(2\alpha a^3/3)/(\pi a^2\sigma/2) = \mathbf{(4/3)\ (a/\pi)}$

So, $y_{cm} = 0$

So, the centre of mass lies at a distance $(4a/3\pi)$ from the flat base along the line of symmetry.

3.2. COLLISIONS

A collision is said to take place when two bodies physically collide with each other. For example when a ball collides with another ball a collision is said to occur.

In some cases, there is no physical contact but *the path of one body is changed due to the influence of the other body.* Even in this case a collision is said to have taken place, e.g., two charged particles separated by a finite distance may collide due to the interaction of their electric fields.

When two bodies collide, each body exerts a force on the other and these forces are due to action and reaction forces on the two bodies. *These forces exist only for a very short time* and are due to the internal forces of the system. Which means that the total momentum of the system is not changed. In every collision *momentum and energy are conserved.*

Elastic collision: *If both the kinetic energy and momentum are conserved in collision, the collision is said to be elastic collision.* The collision between two glass or quartz balls is nearly an elastic collision. The collision between atomic or sub-atomic particle is also an elastic collision.

Inelastic collision: *If the kinetic energy is not conserved in a collision, it is said to be inelastic collision.* Momentum is conserved even in an inelastic collision. The kinetic energy lost in inelastic collision is converted into another form of energy like heat energy, sound energy etc. In daily life most of the collisions are inelastic collisions.

3.3. PERFECTLY ELASTIC COLLISION

Perfect elastic collision in one or two dimension can be understood with the help of Fig. 3.16. In this case ball of mass m_1 travelling with velocity v_1 collides with a ball of mass m_2 at rest. After collision, the balls move in direction θ_1 and θ_2 respctively with velocity v_1 and v_2 respectively. From conservation law of momentum, we have

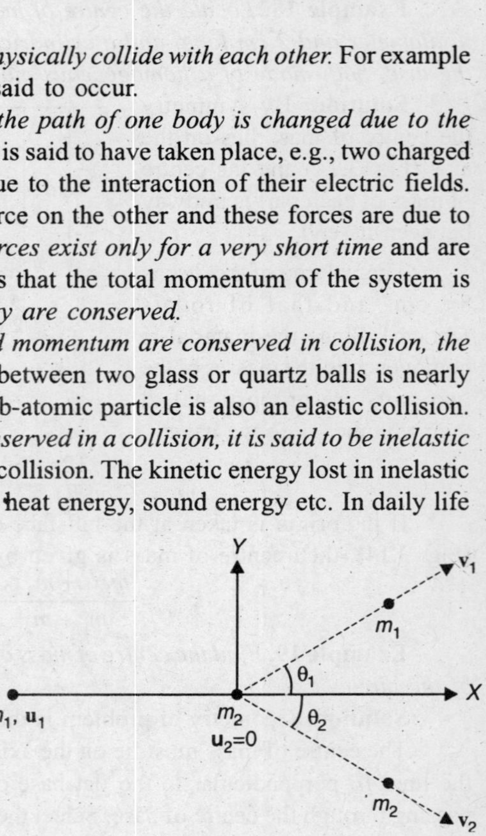

Fig. 3.16

$m_1\mathbf{u}_1 + m_2 \times 0 = m_1\mathbf{v}_1 + m_2\mathbf{v}_2$

x-component of the above equation gives

$$m_1\mathbf{u}_1 = m_1\mathbf{v}_1\cos\theta_1 + m_2\mathbf{v}_2\cos\theta_2 \quad ...(1)$$

and y component is $0 = m_1\mathbf{v}_1\sin\theta_1 - m_2\mathbf{v}_2\sin\theta_2$ ...(2)

For elastic collision, conservation of kinetic energy gives

$$\frac{1}{2}m_1\mathbf{u}_1^2 = \frac{1}{2}m_1\mathbf{v}_1^2 + \frac{1}{2}m_2\mathbf{v}_2^2 \quad ...(3)$$

In case of *one dimension collision,* $\theta_1 = \theta_2 = 0$.

$\therefore$ The above momentum Eqn. 1 reduces to $m_1\mathbf{u}_1 = m_1\mathbf{v}_1 + m_2\mathbf{v}_2$ and K.E. conservation gives

$$m_1\mathbf{u}_1^2 = m_1\mathbf{v}_1^2 + m_2\mathbf{v}_2^2$$

which on simplification gives $\mathbf{v}_1 = \dfrac{(m_1 - m_2)\,\mathbf{u}_1}{(m_1 + m_2)}$ ...(4)

and $$\mathbf{v}_1 = \frac{2m_1\mathbf{u}_1}{(m_1 + m_2)} \qquad ...(5)$$

If mass of m_2 is much heavier than m; *i.e.*, $m_2 >> m_1$, in that case

$$v_1 = -\mathbf{u}_1 \text{ and } v_2 = 0$$

In such a situation heavier ball will not move at all while lighter one rebounds on collision with a final velocity nearly equal in magnitude but opposite in direction to the initial velocity.

3.4. PERFECTLY INELASTIC COLLISION

A collision is said to be perfectly inelastic if two colliding bodies stick to each other after collision. There is maximum loss of kinetic energy in perfectly elastic collision but **momentum is conserved.** It may be noted that the kinetic energy is not lost but changed to some other form. For example (*i*) The collision between a bullet and a block of wood such that the bullet remains embedded in the wood (*ii*) collision between positively and negatively charged particles.

When two elastic bodies moving along the same straight line collide and continue to move in a straight line after the collision, the collision is said to be *one dimensional or head on elastic collision.*

Let us consider a general case when two bodies of masses m_1 and m_2 are travelling with velocities $\mathbf{u}_1$ and $\mathbf{u}_2$ w.r.t. each other before collision. After collision the bodies stick together and move with a velocity $\mathbf{v}$.

Then from the law of conservation of momentum,

$$m_1\mathbf{u}_1 + m_2\mathbf{u}_2 = (m_1 + m_2)\,\mathbf{v}$$

K.E. of the system before collision is

$$k_i = \frac{1}{2}m_1\mathbf{u}_1^2 + \frac{1}{2}m_2\mathbf{u}_2^2$$

K.E. of the system after collision is

$$k_f = \frac{1}{2}(m_1 + m_2)\,\mathbf{v}^2$$

In case $\mathbf{u}_2 = 0$, *i.e.*, the second body was at rest before collision

$$k_i = \frac{1}{2}m_1\mathbf{u}_1^2$$

$$k_f = \frac{1}{2}(m_1 + m_2)\,\mathbf{v}^2$$

and $$m_1\mathbf{u}_1 = (m_1 + m_2)\,\mathbf{v}$$

$$\therefore \qquad \frac{k_f}{k_i} = \frac{\frac{1}{2}(m_1 + m_2)\,\mathbf{v}^2}{\frac{1}{2}m_1\mathbf{u}_1^2} = \frac{m_1}{m_1 + m_2}$$

which means that $k_f << k_i$ *i.e.*, the inelastic collision causes a loss of kinetic energy.

Example 20. *A lead bullet just melts when stopped by an obstacle. Assuming that 25% of the heat is absorbed by the obstacle, find the velocity of the bullet if its initial temperature is 27°C. (Melting point of lead = 327°C, Specific heat of lead = 0.03 calories/gm C., Latent heat of fusion of lead = 6 calories/gm, J = 4.2 joules/calorie).*

Solution: Let m be the mass of the lead bullet. The heat absorbed by the bullet is given by $m \times 0.03 \times 3.00 + m \times 6$

$$= m\,(0.03 \times 300 + 6) = 15 \text{ m cal.} \qquad ...(1)$$

If v be the velocity of the lead bullet, the K.E. dissipated

$$= 1/2\ mv^2 \text{ Joules} = \frac{(1/2)\,mv^2}{4.2 \times 10^7} \text{ cal}$$

As 25% of the heat is absorbed, the heat used to melt the bullet would be given by From equation (1) and (2), v can be calculated as under:

$$v = 4.1 \times 10^2 \text{ ms}^{-1} \textbf{ Ans}$$

$$\text{K.E. dissipated} = \frac{3}{4} \times \frac{(1/2)\, mv^2}{4.2 \times 10^7} = \frac{3}{8} \times \frac{mv^2}{4.2 \times 10^7} \quad ...(2)$$

Example 21. *A ball is thrown vertically downward from a height of 10 meter with an initial velocity V_0. It collides with the ground, loses 50 per cent of its energy in collision and rebounds to the same height. Find (i) the initial velocity v_0 (ii) the height to which the ball would rise after the collision if the initial velocity v_0 were directed upward.*

Solution: (*i*) Let the initial downward velocity be V_0 and v be the velocity just before collision.

We have $h = 10$ m and $g = 9.8$ m/s^2

Using the relation, $v^2 - u^2 = 2gh$

or $v^2 = V_0^2 + 2 \times 9.8 \times 10$

or $v^2 = V_0^2 + 196$

Hence, K.E. of ball $E = \frac{1}{2} m\, (V_0^2 + 196)$

Since, it loses 50% of its energy, so its energy after collision = $(1/4)\, m\, (V_0^2 + 196)$

The velocity after the collision is given by

$$\left(\frac{2E}{m}\right)^{1/2} = \left(\frac{(V_0^2 + 196)}{2}\right)^{1/2}$$

which is the initial velocity for the second case.

Also final velocity = 0, $g = -9.8$ m/s^2, $h = 10$ m.

Using, $v^2 - u^2 = 2gh$, we get

or $0 - [(V_0^2 + 196)/2] = 2 \times (-9.8) \times 10$

or $V_0 = \pm\ 14$ m/sec

(*ii*) When the ball is directed upward, in that case

$$u = +14 \text{ m/sec},$$
$$g = -\ 9.8 \text{ m/sec}^2$$
$$h = -10 \text{ m}$$

$\therefore$ $u^2 - (14)^2 = 2\,(-9.8) \times (-10)$

or $v^2 = 392$

Since, 50% of its energy is lost during collisions, the velocity with which it rises back is

$$v = 196 \qquad \therefore\ u = 14 \text{ ms}^{-1}$$

$\therefore$ initial velocity, $u = 14$ m/sec, final velocity $v = 0$.

Once again using $v^2 - u^2 = 2gh$

or $0 - (14)^2 = 2 \times (-9.8)\, h$

$\therefore$ $h = 10$ m

So, the height upto which ball rises = **10 m.**

Example 22. *A particle of mass m_1 and moving with velocity u_1 is elastically scattered from another particle of mass m_2 at rest. After the collision, the two particles move in opposite directions with the same speed. Find the mass of the target in terms of the mass of incident particle.*

Solution: Momentum of incident particle = $m_1\, u_1$

K.E. of incident particle $= \frac{1}{2} m_1 \mathbf{u}_1^2$

As target is at rest before collision, so its K.E. and mometum are zero before collision.

After the collision, the particle is scattered with velocity $\vec{v}_1$ and target recoils with velocity $\vec{v}_2$ such that they are equal and opposite.

$$\therefore \quad \vec{v}_2 = -\ \vec{v}_1 \quad ...(1)$$

As the collision is elastic, linear momentum and K.E. will be conserved.

Conservation of momentum gives

$$m_1 \vec{u_1} = m_1 \vec{v_1} + m_2 v_2$$

$$= (m_1 - m_2) \vec{v_1} \quad \text{...(2) [From Equation (1)]}$$

For a heavy target $m_2 > m_1$ the R.H.S. of Equation (2) is negative so that v_1 is opposite to $\vec{v_1}$ *i.e.*, the particle is scattered backward while the target moves forward with velocity.

Taking only magnitudes, we have

$$|\vec{v_1}| = \frac{m_1}{(m_1 - m_2)} |\vec{u_1}|$$

Conservation of K.E. gives

$$\frac{1}{2} m_1 \vec{u_1}^2 = \frac{1}{2} m_1 \vec{v_1}^2 + \frac{1}{2} m_2 \vec{v_1}^2$$

or

$$m_1 \vec{u_1}^2 = (m_1 - m_2) \vec{v_1}^2$$

$$= m_1 + m_2 \frac{m_1^2}{(m_1 - m_2)^2} \vec{u_1}^2 \quad \text{[From Equation (3)]}$$

or

$$(m_1 - m_2)^2 = m_1 (m_1 + m_2)$$

or

$$m_1^2 + m_2^2 - 2m_1 m_2 = m_1^2 + m_1 m_2$$

or

$$m_2^2 = 2m_1 m_2$$

or

$$m_2 = \mathbf{3m_1}$$

i.e., target is three times heavier than the incident particle.

Example 23. (*a*) *Two masses of 1 gm and 4 gm are moving with equal kinetic energies. Find out the ratio of the magnitudes of their linear momentum.*

(*b*) *A body of mass 2 kg is being dragged with a uniform velocity of 2 m/sec on a rough horizontal plane. The coefficient of friction between the body and the surface is 0.20 J = 4.2 Joules/cal and g = 9.8 m/ sec². Calculate the amount of heat generated in 5 sec.*

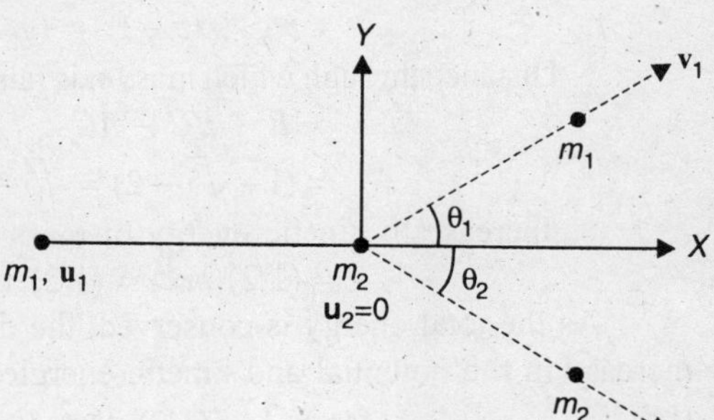

Fig. 3.17

Solution: (*a*) Since K.E.'s are equal so

$$\frac{1}{2} m_1 v_1^2 = \frac{1}{2} m_2 v_2^2$$

where m_1, m_2 are masses of bodies and v_1 and v_2 their respective velocities.

or

$$\frac{v_1^2}{v_2^2} = m_2/m_1 = 4/1$$

or

$$v_1 = 2v_2$$

The ratio of magnitude of their linear momentum

$$= \frac{m_1 v_1}{m_2 v_2}$$

$$= (1 \times 2v_2/4v_2) = \frac{1}{2}$$

$$= \mathbf{1 : 2}$$

(*b*) Now, $F = \mu R$

where F is force of friction, R is normal reaction and μ is coefficient of friction between the body and the surface.

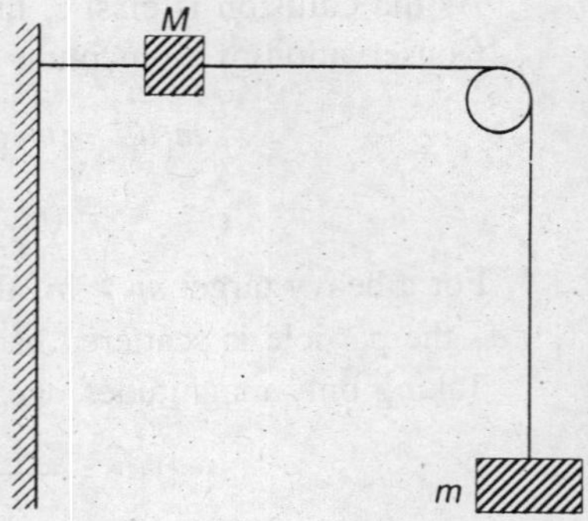

Fig. 3.18

So, work done against the force of friction in one second is given by

$$W = FS = \mu RS = \mu mgS$$
$$= 0.20 \times 2 \times 9.8 \times 10 = 39.2 \text{ J}$$

Thus, work done in 5 seconds

$$= 5 \times 39.2 \text{ J}$$

$\therefore$ Heat generated $= \left(\frac{39.2 \times 5}{4.2}\right) =$ **46.66 cal.**

Example 24. *A string with one end fixed on a rigid wall, passing over a fixed frictionless pulley at a distance of 2 m from the wall, has a point mass M = 2 kg attached to it at a distance of 1 m from the wall. A mass m = 0.5 kg attached at the free end is held at rest so that the string is horizontal between the wall and the pulley and vertical beyond the pulley. What will be the speed with which the mass M will hit the wall when the mass is release?*

Solution: When mass m is released from rest, the heavier mass M loses potential energy and it strikes the wall with velocity v and the string remains stretched as shown in Fig. 3.19. It increases the kinetic energy of mass M and increases the kinetic and potential energies of mass m.

Decrease in potential energy of mass m

$$Mg \times 1 = 2.0 \times g \times 1 = 2g$$

where $g =$ **9.8 ms^{-2}**.

Increase in kenetic energy of mass M

$$= (1/2)\, Mv^2 = (1/2) \times 2.0 \times v^2 = v^2$$

Increase in potential energy of mass m

$$= mg\ (\sqrt{5} - 1) = (g/2)\ (\sqrt{5} - 1)$$

$\therefore$ Distance through which mass m is raised is given by

$$= AB + BC - AC$$
$$= (1 + \sqrt{5} - 2) = \sqrt{5} - 1 \text{ metre}$$

$\therefore$ Increase in kinetic energy of mass m

$$= (1/2)\, mv^2 = (1/2)\, v^2 = v^2/4$$

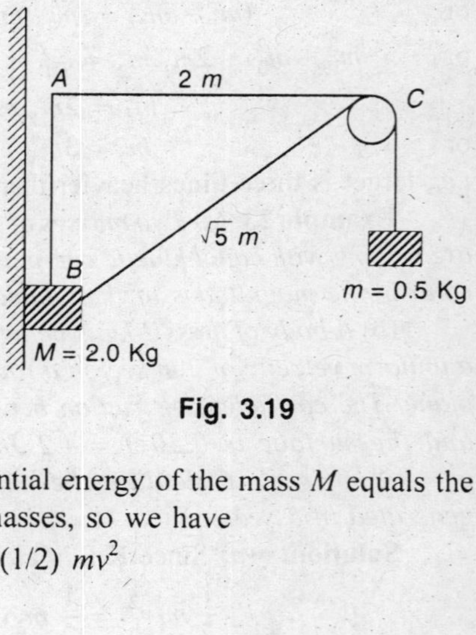

Fig. 3.19

As the total energy is conserved, the decrease in potential energy of the mass M equals the increase in the potential and kinetic energies of both the masses, so we have

$$\therefore \quad Mg \times 1 = (1/2)\, Mv^2 + mg\ (\sqrt{5} - 1) + (1/2)\, mv^2$$

$$\therefore \quad 2g = v^2 + (g/2)\ (\sqrt{5} - 1) + v^2/4$$

or $$5v^2/4 = g\ [2 - (\sqrt{5} - 1)/2)]$$

or $$5v^2/4 = (9.8/2)\ (3 - \sqrt{5})$$

or $$v = \mathbf{3.3\ ms^{-1}}.$$

Example 25. *Particles P and Q of mass 20 g and 40 g respectively are simultaneously projected from points A and B on the ground (Fig. 3.20). The initial velocities of P and Q make angles of 45° and 135° respectively with the horizontal AB as shown in the diagram. The separation AB is 245 m. Both particles have the same initial speed of 49 ms^{-1}. Both particles travel in the same vertical plane and undergo a collision.*

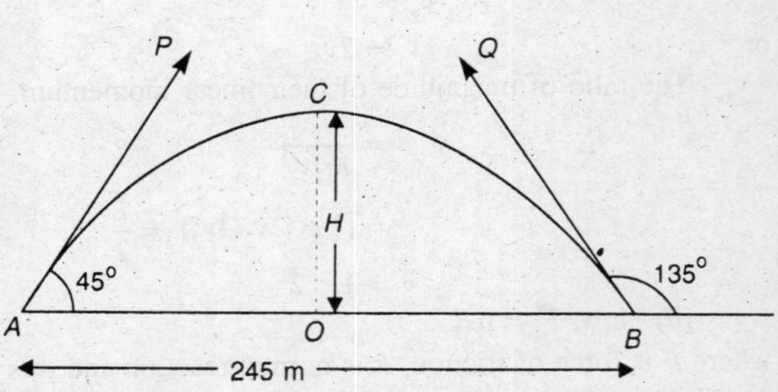

Fig. 3.20

After the collision, P retraces its path. Determine the position of Q when it hits the ground. How much time after the collision does the particle Q take to reach the ground? Take g = 9.8 ms^{-2}.

Solution: As shown in Fig. 3.20 that both P and Q have the same projection velocity and the same inclination of 45° to the horizontal. The range R of both is given by

$$R = \frac{u^2 \sin 2\theta}{g}$$

Since $u = 49$ ms^{-1} and $\theta = 45°$, we have

$$R = \frac{(49)^2 \sin 90°}{9.8} = 5 \times 49$$

$$= 245 \text{ m}$$

Also both P and Q will meet exactly midway between AB. When they meet, the horizontal distance travelled by each is 245/2 m. If the particles meet t_1 seconds after they were projected, then

$$\frac{245}{2} = (u \cos\theta) \times t_1$$

$$= 49 \cos 45° \times t_1$$

$$t_1 = \frac{245}{2} \times \frac{1}{49} \times \sqrt{2} = \frac{5}{\sqrt{2}} \text{ s}$$

At the moment P and Q meet, both are at the highest point in their respective parabolic paths. Their vertical component of velocity is zero and the horizontal component of velocity is $u \cos 45° = \frac{49}{\sqrt{2}}$ ms^{-1}. Also, the directions of the horizontal component of velocities of P and Q are opposite to each other. At this moment P and Q collide, and so we have

∴ Total momentum of P and Q before collision

$$= 20 \times 10^{-3} \times \frac{49}{\sqrt{2}} - 40 \times 10^{-3} \times \frac{49}{\sqrt{2}}$$

$$= 20 \times 10^{-3} \times \frac{49}{\sqrt{2}} \text{ kg ms}^{-1}$$

After collision, article P retraces its path, *i.e.*, the velocity of P reverses. So, the velocity of P after collision will be $\frac{-49}{\sqrt{2}}$ ms^{-1}. Let V be the velocity of Q after collision. In that case

Total momentum of P and Q after collision

$$= -20 \times 10^{-3} \times \frac{49}{\sqrt{2}} + 40 \times 10^{-3} \text{ V}$$

From conservation of momentum law, we get

$$-20 \times 10^{-3} \frac{49}{\sqrt{2}} = -20 \times 10^{-3} \times \left(\frac{49}{\sqrt{2}}\right) + 40 \times 10^{-3} \text{ V}$$

Which gives $V = 0$.

After collision, particle Q is at rest at the highest point of its parabolic path, *i.e.,* at point C. Let H be the maximum height attained by each particle during their parabolic paths. So we have

$$H = (u \sin\theta)\, t_1 - \frac{1}{2} g t_1^2$$

$$H = \frac{49}{\sqrt{2}} \times \frac{5}{\sqrt{2}} - \frac{1}{2} \times 9.8 \times \left(\frac{5}{\sqrt{2}}\right)^2$$

$$= 61.25 \text{ m}$$

If Q reaches ground t_2 seconds after collision, then t_2 is the time taken by it to fall freely through a vertical distance of 61.25 m, starting with initial velocity as zero, so we have

$$61.25 = 0 \times t_2 + 1/2 \times 9.8 \times t_2^2$$

$$\therefore \quad t_2^2 = \frac{61.25 \times 2}{9.8}$$

$$\therefore \quad t_2 = \mathbf{3.535.}$$

Example 26. *A bullet of mass M is fired with a velocity 50 m/s at an angle θ with the horizontal. At the highest point of its trajectory. It collides head-on with a bob of Mass 3 M suspended by a massless string of length 10/3 m and gets embedded in the bob. After the collision, the string moves through an angle of 120°.*

Find:

(i) the angle θ, and

(ii) the vertical and horizontal coordinates of the initial position of the bob with respect to the point of firing of the bullet. (Take g = 10 m/s²).

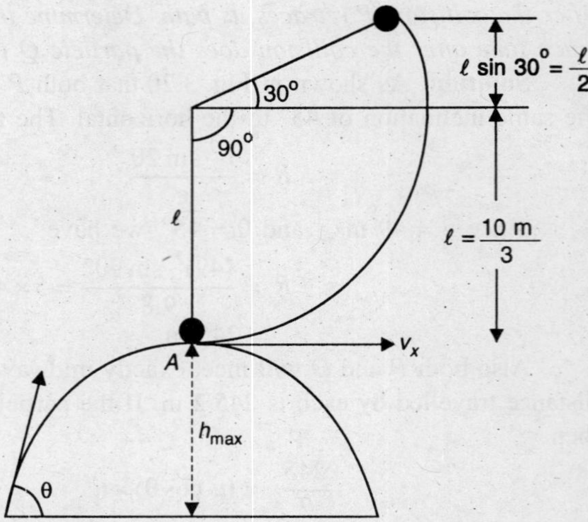

Fig. 3.21

Solution: (*i*) From Fig. 3.21, as the bullet reaches the highest point *A*, its horizontal component of velocity will be given by

$$v_x = v \cos \theta \text{ (where } v = 50 \text{ m/s)}$$

∴ Kinetic energy imparted to the system

$$= (1/2)\ Mv_x^2 = (1/2)\ Mv^2 \cos^2 \theta$$

As a result, the combined system (*i.e.*, bob and bullet) of mass $M + 3M = 4M$ is raised to a vertical height.

$$h = l + l/2 = (3l/2) = (3/2) \times (10/3) = 5 \text{ m}$$

∴ Potential energy gained by the system

$$= 4Mgh = 4Mg \times 5 = 20Mg$$

From the law of conservation of energy, we have

$$(1/2)\ Mv^2 \cos^2 \theta = 20Mg$$

$$\therefore \quad \cos \theta = \frac{\sqrt{40g}}{v} = \frac{\sqrt{40 \times 10}}{v} = 0.4$$

or $\theta = \mathbf{66.4^\circ}$

(*ii*) Vertical component of the initial position *A* of the bob with respect to the initial point of projection.

$$= h_{max} = v^2 \sin \theta/(2g) = [(50)^2 \,.\, \sin^2 66.6^\circ]/(2 \times 10)$$
$$= 105 \text{ m}$$

Horizontal component of the bob

$$= (1/2) \text{ (horizontal range) } v^2 \sin 2\theta/2g$$
$$= (50)^2 \,.\, \sin (132.8^\circ)/(2 \times 10)$$
$$= \mathbf{91.7 \text{ m}}$$

Example 27. *The bob of a simple pendulum of length 980 cm is released from rest with its string making an angle of 60° with the vertical. It collides with a ball of mass 20 gm resting on a smooth horizontal surface just at the position of rest of the pendulum. With what velocity will the bob move immediately after the collision? What will be the maximum angular displacement of the bob after the collision? The mass of the bob is 10g.*

Solution: We have from the Fig. 3.22.

$$h = l(1 - \cos\theta)$$
$$= 980 - (980/2)$$
$$= 490 \text{ cm.}$$

Let the velocity of bob before collision be u_1. Then kinetic energy at the bottom would be same as that of potential energy at the start, which may be written as

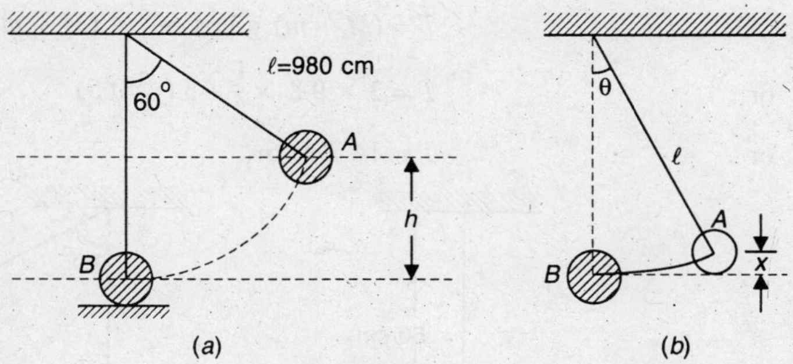

Fig. 3.22

$$\frac{1}{2} m_1 u_1^2 = m_1 gh$$

or $$u_1^2 = 2gh = 2 \times 980 \times 490$$

or $$u_1 = 980 \text{ cm/sec.}$$

Let the velocities of A and B after collision be v_1 and v_2. From the conservation of momentum we have

Momentum before collision = Momentum after collision

or $$m_1 u_1 + 0 = m_1 v_1 + m_2 v_2$$

or $$10 \times 980 = 10v_1 - 20v_2$$

or $$980 = v_1 + 2v_2$$

According to Newton's experimental law

$$v_1 - v_2 = -(u_1 - u_2) = -980$$

From above two equations, one gets

$$v_1 = -(980/3) \text{ cm/sec.}$$

The negative sign shows that the bob gets rebound. Now let θ be the maximum angular displacement of the bob after collision as shown in Fig. 3.22.

In this case x is the maximum vertical height to which the bob is raised.

Now K.E. of bob after impact = P.E. of the bob at top.

$$\frac{1}{2} mv_1^2 = m_1 gx$$

or $$\frac{1}{2} v_1^2 = g(l - l\cos\theta)$$

or $$\frac{1}{2}(980/3)^2 = 980(980 - \cos\theta)$$

$\therefore$ $$\cos\theta = 17/18$$

or $$\boldsymbol{\theta = \cos^{-1}(17/18).}$$

Example 28. *A mass of 2.9 kg is suspended from a string of length 50 cm and is at rest. Another body of mass 100 gm, which is moving horizontally with a velocity of 150 m/sec strikes and sticks to it.*

(*i*) *What is the tension in the string when it makes an angle of 60° with the vertical?*

(*ii*) *Will it complete a vertical circle?*

Solution: (*i*) We have from the Fig. 3.23 the law of conservation of linear momentum, as

$$mu = (m + M)v$$

or, $$v = mu/(m + M) = (0.1 \times 150)/(0.1 + 2.9) = 5 \text{ m/sec.}$$

The centriptal force $= (M + m) v_1^2/l = T - (m + M) g \cos 60^\circ$

where v_1 is the velocity of combined mass at the instant when the string makes an angles 60° with the velocity.

$$T = (M + m)\ g \cos 60^\circ + (M + m)\ (v_1^2/l)$$

or $$T = 3 \times 9.8\ \times \frac{1}{2} + 3\ (v_1^2/0.5)$$

or $$T = 14.7 + 6v_1^2$$

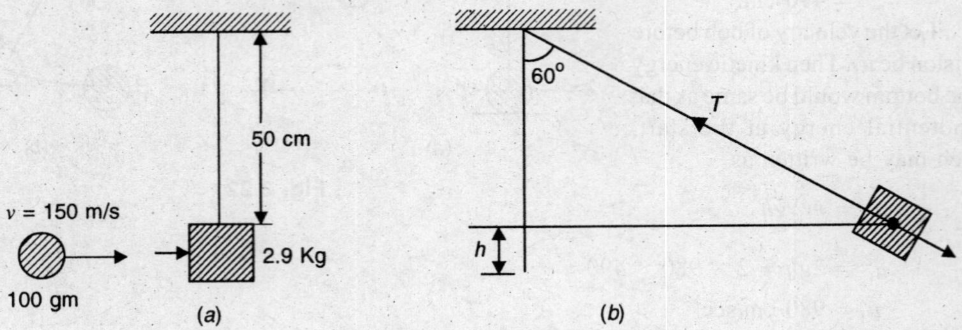

Fig. 3.23

From conservation of mechanical energy we have kinetic energy at the lowest point = (K.E. + P.E.) at given point

or $$\frac{1}{2}\ (m + M)\ v^2 = \frac{1}{2}\ (m + M) + (M + m)\ gh$$

[In this case $h = l\ (1 - \cos\theta)$

or $$\frac{1}{2}\ (m + M) = \frac{1}{2}\ (m + M)\ v_{12} + (M + m)\ gl\ (1 - \cos\theta)$$

Thus, $v_1^2 = 20.1$ (m/sec)

Then, $T = 14.7 + 6 \times 20.1 = 135.3$ N

(*ii*) At the highest point, we have from Fig. 3.24.

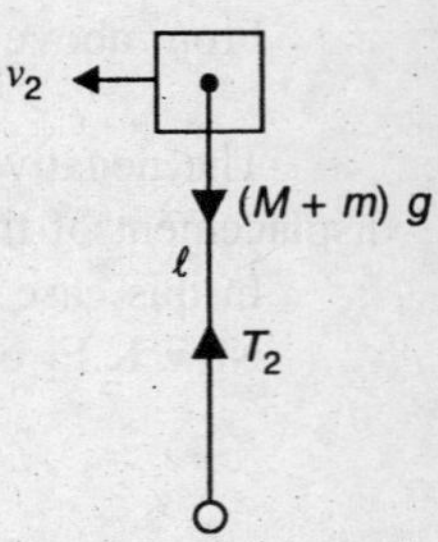

Fig. 3.24

Here centripetal force

$$= T_2 + (M + m)\ g$$

$$= (M + m)\ (v_2^2/l)$$

$$T_2 = (M + m)\left(\frac{v_2^2}{l}\right) - (M + m)\ g$$

If the string is to remain tight at the highest point (condition to complete a circle), $T_2 \geq 0$

i.e., $v_2^2 \geq g.l.$, But $v_2^2 = v_1^2 - 4gl$

or $$v_1^2 = v_2^2 + 4gl$$

Hence, $$v_1^2 \geq 5gl$$

$$v_1 \geq \sqrt{(5gl)} \geq \sqrt{(5 \times 9.8 \times 0.5)}$$

$$= \mathbf{4.949\ m/sec.}$$

As, the calculated value of v_1 is 5 m/sec, so the vertical circle would be completed.

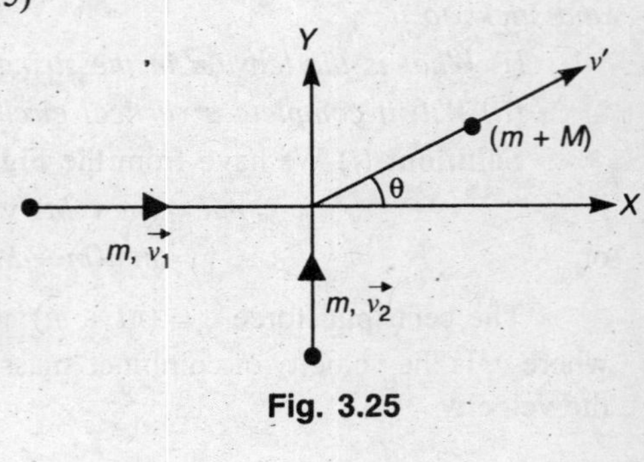

Fig. 3.25

Example 29. *A body of mass m moving with a velocity v_1 in x-direction collides with another body of mass M moving in y-direction with a velocity v_2. They coalesce into one body during collision. Calculate:*

(*i*) *The direction and magnitude of the momentum of the final body.*

(ii) The fraction of the initial kinetic energy transformed into heat during the collision in terms of the two masses.

Solution: We have from Fig. 3.25.

(*i*) Equating the total initial and final momentum along each axis,

$$mv_1 + 0 = (M + m)\ v' \cos\theta \quad ...(1)$$

$$0 + Mv_2 = (M + m)\ v' \sin\theta \quad ...(2)$$

Squaring and adding Equations (1) and (2)

$$(mv_1)^2 + (Mv_2)^2 = (M + m)^2\ v'^2$$

So, the final momentum is

$$(M + m)\ V = \sqrt{(mv_1)^2 + (Mv_2)^2}$$

Also, from Equations (1) and (2)

$$\tan\theta = (Mv_2)/(mv_1) \quad \text{or} \quad \theta = \tan^{-1}\ (Mv_2)/(mv_1)$$

(*ii*) The total kinetic energy is given by the relation

$$= \frac{1}{2}\ mv_1^2 + \frac{1}{2}\ Mv_2^2$$

Final kinetic energy of the system

$$= \frac{1}{2}\ (M + m)\ v$$

$$= \frac{1}{2}\ [1/(M + m)]\ .\ [\ (mv_1)^2 + (Mv_2)^2]$$

Decrease in total kinetic energy will be given by

$$\Delta E = [(1/2)\ mv_1^2\ + (1/2)\ Mv_2^2)]\ - [(1/2)1/(M + m)]\ [(mv_1)^2 + (Mv_2)^2]$$

$$= Mm\ (v_1^2 + v_2^2)\ /(2M + m)$$

so, fraction of kinetic energy transformed into heat is given by

$$= \frac{\Delta E}{(1/2)\ mv_1^2 + (1/2)\ Mv_2^2} = \frac{\boldsymbol{Mm}}{\boldsymbol{(M + m)}}\ \frac{\boldsymbol{v_1^2 + v_2^2}}{\boldsymbol{(mv_1^2 + Mv_1^2)}}.$$

Example 30. *A plastic ball is dropped from a height of one metre and rebounds several times from the floor. If 1.3 seconds elapse from the moment it is dropped to the second impact with the floor what is the coefficient of restitution?*

Solution: Suppose that the ball hits the floor first with a velocity v. After first impact it will rebound with a velocity ev where e is coeff. of restitution. The ball now rises to a certain height where its velocity becomes zero. It again goes down and after another impact with ground its velocity becomes $e^2 v$ and so on.

Now the time before the first impact t_1 and the interval between first and second impact t_2 may be calculated as under

$$v = 0 + gt_1 \quad \text{or} \quad t_1 = (v/g)$$

$$t_2 = (2\ ev/g)$$

$$\therefore \quad t_1 + t_2 = (v/g) + (2ev/g) = (v/g)\ (1 + 2e)$$

Also it is given

$$(v/g)\ (1 + 2e) = 1.3$$

Now the velocity of ball before the first impact is obtained by $v^2 = u^2 + 2gh$ where $u = 0$ and $h = 1$ m, so $v = \sqrt{(2g)}$

$$\sqrt{\frac{2g}{g}}\ (1 + 2e) = 1.3$$

or

$$\frac{\sqrt{2}}{9.8}\ (1 + 2e) = 1.3$$

$$e = \mathbf{0.04}.$$

Example 31. *A ball moving with a speed of 9 m/s strikes an identical stationary ball such that after the collision the direction of each ball makes an angle of 30° with the original line of motion. Find the speeds of the two balls after the collision. Is the kinetic energy conserved in this collision rocess.*

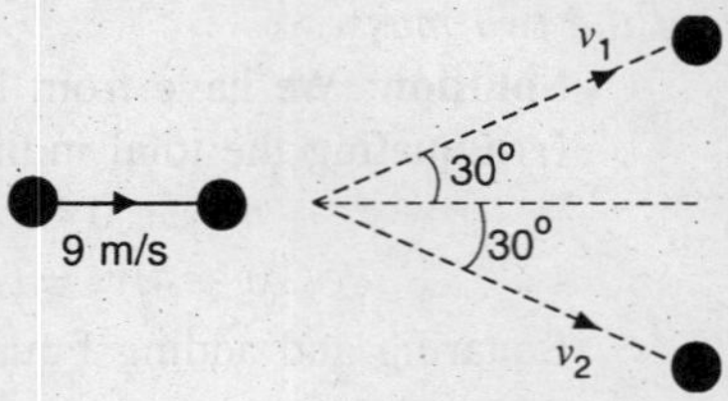

Fig. 3.26

Solution: We can represent the problem by Fig. 3.26.

Let the velocities of balls become v_1 and v_2 respectively after collision. Then the initial momentum of the balls (assuming m be the mass of ball)

$$= m \times 9 \times m \times 0 = 9\text{ m}$$

Final the momentum of system is along horizontal direction.

$mv_1 \cos 30° + mv_2 \cos 30°$

$$= mv_1 \sqrt{3/2} + mv_2 \sqrt{3/2}$$

From, the law of conservation of momentum, we have

$$9m = \frac{mv_1\sqrt{3}}{2} + \frac{mv_2\sqrt{3}}{2}$$

or $$v_1 + v_2 = (18/\sqrt{3})$$

In vertical (perpendicular) direction the initial momentum = 0 and after collision the momentum in this direction would be given by

$mv_1 \sin 30° - mv_2 \sin 30° = (m/2)(v_1 - v_2)$

From the law of conservation of momentum in vertical direction, we get

$(m/2)(v_1 - v_2) = 0$ or $v_1 - v_2 = 0$

So, with the above equations

$$v_1 = 3\sqrt{3}\text{ m/s}, v_2 = 3\sqrt{3}\text{ m/s}.$$

Now from the law of conservation of energy, we have

the energy before collision = energy after collision

or $$(1/2)\ mu_1^2 + (1/2)\ mu_2^2 = (1/2)\ mv_1^2 + (1/2)\ mv_2^2$$

or $$(1/2)\ m\ (9)^2 + 0 = (1/2)\ m\ (3\sqrt{3})^2 + (1/2)\ m\ (3\sqrt{3})^2$$

or $$(81m/2) = (54m/2)$$

As **L.H.S. ≠ R.H.S.**

∴ Energy is not conserved in this process *i.e.*, it is the case of inelastic collision.

Example 32. *A small ball A slides down the quadrant of a circle as shown in Fig. 3.27 and hits the ball B of equal mass which is initially at rest. Find the velocities of the both balls after collision. Neglect the effect of friction and assume the collision to be elastic.*

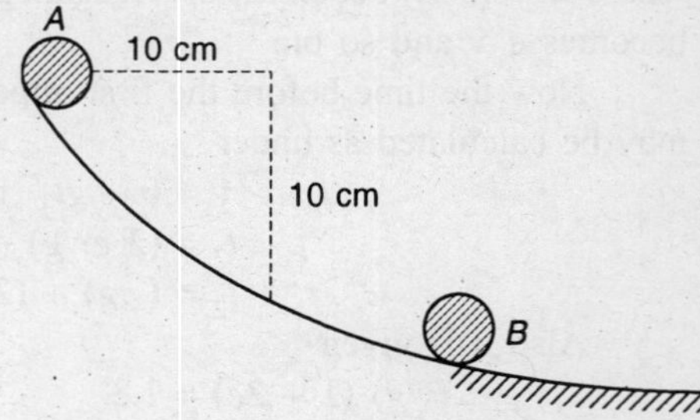

Fig. 3.27

Solution: Let m be the mass of each ball. If u_1 be the velocity of ball A just before collision then since during its slide along a quadrant the P.E. is converted into K.E. so we have

$$mgh = (1/2)\ mv_1^2 \text{ or } u_1^2 = 2gh$$

or $$u_1 = \sqrt{(2gh)} = \sqrt{(2 \times 980 \times 10)}$$

$$= \sqrt{(19600)} = \mathbf{140\ cm/sec.}$$

Now, let after collision, velocities of balls A and B be v_1 and v_2 respectively. Then the conservation of linear momentum will give

$$mu_1 = mv_1 + mv_2$$

or $$u_1 = v_1 + v_2$$

Also $$(v_1 - v_2)/(u_1 - u_2) = -1$$

$\therefore \quad v_1 - v_2 = -u_1$

Then from above two equation we get $v_1 = 0$ and $v_2 =$ **140 cm/sec.**

This shows that the ball A comes to rest while the ball B moves with a velocity of 140 cm/sec.

QUESTIONS AND PROBLEMS

1. What is meant by centre of mass? Give its importance.

2. What is centre of mass? Show that in the absence of external forces, the velocity of centre of mass is constant.

3. What are elastic and inelastic collisions? Explain your answer with suitable examples.

4. Explain the one dimensional elastic collison between two bodies.

5. Three masses 3, 4, and 5 kg are placed at the corners of an equilateral triangle of side 1 metre. Find the centre of mass of the system. **[Ans.** 0.542, 0.361]

6. A ball of mass m makes elastic collison with a ball of mass nm which is initially at rest. Show that the fractional transfer of energy by first ball is $\dfrac{4n}{(1+n)^2}$.

7. A bullet with a speed of 180 ms^{-1} crashes through a wooden plank. After passing through the plank, its speed is 140 ms^{-1}. Another bullet of the same mass and size but travelling at 110 ms^{-1} is fired at the plank. What will be the speed of second bullet after tunnelling through? Assume the resistance of the plank to be independent of bullet's speed. **[Ans.** 26.46 ms^{-1}]

8. A body of mass 1 kg initially at rest explodes and breaks with three fragments of masses in the ratio 1 : 1 : 3. The two pieces of equal mass fly of perpendicular to each other with a speed of 30 ms^{-1} each. What is the velocity of the heavies fragment? **[Ans.** 10 $\sqrt{2}$ ms^{-1}]

9. A ball moving with a speed of 9 ms^{-1} strikes an identical stationary ball such that after the collision the direction of each ball makes an angle of 30° with the original line of motion. Calculate the speed of the two balls after the collision. Also show that the kinetic energy is not conserved in this collision. **[Ans.** 3 $\sqrt{3}$ ms^{-1}, 3 $\sqrt{3}$ ms]

CHAPTER 4

ROTATIONAL MOTION

4.1. INTRODUCTION

In the translatory motion a body moves in a straight line in which case all the constituent particles move along parallel straight lines and undergo equal displacements in equal intervals of time *i.e.*, all the particles have same velocity and acceleration.

In rotatory motion, a body moves about a fixed axis and each particle describes concentric circles about that axis, though different particles at different points of the body will have different linear velocities, they will have same angular velocity. A rotatory motion may be considered as a stationary motion and rotation is caused by a couple acting on the body, its state can be changed only by applying a couple or a set of couples which may be equivalent to a simple couple.

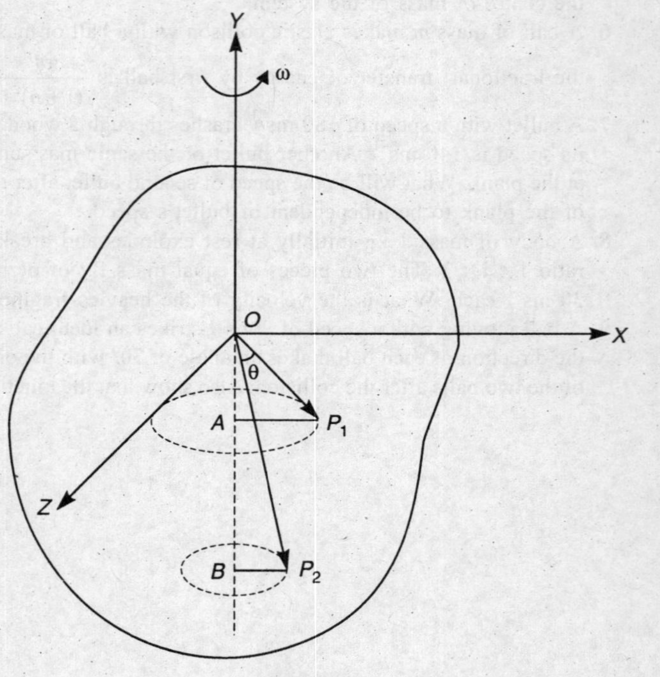

Fig. 4.1

Let a rigid body be taken which is free to revolve around an axis OY through O. All the particles of the body will describe circles in a plane ⊥ to axis of rotation OY. The centres of each circle will lie on OY and radii of these circles will be equal to ⊥ distances from the axis. The radius of each circle sweeps through an angle θ in a small time say δt. The radii of circles swept by a particle P_1 and P_2 will be AP_1 and BP_2.

The angular velocity of all the particles and hence that of the body will be same as in small time δt because all the particles rotate through an angle $\delta\theta$.

The limiting value of $\dfrac{\delta\theta}{\delta t}$ as $\delta\theta \to 0$ is known as angular velocity ω at any instant t, so

$$\omega = \lim_{\delta t \to 0} \frac{\delta\theta}{\delta t} = \frac{d\theta}{dt}$$

and the corresponding angular acceleration will be rate of change of angular velocity *i.e.*,

$$\alpha = \frac{d\omega}{dt}.$$

4.2. MOMENT OF INERTIA

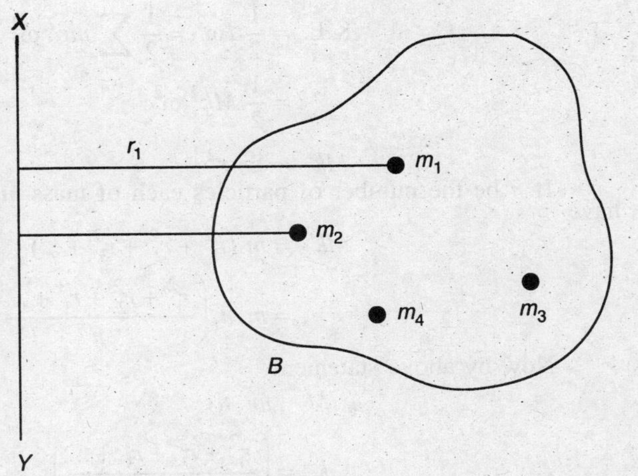

Fig. 4.2

From Newton's first law of motion a body at rest will remain at rest while a body in uniform motion along a stright line will move continuously unless an external force disturbs it. The property due to which a body does not change its state of rest or motion is called 'Inertia'. For the motion in a straight line, inertia depends on the mass of the body, more is the mass of the body, more is the inertia. Also the kinetic energy depends on the mass and linear velocity of the body. However, when a body moves about an axis, the K.E. of its rotation is not dependent on its mass and angular velocity but it also depends on the axis about which the rotation is taking place.

Let a body B rotates about XY axis and let its angular velocity be w, considering that all the particles move with same angular velocities but their linear velocities will be different as they are at different distances from the axis of rotation.

Let the linear velocities of the particles of masses $m_1, m_2, m_3, \ldots$ etc. be $v_1, v_2, v_3, \ldots$ and their distances from the axis of rotation be $r_1, r_2, r_3, \ldots$ The K.E. of these particles will be $\frac{1}{2} m_1 v_1^2$, $\frac{1}{2} m_2 v_2^2$, $\frac{1}{2} m_3 v_2^3$ and the sum total K.E., of all the particles will correspond to the K.E. of the body *i.e.*,

$$\therefore \quad \text{Total K.E.} = \frac{1}{2} m_1 v_1^2, + \frac{1}{2} m_2 v_2^2, + \frac{1}{2} m_3 v_3^2 + \ldots$$

Also $\quad v = r\omega$

$$\therefore \quad \text{Total K.E.} = \frac{1}{2} \mu_1 \rho_1^2 \omega^2 + \frac{1}{2} m_2 r_2^2 \omega^2 + \ldots$$

$$= \frac{1}{2} \left(\sum mr^2\right) \omega^2 = \frac{1}{2} I\omega^2$$

where I is called the moment of inertia of the body about the XY-axis and is given by

$$I = \sum mr^2$$

So, we may define the **moment of inertia of a body about an axis as the sum of the products of the mass and square of the distance of different particles of the body from the axis of rotation.**

Also we see from above

$$\text{K.E. of rotation} = \frac{1}{2} I\omega^2$$

If $\quad \omega = 1, I = 2 \times$ kinetic energy

So moment of inertia may also be defined as **twice the kinetic energy of rotation** of a body if its angular velocity is unity.

4.3. RADIUS OF GYRATION

If we consider the total mass of the body to be concentrated at one point such that the K.E. of rotation is the same as that of the body, then the distance of that point from the axis of rotation is known as radius of gyration of the body about the axis of rotation. If k is considered as radius of gyration and M, the mass of the body, supposed to be concentrated at that point, we have

$$\text{K.E.} = \frac{1}{2} I\omega^2 = \frac{1}{2} \sum mr^2 \omega^2$$

$$= \frac{1}{2} Mk^2 \omega^2$$

$$\therefore \quad Mk^2 = \Sigma mr^2$$

If n be the number of particles each of mass m contained in the given mass M, then we have

$$Mk^2 = m\,(r_1^2 + r_2^2 + r_3^2 + ...)$$

$$= m.n.\left(\frac{r_1^2 + r_2^2 + r_3^2 + ...}{n}\right)$$

Now by above statement

$$M = m\,.\,n$$

$$\therefore \quad k^2 = \left(\frac{r_1^2 + r_2^2 + r_3^2 + ...}{n}\right)$$

or

$$k = \left(\frac{r_1^2 + r_2^2 + r_3^2 + ...}{n}\right)^{\frac{1}{2}}$$

So the radius of gyration is the square root of the mean square distance of the various particles contained in the body from the axis of rotation.

Dimension of k will be the same as the dimensions of length *i.e.*, $[L^1]$ only.

Dimension of moment of inertia can be got from the relation

$$I = Mk^2$$

$$= [M^1\ L^2].$$

4.4. UNITS OF MOMENT OF INERTIA

C.G.S. System. In this system units of moment of inertia are gm-cm^2
and in M.K.S. System, the units are Kilogram – Metre^2

In F.P.S. System, the units are Pounds–Ft^2

It should be noted that moment of inertia is scalar quantity because its value about any axis remains the same even if the direction of rotation is reversed. Also the total moment of inertia of a number of bodies is the sum of their individual moments of inertia about that axis.

4.5. PHYSICAL SIGNIFICANCE OF MOMENT OF INERTIA

Moment of inertia has the same role in rotatory motion as the mass has in linear motion.

From Newton's first law of motion, a body at rest will remain at rest while a body in uniform motion will do so till it is disturbed by an external force. This property is called inertia of the system. Normally, a body has the tendency to resist the external force which tends to change its state of rest or of uniform motion. More is the mass of the body, greater is the force required to produce a given linear acceleration.

Similar to the linear inertia, a body has rotational inertia *i.e.*, a body moving about a fixed axis opposes any change in its state due to an external force. **More is the moment of inertia of a body, the larger is the couple required to produce a given acceleration**. It should be noted that moment of inertia not only depends on the mass of the body but also the distribution of masses about the axis of rotation *e.g.*, If we have a wheel and a disc of the same mass, wheel will have greater moment of inertia as the mass in this case is distributed at larger distances from the axis of rotation. We can summarize some of the similarities of moment of inertia in rotatory motion and mass in linear motion as under.

Linear Motion	*Rotatory Motion*
(*i*) Linear momentum = mv	(*i*) Angular momentum = Iw
(*ii*) Impulse = $m\ (v_2 - v_1)$	(*ii*) Angular Impulse = $I\ (\omega_2 - \omega_1)$
(*iii*) Force = ma	(*iii*) Torque (moment of couple) = I × Angular acceleration
(*iv*) Work = Force × Distance	(*iv*) Work = Couple × Angular placement
(*v*) Kinetic Energy $= \frac{1}{2} mv^2$	(*v*) Rotational Kinetic Energy $= \frac{1}{2} I\omega^2$

4.6. ANGULAR MOMENTUM

The sum of the moment of momentum of all the particles of a rotating rigid body about the axis of rotation is known as its angular momentum. When a body is freely rotating about some axis, the angular velocity of all the particles will be the same, no matter what the distance of these particles may be. However, different particles are at different distances from the axis of rotation, so linear velocities will be different.

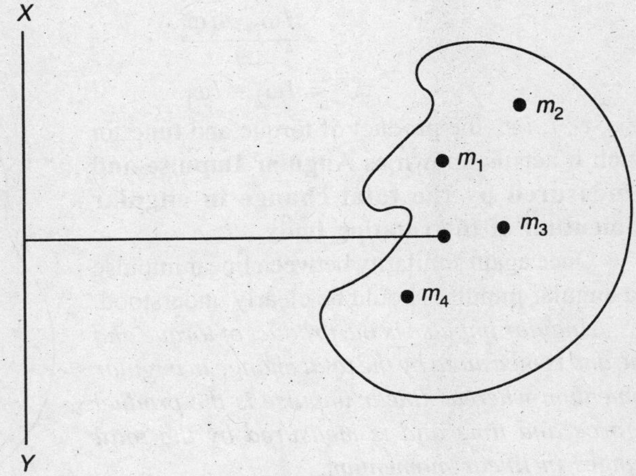

Fig. 4.3

Let us consider various particles of a body having masses as m_1, m_2, m_3, m_4 ...etc., at distance r_1, r_2, r_3, r_4 ... from the axis of rotation and let their linear velocities be $v_1, v_2, v_3, v_4, ...$, respectively.

If w be the angular velocity, then linear velocity of particle $m_1 = v_1 = r_1 \omega$ and corresponding linear momentum $= m_1 v_1$.

So the moment of momentum of m_1 about the axis of rotation

$$= m_1 v_1 r_1 = m_1 r_1^2 \omega$$

Similarly moment of momentum of

$$m_2 = m_2 r_2^2 \omega$$

Total moment of momentum of all the particles

$$= m_1 r_1^2 \omega + m_2^2 \omega + m_3 r_3^2 \omega + ...$$

$$= \sum mr^2 \omega = I\omega$$

So the angular momentum is the product of moment of inertia and angular velocity. It should be clearly understood that whereas linear momentum is the product of mass and linear velocity angular momentum is the product of moment of inertia and angular velocity.

4.7. TORQUE

Torque which is also the moment of couple acting on a rotating body is the rate of change of its angular momentum.

Now Angular Momentum

$$L = I\omega$$

$$\therefore \quad \text{Troque} = I \frac{d\omega}{dt} = I\alpha$$

So $\quad \tau = I\alpha$

where τ is torque and a is angular acceleration.

It should be clearly seen that torque is the product of moment of inertia and angular acceleration and so it has the same significance in rotatory motion as force (mass × linear acceleration) has in linear motion.

4.8. ANGULAR IMPULSE

If an external torque L, is applied to a body which is rotating with an angular velocity w_1 for a time t (say) so that the new angular velocity is w_2, then by definition

$$\text{Angular acceleration} \qquad \alpha = \frac{\text{Change in angular velocity}}{\text{time}}$$

$$= \frac{\omega_2 - \omega_1}{t}$$

Also torque $= I\alpha$

$$\therefore \qquad \tau = \frac{I\omega_2 - I\omega_1}{t}$$

$$\text{or} \qquad \tau . t = I\omega_2 - I\omega_1$$

$\tau . t$, *i.e.*, the product of torque and time for which it acts is known as **Angular Impulse and is measured by the total change in angular momentum of the rotating body.**

Once again similarity between linear impulse and angular impulse should be clearly understood.

Angular impulse is the product of torque and time and is measured by the total change in angular momentum whereas linear impulse is the product of force and time and is measured by the total changes in linear momentum.

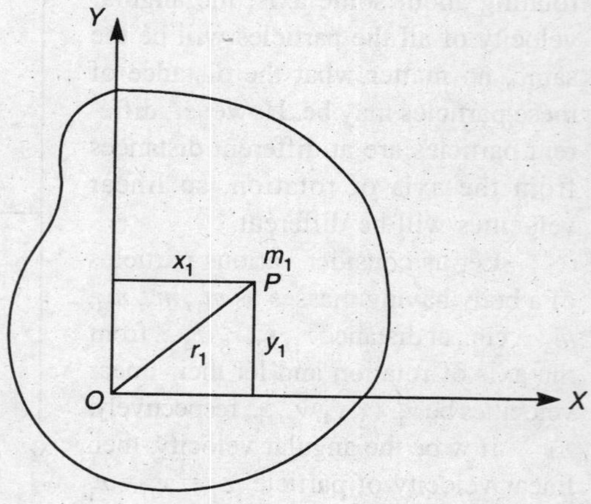

Fig. 4.4

4.9. THEOREM OF PERPENDICULAR AXIS

It states that the moment of inertia of a plane lamina about an axis ⊥ to the plane of lamina is equal to the sum of moments of inertia of the lamina about the two axis at right angles to each other, in its own plane intersecting each other at the point where perpendicular axis passes through it.

The mathematical statement of the above definition can be had by considering I_x and I_y as moment of inertia about two perpendicular axis in the plane of lamina. If I be the moment of inertia about a line passing through the point of intersection and ⊥ to its plane, then

$$I = I_x + I_y$$

Let us have OX and OY as two ⊥ axis in the plane of lamina and let m_1 be the mass of a particle at a distance r_1 from an axis passing through O and ⊥ to the plane XOY. Let the distance of this particle from Y axis be x_1 and that from X-axis be y_1.

So moment of inertia of this particle about X-axis

$$= m_1 \, y_1^2$$

and that about Y-axis $\quad = m_1 \, x_1^2$

If the total lamina is divided into number of particles of masses $m_1, m_2, m_3, ...$, etc. at distances $r_1, r_2, r_3, ...$ So that the corresponding distances are $x_1, x_2, x_3 ...$ from Y-axis and $y_1, y_2\, y_3, ...$ from X-axis.

So we have moment of inertia of the lamina about X-axis

$$I_x = m_1 \, y_1^2 + m_2 \, y_2^2 + m_3 \, y_2^3 + ...$$

$$= \sum my^2$$

and that about X-axis

$$I_y = m_1 x_1^2 + m_2 x_2^2 + m_3 x_3^3 + \ldots$$
$$= \sum mx^2$$

Now moment of inertia of the lamina about a $\perp$ axis passing through O

$$I = m_1 r_1^2 + m_2 r_2^2 + m_3 r_3^3 + \ldots$$
$$= \sum mr^2$$
$$= m_1 (x_1^2 + y_1^2) + m_2 (x_2^2 + y_2^2) + m_3 (x_3^3 + y_3^3) + \ldots$$
$$= m_1 x_1^2 + m_2 x_2^2 + m_3 x_3^2 + \ldots$$
$$= m_1 y_1^2 + m_2 y_2^2 + m_3 y_3^2 + \ldots$$
$$= \sum mx^2 + \sum my^2$$
$$= I_x + I_y$$

So we have mathematically

$$I = I_x + I_y$$

4.10. THEOREM OF PARALLEL AXIS

It states that the moment of inertia of a body about any axes is equal to the sum of its moment of inertia about a parallel axis through its centre of gravity and the product of its mass and the square of the distance between the two axis.

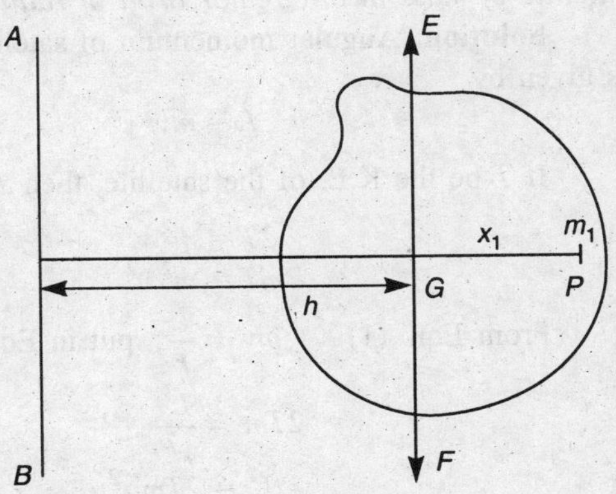

Fig. 4.5

Let AB be an axis in the plane of paper and EF, a parallel axis passing through G, the centre of gravity of the body.

Let the perpendicular distance between the two axis be h and M be the mass of the body and m_1, the mass of an element at P at a distance x_1 from EF.

So moment of inertia of m_1 about

$$AB = m_1 (x_1 + h)^2$$
$$= m_1 (x_1^2 + h^2 + 2x_1 h)$$
$$= m_1 x_1^2 + m_1 h^2 + 2m_1 x_1 h$$

So the moment of inertia of the whole body about AB may be written as

$$I = \sum m_1 x_1^2 + \sum m_1 h^2 + 2 \sum m_1 x_1 h \qquad \ldots(1)$$

Let I_g be the moment of inertia of the body about EF *i.e.*, an axis passing through G, then

$$I_g = \sum m_1 x_1^2 \qquad \ldots(2)$$

Putting in (1) $\quad I = I_g + Mh^2 + 2h \sum m_1 x_1 \qquad \ldots(3)$

Now $m_1 x_1$ correspond to the moments of all the particles about EF passing through G *i.e.*, centre of gravity of the body. Since the body balances about its centre of gravity, so the algebraic sum of all moments about G is zero,

$$\therefore \quad \sum m_1 x_1 = 0$$

$\therefore$ Equation (3) may be writtn as

$$I = I_g + Mh^2 \qquad \ldots(4)$$

Example 1. *A 1 kilogram stone attached to the end of a 60 cm chain is revolving at the rate of 3 rev/sec. What is its angular momentum? If after 30 sec it is making only one rev./sec, find the mean torque.*

Solution: Angular momentum is given by

$$L = mr^2\,\omega$$

Now in this case $m = 1$ kgm, $r = 0.6$ m

$$\omega = 2\pi \times 3 \text{ radians/sec}$$

$$\therefore \quad L = 1 \times (0.6)^2 \times 6\pi$$

$$= 6.785 \text{ Joule-sec.}$$

Now Torque is the rate of change of angular momentum *i.e.*,

$$\tau = \frac{dL}{dt} = \frac{d}{dt}(mr^2\omega) = mr^2\frac{d\omega}{dt} = \mathbf{0.15\ N\text{-}m.}$$

Example 2. *Express angular momentum in terms of kinetic, potential and total energy of a satellite of mass m in circular orbit of radius r.*

Solution: Angular momentum of satellite of mass m moving in a circular path of radius r is given by

$$L = mvr \qquad \text{...(1)}$$

If T be the K.E. of the satellite, then $T = \frac{1}{2}mv^2$

or

$$2T = m^2 v^2$$

$$2mT = m^2 v^2 \qquad \text{...(2)}$$

From Eqn. (1) $mv = \frac{L}{r}$, put in Equation (2)

$$2Tm = \frac{L^2}{r^2}$$

or

$$L^2 = 2Tm.r^2$$

or

$$L = (2Tmr^2)^{1/2} \qquad \text{...(3)}$$

which gives us angular momentum in terms of K.E.

If V be the potential energy of the satellite, then

$$V = -\frac{GMm}{r} \qquad \text{...(4)}$$

where M is the mass of the earth.

Also we have from orbital motion

$$\frac{GMm}{r^2} = \frac{mv^2}{r}$$

or

$$\frac{GMm}{r} = mv^2$$

From Equation (4),

$$V = -\frac{GMm}{r} = -mv^2 \qquad \text{...(5)}$$

From Equations (2), (3) and (5) we have

$$L^2 = -mr^2 V \qquad \text{...(6)}$$

which is an expression of angular momentum in terms of potential energy.

Now total is given as the sum of K.E. and P.E.

$\therefore$ Total Energy $E = T + V$

$$= \frac{1}{2}mv^2 + (-mv^2) = -\frac{1}{2}mv^2 \qquad \text{...(7)}$$

Again from Equations (2), (3) and (7) we have

$$L^2 = \mathbf{2E\ .\ mr^2}$$

which is an expression for angular momentum in terms of total energy.

Example 3. *A flywheel of mass 500 kgs and radius 1 metre make 500 rev/min. Assuming the mass to be concentrated along the rim, calculate the energy of the flywheel.*

Solution: Mass of flywheel $= 500$ kgs

Distance from axis $k = 1$ metre

Angular speed $\omega = 500$ rev/min

$$= \frac{500 \times 2\pi}{60} \text{ rad/sec}$$

Moment of Inertia $I = Mk^2$

$$= 500 \times 1 \times 1 = 500 \text{ kg.m}^2$$

$$= 500 \times 1000 \times 100 \times 100$$

$$= 5 \times 10^9 \text{ gms cm}^2$$

K.E. of flywheel $= \frac{1}{2} I\omega^2 = \frac{1}{2} Mk^2\omega^2$

$$= \frac{1}{2} \times 5 \times 10^9 \times \left(\frac{500 \times 2\pi}{60}\right)^2$$

$$= 6470 \times 10^9 \text{ ergs} = \mathbf{6.870\ joules.}$$

4.11. MOMENT OF INERTIA OF A THIN UNIFORM ROD

Let AB be a thin uniform rod of length l and has a mass M. Let us first find the moment of inertia about an axis passing through its centre and ⊥ to its length *i.e.*, along YY'.

$\frac{M}{l}$ will be the mass/unit length of the rod.

Let us consider an element of thickness dx, its mass will be $\frac{M}{l} \cdot dx$ and its moment of inertia about YY' axis passing through O will be $\frac{M}{l} \cdot dx.x^2$.

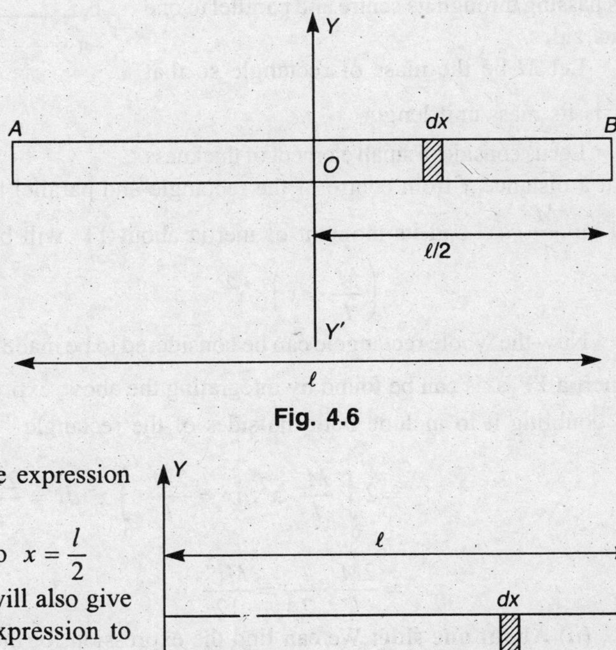

Fig. 4.6

To get the moment of inertia of the whole rod, we integrate the above expression between

$$x = -\frac{l}{2} \text{ to } x = \frac{l}{2}$$

or taking the limits $x = 0$ and $x = l/2$ will also give the same results if we double the expression to include both halves of the rod

$$I = 2\int_0^{l/2} \frac{M}{l}\, dx\,.x^2$$

$$= \frac{2M}{l}\int_0^{l/2} x\,.dx$$

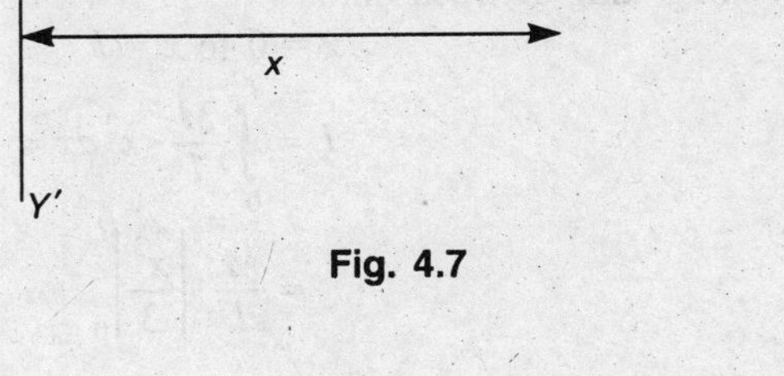

Fig. 4.7

$$= \frac{2M}{l} \cdot \left| \frac{x^2}{3} \right|_0^{l/2} = \frac{2M}{3l} \cdot \frac{l^3}{8} = \frac{Ml^2}{12} \qquad ...(1)$$

To find the moment of inertia of a thin unifrom rod about an axis passing through one end of the rod and ⊥ to its length, as shown in Fig. 4.7, we can have the result by taking integration limits as $x = 0$ to $x = l$.

i.e.,
$$I = \int_0^l \frac{M}{l} x^2 \, dx = \frac{M}{l} \cdot \left| \frac{x^3}{3} \right|_0^l$$

$$= \frac{Ml^3}{3l} = \frac{Ml^2}{3} \qquad ...(2)$$

We can have the same result by the parallel axis theorem as well.

4.12. MOMENT OF INERTIA OF A RECTANGLE

We can find the moment of inertia of a rectangle in the following possible ways.

(*i*) **About an axis passing through its centre and parallel to one of its sides.** Let us consider a rectangle. *ABCD* having length *l* and breadth *b* as shown in the figure and *YY′* be the axis passing through its centre and parallel to one of its sides.

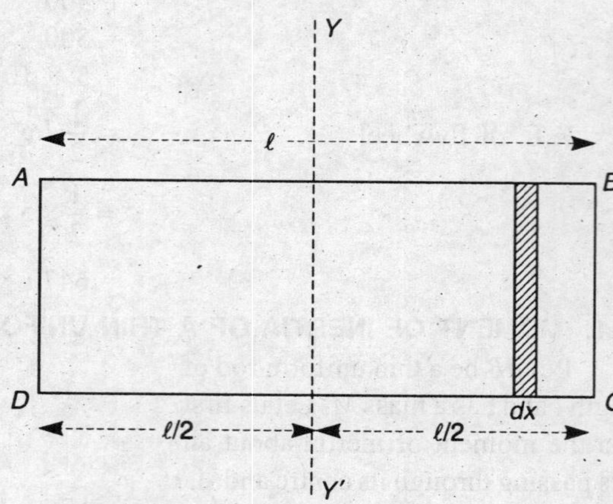

Fig. 4.8

Let *M* be the mass of rectangle so that $\frac{M}{l}$ is its mass/unit length.

Let us consider a small element of thickness dx at a distance x from centre of the rectangle and parallel to *YY′* axis. The mass of this element will be $\frac{M}{l} \cdot dx$ and its moment of inertia about *YY′* will be

$$= \left(\frac{M}{l} \cdot dx \right) \times x^2$$

Now the whole rectangele can be considered to be made up of such small strips and its moment of inertia *YY′* axis can be found by integrating the above expressions between limits $x = 0$ to $x = \frac{l}{2}$ and doubling it to include both the sides of the rectangle

$$= 2 \int_0^{l/1} \frac{M}{l} \cdot x^2 . dx = \frac{2M}{l} \int_0^{l/2} x^2 \, dx = \frac{2M}{l} \cdot \left| \frac{x^3}{3} \right|_0^{l/2}$$

$$= \frac{2M}{l} \cdot \frac{l^3}{24} = \frac{Ml^2}{12} . \qquad ...(1)$$

(*ii*) **About one side:** We can find the expression for moment of inertia along one of its side by integrating between limits

$$x = 0 \text{ to } x = l$$

$$\therefore \qquad I = \int_0^l \frac{M}{l} \cdot x^2 \, dx = \frac{M}{l} \int_0^l x^2 \, dx$$

$$= \frac{M}{l} \cdot \left| \frac{x^3}{3} \right|_0^l$$

$$= \frac{M}{l} \cdot \frac{l^3}{3}$$

$$= \frac{Ml^2}{3} \quad ...(2)$$

(*iii*) **About an axis passing through its centre and ⊥ to its plane:** We can obtain the moment of inertia in this case by applying the principle of perpendicular axis to the case (*i*).

The moment of inertia of the rectangle about an axis passing through its centre and parallel to *AD* will be $\frac{Ml^2}{12}$ and about an axis passing through its centre and parallel to side *AB* will be $\frac{Mb^2}{12}$, where *b* is the breadth of the rectangle.

So from the principle of perpendicular axis

$$I = I_x + I_y$$

$$= \frac{Mb^2}{12} + \frac{Ml^2}{12}$$

$$= \frac{M}{12}(l^2 + b^2) \quad ...(3)$$

However the above relation is true for thick rectangular bars as they can be considered to be built up of a number of small strips.

(*iv*) **About an axis passing through the mid-point of one side and ⊥ to its plane:** When the axis passes through mid point of *BC* or *AD*. We can find moment of inertia by parallel axis theorem *i.e.,*

Moment of inertia about an axis passing through mid point of *BC* or *AD*

= *M.I.* about parallel axis through centre of mass + Mass × (dist. between parallel axis)^2

$$= M\left(\frac{l^2 + b^2}{12}\right) + \frac{Ml^2}{4}$$

$$= M\left[\frac{l^2}{3} + \frac{b^2}{12}\right] \quad ...(4)$$

Similarly we can find *M.I.* if axis passes through the mid-point *AB* or *DC* in exactly the same way as above and it will be *M.I.* about a parallel axis through centre of mass + Mass × (dist. between two axis)^2

$$= M\left(\frac{l^2 + b^2}{12}\right) + \frac{Mb^2}{4}$$

$$= M\left(\frac{l^2}{12} + \frac{b^2}{3}\right)$$

$$= M\left(\frac{b^2}{3} + \frac{l^2}{12}\right).$$

4.13. MOMENT OF INERTIA OF A CIRCULAR RING OR HOOP

(*i*) About an axis passing through its centre and ⊥ to its plane.

Let us consider a very thin circular ring of radius *R* having a mass *M* as shown in Fig. 4.9. Since we consider the ring to the very thin, each particle on the ring may be at a distance *R* from the centre *O* or from an axis passing through *O* and ⊥ to its plane. Let us consider a particle of mass *m*, its moment of inertia about *YY'* will be mR^2 and moment of inertia of total ring will be the sum of moment of inertia of all the particle.

i.e., $$I = \Sigma mR^2$$

$$= MR^2 \qquad [\therefore\ M = \Sigma m] \quad ...(1)$$

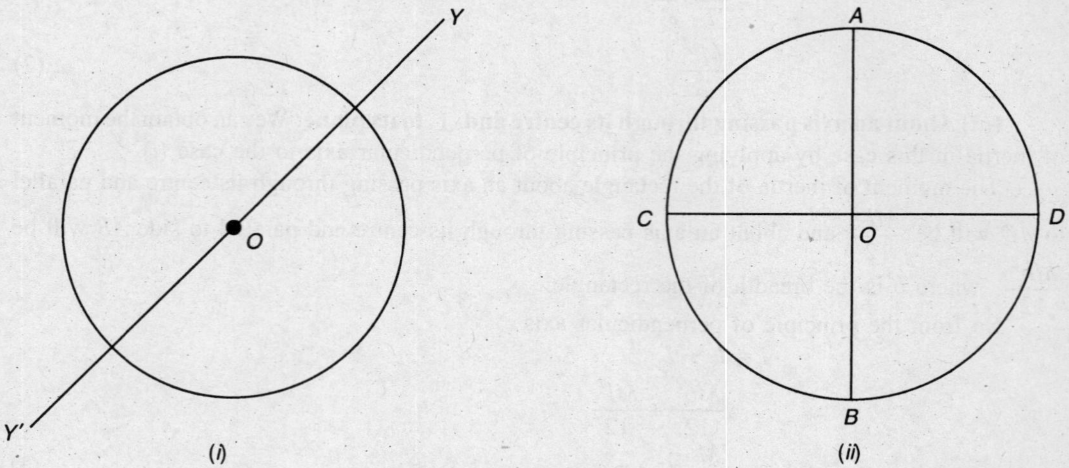

Fig. 4.9

(*ii*) **About its diameter:** Since the moment of inertia of a circular ring is the same about all diameters, so if *I* be M.I. of the ring along *AB*, it will also be *I* along *CD*.

Now by the principle of perpendicular axis theorem, the sum of moments of inertia about *AB* and *CD* will be equal to moment of inertia about an axis passing through its centre and $\perp$ to the ring *i.e.*, it is equal to MR^2.

$$\therefore \qquad I + I = MR^2$$

or
$$I = \frac{1}{2} MR^2. \qquad ...(2)$$

4.14. MOMENT OF INERTIA OF A DISC (OR CIRCULAR LAMINA)

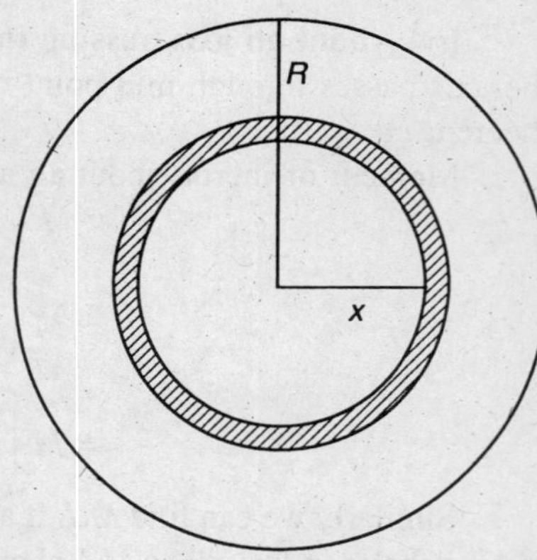

Fig. 4.10 (*i*)

(*i*) **About an axis through its centre and $\perp$ to its plane:** Let *M* be the mass of the disc and *R* be its radius. Consider a very small element of thickness *dx*, a distance *x* from centre. The area of this element will be $2\pi x.\ dx$.

$$\text{Mass of the unit area of disc} = \frac{M}{\pi R^2}$$

$$\text{Mass of the element} = \frac{M}{\pi R^2} \times 2\pi x\ dx = \frac{2Mx}{R^2}\ dx$$

Now moment of inertia of the element about an axis through its centre and $\perp$ to its plane $\frac{2M}{\pi R^2} \cdot x.dx.x^2$

$$= \frac{2M}{\pi R^2} \cdot x^3 \cdot dx$$

Moment of Inertia of the whole disc about this axis

$$I = \frac{2M}{R^2} \int_0^R x^3\, dx = \frac{2M}{R^2} \cdot \frac{R^4}{4} = \frac{1}{2} MR^2 \quad ...(1)$$

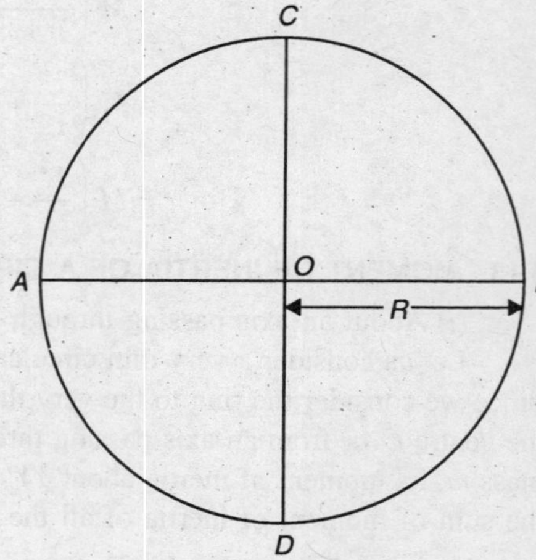

Fig. 4.10 (*ii*)

(*ii*) **About its diameter:** The moment of inertia of a circular disc about an axis $\perp$ to its plane and passing through its centre is given by $I = \frac{1}{2} MR^2$ (as proved above).

Let us consider two ⊥ diameters AB and CD and since all diameters are symmetrical the moment of inertia will be same for all of them.

Let I_1 and I_2 be the moments of inertia of the disc about two ⊥ diameters, then from the principle of perpendicular axis, the moment of inertia I of the disc about an axis ⊥ to the plane of the disc through O will be

$$I = I_1 + I_2$$

Also as stated above,

$$I_1 = I_2$$

$$\therefore \quad I = 2I_1$$

or

$$I_1 = \frac{I}{2} = \frac{1}{2} \cdot \frac{MR^2}{2} = \frac{MR^2}{4}. \quad ...(2)$$

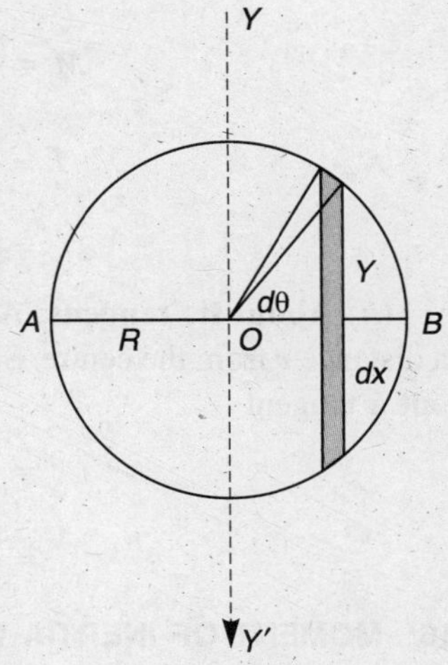

Fig. 4.11

4.15. MOMENT OF INERTIA OF A SOLID SPHERE

(*i*) **About its diameter:** Let us consider a section of a solid sphere of radius R through the centre as shown in the Fig. 4.11.

Let AB be the diameter about which we want to find its moment of interia. Let us consider a thin strip of thickness dx at a distance x from the centre, then

radius of the strip $\quad y = (R^2 - x^2)^{1/2}$

and volume of the strip $\quad = (R^2 - x^2) \,.\, dx$

Let m be the mass/unit volume of the sphere, their mass of the strip

$$M_1 = mp\ (R^2 - x^2)\ dx$$

and moment of inertia of the strip (disc)

$$AB = \frac{1}{2}\ M_1\ Y^2$$

$$= \frac{1}{2}\ m\pi\ (R^2 - x^2)\ dx\ (R^2 - x^2)$$

$$= \frac{1}{2}\ m\pi\ (R^2 - x^2)^2\ dx$$

Considering all such discs, we have the moment of inertia of the sphere about diameter AB as

$$I = 2\int_0^R \frac{1}{2}\ m\pi\ (R^2 - x^2)^2\ dx$$

$$= m\pi \int_0^r (R^4 + x^4 - 2R^2 x^2)\ dx$$

$$= m\pi \left[R^4.x + \frac{x^5}{5} - 2R^2 \cdot \frac{x^3}{3} \right]_0^R$$

$$= m\pi \left[R^5 + \frac{R^5}{5} - 2\,\frac{R^5}{3} \right]$$

$$= m\pi \cdot \frac{8R^5}{15}$$

$$= \frac{8m\pi R^5}{15}$$

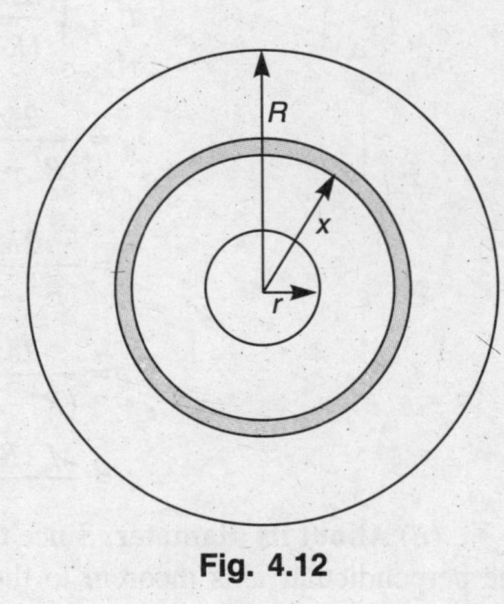

Fig. 4.12

Mass of the solid sphere is given as

$$M = \frac{3}{4}\pi R^3 m$$

$$\therefore \quad I = \frac{8m\pi R^5}{15} + \frac{3}{4}\pi R^3 m \times \frac{2}{5}R^2$$

$$= \frac{2}{5}MR^2$$

(*ii*) **About its tangent:** A tangent to the sphere will be parallel to one of its diameters and at a distance r from the centre. So from the principle of parallel axis, moment of inertia of the sphere about a tangent

$$= \frac{2}{5}MR^2 + MR^2$$

$$= \frac{7}{5}MR^2$$

4.16. MOMENT OF INERTIA OF AN ANNULAR RING

An annular disc is a disc from which a smaller co-axial disc is removed, which in other words means a disc having a concentric hole, a flywheel is an Example of annular ring.

Let the outer and inner radii of the disc be R and r respectively and let M be its mass.

The face area of the annular disc

$$= \pi(R^2 - r^2)$$

and mass/unit area

$$= \frac{M}{\pi(R^2 - r^2)}$$

Let us consider a small ring of thickness dx at a distance x from centre and let the whole disc may be assumed to be made up of a number of such small rings.

Face area of this small ring $= 2\pi x \,.\, dx$

and mass of small ring $= 2\pi x \,.\, dx \times \dfrac{M}{\pi(R^2 - r^2)}$

(*i*) Now the Moment of Inertia of this small ring about an axis through centre and of ⊥ its plane.

$$= \frac{2\pi x\, M dx}{\pi(R^2 - r^2)} \times x^2 = \frac{2Mx^3\, dx}{(R^2 - r^2)}$$

We can get the M.I. of whole annular disc by integrating the above expression between limits $x = r$ to $x = R$

$$\therefore \quad I = \int_r^R \frac{2Mx^3\, dx}{(R^2 - r^2)}$$

$$= \frac{2M}{(R^2 - r^2)} \int_r^R x^3 \,.\, dx$$

$$= \frac{2M}{(R^2 - r^2)} \cdot \left| \frac{x^4}{4} \right|_r^R$$

$$= \frac{2M}{(R^2 - r^2)} \cdot \frac{(R^4 - r^4)}{4}$$

$$= \frac{M(R^2 + r^2)}{2}$$

(*ii*) **About its diameter:** Since the M.I. about any diameter is same, so we can find it by using the perpendicular axis theorem to the above case.

$$\text{M.I. about diameter} = \frac{1}{2}\left[\frac{M\,(R^2 + r^2)}{2}\right]$$

$$= \frac{M\,(R^2 + r^2)}{4}.$$

4.17. FLYWHEEL

A flywheel is a large and heavy wheel through the centre of which passes a long cylindrical axis. The apparatus is such that its centre of gravity lies on its axis of rotation and when it is mounted over ball-bearings, it can be brought to rest at any desired position.

To find its moment of inertia let M be the mass of the disc and m_1 that of the axle. So a flywheel is just a disc of radius R from which a small concentric disc of radius r is cut off. Which in other words mean that it is just an annular ring of outer radius R and inner radius r.

Now the moment of inertia of an annular ring is given by

$$\frac{M\,(R^2 + r^2)}{2}$$

The axle is also a disc whose moment of inertia about an axis through its centre is given by

$$\text{M.I.} = \frac{\text{Mass} \times (\text{Radius})^2}{2}$$

$$\text{M.I. of axle} = \frac{m_1\,r^2}{2}$$

∴ The M.I. of wheel and axle

$$= \text{M.I. of wheel} + \text{M.I. of axle}$$

$$= \frac{M\,(R^2 + r^2)}{2} + \frac{m_1\,r^2}{2}.$$

4.18. Experimental Determination of Moment of Inertia

The flywheel whose M.I. is to be found is mounted on two ball bearings so that the friction is reduced. Also the axle is supported in such a horizontal position or ball bearings that friction is reduced. The whole arrangement is mounted at a convenient distance from ground. One end of a fine cord is attached to a small mass m and the other is slipped on to a peg A in the form of a loop after giving sufficient windings around the axle. This loop is such that it can be easily attached or detached from the peg and the length of the cord is so chosen that it gets detached from the axle as soon as bottom of this mass m is about to reach ground.

When the mass m is allowed to fall, the cord will be unwound under the action of falling mass and the wheel is set into motion. Which in other words means that the potential energy of the falling mass is partly converted into the K.E. of the mass due to the velocity gained by it and partly into the energy of rotation of the flywheel.

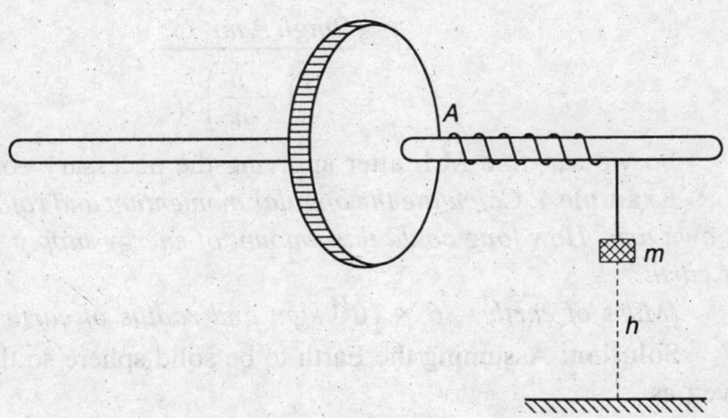

Fig. 4.13

If the friction losses are neglected and we apply the law of conservation of energy, we have

P.E. lost by the mass m

= K.E.

gained by mass m + K.E. gained by flywheel.

If ω be the angular velocity gained by the wheel at the moment when m is detached and r the radius of the axle and h the height through which the mass falls, we have

$$mgh = \frac{1}{2} m (r\omega)^2 + \frac{1}{2} I\omega^2 \quad \text{...(1)}$$

where I is the M.I. of flywheel.

Now when the string gets detached from axle, wheel continues to revolve for sometime, till due to friction after slowing down the wheel comes to stationary position.

If n_1 be the no. of revolutions made by the wheel in time t before coming to rest, then, Average angular velocity

$$\frac{\omega + 0}{2} = \frac{2\pi n_1}{t}$$

or

$$\frac{\omega}{2} = \frac{2\pi n_1}{t}$$

If frictional force is constant throughout, the motion of the wheel is uniformally retarded and finally wheel stops. So we have the angular velocity

$$\omega = \frac{4\pi n_1}{t} \quad \text{...(2)}$$

Putting the value of ω in (1) we can find I.

4.19. CORRECTION FOR FORCE OF FRICTION

If the force of friction is not negligible, then we have to apply correction as under.

Let n be the no. of revolutions made by the wheel during the time the mass m falls and let F be the energy in ergs used/revolution in overcoming the frictional force.

Total energy consumed in n revolutions = $n.F$ ergs

$\therefore$ From (1)
$$mgh = \frac{1}{2} mr^2\omega^2 + \frac{1}{2} I\omega^2 + nF \quad \text{...(3)}$$

Now the energy possessed by the flywheel is $\frac{1}{2} I\omega^2$ and this energy is used in overcoming friction in n_1 revolutions when wheel stops.

$\therefore$
$$n_1 F = \frac{1}{2} I\omega^2$$

or

$$F = \frac{1}{2} \frac{I\omega^2}{n_1}$$

Putting in Equation (3)

$$mgh = \frac{1}{2} mr^2\omega^2 + \frac{1}{2} I\omega^2 \left(1 + \frac{n}{n_1}\right)$$

or

$$I = \frac{2mgh - mr^2\omega^2}{\left(1 + \frac{n}{n_1}\right)}$$

So we can find M.I. after applying the necessary correction due to the frictional force.

Example 4. *Calculate the angular momentum and rotational kinetic energy of the earth about its own axis. How long could this amount of energy supply 1 kW power to each of 4×10^9 persons on earth?*

(Mass of earth = 6×10^{24} kgm and radius of earth = 6.4×10^6 m).

Solution: Assuming the Earth to be solid sphere so that the M.I. about its own axis may be given as

$$I = 2/5MR^2$$
$$= 2/5 \times 6 \times 10^{24} \times (6.4 \times 10^6)^2$$
$$= 9.8 \times 10^{37} \text{ kg m}^2$$

Since earth revolves around its own axis in one day, its angular velocity is given by

$$\omega = \frac{2\pi}{24 \times 60 \times 60} = 7.27 \times 10^{-5} \text{ rad/sec}$$

and corresponding angular momentum

$$= I\omega = 9.8 \times 10^{37} \times 7.27 \times 10^{-5}$$
$$= 7.15 \times 10^{23} \text{ kg–m/sec}^2$$

Its rotational energy will be $= \frac{1}{2} I\omega^2 = \frac{1}{2} \times (9.8 \times 10^{37}) \times (7.27 \times 10^{-5})^2$

$$= 2.61 \times 10^{39} \text{ joules.}$$

The power that can be got from this energy is given by

$$P = \frac{\text{Energy}}{\text{Time}} = \frac{2.6 \times 10^{29}}{t} \text{ watts}$$
$$= \frac{2.61 \times 10^{29}}{t \times 1000} \text{ kW}$$

Now we require 1 kW power to each person, so total power required for 4×10^9 persons $= 4 \times 10^9$ kW

$$\therefore \quad 4 \times 10^9 = \frac{2.61 \times 10^{29}}{t \times 1000}$$

or
$$t = \frac{2.61 \times 10^{29}}{1000 \times 4 \times 10^9}$$
$$= 65 \times 10^{15} \text{ seconds}$$
$$= \mathbf{2.04 \times 10^9 \text{ years.}}$$

Example 5. *A uniform flat disc of mass M and radius R rotates about a horizontal axis through its centre with angular velocity ω. Find (i) Its kinetic energy and angular momentum (ii) If a chip of mass m breaks of the edge of the disc at an instant such that the chip rises vertically above the point at which it broke off. How high above the point does it rise before starting to tall? (iii) What is the final angular momentum and energy of the disc.*

Solution: The kinetic energy of rotation is given by

The Kinetic Energy of rotation is given by

$$T = \frac{1}{2} I\omega^2$$

In this case $I = \frac{1}{2} MR^2$ and $\omega = \omega_0$

$$\therefore \quad T = \frac{1}{2} MR^2 . \omega_0^2 \qquad ...(1)$$

The angular momentum is given by

$$L = I\omega$$
$$\therefore \quad L = \frac{1}{2} MR^2 . \omega_0 \qquad ...(2)$$

(*ii*) We have the relation

$$v = r\omega$$

In this case

$$v = R\omega_0$$

If the particle goes to a height h then,

$$v^2 = 2gh$$

or
$$h = \frac{v^2}{2g} = \frac{\boldsymbol{R^2 \omega_0^2}}{\boldsymbol{2g}} \qquad ...(3)$$

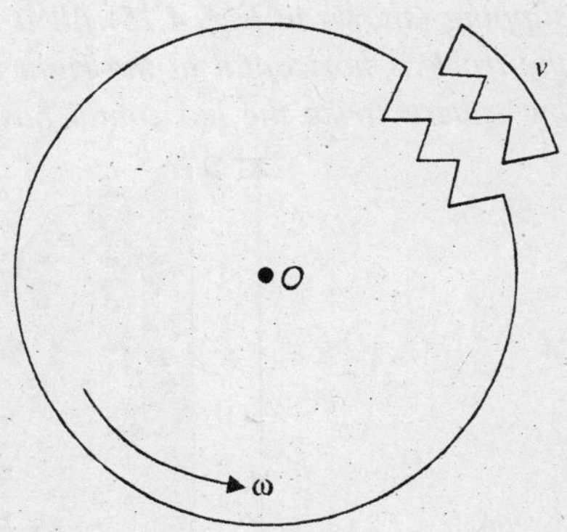

Fig. 4.14

(*iii*) If the angular momentum of the remaining disc $= L_R$ then from the principle of angular momentum

$$L_R + mR^2 \omega_0 = \boldsymbol{\frac{1}{2} MR^2 \omega_0}$$

or $$L_R = \left(\frac{1}{2}M - m\right) R^2 \omega_0$$

Similarly from the principle of conservation of energy we have

$$(T)_R + \frac{1}{2} mR^2 \omega_0^2 = \frac{1}{2}\left(\frac{1}{2} MR^2\right) \omega_0^2$$

or $$(T)_R = \frac{1}{2}\left[\left(\frac{1}{2} M - m\right)\right] R^2 \omega_0^2.$$

Example 6. *A sphere of mass 0.5 kgm and diameter 1 m rolls without slipping with a constant velocity of 5 m/sec. Calculate its total energy.*

Solution: When the sphere is rolling without slipping about its diameter, it will have both translatory and rotatory motion.

So its total K.E. may be written as

$$K_{total} = K_{trans} + K_{rot}$$

For a sphere $I = \frac{2}{5} MR^2$ about its own diameter.

Also $v = R\omega$ or $\omega = v/R$

$$\therefore \quad K_{total} = \frac{1}{2} Mv^2 + \frac{1}{2} I\omega^2$$

$$= \frac{1}{2} Mv^2 + \frac{1}{2}\left(\frac{2}{5} MR^2\right)\left(\frac{v}{R}\right)^2$$

$$= \frac{1}{2} Mv^2 + \frac{1}{5} Mv^2$$

$$= \frac{7}{10} Mv^2$$

In this case $M = 0.5$ kgm and $v = 5$ m/sec

$$\therefore \quad K_{total} = 0.7 \times 0.5 \times (5)^2$$

$$= \mathbf{8.75 \text{ joule.}}$$

Example 7. *A homogeneous sphere starts from rest at the upper end of a track and rolls without slipping (shown in Fig. 4.15) till it rolls of the right hand end. If H = 200 m and h = 60 m and the track is horizontal at the right hand end. Find the distance to the right of point A at which the sphere strike the horizontal base line.*

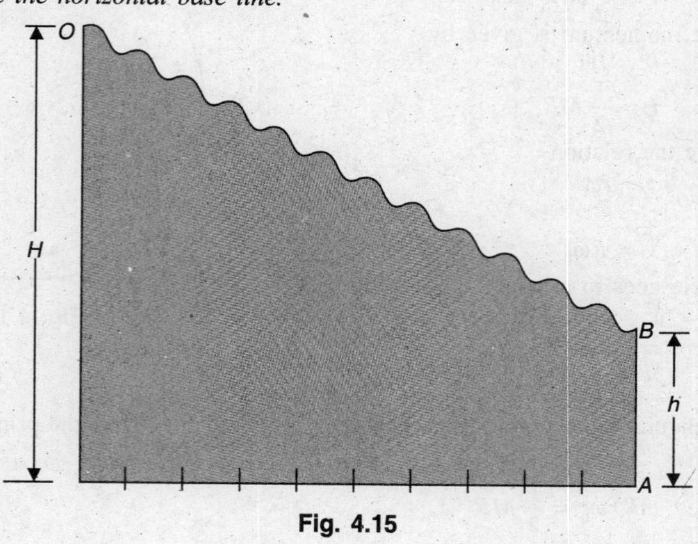

Fig. 4.15

Solution: From the conservation law of energy, the P.E. is lost due to fall and this loss is equal to the gain in K.E. (Rotational K.E. + Translatory K.E.)

$$\therefore \quad mg\,(H-h) = \frac{1}{2} I\omega^2 + \frac{1}{2} mv^2$$

$$= \frac{1}{2} \times \frac{2}{5} mR^2 \times \left(\frac{v}{R}\right)^2 + mv^2$$

$$= \frac{1}{5} mv^2 + \frac{1}{2} mv^2$$

$$= \frac{7}{10} mv^2$$

$$\text{or} \quad g\,(H-h) = \frac{7}{10} v^2 \quad \text{...(1)}$$

Now $H = 200$ m, $h = 60$ m and $v = ?$

$\therefore$ From Equation (1)

$$v = \sqrt{\frac{10}{7} g\,(H-h)}$$

$$= \sqrt{\frac{10}{7} \times 9.8 \times (200-60)}$$

$$= \textbf{44.27 meters/sec.}$$

As the track at B is horizontal, the sphere at B will fall under the gravity and the time of fall is given by

$$S = ut + \frac{1}{2} gt^2$$

$$60 = 0 + \frac{1}{2} \times 9.8 \times t^2$$

$$t = \sqrt{\frac{60 \times 2}{9.8}} = 3.5 \text{ sec}$$

The horizontal distance travelled by sphere in 2 sec. will be

$$2 \times v = 2 \times 44.27 = \textbf{88.54 m}$$

So the sphere will hit the ground at 88.54 m right of A.

Example 8. *Four spheres each of radius r and mass M are placed with their centres on four corners of a square of side l. Calculate the moment of inertia of the arrangement about:*

(i) diameter of the square.

(ii) Any side of the square.

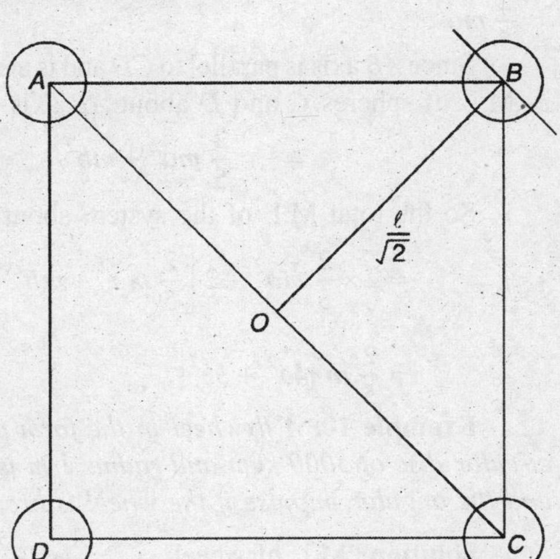

Fig. 4.16

Solution: The moment of inertia of each sphere A and C about diagonal AC of the sphere will be equivalent to that an axis passing through their centres *i.e.*, equal to $\frac{2}{5} Mr^2$. Similarly

M.I. of spheres B and D will be $\frac{2}{5} Mr^2$.

Now since the distance between AC and parallel axis through sphere B or D is $l/\sqrt{2}$, where l is the side of square, the M.I. of sphere B and D about AC can be got from the theorem of parallel axis *i.e.*,

M.I. of B or D about AC

$$= \frac{2}{3} Mr^2 + \frac{Ml^2}{2}$$

So the M.I. of all the spheres about

$$AC = \frac{2}{5} Mr^2 + \frac{2}{5} Mr^2 + 2\left[\frac{2}{5} Mr^2 + \frac{Ml^2}{2}\right]$$

$$= \frac{M}{5}[8r^2 + 5l^2]$$

(*ii*) M.I. of A or B about $AB = \frac{2}{5} Mr^2$

M.I. of D about $AB = \frac{2}{5} Mr^2 + Ml^2$

M.I. of C about $AB = \frac{2}{5} Mr^2 + Ml^2$

$\therefore$ Total M.I. of all the spheres about any side (say AB)

$$= 2\left[\frac{2}{5} Mr^2\right] + 2\left[\frac{2}{5} Mr^2 + Ml^2\right]$$

$$= \frac{\mathbf{M}}{\mathbf{5}}[\mathbf{8r^2 + 10l^2}].$$

Example 9. *Four spheres each of diameter 2a and mass m are placed with their centres on the four corners of a square of side b. Calculate the moment of inertia of the system about one side of the square.*

Solution: Mass of each sphere = m and radius = a

Since the spheres are placed with their centres at the corners of square $ABCD$, the M.I. of each spheres A and B about AB axis passing through their centres $= \frac{2}{5} ma^2$.

Similarly M.I. of spheres C and D about CD axis $= \frac{2}{5} ma^2$.

Since AB axis is parallel to CD and is at a distance b, M.I. of spheres C and D about AB axis

$$= \frac{2}{5} ma^2 + mb^2$$

So the total M.I. of the system about AB axis

$$= 2 \times \frac{2}{5} ma^2 + 2\left[\frac{2}{5} ma^2 + mb^2\right]$$

$$= \frac{\mathbf{2}}{\mathbf{5}} \mathbf{m}[\mathbf{4a^2 + 5b^2}].$$

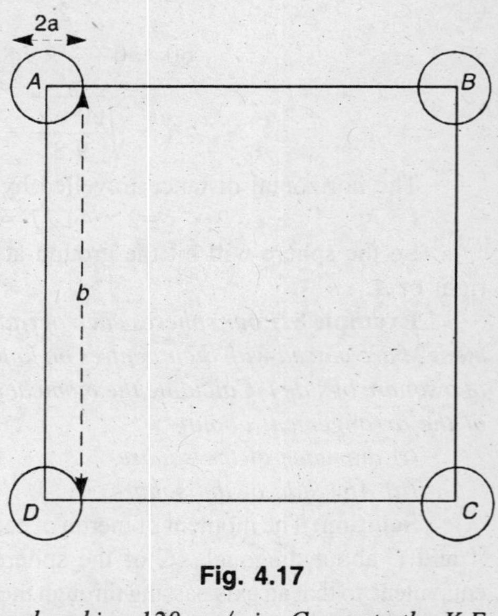

Fig. 4.17

Example 10. *A flywheel in the form of a solid circular disc of 5000 kgm and radius 1 m is rotating and making 120 rev/min. Compute the K.E. and the angular impulse if the wheel is brought to rest in 2 seconds. Friction is to be neglected.*

Solution: M.I. of wheel $= \frac{1}{2} MR^2$

$$= \frac{1}{2} \times 5000 \times 1000 \times 100 \times 100$$

$$= 2.5 \times 10^{10} \text{ C.G.S units.}$$

No. of rotations/sec. $n = 2$

Angular velocity $\omega = 2n = 4$ rad/sec.

$$\text{K.E.} = \frac{1}{2} I\omega^2 = \frac{1}{2} \times 2.5 \times 10^{10} \times 16\pi^2$$

$$= 1.974 \times 10^{12} \text{ ergs.}$$

Angular Impulse = Total change in angular momentum

$$= I\omega = 2.5 \times 10^{10} \times 4\pi$$

$$= \mathbf{3.142 \times 10^{11}} \textbf{ C.G.S. units.}$$

Example 11. *A flywheel has a moment of inertia 70 kg-m^2 and its axle has a diameter 4.4 cm. How much pull should be applied on a thread wound on the axle to produce angular acceleration of 0.5 rad/sec^2.*

Solution: Let F be the force to be applied, so that torque about the axle is given by

$$\tau = F.R \qquad (R \text{ being radius of axle})$$

In this case

$\therefore$ $R = 2.2$ cm $= 0.022$ m

$\therefore$ $\tau = 0.022F$ N-m

Also we have $\tau = I.\alpha$

$\therefore$ $0.22F = I\alpha$

Given $I = 70$ kg-m^2

and $\alpha = 0.5$ rad/sec

$\therefore$ $0.22F = 70 \times 0.5$

$$\therefore \quad F = \frac{70 \times 0.5}{0.22} = \mathbf{1591\ N}$$

4.20. ROLLING DOWN AN INCLINED PLANE

We will consider the case of a rigid body of radius R and mass M rolling down without slipping a smooth inclined place having an angle of repose as shown in Fig. 4.18.

It is clear that as the body rolls down, because of descent it will lose P.E. At the some time it acquire linear and angular speeds and so gains K.E. of *translation and rotation.* As no loss of total energy takes place, so the loss of K.E. compensated by gain of K.E.

Let the body be at rest at A initially and let it reacthes B *i.e.,* it moves through a distance. Let v and ω be the linear and angular velocity acquired by the body respectively.

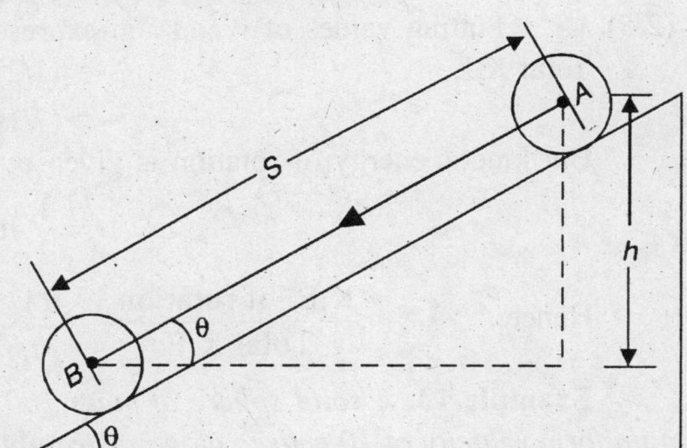

Fig. 4.18

The vertical distance h travelled by the body $h = S \sin\theta$

So, loss in potential energy = $MgS \sin\theta$.

Now gain in kinetic energy of translation = (1/2) Mv^2 and, gain in the kinetic energy of rotation = (1/2) $I\omega^2$ where I is the moment of inertia of the body.

So we have $\quad Mg\,S \sin\theta = (1/2)\,Mv^2 + (1/2)\,Iw^2 \qquad$ ...(1)

Also $\quad v = Rw$

$$Mg\,S \sin\theta = (1/2)\,Mv^2 + (1/2)\,I.\,(v^2/R^2)$$

or $\quad 2Mg\,S \sin\theta = v^2\,[M + (I/R^2)]$

$$v^2 = \frac{2MgS \sin\theta}{[M + (I/R^2)]} \qquad \text{...(2)}$$

Now differentiating Equation (2) with respect to t, we get

$$Mg \sin\theta \frac{dS}{dt} = \frac{1}{2} m.2v \frac{dv}{dt} + \frac{1}{2}\frac{1}{R^2} 2v \frac{dv}{dt}$$

or

$$Mg \sin\theta \frac{dS}{dt} = Mv \frac{dv}{dt} + \frac{l}{R^2} v \frac{dv}{dt}$$

where, (dS/dt) is the magnitude of linear velocity v of the body.

Also, if (dv/dt) is linear acceleration a, then, we may write

$$Mg \sin\theta = M \frac{dv}{dt} + \frac{I}{R^2}\frac{dv}{dt} = Ma + \frac{I}{R^2} a$$

or

$$a = (Mg \sin\theta)/(M + I/R^2)$$

Since, $I = mk^2$ where k is radius of gyration, in that case

$$a = \frac{Mg \sin\theta}{M + (MK^2/R^2)} \quad \text{or} \quad a = \frac{g \sin\theta}{1 + (K^2/R^2)}.$$

Example 12. *A spherical ball rolls on a table. Find the fraction of its total kinetic energy that is rotational.*

Solution: A spherical ball rolling on a table will have both rotation and translation motion simultaneously. So it will have both rotational kinetic energy and translational kinetic energy.

Now, K.E. of rotation $= (1/2)\, I\omega^2$

K.E. of translation $= (1/2)\, Mv^2$

Hence, Total K.E. $= (1/2)\, I\omega^2 + (1/2)\, Mv^2$ ...(1)

During the time the ball completes one round it describes an angle 2π about the axis of rotation (which is diameter). During the same time it covers a linear distance $2\pi r$, where r is its radius. So the common time of motion is

$$(2\pi/\omega) = (2\pi r/v)$$

or,

$$v = r\omega$$

Also, the moment of inertia I of a sphere of mass M and radins r about its diameter is $(2/5)\, Mr^2$. Putting values of v and l in expression of total K.E., *i.e.*, in Equation (1)

$$\text{Total K.E.} = (1/2)\,.\,(2/5)\, Mr^2\,\omega^2 + (1/2)\, Mr^2\,\omega^2 = (7/10)\, Mr^2\,\omega^2$$

But kinetic energy of rotation is given by

$$= \frac{1}{2} I\omega^2 = \frac{1}{2}\frac{2}{5} Mr^2\omega^2 = \frac{1}{5} Mr^2\omega^2$$

Hence,

$$\frac{\text{K.E. of rotation}}{\text{Total K.E.}} = \frac{(1/5)\, Mr^2\omega^2}{(7/10)\, Mr^2\omega^2} = \mathbf{\frac{2}{7}}.$$

Example 13. *A solid sphere of mass 100 gm and radius 2.5 cm rolls without sliding with a uniform velocity of 10 cm/sec along a straight line on a smooth table. Calculate its total energy in ergs.*

Solution: K.E. of the sphere rolling without is given by $= \frac{1}{2} Mv^2 + \frac{1}{2} I\omega^2$

$$\text{Total K.E.} = \frac{1}{2} Mv^2 + \frac{1}{2} \times \frac{2}{5} MR^2 \times \frac{v^2}{R^2} = \frac{7}{10} Mv^2$$

$$= \frac{7}{10} \times 100 \times 10^2 = \mathbf{7000\ ergs.}$$

Example 14. *You are given two sphere of same mass, size and appearance but one of them is hollow and the other is solid? How will you find which is hollow and which is solid? Indicate the theoretical consideration on which your answer is based.*

Solution: Let the two spheres roll down an inclined plane without slipping. The acceleration of either is given by the relation

$$a = \frac{g \sin \theta}{1 + (k^2/R^2)}$$

where g is acceleration due to gravity, q the inclination of the plane to the horizontal. R the radius of sphere (external in case of hollow sphere) and K the radius of gyration about a diameter.

Now the moment of inertia of a solid sphere about a diameter is given by

$$I = Mk^2 = (2/5)\ MR^2 \qquad \text{....(1)}$$

$$\therefore \qquad K^2 = (2/5)\ R^2 \qquad \text{...(2)}$$

Also, the moment of inertia of a hollow sphere about a diameter is given by

$$I = Mk^2 = (2/5)\ M\ [(R^5 - r^5)/(Q^3 - r^3)] \qquad \text{...(3)}$$

where R and r are its external and internal radii respectively.

$$K^2 = \frac{2}{5}\frac{R^5 - r^5}{R^3 - r^3} = \frac{2}{5}\frac{1-(r^5/R^5)}{1-(r^3/R^3)} R^2 \qquad \because (r < R) \text{ ...(4)}$$

Now $\qquad (r^5/R^5) < (r^3/R^3)$

or $\qquad [1-(r^5/R^5)] > [(1-r^3/R^3)]$

or $\qquad \dfrac{\mathbf{1-(r^5/R^5)}}{\mathbf{1-(r^3/R^3)}} > 1$

which means that K^2 for hollow sphere is more than that for solid sphere so the acceleration of hollow sphere is less than that of solid sphere which means that hollow sphere will take more time to roll down the same distance on inclind plane and so can be distinguished from the solid sphere.

Example 4. *A sphere, a disc, a spherical shell and a ring are allowed to roll down simultaneously an inclined plane from the same height without slipping. Prove that the sphere reaches down first, then the disc, then the shell and in the last, the ring.*

Solution: Let M and R represent the mass and radius in each case. The acceleration of sphere, rolling down an inclind plane is given by

$$a = \frac{g \sin \theta}{1 + (K^2/R^2)}$$

where $\qquad \theta$ = angle of inclination, K = radius of gyration.

For sphere: Its moment of inertia about the axis of rotation (diameter) is given by the relation

$$I = MK^2 = (2/5)\ MR^2$$

or $\qquad K^2/R^2 = 2/5$

$$a = \frac{g \sin \theta}{1 + (2/5)} = \frac{5}{7} g \sin \theta \qquad \text{...(1)}$$

For Disc (*or Cylinder*): Moment of inertia about the axis of rotation (axis passing through its centre and perpendicular to its plane) is given by the relation

$$I = MK^2 = (1/2)\ MR^2$$

or $\qquad K^2/R^2 = 1/2$

Hence $\qquad a = \dfrac{g \sin \theta}{1 + (1/2)} = \dfrac{2}{3} g \sin \theta \qquad$ ...(2)

For Spherical Shell: Moment of inertia about the axis of rotation (diameter) is given by the relation

$$I = MK^2 = (2/3)\ MR^2$$

or, $\qquad (K^2/R^2) = (2/3)$

$\therefore \qquad a = \dfrac{g \sin \theta}{1 + (2/3)} = \dfrac{3}{5} g \sin \theta \qquad$ ...(3)

For Ring: Moment of inertia about the axis of rotation (axis passing through its centre and perpendicular to its plane) is given by the relation

$$I = MK^2 = MR^2$$

or, $$K^2/R^2 = 1$$

$$\therefore \quad a = \frac{g \sin\theta}{1+1} = \frac{1}{2} g \sin\theta \qquad ...(4)$$

It is evident that the acceleration for a solid sphere, disc, shell and ring are in a decreasing order. Therefore, if all of them start rolling simultaneously, the sphere will reach down the plane first, then the disc, then the shell and in the last the ring.

It is clear that acceleration for a solid sphere, disc, shell and ring are in descending order, so if all of them roll down an inclined plane simultaneously, sphere will reach first, than disc, then the shell and lastly the ring will reach the ground.

Example 16. *Mass of an electron is 9.0×10^{-31} kg. It revolves around the nucleus of an atom in a circular orbit of radius 4.0Å with a speed of 6.0×10^6 m/s. Calculate the linear kinetic energy and the angular momentum of the electron.*

Solution: Linear K.E. of electron is given by $(K.E.) = \frac{1}{2} mv^2$

where m is mass of electron and v is the linear velocity in the orbit.

Here $m = 9.0 \times 10^{-31}$ kg and $v = 6.0 \times 10^6$ m/sec.

Hence $(K.E.)_{linear} = (1/2) \times (9.0 \times 10^{-31}) \times (6.0 \times 10^6)^2$

$= 1.62 \times 10^{-17}$ Joules

Angular momentum of the electron about the axis of rotation is given by

$$J = I\omega = (mr^2) \times (v/r) = m.v.r$$

$$= (9.0 \times 10^{-31} \text{ kg}) \times (4.0 \times 10^{-10} \text{ m}) \times (6.0 \times 10^6 \text{ m/s})$$

$$= \mathbf{2.16 \times 10^{-33} \text{ kg m}^2/\text{sec.}}$$

Example 17. *A body of mass 20 kg is rotating at a rate of 2000 revolutions/minute in a circular path of diameter 0.20 meter. Find the moment of inertia and the K.E. of the body about the axis of rotation.*

Solution: Mass the body $m = 20$ kg.

The distance of the body from the axis or rotation $r = (0.20/2) = 0.10$ m.

So, moment of inertia of the body about the axis of rotation, is given by

$$I = mr^2 = 20 \text{ kg} \times (0.10 \text{ m})^2 = 0.2 \text{ kg m}^2$$

Angular velocity of the body is given by

$$\omega = \frac{\text{angle turned}}{\text{time}} = \frac{2000 \text{ rev}}{\text{minute}} = \frac{2000 \times 2\pi \text{ radian}}{60 \text{ sec}}$$

$$= (200/3) = 209.52 \text{ rad/sec.}$$

Hence K.E. of body $= (1/2)\, I\omega^2$

$= (1/2) \times 0.2 \times (209.52)^2$

$= \mathbf{4390 \text{ Joules}}$

Questions and Problems

1. Define moment of inertia and give its physical significance. Show that in rotatory motion of inertia plays the same role as the mass does in linear motion.
2. Define moment of inertia and radius of gyration, give the units of moment of inertia. Is it a scalar or a vector quantity?
3. State and prove the principle of parallel and perpendicular axis.
4. Calculate the moment of inertia of a circular disc (*i*) about an axis through its centre and perpendicular to its plane, (*ii*) about a diameter.
5. Calculate the moment of inertia of an annular ring.
6. Calculate the moment of inertia of a thin rod.

7. Calculate the moment of inertia of a rectangular lamina (*i*) about an axis passing through its centre and parallel to one of its sides (*ii*) about an axis passing through its centre and perpendicular to its plane.
8. Obtain an expression for moment of inertia of a solid sphere about (*i*) a diameter and (*ii*) tangent.
9. Describe an experiment to find the moment of inertia of a flywheel. Derive the necessary formula without neglecting friction.
10. A solid sphere of mass 100 gm and radius 2.5 cm rolls without sliding with a uniform velocity of 13 cm along a straight line on a horizontal table. Calculate its total kinetic energy. **[Ans.** 7.0×10^{-4} J]
11. A square frame *ABCD* of side 40 cm and negligible mass has four masses 2, 5, 6 and 3 kg placed at the corners and a mass 10 kg at the centre *O*. It is to be rotated about an axis passing through *O* and perpendicular to the frame compute the torque needed to produce an angular accleration of 20° per sec. **[Ans. 0.446 Nm]**
12. Calculate the moment of inertia of HCl molecule about an axis passing through the centre of mass and perpendicular to the line joining the atoms of HCl molecule. The distance between hydrogen and chlorine atom is 1.275.

    Given $m_H = 1.0$ a.m.u.

    $m_{Cl} = 35.5$ a.m.u.

    and 1 a.m.u. = 1.67×10^{-27} kgm **[Ans.** -2.58×10^{-47} kg-m^2**]**
13. A flywheel is rotated by a 50 kg mass hanging from a string wrapped round the shaft of the flywheel. If the diameter of the shaft is 6.0 cm and if the mass starting from rest falls 3.6 m in 2 seconds, find the moment of inertia. Neglect friction losses. **[Ans.** 0.20 kg m^2**]**
14. A solid cylinder (*i*) rolls (*ii*) slides down an inclined plane. Neglect friction and compare the velocities in the two cases when the cylinder reaches bottom of inclined plane. **[Ans. 1.225]**
15. A thin circular ring of mass *M* and radius *r* is rotating about its axis with a constant angular velocity w. Two objects each of mass *m*, are gently attached to the opposite ends of the diameter of the ring. Calculate the new angular velocity of rotation of the system. $\left[\textbf{Ans. } \omega' = \dfrac{M\omega}{M+2m}\right]$

CHAPTER 5

ELASTICITY

5.1. INTRODUCTION

In the case of a true rigid body, the distance between any two points does not change but in practice no rigid body is perfect and so all bodies get deformed under an applied load. However, when the shape or size of the body is changed under an applied force or load, it has the tendency to regain its original size or shape when the load is removed.

This property of the body *due to which it tends to regain its original shape or size when* **deforming forces are removed is called elasticity.**

If a body quickly regains its original shape or size when the deforming forces are removed, it is called a **perfectly elastic body**. However, if a body retains the deformed shape or size after the removal of deforming forces, it is called a **perfectly plastic body.** In nature no perfect elastic or plastic bodies exist and all the materials are in **between these two extreme limits.**

In the case of fluids, they resist change of volume but not change of shape and so they possess only volume elasticity, solids, on the other hand resist both the change of shape and volume and so possess **rigidity or shear elasticity** as well as volume elasticity.

5.2. STRESS

When some external force acts on a body as stated above, different particles of the body are displaced. These displaced particles try to come back to their original position when the external force is removed.

This restoring force per unit area called into play inside the body is called **stress.**

It should be noted that the reaction set up in the body is equal and opposite to the applied force, so long as no permanent change is produced in the body, so the restoring force is equal to the applied force.

If F is the force applied normally over an area of cross-section a of a body.

Then Stress $= F/a$

5.3. STRAIN

When an external force acts on a body, it displaces various particles of the body and the body is said to be under **strain.**

Strain is normally defined as the **ratio of change in length, volume or shape to the original length, volume or shape.**

Since strain is a pure ratio, it has no units.

5.4. HOOKE'S LAW

Hooke discovered a simple relation between stress and strain. It states that stress is proportional to strain within the elastic limits.

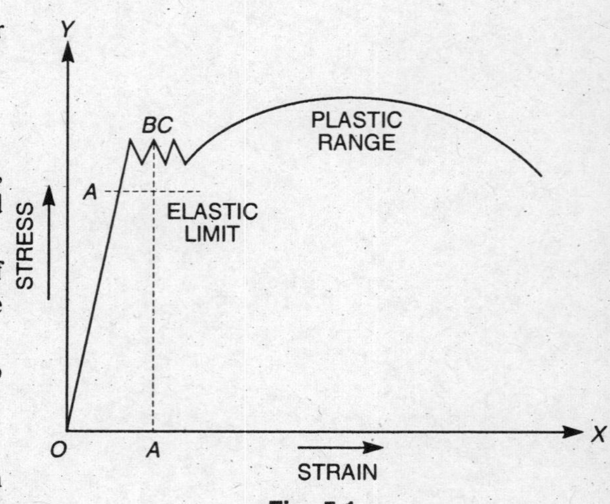

Fig. 5.1

Mathematically,

$$\text{Stress} \propto \text{Strain}$$

or
$$\frac{\text{Stress}}{\text{Strain}} = E \text{ (constant)}$$

This constant is called **coefficient of elasticity or modulus of elasticity**. Its value depends on the nature of the material.

If a graph is plotted between stress and strain, then a curve of the shape shown is obtained. This curve is called **stress-strain curve**.

From the curve it is evident that Hooke's Law is valid in elastic limits only. After the elastic limit, into the change of stress, strain changes more rapidly and after this the body does not come back to its original shape so it is in plastic limits.

Corresponding to the three types of strains, we have three elastic modulus.

(*i*) **Young's Modulus of Elasticity:** It corresponds to linear or tensile strain.

(*ii*) **Bulk Modulus or Volume Modulus:** It corresponds to volumetric strain.

(*iii*) **Shear Modulus or Modulus of rigidity:** It corresponds to shearing strain.

5.5. YOUNG'S MODULUS

It is defined as the ratio of the stress to longitudinal strain within the elastic limits.

Longitudinal strain corresponds to the condition when only change in length takes place under an applied load.

Let us take up a wire of length L and let it changes by l under an applied load, F acting on an area of cross-section a.

$$\text{Longitudinal Strain} = l/L$$
$$\text{Stress} = F/a$$

∴ Young's Modulus of Elasticity

$$Y = \frac{\text{Stress}}{\text{Strain}} = \frac{F/a}{l/L} = \frac{F \cdot L}{a \cdot l}$$

Units of Y are the units of stress as strain has no units. So unit of Y is N/m^2.

5.6. BULK MODULUS

It is defined as the ratio of stress to volumetirc strain.

If a force is applied normally over the surface of a body and only change in volume takes place, the strain is called **volumetric strain.** It is measured by change in volume per unit volume *i.e.*, v/V, where v is change in volume and V, the original volume.

So Bulk Modulus of Elasticity

$$K = \frac{F/a}{v/V} = \frac{FV}{av}$$

Also $F/a = P$ (applied pressure)

If under an applied small pressure dP, a small change in volume dV takes place, we may write Bulk Modulus as

$$K = -\frac{dP}{dV/V}$$

minus sign indicates that as the pressure is increased the volume decreases.

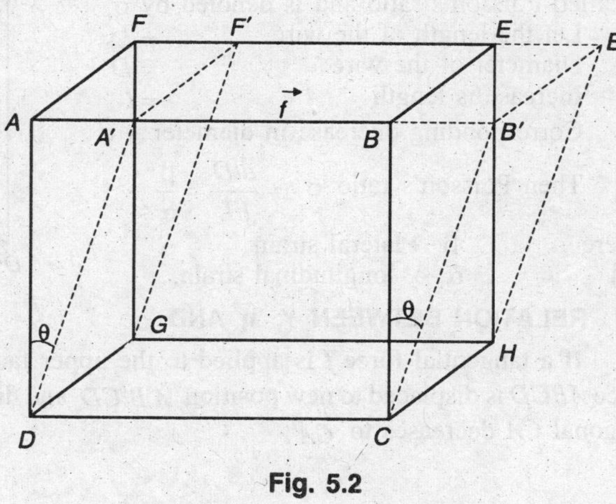

Fig. 5.2

5.7. MODULUS OF RIGIDITY

When a body is put under a force in such a way that its shape changes only without changing the size of the body, the body is said to under **shear stress.**

Normally a force applied tangentially changes the shape of the body without affecting its volume and so modulus of rigidity may be defined as the **ratio of the tangential force per unit area to angular deformation produced in the body.**

In order to find the volume of shearing strain, let us consider a solid cube *ABCDEFGH.*

The lower face *CDGH* is fixed and a tangential force *f* is applied over the face *ABEF* so that it changes to a new position *A'B'E'F'*. Each of the horizontal layer of the cube is displaced and the displacement is proportional to its distance from fixed plane.

Then, $$\text{Shearing strain} = \frac{\text{Displacement of a plane}}{\text{Dist. form the fixed plane}}$$

$$= \frac{AA'}{AD} = \frac{l}{L} = \tan\theta$$

If displacement is small

$$\tan\theta = \theta$$

$\therefore$ $$\text{Shearing strain} = \theta$$

Now Mudulus of rigidity $$= \frac{\text{Tangential stress}}{\text{Shearing strain}}$$

$$\eta = \frac{f/a}{\theta} = \frac{f}{a\theta} = \frac{T}{\theta}$$

where *a* is the area of face *ABEF* of the cube and *T* is the tangential force applied/unit area.

5.8. POISSON'S RATIO

If a body is subjected to some force or a system of forces, then the deformation in the body is not in one direction only but throughout the body. For example if a wire is stretched its length increases but its diameter decreases.

In general if elongation due to an applied force takes place in one direction a contraction results in the lateral dimensions of the body under strain. **This lateral strain is proportional to the longitudinal strain so long as it has a small volume.**

The ratio of lateral strain to longitudinal strain is called Poisson's ratio and is denoted by σ

Let the length of the wire $= L$
Diameter of the wire $= D$
Increase is length $= l$
Corresponding decrease in diameter $= d$

Then Poisson's ratio $$\sigma = \frac{d/D}{l/L} = \frac{\beta}{\alpha}$$

where $\beta \rightarrow$ lateral strain
and $\alpha \rightarrow$ longitudinal strain.

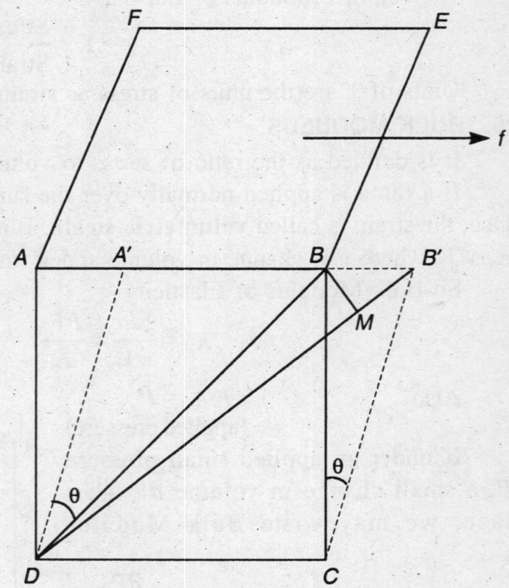

Fig. 5.3

5.9 RELATION BETWEEN *Y*, η AND σ

If a tangential force *f* is applied to the upper face *ABEF* of a cube *ABCDEFGH*, then the force *ABCD* is displaced to new position *A'B'CD* and diagonal *DB* increase in length to *DB'* while diagonal *CA* decreases to *CA'*.

$$\text{Shearing stress} = \frac{f}{\text{Area } ABEF} = T$$

Now the shearing stress along AB is equivalent to the tensile stress along DB and on equal compression stress along CA at right angles.

If α and β are considered to be longitudinal and lateral strains per unit stress respectively, then the increase of diagonal DB due to tensile stress $= DB.T.\alpha$ and increase in diagonal DB due to

$$\text{Compression stress along } AC = DB.T.\beta$$

$$\therefore \quad \text{Total increase along } DB = DB.T(\alpha+\beta)$$
$$= \sqrt{2}L.T(\alpha+\beta) \qquad \text{...(1)}$$

where L is the length of a side of the cube.

Now draw $BM \perp$ on DB'. Then increase in length of diagonal DB can be taken approximately as equal to $B'M$

Since θ is very small, so

$$\angle AB'C = 90' \text{ and } \angle BB'M = 45°$$

$$\therefore \quad B'M = BB'\cos 45° = \frac{BB'}{\sqrt{2}} = \frac{l}{\sqrt{2}}$$

where l is the change in length of the side of the cube.

Putting in Equation (1)

$$\frac{l}{\sqrt{2}} = T(\alpha+\beta)\cdot L\sqrt{2}$$

or
$$T\cdot\frac{L}{l} = \frac{1}{2(\alpha+\beta)}$$

also
$$\eta = \frac{T}{l/L} = \frac{TL}{l} = \frac{T}{\theta}$$

$$\therefore \quad \eta = \frac{1}{2(\alpha+\beta)} = \frac{1}{2\alpha(1+\beta/\alpha)} \qquad \text{...(2)}$$

Now by definition

$$\text{Poisson's Ratio } \sigma = \frac{\text{Lateral strain}}{\text{Longitudinal strain}} = \frac{\beta}{\alpha}$$

and
$$\text{Young's Modulus} = \frac{\text{Stress}}{\text{Longitudinal strain}}$$
$$= \frac{1}{\text{Longitudinal strain per unit stress}} = \frac{1}{\alpha}$$

$\therefore$ Equation (2) may be written as

$$\eta = \frac{Y}{2(1+\sigma)} \quad \text{or} \quad Y = 2\eta(1+\sigma) \qquad \text{...(3)}$$

RELATION BETWEEN *Y*, *K* AND σ

Let us consider a unit cube $ABCDEFGH$ and let forces T_x, T_y and T_z act to the face $BEHC$ and $AFGD$, $EFGH$ and $ABCD$, and $ABEF$ and $CDGH$ respectively as shown in Fig. 5.4.

If it corresponds to the increase per unit length per unit tension along the direction of the force, then the increase produced in the edge AB, BE and BC will be $T_x\cdot\alpha, T_y\cdot\alpha$, and $T_z\cdot\alpha$ respectively.

If β corresponds to the contraction produced per unit length per unit tension in a direction $\perp$ to the force, then the con- traction produced $\perp$ to the edges AB, BE and BC will be $T_x\cdot\beta$, $T_y\cdot\beta$ and $T_z\cdot\beta$ respectively.

So the length's of the edges will be as under:

$$AB = (1 + T_x \cdot \alpha - T_y.\beta - T_z.\beta)$$

$$BE = 1 + T_y \cdot \alpha - T_x \cdot \beta - T_z \cdot \beta$$

$$BC = 1 + T_z \cdot \alpha - T_x \cdot \beta - T_y \cdot \beta$$

So the new volume of the cube will be $AB \times BE \times BC$

$$= (1 + T_x \cdot \alpha - T_y \cdot \beta - T_z \cdot \beta) \times (1 + T_y \cdot \alpha - T_x \cdot \beta - T_z \cdot \beta)$$

$$\times (1 + T_z \cdot \alpha - T_x \cdot \beta - T_y \cdot \beta)$$

Since a, b are very small, so we can neglect their squares and higher powers.

$$\therefore \quad \text{Volume} = 1 + \alpha (T_x + T_y + T_z) - 2\beta (T_x + T_y + T_z)$$

$$= 1 + (\alpha - 2\beta)(T_x + T_y + T_z)$$

If the deforming forces acting on all the three faces are equal

i.e., $T_x = T_y = T_z$

$\therefore$ Volume $= 1 + 3T(\alpha - 2\beta)$

We can take another case also *i.e.,* if instead of applying stretching force T outwardly a pressure P is applied on all the three forces and the cube is compressed, then the contraction in volume will be $= 3P(\alpha - 2\beta)$

$$\therefore \text{Volume strain} = \frac{3P(\alpha - 2\beta)}{1}$$

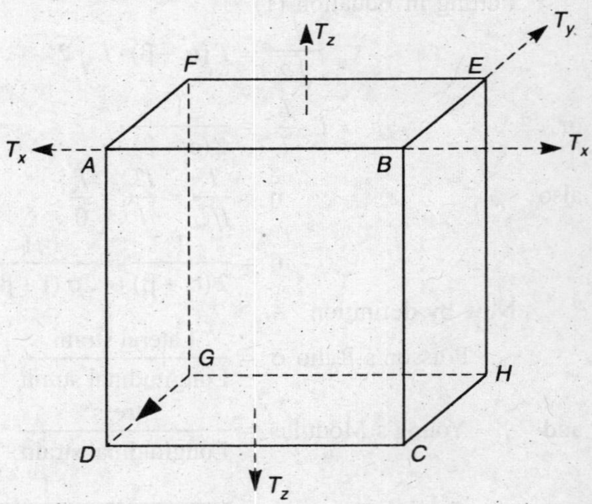

Fig.5.4

Now Bulk Modulus

$$K = \frac{\text{Stress}}{\text{Volume Strain}}$$

$$= \frac{P}{3P(\alpha - 2\beta)}$$

$$= \frac{1}{3(\alpha - 2\beta)}$$

which may be written as

$$K = \frac{\frac{1}{\alpha}}{3(1 - 2\beta/\alpha)}$$

$$= \frac{y}{3(1 - 2\sigma)} \qquad \text{...(4)}$$

(where $y = \dfrac{1}{\alpha}$ and $\beta/\alpha = \sigma$)

5.11. RELATION BETWEEN Y, η AND K

We have the relation between Y, η and σ as

$$\eta = \frac{Y}{2(1 + \sigma)} \qquad \text{...(1)}$$

and that between, Y, K and s as

$$K = \frac{Y}{3(1-2\sigma)} \qquad ...(2)$$

From Equation (1) we have

$$2+2\sigma = \frac{Y}{\eta} \qquad ...(3)$$

Also from Equation (2) we have

$$1-2\sigma = \frac{Y}{3K} \qquad ...(4)$$

Adding Equation (3) and (4) we have

$$3 = \frac{Y}{3K} + \frac{Y}{\eta} = Y\left[\frac{1}{3K} + \frac{1}{\eta}\right]$$

or
$$3\text{ Y} = \frac{(\eta+3K)}{3\eta K}$$

$\therefore$
$$Y = \frac{9\eta K}{\eta+3K} \qquad ...(5)$$

5.12. RELATION BETWEEN K, η AND σ

From Equation (1), we have

$$Y = 2\eta(1+\sigma) \qquad ...(6)$$

and from Equation (2), we have

$$Y = 3K(1-2\sigma) \qquad ...(7)$$

$\therefore$
$$2\eta(1+\sigma) = 3K(1-2\sigma)$$

or
$$2\eta+2\eta\sigma = 3K+6K\sigma$$

or
$$\sigma(2\eta+6K) = 3K-2\eta$$

or
$$\sigma = \frac{3K-2\eta}{2(\eta+3K)} \qquad ...(8)$$

LIMITS OF σ : FORM EQUATION (6) AND (7)

$$3K(1-2\alpha) = 2\eta(1+\sigma)$$

Now since bulk modulus K and coeff. of rigidity η are both positive quantities, for Equation (6) to be true and

(*i*) If σ is to be positive, right hand expression and left hand expression must be positive. This is possible if.

$$2\sigma < 1 \quad \text{or} \quad \sigma < 0.5$$

(*ii*) If σ is to be negative quantity, left hand expression is positive and the right hand expression must also be positive.

This is possible only if $1+\sigma$ is +ve or is not less than –1. So the limiting values of σ are –1 and 0.5.

5.13. TORSION OF A CYLINDER

Let us consider a cylinder of length l and radius a. The cylinder being clamped at the upper end *AB*. Let a twisting couple be applied to the face *CD* in the direction shown by an arrow and in a direction $\perp$ to the length of the cylinder.

The radius $O'P$ is twisted through an angle θ to another position $O'P'$. It should be noted that in this case no change in length or radius is produced due to twist, so this is an example of pure shear :

Now due to the elasticity of the material a restoring couple will be set up in the cylinder which will be equal and opposite to the twisting couple.

As is shown in Fig. 5.5 (*a*) line MP on the rim of the cylinder is || to $O'O'$ is displaced to another position MP' through an angle ϕ due to the application of twisting couple. In this case

ϕ will correspond to the angle of shear. It should be further noted that the displacement PP' is maximum for the points present on the rim and it goes on decreasing as we move towards O' *i.e.*, the centre of the cylinder.

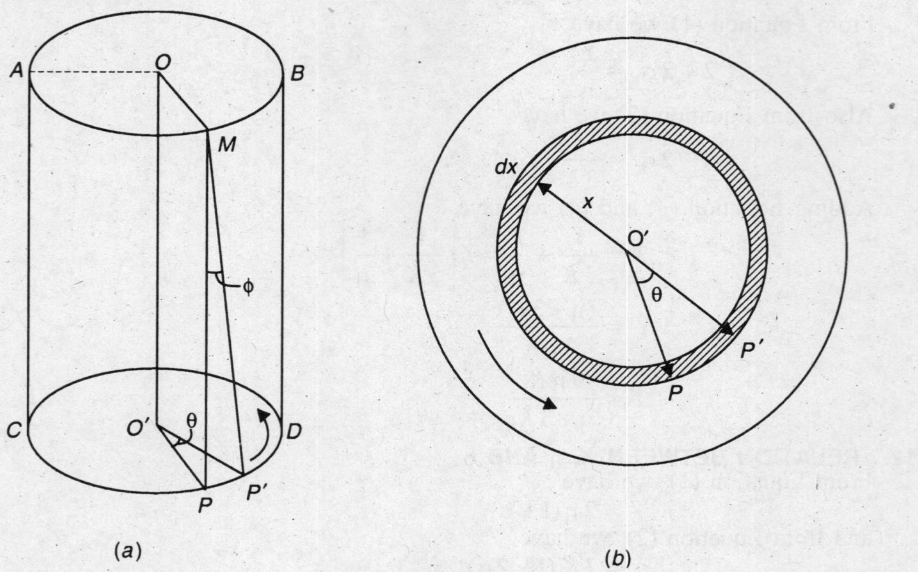

Fig. 5.5

In order to find the value of twisting couple, let the solid cylinder be considered to be divided into large no. of co-axial cylinderical shells and let the radius of one such shell be x and thickness be dx.

Further it should be noted that each radius of the lower end of the cylinder is twisted through same angle θ and the displacement is maximum at the rim. Their displacement decreases as we move from rim towards centre O', where it will be zero. Also angle of shear will be maximum when $x = a$ and minimum at O' which in other words means that shearing strain is not constant throughout the cylinder, it being maximum on the rim and minimum on the innermost side.

The angle of shear will be same for any one of the hollow cylinders, it being greatest for the outermost and least for the innermost cylinder.

Let us say that the points P and P' are on the rim of a hollow cylinder of radius x and let ϕ be the shear angle, then from the Fig. 5.5 (a) we have

$$PP' = l\phi \quad \text{...(1)}$$

Also from Figure 5.5 (b)

$$PP' = x\theta \quad \text{...(2)}$$

$$\therefore \quad l = x\theta$$

or

$$\phi = x \cdot \frac{\theta}{l} \quad \text{...(3)}$$

If η is the coeff. of rigidity, then

$$h = \frac{\text{Shear stress}}{\text{Angle of shear}} = \frac{T}{\phi}$$

$$\therefore \quad T = \eta\phi = \eta \cdot \frac{x\theta}{l} \quad \text{...(4)}$$

The face area of hollow cylinder = $2x. dx$

$\therefore$ Total force on this area

$$= 2\pi x \cdot dx \cdot \eta \frac{x\theta}{l}$$

$$= \frac{2\pi\eta\theta}{l} \cdot x^2 d x$$

and the moment of force about OO'

$$= \frac{2\pi\eta\theta}{l} \cdot x^2 d x \cdot x$$

Now we can get the twisting couple applied to the whole cylinder by integrating the above quantity between the limits $x = 0$ and $x = a$, *i.e.*,

$$\text{Twisting couple} = \frac{2\pi\eta\theta}{l} \int_0^a x^2 \cdot d x$$

$$= \frac{2\pi\eta\theta}{l} \cdot \left| \frac{x^4}{4} \right|_0^a$$

$$= \frac{2\pi\eta\theta}{l} \cdot \frac{a^4}{4} = \frac{\pi\eta\theta a^4}{2l} \qquad \text{...(5)}$$

If in the above relation $\theta = 1$ radian then twisting couple/unit angular

$$\text{twist} = \frac{\pi\eta a^4}{2l} \qquad \text{...(6)}$$

This twisting couple/unit angular twist is called angular twist of the wire or cylinder η is called its torsional rigidity or modulus of torsion.

From Equation (6) it is clear that Twisting couple/unit angular twist $\propto a^4$.

5.14. DETERMINATION OF COEFF. OF RIGIDITY OF A WIRE

The wire W of length l whose coeff. of rigidity is to be found is suspended from a torsion head and a heavy cylinder H is attached to it at lower end.

A couple is applied at the lower end with the help of two cords which are wound over the cylinder in such a way that these leave the cylinder tangentially on either end and then pass over to two frictionless pulleys as shown in Figure 5.6. Two equal masses m and m' are attached to the free ends of these cords.

Now the couple formed by these masses will rotate the cylinder about the wire as axis and thus the wire is twisted through an angle θ which is measured directly by a pointer attached to the wire and the pointer moves on a horizontal circular scales.

Let d be the diameter of the cylinder, then the twisting couple applied to the ends of the wire = $mg.d$

Also the twisting couple when the twist is θ radians form Equation (6)

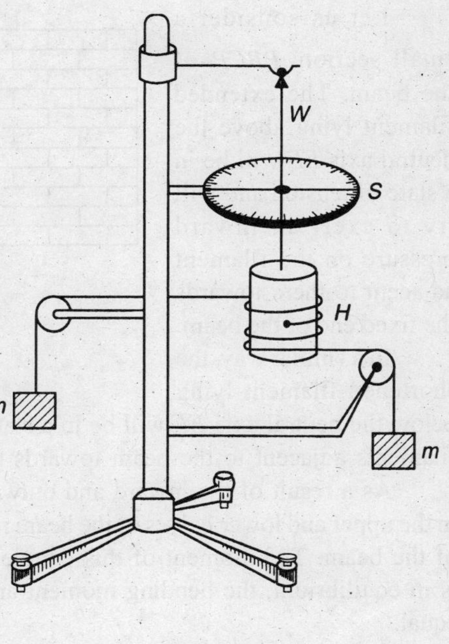

Fig. 5.6

$$= \frac{\pi\eta\theta a^4}{2l}$$

where $a \rightarrow$ radius of the wire

$l \rightarrow$ length of the wire

and $\eta \rightarrow$ coeff. of rigidity.

$\therefore \quad \pi\eta\theta a^4/2l = mgd$

or $$\eta = \frac{2l.m.g.d}{\pi \theta a^4}$$

In this case all the quantities *i.e.*, l, m, d, θ, a etc., can be measured and so η can be found. It should further be noted that radius a should be measured very accurately as in the above equation it occurs as a^4 and a small error in its measurement will cause large error in η.

5.15. BENDING-BEAM

A beam is a rod having uniform cross-section and whose length is very large as compared to the other dimensions, so that the shearing stress at any other section is very small and so negligible.

The beams are normally supported horizontally and are designed in such a way that they can support heavy loads such as in the case of buildings they support the roofs and in case of bridges they support the loads of heavy vehicles passing over them. If a beam is supported at one end and loaded at another end then it is called a Cantilever.

In order to study the bending of beams, we make use of the following assumptions.

(*i*) The beam is very light *i.e.*, its weight is negligible as compared to the load it supports.

(*ii*) The cross-section of the beam remains same so that the geometrical moment of inertia of the beam remains same.

(*iii*) Beam has a very small curvature.

(*iv*) There are no shearing forces.

It should further be noted that if a couple is applied in any direction to bend the beam, longitudinal filament on the convex side of the beam are extended while compression takes place on the concave side. In between these filament, which is neither increased nor decreased but remains constant in length. **This filament is known as neutral filament and the axis of this beam lying on the neutral filament is called the neutral axis.**

5.16. BENDING MOMENT OF THE BEAM

Let us consider a small section $PBCP'$ of the beam. The extended filament lying above the neutral axis EF will be in a state of tension and will try to exert an inward pressure on the filament adjacent to them towards the fixed end of the beam.

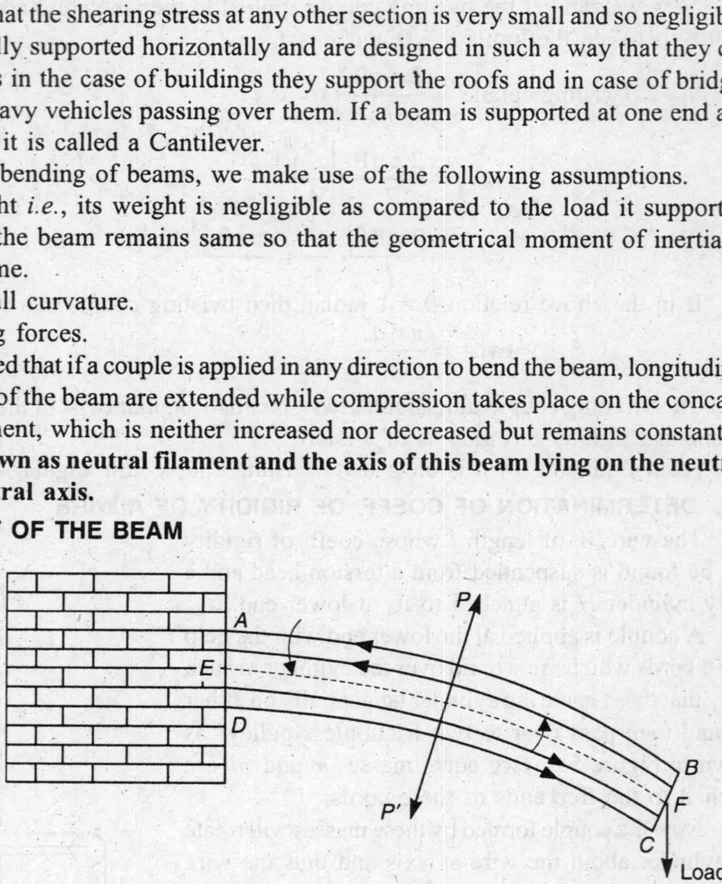

Fig. 5.7

In a similar way, the shortened filament lying below the neutral axis EF will be in a state of compression and will exert outward pressure on the filaments adjacent to the beam towards the loaded end of the beam.

As a result of the inward and outward pressures, tensile and compressive stresses develope in the upper and lower halves of the beam respectively and set up a couple which opposes the bending of the beam. The moment of this couple is called the moment of resistance and when the beam is in equilibrium, the bending moment and restoring moment or moment of resistance should be equal.

In order to get an expression for the moment of restoring couple, consider a fibre DE which is at a distance r from the neutral axis as shown in Fig. 5.8.

Let the radius of curvature of the beam of part PB be R and let ϕ be the angle subtended by it at the centre of curvature; If the beam is not strained, the length of fibre $DC = EF = R\,\phi$ and in the strained condition, length of fibre $DC = (R + r)\,\phi$

The strain in the fibre DC

$$= \frac{\text{Change in length}}{\text{Original length}}$$

$$= \frac{(R+r)\phi - R\phi}{R\phi} = \frac{r\phi}{R\phi} = \frac{r}{R}$$

which shows that strain is proportional to the distance from neutral axis.

Let the area of the fibre be a and the neutral axis of the fibre be at a distance r from the neutral axis of the beam.

Now Young's Modulus

$$Y = \frac{\text{Stress}}{\text{Strain}}$$

or $$\text{Stress} = Y \times \text{ strain} = Y \cdot \frac{r}{R}$$

So the force on area a will be

$$F = Y \cdot \frac{r}{R} \times a$$

and the moment of this force about EF

$$= Y \cdot \frac{r}{R} \times a \times r = Y \cdot \frac{a\,r^2}{R} \quad ...(2)$$

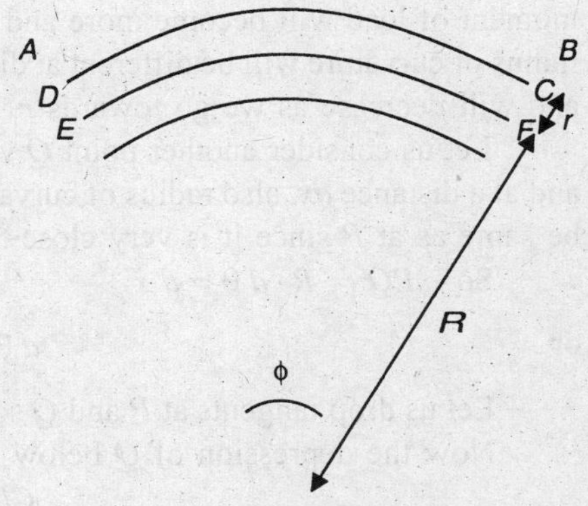

Fig. 5.8

Now since the moment of the force acting on both the upper and lower halves of the section are in the same direction, the total moment of the force acting on the filaments due to the strain will be

$$= \Sigma Y \cdot \frac{a\,r^2}{R} = \frac{Y}{R} \cdot \Sigma a\,r^2$$

$$= \frac{Y}{R} \cdot I_g$$

where I_g corresponds to geometrical moment of inertia and is given by $I_g = Ak^2$, A being total area of cross-section and k, the radius of gyration.

$\therefore$ Moment of force $= \frac{Y}{R} \cdot I_g$

Now in the equilibrium condition, the bending moment of the beam will be equal and opposite to the moment of bending couple due to the load put on one end.

$\therefore$ Bending moment of beam $= \frac{Y}{R} \cdot I_g$

The quantity $Y.\ I_g$ ($= Y.Ak^2$) is known as *Flexural Rigidity* of the beam and is defined as the **bending moment required to produce a unit radius of curvature.**

5.17. CANTILEVER

Cantilever is a beam fixed horizontally at one end and loaded at the other end.

5.18. THEORY OF CANTILEVER LOADED AT ONE END

Let EF represents the neutral axis of the cantilever of length l. The cantilever being fixed at the end E and loaded at F with a load W. Let the end F be deflected to F' under load W.

Now consider a section of the beam at P which is at a distance x from E, then the moment of couple due to load will be given by

$$W \times PF' = W \times (l - x),$$

where we have taken PF' is $\perp$ to FW as F' is very near to F.

Now the beam is in equilibrium, so the bending couple be must equal to the restoring couple

$= \dfrac{YI}{R}$, where R corresponds to the radius of curvature of the neutral axis at P, Y and I correspond to Young's modulus and geometrical M.I. of the beam respectively.

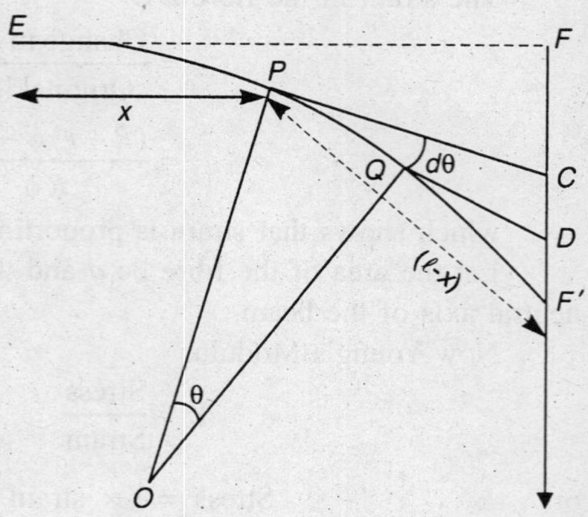

Fig. 5.9

$$\therefore \quad W(l-x) = Y \cdot \frac{1}{R}$$

or
$$\frac{I}{R} = \frac{W(l-x)}{YI} \quad ...(1)$$

Now as we move towards the fixed end r, the moment of load will become more and more. So the radius of curvature will be different at different points and will decrease as we go towards r.

Let us consider another point Q very near to P and at a distance dx, also radius of curvature at Q will be same as at P since it is very close to P.

So $\quad PQ = R \cdot d\theta = dx$

or
$$d\theta = \frac{dx}{R} = \frac{W(l-x)\,dx}{YI}$$

Let us drop tangents at P and Q so that these meet the vertical lines at C and D respectively. Now the depression of Q below P will be given by

$$CD = dy = (l-x)\,d\theta,$$

where $d\theta$ corresponds to the angle between two tangents.

$$\therefore \quad dy = (l-x) \cdot \frac{W(l-x)\,dx}{Y \cdot I}\,dx$$

$$= \frac{W}{YI}(l-x)^2\,dx$$

So total depression
$$FF' = p = \int_0^l \frac{W}{YI}(l-x)^2\,dx$$

$$= \frac{W}{YI}\int_0^l (l^2 + x^2 - 2xl)\,dx$$

$$= \frac{W}{YI}\left[l^2 x + \frac{x^3}{3} - 2\frac{x^2 \cdot l}{2}\right]_0^l$$

$$= \frac{W}{YI}\left[l^3 + \frac{l^3}{3} - \frac{2l^3}{2}\right]$$

$$= \frac{W}{YI}\left[\frac{l^3}{3}\right]$$

$$= \frac{Wl^3}{3YI} \quad ...(2)$$

5.19. CANTILEVER SUPPORTED AT TWO ENDS AND LOADED IN THE MIDDLE

Let us consider a rod which is supported at two knife edges A and B and load at centre D with a load W so that the reactive at the knife edges be $W/2$ in upper direction.

Now since the middle of the rod may be considered as practically horizontal, we may consider the above arrangement as equal to two inverted cantilevers fixed at c and load loaded at A and B with a load $W/2$ in upper direction. If l be the length of the beam or rod, then we have the depression from Equation (2) given by

$$y = \frac{\frac{W}{2}\cdot\left(\frac{l}{2}\right)^3}{3Y\cdot I} = \frac{W\,l^3}{48\,YI} \quad ...(3)$$

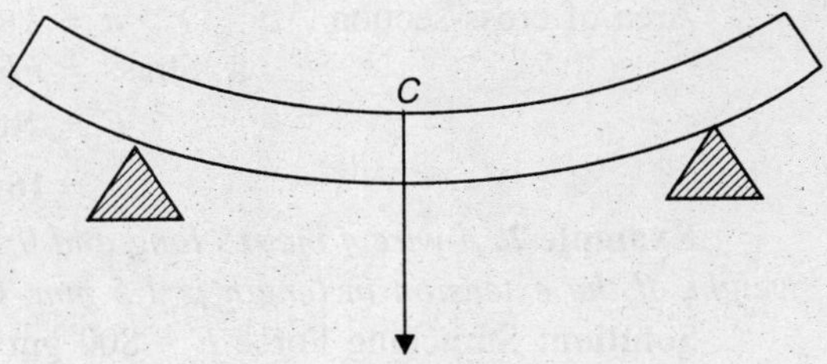

Fig. 5.10

Now we can take up the different cases.

(*i*) If the rod is circular in cross-section with radius r, we have the geometrical M.I. as

$$I = ak^2 = \pi r^2 \cdot \frac{r^2}{4} = \frac{\pi r^2}{4}$$

$$y = \frac{W\,l^3}{48\,Y}\cdot\frac{4}{\pi r^4} = \frac{W\,l^3}{12\,Y\cdot\pi r^4} \quad ...(4)$$

(*ii*) If the rod is rectangular and has b as breadth and d as depth, the geometrical M.I. is given by

$$I = n\,F = b\cdot d\frac{d^2}{12} = \frac{bd^3}{12}$$

$$\therefore \quad y = \frac{W\,l^3}{48\,Y}\cdot\frac{12}{bd^3} = \frac{W\,l^3}{4\,Y\,bd^3}$$

$$\therefore \quad y = \frac{W\,l^3}{4\,Y\,bd^3} \quad ...(5)$$

5.20. DETERMINATION OF *Y*

We can make use of a cantilever for finding Young's modulus in the following way.

The depression y at the centre of the rod can noted with the help of a spherometer or a travelling microscope. To get the mean value of depression, the rod first loaded in steps and depressions are noted and then it is unloaded again depressions are noted. Then mean value of y for a particular load is noted. Knowing the values of b, d, l etc, we can find Young's modulus from the relation

$$Y = \frac{W\,l^3}{4\,ba^3\,y}$$

Example 1. *While at O°C a square steel bar of 20 cm side is fixed at both ends so that it can not expand. If the temperature is raised to 20°C, what force would be exerted on the support at the ends? Given $Y = 2 \times 10^{12}$ dynes/cm^2 and coefficient of linear expansion of steel $= 10^{-5}$°C.*

Solution: Let L be the length of the wire at 0°C.

Coefficient of linear expansion $= 10^{-5}$°C.

$$l = L \times d \times 20$$

$$= L \times 10^{-5} \times 20$$

Longitudinal strain $= \dfrac{\text{Change in length}}{\text{Original length}}$

$$= \frac{L \times 10^{-5} \times 20}{L}$$

$$= 20 \times 10^{-5}$$

Young's Modulus $\quad Y = \dfrac{\text{Stress}}{\text{Strain}}$

$$\text{Stress} = Y \times \text{Strain}$$

$$= 2 \times 10^{12} \times 20 \times 10^{-5}$$

$$= 4 \times 10^{8} \text{ dynes/cm}^2$$

Area of cross-section $a = 2 \times 2 = 4 \text{ cm}^2$

$$\text{Stress} = F/a$$

$$\therefore \quad F = \text{Stress} \times a = 4 \times 10^8 \text{ dynes/cm}^2 \times 4 \text{ cm}^2$$

$$= \mathbf{16 \times 10^8 \text{ dynes}}.$$

Example 2. *A wire 4 metres long and 0.3 mm in diameter is stretched by a force of 800 gm weight. If the extension in length is 1.5 mm. Calculate the energy stored in wire.*

Solution: Stretching Force $F = 800$ gm wt

$$= 800 \times 980 \text{ dynes}$$

$$\text{Change in length} = 1.5 \text{ mm} = 0.12 \text{ cm}$$

Work done in extension of length

or Energy stored $= \frac{1}{2} \times$ streching force $\times$ extension in length

$$= \frac{1}{2} \times 830 \times 980 \times 0.15$$

$$= \mathbf{58.8 \times 10^3 \text{ ergs}}.$$

Example 3. *What force is required to stretch a steel wire to double its length when its area of cross-section 1 cm² and $Y = 2 \times 10^{-11}$ N/m².*

Solution: Let original length be L and change in length be l.

In this case Area of cross-section a 10^{-4} m²

and $$Y = 2 \times 10^{-11} N/m^2$$

Now $$Y = \frac{F/a}{l/L} = 2 \times 10^{-11}$$

$$\therefore \quad F = 2 \times 10^{11} \times 10^{-4}$$

$$= \mathbf{1 \times 10^7 \text{ N}}$$

Example 4. *A steel wire of length 2.3 m and cross-section 1×10^{-6} m is held between rigid supports with a tension 200 N. If the middle of the wire is pulled 5 m.m. side ways, calculate the change in tension. Also calculate, change in tension if temperature changes by 5°C.*

[For steel $Y = 2.2 \times 10^{11}$ N/m² and $\alpha = 8 \times 10^{-6}$]

Solution: Let the change in tension be F

$$\therefore \quad Y = \frac{\text{Stress}}{\text{Strain}} = \frac{F/A}{l/L} = \frac{F\,L}{A \cdot l}$$

or $$F = Y \cdot \frac{Al}{L} \quad ...(1)$$

Now when the wire is pulled 5×10^{-3} m side ways at the middle, the new length of the wire will be

$$L' = 2\sqrt{1^2 + (5 \times 10^{-3})^{-2}}$$

So change in length will be

$$l = L' - L = 2\sqrt{1^2 + (5 \times 10^{-3})^{-2}} \quad ...(2)$$

$$= 25 \times 10^{-6} \text{ m}$$

$\therefore$ From relation (1)

$$F = \frac{2.2 \times 10^{11} \times 10^{-6} \times 25 \times 10^{-6}}{2} = 2.75 \text{ N}$$

When the temp. changes by 5°C, the change in length of the wire is given by

$$l = L \cdot \alpha \cdot t$$

$$= 2 \times 8 \times 10^{-6} \times 5 = 8 \times 10^{-5} \text{ m}$$

$\therefore$ Change in tension

$$F = \frac{2.2\times10^{11}\times10^{-6}\times8\times10^{-5}}{2} = \mathbf{8.8\,N}$$

Example 5. *A material has Poisson's ratio 0.2. If a uniform rod of it suffers longitudinal strain* 4.0 × 10^{-3}, *calculate the change in volume.*

Solution: We have Poisson's ratio

$$= \frac{\text{Lateral strain}}{\text{Longitudinal strain}} = \frac{\Delta r/r}{\Delta l/l}$$

Given $\Delta l/l = 4.0\times10^{-3}$, $\Delta = 0.2$

$\therefore$ $\Delta r/r = 0.2\times4.0\times10^{-3}$

If V is initial volume and $V + \Delta V$ the final volumes then

$$V = \pi r^2\cdot l \qquad \text{...(1)}$$

and

$$V+\Delta V = \pi(r-\Delta r)^2\times(l+\Delta l) \qquad \text{...(2)}$$

Solving the two equations and neglecting the higher powers, we have

$$\Delta V = \pi r^2 \Delta l - 2\pi r l \Delta r$$

$$\therefore \quad \frac{\Delta V}{V} = \frac{\Delta l}{l} - 2\cdot\frac{\Delta r}{r}$$

Putting the values of $\frac{\Delta l}{l}$ and $\frac{\Delta r}{r}$; we have

$$\frac{\Delta V}{V} = 4.0\times10^{-3} - 2\times0.8\times10^{-3}$$

$$= 2.4\times10^{-3}$$

$\therefore$ % of change in volume (2.4 × 10^{-3}) × 100 = **0.24%.**

Example 6. *A wire of length 1 m and diameter 1 mm is clamped at one of its ends. Calculate the couple required to twist the other end by* 90°. Given η = 2.8 × 10^{10} N/m^2.

Solution: The torque required to twist a wire of length l clamped at one end through ϕ radians will be

$$\tau = \frac{\pi\eta r^4}{2l}\cdot\phi$$

Now $= 90° = \frac{\pi}{2}$ radian

$$\eta = 2.8\times10^{10}\ N/m^3,\ l = 1\,\text{m}$$

$$r = \frac{1}{2}\ \text{mm} = 0.0005\ \text{m}$$

$$\therefore \quad = \frac{\pi^2\times2.8\times10^{13}\times(5\times10^{-4})^4}{4}$$

$$= \mathbf{4.31\times10^{-3}\,N\text{-}m.}$$

Example 7. *Find the amount of work done in twisting a steel wire of radius 1 mm and length 25 cm through an angle of* 45°. *The modulus of rigidity of steel is* 8 × 10^{10} N/m^2.

Solution: The work done by some torque τ in twisting through an angle θ is $\tau\,\theta$. So the work done for producing some twist θ_0 is given by

$$W = \int_0^{\theta_0} \tau\cdot\theta\cdot d\theta = \frac{1}{2}\tau\theta_0^2$$

We have the relation

$$r = \frac{\pi\eta r^4}{2l}$$

$$\therefore \quad W = \frac{1}{2}\cdot\frac{\pi\eta r^4}{2l}\cdot\theta_0^2$$

In this case

$$\eta = 8\times10^{10}\ N/m^2, \theta_0 = \frac{\pi}{4}$$

$$r = 10^{-3}\text{ m and } l = 0.25\text{ m}$$

$$\therefore \quad W = \frac{1}{2}\times\frac{\pi\times8\times10^{46}\times(10^{-3})^4}{2\times0.25}\times\left(\frac{\pi}{4}\right)^2$$

$$= \mathbf{15.5\times10^{-2}\ joule}.$$

Example 8. *A circular bar, one metre long and 0.8 cm in diametre is rigidly clamped at one end in a vertical position. A couple of magnitude* 2.5 × 10^7^ *dynes/cm is applied at the other end. As a result a mirror fixed at this end deflects a spot of light by 15 cms on a scale 1 m away. Calculate the modulus of rigidity of the bar.*

Solution: In this case $l = 100$ cm

radius $a = 0.4$ cm

Twisting couple $= 2.5\times10^7$ dynes/cm

$\eta = ?$

Since the spot of light is deflected through 15 cms on a scale 1 m away, the angle 20 through which the reflected ray turns is given by

$$\tan 2\theta = 0.15$$

or $$2\theta = 8°32'$$

or $$\theta = 4°16' = 0.0745\text{ radian}$$

Now using the relation

$$\text{Twisting couple} = \frac{\pi\eta\theta a^4}{2l}$$

$$\therefore \quad 2.5\times10^7 = \frac{\pi\times\eta\times0.0745\times(0.4)^4}{2\times100}$$

or $$\eta = \frac{2.6\times10^7\times200}{3.142\times0.0746\times(0.4)^4}$$

$$= \mathbf{8.34\times10^{11}\ dynes/cm^2}.$$

Example 9. *A gold wire 0.82 mm in diameter elongates by 1 mm when stretched by a force of 330 gms wt and twists through one radian when equal and opposite torques of 145 dynes/cm are applied at its ends. Find the value of Poisson's ratio for gold.*

Solution: In this case

$$F = 330\times980\text{ dynes}$$

$$l = 1\text{ mm} = 0.1\text{ cm}$$

$$\text{Diameter} = 0.32\text{ mm} = 0.032\text{ cm}$$

$$\therefore \quad \text{radius } a = 0.016\text{ cm}$$

Area of cross-section $= \pi\times(0.016)^2$ cm^2

It L is the length of wire, then Young's modulus of elasticity is given by

$$Y = \frac{330\times980\times L}{\pi\times(0.016)^2\times0.1}\text{ dynes/cm}^2$$

Couple/unit angular twist that acts an wire is given by

$$\text{Couple} = 145/1 = 145\text{ dynes cm.}$$

Also we have, Couple/unit angular twist

$$= \frac{\pi\eta a^4}{2L}$$

$$\eta = \frac{145\times2L}{\pi a^4} = \frac{145\times2L}{\pi\times(0.016)^4}$$

Now $$\eta = \frac{Y}{2(1+\sigma)}$$

or $$2(1+\sigma) = \frac{Y}{\eta} = \frac{330 \times 980 \times L}{\pi \times (0.016)^2 \times 0.1} \times \frac{\pi \times (0.016)^2}{145 \times 2L}$$

$$= 2.856$$

$$\therefore \quad 1+\sigma = 1.428$$

$$\boldsymbol{\sigma = 0.428.}$$

Example 10. *In an experiment the diameter of a rod was 1.26 cms and the distance between the knife edges 70 cms. On putting a load of 900 gms at the middle point the depression was 0.025 cm. Calculate the Young's Modulus of the substance.*

Solution: In this case $r = 0.63$ cm

$l = 70$ cm, $y = 0.025$

$w = 900$ gms, $Y = ?$

$$Y = \frac{W l^3}{12 Y \pi r^4}$$

Now depression is given by

$$y = \frac{W l^3}{12 Y \pi r^4} = \frac{900 \times 980 \times (70)^3 \times 7}{12 \times 0.025 \times 22 \times (0.63)^4}$$

$$= \mathbf{20.40 \times 10^{11}\ dynes/cm^2}.$$

Example 11. *One end of a wire of 3 mm radius and 100 cm in length is twisted through 60°. Calculate the angle of shear on its surface.*

Solution: If ϕ is the angle of twist at the free end and θ, the angle of shear then

$$\theta = r \cdot \frac{\phi}{l}$$

In this case $l = 1$ m, $r = 4 \times 10^{-3}$ m

and $\phi = 30°$

$$\therefore \quad \theta = \frac{4 \times 10^{-3} \times 60}{1} = \mathbf{0.24°}.$$

QUESTIONS AND PROBLEMS

1. Explain the terms stress, strain and Young's Modulus of elasticity.
2. Explain Young's Modulus of rigidity and Bulk Modulus.
3. State and explain Hooke's law. What is a yield point?
4. Define Poissions' ratio σ. Derive a relation between them.
5. Prove the relation in elasticity
$$\eta = \frac{Y}{2(1+\sigma)}$$
6. If Y, k and σ represent Young's Modulus, Bulk Modulus and Poisson's ratio respectively, then prove that
$$k = \frac{Y}{3(1-\sigma)}$$
7. Drive the various relations between Y, k, η and σ, where the symbols have usual meanings in elasticity. Also give the limiting values of σ.

8. Show that when a cylinder is twisted the torsional couple/unit angular twist is $\frac{\eta \pi a^4}{2l}$, where the symbols have the usual meaning.
9. Derive an expression for the couple required to bend a uniform straight metallic strip into an arc of a small curvature.
10. Describe with relevent theory how a Cantilever may be used to find the Young's Modulus of Elasticity.
11. A steel wire 2 mm in diameter is just strecthed between two fixed points at a temperature of 20°C. Determine its tension when temperature falls to 10°C. Coeff. of linear expansion for steel $= 11 \times 10^{-6} °C$ and Y for steel $= 2 \times 10^{11}\ N/m^2$ **[Ans.** 69.14 N]
12. A fixed beam of 50 cm is depressed by 15.0 mm at the load end. Calculate the depression a distance 0.3 m from fixed end. **[Ans.** 36.48 mm]
13. The end of a given strip cantilever depresses 10 mm under a certain load. Calculate the depression under the same load for another cantilever of same material, 2 times in length, 2 times in width and 3 times in thickness. **[Ans.** 1.48 mm]
14. A unifrom rigid rod 1.2 m long is clamped horizontally at one end. A weight of 1 kgm is attached to the free end. Calculate the depression of a point 0.9 m distance term the clamped end. Diameter of the rod is 2 cm. Y for material of rod is 1.013×10^{10} N/m^2.**[Ans.** 0.48 mm]

CHAPTER

6

VISCOSITY

6.1. INTRODUCTION

When a liquid moves with a steady speed over a solid and the motion is quite slow then it is observed that the layer of the liquid in contact with the solid is more or less stationary. In other words the velocity of the liquid along the surface of the solid is zero and the velocity of any other layer of the liquid will be **proportional to its distance from the stationary layer and it will be maximum for the most distant layer *i.e.*, the topmost layer.**

If a specific layer of a liquid is taken then it is observed that a layer above it is *moving faster and a layer immediately below is moving slower.*

So we see that an upper layer tends to increase the velocity of the lower layer while a lower layer tends to decrease the velocity of the upper one. In other words the two layers try to destroy their relative motions in a similar way as if some backward force is acting tangentially on the layers. So if a relative velocity is to be maintained between the two layers of a liquid an external force is required to overcome the backward force.

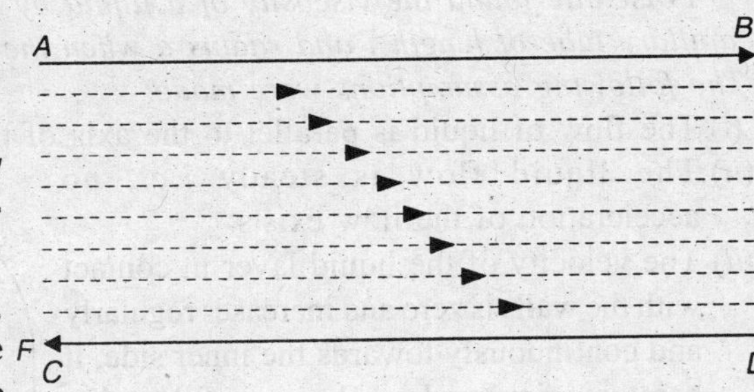

Fig. 6.1

This property by virtue of which a liquid opposes relative motion between its different layers is known as viscosity.

Liquids such as kerosine oil, alcohol, water etc., flow easily and are called **mobile** while liquids like tar, pitch, glycerine etc., move with difficulty and are called **viscous.**

Coefficient of Viscosity. Let us consider a layer AB of a liquid moving with a velocity v w.r.t. a parallel layer CD which is at a distance r from it.

Let us consider that the force required to produce the motion be F acting on an area A and this force is acting along the direction AB *i.e.*, along the direction of motion. An equal force will, therefore, act on it in the backward direction due to viscosity and this backward force F will depend upon the following factors:

(*i*) $$F \propto -v$$

i.e., F is directly proportional to relative velocity and acts in direction opposite to the direction of motion.

(*ii*) $$F \propto A$$

(*iii*) $$F \propto \frac{1}{r}$$

∴ From above

$$F = -\eta A \frac{v}{r}$$

where η is called coefficient of viscosity and it depends on the nature of a liquid.

If the two layers of a liquid are very close to each other, then we may write the above relation as

$$F = -\eta A \frac{dv}{dr}$$

where $\frac{dv}{dr}$ is called *velocity gradient* or rate of change of velocity with distance.

If $A = 1 \text{ cm}^2$ and $\frac{dv}{dr} = 1$

Then $F = \eta$ (In magnitude)

So the coefficient of viscosity may be defined as the tangential force per unit area required to maintain a unit velocity gradient *i.e.*, a unit relative velocity between two layers which are a unit distance apart..

If $F = 1$, then $\eta = 1$ and this unit is called **Poise.**

6.2. DIMENSIONS OF η

From above relation

$$\eta = \frac{Fr}{Av}$$

Dimensions of η are

$$[\eta] = \frac{[M^1L^1T^{-2}][L^1]}{[L^2][L^1T^{-1}]} = [M^1L^1T^{-1}]$$

6.3. POISEUILLE'S EQUATION

Poiseuille found the viscosity of a liquid by measuring the volume of the liquid through a capillary tube of length i and radius a when the liquid is flowing under a constant pressure P. The following assumptions were made:

(*i*) The flow of liquid is parallel to the axis of the tube throughout and it is stream line.

(*ii*) The liquid flow is steady *i.e.*, no acceleration of the flow exists.

(*iii*) The velocity of the liquid layer in contact with the walls is zero and increases regularly and continuously towards the inner side, it being maximum along the axis of the tube.

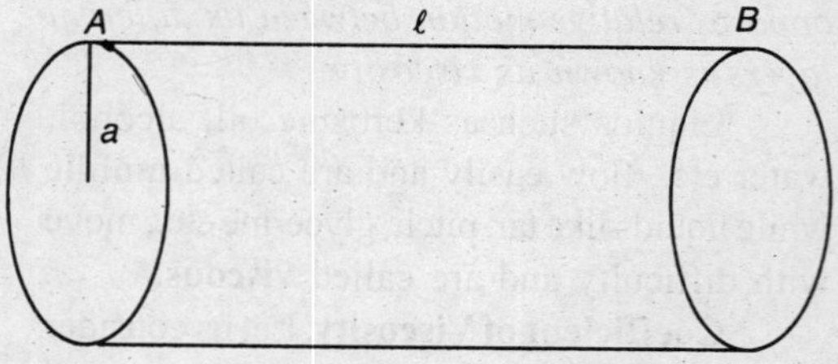

Fig. 6.2

Theory. Let *AB* be a capillary tube of length *l* and radius *a* as shown in Figure 6.2.

Consider a cylindrical layer of the liquid co-axial with the tube of inner radius *r* and outer radius *r* + *dr*.

Now the velocity θ the liquid at a distance *r* from the axis of the tube is *v* and at a distance *r* + *dr* is *v* – *dv*. So that $\frac{dv}{dr}$ is velocity gradient.

Surface area of the cylinder

$$= 2\pi r l$$

As stated above, the liquid on the inner side of this cylindrical layer moves faster while that on the other sides moves slower.

The tangential force exerted by the outer layer on the inner layer is opposite to the direction of motion and is given by

$$F = -\eta \cdot 2\pi r l \cdot \frac{dv}{dr}$$

Now the forward push due to the difference of pressure *P* on two sides of the tube or cylinder of radius *r*

$$= P \times \pi r^2$$

When the motion is steady, *i.e.*, no acceleration is imparted to the liquid

$$= -\eta \cdot 2\pi r \cdot l \cdot \frac{dv}{dr} = P \times \pi r^2$$

or
$$dV = -\frac{P}{2\eta l} \cdot r \cdot dr$$

Integrating both sides

$$\int dV = \int \frac{P}{2\eta l} r \cdot dr$$

$$V = -\frac{P}{2\eta l} \cdot \frac{r^2}{2} + c \quad ...(1)$$

where c is integration constant.

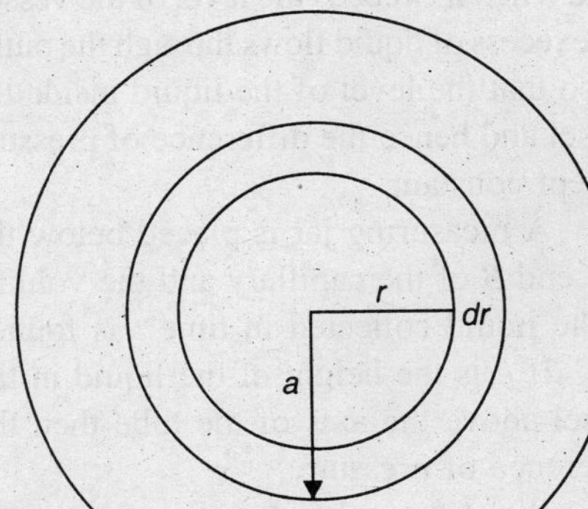

Fig. 6.3

Now the velocity of the liquid along the sides of the tube is zero.

i.e., if $r = a, \quad v = 0$

Putting in Equation (1)

$$0 = -\frac{P}{2\eta l} \cdot \frac{a^2}{2} + c$$

or
$$c = \frac{P}{2\eta l} \cdot \frac{a^2}{2}$$

$\therefore$
$$V = \frac{P}{4\eta l}[a^2 - r^2] \quad ...(2)$$

which gives an average value of the velocity of flow of liquid.

Now cross-sectional area of the cylinder of radius r and thickness dr is $= 2\pi r.dr$ and the volume of the liquid flowing through this area/sec is given by

$$dV = V \cdot 2\pi r \cdot dr$$

So the volume of the liquid flowing through the whole tube/sec is given by

$$\int dV = \int_0^a 2\pi V \cdot r dr$$

or
$$V = \int_0^a \frac{P}{4\eta l}(a^2 - r^2) 2\pi r\, dr \quad \text{[From Equation (2)]}$$

$$= \frac{\pi P}{2\eta l} \int_0^a (a^2 r - r^3)\, dr$$

$$= \frac{\pi P}{2\eta l}\left[\frac{a^2 \cdot r^2}{2} - \frac{r^4}{4}\right]_0^a$$

$$= \frac{\pi P}{2\eta l}\left[\frac{a^4}{2} - \frac{a^4}{4}\right]$$

$$= \frac{\pi P}{2\eta l} \cdot \frac{a^4}{4}$$

$$\eta = \frac{\pi P a^4}{8 V l} \quad ...(3)$$

So if P, a, l and V are known can be found.

6.4. EXPERIMENTAL ARRANGEMENT

A constant level vessel is filled with a liquid and a capillary tube AB of uniform circular bore is fixed horizontally in the side of the vessel.

The liquid enters the vessel through i and when it exceeds the level of the vessel, the execess of liquid flows through the outlet O, so that the level of the liquid inside the vessel and hence the difference of pressure is kept constant.

A measuring jar is placed below the free end B of the capillary and the volume of the liquid collected in time t is found.

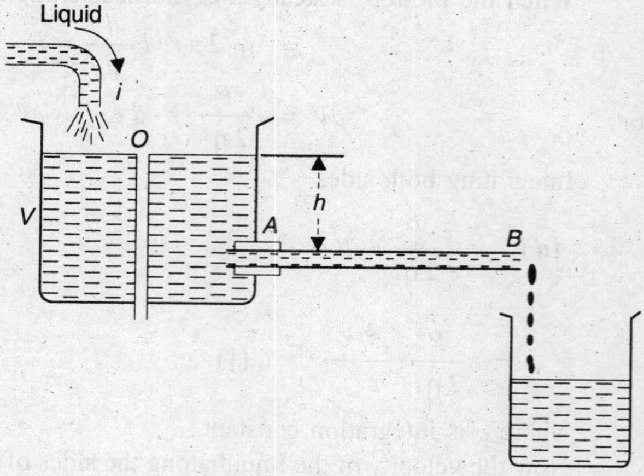

Fig. 6.4

If h is the height of the liquid in the vessel above the axis of the tube then the difference of pressure

$P = h\rho g$ where ρ of in the density the liquid.

The length of the capillary tube is measured by a meter rod and radius of the tube is measured accurately by a high power microscope, knowing these values η can be found from the relation

$$\eta = \frac{\pi P a^4}{8 l V}$$

6.5. OSTWALD'S VISCOMETER

Viscosities of two liquids can be compared with Ostwald's viscometer.

Construction. It consists of two glass bulbs A and B which are joined by a U-shaped capillary tube. One end of the bulb A is connected to a funnel through which liquid can be poured and another end of B is connected through a stop cock S to an exhaust pump. In order to have good thermal insulation, the whole apparatus is placed in a constant temperature bath.

Theory. The flow of liquid from bulb B to bulb A is controlled by the capillary tube DC and the volume passing/sec is given by

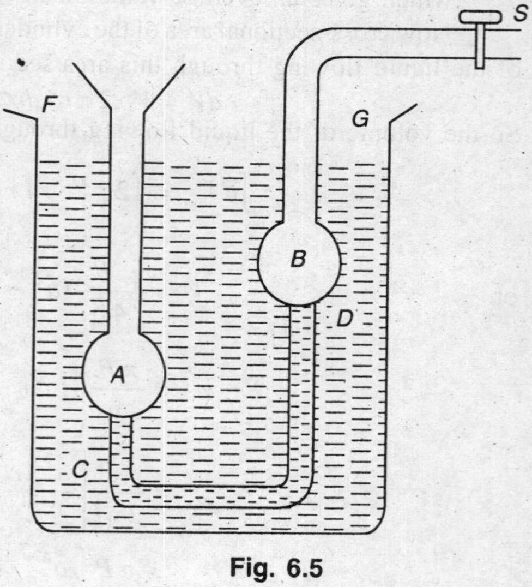

Fig. 6.5

$$V = \frac{\pi a^4}{8 \eta l} \quad ...(1)$$

where $P \rightarrow$ Difference of pressure at two ends

$l \rightarrow$ length of the capillary

$a \rightarrow$ radius of the capillary

For first liquid, we have

$$V_1 = \frac{\pi P_1 a^4}{8 \eta_1 l} \quad ...(2)$$

and for second liquid

$$V_2 = \frac{\pi P_2 a^4}{8 \eta_2 l} \quad ...(3)$$

$$\therefore \quad \frac{V_1}{V_2} = \frac{P_1}{P_2} \cdot \frac{\eta_2}{\eta_1} \qquad ...(4)$$

Now since the pressure is proportional to density and since both the liquids move under the same conditions.

$$\therefore \quad P_1 \propto \rho_1 \text{ and } P_2 \propto \rho_2$$

or
$$\frac{P_1}{P_2} = \frac{\rho_1}{\rho_2} \qquad ...(5)$$

If Q be the quantity of one liquid flowing through capillary in time t_1 and the same quantity of the other liquid flowing in time t_2, then

Rate of flow

$$V_1 = \frac{Q}{t_1} \text{ and } V_2 = \frac{Q}{t_2} \qquad ...(6)$$

or
$$\frac{V_1}{V_2} = \frac{t_2}{t_1} \qquad ...(7)$$

Putting the value of $\frac{P_1}{P_2}$ from Equation (5) and $\frac{V_1}{V_2}$ from Equation (6) in Equation (4)

We have

$$\frac{\eta_2}{\eta_1} = \frac{t_1}{t_2} \times \frac{\rho_2}{\rho_1} \qquad ...(8)$$

Working. A liquid is filled through the funnel so that it occupies volume *GDC* of the apparatus. Points *G, D* and *C* are fixed masses on the apparatus.

Now the stop cork *C* is slightly opened and the pump is put to a small work so that the liquid rises slightly above *G* and then stop cock is closed. At this stage the pump is removed and stop cock again opened so that pressure on the two sides of the tube is now same and the liquid flows towards bulb *A* to equilize level.

A stop watch is started and time noted when the liquid level passes the mark *G* and when it passes the mark *D*, *i.e.*, the time of flow of liquid from *G* to *D* is noted and let it be t_1.

Similarly time t_2 for the second liquid passing through *G* and *D* is noted.

Then from relation

$$\frac{\eta_2}{\eta_1} = \frac{t_1}{t_2} \times \frac{\rho_2}{\rho_1}.$$

the ratio $\frac{\eta_2}{\eta_1}$ can be calculated.

In the above case

$\rho_1 \rightarrow$ density of 1st liquid.

$\rho_2 \rightarrow$ density of 2nd liquid.

6.6. STOKE'S LAW

If a body falls through a fluid say a liquid or a gas then it carries along with it the layer of the fluid in contact and so will tend to produce some relative motion between the layers of the fluid. This relative motion is opposed by forces of viscosity and the opposing force increases as the velocity of the body increases and if the body is small enough then this opposing force soon becomes equal to the driving force that produce motion. **In that case the body moves with a constant velocity called terminal velocity.**

Let us consider a small sphere falling through a viscous medium, then the opposing force . directly proportional to the velocity v of the sphere and also will be depending on

(*i*) coefficient of viscosity of the medium

(*ii*) radius of the sphere

(*iii*) density of the medium.

Combining all the above factors we have

$$F = k.v.\eta^a.r^b.d^c$$

where k is a constant and a, b, c are dimensional coefficients.

Putting the dimensions of various quantities on two sides, we have

$$[M^1L^1T^{-2}] = [L^1T^{-1}][M^aL^{-a}T^{-a}][L^b][M^cL^{-3c}]$$

$$= M^{a+c} \cdot L^{1+b-a-3c} \cdot T^{-1-a}$$

Comparing the coefficients on two sides

$$a + c = 1 \quad ...(1)$$

$$1 + b - a - 3c = 1$$

or $$b - a - 3c = 0 \quad ...(2)$$

and $$-1 - a = -2$$

or $$a = 1 \quad ...(3)$$

Putting $a = 1$ in Equation (1)

we get $c = 0$

Putting $a = 1$ and $c = 0$ in (2)

we get $b = 1$

$$\therefore \quad F = k \cdot v \cdot \eta \cdot r$$

The value of k was calculated by stokes and it come out to be 6π.

$$\therefore \quad F = 6\pi\eta r v \quad ...(4)$$

If P is the density of the material of sphere, then downward force with which the sphere falls under gravity $= \frac{4}{3}\pi r^3 \cdot \rho \cdot g$

Upward thrust on the sphere due to buoyancy = wt. of the fluid displaced $= \frac{4}{3}\pi r^3 d\, g$

Net downward force $= \frac{4}{3}\pi r^3 \cdot (\rho - d)\, g$

Hence we have

$$= \frac{4}{3}\pi r^3 \cdot (\rho - d)\, g = 6\pi\eta r v$$

or $$v = \frac{2}{9} \cdot \frac{r^3(\rho - d)\, g}{\eta}$$

This is the Stoke's formula for velocity of a small sphere falling through a fluid.

6.7. DETERMINATION OF VISCOSITY OF A LIQUID

The viscosity of a viscous liquid like castor oil or glycerine is found by finding the terminal velocity of a small sphere of suitable size which is falling through a liquid.

Let a small sphere of radius r and density ρ be falling freely under the force of gravity through a viscous fluid of density d. After falling through a small distance the sphere attains a constant terminal velocity v. At this stage the viscous force F is equal to the wt. of sphere and upward thrust due to buoyancy.

The apparatus consists of a tall jar of 100 cm long and 10 cm in diameter in which a viscous liquid is filled and a steel ball of 1 cm radius is dropped into the liquid.

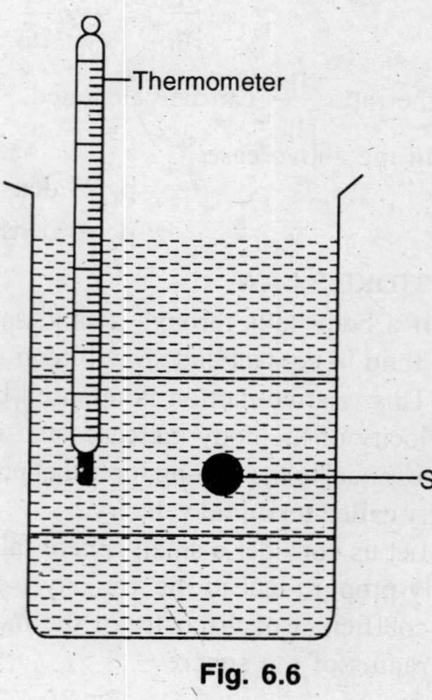

Fig. 6.6

After falling about 10 to 15 cms, it acquires a constant velocity and then time t taken by the ball to pass through distance AB is noted.

Terminal velocity

$$v = \frac{AB}{t}$$

radius of the ball or sphere is noted accurately with a travelling microscope and then η is found from the relation

$$\eta = \frac{2}{9} \cdot \frac{r^3(\rho - d)g}{v}$$

Since the jar has a finite size, so we have to apply a correction for **wall effect and end effect,** which according to Laudenberg are

$$\left(1 + 2.4 \times \frac{r}{R}\right) \quad ...(i)$$

and $$\left(1 + 3.3 \times \frac{r}{l}\right) \quad ...(ii)$$

respectively, where

$R \rightarrow$ radius of jar and $l \rightarrow$ length of the jar.

Applications. Stoke's formula is used to find the charge on an ion which is moving in an electric field as is done in Millikan's oil drop method.

6.8. VISCOSITY OF A GAS ON THE BASIS OF KINETIC THEORY OF GASES

Maxwell explained the viscosity of a gas on the basis of kinetic theory. Let us consider the motion of a gas along horizontal direction and let the gas be having large number of horizontal layers moving over layers of horizontal surface OX as shown in Fig. 6.7. In this case the velocity of the gas molecules moving along the wall of the surface will be zero and the velocity of gas molecules will increase with increase in distance from the fixed layer.

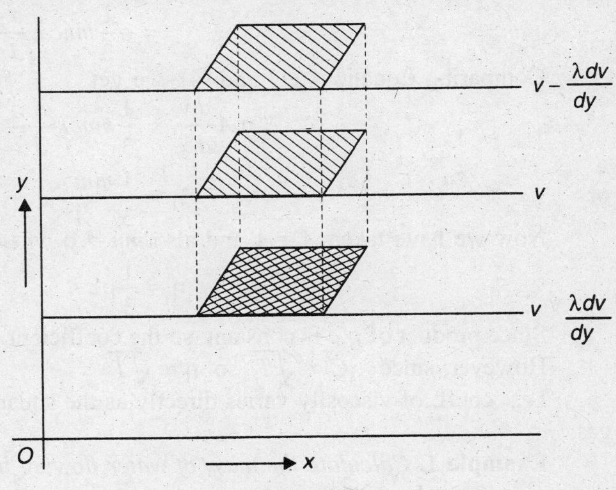

Fig. 6.7

Now from definition viscosity corresponds to *tangential force/unit area required to maintain a unit velocity gradient, i.e.,*

$$F = -\eta A \frac{dv}{dx} \quad ...(1)$$

where

$\eta \rightarrow$ coeff. of viscosity

$A \rightarrow$ area of the layers

and $dv/dx \rightarrow$ velocity gradient

Now dv/dx corresponds to late of change of velocity with distance from fixed layer, so the velocity of a molecule at a distance x above the fixed layer will be $= x \cdot \frac{dv}{dx}$

Let λ be the mean free path of the gas molecules, so the fixed collision will occur after travelling through a distance $(x - \lambda)$ and so the average velocity of the molecule will be

$$= (x - \lambda)\, dv/dx \qquad ...(2)$$

Also the molecules are at random motion so they travel freely in all the three directions. It means that we can consider 1/3 molecules moving in any one direction and out of 1/3 molecules, 1/2 can move in positive direction and remaining half can move in negative direction.

Let $m \rightarrow$ mass of each molecule

$n \rightarrow$ number of molecules

and $c \rightarrow$ r.m.s. velocity

$\therefore$ Number of molecules crossing *per unit area per second* in one direction will be

$$= 1/6\ nc.$$

Mass of all the molecules $= 1/6\ nmc = M$, moving in one direction

Momentum in forward direction

$$= M[(x - \lambda)\, dv/dx]$$

Momentum in backward direction

$$= M[x - (-\lambda)\, dv/dx]$$

$\therefore$ Rate of change of momentum *i.e.*, force responsible for the motion of gas molecules

$$F = M\,(x - \lambda)\frac{dv}{dx} - M\,(x + \lambda)\frac{dv}{dx}$$

$$= -2M\lambda\frac{dv}{dx}$$

$$= -2\left[\frac{1}{6}mnc\right]\lambda\frac{dv}{dx} \qquad \left[\therefore M = \frac{1}{6}mnc\right]$$

$$= -\frac{1}{3}mnc\,\lambda\frac{dv}{dx} \qquad ...(4)$$

Comparing Equations (1) and (4), we get

$$\eta A\frac{dv}{dx} = \frac{1}{3}mnc\lambda\frac{dv}{dx}$$

or

$$\eta = \frac{1}{3}\frac{mnc\lambda}{A}$$

Now we have taken $A = 1$ and also $mn = \rho$ where ρ is the density of the gas.

$$\therefore \qquad \eta = \frac{1}{3}\rho c\lambda$$

Since product of $\rho\, c$ is constant, so the coefficient of viscosity η is independent of pressure. However, since $C \propto \sqrt{T}$ so $\eta \propto \sqrt{T}$

i.e., coeff. of viscosity varies directly as the square root of the absolute temperature of the gas.

Example 1. *Calculate the mass of water flowing in 10 minutes through a tube of 0.1 cm in diameter, 40 cm long, if there is a constant pressure head of 20 cm of water. The coefficient of viscosity for water is 8.9 × 10^{-4} NS m^{-2}.*

Solution: In this case

$$h = 0.2 \text{ m and } l = 2 \text{ m}$$

$$\eta = 8.9 \times 10^{-4}\ Nm^2$$

$$t = 600 \text{ sec.}$$

$$a = \frac{10^{-3}}{2} \text{m} = 5 \times 10^{-4} \text{m}$$

Now

$$P = h.d.g$$

$$= 0.2 \times 10^3 \times 9.8 \text{ N/m}^2 \qquad [\because \text{ density of water} = 10^3 \text{ kgm/m}^3]$$

Now from Poiseuille's relation volume of the liquids flowing/sec.

$$V = \frac{\pi P a^4}{8 \eta l}$$

$$= \frac{3.142 \times 0.2 \times 10^3 \times 9.8 \times (5 \times 10^{-4})^4}{8 \times 8.9 \times 10^{-4} \times 0.4}$$

$$= 1.351 \times 10^{-7} \text{ m}^3$$

Volume of water flowing in 600 sec (10 min)

$$= 1.35 \times 10^{-7} \times 600$$

$$= 81.06 \times 10^{-6} \text{ m}^3$$

Mass of water flowing in 10 minutes

$$= 81.06 \times 10^{-6} \times 10^3 \text{ kg}$$

$$= \mathbf{81.06\ gm.}$$

Example 2. *In a horizontal tube 4 km long and 8 cms in diameter, water flows at the rate of 20 litres/sec against viscous resistance. If the viscosity of water is 0.01 C.G.S. units, calculate the pressure required to maintain the flow.*

Solution: Diameter of tube $= 8.0$ cm

Radius of tube $a = 4.0$ cm

Length of tube $l = 4$ km $= 4 \times 10^5$ cm

Rate of flow $V = 20$ liters/sec

$= 20 \times 10^3$ cm^3/sec

$\eta = 0.01$ C.G. S. units

Now pressure $P = \dfrac{8 \eta l V}{\pi a^4} = 79.56 \times 10^4$

Now $n\, a\, g = l, \ldots h = \dfrac{l}{g}$ [$\therefore d = l$ formula]

$\therefore$ $h = \dfrac{79.56 \times 10^4}{980}$

$= \mathbf{811.8\ cms.}$

Example 3. *A plate of metal 100 cm² in area rests on a layer of castor oil 2 mm thick whose coefficient of viscosity is 15.5 poise. Calculate the horizontal force required to move the plate with a speed of 3 cm/sec.*

Solution: In this case

$A = 100$ cm^2 $\eta = 15.5$

$v = 3$ cm/sec $r = 2$ mm $= 0.2$ cm

Now horizontal viscous force

$$F = -\eta A \frac{v}{r}$$

$$= -\frac{15.5 \times 100 \times 3}{0.2}$$

$$= -23.25 \times 10^3 \text{ dynes}$$

So the external force required

$$= \mathbf{+23.25 \times 10^3\ dynes.}$$

Example 4. *A drop of water of radius 0.01 cm is falling through a medium whose density is 1.21 kg/m³ and $\eta = 1.8 \times 10^{-5}$ NS m^{-2}. Find the terminal velocity of the drop.*

Solution: In this case $r = 10^{-5}$ m

$\rho = 10^3$ kg/m^3

$d = 1.21$ kg/m^3

$$\eta = 1.8 \times 10^{-5} \text{ NS m}^{-2}$$

Now terminal velocity

$$v = \frac{2r^2(\rho - d)g}{9\eta}$$

$$v = \frac{2 \times (10^{-5})^2 \times (10^{-3} - 1.21) \times 9.8}{9 \times 1.8 \times 10^{-5}}$$

$$= \mathbf{1.21 \times 10^{-2} \text{ m/sec.}}$$

Example 5. *In a capillary flow experiment one thousand part of the capillary length was having a radius one fourth of the radius elsewhere. Compute the % error in the result due to ignoring this fact.*

Solution: The rate of flow of liquid through a uniform those of radius a is

$$v = \frac{\pi P a^4}{8\eta l}$$

In this case $\frac{l}{1000}$ part of tube has radius of $\frac{r}{4}$ and remaining $\frac{999}{1000} l$ of the tubes has radius r. The two tubes are joined in series so new rate of flow

$$V' = \frac{\pi P}{8\eta\left[\frac{l/1000}{(r/4)^4} + \frac{999\,l/1000}{r^4}\right]}$$

or

$$V' = \frac{\pi P a^4}{8\eta l}\left(\frac{1000}{1255}\right)$$

So, the error in rate of flow in given by

$$V - V' = \frac{\pi P a^4}{8\eta l} - \frac{\pi P a^4}{8\eta l}\left[\frac{1000}{1255}\right]$$

or the % error is given by $\frac{V - V'}{V} \times 100 = \frac{255}{1255} \times 100 = \mathbf{20.32\%}$

Example 6. *Capillary tubes of length l, 2l and l/2 are connected in series. The radii being r, r/2 and r/3 respectively. If a streamline flow is maintained and the pressure across first capillary is P_l, deduce pressure accross the second and third capallaries.*

Solution: Let P_2 and P_3 be the pressures accross the 2nd and 3rd capillaries respectively. Since the rate of flow of the liquid through the system is same, so

$$V = \frac{\pi P_1 a^4}{8\eta l} = \frac{\pi P^2 (a/2)^4}{8\eta(2l)} = \frac{\pi P^3 (a/3)^4}{8\eta(l/2)}$$

or

$$P_1 = \frac{P_2}{32} = \frac{P_3}{40.5}$$

$\therefore$

$$P_2 = 32 P_1$$

and

$$P_3 = \mathbf{40.5}\, P_1.$$

Example 7. *In giving patient a blood transfusion, the bottle is set up so that the level of blood is 1.3 m above the needle which has an internal diameter of 0.36 mm and is 3.0 cm in length. If 4.5 cm^3 of blood passes through the needle in one minute, calculate the viscosity of blood. The density of blood is 1020 kgm/m^3.*

Solution: In this case

Length of the needle $= 3.0$ cm

Radius of needle $C = \frac{d}{2} = \frac{0.36}{2}$ mm $= 0.18$ mm

$$= 0.018 \text{ cm}$$

Volume of blood flowing in one sec.

$$V = \frac{\text{Total volume}}{\text{Time}} = \frac{4.5}{60}$$

$$= 0.075 \text{ cc/sec.}$$

$$\text{Density of blood} = 1020 \text{ gm/m}^3 = 1020 \times 10^{-3} \text{ gm/c.c.}$$

$$= 1.02 \text{ gm/c.c.}$$

$$\text{Pressure difference} = 1.3 \text{ m column of blood}$$

$$= 1.3 \times 100 \times 1.02 \times 980 \text{ dynes/cm}^2$$

$$V = \frac{\pi Pa^4}{8 \eta l}$$

$$\eta = \frac{\pi Pa^4}{8 Vl} = \frac{3.142 \times 1.3 \times 1300 \times 1.02 \times 980 \times (0.18)^4}{8 \times 0.075 \times 3}$$

$$= \mathbf{0.0236 \text{ poise}}.$$

Example 8. *A vertical wide tube has sealed into its lower end a short length of capillary tube which is horizontal. It is filled with glycerine of density 1.26 gm/c.c and viscosity 8.50 C.G.S. units to a certain mark and it takes 45 sec for the glycerine to fall to a second mark by escaping through the capillary tube with castor oil of density 0.87 gm/c.c. the time taken to fall from one mark to the other is 69 sec. Calculate the viscosity of castor oil.*

Solution: We have the relation,

$$\frac{\eta_2}{\eta_1} = \frac{t_2}{t_1} \times \frac{\rho_2}{\rho_1} \quad ...(1)$$

where η_1 and η_2 are coefficient of viscosities of two liquids t_1 and t_2 time and flow of two liquids between fixed markes and ρ_1 and ρ_2 the densities of the two liquids

In this case $\eta_1 = 8.5$ C.G.S. units

$\rho_1 = 1.26$ gm/c.c., $\rho_2 = 0.87$ gm/c.c.

$t = 45$ sec, $t_2 = 69$ sec.

$\eta_2 = ?$

Putting in (1), we get

$$\frac{\eta_2}{8.5} = \frac{69}{45} \times \frac{0.87}{1.26}$$

or $\eta_2 = \mathbf{8.99}$ **C.G.S. units.**

Example 9. *A capillary tube of 0.1 cm in diameter and 20 cm in length is filled horizontally to a vessel kept full of alcohol. The depth of the centre of capillary tube below the surface of alcohol is 30 cm. If the viscosity and density of alcohol are 0.012 C.G.S units and 0.8 gm/cm³ respectively, find the amount of alcohol that will flow out in 5 minutes.*

Solution: In this case length of tube $l = 20$ cm

Radius of tube $a = \frac{0.1}{2} = 0.05$ cm

Coefficient of viscosity of alcohol $\eta = 0.012$ C.G.S. units

Density of alcohol $\rho = 0.8$ gm/cm^3

time of flow $t = 5$ minutes $= 5 \times 60 = 300$ sec.

Pressure head $h = 2V$ cm of alcohol

Now volume of the liquid flowing/sec.

$$V = \frac{\pi Pa^4}{8 \eta l}$$

and $P = h.\rho.g = 20 \times 0.8 \times 980$

$$\therefore \quad V = \frac{3.182 \times 20 \times 0.8 \times 980 \times (0.05)^4}{8 \times 0.012 \times 20} = 0.160 \text{ cm}^3$$

and mass of the liquid flowing in time t

$$= V \times \rho \times t$$

$$= 0.180 \times 0.8 \times 300 = \mathbf{38.43\ gm}.$$

Example 10. *A capillary tube 1.0 mm in diameter 20 cm in length is fitted horizontally to a vessel kept full of alcohol g density 8 × 10² kgm/cm³. The viscosity of alcohol is 0.0012 M.K.S. units. Find the amount of alcohol that will flow in 5 minutes.*

Solution: The rate of flow of the liquid is given by

$$V = \frac{\pi P a^4}{8 \eta l}$$

In this case

$$P = \rho \cdot g \cdot h = \frac{30}{100} \times 0.8 \times 10^3 \times 9.8 \text{ N/m}^2$$

$$a = \frac{1}{2} = 0.5 \text{ mm} = 5 \times 10^{-4} \text{ m}$$

$$\eta = 0.0012 k \text{ gm/m-sec}$$

$$l = 0.2 \text{ m}$$

$$\therefore \quad V = \frac{\pi \times (0.30 \times 0.8 \times 10^3 \times 9.8) \times (5 \times 10^{-4})^4}{8 \times 0.0012 \times 0.2}$$

$$= \frac{46.18 \times 10^{-8}}{1.92} = 24.05 \times 10^{-8} \text{ m}^3$$

The volume flowing in 5 minutes

$$= 24.05 \times 10^{-8} \times 5 \times 60 = 7.215 \times 10^{-5} \text{ m}^3$$

Now density of alcohol $= 0.8 \times 10^3$ kgm/m³

So mass of alcohol flowing in 5 min

$$= 7.215 \times 10^{-5} \times 0.8 \times 10^3$$

$$= 5.77 \times 10^{-2} \text{ kgm} = \mathbf{57.7\ gm}.$$

Example 11. *Eight rain drops of 1 mm each falling down with a terminal velocity of 5 cm/sec combine to form a bigger drop. Find the velocity of the bigger drops.*

Solution: In this case

$$n = 8$$

Radius of each drop $a = 0.1$ cm

Terminal velocity of each drop $v = 5$ cm/sec.

Let m be the mass of each drop so each drop will attain a terminal velocity v, given by

$$6\pi\eta a v = mg \qquad \text{...(1)}$$

where η is the coefficient of viscosity of air through which rain drops fall. Let R be the radius of the big drop formed by combination of 8 of small drops. Then we have

$$\frac{3}{4}\pi R^3 = 8 \cdot \frac{4}{3}\pi a^3$$

or

$$R = (8a^3)1/3 = 2a = 2 \times 0.1 = 0.2 \text{ cm}$$

Let v_1 be the terminal velocity of big drop, then

$$6\pi\eta R v_1 = 8\, mg \qquad \text{...(2)}$$

Dividing (2) by (1), we get

$$\frac{R v_1}{a v} = 8$$

or

$$v_1 = \frac{8 \times a \times v}{R} = \frac{8 \times 0.1 \times 5}{0.2} = \mathbf{20\ cm/sec.}$$

QUESTIONS AND PROBLEMS

1. Explain viscosity and coefficient of viscosity. Describe an accurate method to find coefficient of viscosity. What are the dimensions of coefficient of viscosity.
2. Describe Ostwald's viscometer and explain how it can be used for comparison of viscosities of two liquids.
3. Derive Poiseuille's formula for the rate of blow of a liquid through a capillary tube. Explain how this formula is used for finding the viscosity of a liquid by capillary flow method.
4. State Stoke's law and apply it to find the terminal velocity of a sphere falling through a fluid.
5. Deduce an expression for the rate of a steady flow of a liquid through a narrow tube.
6. Find the radius of drop of water falling through air if the terminal velocity of drop is 1.2 cm/sec. η for air = 1.8×10^{-5} NS m^{-2} and density of air = 1.51 kgm/m^3.

   [Ans. 9.97×10^{-5} m]
7. A pressure head of 5.6 cm of water causes water to flow at 44 cm^3/min through a tube of length 30 cm and radius 1 mm. Assuming streamline motion, calculate the value of η.

   [Ans. 9.8×10^{-4} NS/m^2]
8. Calculate the mass of water flowing in 10 minutes through a 0.1 cm in diameter and 0.4 m long, if there is a constant pressure head of 0.2 m of water. The coefficient of viscosity of water is 0.00089 M.K.S units. **[Ans.** 81.06×10^{-3} kgm]
9. A horizontal pipeline of constant cross-sectional area, the pressure difference between two points 10 m apart is 3×10^4 Pa. What is the energy loss per cubic metre of oil/unit distance.

   [Ans. 3×10^3 J/m^3]
10. How much water will flow in one minute through 25 cm of a capillary tube of internal diameter 2 mm if the pressure difference across the ends of the tube is 5 cm of Hg. given η for water is 0.80 centipoise. **[Ans.** 1.3 ml/s]
11. Find the Volume of water that will drain out/minute from an open top tank through an opening 2 cm in diameter that is 4m below the water level in the tank. **[Ans.** 0.167 m^3/min]

CHAPTER

7

HIGH VACUUM TECHNOLOGY

7.1. INTRODUCTION

The concept of vacuum is related to the pressure exerted by Earth's atmosphere and this is normally expressed in terms of the weight of a column of atmospheric air of *unit cross-sectional area* measured at a particular altitude above the sea level. With the increase of altitude the pressure decreases, till at a very great height above Earth's surface. where we consider only the empty space exists, it is assumed that the *air pressure is zero.* So the concept of perfect vacuum corresponds to *hypothetical state of zero pressure.*

As vacuum technique is of great use for various experimental investigations and industrial development, so we will study some of the methods for producing high vacuum and some instruments used to measure that high vacuum.

(*i*) ROTARY PUMP. This pump is used to produce moderately low pressures of the order of 10^{-4} torr (1 torr = 1 mm-Hg) and is based on the principle of displacement of a gas. A portion of the gas inside a vessel is *isolated* by a rotating disc and is then compressed till it attains a sufficiently high pressure and gets discharged to the atmosphere. As a gas consists of *fast moving molecules,* more gas comes to occupy that space and, in turn, also removed. This process continues and the pressure of the vessel keeps on decreasing.

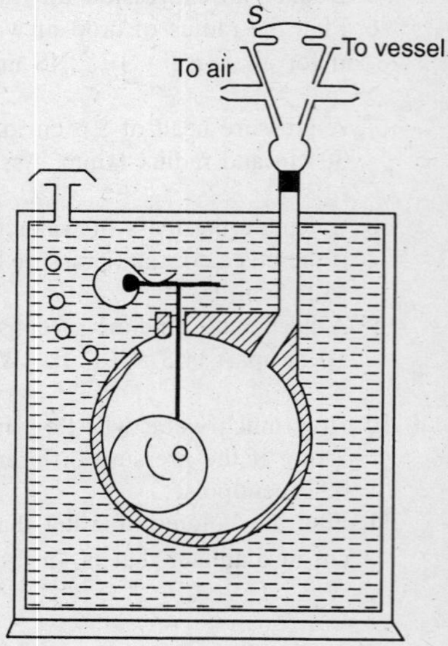

Fig. 7.1

A typical rotary pump is shown in Fig. 7.1. The dark-shaded disc rotates clockwise, thus *tapping* air from the right. *S* is a stop-cock which connects the pump to the vessel to be evacuated. When the pump is stopped, this stop-cock is turned to the left to allow the atmospheric air to enter into the pump. The disc moves in immersed oil. The oil fills the *dead space*, and thus allows a high compression ratio to be obtained. It also seals the moving disc against back leakage, besides *lubricating and cooling the pump.*

The maximum vacuum is limited as the oil gets contaminated when gases and vapours get dissolved in it. However, this problem can be minimized by making the vapours carried by air by the use of 'gas ballasts'.

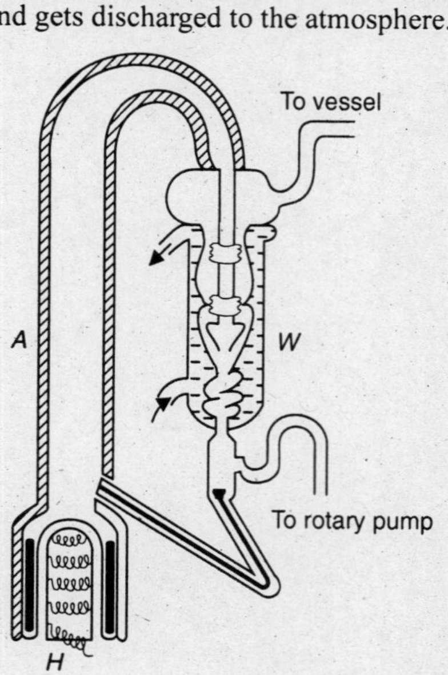

Fig. 7.2

(*ii*) DIFFUSION PUMP. In order to get more vacuum diffusion pumps are used. The vessel is first evacuated by the *rotary pump* and then the diffusion pump is switched on, it is then continuously backed by the rotary pump. A typical diffusion pump is shown in Fig. 7.2. Mercury or a low vapour pressure oil (e.g. silicon oil) is heated by heater H. The vapours go up and steam down at the right. This vapour jet sweeps away along with it the molecules entering the pump from the vessel. In Fig. 7.3, A is the asbestos lining to prevent the cooling of the vapours on their way up. This will give good steam jets without the boiler temperature becoming too high. On the right, the circulation of cold water in the outer jacket condenses to the boiler for further circulation. Baffles are provided to prevent direct back streaming of the vapours into the vessel. The molecules which manage to steam are removed by condensation in the trap shown in Fig. 7.3. If the vessels are clean, a vacuum of the order of 10^{-8} torr can be produced.

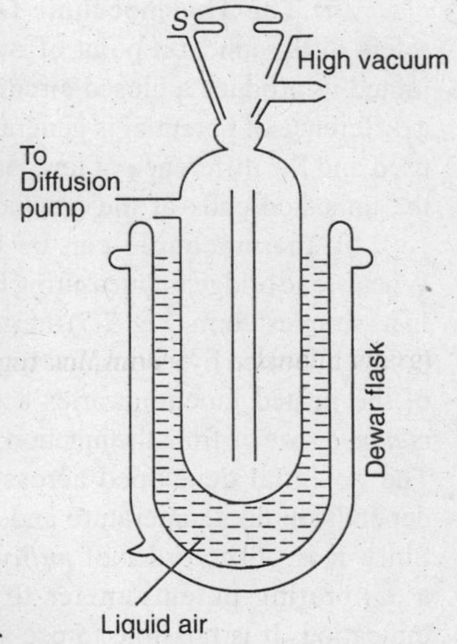

Fig. 7.3

7.2. VACUUM GAUGES

(*i*) The Pirani Gauge: Pirani gauges employ a Wheatstone bridge circuit, as shown in Fig. 7.4, which serves to heat a filament and to balance its resistance against a standard resistor sealed off in high vacuum. A change of pressure causes a change of the filament temperature and, consequently, of the filament resistance, thus unbalancing the bridge. The pressure can then be measured in terms of the *unbalance voltage.* Alternatively, the power required to maintain the filament temperature at a constant level is a measure of pressure. The temperature in this case is kept constant by means of a feedback circuit.

The smallest detectable pressure change is determined either

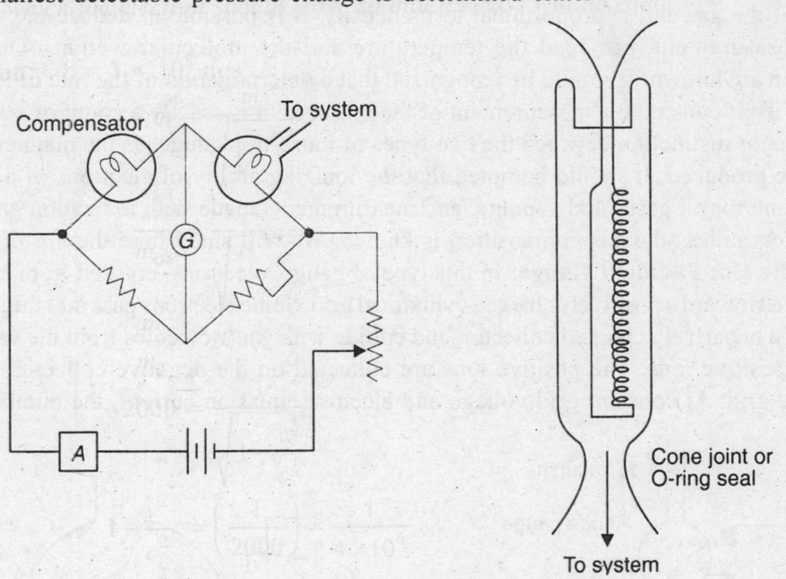

Fig. 7.4

by the *smallest signal* that can be registered on the bridge output meter or by the *background fluctuations* of the signal due to pressure instabilities of the system, whichever is the larger. The errors due to background fluctuations and to changes in the gauge-wall temperature become increasingly more important at higher pressures. They are minimized by operating at constant filament temperature rather than at constant voltage or constant current.

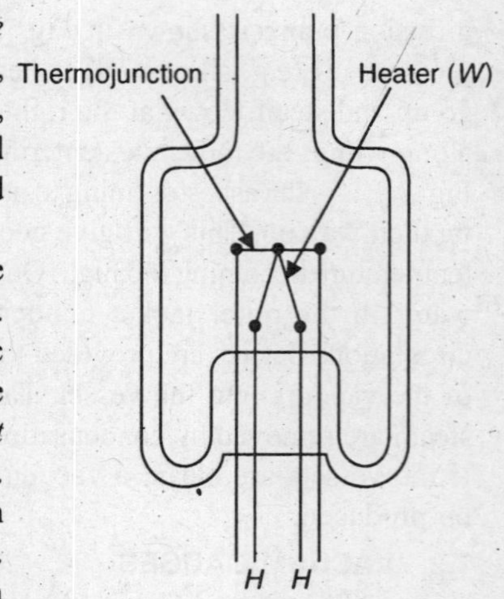

Fig. 7.5

(ii) The Thermocouple Gauge: The term *thermocouple* refers to the junction point of two dissimilar metal conductors, joined to produce a closed circuit. When the junction is heated, a difference of potential is generated which depends on the metal used and the difference of temperature between the junction and the unheated ends of the conductors.

A thermocouple can be used as an alternative to the Wheatstone bridge for measuring the temperature of a hot filament. In its simplest form (Fig. 7.5) the gauge consists of a thermoelectric junction, heated by a constant, regulated current. The temperature of the heated junction varies according to the *amount of heat conducted away* from the junction by the gas in the vaccum system. The potential developed across the terminals of the junction depends on the temperature and, therefore, on the gas pressure. Since it is of the order of *millivolts*, it can be fed directly via a calibrating potentiometer to a simple meter for pressure indication. It is possible to use several such couples in series, in which case the device is known as a "thermopile." Thermocouple and thermopile gauges can be made more rugged than pirani gauges, but they are more limited in *range and sensitivity.* The thermocouple gauge has the advantage of simplicity but the disadvantage of having a nonlinear scale. Its greater usefulness is in systems where *high accuracy is not required.*

(iii) Ionization Gauges: The number of ion pairs produced by an electron of a given energy in a particular gas depends on the probability of an *ionizing collision,* which in turn, depends on the nature of the gas and is proportional to its density. It is possible to deduce the pressure from a density measurement, provided the temperature and the molecular weight of the gas under consideration are known. It should be recognized that a determination of the rate of ion production does not by itself constitute a measurement of the pressure; it is strictly a count of *particle density.*

The major distinction between the two types of ionization gauges is the manner in which the electrons are produced. It should be noted that the ionizing ability of electrons of a given energy is not the same for all gases and vapours, and measurements made with ionization gauges are thus liable to errors unless the gas composition is known. We will study here the *hot filament gauge.*

(iv) The Hot Filament Gauge: In this type of gauge, electrons, emitted from a *hot filament,* are accelerated toward a positively charged cylindrical grid. Some electrons pass into the space between the grid and a negatively charged collector, and collide with gas molecules from the vacuum system to produce positive ions. The positive ions are collected on the negative collector; the electrons return to the grid. At constant grid voltage and electron emission current, the number of positive

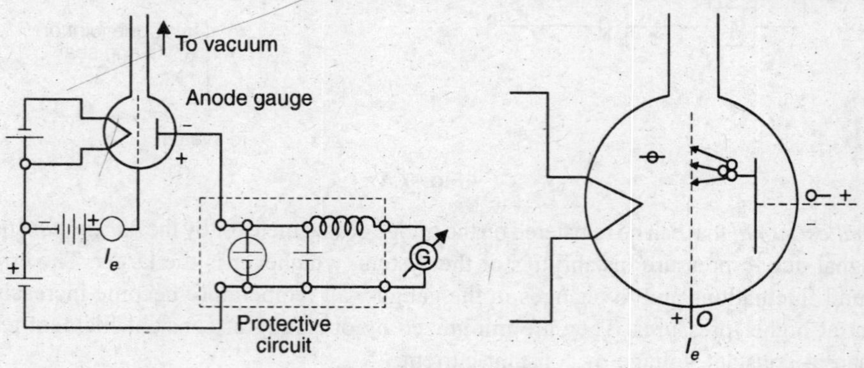

Fig. 7.6

ions formed is proportional to the density below 10^{-3} Torr; the ion current to the collector is, therefore, a measure of the pressure.

The electrical circuits necessary for the reliable operation of the gauge must perform the following functions: (1) keep constant relative potentials between the electrodes, (2) maintain a constant electron current between filament and grid, (3) measure the positive ion current reaching the collector, (4) provide sufficient power in order to outgas the electrodes adequately by heating. For this purpose, the electrode potential are obtained from a stabilized power supply, and the emission current is stabilized by a feedback control of the filament current. The basic circuit of the gauge is shown in Fig. 7.6. For pressures not below 10^{-8} Torr, a structure, as shown in Fig. 7.6, incorporating a central filament surrounded by a concentric grid and outer concentric ion collector is used. performance of this configuration is limited by a constant residual collector current irrespective of pressure. It is found that electrons arriving at the positive electrode produce *X*-rays, which irradiate the negative ion collector and release from its surface photoelectrons which are attracted to the positive electrode. The current of photoelectrons leaving the ion collector is indistinguishable from a current of positive ions arriving, and a residual constant current irrespective of gas pressure results.

To reduce the photoelectric current to as low a level as possible, and hence to extend the range to 10^{-10} Torr or less, the expedient of inverting the structure was adopted by Bayard and Alpert Fig. 7.7 (*b*). The filament *FF* is outside the cylindrical grid *G*, which acts at a positive potential to collect the electrons, whereas the ion collector *C* is at a negative potential and consists of a fine wire suspended centrally within the grid. The volume enclosed by the grid is made comparable with that between grid and plate of the conventional gauge.

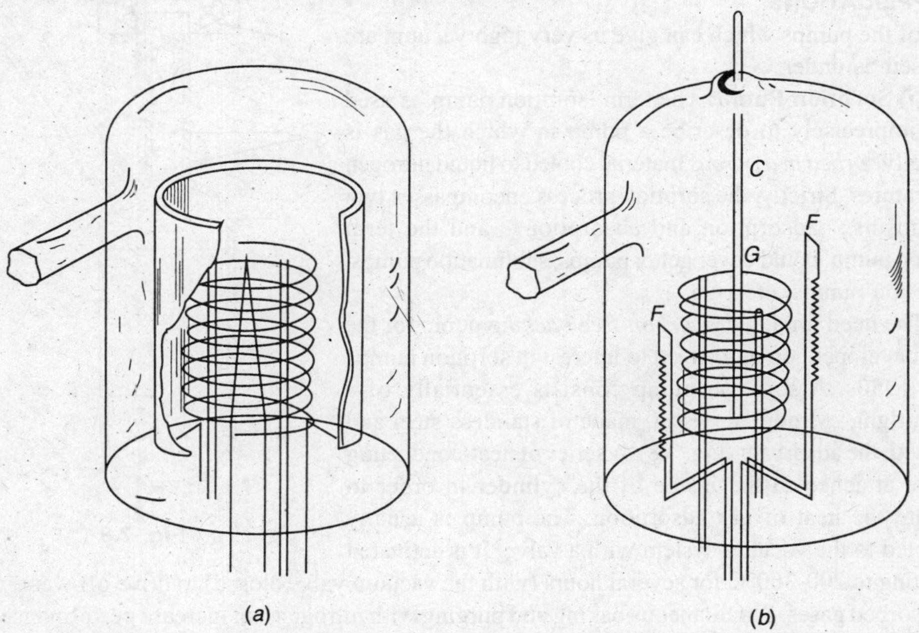

Fig. 7.7 (*a*) Triode ionization gauge. (*b*) Bayard-Alpert ionization gauge.

(*v*) Knudsen Gauge: We can describe the flow of a fluid by its flux (*e.g.* the gas flux). These processes are quite useful in vacuum technology and to get the maximum efficiency in the results, we have to define both the *flow* and *flux rates*.

Flow is generally controlled by pressure and friction (gravity effect is ignored). In some cases friction effect is also ignored. In discussing the gas flow, the value of *Knudsen number* (K_n) is important for consideration and is defined as the ratio of mean free path to the width of flow channel (d), *i.e.*,

$$K_n = \frac{\sigma}{d}$$

In the conditions when $\sigma << d$; $K_n << 1$, many collisions between particles occur and the gas may be considered as a *continuum.* Such conditions are fulfilled in the rough vacuum range. It may be noted that in continuum flow, the *dynamic viscosity* plays an important role. Further flow conditions can be defined by the value of Reynolds number (R_e),

$$R_e = \frac{V \cdot P \cdot d}{\eta}$$

where $V \rightarrow$ free steam velocity

$\rho \rightarrow$ density of fluid

$d \rightarrow$ distance of flow channel

$\eta \rightarrow$ dynamic viscosity

For smooth tubes of circular cross-section, flow in *laminar* when $R_e < 2300$. When R_e is more than 3000, flow becomes *turbulent.*

Under these conditions, in vacuum systems, loosely held deposits on the walls may be swept up and transported towards pumps.

In high and ultra high vacuum regions σ is greater than d. This corresponds to molecular flow conditions and are characterised by *Knudsen number* $K_n > 0.5$.

The transition from $0.5 > K_n > 0.01$ is termed as *Knudsen flow region.*

7.3. PRODUCTION OF ULTRA HIGH VACUUM AND THEIR APPLICATIONS

Some of the pumps which can give us very high vacuum are discussed as under:

(i) Sorption Pump: The term 'sorption pump' is used rather imprecisely to describe a pump in which the gas is physically *sorbed in a porous* material cooled to liquid nitrogen temperatures. Strictly, the sorption process encompasses two mechanisms – adsorption and absorption – and the term 'sorption pump' could cover getter pumps, sublimation pumps, sputter ion pumps, etc.

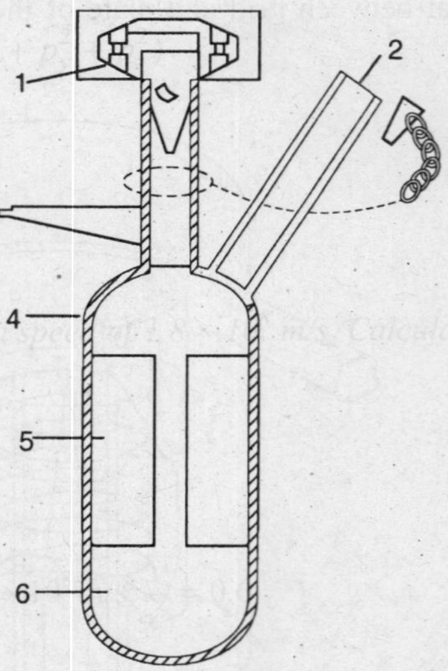

Fig. 7.8

The need for a *hydrocarbon-free backing* pump for the newly developed ion pumps led to interest in sorption pumps in the 1950s. A sorption pump consists essentially of a vacuumtight, cylindrical vessel, made of stainless steel and filled with the adsorbant (Fig. 7.8). A series of heat-conducting fins are attached to the inside of the cylinder in order to dissipate the heat of gas adsorption. The pump is usually connected to the vacuum system with a valve. It is activated by heating to 200-300°C for several hours (with the vacuum valve closed) to drive off water vapour and adsorbed gases. Simultaneous baking and purging with nitrogen can increase performance. After regeneration, the pump is chilled with liquid nitrogen and is then ready for use.

Generally, molecular sieves, in the form of pellets, are used as the adsorbant. Molecule sieves are synthetic zeolites (metal aluminosilicates). Type 5*A* (Ca aluminosilicate) and 13*X* (Na aluminosilicate) are commonly used. These materials have a very large surface area (several hundred $m^2 g^{-1}$) and, therefore, have a considerable gas sorption capability. The pore size depends on the type of molecular sieve. Types 5*A* and 13*X* have pore diameters of 0.5 and 1.3 nm, respectively. Since the amount of gas adsorbed can be large, a safety-valve is provided to prevent a great pressure build-up in an isolated pump if the refrigerant is removed.

The design and performance of adsorption pumps is determined by various factors, including the type of gas being pumped and the capacity of the sieve. Isotherms for a type 5A molecular sieve (Fig. 7.9) (Turner and Feinleib, 1962) show that, at liquid nitrogen temperature, larger quantities of nitrogen are adsorbed than Ne or He. A complete monolayer appears to be formed at about 100 mbar l g^{-1} and, thereafter, adsorption is significantly reduced. The adsorption of gas depends not only on the temperature but also on the pressure above the adsorbate surface. Because of the different adsorption properties, the pumping speed and ultimate pressure of an adsorption pump are different for different gases. The best performance is obtained when pumping N_2, CO_2, water and other vapours. Light noble gases are hardly adsorbed at all.

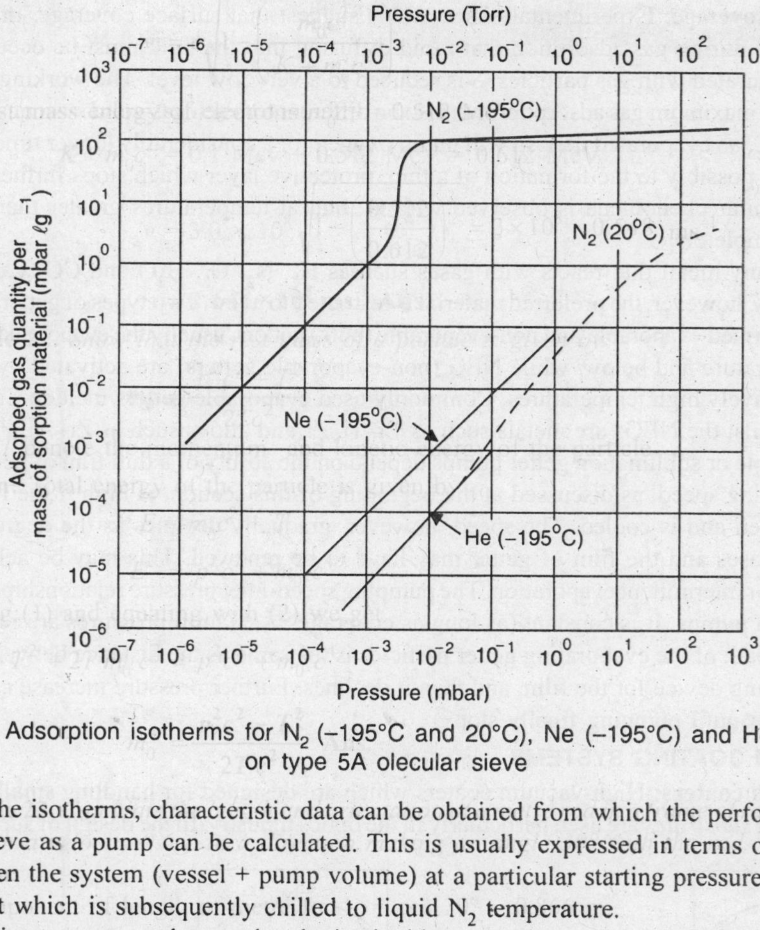

Fig. 7.9 Adsorption isotherms for N_2 (−195°C and 20°C), Ne (−195°C) and He (−195°C) on type 5A olecular sieve

From the isotherms, characteristic data can be obtained from which the performance of the molecular sieve as a pump can be calculated. This is usually expressed in terms of the pressure achieved when the system (vessel + pump volume) at a particular starting pressure is exposed to the adsorbent which is subsequently chilled to liquid N_2 temperature.

The ultimate pressure that can be obtained with an sorption pump evacuating a system from atmospheric pressure is approximately 10^{-2} mbar. By using two pumps, the situation can be improved somewhat. If the pumps are used in parallel, only a sight improvement is obtained. If the pumps are used successively then, depending on the procedure, quite low pressures can be achieved. The most satisfactory method with dual pumps appears to be:

1. With the valves to *both* pumps open, cool one pump down.
2. When the pressure ceases to fall, isolate the cold pump.
3. Chill the second pump.

Unfortunately, evacuation of a vessel initially containing atmospheric air using this technique can achieve only limited success due to the presence of neon and helium (p_{Ne} in air, 2×10^{-2} mbar;

p_{He}, 5×10^{-3} mbar). One method that will reduce the partial pressures of noble gas is to flush or back fill the chamber with a gas such as dry N_2.

***(ii)* Getter pumps:** Getters take up gas by adsorption, absorption (diffusion into the interior of the getter) and chemical compound formation. The initial stage involves the interaction of the particle, with the surface (measured by the 'sticking probability) and, for a getter area *A*, the pumping speed is given by:

$$S = \alpha \overline{Ac}/4$$

where α is the sticking probability and $\bar{c}$ is the thermal speed of the gas molecules. The valuc of α depends on the conditions. It generally increases with *decreasing temperature* and decreases with the degree of coverage. Experimental observations suggest that surface coverage, in some cases, remains so low during gas adsorption that rapid diffusion into the bulk must be occurring. When the getter is saturated with gas particles, *S* is reduced to a very low level. The working temperature of the getter for maximum gas adsorption can also be different for each gas-solid system. For example, with Ba at 50°C, an evaporated deposit will pump oxygen for a considerably longer time than carbon monoxide, due possibly to the formation of a thin, protective layer which stops further interaction. At 40°C, a similar phenomena is observed with O_2 but, at temperatures greater than 80°C, both gases react completely.

Almost any metal that reacts with gases such as H_2, N_2, O_2, H_2O and CO_2 can be used as a getter. Usually, however, the preferred material is restricted to a few. Two types of getter are available and these are termed evaporable and non-evaporable (bulk) getters. usually the evaporated getter works at room temperature and below, while NEG (non-evaporable getters) are activated by heating and operate at relatively high temperatures. Commonly used evaporable getters include Ca, Ba, Ti, Th, Mo and Zr, whilst the NEGs are metals such as Ta, Ti, Zr and alloys such as Zr–Al, Zr–V–Fe, etc.

Evaporable or sublimation getter pumps depend on the ability of a thin film of getter to adsorb gas. The pumping speed, as discussed at the beginning of this section, is high if the film has been freshly deposited and is cooled. The speed, however, gradually declines as the degree of surface coverage increases and the film of getter may have to be renewed. This may be achieved either by continuous or intermittent evaporation. The pumping speed-inlet pressure relationship is interesting for evaporation pumps. It is constant (as long as coverage << 1) until pressures are reached where the mean free path of the evaporating getter particles is the same as the distance between the source and the collecting device for the film, and then it declines. Further pressure increase causes further speed reduction until pumping finally stops.

7.4. VACUUM COATING SYSTEMS

***(a)* Batch coaters:** High-vacuum coaters which are designed for handling small-to medium-sized batches of *substrates* are used particularly in the optics industry. In the design of such equipment,

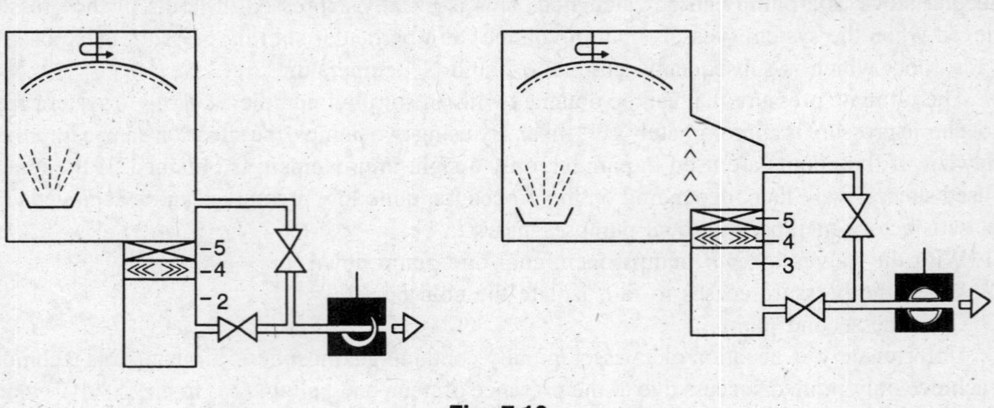

Fig. 7.10

attention has to be paid to the presentation of the substrates and their pretreatment (glow-discharge and ion cleaning, heating) and the source of coating material. Additional factors that have to be considered include case of access and the useable chamber area. Several years ago, batch coaters were usually bell jars which could be raised and lowered onto a base plate which incorporated feed-thoughts for evaporation sources, substrate cleaning facilities, etc. Over a period of time box coaters developed in which access to the chamber was via a hinged door. Unfortunately, the pumping port was connected to the base plate and this could take up approximately 30% of the useable area. In recent development, the preferred position for the pumping port is in the chamber wall thus freeing the base for the accommodation of feedthroughs for sources and monitoring equipment.]

Figure 7.10 shows two designs of box coater. The vacuum chamber is usually made of stainless steel with a diameter in the range 500-1100 mm. Access to the chamber is via door having a width which is approximately 30% of circumference. In modern coaters, the interior of the chamber is fitted with a shield that can be heated or cooled. Heating is often required during loading/unloading in order to minimize adsorption of moisture whilst cooling may be an advantage during evaporation since it inhibits desorption from the surfaces. Depending on the degree of cleanliness required and

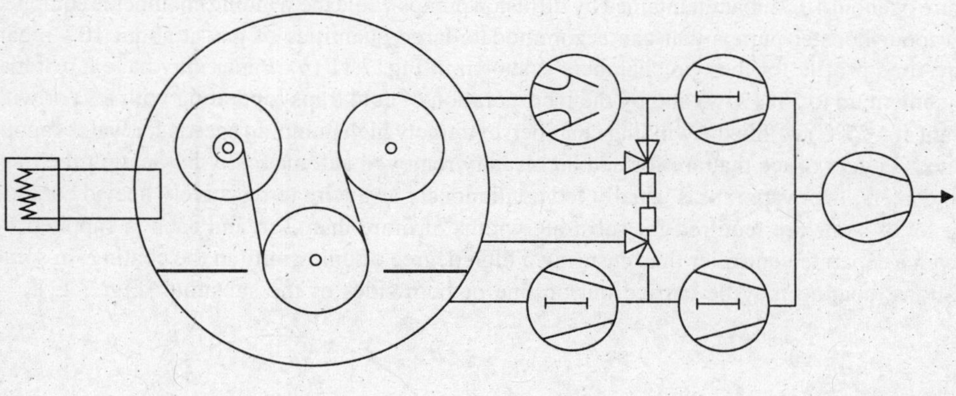

(*a*)

the throughout, box coaters can be evacuated with diffusion, cryo or turbomolecular pumps. If substrates with high outgassing rates (such as PMMA or CR39) are used then additional pumping capacity for water vapour can be incorporated in the form of an LN_2-cooled baffle or Meissner trap. Pressure measurement with such systems usually involves the use of Pirani and hot or cold-cathode ionization gauges whilst control of the deposition process is carried out using a quartz crystal monitor, photometer or one of the devices indicated earlier.

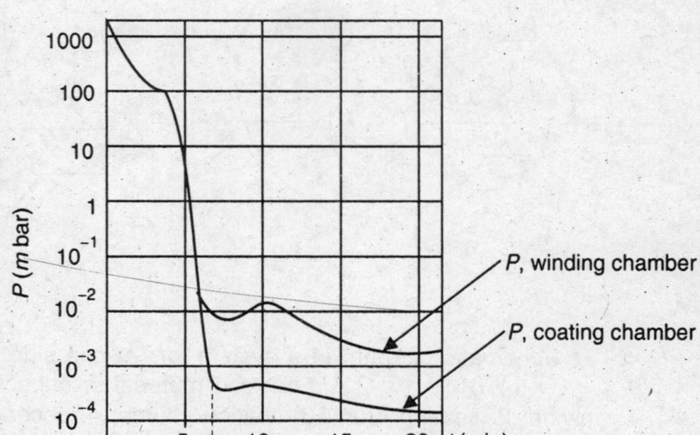

Fig. 7.11 (*b*)

Box coaters are also used extensively in many other areas of *thin–film research and development,* including investigation into metallization by various techniques, ion–assisted deposition an sputtering and some highly specialized areas of reasearch such as the deposition of high T_c superconductors. For such applications, a wide range of substrate conditioning methods, coating sources and process-monitoring equipment must be available. Other facilities may include load-lock connections to external systems for, say, surface analysis by SIMS and Auger electron spectroscopy. Coating systems capable of being evacuated down to 10^{-9} mbar using both turbomolecular and cryopumps are now commercially available.

(*b*) **Web coaters:** In recent years, the market for vacuum *metallized plastic films and papers* has increased significantly. For packaging materials, the vacuum deposition of aluminium onto a film is the preferred method of preparation. The use of aluminium is extensive because it is a relatively inexpensive material which has excellent barrier properties against water, light and oxygen (thereby increasing storage life) and produces a coating with an attractive appearance. A layer of aluminium which is only 0.04µm thick has the required barrier properties.

Commercially available web coaters can accommodate rolls of paper or plastic with widths from 0.2 to 2 m and with thickness from 1.5*µm* to 0.5 mm. They can coat at rates up to 10 ms^{-1}. The winding and coating chambers are separately pumped [Fig. 7.11 (*a*)]. The coating chamber is operated at a pressure of about 10^{-4} mbar maintained by diffusion pumps while the winding chamber is equipped with oil vapour booster pumps that can accommodate large quantities of gas at about 10^{-2} mbar. A pressure-time profile for the two chambers is shown in Fig. 7.11 (*b*). Paper may have a residual moisture content up to 5% (w/w) and by the incorporation of cold traps (operating with LN_2 or with a refrigerant at – 75°C) within the winding chamber, extremely high pumping speeds for water vapour are obtained. Layers of ice that are formed are usually removed automatically by heating the trap. With web coaters, the evaporant is usually fed continuously as a wire to resistively heated ceramic boats. Up to 24 boats are required for substrate widths of more than 2 m and each is supplied by current from a separate controller thus ensuring a high degree of uniformity in the coating. In some configurations, coating may be carried out on one or both sides of the substrate (Fig. 7.12).

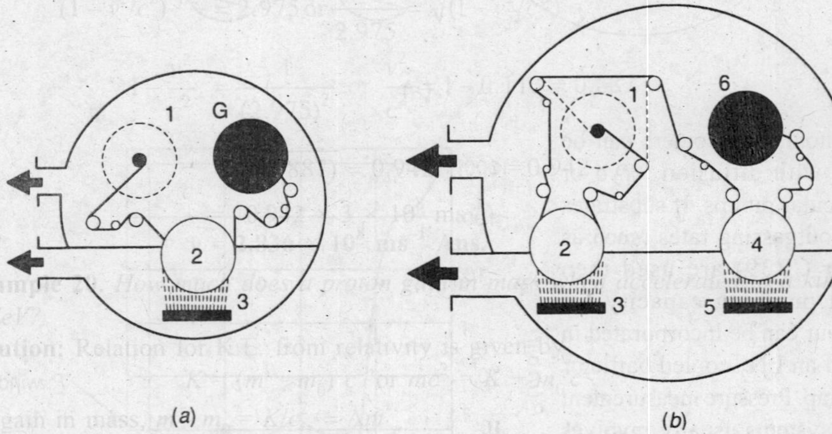

Schematic diagram of a system for (*a*) one side and (*b*) two-side coating of packing material or capacitors.
1. unwind; 2. coating drum; 3. coating source; 4. second coating drum; 5. second coating source; 6. rewind.

Fig. 7.12

A specialized activity in the field of web coating is the preparation of metallized films for use in capacitors. The substrates may be capacitor paper, polyester, polypropylene or polycarbonate ranging in thickness from 1-2µm (ultra-thin polyester and polycarbonate films) up to 10-20µm for power capacitors. Depending on the construction of the capacitor, the metal layers may be aluminium, zinc or a combination, applied to one or both sides of the substrate. Zinc coaters usually involve one-chamber systems and precoating the substrate with silver must be carried out in order to enhance the adhesion process for efficient heat removal from films that would otherwise be damaged. Data storage media such as *floppy disks and magnetic tape* are also produced by coating a polymer web with a multi-layer film produced by sputtering or electron beam evaporation. Details of these systems are given in Table 7.1. A web coater may contain several sputtering cathodes, which must be rigorously separated, together with other stations in which the web is created by glow discharge, plasma-enhanced deposition and other processes. A schematic diagram is shown in Fig. 7.13.

Table 7.1: *Characteristics of vacuum web coaters for the manufacture of data storage media.*

	Floppy disks	*Video tape*	*Substrate*	*Coating*	*Storage density (bits/inch)*
Coater with Magnetron sputtering	U	NU	PET (70–100μm) Polymide (25 – 5μm)	CoNi CoCr	6×10^4 9×10^4
Coater with EB-evaporation	U	U	PET (4 – 20μm) Polymide (10 – 20μm)	CoNi CoCr NiFe	7.5×10^4 9×10^4

U = used for; NU = not used.

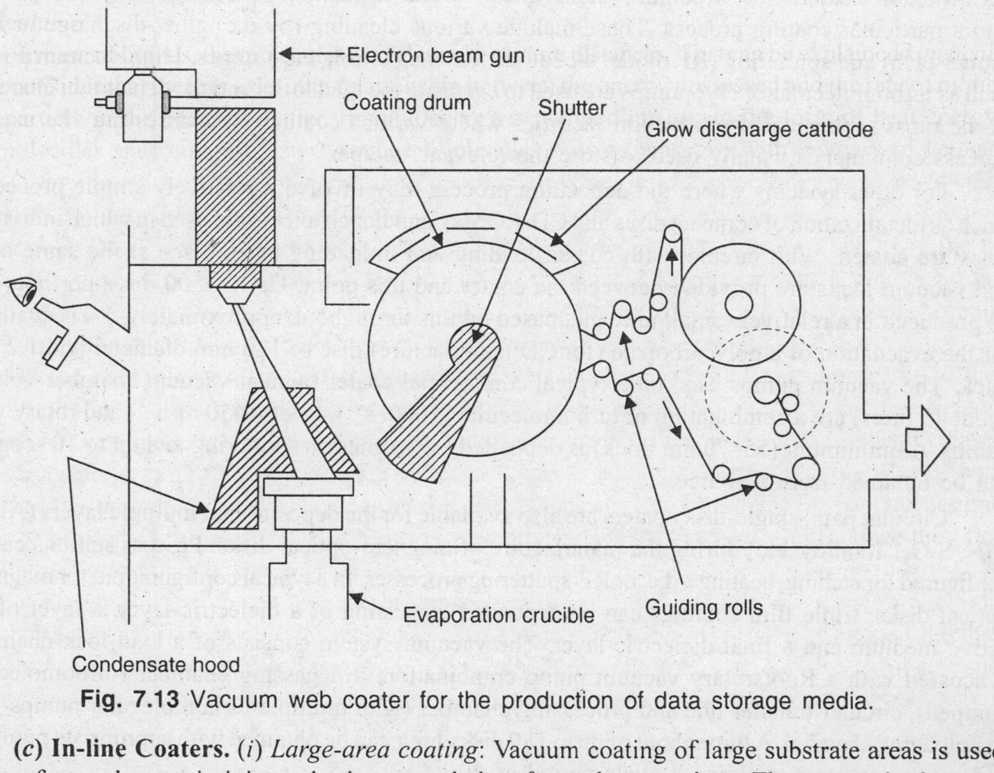

Fig. 7.13 Vacuum web coater for the production of data storage media.

(c) In-line Coaters. (*i*) *Large-area coating*: Vacuum coating of large substrate areas is used to manufacture low-emissivity and solar *control glass for architectural use.* The process is also used to produce transparent, electrically conducting and reflective coatings for use in the automotive industry. For such applications, district control must be exerted on the coating process, and the coatings themselves must have optical properties that are constant to within 0.5%. The largest standard coaters can deal with up to two million m^2 glass/year and usually consist of a series of chambers through which the glass is transported. After a preliminary cleaning at atmospheric pressure, substrates (which may consist of sheets up to 3 × 6 m) are placed into a vacuum-lock entry/glow discharge chamber and the pressure reduced to about 10^{-2} mbar. The ignition of a glow discharge in Ar/O_2 may be necessary for further cleaning of the substrate. From this chamber, substrates pass via a transfer chamber which holds a queue of material, and allows a transition from discontinuous to continuous flow, to the coating chamber. Within this chamber, various sputtering modules are arranged in series. Each is separately pumped (usually by turbomolecular pumps) and has its own gas inlet system for reactive or non-reactive sputtering. Low emissivity films consist of a 10 mm thick silver layer and two layers with high reflectance in the infra-red region but high transmittance in the visible region of the spectrum. For solar control, a 'hard' metal layer-to control the transmission of light and energy is deposited in conjunction with two oxide interference films which affect the colour seen in reflection. After coating, material passes again through a transfer chamber to a vacuum-exit lock.

In line systems are also used for the production of sophisticated data storage media including rigid magnetic, optical and magneto-optical disks. The last type of disk usually consists of two protective layers (AlN or SiO_2), a metal alloy layer (for example, FeCoTb) and a laster-light reflecting coating (Al). For magnetic disks, recording layers of NiFe, CoCr (perpendicular) or Cr. CoNi (Cr) (longitudinal) and C on aluminium are used. For the production of high-quality disks, loading and unloading is performed in clean rooms. Modular systems allow various deposition procedures to be incorporated into a particular coating process. These include various cleaning (by d.c. glow discharge or RF–sputter etch) and sputtering (RF diode, d.c. diode, RF magnetron, etc.) stages. High-vacuum pumps such as turbomolecular or cryopumps are used to evacuate the load locks and the deposition chamber. In the most sophisticated production facilities where various coating procedures may be used, a process computer is usually used to store the relevant 'recipes'.

For other systems where the deposition process may involve a relatively simple procedure (such as metallization of compact discs and CD-ROMs), annular chambers are used in which individual discs are coated. With circular-path coater, loading and unloading takes place at the same place and vacuum locks are provided between the coater and this point. Up to 1500 discs per hour can be produced in a relatively small unit and, based on this throughout, approximately, 1 s is available for the evacuation of a polycarbonate (for CD manufacture) disc of 120 mm diameter and 1.2 mm thick. The vacuum pumps used on a typical commercial coater (annual vacuum chamber volume about 45 litres) are a combination of turbomolecular (1000l s^{-1}), Roots (450 m^3h^{-1}) and rotary vane pumps. Aluminimum (55–70 nm thick) is deposited by magnetron sputtering and up to 10^5 coating can be obtained from a source.

Circular path, single-disk coaters are also available for the deposition of multiple layers (Al_2O_3, AlN, SiO_3, Te-alloy, etc) during the manufacture of magneto-optical disks. Process stations can be configured for etching, heating a d.c. or RF-sputtering processes. In a typical configuration, for magneto-optical disks, triple film coatings can be deposited consisting of a dielectric layer, a layer of the active medium and a final dielectric layer. The vacuum system consists of a load lock chamber, evacuated with a Rootsrotary vacuum pump combination, a degassing chamber (turbomolecular pumped), circular transfer line and processing chamber (with multiple turbomolecular pumps) and an unloading chamber. A throughout of up to 180 disks/hour can be obtained with appropriate pumping and sputtering rates, yielding a cycle period of 20 s.

QUESTIONS AND PROBLEMS

1. Give the importance of high vacuum technology and illustrate the meaning of high vacuum.
2. What is meant by pressure? How is it measured? What do you understand by zero pressure?
3. Give the theory and working of a rotary pump. What is the limit of pressure measured by this system?

4. Give important features of a diffusion pump. What is the order of pressure produced by a diffusion pump?
5. What are the insturment used for measurement of high pressure? Give the theory and working of one of the best instrument and its importance.
6. What is Pirani Gauge? How is it important? Give its merits and demerits.
7. What is the importance of thermocouple gauge? How is it used and give its salient features?
8. How Ionization Gauge is used? Give its theory working and important features.
9. Give the important features of Knudsen Gauge? What is Knudsen number and Knudsen flow region? Explain.
10. How ultra high vacuum is produced? Give the principle and working of a Sorption pump.
11. Give the principle, working and importance of Getter pump.
12. Give an outline of the vacuum coating systems and their importance in thin film technology.
13. What are Web Coaters? How they are useful for coating on two sides of a substrate?

CHAPTER

8

THEORY OF RELATIVITY

8.1. INTRODUCTION

The first theory in Physics was mechanics developed by Newton and so is called classical or Newtonian mechanics. Acutally, the Physics developed before the advent of 20th century is called classical Physics.

The development of 'theory of relativity' by Einstein in 1905 revolutionized the old concepts. It discards the absolute motion through space and deals with objects or observer moving with relative velocities w.r.t. each other. This theory is divided into two parts.

(*i*) *Special theory* (*ii*) *General theory*

The special theory of relativity deals with objects and systems which are either moving at a *constant speed with repect to one another or are at rest.* The **general theory deals with objects or system which are speeding up or slowing down with respect to one another.**

8.2. FRAME OF REFERENCE

We know that the motion of the body has no meaning unless it is described with respect to some well defined coordinate system. The coordinate system is known as *frame of reference.* The velocity is now measured with respect to this **frame of reference.** The frame of reference is selected in such a way that the laws of nature may become fundamentally simpler in that frame of reference. There are two types of frames of reference:

(*i*) Inertial or unaccelerated frames.

(*ii*) Non-inertial or accelerated frames.

(*i*) Inertial frames: A frame of reference is said to be inertial when bodies in this frame obey Newton's law of inertia and other laws of Newtonian mechanics. In this frame, a body not acted upon by external force, is at rest or moves with a constant velocity. Experiments show that all inertial reference frames are equivalent for the measurement of physical phenomenon. Observers in different frames may obtain different numerical values for measured physical quantities but the relationship between the measured quantities *i.e.,* the laws of physics will be same for all observers.

(*ii*) Non-inertial frames: A frame of reference is said to be non-inertial frame when a body, not acted upon by an external force, is accelerated. In this frame, **the Newton's laws are not valid**.

As the motion of Earth around the sun is accelerated motion, ideally it should be non-inertial frame, however since the motion of the Earth during the course of experiment is negligibly small, so for all practical purposes Earth can be considered as *inertial frame of reference.*

8.3. GALILEAN TRANSFORMATIONS

Events and Coordinates. We begin by considering the concept of a physical event. The event might be the striking of a tree by a lightening bolt or

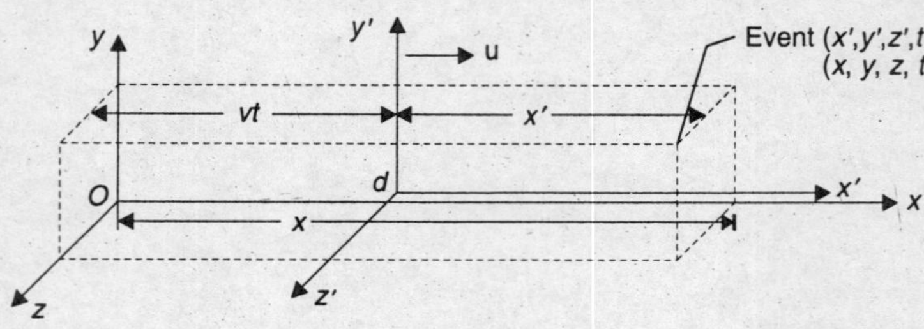

Fig. 8.1

the collision of two particles, and happens at a point in space assigning to it four coordinates: the three position coordinates x, y, z that measure the distance from the origin of a coordinate system where the observer is located, and the time coordinate t that the observers records with his clock.

Consider now two observers, O and O' where O' travels with a constant velocity v with respect to O along their common $X - X'$ axis (Fig. 8.1). Both observers are equipped with metersticks and clocks so that they can measure coordinates of events. Further, suppose both observers adjust their clocks so that when they pass each other at $x = x' = 0$, the clocks read $t = t' = 0$. Any given event P will have eight numbers associated with it, the four coordinates (x, y, z, t) assigned by O and the four coordinates (x', y', z', t') assigned (to the same event) by O'.

8.4. GALILEAN COORDINATE TRANSFORMATIONS

The relationship between the measurements (x, y, z, t) of O and the measurements (x', y', z', t') of O for a particular event is obtained by examining Fig. 8.1.

$$x' = x - vt; \quad y' = y, z' = z$$

In addition, in classical physics it is implicitly assumed that

$$t' = t$$

These four equations are called the *Galilean coordinates transformations.*

8.5. GALILEAN VELOCITY TRANSFORMATIONS

In addition to the coordinates of an event, the velocity of particle is of interest. Observers O and O' will describe the particles' velocity by assigning three components to it, with (u'_x, u'_y, u'_z) being the velocity components as measured by O'.

The relationship between (u_x, u_y, u_z) and (u'_x, u'_y, u'_z) is obtained from the time differentitation of the Galilean coordinate transformations. Thus, from $x' = x - vt$

$$u'_x = \frac{dx'}{dt'} = \frac{d}{dt}(x - vt)\frac{dt}{dt'} = \left(\frac{dx}{dt} - v\right) \times (1) = u_x - v$$

Altogether, the Galilean velocity transformations are

$$u'_x = u_x - v \qquad u'_x = u_y \qquad u'_z = u_z$$

8.6. GALILEAN ACCELERATION TRANSFORMATIONS

The acceleration of particle is the time derivative of its velocity, *i.e.*, $a_x = du_x/dt$. To find the *Galilean acceleration transformations* we differentiate the velocity transformations and use the facts that $t' = t$ and v = constant to obtain

$$a'_x = a_x \qquad a'_y = a_y \qquad a'_z = a_z$$

Thus the measured acceleration components are the same for all observers moving with uniform relative velocity.

8.7. INVARIANCE OF AN EQUATION

By *invariance* of an equation it is mean that the equation will have the same form when determined by two observers. In classical theory it is assumed that space and time measurements of two observers are related by the Galilean transformations. Thus, when a particular form of an equation is determined by one observer, the Galilean transformations can be applied to this form to determine the form for the other observer. If both forms are the same, the equation is invariant under the Galilean transformations.

Example 1. *A passenger in a train moving at 30 m/s passes a man standing on a station platform at* t = *t'* = *0. Twenty seconds after the train passes him, the man on the platform determines that a bird flying along the tracks in the same directions as the train is 800 m away. What are the coordinate of the bird as determined by the passenger?*

Solution: The coordinates assigned to the bird by the man on the station platform are

$$(x, y, z, t) = (800 \text{ m}, 0, 0, 20 \text{ s})$$

The passenger measures the distance x' to the bird as

$$x' = x - vt = 800\ m - (30\ \text{m/s})(20\ \text{s}) = 200\ \text{m}$$

Therefore the bird's coordinates as determined by the passenger are

$$(x', y', z', t) = (200\ \text{m}, 0, 0, 20\ \text{s})$$

Example 2. *A sample of radioactive material, at rest in the laboratory, ejects two electrons in opposite directions. One of the electrons has a speed of 0.6 c and the other has a speed of 0.7 c, as measured by a laboratory observer. According to classical velocity transformations, what will be the speed of one electron as measured from the other?*

Solution: Let observer O be at rest with respect to the laboratory and let observer O' be at rest with respect to the particle moving with speed 0.6 c (taken in the positive direction). Then, from the Galilean velocity transformation.

$$u'_x = u_x - v = -0.7\,c - 0.6\,c = 1.3\,c$$

This problem demonstrates that velocities greater than the speed of light are possible with the Galilean transformation, a result that is inconsistent with Special Relativity.

Example 3. *A train moving with a velocity of 60 km/hr passes through a railroad station at 12.00. Twenty seconds later a bolt of* ***lightning*** *strikes the railroad tracks one km from the station in the same direction that the train is moving. Find the coordinates of the lightning flash as measured by an observer at the station and by the engineer of the train.*

Solution: Both observers measure the time coordinate as

$$t = t' = \left(\frac{1\ \text{hr}}{3600\ \text{s}}\right) = \frac{1}{180}\ \text{hr}$$

The observer at the station measures the spatial coordinate to be $x = 1$ km. The spatial coordinate as determined by the engineer of the train is

$$x' = x - vt = 1\ \text{km} - (60\ \text{km/hr})\left(\frac{1}{180}\ \text{hr}\right) = \frac{2}{3}\ \text{km}.$$

Example 4. *A hunter on the ground fires a bullet in the north east direction which strikes a deer 0.25 km from the hunter. The bullet travels with a speed of 1800 km/hr. At the instant when the bullet is fired, an airplane is directly over the hunter at an altitude of one km and is travelling due east with a velocity of 600 km/hr. When the bullet strikes the deer, what are the coordinates as determined by an observant in the airplane?*

Solution: Using the Galilean transformations,

$$t' = t = \frac{0.25\ \text{km}}{1800\ \text{km/hr}} = 1.39 \times 10^{-4}\ \text{hr}.$$

$$x' = x - vt = (0.25\ \text{km})\cos 45° - (600\ \text{km/hr})(1.39 \times 10^{-4}\ \text{hr})$$

$$y' = y = (0.25\ \text{km})\sin 45° = 0.177\ \text{km}$$

$$z' = z - h = 0 - 1\ \text{km} = -1\ \text{km}$$

8.8. CONCEPT OF ETHER

Newton considered the fixed stars as frame of reference for the explanation of his laws. However in later years a new search for a frame of reference was carried out. Later on Maxwell proved the electromagnetic nature of light waves and also it was proved that unlike sound waves (longitudinal waves), **light waves have transverse character** which was proved by the polarization phenomenon.

Again it was considered that a material medium is necessary for propagation of transverse light waves and Luminiferous ether was supposed to fill the whole space.

Since transverse waves require shearing forces and these forces can occur in solids only, so ether must be a rigid solid filling the whole space and the velocity of propagation of light will depend on the elasticity of the medium. Also the velocity of light being very high so the ether must also have a high elasticity.

Now if the light is propagated the ether medium, then we must expect a change in velocity (drift) depending on the direction of propagation of light w.r.t. the motion of the apparatus.

If we consider earth's motion around the sun in an orbit with a velocity of 3×10^4 m/s and velocity of light is 3×10^8 m/s. The ratio between the two is ~ 10^{-4}. We may consider the motion of the earth through stationary orbit and this led to the experiment of Michelson and Morley in 1881–1887 to verify the concept of ether filling the whole space.

8.9. MICHELSON MORLEY EXPERIMENT

Light from a monochromatic source S is rendered parallel with the lens L and is divided into two portions by the half silver plate A (Fig. 8.2). One portion of the beam travels to the mirror M_1 and is reflected back and the other portion of the beam travels towards M_2 and is reflected back. The two reflected beams interfere and the interference fringes are viewed through the telescope.

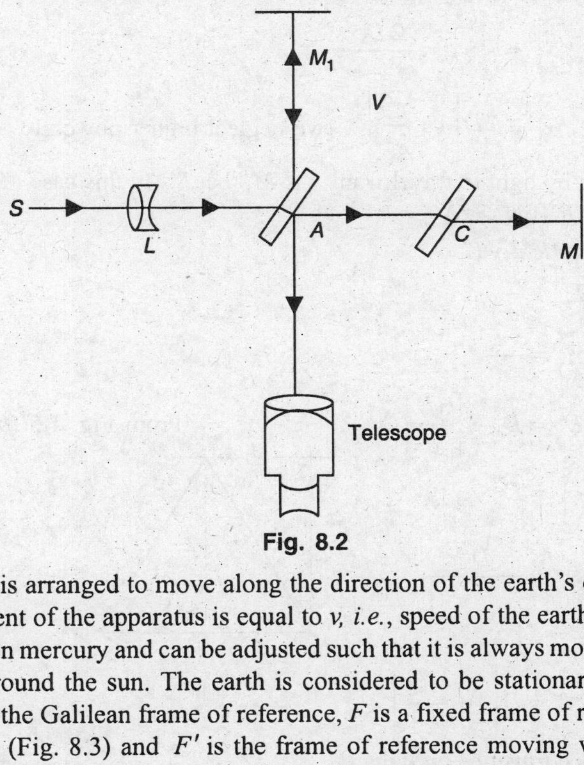

Fig. 8.2

The apparatus is arranged to move along the direction of the earth's orbit round the sun and the speed of movement of the apparatus is equal to v, *i.e.*, speed of the earth in its orbit. The whole apparatus is floated on mercury and can be adjusted such that it is always moving along the direction of the earth's orbit round the sun. The earth is considered to be stationary.

If we consider the Galilean frame of reference, F is a fixed frame of reference corresponding to the ether medium (Fig. 8.3) and F' is the frame of reference moving with a velocity v in the direction of the movement of the apparatus *i.e.*, in the direction of the movement of the earth in

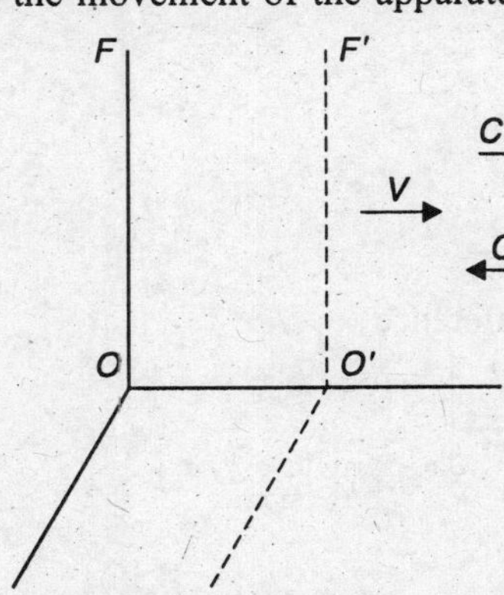

Fig. 8.3

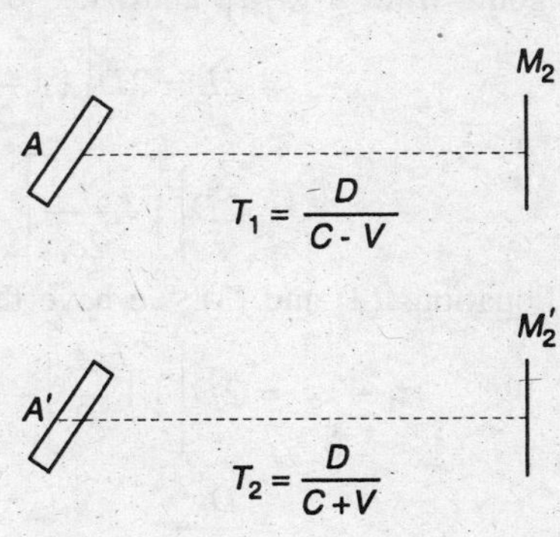

Fig. 8.4

its orbit round the sun. The velocity of light in the direction of the movement of the frame $F' = c - v$ and in the opposite direction it is $= c + v$. Let T_1 be the time taken by light to travel from A to M_2' and T_2 that taken that from M_2' to A' (Fig. 8.4).

In this case $D = AM_1 = AM_2$

So total to and fro times is given by

$$T = T_1 + T_2 = \frac{D}{c - v} + \frac{D}{c + v}$$

$$T = \frac{2\,Dc}{c^2 - v^2} \qquad \text{...(1)}$$

The total distance travelled by light is given by

$$x_1 = T \times c = \frac{2\,Dc^2}{c^2 - v^2}$$

$$x_1 \cong 2D\left[1 + \frac{v^2}{c^2}\right] \quad \text{[we neglect higher powers]} \qquad \text{...(2)}$$

Let the time taken by light to travel from A to M_1' be T'. In this case M_1 is shifted to M_1' in the time A is shifted to R (Fig. 8.5)

The distance $\quad AR = vT'$

Also $\quad T' = \dfrac{L}{c}$

$\therefore \quad AR = \dfrac{vL}{c}$

$$L^2 = D^2 + \left[\frac{vL}{c}\right]^2$$

From Fig. 1.5 (*b*)

$$L^2\left[1 - \frac{v^2}{c^2}\right] = D^2$$

$$L^2 = \frac{D^2}{\left[1 - \frac{v^2}{v^2}\right]}$$

$$L = D\left[1 + \frac{v^2}{2c^2}\right]$$

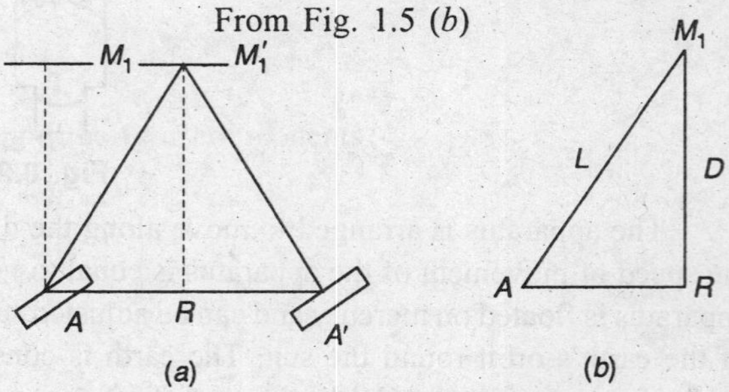

Fig. 8.5

So the total distance travelled by light in time T is going from A to M_1' and back to A'

$$= 2L = 2D\left[1 + \frac{v^2}{2c^2}\right]$$

$$x_2 = 2D\left[1 + \frac{v^2}{2c^2}\right] \qquad \text{...(3)}$$

From Equations (2) and (3), we have the path difference

$$x_1 - x_2 = 2D\left[1 + \frac{v^2}{c^2}\right] - 2D\left[1 + \frac{v^2}{2c^2}\right]$$

$$= \frac{Dv^2}{c^2} \qquad \text{...(4)}$$

If the apparatus is turned through 90°, the path difference will be $\frac{-Dv^2}{c^2}$ *i.e.*, the displacement in the interference fringes is $\frac{2Dv^2}{c^2}$ which will shift the fringes in the telescope T through a number N, given by

$$\Delta N = \frac{2Dv^2/c^2}{\lambda}$$

or
$$\Delta N = \frac{2Dv^2}{\lambda} \cdot \frac{1}{\lambda}$$

where λ corresponds to the wavelength of light used.

In the Michelson-Morley experiment the effective path D was taken = 11 metres and if the velocity of earth through ether is considered to be the orbital velocity of the earth *i.e.*, $v = 3 \times 10^4$ m/s, the fringe shift for a visible light say of wavelength = 5500Å is given by

$$\Delta N = \frac{2Dv^2}{c^2} \cdot \frac{1}{\lambda}$$

or
$$\Delta N = \frac{2 \times 11 \times (3 \times 10^4)^2}{(3 \times 10^8)^2} \times \frac{1}{5.5 \times 10^{-7}} = 0.404$$

This displacement could be very easily detected by Michelson-Morley apparatus. But actually no fringe shift was observed. The experiment was repeated at different times and different places, but all the times no change was observed. **Which concludes that the relative velocity between ether and earth is zero.**

Explanation of the negative results. Following separate interpretations were given to explain the negative results of Michelson-Morely experiment :

(*a*) Ether-drag hypothesis. The moving earth drags the ether along with it and hence there is no relative motion between the earth and the ether. This explanation of Michelson himself, however, is not reasonable because it goes against the observed phenomenon of aberration of starlight.

(*b*) Lorentz-Fitzgerald contraction hypothesis. Fitzgerald and Lorentz independently offered a more radical suggestion that a material body moving through ether is contracted in the direction of motion by a factor $\sqrt{[1-(v^2/c^2)]}$. Thus a length l in the direction of motion is shortened to $l\sqrt{[1-(v^2/c^2)]}$. It would equalize the times t_1 and t_2 in Michelson-Morley experiment and hence no fringe shift can be expected. This contraction phenomenon is purely mathematical without any experimental confirmation.

(*c*) Constancy of speed of light. The proper explanation of zero results of Michelson-Morley experiment was given by Einstein. He proposed that the *speed of light is an invariant, i.e.* the speed of light is constant and does not depend upon the motion of the source, observer or the medium. Then light would take equal times to travel the two paths in Michelson-Morely apparatus *i.e.*,

$$t_1 = t_2 = \frac{2D}{c}$$

and hence, no fringe shift is to be expected.

This negative result suggests that the *velocity of light is invariable and remains constant in all directions.* Also the effects of ether are undetectable. This has been verified by Doppler's principle *i.e.*, the speed of light is independent of the motion of the source or the observer.

So, all attempts to make ether as a fixed frame of reference have failed.

8.10. EINSTEIN'S SPECIAL THEORY OF RELATIVITY

Einstein in 1905 announced the special theory of relativity and in 1915 proposed the general theory of relativity. The special theory deals with the problems in which one frame of reference moves with a *constant linear velocity* relative to another frame of reference. The general theory of relativity deals with problems in which one frame of reference is accelerated with respect to another frame of reference. Einstein assumed that all observers will notice that their motion in space will make no difference in the velocity of light with respect to them. He assumed that a fixed frame of reference cannot be located. All laws must be stated and they are applicable in any frame of reference.

According to Einstein's special theory of relativity:

(*i*) Observation on any particular frame of reference are not preferred to those of the other frames of reference. It means that the laws of physics are applicable equally well for an observer in the frame of reference A moving with a constant linear velocity with respect to an observer in another frame of reference B and *vice versa* and

(*ii*) Velocity of light in free space is constant and is independent of the motion of the observer in any frame of reference.

Einstein proved the following facts based on his theory of relativity. Let v be the velocity of the spaceship with respect to a given frame of reference where an observer makes the observations.

(*a*) All clocks on the spaceship will go slow by a factor

$$\sqrt{1-\frac{v^2}{c^2}}$$

(*b*) The mass of the spaceship increases by a factor

$$\left[\frac{1}{\sqrt{1-\frac{v^2}{c^2}}}\right]$$

(*c*) All objects on the spaceship will have contracted in length by a factor

$$\sqrt{1-\frac{v^2}{c^2}}$$

(*d*) The speed of a material object can never exceed the velocity of light.

(*e*) Mass and energy are interconvertible

$$E = mc^2$$

(*f*) If two objects A and B are moving with velocities u and v with respect to each other along the X-axis, the relative velocity of A with respect to B is given by

$$V_x = \frac{u-v}{1-\frac{uv}{c^2}}$$

Here u and v are comparable with the value of c.

Example 5. *An electron is moving with a speed of 0.85 c in a direction opposite to that of a moving photon. Calculate the relative velocity of the electron with respect to the photon.*

Solution: Here, $u = 0.85\,c,\ v = -c$

$$V_x = \frac{u-v}{1-\frac{uv}{c^2}}$$

$$V_x = \frac{0.85c + c}{1+\frac{0.85 \times c}{c^2}} = \frac{1.85c}{1.85} = c$$

Form this exmaple it is clear that the velocity of any object cannot exceed the velocity of light.

8.11. LORENTZ TRANSFORMATION OF SPACE AND TIME

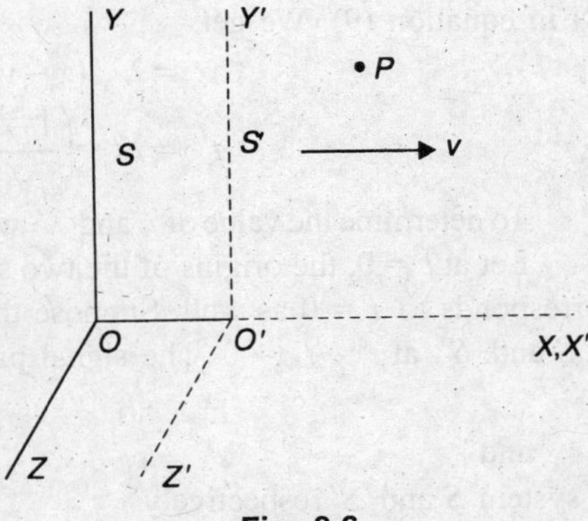

Fig. 8.6

Without making any particular postulates, Lorentz algebraically tried transformation equations which may replace the Galilean transformations to give results in agreement with Michelson-Morley experiment.

Let us consider two inertial system S and S' as shown in Fig. 8.6. The system S' is moving with a velocity v with respect to S in the positive direction of X. For convenience, let us suppose that the axes of two coordinate systems coincide at $t = t' = 0$. Let a pulse of light is generated at time $t = 0$ at the origin which grows in the space. Now consider the situation when the pulse reaches at point P. Let (x, y, z, t) and (x', y', z', t') be the positions and times coordinates of P measured by observers O and O' in frames S and S' respectively. When the pulse is observed from S, then we have

$$x' = x - vt \quad \text{...(1)}$$

$$y = y \quad \text{...(2)}$$

and

$$z' = z \quad \text{...(3)}$$

$$t = t' \quad \text{...(4)}$$

If a velocity is measured in system S and we want to get the components of velocity in system S' in terms of those in system S, we can write

$$v'_x = \frac{dx'}{dt} = v_x - v \quad \text{...(5)}$$

$$v'_y = \frac{dy'}{dt} = v_y \quad \text{...(6)}$$

and

$$v'_z = \frac{dz'}{dt} = v_z \quad \text{...(7)}$$

It is clear that the Galilean transformations and the velocity transformations obtained in equations (5), (6) and (7) violate the postulates of the special theory of relativity. **The equations of physics obtained by these transformations assume totally different forms in frames S and S'.** Moreover, if the velocity of light in frame S is c, that in system S' would be $c' = c - v$. It is, therefore, necessary that different transformation equations must be obtained if the postulates of the special theory of relativity are to be satisfied. These transformation equations–called the **Lorentz transformations** – are developed as under:

Let us assume that possible relation x and x' is of the type

$$x' = \lambda (x - vt) \quad \text{...(8)}$$

where λ is a constant of proportionality. It does not depend upon x and t but it may be a function of v. In choosing the relation between x and x' in the form expressed in equation (8) the following points are considered: (*i*) The relation is linear in x and x'. Hence, a single event in frame S corresponds to a single event in S'. (*ii*) It can be easily reduced to the form $x' = x - vt$ which is known to be correct in Newtonian mechanics if $\lambda = 1$ when $v << c$.

The inverse relation will be written as

$$x = \lambda'(x' + vt') \quad \text{...(9)}$$

in which primed quantities are replaced by unprimed quantities and *vice versa*. The other relations will be as before

$$y' = y \quad \text{...(10)}$$

and

$$z' = z \quad \text{...(11)}$$

Times t and t' are not equal. To verify this, let us substitute the value of x' from equation

(8) in equation (9). We get,

$$x = \lambda\lambda'(x - vt) + \lambda' vt'$$

$$t' = \lambda t + \left(\frac{1-\lambda\lambda'}{\lambda' v}\right)x \qquad ...(12)$$

To determine the value of λ and λ' we use the second postulate of the special theory of relativity.

Let at $t = 0$, the origins of the two system S and S' coincide with each other. Let this instant corresponds to $t' = 0$ as well. Suppose that a signal of light is given from the common or origin of S and S' at $t = t' = 0$. The signal propagates in the two systems satisfying the equations

$$x = ct \qquad ...(13)$$

and

$$x' = ct' \qquad ...(14)$$

in system S and S' respectively.

Substituting the values of x' and t' from equations (8) and (12) into equation (14), we get

$$\lambda(x - vt) = c\lambda t + \left(\frac{1-\lambda\lambda'}{\lambda' v}\right)cx$$

Hence,

$$x = ct\left[\frac{1+v/c}{1-\left\{\frac{1}{\lambda\lambda'}-1\right\}\frac{c}{v}}\right] \qquad ...(15)$$

Comparing equation (15) with equation (13), we have

$$\frac{1+v/c}{1-\left\{\frac{1}{\lambda\lambda'}-1\right\}\frac{c}{v}} = 1$$

or

$$\lambda\lambda' = \frac{1}{1-v^2/c^2} \qquad ...(16)$$

Now, $\lambda\lambda'$ depends on v^2 and not on v, the relative velocity of S' with respect to S. In fact, we cannot choose between the two frames S' and S' except for the sign of v which does not affect the dependence of $\lambda\lambda'$. Therefore, we choose

$$\lambda = \lambda' = \frac{1}{\sqrt{1-v^2/c^2}} \qquad ...(17)$$

Substituting these values in equations (8) and (12), we get

$$\left.\begin{aligned} x' &= \frac{x - vt}{\sqrt{1-v^2/c^2}} \\ y' &= y \\ x' &= z \\ \text{and} \quad t' &= \frac{t - \frac{vx}{c^2}}{\sqrt{1-v^2/c^2}} \end{aligned}\right\} \qquad ...(18)$$

Equations (18) are called the Lorentz transformations.

The inverse Lorentz transformations can now be written as

$$\left.\begin{aligned} x &= \frac{x - vt}{\sqrt{1-v^2/c^2}} \\ y &= y' \\ x &= z' \\ \text{and} \quad t &= \frac{t - \frac{vx'}{c^2}}{\sqrt{1-v^2/c^2}} \end{aligned}\right\}$$

So, the measurements of position and time are found to depend upon the frame of reference of the observer. It should be noted that Lorentz equations reduce to the Galilean transformations when relative velocity v is very small in comparison with velocity c of light. So, the consequences of special theory of relativity will be observable only when velocity v is practically comparable with the velocity of light.

Example 5. *An event occurs at $x = 100$ m, $y = 10$ m, $z = 5$ m and $t = 1 \times 10^{-4}$ second in a frame S. Find the coordinates of this event in a frame S' which is moving with velocity 2.7×10^8 metre per second with respect to the frame S along the common XX' axes using (i) Galilean transformation and (iii) Lorentz transformation.*

Solution : (*i*) According to Galilean transformation, we have

$$x' = x - vt,\ y' = y,\ z' = z \text{ and } t' = t$$

Substituting the given values, we get

$$x' = 100 - 2.7 \times 10^6 \times 1 \times 10^{-4} = -26900 \text{ m}$$

$$y' = y = 10 \text{ m}, \qquad z' = z = 5 \text{ m}$$

and $$t' = t = 1 \times 10^{-4} \text{ second}$$

Thus the coordinates in frame S' are

$$x' = -26900 \text{ m},\ y' = 10 \text{ m},\ z' = 5 \text{ m and } t' = 10^{-4} \text{ second}$$

(*ii*) According to Lorentz transformation, we have

$$x' = \frac{x - vt}{\sqrt{1 - \dfrac{v^2}{c^2}}}$$

and $$t' = \frac{t - vx/c^2}{\sqrt{1 - \dfrac{v^2}{c^2}}}$$

Substituting the given values, we get

$$x' = \frac{100 - 2.7 \times 10^8 \times 1 \times 10^{-4}}{\sqrt{\left\{1 - \left(\dfrac{2.7 \times 20^8}{3 \times 10^8}\right)^2\right\}}} = -61720 \text{ m}$$

$$y' = y = 10 \text{ m}, \quad z' = z = 5 \text{ m}$$

$$t' = \frac{1 \times 10^{-4} - \{(2.7 \times 10^3 \times 100)/(3 \times 10^8)^2\}}{\sqrt{\left\{1 - \left(\dfrac{2.7 \times 20^8}{3 \times 10^8}\right)^2\right\}}}$$

$$= 2.288 \times 10^{-4} \text{ sec.}$$

Thus the coordinates in S' are

$$x' = -61720 \text{ m},\ y' = 10 \text{ m},\ z' = 5 \text{ m and } t' = 2.288 \times 10^{-4} \text{ sec.}$$

Example 6. *Use Lorentz transformation to show that*

$$x^2 + y^2 + z^2 - c^2t^2 = x'^2 + y'^2 + z'^2 - c^2t'^2$$

show that the quantity $x^2 + y^2 + z^2 - c^2t^2$ is invariant under Lorentz transformation.

Solution: According to Lorentz transformation, we have

$$x' = \frac{x - vt}{\sqrt{1 - v^2/c^2}}, \qquad y' = y, \quad z' = z$$

and $$t' = \frac{t - vx/c^2}{\sqrt{1 - v^2/c^2}}$$

$$\therefore \quad x'^2 + y'^2 + z'^2 - c^2t'^2$$

$$= \left(\frac{x - vt}{\sqrt{1 - v^2/c^2}}\right)^2 + y^2 + z^2 - c^2\left(\frac{t - vx/c^2}{\sqrt{1 - v^2/c^2}}\right)^2$$

$$= \frac{(x - vt)^2 - c^2(t - vx/c^2)}{1 - \dfrac{v^2}{c^2}} + y^2 + z^2$$

$$= \frac{(x^2 + v^2t^2 - 2xvt) - c^2\{t^2 + (v^2x^2/c^4) - (2xvt/c^2)\}}{1 - v^2/c^2} + y^2 + z^2$$

$$= \frac{x^2(1 - v^2/c^2) - c^2t^2(1 - v^2/c^2)}{1 - v^2/c^2} + y^2 + z^2$$

$$= \frac{(x^2 - c^2t^2)(1 - v^2/c^2)}{1 - v^2/c^2} + y^2 + z^2$$

$$= x^2 - c^2t^2 + y^2 + z^2 = x^2 + y^2z^2 - c^2t^2$$

This shows that the quantity $x^2 + y^2 + z^2 - c^2t^2$ is same in both frames of reference. So, this is invariant under Lorentz transformation.

8.12. LENGTH CONTRACTION

Consider two coordinate systems S and S', the latter moving with velocity v relative to the former along the positive x-direction. Let a rod be placed in S' along X-axis. If x_1' and x_2' are coordinates of the abscissae of the ends of the rod, then its length l' in system S' is given by

$$l' = x_2' - x_1' \qquad ...(i)$$

Now let an observer in S measure the length of the same rod. If x_1 and x_2 are the abscissae of the ends of the rod at the same time t, then the length l of the rod is given by

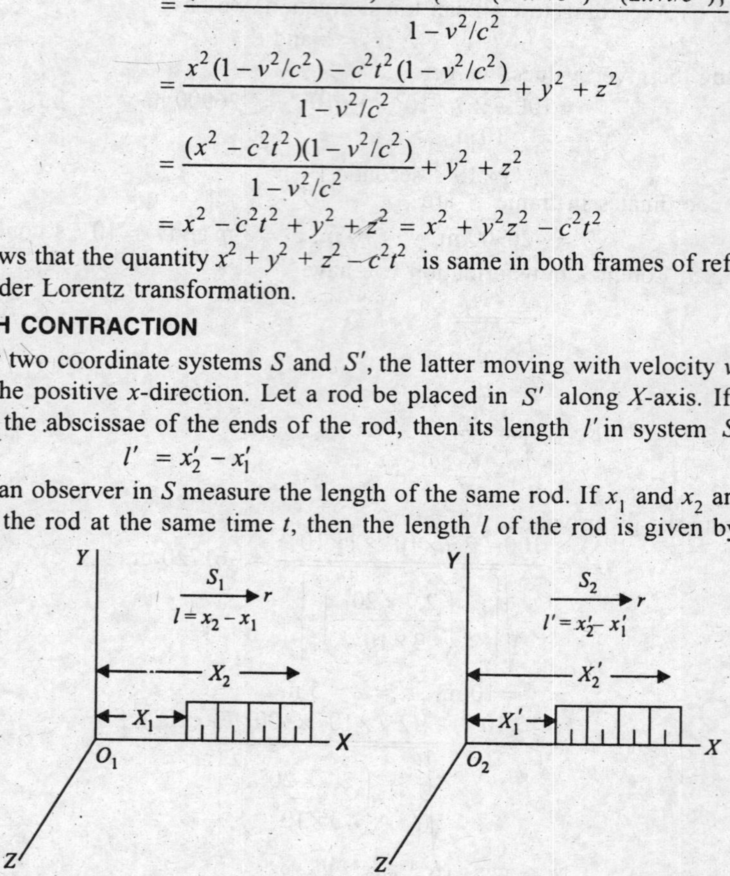

Fig. 8.7

$$l = x_2 - x_1 \qquad ...(ii)$$

From Lorentz transformation equations we may write

$$x_2' = \frac{(x_2 - vt)}{\sqrt{1 - \dfrac{v^2}{c^2}}}$$

and

$$x_1' = \frac{(x_1 - vt)}{\sqrt{1 - \dfrac{v^2}{c^2}}}$$

Putting these values in Equation (*i*) we have

$$l' = \frac{(x_2 - vt)}{\sqrt{1-\frac{v^2}{c^2}}} - \frac{(x_1 - vt)}{\sqrt{1-\frac{v^2}{c^2}}}$$

$$= \frac{(x_2 - x_1)}{\sqrt{1-\frac{v^2}{c^2}}} = \frac{l}{\sqrt{1-\frac{v^2}{c^2}}}$$

or

$$l = l'\sqrt{1-\frac{v^2}{c^2}}$$

Thus the length of the rod moving with velocity v relative to the observer is contracted by a factor $\sqrt{1-\frac{v^2}{c^2}}$ *in the direction of motion.* This is known as *Lorentz-Fitgerald contraction.*

From above we may conclude that

(*i*) When v is very small compared to c, then $\frac{v^2}{c^2}$ will be negligible in comparison to unity, therefore

$$l = l'$$

i.e., the length of the moving rod or object is the same as that, when it was at rest.

(*ii*) When v is comparable to c, $\sqrt{1-\frac{v^2}{c^2}}$ will be less than unity, therefore $l < l'$ *i.e.*, the length of the moving object or rod appears to be less than the length when it was at rest.

(*iii*) When v is equal to c or greater than $c, \frac{v^2}{c^2}$ *will be equal to unity or greater than unity, therefore*

$$l = 0 \text{ or imaginary.}$$

This is impossible. *So no material body can ever attain the velocity of light.*

(*iv*) The contraction takes place only along the direction of motion and remains unchanged in a perpendicular direction.

(*v*) The contraction cannot be visualized, though it really exists.

8.13. TIME DILATION

According to the special theory of relativity, time intervals are also affected by relative motion between two frames of reference and reference frame B is moving with a velocity v along the X-axis.

Let at any instant, the two reference frames coincide at $t = t' = 0$. The observer in B notes the time at any instant in his clock as t_1' whereas an observer in A notes the time as t_1'.

According to in Lorentz transformation,

$$t_1' = \frac{t_1 - \frac{vx}{c^2}}{\sqrt{1-\frac{v^2}{c^2}}} \qquad ...(1)$$

Let t_2 and t_2' be the times measured by the observers in B and A respectively at the same instant,

$$t_2' = \frac{t_2 - \frac{vx}{c^2}}{\sqrt{1-\frac{v^2}{c^2}}} \qquad ...(2)$$

Let t_0 be the interval of time as assumed by an observer in B and t the interval of time measured by an observer in A.

$$\Delta t' = t_2' - t_1' \qquad ...(3)$$

$$\Delta t = t_2 - t_1 \qquad ...(4)$$

$$\Delta t' = \frac{t_2 - \frac{vx}{c^2}}{\sqrt{1-\frac{v^2}{c^2}}} - \frac{t_1 - \frac{vx}{c^2}}{\sqrt{1-\frac{v^2}{c^2}}}$$

$$\Delta t' = \frac{t_2 - t_1}{\sqrt{1-\frac{v^2}{c^2}}} = \frac{\Delta t}{\sqrt{1-\frac{v^2}{c^2}}} \qquad ...(5)$$

Which show that an interval of time observed in a moving frame of reference will be less than the same interval of time observed in a stationary frame of reference. It also show that the clocks in the moving spaceships will appear to go slower than the clocks on the surface of the earth. **It is called time dilation.**

8.14. EXPERIMENTAL VERIFICATION OF TIME DILATION

Time dilation has been verified in case of a nuclear particle meson. There are π meson and μ mesons. The π^+ mesons decay (convert into μ mesons and neutrons) in such a way that in every 1.8×10^{-8} second half of them die away. In the laboratory the π^+ mesons are produced with speed of 0.99 c when high energy particles generated by Synchrotron strike a target. The flux of π^+ mesons was measured at two places separated by 30 metres. The laboratory time interval $\Delta t'$ for this distance is given by

$$\Delta t' = \frac{30 \text{ m}}{0.99 \times 3 \times 10^8 \text{m/sec}} = 10 \times 10^{-8} \text{ sec}.$$

This is nearly 5.6 times of 1.8×10^{-8} sec. Thus the flux of π^+ mesons should decrease to $(2)^{-5.6}$ or or less than 2% of the original flux in travelling a distance 30 metres. It was observed that the flux at the second place was nearly 60% of that of the first place.

This discrepancy can be explained by calculating t given by the relation

$$\Delta t' = \frac{\Delta t}{\sqrt{1-\frac{v^2}{c^2}}}$$

or

$$\Delta t = \Delta t' \times \sqrt{1-\frac{v^2}{c^2}}$$

$$= 10 \times 10^{-8} \sqrt{1 - \left\{\frac{0.99c}{c}\right\}^2} = 1.4 \times 10^{-8} \text{ sec}.$$

This is 0.78 times the half life 1.8×10^{-8} sec. Hence in so much time the flux should fall $(2)^{-0.7} \approx 60\%$.

This is exactly the same, what is observed experimentally. From this experiment, it is obvious that π^+ mesons take a time 10×10^{-8} sec to cover a distance 30 metres in laboratory measurements while a time 1.4×10^{-8} sec in their own frame of reference. Thus a 7-fold time dilation has occurred in this case.

Example 7. *A rocket ship is 100 metre long on the ground. When it is in flight, its length is 99 metres to an observer on the ground. What is its speed?*

Solution: We know that

$$l = l'\sqrt{1 - \frac{v^2}{c^2}}$$

Here $l' = 100 \times 100 = 10000$ cm

and $l = 99 \times 100 = 9900$ cm

$$\therefore \quad 9900 = 10000\sqrt{1 - \frac{v^2}{c^2}}$$

or $$\frac{99}{100} = \sqrt{1 - \frac{v^2}{c^2}}$$

or $$\frac{99 \times 99}{100 \times 100} = 1 - \frac{v^2}{c^2}$$

or $$\frac{v^2}{c^2} = 1 - \frac{99 \times 99}{100 \times 100} = \frac{199}{100 \times 100}$$

or $$v = c\,\frac{\sqrt{199}}{100}$$

or $$v = 3 \times 10^{10}\,\frac{\sqrt{199}}{100}$$

$$= \mathbf{4.23 \times 10^9 \text{ cm/sec.}}$$

Example 8. *A certain process requires 10^{-6} sec to occur in an atom at rest in laboratory. How much time will this process require to an observer in the laboratory, when the atom is moving with a speed of 5×10^7 metres/sec.*

Solution: We have the relation

$$\Delta t' = \frac{\Delta t}{\sqrt{1 - \left(\frac{v^2}{c^2}\right)}}$$

Here $\Delta t = 10^{-6}$ sec and $v = 5 \times 10^9$ cm/sec.

$$\therefore \quad \Delta t' = \frac{10^{-6}}{\sqrt{1 - \left(\frac{5 \times 10^9}{3 \times 10^{10}}\right)^2}}$$

$$= \frac{10^{-6}}{\sqrt{1 - \left(\frac{25}{900}\right)}} = \mathbf{1.013 \times 10^{-6} \text{ sec.}}$$

Example 9. *With what velocity should a space ship fly so that every day spent on it may correspond to three days on earth's surface.*

Solution: We know that

$$\Delta t' = \frac{\Delta t}{\sqrt{1 - \frac{v^2}{c^2}}}$$

According to the given problem, $\Delta t = 1$ day and $\Delta t' = 3$ days

$$\therefore \quad 3 = \frac{1}{\sqrt{1 - \frac{v^2}{c^2}}}$$

or $$\left(1-\frac{v^2}{9\times10^{16}}\right) = \frac{1}{9} \text{ or } v^2 = \frac{9\times10^{16}\times 8}{9}$$

$\therefore$ $$v = \sqrt{8}\times10^8 = \mathbf{2.83\times10^5\ m/s.}$$

Example 10. *A μ meson has a life time 2 × 10^{-6} sec.*

(*a*) *What is the mean life time when the particle is travelling with a speed of* 2.994×10^{10} *cm/sec?*

(*b*) *How far does it go during one mean life?*

(*c*) *What distance would be travelled without relativistic effects?*

Solution: (*a*) We have the relation

$$\Delta t' = \frac{\Delta t}{\sqrt{1-\frac{v^2}{c^2}}}$$

Here $$\Delta t = 2\times10^{-6} \text{ sec}$$

$$v = 2.994\times10^{10} \text{ cm/sec}$$

$\therefore$ $$\Delta t' = \frac{2\times10^{-6}}{\sqrt{1-\left(\frac{2.994\times10^{10}}{3\times10^{10}}\right)}} = 31.7\times10^{-6} \text{ sec}.$$

(*b*) The distance travelled by the particle in one mean life

$$s = v\times\Delta t'$$

$$= 2.294\times10^{10}\times31.7\times10^{-6} \text{cm} = 9510 \text{ metres.}$$

(*c*) Treating the problem non-relativistically, we get

$$s = v\times\Delta t$$

$$= 2.994\times10^{10}\times2\times10^{-6} \text{cm} = \mathbf{598.8\ metres}.$$

8.15. ADDITION OF VELOCITIES

Consider two systems S and S', S' moving with a velocity v relative to system S along the positive direction of X. Let us express the velocity of a body in them. Suppose the body moves a distance dx in time dt in system S and through a distance dx' in time dt' in system S'. Then

$$\frac{dx}{dt} = u \quad \text{and} \quad \frac{dx'}{dt'} = u' \qquad ...(1)$$

From Lorentz transformation equations, we have the relations

$$x = \lambda(x'+vt') \qquad ...(2)$$

and $$t = \lambda\left(t'+\frac{vx'}{c^2}\right) \qquad ...(3)$$

Differentiating Equation (2) and (3), we get

$$dx = \lambda(dx'+vdt') \qquad ...(4)$$

$$dt = \lambda\left(dt'+\frac{vdx'}{c^2}\right) \qquad ...(5)$$

Dividing Equation (4) by Equation (5)

$$\frac{dx}{dt} = \frac{\lambda(dx'+vdt')}{\lambda(dt'+vdx'/c^2)}$$

$$\frac{dx}{dt} = \frac{\frac{dx'}{dt'}+v}{1+\frac{v}{c^2}\frac{dx'}{dt'}}$$

or
$$u = \frac{u' + v}{1 + \frac{u'v}{c^2}} \qquad ...(6)$$

The equation (6) represents the relativistic law of addition of velocities whereas in classical mechanics it is simply $u = u' + v$. Following points are evident from here

(*i*) When u' and v are smaller as compared to c, $\frac{u'v}{c^2}$ can be neglected in comparison to unity, therefore
$$u = u' + v \qquad ...(7)$$
which is classical formula.

(*ii*) When v or $u' = c$, then
$$u = \frac{u + c}{1 + \frac{u'c}{c^2}} = \frac{u' + c}{c + u'} \times c = c \qquad ...(8)$$
i.e., if one object moves with velocity c with respect to other then, their relative velocity is always c, whatever may be the velocity of the other.

(*iii*) When $u' = c = v$, then
$$u = \frac{c + c}{1 + \frac{c^2}{c^2}} = \frac{2c}{c} = c \qquad ...(9)$$

This shows that the addition of velocity of light to the velocity of light merely reproduces the velocity of light.

Example 11. *Two particles came towards each other with speed 0.8 c with respect to laboratory. What is their relative speed?*

Solution: We know that
$$u = \frac{u' + v}{1 + \frac{u'v}{c^2}}$$

Here $u' = v = 0.8\ c$

$$\therefore \quad u = \frac{0.8c + 0.8c}{1 + \frac{0.8c \times 0.8c}{c^2}} = \frac{0.8c + 0.8c}{1 + 0.8 \times 0.8}$$
$$= \frac{1.6c}{1.64} = \mathbf{0.975c}$$

Thus the relative speed is 0.975 c.

Example 12. *An airforce rocket is chasing enemy's spaceship. From earth, it is found that the speed of airforce rocket is 2.55×10^{10} cm/sec, while that of enemy ship is 2.25×10^{10} cm/sec. What is the relative velocity of*

(*i*) Enemy's ship as seen by Airforce rocket?

(*ii*) Airforce rocket as seen by enemy's ship?

(*iii*) Airforce rocket with respect to enemy's ship as seen from earth?

(*iv*) Enemy's ship with respect to airforce rocket as seen from earth?

Solution: We have the relation
$$u = \frac{u' + v}{1 + \frac{u'v}{c^2}}$$

In this case $\quad u' = 2.55 \times 10^{10}$ cm/sec.

$$v = 2.55 \times 10^{10} \text{ cm/sec}$$

$$\therefore \quad u = \frac{2.55 \times 10^{10} - 2.55 \times 10^{10}}{1 - \left(\dfrac{2.55 \times 10^{10} \times 2.25 \times 10^{10}}{3 \times 10^{10} \times 3 \times 10^{10}} \right)}$$

$$= \frac{0.30 \times 10^{10} \times 9}{9 - 2.55 \times 2.55} = 0.82 \times 10^{10} \text{ cm/sec.}$$

$$= \mathbf{0.83 \times 10^{8} ms^{-1}}$$

(*i*) Relative velocity of enemy's ship as seen by airforce rocket will be **0.83×10^8 ms^{-1}** **Ans.**

(*ii*) Similarly the relative velocity of airforce rocket as seen by enemy's ship towards it will be **0.83×10^8 ms^{-1}** **Ans.**

(*iii*) The relative velocity of airforce rocket w.r.t. enemy's ship as seen from Earth is

$$= u' - v = 2.55 \times 10^8 - 2.25 \times 10^8$$

$$= \mathbf{0.3 \times 10^8 \ ms^{-1}} \textbf{ Ans.}$$

(*iv*) Considering airforce rocket to be at rest the observer will see that the enemy's ship is approaching the airforce rocket with a velocity **0.3×10^8 ms^{-1} Ans.**

Example 13. *Show that the motion of one projectile as seen from another projectile will always be a straight line motion.*

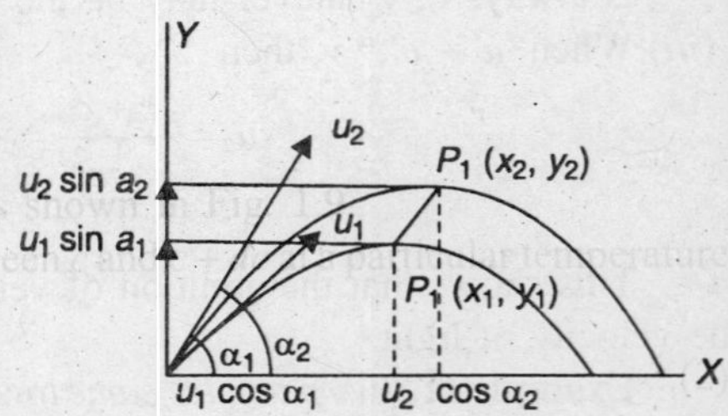

Fig. 8.8

Solution: Let the two projectiles be projected from the origin O with initial velocities u_1 and u_2 at angles of projections α_1 and α_2 respectively as shown in Fig. 8.8. After t sec the position coordinates of points P_1 (x_1, y_1) and P_2 (x_2, y_2) along X-and Y-axes may be written as

$$x_1 = (u_1 \cos \alpha_1)t, \; y_1 = (u_1 \sin \alpha_1)t - \frac{1}{2}gt^2$$

$$x_2 = (u_2 \cos \alpha_2)t, \; y_2 = (u_2 \sin \alpha_2)t - \frac{1}{2}gt^2$$

$$\therefore \quad x_2 - x_1 = (u_2 \cos \alpha_2 - u_1 \cos \alpha_1)t$$

$$\text{and} \quad y_2 - y_1 = (u_2 \sin \alpha_2 - u_1 \cos \alpha_1)t$$

Dividing (2) by (1), we get

$$\frac{y_2 - y_1}{x_2 - x_1} = \frac{u_2 \sin \alpha_2 - u_1 \sin \alpha_1}{u_2 \cos \alpha_2 - u_1 \cos \alpha_1} = m \text{ (constant)}$$

($\because u_1, u_2, \alpha_1$ and α_2 are constants)

$$\therefore \quad (y_2 - y_1) = m\,(x_2 - x_1)$$

Let $y_2 - y_1 = Y$ and $x_2 - x_1 = X$

$$\therefore \quad Y = mX$$

Which is the equation of a straight line. So, **the motion of one projectile as seen from another projectile will always be a straight line motion. Ans.**

Example 14. *A ball has velocity (4i – 5j + 10 k) m/sec w.r.t to a train moving with velocity (3i + 4j) m/sec relative to an observer on the ground. Find the velocity of the ball relative to the ground.*

Solution: Let the train be equivalent to a moving frame of reference S′ and let the ground be stationary frame of reference S. So we have

$\vec{V}$ $(3i + 4j)$ m/sec and $\vec{u'} = (4i - 5j + 10k)$ m/sec

So the velocity of the ball w.r.t. ground will be

$$\vec{u} = u' + \vec{V}$$

$$= (4i - 5j + 10k) + (3i + 4j)$$

or $$\vec{u} = 7i - j + \mathbf{10k \text{ m/sec. Ans.}}$$

Example 15. *If at the time t = t′ = 0, the origins of systems S and S′ just coincide and a spherical pulse of light is produced at the common origin, show that the speed of propagation of the spherical wavefront is the same in both systems-namely c. (where system S′ is moving relative to stationary system S).*

Solution: As per problem the origin of two systems S and S' coincide at $t = t' = 0$. The system S is at rest while S' is moving with fixed velocity v relative to S. Let spherical pulse start from a point where coordinates w.r.t. S and S' are (x, y, z, t) and (x', y', z', t') respectively. We have from invariance of Lorentz transformation equations.

$$x^2 + y^2 + z^2 = c^2 t^2 \text{ and } x'^2 + y'^2 + z'^2 = c^2 t'^2 \quad \text{...(1)}$$

Now the Lorentz Transformation Equations are

$$x' = \frac{x - vt}{\sqrt{(1 - v^2/c^2)}}, y' = y, z' \text{ and } t' = \frac{t - (vx/c^2)}{\sqrt{(1 - v^2/c^2)}} \quad \text{...(2)}$$

From equation (1), we have

$$c = \frac{(x^2 + y^2 + z^2)^{1/2}}{t} \text{ and } c' = \frac{(x'^2 + y'^2 + z'^2)^{1/2}}{t'} \quad \text{...(3)}$$

Putting the values of x', y', z', and t', in equation (3)

Now from equation (2) we have

$$c' = \frac{\{[(x - vt)^2 / (1 - v^2/c^2)] + y^2 + z^2\}^{1/2}}{[(t \; vx/c^2) / \sqrt{(1 - v^2/c^2)}]}$$

or $$c' = \frac{[(x - vt)^2 + (y^2 + z^2)(1 - v^2/c^2)]^{1/2}}{(t - vx/c^2)} \quad \text{...(4)}$$

Adding and subtracting a factor of $x^2\left(1 - \dfrac{v^2}{c^2}\right)$ in the numerator of equation (4) and writing in form

$$c' = \frac{[(x^2 + y^2 + z^2)(1 - v^2/c^2) + \{v^2 t^2 - 2xvt + (v^2/c^2).x^2\}]^{1/2}}{(t - vx/c^2)}$$

But, $x^2 + y^2 + z^2 = c^2 t^2$

$$\therefore \quad c' = \frac{\left[c^2 t^2 \left(1 - \dfrac{v^2}{c^2}\right) + v^2 t^2 - 2xvt + \dfrac{v^2}{c^2} x^2\right]^{1/2}}{\left(t - \dfrac{vx}{c^2}\right)}$$

$$= \frac{\left(c^2 t^2 - 2xvt + \dfrac{v^2}{c^2} x^2\right)^{1/2}}{\left(t - \dfrac{vx}{c^2}\right)} = \frac{\left[c^2 \left(t - \dfrac{vx}{c^2}\right)^2\right]^{1/2}}{\left(t - \dfrac{vx}{c^2}\right)} = c$$

$$\therefore \quad c' = c$$

Which means that the velocity of light is same in both systems or the speed of propagation of the spherical wavefront is same in both systems. Ans.

Example 16. *A light pulse is emitted at the origin of a frame of reference S' at time t' = 0. Its distance x' from the origin after a time t' is given by* $x'^2 = c^2t'^2$. *Use the Lorentz transformation to transform this equation to an equation in x and t and show that this is* $x^2 = c^2t^2$. *Discuss the implication of this result.*

Solution: From given problem, the distance of a light pulse x' from the origin after a time t' in the moving frame S' may be given as

$$x'^2 = c^2t'^2$$

Now Lorentz transformation equations, have the form

$$x' = \frac{x - vt}{\sqrt{1 - v^2/c^2}} \text{ and } t' = \frac{t - xv/c^2}{\sqrt{1 - v^2/c^2}}$$

Putting these values of x' and t' in equation (2) so we have

$$\frac{(x - vt)^2}{1 - v^2/c^2} = \frac{c^2[t - (xv/c^2)]^2}{1 - v^2/c^2}$$

or $$(x - vt)^2 = c^2\left(t - \frac{xv}{c^2}\right)^2$$

or $$x^2 + v^2t^2 - 2xvt = c^2t^2 + \frac{x^2v^2}{c^2} - 2txv$$

$$x^2 + v^2t^2 = c^2t^2 + \frac{x^2v^2}{c^2}$$

or $$x^2 - c^2t^2 + v^2t^2 - \frac{x^2v^2}{c^2} = 0$$

or $$x^2 - c^2t^2 - \frac{v^2}{c^2}(x^2 - c^2t^2) = 0$$

or $$(x^2 - c^2t^2)\left(1 - \frac{v^2}{c^2}\right) = 0$$

Now since $v \neq c$, $\therefore \left(1 - \frac{v^2}{c^2}\right) \neq 0$

$\therefore$ $$x^2 - c^2t^2 = 0$$

or $$x^2 = c^2t^2$$

Which shows that the velocity of light is an absolute constant independent of the frame of reference or the equation is invariant under Lorentz transformation. Ans.

Example 17. *Find the time [as measured by a clock at rest (on the rocket)] taken by a rocket to reach a distant star and return to earth with a constant velocity* $v = \sqrt{(0.9999)}c$, *if the distance to the star is 4 light years (A light year is defined as the distance travelled by a light beam in vacuum in one year).*

Solution: Lorentz transformation equation for time, may be written as

$$t' = \frac{t - vx/c^2}{\sqrt{(1-v^2/c^2)}} \qquad ...(i)$$

In this case $x = 2 \times 4$ light year $= 2 \times 4 \times 365 \times 24 \times 60 \times 60 \times c$ cm. $= 2.523 \times 10^8 c$ cm, where t is the time taken by the rocket to reach the star and back to the earth as measured by an observer in a stationary frame at earth and t' is the time measured by the clock in the moving rocket for the complete journey time is given by $t = x/v$, where $v = c\sqrt{(0.9999)}$

$$\therefore \qquad t' = \frac{[(x/v) - (vx/c^2)]}{\sqrt{(1-v^2/c^2)}} = \frac{\frac{x}{v}\left(1 - \frac{v^2}{c^2}\right)}{\left(1 - \frac{v^2}{c^2}\right)^{1/2}}$$

or
$$t' = \frac{x}{y}\sqrt{(1 - v^2/c^2)} \qquad ...(ii)$$

Putting the values of x and v in equation (*ii*) we have

$$t' = \frac{2523 \times 10^8 c}{c\sqrt{0.9999}}\sqrt{1 - 0.9999} = \frac{2523 \times 10^8 \times (0.0001)^{1/2}}{\sqrt{0.9999}}$$

or
$$t' = \frac{2523 \times 10^8}{(9999)^{1/2}} = \frac{2.523 \times 10^8}{99.995} \simeq \mathbf{2.523 \times 10^6 \text{ sec}}$$ **Ans.**

8.16. VARIATION OF MASS WITH VELOCITY

Einstein's concepts of relativistic change of space and time given an idea that mass also varies with velocity. In fact the mass has to be considered as a function of velocity with which a body moves.

Let us consider two inertial frames S and S' and let S' be moving with velocity V along the positive X-direction.

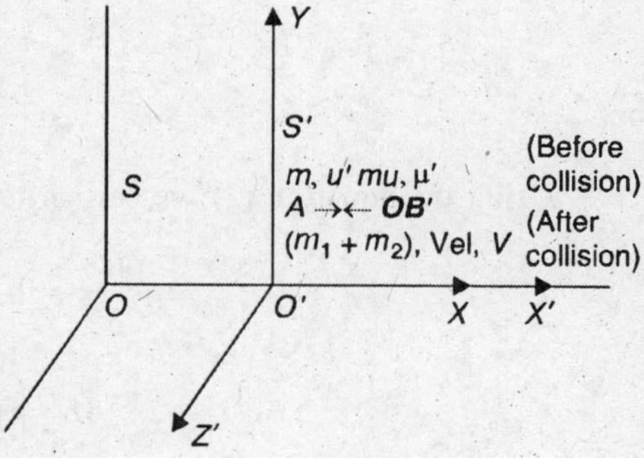

Fig. 8.9

Consider two identical particles A and B of same masses in frame S' approaching each other with a velocity u' in frames (taking positive velocity of particle A and negative velocity of particle B as it is moving in opposite direction).

Let, to an observer in frme S, the bodies be moving with velocities u_1 and u_2 before the collision. Since mass is a function of velocity, so to an observer in S; particles A and B will appear to have different masses. Let these masses be m_1 and m_2 as observed in frame S.

After Collision, the two masses stick to each other as the collision in elastic. The combined mass in frame S' will be 2m and it will be at rest in frame S'.

In frame S, the two bodies will have mass $m_1 + m_2$ and will appear to move with a velocity v, the velocity of frame S'.

From Lorentz inverse velocity transformation equations.

$$u_x = \frac{u'_x + V}{1 + \frac{V}{c^2}u'_x} \qquad ...(1)$$

For body A, but $u_x = u_1$ $u'_x = u'$, so
We have from equation (1)

$$u_1 = \frac{u'+V}{1+\frac{u'V}{c^2}} \qquad ...(2)$$

For body B, put and $u_x = u_2$ and $u'_x = -u'$

$$u_2 = \frac{-u'+V}{1-\frac{u'V}{c^2}} \qquad ...(3)$$

From the conservation law of linear momentum in frame S, we get

$$m_1 u_1 + m_2 u_2 = (m_1 + m_2)V \qquad ...(4)$$

Putting the values of u_1 and u_2 from equations (2) and (3) is equation (4), we get

$$m_1\left[\frac{u'+V}{1+\frac{u'V}{c^2}}\right] + m_2\left[\frac{-u'+V}{1-\frac{u'V}{c^2}}\right] = (m_1 + m_2)\,V$$

or
$$m_1\left[\frac{u'+V}{1+\frac{u'V}{c^2}} - V\right] = m_2\left[V - \left(\frac{-u'+V}{1-\frac{u'V}{c^2}}\right)\right]$$

or
$$\frac{m_1}{m_2} = \left[\frac{1+u'V/c^2}{1-u'V/c^2}\right]^2 \qquad ...(5)$$

Using the equation (2) we can write

$$1-\frac{u_1^2}{c^2} = 1-\frac{1}{c^2}\left[\frac{u'+V}{1+u'V/c^2}\right]^2$$

$$= \frac{\left(1+\frac{u'V}{c^2}\right)^2 - \frac{1}{c^2}(u'+V)^2}{\left(1+\frac{u'V}{c^2}\right)^2}$$

$$= \frac{\left(1-\frac{u'^2}{c^2}\right)(1-V^2/c^2)}{(1+Vu'/c^2)^2} \qquad ...(6)$$

Similary using equation (3) we get

$$1-\frac{u_2^2}{c^2} = \frac{(1-u'^2/c^2)(1-V^2/c^2)}{(1-u'V/c^2)^2} \qquad ...(7)$$

Dividing equation (7) by equation (6), we get

$$\frac{(1-u_2^2/c^2)}{1-u_1^2/c^2} = \left[\frac{1+u'V/c^2}{1-\frac{u'V}{c^2}}\right]$$

$$= \left(\frac{m_1}{m_2}\right)^2 \qquad \text{[From equation (5)]}$$

$$(1 - u_2^2 / c^2)^{1/2} \qquad \ldots (8)$$

Relation (8) is valid for all values of u_1 and u_2. If we select $u' = 0$ for particle B in frame S [This is possible if $u' = V$]

In that case $u_2 = 0$

$m_2 = m_0$ = rest mass of particle B.

Then eq. (8) reduces to $$\frac{m_1}{m_0} = \frac{1}{\left(1 - u_1^2 / c^2\right)^{1/2}}$$

$$\therefore \quad \frac{m_1}{m_2} = \frac{1}{(1 + u_1^2 / c^2)^{1/2}}$$

$$m_1 = \frac{m_0}{(a - u_1^2 / c^2)^{1/2}}$$

As we have taken the two masses exactly identically so the rest mass of other body may also be considered as m_0. This in other words mean that the above equation is valid as if applied to one and same particle.

Replacing m_1 by m and u_1 by v, we can write the general relation

$$m = \frac{m_0}{(1 - v^2/c^2)^{1/2}}$$

Where m_0 is the rest mass and m is the mass when the body is making with a velocity v.

So from relativity theory, the measurements of an object or an event will be different for different observers who are in relative motion. If the rest mass of a body be m_0, then the mass of the body in motion is given by

$$m = \frac{m_0}{\left(1 - \frac{v^2}{c^2}\right)^{1/2}} \qquad \ldots (9)$$

where $v \rightarrow$ velocity of motion of the body and $c \rightarrow$ velocity of light

From (9) it is clear that $m > m_0$ *i.e.*, an object will appear to have more mass when in motion than at rest.

Relative mass of electron in motion moving with different speeds is shown in graph (Fig. 8.10). The relative mass of the electron increases enormously when its velocity approaches the velocity of light.

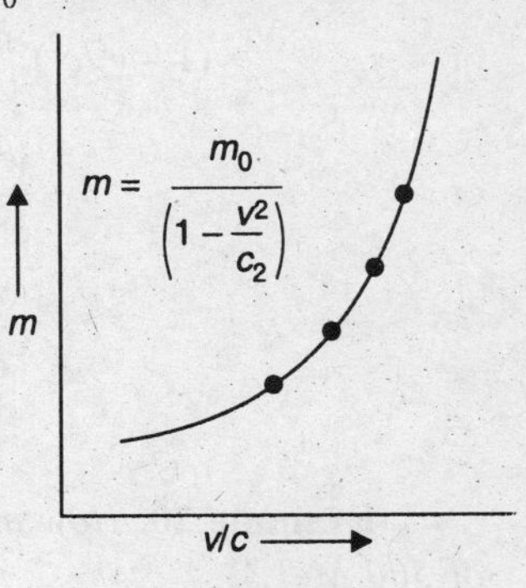

Fig. 8.10

If $v = c$

$$m = \frac{m_0}{\sqrt{1 - \frac{v^2}{c^2}}}$$

$$m = \frac{m_0}{0} = \infty$$

i.e., an electron moving with the speed of light will have an infinite mass which is not possible. **Hence no material particle can have a velocity equal to the velocity of light.**

Example 18. *What is the length of a metre stick moving parallel to its length when its mass is 3/2 of its rest mass.*

Solution: Here $$m = \frac{m_0}{\sqrt{1-\frac{v^2}{c^2}}} \quad ...(i)$$

and $$\frac{m}{m_0} = \frac{1}{\sqrt{1-\frac{v^2}{c^2}}}$$

and $$L = L_0\sqrt{1-\frac{v^2}{c^2}}$$

or $$v = 3\times10^8\times\sqrt{(0.935)} = 3\times10^8\times0.967 = \mathbf{2.9\times10^8\,ms^{-1}}\ \textbf{Ans.}$$

Example 19. *Calculate the velocity of electrons accelerated through a potential of 10^6 volts. Rest mass of an electron = 9×10^{-31} kg.*

Solution. We have the relation for kinetic energy Kinetic Energy K = Total Energy – Rest mass Energy

$$= mc^2 - m_0c^2 = \frac{m_0c^2}{\sqrt{(1-v^2/c^2)}} - m_0c^2$$

$$\therefore \quad K = m_0c^2\,[(1-v^2/c^2)^{-1/2}-1]$$

Now kinetic energy $= 1\text{ MeV} = 10^6\text{ eV} = 10^6 \times 1.6 \times 10^{-19}\text{J} = 1.6 \times 10^{-13}\text{J}$

$$\therefore \quad 1.6\times10^{-13} = 9\times10^{-31}\times(3\times10^8)^2\,[(1-v^2/c^2)^{-1/2}-1]$$

$$(1-v^2/c^2)^{-1/2}-1 = \frac{1.6\times10^{-13}}{9\times10^{-31}\times9\times10^{16}} = 1.975$$

or $$(1-v^2/c^2)^{-1/2} = 2.975 \text{ or } \frac{1}{2.975} = \sqrt{(1-v^2/c^2)}$$

or $$1-\frac{v^2}{c^2} = \frac{1}{(2.975)^2} \text{ or } \frac{v^2}{c^2} = 1-0.113 = 0.887$$

$$\frac{v}{c} = \sqrt{(0.887)} = 0.942 \quad \therefore \quad v = 0.942\times c$$

$$= 0.942 \times 3 \times 10^8\ \text{ms}^{-1}$$

$$\therefore \quad v = \mathbf{2.836 \times 10^8\ ms^{-1}}\ \textbf{Ans.}$$

Example 20. *How much does a proton gain in mass when accelerated to a kinetic energy of 500 MeV?*

Solution: Relation for K.E. from relativity is given by

$$K = (m - m_0)\,c^2 \text{ or } mc^2 = K + m_0c^2.$$

So gain in mass, $m - m_0 = K/c^2 = \Delta m$

In this case K = 500 MeV = $500 \times 10^6 \times 1.6 \times 10^{-19}$ joule, $m_0 = 2.67 \times 10^{-27}$ kg and $c = 3.0 \times 10^8$ m/sec

So $$\Delta m = \frac{500\times10^6\times1.6\times10^{-19}}{(3\times10^8)^2} = \mathbf{8.89\times10^{28}\ kg}\ \textbf{Ans.}$$

Example 21: *Find the speed of 0.1 MeV electron according to relativity mechanics.*

Solution: From classical mechanics the kinetic energy of a particle is given by

$$K = \frac{1}{2}mv^2 \text{ or } v = \sqrt{(2K/m)}$$

In this case $K = 0.1$ MeV $= 0.1 \times 10^6$ eV $= 0.1 \times 10^6 \times 1.6 \times 10^{-19}$J and $m = 9.11 \times 10^{-1}$

$$\therefore \quad v = \sqrt{\left(\frac{2\times 0.1\times 10^6 \times 1.6\times 10^{-19}}{9.11\times 10^{-31}}\right)} = \mathbf{1.87\times 10^8 ms^{-1}}$$

From relativistic mechanics, K.E. of an electron may be written as

$$K = mc^2 - m_0c^2 \quad \text{or} \quad mc^2 = K + m_0c^2$$

$$\frac{m_0c^2}{\sqrt{1-v^2/c^2}} = K + m_0c^2 \quad \text{or} \quad (1 - v^2/c^2) = \frac{m_0c^2}{K + m_0c^2}$$

Hence

$$v = c\sqrt{1-\left(\frac{m_0c^2}{K+m_0c^2}\right)^2}$$

Now rest mass energy of electrons $m_0c^2 = 0.512$ MeV

So $\quad K + m_0c^2 = 0.1$ Mev $+ 0.512$ MeV $= 0.612$ MeV

$$\therefore \quad v = 3.0 \times 10^8\sqrt{1-\left(\frac{0.512}{0.612}\right)^2} = 3\times 10^8 \times 0.548 \ \text{ms}^{-1}$$

$$= \mathbf{1.644\times 10^8\ ms^{-1}\ Ans.}$$

Example 22. *Show that the rest mass of a paricle is given by*

$$m_0 = \frac{p^2c^2 - T^2}{2Tc^2}$$

where P and T denote the momentum and kinetic energy of the particle

Solution: Total energy of the particle is given by

$$E = m_0c^2 + T \quad \text{...(1)}$$

Also

$$E^2 = p^2c^2 + m_0^2 c^4 \quad \text{...(2)}$$

Squaring (1) and equating with (2) we get

$$m_0^2 c^4 + T^2 + 2T m_0 c^2 = p^2c^2 + m_0^2 c^4$$

$$\therefore \quad m_0 = \frac{p^2c^2 - T^2}{2T c^2} \quad \textbf{Ans.}$$

Example 23. *Calculate the length and orientation of a metre rod in a frame of reference which is moving with a velocity 0.6 c in a direction making an angle 30° with the rod.*

Solution:

x – component of length of metre for $1_x = \cos 30° = 0.866$ m.

y – component of meter rod

$Ly = 1. \sin 30° = 0.5$ m.

Component of length along x – axis in moving frame is

$$L_x = Lx\sqrt{1-v^2/c^2}$$

$$= 0.866\sqrt{1-(0.6)^2}$$

$$= 0.693 \text{ m.}$$

Component of length I to the motion in moving frame

$$L'y = Ly = 0.5\,\text{m}$$

Length of rod in moving frame is

$$L = \sqrt{L_x'^2 + L_y'^2} = \sqrt{(0.5)^2 + (0.693)^2}$$

$$= 0.854 \text{ m.}$$

Let v be the angle that the rod appears to make in moving frame

Than $$\tan\theta = \frac{L_x'}{L_y'} = \frac{0.5}{0.693}$$

$$= 0.72$$

$\therefore$ $$v = \tan\theta\ (0.72)$$

$$= \mathbf{35° \ 50 \ Ans.}$$

8.17. ENERGY-MASS RELATION

The increase in kinetic energy E_k when a force F displaces the body through a distance dS is given by

$$dE_k = F \cdot dS = F\left(\frac{dS}{dt}\right)dt = F\,v\,dt \qquad ...(i)$$

Also $$F = \frac{d}{dt}(mv)$$

or $$Fdt = d\,(mc) \qquad ...(ii)$$

$\therefore$ $$dE_k = vd\,(mv)$$

or $$dE_k = vd\left[\frac{m_0 v}{\left(1-\frac{v^2}{c^2}\right)^{1/2}}\right]$$

$$dE_k = vm_0\left[\left(1-\frac{v^2}{c^2}\right)^{1/2} + \frac{v^2/c^2}{\left(1-\frac{v^2}{c^2}\right)^{1/2}}\right]\frac{dv}{\left(1-\frac{v^2}{c^2}\right)}$$

or $$dE_k = \frac{m_0 v dv}{\left(1-\frac{v^2}{c^2}\right)^{3/2}} \qquad ...(iv)$$

Also $$m = \frac{m_0}{\left(1-\frac{v^2}{c^2}\right)^{1/2}}$$

Differentiating, we get

$$dm = \frac{m_0 v dv}{c^2\left(1-\frac{v^2}{c^2}\right)^{3/2}} \qquad ...(v)$$

From Equations (*iii*) and (*v*)

$$dm = \frac{dE_k}{c^2}$$

or $$dE_k = c^2 dm \qquad ...(vi)$$

Now consider that the mass of body at rest is m_0 and when it is moving with a velocity v it is m. Integrating equation (*vi*)

$$\int dE_k = c^2 \int_{m_0}^{m} dm$$

$$E_k = c^2(m - m_0) \quad ...(vii)$$

Total energy = kinetic energy of moving body + energy at rest

$$E = E_k + m_o c^2$$

$$E = c^2(m - m_0) + m_0 c^2$$

$$E = mc^2 \quad ...(viii)$$

which gives us a relation between mass and energy.

8.18. ENERGY-MOMENTUM RELATION

If a body of mass m is moving with a velocity v, its momentum is given by

$$p = mv \quad ...(i)$$

Also energy is given by the relation

$$E = mc^2 \quad ...(ii)$$

and
$$m = \frac{m_0}{\sqrt{1 - \frac{v^2}{c^2}}} \quad ...(iii)$$

$$\therefore \quad E^2 - c^2 p^2 = m^2 c^4 - c^2 m^2 v^2 \quad ...(iv)$$

$$E^2 - c^2 p^2 = \frac{m_0^2 c^4}{\left(1 - \frac{v^2}{c^2}\right)} - \frac{c^2 m_0^2 v^2}{\left(1 - \frac{v^2}{c^2}\right)}$$

$$E^2 - c^2 p^2 = \frac{m_0^2 c^4 \left[1 - \frac{v^2}{c^2}\right]}{\left(1 - \frac{v^2}{c^2}\right)}$$

$$E^2 - c^2 p^2 = m_0^2 c^4$$

$$\mathbf{E^2 = m_0^2 c^4 + c^2 p^2} \quad ...(v)$$

Example 24. *Show that* $\left(\frac{E^2}{c^2}\right) - p^2$ *is invariant.*

Solution: Let us consider two frames of reference A and B and let frame B be moving with a velocity v along the X-axis with respect to frame A.

Let p be the momentum and E the energy of a particle in reference frame A and p' and E' in reference frame B.

$$\therefore \quad p^2 = p_x^2 + p_y^2 + p_z^2 \quad ...(i)$$

and
$$p'^2 = p_x'^2 + p_y'^2 + p_z'^2 \quad ...(ii)$$

$$\lambda = \frac{1}{\sqrt{1 - \frac{v^2}{c^2}}} \quad ...(iii)$$

$$p_y' = p_y, \; p_z' = p_z$$

and
$$E' = (E - vp_x)$$

$$\therefore \quad \frac{E'^2}{c^2} - p'^2 = \frac{\lambda^2}{c^2}(E - vp_x)^2 - [p_{x'}^2 + p_{y'}^2 + p_{z'}^2]$$

$$= \frac{\lambda^2}{c^2}(E - vp_x)^2 - \lambda^2\left(p_x - \frac{vE}{c^2}\right)^2 - p_y^2 + p_z^2$$

$$= \frac{\lambda^2}{c^2}\left[(E - vp_x)^2 - c^2\left(p_x - \frac{vE}{c^2}\right)^2\right] - p_y^2 + p_z^2$$

$$= \frac{\lambda^2}{c^2}\left[E^2 - v^2p_x^2 - 2vEp_x - p_x^2c^2 - \frac{v^2E^2}{c^2} + 2vEp_x - p_y^2 - p_z^2\right]$$

$$= \frac{\lambda^2}{c^2}\left[\left(E^2 - \frac{v^2E^2}{c^2}\right) + (-p_x^2c^2 + v^2p_x^2)\right] - p_y^2 - p_z^2$$

$$= \frac{\lambda^2}{c^2}\left[E^2\left(1 - \frac{v^2}{c^2}\right) - p_x^2c^2\left(1 - \frac{v^2}{c^2}\right)\right] - p_y^2 - p_z^2$$

But $\left(1 - \frac{v^2}{c^2}\right) = \frac{1}{\lambda^2}$

$$\therefore \quad \frac{E'^2}{c^2} - p'^2 = \frac{\lambda^2}{c^2}\left[\frac{E^2}{\lambda^2} - \frac{p_x^2c^2}{\lambda^2}\right] - p_y^2 - p_z^2$$

$$= \frac{E^2}{c^2} - p_x^2 - p_y^2 - p_z^2 = \frac{E^2}{c^2} - (p_x^2 + p_y^2 + p_z^2)$$

$$\frac{E'^2}{c^2} - p'^2 = \frac{E^2}{c^2} - p^2$$

Which shows that

$\frac{E^2}{c^2} - p^2$ is an invariant.

Example 25. *A particle of mass 10^{-24} kg is moving with a speed of 1.8×10^8 m/s. Calculate its mass when it is in motion.*

Solution:
$$m = \frac{m_0}{\sqrt{1 - \frac{v^2}{c^2}}}$$

Here $m_0 = 10^{-24}$ kg; $v = 1.8 \times 10^8$ m/s; $c = 3 \times 10^8$ m/s; $\frac{v}{c} = 0.6$

$$m = \frac{10^{-24}\,\text{kg}}{\sqrt{1 - (0.6)^2}}; \quad m = \mathbf{1.25 \times 10^{-24}\,kg.}$$

Example 26. *A particle is moving with a speed of 0.5 c. Calculate the ratio of the rest mass while in motion.*

Solution:
$$m = \frac{m_0}{\sqrt{1 - \frac{v^2}{c^2}}} \quad \text{or} \quad \frac{m_0}{m} = \sqrt{1 - \frac{v^2}{c^2}}$$

Here $v = 0.5\ c$

or $\frac{v}{c} = 0.5$; $\frac{m_0}{m} = \mathbf{0.866.}$

Example 27. *The mean life of meson is 2×10^{-8} s. Calculate the mean life of a meson moving* *with a velocity of 0.8 c.*

Solution: $$t = \frac{t_0}{\sqrt{1-\frac{v^2}{c^2}}}$$

Here $$t_0 = 2\times10^{-8}\,s;\ v = 0.8\,c$$

$$t = 2\times10^{-8}/\sqrt{1-(0.8)^2} = \frac{2\times10^{-8}}{0.6}$$

$$\mathbf{t = 3.33\times10^{-8}\,s}$$

Example 28. *Calculate the velocity that one atomic mass unit will have if it a kinetic energy equal to twice the mass energy.*

Solution:

Here $$2m_0c^2 = mc^2 - m_0c^2$$

$$3m_0c^2 = mc^2$$

$$m = 3m_0$$

but $$m = \frac{m_0}{\sqrt{1-\frac{v^2}{c^2}}} \quad \therefore \quad 3m_0 = \frac{m_0}{\sqrt{1-\frac{v^2}{c^2}}}$$

or $$1-\frac{v^2}{c^2} = \frac{1}{9} \Rightarrow v^2 = \left(\frac{8}{9}\right)c^2 \Rightarrow v = \mathbf{0.941\,c}$$

Example 29. *The rest mass of a proton is 2000 times the rest mass of an electron. Calculate the speed at which the electron should move so that its mass will be equal to the rest mass of the proton.*

Solution: $c = 3\times10^8$ m/s

$$m = \frac{m_0}{\sqrt{1-\frac{v^2}{c^2}}}$$

$$\frac{m}{m_0} = \frac{1}{\sqrt{1-\frac{v^2}{c^2}}}$$

Here $$\frac{m}{m_0} = 2000$$

$$2000 = \frac{1}{\sqrt{1-\frac{v^2}{c^2}}}$$

$$1-\frac{v^2}{c^2} = \left(\frac{1}{2000}\right) = \frac{1}{4\times10^6}$$

$$\frac{v^2}{c^2} = \left(1-\frac{1}{4\times10^6}\right)$$

$$\frac{v}{c} = \left(1-\frac{1}{4\times10^6}\right)^{1/2} = \left[1-\frac{1}{2\times4\times10^6}\right]$$

$$v = (1-1.25\times10^{-7})\,c$$

$$v = (1-1.25\times10^{-7})\times3\times10^8$$

$$v = \mathbf{3 \times 10^8}\,\text{m/s}$$

Example 30. *The density of gold is 19.3 × 10³ kg/m³ in a frame S which is at rest. Calculate its density that an observer in a frame S' would determine if the frame S' is moving along the x-axis with a speed of 0.9 c.*

Solution: (*i*) $$\text{L} = L_0\sqrt{1-\frac{v^2}{c^2}}$$

Here $$\frac{v}{c} = 0.9$$

$$L = L_0\sqrt{1-(0.9)^2}$$

$$L = 0.4359L_0$$

Volume $$V_0 = L_0 \times y \times z$$

$$V = L \times y \times z$$

$$= 0.4359L_0 \times y \times z$$

or Mass $$m = \frac{m_0}{\sqrt{1-\left(\frac{v^2}{c^2}\right)}}$$

$$m = \frac{m_0}{\sqrt{1-(0.9)^2}} = \frac{m_0}{0.4559}$$

Original density, $$\rho_0 = \frac{m}{V_0}$$

$$\frac{\rho}{\rho_0} = \left(\frac{m}{m_0}\right)\left(\frac{V_0}{V}\right) = \left(\frac{1}{0.4359}\right) \times \left(\frac{1}{0.4359}\right)$$

$$\rho = 19.3 \times 10^3\left(\frac{1}{0.4359}\right)^2 = \mathbf{101.6 \times 10^3\,kg/m^3}$$

Example 31. *Electrons are accelerated upto a kinetic energy of 10⁹ eV. Find* (*i*) *the ratio of their mass to the rest mass* (*ii*) *the ratio of their velocity to the velocity of light and* (*iii*) *the ratio of their energy to the rest mass energy.*

Solution: $$E = 10^9\,\text{eV} = 10^9 \times 1.6 \times 10^{-19}\,\text{J}$$

Also $$E = mc^2$$

$$m = \frac{E}{c^2} = \frac{10^9 \times 1.6 \times 10^{-19}}{(3 \times 10^4)^2} = 1.77 \times 10^{-27}\,\text{kg}$$

Rest mass of election,

$$m_0 = 9 \times 10^{-31}\,\text{kg}$$

(*i*) $$\frac{m}{m_0} = \frac{1.77 \times 10^{-27}}{9 \times 10^{-31}} = 1.95 \times 10^3$$

(*ii*) $$m = \frac{m_0}{\sqrt{1-\left(\frac{v^2}{c^2}\right)}}$$

$$\frac{m}{m_0} = \frac{1}{\sqrt{1-\left(\frac{v^2}{c^2}\right)}}$$

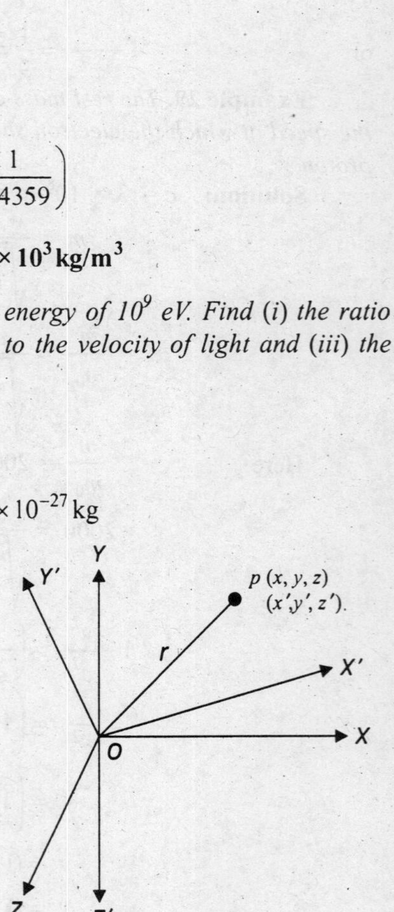

Fig. 8.11

$$1.95 \times 10^3 = \frac{1}{\sqrt{1-\left(\frac{v^2}{c^2}\right)}}$$

$$1-\left(\frac{v^2}{c^2}\right) = \left(\frac{1}{1.95\times10^3}\right)^2$$

$$\frac{v}{c} = 1-\left(\frac{1}{1.95\times10^3}\right)^2 = \left[\left(1-\frac{1}{1.95\times10^3}\right)^2\right]^{1/2}$$

$$= \left(1-\frac{1}{2\times(1.95)^2\times10^6}\right) = [1-1.315\times10^{-7}] = 1$$

(*iii*) Rest mass energy

$$E_0 = m_0c^2$$

$$E_0 = 9\times10^{-31}\times(3\times10^8)^2 = 8.1\times10^{-14}\text{ J}$$

$$\frac{E}{E_0} = \frac{10^3\times1.6\times10^{-19}}{8.1\times10^{-14}} = \mathbf{1.975\times10^{-3}}.$$

Example 32. *The mean life time of a μ meson when it is at rest is 2.2 × 10⁻⁶ sec. Calculate the average distance it will travel in vacuum before decay, if it has velocity 0.9 c.*

Solution: Let t be the mean life period of meson, then the distance travelled by it before decay

$$= \gamma\cdot v\cdot t$$

where $$\gamma = \frac{1}{\sqrt{1-v^2/c^2}} = \frac{1}{\sqrt{1-81/100}} = \frac{10}{\sqrt{19}} = \frac{10}{4.36}$$

∴ Average distance travelled before decay

$$= \frac{10}{4.35}\times0.9\times3\times10^{10}\times2.2\times10^{-6}\text{ cm}$$

$$= \mathbf{1.36\ km}.$$

8.19. MINKOWSKI'S SPACE (FOUR DIMENSIONAL SPACE-TIME CONTINUUM)

The basis of the th eory of relativity is Lorentz transformation equations concerning length, mass and time. Minkowski put forward the concept of time to account for *four dimensional space.*

Minkowski took space and time together and according to him the universe is not made up of points with only x, y and z co-ordinates, but it also consists of *events involving x, y, z and time.* Any event is supposed to be denoted by its position in space and the *time of its occurrence.* So we live in a *four-dimensional* world *i.e.*, in a world of *events* rather than the world of points and this is called a *space-time continuum i.e.*, space and time are *in separable.*

Consider two co-ordinate systems x, y, z *and* x', y', and z' with O as common origin (Fig. 8.9). From three dimensional space, the position of P w.r.t. co-ordinate system is given by

$$r^2 = x^2+y^2+z^2$$

and $$r^2 = x'^2+y'^2+z'^2$$

i.e., the value of r^2 remains same, whatever may be the values of x, y, z for co-ordinate axes.

Minkowski introduced fourth dimension τ, where τ is variable involving time and is given by

$$\tau = ict$$

$$\therefore \quad r^2 = x^2+y^2+z^2+\tau^2 \quad ...(1)$$

$$\text{and} \quad r^2 = x'^2+y'^2+z'^2+\tau'^2 \quad ...(2)$$

From Equation (1) and (2), we have

$$r^2 = x^2 + y^2 + z^2 - c^2t^2 \quad ...(3)$$

and

$$r^2 = x'^2 + y'^2 + z'^2 - c^2t'^2 \quad ...(4)$$

Equations (3) and (4) show that space time continuum differs from the *Euclidean space* in a particular way.

Also as may be seen τ^2 (or τ'^2) has a *negative sign* and r^2 may be zero since $x^2 + y^2 + z^2$ will have positive value. This is however not possible in *three dimensional space.*

It may be clearly noted that **Michelson-Morely** experiment gave a negative result. This negative result can be explained on the basis of four dimensional *space-time* continuum. Minkowski termed the four dimensional-space time continuum as *world* and velocity and acceleration were consequently **world velocity** and **world acceleration.**

8.20. MINKOWSKI'S WORLD VELOCITY

Let u_1, u_2, u_3 and u_4 be the component of velocity in four dimensional space-time continuum.

$$u_1 = \frac{dx}{dt'} = \frac{dx}{dt} \times \frac{dt}{dt'} = u_x \times \alpha \quad ...(1)$$

$$u_2 = \frac{dy}{dt'} = \frac{dy}{dt} \times \frac{dt}{dt'} = u_y \times \alpha \quad ...(2)$$

$$u_3 = \frac{dz}{dt'} = \frac{dz}{dt} \times \frac{dt}{dt'} = u_z \times \alpha \quad ...(3)$$

$$u_4 = \frac{d\tau}{dt'} = \frac{d\tau}{dt} \times \frac{dt}{dt'} = \frac{d\,(ict)}{dt} = ic \times \alpha \quad ...(4)$$

where

$$\alpha = \frac{1}{\left(1 - \frac{v^2}{c^2}\right)^{1/2}}$$

We can get *world velocity u* by compound the four components

$$u = \sqrt{u_1^2 + u_2^2 + u_3^2 + u_4^2}$$

$$= \left[\alpha^2(u_x^2 + u_y^2 + u_z^2 - c^2)\right]^{1/2}$$

Also

$$v^2 = u_x^2 + u_y^2 + u_z^2$$

$\therefore$

$$u = [\alpha^2(v^2 - c^2)]^{1/2}$$

$$= \left(\frac{v^2 - c^2}{1 - v^2/c^2}\right) = ic \quad ...(5)$$

It may be noted that in the space time continuum, there is one *world velocity* $u = i.c$ known as *four velocity* and it is invariant. *In other words;* u_1, u_2, u_3 *may have any magnitude* but u_4 will be such that together, the resultant will always be equal to *ic.*

QUESTIONS AND PROBLEMS

1. Explain the term frame of reference with specific examples.
2. Discuss Galilean transformation for (*a*) position (*b*) velocity (*c*) acceleration.
3. Define Galilean transformations. Prove that Newton's laws of motions are invariable under these transformations.
4. Describe Michelson-Morley experiment and show how it discarded the all prevading ether concept.
5. Explain the postulates of special theory of relativity.
6. Describe Lorentz transformation for (*i*) length (*ii*) time.
7. Describe how length contraction can be explained from Lorentz experiment.
8. Explain time dialation and show its consequences.

9. Show how mass of an object varies with velocity.
10. Derive Einstein's mass energy relation.
11. Derive Energy-momentum relation.
12. Two bodies A and B each of mass 1 kg are moving along X-axis with speeds 12 m/s and –5 m/s in a frame of reference S. Assuming the collision to be elastic, calculate the speeds of two bodies after collision. **[Ans.** (*i*) –5 m/s (*ii*) 12 m/s**]**
13. A particle is moving with speed of 0.6 c. Calculate the ratio of rest mass and mass while in motion. **[Ans.** 0.8**]**
14. What speed must a body have for its mass to be doubled. **[Ans.** 0.866 C**]**
15. A photon is travelling east and another is travelling west. Find the relative velocity of the two photons. **[Ans.** c**]**
16. A rod has length 100 cm. When the rod is in a satellite moving with a velocity 1/2 the velocity of light relative to laboratory, what is the length of the rod as determined by the observer (*a*) in the satellite (*b*) in laboratory. **[Ans.** (*a*) 100 cm (*b*) 86.6 cm**]**
17. A particle with a mean life time of 2 sec. moves through the laboratory with a speed of 0.9 c. Calculate the life time as measured by an observer in the laboratory. **[Ans.** 4.58×10^{-6} sec.**]**
18. The length of a spaceship is measured to be exactly half its actual length. Calculate (*i*) speed of the spaceship (*ii*) Time dialation corresponding to 1 sec. on second spaceship. **[Ans.** (*i*) 2.593×10^{8} m/s (*ii*) 2 sec.**]**
19. Compute three significant places the speed at which the mass of a body becomes 8 times its mass at rest $c = 3 \times 10^8$ m/sec. **[Ans.** 2.98×10^8 m/sec.**]**
20. Calculate to the velocity of electrons accelerated by a potential of 1 million-volts. **[Ans.** 2.8×10^{10} cm/sec.**]**
21. It is assumed that the sun gets energy by fusion of 4 hydrogen atoms to a helium atom. Rest masses of hydrogen and helium are 1.0081 and 4.0039 mass units (a.m.u) respectively. Calculate the energy released in this process.

$$1 \text{ a.m.u.} = 1.66 \times 10^{-27} \text{ kg}$$

[Ans. 26.53 meV**]**
22. From the relativistic concept of mass and energy snow that the kinetic energy of moving mass m with velocity v is $m_0v^2/2$ when $v << c$ (c being the velocity of light).
23. What is the velocity of nuclear particles whose mean life time is 2.5×10^{-7} sec? The proper life time is 2.5×10^{-8} sec. **[Ans.** 3.995 c**]**

PART II
THERMAL PHYSICS

James Prescott Joule (1818-1889)

Joule was born in 1818 at Salford, near Manchester, England, the son of a wealthy owner of a brewery. He was educated at home, and for 3 years had the eminent chemist John Dalton, then 70 years of age, as his tutor. Dalton imparted to Joule his great love for science and his passion for sound numerical data on which to base scientific theories and laws. Unfortunately Joule had little mathematical training, and this prevented him from making even more significant contributions to physics in his later life.

During the years from 1837 to 1847 Joule devoted all his available time to a variety of experiments on the conversion of various forms of energy—mechanical, electrical, chemical—into heat. He developed thermometers capable of measuring temperatures to 1/200 at a Fahrenheit degree, and this gave his work a precision unattainable by other physicists up to that time. In 1840 he proposed for the first time that the rate at which heat was generated by an electric current passing through a wire of resistance R was $Q/t = I^2 R$. This is now called *Joule's law.*

In June 1847 Joule presented a paper to the British Association meeting in Oxford which contained the most accurate experimental value for the mechanical equivalent of heat (Joule's equivalent). This was the turning point in Joule's career.

Joule had excellent laboratory skills and a passion for careful, precise work. Even on his honeymoon he took time out to measure the temperature of the water at the top and bottom of a scenic waterfall to see if the difference agreed with the value predicted by the conversion-of-energy principle! He believed that nature was simple, and strove to find the simple relationships which he was convinced must exist between important physical quantities. His discovery of just two such relationships made a major contribution to the development of the concept of energy as we know it today.

I shall lose no time in repeating and extending those experiments, being satisfied that the grand agents of nature are, by the creator's fiat, indestructable, and that, whenever mechanical force (i.e., work) is expanded, an exact equivalent of heat is always obtaine.

[Sir James Joule (1818-1889)]

Energy will remain in some sense the lord and the giver of life, a reality transcending our mathematical descriptions. Its nature lies at the heart of the mystery of our existence as animate beings in an inanimate universe.

[Freeman Dyson (Born 1924)]

CHAPTER

1

KINETIC THEORY OF GASES

1.1. INTRODUCTION

According to *kinetic molecular theory,* all matter whether solid, liquid or gas consists of molecules which are in *motion* when above a zero activity reference temperature called *absolute zero temperature.* Also, these molecules attract each other with varying degrees. In solids, the molecules are quite close to each other. Which means that the *mean free path i.e.,* the average distance that a molecule travels before colliding with another molecule is generally small as compared to that in case of a liquid or a gas. In fact the molecular motion in a solid may be considered as *vibrational* in nature due to the fact that the molecules are *localized* and remain in one region. In a solid, the forces of attraction are great enough to 'hold the molecules in a set pattern to maintain their *definite shape and volume*'.

In case of liquids, the molecules are relatively at more distances. This can be explained *kinetically* if we consider that there is a definite relationship between the energy possessed and the rate of motion of a particle in the piece of matter. If some energy is pumped in (*i.e.,* in the form of heat energy), the molecules start moving faster. As these molecules possess mass, so with increase of *velocity,* their moment is increased. Due to increase in momentum the molecules colliding with neighbours will clear out more space for themselves and so *mean free path* increases. This explains the property of normal matter which *expands on heating* and *contracts on cooling.* In case of a *liquid* the net result of increased motion is that vibrational motion is changed into translational motion. So in a liquid, molecules may move from one place to another *within the liquid.* So while a liquid maintains a definite volume it assumes the *shape of the container.*

In the case of a gas, the distance between the molecules is very large as compared to their size. If additional energy is *pumped into the gas,* the molecules continue to increase their *velocity and momentum* and create a larger space for themselves and so their mean free path is increased. The forces between these molecules are very small and as a result of this, the molecules of a gas have *random motion.* A gas therefore has neither fixed shape nor fixed volume of its own but assumes those of its container. Liquids and gases are frequently grouped together as fluids, since they flow readily and do not resist shearing stresses.

A class into which a substance falls, depends on the physical conditions surrounding it at that time of observation. Under different conditions, a single substance may be in any one of the three states. We are familiar with water in three phases: ice, water and vapours. Other substances, such as Iron and most other metals are not familiar in liquid or vapour phases. However, they can also exist in liquid or vapour states at *very high temperatures.* Those substances which are commonly observed as gases can all be liquified and solidified if the *temperature is sufficiently reduced* and *pressure is sufficiently raised.*

1.2. PRESSURE EXERTED BY A GAS

Basic assumptions for derivation of pressure exerted by a gas are given as under:

1. All the gases are composed of large number of molecules. The molecules of a gas are all alike but differ from the molecules of another gas.

2. The molecules are continuously moving and their velocities may vary from zero to ∞. The velocities increase with the rise of temperature of the gas.
3. The molecules are just point masses, their dimensions being negligible as compared to their distances.
4. In the steady state the molecules because of their random motions are colliding with each others and with the walls of the vessel containing the gas. Between any two collisions, a molecule moves in a straight line and this distance is called free path of the molecule.
5. The collisions are instantaneous *i.e.*, time of collision is very small as compared to time required to traverse the free path.
6. The no. of molecules/unit volume of the gas is very large and remains constant throughout the time.
7. The molecules are perfect elastic spheres and there is no force of attraction or repulsion between them or with the walls of the vessel. In case of a direct collision, molecules begin to move in opposite direction with the same velocity.

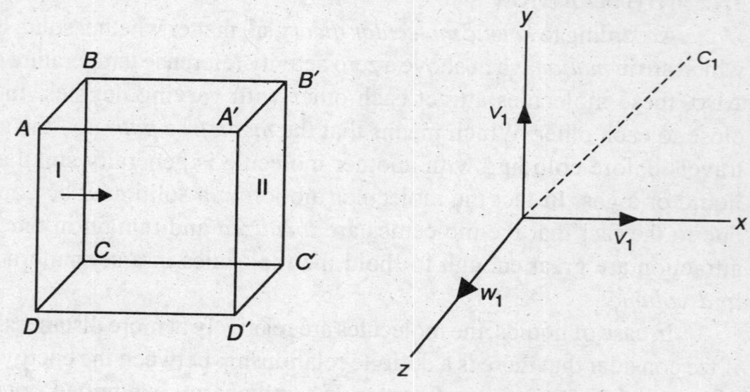

Fig. 1.1

PRESSURE OF A GAS

Let us consider a cubical vessel of each side l and let it have *perfect elastic walls.* Let a large no. of gas molecules (say n) be in the vessel and let the velocity of a gas molecule be c_1.

Resolve the velocity c_1 into three rectangular compo-nents u_1, v_1 and w_1 along the x, y and z-axis respectively, then

$$c_1^2 = u_1^2 + v_1^2 + w_1^2$$

Consider the motion of the molecule in a direction I to the face $ABCD$. The molecule strikes the face $A'B'C'D'$ with velocity u_1 and rebounds with the same velocity as the *collisions are elastic.*

Let m be the mass of the molecule, their change in momentum during collision is

$$mu_1 - (-mu_1) = 2mu_1$$

The molecule moves from face I to face II and then comes back to face I. So it covers a total distance of $2l$. It means that the time taken by the molecule to go from face I to face II and then come back to face I is $\dfrac{2l}{u_1}$

$$\text{Number of collisions/sec.} = \frac{u_1}{2l}$$

$$\therefore \quad \text{rate of change of momentum} = 2mu \times \frac{u_1}{2l} = \frac{mu_1^2}{l}$$

From Newton's second law of motion, rate of change of momentum is equal to the impressed force. Let this force be F_1, so

$$F_1 = \frac{mu_1^2}{l}$$

Similarly force F_2 due to another molecule having velocity c_2 and components u_2, v_2 and w_2 may be written as

$$F_2 = \frac{mu_2^2}{l}$$

$\therefore$ Total force F on face $ABCD$ $= \frac{mu_1^2}{l} + \frac{mu_2^2}{l} + \ldots \frac{mu_n^2}{l}$

$$= \frac{mu}{l}(u_1^2 + u_2^2 + \ldots u_n^2)$$

So pressure on face $ABCD$ $\quad P_1 = \frac{\text{Force}}{\text{Area}} = \frac{m}{l^3}(u_1^2 + u_2^2 + \ldots u_n^2)$

$$= \frac{m}{V}(u_1^2 + u_2^2 + \ldots u_n^2)$$

Similarly pressure on face of the cube I to y-axis.

$$P_2 = \frac{m}{V}(v_1^2 + v_2^2 + \ldots v_n^2)$$

and pressure on face of the cube I to z-axis

$$P_3 = \frac{m}{V}(w_1^2 + w_2^2 + \ldots w_n^2)$$

But pressure is the same in all the directions *i.e.*, $P_1 = P_2 = P_3 = P$ (say)

$$\therefore \quad P = \frac{1}{3}(P_1 + P_2 + P_3)$$

$$= \frac{1}{3}\cdot\frac{m}{v}[(u_1^2 + u_2^2 + \ldots u_n^2) + (v_1^2 + v_2^2 + \ldots v_n^2) + (w_1^2 + w_2^2 + \ldots w_n^2)]$$

$$= \frac{1}{3}\cdot\frac{m}{v}[(u_1^2 + v_1^2 + \ldots w_1^2) + (u_2^2 + v_2^2 + \ldots w_2^2) + \ldots + \ldots (u_n^2 + v_n^2 + \ldots w_n^2)]$$

$$= \frac{1}{3}\cdot\frac{m}{v}(c_1^2 + c_2^2 + \ldots c_n^2)$$

$$= \frac{1}{3}\cdot\frac{mn}{V}c^2$$

where $\quad c^2 = \frac{c_1^2 + c_2^2 + \ldots c_n^2}{n}$

where c is called root mean square (rms) velocity of the molecules and is equal to the *square root of the mean of the squares of the velocities* of the various molecules.

Also $m.n. = M$, the total mass of the gas and $\frac{M}{V} = \rho$, the density of the gas

$$\therefore \quad P = \frac{1}{3}\rho c^2$$

1.3. RELATION BETWEEN PRESSURE AND KINETIC ENERGY

$$P = \frac{1}{3}\rho c^2 = \frac{2}{3}\cdot\frac{1}{2}\rho c^2$$

Also $\frac{1}{2}\rho c^2$ is the mean K.E./unit volume of the gas and let us denote it by E.

$$\therefore \quad P = \frac{2}{3}\cdot E$$

So the pressure of a gas is equal to $\frac{2}{3}$ of the total translational kinetic energy of the molecules in a unit volume.

1.4. KINETIC ENERGY OF A GAS ON THE BASIS OF KINETIC THEORY OF GASES

Since one of the assumptions of the kinetic theory of gases is that there are no forces of attraction or repulsion between gas molecules so no work is required to change the distance between the molecules *i.e.,* the P.E. of the gas is zero and the whole internal energy of the gas is Kinetic Energy.

Now K.E./gm mole of the gas is given by

$$\frac{1}{2}Mc^2 = \frac{3}{2}RT$$

If N is the Avogadro No. *i.e.,* total no. of gas molecules/gm. mole, then

$$\text{K.E./molecule} = \frac{1}{2}\cdot\frac{M}{N}c^2$$

$$= \frac{3}{2}\cdot\frac{R}{N}T = \frac{3}{2}\text{kJ}$$

where k is Boltzmann's constant.

So average translational K.E./molecule

$$\frac{1}{2}mc^2 = \frac{3}{2}kT$$

$$\therefore \quad c = \sqrt{\frac{3kT}{m}}$$

KINETIC INTERPRETATION OF TEMPERATURE

We have the relation $$c^2 = \frac{3R}{m}T = \sqrt{\frac{3kT}{m}}$$

$$\therefore \quad c \propto \sqrt{T}$$

if $$T = 0, c = 0$$

So absolute temperature is that temprature at which linear velocities of all the gas molecules are reduced to zero.

Also $$\text{K.E./molecule} = \frac{3}{2}kT$$

$$\therefore \quad \text{Mean K.E./molecule} \propto T$$

which is the kinetic interpretation of temperature.

1.5. DERIVATION OF GAS LAWS

We can derive the various gas laws from kinetic theory of gases as under:

***(i)* Boyle's law.** From kinetic theory of gases, the pressure of a gas is given by

$$P = \frac{1}{V}.\frac{1}{3}mnc^2$$

where n is the no. of molecules in a volume V.

$$\therefore \quad PV = \frac{1}{3}mnc^2 = \text{constant}$$

Now the heat energy of a gas is due to the motion of molecules and if the temperature of the gas is constant, the K.E. is constant.

So temperature remaining constant

$$P \propto \frac{1}{V} \quad \text{or} \quad PV = \text{constant}$$

This is Boyle's Law.

***(ii)* Gas Equation.** The heat energy of a gas is due to the motion of the molecules. At absolute zero temperature heat energy of a gas is zero and so K.E. and r.m.s. velocity of the molecules must also be zero.

If some heat is added to gas, it will appear as K.E. of motion of its molecules and will be given by

$$\frac{1}{2}mnc^2$$

where c is r.m.s velocity.

Now the rise in temperature is proportional to the amount of heat acquired.

So $$\frac{1}{2}mnc^2 \propto T$$

The pressure of the gas is given by

$$P = \frac{1}{V}\cdot\frac{1}{3}mnc^2$$

or $$PV = \frac{1}{2}mnc^2 \propto T$$

or $$\frac{PV}{T} = \text{constant } (R)$$

or $$PV = RT \text{ Which is gas equation}$$

***(iii)*Charle's law.** We have proved

$$PV = RT$$

If P is constant, $$V \propto T$$

which is Charle's law.

***(iv)* Avogadro's law.** It states that equal volumes of all the gases under some conditions of temperature and pressure contain equal no. of molecules.

Let there be two gases having m_1, n_1 and c_1 mass/molecule, no. of molecules/c.c. and r.m.s. velocity respectively for one gas and m_2, n_2, c_2 the corresponding values for the second gas.

If both gases exert same pressure

$$P = \frac{1}{3}m_1 n_1 c_1^2 = \frac{1}{3}m_2 n_2 c_2^2 \quad ...(1)$$

Also if the two gases have same temperature, there will not be any transfer of heat energy from one gas to other if the two gases are mixed. This is only possible if mean K.E. per molecule of two gases is equal *i.e.*,

$$\frac{1}{2}m_1 c_1^2 = \frac{1}{2}m_2 c_2^2 \quad ...(2)$$

From Equations (1) and (2), we have $n_1 = n_2$, which is Avogadro's law.

***(v)* Dalton's Law of Partial Pressures.** It states that the total pressure of gases which do not react with each other is equal to the sum of the partial pressures which each gas would exert if present alone in the same volume.

Let there be no. of gases having densities $\rho_1, \rho_2, \rho_3, ...$ etc. and let $c_1, c_2, c_3 ...$ etc. be their r.m.s. velocities. If these gases are mixed, the resultant pressure P is given by considering each set of molecules separately as under

$$P = \frac{1}{3}\rho_1 c_1^2 + \frac{1}{3}\rho_2 c_2^2 + \frac{1}{3}\rho_3 c_3^2 + ...$$
$$= p_1 + p_2 + p_3 + ...$$

which is Dalton's law.

***(vi)* Grahm's Law of diffusion.** It states that the rate of diffusion of a gas is inversely proportional to the square root of the density of the gas.

Let us consider two gases having densities ρ_1 and ρ_2 and r.m.s. velocities c_1 and c_2. Let P be the pressure of each gas

$$\therefore \quad P = \frac{1}{3}\rho_1 c_1^2 \quad ...(1)$$

and $$P = \frac{1}{3}\rho_2 c_2^2 \quad ...(2)$$

Comparing Equations (1) and (2)

$$\frac{c_1}{c_2} = \sqrt{\frac{\rho_2}{\rho_1}}$$

or
$$c = \frac{1}{\sqrt{\rho}}$$

Now rate of diffusion is proportional to c.

$$\therefore \qquad \text{Rate of diffustion} \propto \frac{1}{\sqrt{\rho}}$$

Which is Grahm's law of diffusion.

1.6. MEAN VELOCITY, MEAN SQUARE VELOCITY AND R.M.S. VELOCITY

Mean velocity. This is the mean or average value of the velocities possessed by the molecules. If n be the no. of molecules having velocities $c_1, c_2, c_3, \ldots c_n$, then mean or average velocity is given by

$$C_{av} = \frac{c_1 + c_2 + c_3 + \ldots c_n}{n} \qquad \ldots(1)$$

Root Mean Square (r.m.s.) velocity: This velocity is the square root of the mean value of the square of the individual velocities. Let there be n molecules having velocities $c_1, c_2, \ldots, c_n$, then

$$C_{\text{r.m.s}} = \frac{\sqrt{c_1^2 + c_2^2 + c_3^2 + \ldots c_n^2}}{n} \qquad \ldots(2)$$

In the random motion, due to various collisions the individual velocities of different molecules vary but r.m.s. velocity at a constant temperature remains constant. This velocity can be expressed in terms of temperature as

$$C_{\text{r.m.s}} = \frac{\sqrt{3kT}}{m}$$

where k = Boltzman constant, T = Absolute temperature of the gas and m = mass per molecule of the gas.

1.7. MOST PROBABLE VELOCITY

The velocity with which large no. of molecules move is called most probable velocity. If a maxwell distribution curve is plotted between velocity and no. of molecules, then from the curve it is evident that velocity OM is possessed by large no. of molecules and this is called most probable velocity (C_m) and is given by

$$C_m = \frac{\sqrt{2kT}}{m} \qquad \ldots(3)$$

Since the above curve is non-symmetrical, so mean velocity is slightly greater than most probable velocity and r.m.s. velocity is slightly larger than both mean and most probable velocities.

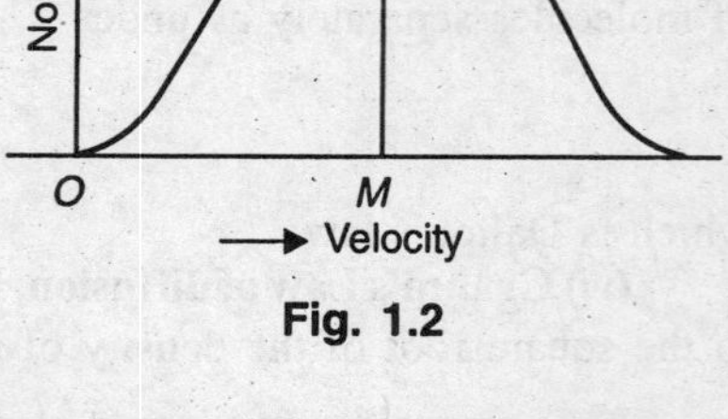

Fig. 1.2

1.8. LAW OF EQUIPARTITION OF ENERGY

To completely describe the motion of a particle in one plane only *two* quantities must be known, say its two rectangular components. Similarly for a particle moving in space *three* independent quantities must be known to describe its motion. A molecule as a rigid body can have three motions of vibration along any of the three co-ordinate axes in addition to its three motions of translation. Hence to completely describe its state of motion *six* independent quantities must be known.

The total number of independent quantities which must be known for describing completely the state of motion of a body are called its **degrees of freedom.**

In the treatment of kinetic theory of gases it has been assumed that in the steady state

$$\overline{u}^2 = \overline{v}^2 = \overline{w}^2$$

where and $\overline{u}, \overline{v}$ are $\overline{w}$ the average values of the component velocities.

$$\therefore \quad \frac{1}{2}m\overline{u}^2 = \frac{1}{2}m\overline{v}^2 = \frac{1}{2}m\overline{w}^2 = \frac{1}{3}$$

$$\frac{1}{2}m\overline{u}^2 = \frac{1}{3}\cdot\frac{3}{2}kT = \frac{1}{2}kT$$

The K.E. per molecule per degree of freedom $= \frac{1}{2}kT$

Hence K.E. per gram molecule per degree of freedom

$$= \frac{1}{2}NkT = \frac{1}{2}RT$$

Thus we find that the total energy is divided equally among all the three degrees of freedom. The result is true for any number of degrees of freedom and is known as the *law of equipartition of energy.* It is stated as follows:

For any dynamical system in thermal equilibrium the total energy is divided equally among all the degrees of freedom, and the energy associated with one molecule per degree of freedom $= \frac{1}{2}kT$.

1.9. RATIO OF SPECIFIC HEATS OF A GAS

The internal energy U of a gas is wholly kinetic as there is no force of attraction between the molecules of an ideal gas. If f be the number of degrees of freedom then according to the law of equipartition of energy

Average kinetic energy per molecule $= \frac{1}{2} f\, kT$

$\therefore$ Total energy/gm. molecule $\quad U = N \times \frac{1}{2} f\, kT$

where N is Avogadro's number.

or $$U = \frac{1}{2} f\, RT \qquad \left[\because \frac{R}{N} = k\right]$$

The molar specific heat of a gas at constant volume

$$C_v = \frac{dU}{dT} = \frac{d}{dt}\left(\frac{1}{2} f\, RT\right) = \frac{1}{2} f\, R$$

The molar specific heat at constant pressure

$$C_p = C_v + R = \frac{1}{2} f\, R + R = R\left(\frac{f}{2}+1\right)$$

$\therefore$ Ratio of specific heats

$$\gamma = \frac{C_p}{C_v} = \frac{\frac{f}{2}+1}{\frac{f}{2}} = 1 + \frac{2}{f}$$

Monoatomic gas. In a monoatomic gas, a molecule is made up of only one atom. It can rotate about polar axis and also can move bodily in the three perpendicular planes. Hence such a molecule has **three** degrees of freedom and the fourth degree of freedom due to rotational motion is ineffective. Such a molecule behaves like a point mass. Hence total energy of such a molecule is divided only among three degrees of freedom due to translational motion.

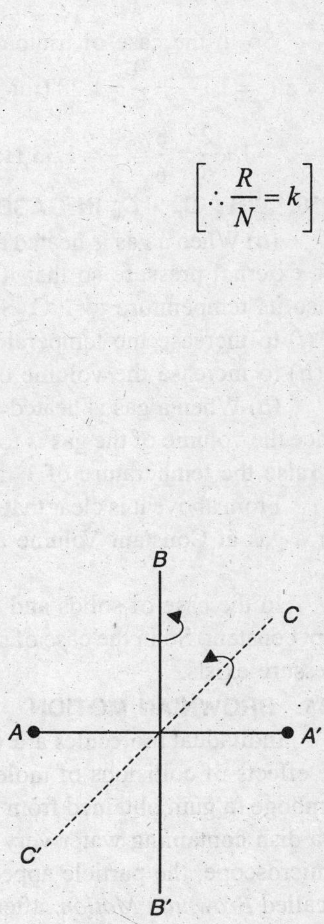

Fig-1.3

So for a monoatomic gas

$$\gamma = \frac{C_p}{C_v} = 1 + \frac{2}{f} = 1 + \frac{2}{3} = 1.67$$

Diatomic gases. The molecule of a diatomic gas like

hydrogen, oxygen, nitrogen etc., has two atoms with axis AA'. The molecule is capable of translatory motion about its centre of mass. So it has three degrees of freedom due to *motion of translation*. The molecule is also capable of rotation about BB' and CC' axis. Rotation about AA' *axis is not possible*. So rotatory motion contributes two more degrees of freedom. So a diatomic molecule has *5 degrees of freedom*.

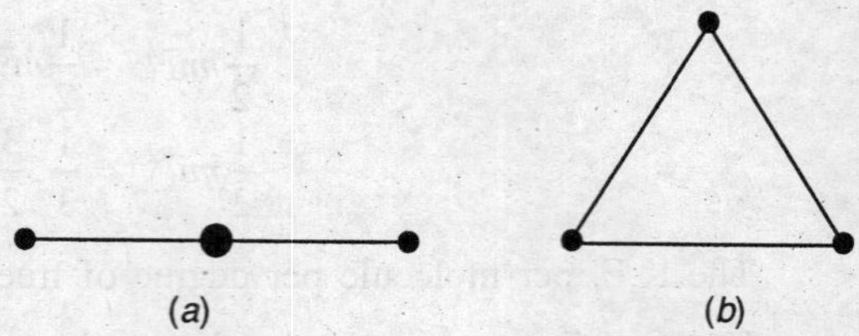

Fig. 1.4

$\therefore$ Ratio of sp. heat for a diatomic molecule is given by

$$r = 1+\frac{2}{f} = 1+\frac{2}{5} = 1.4$$

Triatomic gas. Molecule of a triatomic gas like SO_2, H_2S, CO_2 etc. has three atoms in it. So, there are two independent relations amongst three atoms. In linear molecule [Fig. 1.4 (*a*)], two atoms lie on either side of the central atom. In that case total degree of freedom is $(3 \times 3 - 2) = 7$ (because distance between the atoms from the central atom is fixed).

In *non-linear molecule*, the three atoms are at the vertices of a triangle Fig. 1.4 (*b*), so there are three independent relations between three atoms.

So the total number of degree of freedom

$$= (3 \times 3 - 3) = 6$$

So in the case of triatomic gas

$$r = 1+\frac{2}{7} = \frac{9}{7} = 1.28 \text{ (for a linear molecule)}$$

$$= 1+\frac{2}{6} = \frac{8}{6} = \frac{4}{3} = 1.33 \text{ (for a non-linear molecule).}$$

1.10. WHY $C_P > C_V$ IN GASES?

(*a*) When a gas is heated at constant pressure, it expands and in doing so it has to work against the external pressure so that it gets somewhat cooled. So some more heat has to be supplied to raise its temperature to 1°C. So in case of C_p, heat is required:

(*i*) to increase the temperature of 1 mole of a gas through 1°C.

(*ii*) to increase the volume of the gas against external pressure.

(*b*) When a gas is heated at constant volume, the pressure of the gas increases somewhat and since the volume of the gas is to be kept constant so C_v, corresponds to the amounts of heat required to raise the temperature of 1 mole of a gas through 1°C.

From above it is clear that heat required for a gas at constant pressure is more than that required for a gas at Constant Volume *i.e.*,

$$C_p > C_v$$

In the case of solids and liquids, the change takes place in volume only and the pressure is kept constant. So in the case of solids and liquids only one specific heat *i.e.*, specific heat at constant pressure exists.

1.11. BROWNIAN MOTION

Individual molecules are very small and cannot be visible. However, it is possible to observe the effects of collisions of molecules with small particles with the help of a microscope. If some gamboge (a gum obtained from the trees) is dissolved in alcohol and a small amount of this poured in a dish containing water very fine suspension is formed. If this suspension is observed through a microscope, the particle appear to move *randomly* and the motion is quite *eratic*. This motion is called *Brownian Motion*, after the name of the man who observed it in 1827. The motion is due to collision of the molecules with particles. Each impact imparts some motion to the particle and since the motion can be in any direction after each impact, so this motion is quite *random or eratic.* Brownian motion constitutes one of the direct evidences that the *molecules are always in motion.*

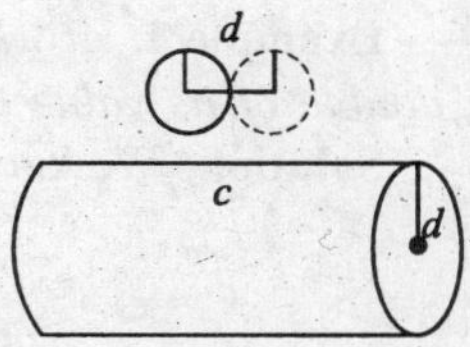

Fig. 1.5

1.12. MEAN FREE PATH

On the basis of the kinetic theory of gases it is assumed that the molecules of a gas are continuously colliding against each other. Between two collisions the molecule travels a certain distance in a straight line. This distance is called the free path.

The average distance travelled by a molecule between two successive collisions is known as the mean free path λ.

If the total path travelled in N collisions is S, then

$$\text{Mean free path } \lambda = \frac{S}{N}$$

Calculation of the mean free path. For getting the required results we consider

(*i*) Only the molecule under consideration is in motion and all other molecules are at rest.

(*ii*) The sphere of influence of the molecule has a diameter d. In other words, it means that the molecule can collide only with those molecules the centres of which lie at a distance d or less.

If c be velocity of the molecule, then in one second it will collide with all the molecules the centres of which lie within a cylinder of radius d and length c.

$\therefore$ The volume of the cylinder $= \pi d^2 c$

Let the number of molecules per c.c. be n, then

Number of molecules in the cylinder $= \pi d^2 cn =$ the number of collisions per second

$\therefore$ Average time between two successive collisions

$$= \frac{1}{\pi d^2 cn} \text{ sec}$$

Hence average distance between two successive collisions

$$= \frac{1}{\pi d^2 cn} \times c$$

$$\therefore \text{ Mean free path } \quad \lambda = \frac{1}{\pi d^2 n} \qquad \text{...(1)}$$

We have considered above all the molecules except one to be at rest.

Claussius on the consideration that all the molecules were moving with a constant speed in all possible directions gave an expression

$$\lambda = \frac{3}{4\pi d^2 n} \qquad \text{...(2)}$$

Maxwell improved upon this result by considering the distribution of molecular speeds and got the result

$$\lambda = \frac{1}{\sqrt{2\pi d^2 n}} \qquad \text{...(3)}$$

All the above three results are similar but differ only by a constant multiplier.

Now pressure $\quad p = nk\text{T}$

$$\therefore \quad n = \frac{p}{kT}$$

$$\text{Hence,} \quad \text{Mean free path } \lambda = \frac{kT}{\sqrt{2 p\pi d^2}}$$

Thus the mean free path varies directly as the absolute temperature and inversely as the pressure.

Example 1. *At what temperature, pressure remaining same will r.m.s. velocity of hydrogen be double of its value at N.T.P.*

Solution: We know

$$\therefore \quad C^2 = \frac{3R}{M} \cdot T$$

$$C \propto \sqrt{T}$$

Let the velocity at $T_0 K$ be C and at TK be $2C$, then

$$\frac{2C}{C} = \sqrt{\frac{T}{T_0}}$$

Squaring

$$4 = \frac{T}{T_0}$$

$$\therefore \quad T = 4 \times 273 = 1092K$$

$$= 1092 - 273 = \mathbf{819°C.}$$

Example 2. *At what temperature the molecular velocity (r.m.s) of oxygen will become half that of hydrogen at N.T.P.?*

Solution: Let T be the absolute temperature at which the r.m.s. velocity of oxygen molecules is half that of hydrogen at N.T.P.

r.m.s. velocity of hydrogen molecules at N.T.P.

$$C_h = \sqrt{\frac{3R}{M_h} \times 273}$$

where M_h is the molecular weight of hydrogen

r.m.s. velocity of oxygen molecules at $T°K$

$$C_0 = \sqrt{\frac{3R}{M_0} \cdot T}$$

where M_0 is the molecular weight of oxygen.

Now

$$\frac{C_0}{C_h} = \frac{1}{2}$$

$$\therefore \quad \sqrt{\frac{3R}{M_0} \cdot T} = \frac{1}{2}\sqrt{\frac{3R \times 273}{M_h}}$$

or

$$T = \frac{M_0}{M_h} \cdot \frac{273}{4}$$

But

$$\frac{M_0}{M_h} = 16$$

$$\therefore \quad T = \frac{16 \times 273}{4} = 1092°K$$

$$= 1092 - 273 = \mathbf{819°C.}$$

Example 3. *The diameter of the molecule of a gas is 2×10^{-8} cm. Calculate the mean free path at N.T.P., given $k = 1.38 \times 10^{-16}$ ergs./°C.*

Solution: Mean free path

$$\lambda = \frac{kT}{\sqrt{2} p \pi d^2}$$

Normal pressure, $p = 76 \times 13.6 \times 981$ dynes/sq. cm

Normal temperature $T = 273K$

$$\therefore \quad \lambda = \frac{1.38 \times 10^{-16} \times 273}{\sqrt{2}\pi \times 76 \times 13.6 \times 981 \times 4 \times 10^{-16}} = \mathbf{2.1 \times 10^{-5}\ cm.}$$

Example 4. *Determine the mean free path of the molecules of hydrogen at N.T.P. given that density of hydrogen is 8.96 × 10^{-5} gm./c.c., coefficient of viscosity 8.6 × 10^{-5} C.G.S units and k = 1.38 × 10^{-16} ergs per degree K.*

Solution: Coefficient of viscosity $\eta = \frac{1}{3} mnc\,\lambda = \frac{1}{3}\rho c\lambda$ where λ is the mean free path and c the average velocity of the molecules. According to Maxwell's law of distribution of velocities

Average velocity $C = \sqrt{\frac{8kT}{\pi m}}$

Mean free path $\lambda = \frac{3n}{\rho c} = \frac{3n}{\rho}\sqrt{\frac{\pi m}{8kT}}$

Mass of hydrogen molecule $m = 2 \text{ a.m.u.} = \frac{2}{N}$

$$= \frac{2}{6.023 \times 10^{23}} = 3.321 \times 10^{-14} \text{ gm}$$

$$\lambda = \frac{3 \times 8.6 \times 10^{-5}}{8.96 \times 10^{-5}} \left\{ \frac{3.321 \times 10^{-24} \times \pi}{8 \times 1.38 \times 10^{-16} \times 273} \right\}^{\frac{1}{2}}$$

$$= \mathbf{1.69 \times 10^{-5} cm.}$$

Example 5. *Calculate the r.m.s. velocity of nitrogen molecule at 0°C. The density of nitrogen at N.T.P. is 1.25 kg/m².*

Solution: In this case $\rho = 1.25 \text{ kg/m}^2$

$$P = 76 \times 10^{-2} \times 13.6 \times 10^3 \times 9.8 \text{ M/m}^2$$

$$C_{\text{r.m.s.}} = \sqrt{\frac{3P}{\rho}}$$

$$= \sqrt{\frac{3 \times 76 \times 13.6 \times 10 \times 9.8}{1.25}}$$

$$= \mathbf{493.05 \text{ m s}^{-1}}.$$

Example 6. *Calculate the mean translational K.E. per molecule of a gas at 727°C, given R = 8.32 Joule/mole K.*

Solution: The K.E. of a molecule of a gas is given by

$$\text{K.E./molecule} = \frac{3}{2}\frac{R}{N}T$$

In this case $R = 8.32$

$T = 727 + 273 = 1000$ K

and $N = 2.073 \times 10^{-2}$

$\therefore$ $\text{K.E./molecule} = \frac{3}{2} \times \frac{8.32}{6.02 \times 10^{23}} \times 1000$

$$= \mathbf{2.073 \times 10^{-2} \text{ Joule.}}$$

Example 7. *Calculate the mean free path of a gas molecule, given that the diameter of the gas molecule is 4Å and no. of molecules is 1.5 × 10^{25}/m³.*

Solution: We have the relation for mean free path

$$\lambda = \frac{1}{\pi d^2 n}$$

In this case $R = 1.5 \times 10^{25}/\text{m}^3$

and $d = 4 \times 10^{-10}$ m

$$\lambda = \frac{1}{\pi(4\times10^{-10})^2\times1.5\times10^{25}\sqrt{2}}$$

$$= \frac{10^{-5}}{\pi\times16\times1.5\times1.412}$$

$$= 947.3 \times 10^{-10} = \mathbf{947.3Å.}$$

Example 8. *Compute the mass of an argon atom and atomic weight of argon if the sp. heat capacity of argon at constant volume is 315 J/kg K.*

Solution: Argon is a monoatomic gas, so we have sp. heat at constant volume as

$$C_v = 3/2\ R = 3/2 \times (8.31 \text{ Joule/mole-K})$$

$$= 12.465 \text{ J/mole-K}$$

Let A be the At. wt. of Argon *i.e.,* a mole of Argon contains A gm. So we have

$$C_v = 12.465 \text{ J/A gm-K}$$

$$= \frac{12.465\times10^{-3}}{A} \text{ J/kg-K}$$

Now, $$C_v = 315 \text{ J kg-K (given)}$$

$\therefore$ $$315 = \frac{12.465\times10^{-3}}{A}$$

or $$A = 0.03957 \times 10^{-3} \text{ kg} = 0.03957 \text{ gm}$$

Now A correspondence to 1 gm mole, *i.e.,* it contains 6.02×10^{23} atoms (Avogadro's No.)

$\therefore$ Mass of an Argon atom $$= \frac{0.03937}{6.02\times10^{23}} = \mathbf{6.57\times10^{-26}\ g.}$$

Example 9. *The gas molecule of a gas have 1.25×10^{25} molecules/m^3 and the mean free path of the molecule is 10^{-7} m. Calculate the diameter of the gas molecule.*

Solution: We have the relation $\lambda = \dfrac{1}{\pi d^2 n\sqrt{2}}$

or $$d = \sqrt{\frac{1}{\pi\lambda n\sqrt{2}}}$$

$$= \frac{1}{\pi\times\sqrt{2}\times10^{-7}\times1.25\times10^{25}} = \mathbf{0.424\times10^{-10}\ m.}$$

Example 10. *An electric bulb of volume 250 cm^3 was sealed off during manufacture at a pressure of 10^{-3} mm of Hg at 27°C. Find the number of molecules in the bulb.*

Solution: Let N be the no. of molecules in the bulb.

We have given, $$P = 10^{-3} \text{ mm of Hg} = 10^{-4} \text{cm of Hg.}$$

$$V = 25 \text{ cm}^3$$

$$T = 273 + 27 = 300 \text{ K.}$$

Also we have the relation $$PV = NkT \quad ...(1)$$

Now at N.T.P., one mole. of a gas occupies a volume V_0 = 22400 cm^3 and contains N_0 = 6.02×10^{23} molecules, where N_0 is the Avogadro's number, P_0 = 76 cm of Hg and T_0 = 273 K. Now from gas equation

$$P_0 V_0 = N_0\ kT_0 \quad ...(2)$$

Dividing Equation (1) by Equation (2), we get

$$N = N_0 \times \frac{T_0}{T} \times \frac{P}{P_0} \times \frac{V}{V_0}$$

$$= 6.02\times10^{23}\times\left(\frac{273}{373}\right)\times\left(\frac{10^{-4}}{76}\right)\times\left(\frac{250}{22,400}\right)$$

$$= \mathbf{8.045 \times 10^{15}\ molecules.}$$

Example 11. *A balloon whose volume is 500 m³ is to be filled with hydrogen at atmospheric pressure. (a) If hydrogen is stored in cylinders of volume 0.05 m³ at an absolute pressure of 15 × 10⁵ Pa, how many cylinders are required? (b) What is the total weight that can be supported by the balloon in air at standard condition?*

$$(\rho_{air} = 1.3\ kg/m^3 \text{ and } \rho_H = 9 \times 10^{-2}\ kg/m^3).$$

Solution: (*a*) The volume of the balloon is 500 m^3 and hydrogen is at a pressure of $1.013 \times 10^5\ N/m^2$.

Hydrogen cylinders are at a pressure of 15×10^5 Pa.

So, we have $\qquad 15 \times 10^5 \times V = 500 \times 1.013 \times 10^5$

or, $\qquad V = 33.77 m^3$

Number of cylinders $\qquad = \dfrac{33.77}{0.05} = \mathbf{675}$

(*b*) Weight that the balloon can support is given by

$$m.g. = V.(\rho - \sigma).g$$

where ρ is the density of air and σ is the density of hydrogen.

$$\therefore \qquad m.g. = 500\ (1.3 - 0.09) \times 1.8$$
$$= 500 \times 1.21 \times 9.8 = \mathbf{5929\ N.}$$

1.13. VAN DER WAAL'S EQUATION OF STATE

No gas in nature obeys the gas laws perfectly. At ordinary temperatures and pressure, the deviation is small but at *high pressure and low temperature,* deviations are quite large. In deriving gas equation, the following assumptions were taken into account.

(*i*) Molecules are mere point masses and do not occupy any space.

(*ii*) There is no force of attraction or repulsion between the molecules.

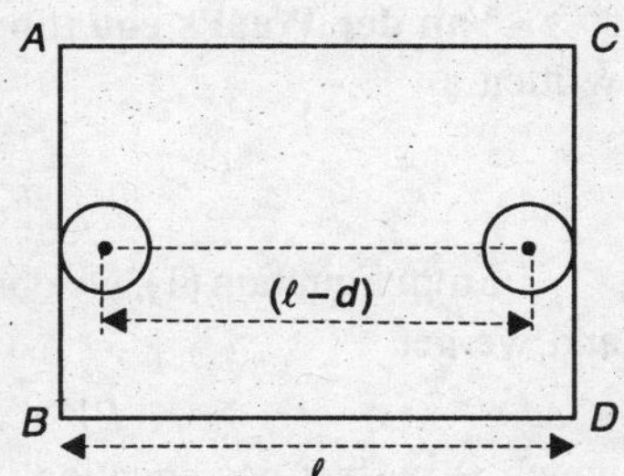

Fig. 1.6

However the experiments by Joule-Thomson on liquifaction of gases and Andrews experiment on carbon dioxide showed the *forces of attraction* between the molecules. Though this force of attraction is very small as compared to that in case of liquids, yet it can not be neglected. Van der Waals modified the gas equation $PV = RT$ by applying the correction for (*i*) finite size of the molecules (*ii*) intermolecular force of attraction.

(*i*) *Finite size of molecule.* Consider a molecule of gas moving parallel to *X*-axis. Let it strike the face *CD* of the vessel. Now if $CD = l$, the distance travelled by the molecule is not $2l$ but is $2\,(l - d)$ where *d* is the diameter of the molecule. If we consider the motion of all the molecules in this way, we find that the effective volume in which the molecule moves is less than the total volume. When pressure is reduced, there is decrease of volume but not the whole volume but some part $(V - b)$, where *b* corresponds to the portion which cannot be compressed. So applying this correction, the net volume of the gas is $(V - b)$.

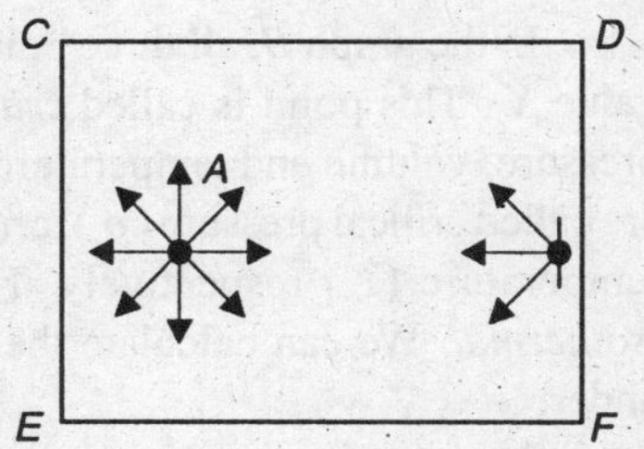

Fig 1.7

(*ii*) *Intermolecular attraction.* If again we consider a molecule (say *A*) well inside the vessel, it is attracted by other molecules equally in all directions and so net force on it is zero.

However when this molecule strikes the wall of the vessel, it is pulled by other molecules. Its *velocity and hence its momentum* with which it strikes the wall would be less than the case when there are no force of attraction between the molecules. So as the *momentum* is less so it will exert less pressure.

It is clear that for *double number of molecules/c.c. of the gas, the decrease in pressure will be four times.* It is due to this reason that decrease in pressure is proportional to

(*a*) number of attracting molecule/unit volume.

(*b*) number of molecules striking a unit area of the walls of the vessel/unit time.

If we combine both the above factor, we observe that the decrease in pressure will be proportional to density of gas in each case.

∴ Decrease in pressure $\propto$ (density of gas)^2

∴ Decrease in pressure $p = \frac{a}{V^2}$

where a is constant of proportionality.

∴ Real pressure = observed pressure + decrease in pressure

$$= p + \frac{a}{V^2}$$

So applying the correction for volume and pressure in the gas equation, we get

$$\left(p + \frac{a}{V^2}\right)(V - b) = RT$$

This is called Van der Waal's equation of state.

It may be noted that $\frac{a}{V^2}$ will have the dimensions of pressure and b will have the dimensions of volume.

Van der Waal's equation and critical constants. Van der Waal's equation of state may be written as

$$P = \frac{RT}{(V - b)} - \frac{a}{V^2} \quad ...(1)$$

From Equation (1), we observe that if V is very large then the terms b and $\frac{a}{V^2}$ can be neglected and we get

$$PV = RT$$

which is perfect gas equation.

For *such volumes,* the P–V diagram is a rectangular hyperbola as shown by the curve D.

At lower temperatures the curve develops maxima and minima, it is due to the fact that Van der Waal's equation is a *cubic equation in V* and same equation is true for three values of V at constant temperature and pressure corresponding to three roots of V. This is shown in curve A of Fig. 1.8 by the points E, F and G.

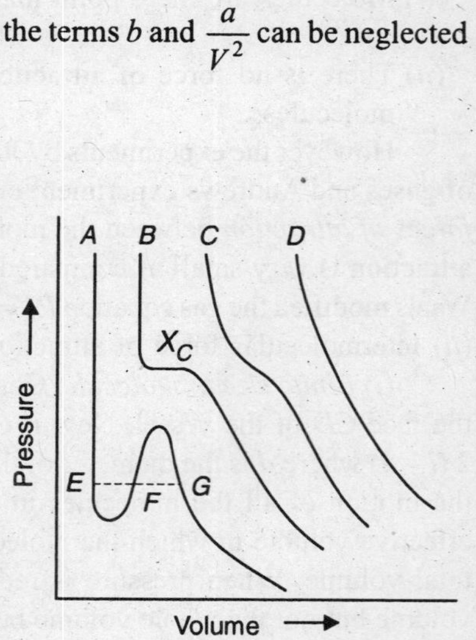

Fig. 1.8

In the graph B, all three points merge to give a common value X_c. This point is called *critical point* and the values of pressure, volume and temperature corresponding to this point are called critical pressure (p_c), critical volume (V_c) and critical temperature (T_c) respectively. This curve is called *critical isothermal.* We can calculate the values critical constants as under:

At critical point X_c, we have

$$\frac{dP}{dV} = 0 \text{ and } \frac{d^2P}{dV^2} = 0 \quad ...(3)$$

We may write equation (1) as $P_c = \frac{RT_c}{(V_c - b)} - \frac{a}{V_c^2}$ ...(4)

Differentiating w.r.t. V, we get $\frac{dP_c}{dV_c} = -\frac{RT_c}{(V_c-b)^2} + \frac{2a}{V_c^3}$ ...(5)

From the condition as shown by equation (3), we get

$$-\frac{RT_c}{(V_c-b)^2} + \frac{2a}{V_c^3} = 0$$

$$\therefore \quad \frac{RT_c}{(V_c-b)^2} = \frac{2a}{V_c^3} \quad ...(6)$$

Again differentiating equation (5), we get

$$\frac{d^2P_c}{dV_c^2} = \frac{2RT_c}{(V_c-b)^3} - \frac{6a}{V_c^4}$$

Again using the condition of equation (3) we get

$$\frac{2RT_c}{(V_c-b)^3} - \frac{6a}{V_c^4} = 0$$

$$\therefore \quad \frac{2RT_c}{(V_c-b)^3} = \frac{6a}{V_c^4} \quad ...(7)$$

From equations (7) and (5), we get,

$$\frac{V_c-b}{2} = \frac{V_c}{3} \text{ or } V_c = 3b$$

Putting this value of V_c in equation (5), we get

$$\frac{RT_c}{(3b-b)^2} = \frac{2a}{(3b)^3}$$

or $$\frac{RT_c}{4b^2} = \frac{2a}{27b^3}$$

$$\therefore \quad T_c = \frac{8a}{27Rb}$$

Putting the value of V_c and T_c in equation (4), we get

$$P_c = \frac{R\cdot\frac{8a}{27Rb}}{3b-b} - \frac{a}{(3b)^2}$$

$$= \frac{8a}{27b\times 2b} - \frac{a}{9b^2}$$

or $$P_c = \frac{a}{27b^2}$$

From the values of critical constants we have

$$\frac{RT_c}{P_cV_c} = \frac{8}{3} = 2.667$$

This relation $\frac{RT_c}{P_cV_c}$ is called **critical coefficient.** It may be noted that its actual experimental value is much more than this theoretical value.

1.14. DEFECTS OF VAN DER WAAL'S EQUATION

(*i*) The values of a and b vary with temperature.

(*ii*) The value of critical volume obtained experimentally is $2b$ while from Van der Waal's equation it comes out to be $3b$.

(*iii*) The calculated values of the critical coefficients do not agree with the experimental values.

1.15. MAXWELL'S DISTRIBUTION LAW

From kinetic theory of gases, we have understood that the velocity of gas molecules is different at different times and it changes after each collision. At any time, molecules are not moving with the same velocity. Some molecules are moving with velocities which are somewhat higher than the r.m.s. velocity while few others are moving with velocities lower than the r.m.s. velocity. The problem of knowing the velocities of different molecules at different times was studied by *Maxwell statistically* and arrived at a law called *Maxwell's distribution law*. According to this law, the number of the molecules having velocities between c and $c + dc$ is given by the relation.

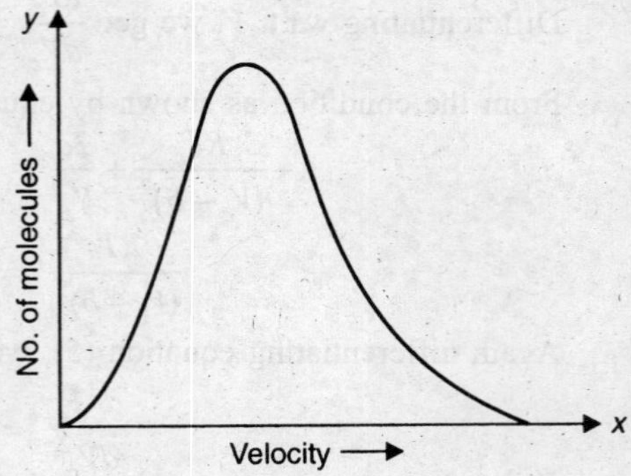

Fig. 1.9

$$dN_0 = 4\pi NA^3 . e^{-bc^2} .c^2 .dc \quad ...(1)$$

where $N \rightarrow$ No. of molecules/unit volume

$$A = \sqrt{\frac{b}{\pi}} \text{ and } b = \frac{m}{2kT}$$

and k is Boltzmann's constant.

A graph showing the Maxwell's distribution law is shown in Fig. 1.9.

Probability that a molecule has its velocity lying between c and $c + dc$ at a particular temperature is given by

$$W = \frac{dN_0}{N} = 4\pi A^3 .e^{-bc^2} .c^2 .dc \quad ...(2)$$

At a particular temperature, the velocity of a gas molecule should vary between 0 and ∞, for which case the probability is maximum. To have this speed, dN_0, the number of molecules having speed between c and $c + dc$ must be maximum or derivative of dN_0 must be zero. So we have from equation (2).

$$4\pi NA^3 \left[e - \frac{mc^2}{2kT} \times \left(\frac{-mc}{kT} \times c^2 + 2c \right) \right] dc = 0$$

for $c = c_m$

$$\therefore \quad \frac{-mc_m^3}{kT} + 2c_m = 0$$

or

$$c_m = \sqrt{\frac{2kT}{m}} \quad ...(3)$$

So the r.m.s. velocity of a gas molecule is constant at a given temperature, though at any time the molecules are moving with different velocities.

1.16. TRANSPORT PHENOMENON

When the gas molecules are colliding with one another or with the walls of the containing vessel, *mass, energy and momentum* are all conserved. A molecule in motion may be considered as a *transporter* of energy and momentum which may be transferred on collision. In a steady state each *transport* would be balanced by an equal and opposite movement in the reverse direction and consequently no *transport* will take place. However, if the gas is not in steady state and concentration of energy and momentum is more in one region and less in the other region, the *transport* will take place from one region to the other region through long chain of collisions. Such a phenomenon is called *transport phenomenon.* Some of the examples of this phenomenon are *thermal conduction* where transfer of *energy* takes place; *viscosity* in which transfer of *momentum* takes place and *diffusion* in which transfer of molecules (mass) takes place.

QUESTIONS AND PROBLEMS

1. Discuss kinetic theory of matter. How will you explain three states of matter on it?
2. Deduce an expression for the pressure of a gas from the assumptions of the kinetic theory. Point out clearly the fundamental hypothesis you adopt.
3. Outline the essential features of kinetic theory of gases and apply it to account for the Boyle's law, Avogadro's law, and Charle's law. What is the interpretation of temperature on this theory?
4. Write a note on degrees of freedom of a mecnanical system.
5. Discuss the caloric theory and modern theory of heat. How the various states of matter can be explained on the kinetic theory of matter?
6. State the law of equipartition of energy. Prove that for a monoatomic gas the ratio between the specific heats is 5/3 and for a diatomic gas it is 7/5.
7. Explain the Maxwell's distribution of velocities.
8. What is degree of freedom? Derive an expression for the determination of ratio of two specific heats.
9. How the specific heats of monoatomic, diatomic and triatomic gases depend on the degree of freedom?
10. How will you explain Brownian movement? Give its importance.
11. Calculate the r.m.s. velocity of a gas of density 1.4 gm/litre at 76 cm. of Hg.
12. The density of helium is 0.178 gram per litre at N.T.P. What is the r.m.s. velocity of helium molecules under these conditions? Compare it with the velocity at 200°C.

    [Ans. 1.306×10^5 cm/sec; 1 : 1.316**]**
13. Find the root mean square velocity of the molecules of nitrogen at 20°C. Molecular weight of nitrogen is 28 and the gas constant for a gram molecule is 8.31×10^7

    [Ans. 5.11×10^4 cm/sec**]**
14. Calculate the K.E. of gm. molecule of helium at 27/C. Given $R = 8.31 \times 10^7$.

    [Ans. 3739.5 Joules**]**
15. Find the molecular kinetic energy of 1 gm. of helium at N.T.P. What will be its energy at 100°C? Molecular weight of helium is 4 and value of $R = 8.31 \times 10^7$.

    [Ans. 8.5×10^9 ergs., 1.16×10^{10} ergs.**]**
16. Calculate the root mean square velocities of air molecules at –100°C and at 1000°C. Given the density of air at N.T.P. to be 0.001296 gm/c.c.

    [Ans. 3.857×10^4 cm/sec; 1.045×10^5 cm/sec**]**

CHAPTER

2

Low Temperture Physics

2.1. INTRODUCTION

For quite some time ice was considered to be coldest and so the temperature of ice was taken as the lowest temperature in the early design of thermometery. Farnheit was the first to experimentally demonstrate that a mixture of ice with common salt gives a lower temperature of the order of –18°C. So in the Farnheit scale of temperature this temperature was taken as the lowest temperature. Later on temperatures lower than this temperature could be attained. Study of the gases shows that the lowest temperature that can be attained theoretically corresponds to 273°C below the melting point of ice and this temperature is used as zero of the absolute scale of temperature.

We list here some of the methods which can be used to produce low temperature. The general principle of production of low temperature is to devise means for removing from a body its heat content.

***(i)* Freezing mixtures of salts in ice:** The basic principle of freezing mixtures is that heat is absorbed when salt dissolves in water. Since different salts absorb different amount of heat, so the low temperature produced by different salts will be different. The lowering of temperature by freezing mixture has similar analogy like the formation of an alloy: *i.e.*, when two metals are mixed to form an alloy, it is found that the melting point is not only always less than that of either constituent, but also varies with the composition of the metals forming alloy. A simple example of freezing mixture is that when a salt say NaCl, which is above 0°C is mixed with ice, some ice melts in cooling the salt at 0°C is produced. But since ice cannot remain in equilibrium with a saturated solution of salt to its own temperature and a saturated solution of salt at 0°C, hence the ice melts taking the necessary latent heat from the solution which in consequence cools down-taking the temperature to about –22°C at which temperature ice attains equilibrium with saturated solution.

A table below shows different temperatures obtained in various freezing mixtures.

Salt	*Amount in 100 gms of mixture*	*Minimum temp. in 0°C*
$MgSO_4$	19.0 gm	–3.9
KCl	19.7 gm	–11.1
NH_4Cl	18.6 gm	–15.8
NaCl	22.4 gm	–21.2
$MgCl_2$	21.6 gm	–33.6
$CaCl_2$,	29.8 gm	–55.0

***(ii)* Cooling due to evaporation of liquids:** Low temperature can also be attained by allowing suitable liquids to evaporate at reduced pressure. When a liquid evaporates, it requires latent heat of vaporisation, if by some process the evaporation is made to take place rapidly and also if the liquid is isolated, the latent heat of vaporisation is obtained at the cost of the latent heat of the liquid itself and so greater cooling results.

***(iii)* Liquefaction of gases by cooling them sufficiently and simultaneously subjecting them to large pressure:** This basic principle was acquired from the knowledge that steam can be liquified either by cooling or by increasing pressure and an analogy was made that substances which are gaseous at ordinary temperatures could be liquified if they were cooled and subjected simultaneously to a large pressure. While cooling reduces the heat motion of the molecules the higher pressure brings the molecules sufficiently close together to coalesce and form a liquid. If the liquid so formed is further allowed to evaporate still lower temperatures could be attained.

***(iv)* Cooling due to Peltier effect:** If suitable external e.m.f. is applied to a thermocouple, one junction is cooled. This is just the reverse of the See-beck Effect.

***(v)* Cooling by desorption:** An example for this process is that charcoal absorbs residual gases and when these gases escape at lower pressure, cooling results due to heat of adsorption, a process similar to the evaporation of liquids.

***(vi)* Cooling due to adiabatic demagnetisation:** Cooling can be attained by demagnetising certain crystals adiabatically the magnetic energy of the system is extracted and extreme cooling is produced under suitable conditions.

2.2. JOULE-THOMSON EFFECT

Van der Waal's equation of state clearly predicted the deviation from an ideal gases and existence of intermolecular forces of attraction between the molecules. Prof. Joule in 1845 while doing experiments on the measurement of mechanical equivalent of heat took into consideration that the work done on the gas would be equivalent to the heat produced, only if no work was done by the mutual attractions of the molecules as they approached each other when the gas was compressed. However, in his experiments, Joule got almost the same results as predicted by other scientists without considering this factor in his experiments. He therefore tried to have more investigations by taking into account the following considerations with regard to a gas which is made to expand without doing external work and without allowing any heat to enter or leave it during expansion:

(*i*) If there is force of attraction between the molecules, internal work will have to be done in pulling them apart and the energy required for this is extracted from the gas, which will, therefore, be cooled by expansion.

(*ii*) If there is force of repulsion between the molecules, the expansion will cause a rise in temperature.

(*iii*) If there is no inter-molecular attraction or repulsion, there will be no change in temperature.

To varify the above predictions, Joule took two small steel vessels joined by a tube and provided with a tap, one of the vessels was exhausted and the other was filled with air at high pressure (About 20 atmosphere). The whole arrangement was placed in a constant temperature enclosure in the form of a water bath whose temperature could be measured with a very sensitive and accurate thermometer to the order of $\frac{1}{200}$th of a degree.

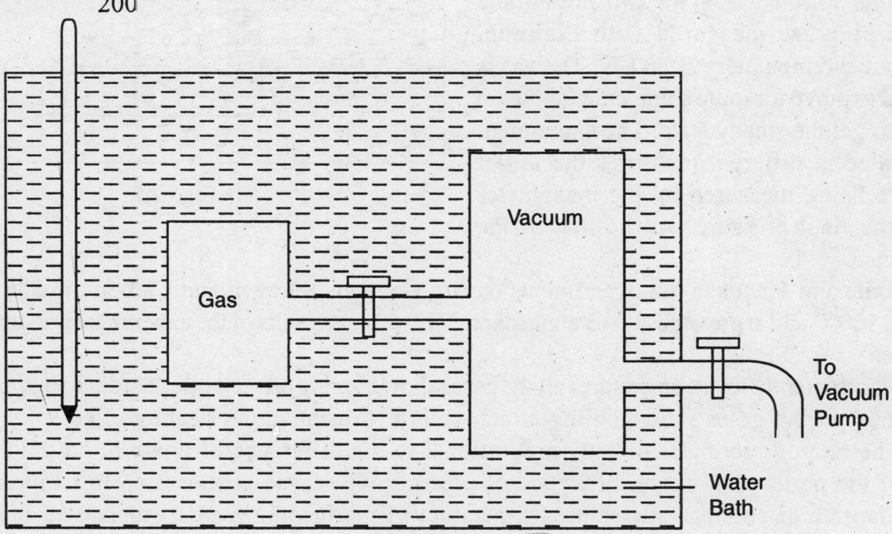

Fig. 2.1

By opening the tap, the air was allowed to expand into the exhausted vessel, thus doing no external work. The expansion was so rapid that no exchange of heat with the surroundings was possible and so a fall in temperature was expected if intermolecular attractions existed. However, in this experiment no fall or rise in temperature could be attained, which indicated that the experimental arrangement is so crude that it could not account for the forces of attraction between the molecules which had clear evidence from other experiments. Later in 1852, Thomson along with Joule devised another more sensitive experiment, known as Porous Plug experiment to establish the existence of intermolecular forces of attraction.

2.3. POROUS PLUG EXPERIMENT

Principle: To measure the difference in temperature of the gas which has been driven by a compressor at constant pressure through a porous plug. Porous plug consists of a porous material, such as cotton wool, silk etc. having a number of holes or pores and so acts like a number of narrow holes placed in parallel. The plug is surrounded by a constant temperature enclosure so that no heat leaves or enters the porous plug. The gas as it flows through narrow orifices under a large difference of pressure, gets throttled *i.e.*, the different molecules are separated from one another and if there are forces of attraction between the gas molecules, some energy is extracted from the gas, which in turn will lower the temperature of the gas coming out of the porous plug. This cooling was experimentally demonstrated by Joule and Thomson and so it is also called Joule-Thomson Effect.

Experiment: It consists of a porous plug P of cotton wool or silk kept in between perforated brass plates and surrounded in a box wood tube B for thermal insulation. A brass case D full of cotton wool or silk surrounds B so that the porous plug is further thermally insulated and also B is not in contact with water of the water bath to ensure constant temperature enclosure.

The box wood tube is connected on both sides with the copper tubes CC and the lower of these tubes is connected through a copper spiral S which is also immersed in water bath so that the gas is at constant temperature before it reaches the plug.

The compressed gas is made to suffer throttle expansion at the porous plug P. The temperature of the gas on the two sides of the porous plug are measured with Platinum resistance thermometers T_1 and T_2. The gas is allowed to move for quite some time (about on hour), to get the steady state. The experiment is repeated at different pressures the initial pressure being measured by the manometer while the final pressure will be that of the atmosphere.

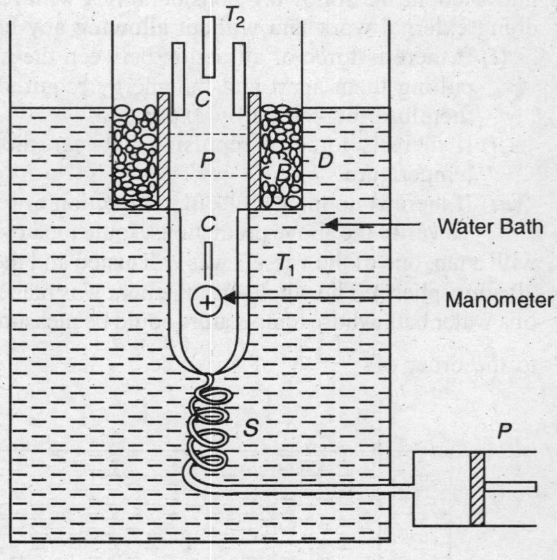

Fig. 2.2

Joule and Thomson did experiments on Air, Oxygen, Nitrogen and Carbon dioxide between 4°C and 100°C and at pressures of 4.5 atmosphere. The various results of the experiment are summarised as under:

(*i*) At sufficiently low temperatures all the gases show cooling effect while at ordinary temperatures, most of the gases show cooling effect while hydrogen shows heating effect.

(*ii*) The rate of cooling is directly proportional to the difference of pressure on the two sides of the porus plug, more difference of pressure will cause greater drop in temperature.

(*iii*) The rate of cooling falls as the temperature of the incoming gas is increased, the cooling becomes zero at a particular temperature, depending on each gas above this temperature instead of cooling, heating begins. The temperature at which Joule-Thomson effect changes sign *i.e.*, cooling effect becomes heating effect is called the *temperature of inversion.*

(*iv*) The magnitude of the effect can be judged from the following results:

At 0°C for Air → cooling of 0.275°C/Atmos.
At 0°C for CO_2 → cooling of 1.39°C/Atmos.
At 0°C for H_2 → heating of 0.03°C/Atmos.

So an important conclusion from the above results can be drawn as: All the gases below their temperature of inversion will cool in passing through a narrow orifice without resultant work.

2.4. THEORY

The above results can be explained from the following simple theory.

Let us consider an arrangement shown in the figure. Gas is filled in chamber A and is made to pass through a narrow orifice *O*. Two imaginary pistons P_1 and P_2 are shown to represent the compressor which drives the gas forward in chamber *A* and the pressure of the atmosphere which drives back the gas in chamber *B*. The positions of the two pistons is such that a constant difference of pressure is maintained.

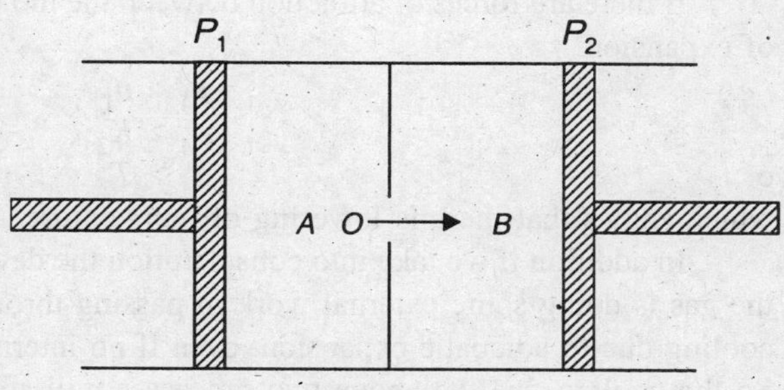

Fig. 2.3

Let p_1, v_1, and T_1 be the values of pressure, volume and temperature in chamber *A* and p_2, v_2, T_2 the corresponding values in chamber *B*. Let U_1 and U_2 be internal energies/unit mass of the gas in chamber *A* and *B* respectively.

Work done by piston P_1 on gas = Force × distance
= Pressure × Area × distance = $p_1 v_1$

Similarly work done by the gas on piston $P_2 = p_2 v_2$

Net work done by a unit mass of the gas in passing through the narrow orifice = $p_2 v_2 - p_1 v_1$. Because of the thermal insulation no heat can be gained or lost by the gas, so the energy spent by the gas in doing the work must be derived from the change in internal energy of the gas.

Hence $$p_2 v_2 - p_1 v_1 = U_1 - U_2$$

or $$U_1 + p_1 v_1 = U_2 + p_2 v_2 \quad \text{... (1)}$$

i.e., the total heat function $i = U + pv$ is constant during throttling process. This is known as is enthalpic process.

Now the internal energy of the gas is the sum total of the kinetic energy due to the translatory motion of the molecules and the potential energy due to the force of attraction between the molecules. If K_1 and ρ_1 correspond to K.E. and P.E. of the gas in chamber *A* and K_2 and ρ_1 the corresponding energies in chamber *B*, then

$$U_1 = K_1 + \rho_1 \text{ and } U_2 = K_2 + \rho_2$$

Also from kinetic theory of gases

$$T_1 \propto K_1 \text{ and } T_2 \propto K_2$$

If $$T_1 > T_2 \text{ then } K_1 > K_2 \quad \text{... (3)}$$

If $$T_1 < T_2 \text{ then } K_1 < K_2$$

We can now discuss the various cases.

2.5. PERFECT GAS

In the case of a perfect gas Boyle's Law is obeyed

∴ $$p_1 v_1 = p_2 v_2$$

Then from (1) $$U_1 = U_2$$

i.e., $$K_1 + \rho_1 = K_2 + \rho_2$$

But from the assumption that there are no forces of attraction between the gas molecules

$$\rho_1 = \rho_2 = 0$$

$$\therefore \quad K_1 = K_2 \text{ or } T_1 = T_2$$

Hence theoretically, no change in temperature is possible in the case of a perfect gas. However as a perfect gas is an ideal gas, rarely found in nature, so we can take up the case of ordinary gases under different conditions.

2.6. ORDINARY GASES

***(i)* Below Boyle's temperature:** Experimentally it is seen that pv decreases with increase of p during the initial stages of compression *i.e.*, $p_1 v_1 > p_2 v_2$

or $\quad p_2 v_2 > p_1 v_1$

i.e., $\quad p_2 v_2 - p_1 v_2$ is + ve quantity.

and $\quad U_1 - U_2$ is + ve quantity.

i.e., $\quad K_1 + p_1 > K_2 + p_2$

If there are forces of attraction between the molecules, the potential energy increases because of expansion

i.e., $\quad p_1 > p_1$

$\therefore \quad K_1 > K_2$

or $\quad T_1 > T_2$

which shows that there is lowering of temperature.

In addition if we take into consideration the deviations from Boyle's Low, since $p_2 v_2 > p_1 v_1$, the gas is doing some external work in passing through the orifice and this will cause additional cooling due to adiabatic expansion, even if no intermolecular forces of attraction are considered. So the cooling due the above two processes will add up to give more cooling.

***(ii)* Above Boyle's temperature.** Since pv increases with p, we have

$$p_1 v_1 > p_2 v_2$$

or $\quad p_2 v_2 < p_1 v_1$

i.e., $\quad p_2 v_2 - p_1 v_1$ is –ve quantity.

and $\quad U_1 - U_2$ is –ve quantity.

i.e., $\quad K_1 + p_1 > K_2 + p_2$

If we consider the intermolecular forces of attraction then P.E. will increase due to expansion at the cost of K.E.

i.e., $p_2 > p_1$ and $K_2 < K_1$

So cooling will be produced. However this does not justify equation (4), but if we consider the derivations from Boyle's Low, since $p_2 v_2 > p_1 v_1$, the gas has some extra work done on it in passing through the narrow orifice, which will produce heating effect.

As the above two effects are opposed, there may be cooling effect, heating effect or no effect depending upon their magnitudes.

So we have Joule-Thomson effect to be divided into two parts:

(*i*) Due to intermolecular forces of attraction.

(*ii*) Some external work done by or on

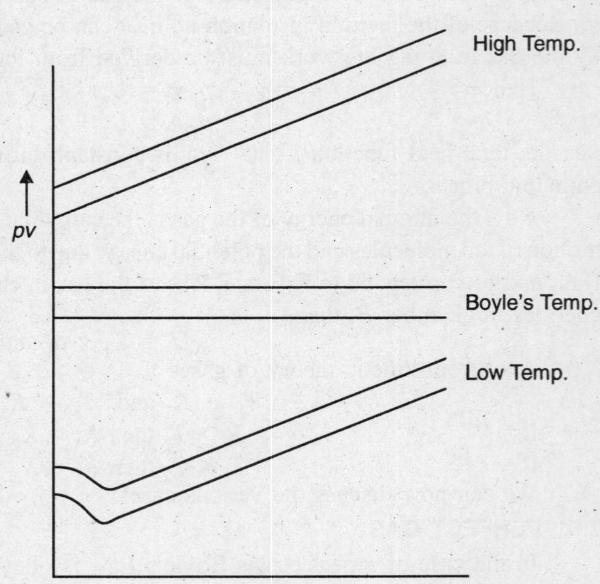

Fig. 2.4

the gas.

At the temperature of inversion which is well above Boyle's temperature, the cooling due to intermolecular forces of attractions is just balanced by the heating due to external work done on the gas. Above the temperature of inversion $T_2 > T_1$ *i.e.*, heating effect results. Below temperature of inversion, the effects due to the two processes gives cooling.

In the case of hydrogen the slight heating at ordinary temperatures is because of the external work done on the gas, since its temperature of inversion is –80°C. In case of hydrogen the deviations from Boyle's Law at ordinary temperatures are known to be small because of the finite size of the molecules.

The above discussion leads to the conclusion that lower the temperature of the incoming gas, more is the cooling for the same difference of pressure. Some of the result are given as under :

Air	→	0.152°C cooling/Atmos.	at 93°C
	→	0.263°C cooling/Atmos.	at 7°C
CO_2	→	0.645°C cooling/Atmos	at 96°C
		1.233°C cooling/Atmos	at 8°C
H2	→	0.039°C heating/Atmos	at 30°C
	→	0.004°C heating/Atmos	–71°C
		no Effect at heating/Atmos	(–80°C) (Temp. of Inversion)

2.7. LIQUEFACTION OF GASES

It is observed that substances which are gaseous at ordinary temperatures can be converted into liquid state if sufficiently cooled and simultaneously subjected to a high pressure. Normally, the methods employed to liquify gases may be classified as under:

1. Cascade Proces: In this case a series of liquids with successively lower boiling points is employed so that the desired low temperature is attained.

2. Regenerative Joule–Thomson Process: Which is based on the principle of regenerative cooling and the Joule–Thomson Effect.

3. Adiabatic Process: It is based on the principle of cooling produced when a gas expands adiabatically doing external work.

2.8. LIQUEFACTION OF CHLORINE

The apparatus used is shown in the Fig. 2.5. It consists of a strong glass tube having a V shape with both the ends closed. One end of the glass tube contains crystals of chlorine and the

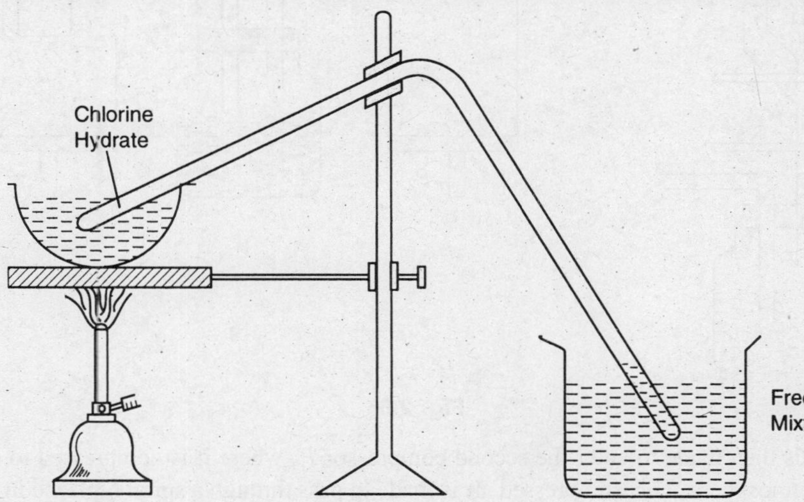

Fig. 2.5

other end is put in a freezing mixture. On heating the end containing chlorine hydrate, a high pressure is developed and chlorine is evolved in large quantities. When it reaches the other end it is liquefied due to the cooling caused by freezing mixture into an oily liquid under its own pressure.

2.9. LIQUEFACTION OF CO_2

The above method can be used to liquify carbon dioxide. In this case a strong cylinder is filled upto $\frac{1}{3}$ with Sodium Bicarbonate. An open copper tube containing sulphuric acid is fitted as above with the cylinder. When the cylinder is inverted, Sodium Bicarbonate reacts with Sulphuric acid to produce CO_2 and also producing very high pressure is 50–80 atmospheres at which pressure the gas is liquified and distils over to the other end. If the liquid CO_2 is allowed to pass through a jet, it is converted into a solid.

2.10. LIQUEFACTION OF AIR

Linde, in 1895 used the Joule–Thomson regenerative process to liquify air. His apparatus and principle is given as under.

It consists of a two stage compress P_1 and P. Air is compressed to about 20 atmospheric pressure and heat of compression is removed by allowing it to pass through a water bath. This air is then made to pass through Caustic potash, Calcium chloride and Phosphorus pentoxide tubes to free it from CO_2, and water vapours because otherwise CO_2 and water vapours will solidify before air is liquified and block the circulation system.

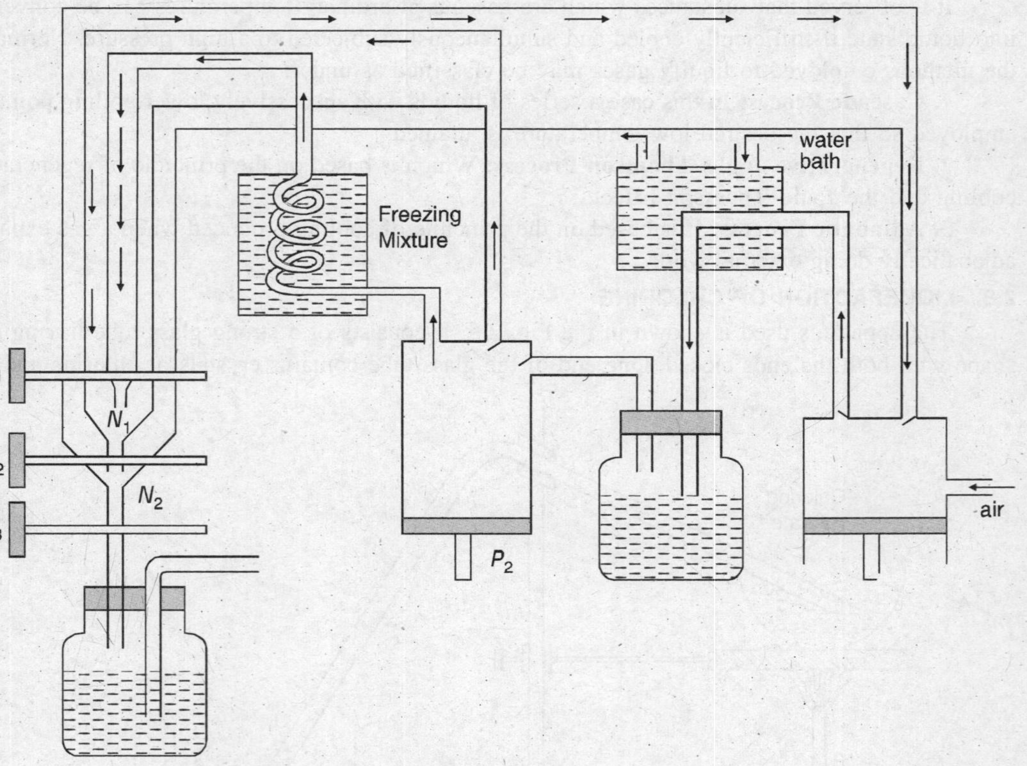

Fig. 2.6

The air is then made to enter the second compressor P_2 where it is compressed to a pressure of about 200 atmosphere. This compressed air is made to pass through a spiral S placed in a freezing mixture where its temperature is lowered to about 20°C. This cooled air is made to enter through the tube T_1 to nozzle N_1 where it suffers a drop in pressure, the drop in pressure can be controlled by valve V_1. The air coming out of the nozzle N_1, suffers throttle expansion and its temperature is lowered to about –70°C. This cooled air is made to pass through another tube T_2 and return to compressure P_2, where it is further compressed to a pressure of 200 atmosphere on its way to P_2, this air further cools the incoming air through the tube T_1, which therefore comes out of the nozzle N_1 at still a lower temperature. After sufficient number of cycles, the temperature of the air is lowered quite appreciably. At this stage the second nozzle N_2 is opened and the already much cooled air is allowed to expand to one atmosphere when it is easily liquified. The liquid air is collected in Dewar flask D from which it can be removed. The unliquified air is led back through the tube T_3 surrounding T_2 to the compressor P_1 for recycling. The whole of the liquifying apparatus is packed with cotton wool to secure thermal insulation as required in the Joule-Thomson Effect.

Linde with the above method using about 3 H.P. could get one litre of liquid air per hour. Later on the above apparatus was improved by Hampson, Dewar and others to produce liquid air in lesser time without preliminary cooling.

2.11. LIQUEFACTION OF HYDROGEN

In the earlier attempts, Oxygen, Nitrogen and Hydrogen could not be liquefied and for quite some time they were called permanent gases. The reason of earlier failures being that these gases have very low temperature of inversion.

Dewar was the first to succeed to liquefy hydrogen by Joule-Thomson Effect. Cascade process failed because of the very low critical temperature (–240°C) of hydrogen. In order to apply Joule Thomson Effect, hydrogen has to be first pre-cooled below its temperature of inversion (–80°C) since above this temperature, it produces heating effect. Moreover, to get appreciable cooling in practice, it is necessary to pre-cool hydrogen below its Boyle's temperature (–177°C). This pre-cooling is achieved by passing the gas through liquid air (B.P. –190°C) evaporating under reduced pressure.

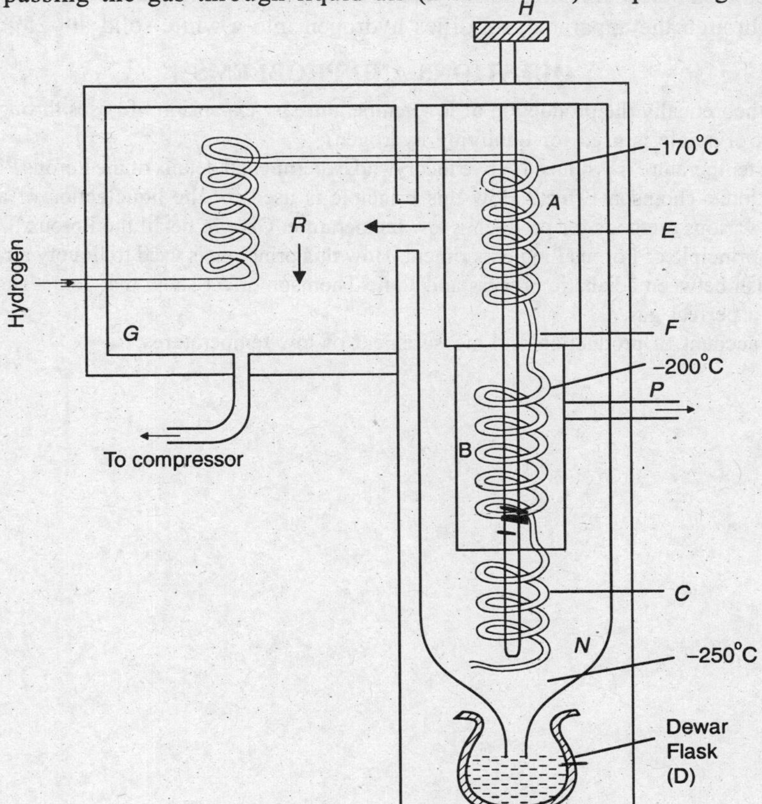

Fig. 2.7

Dewar's apparatus is shown in Fig. 2.7. Hydrogen is compressed to about 150 atmosphere and its heat of compression is removed by making it to pass through cold water bath. Then it is made to pass through tubes containing caustic potash. Calcium chloride, Phosphorus pentoxide to remove CO_2 and water vapours—which otherwise at lower temperatures may solidify and block the passage. Pure, dry hydrogen thus obtained enters the liquefier and passes through the regenerative coil *R*, in the chamber *G* which is cooled by the outgoing cold hydrogen gas that has already suffered a throttle expansion and finally its temperature reduces to about –170°C. This cooled gas passes through a refrigerating coil *A* immersed in a chamber *E* and then through another coil put in liquid air in second chamber *F* and boiling at 100 mm pressure. This is achieved by allowing liquid air from *E* to trickle into *F* and exhausting *F* to a pressure of 10 cms through a pump. The temperature of the gas thus reduces to about –200°C. After this pre-cooling the gas is made to pass through coil *C* and suffers throttle expansion at the nozzle *N*–which operates with a handle *H* placed outside. This gas thus cooled, further passes through *C*, *B* and *A* to chamber *G* and then to compressor. After a few cycles, the temperature of the gas reaching nozzle *N* falls to –250°C and under these conditions, suffering a throttle expansion it is liquified and collected in Dewar flask *D*. Liquid hydrogen is transparent and colourless.

2.12. SOLIDIFICATION OF HYDROGEN

Dewar, an year later was able to solidify hydrogen by taking liquid hydrogen (B.P. –253°C) into a double walled flask and reducing the pressure to around 10 mm and immersing in a bath of liquid hydrogen. Though the liquid hydrogen started evaporating under reduced pressure and the temperature reduced below freezing point, yet no solidification took place. However a trace of air when leaked through the apparatus, solidifies hydrogen into a white solid at –259°C.

QUESTIONS AND PROBLEMS

1. Discuss theoretically the production of low temperature by expansion of gases through porous plug. How this principle is used for liquifying hydrogen?
2. How low temperature is produced? Give theory and experimental details of the Porous Plug experiment.
3. What is Joule-Thomson Effect? How this principle is used for the liquefaction of air?
4. Give the various methods for producing low temperature. Give in detail the Porous Plug experiment.
5. Give the principle of Porous Plug experiment. How this principle is used to liquefy air and hydrogen?
6. Distinguish between adiabatic process and Joule-Thomson effect show that Joule-Thomson effect is zero for a perfect gas.
7. Give an account of production and measurement of low temperatures.

CHAPTER 3

THERMOMETRY

3.1. INTRODUCTION

Thermometers are generally the devices which measure the temperature of a body with scientific precision and that branch of physics that deals with measurement of degree of hotness or coldness of a body is called thermometry.

Normally the basic principle of a thermometer is 'Two bodies in contact will have a thermal equilibrium after sometime'. In practice, thermometry will involve.

(*a*) An instrument called thermometer which is based on certain physical properties of matter *e.g.*, expansion, electrical resistance etc. In designing a thermometer regular change of physical properties for regular changes of temperatures over a long range are required.

(*b*) **Standardizing the instrument.** We have to fix certain points on the thermometers that will account for the range of its measurement *e.g.*, in the case of ordinary mercury thermometer freezing point of pure water (ice point) and boiling point of pure water (steam point) are taken and then the intermediate scale is suitably divided to give arbitrary numerical values of temperature known as thermometric scale *e.g.*, centigrade scale (0°C to 100°C) Farnheit scale (92°F to 212°F) etc.

Requirements of a good thermometer will be that it should have considerable sensitiveness *i.e.*, (*i*) it quickly responds to the temperature of the body it is placed in contact with (*ii*) it responds to even smallest changes of temperature.

A broad outline of the various thermometers which are generally in use are given as under:

(1) Liquid thermometers. These are based on the principle of expansion of liquids *e.g.*, mercury thermometers, alcohol thermometers.

(2) Gas thermometers. They depend on the thermal expansion of gases and are generally considered as accurate thermometers because of their accuracy in results and most of the other thermometers are calibrated with gas thermometers.

(3) Resistance thermometers. These are based on the principle of change electrical resistance with temperature.

(4) Thermoelectric thermometers. Based on the principle of change of thermo e.m.f. due to the difference of temperature of the two functions of a thermocouple.

(5) Vapour pressure thermometers. These depends on the change of vapour pressure with temperatures.

(6) Radiation thermometers. These are based on the principle of radiations emitted by a body and noting the amount of radiations on some instrument and by applying radiation laws, temperature of the body is found.

Here we will concentrate only on resistance thermometers and thermoelectric thermometers.

3.2. RESISTANCE THERMOMETERS

It has been found that the electrical resistance of a material in the form of a wire varies quite uniformly with increase in temperature. This characteristic has been used for measurement of temperatures.

Callender showed that, platinum wire free from impurities like silicon, carbon, tin etc. has got the variation of temperature according to the relation

$$R_t = R_0\ (1 + \alpha t + \beta t^2) \quad ...(1)$$

where α and β are constants of the material of the wire used. For platinum $\alpha = 3.91 \times 10^{-3}$ and $\beta = -5.79 \times 10^{-7}$. The value of α and β are determined by finding the resistance of the platinum coil at:

(*a*) Melting pt. of ice *i.e.*, 0°C (*b*) Boiling pt. of water *i.e.*, 100°C
(*c*) Boiling pt. of sulphur *i.e.*, 444.6°C.

However, for the case of pure platinum, a simple relation of clausius is used, *i.e.*,

$$R_t = R_0\ (1 + \alpha\ t) \quad ...(2)$$

where R_t, R_0 are the resistances at t°C and 0°C, t is unknown temperature and α is constitution of the material.

If R_{100} be the resistance at 100°C then,

$$R_{100} = R_0\ (1 + \alpha\ .\ 100) \quad ...(3)$$

From Equation (3), $d = \dfrac{R_{100} - R_0}{R_0 .100}$ ...(4)

From Equation (2), $t = \dfrac{R_t - R_0}{R_0 \alpha}$ ...(5)

Putting the value of a from Equation (4) in Equation (3)

$$t = \frac{R_t - R_0}{R_0} \times \frac{R_0 \times 100}{(R_{100} - R_0)}$$

$$= \frac{R_t - R_0}{R_{100} - R_0} \times 100 \quad ...(6)$$

which gives the temperature on 'platinum scale'. However, to get direct temperature on a uniform scale, callender showed that the difference of temperatures of platinum resistance thermometer and a standard gas thermometer may be represented over a wide range of accuracy by the relation

$$t' - t = \delta \left[\left(\frac{t'}{100} \right)^2 - \left(\frac{t'}{100} \right) \right] \quad ...(7)$$

where δ is a constant.

So normally in the construction and use of platinum resistance thermometer, two important factors have to be kept in mind *viz.*,

(*i*) Platinum wire should be pure, free from impurities and chemical changes.

(*ii*) Temperature measured on the resistance scale must be changed to the gas scale.

Construction. It consists of a long hard glass or porcelain tube in the bottom of which is placed an insulating mica disc *M*. Pure, platinum wire, which has been well annealed is doubled on itself to avoid induction effects is wound over the mica disc *M*, as shown in Fig. 3.1. The free ends of the resistance wire are attatched to lead wires which pass through number of mica sheets to avoid the variation of temperatures at different points of lead wires. Depending upon the range of temperature for which the instrument is to be used and accuracy, we may have,

(*i*) Copper lead wires and hard glass tube for temperatures upto 700°C.

(*ii*) Platinum lead wires and porcelain tube for temperatures above 700°C.

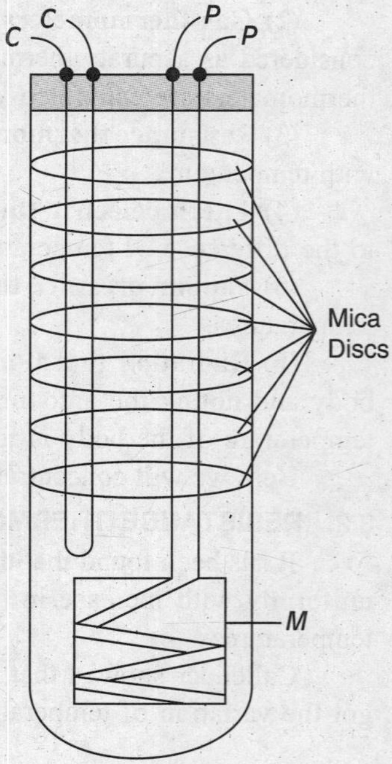

Fig. 3.1

To avoid the variation of resistance of the lead wires with

temperature, a pair of compensating lead wires, C of the same material, length and area of cross-section, pass through the holes of mica discs and are placed close to PP as shown in Fig. 3.1. The tube is evacuated and then scaled so that moisture effect does not effect the insulation of mica.

Working. Callender and Griffith designed, a special bridge, based on the principle of wheat stone bridge for accurate measurement of temperatures. The accuracy of measurement of resistance is $\simeq 10^{-4}$ ohm.

It consists of two ratio arms P and Q of equal resistances, the third arm R consists of a resistance box which can measure resistance of the order of 0.01 ohm. The resistance box and the fourth arm X, which consists of the unknown resistance *i.e.*, the platinum wire of the thermometer are connected to a potentiometer of wire AB of high accuracy in dimensions and whose resistance/cm is $\sim \frac{1}{2} \times 0.1$ ohm.

Another parallel wire LM of the same material as AB is placed to avoid thermoelectric effects and the two are connected together through a button jockey. The platinum leads PP are put in the arm X while compensating leads CC are put in arm R. By adjusting the value of R, a balance point D on the potentiometer wire AB is noted. Then far balance point to occur,

$$P + R + r + \rho r..AD = Q = X + r + \rho . DB$$

Since P and Q are of equal resistance, the above relation may be written as

$$P + r + \rho . AD = X + r + \rho . DB$$

where r is the resistance of lead wire, P of the resistance/cm of potentiometer wire and A, the resistance of the platinum wire of temperature to be measured.

$$\therefore \quad X = R + \rho\,(AD - DB)$$
$$= R + \rho\,[AO + OD - (OB - OD)]$$
$$= R + \rho . 2OD$$

[$\because AO = OB$]

Since we have taken

$$\rho = \frac{1}{2} \times .01 \text{ ohm/cm}$$

$$\therefore \quad X = R + .01OD \quad ...(8)$$

From the Fig. 3.2 it is clear that if the balance point is on the right of O, OD is +Ve and if it is on the left of O, it will be –Ve.

So X will correspond to resistance R_t at the temperature t of the bath. Similarly R_0 and R_{100} can be measured and then using the relation

$$i.e., \quad t = \frac{R_t - R_0}{R_{100} - R_0} \times 100$$

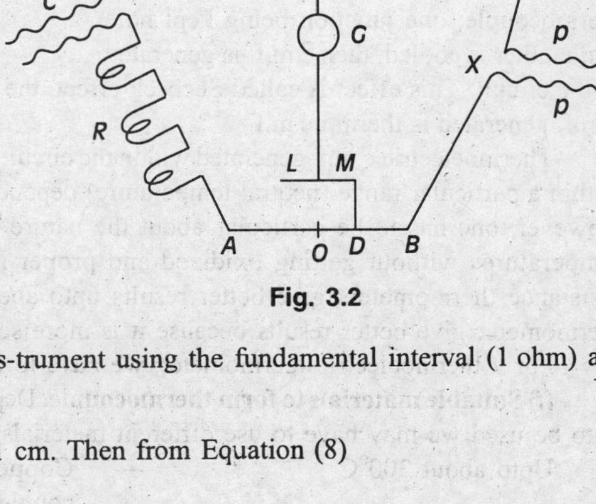

Fig. 3.2

The term $(R_{100} - R_0)$ is called fundamental interval and in the case of platinum resistance thermometer it is 1 ohm.

We can study the accuracy of the ins-trument using the fundamental interval (1 ohm) and $\rho = 0.005$ ohm, we have $\frac{R_{100} - R_0}{200}$

Let R correspond to R_0 and $OD = +1$ cm. Then from Equation (8)

$$\therefore \quad X = R_0 + .01$$

$$\text{or} \quad X - R_0 = \frac{1}{100} \quad ...(9)$$

Since we have the fundamental interval $R_{100} - R_0 = 1$ ohm, so we may write equation (9) as

$$\frac{X - R_0}{R_{100} - R_0} \times 100 = 1$$

which shows that the temperature on the platinum scale is 1°C. If we measure the balance point correct to 1 mm, the accuracy of temperature will be ~ 0.01°C.

To get the accurate temperature on gas scale, we have the relation

$$t' - t = \left[\left(\frac{t'}{100}\right)^2 - \frac{t'}{100}\right]$$

where constant δ can be found by standardizing the thermometer at 0°C, 100°C and 444.6°C (B.P. of sulphur). In practice, each platinum thermometer is provided with a calibration curve drawn from actual comparison with a gas thermometer which can directly give us the temperature on the gas scale corresponding to the resistance scale temperature measured.

Advantage:

(*i*) It has a wide range (From –200°C to 1200°C).

(*ii*) It has good accuracy *i.e.,* it can read accurately upto 0.01°C. At lower temperatures it is as accurate as a gas thermometers, however it should not be used beyond 1000°C on account of the impurity that may come due to contamination of insulating material.

Disadvantages: The container tube has large thermal capacity and low thermal conductivity and so it can not respond to quick changes of temperatures. Moreover it takes quite some time to get a balance point on the bridge.

3.3. THERMOELECTRIC THERMOMETER

Principle. It is based on the principle that when two wires of different materials are soldered together to form to junctions of thermocouple, one junction being kept hole while other is cooled, then e.m.f. is generated in the circuit. This effect is called Seebeck effect, the arrangement is called thermocouple and the e.m.f. generated is thermo e.m.f.

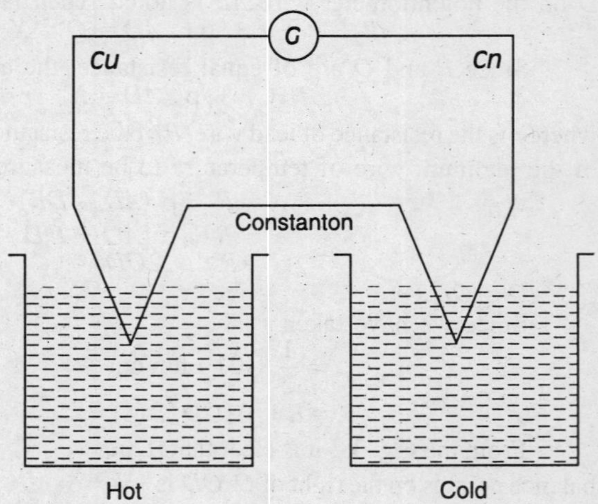

Fig. 3.3

Thermoelectric e.m.f. generated within the circuit is found to be a linear function of temperature within a particular range (neutral temperature) depending on the metals chosen for thermocouple. However, one has to be particular about the nature of metal chosen which can withstand high temperatures without getting oxidized and proper insulations at such temperatures. Platinum resistance thermometers give better results upto about 1000°C, beyond which Thermo-electric thermometer give better results because it is more sensitive and convenient to use.

For a thermoelectric thermometer, we have to consider the following points:

(*i*) **Suitable materials to form thermocouple.** Depending on the range for which a thermocouple is to be used we may have to use different materials forming the two junctions:

Upto about 300°C	—	Copper-constanton or Iron constanton thermocouple is used
Between 300°C – 600°C	—	Iron-Nickel thermocouple is used.
Between 600°C – 1000°C	—	Nickel-Nichrome is used.
Between 1000°C – 1600°C	—	Pt.-Alloy of pt. with Ir or Rh is used.

Between 1600°C – 2000°C — Ir with an alloy of Iridium-Rubidium is used.
Between 2000°C – 3000°C — Tungsten-Molybdenum is used.

(*ii*) Electrical insulation of the wires especially near the hot junctions to avoid e.m.f.'s generated at different points due to variation of temperatures is done by means of fireclay tube or glass beads.

(*iii*) e.m.f. generated in the circuit is measured with a high resistance galvanometer which is calibrated to give direct readings of temperature if small accuracy of measurement of temperature is required. However for good accuracy a potentiomerte is used which can measure e.m.f. ~ 0.1 μV corresponding to a very small difference of temperatures between the junctions.

(*iv*) **Measurement of temperature.** Normally a calibration curve is drawn between e.m.f. and the temperature for each thermocouple. It should be noted that the calibration curve for each thermocouple will depend on the range upto which the thermometer is to be used because different formulas for different thermocouples and ranges of temperatures are to be considered. Then the temperature is directly measured from the calibration curve.

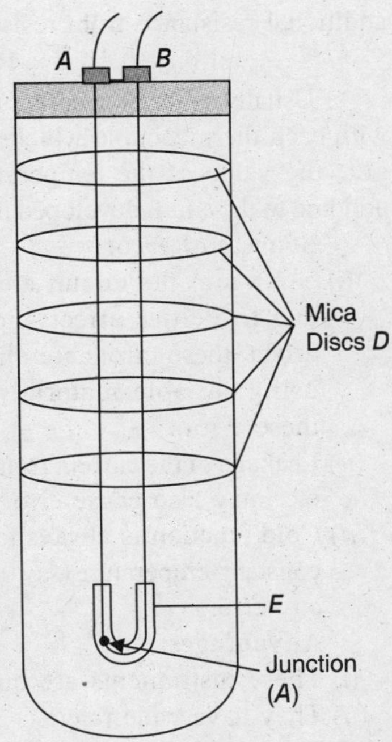

Fig. 3.4

Construction. It consists of a long tube of hard glass, porcelain or quartz depending upon the range of temperature which the instrument is used. Hot junction *A* of a thermocouple is placed in insulator *E* of fireclay or glass. Then the lead wires are passed through mica discs *D* to avoid different amounts of e.m.f.'s in the circuit. The elements of the lead wires are fixed in the termination *A* and *B* on the top of the tube and another junction (cold) being kept at 0°C.

For measuring temperature with thermoelectric thermometer to fairly accuracy, we use a potentiometer which can measure e.m.f.'s in microvolts.

Working. A potentiometer wire *AB* of 100 cm length and resistance of 1 ohm is connected in series with two resistance boxes *P* and *Q*, battery, tap key and a rheostat as shown in Fig. 3.5.

We make use of a standard Cd cell (e.m.f. = 1.0183 V) which is connected with the resistance box *Q* through a galvanometer and key K_2 to the point *B* of the potentiometer. A resistance of 1017.3 Ω is extracted from *Q* and the current with rheostat is so adjusted that there is no deflection in the galvanometer, which means that there is fall of potential of 1.0183 over a resistance of 1018.3Ω. So the fall in potential for

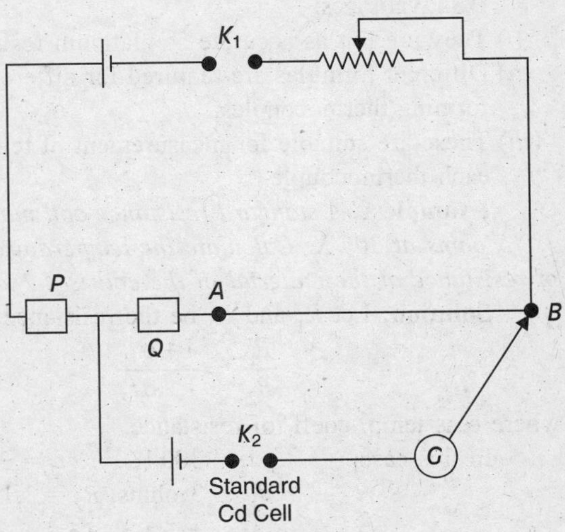

Fig. 3.5

$$1\Omega = \frac{1.018 \text{ V}}{1018.3} = 10^{-3} \text{ V}$$

Since *AB* is 100 cm long and has a resistance of 1 Ω, the fall in potential for 1 cm = 10^{-5} V. Since we can measure accurately 1 mm on the scale, the fall in potential for 1 mm = 10^{-6} V = 1μV.

In order to measure e.m.f., the Cd cell is replaced by thermocouple as shown below, taking care that +ve of the thermocouple is connected to Q. A balance point on AB is got by suitably adjusting resistance in Q but taking out the same resistance from P so that the initial calibration is not disturbed. Let the balance point be at O at a distance l cm and a resistance $\times\ \Omega$ be the additional resistance from resistance box Q, the e.m.f. of the thermocouple will be

$$(X \times 10^{-3} + l \times 10^{-5}) \text{ volts}$$

Usually calibration curves are provided with each thermocouple which can directly give the value of the temperature corresponding to the e.m.f. developed in the circuit.

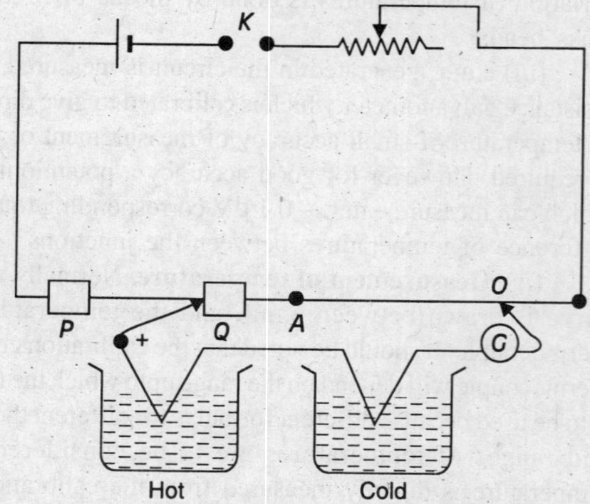

Fig. 3.6

Sources of error:

(*i*) e.m.f.'s in the circuit are developed due to Peltier effect and Thomson effect, these errors are eliminated by using suitable materials free from these errors.

(*ii*) Leakage of the current from the battery etc. may also cause error.

(*iii*) Cold junction is always to be kept at constant temperature (say 0°C) to avoid any error.

Advantages:

(*i*) These instruments are quite cheap and easy to use.

(*ii*) They have wide range (–20°C to 1600°C).

(*iii*) From calibration curves direct reading of temperatures can be got.

(*iv*) Since they have small thermal conductivity, they are quite suitable for measuring the rapidly changing temperatures such as cooling of alloys.

Disadvantages:

(*i*) They are not as accurate as platinum resistance thermometer especially below 1000°C.

(*ii*) Different formulas are required for different ranges of temperatures and different materials forming thermocouples.

(*iii*) These are suitable for measurement of temperatures upto only neutral point in the case of each thermocouple.

Example 1. *A standard resistance coil marked 3 ohm is found to have a true resistance of 3.115 ohms at 300°K. Calculate the temperature at which the marking is correct. Temp. Coeffi. of resistance of the material of the coil is 4.2 × 10^{-3}/°C.*

Solution: Let R_1 and R_2 be the resistance of the coil at t_1°C and t_2°C respectively

$$\therefore \qquad \frac{R_1}{R_2} = \frac{1+\alpha t_1}{1+\alpha t_2}$$

where α is temp. coeff. of resistance.

In this case $t_2 = 300$ K $\quad \therefore\ t_2 = 300 - 273 = 27$°C

$R_1 = 3$ ohms, $R_2 = 3.115$ ohms

$$\therefore \qquad \frac{3}{3.115} = \frac{1+4.2\times10^{-3}\times t_1}{1+4.2\times10^{-3}\times t_2}$$

$$= \frac{1+0.0042\ t_1}{1+0.0042\ t_2}$$

$$\therefore \qquad 1 + 0.0042\ t_1 = \frac{3.115}{3}\ [1 + .0042 \times 27] = 1.0723$$

$$\therefore \quad t_1 = = \frac{0.0723}{0.0042} = 7.21^\circ\text{C}$$

$$\therefore \quad t_1 = 273 + 17.21 = \mathbf{290.21\ K.}$$

Example 2. *The resistance of a platinum wire of platinum resistance thermometer at 0°C is 5Ω and at 100°C it is 5.39Ω. When this thermometer is put in a hot bath, the resistance of platinum wire is found to be 5.795Ω. Calculate the temperature of the bath.*

Solution: In this case

$$R_0 = 5\Omega$$

$$R_{100} = 5.23\Omega$$

and

$$R_t = 5.795\Omega$$

We have the relation

$$t = \frac{R_t - R_0}{R_{100} - R_0} \times 100$$

$$= \frac{5.795 - 5}{5.25 - 5} \times 100$$

$$= \frac{0.795}{0.25} \times 100 = \mathbf{345.652^\circ C}$$

Example 3. *The resistance of platinum wire at 0°C, 100°C and 444.6°C is found to be 5.5, 7.5 and 14.5 ohms respectively. The resistance of a wire is given by the relation*

$$R_t = R_0\,(1 + \alpha t + \beta t^2)$$

find the values of α and β.

Solution: We have $R_0 = 5.5\Omega$, $R_{100} = 7.5\Omega$, $R_{444.6} = 14.5\Omega$

$$\therefore \quad R_{100} = R_0\,[1 + \alpha\,.\,(100) + \beta\,.\,(100)^2] \qquad \text{...(1)}$$

and

$$R_{444.6} = R_0\,[1 + \alpha\,.\,(444.6) + \beta\,.\,(444.6)^2] \qquad \text{...(2)}$$

Putting the values of R_0, R_{100} and $R_{444.6}$, we get

$$7.5 = 5.5\,[1 + 100\,\alpha + (100)^2\,.\,\beta] \qquad \text{...(3)}$$

$$14.5 = 5.5\,[1 + 444.6\,\alpha + (444.6)^2\,\beta] \qquad \text{...(4)}$$

From Equation (3), we have

$$\alpha + 100\,\beta = \frac{2}{5.5 \times 100} \qquad \text{...(5)}$$

From Equation (4), we have

$$\alpha + 444.6\beta = \frac{9}{5.5 \times 444.6}$$

Subtracting Equation (5) from Equation (6), we get

$$344.6 = \frac{900 - 2 \times 444.6}{5.5 \times 444.6 \times 100}$$

or

$$\beta = \mathbf{1.281 \times 10^{-7}/^\circ C}$$

Putting the value of β in Equation (5), we get

$$\alpha + 100\,(1.281 \times 10^{-7}) = \frac{2}{5.5 \times 100}$$

$$\alpha = \mathbf{3.62 \times 10^{-3}/^\circ C.}$$

Example 4. *If the platinum temperature corresponding to 50°C on the gas scale is 50.25°C, what will be the temperature on the platinum scale corresponding to 150°C on the gas scale?*

Solution: Let t' be the temperature on gas scale and t that on platinum scale, then we have

$$t' - t = \delta\left[\left(\frac{t'}{100}\right)^2 - \frac{t'}{100}\right]$$

(*i*) In this case

$$t' = 50^\circ, \quad t = 50.25^\circ$$

$$\therefore \quad 50 - 50.25 = \delta\left[\left(\frac{50}{100}\right)^2 - \left(\frac{50}{100}\right)\right] \qquad \text{...(1)}$$

(*ii*) In this case

$$t' = 150^o \quad t = x$$

$$\therefore \quad 150 - x = \delta\left[\left(\frac{150}{100}\right)^2 - \frac{150}{100}\right] \quad ...(2)$$

Now from Equation (1), we have

$$-0.25 = \delta\,[-0.25]$$

$$\therefore \quad \delta = 1$$

Put in Equation (2) $\quad 150 - x = 1\cdot\left[\frac{9}{4} - \frac{3}{2}\right]$

or $\quad 150 - x = \frac{3}{4}$

$\therefore \quad 600 - 4x = 3$

or $\quad 4x = 597$

or $\quad x =$ **149.25°C.**

Example 5. *An e.m.f. of 0.9 V is generated when the temperature difference between the hot and cold junctions of a thermocouple is 75°C. Assuming that the cold junction is heated by 15°C, determine the extent by which thermo e.m.f. will change.*

Solution: In this case the temperature difference being small, so variation of e.m.f. with temperature may be considered linear, *i.e.*, we may write

$$E \propto t$$

$$= \alpha t \text{ where } \alpha \text{ is constant}$$

Now $\quad t = 75°C,\ E = 0.9$ V

$$\therefore \quad \alpha = \frac{E}{t} = \frac{09}{75} \text{ volt/°C.}$$

When the cold junction is heated by 15°C, the difference of temperature between hot and cold junction will become 75 – 15 = 60°.

Now thermo e.m.f. generated will be

$$E' = \alpha t' = \frac{0.9}{75} \times 60 = 0.72 \text{ Volt}$$

$\therefore$ Change in e.m.f. $= E - E' = 0.90 - 0.72$

$= 0.18$ V

$\therefore$ % change of e.m.f. $= \frac{0.18}{0.9} \times 100 =$ **20%.**

QUESTIONS AND PROBLEMS

1. Describe the basic principle of measurement of temperature. Give construction and use of platinum resistance thermometer.
2. What is a thermoelectric thermometer? How it is used for measurement of temperature?
3. Give in details the relative merits of thermoelectric and platinum resistance thermometer. In which respect thermoelectric and thermometer is more useful than the platinum resistance thermometer?
4. Give in details the measurement of temperature by platinum resistance thermometer. How the apparatus is calibrated and how the temperature is measured on the gas scale thermometer?
5. Give the importance of thermoelectric thermometer. How this thermometer is calibrated and what is the range of accuracy of this thermometer?
6. If the platinum temperature corresponding to 60°C on the gas scale is 60.36°C, what is the platinum scale temperature corresponding to 151.7°C on the gas scale? **[Ans.** 150°C]
7. If the platinum temperature corresponding to 60°C on the gas scale is 60.25°C what will be the temperature on the platinum scale corresponding to 120°C on the gas scale? **[Ans.** 119.25°C]
8. The resistance R_t of a platinum wire at temperature t°C measured on the gas scale is given by $R_t = R_0(1 + \alpha t + \beta t^2)$ where $\alpha = 3.8 \times 10^{-3}$ and $\beta = -5.6 \times 10^{-7}$. What temperature will the platinum thermometer indicate when temperature on the gas scale is 200°C? **[Ans.** 197°C]

CHAPTER

4

TRANSMISSION OF HEAT

4.1. INTRODUCTION

There are normally three modes of transmission of heat *viz.,* conduction, convection and radiation. When the temperature of a body is increased, the energy supplied to the body is being stored in the form of thermal or heat energy. In the normal process, the transmission of heat takes place from a region of higher temperature to a region of lower temperature.

The various modes of transmission of heat are given as under.

4.2. CONDUCTION

If a hot body is placed in contact with a cold body, we observed that after some time the cold body becomes warmer while the hot body becomes colder. This is due to the fact that some of the energy is transferred from the hot body to the cold body. If a metallic rod is heated from one end, we see that the temperature of the other end also rises, which indicates that the heat has travelled to the other end of the rod. This gradual flow of heat from hot end to cold end is called thermal conduction. The explanation for this can be given as under.

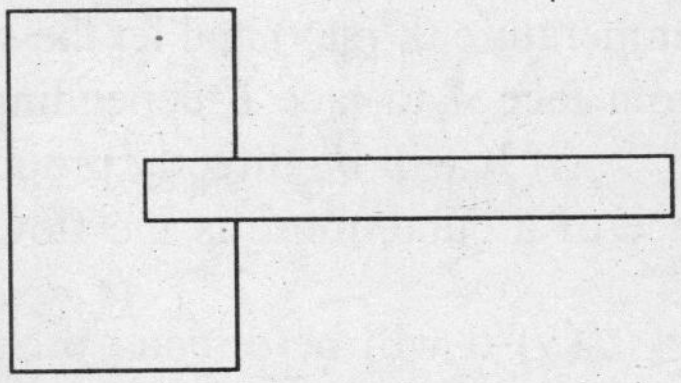

Hot Bath

Fig. 4.1

The molecules of the rod near the hot end start vibrating vigorously due to thermal energy and this vibration is handed over to neighbouring molecules and thereby some energy transfer takes place. As a result the neighbouring molecules also start vibrating and so the heat spreads through the body. In this process the transfer of energy is due to molecular impact and molecules themselves do not move. Since the transfer of energy is due to molecular movements, so this mode of transmission cannot take place in vacuum. So conduction may be defined as the mode of transmission of heat from one part to another part without the actual movement of the particles.

4.3. CONVECTION

In this mode of transmission the particles themselves to propagate the heat. Convection is the normal mode of propagation of heat in liquids and gases. The regions and fluids which are heated, expand and become less dense thereby carrying heat to the other regions and the walls of the vessel containing fluid.

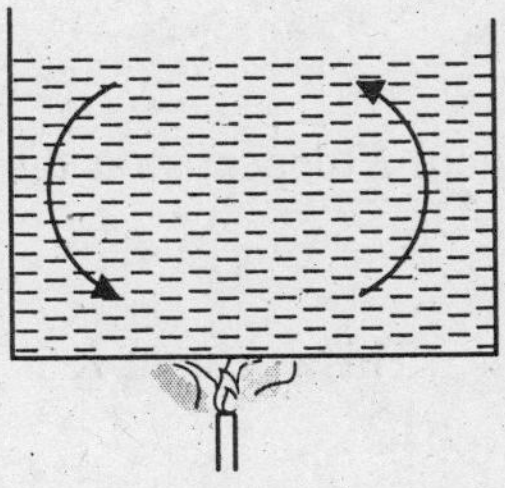

Fig. 4.2

An illustration of convection process can be had from a very simple experiment. If some water is put into a beaker and few grams of potassium permanganate are added into water. On heating, we find that the heat is conducted through glass to water in the lower region of the beaker. When the lower region becomes warm it expands and so becomes less dense. Colder water at the surface of water contained in the beaker, being less dense, flows under the warm water and displaces it upward. The colder layers which go to the bottom of the beaker become warm and are in turn displaced upward by colder water. So sort of convection current is maintained and the actual motion of the grains of potassium permanganate can be seen. From the above experiment it is clear that heat is transferred by the bodily movement of the heated particles of the medium and the convection may be defined as the process of transmission of heat from one part of the medium to the other by the actual movement of the heated particles.

4.4. RADIATION

In this mode of propagation of heat, no material medium is required for the transmission of heat. If we are near an open fire or we hold a hand near an electric bulb, we feel some hotness. Also the heat from the sun reaches us by this process. So radiation may be defined as the process by which heat is transmitted from one body to another without heating the intervening medium.

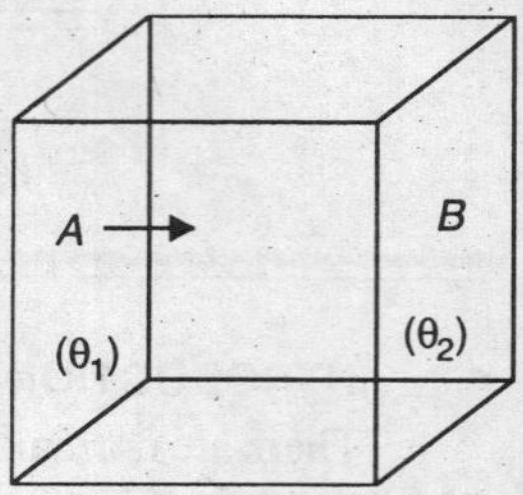

Fig. 4.3

4.5. THERMAL CONDUCTIVITY

Different materials conduct heat differently and so it is essential to know about the conducting power of a material.

Let us consider a cubic section of a material whose face A is at a higher temperature θ_1 (say) and let the opposite parallel face B be at a temperature θ_2. Now heat will flow from face A to face B depending on the following factors.

(*i*) It will be directly proportional to the area of cross-section, since more is the area of the face of a cube, more is the flow of heat *i.e.*,

$$Q \propto A \quad \text{...(1)}$$

(*ii*) It will be directly proportional to the time of flow of heat, *i.e.*,

$$Q \propto t \quad \text{...(2)}$$

(*iii*) It will be directly proportional to the difference of temperature of the two faces of the cube. The more is the difference of temperature, the more rapid is the flow of heat, *i.e.*,

$$Q \propto (\theta_1 - \theta_2) \quad \text{...(3)}$$

(*iv*) It will be inversely proportional to the thickness of the cube, *i.e.*, more is the distance between the two faces of the cube, less is the flow of heat *i.e.*,

$$Q \propto \frac{1}{d} \quad \text{...(4)}$$

Combining the above equations, we have

$$Q \propto \frac{A(\theta_1 - \theta_2)\,t}{d}$$

Introducing the proportionality constant K, we have

$$Q = \frac{KA(\theta_1 - \theta_2)\,t}{d} \quad \text{...(5)}$$

where K depends on the material in which the heat is flowing. From Equation (5)

$$K = \frac{Q.d}{A(\theta_1 - \theta_2)\,t}$$

This constant K is known as coefficient of thermal conductivity.

If $\quad A = 1 \text{ cm}^2,\ \theta_1 - \theta_2 = 1^\circ\text{C},\ t = 1 \text{ sec},\ d = 1 \text{ cm}$

then $\quad Q = K$

So K can be defined as the quantity of heat that flows for one second through a cm cube whose opposite faces are maintained at a difference of temperature of 1°C.

4.6. UNITS OF *K*

From Equation (6) we can derive the units of K.

$$\text{Unit of } K = \frac{\text{Unit of length} \times \text{unit of heat}}{\text{unit of area} \times \text{unit of temperature} \times \text{unit of time}}$$

$$= \frac{\text{Unit of heat}}{\text{Unit of length} \times \text{unit of temperature} \times \text{unit of time}} \quad \text{...(7)}$$

= cal/cm/°C/sec. (C.G.S. System)

In M.K.S. or S.I. unit of thermal conductivity is K cal/m/K/sec.

Also heat is a form of energy. So it can be expressed in terms of units of power., *i.e.*, K in S.I. units is watt/m/K.

4.7. DIMENSIONS OF *K*

We can write the dimensions of K from Equation (7).

$$K = \frac{M^1 L^3 T^{-2}}{L^2 . \theta . T} = M^1 L^1 T^{-3} \theta^{-1}.$$

4.8. THERMAL DIFFUSIVITY

An equally important term used for the flow of heat is thermal diffusivity or thermometric conductivity. It is given by the ratio of the thermal conductivity to the thermal capacity per unit volume of the material of the rod. If s is the specific heat and ρ the density of the material of the rod, then diffusivity which is denoted by h is given by

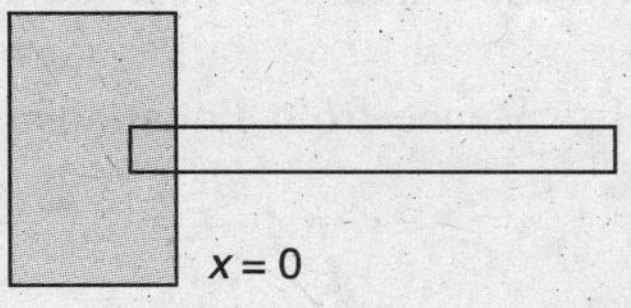

Fig. 4.4

$$h = \frac{K}{\rho s}.$$

4.9. RECTILINEAR FLOW OF HEAT THROUGH A ROD

Let us consider a long metal bar heated at one end and let us consider the flow of heat along X-axis.

Let us consider two parallel planes $\perp$ to the X-axis and at a distance x and $x + \delta x$ from the hot end. Let θ be the excess of temperature at the plane x and $\frac{d\theta}{dx}$ be the temperature gradient. (Rate of change of temperature with distance). Then the excess of temperature at the plane $x + \delta x$ will be $\left(\theta + \frac{d\theta}{dx} \cdot \delta x\right)$ and the temperature gradient will be $\frac{d}{dx}\left(\theta + \frac{d\theta}{dx} \cdot \delta x\right)$.

If Q_1 and Q_2 be the quantities of heat that enter the plane at x and leave the plane at $(x + \delta x)$/sec. respectively.

We have
$$Q_1 = -KA \cdot \frac{d\theta}{dx} \qquad ...(1)$$

where K is coefficient of thermal conductivity and A the area of cross-section of the bar.

$$\therefore \quad Q_2 = -KA \frac{d}{dx}\left(\theta + \frac{d\theta}{dx} \cdot \delta x\right)$$

$$= -KA \left[\frac{d\theta}{dx} + \frac{d^2\theta}{dx^2} \cdot \delta x\right] \qquad ...(2)$$

So the net amount of heat gained per second by the element of thickness δx, Sandwitched between the two planes will be

$$Q_1 - Q_2 = -KA \frac{d\theta}{dx} - \left[-KA\left(\frac{d\theta}{dx} + \frac{d^2\theta}{dx^2}\delta x\right)\right]$$

$$= KA \frac{d^2\theta}{dx^2} \cdot \delta x \qquad ...(3)$$

Before we get the steady state, this heat will be partly used in raising the temperature of the element of the rod and partly it will be radiated.

If $\frac{d\theta}{dt}$ is the change of temperature per unit time, ρ the density of the material, *s* the specific heat of the material, then the quantity of heat required to raise the temperature of the element per unit time will be

$$= A \cdot \delta x \cdot s \cdot \frac{d\theta}{dt} \qquad ...(4)$$

If *p* is the perimeter of the element, the surface area of the element = $p \,.\, \delta x$. Let *E* be the emissivity or the radiative power of the surface and θ the average excess temperature of the element, then the amount of heat radiated by the element/sec

$$= E \,.\, p \,.\, \delta x . \theta \qquad ...(5)$$

Hence from Equations (3), (4) and (5) we have

$$KA \frac{d^2\theta}{dx^2} \cdot \delta x = A \cdot \delta x \cdot \rho \cdot s \cdot \frac{d\theta}{dt} + E \cdot p \cdot \delta x \cdot \theta$$

or

$$\frac{d^2\theta}{dx^2} = \frac{\rho s}{K} \cdot \frac{d\theta}{dt} + \frac{E_p}{KA} \cdot \theta \qquad ...(6)$$

This is the standard equation for the flow of heat in one direction, and this equation can be solved by taking into consideration the actual conditions of the problem.

(*i*) Let the amount of heat lost by radiation be negligibly small *i.e.*, Equation (6) may be written as

$$\frac{d^2\theta}{dx^2} = \frac{\rho s}{K} \cdot \frac{d\theta}{dt}$$

or

$$\frac{d\theta}{dt} = \frac{K}{\rho s} \cdot \frac{d^2\theta}{dx^2}$$

$$= h \cdot \frac{d^2\theta}{dx^2} \qquad ...(7)$$

where *h* is diffusivity and it determines the rate at which temperature changes.

(*ii*) At the steady state, no amount of heat is taken up by any part of the rod *i.e.*, $\frac{d\theta}{dt} = 0$. So relation (6) may be written as

$$\frac{d^2\theta}{dx^2} = \frac{E_p}{KA} \cdot \theta$$

$$= \mu^2 . \theta \qquad ...(8)$$

where $\mu^2 = Ep/KA$.

Now Equation (8) is a differential equation of second order, so its general solution may be written as

$$\theta = Ae^{\mu x} + Be^{-\mu x}. \qquad ...(9)$$

where *A* and *B* are constants and their values can be found from the initial and final conditions. Since above relations have the exponential form, so in order to have a viable solution we may taken a very long bar so that the other end of the bar may approximately be taken at the room temperature. Then

$$x = 0, \quad \theta = \theta_0$$

and when $x = \infty, \quad \theta = 0$

Putting these in Equation (9)

$$\theta_0 = A + B$$

and $0 = Ae^{\infty}$

Now $e^{\infty} \neq 0$

So $A = 0$

∴ $B = \theta$

and $\theta = \theta_0 \, e^{-\mu x}$

Equation (10) gives a general relation for the excess of temperature at any point at a distance x from the hot end after the steady state is reached. Equation (10) is an exponential function of the distance x but independent of time.

(*iii*) If the bar is covered, so that amount of heat lost due to radiation is negligible, $E = 0$ and hence $\mu^2 = 0$.

∴ Relation (6) becomes

$$\frac{d^2\theta}{dx^2} = 0$$

Integrating twice $\quad q = A'x + B' \qquad ...(11)$

where A' and B' are constants and their values can be found from the initial and final conditions.

When $\quad x = 0,\ \theta = \theta_0 = \theta_0 = B'$

When $\quad x = l,\ \theta = \theta_1$

∴ Equation (11) may be written as

$$\theta_1 = A'\,l + B'$$

or

$$A' = \frac{\theta_0 - \theta_1}{l}$$

Putting the values of A' and B' we get

$$\theta = \theta_0 - \frac{\theta_0 - \theta_1}{l} \cdot x \qquad ...(12)$$

Equation (12) gives the excess of temperature at any point x along the bar, the bar being covered so that amount of heat lost by radiation is zero.

4.10. THREE DIMENSIONAL FLOW OF HEAT

The above relation may be extended to three dimensions. The general case in which we consider that the heat flows in one direction may be treated in a similar way and then flow of heat may be found in there directions. The general equation of conduction may be written as

$$K\left(\frac{d^2\theta}{dx^2} + \frac{d^2\theta}{dy^2} + \frac{d^2\theta}{dz^2}\right) = \rho \cdot \frac{d\theta}{dt}$$

$$\frac{K}{\rho S}\nabla^2\theta = \frac{d\theta}{dt}$$

or

$$h\nabla^2\theta = \frac{d\theta}{dt} \qquad ...(13)$$

where h is diffusivity and ∇^2 the Laplacian operator. After the steady state

$$\frac{d\theta}{dt} = 0,$$

so we have

$$\nabla^2\theta = 0 \qquad ...(14)$$

The simplest case of conduction in three dimensions is the flow of heat in a sphere where the heat is supplied from the centre.

4.11. FLOW OF HEAT THROUGH A COMPOUND MEDIA

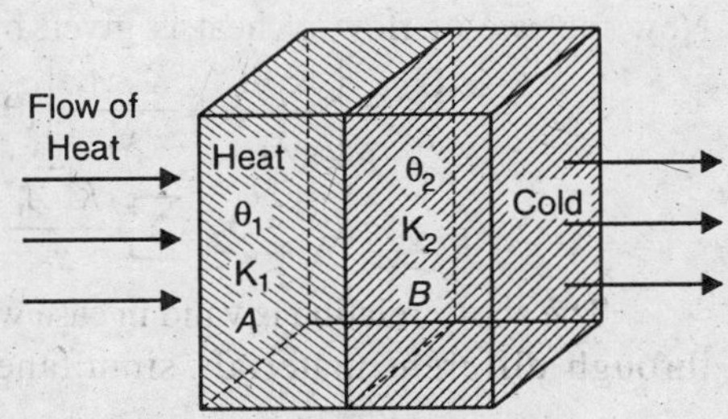

Fig. 4.5

(*i*) Bodies in Series. Let us consider a compound media *e.g.* a slab or a wall made up of two different materials A and B having thickness x_1 and x_2. (Shown in Fig. 4.5). Let the temperatures of two end faces be θ_1 and θ_2 and let the temperature of the common surface be θ. Let K_1 and K_2 be the thermal conductivities of A and B respectively. If we consider $\theta_1 > \theta_2$, so that heat will flow from A to

B and when the steady state is reached,

$$\theta = \frac{K_1 A (\theta_1 - \theta)}{x_1} \quad ...(1)$$

and for B medium,
$$\theta = \frac{K_2 A (\theta - \theta_2)}{x_2} \quad ...(2)$$

Now at steady state, we have

$$\frac{K_1 A (\theta_1 - \theta)}{x_1} = \frac{K_2 A (\theta - \theta_2)}{x_2}$$

or
$$\frac{K_1}{x_1}\theta_1 - \frac{K_1}{x_1}\theta = \frac{K_2}{x_2}\theta - \frac{K_2}{x_2}\theta_2$$

or
$$\left(\frac{K_1}{x_1} + \frac{K_2}{x_2}\right)\theta = \left(\frac{K_1}{x_1}\theta_1 + \frac{K_2}{x_2}\theta_2\right)$$

or
$$\theta = \frac{\frac{K_1 \theta_1}{x_1} + \frac{K_2 \theta_2}{x_2}}{\left(\frac{K_1}{x_1} + \frac{K_2}{x_2}\right)}$$

or
$$\theta = \frac{\sum \frac{K_1}{x_1}\theta_1}{\sum \frac{K_1}{x_1}} \quad ...(3)$$

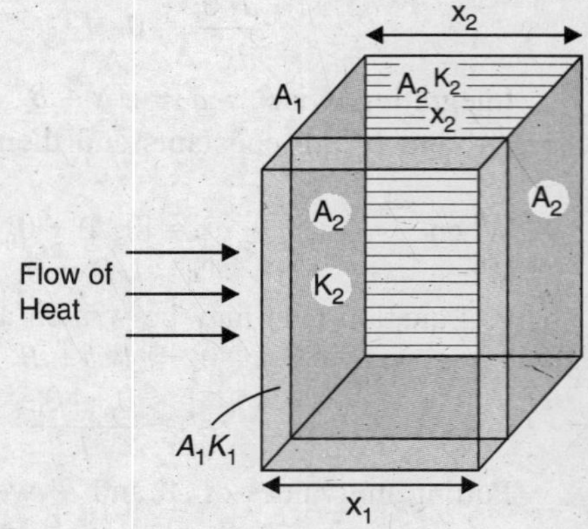

Fig. 4.6

Now putting the value of θ in Equations (1) or (2) we can get the quantity of heat flowing through the compound media.

***(ii)* Bodies in Parallel.** Let a compound wall or slab be made up of two different materials which are arranged parallel to each other (Fig. 4.6).

Again let K_1 and K_2 be the coefficient of thermal conductivities and x_1, x_2 their respective thickness of the two materials. Let the two faces have temperatures θ_1 and θ_2 and let A_1 and A_2 be their respective area of cross-section. If Q_1 be the quantity of heat flowing through the first material having area of cross-section A_1 and Q_2, the quantity of heat flowing through the second material having area of cross-section A_2, then we have

$$Q_1 = \frac{K_1 A_1 (\theta_1 - \theta_2)\, t}{x_1} \quad ...(1)$$

and
$$Q_2 = \frac{K_2 A_2 (\theta_1 - \theta_2)\, t}{x_2} \quad ...(2)$$

The total amount of heat flowing will be the sum of the two heats,

i.e.,
$$Q = Q_1 + Q_2$$
$$= \frac{K_1 A_1 (\theta_1 - \theta_2)\, t}{x_1} + \frac{K_2 A_2 (\theta_1 - \theta_2)\, t}{x_2} \quad ...(3)$$

Now the rate of flow of heat is given by

$$\frac{Q}{t} = \left[\frac{K_1 A_1}{x_1} + \frac{K_2 A_2}{x_2}\right][\theta_1 - \theta_2]$$
$$= (\theta_1 - \theta_2) \sum \frac{K_1 A_1}{x_1} \quad ...(4)$$

The above relation is valid in case where heat is conducted through different materials simultaneously under similar

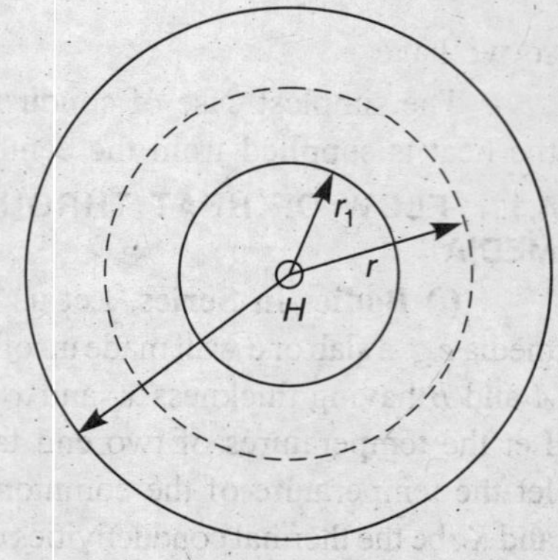

Fig. 4.7

conditions of temperatures. For example, in case the heat is flowing through a room, then the inner temperature of the walls, doors and glass panes is same and also the outer temperatures are also the same.

4.12. RADIAL FLOW OF HEAT THROUGH A SPHERICAL SHELL

Consider that a source of heat H is placed in the centre of a spherical shell having radius of the inner sphere as r_1 and that of the outer as r_2. Let us take a surface of radius r, then the flow of heat through this surface/sec.

$$= -K.A.\frac{d\theta}{dr} = -K.4\pi r^2 \cdot \frac{d\theta}{dr}$$

If Q be the quantity of heat supplied at the centre, then

$$Q = -K.4\pi r^2 \cdot \frac{d\theta}{dr}$$

or $$\frac{Q}{4\pi K} \cdot \frac{1}{r^2} dr = -d\theta$$

Integrating both sides with proper limits

$$\frac{Q}{4\pi K} \cdot \int_{r_1}^{r_2} \frac{1}{r^2} \cdot dr = \int_{\theta_1}^{\theta_2} -d\theta$$

$$-\frac{Q}{\pi K} \cdot \left(\frac{1}{r_2} - \frac{1}{r_1}\right) = -[\theta_2 - \theta_1]$$

or $$\frac{Q}{4\pi K} \cdot \frac{r_2 - r_1}{r_2 \cdot r_1} = -[\theta_1 - \theta_2]$$

or $$Q = \frac{4\pi K \cdot (\theta_1 - \theta_2) \cdot r_1 r_2}{(r_2 - r_1)} \quad ...(2)$$

From this relation we can find K if Q, θ_1, θ_2, r_1, r_2 are known.

4.13. FLOW OF HEAT THROUGH A CYLINDERICAL TUBE

Let us consider a cylinderical tube of internal radius r_1 and external radius r_2 and length l. The tube being heated electrically from the centre with a heating coil H.

We assume that the heat is flowing radially through the tube and at the steady state θ_1 and θ_2 are the temperatures of the inner and outer surfaces of the tube respectively.

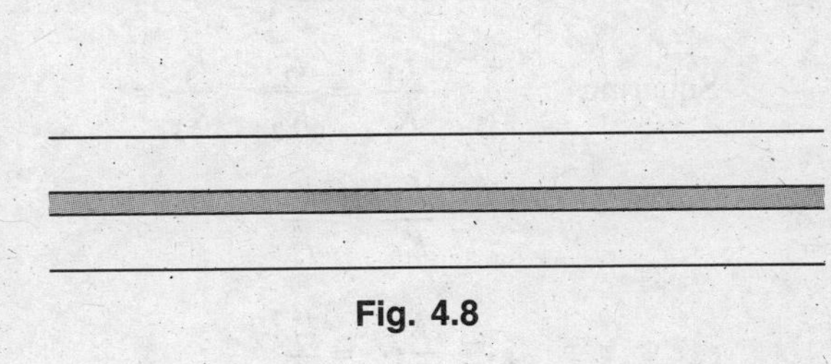

Fig. 4.8

Let there be a thin cylindered ring of thickness dr at a distance r_1 from the centre and let the temperature at that point be θ. The amount of heat Q flowing through this shell/sec will be

$$Q = -KA\frac{d\theta}{dr} = -K.2\pi r.l \cdot \frac{d\theta}{dr} \quad ...(1)$$

or $$\frac{Q}{2\pi Kl} \cdot \frac{1}{r} dr = -d\theta$$

Integrating both sides with proper limits

$$\frac{Q}{2\pi Kl}\cdot\int_{r_1}^{r_2}\frac{1}{r}\,dr = \int_{\theta_1}^{\theta_2} d\theta$$

or $$\frac{Q}{2\pi Kl}\log_e\frac{r_2}{r_1} = (\theta_1-\theta_2)$$

or $$K = \frac{Q\cdot\log_e r_2/r_1}{2\pi l\cdot(\theta_1-\theta_2)} \quad ...(2)$$

Hence we can find K if Q, l, θ_1, θ_2, r_1 and r_2 are known.

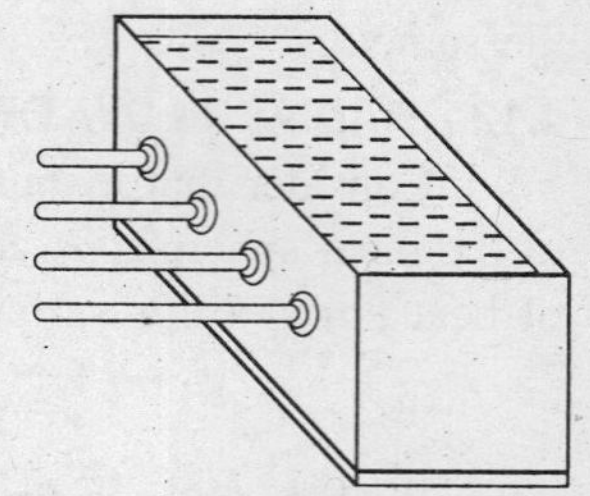

Fig. 4.9

4.14. DETERMINATION OF THERMAL CONDUCTIVITY

Metals are generally good conductors of heat, while non-metals like glass, rubber, wood etc. are bad conductors. Liquids in general are less good conductors of heat than solids, gases are the worst conductors. We will study some of the methods for finding thermal conductivity in the case of good and bad conductors.

4.15. DETERMINATION OF THERMAL CONDUCTIVITY IN THE CASE OF GOOD CONDUCTORS

(*i*) Ingen Hausz's experiment. This is one of the earliest experiments to demonstrate the flow of heat by conduction. Rods of different metals but exactly of the same size and shapes and also polished similarly are fitted into the holes on the sides of a vessel. The portions of the rods which are outside the vessel are coated with wax and the vessel is filled with hot water or some oil.

When steady state is reached, the wax will melt upto different lengths in the various rods. If l_1, l_2, l_3, ... are lengths of the rods upto which the wax has melted and θ_0, the temperature of the hot bath. If θ is the melting point of wax, then

$$\theta = \theta_0\, e^{-\mu_1 l_1} = \theta_0\, e^{-\mu_2 l_2} = \theta_0\, l^{-\mu_3 l_3} = ...$$

Hence, $$\mu_1 l_1 = \mu_2 l_2 = \mu_3 l_3$$

Putting the values of μ_1, μ_2, μ_3 etc.

$$\left(\frac{Ep}{K_1 A}\right)^{\frac{1}{2}} l_1 = \left(\frac{Ep}{K_2 A}\right)^{\frac{1}{2}} l_2 = \left(\frac{Ep}{K_3 A}\right)^{\frac{1}{2}} l_3 = ...$$

where K_1, K_2, K_3 etc. are the conductivities of the different lengths and E the emissivity, p the perimeter and A the cross-sectional area of the rods which gives

$$\frac{l_1}{K_1^{1/2}} = \frac{l_2}{K_2^{1/2}} = \frac{l_3}{K_3^{1/2}}$$

Squaring $$\frac{l_1^2}{K_1} = \frac{l_2^2}{K_2} = \frac{l_3^2}{K_3} = ...$$

or $$\frac{K_1}{K_2} = \frac{l_1^2}{l_2^2}$$

$$\frac{K_2}{K_3} = \frac{l_2^2}{l_3^2}$$

which means that if the coefficient of thermal conductivity of one rod is known, that of all others can be found. Alternately this method gives a relative values of the thermal conductivities of the different metals. However, this method is only an indirect method and does not give every accurate results.

(*ii*) Searle's Method. This method is ordinarily used to measure the coefficient of thermal conductivity in laboratory. A long rod of the metal under test (About 30 cm long and 3 cm in diameter) is used. One end A of the rod is enclosed in a steam chamber S while the other end B is enclosed in a copper coil through which water is constantly circulated. Two thermometers T_1 and T_2 are put in the cavities of the rod and two other thermometers T_3 and T_4 measure the temperatures of the incoming and outgoing water. The whole arrangement is properly insulated by some insulating material and placed in a wooden box as shown in Fig. 4.10.

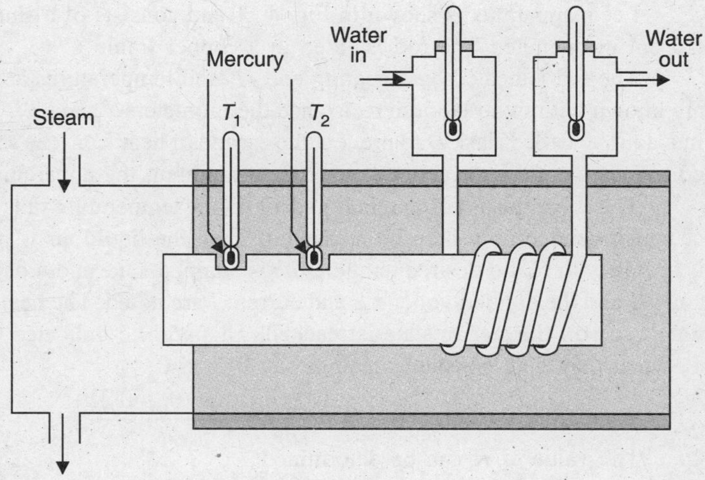

Fig. 4.10

The steam is passed into the steam chamber and when thermo-meters T_1 and T_2 give constant reading and also the flow of water is adjusted so that T_3 and T_4 give constant reading. The temperature of the incoming and outgoing water is noted and also water is collected for some known time.

Let the mass of water collected $= M$ gm

Time taken for water collection $= t$ sec

Specific heat of water $= s$

Let the temperature of the thermometer $T_3 = \theta_3$

and the temperature of the thermometer $T_4 = \theta_4$

Also temperature of the thermometer $T_1 = \theta_1$

and temperature of the thermometer $T_2 = \theta_2$

Let the area of cross-section of the rod $= A$

Distance between two thermometers T_1 and $T_2 = d$ cm

$\therefore$ Heat gained by water $= M.s.\ (\theta_1 - \theta_2)$ ...(1)

Also by definition the amount of heat passing through distance d

$$= \frac{K \cdot A\,(\theta_1' - \theta_2) \times t}{d}$$

At the steady state $M.s.\ (\theta_3 - \theta_2) = \dfrac{KA\,(\theta_1 - \theta_2)\,t}{d}$

or
$$K = \frac{M.s.\ (\theta_3 - \theta_4)\,d}{A\,(\theta_1 - \theta_2)\,t}$$

Knowing the various parameters, the value of K can be found.

(*iii*) Lee's Method. In case we are interested to find the coefficient of thermal conductivity at various temperatures, then the above method is not suitable since it gives an average value of coefficient of thermal conductivity. Lee's method is quite suitable for finding K at very low temperature of the order of liquid air temperatures and also K can be found at various

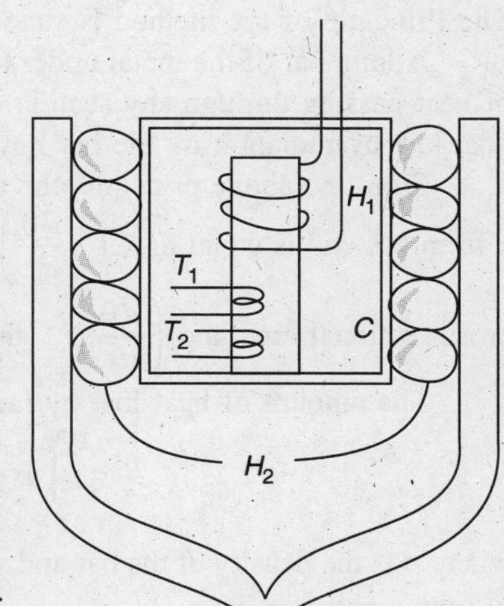

Fig. 4.11

temperatures.

Lee's apparatus is shown in Fig. 4.11 and consists of a small rod of about 6 cm length and 0.5 cm in diameter. The rod is fitted in a copper frame *C*.

The rod is heated by a heating coil H_1 and temperatures at two points which are *d* cm apart are known with two Platinum resistance thermometers T_1 and T_2. The whole arrangement is placed in a double walled flask *D* to prevent the escape of heat from the surface of the rod. Another heating coil H_2 is wound over copper frame *C* to maintain the apparatus at any desired temperature.

The apparatus is first put in liquid air till the temperature falls somewhat below the temperature at which conductivity is to be measured. Then the liquid air is poured out and the copper frame is heated with H_2 till desired temperature is attained. One end *A* of the rod is heated with the heating coil H_1 and the applied voltage *E* and current *I* are noted. The heating is continued for some known time *t* sec till the steady state is reached. This will be indicated by the two thermometers T_1 and T_2 when they give constant readings say θ_1 and θ_2.

Then from the relation $Q = \frac{EIt}{J} = \frac{KA(\theta_1 - \theta_2)}{d} \cdot t$

The value of *K* can be determined

i.e.,
$$K = \frac{EId}{JA(\theta_1 - \theta_2)} \qquad ...(1)$$

It should be noted that the value of *K* can be found at different temperatures.

Corrections. When the end *A* of the rod is heated by coil H_1, the copper frame is also heated and so there is a backward flow of heat from end *B* to *A*. So the reading $(\theta_1 - \theta_2)$ is not the correct reading. In order to remove this defect, the difference in temperature of the two thermometers $d\theta_1$ and $d\theta_2$ are noted for the same currents and voltages in coil H_1 and H_2 one after the other after the steady state is reached. The average of these difference of temperatures give the necessary correction which is added into $(\theta_1 - \theta_2)$ to give the correct value of difference of temperatures.

So the relation (1) may be written as

$$K = \frac{EId}{JA \cdot \left[(\theta_1 - \theta_2) + \frac{d\theta_1 + d\theta_2}{2}\right]} \qquad ...(2)$$

Lee was able to perform experiment over a wide range of temperatures (from 20°C to –170°C) in the case of different materials and thus some valuable information regarding the variation of thermal conductivity with temperature was provided.

***(iv)* Forbe's Method.** This is the absolute method for finding the value of *K* of a good conductor. The Principle of the method is stated as under:

A long bar of the metal under test is heated till the steady state is reached, when the amount of heat passing through any section *B* of the bar at a distance *x* from the hot end is equal to the heat lost by radiation by the bar beyond the section *B* chosen.

From the above principle, the amount of heat passing/sec through a section *B* at a distance *x* from hot end is equal to $KA\left(\frac{d\theta}{dx}\right)_B$, where *K* is the coefficient of thermal conductivity, *A* the cross-sectional area and $\left(\frac{d\theta}{dx}\right)_B$, the temperature gradient at *B*.

The amount of heat lost by radiation by the bar beyond the section

$$B = \int_B^{\text{end}} \rho \cdot A \cdot s \cdot dx \cdot \frac{d\theta}{dt}$$

where ρ is the density of the bar and *s* is the specific heat of the bars, and $\frac{d\theta}{dt}$ is the rate of change of temperature.

$$\therefore \quad KA\left(\frac{d\theta}{dx}\right)_B = \int_B^{\text{end}} \rho As \frac{d\theta}{dt}\cdot dx$$

or

$$K = \frac{\rho s \int_B^{\text{end}} \frac{d\theta}{dt}\cdot dx}{\left(\frac{d\theta}{dx}\right)_B}$$

So in order to find K we have to find $\int_B^{\text{end}} \frac{d\theta}{dt}\cdot dx$ and $\left(\frac{d\theta}{dx}\right)_B$. The experiment is done in two parts:

(*i*) Static experiment to find $\left(\frac{d\theta}{dx}\right)_B$ (*ii*) Dynamic experiment to find $\left(\frac{d\theta}{dt}\right)$ and hence to find $\int_B^{\text{end}} \frac{d\theta}{dt}\cdot dx.$

Experimental Arrangement. The apparatus consists of a long metallic bar of constant area of cross-section. The rod is sufficiently long so that one end is heated the other end is practically at the room temperature.

Forbe used a bar of iron about 8 ft. long and 1 ¼ inch sq. section. One end of the bar is placed at a very high temperature bath of molten lead. However in a laboratory a steam chamber is used to heat the bar. The bar has number of holes in which various thermometers are fitted with a little mercury in each hole to have good thermal contact. A non-conducting screens protects the thermo-meters from the direct heat of the bath.

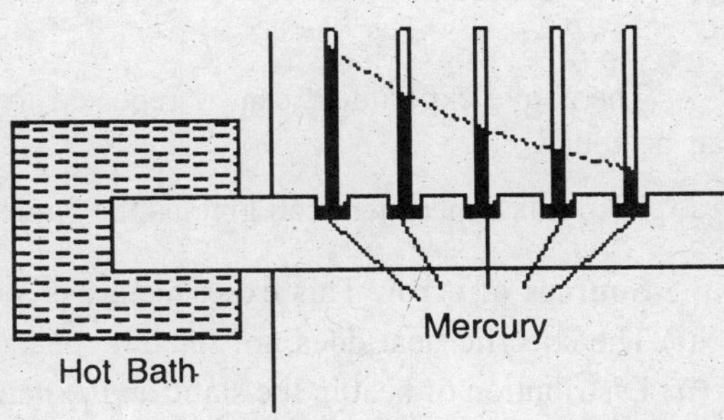

Fig. 4.12

(*i*) Static experiment. After sufficient heating, a steady state is reached in the various thermometers which will indicate constant temperatures. Then a curve is plotted between the temperatures of the various thermometers and their distances from the hot end.

A tangent is drawn on the curve corresponding to the point B chosen and the value of $\left(\frac{d\theta}{dx}\right)_B$ is got which may be written as

$$\left(\frac{d\theta}{dx}\right)_B = \frac{AB}{BC} \quad \text{...(2)}$$

(*ii*) Dynamic experiment. A short piece of the bar with same area of cross-section as the original one is heated directly till it has the temperature approximately the temperature of the hot end in the static experiment. This bar is then suspended freely

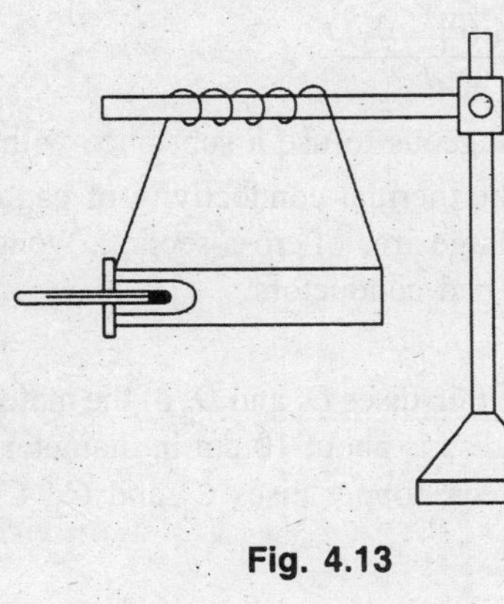
Fig. 4.13

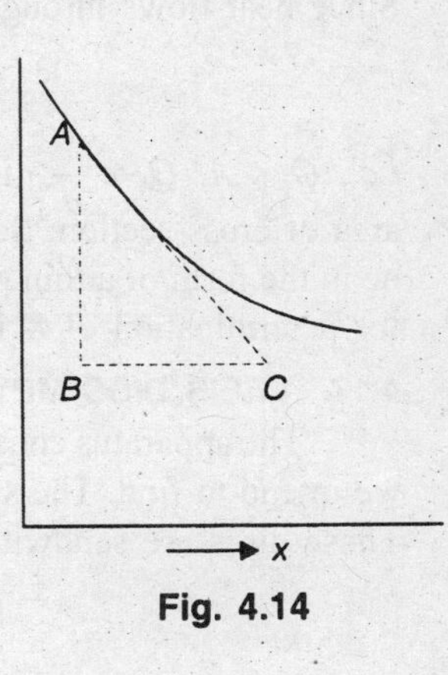

Fig. 4.14

in air and allowed to cool. At the same time a stop water is started and temperatures at different intervals of time are noted till we get temperature somewhat below the temperature of the section B of the bar chosen. This is dynamic experiment since we are studying the variation of time as well different values of $\frac{d\theta}{dt}$ are noted from the curve at different values of x beyond the section B chosen.

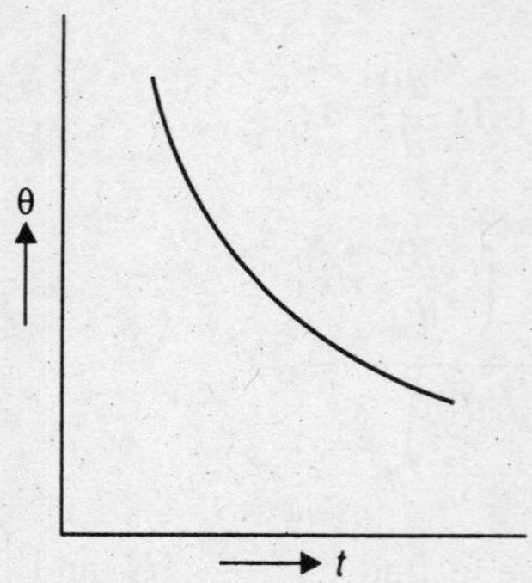

Fig. 4.15

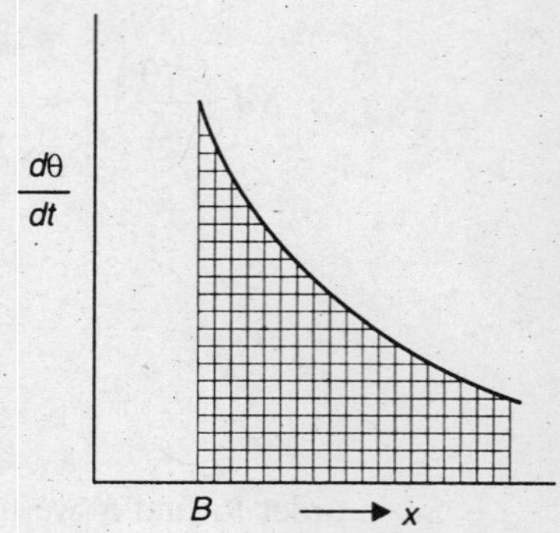

Fig. 4.16

With the above data, a third curve is plotted with $\frac{d\theta}{dt}$ against distance from the hot end beyond the section B.

The area of the shaded portion of the curve, where B is the point chosen in the static experiment, will correspond to $\int_B^{\text{end}} \frac{d\theta}{dt} \cdot dx$, since $\frac{d\theta}{dt} \cdot dx$ corre-sponds to the area of small square on the graph.

So K can be got from $$K = \frac{\rho s \times \text{area of shaded curve}}{\frac{AB}{BC}}$$

The above experiment can be repeated for different values of B and so average value of K can be found.

Also this experiment can give us the value of diffusivity $\left(h = \frac{K}{\rho s}\right)$ at various temperatures.

Sources of error. This experiment has some sources of error which are given as under:

(*i*) The specific heat does not remain constant at various temperatures.

(*ii*) Distribution of heat in the static and dynamic experiments is not the same, since in the static experiment, the bar is heated from one end and in the dynamic experiment it is heated directly.

However, inspite of the above errors this is a very good method for finding the absolute value of K without any theoretical assumptions.

4.16. THERMAL CONDUCTIVITY OF BAD CONDUCTORS

All the non-metallic bodies like cork, wood, glass, rubber etc. are bad conductors of heat. Since heat flows through bad conductors with great difficulty and from the relation

$$Q = \frac{KA(\theta_1 - \theta_2)\,t}{d}$$

i.e., $Q \propto A$, $Q \propto \frac{1}{d}$, it is advantageous to use a substance with very small thickness and large area of cross-section. So to find the thermal conductivity of bad conductors, the material should be in the form of a thin plate with large area of cross-section. We will discuss some of the methods for determination of K in case of bad conductors.

4.17. LEE'S DISC METHOD

The apparatus consists of two thin discs D_1 and D_2 of the material whose coefficient of thermal we intend to find. The size of the disc is about 10 cm in diameter and 0.2 to 0.3 cm in thickness. These discs are sandwitched between copper discs C_1 and C_2, C_3 and C_4.

In order to have good thermal contact of all the surfaces, some glycerine is put on the faces of the discs. The heat is supplied by a heating coil placed in the centre and the temperatures at various faces are noted.

Let a current I amperes at a p.d. of E volts be passed through the heating coil. When the steady state is reached, let the various temperatures be θ_1, θ_2, θ_3 and θ_4. If we neglect the radiation losses then the amount of heat supplied by the heating coil should be equal to the amount of heat passing through the two discs D_1 and D_2.

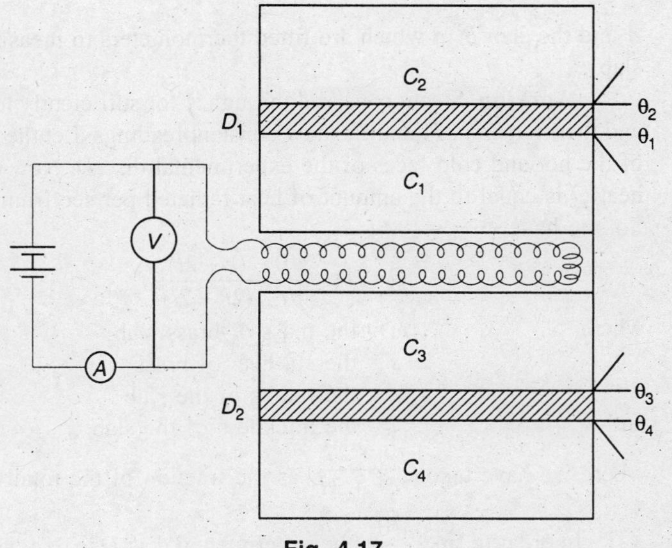

Fig. 4.17

Using the general relation

$$Q = \frac{KA(\theta_1 - \theta_2)}{d}$$

Let the thickness of disc D_1 be x_1 and that of D_2 be x_2, then heat conducted/sec through

$$D_1 = \frac{KA(\theta_1 - \theta_2)}{x_1}$$

and heat conducted/sec through

$$D_2 = \frac{KA(\theta_3 - \theta_4)}{x_2}$$

The heat supplied by the heating coil/sec $= \dfrac{EI}{J}$

So we have $\dfrac{EI}{J} = KA\left[\dfrac{\theta_1 - \theta_2}{x_1} + \dfrac{\theta_3 - \theta_4}{x_2}\right]$

From the above relation, the value of K is determined.

4.18. LEE'S AND CHARLTON'S DISC METHOD

Another laboratory method for finding the coeff. of thermal conductivity in laboratory is by Lee and Charlton's disc method.

Construction. In this method, a hollow metal box A with a flat and horizontal upper surface serves as a steam chamber.

A thin disc D of the material whose coeff. of thermal conductivity is to be found (say ebonite, glass, wood etc.) is placed on A and over it is placed a polished cylinderical brass slab B. There are holes in the steam chamber

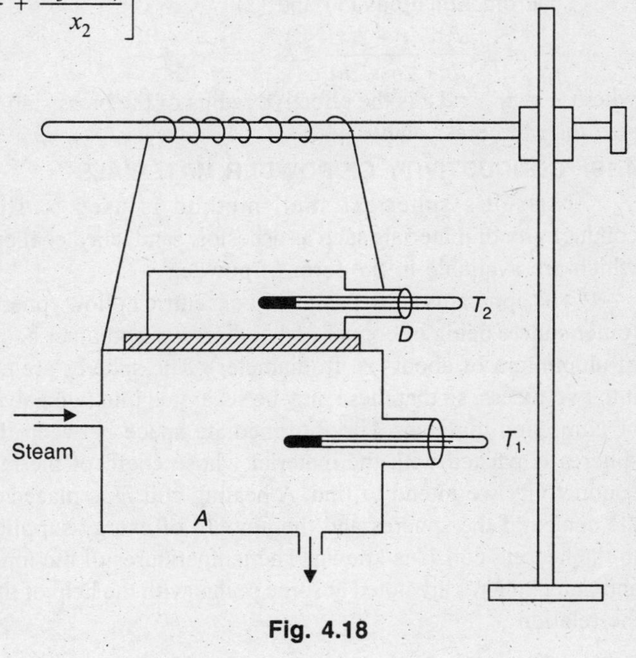

Fig. 4.18

A and the slab B in which are fitted thermometers to measure the tempera-tures of steam and the slab B.

Working. Steam is passed through A for sufficiently long time till the steady state is reached and thermometers T_1 and T_2 show constant readings. Let the temperatures be θ_1 and θ_2 respectively of the hot and cold faces of the experimental disc D. Now at the steady state, the rate of flow of heat Q is equal to the amount of heat radiated per sec from the top and bad surface of he slab B. So, we have

$$Q = ms.\frac{d\theta}{dt} \cdot \frac{(r+2l)}{(2r+2l)} \qquad ...(1)$$

where m = the mass of brass slab

s = the sp. heat of brass

r = the radius of the slab

and l = the thickness of the slab

where we have taken $\frac{r+2l}{2r+2l}$ as the fraction of the total radiation from the slab.

In order to find $\frac{d\theta}{dt}$, the experimental disc D is removed and the brass slab is heated directly with the steam chamber till its temperature is above 10°C above the temperature of the slab at steady state. Then the brass slab is suspended freely and its temperature is noted at regular intervals of time. With this a cooling curve is plotted for the brass slab. A tangent is dropped on the curve at the steady state, temperature θ_2 and $\frac{d\theta}{dt}$ at θ_2 is noted.

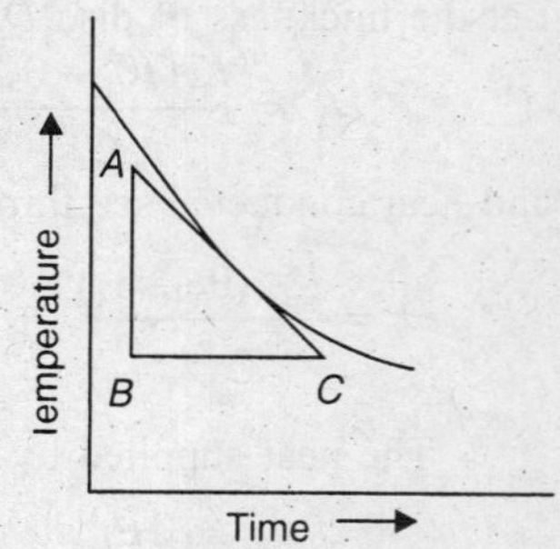

Fig. 4.19

From the definition, we have

$$Q = \frac{KA(\theta_1 - \theta_2)}{d} \qquad ...(2)$$

where d is the thickness of the experimental disc and A the effective area of the disc.

∴ From Equations (1) and (2)

$$m.s.\frac{d\theta}{dt} \cdot \frac{r+2l}{2r+2l} = K \cdot \pi r^2 \frac{\theta_1 - \theta_2}{d}$$

where $A = \pi r^2$ and r is the effective radius of the brass slab. (Assuming disc to be of slightly larger size than the brass slab)

4.19. CONDUCTIVITY OF POWDER MATERIALS

Nusselt's spherical shell method is used to find conductivity of materials such as asbestos, sand, clay, coal etc. which are available in the form of powders.

The apparatus consists of two concentric hollow spheres (outer sphere being of copper of 1 ft diameter and inner being of aluminium of about ½ ft. diameter). The spheres are cut into two pieces, so that these may be separated into two halves or combined together: The intermediate space between the spheres is packed with the material whose coeff. of thermal conductivity we intend to find. A heating coil H is placed in the centre of the spheres and the amount of energy supplied to the heating coil H is known. The temperatures of the inner and outer spheres are noted at some points with the help of thermoelectric thermometers. Then from the relation

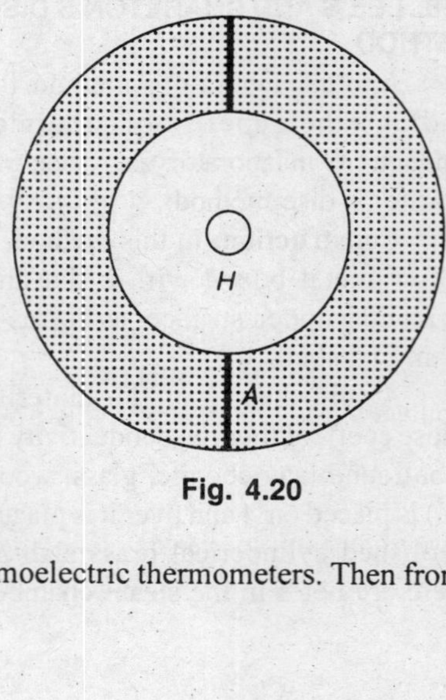

Fig. 4.20

$$Q = \frac{EI}{J} = \frac{4\pi K(\theta_1 - \theta_2) r_1 r_2}{r_2 - r_1}$$

Coefficient of thermal conductivity, K, can be found

where E = voltage given to the heating coil

I = the current supplied to heating coil

r_2 and r_1 = the radii of the two spheres

$\theta_1 - \theta_2$ = the temperatures of inner and outer spheres.

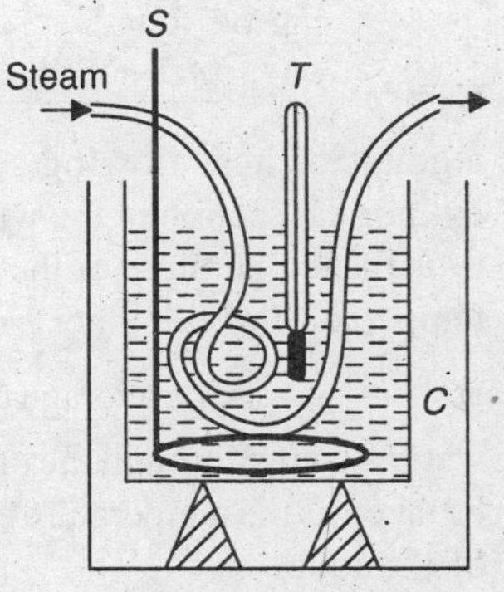

Fig. 4.21

4.20. CONDUCTIVITY OF MATERIALS SUPPLIED IN THE FORM OF A CYLINDER (CONDUCTIVITY OF RUBBER)

In this experiment a known length of a rubber tube is immersed in water placed in a double walled calorimeter. The mass of water is known. Steam is passed till the temperature of the water is raised by about 10°C. Let the time for which steam is passed be t sec and the initial and final temperatures of water be θ_1 and θ_2. Let r_1 and r_2 be the internal and external radii of the tube with the above data, we have the relation

$$K = \frac{Q \cdot \log_e \frac{r_2}{r_1}}{2\pi l \left(\theta_3 - \frac{\theta_1 + \theta_2}{2}\right)}$$

where Q is the quantity of heat flowing per unit time across any cylinderical section of the wall of the tube, which may be given as

$$Q = \frac{(m + \omega)(\theta_2 - \theta_1)}{t}$$

where m = mass of water

ω = water equivalent of the calorimeter.

θ_1 and θ_2 are the initial and final temperatures of the water, θ_3 is the temperature of the steam, *i.e.*, the temperature of the hot side of the tube and $\frac{\theta_1 + \theta_2}{2}$ gives the average temperature of the cold side of the tube.

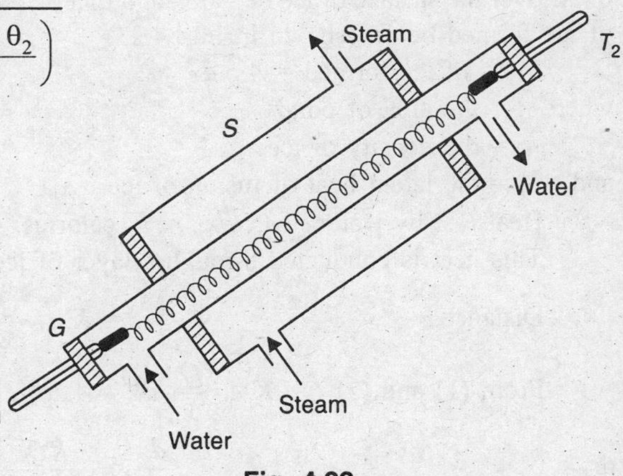

Fig. 4.22

4.21. CONDUCTIVITY OF GLASS

A glass tube G is enclosed in a steam jacket S. Two thermometers T_1 and T_2 are put in the glass tube to measure the temperatures of the incoming water and outgoing water. In order that there may be a steady stream of flow of water, a spiral is put into the glass tube and water is made to flow from lower side to the upper side so that it is in contact with all the sides of the glass tube.

Steam is passed through S and flow of water is adjusted through G so that there is a good difference of temperature at T_1 and T_2. At the steady state let the temperatures be θ_1 and θ_2. The outgoing water is collected for some known time t and let m be the mass of water collected in time t. The length l of the glass tube exposed to steam is measured and the internal and external radii of the tube r_1 and r_2 are measured accurately. Then using the relation

$$K = \frac{Q \cdot \log_e \frac{r_2}{r_1}}{2\pi l \left(\theta_3 - \frac{\theta_1 + \theta_2}{2}\right)}$$

K can be found,

where $Q = \frac{(m + \omega)(\theta_2 - \theta_1)}{t}$ is the amount of heat flowing/sec across any cylindrical section in the wall of the tube, θ_3 is the temperature of the steam *i.e.*, the temperature of the inner side of the tube and $\frac{\theta_1 + \theta_2}{2}$ is the average temperature of water flowing through the tube which may be taken as the temperature of the cold side of the tube.

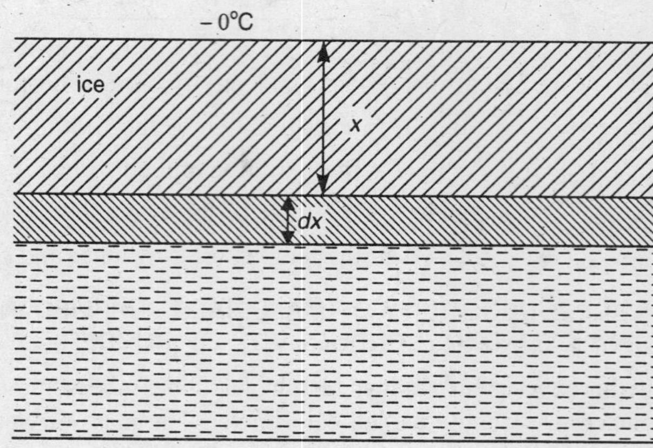

Water at 0°C

Fig. 4.23

4.22. FORMATION OF ICE ON PONDS

Let us consider a layer of ice x cm thick on the surface of a pond and let the temperature of air over the surface of ice be – 0°C and that of water below ice be 0°C. Let in time dt a thickness of ice formed be dt. (shown in Fig. 4.23)

Mass of ice formed = $A.\rho.dx$

where A = the area of pond

ρ = the density of ice

and L = the latent heat of fusion of ice.

Heat lost by water = $A \,.\, dx \,.\, r \,.\, L.$ calories ...(1)

This heat is conducted through a layer of ice of thickness x in the upward direction.

Distance $x = K.A.\ \frac{Q}{x} \cdot dt$ calories ...(2)

From (1) and (2) $K.A.\ \frac{Q}{x} \cdot dt = A \,.\, dx \,.\, \rho \,.\, L$

or $$\frac{dx}{dt} = \frac{KQ}{\rho.L.x} \quad ...(3)$$

where $\frac{dx}{dt}$ corresponds to the rate of growth of the thickness of ice.

$$\therefore \quad dt = \left(\frac{\rho L}{KQ}\right) \cdot x.dx$$

So total time taken by the layer of ice to increase in thickness by an amount x is given by

$$\int dt = \frac{\rho L}{KQ} \int x.dx$$

or $$t = \frac{\rho L}{KQ} \frac{x^2}{2} + C \text{ (constant)}$$

when $t = 0;\ x = 0;\quad \therefore\ c = 0$

So we have $$t = \frac{\rho L}{2KQ} \cdot x^2 \quad ...(4)$$

which gives us a relation between time and thickness of ice formed.

Example 1. *A copper rod 19 cm long and of 0.785 cm^2 area of cross-section which is thermally insulated is heated at one end through 100°C while the other end is kept at 30°C. Calculate the* *amount of heat that will flow in 10 minutes along the way. K of copper is 380 W/m/K.*

Solution: Area of cross-section of rod $A = 0.785\ cm^2 = 785 \times 10^{-7}\ m^2$

Distance between two ends $d = 19\ cm = 0.19\ m$

Temperature difference $= \theta_1 - \theta_2 = 100 - 30 = 70°C$

Time $t = 10 \times 60 = 600$ sec.

$K = 380$ W/m/°K

Now using relation $$Q = \frac{KA(\theta_1 - \theta_2)t}{d}$$

$$= \frac{380 \times 785 \times 10^{-7} \times 70 \times 600}{0.19}$$

$= 6.6 \times 10^3$ Joules

$=$ **6.6 kJ.**

Example 2. *The total area of glass window pane is 0.5 m². Calculate how much heat is conducted/hr through the pane if thickness of glass is 0.6 cm, the inside temperature is 23 °C and outside temperature is 2°C. K for glass is 1.0 W/m/K.*

Solution: Thickness of glass pane

$d = 0.6\ cm = 6 \times 10^{-3}\ m$

$\theta_1 - \theta_2 = 23 - 2 = 21°C$

Time for flow of heat $t = 1$ hr $= 300$ sec.

Total area of glass $A = 0.5\ m^2$

Thermal conductivity $K = 1.0$ W/m/K

∴ From relation $$Q = \frac{KA(\theta_1 - \theta_2)t}{d}$$

$$= \frac{1 \times 0.5 \times 21 \times 3600}{6 \times 10^{-3}}$$

$=$ **63 × 10⁵ joules.**

Example 3. *Calculate the rate of increase of thickness of ice layer on a lake when thickness of ice is 20 cm and air temperature is – 40°C. K for ice = 0.004 cal/cm/sec/°C density of ice = 0.92 gm/cc and latent heat of ice = 80 cal/gm. After what time the thickness will be doubled?*

Solution: Temperature of atmosphere

$q^2 = -40°C$

Temperature of water $\theta_1 = 0°C$

K of ice $= 0.004$ cal/cm/sec/ °C

Thickness of ice $d = 20$ cm

Let A be the area of the surface of ice, then the quantity of heat given in time dt

$$Q = \frac{KA(\theta_1 - \theta_2)\,dt}{x} \qquad ...(1)$$

Since the outer temperature is below the freezing point of water so heat is slowly conducted away and water freezes gradually. Let the increase in thickness of ice be dx in a time dt.

Also density of ice $\rho = 0.92$ gm/cc

and $L = 80$ cal

∴ Mass of ice formed $= A \cdot dx \cdot \rho$

and Quantity of heat lost $= A \cdot dx\, \rho \cdot L \qquad ...(2)$

∴ From Equations (1) and (2)

$$\frac{KA(\theta_1 - \theta_2)}{x} \cdot dt = A \cdot dx \cdot \rho \cdot L$$

or Rate of increase of ice layer $$\frac{dx}{dt} = \frac{KA(\theta_1 - \theta_2)}{x \cdot \rho \cdot L} \qquad ...(3)$$

$$= \frac{0.004 \times 40}{20 \times 0.92 \times 80} = \frac{1}{9200} \text{ cm/sec}$$

Now we are to find the time when ice layer is doubled *i.e.,* it increases from 20 cm to 40 cm. We have from Equation (3)

$$\frac{K(\theta_1 - \theta_2)\,dt}{\rho L} = x\,.\,dx$$

Integrating both sides, we have

$$\frac{K(\theta_1 - \theta_2)}{\rho L}\int dt = \int_{20}^{40} x \cdot dx = \left[\frac{x^2}{2}\right]_{20}^{40}$$

$$\therefore \quad \frac{K \cdot (\theta_2 - \theta_1)}{\rho L}\, t = \frac{40^2 - 20^2}{2}$$

$$\therefore \quad t = \frac{(800 - 200) \times 0.92 \times 80}{0.004 \times 40 \times 60 \times 60}$$

$$= \textbf{76 hrs 40 minutes.}$$

Example 4. *A glass vessel with an insulating cover and with a surface area of 100 cm² and 0.15 cm thick is filled with ice at 0 °C and placed in a second vessel maintained at a temperature of 100 °C. Find the mass of ice that melts/minute when steady state is reached. Latent heat of ice = 3.3 × 10⁵ J/kg m and K for glass = 1.0 W/m/K.*

Solution: Thickness through which heat flows

$$d = 0.15 \text{ cm} = 1.5 \times 10^{-3} \text{ m}$$

Area through which heat flows $A = 100 \text{ cm}^2 = 10^{-2} \text{ m}^2$

Difference of temperature $\theta_1 - \theta_3 = 100 - 0 = 100°\text{C}$

Time of flow of heat $= 1 \text{ min} = 600$ sec.

Thermal conductivity of glass $K = 1.0$ W/m/K

Then from the relation $Q = \dfrac{KA(\theta_1 - \theta_2)\,t}{d}$

We have $Q = \dfrac{1.0 \times 10^{-2} \times 100 \times 60}{1.5 \times 10^{-3}}$

$$= \frac{60}{15} \times 10^5 = 4 \times 10^5 \text{ joules}$$

If M kgm of ice melts, the heat required to melt ice *i.e.,* $M \times L$ must be equal to Q.

$$\therefore \quad 4 \times 10^5 \text{ joules} = M \times 3.30 \times 10^3 \text{ joules/kgm}$$

or

$$M = \frac{4}{3.3} = 1.12 \text{ kgm}$$

$$= \textbf{1190 gm.}$$

Example 5. *A slab of compressed cork 5 cms thick and 2 sq m area beas a heating coil at one face. A current of 1.18 amps. at 20 volts passes through the coil. The two faces of the coil attain steady temperatures of 12.5°C and 0°C. Assuming that the whole of heat developed in the coil is conducted through the slab. Calculate the thermal conductivity of the cork.*

Solution: Thickness of the slab $d = 5$ cms

Area of slab $= 2 \text{ cm}^2 = 2 \times 100 \times 100 \times\ 2 \times 10^4 \text{ cm}^2$

$$\theta_1 - \theta_2 = (12.5 - 0) = 12.5°\text{C}.$$

Heat produced by the coil/sec

$$Q = \frac{EI}{J}$$

Also

$$Q = \frac{KA(\theta_1 - \theta_2)}{d}$$

$$\therefore \qquad \frac{EI}{J} = \frac{KA(\theta_1 - \theta_2)}{d}$$

$$\therefore \qquad K = \frac{EI}{J} \cdot \frac{d}{A(\theta_1 - \theta_2)}$$

$$= \frac{20 \times 11.8}{4.2} \times \frac{5}{2 \times 10^4 \times 12.5}$$

$$= \mathbf{1.114 \times 10^{-4}\ cal/sec/cm/^\circ C.}$$

Example 6. *A metal pipe having an external diameter 15 cm carries steam at 200°C. This is covered by a layer of insulating material of 2.5 cm thickness for which K = 0.21 W/m/K. If the outer surface is at 100 °C, calculate the heat lost per metre/min. Neglect the temperature drop in metal. Also calculate the amount of steam condensed/min. Take latent heat of steam = 2.4 × 10⁶ J/kgm.*

Solution: We have the relation

$$Q = \frac{2\pi l K\left(\theta_3 - \frac{\theta_1 + \theta_2}{2}\right)}{2.303 \log_{10}\left(\frac{r_2}{r_1}\right)} \times 60 \text{ J/min}$$

Here $l = 1$ m, $r_1 = 7.5 \times 10^{-2}$ m

$r_2 = r_1$ + thickness of layer $= (7.5 + 2.5) \times 10^{-2}$

$= 10 \times 10^{-2}$ m

$\theta_3 = 200°C, \quad \frac{\theta_1 + \theta_2}{2} = 100°C$

$K = 0.21$ W/m/K

$$Q = \frac{2 \times 3.142 \times 1 \times 0.21 \times (200 - 100)}{2.303 \log_{10}\left(\frac{10}{7.5}\right)}$$

$$= \frac{3437.6}{0.1249} = 27522.82 \text{ J}$$

Amount of steam condensed $m = Q/L$

$$= \frac{27522.82}{24 \times 10^6} = 0.01146 \text{ kgm}$$

$$= \mathbf{11.46\ gm.}$$

Example 7. *Water is pumped through an iron tube 2 m long at the rate of 100 litres/sec. The inner and outer diameter of the tube are 5.0 and 6.0 cm respectively. The outside of the tube is surrounded by hot gases at a temperature of 500°C. If the temperature of the water as it enters is 30°C. Calculate the rise in temperature as it leaves the tube. K for iron = 65 W/m/K.*

Solution: We have the relation for flow of heat through a tube

$$Q = \frac{2\pi K l\left(\theta_3 - \frac{\theta_1 + \theta_2}{2}\right)}{2.303 \log_{10}\left(\frac{r_2}{r_1}\right)}$$

Here $l = 2$ m, $r_1 = 2.5 \times 10^{-2}$ m

$r_2 = 3 \times 10^{-2}$ m, $K = 65$ W/m/K.

Temperature of water entering the tube $\theta_1 = 30°C$

Let the temperature of the water leaving the tube = θ_2

Average temperature of water $= \dfrac{30 + \theta_2}{2}$

Temperature of hot gases $= \theta_3 = 500^\circ C$

$$\therefore \quad Q = \frac{2 + 3.142 \times 65 \times 2 \times \left(500 - \dfrac{30+\theta_2}{2}\right)}{2.303 \log_{10}\left(\dfrac{3 \times 10^{-2}}{2.5 \times 10^{-2}}\right)}$$

$$= \frac{816.8 \times (1000 - 30 - \theta_2)}{4.606 \times \log_{10} \dfrac{3}{2.5}}$$

$$= \frac{816.8 \times (970 - \theta_2)}{4.606 \times 0.0792}$$

$$= 22391 \times (970 - \theta_2) \qquad ..(1)$$

Heat taken by water $= m.S. (\theta_2 - \theta_1)$ [$S = 4200$ J/kgm/K]

$= 100 \times 4200 \times (\theta_2 - 30)$ ...(2)

From Equations (1) and (2)

$$42 \times 10^4 (\theta_2 - 30) = 2239.1 \times (970 - \theta_2)$$

or $187.575\theta_2 - 5627.26 = 970 - \theta_2$

or $\theta_2 = 35.36$

$\therefore$ Rise in temperature $= \theta_2 - \theta_1 = (35.36 - 30)$

$= \mathbf{5.36^\circ C}$.

Example 8. *A steady stream of water flowing at the rate of 500 gm/min through a glass tube 30 cm long and 1 cm in external diameter and 0.8 cm in bore, outside of which is surrounded by steam at a pressure of 76 cm of mercury is raised in temperature from 20°C to 30°C as it passes through the tube. Calculate the conductivity of the glass.*

Solution: Mass of water flowing/sec $= \dfrac{500}{60} = 8.33$ gm

Initial temperature of water $\theta_1 = 20^\circ C$

Final temperature of water $= 30^\circ C$

Average temperature of water $= \dfrac{20 + 30}{2} = 25^\circ C$

Temperature of steam $\theta_3 = 100^\circ C$

Length of the tube $l = 30$ cm

External radius of the tube $r_2 = \dfrac{1.00}{2} = 0.5$ cm

Internal radius of the tube $r_1 = \dfrac{0.8}{2} = 0.4$ cm

$$Q = 8.33 \times (30 - 20) = 83.3 \text{ cal}$$

Now
$$K = \frac{Q \times 2.3026 \times \log_{10} \dfrac{r_2}{r_1}}{2\pi l \left(\theta_3 - \dfrac{\theta_1 + \theta_2}{2}\right)}$$

$$= \frac{83.3 \times 2.3026 \times \log\left(\dfrac{0.5}{0.4}\right)}{2\pi \times 30 \times (100 - 25)}$$

$= \mathbf{0.00132\ cal/cm/sec/^\circ C.}$

Example 9. *A window glass pane consists of two similar panes separated by 6.00 mm of air. Calculate the heat which would be lost by conduction per hour. The glass panes are of size 50 × 50 cm and 6.00 mm thick. The temperature of the inside surface of glass is 24°C and of the outer surface is 3°C. Thermal conductivity of glass = 1.00 W/m/K and of air 0.025 W/m/K.*

Solution: In this case heat will be conducted through two identical glass panes and from air gap of 6 mm width and 50 cm length. The total heat conducted will be the sum of the heat flowing through glass panes and air gap.

Now total area of glass panes $A_1 = 2 \times (50 \times 50) = 5000 \text{ cm}^2 = 0.5 \text{ m}^2$

Difference of temperature on two sides

$$\theta_1 - \theta_2 = 24 - 3 = 21°\text{C}$$

Thermal conductivity for glass $K_1 = 1.0$ W/m/K

Time of conduction of heat $t = 1 \text{ hr} = 3600$ sec

Total area of air gap $= 0.6 \times 50 \text{ cm}^2 = 30 \text{ cm}^2 = 3 \times 10^{-3} \text{ m}^2$

Thermal conductivity of air gap $K_2 = 0.025$ W/m/K.

Thickness of glass panes $x_1 = 6 \text{ mm} = 6 \times 10^{-3}$ m

Thickness of air gap $x_2 = 6 \text{ mm} = 6 \times 10^{-3}$ m

If Q be the heat conducted/hour.

Then
$$Q = (\theta_1 - \theta_2) \sum \frac{K_1 A}{x} \times 3600$$

$$= 21 \left[\frac{1 \times 0.5}{6 \times 10^{-3}} + \frac{0.025 \times 3 \times 10^{-3}}{6 \times 10^{-3}}\right] \times 3600$$

$$= 21\ (83.3 + 0.0125) \times 3600 = \mathbf{63 \times 10^6 \text{ joules}}$$

QUESTIONS AND PROBLEMS

1. Distinguish between thermal conductivity and thermometric conductivity and give a relation between the two. Give Searle's method for finding the thermal conductivity of a solid.
2. Describe Ingen Hausz's experiment and prove from mathematical theory that the ratio of thermal conductivities is equal to the ratio of the square of the length upto which wax has melted.
3. Show that in the steady state of a metal bar heated at one end is $\frac{d^2\theta}{dx^2} = \mu^2 \theta$, where symbols have the usual meaning.
4. Describe Forbe's method for finding thermal conductivity of metals. In what respects this method is superior to the other conventional methods.
5. Describe Lee's method for finding coefficient of thermal conductivity of solids.
6. Define thermal conductivity and give general principles for measuring coefficient of thermal conductivities of good and bad conductors.
7. Give Lee's method for finding thermal conductivity of a bad conductor.
8. Give a method for finding thermal conductivity of a material supplied in the powder form.
9. How the conductivity can be found for a material supplied in the form of a cylinder? Give methods for finding thermal conductivity in the case of glass and rubber.
10. In an experiment performed to find thermal conductivity of cardboard by Lee's method, the following observations are made.

    Steady state temperature of upper disc = 99.5°C
    Steady state temperature of lower disc = 83.5°C
    Time taken for upper disc to cool from 86°C to 81°C = 4 min
    Thickness of cardboard disc = 4.8 mm
    Thickness of upper disc = 1 cm

Material of the upper disc is Cu. Find the thermal conductivity of cardboard.

[Ans. 0.000566 cal/cm/°C/sec]

11. Calculate the heat passing/hr through the walls and windows of a room (5 × 25 metres). If the walls are of bricks of thickness 30 cms and have windows of glass 0.3 cm thick and total area 5 m^2. Temperature of the room is 30°C below that of the outside and the thermal conductivities of bricks and glass are 12×10^{-4} and 25×10^{-4} C.G.S. units respectively. **[Ans.** 4.91×10^7 cal]

12. In Searle's experiment, the readings of thermometers in the holes of bar at a distance 5.0 cm from each other are 80°C and 70°C. If a litre of water flows through the coil in 7 min, and the radius of the bar is 2 cms, calculate the conductivity of the bar. Inflow temperature = 30°C and outflow temperature = 40°C. **[Ans.** 0.947]

13. Find the time in which a layer of ice 3 cm thick on the surface of a tank will increase in thickness by 0.1 cm when the temperature of the air above is 20°C below zero. K of ice = 2.1 W/m/K, *L* for ice = 3.3×10^5 J/kgm and density of ice at 0°C = 0.91×10^3 kgm/m^3. **[Ans.** 218 sec]

14. If the air temperature remains constant at – 10°C find the time required for (*i*) first cm (*ii*) second cm. thickness of ice on a pond. Assume water to be at 0°C. Density of ice = 0.91×10^3 kgm/m^3. Latent heat of ice = 3.3×10^5 J/kgm and *K* of ice = 2.1 W/m/K. **[Ans.** 2145 sec]

15. How much time will it take for a layer of ice of thickness 10 cm to increase by 5 cm on the surface when the temperature of the surroundings is –10°C? **[Ans.** 9×10^4 S]

CHAPTER 5

RADIATION

5.1. INTRODUCTION

Radiation is the phenomenon of propagation of heat through vacuum or a material medium without the medium playing an active part. However, it does not mean that the intervening medium will not be heated when the heat radiations fall on it or pass through it. Some materials like glass, wood, metal *etc.* get heat to less or more extent while materials like quartz, rock salt *etc.* allow the radiations to pass through them without being appreciably heated. It must also be understood that the molecules of intervening material owing to the heat vibrations act as source of radiations which are in the form of waves that carry the thermal energy.

The statement that 'no intervening medium is required for the radiation phenomenon' simply means that unlike conduction and convection, in radiation phenomenon it is not necessary for the intervening medium to get itself heated. Also since heat is a form of energy, the mode of propagation is in the form of waves just as light waves travel even in empty space.

5.2. SOURCES OF THERMAL RADIATION

A body which can be maintained at a constant temperature can be a source of thermal radiation, however two special devices which are normally employed as radiation sources are given below:

(*a*) Leslie Cube. This was devised by Leslie and consists of a hollow metal cube whose four vertical faces are coated with different materials *e.g.,* one face is coated with lamp black, the second face is coated with white paint, the third is polished and the fourth is left in original form. The cube when filled with hot water becomes a source of radiation, giving radiations of different emissive powers from different faces. The cube can be rotated along a vertical axis, so that different radiating powers can be got depending upon the face of the cube giving out the radiations.

(*b*) Black Body. A perfectly black body is that which can absorb radiations of all the wavelengths which fall on it. It neither reflects nor refracts and so it appears black. Whatsoever form of radiations may fall on it. The property of a black body is that when hot, it gives out radiations of all wavelengths or colours because it has the characteristic of emitting full or total radiations at high temperature. A perfect black body being a good absorber, is also a perfect radiator *i.e.,* it gives out maximum emission of radiation for every wavelength that falls on it at any temperatures.

5.3. BLACK BODY IN PRACTICE

A surface coated with lamp black may be considered as a black body, however normally a closed chamber having a narrow hole at one point O, is a perfect absorber and is used as a perfect black body. Any radiation entering through O is lost inside the blackened sphere because of successive reflections inside. As no direct reflection may take place from the sphere, a projection just opposite to the opening O is made. At uniform high temperatures such a body gives out full radiations.

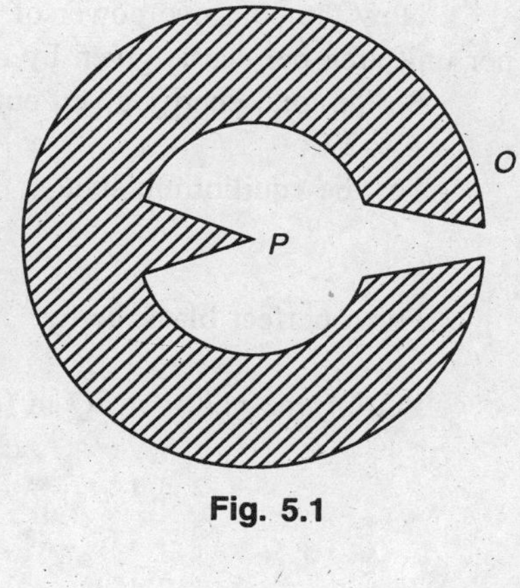

Fig. 5.1

A very important property of such a body is that if some substance is placed in such a body, it will also emit black body radiations, because any substance that absorbs any radiation from the black body will emit the same radiation (Kirchhoff's Law) and so that total radiations from O must be identical with those from the walls of the black body in Fig 5.1.

5.4. WIEN'S BLACK BODY

Wien constructed a more convenient cylinderical type of black body based on the Fery's principle Fig. 5.2.

It constitutes a hollow cylindrical metallic chamber C, heated electrically by a coil of thin platinum coil wound over it. The material of chamber may be either brass or platinum depending upon the temperature falling on it, the inner side of the chamber is blackened so that the equilibrium state is reached quickly. The radiations from the chamber can be restricted by a number of blackened diaphragms D and can be taken out from hole H. In order that outer radiations do not interfere with the radiations from the chamber, the cylinder is protected by a coaxial procelain tube P with an air space. The cylinder can be heated to desired temperature by means of the platinum heating coil. To measure accurately the temperature, a thermoelectric thermometer T is used. This type of black body, because of its accuracy of results is now commonly used.

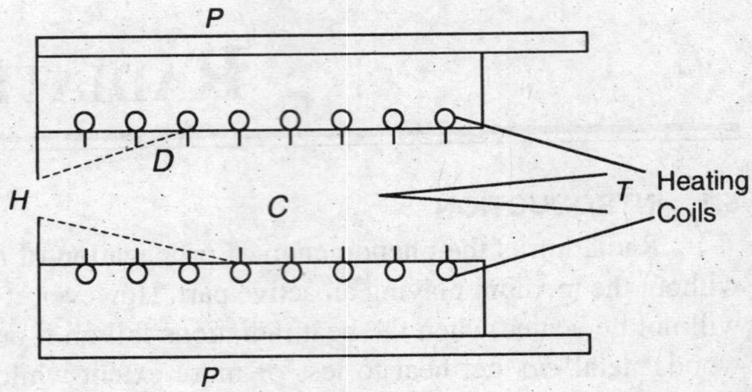

Fig. 5.2

5.5. EMISSIVE AND ABSORTIVE POWER

Some of the terms which are frequently used in the black body radiation phenomenon must be clearly understood. The following terms are normally used.

Emissive Power. The emissive power of a body at temperature corresponding to a wavelength λ is defined as the energy radiated by the body/sec per sq cm of the body for a unit range of a wavelength. Normally emissive power is denoted by e_λ. The energy radiated by a body/sec/cm^2 within a range of wavelength, λ and $\lambda + d\lambda$ is given by $e\lambda \,.\, d\lambda$. For a prefect black body, emissive power is maximum and is given by E_λ.

Absorptive Power. It is ratio of the amount of energy absorbed by a body to the amount of energy incident on it. It is normally denoted by a_λ.

5.6. KIRCHHOFF'S LAW

Let dQ be the amount of radiation within the incident wavelength λ and $\lambda + d\lambda$ by unit area of the surface. So the amount absorbed is given by $a_\lambda \,.\, dQ$. The remaining energy $(dQ - a_\lambda \,.\, dQ)$ is either reflected or transmitted.

Now the emissive power of the body giving radiations between wavelength λ and $\lambda + d\lambda$ per unit area per sec is given by $e_\lambda \,.\, d\lambda$.

So the total energy given out per sec. per unit area of the surface will be given as $(1 - a_\lambda)\, dQ + e_\lambda \,.\, d\lambda$.

For the equilibrium state, $$dA = (1 - a_\lambda)\, dQ + e_\lambda\, d\lambda$$
$$= dQ - a_\lambda \,.\, dQ + e_\lambda \,.\, d\lambda$$

$\therefore$ $$a_\lambda \,.\, dQ = e_\lambda \,.\, d\lambda$$

For a perfect black body, $$a_\lambda = 1,\ e_\lambda = E_\lambda$$
$$dQ = E_\lambda \,.\, d\lambda$$

Putting the value of dQ in (i)

$$a_\lambda \,.\, E_\lambda \,.\, d\lambda = e_\lambda \,.\, d\lambda$$

or $$E_\lambda = \frac{e_\lambda}{a_\lambda}$$

which shows that the ratio of the emissive power of a body to its absorptive power at any temperature and wavelength is equal to the emissive power of a perfectly black body and so is constant. This is known as Kirchhoff's law.

The emissive and absorptive powers of some of the sources are given as under:

Emissive power	
Lamp Black	98%
Platinum	11%
Copper	5%
Silver	2%
Indian Ink	88%
Absorptive Power	
Lamp Black	98%
Shellac	72%
Copper	13%
Indian Ink	85%

When an emitted radiation falls on matter, some of it is reflected, some absorbed and some is transmitted. The total radiation is given by sum of the reflected, absorbed and transmitted coefficients by the relation

$$r + a + t = 1$$

where r is the fraction of radiation reflected, a is the fraction of radiation absorbed and t is the fraction of radiation transmitted.

If we put $r = 0$, $t = 0$ in the above relation, we get $a = 1$ and this defined a perfectly black body which absorbs all the radiations and no part is reflected or transmitted. From above table lamp black is a nearest approach to a perfectly black body as it absorbs 98% of incident radiation and also when hot, emits the same amount of radiation.

5.7. STEFAN-BOLTZMANN'S LAW

Newton was the first scientist who studied systematically the relation between the amount of heat radiated from a body in a given enclosure and its temperature. The rate of fall in temperature for given body was found to be directly proportional to the difference of temperature between the body and the surrounding by number of experiments which were conducted by Newton, *i.e.*, when a graph is plotted between the fall in temperature $\left(-\frac{d\theta}{dt}\right)$ and the temperature difference $(\theta - \theta_0)$ between the body and the surroundings – a straight line is obtained

i.e., $$-\frac{d\theta}{dt} = K(\theta - \theta_0)$$

which is Newton's Law of cooling.

However, it was found to be valid for a small difference of temperature between the body and the surroundings. Furthermore, for higher difference of temperatures some other corrections due to convection phenomenon are also to be taken into account. With all the corrections and precautions, Newton's law could be found to give correct results upto 40°C difference of temperature between the body and the surroundings.

In order to get more general results, Dulong and Petit conducted number of experiments for different ranges of temperatures and got an other improved form of cooling law which may be expressed by the relation

$$V = K\{f(\theta) - f(\theta_0)\} \quad \text{...}(i)$$

where V is the rate of cooling of a body at temperature θ and θ_0 is the temperature of the surroundings.

Later on, it was further found that $f(\theta)$ may have an exponential form and in order to have a co-relation of theoretical and experimental results, the above relation is expressed as

$$V = k\,[a^{\theta} - a^{\theta_0}] \qquad ...(ii)$$

where a is a constant.

In 1879, Stefan was studying the experiments done by Tyndall for the radiations emitted by a platinum wire when it is white hot (1200°C) and those emitted when it is red hot (525°C). The ratio as verified by Tyndall came out to be 11.7.

Stefan tried to have a co-relation between temperatures and radiations and came to the result:

$$\frac{(1200+273)^4}{(525+273)^4} = 11.6 \qquad ...(iii)$$

i.e., the ratio of the radiations from white hot platinum to red hot platinum very nearly depended on the ratio of 4th power of the absolute temperatures. So he enunciated the formula

$$R = \sigma T^4 \qquad ...(iv)$$

which means that the total amount of heat radiated from a perfectly black body per sec per unit area is directly proportional to the fourth power of its temperature. Where σ is Stefan's constant.

On testing the formula by the results of Dulong and Petit, the formula fitted perfectly with the experimental results. So, Dulong and Petit formula may be written in the modified form as

$$V = K\,[T^4 - T_0^4] \qquad ...(v)$$

where T is the absolute temperature of the hot body and T_0 that of the enclosure.

This law had certain deviations because Dulong and Petit did not use the relations for a Black body. Lummer and Pringsheim using a black body enunciated that this law is strictly valid for a black body. Later on, Boltzmann also came to the same results theoretically from the thermodynamic treatment and so this law is also called Stefan-Boltzmann's Law.

This law is also sometimes called fourth power law of radiation and it deals with the amount of heat radiated by a black body by virtue of its temperature and is different from the cooling law in respect that in this case radiations take place and exchange of heat is not valid. Moreover, it is a law of total radiation which applies to the whole range of wavelength.

The above law can, however, be extended to the heat exchange law with the surroundings as:

If a black body at absolute temperature T is surrounded by another black body at absolute temperature T_0, the amount of heat lost by the former per sec per square cm is given by

$$R = \sigma (T^4 - T_0^4)$$

5.8. NEWTON'S LAW OF COOLING DEDUCED FROM STEFAN-BOLTZMANN'S LAW

Let the body at absolute temperature T be placed in another body at absolute temperature T_0. From Stefan-Boltzmann's law, the rate of loss of heat is given by

$$R = \sigma\,(T^4 - T_0^4)$$

$$= \sigma (T^2 + T_0^2)(T^2 - T_0^2)$$

$$= \sigma (T^2 + T_0^2)(T + T_0)(T - T_0)$$

If the temperature difference is very small

i.e., $T + T_0 \cong 2T_0$

$$T^2 + T_0^2 \cong 2T_0^2$$

Then, $R = \sigma \,.\, 4T_0^3\ (T - T_0)$

$K\,(T - T_0)$ where $K = 4T_0^3 \,.\, \sigma$ is constant.

which shows that rate of loss of heat from a body to enclosure is proportional to the temperature difference between the body and the enclosure–that is the Newton's Law of cooling. So, Stefan-Boltzmann's law can also lead to Newton's law of cooling for **small difference of temperature between the body and the surroundings.**

5.9. BOLTZMANN'S PROOF OF STEFAN'S LAW

According to quantum theory, energy carried by a photon of frequency ν is given by $h\nu$, where h is Planck's constant. Also from Relativity theory, mass and energy are interconvertible *i.e.*, $E = mc^2$.

So we may write $E = h\nu = mc^2$

or $$mc = \frac{h\nu}{c} \qquad ...(1)$$

which gives momentum of a photon of frequency ν.

If n number of such photons are incident normally on a unit area, the total momentum imparted/sec $= n\frac{h\nu}{c}$.

Now pressure is force per unit area, so pressure $= n\frac{h\nu}{c}$. ...(2)

The volume of a cylinder containing n photons and having length c and unit area of cross-section $= c \times 1 = c$.

$$\therefore \quad \frac{n}{c} = \frac{\text{No. of photons}}{\text{Unit volume}}$$

So $$\text{Pressure} = \frac{n}{c} \cdot h\nu$$

= Total energy of the photons per unit volume

= E (Energy density)

If we consider that the radiations are incident in all the directions, then the pressure p is equally distributed along three axes

$$\therefore \quad p = \frac{1}{3}E \qquad ...(3)$$

In order to find the work done, when a reflecting surface is moved against radiations, let us consider a cylinder of unit area of cross-section, having perfectly reflecting walls and piston P, the base B of the cylinder being perfectly black. Let the cylinder be filled with radiations at uniform tem-perature T and let a small quantity of radiation dQ be incident on base B from outside which raises the temperature by an amount dT which inturn raises the internal energy.

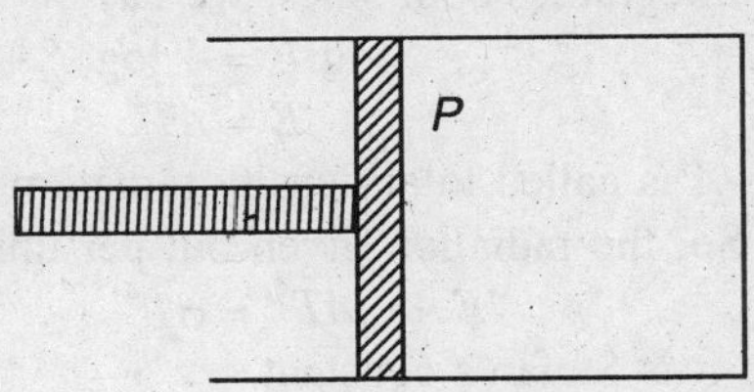

Fig. 5.3

If v be the volume and E the energy density of the system, then Internal Energy $= vE$.

Change in Internal Energy

$$= d\,(vE) = v dE + E\,dv \qquad ...(4)$$

If some work is done on the piston, so that the volume increases by a small value dv against a constant, pressure p of the radiations inside the cylinder, then from Equation (3) we have

$$\text{External work} = p.dv = \frac{1}{3}E.\,dv \qquad ...(5)$$

The energy dQ is used partly in increasing the internal energy and partly is doing the external work and the change is reversible.

$$\therefore \quad dQ = (v\,.\,dE + E\,.\,dv) + \frac{1}{3}E.dv \qquad ...(6)$$

$$= v\,.\,dE + \frac{4}{3}E.dv$$

If ds is the change in entropy of the system, then

$$dQ = T\,.\,ds \qquad ...(7)$$

$\therefore$ From Equations (6) and (7),

$$T \, . \, ds = v \, . \, dE + \frac{4}{3} \, E \, . \, dv$$

or $$ds = \frac{v}{T} dE + \frac{4}{3} \frac{E}{T} dv \qquad ...(8)$$

Also ds being a perfect differential, may be written as

$$ds = \frac{\delta s}{\delta E} \cdot dE + \frac{\delta s}{\delta v} dv \qquad ...(9)$$

From Equations (8) and (9),

$$\frac{v}{T} = \frac{\delta s}{\delta E} \text{ and } \frac{\delta s}{\delta v} = \frac{4}{3} \frac{E}{T}$$

Also, $$\frac{\delta}{\delta v}\left(\frac{\delta s}{\delta E}\right) = \frac{\delta}{\delta E}\left(\frac{\delta s}{\delta v}\right)$$

$\therefore$ $$\frac{\delta}{\delta v}\left(\frac{v}{T}\right) = \frac{\delta}{\delta E}\left(\frac{4}{3} \frac{E}{T}\right)$$

Since temperature T is independent of *i.e.*, and is a function of E alone

$\therefore$ $$\frac{1}{T} = \frac{4}{3}\left(\frac{1}{T} - \frac{1}{T^2} . E \cdot \frac{\delta T}{\delta E}\right)$$

or $$1 = \frac{4}{3} - \frac{4}{3} \cdot \frac{E}{T} \cdot \frac{\delta T}{\delta E}$$

or $$\frac{1}{3} = \frac{4}{3} \cdot \frac{E}{T} \cdot \frac{\delta T}{\delta E}$$

or $$\frac{\delta E}{E} = 4 \, \frac{\delta T}{T}$$

Integrating both sides, we have

$$\log E = 4 \log T + C$$

or $$E = AT^4 \qquad ...(10)$$

where A is called total density radiation constant.

So, the radiation given out per unit area per unit time is proportional to *i.e.*,

$$E \propto AT^4 = \sigma T^4 \qquad ...(11)$$

where σ is Stefan's constant.

So Stefan's law is proved from thermodynamics.

5.10. EXPERIMENTAL VERIFICATION OF STEFAN'S LAW

Lummer and Pringsheim experimentally demonstrated the validity of Stefan's law. The experiment was done over wide range of temperature (from 100°C to 1300°C) and Stefan's fourth power law was found to hold good.

Depending upon the range of temperature over which the experiment is to be verified, the black body and the hot bath is chosen. From 200°C to 600°C, a hollow copper sphere blackened from inside with platinum black was used as black body, while fused nitrates of sodium and potassium melting at 219°C

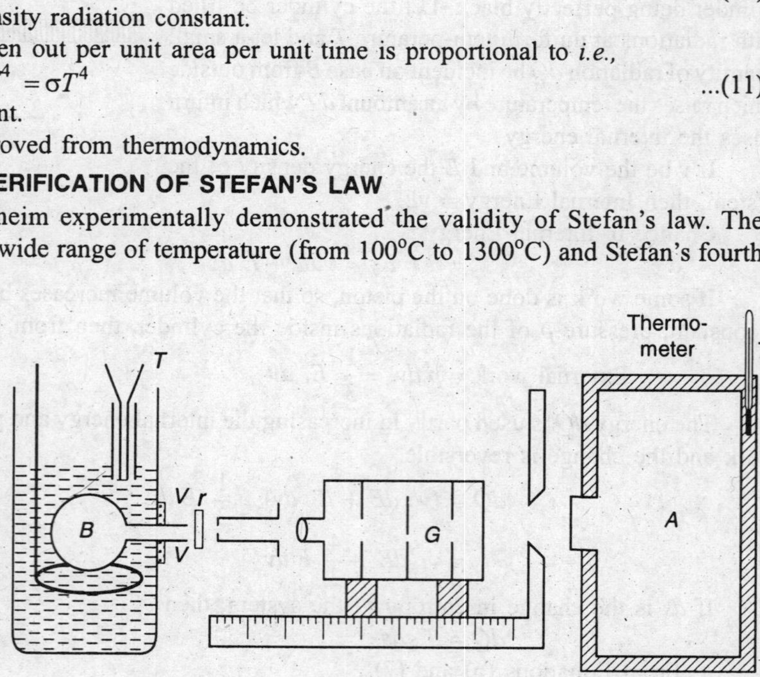

Fig. 5.4

were used as hot bath. In the higher ranges of temperature *e.g.*, between 900°C to 1300°C, an iron cylinder blackened from side with platinum black was used as black body and gas furnance as hot bath was used. The temperatures of the black body and the surroundings were kept at desired values. The temperature of the black body was measured by a thermoelectric thermo-meter. The opening of the black body was surrounded by a vessel *VV* through which a constant stream of water was kept at 17°C.

The radiations from the black body *B* were made to pass through water cooled shutter *r*, so that these may be cut off when required. These radiations were made to fall on bolometer *G* which can measure the intensity of radiations. The bolometer was standardized by means of Fery's black body *A*, which consisted of a double-walled vessel of copper blackened from inside, the inbetween space of the walls contained boiling water kept at 100°C. The bolometer was made to face *A* and deflections for different distances were noted. The deflection was found to be inversely proportional to the square of distance from the aperture of the Fery's black body which meant that the deflection of bolometer was proportional to incident radiation. This condition was checked during the further course of the experiment.

In the experiment, different deflections of the galvanometer of the bolometer were noted after heating the bath surrounding *B* at desired temperatures and allowing the radiations to fall on bolometer after removing the shutter *r*. With the black body at different steady temperatures, different readings of the galvanometer were noted. Then the data was reduced to some arbitrary unit depending on the radiations from the black body *A* at 100°C and kept at 63.3 cm. If θ is the deflection of galvanometer, *T* the absolute temperature of the black body *B*, and 17°C or 290 K, the temperature of the surroundings, then it was found that

$$\frac{\theta}{(T^4 - 290^4)} = a \text{ (constant)}$$

thereby, establishing the validity of the Stefan's Law.

The constant a is however not Stefan's constant σ, but some other constant which depends upon the calibration of the bolometer and whose value came out to be 123.8×10^{10}. When this value was reduced to Stefan's constant by experimental data, the value of σ was found to be 5.32×10^{-5} ergs/cm^2/sec/K^4.

Later on, the experiment was checked by Coblentz with greater accuracy by using an electrically heated black body, the temperature of which was measured by thermoelectric thermometer. The radiation were measured by a special type of bolometer called 'absolute bolometer' devised by Coblentz. The experiment was successfully performed for temperatures upto 1600°C and the value of Stefan's constant came out to be 5.72×10^{-5} ergs/cm^2/sec/K^4.

5.11. DETERMINATION OF STEFAN'S CONSTANT

The experimental arrangement is shown in Fig. 5.5. It consists of a hollow copper

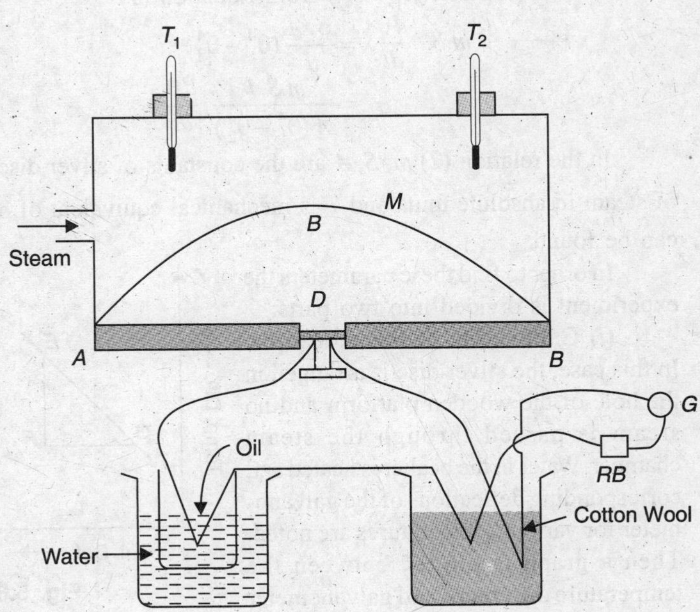

Fig. 5.5

hemisphere B having a diameter of about 25 cm with its inner surface uniformally blackened and fitted in a metal box M, which serves as a steam chamber. The arrangement is fitted on a wooden platform AB, with a small hole of 2.5 cm at the centre. B is heated to a steady temperature which can be measured accurately by two thermometers T_1 and T_2 inserted in M. The blackened face of B, thus constitutes a black body at constant temperature say θ_1 K (temperature of steam). A silver disc D is filled in the hole and is supported by an ebonite handle. The heat radiated by the black body is received by silver disc D blackened on the top surface and placed in the hole. The ebonite handle can be used to put the silver disc in the hole or taking it out when required. One junction of a thermocouple is fixed at the bottom of the disc while the other is placed in a test tube containing oil and placed in a beaker W filled with water. A galvanometer is introduced in the circuit and in order that different parts of the lead wires may not be at different temperatures, these are packed in cotton wool in vessel C. A resistance box R is introduced in the circuit to check the deflection of the galvanometer.

When the black body is at constant temperature θ_1, and silver disc is introduced in the hole, it will absorb radiations emitted by black body. Let θ_2 be the temperature of silver disc at any time. The disc will also give radiations and let R_1 be the amount of radiations absorbed by the silver disc/unit area/sec at temperature θ_1 K and R_2 be the amount of radiations emitted/unit area/sec at temperature θ_2 K, then from Stefan's Law

$$R_1 = \sigma\theta_1^4$$

$$R_2 = \sigma\theta_2^4$$

$$\therefore \quad R_1 - R_2 = \sigma(\theta_1^4 - \theta_2^4) \quad \text{...(1)}$$

Since some of the heat will be absorbed by silver disc, it will gain energy equal to $(R_1 - R_2) A$, where A is the area of the disc. Again if m is its mass and S specific heat, $\frac{d\theta}{dt}$, the rate of change of temperature,

then $m.S.$ $$\frac{d\theta}{dt} = \frac{(R_1 - R_2)\,A}{J}.$$

Substituting for $(R_1 - R_2)$ from Equation (1)

$$m.S.\frac{d\theta}{dt} = \frac{\sigma.A}{J}(\theta_1^4 - \theta_2^4)$$

$$\therefore \quad \sigma = \frac{m.S.J.}{A(\theta_1^4 - \theta_2^4)}\cdot\frac{d\theta}{dt} \quad \text{...(2)}$$

In the relation (2) m, S, A are the constants of silver disc. θ_1 corresponds to the temperature of steam in absolute units and T is mechanical equivalent of heat. So, if θ_2 and $\frac{d\theta}{dt}$ are found, σ can be found.

In order to find these parameters the experiment is divided into two parts:

(i) Calibration of thermocouple. In this case, the silver disc is arranged in the hole of the wooden platform and no steam is passed through the steam chamber. Water in the beaker is heated and corresponding deflections of the galvanometer for various temperatures are noted. Then a graph is plotted between the temperature difference and galvanometer deflection, from the graph, slope of the curve is

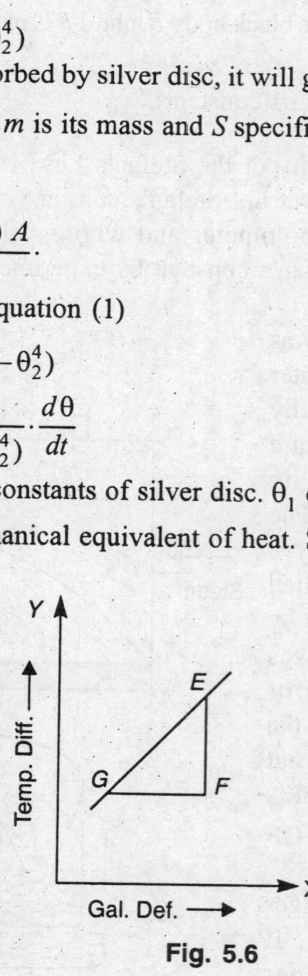

Fig. 5.6

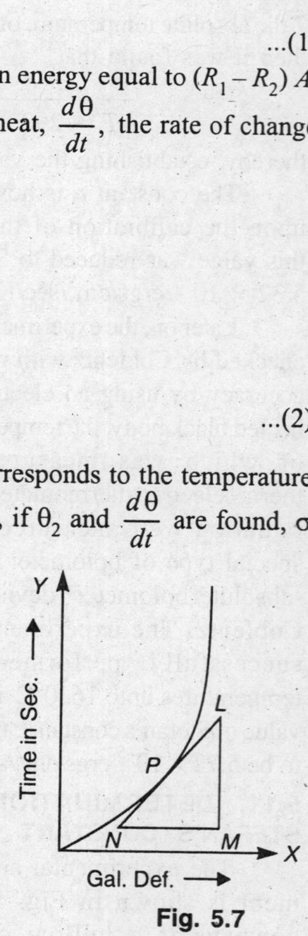

Fig. 5.7

$$\frac{EF}{GF} = \frac{d\theta}{dx}$$

Determination of $\frac{d\theta}{dt}$. The silver disc is removed from the hole and steam is made to pass through the steam chamber till two thermometers T_1 and T_2 give constant readings. At this instant, the silver disc is inserted into the hole and at the same time a stopwatch is started. Reading of the galvanometer at different intervals of time is noted and then a graph is plotted between time and the galvanometer deflection. A tangent is drawn at most curved part of the curve and then $\frac{NM}{LM}$ gives the rate of change of deflection $\frac{dx}{dt}$ and using the first result, we get

$$\frac{d\theta}{dt} = \frac{d\theta}{dx} \times \frac{dx}{dt} = \frac{EF}{GF} \times \frac{NM}{LM}$$

and so knowing $\frac{EF}{GF}$ and $\frac{NM}{LM}$, $\frac{d\theta}{dt}$ can be obtained.

Measurement of T_2. From the deflection of the galvanometer corresponding to the point P chosen, we can find the difference of temperature from the first graph. This added to the initial temperature of the disc gives the temperature in °C and adding 273 to it we get θ_2 K.

Example 1. *A sphere of radius 6.0 cm at 1200°C is suspended in an enclosure at 500°C. Find the rate of loss of heat of the sphere assuming it to behave as 60% of a black body ($\sigma = 5.7 \times 10^{-5}$ ergs/cm²/sec/K⁴).*

Solution: From Stefan's Law

$$R = e \, . \, A \, . \, \sigma \, (T_1^4 - T_2^4)$$

[where e is the emissivity and A is the area of the sphere]

Now $A = 4\pi r^2 = 4\pi . (6)^2$

$= 4 \times 3.142 \times 36 = 452.4$ sq cm.

$e = 0.60$

$\therefore$ $R = 0.60 \times 452.4\,(\text{cm}^2) \times 5.7 \times 10^{-12}\,\text{joules/sec/cm}^2/\text{K}^4 \times (1200^4 - 500^4)\,\text{K}^4$

$= \mathbf{5.6 \times 10^3}$ **watt.**

Example 2. *The surface of a furnace is at 1700°C. Assuming it to be a black body, how much heat is radiated by 1.0 m² of this surface in one hour. ($\sigma = 5.7 \times 10^{-8}$ W/m²/K⁴).*

Solution: From Stefan's Law $R = A.\sigma.T^4 \times 60 \times 60$

$A = 1 \text{ m}^2$

$T = (1700 + 273) = 1973$ K

$\therefore$ $R = 1 \times 5.7 \times 10^{-8} \times (1973)^4 \times 3600 = \mathbf{3.1 \times 10^9}$ **joules.**

Example 3. *Calculate the amount of heat lost per sec by radiation from a sphere of 10.0 cm diameters at a temperature of 227°C when placed in an enclosure having temperature 27°C. Given $\sigma = 5.7 \times 10^{-12}$ joules/sec/cm²/K⁴.*

Solution: From Stefan's Law $R = As\,(T^4 - T_0^4)$

Here $A = 4\pi r^2$ $\left(\because r = \frac{d}{2} = \frac{10}{2} = 5 \text{ cm}\right)$

$= 4 \times 3.142 \times 25$

$= 314.2$ sq cm

$T_1 = 227 + 273 = 500$ K

$T_0 = 27 + 273 = 300$ K

$$\therefore \quad R = \frac{314.2 \times 5.7 \times 10^{-12} \times [(500)^4 - (300)^4]}{4.2 \times 10^7}$$

Heat lost/sec = **23.2 calories.**

Example 4. *What power is radiated from a tungsten filament 20 cm long and 0.010 mm in diameter when the filament is kept at 2500 K in an evacuated bulb? The tungsten radiates at 30% of the rate of a black body at the same temperature. Neglect conduction losses.*

$$(\sigma = 5.72 \times 10^{-5} \text{ ergs/cm}^2\text{/sec/K}^4).$$

Solution:
$$A = \pi \,.\, d \,.\, l = (3.142 \times 0.001 \text{ cm} \times 20 \text{ cm})$$
$$= 0.0628 \text{ cm}^2$$
$$T^4 = (2500 \text{ K})^4 = 39.1 \times 10^{12} \text{ (K}^4)$$

From Stefan's Law, $R = e \,.\, \sigma \,.\, A \,.\, T^4$

$$= 0.30 \times 5.70 \times 10^{-12} \text{ joules/cm}^2\text{/sec/K}^4 \times 0.0628 \text{ cm}^2 \times 39.1 \times 10^{12} \text{ (K)}^4$$
$$= 4.2 \text{ joules/sec}$$
$$= \mathbf{4.2 \text{ watts.}}$$

Example 5. *A black body with an initial temperature of 300°C is allowed to cool inside an evacuated enclosure which is surrounded by melting ice at the rate of 0.35°C/sec. If the surface area of the body is 8 sq cm and its mass and specific heat are 32 gms and 0.10 respectively. Find the Stefan's constant.*

Solution: We have the relation for Stefan's constant

$$\sigma = \frac{J.m.S}{A\,(\theta_1^4 - \theta_2^4)} \cdot \frac{d\theta}{dt}$$

Here
$$\theta_1 = 300°\text{C} = 573 \text{ K}$$
$$\theta_2 = 0°\text{C} = 273 \text{ K}$$
$$m = 32 \text{ gms}$$
$$A = 8 \text{ sq cm}$$
$$S = 0.10, \quad \frac{d\theta}{dt} = 0.35°\text{C}$$

$\therefore$
$$\sigma = \frac{4.2 \times 10^7 \times 32 \times 0.1 \times 0.35}{8 \times (573^4 - 273^4)}$$
$$= \mathbf{5.72 \times 10^{-5} \text{ ergs/cm}^2\text{/sec/K}^4.}$$

Example 6. *The element of an electric fire attains a steady temperature of 800°C when connected to its normal power supply. If the power supply is increased by 20%, calculate the new steady temperature of the element. Assuming the element to be black body and*

$$\sigma = 5.72 \times 10^{-} \text{ ergs/cm}^2\text{/sec/K}^4.$$

Solution: From Stefan's Law

$$R = \sigma . T_1^4$$
$$= \sigma\,(803 + 273)^4$$
$$= \sigma \,.\, (1073)^4 \qquad \text{...(1)}$$

Let the temperature be T_2 when the power supply is increased by 20%.

$$\therefore \quad 1.2 \times R = \sigma \times T_2^4 \qquad \text{...(2)}$$

From Equations (1) and (2)

$$1.2 = \frac{T_2^4}{(1073)^4}$$

or
$$T_2 = 1073 \times (1.2)^{1/4}$$
$$= \mathbf{1123 \text{ K or } 850°\text{C.}}$$

5.12. CONCEPT OF HEAT AND WORK

Heat is defined as the *energy in motion*. Since we cannot speak of work in a body, so we cannot talk of heat in a body.

Work is either done on a body or by a body. Similarly, heat can flow from a body or to a body.

If a body is at a constant temperature, it will have both **mechanical and thermal energies** due to the molecular movements and *it is not possible to separate them.* In this case, we cannot talk about heat energy. This in other words may be stated as that in case no flow of heat takes place, the word *heat* cannot be used. **Heat** can be used only when transfer of energy takes place between two or more systems.

COMPARISON OF HEAT AND WORK

1. Heat is defined as form of energy which is transferred across a boundary due to the temperature gradient, *i.e.*, rate of change of temperature with distance.

   Work is said to be done by the system if the sole effect on things external to system is the change of position.

   Normally, work done by the system is positive and work done on the system is negative. Similarly, heat flowing into the system is positive and heat flowing out of system is negative.
2. Heat and work are both *transient phenomenon*, systems do not possess heat or work.
3. Heat and work are boundary phenomenon, they are observed at the boundary of the system. Heat and work represent the *energy* crossing the boundary of the system.
4. Heat and work depend on the path followed, *i.e.*, they are path functions and so they are *exact differentials* and may be written *as* S_H and S_W.
5. When a system undergoes a change, heat transfer or work done may occur.

5.13. STATISTICS AND DISTRIBUTION LAWS

Statistics corresponds to that branch of science which deals with the *collection, classification and interpretation* of numerical data or facts.

Basically, two types of statistics exist:

(*i*) Classical Statistics

(*ii*) Quantum Statistics.

Classical statistics (also known as Maxwell-Boltzman Statistics) treats particles as *distinguishable* while Quantum statistics treats them *undistinguishable.*

PHASE SPACE

In a static system, all the particles stay fixed at various points in space. If we consider a particle in space, it can be completely located by knowing the three space co-ordinates (x, y, z). In other words, a knowledge of $3n$ position co-ordinates of n particles can completely define the static system.

However, in actual systems of nature which are mostly dynamic, *i.e.*, particles are *moving about and hence possess momenta.*

In this case, in addition to space co-ordinates, momenta co-ordinates need to be mentioned to completely define the state of particles. So, in addition to position space (x, y, z), momentum space (p_x, p_y, p_z) has to be mentioned. Just as space co-odinates, momentum co-ordinates p_x, p_y and p_z may be assigned three mutually perpendicular axes. If we combine position and momentum space, we get what is called **phase space**. A point in phase space is fixed by six co-ordinates x, y, z; p_x, p_y and p_z. It means that a knowledge of $6n$ co-ordinates in phase *space for all n particles of the system completely defines the position and momenta of these particles.*

So, the concept of *phase space* is very important in dealing with actual dynamic systems. The phase space can be further sub-divided into small elements of sides Δx, Δy, Δz, Δp_x, Δp_y, Δp_z.

The volume of each state is may be written as

$$H = \Delta x\, \Delta y\, \Delta z\, \Delta p_x\, \Delta p_y\, \Delta p_z \qquad ...(1)$$

From Hiesenburg's uncertainty principle, we have

$$\Delta x\, \Delta p_x \approx h$$
$$\Delta y\, \Delta p_y \approx h$$

$$\Delta z \, \Delta p_z \approx h$$

So, $\Delta x \, \Delta p_x \, \Delta y \, \Delta p_y \, \Delta z \, \Delta p_z = h$ ...(2)

Comparing Equations (1) and (2)

Volume of each state may be given as

$$H = h^3 \quad ...(3)$$

So, from uncertainty principle, in phase space, the co-ordinates of a particle can be specified only to the extent that the particle under consideration has the position and momentum lying within an element of phase space of volume h^3. The unit of phase space volume is given by (Joule-sec)^3.

5.14. MAXWELL-BOLTZMAN DISTRIBUTION LAW

Assemblies in nature generally consist of three kinds of particles.

(*i*) Identical but distinguishable particles of any spin. *Molecules of a gas* constitute this kind of particles.

(*ii*) Identical and indistinguishable particles of spin $\frac{1}{2}$. These particles obey Pauli's exclusion principle and are called the Fermi particles (or Fermions). Examples of this type of particles are electrons, protons, neutrons etc.

(*iii*) Identical and indistinguishable particles of zero or integral spin. Photons, phonons and helium atoms at low temperatures are particles of this type. These particles are called Bose particles (or Bosons) and do not obey the Pauli's exclusion principle.

Particle of first kind obey Maxwell-Boltzman Distribution law.

Let us consider N similar but distinguishable molecules of a gas. Let the phase space be divided into large number of cells 1, 2, 3, 4, ..., i, ..., k.

Let $n_1, n_2, n_3, n_4, ..., n_i, ..., n_k$ be the number of gas molecules whose phase space lie in cell 1, 2, 3, 4, ..., i, ..., k respectively in the equilibrium state. As the gas molecules are moving continuously, number of molecules will change in different ways but always keeping the position close to the equilibrium one.

Let us consider the number n_i's change but the *fundamental postulates* of statistical mechanics are obeyed, *i.e.*,

1. The total number of molecules is constant, which means

$$N = n_1 + n_2 + n_3 + ... + n_i + ... + n_k = \text{constant}$$

$$\delta N = \delta n_1 + \delta n_2 + \delta n_3 + ... + \delta n_i + ... + \delta n_k = 0$$

i.e., $$\sum_i \delta n_i = 0 \quad ...(1)$$

2. The total energy of the system is constant, and may be written as

$$E = \varepsilon_1 n_1 + \varepsilon_2 n_2 + \varepsilon_3 n_3 + ... + \varepsilon_i n_i + ... + \varepsilon_k n_k = \text{constant}$$

$$\delta E = \varepsilon_1 \delta n_1 + \varepsilon_2 \delta n_2 + \varepsilon_3 \delta n_3 + ... + \varepsilon_i \delta n_i + ... + \varepsilon_k \delta n_k = 0$$

i.e., $$\Sigma \varepsilon_i \delta n_i = 0 \quad ...(2)$$

3. When gas is in equilibrium, the probability is maximum, which means

$$\delta P = 0$$

Also when P is maximum, $\log P$ is maximum, so

$$\delta (\log P) = 0$$

But $$P = \frac{N!}{n_1! n_2! n_i! ... n_k!} (g_1)^{n_1} (g_2)^{n_2} ... (g_i)^{n_i} ... (g_k)^{n_k} \quad ...(3)$$

where g_i is the **priori probability** for a molecule to have the energy ε_i.

Taking natural logarithm of above equation, we have the relation

$$\log P = \log N! = \sum_i \log n_i! + \sum_i n_i \log g_i$$

Now, using **Stirling's formula**, we get

$$\log P = N \log n - N - \sum_i n_i \log n_i + \sum_i n_i! + \sum_i n_i \log g_i$$

But $\sum_i n_i = N$, so we have

$$\log P = N \log N - \sum_i n_i \log n_i + \sum_i \log g_i$$

So, from Equation (3), we get

$$(\log P) = -\sum_i n_i\, \delta(\log n_i) - \sum_i \log n_i\, \delta n_i + \sum_i \log g_i\, \delta n_i = 0 \quad \text{...(4)}$$

($\because N \log N$ is constant)

Now, $\delta(\log n_i) = \frac{1}{n_i}\delta n_i$

or $n_i\, \delta(\log n_i) = \delta n_i$

or $\sum_i n_i\, \delta(\log n_i) = \sum_i \delta n_i = 0$ [using Equation (1)]

So, Equation (4) may be written as

$$\sum_i \log n_i\, \delta n_i - \sum_i \log g_i\, \delta n_i = 0 \quad \text{...(5)}$$

Let us now use the method of *Lagrangian undermined multiplier*. For this purpose, multiplying Equations (1) and (2) by α and β respectively, where α and β are undermined quantities independent of n_i's and then adding the resulting expression to Equation (5), we get

$$\sum_i \log n_i\, \delta n_i - \sum_i \log g_i\, \delta n_i + \sum_i \alpha\, \delta n_i + \sum_i \beta\varepsilon_i\, \delta n_i = 0$$

i.e.,

$$\sum_i [\log n_i - \log g_i + \alpha + \beta\varepsilon_i]\, \delta n_i = 0 \quad \text{...(6)}$$

Since, δn_i's are independent variables, so for the validity of Equation (6), the quantity within brackets must always be zero regardless of the value of i. So, we can write

$$\log n_i - \log g_i + \alpha + \beta\varepsilon_i = 0$$

which may be written as

$$n_i = g_i\, e^{-\alpha}\, e^{-\beta\varepsilon_i}$$

This result determines the most probable distribution of molecules among the various possible individual states and is called *Maxwell-Boltzman distribution law*.

β can be calculated by considering continuous distribution of energy of molecules and we find

$$\beta = \frac{1}{kT}$$

$\therefore$ Maxwell-Boltzman distribution law may be written as

$$n_i = g_i\, e^{-\alpha} . e^{-\varepsilon_i/kT}$$

5.15. PARTITION FUNCTIONS

In order to develop the concept of *partition function*, consider an ideal gas molecules obeying *classical statistics* and let the distribution of gas molecules be such that n_i molecules occupy the ith state between energies ε_i and $\varepsilon_i + d\varepsilon_i$ and let the degeneracy (*i.e.*, number of energy levels be g_i).

Then from Maxwell-Boltzman ($M - B$ statistics) distribution law, we have

$$n_i = g_i . e^{-\alpha} . e^{-\varepsilon_i/kT}$$

Put $e^{-\alpha} = A$. Then, we have

$$n_i = g_i\, A e^{-\varepsilon_i/kT} \quad \text{...(1)}$$

Let total number of gas molecules in the assembly be N, then

$$N = \sum_i n_i = \sum_i g_i\, A e^{-\varepsilon_i/kT}$$

$$= A \sum_i g_i\, e^{-\varepsilon_i/kT} \quad \text{...(2)}$$

The term A $(= e^{-\alpha})$ has been taken out of the summation sign as α is a constant, called the *Lagrangian undetermined multiplier*.

From Equation (2), we have

$$\frac{N}{A} = \sum_i g_i \, e^{-\varepsilon_i/kT} = Z \text{ (say)} \qquad ...(3)$$

The quantity Z is called the *Boltzman partition function* or simply the *partition function*. In Equation (3), the expression for Z represents the sum of all the $g_i \; e^{-\varepsilon_i/kT}$ terms for every energy state of the given molecule. So, *the quantity Z indicates how the gas molecules of an assembly are distributed or partitioned among the various energy levels.*

If the energy of the ith level is ε_i, then keeping in mind that the weight of an individual level is unity; Equation (3) may be expressed as

$$Z = \sum_i e^{-\varepsilon_i/kT} \qquad ...(4)$$

In Equation (4), the term Z is the summation of all the levels of a system in the assembly. It is due to this reason, it is customary to call Z as the *partition function* for a system in the assembly. The energy term in the expression for partition function does not mean only the translational component, as is the case for *structureless molecules*, but in general, it may contain the components corresponding to other degrees of freedom too, *e.g.*, rotational, vibrational and electronic etc.

5.16. PARTITION FUNCTION AND THERMODYNAMICAL QUANTITIES

Let us establish relations between the partition functions and the *thermodynamical quantities*. The **entropy** of an assembly and the weight of the most probable configuration are given by the equation

$$S = k \log W_{\max} \qquad ...(1)$$

where weight of the configuration for a classical system is given by the relation

$$W = N! \prod_i \frac{g_i^{n_i}}{n_i!} \qquad \text{[From probability theory]}$$

So,
$$\log W = \log N! + \sum_i (n_i \log g_i - \log n_i!)$$

From Stirling approximation, $\log n! = n \log n - n$, we have

$$\log W = N \log N - N + \sum_i (n_i \log g_i - n_i \log n_i + n_i) \qquad ...(2)$$

According to Maxwell-Boltzman distribution law, for the most probable configuration, we have the relation

$$n_i = g_i \, e^{-\alpha} \; e^{-\beta\varepsilon_i} \qquad ...(3)$$

Since, total number of molecules is constant, so we have

$$N = \sum_i n_i \qquad ...(4)$$

Put these values in Equation (2), we get

$$\log W_{\max} = N \log N - N + \sum_i [n_i \log g_i - n_i \log g_i \, e^{-\alpha} \, e^{-\beta\varepsilon_i} + n_i]$$

$$= N \log N - N + \sum_i n_i \log g_i - \sum_i n_i \log g_i + \sum_i n_i \alpha + \sum_i n_i \beta\varepsilon_i + \sum_i n_i$$

$$= N \log N + \alpha N + \beta \sum_i n_i \varepsilon_i \qquad \left(\because N = \sum_i n_i\right) \qquad ...(5)$$

Let the total energy of the gas molecules of the assembly be E, then

$$E = \sum_i n_i \varepsilon_i$$

So, Equation (5) may take the form

$$\log W_{\max} = N \log N + \alpha N + \beta E \qquad ...(6)$$

Now, put $A = e^{-\alpha}$, *i.e.*, $\alpha = -\log A$.

So, Equation (6) may be written as

$$\log W_{\max} = N \log N - N \log A + \beta E$$

$$= N \log \frac{N}{A} + \beta E$$

$$= N \log Z + \beta E \qquad \left(\because \text{partition function, } Z = \frac{N}{A}\right) \qquad ...(7)$$

Also, $$\beta = \frac{1}{kT},$$

$$\therefore \quad \log W_{max} = N \log Z + \frac{E}{kT}$$

So, the entropy of the ensemble (or assembly), is given by the relation

$$S = k \log W_{max}$$

$$= Nk \log Z + \frac{E}{T} \quad ...(9)$$

In the case of an ideal gas, we have

$$E = \frac{3}{2} NkT$$

So, Equation (9) becomes

$$S = Nk \log Z + \frac{3}{2} Nk \quad ...(10)$$

The equation gives the expression for the *entropy of the assembly of ideal gas molecules. We can also evaluate the other thermodynamic functions as under:*

Helmholtz Free Energy *F*. The Helmholtz free energy F is given by the relation

$$F = E - TS$$

With Equation (9), we get

$$F = E - T\,(Nk \log Z + E/T)$$

$$= NkT \log_e Z \quad ...(11)$$

This is the required expression for *Helmholtz free energy E in terms of partition function.*

Total Energy *E*. Let $\overline{\varepsilon}$ be the average energy of a system of an assembly of N such independent systems, so we have

$$\overline{\varepsilon} = \frac{E}{N} = \frac{\sum_i n_i \varepsilon_i}{\sum_i n_i}$$

From Maxwell-Boltzman distribution law, we have

$$\overline{\varepsilon} = \frac{\sum_i g_i e^{-\alpha} e^{-\beta\varepsilon_i} \varepsilon_i}{\sum_i e^{-\alpha} e^{-\beta\varepsilon_i}}$$

$$= \frac{\sum_i g_i \varepsilon_i e^{-\beta\varepsilon_i}}{\sum_i g_i e^{-\beta\varepsilon_i}} = \frac{\sum_i g_i \varepsilon_i e^{-\varepsilon_i/kT}}{\sum_i g_i e^{-\varepsilon_i/kT}} \quad \left(\because \beta = \frac{1}{kT}\right)$$

$$= \frac{\sum_i g_i \varepsilon_i e^{-\varepsilon_i/kT}}{Z} \quad ...(12)$$

Now, partition function $Z = \sum_i g_i e^{-\beta\varepsilon_i} = \sum_i g_i e^{-\varepsilon_i/kT},$

so, we have $$\left(\frac{\partial Z}{\partial T}\right)_V = \sum_i g_i \frac{\varepsilon_i}{kT^2} e^{-\varepsilon_i/kT}$$

$$= \frac{1}{kT^2} \sum_i g_i \varepsilon_i e^{-\varepsilon_i/kT}$$

for isothermal isochoric transformation.

So, for *isothermal isochoric transformation*, we have the relation

$$\overline{\varepsilon} = \frac{kT^2}{Z}\left(\frac{\partial Z}{\partial T}\right)_V$$

or $$\bar{\varepsilon} = kT^2 \left[\frac{\partial (\log Z)}{\partial T}\right]_V$$

Therefore, $$E = N\bar{\varepsilon} = NkT^2 \left[\frac{\partial (\log Z)}{\partial T}\right]_V \quad ...(13)$$

which is the expression for total *energy of an ensemble in terms of partition function.*

Enthalpy *H*. The enthalpy H is given by the relation

$$H = E + pV$$

In case of ideal gas, we have

$$pV = RT$$

So, $$H = E + RT$$

With Equation (13), we get

$$H = NkT^2 \left[\frac{\partial (\log Z)}{\partial T}\right]_V + RT \quad ...(14)$$

Gibb's potential *G*. The Gibb's potential G is given by the relation

$$G = H - TS$$

With Equations (10) and (14) we have

$$G = NkT^2 \left[\frac{\partial (\log Z)}{\partial T}\right]_V + RT - T\left(Nk \log Z + \frac{3}{2} Nk\right)$$

$$= NkT^2 \left[\frac{\partial (\log Z)}{\partial T}\right]_V + RT - NkT \log Z - \frac{3}{2} NkT \quad ...(15)$$

Pressure of a gas *p*. The pressure of a gas p is given by the relation

$$p = -\left(\frac{\partial F}{\partial V}\right)_T$$

From thermodynamic relation for pressure (p) and F (Helmholtz function),

$$= NkT \left[\frac{\partial (\log Z)}{\partial T}\right]_T \quad \text{[use Equation (11)]} \quad ...(16)$$

Specific heat at constant volume C_V. For an isothermal isochoric transformation, the internal energy U of the assembly is equal to its total energy E. So, we have

$$C_V = \left(\frac{\partial E}{\partial T}\right)_V$$

which may be written as $$C_V = \frac{\partial}{\partial T}\left\{NkT^2 \left(\frac{\partial \log Z}{\partial T}\right)_V\right\}, \quad \text{[Using Equation (13)]}$$

$$= Nk\left[2T \frac{\partial (\log Z)}{\partial T} + T^2 \frac{\partial^2 (\log Z)}{\partial T^2}\right]_V \quad ...(17)$$

Example 7. *Show that if the partition function is given by Z, the mean energy $\bar{\varepsilon}$ is given by*

$$\bar{\varepsilon} = -\frac{\partial \log Z}{\partial \beta} \text{ where } \beta = \frac{1}{kT}.$$

Solution: Let $\bar{\varepsilon}$ be the mean energy of a system of an ensemble consisting *of N* such independent systems, so we have

$$\bar{\varepsilon} = \frac{E}{N}, \text{ where } E \text{ is the total energy of the ensemble.}$$

$$= \frac{\sum_i \varepsilon_i n_i}{\sum_i n_i} \quad ...(1)$$

From Maxwell-Boltzman distribution law, we have

$$n_i = g_i\, e^{-\alpha}\, e^{-\beta\varepsilon_i} \text{ where } \beta = \frac{1}{kT}$$

So, Equation (1) takes the form

$$\overline{\varepsilon} = \frac{\sum_i \varepsilon_i\, g_i\, e^{-\alpha}\, e^{-\beta\varepsilon_i}}{\sum_i g_i\, e^{-\alpha}\, e^{-\beta\varepsilon_i}} \qquad ...(2)$$

As α is a constant, being Lagrangian undetermined multiplier, so relation (2) may be written as

$$\overline{\varepsilon} = \frac{\sum_i g_i\, \varepsilon_i\, e^{-\beta\varepsilon_i}}{\sum_i g_i\, e^{-\beta\varepsilon_i}}$$

$$= \frac{\sum_i g_i\, \varepsilon_i\, e^{-\beta\varepsilon_i}}{Z} \qquad ...(3)$$

where Z (partition function) $= \Sigma\, g_i\, e^{-\beta\varepsilon_i}$. ...(4)

Differentiating Equation (4) *w.r.t.* β we get

$$\frac{\partial Z}{\partial \beta} = -\sum_i g_i\, \varepsilon_i\, e^{-\beta\varepsilon_i} \qquad ...(5)$$

Put in Equation (1), we get

$$\overline{\varepsilon} = -\frac{(\partial Z/\partial\beta)}{Z},$$

or mean energy $$\overline{\varepsilon} = -\frac{\partial (\log Z)}{\partial \beta} \qquad ...(3)$$

where $\beta = \dfrac{1}{kT}$ which is the required result.

5.17. QUANTUM STATISTICS

In Maxwell-Boltzman distribution law, it is assumed that all the energy levels are accessible to all particles of the system. However, some energy levels may be *prohibited* for certain *group of particles* as each state is associated with a certain available *wave function* and the probabiliy of a particular distribution is restricted by the available *wave functions of any state*.

These restrictions are taken into account in **quantum statistics** and there are two types of quantum statistics.

(*i*) Fermi-Dirac statistics (F-D statistics)

(*ii*) Bose-Einstein statistics (B-E statistics)

(*i*) In F-D statistics, particles are assumed to obey Pauli's exclusion principle and they are characterized by *antisymmetric wave functions*.

Particles under this category are called *Fermions*. Protons, neutrons, electrons etc. are *Fermions*.

(*ii*) In B-E statistics, it is assumed that particles are not restricted by **Pauli's exclusion** principle and they are characterized by *symmetric wave functions*. These particles are called *Bosons*. Experimentally, these particles are found to have **zero** or **integral multiple of 1 spin**.

Helium nuclei, photons, mesons etc. are **Bosons**.

In both kinds of Quantum Statistics, the particles are *identical and indistinguishable*.

At high temperatures and low pressures, all the three statistics give practically the same result.

5.18. FERMI-DIRAC DISTRIBUTION LAW

Let us consider a large number of particles which are

(*i*) identical and indistinguishable

(*ii*) these particles obey Pauli's exclusion principle.

It means that no two particles can have same dynamical state and the wave function of the whole system must be antisymmetric. Particles satisfying these conditions are called *Fermions*. **In general, all fundamental particles with spin 1/2 are Fermions.**

In quantum statistics, the intrinsic probability g_i is governed by different quantum states relating to a given energy, *i.e.*, *the degeneracy of energy state.*

Each quantum state corresponds to a particle wave function. Wave functions of various particles are determined by each of possible arrangements of quantum numbers corresponding to a particular energy level. For a particle with spin 1/2, in the absence of magnetic field, each particle may have energy state corresponding to spin 1/2 or – 1/2. **So, the intrinsic probability in this case is 2.**

For motion in a central field, the energy of a particle is independent of the orientation of the orbital angular momentum. Due to this, a degeneracy of $(2l + 1)$ is introduced and it is the value of g for a particular state. If the particle has spin, the total degeneracy = $2\ (2l + 1)$.

Since no two particles can be in the same energy state with same quantum numbers, the intrinsic probabilities g_i's give the maximum number of particles that can be accommodated in a particular energy level without violating the Pauli's exclusion principle, *i.e.*,

$$n_i \leq g_i$$

which means that the n_i *value for a given distribution does not exceed the corresponding value of* g_i.

Let n_i be the number of particles for energy level E_i. The first particle can be placed in any one of the available g_i states, *i.e.*, this particle can be assigned to any of g_i sets of quantum numbers. So, the first particle can be distributed in g_i different ways. Similarly, the second particle can be arranged in $(g_i - 1)$ different ways and so on.

The total number of different ways of arranging n_i particles among the available g_i states with energy level E_i will be

$$= g_i\ (g_i - 1)\ (g_i - 2)\ ...\ [g_i - (n_i - 1)]$$

$$= \frac{g_i\,!}{(g_i - n_i)!} \qquad ...(1)$$

Now, if the particles are taken to be indistinguishable, it will not be possible to detect any difference when n_i particles are reshuffled into different states occupied by them in the energy level E_i. So, the total number of different and distinguishable ways will be

$$= \frac{g_i\,!}{n_i\,!(g_i - n_i)!}$$

The total number of different and distinguishable ways of getting the distribution n_1, n_2, n_3, etc., among the various energy levels, E_1, E_2, E_3, ... etc. can be got by multiplying the various factors. Thus, writing the probability as

$$\therefore \qquad P = \frac{g_1\,!}{n_1\,!(g_1 - n_1)!} \cdot \frac{g_2\,!}{n_2\,!(g_2 - n_2)!} \cdots$$

$$P = \prod_i \frac{g_i\,!}{n_i\,!(g_i - n_i)!} \qquad ...(3)$$

The most probable distribution can be bad by finding the maximum value of $\log_e P$ in Equation (4).

This should also satisfy the condition

$$\sum_i n_i = N \text{ and } \sum_i n_i\ E_i = U \qquad ...(4)$$

From **Stirling's approximation,**

$$\log_e N! = N \log_e N - N$$

In Equation (3), applying **Stirling's approximation,** we get

$$\log_e P = \sum_i [(g_i \log_e g_i - g_i) - (n_i \log_e n_i - n_i)] - [(g_i - n_i) \log_e (g_i - n_i) - (g_i - n_i)]$$

$$\log_e P = \sum_i [g_i \log_e g_i - n_i \log_e n_i - (g_i - n_i) \log_e (g_i - n_i)] \quad ...(5)$$

Differentiating Equation (5), we get

$$-d (\log_e P) = \sum_i [\log_e n_i - \log_e (g_i - n_i)]\, dn_i$$

For P to be maximum,

$$d (\log_e P) = 0$$

$$\therefore \quad \sum_i [\log_e n_i - \log_e (g_i - n_i)]\, dn_i = 0 \quad ...(6)$$

Now, $$\sum_i dn_i = 0 \quad ...(7)$$

and $$\sum_i E_i\, dn_i = 0 \quad ...(8)$$

Multiplying (7) by α and (8) by β and adding to Equation (6), we get

$$\sum_i [\log n_i - \log_e (g_i - n_i) + \alpha + \beta E_i]\, dn_i = 0$$

Now, equilibrium distribution is possible if

$$\therefore \quad \log_e n_i - \log_e (g_i - n_i) + \alpha + \beta E_i = 0$$

$$2al \log_e \left(\frac{n_i}{g_i - n_i}\right) = -\alpha - \beta E_i$$

or $$\frac{n_i}{g_i - n_i} = e^{-\alpha - \beta E_i}$$

or $$\frac{g_i - n_i}{n_i} = e^{\alpha + \beta E_i}$$

$$\frac{g_i}{n_i} - 1 = e^{\alpha + \beta E_i}$$

or $$\frac{g_i}{n_i} = (e^{\alpha + \beta E_i}) + 1$$

or $$n_i = \frac{g_i}{(e^{\alpha + \beta E_i}) + 1} \quad ...(9)$$

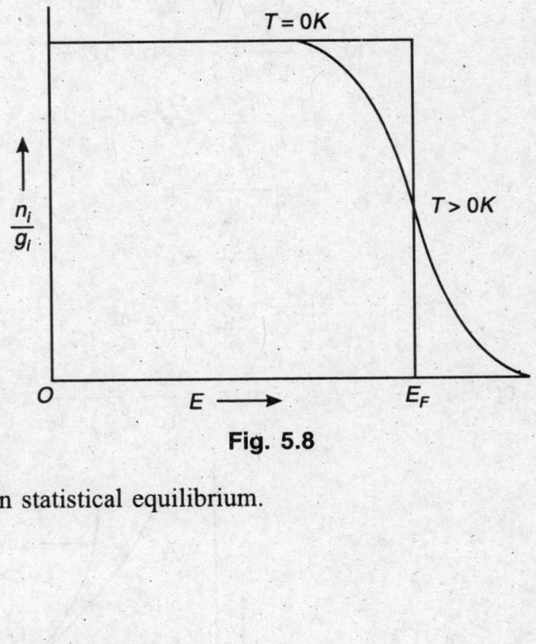

Fig. 5.8

Equation (9) represents the *Fermi-Dirac distribution law*. The parameter has the same role as in the case of Maxwell-Boltzman distribution law, *i.e.*, for the system consisting of fermions in statistical equilibrium.

Now, $$\beta = \frac{1}{kT}$$

$$\therefore \quad n_i = \frac{g_i}{[e^{(\alpha + E_i/kT)}] + 1}$$

In most cases, the value of α is negative and is given by

$$\alpha = -E_F/kT$$

$$\therefore \quad n_i = \frac{g_i}{[e^{(E_i - E_F)/kT}] + 1} \quad ...(10)$$

The value of E_F is positive and is found to be independent of temperature.

If a graph is plotted between n_i/g_i and E, we get a curve as shown in Fig. 5.8.

At high temperatures $T >> 0$ K, the shape of the curve is shown by dotted line.

At $T = 0$ K, there is complete cut off beyond $E = E_F \cdot n_i$ becomes zero for $E \geq E_F$. At higher temperatures, the cut off is not sharp.

So, we observe that in case of F-D statis-tics, accumulation of particles at ground level is not allowed and at temperature $T = 0$ K, the particle occupy lowest energy levels upto E_F and E_F

gives the indication of maximum energy of fermions in the system. E_F is called **Fermi Energy**.

For P to be maximum, $d(\log_e P) = 0$.

So, Equation (2), may be written as

$$d(\log_e P) = \sum_i [\log_e (n_i + g_i - 1)\, dn_i - \log_e n_i\, dn_i] = 0$$

$$\therefore \quad -d(\log_e P) = \sum_i [-\log_e (n_i + g_i - 1)\, dn_i + \log_e n_i\, dn_i] = 0$$

or $$\sum_i [-\log_e (n_i + g_i - 1) + \log_e n_i]\, dn_i = 0 \qquad ...(3)$$

As the total number of particles and total energy are constants, so

$$\sum_i dn_i = 0 \qquad ...(4)$$

$$\sum_i E_i\, dn_i = 0 \qquad ...(5)$$

Multiplying Equation (4) by α and Equation (5) by β and adding to Equation (3), we have

$$\sum_i [-\log_e (n_i + g_i - 1) + \log_e n_i + \alpha + \beta E_i]\, dn_i = 0$$

$$\therefore \quad -\log_e (n_i + g_i - 1) + \log_e n_i + \alpha + \beta E_i = 0$$

Taking $n_i + g_i$ very large as compared to 1, so that 1 may be neglected.

$$\therefore \quad -\log_e (n_i + g_i) + \log_e n_i + \alpha + \beta E_i = 0$$

$$\log_e \left(\frac{n_i}{n_i + g_i}\right) = -\alpha - \beta E_i$$

$$\frac{n_i}{n_i + g_i} = e^{-\alpha - \beta E_i}$$

or $$\frac{n_i}{n_i + g_i} = e^{\alpha + \beta E_i}$$

or $$1 + \frac{g_i}{n_i} = e^{\alpha + \beta E_i}$$

$$\frac{g_i}{n_i} = (e^{\alpha + \beta E_i}) - 1$$

or $$n_i = \frac{g_i}{(e^{\alpha + \beta E_i}) - 1} \qquad ...(6)$$

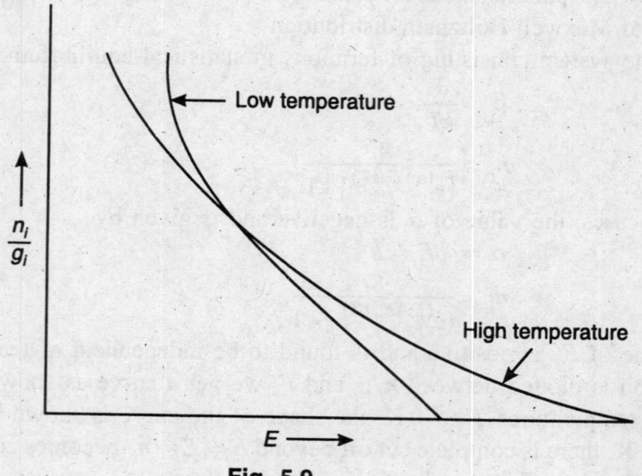

Fig. 5.9

Taking $$\beta = \frac{1}{kT}$$

$$n_i = \frac{g_i}{[e^{(\alpha + E_i/kT)}] - 1} \quad ...(7)$$

Equation (7) is referred to Bose-Einstein's distribution law.

The value of the constant α is governed by the equation $\Sigma\, n_i = N$. Since, n_i cannot be negative, so α must always have a positive value. The distribution of the particles for different energy levels is shown in Fig. 5.9 for low and high temperatures.

5.20. DISTRIBUTION OF ENERGY IN A BLACK BODY SPECTRUM

The intensity of radiation emitted by a black body is not uniformly distributed over the whole range of spectrum. We have to see how the radiations from a black body depend on wavelength and temperature, *e.g.*, in the case of blackened platinum wire which is electrically heated its appearance varies as under:

At 525°C it appears dull red
At 900°C it changes to cherry red
At 1100°C it appears orange red
At 1250°C it appears yellow
and At 1600°C it becomes white hot.

From above it is clear that with the increase in temperature the wavelength decreases *i.e.*, *The maximum intensity of emission shifts towards the shorter wavelength side as the temperature is increased.*

Wien's law of energy distribution. Wien, in 1893 applied thermodynamic laws to the black body radiations and got result like the Stefan's Law. His two results are given as:

$$\lambda_m . T = \text{constant} \quad ...(1)$$

$$E_m . T^{-5} = \text{constant} \quad ...(2)$$

where λ_m corresponds to the wavelength at which emissive power E_λ is max. (E_m) at the absolute temperature T.

The first relation shows that $\lambda_m \propto \frac{1}{T}$ *i.e.*, with the increase of temperature, the max. intensity of radiation shifts towards shorter wavelengths and so is called Wien's displacement law.

The second relation shows that $E_m \propto T^5$ and is sometimes called fifth power law.

If the two above relations are combined, then a general relation may be expressed as

$$E = K . \lambda^{-5} f(\lambda T) \quad ...(3)$$

where K is constant and $f(\lambda T)$ is some function of (λT).

Wien further applied some valid assumptions and got the following expression for the energy distribution

$$dE = K . \lambda^{-5} \; e^{\frac{-a\lambda}{T}} . d\lambda \quad ...(4)$$

where K and a are constants and dE corresponds to the energy contained in the spectral region between wavelength λ and $\lambda + d\lambda$ of a black body at temperature T. This expression is called Wien's Law of energy distribution.

Experimental Study. The above theoretical results were tested experimentally by Paschen, Lummer, Pringshein, Rubens etc. between the years 1893 to 1899.

We can here take up some of the results of these scientists in the case of a black body in the form of a carbon tube with a small opening which was heated electrically for different temperatures between 600 K to 1650 K. The results

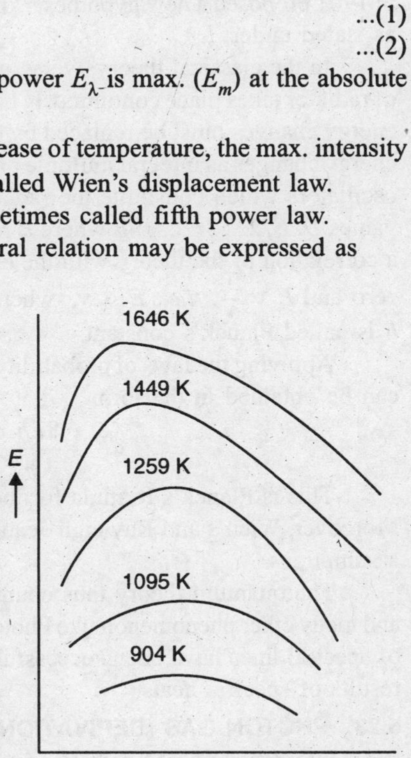

Fig. 5.10

are represented graphically in the Fig. 5.10, where amount of energy emitted in a narrow wavelength range is plotted against mean wavelength of range. Different curves are shown for different ranges of temperatures. The total energy of radiation at any given temperature given by the area between the curve corresponding to that temperature and the horizontal. This area is found to vary according to fourth power of absolute temperature, which verifies Stefan's Law.

The wavelength corresponding to max. energy repre-sented by the peak of the curve shifts to shorter wavelength side as the temperature increases, which verifies Wien's Law. $\lambda_m T$ = const. Wien's relation (4) is found to be valid for shorter wavelength but some deviations are there for longer wavelengths. This defect was removed to some extent by Rayleigh and Jean's by using different treatment of energy distribution in the case of a black body.

5.21. RAYLEIGH-JEAN'S LAW

From E.M. theory, a black body radiator gives radiations of continuous variable wavelengths from zero to infinity. The mode of vibrations of number of wave trains can be got from the principle of probability. The energy carried by each mode of vibration and so the energy distribution can be got from the expression

$$dE = B.\lambda^{-4}\ .\ T.e^{-a/\lambda T}\ .\ d\lambda \qquad ...(1)$$

where B and a are constants. This is known as Rayleigh-Jean's Law of energy distribution.

5.22. PLANCK'S LAW

Both the theoretical results of Wien's Law and Rayleigh-Jean's law were found not to be strictly valid over the total range of spectrum. Moreover, these laws sometimes lead to absurd results, *e.g.*, in Wien's formula even when $T = \infty$, E is still finite, which contradicts Stefan's law.

In order to overcome the errors in the theoretical results, Max Planck suggested that the classical assumptions on which Wien's and Rayleigh-Jeans laws were based might have some flaws. So, Planck in 1901 proposed a new hypothesis known as quantum theory which differs from the classical theory as stated under:

In the classical theory, we assume that the radiations are continuous *i.e.*, the energy changes of radiator takes place continuously has led to inconsistent results. So the classical idea of continuous energy changes must be replaced by discrete or discontinuous energy changes. We may take up the energy changes as integral multiple of a small unit of energy known as quantum. The electrical linear oscillators which constitute the radiations of a black body may vibrate only with integral energy values $O, E, 2E, 3E, ..., nE$ where E represents the elementary quantum. Moreover, in order to have a correlation of the theory with the experimental results, E should be a finite quantity greater than zero and $E \propto \frac{1}{\lambda}$, *i.e.*, $E \propto \nu$, where ν represents the frequency of radiation. So $E = h\nu$, where h is called Planck's constant.

Applying the laws of probability to such oscillators, an expression for the black body radiations can be obtained in the form

$$dE = \frac{8\pi h.c\lambda h^{-5}\ d\lambda}{e^{hc/k\lambda.T} - 1}$$

This is Planck's formula for the thermal radiations and fits well in the experimental results. Moreover, Wien's and Rayleigh-Jean's laws can be derived from this formula by using some valid assumptions.

The quantum theory thus enumerated by Max Planck has revolutionised the classical ideas and many other phenomenon like Photoelectric effect, Crompton effect, atomic structure, fine structure of spectral lines have been successfully explained by it. Einstein was able to use it to give correct results of specific heats.

5.23. PHOTON GAS (DERIVATION OF PLANCK'S LAW FOR BLACK BODY RADIATIONS)

Interaction of E.M. radiations with matter led to the conclusion that electro-magnetic radiations are composed of discrete energy particles called *photons*. Each photon carries an energy quanta

$h\nu$ and momentum $\frac{h}{\lambda}$, where ν is the frequency and λ, the wavelength of radiations. The E.M. radiations trapped in a cavity and in thermal equilibrium with the walls of the cavity are termed as *black body radiations.*

In the equilibrium conditions, black body radiations may be considered as a *photon-gas.* It is further assumed that the photons do not interact with each other but interact only with the atoms of the walls of the cavity. Further, it is assumed that the photons are indistinguishable and number of photons may have same energy.

These photons are considered as *Bosons* since they obey Bose-Einstein Statistics.

Since in this case photons can either be emitted or absorbed, the total number of photons is not constant *i.e.*, $\sum dn_2 \neq 0$.

In Bose-Einstein statistics, the values of g_i refer to *degeneracy* of each energy level. We have from Bose-Einstein Statistics the number of particles n_i distributed in a row as

$$n_i = \frac{g_i}{e^{(\alpha + E_2/kT)} - 1} \quad ...(1)$$

Taking $\alpha = 0$, as the particles are indistinguishable.

$$n_i = \frac{g_i}{e^{E_i/kT} - 1} \quad ...(2)$$

Now, if we consider the cavity to be quite large as compared to the wavelength of radiations, then the energy spectrum of photons can be considered to be continuous. So in this case the energy difference between successive allowed energy values will be quite small. So we can replace g_i by $g(E)\, dE$ and n_i by dn and write Equation (2) as

$$dn = \frac{g(E)\, dE}{e^{E/kT} - 1} \quad ...(3)$$

Since the energy of a photon is given by $E = h\nu$, we can take $g(\nu)\, d\nu$ instead of $g(E)\, dE$ where factor $g(\nu)\, d\nu$ corresponds to the *number of modes of oscillations* in the frequency range $d\nu$ relating to energy range dE.

We can calculate the number of states in black body radiations in the frequency range of ν and $\nu + d\nu$ by finding the spherical volume bound by spheres of radii $\frac{h(\nu + d\nu)}{c}$ and $\frac{h\nu}{c}$.

The volume of spherical shell will be

$$= \frac{4}{3}\pi \frac{h^3}{c^3}(\nu + d\nu)^3 - \frac{4}{3}\pi \frac{h^3}{c^3}\nu^3$$

$$= \frac{4}{3}\pi \frac{h^3}{c^3}[\nu^3 + 3\nu^2\, d\nu + ... - \nu^3]$$

$$= \frac{4}{3}\pi \frac{h^3}{c^3} \cdot 3\nu^2\, d\nu$$

$$= 4\pi \frac{h^3}{c^3} \cdot \nu^2\, d\nu \quad ...(4)$$

The phase space has a volume $V = h^3$ and there are two states of polarization of the radiation. Number of states in the black body radiation in the frequency range ν and $\nu + d\nu$ are given by

$$g(\nu)\, d\nu = \frac{8\pi V}{c^3}\nu^2\, d\nu$$

$$\therefore \quad dn = \frac{8\pi V}{c^3} \cdot \frac{\nu^2\, d\nu}{e^{h\nu/kT} - 1} \quad ...(5)$$

For dn photons in frequency range $\nu + d\nu$ and ν, the energy is equal to $(h\nu)\, dn$ and the energy/unit volume

$$= \frac{(h\nu)\, dn}{V} \quad ...(6)$$

So the energy density distribution for black body radiation is given by

$$E(\nu) = \frac{h\nu}{V} \cdot \frac{dn}{d\nu} \qquad ...(7)$$

Putting the value of *dn* from Equation (5) in Equation (7) we get

$$E(\nu) = \frac{8\pi\nu}{c^3} \cdot \frac{1}{V} \cdot \frac{h\nu^3\, d\nu}{(e^{h\nu/kT} - 1) \cdot d\nu}$$

$$= \frac{8\pi h\nu^3}{c^3} \cdot \frac{1}{(e^{h\nu/kT} - 1)} \qquad ...(8)$$

Equation (8) represents the *Planck's law for black body radiations.*
From Equation (8), we have

$$E(\nu)\, d\nu = \left[\frac{8\pi h\nu^3}{c^3} \cdot \frac{1}{(e^{h\nu/kT} - 1)}\right] d\nu \qquad ...(9)$$

Also, $\nu = \dfrac{c}{\lambda}$

$\therefore$ $d\nu = -\dfrac{c}{\lambda^2}\, d\lambda$

Neglecting the negative sign, we may write Equation (9) as

$$E(\lambda)\, d\lambda = \frac{8\pi h\nu^3}{c^3} \cdot \frac{c}{\lambda^2} \cdot \left[\frac{1}{e^{h\nu/kT} - 1}\right] d\lambda$$

$$= \frac{8\pi h\nu^3}{\nu^3\lambda^3} \cdot \frac{c}{\lambda^2} \cdot \left[\frac{1}{e^{h\nu/kT} - 1}\right] d\lambda$$

$$\therefore \quad E(\lambda) = \frac{8\pi hc}{\lambda^5} \cdot \left[\frac{1}{e^{h\nu/kT} - 1}\right] \qquad ...(10)$$

Equation (10) represents the energy density for wavelength λ of the black body.
The above Equations (9) and (10) represent Planck law of black body radiations.
We can derive Wien's law and Rayleigh-Jean's Law from Planck Law (Equation 10)
(*i*) For small temperatures, λT is small
$\therefore$ $e^{hc/\lambda kT} >> 1$ and so can be neglected in the denominators of Equation (10)

$$\therefore \quad E(\lambda) = \frac{8\pi hc}{\lambda^5} e^{-hc/\lambda kT} \qquad ...(11)$$

which is Wien's law.
(*ii*) For large temperatures, λT is large.

$$\therefore \quad e^{hc/\lambda kT} = 1 + \frac{hc}{\lambda kT} + \left(\frac{hc}{\lambda kT}\right)^2 \cdot \frac{1}{2!} + ...$$

$$= 1 + \frac{hc}{\lambda kT}$$

or

$$E\lambda . d\lambda = \frac{8\pi kT}{\lambda^4} \cdot d\lambda \qquad ...(12)$$

which is Rayleigh-Jean's law. From above, it is clear that **Plank's law is the perfect law while all other laws follow from it.**

Also from Planck's law, energy changes take place discontinuously and discretely as an integral multiple of small unit of energy *hv* which is called **quantum or a photon.**

Example 8. *A furnace is at a temperature of 2000 K. At what wavelength would you expect the greatest intensity of radiation? Is it in visible region? (Given b = 2.898 × 10^{-3} mK).*

Solution: We have the relation from Wien's Law

$$\lambda_m . T = \text{Constant} = b \text{ (say)}$$

$$\therefore \quad \lambda_m = \frac{b}{T} = \frac{2.898 \times 10^{-3} \text{ mK}}{2000 \text{ K}}$$
$$= 1.449 \times 10^{-6} \text{ m}$$
$$= 1.449 \text{ micron.}$$

Visible region is between 4000 Å to 800 Å (0.4 micron to 0.8 micron). So the above wavelength is beyond red colour of visible spectrum.

5.24. RADIATION-PYROMETRY

The measurement of high temperatures is called pyrometry and the instruments used are called **pyrometers.** Because of the basic properties of the instrument used for the measurement of temperatures, *e.g.*, gas thermometers, resistance thermometers, thermoelectric thermometers etc. have got their restricted use only upto certain range of temperatures. For measuring high temperatures, phenomenon of radiation because of its dependence on temperature is used and such instruments are called **Radiation Pyrometers.**

The pyrometers have got the advantage over other thermometers because they can be used to measure very high temperatures at whatsoever distance the body may be. In other instruments, the body whose temperature is to be measured is to be placed in contact with the instrument, that restricts the use of the instruments which in practice, is about 1600°C. But in the case of 'radiation pyrometers there is no upper limit of temperature as the body is not to be placed in contact with the pyrometer. Moreover, radiation formulas are valid at all temperatures and so no extrapolation is required for very high temperatures.

However, these instruments suffer from the following drawbacks:

(*i*) Since they are based on the principles of black body radiations, when they are used to measure the temperature of a body, they give that temperature only at which a perfectly black body gives intensity of radiation falling on the instrument. However, black body radiations being more intense for the same temperature, there will be greater error in temperature measurement, if there is more departure from the perfect black body.

(*ii*) Below 600°C, the radiation pyrometers cannot be used as the radiations from a body are too small to be measured by these instrument.

5.25. TYPES OF RADIATION PYROMETERS

For measurement of high temperatures, two types of instruments are used:

(*i*) Total radiation pyrometers. In this case, the total radiations from a source are measured and then the temperature is deduced from the application of Stefan's Law.

(*ii*) Optical pyrometers. In this type of instrument the intensity of radiations from a source are compared with the intensity from a standard temperature which gives intensity of emission of the same wavelength and then the temperature of the source is deduced using Planck's Law.

5.26. TOTAL RADIATION PYROMETERS

This device was first designed by Fery and so is sometimes called Fery's total radiation pyrometer. It consists of a concave mirror M, which is made of copper and polished with Nickel on the front surface with a small hole in the centre, where eye piece E is fitted. In front of M, a small diaphragm D is put and immediately behind it is metallic strip S, whose front face is blackened and on the back surface is fitted one junction of a thermocouple T. The other junction of T as well as S are protected by suitable screens from the direct radiations. The leads from the thermocouple are connected to a millivoltmeter mV which measures e.m.f. developed in the circuit. The radiations are allowed to fall on the mirror M and focussing of radiations from M on S can be controlled by the movement of the mirror M.

For accurate focussing, the diaphragm D consists of two semi-circular mirrors inclined to each other at an angle of about 5° with a central hole of about 1.5 mm for the incident radiations to pass. When the diaphragm is viewed through eye piece E, if the focussing is correct the two halves will form a complete circle and if the focussing is not correct, the two halves will appear displaced (as shown in Fig. 5.11). The concave mirror is moved till the focussing becomes accurate and relative displacement of the two halves disappears.

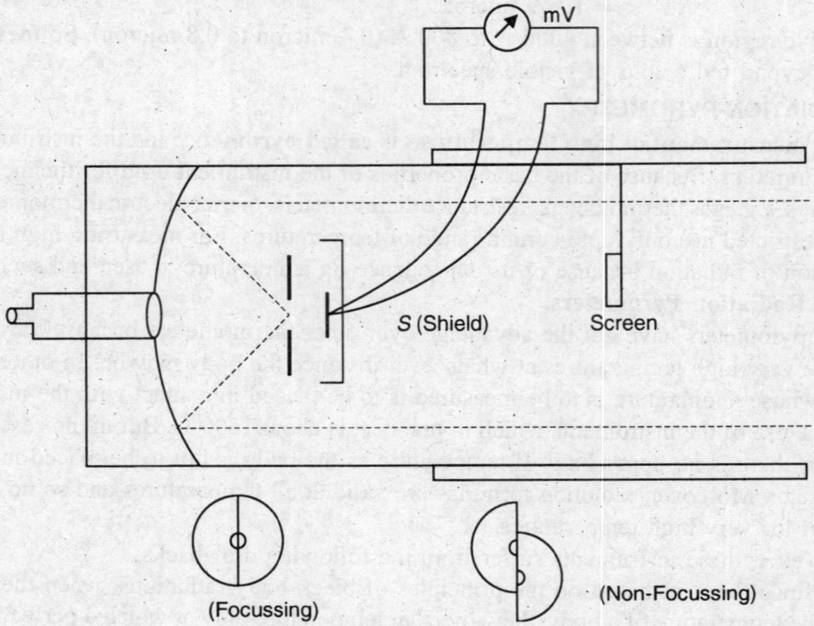

Fig. 5.11

It is always advisable to form heat image by concave mirror which should be larger than the size of diaphragm D, because in that case the reading of pyrometer is independent of its distance from source. It is because of the fact that inverse square law is observed by radiations and when the size of image is greater than the size of diaphragm, the amount of radiations received/sec/cm^2 is constant.

If V corresponds to the millivoltmeter reading, to the temperature of the surroundings, then unknown temperature T is given by $V = a\ .\ (T^b - T_0^b)$ where b is constant and varies between 3.8 to 4.2, showing some deviation from fourth power law. In order to get accurate results to account for the said deviations, the instrument is standardized by means of some standard sources or standard thermometers. However, because of direct calibration of the instrument, its use is restricted within the range of calibration.

The above defect can be surmounted by a rotating sector device, which allows only part of radiations to fall on the instrument and so very high temperatures can be measured as follows:

An opaque disc making an angle θ at the centre is cut off as shown in Fig. 5.12. This disc is rotated in the path of radiations, so that only a fraction of radiations enter the pyrometer per revolution. If T_1 is the temperature now measured, then the actual tem-perature T of the source is given according to Stefan's Law as

$$\frac{T_1^4}{T^4} = \theta/360^\circ$$

Let $\theta = 22.5^\circ$ and $T_1 = 2000$ K.

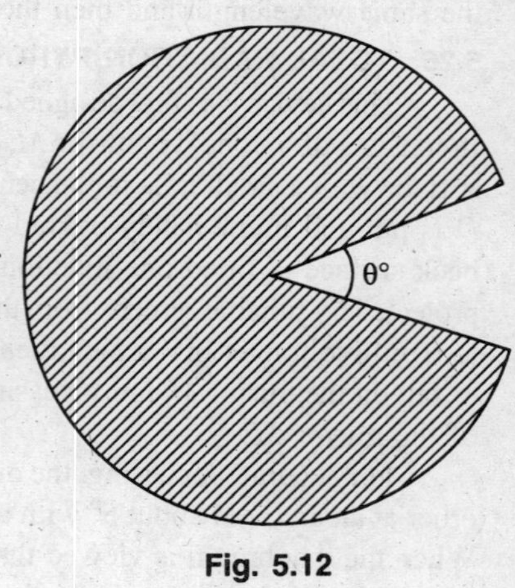

Fig. 5.12

$$\therefore \quad T = 2000 \times \left(\frac{360}{22.5}\right)^{\frac{1}{4}} = 2000 \times (16)^{\frac{1}{4}}$$
$$= 2000 \times 2 = 4000 \text{ K}.$$

So with the arrangement, as shown in the above example, the temperature as measured is very less as compared to the actual temperature of the source but the actual temperature can be measured accurately by knowing the sector angle.

5.27. OPTICAL PYROMETER

In this type of pyrometer, the most important one is the disappearing filament pyrometer. The construction and working of the instrument is given as under:

It essentially consists of a telescope, the crosswire of which is replaced by an electric lamp L. There are two diaphragms on either side of the lamp, the lamp can be heated to different high temperatures by adjusting the current from the battery circuit with the help of a rheostat R, the current being measured with ammeter A.

Radiations from the source whose temperature is to be measured are made to fall on the objective O and the position is so adjusted that these are focussed on the lamp where the heat image of the source is formed. In order to have a good comparison of the brightness, lights of only one wavelength *e.g.*, red, are compared by putting a red glass filter is placed behind eye piece. If the current is more, the filament will appear brighter than the image of the heat source and if the current is less, the filament will appear darker than the heat source. The current is so adjusted, so that the filament disappears. The red glass filter is used to match the monochromatic lights from the source and the lamp which is easier for the eye for comparison.

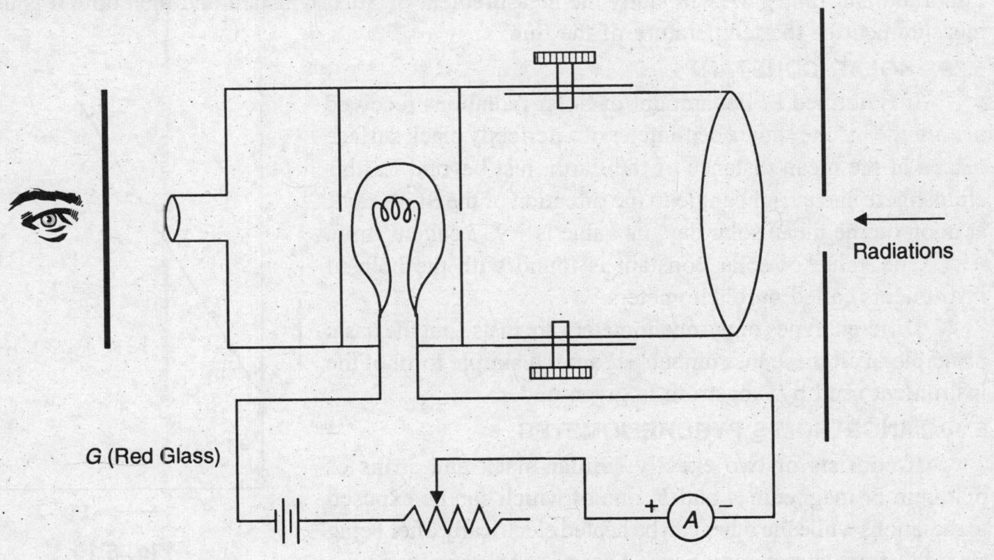

Fig. 5.13

When the matching is acheived, the temperature of the source can be obtained by knowing the temperature of the filament of the lamp which is given by

$$I = a + bT + cT^2$$

where a, b, c are constants and can be found by standardizing the instruments by three known sources and T is the unknown temperature, then according to Planck's law the temperature of the source is equal to the temperature of the filament as the energy emitted/sec/cm^2 in both the cases is same.

For using the instrument to measure very high temperatures, sector type device may be used as discussed earlier.

5.28. RADIATIONS FROM THE SUN (SOLAR RADIATIONS)

The sun which forms the centre of our planetary system is the source which gives out radiations to the number of planets around it and in the absence of such radiations life at our planet, Earth, would have been impossible. It must be understood that Earth gets only a fraction of radiations from the Sun as major part of the radiations is lost by reflection and absorptions in the Earth's atmosphere. Moreover, water vapours, clouds, snow and dust particles scatter the radiations and send them back into the space.

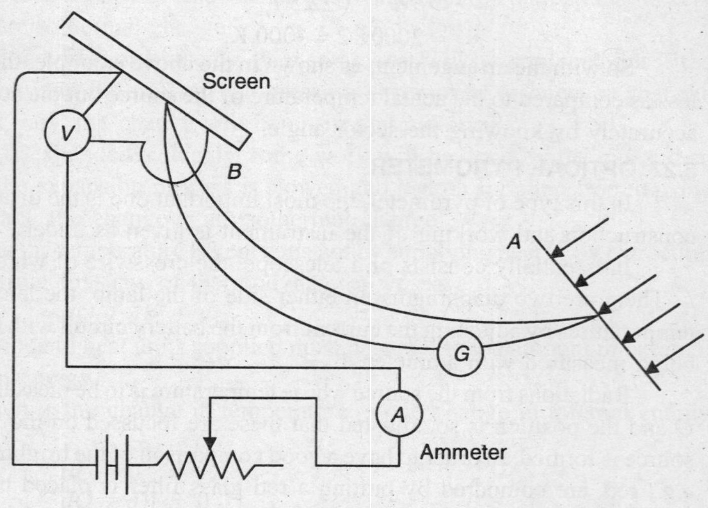

Fig. 5.14

Even from small amount of radiations that we get on Earth, it is possible to measure the temperature of the Sun from the laws of radiations. For this purpose, a term *Solar constant* is very important and firstly we will study the measurement of Solar constant and then how it is used for measurement of the temperature of the Sun.

5.29. SOLAR CONSTANT

It is defined as the amount of Solar radiations received per minute by one square centimeter of a perfectly black surface placed at the mean distance of the Earth, just beyond Earth's atmosphere and at right angle to the direction of the Sun's rays, at noon on the mean solar day. Its value is 1.937 cal/cm^2/min.

The value of solar constant is found with the help of instruments called pyroheliometers.

Different types of pyroheliometers are in use but the basic principle in all these instruments is same. A simple form of the instrument used by Angstrom is given under.

5.30. ANGSTROM'S PYROHELIOMETER

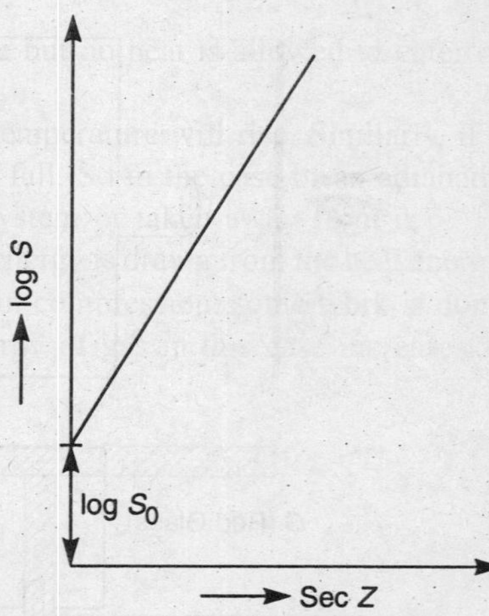

Fig. 5.15

It consists of two exactly similar black and strips of platinum or magneuin A and B, one of which can be exposed to radiations while the other can be heated electrically after being shielded from direct radiations. Thermocouple junctions are fixed on the backs of A and B. As the radiations fall on strip A, it gains solar energy and so the galvanometer C, gives some deflection. Some current is passed through B and its value is so adjusted that there is zero deflection in the galvanometer. In this condition, both A and B are at the same temperature and so from the known values of current and voltage, the heat supplied to B and consequently solar energy absorbed by A can be obtained.

So the above instrument can be used for determining the energy absorbed per unit area of a given surface. In order to eliminiate the error due to absorption by atmosphere, solar constant is found at different positions of the Sun on the same day when the sky is under same conditions.

If V is the applied voltage to strip B and I is the current, then heat supplied to it per sec.

will be given by

$$\frac{VI}{J} \quad ...(1)$$

If a be the area of the exposed surface, a is the absorptive power and S is solar constant, then we have the heat energy absorbed by A

$$= \alpha \,.\, S \,.\, a \quad ...(2)$$

From Equations (1) and (2),

$$\frac{VI}{J} = \alpha \; S \,.\, a$$

$$\therefore \quad S = \frac{VI}{\alpha . J . a}$$

Let S be the observed value of solar constant, the true value S_0 can be obtained from the relation,

$$S = S_0 \; a^{\sec Z}$$

where Z is the zenith distance of the angular altitude of the Sun, a is transmission coefficient, its value varies between 0.55 to 0.85.

Taking log S on both sides

$$\log S = \log S_0 + \sec Z \log a$$

Now if a graph is plotted between log of S along Y axis and sec Z along X-axis it will be a straight line and the intercepts on Y-axis will give log S_0, from which S_0 can be found. The most accurate values of S_0 vary between 1.9 to 2.6 cal/cm^2/min.

5.31. TEMPERATURE OF THE SUN

The Sun consists of a central hot core called corona, surrounded by a thick shell called photosphere. The temperature of the central core of the Sun is very high (~ 20,000,000 K) where most of the gases are burning. However, we are concerned about the temperature of the photosphere of the Sun which may be considered to constitute a black body. Above photosphere lies *chromosphere*, an irregular layer of gases. It is in *chromosphere* that Sun spots, flares and prominences occur.

Let R be the amount of radiations emitted/sec/cm^2 surface of the Sun.

Let a be the radius of the Sun.

So the total amount of radiant energy emitted by the Sun/min = $4\pi a^2 \times R \times 60$.

This energy is received by a sphere whose radius is equal to the radius of the Earth's orbit.

If b is the radius of the earth's orbit, the total amount of the energy of radiation received by the surface of a sphere of radius b/min. = $4\pi b^2 \times S$, where S is solar constant.

So we have $\quad 4\pi a^2 \times R \times 60 = 4\pi b^2 \times S$

$$\therefore \quad R = \frac{b^2}{a^2} \frac{S}{60}$$

From Stefan's Law $\quad R = \sigma T^4$

$$\therefore \quad T = \left(\frac{R}{\sigma}\right)^{\frac{1}{4}} = \left(\frac{b^2}{a^2} \cdot \frac{S}{60} \cdot \frac{1}{\sigma}\right)^{\frac{1}{4}}$$

Given

$$S = 1.9 \text{ cal/cm}^2\text{/min.}$$

$$\sigma = 1.37 \times 10^{-12} \text{ cal/cm}^2\text{/sec/K}^4$$

$$a = 6.88 \times 10^5 \text{ km.}$$

$$b = 1.50 \times 10^8 \text{ km.}$$

$$\therefore \quad \mathbf{T = 5760\ K.}$$

5.32. TEMPERATURE OF STARS

The stars, which are at far off distances than the sun, may be considered to constitute number of Suns. However, since diameters of most of the stars are not accurately known. So, Stefan's Law cannot be applied in these cases. However, with the application of Wien's Law ($\lambda_m . T$ = const.), their temperature can be measured accurately. Temperatures of these stars vary between 2800 K red star to 30,000 K (Bluish Stars). It should be clearly noted that these temperatures refer to surface temperatures only while the interiors will have several million degrees of temperatures.

Example 9. *Calculate the temperature of the sun treated as a black body from the following data:*

Ratio of the radius of the sun to the radius of the earth orbit around Sun = 4.649 × 10^{-3}, $\sigma = 5.77 \times 10^{-5}$ ergs/sec/cm²/K⁴, S = 1.937 cal/cm²/min.

Solution: In this case, ratio of the radius of the Sun (a) to the radius of the Earth's orbit (b) *i.e.,* $\frac{a}{b} = 4.649 \times 10^{-3}$.

Stefan's const. $\sigma = 5.77 \times 10^{-5}$ ergs/cm²/sec/K⁴

$= 1.37 \times 10^{-12}$ cal/cm²/sec/K⁴,

$S = 1.937$ cal/cm²/min.

$$\text{Temp. of the Sun, } T = \left(\frac{b^2}{a^2}\cdot\frac{S}{60}\cdot\frac{1}{\sigma}\right)^{\frac{1}{4}}$$

$$= \left(\frac{1}{4.649\times10^{-3}}\times\frac{1.937}{60}\times\frac{1}{1.37\times10^{-12}}\right)^{\frac{1}{4}}$$

$$= \mathbf{5742\ K.}$$

Example 10. *If the Sun's surface radiates heat at the rate of 6.3×10^7 watts/m². Calculate the temperature of the Sun's surface, assuming it to be a black body ($\sigma = 5.7 \times 10^{-8}$ W/m²/K⁴).*

Solution: From Stefan's Law,

$$R = \sigma T^4$$

$$R = 6.3 \times 10^7 \text{ W/m}^2$$

$$\sigma = 5.7 \times 10^{-8} \text{ W/m}^2/\text{K}^4$$

$$\therefore \quad 6.3 \times 10^7 = 5.7 \times 10^{-8} \times T^4$$

$$\therefore \quad T^4 = \frac{6.3\times10^7}{5.7\times10^{-8}} = \frac{6.3}{5.7}\times10^{15}$$

$$\therefore \quad T = \mathbf{5500\ K.}$$

Example 11. *Calculate the temperature of the sun from the following data, S = 1.34 kW/m², radius of the Sun = 6.92×10^5 km. Distance of the Sun from the Earth = 1.5×10^8 km and $\sigma = 5.7 \times 10^{-8}$ W/m²/K⁴.*

Solution:

$$S = 1.34 \text{ kW/m}^2$$

$$a = 6.92 \times 10^5 \text{ km}$$

$$b = 1.5 \times 10^8 \text{ km}$$

$$\sigma = 5.7 \times 10^{-8} \text{ W/m}^2/\text{K}^4$$

$$\therefore \text{ Temperature of the Sun} \quad T = \left(\frac{b^2}{a^2}\cdot\frac{S}{60}\cdot\frac{1}{\sigma}\right)^{\frac{1}{4}} = \left[\left(\frac{1.5\times10^8}{6.92\times10^5}\right)^2\cdot\frac{1.34}{5.7\times10^{-8}}\right]^{\frac{1}{4}}$$

$$T = \mathbf{5998\ K.}$$

Example 12. *Calculate the surface temperature of the sun and moon, given $\lambda_m = 4753 Å$ and 14μ respectively, where λ_m corresponds to maximum intensity of emission.*

Solution: From Wien's Displacement Law,

$$\lambda_m T = \text{const.} = 0.2898$$

(*i*) For Sun $\lambda_m = 4753 \times 10^{-8}$ cm

$$\therefore \quad T_s = \frac{0.2898}{4753\times10^{-8}} = 6097 \text{ K}$$

(*ii*) For Moon $\lambda_m = 14\mu = 14 \times 10^{-4}$ cm

$$\therefore \qquad T_m = \frac{0.2898}{14 \times 10^{-4}} = \mathbf{207\ K.}$$

QUESTIONS AND PROBLEMS

1. What is a black body? What are the characteristics of a black body radiations and how these are realized in practice? How Stefan's law has been verified by it?

2. Explain the terms 'emmissive power' and absorptive power. Deduce that at any temperature the ratio of the emissive power to the absorptive power of a substance is constant and is equal to the emissive power of a perfectly black body.

3. State and explain the laws relating to the radiation and temperature of a radiating body. What do you mean by (*a*) perfectly black body (*b*) coeff. of absorption (*c*) coeff. of emission?

4. Explain Stefan's law of radiation and give Boltzman's theoretical proof. How Stefan's constant can be determined in laboratory?

5. What is Stefan's fourth power law? How it can be verified?

6. Describe a method for measurement of high temperatures.

7. Define Solar constant. Explain with necessary theory how Solar constant is used to find the temp. of the Sun?

8. What is pyrometry? Give in details the different types of pyrometers and how they are used for measurement of high temperatures?

9. The black body at 0°C radiates heat at the rate of 315 W/m^2. How much heat is radiated in 1 minute by a black sphere of diameter 10 cm. at 1000°C? **[Ans.** 2.8×10^5 J]

10. Calculate the effective temp. of the sun from the following data:
Solar const. = 2.279 cal/cm^2/min. Radius of the Sun = 4.3×10^5 miles, Radius of Earth's orbit 9.3×10^7 miles, Stefan's const. = 1.37×10^{-12} cal/cm^2/sec/K^4. **[Ans.** 6001 K]

11. By experiment, it is found that the amount of heat energy from the Sun falling on 1.0 m^2 of the earth's surface/sec is 1.8×10^3 J. The distance of the Sun from the Earth is 1.5×10^8 km. and radius of the Sun is 6.9×10^5 km. Assuming sun to be a black body, calculate its temperature. **[Ans.** 6200 K]

12. An iron furnace radiates 1.53×10^5 cal/hr through an opening of cross-section 10^{-4} m^2. If the emissivity of furnace is 0.8, calculate the temperature of the furnace. (Given $\sigma = 1.36 \times 10^{-8}$ cal/m^2/s/K^4). **[Ans.** T = 2500 K]

CHAPTER

6

LAWS OF THERMODYNAMICS

6.1. INTRODUCTION

The branch of science in which we study the heat in motion is called **thermodynamics.** Also thermodynamics may be defined as the science that shows the relation between heat and mechanical work. The principles of thermodynamics are very general and give a relation between heat and various forms of energy like electrical, chemical, optical etc.

6.2. ZEROTH LAW OF THERMODYNAMICS

When two bodies are in thermal contact with each other and no exchange of heat takes place between them, they are said to be in thermal equilibrium. Also, it is noticed that when two bodies are in **thermal equilibrium** with a third body then these two bodies are also in **thermal equilibrium** with each other. This statement is called **zeroth law of thermodynamics.** The term **zero** is used as this law logically provides the basis for **first and second laws of thermodynamics** and is assumed in these two laws. Also **zeroth law of thermodynamics** forms the basis of **temperature.** In fact, three bodies which are in thermal equilibrium with one another ensure **equal thermal activity** in them and this property is called **temperature.** So temperature of body is the **property which determines whether the body is in thermal equilibrium with neighbouring bodies or not.** It is clear that if the two bodies are not in thermal equilibrium, they will be at **different temperatures.**

6.3. FIRST LAW OF THERMODYNAMICS

According to the first law of thermodynamics, whenever work is converted into heat or heat into work, the quantity of work is mechanically equivalent to the quantity of heat, *i.e.*,

$$W \propto H$$

or

$$\frac{W}{H} = J \text{ (constant)}$$

where J is called mechanical equivalent of heat and has value 4.2×10^7 ergs/cal in C.G.S. units and 4.2 Joule/cal in M.K.S. units.

The above law can also be stated as under.

To produce a definite amount of heat, mechanical work is needed and *vice versa i.e.,* ratio of amount of work done to heat produced is always constant.

The first law, as seen above, is based on the principle of conservation of energy *i.e.*, energy can neither be created nor destroyed but changes its form. In fact first, law deals with the equivalence of work with the amount of heat produced.

6.4. SECOND LAW OF THERMODYNAMICS

The first law of thermodynamics shows an equivalence of mechanical work with the quantity of heat produced but does not give an idea about the conditions under which work is transformed into heat and *vice versa*. It also does not give any idea about the direction in which the heat transfer takes place. In order to explain these two conditions, second law of thermodynamics was put forward which may be stated in the following two ways.

6.5. CLAUSSIUS STATEMENT

It is impossible for a self-acting machine which is working in a cyclic process and is unaided by any external energy to convey heat from a body at a lower temperature to a body at a higher temperature.

The above statement may be given in other words as-heat cannot flow from a cold body to a hot body without the aid of some external agency.

6.6. KELVIN STATEMENT

It is impossible to get a continuous supply of work by cooling a body to a temperature lower than that of the coldest of its surroundings.

From the above statements, it is clear that no heat engine can convert whole of heat energy into useful work which may be restated as no actual or ideal engine that is operating in cycles can convert the total heat energy supplied to it into work. The ratio of the amount of heat converted into useful work and the heat taken by it is a fraction and is known as thermal efficiency of the engine. From above statements, it is clear that thermal efficiency of an engine is always less than unity.

6.7. THERMODYNAMIC PROCESSES

In order to study the transformation of heat energy into work, all the thermodynamic processes have been classified into two groups:

(*i*) Reversible process

(*ii*) Irreversible process.

6.8. REVERSIBLE PROCESS

A reversible process is that which can be retraced in the reverse direction *i.e.*, if a minute change is made in the reverse direction, the process is reversed completely and the working substance undergoes exactly, the same conditions as it does in the direct process.

Examples of reversible process. All the isothermal and adiabatic changes which are performed very slowly are examples of reversible process if in these changes it is assumed that no heat is lost in friction or to surroundings. In that case the total heat energy supplied say, to a given mass of gas at constant pressure will be used for some external work or for expansion of the gas. Now if the same work is done on the gas in compression, the same quantity of heat will be ejected.

For example, let us consider the conversion of water at 0°C to ice at 0°C by removing same quantity of heat. Now if same quantity of heat is given at 0°C, ice will melt and will be converted into water at 0°C. However, it should be noted that a reversible process must be in thermodynamic equilibrium at all the times.

6.9. IRREVERSIBLE PROCESS

A process which cannot be retraced is called irreversible process.

Examples of irreversible process. 1. Work done against friction is irreversible because even if the direction of motion is reversed, work has to be done against friction.

2. Joule-Thomson effect is irreversible because if the flow of gas is reversed, similar heating and cooling effect are not observed. In the same way conduction, radiation, diffusion etc. are irreversible process.

3. Heat due to flow of current through a conductor is irreversible because when the direction of current is reversed, heating is produced.

6.10. CYCLES

If a series of operations are taking place in some particular order such that after such set of operations, initial conditions are restored, the process is known as a cycle.

In the case of engines, the working fuel may be in the form of a gas and after undergoing series of changes in pressure, volume and temperature returns to its original form.

6.11. ISOTHERMAL CHANGE

When a change in pressure and volume of a substance takes place but the temperature is kept constant, the change is known as isothermal change.

As an example, if a gas is compressed suddenly, then some heat will be produced but if the compression process is very slow and the heat of compression is removed at once, so that the temperature of the gas does not vary, the change will be an isothermal one.

In the same, way if a gas is expanded suddenly, some work will be done by the gas and so some heat will be extracted. If the expansion process is slowed and heat is supplied from outside so that the temperature is constant, the change is an isothermal change.

So, in the isothermal change the temperature is kept constant by supplying heat or by extracting heat from the substance. Now, from first law of thermodynamics we have

$$dQ = dW + dU \quad ...(1)$$

i.e., All transformation of energy due to heat units supplied must be equal to the amount of external work plus the change in internal energy.

Now since in this case, there is no change in temperature, so no change in internal energy takes place, *i.e.,*

$$dU = 0$$

∴ From Equation (1) $\quad dQ = dW + 0$

or $\quad dQ = dW \quad ...(1)$

So in an isothermal change, we have heat added (or subtracted)

= Work done by (or on) the gas

In the case of a perfect gas, isothermal change is represented by Boyle's law equation,

i.e., $\quad pV = \text{constant}.$

6.12. ADIABATIC CHANGE

If a substance undergoes a change in pressure or volume but no heat is allowed to enter or leave it, the change is called an adiabatic change.

If a gas is compressed and no heat is taken from it, its temperature will rise. Similarly, if a gas expands and no heat is supplied to it, its temperature will fall. So in the case of an adiabatic change, temperature changes and no heat is supplied to the system or taken away from it.

Now if a gas expands, it does some external work and the energy is drawn from the heat energy of the gas. So the heat energy of the gas decreases. Similarly in compression, some work is done on gas and heat energy is gained by the gas. So the heat energy of gas in this case increases.

From first law of thermodynamics

$$dQ = dU + dW$$

Since no heat leaves or enters the gas in adiabatic change,

∴ $\quad dQ = 0$

∴ $\quad dU + dW = 0$

or $\quad dU = -dW \quad ...(2)$

i.e., in the adiabatic process, increase (or decrease) in internal energy

= Work done on (or by) the gas

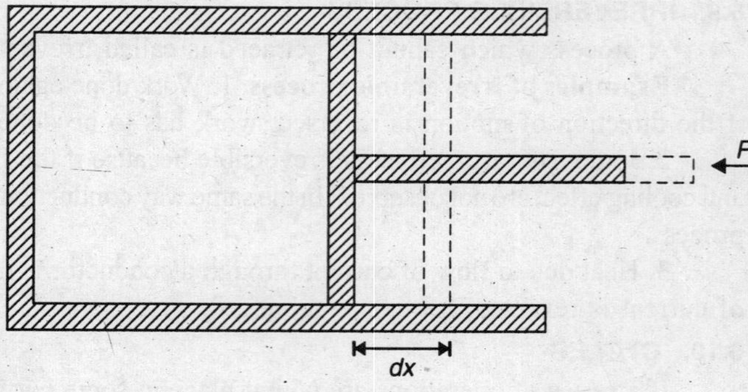

Fig. 6.1

In order to get an equation for the adiabatic change, let us consider a gram-molecule of a perfect gas put in a perfectly non-conducting cylinder which is fitted with a non-conducting piston. Let the pressure, temperature and volume of the gas be P, T and V respectively. Let the gas be compressed adiabatically, so that the piston moves towards inside the cylinder through a distance dx, as shown in Fig. 6.1.

If A be the area of cross-section of the piston, then total force applied will be $P \times A$ and the work done by the piston

$$dW = \text{Force} \times \text{distance}$$
$$= P \times A \times dx$$
$$= P.dV$$

where $A.dx = dV$ corresponds to change in volume.

Now due to compression, there will be rise of temperature dT and the heat of compression will be equal to $C_v \,.\, dT$ where C_v is gram molecular specific heat.

From the first law of thermodynamics, we have

$$dQ = dU + dW = dU + P.dV$$

and for an adiabatic change, $\quad dQ = 0$

$$\therefore \quad dU + PdV = 0$$

Also by definition $\quad C_v = \dfrac{dU}{dT}$

or $\quad dU = C_v.dT$

$$\therefore \quad C_v.dT + PdV = 0 \quad \text{...(1)}$$

In the case of a perfect gas, $\quad PV = RT$

Differentiating this equation, we have

$$PdV + VdP = R.dT$$

or $\quad dT = \dfrac{PdV + VdP}{R}$

Putting the value of dT in Equation (1)

$$C_v \cdot \frac{PdV + VdP}{R} + P.dV = 0$$

or $\quad C_v.P.\,dV + C_v.VdP + R.P.dV = 0$

or $\quad (C_v + R)\,.\,P.\,dV + C_v.V.\,dP = 0 \quad \text{...(2)}$

Also we have $\quad C_p - C_v = R$

or $\quad C_p = C_v + R$

$\therefore$ Equation (2) may be written as

$$C_p.\,P.\,dV + C_v.V.\,dP = 0$$

Dividing by $C_v.P.\,V$, we get

$$\frac{C_p}{C_v} \cdot \frac{dV}{V} + \frac{dP}{P} = 0$$

Now, $\quad \gamma = \dfrac{C_p}{C_v}$

$$\therefore \quad \gamma \cdot \frac{dV}{V} + \frac{dP}{P} = 0$$

Integrating both sides, we get

$$\gamma.\log_e V + \log_e P = \text{constant}$$

or $\quad \log_e PV^\gamma = \text{constant}$

or $\quad PV^\gamma = \text{constant}$

which in other words mean that if P_1 and V_1 are the initial values of pressure and P_2, V_2, the corresponding final values, then

$$P_1V_1^\gamma = P_2V_2^\gamma \quad \text{...(3)}$$

From above relation, we can derive a relation corresponding to adiabatic change in terms of volume and temperature also.

We have derived the relation $PV^\gamma = K$ (constant) ...(4)

Also for a perfect gas, $\quad PV = RT$

$$\therefore \quad P = \frac{RT}{V}$$

Putting in Equation (4), we have $\frac{RT}{V}.V^{\gamma} = K$

or $$T.V^{\gamma-1} = \frac{K}{R} = K_1 \text{(say) (another constant)} \quad ...(5)$$

Similarly, we can get an expression for pressure and temperature also.

From gas equation $$PV = RT$$

or $$V = \frac{RT}{P}$$

Putting in Equation (4), $P\left(\frac{RT}{P}\right)^{\gamma} = K$

or $$\frac{T^{\gamma}}{P^{\gamma-1}} = \frac{K}{R^{\gamma}} = K_2 \text{ (another constant)}$$

$$\therefore \quad T^{\gamma}.P^{1-\gamma} = K_2 \text{ (constant).}$$

6.13. WORK DONE DURING ISOTHERMAL EXPANSION

Let us consider a gram molecule of a perfect gas placed in a cylinder having perfectly non-conducting walls and perfectly conducting bottom, fitted with a non-conducting piston.

Let P and V be the pressure and volume of the gas at temperature T K and let the cylinder be placed on a source of heat at the same temperature *i.e.,* at T K.

Now if the piston is allowed to move slowly outwards, expansion of the gas takes place and so the gas does some work and thereby its temperature will fall. Heat will be absorbed by the gas from the source so that it has the same constant temperature TK. Since the temperature is not allowed to vary, so the change is isothermal.

Let A be the area of cross-section of the piston and let it move through a small distance dx, so the small amount of work done by the gas is given by

$$dW = P \times A \times dx$$
$$= P.dV$$

We can represent a relation between pressure and volume under isothermal conditions graphically by line AB depicting isothermal change as shown in [Fig. 6.2 (*a*)], where a small change in volume dV corresponding to some pressure P is given by an area of the strip *abcd* and will be equal to $P.dV$.

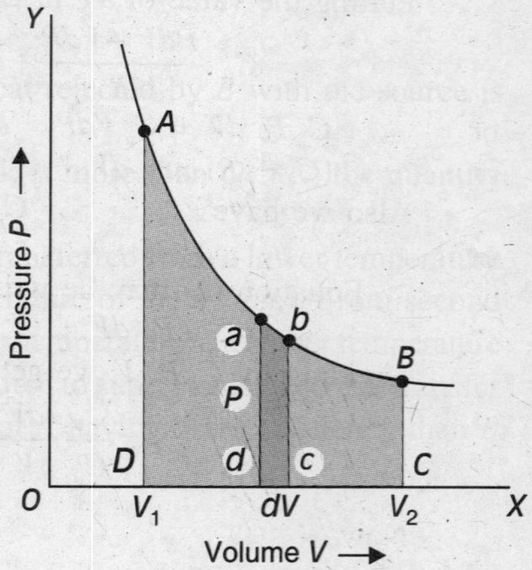

Fig. 6.2 (*a*)

From gas Equation we have

$$PV = RT$$

$$\therefore \quad P = \frac{RT}{V}$$

so amount of work done is given by

$$dW = P.dV = \frac{RT}{V} \cdot dV$$

Now, total work done when the gas expands isothermally from an initial volume V_1 to final volume V_2 is given by the area $ABCD$.

$\therefore$ Total work done $$W = \int_{V_1}^{V_2} RT \cdot \frac{dV}{V} = RT \log_e \frac{V_2}{V_1}$$

Now, from first law of thermodynamics

$$dQ = dU + dW$$

Since in an isothermal change, temperature remains constant, so there is no change in internal energy of the gas, *i.e.,* $dU = 0$

$\therefore \qquad dQ = dW$

So in the isothermal change, the total heat supplied is changed into work. If W be the total work done and Q be the amount of heat supplied, then

$$Q = \frac{W}{J} \text{ calories}$$

where J is Joule's mechanical equivalent of heat.

6.14. WORK DONE IN ADIABATIC EXPANSION

Let us consider a gram molecule of a gas contained in a perfectly non-conducting cylinder fitted with a non-conducting piston. Let P, V and T be the pressure, volume and temperature of the gas respectively. If suddenly the piston is moved outward, the gas expands and so does some external work. Since we are considering the adiabatic expansion, so in this case no heat is supplied from outside and the energy is taken from the gas itself so that its temperature will be lowered.

If A be the area of cross-section of the piston and if it moves through a distance dx, then

$$\text{Work done} = P \times A \times dx$$

$$dW = P \times dV \qquad \text{...(1)}$$

We can represent the relation between pressure and volume in the adiabatic change graphically and AB will correspond to this adiabatic change as shown in [Fig. 6.2 (b)].

Now the work done for a small change in volume dV, corresponding to some pressure P is given by the area of the small strip abcd and its value is equal to $P.dV$.

Now for the case of adiabatic change

$$PV^\gamma = K$$

$$\therefore \quad P = \frac{K}{V^\gamma}$$

$\therefore$ From Equation (1), $dW = \frac{K}{V^\gamma} \cdot dV$

Total work done when the gas expands adiabatically from volume V_1 to volume V_2 will be given by area $ABCD$.

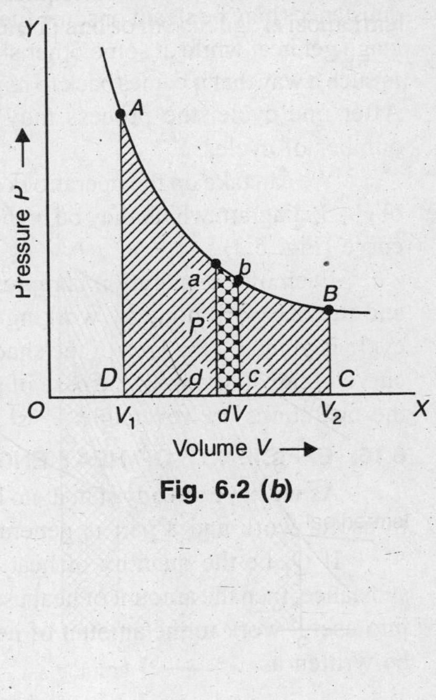

Fig. 6.2 (b)

$$\therefore \text{Total work done } W = K \int_{V_1}^{V_2} \frac{dV}{V^\gamma}$$

$$= \frac{K}{1-\gamma}\left[\frac{1}{V_2^{\gamma-1}} - \frac{1}{V_1^{\gamma-1}}\right]$$

$$= \frac{K}{\gamma-1}\left[\frac{1}{V_1^{\gamma-1}} - \frac{1}{V_2^{\gamma-1}}\right]$$

Now, we have for an adiabatic change

$$P_1V_1^\gamma = P_2V_2^\gamma = K$$

$$\therefore \quad W = \frac{1}{\gamma-1}\left[\frac{P_1V_1^\gamma}{V_1^{\gamma-1}} - \frac{P_2V_2^\gamma}{V_2^{\gamma-1}}\right]$$

$$= \frac{1}{\gamma-1}[P_1V_1 - P_2V_2]$$

If T_1 and T_2 are the initial and final temperatures of the gas then $P_1V_1 = RT_1$ and $P_2V_2 = RT_2$

$$\therefore \quad W = \frac{R}{\gamma-1}[T_1 - T_2]$$

So, the work done in an adiabatic expansion is proportional to the fall in temperature. In the case of adiabatic compression, the work done will be proportional to the rise in temperature.

6.15. HEAT ENGINE

A device used to convert heat energy into mechanical work through a medium which is called a working substance (normally a gas or vapours) is called a **heat engine.**

It should be noted that the temperature at which heat is supplied to the working substance is always greater than that at which it is rejected. Hot body which supplies heat is called source while the cold body into which heat is rejected is called sink.

Heat engines in which water or steam is used as working substance are called steam engines such as locomotive engines or steam turbines while if the working substance is air, then these are called Internal combustion engines. Such engines are used in Aeroplanes, cars, motor cycles etc.

In actual practice, a heat engine consists of three parts:

(*i*) Working substance

(*ii*) Heat source

(*iii*) Sink.

Engines which require one complete revolution of the crank are called two stroke engines while those which require two complete revolutions of the crank are called four stroke engines and the working substance is said to perform work in a closed cycle. In the process of a cycle, the working substance (may be steam or some fuel vapours) at some stage get heat while at some other stage gives out heat in such a way that it comes back to its original condition. After one cycle, the process may be repeated into number of cycles.

We can take up the operations of a cycle in terms of P—V diagram which may be in the form of a closed curve (Fig. 6.3).

Each of the operation is represented by a curve and the net work done by working substance in one cycle is given by the area of the shaded portion of the curve in the diagram. In the case of an ideal cycle, all the operations are reversible.

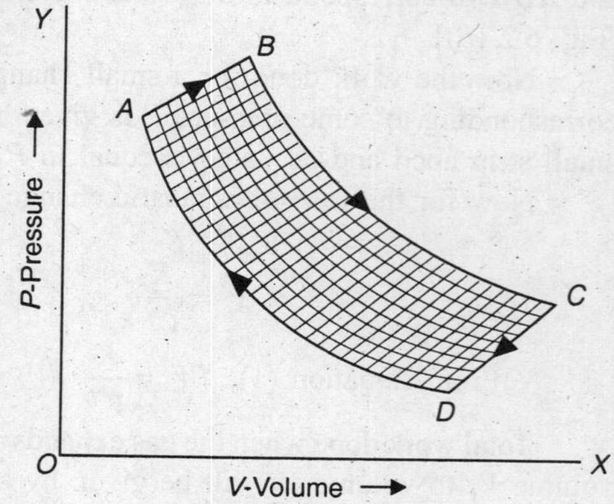

Fig. 6.3

6.16. EFFICIENCY OF HEAT ENGINES

As we have already stated, no heat engine can completely convert the amount of heat supplied to useful work and a part is generally rejected as used heat.

If Q_1 be the quantity of heat supplied and Q_2 be the quantity of heat rejected by working substance, then the amount of heat used in useful work will be $Q_1 - Q_2$. The ratio of the heat converted into useful work to the amount of heat supplied is called thermal efficiency of the engine and may be written as

$$\text{Efficiency} = \frac{\text{heat converted into useful work}}{\text{heat supplied}}$$

i.e.,

$$\eta = \frac{Q_1 - Q_2}{Q_1}$$

or

$$\eta = 1 - \frac{Q_2}{Q_1} \quad ...(1)$$

6.17. CARNOT'S CYCLE

In theory, Carnot devised an ideal cycle, free from the normal imperfections we see in engines, though such an engine cannot be realized in practice.

Carnot's engine consists of:

(*i*) Cylinder which is having a perfectly non-conducting walls and a perfectly conducting bottom. In the cylinder, moves a perfectly nonconducting piston in horizontal direction and the motion is frictionless. The cylinder has a perfect gas as the working substance and is placed on a perfectly insulated stand.

(*ii*) A source of heat that supplies heat at constant temperature say T_1 K.

(*iii*) A sink which gets rejected heat at constant temperature say T_2 K.

(*iv*) An insulating cap that can be connected to the bottom of the cylinder whenever it is desired.

Normally, the source and sink are large reservoirs of heat so that their temperatures are practically constant during the transfer of heat to or from the cylinder.

The Carnot's engine is supposed to work in the following four stages known as Carnot's cycle. A working substance gets some heat from a source, converts a part into mechanical work and gives the rejected heat to sink. Let m be the mass of air enclosed in cylinder and let the point A in Fig. 6.4 (*b*) represent its condition on *P-V* diagram.

(*iii*) A cold body to serve as a sink at a constant lower temperature T_2 absolute.

Let the volume, pressure and temperature at this stage be V_1, P_1 and T_1 respectively.

First operation. Let the initial temperature of the gas in the cylinder be T_1 K and is placed in contact with a source which is also at temperature T_1 K. Let the piston move outward very slowly so that the expansion of the gas takes place. Consequently, the temperature of the gas tends to fall. Heat is therefore, absorbed from the source at constant temperature T_1 and so the change is isothermal as shown by curve *AB* in Fig. 6.4 (*b*).

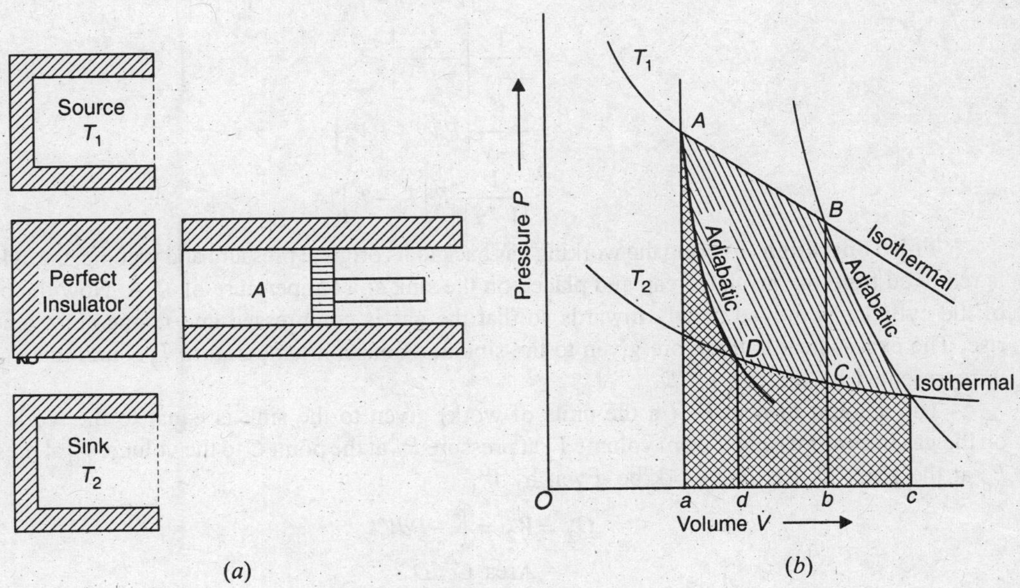

Fig. 6.4

Since the change is isothermal, so there is no change in internal energy of the gas *i.e.*,

$$dU = 0$$

Also $$dQ = P.dV$$

∴ The amount of heat Q_1 absorbed (in units of work) from source is equal to the work done by the gas when it expands from pressure P_a and volume V_a at A to a pressure P_b and volume V_b at B.

i.e., $$Q_1 = W_1 = \int_a^b P.dV = \text{Area } A\ ab\ B$$

We can find the value of the integral from the gas equation,

$$PV = RT_1$$

or $$P = \frac{RT_1}{V}$$

$$\therefore \quad Q_1 = RT_1 \int_a^b \frac{dV}{V}$$

$$= RT_1 \log_e \frac{V_b}{V_a} \qquad ...(1)$$

Second operation. In the second stage, the cylinder is removed from the source and is placed on the perfectly insulated cap and the gas is allowed to expand adiabatically till the temperature falls to T_2 *K*. This adiabatic change is represented by curve *BC*. In this case, $dQ = 0$ and from the first law of thermodynamics

$$dQ = dU + dW$$

$$\therefore \quad dW = -dU$$

which means that the gas is doing external work at the cost of decrease in internal energy and so its temperature will fall.

Now, the work done by the gas is given by relation

$$W_2 = \int_b^c P.dV = \text{area } BbcC$$

The value of integral is found from the adiabatic relation,

$$PV^\gamma = K \text{ or } P = \frac{K}{V^\gamma}$$

$$\therefore \quad W_2 = K\int_b^c \frac{dV}{V^\gamma} = \frac{K}{1-\gamma}\left[\frac{1}{V_c^{\gamma-1}} - \frac{1}{V_b^{\gamma-1}}\right]$$

$$= \frac{1}{1-\gamma}\left[\frac{K}{V_c^{\gamma-1}}\right]$$

$$= \frac{1}{1-\gamma}\left[P_cV_c - P_bV_b\right]$$

$$= \frac{1}{[1-\gamma]}.R\left[T_2 - T_1\right] \qquad ...(2)$$

Third operation. To bring the working gas back to its original pressure and volume, the cylinder is removed from the insulated cap and placed on the sink at a temperature of T_2 Å. Now the piston of the cylinder is moved slowly inwards so that the gas is compressed and temperature tends to rise. The excess heat is therefore given to the sink at a constant temperature T_2 *K* and this change is represented by the curve *CD*.

The amount of heat Q_2 (in the units of work) given to the sink is equal to the work done on the gas in compressing it from volume V_c at pressure P_C at the point *C* to the volume V_d at pressure P_d at the point *D*. Let this work be given by W_3

$$\therefore \quad Q_2 = W_3 = \int_c^d -PdV$$

$$= \text{Area } CcdD$$

$$= RT_2 \log_c \frac{V_c}{V_d} \qquad ...(3)$$

Fourth operation. Cylinder is removed from the sink and placed on the insulated cap. The piston is allowed to move inwards so that the gas is compressed adiabatically till its temperature rises to T_1 and the gas gets back to its original value of pressure and volume. This change is represented by *DA* curve, and the work done W_4 on the gas is given by

$$W_4 = \int_d^a -PdV = \text{Area } DdaA$$

$$= \frac{R(T_2 - T_1)}{1-\gamma}$$

From Equations (2) and (4), we have

$$W_2 = W_4$$

So net work done $= W_1 + W_2 - W_3 - W_4$

$$= \text{Area } ABCD = W_1 - W_3$$

Net amount of heat absorbed in units of work $= Q_1 - Q_2$

$$\therefore \quad Q_1 - Q_2 = W_1 - W_3$$

$$= RT_1 \log_c \frac{V_b}{V_a} - RT_2 \log_e \frac{V_c}{V_d}$$

$$= R\left[T_1 \log_e \frac{V_b}{V_a} - T_2 \log_e \frac{V_c}{V_d}\right] \quad ...(5)$$

6.18. EFFICIENCY OF CARNOT'S CYCLE

Since points B and C lie on the same adiabatic, so we have

$$P_b V_b^{\gamma} = P_c V_c^{\gamma} \quad ...(6)$$

Also points A and D are on the same adiabatic

$$\therefore \quad P_d V_d^{\gamma} = P_a V_c^{\gamma} \quad ...(7)$$

Similarly points A and B lie on the smae isothermal, at temperature T_1 K so we have

$$P_a V_a = P_b V_b \quad ...(8)$$

Also points C and D lie on the same isothermal at temperature T_2 K.

$$P_c V_c = P_d V_d \quad ...(9)$$

Multiplying the corresponding sides of Equations (6), (7), (8) and (9) we have

$$V_a . V_b^{\gamma} . V_c . V_d^{\gamma} = V_a^{\gamma} . V_b . V_c^{\gamma} . V_d$$

or

$$V_a^{\gamma - 1} . V_c^{\gamma - 1} = V_b^{\gamma - 1} . V_d^{\gamma - 1}$$

or

$$\frac{V_b}{V_a} = \frac{V_c}{V_d}$$

Putting in Equation (5), we get $Q_1 - Q_2 = R.(T_1 - T_2) \log_e \dfrac{V_b}{V_a}$

Now efficiency is given by $\quad \eta = \dfrac{\text{heat converted into useful work}}{\text{Total heat supplied}}$

$$\eta = \frac{Q_1 - Q_2}{Q_1} = \frac{R.(T_1 - T_2) \log_e \frac{V_b}{V_a}}{RT_1 \log_e \frac{V_b}{V_a}}$$

$$= \frac{T_1 - T_2}{T_1} = 1 - \frac{T_2}{T_1}$$

From above relation, we observe that

(*i*) Efficiency depends on the temperature of the hot and cold bodies.

(*ii*) Efficiency is independent of the nature of the working substance.

(*iii*) Reversible engines working between the same temperatures will have same efficiency.

(iv) From efficiency to be maximum, either T_2 should be zero or T_1 should be infinite. As none of these conditions are possible in practice so it is impossible to get 100% efficiency.

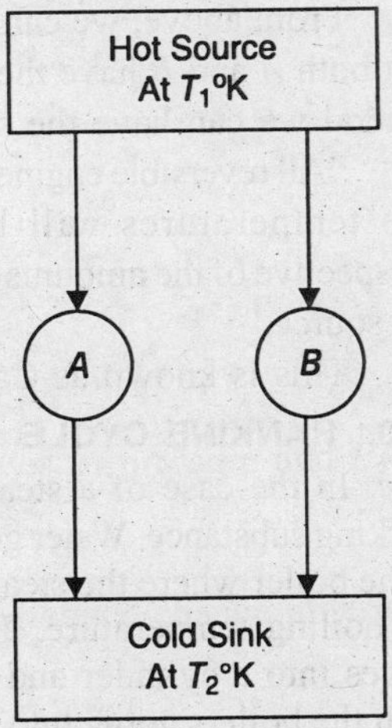

Fig. 6.5

6.19. CARNOT'S THEOREM

A reversible engine working between the same limits of temperatures has maximum efficiency and Carnot's theorem states that no heat engine which is working between the same temperature limits has more efficiency than a Carnot's engine. The proof of this theorem can be given as under:

Let two engines A and B be working between the same limits of temperature say T_1 K and T_2 K. The heat being received at temperature T_1 from the source is rejected to sink at temperature T_2. Let us consider A to be more efficient than B and let the length of the stroke of piston B is so adjusted that both the engines perform work under similar conditions.

Let A and B be coupled in such a way that when A is working directly, it drives B in backward direction as shown in Fig. 6.5.

Let A receive heat Q_1 at temperature T_1 K and rejects heat Q_2 at temperature T_2 K during each cycle. It is clear that B will travel the cycle in reverse direction. Let Q'_2 be the amount of heat absorbed at T_2 K and Q_1' the amount of heat rejected at T_1 K by B, the work performed in both the cases being same.

Now, heat energy converted into work by engine $A = Q_1 - Q_2$

and heat energy converted into work by engine $B = Q'_1 - Q'_2$

$$\therefore \quad Q_1 - Q_2 = Q_1' - Q_2'$$

We have assumed that efficiency of A is more than of B

$$\therefore \quad \frac{Q_1 - Q_2}{Q_1} > \frac{Q_1' - Q_2'}{Q_1'}$$

or
$$\frac{1}{Q_1} > \frac{1}{Q_1'}$$

or
$$Q_1' > Q_1$$

Similarly, we have
$$Q_2' > Q_2$$

From above, it is evident that Q'_1 *i.e.,* the quantity of heat rejected by B with the source is more than Q_1, the quantity of heat which A absorbs.

Also Q'_2, the quantity of heat which B absorbs from sink is more than Q_2 *i.e.*, the quantity which A rejects to sink.

So we see that quantity of heat $Q'_1 - Q_1$, $Q'_2 - Q_2$ will be transferred from a lower temperature T_2 (temperature of the sink) to a higher temperature T_1 (temperature of the source). From second law of thermodynamics, it is not possible to transfer heat from lower temperature to a higher temperature without the aid of an external agent. Since there is no external source to supply energy so this transfer is not possible. This in other words mean that the assumption of A having more efficiency than B is not correct and so not possible.

From above, we can come to the conclusion that both A and B have the same efficiency and in general we can have the statement.

"All reversible engines working between same two temperatures will have same efficiency irrespective of the amounts of heat they derive from the source."

This is known as Carnot's theorem.

6.20. RANKINE CYCLE

In the case of a steam engine, water is the working substance. Water gets converted into *steam* in the boiler where the steam is superheated above the boiling temperature. This superheated steam moves into a cylinder and in the first part, steam from the boilers going into the cylinder, moves the piston. The inlet valve remains open and the energy of the steam at constant pressure continues.

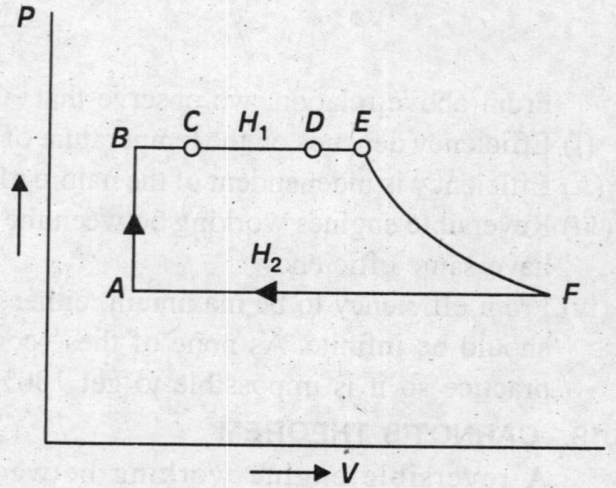

Fig. 6.6

In the second part, the inlet valve closes and the steam expands *adiabatically*. The piston goes on moving throughout the rest of the working stroke.

As soon as the piston moves to the end of the *stroke*, the exhaust value opens and steam goes to the condenser where it gets condensed.

This cycle of operation is called Rankine Cycle and corresponds to an ideal case. We can have the Rankine Cycle from the *P-V* diagram as under:

(*i*) *A* to *B* in the *P-V* diagram corresponds to compression of water and the boiler pressure is reached to *B*.

(*ii*) *B* to *C* corresponds to the heating of the water in boiler at constant pressure and boiling point is reached at *C*.

(*iii*) *C* to *D* represents water conversion into steam, pressure being constant.

(*iv*) *D* to *E* corresponds to superheating of steam beyond the boiling point of water.

(*v*) *E* to *F* represents the adiabatic expansion of steam in doing some work, thereby it gets cooled and so condenses.

(*vi*) *F* to *A* corresponds to rejection of steam to the condenser where total steam gets condensed and finally the original point *A* is reached.

Efficiency of Rankine Cycle. As is evident from the *P-V* diagram, H_1 corresponds to the amount of heat absorbed from *B* to *E* and H_2 the amount of heat rejected from *F* to *A*.

$$\therefore \qquad \text{Efficiency } \eta = \frac{H_1 - H_2}{H_1}$$

$$= 1 - \frac{H_2}{H_1}$$

It is also known as standard or ideal efficiency of a steam engine.

6.21. OTTO AND DIESEL CYCLES

Gas and petrol engines work on the principle of Otto-cycle. In this case, since heat is received *and rejected at constant volume*, so it is also called *constant volume cycle*. Though original cycle was due to a **Frenchman Beau-de-Rochas,** it was first successfully built by **Nicholas A. Otto** in 1876.

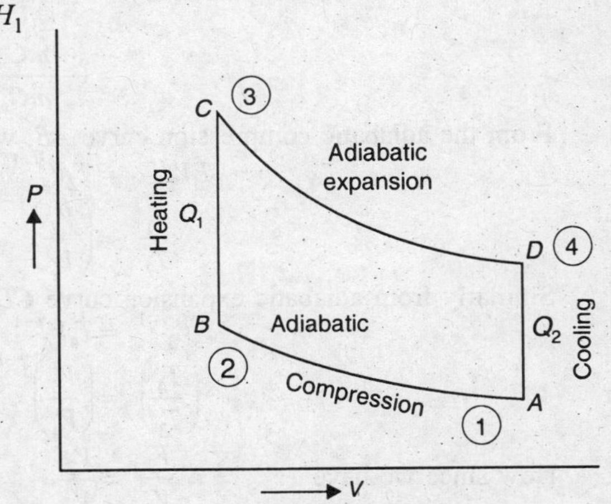

Fig. 6.7

We will consider *air theoretical standard cycle* where air is assumed to be working substance. It is supposed that the air is contained in an imaginary cylinder whose walls are perfectly non-conducting but whose base is a perfect conductor of heat (As in case of Carnot's cycle). Similarly, we have a hot source, a cold sink and an insulating cap, which are alternately brought in contact with the bottom of the imaginary cylinder.

6.22. THEORETICAL OTTO CYCLE

P–V diagram of the cycle is given below. The cycle is constituted by two constant volume lines and two *adiabatic* curves. At *A*, the cylinder is full of air of mass *m* gm and has a volume V_1 pressure P_1 and temperature T_1. The cycle has the following four stages:

(*i*) **First Stage.** The bottom of the cylinder is covered with an insulating cap and the piston is moved inwards compressing the air adiabati-cally till end of the stroke reaches at *B*. The pressure and temperature of the air rises to P_2 and T_2 respectively while the volume is reduced to a value V_2.

Work is done adiabatically during this adiabatic compression.

(*ii*) **Second Stage.** Insulating cap is removed at point B and the bottom of the cylinder is brought in contact with hot source and then the piston is *held stationary*. The air gets heated at constant volume till the pressure and temperature are raised to P_3 and T_3 respectively.

In this case, heat taken in is given by

$$Q_1 = mC_v (T_3 - T_2)$$

(*iii*) **Third Stage.** At the point C, the hot body is removed and again the insulating cap is put on the bottom of the cylinder. Air is now allowed to expand adiabatically till point D is reached. During this expansion, work is done *by the air* thereby its pressure and temperature fall to P_4 and T_4 respectively.

(*iv*) **Fourth Stage.** At the point D, cylinder cap is removed and the cylinder is brought into contact with the cold sink. *Piston is kept stationary* and the air is allowed to reject heat at *constant volume* till its pressure and temperature drop to the original values P_1 and T_1 respectively. In this part, the amount of heat rejected to the sink is given by

$$Q_2 = mC_v (T_4 - T_1)$$

So the cycle is completed as the air comes back to its initial condition of temperature and pressure.

Efficiency of Otto Cycle. In this case:

heat received, $Q_1 = mC_v (T_3 - T_2)$

heat rejected, $Q_2 = mC_v (T_4 - T_1)$

Heat converted to work $= Q_1 - Q_2$

$$\text{Efficiency} = \frac{Q_1 - Q_2}{Q_1} = 1 - \frac{Q_2}{Q_1}$$

$$= 1 - \frac{mC_v(T_4 - T_1)}{mC_v(T_3 - T_2)} = 1 - \frac{(T_4 - T_1)}{(T_3 - T_2)}$$

From the adiabatic compression curve AB, we have

$$T_1 V_1^{\gamma-1} = T_2 V_2^{\gamma-1}$$

or $$\frac{T_1}{T_2} = \left(\frac{V_2}{V_1}\right)^{\gamma-1} \quad ...(2)$$

Similarly from adiabatic expansion curve CD, we have

$$T_3 V_3^{\gamma-1} = T_4 V_4^{\gamma-1}$$

or $$\left(\frac{T_4}{T_3}\right) = \left(\frac{V_3}{V_4}\right)^{\gamma-1} \quad ...(3)$$

Now since we have $$\frac{V_3}{V_4} = \frac{V_2}{V_1}$$

∴ From Equations (2) and (3), we get

$$\frac{T_1}{T_2} = \frac{T_4}{T_3}$$

or $$\frac{T_3}{T_2} = \frac{T_4}{T_1}$$

or $$\frac{T_3}{T_2} - 1 = \frac{T_4}{T_1} - 1$$

or $$\frac{T_3 - T_2}{T_2} = \frac{T_4 - T_1}{T_1}$$

or $$\frac{T_1}{T_2} = \frac{T_4 - T_1}{T_3 - T_2} \quad ...(4)$$

Also from (1) $$\eta = 1 - \frac{T_4 - T_1}{T_3 - T_2}$$

$\therefore$ *Air standard efficiency*, from Equation (4) may be written as

$$\eta = 1 - \frac{T_1}{T_2} \qquad ...(5)$$

Now $\frac{V_1}{V_2}$ = adiabatic compression ratio, ρ, so we may write from Equation (2)

$$\frac{T_1}{T_2} = \left(\frac{1}{\rho}\right)^{\gamma-1}$$

Equation (5) may be written as $\eta = 1 - \left(\frac{1}{\rho}\right)^{\gamma-1}$ ...(6)

From Equation (6) above, we observe that

(*i*) efficiency increases with increase of compression ratio ρ. However in practice, we cannot increase ρ beyond a value of around 7.

(*ii*) efficiency is dependent only on ρ and is independent of the heat input.

We can have expression for the clearance volume from the cylinder as under:

By definition, compression ratio is given as

$$\rho = \frac{\text{Total cylinder volume}}{\text{clearance volume}}$$

$$= \frac{\text{Max. cylinder capacity}}{\text{Min. cylinder capacity}}$$

$$= \frac{\text{Clearance volume + Stroke volume}}{\text{Clearance volume}}$$

$$= 1 + \frac{\text{Stroke volume}}{\text{Clearance volume}}$$

or Clearance volume $= \frac{\text{Stroke volume}}{\rho - 1}$

6.23. THEORETICAL DIESEL CYCLE

The theoretical form of the diesel cycle is shown in the *P–V* diagram.

As is evident from the Fig. 6.8, it consists of two adiabatics, one constant pressure line and one constant volume line.

In order to get the *air standard efficiency* in this cycle, let us consider that the cylinder is having air as the working substance. Also, we consider the cylinder to be perfectly *non-conducting* except at the *bottom*.

As we considered previously, there is a hot source, a cold source (or sink) and an insulating cap that can be put at the bottom of the cylinder whenever we want. This is also called a *constant pressure cycle* as heat is taken in at a constant pressure.

We can study the different stages of the *P–V* diagram.

Let us consider that the cylinder is full of air of mass *m* gram at the beginning. Let the initial pressure, volume and temperature of the air at *A* be P_1, V_1 and T_1 respectively.

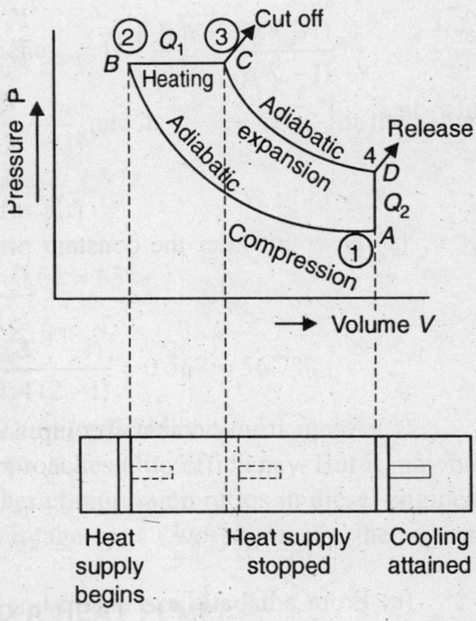

Fig. 6.8

(*i*) **First Stage.** First let us apply the insulating cap to the cylinder and let the piston be moved inward so as to compress the air adiabatically (In this case, from adiabatic law PV^γ = constant) upto the point

B. Now, the air occupies the clearance space and let it have pressure, volume and temperature as P_2, V_2 and T_2 respectively.

(*ii*) **Second Stage.** At the point *B*, insulating cap is removed and the bottom of the cylinder is brought in contact with a hot body. Now, absorption of heat by air takes place at constant pressure till we reach point *C*. The amount of heat absorbed is given by

$$Q_1 = mC_p (T_3 - T_2)$$

(*iii*) **Third Stage.** At the point *C*, heat supply is stopped and the point *C* is called *cut off point*. Now, hot source is removed and the insulating cap is put on the bottom of the cylinder. Now, air is allowed to expand adiabatically till point *D* is reached (Again from Adiabatic law PV^γ constant).

(*iv*) **Fourth Stage.** On reaching the point *D*, the insulating cap is removed and the bottom of the cylinder is brought in contact with the cold source. Now rejection of heat takes place from air at constant volume. Piston is kept stationary during this process till pressure falls to P_1. Also since the air comes back to the original conditions, the cycle is completed.

The amount of heat rejected

$$Q_2 = mC_v (T_4 - T_1)$$

6.24. AIR STANDARD EFFICIENCY OF DIESEL CYCLE

Heat received $Q_1 = mC_p (T_3 - T_2)$

Heat rejected $Q2 = mC_p (T_4 - T_1)$

Heat converted to work $= Q_1 - Q_2$

∴ **Ideal or air-standard efficiency**

or $$\eta = \frac{Q_1 - Q_2}{Q_1} = 1 - \frac{Q_2}{Q_1}$$

$$= 1 - \frac{mC_v (T_4 - T_1)}{mC_p (T_3 - T_2)} = 1 - \frac{1}{\gamma} \frac{(T_4 - T_1)}{(T_3 - T_2)}$$

or $$\eta = 1 - \frac{1}{\gamma} \frac{(T_4 - T_1)}{(T_3 - T_2)} \quad ...(1)$$

Now let $\frac{V_1}{V_2}$ = adiabatic compression ratio ρ

and $\frac{V_4}{V_3}$ = adiabatic expansion ratio *r*.

also let $\frac{\rho}{r} = K$

∴ $$K = \frac{V_1/V_2}{V_4/V_3} = \frac{V_1}{V_2} \times \frac{V_3}{V_4} = \frac{V_3}{V_2} \qquad [\because V_1 = V_4]$$

(*a*) Now consider the constant pressure line *BC*, we have, from Charle's law

$$\frac{V_2}{T_2} = \frac{V_3}{T_3}$$

∴ $$\frac{V_3}{V_2} = K = \frac{T_3}{T_2} \quad ...(2)$$

(*b*) Again from adiabatic compression *AB*, we have

$$T_1 V_1^{\gamma-1} = T_2 V_2^{\gamma-1}$$

or $$\frac{T_2}{T_1} = \left(\frac{V_1}{V_2}\right)^{\gamma-1} = \rho^{\gamma-1} \quad ...(3)$$

(*c*) From adiabatic expansion line *CD*, we have

$$T_3 V_3^{\gamma-1} = T_4 V_4^{\gamma-1}$$

or $$\frac{T_3}{T_4} = \left(\frac{V_3}{V_4}\right)^{\gamma-1} = \left(\frac{1}{r}\right)^{\gamma-1} \quad ...(4)$$

Also from Equation (1), we have $\eta = 1 - \frac{1}{r} \cdot \frac{T_1(T_4/T_1 - 1)}{T_2(T_3/T_2 - 1)}$ ...(5)

Putting the value of $T_2/T_1 = \rho^{\gamma-1}$ from Equation (3) and also we have

$$\frac{T_4}{T_1} = \frac{T_4}{T_3}.\frac{T_3}{T_2}.\frac{T_2}{T_1}$$

$$= \left(\frac{1}{r}\right)^{\gamma-1}.K.\rho^{\gamma-1}$$

$$= \left(\frac{\rho}{r}\right)^{\gamma-1}.K$$

$$= K^{\gamma-1}.K = K^{\gamma}$$

So, we have the air standard efficiency

$$\eta = 1 - \frac{1}{r} \cdot \left(\frac{1}{\rho}\right)^{\gamma-1} \cdot \frac{(K^{\gamma}-1)}{(K-1)}$$

$$= 1 - \left(\frac{1}{\rho}\right)^{\gamma-1} \frac{1}{r} \cdot \frac{(K^{\gamma}-1)}{(K-1)}$$

where we have taken $K = \frac{\text{Adiabatic compression ratio}}{\text{Adiabatic expansion ratio}}$

$$\frac{V_3}{V_2} = \frac{\text{Volume cut off}}{\text{Clearance volume}}$$

$\rho \rightarrow$ compression ratio

$r \rightarrow$ expansion ratio

6.25. COMPARISON OF OTTO AND DIESEL CYCLE

In case of **Otto Cycle,** we have air standard efficiency $= 1 - \left(\frac{1}{\rho}\right)^{\gamma-1}$

In case of Diesel Cycle, we have air standard efficiency $= 1 - \left(\frac{1}{\rho}\right)^{\gamma-1} \cdot \frac{(K^{\gamma}-1)}{\gamma(K-1)}$

In case of diesel cycle, we have additional factor $\frac{K^{\gamma}-1}{\gamma(K-1)}$ and as can be seen, for the same compression ratio, *Otto Cycle* has **more** efficiency **than** *diesel cycle.*

For example: For $\rho = 12$ and $r = 6$ and $\gamma = 1.4$ (for air)

$$\eta_{\text{Otto}} = 1 - \left(\frac{1}{12}\right)^{1.4-1} = 0.63 = 63\%$$

$$\eta_{\text{Diesel}} = 1 - \left(\frac{1}{12}\right)^{1.4-1} \cdot \frac{(2^{1.4}-1)}{1.4\,(2-1)} = 0.567 = 56.7\%$$

It is clear that *Otto-efficiency* is more than that of *Diesel efficiency*.

However as K approaches unity, Diesel efficiency approaches Otto efficiency. But it may be noted that practically it is possible to use considerable higher compression ratios in diesel engines than in Otto engines so *diesel engines* have practical advantage over *Otto engines* in the values of *operating efficiency*.

6.26. DIFFERENCE BETWEEN A REFRIGERATOR AND A HEAT PUMP

As is evident from the principle of Carnot's cycle, if all the operations in a Carnot's cycle are performed in the reverse direction, there will be an absorption of heat Q_2 from the sink or the cold body at temperature T_2 and transfer of an amount Q, at higher temperature T_1. Now, from the second law of thermodynamics, no transfer of heat can take place from a cold to hot body without some internal agency that supplies the requisite amount of energy. So some mechanical energy has to be supplied to the engine to make it work in the reverse direction. Then we have the relation

$$Q_1 = Q_2 + W$$

The Carnot's cycle working in the opposite direction can perform the following two functions:

(*i*) It can further cool the colder body.

(*ii*) It can further heat the hot body.

When a reversed heat engine is used to further cool the colder system the arrangement is called a refrigerator.

When the reversed heat engine is used to further heat the hotter system it is known as a heat pump.

6.27. COEFFICIENT OF PERFORMANCE OF A REFRIGERATOR

It is defined as the ratio of the heat drawn from the cold body to the work done on the machine.

In order to have a relation between coeff. of performance and the efficiency of an engine, let us assume a machine performing Carnot's cycle in the reverse direction. If Q_2 be the amount of heat absorbed from the cold body at temperature T_2 and Q_1 be quantity of heat given to the hot body at temperature T_1 then the work done by the machine will be equal to $Q_1 - Q_2$.

In the case of an ideal refrigerating machine, let W be the amount of work done on the machine then

$$W = Q_1 - Q_2$$

$\therefore$ By definition, coeff. of performance

$$\beta = \frac{Q_2}{W} = \frac{Q_2}{Q_1 - Q_2}$$

We have, from Carnot's cycle

$$\frac{Q_1}{T_1} = \frac{Q_2}{T_2}$$

or $$\frac{Q_1}{Q_2} = \frac{T_1}{T_2}$$

or $$\frac{Q_1}{Q_2} - 1 = \frac{T_1}{T_2} - 1$$

or $$\frac{Q_1 - Q_2}{Q_2} = \frac{T_1 - T_2}{T_2}$$

$\therefore$ Coeff. of performance $$\beta = \frac{Q_2}{Q_1 - Q_2} = \frac{T_2}{T_1 - T_2} \quad \text{...(1)}$$

Now we can have a relation between efficiency η and coeff. of performance β.

From Equation (1) $$\beta = \frac{T_2}{T_1 - T_2} = \frac{\frac{T_2}{T_1}}{1 - \frac{T_2}{T_1}} \quad \text{...(2)}$$

Also efficiency of an engine $$\eta = \frac{T_1 - T_2}{T_1} = 1 - \frac{T_2}{T_1}$$

$\therefore$ $$\frac{T_2}{T_1} = 1 - \eta \quad \text{...(3)}$$

Put in Equation (2) $$\beta = \frac{1-\eta}{1-1+\eta} = \frac{1-\eta}{\eta} \quad \text{...(4)}$$

We can see from above that as the temperature T_2 of the cold body decreases, $T_1 - T_2$ increases. So, from Equation (2), β decreases due to decrease in T_2 or increase in $T_1 - T_2$.

We also know that $\beta = \dfrac{Q_2}{W}$, so for a given value of Q_2, W will increase for β to decrease.

So the mechanical energy W required to extract a given amount of heat Q_2 from the cold body increases as the temperature T_2 of the colder body decreases.

6.28. THERMODYNAMIC SCALE OF TEMPERATURE

In the measurement of temperature, different scales have been devised. However, these measurements depend on the properties of the substance used to measure the temperature, *e.g.* in case of liquid thermometers, expansion of the liquid forms the basic property for measurement of temperature. Similarly, in case of resistance thermometer, change of resistance with temperature forms the basis for measurement of temperature.

Kelvin, from the study of the efficiency of the reversible heat engine got a temperature scale which is independent of any particular substance and such a scale is called thermodynamic scale or absolute scale of temperature.

We know that the efficiency of an engine working between two temperatures is independent of the nature of the working substance but is a function of the two temperatures only.

Let Q_1 be the quantity of heat absorbed at higher temperature (say θ_1) and Q_2 be the quantity of heat rejected at lower temperature (say θ_2).

where Q_1 and Q_2 may be temperatures measured on any arbitrary scale, then efficiency may be written as

$$\eta = \frac{Q_1 - Q_2}{Q_1} = f(\theta_1, \theta_2)$$

$$= 1 - \frac{Q_2}{Q_1} = f(\theta_1, \theta_2)$$

$$\frac{Q_1}{Q_2} = \frac{1}{1 - f(\theta_1, \theta_2)} = F(\theta_1, \theta_2) \quad ...(1)$$

Similarly, for another reversible heat engine working between temperatures θ_2 and θ_3 $(\theta_2 > \theta_3)$ we have

$$\frac{Q_2}{Q_3} = F(\theta_2, \theta_3) \quad ...(2)$$

Multiplying Equations (1) and (2) we get

$$\frac{Q_1}{Q_3} = F(\theta_1, \theta_3) = F(\theta_1, \theta_2) \times F(\theta_2, \theta_3) \quad ...(3)$$

Now if we take a reversible engine working between temperatures θ_1 and θ_2 such that it absorbs an amount of heat Q_1 at temperature θ_1 and rejects an amount of heat Q_3 at temperature θ_3, then

$$\frac{Q_1}{Q_3} = F(\theta_1, \theta_3) \quad ...(4)$$

From Equations (3) and (4)

$$F(\theta_2, \theta_3) = F(\theta_1, \theta_2) \quad ...(5)$$

Now in Equation (5), θ_2 exists on L.H.S. in both factors, while it is not present in R.H.S. so we may have the form of $F(\theta_1, \theta_2)$ as

$$F(\theta_1, \theta_3) = \frac{\phi(\theta_1)}{\phi(\theta_3)} \quad ...(6)$$

and

$$F(\theta_2, \theta_3) = \frac{\phi(\theta_2)}{\phi(\theta_3)} \quad ...(7)$$

where ϕ is another function of θ. The proof for this can be had by multiplying Equations (6) and (7) we get

$$F(\theta_1, \theta_2) \times F(\theta_2, \theta_3) = \frac{\phi(\theta_1)}{\phi(\theta_3)} = F(\theta_1, \theta_3)$$

which is Equation (5).

Let $\phi(\theta)$ be represented by τ, the temperature on absolute scale,

$$\therefore \quad \frac{Q_1}{Q_2} = \frac{\tau_1}{\tau_2}$$

In terms of efficiency we have

$$\eta = \frac{Q_1 - Q_2}{Q_1} = \frac{\tau_1 - \tau_2}{\tau_1} \quad \text{...(8)}$$

As on this scale, the temperatures τ_1 and τ_2 are measured in units of work, so this scale is also known as work scale of temperature.

As is clear from above, the efficiency η on thermodynamic scale is given by

$$\eta = 1 - \frac{Q_2}{Q_1} = 1 - \tau_2/\tau_1 \quad \text{...(9)}$$

Now for η to be equal to unity, τ_2 should be zero. A, temperature $\tau_2 < 0$ is not possible because in that case, efficiency of the heat engine will be more than one which is not possible. So the efficiency $\eta = 1$, only when all the heat taken by engine is converted into work *i.e.* $Q_1 = W$ or $Q_2 = 0$ which will correspond to $\tau_2 = 0$, so this will represent zero on thermodynamic scale.

Having fixed zero on this scale, we can determine the size of each degree on it. Let the reversible engine work between the boiling and freezing points of water at normal pressure and let *ABCD* represent Carnot's cycle of the engine working between these temperatures. The work done by the engine will be numerically equal to the area *ABCD* and let this area be divided into 100 equal parts by isothermals parallel to *AB* or *CD*. The area of each part will correspond to one degree on absolute scale or Kelvin scale.

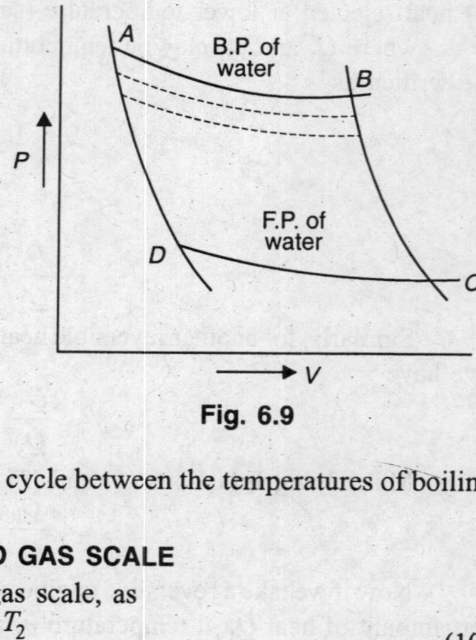

Fig. 6.9

A degree on the scale may be defined as the difference in temperatures between which a reversible engine should work to get energy equal to $\frac{1}{100}$th of the area of the indicator diagram that represents the Carnot's cycle between the temperatures of boiling point of water and melting point of ice.

6.29. RELATION BETWEEN THERMODYNAMIC AND GAS SCALE

We have the efficiency of an engine on perfect gas scale, as

$$\eta = \frac{T_1 - T_2}{T_1} = 1 - \frac{T_2}{T_1} \quad \text{...(10)}$$

Now on this scale if $T_2 = 0$, $\eta = 1$ and also temperatures less than $T_2 = 0$ are not possible because the motion of the molecules is zero at absolute zero temperature, which in other words mean that thermal energy of the gas is zero at absolute zero temperature. So the zero on gas scale and thermodynamic scale is same.

Also, since efficiencies in both the cases have same value, so

$$\frac{T_1}{T_2} = \frac{\tau_1}{\tau_2}$$

On the gas scale, if T_1 is the temperature of boiling water and T_2 is the temperature of melting ice, then

$$T_1 - T_2 = 100$$

Also as we have seen $\tau_1 - \tau_2 = 100$

Using the relation for a perfect gas

$$\eta = \frac{Q_1 - Q_2}{Q_1} = \frac{T_1 - T_2}{T_1} = \frac{100}{T_1}$$

And for thermodynamic scale

$$\eta = \frac{Q_1 - Q_2}{Q_1} = \frac{\tau_1 - \tau_2}{\tau_1} = \frac{100}{\tau_1}$$

$$\therefore \quad \frac{100}{\tau_1} = \frac{100}{T_1}$$

$$\tau_1 = T_1$$

So, we see that the temperatures on the gas scale and thermodynamic scale are the same which in other words mean that the two scales are identical.

8.30. CLAPEYRON'S EQUATION

Clapeyron's equation gives a relation between the **boiling point** or **the freezing point** of a liquid with pressure. This relation can be obtained by the application of **second law of thermodynamics.**

Let us consider a **Carnot's engine** having a unit mass of a **liquid** as a working substance. Let *A' ABB'* and *D'DCC'* be the two isothermals for the subtance at temperatures $T°$ and $(T° - dT°)$ in absolute scale of temperature. Upto the point *A*, the substance is wholly liquid and along *AB* it is partly in **liquid state** and partly in **vapour state**. It means that the temperature at *T* is, therefore, the boiling point of a liquid at a pressure corresponding to the point *A*.

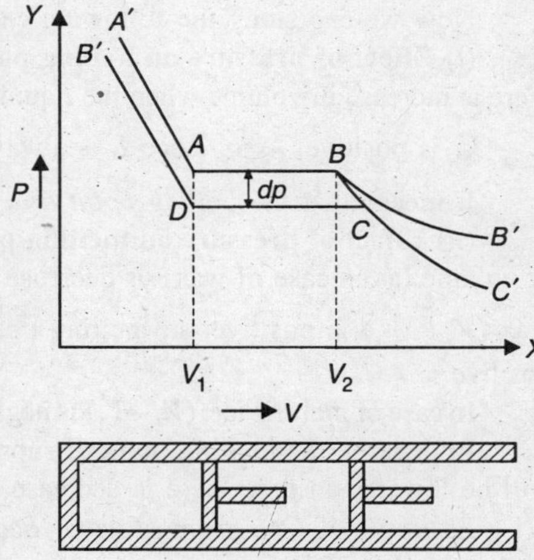

Fig. 6.10

Similarly, along *CD isothermal,* the material will be partly in liquid state and partly in vapour state at temperature $(T° - dT°)$. So the temperature $(T° - dT°)$ represents the B.P. of the liquid at pressure corresponding to the point *D*.

Let the volume of the liquid be v_1 c.c. at *A*, so we have the following conditions :

(*i*) If the cylinder is placed on the source at tem-perature *T*, the liquid slowly gets heated and changes partly into vapour state at constant pressure. At *B* the total liquid is converted into vapour and occupies a volume V_2 *c.c.* If *L* be the latent heat of the vaporization of liquid, then amount of heat absorbed at $T° = L$.

(*ii*) Now place the cylinder on the non-conducting stand and let the vapour expand adiabatically, so that the pressure decreases by a very small value *dp* and the temperature to $(T° - dT°)$. This is represented by the adiabatic curve *BC*.

(*iii*) Now place the cylinder on the sink at a temperature $(T° - dT°)$ and compress the vapour very slowly, so that the total vapour is liquified at temperature $(T° - dT°)$ and we reach the point *D* shown in the curve. This corresponds to isothermal process and the quantity of heat $L - dL$ is given out at temperature $(T° - dT°)$.

Again place the cylinder on a non-conducting stand and compress the liquid so that the initial *temperature and volume* of the liquid at *A* are attained.

As is evident, this is a reversible cycle as we can perform the cycle in opposite direction and each change in it can be reversed.

Now, the efficiency of a reversible cycle is given by

$$\frac{Q_1 - Q_2}{Q_1} = \frac{T_1 - T_2}{T_1}$$

which may be written as

$$\frac{L-(L_1-dL)}{L} = \frac{T-(T_1-dT)}{T}$$

or
$$\frac{dL}{L} = \frac{dT}{T} \quad ...(i)$$

Also quantity of heat converted into work

$$= L - (L - dL) = dL \quad ...(ii)$$

Work done in one cycle = Area $ABCD = dP\,(V_2 - V_1)$ ...(iii)

So from (ii) and (iii), we have $dL = dP\,(V_2 - V_1)$

Putting the value of dL from Equation (i), we get

or
$$\frac{dP}{dT} = \frac{L}{T(V_2 - V_1)} \quad ...(iv)$$

This is **Clausius Clapeyron's** equation.

Now we can study the following cases :

(*i*) **Effect of pressure on boiling point.** In Equation (*iv*), if $V_2 > V_1$, *i.e.*, the case in which there is increase in volume when the liquid is converted to vapour state (as is in most of the cases), $V_2 - V_1$ is positive. Also, since L is always positive, so $\frac{dP}{dT}$ is positive.

It means that the *boiling point of a liquid is raised due to increase in pressure.*

(*ii*) **Effect of pressure on melting point.** When a solid melts, then there can be an increase in volume (as in case of wax) or decrease in volume (as in case of ice). So, in case of melting of wax $(V_2 - V_1)$ is positive, so melting point is raised with increase of pressure because $\frac{dP}{dT}$ is positive.

In case of melting ice, $(V_2 - V_1)$ is negative and melting point is lowered by increase of pressure.

In general, the melting point of substances which expand on melting will be raised while it will be lowered in case there is decrease in volume on melting.

Example 1. *One gram of water becomes 1671 c.c. of steam when boiled at a pressure of one atmosphere. The heat of vaporisation at this temperature is 540 cals/gm. Compare the external work and the increase in internal energy.*

Solution: Let dV be the change in volume, then the work done against the pressure P is given by

$$dW = PdV$$

In this case,
$$P = 1 \text{ atm} = 76 \times 13.6 \times 980 \text{ dynes/cm}^2$$

and
$$dV = 1671 - 1 = 1670 \text{ c.c.}$$

$\therefore$
$$dW = 76 \times 13.6 \times 980 \times 1670$$
$$= 169.2 \times 10^7 \text{ ergs}$$
$$= \frac{169.2 \times 10^7}{4.2 \times 10^7} = 40.28 \text{ cal}$$

Now, we have the relation

$$dQ = dU + dW$$

$\therefore$ Change in thermal energy

$$dU = dQ - dW = 570 - 40.28 = \mathbf{499.72 \text{ cals.}}$$

Example 2. *A car is brought to halt by applying brakes in 50 metres. If the average friction force, which stops the car is 7.5 kN, how much heat is produced.*

Solution: Distance moved = 50 metres

Frictional Force = 7.5×10^3 N

$\therefore$ Work done = $7.5 \times 10^3 \times 50$

$= 375.0 \times 10^3$ Joule

$\therefore$ Heat produced $= \frac{375 \times 10^3}{J}$

$$= \frac{375 \times 10^3}{4.2} \text{ cal.}$$
$$= 89.29 \times 4.2 \text{ kJ}$$
$$= \mathbf{375 \ kJ.}$$

Example 3. *Calculate the final temperature when a given mass of gas at 0°C is suddenly compressed to a pressure 20 times the initial pressure. Given γ = 1.42.*

Solution: Expression for the pressure and temperature during an adiabatic change is given by the relation

$$\left(\frac{P_2}{P_1}\right)^{1-\gamma} = \left(\frac{T_1}{T_2}\right)^{\gamma}$$

Putting the values, we have

$$(20)^{1-1.42} = \left(\frac{273}{T_2}\right)^{1.42}$$

or
$$(20)^{-0.42} = \left(\frac{273}{T_2}\right)^{1.42}$$

Taking log on both sides, we have

$$-0.42 \log 20 = 1.42 \log 273 - 1.42 \log T_2$$

or
$$1.42 \log T_2 = 1.42 \log 273 + 0.42 \log 20 = 4.0058$$

∴
$$\log T_2 = 2.821$$

or
$$T_2 = 662.2\text{Å} = \mathbf{389.2°C.}$$

Example 4. *In wilson chamber, the volume of the gas is increased in the ratio 1.375 to 1. If the initial temperature of the gas is 20°C, find the final temperature. Given γ = 1.4.*

Solution: Let V_1 and T_1 be the initial volume and temperature of the gas and V_2 and T_2 be the final volume and temperature of the gas. Since the change is adiabatic, we have the relation

$$\frac{T_2}{T_1} = \left(\frac{V_1}{V_2}\right)^{\gamma-1}$$

Now
$$\frac{V_1}{V_2} = \frac{1}{1.375}$$

and
$$\gamma = 1.4, \ T = 273 + 20 = 293 \text{ K.}$$

$$T_2 = 293 \times \left(\frac{1}{1.375}\right)^{1.4-1}$$
$$= 293 \times \left(\frac{1}{1.375}\right)^{0.4}$$
$$= 258°\text{A} = \mathbf{-15°C.}$$

Example 5. *A quantity of air at N.T.P. is suddenly compressed to 1/5th of its volume. Calculate the temperature attained by air, γ for air = 1.40.*

Solution: Let V_1, T_1 be the original volume and temperature of air respectively and V_2, T_2 be the final volume and temperature of air respectively. Since the change is adiabatic, so we have the relation

$$\frac{T_2}{T_1} = \left(\frac{V_1}{V_2}\right)^{\gamma-1} \quad ...(1)$$

Now
$$\frac{V_2}{V_1} = \frac{1}{5}$$

$\therefore$ $$\frac{V_1}{V_2} = 5$$

and $$\gamma = 1.40,\ T_1 = 273°K$$

Putting in Equation (1)

$$\frac{T_2}{273} = (5)^{1.40-1} = (5)^{0.40}$$

$\therefore$ $$T_2 = 273 \times (5)^{0.40}$$

Taking log on both sides

$$\log T_2 = \log 273 + 0.40 \log 5$$
$$= 2.176$$

$\therefore$ $$T_2 = 519.7 \text{ K} = \mathbf{246.7°C.}$$

Example 6. *A certain volume of dry air at N.T.P. is expanded to three times its volume under (a) isothermal conditions (b) adiabatic conditions. Find in each case, the final pressure and temperature. Given $\gamma = 1.40$ for air.*

Solution: Let V_1 be the volume of dry air at N.T.P.

(*a*) **Isothermal conditions**

$$\text{Initial pressure } P_1 = 76 \text{ cm}$$
$$\text{Initial volume} = V_1$$
$$\text{Final volume} = 3V_1$$

If P_2 be the final pressure under isothermal conditions, then we have from the relation

$$P_1 V_1 = P_2 V_2$$

or $$76 \times V_1 = P_2 \times 3V_1$$

or $$P_2 = \frac{76}{3} = 25.33 \text{ cm}$$

(*b*) **Adiabatic condition**

Determination of temperature. We have the relation

$$\frac{T_1}{T_2} = \left(\frac{V_2}{V_1}\right)^{\gamma-1}$$

In this case $$T_1 = 273 \text{ K},\ \gamma = 1.4,\ V_2 = 3V_1$$

$\therefore$ $$\frac{273}{T_2} = \left(\frac{3V_1}{V_1}\right)^{1.4-1} = (3)^{0.4}$$

or $$T_2 = \frac{273}{(3)^{0.4}}$$

Taking log on both sides

$$\log T_2 = \log 273 - 0.4 \log 3$$
$$= 2.4362 - 0.4 \times 0.4771$$
$$= 2.2544$$

$\therefore$ $$T_2 = 176.0 \text{ K} = -97°\text{C}.$$

(*ii*) **Determination of pressure.** We have the relation

$$\frac{P_1}{P_2} = \left(\frac{V_2}{V_1}\right)^{\gamma}$$

$\therefore$ $$\frac{76}{P_2} = \left(\frac{3V_1}{V_1}\right)^{1.4} \text{ or } P_2 = \frac{76}{(3)^{1.4}}$$

Taking log on both sides

$$\log P_2 = \log 76 - 1.4 \log 3$$
$$= 1.8808 - 1.4 \times 4.71 = 1.2129$$
$$\boldsymbol{P = 16.39 \text{ cm.}}$$

Example 7. *A gram molecule of a gas at 127°C expands isothermally until its volume is doubled. Find the amount of work done and heat absorbed. (Given R = 8.3 × 10⁷ ergs per gm mole/°C).*

Solution: In this case,

$$\frac{V_1}{V_2} = 2$$

$$T = 127°C = 273 + 127 = 400 \text{ K}$$

$$R = 8.3 \times 10^7 \text{ ergs/gm mole/°C}$$

Since the change is isothermal, we have work done given by

$$W = RT \log_e \frac{V_2}{V_1}$$

$$= 8.3 \times 10^7 \times 400 \times 2.3026 \log_{10} 2$$

$$= 8.3 \times 10^7 \times 400 \times 2.3026 \times 0.3010$$

$$= 2.302 \times 10^{10} \text{ ergs}$$

∴ Amt. of heat absorbed

$$= \frac{2.302 \times 10^{10}}{4.2 \times 10^7} = \mathbf{548\ cal.}$$

Example 8. *Calculate the work done in compressing 1 gm helium at S.T.P. adiabatically; so that its pressure is doubled. Given R = 8.3 × 10⁷ ergs/°C/mole.*

Solution: Let P_1, T_1 be the original pressure and temperature and P_2, T_2, the final pressure and temperature

$$\therefore \quad \frac{P_2}{P_1} = 2$$

Now from the adiabatic relation

$$\left(\frac{T_2}{T_1}\right)^{\gamma} = \left(\frac{P_1}{P_2}\right)^{1-\gamma} = \left(\frac{1}{2}\right)^{1-\gamma}$$

$$\therefore \quad T_2 = T_1 \times \left(\frac{1}{2}\right)^{1-\gamma/\gamma}$$

$$= 273 \times (2)^{\frac{1-\gamma}{\gamma}}$$

$$= 273 \times (2)^{0.4/1.4}$$

$$= 273 \times (2)^{0.286}$$

$$= 333.5 \text{ K}$$

Now gas constant for 1 gm at helium is given by

$$R_1 = \frac{R}{4}$$

Work done

$$= \frac{R_1}{\gamma - 1}(T_1 - T_2)$$

$$= \frac{8.3 \times 10^7}{4 \times 0.4} \times (273 - 333.5)$$

$$= -\frac{8.3 \times 10^7}{1.6} \times 60.5$$

$$= \mathbf{-\ 313.9 \times 10^7\ ergs}$$

the minus sign indicates that the work is done on the gas.

Example 9. *A reversible engine takes in a mixture of fuel and air at 27°C and the highest temperature after combustion is 427°C. Calculate the Carnot's efficiency of the engine working between these two limits of temperature.*

Solution: Carnot's efficiency is given by

$$\eta = \frac{T_1 - T_2}{T_1} \times 100\%$$

In this case $T_1 = 273 + 427 = 700$ K

$T_2 = 273 + 27 = 300$ K

$$\therefore \quad \eta = \frac{700 - 300}{700} \times 100$$

$$= \frac{400}{700} \times 100 = \mathbf{57.14\%}$$

Example 10. *Calculate the maximum efficiency of a heat engine working between 30°C and 300°C.*

Solution: Efficiency $\eta = \frac{T_1 - T_2}{T_1}$

In this case $T_1 = 300 + 273 = 573$ K

$T_2 = 30 + 273 = 303$ K

$$\therefore \quad \eta = \frac{573 - 303}{573} \times 100 = \mathbf{47.1\%}$$

Example 11. *A reversible engine converts $\frac{1}{6}$ th of heat input into work. When the temperature of the sink is reduced by 62°C, its efficiency is doubled. Find the temperature of the source and the sink.*

Solution: Efficiency is given by

$$\eta = \frac{T_1 - T_2}{T_1} = \frac{1}{6}$$

$$\therefore \quad 6T_1 - 6T_2 = T_1 \text{ or } T_1 = \frac{6}{5} T_2 = 1.2\, T_2$$

In second case

$$\eta = \frac{T_1 - (T_2 - 62)}{T_1} = 2 \times \frac{1}{6}$$

$$= \frac{T_1 - T_2 + 62}{T_1} = \frac{1}{3}$$

$$= 3T_1 - 3T_2 + 186 = T_1$$

or $2T_1 = 3T_2 - 186$

Now $T_1 = 1.2\, T_2$, so we have

$2 \times 1.2T_2 = 3T_2 - 186$

or $2.4T_2 = 3T_2 - 186$

or $0.6T_2 = 186$

or $T_2 = \frac{186}{0.6} = 310\text{K}$

or $T_2 = 310 - 273 = 37°\text{C}$

$\therefore$ $T_1 = 1.2 \times 310 = 372 \text{ K} = \mathbf{99°C.}$

Example 12. *A reversible engine takes in heat from a reservoir of heat at 527°C and gives out heat to sink at 127°C. How many calories per second must it take from the reservoir in order to produce useful work at the rate of 750 watts?*

Solution: Efficiency is given by

$$\eta = \frac{T_1 - T_2}{T_1}$$

In this case, $T_1 = 527°\text{C} = 527 + 273 = 800$ K

$$T_2 = 127°C = 127 + 273 = 400 \text{ K}$$

$\therefore$ $$\eta = \frac{800-400}{800} = \frac{1}{2}$$

Useful work $$W = 750 \text{ watts}$$
$$= 750 \text{ joules/sec.}$$

Now, efficiency $$\eta = \frac{\text{work done}}{\text{heat taken from source}} = \frac{W}{Q_1}$$

$\therefore$ $$\frac{1}{2} = \frac{750}{Q_1}$$

or $$Q_1 = 1500 \text{ joule/sec.}$$
$$= \frac{1500}{4.2} = \mathbf{357.17 \text{ cal/sec.}}$$

Example 13. *A Carnot's engine takes 1000 Kilo calories of heat from a reservoir at 627°C and exhausts to the sink at 27°C. What is its efficiency?*

How much work does it perform? Express it in (i) Ergs (ii) Kilowatt hours (iii) Electron volts.

Solution: Efficiency is given by the relation

$$\eta = \frac{W}{Q_1} = \frac{Q_1 - Q_2}{Q_1} = \frac{T_1 - T_2}{T_1}$$

In this case $$Q_1 = 1000 \text{ Kcal.} = 10^6 \text{ cal.}$$
$$T_1 = 627 + 273 = 900 \text{ K}$$
$$T_2 = 27 + 273 = 300 \text{ K}$$

$\therefore$ $$\eta = \frac{900-300}{900} = \frac{6}{9} = \frac{2}{3} = 66.67\%$$

Now work done by the engine is given by ηQ_1

(*i*) $$W = \eta Q_1$$
$$= 66.67 \times 10^6 \text{ cal.}$$
$$= 66.67 \times 10^6 \times 4.2 \times 10^7 \text{ ergs}$$
$$= 2.8 \times 10^{13} \text{ ergs.}$$

(*ii*) $$\text{One kWh} = 1000 \text{ watts} \times 3600 \text{ sec}$$
$$= 1000 \text{ joule/sec} \times 3600 \text{ sec}$$
$$= 1000 \times 10^7 \text{ ergs} \times 3600$$
$$= 36 \times 10^{12} \text{ ergs.}$$

$\therefore$ $$W = \frac{2.8 \times 10^{13}}{36 \times 10^{12}} = \frac{7}{9} = 0.778 \text{ kWh.}$$

(*iii*) One electron volt *i.e.*,

$$1 \text{ eV} = 1.6 \times 10^{-12} \text{ ergs.}$$

$\therefore$ $$W = \frac{2.8 \times 10^{13}}{1.6 \times 10^{-12}} = \mathbf{1.75 \times 10^{25} \text{ eV.}}$$

Example 14. *An engineer claims his engine to develop 5 H.P. on testing, the engine consumes 0.44 kgm. of fuel/hour having a calorific value of 10,000 k.cal/kg. The maximum temperature recorded in the cycle is 1400°C and minimum is 350°C. Find whether the engineer is justified in his claim.*

Solution. Efficiency is given by

$$\eta = \frac{T_1 - T_2}{T_1}$$

In this case $$T_1 = 1400 + 273 = 1673 \text{ K}$$
$$T_2 = 350 + 273 = 623 \text{ K}$$

$\therefore$ $$\eta_{max} = \frac{1673 - 623}{1673} \times 100 = 62.76\%$$

Efficiency of the engine developed by the engineer is given by

$$\eta = \frac{Q_1 - Q_2}{Q_1}$$

Here Q_1 = heat consumed in 1 hour
= 0.44 × 10,000 = 4400 k.cal.
= heat input

Power developed = 5 H.P.

So equivalent heat output

$$= \frac{5 \times 45000 \times 60}{4200} \text{ k.cal}$$

$$\therefore \quad \text{Efficiency} = \frac{\text{Output}}{\text{Input}}$$

$$\therefore \quad \eta = \frac{5 \times 4500 \times 60}{4400 \times 420} \times 100$$

$$= \mathbf{7305\%}$$

Since, the efficiency of the engine is more than the maximum efficiency of Carnot's cycle, so the claim of the engineer is not justified.

Example 15. *Two Carnot engines A and B are operated in series. The first engine A receives heat at 900°A and rejects to the reservoir at a temperature T°A. The second engine B receives the heat rejected by A and inturn rejects to a reservoir at 400°A. Calculate the temperature T for the situation, when*

(i) Work output of the two engines are equal.

(ii) Efficiencies of the two engines are equal.

Solution: We have the relation for efficiency

$$\eta = \frac{W}{Q_1} = \frac{Q_1 - Q_2}{Q_1} = \frac{T_1 - T_2}{T_1}$$

For engine A

$$\eta_1 = \frac{W}{Q_1} = \frac{900 - T}{900}$$

and for engine B

$$\eta_2 = \frac{W'}{Q_1'} = \frac{T - 400}{T}$$

(*i*) When the work output of the two engines are equal *i.e.*,

$$W = W'$$

$$\therefore \quad 900 - T = T - 400$$

or $$2T = 1300$$

or $$T = 650 \text{ K}$$

(*ii*) When efficiencies of A and B are equal

i.e., $$\eta_1 = \eta_2$$

$$\therefore \quad \frac{900 - T}{900} = \frac{T - 400}{T}$$

or $$900 - T^2 = 900T - 360000$$

$$\therefore \quad T^2 = 36 \times 10^4$$

or $$T = \mathbf{600 \text{ K}}.$$

Example 16. *A carnot's refrigerator takes heat from water at 0°C and discards it to room at 27°C. 1 kgm of water at 0°C is to be changed into ice at 0°C. How many calories of heat are discharged to the room? What is the work done by the refrigerator in the process?*

Solution: In this case

$$T_1 = 300 \text{ K}$$
$$T_2 = 273 \text{ K}$$
$$Q_1 = ?, \quad Q_2 = 1000 \times 80 = 80 \times 10^3 \text{ cal.}$$

Now we know
$$\frac{Q_1}{Q_2} = \frac{T_1}{T_2}$$
$$Q_1 = Q_2 \times \frac{T_1}{T_2}$$
$$= 80 \times 10^3 \times \frac{300}{273}$$
$$= 87.91 \times 10^3 \text{ cal}$$
$$= 87.91 \text{ k.cal.}$$

Now the work done by the refrigerator
$$W = J(Q_1 - Q_2)$$
$$= 4.2\,(87910 - 80000)$$
$$= 4.2 \times 7910$$
$$= 3.322 \times 10^4 \text{ joule}$$

Coefficient of performance
$$B = \frac{Q_2}{Q_1 - Q_2} = \frac{80000}{87910 - 80000}$$
$$= \frac{80000}{7910} = \mathbf{10.11}.$$

Example 17. *(a) A Carnot's engine operates between a hot reservoir at 320 K and a cold reservoir at 260 K. If it absorbs 500 joules of heat at the hot reservoir, how much work does it deliver? (b) If the same engine working in the reverse functions as a refrigerator between the same two reservoirs, how much work must be supplied to remove 100 joules of heat from the cold reservoir.*

Solution: Let η be the efficiency and T_1, T_2 the corresponding temperatures of hot and cold reservoirs

then
$$\eta = \frac{T_1 - T_2}{T_1} = \frac{320\text{ K} - 260\text{ K}}{320} = \frac{3}{16}$$

Now work done $W = \eta Q_1$ where Q_1 is the heat absorbed at hot reservoir.

$\therefore$
$$W = \frac{3}{16} \times 500 \text{ joules}$$
$$= 93.75 \text{ joules.}$$

(*b*)
$$\eta = 1 - \frac{Q_2}{Q_1}$$

where
$$\eta = \frac{3}{16} \text{ and } Q_2 = 1000 \text{ joules.}$$

$\therefore$
$$\frac{3}{16} = 1 - \frac{1000}{Q_1} \text{ joules}$$

or
$$Q_1 = 1000 \times \frac{16}{13} \text{ joules}$$

The work that must be supplied to remove 1000 joules of heat from cold reservoir will be
$$W = Q_2 - Q_1 = \left(1000 \times \frac{16}{3} - 1000\right) \text{ joules}$$
$$= \mathbf{230.7 \text{ joules.}}$$

Example 18. *The motor in a refrigerator has a power output of 200 watts. If the freezing compartment is at 270 K and the outside air is at 300 K, assuming ideal efficiency, what is the maximum amount of heat that can be extracted from the freezing mixture?*

Solution: If the quantity of heat Q_2 is taken in at a lower temperature T_2 and a quantity of heat Q_1 is given out at the higher temperature T_1, then the difference $W = Q_1 - Q_2$ will be the work supplied to the refrigerator.

Now we have the relation

$$\frac{Q_1 - Q_2}{Q_2} = \frac{T_1 - T_2}{T_2}$$

In this case $T_1 = 300 \text{ K}, T_2 = 270 \text{ K}$

and
$$Q_1 - Q_2 = 200 \text{ watts} \times 600 \text{ sec.}$$
$$= 200 \text{ joules/sec} \times 600 \text{ sec.}$$
$$= 12 \times 10^4 \text{ joules}$$

$\therefore$
$$Q_2 = \frac{Q_1 - Q_2}{T_1 - T_2} \times T_2$$
$$= \frac{12 \times 10^4 \text{ joules}}{(300 \text{ K} - 270 \text{ K})} \times 270 \text{ K}$$
$$= \mathbf{1.08 \times 10^6 \text{ joules.}}$$

QUESTIONS AND PROBLEMS

1. State and explain the first law of thermodynamics.
2. State Second law of thermodynamics. Illustrate your answer with suitable examples.
3. What are reversible and irreversible processes? Give the conditions for a reversible cycle.
4. Give the different operations of Carnot's cycle. How does it account for the efficiency of a heat engine.
5. Show that in a reversible engine $\frac{Q_1}{T_1} = \frac{Q_2}{T_2}$.
6. If a Carnot's engine is independent of the working substance, then similarly real engines should also be independent of the working substance to some extent. Then why do we show much concern for a suitable fuel like oil, gasoline etc? Why not use stone as a fuel?
7. What factors reduce the efficiency of a heat engine from its ideal value?
8. In order to increase the efficiency of a Carnot's engine most effectively, would you increase T_1 keeping T_2 constant or would you decrease T keeping T_1 constant.
9. Explain Carnot's cycle on a $P - V$ diagram for an ideal gas and calculate its efficiency. What are the characteristics of Carnot's engine?
10. Differentiate between a refrigerator and a heat pump. Derive a relation between efficiency η of an ideal engine and coefficient of performance β of refrigerator.
11. Explain what do you understand by thermodynamic scale of temperature. How does it agree with ideal gas scale?
12. A motor tyre containing air under pressure of 2 atmosphere bursts. Calculate the lowering in temperature of air, initial temperature being 27°C and $\gamma = 1.4$ **[Ans.** 26.9°C]
13. An ideal heat engines operates in Carnot's cycle between 227°C and 127°C. It absorbs 6×10^4 cal at higher temperature. How much work per cycle is this engine capable of performing? **[Ans.** 5×10^4 joules]
14. A combination mercury-steam turbine takes saturated mercury vapour from boiler at 876°F and exhausts it to heat a steam boiler at 460°F. The steam turbine receives steam at this temperature and exhausts it to condenser at 100°F. What is the maximum efficiency of the combination? **[Ans.** 58%]
15. In Carnot's cycle, the isothermal expansion of the gas takes place at 400 K and the isothermal compression at 300 K. During the expansion 500 cal of heat are transferred to gas. Find
    (*a*) The work performed by the gas during isothermal expansion.
    (*b*) The heat rejected from the gas during the isothermal compression.
    (*c*) The work done on the gas during the isothermal compression.
    [Ans. (*a*) 2090 joules (*b*) 380 cal. (*c*) 1570 joules]
16. Find the thermal efficiency of a Carnot's engine whose hot and cold bodies have temperatures 154°C and 15°C respectively.
17. A reversible heat engine of efficiency 40% has its efficiency decreased to 20% when the temperature of the sink is lowered by 50°C. Find the temperature of the source. **[Ans.** 227°C]

18. Calculate the rise in temperature of a gas at 30°C if the pressure on it is doubled adiabatically ($\gamma = 1.4$). **[Ans.** 62.7°C]

19. Assuming a domestic refrigerator to be a reversible engine working between temperatures of melting ice and room temperature 17°C. Find the efficiency and the energy that must be supplied to freeze 10 gm of water which is already at 0°C. **[Ans.** 5.8%, 2.092×10^{11} ergs]

20. A refrigerator working between 200 K and 100 K and works at 60% coefficient. Calculate the power for refrigeration effect at 100 watts. **[Ans.** 333.3 watts]

21. In a mechanical refrigerator, the low temperature coils are at a temperature of – 13°C and the compressed gas in a condenser has a temperature of 27°C. What is the theoretical coefficient of performance? **[Ans.** 6.5]

22. A Carnot's engine whose temperature of source is 400 K takes 200 calories of heat at this temperature and rejects 150 calories of heat to the sink. What is the temperature of the sink? Also calculate the efficiency of the engine. **[Ans.** (*i*) 300 K (*ii*) 25%]

23. Find the efficiency of Carnot's engine working between steam point and ice point. **[Ans.** 26.81%]

CHAPTER 7

ENTROPY

7.1. INTRODUCTION

We know that in the case of an isothermal process, the temperature of the body remains constant while in the case of an adiabatic process, no external heat is supplied when the gas expands adiabatically and no heat is extracted when it is compressed adiabatically. So in the first case, temperature of the gas is lowered while in the second case it is raised.

From above it is evident that neither heat energy nor the temperature remains constant and these two increase or decrease together. However, there is something in the adiabatic process that remains constant just as temperature remains constant in an isothermal process.

This thermal property of the body which remains constant during an adiabatic process when no heat energy is supplied to or extracted from the body is known as Entropy.

To get a mathematical expression for entropy, let us consider isothermals T_1, T_2, T_3, ..., etc. drawn at very small but equal temperature differences. Let A and B be two adiabatics which cut these isothermals T_1 at a and b, T_2 at c and d and T_3 at e and f respectively as shown in the Fig. 7.1.

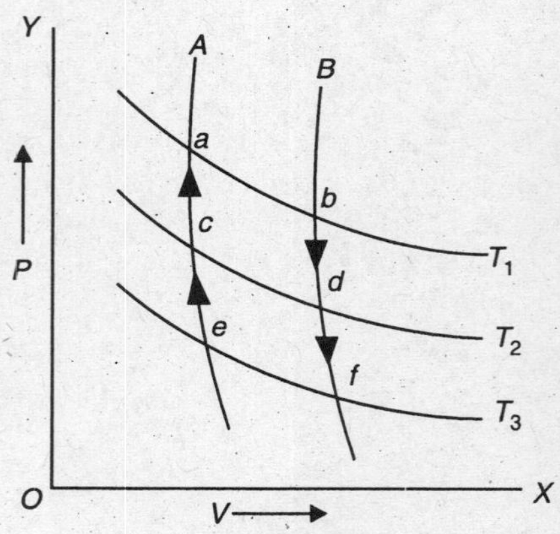

Fig. 7.1

Let us consider Carnot's Cycle in the direction *abdc*. In going from a to b some amount of heat say Q_1 is absorbed at temperature T_1 and in going from d to c, an amount say Q_2 is given at a temperature T_2. Then we have from Carnot's theorem

$$\frac{Q_1 - Q_2}{Q_1} = \frac{T_1 - T_2}{T_1}$$

or $$1 - \frac{Q_2}{Q_1} = 1 - \frac{T_2}{T_1}$$

or $$\frac{Q_2}{Q_1} = \frac{T_2}{T_1}$$

or $$\frac{Q_1}{T_1} = \frac{Q_2}{T_2}$$

Similarly, let us consider cycle *cdfe*, where Q_2 amount of heat is absorbed at temperature T_2 and Q_3 is given out at temperature T_3, then we have

$$\frac{Q_2}{T_2} = \frac{Q_3}{T_3}$$

which means that $$\frac{Q_1}{T_1} = \frac{Q_2}{T_2} = \frac{Q_3}{T_3} = \text{constant.}$$

So in general, if Q be the quantity of heat absorbed at temperature T, then in going from adiabatic A to B, we have

$$\frac{Q}{T} = \text{constant}$$

Now, if we consider the two adiabatics to be very near each other, so that δQ is the quantity of heat absorbed at temperature T in going from one adiabatic to another one, then the change in entropy δS is given by

$$\delta S = \frac{\delta Q}{T}$$

7.2. CHANGE IN ENTROPY

***(i)* In case of a reversible cycle.** We know by definition that a reversible process is one in which the system successively changes from one equilibrium state to the next equilibrium state differing only by infinitely small value from the previous state.

As is clear from above and from the definition of entropy, change in entropy δS for an equilibrium state is zero. So, if the system changes from one equilibrium state to another infinitesimally equilibrium state, $\delta S \to 0$ which means that the system can go from one equilibrium state to another equilibrium state at a very small distance in very small steps for infinitely long time. This type of infinitely slow process is called a quasi-static process and a reversible process is invariably a quasi-static process.

So in the case of a reversible process, δS approaches zero and we can prove this result from the Carnot's theorem as well.

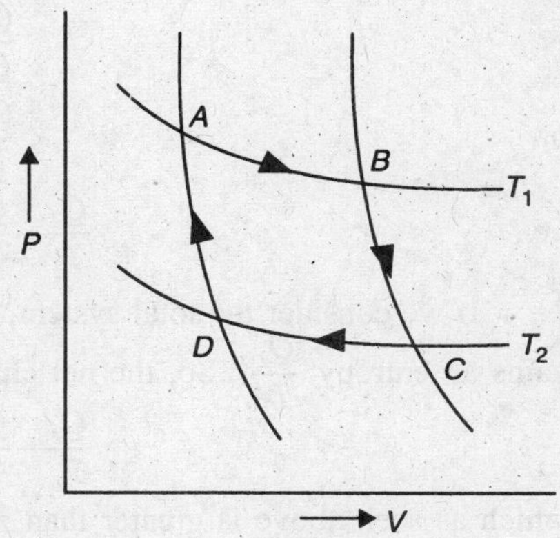

Fig. 7.2

Let us consider a reversible cycle $ABCD$ formed by two adiabatics BC and DA and isothermals AB and CD.

(*a*) Let Q be the quantity of heat absorbed in going from A to B along the isothermal AB at temperature T_1 then we have the entropy of the system as $\frac{Q_1}{T_1}$.

(*b*) In going from B to C along the adiabatic BC, there is no change in entropy but temperature falls from T_1 to T_2 because of the expansion of the working substance.

(*c*) Again in moving from C to D in isothermal CD at temperature T_2, Q_2 quantity of heat is given out. So the entropy of the working substance decreases and is given by $\frac{Q_2}{T_2}$.

(*d*) In moving from D to A along the adiabatic DA, there will not be any change in entropy but the working substance regains its original conditions.

So from above, the net change in entropy

$$\delta S = \frac{Q_1}{T_1} - \frac{Q_2}{T_2}$$

Since the process is reversible, so $\quad \frac{Q_1}{T_1} = \frac{Q_2}{T_2}$

Hence net change in entropy $\quad \delta S = \frac{Q_1}{T_1} - \frac{Q_2}{T_2} = 0.$

***(ii)* Irreversible process.** We can study the change in entropy in an irreversible process by considering a simple irreversible cycle in which Q_1 is the quantity of heat absorbed at temperature T_1 and Q_2 is the quantity of heat rejected to the sink at a temperature T_2.

Efficiency of such a cycle is given by $\frac{Q_1 - Q_2}{Q_1}$

From Carnot's theorem, we have the efficiency

$$\eta = \frac{T_1 - T_2}{T_1}$$

and by definition, efficiency of an irreversible cycle is always less than that of a reversible cycle *i.e.*,

$$\frac{Q_1 - Q_2}{Q_1} < \frac{T_1 - T_2}{T_1}$$

or
$$1 - \frac{Q_2}{Q_1} < 1 - \frac{T_2}{T_1}$$

or
$$\frac{Q_2}{Q_1} > \frac{T_2}{T_1}$$

or
$$\frac{Q_2}{T_2} > \frac{Q_1}{T_1}$$

i.e.,
$$\frac{Q_2}{T_2} - \frac{Q_1}{T_1} > 0$$

If we consider the total system, we observe that the source loses an entropy $\frac{Q_1}{T_1}$ while sink gains an entropy $\frac{Q_2}{T_2}\cdot$ So, the net change in the process is given by

$$\frac{Q_2}{T_2} - \frac{Q_1}{T_1}$$

which as seen above is greater than zero and hence positive.

Thus we observe that in the case of an irreversible cycle, there is a net gain of entropy, *i.e.*, change in entropy is given by $\delta S > 0$.

7.3. UNITS OF ENTROPY

In C.G.S. units, entropy is measured in ergs/K while in M.K.S. units it is measured as joules/K.

7.4. PROOF OF CARNOT'S THEOREM ON THE BASIS OF ENTROPY

Carnot's theorem can also be proved from the entropy considerations. Let a reversible engine operating in a cyclic way absorb an amount of heat Q_1 from the source at temperature T_1 and let a part of it may be converted into mechanical work and the rest say Q_2 be delivered to sink at temperature T_2. Then we have

Increase in entropy of sink $= \frac{Q_2}{T_2}$

and decrease in entropy of source $= \frac{Q_1}{T_1}$.

From definition, in the case of a reversible cycle, change in entropy is zero.

$\therefore$ Increase in entropy of the sink

= Decrease in entropy of the source

$\therefore$
$$\frac{Q_1}{T_1} = \frac{Q_2}{T_2} = \delta S$$

or
$$Q_1 = T_1 \,.\, \delta S$$

and
$$Q_2 = T_2 \,.\, \delta S$$

Also work done is given by

$$W = Q_1 - Q_2 = (T_1 - T_2)\,\delta S$$

$$= (T_1 - T_2)\,\frac{Q_1}{T_1}$$

Hence, efficiency
$$\eta = \frac{W}{Q_1} = \frac{T_1 - T_2}{T_1}$$

$$=1-\frac{T_2}{T_1}$$

which is the same as the Carnot's cycle.

So, we see that the efficiency of all the reversible engines working between the same temperatures is the same and is equal to the efficiency of the Carnot's engine.

From entropy concept, we can also prove that no heat engine can be more efficient than a reversible engine working between the same limits of temperatures.

We have already proved that in the case of an irreversible engine, $\delta S > 0$ *i.e.*, there is net increase in entropy. So for such an engine, increase in entropy at sink $\delta S'$ is greater than the decrease in entropy at the source δS.

Let us consider both reversible and irreversible heat engines to absorb same amount of heat say Q_1 from the source at temperature T_1, then we have for both the cases

$$\delta S = \frac{Q_1}{T_1} \text{ or } Q_1 = T_1 . \delta S$$

Let us say that the irreversible engine reject Q_2' quantity of heat to sink and reversible engine rejects Q_2 quantity of heat to sink at temperature T_2. Then we have

$$Q_2' = T_2 . \delta S'$$

and

$$Q_2 = T_2 . \delta S$$

From definition

$$Q_2' > Q_2$$

Let η and η' be the efficiencies of reversible and irreversible heat engines respectively, then we have

$$\eta = \frac{Q_1 - Q_2}{Q_1} \text{ and } \eta' = \frac{Q_1' - Q_2'}{Q_1'}$$

Since

$$Q_2' > Q_2$$

$\therefore$

$$Q_1' - Q_2' > Q_1 - Q_2$$

or

$$\eta' < \eta \text{ or } \eta > \eta'$$

$\therefore$ The efficiency of a reversible cycle is greater than that of an irreversible heat engine.

7.5. ENTROPY OF A PERFECT GAS

Let us consider a gram mole of a perfect gas at temperature T and pressure P having a volume V.

Let a quantity dQ of heat be given to the gas so that a part of this energy is used in raising the temperature of the gas through dT at constant volume and rest in doing work for the expansion of the gas by an amount dV against the pressure P. If C_v be the gram molecular specific heat of the gas at constant volume, then we may have

$$dQ = C_v . dT + P . dV$$

From the perfect gas equation

$$PV = RT$$

or

$$P = \frac{RT}{V}$$

$\therefore$

$$dQ = C_v . dT + \frac{RT}{V} \cdot dV$$

So, Increase in entropy

$$\delta S = \frac{dQ}{T} = C_v \cdot \frac{dT}{T} + \frac{R}{V} \cdot dV$$

In order to find the change in entropy from temperature T_a and volume V_a to a temperature T_b and volume V_b, we can integrate the above expression between these limits.

$\therefore$ Change in entropy

$$S_b - S_a = C_v \int_{T_a}^{T_b} \frac{dT}{T} + R \int_{V_a}^{V_b} \frac{dV}{V}$$

$$= C_v \log_e \frac{T_b}{T_a} + R \log_e \frac{V_b}{V_a}$$

If M is the molecular weight of the gas, then the entropy of m gm of the perfect gas may be given as

$$\frac{m}{M}(S_b - S_a) = \frac{m}{M}\left[C_v \log_e \frac{T_b}{T_a} + R \log_e \frac{V_b}{V_a}\right]$$

7.6. CARNOT'S CYCLE ON TEMPERATURE-ENTROPY DIAGRAM

We can represent the Carnot's cycle on temperature-entropy diagram by considering the Carnot's cycle undergoing the following operations.

(*a*) Isothermal expansion represented by AB of the curve at temperature T.

(*b*) Adiabatic expansion represented by curve BC so that temperature falls from T_1 to T_2.

(*c*) Isothermal compression represented by curve CD at temperature T_2.

(*d*) Adiabatic compression represented by curve DA when temperature rises from T_1 to T_2.

On the temperature-entropy diagram, an isothermal is represented by a straight line AB parallel to the entropy axis (X-axis in Fig. 7.3 (*ii*) because the temperature in this case is constant. The adiabatic curve is represented by a straight line parallel to temperature-axis (Y-axis in Fig. 7.3 (*ii*), since in the adiabatic change, entropy of the system remains constant and the process is called entropic process. So, on the T–S diagram, Carnot's cycle can be represented by isothermal AB at temperature T_1, adiabatic BC, isothermal CD at temperature T_2 and adiabatic DA.

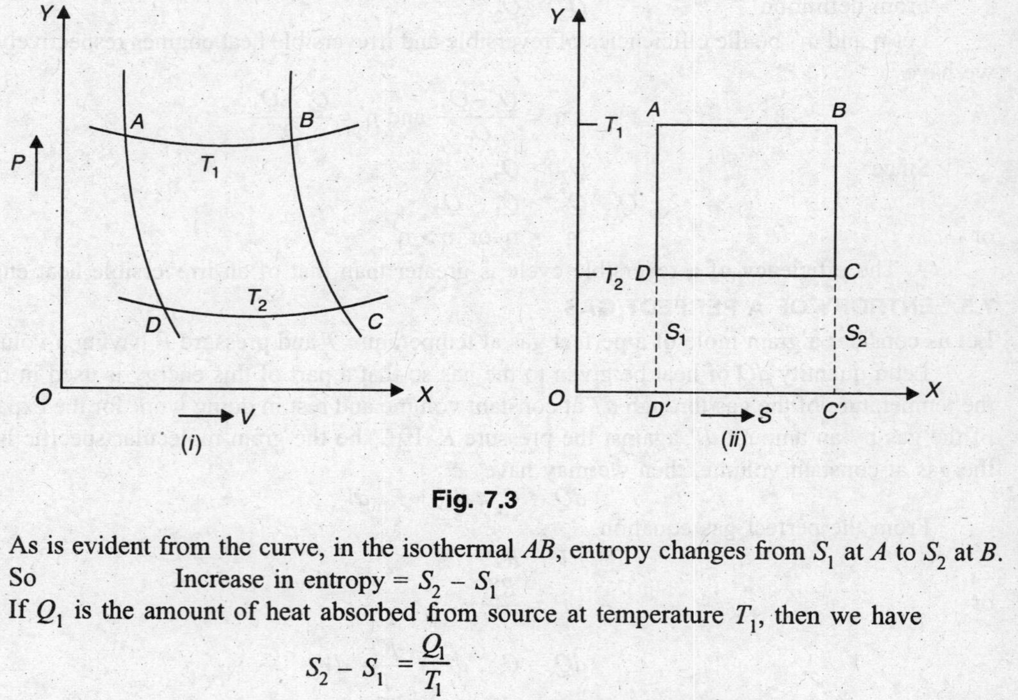

Fig. 7.3

As is evident from the curve, in the isothermal AB, entropy changes from S_1 at A to S_2 at B.

So Increase in entropy $= S_2 - S_1$

If Q_1 is the amount of heat absorbed from source at temperature T_1, then we have

$$S_2 - S_1 = \frac{Q_1}{T_1}$$

$$\therefore \quad Q_1 = (S_2 - S_1) \times T_1 = \text{Area } ABC'D' \quad ...(1)$$

In the adiabatic BC, since there is no change of entropy, so no exchange of heat takes place.

Again in the isothermal change CD, the entropy changes from S_2 at C to S_1 at D.

$\therefore$ Decrease in entropy $= S_2 - S_1$

Now if Q_2 is the amount of heat given to the sink at temperature T_2,

then
$$S_2 - S_1 = \frac{Q_2}{T_2}$$

or
$$Q_2 = (S_2 - S_1) \times T_2 \quad ...(2)$$
$$= \text{Area } CDC'D'$$

From Equations (1) and (2) energy available for useful work

$$\begin{aligned} &= (Q_1 - Q_2) \\ &= (T_1 - T_2) \times (S_2 - S_1) \\ &= \text{Area } ABC'D' \ - \text{Area } CDC'D' \\ &= \text{Area } ABCD \end{aligned}$$

Hence we can see that the area of a Carnot's cycle on temperature-entropy curve will represent the energy available for useful work.

We can define efficiency from above concepts given by

$$\eta = \frac{Q_1 - Q_2}{Q_1} = \frac{(T_1 - T_2)(S_1 - S_2)}{T_1 (S_1 - S_2)}$$

$$= \frac{T_1 - T_2}{T_1} = 1 - \frac{T_2}{T_1}.$$

Example 1. *Find the difference of entropy between 1 gm of ice at 0°C and 1 gm of steam at 100°C. Latent heat of fusion of ice = 80 calories, Sp. heat of water = 1 and latent heat of steam at 100°C = 540 cals.*

Solution: We can also get the total change of entropy in the following three steps:

(*i*) When 1 gm of ice at 0°C changes to water at 0°C,

$$\text{Change in entropy } \delta S = \frac{\delta Q}{T}$$

Now latent heat of ice = 80 cals *i.e.*, 80 calories of heat are required to convert ice at 0°C = 273 Å into water without any change in temperature.

$$\therefore \qquad \delta S = \frac{80}{273} = 0.293 \text{ cal/°C.}$$

(*ii*) When 1 gm of water at 0°C is heated to 100°C.

$$\text{Change in entropy, } S_2 - S_1 = \int_{T_1}^{T_2} \frac{dQ}{T}$$

Also $\qquad dQ = m \, . \, s \, . \, dT$

where m is the mass of the substance, and its sp. heat and dT, a very small change in temperature of the initial temperature T.

$$\therefore \qquad S_2 - S_1 = ms \int_{T_1}^{T_2} \frac{dT}{T}$$

$$= ms \log_e \frac{T_2}{T_1}$$

$$= ms \times 2.3026 \log_{10} \frac{T_2}{T_1}$$

In this case

$$\begin{aligned} T_1 &= 273 \text{ K} \\ T_2 &= 273 + 100 = 373 \text{ K} \\ m &= 1 \text{ gm and } s = 1 \end{aligned}$$

$$\begin{aligned} \therefore \qquad S_2 - S_1 &= 2.3026 \times 1 \times 1 [\log_{10} 373 - \log_{10} 273] \\ &= 2.3026 [2.5717 - 2.4362] \\ &= 2.3026 \times 0.1355 \\ &= 0.312 \text{ cal/°C.} \end{aligned}$$

(*iii*) When 1 gm of water at 100°C is converted into steam at 100°C.

$$\text{Change in entropy} \qquad \frac{\delta Q}{T} = \frac{540}{273 + 1000}$$

$$= 1.448 \text{ cal/°C.}$$

$$\therefore \text{ Total change in entropy} \qquad = 0.293 + 0.31 + 1.448$$

$$= \mathbf{2.053 \text{ cal/°C.}}$$

Example 2. *A brass rod is in contact thermally with a heat reservoir at 127°C at one end* *and a heat reservoir at 27°C at the other end. Compute the total change in entropy arising from the process of conduction of 1200 cal. of heat through the rod. Does the entropy of the rod change in the process?*

Solution: Q calories of heat are conducted through brass rod from hot reservoir at $T_1 = 127°\text{C} = 400$ K, to a cold reservoir at $T_2 = 27°\text{C} = 300$ K, so the change in entropy in each case is given by

For hot reservoir,
$$\Delta S_1 = -\frac{Q_2}{T_1} = -\frac{1200 \text{ cal}}{400 \text{ K}} = -3 \text{ cal/K.}$$

For cold reservoir
$$\Delta S_2 = +\frac{Q}{T_2} = \frac{1200}{300} = 4 \text{ cal/K.}$$

So total change in entropy during the conduction process
$$\Delta S = \Delta S_1 + \Delta S_2 = -3 + 4 = \mathbf{1.0 \text{ cal/K.}}$$

The entropy of the rod remains constant, because for any portion of rod, as much heat is given out as is received by it at a given temperature.

Example 3. *Calculate the change in entropy when 10 gms of ice at 0°C is converted into water at the same temperature.*

Solution: Change in entropy is given by
$$S_2 - S_1 = Q/T$$

In this case
$$Q = M.L = 10 \times 80 \text{ cal}$$
$$T = 273 \text{ K}$$

$\therefore$ Change in entropy
$$\frac{800}{273} = \mathbf{2.93.}$$

Example 4. *50 grams of water at 0°C is mixed with an equal mass of water at 80°C. Calculate the resultant increase in entropy.*

Solution: In this case, $m_1 = 50$ gm, $m_2 = 50$ gm
$$T_1 = 273 \text{ K}, T_2 = 273 + 80 = 353 \text{ K.}$$

Let the final temperature of the mixture be T, then we have
$$m_1 s (T - T_1) = m_2 s (T_2 - T)$$

or
$$50 \times 1 (T - 273) = 50 \times 1 (353 - T)$$

or
$$T = \mathbf{373 \text{ K.}}$$

Now change in entropy of 50 gm of water when temperature changes from 313 K to 273 K
$$= ms \int_{T_1}^{T} \frac{dT}{T}$$
$$= 50 \times 1 \log_e \frac{313}{273}$$
$$= 50 \times 2.3026 \log_{10} \frac{313}{273}$$
$$= 6.829 \text{ cal/K (gain is entropy)}$$

Also change in entropy of 50 gm of water when its temperature falls from 353 K to 313 K.
$$= ms \int_{T_1}^{T} \frac{dT}{T}$$
$$= 50 \times 1 \times \log_e \frac{313}{353}$$
$$= 50 \times 1 \times 2.3026 \log_{10} \frac{313}{353}$$

$= -\ 6.023$ cal/K (loss in entropy)

$\therefore$ Total gain in entropy $= 6.829 - 6.023$

$=$ **0.806 cal/K.**

Example 5. *Four moles of an ideal gas are caused to expand from a volume V_1 to a volume V_2 $(= 2V_1)$ (a) if the expansion is isothermal at the temperature $T = 400$ K, deduce an expression for the work done by the expanding gas (b) for the isothermal expansion just described, deduce an expression for the change in entropy, if any (c) if the expansion were reversible adiabatic instead of isothermal, would the change in entropy be positive, negative or zero?*

Solution: (*a*) Work done by an ideal gas expanding isothermally from volume V_1 to V_2 is given by

$$W = \int_{V_1}^{V_2} P \cdot dV$$

For an ideal gas we have

$$p = \frac{nR}{V}$$

So, the work done by n molecules is

$$W = \int_{V_1}^{V_2} \frac{nRT}{V} dV = nRT \int_{V_1}^{V_2} \frac{dV}{V}$$

$$= nRT \log \frac{V_2}{V_1}$$

where T is taken outside the integral because it is constant in an isothermal process.

Now $n = 4$ moles, $V_2 = 2V_1$

$T = 400$ K

$\therefore$ $W = 4RT \log 2$

$= 4 \times (1.986$ cal/mole K$) \times (400$ K$) \times (0.693)$

$=$ **2202 calories.**

(*b*) From the law of thermodynamics, we have

$$\Delta U = Q - W$$

Since temperature is constant in isothermal expansion, change in internal energy is zero.

i.e., $\Delta U = 0$

$\therefore$ $Q = W = 4RT \log 2$

$\therefore$ Change in entropy $\Delta S = \frac{Q}{T} = 4R \log 2$

$= 4$ mole $\times$ 1.986 cal/mole K $\times$ 0.698

$=$ **5.5 cal/K.**

(*c*) Change in entropy for a reversible adiabatic compression or expansion is

$$\Delta S = \int \frac{dU + PdV}{T}$$

Also from first law of thermodynamics

$$dQ = dU + PdV = \mathbf{0}$$

Since no heat transfer takes place change in entropy $\Delta S = 0$, in a reversible adiabatic process.

Example 6. *Heat can be removed from water at 0^oC at atmospheric pressure without causing the water to freeze, if done with little disturbance of the water. Suppose the water is cooled to -5^oC, before ice begins to form. What is the change in entropy per unit mass during the sudden freezing that takes place?*

Solution: Freezing takes place at $T = -5^oC = 268$ K.

Heat required to freeze 1 gm of water

$dQ = 80$ cal/gm. So change in entropy

$$dS = \frac{dQ}{T} = \frac{80 \text{ cal/gm}}{268 \text{ K}}$$
$$= \mathbf{0.3 \text{ cal/gm K}}.$$

Example 7. *A 20 gm piece of ice at 0°C is dropped into a thermous bottle containing 200 gms of water at 30°C. Calculate the net entropy change of the system of water and ice when a final equilibrium state is reached. L = 80 cals.*

Solution: Let the final temperature of the mixture be t, then we have

$$20 \times 80 + 20 \times t = 200 \times (30 - t)$$

or $$1600 + 20t = 6000 - 200t$$

or $$220t = 6000 - 1600 = 4400$$

or $$t = 20°\text{C}.$$

(*a*) **Increase in entropy when the ice melts:** Increase in entropy to convert 20 gm of ice at 0°C into water at 0°C is given by

$$\delta S = \frac{\delta Q}{T} = \frac{20 \times 80}{273} = \mathbf{5.86 \text{ cal/°C}}$$

(*b*) Increase in entropy of the melted ice when temperature changes from 0°C to 20°C.

$$\delta S = ms. \log_e \frac{T_2}{T_1} = ms \times 2.3026 \log_{10} \frac{T_2}{T_1}$$
$$= 20 \times 2.3026 \times 1 \log_{10} \frac{273+20}{273}$$
$$= \mathbf{20 \times 2.3026 \times 0.0307}$$

(*c*) Decrease in entropy of 200 gm of water when temperature changes from 30°C to 20°C.

$$\delta S = ms \log_e \frac{T_2}{T_1}$$
$$= 200 \times 1 \times 2.3026 \log_{10} \frac{273+30}{273+20}$$
$$= 200 \times 2.3026 \times 0.0145$$

∴ Net increase in entropy

$$= 5.86 + 2.3026 (20 \times 0.0307 - 200 \times 0145)$$
$$= 5.86 - 5.264 = \mathbf{0.596 \text{ cal/°C}}.$$

Example 8. *In a specific heat experiment, 100 gms of lead (C_p = 0.0345 cal/gm °C) at 100°C is mixed with 200 gms of water at 20°C. Find the difference in entropy of the system at the end from its value before mixing.*

Solution: Let the resultant temperature be T°C.

We have the relation

Heat lost by lead = heat gained by water

i.e., $$m_1 s_1 (T_1 - T) = m_2 s_2 (T - T_2)$$

where
m_1 = mass of lead = 100 gm
s_1 = sp. heat of lead = 0.0345 cal/gm °C
T_1 = initial temperature of lead = 100°C = 373 K
m_2 = mass of water = 200 gm.
s_2 = sp. heat of water = 1 cal/gm °C
T_2 = initial temperature of water = 20°C = 293 K

∴ We have

$$100 \times 0.0345 \times (100 - T) = 200 \times 1 \times (T - 20)$$

or $$T = 21.36°\text{C} = 294.36 \text{ K}$$

Change in entropy of lead

$$\Delta S_1 = -m_1 s_1 \log\left(\frac{T_1}{T}\right)$$

where negative sign shows that there is decrease in entropy when the substance is cooling.

$$\therefore \quad \Delta S_1 = -100 \times 0.0345 \times \frac{373}{294.36}$$

$$= -\,0.817 \text{ cal/K}$$

Change in entropy of water

$$\Delta S_2 = +\, m_2\, s_2 \log\left(\frac{T_1 \text{ K}}{T_2 \text{ K}}\right)$$

$$= 200 \times 1 \times \left(\frac{294.36 \text{ K}}{293 \text{ K}}\right)$$

$$= 0.926 \text{ cal/K}$$

So net change in entropy $\quad \Delta S = \Delta S_1 + \Delta S_2$

$$= -\,0.817 + 0.926$$

$$= \mathbf{0.11 \text{ cal/K.}}$$

Example 9. *Calculate the increase in entropy of 2 gm of O_2 when its temperature is raised from 20°C to 100°C and its volume is doubled.* [C_v *for* O_2 = *0.19 kcal/kgm/K, R = 0.062*]

Solution: We have the change in entropy for a perfect gas given by

$$S_b - S_a = m\left[C_v \log_e \frac{T_b}{T_a} + R \log_e\left(\frac{V_b}{V_a}\right)\right]$$

$$= 2.3026 \times m\left[C_v \log_{10} \frac{T_b}{T_a} + R \log_{10}\left(\frac{V_b}{V_a}\right)\right]$$

In this case, $\quad C_v = 0.19$ k.cal

$$R = 0.062$$

$$\frac{V_2}{V_1} = 2,\ m = 2 \times 10^{-3} \text{ kg}$$

$$T_a = 293\ K,\ T_b = 373 \text{ K}$$

$$\therefore \quad S_b - S_a = 2.3026 \times 2 \times 10^{-3} \times \left[0.19 \log \frac{373}{293} + 0.062 \log_{10} 2\right]$$

$$= 2.3026 \times 2 \times 10^{-3}\, [0.19 \times 0.1048 + 0.062 \times 0.301]$$

$$= \mathbf{0.0423 \text{ cal/K.}}$$

Example 10. *An 8 gm ice cube at – 10°C is dropped into thermous flask containing 100 cc of water at 20°C. What is the change in entropy of the system when final equilibrium is reached?*

Solution: Let the final temperature when equilibrium reaches be T°C.

Now heat lost by 10 c.c. of water *i.e.*, 100 gm of water = heat gained by 8 gm of ice when its temperature rises from –10°C to 0°C + heat gained by ice when it melts + heat gained by melted ice to raise its temperature from 0°C to T°C.

i.e., $\quad 100 \times 1 \times (20 - T) = 8 \times 0.5 \times 10 + 8 \times 80 + 8 \times 10 \times T$

$$\therefore \quad T = 12.2°\text{C}$$

Now change in entropy of ice from temperature – 10°C to 0°C.

$$\Delta S_1 = \frac{\text{heat absorbed by ice}}{\text{average temperature K}}$$

$$= \frac{\text{Mass} \times \text{sp. heat} \times \text{rise in temperature}}{\text{Average temperature K}}$$

$$= \frac{8 \times 0.5 \times 10}{268} = \mathbf{0.15 \text{ cal/K.}}$$

(*ii*) Change in entropy when ice melts at 0°C.

$$\Delta S_2 = \frac{\text{Heat absorbed to melt ice}}{273}$$

$$= \frac{\text{Mass of ice} \times \text{latent heat of ice}}{273}$$

$$= \frac{8 \times 80}{273} = \mathbf{2.34\ cal/K.}$$

(*iii*) Change in entropy when melted ice gets heated up from 0°C to 12.2°C.

$$\Delta S_3 = \frac{\text{Heat absorbed by melted ice}}{\text{Average temperature K}}$$

$$= \frac{\text{Mass of melted ice} \times \text{sp. heat of water} \times \Delta T}{\text{Average temperature K}}$$

$$= \frac{8 \times 1 \times 12.2}{279 \text{ K}} = \mathbf{0.35\ cal/K.}$$

(*iv*) Change in entropy of 100 c.c. of water when it cools from 20°C to 12.2°C

$$\Delta S_4 = + \frac{\text{Heat lost by water}}{\text{Average temperature}}$$

$$= - \frac{(\text{Mass of water}) \times \text{sp. heat} \times \Delta T}{\text{Average temperature K}}$$

$$= - \frac{100 \times 1 \times (20 - 12.2)}{290 \text{ K}}$$

$$= -2.69 \text{ cal/K.}$$

So, total change in entropy

$$\Delta S = \Delta S_1 + \Delta S_2 + \Delta S_3 + \Delta S_4$$

$$= 0.15 + 2.34 + 0.35 - 2.69$$

$$= \mathbf{0.15\ cal/K.}$$

Example 11. *A volume of a perfect gas expands isothermally to four times its initial volume. Calculate the change in entropy in terms of gas constant.*

Solution: Let a gram. molecule of gas expands isothermally at temperature T from volume V to $4V$.

Let Q be the quantity of heat extracted so that the amount of work done W in this expansion process is given by

$$W = Q = = \int_{V_2}^{V_1} P \cdot dV$$

where V_1 and V_2 correspond to initial and final volume of the gas.

In the case of a perfect gas

$$PV = RT$$

$$\therefore \quad P = \frac{RT}{V}$$

So we have
$$Q = RT \int_{V_1}^{V_1} \frac{dV}{V} = RT \int_{V}^{4V} \frac{dV}{V}$$

$$= RT \log_e \frac{4V}{V} = RT \log_e 4$$

So change in entropy
$$\frac{Q}{T} = R \log_e 4$$

$$= R \times 2.3026 \log_{10} 4$$

$$= R \times 2.3026 \times 0.6021$$

$$= \mathbf{1.386}\ \boldsymbol{R}.$$

QUESTIONS AND PROBLEMS

1. Explain what do you know about entropy of a substance. Show that for any reversible cyclic change of a system the total change in entropy is zero. Show why this statement is untrue for irreversible change.
2. From entropy considerations, state and prove Carnot's theorem.
3. Represent Carnot's cycle on a temperature entropy diagram and prove that its area represents available energy. Establish from above that the efficiency of an engine is $\frac{T_1 - T_2}{T_1}$ where T_1 and T_2 are the temperatures between which engine works.
4. Derive an expression for the entropy of *m* gm of a perfect gas.
5. (*a*) Show that when a substance of mass *m* having a constant specific heat *s* is heated from T_1 to T_2, the change in entropy is $S_2 - S_1 = ms \log_e \frac{T_2}{T_1}$.
   (*b*) Does the entropy of a substance decrease on cooling? If so, does the total entropy decrease in such a process? Explain.
6. Two samples of a gas initially at the same temperature and pressure are compressed from a volume V to $\frac{V}{2}$, one isothermally and the other adiabatically. In which sample, the final pressure is greater? Does entropy change in either process?
7. In a specific heat experiment 106 gm of lead (C_p = 0.0345 cal/gm °C) at 100°C is mixed with 200 gm of the system at the end from its value before mixing. **[Ans.** + 0.20 cal/K]
8. A molecule of a monatomic ideal gas is taken from an initial state of pressure *p* and volume *V* to a final state of pressure 2*p* and volume 2*V* by two different processes. (*a*) It expands isothermally until its volume is doubled and then its pressure is increased at constant volume to the final state. (*b*) It is compressed isothermally until its pressure is doubled and then its volume is increased at a constant pressure to the final state.
   Show the path of each process on *P–V* diagram. For each process, calculate the terms of *P* and *V* or *T* (*i*) The heat absorbed by the gas in each part of the process (*ii*) The work done on gas in each process (*iii*) The change in internal energy of gas $U_f - U_i$ (*iv*) change in entropy of the gas $S_f - S_i$.
9. Calculate the change in entropy when 20 gms of steam at 100°C is passed through 200 gms of water at 13°C. Latent heat of steam is 540 calories. **[Ans.** 5.763 cal/°C]
10. Calculate the change in entropy when 10 gms of ice is converted into water at the same temperature. **[Ans.** 2.93 cal/°C]

PART III
OPTICS

Christian Huygens (1629-1695)

"Experiment and observation provide the only way of arriving at the knowledge of the causes of all that one sees in *nature*."

—Christian Huygens (1629–1695)

Huygens was born on April 14, 1629, in The Hague, the son of the famous Constantine Huygens, who was a poet, philosopher, classical scholar, and diplomat. He was taught by a private tutor until he was 16, including Latin, music, and drawing. He grew to love mathematics more than all other subjects, however, this interest dominated his later life.

Huygens lived at home from 1650 to 1666, except for journeys to Paris and London, and these 16 years were the most productive of his scientific career. His first published work came out in 1651; it was followed by a constant stream of important books on a great variety of scientific subjects.

He remained in Paris from 1666 until 1681, except for occasional visits to his native Holland for reasons of health. In 1672, when the French armies were attacking the Dutch Republic, he stayed on in Paris. He even dedicated a celebrated book on his invention of the pendulum clock to Louis XIV, who had vowed to destory the Dutch Republic.

In 1681 Huygens was forced to leave Paris because of ill health, and he never returned. The years after 1683 were hard ones for Huygens. He lived in The Hague, cut off from the scientific centers of the world. He remained single, living alone at his family's estate most of the time, and the income from his family's property was barely enough to support him. In 1689 he visited England, met Newton and Boyle, and returned home even more unhappy about his intellectual isolation in the Netherlands.

He died in The Hague in 1695, survived by no disciples and relatively forgotten by the swiftly moving world of science, which owed so much to his ideas.

CHAPTER 1

NATURE OF LIGHT

1.1. INTRODUCTION

When we say that light travels from one point to another point, the natural question is 'what' is the thing that constitutes light. The case of sound is different because in that 'case the vibrations from one particle to another travel as sound is made up of **vibrations of the medium.** The medium as a whole does not travel but the vibrations move..

Many theories were put forward to explain the nature of light, the most important amongst these are given as under:

(*i*) Newton's Corpuscular theory (*ii*) Huygen's wave theory (*iii*) Electromagnetic theory (*iv*) Quantum theory.

We can study the various theories one by one.

(*i*) Corpuscular theory: Newton in 1675 put forward this theory, which states that light consists of very small particles called Corpuscles which are emitted from a luminous source and travel with very large velocities. When they reach the ratina of the eye they produce the sensation of vision.

This theory was able to explain the rectilinear propagation of light, because size of the particles or corpuscles is considered to be so small that Earth's gravitation has any little effect on them.

This theory was also able to explain the laws of Reflection (Angle of Incidence = Angle of Reflection) and the laws of Refraction ($\sin i/\sin r = \mu$, Refractive index of the medium) by considering another medium more rarer and more elastic than the air prevading all space. But this theory was not able to explain **Partial Reflection and Partial Refraction.**

Later on when the phenomenon of interference, diffraction and polarization also came into being, Newton's theory was unable to explain them.

(*ii*) Wave Theory of Light: Huygens in 1678, said that light may be considered to consist of periodic disturbance transmitted through a medium in the form of waves. He considered the waves to be of **longitudinal type** and said that a source of light created periodic disturbances in other and travelled as wave in a way similar to that of sound which travels in air.

Huygens argued that because light beam can travel in all directions independently, so it cannot be considered to consist of corpuscles, because due to collision between the various corpuscles, intensity of light should be different in different directions (this objection against corpuscular theory can be removed if the corpuscles are considered to be of very very small size). The waves can easily explain the independence of the light beams travelling in the same medium in all the directions. Huygens was able to explain the phenomenons of reflection, refraction and **simultaneous occurrence of reflection and refraction** although it was only after more than 100 years of the prediction of wave theory that the various results could be verified by (Focault, 1850 and Fresnel, 1822).

Huygens was able to discover polarization of light in around 1690, even before the interference and diffraction phenomenon came into being. This instead of any help to the wave theory of light, proved to be a hinderance bacause Huygens longitudinal waves could not explain polarization. It was **Fresnel** 140 years later, who introduced the idea of transverse waves in ether, considering **ether to be an elastic solid.**

Diffraction was first reported by Grimaldi in 1665, but due to lack of experimental evidence this phenomenon could not be established at that time. The firm proof of this phenomenon also came from Fresnel. Fresnel was also able to explain the interference phenomenon and rectilinear propagation of light.

It seems as if Huygens was able to give wave theory of light by some intution, the firm mathematical concepts of wave theory came from Fresnel.

Discarding ether concept: The concept of ether filling the whole space had to be discarded because it involved various experimental difficulties. Einstein's theory of Relativity and later on experiment by Michelson and Morley proved that concept of ether is superfluous. As gravitational forces can act across space without a medium, electric and magnetic fields can also act without a medium, so *light can also travel without any medium.*

e.m. theory of light: A moving electric charge produces magnetic field and a changing magnetic field produces an electric field. **Maxwell** was able to explain the intimate relation between the electric and magnetic fields. If we study the effect at some distance from the first point then we have the same changing electric and magnetic fields, though they are not in phase because of distance between the two points. The disturbance is found to travel in waves of electric and magnetic intensity. These are called **electromagnetic waves** and the ratio of e.m.u. of current, to e.s.u. of current gives the speed of these waves. This speed came out to be equal to the speed of light.

Hertz actually detected electromagnetic waves from an oscillating electric circuit and was able to explain that these waves could be reflected, refracted and polarized. Though Hertz did experiments with waves of several meters wavelength, (visible light has wavelength 6×10^{-7} meters) e.m. theory of light was accepted because it simplified many results and was considered to be an extension of the wave theory. In this concept transverse waves come up naturally and their speed in vacuum can be found by any method and the speed in other transparent medium can be found from $c\sqrt{\varepsilon}$, where ε is dielectric constant which can be measured by an electrical experiment.

Quantum Theory of Light: Upto the beginning of 20th century, it was thought that perhaps all the characteristics of light could be explained by the e.m. theory. But the emission spectrum of light and why light from a particular source gave same set of wavelengths remained a problem. To this problem were added two more prominent ones, photo-electric effect and Compton effect. Photo-electric effect showed that when light of some frequency fell on a surface, it did not give energy to the surface continuously but in units of $h\nu$. No transfer of energy less than $h\nu$ units could take place. Compton effect showed that interaction between electrons and light can be treated like an elastic collision between two balls, with light to be considered to consist of number of balls each of energy $h\nu$ and momentum $h\nu/c$ and conservation laws of energy and momentum holding good for the collisions. So light in addition to showing all the properties of waves also showed the characteristics of units of **energies and momenta** carried by these waves. These carriers of energies and momenta and associated with light waves are given the name of **photons.**

Planck in 1900 said that an atomic oscillator may take in or give out energy not continuously but in units of $h\nu$, where ν is the frequency of the absorbed or emitted radiation and h is Planck's constand ($h = 6.626 \times 10^{-27}$ erg. sec).

Einstein in 1905 explained the Photo-electric effect by assuming that light of frequency ν carried energies in quantas of $h\nu$ and not continuously. The thing that carried energy $h\nu$ and momentum $h\nu/c$ is called **Photon.**

Bohr in 1912 made another remarkable advance by explaining that atomic systems could exist only in discrete states of energy. These could stay either in one of these states or jump to some other state, but could not exist in any in between states. Whenever there is change of energy $-E$ due to the jump of the states, light of frequency $-E/h$ is emitted or absorbed so that $\Delta E = h\nu$. Due to tremendous success of the discreteness of states, this concept has been extended to explain all kinds of characteristics observed in atomic and Molecular sysums.

Modern Ideas. We are now in a position where physicists often find themselves, from time to time, many theories were put forward which could explain some characteristics while others posed problems. It is almost always true that the efforts spent never go waste. The old corpuscular model which was once discarded has once again started playing part and modern concepts of light are that it has dual character *i.e.*, some times it behaves like a wave and sometimes like a particle. With the advance of wave theory and the quantum theory it is proved beyond doubt that not only light but other particles like electrons, protons etc. have double character. We can elaborate the above mentioned advances to explain the various theories of light as under.

1.2. WAVE THEORY OF LIGHT AND HUYGEN'S PRINCIPLE

Whenever we mention the term light, it means that part of radiant energy which can give us the sensation of vision by stimulating the ratina of the eye. From the experiments, it has been proved that light travels with a finite velocity $\sim 3 \times 10^8$ ms^{-1}. From the various experiments it has been further proved that light produces heat, affects photographic plates, influences the growth of plants etc., which shows that just like heat, light is also a form of energy. Now since light is a form of energy, so we can see how this energy is transported from one point to another point. For example, when light energy comes from the sun, there are two modes of transportation.

(*i*) This energy may be carried by small material particles travelling with finite velocity, and each particle possesses K.E. = mv^2 and no material medium is required for propagation of these particles. This is the basic form of the Newton's corpuscular theory.

(*ii*) The second mode of transportation may be considered that the light energy moves by wave motion where the actual matter does not travel. This process requires some intevening medium for transference of energy. This interpretation was further supplemented by wave theory of light given by Huygen's in 1678.

1.3. NEWTON'S CORPUSCULAR THEORY OR EMISSION THEORY

According to Newton's Corpuscular Theory, a luminous body emits very minute and weightless particles called corpuscles which travel through a homogeneous medium in straight lines and in all directions with the velocity of light *i.e.,* 3×10^8 metres/sec. and each particle carries with it kinetic energy. The particles move in straight lines due to the property of inertia and when these particles fall on eye, they give the sensation of vision. Different colours of light were thought to be associated with these particles and the size of the particles is so small that they pass through the interspace of the particles of the matter. Also due to their very high speed these particles are not affected by gravity. This theory could easily explain laws of reflection, refraction and rectilinear propagation of light.

1.4. WAVE THEORY OF LIGHT

This theory was put forward by Huygens in 1678. According to this theory light is a form of wave motion which is produced due to the mechanical vibrations of an all pervading homogeneous medium known as ether, hypothetical medium is supposed to possess the properties of **elasticity and inertia** and is able to transmit wave motion : Huygens supposed these waves to be longitudinal and he was able to explain the phenonmenon of reflection, refraction, interferences diffraction. But the phenomenon of rectilinear propagation of light and polarization could not be properly explained.

Fresnel and Young overcame the above defects by supposing the light waves to be transverse in nature. Fresnel explained the phenomenon of rectilinear propagation of light by making use of secondary wave fronts and dividing these wave fronts into half period zones or half period elements. Now if V be the velocity of a wave through an elastic medium of elasticity E and density ρ, then we have

$$V = \sqrt{\frac{E}{\rho}}$$

which means that Ether must have very high elasticity and low density. These two properties are just opposite to each other, so it is very difficult to conceive a medium with contracting properties.

Maxwell in 1860 introduce the idea of e.m. waves and concluded that light waves are just like e.m. waves. According to this theory, light wave may be considered as similar to an e.m. wave. The velocity of e.m. wave is given by

$$c = \frac{1}{\sqrt{\varepsilon\mu}}$$

where ε is dielectric constant and μ, the permeability of the medium. Now since these quantities have definite values for vacuum, so there is no necessity of assuming a hypothetical medium like ether.

1.5. HUYGEN'S PRINCIPLE

According to this principle a source of light in a hypothetical ether medium sends out waves in all directions and these waves carry energy.

Let S be a source of light that sends waves in all directions. After and interval of time t, all the particles of the medium lying on AB are vibrating in same phase where AB is drawn by taking S as centre and radius equal to SA, where $SA = ct$ (c being velocity of pro-pagation of wave).

AB is called primary wave-front. So we may define a wavefront as the locus of all such points which are in same phase *i.e.*, which are in the same state of vibration.

The wave front will be spherical for a point source of light if the distance is small, as shown in Fig. 1.1 (*a*). For large distance, the wave-front can be considered as a plane wave as shown in Fig. 1.1 (*b*).

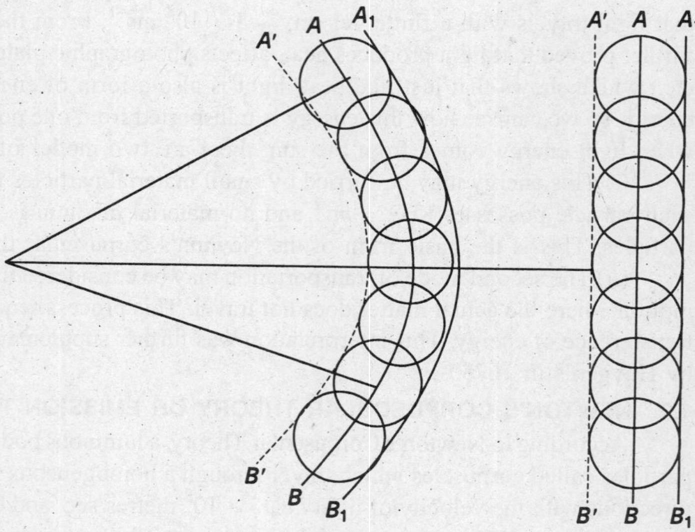

Fig. 1.1

From above it is clear that a beam of light converging to or diverging from a point will give rise to spherical wave front while a parallel beam of light will give rise to a plane wavefront.

A ray is defined as the direction in which the disturbance is propagated into the medium and it is always normal to wavefront.

From Huygen's Principle we may consider all the points on primary wavefronts as **centres of disturbance** which send out disturbance in all the directions which will travel through the space with the same velocity in an isotropic medium.

In order to find the position of the new wavefront after a time say t seconds, number of points on AB are taken as centres and radii equal to ct, various spheres are drawn. These spheres will represent secondary waves and a surface, $A_1 B_1$ touching these spheres in the forward direction is the new wavefront is the new wavefront.

In similar way propagation of light due to numerous wavefronts can be explained.

To establish a connection between Newton's corpuscular theory and e.m. theory of light, let us consider an e.m. wave of frequency ν incident on a screen.

The intensity I of the wave which is the rate at which it transports energy/unit area, depend on E and B *i.e.*, the magnitude of electric and magnetic fields. Also, we have the relation $E = cB$, so either E or B is sufficient to describe the wave intensity. It is customary to choose E and so the intensity I of the wave at the screen is given by

$$I = \varepsilon_0 c \overline{E}^2 \qquad ...(1)$$

where $\overline{E}^2$ correspond to the coverage of the *square of instantaneous magnitude of the wave's electric field over a complete cycle.*

In the *corpuscular theory* (photon model) for the same e.m. wave, energy is transported by n photons/unit area/sec. From quantum theory ~~h~~ each photon has energy $h\nu$, so the intensity at the screen is given by

$$I = nh\nu \qquad ...(2)$$

Compare Equations (1) and (2), we have

$$n = \frac{\varepsilon_0 c}{h\nu} \overline{E}^2 \qquad ...(3)$$

If n is very large and we look at the screen, we will find a continuous distribution of light whose pattern corresponds to distribution of $\overline{E}^2$ *and so will lead us to wave theory of light.*

If n is very small, *i.e.*, so small that only one photon reaches the screen at a time, we will get random flashes, which will account for the *quantum nature of light.*

If we keep track of the flashes for sufficient long time on the screen in the same form, we may come to the conclusion that *probability* of finding photon at a certain place and time depends on the value of $\overline{E}^2$ at that place.

The above argument may lead to an interesting situation as in the case of a double slit experiment. Let the light be so weak that only *one photon at a time moves.* How the interference phenomenon can be explained? How the photon decides to move through one slit or the other? Which means how can one photon interfere with itself. So the idea of a wave which spreads out in space and a photon which is *localized* in small region would *contradict each other.*

This contradiction can be overcome by regarding a photon to be associated with a wave. When it passes through the slits, light is behaving as a *wave* and when it strikes the screen it behaves as a *particle.*

So we conclude that light has dual characters: *wave theory of light and quantum theory of light compliment each other.* Either of the two theories by themselves cannot give the complete nature of light. It may be noted that this dual character *wave and particle* is exhibited by all moving objects.

QUESTIONS AND PROBLEMS

1. Illustrate the nature of light. What are the various theories to explain the different laws?
2. How wave theory of light explains the interference and polarization phenomenon?
3. What is a wave-front? How it explains the various characteristic of light?
4. How the quantum theory explains the nature of light?
5. What are the modern ideas about the nature of light?

CHAPTER

2

DISPERSION AND ABERRATIONS

2.1. DISPERSION

When white light is passing through a prism, it gets deviated through different extent because refractive index of a material is different for different wavelengths, and since white light consists of seven colours, so all the colours or wavelengths will be deviated through different degrees. The deviation for violet light being more than red light as the refractive index for violet light will be more than that for red light.

So we see that a ray of white light after passing through a prism splits up into its constituent colours and this phenomenon is called *dispersion.*

If the angle of the prism is very small then angle of minimum deviation δ is also very small and so we have

$$\mu = \frac{\sin \frac{A+\delta}{2}}{\sin A/2} = \frac{A+\delta}{A} = 1 + \frac{\delta}{A}$$

or

$$\delta = (\mu - 1)A$$

If δ_v and δ_r are the deviations for the extreme violet and extreme red radiations respectively and μ_v and μ_r their corresponding values of refractive indices,

then $$\delta_v = (\mu_v - 1)A$$

and $$\delta_r = (\mu_r - 1)A$$

$$\therefore \quad \delta_v - \delta_r = (\mu_v - \mu_r)A \qquad ...(1)$$

This difference in deviations for the extreme violet and extreme red radiations is called *angular dispersion.*

2.2. DISPERSIVE POWER

If we consider δ and μ as the deviation and refractive index for some mean wavelength light (say yellow in spectrum of white light), then

$$\delta = (\mu - 1)A \qquad ...(2)$$

Dividing Equations (1) by (2), we have

$$\frac{\delta_v - \delta_r}{\delta} = \frac{\mu_v - \mu_r}{(\mu - 1)} \qquad ...(3)$$

The factor $\frac{\delta_v - \delta_r}{\delta}$ is known as **dispersive power** of the material and is denoted by ω.

So, we may define the dispersive power of a material as the ratio of the **angular dispersion to the mean deviation**, Equation (3) may be written as

$$\omega = \frac{\mu_v - \mu_r}{\mu - 1} = \frac{d\mu}{\mu - 1}$$

where $d\mu = \mu_v - \mu_r$ *i.e.,* the change in refractive index for violet and red light.

Our eye is sensitive only to the regions of spectrum between C and F lines that lie in the red and blue portion of hydrogen spectrum, so it means that μ does not differ from D-line of the sodium spectrum. The internationally accepted expression for dispersive power is

$$\omega = \frac{\mu_F - \mu_C}{\mu_D - 1}. \qquad ...(4)$$

2.3. ACHROMATIC PRISM

As is evident from the different diviations of light from different prisms, the dispersive power of prisms of different materials will be different. It is seen that the dispersive power of flint glass is greater than that of the crown glass. So it is possible to have a combination of two prisms, so that a ray of white light after passing through the combination of prisms may suffer deviation without dispersion such a combination of prisms is called an achromatic combination of prisms; as shown in Fig. 2.1.

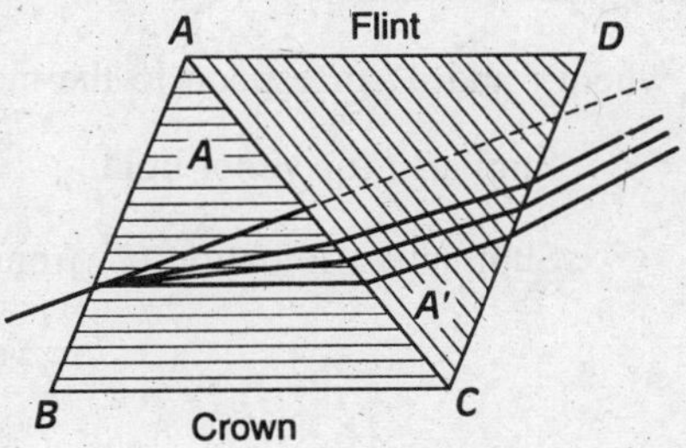

Fig. 2.1

As the light comes out of the combination of prisms, the constituent colours are parallel and so combine to give white light. In the strict sense of words, combination is a chromatic only for extreme colours (C and F) while for others it is only approximate.

2.4. CHROMATIC ABERRATION

A lens may be considered to consist of large number of prisms of different refracting angles placed one above the other.

In the case of a convex lens, the refracting angle of the prism goes on decreasing at a uniform rate as we move from centre towards periphery. In case white light is allowed to fall on a lens, then it will be dispersed in its constituent colours and since diviation are different for different colours, so dispersed rays will be focused at different points on the axis.

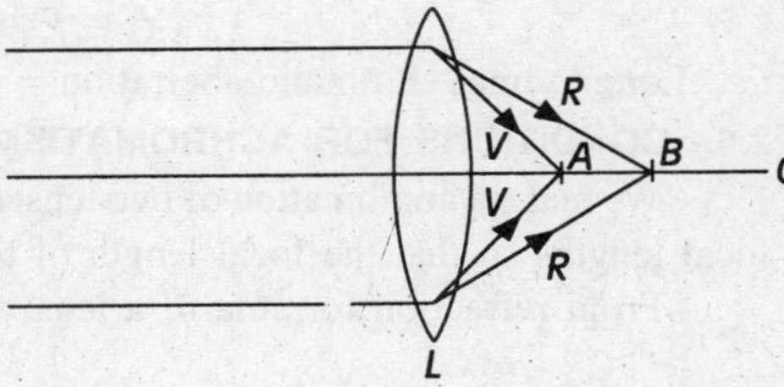

Fig. 2.2

Since μ for violet rays is greater than that for red rays, so focal length of the lens for violet rays is less than that for red rays. So the violet light is brought to focus at a point nearer to lens than red light. The various colours are brought to focus in between A and B as shown in Fig. 2.2.

If f_v is the focal length of the lens for extreme violet rays and f_r that for extreme red rays, then $f_v - f_r$ is the longitudinal chromatic aberration. Also since the focal length depends on different colours (or wavelengths), so the images of different colours are magnified to different degrees. This defect is called the **lateral chromatic aberration.**

The chromatic aberration can be minimized by making use of a combination of two or more lenses of different materials. However, it should be noted that chromatic aberration can be corrected only for two colours. Such a combination is called an **achromatic combination.**

2.5. ACHROMATIC COMBINATION OF TWO THIN LENSES IN CONTACT

An achromatic doublet generally consists of two thin lenses, a convex lens of crown glass having small focal length (or large positive power) and a concave lens of flint glass of large focal length (or small negative power). The combination is so designed that these two lenses have equal and opposite **dispersions** and the combination has net positive power.

We can find the chromatic aberration for a lens as under:

Let f_r and μ_r be the focal length and refractive index for red light f_v and μ_v the focal length and refractive index for violet light then we have for red light.

$$\frac{1}{f_r} = (\mu_r - 1)\left(\frac{1}{R_1} - \frac{1}{R_2}\right) \qquad ...(1)$$

Also from relation $\quad \frac{1}{f} = (\mu - 1)\left(\frac{1}{R_1} - \frac{1}{R_2}\right)$

We have $\quad \left(\frac{1}{R_1} - \frac{1}{R_2}\right) = \frac{1}{f} \cdot \frac{1}{(\mu - 1)}$

Put in Equation (1) $\quad \frac{1}{f_r} = \frac{(\mu_r - 1)}{(\mu - 1)} \cdot \frac{1}{f} \qquad ...(2)$

where f and μ correspond to the mean values of focal length and refractive index respectivly.

Similarly for violet light $\quad \frac{1}{f_v} = \frac{(\mu_v - 1)}{(\mu - 1)} \cdot \frac{1}{f}$

Subtracting Equation (2) from Equation (3)

$$\frac{1}{f_v} - \frac{1}{f_r} = \frac{(\mu_v - \mu_r)}{(\mu - 1)} \cdot \frac{1}{f}$$

or $\quad \frac{f_r - f_v}{f_v \cdot f_r} = \frac{\omega}{f}$

$$f_r \cdot f_v = f^2$$

Now $\quad f_r - f_v = \text{longitudinal chromatic aberration}$

$\therefore \quad f_r - f_v = \omega \cdot f$

i.e., Longitudinal chromatic aberration = dispersive power × mean focal length.

2.6. CONDITIONS FOR ACHROMATISM

We make a combination of two lenses of different materials placed in contact and choose their focal lengths so that the focal length of the combination is same for the both colours.

From refraction formula of a lens we have

$$\frac{1}{f} = (\mu - 1)\left(\frac{1}{R_1} + \frac{1}{R_2}\right)$$

Differentiating $\quad -\frac{df}{f^2} = d\mu\left(\frac{1}{R_1} + \frac{1}{R_2}\right) = \frac{d\mu}{(\mu - 1)} \cdot \frac{1}{f} \qquad ...(1)$

where $d\mu$ will correspond to the change in refractive indices. If violet and red colours are taken then

$$d\mu = (\mu_v - \mu_r)$$

Also dispersive power $\quad \omega = \frac{d\mu}{\mu - 1}$

$\therefore$ Equation (1) may be written as

$$-\frac{df}{f^2} = \frac{\omega}{f} \qquad ...(2)$$

When the two lenses of focal lengths f and f' are placed in contact, then their combined focal length is given by

$$\frac{1}{F} = \frac{1}{f} + \frac{1}{f'}$$

Differentiating $\quad -\frac{dF}{F^2} = -\frac{df}{f^2} - \frac{df'}{f'^2} \qquad ...(3)$

In order that the different colours reach the focus at the same point, df *i.e.*, the change in focal length of the combination should be zero.

i.e., from Equation (3) $\quad -\frac{df}{f^2} - \frac{df'}{f'^2} = 0$

or from Equation (2) $$\frac{\omega}{f}+\frac{\omega}{f'} = 0$$

where ω and ω' are dispersive powers of the materials of two lenses.

Equation (4) shows that f and f' must be of opposite signs *i.e.,* the combination consists of one concave lens and other convex lens of different materials.

2.7. CONDITION FOR ACHROMATISATION OF TWO LENSES SEPARATED BY A DISTANCE

Let f_1 and f_2 be the focal lengths of two lenses which are separated by a distance d and let F be focal length of the combination, then

$$\frac{1}{F} = \frac{1}{f_1}+\frac{1}{f_2}-\frac{d}{f_1 f_2}$$

Differentiating both sides, we have

$$-\frac{dF}{F^2} = -\frac{df}{f_1^2}-\frac{df_2}{f_2^2}+d\left[\frac{df_1}{f_1^2 f_2}+\frac{df_2}{f_2^2 f_1}\right] \quad ...(1)$$

Also in the case of single lens, we have

$$\frac{1}{f} = (\mu-1)\left(\frac{1}{R_1}+\frac{1}{R_2}\right) \quad ...(2)$$

Differentiating both sides, we have

$$-\frac{df}{f^2} = d\mu\left(\frac{1}{R_1}+\frac{1}{R_2}\right) = \frac{d\mu}{(\mu-1)}\cdot\frac{1}{f} \quad \text{[from Equation (2)]}$$

$$= \frac{\omega}{f}$$

or we have $$-\frac{df}{f^2} = \frac{\text{Dipersive power}}{\text{focal length}}$$

Now if ω_1 and ω_2 be the dispersive powers of the two lenses and f_1 and f_2 their corresponding focal lengths, we have from Equation (1)

$$\frac{\omega}{F} = \frac{\omega_1}{f_1}+\frac{\omega_2}{f_2}-d\left[\frac{\omega_1}{f_1 f_2}+\frac{\omega_2}{f_1 f_2}\right] \quad ...(3)$$

If the different colours are to be brought to focus at one point, ω *i.e.,* the dispersive power of combination should be zero. So Equation (3) may be written as

$$0 = \frac{\omega_1}{f_1}+\frac{\omega_2}{f_2}-d\left[\frac{\omega_1}{f_1 f_2}+\frac{\omega_2}{f_1 f_2}\right]$$

or $$\frac{\omega_1}{f_1}+\frac{\omega_2}{f_2} = d\,\frac{(\omega_1+\omega_2)}{f_1 f_2} \quad ...(4)$$

which gives us a relation for the two lenses separated by a distance d.

If, however, the two lenses are in contact, then $d = 0$

$\therefore$ Equation (4) may be written as

$$\frac{\omega_1}{f_1}+\frac{\omega_2}{f_2} = 0 \quad ...(5)$$

If we have the two lenses of the same material and are separated by distance d, *i.e.,* $\omega_1 = \omega_2 = \omega$. Then equation (4) may be written as

$$\frac{\omega_1}{f_1}+\frac{\omega_2}{f_2} = \frac{2d\cdot\omega}{f_1 f_2}$$

or $$\omega\frac{[f_1+f_2]}{f_1 f_2} = 2\cdot\frac{d\cdot\omega}{f_1 f_2}$$

or $$f_1+f_2 = 2d$$

or $$d = \frac{f_1 + f_2}{2} \qquad ...(6)$$

which means that if the two lenses are of the same material and they are seperated by a distance equal to 1/2 the sum of focal lengths of the two lenses, then they form an achromatic combination.

2.8. ADVANTAGE OF ACHROMATIC COMBINATION OF LENSES SEPARATED BY A DISTANCE

It should be noted that the above chromatic combination of lenses is completely free from **achromatic aberrations** of all colours since the above expression does not contain **dispersive power.** Also as seen from Equation (6), d is positive, so either both the lenses should be convex or at least lens of higher focal length should be convex.

2.9. SPHERICAL ABERRATION

In the derivation of the formula for mirror, it is assumed that the aperature of the mirror is small so that a beam of light parallel to the principal axis and incident on the surface of the mirror, after suffering reflection will be brought to focus at one point known as Principal Focus.

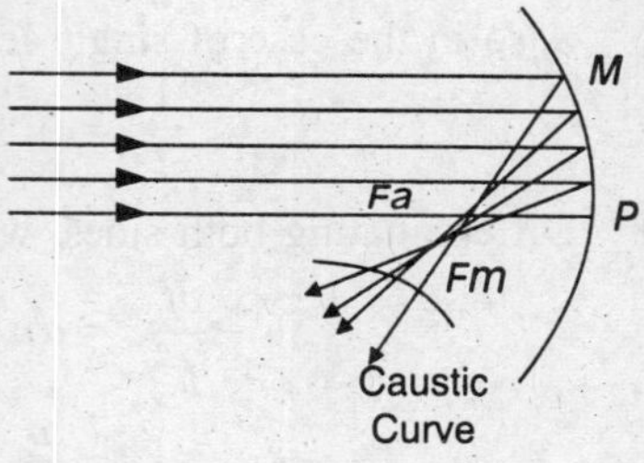

Fig. 2.3

However, if the aperture of the mirror is very large, then all the rays incident parallel to Principal axis and falling on the mirror do not come to a single point F_a as shown in Fig. 2.3. The rays reflected from points lying close to pole P will meet at F_a, which is lying away from P, while the rays reflected from outer portions of the mirror like M will come to a point F_m on the principal axis, which is near the pole. All the rays of light rerflected from points between P and M will come to foucs between F_a and F_m as shown in figure, which mean that the image formed is not a point image, but a blurred one.

The failure of the rays to come to a single point after reflection from a curved spherical surface is known as spherical aberration and the distance $F_a\ F_m$ will be a measure of axial or longitudinal spherical aberration.

If we place a screen ⊥ to axis to receive light, we get a circular patch of light. If the screen is at F_a, the circular patch is formed to have a bright centre and a faint periphery but if it is placed at F_m, we get a faint, centre and bright periphery. At a point in between F_a and F_m lying in the middle of $F_a\ F_m$, the two circles will overlap and same illumination throughout the point will be observe. This circle known as the circle of least confusion and may be considered as close to the true image of the object. The radius of this circle of least confusion will be measure of lateral spherical aberration.

2.10. CAUSTIC CURVES

As is evident from the above Fig. 2.3 that the two rays from the neighbouring points of the mirror interesect each other before reaching the axis each point of intersection of neighbouring reflected rays will form an image. A curve joining all such images is called caustic curve and is shown in the Fig. 2.3.

2.11. SPHERICAL ABERRATION IN CASE OF LENSES

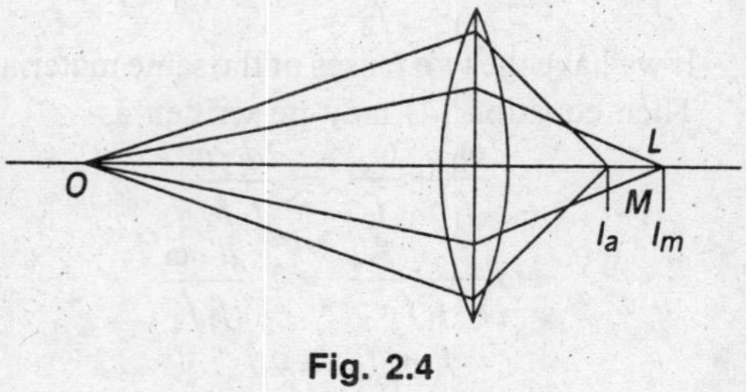

Fig. 2.4

In case of a large lens if an object is placed on the axis of the lens, then it is observed that the rays which are incident from the point farther away from the axis are brought to focus at a point nearer to the lens (shown at I_a in figure) and the rays which are coming from the central portion of the lens are brought to focus at a point away from the lens (shown at I_m in Fig. 2.4).

We can see from the figure that since central rays and outer rays form images at different positions so the image will not be a sharp on at any point on the axis. If a screen is placed ⊥ to principal axis at I_a, we observe, the outer contours of the object lie in focus while if it is placed at I_m, the central part of the object are in focus. If the screem is placed at LM, the image looks to be a circular patch of diameter LM and this circle of diameter LM is called **circle of least confusion** and will give the best position for the image.

The distance, I_a I_m measures longitudinal or axial spherical aberration.

As can be seen from above that an object O gives an extended form of image from I_a to I_m and this defect is known as spherical aberration and is due to the fact that different portions of the lens have different focal lengths.

In the case of concave lens spherical aberration is shown in Fig. 2.5. It should be noted that spherical aberration is positive for convex lens and negative for concave lens.

As can be seen in Fig. 2.5, if a parallel beam is incident on a concave lens, the central rays will appear to diverge from F_m *i.e.*, a point away from lens and outer rays will appear to diverge from F_a *i.e.*, a point nearer the concave lens.

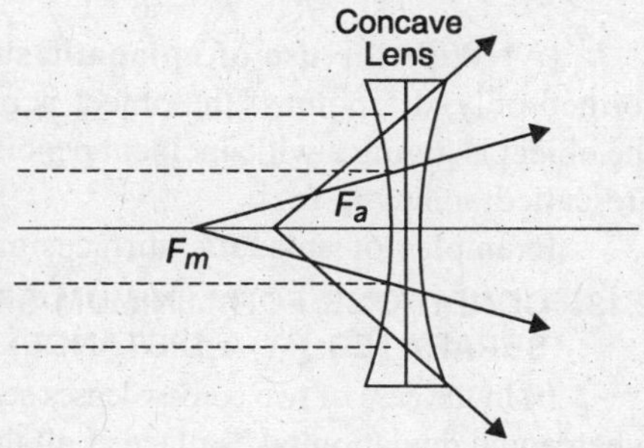

Fig. 2.5

2.12. CONDITION FOR MINIMUM SPHERICAL ABERRATIONS (WHEN LENSES ARE IN CONTACT)

In the case of a single lens we can reduce the spherical aberrations in the following ways:

(*i*) Since spherical aberration depends on the aperture of the lens, so we can reduce it by reducing the lens aperature with the help of co-axial aperture stop which allows only the central portion of the lens to be used for the rays.

(*ii*) By making use of a plano convex lens with convex surface towards the incident or the refracted light, whichever is more, parallel to the principal axis of the lens.

We know that the spherical aberration is due to the more deviations from outer rays than the central rays. If by some method we make the outer rays deviations to be minimum, the focal point F_a will come near to F_m *i.e.*, the focal point for central rays and so the spherical aberrations will be minimum.

As in the case of a prism, we have the minimum deviation when the incident and the refracted rays make equal angles with its faces, similarly in case of a lens, the deviations from outer rays will be minimum when these rays are falling on one surface and leaving the other surface at nearly equal angles. So we see that spherical aberration is minimum when the lens is so designed in such a way.

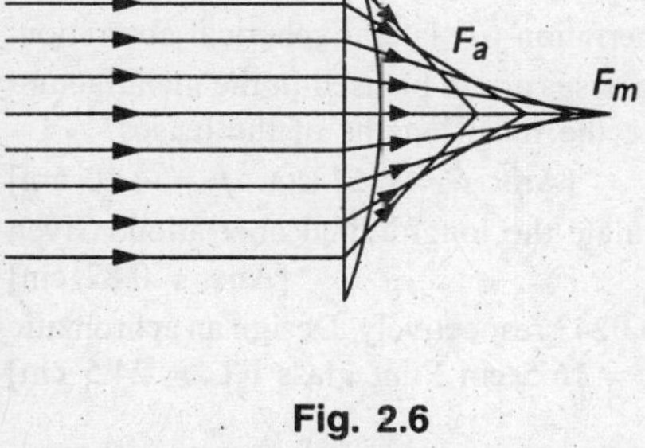

Fig. 2.6

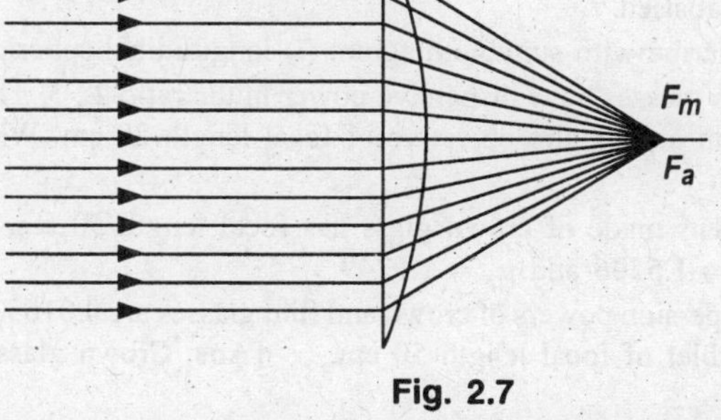

Fig. 2.7

If we make use of a plano convex lens than as can be seen from above figures spherical aberration is very much reduced in the second case, deviation partly takes place from convex surface and partly from plane surface, while in the first case the total deviation is from the convex surface.

(*iii*) Spherical aberration is minimum for some special class of lenses called crossed lens which satisfy the condition

$$\frac{R_2}{R_1} = \frac{\mu(2\mu+1)}{2\mu^2-\mu-4}$$

where R_1 and R_2 correspond to the radius of curvature of the two surfaces of the lens.

Taking the case of a lens having

$$\mu = 1.5$$

$$\frac{R_2}{R_1} = \frac{1.5(3+1)}{2\times(1.5)^2-1.5-4} = \frac{6}{1}$$

$$\therefore \quad R_2 : R_1 = 6 : 1.$$

(*iv*) **With the use of aplanatic surfaces**. In the case of an aplanatic surface, the image is formed only at a point if the object is placed at a specific point on the principal axis even when the object is giving a wide incident pencil of rays. These particular positions of the object and image are called aplanatic foci.

Examples of aplanatic surfaces are ellipsoidal or aparaboloidal surface.

2.13. CONDITIONS FOR MINIMUM SPHERICAL ABERRATIONS (WHEN THE LENSES ARE SEPARATED BY A DISTANCE)

(*i*) In the case of two convex lenses separated by some distance, spherical aberration will be minimum, when equal deviation takes place at all the four surfaces. This condition is achieved if

$$f_2 - f_1 = d$$

i.e., the two lenses are separated by a distance equal to the difference of their focal lengths. This particular arrangement is used in eye pieces.

(*ii*) It can also be reduced if a suitable concave and convex lens combination is used so that the focus F_a shifts to F_m. However, this combination has a drawback that it can be used only for particular object for which it is constructed.

(*iii*) If a stop at some position on the lens is used to restrict the rays to pass through a specific part of the lens, spherical aberration can be reduced. However, with this arrangement the intensity of light will also be reduced because some portion of the light passing through the lens will be cut off.

QUESTIONS AND PROBLEMS

1. What is meant by dispersion, give its definition and derive dispersive power for a material?
2. What is chromatic aberration? Give the conditions of achromatism for two thin lenses in contact.
3. Discuss the condition for achromatism of two lenses separated by a distance.
4. Explain the spherical aberration. How this defect can be minimised in lenses?
5. If f_1 and f_2 are the focal lengths of two lenses separated by a distance *d*. Show that spherical aberration for such a combination is minimum if $d = f_2 - f_1$. Mention a useful combination in which this condition is satisfied.
6. Describe with suitable diagram (*i*) longitudinal spherical aberration (*ii*) lateral spherical aberration.
7. Two glasses have dispersive power in the ratio 2 : 3. These glasses are to be used in the manufacture of an achromatic objective of focal length 20 cm. What are the focal lengths of the lenses? [**Ans.** f_1 = 6.67 cm., f_2 = – 10 cm]
8. A lens made of crown glass has focal length 50 cm. Calculate the longitudinal aberration. Given μ_C = 1.5206 and μ_F = 1.5249. [**Ans.** = 0.82 cm]
9. Dispersion powers of crown and flint glasses are 0.0163 and 0.0243 respectively. Design an achromatic doublet of focal length 50 cm. [**Ans.** Crown glass F.L. = 16.5 cm Flint glass F.L. = 24.5 cm]

CHAPTER

3

EYEPIECES

3.1. INTRODUCTION

If we make use of a simple convex lens as an eyepiece, then as is clear from the Fig. 3.1, in addition to the spherical and chromatic aberrative, another drawback exists, *i.e.*, when the eye is placed very close to eye lens, the rays from the outer parts of the image I would cross over the eye lens and so will fail to enter the eye along with paraxial rays. So we will only be able to see part of the field near the axis and thus field of view will be limited.

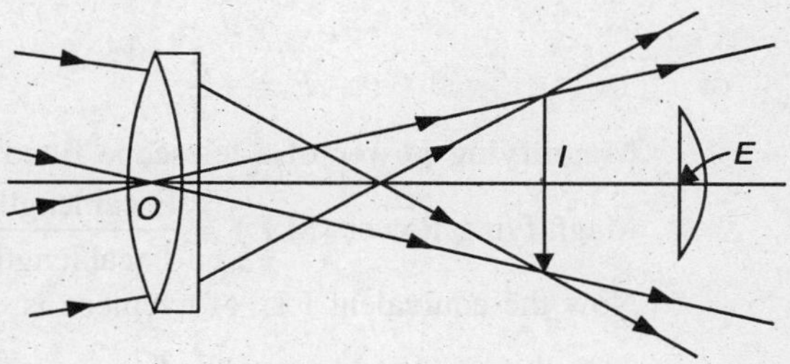

Fig. 3.1

Now in order to make the marginal rays from the periphery to pass through the eyepiece or in other words to increase the field of view, another lens F is introduced between the eye lens and the objective as shown in Fig. 3.2.

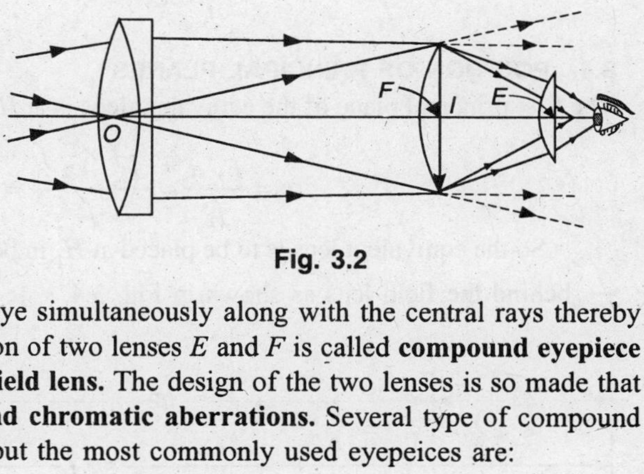

Fig. 3.2

This arrangement deflects the outer rays towards the principal axis which after entering the eye lens will be seen by the eye simultaneously along with the central rays thereby enlarging the field of view. This combination of two lenses E and F is called **compound eyepiece** and E is called **eye lens** while F is called **field lens.** The design of the two lenses is so made that this combination is free from **spherical and chromatic aberrations.** Several type of compound eyepieces are used in optical instruments but the most commonly used eyepeices are:

(*i*) Ramsden's eyepiece

(*ii*) Huygen's eyepiece.

3.2. RAMSDEN'S EYEPIECE

This type of eyepiece consists of two plano convex lenses each having focal length f and separated by a distance equal to 2/3 of the focal length of either. The lenses are placed so that their curved surfaces face each other and the eyepiece is placed just beyond the image formed by the objective lens. The eyepiece is provided with cross wires so that accurate measurements in case of optical instruments can be made.

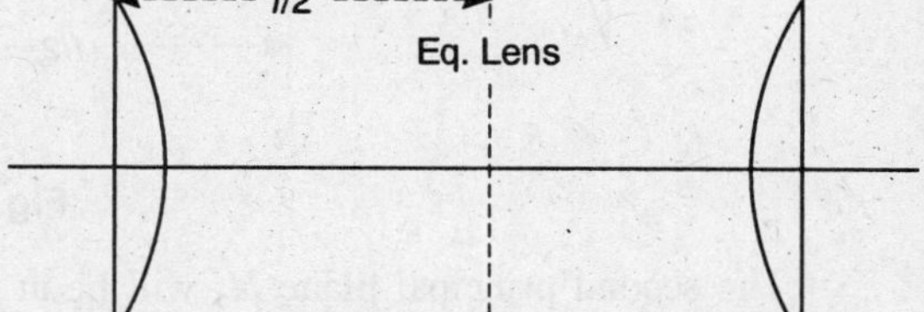

Fig. 3.3

3.3. FOCAL LENGTH OF EQUIVALENT LENS

Let F be the focal length of the equivalent lens, so that we have,

$$\frac{1}{F} = \frac{1}{f_1} + \frac{1}{f_2} - \frac{d}{f_1 f_2}$$

In this case $f_1 = f_2 = f$

and $d = \frac{2}{3} f$

$$\therefore \quad \frac{1}{F} = \frac{1}{f} + \frac{1}{f} - \frac{\frac{2}{3} f}{f^2}$$

$$= \frac{2}{f} - \frac{2}{3f} = \frac{4}{3f}$$

or $F = +\frac{3}{4} f$

Magnifying power of a telescope fitted with Ramsden's eyepiece is given as

$$\text{Magnifying Power (M.P.)} = \frac{\text{Focal length of objective}}{\text{Focal length of eyepiece}}$$

Now the equivalent F.L. of eyepiece is

$$= +\frac{3}{4} f$$

$$\therefore \quad \text{M.P.} = \frac{F}{\frac{3}{4} f} = \frac{4}{3} \frac{F}{f}$$

3.4. POSITION OF PRINCIPAL PLANES

The first principal plane of the equivalent lens is at H_1 behind the field lens at a distance α given by

$$\alpha = \frac{F \cdot d}{f_2} = \frac{\frac{3}{4} f \times \frac{2}{3} f}{f} = \frac{1}{2} f$$

So the equivalent lens is to be placed at H_1 in between the field lens F and eye E at a distance $\frac{f}{2}$ behind the field lens as shown in Fig. 3.4.

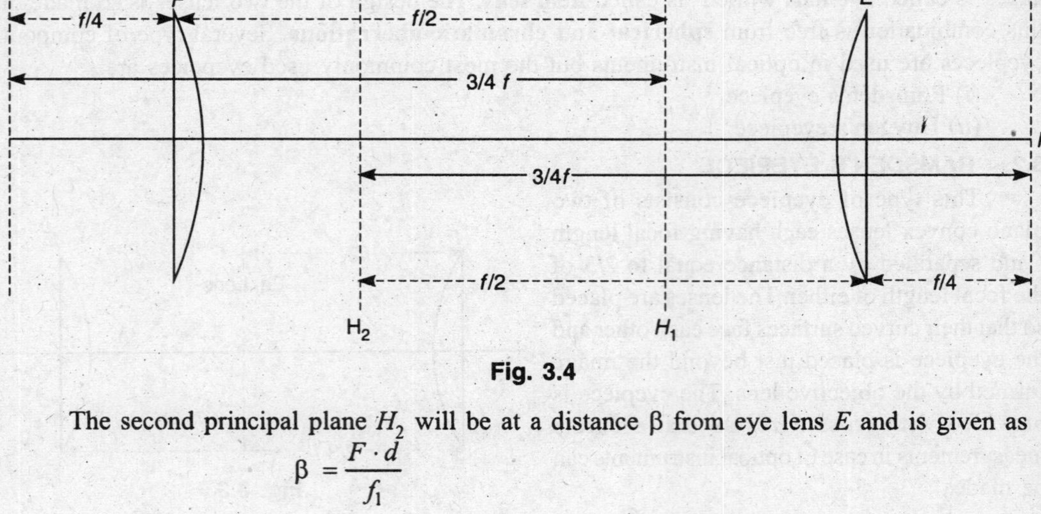

Fig. 3.4

The second principal plane H_2 will be at a distance β from eye lens E and is given as

$$\beta = \frac{F \cdot d}{f_1}$$

$$= \frac{\frac{3}{4}f \times \frac{2}{3}f}{f} = \frac{1}{2}f$$

So the two principal planes lie at distance $f/2$ from either of the two lenses E and F the nodal points will coincide with the principal points since the arrangement is placed in air.

3.5. POSITION OF FOCAL PLANES

The first focal plane is at F which is at a distance $\frac{3}{4}f$ from the first principal plane *i.e.*, at a distance $\frac{3}{4}f - \frac{1}{2}f = \frac{1}{4}f$ from the field lens on the side which is away from eye lens.

The second focal plane is at F_2 which is at a distance $f/4$ behind eye lens on the side away from field lens.

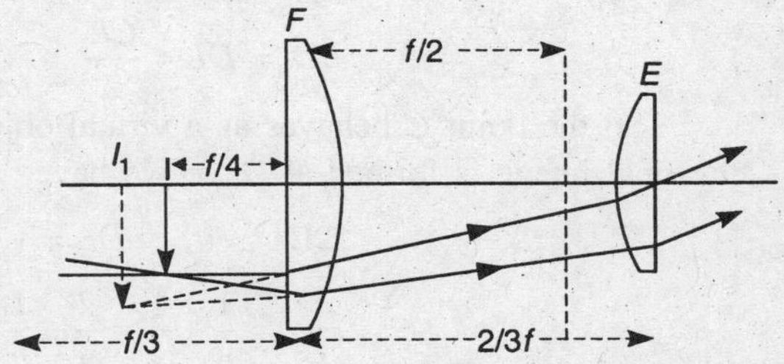

Fig. 3.5

3.6. CROSSWIRE POSITION

Since the focal length of the equivalent lens is $\frac{3}{4}f$ and the first focal point F is at a distance $\frac{3}{4}f$ from the first principal plane *i.e.,* at a distance $\frac{3}{4}f - \frac{f}{2} = \frac{f}{4}$.

On the left side of field lens, so the position I of the cross-wire is fixed at $f/4$, where the image of the object is formed by the objective lens. This image will act as the real object for eye lens and the final image will be formed at ∞. The Ramsden's eyepiece is called a positive eyepiece due to this reason.

In fact the eyepiece E produces the final image in two steps. As we have seen above, the real image is formed at a distance $\frac{1}{4}f$ in front of field lens F.

So for field lens $u = \frac{1}{4}f$

and from relation
$$\frac{1}{f} = \frac{1}{u} + \frac{1}{v}$$

$$\therefore \quad \frac{1}{v} = \frac{1}{f} - \frac{1}{u} = \frac{1}{f} - \frac{4}{f} = -\frac{3}{f}$$

$$\therefore \quad v = -\frac{f}{3}$$

which means that the field lens F will form a virtual image I_1 at a distance $\frac{f}{3}$ in front of it or at distance $\frac{1}{3}f + \frac{2}{3}f = f$ in front of eye lens E. The lens will, therefore, produce the final image at as shown in Fig. 3.5.

3.7. HUYGEN'S EYEPIECE

It consists of two plano convex lenses of the same kind of glass and the lenses are so arranged that both **spherical and chromatic** aberrations are eliminated.

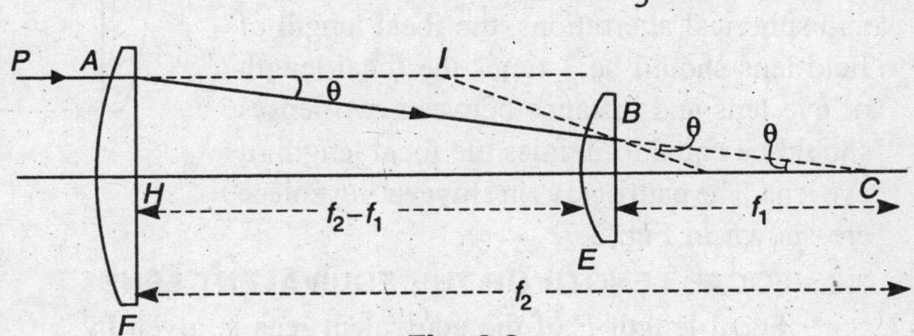

Fig. 3.6

In order to have minimum spherical aberration, the lenses are so arranged that a ray which is parallel and close to the axis suffers equal deviations from both the lens surfaces as shown in Fig. 3.6.

Now in order to see the condition for equal deviation, let a ray *PA* parallel to axis be diviated through an angle θ on suffering refraction at the surface of field lens *F*. This ray will further deviate through an angle θ from the surface of eye lens *E* and will finally come to point *D* instead of *C*.

Now $\angle IAB = \angle BCD = \angle CBD = \theta$

$\therefore \quad BD = CD$

If the point *B* is very close to *L*, then,

$$BD = LD$$

$$\therefore \quad BD = DL = \frac{CL}{2}$$

So the point *C* behaves as a virtual object for the eye lens *E* having a focal length f_1 and the final image is formed at *D*, so we have

$$CL = -u, \quad DL = v = \frac{u}{2}$$

$$\therefore \quad \frac{1}{f_1} = -\frac{1}{u} + \frac{1}{v} = \frac{2}{u} - \frac{1}{u} = \frac{1}{u}$$

$$\therefore \quad f_1 = u = CL$$

As the ray *PA* is parallel to the principal axis and would have converged at *C* if there were no eye lens, so

$$HC = f_2$$

or $$HL = f_2 - f_1 = d \qquad ...(1)$$

So we see that there will be equal deviations produced by field and eye lens if the distance between two lenses is equal to the difference between their focal lengths. Normally the field and eye lenses are plano-convex lenses with their convex surfaces towards incident light because in this way the total deviation due to this combination is divided into four parts which will have minimum spherical aberration.

Now for minimum chromatic aberration

$$d = \frac{f_1 + f_2}{2} \qquad ...(2)$$

If both spherical and chromatic aberrations are to be simultaneously reduced, we have from Equations (1) and (2)

$$\frac{f_2 + f_1}{2} = f_2 - f_1$$

$$\therefore \quad f_2 = 3f_1$$

$$\therefore \quad d = \frac{3f_1 + f_1}{2} = 2f_1$$

So, we see that in order to satisfy the conditions for minimum chromatic aberration and spherical aberrations, the focal length of field lens should be 3 times the focal length of eye lens and distance between two lenses should be equal to 2 times the focal length of eye lens. The path of rays in Huygen's eyepiece are shown in Fig. 3.7.

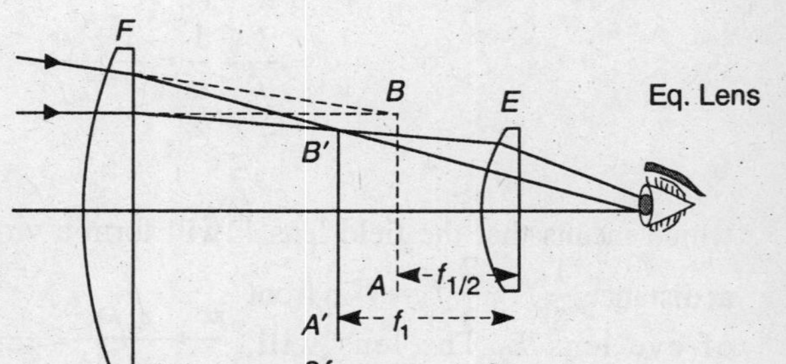

Fig. 3.7

3.8. FOCAL LENGTH OF THE EQUIVALENT LENS

Focal length *F* of the equivalent lens is given by

$$\frac{1}{F} = \frac{1}{f_1} + \frac{1}{f_2} - \frac{d}{f_1 f_2}$$

Now $\quad f_2 = 3f_1$ and $d = 2f_1$

$$\therefore \quad \frac{1}{F} = \frac{1}{f_1} + \frac{1}{3f_1} - \frac{2f_1}{f_1 \cdot 3f_1} = \frac{2}{3f_1}$$

$$\therefore \quad F = \frac{3}{2} f_1 \qquad \text{...(3)}$$

3.9. POSITIONS OF PRINCIPAL POINTS

The position of equivalent lens which will be at distance α from field lens is given by

$$\alpha = \frac{F \cdot d}{f_2} = \frac{\frac{3}{2} f_1 \times 2f_1}{f_1} = 3f_1$$

Since d, the distance between field lens and eye lens is given by $d = 2f_1$ so the position of equivalent lens is $2f_1 - 3f_1 = -f_1$ *i.e.*, it should be placed behind eye lens as shown by dotted line in Fig. 3.8.

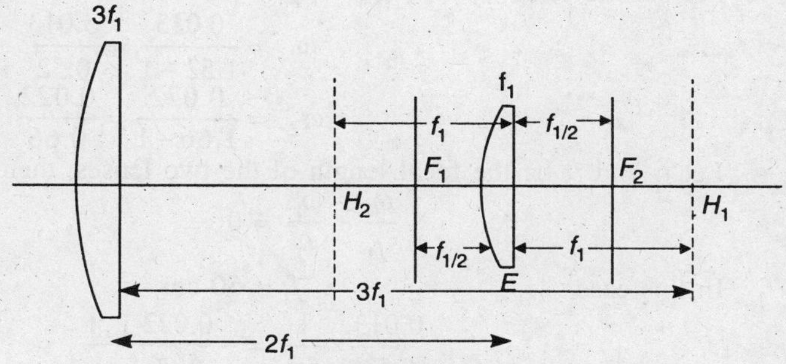

Fig. 3.8

Now the second principal point will be at a distance β from eye lens towards field lens and is given by

$$\beta = \frac{F \cdot d}{f_2} = \frac{\frac{3}{2} f_1 \cdot 2f_1}{3f_1} = f_1 \qquad \text{..(4)}$$

3.10. POSITION OF FOCAL POINTS

First Focal Point. The first focal point F_1 will lie at a distance $\frac{3}{2} f_1$ from the first principal point f_1, *i.e.*, at a distance $\left(\frac{3}{2} f_1 - f_1\right) = \frac{1}{2} f_1$ from the eye lens towards the side of the field lens.

Second Focal Point. It will be at distance $\frac{3}{2} f_1$, from second principal point H_2 *i.e.*, at distance $\left(\frac{3}{2} f_1 - f_1\right) = \frac{1}{2} f_1$ from the eye lens and at a point away from field lens.

3.11. CROSS WIRE POSITION

The first principal focus of equivalent lens is at a distance $\frac{3}{2} f_1 - f_1 = \frac{1}{2} f_1$ in front of eye lens and this point gives the position at which image AB of the object must be formed so that the final image lies at ∞. As seen in Fig. 3.7, the rays coming from objective are intercepted by the field lens and the image AB will be displaced to $A'B'$.

So for field lens of focal length $3f_1$, AB at a distance $\frac{3}{2} f_1$ will act as virtual object. So the distance v of the real image $A'B'$ is given by

$$\frac{1}{v} = \frac{1}{3f_1} + \frac{2}{3f_1} = \frac{1}{f_1}$$

$$\therefore \quad v = f_1$$

That means AB lies mid way between the two lenses and this will be the position of cross-wire.

As is clear from above, the image due to the objective is formed on negative side v behind the field lens, this eyepiece is called a negative eyepiece.

It should be noted that in this case since the cross wires are seen only through the eye lens so we have spherical and chromatic aberrations and so measurements are not very reliable.

Example 1. *Mean refractive indices of two specimens of glass are 1.52 and 1.66 respectively. The difference in indices for the same two lines of the spectrum are 0.013 for the first lens and 0.022 for the second. Find the focal length of a lens of the second glass which when combined with a convex lens of 50 cm focal length will make an object glass achromatic for these two lines.*

Solution: Let ω_1 and ω_2 be the dispersive powers of the two specimens of glass and let the two lights be violet and red, then we have

$$\omega_1 = \frac{\mu_v - \mu_r}{\mu - 1} \quad \text{and} \quad \omega_2 = \frac{\mu'_v - \mu'_r}{\mu' - 1}$$

For Ist lens we have $\mu_v - \mu_r = 0.013$ and $\mu = 1.52$

For second lens $\mu'_v - \mu'_r = 0.022$ and $\mu' = 1.66$

$$\therefore \quad \omega_1 = \frac{0.013}{1.52 - 1} = \frac{\sigma.013}{0.52}$$

$$\omega_2 = \frac{0.022}{1.66 - 1} = \frac{0.022}{0.66}$$

Let f_1 and f_2 be the focal length of the two lenses, then for an achromatic combination

$$\frac{\omega_1}{f_1} + \frac{\omega_1}{f_2} = 0$$

In this case $f_1 = 50$ cm

$$\therefore \quad \frac{0.013}{0.52} \times \frac{1}{50} = -\frac{0.022}{0.66} \times \frac{1}{f_2}$$

or
$$\frac{1}{f_2} = \frac{-1}{40} \times \frac{1}{50} \times 30$$

or
$$f_2 = -\frac{200}{3} = \mathbf{-66.67\,cm}$$

negative sign shows that the second lens is concave lens.

Example 2. *Given the dispersive powers of crown and flint glass as 0.2 and 0.04 respectively, find the focal lengths of the two components of an achromatic doublet of focal length 20 cm.*

Solution: Let f and f' be the focal lengths of the crown and flint glasses and let F be the focal length of the combination, then

$$\frac{1}{F} = \frac{1}{f} + \frac{1}{f'} \qquad \text{...(1)}$$

Also
$$\frac{\omega}{f} + \frac{\omega'}{f'} = 0 \qquad \text{...(2)}$$

In this case $\omega = 0.02$ and $\omega' = 0.04$, and

$\therefore$ From Equation (2)

$$f = \frac{\omega}{\omega'} f' = \frac{0.02}{0.04} \times f' = \frac{-1}{2} f'$$

or
$$f' = -2f$$

Now $F = 20$ cm.

From Equation (1)

$$\frac{1}{20} = \frac{1}{f} - \frac{1}{2f} \qquad [\because f' = -2f]$$

or
$$\frac{1}{2f} = \frac{1}{20}$$

or
$$f = 10 \text{ cm}$$

From Equation (1) $\dfrac{1}{20} = \dfrac{1}{10} + \dfrac{1}{f'}$

or
$$\frac{1}{f'} = \frac{1}{20} - \frac{1}{10} = -\frac{1}{20}$$

or $$f' = -20 \text{ cm}$$

So in this case we have an achromatic combination of two lenses (*i*) convex lens of focal length = 10 cm (*ii*) concave lens of focal length = 20 cm.

Example 3. *An achromatic objective of focal length 30 cm is to comprise two thin lenses cemented together with canada balsam and made from head crown and dense flint respectively. The free surface of the divergent lens is plane, calculate the radii of curvature of both lenses. Given*

	Crown	*Flint*
μ red	*1.514*	*1.644*
μ blue	*1.524*	*1.664*

Solution: We have the relations

$$\frac{1}{F} = \frac{1}{f} + \frac{1}{f'} \quad ...(1)$$

$$\frac{\omega}{f} + \frac{\omega'}{f'} = 0 \quad ...(2)$$

$$\frac{1}{f} = (\mu - 1)\left(\frac{1}{R_1} - \frac{1}{R_2}\right) \quad ...(3)$$

$$\frac{1}{f'} = (\mu - 1)\left(\frac{1}{R_1'} - \frac{1}{R_2'}\right) \quad ...(4)$$

For crown glass, average refractive index (μ) is given by

$$\frac{1.514 + 1.524}{2} = 1.519$$

and dispersive power is given by

$$\omega = \frac{\mu_v - \mu_r}{\mu - 1} = \frac{1.524 - 1.514}{1.519 - 1}$$

$$= \frac{0.01}{0.519} = \frac{10}{519} \quad ...(5)$$

and for flint glass μ' the average refractive index is given by

$$\mu' = \frac{1.664 + 1.644}{2} = 1.654$$

and dispersive power is $$\omega' = \frac{\mu_v' - \mu_r}{\omega - 1} = \frac{1.664 - 1.644}{1.654 - 1}$$

$$= \frac{0.02}{0.654} = \frac{20}{654} = \frac{10}{327} \quad ...(6)$$

Now from Equation (2)

$$\frac{f}{f'} = \frac{-\omega}{\omega'}$$

$$= -\frac{\frac{10}{519}}{\frac{10}{327}} = \frac{-327}{519} = \frac{-109}{173} \quad ...(7)$$

Also from Equation (1)

$$\frac{1}{30} = \frac{1}{f} + \frac{1}{f'}$$

or $$\frac{f}{30} = 1 + \frac{f}{f'}$$

$$= 1 - \frac{109}{173} = \frac{64}{173} \quad \left[\text{From (7) } \frac{f}{f'} = -\frac{109}{173}\right]$$

or $$f = \frac{64}{173} \times 30 = 11.1 \text{ cm}$$

and $$f' = \frac{-173}{109} \times 11.1 = -17.6 \text{ cm}$$

Since the free surface of the divergent lens is plane, so $R_2' = \infty$

$\therefore$ From Equation (4)

$$\frac{1}{f'} = (\mu' - 1)\left(\frac{1}{R_1'} - \frac{1}{R_2'}\right)$$

Putting the various values we get

$$-\frac{1}{17.6} = (1.654 - 1)\left(\frac{1}{R_1'} - \frac{1}{\infty}\right)$$

$$= \frac{0.654}{R_1'}$$

or $$R_1' = -0.654 \times 17.6 = \mathbf{-11.51\ cm}$$

For crown glass $$\frac{1}{f} = (\mu - 1)\left(\frac{1}{R_1} - \frac{1}{R_2}\right)$$

$$f = \frac{1920}{173},\ \mu = 1.519, \qquad R_2 = -R_1' = -\frac{288}{25}$$

Putting in above equation

$$\frac{173}{1920} = 0.519\left(\frac{1}{R_1} + \frac{25}{288}\right)$$

or $$\frac{1}{R_1} + \frac{25}{288} = \frac{173}{1920} \times \frac{1}{0.519}$$

$$\frac{1}{R_1} = \frac{173}{1920} \times \frac{1}{0.519} - \frac{25}{288} = 0.11$$

$\therefore$ $$R_1 = \frac{1}{0.11} = \mathbf{9.1\ cm}.$$

Example 4. *An achromatic lens having a focal length of 10 cm is to be made as a cemented doublet using crown and flint glasses having the following indices:*

Glass	μ_C	μ_D	μ_F
Crown	*1.50868*	*1.51100*	*1.51673*
Flint	*1.61611*	*1.62100*	*1.63327*

Find the radii of curvature of both lenses if the crown glass lens to be equi-convex and the combination is to be corrected for C and F lines.

Solution: For crown glass, dispersive power is given by

$$\omega_1 = \frac{\mu_F' - \mu_C}{\mu_D} = \frac{1.51673 - 1.50868}{0.511} = 0.05175$$

and for flint glass $$\omega_2 = \frac{\mu_F' - \mu_C'}{\mu_D' - 1} = \frac{1.63327 - 1.61611}{0.56200} = \sigma.02764$$

Let f_1 and f_2 be the focal length of crown and flint glasses

$\therefore$ $$\frac{\omega_1}{f_1} + \frac{\omega_2}{f_2} = 0$$

or $$\frac{1}{f_1} = -\frac{\omega_2}{\omega_1} \cdot \frac{1}{f_2}$$

$$\therefore \quad \frac{1}{f_1} = -\frac{0.02764}{0.01575}\cdot\frac{1}{f_2}$$

Now we have the relation for combination

$$\frac{1}{F} = \frac{1}{f_1}+\frac{1}{f_2} \quad \text{and} \quad F = 10 \text{ cm}$$

$$\therefore \quad \frac{1}{10} = \left[1-\frac{0.2764}{0.1575}\right]\cdot\frac{1}{f_2}$$

$$= -\frac{1189}{1575}\cdot\frac{1}{f_2}$$

or $$\frac{1}{f_2} = \frac{1575}{11890}$$

$$\therefore \quad \frac{1}{f_1} = \frac{0.2764}{0.1575}\times\frac{1575}{11890}$$

$$= \frac{2764}{11890}$$

Now since crown glass lens is equi convex, we have

$$\frac{1}{f_1} = (\mu_D - 1)\left(\frac{1}{R}+\frac{1}{R}\right)$$

where R is the radius of curvature of either surface of the lens

$$\therefore \quad \frac{2764}{11890} = 0.511\times\frac{2}{R}$$

or $$R = 0.511\times 2\times\frac{11890}{2764} = 4.396 \text{ cm}$$

Let R_1 be the radius of curvature of the uncemented surface of the concave lens, then we have

$$\frac{1}{f_2} = (\mu'_D - 1)\left(\frac{1}{R}+\frac{1}{R_1}\right)$$

$$-\frac{1575}{11890} = 0.621\left(-\frac{1}{4.396}+\frac{1}{R_1}\right)$$

or $$\frac{1}{R_1} = -\frac{1575}{11890\times 0.621}+\frac{1}{4.396}$$

$$= -0.2133 + 0.2275$$

$$= 0.0142$$

or $$R_1 = 70.42 \text{ cm.}$$

So the radii of each surface of crown glass convex lens is 4.396 cm and those for flint glass convex-concave lens are – 4.396 cm for cemented surface and + 70.42 cm for the other surface.

Example 5. *Two thin lenses of focal lengths f_1 and f_2 separated by a distance d have an equivalent focal length 50 cm. The combination satisfies the conditions for no chromatic aberration and minimum spherical aberration. Find the value of f_1, f_2 and d. Assume that both the lenses are of the same material.*

Solution: The relation for the two lenses separated by a distance is given as

$$\frac{1}{F} = \frac{1}{f_1}+\frac{1}{f_2}-\frac{d}{f_1 f_2} \quad \text{...(1)}$$

Now we have for no chromatic aberration

$$d = \frac{f_1+f_2}{2} \quad \text{...(2)}$$

and for minimum spherical aberration

$$d = f_1 - f_2 \qquad ...(3)$$

From Equation (1) and (3) $\quad f_1 = \frac{3d}{2} \quad$ and $\quad f_2 = \frac{d}{2}$

Putting in Equation (1)

$$\frac{1}{50} = \frac{2}{3d} + \frac{2}{d} - \frac{d}{\frac{3d}{2} \times \frac{d}{2}}$$

$$= \frac{2}{3d} + \frac{2}{d} - \frac{4}{3d} = \frac{4}{3d}$$

or $\quad 3d = 200, \quad$ or $\quad d = 66.67$ cm

$$\therefore \quad f_1 = \frac{3}{2} \times 66.67 = 100 \text{ cm}$$

and

$$f_2 = \frac{66.67}{2} = \mathbf{33.33\ cm.}$$

Example 6. *A double convex lens has radii of curvature of 40 cm and 10 cm. Find the longitudinal chromatic aberration for an object at* ∞. *Given* $\mu_v = 1.5230$ *and* $\mu_r = 1.5145$.

Solution: The relation for red ray is

$$\frac{1}{f_r} = (\mu_r - 1)\left(\frac{1}{R_1} + \frac{1}{R_2}\right)$$

We have $\quad \mu_r = 1.5145$

$R_1 = 40$ cm $\quad$ and $\quad R_2 = 10$ cm

$$\therefore \quad \frac{1}{f_r} = (1.5145 - 1)\left(\frac{1}{40} + \frac{1}{10}\right)$$

$$= 0.5145\left(\frac{5}{40}\right) = 0.5145 \times \frac{1}{8}$$

$$\therefore \quad f_r = \frac{8}{0.5145} = 15.549 \text{ cm.}$$

Relation for violet ray is

$$\frac{1}{f_v} = (\mu - 1)\left(\frac{1}{R_1} + \frac{1}{R_2}\right)$$

$$= (1.5230 - 1)\left(\frac{1}{40} + \frac{1}{10}\right)$$

$$\therefore \quad \frac{1}{f_v} = \frac{0.5230}{8}$$

or $\quad f_v = 15.296$ cm

So longitudinal chromatic aberration for an object at ∞ is

$$f_r - f_v = 15.549 - 15.269 = \mathbf{0.253\ cm.}$$

Example 7. *Light from the sun is falling on Ramsden's eyepiece. Locate the position of the image thus formed and also find the point from which the distance is to be measured. The focal length of each lens of the eye piece is 3.0 cm.*

Solution: Focal length of each lens $f = 3.00$ cm

Distance between the lens $d = \frac{2}{3} f = \frac{2}{3} \times 3 = 2.0$ cm

Focal length of equivalent lens

$$F = \frac{3}{4} f = \frac{3}{4} \times 3 = 2.25 \text{ cm}$$

Since the object *i.e.*, the sun is at ∞, so the image will be formed at second focal plane of eyepiece *i.e.*, at distance 2.25 cm from the second principal plane.

Distance of 2nd principal plane from eye lens F_2 is

$$= \frac{F \cdot d}{f_1} = \frac{2.25 \times 2}{3} = \mathbf{1.5\,cm.}$$

Example 8. *Two thin convex lenses of focal length 12.0 cm and 4.0 cm are kept separated by a distance of 8.0 cm. Plot the positions of the cardinal points for the combination.*

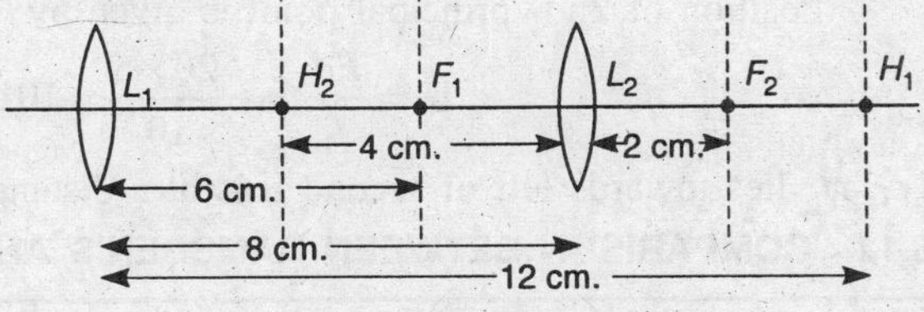

Fig. 3.9

Solution: Focal length of the equivalent lens is given by

$$\frac{1}{F} = \frac{1}{f_1} + \frac{1}{f_2} - \frac{d}{f_1 f_2}$$

In this case $f_1 = 12$ cm, $f_2 = 4$ cm and $d = 8$ cm.

$$\frac{1}{F} = \frac{1}{12} + \frac{1}{4} - \frac{8}{12 \times 4} = \frac{2}{12}$$

$$F = 6 \text{ cm}$$

First Principal point. The distance of first principal point is given by

$$L_1 H_1 = \alpha = \frac{F \cdot d}{f_2} = \frac{6 \times 8}{4} = 12 \text{ cm.}$$

2nd Principal point. The distance of the second principal point is given by

$$L_2 H_2 = \beta = -\frac{F \cdot d}{f_1} = -\frac{6 \times 8}{12} = -4 \text{ cm.}$$

Nodal points. When the system is in air, principal points are also the nodal points.

First focal point. The first focal point is given by

$$L_1 F_1 = -F\left(1 - \frac{d}{f_2}\right) = -6\left(1 - \frac{8}{4}\right) = \mathbf{+6\,cm.}$$

Second focal point is given by

$$L_2 F_2 = F\left(1 - \frac{d}{f_1}\right) = 6\left(1 - \frac{8}{12}\right) = \mathbf{+2\,cm.}$$

The position of the cardinal points are given in Fig. 3.9.

Example 9. *A thin converging lens and a thin diverging lens are placed co-axially at a distance 5 cm. If the focal length of each lens is 10 cm, find for the combination (i) focal length (ii) power (iii) position of principal points.*

Solution: In this case $f_1 = 10$ cm $f_2 = -10$ cm and $d = 5$ cm

We have the relation

$$\frac{1}{F} = \frac{1}{f_1} + \frac{1}{f_2} - \frac{d}{f_1 f_2}$$

$$\therefore \quad \frac{1}{F} = \frac{1}{10} - \frac{1}{10} + \frac{5}{10 \times 10}$$

$$\therefore \quad F = \frac{100}{5} = 20 \text{ cm}$$

(*ii*) Power $= \frac{100}{F} = \frac{100}{20} = 5$ diapters

(*iii*) Position of first principal points is given by

$$\alpha = \frac{Fd}{f_1} = \frac{20 \times 5}{-10} = -10\,\text{cm}$$

i.e., H_1 lies towards left of the first lens at a distance 10 cm.

Position of 2nd principal point is given by

$$\beta = \frac{Fd}{f_1} = -\frac{20 \times 5}{10} = 10\,\text{cm}$$

i.e., H_2 lies towards left of second lens at a distance of 10 cm.

3.12. COMPARISON BETWEEN RAMSDEN'S AND HUYGEN'S EYEPIECE

Ramsden's eyepiece	*Huygen's eyepiece*
1. It has quite large field of view.	1. It also has large field of view.
2. It has minimum spherical aberrations.	2. It has also minimum spherical aberrations.
3. It is not free from chromatic aberrations and show lateral colours.	3. It is free from chromatic aberrations as in this case $d = \frac{f_1 + f_2}{2}$
4. It is achromatic only for same colours.	4. It is achromatic for all colours.
5. The cross-wires in this case are placed outside and so it is an easier process.	5. Cross-wires are to be placed in between the lenses and is a difficult process.
6. The cross-wires can be adjusted for different persons as they are outside the eyepiece.	6. In this case the cross-wires are to be fixed for a normal person as these are in between the lenses.
7. Size measurements are accurate since the image and the scale used are equally magnifield.	7. Size measurements are not accurate as final image will undergo refraction through both the lenses, while the scale on which measurements are to be made will be refracted from one lens.
8. It is used in microscopes and telescopes for various optical measurements.	8. It is used only in microscopes which is used for biological work where no measurements are required.
9. Final image is flat.	9. Final image is convex towards eye.
10. The distance between eye lens and eye ring is greater, so it is not taxing on eye.	10. The distance between eye lens and eye ring is comparatively less and so it will be somewhat taxing on eye.

QUESTIONS AND PROBLEMS

1. Why does an eyepiece system consist of two lenses instead of one? Give the functions of the field lens.

2. Show with neat diagram the construction and working of Ramsden's eyepiece. Give the position of principal points and focal points.

3. Describe the construction and theory of Huygen eyepiece. Show that Huygen's eyepiece satisfies the condition of minimum spherical and chromatic aberration.

4. Calculate the focal length and positon of cross-wire in Huygen's eyepiece and show in diagram the position of focal points.

5. Give the theory and working of Huygen's eyepiece. Discuss the comparison between Huygen's and Ramsden's eyepieces.

6. Calculate the focal length of the components of an achromatic telescope objective having a focal length of 30 cm made from crown glass and flint glass whose dispersive powers are 0.012 and 0.02 respectively.

[**Ans.** 12 cm and – 20 cm]

7. Two glasses of dispersive power in the ratio 2 : 3 are to be used in the manufacture of an achromatic objective of focal length 20 cm. What are the focal lengths of the lenses?
[**Ans.** f_1 = 6.67 cm and f_2 = – 10 cm]

8. Two lenses of focal lengths 4.5 cm and 1.5 cm are placed at a certain distance apart. Calculate the distance between the lenses if they form an achromatic combination. [**Ans.** d = 3 cm]

9. A convex lens of focal length 35 cm and achromatic for C and F lines is to be made from two glasses, data for which are as under:

	μ_D	$\mu_F - \mu_C$
Hard crown	1.5175	0.00856
Dense flint	1.6214	0.01722

Find the focal length of the components. [**Ans.** 14.1 cm]

10. An achromatic converging lens combination has to be made from two thin crown glass and flint glass lenses in contact, the surface in contact having a common radius of curvature of 25 cm. If the focal length of the combination is to be 40 cm, calculate the radii of curvature of the second face of two lenses. Dispersive powers for crown and flint glasses are 0.021 and 0.045 and corresponding refractive indices are 1.50 and 1.60 respectively.
[**Ans.** f_1 = 21.33, f_2 = – 45.7 cm, r_1 = 13.8 cm, r_2 = – 283.2 cm]

11. An eye piece is formed by two convex lenses of focal length 3 and 8 cm placed 6 cm apart. Find the focal length of eye piece. Also find (*i*) position of the object for which image is formed at infinity (*ii*) the position of the image if the object is placed at infinity.
[**Ans.** F = 4.8 cm (*i*) F_1 = 1.2 cm at left to first lens (*ii*) F_2 = 1.2 cm at right to first lens]

12. Two convex lenses each of 20 cm focal length are placed 5 cm apart. A tower 100 metres high and 200 metres distant is seen through them. Find (*i*) the position of unit focal planes (*ii*) position and size of image. [**Ans.** (*i*) 100/35 (*ii*) –100/35, F = 400/35 cm,
V = 11.4 cm and I = –5.67 cm]

13. A convex lens of 25 cm focal length is placed at 15 cm from a convex lens of 15 cm focal length. Locate the cardinal points of the system.
[**Ans.** 1st Principal pt.= – 75 cm, 2nd principal pt.= – 45 cm
1st focal pt. = – 150 cm, 2nd focal pt. = – 30 cm]

14. A convex lens of focal length 40 cm is in contact with a concave lens of focal length 25 cm. Find the power of combination. [**Ans.** 1.5 diopters]

15. A convex lens A of focal length 20 cm and concave lens B of focal length 5 cm are kept along the same axis with a distance d between them. If a parallel beam of light falling on A leaves B as a parallel beam what will be the value of d? [**Ans.** 15 cm]

Thomas Young (1773-1829)

Few men or women have ever contributed so much first-rate scholarly work to so many different fields of learning and did Thomas Young.

Born in Somerset, England, in 1773, Young learned to read at the age of 2 and by the age of 6 is said to have read the complete Bible twice.

From 1792 to 1799 Young was engaged in his medical studies, receiving his M.D. degree from Gottingen in 1796. For most of his life he practiced medicine, but without great success. Constantly torn between the practice of medicine and his love for pure research. Young lacked sufficient interest in the routine problems of patients to build up a successful medical practice. Fortunately he had inherited a large fortune from his uncle, and did not have to rely on medicine for a livelihood.

During his medical studies Young became interested in the physiology of the human eye. He was the first to explain how the human eye can focus on objects at different distances. He also first explained, by experimenting with his own eyes, the nature of astigmatism. Finally, research on the nature of colour blindness led him to put forth his three-colour theory of colour vision, which proposed that only three different kinds of colour receptors are present in the human eye, one for blue, one for green,, and one for red light. All other colours are merely combinations of these three basic colours. This theory, now called the Young-Helmholtz theory, is the basis of modern *colour photography* and *colour television.*

Young's work on vision naturally led him to explore the nature of light. His experiments with double-slit interference, Newton's rings, and thin films convinced him that light was a wave motion of some kind. Young never developed the mathematics to explain his wave model of light properly, and hence the French physicist Fresnel, who did, is often given more credit than Young for the wave theory.

Young was brilliant intellectual, but had little understanding of human sensitivities. He was honest—sometimes too honest—in appraising the scholarly work of his contemporaries, and made enemies as a result. Much of his scientific work was published anonymously because he did not want his patients to think that his heart was not in his medical practice.

Thomas Young was one of the most ***acute*** *man ever lived, but had the misfortune to be too far in advance of his contemporaries. They looked on him with astonishment, but could not follow his bold speculations, and thus a mass of his most important thought remains buried and forgotten in the Transactions of the Royal Society until a later generation arrived at the rediscovery of his discoveries, and came to appreciate the force of his arguments and accuracy of his conclusions.* **[Hermann Von Helmholtz]**

All the compliments that I have received from Arago, Laplace and Biot never gave me as much pleasure as the discovery of a theoretic truth, or the confirmation of a calculation of experiment.

[Augustin Fresnel (1788–1827)]

CHAPTER

4

INTERFERENCE

4.1. INTRODUCTION

When light from two sources move in a direction then the light wave trains from these sources **superimpose** upon each other resulting in the modification of distribution of intensity *i.e.*,we get **position of maximum intensity and positions of minimum intensity.** This modification of light energy due to super position of two or more wave trains is called **interference.** When the amplitudes of the two wave trains and thereby increasing the intensity of the resulting wave (Intensity and Amplitude), it is called **Constructive Interference** and when the resulting amplitude is equal to the difference of two amplitudes, it is called **destructive interference.**

4.2. SUPERPOSITION PRINCIPLE

According to this principle the resultant displacement of a particle of the medium acted upon by two or more waves simultaneously is equal to the algebraic sum of the displacement of the particle due to individual waves.

Let the displacement due to a single wavetrain at a point be x_1 in a given direction and let the displacement due to another wavetrain in the same direction be x_2. Then from the above principle, the resultant displacement R_1 may be given as

$$R_1 = x_1 + x_2$$

If the two displacements are in the opposite directions, the resultant displacement due to two waves may be given as

$$R_2 = x_1 - x_2.$$

4.3. YOUNG'S EXPERIMENT

In 1801, Young first demonstrated experimentally the phenomenon of interference. His experiment is shown as under:

Light from Sun was allowed to fall on a slit S and then through a double slit S_1 and S_2. The interference pattern was observed on a screen XY–where few coloured bright and dark bands could be seen.

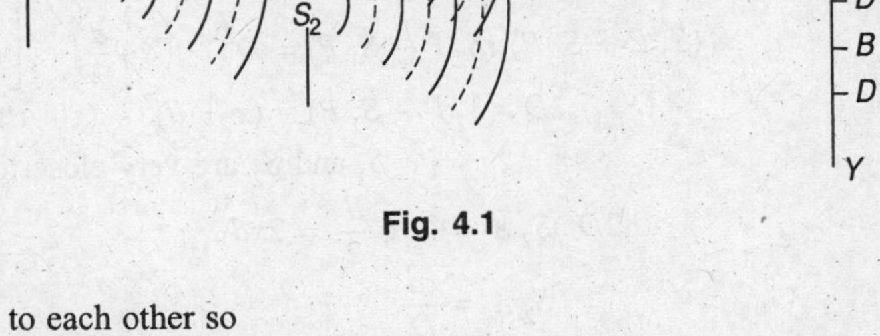

Fig. 4.1

In the present day experiments, the white light is replaced by monochromatic light and the double slits are very fine narrow and close to each other so that the dark and bright fringes are observed at equal distances.

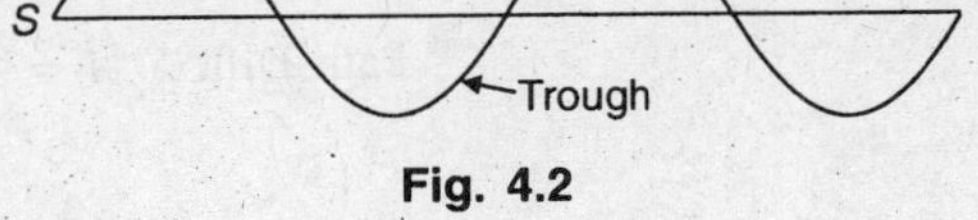

Fig. 4.2

4.4. FORMATION OF FRINGES

When light from the slit S passes, then from

Huygen's principle, we get wave fronts. The bold lines show the amplitude in positive direction (crests) and dotted lines show the amplitude in negative direction (Troughs). These wavelets on entering into slits S_1 and S_2 will look to be coming from two sources and the radii of these wave fronts increase as they move away from S_1 and S_2 and so will superimpose upon each other.

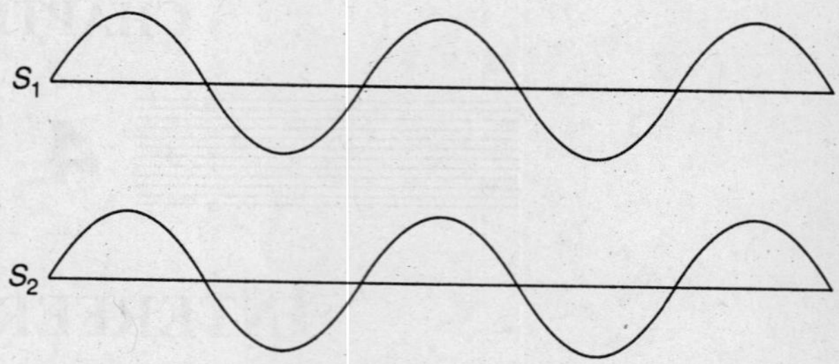

Fig. 4.3

At points where crests fall on crests or troughs fall on troughs the amplitudes add up and the result is intensity increases ($I \propto A^2$) and at points where crests fall on troughs or troughs fall on crests, the amplitudes are reduced (the resultant amplitude is zero). This is known as destructive interference. Thus on the screen alternate bright and dark regions of equal width, called **interference fringes** are observed.

4.5. THEORY OF FRINGES

Let S be a monochromatic source of light and let S_1 and S_2 be double slits, equidistant from S. The waves on reaching S_1 and S_2 will have the same phase at all the times.

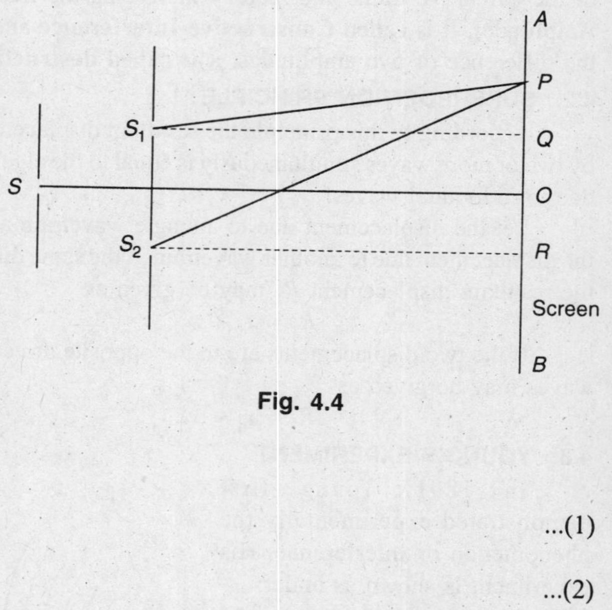

Fig. 4.4

Let the distance between S_1 and S_2, be d and D be the distance of the screen AB from S_1 S_2. The point O on the screen being at equal dist. from S_1 and S_2 will have zero path difference due to two lights and so will be centre of brightness. Let us consider a point P at a distance x from the centre O. We will consider the conditions for bright or dark fringes at this point.

From the right angled $\Delta S_1 QP$

$$S_1 P^2 = S_1Q^2 + QP^2 \quad ...(1)$$

Similarly from right angled $\Delta S_2 RP$

$$S_2 P^2 = S_2 R^2 + PR^2 \quad ...(2)$$

From Equations (1) and (2)

$$S_2 P^2 - S_1 P^2 = S_2 R^2 + PR^2 - (S_2 Q^2 + QP^2)$$

$$(S_2 P + S_1 P)(S_2 P - S_1 P) = D^2 + \left(x+\frac{d}{2}\right)^2 - \left[D^2 + \left(x-\frac{d}{2}\right)^2\right]$$

or

$$2D \,.\, (S_2 P - S_1 P) = (x + d)^2 - (x - d)^2$$

[$\because$ S_1 and S_2 are very close to each other so $S_1 P + S_2 P \approx 2D$]

$$2D \,.\, S_2 A = 4 \,.\, x\, \frac{d}{2} = 2xd$$

or

$$S_2 A = \frac{xd}{D}$$

where, we have taken $S_2 P + S_1 P = 2D$ because the two sources S_1 and S_2 are very close to each other and D is thousand times greater than x or d.

$$\therefore \quad \text{Path Diff. } S_2 A = \frac{xd}{D} \quad ...(3)$$

4.6. CONDITIONS FOR MAXIMUM INTENSITY

The point P will be centre of maxima if the path difference is an integral multiple of λ— because then the crest of one light will fall on the crest of the other and trough of one light will fall on the trough of the other

i.e., $$S_2 A = \frac{xd}{D} = n\lambda$$

$\therefore$ $$x = \frac{nD\lambda}{d} \quad ...(4)$$

Equation (4) gives the positions of bright fringes. At O, the path difference is zero and hence it will be position of maxima. The next bright fringes are formed when $n = 1, 2, 3, ...$

When $n = 1$, $x_1 = \frac{D}{d}\lambda$

When $n = 2$, $x_2 = \frac{D}{d^2}\lambda$

When $n = 3$, $x_3 = \frac{D}{d}3\lambda$

....

....

When $n = n - 1$, $x_{n-1} = \frac{D}{d}(n-1)\lambda$

When $n = n$, $x_n = \frac{D}{d}n\lambda$

Now $$x_2 - x_1 = \frac{D}{d}\cdot\lambda$$

$$x_3 - x_2 = \frac{D}{d}\cdot\lambda$$

....

....

$$x_n - x_{n-1} = \frac{D}{d}\cdot\lambda$$

i.e., the distance between two consecutive bright fringes is constant.

4.7. CONDITIONS FOR DARK FRINGES

The point P will be centre of darkness if the path difference is an odd multiple of $\frac{\lambda}{2}$.

i.e., $$S_2 A = x\frac{d}{D} = (2n+1)\frac{\lambda}{2}$$

or $$x = \frac{D}{d}(2n+1)\frac{\lambda}{2} \quad ...(5)$$

Equation (5) gives the distances of the dark fringes from point O. The dark fringes will be as under.

When $n = 1$ $x_1 = \frac{D}{d}\cdot\frac{3\lambda}{2}$

$n = 2$ $x_2 = \frac{D}{d}\cdot\frac{5\lambda}{2}$

$n = 3$ $x_3 = \frac{D}{d}\cdot\frac{7\lambda}{2}$

....

....

$n = (n-1)$, $x_{n-1} = \frac{D}{d}(2n-1)\frac{\lambda}{2}$

$n = n$ $x_n = \frac{D}{d}(2n+1)\frac{\lambda}{2}$

So we have for dark fringes

$$x_2 - x_1 = \frac{D}{d}\cdot\lambda$$

$$x_3 - x_2 = \frac{D}{d}\cdot\lambda$$

....

....

$$x_n - x_{n-1} = \frac{D}{d}\cdot\lambda$$

Hence, the distance between any two consecutive dark fringes or bright fringes is the same. This is expressed by β, $\left(\beta = \frac{D}{d}\cdot\lambda\right)$ and is called **Fringe width**. It is evident that the fringe width β is **directly** proportional to D and λ and **inversely** proportional to d.

4.8. CONDITIONS OF INTERFERENCE

We are now in a position to state the conditions for interference of light. These are given as under:

1. The beams from two sources must be propagated along the same line, otherwise the vibrations cannot be along a common line.

2. The two sources must be very narrow because a broad source is equivalent to number of sources and so positions of darkness and brightness cannot be well defined.

3. The two sources must have equal intensities.

4. The two sources should have equal amplitudes and frequencies.

5. The common original source must be monochromatic *i.e.*, emitting light of single wavelength.

6. The light from two sources must have either zero phase or a constant difference of phase. This condition is very well observed in the case of coherent sources.

7. The condition for maximum intensity is that path difference between two waves should be equal to integral multiple of λ or phase difference should be equal to an integral multiple of 2π.

i.e., path diff. $= n\lambda$

or phase diff. $= 2n\pi$

Similarly for minimum intensity, the path difference should be an odd multiple of $\frac{\lambda}{2}$ or an odd multiple of π

i.e., path diff. $= (2n + 1)\ \frac{\lambda}{2}$

or phase diff. $= (2n + 1)\ \pi$

4.9. COHERENT SOURCES

Two narrow light sources are said to be coherent if they emit light waves of the same wavelength and frequency and also there is a constant phase relation between them.

Two independent sources can never act as coherent sources as a source of light consists of large number of atoms and from the atomic theory, each atom consists of a central nucleus around which revolve electrons in various definite orbits. In an excited state an electron jumps to a higher orbit and so the atom becomes unstable. The electron spontaneously fall back to inner orbit within 10^{-8} seconds and in doing so will give light pulses. The emission of light pulses from various atoms is random and so there is no constant phase relationship from two pulses. So independent sources or different parts of the same sources cannot act as coherent sources.

A real source and its virtual image or two virtual images of a single source can act as coherent sources. So in the interference experiments a double slit is used to produce two virtual sources from a single slit, so that the conditions of coherence is satisfied.

4.10. ANALYTICAL TREATMENT OF INTERFERENCE

Let us consider a narrow slit S illuminated by a monochromatic source of light of wavelength λ. Let S_1 and S_2 be two parallel slits at equal distances from S. The two sources S_1 and S_2 will act as coherent sources.

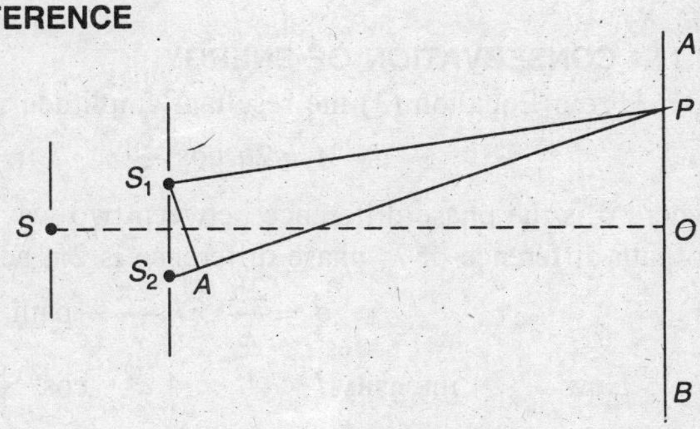

Fig. 4.5

To find the intensity at any point P on the screen AB. Let us consider a general wave equation

$$y_1 = a \sin \frac{2\pi}{\lambda} vt \quad ...(1)$$

where y_1 is the displacement of a particle from its mean position at any time t, v the velocity of propagation of the wave, a and λ the amplitude and wavelength of the light wave respectively.

Let y_2 be displacement of the wave reaching P from S_2 and x the path difference w.r.t. the first wave from S_1, then the general equation for the second wave is given by

$$y_2 = a \sin \frac{2\pi}{\lambda}(vt + x) \quad ...(2)$$

When the two waves superimpose, the resultant displacement y is given by

$$y = y_1 + y_2$$

$$= a \sin = \frac{2\pi}{\lambda} vt + a \sin \frac{2\pi}{\lambda}(vt + x)$$

Using the relation

$$\sin A + \sin B = 2 \sin \frac{A+B}{2} \cdot \cos \frac{A-B}{2}$$

We get $$y = 2a \sin \frac{2\pi}{\lambda}\left(vt + \frac{x}{2}\right) \cdot \cos\left(\frac{2\pi}{\lambda} \cdot \frac{x}{2}\right)$$

which is an *e.g.*, of S.H.M. with an amplitude

$$A = 2a \cos\left(\frac{2\pi}{\lambda} \cdot \frac{x}{2}\right) = 2a \cos\left(\frac{\pi}{\lambda} \cdot x\right) \quad ...(3)$$

For amplitude to be minimum

$$\cos\left(\frac{\pi}{\lambda} \cdot x\right) = 0$$

or $$\frac{\pi}{\lambda} \cdot x = \frac{\pi}{2} \cdot \frac{3\pi}{2}, ... (2n+1)\frac{\pi}{2}$$

or $$x = (2n + 1)\frac{\lambda}{2} \quad ...(4)$$

which shows that the intensity is minimum when the path difference between two waves is equal to an odd multiple of $\frac{\lambda}{2}$.

For amplitude to be maximum

$$\cos \frac{\pi}{\lambda} x = 1$$

$\therefore$ $$\frac{\pi x}{\lambda} = n\pi$$

or $$x = n\lambda \quad ...(5)$$

Thus the intensity is maximum if the path difference between two waves is an even multiple of $\frac{\lambda}{2}$.

4.11. CONSERVATION OF ENERGY

From Equation (3) the resultant amplitude is given by

$$A = 2a \cos \frac{\delta}{2}$$

where δ is the phase difference between two waves reaching P at any instant. As corresponding to path difference of λ, phase difference is 2π, so the phase difference for a path difference x is

$$\delta = \frac{2\pi}{\lambda} \cdot x = \frac{2\pi}{\lambda} \times \text{ path difference}$$

Now $\quad$ intensity $I \propto A^2 \propto 4\,a^2 \,.\, \cos^2 \frac{\delta}{2}$

So $\quad$ intensity $\propto \cos^2 \frac{\delta}{2}$

Let us examine the different cases.

(*i*) When phase difference $= 0, \pm 2\pi, \pm 2\,(2\pi) \ldots \pm n\,(2\pi)$

or path diff. $x = 0, \lambda, 2\lambda, \ldots\, n\lambda$, the intensity is given by

$$I = 4a^2$$

So the intensity is maximum when the phase difference is an integral multiple of 2π or path difference is an integral multiple of λ.

(*ii*) When phase diff. $\delta = \pm\pi, \pm 3\pi, \ldots, \pm(2\pi + 1)\,\pi$

or $\quad x = \frac{\lambda}{2}, \frac{3\lambda}{2}, \frac{5\lambda}{2}, \ldots, (2n+1)\frac{\lambda}{2}$

$$A = 0 \qquad \therefore\ I = 0$$

Hence, intensity is minimum when the phase difference is an odd multiple of π or path difference is an odd multiple of $\frac{\lambda}{2}$.

If we plot the values of δ along X-axis and those of I along Y-a`:s then the graph is shown in Fig. 4.6.

From the graph it is clear that the intensity varies between 0 and $4a^2$ according to the difference of phase when it is *odd or even multiple of π*. Now from the principle of conserva- tion of energy, since the energy cannot be destroyed, but, is transferred from the points of minimum intensity to points of maximum intensity. If there is no interference, the intensity due to two waves should be $a^2 + a^2 = 2a^2$. But due to interference it is $4a^2$ at bright points and 0 at dark points. So the energy which apparently disappears at minima actually appears at maxima. So there is actually transfer of energy and the total energy of the system is conserved, which is in accordance with the law of conservation of energy.

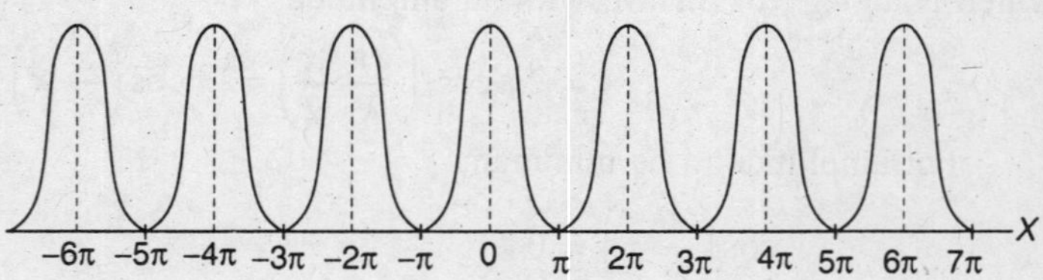

Fig. 4.6

4.12. METHODS OF GETTING INTERFERENCE PATTERN

Two methods are most commonly used to get interference pattern. These methods are given as under:

(*i*) Division of wave front: In this method a narrow source and its virtual image or two virtual images of a narrow source can be used as coherent sources. Two slits illuminated by light from a single slit can also be used as coherent source as was done by *Thomas Young for the first experiment on interference.* In the Llyod's single mirror method, a source and *its virtual image* are used as coherent sources. In Fresnel biprism two virtual images of a narrow slit are used as coherent sources.

In all these cases any random phase change produces as similar phase change in the secondary source and so interfering rays will always have **constant phase difference.**

(*ii*) **Division of Amplitude:** In this method, the same beam of light is partially reflected from two surfaces and the two reflected beams so produced **interfere** and because of the different distance travelled by these two beams, an additional path difference is created between the secondary waves. However, the random phase change in primary wave front is equally reflected in two beams and so the two beams satisfy the condition of **coherence.** As in this case coherence is achieved because the same beam of light is reflected from different surfaces, so in this method, the interference of light is due to division of amplitude (or intensity).

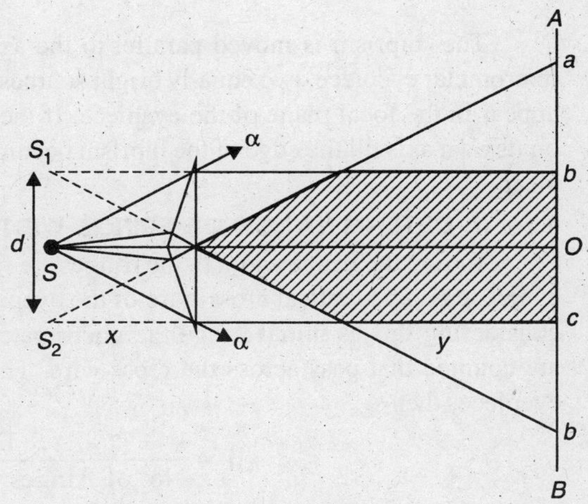

Fig. 4.7

Colour in thin films, Newton rings, interference pattern in Michelson Morley interferometer involve **interference by division of amplitude.**

4.13. FRESNEL BIPRISM (METHOD FOR FINDING WAVE LENGTH OF LIGHT)

It consists of two acute angle prisms with their bases in contact forming a single obtuse angle. In practice, however, a single glass plate is grounded till we have two side angles of the order of 30′ to 1° each and the angle at the vertex is very large. The prism is placed with its refracting edge parallel to the line joining the source and the point *O* on the screen placed ⊥ to the plane of paper.

When the light from source *S* falls on the lower portion of the prism, after refraction it appears to come from virtual source S_2. Similarly light falling on the upper portion of the prism will appear to come from the virtual source S_1. So S_1 and S_2 will act as two coherent sources. The distance *x* between the biprism and the source *S* is so adjusted that the source S_1 and S_2 are very close to each other. Interference fringes are observed within the region *bc* on the screen. If *d* be the distance between the sources S_1 and S_2, *D* the distance between the source and the screen and β the fringe width, then

$$\beta = \frac{D}{d}\lambda$$

or

$$\lambda = \frac{d}{D}\beta = \frac{d}{(x+y)}\beta$$

Knowing *d*, *D* and β, wavelength of light can be known.

Experimental Set up. A narrow adjustable slit *S*, a biprism and a micrometer eyepiece *E* are put on uprights of an optical bench and they are adjusted at the same height and same straight line.

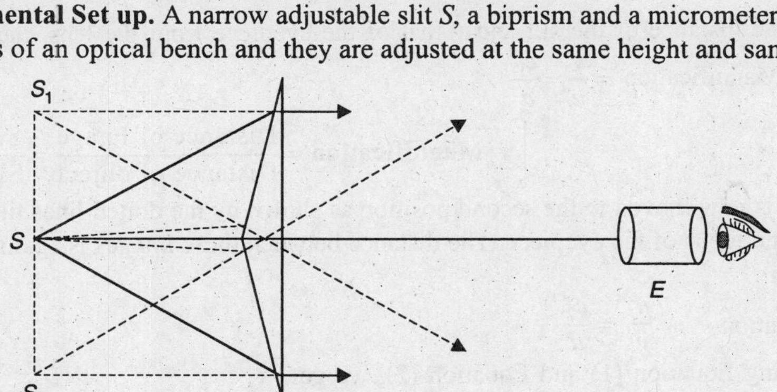

Fig. 4.8

The biprism is moved parallel to the axis of the optical bench till on looking through the micrometer eyepiece, two equally bright sources S_1 and S_2 are seen. The slit is adjusted till the fringes appear in the focal plane of the eyepiece. If the fringes are not well defined, the biprism is rotated on its own axis till the edge of the biprism becomes exactly parallel to the slit and the fringes become clear.

4.14. (*I*) DETERMINATION OF FRINGE WIDTH

To find the fringe width β, the fringes are firstly observed in the field of view of the micrometer eyepiece *E*. The vertical cross-wire of the eyepiece is made to coincide with one of the fringes and considering this as initial fringe the micrometer screw is moved sideways and number of fringes are counted that pass across the cross-wire. The second position is also noted on the scale of the eyepiece, then

$$\beta = \frac{\text{Distance moved}}{\text{No. of fringes moved across the cross-wire}}.$$

***(ii)* Determination of *D*:** The distance between the slit and the focal plane of the eyepiece is noted on the scale of the optical bench and the usual correction is applied by taking the distance by a needle between the slit and the eyepiece and correct value of *D* is noted.

***(iii)* Determination of *d*:** To find *d*, the distance between S_1 and S_2, a convex lens whose focal length is less than 1/4 of the distance between slit and eyepiece *i.e.*, *D*, is mounted between the biprism and the eyepiece. The position of the lens is adjusted till we have the view of the two sources S_1 and S_2 through eyepiece *E*. The distance between the two images of S_1 and S_2 is measured and let it be c_1.

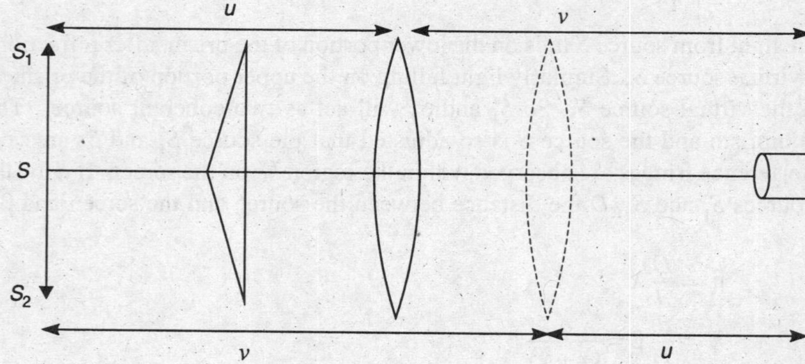

Fig. 4.9

If *u* be the distance of the slit and *v* that of the eyepiece from the lens, then

$$\text{Magnification} = \frac{v}{u} = \frac{c_1}{d} \qquad ...(1)$$

$$\left[\because \text{Magnification} = \frac{\text{Distance of Image}}{\text{Distance of object}} = \frac{\text{Size of image}}{\text{Size of object}}\right]$$

The lens is now moved to the second position as shown by the dotted lines till again the two images are in the focus of the eyepiece. The distance between these images is again measured. Let this be c_2, then

$$\text{Magnification} \quad \frac{u}{v} = \frac{c_2}{d} \qquad ...(2)$$

Multiplying Equation (1) and Equation (2), we get,

$$\frac{c_1 c_2}{d^2} = 1 \quad \text{or} \quad d = \sqrt{c_1 c_2}$$

So knowing β, *D* and *d*, wavelength of light λ can be known.

4.15. WHAT HAPPENS IF WHITE LIGHT IS USED?

Since white light consists of seven colours, the central fringe O will be achromatic as light of all wavelengths will meet there in phase. On either side of O, the path difference gradually increases from zero and will become equal to half the wavelength of the component corresponding to smallest wavelength *i.e.*, violet at some point. This corresponds to the first dark fringe for the violet light. Beyond this point the first dark fringe from various colours like indigo, blue, green etc., and lastly for red will lie. So figure will be violet on inner side and red on outer side. As $\beta \propto \lambda$, so the fringe width of red will be nearly double that of violet. As a result there may be some overlapping of the various fringes and this overlapping will increase with the order of fringes till we get a general illumination.

4.16. LOCATION OF ZERO ORDER FRINGE

If monochromatic light is used then all the fringes in the biprism experiment are alike. However, with white light, the central fringe will be white. So if a monochromatic source of light is replaced by white light the central fringe is located and by putting the cross-wire of the eyepiece on the central fringe its position can be noted then white light may be replaced by the monochromatic light and position of other fringes may be noted with reference to the central fringe.

Example 1. *Two straight and narrow parallel slits 0.6 mm apart are illuminated by a monochromatic light of wavelength 6000 Å. Calculate the width of the fringes obtained on a screen placed at distance of 60 cm from the slit.*

Solution: Here $D = 60$ cm, $d = 0.06$ cm

$$\lambda = 6000\text{Å} = 6000 \times 10^{-8} \text{ cm.}$$

$$\therefore \quad \text{Fringe width } \beta = \frac{D}{d}\cdot\lambda = \frac{60 \times 6000 \times 10^{-8}}{0.06} = \mathbf{0.06\ cm.}$$

Example 2. *A double slit of 0.5 mm separation is illuminated by light of blue cadmium line. How far behind the slits must one go to obtain fringes that are 0.1 cm apart? The wave-length of blue cadmium light is 4800Å.*

Solution: $d = 0.5$ mm $= 0.05$ cm.

$D = ?$ $\quad \beta = 0.1$ cm, $\quad \lambda = 4800 \times 10^{-8}$

Now $$\beta = \frac{D}{d}\cdot\lambda$$

or $$D = \frac{D}{d}\cdot\beta = \frac{0.05 \times 0.1}{4800 \times 10^{-8}} = \mathbf{104.17\ cm.}$$

Example 3. *In Young's experiment the fringe width obtained is 0.60 cm by using a light of wavelength 5000Å. If the distance between screen and slit is reduced to half, what should be the wavelength of light source to get fringes 0.40 cm wide?*

Solution: $\lambda = 5000 \times 10^{-8}$ cm; $\quad \beta = 0.60$ cm

Now $$\beta = \frac{\lambda D}{d}$$

or $$D = \beta\cdot\frac{d}{\lambda} = \frac{0.60 \times d}{5000 \times 10^{-8}}$$

Let the new distance between the slit and the source be x, then

$$x = \frac{1}{2}D = \frac{1}{2}K\cdot\frac{0.60 \times d}{5000 \times 10^{-8}} \text{ cm}$$

Here $\beta' = 0.40$ cm, $\quad \lambda' = ?$

Now $$\beta' = \frac{\lambda' x}{d}$$

$$\lambda' = \beta' \cdot\frac{d}{x}$$

$$= \frac{0.40 \times d \times 2 \times 5000 \times 10^{-8}}{0.60 \times d}$$

$$= 6667 \times 10^{-8} \text{ cm} = \mathbf{6667Å.}$$

Example 4. *In a two slit interference pattern at a point we observe 10th order maximum for λ = 7000 Å. What order will be visible here if the source of light is replaced by light of wavelength 5000 Å?*

Solution: Let x_{10} be the distance of the 10th bright fringe from the central fringe, then we know

$$x_n = n \cdot \frac{D}{d} \cdot \lambda$$

$$x_{10} = 10 \cdot \frac{D}{d} \times 7000 \times 10^{-8}$$

If the nth bright fringe is formed at that place, when the new wavelength is 5000 A.U. then

$$x_n = n \cdot \frac{D}{d} \cdot \lambda$$

$$= n \cdot \frac{D \times 5000 \times 10^{-8}}{d}$$

Now $x_{10} = x_n$

$$\therefore \frac{10.D \times 7000 \times 10^{-8}}{d} = n \times \frac{D \times 5000 \times 10^{-8}}{d}$$

or $n = 14$

So the 14th bright fringe will be visible at the place of 10th bright fringe.

Example 5. *The inclined faces of a biprism of μ = 1.50 make an angle of 2° with base. A slit illuminated by a monochromatic light is placed at a distance of 10 cm from the biprism. If the distance between two dark fringes observed at a distance of 1 metre from biprism is 0.018 cm, find the wavelength of light used. Also derive the formula used.*

Solution: In the case of biprism, the refracting angle α is very small so the deviation of the rays is given by (μ – 1) α where μ is the refractive index of the material of the prism. If θ be the total deviation produced in a ray that strikes the centre of the biprism.

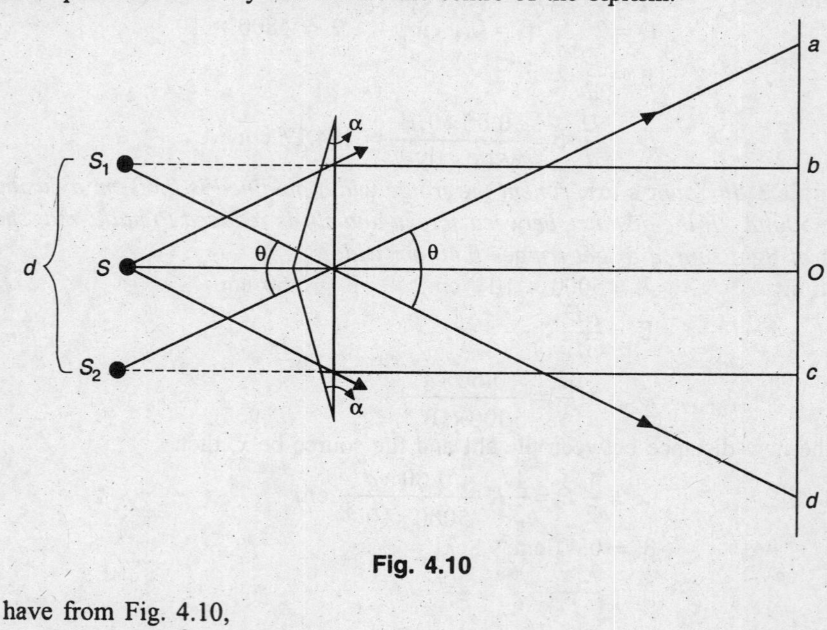

Fig. 4.10

We have from Fig. 4.10,

$$\frac{\theta}{2} = (\mu - 1)\,\alpha$$

Let x be the distance of the biprism from the source, then

$$d = x \,.\, \theta = 2x\,(\mu - 1)\,\alpha$$

Also $$\lambda = \beta \cdot \frac{d}{D}$$

$\therefore$ $$\lambda = \frac{2\beta . x\,(\mu - 1)\,\alpha}{D}$$

Here $\mu = 1.50,\quad x = 10 \text{ cm},\quad \beta = 0.018 \text{ cm}$

$$D = 100 + 10 = 110 \text{ cm}, \ \alpha = 2^\circ = \frac{\pi}{90} \text{ radian}$$

$\therefore$ $$\lambda = \frac{2 \times 0.018 \times 10 \times 0.5 \times \pi}{110 \times 90}$$

$$= 5714 \times 10^{-8} \text{ cm} = \mathbf{5714 Å.}$$

Example 6. *In a biprism experiment, the eyepiece was placed at a distance of 120 cm from the source. The distance between the two virtual images was found to be 0.075 cm. Find the wavelength of the light used if the eyepiece is moved through a distance of 1.888 cm for 20 fringes to cross the field of view.*

Solution: $D = 120 \text{ cm},\quad d = 0.075 \text{ cm}$

$$\beta = \frac{1.888}{20} = 0.0944 \text{ cm}$$

Now $$\lambda = \beta \cdot \frac{d}{D} = \frac{0.0944 \times 0.075}{120}$$

$$= 5900 \times 10^{-8} \text{ cm} = \mathbf{5900Å}$$

Example 7. *Fringes are produced by a Fresnal's biprism in the focal plane of reading microscope which is 100 cm. from the slit. A lens inserted between the biprism and microscope gives two images of the slit are 4.05 mm and in the other 2.09 mm apart. If wave length of light used is 5893Å. Find the distance between interference bands.*

Solution: $c_1 = 4.05 \text{ mm} = 0.405 \text{ cm}$

$$c_2 = 2.09 \text{ mm} = 0.209 \text{ cm}$$

Now $$d = \sqrt{c_1\, c_2}$$

$$= \sqrt{0.405 \times 0.209}$$

$$= 0.3428 \text{ cm.}$$

Now $D = 100 \text{ cm}$ and $\lambda = 5893 \times 10^{-8}$ cm

From the relation $$\beta = \lambda \frac{D}{d}$$

We have $$\beta = \frac{5893 \times 10^{-8} \times 100}{0.3428} = \mathbf{0.172 \text{ cm.}}$$

Example 8. *Calculate the separation between two coherent sources formed by a prism whose inclined faces make angles of 2° with the base. The slit source being 10 cm away from the biprism and refractive index of prism = 1.5.*

Solution: If μ is the refractive index of the biprism and α the angle which inclined faces make with the base, then

Distance between coherent sources

$$d = 2x\,(\mu - 1)\,\alpha$$

where x is the distance of the biprism from the source.

Now $x = 10 \text{ cm};\quad \alpha = 2^\circ = \frac{\pi}{90} \text{ radian};\quad \mu = 1.5$

$\therefore$ $$d = \frac{2 \times 10 \times (1.5 - 1) \times \pi}{90} = \mathbf{0.349 \text{ cm.}}$$

4.17. DISPLACEMENT OF FRINGES

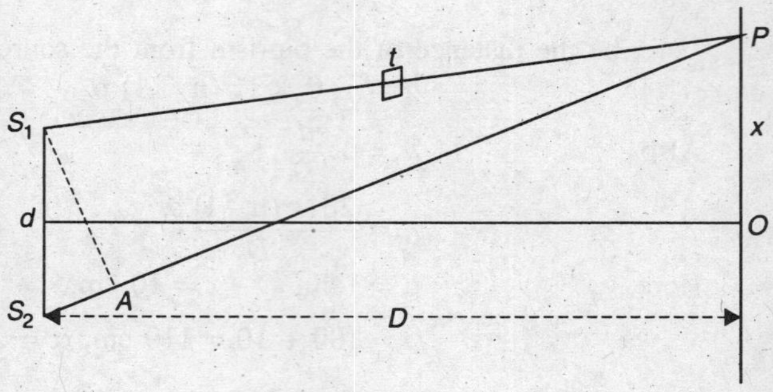

Fig. 4.11

When a transparent material of uniform thickness and having refractive index m is placed in the path of one of the lights, then we observe that shifting of fringes make place. If the shift of the fringes is measured then we can use this method to find the thickness of the material or its refractive index or the wavelength of light used.

Let a thin glass plate of thickness t and refractive index μ be placed in the path $S_1 P$. Since the velocity of light will change in the medium of glass, so, we can note the path difference of two waves $S_1 P$ and $S_2 P$ as

$$\text{Path differential} = S_2 P - [(S_1 P - t) + \mu t]$$

$$\therefore \quad \text{Path differential} = S_2 P - S_1 P - (\mu - 1)\, t = S_2 A - (\mu - 1)\, t$$

Now P will be centre of maxima if path difference $= n\lambda$,

i.e.,
$$S_2 A - (\mu - 1)\, t = n\lambda$$

$$x\frac{d}{D} - (\mu - 1)\, t = n\lambda \qquad \left[\because S_2 A = x\frac{d}{D}\right]$$

or
$$x = \frac{D}{d}[n\lambda + (\mu - 1)\, t]$$

We can note down the positions of the various fringes as

$$x_0 = \frac{D}{d}(\mu - 1)\, t \qquad n = 0$$

$$x_1 = \frac{D}{d}[\lambda + (\mu - 1)\, t] \qquad n = 1$$

$$x_2 = \frac{D}{d}[2\lambda + (\mu - 1)\, t] \qquad n = 2$$

....

....

$$x_{n-1} = \frac{D}{d}[2\,(n - 1)\,\lambda + (\mu - 1)] \qquad n = n - 1$$

$$x_n = \frac{D}{d}[n\lambda + (\mu - 1)\, t] \qquad n = n$$

From where we get

$$x_1 - x_0 = \frac{D}{d}\lambda$$

$$x_2 - x_1 = \frac{D}{d}\lambda$$

....

....

$$x_n - x_{n-1} = \frac{D}{d}\lambda$$

i.e., the fringe width $\beta = \dfrac{D}{d}\lambda$ and is not changed because of the introduction of a thin plate.

Also we observe that $x_0 = \dfrac{D}{d}(\mu - 1)\, t$ and since $\mu > 1$, so x_0 is positive *i.e.,* the central fringe is shifted upwards and correspondingly all other fringes are shifted.

From above we have

$$x_0 = \frac{D}{d}(\mu - 1)\,t = \frac{\beta}{\lambda}(\mu - 1)\,t$$

The fringe width β and x_0 shift of the central fringe can be measured experimentally, we can locate the position of the central fringe by using white light when except the central fringes, all the fringes are coloured and the shift can be noted by making use of a monochromatic light.

So the above experimental set up can provide a method for finding μ, t or λ depending upon the availability of the unknown parameters.

Example 9. *When a thin sheet of transparent material of thickness 7.2×10^{-4} cm is introduced in the path of one of the interfering beams, the central fringe shifts to a position occupied by the sixth fringe. If $\lambda = 6000$Å, find the refractive index of the sheet.*

Solution: $\lambda = 6000 \times 10^{-8}$ cm.

$t = 7.2 \times 10^{-4}$ cm.

Central fringe shift $x_0 = 6\beta$

Now $x_0 = \frac{\beta}{\lambda}(\mu - 1)\,t$

$\therefore$ $6\beta = \frac{\beta}{\lambda}(\mu - 1)\,t$

$$\mu = \frac{6\lambda}{t} + 1 = \frac{6 \times 6000 \times 10^{-8}}{7.2 \times 10^{-4}} + 1$$

$\therefore$ $\mu = \mathbf{1.5.}$

Example 10. *Fringes are produced with monochromatic light of wavelength 5.800×10^{-8} cm. A thin glass plate of refractive index 1.5 is kept in the path of one of the interfering beams and the central bright band of the fringe system is found to move into position previously occupied by third bright band from the centre calculate the thickness of glass plate.*

Solution: $\mu = 1.5$, $\lambda = 5800 \times 10^{-8}$ cm

$n = 3$, $t = ?$

Using the relation

$$(\mu - 1)\,t = n\lambda$$

$$t = \frac{n\lambda}{\mu - 1} = \frac{3 \times 5800 \times 10^{-8}}{1.5 - 1} = \frac{3 \times 5800 \times 10^{-8}}{0.5} = \mathbf{3.48 \times 10^{-4}\ cm.}$$

4.18. PHASE CHANGE ON REFLECTION

Consider a ray of light incident on a medium which can partly reflect and partly transmit the light. Let a be the amplitude of the incident light and μ_1 and μ_2 be the refractive indices of two media ($\mu_2 > \mu_1$) I and II as shown.

Let r be the fraction of amplitude reflected as light travels from medium I to II. And let t be the fraction of amplitude transmitted for light from medium I to II.

Also let r' be the fraction of amplitude reflected as light travels from medium II to I. And Let t' be the fraction of amplitude transmitted as light travels from medium II to I.

Now the amplitude reflected in the medium I will be ar.

And amplitude transmitted in the medium I to II will be at.

If we retrace the reflected and transmitted lights, we should expect to get the original light if there is no absorption of light in the medium. This is from the principle of reversibility.

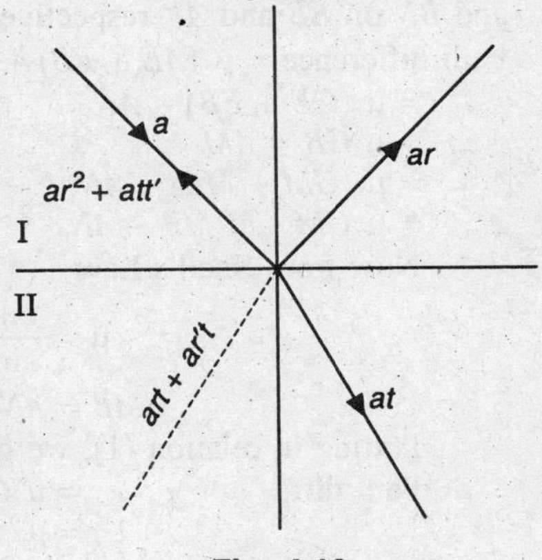

Fig. 4.12

We will have the following relations

$$ar^2 + att' = a$$

or $\qquad r^2 + tt' = 1$

or $\qquad tt' = (1 - r^2) \qquad$ ...(1)

and since there is no light along the dotted line, so

$$art + ar't = 0$$

or $\qquad r' = -r \qquad$ ...(2)

Equations (1) and (2) are known as **Stoke's relations.**

The second relation $r = -r'$ shows that intensity is reflected from medium I or medium II. However, a phase change of π or a path difference of $\lambda/2$ occurs due to light reflected from medium I to medium II.

From above a very important result is deduced *i.e.,*

When a ray of light is reflected at the surface of an optically denser medium while it is travelling through a rarer medium, a phase change of π or a path difference of $\lambda/2$ is introduced. When reflection takes place at the surface of a rarer medium no change in phase or path difference takes place.

4.19. COLOUR IN THE FILMS (PARALLEL FILM)

Colour in the case of thin flims such as soap solution or oil spread on a water surface can be explained from the interference phenomenon. This is the case of multiple reflections from thin films, the reason being due to the division of amplitudes we get interference patterns. The explanation of formation of colours was given by Thomas Young that the interference is between light reflected from the two surfaces of a thin film.

Consider a thin film of uniform thickness t and refractive index μ bounded by two plane surfaces XY and RZ.

A ray of light SA incident on the surface XY at an angle i is partly reflected along AP and partly transmitted into the medium along AE and angle of refraction being r. At E it is again partly reflected and partly transmitted (refrac-ted). Similar reflections and refractions take place at B, K etc., as shown in the Fig. 4.13.

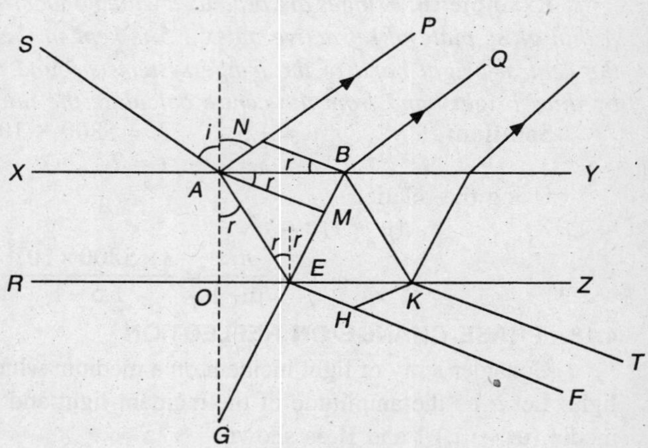

Fig. 4.13

Now we can study the interference patterns in the reflected and refracted systems.

Reflected System. To find the path difference between AP and BQ in the reflected system, drop perpendiculars AM and BN on EB and AP respectively.

Path difference $= \mu\,(AE + EB) - AN$

$= \mu\,(GE + EB) - AN$

$= \mu\, GB = AM$

$= \mu\,(GM + MB) - AN$

$= \mu\, GM + \mu\, MB - AN \qquad$ ...(1)

Now from Snell's Law

$$\mu = \frac{\sin i}{\sin r} = \frac{AN/AB}{MB/AB}$$

$\therefore \qquad \mu MB = AN$

Putting in relation (1), we get

Path diff. $\qquad = \mu\, GM$

$\qquad = \mu\, .\, GA \cos r$

$$= 2\mu\, AO.\cos r$$
$$= 2\mu\, t\cos r$$

Now since the ray AP suffers reflection from optically denser medium while the ray BQ is transmitted into rarer medium, so an additional path difference of $\lambda/2$ will be introduced.

$\therefore$ The total path difference

$$= 2\mu t\cos r + \frac{\lambda}{2} \qquad ...(1)$$

we can study some important cases from the above results.

(*i*) When the film is very very thin as compared to λ, *i.e.*, $t << \lambda$, then $2\mu t\cos r$ can be neglected as compared to $\lambda/2$, the net result will be that the film will appear dark and this will be the case of destructive interference.

(*ii*) If the thickness of the film cannot be neglected as compared to λ, we will have condition for brightness as

$$\text{Path diff.} = 2\mu t\cos r + \frac{\lambda}{2} = n\lambda$$

or $$2\mu t\cos r = (2n-1)\frac{\lambda}{2} \qquad ...(2)$$

where $n = 1, 2, 3, ...$

(*iii*) Condition for darkness will be

$$\text{Path diff.} = 2\mu t\cos r + \frac{\lambda}{2} = (2n+1)\frac{\lambda}{2}$$

or $$2\mu t\cos r = n\lambda \qquad ...(3)$$

where $n = 1, 2, 3, 4, ...$

Transmitted System. The path difference between two rays EF and KT can similarly be given as

$$= \mu\,(EB + BK) - EH$$
$$= 2\mu t\cos r$$

In this case there will not be any additional path difference as both the rays are transmitted from denser to rarer medium.

We can again study the different cases.

(*i*) When $t < \lambda$, $2\mu t\cos r$ can be neglected and the path difference between the two rays will be zero *i.e.*, the two rays will reinforce each other and this will be the case of constructive interference and so the pattern will appear bright.

(*ii*) If t can be neglected as compared to λ, we will have condition for brightness as

$$\text{Path diff.} = 2\mu t\cos r = n\lambda \qquad ...(4)$$

where $n = 1, 2, 3, ...$

(*iii*) The condition for darkness will be

$$\text{Path diff.} = 2\mu t\cos r = (2n+1)\frac{\lambda}{2} \qquad ...(5)$$

where $n = 1, 2, 3, ...$

4.20. STUDY OF COLOURS

We can see from above relations that the conditions of interference are complementary *i.e.*, the conditions of interference in the case of reflected and transmitted systems are opposite. For a certain path difference if a bright fringe is observed in reflected system, a dark fringe will be observed in the transmitted system.

With monochromatic light alternate dark and bright fringes are observed because λ depends on μ, t and r.

From above relations we see different colours if either μ, t or r are changed.

(*i*) If t and r are constants and μ is varied we will get various colours because their λ will be varied. Since white light consists of seven colours, so we will observe colours in the order of violet, indigo... red etc. because $\lambda \propto \mu$.

(*ii*) If t and μ are constants and r is varied *i.e.*, the angle of incidence is changed so that the angle of refraction is also changed, we observe different colours with white light because $\lambda \propto r$.

(*iii*) If μ and r are constants and t is changed we still get different colours with white light because $\lambda \propto t$.

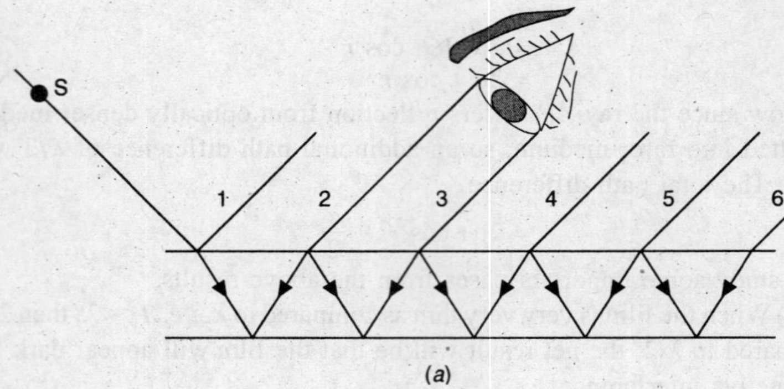

Fig. 4.14 (*a*)

So it is clear from above that we get different colours in the reflected and transmitted system if one of the three parameters μ, t and r is changed.

4.21. CONDITIONS FOR GETTING COLOURS IN THIN FILMS

1. For getting colours in thin films, a broad source of light is necessary as can be seen as under.

If we use a point source, then we observe that different pairs of reflected light cannot reach the eye due to its small aperture. Only some pairs reach and so a small portion of the film will be visible, and since white light consists of seven colours, so only some colours, can be seen.

However, if a broad source of light is used then the field of view can be sufficiently increased and so more colours can be seen from Fig. 4.14 (*b*).

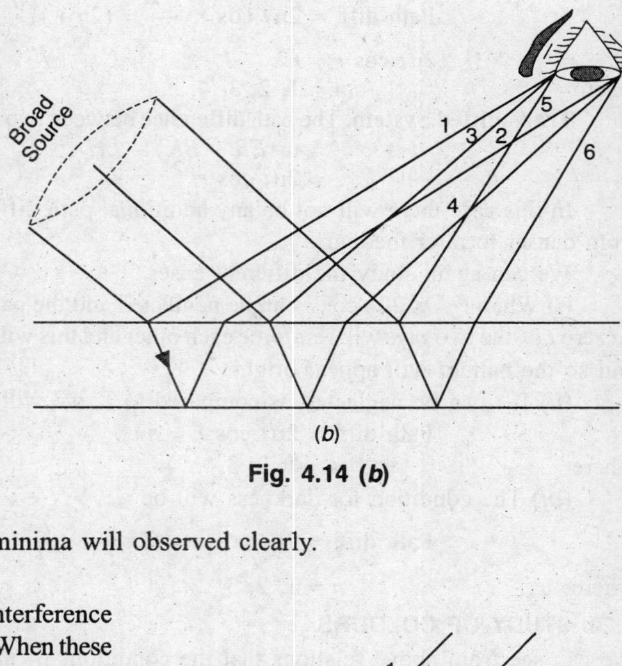

Fig. 4.14 (*b*)

2. Thin film is necessary for seeing colours because if a thick film is used then the path difference between pairs of interfering beams will be quite large as compared to wavelength of light used and so the conditions of maxima and minima will observed clearly.

4.22. NON-REFLECTING FILMS

A very important application of interference in thin films is in the non-reflecting films. When these films are coated on lens or prism surfaces, the reflectivity of these surfaces is very much reduced.

With non-reflecting films, the loss of light by reflections at various surfaces of a system of lenses or prism is appreciably reduced. So we can reduce the stray light reaching the image as a result of reflections. In these days all optical parts which require good quality are coated with non-reflecting films. Previously many monomolecular layer

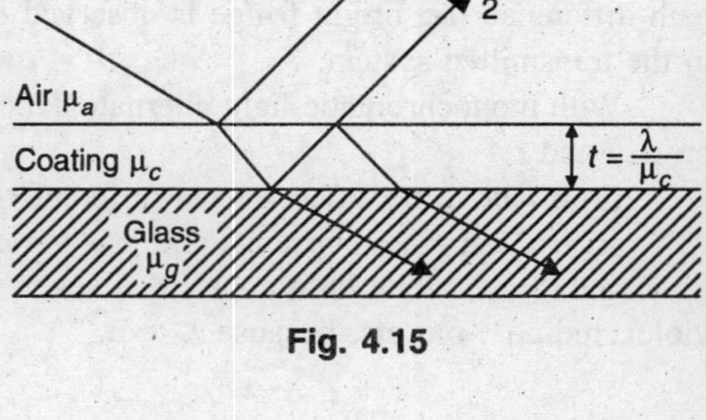

Fig. 4.15

coatings of some organic substance were made on glass plates but these were not durable. These have been replaced now by either (*a*) evaporating calcium or magnesium flouride on surface in vacuum or (*b*) by chemical treatment of the surface with acids which leave a thin layer of silica on the surface of glass. It should be noted that no light is destroyed by non-reflecting films. There is merely a redistribution *i.e.*, decrease of reflection appears as an increase of transmission.

The material chosen for coating on glass should have a refractive index in between that of air and gas.

If μ_c is the refractive index of transparent material

$\mu_a \rightarrow$ refractive index of air

$\mu_g \rightarrow$ refractive index of glass

then $\mu_g > \mu_c > \mu_a$

If λ be the wavelength of incident light, then the path difference between two rays in $2\mu_c t \cos r$.

For normal incidence, $r = 0$

$\therefore \quad \cos r = 1$

$\therefore \quad$ path difference $= 2\mu_c t$

For destructive interference, path difference should be $= \dfrac{\lambda}{2}$

i.e., $$2\mu_c t = \frac{\lambda}{2}$$

or $$t = \frac{\lambda}{4\mu_c}$$

Thickness of the film is adjusted to be $\dfrac{\lambda}{4\mu_c}$ for a particular light of wavelength λ. In case of white light, λ is generally chosen near the middle of the visible spectrum.

At angles other than normal incidence, the path difference will change because of the factor $\cos r$. However, since $\cos r$ does not change very much near 0°, so reflection remains quite low over a fair range of small angles about the normal.

4.23. WEDGE SHAPED FILMS

Let a wedge be there between two glass plates slightly inclined to each other at an angle of θ and let the refractive index of the enclosing film in between the two plates be μ. Let the wedge be illuminated by a monochromatic light from a slit held parallel to the edges of the wedge in contact. Interference occurs between the rays reflected from the upper and lower surface of the film respectively. If t be the thickness of the film at a distance x from the edge, for normal incidence, the path difference $= 2\mu t = 2\mu x\theta$ (If θ is very small).

The condition for bright fringe of the reflected light is

$$2\mu t = (2n + 1) \qquad \left[\therefore \text{ additional path difference of } \frac{\lambda}{2} \text{ is introduced}\right]$$

or $$2\mu x\theta = (2n + 1)\frac{\lambda}{2}$$

and the condition for darkness is

$$2\mu t = n\lambda$$
$$2\mu x\theta = n\lambda$$

If x is the distance of the nth dark fringe from the edge and x_m is that of $(n + m)$th dark fringe, then

$$x_n = \frac{n\lambda}{2\mu\theta} \text{ and } x_{n+m} = \frac{(n+m)\lambda}{2\mu\theta}$$

$$\therefore \quad x_{m+n} - x_n = \frac{(m+n)\lambda}{2\mu\theta} - \frac{n\lambda}{2\mu\theta}$$

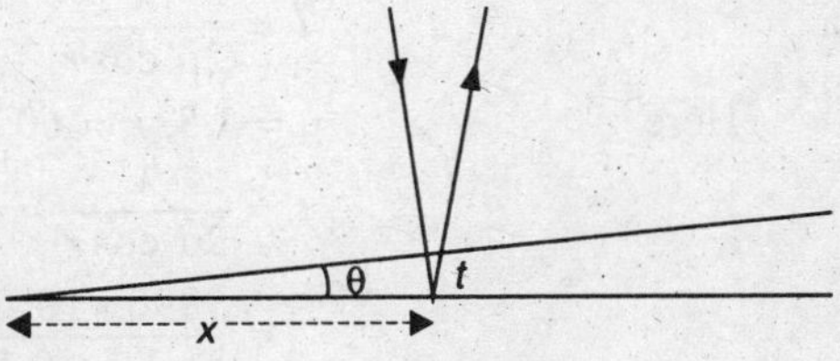

Fig. 4.16

$$= \frac{m\lambda}{2\mu\theta} \quad ...(1)$$

which affords us a method

(*i*) to find λ if μ and θ are known
(*ii*) to find μ if λ and θ are known
(*iii*) to find θ if λ and μ are known.

Also from (1) the fringe width

$$\beta = \frac{\lambda}{2\mu\theta}$$

If the two surfaces forming wedge shaped film are optically plane, then fringes of equal thickness will be straight. Each fringe will give locus of equal thickness of the film. This method is used to test the flatness of a surface.

Example 11. *A soap film of μ = 1.333 is illuminated by white light incident at an angle of 45°. The light refracted by it is examined by a spectrometer and a bright band is found corresponding to a wavelength of 6000Å. Find the thickness of the film.*

Solution: Here Angle of Incidence $i = 45°$, $\mu = 1.333$

$$\lambda = 6000 \times 10^{-8} \text{ cm}, \quad n = 1$$

For a bright band in the refracted system, we have

$$2\mu t \cos r = n\lambda$$

where $t \rightarrow$ thickness of the film
and $r \rightarrow$ Angle of refraction

Now
$$\mu = \frac{\sin i}{\sin r} = \frac{\sin 45°}{\sin r}$$

$$\therefore \quad \sin r = \frac{0.7071}{1.333}$$

$$\cos r = \sqrt{1-\sin^2 r} = \sqrt{1-\left(\frac{0.7071}{1.333}\right)^2}$$

$$= 0.8476$$

$$\therefore \quad t = \frac{n\lambda}{2\mu t \cos r} = \frac{1 \times 6000 \times 10^{-8}}{2 \times 1.333 \times 0.8476}$$

$$= \mathbf{2.26 \times 10^{-5} \text{ cm}}$$

Example 12. *A parallel beam of light λ = 5890Å, is incident on a glass plate (μ = 1.5) such that angle of refraction into plate is 60°. Calculate the smallest thickness of the plate which will make it appear dark by reflection.*

Solution: Let the smallest thickness of the plate = t cm.

For least thickness of the plate,

$$\text{Path difference} = 2\mu t \cos r = 2 \cdot \frac{\lambda}{2} = \lambda$$

or
$$t = \frac{\lambda}{2\mu \cos r}$$

Here
$$\mu = 1.5, r = 60°, \lambda = 5890 \times 10^{-8} \text{ cm}$$

$$\therefore \quad t = \frac{\lambda}{2\mu \cos r} = \frac{5890 \times 10^{-8}}{2 \times 1.5 \cos 60°}$$

$$= \frac{5890 \times 10^{-8}}{2 \times 1.5 \times 0.5} = \mathbf{3927 \text{ A.}}$$

Example 13. *On placing a thin sheet of mica of thickness 12×10^{-5} cm in the path of one of the interfering beams in a biprism experiment, the central fringe shifts a distance equal to the width of the bright fringe.*

Calculate the refractive index of mica (λ = 6000Å).

Solution: $t = 12 \times 10^{-5}$ cm, $\lambda = 6000 \times 10^{-8}$ cm
$n = 1$, $\mu = ?$

Now $(\mu - 1)\, t = n\lambda$

$$\therefore \quad \mu = \frac{n\lambda}{t} + 1 = \frac{1 \times 6000 \times 10^{-8}}{12 \times 10^{-5}} + 1 = \mathbf{1.50.}$$

Example 14. *A glass plate 0.40 micron thick is illuminated by a beam of white light normal to the plate. m for glass is 1.50. Calculate the wavelengths within the limits of visible spectrum ($\lambda = 4.0 \times 10^{-5}$ cm to 7.0×10^{-5} cm) which will be intensified in the reflected beam.*

Solution: Condition for the constructive interference in reflected system is

$$2\mu t \cos r = (2n + 1)\frac{\lambda}{2}$$

Here $t = 0.40$ micron $= 40 \times 10^{-4}$ cm
$\mu = 1.50$, $\lambda = ?$

Since incidence is normal, so $r = 0$

$\therefore$ $\cos r = 1$

So we get
$$\lambda = \frac{2 \times 2 \times 1.50 \times 0.40 \times 10^{-4}}{2n+1}$$
$$= \frac{24 \times 10^{-5}}{2n+1} \text{ cm}$$

For $n = 0$, $\lambda = 24 \times 10^{-5}$ cm.

This wavelength does not lie in visible region.

For $n = 1$, $\lambda = \frac{24 \times 10^{-5}}{3} = 8 \times 10^{-5} = 8000$Å

This wavelength also does not lie in visible region

For $n = 2$, $\lambda = \frac{24 \times 10^{-5}}{5} = 4.8 \times 10^{-5} =$ **4800Å**

This waveleng is in visible region.

Example 15. *White light is incident on a soap film at an angle $\sin^{-1}$ (4/5) and the reflected light on examination by a spectromet r shows dark bands. The consecutive dark bands correspond to wavelength 6.1×10^{-5} cm and 6.0×10^{-5} cm. If m = 1.333 for the film, calculate the thickness.*

Solution: $i = \sin^{-1}\left(\frac{4}{5}\right)$ or $\sin i = \frac{4}{5}$

$\lambda_1 = 6.1 \times 10^{-5}$ cm $\quad \lambda_2 = 6.0 \times 10^{-5}$ cm

$\mu = 1.333$

The condition for dark band in reflected light is

$$2\mu t \cos r = n\lambda \quad \text{...(}i\text{)}$$

For two consecutive dark bands corresponding to wavelength λ_1 at λ_2, the conditions are given by equations,

$$2\mu t \cos r = n\lambda$$
$$2\mu t \cos r = (n + 1)\, \lambda_2$$
$$\therefore \quad n\lambda_1 = (n + 1)\, \lambda_2$$
$$n\lambda_1 = n\lambda_2 + \lambda_2$$

or
$$n\,(\lambda_1 - \lambda_2) = \lambda_2$$

or
$$n = \frac{\lambda_2}{\lambda_1 - \lambda_2}$$

Putting in Equation (1)

$$2\mu t \cos r = \frac{\lambda_1 \lambda_2}{\lambda_1 - \lambda_2} \quad \text{...(2)}$$

Also $\cos r = \sqrt{1 - \sin^2 r}$

$$= \sqrt{1 - \frac{\sin^2 i}{\mu^2}} \qquad \left[\because \mu = \frac{\sin i}{\sin r}\right]$$

$$= \sqrt{1 - \frac{4}{5} \cdot \frac{1}{(1.33)^2}}$$

$$= \sqrt{1 - \frac{9}{25}} = \sqrt{\frac{16}{25}} = \frac{4}{5}$$

From Equation (2)

$$2 \times 1.333 \times t \times \frac{4}{5} = \frac{6.1 \times 10^{-5} \times 6.0 \times 10^{-5}}{(6.1 \times 10^{-5} - 6.0 \times 10^{-5})}$$

or $$\frac{32t}{15} = \frac{6.1 \times 6 \times 10^{-5} \times 15}{1 \times 10^{-5}}$$

or $$t = \mathbf{0.00172\ cm.}$$

Example 16. *Monochromatic light emitted by a broad source of wavelength 5800Å, falls normally on two plates of glass enclosing a wedge shaped film. The plates touch at one end and are separated at a point 15 cm from that end by a wire of 0.05 mm diameter. Find the fringe width.*

Solution: $$\lambda = 5800 \times 10^{-8} \text{ cm}$$

$$\theta = \frac{t}{x} = \frac{0.005}{15} = 33 \times 10^{-4} \text{ radians}$$

$$\mu = 1 \text{ (for air)}$$

Now $$\beta = \frac{\lambda}{2\mu\theta}$$

$$= \frac{5800 \times 10^{-8}}{2 \times 1 \times 33 \times 10^{-5}}$$

$$= \mathbf{0.0878\ cm.}$$

Example 17. *A non-reflecting layer is to be deposited. What would be the necessary thickness for zero reflection at 5500Å? μ for the layer is 1.334.*

Solution: $$\lambda = 5500 \times 10^{-8} \text{ cm}, \quad t = ?$$

$$\mu_c = 1.334$$

Now $$t = \frac{\lambda}{4\mu_c} = \frac{5500 \times 10^{-5}}{4 \times 1.334}$$

$$= \mathbf{1.031 \times 10^{-5}\ cm}$$

Example 18. *Two pieces of plane glass are placed together with a piece of paper between the two at one edge. Find the angle in seconds of the wedge shaped air film between the plates, if on viewing the film normally with monochromatic light of wavelength 4800Å. Given β = 0.0555.*

Solution: $$\theta = ?, \qquad \lambda = 4800\text{Å}$$

$$\mu = 1 \text{ (for air)}, \quad \beta = 0.0555$$

Now $$\theta = \frac{\lambda}{2\mu\beta} = \frac{4800 \times 10^{-8}}{2 \times 1 \times 0.0555}$$

$$= 432 \times 10^{-6} \text{ radians}$$

$$= \frac{432 \times 10^{-6} \times 180}{\pi} \text{ degrees}$$

$$= \frac{432 \times 10^{-6} \times 180 \times 60'}{3.142} = 1.49'$$

$$= \mathbf{1'29''\ (Approx.)}$$

4.24. TESTING OF FLAT SURFACES

We call a surface to be optically flat if it is plane upto $\frac{1}{10}$ th of wavelength of light used. In order to test a flat surface, we may use interference fringes obtained from a film of varying thickness.

The surface which is to be tested is placed in contact with an optically flat glass plate and the fringes of equal thickness formed are viewed. Normally, an extended source of light may be used and suitable filter is placed in between the source and the surface to select a suitable colour for study of fringes.

If the surface is not flat, then the fringes will have irregular shape while for a flat surface, straight fringes will be observed.

From the relation

$$2\mu t \cos\theta = (2n + 1)\lambda/2$$

We have for normal incidence $\theta = 0$

$$\therefore \quad 2\mu t = (2n + 1)\lambda/2$$

It is evident that a fringe will be characterised by a particular value of n and where the thickness of the air film or flatness of the surface is constant, we will have straight fringes. In other words, the colours of the fringes will indicate the varying thickness of the air film. With this type of arrangement we are able to test the flatness of a surface upto $\frac{1}{10}$ th of the wavelength of the light used and any irregularity in the flatness of the surface can be easily detected.

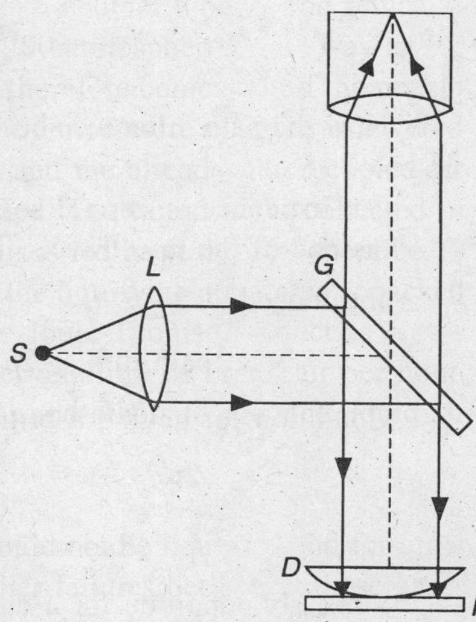

Fig. 4.17

4.25. NEWTON'S RINGS

Introduction: Circular interference fringes can be observed if a very thin film of air or some other transparent medium of varing thickness is enclosed between a plane glass plate and a plano-convex lens of large focal length. Such fringes were first observed by Newton and so are called Newton's Rings.

Experimental Arrangement: Light from a monochro- matic source S is rendered parallel bya convex lens L and then it is made to fall on a glass plate G inclind at an angle of 45° to the incident beam. This beam is reflected normally on to a plano-convex lens placed on a glass plate P as shown in Fig. 4.17.

Light rays reflected from the top and bottom surfaces of the air film formed between the lens D and glass plate P superimpose upon each other and depending upon the path difference between these rays, circular bright and dark rings are observed with a monochromatic light. The fringes are circular because the air film is symmetrical about the point of contact of the lens) and the plane glass plate P. These fringes can be observed by a travelling microscope which can also measure the diameter of the various rings.

Theory of Newton's rings: Let CME be the vertical section of the lens surface and let R be the radius of curvature of the lens. The lens is contact with plate at M in such a way that the film thickness at CL and EN is equal, *i.e.*, $CL = EN = t$

$$\begin{aligned} CD \times DE &= AD \times DM \\ &= (AM - DM) \times DM \\ &= (2R - t) \times t \\ &= 2Rt - t^2 \end{aligned}$$

Since air film thickness is very small, so t^2 can be neglected as compared t^2 Rt.

If d_n be the diameter of some (say nth) ring, then

$$\frac{d_n^2}{4} = 2Rt$$

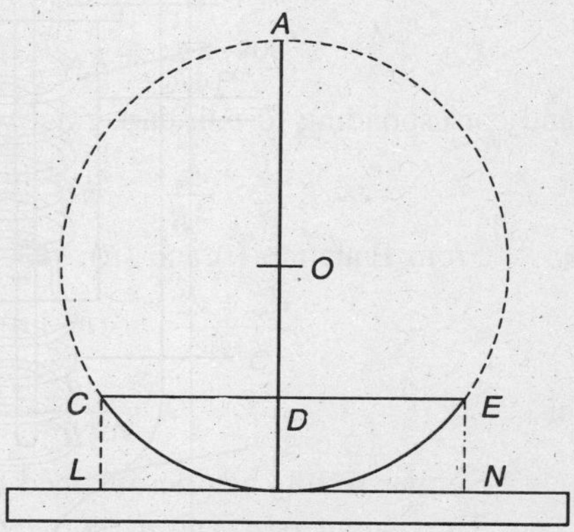

Fig. 4.18

or $$2t = \frac{d_n^2}{4R} \qquad ...(1)$$

Now the path difference between the two rays, one reflected from C and other from $L = 2\mu t \cos r = 2\mu t$

[$\because$ Incidence is normal so $r = 0$ and $\cos 0^\circ = 1$]

As one of the rays travels from denser to rare medium, so an additional path difference of $\frac{\lambda}{2}$ is introduced.

$\therefore$ Total path diff. $= 2\mu t + \frac{\lambda}{2}$.

Condition for getting a bright ring is

$$2\mu t + \frac{\lambda}{2} = n\lambda$$

or $$2\mu t = (2n - 1)\frac{\lambda}{2} \qquad ...(2)$$

Similarly condition for a dark ring is

$$2\mu t + \frac{\lambda}{2} = (n+1)\frac{\lambda}{2}$$

$$2\mu t = n\lambda \qquad ...(3)$$

Putting the values of $2t$ from (1) in (2) and (3), we get for nth bright ring

$$\frac{d_n^2}{4R\mu} = (2n-1)\frac{\lambda}{2}$$

or $$\frac{d_n^2}{4R} = (2n-1)\frac{\lambda}{2} \qquad \text{[For air } \mu = 1]$$

$\therefore$ $$d_n \propto \sqrt{\left(n - \frac{1}{2}\right)} \text{ for a bright ring.} \qquad ...(4)$$

and for nth dark ring in air

$$\frac{d_n^2}{4R} = n\lambda$$

$\therefore$ $$d_n \propto \sqrt{n} \text{ for a dark ring.}$$

4.26. FORMULA FOR DETERMINATION OF λ

Considering the condition for nth dark ring

$$\frac{d_n^2}{4R} = n\lambda \qquad ...(i)$$

and corresponding to mth dark ring, we have

$$\frac{d_m^2}{4R} = m\lambda \qquad ...(ii)$$

From Equation (*i*) and (*ii*),

$$\frac{d_m^2 - d_n^2}{4R} = (m - n)\lambda$$

or $$\lambda = \frac{d_m^2 - d_n^2}{4R(m-n)} \qquad ...(iii)$$

Similar results can be obtained by taking bright rings.

The diameters d_n and d_m corresponding to nth and mth dark rings are noted with travelling microscope and radius of curvature of the lens is measured with a spherometer. So the wave length of light can be found out.

4.27. DETERMINATION OF REFRACTIVE INDEX OF A LIQUID BY NEWTON'S RINGS

If the wavelength of monochromatic light used is known then, the refractive index of a liquid torming film between lens and glass plate is got in the following way:

For an nth dark ring, we have

$$\frac{d_n^2}{4R} = \frac{n\lambda}{\mu} \quad ...(1)$$

and for an mth dark ring, we have

$$\frac{d_m^2}{4R} = \frac{m\lambda}{\mu}$$

Subtracting Equation (1) from (2)

$$\frac{d_m^2 - d_n^2}{4R} = \frac{(m-n)\,\lambda}{\mu}$$

So knowing d_m, d_n R and λ; μ can be found.

However, if the wavelength of light is not known, then diameter corresponding to some ring say nth dark ring is noted with air as film and then with the liquid as film, then we have

$$\frac{d_{\text{air}}^2}{4R} = n\lambda \quad ...(3)$$

and

$$\frac{d_{\text{liq}}^2}{4R} = \frac{n\lambda}{\mu} \quad ...(4)$$

Dividing Equation (3) by (4), we get

$$\frac{d_{\text{air}}^2}{d_{\text{liq}}^2} = \mu$$

$\therefore$ μ can be found by simply knowing the diameters of a ring with air as film and then with the liquid as the film.

4.28. IF GLASS PLATE UNDER THE LENS IS REPLACED BY PLANE MIRROR

In the above case we will not get interference fringes because the intensity of light reflected from the plane mirror will be so large that the different interference fringes will simply be not visible and we will get general illumination.

4.29. IF WHITE LIGHT IS USED INSTEAD OF MONOCHROMATIC LIGHT

We know that the diameter of a ring depends on the wavelength of light used. If we use white light then because it consists of seven colours, we will observe coloured rings but because of so much super-positions only few rings will be observed, however, in this case also the central fringe will be dark.

4.30. NEWTON'S RINGS IN TRANSMITTED SYSTEM

Newton's rings can be observed in transmitted system also.

However, in this case since both the lights travel from denser to rarer medium so no additional path difference is introduced between the two rays *i.e.*, for bright ring in air

$$2t = n\lambda$$

and for dark ring

$$2t = (2n-1)\,\frac{\lambda}{2}$$

and correspondingly, the diameters will be given as

$$d_n^2 = 4Rn\lambda$$

i.e.,

$$d_n \propto \sqrt{n} \text{ for bright ring}$$

and

$$d_n^2 = 4R\lambda\left(n - \frac{1}{2}\right)$$

i.e.,

$$d_n \propto \sqrt{n - \frac{1}{2}} \text{ for dark ring.}$$

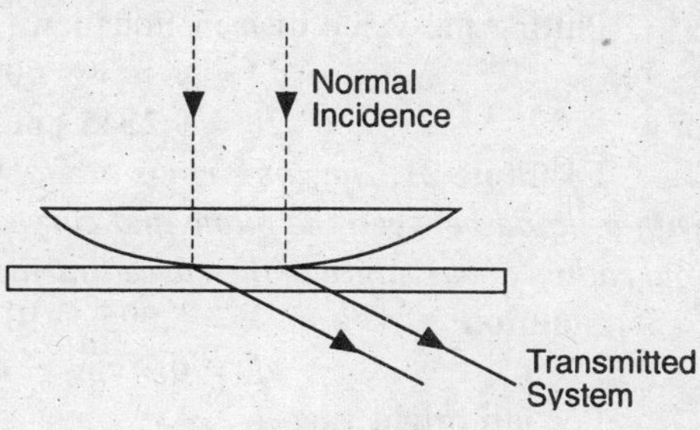

Fig. 4.19

So comparing the two we see that the reflected and transmitted systems are complementary to each other. In the reflected system, the central fringe is dark while in the transmitted system it is bright.

Example 19. *Newton's rings are observed in reflected light of wavelength*

= 5900Å. The diameter of 10th dark ring is 0.50 cm. Find the radius of curvature of the lens and the thickness of the air film.

Solution: $\lambda = 5900 \times 10^{-8}$ cm

$$d_n = d_{10} = 0.5 \text{ cm}$$

Now we have the relation

$$d_n^2 = 4Rn\lambda$$

$\therefore$

$$R = \frac{d_n^2}{4n\lambda} = \frac{0.5 \times 0.5}{4 \times 10 \times 5900 \times 10^{-8}}$$
$$= 105.93 \text{ cm}$$

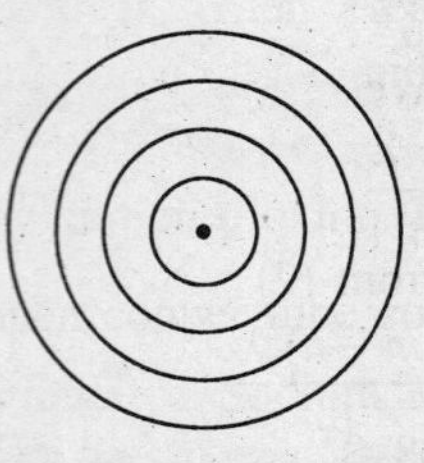

Reflected System

Fig. 4.20

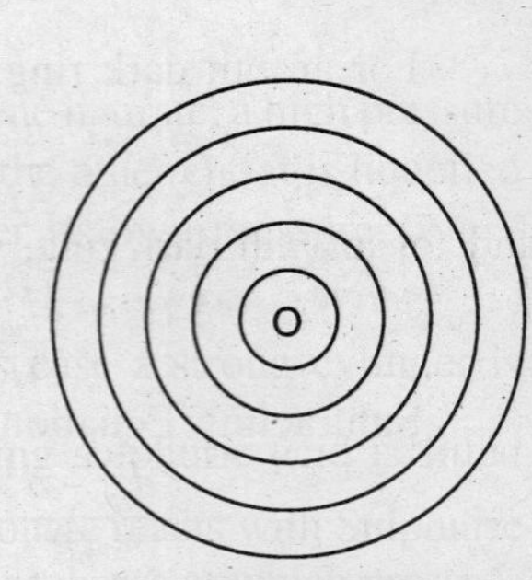

Transmitted System

Fig. 4.21

For thickness of the film at 10th dark ring we have,

$$2t = \frac{d_n^2}{4R}$$

or

$$t = \frac{d_n^2}{8R} = \frac{0.5 \times 0.5}{8 \times 105.93}$$
$$= \mathbf{0.000295 \text{ cm}}$$

Example 20. *A Newton's ring arrangement is used with a source emitting two wavelengths $\lambda_1 = 6000Å$. and $\lambda_2 = 4500$ A and it is found that the nth dark ring due to λ_1 coincides with (n + 1)th dark ring for λ_2. If the radius of curvature of the curved surface is 90 cm, find the diameter of the nth dark ring of λ_1.*

Solution: Let d_n be the diameter of nth ring corresponding to wavelength λ_1, then diameter of $(n+1)$th dark ring corresponding to wavelength λ_2 will also be d_n

$\therefore$ $$d_n^2 = 4n\,\lambda_1\,R \quad \text{...(1)}$$

and $$d_n^2 = 4\,(n+1)\,\lambda_2\,R \quad \text{...(2)}$$

Dividing Equation (2) by (1)

$$\frac{n+1}{n} = \frac{\lambda_1}{\lambda_2} = \frac{6000 \times 10^{-8}}{4500 \times 10^{-8}}$$

or $$1 + \frac{1}{n} = \frac{60}{45}$$

or $$\frac{1}{n} = \frac{60}{45} - 1 = \frac{15}{45} = \frac{1}{3}$$

or $$n = 3$$

Putting the value of n in Equation (1)

$$d_n^2 = 4 \times 3 \times 6000 \times 10^{-8} \times 90$$

or $$d_n = \mathbf{0.2545 \text{ cm.}}$$

Example 21. *Newton's rings are formed with reflected light of wavelength 5.095×10^{-5} cm with a liquid between the plane and curved surface. The diameter of the 5th bright ring is 0.3 cm and radius of curvature of the curved surface is 100 cm. Calculate the refractive index of the liquid.*

Solution: $\lambda = 5.095 \times 10^{-5}$ cm, $n = 5$

$d_5 = 0.3$ cm, $R = 100$ cm, $\mu = ?$

For nth bright ring

$$2\mu t = (2n - 1)\,\frac{\lambda}{2}$$

Also $$2t = \frac{d_n^2}{4R}$$

$$\therefore \qquad \mu = \frac{(2n-1)\,\lambda . R}{2r_n^2} \qquad \left[\because\ r_n = \frac{d_n}{2}\right]$$

$$= \frac{2 \times (5-1) \times 5.095 \times 10^{-5}}{2 \times 0.15 \times 0.15} = 1.019$$

Example 22. *Newton rings are formed with reflected light of wavelength 5.895 × 10⁻⁵ cm with a liquid between the plane and the curved surface. The diameter of the 5th dark ring is 0.3 cm and the radius of curvature of the curved surface is 100 cm. Calculate the refractive index of the liquid.*

Solution: $\lambda = 5.895 \times 10^{-5}$ cm, $R = 100$ cm, $d_5 = 0.3$ cm

For nth dark ring, we have

$$\frac{d_n^2}{4R} = \frac{n\lambda}{\mu}$$

or

$$\mu = \frac{4R}{d_n^2} \times n\lambda$$

$$= \frac{4 \times 100 \times 5 \times 5.895 \times 10^{-5}}{0.3 \times 0.3} = \mathbf{1.31.}$$

4.31. MICHELSON'S INTERFEROMETER

Principle. The amplitude of light beam from a source is divided into two parts of equal intensities by partial reflection and transmission. These beams are then sent in two directions at right angles and are brought together after they suffer reflection from plane mirrors to produce interference fringes.

Construction. It consists of two highly polished plane mirrors M_1 and M_2 and two plane parallel glass plates G_1 and G_2 of exactly the same thickness. The plate G_1 is half silvered at the back so that the incident beam is divided into a reflected and transmitted beam of equal intensity. The plates G_1 and G_2 are parallel to each other and are placed 45° to the mirrors M_1 and M_2 which are mutually at right angles to each other. The mirrors M_1 and M_2 are provided with screws on their backs, so that they can be adjusted exactly ⊥ to each other. The mirror M_1 is mounted on a carriage and can be moved exactly parallel to itself with a

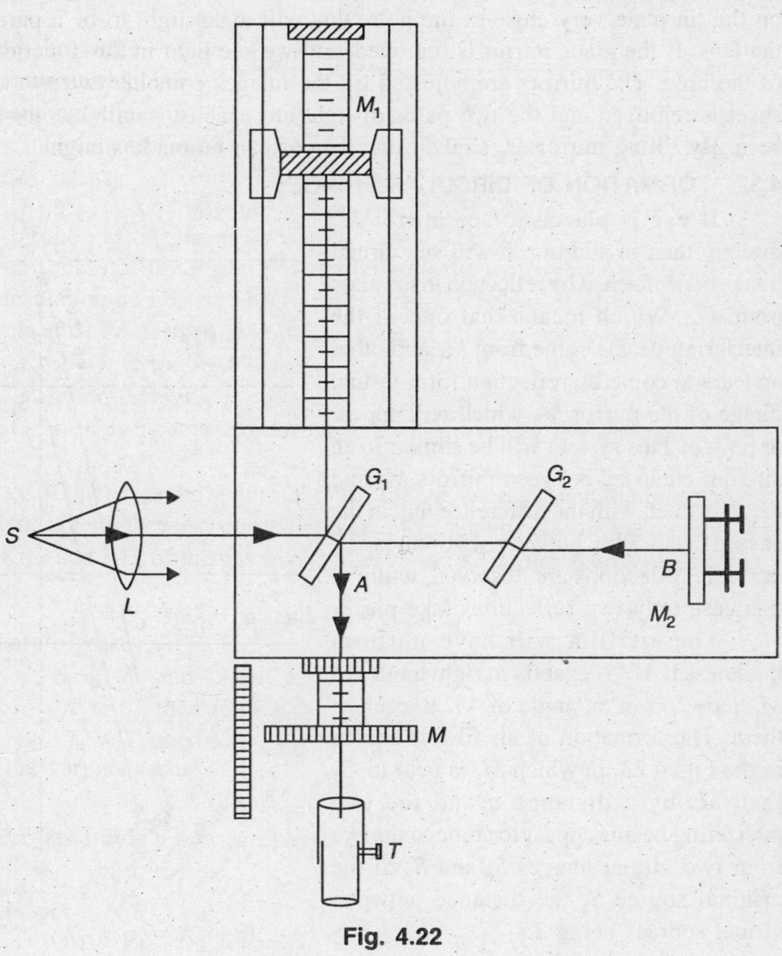

Fig. 4.22

micrometer screw filted with a graduated drum which can read a displacement ~ 10^{-5} cm. The interference bands are observed with the help of telescope T.

Working. Light from the source S is made parallel with a collimating lens L and then it is made to fall on the glass plate G_1. It is partly reflected at the back surface of G_1 along AC and partly transmitted along AB. The ray AC travels normally towards plane mirror M_1 and so it is reflected along the same path and comes out along AT. The refracted or transmitted light is received by the mirror M_2 normally, reflected along the same path and then moves along AT after reflection from the back surface of G_1. So the two beams received along AT are produced from a single source by the division of amplitude and we may get interference patterns depending upon the distances travelled by them.

It is clear from Fig. 4.22 that a ray starting from source S and suffering reflection at the mirror M_1 travels the glass plate G_1 twice where as the ray reflected from mirror M_2 travels through the glass plate G_1 only once. To compensate it a second glass plate G_2 is introduced in the path AB and is made exacty || to G_1.

Adjustment. (*i*) Dis-tances of the mirrors M_1 and M_2 are adjusted to be nearly equal from glass plate G_1.

(*ii*) In order to make the incident light parallel, a tin sheet with a small hole is placed in front of source S. The hole and the light from the source are adjusted in line with the centre of glass plates and mirror M_2. A lens is then placed between G_1 and tin sheet and a plane mirror is placed between G_1 and lens normally. The lens position is adjusted till the image of the hole falls back on the tin plate very close to the hole, this will make light to be a parallel beam when it leaves the lens. If the plane mirror is removed, and we see light in the direction AT, we see four images of the hole. The mirrors are adjusted till the images coincide two by two. If at this stage the tin sheet is removed and the two paths of light are exactly, parallel to itself, circular fringes can be seen. By tilting mirror M_2 slightly, the fringes can be made straight.

4.32. FORMATION OF CIRCULAR FRINGES

If eye is placed to see mirror M_1 directly then in addition it will see virtual image of M_2 formed by reflection in the glass plate G_1. Which means that one of the interfering beams come from M_1 and other appears to come by reflection form virtual image of the mirror M_2 which will appear to be M_2'. This system will be similar to an air film enclosed between mirrors M_1 and M_2' however, with the difference that in the case of real film between two surfaces, multiple reflections are observed, while in this case only two reflections take place.

The air film will have uniform thickness if M_2' is exactly at right angles to M_1 and G_1 is at an angle of 45° to each of them. The formation of air film is shown in the Fig. 4.23, in which M_2 is near to G_1 than M_1 by a distance d and the two interfering beams appear to come to the eye from two virtual images S_1 and S_2 of the original source S_1 the distance between virtual sources being $2d$.

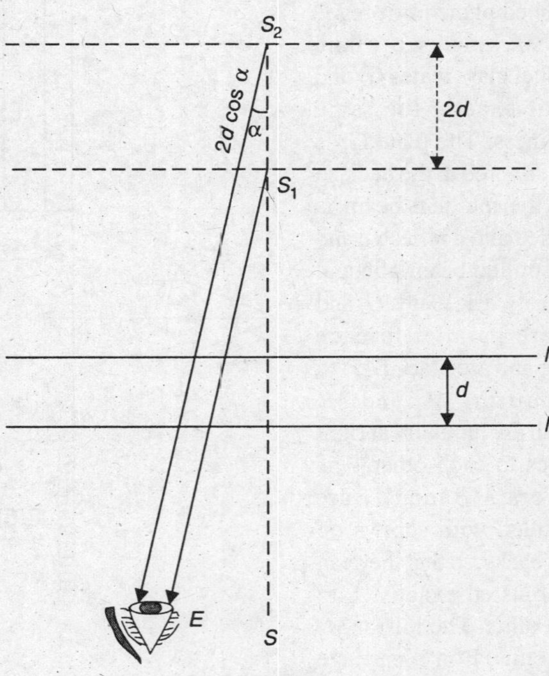

Fig. 4.23

Now since the ray AB coming after reflection from M_2' suffers reflection at the silvered surface of the glass plate G_1, so an additional path difference of $\lambda/2$ is introduced.

$\therefore$ Total path diff. $= 2d + \dfrac{\lambda}{2}$

If the eye is placed at an angle a w.r.t. the normal to the mirrors

Total path diff. $= 2d \cos \alpha + \dfrac{\lambda}{2}$

Fig. 4.24

Fig. 4.25

The condition for getting brightness will be

$$2d \cos \alpha + \frac{\lambda}{2} = n\lambda \qquad \text{...}(i)$$

where $\qquad n = 0, 1, 2, \ldots$ etc.

In case M_2 coincides with M_1 *i.e.*, the air paths AC and AB are exactly equal, the total path difference will be $\dfrac{\lambda}{2}$, *i.e.*, the case of destructive interference and we will get total darkness, as shown in Fig. 4.24.

When M_2 is moved either forward or backward parallel to itself, we observe suffi-ciently spaced circular fringes as shown in Fig. 4.25.

(*ii*) In case the path difference is increased, the distance between the fringes decreases and these fringes can be seen even for a very large path difference.

In case the two mirrors are exactly at right angles to each other, then the path difference = $2d \cos \alpha + \dfrac{\lambda}{2}$.

In the case of an extended monochromatic light source, for a given value of n, the value of a is constant and so the locus of the fringes is circle. So we see circular bright rings and circular dark rings through the telescope.

4.33. LOCALIZED FRINGES

In case M_2 and M_2' are not exactly parallel, and the path difference is very small, we observe fringes as in the case of a wedge shaped film. In this case, the locus of points of equal thickness is a straight line parallel to the edge of the wedge.

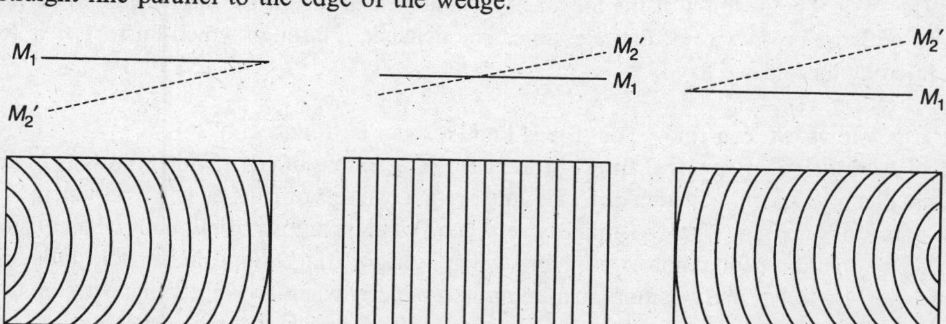

(*i*) (*ii*) (*iii*)

Fig. 4.26

For small path difference, the fringes are nearly straight but for larger path difference the fringes are generally curved with the convex surface of the film towards the edge of the wedge.

For a given value of d, the fringes observed are as shown in Fig. 4.26 (*i*) with the positions of mirror M_1 and M_2' shown. If the separation of the mirrors is reduced, the fringes will move towards left of the field of view and few fringes cross the centre each time as d is changed by $\lambda/2$. When path difference becomes zero, the fringes will look to be straight till the point when M_1 intersects

M_2'. Beyond this point, the fringes begin to show curves in the opposite direction as shown in Fig. 4.26 (*iii*).

4.34. WHITE LIGHT FRINGES

In case monochromatic light is replaced by white light, a few coloured fringes with the central dark fringe can be seen. To observe these fringes, the mirrors are slightly tilted to get localized fringes and also position of intersection of M_1 and M_2' is noted. This position is often difficult with white light, however, if monochromatic light is used, the fringes become nearly straight and after locating the position of M_1 where it intersects M_2', white light is used instead of monochromatic light when a central dark ring is seen being surrounded by about 8-10 coloured fringes. These fringes are very much useful to find zero path difference which is used for the standardization of metre.

4.35. APPLICATIONS OF MICHELSON'S INTERFEROMETER

(*i*) Determination of wavelength. Mirrors M_1 and M_2 are adjusted exactly ⊥ to each other. The position of mirror M_1 is adjusted till some bright fringe appears in the field of view of telescope. Then with the help of a micrometer screw, the number of fringes that pass through the cross-wire of the telescope, the distance is noted for some number (say *n*) of the fringes.

Let the distance through which *n* number of fringes are displaced be *d*, then

$$2d = n\lambda$$

or

$$\lambda = \frac{2d}{n}$$

with the help of micrometer, *d* is measured correctly upto 10^{-4} mm.

(*ii*) Determination of refractive index or the thickness of a transparent medium. If a transparent material is placed in the path of one of the interfering beams, the optical path in this beam will increase as light travels slowly in the medium. If μ be the refractive index and *t* the thickness of the medium, the optical path will become m*t* in the medium and the increase in optical path will be $(\mu - 1)\, t$ passing through the medium.

To find either *t* or μ, the two paths *AB* and *AC* are made exactly equal by setting up the apparatus with white light so that a straight dark fringe coincides with the cross-wire of the telescope.

The thin transparent material is now introduced between glass plate G_2 and mirror M_2 and the adjustments are made so that light passes through it normally. In this position the path difference will be $(\mu - 1)\, t$. The position of the mirror M_1 is adjusted till again the dark fringe of zero path coincides with the cross-wire of the telescope. The distance *d* through which mirror M_1 is found on the micrometer and we have

$$(\mu - 1)\, t = d$$

If *t* is known we can find μ, or if μ is known *t* can be found out.

(*iii*) Resolution of spectral lines. In the case of a source of light which consists of lights of two wavelengths λ_1 and λ_2 which differ slightly, we get two sets of fringes which on superposition give positions of maxima (where visiblity is very good) and positions of minima (where visibility is poor). The position of maxima is produced when a bright fringe from one system falls on the bright fringe of another and position of minima is produced when a bright fringe of one system falls on a dark fringe of another system. So if mirror M_1 is moved so that the position of one bright fringe is replaced by the next bright fringe in the field of view of telescope, then the number of fringes contained in it of smaller wavelength, λ_1 is one more than those for wavelength λ_2 $(\lambda_1 < \lambda_2)$. Let *d* be the distance through which the mirror is moved from position of one maxima to that of next maxima, then

$$\frac{2d}{\lambda_1} - \frac{2d}{\lambda_2} = 1$$

or

$$2d\left(\frac{\lambda_2 - \lambda_1}{\lambda_1 \lambda_2}\right) = 1$$

Since the difference between λ_1 and λ_2 is very small, so let $\lambda_1 \lambda_2 \sim \lambda^2$ and this very small difference is given by

$$\delta\lambda = \frac{\lambda^2}{2d}$$

To find d the interferometer is adjusted for circular fringes and the mirror M_1 is so adjusted that we get positions of maximum intensity. In this position maxima of λ_1 falls on maxima of λ_2. When mirror M_1 is moved, the two sets of fringes get out of step and we will get a position of minimum intensity when a bright fringe of one set falls on the dark fringe of another. Again the mirror M_1 is moved till we get again a position of maximum intensity. The distance between the two positions is noted on the micrometer screw, which gives the value of d and so $\delta\lambda$ is known from the relation

$$\delta\lambda = \frac{\lambda^2}{2d}$$

(iv) Calibration of standard metre. Michelson used this interferometer to standardize the metre. He used a monochromatic light like *Cd*, red, blue and green and noted the wavelengths of these light contained in a metre. However, to count the number of fringes that cross the field of view of telescope when the movable mirror is displaced through 1 metre is not practically possible because we can get interference fringes for distances greater than 24 cm *i.e.*, the mirror displacement should not be more than 12 cm. Also a displacement of 10 cm involves a movement of about 3 $\times$ 10^5 fringes, which are very difficult to count.

In order to overcome the above difficulties Michelson devised nine sub-standards known as etalons, each approximately twice of the other. The longest etalon was 10 cm and the shortest was $\frac{10}{108} = 0.093$ cm. An etalon con-sists of a bronze bar B (shown in Fig. 4.27) on which two front silvered plane mirrors M_1 and M_2 are fixed and the planes of these mirrors are so adjusted that they are exactly vertical and parrallel to each other. Firstly the number of wavelengths between the two mirrors of shortest etalon were counted. The two shortest etalons (I and II) were mounted in a specially designed and modified version of interferometer whose field of view covered all the four mirrors M_1, M_2, M_3 and M_4.

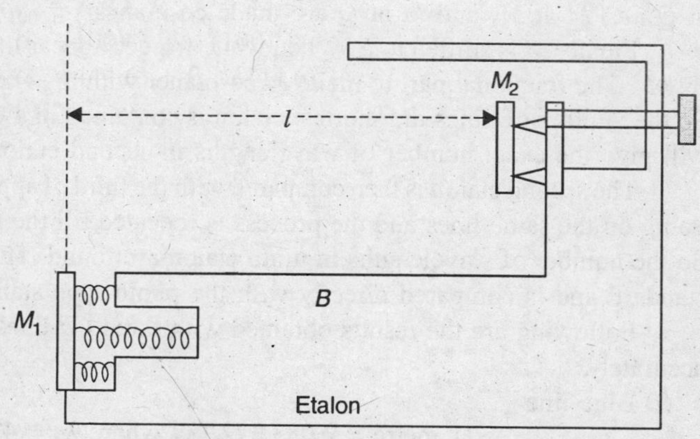

Fig. 4.27

For this purpose an extra compensating glass plage G_3 is employed as shown in Fig. 4.28.

The etalon and mirror should move parallel to each other, with their faces parallel and vertical.

Firstly all the mirrors M, M_1 and M_3 are made co-planar using white light, this will also make distances of M, M_1 and M_3 from the eye equal.

White light is replaced by one of the *Cd* lines and mirror M is moved slowly from position P to Q and the number of fringes passing through the cross-wire are counted till mirror M reaches Q when M becomes co-planer with M_2. This can be judged by the appearance of white fringes in the mirror M_2, the upper mirror of first etalon. The fraction of excess fringes can also be counted by the rotation of compensating plate G_3. So, the distance $d = M_1 M_2$ is determined in terms of wavelengths.

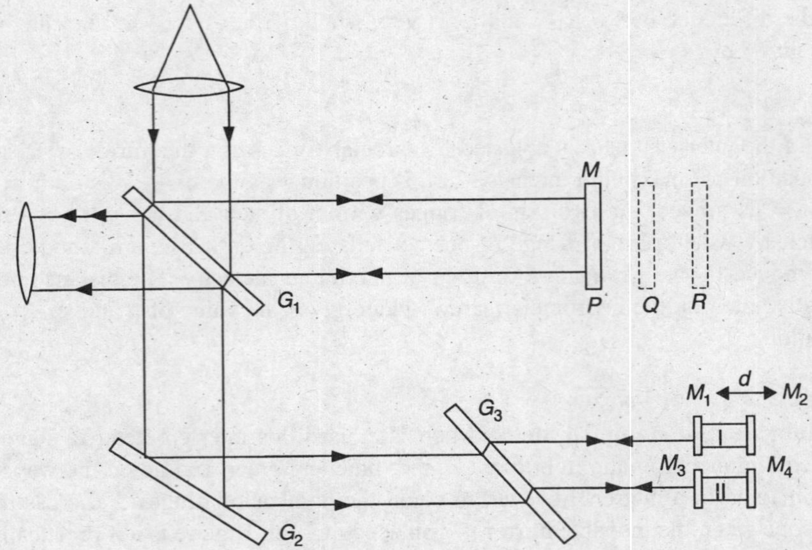

Fig. 4.28

The etalon I is then moved backward till white light fringes again appear when M_1 and M at Q *i.e.*, M at M_2 and M at Q are made co-planes.

Finally M is shifted to R. When white fringes appear in M_4 as well as in M_2 the position taken by M_2. The fractional part to make M co-planer with M_4 is determined in terms of fringes shifted. So the number of times the shortest etalon is contained in the next sub-standard is determined, that will give the exact number of wavelengths in second etalon.

The second etalon is then compared with the third of approximately double the length of second going on the same lines and the process is repeated till the ninth etalon which is 10 cm in length. So the number of wavelengths in ninth etalon are found. The 10 cm long etalon is taken as a sub-standard and is compared directly with the photo type standard metre.

Following are the results obtained with three *Cd* lines whose wavelengths have been found accurately.

(*i*) Blue line

$$1 \text{ metre} = 2{,}083{,}372.1\lambda_b \text{ where } \lambda_b = 4799.9107\text{Å}$$

(*ii*) Green line

$$1 \text{ metre} = 1{,}966{,}249.7\lambda_g \text{ where } \lambda_g = 5085.8240\text{Å}$$

(*iii*) Red line

$$1 \text{ metre} = 1{,}553{,}163.5\lambda_r \text{ where } \lambda_r = 6438.4722\text{Å}$$

Nowadays a more satisfactory result is got with green line of a single isotope of Hg^{198}. This is very sharp line having better intensity than *Cd* lines and has a wavelength of 5460.7532Å.

Example 23. *When the movable mirror of the interferometer is shifted through 0.0589 mm, 200 fringes move across the field. Calculate the wavelength of light used.*

Solution: No of fringes $n = 200$

Distance through which the mirror is moved

$$d = 0.0589 \text{ mm} = 0.00589 \text{ cm}$$

Now $$\lambda = \frac{2d}{n} = \frac{2 \times 0.00589}{200} = 5890 \times 10^{-8} \text{ cm} = \mathbf{5890\text{Å}}.$$

Example 24. *In Michelson interferometer, the scale readings for a pair of positions of maximum restrictness were found to be 0.6025 mm and 0.8970 mm. If the mean wavelength of the two components of sodium light be 5893Å. Find the difference between the wavelengths of the components.*

Solution: $\lambda = 5893 \times 10^{-8}$ cm

$d = (0.8970 - 0.6025)$ mm $= 0.2945$ mm $= 2945 \times 10^{-5}$ cm

Now from the relation

$$\lambda_2 - \lambda_1 = \frac{\lambda^2}{2d}$$

We have $$\lambda_2 - \lambda_1 = \frac{(5893 \times 10^{-8})^2}{2 \times 2945 \times 10^{-5}} = \mathbf{5.896 \times 10^{-8}\ cm.}$$

Example 25. *Michelson's Interferometer is set for the white light straight fringes. When a mica sheet of thickness 0.005 cm is put in front of the fixed mirror, then in order to bring back the coloured fringes to the original position, the movable mirror is moved by 0.0025 cm. Calculate the refractive index of mica.*

Solution: Using the relation

$$(\mu - 1)\, t = d$$

$\mu = ?$ $\quad t = 0.005$ cm

Distance through which mirror is moved

$d = 0.0025$ cm

$$\therefore \quad \mu - 1 = = \frac{d}{t} = \frac{0.0025}{0.005} = 0.50$$

$$\therefore \quad m = \mathbf{1.50.}$$

4.36. THEORY OF COHERENCE

We have been considering monochromatic waves, the waves that exist *everywhere at all times*. The superposition of waves of same frequencies give *interference patterns*. While that due to waves of different frequencies give *beat phenomenon.*

In the case of a monochromatic plane wave, **the amplitude is constant while phase varies linearly with time**. It means that the phase difference between any two points is *time dependent*. It may be noted that a monochromatic wave can not be achieved in practice, it is only theoretically possible *i.e.*, a monochromic wave is an *abstract concept*. Also, the quantity of interest in a wave field is amplitude which is a function of both *space and time*. For points very close to each other, the *amplitudes are strongly co-related*; while for points which are very far away from each other, the amplitudes are *weakly co-related or unrelated.*

Unlike amplitudes, the *coherence degree* can be experimentally measured. If the degree of coherence is unity, we have *perfect coherence* while for zero degree of coherence, we have *complete incoherence*. Again *complete coherence and incoherence* are possible theoretically only while in practice we have coherence between 0 and 1 and is refereed to as *partially coherent light.*

4.37. SPATIAL COHERENCE

As stated above, the amplitude of a wave depends both on *space and time*. If we consider amplitudes of a wave at two points on a plane normal to the direction of propagation and study the co-relation, we are dealing with *spatial coherence*. It may be noted that the spatial coherence is governed by the *size of the source.*

4.38. TEMPORAL COHERENCE

If we study the co-relation between two points in the direction of wave propagation and if the wave is a monochromatic wave of *circular frequency* ω and the points are separated by a distance z along direction of propagation, phase difference between these two points will be $\frac{\omega z}{c}$ where c is the velocity of the wave. This phase difference is independent of time. Again, on a monochromatic wave, any two points are co-related. A wave due to superposition of a large number of monochromatic waves of slightly different frequencies is localized in space and a co-relation exists between

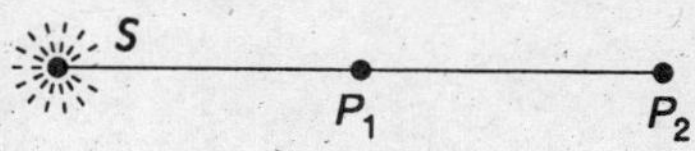

Fig. 4.29

any two points say P_1 and P_2 in space. These co-relations are interpreted in terms of *time coherence or temporal coherence.*

Temporal coherence is governed by the monochromaticity of the wave. If the line width of the wave is $\Delta\nu$ in frequency, then coherence exists over a distance $c/\Delta\nu$ and this called the *coherence length* of the wave.

4.39. PURITY OF A SPECTRAL LINE

Finite purity, Q of a spectral line is defined on the basis of quantum theory as

$$Q \simeq \frac{\lambda}{\Delta\lambda} \quad \text{...(1)}$$

The purity of Cd red light $\simeq 10^6$ and that of sodium yellow light $\simeq 10^5$.

If the radiations emitted from a source of frequency ν has a frequency spread $\Delta\nu$ and its coherence time is τ, then the fading of amplitude is explained by putting the difference of number of oscillations of frequency ν and $\nu + \Delta\nu$ in *time t by unity.*

i.e., $$\tau(\nu + \Delta\nu) - \tau\nu = 1$$

$\therefore$ $$\tau\,\Delta\nu = 1 \quad \text{...(2)}$$

also $$\nu = \frac{c}{\lambda}$$

$\therefore$ $$\Delta\nu = \frac{c\Delta\lambda}{\lambda^2} \quad \text{(in magnitude)}$$

Putting in Equation (2) $$\tau \cdot \frac{c\Delta\lambda}{\lambda^2} = 1 \quad \text{or} \quad \Delta\lambda \simeq \frac{\lambda^2}{cz} \quad \text{...(3)}$$

Also coherence length L is given by

$$L = \tau c$$

$\therefore$ Equation (3) may be written as

$$\Delta\lambda = \frac{\lambda^2}{L}$$

$\therefore$ Coherence length $$L = \frac{\lambda^2}{\Delta\lambda} \quad \text{...(4)}$$

which may be written in terms of purity Q of spectral line as

$$L = \lambda \cdot \frac{\lambda}{\Delta\lambda} \simeq \lambda Q \quad \text{...(5)}$$

i.e., Coherence length $\cong$ wave length of a spectral line $\times$ finite purity.

4.40. RELATION BETWEEN COHERENCE LENGTH AND LINE WIDTH

We have $$c = \nu\lambda$$

$\therefore$ $$\lambda = \frac{c}{\nu}$$

Differentiating $$\Delta\lambda = -c\,\frac{\Delta\nu}{\nu^2} = -c \cdot \frac{\Delta\nu}{c^2}\,\lambda^2$$

$$= -\frac{\lambda^2\,\Delta\nu}{c} \quad \text{...(1)}$$

Negative sign shows that change in $\Delta\lambda$ is opposite to change in frequency.

Writing the above expression in magnitude.

$$\Delta\lambda = \lambda^2\,\frac{\Delta\nu}{c}$$

$$= \frac{\lambda^2}{c} \cdot \frac{1}{\Delta t} \quad \left[\because \Delta\nu = \frac{1}{\Delta t}\right]$$

$\therefore$ $$\Delta t = \frac{\lambda^2}{c.\Delta\lambda} \quad \text{...(2)}$$

and coherence length is given by

$$\Delta l = c \cdot \Delta t$$

$$= c \cdot \frac{\lambda^2}{c \cdot \Delta\lambda}$$

or

$$\Delta l = \frac{\lambda^2}{\Delta\lambda} \quad \text{...(3)}$$

which gives us a relation between *coherence length* Δl *and line width* $\Delta\lambda$.

Example 26. *A nearly monochromatic source s of mean wavelength 5893Å illuminates a circular hole of diameter 0.1 mm in Young's double slit experiment. The distance between the source S and the plane of the double slit is 2 m. Calculate the separation between the two slits at which the fringes first disappear.*

Solution: We have the relation

$$a_0 = \frac{1.22\lambda_0 R}{D}$$

In this case $\lambda_0 = 5893\text{Å}$

$$R = 2 \text{ m},\ D = 0.1 \text{ mm} = 10^{-4} \text{ m}$$

$$a_0 = \frac{1.22 \times 5893 \times 10^{-10} \times 2}{10^{-4}} = \mathbf{1.437\ cm.}$$

Example 27. *An optical filter has a line width of 1.5 nm and mean wavelength 550 nm. With white light incident on the filter, calculate (i) coherence length (ii) number of wavelengths in wave train.*

Solution: In this case $\Delta\lambda = 1.5 \text{ nm} = 1.5 \times 10^{-9} \text{ m}$

$$\lambda = 550 \text{ nm} = 550 \times 10^{-9} \text{ m}$$

(*i*) Coherence length $\Delta l = \frac{\lambda^2}{\Delta\lambda} = \frac{(550 \times 10^{-9})^2}{1.5 \times 10^{-9}}$

$$= \mathbf{2.017 \times 10^{-4}\ m}$$

(*ii*) Number of waves in wave train

$$n = \frac{\Delta l}{\lambda} = \frac{2.017 \times 10^{-4}}{550 \times 10^{-9}} = \mathbf{367\ (approx).}$$

Example 28. *Cadmium led light of mean wave length 6438Å is used in Michelson interferometer line width of the source is 0.013Å for the initial setting of the interferometer for zero optical path difference. One of the mirrors is slowly moved until the fringes first disappear. Calculate the distance by which the mirror is moved. Also calculate the number of wavelengths corresponding to this distance.*

Solution: In this case

$$\lambda = 6438\text{Å} = 6438 \times 10^{-10} \text{ m}$$

$$\Delta\lambda = 0.013\text{Å} = 0.013 \times 10^{-10} \text{ m}$$

Coherence length $\Delta l = \frac{\lambda^2}{\Delta\lambda} = \frac{(6438 \times 10^{-10})^2}{0.013 \times 10^{-10}} = 0.319 \text{ m}$

Now the condition for first disappearance of fringes is that the optical path must exceed the coherence length Δl.

$$\therefore \quad 2d = 0.319 \text{ m}$$

or

$$d = 0.1595 \text{ m} = 15.95 \text{ cm}$$

which means that the mirror has to be moved through a distance 15.95 cm

Also $d = n\lambda$

$$\therefore \quad 0.1595 = n \times 6438 \times 10^{-10}$$

$$\therefore \quad n = \mathbf{24770\ waves.}$$

Example 29. *A quasi monochromatic source emits radiations of mean wavelength* $\lambda = 5461\text{Å}$ *and has a band width* $\Delta\nu = 10^9$ *Hz. Calculate (i) coherence time (ii) coherence length (iii) frequency stability.*

Solution: In this case $\lambda = 5461\text{Å} = 5461 \times 10^{-10}$ m

$$\Delta\nu = 10^9 \text{ Hz}$$

∴ (*i*) $$\Delta t = \frac{1}{\Delta\nu} = \frac{1}{10^9} = 10^{-9} \text{ sec.}$$

(*ii*) $$\Delta l = c \,.\, \Delta t$$
$$= 3 \times 10^8 \times 10^{-9} = \mathbf{0.3\ m}$$

(*iii*) We have $$\nu = \frac{c}{\lambda} = \frac{3\times10^8}{5461\times10^{-10}}$$
$$= 5.5 \times 10^{14} \text{ Hz}$$

∴ Frequency stability $$\frac{\Delta\nu}{\nu} = \frac{10^9}{5.5\times10^{14}} = \mathbf{1.82 \times 10^{-6}}.$$

4.41. HOLOGRAPHY

Though there have been great changes, the basic principles of the photography have not changed for the past many years. Essentially, the photographic process consists of recording an illuminated three dimensional scene on a two dimensional light sensitive surface. So the light reflected from the objects in the scene is focussed on the light sensitive surface by some kind of image forming device, which can be a complex series of lenses as in the case of expensive cameras or may be a pinhole in an opaque screen.

A different technique of photography invented in 1947 by Dennis Gabor of the Imperial College of Science and Technology, London, which can also be called *photography by wavefront reconstruction*, does not record an image of the object being photographed but rather records the reflected light waves themselves. The photographic record is called *hologram*; it bears no resembles to the original object, but contains in some kind of optical code all the information about the object that would be contained in an ordinary photograph and in addition some more information which cannot be recorded by ordinary photography.

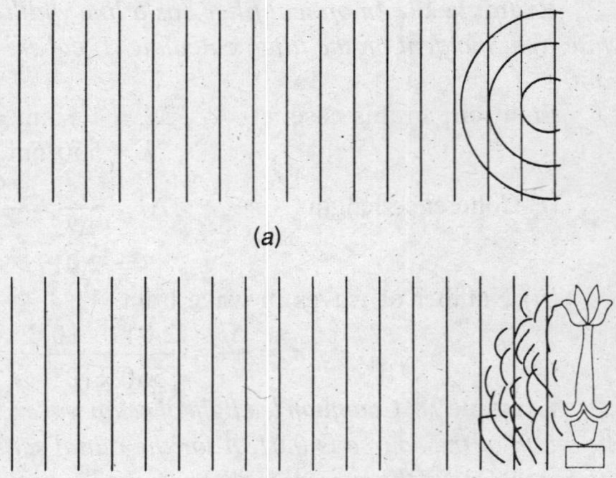

(*a*)

(*b*)

Fig. 4.30

The above mentioned process is called **Holography**. Though this process was known in 1947, the actual technique of Holorgaphy could be had only after the invention of LASERS in 1960 because previously it was difficult to get true coherent sources. So with lasers in 1962, many scientists got very high quality Holograms and a new chapter of three dimensional photography started.

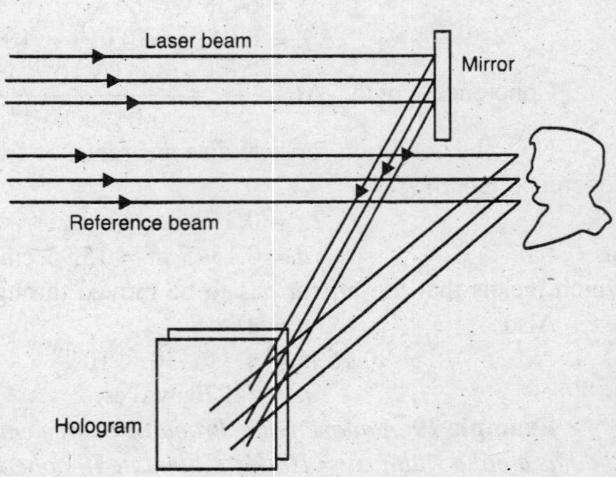

Fig. 4.31 (*a*)

Principle of Holography. We know that light waves reflected by an object are characterised by their amplitude and phase. In the case of a point source, the reflected wave fronts are spherical and concentric around the point of origin Fig. 4.30 (*a*).

However, if the reflecting object is not a simple point but a complex object, we may consider it to consist of number of points and the resulting wave pattern reflected from the surface of the objects can be taken as sum of many such sets of spherical waves, each set concentric about its point of origin [Fig. 4.30 (*b*)]. The exact form of the wave patterns from a very broad source will be quite complex. The basic problem of wave front reconstruction photography is to record this complex wave pattern at a given plane at some instant of time. Such a record may be just as 'freezing' of the wave pattern; the pattern remains frozen till the reconstruction process. When the waves are 'read out' of the recording medium. An important consideration for capturing of the wave pattern completely is that both the amplitude and the phase of the waves must be recorded at each point on the recording surface.

For recording the wave pattern, only photographic plate is not sufficient, since phtographic plate is sensitive to intensity *i.e.*, the square of amplitude and the phase information cannot be recorded.

So the problem is to have such apparatus that can convert these phase relations into intensity variations which can then be recorded photographically. Gabor solved the problem by superposing to this wave another known wave thereby producing interference phenomenon. Due to interference the intensity will increase at the points where phases are in concordance and decreases where they are opposite.

The experimental arrangement Fig. 4.31 (*a*) consists of illuminating the object by Laser beam (because of its uniform intensity at large distances and very good coherent characteristic laser beam sources are used). Each point on the object reflects light to the total photographic plate or alternately each point of the photographic plate gets light form the entire object. A part of the laser beam reflected by a plane mirror and produces another beam called **reference beam**. This reference beam interferes with the light reflected from the object and produces an interference pattern called **Hologram on the photographic plate.**

In the reproduction [Fig. 4.31 (*b*)] process, the hologram is illuminated by a collimated laser beam and two images are produced by diffracted waves coming out of the Hologram. One of them, projected towards the illuminating sources, seem to originate from a virtual object located at position where orignal object was present. These waves produce virtual image and the other wave produces a real image, that can be photographed directly, without using a lens *i.e.*, by simply placing a photographic plate at the position of the image [Fig. 4.31 (*b*)].

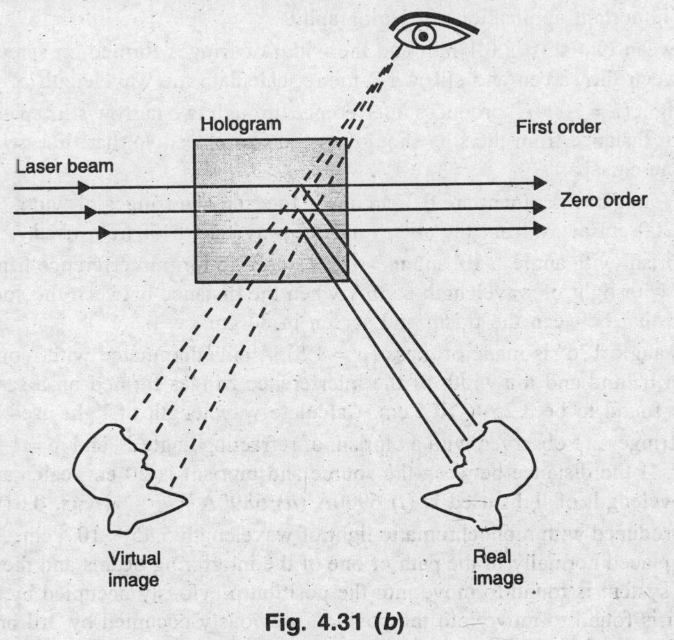

Fig. 4.31 (*b*)

QUESTIONS AND PROBLEMS

1. Explain the concepts of light. What are the modern ideas about the nature of light?
2. Explain the phenomenon of interference of light. What are the conditions necessary to get clear and distinct interference fringes?
3. What is interference? Can interference be observed with two independent sources of light?
4. Give the theory and experimental details of biprism method for finding wavelength of light. Explain what happens when (*i*) White light is used (*ii*) the width of the slit is increased continuously.
5. Explain the formation of colours in thin films. What are the parameters on which these colours depend?
6. Explain the formation of colours in thin films and show that films which appears bright in reflected light appear dark in transmitted light.
7. Give the theory and formation of wedge shaped films. How these can be used to find the wavelength of light used?
8. Give the theory of the Newton's rings. How rings can be used to find
   (*i*) wavelength of light?
   (*ii*) refractive index of a liquid?
9. Give the formulation of Newton's rings. How their formation differs in reflected and transmitted system? What happens if monochromatic source is replaced by white light?
10. What are coherent sources? Why are they necessary for interference fringes? Give a method of their formation.
11. Describe the principle, construction and some of the applications of Michelson's interferometer.
12. Describe Michelson's interferometer and explain the formation of fringes in it. How this interferometer is used for the standardisation of the metre?
13. Account for the colours observed when a little oil spreads over water. Give the theory of the formation of these colours. A parallel beam of light λ = 5890Å is incident on a thin glass plate (μ = 1.5) such that the angle of refraction into plate is 60°. Calculate the smallest thickness of the plate which will make it appear dark by reflection. **[Ans.** 4000Å]
14. Discuss the formation of colours in thin transparent films due to multiple reflection and show that with monochromatic light the interference patterns of the reflected and the transmitted light are complementary.
15. Give simple principles of Holography. What is the importance of coherence in this phenomenon? State some of the important applications of Holography.
16. Distance between two slits is 0.1 mm and the width of fringes formed on screen is 0.6 mm. If the distance between the screen and slit is 1.2 metre, calculate the wavelength of light. **[Ans.** 5000Å]
17. A source of light (λ = 7000Å) produces interference through two narrow slits spaced 0.01 cm distance apart. At what distance from the slits should a screen be placed so that first two interference bands are spaced one cm apart? **[Ans.** 142.86 cm]
18. Two slits in Young's experiment are 0.2 cm apart. Interference fringes of width 0.2 cm are observed on a screen 200 cm away from the slits. Find the wavelength of light used. **[Ans.** 5100Å]
19. A Fresnel biprism with angle 1°80′ and m = 1.52 is used to form interference fringes. Find the width of the fringe with light of wavelength 6536Å when the distance between the sources and the prism is 20 cm and that between the prism and screen in 80 cm. **[Ans.** 0.0149 cm]
20. A biprism of angle 176° is made of glass (μ = 1.5). A slit illuminated with monochromatic light is placed 20 cm behind and the width of the interference fringes formed on a screen 50 cm in front of biprism is found to be 8.25×10^{-3} cm. Calculate wavelength of light used. **[Ans.** 57556Å]
21. Interference fringes are observed with a biprism of refracting angle 1° and m = 1.5 on a screen 20 cm away from it. If the distance between the source and biprism is 20 cm, calculate the fringe width, when the wavelength of light used is (*i*) 6900Å (*ii*) 5890Å. **[Ans.** 0.0198 cm, 0.0169 cm]
22. Fringes are produced with monochromatic light of wavelength 5.45×10^{-5} cm. A thin plate of glass of μ = 1.5 is placed normally in the path of one of the interfering beams and the central bright band of the fringe system is found to move into the position previously occupied by 3rd brigth band the fringe system is found to move into the position previously occupied by 3rd bright band form the centre, calculate thickness of the glass plate. **[Ans.** 32.7×10^{-5} cm]

23. A soap film 5×10^{-3} cm thick is viewed at an angle of 35° to the normal. Find the wavelength of light in the visible spectrum which will be absent from reflected light m of soap film = 1.33. (Visible spectrum is in the range of 3900Å to 7800Å). **[Ans.** 6000Å, 4000Å]

24. A wedge shaped air film having an angle of 45° is illuminated by monochromatic light and fringes are observed normally. If the fringe width is 0.12 cm, calculate wavelength of light used. **[Ans.** 5238Å]

25. Monochromatic light is incident normally on a thin wedge shaped film of a transparent material ($\mu = 1.42$). Interference fringes are observed when viewed in reflected light with fringe width 0.26 mm. If $\lambda = 5900$Å. Find the angle of the wedge. **[Ans.** 7.9×10^{-14} radians]

26. A convex lens is placed on a slab of plane glass and illuminated by monochrômatic light. The diameter of the 10th dark ring is found to be 0.453 cm. Find the wavelength of light used if radius of curvature of the lower surface of the lens is 70 cm. **[Ans.** 7320Å]

27. In a Newton's ring experiment, the radii of the nth and $(n + 5)$th dark rings were found to be 6 mm and 8 mm respectively. Calculate the wavelength of the light used, if the radius of curvature of the curved surface is 10 meters. **[Ans.** 5600Å]

28. Newton's rings are formed by the light reflected normally from a plane convex lens and plane glass plate with a liquid between them. The diameter of the nth ring is 2.18 mm and that of $(n + 10)$th is 4.51 mm. Calculate refractive index of the liquid. (Given $\lambda = 5893$Å and radius of curvature of lens = 90 cm). **[Ans.** 1.7]

29. As the movable mirror of Michelson's interferometer is moved back 1 mm, 3663 fringes grow out from the centre of the pattern. Calculate the wavelength of light used. **[Ans.** 5460Å]

30. Calculate the distance through which the mirror of Michelson's interferometer has to be moved between two consecutive positions of maximum distinctness of D_1 and D_2 of sodium lines. (λ of D_2 line = 5890Å and λ of D_1 line = 5896Å. **[Ans.** 0.2894 m.m.]

31. In moving the mirror M_1 of Michelson interferometer through a distance of 0.4220 mm, 1500 fringes are counted, calculate the wavelength of light used. **[Ans.** 5626.7Å]

CHAPTER

5

DIFFRACTION

5.1. INTRODUCTION

Newton's corpuscular theory was able to explain the rectilinear propagation of light. However, it was observed that some bending of light takes place around the corners of some obstacles placed in the path of light. In a normal case though light looks to travel in a straight line because of its very short wavelength (of the order of 6×10^{-5} cm) unlike sound waves which have wavelengths of the order of 10^7 times that of light and bending can be observed easily. This effect was first observed by Grimaldi of Italy (1665). He observed that with a small source of light the shadow of an obstacle was larger than that given by geometrical construction. It was also observed that the shadow was not well defined but consisted of some coloured fringes near the outer sides of the shadow. These bands outside the geometrical shadow were observed to become closer till they faded into uniform illumination.

This phenomenon of bending of light round the corners and spreading of light waves into the geometrical shadow of an obstacle placed in the path of light is called diffraction.

Newton also observed this phenomenon and gave some explanation on the basis of corpuscular theory. Thomal Young later tried to give explanation from interference phenomenon between the direct light near the edge of the obstacle and the light reflected from edge at grazing angle. Fresnel, later on, observed that this phenomenon did not depend on the nature or shape of the obstacle *i.e.*, a round edge or a sharp edge gave the same effect and so he did not agree with Young's idea of reflection. He proved experimentally that the phenomenon was due to mutual interference between the secondary wavelets from the some wave front. It may be explained in another way as well that it is the interference between the light coming from different parts of any aperture.

Huygens had already predicted that these secondary wavelets could be considered as secondary sources whose wavetrains could interfere with each other. But in this case since large number of secondary wavelets are to be considered to interfere, so mathematical treatment is quite difficult.

Modern ideas about the diffraction phenomenon are based on the Huygen's principle according to which every point in any surface through which light passes may be considered as a secondary source and subsequent effects beyond the surface may be considered to be due to the action of secondary wavelets alone.

The validity of the above explanation is got by experimental facts in which light from a small aperture is found to spread in large directions surrounding the aperture.

Basically, the diffraction phenomenon is divided into two main groups:

(*i*) Fresnel Diffraction. (*ii*) Fraunhoffer Diffraction.

5.2. FRESNEL DIFFRACTION

In this phenomenon the interference is considered to take place from different parts of an aperture when either the source or screen or both are at finite distance from the obstacle.

5.3. FRESNEL'S FORMATION OF WAVEFRONTS

Consider a source of light S which sends spherical waves in all directions in the medium surrounding it. Let AB be the position of spherical wave front at any instant of time.

Then from Huygen's Principle.

(*a*) Every point on this wave front *AB* will become a source of secondary waves spreading wavelets in all directions. At any other instant, *A'B'* may be considered as secondary wavefront. The effect at *O* on the screen may be considered due to the resultant secondary wavelets reaching *O* from various points of the wave-front, the disturbance from *P* will reach *O* first and from other points a bit later.

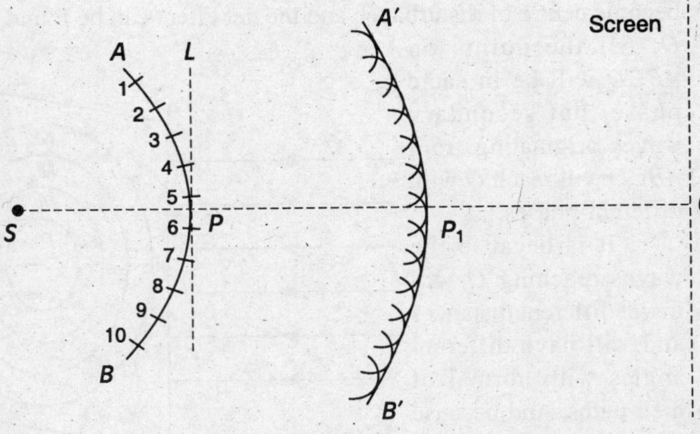

Fig. 5.1

(*b*) The amplitude of the disturbance will be maximum along the normal PP_1 to the wavefront, but it will decrease with the increase of angle of inclination θ *i.e.*, the amplitude of disturbance varies with the angle, in fact the effect of disturbance will be proportional to (1 + cos θ). It is maximum along *PO*, half along *PL* and zero along *PS* which means that backward propagation of energy is not possible.

(*c*) The total effect at any point is due to the combined effect of all amplitudes of the waves reaching that point.

5.4. FRAUNHOFFER DIFFRACTION

It consists of all those cases in which either the source or screen or both are at infinite distance from the obstacle. This may be achieved by using two convex lenses, one to make the light from the source parallel before it falls on the aperture and the other to focus the light after diffraction on the screen. This arrangement in fact makes the source and the screen to be at infinity.

The incident wavefront will be a plane one and the secondary wavelets which originate from the unblocked portions of the wavefront are in the same phase at every point in the plane of aperture. So we get the diffraction because of the interference between parallel rays which are brought to focus with a convex lens.

5.5. RECTILINEAR PROPAGATION OF LIGHT (HALF PERIOD ZONES OR ELEMENTS)

Consider a point source *S* of monochromatic light which is at sufficiently large distance, so that for all practical purposes, we may consider plane wavefronts only.

We can find the resultant amplitude at a distant point *O* due to an aperture limiting the exposed part of the wave-front *XY*. If *O* is joined with *S*, the point *P* on the wave front is called the pole of the wavefront w.r.t. *O*.

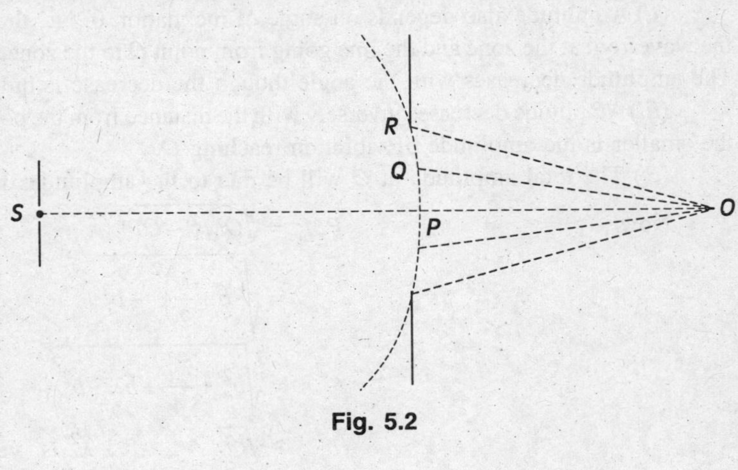

Fig. 5.2

To find the effect at O. Let $ABCD$ be the position of the plane wave front of the monochromatic light of wavelength λ at any time. To find its effect at O, we consider every point on $ABCD$ to become centre of disturbance and the net effect can be found due to all these disturbances reaching O. All the points on $ABCD$ will be in same phase, but secondary waves originating from $ABCD$ will reach O with different phases.

It is because the waves reaching O will travel different distances and will have different angles with normal of their paths. And because of unequal amplitudes the interference will be partial.

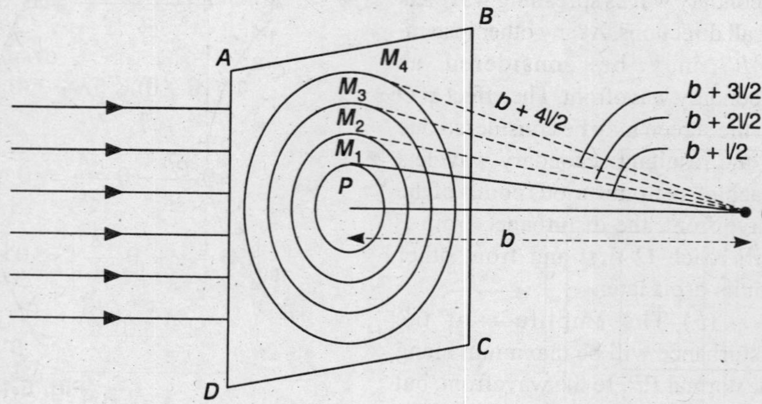

Fig. 5.3

To find the net effect at O, wavefront $ABCD$ may be divided into a number of half period zones. To get these zones or elements, drop a $\perp$ from O to P and net $OP = b$, with O as centre and radii

$$OM_1 = \left(b+\frac{\lambda}{2}\right) \qquad OM_2 = \left(b+\frac{2\lambda}{2}\right)$$

$$OM_3 = \left(b+\frac{3\lambda}{2}\right) \qquad OM_4 = \left(b+\frac{4\lambda}{2}\right) \ldots\ldots$$

Construct a series of concentric spheres that cut the wavefront $ABCD$ in circles of radii PM_1, $PM_2, PM_3, \ldots$ as shown in Fig. 5.3. The areas enclosed between PM_1, M_1M_2, M_2M_3 etc. ... are known as half period elements or half period zones. The area corresponding to PM_1, M_1M_2, M_2M_3 etc. are called first, 2nd, 3rd etc. half period elements respectively.

We can find the resultant effect of the whole wavefront at O by finding the resultant effect of these half period elements by taking into considerations.

(*i*) The amplitude of the disturbance reaching O is directly proportional to area, greater the area, more is the disturbance.

(*ii*) Amplitude also depends on angle of inclination θ, *i.e.*, the angle between the normal to the wavefront at the zone and the line going from point O to the zone. This is called obliquity factor. The amplitude decreases with the angle though the decrease is quite slow but is steady.

(*iii*) Amplitude decreases inversely with the distance from the point O. The more is the distance, the smaller is the amplitude of vibration reaching O.

(*iv*) The total amplitude at O will be due to the amplitudes from the different zones.

Now
$$PM_1 = \sqrt{OM_1^2 - OP^2}$$
$$= \sqrt{\left(b+\frac{\lambda}{2}\right)^2 - b^2}$$
$$= \sqrt{b^2 + \frac{\lambda^2}{4} + b\lambda - b^2}$$
$$\approx \sqrt{b\lambda} \qquad [\because \lambda^2 \text{ is very small as compared to } b\lambda]$$

Similarly,
$$PM_2 = \sqrt{OM_2^2 - OP^2}$$
$$= \sqrt{(b+\lambda)^2 - b^2}$$
$$= \sqrt{b^2 + \lambda^2 + 2b\lambda - b^2}$$
$$= \sqrt{2b\lambda} \text{ approx}$$
....

....

$$PM_{n-1} = \sqrt{\left(b+(n-1)\frac{\lambda}{2}\right) - b^2} = \sqrt{(n-1)b\lambda}$$
$$PM_n = \left(b + n\frac{\lambda}{2}\right)^2 - b^2 = nb\lambda$$

$\therefore$ Area of 1st half period zone $= PM_1^2 = \pi b\lambda$

Area of 1st two half period zone $= \pi PM_2^2 = 2\pi b\lambda$

Area of 2nd half period zone $= 2\pi b\lambda - \pi b\lambda = \pi b\lambda$

Area of nth half period zone

= Area of n half period zones – Area of (n – 1) half period zones
$$= \pi PM_n^2 - \pi PM_{n-1}^2$$
$$= \pi\,(nb\lambda - \overline{n-1}\,b\lambda) = nb\lambda$$

which shows that area of all the half period zones are approximately equal.

Now since the distance of these half period elements from O are different, so the secondary waves originating from these elements will reach O in different phases. Since path difference between each consecutive half period zones is $\lambda/2$, phase difference between successive zones will be π.

Now we have to find the resultant amplitude of n S.H.M. all in the same straight line, with slowly decreasing amplitudes and phase difference π for each element.

Let m_1, m_2, m_3, ... be the amplitudes at O due to the first, 2nd 3rd etc. half period elements respectively.

The resultant amplitude at O due to whole wavefront is given by
$$A = m_1 - m_2 + m_3 - m_4 + m_5 - m_6 + \ldots\ldots$$
$$= \frac{m_1}{2} + \left(\frac{m_1}{2} + \frac{m_3}{2} - m_2\right) + \left(\frac{m_3}{2} + \frac{m_5}{2} - m_4\right) + \ldots\ldots \frac{m_n}{2}$$

Now since with the increase of order of the zone, m_1, m_2, m_3, ... go on decreasing regularly *i.e.*, m_1 is slightly greater than m_2, m_2 slightly greater than m_3 ... m_{n-1} slightly greater than m_n.

So we can take to a first approximation
$$m_2 = \frac{m_1 + m_3}{2}, \qquad m_4 = \frac{m_3 + m_5}{2} \ldots$$

So that the various terms in brackets cancel and we have the resultant amplitude
$$A = \frac{m_1}{2} + \frac{m_n}{2}$$

So the resultant amplitude at O is the sum of the half displacement due to first half displacement due to the last zone. If we consider n to be quite large, then the effect due to nth zone is almost negligible and the total effect of whole wavefront reduces to $m_1/2$.

If I be the intensity and since $I \propto A^2$, so we have $I = m_1^2/4$ *i.e.*, intensity at O due to the whole wavefront is 1/4 of the intensity due to the first half period element.

If we plot a curve between Intensity and exposed half period zones the graph is shown in Fig. 5.4.

As shown in Fig. 5.4 since $\lambda \cong 6 \times 10^{-5}$ cm is very small and $\pi b\lambda$ is also very small, for simplicity, let $b = 30$ cm.

$$\therefore \qquad \pi\, b\, \lambda = 3.142 \times 30 \times 6 \times 10^{-3} \text{ cm}$$
$$= 5.65 \times 10^{-3} \text{ cm}$$

which is very small, so even a small obstacle placed at P will cut off considerable number of half period elements and the resultant intensity at O being 1/4 of the intensity of the first exposed zone, so it will be negligibly small. This explains why propagation of light is considered to be nearly rectilinear, so rectilinear propagation of light is due to the short wavelength of light waves.

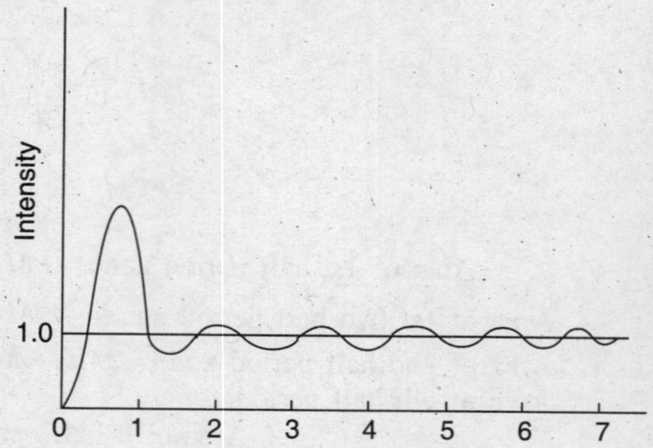

Fig. 5.4

Since sound waves of ordinary hearing range have wavelengths approximately 10^6 times larger than those of light waves, so the area of the Ist half period zone of sound waves will be very large to stop sound waves, an obstacle which should be 10^6 times the size required for light waves will stop the first half period zone. Thus for an ordinary obstacle sound waves do not appear to travel in straight line, which agrees with the wave theory for its explanation.

5.6. DIFFERENCE BETWEEN INTERFERENCE AND DIFFRACTION

1. In the interference phenomenon, the superposition is due to two separate wavefronts originating from two coherent sources, while in the diffraction phenomenon the superposition is due to secondary wavelets originating from different parts of the same wavefront.

2. In the interference, the regions of minimum intensity are perfectly dark while in the diffraction, they are not perfectly dark.

3. The width of the fringes in the interference phenomenon is normally equal and uniform while in diffraction, the width between the fringes is never equal.

4. In the interference phenomenon, all the positions of maxima are of the same intensity, but in the diffraction phenomenon they are of varying intensity.

5.7. DIFFRACTION AT A STRAIGHT EDGE

Let A be a sharp and straight edge of some opaque obstacle AB (tip of a needle or edge of a blade) and S be a narrow slit placed parallel to each other and ⊥ to the plane of paper. Let, the slit be illuminated by a monochromatic source of lights of wavelength λ. Join SA and produce it to meet XY placed ⊥ to the plane of paper at O. Let us consider the effect of light at O.

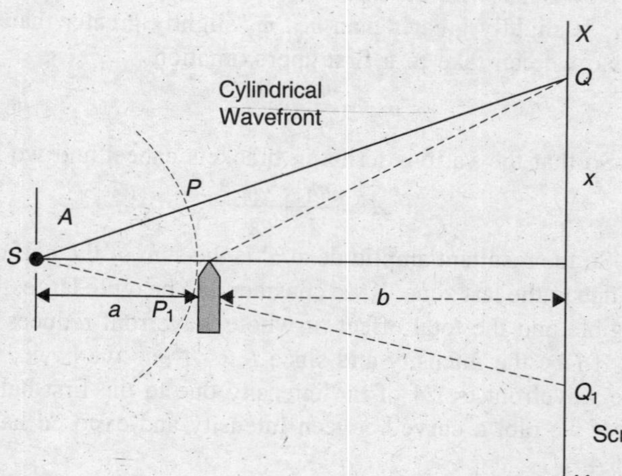

Fig. 5.5

According to the simple laws of the geometrical optics, the shadow should start from O

in the downward direction while light should be present above the point O *i.e.*, below O there should be total darkness and above O total illumination. But actually we observe:

(*i*) Below O the intensity falls off gradually as we move into the geometrical shadow.

(*ii*) Just above O, alternate dark and bright bands of light of unequal width and different intensities are observed. These bands become closer to each other as we move away from O and ultimately disappear and we get uniform intensity.

The explanation for above can be given as:

(*i*) For the point O on the screen, the pole of the wavefront is at A, and only 1/2 of the wavefront is exposed so that the displacement at O is only due to the upper half on the wavefront *i.e.*, $m_1/2$. As we move below O, into the geometrical shadow, say to Q_1, the pole also moves from A to P_1 and gradually, the first, two, first three etc. half period elements of the upper half are blocked, so that the displace- ments at various points below O become $= m_1/2, m_2/2, -m_3/2, ...$ As m_1, m_2, m_3 etc., are in the descending orders, so that as the point moves into the geometrical shadow, the intensity of light reduces to zero.

(*ii*) If we move from O in the upward direction to the point Q distance x from O, the pole of the wavefront also moves up towards P and one after another the first, the first two, the first three etc. half period of zones of the upper half of the wavefront are exposed so that the illumination at the point Q will be due to the complete upper half of the wavefront above P together with the half period elements contained in the wavefront PA. If the no. of half period elements in PA is even, the displacement from the alternate zones cancel each other out inpairs and the intensity at Q is due to only the upper half of the wavefront above P and so will be less. If the no. of half period zones contained in PA is odd, after cancellation, one zone will be left and the effect of this at Q will be added to that of the half wavefront above P so that the intensity will increase. So as the point Q moves up from O it alternately passes from positions of maximum intensity and minimum intensity according as PA contains odd or even number of half period elements. This is how we get the diffraction bands. The intensity of maximum goes on decreasing and that of minimum goes on increasing and finally Q goes into uniform intensity.

To consider the effect at Q, join SQ cutting the wavefront at P and join AQ also. So the number of half period elements contained in PA is equal to half wavelengths in the path difference $(AQ - PQ)$

Let $SA = a$, $AO = b$, we have

$$AQ = \sqrt{AO^2 + OQ^2}$$

$$= \sqrt{b^2 + x^2} = b\left(1 + \frac{x^2}{b^2}\right)^{\frac{1}{2}}$$

$$= b + \frac{1}{2}\frac{x^2}{b^2} \qquad \text{(from binomial to 1st approx.) ...(1)}$$

and

$$PQ = SQ - SP$$

$$= \sqrt{(a+b)^2 + x^2} - a$$

$$= (a+b)\left[1 + \frac{x^2}{(a+b)^2}\right]^{\frac{1}{2}} - a$$

$$= (a+b) + \frac{1}{2}\frac{x^2}{(a+b)^2} - a$$

$$= b + \frac{1}{2}\frac{x^2}{(a+b)} \qquad \text{...(2)}$$

So the path difference

$$AQ - PQ = \left(b + \frac{x^2}{2b}\right) - \left(b + \frac{x^2}{2(a+b)}\right)$$

$$= \frac{x^2}{2}\left[\frac{1}{b} - \frac{1}{(a+b)}\right]$$

$$= \frac{x^2}{2}\left[\frac{1}{b(a+b)}\right]$$

$$= \frac{ax^2}{2b(a+b)} \quad \text{...(3)}$$

For Q to be centre of maxima, this path difference should be odd multiple of $\lambda/2$ and for Q to be centre of minima, it should be even multiple of $\lambda/2$.

For maximum brightness $\dfrac{ax^2}{2b(a+b)} = (2n+1)\dfrac{\lambda}{2}$

or $$x = \sqrt{\frac{b(a+b)(2n+1)\lambda}{a}}$$

and for minimum brightness $\dfrac{ax^2}{2b(a+b)} = n\lambda$

$$= \sqrt{\frac{b(a+b)2n\lambda}{a}}$$

where $n = 0, 1, 2, 3$ ....

Since $x \propto \sqrt{2n+1}$ for maximum intensity

and $\propto \sqrt{2n}$ for minimum intensity

So the width of the band goes on decreasing with the order of the maximum number of bands *i.e.*, as Q moves away from O.

Since $x \propto \sqrt{2n}$ for minimum intensity

and $x \propto \sqrt{2n+1}$ for bright band, we can have an example

For dark fringes $x_1 = K\sqrt{1} = K$, when $n = 1$

$x_2 = K\sqrt{2} = 1.414\,K$, when $n = 2$

$x_3 = K\sqrt{3} = 1.732\,K$, when $n = 3$

$x_4 = K\sqrt{4} = 2\,K$, when $n = 4$

width of 1st dark fringe + 1.414 – $K = 0.441K$

width of 2nd dark fringe = 1.732 K – 1.414 $K = 0.318K$

width of 3rd dark fringe = $2K$ – 1.732 $K = 0.268K$

which shows that the width of the fringes goes on decreasing *i.e.*, the bands are not equally spaced.

5.8. DETERMINATION OF λ

Straight edge experiment can be used to find the wavelength of light used.

The position of the nth maxima is given be

$$x_n = \sqrt{\frac{b(a+b)(2n+1)\lambda}{a}}$$

Let the position of 1st maxima be corresponding to $n = 0$

$$\therefore \quad x_1 = \sqrt{\frac{b(a+b)\lambda}{a}}$$

$$\therefore \quad x_n - x_1 = \sqrt{\frac{b(a+b)\lambda}{a}}(\sqrt{2n+1} - 1)$$

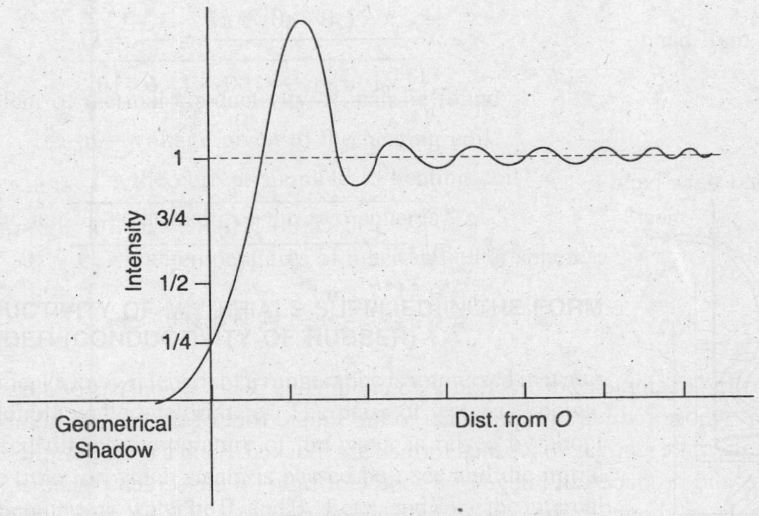

Fig. 5.6

Knowing a, b and $(x_n - x_1)$, the value of λ can be found.

Figure 5.6 shows the variation of intensity above the point O and into the geometrical shadow of the obstacle.

Example 1. *A narrow slit illuminated by light of wavelength 6.4 × 10⁻⁵ cm is placed at a distance of 3 metres from a straight edge. If the distance between the straight edge and the screen is 6 metres, calculate the distance between the first and fourth dark band.*

Solution: The position of the nth dark band in case of straight edge is

$$x_n = \sqrt{\frac{b(a+b)\,2n\lambda}{a}}$$

Distance of the first dark band will be for $n = 1$

$$\therefore \quad x_1 = \sqrt{\frac{b(a+b)\,2\lambda}{a}}$$

Distance of the fourth dark band will be for $n = 4$

$$x_4 = \sqrt{\frac{b(a+b)\,8n\lambda}{a}}$$

$$\therefore \quad x_4 - x_1 = \sqrt{\frac{b(a+b)\,2\lambda}{a}}\,[\sqrt{4}-1]$$

$$= \sqrt{\frac{b(a+b)\,2\lambda}{a}}\,[2-1] = \sqrt{\frac{b(a+b)\,2\lambda}{a}}$$

Now $a = 3$ m $= 300$ cm

$b = 6$ m $= 600$ cm

$\lambda = 6.4 \times 10^{-5}$ cm

$$\therefore \quad x_4 - x_1 = \sqrt{\frac{600(600+300)\times 2\times 6.4\times 10^{-5}}{300}} = \mathbf{0.48\ cm}$$

Example 2. *A narrow slit illuminated by light of wavelength 6 × 10⁻⁵ cm is placed at a distance of 10 cm from a straight edge. If measurements are done at a distance of 100 cm from the straight edge, calculate distance between the first and the second dark bands.*

Solution:

$$x_n = \sqrt{\frac{2b(a+b)n\lambda}{a}}$$

$a = 10$ cm, $b = 100$ cm, $\lambda = 6 \times 10^{-5}$ cm

For first dark band $n = 1$

$$\therefore \quad x_1 = \sqrt{\frac{2\times 100\times(10+100)\times 1\times 6\times 10^{-5}}{10}}$$

$$= 0.3634 \text{ cm}$$

For second dark band $n = 2$,

$$\text{so,} \quad x_2 = \sqrt{\frac{2\times 100\times(10+100)\times 2\times 6\times 10^{-5}}{10}}$$

$$= 0.5137 \text{ cm}$$

$$\therefore \quad x_2 - x_1 = 0.5137 - 0.3634 = \mathbf{0.1503\ cm.}$$

5.9. ZONE PLATE

An interesting example of diffraction is found in zone plate. It consists of a transparent plate on which circles whose radii are proportional to the square root of the natural numbers 1, 2, 3, ... are drawn. The alternate annular zones thus formed are blocked. Such a plate gets the characteristics of a convex lens and produces an image of a source of light on the screen placed at some suitable distance. As has been shown in the case of half period elements, if b be the distance of the point O from the pole P of the wavefront and λ the wavelength of the monochromatic light, then the radii of the various zones are given by

$$r_1 = \sqrt{1b\lambda}; \quad r_2 = \sqrt{2b\lambda}$$
$$r_3 = \sqrt{3b\lambda}; \quad r_4 = \sqrt{4b\lambda}$$

So the radii are proportional to the square root of the natural numbers *i.e.*, $\sqrt{1}$, $\sqrt{2}$, $\sqrt{3}$, $\sqrt{4}$ ...., etc. It is on this fact that the construction of zone plate depends.

5.10. THEORY OF ZONE PLATE

Consider an imaginary transparent plate $AB \perp$ to the plane of paper and let S be a point source of light giving out monochromatic light with spherical wavefronts. SP and to the plane AB and produce it to O. Let us now try to find the resultant displacement at O on the screen at right angles to the plane of paper.

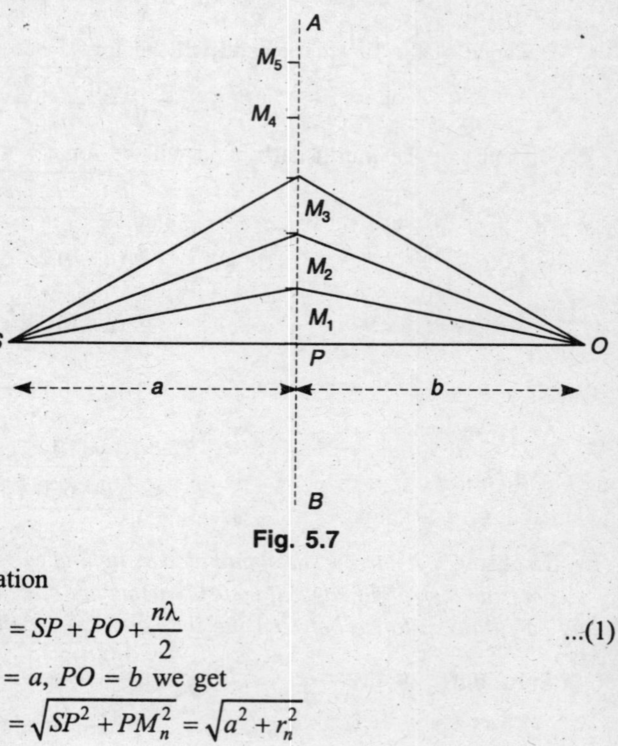

Fig. 5.7

With P as centre and radii equal to PM_1, PM_2, PM_3 ..., divide plate AB into circular zones, such that

$$SM_1 + M_1O = SP + PO + \frac{\lambda}{2}$$

$$SM_2 + M_2O = SP + PO + \frac{2\lambda}{2}$$

$$SM_3 + M_3O = SP + PO + \frac{3\lambda}{2}$$

$$SM_n + M_nO = SP + PO + \frac{n\lambda}{2}$$

To find the radius of any circle, ray r_n of the nth circle, consider the relation

$$SM_n + M_nO = SP + PO + \frac{n\lambda}{2} \qquad ...(1)$$

Put $SP = a$, $PO = b$ we get

$$SM_n = \sqrt{SP^2 + PM_n^2} = \sqrt{a^2 + r_n^2}$$

$$= a\left(1+\frac{r_n^2}{a^2}\right)^{\frac{1}{2}} = a+\frac{1}{2}\cdot\frac{r_n^2}{a} \qquad ...(2)$$

Since $r_n < a$, so we may neglect higher powers of r_n.

Similarly $$M_nO = \sqrt{PO^2 + PM_n^2} = \sqrt{a^2 + r_n^2}$$

$$= b + \frac{r_n^2}{2b} \qquad ...(3)$$

Putting these relations in (1), we get

$$a+\frac{r_n^2}{2a}+b+\frac{r_n^2}{2b} = a+b+\frac{n\lambda}{2}$$

or $$r_2^2\left[\frac{1}{2a}+\frac{1}{2b}\right] = \frac{n\lambda}{2}$$

or $$r_n^2\left[\frac{1}{a}+\frac{1}{b}\right] = n\lambda$$

or $$r_n^2 = n\lambda\left[\frac{ab}{a+b}\right] \qquad ...(4)$$

Since a, b, λ are constants, $\therefore\ r_n \propto \sqrt{n}$

The area of nth zone $= \pi(r_n^2 - r_{n-1}) = \pi\left[\frac{n\lambda\, ab}{(a+b)} - \frac{(n-1)\,\lambda ab}{a+b}\right] = \pi\frac{\lambda ab}{(a+b)}$

which is independent of order n, thereby showing that for the given position of object and image the areas of all the zones is constant. As the objects and the image O approach the zone plate, the areas of the zones, reduce rapidly but the magnitudes of $m_1, m_2, m_3, ...,$ the displacements at O due to the various zones decrease very slowly with the order of the zone, so that

$$A = m_1 - m_2 + m_3 - m_4 + ...$$

If we allow monochromatic light to fall on this zone plate and the emergent value of b, hypothetical half period elements may coincide with the actual elemental areas of the plate. In that case if the even number of elements are made opaque, the resultant displacement

$$A = m_1 + m_3 + m_5$$

which is many times greater than the zones and this accounts for the focussing action of the zone plate.

Putting Equation (4) in the form

$$\frac{1}{a}+\frac{1}{b} = \frac{n\lambda}{r_n^2}$$

and comparing with convex lens formula

$$\frac{1}{v}+\frac{1}{u} = \frac{1}{f}, \text{ we have}$$

$$\frac{1}{f} = \frac{n\lambda}{r_n^2}$$

or $$f = \frac{r_n^2}{n\lambda} \qquad ...(5)$$

which gives the focal length of the zone plate.

So the zone plate behaves like a convex lens. However, the difference between a convex lens and a zone plate is that a convex lens has one focus given by the relation

$$\frac{1}{f} = (\mu-1)\left(\frac{1}{r_1}-\frac{1}{r_2}\right) \qquad ...(6)$$

whereas a zone plate has a number of foci between it and the point O, where S and O represent the position of object and image respectively. For source at S, we may have series of images O_1,

O_2, O_3 etc. determined by the condition that for each of the positions a zone contains an odd number of half period zones. The positions of these foci are given by

$$f_1 = \frac{r_n^2}{n\lambda},\ f_2 = \frac{r_n^2}{3n\lambda},\ f_3 = \frac{r_n^2}{5n\lambda} \ldots\ldots$$

The intensity of the image decreases as the focal length decreases.

Another important point is that in zone plate, each colour has its own separate focus, the red being near P than violet. In case of a lens, the case is just opposite; red comes to focus away from the violet.

5.11. CONSTRUCTION OF ZONE PLATE

To construct a zone plate, concentric circles whose radii are proportional to the square root of the natural numbers are drawn on a sheet of white paper as shown in Fig. 5.8.

The alternate zones are painted black and a very much reduced photograph of the drawing is got on a glass plate. The negative when held in the light path from a distant point source, produces a large intensity at a point on its axis at a distance corresponding to the size of the zone and the wavelength of the light used.

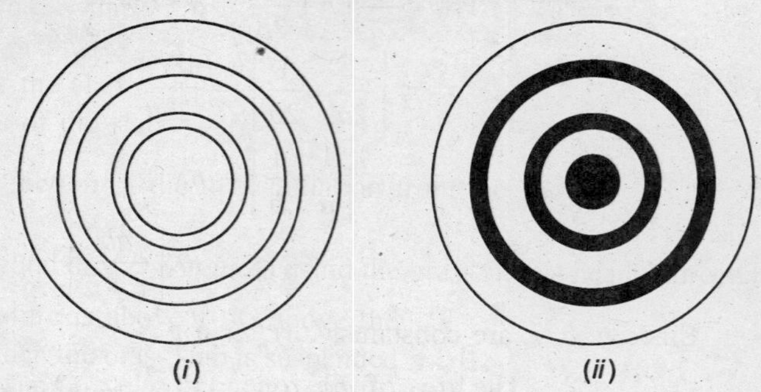

Fig. 5.8

Example 3. *Calculate the radii of the first three clear elements of a zone plate which is designed to bring a parallel beam of light of wavelength 6000 Å to the first focus at a distance of 2 metres.*

Solution: As the incident beam is parallel, $u = \infty$ and, therefore, $f = v$

Now $$f = \frac{r_n^2}{n\lambda} \text{ or } r_n = \sqrt{nf\lambda}$$

In this case $v = f = 200$ cm, $\lambda = 6000$ Å $= 6 \times 10^{-5}$ cm

(*i*) For first clear zone, $n = 1$

$\therefore$ $$r_1 = \sqrt{\lambda f} = \sqrt{6 \times 10^{-5} \times 200} = \mathbf{0.1095\ cm}$$

(*ii*) For 2nd clear zone, $n = 3$

$$r_2 = \sqrt{3 \times 6 \times 10^{-5} \times 200} = \mathbf{0.1897\ cm}$$

(*iii*) For third clear zone, $n = 5$

$$r_5 = \sqrt{5 \times 6 \times 10^{-5} \times 200} = \mathbf{0.2450\ cm.}$$

Example 4. *A zone plate is constructed in such a way that the radii of the circles which define the zones are the same as the radii of Newton rings formed between a plane surface and surface having radius of curvature of 1 metre. Find the primary focal length of zone plate.*

Solution: The diameter of d_n of nth Newton's dark ring is given by

$$\frac{d_n^2}{4} = n\lambda R$$

or $r_n^2 = n\lambda R$, where r_n is the radius of nth ring

For $n = 1$, $r_1^2 = \lambda R$ ...(1)

Focal length of zone plate $$= \frac{r_n^2}{n\lambda}$$

For primary focal length $n = 1$

$\therefore$ $$f_1 = \frac{r_1^2}{\lambda} \quad \ldots(2)$$

Putting the value of r_1^2 from Equation (1) in (2)

$$f_1 = \frac{\lambda R}{\lambda} = R = 1\text{ m} = \mathbf{100\ cm.}$$

Example 5. *With a zone plate, for a point source of light on the axis, the strongest and the next strongest images are formed at 30 cm and 6 cm respectively from the zone plate. Both the images are on the same side and on the other side of the source. Calculate:*

(i) Distance of the source from the zone plate

(ii) The radius of the first zone

(iii) Principal focal length.

Given λ = 6000Å

Solution: For a zone plate

$$\frac{1}{v} - \frac{1}{u} = \frac{n\lambda}{r_n^r}$$

where u and v are the distances of object and image from the zone plate.

Here $v = 30$ cm for first strongest image

and $v = 6$ cm for second strongest image

$$\therefore \quad \frac{1}{30} - \frac{1}{u} = \frac{\lambda}{r_1^2} \text{ for first image} \quad ...(1)$$

$$\frac{1}{6} - \frac{1}{u} = \frac{3\lambda}{r_1^2} \text{ for second image} \quad ...(2)$$

$$\therefore \quad \frac{1}{30} - \frac{1}{u} = \frac{1}{3}\left(\frac{1}{6} - \frac{1}{u}\right)$$

or $u = -30$ cm

Negative sign shows that the point source lies towards left of the zone plate at a distance of 30 cm.

Putting the value of u in equation (1) we get

$$\frac{1}{30} + \frac{1}{30} = \frac{6000 \times 10^{-8}}{r_1^2}$$

or $r_1 = 0.03$ cm

Again

$$f_1 = \frac{r_1^2}{\lambda} = \frac{(3 \times 10^{-2})^2}{6000 \times 10^{-8}}$$

$$= \frac{9 \times 10^{-4}}{6 \times 10^{-5}} = \frac{90}{6} = 15 \text{ cm}$$

5.12. DIFFERENCE BETWEEN A ZONE PLATE AND A CONVEX LENS

Similarities. (*i*) Both form real images of an object on the other side of the object and the distances of object and images are connected by similar relations.

(*ii*) In both cases focal length depends on λ and so both show chromatic aberrations.

Differences. (*i*) In the case of lens the rays are brought to focus by refraction while in the zone plate, the image is formed by diffraction.

(*ii*) In the case of convex lens, all the rays reaching the point where the image is formed travel the same optical path while in zone plate, the waves reaching the image through two alternate zones have a path difference of λ.

(*iii*) Image formed by a convex lens is more intense than that due to a zone plate.

(*iv*) For convex lens $f_v < f_r$ while for zone plate $f_r < f_v$.

(*v*) A convex lens has only one focus on one side of it while a zone plate has multiple foci at distances

$$\frac{r_n^2}{n\lambda}, \frac{r_n^2}{3n\lambda}, \frac{r_n^2}{5n\lambda} \cdots \frac{r_n^2}{(2m+1)\,n\lambda}$$

The intensity reduces as the focal length decreases.

5.13. FRAUNHOFER DIFFRACTION AT SINGLE SLIT

In order to get this class of diffraction at a single slit, the wavefront must be plane and the diffracted light be focussed on the screen using a convex lens. To obtain the diffraction pattern either the light source should be at large distance from the slit or a collimating lens be used.

In Fig. 5.9, S is a narrow source of mono-chromatic light. Let XY be the incident spherical wavefront, L_1 the collimating lens and AB be the slit of width a.

Now as the light passes through the slit AB, after getting diffracted from single slit, it is collimated on screen MN with the help of lens L_2 where L_1 and L_2 are both achromatic lenses.

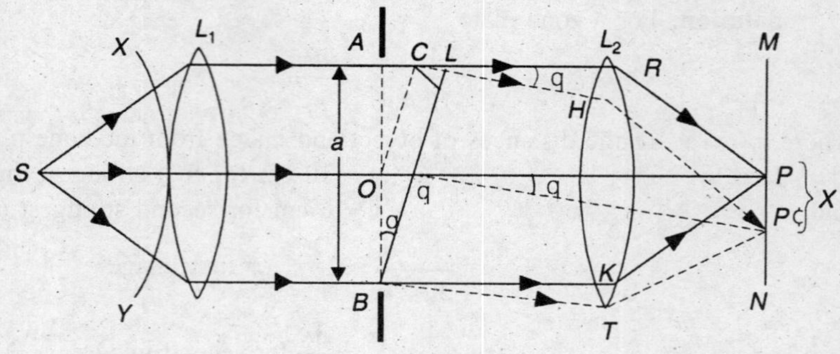

Fig. 5.9

As the plane wave is incident on the slit AB, each point on the wavefront is a source of secondary disturbance, the secondary waves travelling in the direction parallel to OP *i.e.,* along HA and BK come to focus at O and a central bright fringe is observed. The secondary waves which are at equidistant from O and situated in the upper and lower halves OA and OB of the wavefront move through the same distance to reach P and thereby have zero path difference SOP *will be the position of maximum intensity.*

Let us now consider the secondary wave moving along AHP inclined at angle θ to the horizontal. All the waves travelling along this direction will reach P' on the screen and point P will be position of maxima or minima depending upon the path difference between the secondary waves that originate from the corresponding points of the wavefront. Draw OC and BL to AH,

Now from $\Delta\, ABN$, we have

$$\sin\theta = \frac{AL}{AB} = \frac{AL}{a}$$

where a is the width of the slit and AL is the path difference between the secondary waves coming from A and B.

Now if the path difference is equal to λ, the wavelength of light used the P' will be a point of minimum intensity. It is because, the path difference between two halves OA and OB can be considered as $\lambda/2$, thereby each point on upper half of wavefront will interface destructively with corresponding point on the lower half of wavefront.

Similarly if the direction of secondary waves is such that $AL = 2\lambda$, then also the corresponding points on screen will interface destructively and so we will get minimum intensity at P'.

In general, for minimum intensity

$$a \sin\theta_n = n\lambda$$

or

$$\sin\theta_n = \frac{n\lambda}{a} \qquad ...(1)$$

where θ_n corresponds to the direction of the nth minima and n is an integer.

In case the path difference is odd multiple of $\lambda/2$, we get position of secondary maxima and in that case

$$a \sin\theta_n = (2n+1)\,\lambda/2 \qquad ...(2)$$

or $$\sin\theta_n = \frac{(2n+1)\lambda}{2a}$$

where $n = 1, 2, 3, \ldots$ etc.

In Fig. 5.10 intensity distribution on the screen is given. It is evident that P corresponds to central maxima and points on the screen for which path difference is $\lambda, 2\lambda, 3\lambda, \ldots, n\lambda$ correspond to position secondary minima. As is evident the secondary maxima are of lesser intensity and the intensity falls off rapidly on either side of P.

Now in the case, when lens L_1 is very near the slit or the screen is away from lens L_2, then we have

$$\sin\theta = \frac{x}{f}$$

where f is the focal length of lens L_2.

Also from Equation (1)

$$\sin\theta = \frac{\lambda}{a}$$

$$\therefore \quad \frac{x}{f} = \frac{\lambda}{a}$$

or $$x = f\frac{\lambda}{a}$$

where x is the distance of secondary minimum from the point P. So the width of central maxima = $2x$

$$\therefore \quad 2x = 2f\frac{\lambda}{a}$$

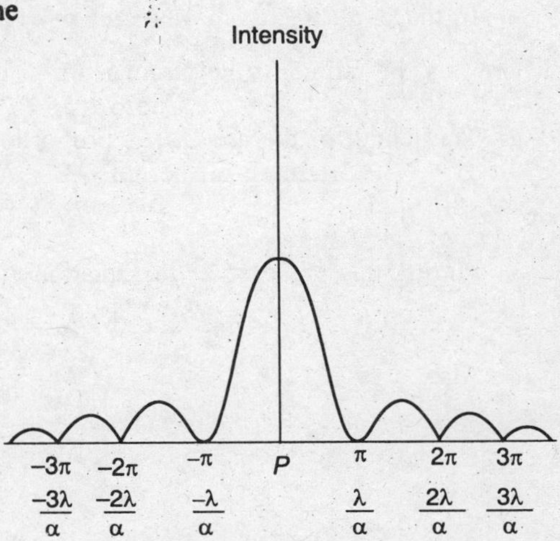

Fig. 5.10

It means that the width of the central maxima is proportional to λ. For light of longer wavelength (say red light) the width will be more than that for light of shorter wavelength (say violet light).

Also with small a, *i.e.*, with narrow slit the width of central maxima will be more.

In case white light is used, we will observe central maxima as white while all other diffraction patterns are coloured.

Also from Equation (1), if the slit width is large, $\sin\theta$ and thereby θ is small. In that case the maxima are very close to the central maxima at the point P. In case a is small, $\sin\theta$ is large and so we get distinct diffraction maxima.

5.14. INTENSITY DISTRIBUTION IN THE DIFFRACTION PATTERN DUE TO A SINGLE SLIT

We can study the intensity variation in the diffraction pattern due to a single slit as follows. The incident wavefront on the slit AB (Fig. 5.11) can be imagined to be divided into a large number of infinitesimally small strips. The path difference between the secondary waves originating from the extreme points A and B $a\sin\theta$ where a is the width of the slit and $\angle ABN = \theta$. For a parallel beam of incident light, the amplitude of vibration of the waves from each strip can be taken to be the same. If we consider the secondary waves in a direction inclined at angle θ from the point B upwards, with change of path difference, phase difference also increases. Let α be the phase difference between the secondary waves from the points B and A of the slit (Fig. 5.11).

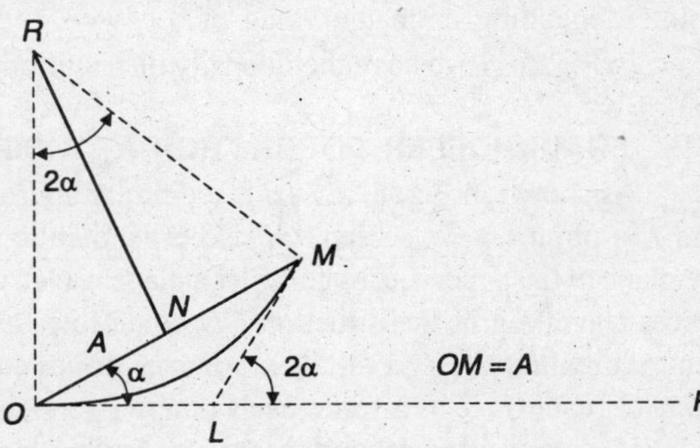

Fig. 5.11

Since we have divided the wavefront into a large number of strips, the resultant amplitude due to all the individual small strips can be got by the vector polygon method. In case, the amplitudes are small and the phase difference increas by **infinitesimally small** amounts from strip to strip. So the vibration polygon coincides with the circular are *OM* (Fig. 5.11). *OP* gives the direction due to the secondary waves from *A*. *R* is the centre of the circular are, from geometry we have

$$\angle MLP = 2\alpha$$

$$\therefore \quad \angle ORM = 2\alpha$$

In the ΔORN $\quad \sin\alpha = \frac{ON}{r}; \; ON = r\sin\alpha$

where r is the radius of the circular arc

$$\therefore \quad \text{Chord } OM = 2ON = 2r\sin\alpha \qquad ...(1)$$

The length of the arc *OM* is proportional to the width of the slit

$\therefore$ Length of the arc $OM = K\alpha$ (K being proportionally constant)

or $$2r = \frac{K\alpha}{\alpha} \qquad ...(2)$$

Putting this value of $2r$ in Equation (1), we get

$$\text{Chord } OM = \frac{K\alpha}{\alpha}$$

Also, $OM = A$ is the amplitude of the resultant, so we have

$$A = (K\alpha)\frac{\sin\alpha}{\alpha}$$

$$A = A_0\frac{\sin\alpha}{\alpha}$$

which means that the resultant amplitude of vibration at a point on the screen is given by

$$I = A^2 = A_0^2\frac{\sin^2\alpha}{\alpha^2} = I_0\left(\frac{\sin\alpha}{\alpha}\right)^2$$

i.e., the intensity at any point on the screen is proportional to $\left(\frac{\sin\alpha}{\alpha}\right)^2$

Now phase difference of 2π corresponds to a path difference of λ. So as phase difference of 2α will be given by

$$2\alpha = \frac{2\pi}{\lambda}\cdot a\sin\theta$$

where $a\sin\theta$ is the path difference between the secondary waves from *A* and *B* (Fig. 5.11)

or $$\alpha = \frac{\pi}{\lambda}\cdot a\sin\theta$$

It means that α depends on the angle of diffraction θ. Also $\frac{\sin^2\alpha}{\alpha^2}$ will give intensity at different points depending upon the value of θ.

In Fig. 5.11 we have the intensity distribution curve for α or $\sin\theta$ along *X*-axis *v/s* $\frac{\sin^2\alpha}{\alpha^2}$ along *Y*-axis.

5.15. FRAUNHOFFER DIFFRACTION AT A CIRCULAR APERTURE

As shown in Fig. 5.12, *AB* is a circular aperture of diameter *d*. Let *C* be the centre of aperture and *P* a point on the screen. *OP* is perpendicular to the screen and the screen is perpendicular to the plane of the paper. Let us consider a plane wavefront incident on the circular aperture. The secondary waves travelling in the direction *CO* come to focus at *P*. So *P* corres-ponds to the position of the central maximum. Here, all the secondary waves coming from points which are at the same distance from *O* reach *P* to reinforce each other.

If we consider, secondary waves moving in a direction inclined at an angle θ with *CP*, they will be reaching at point *P* that $PP' = x$

Now the path difference between the waves coming from points A and B (*i.e.*, from extreme positions) is AD, where $AD = d \sin \theta$.

Now as before, we will have the point P; to be centre of minima if path difference is equal to integral multiple of λ *i.e.*,

$$d \sin \theta = n\lambda \quad ...(1)$$

and P' for to be centre of maxima path difference must be odd multiple of $\lambda/2$ *i.e.*,

$$d \sin \theta = (2n + 1)\, \lambda/2 \quad ...(2)$$

if P is at minimum intensity point, then all the points which are at equal distance from P and P' lying on a circle with radius x will be positive of minima.

So the diffraction pattern due to a circular aperture consists of a central bright disc called **Airy's disc** and it is surrounded by alternate dark and bright rings known as **Airy's rings.** It may be noted that the intensity of dark rings is zero while that of the bright rings decrease gradually as we move outward from P.

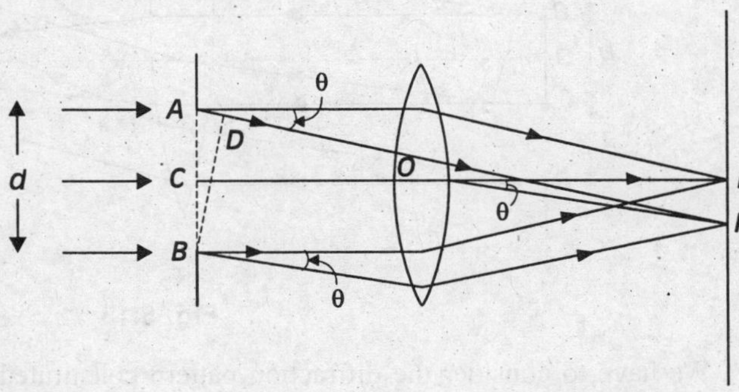

Fig. 5.12

Again, if we place the converging lens very near the slit, or when the distance of the screen in very large from the lens

$$\sin \theta = \theta = \frac{x}{f} \quad ...(3)$$

Also, for the first secondary minima, we have

$$d \sin \theta = \lambda$$

$$\sin \theta = \theta = \frac{\lambda}{d} \quad ...(4)$$

So from Equations (3) and (4)

$$\frac{x}{f} = \frac{\lambda}{d}$$

or

$$x = \frac{f\lambda}{d} \quad ...(5)$$

where x is the radius of the Airy's disc. But experimentally, the radius of the first dark ring is slightly more than given by Equation (5), which according to Airy is given as

$$x = \frac{1.22 f\lambda}{d} \quad ...(6)$$

The intensity distribution in this case is similar to the one given in case of a rectangular slit in which case with the increase in diameter of the aperture, in central bright ring has its radius reduced.

5.16. FRAUNHOFER DIFFRACTION AT A DOUBLE SLIT

As shown in Fig. 5.13, AB and CD are two rectangular slits with each having width a and b as the width of the opaque portion. Both the slits are parallel to one another and $\perp$ to the plane of paper. Let L be the colliating lens and SS' be the screen to the plane of paper. Let P be a point on the screen such that OP is $\perp$ to both screen and the slits.

Now let a plane wavefront XY be incident so that all secondary waves moving along parallel direction to OP come to focus at O *i.e.*, P is the position of central maxima.

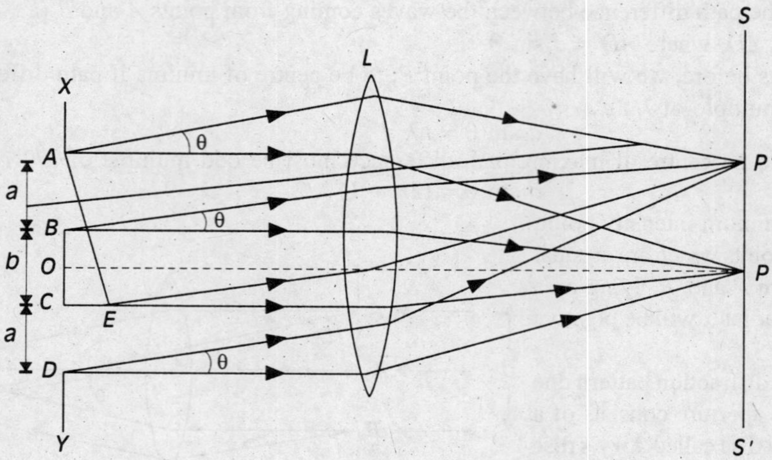

Fig. 5.13

We have to consider the diffraction pattern constituted by the following parts:

(*i*) Interference fringes due to secondary waves from the two slits taken separately.

(*ii*) Diffraction fringes due to the secondary waves from the two slits taken separately.

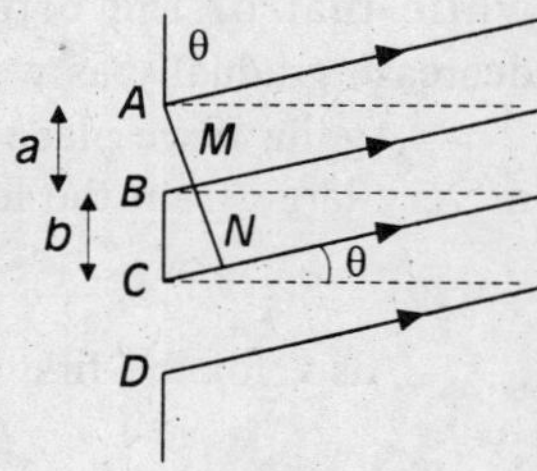

Fig. 5.14

Let us denote the angle of diffraction θ, for calculating the interference pattern and angle ϕ for calculating the diffraction pattern. Both θ and ϕ refer to the direction of secondary waves of direction of incident light.

(*i*) Interference maxima and minima. Let the secondary waves travelling in a direction inclined at an angle θ with the initial direction, then we have

In the Δ *ACN* $\quad \sin\theta = \dfrac{CN}{AC} = \dfrac{CN}{a+b}$

or $\quad ON = (a+b)\sin\theta$

Now if this path difference is equal to odd multiples of $\dfrac{\lambda}{2}$, θ gives the direction of minima due to **interference of the secondary waves** from the two slits.

$$\therefore \quad CN = (a+b)\sin\theta_n = (2n+1)\frac{\lambda}{2} \quad \text{...(1)}$$

For n = 1, 2, 3, etc. the values of $\theta_1', \theta_2', \theta_3'$ etc. corresponding to the directions of minima can be known

Again from Equation (1) $\quad \sin\theta_n = \dfrac{(2n+1)\lambda}{2(a+b)} \quad$ ...(2)

Again, if the secondary waves travel in a direction θ′ such that the path difference is even multiples of λ/2. The θ′ gives the direction of the maxima due to interference of light waves coming from the two slits.

$$CN = (a+b)\sin\theta_n' = 2n\cdot\frac{\lambda}{2}$$

$$\sin\theta_n' = \frac{\lambda}{(a+b)} \quad \text{...(3)}$$

For n =1, 2, 3, etc. the values of $\theta_1', \theta_2', \theta_3'$ etc., corresponding to the directions of minima can be known.

Now from Equation (3) $\quad \sin\theta_1 = \dfrac{3\lambda}{2\,(a+b)}$

and $$\sin\theta_2 = \frac{5\lambda}{2(a+b)}$$

$$\sin\theta_2 - \sin\theta_1 = \frac{\lambda}{a+b} \quad ...(4)$$

which means that angular separation between any two consecutive maxima or minima is $\frac{\lambda}{(a+b)}$.

It means that angular separation $\propto \frac{\lambda}{(a+b)}$, where $(a + b)$ is the distance between the two slits.

***(ii)* Diffraction maxima and minima.** Now consider the secondary waves travelling along a direction at an angle ϕ with the initial direction of incident light.

In this case if the path difference BM is equal to λ, the wavelength of light used, then the angle of diffraction will be ϕ for getting **diffraction minima.**

It means that the path difference between the secondary waves, originating from the points A and B *i.e.*, extremities of a slit will be equal to λ. If we consider the wavefront on AB to be made up of two halves, then $\lambda/2$ will be the path difference between corresponding points of upper and lower half of the wavefront.

So the effect at P' due to these two halves of the wavefront will be zero. Similarly, the affect at P' for same direction of secondary waves, due to wavefront incident on the slit CD is also zero.

In general, for getting diffraction minima, we have

$$a\sin\theta_n = n\lambda$$

when $n = 1, 2, 3, ...$ we get the corresponding order of diffraction minima.

Difference between single slit and double slit diffraction pattern

Single slit pattern	*Double slit pattern*
1. It consists of central bright maxima with secondary maxima and minima of gradual lower intensity	1. It consists of diffraction pattern of equally spaced interference maxima and minima within the central maxima.
2. Intensity of this pattern is quite less.	2. Intensity pattern of double slit diffraction is four times that of single slit pattern.
3. The spacing of single slit diffraction pattern depends on a the width of the slit.	3. Spacing of double slit pattern depends on a and b where b is the width of opacity.
4. Intensities of maxima are not constant but decrease to zero on either side of the central maxima.	4. The maxima appear on either side, but the intensity is too weak to observe them after two or three maxima.

MISSING ORDER IN A DOUBLE DIFFRACTION PATTERN

We have taken the slit width as a and double slit separation as b. If a is kept constant, we will observe the sam diffraction pattern. However, if a is kept constant and b is varied, the spacing between interference maxima changes, depending upon the relative values of a and b. Some order of interference maxima will be missing in the resultant pattern.

We have direction for in interference maxima as

$$(a + b)\sin\theta = n\lambda \quad ...(1)$$

and the direction for diffraction minima as

$$a\sin\theta = m\lambda \quad ...(2)$$

where n and m are integers.

If the values of a and b are such that both the Equations (1) and (2) are satisfied simultaneously, for same value of θ, in that case position of **interference maxima** correspond to that of **diffraction minima.**

Let us take up some cases

(*i*) If $a = b$

then $$2a\sin\theta = n\lambda \quad ...(3)$$

and $$a \sin \theta = m\lambda \quad \text{...(4)}$$

∴ From Equations (3) and (4)

$$\frac{n}{m} = 2$$

or $$n = 2m$$

if $$m = 1, 2, 3, \ldots$$

$$n = 2, 4, 6, \ldots$$

i.e., 2nd, 4th, 6th etc., orders of the interference maxima will be missing in the diffraction pattern.

(*ii*) If $$2a = b$$

$$3a \sin \theta = n\lambda$$

and $$a \sin \theta = m\lambda$$

∴ $$\frac{n}{m} = 3$$

or $$n = 3m$$

Putting $$m = 1, 2, 3 \ldots \text{ etc.}$$

∴ $$n = 3, 6, 9, \ldots \text{ etc.}$$

So the 3, 6, 9, etc., of the interference maxima will be missing in the diffraction pattern.

(*iii*) If $a + b = a$ *i.e.,* $b = 0$

In that case we will have a single slit, so all interference pattern will be missing. In this case diffractions pattern observed on the screen is similar to that due to single slit with width $2a$.

5.17. CORNU'S SPIRAL

In order to find the effect at a point due to an incident wavefront. Fresnel's method consists in dividing the *wavefront* into *half period zones.* The path difference between the secondary waves from two corresponding points of neighbouring zones being $\lambda/2$.

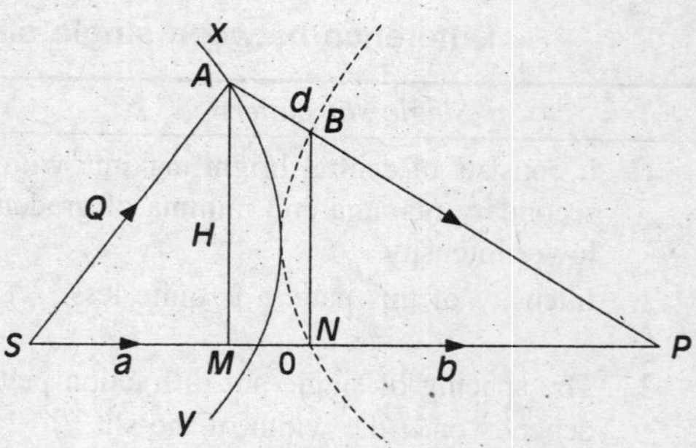

Fig. 5.15

In Fig. 5.15, S is a point source of light and XY is the incident spherical wavefront. With reference to the point P, O is the pole of the wavefront. Let a and b be the distances of the points S and P from the pole of the wavefront. With P as centre and draw a sphere touching the incident wavefront at O. The path difference between the waves travelling in the directions SAP and SOP is given by

$$\begin{aligned} d &= SA + AP - SOP \\ &= SA + AP - (SO - OP) \\ &= a + AB + b - (a + b) = AB \end{aligned}$$

For large distances of a and b, AM and BN can be taken to be approximately equal and the path difference d can be written as

$$d = AB = MO + ON$$

From the geometry we have for a circle,

$$MO = \frac{AM^2}{2SO} = \frac{h^2}{2a}$$

and $$ON = \frac{BN^2}{2OP} = \frac{h^2}{2b}$$

∴ $$d = \frac{h^2}{2b} + \frac{h^2}{2b} = \frac{h^2(a+b)}{2ab} \quad \text{...(1)}$$

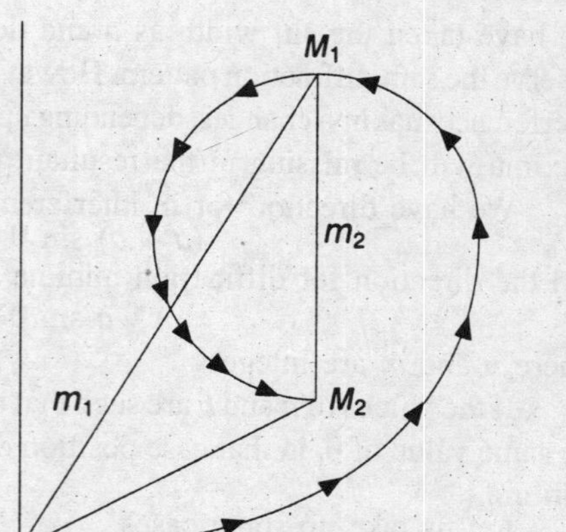

Fig. 5.16

Let AM be the radius of the nth half period zone, then this path difference is equal to $n\lambda/2$. According to Fresnel's method of constructing the half period zones, we have

$$\therefore \quad \frac{h^2(a+b)}{2ab} = \frac{n\lambda}{2} \qquad ...(2)$$

We can obtain the resultant amplitude at an external point due to the wavefront by the following method. Let the first half period strip of the Fresnel's zones and these vectors be represented from O to M_1 (Fig. 5.16).

The continuous phase changes is due to the continuous increase in the obliquity factor O to M_1. The resultant amplitude at the external point due to first half period strips is given by OM_1 (= m_1). Similarly if the process is continued, we obtain the vibration curve M_1 M_2. The portion M_1M_2 corresponds to the second half period strip. The resultant amplitude at the point due to the first two half period strips is given but OM_2 (= A.) If instead of eighth substrips, each half period zone is divided into substrips of infinitesimal width, we get a smooth curve. The complete vibration curve for the whole wavefront will be a spiral as shown in Fig. 5.17. X and Y correspond to the two extremities of the wavefront and M_1 M_2 etc., refer to the edges of the first, second etc., half period strips. Similarly $M_1{}'M_2{}'$ etc. refer to the edges of the first, second etc. half period strips of the lower portion of the wavefront. This is known as *Cornu's spiral.* The characteristic of this curve is that for any point P on the curve, the phase lag δ is directly proportional to the *square of the distance v.* The distance is measured along the length of the curve from the point O. For a path difference of λ by phase difference is 2π. So, for a path difference of d, the phase difference δ will be given by

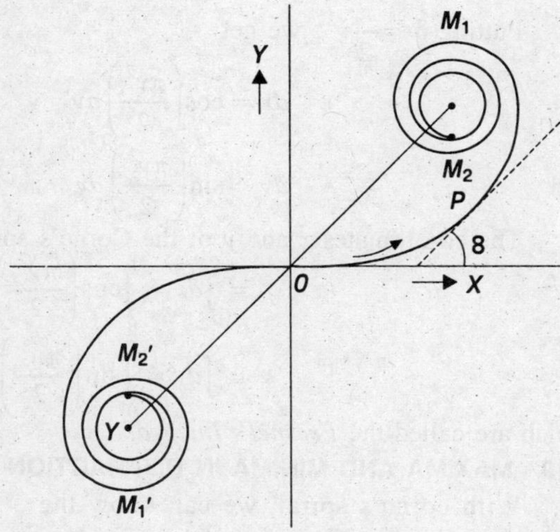

Fig. 5.17

$$\delta = \frac{2\pi}{\lambda} \cdot d$$

Putting the value of d from Equation (1), we get

$$\delta = \left[\frac{2\pi h^2(a+b)}{\lambda\, 2ab}\right] \qquad ...(3)$$

$$\therefore \quad \delta = \frac{\pi}{2}\left[\frac{2h^2(a+b)}{2b\lambda}\right]$$

$$\delta = \frac{\pi}{2} \cdot v^2 \qquad ...(4)$$

where we have introduced a new dimensionless variable v given by

$$v^2 = \frac{2h^2(a+b)}{ab\lambda}$$

$$v^2 = h\sqrt{\frac{2(a+b)}{ab\lambda}} \qquad ...(5)$$

we can use Cornu's spiral for any diffraction problem irrespective of the values of a,b and λ.

5.18. FRESNEL INTEGRAL

For any point of Cornu's spiral, we introduce two integrals for x and y co-ordinates, which are called Fresnel Integrals. Let us consider a point P on the spiral which has a distance v from the origin. Let the length to the curve of P made an angle S with X-axis, where δ will correspond to the phase change from O to P. Let the change in co-ordinates of the point be dx and dy for a small displacement dv of the point along the curve.

$$dx = dv \cos \delta \qquad ...(1)$$

and

$$dy = dv \sin \delta \qquad ...(2)$$

Putting $\delta = \frac{\pi}{2} v^2$, we get

$$dx = \cos\left(\frac{\pi v^2}{2}\right) dv \qquad ...(3)$$

and

$$dy = \sin\left(\frac{\pi v^2}{2}\right) dv \qquad ...(4)$$

The co-ordinates x and y of the Cornu's spiral are thereby given as

$$x = \int dx = \int_0^v \cos\left(\frac{\pi v^2}{2}\right) dv \qquad .(5)$$

$$y = \int dy = \int_0^\pi \sin\left(\frac{\pi v^2}{2}\right) dv \qquad ...(6)$$

which are called the *Fresnel's Integral.*

5.19. MAXIMA AND MINIMA IN DIFFRACTION PATTERNS (CORNU'S SPIRAL)

With cornu's spiral, we can know the position of maxima and minima of the diffraction patterns.

In Fig. 5.18, O is the origin of co-ordinates. OX is the vibration curve for the upper half of the wavefront and OY corresponds to the vibration curve for the lower half of the wavefront. If no part of the wavefront is obstructed, then the resultant amplitude at a point is given by XY.

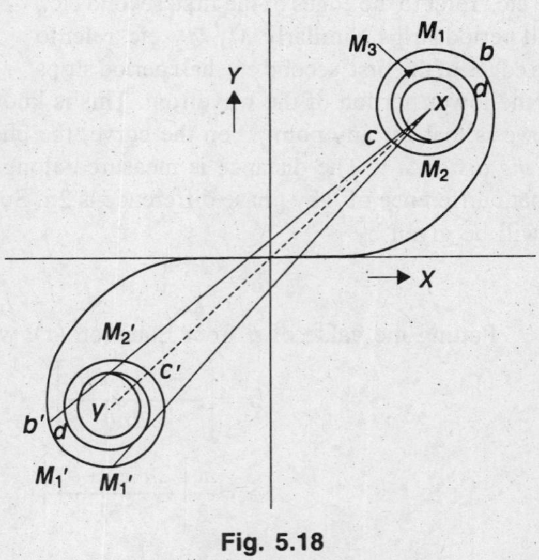

Fig. 5.18

If we consider a cylindrical wavefront incident on straight edge, the amplitude at the point p on the edge to the geometrical shadow (see diffraction at straight edge) is given by OX. If we consider point above P in the illuminated portion gradually more part of the lower half of the wavefront gets exposed and amplified will pass from maxima and minima. In Fig. 5.16 xb'; xd' etc., correspond to maxima xc' and corresponds to minima. So we get on screen in the illuminated portion, alternate bright and dark bands parallel to the length of the slit.

Now if we consider below point P in the geometrical shadow, the lower half of the wavefront and a portion of upper half of the wavefront are cut off and *tail* of the amplitude vector moves to the right of O. The amplitude slowly decreases and becomes zero when the tail reaches X. So, we see that in the region of geometrical shadow, the intensity falls slowly.

We can get the quantitative values of intensity for different points on screen by knowing the different values of amplitude A for various values of v. The intensity at any point will be square of the amplitude at that point, the point M_1'; M_2' point etc., represent the edges of the first, 2nd etc. half period strips in the upper half of the wavefront and the points M_1 M_2 etc. refer to the lower half of the wavefront. The points b', d' etc., on the spiral corresponding to maximum intensity will be a little earlier than the M_1' and M_2' point are obtained.

We can give the *x*, *y* co-ordinates of Cornu's spiral as

$$x = \int dx = \int_0^v \cos\left(\frac{\pi v^2}{2}\right) dv \qquad ...(1)$$

$$y = \int dy = \int_0^\pi \sin\left(\frac{\pi v^2}{2}\right) dv \qquad ...(2)$$

We can find the values of these integrals for different values of *v*.

As shown in the curve the two integrals represent the horizontal and vertical components of the resultant amplitude. The intensity is proportional to the square of the resultant amplitude, so we have

$$I_P = k\,[x^2 + y^2] \qquad ...(3)$$

When the whole of the wavefront is exposed to the point, we get

$$v \to \infty$$

and the values of the integrals will be given by

$$x = \int_0^\infty \cos\left(\frac{\pi v^2}{2}\right) dv = \frac{1}{2} \qquad ...(4)$$

and $$y = \int_0^\infty \sin\left(\frac{\pi v^2}{2}\right) dv = \frac{1}{2} \qquad ...(5)$$

So, for the point *x* in Fig. 5.18, the *x* and *y* coordinates are $\left(\frac{1}{2}, \frac{1}{2}\right)$, similarly for the point *y* on the lower half of the spiral the co-ordinates are $\left(-\frac{1}{2}, \frac{1}{2}\right)$.

At the origin *i.e.,* when $v = 0$, $x = 0$ and $y = 0$ the spiral will pass through the origin and it is symmetrical with the origin. At any point on the spiral, the tangent to the curve makes an angle ϕ with the *X*-axis and we have

$$\tan\phi = \frac{dy}{dx} = \frac{\sin(\pi v^2/2)\,dv}{\cos(\pi v^2/2)\,dv}$$

$$\tan\phi = \tan\left(\frac{\pi v^2}{2}\right)$$

$$\therefore \qquad \phi = \frac{\pi v^2}{2} \qquad ...(6)$$

when $v = 0$, $\phi = 0$.

Which implies that the curve is parallel to the *X*-axis at the origin.

The element of length *dv* along the spiral is given by the relation

$$dv = \sqrt{(dx)^2 + (dy)^2} \qquad ...(7)$$

Differentiating Equation (6), we get

$$d\phi = \frac{2\pi v}{2} \cdot dv$$

or $$d\phi = \frac{1}{\pi v} \qquad ...(8)$$

where, $\frac{dv}{d\phi}$ measures the radius of curvature of the spiral at the point we have considered

Also from Equation (7)

$$\frac{dv}{d\phi} \propto \frac{1}{v}$$

which shows that with the increase in the value of v, the radius of curvature of the curve gradually decreases and takes the shape of a spiral. Finally when $v \to \infty$, the curve will end at a point x or y.

5.20. PLANE DIFFRACTION GRATING

An arrangement which is equivalent in action to a number of parallel, equidistant, small rectangular slits of equal width placed side is called diffraction grating.

Since the slits are arranged in a plane, so diffraction grating is also called plane diffraction grating and is a very good and accurate device for study of spectra.

In diffraction grating, the slits are very narrow and are separated by opaque spaces, which do not allow light to pass through them and are so called 'opacities'. Between the opacities, we have the slits which allow the light to pass through and so are called transparencies.

Fraunhofer in 1821 was able to prepare the first grating by making 310 narrow slits in an inch, which he got by ruling lines on glass with fine diamond point. Using the grating, he was able to study the solar spectrum. It was because of the original formulations by Fraunhofer in the making of parallel slits of study the diffraction phenomenon that his name is associated with this type of diffraction.

So the gratings are made by ruling equidistant parallel lines on an optically transparent sheet of material, the line portion being opaque and the space between the lines will be transparent to light. In the construction of a good quality grating, the following considerations must be taken into account.

(*i*) It should have large number of slits *i.e.,* the number of lines should be very large. Normally this number may be between 12,000 to 30, 000 lines/inch.

(*ii*) The spacings between the lines should be equal.

Initially diamond point was use for getting large number of lines/inch on glass with the help of a precision machine. After number of experiments, R.W. Wood showed that luminised pyrex glass is quite suitable for preparing large gratings.

Original gratings are quite expensive. So their photographic reproductions or replicas are got. A thin layer of collodion solution is poured uniformally over the surface of an original grating and the solution is allowed to harden to form a collodion film. This film is them detached from the grating and the impressions of the rulings of the original are preserved by mounting the collodion film between glass plates. This film is mounted against a plane silvered surface and not between glass plates. So in this way a large number of replicas can be produced from the single original grating.

In these days, because of the advances in the holography, holographic grating are made which have a large number of lines/inch than the ruled grating.

5.21. PLANE TRANSMISSION GRATING

The diffraction grating as stated above consists of a large number of fine equidistant lines of large number (between 12,000*LPI* to 30,000*LPI*) marked on a polished glass plate. The spaces between the lines act as transparencies while the lines act as opacities. If the width of the transparency and opacity be 'a' and 'b' respectively the distance $(a + b)$ is called grating constant or grating element and is generally denoted by g. The points between adjoining transparencies separated by a distance, $(a + b)$ are known as corresponding points. In Fig. 5.19, $ABC... H$ represents the section of a plane transmission gratin having N number of clear spaces placed $\perp$ to the plane of paper. Let a parallel beam of monochromatic light of wavelength λ be incident normally on the grating surface.

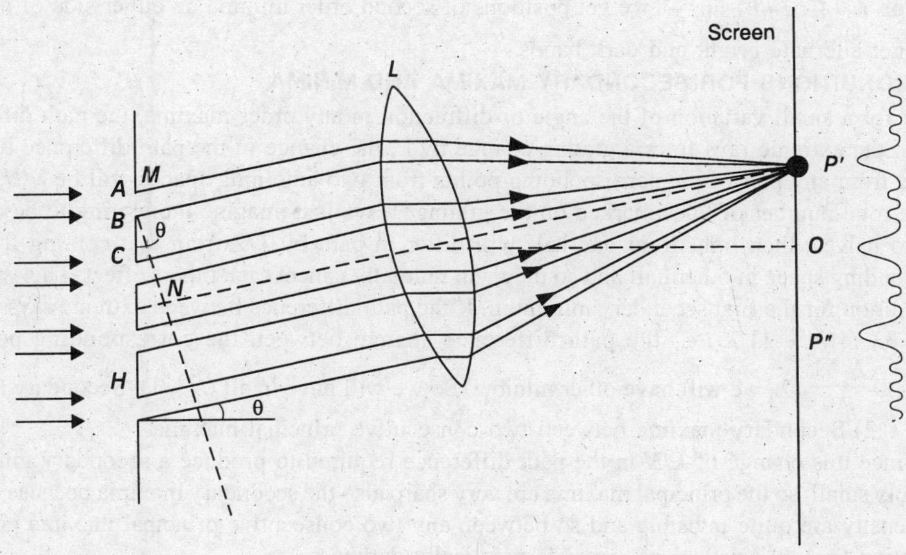

Fig. 5.19

Most of the light from the spaces go straight, but a part of it gets diffracted in various directions. This light is converged by convex lens L on the screen where alternate dark and bright bands are formed on both sides of the central maximum.

All the rays starting from AH reach O in phase with each other and so reinforce each other producing an image of the slits, corresponding to central maximum. The rays which are diffracted at an angle θ with the normal reach P_1 on passing through the lens in different phases. If AK is drawn $\perp$ to the direction of the diffraction light, then CN will correspond to the path difference between the two diffracted lights from the corresponding points A and C at an angle θ.

So path difference $\quad CN = AC \sin\theta = (a+b)\sin\theta$

Now the condition for maxima

$$(a+b)\sin\theta = n\lambda \qquad ...(1)$$

And condition for minima is

$$(a+b)\sin\theta = (2n+1)\frac{\lambda}{2}$$

where $\quad n = 0, 1, 2, 3, ...$

So the diffracted rays from any pair of corresponding points of the slits will produce constructive or destructive interference at a point P according as the path difference is even or odd multiple of $\lambda/2$. This condition holds for all the pairs of rays from the corresponding points of the total grating surface.

For $\quad n = 0, \quad \theta = 0, \quad (a+b)\sin\theta = 0$

which corresponds to the central maxima at 0

For $\quad n = 1, \quad (a+b)\sin\theta = \lambda$

This corresponds to the first order principal maxima on either side of the central maxima *i.e.*, at P' and P''

For $\quad n = 2, (a+b)\sin\theta_2 = 2\lambda$

This gives the position of the second order principal maxima at P_1' and P_1'' on either side of 0.

Also for $n = 0$, $(a+b)\sin\theta = \lambda/2$, which will correspond to the first order minima on either side of 0.

For $n=1, (a+b)\sin\frac{3\lambda}{2}$ we get positions of second order minima on either side of point O. So we get alternate bright and dark bands.

5.22. CONDITIONS FOR SECONDARY MAXIMA AND MINIMA

If for a small variation of the angle of diffraction in any order maxima, the path difference between the extreme rays from a grating change by λ, the change in the path difference between the rays from any pair of the corresponding points from two adjoining spaces will be λ/N, where N is the total number of lines marked on the grating. If we just imagine the grating to be divided into two halves, then light from one half will differ in path by $\lambda/2$ from that coming from the corresponding space in other half and so they will mutually cancel each others effect. This will give the condition for the first secondary minimum. If the path difference between extreme rays change by 2λ, 3λ... $(N-1)\lambda$ *i.e.,* the path difference change between the corresponding points is $\frac{2\lambda}{N}, \frac{3\lambda}{N} \ldots \frac{N-1}{N}$, λ, we will have other minima. So we will have in all $(N-1)$. Secondary minima and $(N-2)$ Secondary maxima between two consecutive principal maxima.

Since this change of λ/N in the path difference required to produce a secondary minima is negligibly small, so the principal maxima are very sharp also the secondary maxima because of their low intensity are quite invisible and so between any two consecutive principal maxima given by the relation $(a+b)\sin\theta_n = n\lambda$, there is practically darkness.

So after the principal maxima the condition for the secondary minima is

$$(a+b)\sin(\theta - d\theta) = n\lambda + \frac{n}{\lambda}$$

5.23. ABSENT SPECTRA

Some times, for a particular angle of diffraction θ satisfying the relation $(a+b)\sin\theta_n = n\lambda$, there is no visible spectrum. It is because for a given direction θ, the path difference between the diffracted rays from two extreme edges of one space is equal to an even multiple of $\lambda/2$ *e.g.,* if the path difference is λ, the slit can be considered to be divided into two halves and the path difference between the two pairs of corresponding points in the two halves will be $\lambda/2$, which will give zero intensity in that direction. So the condition for a minima in a single slit is given by

$$a\sin\theta = m\lambda \qquad \text{...(1)}$$

where m is an integer and is not equal to zero.

The condition for the principal maxima in the nth order spectrum is given by

$$(a+b)\sin\theta = n\lambda \qquad \text{...(2)}$$

If the conditions (1) and (2) are simultaneously satisfied, the diffracted rays from all the transparencies superimpose upon each other, but the resultant intensity is zero, *i.e.,* the spectrum is absent.

From Equation (1) and (2) we have

$$\frac{a+b}{a} = \frac{n}{m}$$

or

$$\frac{b}{a} = \left(\frac{n}{m} - 1\right) \qquad \text{...(3)}$$

If width of opacity is equal to that of transparency *i.e.,* $a = b$,

then from (3) $\quad n = 2m$

So that 2nd, 4th, 6th etc., orders of the spectra will be missing corresponding to the minima due to a single slit given by $m = 1, 2, 3, \ldots$ etc.

If width of opacity is twice the width of transparency *i.e.,* $b = 2a$, then from (3) $n = 3m$, so that 3rd, 6th, 9th etc., order of the spectra will be missing corresponding to the minima due to a single slit given by $m = 1, 2, 3, \ldots$ etc.

5.24. DISPERSIVE POWER OF GRATING

The change in angle of diffraction corresponding to a unit change in wavelength of light used measures the dispersive power of the grating.

Now from the condition of the nth order maxima, we have

$$(a+b)\sin\theta = n\lambda \qquad ...(1)$$

which shows that for a given order n, and grating element $(a+b)$, θ changes with λ. So, if wavelength changes from λ to $\lambda + d\lambda$, the corresponding angle of diffraction will change from θ to $(\theta + d\theta)$. The factor $\frac{d\theta}{d\lambda}$ is called dispersive power of the grating and may be got by differentiation of Equation (1)

$$(a+b)\cos\theta\frac{d\theta}{d\lambda} = n$$

or

$$\frac{d\theta}{d\lambda} = \frac{n}{(a+b)\cos\theta} = \frac{n}{(a+b)} \times \frac{1}{\cos\theta}$$

$$= \frac{m \times n}{\cos\theta}$$

where $m = \frac{1}{a+b}$ number of lines/cm of length. Since for small values of θ, $\cos\theta$ is practically constant, so that $d\theta$ is directly proportional to $d\lambda$. Such a spectrum is called Normal spectrum. It is also clear from above that dispersive power varies directly with order of spectrum n and number of lines/unit length m.

5.25. NORMAL SPECTRUM

From above we have seen that in the grating the spectrum formed is pure *i.e.,* there is no overlapping or mixture of colours unlike the prism spectrum where there is overlapping of different colours. Another important fact is that the dispersion depends in λ and $(a+b)$. So that the spectrum formed by one grating is exactly similar to that formed by another grating. The ratio of their lengths being $g : g'$, since the dispersion $\propto \frac{1}{g}$. So the widths occupied in the spectrum by any two colours is invariable. In case of prisms, we have different spectra because the relative dispersion of two colours will be different prisms because of the change of glass. This is also known as the irrationality of dispersion. Because of the fact that all grating spectra are similar, they are, therefore, called standard or Normal Spectra.

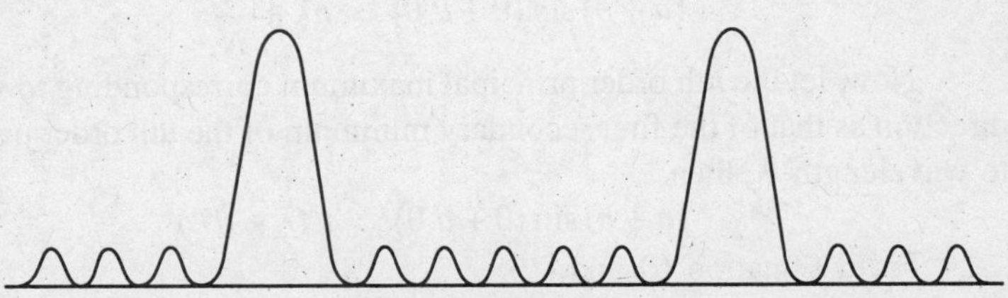

Fig. 5.20

5.26. RESOLVING POWER

The objects lying very close to each cannot be seen as separate by a naked eye but if these are enlarged by magnifying them then these can be seen as separate. This is because an eye can see two objects as separate only if the angle subtended by them at the eye is greater than one minute. Different optical instruments such as telescope, microscope, prism, grating etc., can be used to increase the angle subtended at the eye so that these objects appear as separate.

So this method of seeing two objects or images which are very close to each other as separate using some optical instrument is called resolution:

In the case of circular aperture of a telescope or a microscope, with a point object, we get diffraction patterns consisting of central bright disc surrounded by dark and bright concentric circular rings of rapidly decreasing intensity.

The diffraction patterns of two widely separated sources may be given as under.

If the two point sources are very near, then the diffraction patterns of these sources will overlap and they will appear as one, optical instruments are used to see them separate. So the capacity of an instrument to produce two separate images of very close objects is called resolving power.

In the case of a telescope, the reciprocal of the smallest angle subtended between two objects which can be seen as separate through the telescope measures its resolving power.

In case of microscope, the reciprocal of minimum distance between two point objects which can be seen separately through the microscope measures its resolving power.

The resolution in case of two very close spectral lines is known as spectral resolution.

$$\text{Spectral resolving power} = \frac{\lambda}{d\lambda}$$

where $d\lambda$ is the smallest difference in wavelengths than can be resolved by the instrument from wavelength λ.

5.27. RESOLVING POWER OF A DIFFRACTION GRATING

The resolving power of a grating is its ability to show two neighbouring lines in a spectrum as separate.

It may also be defined as the ratio of the wavelength of any spectral line to its difference of wavelengths between this line and a neighbouring line such that the two spectral lines can be just seen as separate.

Two lines of wavelengths λ and $\lambda + d\lambda$ are said to be just resolved if the central maxima due to $\lambda + d\lambda$ falls on the first minima of λ. The resolving power of diffraction grating is given by $\lambda/d\lambda$.

Consider two wavelets of wavelengths λ and $\lambda + d\lambda$ to be incident normally on the surface of the grating. The two wavelengths will give their own diffraction patterns. According to Rayleigh's criteria, the two patterns would be just resolved if the principal maxima of one falls on the first secondary minima of the other in any order.

The condition for the first secondary minimum after the nth order maximum is given by

$$(a+b)\sin(\theta + d\theta) = n\lambda + \frac{\lambda}{N} \qquad ...(i)$$

Now, let the nth order principal maximum corresponding to wavelength $\lambda + d\lambda$ has the same direction as that of the first secondary minimum of the nth order principal maximum corresponding to wavelength λ, then

$$(a+b)\sin(\theta + d\theta) = n(\lambda + d\lambda) \qquad ...(ii)$$

From Equation (i) and (ii)

$$n\lambda + \frac{\lambda}{N} = n\lambda + nd\lambda$$

$$\therefore \qquad nd\lambda = \frac{\lambda}{N}$$

or

$$\frac{\lambda}{d\lambda} = N \times n = (\text{Total no. of lines on grating} \times \text{order of spectrum})$$

which shows that resolving power depends on the order of spectrum and the number of lines on the grating but does not depend on the grating element.

5.28. DETERMINATION OF WAVELENGTH BY DIFFRACTION GRATING

With the help of diffraction grating the wavelength of any light can be found very accurately because the expression $(a+b)\sin\theta = n\lambda$ involves the determination of angle of diffraction only and this can be found quite accurately with the help of spectrometer. The grating constant $(a + b)$ is supplied by the manufacturer. The measurement of θ is done quite accurately as compared to the measurement of fringe width in case of interference phenomenon where the measurement requires lot of precision. For example in the biprism method $\lambda = \beta \frac{d}{D}$, where both β and d being very small cannot be measured with much accuracy.

For measurement of λ of a given source of light a spectrometer is used on the prism table of which is mounted the grating. The collimator slit is illuminated first by a sodium light source and the spectrometer is levelled and adjusted for parallel rays. The telescope is turned around and fixed at 90° with respect to collimator. The grating is fixed vertically on the prism table and the table is rotated till an image of the slit reflected from the grating surface is formed on the cross-wire of the telescope. Prism table is suitably levelled to get the image of the slit in the proper position. The grating table is next rotated through exactly 45° in backward direction so that its plane is ⊥ to the axis of the collimator. The telescope is then rotated and the positions of the diffracted images on either side of the central direct image are observed.

Sodium light is then replaced by the source whose wavelength is to be found. If the light contains more than one wavelengths we will get a spectrum. The telescope is rotated and fixed at the positions of diffracted images on either side of the central image are observed and their positions noted. From this, angle of diffraction is calculated from the central direct reading and the readings on either side of the central image.

Then knowing $(a + b)$, the grating element, the order n of the spectra and the corresponding angle of diffraction θ, the value of wavelength λ of the light used can be got from the expression

$$\lambda = \frac{(a+b)\sin\theta}{n}$$

Normally the number of lines on the grating are given and if there are N lines/inch, then grating element

$$(a + b) = \frac{2.54}{N} \text{ cm.}$$

Example 6. *A parallel beam of monochromatic light is allowed to be incident on a plane grating having 5000 lines/cm and the second order spectral line is found to be diffracted through 30°. Calculate the wavelength of light used.*

Solution: $(a + b) = \frac{1}{5000}, n = 2, \theta_2 = 30°$

Using the relation $(a+b)\sin\theta_2 = n\lambda$

We have $\frac{1}{5000}\sin 30° = 2\lambda$

$$\lambda = \frac{\sin 30°}{2\times 5000} = \frac{1}{2}\times\frac{1}{1000}$$

$$= 5\times 10^{-5} \text{ cm}$$

$$= \mathbf{5000\ Å}$$

Example 7. *Calculate the angle between the lamp filament and its first diffracted image produced by a fabric with 160 threads/cm. λ = 6000Å.*

Solution: Here $a + b = \frac{1}{160}, n = 1, \lambda = 6000\times 10^{-8}$ cm

Using the relation $(a+b)\sin\theta = n\lambda$

$$\sin\theta = \frac{n\lambda}{a+b} = \frac{1\times 6000\times 10^{-8}}{\frac{1}{160}}$$

$$= 160\times 6\times 10^{-5}$$

$$= 96\times 10^{-4} = 0.0096$$

$$\therefore \quad \theta = \sin^{-1} 0.0096 = \mathbf{33\ min.}$$

Example 8. *A plane transmission grating having 6000 lines/cm is used to obtain a spectrum of light from a sodium lamp in the second order. Calculate the angular separation between the two sodium lines whose wavelengths are 5890Å and 5896Å.*

Solution: Here $a + b = \frac{1}{6000}$, $n = 2$, $\lambda_1 = 5890 \times 10^{-8}$ cm

and $\lambda_2 = 5896 \times 10^{-8}$ cm

Now $(a+b)\sin\theta = n\lambda$

If θ_1 and θ_2 are the angles of diffraction for the two sodium lines of wavelengths λ_1 and λ_2 respectively in the second order,

$$(a+b)\sin\theta_1 = 2\lambda_1$$
$$(a+b)\sin\theta_2 = 2\lambda_2$$

Now $\sin\theta_1 = \frac{1}{(a+b)} \times 2\lambda_1 = 6000 \times 2 \times 5890 \times 10^{-8}\text{ cm} = 0.7068$

$\therefore$ $\theta_1 = 44°58'$

Again $\sin\theta_2 = \frac{1}{(a+b)} \times 2\lambda_2$

$= 6000 \times 2 \times 5896 \times 10^{-8} = 0.7074$

$\therefore$ $\theta_2 = 45°1'$

So the angular separation

$$\theta_2 - \theta_1 = 45°1' - 44°58' = 3'.$$

Example 9. *Light is incident normally on a grating 1/2 cm wide with 2,500 lines. Find the angular separation of the two sodium lines in the first order spectrum. Can they be seen distinctly, wavelength of the lines are 5890 and 5896Å.*

Solution: Width of grating = 1/2 cm = 0.5 cm

Number of lines $N = 2500$

Grating element $a + b = \frac{0.5}{2500} = \frac{1}{5000}$

Mean wavelength of two lines $\lambda = \frac{5890 + 5896}{2} = 5893\text{ Å}$

Also difference $d\lambda$ in wavelengths

$= 6\text{ Å} = 6 \times 10^{-8}\text{ cm}$

If θ be the mean angle of diffraction for the first order spectrum, then

$$\sin\theta = \frac{\lambda}{a+b} = 5000 \times 5893 \times 10^{-8} = 0.2946$$

$\therefore$ $\theta = 17°8'$

We know dispersive power $\frac{d\theta}{d\lambda} = \frac{n}{(a+b)\cos\theta}$

$\therefore$
$$d\theta = \frac{n \cdot d\lambda}{(a+b)\cos\theta}$$
$$= \frac{1 \times 6 \times 10^{-8} \times 5000}{\cos 17°8'}$$
$$= \frac{6 \times 10^{-8} \times 5000}{0.9556} \times \frac{180}{\pi}\text{ radians}$$
$$= \frac{6 \times 10^{-8} \times 5000 \times 180}{0.9556 \times \pi} \times 3600\text{ seconds} = 1'47.5''$$

The two lines will be seen distinctly if they are resolved by grating.

Now resolving power $\frac{\lambda}{d\lambda} = \frac{5893 \times 10^{-8}}{6 \times 10^{-8}} = 982$ N

Actual power of the diffraction grating

$$n \times N = 1 \times 2500 = \mathbf{2500}$$

The actual power is greater than the required resolving power, so the lines will be distinctly visible.

Example 10. *Sodium yellow light of λ = 5893Å, is a doublet of 6Å. What is the minimum number of lines a grating should have to resolve the doublet in the third order spectrum?*

Solution: $\lambda = 5893$Å, $d\lambda = 6 \times 10^{-8}$ cm, $N = ?$, $n = 3$

Resolving power $\frac{\lambda}{d\lambda} = N \cdot n$

$$\therefore \quad N = \frac{1}{n} \cdot \frac{d}{d\lambda} = \frac{5893 \times 10^{-8}}{3 \times 6 \times 10^{-8}} = \mathbf{327.4}$$

QUESTIONS AND PROBLEMS

1. What are the various diffraction phenomenon? How these can be explained from the wave theory, give the difference between interference and diffraction?
2. Explain the formation of half period elements and hence explain the rectilinear propagation of light. How this phenomenon differs from the rectilinear propagation of sound?
3. What is a zone plate? Give its construction and theory and explain its similarities and dissimilarities with a convex lens.
4. Explain the phenomenon of diffraction at a straight edge. How will you explain the bands with the geometrical shadow of the obstacle?
5. Give the theory of plane transmission grating. How it can be used to find wavelength of light and in what respect the determination of wavelength gives better results than those obtained by interference phenomenon?
6. Explain the terms:
   (*i*) Dispersive power (*ii*) Resolving power
   (*iii*) Absent spectra in the case of a diffraction grating.
7. A zone plate has the radius of the first ring 0.025 cm. If plane wave (λ = 5000Å) falls on the plate, find where the screen must be placed so that light is focussed to a bright spectral. **[Ans.** 12.5 cm]
8. If the diameter of the central zone of a zone plate is 2.5 mm and a point source of light of wavelength 5×10^{-5} cm is placed 5 meters away from the zone plate, find the position of the primary and secondary images.
9. In a particular experiment λ = 6000Å, the distance between the slit and the edge is 6 meters and between the edge and the eye piece is 4 meters. Find the positions of the first three maxima and their separations. **[Ans.** 0.2 cm, 0.346 cm, 0.447 cm, 0.146 cm and 0.101]
10. Calculate the least width that a grating must have to resolve two components of sodium D lines in the second order, the grating having 800 lines/cm. (The wavelengths of the two sodium lines are 5890Å and 5896Å). **[Ans.** 0.614 cm]
11. A beam of monochromatic light ($\lambda = 6.28 \times 10^{-5}$ cm) incident normally on a grating produces first order diffracted image 55′ from the direction of incidence. Calculate the grating constant. **[Ans.** 0.004064]
12. The separation of Na – lines (**Mean** λ = 5893Å) in the second order spectrum of a transmission grating containing 5000 lines per cm is 12.5 minute for normal incidence. What is the difference between two yellow lines? **[Ans.** 5.9Å]
13. Light of wavelength 5000Å and 5200Å falls normally on a plane transmission grating having 5000 lines/cm. If a lens of 200 cm focal length is used to form spectrum on a screen, find the distance between two lines in the first order. **[Ans.** 0.4 cm]

CHAPTER

6

POLARIZATION

6.1. INTRODUCTION

The interference and diffraction phenomenon have been able to establish firmly the wave nature of light, however, both these phenomenon fail to establish whether light waves are longitudinal or transverse.

The exact nature of light waves that they possess the transverse character has been established by polarization phenomenon. This phenomenon was studied by many scientists like Malus, Huygens and others but it was Fresnel who was able to satisfactorily explain this phenomenon.

We know that in the longitudinal wave motion, the particles of the medium execute periodic motion about their mean positions along the direction of propagation of waves and in the transverse waves, the motion of the particles is perpendicular to the motion of the wave. So in case of longitudinal waves, they are symmetrical about their direction of propagation and two waves can be similar in all their properties.

In the case of transverse waves, there can be number of directions in which the particles can execute periodic vibration perpendicular to the direction of propagation. So two similar transverse waves may differ from one another because of their different directions of vibrations.

Transverse waves in which the vibrations are restricted to one particular direction cannot be symmetrical and must exhibit some characteristics having some relation with the direction of propagation. This dependence of certain properties on the direction or this property of asymmetry or one sidedness is called polarity. A wave having such a characteristics is said to be polarized and the phenomenon is called polarization. So the polarization is the characteristic of a transverse wave and this can be explained by a simple analogy. Take a stretched rubber cord *AB* passing through two slits, S_1 and S_2 parallel to each other in the vertical plane as shown in Fig. 6.1 (*i*). Set up a longitudinal wave by moving the end *A* to and for || to the length of the cord. Rotate any side about *AB* as axis, it will be seen that the passage of wave is not affected. However, in the case of a transverse wave *i.e.*, when the end *A* is moved up and down in vertical plane, it will be seen that the waves pass undisturbed as in Fig. 6.1 (*i*).

If now the slit S_2 is rotated through 90° about *AB* as axis a transverse wave will pass through slit S_1 undisturbed as before but not through S_2 so beyond S_2, the cord remains practically undisturbed as in Fig. 6.1 (*ii*). So the transverse waves can pass through the slits when they are parallel to each other and lie in the same plane in which the wave lies. So the transverse waves have different properties with respect to vertical plane as compared to the other plane. This is an important difference between a longitudinal and a transverse wave.

Tourmaline crystals have the same behaviour with respect to light waves as two slits with respect to transverse waves in the string.

Consider a tourmaline crystal *A* cut parallel to its crystallographic axis. If the crystal is held in the path of light as shown in Fig. 6.2 (*a*) the emergent beam will be slightly coloured. If the crystal is rotated about *SO* as axis, there will be no change in the colour or the intensity of the emergent beam. If a second crystal *B* is held in the path of light and is cut as the crystal *A*, so that its axis is parallel to the axis of the crystal *A*, the beam will pass through both crystals and will still be coloured.

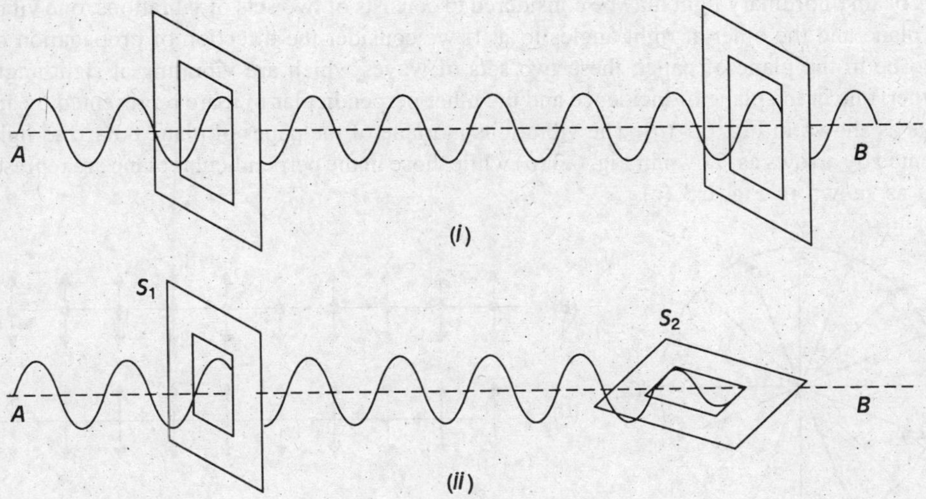

Fig. 6.1

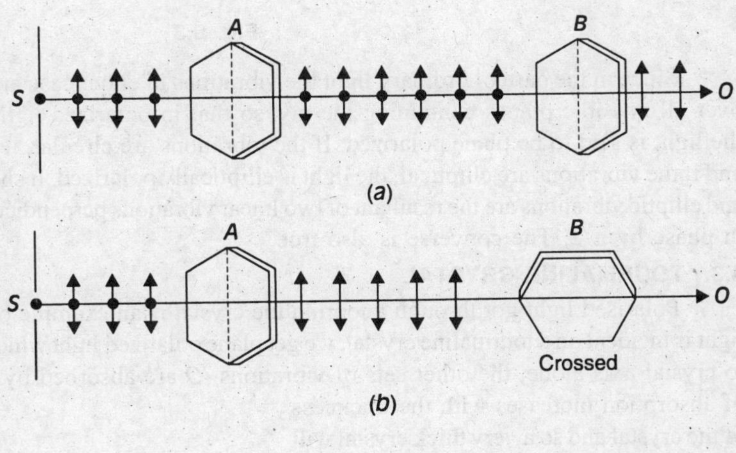

Fig. 6.2

Now if the crystal B is rotated about the beam axis, the intensity of light will appear to decrease and when both axis of the crystal are at right angles to each other, no light will come out of B as shown in Fig. 6.2 (*b*). If the crystal B is further rotated light, re-appears and intensity becomes maximum again when the axes are parallel. This will occur after a further rotation of B through 90°. This experiment firmly confirms the transverse nature of light *i.e.*, their oscillations are perpendicular to the direction of propagation.

In the case of light coming out of tourmaline crystal A, the oscillations are confined to only one direction *i.e.*, in a direction ⊥ to the direction of propagation of light, such a light which acquires the property of one sidedness is called *polarized light*. If the oscillations constituting the beam of light are confined only to one plane, the light is called *plane polarised light*.

The properties of a plane polarized beam differ with respect to two planes, one containing oscillations and the other at right angles to it.

The plane containing oscillations is called *plane of oscillation* or plane of vibration and a plane ⊥ to the plane of vibration of the light wave is called the *plane of polarization*.

6.2. REPRESENTATION OF VIBRATIONS

Light waves are transverse waves and so have vibrations at right angles to the direction of propagation. An ordinary beam of light which consists of millions of such waves, each with its own plane of vibration, will contain waves vibrating in all possible planes with equal probabilities. Figure 6.3 shows that the direction of propagation of a beam of light is perpendicular to the plane of paper while its vibrations are in the plane of paper.

A beam of ordinary light may be considered to consists of two sets of vibrations, one vibrating in one plane and the other at right angles to it. If we consider the direction of propagation of the beam to be in the plane of paper, these two sets of waves which are vibrating at right angles to each other, one in the plane of incidence and the other perpendicular to it are represented by arrows and dots as shown in Fig. 6.3 (*a*). The vibrations in plane of the paper of plane polarized light are represented by arrows as shown in Fig. 6.3 (*b*) while those in the perpendicular plane are represented by dots as shown in Fig. 6.3 (*c*).

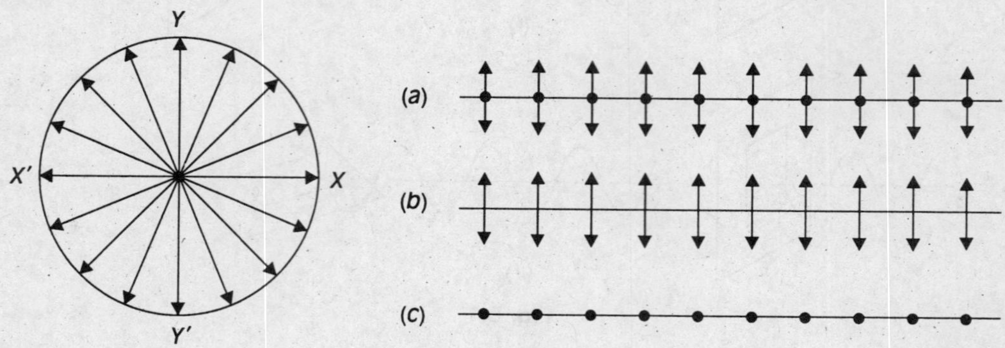

Fig. 6.3

Since in the case of ordinary light the vibrations in either case are considered to be distributed over all possible planes containing the ray, so that in polarized light, if the vibrations are linear the light is said to be plane polarized. If the vibrations are circular, it is called circularly polarized and if the vibrations are elliptical, the light is elliptically polarized. It should be noted that the circular and elliptic vibrations are the resultant of two linear vibrations perpendicular to each other and differing in phase by $\pi/2$. The converse is also true.

6.3. TOURMALINE CRYSTAL

Polarised light got through a tourmaline crystal is an example of selective absorption. When light is incident on a tourmaline crystal, we get plane polarized light which have displacements parallel to crystal axis alone, the other sets of vibrations, *O* are absorbed by the crystal, and this amount of absorption increases with the thickness of the crystal and so a very thick crystal will allow only the extraordinary component (*E*) to pass through it.

6.4. PRODUCTION OF PLANE POLARIZED LIGHT

Some of the methods for producing plane polarized light are given under:

(*i*) By reflection.
(*ii*) By transmission through a pile of plates.
(*iii*) By double refraction.
(*iv*) By selective absorption in crystals.
(*v*) By scattering.

6.5. REFLECTION METHOD

This is the simplest method of getting plane polarized light and was first given by Malus, a French scientist, in 1808. After a

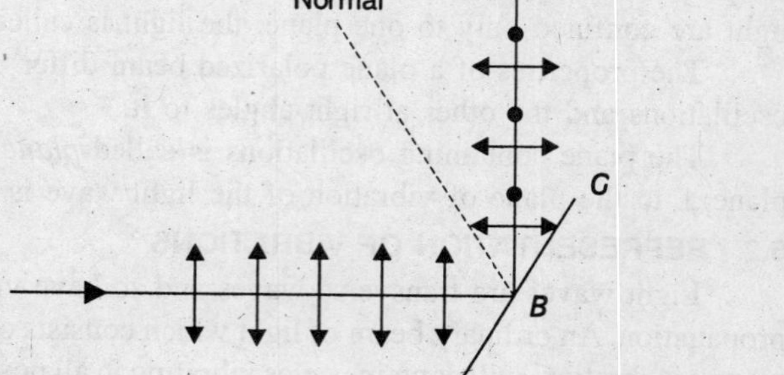

Fig. 6.4

set of experiments, Malus, was able to see that if an ordinary beam of light is reflected from some transparent medium like water, glass etc. the reflected light is partially polarized and the degree of polarization varies with the angle of incidence, and this percentage of polarized light in the reflected beam is greatest when the angle of incidence is equal to the angle of polarization. For an ordinary glass, the angle of maximum polarization is 57.5° which varies slightly with the wavelength of incident light, since refractive index depends on wavelength. So, we can get complete polarization with monochromatic light. This reflected light is said to be plane polarized in the plane of incidence and the vibrations of this plane polarized reflected light are found to be perpendicular to the plane of incidence.

Experimentally we can demonstrate the phenomenon by considering an ordinary beam of light incident along AB on a reflecting surface G of the transparent medium. The beam will be reflected along BC.

If a tourmaline crystal T is held in the path of reflected light and rotated along the BC axis, we find that the transmitted light shows variation in intensity showing that the reflected light is partially polarized. If the crystal is rotated till we get a position where transmitted light has minimum intensity and vary the angle of incidence till the intensity of the reflected beam after transmission through crystal is almost zero. At this stage the percentage of polarized light in the reflected beam will be maximum and this angle at which reflected polarized light has maximum intensity is called 'Angle of Polarization'. The value of this angle for glass is 57.5°.

Further if the crystal is rotated along BC axis, we have the intensity of the reflected beam after passing through the crystal as maximum or minimum according as the longer axis of the crystal is perpendicular or lies in the plane of incidence and also the intensity will be maximum and minimum twice in one complete rotation. This means that the light vibrations in the plane polarized reflected light are perpendicular to the plane of incidence. This is a very easy method for production and detection of plane polarized light. We can detect the plane polarized reflected light by using another mirror instead of tourmaline crystal. This can be got by 'Biot's Polariscope'.

6.6. BIOT'S POLARISCOPE

Monochromatic light from a source S is allowed to fall at an angle of 57.5° on a glass plate G_1 and on reflection along BC, it is again reflected at 57.5° by second glass plate G_2 placed parallel to G_1 as shown in Fig. 6.5.

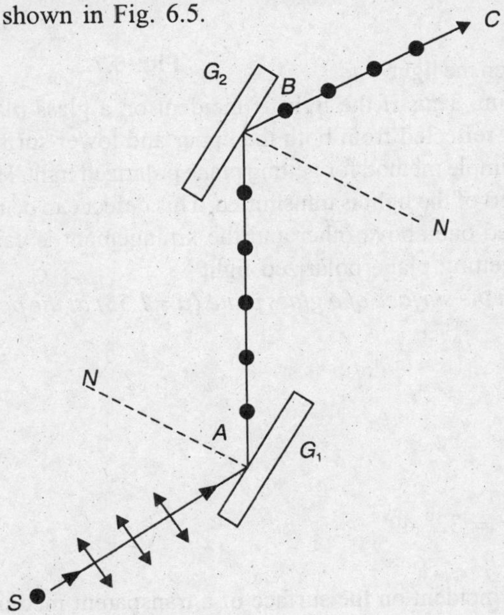

Fig. 6.5

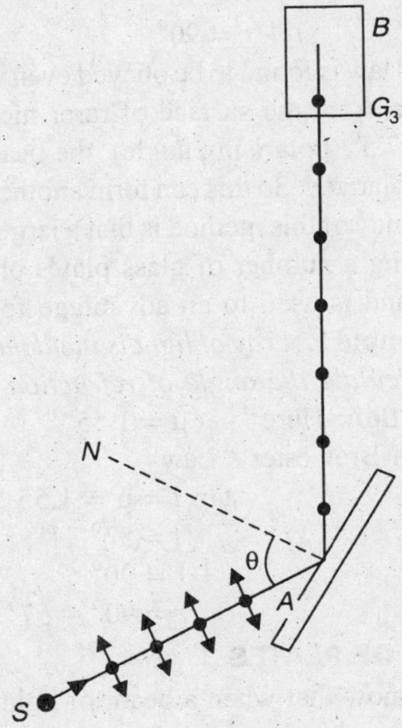

Fig. 6.6

Both the plates G_1 and G_2 are held at an equal angle to the horizontal, so that the light always falls on them at the same angle of incidence. In this position we get the intensity of the reflected beam along *BC* as maximum.

If now the upper plate is rotated about *BA* axis, the intensity along *BC* will be found to decrease and it will become zero when the plate has been rotated through 90°. The experiment can be best performed if the upper plate is painted black to avoid internal reflections.

In the above experiment, since plate G_1 polarizes light it is called a 'polarizer' and since G_2 analyses light, it is called 'analyser'.

6.7. BREWESTER'S LAW

In 1811, Brewester found that a simple relation between angle of polarization and refractive index μ of the medium exists. He got this relation as $\mu = \tan i$, *i.e.*, refractive index of the material medium is equal to tangent of the angle of polarization. This is called Brewester's law.

An important conclusion of the above law is that the reflected rays are at right angles to the refracted rays.

Proof: Let *AB* be the surface of separation of a transparent medium, say glass and *PQ* the incident beam, *QR* the reflected beam and *QS* the transmitted or refracted beam.

Angle *i* and *r* are the incident and refracted angles.

Now from Brewester's Law

$$\mu = \tan i = \frac{\sin i}{\cos i} \quad ...(1)$$

Also from Snell's Law

$$\mu = \frac{\sin i}{\cos i} \quad ...(2)$$

From (1) and (2)

$$\frac{\sin i}{\cos i} = \frac{\sin i}{\sin i}$$

or
$$\frac{\sin i}{\sin (90 - i)} = \frac{\sin i}{\sin r}$$

∴
$$\sin r = \sin (90^\circ - i)$$

or
$$r = 90^\circ - i$$

or
$$i + r = 90^\circ$$

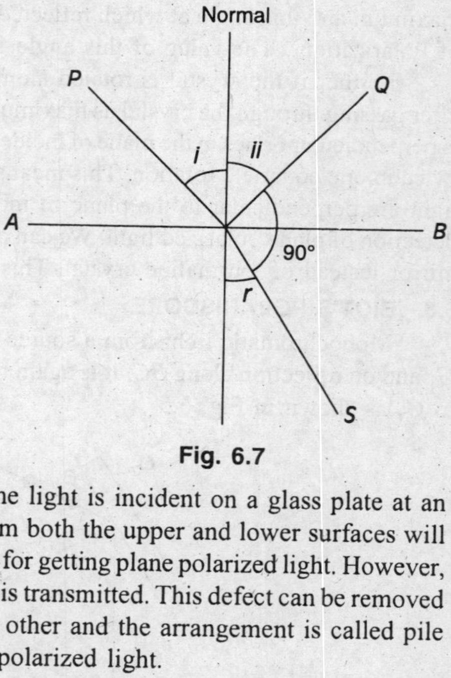

Fig. 6.7

This law is found to be obeyed even when the light is reflected from the surface of rarer medium. Thus if the light is incident on a glass plate at an angle of 57.5° (polarising angle), the beams reflected from both the upper and lower surfaces will be plane polarized. So this can form another simple method for getting plane polarized light. However, a descrepancy of this method is that a large part of the light is transmitted. This defect can be removed by arranging a number of glass plates placed one above other and the arrangement is called pile of plates and is used to an advantage for getting plane polarized light.

Example 1. *A ray of light is incident on the surface of a glass plate (μ = 1.53) at the polarizing angle. Calculate the angle of refraction.*

Solution: Here $\mu = 1.55$

From Brewester's Law

∴ $\tan i = \mu = 1.55$

or $i = 57^\circ\ 11'$

Also $i + r = 90^\circ$

∴ $r = 90^\circ - 57^\circ\ 11' = 32^\circ\ 49'$

6.8. PILE OF PLATES

We know that when a beam of light is incident on the surface of a transparent medium, part of it is reflected and rest transmitted through the medium. The reflected light is partially polarized for all the angles of incidence, but is completely plane polarized in the plane of incidence at the polarizing angle.

The transmitted beam will have components of vibrations in the plane of paper and some components perpendicular to the plane of paper and so will be unpolarized.

If the incident light at the polarizing angle on a pile of thin parallel faced plates placed one another, then at each reflection more of the components having vibrations ⊥ to the plane of incidence will be filtered out with the result that the transmitted light will have more and more plane polarized light having vibrations in the plane of incidence. It is found that between 10 – 15 plates are able to completely polarise the vibrations in the transmitted beam.

The apparatus consists of about 10 – 15 thin glass plates placed in contact with each other making an angle of 32.5° (90° – 57.5°) with the axis of the wooden tube as shown in Fig. 6.8. We will get reflected and transmitted polarized light. This arrangement can further be extended by another set of plates which can analyse the transmitted light.

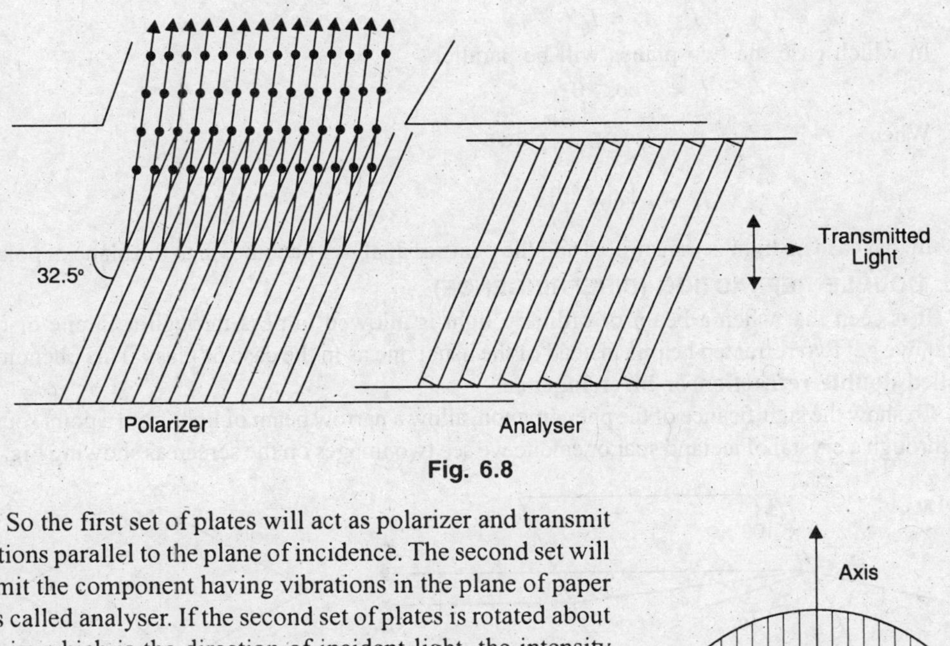

Fig. 6.8

So the first set of plates will act as polarizer and transmit vibrations parallel to the plane of incidence. The second set will transmit the component having vibrations in the plane of paper and is called analyser. If the second set of plates is rotated about the axis, which is the direction of incident light, the intensity of transmitted beam gradually decreases and becomes practically zero when, it is rotated through 90°. It is because of this position, the vibrations now become perpendicular to the plane of incidence of the analyser, and are now reflected instead of transmission. Further rotation through 90° will bring back the normal intensity. The intensity, I_1 of the polarized light transmitted through the analyser is given by Malus law,

$$I_1 = I \cos^2 \theta$$

where I is the original intensity, and θ is the angle between planes of the polariser and the analyser.

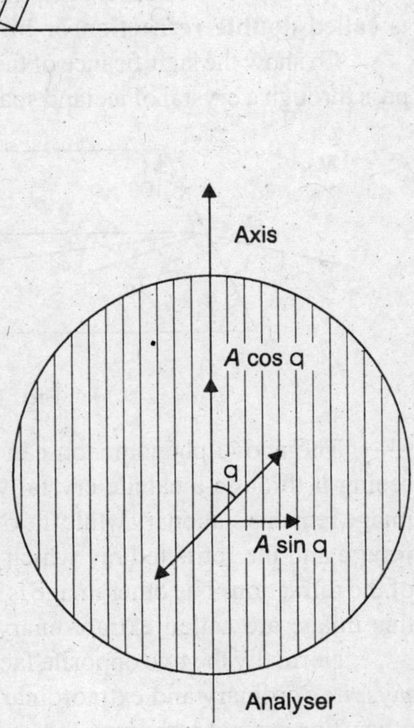

Fig. 6.9

6.9. MALUS LAW

We have seen that when the light is incident on a polariser, the transmitted light is plane polarized and if this light is made to pass through the analyser, the intensity varies with the angle between the planes of polariser and analyser. Malus stated that the *intensity of the polarized light transmitted through the analyser varies as the square of cosine of the angle between the plane of transmission of the analyser and the plane of polariser.*

The same statement is true for the reflected polarized light.

Proof: The proof is based on the fact that any polarized light can be split into rectangular components (*i*) parallel to the plane of transmission of the analyser (*ii*) perpendicular to the plane of analyser.

If A be the amplitude of the incident plane polarized light and θ the angle that this vibration makes with the axis of analyser, we have the amplitude component || to the optic axis = $A \cos \theta$.

This component will be transmitted through the analyser. The intensity I_1 of this component is given by

$$I_1 = A^2 \cos^2 \theta$$

When the axis of analyser makes any angle of 0° or 180° with the incident vibration, then

$$I = A^2 = I_0$$

In which case the two planes will be parallel

$$\therefore \quad I_1 = I \cos^2 \theta$$

When $\theta = \frac{\pi}{2}$ or $\frac{3\pi}{2}$ then

$$I_1 = I\left(\cos \frac{\pi}{2}\right)^2 = 0$$

In that case no light is transmitted and the plane of analyser becomes ⊥ to the plane of polarizer.

6.10. DOUBLE REFRACTION (BIREFRINGENCE)

It is seen that when a beam of ordinary light is allowed to pass through a calcite or quartz crystal, we get two refracted beams instead of the usual one as in the case of glass. This phenomenon is called **double refraction** or **birefringence.**

To show the significance of the phenomenon, allow a narrow beam of light from a point source to pass through a crystal of iceland spar or calcite we see two images on the screen as showing Fig. 6.10.

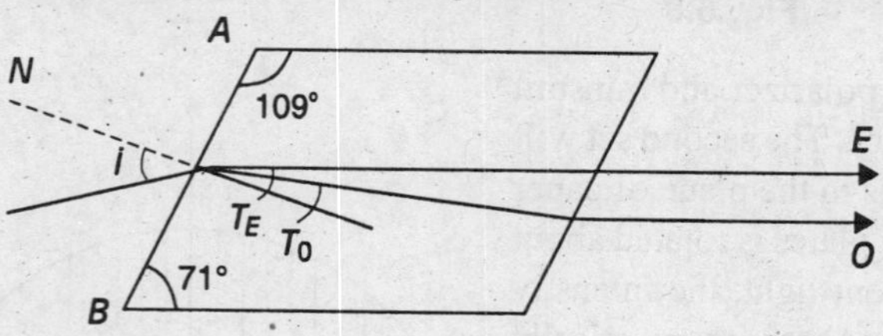

Fig. 610

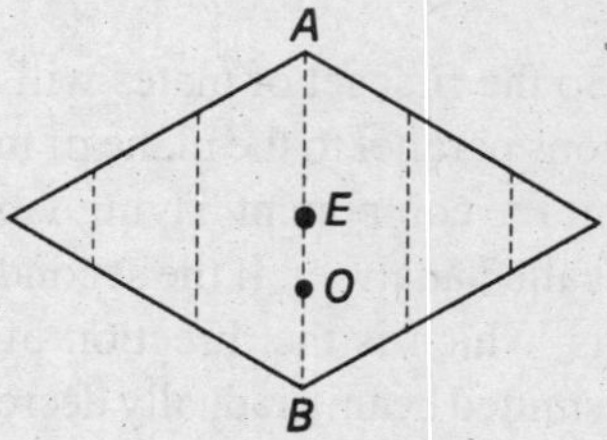

Fig. 6.11

The above phenomenon can be observed in a simple way by putting a dot on a paper and seeing it through a calcite crystal when two images will be observed. If the crystal is rotated one image remains stationary while the other rotates round the first. The stationary image is called **ordinary image** and the refracted ray which produces it is called ordinary ray and these rays obey the laws of the refraction. The other image is known as **extraordinary image** and the refracted rays producing this image are called extraordinary rays because they do not obey the laws of refraction.

Normally the two opposite faces of a calcite crystal are always parallel so that both the refracted rays, *i.e.*, ordinary and extraordinary rays emerge parallel to the incident beam and hence || to each other. For normal incidence the ordinary ray will pass straight through without deviation whereas the extraordinary ray will be transmitted at some angle and come out || to and displaced from the incident ray. As shown above on rotating the crystal about the beam as axis, the E ray rotates about the fixed *O* ray, the line joining the shorter diagonal of the emergent crystal face.

It may further be noticed that both ordinary, and extraordinary rays are plane polarized *i.e.*, their planes of polarization are perpendicular to each other. If a tourmaline plate is rotated, the intensities of both the images undergo change. It is seen that if the intensity of ordinary image is increasing that of extraordinary is decreasing and vice versa *i.e.*, if one light is very bright the other is not visible which means that ordinary and extraordinary images are brightest twice in each complete rotation.

6.11. CALCITE CRYSTAL

Iceland spar which is also known as calcite is transparent to visible and ultraviolet light. Chemically this is calcium carbonate ($CaCO_3$) and in nature they occur in different forms, all of which have rhombohedron shape on cleavage as shown in Fig. 6.12. Each of the six faces of the crystal form a parallelogram having angles of 78° and 102° approximately. At the two opposite corners *B* and *H*. The angles of three faces meeting there are obtuse while at the remaining six corners one angle is obtuse and two are acute. The corners *B* and *H* are called Blunt corners.

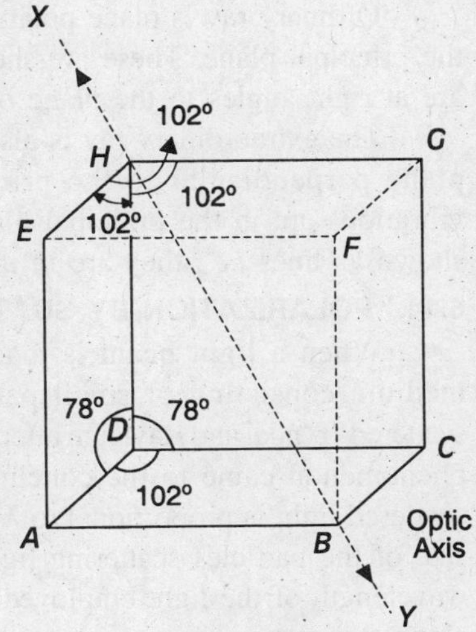

Fig. 6.12

6.12. OPTIC AXIS

In the case of calcite and quartz crystals, there is an important characteristic *i.e.*, there is only one direction through the crystal along which no double refraction takes place. This particular direction is called optic axis.

In the case of calcite crystal it is the line through which one of the blunt corners (*B* or *H*) making equal angles with the three edges meeting there or any other line || to it.

If the Rhombohedron is cut in such a way that all the edges are equal as shown in the figure above, then the line joining the opposite blunt corners *BH* or any line parallel to it gives direction of optic axis. It should be noted that optic axis is direction and not a particular line and the crystal is symmetrical about the optic axis. Also a ray passing along optic axis does not break into ordinary (*P*) and extra-ordinary (*E*) rays *i.e.*, both have the same velocities along the optic axis.

6.13. PRINCIPAL SECTION AND PRINCIPAL PLANE

A plane containing the optic axis and ⊥ to the face of the crystal is called the principal plane or principal section for that face. In the Fig. 6.13 below, *AA′ BB′* is the principal plane for the top and bottom faces of calcite crystal.

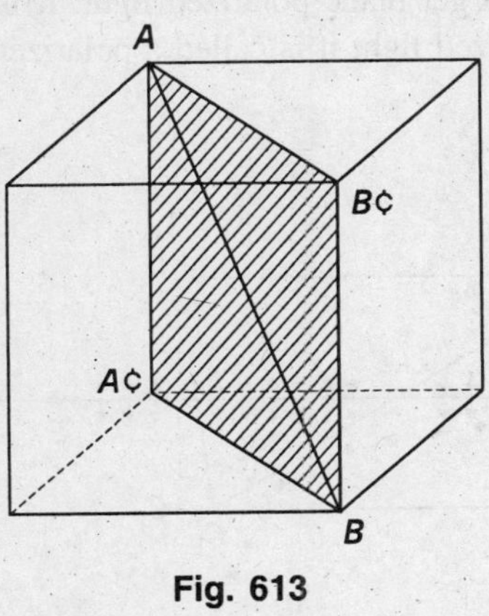

Fig. 613

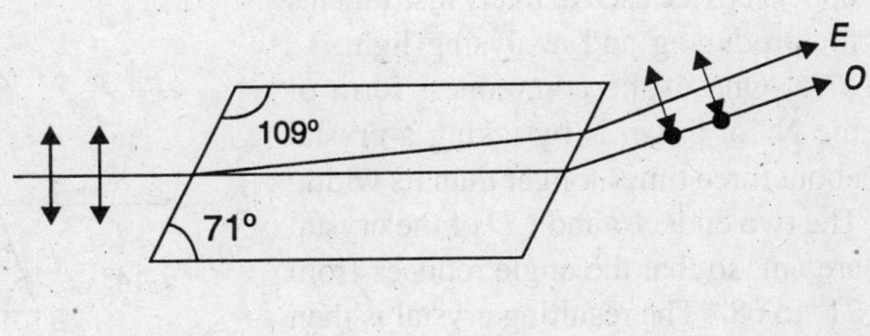

Fig. 6.14

For each point there are three principal planes one for each pair of opposite faces of the crystal.

The principal plane for calcite crystal is a parallelogram with angles 71° and 109°. Since the opposite faces of the calcite crystal are always parallel the two refracted rays come out parallel to each other and the incident ray.

Ordinary ray is plane polarized in the principal plane *i.e.,* its vibrations are perpendicular to the principal plane. These are shown by dots *i.e.,* they are at right angles to the plane of paper.

The extraordinary ray is also plane polarized in a plane perpendicular to the principal palne *i.e.,* the vibrations are in the principal plane. These have been shown as lines *i.e.*, they are in the plane of paper.

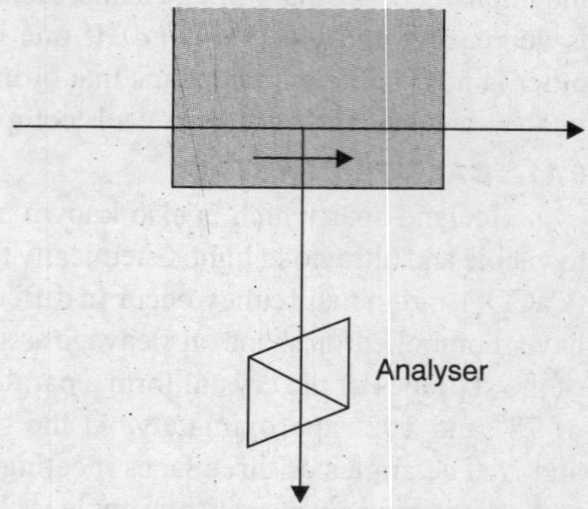

Fig. 6.15

6.14. POLARIZATION BY SCATTERING

When a light beam is made to pass through a medium consisting of small particles, some of it is scattered. Tyndal and Rayleigh after studying the scattering phenomenon came to the conclusion that intensity of scattered light is proportional to λ^{-4} *i.e.,* $1/\lambda^4$. When the size of the particles scattering light is smaller than the wavelength of the light employed. If the scattered light along the perpendicular path to the original light is examined by an analyser crystal, it is found that intensity changes from maximum to minimum if the caystal is rotated through 90°, which shows that the scattered light is polarized. Brewester discovered that blue light from the sky is also polarized because of scattering.

6.15. POLARIZATION BY SELECTIVE ABSORPTION (DICHROISM)

Some crystals have the charateristic of selective absorption and the most notable example being the tourmaline crystal. This crystal is doubly refracting crystal and transmits only *E*-rays whose vibrations are parallel to the crystal axis and absorbs *O* rays. The absorption depends on the thickness of the crystal and few mm thick crystal is able to absorb *O*-rays whose vibrations are ⊥ to its axis.

This phenomenon of selective absorptions (Dichroism) is used in the polaroids, which are large sized polarizing films of suitable cross-section that helps in getting polarized light beams. They can be made by deposition of ultra microscopic polarizing crystals, with their optic axis all parallel, on a thin sheet of nitrocellulose. Each individual crystal will be transmitting *E*-rays and absorb the *O*-rays.

6.16. NICOL'S PRISM

This is one of the commonest and most important device to get plane polarized light. It is made from calcite crystal and when it is used for producing polarized light it is called a polarizer and when it is used for analysing light, it is called an analyser. It is a useful optical device used in many instruments for producing and analysing light.

One of the commonest form of the Nicol prism is by taking a crystal about three times longer than its width. The two ends *AB* and *CD* of the crystal are cut, so that the angle reduces from 71° to 68°. The resulting crystal is then cut into two parts along the plane A_1 C_1 passing through the blunt corners and ⊥ to both the principal section and end forces, so that A_1 C_1 makes and

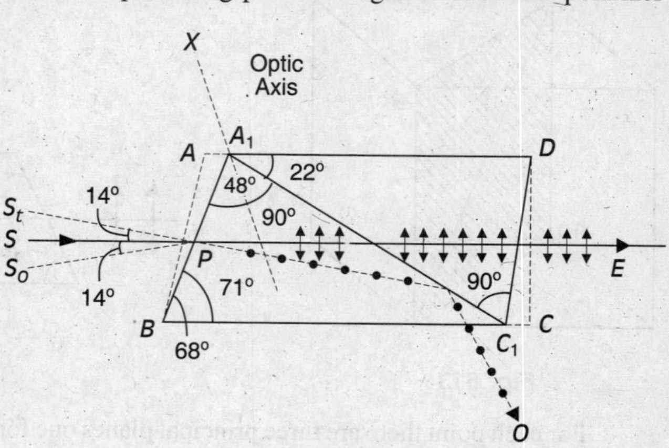

Fig. 6.16

angle of 90° with $C_1 D$ and $A_1 B$ as shown in Fig. 6.16.

The two cut faces are grinded and polished to make very smooth surfaces and they are optically flat. Then these surfaces are cemented by a thin layer of canda balsam, a clear transparent material having refractive index midway for *O* and *E*-ray. For example, for sodium light λ = 5893 A.U.

$$\mu_G = 1.6584$$
$$\mu_{CB} = 1.55$$
$$\mu_E = 1.4964$$

It should be noted that the principal section of the crystal coincides with the principal planes of both *O* and *E*-rays. The two end faces of the crystal are left open while its sides are coated with lamp black and kept in a brass tube. A ray *SP* incident on one of the end faces of the Nicol prism and moving nearly parallel to its length is divided into *O* and *E* rays which have their respective vibrations ⊥ and || to the principal section of the Nicol Prism. The ordinary ray going from denser to rarer medium will suffer total internal reflection provided the angle of incidence at balsam layer is greater than the critical angle for the two media *i.e.*, its value is greater than $\sin^{-1}\left(\frac{1.55}{1.658}\right) = 69^\circ$

This totally reflected ray will be absorbed by lamp black layer on the side of the prism. As is clear from above, the angle of incidence on the face $A_1 B$ is limited to a very narrow range of about 14°. For angle of incidence greater than this value, the *O*-ray will strike the balsam layer at angles less than the critical angle and so will be transmitted. It should be noted that optically, Canada balsam is more dense than calcite for the *E*-ray and less dense for *O*-ray.

This is why the *E*-ray travels from an optically rarer to a denser medium and will emerge out of the crystal, whereas *O*-ray for large angles of incidence will be totally reflected.

Also, it is only along the optics axis that *E*-ray and *O*-ray have same velocities and in all other directions they have different velocities, *E*-ray having maximum value of velocity at right angles to the optic axis, the value of μ_E for sodium light is 1.486 or 1.658 according as *E*-ray leaves ⊥ or along the optic axis.

Along optic axis $\mu_E = \mu_O = 1.658$. So for intermediate angles, the effective value of μ_E lies between 1.486 and 1.658, so that when the angle of incidence at the end faces exceeds certain limit, the Canada balsam becomes optically less dense than calcite and *E*-ray is totally reflected by balsam. This limits the incident angle for *E*-ray to about 14°. So to avoid the transmission of *O*-ray and totally internal reflection of *E*-ray the angle between the extreme incident rays is limited to about 20°. However must be noted that Nicol prism cannot be used for highly convergent or divergent lights.

Example 2. *The critical angle for certain wavelength of light in the case of a piece of glass is 40°. Find the polarising angle for glass.*

Solution: Critical angle $C = 40^\circ$

Refractive index for glass $\mu = \frac{1}{\sin C} = \frac{1}{\sin 40^\circ} = \frac{1}{0.6427}$

$= 1.556$

Using Brewester's law $\mu = \tan i$, i being polarizing angle.

$\therefore$ $1.556 = \tan i$

or $i = $ **57.3°**.

Example 3. *Intensity of light through a polariser and analyer is maximum when their principal planes are parallel. Through what angle the analysing Nicol must be rotated so that the intensity gets reduced to 1/4 of the maximum value?*

Solution: From Malus Law $I_1 = I\cos^2\theta$

$\therefore$ $I_1 = \frac{I}{4} = I\cos^2\theta$

or $$\cos^2 \theta = \frac{1}{4}$$

$$\cos \theta = \pm \frac{1}{2}$$

$$\theta = \mathbf{60^o} \text{ or } \mathbf{120^o}.$$

Example 4. *Two Nicol prisms are so arranged that the amount of light transmitted through them is maximum. What will be the percentage-reduction in intensity of the incident light when analyser is rotated through (i) 30° (ii) 45° (iii) 60° (iv) 90°?*

Solution: From Malus Law $I_1 = I \cos^2 \theta$

$\therefore$ $$\frac{I_1}{I} = \cos^2 \theta$$

where $I \rightarrow$ original intensity

$I_1 \rightarrow$ Intensity of transmitted light

$\theta \rightarrow$ the angle through which analyer is rotated

% of reduction in intensity

$$\left(\frac{I - I_1}{I}\right) \times 100 = \left(1 - \frac{I_1}{I}\right) \times 100$$

$$= (1 - \cos^2 \theta) \times 100$$

(*i*) When $\theta = 30^o$, $\cos 30^o = 0.866$

% reduction of intensity $= [1 - (0.866)^2] \times 100$

$= 25\%$

(*ii*) When $\theta = 45^o$, $\cos 45^o = 0.777$

% reduction of Intensity $= (1 - 0.5^2) \times 100 = 75\%$

(*iii*) When $\theta = 60^o$, $\cos 60 = 0.50$

% reduction of Intensity $= (1 - 0.5^2) \times 100 = 75\%$

(*iv*) When $\theta = 90^o$, $\cos \theta = 0$

% reduction of Intensity $= [1 - 0^2] \times 100 = 100\%$.

6.17. HUYGEN'S THEORY OF DOUBLE REFRACTION

Huygens extended his theory of secondary wavefronts to explain the phenomenon of double refraction in uniaxial crystals. According to this theory

(*i*) A point source of mono-chromatic light in a doubly refracting medium is the origin of two wavefronts.

(*ii*) The ordinary wave travels with same velocity in all directions and so the corres-ponding wavefront will spherical.

(*iii*) Extraordinary waves have different velocities in different directions, so the corresponding wavefront will be elliptical.

Now since along the optic axis both *O*-ray and *E*-ray travel with the same velocity, the two wavefronts touch each other at two points. The line joining these points will give the direction of optic axis as shown in Fig. 6.17 (*a*). In general, since both *O*-ray

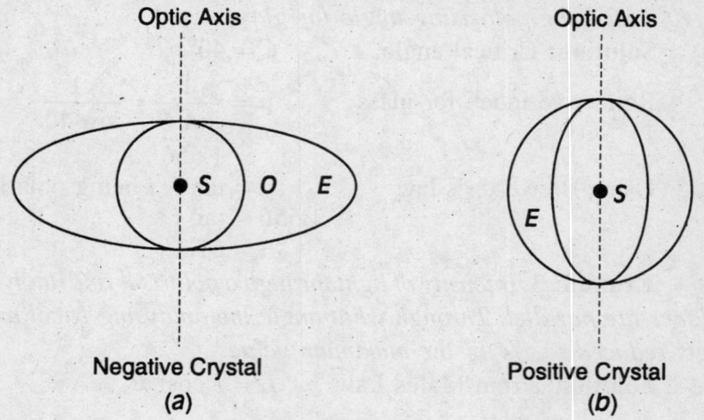

Fig. 6.17

and E-ray travel with different velocities in different directions, these rays separate out in passing through such crystals and double refraction takes place.

6.18. NEGATIVE CRYSTALS

The wavefronts surrounding a point source of light S in such crystals is given in Fig. 6.17 (a). The figure shows a section in the plane of paper and sphere gives the wave surface for O-ray and ellipse that for E-ray. It is evident that

(*i*) O-ray surface lies inside E-ray surface.

(*ii*) The velocity of O-ray is constant in all directions.

(*iii*) The velocity of E-ray varies with direction. It is minimum along the direction of optic axis where it will be equal to the velocity of O-ray and maximum $\perp$ to the direction of optic axis.

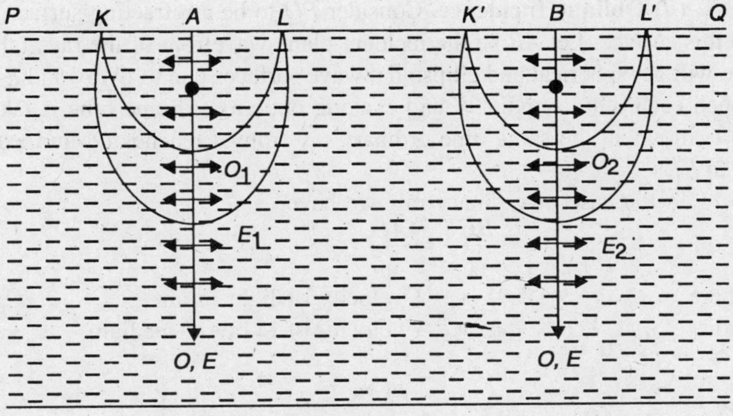

Fig. 6.18

(*iv*) Refractive index for E-ray is less than the refractive index for O-ray ($\mu_0 << \mu_e$).

This type of crystals are called negative crystals *e.g.*, calcite is a negative crystal.

6.19. POSITIVE CRYSTALS

The wavefronts surrounding a point source S in such a crystal are shown in Fig. 6.17 (b). It is evident that

(*i*) O-ray surface lies outside E-ray surface.

(*ii*) Velocity of O-ray is constant in all directions.

(*iii*) The velocity of E-ray varies with direction. It is max. along optic axis where it will be equal to velocity of O-ray and minimum $\perp$ to the direction of optic axis.

(*iv*) Refractive Index for E-ray is more than that for O-ray ($\mu_e >> \mu_0$).

Such crystals are called positive crystals *e.g.*, quartz is a positive crystals.

We can now study the Huygen's construction for double refraction in certain special cases.

Case 1.

6.20. OPTIC AXIS PARALLEL TO THE REFRACTING SUFRACE AND IN THE PLANE OF INCIDENCE

(*i*) Normal Incidence. Let PQ be the refracting source of calcite crystal and optic axis is parallel to PQ. Let AB be the incident wave front. Now points A and B will act as secondary sources of dis- turbance. Since the crystal has the characteristics of double refraction,

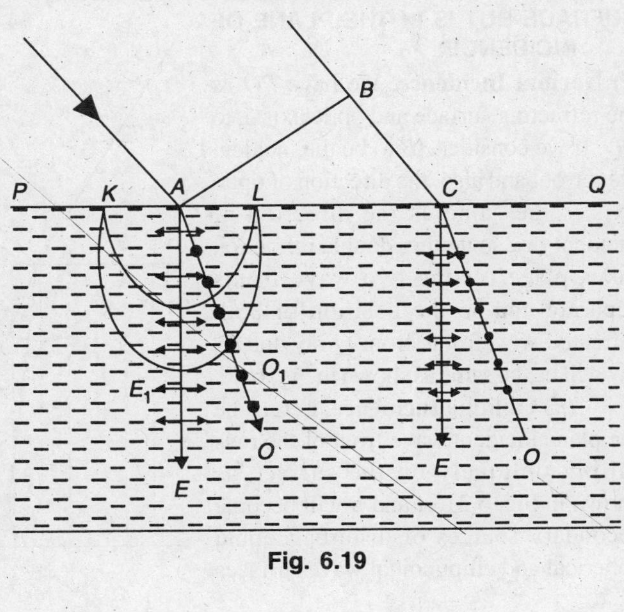

Fig. 6.19

we will have spherical and elliptical waves due to *O*-rays and *E*-rays respectively. These rays touch each other at *K, L, K′* and *L′* along the optic axis. Since this is the case of a negative crystal, So *E*-ray will travel faster than *O*-ray along ⊥ to the optic axis but both the rays travel in the same direction.

Tangent planes $O_1\, O_2$ and $E_1\, E_2$ represent the positions of refracted wavefronts due to *O*-rays and *E*-rays respectively. This fact is used in the construction of quarter and half wave plates.

***(ii)* Oblique Incidence.** Consider *PQ* to be a refracting surface and the optic axis is parallel to this surface. Let *AB* be the incident plane wavefront of the rays falling obliquely on the surface we will get spherical and ellipsoidal wave surfaces due to *O*-ray and *E*-rays respectively. These rays touch each other at point *K* and *L* along optic axis. Some time is taken by a ray tc reach form *B* to *C* in air, in the same time ordinary ray from *A* reaches O_1 and extraordinary ray reaches from *A* to E_1.

In the case of ordinary ray, we have

$$\frac{BC}{v_c} = \frac{AO}{v_0}$$

where $v_c \rightarrow$ velocity of light in air

and $v_0 \rightarrow$ velocity of *O*-ray in medium.

or $$BC = \frac{v_c}{v_0},$$

$AO_1 = \ m_0\, AO_1$

where μ_0 is the refractive index for ordinary rays.

Similarly $$BC = \mu_e\, AE_1$$

where μ_e is the refractive index of extraordinary rays.

The tangent planes to the two surfaces from *C i.e.*, $O_1\, C$ and $E_1\, C$ represent the refracted plane wavefronts for *O*-rays and *E*-rays respectively.

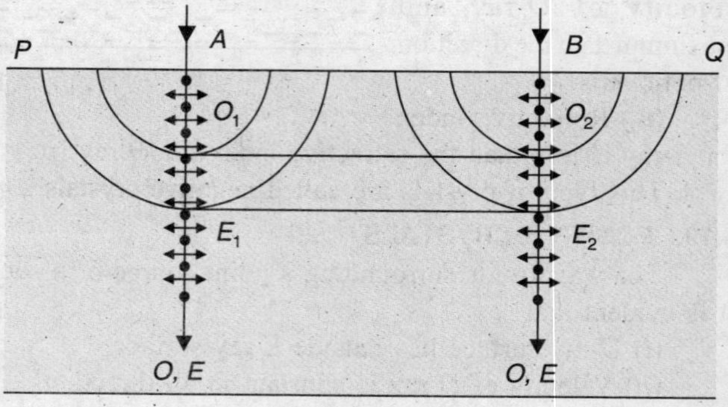

Fig. 6.20

Case 2

6.21. OPTIC AXIS ⊥ TO THE REFRACTING SUFRACE BUT IS IN THE PLANE OF INCIDENCE

***(i)* Normal Incidence.** We have *PQ* as the refracting surface and optic axis ⊥ to *PQ*. If we consider *AB* to be the incident wavefront and also, the direction of optic axis is the same as the direction of incident ray, so neither double refraction takes place for the two wave fronts separate and so the position of both refracted wavefronts from *O*-ray and *E*-ray will be the same as shown in Fig. 6.20.

***(ii)* Oblique Incidence.** Let *AB* be the plane incident wave front of the rays falling obliquely on the surface as shown in Fig. 6.21. *A* and *B* will become secondary sources of disturbance and spherical and ellipticoidal wave surfaces

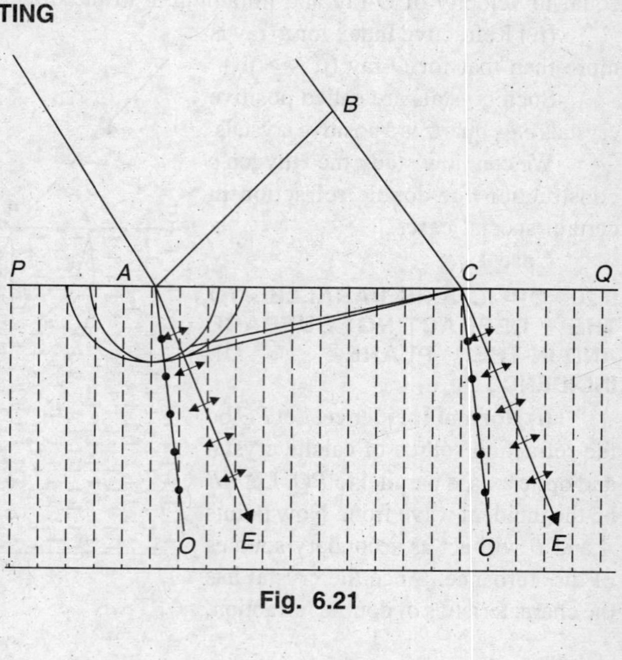

Fig. 6.21

due to O-rays and E-rays touch each other at the point L along the direction of optic axis. The time taken by the secondary waves to travel from B to C in air will be equal to the time during which O-ray reaches from A to O_1 and E-ray from A to E_1

where $BC = \mu_0 AO_1$

and also $BC = \mu_e AE_1$

where μ_0 and μ_e are refractive indices for O-ray and E-ray respectively.

Tangent places O, C and E, C represent the refracted wavefronts for O-rays and E-rays respectively.

From above it is clear that O-rays and E-rays travel in different directions with different velocities are so double refraction will be observed.

Case 3

6.22. OPTIC AXIS IN THE PLANE OF INCIDENCE AND INCLINED TO THE REFRACTING SURFACE

***(a)* Normal Incidence.** PQ is the refracting surface and AB the incident wavefront where A and B will act as secondary sources of disturbance. The spherical and ellipsoidal waves due to O-rays and E-rays touch each other at points along the direction of optic axis. Now during the time the secondary waves for O-rays reach from A to O_1 or B to O_2, that for E-rays reach from A to E_1 and B to E_2

where $\mu_0 AO_1 = \mu_e AE_1$

and $\mu_0 BO_2 = \mu_e BE_2$

where μ_0 and μ_e are refractive indices for O-rays and E-rays.

Tangential planes $O_1 O_2$ and $E_1 E_2$ will represent the position of refracted plane wave fronts for O-rays and E-rays respectively.

As is clear O-rays and E-rays travel with different velocities in different directions, so we will observe double refraction phenomenon.

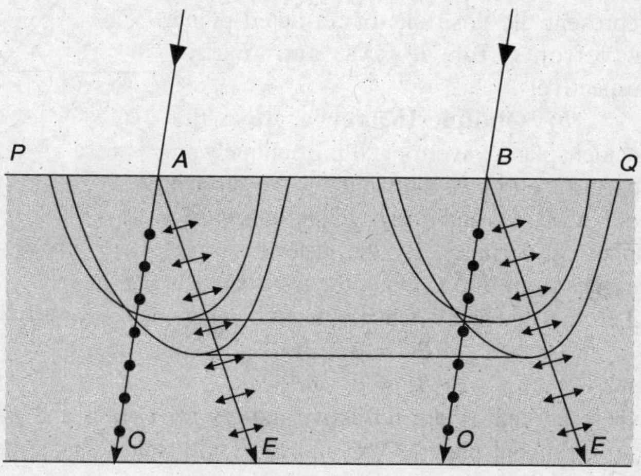

Fig. 6.22

***(b)* Oblique Incidence.** AB the incident wave front of the rays is shown to fall obliquely on the sufface PQ. A and B will become secondary sources of disturbance. Spherical and ellipsoidal wave fronts due O-ray and E-ray touch each other at the point which is along the direction of optic axis. During the time the secondary wave from B reaches C in air the secondary rays reach from A to O_1 and A to E_1 for O-rays and E-rays respectively,

where $BC = \mu_0 AO_1$

and $BC = \mu_e AE_1$

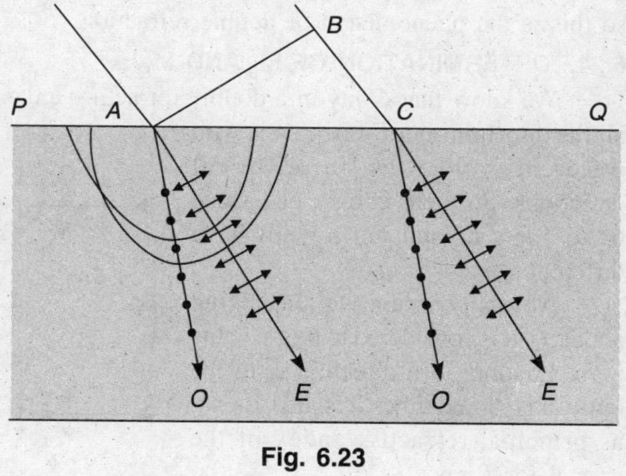

Fig. 6.23

where μ_0 and μ_e are refractive indices for O-ray and E-rays respectively.

Tangent planes $O_1 C$ and $E_1 C$ represent the positions of refracted plane wave fronts due to O-rays and E-rays. As is clear from above, O-rays and E-rays travel in different directions with different velocities. So double refraction is observed.

Case 4

6.23. (*A*) OPTIC AXIS PARALLEL TO REFRACTING SURFACE BUT ⊥ TO THE PLANE OF INCIDENCE

PQ is the refracting surface and *AB* is the incident plane wavefront. The optic axis is perpendicular to plane of incidence or the plane of paper, as shown in Fig. 6.24.

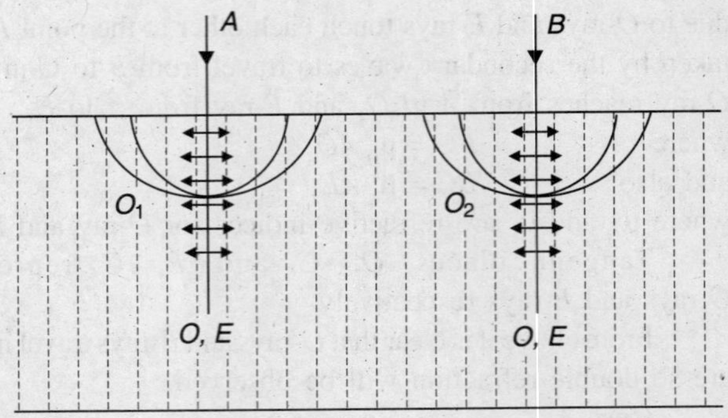

Fig. 6.24

A and *B* will act as secondary sources of disturbance. Both *E*-rays and *O*-rays travel in the same direction but with different velocities and no separation of *O*-rays and *E*-rays will take place *i.e.*, double refraction will not be observed. Tangent planes $O_1\,O_2\,E_1\,E_2$ will represent the positions of refracted plane wavefronts for *O*-rays and *E*-rays respectively.

(***b***) **Oblique Incidence.** *AB* is the incident plane wavefront falling obliquely on the surface *PQ*. As shown the wave surface for both *O* rays and *E*-rays will be concentric spherical surfaces. By the time the wave reaches from *B* to *C* in air, the *O*-rays reach O_1 from *A* and *E*-rays reach E_1 from A_1 where

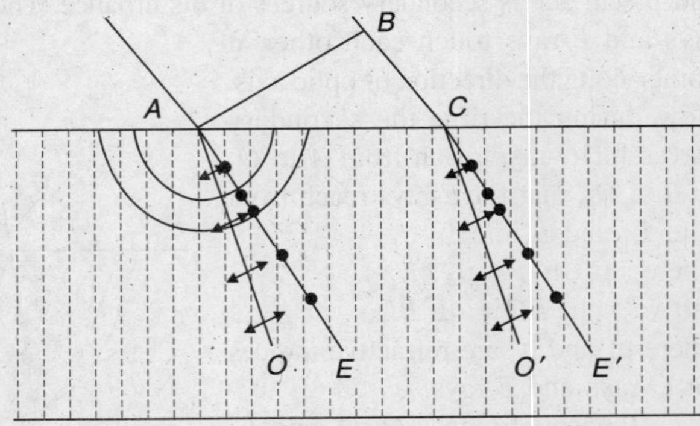

Fig. 6.25

$$BC = \mu_0 \,.\, AO_1$$

and
$$BC = \mu_e \,.\, AE_1$$

where μ_0 and μ_e are refractive indices for *O*-rays and *E*-rays respectively.

Tangent planes *O*, *C* and *E*, *C* will represent the positions of refracted wave fronts for *O*-rays and *E*-rays respectively.

As is seen from above, *O*-rays and *E*-rays travel in different direction with different velocities, so this is the phenomenon of double refraction.

6.24. DETERMINATION OF M_0 AND M_E

We know that *O*-ray in a doubly refracting uniaxial crystal travels with the same velocity in all the directions but *E*-ray travels with different velocities in different directions. So there is only one value of μ_0 for a crystal but μ_e will have different values.

Normally, when we are talking about μ_e, it is considered to be the value corresponding to a direction ⊥ to the optic axis. Some times it is also known as principal refractive index of the crystal for *E*-ray.

To determine μ_0 and μ_e we take the prism of the material for which these values are required. Then, by measuring

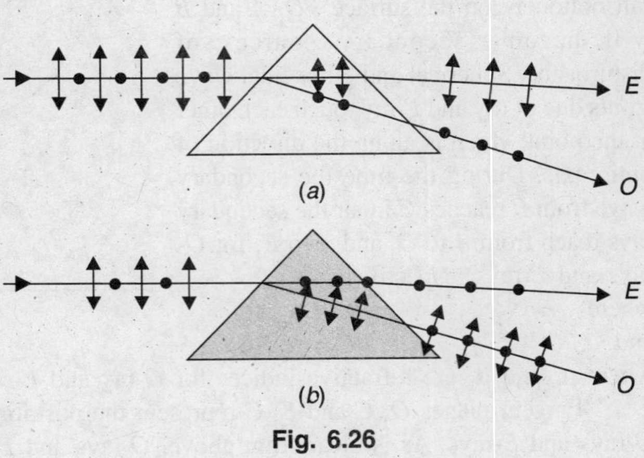

Fig. 6.26

the angles of minimum deviation for O-rays and E-rays with the help of a spectrometer μ_0 and μ_e can be found from the relation

$$\mu_0 = \frac{\sin \frac{A + D_0}{2}}{\sin \frac{A}{2}}$$

$$\mu_e = \frac{\sin \frac{A + D_e}{2}}{\sin \frac{A}{2}}$$

where D_0 and D_e correspond to angles of minimum deviations for O-rays E-rays respectively.

The prisms of doubly refracting material like quartz and calcite are cut with optic axis either (*i*) perpendicular to the refracting edge and parallel to the principal section as shown in Fig. 6.26 (*a*) or (*ii*) parallel to the refracting edge and perpendicular to the principal section of the prism as shown in Fig. 6.26 (*b*).

In both the cases, E-ray will travel parallel to the base of the prism in the position of minimum deviation and so it will be perpendicular to the direction of optic axis, the condition required for determination of principal refractive index.

Quarter and Half-wave Plates

6.25. QUARTER WAVE PLATE

It is a crystal wave plate of a doubly refracting material of a uniaxial crystal, cut in a way that the optic axis is parallel to the refracting faces. It can produce a phase difference of $\pi/2$ or path difference of $\lambda/4$ between O-rays and E-rays when a monochromatic polarized light of wavelength λ is incident normal to the surface. Since, incidence is normal to the optic axis, so both E and O-rays will travel in the same direction but with different velocities.

In case of a negative crystal (calcite) the velocity of E-rays is more than that for O-rays so $\mu_0 > \mu_e$. If t be the thickness of the crystal plate, then the path difference between the two components on coming out will be

$$t\,(\mu_0 - \mu_e)$$

So to get a path difference of $\lambda/4$, we have

$$t\,(\mu_0 - \mu_e) = \lambda/4$$

or
$$t = \frac{\lambda}{4\,(\mu_0 - \mu_e)} \qquad ...(1)$$

For a positive crystal (quartz), $\mu_e > \mu_0$

$\therefore$ Thickness of plate
$$t = \frac{\lambda}{4\,(\mu_e - \mu_0)} \qquad ...(2)$$

As thickness of the plate depends on the wavelength, so it should be used only for the wavelength for which it is constructed.

A quarter wave plate is used for the production and detection of **circularly polarized light**, however, with a Nicol prism it may be used for analysing any kind of light.

6.26. HALF-WAVE PLATE

It is also made from a doubly refracting uniaxial crystal (like quartz or calcite) with its refracting faces cut parallel to the optic axis. The thickness of this plate is such that it introduces a path difference of $\lambda/2$ or a phase difference of π between the O-ray and E-ray passing through it when the light is incident normally on the face of the crystal.

For a negative crystal, thickness of the plate

$$t = \frac{\lambda}{2\,(\mu_0 - \mu_e)} \qquad ...(3)$$

For a positive crystal, thickness of the plate

$$t = \frac{\lambda}{2\,(\mu_e - \mu_0)} \qquad ...(4)$$

Quarter wave plates or half-wave plates are normally made from thin sheets of mica or quartz cut parallel to the optic axis. The thickness can be calculated from relations Equations (2) and (4) respectively using the values of μ_e and μ_0 for sodium light of wavelength 5893Å.

Reasonably good quarter wave plates can be prepared by spliting good, fine and clear mica into very thin sheets of thickness of about 0.035 mm with the help of a fine knife or a needle. The uniform thickness can be checked with an accurate vernier calliper.

In the case of quartz, the crystal is cut to the calculated thickness and the faces are polished and made optically flat.

Example 5. *Calculate the thickness of a quarter waveplate for light of wavelength 5896Å. The refractive index for ordinary and extraordinary rays being 1.54 and 1.55 respectively.*

Solution: $\mu_e = 1.55, \quad \mu_0 = 1.54, \quad \lambda = 5896 \times 10^{-8}$ cm

Since $\mu_e > \mu_0$ for a positive crystal, the required thickness t is given by

$$t = \frac{\lambda}{4(\mu_e - \mu_0)} = \frac{5896 \times 10^{-8}}{4(1.56 - 1.54)} = \mathbf{0.001474\ cm.}$$

Example 6. *Calculate the thickness of a doubly refracting crystal required to introduce a path difference of λ/2 between the O and E-rays when λ = 6000Å, $\mu_0 = 1.55$, and $\mu_e = 1.54$.*

Solution: $\mu_0 = 1.55, \quad \mu_e = 1.54, \quad \lambda = 6000 \times 10^{-8}$ cm

Since, $\mu_0 < \mu_e$, the crystal is negative and the required thickness, t is given by

$$t = \frac{\lambda}{2(\mu_0 - \mu_e)} = \frac{6000 \times 10^{-8}}{2(1.55 - 1.54)} = \mathbf{0.003\ cm.}$$

6.27. POLARIMETRY

If plane polarized light is made to pass through certain substances, it is found that the plane of polarization of the emergent light is not the same as that of the incident light and it is rotated through certain angle. This property of crystals or substances is called optical activity and the substances showing the said characteristic are said to be optically active and the phenomenon is called optical rotation.

There are two types of optically active substances (*i*) those which produce clockwise rotation are known as Dextro-rotatory (right handed) substances and others which rotate the plane of polarization in anticlockwise direction are called Laevo-rotatory (left-handed) substances.

If a monochromatic light from source *S* is made to pass through two nicols *PN* and *AN* but in crossed positions, no emergent light can be received by the eye.

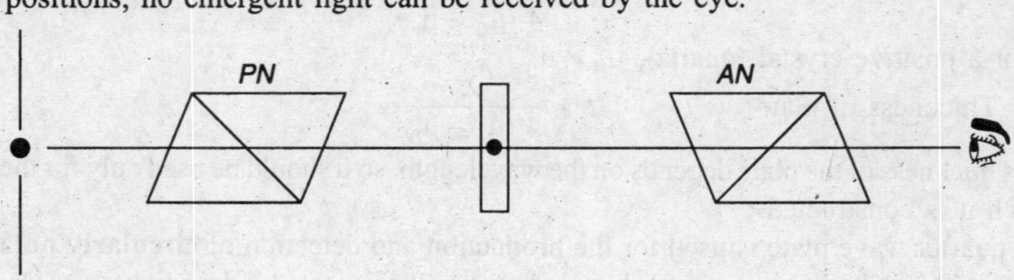

Fig. 6.27

In this case polarizing nicol *PN* renders the beam polarized with its vibrations in principal plane. In the analysing prism, the position is set with its principal plane ⊥ to the direction of these vibrations, so that no light passes through *AN*. If a quartz crystal cut with its refracting faces ⊥ to the optic axis is placed in between *PN* and *AN*, the two crystals still being crossed, it is seen that the light is no longer cut by analysing prism. In order to cut off the light, the analysing prism is required to be rotated through some angle. So the emergent light is still plane polarized, but its plane of polarization has been rotated.

It has been experimentally established that depending upon the arrangement of molecules, some quartz crystals are Dextero-rotatory and some are Laevorotatory.

The amount of rotation is found to depend on:

(*i*) thickness and density of crystals.

(*ii*) temperature and wavelength of light used.

$\left(\text{Biot found that rotation} \propto \frac{1}{\lambda^2}\right)$

(*iii*) concentration in case of solutions.

6.28. SPECIFIC ROTATION

The term specific rotation is used to bring rotations of all optically active substances in a comparable form.

Specific rotation for a given temperature t and for a light of given wavelength λ is defined as the rotation (in degrees) produced by a path of one decimeter length in a substance of unit density.

If θ is the rotation produced by l decimeter length of a solution of density d/cc; then specific rotation S corresponding to some temperature t and light of wavelength λ is given by

$$S_{e\lambda} = \frac{\theta}{l \times d}$$

$$= \frac{\text{Rotation in degrees}}{\text{length in decimeter} \times \text{concentration in gm/cc}}$$

Sugar is the most commonly used optically active substance and the instruments used for measuring the optical rotation produced by a substance are called polarimeters or sacchrimeters. Basically, they consist of two Nicols capable of rotation about the incident beam as axis. The optically active substance is placed between the two Nicols. Generally, it is very difficult to locate the position of analysing nicol, when no light is received but with the use of Laurent's half shade device, this difficulty is overcome.

6.29. LAURENT'S POLARIMETER

It is an instrument used for finding optical rotation of certain solutions. When used for finding specific rotation of sugar, it is called a sacchrimeter. Also if the specific rotation of sugar is known, the concentration of the solution can be found.

Construction. It consists of two Nicols P and A mounted in brass tubes and capable of rotation about a common axis. A glass tube D containing optically active solution is mounted in between the two Nicols as shown in Fig. 6.28. All the tubes are in line. Monochromatic light from a sources is rendered parallel by a convex lens L and is made to fall on polarizing prism P which renders the light plane polarized with its vibrations in the principal plane of the Nicol. This plane polarized light is then made to pass through half shade device H and the tube D containing the active solution and then it passes through analysing Nicol prism A.

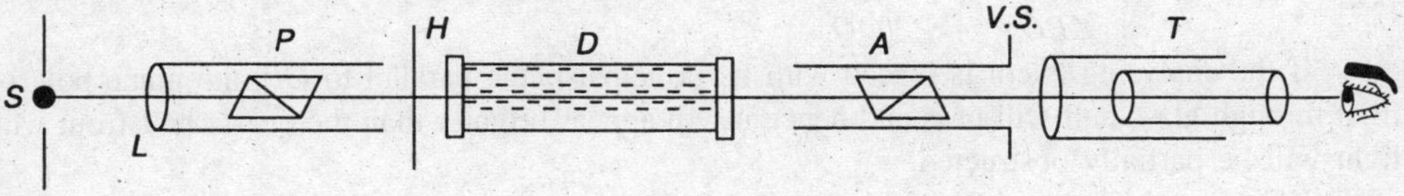

Fig. 6.28

The emergent light is viewed through telescope T. The analysing Nicol A can be rotated about its axis and the rotation can be measured on a graduated circular scale in terms of degrees with the use of Vernier Scale *V.S.*

Experiment. The position of the analysing Nicol A is adjusted so that the field is completely dark when the tube is empty. Now the tube is filled with the solution and placed in position, with a precaution that it does not contain any air bubble. On seeing through the telescope, the field will no longer look dark. To get the position of darkness, the Nicol A is rotated through certain angle which will be the measure of optical rotation of the given solution. However, it is very difficult to get the position of total darkness of the field of view as this position is achieved only gradually. To overcome this difficulty Laurent's half shade device is employed which is nothing but a half-wave plate and gives the measurement of rotation very accurately.

6.30. LAURENT'S HALF SHADE DEVICE

It consists of two semi-circular plates, one made of glass and other of quartz. Both glass and quartz are cemented together as shown in Fig. 6.29.

The quartz plate is cut parallel to its axis which being parallel to the line joining two plates YY' quartz plate is made as half-wave plate, *i.e.,* its thickness is so adjusted that it introduces a path difference of $\lambda/2$ between E-ray and O-ray for sodium light. In other words, in passing through the quartz plate O-ray gains $\lambda/2$ on E-ray. Thickness of glass plate is so adjusted that it transmits the same amount of light as the quartz plate.

Let the light after passing through the polariser P be incident normally on half shade plate and has vibrations along OP, on passing through the glass half, the vibrations will remain along OP, but on passing through the quartz half these will split into E and O components. The vibrations of E-component are along OY and those of O-component along OX.

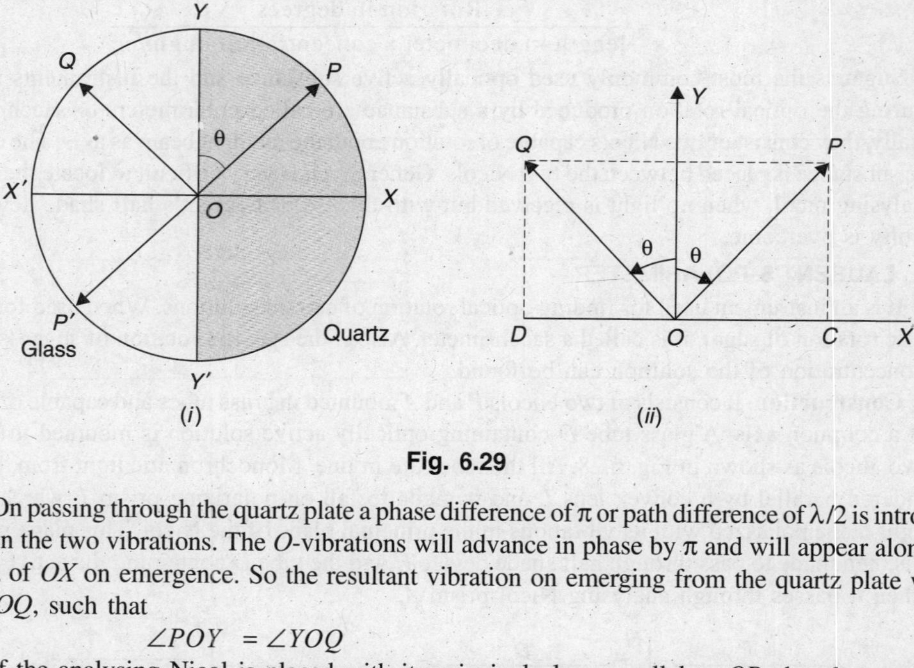

Fig. 6.29

On passing through the quartz plate a phase difference of π or path difference of $\lambda/2$ is introduced between the two vibrations. The O-vibrations will advance in phase by π and will appear along OX' instead of OX on emergence. So the resultant vibration on emerging from the quartz plate will be along OQ, such that

$$\angle POY = \angle YOQ$$

If the analysing Nicol is placed with its principal plane parallel to OP, the plane polarized light through glass half will pass and hence it will appear brighter than the quartz half from which light will be partially obstructed.

If the principal plane of Nicol is parallel to OQ, the quartz half will appear brighter than glass due to the above reason.

When the principal plane of analysing Nicol is parallel to YOY', the two halves will appear equally bright. It is because the two vibrations coming out of the two halves are equally inclined to its principal plane and so the two components will have equal intensity.

Again if the principal plane of the analyser is $\perp$ to YOY', we will have components of OP and OQ equal. The two halves are again equally illuminated but since the intensity of the two components passing through is small as compared to that in the first case, the two halves are equally dark. Since the eye can easily detect the change when the two halves are equally dark, so the readings are taken for this position.

To find the concentration of sugar solution. The polarimeter tube is filled with plain water and the reading on the circular r scale corresponding to equally dark position of half shade device is noted.

Now the tube is filled with given sugar solution and again the reading on the circular scale corresponding to equally dark position of half shade device is noted. The difference between the scale reading gives the optical rotation θ produced by the given length l in decimeter of sugar solution. If S is the specific rotation of sugar for a particular wavelength and temperature then,

$$\text{Concentration } c = \frac{\theta}{l \times S} \text{ gm/cc.}$$

Example 7. *A 20 cm long tube containing sugar solution is placed between crossed Nicols and illuminated by light of wavelength λ = 6000Å. If the specific rotation is 60° and optical rotation produced is 12°. What is the strength of the solution?*

Solution: The specific rotation, at a given temperature t and for a given wavelength λ is given by

$$S_{e\lambda} = \frac{\theta}{l \times d}$$

where θ is the rotation, d the density of solution in gm/c.c. and l its length in decimeters.

Here $S = 60^\circ$

$l = 2$ decimeters

$\theta = 12^\circ$

and $\lambda = 6000 \times 10^{-8}$ cm

$$\therefore \quad d = \frac{\theta}{l \times S} = \frac{12}{2 \times 60} = \frac{1}{10}$$

So it is 10% solution of sugar *i.e.,* 1 gm sugar is dissolved in 10 gm of water.

6.31. PHOTO ELASTICITY

Photo elasticity deals with the analysis of stress distribution in structural and mechanical components with the help of polarized light.

It is observed that certain transparent materials like glass, bakelite etc., are not doubly refracting under normal conditions, but show the phenomenon when subjected to mechanical stresses. If such a specimen is placed between two crossed Nicol prisms, no change in field of view is observed but if some mechanical stress is applied the emergent light is in general elliptically polarized and hence partly transmitted through the analysing. Nicol, thereby changing the field of view. In the cases of complex structures like bridges, boiler plates, gear teeth etc., a transparent model of the structure is studied by photo elasticity and the scale of these structures are constructed by materials like bakelite or zylonite and these structures are subjected to loads or stresses and from the study of interference patterns, the effect of these stresses on the various parts of the models are studied. From this the actual effect of stresses for the materials can be calculated which helps in design of the structures.

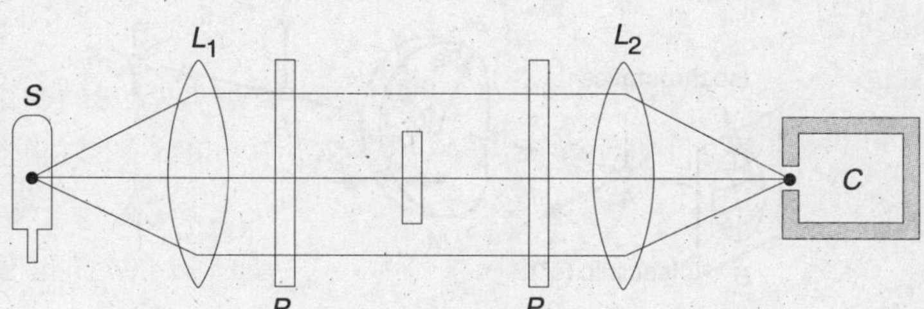

Fig. 6.30

6.32. PHOTO ELASTIC BENCH

As stated earlier, when a polarized light is passed through a *transparent elastic object* such as glass or bakelite, colours are observed when a stress is applied to the object. This fact has been developed with a full fledged and a accurate technique for measurement of complicated stress distributions.

Figure 6.30 shows the experimental set up for this technique. *S* is a source of monochromatic light (usually blue-green mercury are lamp). Lens L_1 reduces the beam of light parallel and this parallel light passes through two *polaroids* P_1 *and* P_2. Light converging from second polaroid P_2 is converged by the lens L_2. *C* is a photographic camera. However, for visual observation the second lens L_2 and camera are removed and a milk glass screen is placed behind P_2. The two polaroids are kept crossed, so that field of view is completely dark.

Test piece *T* of the material is made of a special yellowish sheet of *transparent bakelite*. This test piece is cut from a flat plate of thickness between 0.6 cm to 1.2 cm and given the proper form. This sheet is placed between the two crossed polaroids. When no stress is applied to the plate, the field of view is completely dark. As the stress is applied on the test piece, some coloured patterns of fringes will appear. From the pattern, the information of the stress distribution at various points can be known.

It may be noted that stress analysis of various complicated structures and machines can be done using the *photoelasticity technique*. Also the stress distribution in a railway wagon wheel and aircraft components can be studied. Photoelasticity is also being used in solving thermal problems, high speed photography, computer technology and lasers.

Some important terms used for stress analysis are given as under.

6.33. ISOCLINICS AND ISOCHROMATICS

The locus of points where the directions of principal stresses coincide with a particular orientation of the *polarizer-Analyzer* combination is called isoclinic (Meaning same angle of inclination). For example, if polarizer is kept vertical and analyzer is kept horizontal, then on the Screen a dark band will be observed which will correspond to locus of points where two directions say σ_1 and σ_2 happen to be vertical and horizontal.

If the measurement is made from vertical reference axis, the isoclinic will be called 0° isoclinic. If the polarizer is turned through some angle say 45° and analyzer is also rotated through some angle (to make them always in crossed position), then the previous 0° and a new dark band is observed. This is the 45° isoclinic that represents locus of points in the model where principal axes are oriented at 45° and $45° + \pi/2$ w.r.t the vertical axis.

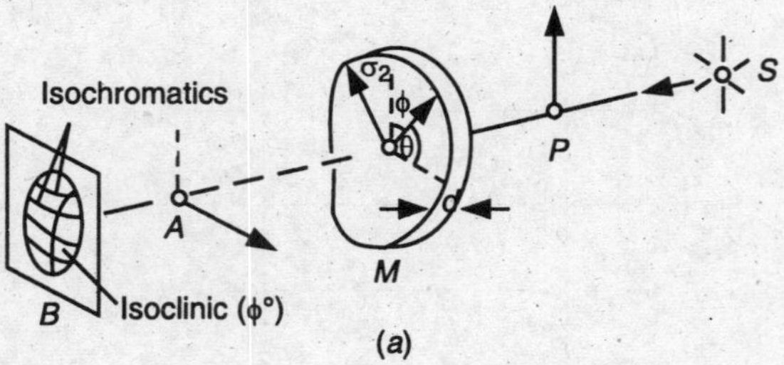

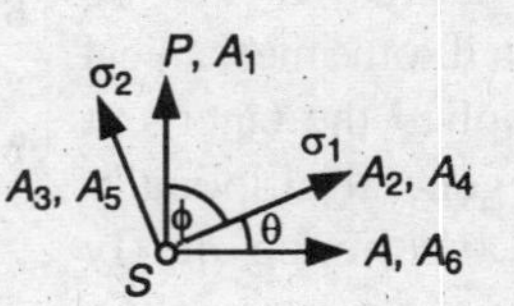

Fig. 6.31

If we consider a situation where σ_1 and σ_2 have such a magnitude that causes a relative phase difference of $2n\pi$, where *n* is an integer. As stated in case of a full wave plate, the model will be like a full wave plate and in the case of points where relative phase is $2n\pi\,(n = 0, 1, 2, 3, ...)$ the intensity of light will be zero. (As the polarizer and analyzer are crossed).

On the screen we get a series of dark fringes corresponding to the locus of the respective points. *These dark bands or fringes are called isochromatics.* The locus of points where the values of σ_1 and σ_2 are such that the relative phase difference $\sigma_1 - \sigma_1$ is zero, then we get *zero order* fringe. In case $\sigma_1 - \sigma_2 = 2\pi$ (which will correspond to path difference λ) we get *first order fringe.* In the same way higher order fringes are observed.

6.34. MATHEMATICAL ANALYSIS OF ISOCLINICS AND ISOCHROMATICS

Let the light component coming from polarizer be given by

$$A_1 = a\cos\omega t \quad ...(1)$$

On entering the Photo Elastic Model *M*, the light vector gets resolved along principal stress axes [shown in Fig. 6.31(*b*)]. So we have

$$\left.\begin{aligned} A_2 &= a\cos\phi\cos\omega t \\ A_2 &= a\sin\phi\cos\omega t \end{aligned}\right\} \quad ...(2)$$

In travelling through thickness *x* of the model, two components acquire a relative phase difference of θ. So on leaving the model, two components are

$$A_4 = a\cos\phi\cos(\omega t+\theta)$$

$$A_5 = a\sin\phi\cos\omega t$$

In general A_4 and A_5 should have a certain net amount of phase difference; however in the final analysis a relative phase difference is observed. On entering the analyzer only *components along A* are allowed to come out.

So we have

$$\begin{aligned} A_6 &= A_4\sin\phi - A_5\cos\phi \\ &= a\cos\phi\sin\phi\cos(\omega t+\theta) - a\cos\phi\sin\phi\cos\omega t \\ &= \frac{a}{2}\sin 2\phi\cos(\omega t+\theta) - \cos\omega t \\ &= \frac{a}{2}\sin 2\phi[\cos\omega t(\cos\theta-1) - \sin\omega t.\sin\theta] \\ &= \frac{a}{2}\sin 2\phi\left[-2\cos\omega t.\sin^2\frac{\theta}{2} - 2\sin\omega t.\sin\frac{\theta}{2}.\cos\frac{\theta}{2}\right] \\ &= -a\sin 2\phi\sin\frac{\theta}{2}\left[\cos\omega t\sin\frac{\theta}{2} + \sin\omega t.\cos\frac{\theta}{2}\right] \\ &= -a\sin 2\phi\cdot\sin\frac{\theta}{2}\cdot\sin\left(\omega t+\frac{\theta}{2}\right) \\ &= -b\sin\left(\omega t+\frac{\theta}{2}\right) \end{aligned}$$

where $b = a\sin 2\phi.\sin\frac{\theta}{2}$ corresponds to the *amplitude* of the emerging light from analyzer. Since intensity is proportional to the square of amplitude, so the intensity of emerging light will be

$$I = a^2\sin^2 2\phi\sin^2\frac{\theta}{2}$$

The intensity of emerging light will be zero

(*i*) When $\phi = 0$ or $\pi/2$

(*ii*) When $\phi = 2n\pi\ (n = 0, 1, 2...)$

Condition (*i*) corresponds to the point where direction of principal stress coinicide with the direction of polariser and analyser. Condition (*ii*) corresponds to the point where relative phase is $2n\pi$.

The locus of points where the direction of principal stress has a common orientation w.r.t. a given axis is called *isoclinic*. The locus of points satisfying condition (*ii*) is called as *isochromatic.*

6.35. MOIRE FRINGES

The word 'Moire' is derived from *silk fabric which when superimposed on itself exhibit light and dark bands.*

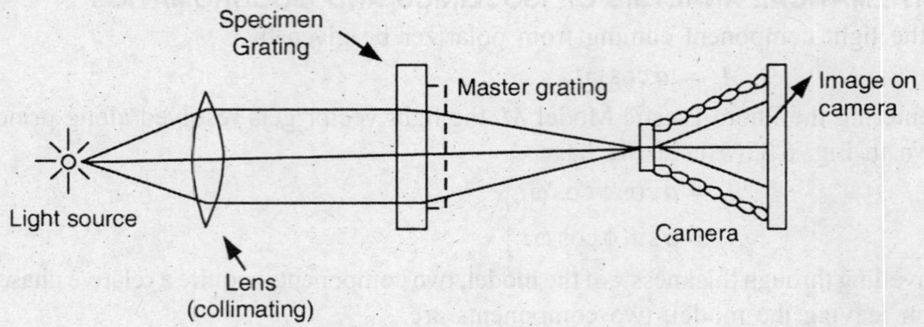

Fig. 6.32

These days Moire Fringes are produced by superposition of two sets of gratings under some specific conditions. *The gratings could be linear, circular or elliptical* and beautiful. Moire patterns are generated using such gratings. The idea of *using* such Moire pattern to study strain was first introduced by Faucault and in 1874 Lord Rayleigh suggested that Moire pattern could be used to study the perfection of diffraction grating. By placing the plastic replace over original grating, any error due to ruling on the grating could be studied.

The basic principle of analysis of grating using Moire Fringes is by putting one grating which is either bonded to or etched or printed on the specimen being analyzed. This is termed as *Model or Specimen grating*. The specimen grating undergoes deformation depending on the state of strain on the surface and this is shown by change of spacing (Pitch) between the lines. *The master grating does not undergo any change*. The strained specimen and the unstrained specimen grating interfere optically to form *Moire Fringes*. The experimental set up of the Moire Fringes is shown in the Fig. 6.32.

It is observed that the Moire Fringe is the locus of points having same component of displacement in the principal direction of the master grating. Such a locus is called *isothetic*. From the study of the Moire fringes, the nature of the strains can be known.

6.36. POLAROIDS

Polaroids are commercially available *polarizing* devices. Recent polaroids are manufactured using a sheet of *polyvinyl alcohol* and making it undergo a large strain. By doing so, the molecules will get oriented parallel to the strain and sheet becomes *double refracting*. If this sheet is strained with iodine then the *ordinary ray* is absorbed. So that the sheet become a *polaroid* by transmitting completely the *extraordinary ray* only. These sheets are sandwitched between glass films so that these are not spoiled.

Polaroids find use in Sun glasses. They are used in making 3-dimensional films and photos. Audience use *polaroid glasses* to view the three dimensional films. Car head lamps and wind screens are fitted with polaroids at an angle of 45° to the left, to make the driver see the light of his car and the glare of vehicles from opposite direction is cut off as it is at angle of 90° to his wind screen. If we view the aerial pictures taken at different angles through polaroids, we get better details of the depth. Also polaroids are highly useful to improve the colour contrast in old pictures.

6.37. SUPER POSITION OF TWO PLANE POLARIZED LIGHTS HAVING VIBRATIONS PERPENDICULAR TO EACH OTHER

Let us consider a plane polarized light emerging from a polarizer and incident in perpendicular direction on a thin calcite crystal with faces parallel to the optic axis of the polarizer and oriented in such a way so that the vibrations of plane polarized light make θ angle with optic axis (as shown in Fig. 6.33).

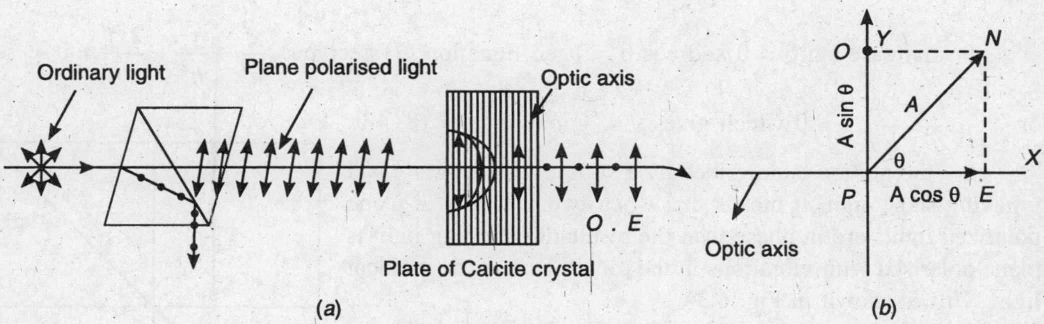

Fig. 6.33

Now amplitude A of incident light may be resolved into two components corresponding to E-ray and O-ray with E-component of amplitude as $A \cos \theta$ and O-component as $A \sin \theta$ as shown in Fig. 6.31.

Since, both the vibrations move with different velocities in the same direction. So, a phase difference δ in introduced between E-ray along PE and O-ray along PO for travelling the same distance.

In case of calcite, since $V_e > V_0$. So, E-ray leads the O-ray. So, we can express the displacement of E and O waves *along optic axis* and perpendicular to optic axis by the equations

$$x = A \cos \theta \sin (\omega x + \delta) \qquad ...(1)$$

and

$$y = A \sin \theta \sin \omega t \qquad ...(2)$$

Let $A \cos \theta = a$ and $A \sin \theta = b$ put in Equation (1) and (2), we have

$$x = a \sin (\omega t + \delta) \qquad ...(3)$$

and

$$y = b \sin \omega t \qquad ...(4)$$

Now, from Equation (3),

$$\frac{x}{a} = \sin \omega t \cos \delta + \cos \omega t \sin \delta \qquad ...(5)$$

and from Equation (4)

$$\frac{x}{b} = \sin \omega t \text{ which gives } \left(1 - \left(\frac{y}{b}\right)^2\right) = \cos \omega t \qquad ...(6)$$

Put these values of sin ωt and cos ωt in Equation (5) and writing in the form.

$$\frac{x}{a} = \frac{y}{b} \cos \delta + \sqrt{\left(1 - \frac{y^2}{b^2}\right)} \cdot \sin \delta$$

or

$$\frac{x}{a} - \frac{y}{b} \cos \delta = \sqrt{\left(1 - \frac{y^2}{b^2}\right)} \cdot \sin \delta$$

Which on squaring gives $\left(\frac{x}{a} - \frac{y}{b} \cos \delta\right)^2 = \left(1 - \frac{y^2}{b^2}\right) \sin^2 \delta$

which may be written as

$$\frac{x^2}{a^2} + \frac{y^2}{b^2} - \frac{2xy}{ab} \cos \delta = \sin^2 \delta \qquad ...(7)$$

which is the equation of an **oblique ellipse**. So, in general the emergent light is elliptically **polarised**. The exact nature of the resultant motion or the light coming out from calcite plate depends upon the value of δ.

Let us take some special cases.

Case I. If the thickness of the plate is such that $\delta = 0, 2\pi, 4\pi, ..., 2n\pi$.

In that case $\sin \delta = 0$ and $\cos \delta = 1$, so Equation (7) becomes $\frac{x^2}{a^2} + \frac{y^2}{b^2} - \frac{2xy}{ab} = 0$

or $\left(\frac{x}{a} - \frac{y}{b}\right)^2 = 0$ which gives $y = \pm (b/a)^x$ ...(8)

So, the motion is described by a *st. line* with length x and y having same sign. It means that when two mutually ⊥ plane polarized lights are in phase then the resultant emerging light is plane polarised with vibrations in the same plane as the incident light. This is shown in Fig. 6.34.

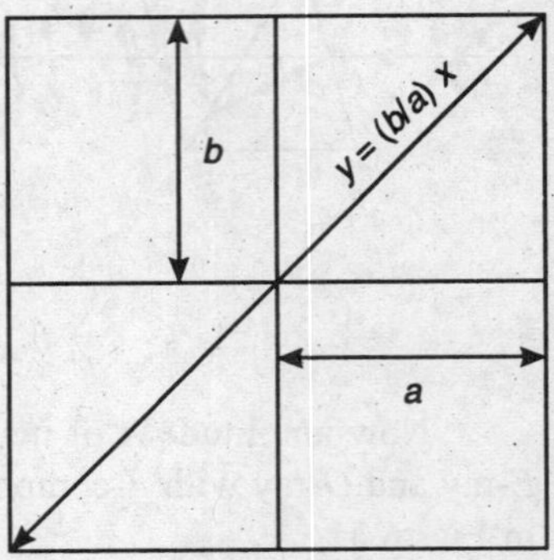

Fig. 6.34

Case 2. In case the thickness of the plate is such that

$$\delta = \pi/2, 3\pi/2, 5\pi/2, \ldots, (2n+1)\,\pi/2$$

where $n = 0, 1, 2, 3, \ldots$

In that case, $\sin \delta = 1$ and $\cos \delta = 0$

So, Equation (7) takes the form

$$\frac{x^2}{a^2} + \frac{y^2}{b^2} = 1 \qquad \ldots(9)$$

which is the equation of a simple ellipse. So, when two mutually ⊥ plane polarized lights are $(2n + 1)\,\pi/2$ phase, then emerging light is elliptically polarized (with the condition $a \neq b$). The plane of ellipse being ⊥ to the direction if propagation. (shown in Fig. 6.35).

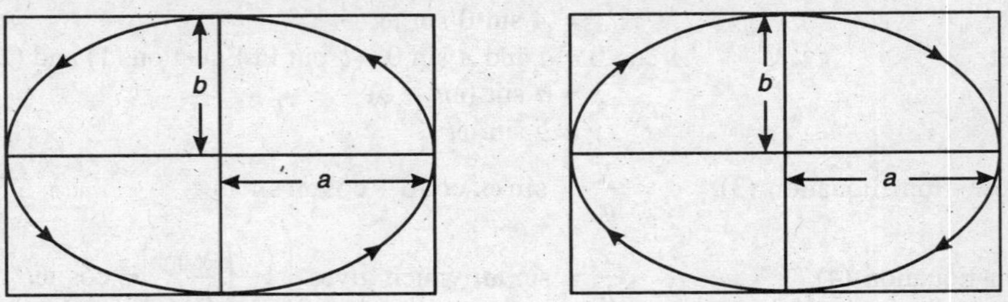

$\delta = \frac{\pi}{7}, \frac{5\pi}{2}, \ldots$ $\qquad$ $\delta = \frac{3\pi}{2}, \frac{7\pi}{2}, \ldots$

Fig. 6.35

Case 3. Increase the thickness of plate is such that $\delta = \pi, 3\pi, 5\pi, \ldots, (2n + 1)\,\pi$ then $\sin \delta = 0$ and $\cos \delta = -1$, so Equation (7) becomes

$$\frac{x^2}{a^2} + \frac{y^2}{b^2} + \frac{2xy}{ab} = 0$$

$$\left(\frac{x}{a} + \frac{y}{b}\right)^2 = 0 \text{ or } y = -\frac{b}{a}x \qquad \ldots(10)$$

which is again the equation of a straight line passing through the origin P. So, the emerging light is once again plane polarized with vibrations making an angle $2\theta = 2\tan^{-1}(b/a)$ with the direction of incident light.

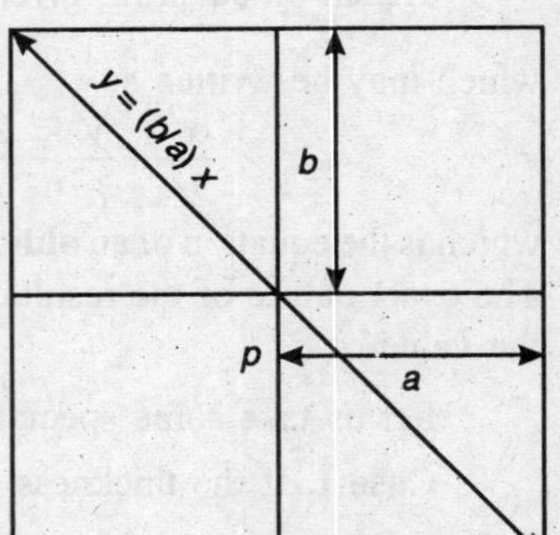

Fig. 6.36

Case 4. $a = b$ and $\delta = (2n + 1)\,\pi/2$, $n = 0, 1, 2$, then we have $\sin \delta = 1$ and $\cos \delta = -1$. So, Equation 7 takes the form $x^2 + y^2 = a^2$ which is the equation of a circle of radius a. So, **the emergent light is circularly polarised** as shown in Fig. 6.37.

From above discussion we may conclude that when two mutually ⊥ light waves super impose then the result and may be plane polarized light, elliptically polarized light or circularly polarized light depending upon the value of phase angle δ.

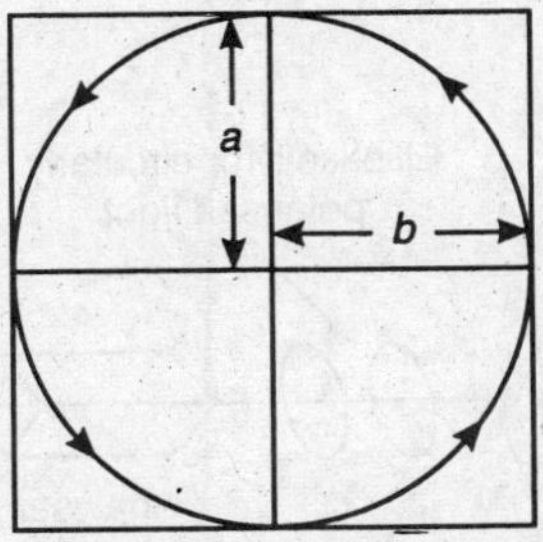

$\delta = \frac{\pi}{7}, \frac{5\pi}{2}, ...$

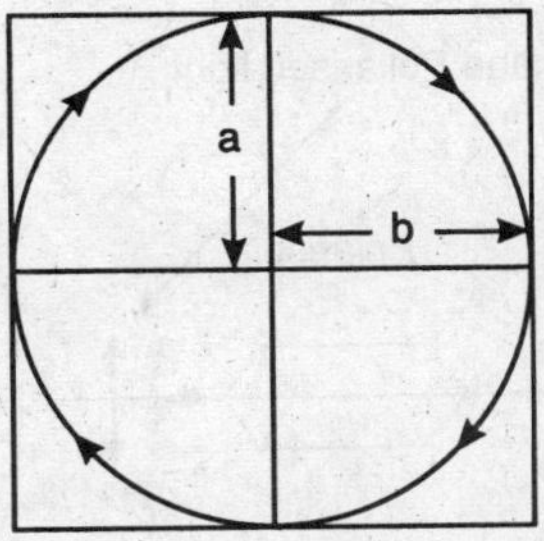

$\delta = \frac{3\pi}{2}, \frac{7\pi}{2}, ...$

Fig. 6.37

6.38. PRODUCTION AND DETECTION OF PLANE CIRCULAR AND ELLIPTICAL POLARIZED LIGHT

1. Plane polarized light: A beam of monochromatic light is incident on a nicol prism which splits it up to *O*-ray and *E*-ray. The *O*-ray is *totally internally reflected* by the canada balsom layer while *r*-ray passes through the nicol prism. So, the emerging beam is plane polarized.

2. Circularly polarized light: In this case, two waves of equal amplitudes and vibrations $\perp$ to each other having a phase angle of $\pi/2$ or path difference of $\lambda/4$ are superimposed. In order to get a circularly polarized light, plane polarized light must fall normally on a quarter wave plate in such a way that the *plane of incident wave vibrations make an angle of 45° with the direction of optic axis.*

If polarized light is made to pass through nicol prism *P* (called polarizer) to obtain plain polarized light. Another nicol prism *A* which acts as *analyser* is placed in a crossed position at some distance from *P*. In these position no light appears after *A* and so the field of view appears dark.

A quarter wave plate *Q* which is mounted on a tube *c* is placed in between *P* and *A*, so that the polarized beam falls normally on the quarter wave plate. Amplitude of the plane polarized light while entering the quarter wave plate splits into two mutually $\perp$ components having a phase angle of $\pi/2$, *i.e.*, we get the *ordinary and extraordinary components*. Now, the field of view becomes bright. The quarter wave plate *Q* is again rotated about the outer fixed tube till the field again becomes dark. This will happen when the vibrations of incident light are along optic axis of the quarter wave plate.

Again, keeping fixed, tube *C* is rotated that the mark *M* on *Q* coincides with mark 45° on the tube *C*. Now, the position of quarter plate *Q* is such that the incident plane polarized light make an angle of 45° with optic axis of the plate. In this condition, plane polarized light on entering the quarter wave plate *Q* splits into ordinary extra-ordinary components of equal amplitude and we get emergine light as circularly polarized. If analysing nicol *A* is rotated, no intensity change is observed.

Experimental arrangement is shown in Fig. 6.38.

3. Elliptically Polarized Light: Elliptically polarized light is obtained by superposition of *two waves of inequal amplitudes* vibrating at right angles to each other and having a phase difference of $\pi/2$ or path difference of $\frac{\lambda}{4}$ between them.

To get elliptically polarized light, plane polarized light must fall normally on a quarter wave plate in such a way that the plane of vibration of incident light makes an angle other *than 45°* with the direction of optic axis.

Experimental arrangement for elliptically polarized light as the same as explained in case of circularly polarized light.

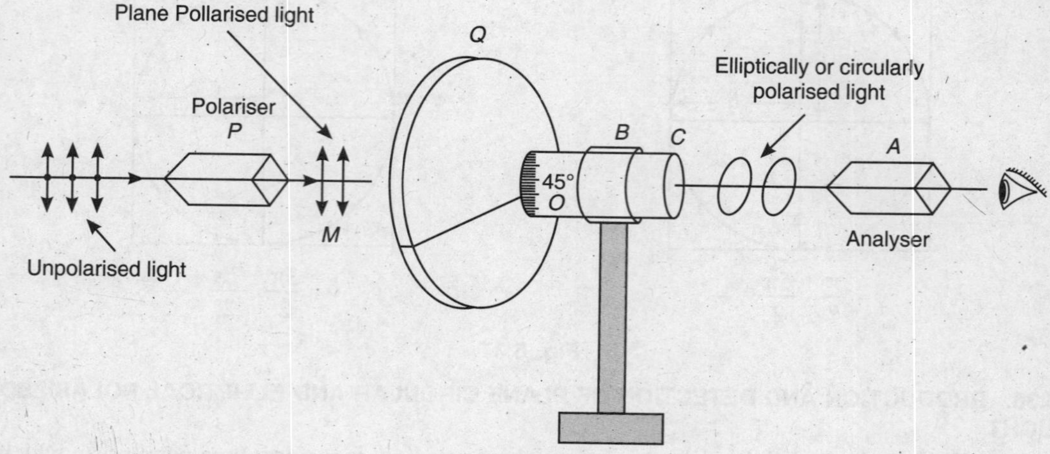

Fig. 6.38

Initial experimental setting is done in the same way. However, in this case when incident polarized light is made to fall on C at angle θ (other than 0°, 45° and 90° with optic axis of the plate as at these angles the emerging light is either circularly polarized or plane polarized). Plane polarized light on entering the quarter wave plate splits up into ordinary and extra-ordinary components of *unequal amplitudes*. So, the light coming out of quarter wave plate is elliptically polarized.

In this case, if the analysing nicol A is rotated the change in intensity is observed between a *maximum and minimum* value in the field of view, *i.e.*, minimum intensity is never zero.

6.39. DETECTION OF PLANE, CIRCULARLY AND ELLIPTICALLY POLARIZED LIGHT

I. Plane Polarized Light

In order to detect the plane polarized light, it is made to pass through a nicol prism rotating about the direction of propagation of light. If the intensity of the emergent light from the rotating nicol changes from maximum to zero minimum twice in each rotation, then the emerging light is plane polarised (Fig. 6.39).

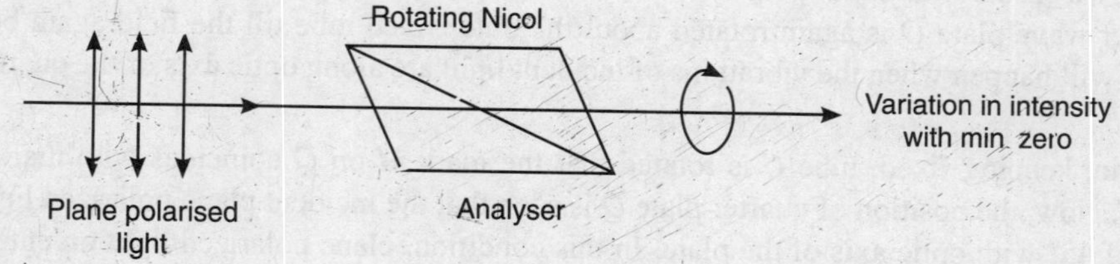

Fig. 6.39

II. Circularly Polarized Light

If we allow a circularly polarized light to pass through a rotating nicol, we observe no change in the intensity of transmitted light. Also in case of unpolarised light which is allowed to pass through a rotating nicol, no change in intensity is observed so ordinary, unpolarised light and circularly polarised light have the same behaviour when seen through a rotating nicol and so they cannot distinguished by a nicol prism alone.

So, to detect a circularly polarised light, it is first passed through a quarter wave plate and then viewed through a rotating nicol. It the original beam is circularly polarised, then it is first converted to a plane polarised light by quarter wave plate. So, on emerging from quarter wave plate, if this beam is examined by a rotating nicol, it will show variation of intensity from maximum to zero minimum twice during each rotation [shown in Fig. 6.40 (*a*)].

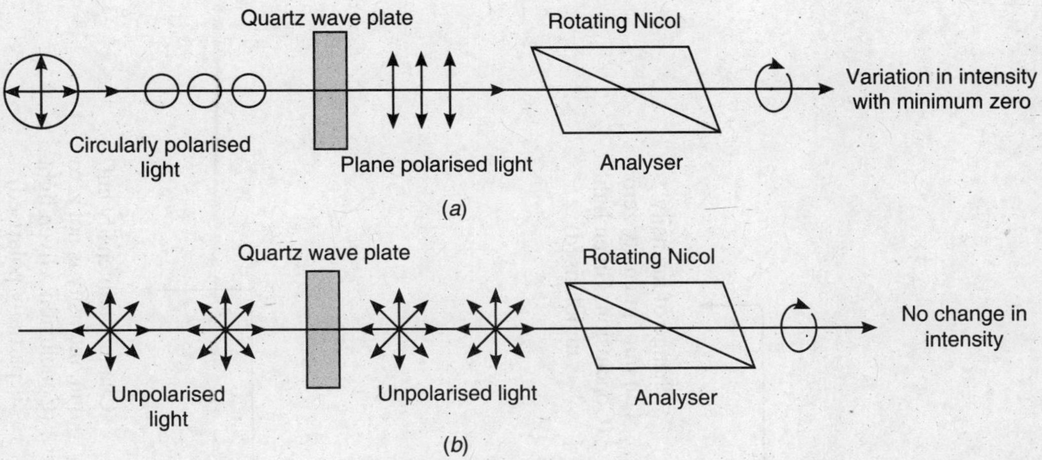

Fig. 6.40

III. Elliptically Polarized Light

It elliptically polarised light beam is passed through a rotating nicol, the intensity of emergent light varies from maximum to minimum light the *minimum intensity is never zero*. Partial plane polarised light behaves in the same way if it is passed through a rotating nicol. So, elliptical polarised light and partial polarised light cannot be distinguished by rotating nicol alone.

To detect elliptical polarised light, the beam is first passed through a rotating nicol. In case the beam is elliptically polarised, it will be converted to plane polarised light by *quarter wave plate*. If the emerging beam is studied by a rotaing nicol, a variation of intensity is observed twice from maximum to zero minimum on each rotation.

In case the original beam is partial polarised, its intensity will not change on passing through quarter wave plate [Fig. 6.41 (*b*)].

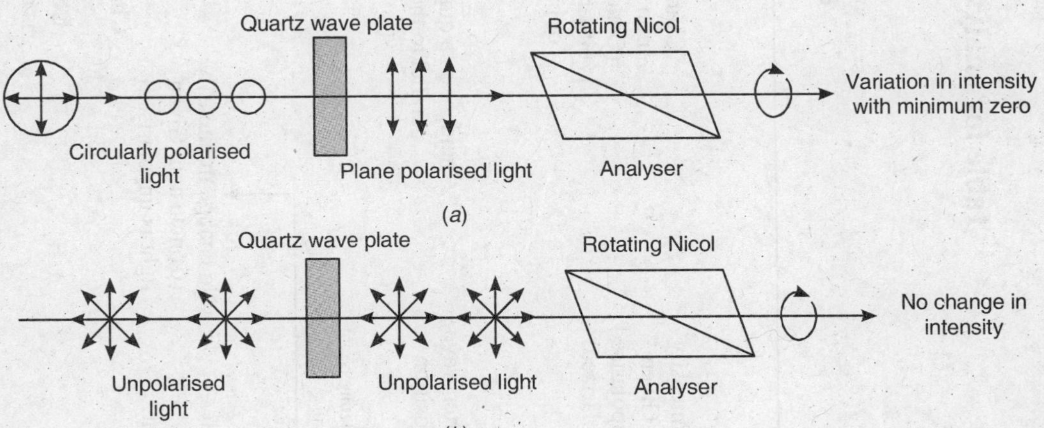

Fig. 6.41

Example 8. *Calculate the specific rotation of sugar from the following data:*

Length of the tube containing solution	*= 20 cm*
Volume of the solution	*= 80 cc*
Angle of rotation	*= $9^o\ 54'$*
Amount of sugar in solution	*= 6 gm*

Table for analysis of a given beam of light

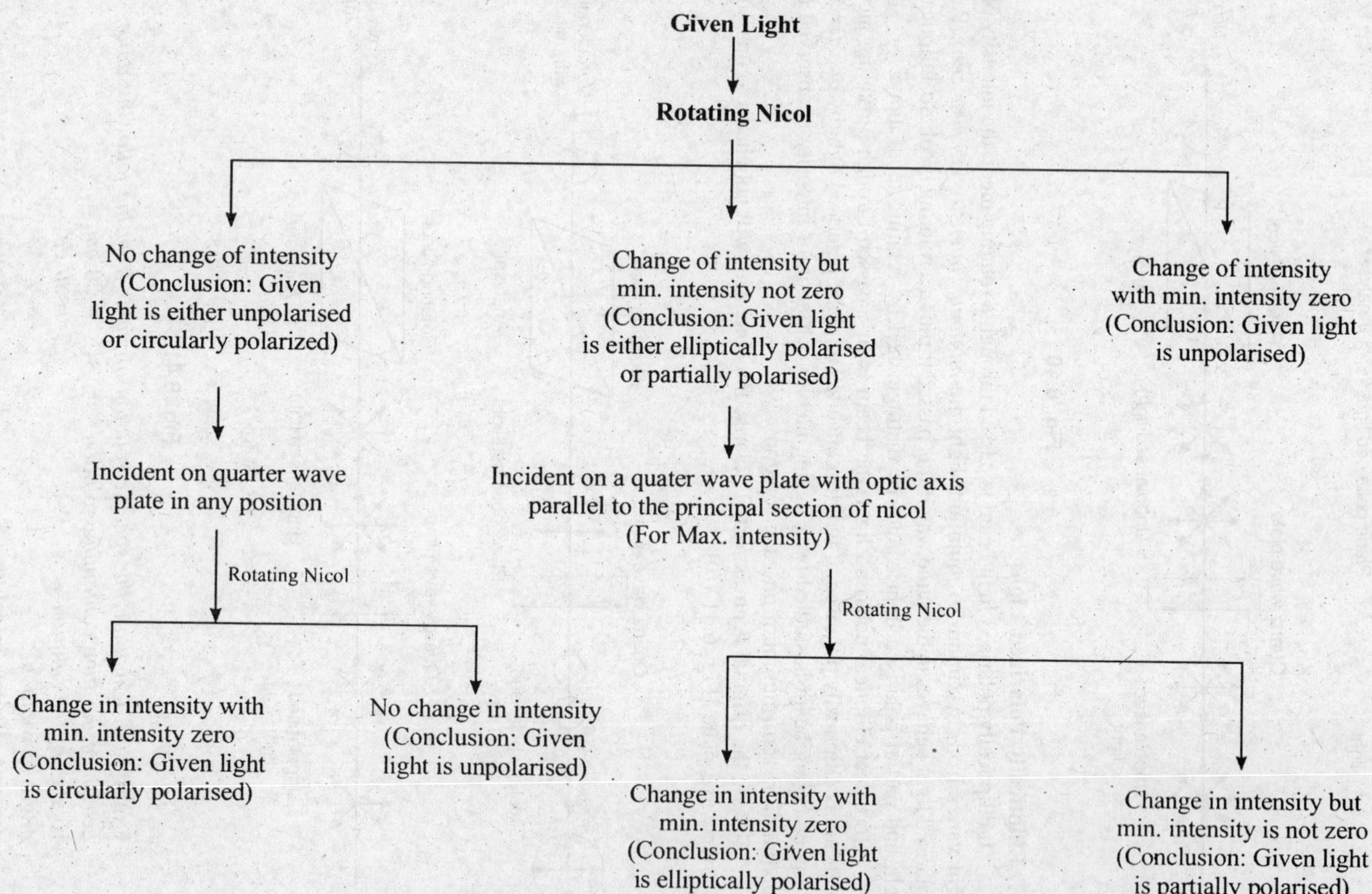

Solution: Specific rotation is given by

$$S = \frac{\theta}{l.d}$$

In this case $\theta = 9^\circ\ 54' = 9.9^\circ$

$l = 20$ cm $= 2$ decimeter

$$d = \frac{6 \text{ gm}}{88 \text{ cm}}$$

$$\therefore \quad S = \frac{9.9 \times 88}{2.2 \times 6} = 66^\circ \quad \textbf{Ans.}$$

6.40. FARADAY EFFECT

Faraday effect was discovered in 1845 and corresponds to the effect of a strong magnetic field on an **isotropic substance of high refractive index**. It is observed that if an isotropic substance of high refractive index is placed in a strong magnetic field it becomes temporarily an *optical active substance.* In case a plane polarized light is passed through such a substance such that the direction of propagation is along the direction of the field, then it is observed that the plane of vibration gets rotated through certain angle θ which depends on the strength of magnetic field H and the length l of the path of light through the substance. This can be written as

$$\theta = V.H.\ l$$

where V is called *Verdet Constant* and is dependent on nature of substance. This effect called *Faraday Effect* is shown by substances like carbon bisulphide, glass and water. It is experimentally observed that rotation angle is doubled if the light after passing through the *temporarily optical active* substance is reflected back in the opposite direction.

In case of ordinary optical active substance like quartz, calcite etc. the effect is annulled when light is sent back after reflection through them.

6.41. KERR EFFECT

Kerr in 1875 discovered experimentally the effect of *strong electric field* on some transparent substances like *Nitrobenzene.* It was observed that the electric field rotated the plane of polarization of light when the transparent substance is placed between a strong electric field. *The effect has been given the name of* **Kerr Effect**.

Fig. 6.42 shows a Kerr cell which is a small container with two electrodes immersed in the transparent insulating liquid (ortho-Dichlorobenzene or Nitro Benzene). A very high p.d. is applied between the two electrodes. The cell behaves like a *uniaxial crystal* with optic axis parallel to the line of force of the electric field. Cell is placed in between two crossed Nicols N_1 and N_2.

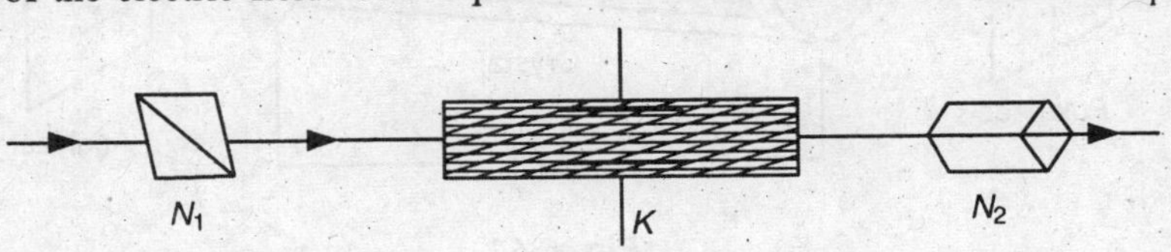

Fig. 6.42

If a plane polarized light from Nicol N_1 is made to pass through *Kerr Cell*, it is split into two components, one component is parallel to the electric field while the other is perpendicular to the electric field. The two components move with different velocities inside the cell and hence some *phase difference* is introduced between them.

It is observed that the emergent light is *elliptically polarized* which now enters the analyzing Nicol N_2. Again the light is broken up into two components in Nicol N_2, one parallel and other perpendicular to the principal section of N_2. Perpendicular component is lost in the Nicol Prism N_2 due to internal reflections while the other emerges through N_2.

So when a high voltage is applied to the electrodes of Kerr Cell, light is transmitted through the system. However, as the electric field is switched off, no light is available after the Nicol N_2 as the two Nicols are crossed. Thus, justifying the effect of electric field in creating optical activity in certain transparent substances.

6.42. COTTON-MOUTON EFFECT

Cotton-Mouton Effect is magneto-optic effect.

It is observed that certain isotropic materials acquire the optical behaviour of a uniaxial crystal under an applied external magnetic field.

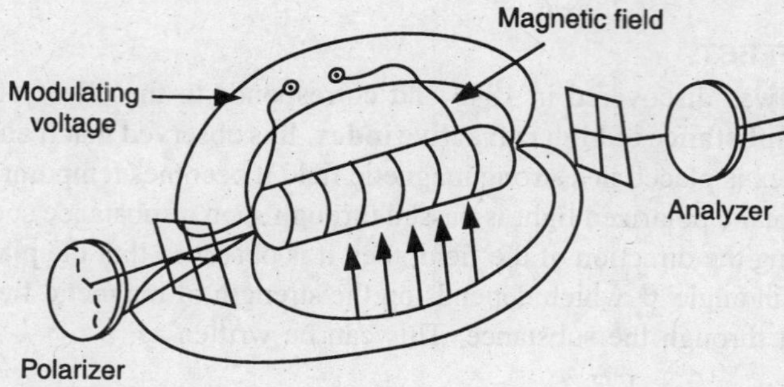

Fig. 6.43

The induced birefringence is given by the relation.

$$\Delta\mu = c.\lambda.B$$

The magnitude of induced birefringence is found to be very small. In the above relation C is a constant, B the applied magnetic field and λ the wavelength of light.

6.43. POCKETS EFFECT

Pocket effect is an *opto-electric effect.* F. Pockets in 1883 found that when an electric field is applied to some *piezo-electric* crystals, birefringences are produced.

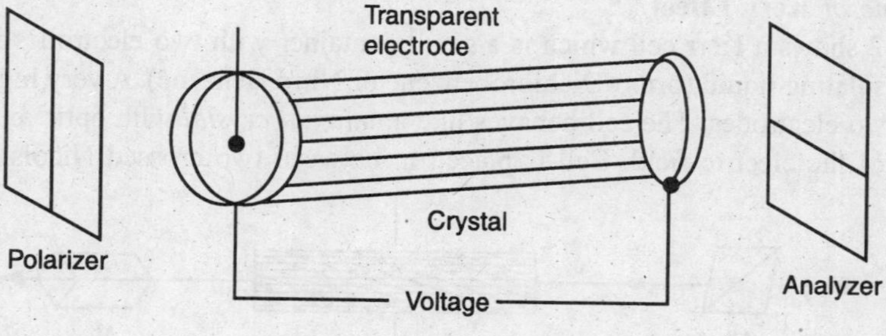

Fig. 6.44

Normally piezo-electric crystals are *birefringent,* but birefringence is restricted to only in some preferred directions. However, if an external electric field is applied on the piezo-electric crystals, birefringence (Double refraction) is observed in other directions as well.

A pocket cell consists of a piezo-electric crystal with suitable orientation such that the transparent electrodes of *trioxide* or *Indium* are put on the opposite faces.

The birefringence induced in the crystal is proportional to the strength of the applied electric field

i.e.,

$$\Delta\mu \propto E$$
$$= KE \quad \text{...(1)}$$

where k is a constant of the nature of material. Above equation shows that Kerr effect is a linear effect.

A *pocket cell* is simple in construction and requires very small applied voltage ~ 1.5 kV unlike Kerr cell which requires a very high voltage ~15 kV. Piezo-electric crystals like Potassium-Di-hydrophosphate (KDP) and Ammonium-Di-Hydrophosphate (ADP) are extensively being used as Pockets Cells.

Because of their optoelectric properties *Pockets cells* and *Kerr cells* are widely used as *optoelectric shutters* in *Q*-Switching in Lasers.

QUESTIONS AND PROBLEMS

1. What is polarization? How the polarized light can be differentiated from ordinary light?

2. Give various methods for producing plane polarized light.

3. Give various methods for production and detection of polarized light.

4. Explain the terms:

(*i*) Polarized light. (*ii*) Plane of polarization.

(*iii*) Plane of vibration. (*iv*) Double refraction.

(*v*) Optic axis. (*vi*) Principal plane.

5. Describe the construction and working of a Nicol's prism.

6. What is Brewester's law? How it can be used to find polarizing angles in crystals?

7. What is double refraction in uniaxial crystals? Explain about the positive and negative crystals.

8. What do you understand by a quarter and half-wave plate?

9. Thin plates are cut from a negative doubly refracting crystal with the position of the axis as described under. Explain giving diagrams, the nature of refraction observed in each case.

(*i*) Optic axis inclind in the upper face but lying in the plane of incidence.

(*ii*) Optic axis parallel to the upper face but lying in the plane of incidence.

(*iii*) Optic axis parallel in the unper face but perpendicular to the plane of incidence.

(*iv*) Optic axis perpendicular to upper face.

10. What is polarimetry? Explain the terms Laevorotatory and Dextrorotatry as used in polarimetry.

11. Explain briefly:

(*i*) Opticallyactive substance. (*ii*) Specific rotation.

12. Give theory and construction of Laurant's polarimeter. How half shade device is formed and used in the instrument to find the strength of sugar solution?

13. Explain Faraday effect? How is it different from Kerr effect.

14. What is Kerr effect? How it explains the optical activity due to electric field.

15. Explain Cotton Mouton effect and Packets effect. Give their importance.

16. Calculate the thickness of a quarter waveplate for light of wavelength 4890Å. (Given $\mu_0 = 1.55$ and $\mu_e = 1.60$). **[Ans.** 0.002945 mm**]**

17. Calculate the thickness of a quartz half waveplate for Fraunhofer's *c*-line (λ = 6563Å) for which $\mu_e = 1.5508$ and $\mu_0 = 1.5533$). **[Ans.** 0.00363 cm**]**

18. A beam of linearly polarized light is changed into circularly polarized light by passing it through a thin sheet of crystal 0.003 cm thick. Calculate the difference of refractive indices of the two rays, assuming it to be of minimum thickness and the wavelength is 6000Å. **[Ans.** 0.005**]**

19. A 10 cm long tube containing 10% sugar solution produces optical rotation of 13.2°. Find the specific rotation of sugar under given experimental conditions. **[Ans.** 66°**]**

CHAPTER

7

PHOTOMETRY AND OPTICAL INSTRUMENTS

7.1. PHOTOMETRY

Photometry corresponds to that branch of optics which deals with the comparison of *illuminating process* of different sources of light.

Our eye can give quantitative comparsion between two illuminating sources and this forms the basic principle of the visual photometric.

To compare the illuminating process of two sources, they are made to illuminate the adjacent portions of the same screen. Their distances from the screen are adjusted till the surfaces appear to be equally bright. Let L_1 and L_2 be the *illuminating powers* of two sources and let r_1 and r_2 be their corresponding distances from the screen. If the two portions are equally illuminated then we have

$$\frac{L_1}{r_1^2} = \frac{L_2}{r_2^2} \quad \text{or} \quad \frac{L_1}{L_2} = \frac{r_1^2}{r_2^2}$$

So the *illuminating powers of two sources are directly proportional to the square of their distance from the screen for producing the same illumination.*

The instruments which are used to compare the illuminating powers of two sources are called **photometers.**

7.2. LUMMER–BRODHUN PHOTOMETER

This is one of the most accurate photometers. Its construction and working is given as under:

Construction. 1. *Screens.* It is made up of some white diffusing material of high reflecting power *like magnesium carbonate* or *plaster of paris.*

2. *Total reflecting prisms P_1 and P_2.* P_1 and P_2 are two isosceles right angled prisms whose base angles are 45° – 45°. When a ray of light falls normally, on one of the faces, as shown in Fig. 7.1, the ray enters the prism without any change in direction, it is totally internally reflected and emerges normally from the other face.

3. *Lummer Brodhun Cube D_1 D_2.* It is the most important part of the photometer and consists of two right angled isosceles prisms D_1 and D_2 in contact with each other. Edges of prism D_2 are cut in such a way that it is slightly curved at the outer side while plane at the centre. The prisms are cemented together at the centre with a transparent material *Canada balsam* having the same refractive index as the material of the prism. So the two prisms are in optical contact with each other at the centre and enclose a thin air film at the periphery.

4. *Telescope.* It is used to receive the light from Lummer–Brodhun Cube, the whole apparatus is placed in a box blackened from inside.

Working. Light from the given sources L_1 and L_2, whose illuminating powers have to be compared, falls on the opposite faces of screen S. The light scattered from the two surfaces of the screen is reflected by the two identical, totally reflecting prisms, P_1 and P_2. The light reflected from the prisms then falls on the "Lummer–Brodhun Cube". The ray bb from the source L_1 is transmitted through the face D_1. The outer rays aa and cc are totally reflected from the face D_1. Similarly, the central ray BB and source L_2 is transmitted through the hypotenuse face D_2, whereas the outer rays AA and CC are totally reflected. A telescope T is suitably adjusted to view the light coming from both sides of the screen. To avoid complications due to external stray light, the whole apparatus is enclosed in blackened box.

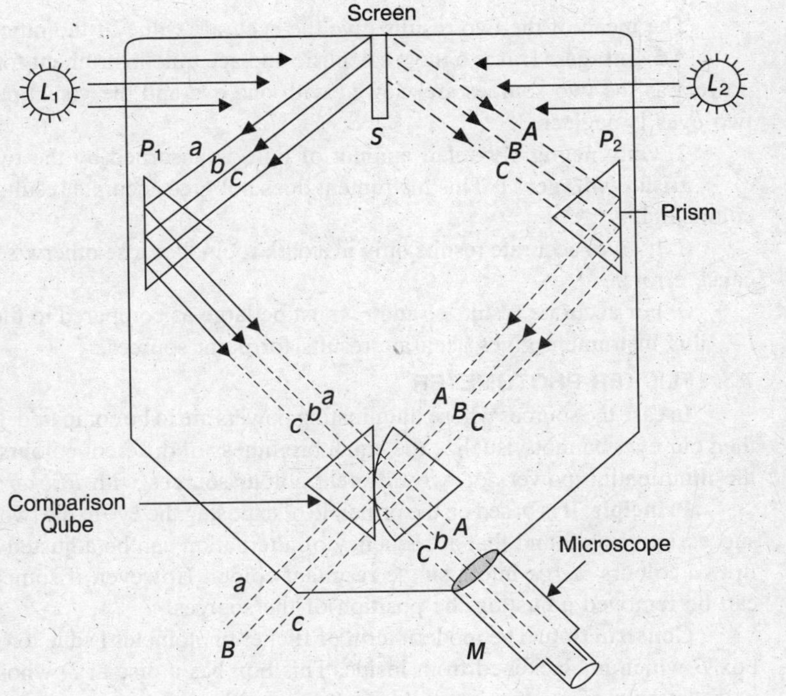

Fig. 7.1

The central part of the field of view through the telescope is thus illuminated only by light from L_1, while the outer parts are illuminated only by light from L_2, Fig. 7.2 (*a*). One of the sources is moved towards or away from the screen till the field of view is uniformly bright, *i.e*, no dividing line is visible between the central portion and the surroundings, Fig. 7.2 (*b*). A "photometric balance" thus exists between the brightness of the two sides of the screen. The distances of the sources from the screen are then measured. If L_1, L_2 are the illuminating powers of the two sources, and r_1, r_2 their respective distances from the screen, then for photometric balance, we have

$$\frac{L_1}{L_2} = \frac{r_1^2}{r_2^2}$$

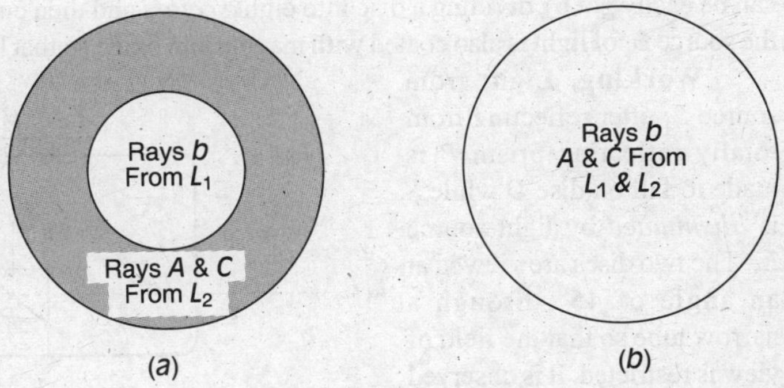

Fig. 7.2

Theory. Let I_1 and I_2 be the intensity of illumination on the two faces of the screen due to the sources L_1 and L_2 respectively.

Then $I_1 = \frac{L_1}{r_1^2}$ and $I_2 = \frac{L_2}{r_2^2}$

If β = *Reflecting power* of both sides of the screen, the brightness B_1 of the face towards L_1 is given by

$$B_1 = \beta . I_1$$

Similarly the brightness B_2, of the face towards I_2 is given by

$$B_2 = \beta . I_2$$

When the field of view is uniformly bright,

$$B_1 = B_2$$

or $$\beta . I_1 = \beta . I_2 \quad \text{or } I_1 = I_2$$

or $$\frac{L_1}{r_1^2} = \frac{L_2}{r_2^2} \quad \text{or } \frac{L_1}{L^2} = \frac{r_1^2}{r_2^2}$$

which gives us the ratio of candle powers of two sources.

In order to eliminate the error because of unequal reflecting power on the two sides of the screen, two sources are interchanged and the experiment is repeated.

The mean of the two results give the accurate value of the ratio of the illuminating powers.

Advantages. 1. It is a most scientific and accurate instrument for comparison of illuminating powers as the two sources are viewed with one eye and the visual power difference between the two eyes is neglected.

2. Only negligibly small amount of light is absorbed by the two totally reflecting prisms.

Disadvantages. 1. This instrument does not give accurate results for two sources of different colour light.

2. It gives accurate results only in a dark room, because otherwise multiple reflections of light cause errors.

3. For accurate results r_1 and r_2 must be large as compared to the dimensions of the sources, *i.e.,* this instrument gives accurate results for point sources.

7.3. FLICKER PHOTOMETER

In case the sources whose illuminating powers are to be compared, give light of different colours, then our eyes cannot visualize the equal brightness of different colours. However, we can compare the illuminating powers of two different colour sources with *Flicker Photometer.*

Principle. It is based on the principle of exposing the eye to the two different sources in alternate succession. It is found that a frequency of alternation can be adjusted till *flicker* due to difference in two colours *merge* into a single resultant colour. However, if some *flicker* still exists the same can be removed adjusting the position of the sources.

Construction. The modern term of flicker photometer is due to Guild and consists of a metal box *B* which is blackened from inside. This box has a disc at *D* whose surface is made diffusing with the help of *magnesium oxide* coating on it.

S is a rotating sector disc which is cut in the form of a *maltese cross* (also shown separately) and can be easily got by dividing a disc into eight *sectors* and then cutting off alternate sectors of *S* facing the source L_1 of light is also coated with magnesium oxide so that both discs are made equally diffusing.

Working. Light from source L_1 after reflecting from totally reflecting prism *P* is made to fall on disc *D* while *S* is *illuminated* by light source L_2. The two discs are viewed at an angle of 45° through a narrow tube so that the field of view is restricted. It is observed that if the outerfield is kept at a steady brightness then the results obtained have good accuracy. For this purpose, a *flash lamp* is provided in the side of the tube and the tube is covered by a diffusing disc *M*.

The Disc *S* is rotated about an axle at right angles to its plane. An electric motor can be used to provide different speeds. However, at one speed the frequency of rotation is constant. Due to rotation of *S*, the eye *E* will observe *D* and *S* alternately. In case two surfaces *D* and *S* are equally bright then at certain low speed no *flickering* will occur. However, if there is slight flickering the same can be removed by adjusting the distance of one of the sources.

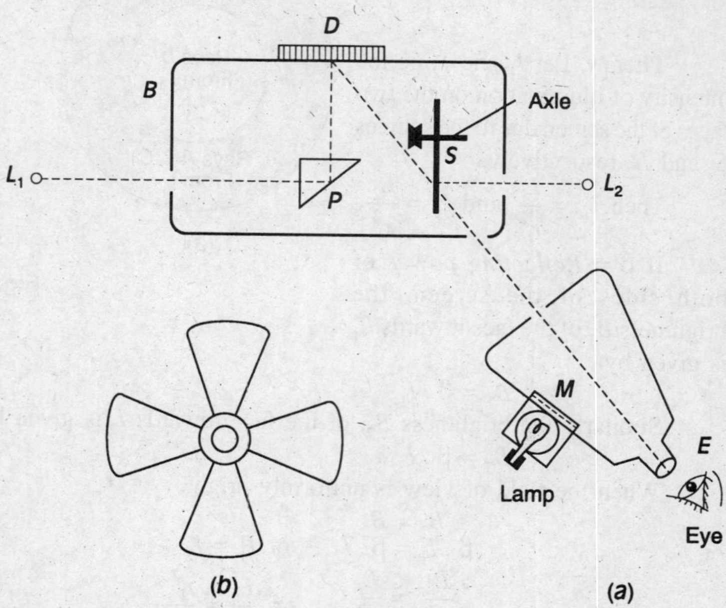

Fig. 7.2

If r_1 and r_2 are the distances of the two sources L_1 and L_2 from diffusing surfaces, then for no flickering

$$\frac{L_1}{r_1^2} = \frac{L_2}{r_2^2}$$

So the illuminating powers of two sources can be compared.

For getting accurate results, the following few conditions must be observed:

(*i*) The two surfaces must be brightly lit.

(*ii*) Rotation of sector disc should be uniform and its speed should be controllable.

(*iii*) The field of view should be kept small so that light falls on central part of retina.

(*iv*) Background should be brightly illuminated.

(*v*) Two diffusing surfaces must be free from impurities as otherwise positions of equal brightness will give inaccurate results.

If the above precautions are observed then 99% accuracy may be attained.

7.4. METALLURGICAL MICROSCOPE

It is a form of compound microscope and is useful for the study of crystal structure.

Construction. It consists of a set of two convex lenses placed co-axially within two sliding tubes. The lens *O* which is towards the object is called *objective lens,* while that for observing the object through eye is called eyepiece. Objective lens is of *short focal length* while eyepiece has long focal length. The eyepiece is fitted in a *draw tube* and the distances between objective and eyepiece can be adjusted by *rack and pinion* arrangement.

Working. The object *AB* to be magnified is placed just beyond the principal focus of the objective lens and a real, inverted and magnified image *A'B'* is formed beyond 2*f* of the objective lens. The position of eyepiece is adjusted so that the image *A'B'* falls with the focal length. The eyepiece then merely acts as a magnifying lens with which *A'B'* serves as the real object. A large, virtual image *A''B''* is formed on the same side of *A'B'* and is normally seen at the distance of distinct vision from the eye placed close behind the eyepiece.

Magnifying Power. *The magnifying power of a microscope is defined as the ratio of the angle subtended at the eye by the final virtual image to the angle subtended at the eye by the object when both are placed at the least distance of distinct vision.*

Let the final image *A''B''* formed at the distance of distinct vision *D* subtend and angle β at the eyepiece or eye (Fig. 7.4). Let the object *AB* be displaced to the position *ab* at the distance of distinct vision and let *ab* subtend an angle α at the eyepiece. Then magnifying power

$$m = \frac{\beta}{\alpha}$$

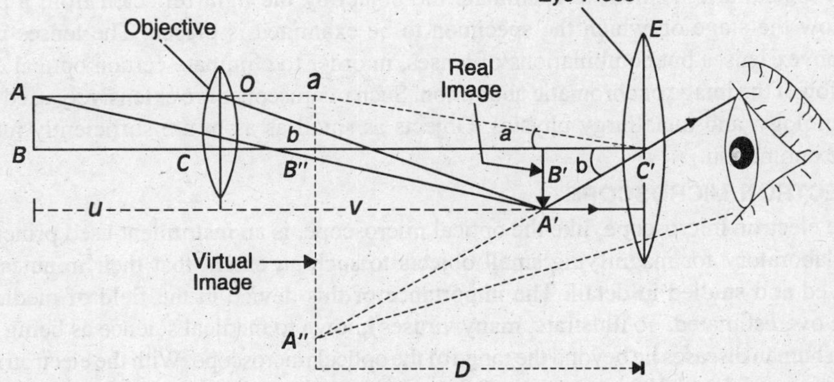

Fig. 7.4

Since α and β are small, they can be replaced by their tangents. Hence

$$m = \frac{\tan\beta}{\tan\alpha}$$

Now $$\tan\beta = \frac{A''B''}{B''C''} = \frac{A''B''}{D}$$

and $$\tan\alpha = \frac{ab'}{bC'} = \frac{AB}{D} \qquad (\because ab = AB)$$

$$\therefore \quad m = \frac{\tan\beta}{\tan\alpha} = \frac{A''B''}{D} \times \frac{D}{AB} = \frac{A''B''}{AB}$$

$$= \frac{A''B''}{A'B'} \times \frac{A'B'}{AB} = m_2 \times m_1$$

where m_1 is the magnification produced by the objective and m_2 that by the eyepiece.

Now $$m_1 = \frac{A''B''}{AB} = \frac{u}{v} \qquad ...(1)$$

where $$CB = u \text{ and } CB' = v$$

As the image formed by the eyepiece is at D, we have

$$m_1 = \frac{A''B''}{A'B'} = 1 + \frac{D}{f_c} \qquad ...(2)$$

where f_c is the focal length of the eyepiece.

Hence $$m = \frac{v}{u}\left(1 + \frac{D}{f_c}\right) \qquad ...(3)$$

Approximate formula for mangification. As the focal length of the objective is very small, the object lies very close to the principal focus of the objective lens and, therefore, u will be nearly equal to f_o, the focal length of the objective. If the focal length of the eyepiece is also short, the image $A'B'$ will be formed very near to the eyepiece so that v will be nearly equal to the distance between the two lenses, *i.e.*, equal to the length L of the microscope tube.

∴ Magnifying power can be approximately written as

$$m = \frac{L}{f_0}\left(1 + \frac{D}{f_0}\right) \qquad ...(4)$$

It is clear from the above expression that the smaller the focal lengths of the objective and the eyepiece, the larger is the magnifying power.

A good microscope gives magnification upto 2000 times or more. The object must be quite strongly illuminated if a bright image is to be seen through the eyepiece. In a high power microscope, there is a special arrangement to illuminate the object by the light reflected from a plane mirror fitted below the stage on which the specimen to be examined is placed. The lenses used are not simple convex lenses but combinations of lenses, in order to eliminate certain optical defects such as distortion of the image or chromatic aberration. Such microscopes are extensively used for research work in biology, and metallurgy physics. Objects as small as atom are sufficiently magnified for detailed examination.

7.5. ELECTRON MICROSCOPE

The electron microscope, like the optical microscope, is an instrument used principally in the research laboratory for magnifying small objects to such an extent that their minutest parts may be observed and studied in detail. The importance of this device in the field of medical research cannot be overestimated. To illustrate, many viruses known to medical science as being responsible for certain human diseases lie beyond the range of the optical microscope. With the electron microscope, magnifications of 10 to 100 times that of finest optica microscopes make many of these viruses, and some of their detailed structure, visible to the eye. While the highest magnification obtained with the best optical microscope is about 2000 ×, electron microscopes have already been made that give magnifications as high as 100,000 ×.

Principle. It is based upon the principle that

(1) particles such as electrons exhibit wave properties like light rays but have much shorter wavelength; and

(2) electrons can be focussed by suitable electric and magnetic fields just as light rays can be focussed with the help of lenses.

By Definitions. *The resolving power of an optical instrument like a microscope is directly proportional to the aperture (diameter) of a lens used and inversely proportional to the wavelength of the light used.* So, the smaller the wavelength, the greater shall be the magnifying power. If we accelerate electrons through a potential difference of say, 60000 volts; the velocity v attained by them can be calculated from the relation $eV = 1/2mv^2$ where m, the electron mass = 9×10^{-31} kg. Calculating λ from de Brogile wave equation ($\lambda = h/mv$), the wavelength come out to be about 5×10^{-12} m which is 10^5 times smaller than that of visible light.

Construction. In Fig 7.5, the schematic diagram of electron microscope has been shown by the side of an optical microscope for comparison. In an optical microscope, Fig. 7.5 (*b*), the light from a source *S* such as sunlight or an electric lamp is made to pass through a narrow slit and then made to fall on a condeser lens. The condensed light is then made to illuminate the object to be seen magnified. An objective lens gives the first magnified image of the object. This image is again magnified by a projection lens. The final magnified image is either seen directly or is obtained on a photographic plate for permanent record.

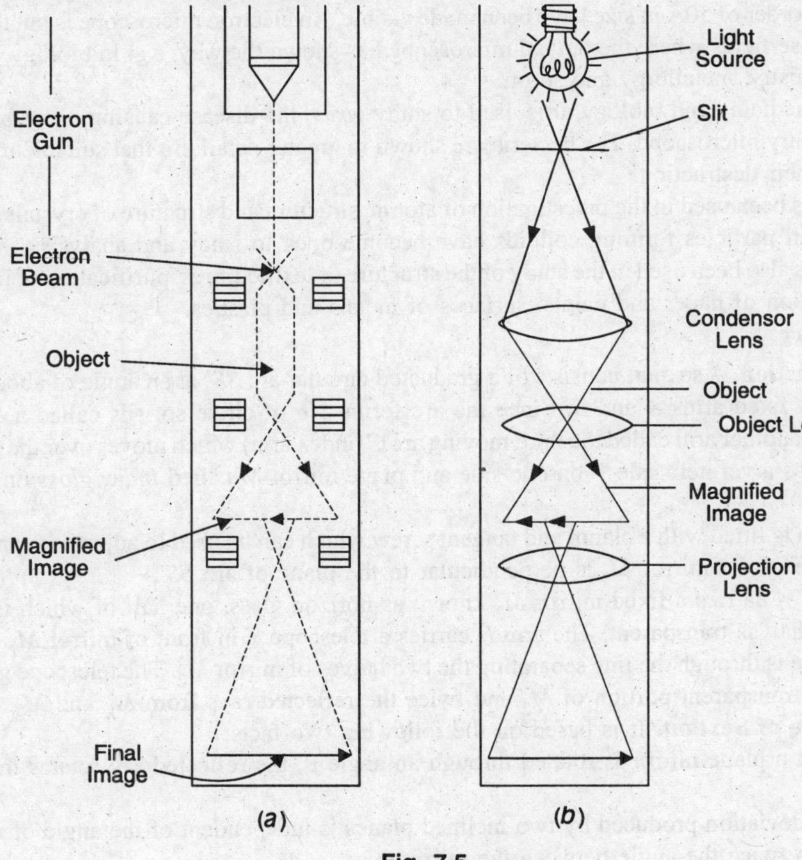

Fig. 7.5

An electron microscope shown in Fig. 7.5 (*a*) is similar in principle to an optical microscope. The modern electron microscope is ordinarily of transmission type in which magnetic lenses of short focal length are used to obtain large magnification. An electron beam emitted by a filament is accelerated through a large potential difference in a device called electron gun. The electron beam got from the electron gun, is made to pass through the centre of the doughnut shaped electromagnet '*a*'. The electrons get deflected to form a parallel beam which strikes the object to be magnified. It should be noted that the electrons will be transmitted more through the transparent parts of the object and transmitted in less number through comparatively denser portions. The transmitted beam will thus have the likeness of the object transversed by it. The second electromagnet '*b*' causes the electron beam to diverge to produce enlarged image of the object. The electromagnet '*c*' focuses the electron beam from part of the enlarged image on the fluorescent screen producing still greater magnification. The image obtained on the fluorescent screen is made visible by scintillation for direct view. It can also be obtained on a suitable photographic plate for a permanent record. Sharp focussing is obtained by adjusting the intensity of magnetic fields produced by electromagnets.

The object can be given fine micro-movement to bring different parts of object under the incident beam for detailed examination. Since the electron beam operates in vacuum, the apparatus is mounted in a glass tube, completely evacuated. Accelerating voltage of the order of 3,00,000 volt are usually employed for the examination of thicker objects. With the electron microscope, it has been possible to magnify the objects 100,000 times with adequate resolving power.

Uses. The resolving power of an optical microscope is limited by the wavelength of visible light and the smallest size of the object that can be examined with it is of the order of 2000Å. The electron microscope has a very large magnifying power and with the help of the best type of instrument objects of the order of 50Å in size have been made visible. An electron microscope is put to valuable uses in all those fields where the optical microscope has shown the way, *e.g.* in biology, medicine, physics, chemistry, metallurgy and so on.

(1) In medicine and biology, it is used to study *virus*, the disease causing agent beyond the reach of ordinary microscope. The bacteria are shown in greater detail. So that suitable means may be used for their destruction.

(2) It has been used in the investigation of atomic structure and structure of crystals in details.

(3) Small particles forming colloids have become open to study and analysis.

(4) It has also been used in the study of the structure of textile fibres, purification of lubricating oils, composition of paper and paints, surfaces of metals and plastics.

7.6. SEXTANT

Construction. A sextant consists of a graduated circular arc SS' at an angle of about 60° and has two radial fixed arms A and B. Since the arc forms 1/6 of circle so it is called a *Sextant*.

There is another arm called the third moving arc C (index arm) which moves over the graduated scale SS'. It has a vernier scale V on one side and plane mirror M called *index glass* on the other side.

This arm is fitted with a clamp and tangent screw which can be used to adjust it to any desired position. Plane of the mirror M_1 is perpendicular to the plane of arc SS'.

The arm A carries a fixed mirror M_2 known as horizon glass, one half of which is silvered and the other half is transparent. The arm B carries a telescope T in front of mirror M_2. The axis of the telescope is through the line separating the two halves of mirror M_2. The telescope gets direct rays from the transparent portion of M_2 and twice the reflected rays from M_1 and M_2.

Principle of Sextant. It is based on the following two facts:

(*i*) When a plane mirror is rotated through an angle θ, the reflected ray rotates through an angle 2θ.

(*ii*) The deviation produced by two inclined planes is independent of the angle of incidence and is equal to twice the angle between the mirrors.

Working. A sextant is used to measure the angle subtended by two distant points in the same vertical or horizontal line.

First let us consider that the two points lie in same vertical line. Now the sextant is fixed in a stand, and telescope and the horizon glass are put at the same height as the lower point, and a direct image of the lower point is seen directly through the clean part of the mirror M_2 (Fig. 7.6).

Again the movable arm is moved and mirror M_1 is brought to a position that the image of the lower point after reflection at M_1 and M_2 coincides with the direct image. In this position *zero of the main scale should coincide with the zero of the vernier scale.* If it does not happen then the zero error with proper sign should be carefully noted.

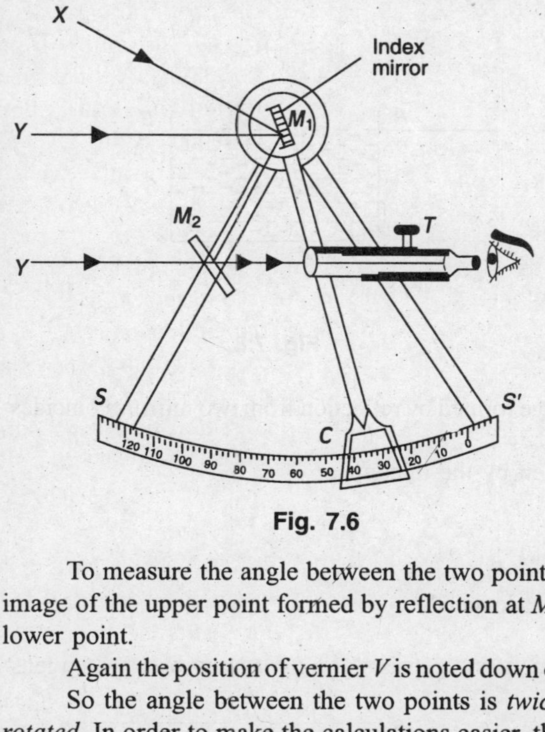

Fig. 7.6

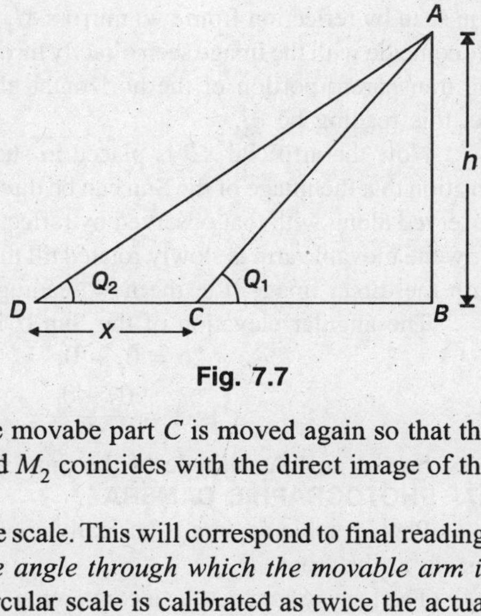

Fig. 7.7

To measure the angle between the two points, the movabe part C is moved again so that the image of the upper point formed by reflection at M_1 and M_2 coincides with the direct image of the lower point.

Again the position of vernier V is noted down on the scale. This will correspond to final reading.

So the angle between the two points is *twice the angle through which the movable arm is rotated.* In order to make the calculations easier, the circular scale is calibrated as twice the actual degrees. The difference between two vernier reading is the *required angle between the two points.*

Applications. (1) It is useful for measuring angle subtended at the eye by two distant points objects situated in a vertical or horizontal line.

(2) To measure the height of inaccessible towers etc.

(3) To measure altitude of the sun.

(*i*) *Measurement of height of an object.* Let AB be a distance object whose height is to be determined.

Let the angle subtended by A be θ_1 and θ_2 at two different positions 'C' and 'D' respectively.
Now in ΔBCA

$$\tan\theta_1 = \frac{AB}{BC} = \frac{h}{BC}$$

$$\therefore \quad BC = \frac{h}{\tan\theta_1}$$

$$= h\cot\theta_1 \qquad ...(1)$$

Again in ΔBDA,

$$\tan\theta_2 = \frac{AB}{BD} = \frac{h}{BD}$$

$$\therefore \qquad BD = \frac{h}{\tan \theta_2} = h \cot \theta_2 \qquad ...(2)$$

Now $\qquad BD - BC = h\,(\cot \theta_2 - \cot \theta_2)$

or $\qquad x = h\,(\cot \theta_2 - \cot \theta_1)$

$$\therefore \qquad h = \frac{x}{\cot \theta_2 - \cot \theta_1} \qquad ...(3)$$

So the height of an inaccessible object can be measured using the relation (3).

(*ii*) *Measurement of Sun's altitude.* Angular altitude corresponds to the angle between the *horizon* and the direction of the Sun at place. Since it is difficult to know about the position of *horizon* so an artificial horizon is used in the form of a horizontal surface of mercury taken in a dish as shown in the Fig 7.8.

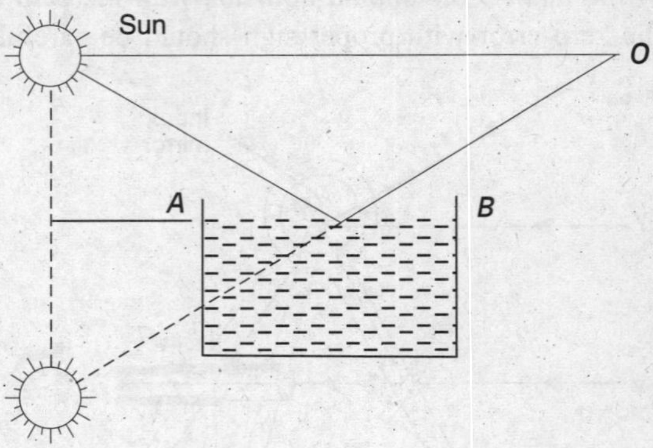

Fig. 7.8

In order to note the zero error of sextant, the movable arm C is turned till the image of the Sun seen by reflection from two mirrors M_1 and M_2 coincide with the image seen directly through the transparent portion of the horizontal glass. Let this reading be θ_1.

Now the artificial AB is placed in such a position that the image of the Sun can be directly observed along with that observed by reflection. Now the movable arm is slowly rotated till the image formed by reflection from two mirrors coincides with the direct image. Let the new reading be θ_2.

The angular elevation of the Sun α is given by the relation

$$2\alpha = \theta_2 - \theta_1$$

or $$\alpha = \frac{\theta_2 - \theta_1}{2}$$

So the angular altitude of the Sun can be calculated.

7.7. PHOTOGRAPHIC CAMERA

Photographic camera is basically a lens camera and consists of a *light-proof* base with a lens system at the front.

The lens system acts as a converging lens and light from an object is *focussed* on to the film F by the lens system. There is a *shutter* between the lens and the film.

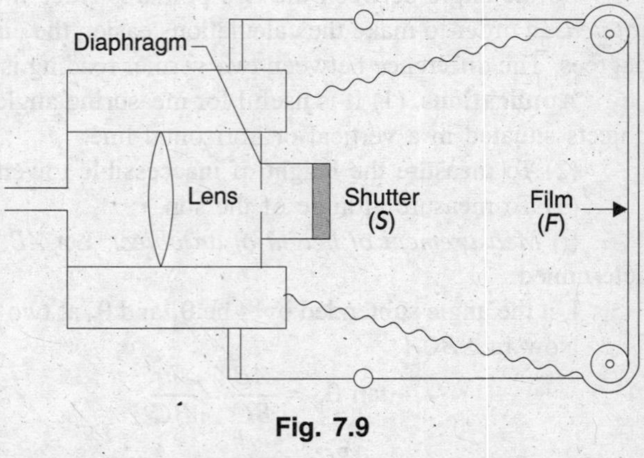

Fig. 7.9

In order to take a photograph the shutter is opened and closed very rapidly so that only desired amount of light after reflection from the object enters the camera and the film is exposed for a very short time. An arrangement fitted in the camera can move the lens forward or backward to get a sharp picture of the object through the lens.

An important point to note is that the image produced by the lens system has to be *bright* enough so that during a very short time of exposure, a sharp image of the object is got on the film. The brightness and sharpness of the image depends on the diameter of the *aperture,* which is a circular opening, through which light passes. The amount of light falling on the film is controlled by the *aperture* while the focal length of the lens controls the *area of the image on the film*. The *exposure time will determine the brightness of the image.*

Diameter of the aperture is given in terms of fraction of the focal length of the lens and is called *f-numbers*.

As may be evident the exposure time will be less on a bright sunny day than that on a dull day. Also for fast moving objects the exposure time should be small otherwise the image will be *blurred.*

An other important point is that the speed of the film has also to be considered for a good photograph. A fast film needs comparatively a short exposure time while a slow film needs a longer exposure time.

7.8. PHOTOGRAPHIC OBJECTIVE LENS (CAMERA LENS)

An elementary objective camera lens is a convex lens which forms the image of an object with quite accuracy and over a reasonable specified field.

However, as explained earlier, a single lens suffers from many defects. In order to overcome these defects, many improvements have been made and in modern cameras a complicated system of lenses is used. By selecting suitable communication of convex and concave lenses, *chromatic aberration* is eliminated. Suitable diaphragms are used to eliminate *spherical aberration and distortions.* A modern camera may consist of more than 10 lenses with good accuracy using computers.

7.9. *F*-NUMBER OF A LENS (*F*-RATIO)

It is the ratio of the focal length of the lens (f) and the diameter of the lens aperture (d).

f-number (*f*-ratio) = focal length of lens (f)/diameter of the lens aperture (d).

f-number is also called the speed of the lens and determines the amount of energy transmitted by the lens.

$\frac{d}{f}$ is known as *aperture ratio or relative aperture* of the lens.

If the focal length of a lens is 11 cm and the diameter of the aperture is 1 cm, then *f*-ratio of *f*-number is 11, and $d = \frac{f}{11}$.

Normally, the diameters of lens aperture (stop) of a camera are marked in such a way that exposure time is doubled between each position.

For example, in case of $\frac{f}{5.6}, \frac{f}{8}, \frac{f}{11}, \frac{f}{16}, \frac{f}{22}$; the *f*-ratios are 5, 6, 8, 11, 16 and 22 while their *squares* are in the ratio of 1 : 2 : 4 : 8 : 16 etc.

So, if the correct exposure for $\frac{f}{5.6}$ is 0.5 sec., then for $\frac{f}{8}$ it is 1 sec. and for $\frac{f}{11}$ it is 2 sec. and so on.

7.10. DEPTH OF FOCUS OF A LENS

The depth of focus of a lens is *small range of distances* on either side of the focal plane of the lens in which no appreciable *deformation* of the image quality is detectable. It is formed by the *relative aperture and by the size of the point image.*

7.11. TYPES OF CAMERA LENSES

Following are the various types of camera lenses which find use in the *cameras:*

***(i)* Meniscus Lens (Single Lens System).** This is the simplest form of lens which find use in box cameras. The concave surface is facing the object and the diaphragm is placed in front of the meniscus lens to overcome the *lens aberrations.* Lens aberrations can be controlled by changing the shape of the lens and the position of the diaphragm (Fig. 7.10).

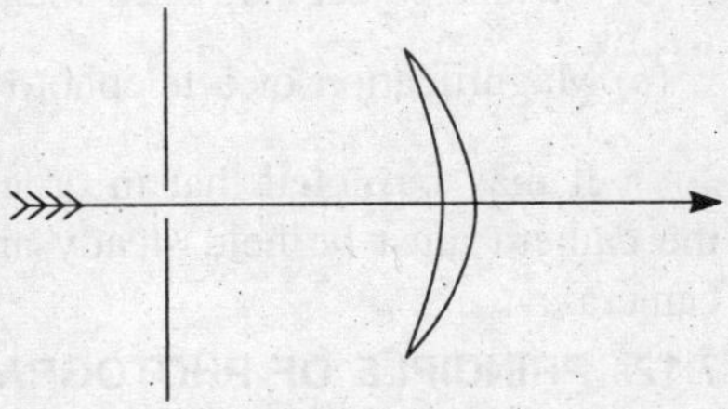

Fig. 7.10 Single lens system.

***(ii)* Achromat (Double Lens System).** In this system two *achromatic* lenses are placed at a certain distance apart from each other with a diaphragm between them. This arrangement of lenses is called *aplanats* (aperture is ~ $f/8$ and the field is ~ 50°). Also the rapid rectilinears (with large apertures and small fields) are useful for taking group or full figure photographs. The achromat lenses are formed using crown glass which acts as positive element and fluit glass which acts as negative element. Using suitable glasses with correct indices of refraction, simultaneous corrections are obtained for spherical aberrations, coma. The achromatic doublets separated by diaphragm gives simultaneous correction of field curvature and astigmatism.

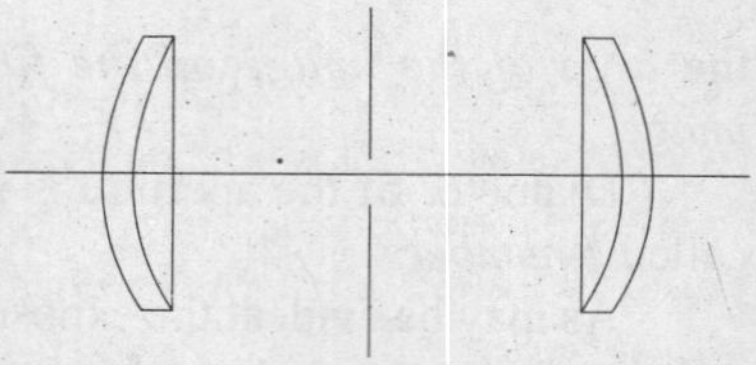

Fig. 7.11 Double lens system.

***(iii)* Zeiss Sonnar (Triple Lens System).** Zeiss Sonnar is an improved form of triple lens system and is free from corrections for lens aberrations. It permits a larger angular field and the defects of coma and a stigmatisms are greatly reduced.

***(iv)* Gauss Type Lenses (Four Lens System).** The four lens system is quite useful designing wide angle lenses and process lenses.

Wide angle lenses are quite because of the following important properties:

(*a*) A good impression of depth is obtained with them.

(*b*) Object seems to be smaller and more away from the eye.

***(v)* Zoom Lenses.** These lenses are quite useful for amataur photography used in movie and T.V. cameras. The variation of focal length is achieved by moving one or more lens components longitudinally.

***(vi)* Telephoto Lenses.** Lenses having focal length more than the normal lenses are called *telephoto lenses.* Because of larger focal length of the lens, the image is focussed at a large distance in the lens and to achieve it the size of the camera has to be greatly increased, which is impractical and inconvenient. In order to avoid it, a special lens system called *Telephoto lens* is used in which *effective focal* is increased.

It consists of two lenses (one convex and one concave) mounted in a tube. In place of a single lens, the tube is fitted. Each of the lenses is made achromatic.

The concave lens L_2 is placed in original position of the lens in camera, while lens L_1 (convex lens) is placed infront of lens L_2. The combination of these two lenses acts as a single convex lens of large focal length.

If f_1 and f_2 are the focal lengths of lenses L_1 and L_2 and d is the distance between them, then

$$\frac{1}{f} = \frac{1}{f_1} + \frac{1}{f_2} - \frac{d}{f_1 f_2}$$

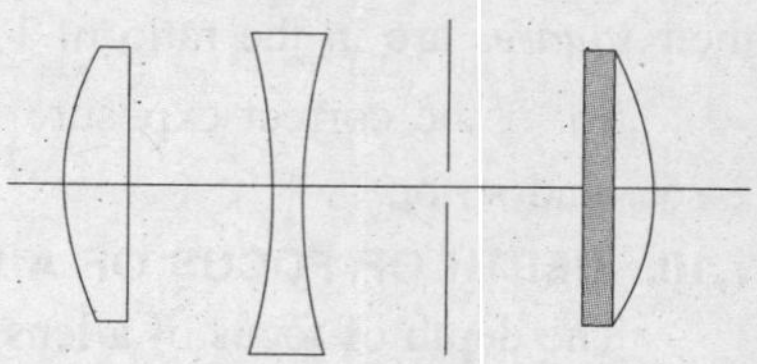

Fig. 7.12 Zeiss sonnar—Triple lens system

Some important features of telephoto lense are:

(*a*) Angle of view is very small ~ 5°.

(*b*) Distant objects are seen magnified and nearer.

(*c*) Magnification of a telephoto lens $= \dfrac{\text{Focal length of telephoto lens}}{\text{Focal length of normal lens}}$.

It may be noted that in order to have sharp and clear pictures with a telephoto camera, the camera must be held steady and normally a *tripod* is used to avoid any movement of the camera.

7.12. PRINCIPLE OF PHOTOGRAPHY

Consists of the following steps:

(*a*) Exposure, (*b*) Developing, (*c*) Fixing, (*d*) Printing.

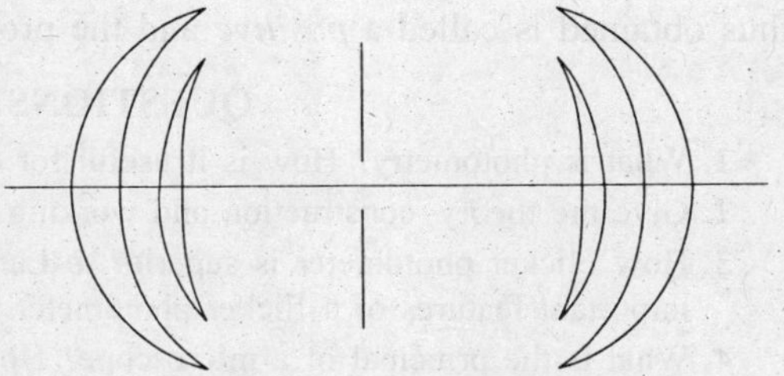

Fig. 7.13 Four lens system.

(*a*) Exposure. When we want to take the photograph of an object, first of all a *clear and sharp* image of the object has to be got on the focus of the film of the camera. This can be done by adjusting the lens. Only the required amount of light has to be admitted into the camera and if the light is dull then a *flash* can be used. An unexposed photographic film is inserted into the holder of the camera and the shutter is kept closed so that no light enters. Shutter speed is adjusted to have a favourable value of exposure time, which depends on the aperture of the lens, intensity and colour of light and nature of the film. During the exposure halides of silver grains on the film are exposed to the *reflected* light from the object and there is a formation of an invisible or a latent image of the film.

(*b*) Developing. The exposed film is taken out of the camera and is immersed in a *developer* in a dark room let with red light. Developer is a reducing agent which is a capable of reducing the exposed grains of *silver halide* in the film to silver without reducing the unexposed grains at the same time. Normally a developing agent has the following components.

(*i*) Reducing agent. It mostly consists of organic reducing agents like pyrogallol, amidol, adurol, metol glycin etc. Mostly a combination of reducing agents is used to get optimum results for various temperatures.

(*ii*) Accelerator. It is an *alkali* and controls the pH of the developer. In order that the pH of the developer is not changed due to repeated uses, the original pH of the developer is always maintained. For this an Alkali with high buffer capacity is required and normally *sodium carbonate* or *borax* fulfils the requirement.

(*iii*) Preservative. In order to protect the reducing agent from atmosphere, preservative is required. A preservative protects the oxidation and transforms the oxidation products with colourless components. Sodium sulphite or sodium bisulphite can act as preservatives.

(*iv*) Restrainers. Restrainers are required to lower the degree of ionization of silver bromide and retain the concentration of silver cations. Potassium bromide is used as *restrainer.* It also reduces the fog on negative of the photograph.

So a solution comprising of the above four components along with water is used for developing of the film.

When the exposed film is put in the developer, white part of the films turns black and the amount of blackness will depend upon the intensity of light received at any point. Time required for developing a film depends on the nature of the film.

After developing the film is washed in water with a precaution that no finger prints or scratches are put on the film. After this the film is put in a *fixer.*

(*c*) Fixer. The purpose of the fixer is to dissolve the remaining halides from the unaffected parts of the film and fix the developed image on the film. So it converts silver halides into water soluble compounds.

Again the film is thoroughly washed with water and dried. Now the film is called *negative* as the bright portions of the object appears black and the black portions of the object appear white in it.

(*d*) Printing. Now the negative is put in contact with a piece of light sensitive paper and this combination is exposed to light from a very powerful lamp. The required size of the image on the photo sensitive paper can be obtained by adjusting the distance between photo-sensitive paper and the negative and consequently the image can be enlarged as per our requirement. Again the exposed paper is developed and fixed as explained earlier and a true photograph of the object is obtained

In this way from a negative many photographs can be obtained by varying enlargements. The photograph thus obtained is called a *positive* and the process of getting a photograph is called *printing*.

QUESTIONS AND PROBLEMS

1. What is photometry? How is it useful for comparison of illuminating powers of two sources?
2. Give the theory, construction and working of Lummer Brodhun Photometer.
3. How Flicker photometer is superior to Lummer-Brodhum Photometer? Give the construction and important features of a flicker photometer.
4. What is the principal of a microscope? How metallurgical microscope is useful?
5. Give the theory, working and applications of a sextant.
6. What is the reason of calling the instrument *Sextant* for measuring heights? Give its use for measuring the height of a tower.
7. How sextant can be used for measuring the altitude of the sun?
8. What is the principle of an electron microscope? How is it useful? In what way is it superior to conventional microscope?
9. Give the various methods to get a good photograph.
10. What is the principle of photography?
11. Give in details the construction and working of a photographic camera.
12. Give the various types of photographic camera lens systems. State their important applications.
13. Give the principle of photography? What their important steps to get a good photograph?
14. Explain the *f*-number of a photographic lens.

CHAPTER

8

LASERS AND FIBRE OPTICS

8.1. INTRODUCTION

In interference of light, we use a term coherence between two sources of light. The two sources are coherent, when they vibrate in the same phase or there is a constant phase difference. We know that light from a source comes as the sum total of radiations by billions and billions of atoms or molecules in the source. The phase is different at different times. Now the question is that to what extent may the radiations from different atoms of a given source be related in phase, in direction of emission and in polarization *i.e., the coherence of a given source.* In recent years, some sources are developed which are highly coherent *i.e.*, the radiation given out by all the emitters in the source is in mutual agreement not only in phase but also in the direction of emission and polarization. *These coherent sources are called lasers.* The light beams from lasers, with coherence lengths of upto 10^{14} Hz will make possible many experimetnts and practical applications which would not be possible with the previously available light sources of coherence lengths extending alomst upto 10^7 Hz.

Before the description of lasers and their applications, we may underline some of the basic concepts of the various energy states.

8.2. ATOMIC EXCITATION

An atom can be excited to a state higher than its ground state by the following two principal methods.

(*i*) By collision with another atom during which part of the kinetic energy is absorbed by the atom and an atom in this way returns to ground state in time ~ 10^{-8} sec by emitting one or more photons (Fig. 8.1).

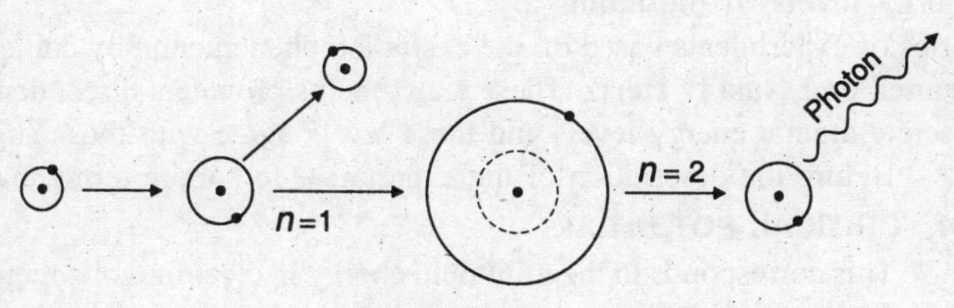

Fig. 8.1

(*ii*) By establishing an electric field that accelerates electrons and atomic ions till kinetic energies are sufficient to excite atoms they collide with. Since the energy transfer is maximum when the colliding particles are of the same mass, so electrons in luminous discharge in a rarefied gas are more effective than the ions in providing energy in atomic electrons.

Neon signs and mercury vapour lamps are some of the familiar examples of the characteristic colours in spectral emission when strong electric fields are applied between electrods in a gas filled tube. Neon gives reddish light while mercury vapour give bluish light.

8.3. EXCITATION BY ABSORPTION

A different kind of excitation is involved when an atom absorbs a photon of light whose energy is just the right amount of energy to raise an atom to higher energy level. For example, a photon of $\lambda = 121.7$ nm is emitted when a hydrogen atom in $n = 2$ state comes to $n = 1$ state. The absorption of a photon of $\lambda = 121.7$ nm by hydrogen at ground state $n = 1$ will bring it to $n = 2$ state. Figure 8.2 explains the absorption spectrum.

We can explain the absorption and emission spectrum as under:

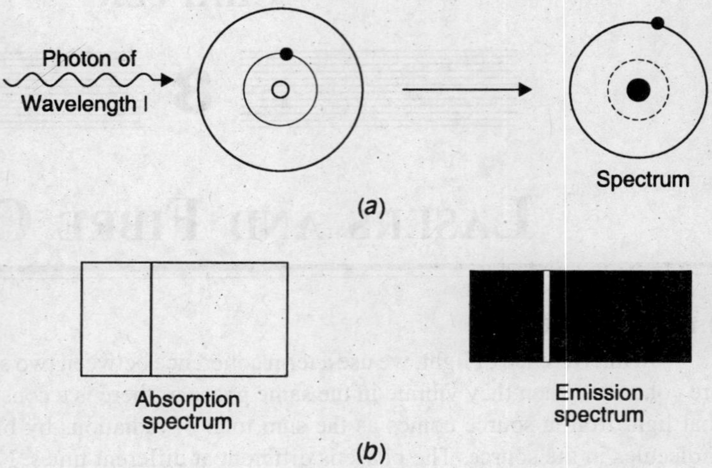

Fig. 8.2

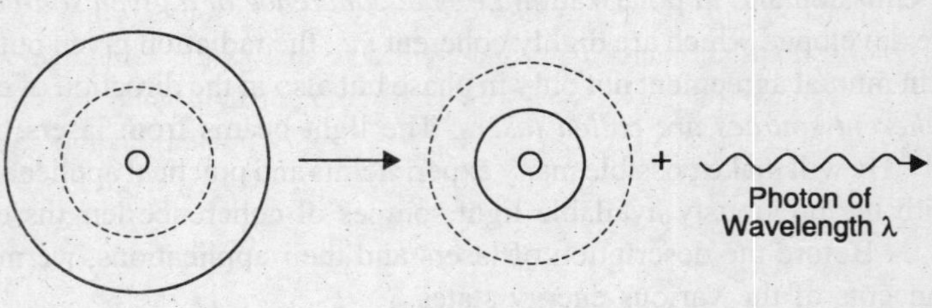

Fig. 8.3

White light consisting of all the colours (wavelengths) when passes through hydrogen gas, then photons of those wavelengths that correspond to the transitions between energy levels are absorbed resulting in the excitation of the hydrogen atom. The excited hydrogen atom radiates photons almost (instantly at once) but these photons come in random directions and only few of them in the same direction as the original beam of white light. So the dark lines in an absorption spectrum are never completely dark but only appear to be **black because of bright background.**

The dark lines in the absorption spectrum are at the same position as the bright lines in the emmission spectrum.

To explain the discrete energy levels of the atoms a series of experiments based on the excitation phenomenon by collision were performed by **James Franck and Gustav Hertz.** These experiments provide a direct demonstration of the existence of discrete atomic energy levels and these levels agree with those suggested by the line spectra.

Before the description of the experiment few more terms may also be clearly understood.

8.4. CRITICAL POTENTIAL

This corresponds to the minimum energy in electron volts required to excite an electron from ground state to a higher state.

8.5. IONIZATION POTENTIAL

The critical energy is known as ionization potential when the absorption of the energy just removes an electron from the atom *i.e.*, the electron is raised from level $n = 1$ to $n = \infty$.

8.6. EXCITATION POTENTIAL

This corresponds to the raising of energy level of an unexcited atom to a higher energy state (but not ∞).

From Bohr's theory, we have the energy of hydrogen atom given by

$$E_n = \frac{13.6 \text{ eV}}{n^2}$$

when $n = 1$, $E_1 = -13.6$ eV

$n = 2$, $E_2 = -3.4$ eV

$n = 3$, $E_3 = -1.51$ eV

$n = \infty$ $E_\infty = 0$ eV

So we have the energy required for hydrogen as

For first excited state = 13.6 – 3.4 = 10.2 eV

2nd excited state = 13.6 – 1.51 = 12.09 eV

For ionization = 13.6 – 0 = 13.6 eV

So the discrete amounts of energy are required to excite hydrogen atom to various energy states which are thereby quantized.

Now the Franck-Hertz experiment can be described to illustrate about these energy states.

8.7. THE FRANCK-HERTZ EXPERIMENT

As is clear from above, the excitation of an atom, by any of the two mechanisms from one state to another involve transfer of energy in quanta. This quanta either absorbed or emitted corresponds to some line in the spectrum of that atom. It was with these considerations that Franck and Hertz in 1914 performed the experiment using **inelastic collision of accelerated electrons with the atoms as the source of excitation**. This experiment provides a direct demonstration that **atomic energy levels are present in the atom.**

Experimental Arrangement. A tube filled with the gas or vapours of the atoms under study is fitted with an indirectly heated cathode C, grid G and anode A.

A sensitive milliammeter and voltmeters connected along with batteries, as shown in Fig. 8.4. The emitted electrons from C are accelerated by the potential difference V_a towards G, so every electron reaching G must have acquired a kinetic energy eV_a. After crossing G the electrons pass through a retarding potential V_r and only those electrons reach 'A' which have sufficient kinetic energy to overcome V_r. As a result a current I_p flow inside the grid-anode circuit, which is measured by the milliammeter mA. A typical plot of I_p and V_a is shown in Fig. 8.5.

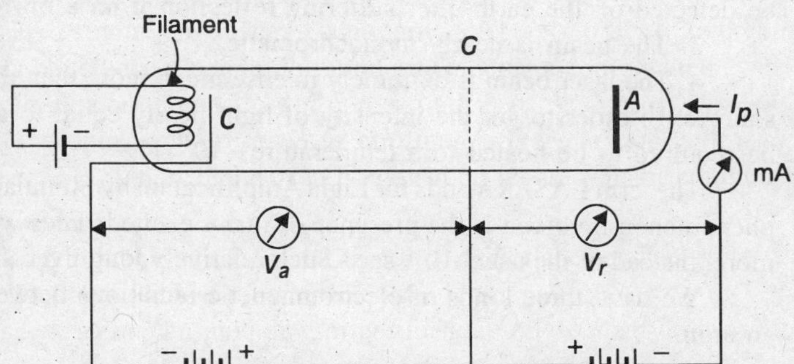

Fig. 8.4

Results of the Experiments. As V_a is increased from zero, the kinetic energy eV_a of electron is increased so that more of them cross G and reach A, increasing I_p. The rise from O to K in the curve in figure indicates that upto the corresponding value of the accelerating voltage, the collision of the electrons with the atoms are elastic and number of electrons reaching A increases with V_a. The abrupt fall in I_p form M to N or $V_a = V_{a_1}$ indicates that when the kinetic energy of the electron is eV_a these undergo inelastic collisions with the atoms close to G so that their kinetic energy after crossing G is smaller than eV_r. Further rise from N to O shows that the collisions of the electrons are again elastic. If they undergo an inelastic collision it involves energy eV_{a_2} and after that these have gained kinetic energy sufficient to overcome V_r. Thus the collisions continue to be upto V_{a_2} which again most of the collisions are inelastic as there is sudden drop in I_p from O to P.

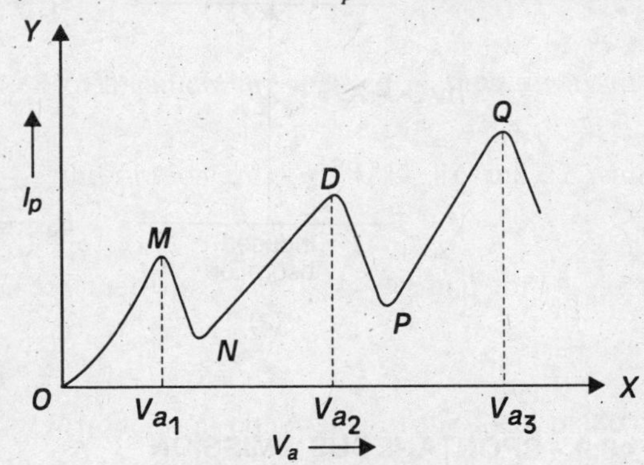

Fig. 8.5

From the above discussion it is clear that the collisions of the electrons with the atoms are inelastic only for some critical potentials *i.e.*, $V_{a_1}, V_{a_2}, V_{a_3}, \ldots$ The direct evidence from this is that atoms absorb discrete energies $eV_{a_1}, eV_{a_2}, eV_{a_3}, \ldots$ *to get excited. The energies other than these are not transferred. To check the interpretation of critical potentials as being due to discrete atomic energy levels. Franck and Hertz observed the emission spectra of vapour during electron bombardment. In the case of* **mercury vapours**, they found that a minimum electron energy of 4.9 eV was required to excite 2536Å spectral line of mercury. A photon of 2536Å has an energy of just 4.9 eV. It was concluded that energy levels in the atoms are **quantized** and the absorption and emission of radiation by these are in **quanta and not continuous.** It is observed that mercury accepts only and exactly 4.86 eV, 6.67 eV and 8.84 eV which means that mercury atoms have a set of discrete energy states above ground state and continuous states above 10.4 eV.

Some of the following terms must be clearly understood to know the phenomenon of production of laser beams:

Laser is a device that produces light beam with the following properties:

1. The light is coherent with all the waves exactly in phase with each other.
2. Laser beam hardly diverges. Such a light if sent from Earth to Moon remains narrow to be detected on the earth after suffering reflection from a mirror at Moon's surface.
3. The beam is nearly monochromatic.
4. The laser beam is extremely intense much more than the light from ordinary conventional sources. In order to get the intensity of light nearly equal to that of laser beam an object would be required to be heated to a temperature ~10^{30} K.

The term LASER stands for Light Amplification by Stimulated Emission of Radiation. The basic phenomenon involved is the presence of many excited states whose life times may be 10^{-3} sec. or more, instead of the usual 10^{-8} sec. Such relatively long lived states are called *'Metastable States'*.

We have three kinds of electromagnetic radiations between two energy levels E_0 and E_1 in an atom.

8.8. INDUCED ABSORPTION

If the atom is initially in the lower state E_0 it can be raised to E_1 by absorption of a photon of energy $E_1 - E_0 = h\nu$. This is called 'induced absorption' [Fig. 8.6 (*a*)].

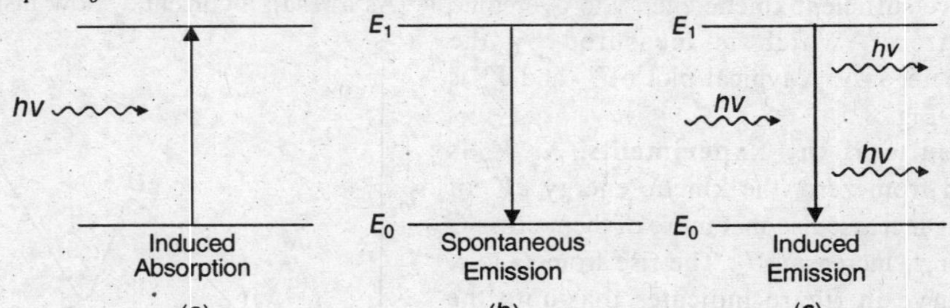

Fig. 8.6

8.9. SPONTANEOUS EMISSION

If the atom is initially in the upper state E_1, it can drop to E_0 by emmition of a photon of energy $h\nu$. This process is known as **Spontaneous Emission** [Fig. 8.6 (*b*)].

8.10. INDUCED EMISSION

In this case an incident photon of energy $h\nu$ causes a transition from E_1 to E_0. In the induced emission, the radiated light waves are exactly in phase with the incident waves, so that we get an enhanced beam of coherent light. In 1917 Einstein showed that induced emission and induced absorption have the same probabilities. Which means that a photon of energy $h\nu$ when incident on an atom has the same likelihood of induced emission as that of induced absorption.

The concept of induced absorption forms the basis of LASERS.

8.11. ACTIVE MEDIUM

A medium in which light gets amplified is called **active medium** and this medium may be solid, liquid or a gas. It is important to note the only a small fraction of a particular medium is responsible to **stimulated emission** and the remaining part of the medium merely supports the active centres.

Population

Number of active atoms occupying an energy state is known as the *population* of that state.

Let N_1 and N_2 be the population of the lower (E_1) and upper (E_2) energy states respectively. At the thermal equilibrium, population of energy levels E_2 and E_1 are given by the Boltzmann's relation as:

$$N_2 = N_1\, e^{-(E_2-E_1)/kT}$$

$$\therefore \quad N_1 = N_2\, e^{(E_2-E_1)/kT}$$

$$= N_2\, e^{-(h\nu/kT)} \qquad \text{... (1)}$$

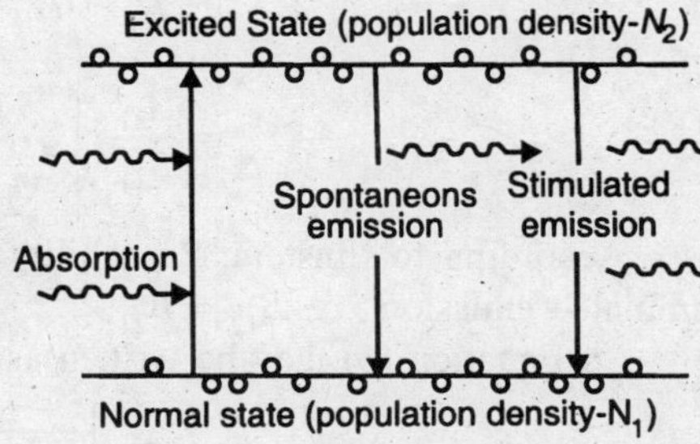

Fig. 8.7

where negative exponent factor in Equation (1) shows that $N_2 << N_1$, at the equilibrium, which in other words mean that there are more atoms in the tower energy state E_1 and this state is called **Normal state.**

If an external radiation is incident on the medium under normal state condition, the three processes, *viz.*, **absorption, spontaneous emission and stimulated emission** occur simultaneously as shown in Fig. 8.7.

As is clear from the figure, atoms in the lower level E_1 will **randomly absorb radiation** to make a transition of E_2 level, which may again randomly give radiation and come to the lower level E_1.

To make radio of N_1 and N_2 constant, number of upward transitions must be equal to the number of downward transition, *i.e.*,

$$N_{ab} = N_{sp} + N_{st} \qquad \text{...(2)}$$

which may be written in terms of **Einstein coefficient** as

$$B_{12}\, N_1\, Q = A_{21}\, N_2 + B_{21}\, N_2\, Q \qquad \text{...(3)}$$

where B_{12}, A_{21} and B_{21} are called Einstein's coefficient and Q is the radiation density of light.

At thermal equilibrium, ratio of stimulated and spontaneous emission is generally very small and stimulated emission is negligible may write the ratio as

$$\frac{\text{Stimulated emission}}{\text{Spontaneous emission}} = \frac{B_{21}\, N_2\, Q}{A_{21}\, N_2} = \frac{B_{21}}{A_{21}}\, Q \qquad \text{...(4)}$$

also the ratio of stimulated emission and asborption transition is given by

$$\frac{\text{Stimulated emission}}{\text{Absorption transition}} = \frac{B_{21}\, N_2\, Q}{B_{21}\, N_1\, Q} = \frac{N_2}{N_1} \qquad \text{...(5)}$$

where $B_{21} = B_{12}$, since the probability of emission transition is equal to that of **absorption transition.**

As is clear from Equation (4), to get more stimulated transitions, *radiation density Q* must be large.

Also from Equation (5) stimulated emission will be more than the absorption transition only when $N_2 > N_1$. If $N_1 > N_2$, *absorption dominates stimulated emission. i.e.*, the medium will only absorb light rather than amplifying it.

When the above two conditions are satisfied, the light passing through the medium gets amplified.

From Equation (3)

$$N_1\, B_{12}\, Q = N_2\, (A_{21} + B_{21}\, Q)$$

which gives $$\frac{N_1}{N_2} B_{12}\, Q = A_{21} + B_{21}\, Q$$

or $$\left(\frac{N_1}{N_2} B_{12} - B_{21}\right) Q = A_{21}$$

$$\therefore \quad Q = \frac{A_{21}}{\frac{N_1}{N_2} B_{12} - B_{21}}$$

$$= \frac{A_{21}}{B_{21}} \left[\frac{1}{\frac{N_1}{N_2}\left(\frac{B_{12}}{B_{21}}\right) - 1} \right] \quad ...(6)$$

According to Einstein, Probability of stimulated absorption is equal to the probability of stimulated emission *i.e.,* $B_{12} = B_{21}$.

So Equation (6) may be written as

$$Q = \frac{A_{21}}{B_{21}} \left[\frac{1}{\left(\frac{N_1}{N_2} - 1\right)} \right]$$

From Boltzmann's law the population of two levels at temperature T is expressed as

$$\frac{N_1}{N_2} = e^{hv/kT} \quad [(\text{Equation } (1)]$$

where K is Boltzmah's constant and h is Planck's contant. So Equation (7) takes the form

$$Q = \frac{A_{21}}{B_{21}} \left[\frac{1}{e^{hv/kT} - 1} \right] \quad ...(8)$$

From Planck's law, energy density of radiation is given by

$$Q = \frac{8\pi h v^3}{c^3} \left[\frac{1}{e^{hv/kT} - 1} \right] \quad ...(9)$$

comparing Equation (8) and Equation (9) we get

$$\frac{A_{21}}{B_{21}} = \frac{8\pi h v^3}{c^3} \quad ...(10)$$

From above equation it is clear that ratio of spontaneous emission coeff. and stimulated emission coeff. is proportional to V^3 which means that at **thermal equilibrium**, the probability of spontaneous emission increases rapidly with energy differences between the two states.

Units of Einstein Coefficients

From Equation (10)

$$\frac{A_{21}}{B_{21}} = \frac{8\pi h v^3}{c^3}$$

$$= \frac{\text{Joule} - \text{sec} \times \text{sec}^{-3}}{(\text{meters sec}^{-1})^3}$$

$$= \text{Joule} - \text{Sec} - \text{meters}^{-3}$$

$$= \text{J} - \text{s} \quad \text{m}^{-3}$$

These are the units of Einstein Coefficients.

8.12. LASER ACTION

Let us consider a collection of atoms all in the same excited state. A passing photon may cause stimulated emission in one of these atoms. It will cause emission of *two photons.*

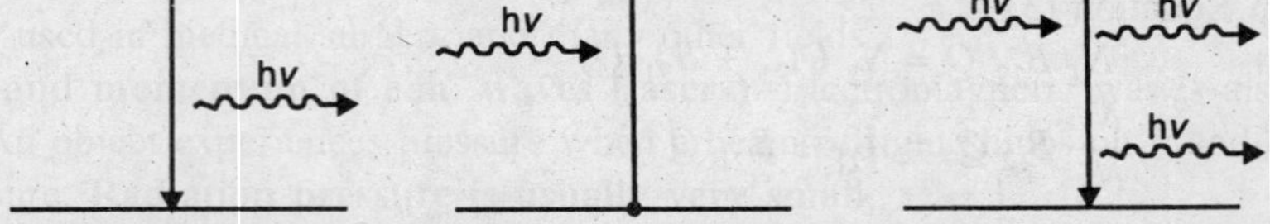

Fig. 8.8

Each of these photons may cause induced emission in two other excited atoms. This will result into induced emission in number of atoms and give an avalanche of an intense beams of photons moving of the same direction and all coherent. This is the basic operation of *laser.*

8.13. POPULATION INVERSION

The process described above is not workable due to the fact that the life time of an atom in an excited state is very small ~ 10^{-8} sec. Which means that before an atom can be stimulated to exit a photon it may make a spontaneous emission. So the photons resulting from the spontaneous emission are not coherent. The ratio of number N^* of excited atoms to that N at ground state is given by

$$\frac{N^*}{N} = e^{-E/kT}$$

where $E \rightarrow$ Energy difference between excited state and ground state.

$k \rightarrow$ Boltzmann's constant.

$T \rightarrow$ Temperature on celcius scale.

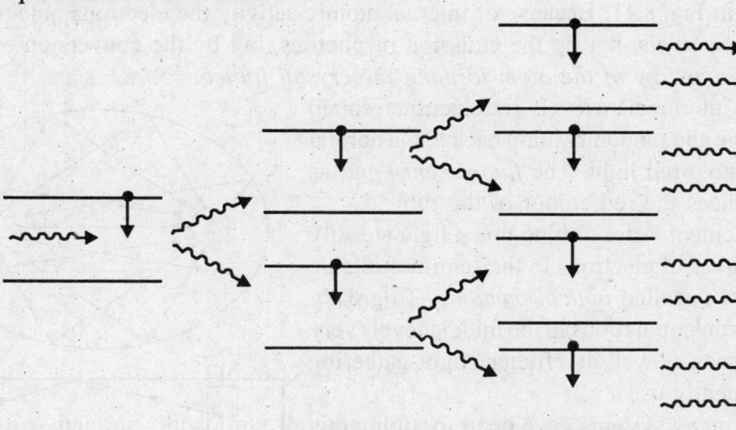

Fig. 8.9 (*a*)

For atomic radiation E is much greater than kT and so in thermal equilibrium the population of higher state is smaller than that of the ground state, with the result that there will be few stimulated emission of radiations.

A solution of this problem is found if an atom has there energy levels. The atoms that are originally in the ground state G are raised to higher short lived states by some external source of energy. However, the atom can not remain in the excited state S for long and rapidly decay into an intermediate state is 10^{-3} sec. Which is very much higher than the life time (10^{-8} sec.) of the atom in the excited state.

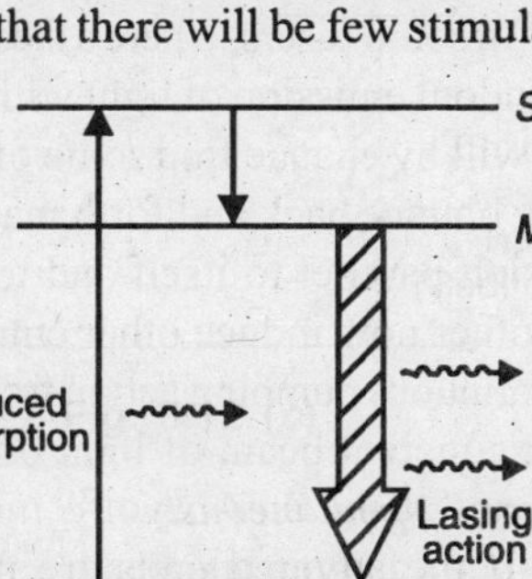

Fig. 8.9 (*b*)

A passing photon produces the lasing transition from on state to two ground state. The above process in which the ratio of the Population in highers state M to the ground state G is appreciably increased is called **Population Inversion.**

8.14. OPTICAL PUMPING

A convenient method of describing laser action is to refer to an energy level diagram of the electronic states involved in light absorption and emission. An energy-level diagram for the electrons in the atomic lattice of a ruby crystal is given in Fig. 8.10. Here there are three sets of levels, the *normal state*, the *semi-metastable states* near the middle and the top *wide-band energy levels.*

(Pink ruby is a crystal lattice of aluminum and oxygen atoms. Al_2O_3, with a small amount of chromium ions as an impurity, 0.04% of Cr^{3+}).

Semi metastable states are states in which electron transitions to lower levels are not entirely forbidden, but there is a delay in the electrons' jumping down to a lower level.

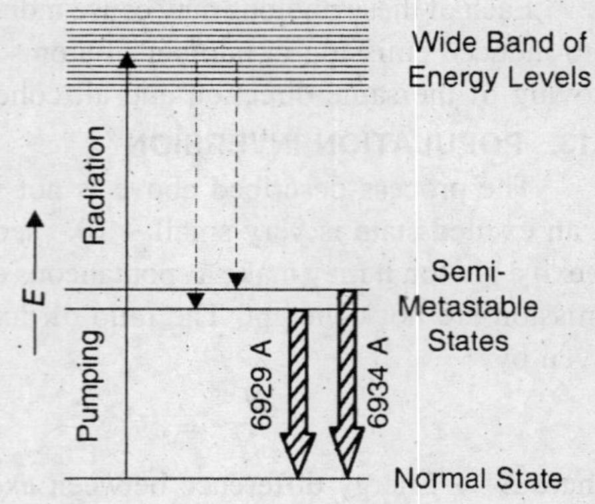

Fig. 8.10

When a beam of white light enters a ruby crystal, strong absorption occurs in the blue and green part of the spectrum and the transmission of only the red region is what gives the ruby its red colour.

On shining a strong beam of blue-green light into a ruby crystal, the absorption that takes place raises many electrons to the wide-band levels, as shown by the up arrow at the left in Fig. 8.11. Because of internal atomic activity the electrons quickly drop down to the intermediate levels, not by the emission of photons, but by the conversion of energy into *vibrational kinetic energy of the atom forming the crystal lattice.*

Once in the intermediate levels, the electrons remain there for some time and randomly jump back to the normal state emitting visible red light. The *fluorescent light*, as it is called, enhances the red colour of the ruby.

Since an incident beam of blue-green light steadily increases the number of electrons in the semi-metastable states, the process is called *optical pumping*. To greatly increase the electron populations in the middle levels very intense light sources as well as efficient light-gathering systems are frequently used.

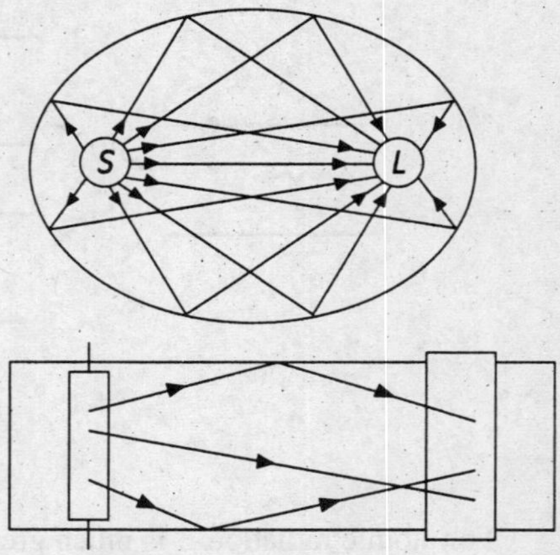

Fig. 8.11

One of the many systems developed for doing this is shown in Fig. 8.11. By placing the exciting light source at one focus of an elliptical reflector and the laser at the other focus, **high efficiency can be obtained.**

In some lasers, where a steady light beam is required, the random emission of light within the gas, liquid or solid mass will by chance find some of the waves emitted along the axis. See Fig. 8.11. This light, because it may bounce back and forth many times between the highly reflective ends, will stimulate coherent emission parallel to itself and to the axis. These two waves travelling back and forth in step with each other now induce other emission, thereby giving rise to a rapid growth or chain reaction. Thus by continuous pumping action from a separate light source, a large part of the stored energy is converted into a coherent beam of light of a different wavelength. *Under these conditions the device is said to "laze" spontaneously.*

If the silvered ends are not highly reflecting, too much light escapes from the ends and spontaneous lazing cannot occur. Under these conditions a beam of light of the stimulating frequency from another laser can be sent into the crystal where it is intensified. By modulating the input beam the greatly intensified output beam will be modulated accordingly. This modulation capability is one of the important properties that gives lasers their promising future applications to the field of communications.

8.15. RUBY LASER (THREE LEVEL LASER SYSTEM)

Its main part is a ruby rod called the laser rod (Fig. 8.12). Ruby is a crystal of aluminium oxide (Al_2O_3) with an addition of 0.03 to 0.05% of chromium oxide (Cr_2O_3) in which Cr^{+3} ions replace some of the atoms in the crystal lattice. Ruby crystals are grown in special furnaces and then shaped into rods. These rods are *2 to 30* cm in length and about 0.5 to 2 cm in diameter. Flat end faces of the rod are made strictly parallel, ground and silvered, so that the surface of one face *A* becomes fully reflecting and that of the other face *B* partially reflecting.

The ruby rod is placed along the axis of the *Xenon flash* tube in such a way that it is surrounded by the coils of the tubes through which electric current flows. A flash is sent through the tube in a millisecond, the major part of the energy emitted is used in heating the apparatus, while some part of the energy in the form of blue and green radiation is absorbed by ruby which in turn excites the chromium ions of the rod.

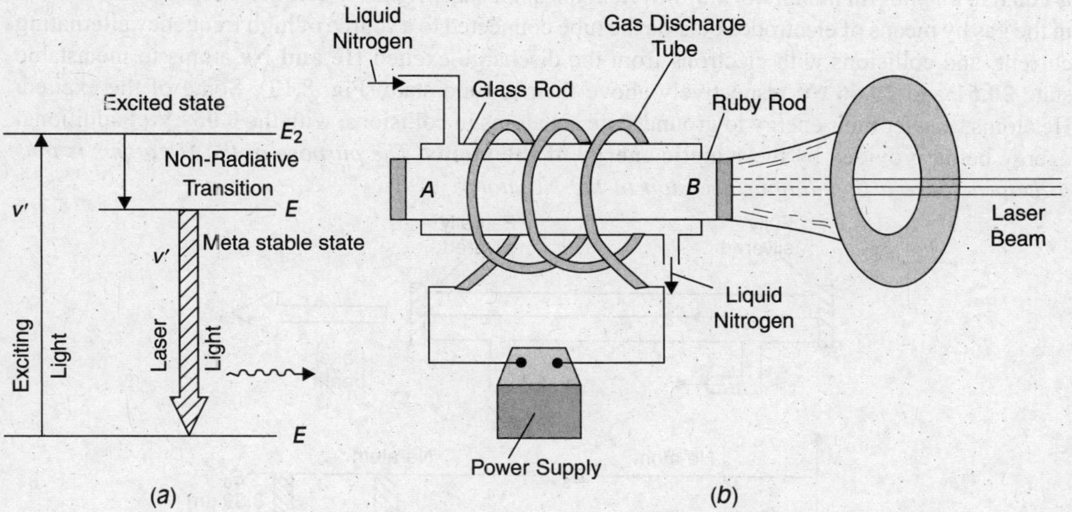

Fig. 8.12 Design of Ruby laser.

The energy diagram illustrating the operating principle is shown in Fig. 8.12 (*b*). The lines *A, B* and M correspond for energy levels of chromium ions. When light from the flash tube is made to fall upon the ruby rod, the chromium atom absorbs the green component and gets excited from the energy level *A* to the level *B*. The chromium ions have two wide energy bands at *B* and also a double level (Metastable state). From level *B*, part of the excited atoms return to the ground level *A*, whereas the other part moves to the level *M*. From selection rules the probability of transition from *B* to *M* is greater than that from *B* to *A*, whereas the probability of transition from *M* to *A* is much smaller than that from *B* to *A*. As a result, the level *M* becomes more populated than *A*. Population inversion thus takes place and proper conditions for intensive induced emission are created.

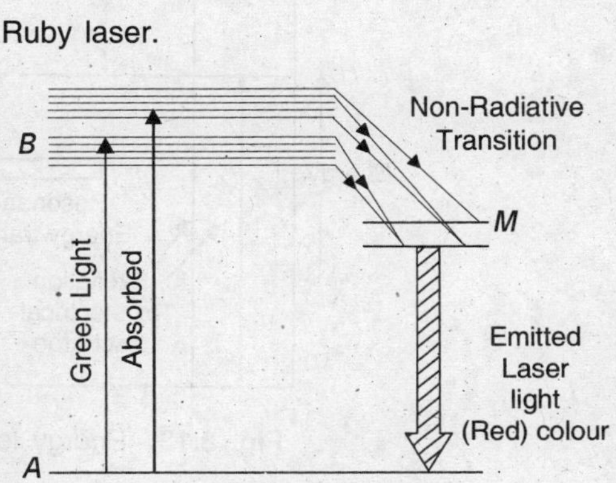

Fig. 8.12 (*c*)

Such a system with population inversion is very unstable and any moment, the probability of spontaneous emission is very high. The first photon of right frequency appearing during such spontaneous emission produces a second identical photon from a neighbouring atom, with the latter moving from the level *M* to the ground level *A*. The emitted photon correspond to red light of wavelength 6034Å. The two photons produce two more photons and so on and the action quickly builds up. The function of the mirror is to reflect the stimulated photons back and forth from the ends of the rod and generate more coherent photons from the population all moving in one direction. A narrow intense flash emerges from the partially silvered end which lasts for about a microsecond. The power output of this emitted flash reaches the order of the megawatts. Several such flashes are emitted in a millisecond. As very high temperature is produced during the operation of the laser, the rod is surrounded by liquid nitrogen to cool the apparatus.

8.16. HELIUM-NEON LASER (FOUR LEVEL LASER SYSTEM)

The helium-neon gas laser achieves a population inversion in a different way. Mixture of about 7 **parts of helium and 1 part of neon** at a low pressure (~ 1 mm mercury) is placed in a glass tube that has parallel mirrors. One of them partly transparent at both ends. The spacing of the mirrors is equal to *an integral number of half wavelengths of the laser light.* An electric discharge is produced in the gas by means of electrodes outside the tube connected to a source of high frequency alternating current, and collisions with electrons from the discharge excited **He and Ne** atoms to metastable state **20.61** and **20.66 eV** respectively above their ground state (Fig. 8.13). Some of the excited. He atoms transfer their energy to ground state Ne atoms in collisions, with the **0.05 eV** of additional energy being provided by the **Kinetic energy of the atoms.** *The purpose of the* ***He*** *atoms is thus to help achieve a population inversion in the Ne atoms.*

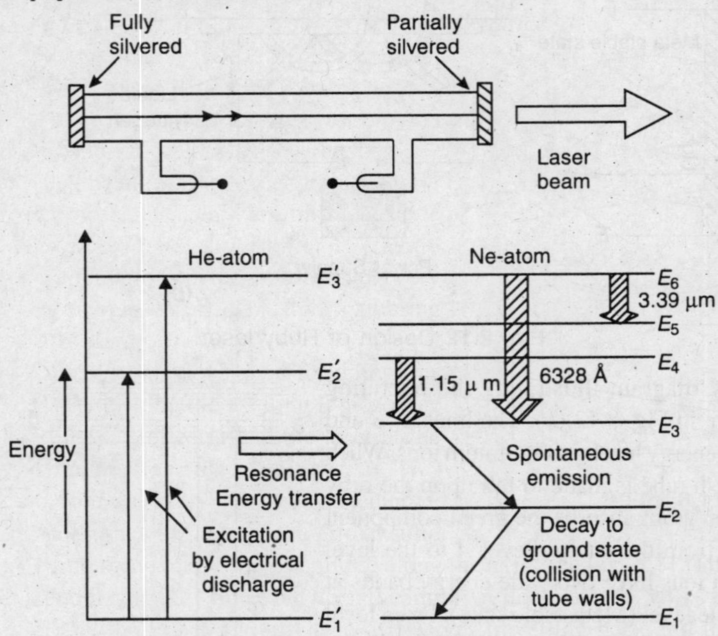

Fig. 8.13 Energy level diagram of He-Ne laser.

Working: The electrical discharge accelerates the electrons. These electrons collide with He atoms and these atoms are raised to level E_2' and E_3'. These excited He atoms cannot dexcite spontaneously rather they collide with Ne atoms and transfer their excited energies through the process of resonance energy transfer. Ne-atoms are excited to level E_6 and E_7. Reverse energy transfer also takes place, but due to small concentration of Ne-atoms, it is neglegible. As collisions continues population of Ne-atoms in these two excited states increases. Three stimulated transitions occur which produce laser light of three wavelengths (Fig. 8.13).

(*i*) $E_6 \rightarrow E_3$ transition $\lambda = 6328Å$ Red colour

(*ii*) $E_4 \rightarrow E_3$ transition $\lambda = 1.15 \mu m$ (far IR region)

(*iii*) $E_6 \rightarrow E_5$ transition $\lambda = 3.39 \mu m$ for (far IR region)

After these transitions, Ne-atoms are rapidly decay to level E_2 which is again metastable level. In order to dexcite them to ground state, tube is made narrow so that they can collide with inner wall of the tube frequently.

Conclusions: The laser transition in Ne is from the excited state at 20.66 eV to meta stable state at 18.70 eV with the emission of 632.8 nm photon. Another photon is *spontaneously* emitted in a transition to a lower metastable state which yields only in coherent light (From E_3 state to E_2 state). The remaining energy is lost in collisions with the tube walls.

As the electron impacts that excite the He and Ne atoms occur all the time unlike the pulsed excitation from xenon flash lamp in a ruby laser a He–Ne laser operates continuously.

In ruby laser, at least the Cr^{3+} *ions* must be in the metastable state otherwise absorption rather than induced emission will predominate and laser transition (At 18.70 level of Ne) is above the ground state and hence *largely* unpopulated at room temperature, which means that few Ne atoms will be able to absorb photons with laser frequency. For this reason the fraction of *Ne atoms* must be excited in order that light amplifications to occur are much less than half the total, an advantage shared by other four level lasers.

8.17. SOLID STATE LASERS

Solid state lasers that operate in the **infrared** are the most common because of their use in compact disc player. Lasers of the kind are also ideal for fiberoptic telephone lines in which the electric signals that works normally to be send along copper wires are first converted into a series of pulses according to a standard code and then sent as flashes of light down in thin (0.005 to 0.1 mm) glass fibre. At the other end of flashes are converted back into electric signals. Thousands of telephone conversations can be carried by a single fibre. Fiber-optic telephone lines already 'line' many cities and exchanges within the cities all over the world, and transoceanic fiber-optic cables will also become operational in near future.

Chemical Lasers. These are based on the production by chemical reactions of molecules in **metastable excited** states. These lasers are quite efficient and useful. For example, in the case of a chemical laser involving hydrogen and fluorine to form *hydrogen fluoride* has been found to generate infrared beam of the order of 2 MW.

CO_2 Gas Lasers. These lasers of about 100 W output are finding many uses in surgery as they seal small blood vessels while cutting through tissue by vaporizing water in the path of infrared beams. CO_2 lasers of higher power have tremendous use in industry especially for the precise cutting of all materials which also include metals.

Recently lasers have been developed using **solids, liquids and gases** to get radiations in a wide range *i.e.,* from microwaves to ultraviolet regions. Some lasers like, *He-Ne* laser or *YAG* (Yittrium Aluminium Garnet) laser can be used continuously while in the other cases light is received in pulses. An interesting gas laser as discussed above is the CO_2 laser developed in 1964 by ***C.K.N. Patel*** which operates continuously at quite higher power.

8.18. ND: YAG LASER

Yittrium Aluminium Garnet ($Y_3A_{15}O_{12}$) which is commonly called YAG is an optically **isotropic crystal**. Some of the Y^{3+} ions in the crystal are replaced by Nd^{3+} (neodymium ions) ***Nd: YAG*** laser is one of the most popular **solid state laser** and is a **four level laser system.** In *Nd: YAG* laser, crystal atoms do not participate in losing action but act as host lattice in which Nd^{3+} ions (active centres) are present.

Construction and Working. *Nd: YAG* system consists of an **elliptically cylindrical** reflector which comprises of a laser rod along one of its focus lines and flash lamp along the other focus line. Light coming from one focus will reach the other focus at free suffering multiple reflections from the silvered surfaces of the reflector. So the total light is received by the *Nd: YAG* rod, the two ends of which are polished and silvered so as to constitute and **optical resonator**.

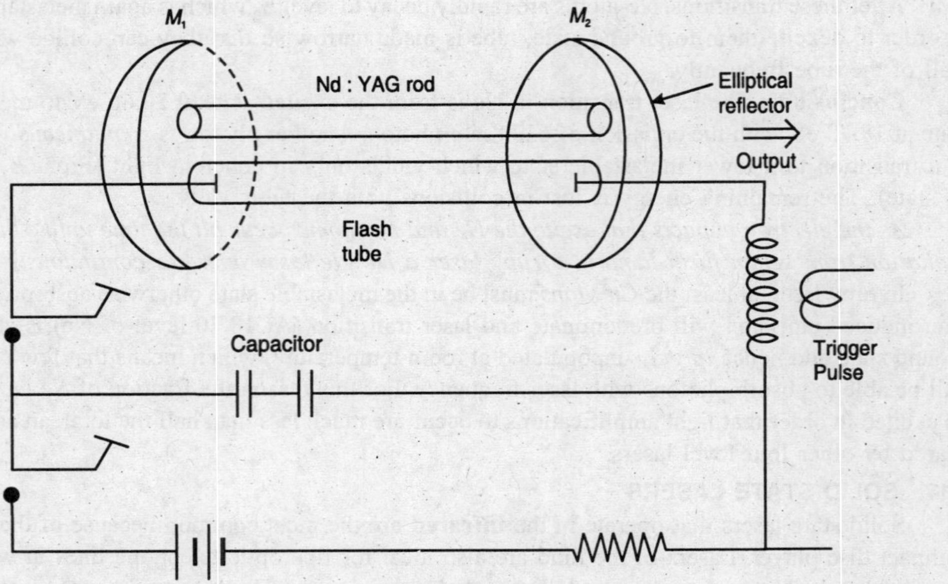

Fig. 8.14 (*a*)

As shown in the energy level diagram, the energy levels of this laser system are complex. The pumping of Nd^{3+} ion to the upper states is done using **krypton arc lamp.** Optical pumping with light of wavelength 5000 to 8000Å excites the ground state of Nd^{3+} ions to higher states. E_3 corresponds to **the metastable state** and E_2 corresponds to the **lower laser level.** Upper laser level E_3 will be rapidly populated as Nd^{3+} ions quickly **transit** down-ward from upper levels. E_2 level is quite above the ground level and so it cannot be pupulated by nd^{3+} ions through normal transi-tions from the ground level. So the population inversion is achieved by between E_3 and E_2 level and laser emission occurs in the infrared (*IR*) range of spectrum, *i.e.,* about 10600Å (1.06μ*m*).

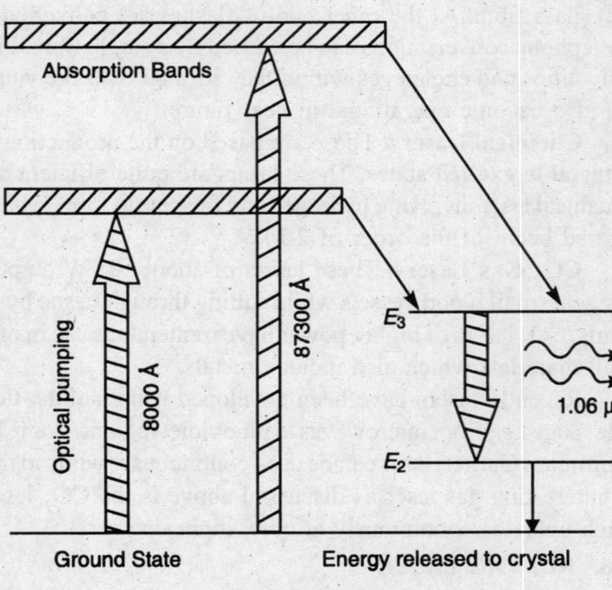

Fig. 8.14 (*b*)

Nd : YAG laser finds applications in many machine operations like drilling, welding, etc. Also they are being used in medical surgery and many other fields.

Energy and momentum of e.m. waves (lasers). Electromagnetic waves also **transport momentum.** An object experiences pressure when a beam of light shines on it and this is called radiation pressure. **Radiation pressure is usually very small.**

Consider a plane *e.m.* wave or laser incident on an object for a very small time *t*. Let this wave completely absorbed during this time so the momentum *p* delivered to the object due to absorption of energy (say *E*) may be written as:

$$p = \frac{E}{c} \quad ...(1)$$

Now if the energy E is entirely reflected, the momentum imparted to the perfectly reflecting object will be twice the above value, *i.e.,*

$$p = \frac{2E}{c}.$$

8.19. CARBON DIOXIDE LASER

A molecule consists of two or more atoms bound together, so in addition to electronic **rotations, atoms in molecules may vibrate in different modes or rotate about various axes.**

For example Fig. 8.15 (*a*) shows three modes of vibration of a CO_2 molecule, so the molecule is characterized not only by **electronic levels** but also by **vibrational and rotational levels**. Each electronic level is split into various vibrational sub-levels (because of vibrational motion) and each of vibrational level is further sub divided into *rotational sub-levels.*

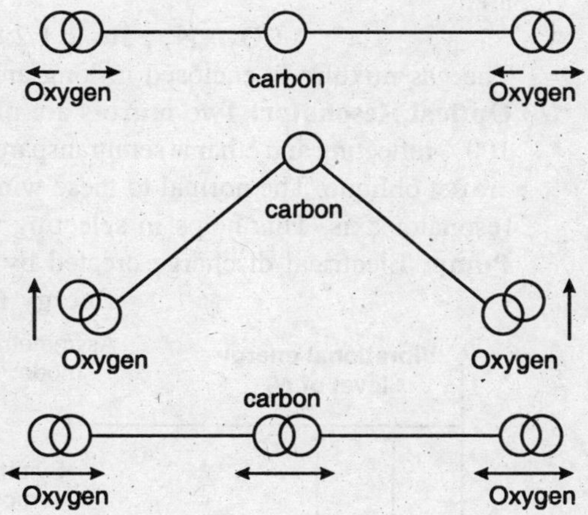

Fig. 8.15 (*a*)

The energy difference between the various electronic levels corresponds to **visible and ultraviolet region** while the energy difference between various vibrational levels correspond to **infra red region**.

The energy difference between various rotational levels corresponds to the far-infra red region of the spectrum.

In CO_2 laser, use is made of **transition** between *rotational sub level* of a vibrational level and a *rotational sub level* of a vibrational level and a rotational sub level of a vibrational level.

Population inversion is achieved through a gas discharge and the lasing action produces output *at 10.6μm and at 9.6μm.*

Addition of N_2 gas to CO_2 gas is found to increase the efficiency of CO_2 laser. So the nitrogen molecules are excited to the upper vibrational level and it is seen that the energy difference between this level and vibrational ground level of the nitrogen molecule matches very well with the energy required to excite CO_2 molecule to upper laser level. So very *efficient* energy transfer between nitrogen molecule and CO_2 molecule takes place resulting in excitation of CO_2 molecule as shown in Fig. 8.15 (*c*). Hence, efficiency of CO_2 is also enormously increased. CO_2 laser is more **efficient** than other gas lasers due to the fact that in gas lasers, sufficient energy is involved in energy transfer and this does not contribute to the output laser beam. In CO_2 laser, levels taking part in energy transfer are the *vibrational rotational levels of the lowest electronic level.* These levels being very close to the ground level, so a large part of the input energy is converted into output laser energy, thereby enhancing the output laser efficiency.

Constraction and working of CO_2 Laser.

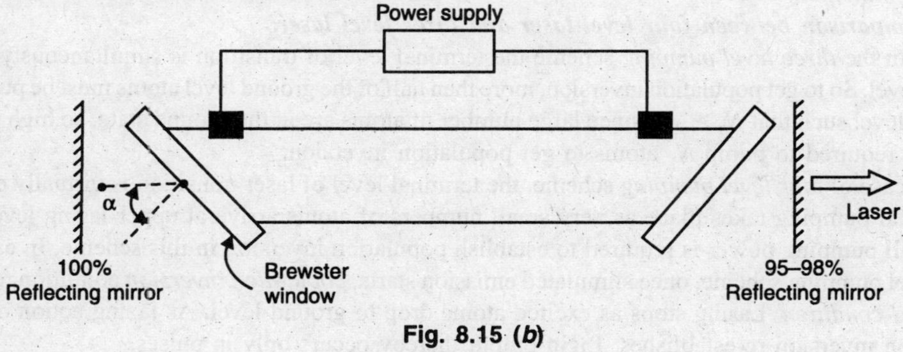

Fig. 8.15 (*b*)

(*a*) **Active Medium:** It is a gas mixture of CO_2, N_2, and additives such as He and water vapours. CO_2 molecules act as **active centres**. The proportion of these three gases in the mixture depends on the laser system and the excitation mechanism. In general, for CW mode the proportions are:

$$CO_2 : N_2 : He :: 1:2:8$$

The gas mixture is enclosed in long and narrow discharge tube.

(*b*) **Optical Resonator:** Two mirrors are placed across the ends of the discharged tube One is 100% reflecting and other is semitransparent from where laser in emitted. The end glass windows are set oblique. The normal to these windows makes angle equal to Brewster angle with the resonator axis. This helps in selecting the laser light having one plane of polarization. **Pump:** Electrical discharge created by dc voltage act as pumping agent [Fig. 815 (c)].

Energy Level Diagram

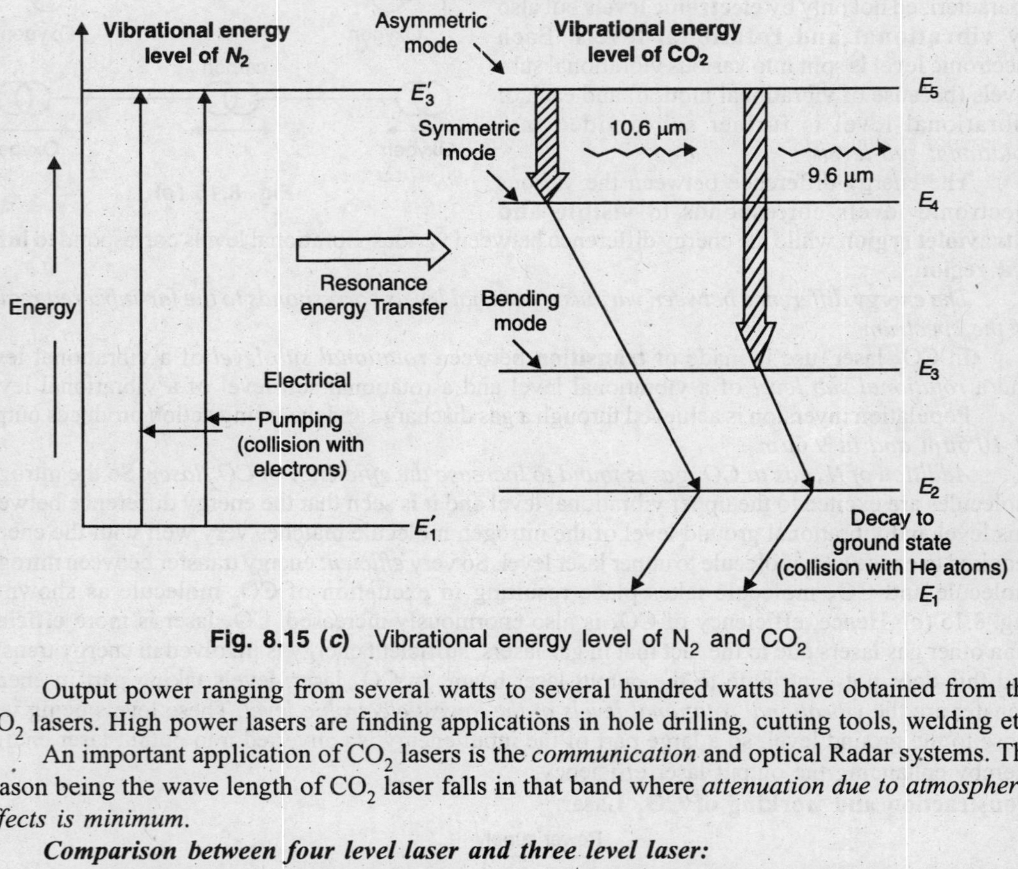

Fig. 8.15 (*c*) Vibrational energy level of N_2 and CO_2

Output power ranging from several watts to several hundred watts have obtained from the CO_2 lasers. High power lasers are finding applications in hole drilling, cutting tools, welding etc.

An important application of CO_2 lasers is the *communication* and optical Radar systems. The reason being the wave length of CO_2 laser falls in that band where *attenuation due to atmospheric effects is minimum.*

Comparison between four level laser and three level laser:

1. In the *three level pumping* scheme the terminal level of transition is simultaneously from ground level. So to get population inversion, more than half of the ground level atoms must be pumped to upper level such that $N_2 > N_1$ –since large number of atoms are in the ground state, so high pump power is required to pump N_1 atoms to get population inversion.

In case of *four level pumping* scheme, the terminal level of laser transition is virtually empty and optical pumping takes place as very small numbers of atoms arrive at upper lasing level. So very small pumping power is required to establish population inversion in this scheme. In case of three level pumping scheme, once stimulated emission starts, *population inversion* condition reverts to *normal condition.* Lasing stops as excited atoms drop to ground level. As lazing action occurs population inversion re-establishes. Light output thereby occurs only in pulses.

In the case of four level scheme, condition of population inversion can be held without interruption and so light output is continuous. So in this case lasing action has a continues mode. **Four level laser is preferred over three level laser where uninterrupted continuous laser beam is required.**

8.20. SEMI CONDUCTOR LASERS

Semi conductors are materials having conductivity between conductors and insulators.

As explained in the chapter on *semi conductors*, *p*-type semi conductor and *n*-type semi conductor form a junction where *n*-type region consists of excess of electrons and *p*-type region consists of excess of holes. So an electric field is created in the region resulting in the formation of a depletion layer.

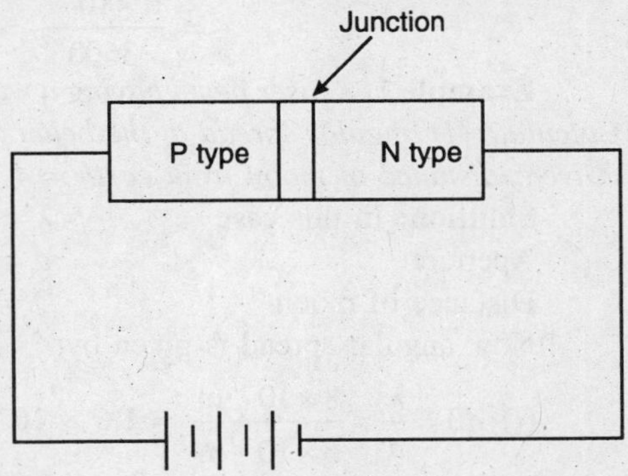

Fig. 8.16

If now *p*-type semi conductor is connected to the positive of a direct current source and *n*-type semiconductor is connected to the negative terminal, then flow of electrons (from *n*-region) and flow of holes (from *p*-region) takes place. These *electrons and holes may recombine to produce radiation.*

This amounts to shifting of an electron from conduction band to valence band. *In case the current is sufficiently large laser action may take place.*

A resonant cavity can be created by just clearing the junction ends. Since the refractive index of a semiconductor is quite large ~ 3.6, the reflectivity at the *Crystal-Air* interface is about 32%.

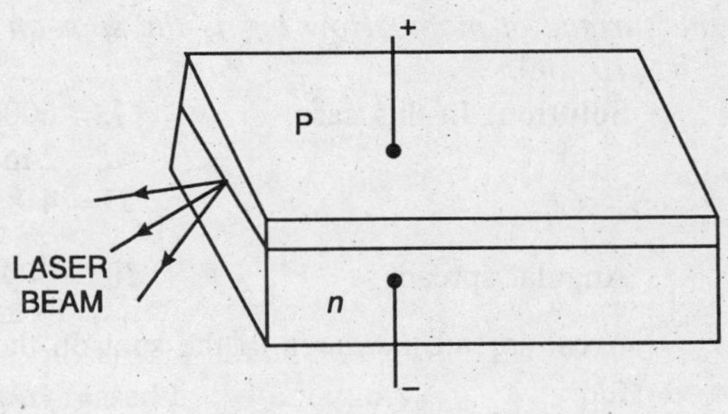

Fig. 8.17

Laser emission is restricted to a very thin region ($\sim / \mu m$) around the junction (as shown in Fig. 8.17).

Hence the laser beam may come out over a large angle. Also as compared to gas lasers, the emission is between two bands of energies and so the emission is not as monochromatic as in gas lasers.

Semi conductor laser is very small in Physical dimensions as direct conversion of electric current to light energy takes place and also it is very efficient.

Because of the above properties Semi conductor lasers are finding many applications. They are useful in light wave communications and also since the wave length of some semi conductor lasers (like GaAs and GaIn AsP) match well with the regions where optical fibres shown least attenuation, so they are useful in *portable optical radar systems, space Communications* etc.

As from the basic principle of semi conductor lasers, it is evident that *p-n* junction lasers can be operated continuously only at low temperatures – 77°. (typically liquid Nitrogen temperature. There operation at room temperature generates only pulsed form. However recently *Later-ojuction lasers* have been designed which give continuous lasers even at room temperatures. This is achieved by combined effect of population inversion. The heat produced is dissipated efficiently by cementing the dioxide to a tin plate which acts as a sink.

Example 1. *A laser beam of intensity 50 w/m² falls on a perfectly reflection plane mirror for an hour, the area of the mirror is 0.5 cm². Calculate:*

(*i*) *Momentum imparted to the mirror in this time.*

(*ii*) *Average force acting on this mirror.*

Solution: Energy falling on the mirror in one hour is given by

$$E = 50 \times .05 \times 3600 = 9 \times 10^4 \text{ J}$$

∴ (*i*) Momentum imparted to the mirror

$$P = \frac{2E}{c} = \frac{2 \times 9 \times 10^4}{3 \times 10^8} = \mathbf{6 \times 10^{-4}\ kgms^{-1}}$$

(*ii*) Average force $F = \frac{\text{Change of momentum}}{\text{time}} = \frac{p}{t}$

$$= \frac{6 \times 10^{-4}}{3600} = \mathbf{1.67 \times 10^{-7}\ N.}$$

Example 2. *A laser beam having a wavelength of 8000Å and aperture 0.5 cm is sent to moon. Calculate (1) angular spread of the beam (ii) areal spread of the beam when it reached to moon. (Given: distance of moon from earth = 4 × 10^8 m).*

Solution: In this case $\lambda = 8000 = 8 \times 10^{-7}$ m

Aperture $d = 0.5$ cm $= 5 \times 10^{-3}$ m

Distance of moon $D = 4 \times 10^8$ m

Now angular spread is given by:

(*i*) $d\theta = \frac{\lambda}{d} = \frac{8 \times 10^{-7}\ \text{m}}{5 \times 10^{-3}\ \text{m}} = \mathbf{1.6 \times 10^{-4}\ radian}$

(*ii*) Areal spread $= (D.d\theta)^2 = (4 \times 10^8 \times 1.6 \times 10^{-4})^2 = \mathbf{4.096 \times 10^9\ m^2}$.

Example 3. *A laser beam λ = 6000Å on earth is focussed by a lens of diameter 2 m on to the surface of moon. How big is the spot on the moon? (Given distance of moon from earth = 4 × 10^5 km)*

Solution: In this case, $\lambda = 6000\ \text{Å} = 6 \times 10^{-7}$ m

$d = 2$ m

$D = 4 \times 10^5$ km $= 4 \times 10^8$ m

Angular spread, $d\theta = \frac{\lambda}{d} = \frac{5 \times 10^{-7}}{2} = 3 \times 10^{-7}$ radian

Areal spread, *i.e.*, area of the spot on the moon's surface

$$= (D \times d\theta)^2 = (4 \times 10^8 \times 3 \times 10^{-7})^2$$

$$= \mathbf{1.44 \times 10^4\ m^2}.$$

Example 4. *A laser beam has a power of 50 mW. It has an aperture of 5 × 10^{-3} m and wavelength 7200Å. A beam is focussed with a lens of focal length 0.1 m. Calculate the area and the intensity of the image.*

Solution: In this case $\lambda = 7200\text{Å} = 7.2 \times 10^{-7}$ m

$d = 5 \times 10^{-3}$ m

Focal length of the lens $= 0.1$ m

∴ $d\theta = \frac{7.2 \times 10^{-7}}{5 \times 10^{-3}} = 1.44 \times 10^{-4}$ radian

(*i*) Areal spread $= (d\theta.f)^2 = (1.44 \times 10^{-8} \times 0.1)^2 = 2.074 \times 10^{-10}\ \text{m}^2$

(*ii*) Intensity $= \frac{\text{Power}}{\text{Area}} = \frac{\text{Power}}{\text{Areal spread}}$

Power $= 50 \times 10^{-3}$ watt

Areal spread $= 2.074 \times 10^{-10}\ \text{m}^2$

∴ $I = \frac{50 \times 10^{-3}}{2.074 \times 10^{-10}} = \mathbf{2.411 \times 10^8\ watt/m^2}$.

8.21. APPLICATION OF LASERS

Lasers have found use and applications in almost all the fields of science and technology. Some of the important applications may be described us under:

1. Because of the intense energy, these have been used in drilling holes in hard metals and diamonds.
2. In surgery, these are used because due to the production of localized heat, they cut and seal all the blood vessels simultaneously. They are infact used in 'blood less surgery'. These are also used for welding the detached retina of the eye and also in the treatment of malignant tumours.
3. These are used in radios and televisions because being coherent, these can be modulated to send hundred of messages simultaneously. Also since these are not absorbed, by water, so for under waters transmissions, these are highly useful.
4. These have also been used for automatic control of rockets and satellites which may be used for determination of distant stars and planets.
5. These have also found used in warfare for detecting and destroying aeroplanes, missiles and tanks.
6. In scientific research, these are being used for chemical and crystalline structures of various molecules. In Raman spectroscopy and various other related phenomenon these are being extensively used.
7. In high speed photography, Lasers are useful because these can detect the fast moving bullets and missiles. Holography, *i.e.*, the three dimensional photography has been possible because of lasers. Also in fibre optics lasers are being extensively used.

MASERS

The term Maser stands for "*Molecular* (previously microwave) *amplification by stimulated emission of radiation*". It is extremely useful as a source for carrier of microwaves.

The working of maser is similar to that of laser. Actually the principle of stimulated emission on which laser is based was first employed in microwave region (in frequency range ~ 10^{10} Hz).

In the same way, the *principle of maser* can be stated as under:

The maser action is based on the principle of '*population inversion*' *followed by* '*stimulated emission.*' For stimulated emission to predominate, most of the atoms should be in the excited state. So population of atoms in higher energy level is increased by pumping the matter by photons of appropriate size. *If during the excited state of the atoms, it is struck by a matching photon having precisely the energy as that emitted by the atom in spontaneous emission, then the excited atoms emit photons that fall precisely in phase with the stimulating photons.* So the excited atom emits waves in step with incoming wave and increases its intensity. Finally resulting emergent wave is extremely intense and diverges the least.

8.22. WAVE PACKET

Light is emitted by an excited atom when it comes to lower state and difference of energy is given in the form of a photons. So a photon corresponds to a light wave. As the process of transition from higher to lower state takes about 10^{-8} s, the atom will emit light wave in about 10^{-8} second. It means that the light emitted from an atom is not a continuous sine wave, **but a train of wave oscillations, having a finite length.**

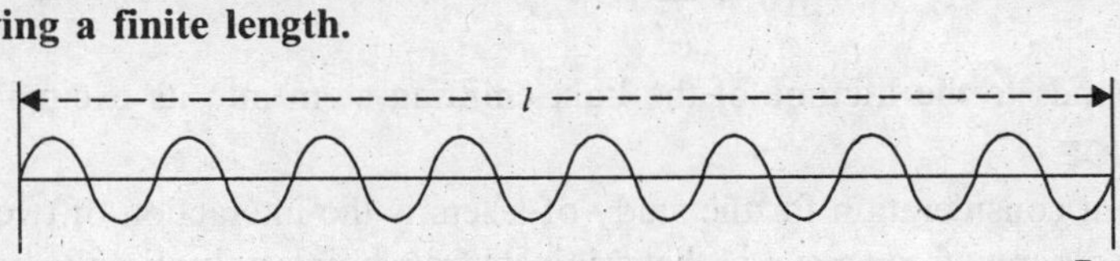

Fig. 8.18

A wave train is given by $E = E_0 \sin(kx - \omega t)$

Such a wave representation is called a **wave packet** or a **wavetrain**.

Let the length of the wavetrain in vacuum be l, then $l = x_2 - x_1 = ct$.

If $$t = 10^{-8} \text{ s, and } c = 3 \times 10^8 \text{ m s}^{-1}$$

$$l = 3 \times 10^8 \text{ m s}^{-1} \times 10^{-8} \text{ s} = 3 \text{ m}$$

∴ No. of wave oscillation present in a wavetrain of mean wavelength 5000 Å (say) will be given by

$$N = \frac{l}{\lambda} = \frac{3\,\text{m}}{5 \times 10^{-7}} = 6 \times 10^6$$

which means that a wavetrain comprising of a single photon contains about a million wave oscillations.

8.23. BAND WIDTH

It may be clearly understood that a wave packet is not a **harmonic** wave and so can not be represented by simple sine functions. A mathematical representation of a wave packet is done in terms of **Fourier Integrals** and is a **complex identity**. This representation we will not take up here but will describe some important features of wave packets.

If light given out by a source is analysed by a spectrograph, we observe it to be made up of **discrete spectral line** which are formed by wave packets emitted by atoms. So a **spectral line and a wave packet corresponds to the same phenomenon** referred in different ways.

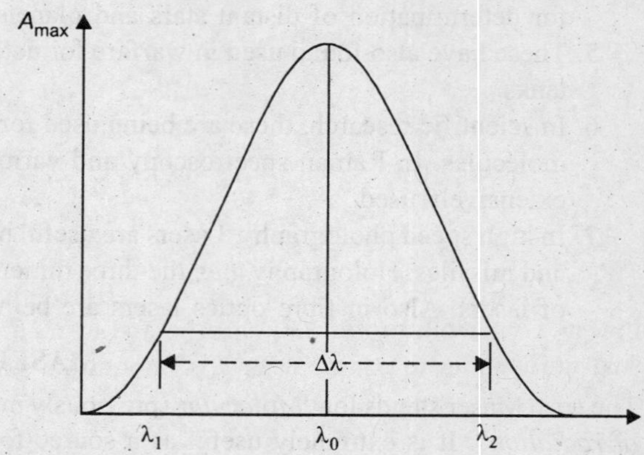

Fig. 8.19

As stated, wavelength of a **wave packet of a spectral line is not precisely** defined. It means that there is a continuous spread of wavelengths over a finite range $\Delta\lambda$, centred around a wavelength λ_0. A wave packet may be considered as comprising of a number of harmonic waves which differ only by very **very small wavelengths**. Maximum intensity of the wave packet will occur at λ_0 and the intensity falls off rapidly on either side of l0 as shown in Fig. 8.19.

The spread of wavelength is called line width or band width. As shown in the figure, the band width is the wavelength interval from $\lambda_0 - \frac{\Delta\lambda}{2}$ to $\lambda_0 + \frac{\Delta\lambda}{2}$ that contains the major part of the energy possessed by the wave packets.

Fourier analysis of the **wave packet** gives a relation between **band width and frequency of oscillation** as

$$\Delta\nu = \frac{1}{\Delta t}$$

where Δt corresponds to the lifetime of the light emitting atom into its excited state.

8.24. COHERENCE

An important consideration for the study of lasers is the interaction of two e.m. waves that have only *slightly different frequencies* or that originate from points only slightly separated. Spatially, *e.g.*, two closely located but separate laser beams or a single beam illuminating two *closely positioned apertures.* In these cases, the two beams will interfere and give some important effects like longitudinal modes, mode locking, phase matching and frequency multiplication etc.

When complete coherence is not achieved but still some interference takes place, the condition is called *partial interference.* Partial interference condition can be grouped into two categories.

(*i*) Temporal coherence associated with difference in frequencies of two waves. It is also referred to as *longitudinal coherence*.

(*ii*) Spatial coherence associated with location of two waves. It is also referred to the *transverse coherence*.

8.25(a). TEMPORAL COHERENCE

It refers to the relative phase or coherence of the two waves at two separate locations along the propagation direction of the two beams.

If we assume that the two beams are exactly in phase at *initial* (say first) location then they will be at least partially in phase in some other (say 2nd) location upto some length say l_c which is called coherence length and may be found as

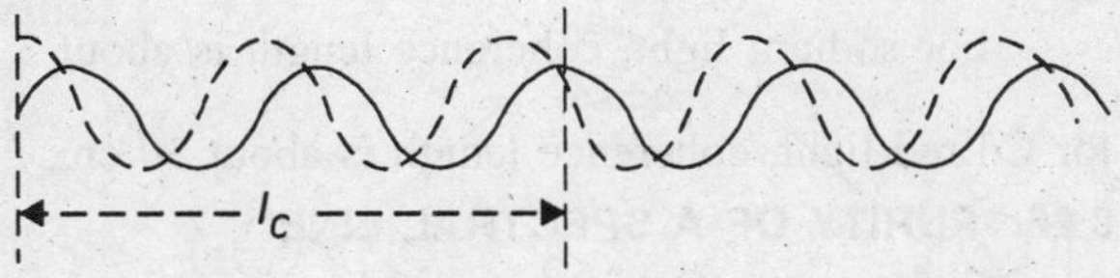

Fig. 8.20

$$l_c = \lambda.\left(\frac{\lambda}{\Delta\lambda}\right) = \frac{\lambda^2}{\Delta\lambda} \qquad ...(1)$$

where $\Delta\lambda$ represents the difference in wavelength between the two waves and λ corresponds to their average wavelength.

It may be seen from eq. (1) that l_c will have significant value only when $\Delta\lambda << \lambda$ *i.e.*, when the wavelength of both waves is nearly same.

8.25 (b). SPATIAL COHERENCE

Spatial coherence is also referred to as *transverse coherence* and it describes, how far apart two sources or two portion of the same source can be located in a direction transverse to the direction of observation and still exhibit coherent properties over a range of observation points. This is sometimes called lateral coherence.

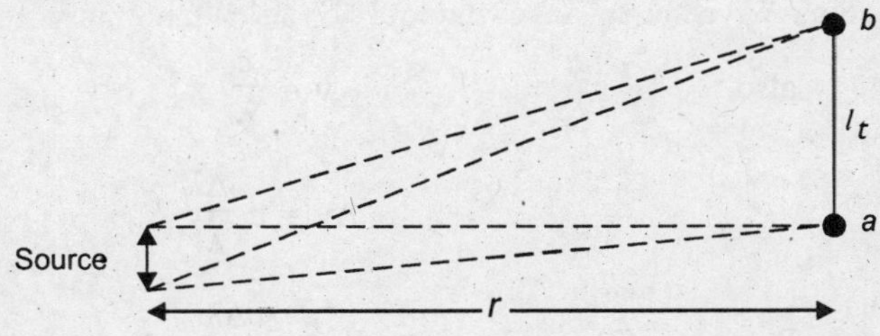

Fig. 8.21

In other words, it means that the distance l_t by which two points are separated in transverse direction and still interference effects are observed at *Specific distance from the source.*

Let two sources be separated by a distance *S* in *transverse direction* to the direction of observation and distance *r* from point of observation. If the two sources exhibit *interference phenomenon* at point *a* then transverse length l_t is the transverse distance from *a* to *b*, where two sources show interference effects. This distance may be calculated from

$$l_t = \frac{\lambda}{\theta_S} = \frac{r\lambda}{S} \qquad [\because \theta_S = S/r]$$

Temporal coherence can be studied from the *Michelson's interferometer* by moving one of the mirrors. In the case of an interferometer, the visibility *V* of the fringes is a measure of *degree of coherence*. In the case of Michelson interferometer, path difference is adjusted to be zero to get *maximum visibility*. From this position, one of the mirror is moved to a position where no fingers are observed. Then distance between these positions give us the *maximum range over which fringes can be observed*. The time $\tau = \dfrac{2d}{c}$ is the time taken by the wave to travel a distance $2d$ and V_1, V_2 are the frequencies of the two waves. The distance l for which the field remains sinusoidal is called coherence length and is given by

$$l = \tau.c$$

where c is the velocity of light.

Now if $2d < l$, there will be a definite phase relationship between two interfering beams and if $2d < 1$, no definite phase relationship between the beams will be observed, so starting with equal separation of mirrors from the silvered glass plate, (*i.e., zero* path difference between interfering beams), if now the path difference is continuously increased, the dark and bright fringe contrast goes on reducing, till the fringes disappear. The path difference at disappearence of fringes gives us an estimate of *coherence length.*

For sodium light, coherence length is about 3 cm, so coherence time is $\frac{l}{c} = \frac{3}{3 \times 10^{10}}$ and for Cd red light, coherence length is about 30 cm. *i.e.,* 10 times more than that of sodium light.

8.26. PURITY OF A SPECTRAL LINE

Finite purity Q of a spectral line is defined on the basis of quantum theory as

$$Q = \frac{\lambda}{\Delta\lambda} \qquad ...(1)$$

The purity of Cd red light $\simeq 10^6$ and that of sodium yellow light $\simeq 10^5$.

If the radiations emitted from a source of frequency V has a frequency spread $\Delta\nu$ and its coherence time τ, then the radius of amplitude is explained by putting the difference of number of oscillations of frequency ν and $\nu + \Delta\nu$ in time by unity.

i.e., $\qquad \tau(\nu + \Delta\nu) - \tau\nu = 1$

$\therefore \qquad \tau\,\Delta\nu = 1 \qquad ...(2)$

also $\qquad \nu = \frac{c}{\lambda}$

$\therefore \qquad \Delta\nu = c\,\frac{\Delta\lambda}{\lambda^2} \qquad$ (in magnitude)

Put in eq. (2) $\qquad \tau\,\frac{c\Delta\lambda}{\lambda^2} = 1 \quad \text{or} \quad \Delta\lambda \simeq \frac{\lambda^2}{c\tau} \qquad ...(3)$

Also coherence length l is given by

$$l = \tau\, c$$

$\therefore$ eq. (3) may be written as $\Delta\lambda = \frac{\lambda^2}{l}$

$\therefore$ coherence length $\qquad l = \frac{\lambda^2}{\Delta\lambda} \qquad ...(4)$

which may be written in terms of purity Q of spectral line as

$$l = \lambda.\ \frac{\lambda}{\Delta\lambda} = \lambda Q \qquad ...(5)$$

i.e., coherence length = wavelength of a spectral line × finite purity.

8.27. RELATION BETWEEN COHERENCE LENGTH AND LINE WIDTH

We have $\qquad c = \nu\lambda$

$\therefore \qquad \lambda = \frac{c}{\nu}$

Differentiating $$\Delta\lambda = -c.\ \frac{\Delta\nu}{\nu^2} = -c.\ \frac{\Delta\nu}{c^2}\lambda^2$$

$$= -\frac{\lambda^2 \Delta\nu}{c} \quad ...(1)$$

Negative sign shows that change in $\Delta\lambda$ is opposite to change in frequency.

Writing the above expression in magnitude

$$\Delta\lambda = \lambda^2 \frac{\Delta\nu}{c} = \frac{\lambda^2}{c}.\frac{1}{\Delta t} \qquad \left[\because \Delta\nu = \frac{1}{\Delta t}\right]$$

$$\therefore \quad \Delta t = \frac{\lambda^2}{c\,.\,\Delta\lambda} \quad ...(2)$$

and coherence length is given by

$$l_c = c.\ \Delta t$$

$$= c.\ \frac{\lambda^2}{c\,.\,\Delta\lambda}$$

or $$l_c = \frac{\lambda^2}{\Delta\lambda} \quad ...(3)$$

which gives us a relation between coherence **length l_c and line width $\Delta\lambda$.**

Example 5. *An optical filter has a line width of 1.5 nm and mean wavelength 550 nm with white light incident on the filter, calculate:*

(i) coherence length (ii) number of wavelengths in wavetrain.

Solution.

In this case $$\Delta\lambda = 1.5 \text{ nm} = 1.5 \times 10^{-9} \text{ m}$$
$$\lambda = 550 \text{ nm} = 550 \times 10^{-9} \text{ m}$$

(*i*) coherence length

$$l_c = \frac{\lambda^2}{\Delta\lambda} = \frac{(550\times10^{-9})^2}{1.5\times10^{-9}}$$

$$= \mathbf{2.017 \times 10^{-4}\ m.} \quad \textbf{Ans.}$$

(*ii*) number of waves in wavetrain

$$n = \frac{l_c}{\lambda} = \frac{2.017\times10^{-4}}{550\times10^{-9}}$$

$$= \mathbf{367\ (approx.)} \quad \textbf{Ans.}$$

Example 6. *Cadmium red light of mean wavelength 6438 Å is used in Michelson interferometer. Line width of the source is 0.013 Å for the initial setting of the interferometer for zero optical path difference. One of the mirrors is slowly moved until the fringes first disappear. Calculate the distance by which the mirror is moved, also calculate the number of wavelength corresponding to the distance.*

Solutions. In this case

$$\lambda = 6428 \text{ Å} = 6438 \times 10^{-10} \text{ m}$$
$$\Delta\lambda = 0.013 \text{ Å} = 0.013 \times 10^{-10} \text{ m}$$

(*i*) coherence length

$$l_c = \frac{\lambda^2}{\Delta\lambda} = \frac{\left(6438\times10^{-10}\right)^2}{0.013\times10^{-10}}$$

$$= \mathbf{0.319\ m} \quad \textbf{Ans.}$$

Now the condition for first disappearance of fringes is that the optical path must exceed the coherence length $\Delta\lambda$.

$\therefore$ $$2d = 0.319 \text{ m}$$

or $$d = 0.1595 \text{ m} = 15.95 \text{ cm}.$$

which means that the mirror has to be moved through a distance 15.95 cm.

$$\text{also } d = n\lambda$$

$\therefore$ $$0.1595 = n \times 6438 \times 10^{-10}$$

$\therefore$ $$n = \mathbf{24770 \text{ waves}} \quad \textbf{Ans.}$$

Example 7. *Quasi monochromatic source emits radiations of mean wavelength l = 5461 Å and has a band width $\Delta\nu = 10^9$ Hz. Calculate:*

(i) coherence time (ii) coherence length (iii) frequency stability.

Solution. In this case

$$\lambda = 5461 \text{ Å} = 5461 \times 10^{-10} \text{ m}$$

$$\Delta\nu = 10^9 \text{ Hz}$$

(*i*) $$\Delta t = \frac{1}{\Delta\nu} = \frac{1}{10^9} = 10^{-9} \text{ sec.}$$

(*ii*) $$l_c = c.\ \Delta t = 3 \times 10^8 \times 10^{-9} = 0.3 \text{ m.}$$

(*iii*) We have

$$\nu = \frac{c}{\lambda} = \frac{3\times10^8}{5461\times10^{-10}} = 5.5 \times 10^{14} \text{ Hz.}$$

$\therefore$ Frequency stability

$$\frac{\Delta\nu}{\nu} = \frac{10^9}{5.5\times10^{14}} = \mathbf{1.82\times10^{-6}} \quad \textbf{Ans.}$$

Example 8. *Calculate temporal coherence length l_c for*

(i) mercury vapour lamp emitting in green portion of spectrum at wavelength of 546.1 nm with emission band width of $\Delta\nu = 6 \times 10^8$ Hz.

(ii) A helium neon laser operating at wavelength of 632.8 nm.

Solution. We have the relation

$$c = \nu\lambda$$

or $$\nu = \frac{c}{\lambda} \quad \therefore \quad \Delta\nu = -\frac{c}{\lambda^2}\Delta\lambda.$$

(*i*) For mercury lamp, we have

$$\Delta\lambda = \frac{\lambda^2\Delta\nu}{c} \text{ (in magnitude)}$$

$$= \frac{(546.1\times10^{-9}\text{ m})^2\times(6\times10^8\text{ s}^{-1})}{3\times10^8\text{ m s}^{-1}} = 5.96\times10^{-13} \text{ m}$$

and temporal length is given

$$l_c = \frac{\lambda^2}{\Delta\lambda} = \frac{(546.1\times10^{-9}\text{ m})^2}{5.96\times10^{-13}\text{ m}} = \mathbf{0.50 \text{ m}} \quad \textbf{Ans.}$$

(*ii*) For helium-neon lasers, we have

$$\Delta\lambda = \frac{\lambda^2 \Delta\nu}{c} = \frac{(632.8\times10^{-9}\,\text{m})^2\,(1\times10^{6}2^{-1})}{3\times10^{8}\,\text{m s}^{-1}}$$

$$= 1.33 \times 10^{-15}\ \text{m} = 1.33 \times 10^{-6}\ \text{nm}$$

Temporal length is thus given by

$$l_c = \frac{\lambda^2}{\Delta\lambda} = \frac{(632.8\times10^{-9}\,\text{m})^2}{1.33\times10^{15}\,\text{m}} = \mathbf{30\,m}$$

So temporal coherence length of laser beam is much longer than that of mercury lamp.

Example 9. *A laser produced plasma consisting of a 100 mm diameter ball radiates very strongly at a wavelength of 10 nm. Calculate spatial coherence at a distance 0.5 m from the source.*

Solution. In this case

$$r = 0.5\ \text{m}$$
$$l = 10\ \text{nm}$$
$$S = 100\ \mu\text{m} = 10^{-4}\ \text{m}.$$

We have the relation for spatial coherence as

$$l_t = \frac{r\lambda}{S} = \frac{(0.5\text{m})\,(1\times10^{-8}\,\text{m})}{1\times10^{-4}\,\text{m}} = \mathbf{5\times10^{-5}\,m} \quad \textbf{Ans.}$$

Example 10. *Find the average number of photons/m^3 in a monochromatic beam of radiation having frequency 10 kHz and intensity 10^4 W/m^2.*

Solution. *Intensity of a beam of radiation is the total energy of the no. of photons passing normally through unit area a cross-section of beam/sec.*

Let x be the no. of photons/unit volume each of energy $h\nu$. Let us consider a length l of the beam having cross-section area A.

So in volumne Al, the total photon energy = $nAl\ h\nu$ and time taken by this energy to pass through the length l is given by

$$t = l / c$$

where c is the velocity of light in free space.

∴ Intensity I of the beam is given by

$$I = \frac{nAlh\nu}{A.t} = \frac{hl.h\nu}{t}$$

$$= \frac{nl\ h\nu}{l/c} = nh\nu c$$

$$\therefore \quad n = \frac{I}{h\nu c}$$

Putting the various value, we get :

$$n = \frac{10^4}{6.62\times10^{-34}\times10^4\times3\times10^8}$$

$$= \frac{10^{26}}{19.88} = 5.03\times10^{24}\ \text{photons/m}^3. \quad \textbf{Ans.}$$

Example 11 . *A certain ruby laser units 1.0 J pulses of light having wavelength 6940Å. Find the minimum number of Cr^{3+} ions in the ruby.*

Solution. We have the relation for power of light pulses;

$$P = \frac{nhc}{\lambda}$$

$$\therefore \quad 1.0 \text{ J} = \frac{n \times 6.62 \times 10^{-34} \text{ J.S} \times 3 \times 10^{8} \text{ ms}^{-1}}{6940 \times 10^{-10} \text{ m}}$$

$$\therefore \quad n = \frac{1.0\, J \times 6940 \times 10^{-10} \text{ m}}{6.62 \times 10^{-34}\, J\text{-}S \times 3 \times 10^{8} \text{ m s}^{-1}}$$

$$= \mathbf{3.4945 \times 10^{18} \text{ ions}} \quad \textbf{Ans.}$$

Example 12 . *Find the intensity of laser beam of 20 mw and diameter 1.5 mm. Assume the intensity to be uniform throughout the beam.*

Solution. Power $P = 20\ mw = 20 \times 10^{-3}\ w$ diameter $d = 1.5$ m m $= 1.5 \times 10^{-3}$ m.

$\therefore$ Intensity
$$I = \frac{Power}{Area} = \frac{P}{\pi d^2/4} = \frac{4P}{\pi d^2}$$

$$I = \frac{4 \times 20 \times 10^{-3} w}{3.142 \times (1.5 \times 10^{-3} \text{ m})^2}$$

$$= \mathbf{11.32 \text{ kw / m}^2} \quad \textbf{Ans.}$$

Example 13. *Imagine that we chop a continuous laser beam assumed to be perfectly monochromatic* (λ_0 = 623.8 nm) *with* 0.05 ns *pulses using some shultee. Compute the resultant line width* $\Delta\lambda$, *band width and coherence length.*

Solution. In this case $\Delta t = 0.05 \times 10^{-9}$s; $\lambda_0 = 623.8 \times 10^{-9}$ m

$\Delta\lambda = ?$

Coherence length is given by

$$l_c = c.\ \Delta t = 3 \times 10^{8} \times 0.05 \times 10^{-9} = 15 \times 10^{-3} \text{ m.}$$

Band width
$$\Delta\nu = \frac{1}{\Delta t} = \frac{1}{0.05 \times 10^{-9}} = 2 \times 10^{10} \text{ Hz.}$$

$$\text{line width } \Delta\lambda = \left(\frac{\lambda^2}{c}\right)\Delta\nu$$

$$= \frac{(623.8 \times 10^{-9})^2 \times 2 \times 10^{10}}{3 \times 10^{8}}$$

$$= \mathbf{0.026 \text{ nm.}} \quad \textbf{Ans.}$$

Example 14. *Calculate the energy difference in e.v. between two energy levels of Ne atoms of a He–Ne laser, the transition between which result in the emission of a light of wavelength 632.8 nm. Also find the number of photons emitted/sec, if the optical power output is 2 mw.*

Solution. In this case $\lambda = 632.8 \times 10^{-9}$ m

$P = 2 \times 10^{-3}\ w$, $\Delta E = ?$ $N = ?$

Energy difference is given by

$$h\nu = \frac{hc}{\lambda} \text{ J}$$

$$= \frac{6.63 \times 10^{-34} \times 3 \times 10^{8}}{632.8 \times 10^{-9}} = 3.143 \times 10^{-19} \text{ J}$$

$$= \frac{3.143 \times 10^{-19}}{1.6 \times 10^{-19}} \text{ e.v.} = \mathbf{1.96 \text{ e.v.}}$$

So energy difference between two energy levels of Ne atoms = 1.96 e.V. Ans.

Let Ne be the number of photons emitted per sec. to give a power out put of 2 mw;

So $$P = N \times \Delta E = 2 \times 10^{-3}$$

$$\therefore \quad N = \frac{2\times10^{-3}}{\Delta E} = \frac{2\times10^{-3}}{3.143\times10^{-19}}$$

$$N = 6.364 \times 10^{15}. \quad \textbf{Ans.}$$

Example 15. *Calculate the ratio of population of two energy states of Ruby laser, the transition between which is responsible for emission of photons of wavelength 6928 Å. Assume the transition temperature to be 18 K.*

Solution. In the case $\lambda = 6928$Å, $T = 273 + 18 = 291$ K

$$\frac{N_2}{N_1} = ?$$

We have the ratio of population of two energy states as

$$\frac{N_2}{N_1} = e^{-(E_2-E_1)/KT} = e^{-\Delta E/KT}$$

$$E = \frac{hc}{\lambda} = \frac{6.62\times10^{-34}\times3\times10^8}{6928\times10^{-10}} = 2.871 \times 10^{-19}.$$

$$\frac{N_2}{N_1} = \text{Exp.}\left(\frac{-2.871\times10^{-19}}{1.38\times10^{-23}\times291}\right)$$

$$= e^{-71\cdot49} = \mathbf{8.85 \times 10^{-32}}. \quad \textbf{Ans.}$$

Example 16. *In a Helium-neon laser the two plane mirrors forming resonant cavity are at a distance of 0.5 m. What is the mode separation of longitudical cavity in terms of frequency.*

Solution. In the case $L = 0.5$ m $\Delta\nu = ?$

Difference between frequencies of consecutive modes is

$$\Delta\nu = \frac{c}{2L} = \frac{3\times10^8}{2\times0.5} = \mathbf{3 \times 10^8\ Hz.} \quad \textbf{Ans.}$$

Example 17. *At what temperature are the rates of spontaneous and stimulated emission equal ? Take $\lambda = 5000$ Å.*

Solution. Ratio of spontaneous to stimulated emission is given by

$$R = e^{\frac{h\nu}{kT}} - 1$$

$$= \text{Exp.}\left[\frac{6\cdot6\times10^{-34}\times3\times10^8}{1\cdot38\times10^{-23}\times T\times5000\times10^{-8}}\right] - 1$$

Put $R = 1$, and Solving

$$T = \mathbf{41532\ K.}$$

So at temp. 41532 K spontaneous and stimulated emission are equal. **Ans.**

Example 18. *Calculate the ratio of spontaneous to stimulated emission by an incandescent lamp at 2000 K. (take $\nu = 6 \times 10^{14}$ Hz optical region).*

Solution. Ratio of spontaneous emission to stimulated emission is given by

$$R = e^{h\nu/kT} - 1$$

$$= \text{Exp.}\left[\frac{6\cdot6\times10^{-34}\times6\times10^{14}}{1\cdot38\times10^{-23}\times2000}\right]-1$$

$$= 1\cdot7\times10^{6}$$

which shows that only one stimulated emission exists in about 1·7 million spontaneous emissions.

Example 19. *Calculate the coherence time and coherence length of white light of wavelength range from 3500 Å to 6500 Å.*

Solution. Frequency $\nu = c/\lambda$

$$\therefore \quad \nu_1 = \frac{c}{\lambda_1} = \frac{3\times10^8}{3500\times10^{-10}} = 8.57\times10^{14} \text{ Hz}$$

and
$$\nu_2 = \frac{c}{\lambda_2} = \frac{3\times10^8}{6500\times10^{-10}} = 4.61\times10^{14} \text{ Hz}$$

Frequency spread $\quad \Delta\nu = \nu_1 - \nu_2$

$$\therefore \quad \Delta\nu = (8.57-4.61)\times10^{14} = \mathbf{3.96\times10^{14}\ Hz}$$

So coherence time $= \dfrac{1}{\Delta\nu} = \dfrac{1}{3.96\times10^{14}} = \mathbf{2.52\times10^{-15}\ s}$

and coherence length, $i = c\tau_c = 3\times10^8\times2.52\times10^{-15}$

$= \mathbf{7.56\times10^{-7}\ m}$ **Ans.**

Example 20. *Calculate the coherence length for CO_2 laser whose line width is 1×10^{-5} xm at IR emission wavelength of 10.6 μm.*

Solution. Coherence length $= \lambda^2/(\Delta\lambda)$

$$\therefore \quad \text{Coherence length} = \frac{(10.6\times10^{-6})^2\,\text{m}^2}{10^{-5}\times10^{-9}\,\text{m}} = 11.2\times10^3 = \mathbf{11.2\ km}$$ **Ans.**

Example 21. *The coherence length for sodium D_2 line is 2.5 cm. Deduce (i) the coherence time τ_c, (ii) the spectral width of the line and (iii) the purity factor Q.*

Solution. (i) $\tau_c = \dfrac{l}{c} = \dfrac{2.5}{3\times10^{10}} = 0.8\times10^{-10}$ s

(*ii*) For sodium D_2 line $\lambda = 5896$ Å $= 5.896\times10^{-5}$ cm.

The spectral width of the line is given by the relation

$$\Delta\lambda = \frac{\lambda^2}{l} = \frac{(5.896\times10^{-5})^2}{2.5} = 13.90\times10^{-10} \text{ cm}$$

(*iii*) The purity factor Q is given by

$$Q = \frac{\lambda}{d\lambda} = \frac{5.896\times10^{-5}}{13.90\times10^{-16}} = 4.24\times10^4$$ **Ans.**

8.28. FIBRE OPTICS (GENERAL INTRODUCTION)

In case of electronic communication, the information is normally carried over in the form of **radio waves and microwaves**, through copper wires and co-axial cables. However, the information carrying capacity of these wires is restricted, due to their limited band-width and is not sufficient, as per the needs of the modern communication techniques.

If **light waves** are used instead of radio waves, or microwaves, the number of signals transmitted can be increased enormously. Technically, **Grahm Bell** in 1880 invented **Photophone,** which transmitted speech on a beam of light. By 1960, it was established that light could be guided by glass fibre. However, the glass available at that time *heavily attenuated light propagating through them.*

It was due to the invention of **solid state lasers** in 1970 and fabrication of **low-loss glass fibres**, that made the optical communication practical. With this the era of **photonics** started.

In addition to the use of optical fibre in communication, these have been extensively used as **sensors** to detect **electrical, mechanical and thermal energies.** Also they are buying used in copying machines and in medical diagnostics (Endoscopy).

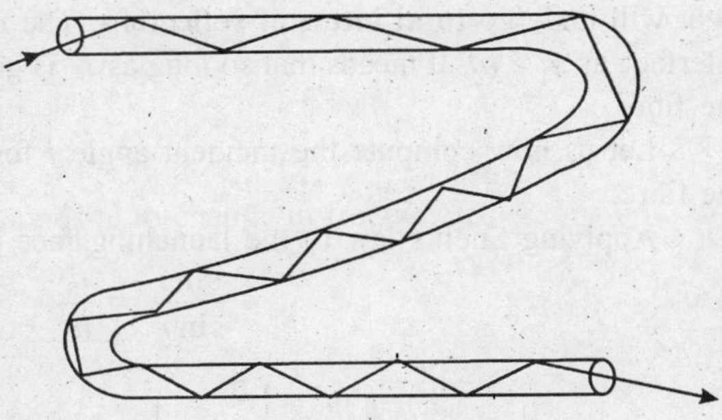

Fig. 8.22

OPTICAL FIBRES

These are made up of **glass or plastic conduicts** as thin as a **human hair** which is designed to guide light waves along the length of the fibre with the help of successive **total internal reflections** from side walls of the fibre. The movement of light is in a *zig-zag* path and some fraction of light may escape through side walls but major fraction comes out from the other end of the fibre. Fig. 8.22.

Construction:

An optical fibre consists, in general, the following three regions:

(*i*) Inner-most region, which is the light guiding region is called core.

(*ii*) Core region is surrounded by a middle region, called cladding. The refractive index of cladding is always lower than that of the core.

(*iii*) The outermost region is called sheath. The sheath protects the core and cladding from moisture, abrasion contamination and to give mechanical strength to fibre.

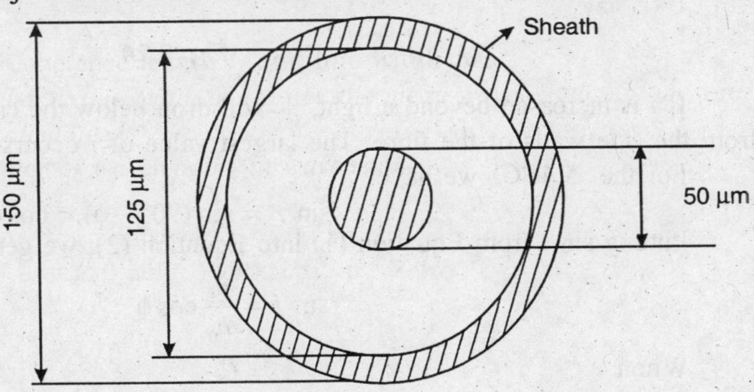

Fig. 8.23

Optical fibres are either made as a single fibre or a flexible bundle of cable. A bundle fibre is a **number of fibres in a single jacket.**

Basic propose of cladding is to confine the light to the core as the light falling on core and striking core cladding interface at angle greater than the *critical angle* will be reflected back to the core.

8.29. PROPAGATION OF LIGHT THROUGH FIBRE

Main function of the optical fibre is to accept maximum light and transmit the same with minimum **attenuation.**

Light gathering ability of a fibre depends on two factors:

(*i*) Core size

(*ii*) Numerical aperture

Numerical aperture of fibre is determined by **acceptance angle** and **fractional refraction** index **change.**

8.30. ACCEPTANCE ANGLE AND ACCEPTANCE CONE

Consider an optical fibre through which light is being sent. The end at which light enters is called **launching end.** Let the refractive indices of the core and cladding be n_1 and n_2 respectively; $n_2 < n_1$. Let the refractive index of the medium k through which light is launched be n_0.

Let the light beam enter at an angle i to the axis of the fibre. The ray gets refracted at an angle r and strikes the core cladding interface at an angle ϕ. For angle ϕ more than the critical angle C, } light will undergo **total internal reflection**. The ray will also undergo total internal reflection at interface as $n_1 > n_2$. It means that so long as ϕ is greater than critical angle C, light will stay within the fibre.

Let us now compute the incident angle i for which $\phi \le C$ such that light rebounds within the fibre.

Applying Snell's law to the launching face of the fibre, we get

$$\frac{\sin i}{\sin r} = \frac{n_1}{n_0} \quad ...(1)$$

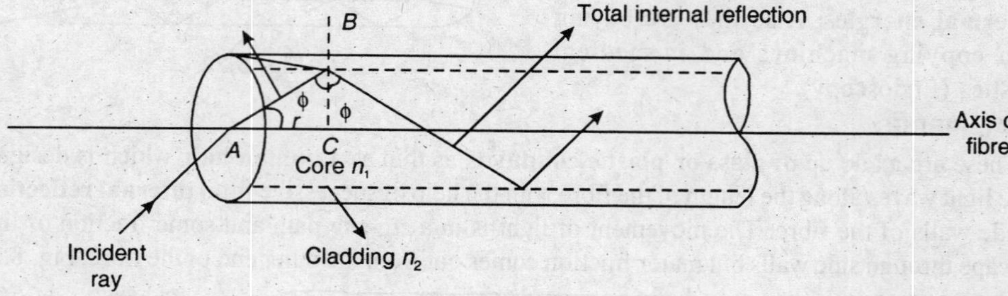

Fig. 8.24

If i is increased beyond a light, ϕ will drop below the critical value C and the ray escapes from the side walls of the fibre. The largest value of i occurs when $\phi = C$.

For the ΔABC, we have

$$\sin r = \sin(90° - \phi) = \cos\phi \quad ...(2)$$

Putting sin i from Equation (1) into Equation (2), we get

$$\sin i = \frac{n_1}{n_0}\cos\phi$$

When $$\phi = C$$

$$\sin i\,(\text{max}) = \frac{n_1}{n_0}\cos C \quad ...(3)$$

But $$\sin C = \frac{n_2}{n_1}$$

$\therefore$ $$\cos C = \frac{\sqrt{n_1^2 - n_2^2}}{n_1} \quad ...(4)$$

Putting the expression (4) into (3), we get

$$\sin i\,(\text{max}) = \frac{\sqrt{n_1^2 - n_2^2}}{n_0} \quad ...(5)$$

When incident ray is launched from air medium, we have $n_0 = 1$.

Designing i (max) = θ_0, Equation (5) may be simplified to

$$\sin\theta_0 = \sqrt{n_1^2 - n_2^2} \quad ...(6)$$

$\therefore$ $$\theta_0 = \sin^{-1}[\sqrt{n_1^2 - n_2^2}] \quad ...(7)$$

The angle θ_0 is called the acceptance angle of the fibre. Acceptance angle may be *defined as the maximum angle that a light ray can have relative to the axis of the fibre and propagate down the fibre.*

The light rays contained within the cone having a full angle $2\theta_0$ are accepted and transmitted along the fibre. The cone is called the acceptance cone (Fig. 8.25).

Light incident at an angle beyond θ_0 refracts through the cladding and the corresponding opti- cal energy is lost. *It is clear that the greater the diameter of the core, the larger the acceptance angle.*

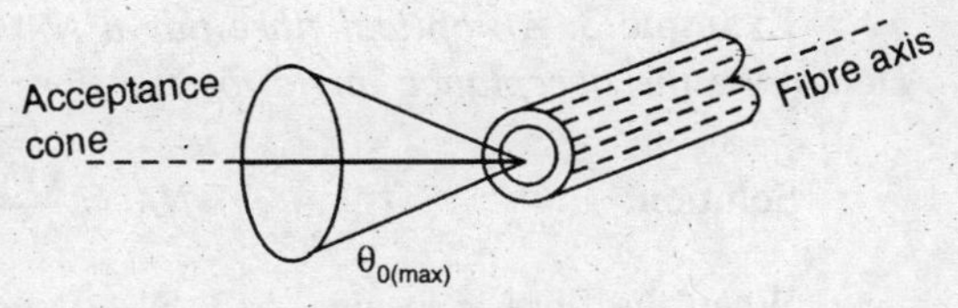

Fig. 8.25

8.31. FRACTIONAL REFRACTIVE INDEX CHANGE

The fractional difference Δ between the refractive indices of the core and the cladding is called *fractional refractive index change.* It may be expressed as

$$\Delta = \frac{n_1 - n_2}{n_1} \quad ...(8)$$

This parameter is always positive as $n_1\ n_2$ for the total internal reflection conduction condition. In order to guide light rays effectively through a fibre, $\Delta < 1$. Typically, Δ is of the order of 0.01.

8.32. NUMERICAL APERTURE

The numerical aperture (*NA*) is defined as the sine of the acceptance angle. So we may write

$$NA = \sin \theta_0$$

$$\therefore \quad NA = \sqrt{n_1^2 - n_2^2} \quad ...(9)$$

$$n_1^2 - n_2^2 = (n_1 + n_2)(n_1 - n_2) = \left(\frac{n_1 + n_2}{2}\right)\left(\frac{n_1 - n_2}{n_1}\right) 2n_1$$

Approximating $\frac{n_1 + n_2}{2} \approx n_1$, we can express the above may be written as

$$(n_1^2 - n_2^2) = 2n_1^2 \Delta$$

$$\therefore \quad NA = n_1\sqrt{2\Delta} \quad ...(10)$$

Numerical aperture accounts for the light gathering ability of the fibre and it measures the amount of light accepted by the fibre. As is clear from Equation (9) numerical aperture depends only on the refractive indices of the core and cladding materials.

Numerical aperture ranges between 0.13 to 0.50 and a larger value of numerical aperture will mean that the fibre can accept more light from the source.

Example 1. *Calculate the refractive indices of the core and cladding material of a fibre from following data: NA = 0.22 and* Δ *= 0.012.*

Solution: Given, $\Delta = \frac{n_1 - n_2}{n_1} = 0.012 \quad ...(1)$

and $NA = n_1\sqrt{2\Delta}$

$$\therefore \quad n_1 = \frac{NA}{\sqrt{2\Delta}} = \frac{0.22}{\sqrt{2 \times .012}} = \mathbf{1.42}$$

Putting in Equation (1), we get

$$\frac{1.42 - n_2}{1.42} = 0012$$

$$\therefore \quad n_2 = 1.40.$$

Example 2. *Calculate the numerical aperture, acceptance angle and the critical angle of a fibre having core refractive index = 1.50 and the cladding refractive index = 1.45.*

Solution: We have the relation, $\Delta = \frac{n_1 - n_2}{n_1} = \frac{1.5 - 1.45}{1.5} = 0.033$

Numerical aperture $(NA) = n_1\sqrt{2\Delta} = 1.5 \times \sqrt{2 \times 0.033} = 0.387$

Acceptance angle $\theta_0 = \sin^{-1} NA = 22.78°$

and critical angle $C = \sin^{-1}\frac{n_2}{n_1} = \sin^{-1}\frac{1.45}{1.50} = \mathbf{75.2°.}$

Example 3. *An optical fibre has a NA of 0.20 and a cladding refractive index of 1.59. Determine the acceptance angle for the fibre in water which has a refractive index of 1.33.*

Solution: $$NA = \frac{\sqrt{n_1^2 - n_2^2}}{n_0}$$

When the fibre is in air $\quad n_0 = 1$ and $NA = \sqrt{n_1^2 - n_2^2} = 0.20$

$$\therefore \quad n_1 = \sqrt{(NA)^2 + n_2^2} = \sqrt{(0.20)^2 + (1.59)^2} = \mathbf{1.6025}$$

When the fibre is in water $\quad n_0 = 1.33.$

$$\therefore \quad NA = \frac{\sqrt{n_1^2 - n_2^2}}{n_0} = \frac{\sqrt{(1.6025)^2 - (1.59)^2}}{1.33} = \mathbf{0.15}$$

$$\therefore \quad \theta_0\ (\max) = \sin^{-1}(NA) = \sin^{-1}(0.15) = 8.6°.$$

Example 4. *A glass clad fibre is made with the core glass of refractive index 1.5 and the cladding is doped to give a fractional index difference of 0.0005. Determine (i) the cladding index, (ii) the critical internal reflection index, (iii) the external critical acceptance angle, (iv) the numerical aperture.*

Solution: Given, $n_1 = 1.5, \Delta = 0.0005$

(*i*) Let the refractive index of cladding be n_2. So we have

$$\Delta = \frac{n_1 - n_2}{n_1}$$

$$\therefore \quad 0.0005 = \frac{1.5 - n_2}{1.5}$$

or $$n_2 = 1.5 - 1.5 \times 0.005 = \mathbf{1.4925}$$

(*ii*) Let the critical internal reflection angle be C

$$\sin C = \frac{n_2}{n_1}$$

$$\therefore \quad C = \sin^{-1}\left[\frac{n_2}{n_1}\right] = \sin^{-1}\left[\frac{1.4925}{1.5}\right] = \sin^{-1}(0.9995)$$

$$= 88.2°$$

(*iii*) Let the external critical acceptance angle be θ_0, so we have,

$$\sin C = \frac{\sqrt{n_1^2 - n_2^2}}{n_0}; \text{ where } n_0 = 1$$

$$\therefore \quad \theta_0 = \sin^{-1}[\sqrt{n_1^2 - n_2^2}]$$

$$= \sin^{-1}(1.5^2 - 1.4925^2)^{1/2}$$

$$= \sin^{-1}(2.25 - 22.2764625)^{1/2}$$

$$= \sin^{-1}(0224) = 112°$$

(*iv*) Numerical aperture $\quad NA = n_1\sqrt{2\Delta}$

$$\therefore \quad NA = 1.5\sqrt{2 \times 0.005} = 1.5\ (0.03162) = \mathbf{0.0474}.$$

8.33. MODES OF PROPAGATION

In an optical fibre light travels as an e.m. wave and all the waves moving in directions above the critical angle will be trapped in the fibre due to total internal reflections. However, all such waves do not propagate through the fibre, and only certain ray directions are allowed for propagations. These allowed directions corresponds to **modes of the fibre.**

In simple form, we can understand the modes as the **number of paths of light** in an optical fibre. The paths are all **zig-zag** with the exception of light moving along axial directions. So the light rays moving through a fibre may be classified as (*i*) axial rays (*ii*) zig-zag rays.

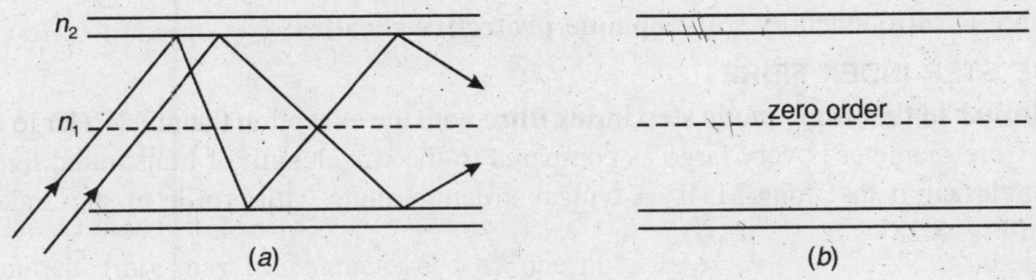

Fig. 8.26

Since the rays get repeatedly reflected at the walls of the fibre, there occurs a phase shift. So wave along certain zig-zag path be in phase and get intensified while some other waves moving in other paths will be out of phase and may fade out due to **destructive interference.**

The light ray paths along which the waves are in phase inside the fibre are called modes. Number of modes, a fibre can support depends on the ratio $\frac{d}{\lambda}$ where ***d* is the diameter of the core and λ is the wavelength of the wave transmitted.**

In general, modes are represented by an order number *m*. In a fibre of fixed thickness, the higher order propagates at smaller angles than the lower order modes.

Axial ray that travels along the axis of the fibre is called **zero order ray.**

8.34. TYPES OF OPTICAL FIBRE

Optical fibres are in general of two types:

(*i*) Single Mode fibre (**SMF**) (*ii*) Multimode Fibre (**MMF**)

A single fibre has a smaller core diameter and can support only one mode of propagation while a **multimode fibre** has a large core diameter and can support a large number of modes.

Multimode fibres are further classified on the basis of index profile. **An index profile** is graph of refractive index (along *X*-axis) and distance from the core (along *Y*-axis). Index profit of a MMF can be either step index (SI) type or graded index (GRIN) type. Index profile of SMF is usually a step index (SI) type.

SINGLE MODE STEP INDEX FIBRE

A single mode step index fibre consists of a very fine thin core of uniform refractive index surrounded by a cladding of refractive index lower than that of the core. Since the refractive index abruptly changes at the core cladding boundary, it is known as step index fibre. A typical SMF has a core diameters of 4μ*m* which corresponds to some of the wavelengths of light waves. Light travels along a single path, *i.e.*, along the axis only and so zero order mode is supported by SMF. Usually SMF is characterised by a very small value of Δ (~0.002).

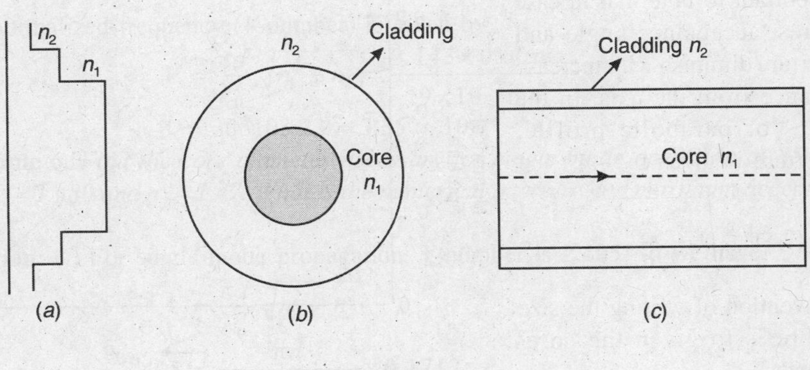

Fig. 8.27

The fibre is surrounded by some **opaque protective sheath.**

MULTIMODE STEP INDEX FIBRE

It is similar to the single mode step index fibre with the exception that it has a large diameter (~100 m*m*). Core diameter is very large as compared to the wavelength of transmitted light. Light moves along zig-zag paths along MMF. A typical structure along with profile of step index MMF are shown in Fig. 8.28.

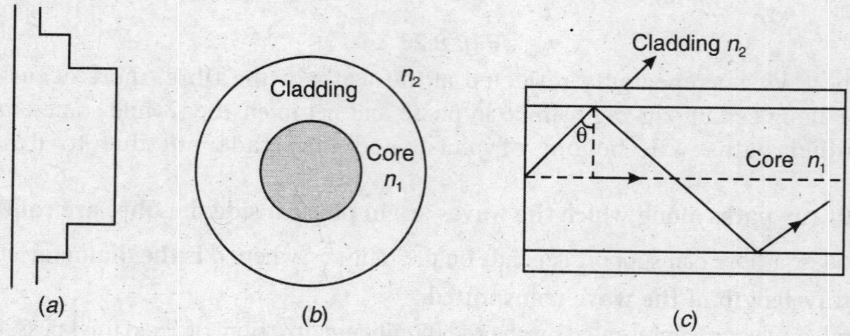

Fig. 8.28

NA in case of a MMF is quite large as core diameter is larger and is ~ 0.3.

GRAIN INDEX FIBRE (GRIN)

A GRIN is a multimode fibre which has concentric layers of refractive indices which means that the refractive index of the core varies with distance from the fibres axis, *i.e.,* it has high value at the centre and falls of rapidly as the radial distance increases from the axis.

As is shown in profile (Fig. 8.29) such a profile causes typical periodic focussing of the light moving through the fibre.

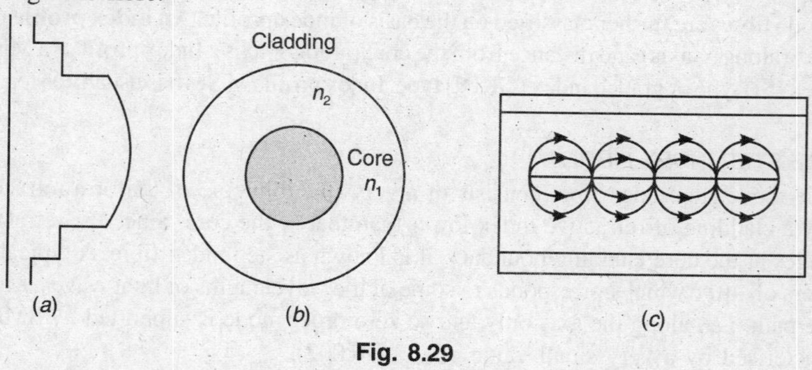

Fig. 8.29

It is important to note that in case of GRIN fibres, acceptance angle and numerical aperture diminish with increase of radial distance from the axis. In the case of fibres of parabolic profile, numerical aperture is given by:

$$NA = n_1 (2\Delta)^{1/2} \sqrt{1 - \left(\frac{r}{a}\right)^2}$$

The convention of writing the size of optical fibre is to given the core/cladding diameter.

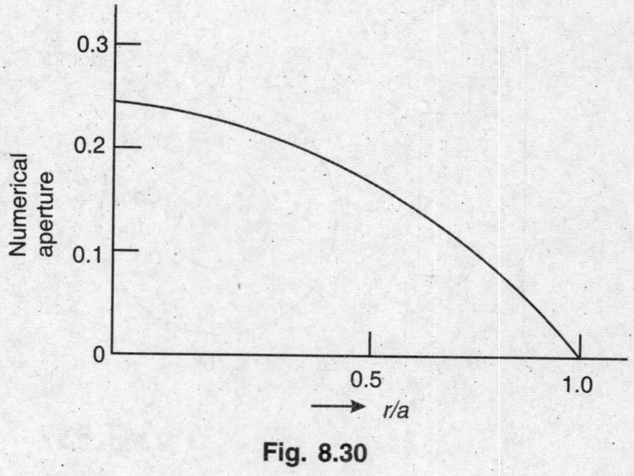

Fig. 8.30

Sizes of SI fibre are $\frac{50}{125}, \frac{10}{140}, \frac{200}{230}$ etc.

Size of GRIN fibres are $\frac{50}{125}, \frac{62.5}{125}, \frac{85}{125}$, etc. For example, a $\frac{50}{125}$ fibre has 50μ*m* core and 125μ*m* cladding.

8.35. NORMALIZED FREQUENCY (V-NUMBER) OF A FIBRE

An optical fibre may be characterised by one more or parameter called V-number or the normalized frequency of the fibre.

V-number is denoted by the relation

$$V = \frac{2\pi a}{\lambda} \cdot \sqrt{n_1^2 - n_2^2} \qquad ...(1)$$

where $a \to$ Radius of the core
and $\lambda \to$ Free space wavelength.

Also we have Numerical Aperture given as:

$$NA = \sqrt{n_1^2 - n_2^2} = n_1 \sqrt{2\Delta} \qquad ...(2)$$

$$V = \frac{2\pi a}{\lambda}(NA) = \frac{2\pi a}{\lambda} n_1 \sqrt{2\Delta} \qquad ...(3)$$

Maximum number of modes N_m supported by an SI fibre may be got using the relation

$$N_m = \frac{1}{2} V^2$$

e.g., for $V = 10, N_m = 50$.

In case $V < 2.405$, fibre can support only one mode and is called SMF.

For $V > 2.405$, it is classified as MMF and it can support number of modes simultaneously.

Wavelengths corresponding to value of $V = 2.405$ is called the *cut off wavelength* of the fibre and is given as

$$\text{Cut off wavelength } \lambda_C = \frac{\lambda V}{2.405} \qquad ...(4)$$

In the case of GRIN fibres, for large values of *V*, we have

$$N_m = \frac{V^2}{4} \qquad ...(5)$$

Example 5. *A step index fibre in air has a NA of 0.16, a core refractive index of 1.45 and a core diameter of 60 cm. Determine the normalized frequency for the fibre when light at a wavelength of 0.9μm is transmitted.*

Solution: Given, $NA = 0.16$

$n_1 = 1.45$

$d = 60 \text{ cm} = 0.6 \text{ m}$

$\lambda_0 = 0.9\mu m = 9 \times 10^{-7}$ m

The normalized frequency (*V*-number) is given by

$$V = \frac{\pi d}{\lambda_0} \sqrt{n_1^2 - n_2^2} = \frac{3.143 \times 0.60 \text{ m}}{9 \times 10^{-7} \text{ m}}$$

$\therefore \quad V = 335103.22 = 3.35 \times 10^5$.

Example 6. *Find the core diameter necessary for single mode operation at 850μm in S.I. fibre with $n_1 = 1.480$ and $n_2 = 1.47$. What is the numerical aperture and maximum acceptance angle of this fibre?*

Solution: (*i*) For single mode propagation, *V*-number is 2.405, so we have,

$$V = \frac{\pi d}{\lambda_0} \sqrt{n_1^2 - n_2^2}$$

$$\therefore \quad 2.405 = \frac{\pi d}{450 \times 10^{-9} \text{m}} \times 0.1717$$

$$d = \frac{2.405 \times 850 \times 10^{-9}}{3.143 \times 0.1717}\ \text{m} = 3.79\mu m$$

(*ii*) $$\text{NA} = \sqrt{n_1^2 - n_2^2} = \sqrt{1.48^2 - 1.47^2} = 0.1717$$

(*iii*) $$\sin\theta_0 = NA$$

$$\theta_0 = \sin^{-1}(NA) = \sin^{-1}[n_1^2 - n_2^2]^{1/2}$$

$$= \sin^{-1}(0.1717) = 953'12''.$$

Example 7. *A step index fibre is made with a core of index 1.52, a diameter of 29μm and fractional difference index of 0.007. It is operated at a wavelength of 1.3μm: find (i) the fibre V-number, and (ii) the number of modes the fibre will support.*

Solution: Core refractive index $n = 1.52$

$$d = 29\,\mu m = 29 \times 10^{-6}$$

$$\Delta = 0.007$$

$$\lambda_0 = 1.3\mu m = 1.3 \times 10^{-6}$$

Using the relation, $\Delta = \dfrac{n_1 - n_2}{n_1}$

We get, $0.007 = \dfrac{1.52 - n_2}{n_1}$

$\therefore$ $n_2 = 1.52 - 1.52 \times 0.007 = 1.5189$

(*i*) $$V = \frac{\pi d}{\lambda_0}\sqrt{n_1^2 - n_2^2}$$

$\therefore$ $$V = \frac{3.143 \times 29 \times 10^{-6}}{1.3 \times 10^{-6}}[1.52^2 - 1.5189^2]^{1/2} = \mathbf{4.049}$$

(*ii*) Number of modes $N = \dfrac{1}{2}V^2 = \dfrac{1}{2}(4.049)^2 = 8.19 = \mathbf{8\ modes.}$

Example 8. *Calculate the maximum radius allowed for a fibre having core refractive index = 1.47 and cladding refractive index = 1.46. Fibre has a support of only one mode at a wavelength of 1300 nm.*

Solution: Condition for single mode is $V < 2.405$

or $$\frac{\pi d}{\lambda}\sqrt{n_1^2 - n_2^2} < 2.405$$

or $$d < \frac{2.405\lambda}{\pi\sqrt{n_1^2 - n_2^2}} < \frac{2.405 \times 1300 \times 10^{-9}}{3.142 \times \sqrt{(1.47)^2 - (1.46)^2}} < \mathbf{5.82\mu m}$$

$\therefore$ Maximum diameter = **5.82μm.**

Example 9. *Calculate the maximum value of Δ in the case of a single mode fibre core of diameter 10μm and core refractive index 1.5. Fibre is coupled to a light source of wavelength 1.3μm. Also calculate the cladding refractive index and the acceptance angle. Given V cut off for a single mode propagation = 2.405.*

Solution: We have the relations,

$$V_{\text{cut-off}} = \frac{\pi d}{\lambda} NA \text{ (cut off)}$$

$\therefore$ $$NA \text{ (cut off)} = \frac{V \times \lambda}{\pi d} = \frac{2.405 \times 1.3 \times 10^{-6}}{3.142 \times 10} = 0.09957$$

$\therefore$ Acceptance angle $= \sin^{-1}(0.09957) = 5.71°$

Also numerical aperture, $NA = n_1\sqrt{2\Delta}$

$\therefore$ $$\Delta = \frac{NA^2}{n_1^2} \times \frac{1}{2} = \frac{(0.09957)^2}{(1.5)^2} \times \frac{1}{2} = 2.2 \times 10^{-3}$$

Also $$\Delta = \frac{n_1 - n_2}{n_1}$$

$$\therefore \quad n_2 = n_1 - \Delta n_1 = n_1\ (1 - \Delta) = 1.5\ (1 - 2.2 \times 10^{-3}) = 1.496.$$

8.36. FABRICATION OF OPTICAL FIBRES

Materials which are used for optical fibres are transparent to optical frequencies. Normally, step index fibres are produced in the following three forms:

(*i*) Glass core cladded with glass which has lower refractive index than the glass-core.
(*ii*) Silica glass core cladded with plastic.
(*iii*) Plastic core cladded with another plastic.

In case of all glass fibres, **refractive step index** is smallest, slightly more for plastic cladded silica (PCS) and largest for plastic core and plastic cladded fibre.

(*i*) **Glass Fibre**

Most abundant and inexpensive optical fibres are made from silica (SiO_2) which has refractive index = 1.458 for λ = 8500Å.

We get material for different refractive index by dropping the basic material. If basic material is doped with Germenia (GeO_2) or Phosphorus pentoxide (P_2O_5), value of refractive index increases and **such materials can be used as core while pure silica is used as cladding.**

If pure silica is doped with B_2O_3 or fluorine, its refractive index reduces and so **it can be used as cladding while pure silica is used as core material.**

Glass optical fibres exhibit very low losses and are used for long distance communication.

(*ii*) **PCS Fibres**

Core is normally of pure quartz while cladding and sheath are of some transparent polymer such as silicon resin of low refractive index. These fibres are less expensive but transmission losses are very high. So these are used for short distances.

(*iii*) **Plastic Fibres**

In this case both core and cladding are of plastics. Using approximate polymers, a high refractive index difference can be achieved between core and cladding. So this type of fibre will have large **Numerical Aperture.** The main advantage of all plastic fibres is low cost and high mechanical flexibility. However, since those fibres are temperature sensitive. So they are limited in application between 80°C to 100°C temperatures. Also they have high transmission losses and so have limited use because of this limitation.

8.37. ATTENUATION

An optical signal passing through a fibre will get progressively reduced. This reduction or attenuation of signal may be defined as the **ratio of the optical output power from a fibre of length *L* to the input optical power.** It is expressed in terms of decibel/km (dB/km)

$$\alpha = \frac{10 \log P_1/P_2}{L}$$

where $P_1 \rightarrow$ Power of optical signal at launching end
$P_2 \rightarrow$ Power of optical signal at the receiving end.

In an ideal fibre $P_1 = P_2$ and attenuation will be 0 dB/km. However an attenuation upto 3 dB is within permissible limit. It is important to note that attenuation is wavelength dependent and so the wavelength must also be specified.

8.38. FIBRE LOSSES

The following three losses may occur in optical fibres:

(*i*) Absorption
(*ii*) Geometric effects
(*iii*) Rayleigh scattering

(*i*) **Absorption**

Even very pure glass absorbs light of a specific wavelength. Strong **electronic absorption** occurs in UV region and **vibrational absorption** occurs in IR region of wavelength $7\mu m$ to $12\mu m$. These losses are attributed due to inherent property of the glass and is called **intrinsic absorption.** However, this loss is insignificant.

Impurities are major extrinsic source of losses in fibre. Hydroxyl radical ions (OH) and transition metals like Nickel, Chromium, Copper, Manganese etc. have electronic losses in **near visible range of spectrum.** These impurities should be kept away as far as possible from the fibre. Intrinsic as well as extrinsic losses are found to be minimum at about $1.3\mu m$.

(*ii*) **Geometric Effects**

These may occur due to manufacturing defects like irregularities in fibre dimensions during drawing process or during coating, cabling or insulation processes.

(*iii*) **Rayleigh Scattering**

As glass has disordered structure having local microscopic variation in density which may also cause variation in refractive indices. So light travelling through these structures may suffer scattering losses due to Rayleigh, *i.e.*, scattering $\propto \frac{1}{\lambda^4}$. It means Rayleigh scattering sets a lower limit on wavelength that can be transmitted by a glass fibre at $0.8\mu m$ below which scattering loss is appreciably high.

8.39. TIME DELAY BETWEEN HIGHEST AND LOWEST ORDER MODE

Two rays of different modes are given in the Fig. 8.31.

If $AD = x$, then from Fig. 8.31.

$$AB = \frac{x}{\sin \phi_2}$$

So the total zig-zag path for L km fibre may be written as

$$L = \frac{L}{\sin \phi_2} \quad ...(i)$$

In general, we may write, Length $= \dfrac{L}{\sin \phi}$

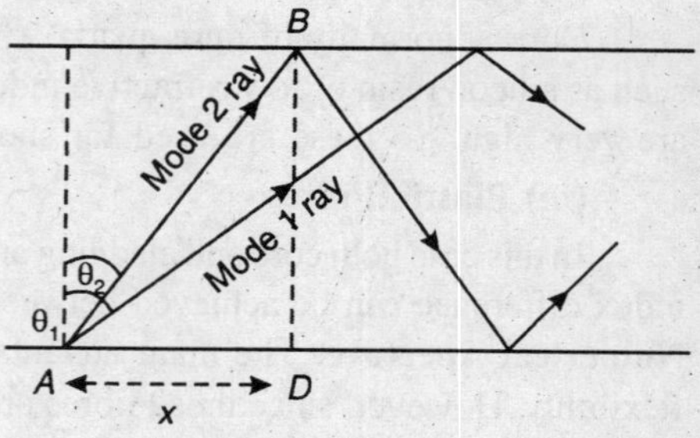

Fig. 8.31

where ϕ corresponds to the angle of incidence in the mode in question, with the normal to the wall of the fibre. In the highest order mode, the angle of incidence has minimum value and so is given by the critical angle C.

$$\phi_{min} = C = \sin^{-1} \frac{n_1}{n_2} \quad ...(2)$$

where n_2 is the refractive index of cladding material and n_1 is the refractive index of core material. In the lowest order mode, path length is given by

$$L_{min} = \frac{L}{\sin 90°} = L \quad [\because = \phi_{max} = 90°]$$

So the highest mode path length

$$L_{max} = \frac{L}{\sin \phi_{min}} \quad ...(3)$$

$$L_{max} = \frac{L}{\sin C} = L \cdot \frac{n_1}{n_2} \quad \left[\because \sin C = \frac{n_1}{n_2}\right] \quad ...(4)$$

Let Δx be the path-difference between the two rays,

$$\therefore \quad \Delta x = L_{max} - L_{min} = L\left[\frac{n_1}{n_2} - 1\right]$$

also $$\Delta = \frac{n_1 - n_2}{n_1}$$

$$\therefore \quad \Delta x = L \cdot \left(\frac{\Delta}{1-\Delta}\right) \qquad ...(5)$$

As the fibre acts like a dielectric medium of relative dielectric constant ε_r (say) which is normally more than one, so the rays move with a speed less than that in free spaces.

Also phase velocity V_p is given by

$$V_p = \frac{1}{\sqrt{\mu_r \mu_0 \varepsilon_r \varepsilon_0}} \qquad ...(6)$$

where μ_0 and ε_0 are free space permeability and permittivity respectively.

Velocity of light is given by

$$c = \frac{1}{\sqrt{\mu_0 \varepsilon_0}}$$

$$\therefore \quad V_p = \frac{c}{\sqrt{\varepsilon_r \mu_r}}$$

Also since $\mu_r = 1$

$$\therefore \quad V_p = \frac{c}{\sqrt{\varepsilon_r}}$$

This phase velocity is the same as that of transmission line. Also we have the relation for refractive index and dielectric constant as

$$n^2 = \varepsilon r$$

$$\therefore \quad V_p \text{ (glass)} = \frac{c}{n_1}$$

So we may represent the time delay between highest and lowest order mode as

$$\Delta t = \frac{\Delta x}{V_p \text{ (glass)}} = L \cdot \frac{\Delta}{1-\Delta} / V_p$$

$$= L \cdot \frac{\frac{\Delta}{1-\Delta}}{c/n_1} = \frac{L.\Delta.n_1}{c\,(1-\Delta)} \qquad ...(9)$$

$$= \frac{L}{c} \cdot \frac{\Delta}{1-\Delta} \cdot n_1$$

This time delay is the characteristic of the fibre and **independent of the wavelength of light and** is of the order of **neno second per kilometer.**

It may be noted that this type of delay is absent in single mode fibre as it has only **one mode of propagation.**

In the case of GRIN fibre, the time delay due to intermodal dispersion is given by,

$$\Delta t = \frac{n_1 L \Delta^2}{8c} \qquad ...(10)$$

In this case n_1 in the refractive index of glass at the core centre and Δ is the fractional refraction index change from *core centre to cladding.*

It is observed that the graded index fibres (GRIN fibre) have much lower intermodal dispersion than that of S.I. fibre, as in the case of GRIN fibre, the ray on its passage from centre of fibre core to outer edge in zig-zag path moves through regions of decreasing refractive index. As phase velocity is given by

$$V_p = \frac{c}{n}$$

So the phase velocity increases with distance, measured from fibre core. So the group velocity is also faster at the outer edge.

Reverse phenomenan occurs when the ray is refracted back to the axis of fibre core. But again it will increase on the other side of the core axis. So we will get alternatively rising and falling group velocity with an average higher than that in the lower modes which propagate straight down the fibre, with the result that a much lower intermodal dispersion is left out in the graded index.

Time delay in this type of fibre is much less than 1 ns/km.

Example 10. *A certain optical fibre has an attenuation of 3.5 dB/km at 850 nm. If 0.5 mw of optical power is initially launched with fibre, calculate the power level after 4 km.*

Solution: In this case

$$\text{Attenuation } \alpha = 3.5 \text{ dB/km}$$
$$\text{Initial power level } P_1 = 0.5 \text{ mW}$$
$$\text{Final power level } P_2 = ?$$
$$\text{length of cable} = 4 \text{ km}$$

We have the relation $\alpha = \dfrac{10}{L} \log \dfrac{P_1}{P_2}$

$$\therefore \quad 10 \log_{10} \frac{P_1}{P_2} = \alpha . L = 3.5 \times 4 = 14 \text{ dB}$$

$$\log_{10} \frac{P_1}{P_2} = \frac{14}{10} = 1.4 \text{ or } \log_{10} \frac{0.5}{P_2} = 1.4$$

Taking antilog on both sides

$$\therefore \quad \frac{0.5}{P_2} = 25.11 \text{ or } P_2 = \frac{0.5}{25.11} \text{ mw} = 19.9 \mu\text{W}.$$

Example 11. *A 15 km optical fibre link uses fibre with a loss of 1.5 dB/km. The fibre is joined every km with connectors, which give attenuation of 0.80 dB each. Find the minimum mean optical power which must be launched with the fibre to maintain a mean optical power level of 0.3 μω at the detector.*

Solution: (*i*) Connector loss = 0.8 dB/km

∴ Connector loss for 15 km length = 0.8 × 15 = 12 dB.

(*ii*) Fibre loss = 1.5 dB/km

∴ Fibre loss for 15 km = 1.5 × 15 = 22.5 dB

Total loss = 12 dB + 22.5 dB = 34.5 dB

Optical power level at detector $P_2 = 0.3\ \mu\text{W}$

Optical power at launch position $= P_1$

Now we have the relation $\alpha = 10 \log \dfrac{P_1}{P_2}$

$$34.5 = 10 \log \frac{P_1}{0.3 \mu\text{W}} \text{ or } \log \frac{P_1}{0.3 \mu\text{W}} = 3.45$$

Taking antilogs on both sides

$$\frac{P_1}{0.3 \mu\text{W}} = 2818.3$$

$$\therefore \quad P_1 = 8.455 \times 10^{-4} \text{ W}$$
$$= \mathbf{0.846 \text{ mW}}.$$

8.40. OPTICAL WINDOWS

A typical curve of fibre attenuation v/s wave length for a silica based optical fibre is given in Fig. 8.32.

As may be observed from the above curve, there are minima at a particular optical wavelengths. The band of wavelengths, at which the attenuation is minimum, is called an optical window or transmission window. Optical signals, having wave-lengths in this region travel through the fibre with least attenuation. Because of the said property, such windows are quite suitable for transmission of information.

In fibre optical communication, two such windows are chosen:

(*i*) First window lies between $0.8 \mu m$ to $0.9 \mu m$.

(*ii*) Second window lies between $1.3 \mu m$ to $1.6 \mu m$.

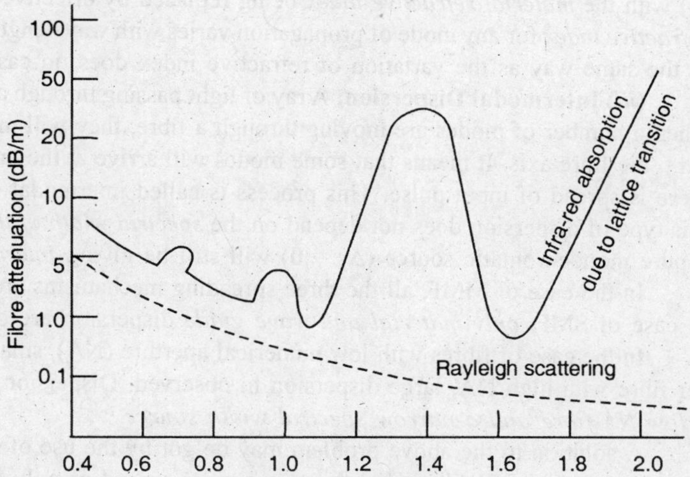

Fig. 8.32

8.41. DISPERSION

As stated earlier a pulse launched with a fibre gets attenuated due to losses in fibre. Moreover the incoming pulse also spreads during the transit through the fibre. So a pulse at the output is wider than the pulse at the in put *i.e.*, the pulse gets distorted as it moves through the fibre (Fig. 8.33). This distortion of pulse is due to dispersion effects which is measured in terms of nanoseconds per km (ns/km).

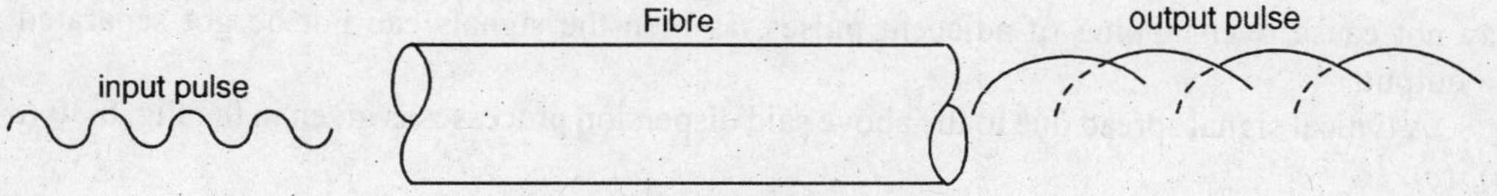

Fig. 8.33

There are three phenomenan that may contribute towards the distortion effect:

(*i*) Material dispersion

(*ii*) Wave guide dispersion.

(*iii*) Intermodal dispersion.

(*i*) Material Dispersion: Light waves having different wavelengths will travel with different speeds through a medium. Light waves having *shorter wavelengths* will move slower than those of longer wavelengths. So narrow pulses of light tend to broaden as they move through an optical fibre. This process is called material dispersion. As may be evident, the spectral width of the source will determine the extent of dispersion and the material dispersion is given by the equation

$$D_m = \frac{\lambda(\Delta\lambda)}{C} \cdot L \frac{d^2 n}{d\lambda^2} \qquad \ldots(1)$$

where λ = corresponds to peak wavelength

$\Delta\lambda$ = is the spectral width

L = is the angle; of the fibre

and n = corresponds to core refractive index

It may be mentioned that the material dispersion can be very much reduced by a using a pure monochromatic source.

To give an example of material dispersion we observe that a LED operating at 820 nm and having a spectral spread of 38 nm gives dispersion of 3 ns/km. This dispersion can be reduced of to about 0.3 ns/km with the help of a laser diode that operates at 1140 nm with a spectral width 3 nm.

(*ii*) Wave-Guide Dispersion: Wave guided dispersion arises due to the *guiding properties of the fibre*. Amount of wave guide dispersion may also be expressed in a way similar to the Equation

(1) with the *material refractive index* being replaced by effective refractive index as the *effective refractive index* for any mode of propagation varies with wavelength that may cause pulse spreading in the same way as the variation of refractive index does, in case of material dispersion.

***(iii)* Intermodal Dispersion:** A ray of light passing through a fibre follows a zig-zag path and when a number of modes are moving through a fibre, they will move with different net velocities w.r.t. the fibre axis. It means that some modes will arrive at the out put earlier than the others *i.e.*, there is spread of input pulse. This process is called intermodal dispersion. It may be noted that this type of dispersion does not depend on the *spectral width of the source i.e.*, a light pulse from a pure monochromatic source ($\Delta\gamma = 0$) will still be giving *intermodal dispersion.*

In the case of MMF, all the three spreading mechanisms are observed simultaneously while in case of SMF, *only material and wave guide* dispersion are observable.

In the case of fibres with low numerical aperture (NA), smaller dispersion is observed while for fibre with high NA, large dispersion in observed. Dispersion may be restricted by the use of a *low NA fibre and a narrow spectral width source.*

A solution to the above problem may be got by the use of GRIN fibres which produce less distortion than the SI fibres and they have pulse spread of only few ns/km, which is guide small as compared to the spread in case of SI fibres. However, GRIN fibres are more expensive. Also in the case of a short length of fibre, dispersion is proportional to the length of the fibre and so may not pose any problem but in case of large lengths, dispersion is proportional to the square root of fibre length and so may limit the use of *rate of data transmission.*

Another important aspect to be kept in mind is that the *dispersion limits the bandwidth of fibre*. Information flow, which means the rate of varies pulses must be quite slow so that the dispersion may not cause over-lapping of adjacent pulses, as then the signals can not be got separated at the output.

A typical signal spread due to the above said dispersion processes is given in the Fig. 8.30 (*a*), (*b*), (*c*).

8.42. RELATION BETWEEN BAND WIDTH, DISTANCE AND DISPERSION

Band width in the case of optical fibres can be known from the knowledge of dispersion using the relation

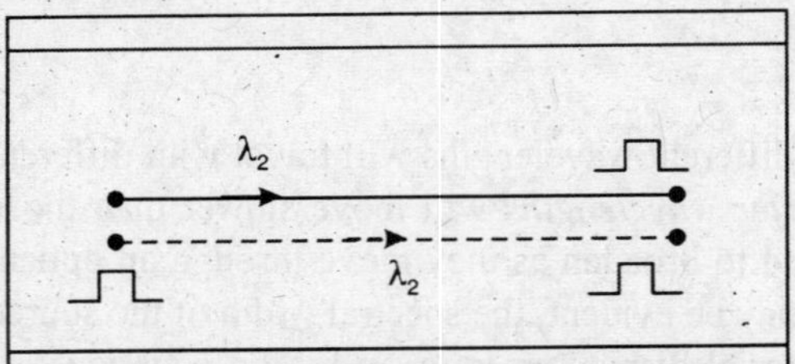

(*a*) wavelengths move with different speeds

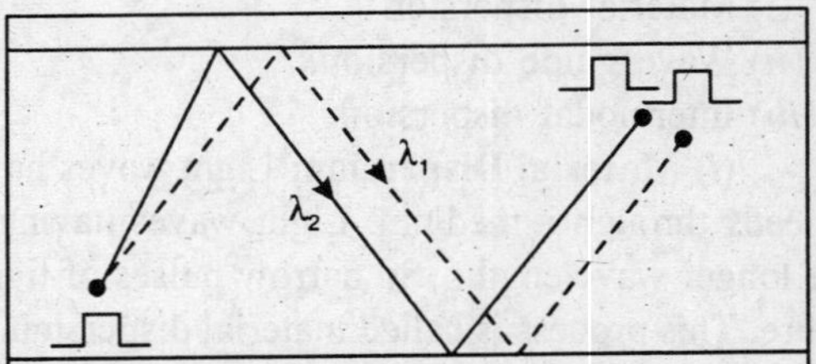

(*b*) (Wave guide dispersion, though propagating with same mode, travel at different angles.)

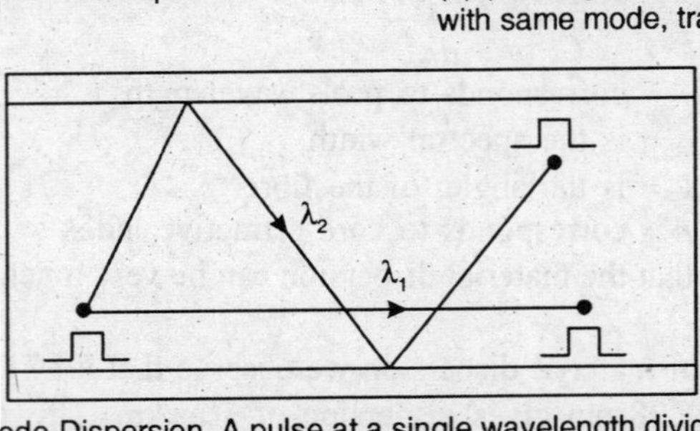

(*c*) Mode-Dispersion. A pulse at a single wavelength divides itself into modes that wave at different axial velocities.

Fig. 8.34

$$\text{Bandwidth (MHZ.km)} = \frac{310}{\text{(Dispersion ns/km)}} \quad ...(2)$$

Bandwidth is a very important property of fibre cable and as is evident, cables with low value of NA will have the widest band width.

Band width-distance product will specify the information capacity of the optical fibre. Another important characteristic of the cable is the attenuation/km. Bandwidth-kilometer product and attenuation/km may be used to find the *bit rate for transmission of information.*

Maximum bit rate B_{max} may be calculated from the relation

$$B_{max} = \frac{1}{5\,\text{(Dispersion)}} \quad ...(3)$$

Example 12. *A step index multimode fibre has core index of 1.5 and cladding index of 1.498. Calculate (i) intermodal factor for the cable; (ii) Total dispersion in a 18 km length and (iii) Maximum bitrate allowed.*

Solution: In this case,

$$n_1 = 1.5;\ n_2 = 1.498;\ l = 18 \text{ km}$$

$$\therefore \quad \Delta = \frac{n_1 - n_2}{n_1} = \frac{1.5 - 1.498}{1.5} = 0.00133$$

(*i*) For 1 km length fibre $\quad \Delta t = \frac{n_1 . L}{c}\left(\frac{\Delta}{1-\Delta}\right) = \frac{1.5 \times 1000 m}{3 \times 10^8 \text{ ms}^{-1}} \times \frac{0.00133}{1 - 0.00133}$

$$= 0.6659 \times 10^{-8} \text{ sec} = \mathbf{6.66 \text{ ns/km.}}$$

(*ii*) Total dispersion for 18 km length

$$\Delta t\,\text{(total)} = (\Delta t) 1 \text{ km} \times 10 \text{ km}$$

$$= 6.66 \text{ ns/km} \times 18 \text{ km} = 119.9 \text{ ns}$$

(*iii*) $\quad B_{max} = \frac{1}{5\Delta_t\,\text{(total)}} = \frac{1}{5 \times 119.9 ns} = 0.00166 \times 10^9 \text{ bits/sec}$

$$= 1.66 \times 10^6 \text{ bits/sec} = \mathbf{1.66 \text{ M bits/sec.}}$$

Example 13. *Compute the dispersion/km of length and total dispersion in a 10 km length of SI fibre from the following data:*

Refractive index of core = 1.558 and Δ = 0.026.

Solution: Maximum dispersion $= \frac{n_1}{c} \cdot \frac{\Delta}{1 - \Delta}$

∴ Max. dispersion/km $= \frac{1.558}{3 \times 10^8} \times \frac{.026}{1 - 0.026} \times 1000 \text{ m} \quad [\because L = 1 \text{ km} = 10^3 m]$

$$13.8 \times 10^{-9} \text{ sec} = 13.8\ n \text{ sec.}$$

∴ Dispersion in 10 km $= 13.8 \times 10 = \mathbf{138}$***n*** **sec.**

Example 14. *Compute the maximum dispersion for an optically graded fibre from the following data refractive index at the centre of core = 1.5, Δ = 0.0260 and length of fibre = 1 km.*

Solution: In case of GRIN fibre,

We have the relation, $\quad \Delta t = \frac{n_1 . L . \Delta^2}{8c} = \frac{1.5 \times 1000 m \times (0.026)^2}{8 \times 3 \times 10^8 \text{ ms}^{-1}} = \mathbf{0.42 \text{ ns/km.}}$

8.43. COUPLING COMPONENTS OF AN OPTICAL FIBRE (SPLICES, CONNECTORS AND COUPLERS)

1. Fibre Splices: *Fibre splices are permanent joints in fibres and are analogous to soldered joints in electrical system.* A splice in a fibre which produces a localised optical distortion that may increase the signal attenuation and some of the signal may get scattered at splices. *Mode conversion* generally from low to high order modes may occur thereby increasing the attenuation beyond splices. Combination of all these effects may *attenuate* a signal significantly.

So, for less distortion and attenuation of the signal, proper matching of the front ends of the fibres to be joined is required. These losses may be due to the *misalignment* of core and cladding of the fibres to be joined. The joining process generally introduces a loss of 0.2 to 0.3 db for a 50 micron *core index fibre*. Fibre mismatches of core diameter, cladding diameter, index profile and numerical aperture between the adjoining fibres increase this loss to about 0.5 db.

In view of the above discussion, it is important that the fibre splices must have low *insertion loss* and sufficient mechanical strength. A number of techniques have been developed, *e.g.*, *V*-groove splices, fusion splices and sleeve splices etc. It should be noted that in designing a splice system, simplicity of the operation must have high priority because of the following reasons.

1. Working conditions in the fields do not allow complicated splicing procedures.
2. Operating companies do not want to increase cost factor by employing more persons to maintain the fibres.

8.44. V-GROOVE TECHNIQUE

This technique is usually employed during testing of optical fibres in the laboratory or in field or for quick restoration of fibre fault in the communication link. The V-grooved system or *Jig* consists of a central pillar and two side alignment blocks called magnetic clamps.

On the central pillar V-grooves impressed by mandrel into copper substrate measuring about 6 × 2 mm are mounted. Figure 8.31 shows the V-groove arrangement in which the fibres to be joined are held in alignment blocks by magnetic clamps. Central pillar is raised so that the fibres are forced down into V-grooves under there own spring force. Some of epoxy resin or glue is applied over the fibre ends which are brought near by moving the right hand fibre clamp towards *V-groove jig*. As soon as the fibre ends touch, a small heater is switched on in the central pillar so that the *epoxy resin* fixes the fibre in place. The fibres with V-groove substrates are then taken out and mounted in a joint box. Joints made with this technique for identical fibres show losses of the order of 0.1 db or even lesser than this value.

In *sleeve splice* techni-que, a glass sleeve is used which has lower softening temperatures than the fibres to be joined. The basic principle in this method is that when a glass tube is heated to its softening point, surface tension forces tend to collapse the tube into a solid rod.

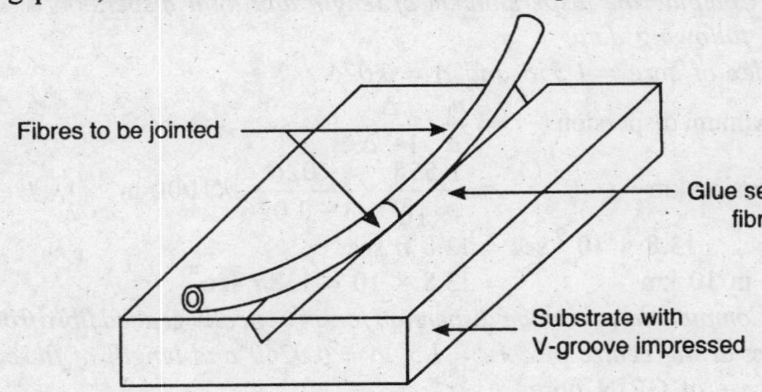

Fig. 8.35

Fusion splices give lowest losses, however, they involve difficult techniques and are not suitable to be used in the fields. Fibre ends are first cleansed and then butted and aligned using a *binocular microscope* which requires a considerable skill. Once aligned the ends are heated either by a **microtorch** or using *electric arc* till the ends are fused. This is a very sophisticated step as more heat melts the glass and so the fibre may be deformed. It properly used, this technique reduce the fibre losses to less than 0.1 db.

8.45. FIBRE CONNECTORS

These are essential components of an optical fibre which allow the direct **transition** of optical power from one core to another. Fibre connectors are used at points where fibres terminate in transmitters and receivers either at the end of line or at the points where repeaters are put connectors must be dismountable so that the equipment may be removed for servicing. A connector must not have any angular or misalignment effects as these may increase the fibre losses.

A variety of fibre connectors have been develope(but all these comprise three main parts: *central alignmen section and two normally identical insertion member.* attched to the two fibres to be connected.

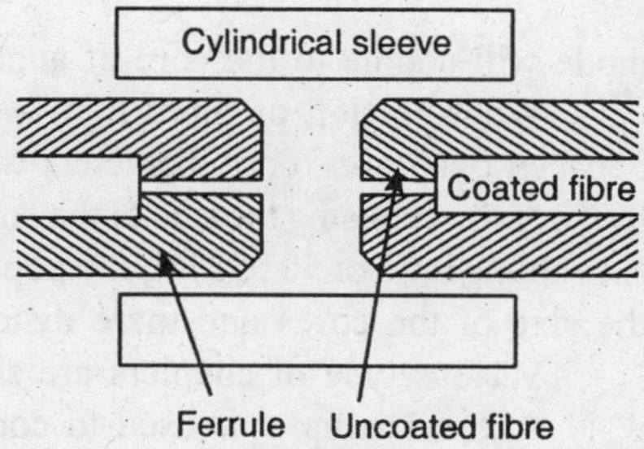

Fig. 8.36

Ferrule type connectors are quite in use which are made by precisely centering the uncoated fibres into ferrules and fixing them with some *epoxy resin*. The terrule and fibre surfaces are and polished. Then the two ferrules are brought to touch ech other inside a precisely guiding cylinder sleeve. Ferrule type connectors are made in different dimensions and have fibre losses in the range of 0.5 to 1 db.

8.46. COUPLERS

Couplers are needed for signal distributions while permanent splices and demountable connectors are used for connecting the fibres.

Main significance of these coupling components is that they do not require optical equipments like *lenses, mirrors, reflectors* etc.

The basic parameters for couplers are:

1. **Coupling ratio** defined as the optical power of a simple output divided by the total output power.
2. **Excess loss:** Defined as the total output power divided by total input power. Excess loss must be quite small and in modern couplers; losses ~ 2 db have been realized for couplers with more than 20 fibres.

For coupling ratio, it is essential that it should be very less and also independent of the fibre materials.

Functionally, there are three types of couplers such as directional, distributive and wavelength dependent couplers. However, depending upon their mechnism they are put as

(1) Diffusion type
(2) Arca-splitting type
(3) Beam-splitting type

A coupler is manufactured by placing two or more fibres adjacent to each other, then fusing and stretching them, thereby creating a coupling region. Heated area is then stretched till required coupling properties are attained.

This device is called *fused biconical taper* (FBT) coupler.

When heated fibre is elongated, the diameter of the core is decreased which means V number is decreased as V-number is given by

$$V = \left[\frac{2\pi\ an}{\lambda}\right]\sqrt{2\Delta}$$

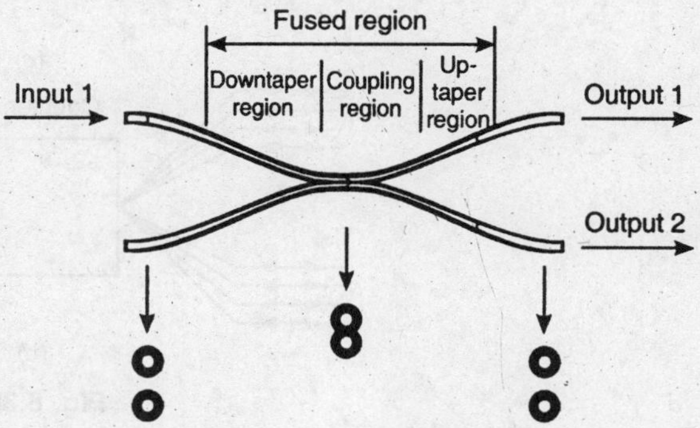

Fig. 8.37

Since $2a = d$ corresponds to diameter of core and when it decreases, V decreases. l is the operating wavelength, n and D are average and relative refractive indices respectively.

Now, V number determines how tightly light power is confined within a core. It means that smaller the V-number, more the mode field diameter (MFD) exceeds the fibre core diameter. So, by decreasing the core diameter, we allow a greater portion of an optical mode to propagate in the cladding where this mode will couple to the core of another fibre. This is illustrated in the Fig. 8.37.

What portion of input light will be coupled to the second fibre depends on the distance between two cores, core diameters within the coupling region and the operating wavelengths. So, by carefully choosing the size of the cores, output power ratios called *coupling ratio* can be controlled. A coupling ratio of 50 : 50 is quite popular, however, any coupling ratio can be attained by controlling the size of the cores and there distances.

Various type of couplers are shown in the Fig. 8.38

A 2 × 1 coupler is used to combine two lights into a single fibre Fig. 8.38 (*a*). When the direction of propagation of light is changed, this will *split* the light signal into two as shown in Fig. 8.38 (*b*). In this case, a coupler is called a splitter due to the function it performs.

There are couplers that couple or split 1 × *N* or *N* × 1 ports as shown in Fig. 8.38 (*c*) and they are called *tree couplers* and may have *N* × *m* configuration.

An important coupler type for wavelength-division multiplexing (WDM) network is a star coupler shown in Fig. 8.38 (*d*). In this case, the same number of ports serve as input and output.

In a sense, a star coupler is an *N* × *m* bidirectonal coupler. However, a star coupler can also be built as an *N* × *m unidirectional coupler*.

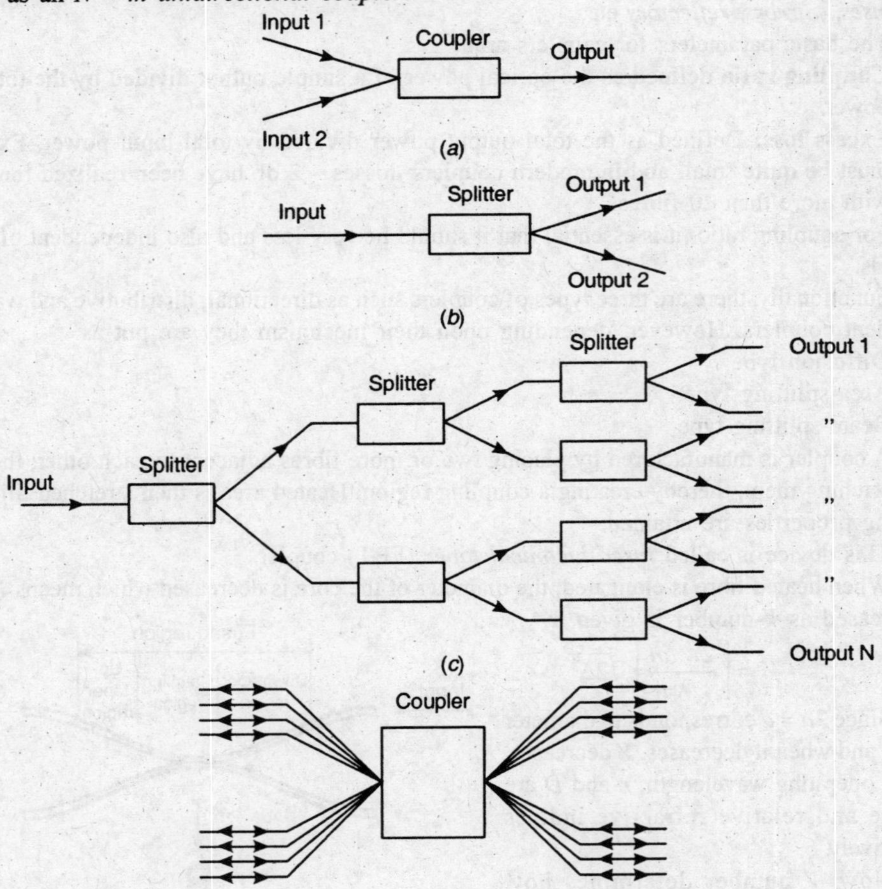

Fig. 8.38

8.47. FABRICATION METHODS OF OPTICAL FIBRE

The fabrication methods used upto sixties were only two: *rod in tube method* and *double crucible method*. These days many other methods are being used, however, we will restrict ourselves to only *double crucible method*.

DOUBLE CRUCIBLE METHOD

In this method, two very high purity glass rods with different refractive indices are prepared using the doping techniques. These rods consist of different compositions for *core and cladding*.

In order to draw cladded core fibres from these glass rods, two cylindrical crucibles are arranged concentrally as shown in Fig. 8.39. The rods are allowed to move slowly downward where they melt due to the heated crucibles, thus filling up the crucibles with glass melt.

The height of the melt is kept constant and the crucible temperature is maintained between 850°C to 1100°C depending upon the composition of the glass.

At the base, the crucibles are tapered to concentric nozzles which mold the glass into a fibre. Pulling speed of the fibre is controlled by the *drum* on which the fibre is wound. So, in this method fabrication of the fibre is achieved only in one step and fibres of unlimited lengths can be manufactured.

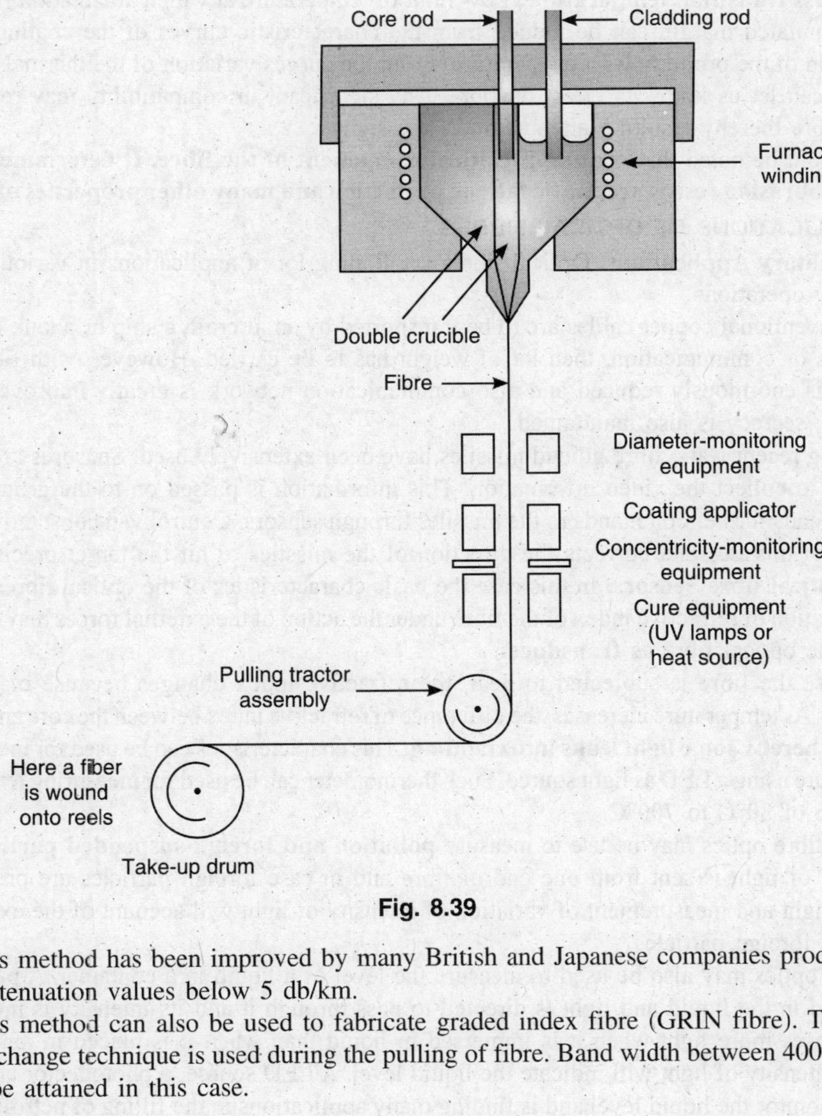

Fig. 8.39

This method has been improved by many British and Japanese companies producing fibres having attenuation values below 5 db/km.

This method can also be used to fabricate graded index fibre (GRIN fibre). To achieve it, an ion exchange technique is used during the pulling of fibre. Band width between 400 to 900 MHz km can be attained in this case.

COATINGS

As we know that an optical fibre has to be coated, the Fig. 8.39 shows how the coating on line can be achieved. A liquid polymer is applied by the coating applicator as the fibre passes through the coating line. The liquid gets solidified by heat or using ultraviolet light.

Main function of the coating is to protect the fibre from external damage. However, a close look reveals the following important functions of the coating.

1. Adhesion: It firmly sticks to the glass due to adhesion property.

2. Toughness: This property is important to provide enhanced abrasion protection and fibre handling and cabling without the loss of strength. Toughness is judged by the elastic (Young's Modulus) testing which shows whether the coating is hard or soft.

3. Moisture resistance: This coating protects the glass from moisture. Moisture resistance can be judged from the aging of fibre when it is exposed to water.

4. Glass transition temperature: Low limit of temperature at which attenuation is observed within a stipulated margin can be judged from the characteristic curves of the coating on fibre, *i.e.*, variation of the property with temperature. From the curves variation of the thermal expansion coefficient can let us know its effect on fibre. Any significant in compatibility may result in the stress on fibre thereby resulting attenuation of the signals.

It should be noted that **coating is critical component of the fibre. It determines bending sensitivity, abrasion resistance, static fatigue protection and many other properties of the fibre.**

8.48. APPLICATIONS OF OPTICAL FIBRES

1. Military Applications. Optical fibres are finding lot of applications in various warfares and military operations.

If conventional copper cables are to be transported by an aircraft, a ship or a tank for control mechanisms or communication, then lot of weight has to be carried. However, with fibre optics, the weight is enormously reduced and also communication network is greatly improved, as with fibre optics; secrecy is also maintained.

During recent wars, fibre guided missiles have been extensively used. Sensores are mounted on missiles to collect the video information. This information is passed on to the ground control van which sends further commands to the missiles through sensors. Control van constantly monitors the course of missiles and corrects the direction of the missiles to hit the target precisely.

2. Optical fibre Sensors. In this case the basic characteristics of the optical fibres are used, viz, the variation of refractive index of the fibre under the action of the external forces may be utilized for using the optical fibre as **transducer.**

In case the fibre is subjected to heat, the refractive index changes because of change of temperature. As temperature increases, the difference of refractive index between the core and cladding is reduced, **thereby some light leaks into cladding**. This characteristics can be used for measurement of temperature using a LED as light source. Such thermometer can be used for measuring temperatures in the range of 80°C to 700°C.

Also fibre optics may be use to measure **pollution and foreign suspended particles** in the air. A beam of light is sent from one end of fibre and in case foreign particles are present, they scatter the light and measurement of variation of intensity of light will account of the extent of the presence of foreign particles.

Fibre optics may also be used to measure the level of a liquid in a container. A part of fibre is suspended in the liquid and light is directed to pass through it and its intensity is measured. A bare core loses more light when it is immersed in liquid than when it is placed in air. A sudden change in intensity of light will indicate the liquid level. A LED source, a photodector and a MMF is used to monitor the liquid level and is finding many applications in the filling of petroleum tanks.

In addition, fibre optics can be used in the fabrication of various sensors using the phenomenon of **double refraction and birefringes.** Due to the application of external forces, the optical properties of the fibre are varied. For example, we can have a **pressure sensor** using the induced birefringes,

current sensor using the Faradays rotation, **mag. sensor** using the variation of magnetic fields and **voltage sensors** to account for the variation of voltages.

3. Entertainment Applications. A coherent optical fibre bundle can greatly enhance the size of the image on a TV screen. In case we have to use the conventional method of getting larger images then the projection system becomes quick bulky and unmanageable.

4. Medical Applications. Fibre optics are finding large number of applications in medical fields. Many parts of the human body which are otherwise inaccessible can be studied by illumination with fibre optics. A bundle of MMF is used to illuminate a part of human organ and the other part to collect the reflected light. This *fibrescopic* technique is employed for endoscopic applications. Fibre optics can be employed to attach a detached retina or rectify other eye defects using lasers.

In heart problems also **lasers angioplasty** is replaced the **balloon angioplasty** and bypass surgery. A special catheter is used which consists of three channels, viz. (*i*) to visualize the various viens (*ii*) to supply power for lasers (*iii*) an open tube to collect and suck the debris and other blocking materials.

5. In Communications. A fibre optic communication system is similar to the **traditional communication** system with the exception that this system is highly efficient.

A block diagram of the typical communication system is shown in Fig. 8.36. A transducer changes a non-electrical message to an electrical signals and in case of fibre communication **light waves act as carrier waves.** Light waves are generally generated by **a laser diode or a light emitting diode (LED)** and the intensity of light given by these sources is **proportional to the current passing through them.**

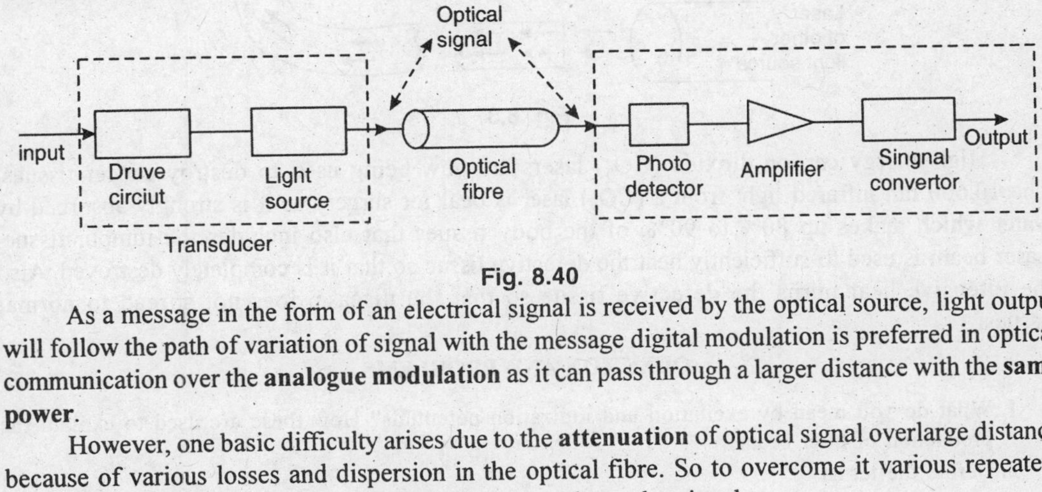

Fig. 8.40

As a message in the form of an electrical signal is received by the optical source, light output will follow the path of variation of signal with the message digital modulation is preferred in optical communication over the **analogue modulation** as it can pass through a larger distance with the **same power**.

However, one basic difficulty arises due to the **attenuation** of optical signal overlarge distance because of various losses and dispersion in the optical fibre. So to overcome it various repeaters are used in the transmission line to amplify and reshape the signal.

At the receiver end an optical coupler is used to direct the light coming from the fibre into a **photoductor** or, which converts light into **electric current**, which will contain the message fed into the optical fibre. In the next step, the message is extracted from this current and is amplified if the nearby. Then this message is put into a transducer to get it either in video or radio form depending upon the nature of the **original message**.

8.49. USE OF FIBRE OPTICS AND LASER IN MEDICAL (ENDOSCOPY)

Every doctor wants to see through the body of a patient without cutting any part of the body and then perform the surgery. This dream of the doctors has come true with the lasers and fibre optics.

For diagnostic purpose, an endoscopy (meaning to see within) is extensively used. An endoscope consists of a bunch of optical fibres which carries light to the inside of body and then transfer an image of the minor parts of the body to be viewed on the screen by the doctor. Endoscopy can better be done using laser beams because of the coherence characteristics.

Endoscope can be inserted in the body through the mouth or rectum and moved along any part of the body, where a particular defects has to be studied by a doctor.

Pulsed ruby lasers have been extensively used to the detached retina as the laser beam passes through transparent outer parts of the eye without causing any damage and it is then focussed on the defective part of the retina.

A burst of photons then generate sufficient energy to 'weld' the retina to the choroid from which it has become detached.

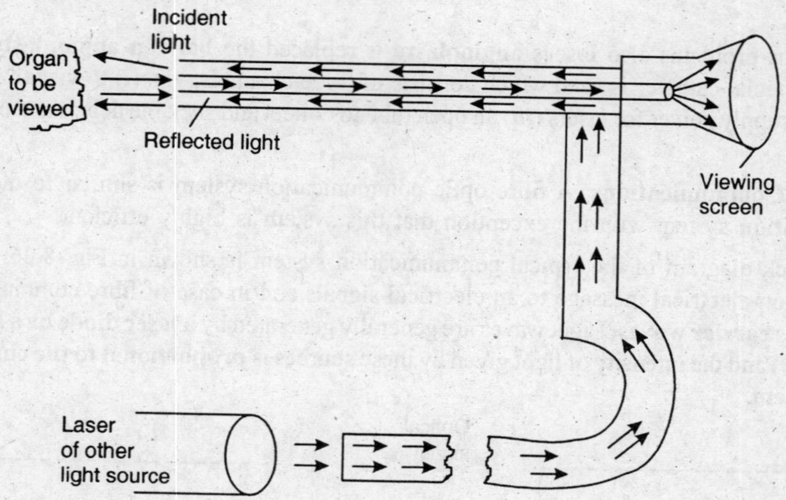

Fig. 8.37

High energy carbon dioxide (CO_2) lasers are now being used to destroy cancer tissues. The 10,600 nm infrared light from a (CO_2) laser is deal for surgery as it is strongly absorbed by water which makes up 80% to 90%, of the body tissues that also includes the tumour tissue. Laser beam is used to sufficiently heat the defective tissue so that it is completely destroyed. Also the intensive heat burns the defective tissue so that the disease does not spread to normal tissues.

QUESTION AND PROBLEMS

1. What do you mean by excitation and ionization potentials? How these are used to explain the production of laser beams?
2. Define the terms:
   (*i*) Induced absorption (*ii*) Spontaneous emission
   (*iii*) Induced emission.
3. How induced emission explain the lasers? Give Frank-Hertz experiment to explain the various energy states of atoms.
4. Explain the terms:
   (*i*) Life time of an energy states (*ii*) Metastable state
   (*iii*) Optical pumping (*iv*) Population inversion.
5. Explain the production of lasers by Ruby crystal.
6. Explain how lasers can be produced by Helium-Neon gas?
7. Give the various uses of lasers in medical, engineering and scientific fields.
8. What does masers stand for? Give their field of applications.

9. Give the principle of optical fibres and give the importance of the terms : Acceptance Angle, Numerical Aperture, V-number, Cone of Acceptance used in optical fibres.
10. How the light gets attenuated in an optical fibre? Give an expression for fibre attenuation in terms of db/km.
11. Give the important components of a fibre optic system with specific use of each component.
12. Give the reasons of attenuation and distortion of light through the optical fibre.
13. How optical fibres can be used as sensors?
14. What are the main laser sources used for optical fibre system?
15. Define V-number as used in optical fibres and give its importance.
16. How optical fibres can be used in medical and communication fields?
17. Calculate the ratio of spontaneous emission to stimulated emission if the wavelength of radiation emission is 5500Å at 2000 K (Given $h = 6.602 \times 10^{-34}$ JS and K = 1.38×10^{-23} SI units). (**Ans.** 4.62×10^5)
18. Find the coherence length of He-Ne line (1150Å) if coherence time is 26.7 ns. (**Ans.** 8 m)
19. Find the intensity of laser beam of 20 mW power having a diameter of 0.13 cm. Consider the intensity to be uniform across the beam. (**Ans.** 1.5 kW/m^2)
20. The ratio of population of two energy levels out of which upper one corresponds to metastable state is 1.059×10^{-30}. Calculate the wavelength of light emitted at 330 K. (**Ans.** 6320 Å)
21. In a He-Ne laser beam, the two plane mirrors form the resonant cavity are at a distance 0.5 m. Find the mode separation of longitudinal cavity in terms of frequency. (**Ans.** 3×10^8 Hz)
22. Calculate the energy difference in eV between two energy levels of Ne atoms of He-Ne gas laser, the transitions between which results in the emission of a light of wavelength 6328 Å. Also calculate the number of photons emitted/sec if optical power output is 1 mW. (**Ans.** 3.182×10^{15})
23. Calculate Δ for a step index fibre having core and cladding refractive indices 1.48 and 1.46 respectively. [**Ans.** 1.351%]
24. Calculate the number of modes in the graded index fibre at an operating wavelength of 850 nm. [**Ans.** 504 modes]
25. Calculate the NA, acceptance angle and the critical angle of a fibre having μ1 (core) = 1.50 and μ_2 (cladding) = 1.45. [**Ans.** (*i*) 0.387; (*ii*) 22.78°; (*iii*) 75.2°]
26. At a distance of 1 km, spot size = 9800 cm; calculate the cut-off parameter and also the number of modes for a fibre of core diameter = 50 μm. μ1 = 1.47 (core), μ = 1.45 (cladding) and operating wavelength = 8500 Å. [**Ans.** 996 modes]
27. The core of a glass fibre has a refractive index of 1.5 while its cladding is doped to give a fractional refractive index change of 0.005.

    Find. (*a*) Reflective index of cladding
    (*b*) Critical internal reflecting angle
    (*c*) Acceptance angle
    (*d*) Numerical Aperture [**Ans.** (*a*) 1.4925 (*b*) 84.26 (*c*) 8.62° (*d*) 1.498]
28. Find the N.A. and acceptance angle for an optical fibre having core and cladding refractive indices 1.45 and 1.40 respectively [**Ans.** 0.377; 22.15°]
29. Find the core radius necessary for a single mode fibre operating at 850 nm is a step index fibre. The core and cladding refractive indices being 1.48 and 1.47 respectively. [**Ans.** 1.91 μm]
30. A step index single mode fibre has a core radius of 30 μm ; $n_1 = 1.5$ and $n_2 = 1.47$. If the operating wavelength is 800 nm, find the number of modes propagating through the fibre. [**Ans.** 2474]
31. A step index single mode fibre has a length of 5 km. Ratio of output to input power is 0.95. Find the alternation coefficient. [**Ans.** 0.045 dB/km]
32. A step index fibre has a core of refractive index 1.3725 and cladding of refractive index 1.3274. If the cut off parameter is 2.405, find the N.A. and radius of core. [**Ans.** 0.349; 1.43 μm]

❑❑❑

CHAPTER

9

ELECTROMAGNETIC WAVES

9.1. INTRODUCTION

Maxwell's equations bear the same relation to electromagnetism that Newton's laws of motion do to *mechanics*. There is, however, an important difference. Einstein presented his special theory of relativity in 1905, about 200 years after Newton's laws were predicted and about 40 years after Maxwell's equation. However Newton's laws had to be drastically modified for cases in which relative speeds approached the speed of light while no changes were required for Maxwell's equations. Maxwell's equations were totally consistent with special theory of relativity. In fact, Einstein's theory grew out of the deep and careful study of Maxwell's e.m. equations.

Maxwell's equations predict the existence of e.m. waves (travelling electric and magnetic fields) propagating in free space with the speed of light. These equations show how the e.m. waves are radiated by accelerating charges. Some of the important properties of e.m. wave are gives as under:

1. Electromagnetic waves are propagated in the form of electric and magnetic fields such that both the fields are at right angles to each other and also to the direction of propagation of waves. It means that the e.m. waves are transverse in nature.
2. In vacuum, the e.m. waves move with a definite velocity given by
$$c = \frac{1}{(\mu_0\ \varepsilon_0)^{1/2}}$$
which is the velocity of light in free space.
3. As the waves spread out in free space, the amplitude of the electric and magnetic fields viz. $\vec{E}$ and $\vec{B}$ decreases steadily. $\vec{E}$ and $\vec{B}$ in steady state are related by $c = \frac{\vec{E}}{\vec{B}}$.
4. When two desturbances are superimposed, they observe the laws of superposition *i.e.*, at any instant, the resultant electric field or magnetic field is the vector sum of the *individual field*.
5. E.M. waves of different frequencies can exist. However, all such waves would be propagated with the speed of light.
6. Unlike other types of waves *i.e.*, water waves, sound waves, stretched string waves etc., no material particle moves with the e.m. waves. The only changes are in the field intensities.
7. Wiener's experiment on stationary waves have shown that it is the electric vector which *plays the role of light vector*.

As is clear from the discussions on wave propagation in medium (e.g. string, rod or a transmission line), a wave carries energy and the physical medium is visualized as a continuum of coupled oscillator. We know that a simple oscillator is characterised by three parameters, two of them being responsible for string and exchanging energy (inertia and elasticity in case of mechanical medium while L and C in an electrical circuit. The third parameter governs the mechanism of energy dissipation (friction in a mechanical oscillator and resistor in the case of an electrical oscillator), propagation of e.m. waves, however, is constituted by the periodic variation of *electric and magnetic fields*. Though, these are separate, yet they are closely interrelated. These vector fields can exist in empty space and no medium is *compressed or rarefied* during their propagation. The *inertia* or the *inductive* property of the medium is provided by the *permeability* μ, the *elasticity or capacity parameters* by the *permittivity* ε. The medium stores the magnetic energy through permeability and electric energy through permitivity. A dielectric medium has finite magnitude of μ and ε and its conductivity $\sigma = 0$. However, in the case of a conductor, all the three parameters have to be accounted for to describe the e.m. wave.

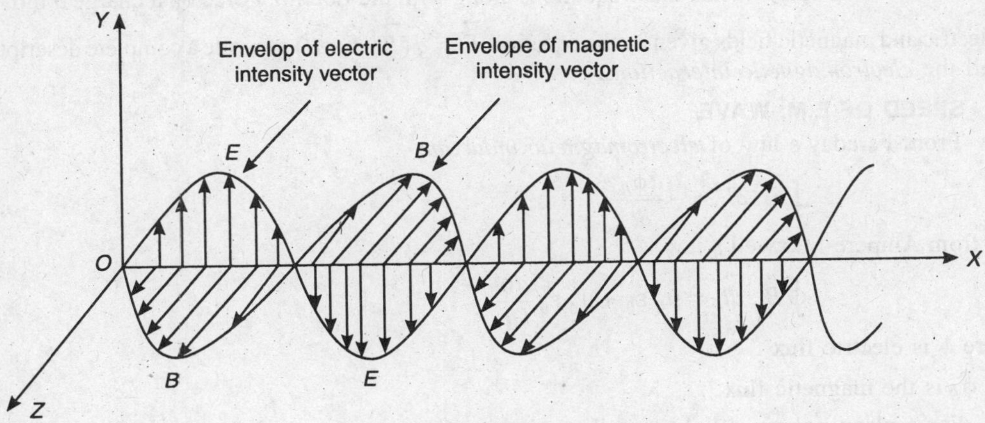

Fig. 9.1

9.2. IMPORTANCE OF MAXWELL'S EQUATIONS

1. Maxwell's first equation is the statement of Gauss's law in electrostatics i.e.

$$\oint \vec{E} \cdot \vec{ds} = \frac{q}{\varepsilon_0} \quad \text{...(1)}$$

which states that the total electric flux through any closed surface equals the net charge inside that surface divided by ε_0. This law gives the relation between *electric field and charge distribution*. Electric field lines start from positive charges and end at negative charges. In fact, Gauss's law is equivalent to Coulomb's inverse square law in electrostatics which has been found to be experimentally true thereby confirming the validity of the equation.

2. Maxwell's second equation describes Gauss's law of electromagnetism *i.e.*, $\oint \vec{B} \cdot \vec{ds} = 0$. It states that the net magnetic flux through a closed surface is zero. It means that the number of magnetic lines entering a closed surface is equal to the number of field lines leaving the surface. It also implies that isolated monopoles do not exit. Since isolated magnetic monopoles are not observed in nature so this equation also stands confirmed.

3. The Maxwell's third equation describes Faraday's law of electromagnetic induction *i.e.*,

$$\oint \vec{B} \cdot \vec{dl} = -\frac{d\,\phi_B}{d\,t} \quad \text{...(3)}$$

This equation relates on electric field with changing magnetic flux (ϕ_B). According to this law, the line integral of electric field around any closed path (*i.e.* e.m.f.) is equal to the rate of change of magnetic flux through any surface bounded by that. So Equation (3) may be written as

$$e = -\frac{d\,\phi_B}{d\,t}$$

which is Faraday's law of electromagnetic induction. This law has been verified experimentally thereby confirming the validity of the Maxwell's third equation.

4. Maxwell's fourth equation is the statement of Ampere-Maxwell's law *i.e.*,

$$\oint \vec{B} \cdot \vec{dl} = \mu_0\, \varepsilon_0 + \varepsilon_0\, \mu_0 \frac{a\varphi}{dt} \quad \text{...(4)}$$

This law states that the line integral of the magnetic field around any closed path is determined by the sum of the net conduction current (I) through that path and the rate of change of flux (ϕ) through any surface bounded by the path. This law describes how a magnetic field can he produced by both changing electric flux and a conduction current.

A very important consequence of Maxwell's equations is that these can be used to derive law of conservation of charge. Also these equations along with the Lorentz Force on a charge q moving in electric and magnetic fields given by the equation $\vec{F} = q(\vec{E} + v \times \vec{B})$ provide a complete description of all the *electromagnetic interactions.*

9.3. SPEED OF E.M. WAVE

From Faraday's law of *electromagnetic induction*

$$\oint \vec{E} \cdot \vec{dl} = -\frac{d\phi_B}{dt} \quad ...(1)$$

and from Ampere-Maxwell's law

$$\oint \vec{B} \cdot \vec{dl} = \mu_0 \varepsilon_0 + \mu_0 \varepsilon_0 \frac{d\phi}{dt} \quad ...(2)$$

where ϕ is electric flux

and ϕ_B is the magnetic flux.

For a plane polarized light travelling along x-direction equations (1) and (2) give

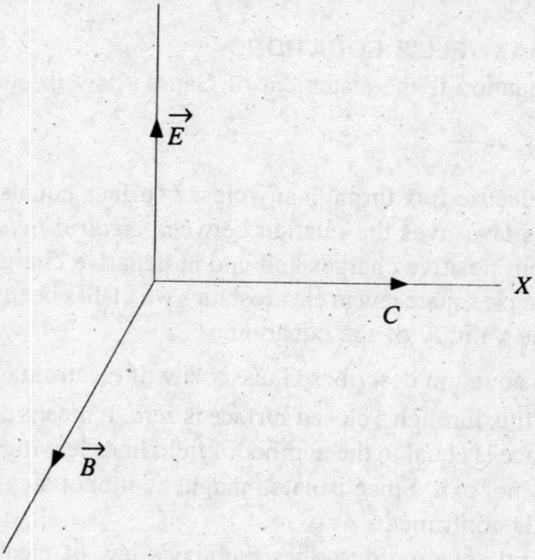

Fig. 9.2

or $$\frac{\partial \vec{E}}{\partial x} = -\frac{\partial \vec{B}}{\partial t} \quad ...(3)$$

and $$\frac{\partial \vec{B_z}}{\partial x} = -\mu_0 \varepsilon_0 \frac{\partial \vec{E_y}}{dt}$$

or $$\frac{\partial \vec{B}}{\partial x} = -\mu_0 \varepsilon_0 \frac{\partial \vec{E}}{\partial t} \quad ...(4)$$

where we have dropped the subscript of components E_y and B_z for simplification.

Taking partial derivative of Equation (3), we have

$$\frac{\partial}{\partial x}\left(\frac{\partial \vec{E}}{\partial x}\right) = -\frac{\partial}{\partial x}\left(\frac{\partial \vec{B}}{\partial t}\right) = -\frac{\partial}{\partial t}\left(\frac{\partial \vec{B}}{\partial x}\right)$$

$$\therefore \quad \frac{\partial^2 \vec{E}}{\partial x^2} = -\frac{\partial}{\partial t}\left(-\mu_0 \varepsilon_0 \frac{\partial \vec{E}}{\partial t}\right) = \mu_0 \varepsilon_0 \frac{\partial^2 \vec{E}}{\partial t^2} \quad ...(5)$$

Similarly $$\frac{\partial^2 \vec{B}}{\partial x^2} = \mu_0 \varepsilon_0 \frac{\partial^2 \vec{B}}{\partial t^2}$$

Also we have the general wave equation

$$\frac{\partial^2 f}{\partial x^2} = \frac{1}{v^2}\frac{\partial^2 f}{\partial t^2}$$

where v is the speed of wave and f is the wave amplitude, we get,

$$\frac{1}{v^2} = \mu_0 \varepsilon_0$$

or $$v = \frac{1}{\sqrt{\mu_0 \varepsilon_0}}$$

Now $$\mu_0 = 4\pi \times 10^7 \text{ Wb A}^{-1} \text{ m}^{-1}$$

and $$\varepsilon_0 = 8.85 \times 10^{-12} \text{ C}^2 \text{ N}^{-1} \text{ m}^{-2}$$

$$\therefore \quad v = 3 \times 10^8 \text{ ms}^{-1}$$

which is the speed of light. So we can replace v by c and write the speed of e.m. wave as

$$c = \frac{1}{\sqrt{\mu_0 \varepsilon_0}}. \quad ...(7)$$

9.4. RATIO OF ELECTRIC AND MAGNETIC FIELDS

The sinusoidally varying electric and magnetic fields of a plane e.m. wave are given as

$$\vec{E} = \vec{E}_0 \cos (kx - \omega t) \quad ...(i)$$

$$\vec{B} = \vec{B}_0 \cos (kx - \omega t) \quad ...(ii)$$

where E_0 and B_0 are the maximum values of the electric and magnetic fields associated with the e.m. waves. Also k, the propagation constant $= \frac{2\pi}{\lambda}$, where λ is the wavelength of e.m. wave and ω $(2\pi v)$ is angular frequency of the wave.

$$\therefore \quad \frac{\omega}{k} = \frac{2\pi v}{2\pi/\lambda} = v\lambda = c \quad ...(iii)$$

Taking partial differentiation of Equations (*i*) and (*ii*) we have

$$\frac{\partial \vec{E}}{\partial x} = -k\vec{E}_0 \sin (kx - \omega t)$$

and $$\frac{\partial \vec{B}}{\partial t} = \omega \vec{B}_0 \sin (kx - \omega t)$$

Now since $$\frac{\partial \vec{E}}{\partial x} = -\frac{\partial \vec{B}}{\partial t} \quad \text{[From Faraday's law]}$$

$$\therefore \; -k\vec{E}_0 \sin (kx - \omega t) = -\omega \vec{B}_0 \sin (kx - \omega t)$$

or $$k\vec{E}_0 = \omega \vec{B}_0$$

or $$\frac{E_0}{B_0} = c \quad \text{[From Equation (iii)]}$$

$$\therefore \quad \frac{E_0}{B_0} = c \qquad ...(iv)$$

From Equations (*i*) and (*ii*)

$$\frac{E}{B} = \frac{E_0}{B_0} = c$$

So at every instant the *ratio of the electric field to the magnetic field of an e.m. wave is equal to the speed of light.*

9.5. POYNTING VECTOR

E.M. waves carry energy with them, when these waves pass through space, they transfer energy to the objects in their path. The rate of flow of energy in an e.m.f. wave is given by a vector $\vec{S}$, called Poynting vector, (named after J.H. Poynting). It is given by

$$\vec{S} = \frac{\vec{E} \times \vec{B}}{\mu_0} \qquad ...(1)$$

For a plane e.m. wave, $\vec{E} \perp \vec{B}$

$$\therefore \quad S = \frac{EB}{\mu_0} \qquad ...(2)$$

Units of *S*

$$[S] = |E|\,|H| = [\text{Vm}^{-1}][\text{Am}^{-1}] = \text{VAM}^{-2} = \text{Wm}^{-2} = \text{Js}^{-1}\,\text{m}^{-2}$$

So units of S are J/sec/m^2

Also $\quad c = \dfrac{E}{B}$

$$\therefore \quad S = \frac{E^2}{\mu_0 c} = \frac{cB^2}{\mu_0} \qquad ...(3)$$

We have the equation for electric and magnetic fields as

$$\vec{E} = \vec{E_0} \cos (kx - \omega t)$$

and

$$\vec{B} = \vec{B_0} \cos (kx - \omega t)$$

$$\therefore \quad S = \left[\frac{E_0 B_0}{\mu_0}\right] \cos^2 (kx - \omega t)$$

If S_{av} is the average value of S taken over a complete cycle,

$$S_{av} = \frac{1}{T}\int_0^T S\,.dt = \frac{1}{T}\int_0^T \frac{E_0 B_0}{\mu_0} \cos^2 (kx - \omega t)\, dt$$

$$= \frac{1}{T} \times \frac{E_0 B_0}{\mu_0} \times \frac{T}{2} \qquad \left[\because \int_0^T \cos^2 (kx - \omega t)\, dt = \frac{T}{2}\right]$$

$$= \frac{E_0 B_0}{2\mu_0} = \frac{E_0^2}{2\mu_0 c} = \frac{cB_0^2}{2\mu_0} \qquad \left[\because \frac{E_0}{B_0} = c\right] \qquad ...(4)$$

where S_{av} is called the wave intensity. The instantaneous energy density associated with an electric field is given by

$$U_e = \frac{1}{2}\varepsilon_0 E^2 \qquad ...(5)$$

Because Energy Stored in magnetic field

$$= \frac{1}{2}LI^2 = \frac{1}{2}\left[\frac{\mu_0 N^2 A}{l}\right]\left[\frac{Bl}{\mu_0 N}\right]^2 = \frac{1}{2}\frac{B^2}{\mu_0} A.l.$$

$\therefore$ Energy Stored/unit volume $= \dfrac{1}{2}\dfrac{B^2}{\mu_0}$.

$$= \frac{1}{2}\frac{E^2}{c^2}\cdot\frac{1}{\mu_0} \qquad \left[\because \frac{E}{B} = c.\right]$$

$$= \frac{1}{2}\epsilon_0 E^2 \qquad \left[\because c = \frac{1}{\sqrt{\mu_0 \epsilon_0}}\right]$$

and instantaneous energy density associated with a magnetic field is given by

$$U_m = \frac{B^2}{2\mu_0}$$

$$\therefore \quad U_m = \frac{(E/c)^2}{2\mu_0} = \frac{E^2/c^2}{2\mu_0} = \frac{E^2}{2\mu_0}\times(\mu_0\,\varepsilon_0)$$

$$= \frac{1}{2}\,\varepsilon_0\,E^2 = U_e \qquad \left[\because \frac{E}{B} = c \text{ and } c = \frac{1}{\sqrt{\varepsilon_0\,\mu_0}}\right] \qquad ...(6)$$

which means that for an e.m. wave, *the instantaneous energy density associated with the magnetic field is equal to the instantaneous energy density associated with the electric field.*

Which is other words mean that in a given volume, the energy is shared equally by two fields.

Relation between wave Intensity and total Energy density

If U be the total energy density,

$$\text{Then} \qquad U = U_e + U_m = \varepsilon_0 E^2 = \frac{B^2}{\mu_0} \qquad ...(7)$$

If U_{av} is the total energy density (*i.e.*, U_{av} is the average value of U taken over a complete cycle), then

$$U_{av} = \frac{1}{T}\int_0^T U\cdot dt = \frac{1}{T}\int_0^T \varepsilon_0 E^2\,dt$$

$$= \frac{\varepsilon_0}{T}\int_0^T E_0^2\cos^2(kx-\omega t)\,dt$$

$$= \frac{\varepsilon_0}{T}\cdot E_0^2\cdot\frac{T}{2} \qquad \left[\int_0^T \cos^2(kx-\omega t)\,dt = \frac{T}{2}\right]$$

$$\therefore \quad U_{av} = \frac{1}{2}\varepsilon_0 E^2$$

$$\text{Also} \qquad \frac{E_0}{B_0} = c$$

$$\therefore \quad U_{av} = \frac{B_0^2}{2\mu_0}$$

$$\therefore \quad U_{av} = \frac{1}{2}\varepsilon_0 E^2 = \frac{1}{2}\frac{B_0^2}{\mu_0}$$

From Equation (4)

$$S_{av} = c\cdot\frac{1}{2}\frac{B_0^2}{\mu_0}$$

$$\therefore \quad S_{av} = c\cdot U_{av}$$

which means that the intensity of an e.m. wave is equal to the product of average energy density with the speed of light.

9.6. INTERPRETATION OF DIVERGENCE AND CURL OF A VECTOR POINT

Divergence. *Divergence of a continuously differentiable vector point function F is denoted by div F and is defined by the equation.*

$$\text{Div } \vec{F} = \vec{\nabla}\cdot\vec{F} = \hat{i}\frac{\partial F}{\partial x} + \hat{j}\frac{\partial F}{\partial y} + \hat{k}\frac{\partial F}{\partial z}$$

If we consider $\vec{F} = f\hat{i} + \phi\hat{j} + \psi\hat{k}$

then
$$\text{div}\,\vec{F} = \vec{\nabla}\cdot\vec{F} = \left(\hat{i}\frac{\partial}{\partial x} + \hat{j}\frac{\partial}{\partial y} + \hat{k}\frac{\partial}{\partial z}\right)(f\hat{i}\,\phi\hat{j} + \psi\hat{k})$$
$$= \frac{\partial f}{\partial x} + \frac{\partial \phi}{\partial y} + \frac{\partial \psi}{\partial z}$$

Curl. The curl of a continuously differentiable vector point function F is defined by the equation

$$\text{Curl } F = \vec{\nabla}\times\vec{F} = \hat{i}\frac{\partial F}{\partial x} + \hat{j}\frac{\partial F}{\partial y} + \hat{k}\frac{\partial F}{\partial z}$$

if $\vec{F} = f\hat{i} + \phi\hat{j} + \psi\hat{k}$

then
$$\text{curl } \vec{F} = \vec{\nabla}\times\vec{F}$$
$$= \left(\hat{i}\frac{\partial}{\partial x} + \hat{j}\frac{\partial}{\partial y} + \hat{k}\frac{\partial}{\partial z}\right)(f\hat{i} + \phi\hat{j} + \psi\hat{k})$$
$$= \begin{vmatrix} \hat{i} & \hat{j} & \hat{k} \\ \frac{\partial}{\partial x} & \frac{\partial}{\partial y} & \frac{\partial}{\partial z} \\ f & \phi & \psi \end{vmatrix}$$
$$= \hat{i}\left(\frac{\partial y}{\partial y} - \frac{\partial \phi}{\partial z}\right) + \hat{j}\left(\frac{\partial f}{\partial z} - \frac{\partial \phi}{\partial x}\right) + \hat{k}\left(\frac{\partial \phi}{\partial z} - \frac{\partial f}{\partial y}\right)$$

9.7. PHYSICAL INTERPRETATION OF DIVERGENCE

Consider the motion of a fluid having velocity $\vec{U} = U_x\hat{i} + U_y\hat{j} + U_z\hat{k}$ at a point P (x, y, z). Consider a small parallelopiped with edges S_x, S_y, S_z parallel to the axes in the mass of the fluid, with one of it corners at p,

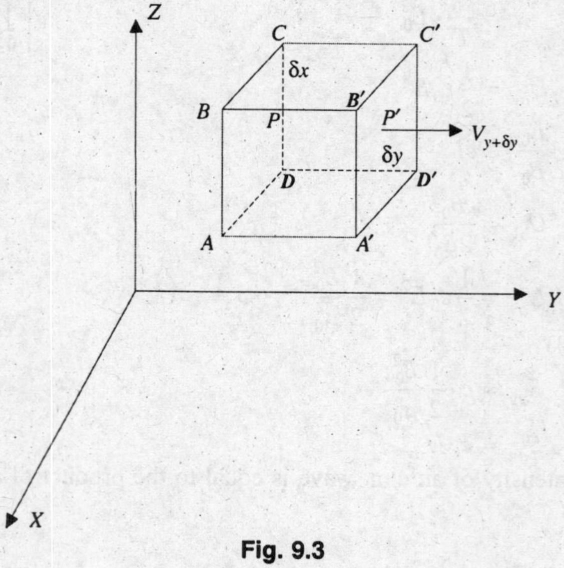

Fig. 9.3

So the amount of fluid entering the face P per unit time

$$= U_y \cdot \delta_z \cdot \delta_x$$

and the amount of fluid leaving the face P' in unit time

$$= U_{y+\delta y}\, dz,\, dx$$

Expanding by Taylor's theorem $= \left(U_y + \frac{\partial U_y}{\partial y} \cdot \delta y\right) \cdot \delta z \cdot \delta x$

So the net decrease in the amount of fluid due to the flow across these two faces

$$= \frac{\partial U_y}{\partial y} \cdot \delta x \cdot \delta y \cdot \delta z$$

Going an the same lines, we have the contribution due to the other two pairs of faces. So the total decrease of amount of fluid inside the parallelopiped per unit time

$$= \left(\frac{\partial U_x}{\partial x} + \frac{\partial U_y}{\partial_y} + \frac{\partial U_z}{\partial_z}\right) \cdot \delta x \cdot \delta y \cdot \delta z$$

Hence the rate of loss of fluid/unit volume

$$= \frac{\partial U_x}{\partial x} + \frac{\partial U_y}{\partial y} + \frac{\partial U_z}{\partial z} = \text{div}\,\vec{V}$$

i.e., div V gives the rate at which the fluid is originating per unit Volume.

In the same way if V represents an electric flux, div V as the amount of flux which diverges per unit volume per unit time. Again if V represents heat flux, div V will represent the amount of heat issuing from a point per unit volume per unit time.

In general, divergence of a vector point function representing any physical quantity gives at each point the rate/units volume at which the physical quantity is issuing from the point.

It is due to this fact that it is called divergence of a vector point function.

If a fluid is incompressible, there will not be any gain or loss in volume of the element. In that case div. $V = 0$. This in hydrodynamics is called equation of continuity for an incompressible field.

If the flux entering any element of space is the same as leaving it *i.e.*, div. $V = 0$ every where, then such a function is called a solenoidal vector function.

9.8. PHYSICAL INTERPRETATION OF CURL

Consider the motion of a rigid body rotating about a fixed axis through O. If Ω be the angular velocity then the velocity $\vec{V}$ of any particle P (R) of the body is given by

$$\vec{V} = \vec{\Omega} \times \vec{R}$$

if $\vec{\Omega} = \hat{i}\,\omega_1 + \hat{j}\,\omega_2 + \hat{k}\,\omega_3$

and $\vec{R} = \hat{i}\,x + \hat{j}\,y + \hat{k}\,z$

then
$$\vec{V} = \vec{\Omega} \times \vec{R} = \begin{vmatrix} \hat{i} & \hat{j} & \hat{k} \\ \omega_1 & \omega_2 & \omega_3 \\ x & y & z \end{vmatrix}$$

$$= \hat{i}(\omega_2 z - \omega_3 y) + \hat{j}(\omega_3 x - \omega_1 z) + \hat{k}(\omega_1 y - \omega_2 x)$$

$$\text{Curl}\;\vec{V} = \begin{vmatrix} \hat{i} & \hat{j} & \hat{k} \\ \frac{\partial}{\partial x} & \frac{\partial}{\partial y} & \frac{\partial}{\partial z} \\ \omega_2 z - \omega_3 y & \omega_3 x - \omega_1 z & \omega_1 y - \omega_2 x \end{vmatrix}$$

$$= \hat{i}(\omega_1 + \omega_1) + \hat{j}(\omega_2 + \omega_2) + \hat{k}(\omega_3 + \omega_3)$$

$$= 2\omega_1 \hat{i} + 2\omega_2 \hat{j} + 2\omega_3 \hat{k}$$

$$= 2(\omega_1\hat{i} + \omega_2\hat{j} + \omega_3\hat{k})$$
$$= 2\vec{\Omega}$$
$$\vec{\Omega} = \frac{1}{2}\text{ curl }\vec{V}$$

So the angular velocity of rotation at any point is equal to half the curl of velocity vector. This gives the justification of using *rotation* for the curl of a vector.

In general the *curl of any vector point function gives the measure of the angular velocity at any point of the vector field.*

Also any motion in which the curl of the velocity vector is not zero is said to be rotational, otherwise it is irrotational.

9.9. DIFFERENTIAL FORM OF MAXWELL EQUATIONS

As described above, Maxwell's equations are a set of four equations. We can have the differential form of the above equations in terms of spatial derivatives (curl or divergence) of the vector fields $\vec{E}$ and $\vec{B}$ and get a relation with their sources viz charge density ρ and current density $\vec{J}$ and express the experimental laws of *Guass, Ampere and Faraday.*

Maxwell equations in SI system for the fields in vacuum are summarized as below, along with the laws which can be expressed from them.

	Equation	*Corresponding law*
(*i*)	$\nabla \cdot \vec{D} = \rho$	(1) Gauss's law
(*ii*)	$\nabla \cdot \vec{B} = 0$	(2) Absence of isolated magnetic fields
(*iii*)	$\nabla \times \vec{E} = -\dfrac{\partial \vec{B}}{\partial t}$	(3) Faraday's law
(*iv*)	$\vec{\nabla} \times \vec{H} = \vec{J} + \dfrac{\partial \vec{D}}{\partial t}$	(4) Ampere's circuital law.

where ρ and $\vec{J}$ are charge and current densities and constitute the sources for the field vector $\vec{E}$ and $\vec{B}$. The *microscopic vector* E and $\vec{B}$, on the introduction of a material medium becomes *macroscopic* vectors viz the electric displacement vector $\vec{D}$ and magnetic field $\vec{H}$ given by the relation;

$$\vec{D} = \varepsilon_r \cdot \varepsilon_0 \vec{E} = \varepsilon \vec{E} \qquad ...(5)$$

and
$$\vec{B} = \mu_r \mu_0 \vec{H} = \mu \vec{H} \qquad ...(6)$$

where ε_0 and ε_r are the permittivity of vacuum and relative permittivity of the dielectric medium, while ε corresponds to the absolute permittivity of the medium.

Similar comments can be applied for μ.

The conduction current density $\vec{J}$ is related to vector $\vec{E}$ though *Ohm's law* having the form

$$\vec{J} = \sigma \vec{E} \qquad ...(7)$$

where σ is the conductivity of the medium.

The importance of the above equations lie from *Helmholtz's theorem of vector analysis*. The field vectors $\vec{E}$ and $\vec{B}$ are cross coupled with each other, which in other words mean that *each is generated by the time variation of the other*. From Helmholtz's theorem, vector field is known completely if its curl and *divergence are given.*

As may be observed, the Equations (1) and (2) are steady state equations while Equations (3) and (4) are time varying and together, they explain all the known properties of $\vec{E}$ and $\vec{B}$.

Maxwell showed that with some manipulation, these equations lead to wave equation. He again showed that these waves could propagate in free space with wave velocity which could be deduced from *electrical measurements* which has no connection with the wave phenomenon. The value thus calculated is the same as the value of velocity of light in vacuum. This establishes *the electromagnetic nature of light*. The velocity of light c in vacuum is the maximum limit of velocity of propagation of interactions and is called *signal velocity*.

9.10. HISTORICAL BACKGROUND

In the early day, electricity and magnetism were considered to be totally separate. However *Oersted* in 1820, noted that an *electron current* could *deflect* a magnetic needle and later on Ampere showed that all magnetic phenomenan are due to *moving electric charges*.

In 1831, Faraday discovered that moving magnet can generate electric current. Thus *time varying electric field* may produce *magnetic field* and vice-versa. *Maxwell* and *Lorentz* gave the idea that electricity and magnetism are inter-related and the effect of combined electric and magnetic field was called *electromagnetism*.

For understanding the four basic Maxwell equations, we will first review the basic laws of *Electricity and Magnetism* which deal with *electrostatics, magnetostatics, varying electric and magnetic fields* and see how Maxwell's equations can be expressed by the four basic laws of electromagnetism.

***(i)* Gauss's law for electric fields:** According to this law the net outward electric flux through a closed surface is equal to the total charge enclosed by the surface which may be written in mathematically form as

$$\int \vec{D}\cdot \vec{dS} \equiv q \qquad ...(1)$$

where $\vec{D}$ is the electric displacement vector measured in *coulomb/m²* and q, the enclosed charge measured in *coulombs*.

***(ii)* Gauss's law for Magnetic fields:** It states that the net outward magnetic flux through a closed surface is zero

$$\int \vec{B}\cdot \vec{dS} = 0 \qquad ...(2)$$

where $\vec{B}$ is the *magnetic induction* and measured in weber/m^2

***(iii)* Faraday's law of electromagnetic Induction:** According to this law the e.m.f. induced in a closed path is equal to the rate of change of magnetic flux linking the path. It may be stated mathematically as:

$$\int \vec{E}\cdot \vec{dl} = \frac{-\partial \varphi}{\partial t} \qquad ...(3)$$

where $\vec{E_a}$ is the electric field vector and measured in volts/m. The integral is over closed path (line integral) and φ corresponds to the magnetic flux linking the path.

***(iv)* Ampere's law:** It states that the line integral of the magnetic field around a closed path is equal to the current linking the path. Mathematically it may be written as

$$\int \vec{H}\cdot \vec{dl} = I \qquad ...(4)$$

where H is the magnetic field intensity and is measured in ampere/meter. The current can be expressed as the surface integral of current density $\vec{J}$. It may be written in the form

$$\int_l \vec{H}\cdot \vec{dl} = \int_S \vec{J}\cdot \vec{dS} \qquad ...(5)$$

where J is the current density and is measured in ampere/m^2.

9.11. CURRENT DENSITY AND CONTINUITY EQUATION

Let there be N charges/unit volume at a point each having a charge q and moving with a mean velocity v.

The amount of charge crossing per unit area perpendicular to mean drift velocity, per unit time may be given as

$$\vec{J} = N \cdot q v \qquad ...(6)$$

where $\vec{J}$ is the current density and is a *vector quantity*. Direction of $\vec{J}$ will depend on the velocity of the charge carriers.

Let $\vec{dS}$ be the small area element of the material. The amount of charge flowing per unit time through that area is $\vec{J}\cdot\vec{dS}$ and it corresponds to the flux of $\vec{J}$ across area $d\vec{S}$. The net charge passing through the surface in unit time is called *electric current* and is given as

$$I = \frac{dq}{dt} = \int_S \vec{J}\cdot\vec{dS}$$

Now let us consider a closed surface enclosing a volume V. If ρ is *volume density* of the charge, total charge within the volume V is $\int \rho \cdot dV$.

From the law of conservation of charge, the net outward flow of charge through this volume must be equal to the rate of decrease of total charge in the said volume *i.e.*

$$\int_S \vec{J}\cdot\vec{dS} = -\frac{d}{dt}\int_V \rho\, dv$$

As the surface s is fixed, the rate of change of charge within the volume is mainly due to the time rate of change of ρ. So we may write

$$\int_S \vec{J}\cdot\vec{dS} = -\int_V \frac{\partial \rho}{\partial t} dV$$

Using *Gauss's divergence theorem*, we can change the surface integral to volume integral as

$$\int_S \vec{J}\cdot d\vec{S} = \int_V \text{div}\,\vec{J}\, dV = -\int_V \frac{\partial \rho}{\partial t} dV$$

which may be written as

$$\int_V \left(\text{div}\,\vec{J} + \frac{\partial \rho}{\partial t}\right) dV = 0$$

As the integral must be zero for any arbitrary value of volume V, so integrand must be zero every where *i.e.*

$$\text{div}\,\vec{J} + \frac{\partial \rho}{\partial t} = 0$$

This is the mathematical expression of conservation of charge and is the continuity equation. According to the equation:

The total current flowing out of some volume should be equal to the rate of decrease of charge within the volume which is other words mean that charge can neither be *created nor be destroyed.*

In case of a steady state, the charge density at any point within the region remains constant *i.e.*

$$\frac{\partial \rho}{\partial t} = 0$$

$$\therefore \quad \text{div}\,\vec{J} = 0$$

or

$$\vec{\nabla}\cdot\vec{J} = 0$$

which means that there is no net outward flux of current density $\vec{J}$ i.e. lines of electric *current are continuous*. This equation is valid for the regions having no *source or sink* for the current:

Again experimentally, it is observed that for a given material, current dinsity in steady state is proportional to the applied field $\vec{E}$. So we may write

$$\vec{J} \propto \vec{E}$$

or

$$\vec{J} = \sigma \vec{E}$$

where is σ is called *electrical conductivity* and is constant for a given material at a constant temperature. Reciprocal of conductivity is called *resistivity* which is measured in ohm-meters.

9.12. LORENTZ FORCE

Let us consider a point charge q moving in an *electric and magnetic field.*

The force $\vec{F}$ acting on the charge in addition to the position of the charge depends on the velocity of the charge. The force $\vec{F}$ may be considered to be composed of two components

(*i*) *Electrostatic force* $\vec{F_e}$ which does not depend on the velocity of the charge.

(*ii*) *Magnetic force* $\vec{F_m}$ which depends on the velocity of the charge.

Let $\vec{F_e}$ be the intensity of the electric field at any point in the field, the electrostatic force experienced by a test charge placed in the electric field is given by

$$\vec{F_e} = q\vec{F} \qquad ...(1)$$

If the test charge moves with a velocity v through a magnetic field, it experiences a deflecting force F_m given by

$$\vec{F_m} = q(\vec{v} \times \vec{B}) \qquad ...(2)$$

where $\vec{B}$ is the *magnetic field induction* or magnetic flux density.

$\vec{B}$ is the characteristic of a given magnetic field and is defined as:-

Force acting on a unit charge moving with a unit velocity in a direction at right angles to the magnetic field.

As is clear from Equation (2) F_m is $\perp$ to both $\vec{v}$ and $\vec{B}$.

Total force on a unit charge q may be written as

$$\vec{F} = \vec{F_e} + \vec{F_m}$$

or

$$\vec{F} = q\vec{E} + q(\vec{v} \times \vec{B})$$

$$= q[\vec{E} + (\vec{v} \times \vec{B})] \qquad ...(3)$$

This is called *Lorentz force* and the above equation is called *Lorentz Force Equation.* This is a universal equal and is valid for steady as well as changing electric and magnetic fields.

Units of flux density or magnetic field induction can be obtained from Equation (2). The maximum value of force F_m is given by

$$F_m = qv\vec{B}$$

i.e. when

$$\vec{v} \times \vec{B} = \vec{v}\,\vec{B} \sin 90°$$

$$\therefore \quad \vec{B} = \frac{F_m}{v \cdot q}$$

If $F_m = 1$ N, $q = 1$ coulomb and $v = 1\ \text{ms}^{-1}$

Then $B = 1$ Tesla

Thus in MKS system unit of $\vec{B}$ ***is Tesla or weber/m²*. Tesla can also be expressed as *Newton/Ampere/meter*.**

9.13. MAGNETIC FLUX AND ELECTROMOTIVE FORCE

Magnetic flux is defined as the number of magnetic lines of force passing normally through an area. Magnetic flux is represented by φ.

Let $\vec{B}$ be the magnetic field at a point surrounded by an area element ds, the mag. flux through the area element may be given as

$$d\varphi = \vec{B} \cdot d\vec{S}$$

and magnetic flux through an area S may be written as

$$\varphi = \oiint_S \vec{B} \cdot d\vec{S}$$

ELECTROMOTIVE FORCE

If a steady current flows through a conductor, *potential difference must exist between the ends of the conductor*. This potential difference is maintained by some source of electromotive force (e.m.f.). Many types of devices supply the *e.m.f.* For example a battery, a dynamo etc. The basic principle in all these cases in conversion of some type of energy into electrical energy.

Let us take the case of a battery which uses the stored chemical energy to drive current in the circuit. In case the battery maintains a steady potential difference of V volts, between its terminals its *e.m.f. is V volts*. If the terminals of the battery are connected to a conductor, an electric field which is static is set up in the conductor.

Line integral of the field over the path between two points say A and B is equal to the potential difference between the points. In case these points are the terminals of the battery, e.m.f. of the battery will be

$$V = \int_A^B \vec{E} \cdot d\vec{l} \qquad ...(1)$$

Now, as the current flows through the conductor, the battery must do some work to keep potential difference constant between its terminals.

Let q charge move from one terminal of the battery to the other through the conductor. The work done by the battery is given by

$$V.q = \int_A^B q\, \vec{E} \cdot d\vec{l} \qquad ...(2)$$

So the e.m.f. of the battery is given by

$$V = \frac{1}{q} \int_A^B q\, \vec{E} \cdot d\vec{l}$$

$$= \int_A^B \vec{E} \cdot d\vec{l} \qquad ...(3)$$

So we may define the electromotive force (e.m.f.) as the *closed line integral of the non-conservative electric field produced by the source* (in this case a battery). So a source that produces such a non-conservative electric field so that there is a flow of current in the circuit is called *e.m.f. Source*.

Unit of e.m.f. is volt.

9.14. FARADAYS LAWS OF ELECTROMAGNETIC INDUCTION (DIFFERENTIAL FORM)

Faraday observed from the various experiments that

(*i*) In case a magnet is moved near a wire in the form of a closed circuit without a battery in the circuits, then a current is produced in the circuits so long as the movement of the magnet lasts and as the movement of the magnet is stopped, the current also disappears.

(*ii*) Same effect is also observed if the magnet is kept still and current is moved. From above it is clear that for the production of current in the wire, relative movement of either the wire or the magnetic is required.

(*iii*) Faraday also observed that a current is induced in a loop of wire if the stationary current in a adjacent circuit is turned on or turned off.

In other words a transient current flows when the *flux through a circuit is altered*. The changing flux induces an electric field and consequently an e.m.f. in the circuit which helps the flow of current.

Faraday called this phenomenon as *electromagnetic induction* which may be stated as:

(*i*) When a magnetic flux through a circuit changes an e.m.f. in the circuit is induced which makes the induced current flow in the circuit till the change in flux persist.

(*ii*) Magnitude of the induced e.m.f. is proportional to the rate of change of magnetic flux in the circuit.

Let e be the e.m.f. and φ, the magnetic flux, then

$$e \propto \frac{d\varphi}{dt}$$

The direction of e.m.f. may be got from Lenz's law, which states that the *direction of e.m.f. produced is always opposite to the e.m.f. produced due to changing flux*. So we may write

$$e = -k\frac{d\varphi}{dt}$$

where k is proportional constant and is unity in S.I. units

So we may write

$$e = -\frac{d\varphi}{dt} \qquad ...(1)$$

This equation is called the *Faraday's law of induction.*

Induced e.m.f. is also called the back e.m.f. It is independent of the way of change of flux.

If φ be the instantaneous rules of magnetic flux, then

$$\varphi = \oiint \vec{B}\cdot d\vec{S} \qquad ...(2)$$

Also the induced e.m.f. around a closed path is given by the line integral of the induced electric field $\vec{E}$ around the coil *i.e.*

$$e = \int_l \vec{E}\cdot d\vec{l} \qquad ...(3)$$

If the circuit path l is bounded by an open surface S, in that case

$$e = \int_l \vec{E}\cdot d\vec{l} = -\frac{d}{dt}\int\oint_S \vec{B}\cdot d\vec{S} \qquad ...(4)$$

So the Faraday's law states that the line *integral of the electric intensity around any fixed closed path is equal to the time rate of decrease of flux of magnetic induction through the surface enclosed by the curve.*

We can write the differential form of Faraday's law by using the Stoke's law to replace the line integral to surface integral.

$$\int_l \vec{E}\cdot d\vec{l} = \oiint_S (\vec{\nabla}\times\vec{E})\cdot d\vec{S} = -\frac{\partial}{\partial t}\oiint_S \vec{B}\cdot d\vec{S}$$

$$\oiint\left(\vec{\nabla}\times\vec{E}+\frac{\partial\vec{B}}{\partial t}\right)d\vec{S} = 0$$

where we have chosen S to be an arbitrary surface, so the integrand and must be zero. Thus

$$\vec{\nabla}\times\vec{E}+\frac{\partial\vec{B}}{\partial t} = 0$$

or

$$\vec{\nabla}\times\vec{E} = -\frac{\partial\vec{B}}{\partial t} \qquad ...(5)$$

which is the differential form of Faraday's law.

9.15. MAXWELL'S FIELD EQUATIONS

Maxwell made certain modifications in the basic four equations of electromagnetism and presented four fundamental differential equations of electromagnetic fields. **These are known as Maxwell's** field equations. These equations may be written as:

(*i*) $\nabla\cdot\vec{D} = \rho$

This results from the Gauss's law of electrostatics. $\vec{D}$ in the *electric displacement vector* and is measured in coulomb/m^2, ρ is the free charge density, measured in coulombs/m^2.

(*ii*) $\vec{\nabla} \cdot \vec{B} = 0$

It results from the application of *Gauss's law to magnetism.*

(*iii*) $\nabla \times \vec{E} = -\dfrac{\partial \vec{B}}{\partial t}$

It results from *Faraday's law of electromagnetic induction*

(*iv*) $\vec{\nabla} \times \vec{H} = \vec{J} + \dfrac{\partial \vec{D}}{\partial t}$

It corresponds to the Maxwell's modified *ampere's law in a circuital form* or magnetic field accompanying an electric current.

All these equations are equally valid for *static as well as varying fields.*

(*i*) **Derivation of $\vec{\nabla} \cdot \vec{D} = \rho$.**

From the Gauss's law, the integral of normal component of electric field *i.e.*, outward electric flux over any closed surface is equal to the charge within the surface *i.e.*,

$$\int \vec{E} \cdot d\vec{S} = q/\epsilon_0$$

where ϵ_0 is the permittivity of the free space. As ρ is the charge density so $\rho\, dV$ will be charge of volume element dv which contributes a charge of $\rho \dfrac{dV}{\epsilon_0}$ to the surface integral.

If surface S encloses a volume V, in that case

$$\int_S \vec{E} \cdot d\vec{S} = \frac{1}{\epsilon_0} \int_V \rho\, dV \qquad \text{...(1)}$$

If S correspond to the surface bounding volume V of the *dielectric*, in that case total charge must be constituted by both *free and polarizing charges.* In that case the total charge density at a point in a small volume element dV must be $(\rho + \rho')$, where ρ' corresponds to the *polarization charge density* and ρ is the *free charge density.*

So the above equation may be written as

$$\int_S \vec{E} \cdot d\vec{S} = \frac{1}{\epsilon_0} \int_V (\rho + \rho')\, dV \qquad \text{...(2)}$$

Now polarization charge density is related to *Polarization vector.* $\vec{P}$ by the relation

$$\rho' = -div\, \vec{P}$$
$$= -\vec{\nabla} \cdot \vec{P}$$

$\therefore$ e.g. (2) may be written as

$$\int_S \vec{E} \cdot d\vec{S} = \frac{1}{\epsilon_0} \int_V (\rho - \vec{\nabla} \cdot \vec{P})\, dv$$

or

$$\int_S \epsilon_0 \vec{E} \cdot d\vec{S} = \int_V \rho\, dV - \int (\vec{\nabla} \cdot \vec{P})\, dV$$

Using Gauss's divergence theorem, we can change surface integral to volume integral and get

$$\int_S \epsilon_0 \vec{E} . d\vec{S} = \int \text{div}\, (\epsilon_0 \vec{E})\, dV = \int_V (\nabla . \epsilon_0 E)\, dV$$

$$\therefore \quad \int_V (\nabla . \epsilon_0 E)\, dV = \int_V \rho\, dV - \int (\vec{\nabla} \cdot \vec{P})\, dV$$

or

$$\int_V \rho\, dV = \int_V \nabla . (\epsilon_0 \vec{E} + \vec{P})\, dV$$

$\epsilon_0 \vec{E} + \vec{P} = \vec{D}$ is the electric displacement vector.

so $$\int_V \rho\, dV = \int_V (\vec{\nabla}\cdot\vec{D})\, dV$$

or $$\int_V (\vec{\nabla}\cdot\vec{D}-\rho)\, dV = 0$$

This equation being true for any volume since V is chosen arbitrarily, so the integral must be zero *i.e.*

$$\vec{\nabla}\cdot\vec{D}-\rho = 0$$

or $$\vec{\nabla}\cdot\vec{D} = \rho \qquad ...(3)$$

In the case of an isotropic medium $\vec{D}, \vec{E}$ and $\vec{P}$ are in the same direction. Also for small applied field, $\vec{D}$ is proportional to $\vec{E}$ *i.e.*

$$\vec{D} = \epsilon\, \vec{E}$$

where ϵ is permittivity of the dielectric medium.

In the case of free space

$$\vec{D} = \epsilon_0\, \vec{E}$$

where ϵ_ρ is the permittivity of free space.

(ii) DERIVATION OF $\vec{\nabla}\cdot\vec{B} = 0$

In case of magnetic lines of force, the number of lines of force entering any close arbitrary urface is equal to the number of lines of force leaving the surface. This in others words mean hat total outward flux of magnetic induction $\vec{B}$ through any closed surface S is equal to zero i.e

$$\int_0 \vec{B}\cdot d\vec{S} = 0 \qquad ...(1)$$

Using Gauss's divergence theorem, we can convert surface integral to volume integral and get

$$\int_V (\vec{\nabla}, \vec{B})\, dV = 0$$

Since volume chosen is arbitrary so the integrand must be zero

$$\therefore \qquad \vec{\nabla}\cdot\vec{B} = 0 \qquad ...(2)$$

This is the Gauss's law of magnetism.

(iii) DERIVATION OF $\vec{\nabla}\times\vec{E} = -\frac{\partial \vec{B}}{\partial t}$

From Faraday's law of electromagnetic induction when the magnetic flux through a circuit changes an e.m.f is induced whose direction is always Opposite to the charge producing it (Lenz's law)

Let e be the induced e.m.f. and φ the magnetic flux then

$$e = -\frac{d\varphi}{dt} \qquad ...(1)$$

Now induced e.m.f. is equal to the induced electric field $\vec{E}$ around the circuit, mathematically

$$e = \int_l \vec{E}\cdot d\vec{l} \qquad ...(2)$$

and magnetic flux through the circuit is

$$\varphi = \int_S \vec{B}\cdot d\vec{S} \qquad ...(3)$$

where integral is over a bound surface of area S.

From the above equation, we have

$$\int \vec{E}\cdot d\vec{l} = -\frac{d}{dt}\int_S \vec{B}\cdot d\vec{S}$$

Since surface integral does not change with time, we may write the above equation.

$$\int_l \vec{E}\cdot d\vec{l} = -\int_S \frac{d\vec{B}}{dt}\cdot d\vec{S}$$

$$= -\int_S \frac{\partial \vec{B}}{\partial t}\cdot d\vec{S} \qquad ...(4)$$

where we have changed the total derivative into partial derivative as we are interested only in the change of $\vec{B}$ over a small area element *ds* Equation (4) shows that *e.m.f. around a closed path is equal to the time derivative of the magnetic displacement through any surface bounded by the path.*

Now using Stoke's theorem, we can transform the line integral into surface integral *i.e.*

$$\int_l \vec{E}\cdot d\vec{l} = \int_S (\nabla \times \vec{E})\cdot d\vec{S} = -\int \frac{\partial \vec{B}}{\partial t}\cdot d\vec{S}$$

$$= \int_S \left(\vec{\nabla}\times\vec{E} + \frac{\partial \vec{B}}{\partial t}\right)\cdot d\vec{S} = 0$$

This equation is valid for any surface field, so we may write

$$\left(\vec{\nabla}\times\vec{E} + \frac{\partial \vec{B}}{\partial t}\right) = 0$$

or

$$\vec{\nabla}\times\vec{E} = -\frac{\partial \vec{B}}{\partial t}$$

which is the differential form of Faraday's law of e.m. induction.

(*IV*) DERIVATION OF $\vec{\nabla}\times\vec{H} = \vec{J} + \frac{\partial \vec{D}}{\partial t}$

This equation correspond to the Ampere's law in circuital form for a magnetic field accompanying an electric current.

This law states that the line integral of magnetic field around a closed path is equal to the total current crossing any surface bounded by the path of line integral.

Mathematically

$$\int_l \vec{H}\cdot d\vec{l} = \int_S \vec{J}\cdot d\vec{S} \qquad ...(1)$$

where we taken the surface integral corresponding to the charge flow that constitute the current I.

Changing line integral into surface integral using Stoke's theorem, we get

$$\int_l \vec{H}\cdot d\vec{l} = \int_S (\vec{\nabla}\times\vec{H})\cdot d\vec{S} = \int \vec{J}\cdot d\vec{S}$$

$$\int_S [(\vec{\nabla}\times\vec{H}) - \vec{J}]\cdot d\vec{S} = 0$$

For any arbitrary surface, we have

$$(\vec{\nabla}\times\vec{H}) - \vec{J} = 0$$

$$\therefore \qquad \vec{\nabla}\times\vec{H} = \vec{J} \qquad ...(2)$$

The above equation is valid for steady state current, however it can be modified for the time varying fields.

Which current density has to be modified.

Taking divergence on both sides, we get

$$\vec{\nabla}\cdot(\vec{\nabla}\times\vec{H}) = \nabla\cdot J = 0$$

As the divergence of a curl of a vector is zero. This is valid for steady state phenomenon. However, when the currents are changing with time, we have from the continuity equation.

$$\vec{\nabla}\cdot\vec{J}+\frac{\partial\rho}{\partial t} = 0$$

or
$$\vec{\nabla}\cdot\vec{J} = -\frac{\partial\rho}{\partial t}$$

where ρ is the volume charge density. It shows that $\nabla\cdot J \neq 0$ *for time varying currents.*

Maxwell improved upon the above equation by introducing the concept that the current density is incomplete and suggested that some factor must be added into $\vec{J}$ so that e.g. (2) becomes

$$\nabla\times\vec{H} = \vec{J}+\vec{J'} \qquad ...(3)$$

Taking divergence of this equation, we get

$$\nabla\cdot(\nabla\times\vec{H}) = [\nabla\cdot(\vec{J}\times\vec{J})]$$

Since $\nabla\cdot(\vec{\nabla}\times\vec{J}) = 0$

$$\nabla\cdot\vec{J} = -\nabla\cdot\vec{J'}$$

But from equation of continuity

$$\nabla\cdot\vec{J} = -\frac{\partial\rho}{\partial t}.$$

Also we have $\vec{\nabla}\cdot\vec{D'} = \rho$, So we have

$$\vec{\nabla'}\cdot J' = \frac{\partial}{\partial t}(\vec{\nabla}\cdot\vec{D}) = \vec{\nabla}\cdot\left(\frac{\partial\vec{D}}{\partial t}\right)$$

$$\vec{J'} = \frac{\partial\vec{D}}{\partial t}$$

Putting the value of J' in e.g. (3) we get the modified form of Ampere's law as

$$\vec{\nabla}\times\vec{H} = \vec{J}+\frac{\partial\vec{D}}{\partial t} \qquad ...(4)$$

where term $\dfrac{\partial\vec{D}}{\partial t}$ is due to the change of electric displacement $\vec{D}$ with time and is called *displacement current density.* $\vec{J}$ is *called conduction current density.*

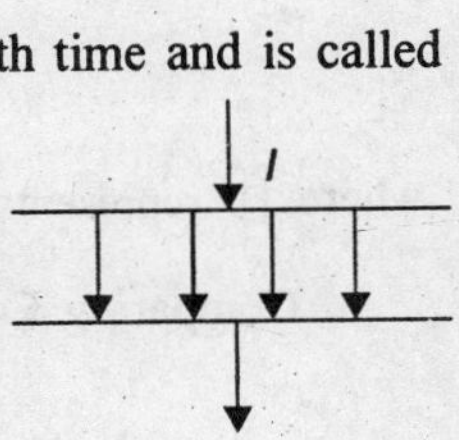

Magnetic flux $\vec{B}$ is related to magnetic field intensity $\vec{H}$ by the relation $\vec{B} = \mu\vec{H}$ where μ is the permeability. $\vec{B}$ In case of free space $\mu_0\vec{H}$ where μ_0 is the permeability of free space and has the value $4\pi\times10^{-7}$ henry/meter. In the same way as above, the equation may be written in integral form

$$\oint\vec{H}.d\vec{l} = \int_S\left(J+\frac{\partial\vec{D}}{\partial t}\right)d\vec{S} \qquad ...(5)$$

This equation signifies that the magnetomative force around a closed path is equal to the conduction current plus the time derivative of the current displacement through any surface bounded by that path.

9.16. DISPLACEMENT CURRENT

Displacement current has not the same significance as the actual current because in this case the charges actually do not move. To understand the displacement current concept, let us consider a simple circuit in the case of a capacitor joined by a conducting wire.

Current in the circuit is equal to the rate of flow of charge i.e.

$$I = \frac{dq}{dt} \quad ...(1)$$

where q corresponds to the positive charge on a plate of capacitor and is related to electric field as

$$\vec{E} = \frac{\sigma}{\epsilon_0} = \frac{q}{\epsilon_0 A} \quad ...(2)$$

where A is the area of the plate, σ is the surface charge density on the plate and ϵ_0 is the permittivity of the medium between the plate. In case of air $\epsilon_0 = 1$

Now $$I = \frac{dq}{dt} = A\epsilon_0\frac{d\vec{E}}{dt} = A\frac{\partial \vec{D}}{\partial t}$$

$$\therefore \quad \frac{I}{A} = \frac{\partial \vec{D}}{\partial t} \quad ...(3)$$

Now I/A corresponds to current density. So the quality $\frac{\partial \vec{D}}{\partial t}$ may be interpreted as the density of some *current that flows in the Space between the plates of the capacitor*. When the circuit is completed by the connecting wires connected to the two capacitor plates.

9.17. PHYSICAL SIGNIFICANCE OF MAXWELL'S EQUATIONS: (CORRELATION WITH ELECTRICITY AND MAGNETISM LAWS)

1 steady state equation

I. $$\vec{\nabla}\cdot\vec{D} = \rho \quad ...(1)$$

This equation is a restatement of Gauss's law in electrostatics. Putting Equation (1) in the form $\vec{\nabla}\cdot\vec{E} = \frac{\rho}{\varepsilon}$ (as $\vec{D}$ is the displacement vector given by $\vec{D} = \varepsilon\vec{E}$ and ρ is the charge density and integrating over volume V, we get

$$\iiint \vec{\nabla}\cdot\vec{E}\,dV = \frac{1}{\varepsilon}\iiint \rho\,dV \quad ...(2)$$

From the divergence theorem of vector analysis,

$$\iiint \vec{\nabla}\cdot\vec{E}\,dV = \oiint_S \vec{E}\cdot d\vec{S} \quad ...(3)$$

where $\oiint_S$ represents integration over a closed surface

Comparing Equation (2) and (3), we have

$$\oiint \vec{E}\cdot d\vec{S} = \frac{1}{\varepsilon}\iiint \rho\,dV.$$

Now $\oiint \vec{E}\cdot d\vec{S}$ represents the electric flux through a closed surface and $\iiint \rho\,dv = q$ represent the charge in the closed surface.

$$\therefore \quad \oiint \vec{E}\cdot d\vec{S} = q/\varepsilon$$

which means that the total flux through a closed surface is equal to the total charge inside the surface divided by ε.

This is the Gauss's law of electrostatics.

II. Second steady state equation: In vectors form the second steady state equation is given by

$$\vec{\nabla}\cdot\vec{B} = 0 \qquad ...(4)$$

where $\vec{B}$ is the magnetic induction vector. This equation is the restatement of Gauss's law in magnetism and proves that *magnetic poles exist only in pairs. i.e.* isolated magnetic poles can not be found.

For a closed surface, we have

$$\iiint \nabla\cdot\vec{B}\cdot dv = 0$$

Applying the divergence theorem in vector analysis, we may have

$$\iiint \vec{\nabla}\cdot\vec{B} = \oiint \vec{B}\cdot d\vec{S}$$

$$\therefore \quad \oiint \vec{B}\cdot d\vec{S} = 0 \qquad ...(5)$$

Now $\oiint \vec{B}$ represents the magnetic flux through a closed surface, so the magnetic flux through a closed surface is zero. In other words it means that a magnetic flux entering into the volume is equal to the magnetic flux leaving the volume. So there is no source of sink in the volume, which in other words mean that *isolated magnetic poles (monopoles) cannot exist.*

III. First time varying equation: This equation is the restatement of Faraday's law of electromagnetic induction.

The equation is given by

$$\vec{\nabla}\times\vec{E} = \frac{-\partial\vec{B}}{\partial t} = -\mu\frac{\partial\vec{H}}{\partial t} \qquad ...(6)$$

Integrating Equation (5) over a bounded surface bounding a path *c*, we get

$$\int_S (\vec{\nabla}\times\vec{E})\cdot d\vec{S} = \oint_S \frac{\partial\vec{B}}{\partial t}\cdot d\vec{S} \qquad ...(7)$$

From Stokes's theorem of vector analysis

$$\int_S (\vec{\nabla}\times\vec{E})\cdot d\vec{S} = -\oint_c \vec{E}\cdot d\vec{l}$$

So Equation (7) may be written as

$$\oint \vec{E}\cdot d\vec{l} = -\oint_s \frac{\partial\vec{B}}{\partial t}\cdot d\vec{S}$$

where $\oint \vec{E}\cdot d\vec{l}$ represent as *induced e.m.f.* in a closed circuit and $-\oint_s \frac{\partial\vec{B}}{\partial t}\cdot d\vec{S}$ represents *negative rate of change of magnetic flux through the circuit.*

This is Faraday's law of electromagnetic induction. It shows that time variation of magnetic induction $\vec{B}$ generates the electric field $\vec{E}$

IV. Second time varying equation: In vector form second time varying equation is given by

$$\vec{\nabla}\times\vec{H} = \vec{J} + \frac{\partial\vec{D}}{\partial t}$$

where J is the conduction corrent density given by $\vec{J} = \sigma\vec{E}, \sigma$ being the conducting of the medium and $\frac{\partial \vec{D}}{\partial t}$ is the displacement current density.

PHYSICAL SIGNIFICANCE

This equation is the restatement of Ampere's law combined together for *displacement current and conduction current,* and states that a *changing electric field* produces a *changing magnetic field.*

Also we know that changing magnetic field produces a changing electric field. So we may conclude that:

A changing electric field produces a changing magnetic field and a changing magnetic field produces a changing electric field. *Alternate production of electric and magnetic fields give rise to the propagation of electromagnetic waves.*

9.18. DISTINCTION BETWEEN CONDUCTION AND DISPLACEMENT CURRENT

Conduction current is the actual current that flows through the conducting medium and so obeys ohm's law *i.e.* $I = V/R$ as well as its vector form $\vec{J} = \sigma \vec{E}$ where J is the *conduction current density,* σ the *conductivity* of the medium and $\vec{E}$ the *electric field intensity.*

Displacement current is the current which is set up in the dielectric medium (*i.e.* $\sigma = 0$) due to the variation of induced displacement charge produced by the varying electric field applied across the dielectric. Displacement charge density is given by $\frac{\partial \vec{D}}{\partial t} = \varepsilon \frac{\partial \vec{E}}{\partial t}$ and it has the units of current/unit area as shown under.

Maxwell's second time varying equation for e.m. field is given by

$$\vec{\nabla} \times \vec{H} = \vec{J} + \frac{\partial \vec{D}}{\partial t}$$

where $\vec{J} = \sigma \vec{E}, \sigma$ being the conductivity of the medium.

$\vec{D} = \varepsilon \vec{E}, \varepsilon$ being the permittivity of the medium having units. Farad/meter $\vec{E}$ being the electric field vector having units volt/meter.

$\frac{\partial \vec{D}}{\partial t}$ is called displacement term and $\vec{D}$ is the electric displacement vector.

$$\therefore \quad \text{Dimension of } D = \frac{\text{Farad}}{\text{meter}} \times \frac{\text{Volt}}{\text{meter}} = \frac{\text{Coulomb}}{(\text{meter})^2} \text{ and}$$

$$\text{dimensions of } \frac{\partial D}{\partial t} = \frac{\text{Coulomb}}{\text{meter}^2} \times \frac{1}{\text{sec}} = \text{Ampere/meter}^2$$

So the dimensions of *displacement term* are that of current/unit area *i.e.,* **current density.**

9.19. MAJOR CONTRIBUTION MAXWELL EQUATIONS TO E.M. WAVE THEORY

Maxwell introduction of *displacement vector* in a polarizable dielectric medium corresponding to the electric vector in vacuum or free space explained many relevant phenomenon as shown under:

Displacement vector D when related to electric field vector E is given by

$$\vec{D} = \varepsilon \vec{E}$$

where ε is the permittivity of the medium.

Now from the Maxwell's idea of *displacement current,* we have

$$\frac{\partial \vec{D}}{\partial t} = \frac{\partial}{\partial t}(\varepsilon \vec{E}) = \varepsilon \frac{\partial \vec{E}}{\partial t}$$

So Maxwell's second time varying equation for e.m. wave in modified form will be

$$\vec{\nabla} \times \vec{H} = \vec{J} + \frac{\partial \vec{D}}{\partial t} = \sigma \vec{E} + \varepsilon \frac{\partial \vec{E}}{\partial t}$$

The term $\frac{\partial \vec{D}}{\partial t} = \varepsilon \frac{\partial \vec{E}}{\partial t}$ accounts for all radiative processes from a charge (positive or negative undergoing an acceleration. Various ways for this to happen may be when an electron:

(*i*) jumps from high energy level to low energy level due to the de-excitation of atom.

(*ii*) suffers collision and its path changes and so gets accelerated.

(*iii*) undergoes oscillatory motion in an alternating field as in the case of a wireless antina.

9.20. REPRESENTATION OF LIGHT VECTOR

Light is a form of e.m. wave and the force exerted by its vector $\vec{E}$ on a particle of charge q is given by $q\vec{E}$ (magnitude is qE). If the particle is moving with velocity v, then the force exerted by magnetic vector $\vec{B}$ is given by $q(v \times B)$ and its maximum value is given by the magnitude qvB.

For an e.m. wave

$$\frac{|\vec{E}|}{|\vec{H}|} = \frac{\sqrt{\mu}}{\varepsilon} = \frac{\sqrt{1}}{\mu\varepsilon} = \mu \cdot c.$$

or

$$\frac{|\vec{E}|}{|\vec{B}|} = c \text{ or } B = E/c.$$

$$\therefore \qquad qvB = q \cdot v \cdot \frac{E}{c} = q\,E \cdot v/c.$$

which means that *magnetic force* is equal to $\frac{v}{c}$ times *electric force*. Since c is very large as compared to v, so *electric force* is much greater than the *magnetic force*. It is due to this reason that *light wave is characterised by its electric field vector*.

It may be noted that although a wave cannot exist unless both $\vec{E}$ and $\vec{B}$ are present, yet $\vec{E}$ is more important than $\vec{B}$ from the point of view of effects due to $\vec{B}$ and $\vec{B}$. We can ignore the effects of $\vec{B}$ as compared to those of $\vec{E}$ as the wave passes through matter.

As an example, let us take the intensity of light from bright sun on earth which is 4 kw/m^2 which means electric vector has an amplitude of 1 KV/m; which is quite measurable quantity.

On the other hand the amplitude of corresponding magnetic field is about 3 micro tesla (3 μT). Which is $\frac{1}{10}$th of the Earth's magnetic field and so is very difficult to measure.

The above example clearly shows that importance of electric field vector $\vec{E}$ over magnetic field vector $\vec{H}$. It is due to this reason that in *polarization of light*. (*i.e.* when light wave passes through some polarizing materials like calcite tourmaline etc., one of its components is selectively absorbed). As the *electric field vector* $\vec{E}$ has much greater magnitude than the magnetic field vector $\vec{H}$, *polarization is determined by the electric field vector*.

9.21. E.M. WAVES IN A MEDIUM OF FINITE PERMEABILITY μ PERMITTIVITY ε BUT CONDUCTIVITY σ = 0

As $\sigma = 0$, let us consider the case of a non-coducting medium (insulator or dielectric) having finite permeability μ and permittivity ε. In such a case as there is no charge in the medium, so charge density $\rho = 0$. In that case Maxwell equations for vectors $\vec{E}$ and $\vec{H}$ may be written as

$$e\vec{\nabla} \cdot \vec{E} = 0 \quad \text{or} \quad \vec{\nabla} \cdot \vec{E} = 0 \qquad ...(1)$$

$$\vec{\nabla} \cdot \vec{B} = 0; \ \mu\vec{\nabla} \cdot \vec{H} = 0 \quad \text{or} \quad \vec{\nabla} \cdot \vec{H} = 0 \qquad ...(2)$$

$$\vec{\nabla} \times \vec{E} = -\mu \frac{\partial \vec{H}}{\partial t} \qquad ...(3)$$

$$\vec{\nabla}\times\vec{H} = \vec{J}+\varepsilon\frac{\partial \vec{E}}{\partial t} \text{ or } \vec{\nabla}\times\vec{H} = \sigma\vec{E}+\varepsilon\frac{\partial \vec{E}}{\partial t} \quad ...(4)$$

As $\sigma = 0$, the above equation for a non-conducting medium may be written as

$$\vec{\nabla}\times\vec{H} = \varepsilon\frac{\partial \vec{E}}{\partial t} \quad ...(5)$$

For electric field $\vec{E}$. Take the curl of Equation (3), and putting the value of $\vec{\nabla}\times\vec{H}$ from Equation (5), we get

$$\vec{\nabla}\times(\vec{\nabla}\times\vec{E}) = \vec{\nabla}\times\left(-\mu\frac{\partial \vec{H}}{\partial t}\right) = -\mu\left(\vec{\nabla}\times\frac{\partial \vec{H}}{\partial t}\right) = -\mu\frac{\partial}{\partial t}(\vec{\nabla}\times\vec{H})$$

$$= -\mu\frac{\partial}{\partial t}\left(\varepsilon\frac{\partial \vec{E}}{\partial t}\right) = -\mu\varepsilon\frac{\partial^2 \vec{E}}{\partial t^2} \quad ...(6)$$

Let us use the vector triple identity

$\vec{A}\times(\vec{B}\times\vec{C}) = \vec{B}(\vec{A}\cdot\vec{C})-\vec{C}(\vec{A}\cdot\vec{B})$ to the case of $\vec{\nabla}\times(\vec{\nabla}\times\vec{E})$, we get

$$\vec{\nabla}\times(\vec{\nabla}\times\vec{E}) = \vec{\nabla}(\vec{\nabla}\cdot\vec{E})-\vec{E}(\vec{\nabla}\cdot\vec{\nabla}) = -\nabla^2\vec{E} \quad ...(7)$$

As $\vec{\nabla}\cdot\vec{E} = 0$ according to Equation (1) and $-\vec{E}(\vec{\nabla}\cdot\vec{\nabla}) = -\nabla^2\vec{E}$

So we may have from Equation (6) and (7)

$$-\mu\varepsilon\frac{\partial^2 \vec{E}}{\partial t^2} = -\nabla^2\vec{E}$$

or

$$\nabla^2\vec{E} = \mu\varepsilon\frac{\partial^2 \vec{E}}{\partial t^2} \quad ...(8)$$

Equation (8) is the general equation for the vector field E in a non-conducting medium.

For magnetic field $\vec{H}$. Let us take curl of Equation (5) and putting the value of $\vec{\nabla}\times\vec{E}$ in Equation (3) we get

$$\vec{\nabla}\times(\vec{\nabla}\times\vec{H}) = \vec{\nabla}\times\left(\varepsilon\frac{\partial \vec{E}}{\partial t}\right) = \varepsilon\left(\vec{\nabla}\times\frac{\partial \vec{E}}{\partial t}\right) = \varepsilon\frac{\partial}{\partial t}(\vec{\nabla}\times\vec{E})$$

$$= \varepsilon\frac{\partial}{\partial t}\left(-\mu\frac{\partial \vec{H}}{\partial t}\right) = -\mu\varepsilon\frac{\partial^2 \vec{H}}{\partial t^2} \quad ...(9)$$

Again from vector identity

$$\vec{A}\times(\vec{B}\times\vec{C}) = \vec{B}\times(\vec{A}\cdot\vec{C})-\vec{C}(\vec{A}\cdot\vec{B})$$

Applying to $\vec{\nabla}\times(\vec{\nabla}\times\vec{H})$, we get

$$\vec{\nabla}\times(\vec{\nabla}\times\vec{H}) = \vec{\nabla}(\vec{\nabla}\cdot\vec{H})-\vec{H}(\vec{\nabla}\cdot\vec{\nabla}) = -\nabla^2\vec{H} \quad ...(10)$$

As $\vec{\nabla}\cdot\vec{H} = 0$ [From Equation (2)]

$$\therefore \quad -\vec{H}(\vec{\nabla}\cdot\vec{\nabla}) = -\nabla^2\vec{H}$$

From Equation (9) and (10), we get

$$-\mu\varepsilon\frac{\partial^2 \vec{H}}{\partial t^2} = -\nabla^2\vec{H}$$

or

$$\nabla^2\vec{H} = \mu\varepsilon\frac{\partial^2 \vec{H}}{\partial t^2} \quad ...(11)$$

Equation (11) is the general wave equation for the vector field $\vec{H}$ in case of a non-conducting medium.

9.22. E.M. WAVES IN A CONDUCTING MEDIUM.

Consider a conducting medium having finite permeability μ, permittivity ε and conductivity σ.

If the medium is free from a charge source, in that case ρ = 0 and Maxwell's equations for vectors $\vec{E}$ and $\vec{H}$ may be written as

$$\varepsilon\vec{\nabla}\cdot\vec{E} = 0 \quad \text{or} \quad \vec{\nabla}\cdot\vec{E} = 0 \qquad ...(1)$$

$$\mu\vec{\nabla}\cdot\vec{H} = 0 \quad \text{or} \quad \vec{\nabla}\cdot\vec{H} = 0 \qquad ...(2)$$

$$\vec{\nabla}\times\vec{E} = -\mu\frac{\partial\vec{H}}{\partial t} \qquad ...(3)$$

$$\vec{\nabla}\times\vec{H} = \vec{J}+\varepsilon\frac{\partial\vec{E}}{\partial t}$$

$$= \sigma\vec{E}+\varepsilon\frac{\partial\vec{E}}{\partial t} \quad [\because \vec{J} = \sigma\vec{E}] \qquad ...(4)$$

For electric field $\vec{E}$. Take curl of Equation (3) and put the value of $(\vec{\nabla}\times\vec{H})$ from Equation (4), we get

$$\vec{\nabla}\times(\vec{\nabla}\times\vec{E}) = \vec{\nabla}\times\left(-\mu\frac{\partial\vec{H}}{\partial t}\right) = -\mu\left(\vec{\nabla}\times\frac{\partial\vec{H}}{\partial t}\right)$$

$$= -\mu\frac{\partial}{\partial t}(\vec{\nabla}\times\vec{H}) = -\mu\frac{\partial}{\partial t}\left(\sigma\vec{E}+\frac{\partial\vec{E}}{\partial t}\right) = -\mu\sigma\frac{\partial\vec{E}}{\partial t}-\mu\varepsilon\frac{\partial^2\vec{E}}{\partial t^2} \qquad ...(5)$$

Again from triple vector identity

$$\vec{A}\times(\vec{B}\times\vec{C}) = \vec{B}(\vec{A}\cdot\vec{C})-\vec{C}(\vec{A}\cdot\vec{B}) \quad \text{applied to } \vec{\nabla}\times(\vec{\nabla}\times\vec{E}), \text{ we get}$$

$$\vec{\nabla}\times(\vec{\nabla}\times\vec{E}) = \vec{\nabla}(\vec{\nabla}\cdot\vec{E})-\vec{E}(\vec{\nabla}\cdot\vec{\nabla}) = -\nabla^2\vec{E} \qquad ...(6)$$

As $\vec{\nabla}\cdot\vec{E} = 0$ from Equation (1) and $-\vec{E}(\vec{\nabla}\cdot\vec{\nabla}) = -\nabla^2\vec{E}$

So from Equation (5) and (6) we have

$$-\nabla^2\vec{E} = -\mu\sigma\frac{\partial\vec{E}}{\partial t}-\mu\varepsilon\frac{\partial^2\vec{E}}{\partial t^2} \quad \text{or} \quad \nabla^2\vec{E} = \mu\sigma\frac{\partial\vec{E}}{\partial t}+\mu\varepsilon\frac{\partial^2\vec{E}}{\partial t^2} \qquad ...(7)$$

In case of a non-conducting medium

$$\sigma = 0, \text{ so we have } \nabla^2\vec{E} = \mu\varepsilon\frac{\partial^2 E}{\partial t^2} \qquad ...(8)$$

Comparing Equation (7) and (8), we see that $\mu\sigma\frac{\partial\vec{E}}{\partial t}$ is the *dissipative term which allows the current to flow through the medium.*

For magnetic field $\vec{H}$. Taking curl of Equation (4) and putting the value of $\vec{\nabla}\times\vec{E}$ from (3), we get

$$\vec{\nabla}\times(\vec{\nabla}\times\vec{H}) = \vec{\nabla}\times\left(\sigma\vec{E}+\varepsilon\frac{\partial\vec{E}}{\partial t}\right) = \sigma(\vec{\nabla}\times\vec{E})+\varepsilon\frac{\partial}{\partial t}(\vec{\nabla}\times\vec{E})$$

$$= -\mu\sigma\frac{\partial\vec{H}}{\partial t}-\varepsilon\mu\frac{\partial^2 H}{\partial t^2} \qquad ...(9)$$

Applying the triple vector identity

$$\vec{A}\times(\vec{B}\times\vec{C}) = \vec{B}(\vec{A}\cdot\vec{C})-\vec{C}(\vec{A}\cdot\vec{B}) \text{ to the case } \vec{\nabla}\times(\vec{\nabla}\times\vec{H}), \text{ we get}$$

$$\vec{\nabla}\times(\vec{\nabla}\times\vec{H}) = \vec{\nabla}(\vec{\nabla}\cdot\vec{H})-\vec{H}(\vec{\nabla}\cdot\vec{\nabla}) = -\nabla^2\vec{H} \quad ...(10)$$

Since $\vec{\nabla}\cdot H = 0$ [from Equation (2)]

$$\therefore -\vec{H}(\vec{\nabla}\cdot\vec{\nabla}) = -\nabla^2\vec{H}$$

From Equation (9) and (10), we have

$$-\nabla^2\vec{H} = -\mu\sigma\frac{\partial\vec{H}}{\partial t}-\varepsilon\mu\frac{\partial^2\vec{H}}{\partial t^2}$$

or

$$\nabla^2\vec{H} = \mu\sigma\frac{\partial\vec{H}}{\partial t}+\mu\varepsilon\frac{\partial^2\vec{H}}{\partial t^2} \quad ...(11)$$

In case of non-conducting medium $\sigma = 0$, so we have

$$\nabla^2\vec{H} = \mu\varepsilon\frac{\partial^2\vec{H}}{\partial t^2} \quad ...(12)$$

Comparison of Equation (11) and (12) shows that $\mu\sigma\frac{\partial\vec{H}}{\partial t}$ is the *dissipative term* which allows the current to flow through the medium.

VELOCITY OF E.M. WAVES FROM MAXWELL'S EQUATION

Wave equation for an e.m. wave in case of a non-conducting medium ($\sigma = 0$) having finite magnetic permeability μ and electric susceptibility ε in terms of vector field $\vec{E}$ is given by

$$\nabla^2\vec{E} = \mu\varepsilon\frac{\partial^2\vec{E}}{\partial t^2} \quad ...(1)$$

and in terms of magnetic field vector $\vec{H}$, it is given by

$$\nabla^2\vec{H} = \mu\varepsilon\frac{\partial^2\vec{H}}{\partial t} \quad ...(2)$$

The general wave equation in three dimensions has the form

$$\nabla^2\psi = \frac{1}{c^2}\frac{\partial^2\psi}{\partial t^2} \quad ...(3)$$

where ψ is the '*disturbance*' or '*wave function*' and is a function of *x*, *y*, *z* and *t*, *c* is the wave velocity.

Comparasion of Equation (1), (2) and (3) gives

Velocity of the electromagnetic wave $c = \frac{1}{\sqrt{\mu\varepsilon}}$

Equation (1) gives the wave equation for vector $\vec{E}$ and Equation (2) for the vector $\vec{H}$. So the velocity of the vector $\vec{E}$ or $\vec{H}$ in electro magnetic waves propagating through free space is given by $\frac{1}{\sqrt{\mu_0\varepsilon_0}}$

It can be seen that $\frac{1}{\sqrt{\mu\varepsilon}}$ has dimensions of velocity.

Taking the dimensions of μ and ε we have

$$\mu\varepsilon = \frac{\text{kg metre}}{\text{coulomb}^2}\times\frac{\text{coulomb}^2\ \text{sec}^2}{\text{kg. metre}^2} = \frac{\text{sec}^2}{\text{m}^2}$$

$$\frac{1}{\sqrt{\mu\varepsilon}} = \frac{\text{metre}}{\text{sec}} = \text{m/sec}$$

which corresponds to the dimensions of velocity.

Example 1. *Calculate the value poynting vector for a 60 w bulb at a distance of 0.5 m from it.*

Solution: Powers of the bulb = 60 w

Since the light from the bulb is spread in a sphere at radius $r = 0.5$ m, so the poynting vector S_{Av} (average value of poynting vector over the surface of sphere) is given by

$$S_{Av} = \frac{P}{4\pi r^2} = \frac{60}{4\pi(0.5)^2} = 19.1\,\text{w/m}^2 \text{ Ans.}$$

Example 2. *From the sun energy received on the surface of earth is found to be 1.33 kw/m². Find the electric field associated with sunlight as received on the earth's surface. Assume the sun's light to be monochromatic (λ = 6000 Å). Take the value of ε = 9 × 10⁻¹² S.I. units.*

Solution: Energy received/sec/m² on the surface of Earth = Average value of poynting vector on the Earth's surface.

$$S_{Av} = \frac{1}{2} c\varepsilon E_0^2$$

$$\therefore \quad E_0 = \left(\frac{2S_{Av}}{c\varepsilon}\right) = \left(\frac{2\times 1.33\times 10^3}{3\times 10^8 \times 9\times 10^{-12}}\right)^{1/2} = 992.5 \text{ volt/meter Ans.}$$

Example 3. *Calculate the poynting vector and the amplitude of the electric field if a laser beam of 500 w is concentrated using a bust on a cross-sectional area of 10^{-10} m².*

Given $\varepsilon_0 = 9 \times 10^{-12}$ F/m

Solution:

Poynting vector is given by

$$\vec{S} = \frac{\text{Power}}{\text{Area}} = \frac{500}{10^{-10}} = 5\times 10^{12} \text{ w/m}^2$$

Average value of Poynting vector is given by the relation

$$S_{Av} = \frac{1}{2} c\cdot\varepsilon\cdot E_0^2$$

$$\therefore \text{ Electric field } \quad E_0 = \left(\frac{2S_{Av}}{c\varepsilon}\right)^{1/2} \qquad \text{...(1)}$$

Now $t = 9 \times 10^{-12}$ F/m

$c = 3 \times 10^8$ m/sec

$$\therefore \quad E_0 = \left(\frac{2\times 5\times 10^{12}}{3\times 10^8 \times 9\times 10^{-12}}\right)^{1/2} = 60.9\times 10^6 \text{ V/m}$$

[**Note:** Units of electric vector E may be got from Equation (1) as

$$E_0 = \left(\frac{W}{m^2}\times\frac{S}{m}\times\frac{m}{\text{farad}}\right)^{1/2} = \left(\frac{w\times\text{sec}}{m^2}\times\frac{\text{volt}}{\text{coulomb}}\right)^{1/2}$$

$$= \left[\frac{\text{Coulomb}\times\text{volt}\times\text{volt}}{\text{m}^2\times\text{coulomb}}\right]^{1/2} = \text{volt/meter]}$$

Example 4. *Calculate the magnitude of the poynting vectors and the amplitude of the electric and magnetic fields at a distance of 10 km from a radio station which is radiating power of 10^5 watt uniformly over a hemisphere with radio station as centre.*

Given $\mu_0 = 4\pi \times 10^{-7}$ henry/meter and $\varepsilon = 9 \times 10^{-12}$ farad/meter

Solution: Total average power radiated by radio station $\rho = 10^5$ w

Radius of hemisphere = 10 kms = 10^4 m

Let S_{Av} be the average poynting vector over the surface of hemisphere, then

$$S_{Av} = \frac{\rho}{2\pi r^2} = \frac{10^5}{2\times 3.142\times(10^4)^2} = 1.59\times10^{-4}\ \text{w/m}^2$$

For free space, average poynting vector is $S_{Av} = \frac{1}{2}c\,\varepsilon_0 E_0^2$

$$\therefore \quad E_0 = \left[\frac{2S_{Av}}{c\,t_0}\right]^{1/2} = \left[\frac{2\times1.59\times10^{-4}}{3\times10^8\times9\times10^{-12}}\right]^{1/2} = 0.343\ \text{V/m}$$

$$\text{Now } H_0 = \frac{B_0}{\mu_0} = \frac{E_0}{\mu_0 c} = \frac{0.343}{4.7\times10^{-7}\times3\times10^8} = 9.1\times10^{-4}\ \text{ampere turns/m Ans.}$$

Example 5. *If the earth receives 2 cal/min/cm² soler energy, what are the amplitudes of electric and magnetic fields of radiation.*

Solution: Poynting vector $\vec{S}$ gives the energy transported by e.m. waves/unit area/sec. So from equation

$$\vec{S} = (\vec{E}\times\vec{H})$$

Energy flux/unit area/sec is given by

$$|\vec{S}| = |\vec{E}\times\vec{H}| = EH\sin 90°$$

$$= EH \qquad [\because E\perp H]$$

The energy flux per unit area per second at the earth surface is given in the problem as:

$$|\vec{S}| = 2\ \text{cal min}^{-1}\ \text{cm}^{-2} = \frac{2\times4.2\times10^4}{60}\ \text{Joule m}^{-2}\ \text{sec}^{-1}$$

$$\text{Also} \qquad \frac{E}{B} = c$$

$$\therefore \qquad \frac{E}{H} = \mu_0 c$$

$$= \frac{\mu_0}{\sqrt{\mu_0\varepsilon_0}} = \sqrt{\frac{\mu_0}{\varepsilon_0}}$$

$$\frac{E}{H} = \sqrt{\frac{\mu_0}{\varepsilon_0}} = \sqrt{\frac{4\pi\times10^{-7}}{8.85\times10^{-12}}} = 376.72\ \text{ohm}$$

$$\therefore \qquad |S| = EH = \frac{2\times4.2\times10^4}{60}\ \text{J/m}^2/\text{sec}$$

Product of Equation (1) and (2) gives

$$EH\times(E/H) = 1400\times376.72 \text{ or } E^2 = 5.274\times105$$

$$\text{or} \qquad E = 726.6\ \text{volt/m}$$

Put this value of E in Equation (1), we get

$$(726.2)H = 1400 \text{ or } H = \frac{1400}{726.2} = 1.928\ \text{amp-turn/m}$$

So, the amplitudes of electric and magnetic fields of radiation are

$E_0 = E\sqrt{2} = 726.2\sqrt{2} =$ **1026.8 volt/m** and $H_0 = H\sqrt{2} = 1.928\sqrt{2} = 2.726$ amp-turn/m

Example 6. *The maximum electric field in a plane electromagnetic wave is 10² Newton/coul. The wave is going in the X-direction and the electric field is in the Y-direction. Find the maximum magnetic field in the wave and its direction.*

Solution: We have the relation for maximum electric field and maximum magnetic field as

$$E_0 = cB_0, \text{ where } c \text{ is the velocity of light}$$

$$\therefore \quad B_0 = \frac{E_0}{c} = \frac{10^2}{3 \times 10^8} = 3.33 \times 10^{-7} \text{ Weber/m}^2 \text{ or } 3.33 \times 10^{-7} \text{ tesla Ans.}$$

Example 7. *The relative permittivity of distilled water is 81. Calculate refractive index and velocity of light in it.*

Solution: We have the relation for refractive index $n = \sqrt{(\mu\varepsilon/\mu_0\varepsilon_0)}$ and $\upsilon = c/n$

In this case, $\varepsilon = 81\varepsilon_0$ and for distilled water $\mu = \mu_0$

So,
$$n = \sqrt{(81)} = 9 \text{ and } \upsilon = \frac{3 \times 10^8}{9} = \mathbf{3.33 \times 10^7 \text{ m/sec}}$$

Example 8. *Show that equation of continuity div* $\vec{J} + \frac{\partial \rho}{\partial t} = 0$ *is contained in Maxwell's equations.*

Solution: Maxwell's fourth equation has the differential form.

$$\vec{\nabla} \times \vec{H} = \vec{J} + \frac{\partial \vec{D}}{\partial t}$$

Taking divergence of both sides, we have

$$\vec{\nabla} \cdot (\vec{\nabla} \times \vec{H}) = \vec{\nabla} \cdot \left(\vec{J} + \frac{\partial \vec{D}}{\partial t} \right)$$

But $(\vec{\nabla} \cdot \vec{\nabla} \times \vec{H}) = 0$

$$\therefore \quad \vec{\nabla} \cdot \left(\vec{J} + \frac{\partial \vec{D}}{\partial t} \right) = 0$$

or
$$\vec{\nabla} \cdot \vec{J} + \vec{\nabla} \cdot \left(\frac{\partial \vec{D}}{\partial t} \right) = 0$$

or
$$\vec{\nabla} \cdot \vec{J} + \frac{\partial}{\partial t} (\vec{\nabla} \cdot \vec{D}) = 0$$

Now, from well's first equation $\vec{\nabla} \cdot \vec{D} = \rho$ (where ρ is surface charge density)

So,
$$\vec{\nabla} \cdot \vec{J} + \frac{\partial \rho}{\partial t} = 0$$

Hence, the result. Ans.

Example 9. *Assuming that all the energy from a 1000 watt lamp is radiated uniformly, calculate the average the intensities of electric and magnetic fields of radiation at a distance of 2 m from the lamp.*

Solution: Let ρ_0 be the power radial uniformly in all directions, then the power or energy flux per unit, per second at a distance r from the point source (in this case lamp) will be

$$S_{av} = \frac{P_0}{4\pi r^2} = \frac{1000}{4\pi (2)^2} \text{ Watt/m}^2$$

Now, from Poynting theorem,

$$|\vec{S}| = |\vec{E} \times \vec{H}| = EH \sin 90° \qquad (\because \text{ as } E \perp H)$$

$$\therefore \quad EH = \frac{1000}{16\pi} \qquad \ldots(1)$$

Now, $$\frac{E}{H} = \sqrt{\frac{\mu_0}{\varepsilon_0}} = 376.72 \text{ ohm} \qquad ...(2)$$

Product of (1) and (2) equation will be

$$EH \times \frac{E}{H} = \frac{376.72 \times 1000}{16\pi}$$

or $$E = \left(\frac{376.72 \times 1000}{16 \times 3.14}\right)^{1/2} \text{ or } \mathbf{E = 86.59 \text{ volt/m Ans.}}$$

Put in Equation (1), we get

$$H = \frac{1000}{16 \times 3.14 E} = \frac{1000}{16 \times 3.14 \times 86.59} = \mathbf{0.23 \text{ amp - turn/m Ans.}}$$

Example 10. *If the magnitude of $\vec{H}$ in a plane wave is 1 amp/metre, find the magnitude of for plane wave in free space.*

Solution. We have the relation,

$$\frac{H_0}{E_0} = \sqrt{\frac{\varepsilon_0}{\mu_0}} \qquad \therefore \qquad E_0 = \sqrt{\frac{\mu_0}{\varepsilon_0}} H_0$$

In this case $H_0 = 1$ amp/m, $\mu_0 = 4\pi \times 10^{-7}$ w/-Am and $\varepsilon_0 = 8.85 \times 10^{-12}$ C/N-m^2

$$E_0 = 1 \times \sqrt{\frac{4\pi \times 10^{-7}}{8.85 \times 10^{-12}}} = \mathbf{376.72 \text{ V/m Ans.}}$$

FORMS AND IMPORTANCE OF MAXWELL'S EQUATIONS INTEGRAL FORM

Name	*Equation*	*Importance*
1. Gauss's law of electricity	$\oint \vec{D} \cdot d\vec{s} = \int_V \rho dV$	It describes charge and electric field.
2. Gauss's law magnetism	$\oint \vec{B} \cdot d\vec{s} = 0$	It describes magnetic field.
3. Faraday's law	$\oint \vec{E} \cdot d\vec{l} = -\frac{\partial}{\partial t}\int \vec{B} \cdot d\vec{s}$	Describes an electric field produced by a magnetic field.
4. Maxwell's Law Ampere	$\int H \cdot \vec{dl} = \int \left(J + \frac{\partial \vec{D}}{\partial t} \right) \cdot \vec{ds}$	A magnetic field produced by a changing electric field or by a current or both.

DIFFERENTIAL FORM

1. Gauss's law of electricity $\qquad \vec{\nabla} \cdot \vec{D} = \rho$

2. Gauss's law of magnetism $\qquad \vec{\nabla} \cdot \vec{B} = 0$

3. Faraday's law $\qquad \vec{\nabla} \times \vec{E} = -\frac{\partial \vec{B}}{\partial t}$

4. Ampere Maxwell's law $\qquad \vec{\nabla} \times \vec{H} = \vec{J} + \frac{\partial \vec{D}}{\partial t}$

QUESTIONS AND PROBLEMS

1. Give the significance and importance of Maxwell's equations.
2. What are e.m. waves? How they propagate?
3. Give the importance of displacement current.
4. How e.m. waves propagate through a dielectric?
5. How will you calculate the speed of e.m. waves?
6. How the permittivity and permeability are related to the speed of e.m. waves?
7. Show that the ratio of electric build to the magnetic field is equal to the speed of light.
8. What is Poynting vector? Give its importance.
9. How will you signify the divergence and curl of a vector field?
10. Give the various forms of Maxwell's equations stating their importance.
11. If the electric amplitude of wave in 5 v/m, what is the magnetic amplitude of the wave.

    [Ans. 1.675×10^{-8} wb/m^2**]**
12. A place monochromatic linearly polarized light wave is travelling east ward, the wave in polarized with *e* directed vertically up and down write the expression for E, H and B if $E_0 = 0.1$ V/m and frequency = 20 MHZ.

    [Ans. $E_z = 0.1 \sin 4\pi \times 10^7\, t - 0.419\, x$; $By = -E_z/C$; $Hy = By/\mu_0$**]**
13. A mercury lamp is radialing monochromatic hight of 10 watt power. Find the electric field strength at a distance 5 m from the lamp. **[Ans.** 4.9 v/m**]**

PART IV
SOUND

Hermann von Helmholtz (1821-1894)

Hermann von Helmholtz was one of the most versatile scientists who ever lived. He was born in Potsdam, Germany, in 1821, the son of a high school teacher. From an early age he wanted to be a physicist but his family could not afford the money required for his studies. Instead, his father persuaded him to take up medicine since he could receive a free medical education on condition that he serve as a doctor in the Prussian army after receiving his medical degree.

In 1847 Helmholtz published a mathematical formulation of the principle of conservation of energy which caused quite a stir in scientific circles.

In 1858 he was appointed to a chair in anatomy at Heidelberg, and while there, devoted most of his time to studies in optics and sound. In 1867 he published his famous *Treatise on Physiological Optics* in three volumes.

In 1871 Helmholtz abandoned physiology in favour of physics. He accepted the chair in Physics at the University of Berlin, and spent the rest of his life there.

As he grew older, Helmholtz became more and more interested in the mathematical side of physics, and made noteworthy contributions to fluid mechanics, thermodynamics, and electrodynamics. He devoted the last decade of his life to an attempt to unify all of physics under one fundamental principle, the principle of least action.

Helmholtz was a sensitive and sickly man all his life, plagued by severe migraine headaches and fainting spells. He sought relief from his pain in music and the arts, and in mountain climbing in the Alps. It is intriguing to imagine what he might have accomplished if he had been in good health for all his 73 years! On Helmholtz's death Lord Kelvin summed up his accomplishments: "**In the historical record of science the name of Helmholtz stands unique in grandeur, as a master and leader in mathematics, and in biology, and in physics.**"

CHAPTER 1

SIMPLE HARMONIC MOTION

1.1. INTRODUCTION

Let a particle P move around a circle with a uniform speed v and let N be the foot of perpendicular drawn from P on YY', *i.e.*, on the vertical diameter of the circle. Let us consider the movement of N as P moves around the circle say in anticlockwise direction starting from X. When P is at X, N is at O, when P reaches Y, N moves along OY and coincides with Y. As P moves along the circle and reaches from Y to X', N moves from Y and reaches O. As P further travels from X' to Y', N travels from O to Y' and finally when P travels from Y' to X to complete one revolution N travels from Y' to reach O. So during the time P completes one revolution, N travels through diameter YY' twice.

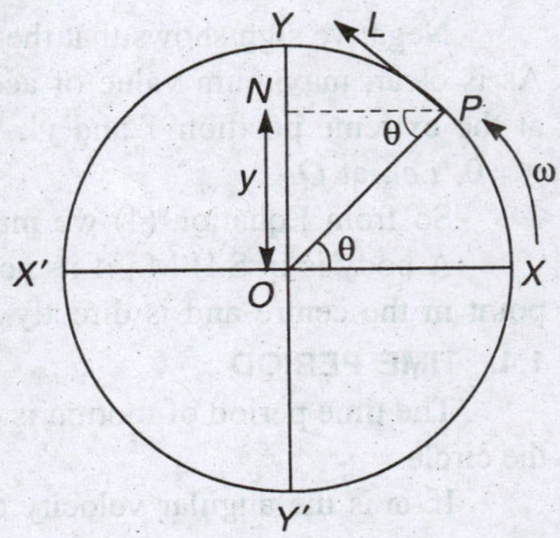

Fig. 1.1

This oscillatory motion of N about centre O is called Simple Harmonic Motion and it may be defined as the projection of a uniform circular motion on the diameter of a circle. This circular may be called circle of reference and the point P is called reference point.

If a be the radius of the circle and θ, the angle which the rotatory vector OP makes with OX, then the angular velocity ω of P is v/a and the displacement of N is given by

$$y = a \sin \theta \quad \text{...(1)}$$

if t be the time during which OP travels through an angle θ, then $\theta = \omega t$ and we have

$$y = a \sin \omega t$$

The radius a, *i.e.*, the maximum displacement on either side of O, is called the amplitude of motion.

1.2. VELOCITY OF MOTION

Velocity of N is given by a component parallel to diameter YY'. As is clear from the figure, the velocity of P is v along PL, *i.e.*, along the tangent of the circle. Its component parallel to YY', *i.e.*, $\perp$ to PN, is $v \sin \angle LPN = v \sin (90 - \theta) = v \cos \theta$.

So velocity of

$$\begin{aligned} N &= v \cos \theta \\ &= v \cos \omega t \\ &= v (1 - \sin^2 \omega t)^{1/2} \quad \text{...(2)} \end{aligned}$$

Now from Equation (2), $\sin \omega t = \dfrac{y}{a}$

and $$v = a\omega$$

$\therefore$

$$\begin{aligned} \text{Velocity of } N &= a\omega\sqrt{1 - y^2/a^2} \\ &= \frac{a\omega\sqrt{a^2 - y^2}}{a} \\ &= \omega\sqrt{a^2 - y^2} \quad \text{...(3)} \end{aligned}$$

So, we see that velocity of N varies along the path depending upon the displacement, *i.e.*, on the distance from O. It becomes maximum when y is minimum and minimum when y is maximum.

As is clear from Fig 1.1, the maximum value of v occurs when particle passes through the mean position, *i.e.*, O, and the velocity is zero when displacement is maximum on either side of O.

1.3. ACCELERATION

Acceleration of N will be the component along NO of the centripetal acceleration of P. As P is moving in a circle, so acceleration of P is $\frac{v^2}{a} = \frac{a^2\omega^2}{a} = a\omega^2$ directed towards centre, so we have

$$\text{Acceleration of } N = -a\omega^2 \sin\theta$$
$$= -\omega^2 . a \sin\theta$$
$$= -\omega^2 . y \quad \text{...(4)}$$

Negative sign shows that the acceleration acts opposite to the direction in which y increases. As is clear, maximum value of acceleration $= a\omega^2$, *i.e.*, the same as that of P and it will occur at the extreme position Y and y', when $y = a$ and minimum value of acceleration is zero when $y = 0$, *i.e.*, at O.

So from Equation (4) we may define S.H.M. (Simple Harmonic Motion) as under:

A body is in S.H.M., if its acceleration of motion $(\omega^2 y)$ is always directed towards a fixed point in the centre and is directly proportional to its distance from it.

1.4. TIME PERIOD

The time period of motion is equal to the time taken by P to complete one revolution around the circle.

If ω is the angular velocity of motion, then by definition

$$\omega = \frac{\text{Angle described in one revolution}}{\text{Time taken to complete one revolution}} = \frac{2\pi}{t}$$

or

$$T = \frac{2\pi}{\omega} \quad \text{...(5)}$$

Also we have from Equation (4)

$$\omega = \sqrt{\frac{\text{Acceleration}}{\text{Displacement}}}$$

$\therefore$ Equation (5) may be written as

$$T = 2\pi\sqrt{\frac{\text{Displacement}}{\text{Acceleration}}} \quad \text{...(6)}$$

1.5. FREQUENCY

Number of revolutions made by a body in one second is called the frequency of motion of the body. Which means that frequency is the reciprocal of time period, *i.e.*,

$$\text{Frequency } n = \frac{1}{T} \quad \text{...(7)}$$

1.6. PHASE

Phase of a vibrating body tells us about the position of the body and at any instant it may be defined as the position and direction of motion of the body, since it last passed through its mean position. The phase is either measured as fraction of time period or fraction of angle that has elapsed since the particle last passed through its mean position in the positive direction.

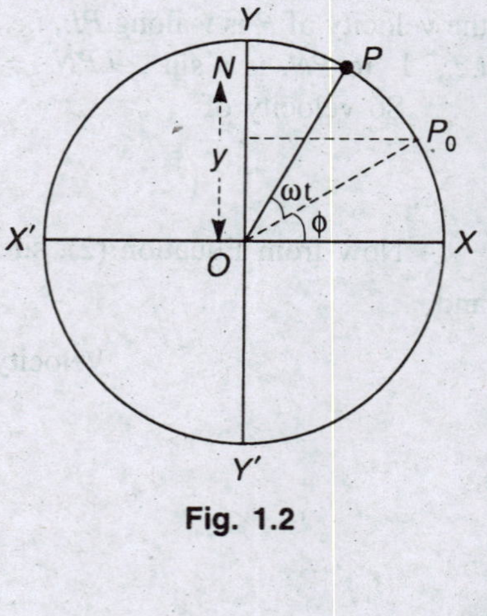

Fig. 1.2

The phase difference between two simple harmonic motions indicates, how much the two motions are out of step from each other, which in other words means how much a motion is ahead of other by time or angle.

Let us consider Fig. 1.2 alongside to illustrate the concept of phase further

Let a point P start its motion from X-axis as $t = 0$, and let θ be the angle it traces in time t with an angular velocity ω.

Displacement, y, along Y-axis is given by

$$y = ON = a \sin \theta = a \sin \omega t \quad ...(1)$$

If, however, time is measured when P is at P_0, then $\angle P_0OX = \phi$ and the displacement ON at any time t is given by

$$y = a \sin (\omega t + \phi) \quad ...(2)$$

where ωt is the angle traced in time t from position P_0.

At $t = 0$, phase is ϕ which is known as initial phase or epoch.

If the particle travels through an angle ϕ below X-axis, then the displacement is given by

$$y = a \sin (\omega t - \phi)$$

So we see that $y = a \sin (\omega t + \phi)$

represents the general equation for displacement in S.H.M. The angle ϕ is called the initial phase.

1.7. PERIOD OF VIBRATION OF A BODY UNDER THE ACTION OF ELASTIC FORCES

Let a mass m of a body be vibrating under the action of an elastic force. Let Fy be the force acting on M to displace it through a distance y, then from Newton's second law of motion

$$Fy = M \cdot \frac{d^2 y}{dt^2}$$

where $\frac{d^2 y}{dt^2}$ is the acceleration of the body. Since the motion is Simple Harmonic, so acceleration is proportional to displacement, and is given by

$$Fy \propto y \qquad \text{or} \qquad Fy = -F.y$$

where F corresponds to restoring force/unit displacement, which sometimes is also termed as stiffness. So we have

$$M \frac{d^2 y}{dt^2} = -F.y \qquad \text{or} \qquad \frac{d^2 y}{dt^2} = \frac{F}{M}.y$$

$\therefore$ Time period is given by

$$T = \frac{2\pi}{\sqrt{F/M}} = 2\pi \sqrt{M/F}$$

$$= 2\pi \sqrt{\frac{\text{Mass}}{\text{Restoring force/unit displacement}}}$$

1.8. COMPOUND PENDULUM

A compound pendulum is a rigid body which is capable of oscillating freely about a horizontal axis passing through it.

The above statement means that a rigid body of any shape or internal structure will constitute a pendulum if it is capable of oscillating due to gravity about a fixed horizontal axis.

The point where the axis of vibration meets the vertical plane through the centre of gravity perpendicular to the axis of oscillation is known as centre of suspension. In Fig. 1.3 alongside, such a pendulum is shown, where m is considered to be the mass of the pendulum and it is oscillating about an axis through O under the action of gravity.

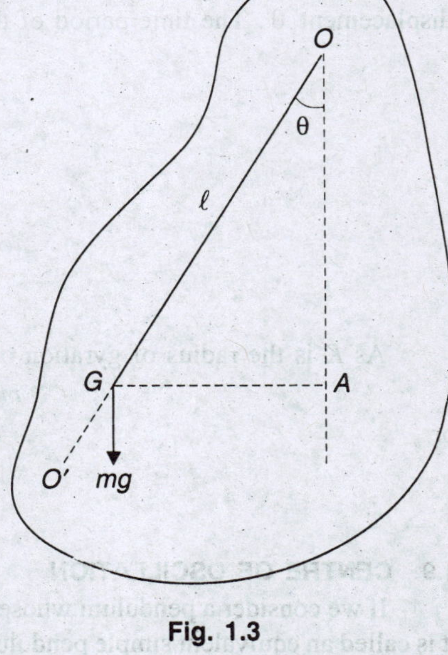

Fig. 1.3

Let G be the centre of gravity and l be the distance OG.

When the pendulum is displaced from its position of rest so that OG makes an angle θ with vertical OA, the weight of the pendulum mg is acting vertically downward and will produce an equal and opposite reaction at O. These two forces will form a couple and try to bring the pendulum back to initial position. The moment of this restoring couple is given by

$$C = mg \times OG = mg \times l \sin\theta \quad ...(1)$$

When the pendulum tries to come back to the position of rest, it will have an angular acceleration produced due to the couple given by

$$-I \cdot \frac{d\omega}{dt}$$

$$\therefore \quad C = -I \cdot \frac{d\omega}{dt} = mgl \sin\theta$$

or

$$mgl \sin\theta = -I \cdot \frac{d^2\theta}{dt^2} \quad ...(2)$$

where I is the moment of inertia of the pendulum and $\frac{a\omega}{dt} = \frac{d^2\theta}{dt^2}$ is its angular acceleration. The negative sign shows that as θ increases, the couple tries to reduce it.

Also we know that $I = mk^2$, where k is the radius of gyration about O

$$\therefore \quad mk^2 . \frac{d^2\theta}{dt^2} + mgl \sin\theta = 0$$

When θ is small $\sin\theta \simeq 0$

$$\therefore \quad mk^2 . \frac{d^2\theta}{dt^2} + mgl . \theta = 0$$

or

$$k^2 . \frac{d^2\theta}{dt^2} = -gl\,\theta$$

or

$$\frac{d^2\theta}{dt^2} = -\frac{g}{k^2} . l . \theta$$

From the above it is clear that the motion is S.H.M. as $\frac{d^2\theta}{dt^2}$ is proportional to angular displacement θ. The time period of the pendulum is given by

$$T = 2\pi \sqrt{\frac{\text{Displacement}}{\text{Acceleration}}}$$

$$= 2\pi \sqrt{\frac{\theta}{\frac{l.g.}{k^2}.\theta}}$$

$$= 2\pi \sqrt{\frac{k^2}{l.g.}}$$

As K is the radius of gyration of the pendulum about the parallel axis theorem,

$$mk^2 = mK^2 + ml^2$$

or

$$k^2 = K^2 + l^2$$

$$\therefore \quad T = 2\pi \sqrt{\frac{K^2 + l^2}{lg}} \quad ...(3)$$

1.9. CENTRE OF OSCILLATION

If we consider a pendulum whose time period is equal to that of given simple pendulum, then it is called an equivalent simple pendulum and if L be the length of the equivalent simple pendulum, then time period is given by

$$T = 2\pi\sqrt{\frac{L}{g}} \qquad \text{...(4)}$$

Comparing Equations (3) and (4),

$$L = \frac{K^2 + l^2}{l} = \frac{K^2}{l} + l$$

which shows that the length of the equivalent simple pendulum is more than l by an amount $\frac{K^2}{l}$. If a $\perp$ is dropped on the line OG, produced from A, then from the figure we have

$$OO' = \frac{K^2}{l} + l$$

which means that distance OO' is equal to the length of an equivalent simple pendulum. The point O' is called the centre of oscillation.

1.10. INTERCHANGEABILITY OF CENTRES OF SUSPENSION AND OSCILLATION

If we suspend the pendulum from O, the time period is given by

$$T = 2\pi\sqrt{\frac{K^2 + l^2}{lg}} \qquad \text{...(5)}$$

Now if the pendulum is suspended about a parallel axis through O', then the time period will be given by

$$T' = 2\pi\sqrt{\frac{K^2 + l_1^2}{l_1 g}}$$

which may be written as

$$T' = 2\pi\sqrt{\frac{K^2 + \left(\frac{K^2}{l}\right)^2}{\frac{K^2}{l} \times g}}$$

$$= 2\pi\sqrt{\frac{\frac{K^2}{l}\left[l + \frac{K^2}{l}\right]}{\frac{K^2}{l} \times g}}$$

$$= 2\pi\sqrt{\frac{K^2 + l^2}{gl}} \qquad \text{...(6)}$$

From Equations (5) and (6) it is clear that the time period of the pendulum in both the cases is same. So the centres of suspension and oscillation are interchangeable.

1.11. MAXIMUM TIME PERIOD

From Equation (5), we see that $T = \infty$, when $l = 0$ or $l = \infty$, which means that time period is maximum about the centre of gravity or about a point at infinite distance from centre of gravity.

1.12. MINIMUM PERIOD

Equation (5) can be written in another form as

$$T = 2\pi\sqrt{\frac{(K-l)^2 + 2Kl}{lg}}$$

From which it is clear that when $K = l$, T is minimum, which in other terms means that time period will be minimum when the distance between axis through the point of gravity is equal to radius of gyration of the pendulum about a parallel axis.

1.13. POINTS ABOUT WHICH TIME PERIOD OF A COMPOUND PENDULUM IS SAME

Time period of a compound pendulum is given by

$$T = 2\pi\sqrt{\frac{K^2 + l^2}{lg}} \qquad ...(1)$$

where l is the distance of point of suspension form centre of gravity G and K is the radius of gyration about G.

Squaring Equation (1), we have

$$T^2 = 4\pi^2\left(\frac{K^2 + l^2}{lg}\right)$$

or $$l^2 - \frac{gT^2}{4\pi^2}\cdot l + K^2 = 0$$

which is a quadratic equation, and two values of l are given by

$$l = \frac{\frac{gT^2}{4\pi^2} \pm \sqrt{\frac{g^2T^4}{16\pi^2} - 4K^2}}{2}$$

Let the two values be l_1 and l_2, then

$$l_1 = \frac{gT^2}{8\pi^2} + \sqrt{\frac{g^2T^4}{16\pi^2} - K^2} \qquad ...(1)$$

and $$l = \frac{gT^2}{8\pi^2} - \sqrt{\frac{g^3T^4}{16\pi^2} - K^2} \qquad ...(2)$$

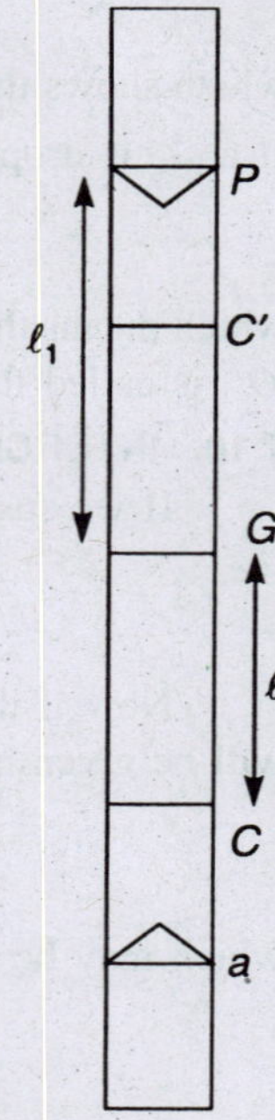

Fig. 1.4

So we see that there are two points on either side of G at a distance l_1 and l_2 respectively about which the compound pendulum has the same time periods. So in all there are four points collinear with $C.G.$ about which the periods of vibration are equal.

In Fig. 1.4, P and C' are two points at a distance l_1 and l_2 above $C.G.$ Q and C are corresponding points below $C.G.$

If L be the length of the equivalent simple pendulum, which will correspond to distance between two points lying on either side of $C.G.$ but not equidistant from it then

$$L = PC = QC'$$

But $$PC = PG + GC = l_1 + l_2$$

$\therefore$ $$L = l_1 + l_2$$

Adding Equations (1) and (2), we get

$$l_1 + l_2 = \frac{2gT^2}{8\pi^2} = L$$

or $$L = \frac{gT^2}{4\pi^2}$$

which will give us the length of the equivalent simple pendulum which will have the same time period.

Example 1. *A hydrogen atom has a mass of 1.68×10^{-27} kgm. When attached to a certain massive molecule, it oscillates as a classical oscillator with a frequency of 10^{14} cycles/sec and with an amplitude of 10^{-10} m. Calculate the force acting on hydrogen atom.*

Solution: In this case amplitude of hydrogen atom executing S.H.M. $a = 10^{-10}$ m and angular velocity is given by

$$\omega = 2\pi n$$
$$= 2\pi \times 10^{14} \text{ radians/sec}$$

Mass of hydrogen atom $m = 1.68 \times 10^{-27}$ kgm

Force acting on hydrogen atom $= \dfrac{mv^2}{a} = ma\omega^2$

$$= 1.68 \times 10^{-27} \times 10^{-10} \times (2\pi \times 10^{14})^2$$

$$= 1.68 \times 4 \times \pi^2 \times 10^{-9}$$

$$= \mathbf{66.29 \times 10^{-9}\ Newton}$$

Example 2. *A body executing S.H.M. describes 120 vibrations per minute and has a velocity of 5 m/sec. What is the length of its path? What is its velocity when it is half way between its mean position and an extremity of its path?*

Solution: Time period $T = \dfrac{1}{120}\text{min} = \dfrac{60}{120} = 0.5\,\text{sec}$

Velocity at mean position, when $y = 0 = 5$ m

Now we have the relation $v = \omega\sqrt{a^2 - y^2}$ ...(1)

and at mean position, $y = 0$

$\therefore$ $v = a\omega = 5$

or $a \cdot \dfrac{2\pi}{T} = 5$

or $a = \dfrac{5 \times T}{2\pi} = 0.394$ m

Since the total path $= 2 \times$ amplitude

$\therefore$ length of path $= 2 \times 0.394$
$= 0.788$ m.

Again from Equation (1), when the body is half way between mean position and an extremity, *i.e.*

$y = a/2$, we have

$$v = \omega\sqrt{a^2 - \frac{4}{a^2}}$$

$$= \frac{3}{4}\cdot a\omega = \frac{\sqrt{3}}{4}a \times \frac{2\pi}{T}$$

$$= \frac{\sqrt{3}}{4} \times 0.094 \times \frac{2\pi}{0.5}$$

$$v = \mathbf{4.286\ m/sec.}$$

So the length of the path of a body is 0.788 m and its velocity half way between an extreme position and mean position is 4.286 m/sec.

Example 3. *A body executes S.H.M. such that its velocity at mean position is 1 m/sec and acceleration at one extremity is 1.57 m/sec². Calculate the time period of vibration.*

Solution: We have the velocity at any point given by

$$v = \omega\sqrt{a^2 - y^2}$$

At mean position $v = 1$ m/sec

and $y = 0$

$\therefore$ $1 = a.\omega$ ...(1)

The acceleration of a body executing S.H.M. is given by

Acceleration $= \omega^2 y$

At extremity $y = a$.

and acceleration at extremity is given $= 1.57$ m/sec^2

$\therefore$ $1.57 = \omega^2.a$ ...(2)

Dividing Equation (2) by (1)

$\therefore$ $\omega = 1.57$

Now Time period is given by

$$T = \frac{2\pi}{\omega} = \frac{2 \times 3.142}{1.57}$$
$$= \mathbf{4 \cdot 5\ sec}$$

Example 4. *A thin uniform bar of length 120 cm is made to oscillate about an axis through its end. Find the period of oscillation and other points about which it can oscillate with same period.*

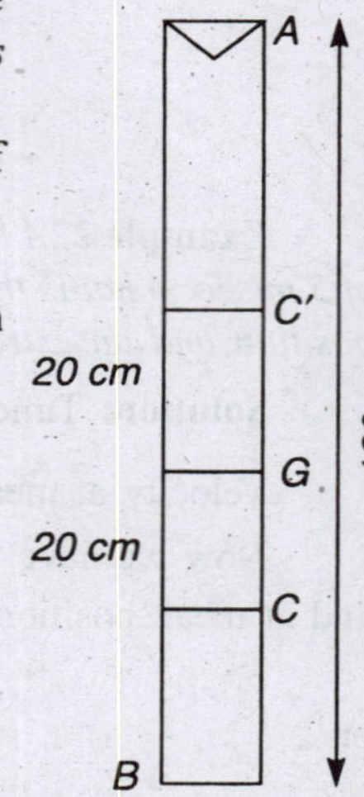

Fig. 1.5

Solution: Let AB be the bar of length a = 120 cm and let the point of oscillation be A.

Let M be the mass of the bar and G be its centre of gravity.

Moment of Inertia about an axis through G and $\perp$ to the bar is given by

$$I = Mk^2 = \frac{Ma^2}{12} = \frac{M \times 120^2}{12}$$

$$\therefore \quad k^2 = \frac{120 \times 120}{12} = 1200$$

Also $\quad AG = 60 \text{ cm} = l$

Now time period of the pendulum about A is given by

$$T = 2\pi\sqrt{\frac{k^2 + l^2}{lg}} = 2\pi\sqrt{\frac{1200 + (60)^2}{60 \times 980}}$$
$$= \frac{4\pi}{7} = 1.796 \text{ second}$$

Let L be the length of the equivalent pendulum, then

$$T = 2\pi\sqrt{L/g}$$

or
$$T^2 = \frac{4\pi^2 L}{g}$$

$$\therefore \quad L = \frac{gT^2}{4\pi^2} = \frac{980 \times 16\pi^2}{4 \times 49 \times \pi^2} = 80 \text{ cm}$$

So there is another point C which is at a distance of 80 – 60 = 20 cm from G about which pendulum will have same time period.

In case the pendulum is suspended from the end B, it will have the same time period as at A. Similarly the time period will be same at C', at a distance of 20 cm from G on the other side.

So there will be same time period at four points A, B, C and C' of the pendulum.

Example 5. *A uniform thin rod of length 120 cm and width 6 cm is swinging in a vertical plane as pendulum about a point A at some distance from one end. If the time of swing is minimum, find the distance of A from the end of the rod.*

Solution: Length of the rod

$$l = 120 \text{ cm}$$

Breadth of rod $\quad b = 60$ cm

Moment of Inertia $\quad I = M\frac{l^2 + b^2}{12} = Mk^2$

where k is the radius of gyration.

$$\therefore \quad k^2 = \frac{l^2 + b^2}{12} = \frac{(120)^2 + (6)^2}{12} = 1203$$

or $\quad k = 34.70$ cm.

For time period to be minimum about A, its distance from C.G. of the bar is equal to the radius of gyration *i.e.*, $l = k$.

∴ Distance from one end

$$= \frac{1}{2} \times 120 - 34.70 = \mathbf{25.30\ cm}$$

Vibrations and Resonance

1.14. FREE VIBRATION

If a simple pendulum is observed which is undergoing vibrations freely, then its time of vibrations is given by

$$T = 2\pi\sqrt{\frac{l}{g}}$$

and the frequency of vibration is given by

$$n = \frac{1}{2\pi}\sqrt{\frac{g}{l}}$$

where n will correspond to the natural frequency of the pendulum. A pendulum which is set to vibrate freely will always have the same fixed natural frequency and time period.

In a similar way if a simple spring mass system is set to vibrate freely, then its natural frequency is given by

$$n = \frac{1}{2\pi}\sqrt{\frac{k}{m}}$$

If no frictional force or resistance is offered to such a system, the body will keep on vibrating indefinitely and such vibrations are called free vibrations. So we may define free vibrations as the vibrations in which a body vibrates with its own natural frequency when left freely.

However, in practice free vibrations are not possible because friction of air cannot be removed completely. So the amplitude of vibrating body will go on decreasing till it becomes zero and body stops vibrating.

1.15. DAMPED VIBRATIONS

As we have seen above, a body set into vibrations will have its amplitude continuously decreasing due to frictional resistance and so the vibrations will die down after some time. This motion is said to be damped by friction and vibrations are called damped vibrations. The resisting frictional force will be proportional to the velocity of motion of the body.

An example of damped vibrations can be had from a simple experiment. A mass is suspended from the spring and set to vibrate in air will vibrate for longer time as compared to the mass which vibrates partly in air and partly in a liquid kept below the mass as shown in Fig. 1.6. The damping force will be more when the mass moves in liquid and so the vibrations will stop quickly in liquid than in air.

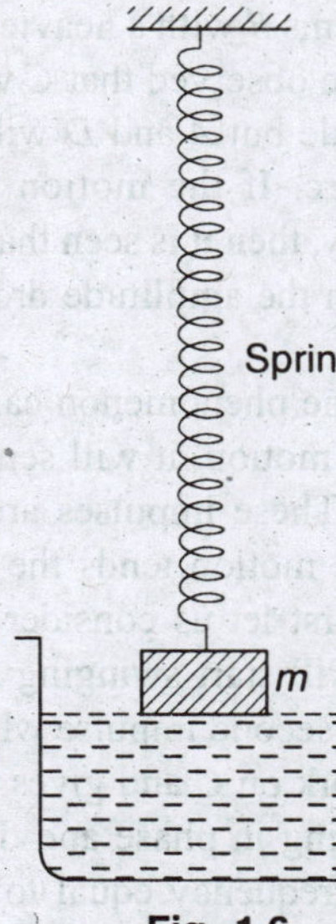

Fig. 1.6

1.16. FORCED VIBRATIONS

If a body is subjected to an external force while it is vibrating, then we have the case of forced vibrations. For example, if a bob of simple pendulum is held in hand and then given number of swings by the motion of hand. In this case the pendulum vibrates with frequency of the hand and not due to its natural frequency.

So forced vibrations may be defined as the vibrations in which a body vibrates with a frequency other than its natural frequency by some other external periodic force.

An example of the forced vibrations can be had by taking a tuning fork which is set into vibrations. If the stem of the tuning fork is kept in hand, very low sound intensity is produced, because the prongs of the tuning fork will set very little air around it into vibrations. However, if the stem is pressed on a table, the table is set into forced vibrations and the sound becomes very loud because a large volume of air is set into vibrations by table. However, in second case the intensity of sound will die down quickly as dissipation of energy is quick unlike the first case where energy dissipation is very slow.

1.17. RESONANCE

If a vibrating tuning fork is brought near a stationary tuning fork having frequency as that of the vibrating tuning fork, then we observe that stationary tuning fork also sets into vibrations. This phenomenon of making a body vibrate with its natural frequency equal to that of another body vibrating with the same frequency is called resonance

An example of resonance can be had by striking a tuning fork and holding it over a tube containing water. If the water level of the tube is slowly lowered so that the length of the air column is increased, then at some stage we hear a loud sound when the natural frequency of the air column is equal to the frequency of vibration of the tuning fork. This is the case of resonance.

Another example of resonance and forced vibrations can be had from a simple demonstration experiment.

Let four pendulums be suspended from a flexible support and let pendulums *B* and *C* be of equal lengths and *A* and *D* be of different lenghts. Let one of the pendulums *B* with a heavier bob be set into vibrations, it will be observed that *C* will also start gaining equal amplitude but *A* and *D* will never swing more than a small arc. If the motion of *A* and *D* are observed carefully, then it is seen that they gain some amplitude and then the amplitude drops to zero and the process goes on.

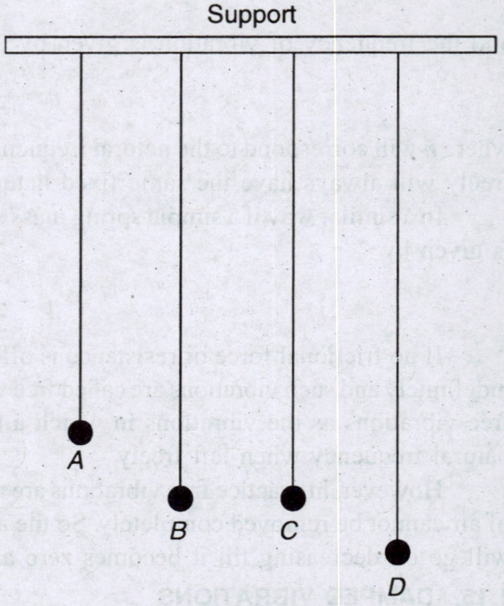

Fig. 1.7

The phenomenon can be explained that as *B* is set into motion, it will send periodic impulses along the flexible support which pass then to *A, C* and *D*. These impulses are due to the sudden motion towards either side of the support and the periodic motion tends the other pendulums into vibration.

First let us consider the effect on *C*. The first impulse will disturb it from position of rest and it will start swinging in forward and backward. Just when it has completed one vibration, it gets the second impulse which tends to move it in the direction of its motion. So the second impulse does work on *C* and gives it energy to increase its amplitude. The successive impulses will always be arriving in phase and due to cumulative effect, the pendulum *C* is set into resonant vibrations with a frequency equal to the applied frequency.

Now let us consider the effects on *A* and *D*. As stated above, the first impulse will disturb *A* and *D* and they will start vibrating. As the second impulse comes to *D*, it has not yet completed first vibration, because it has longer time period than *B*. If, however, the difference of time period is small, the second impulse is only slightly out of phase and still gives some energy to increase its amplitude. This will continue for some vibrations during which the impulses will go on out of phase more and more till time when *B* and *D* are moving in opposite directions. So the impulse from *B* will reach *D* in opposite phase and take away some energy from it, thereby reducing its amplitude of motion. So during the next few vibrations, the amplitude of *D* goes on decreasing till it drops to zero. The rise or fall of amplitude will go on till the pendulum *D* is forced to vibrate with a frequency equal to that of *B*, which is not the natural frequency of *D*. Similarly pendulum *A* will be set into forced vibrations due to the pendulum *B*.

Other examples of resonance can be given as under:

1. Musical instruments can be set to vibrate by bringing them in contact with instruments which have the frequency equal to the natural frequency of the instruments.

2. Soldiers marching on a suspension bridge are not allowed to march because the rhythm of their march and the natural frequency of the bridge may produce resonance and the result may be the collapse of the bridge.

3. In the case of two clocks which are placed near each other, tend to have the same time period because of the rhythm of one clock synchronizes with the other one and two time periods tend to be equal.

4. Tuning of the radios and transistors is also based on the principle of resonance between the incoming waves and the frequency of the receiver, and when the matching is suitably amplified we can listen to the sound of the incoming waves from the transmitter.

5. This phenomemon has also practical significance in the construction of machinery. In case the period of vibrations applied to the machinery match with its natural frequency, resonance takes place resulting in the breakdown of the machinery. So a precaution is to be taken so that the period of applied force does not match with the natural period of the machinery.

In addition to above examples, a number of other examples can be quoted from daily life for the resonance phenomenon.

1.18. SHARPNESS OF RESONANCE

The amplitude in case of forced vibration when resonance occurs is given by the relation

$$a = \frac{f_0}{2K\omega} \qquad ...(1)$$

where f_0 corresponds to applied force per unit mass, K is the damping coefficient and $\omega / 2\pi$ is the frequency of the applied force.

When resonance occurs, we have the relation

$$\omega_0^2 - \omega^2 = 0$$

or

$$\omega_0 = \omega$$

i.e., the frequency of the applied force is equal to the natural frequency of the body. Now we can take up the different cases.

(*a*) When $K = 0$ *i.e.*, no damping is present, from Equation (1) we have $a = \infty$ *i.e.*, the amplitude of resonance becomes infinite as shown in Fig. 1.8 (*i*).

(*b*) When damping is very small, the amplitude is quite large but falls off rapidly as the frequency of applied force is slightly different from the natural frequency of the body.

(*c*) When *K* increases, the amplitude at resonance decreases, but as shown in the figure, curves (*iii*) and (*iv*) fall off gradually and even when the frequency of applied force is slightly different from natural frequency of the body, the amplitude is nearly the same. Which in other words means that when damping is very small, resonance is quite sharp and when damping is large, resonance is flat.

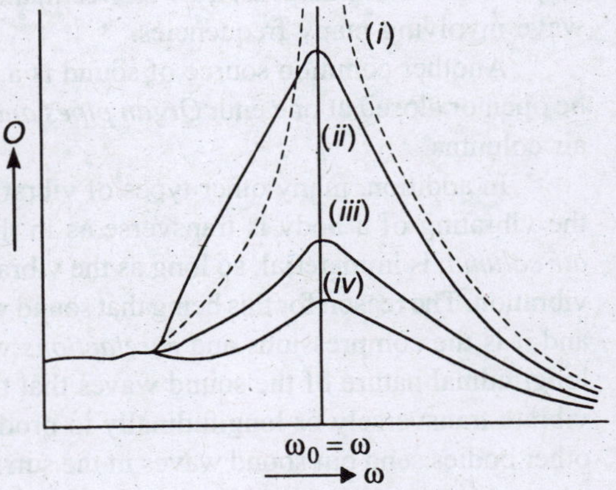

Fig. 1.8

We can take the examples of effect of dampness in case of air column and tuning fork and how the resonance is affected in these cases.

We know that the resonance is accumulated effect of a number of successive impulses. In case of an air column, the natural vibrations die out very quickly because the damping is large and by the time the effect of say 8th impulse is coming, that of the first almost disappears. Which in other words means that the first impulse has no effect on 8th impulse and there is no chance of the impulse to go out of phase and thereby reduce the amplitude.

It means that in such a case slight mistuning does not effect.

However, the case of a tuning fork is different. In this case the damping is small and the natural vibrations of the fork die out quite slowly. So when the 8th impulse comes, the first one still exists and there is every chance of impulses to go out of phase and destroy each other's effect even when the frequency difference is very less. So in case of tuning forks, the frequency of the applied force and the natural frequency of the tuning fork should perfectly match, so that the natural frequency of vibration and the impressed force vibrations are always in phase. In case of two tuning forks, for resonance to occur, both should be in unison because of the above reason, while in case of a tuning fork and an air column slight mistuning we also show resonance phenomenon.

1.19. NATURE OF SOUND

Sound may be defined *Psychologically or Physically*. In the first case, it is a sensation produced by a body and received by an ear. In the second definition, it is some stimulus which is capable of producing the sensation of sound. Whether an observer is necessary for sound to exist depends on which of the above definitions is used.

1.20. VIBRATING SYSTEMS

Any body that has the properties of inertia and elasticity may be sent into vibrations. Most of the bodies vibrate in more than one manner and with each of these modes, there is associated some frequency. In each of these modes, a set of *standing waves* is set up in the body and such standing waves may be *longitudinal* or *transverse*, depending on the shape and characteristics of the vibrating body.

A common source of sound is the **vibrating string**. A string, when plucked or disturbed, sends transverse waves in both the directions along the string and are reflected at the ends which are the fixed points for the string. As a result, the string vibrates in any of its natural harmonic frequencies. In the case of a string these characteristic frequencies are given by

$$f = n\frac{v}{f} = n.\frac{v}{2l}$$

where $n \rightarrow$ is an integer

$l \rightarrow$ the length of the string

and $v \rightarrow$ the speed of the transverse waves in the string.

It may be noted that in practice a string when plucked may vibrate with many of *characteristic frequencies*. Simultaneously, so the resultant wave may not be a simple *sine wave* but a complex wave involving many frequencies.

Another common source of sound is a *vibrating or resonating* air column. This column may be open or closed at one end. *Organ pipes and wind instruments* are common examples of vibrating air columns.

In addition, many other types of vibrating bodies are able to produce sound waves. Whether the vibrating of a body is transverse as in the case of a string or a longitudinal as in the case of *air column*, is immaterial, so long as the vibration can set the surrounding medium into longitudinal vibration. The reason for this being that sound waves are constituted by *longitudinal mode of vibrations* and it is the compressions and *rarefactions* which travel through a medium. Again it is due to the longitudinal nature of the sound waves that these waves cannot travel through *vacuum*. Rods may vibrate transversely or longitudinally to produce sound waves. Plates, bells, diaphragm and many other bodies send out sound waves in the surrounding medium. A string stretched between the fixed points may send very little energy to the surrounding air. Many *primary* vibrator are *coupled* to another body so that when the second body is set into vibrations, more energy is transmitted efficiently. The second body set into vibration may acquire *forced vibrations* or *resonance*. Under these conditions, the vibrators releases more energy per unit time and the sound is greatly amplified.

A loud speaker is used to increase the intensity of sound either by **electrical amplification or by resonance.** Two general types of loud speakers are in normal use, the direct radiator, such as a *cone loud speaker used in radios or horn type used in public address systems.* The direct radiator is more commonly used because of its simple design and less space it occupies. A horn type loud speaker consists of an electrically or mechanically driven diagram coupled to a horn. The air column of the horn produces resonance for a very wide range of frequencies and so the intensity of sound emitted is amplified quite appreciably. Horn type loud speaker is very much in use for public address systems or in cases where sound has to be transmitted over a long distance.

1.21. ATTENUATION COEFFICIENTS OF A VIBRATING SYSTEM

By definition, the energy of an oscillator is proportional to the square of its amplitude. In case of a *damped oscillator,* the amplitude decays exponentially with time as e^{-rt}, which means that energy will decay as $(e^{-rt})^2 = e^{-2rt}$. So the decay rate of energy depends on 2p *which is the damping force/unit mass* at an instant when the vibrating body is moving with a unit velocity. The following three characteristics may give the *attenuation of a vibrating system completely.*

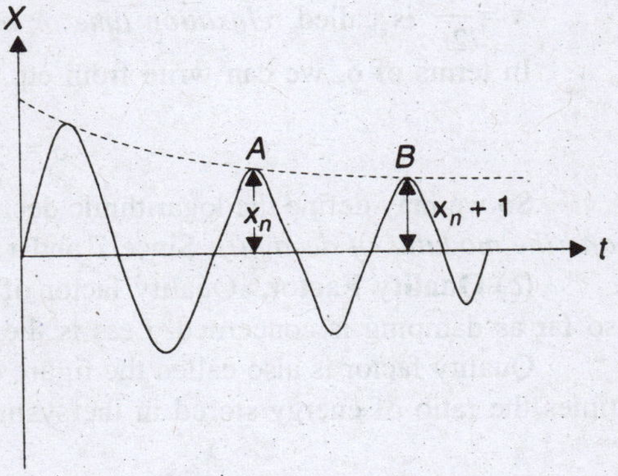

Fig. 1.9

(a) **Logarithmic Decrement.** It measures the rate at which the amplitude of vibration decays, as shown in Fig. 1.9. Let A and B be two successive maxima corresponding to displacements x_n and x_{n+1} and separated by a time period $\frac{2\pi}{q}$, where q corresponds to the angular frequency. So if the maxima at A occurs at time t, then at B it will occur, at time t such that

$$t = t_1 + \frac{2\pi}{q}$$

$$\text{also} \quad x_n = a_0\, e^{-rt} \qquad \ldots (1)$$

and

$$x_{n+1} = a_0\, e^{r(t_1 + 2\pi/q)} \qquad \ldots (2)$$

Dividing (2) by (1), we get

$$\frac{x_{n+1}}{x_n} = e^{-rT} \quad (\text{where } T = 2\pi/q)$$

Taking log on both sides

$$-\log \frac{n+1}{x_n} = rT \qquad \ldots (3)$$

Let us define logarithmic decrement (δ) of the damped motion as the natural logarithm of the ratio of amplitude of vibration at the instant of time t_1 and $t_1 + T$, we get

$$\delta = rT = \frac{2\pi r}{q} = \frac{\pi C}{mq} \qquad \ldots (4)$$

where,

$$2r = \frac{C}{m}\left(\frac{\text{Damping coeff.}}{\text{Mass of vibrating system}}\right)$$

As is evident, δ gives us a method for finding daming coefficient C, as all other parameters can be found either experimentally or from the displacement curve of damped vibrations.

(*b*) **Relaxation time (Modulus of decay).** An alternate method of describing the attenuation of damped vibration is by means of time required to elapse for the amplitude to decay to $\frac{1}{e}$ of its original value a_0.

At time t, the amplitude is given by

$$a_t = a_0 e^{-rt}$$

$$\text{for } r = \frac{1}{t}$$

$$= a_0 e^{-1} = \frac{a_0}{e}$$

$\tau = \frac{1}{2r}$ is called *relaxation time or modulus of decay.*

In terms of δ, we can write from eq., $\delta = rT$ as

$$\delta = \frac{T}{2\tau}$$

So we may define the logarithmic decrement as *half the ratio of time period of vibration (T) and the modulus of decay (t).* Since T and τ can be found experimentally, so δ can be calculated.

(*c*) **Quality Factor.** Quality factor of a damped oscillator defines the quality of oscillator so far as damping is concerned. Less is the damping higher is the quality factor Q.

Quality factor is also called the figure of merit of a harmonic oscillator and is defined as 2π times the ratio of energy stored in the systme to the energy lost per cycle. So,

$$Q = \frac{\text{Energy stored}}{\text{Energy lost per cycle}} = \frac{2\pi E}{-dE}$$

where – dE is the energy lost over a period.

Now $\frac{2\pi}{T} = 2\pi\nu = \omega$, so we have

$$Q = \frac{E\omega}{\frac{-dE}{T}} = \frac{E\omega}{\frac{-dE}{dt}} = \omega.\tau \qquad \text{... (6)}$$

where t is the relaxation time of the oscillator.

When damping is low, $\omega = \omega_0$, So

$Q = \omega_0.\tau$, which is a constant of damped motion

Also

$$w_0 = \sqrt{\frac{S}{m}}$$

where S → Force constant

and m → mass of vibrating system

and $\tau = \frac{1}{2\pi}$

$$\therefore \qquad Q = \frac{1}{2\pi}\sqrt{\frac{S}{m}}$$

which shows that for low damping quality factor is higher. In fact *lower* the damping *higher* is the quality factor. The time for energy E to $E_0 e^{-1}$ is given by $\tau = \frac{1}{2\pi}$.

In this interval of time, the oscillator executes $\frac{\omega_0 \tau}{2\pi}$ or $\frac{Q}{2\pi}$ oscillations. So *the phase of oscillator shifts by Q.*

It is evident that the quality factor has an important significance in the sense that Q of a damped oscillator is the phase change brought about in the oscillaor during the time interval when energy of oscillator decays to $\frac{1}{e}$ of its initial value. So the quality factor can given us an idea of the speed of *attenuation of amplitude* of a vibrating system.

Example 6. *If the quality factor of an undamped tuning fork of frequency 256 is 10^3, calculate the time in which its energy is reduced to $\frac{1}{e}$ of its energy in the absence of damping. How many oscillations the turning fork will make in this time?*

Solution. In the case of a damped harmonic oscillator, the energy at any time t is given by

$$E = E_0 \, e^{-2rt} = E_0 \, e^{-t/\tau}$$

where E_0 = initial energy in the absence of dampoing and τ is the relaxation time.

When
$$t = \tau$$
$$E = E_0 / e$$

Also quality factor Q is given by

$$Q = \omega_0 . t$$

$$\therefore \qquad t = \frac{Q}{\omega_0} = \frac{10^3}{2\pi \times 256} = \frac{10^3}{512} = 0.62 \text{ sec.}$$

So the energy is reduced by $\frac{1}{e}$ to the value in 0.62 second. The number of oscillations made by turning fork in time t will be

$$= \left(\frac{\omega_0}{2\pi}\right)\tau = \frac{Q}{2\pi} = \frac{10^3}{2\pi} = 159 \textbf{ Ans.}$$

QUESTIONS AND PROBLEMS

1. Define amplitude, time period, frequency and phase and explain Simple Harmonic Motion.

2. Derive an expression for time period and frequency of S.H.M.

3. What is S.H.M.? How can it be applied to a compound pendulum? Illustrate that centre of suspension and oscillation of acompound pendulum can be interchanged.

4. Differentiate between free, force and damped motion. Give an example of each.

5. Distinguish between free vibrations and forced vibrations. What is resonance? Give some practical examples of resonance.

6. Does the hammer of an electric bell make free or forced vibrations when the bell is ringing?

7. Discuss the effect of damping on sharpness of resonance. Why should two tuning forks be very accurately in unison to show resonance, while a tuning fork and an air column required only to be approximately tuned?

8. A particle executing S.H.M. of amplitudes 5 cm, when the particle is 3 cm from its mean position, its acceleration is found to be 48 cm/sec. Find

(*i*) its velocity at that instant.

(*ii*) its time period.

(*iii*) its maximum velocity. **[Ans.** (*i*) 16 cm/sec (*ii*) 1.63 (*iii*) 20 cm/sec]

9. Find the acceleration of a body excuting S.H.M. at 20 cm from its mean position. The time pe;riod of the body is 10 sec and amplitude is 1 metre. How long will it take to travel 70 cm from one of its extremities? $\left[\textbf{Ans. } \frac{7}{9} \text{ cm/sec}^2 \text{ and 2 sec.}\right]$

CHAPTER 2

WAVE MOTION

2.1. WAVE MOTION

Wave motion is a form of disturbance which travels through the medium due to the repeated periodic motion of the particles of the medium about their mean positions, the disturbance being handed over from one particle to the next. For example, when a stone is dropped into a pond of water, waves are produced at the point where the stone strikes the water. These waves travel outward, the particles of water vibrate only up and down about their mean positions. In the same way when a tuning fork is set into vibration, it produces waves in air, and the wave travels from one particle to the next but the particles of air vibrate about their mean positions.

Wave motion usually refers to the *transfer* of energy from one point to another point of the medium, *e.g.*, transfer of energy like sound, heat, light, *X*-rays, γ -rays, radio waves etc., takes place in the form of wave motion.

In order that the transfer of energy may take place, the medium must possess the following characteristics:

(*i*) Negligible frictional resistance.

(*ii*) Elastic property.

(*iii*) Property of inertia.

2.2. CHARACTERISTICS OF WAVE MOTION

1. Wave motion is a disturbance produced in the medium by the repeated periodic motion of the particles of the medium.

2. Only the wave travels forward while the particles of the medium vibrate about their mean positions.

3. There is a regular phase change between the various particles of the medium which means that the particle ahead starts vibrating a little later than a particle just preceding it.

4. The velocity of the wave is different from the velocity with which the particles of the medium are vibrating about their mean positions. The basic difference being that the wave travels with a uniform velocity while the velocity of the particles is different at different positions. It is maximum at the *mean position* and zero at the *extreme position* of the particles.

The following two types of wave motion exist:

***(i)* Transverse wave motion.** It is that wave motion in which the particles of the medium vibrate about their mean position in the direction at right angles to the direction of propagation of the wave.

Example: Light waves.

***(ii)* Longitudinal wave motion.** It is that motion in which the particles of the medium vibrate about their mean position in the direction of the wave.

Example: Sound waves.

2.3. EQUATION OF A SIMPLE HARMONIC WAVE

The displacement of a particle *P* in S.H.M. is given by

$$y = a \sin \omega t \qquad ...(1)$$

If we consider another particle Q at a distance x from P (towards right of P) and assume that the wave is travelling with a velocity v from left to right *i.e.*, from P to Q, the displacement of Q is given by

$$y = a\sin(\omega t - \phi) \quad \text{...(2)}$$

where ϕ is the phase difference between the particles P and Q. As phase difference corresponding to path difference λ is 2π, so we have

$$\frac{2\pi}{\lambda} = \frac{\phi}{x} \quad \text{...(3)}$$

where x is the path difference between P and Q.

$$\therefore \quad \phi = \frac{2\pi x}{\lambda}$$

Now $$\omega = \frac{2\pi}{T} = \frac{2\pi v}{\lambda}$$

Putting the values of ϕ and ω in Equation (2), we get

$$y = a\sin\left(\frac{2\pi vt}{\lambda} - \frac{2\pi x}{\lambda}\right)$$

$$y = a\sin\frac{2\pi}{\lambda}(vt - x) \quad \text{...(4)}$$

Equation (4) represents the general equation for a simple harmonic wave.

Similarly for a particle at a distance x in the negative direction (*i.e.*, to the left of ϕ, the equation for displacement may be written as)

$$y = a\sin\frac{2\pi}{\lambda}(vt + x) \quad \text{...(5)}$$

2.4. DIFFERENTIAL EQUATION OF WAVE MOTION

We have the general equation for S.H. wave as

$$y = a\sin\frac{2\pi}{\lambda}(vt - x) \quad \text{...(1)}$$

Differentiating Equation (1) w.r.t. time, we get

$$\frac{dy}{dt} = \frac{2\pi av}{\lambda}\cos = \frac{2\pi}{\lambda}(vt - x) \quad \text{...(2)}$$

Again differentiating Equation (2) w.r.t. time, we have

$$\frac{d^2y}{dt^2} = -\frac{4\pi^2 av^2}{\lambda^2}\sin\frac{2\pi}{\lambda}(vt - x) \quad \text{...(3)}$$

In order to get the value of compression, differentiate Equation (1) w.r.t. x. We get

$$\frac{dy}{dx} = -\frac{2\pi a}{\lambda^2}\cos\frac{2\pi}{\lambda}(vt - x) \quad \text{...(4)}$$

To get the value of compression in terms of distance, differentiate Equation (4) w.r.t x. We get

$$\frac{d^2y}{dx^2} = -\frac{4\pi^2 a}{\lambda^2}\sin\frac{2\pi}{\lambda}(vt - x) \quad \text{...(5)}$$

From (2) and (4), we have $$\frac{dy}{dt} = -v\cdot\frac{dy}{dx} \quad \text{...(6)}$$

Also from Equations (3) and (5), we have

$$\frac{d^2y}{dt^2} = v^2\frac{d^2y}{dx^2} \quad \text{...(7)}$$

which represents the *differential equation of wave motion.*

So. we may write the general wave motion equation as

$$\frac{d^2 y}{dt^2} = K\frac{d^2 y}{dx^2} \quad ...(8)$$

where $K = v^2$ or $\quad v = \sqrt{K}$

which shows that the value of *wave velocity* can be found if K is known.

2.5. PARTICLE VELOCITY AND WAVE VELOCITY

The general equation of S.H. wave is given by

$$y = a\sin\frac{2\pi}{\lambda}(vt - x) \quad ...(1)$$

where v is *the velocity of the wave and y is the displacement of the particle.*

The velocity of the particle U will be given by

$$U = \frac{dy}{dt}$$

∴ Differentiating Equation (1) w.r.t. time we have,

$$U = \frac{dy}{dt} = \frac{2\pi av}{\lambda}\cos\frac{2\pi}{\lambda}(vt - x) \quad ...(2)$$

The maximum value of the particle velocity is

$$U_{\max} = \frac{2\pi av}{\lambda} \quad ...(3)$$

when $\quad \cos\frac{2\pi}{\lambda}(vt - x) = 1$

∴ $\quad$ [Maximum particle velocity] $= \frac{2\pi a}{\lambda} \cdot$ wave velocity

The particle acceleration is given by

$$f = \frac{d^2 y}{dt^2} = -\frac{4\pi^2 av^2}{\lambda}\sin\frac{2\pi}{\lambda}(vt - x) \quad ...(4)$$

$$f = -\frac{4\pi^2 v^2}{\lambda^2}\left[a\sin\frac{2\pi}{\lambda^2}(vt - x)\right] \quad \text{or} \quad f = -\left(\frac{4\pi^2 v^2}{\lambda}\right)y$$

For maximum acceleration

$$y = a$$

$$\therefore \quad f_{max} = -\left(\frac{4\pi^2 v^2}{\lambda^2}\right)a \quad ...(5)$$

The negative sign shows that the acceleration of the particle is directed towards its mean position.

Differentiating Equation (1) w.r.t. x,

$$\frac{dy}{dt} = -\frac{2\pi a}{\lambda}\cos\frac{2\pi}{\lambda}(vt - x) \quad ...(6)$$

where dy/dx represents the slope of the displacement curve.

From Equations (2) and (6),

$$U = \frac{dy}{dt} = -v\left(\frac{dy}{dx}\right) \quad ...(7)$$

∴ Particle velocity at any instant = wave velocity × slope of the displacement curve at that instant.

2.6. DISTRIBUTION OF VELOCITY AND PRESSURE IN A PROGRESSIVE WAVE

For a plane Progressive wave, we have the equation

$$y = a\sin\frac{2\pi}{\lambda}(vt - x) \quad ...(1)$$

The particle velocity is given by,

$$U = \frac{dy}{dt} = \frac{2\pi av}{\lambda}\cos\frac{2\pi}{\lambda}(vt - x) \quad ...(2)$$

The strain in the medium is dy/dx. Which means that if dy/dx is positive, it represents a region of rarefaction. If dy/dx is negative, it represents a region of compression. By definition the modulus of elasticity of the medium

$$K = \frac{\text{Change in pressure}}{\text{Volume strain}} = \frac{-dp}{(dy/dx)}$$

or
$$dP = -K.\left(\frac{dy}{dx}\right)$$

or
$$dP = K.\left(-\frac{dy}{dx}\right) \quad ...(3)$$

which means if dy/dx is negative, dP is +ve *i.e.*, it is a region of *compression*. If dy/dx is positive, dP is negative, *i.e.*, it is a region of *rarefaction*.

Differentiating Equation (1) w.r.t. x, we get

$$\frac{dy}{dx} = -\frac{2\pi a}{\lambda}\cos\frac{2\pi}{\lambda}(vt - x) \quad ...(4)$$

From Equation (3)

or
$$dP = K\left(-\frac{dy}{dx}\right)$$

$$\therefore \quad dP = \frac{2\pi Ka}{\lambda}\cos\frac{2\pi}{\lambda}(vt - x) \quad ...(5)$$

The graphs for displacement, velocity and change in pressure are given in Fig. 2.1.

Where dP corresponds to the change in pressure, P_0 is the normal pressure of the medium when the wave is not propagating through it.

2.7. ENERGY OF A PROGRESSIVE WAVE

In the case of progressive wave, new waves are continuously formed *i.e.*, there is continuous transfer of energy in the direction of propagation of the wave. This energy is supplied from the source. The energy transferred per second also corresponds to the energy possessed by the particles in a length v, where v represents the velocity of the wave. The energy of the particles is partly *kinetics* and partly *potential.* The kinetic energy being due to the velocity of the vibrating particles and for a particle executing simple harmonic motion, the velocity is maximum at the mean position and it is zero at the extreme positions. So the kinetic energy of the particle at the mean position is *maximum* and *zero* at the extreme positions. Similarly the particles also possess potential energy due to their displacements from their mean positions.

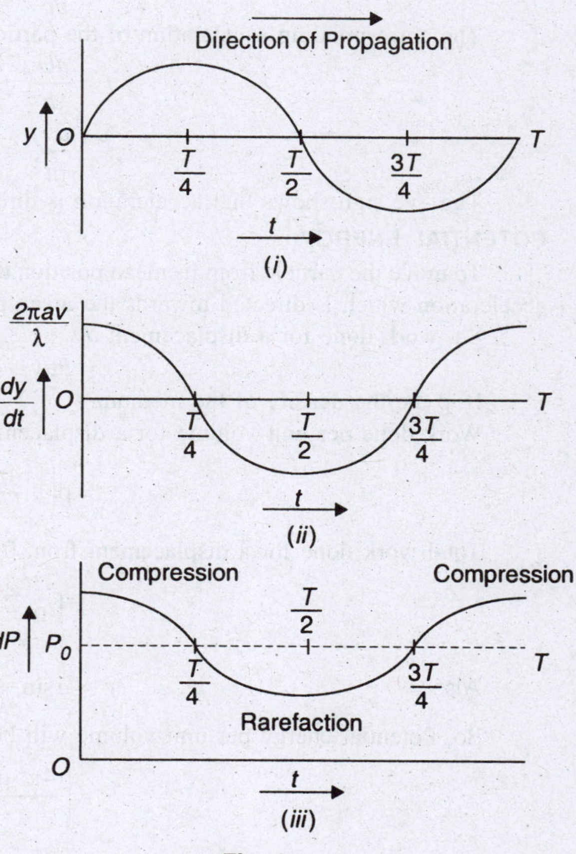

Fig. 2.1

The potential energy being *maximum* at the extreme positions and *minimum* at the mean positions.

In the case of longitudinal wave motion, because the compressions and rarefactions are formed so the energy distribution is not uniform over the wave. At the points where particles prossess maximum velocity, the compressions will be *maximum* and so the energy will be *maximum*. At points of zero velocity, no compression will be there and so the particles do not possess any energy at these points.

However, in the case of progressive wave motion, *there is no transfer of the medium in the direction of propagation of the wave, but there is always transfer of energy in the direction of propagation of the wave.*

ANALYTICAL TREATMENT FOR ENERGY TRANSFER

The equation of a simple harmonic wave is given by

$$y = a\sin\frac{2\pi}{\lambda}(vt - x) \quad ...(1)$$

The particle velocity U at any instant can be got by differentiating Equation (1) w.r.t. time

$$\therefore \quad U = \frac{dy}{dt} = \frac{2\pi av}{\lambda}\cos\frac{2\pi}{\lambda}(vt - x) \quad ...(2)$$

The corresponding acceleration of the particle at that instant,

$$f = \frac{dU}{dt} = \frac{d^2y}{dt^2} \quad ...(3)$$

$$\therefore \quad f = \frac{d^2y}{dt^2} = -\frac{4\pi^2 av^2}{\lambda^2}\sin\frac{2\pi}{\lambda}(vt - x) \quad ...(4)$$

The –ve sign shows that acceleration is directed towards the mean position.

POTENTIAL ENERGY

To move the particle from its mean position through a distance y, work has to be done against acceleration which is directed towards the mean position

So, work done for a displacement dy

$$= Fdy$$

If ρ be the density of the medium.

Work done per unit volume for a displacement dy is given by

$$= \rho\left(+\frac{4\pi^2 av^2}{\lambda^2}\sin\frac{2\pi}{\lambda}(vt - x)\right)dy$$

Total work done for a displacement from 0 to y will be

$$= \int_0^y \rho\left(\frac{4\pi av^2}{\lambda^2}\sin\frac{2\pi}{\lambda}(vt - x)\right)dy$$

Also $$y = a\sin\frac{2\pi}{\lambda}(vt - x)$$

So, Potential energy per unit volume will be

$$= \left(\frac{4\pi^2 \rho v^2}{\lambda^2}\right)\int_0^y ydy$$

$$= \frac{4\pi^2 \rho v^2 y^2}{2\lambda^2}$$

$$= \frac{2\pi^2 \rho v^2 y^2}{\lambda} = \frac{2\pi^2 \rho v^2}{\lambda} a^2 \sin^2\left[\frac{2\pi}{\lambda}(vt - x)\right] \quad ...(5)$$

Kinetic energy per unit volume $$= \frac{1}{2}\rho U^2$$

$$= \frac{1}{2}\rho\left[\frac{2\pi av}{\lambda}\cos\frac{2\pi}{\lambda}(vt - x)\right]^2$$

$$= \frac{2\pi^2 \rho v^2 a^2}{\lambda} \cos^2\left[\frac{2\pi}{\lambda}(vt - x)\right] \quad ...(6)$$

Total energy per unit volume,

$$E = PE + KE$$

$$= \frac{2\pi^2 \rho v^2 a^2}{\lambda^2}\left[\sin^2 \frac{2\pi}{\lambda}(vt - x) + \cos^2 \frac{2\pi}{\lambda}(vt - x)\right]$$

or

$$E = \frac{2\pi^2 \rho v^2 a^2}{\lambda^2} \quad ...(7)$$

Also $v = n\lambda$

$\therefore$

$$E = 2\pi^2 \rho n^2 a^2 \quad ...(8)$$

Which shows that average kinetic energy per unit volume and the average potential energy per unit volume *are equal* and each is equal to *half the total energy per unit volume.*

Average K.E. per unit volume $= \pi^2 \rho n^2 a^2$ ...(9)

Average P.E. per unit volume $= \pi^2 \rho n^2 a^2$ ...(10)

Let the area of cross-section of a parallel beam of radiation be 1 unit and the velocity of the wave be v.

$\therefore$ Volume $= 1 \times v = v$

So, Energy transfer per unit area per second

$$= E \times v = 2\pi^2 \rho n^2 a^2 v \quad ...(11)$$

Energy transfer per second is also called *energy current per unit area of cross section.*

It is to be noted that the P.E. and K.E. of every particle will change with time, but the average kinetic energy per unit volume and the average potential energy per unit volume remain constant. The current density per unit area of cross section and the total energy per unit volume also remain constant.

Example 1. *At a time t = 0, a train of waves has the form* $y = 4 \sin 2\pi\left(\frac{x}{100}\right)$. *The velocity of the wave is 30 cm/s. Find the equation giving the wave form at a time t = 2s.*

Solution: The general wave equation is given by

$$y = a \sin 2\pi\left[\frac{t}{T} - \frac{x}{\lambda}\right] \quad ...(1)$$

when $t = 0$, Equation (1) may be written as

$$y = a \sin 2\pi\left(\frac{x}{\lambda}\right) \quad ...(2)$$

We have the equation

$$y = 4 \sin 2\pi\left(\frac{x}{100}\right) \quad ...(3)$$

Comparing Equations (2) and (3)

$$a = 4$$

and $\lambda = 100$ cm

At time $t = 2$ s, $x = 2 \times 30 = 60$ cm

$\therefore$

$$y = a \sin 2\pi\left[\frac{x}{100} - \frac{60}{100}\right]$$

$$\mathbf{y = 4 \sin 2\pi\left[\frac{x}{100} - 0.6\right]} \quad \textbf{Ans.}$$

Example 2. *When a simple harmonic wave is propagated through a medium, the displacement of a particle in cm at any instant of time is given by* $y = 10 \sin \frac{2\pi}{100}(36000t - 20)$. *Calculate the amplitude of the vibrating particle, wave velocity, wavelength, frequency and time period.*

Solution: The general equation of a simple harmonic wave is given by

$$y = a\sin\frac{2\pi}{\lambda}(vt - x) \qquad ...(1)$$

We have the equation

Here $$y = 10\sin\frac{2\pi}{100}(36000t - 20) \qquad ...(2)$$

Comparing Equations (1) and (2)

$a = 10$ cm, $\lambda = 100$ cm

Wave velocity, $v = 36000$ cm/s, Frequency $n = \frac{v}{\lambda} = \frac{36000}{100}$

or $n = 360$ hertz

and time period $T = \frac{1}{n} = \frac{1}{360}$ **second. Ans.**

Example 3. *What is amplitude, the wavelength and the velocity of the wave represented by the formula, $\phi(x, t) = 5\sin(6\pi t - 4x)$ where x, the distance and t, the time are measured in S.I. units?*

Solution: We have

$$\phi(x, t) = 5\sin(6\pi t - 4x)$$

$$= 5\sin 2\pi\left(3t - \frac{2x}{\pi}\right) \qquad ...(1)$$

The general equation of S.H.M. is

$$\phi(x, t) = a\sin 2\pi\left(\frac{t}{T} - \frac{x}{\lambda}\right) \qquad ...(2)$$

Comparing Equations (1) and (2), we have

$$a = 5\text{m}$$

$$\frac{1}{T} = 3 \qquad \therefore \quad \text{Frequency } n = 3$$

and $$\lambda = \frac{\pi}{2}\text{m}$$

Now velocity $c = n\lambda = 3 \times \frac{\pi}{2} = \frac{3\pi}{2}$ ms^{-1} **Ans.**

Example 4. *Show that in simple harmonic wave $\dot{y}$ is ahead of the phase by $\pi/2$ w.r.t. y and $\ddot{y}$ is further ahead of the phase by $\pi/2$.*

Solution: We have the general S.H. wave as

$$y = a\sin 2\pi\left(\frac{t}{T} - \frac{x}{\lambda}\right)$$

Differentiating, we get $$\dot{y} = \frac{2\pi}{T}a\cos 2\pi\left(\frac{t}{T} - \frac{x}{\lambda}\right)$$

$$= \frac{2\pi}{T}\cdot a\sin\left[2\pi\left(\frac{t}{T} - \frac{x}{\lambda}\right) + \frac{\pi}{2}\right] \qquad ...(1)$$

i.e., $\dot{y}$ is ahead of phase by $\pi/2$.

Again differentiating, we have

$$\ddot{y} = -\left(\frac{2\pi}{T}\right)^2 a\sin 2\pi\left(\frac{t}{T} - \frac{x}{\lambda}\right)$$

$$= \left(\frac{2\pi}{T}\right)^2 \sin 2\pi\left[\left(\frac{t}{T} - \frac{x}{\lambda}\right) + \pi\right]$$

i.e., $\ddot{y}$ is ahead of phase by π w.r.t. y and by $\frac{\pi}{2}$ w.r.t. $\frac{\pi}{\dot{y}}$ **Ans.**

Example 5. *A simple harmonic wave of amplitude 8 units traverses a line of particles in the direction of the positive X-axis. At any given instant of time, for a particle at a distance of 10 cm from the origin, the displacement is +6 units, and for a particle at a distance of 25 cm from the origin, the displacement is +4 units. Calculate the wavelength*

$$y = a \sin \frac{2\pi}{\lambda}(vt - x).$$

Solution: We have the general wave equation as

or $$\frac{y}{a} = \sin 2\pi\left(\frac{t}{T} - \frac{x}{\lambda}\right)$$

We have the relation

$$\frac{y_1}{a} = \sin 2\pi\left(\frac{t}{T} - \frac{x_1}{\lambda}\right)$$

In this case $y_1 = +6,\ a = 8,\ x_1 = 10$ cm

$$\therefore \quad \frac{6}{8} = \sin 2\pi\left(\frac{t}{T} - \frac{10}{\lambda}\right) \quad ...(1)$$

Again we have the relation

$$\frac{y_2}{a} = \sin 2\pi\left(\frac{t}{T} - \frac{x^2}{\lambda}\right)$$

In this case $y_2 = +4,\ a = 8,\ x_2 = 25$

$$\therefore \quad \frac{4}{8} = \sin 2\pi\left(\frac{t}{T} - \frac{25}{\lambda}\right) \quad ...(2)$$

From Equation (1), we have

$$0.75 = \sin 2\pi\left(\frac{t}{T} - \frac{10}{\lambda}\right)$$

Now $$\sin\left(\frac{48.6\pi}{180}\right) = 0.75$$

$$\therefore \quad 2\pi\left(\frac{t}{T} - \frac{10}{\lambda}\right) = \frac{48.6\,\pi}{180}$$

or $$\frac{t}{T} - \frac{10}{\lambda} = \frac{48.6}{360} \quad ...(3)$$

From Equation (2), we have

$$0.5 = \sin 2\pi\left(\frac{t}{T} - \frac{25}{\lambda}\right)$$

Now $$\sin\frac{\pi}{6} = 0.5$$

$$\therefore \quad 2\pi\left(\frac{t}{T} - \frac{25}{\lambda}\right) = \frac{\pi}{6}$$

$$\frac{t}{T} - \frac{25}{\lambda} = \frac{1}{12} \quad ...(4)$$

Subtracting Equation (4) from Equation (3), we get

$$\frac{25}{\lambda} - \frac{10}{\lambda} = \frac{48.6}{360} - \frac{1}{12}$$

or λ = **290.8 cm Ans.**

Example 6. *Consider the wave y = 2.2 cos (300t – 0.24x). If the units of y, t and x are mm, sec and metres respectively, deduce (i) the amplitude (ii) frequency (iii) wave velocity (iv) the amplitude of particle velocity.*

Solution: We have the general wave equation as

$$y = a\cos 2\pi\left(\frac{t}{T} - \frac{x}{\lambda}\right) \quad ...(1)$$

In this case, the equation given is

$$y = 2.2 \cos (300t - 0.24x) \quad ...(2)$$

Comparing Equations (1) and (2), we have

(*i*) amplitude a = **2.2 mm.**

(*ii*) $\frac{2\pi}{T} = 300$

$\therefore$ $T = \frac{2\pi}{300}$

and frequency $n = \frac{300}{2\pi} = \mathbf{\frac{150}{\pi}\ Hz}$

(*iii*) $\frac{2\pi}{\lambda} = 0.24$

$\therefore$ $\lambda = \frac{2\pi}{0.24}$

wave velocity $c = v\lambda = \frac{300}{2\pi} \times \frac{2\pi}{0.24} = \mathbf{1250\ ms^{-1}}$

(*iv*) Amplitude of particle velocity

$$= \frac{2\pi}{T} \cdot a = 300 \times 2.2 \text{ m/s}$$

$$= \mathbf{0.66\ ms^{-1}}\ \textbf{Ans.}$$

Example 7. *A simple harmonic wave is represented by*

$$y = 10\sin\left(\frac{2\pi t}{T} + \alpha\right)$$

The time period is 30 s. At time t = 0, the displacement is 5 cm. Calculate (*i*) *phase angle at t = 7.5 s, and* (*ii*) *phase difference between two positions at a time interval of 6 s.*

Solution: In this case

$$y = 10\sin\left(\frac{2\pi t}{T} + \alpha\right), \quad T = 30 \text{ s}$$

(*i*) At $t = 0$

$y = 5$ cm

$\therefore$ $5 = 10\sin\alpha$

or $\sin\alpha = 0.5$

$\therefore$ $\alpha = \mathbf{\frac{\pi}{6}}$ **radian**

(*i*) When $t = 7.5$ s

The phase angle $\phi = \left(\frac{2\pi t}{T} + \alpha\right)$

$$\phi = \frac{2\pi \times 7.5}{30} + \frac{\pi}{6}$$

$\therefore$ $\phi = \mathbf{\frac{2\pi}{3}}$ **radian**

(*ii*) The phase difference is given by

$$\theta = \frac{2\pi t}{T}$$

When $t = 6$ s

$$\theta = \frac{2\pi \times 6}{30} = \mathbf{\frac{2\pi}{5}}\ \textbf{radians Ans.}$$

Example 8. *The equation of a transverse wave travelling along a coil spring is*

$$y = 4 \sin \pi (0.010x - 2.0t)$$

where y and x are in cm and t in s. Find (i) amplitude (ii) wavelength (iii) initial phase at the origin (iv) speed (v) frequency of the wave.

Solution: We have the general wave equation

$$y = a \sin \frac{2\pi}{\lambda}(x - vt + \phi) \qquad \text{... (1)}$$

In the present case $y = 4 \sin \pi (0.010x - 2.0t)$ ... (2)

Comparing Equations (1) and (2), we have

(*i*) Amplitude $a = 4.0 \text{ cm} = \mathbf{4 \times 10^{-2}\ m}$

(*ii*) $$\frac{2\pi}{\lambda} = 0.010\pi$$

$$\therefore \qquad \lambda = \frac{2\pi}{0.010\pi} = 200 \text{ cm} = \mathbf{2.0\ m}$$

(*iii*) Initial phase at origin

$$\phi = 0$$

(*iv*) $$\frac{2\pi}{\lambda} v = 2.0\ \pi$$

$$\therefore \qquad v = \frac{2\lambda}{2} = \frac{2 \times 2}{2} = \mathbf{2.0\ ms^{-1}}$$

(*v*) Frequency $$n = \frac{v}{\lambda} = \frac{2.0}{2.0} = \mathbf{1.0\ Hz.\ Ans.}$$

Example 9. *A source of sound has a frequency of 512 Hz and an amplitude of 0.25 cm. What is the flow of energy across a square cm per second, if the velocity of sound in air is 340 m/s and the density of air is 0.00129 g/cm³?*

Solution: In this case

$$n = 512 \text{ Hz}, \qquad a = 0.25 \text{ cm}$$

$$\rho = 0.00129 \text{ g/cm}^3, \qquad v = 340 \text{ m/s} = 34000 \text{ cm/s}$$

Total energy per unit volume is given by

$$= 2\pi^2 \rho n^2 a^2$$

$\therefore$ Energy transferred per sq. cm. in one sec

$$= 2\pi^2 \rho n^2 a^2 v$$

$$= 2 \times (3.14)^2 \times 0.00129 \times (512)^2 \times (0.25)^2 \times 34000$$

$$= \mathbf{1.417 \times 10^7\ ergs/cm^2/s\ Ans.}$$

QUESTIONS AND PROBLEMS

1. Explain the term wave motion. Distinguish clearly between transverse and longitudinal wave motion.

2. Give the characteristics of wave motion.

3. Derive the equation for a plane simple harmonic wave.

4. Show that $U = -v \dfrac{dy}{dx}$ where U is the particle velocity and v is the wave velocity.

5. Derive the relation $\dfrac{d^2y}{dt^2} = v^2 \cdot \dfrac{d^2y}{dx^2}$.

6. Show that the energy of a plane progressive wave is given by $E = 2\pi^2 \rho n^2 a^2$.

7. Derive the equation of wave motion in the form $y = a \sin \frac{2\pi}{\lambda}(vt - x)$.

8. Discuss the distribution of energy in plane progressive wave.

9. Show that for a plane progressive wave, on the average, half the energy is kinetic and half potential. Hence derive the expression for intensity of sound in terms of pressure amplitude.

10. For a plane waves in air of frequency 1000 Hz and displacement amplitude 0.2×10^{-7} m, deduce (*i*) the velocity amplitude (*ii*) the intensity.
(Given $\rho = 1.3$ kg/m^2, $v = 340$ ms^{-1}). **[Ans.** (*i*) 1.3×10^{-4} ms^{-1}. (*ii*) $3.7 - 10^{-6}$ Wm^{-1}]

11. Find the displacement of an air particle, 3.5 m from origin of disturbance at $t = 0.055$ when a wave of amplitude 0.2 mm and frequency 500 Hz travels along it with a velocity of 350 ms^{-1}.
[Ans. 0 cm]

12. The equation of a transverse wave in a stretched wire is $y = 2 \sin 271\left(\frac{t}{0.01} - \frac{x}{30}\right)$ cm. Find (*i*) amplitude (*ii*) wavelength (*iii*) frequency (*iv*) speed of the wave.
[Ans. (*i*) 2 cm (*ii*) 30 cm (*iii*) 100 Hz (*iv*) 3×10^3 cms^{-1}]

CHAPTER

3

INTERFERENCE AND BEATS

3.1. INTERFERENCE

When two wave trains of sound are moving along the same straight line in a medium, then every particle of the medium is simultaneously acted upon by both these waves. If the two waves arrive at a point in the same phase *i.e.*, the *two crests* or *two troughs* (*two condensations* or *two rarefactions*) superimpose, then the resultant amplitude is equal to the sum of the two amplitudes of the respective waves, while at some other points where they arrive in opposite phase, *i.e.*, *crest* of one falls over the *trough* of the other or the *condensation* due to one falls over the rarefaction due to the, other the resultant amplitude is equal to the difference of the individual amplitudes. The points where two waves meet in phase will give maximum amplitude and so the sound intensity will be maximum, while the points where the two waves meet in opposite phase, those points will have minimum amplitudes and so the intensity of sound will also be minimum.

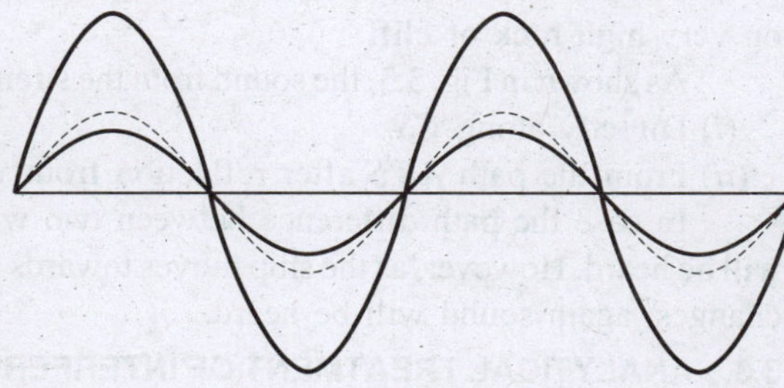

Fig. 3.1

This redistribution of sound energy when two sound waves superimpose upon each other is called interference of sound waves.

Examples. (*i*) Constructive Interference. One case in which the two waves of nearly the same amplitude superimpose is represented in Fig. 3.1 The resultant form of the waves is shown by a thick line. In this case the crest of one falls over the crest of the other and the trough falls over the trough of the other. This is case of *constructive interference* as the resultant amplitude of the two waves, the sum of the amplitudes of the component waves.

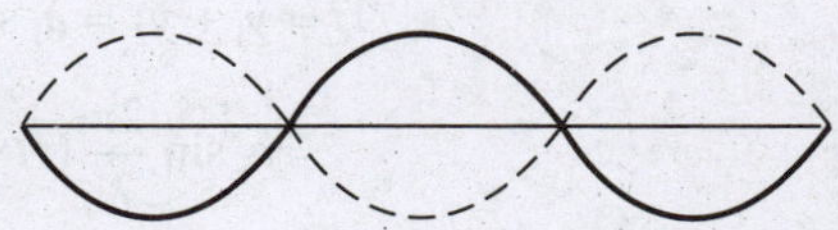

Fig. 3.2

(*ii*) Destructive Interference. When crest of one wave falls on the trough of the other *i.e.*, the two waves reach at a point with opposite phase, the resultant amplitude will be minimum (zero, if the two waves have equal amplitudes). This is the case of *destructive interference* (Fig. 3.2).

3.2. EXPLANATION OF SILENCE ZONES

The interference phenomenon explains the existence of silence zones.

Silence zones correspond to the positions where no sounds are heard. If a powerful source of sound is at some distance from an observer, then the observer will receive the direct sound waves as well as reflected sound waves from big surfaces, the intensity of both waves being nearly the same. However, since these waves have to travel different distances so these may reach

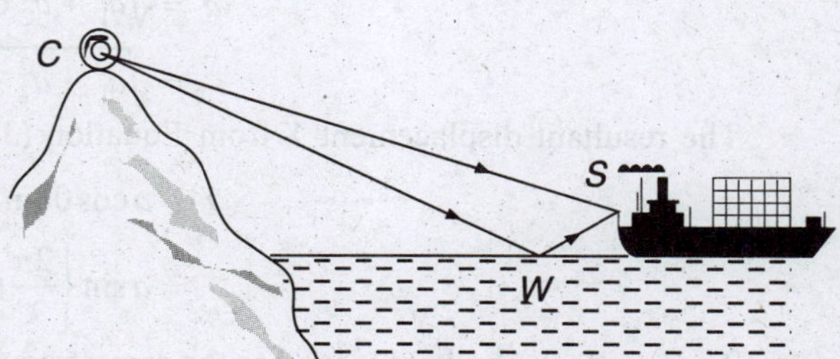

Fig. 3.3

the observer in the same phase or in opposite phase. If these waves reach the observer in same phase, the resultant amplitude will be maximum and so the sound heard is maximum. If these waves reach in opposite phase, they will cancel the amplitudes of each other and practically no sound will be heard. This will happen *if the path difference between the two waves is equal to an odd multiple of half a wavelength.* This explains the existence of *silence zones.* If the observer moves towards or away from the source, the sound is again heard. Its intensity increases and becomes maximum when the path difference is equal to an *even multiple of half a wavelength.* It again becomes minimum or zero when the path difference is an *odd multiple of half a wavelength.*

The presence of silence zones is observed when a ship approaches a fog siren which is installed on very high rock or cliff.

As shown in Fig. 3.3, the sound from the siren C may reach the ship S in the following two ways.

(*i*) Directly along CS.

(*ii*) From the path AWS after reflection from water surface at ω.

In case the path difference between two waves is $\lambda/2$ or odd multiple of $\lambda/2$ no sound will be heard. However, as the ship moves towards the cliff or away from it, so that the path difference changes, again sound will be heard.

3.3. ANALYTICAL TREATMENT OF INTERFERENCE PHENOMENON COMBINATION OF TWO SIMPLE HARMONIC MOTIONS OF THE SAME PERIOD BUT DIFFERING IN PHASE AND AMPLITUDE

Let the two waves having amplitudes a_1 and a_2 be represented by the equations

$$y_1 = a_1 \sin \frac{2\pi}{\lambda}(vt - x) \qquad \text{...(1)}$$

and

$$y_2 = a_2 \sin \left\{\frac{2\pi}{\lambda}(vt - x) + \phi\right\} \qquad \text{...(2)}$$

where ϕ is the phase difference between the two waves, not at the same time but after a time.

The resultant displacement is given by the relation

$$Y = y_1 + y_2 = a_1 \sin \frac{2\pi}{\lambda}(vt - x) + a_2 \sin\left\{\frac{2\pi}{\lambda}(vt - x) + \phi\right\}$$

$$= a_1 \sin \frac{2\pi}{\lambda}(vt - x) + a_2 \sin \frac{2\pi}{\lambda}(vt - x) + \cos\phi + a_2 \cos \frac{2\pi}{\lambda}(vt - x) + \sin\phi$$

$$= \sin \frac{2\pi}{\lambda}(vt - x)\{a_1 + a_2 \cos\phi\} + \cos \frac{2\pi}{\lambda}(vt - x)\{a_2 \sin\phi\} \qquad \text{...(3)}$$

Now put $a_1 + a_2 \cos\phi = a\cos\theta$ ...(4)

and $a_2 \sin\phi = a \sin\theta$ ...(5)

Dividing Equation (5) by (4), we get,

$$\tan\theta = \frac{a_2 \sin\phi}{a_1 + a_2 \cos\phi}$$

Squaring and adding Equations (4) and (5), we have the resultant amplitudes as

$$a = \sqrt{a_1^2 + a_2^2 \cos^2\phi + 2a_1 a_2 \cos\phi + a_2^2 \sin^2\phi}$$

$$= \sqrt{a_1^2 + a_2^2 + 2a_1 a_2 \cos\phi} \qquad \text{...(6)}$$

The resultant displacement Y from Equation (3) is given by

$$Y = a\cos\theta \sin\frac{2\pi}{\lambda}(vt - x) + a \sin\theta \cos\frac{2\pi}{\lambda}(vt - x)$$

$$= a \sin\left\{\frac{2\pi}{\lambda}(vt - x) + \theta\right\} \qquad \text{...(7)}$$

As is evident, resultant wave has the same frequency but has a different amplitude and phase from the amplitude and phase of the component wave trains.

Special cases

(*i*) When $\phi = 0$, $a = a_1 + a_2$ and $\tan \theta = 0$

i.e., the resultant vibration is in phase with the component vibrations.

(*ii*) When $\phi = 180°$ $a = a_1 - a_2$

If $a_1 = a_2$ then $a = 0$

Which means that phase difference between the two waves is zero, they reinforce each other. The resultant wave thus has the same period and its amplitude is the sum of the amplitudes of the component waves.

However, *i.e.,* the phase difference between the two waves is 180°, they destroy each other's effect, and the resultant amplitude is the difference between the amplitude of the component waves.

3.4. INTERFERENCE FRINGES PRODUCED BY A TUNING FORK WHICH STRIKES A LIQUID SURFACE

When the prongs of a vibrating tuning fork are made to touch the surface of water, they form the centres of two series of spherical waves of the same wavelength and amplitude. The waves consist of alternate crests and troughs as shown in Fig. 3.4. The crests are shown by bold lines and the troughs by dotted lines. When the crest of one wave falls over the crest of the other or the trough of one falls over the trough of the other, we get points, but when the crest of one falls over the trough of the other and vice-versa, we get points of minimum. If we denote the distance of any point from the source A by x_a and from the source B by x_b, then for positions of maxima

$$x_a - x_b = n\lambda$$

And for destructive interference phenomenon

$$d_a - d_b = (2n+1)\frac{\lambda}{2}$$

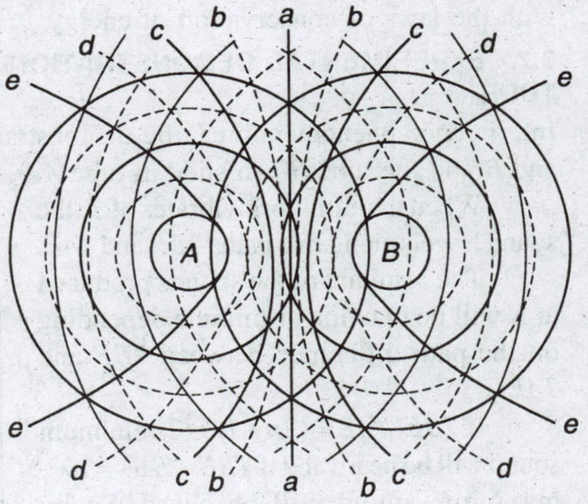

Fig. 3.4

As is clear from above, the locus of all such points for which there is a constant path difference between two fixed points is constant, is a *hyperbola,* which means that the interference fringes observed will have the hyperbolic representation.

3.5. CONDITIONS FOR INTERFERENCE IN THE CASE OF SOUND WAVES

1. The two wave trains must move in the same direction because if the two wave trains move in some obliguity, then the resultant intensity at maxima will be very much reduced. Also the positions of minima will not be well defined.

2. The two sources of sound must give waves of same frequency and amplitude so that the positions of maxima and minima are quite distinct.

3. The two sources must give sounds of same phase or constant difference of phase, *i.e.,* the two sources must be coherent so that there is a sustained interference *i.e.*, maxima and minima positions are fixed over a wider range of interference fringes.

3.6. ENERGY DISTRIBUTION DUE TO INTERFERENCE OF SOUND WAVES

From Equation (6) of analytical treatment of interference phenomenon, we have:

$$a = \sqrt{a_1^2 + a_2^2 + 2a_1\, a_2 \cos\phi}$$

If $a_1 = a_2$

then $a = 2a_2(1+\cos\phi)$

$$= 2a_2 . 2\cos^2 \phi/2$$

$$= 4a_2 \cos^2 \phi/2 \quad ...(1)$$

which gives the conditions for maxima and minima of amplitudes,

(*i*) When $\phi = 0$

$a_{max} = 4a^2$

(*ii*) When $\phi = \pi$

$a_{min} = 0$

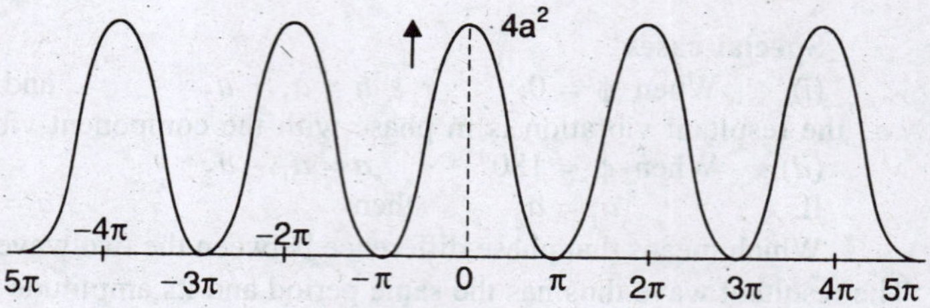

Fig. 3.5

From the laws of conservation of energy, since energy can neither be created nor destroyed, so interference phenomenon gives only redistribution of energy.

The above curve shows that the maximum intensity is $4a^2$ and minimum is 0 and the average is $2a^2$ *i.e.*, it is equal to the uniform intensity of $2a^2$ in the absence of interference between the two waves.

So we observe that interference phenomenon showing maxima and minima is in accordance with the laws of conservation of energy.

3.7. EXPERIMENTAL DEMONSTRATION OF INTERFERENCE PHENOMENON—QUINCKE'S TUBE

Interference phenomenon can be demonstrated by Quincke's tube which consists of two tubes *A* and *B* and the tube *B* can slide in tube *A*.

When a tuning fork vibrates at *T*, the sound travels along two paths *TBF* and *TAE*.

The resultant of the sounds produced at *E* will be maxima or minima depending on the path difference between *TBE* and *TAE*.

If $TBE–TAE = (2n + 1)\ \lambda/2$ minimum sound will be heard and if $TBE - TAE = n\lambda$, maximum sound will be heard. So by moving tube *B* inside *A* alternately maximum and minimum sounds will be heard, thereby demonstrating the interference phenomenon.

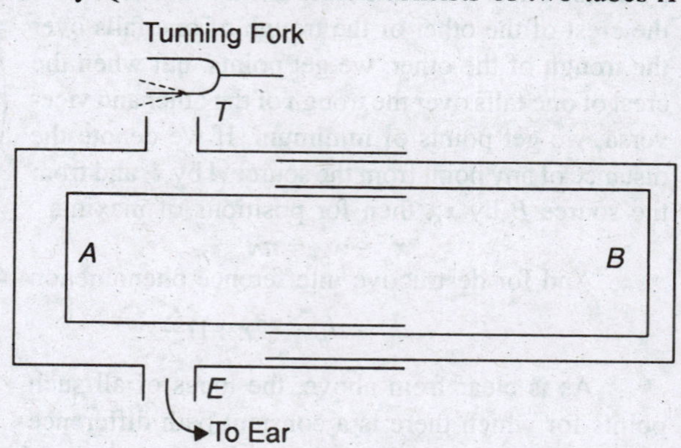

Fig. 3.6

3.8. BEATS

When two wave-trains of nearly the same frequency travel along the same straight line in the same direction, the resultant displacement at a point alternately waxes and wanes in amplitude as many times per second as the difference in their frequencies. This phenomenon is known as beats. The basic theory of the phenomenon is given as under:

The displacement *y* of a particle at any time *t* when a harmonic wave of frequency *n* is progressing with a velocity *v*, is given by the relation

$$y = a \sin 2\pi\ nt \quad ...(1)$$

where *a* is the amplitude of the vibration and *n* the frequency.

If we consider two wave motions of slightly different frequencies of values n_1 and n_2 per second, then the displacements of a particle due to these waves are given by the following equations

$$y_1 = a \sin 2\pi\ nt \quad \text{and} \quad y_2 = a \sin 2\pi\ nt$$

Due to superposition, the resultant displacement *Y* will be given by

$$Y = y_1 + y_2 = a \sin 2\pi n_1 t + a \sin 2\pi n_2 t$$

$^*\sin A + \sin B = 2 \cos \dfrac{A-B}{2} \sin \dfrac{A+B}{2}$

$$= {}^{*}2a\cos 2\pi\left(\frac{n_1-n_2}{2}\right)t\sin 2\pi\left(\frac{n_1+n_2}{2}\right)t$$

$$= A\sin 2\pi\left(\frac{n_1+n_2}{2}\right)t$$

Which represents a periodic vibration of amplitude $A = 2a\cos 2\pi\left(\frac{n_1-n_2}{2}\right)t$ and frequency $\frac{n_1+n_2}{2}$, which is the arithmetic mean of the individual frequencies. The amplitude is, however, not constant as it varies with time.

The maximum value of the amplitude $A = \pm 2a$ when

$$\cos 2\pi\left(\frac{n_1+n_2}{2}\right)t = \pm 1$$

or $$\pi(n_1+n_2)\,t = k\pi \text{ where } k = 0,1,2,3 \text{ etc.}$$

or $$t = \frac{k}{n_1-n_2} = 0, \frac{1}{n_1-n_2}, \frac{2}{n_1-n_2}, \frac{3}{n_1-n_2}$$

and so on, *i.e.*, t is an intergral multiple of $\frac{1}{n_1-n_2}$

Time interval between two consecutive maxima will be $\frac{1}{n_1-n_2}$

$\therefore$ Frequency of maxima $= n_1 - n_2$

Amplitude $A = 0$, *i.e.*, it will be minimum, when

$$\cos 2\pi\left(\frac{n_1-n_2}{2}\right)t = 0$$

or $$\pi\,(n_1-n_2) = (2k-1)\,\pi/2, \text{ where } k = 0, 1, 2, 3...$$

$\therefore$ $$t = \frac{k}{n_1-n_2} + \frac{1}{2(n_1-n_2)}$$

$$= \frac{1}{2(n_1-n_2)} \text{ when } k = 0$$

$$= \frac{3}{2(n_1-n_2)} \text{ when } k = 1$$

$$= \frac{5}{2(n_1-n_2)} \text{ when } k = 2$$

and so on *i.e.*, t is odd multiple of $\frac{1}{2\,(n_1-n_2)}$. So the minima are regularly timed between maxima and the interval between consecutive minima is $= \frac{1}{(n_1-n_2)}$

$\therefore$ Frequency of minima $= n_1 - n_2$

Now one beat is constituted by one maxima and one minima of sound, so number of beats $= n_1 - n_2$, *i.e.*, the resultant intensity of sound rises and falls $(n_1 - n_2)$ times/sec.

So, we have the resultant waves as Simple Harmonic wave whose frequency is equal to the mean of the component frequencies and whose amplitude varies between $2a$ *i.e.*, the sum of two amplitudes of individual waves and zero *i.e.*, their difference $[a - (-a) = 2a]$. Also the variation has a frequency equal to the difference of the constituent frequencies.

3.9. DETERMINATION OF FREQUENCY BY METHOD OF BEATS

Let a tuning fork A have known frequency N and let B tuning fork is of unknown frequency. Both A and B have nearly the same frequencies. Sound the two forks simultaneously and count the number of beats. Let these be n.

Now the frequency of B will be $N \pm n$.

To get the correct value of B *i.e.*, whether it is $N + n$ or $N - n$, load the fork B with wax, so that its frequency decreases. If the number of beats/sec increases, the frequency of B will be $N - n$. However, if the number of beats per second decreases, the original frequency of B will be $N + n$.

It should be noted that

(*i*) When a tuning fork is loaded, its frequency decreases.

(*ii*) When a tuning fork is filed, its frequency increases.

Example 1. *When two tuning forks were sounded together, 20 beats were produced in 8 seconds. After loading one of the tuning forks with a little wax, they produced 32 beats in 8 seconds. If the unloaded fork had frequency 512, calculate the frequency of the other.*

Solution: In this case we have

$$\nu_1 = 512$$

The frequency of the beats before loading is

$$= \frac{10}{8} = 2.5 \text{ s}^{-1}$$

So the frequency of the other tuning fork will be

$$\nu_2 = \nu_1 \pm 2.5 = 512 \pm 2.5$$
$$= 514.5 \text{ or } 509.5$$

After loading, the frequency of the beats is

$$= \frac{32}{8} = 4 \text{ s}^{-1}$$

which means that the frequency of the beats after loading is increased, so the frequency of the other fork should be less than that of the first fork before loading also.

$\therefore$ the frequency of other fork = **509.5 Hz**

Example 2. *When two tuning forks were sounded together, 20 beats were produced in 8 seconds. After loading one of the tuning forks with a little wax, they produce 32 beats in 8 seconds. If the unloaded fork had a frequency of 512 Hz, calculate the frequency of the other.*

Solution: Frequency of A = 512 Hz

Frequency of B = ?

Before loading beats per second

$$= \frac{20}{8} = 2.5$$

Frequency of B before loading will be

either $512 + 2.5 = 514.5$

or $512 - 2.5 = 509.5$

After loading, beats per second

$$\frac{32}{8} = 4$$

Frequency of B after loading will be

either $512 + 4 = 516$

or $512 - 4 = 508$

Since, after loading the frequency of a tuning fork becomes less the frequency of B, before loading is 509.5 Hz as after loading with a *little* wax, its frequency is *slightly* lowered to 508.

$\therefore$ Frequency of B before loading = **509.5 Hz.**

Example 3. *Two sitar strings A and B playing the note 'Ga' are slightly out of tune and produce beats of frequency 6 Hz. The tension in the string A is slightly reduced and the beat frequency is found to reduce to 3 Hz. If the original frequency of A is 324 Hz, what is the frequency of B?*

Solution: Frequency of A = 324 Hz

Beat frequency = 6 Hz

$\therefore$ Possible frequencies of $B = 324 \pm 6 = 330$ or 318 Hz

On reducing the tension on the string A its frequency should is reduced. If the frequency of B is 330, on reducing the tension on the string A, the beat frequency will increase. But according to the question, the beat frequency decreases from 6 Hz to 3 Hz. So we have

Frequency of B = **318 Hz.**

Example 4. *A tuning fork A of frequency 384 Hz gives 6 beats per second when sounded with another tuning fork B. On loading B with a little wax, the number of beats per second becomes 4. What is the frequency of B?*

Solution: Frequency of A = 384 Hz

Beats per second = 6

Frequency of B before loading should be

either 384 + 6 = 390 Hz

or 384 – 6 = 378 Hz

After loading, beats per second = 4

Frequency of B after loading should be

either 384 + 4 = 388 Hz

or 384 – 4 = 380 Hz

Since after loading frequency is reduced, the frequency of B before loading is **390 Hz.**

Example 5. *Two sitar strings A and B playing the note 'Dha' are slightly out of tune and produce beats of frequency 5 Hz. The tension of the string B is slightly increased and the beat frequency is found reduced to 3 Hz. What is the original frequency of B if the frequency of A is 427 Hz?*

Solution: Frequency of A = 427 Hz

Beat frequency = 5 Hz

$\therefore$ Possible frequencies of $B = 427 \pm 5 = 432$ or 422 Hz

On increasing the tension of the string B, its frequency should increase. If its original frequency were 432, on increasing the tension, it will further increase and hence the beat frequency will also increase, which is not the case as the beat frequency decreases, 5 Hz to 3 Hz.

Original frequency of B = **422 Hz.**

Example 6. *Two tuning forks P and Q give 6 beats per second. P resounds with a closed air column 15 cm long and Q with a similar open column 30.5 cm long. Calculate the frequencies of the forks.*

Solution: Let n_1 and n_2 be the frequencies of forks P and Q

$$v_1 - v_2 = 6 \qquad ...(1)$$

Now for P,
$$n_1 = \frac{V}{4l} = \frac{V}{4 \times 15}$$

And for Q,
$$n_2 = \frac{V}{2l} = \frac{V}{2 \times 30.5}$$

Dividing,
$$\frac{n_1}{n_2} = \frac{30.5}{30} \qquad ...(2)$$

From Equations (1) and (2) we get

$$n_1 = \mathbf{366\ Hz}$$

and
$$n_2 = \mathbf{360\ Hz.}$$

Example 7. *Calculate the velocity of sound in a gas in which two waves of lengths 1 metre and 1.01 metres produce 10 beats in 3 seconds.*

Solution: Number of beats in 3s = 10

$\therefore$ Beat/second = 10/3

In this case λ_1 = 100 cm

$$\lambda_2 = 1.01 \text{ metres} = 101 \text{ cm}$$

Let the velocity of sound be V.

Frequency of the first wave = n_1

Frequency of the second wave = n_2

$$\therefore \quad n_1 - n_2 = \text{beat/s} = 10/3$$

Also $$n_1 = \frac{V}{\lambda_1} = \frac{V}{100} \text{ and } n_2 = \frac{V}{\lambda_2} = \frac{V}{100}$$

Putting in Equation (1) the values of n_1 and n_2

$$\frac{V}{100} - \frac{V}{101} = \frac{10}{3} \text{ or } \frac{V}{10100} = \frac{10}{3}$$

$$V = \mathbf{336.67\ ms^{-1}.}$$

Example 8. *A tuning fork produces 6 beats per second when sounded with a tuning fork of frequency 256 Hz. The same tuning fork when sounded with another tuning fork of frequency 252 Hz produces 2 beats per second. Find the frequency of the tuning fork.*

Solution: Let the frequency of the tuning fork = n

It produces 6 beats/second with a tuning fork of frequency 256 Hz.

$$\therefore \quad n = 256 \pm 6$$

$$= 250 \text{ or } 262 \text{ Hz} \qquad ...(1)$$

It also produces 2 beats/second with a tuning fork of frequency 252 Hz.

$$\therefore \quad n = 252 \pm 2$$

$$= 250 \text{ or } 254 \text{ Hz} \qquad ...(2)$$

Equations (1) and (2) show that the common frequency is 250.

$$n = \mathbf{250\ Hz.}$$

Example 9. *Two tuning forks A and B give 5 beats/second. The frequency of A = 512. When B is filled, 5 beats/second are again produced. Find the frequency of B before and after filling.*

Solution: Frequency of $A = 512$

Beats per second = 5

Frequency of B before filling will be either

$$512 + 5 = 517$$

or $$512 - 5 = 507$$

Since after filling 5 beats/second are again produced, so frequency of B after filling is either

$$512 + 5 = 517$$

or $$512 - 5 = 507$$

Let us consider the frequency of B before filling as 517. After filling the frequency increases. So after filling the frequency of B cannot be equal to either 517 or 507. So, frequency of B cannot be equal to 517.

Consider the frequency of B before filling 507

After filling its frequency can be equal to 517

Which will be the possible value.

So, frequency of B before filling = **507**

after filling = **517.**

Example 10. *A note produces 4 beats/second with a tuning fork of frequency 512 and 6 beats/second with a fork of frequency 514. Find the frequency of the note.*

Solution:

(*i*) Frequency of the tuning fork = 512

Beats/second = 4

So, possible frequencies of the note are

$$\left.\begin{matrix} 512+4 \\ 512-4 \end{matrix} = \begin{matrix} 516 \\ 508 \end{matrix}\right] \quad ...(1)$$

(*ii*) Frequency of the tuning fork = 514

Beats/second = 6

Possible frequencies of the note are

$$\left.\begin{matrix} 514+6 \\ 514-6 \end{matrix} = \begin{matrix} 520 \\ 508 \end{matrix}\right] \quad ...(2)$$

Since the common frequency in (1) + (2) is 508

So the frequency of the note is **508.**

QUESTIONS AND PROBLEMS

1. Obtain an expression for the resultant of two wave trains travelling along the same direction with the same velocity and amplitude, but of slightly different frequencies and explain its meaning.

2. Explain the phenomenon of Beats and distinguish it from Combination tones.

3. Investigate mathematically the phenomenon of 'beats'. Differentiate carefully between the phenomenon of 'beats' and that of interference. Give examples of each.

Write short notes on the following:

(*a*) Interference of sound waves, (*b*) Beats, (*c*) Silence zones in the neighbourhood of a fog-siren.

How can the existence of these phenomena be shown?

4. Under what conditions can two sound waves produce interference? Describe some methods of experimentally demonstrating the phenomenon.

5. A tuning fork is struck and then slowly rotated about its stem. Explain the changes in the sound heard during one complete rotation of the fork.

6. When two waves interfere

(*i*) does one alter the progress of the other? (*ii*) is there a loss of energy?

7. Explain mathematically the phenomenon of beats. How does it differ from that of interference?

8. A man is travelling in a straight line joining two identical sources each of frequency 1000 and vibrating in phase. How fast would he move from one towards the other so that he may hear 10 beats per second? Velocity of sound = 1120 ft/sec. **[Ans.** 5.6 ft/sec.]

9. Two organ pipes when sounded together give 5 beats per second. If their lengths are in the ratio of 50 to 51, calculate their frequencies. **[Ans.** n_1 = 250, n_2 = 255]

10. Two tuning forks A and B give 4 beats per second. On loading A slightly, we get 3 beats in 5 seconds. What is the frequency of A before and after loading, if that of B is 256?

[Ans. 260 before loading and 256.6 after loading]

11. Two tuning forks make 10 beats in 3 seconds when sounded simultaneously. One fork makes 384 vibrations per second and the beats cease when the other fork is filed a bit at the end of the prong. What is the frequency of the second fork? **[Ans.** $380\frac{2}{3}$ Hz]

CHAPTER

4

VIBRATION IN STRINGS

4.1. VELOCITY OF TRANSVERSE WAVES ALONG A STRETCHED STRING

Consider one portion *APQB* of a string through which a transverse wave is progressing from left to right. In this case, the particles vibrate ⊥ to its length and the movement is handed from one particle to another particle. The velocity of transverse wave will depend on certain constants of the string.

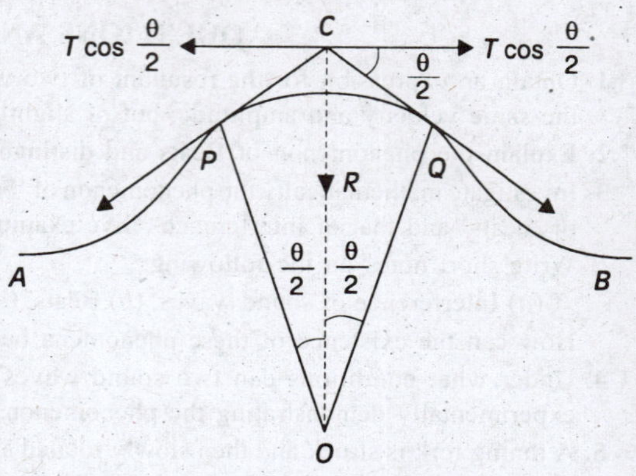

Fig. 4.1

Consider a small element *PQ* of the string and let its centre of curvature be at *O*. Let $\angle POQ = \theta$. Also consider that the curvature *PQ* is very small so that angle θ is also very small. Let *T* be the tension at *P* and *Q*, the direction of the tensions being tangential to the element *P* and *Q*.

Resolve the tensions into horizontal and vertical components. The horizontal components at *O* are equal and opposite while the ⊥ components are along *CO*.

∴ The resultant tension along

$$CO \text{ will be } = 2T\sin\frac{\theta}{2}$$

$$= 2T\cdot\frac{\theta}{2} = T.\theta \quad [\text{if } \theta \text{ is small, } \sin\theta \cong \theta] \qquad ...(1)$$

For equilibrium position, this resultant tension provides the necessary centripetal force given by

$$= \frac{(m\,\delta x)v^2}{R} \qquad ...(2)$$

$$\therefore \quad \frac{m\,\delta x\,v^2}{R} = T\theta$$

$$\text{Also} \quad \theta = \frac{\delta x}{R}$$

$$\text{So} \quad \frac{m.\delta x.v^2}{R} = \frac{T.\delta x}{R} \text{ or } v^2 = \frac{T}{m}$$

$$\therefore \quad v = \sqrt{\frac{T}{m}} \qquad ...(3)$$

where *m* is the mass per unit length of the string.

Also, we have the relation

$$v = n\lambda \qquad ...(4)$$

If the string of length *l* vibrates in *p* segments, length of each segment being $= \frac{l}{p}$ and each segment will correspond to $\frac{\lambda}{2}$.

$$\therefore \quad \frac{l}{p} = \frac{\lambda}{2} \quad \text{or} \quad \lambda = \frac{2l}{p}$$

Putting in Equation (4), we get,

$$v = \frac{n(2l)}{p} \quad \text{...(5)}$$

Comparing Equations (2) and (4) we have,

$$\frac{n(2l)}{p} = \sqrt{\frac{T}{m}}$$

or

$$n = \frac{p}{2l}\sqrt{\frac{T}{m}} \quad \text{...(6)}$$

In case the string is vibrating in one segment,

then $p = 1$

$$\therefore \quad n = \frac{1}{2l}\sqrt{\frac{T}{m}} \quad \text{...(7)}$$

4.2. LAWS OF TRANSVERSE VIBRATION OF STRINGS

We have the following fundamental laws of vibrating string.

(*i*) The fundamental frequency is inversely proportional to the length of the string, *i.e.*,

$$n \propto \frac{1}{l}$$

(*ii*) The fundamental frequency is directly proportional to the square root of the stretching force or tension, *i.e.*,

$$n \propto \sqrt{T}$$

(*iii*) The fundamental frequency is inversely proportional to the square root of the mass per unit length, *i.e.*,

$$n \propto \frac{1}{\sqrt{m}}$$

If we combine the above three laws, we may write

$$n \propto \frac{1}{l}\sqrt{\frac{T}{m}}$$

or

$$n = \frac{k}{l}\sqrt{\frac{T}{m}}$$

The value of constant k comes out to be 1/2.

$$\therefore \quad n = \frac{1}{2l}\sqrt{\frac{T}{m}} \quad \text{...(1)}$$

If D be the diameter of the string or wire and d is the density of the material of the wire, then we may write

$$m = \frac{\pi D^2}{4} \times 1 \times d$$

$$\therefore \quad n = \frac{1}{2l}\sqrt{\frac{4T}{\pi D^2 d}}$$

or

$$n = \frac{1}{l\,D}\sqrt{\frac{T}{\pi d}} \quad \text{...(2)}$$

4.3. VERIFICATION OF THE LAWS OF TRANSVERSE VIBRATION OF STRINGS

(*i*) $n \propto \frac{1}{l}$ or nl = constant

To verify this law,

Take a sonometer and a tuning fork of frequency 256 (say). Keeping a load of 1 kg, note the position of the two bridges when the wire is in unison with the tuning fork. This can be obtained by using a small paper rider on the sonometer wire, let the length be $= l_1$. Now, take a fork of frequency 512, find the corresponding resonant length, keeping tension the same. Let this length be l_2. It is found that l_2 is $P\frac{1}{2}$,

$$\therefore \quad n_1 l_1 = n_2 l_2$$

i.e., $$n \propto \frac{1}{l}$$

or nl = constant

∴ Law of length is verified.

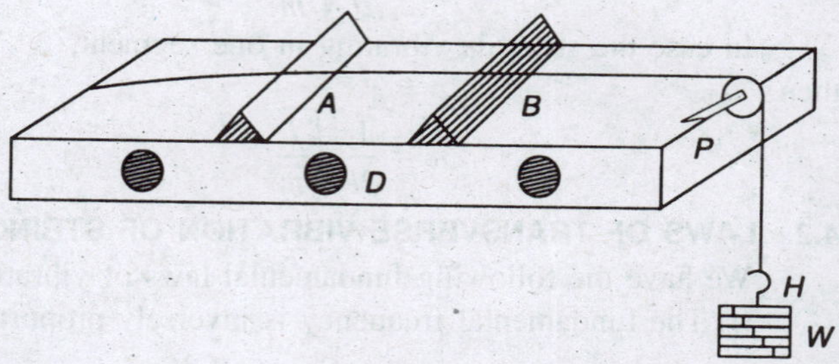

Fig. 4.2

(*ii*) $n \propto \sqrt{T}$

For a tuning fork of frequency 256, note the resonating length l_1 for a load of 1 kg.

Now, increase the load to 4 kg, you will note that the tuning fork of frequency 512 is in resonance with the same length l_1 of the wire

$$\frac{n_2}{n_1} = \sqrt{\frac{T_2}{T_1}}$$

or $$n \propto \sqrt{T}$$

So the law of tension is verified.

(*iii*) Take two wires of the same material but having diameters in the ratio of 1 : 2.

Stretch one wire and find the resonating length with a tuning fork of frequency 512 when a load of 1 kg is attached. Now, stretch the second wire and using the same tension, find the resonating frequency with turning fork of frequency 256. It will be found that,

$$\frac{n_1}{n_2} = \frac{D_2}{D_1}$$

Also $$m_1 = \frac{\pi D_1^2}{4} \cdot d \quad \text{and} \quad m_2 = \frac{\pi D_2^2}{4} \cdot d$$

$$\therefore \quad \frac{m_2}{m_1} = \frac{D_2^2}{D_1^2} \quad \text{or} \quad \frac{D_2}{D_1} = \sqrt{\frac{m_2}{m_1}}$$

$$\therefore \quad \frac{n_1}{n_2} = \sqrt{\frac{m_2}{m_1}}$$

or $$n \propto \frac{1}{\sqrt{m}}$$

so law of mass is verified.

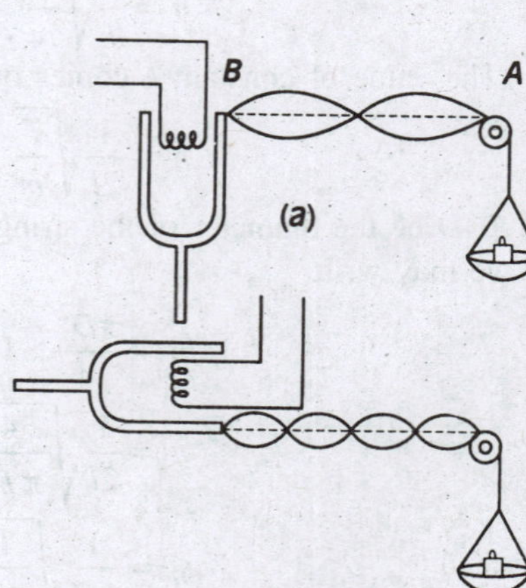

Fig. 4.3

4.4. MELDE'S EXPERIMENT

Melde's experiment is very convenient method for determining the resonant vibrations of a stretched wire and investigating the laws which govern such vibrations. It consists of light string about 1.5 meter long fixed at one end to the prong of the tuning fork which is maintained in vibration electrically while the other end passes over an adjustable pulley and is fastened to a small scale pan. The tension in the string can be changed by putting weights in it. The vibrations of the form may be either along the string as in Fig. 4.3 (*a*) (*the longitudinal mode of vibration*), or at right angles to the string as shown in Fig. 4.3 (*b*) (*the transverse mode of vibration*).

Construction of electrically maintained tuning fork. The chief parts of the electrically maintained tunning fork are the force T, the electromagnet M, the steel spring S and the platinum tip P, at which contact is made. The vibrations of the fork are maintained by the intermittent current that flows through the electromagnet placed between the prongs of the tuning fork. When current flows through the circuit, the electromagnet is magnetised and the prongs are pulled inwards and the steel spring S fails to make any contact with P. So the circuit is broken and consequently, the electromagnet fails to pull the prong, which falls back and the contact is remade. In this way the fork makes the vibration.

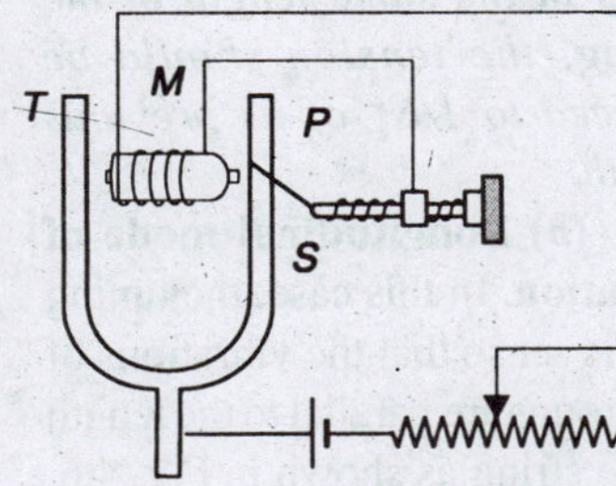

Fig. 4.4

It should be noted that when the circuit is broken, a spark is produced. To reduce this sparking, and thereby increasing the efficiency of the electrically maintained vibrations of the tuning fork, a condenser of suitable capacitance can be placed in parallel with the spark gap as is done in the case of an induction coil.

Construction and Theory. The apparatus consists of a light string AB of uniform diameter and about one metre long, one end of which is connected to the prong of an electrically maintained tuning fork and the other end after passing over a verification less pulley is connected to a light scale pan P. The tension in the string can be varied by placing different weights in the pan. Stationary waves are produced due to the superimposition of the direct waves sent by the tuning fork and the reflected waves from the pulley. There are two modes of vibrations.

(*a*) Transverse mode of vibration. In this arrangement, the vibrations of the prong of the tuning fork are in a direction at right angles to the length of the string as shown in Fig. 4.5. The time during which the tuning fork executes one vibration, the string also completes one vibration. Now, if the thread of length l vibrates in one loop, the frequency of the fundamental note under a tension T is given by the relation

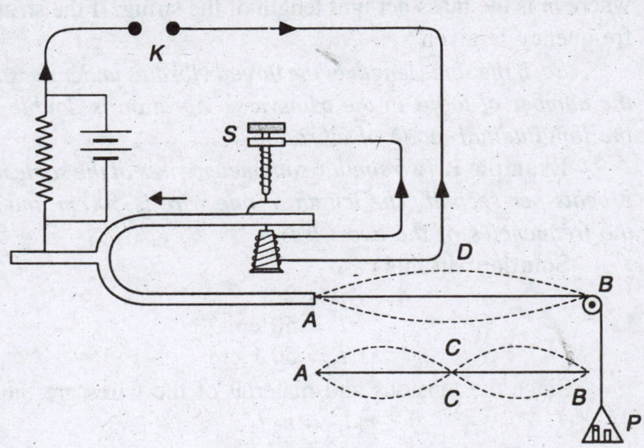

Fig. 4.5

$$n = \frac{1}{2l}\sqrt{\frac{T}{m}} \quad ...(1)$$

where m is the mass per unit length of the thread. If the same length of the thread vibrate in N loops under a tension T_N, we have

$$n = \frac{N}{2l}\sqrt{\frac{T_N}{m}} \quad ...(2)$$

From Equations (1) and (2) we have

$$\frac{T_N}{T} = \frac{1}{N^2} \quad \text{or} \quad T_N = \frac{T}{N^2}$$

So, if we wish to have N loops in the same length of the string, the tension should be reduced to $1/N^2$ of its previous length.

(b) Longitudinal mode of vibration. In this case, the tuning fork is set so that the vibrations of the prongs are parallel to the length of the string as shown in Fig. 4.6, and the time during which the tuning fork completes one vibration, string completes *half* a vibration. So the frequency of the thread is half that of the tuning fork. If the length l of the thread vibrates in one loop under a tension T, then we have

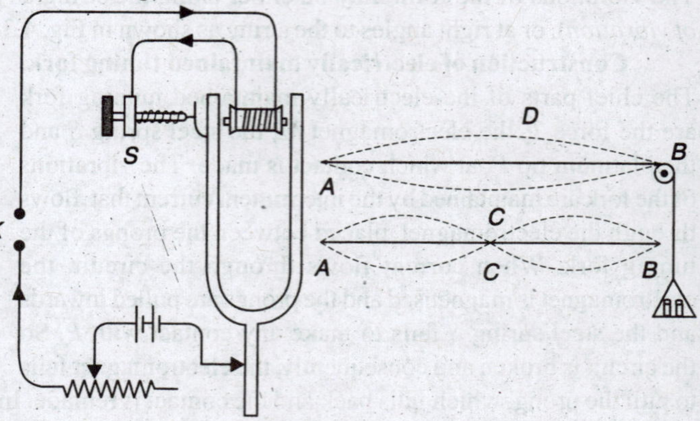

Fig. 4.6

Frequency of fork $n = 2 \times$ frequency of the string

$$= 2 \times \frac{1}{2l}\sqrt{\frac{T}{m}} = \frac{1}{l}\sqrt{\frac{T}{m}}$$

where m is the mass per unit length of the string. If the string vibrates in N loops, the corresponding fre-quency is given?

So, if the same length of the thread vibrates under the same tension in the two modes of vibration, the number of loops in the transverse vibration is double as compared to the number of loops in the longitudinal mode of vibration.

Example 1. *Two similar sonometer wires of the same material, under the same tension produce 2 beats per second. The length of one wire is 50 cm and that of the other is 50.1 cm. Calculate the frequencies of the two wires.*

Solution: In this case

$$n_1 - n_2 = 2 \quad \text{...(1)}$$

$$l_1 = 50 \text{ cm} \quad \text{...(2)}$$

$$l_2 = 50.1 \text{ cm}$$

Since the tensions and material of the wires are the same, So we have

$$n_1 l_1 = n_2 l_2$$

or

$$\frac{n_1}{n_2} = \frac{l_2}{l_1}$$

$$\frac{n_1}{n_2} = \frac{50.1}{50}$$

or

$$n_1 = \left(\frac{50.1}{50}\right) n_2 \quad \text{...(3)}$$

Putting value of n_1 in Equation (1), we get

$$\left(\frac{50.1}{50}\right) n_2 - n_2 = 2$$

$$n_2 = \mathbf{1000} \text{ Hz}$$

From (1) $\quad n_1 = n_2 + 2$

$\therefore \quad n_1 = 1000 + 2$

or $\quad n_1 = \mathbf{1002}$ Hz

Example 2. *A wire of specific gravity 7, one meter long and 1 mm in diameter is stretched by a weight of 11 kg. Calculate the pitch of the fundamental note.*

Solution: We have

$$d = 7, \quad l = 100 \text{ cm}$$
$$D = 1 \text{ mm} = 0.1 \text{ cm}$$
$$T = 11 \text{ kg} = 11 \times 1000 \times 980 \text{ dynes}$$
$$n = ?$$

Using the relation $n = \frac{1}{lD}\sqrt{\frac{T}{\pi d}}$

$$\therefore \quad n = \frac{1}{100 \times \frac{1}{10}}\sqrt{\frac{11 \times 1000 \times 980 \times 7}{22 \times 7}} = \mathbf{70}$$

Example 3. *A string when stretched by a weight of 4 kg gives a note of frequency 256. What weight will produce an octave of this note?*

Solution: We have

$$n_1 = 256, \quad n_2 = 2 \times 256 = 512 \qquad \text{(given)}$$
$$T_1 = 4 \text{ kg wt}; \quad T_2 = ?$$

Now $\quad \frac{n_2}{n_1} = \sqrt{\frac{T_2}{T_1}}$ or $\frac{512}{256} = \sqrt{\frac{T_2}{4}}$

or $\quad \sqrt{\frac{T_2}{4}} = 2, \quad T_2 = \mathbf{16 \text{ kg wt}}$

Example 4. *A sitar wire is under tension 30 N and the length of the bridges is 80 cm. A 10 metre sample of that wire has mass 2.2 g. Deduce:*

(i) the speed of the transverse waves on the wire and

(ii) the frequency of the fundamental.

Solution: We have

$$T = 30 \, N$$
$$l = 80 \text{ cm} = 0.8 \text{ m}$$

and $\quad m = \frac{2.2}{10} \text{ g/m} = \frac{2.2}{10} \times 10^{-3} \text{ kg/m}$

$$= 2.2 \times 10^{-4} \text{ kg/m}$$

(*i*) The speed of the transverse waves on the wire is given by the relation

$$v = \sqrt{\frac{T}{m}}$$
$$= \sqrt{\frac{30}{2.2 \times 10^{-4}}} = \mathbf{369.8 \text{ ms}^{-1}}$$

(*ii*) The frequency of the fundamental is given by,

$$n = \frac{v}{2l} = \frac{369.27}{2 \times 0.80}$$
$$= \mathbf{230.8 \text{ s}^{-1}}$$

Example 5. *Calculate the frequency of the fundamental note of a string, 1 metre long and weighing 2 grams when stretched by a weight of 400 kg.*

Solution: We have

$$l = 1 \text{ metre}$$

Total mass $\quad = 2 \text{ g} = 0.002 \text{ kg}$

Mass/unit length $m = \frac{0.002}{1} = 0.002$ kg/m

$$T = 400 \text{ kg} = 400 \times 9.8 \text{ N}$$

$$\therefore \quad n = \frac{1}{2l}\sqrt{\frac{T}{m}} = \frac{1}{2 \times 1}\sqrt{\frac{400 \times 9.8}{0.002}}$$

$$n = \frac{1}{2}\sqrt{400 \times 4900} = \mathbf{700\ Hz}$$

Example 6. *A steel wire has a length of 12.0 m and a mass of 2.10 kg. What should be the tension in the wire so that the speed of a transverse wave on the wire equals the speed of sound in dry air at 20°C (= 343 m/s)?*

Solution: The speed of a transverse wave is given by the relation

$$v = \sqrt{\frac{T}{m}}$$

$$\therefore \quad v^2 = \frac{T}{m}$$

or $$T = v^2 \text{ m}$$

In this case

$$v = 343 \text{ m/s}$$

and $$m = \frac{2.10}{12.0} \text{ kg/m}$$

$$\therefore \quad T = (343)^2 \times \frac{2.10}{12.0}$$

$$= \mathbf{2.06 \times 10^4\ N.}$$

Example 7. *Two strings each of length 60 cm are stretched, one by a force of 4 kg wt. and the other by a force of 9 kg wt. What is the interval between the two notes that are produced?*

Solution: We have

$$\frac{n_2}{n_1} = \sqrt{\frac{T_2}{T_1}} = \sqrt{\frac{9}{4}} = \frac{3}{2}$$

The interval between the two notes is the fifth.

Example 8. *Two tuning forks produce 4 beats per sec when sounded together. One of them is in unison with a length of 120 cm of a sonometer wire and other with 125 cm of the same wire under the same tension. Calculate the frequencies of the forks.*

Solution: The frequencies of the two wires are given by the relation

$$n_1 = \frac{1}{2l_1}\sqrt{\frac{T}{m}} \quad \text{...(1)}$$

and $$n_2 = \frac{1}{2l_2}\sqrt{\frac{T}{m}} \quad \text{...(2)}$$

Dividing Equation (1) by (2), we get

$$\frac{n_1}{n_2} = \frac{l_2}{l_1}$$

In this case, $l_1 = 120$ cm

$$l_2 = 125 \text{ cm}$$

$$\frac{n_1}{n_2} = \frac{125}{120} = \frac{25}{24} \quad \text{...(3)}$$

As the frequency of beats is 4,

i.e., $n_1 - n_2 = 4$ ...(4)

Solving Equation (3) and (4), we get

$$n_1 = \mathbf{100\ Hz} \quad n_2 = \mathbf{96\ Hz.}$$

Example 9. *A steel wire, 60 cm long and 0.05 cm diameter gives a note 240 vibrations per sec when stretched with a certain weight. Another steel wire 40 cm long and 0.06 cm diameter is stretched with the same force. Find the frequency of the fundamental note.*

Solution: In this case

$l_1 = 60$ cm $l_2 = 40$ cm

$r_1 = 0.025$ cm $r_2 = 0.03$ cm

$n_1 = 240$ vib/s $n_2 = ?$

For the first wire, the frequency is given by the relation

$$n_1 = \frac{1}{2l_1}\sqrt{\left(\frac{T}{\pi r_1^2 \rho}\right)}$$

$$240 = \frac{1}{2 \times 60}\sqrt{\frac{T}{\pi (0.025)^2 \rho}} \quad ...(1)$$

For the second wire, the frequency is given by the relation

$$n_2 = \frac{1}{2l_2}\sqrt{\left(\frac{T}{\pi r_2^2 \rho}\right)}$$

$$= \frac{1}{2 \times 40}\sqrt{\frac{T}{\pi (0.03)^2 \rho}} \quad ...(2)$$

Dividing Equation (2) by (1), we get

$$\frac{n_2}{240} = \frac{3}{2}\sqrt{\frac{(0.05)^2}{(0.03)^2}}$$

$$n_2 = 240 \times \frac{3}{2} \times \frac{0.025}{0.03}$$

$$= \mathbf{300\ Hz.}$$

Example 10. *A wire of length 108 cm produces a fundamental note of frequency 256 when stretched by a weight of 1 kg. By how much its length should be increased so that its pitch is raised by a major tone, if it is now stretched by a weight of 4 kg (Given: $n_2/n_1 = 9/8$)?*

Solution: We have

$l_1 = 108$ cm, $T_1 = 1$ kg

$l_2 = ?$ $T_2 = 4$ kg

$n_1 = 256$

Now $$\frac{n_2}{n_1} = \frac{9}{8} \quad \text{(given)}$$

$\therefore$ $$n_2 = \frac{9}{8} \times 256 = 288$$

Also $$\frac{n_2}{n_1} = \frac{l_1}{l_2}\sqrt{\frac{T_2}{T_1}}$$

$$\frac{9}{8} = \frac{108}{l_2}\sqrt{\frac{4}{1}}$$

$$l_2 = 192 \text{ cm}$$

So the increase in length = 192 – 108 = **84 cm.**

Example 11. *A tuning fork of frequency 160, is sounded along with a sonometer wire of length 25 cm, stretched to a tension of 1.25 kg weight. Calculate the number of beats per second.*

(*Mass per unit length = 0.025 g/cm*)

Solution: We have

$$l = 25 \text{ cm}, m = 0.025 \text{ g/cm}$$

$$T = 1.25 \text{ kg.wt} = 1.25 \times 1000 \times 980 \text{ dynes}$$

Now $$n = \frac{1}{2l}\sqrt{\frac{T}{m}} = \frac{1}{2 \times 25}\sqrt{\frac{1.25 \times 1000 \times 980}{0.025}}$$

$$n = \frac{7000}{50} = 140$$

Frequency of the tuning fork $N = 160$

∴ Beats/s = $N - n = 160 - 140 =$ **20.**

Example 12. *A rope, 5 metre long, has a total mass of 245 g. It is stretched with a constant tension of 1 kg weight. If it is fixed at one end and shaken by hand at the other, what frequency of shaking will make it break up into three vibrating segments? (g = 980 cm/s²)*

Solution: The frequency in transverse mode is given by the relation

$$n = \frac{p}{2l}\sqrt{\frac{T}{m}}$$

In this case $$l = 5 \text{ m} = 500 \text{ cm}, \quad m = \frac{245}{500} = 0.49 \text{ g/cm}$$

$$T = 1 \text{ kg.wt.} = 1000 \times 980 \text{ dyne}$$

and $$p = 3$$

∴ $$n = \frac{3}{2 \times 500}\sqrt{\frac{1000 \times 980}{0.49}}$$

$$= \mathbf{4.24 \text{ Hz.}}$$

Example 13. *A rope of mass 0.1 kg/m is held with a tension of 1,000 N. As a wave passes along the rope, every point on the rope moves with SHM with a frequency of 10 Hz and an amplitude 0.1 m. For this wave, what is (i) the velocity, (ii) the wavelength.*

Solution: Given that,

$$m = 0.1 \text{ kg/m}, \quad T = 1000 \text{ N}$$

$$\nu = 10 \text{ Hz}, \quad a = 0.1 \text{ m}$$

(*i*) Velocity of the wave, is given by

$$v = \sqrt{\frac{T}{m}} = \sqrt{\frac{1000}{0.1}}$$

$$= \mathbf{100 \text{ ms}^{-1}}$$

(*ii*) Now $$v = n\lambda$$

∴ $$\lambda = \frac{v}{n} = \frac{100}{10} = \mathbf{10 \text{ m.}}$$

Example 14. *In an experiment, it was found that a tuning fork and a sonometer wire gave 5 beats per second, both when the length of the wire was 1 m and 1.05 m. Calculate the frequency of the fork.*

Solution: Let the frequency of the tuning fork be *n*. Since with the two lengths of the sonometer, the number of beats heard everytime is 5 per second, the two frequencies of the sonometer wire for its two lengths must be,

$$n_1 = n + 5$$

and $$n_2 = n - 5$$

Also, $$n_1 = \frac{1}{2l_1}\sqrt{\frac{T}{m}}$$

and $$n_2 = \frac{1}{2l_2}\sqrt{\frac{T}{m}}$$

Dividing the two, $$\frac{n_1}{n_2} = \frac{l_2}{l_1}$$

In this case
$$n_1 = n + 5$$
$$n_2 = n - 5$$
$$l_1 = 1 \text{ m}$$
$$l_2 = 1.05 \text{ m}$$

Substituting these values,

$$\frac{n+5}{n-5} = \frac{1.05}{1}$$

or $$n + 5 = 1.05n - 5.25$$

or $$0.05n = 10.25$$

or $$n = \mathbf{205.}$$

Example 15. *(a) Compute the fundamental frequency of a sonometer wire of the length 20.0 cm, $T = 20$ N, $m = 5.2 \times 10^{-3}$ kg/m.*

(b) A resonance air column resonates with a tuning fork of the frequency 512 Hz at the length 17.4 cm. Neglecting the end correction, deduce the speed of the sound in air. Is your answer unique for the given data?

Solution: (*a*) The fundamental frequency of a sonometer wire is given by the relation

$$n = \frac{1}{2l}\sqrt{\frac{T}{m}}$$

In this case
$$l = 20 \text{ cm} = 0.2 \text{ m}$$
$$T = 20 \text{ N}$$

and $$m = 5.2 \times 10^{-3} \text{ kg/m}$$

$$\therefore \quad n = \frac{1}{2 \times 0.2}\sqrt{\frac{20}{5.2 \times 10^{-3}}}$$
$$= \mathbf{155\ Hz}$$

(*b*) Speed of sound in air is given by,

$$c = n.4l$$

Here,
$$n = 512 \text{ Hz}$$
$$l = 17.4 \text{ cm} = 0.174 \text{ m}$$

$$\therefore \quad c = 512 \times 4 \times 0.174$$
$$= \mathbf{365.35\ ms^{-1}}$$

This answer is **unique** for the fundamental mode.

Example 16. *A string of mass 2.50 kg is under a tension of 200 N. The length of the stretched string is 20.0 m. If a transverse jerk is struck at one end of the string, how long does the disturbance take to reach the other end?*

Solution: We have

$$m = \frac{2.50}{20.0} \text{ kg/m}$$
$$T = 200 \text{ N}$$

The speed of the transverse jerk is given by the relation

$$v = \sqrt{\frac{T}{m}}$$
$$= \sqrt{\frac{200 \times 20.0}{2.50}}$$

$$= \sqrt{1600} = 40 \text{ ms}^{-1}$$

$\therefore$ Time taken by the jerk to reach the other end $= \dfrac{\text{Distance}}{\text{Speed}}$

$$= \frac{20}{40} = \mathbf{5.0\,s}$$

Example 17. *Calculate the fundamental frequency of a sonometer wire of length = 20 cm, tension 25 N, cross-sectional area = 10^{-2} cm^2 and density of the material of wire = 10^4 kg m^{-3}.*

Solution: We have

$$l = 20 \text{ cm} = 0.20 \text{ m}$$
$$T = 25 \text{ N}$$
$$a = 10^{-2} \text{ cm}^2 = 10^{-6} \text{ m}^2$$
$$\rho = 10^4 \text{ kg m}^{-3}$$

Now, mass per unit length of the wire

$$m = a.\rho = 10^{-6} \times 10^4 \text{ kg m}^{-1} = 10^{-2} \text{ kg m}^{-1}$$

Fundamental frequency of the sonometer wire is given by the relation

$$n = \frac{1}{2l}\sqrt{\frac{T}{m}} = \frac{1}{2 \times 0.20}\sqrt{\frac{25}{10^{-2}}} = \frac{50}{2 \times 0.20}$$
$$= \mathbf{125\ Hz.}$$

Example 18. *A wire 75 cm long and of mass 1 g is making 256 vibration/sec under a tension supplied by a brass weight hanging vertically. On immersing the weight in water the vibrating length of the wire has to be shortened by 4.45 cm to regain its original pitch. What is the density of brass?*

Solution: In this case

$$l_1 = 75 \text{ cm}$$
$$l_2 = 75 - 4.45 = 70.55 \text{ cm}$$
$$m = \frac{1}{75} \text{ gm/cm}$$
$$n = 256 \text{ Hz}$$

The frequency, in the first case, is given by the relation

$$n = \frac{1}{2l_1}\sqrt{\frac{T_1}{m}}$$

$$\therefore \quad T_1 = 4 l_1^2 n^2 m = 4 \times 75 \times 75 \times 256 \times 256 \times \frac{1}{75}$$
$$= 1965 \times 10^4 \text{ dyne}$$

Similarly, in the second case, we have the relation

$$T_2 = 4 l_2^2 n^2 m = 4 \times 70.55 \times 70.55 \times 256 \times 256 \times \frac{1}{75}$$
$$= 1739 \times 10^4 \text{ dyne}$$

So loss of weight on immersing in water, we have

$$= T_1 - T_2$$
$$= (1965 - 1739) \times 10^4 \text{ dyne}$$

$$= 226 \times 10^4 \text{ dyne}$$

$$\therefore \quad \text{Density of brass} = \frac{T_1}{T_1 - T_2} = \frac{1965 \times 10^4}{226 \times 10^4}$$

$$= \mathbf{8.7\ g/cm^3}.$$

Example 19. *Find the tension needed to produce stationary waves with 4 loops in a string 1 metre long, 0.5 g in weight, fixed to a tuning fork vibrating 200 times per second.*

Solution: We can calculate the tension in both the modes of vibrations.

(*i*) For longitudinal mode of vibration

$$\frac{1}{2}N = \frac{p}{2l}\sqrt{\frac{T_1}{m}}$$

$$T_1 = \frac{N^2 l^2 m}{p^2} \qquad \text{...(1)}$$

In this case, $p = 4$

$$l = 1 \text{ metre} = 100 \text{ cm.}$$

$$m = \frac{0.5}{100} = 0.005 \text{ g/cm}$$

$$N = 200 \text{ Hz}$$

$$\therefore \quad T_1 = \frac{(200)^2 \times (100)^2 \times 0.005}{(4)^2}$$

$$= 125000 \text{ dyne}$$

$$= \frac{125000}{981} \text{ g.wt.}$$

$$= \mathbf{127.4\ g\ wt.}$$

(*ii*) For transverse mode of vibration

$$N = \frac{p}{2l}\sqrt{\frac{T_2}{m}}$$

$$T_2 = \frac{4N^2 l^2 m}{p^2}$$

$$= 4T_1 \qquad \text{[from (1)]}$$

$$= 4 \times 127.4 \text{ g.wt.}$$

$$= \mathbf{509.6\ g\ wt.}$$

Example 20. *A tuning fork gives 15 beats per sec., when sounded with a sonometer wire of length 20 cm and 20 beats with that of a length of 25 cm. Calculate the frequency of the fork. The tension and mass per unit length of the wire are 1.25 kg and 0.025 g/cm respectively. Given g = 980 cm/s².*

Solution: In this case,

$$l_1 = 20 \text{ cm}, \qquad l_2 = 25 \text{ cm}$$

$$T = 1.25 \text{ kg} = 125 \times 1000 \times 980 \text{ dyne}$$

$$m = 0.025 \text{ g/cm}$$

The frequency of the first wire is given by the relation

$$n_1 = \frac{1}{2l_1}\sqrt{\frac{T}{m}}$$

$$= \frac{1}{2 \times 20}\sqrt{\frac{1.25 \times 1000 \times 980}{0.025}} = 175$$

As it produces 15 beats per sec, so the frequency of the fork

$$= 175 \pm 15 = 190 \text{ or } 160 \text{ Hz} \qquad ...(1)$$

Again the frequency of the second wire is given by the relation

$$n_2 = \frac{1}{2l_2}\sqrt{\frac{T}{m}}$$

$$= \frac{1}{2 \times 25}\sqrt{\frac{1.25 \times 1000 \times 980}{0.025}} = 140$$

Since it gives 20 beats/sec, so the frequency of the fork

$$= 140 \pm 20 = 160 \text{ or } 120 \text{ Hz} \qquad ...(2)$$

Comparing Equation (1) and (2), the common frequency is

$$= \mathbf{160\ Hz.}$$

Example 21. *In a Melde's Experiment, the string is stretched by a load of 10 grams. When the fork is bowed, the string vibrates in four loops. By what weights must the string be stretched for the number of loops to be 5, 3 and 1 respectively?*

Solution: Using the relation, $T_p \times p^2$ = a constant. In this ,we have for 5 loops,

$$T_5 \times 5^2 = 10 \times 4^2$$

or

$$T_5 = \frac{160}{25} = \mathbf{6.40\ gm\ wt}$$

For 3 loops, $T_3 \times 3^2 = 10 \times 4^2$

$$\therefore \quad T_3 = \frac{160}{9} = \mathbf{17.77\ gm\ wt}$$

and for one loop, $T_1 \times 1^2 = 10 \times 4^2$

or $T_1 = \mathbf{160\ gm\ wt.}$

QUESTIONS AND PROBLEMS

1. Derive an expression for the velocity of propagation of a transverse disturbance in a stretched string and hence deduce the laws of transverse vibrations of strings.
2. Write down the equation of motion of a stretched string and find an expression for the velocity of a transverse wave travelling along the string.
3. Show that the velocity of waves in a stretched string is given by the expression $\frac{T}{\rho}$, where T is the tension and ρ the mass of unit length of the string. Hence deduce an expression for the frequency of a vibrating string.
4. Obtain an expression for the velocity of transverse vibrations along a stretched string, and thence deduce the frequency of a string vibrating in p segments.
5. Work out completely the laws of transverse vibrations of a string, and show how the diameter of two wires of the same material can be compared with the help of a sonometer.
6. Describe and explain Melde's Experiment. In an experiment it was found that the string vibrated in 5 loops when 10 grams were placed in the scale pan. What mass must be placed in the pan to make the string vibrate in 7 loops? (Neglect the wt. of the scale pan.) [**Ans.** 5.1 gm.]
7. Four violin strings all of the same length and material but of diameters in the ratios 5 : 4 : 3 : 2 are to be stretched so that each gives a note a two fifth above the preceding. Compare the stretching forces necessary. [**Ans.** 400 : 576 : 729 : 729]
8. A piece of wire 50 cm long is stretched by a load of 2.5 kg and has a mass of 1.44 gm. Find the frequency of the second harmonic. [**Ans.** 184.4 Hz]
9. A metal wire of mass 9.8 gm/metre is stretched with a tension of 10 kg wt. between two rigid supports one metre apart. The wire passes at its middle point between the poles of a permanent horse-shoe magnet and it vibrates in resonance when carrying an alternating current of frequency f cycles per second. Calculate f and mention the effect of using an iron wire and replacing the permanent magnet by an electromagnet energised by the AC supply. [**Ans.** 50 Hz]

10. In performing Melde's experiment in the longitudinal mode of vibration the string divides into 4 segments with a weight of 40 grams. What weight must be carried to cause the string to vibrate in 5 and 6 segments? Explain what effect will be produced by tuning the fork so that the plane of vibration of the fork is at right angles to the string.

[**Ans.** $25\frac{3}{5}$ gm; $17\frac{7}{9}$ gm; Tension must be increased four times to produce the same number of loops]

11. The E string of a violin is 36 cm long and has a mass of 0.15 gm. With what tension must it be stretched if tuned to 660 vibrations for its fundamental?
[**Ans.** 9.50 kg]

12. Prove that the velocity of wave in a stretched string is $\sqrt{\frac{T}{m}}$. Hence find the period of vibration of a string when giving out its fundamental note. If an addition of 2.5 kg to the tension of a string raises its pitch a fifth, what was original tension?
[**Ans.** $T = 2.0$ kg]

13. In a Melde's experiment, a string vibrated in 3 loops when 8 grams were placed in the scale pan. What mass must be placed in the scale pan to make the string vibrate in 5 loops ? (neglect the weight of the scale pan).
[**Ans.** $2\frac{23}{25}$ grams]

14. Calculate the tension to give standing waves with three loops in a string 90 cm long, weighing 0.4 gm and hanging from a fork vibrating 150 times per second.
[**Ans.** 91.7 gm wt.]

CHAPTER

5

DOPPLER EFFECT

5.1. INTRODUCTION

It is a matter of common experience that the pitch of a note appears to change when either the source or observer or both are in motion relative to each other. When either the source or the observer or both move towards each other, the apparent pitch produced by a sounding body appears to be higher than the actual pitch. In the same way, when the source is moving away from the observer or the observer is moving away from the source or both are moving away from each other, the apparent pitch will appear to be lower than the actual pitch of the sounding body.

The apparent change in pitch due to a relative motion between the source and the observer is known as **Doppler's principle.**

Example 1. If a railway engine moving fast with its whistle blowing is approaching an observer, the pitch appears to become more. The pitch of the note appears to become less just as the engine moves away from the observer on the platform.

Example 2. If a tunning fork mounted on a resonance box is made to move in a circle, then to a listener who is in the plane of the fork, pitch appears to be higher as the fork is moving towards him than when it is moving away from him.

It should be noted that Doppler effect in *sound* is *asymmetric i.e.*, when the source moves towards an observer with some velocity, the apparent pitch is different than that observed when the observer moves towards source with the same velocity. However, in *light*, Doppler effect is *symmetric i.e.*, in this case the apparent pitch or frequency does not change with the motion of source or observer moving with same velocity.

5.2. EXPRESSION FOR CHANGE IN FREQUENCY

(*i*) When the source moves towards the stationary observer and the medium is at rest. Let the frequency of the note given by the source be n and let V be the velocity of sound. The number of waves emitted by the source per second will be n' and these would occupy a length V, if the source is at rest at S. If *the source now moves with velocity a towards the observer* (in the same direction in which the sound is propagating), then after one second it will be at S. The n waves produced per second will now be crowded in the space $S'D$ instead of SO and occupy a length $(V - a)$ as shown in Fig. 5.1.

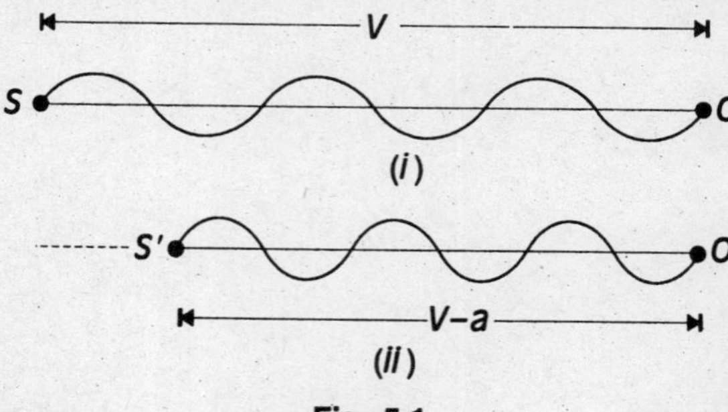

Fig. 5.1

$$\therefore \quad \text{Reduced wavelength} = \frac{(V-a)}{n}$$

So the apparent pitch of the sound heard by the listener

$$n' = \frac{\text{Velocity}}{\text{Wavelength}} = \frac{V}{(V-a)}n \qquad ...(i)$$

Since $(V - a) < V$, so $n' > n$. This shows that the apparent frequency, as heard by the observer, is **higher** than the actual frequency of the note when the source moves towards the observer.

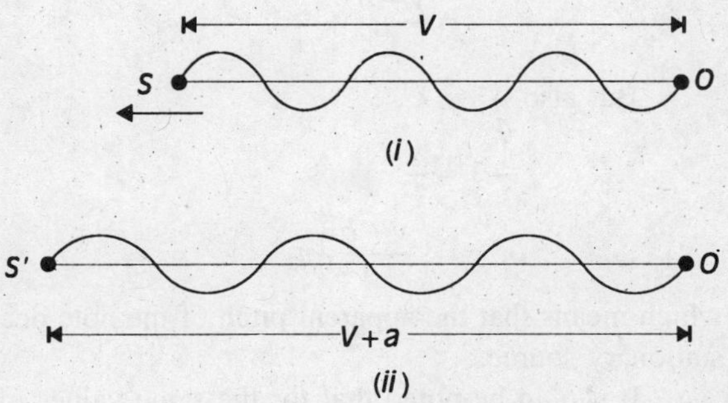

Fig. 5.2

If the source instead of moving towards the observer moves away from it (in a direction opposite to that of the sound), then a is negative and the new frequency n' is given by the relation

$$n' = \frac{V}{(V + a)} n \qquad ...(ii)$$

Since $V + a >> V$, so $n' << n$. This shows that the apparent frequency, as heard by the observer, is lower than the actual frequency of the note when the source moves away from the observer (see Fig. 5.2).

(*ii*) Source at rest, medium also at rest and observer in motior

(*a*) When the observer moves towards a stationary source

Let a source S is producing sound of ptich n and wavelength λ. The velocity of sound is v (Fig. 5.3). Let the observer move with a velocity b towards stationary source S. In this case, the observer receives more number of waves in one second. So the apparent wavelength remains the same. The apparent frequency will be

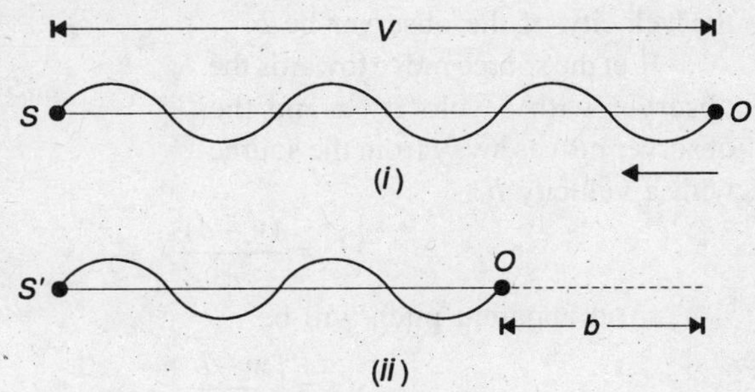

Fig. 5.3

$$n' = n + \frac{b}{\lambda} = \frac{v}{\lambda} + \frac{b}{\lambda}$$

$$\therefore \quad n' = \left(\frac{v + b}{\lambda}\right)$$

$$\text{Also} \quad \lambda = \frac{v}{n}$$

$$\therefore \quad n' = \left(\frac{v + b}{\lambda}\right) n \qquad ...(1)$$

So the apparent pitch of the note is more when the observer moves towards the stationary source.

(*b*) When the observer moves away from a stationary source (Medium at rest)

Let a source S is producing sound of pitch n and wavelength λ (Fig. 5.4). The velocity of sound is v. Let the observer move with a velocity b away from a stationary source. In this case, the observer gets less number of waves in one second. So the apparent wavelength remains the same. The apparent frequency will be given by

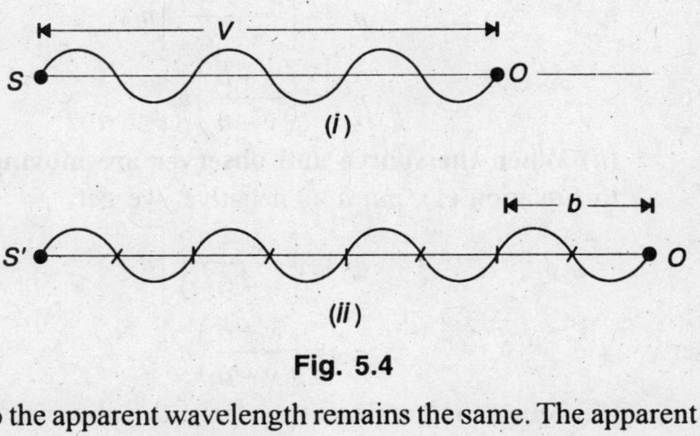

Fig. 5.4

$$n' = n - \frac{b}{\lambda} = \frac{v}{\lambda} - \frac{b}{\lambda}$$

$$= \frac{v-b}{\lambda}$$

But also $\frac{v}{n} = \lambda$

$$\therefore \quad \frac{1}{\lambda} = \frac{n}{v}$$

$$\therefore \quad n' = \left(\frac{v-b}{v}\right) n \qquad \text{...(2)}$$

which means that the apparent pitch of the note decreases when the observer moves away from a stationary source.

It should be noted that for the same values of v, n and b, the values of n' as calculated in Equations (1) and (2) will be different.

(*iii*) Source and observer both in motion and medium at rest

(*a*) When the source is moving towards the observer and the observer moves away from the source

Let a source S be producing sound of pitch n and wavelength λ. The velocity of sound being v (Fig. 5.5). Let the velocity of the source be a and the velocity of the observer be b.

Let the source move towards the observer with a velocity a and the observer moves away from the source with a velocity b.

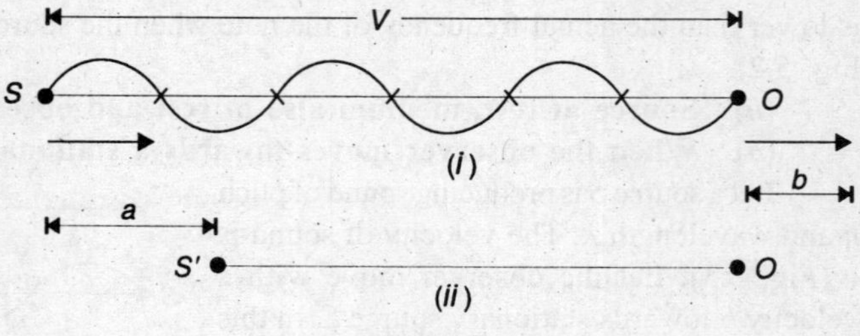

Fig. 5.5

$$\lambda' = \frac{(v-a)}{n}$$

And apparent pitch will be

$$n' = \left(\frac{v-b}{\lambda'}\right)$$

or
$$n' = \left(\frac{v-b}{v-a}\right) n \qquad \text{...(1)}$$

Special cases

(*i*) When the source and observer are moving towards each other

In Equation (1), put b as negative, we get,

$$n' = \left[\frac{v-(-b)}{v-a}\right] n$$

or
$$n' = \left(\frac{v+b}{v-a}\right) n \qquad \text{...(2)}$$

(*ii*) When the source and observer are moving away from each other

In Equation (1), put a as negative, we get

$$n' = \left[\frac{v-b}{v-(-a)}\right] n$$

or
$$n' = \left(\frac{v-b}{v+a}\right) n \qquad \text{...(3)}$$

(*iii*) Source is moving away from the observer and the observer is moving towards the source

In Equation (1), putting both a and b negative, we get

$$n' = \left[\frac{v-(-b)}{v-(-a)}\right] n$$

or
$$n' = \left(\frac{v+b}{v+a}\right) n \quad ...(4)$$

It should be noted that the general relation for apparent frequency is

$$n' = \left(\frac{V-b}{V-a}\right) n \quad ...(5)$$

Which refers to the case when the source is moving towards the observer and the observer is moving away from the source. Depending upon the motion of observer or source, appropriate sign should be put for a and b to get the value of the apparent pitch.

If $a = b$, *i.e.*, the source and the observer move with the same velocity in the same direction, then

$$n' = n$$

which means that there is no apparent change of frequency as there is no relative motion between the source and the observer.

5.3. EFFECT OF MOTION OF THE MEDIUM

When the medium carrrying sound waves is in motion say, there is a wind blowing from S to O with velocity w, the apparent velocity of sound increases from V to $(V + w)$.

$$\therefore \quad n' = \frac{(V+w)-c}{(V+w)-b} n$$

If the wind is blowing with the same velocity in the opposite direction, *i.e.*, from O to S, w will be negative.

$$\therefore \quad n' = \frac{(V-w)-c}{(V-w)-b} n$$

IMPORTANT NOTE

When the source is approaching the observer with a velocity b, the apparent frequency

$$n'' = \frac{V}{V-b} n$$

When the observer is approaching the source with the same velocity, b being negative

$$n' = \frac{V+b}{V} n$$

$$\therefore \quad n' - n'' = \left(\frac{V}{V-b} - \frac{V+b}{V}\right) n = \left\{\frac{b^2}{V(V-b)}\right\} n$$

$$= \text{a positive quantity}$$

or
$$n' = n''$$

which means that when the observer is at rest and the source is in motion towards the observer, the frequency of the note observed by the observer is more than the case when the source is at rest and the observer is moving with the same velocity towards the source.

5.4. DOPPLER EFFECT IN LIGHT

Due to Doppler's principle, we observe a change in frequency of light radiation when the source of light or observer move relative to each other. This principle in similar to the apparent change in pitch of sound due to relative motion of source of sound or observer.

However, there is a difference, in the case of light, no material medium is required for its propagation, while in the case of sound, material medium is necessary for its propagation. In sound as we have seen, the apparent frequencies are different when a source is moving towards a stationary observer and when the observer is moving towards stationary source, *i.e.*, Doppler effect is **asymmetric** in sound.

We have the following relations in the case of sound.

1. When the source is moving towards stationary observer, apparent pitch is given by

$$n' = \left[\frac{v}{(v-a)}\right]n \qquad ...(1)$$

2. When the observer is moving with same velocity towards stationary source,
We have the apparent pitch as

$$n' = \left(\frac{v+a}{v}\right)n \qquad ...(2)$$

However, in the case of light, there is no apparent change of frequency when either the source moves towards the stationary observer or observer moves towards the stationary source with the same velocity, *i.e.*, in the case of light, Doppler effect is **symmetric.**

(*i*) Let the observer be stationary and the source is moving towards the observer with a velocity v. Let the velocity of light be c.

$$\therefore \qquad c = \nu\lambda$$

where, ν is the frequency and λ is the wavelength of light in no medium for propagation, the observer will receive more waves due to the motion of the source. So the wavelength is not changed.

The apparent frequency will be

$$\nu' = \nu + \frac{v}{\lambda} \qquad \nu' = \nu + \frac{\nu v}{c}$$

$$\nu' = \nu\left[1+\frac{v}{c}\right] \qquad ...(1)$$

(*ii*) When the source is stationary and the observer is moving towards the source with a velocity v, we have

$$\nu' = \nu + \frac{v}{\lambda} \qquad \nu' = \nu + \frac{\nu v}{c}$$

$$\nu' = \nu\left[1+\frac{v}{c}\right] \qquad ...(2)$$

which shows Equations (1) and (2) are similar.

(*iii*) When the source moves away from the stationary observer or the observer moves away from the stationary source with a velocity v. We have

$$\nu' = \nu - \frac{v}{\lambda} \qquad \nu' = \nu - \frac{\nu v}{c}$$

$$\nu' = \nu\left[1-\frac{v}{c}\right] \qquad ...(3)$$

(*iv*) When the source and the observer move towards each other and each is moving with a velocity v, we have the relation

$$\nu' = \nu\left[1+\frac{v}{c}\right]^2 \qquad ...(4)$$

(*v*) When the source and the observer move away from each other and each is moving with a velocity v, we have

$$\nu' = \nu\left[1-\frac{v}{c}\right]^2 \qquad ...(5)$$

Doppler effect has great significance in light and astronomy. Some of the important applications of Doppler effect in light are given as under:

5.5. APPLICATIONS OF DOPPLER EFFECT IN LIGHT

1. Red shift. It is observed that some distant nebulae are receding from us with a velocity more than 20×10^3 km/s and the important spectral lines appear to shift to the red end of spectrum, confirming the theory of the **expanding universe.**

2. Saturn Rings. Planet Saturn is found to have concentric rings. By the application of Doppler effect, it is observed that these rings are not of solid matter, but infact consist of small 'Satellites' revolving in orbits around Saturn. Also, if these rings were of solid matter, the outer edges of rings should have more velocity than the inner ones.

From the principle of motion of satellites

$$\frac{mv^2}{R} = \frac{GMm}{R^2}$$

or $$v^2 = \frac{GM}{R}$$

or $$v = \sqrt{\frac{GM}{R}} = \sqrt{\frac{GM}{R}}$$

which means that the velocity of inner orbit is more than that in the outer orbit. The same fact has been established by Doppler's shift, confirming the fact that the rings of Saturn are not solids but satellites moving in orbits of Saturn.

3. Velocity of Sun's rotation. By the application of Doppler's shift to sun's light, it is observed that there is a shift due to velocity of about 2 kms^{-1}, if the observation is made on the light received from the western and eastern edges of the sun. If the observation is made on the north and south edges, no such shift is observed showing that the sun is rotating about the north-south axis.

4. Explanation and discovery of Binary stars. If the sky is observed constantly, it is found that some stars which appear to be single are actually double stars and known as **spectroscopic binaries.** These stars are revolving about each-other, with one approaching the Earth and the other receding from it. There is a shift in their spectral lines and a single spectral line splits into two spectral lines whose separation depends on the time period which is equal to the time period of revolution of the stars, with this method, number of binary stars have been detected.

Example 1. *The pitch of the whistle of an engine appears to drop to 5/6th of the original value when it passes a stationary observer. Assuming the velocity of sound in air to be 350 m/sec-calculate the speed of the engine.*

Solution: Let b be the speed of the engine relative to the observer at rest.

Velocity of sound $V = 350$ m/sec.

Let n be the original frequency of the sound emitted by the engine and n' the observed frequency, then we have

$$\frac{n'}{n} = \frac{5}{6}$$

As the listener is at rest, so from Doppler's principle, we have

$$n' = \frac{V}{V+b}n$$

or $$\frac{n'}{n} = \frac{V}{V+b} = \frac{5}{6}$$

or $$5b = V = 350 \text{ m/sec.}$$

$$\therefore \text{ Speed of engine } b = \frac{350}{5} = \mathbf{70 \text{ m/sec.}}$$

Example 2. *Two aeroplanes pass each other in opposite directions and one of them is blowing a whistle of frequency 540 hertz. Calculate the frequencies of the notes heard in the other aeroplane (i) before and (ii) after they have passed each other. Velocity of either of the aeroplanes is 540 km hr^{-1} and the velocity of sound = 350 ms^{-1}.*

Solution: We have

$$V = 350 \text{ m/s}$$

$$a = 540 \text{ km/hr} = \frac{540 \times 1000}{60 \times 60} = 150 \text{ ms}^{-1}$$

$$b = 540 \text{ km/hr} = 150 \text{ ms}^{-1}$$
$$n = 540 \text{ Hz}$$

(*i*) The source and the observer are moving towards each other. So we have

$$n' = n\left[\frac{V+b}{\text{v}-a}\right]$$

$$n' = 540\left[\frac{350+150}{350-150}\right]; \qquad n' = \mathbf{1350\ Hz}$$

(*ii*) The source and the observer are moving away from each other. So we have

$$n' = n\left[\frac{V-b}{V+a}\right]$$

$$n' = 540\left[\frac{350-150}{350+150}\right]; \qquad n' = \mathbf{216\ Hz}$$

Example 3. *Two observer A and B carry identical sound sources of frequency 500 hertz. If the observer A is stationary and B moves away from A at a speed of 3.6 km/hr, how many beats per second are heard by A and B respectively. Velocity of sound in air = 350 m/s.*

Solution: (*i*) Let *A* be source and *B* be observer. So when the source is at rest and the observer is moving away from it, the apparent pitch will be given by

$$n' = \left(\frac{v-b}{v}\right)n$$

Here
$$v = 350 \text{ m/s}$$
$$b = 3.6 \text{ km/hr} = 1 \text{ m/s}$$
$$n = 500$$
$$n' = \left(\frac{350-1}{350}\right)500$$
$$n' = \frac{349 \times 500}{350}$$

Number of beats per second heard by $B = n - n'$

$$= 5000 - \frac{500 \times 349}{350} = \frac{500}{350} = \mathbf{1.43}$$

(*ii*) Now consider *B* as a source and *A* as an observer, so that the source is moving away from a stationary observer. The apparent pitch will be given by

$$n' = \left(\frac{v}{v+a}\right)n$$

Here
$$v = 350 \text{ m/s}$$
$$a = 1 \text{ m/s}$$
$$n = 500 \text{ hertz}$$
$$n' = \left(\frac{350}{350+1}\right)500$$
$$n' = \left(\frac{350}{351}\right)500$$

Number of beats per second heard by $A = n - n'$

$$= 500 - \left(\frac{500 \times 350}{351}\right) = \frac{500}{351} = \mathbf{1.424.}$$

Example 4. *A motor car fitted with two sounding horns which have a difference in frequency by 320 vibrations per second is speeding at the rate of 36 km/hr towards a person standing. Calculate the difference in the frequencies of the notes heard by him. Velocity of sound in air = 330 m/s.*

Solution: Let n_1 and n_2 be the two frequencies, we have

$$n_1 - n_2 = 320$$

$$n_1' = \left(\frac{v}{v-a}\right) n_1 \quad ...(i)$$

$$n_2' = \left(\frac{v}{v-a}\right) n_2 \quad ...(ii)$$

Subtracting Equation (*ii*) from (*i*) we have

$$(n_1' - n_2') = \left(\frac{v}{v-a}\right)(n_1 - n_2)$$

Here $a = 36 \text{ km/hr} = 10 \text{ m/s}$

also $v = 330 \text{ m/s}$

$$\therefore \quad n_1' - n_2' = \left(\frac{330}{330-10}\right)(320)$$

or $n_1' - n_2' =$ **330 vibrations/second**

Example 5. *A person is standing near a railway track and a train moving with a speed of 36 km/hr is approaching him. The apparent pitch of the whistle as heard by the person is 700 hertz. Calculate the actual frequency of the whistle. Velocity of sound = 350 m/s.*

Solution: We have

$$V = 350 \text{ m/s}, \qquad a = 36 \text{ km/hr} = 10 \text{ m/s}$$

$$n' = 700 \text{ Hz}, \qquad n = ?$$

The source is moving towards a stationary observer, from Doppler's principle,

$$n' = \left(\frac{V}{V-a}\right) n$$

$$\therefore \quad n = \left(\frac{V-a}{V}\right) n'$$

$$n = 700\left(\frac{350-10}{350}\right); \; n = \mathbf{680 \ Hz}$$

Example 6. *Two aeroplanes A and B are approaching each other with a speed of 360 km/hr. The frequency of the whistle emitted by A is 1000 hertz. Calculate the apparent pitch of the whistle as heard by the passengers of aeroplane B. Velocity of air = 350 m/s.*

Solution: In this case

$$V = 350 \text{ m/s}$$

$$a = 360 \text{ km/hr} = 100 \text{ m/s}$$

$$b = 360 \text{ km/hr} = 100 \text{ m/s}$$

$$n = 1000 \text{ Hz}$$

The source and the observer are moving towards each other, from Doppler's Principle

$$\therefore \quad n' = n\left[\frac{V+a}{V-a}\right]$$

$$\therefore \quad n' = 1000\left[\frac{350+100}{350-100}\right]; \; n' = \mathbf{1800 \ Hz}$$

Example 7. *Two aeroplanes are approaching each other and their velocities are 100 and 150 miles/hr. The frequency of a note emitted by the first as heard by the passengers in the other is 1000. Calculate the true frequency of the note as heard by its own passengers. Take the velocity of sound as 750 mile/hour.*

Solution: The apparent frequency, when both source and the listener are moving in the direction of the sound is given by the relation

$$= \frac{c - V_0}{c - V_s}$$

But as both the aeroplanes are approaching each other, the direction of the second aeroplane (listener) is opposite to the direction of the sound *i.e.*, in this case V_0 is negative.

$$\therefore \quad n' = \frac{c + V_0}{c - V_s}$$

or $$n = \frac{c - V_s}{c + V_0}$$

Also we have

$$V_0 = 100 \text{ m.p.h.}$$
$$V_s = 150 \text{ m.p.h.}$$
$$c = 750 \text{ m.p.h.}$$
$$\nu' = 1000 \text{ Hz}$$

Hence, $$\nu = \frac{750 - 100}{750 + 150} \times 1000$$
$$= \mathbf{722 \ Hz}$$

Example 8. *At each of two stations A and B, a siren is sounding with a constant frequency of 250 cycles/s. A cyclist from A proceeds straight towards B with a velocity of 12 km/hr and hears 5 beats/s. Calculate the velocity of sound.*

Solution: (*i*) The cyclist is moving away from the source *A*. From Doppler's Principle, we have,

$$\therefore \quad n_1 = \left(\frac{v - b}{v}\right) n$$

Here $$b = 12 \text{ km/hr} = \frac{10}{3} \text{ m/s}$$
$$n = 250$$

$$\therefore \quad n_1 = \left(\frac{v - \frac{10}{3}}{v}\right) 250 \qquad \text{...(1)}$$

(*ii*) The cyclist is moving towards the source *B*.

$$\therefore \quad n_2 = \left(\frac{v + b}{v}\right) n$$

or $$n_2 = \left(\frac{v + \frac{10}{3}}{v}\right) 125 \qquad \text{...(2)}$$

Subtracting Equation (1) from (2),

$$n_2 - n_1 = \frac{250}{v} = \frac{20}{3}$$

Also $$n_2 - n_1 = 5$$

$$\therefore \quad v = \frac{250 \times 20}{5 \times 3}$$
$$v = \mathbf{333.33 \ m/s.}$$

Example 9. *A spectral line of wavelength 4000×10^{-8} cm in the spectrum of light from a star is found to be displaced from its normal position towards the red end of the spectrum by an amount 10^{-8} cm. What velocity of the star in the line of sight would account for this? (Velocity of light = 3×10^{10} cm/s).*

Solution: Let λ be the actual wavelength and λ', the changed wavelength of the spectral line. Let V be the velocity of the star.

Then the changed wavelength of the spectral line is given by the relation

$$\lambda' = \frac{c - V}{v}$$

where $$v = \text{Actual frequency} = \frac{c}{\lambda}$$

$$\therefore \quad \lambda' = \frac{c - V}{c} \quad \text{...for approaching star}$$

and $$\lambda' = \frac{c + V}{c}\lambda \quad \text{...for receeding star}$$

or $$\lambda' = \frac{c \pm V}{c}\lambda$$

$$\frac{\lambda'}{\lambda} = \frac{c \pm V}{c}\lambda$$

or $$\frac{\lambda \pm d\lambda}{\lambda} = \frac{c \pm V}{c} \quad (\text{Taking } \lambda' = \lambda \pm d\lambda)$$

or $$\frac{d\lambda}{\lambda} = \frac{V}{c}$$

or $$V = \frac{d\lambda}{\lambda}c$$

Now we have $$\lambda = 4000 \times 10^{-8} \text{ cm}$$

$$d\lambda = 10^{-8} \text{ cm}$$

$$c = 310^{10} \text{ cm/s}$$

$$V = \frac{10^{-8}}{4000 \times 10^{-8}} \times 3 \times 10^{10}$$

$$= 75 \times 10^{-8} \text{ cm/s}$$

$$= \mathbf{75 \text{ km/s}}$$

Since the shift is towards longer wavelength side (red end), so the star is receeding.

Example 10. *A vibrating tuning fork tied to the end of a string 6 ft long, is whirled round in a circle. It makes 120 rev/min. Calculate the difference of frequency between the highest and the lowest note heard by an observer situated in the plane of rotation.*

(*Velocity of sound = 1100 ft/s.*)

Solution: The apparent frequency, when the listener is at rest, is given by the relation

$$n' = \frac{c}{c - V_s}n \quad \text{— in the highest position}$$

and $$n'' = \frac{c}{c + V_s}n \quad \text{— in the lowest position}$$

So the difference of the frequencies between the highest and the lowest note heard by the listener is given by,

$$n' - n'' = \left(\frac{c}{c - V_s} - \frac{c}{c + V_s}\right)$$

$$= \frac{c(2V_s)}{(c - V_s)(c + V_s)}$$

we have $$V_s = r\,\omega = r.2\pi\, n$$

$$= \frac{6 \pm 2\pi \times 120}{60} = 24\pi \text{ ft/s}$$

and $= 1100$ ft/s

$$n' - n'' = \frac{1100 \times 2 \times 24\pi \times n}{(1100 - 24\pi)(1100 + 24\pi)}$$

$$= \mathbf{0.31377n.}$$

Example 11. *An observer on a railway platform noticed that when a train passed through the station, at a speed of 72 km/hr, the frequency of the whistle appeared to drop by 500 Hz. Calculate the actual frequency of the note given by the whistle.*

(*Velocity of sound in air = 340 m/s*).

Solution: We have $V = 340$ m/s

$a = 72$ km/hr $= 20$ m/s

$n = ?$

$n_1 - n_2 = 500$

where n_1 is the apparent pitch when the train is approaching the observer and n_2 is the apparent pitch when the train is receeding from the observer. From Doppler's Principle

$$n_1 = n\left[\frac{V}{V - a}\right]$$

and

$$n_2 = n\left[\frac{V}{V + a}\right]$$

$$\therefore \quad n_1 - n_2 = n\left[\frac{V}{V - a}\right] - n\left[\frac{V}{V + a}\right]$$

or

$$n_1 - n_2 = n\left[\left(\frac{V}{V - a}\right) - \left(\frac{V}{V + a}\right)\right]$$

$$\therefore \quad 500 = n\left[\left(\frac{340}{340 - 20}\right) - \left(\frac{340}{340 + 20}\right)\right]$$

$$n = \mathbf{4235 \text{ Hz.}}$$

Example 12. *A policeman on duty detects a drop of 10% in the pitch of the horn of a motor car as it crosses him. If the velocity of sound is 330 m/s, calculate the speed of the car.*

Solution: In this case

$v = 330$ m/s

Let the frequency of the horn $(n) = 100$ Hz

$\therefore$ Apparent frequency $(n') = 90$ Hz

We have the relation

$$n' = \frac{v}{v + V} n$$

where $V =$ speed of the car.

$$\therefore \quad 90 = \frac{330}{330 + V} \times 100$$

or

$$2970 + 9V = 3300$$

or

$$V = \frac{330}{9} \text{ m/s} = \mathbf{36.67 \text{ ms}^{-1}.}$$

Example 13. *A train travelling at 45 miles per hour is whistling, the frequency of the whistle is 600 Hz. Another train is approaching this train on parallel rails and the person in the second train hears this whistle as a sound of 800 Hz. What is the speed of the second train? Velocity of sound = 1100 ft/s.*

Solution: The observed frequency, when both source and listener are moving in the direction of sound is given by the relation

$$n = \frac{v - V_0}{v - V_s} n$$

In this case, the listener is moving in the opposite direction of the sound *i.e.*, V_s is negative, so we have

$$n = \frac{v + V_0}{v - V_s} n$$

or
$$V_s = \frac{n'}{n}(v - V_s) - v$$

Now
$$V_0 = 45 \text{ mile/hour}$$
$$n = 600 \text{ Hz}$$
$$n' = 800 \text{ Hz}$$
$$v = 1100 \text{ ft/s} = 750 \text{ mile/hour}$$

$$\therefore \quad V_0 = \frac{800}{600}(750 - 45) - 750$$
$$= \mathbf{190 \text{ mile/hour.}}$$

Example 14. *A motor car is fitted with two horns, which differ in frequency by 288 vibrations per sec. If the car sounding the horns is moving at 30 m.p.h. towards a person, who is at rest, calculate the change in the difference of frequencies of the notes heard by him. (Velocity of sound in air = 1120 ft/s).*

Solution: When the source is moving, but listener is at rest, the apparent frequency is given by

$$n_1' = \frac{V}{v - V_s} n_1 \quad \text{...for one horn}$$

$$n_2' = \frac{v}{v - V_s} n_2 \quad \text{...for other horn}$$

So, the difference of frequencies of the notes heard by the listener will be

$$n_1' - n_2' = \frac{v}{v - V}(n_1 - n_2)$$

In this case
$$V = 30 \text{ m.p.h.} = 44 \text{ ft/s}$$
$$v = 1120 \text{ ft/s}$$
$$n_1' - n_2' = 288 \text{ Hz}$$

Hence,
$$n_1' - n_2' = \frac{1120}{1120 - 44} \times 288$$
$$= \mathbf{299 \text{ Hz.}}$$

Example 15. *A train is approaching a vertical cliff with a velocity v km/hour. The driver sounds a whistle of pitch n and finds the frequency of the echo equal to n_2 where*

$$n_2 = n\frac{V + v}{V - v}$$

Explain how the above result is obtained if V is the velocity of sound in km/hour?

Solution: As the whistle is sounded by the moving engine, so apparent frequency of the echo as heard by a *stationary observer* can be calculated by considering the source to be moving towards the observer with a velocity v. This is because when the engine approaches the cliff, the distance of the path by which the echo reaches the listener decreases. So we have

Apparent frequency
$$n_1 = \frac{V}{V - v} n \quad \text{...(1)}$$

where V is the velocity of the sound.

Again the listener (driver) moves towards the cliff with a velocity v which is equivalent the *listener moving towards the source.* So frequency of sound as heard by the driver (listener) is given by the relation

$$n_2 = \frac{V - v}{V} n_1 = \frac{V + v}{V} n_1 \quad ...(2)$$

The value of v is taken as **positive** when the listener is moving **away** from the source and it is **negative** when the listener is moving **towards** the source.

Substituting the value of n_1 from Equation (1) in (2) we get

$$n_1 = \frac{V + v}{V} \times \frac{V}{V - v} n = \frac{V + v}{V - v} n$$

Example 16. *An observer on a platform observed a train passing through the station at a speed b. Show that the frequency of the whistle changes by*

$$nV \left[\frac{2b}{V^2 - b^2}\right]$$

where V is the velocity of sound and n the frequency of the whistle.

Ans. (*a*) Let n_1 be the apparent frequency, when the train is approaching the observer on the platform, and n_2 the apparent frequency, when the train is going away from the observer, then we have

(*i*) when the source is approaching the observer

$$n_1 = \frac{V}{V - b} n \quad ...(1)$$

(*ii*) when the source is moving away from the observer

$$n_2 = \frac{V}{V + b} n \quad ...(2)$$

$$\therefore \quad \text{Change in frequency } n_1 - n_2 = n\left[\frac{V}{V - b} - \frac{V}{V + b}\right]$$

$$= \boldsymbol{nV\left[\frac{2b}{V^2 - b^2}\right]}$$

Example 17. *Find the velocity of the source when the frequency appears to be (a) double and (b) half the original frequency to a stationary observer.*

Solution: The apparent frequency when the source is in motion and listener at rest is given by the relation:

$$n' = \frac{v}{v - V_s} n$$

(*a*) In this case

$$n' = 2n$$

$$2n = \frac{v}{v - V_s}$$

$$v = 2V_s$$

i.e., the velocity of the source should be half of the velocity of sound.

(*b*) Here,

$$n' = \frac{n}{2}$$

$$\frac{n}{2} = \frac{v}{v - V_s} n$$

or

$$V_s = -v$$

i.e, the velocity of the source should be equal to the velocity of sound and it should move in the direction opposite to the sound.

Example 18. *A tuning fork of frequency 440 approaches a wall with a velocity of 4 metres per second. What will be the number of beats heard between the direct and the reflected sounds, if the velocity of sound is equal to 332 metres/sec?*

Solution: We have the frequency of the tuning fork $n = 440$

Velocity of sound $V = 332$ m/sec.

Let the tuning fork lie between the listener and the wall. When the tuning fork approaches the wall, it moves away from the listener. The listener receives two sounds of different frequencies, one directly, from the receeding tuning fork and the other, after reflection from the wall. So, we will have beats due to these two notes.

(*i*) *Frequency of sound received directly,*

Velocity of the source $b = 4$ meters/sec.

Since the source is moving away from the listener so b will be negative.

$$\therefore \text{ Apparent frequency } \quad n_1 = \frac{V}{V+b}n = \frac{332}{332+4} \times 440 = 434.8 \text{ Hz}$$

(*ii*) *Frequency of sound after reflection*

As the source is moving towards the wall and the listener is at rest w.r.t. the wall. It is equivalent to source moving towards the listener, so from Doppler's Principle

$$\therefore \text{ Apparent frequency } \quad n_2 = \frac{V}{V-b}n = \frac{332}{332-4} \times 440$$

$$= 445.4 \text{ Hz}$$

Hence, number of beats heard $= 445.4 - 434.8 =$ **10.6 /sec.**

Example 19. *Two trains travelling in opposite directions at 100 km/hr each, cross each other while one of them is whistling. If the frequency of the note is 800 Hz, find the apparent pitch as heard by an observer in the other train.*

(a) before the trains cross each other

(b) after the trains have crossed each other.

(Velocity of sound in air = 340 m/s)

Solution: We have $V = 340$ m/s

$$a = 100 \text{ km/hr} = \frac{250}{9} \text{ m/s}$$

$$b = 100 \text{ km/hr} = \frac{250}{9} \text{ m/s}$$

$$n = 800 \text{ Hz}$$

(*a*) The source and the observer are moving towards each other. So we have,

$$n' = n\left[\frac{V+b}{V-a}\right]$$

$$n' = 800\left[\frac{340 + \left(\frac{250}{9}\right)}{340 - \left(\frac{250}{9}\right)}\right] = \mathbf{942.3 \text{ Hz}}$$

(*b*) In this case, the source and observer are moving away from each other, so we have

$$n' = n\left(\frac{V-b}{V+a}\right)$$

$$= 800\frac{\left(340 - \frac{250}{9}\right)}{\left(340 + \frac{250}{9}\right)}$$

$$= \mathbf{679.2 \text{ Hz.}}$$

Example 20. *A train moving at 30 m.p.h. approaches another train, also moving at 30 m.p.h. in the opposite direction and its driver sounds a whistle of frequency 1000 Hz. Calculate the apparent frequencies heard by the second driver (a) when the trains are approaching (b) when they are receeding. (Velocity of sound = 1100 ft/s.).*

Solution: The apparent frequency, when both source and the listener are moving in the direction of sound is given by the relation,

$$n' = \frac{v - V_0}{v - V_s} \times n$$

In this case $V_s = V_s = 30$ m.p.h.

$$= 44 \text{ ft/s}$$

$$n = 1000 \text{ Hz}$$

Here $v = 1100$ ft/s

(*a*) Here, the direction of the listener is opposite to the direction of the sound *i.e.,* V_o is negative, so we have

$$n_1 = \frac{v + V_0}{v - V_s} \times n$$

$$= \frac{1100 + 44}{1100 - 44} \times 1000$$

$$= \mathbf{1083\ Hz}$$

(*b*) In this case, V_s is negative, so we have

$$n' = \frac{v - V_0}{v + V_s} \times n$$

$$= \frac{1100 - 44}{1100 + 44} \times 1000$$

$$= \mathbf{923\ Hz.}$$

Example 21. *An observer on the railway platform observed that as the train passed through the station at 90 km/hour, the frequency of the whistle appeared to drop by 400 hertz. Find the frequency of the whistle. (Velocity of sound in air = 350 m/sec)*

Solution:

Velocity of sound $V = 350$ m/sec

Velocity of the train $b = 90$ km/hour $= 25$ m/sec

Change in frequency $= 400$ hertz

Let n be the frequency of the whistle, then from Doppler's principle,

$$\text{Change in frequency} = nV\left[\frac{2b}{V^2 - b^2}\right]$$

$$\therefore \quad 400 = n \times 350\left[\frac{2 \times 25}{350^2 - 25^2}\right]$$

$$= 350\left[\frac{50}{375 \times 325}\right]$$

or

$$n = \frac{400 \times 375 \times 325}{350 \times 50} = \mathbf{278.7\ Hz.}$$

Example 22. *A motor car sounding a horn at a frequency of 100 hertz moves away from a stationary observer towards a rigid flat wall with a velocity of 36 km/hr. How many beats per second will be heard by the observer? The velocity of sound in air at room temperature = 350 m/s.*

Solution: We have in this case

$$V = 350 \text{ ms}^{-1}$$

$$a = 36 \text{ km/hr} = 10 \text{ ms}^{-1}$$

$$n = 100 \text{ Hz}$$

Now, the observer receives sound of apparent frequency n_1 from the moving source directly and also sound of apparent frequency n_2 from the sound waves reflected from the wall, so we have

$$n_1 = n\left[\frac{V}{V + a}\right]$$

$$n_1 = 100\left[\frac{350}{350+10}\right]$$

$$n_1 = 97.23 \text{ hertz}$$

Also $$n_2 = n\left[\frac{V}{V-a}\right]$$

$$n_2 = 100\left[\frac{350}{350-10}\right]$$

$$n_2 = 102.9 \text{ hertz}$$

Number of beats/sec

$$n_2 - n_1 = 102.9 - 97.23 = \mathbf{5.67.}$$

Example 23. *A policeman on duty at a crossing challanges a motor driver for crossing the speed limit of 100 kilometres per hour by detecting a change of 20 vibrations in the horn note of frequency 128 as the car passes by him. Is he correct?*

(Velocity of sound = 350 m/sec.)

Solution: As the car is approaching the policeman, the apparent frequency n_1 is given by the relation

$$n_1 = \frac{V}{V-b}.n$$

where b is the velocity of the car (in this case, source). When the car crosses the policeman the apparent frequency is given by the relation

$$n_2 = \frac{V}{V+b}.n$$

∴ Change in frequency

$$n_1 - n_2 = nV\left[\frac{1}{V-b} - \frac{1}{V+b}\right] = nV\left(\frac{2b}{V^2-b^2}\right) \quad ...(1)$$

Also, velocity of sound $V = 300$ m/sec

Frequency of horn $n = 128$ Hz

Change in frequency $(n_1 - n_2) = 20$ Hz

Putting these values in Equation (1), we have

$$20 = \frac{128 \times 350 \times 2b}{350^2 - b^2}$$

or $$b_2 + 4480b - 350 \times 350 = 0$$

∴ $$b = \frac{-4480 \pm \sqrt{4480^2 + 4 \times 350 \times 350}}{2}$$

$$= 27.5 \text{ m/sec.} = \mathbf{99\,km/hour}$$

which means that the motor driver has not crossed the speed limit of 100 km/hour.

Example 24. *The apparent frequency of the whistle of an engine changes in the ratio 6 : 5 as the engine passes a stationary observer. If the velocity of sound is 352 m/s, calculate the velocity of the engine.*

Solution: (*i*) When the source moves towards the stationary observer, we have the relation

$$n_1 = \left(\frac{v}{v-a}\right) \quad ...(i)$$

(*ii*) When the source moves away from the stationary observer, we have

$$n_2 = \left(\frac{v}{v+a}\right) \quad ...(ii)$$

Dividing Equation (*i*) by (*ii*), we get

$$\frac{n_1}{n_2} = \frac{v+a}{v-a}$$

In this case $\frac{n_1}{n_2} = \frac{6}{5}$

$v = 352$ m/s

$a = ?$

$\therefore \quad \frac{6}{5} = \frac{353 + a}{353 - a}$ or $a = \mathbf{32\ ms^{-1}}$

Hence, the velocity of the engine is 32 ms^{-1}.

Example 25. *A vibrating fork tied to the end of a string 1.988 metres long is whirled round in a circle. If it makes two revolutions in a second, calculate the ratio of the frequencies of the highest and the lowest notes heard by an observer situated in the plane of the tuning fork. Velocity of sound is 350 metres/sec.*

Solution: We have

Radius of the circle = 1.988 m

Number of rotations per second = 2

So, linear velocity of tuning fork$b = 2 \times 2\pi r$

$$= 4 \times \frac{22}{7} \times 1.988 = 25\ ms^{-1}$$

As the tuning fork is revolving in a circle, the apparent pitch rises and will be the highest when the tuning fork is approaching the listener and will be the lowest when the tuning fork is moving away from the listener.

(*i*) Apparent frequency when the tuning fork is approaching the listener will be

$$n' = \frac{V}{V - b}.n = \frac{350}{350 - 25}n = \frac{14}{13}n$$

(*ii*) Apparent frequency when the tuning fork is receeding from the listener

$$n'' = \frac{V}{V + b}.n = \frac{350}{350 + 25}n = \frac{14}{15}n$$

$$\therefore \quad \frac{n'}{n''} = \frac{14}{13} \times \frac{15}{14} = \mathbf{1.154.}$$

Example 26. *An engine blowing a whistle of frequency 130, moves with a velocity of 90 km/ hour towards a hill from which a well-defined echo is heard. Calculate the frequency of the echo heard by the driver. Velocity of sound in air is 350 m/sec.*

Solution: Let V be the velocity of sound and v that of the engine moving towards the hill, then the apparent frequency n'' is given by the relation

$$n'' = \frac{V + v}{V - v}n$$

where n is the frequency of the whistle.

Now, we have frequency of the whistle $n = 30$

Velocity of sound $V = 300$ m/sec.

Velocity of the engine $v = 90$ km/hour = 25 m/sec.

$\therefore$ Apparent frequency of the echo as heard by the driver

$$n'' = \frac{V + v}{V}n = \frac{350 + 25}{350 - 25} \times 130$$

$$= \mathbf{150\ Hz.}$$

QUESTIONS AND PROBLEMS

1. Explain Doppler's principle. Calculate the apparent pitch of a note due to the motion of the source and the listener.
2. Explain Doppler's effect. A source produces a note of frequency n and is moving towards a stationary observer with a uniform speed a. Show that the apparent pitch is

$$n' = n\left[\frac{V}{V-a}\right].$$

3. State and Explain Doppler's effect. Derive an expression for the change in frequency of a note when both the source of sound and the observer are in motion.

4. Explain Doppler effect. Explain the apparent frequency of a note when both the source and the listener are in motion, the medium being at rest.

5. Discuss Doppler's effect in sound and obtain an expression for the apparent frequency of the note when the source and the listener are (*i*) moving towards each other, and (*ii*) moving away from each other.

6. An observer on a railway platform observed a train passing through the station at a speed a. Show that the frequency of the whistlen changes by

$$nV\left[\frac{2a}{V^2 - a^2}\right]$$

7. Explain Doppler's effect due to source and listener in motion. How do you show that in the absence of wind, a given speed of approach of source raises the apparent pitch more than the same speed of approach of the listener? What change will be observed if the source of sound travels in a straight line which does not pass through the position of the observer?

8. Show that if the source moves away with the velocity of sound from an observer who is at rest, the frequency of the note heard is halved.

9. A railway engine while whistling moves past an observer with a speed of 30 miles per hour. Find the ratio of the pitches when the source is first approaching and then receeding (velocity of sound in air = 1088 ft/sec). **[Ans.** 1.08]

10. A train approaches a stationary observer, the velocity of the train being 1/20th of the velocity of sound, and a sharp blast is blown with the whistle of the engine, at equal intervals of a second. Find the interval between the successive blasts as heard by the observer. **[Ans.** 0.95 sec.]

11. A train is approaching a vertical cliff at 60 miles per hour. The driver sounds a whistle of pitch 500. What will be the pitch of the echo as heard (1) by the engine driver, (2) by an observer who is stationary? **[Ans.** (1) 585.2 Hz (2) 542.6 Hz]

12. The whistle of an engine moving at 30 miles/hr is heard by a motorist driving at 15 miles/hr and estimated to have a pitch of 500. What must be the actual pitch of the whistle to the nearest whole number when,

(*a*) the two are moving in opposite directions but approaching each other;
(*b*) the two are moving in opposite directions but moving away from each other;
(*c*) the two are moving in the same directions, the motorist being behind the engine;
(*d*) the two are moving in the same direction, the motorist being in front of the engine?

(Velocity of sound = 1200 ft/sec.) **[Ans.** (*a*) 473 Hz (*b*) 508 Hz (*c*) 508 Hz (*d*) 491 Hz]

13. Describe Doppler effect in sound. Show how the frequency of a note is affected by the motion of an observer towards or away from the source.

A motor-car travelling 30 miles per hour, is proceeding towards a high building. If the frequency of the note produced by its horn is 600, what would be the pitch of the note heard by the driver after reflection from the building? **[Ans.** 649.2]

14. Find the velocity of the car with respect to a stationary observer if the latter observes a drop in pitch of 100 vibrations per sec, when the car passes him. The car emits a note with a frequency of 1000 vibrations/sec and the velocity of sound is 1142 ft./sec. **[Ans.** 57.1 ft/sec.]

15. A spectral line, of wavelength 4×10^{-5} cm in the spectrum of light from a star is found to be displaced from its normal position towards the red end of the spectrum by an amount equivalent to 10^{-8} cm. What velocity of the star in the line of sight would account for this? (Velocity of light = 3×10^5 km/sec.) **[Ans.** A velocity of recession of 75 km/sec.]

16. A man standing by a railway track notices the change of pitch of the note due to the whistle of an engine as it passes him. If the frequency of the whistle be 256 vibrations/sec. and the velocity of the engine be 1/20th of that of sound, what will be the frequencies of the notes heard by the man before and after the engine passes by him? **[Ans.** 269.47; 243.8]

17. A locomotive whistle emitting 2000 waves/sec is moving towards you at the rate of 60 miles/hr on a day when the thermometer stands at 15°C. Calculate the apparent pitch of the whistle (velocity of sound at 0°C = 1090 ft/sec). **[Ans.** 2157]

CHAPTER 6

ULTRASONICS

6.1. INTRODUCTION

Sound waves in which the frequencies are above the limits of human audibility, *i.e.*, greater than 20,000 cycles/sec, are called **ultrasonics or supersonics**. However, supersonic term is nowadays used only to velocities greater than that of sound, *i.e.*, about 700 miles/hr, which is a common term in aviation. Sounds of frequency less than 20 cycles/sec are called **infrasonics.**

6.2. PIEZO ELECTRIC EFFECT

In 1880, J. and P. Curie discovered that if mechanical pressures are applied to the opposite faces of certain crystal slices cut suitably, then there would be developed equal and opposite electric charges on other faces resulting in a difference of potential, the magnitude of the potential developed being proportional to the pressure. However, if the pressure is replaced by tension, the sign of the charges is reversed. This phenomenon is called **piezo electric effect**.

The converse effect is also possible, *i.e.*, if a potential is applied to the opposite forces of a crystal, a change in dimensions would take place, the change or increase and decrease being according to the direction of the potential difference.

This effect was best observed in a number of crystals like Quartz Tourmaline, Rochille salt etc., which has been cut into slices in suitable directions.

6.3. PRODUCTION OF ULTRASONICS

1. Mechanical Method. This is one of the earliest methods for producing frequencies up to 100 K. cycles/sec with the help of Galton's whistle.

2. Piezoelectric Generator. Langevin in 1917 devised a method for producing ultrasonics by the use of piezo electric effect. He used an electric oscillatory circuit to provide e.m.f. and the tuning is achieved by a variable condenser.

Figure 6.1 shows a simple piezoelectric generator. Q is a slice of a crystal placed between metal plates A and B so as to form a parallel plate condenser with crystal Q as dielectric. The plates are connected to the primary of a transformer which is coupled inductively to the oscillatory circuit of a valve as shown in Fig. 6.1. If the natural frequency of oscillations of the valve circuit is equal

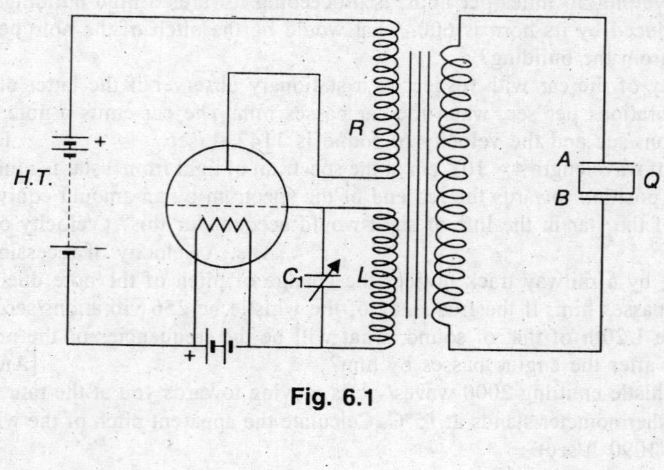

Fig. 6.1

to the frequency of the crystal mode to which we intend it to excite, resonance will occur and the crystal will undergo linear expansions and contractions at exactly the same rate *i.e.,* the crystal is set into mechanical vibrations. This arrangement is similar to an electrically maintained vibrator emitting sound waves of definite frequency. With a moderate size quartz crystal, ultrasonics of frequencies ~ 5,40,000 cycles/sec can be produced. We may get higher frequency as well but then the plate has to be very thin and strong enough to stand the strain. Tourmaline plates can be prepared to give frequencies ~ 1.5×10^8 cycles/sec.

3. Magnetostriction Oscillator. Before actually taking up the method for production of ultrasonics, first we may study the magnetostriction effect.

Magnetostriction Effect. We can get very high frequency oscillations with Magnetostriction. According to this phenomenon a bar of a ferromagnetic material like iron or nickel changes in length when it is placed in a strong magnetic field parallel to its length.

A nickel tube or bar placed in a rapidly varying magnetic field alternately expands and contracts with twice the frequency of the applied magnetic field. The longitudinal expansions and contractions produce ultrasonic sounds in the medium surrounding the nickel tube with frequencies between 8,000 and 20,000 cycles/sec. The range of frequencies depends on the mode of vibrations with magnetostrictive material and varies between few hundred to 3,00,000 cycles/sec.

Since the rod will vibrate longitudinally, so we can use the fundamental mode of vibration as

$$N = \frac{1}{2l}\sqrt{\frac{Y}{\rho}},$$

where Y is the Young's modulus of the material and ρ its density.

6.4. EXPERIMENTAL PRODUCTION OF ULTRASONICS

Figure 6.2 shows the magnetostriction oscillator, which involves the interaction of a magnetostrictive bar and the oscillations of a valve circuit.

The bar normally invar 36% nickel and 64% Fe or "monel" 65% Ni and 31% Cu and 4% (Fe, Si, Mn and C) is placed inside a coil parallel to its axis and a high frequency current is passed through the coil, the rod is magnetised and demagnetised with the current. The length of the rod varies accordingly and its free ends produce ultrasonics. If the length of the rod is such that the frequency of its vibrations is equal to the frequency of the applied current, resonance occurs and the amplitude of vibrations becomes very high.

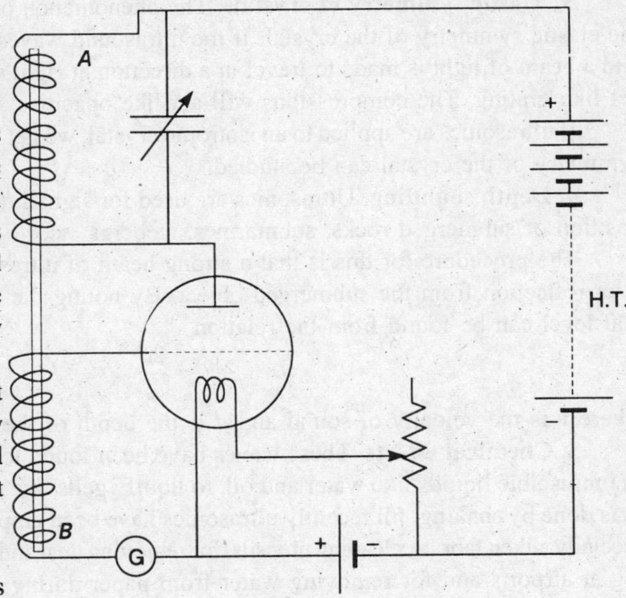

Fig. 6.2

It should be noted that the best results are got if the bar is previously magnetised by means of a second coil wound over it and carrying a direct current.

6.5. DETECTION OF ULTRASONIC

1. Kundt's tube method. Kundt's tube can be used to detect ultrasonics in the same way as it is used for detection of ordinary sound waves.

2. Thermal detectors. If a probe of fine platinum wire is placed in the region of ultrasonics, we set compressions and rarefactions very rapidly at nodes and it causes adiabatic changes, so the platinum probe is alternately heated and cooled resulting in the change of resistance which may be detected by suitable experiments. However, no such change occurs at an antinode.

3. Sensative Flame. These can be used to detect ultrasonics since very high frequency sounds will change the intensity of the flame.

4. Piezoelectric detectors. If a pair of faces of a piezoelectric crystal like quartz is subjected to waves, opposite charges develop on the other pair perpendicular to the first. The changes are amplified by an amplifier and then detected by suitable means.

6.6. APPLICATION OF ULTRASONICS

1. Velocity in gases and liquids. To find the velocity of sound in gases and liquids, stationary wave method is applied. Stationary waves are set up in a gas or a liquid analogous to Kundt's tube method. Then the velocity of ultrasonics in the medium can be calculated from the relation $v = n\lambda$, where n is the frequency of oscillation and λ is the wavelength of the sound waves. This method is suitable for finding the velocity of sound in gases and liquids which are available in small quantities. Changes of velocity with frequency and temperature can also be measured with this arrangement.

2. Velocity of ultrasonics by acoustic grating. Brillouin in 1925 gave an alternative method for finding the velocity of ultrasonics. In this case a liquid or solid through which ultrasonics travel acts like an acoustic grating. Then using the phenomenon of diffraction from optics we have the relation

$$d \sin\theta = n\lambda$$

where d is the distance between consecutive nodal points, θ the angle of diffraction, n the order of spectrum and λ the wavelength of light used. Knowing θ and λ, d can be found and it will be equal to $\lambda a/2$ where λa corresponds to wavelength of ultrasonic sound. Now by the method of stationary waves using a piezo electric crystal, if the frequency of the crystal oscillator is known, the velocity of ultrasonics can be calculated.

3. Elastic symmetry of crystals. The phenomenon of acoustic grating can be used to study the elastic symmetry of the crystal. If the ultrasound waves are passed through a solid or a liquid and a beam of light is made to travel in a direction at right angles, the ultrasonic wave system will act like grating. The compressions will act like opacities and rarefactions as transparencies.

If ultrasonics are applied to an isotropic crystal, we get interference fringes from which electric symmetry of the crystal can be studied.

4. Depth sounding. Ultrasonics are used for signalling to find the depth of sea, to detect the position of submerged rocks, submarines, icebergs etc.

The procedure for this is that a strong beam of ultrasonics is transmitted and then received after reflection from the submerged object. By noting the time of travel, the depth of the sea to that level can be found from the relation

$$t = \frac{2d}{v}$$

where v is the velocity of sound and d is the depth of the sea.

5. Chemical effects. These waves have been found to be useful for forming stable emulsion of immiscible liquids like water and oil, to liquify gells like Aluminium hydroxide in the same way as is done by shaking, till recently ultrasonics have been employed industrially for washing clothes, specially silken fabrics, cleaning utensils, for removing dust and soot from chimney gases, for dispersing fog at airports and for removing water from paper during manufacturing process. It can also be used for drilling holes, *e.g.*, a glass rod oscillating with ultrasonic frequency can make a hole in a piece of steel or glass plate.

6. Sound signalling. These have been used for directional signalling on account of their high frequency, these waves can be sent out in the form of a beam and as such they have great advantage over audible sound waves, which spread out in all directions. It is thus possible to radiate out a large amount of energy without the amplitude of the vibrating crystal being large while a much larger amplitude will be required for radiating out energy at the same rate of ordinary frequencies. Experimentally, it is seen that if a vibrating crystal in the form of a circular disc of radius r be used for transmission, the angle of the cone containing the waves is given by $\sin\theta = \frac{0.61\lambda}{r_1}$ which is quite small for short wavelengths.

7. Thermal effects. If a fluid is subjected to ultrasonic waves, its temperature will increase due to the absorption of waves.

8. Biological effects. Small animals like frogs, fish etc. are killed or maimed because of exposure to ultrasonics; yeast cells when exposed to them lose their power of reproduction.

These waves are put to useful purposes in medical sciences as well. The vibration of these waves has been found to have good massage action and can be used for treatment of muscular pains. These waves have been found to be quite useful for neuro patients where other conventional methods of electric shocks, psychotherapy or drugs fail. These waves have also been found useful to locate the position of eye tumour.

9. Other effects. These waves are useful to detect cracks or other defects in the homogeneity by reflection or absorption.

6.7. DETERMINATION OF VELOCITY OF SOUND IN A LIQUID

We can find the velocity of sound in a liquid (say kerosene oil) using ultrasonics.

A quartz crystal Q is placed between two metal plates. These two plates are connected to an audio frequency oscillator whose frequency is so adjusted that the crystal vibrates in resonance with the oscillator. Due to longitudinal compression and rarefactions of the waves, *ultrasonics are produced*. Those waves are reflected by the reflector R (Fig. 6.3). Due to superposition of the forward and reflected waves, longitudinal stationary waves are produced in the medium and consequently fixed values of *nodes* and *antinodes* can be assigned. At *nodes* density is maximum and at *antinodes* density is minimum. The arrangement is just similar to diffraction grating and so is called *acoustic grating*.

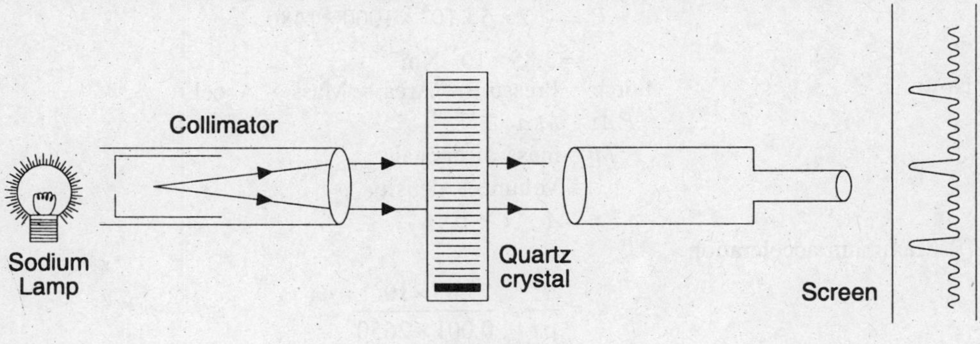

Fig. 6.3

The above arrangement is suitable for measurement of wavelength and velocity of ultrasonics in a medium.

As shown in Fig. 6.3, the acoustic grating is mounted on the prism table of a spectrometer. A parallel beam of sodium light from collimator is incident normally on the acoustic grating and the diffracted light is viewed through the telescope. As shown, the diffraction pattern consists of a *central maximum and principle maxima* on either side. The position of the principle maxima is given by the relation

$$d\sin\theta_n = n\lambda \qquad \text{...(1)}$$

where $\lambda \rightarrow$ wavelength of sodium light

$d \rightarrow$ distance between two nodal or antinodal waves

$\theta_n \rightarrow$ angle of diffraction for nth order

and $n \rightarrow 1, 2, 3$, etc. *i.e.*, order of spectrum.

If n, θ_n and λ are known, then d can be found from Equation (1).

If λ_m is the wavelength of ultrasonics through the medium, then

$$d = \lambda_m/2$$

or

$$\lambda_m = 2d \quad \text{...(2)}$$

If the resonant frequency of the piezoelectric oscillator is N, the velocity of ultrasonic waves is given by

$$v = N\lambda_m = 2Nd \quad \text{...(3)}$$

The above method is useful for determination of velocity of ultrasonic waves through liquids and gases at various temperatures.

Example 1. *A quartz crystal of thickness 0.001 m radiates ultrasonics into water of frequency 20×10^3 Hz and intensity 50 kW/m². Calculate (i) maximum acceleration (ii) maximum displacement.*

Velocity of sound in water $= 1480\ ms^{-1}$

Density of water $= 1000\ kg/m^3$

Density of crystal $= 2650\ kg/m^3$.

Solution: Sound intensity is given by

$$I = \frac{p^2}{2\rho v} \text{ watt m}^{-2}$$

where P = sound pressure

ρ = density

and v = velocity of sound

$\therefore$ Sound pressure $P = \sqrt{2I\rho v}$

In this case $I = 50$ kW/m² $= 5 \times 10^4$ W/m²

$\rho = 1000$ kg/m³

$v = 1480$ ms⁻¹

$$\therefore \quad P = \sqrt{2 \times 5 \times 10^4 \times 1000 \times 1480}$$

$$= 3.85 \times 10^5 \text{ Nm}^{-2}$$

Now Force = Pressure × Area = Mass × Accel.

$\therefore$ $P.A. = M.a$

M = mass of crystal

= Volume × density

$= A \times t \times \rho$

(*i*) Maximum acceleration

$$a = \frac{P}{\rho.t} = \frac{3.85 \times 10^5}{0.001 \times 2650}$$

$$= \mathbf{1.45 \times 10^5\ ms^{-2}}$$

(*ii*) Maximum displacement

$$x = \frac{a}{\omega^2} = \frac{1.45 \times 10^5}{(2\pi n)^2}$$

Now $n = 20 \times 10^3$ Hz

$$\therefore \quad x = \frac{1.45 \times 10^5}{(2\pi \times 20 \times 10^3)^2}$$

$$= \mathbf{9.4 \times 10^{-6}\ m.}$$

Example 2. *A quartz crystal of thickness 0.001 m is vibrating at resonance. Calculate the fundamental frequency.*

Given Y, Young's modulus for quartz $= 7.9 \times 10^{10}\ Nm^{-2}$

and ρ, *density of quartz* = *2.650* × *10*^{3} *kg/m*^{3}.

Solution: We have the relation for longitudinal vibrations

$$v = \sqrt{\frac{Y}{\rho}}$$

In this case, $Y = 7.9 \times 10^{10}\ \text{Nm}^{-3}$

$$\rho = 2.650 \times 10^3\ \text{kg/m}^3 = 2650\ \text{kg/m}^3$$

$$\therefore \quad v = \sqrt{\frac{7.9 \times 10^{10}}{26.50}} = 5460\ \text{ms}^{-1}.$$

For fundamental mode of vibration, thickness

$$= \lambda/2$$

$$\therefore \quad \lambda = 2 \times \text{thickness} = 2 \times 0.001 = 0.002\ \text{m}$$

$$\text{Frequency } n = \frac{5460}{0.002} = 2730 \times 10^3\ \text{Hz} \doteq \mathbf{2730\ kHz.}$$

6.8. ACOUSTICS OF BUILDINGS

Reverberation. The sound produced in a room or a hall suffers multiple reflection from the various walls, ceilings and floor of the hall so that in addition to the direct sound, a series of sound waves are heard by a listeners in the hall. The listener feels the persistence of sound for some time even when the original sound has ceased. This is known as reverberation.

The various causes for reverberation are:

(*i*) The fall in intensity of the sound produced in the room is exponential and so it will take a longer time to become zero.

(*ii*) Due to multiple reflections from walls, ceilings and floor, the sound reverberates and so persists inside the wall or room for longer time.

(*iii*) As the various sounds reaching a point in the room are not in phase, so these may produce interference, thereby causing reinforcement or destruction of sound wave accordingly as it may be a case of constructive or destructive interference.

Reverberation Time. The interval of time taken by a sustained or continuous sound to fall to an intensity equal to one millionth of its original value, *i.e.,* fall by 60 decibels in loudness, is called reverberation time.

Sabine's Formula. The relation between reverberation time, volume of the hall, area of the surface and its absorption coefficient is called Sabine's formula.

6.9. ABSORPTION

Coefficient for the material of the surface is defined as the ratio of the sound energy absorbed by the surface to the sound energy absorbed by a perfect absorber such as an open window.

The unit of absorption is called Sabine, and it is the sound energy absorbed by 1 sq. ft. of an open window.

So the effective absorbing area A for a surface having total area S and absorption coefficient α is given by

$$A = \alpha.S$$

If α_1, α_2, α_3...etc., are absorption coefficients for each reflecting surface and S_1, S_2, S_3.. etc. the various corresponding areas, then the average value of the absorption coefficient is

$$\alpha = \frac{\alpha_1 S_1 + \alpha_2 S_2 + \alpha_3 S_3 + ...}{S_1 + S_2 + S_3 + ...} = \frac{\Sigma \alpha_1 S_1}{S}$$

where S is the total area of all the surfaces.

If we consider I to be the average intensity, *i.e.,* Intensity/unit volume of the room, at any instant of time and δI the fall in intensity due to absorption in a small instant of time δt, then

$$\delta I = -\alpha . n . I . \delta t \quad ...(1)$$

where n corresponds to the number of reflections of sound waves/sec and α is the *absorption coefficient* of all absorbing surfaces.

By statistical method, it was shown by Jaeger that sound travels an average distance $\frac{4V}{S}$ between two successive reflections where V is the volume of the hall and S the total area of the reflecting surfaces.

Now if v is the velocity of sound, the time between two successive reflections $= \frac{4V}{vS}$.

$$n = \frac{Sv}{4V} \quad ...(2)$$

Putting the value of n in Equation (1),

$$\delta I = \alpha \cdot \frac{Sv}{4V} \cdot I \delta t$$

or

$$\frac{\delta I}{\delta t} = -\alpha \cdot \frac{Sv}{4V} \cdot I$$

which may be written in the limit as

$$\frac{dI}{dt} = -\alpha \cdot \frac{Sv}{4V} \cdot I$$

or

$$\frac{dI}{I} = -\alpha \cdot \frac{Sv}{4V} \cdot dt$$

If I_0 corresponds to the steady intensity at the instant when the source of sound is cut off and I_t its value after t seconds, then,

$$\int_{I_0}^{I_t} \frac{dI}{I} = -\int_0^t \frac{\alpha S v}{4V} \cdot dt$$

or

$$\log_e \frac{I_t}{I_0} = -\frac{\alpha S v}{4V} \cdot t$$

$\therefore$

$$\frac{I_t}{I_0} = e^{-\frac{\alpha S v}{4V} \cdot t}$$

Now according to the definition of reverberation time

$$\frac{I_t}{I_0} = 10^{-6}$$

$\therefore$

$$e^{-\frac{\alpha S v}{4V} \cdot t} = 10^{-6}$$

or

$$\frac{\alpha S v}{4V} \cdot t = 6 \log_e 10$$

$$= 6 \times 2.3026 \times 1$$

If we take the velocity of sound approximately 350 m/sec at room temperature, then reverberation time T is given by

$$T = \frac{6 \times 2.3026 \times 4V}{\alpha . S \times 350} = \frac{0.158 V}{\alpha . S} \quad ...(3)$$

In (3) we get the relation in M.K.S. system of units. If we take v approximately equation 1120 ft/sec, then the above relation in F.P.S. system will be

$$T = \frac{0.05 V}{\alpha \cdot S} \quad ...(4)$$

6.10. IMPORTANCE OF REVERBERATION TIME

For good acoustics of a hall, the reverberation time should have some optimum value, *i.e.,* it should not be too small or too large. It is because, if the reverberation time is too small, the sound vanishes almost instantaneously and gives the hall 'dead' effect. If it is too large, there may be multiple reflections and overlappings, thereby causing confusion to the listener of the sound and making sound inaudible.

So, the reverberation time is maintained at some optimum value τ which is approximately 1.03 sec, for speech for a hall of about 10,000 cu. ft. capacity. In case of music, this time is adjusted, slightly larger, so that the sound is distinctly heard.

Example 3. *A lecture hall with a volume of 45,000 cu. ft. is formed to have reverberation time of 1.5 sec. What is the total absorbing power of all the surfaces in the hall? If the area of the sound absorbing surface is 8000 sq. ft., calculate the average absorption coefficient.*

Solution: Volume of the hall $V = 45,000$ cu. ft

Reverberation time $T = 1.5$ sec

Now using the relation

$$T = \frac{0.005V}{\alpha \cdot S}$$

where $$\alpha S = \alpha_1 S_1 + \alpha_1 S_2 + ...$$

i.e., the total absorbing power of all the surface we have

Total absorbing power $$= \frac{0.05V}{T} = \frac{0.05 \times 45000}{1.5}$$

$$= 1500$$

Area of absorbing surface $S = 1,000$ sq. ft.

Average absorption coefficient

$$\alpha = \frac{1500}{8000} = \mathbf{0.1875.}$$

Example 4. *Entry of people in an auditorium of volume 1,60,000, cu. ft. and total absorption of 1000 sq. ft. of open window raises the absorption by 600 sq. ft. Find the change in reverberation time.*

Solution: Volume of Auditorium $V = 1,60,000$ cu. ft.

Total absorption $\alpha S = 1000$ sq. ft.

Now Absorption time $$T = \frac{0.05V}{\alpha \cdot S} = \frac{0.05 \times 160000}{1000} = 8 \text{ sec}$$

Increase in total area due to audience = 600 sq. ft.

$\therefore$ Total absorption with audience = 1000 + 600 = 1600 sq. ft.

Hence Reverberation time

$$T = \frac{0.05 \times 160000}{1600} = 5 \text{ sec}$$

Change in reverberation time

$$= 8 - 5 = \mathbf{3\ sec.}$$

Example 5. *Calculate the reverberation time in a hall measuring 40 × 100 × 20 ft. with the following parameters: (i) 7500 sq. ft. of plaster, $\alpha_1 = 0.03$. (ii) 6000 sq. ft. of wood and floor etc. $\alpha_2 = 0.06$, (iii) 400 sq. ft. of glass $\alpha = 0.025$, (iv) 600 seats, $\alpha_4 = 0.3$, (v) Audience of 500 persons $\alpha_5 = 4.0$ per person.*

Solution: Total absorption is calculated from Sabine's formula

Plaster $(\alpha_1 S_1) = 7500 \times 0.03 = 225$

Wood (doors etc.) and floor $(\alpha_2 S_2) = 6000 \times 0.06 = 360$

Glass $(\alpha_3 S_3) = 400 \times 0.025 = 10$

Seats $(\alpha_4 S_4) = 600 \times 0.3 = 180$

Audience $(\alpha_5 S_5) = 500 \times 4.0 = 2000$

(Subtracting the absorption by seats)

∴ Total absorption $\Sigma \alpha S = 2775$ Sabines

Now from Sabine's formula

$$T = \frac{0.05 \times V}{\alpha S} = \frac{0.05 \times 40 \times 100 \times 20}{2775} = \textbf{1.44 seconds}$$

which shows that the hall is quite satisfactory for performance with light music.

However, if we consider the hall to be without audience, then from above, the total absorption will be (2775 – 200) = 775 Sabines only. The corresponding reverberation time will be

$$T = \frac{0.05 \times 40 \times 1000 \times 20}{775}$$

$$= \textbf{5.16 sec} \text{ approximately}$$

So, we see that the presence or absence of audience affects the reverberation time considerably.

6.11. UNITS OF LOUDNESS

From Weber-Fechner Law in Psychology, the loudness of sound judged by the ear is proportional to logarithm of intensity. Experimentally also it is found to be correct. If I_1 and I_0 correspond to the intensities of two sounds of a particular frequency and L_1 and L_0 the corresponding measurements of loudness, then

$$L_1 \propto \log_{10} I_1$$

$$= K \log_{10} I_1$$

and

$$L_0 = K \log_{10} I_0$$

The difference in intensity, known as intensity level in technical terms, is given by

$$L = L_1 - L_0 = K \log_{10} I_1 - \log_{10} I_0$$

or

$$L = K \log_{10} \frac{I_1}{I_0} \qquad ...(1)$$

where K is a constant depending on the units taken, I_0 some standard reference intensity arbitrarily taken as 10^{-16} watt/cm^2, which corresponds to sound that can be just heard at a frequency of about 1000 cycles/sec. This is taken as the threshold of audibility of a normal ear.

6.12. DECIBEL

If K in Equation (1) is taken as 1, the difference in loudness is expressed in bels, a unit named after Graham Bell, the inventor of telephone. This unit is however quite large, and its 1/10 called decibel (dB) has become a standard. So in terms of decibels, Equation (1) may be written as

$$I = 10 \log_{10} \left(\frac{I_1}{I_0}\right) \text{dB} \qquad ...(2)$$

or

$$\log_{10} \frac{I_1}{I_0} = \frac{1}{10}$$

or

$$\frac{I_1}{I_0} = 1.26$$

i.e., about 26% change in intensity, changes the level by 1 dB. This is practically the smallest change in intensity that the ear can ordinarily detect. It will also be clear from equation (2).

When

$$I_1 = 100\, I_0$$

$$L = 10 \log_{10} 100 = \log_{10} 10^2 = 20 \text{ dB}$$

and when

$$I_1 = 1000\, I_0$$

$$L = 10 \log_{10} 1000 = 10 \log_{10} 10^3 = 30 \text{ dB}$$

In other words, it means that when the two sounds differ by 20 dB, the louder of them is 100 times more intense and when they differ by 30 dB, the louder one is 1000 times more intense.

To build up a scale of loudness, zero on the scale is taken as the threshold of hearing. This corresponds to the intensity of sound $I_0 = 10^{-16}$ watt/cm^2. The maximum intensity that ear can tolerate without sensation of pain is about 10^{-4} watt/cm² and corresponds to intensity level.

$$L = 10\log_{10}\frac{10^{-4}}{10^{-16}} = 10\log_{10}10^{12} = 120 \text{ dB}$$

The following Table gives approximate value of sounds measured in decibels. The table is quite useful for an engineer interested in noise reduction and design of various devices.

Source	*Intensity Level in dB*
1. Threshold of hearing	0
2. Rustle of leaves	10
3. Whisper	15 – 20
4. Ordinary conversation	60 – 65
5. Motors or Heavy traffic trucks	70 – 80
6. Roaring of lion at 20 ft	90
7. Thunder	100 – 110
8. Painful sounds	120 and above

6.13. PHON

In the above table we have expressed the loudness in dB on the assumption that the threshold of audibility is the same for all frequencies. However, the sensitivity of the ear and the limits of audibillity vary over a wide range of intensity and frequencies. So, the sounds of same intensity but of different frequencies will seem to differ in loudness. So, a different unit of measuring loudness is used. It is known as PHON. The measure of loudness in Phons of any sounds is equal to the loudness in decibels of an equally loud pure tone of frequency 1000 c.p.s. The two sounds are heard alternately with both ears, while the sound comes from a position directly in front of the listener. So the Phon scale and decibel scale agree at frequency of 1000 c.p.s. but differ at other frequencies.

Let us take an example. Suppose we want to find loudness of a tone of frequency 512/sec. The standard source of frequency 1000/sec is sounded and its loudness adjusted till it is equal to the loudness of the given tone. The intensity level of the standard source in 'dB' measures the loudness of given sound in 'Phons'.

A table below gives some sound contents of different frequencies which appear equally loud, *i.e.*, are around 70 phons.

Frequency	*Decibels*
100	80
1000	70
3000	67
10000	63

6.14. CONDITIONS FOR GOOD ACOUSTICS

Most of the big halls or rooms are acoustically poor, unless some precautions are taken for the design of these halls. The main causes of bad acoustics are:

(*i*) When sound is produced in a closed room, it reaches a point directly as well as after reflections from walls, roof etc. The intensity of the sound depends on the distance of the speaker and the distance of wall and the roof. These sounds are generally out of phase and due to interference the distribution of intensity in the room is not uniform.

(*ii*) It is also possible that the different frequency sounds may interfere differently at the some point and so a change in quality of the sound may occur. This produces a very unpleasant effect, especially in the case of music.

(*iii*) In a hall, the sound persists for some time due to multiple reflections, even when the original sound has ceased. This is called reverberation time. Ordinarily a syllable takes about 0.2 seconds to decay. If during this time another syllable is received after some reflections, the two superimpose and cause indistinctness in audibility.

6.15. METHODS OF DESIGNS FOR GOOD ACOUSTICS

For getting good acoustic properties, a hall should have the following features:

(*i*) The distribution of intensity throughout the hall should be uniform. A usual design of the hall should have a parabolic shape at the speaker's end as shown in Fig. 6.4.

As the speaker speaks, the sound is reflected from the back almost as a parallel beam. So there is uniform distribution of sound intensity at all the positions of the hall.

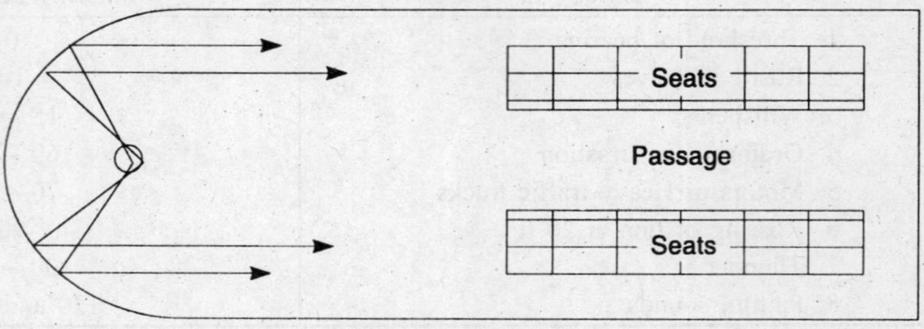

Fig. 6.4

(*ii*) Sound of each syllable should be distributed in such a way that each word is heard distinctly and there are no reinforcements causing the change in quality.

(*iii*) Reverberation time should be small but not zero, as then the hall walls given dead effect. The time of reverberation is give by

$$t = \frac{0.05V}{\alpha S}$$

where V is the volume of the hall in cu. ft, S, the area of reflecting surfaces and α the absorption coefficient of surfaces. We consider an open window as a perfect absorber and the number of windows are chosen to have the required reverberation time. To get an optimum value of reverberation time, we should have

(*i*) a few open windows;

(*ii*) some porous materials like cloth, asbestos, cushions etc., are good absorbers of sound and they are put at various places to have the required value of reverberation time;

(*iii*) a good audience also reduces the reverberation because a listener is equal to 5 sq. ft. of an open window;

(*iv*) curved walls or corners should be bare minimum, because then there may be unduly concentration of sound at some place and some other places may become zones of silence.

6.16. DETERMINATION OF ABSORPTION COEFFICIENT

A source of known frequency 512 Hz (say) is taken and the time of reverberation is found using a chronographic the following two steps:

(*i*) Without any absorbing material in the hall

(*ii*) With absorbing material in the hall.

Let the two times be t_1 and t_2 respectively.

From Sabine's formula $$t_1 = \frac{0.158V}{\Sigma\alpha_1 S_1}$$

and $$t_2 = \frac{0.158V}{\Sigma\alpha S_1 + \alpha_2 S_2}$$

where α_2 is the absorption coeff. of material of area A_2.

$$\therefore \quad \frac{1}{t_1} = \frac{\Sigma\alpha_1 S_1}{0.158V} \qquad ...(1)$$

and $$\frac{1}{t_2} = \frac{(\Sigma\alpha_1 S_1) + \alpha_2 S_2}{0.158V} \qquad ...(2)$$

Subtracting (1) from (2), we get

$$\therefore \qquad \frac{1}{t_2} - \frac{1}{t_1} = \frac{\alpha_2 S_2}{0.158\,V}$$

or

$$\alpha_2 = \frac{0.158\,V}{S_2}\left[\frac{t_1 - t_2}{t_1 t_2}\right] \qquad ...(3)$$

From Equation (3), if t_1, t_2, S_2 and V are known, then α_2 can be calculated.

It may be noted that the absorption coefficient of a given material is different at frequencies. It is *normally higher for higher frequencies.*

6.17. NOISE POLLUTION

Noise is undesirable sound and is recognized all over the world as a slow and insidious killer. Noise due to loudspeakers and public address systems is a social crime and many organizations are active in India and abroad to educate the public about long-term ill effects of noise.

EFFECTS OF NOISE POLLUTION:

Various noise pollution effects on human beings can be broadly classified as

(*i*) Auditory effects

(*ii*) Non-auditory effects

(*i*) Auditory effects

The most acute and immediate auditory effect of noise pollution in impairing of hearing which may further lead to total deafness.

(*ii*) Non-Auditory effects

It includes interference with speech communication, annoyance leading to ill temper, mental disorder and violent behaviour. This effect can reduce working efficiency and may cause neurosis, insomnia, hypertension, cardio vascular diseases, fatigue, change in skin temperature, nervous disorder and many other problems.

Auditory fatigue appears at 90 dB and leads to whistling and buzzing in ears. Temporary deafness occurs at 4000 to 6000 Hz and the effect is called *Temporary Threshold Shift* (TTS). Permanent loss of hearing occurs at 100 dB due to continuous exposure and the corresponding *auditory Threshold Shift* is called *Permanent Threshold Shift* (PTS).

In addition, chronic loss of hearing may occur due to sudden explosion or sudden excessive noise at 150 dB.

Reverberating noise experienced by boiler makers and structural rivets have resulted in *deafness* and in the former case it has been named as '*Boiler Makers Deafness Syndrome*'. Hearing loss has been observed in many cases where people are working on machines making huge noise and constant exposure to these noises have rendered many people partially or totally deaf.

Actually hearing loss is of two types : one is *acoustic trauma* which is due to severe damage to the internal structure of the ear caused by sudden explosion. It may be corrected by surgery depending upon the extent of damage.

Another type, which is a more serious one, is due to the *shift of the hearing threshold*. It can lead to many other faults in the body and the person may not even be able to hear loud sounds like sirens or alarms. The chances of recovery of the person to hear sounds are very rare.

Brain is also adversely affected by loud and sudden sounds like that from a jet or an aeroplane. These sounds may affect the pregnant women and the foetus.

Recently it has been observed that blood is thickened by excessive noise. Noise is also responsible for biological disorders. Even non-living things like buildings undergo cracks due to sudden sounds or shock waves which may also lead to breaking of windowpanes, doors etc.

Recent research has shown that more people die due to noise pollution than due to road accidents.

Prevention and control of noise pollution

It is almost impossible to curb completely all the annoying sounds. However, noise can be minimized using the following measures:

1. Reduction of noise at source and reduction noise expore time and also the exposure of noise to minimum people.

2. Using ear protective acids for workers who are constantly exposed to noise of machinery.

3. Proper design of doors and windows so that they absorb the impact of loudness of sound without impairing the quality of sound.

4. Improving working methods so that noise has minimum effect. For example welding makes less noise, so it should be preferred over riveting.

5. **Tree planting:** It is observed that thick trees like Neem etc., absorb sound to the extent of 6-8 dB. So they should be planted near schools, hospitals etc.

6. Good enclosures also reduce the impact of sounds. For example, thick lead enclosure reduces the noise level considerably.

7. Walls, floors etc., may be made of materials like felt, esbestos to reduce the noise levels.

8. In factories, silencers may be fitted to the machines to reduce noise levels.

9. Vibration damping material may be used to reduce the noise due to vibration of machines.

10. City zones may be divided in such a way that residential areas, hospitals, educational institutes etc., are away from the industrial or noise producing units.

11. Law enforcing agencies to be particular about the noise levels of blowing horns or marriage functions and strict action to be taken against defaulters.

Example 7. *A room has dimensions 6 × 4 × 5 m. Find (i) the mean free path of the sound wave in the room. (ii) No. of reflections made/sec. by sound waves with the walls of the room. Given velocity of sound in air = 350 ms^{-1}.*

Solution: (*i*) The mean free path of sound waves may be defined as the *average distance covered by a sound wave through air between two consecutive reflections with the walls of the room.*

$$\therefore \quad \text{Mean free path } \sigma = \frac{4 \times \text{Volume of the room}}{\text{Total surface area}}$$

In this case volume of the room $= 6 \times 4 \times 5 = 120 \text{ m}^3$

Total surface area $= 2[6 \times 4 + 4 \times 5 + 6 \times 5]$

$= 148 \text{ m}^2$

$$\therefore \quad \sigma = \frac{0.4 \times 120}{148} = 3.242 \text{ m.}$$

(*ii*) No. of reflection/sec.,

$$n = \frac{\text{Velocity of Sound}}{\text{Mean free path}}$$

$$= \frac{350}{3.243} \simeq \mathbf{108}$$

Example 8. *The volume of a room is 600 m^3 and wall area of the room is 220 m^2, the floor area is 120 m^2, the ceiling area is 120 m^2. Given the average absorption coefficient (i) For walls = 0.03, (ii) for ceiling = 0.80 and (iii) for floor = 0.06. Calculate the average absorption coeff. and reverberation time.*

Solution: We have the average absorption coeff.

$$\alpha = \frac{\Sigma \alpha S}{\Sigma S}$$

$$= \frac{\alpha_1 S_1 + \alpha_2 S_2 + \alpha_3 S_3}{S_1 + S_2 + S_3}$$

In this case, $S_1 = 220 \text{ m}^2, \alpha_1 = 0.03$

$S_2 = 120 \text{ m}^2, \alpha_2 = 0.80$

$$S_3 = 120 \text{ m}^2, \alpha_3 = 0.06$$

$$\therefore \quad \alpha = \frac{0.3 \times 220 + 0.8 \times 120 + 0.06 \times 120}{220 + 120 + 120}$$

$$= \frac{109.9}{460} \simeq 0.24$$

Total sound absorption of the room

$$= \alpha \Sigma S$$

$$= 0.24 \times 460$$

$$= 110.40 \text{ metric Sabine}$$

Reverberation time $t = \dfrac{0.158 V}{\alpha \Sigma S} = \dfrac{0.158 \times 600}{110.4}$

$$= \mathbf{0.86\ S}$$

Example 9. *A blast gives a sound of intensity 0.80 w/m² and the frequency is measured as 1000 Hz. If the density of air is 1.29 kg/m³, and the speed of sound is 330 ms⁻¹, find the amplitude of the sound wave.*

Solution: Intensity $I = 2\pi^2 n^2 \rho v a^2$

$$\therefore \quad a = \frac{1}{\pi n}\sqrt{\frac{1}{2\rho v}} = \frac{1}{\pi \times 1000}\sqrt{\frac{0.80}{2 \times 1.29 \times 330}}$$

$$= \mathbf{9.7 \times 10^{-5}\ m.}$$

Example 10. *A sound has intensity of 5 × 10⁻⁸ w/m². What is the sound level in decibels? (Given intensity $I_0 = 10^{-12}$ w/m²).*

Solution: Intensity in dB $= 10 \log \dfrac{I}{I_0}$

$$= 10 \log \left(\frac{5 \times 10^{-8}}{1 \times 10^{-12}}\right)$$

$$= 10 \log (5 \times 10^4)$$

$$= 10 \times 4.699$$

$$= \mathbf{46.99\ dB.}$$

QUESTIONS AND PROBLEMS

1. What are supersonic or ultrasonic waves? How are they produced and detected? Describe some of their applications.
2. What is piezoelectric effect? How is it used for the production of ultrasonics? Give some of the applications of ultrasonics.
3. Explain the causes of reverberation in a hall and how it can be minimised? Derive Sabine's expression for reverberation time.
4. What is absorption coefficient? How reverberation time is related to absorption coefficient? Give the importance of reverberation time for good acoustics.
5. Given an account of bad acoustic properties of a hall and discuss the methods to improve these defects.
6. Explain why the acoustic properties of a concert hall may vary with the size of audience.
7. A lecture hall with a volume of 45,000 cu. ft. is found to have a reverberation time of 1.5 sec. What is the total absorbing power of all the surfaces in the hall? If the area of the sound absorbing surface is 8000 sq. ft. calculate the average absorption coefficient. [**Ans.** Total absorbing power = 1500; Average absorption coefficient = 0.19]
8. Deduce Sabine's formula for reverberation of time. Discuss its importance. A hall has a volume of 80,000 cu. ft. Its total absorption coefficient is equivalent to 1000 sq. ft. of an open window. What will be the effect on reverberation time?

PART V
ELECTROSTATICS

Benjamin Franklin (1706-1790)

Franklin was born in Boston in 1706, the fifteenth of 17 children. When he was 17 years old, he went to Philadelphia to work as a printer. Franklin had only 2 years of formal schooling, but he loved books and studied foreign languages, philosophy, and science on his own. He became so involved in scientific study that in 1758 he turned over to his foreman, what was then his own printing business, intending to devote the best of his life to science.

During the years from 1747 to 1753 Franklin carried out his most important work on the nature of electricity, *the conservation of charge,* and the properties of *Leyden jars.*

One of Franklin's discoveries in electricity was the important role played by pointed metal objects. He found that a grounded metal rod with a pointed end could cause a charged conductor to lose its charge, if the point was brought within 6 inches of the charged object, but that a flat or rounded conductor had no such effect. Franklin used this discovery in the design of lightning rods to protect buildings. A lightning rod is simply a sharp-pointed rod which is connected by a very good conducting path from the roof of a building to the ground. A lightning rod allows charge to leak off thunder clouds before it builds up sufficiently to create destructive lightning bolts. Also, if lightning should strike, it would hit the lightning rod, not the house, and the charge would pass through the lightning rod to the ground without damaging the house.

Among Franklin's many inventions were the rocking chair, the Franklin stove, and bifocal glasses. Franklin had an attitude toward life which reflected the youthfulness of his spirit. In 1787, at the age of 81, after 50 years of public service, Franklin wrote to a Dutch friend asking him to come to America, where "in the little remainder of my life... we will make plenty of experiments together."

In going on with these experiments, how many pretty systems do we build, which we soon find ourselves obliged to destroy! If there is no other use of discovery of electricity; this, however, is something considerable, that it may help to make a ***vain man*** *humble.*

[Benjamin Franklin (1706-1790)]

If we accept the hypothesis that the elementary substances are composed of atoms, we can not avoid concluding that electricity also is divided into definite elementary portions, which behave like atoms of electricity. **[Hermann Von Helmboltz (1821-1894)]**

CHAPTER 1

ELECTRIC FIELDS

1.1. INTRODUCTION

We can show from the following experiment that there are *two kinds* of charges: if a glass rod is rubbed with silk and hanged from a long silk thread as shown in Fig. 1.1, and if a second glass rod is rubbed with silk and held near the rubbed end of the first rod, the rods will repel each other. On the other hand, a hard-rubber rod rubbed with fur will *attract* the glass rod. Two hard-rubber rods rubbed with fur will repel each other. We explain these facts by saying that rubbing a rod gives it an *electric charge* and that the charges on the two rods exert forces on each other. Clearly the charges on the glass and on the hard rubber must be *different in nature.*

Benjamin Franklin (1706–1790), the first American physicist, who named the kind of electricity that appears on the glass *positive* and the kind that appears on the hard rubber *negative;* these names have been retained to this day. We can sum up these experiments by saying that *like charges repel and unlike charges attract.*

Electric effects are not limited to glass rubbed with silk or to hard rubber rubbed with fur. Any substance rubbed with any other substance under suitable conditions will become *charged* to some extent; by comparing the known charge with a charged glass rod or a charged hard-rubber rod, it can be labelled as either *positive or negative.*

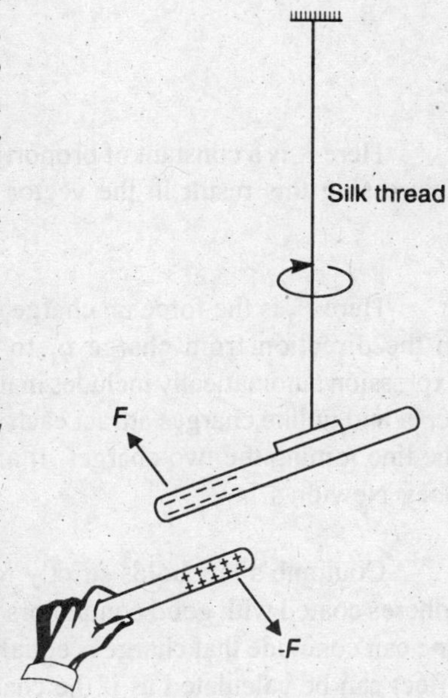

Fig. 1.1

The modern view of bulk matter is that, in its normal or neutral state, it contains equal amounts of positive and negative electricity. If two bodies, like glass and silk are rubbed together, a small amount of charge is transferred from one to the other, upsetting the electric neutrality of each. In this case the glass would become positive, and the silk negative.

1.2. CONDUCTORS AND INSULATORS

A metal rod held in the hand and rubbed with fur will not seem to develop a charge. It is possible to charge such a rod, however, if it is furnished with a glass or hard-rubber handle and if the metal is not touched with the hands while rubbing it. The explanation is that metals, the human body, and the earth are *conductors* of electricity and that glass, hard rubber, plastics, etc., are *insulators* (also called *dielectrics*).

In conductors electric charges are free to move through the material, whereas in insulators they are not. Although there are no perfect insulators, the insulating ability of fused quartz is about 10^{25} times as great as that of copper, so that for many practical purposes some materials behave as if they were perfect *insulators.*

A class of materials called *semiconductors* is intermediate between *conductors* and *insulators* in its ability to conduct electricity. Among the elements, *silicon and germanium* are well-known examples. In semiconductors the electrical conductivity can often be greatly increased by adding very small amounts of other elements; traces of arsenic or boron are often added to silicon for this purpose. Semiconductors have many practical applications, among which is their use in the construction of transistors.

1.3. COULOMB'S LAW

It can be easily demonstrated that like charges repel each other, and unlike charges attract each other. Quantitative investigations regarding the dependence of forces of attraction and repulsion on the magnitude of the charges and the distance between them were carried out by Coulomb. He found that for a given pair of charges, this force varied inversely as the square of the distance r_{12}, between them. He further found that the force between two charges q_1 and q_2, held at a given distance, varied directly as the product $q_1 q_2$. Thus mathematically, Coulomb's law may be described as

$$F \propto \frac{q_1 q_2}{r_{12}^2}$$

or

$$F = K\frac{q_1 q_2}{r_{12}^2} \quad \text{... (1)}$$

Here K is a constant of proportionality and its value depends upon the system of unit we select. Expressing this result in the vector form, we get,

$$F_2 = K\frac{q_1 q_2}{r_{21}^2}\hat{r}_{21} \quad \text{... (2)}$$

Here F_2 is the force on charge q_2, and r_{21} is the unit vector in the direction from charge q_1 to charge q_2 (Fig. 1.2). This expression automatically includes in itself the fact that like charges repel and unlike charges attract each other, and the force is along the line joining the two charges. It also reveals that these forces obey Newton's law

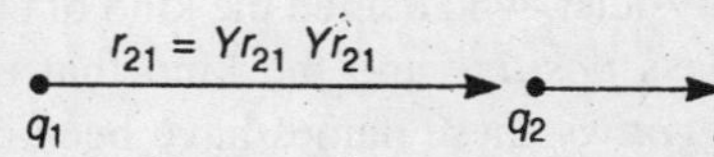

Fig. 1.2

$$F_2 = -F_1 \quad \text{... (3)}$$

Coulomb's law holds strictly for point charges. But, for experimental investigations, charged spheres coated with good conductors of electricity can be employed. From the argument of symmetry one can conclude that charge is equally distributed at the surface of such a sphere and thus its overall effect can be calculated as if the charge were located at its centre. But this assumption is true only if the distance r is much larger than the size of the spheres. If the spheres are quite close, then due to mutual forces, the charge distribution will not be uniform and, hence, we cannot take it to be located at the centre.

In S.I. units (rationalised M.K.S. units)

$$\frac{1}{4\pi\varepsilon_0}\frac{q_1 q_2}{r_{12}^2} \quad \text{... (4)}$$

The unit of charge is coulomb with symbols C. *A coulomb is defined as that amount of charge which flows through a given cross-section of a wire in one second when the wire is carrying a current of one ampere.* Thus in Equation. (4) everything has been defined except the constant to ε_0 known as the *permittivity* of free space, which is adjusted in such a way that Equation. 4 fits in the experiment. The experimental value of ε_0 is :

$$\left.\begin{aligned} \epsilon_0 &= 8.854 \times 10^{-12}\,\text{C}^2/\text{N m}^2 \\ &= 8.854 \times 10^{-12}\,\text{farad/m} \end{aligned}\right] \quad \text{... (5)}$$

The unit farad/m for ε_0 will be explained in the chapter of condensers. The form of the Coulomb's law equation, and the method of evaluation of the constant of proportionality, are similar to that being done in *gravitation.* But the *electric force* is much stronger than the *gravitational forces. A charged glass rod can lift a piece of paper against the gravitational pull of such a huge mass i.e., the earth.*

Example 1. *Calculate the force between two charges of one coulomb each, when they are at a distance of 1 m.*

Solution:
$$F = \frac{1}{4\pi\varepsilon_0} \times \frac{1 \times 1}{(1)^2} \text{N}$$
$$= \frac{1}{4\pi \times 8.854 \times 10^{-12}} \text{N} = 8.986 \times 10^9 \text{N}$$

This force is equal to the force with which $\mathbf{9.2 \times 10^5}$ (about one million) metric tons of mass are attracted by the earth.

Example 2. *Find the radius of the water drops which will remain just suspended with a charge of 1 electron in the Earth's field of 3 volt/cm. ($e = 4.8 \times 10^{-10}$ e.s.u)*

Solution: We have the relation
$$eE = mg$$
$$4.8 \times 10^{-10} \times \frac{3}{300} = \frac{4}{3}\pi r^3 \times 981$$
$$r = \mathbf{1.04 \times 10^{-5}\ cm.}$$

Example 3. *The distance between an electron and a proton in hydrogen atom is 5.3×10^{-10} metres, calculate the magnitude of (a) electrical force (b) gravitational force between them (elementary charge = 1.602×10^{-19} coulomb, mass of electron = 9.11×10^{-31} kgm, mass of Proton = 1.67×10^{-27} kgm, $G = 6.67 \times 10^{11}$ M.K.S. units.*

Solution: Electrical force between the two charges
$$= \frac{1}{4\pi\varepsilon_0} \times \frac{q_1 q_2}{r^2}$$
$$= 9 \times 10^9 \times \frac{1.602 \times 10^{-19} \times 1.602 \times 10^{-19}}{(5.3 \times 10^{-10})^2}$$
[$\because q_1 = q_2$ = charge on electron or proton *i.e.*, the elementary charge]
$$= 8.212 \times 10^{-10} \text{ Newton}$$

Gravitational Force
$$= G\frac{M.m}{d^2}$$
$$= 6.67 \times 10^{-11} \times \frac{1.67 \times 10^{-27} \times 9.11 \times 10^{-31}}{(5.3 \times 10^{-11})^2}$$
$$= \mathbf{3.6 \times 10^{-47}\ Newtons}$$

which shows that electrostatic force is 10^{37} times the gravitational force.

Example 4. *Two parallel plates are kept in air 1 cm apart and a p.d. of 100 volts is applied. If there be an electron resting on the plate at lower potential, find (i) the force on the electron in newtons (ii) K.E. and velocity of this electron when it reaches the other plate ($e = 1.6 \times 10^{-19}$ coulomb, $m = 9.1 \times 10^{-31}$ kgm)*

Solution: Electric intensity $\quad E = \frac{V}{d} = \frac{100}{0.01} = 10^4$ V/m

Force $\quad F = \frac{1}{4\pi\varepsilon_0} \times eE = 9 \times 10^9 \times 1.6 \times 10^{-19} \times 10^4$
$$= 1.44 \times 10^{-5} \text{ N}$$

Acceleration $\quad a = \frac{F}{m} = \frac{1.44 \times 10^{-5}}{9.1 \times 10^{-31}} = 1.582 \times 10^{25}$ m/sec^2

Velocity $\quad v = \sqrt{2ad} = \sqrt{2 \times 1.582 \times 10^{25} \times 0.01}$

$$= 5.536 \times 10^{11} \text{ m/sec}$$

$$\text{K.E.} = \frac{1}{2}mv^2 = \frac{1}{2} \times 9.1 \times 10^{-31} \times (5.536 \times 10^{11})^2$$

$$= \mathbf{1.314 \times 10^{-7} \text{ joules}}$$

Example 5. *If the total positive and the total negative charges in a copper penny each of value 1.3×10^5 coulomb are separated to a distance such that their force of attraction is 4.5 N. How far apart must they be?*

Solution: We have

$$F = \frac{1}{4\pi\varepsilon_0} \frac{q_1 q_2}{r^2}$$

Putting $q_1 q_2 = q$ and solving for r we get

$$r = q\sqrt{\frac{1/4\pi\varepsilon_0}{F}}$$

$$= 1.3 \times 10^5 \text{ coul} \sqrt{\frac{9.0 \times 10^9 \text{N-m}^2/\text{coul}^2}{4.5 \text{ nt}}}$$

$$= \mathbf{1.8 \times 10^8 \text{ m}}$$

If more than two charges are present, Equation (4) holds for every pair of charges. Let the charges be q_1, q_2 and q_3, etc., we calculate the force exerted on any one (say q_1) by all the others from the vector equation

$$\vec{F_1} = \vec{F_{12}} + \vec{F_{13}} + \vec{F_{14}} + \ldots\ldots\ldots$$

where F_{12} , for example, is the force exerted on q_1 by q_2

1.4. CHARGE IS QUANTIZED

Earlier electric charge was thought of as a *continuous fluid,* an idea that was useful for many purposes. The atomic theory of matter, however, has shown that fluids themselves, such as water and air, are not continuous but are made up of atoms. Experiment shows that the 'electric fluid" is not continuous either, but that it is made up of integral multiples of a certain minimum electric charge. This *fundamental charge,* to which we give the symbol e, has the magnitude 1.60210×10^{-19} coul. Any physically existing charge q, no matter what its origin, can be written as ne where n is a positive or a negative integer.

When a physical property such as charge exists in discrete "packets" rather than in continuous amounts, the property is said to be *quantized.* Quantization is basic to modern physics. The existence of atoms and of particles like the electron and the proton indicates that *mass* is quantized also.

The quantum of charge e is so small that the "graininess" of electricity does not show up in large-scale experiments, just as we do not realize that the air we breathe is made up of atoms. In an ordinary 220 volt, 100 watt light bulb, for example, 3×10^{18} *elementary charge* enter and leave the filament every second.

1.5. MATTER AND CHARGE

Matter as we ordinarily experience it can be regarded as composed of three ends of elementary particles, *proton, neutorn,* and *electron,* then masses and charges are given below. Note that the masses of the *neutron* and the *proton* are approximately equal but that the electron is $\frac{1}{1840}$ times smaller.

Particle	*Symbol*	*Charge*	*Mass*
Proton	p	$+e$	1.67252×10^{-27} kg
Neutron	n	0	1.67482×10^{-27} kg
Electron	e	$-e$	9.1091×10^{-31} kg

Atoms are made up of a dense, positively charged *nucleus,* surrounded by a cloud of electrons; The nucleus varies in radius from about 1×10^{-15} meter for hydrogen to about 7×10^{-15} meter for the heaviest atoms. The outer diameter of the electron cloud, that is, the diameter of the atom, lies in the range $1 \text{ to } 3 \times 10^{-10}$ meter, about 10^5 times larger than the nuclear diameter.

Example 6. *A point charge of* $+3.0 \times 10^{-6}$ *coul is 12 cm distant from a second charge of* -1.5×10^{-6} *coul. Calculate the magnitude and direction of the force of each charge.*

Solution: Magnitude of the force on each charge is

$$F = \frac{q_1 q_2}{4\pi\varepsilon_0 r^2}$$

$$= \frac{(3 \times 10^{-6}\text{C})\,(1.5 \times 10^{-6}\text{C})}{(0.12\text{m})^2} \times 9 \times 10^9 \text{ N-m}^2/\text{C}^2$$

$$= \mathbf{2.8\ N}$$

The force will be attractive.

Example 7. *How far apart must two protons be if the electrical repulsive force on either is equal to its weight?*

Solution: Let the charge on each proton be q. Coul and let d be the distance between them. Then electrical repulsive force on either is given by

$$F = \frac{q^2}{4\pi\varepsilon_0 d^2} = mg$$

or

$$d = \frac{q}{\sqrt{4\pi\varepsilon_0\ mg}}$$

$$= \frac{1.6 \times 10^{-19}}{\sqrt{\dfrac{1.67 \times 10^{-27} \times 9.8}{9 \times 10^9}}}$$

$$= \mathbf{0.119\ m = 11.9\ cm}$$

Example 8. *In the radioactive decay of* U^{238}, *the centre of emerging* α*-particles is at a certain instant* 9×10^{-15} *m from the centre of the residual nucleus* Th^{234}. *At this instant*

(*a*) *what is the force on the* α*-particle*

(*b*) *what is the acceleration?*

Solution: (*a*) We have the coulomb force given by the relation

$$F = \frac{Z_1 Z_2 e^2}{4\pi\varepsilon_0 r^2}$$

where $Z_1 \rightarrow$ At. No. of α-particle

and $Z_2 \rightarrow$ At. No. of Thorium nucleus

$r \rightarrow$ distance of separation

$$\therefore \quad F = \frac{2 \times 90 \times (1.6 \times 10^{-19}\text{C})^2}{(9 \times 10^{-15}\text{m})^2} \times 9 \times 10^9 \text{ N-m}^2/\text{C}^2$$

$$= \mathbf{512\ N.}$$

(*b*) Mass of α-particle $= 4 \times 1.67 \times 10^{-27}$ kg

$$M = 6.69 \times\times 10^{-27} \text{ kg}$$

$$\therefore \quad \text{Acceleration } a = \frac{F}{M} = \frac{512\text{N}}{6.69 \times 10^{-27}\text{kg}} = \mathbf{7.7 \times 10^{-28}\ ms^{-2}}$$

Example 9. *Find the ratio between the electrical force* F_e *and gravitational force* F_g *between two electrons.*

Solution:

$$\frac{F_e}{F_g} = \frac{Q^2 / 4\pi\varepsilon_0 r^2}{G\dfrac{m^2}{r^2}}$$

$$= \frac{Q_2}{4\pi\varepsilon_0\ Gm^2}$$

$$= \frac{(9\times10^9\ \text{Nm}^2\text{C}^{-2})\times(1.60\times10^{-19}\text{C}^2)}{(6.67\times10^{11}\ \text{Nm}^2\text{kg}^{-2})\times(9.1\times10^{-31}\text{kg})^2}$$

$$= 4.16\times10^{42}$$

$$\therefore \qquad F_e = \mathbf{4.16\times10^{42}}\,F_g$$

Which shows that electrical force is about 10^{42} times stronger than the gravitational force for two electrons.

The significance of Coulomb's law goes far beyond the description of the forces acting between charged balls or rods. This law, when incorporated into the structure of quantum physics, correctly describes

(*i*) the electric forces that bind the electrons of an atom to its nucleus,

(*ii*) the forces that bind atoms together to form molecules, and

(*iii*) the forces that bind atoms or molecules together to form nuclids or liquids. Thus most of the forces of our daily experience that are not gravitational in nature are electrical.

In the atomic *nucleus* we encounter a new force which is neither gravitational nor electrical in nature. This strong attractive force, which binds together the *protons* and *neutrons* that make up the nucleus is called simply *the nuclear force.* If this force were not present, the nucleus would fly apart at once because of the strong Coulomb repulsion force that acts between its *protons.*

Example 10. *What repulsive Coulomb force exists between two protons in a nucleus of copper? Assume a separation of* 4.0×10^{-15} *meter.*

Solution: From Coulomb's law,

$$F = \frac{1}{4\pi\varepsilon_0}\frac{q_1 q_2}{r^2}$$

$$= \frac{(9.0\times10^9\,\text{N-m}^2/\text{coul}^2)\ (1.6\times10^{-19}\,\text{coul})^2}{(4.0\times10^{-15})^2}$$

$$= \mathbf{14\ N}$$

This enormous repulsive force must be more than compensated for by the strong attractive nuclear forces. This example, combined with Example 4, shows that *nuclear binding, forces* are much stronger than *atomic binding forces. Atomic binding forces* are, in turn, much stronger than *gravitational forces* for the same particles separated by the same distance.

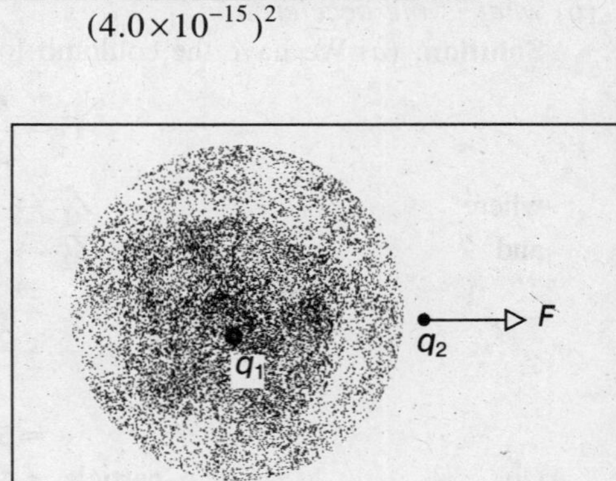

Fig. 1.3

1.6. CONSERVATION OF CHARGES

When a glass rod is rubbed with silk, a positive charge appears on the rod. Measurement shows that a negative charge of equal magnitude appears on the silk. This suggests that rubbing does not create charge but merely transfers it from one object to another, disturbing slightly the electrical neutrality of each. This hypothesis of the *conservation of charge* has been verified experimentally both for large scale events and at the atomic and nuclear level; no exceptions have ever been found.

An example of charge conservation is found in radioactive decay, of which the following process is typical:

$$U^{238} = Th^{234} + He^{4} \qquad (26-5)$$

The radioactive "parent" nucleus, U^{238}, contains 92 protons (that is, its atomic number Z = 92), It disintegrates spontaneously by emitting an α-particle (He^4; $Z = 2$) transmitting itself into the nucleus Th^{234} with $Z = 90$. Thus the amount of charge present before disintegration (+ 92e) is the same as that present after the disintegration.

Another example of charge conservation is found in nuclear reactions, of which the bombardment of Ca^{44} with cyclotron accelerated protons is typical. In a particular collision a neutron may emerge from the nucleus, leaving Sc^{44} as a "residual" nucleus:

$${}_{20}Ca^{44} + p = {}_{21}Sc^{44} + n$$

The sum of the atomic numbers before the reaction (20 + 1) is exactly equal to the sum of the atomic numbers after the reaction (21+0). So in this case also charge is conserved.

1.7. THE ELECTRIC FIELDS

With every point in space near the earth is associated with a *gravitational field strength* vector g. This is the gravitational acceleration that a test body, placed at that point and released, would experience. If m is the mass of the body and F the gravitational force acting on it, g is given by

$$g = F / m \qquad \text{... (1)}$$

This is an example of a *vector field.* For points near the surface of the earth the field is often taken as *uniform;* that is, g is constant at all points.

The flow of water in a river provides another example of a vector field called a *flow field.* Every point in the water has associated with it a vector quantity, the velocity v with which the water flows past the point. If g and v do not change with time, the corresponding field is described as *stationary.* In the case of the river, note that even though the water is moving the vector v at any point does not change with time for steady flow conditions.

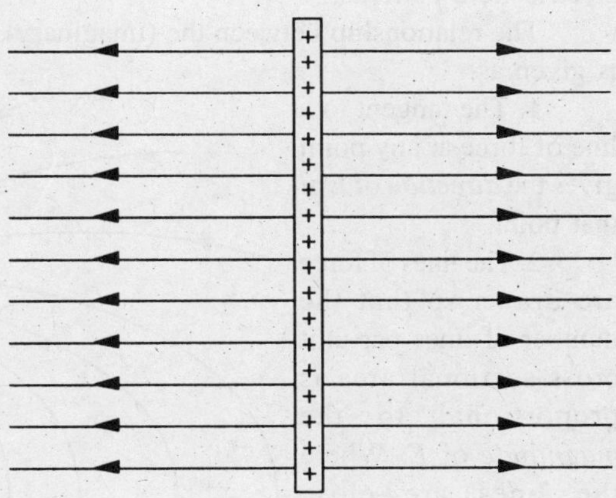

Fig. 1.4

The space surrounding a charged rod is affected by the presence of the rod, and we speak of an *electric field* in this space. In the same way we speak of a *magnetic field* in the space around a *bar magnet.*

We can explain the electric fields as under :

1. Charge q_1 in Fig. 1.3 sets up an electric field in the space around itself. This field is represented by the shading in the figure; later we shall show how to represent electric fields more effectively.
2. The field acts on charge q_2; this shows up in the force F that q_2 experiences.

1.8. THE ELECTRIC FIELD STRENGTH *E*

To define the electric field we place a small test body carrying a test charge q_0 (assumed positive for convenience) at the point in space that is to be examined, and we measure the electrical force F (if any) that acts on this body. The *electric field strength* E at the point is defined as

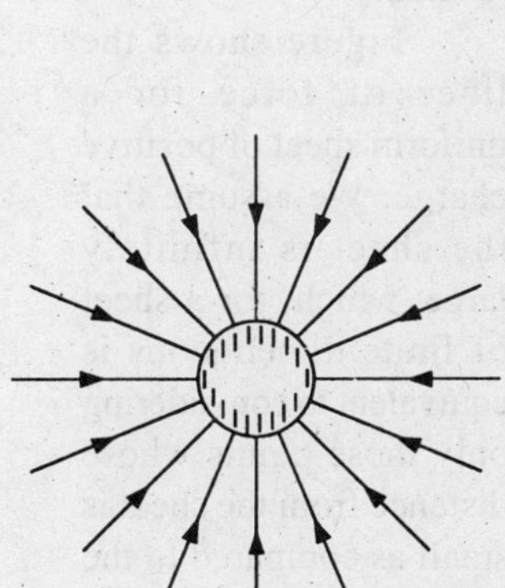

Fig. 1.5 (*a*)

$$E = F / q_0 \qquad \text{... (2)}$$

Here E is a vector, F is also vector, q_0 being a scalar. The direction of E is the direction of F, that is, it is *the direction in which a resting positive charge placed at the point would tend to move.*

The definition of gravitational field strength g is much like that of electric field strength except that the mass of the test body rather than its charge is the property of interest. Although the units of g are usually written as meters/sec^2, they could also be written as N/kg (Equation 1) ; those for E are N/coul. Thus both g and E are expressed as a force divided by a property (mass or charge) of the test body.

In applying Equation (2), we must use a test charge as small as possible. A large test charge might disturb the primary charges that are responsible for the field, thus changing the very quantity that we are trying to measure. Equation (2) should, strictly be replaced by

$$E = \lim_{q_0 \to 0} \frac{F}{q_0} \qquad \text{... (3)}$$

This equation tells us to use a smaller and smaller test charge calculating the ratio F/q_0 at every step. The electric field E is then the limit of this ratio as the size of the test charge tends to zero.

1.9. LINE OF FORCE

The concept of lines of force given by Michael Faraday, form a convenient way of visualizing electric field patterns.

The relationship between the (imaginary) lines of force and the electric field strength vector is given as:

1. The tangent to a line of force at any point gives the *direction* of E at that point.

2. The lines of force are drawn so that the number of lines per unit cross-sectional area is proportional to the *magnitude* of E. Where the lines are close together E is large and where they are far apart E is small.

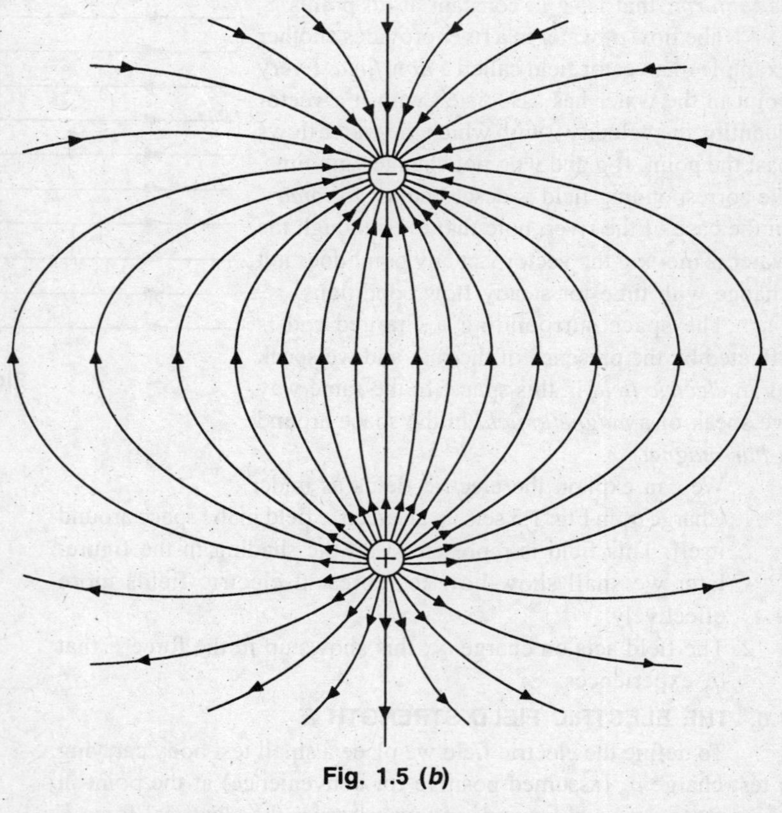

Fig. 1.5 (*b*)

Figure shows the lines of force for a uniform sheet of positive charge. We assume that the sheet is infinitely large, which, for a sheet of finite dimen-sions is equivalent to con-sidering only those points whose distance from the sheet is small as compared to the distance to the nearest edge of the sheet. A positive test charge, released in front of such a sheet, would move away from the sheet along a perpendicular line. So the electric field strength vector at any point near the sheet must be at right angles to the sheet. The lines of force are uniformly spaced, which means that E has the same magnitude for all points near the sheet.

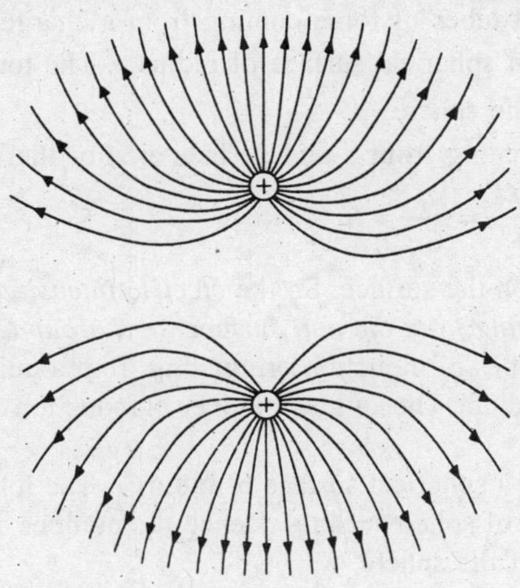

Fig. 1.6

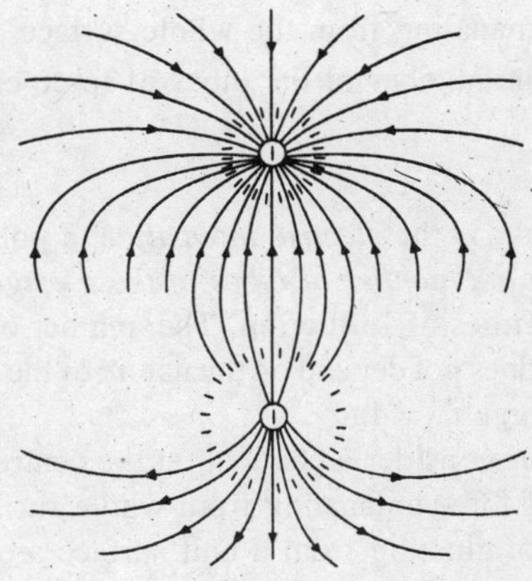

Fig. 1.7

Figure 1.5 shows the lines of force for a negatively charged sphere. From symmetry the lines must be along radii. They point inward because a free positive charge would be accelerated in this direction. The electric field E is not constant but decreases with increasing distance from the charge. This is evident in the lines of force, which are farther apart at greater distances. From symmetry, E is the same for all points that lie at the same distance from the centre of the charge.

Figures 1.6 and 1.7 show the lines of force for two equal like charges and for two equal unlike charges, respectively.

Example 10. *What is the magnitude of the electric field strength E such that an electron, placed in the field, would experience an electrical force equal to its weight?*

Solution: From Equation (2), replacing q_0 by e and F by mg, where m is the electron mass, we have

$$E = \frac{F}{q_0} = \frac{mg}{e}$$

$$= \frac{(9.1 \times 10^{-31} \text{ kg})\,(9.8 \text{ meters/sec}^2)}{1.6 \times 10^{-19} \text{ coul.}}$$

$$= 5.6 \times 10^{-11} \text{ N/coul.}$$

This is a very weak electric field.

Example 11. *Calculate the intensity at a point due to a charge of 100 micro coulomb at a distance of 9 metres.*

Solution. $q = 100\ \mu \text{ coulomb} = 100 \times 10^{-6} \text{ coulomb}$
$= 10^{-4} \text{ coulomb}$

and $$\frac{1}{4\pi\varepsilon_0} = 9 \times 10^9 \text{ N-m}^2 \text{ C}^{-2}$$

Using the relation $$E = \frac{q}{4\pi\varepsilon_0} \cdot \frac{1}{r^2}$$

We get $$= 10^{-4} \times 9 \times 10^9 \frac{1}{9 \times 9}$$

$$= \frac{9 \times 10^5}{81} = \mathbf{1.1 \times 10^4} \text{ N/coulomb}$$

1.10. TUBES OF FORCE AND INDUCTION

(*a*) **Maxwell Tubes of Force.** According to Maxwell the number of tubes of force originating from a unit charge $= \dfrac{4\pi}{k}$ and so the number of tubes of force coming from a charge $q = \dfrac{4\pi q}{k}$

Let us consider a charge q at the centre of spherical surface of radius r. The total number of force emanating from the whole surface of the sphere $= \dfrac{4\pi q}{k}$.

So the number of the tubes of force emanating from a unit surface area of the sphere

$$= \frac{4\pi q / k}{4\pi r^2} = \frac{q}{kr^2} = E$$

Here E is the *electric intensity* at a point on the surface. So the *electric intensity at a point is equal to the number of tubes of these originating from the unit surface area around that point.*

(*b*) **Tubes of Induction.** The number of tubes of induction originating from a unit charge = 4π and it does not depend on the nature of the medium. The number of tubes of induction originating from a charge $q = 4\pi q$.

Let us consider a charge q at the centre of a spherical surface of radius r. The total number of tubes of force emanating from whole surface of sphere $= 4\pi q$. Hence the number of tubes of induction originating from a unit surface area of the sphere

$$= \frac{4\pi q}{4\pi r^2} = \frac{q}{r^2} = kE$$

i.e. Induction $= k \times$ (no. of maxwell tubes/unit area)

$= k \times$ electric intensity.

(*c*) **Faraday Tubes.** Unlike Maxwell tubes of force, Faraday tubes are grouped considering the reference to charge and not the medium. A unit positive charge is supposed to give one tube of force and q charge will give q tubes of force. These are called *Faraday unit tubes.*

If a charge q is kept at the centre of an imaginary sphere of radius r the number of Faraday tubes that originate from the surface of the sphere $= q$. Hence, the number of Faraday tubes that originate from unit area of the spherical surface $= \dfrac{q}{4\pi r^2} = \dfrac{q}{kr^2} \cdot \dfrac{k}{4\pi} = E \cdot \dfrac{k}{4\pi}$

[Since $E = \dfrac{q}{k\,r^2}$, electric intensity of a point on the surface of sphere]

So the Electric Intensity at a point is equal to $4\pi / k$ times the number of Faraday's tubes surrounding that point.

No. of Maxwell tubes of force crossing unit area around a point

$$= \frac{q}{k\,r^2}$$

And no. of Faraday tubes crossing unit area around a point

$$= \frac{q}{4\pi r^2}$$

1 Faraday tube $= \dfrac{q}{k\,r^2} \cdot \dfrac{4\pi r^2}{q}$

$= \dfrac{4\pi}{k}$. Maxwell tubes of force

Further the no. of Faraday tubes/unit area

$$= \frac{q}{4\pi r^2} = \frac{1}{4\pi} \cdot \frac{q}{r^2} = \frac{1}{4\pi}\left[k.\frac{q}{k\,r^2}\right]$$

$= \dfrac{1}{4\pi} \times$ no. of tubes of induction/unit area.

$\therefore$ No. of tubes of induction per unit area $= 4\pi$ times the Faradays tubes per unit area.

1.11. DETERMINATION OF *E*

Let a test charge q_0 be placed at a distance r from a point charge q. The magnitude of the force acting on q_0 is given by Coulomb's law

$$= \frac{1}{4\pi\varepsilon_0} \cdot \frac{qq_0}{r^2} \quad \text{... (1)}$$

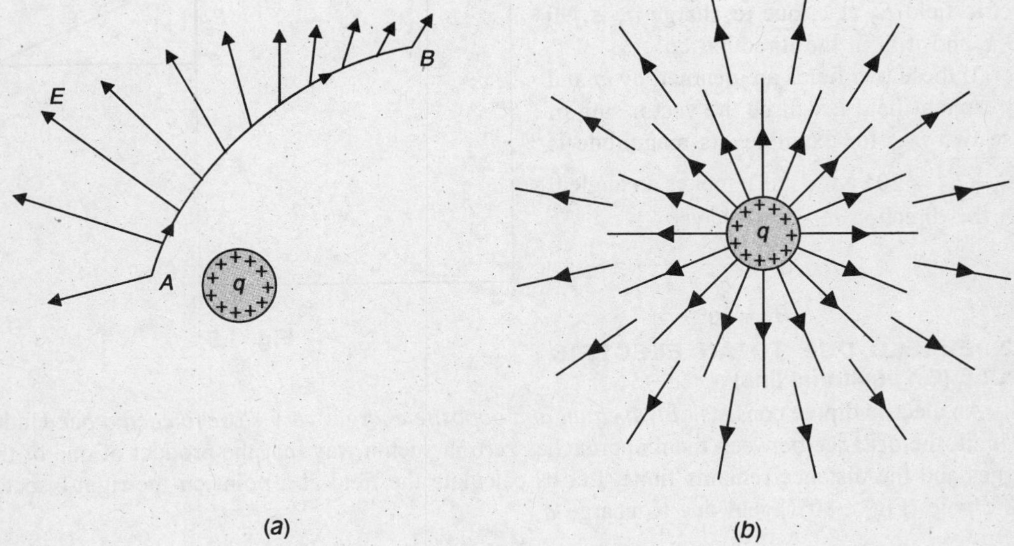

Fig. 1.8

The electric field strength at the site of the test charge is given by

$$E = \frac{F}{q_0} = \frac{1}{4\pi\varepsilon_0} \frac{q_0}{r^2} \quad \text{... (2)}$$

The direction of E is on a radial line from q, pointing outward, if q is positive, and inward if q is negative (Fig. 1.8).

To find E for a group of point charges : (*a*) Calculate E_n due to each charge at the given point *as if it were the only charge present.* (*b*) Add these separately calculated fields vectorially to find the resultant field E at the point. In equation form

$$E = E_1 + E_2 + E_3 + ... = E_n \qquad n = 1, 2, 3, 4 \quad \text{... (3)}$$

The sum of a vector sum, taken over all the charges.

If the charge distribution is a continuous one, the field it sets up at any point P can be computed by dividing the charge into infinitesimal element dq. The field dE due to each element at the point is then calculated, treating the elements as point charges. The magnitude of dE is given by

$$dE = \frac{1}{4\pi\varepsilon_0} \frac{dq}{r^2} \quad \text{... (4)}$$

where r is the distance from the charge element da to the point P. The resultant field at P is then found by adding the field contributions due to all the charge element and is given by

$$E = \int dE \quad \text{... (5)}$$

The integration, like the sum in Equation (3) is a vector operation.

Example 12. *Two point charges q_1 and q_2 of $+5 \times 10^{-9}$ C and 2.5×10^{-9} C respectively and placed at two opposite corners of a square whose side is 0.50 m. What are the magnitude and the direction of the electric field at point C?*

Solution: Electric field at c due to q_1 is given by.

$$E_1 = \frac{5\times10^{-9}}{4\pi\times8.85\times10^{-12}\times(0.5)^2}\ \text{N C}^{-1}$$
$$= \mathbf{180\ N\,C^{-1}}$$

It is in the direction *cf*. Similary the electric field E_2 at c, due to charge q_2 is, 90 NC^{-1}, and it is in the direction *cb*.

If these two fields are denoted by *cf* and *ce*, resultant field E will be the vector sum of these two *i.e.*, for example, Its magnitude is $\sqrt{E_1^2+E_2^2} = 201\ \text{NC}^{-1}$, if it makes an angle θ with the direction *d.c.f.*, we have

$$\tan\theta = \frac{fg}{cf} = \frac{90}{180} = \frac{1}{2}$$
$$\therefore \quad \theta = \mathbf{26°}$$

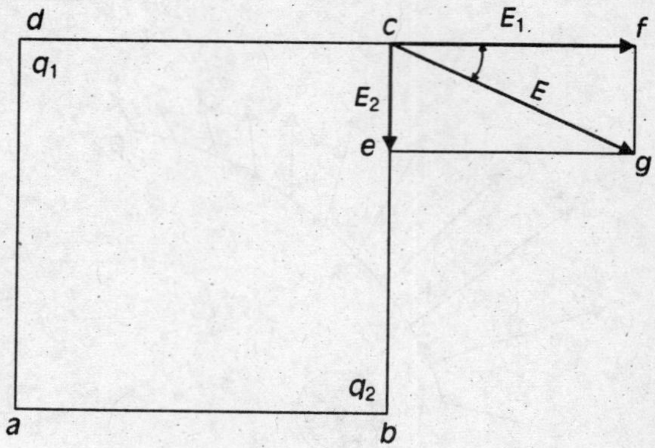

Fig. 1.9

1.12. *E*-FIELD DUE TO AN ELECTRIC DIPOLE (On equitorial line)

An electric dipole consists of *two equal and opposite charges very close to each other.* Under the limit, the distance between them approaches zero in such a way that the product of one of the charges and the distance remains finite. Let us calculate the field at a point on the right-bisector of a dipole (Fig. 1.10). Field due to charge q,

$$E_1 = \frac{q}{4\pi\varepsilon_0 AP^2}$$
$$= \frac{q}{4\pi\varepsilon_0(r^2+a^2)},\ \text{in the direction } AP$$

Field due to charge $-q$,

$$E_2 = \frac{1}{4\pi\varepsilon_0}\frac{-q}{(r^2+a^2)}$$
$$= \frac{1}{4\pi\varepsilon_0}\frac{q}{(r^2+a^2)},\ \text{in the direction } PB$$

Resolving both E_1 and E_2 along OP and perpendicular to it, the components along OP will be cancelled and the resultant field will be in the direction PQ, parallel to AB given by

$$\mathbf{E} = 2\frac{1}{4\pi\varepsilon_0}\frac{q}{(r^2+a^2)}\cos\theta$$
$$= 2\frac{1}{4\pi\varepsilon_0}\frac{q}{(a^2+r^2)}\frac{a}{(a^2+r^2)^{1/2}},\ \text{or}$$
$$\mathbf{E} \cong \frac{1}{4\pi\varepsilon_0}\frac{\mathbf{P}}{r^3}$$

In vector notation we may write

$$\mathbf{E} = \frac{1}{4\pi\varepsilon_0}\cdot\frac{\mathbf{P}}{r^3} \qquad ...(1)$$

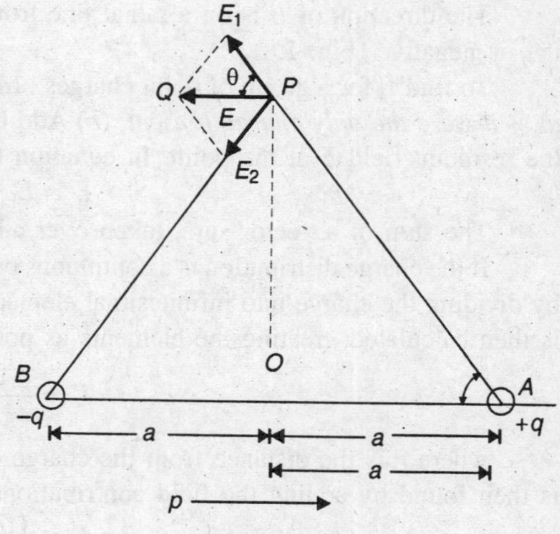

Fig. 1.10

Here we have neglected a because $a << r$ in general. The term 2 aq is called the *electric dipole moment* and denoted by a symbol **p**. It is vector quantity with its direction from the negative charge. Similarly, the electric field at a point P' at a distance r from the centre of the dipole and on its axis is

$$\mathbf{E} = \frac{1}{2\pi\varepsilon_0}\,\frac{\mathbf{P}}{r^3} \qquad \text{... (2)}$$

Example 13. *Two charges q_1 and q_2 of 0.1 μμC (micro-micro coulomb) and −0.1 μμC respectively are 10 Å apart. What is the electric field at a point on the line joining them at a distance 10 cm from their mid point?*

Solution: In this case 10 cm is a much larger distance as compared to 5 Å. Therefore, using Equation. (2) of 1.12, we have

$$E = \frac{1}{2\pi\varepsilon_0}\cdot\frac{\mathbf{P}}{r^3}$$

$$= \frac{0.1\times10^{-12}\times10\times10^{-10}}{2\pi\times8.85\times10^{-12}\times\left(\frac{10}{100}\right)^3}$$

$$= \mathbf{1.80\times10^{-9}\ N\,C^{-1}}$$

Example 14. *Figure shows a ring of charge q and of radius. Calculate E for a point on the axis of the ring at a distance x from centre.*

Solution: Consider a differential element of the ring of length *ds*, located at the top of the ring in Fig. 1.11. It contains an element of charge given by

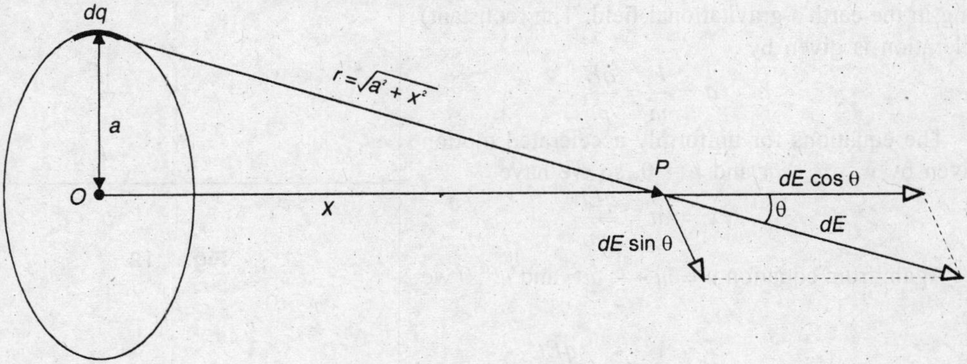

Fig. 1.11

$$dq = q\frac{ds}{2\pi a}$$

where $2\pi a$ is the circumference of the ring. This element sets up differential electric field dE at a point P.

The resultant field E at P can be found by integrating the effects of all the elements that make up the rings. From symmetry this resultant field must be along the ring axis. So only the component of dE parallel to this axis contributes to the final result. The component perpendicular to the axis is cancelled out by an equal but opposite component established by the charge element on the opposite side of the ring.

Thus the general vector integral

$$E = \int dE$$

becomes a scalar integral $\quad E = \int dE\cos\theta$

Now $\quad dE = \dfrac{1}{4\pi\varepsilon_0}\cdot\dfrac{dq}{r^2}$

$$= \frac{1}{4\pi\varepsilon_0} \cdot \frac{dq}{(a^2 + x^2)}$$

$$\therefore \quad E = \int dE \cos\theta$$

$$= \int \frac{1}{4\pi\varepsilon_0} \cdot \frac{dq}{(a^2 + x^2)} \cdot \frac{x}{(a^2 + x^2)^{1/2}} \qquad \left[\because \cos\theta = \frac{x}{(a^2 + x^2)^{1/2}}\right]$$

$$= \frac{1}{4\pi\varepsilon_0} \cdot \frac{x}{(a^2 + x^2)^{3/2}} \int dq$$

$$= \frac{1}{4\pi\varepsilon_0} \cdot \frac{x}{(a^2 + x^2)^{3/2}} \cdot q$$

For far off axial points $x >> a$, so a^2 can be neglected as compared to x^2.

$$\therefore \quad E = \frac{1}{4\pi\varepsilon_0} \cdot \frac{q}{x^3}$$

This is an expected result as at large distance, the ring behaves like a point charge q.

Example 15. *A particle of mass m and charge q is placed at rest in a uniform electric field (Fig. 1.12) and released. Describe its motion.*

Solution: The motion resembles that of a body falling in the earth's gravitational field. The (constant) acceleration is given by

$$a = \frac{F}{m} = \frac{qE}{m}$$

The equations for uniformly accelerated motion is given by $v = u + at$ and $u = 0$, so we have

$$v = at = \frac{qEt}{m}$$

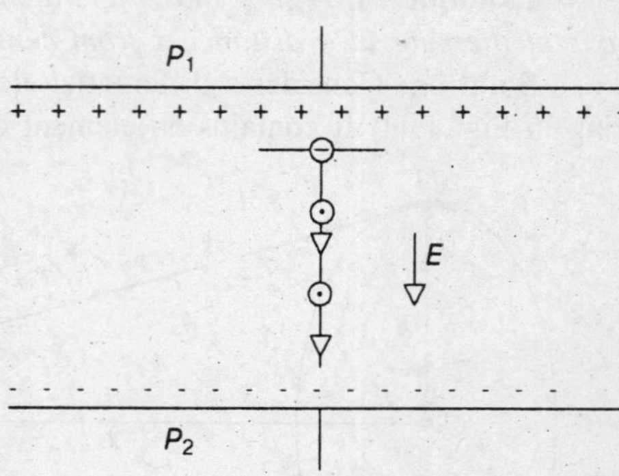

Fig. 1.12

Again from equation $y = ut + \frac{1}{2}at^2$ and $u = 0$ we have

$$y = \frac{1}{2}at^2 = \frac{qEt^2}{2m}$$

and $$v^2 = 2ay = \frac{2qEy}{m}$$

The kinetic energy attained after moving a distance y is given by

$$K = \frac{1}{2}mv^2 = \frac{1}{2}m\left(\frac{2qEy}{m}\right) = qEy$$

This result also follows directly from the work-energy theorem because a constant force qE acts over a distance y.

1.13. A DIPOLE IN AN ELECTRIC FIELD

An electric dipole moment can be regarded as a vector **p** whose magnitude p, for a dipole like that described is the product $2aq$ of the magnitude of either charge q and the distance $2a$ between the charges. The direction of **p** for such a dipole is from the negative to the positive charge.

Figure 1.13 shows an electric dipole formed by placing two charges $+q$ and $-q$ at a fixed distance $2a$ apart. The arrangement is placed in a uniform *external* electric field E, its dipole moment **p** making an angle θ with this field. Two equal and opposite forces F and $-F$ act as shown in figure where

$$F = qE \qquad \text{... (1)}$$

The net force will be zero, but there is a net torque about an axis through O given by

$$\tau = 2F\,(a\sin\theta) = 2aF\sin\theta \qquad \text{... (2)}$$

Combining these two equations and putting $p = 2aq$ we get

$$\tau = 2aq\, E\sin\theta = pE\sin\theta \qquad ...(3)$$

So an electric dipole placed in an external electric field E experiences a torque tending to align it with the field. Equation (3) can be written in vector form as

$$\tau = p \times E \qquad ...(4)$$

the appropriate vectors being shown in his Fig. 1.13.

Work (positive or negative) must be done by an external agent to change the orientation of an electric dipole in an external field and this work is stored as potential energy U in the system consisting of the dipole and the arrangement used to set up the external field. If θ in Fig. 1.13(a) has the initial value θ_0, the work required to turn the dipole axis to an angle θ is given by

$$W = \int dW = \int_{\theta_0}^{\theta} \tau\, d\theta = U \qquad ...(5)$$

where τ is the torque exerted by the agent that does the work. From equation (3) and (5) we get

$$U = \int_{\theta_0}^{\theta} pE\sin\theta\, d\theta = pE\int_{\theta_0}^{\theta} \sin\theta\, d\theta$$

$$= pE\,|-\cos\theta|_{\theta_0}$$

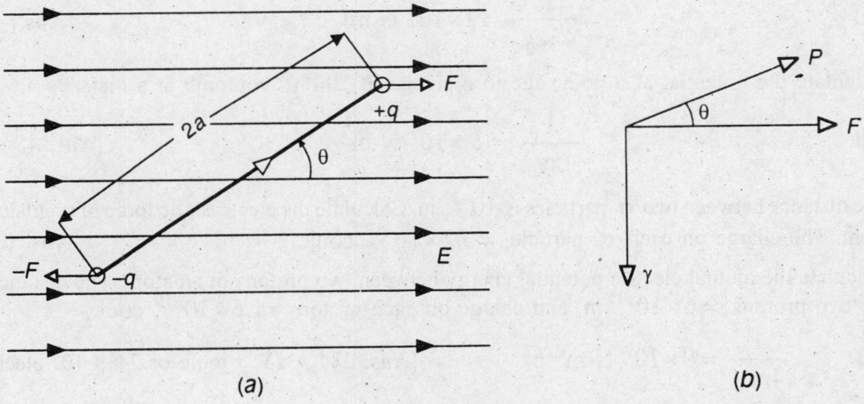

Fig. 1.13

Since we are interested only in *changes* in potential energy, we choose the reference orientation θ_0 to have any convenient value, say 90°, so we have

$$\mathbf{U = -pE\cos\theta} \qquad ...(6)$$

or, in vector notation

$$\mathbf{U = -p.E} \qquad ...(7)$$

Example 15. *An electric dipole consists of two opposite charges of magnitude* $q = 1.0\times10^{-6}$ *coul separated by* $d = 2.0$ *cm. The dipole is placed in an external field of* 1.0×10^{5} *N/coul.*

(a) What maximum torque does the field exert on the dipole?

(b) How much work an external agent must do to turn the dipole and to end?

Solution: Max torque is found by putting $\theta = 90°$ in equation (3)

$$\tau = pE\sin\theta = qd\, E\sin\theta$$

$$= (1.0\times10^{-6}\text{ coul})\,(0.020\text{ meter})\,(1.0\times10^{5}\text{ N/coul})\,(\sin 90°)$$

$$= 2.0\times10^{-3}\text{ N-m}$$

(b) The work is the difference in potential energy U between the positions $\theta = 180°$ and $\theta = 0$. From equation (6)

$$W = U_{180°} - U_{0°} = (-pE.\cos180°) - (-pE\cos0°)$$

$$= 2pE = 2qdE$$
$$= (2)\,(1.0 \times 10^{-6} \text{ coul})\,(0.020 \text{ meter})\,(1.0 \times 10^{5} \text{ N/coul})$$
$$= \mathbf{4.0 \times 10^{-3}\ J.}$$

QUESTIONS AND PROBLEMS

1. Explain the terms electric field and lines of force.
2. Derive an expression for a dipole in an electric field.
3. Compare electrostatic and gravitational field and show how they differ from each other.
4. Describe an experiment to show that the electric intensity inside a charged hollow sphere is zero.
5. What is meant by Maxwell tube of force? How is the intensity of an electric field represented in terms of these tubes?
6. Show theoretically that there is no electric field inside a hollow charged spherical conductor. How may this be tested experimentally?
7. Distinguish between unit tubes, Maxwell unit tubes and unit tubes of induction. How are they related to one another? What do you understand by electrical displacement?
8. Calculate the electric potential energy of an electron-proton system of an atom. The radius of the orbit of the electron $= 21.16 \times 10^{-11}$ m. The charge on the electron $= 1.6 \times 10^{-19}$ coul.

   and $\dfrac{1}{4\pi\varepsilon_0} = 9 \times 10^{9}$ N-m^2 c^{-2} **[Ans.** –7.275 eV]

9. Calculate the potential at a point due to a charge of 200 μ coulomb at a distance 4.6 m.

   and $\dfrac{1}{4\pi\varepsilon_0} = 9 \times 10^{9}$ N-m^2 c^{-2} **[Ans.** 4 × 10^5 volts]

10. The distance between two α particles is 10^{-13} m. Calculate the electrostatic force of repulsion between them. The charge on each α particle $= 3.2 \times 10^{-19}$ coul. **[Ans.** 0.09216 N]
11. Calculate the mutual electric potential energy between two protons of an atom. The distance between the two protons $= 6 \times 10^{-15}$ m and charge on each proton $= 1.6 \times 10^{-10}$ coul.

    and $\dfrac{1}{4\pi\varepsilon_0} = 9 \times 10^{9}$ N-m^2 c^2 **[Ans.** 384 × 13^{-10} joule or 24 × 10^4 electron volts]

12. In an atom the distance between an electron and a proton is 2.12×10^{-10} m. Calculate the electrical force of attraction between them. Charge on the electron $= 1.6 \times 10^{-10}$ coul.

    and $\dfrac{1}{4\pi\varepsilon_0} = 9 \times 10^{9}$ N-m^2 c^{-2} **[Ans.** 5.06 × 10^{-9} N]

CHAPTER

2

GAUSS'S LAW AND ELECTRIC POTENTIAL

2.1. FLUX OF THE ELECTRIC FIELD

Flux denoted by Φ is a property of any vector field; it refers to a hypothetical surface in the field, which may be closed or open. For a *flow field* the flux (Φ_v) is measured by the number of streamlines that cut through such a surface. For an *electric field* the flux (Φ_E) is measured by the number of lines of force that cut through such a surface.

For closed surfaces we shall see that Φ_E is positive if the lines of force point outward everywhere and negative if they point inward. Figure 2.1 shows two equal and opposite charge and their lines of force. Curves S_1, S_2, S_3 and S_4 are the intersections with the plane of the figure of four hypothetical closed surfaces. From the statements just given, Φ_E is positive for surface S_1 and negative for S_2. The flux of the electric field is important because Gauss's law, *one of the four basic equation of electromagnetism,* is expressed in terms of it.

To define Φ_E precisely, consider Fig. 2.2 which shows an arbitrary closed surface immersed in electric field. Let the surface be divided into elementary squares Δ**S**, each of which is small enough so that it may be considered to be plane. Such an element of area can be represented as a vector Δ**S**, whose magnitude is the area Δ**S**; the direction of Δ**S** is taken as the *outward-drawn normal* to the surface.

At every square in Fig. 2.2 we can also construct an electric field vector **E**. Since the squares have been taken to be arbitrarily small, **E** may be taken as constant for all points in a given square.

The vectors **E** and Δ**S** that characterize each square make an angle θ with each other. Figure 2.2(*b*) shows an enlarged view of the three squares on the surface of Fig. 2.2(*a*) marked *x*, *y* and *z*. Note that at *x*, $\theta > 90°$; at *y*, $\theta = 90°$; and at *z*, $\theta < 90°$.

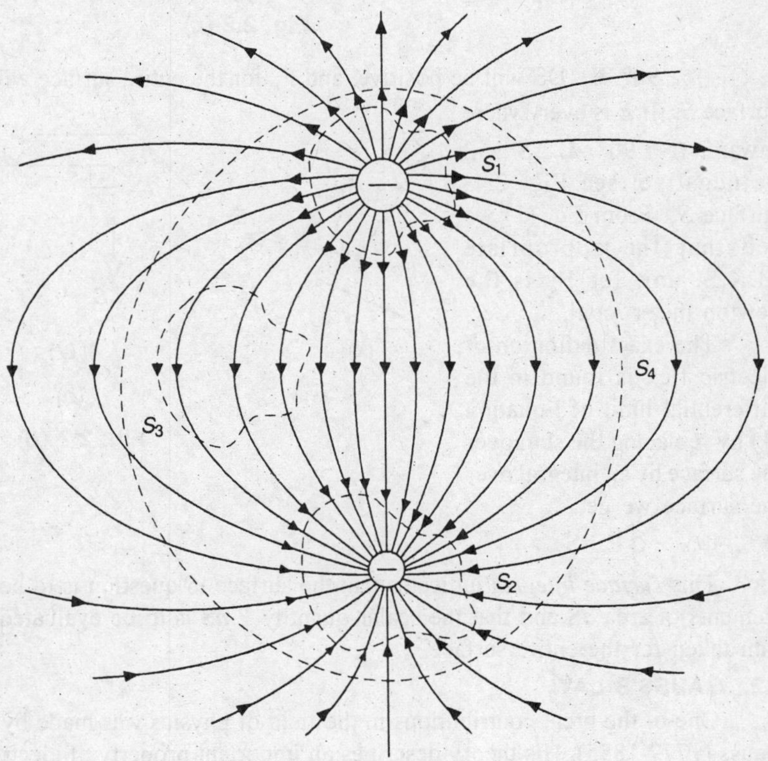

Fig. 2.1

A semiquantitative

definition of flux is $\Phi_E \cong \Sigma \mathbf{E}.\Delta\mathbf{S}$ ...(1)

which tells us to add up the scalar quantity $\mathbf{E}.\Delta\mathbf{S}$ for all elements of area into which the surface has been divided. For points such as x in Fig. 2.2 the contribution to the flux is negative; at y it is zero and at z it is positive. Thus if $\mathbf{E}$ is everywhere outward,

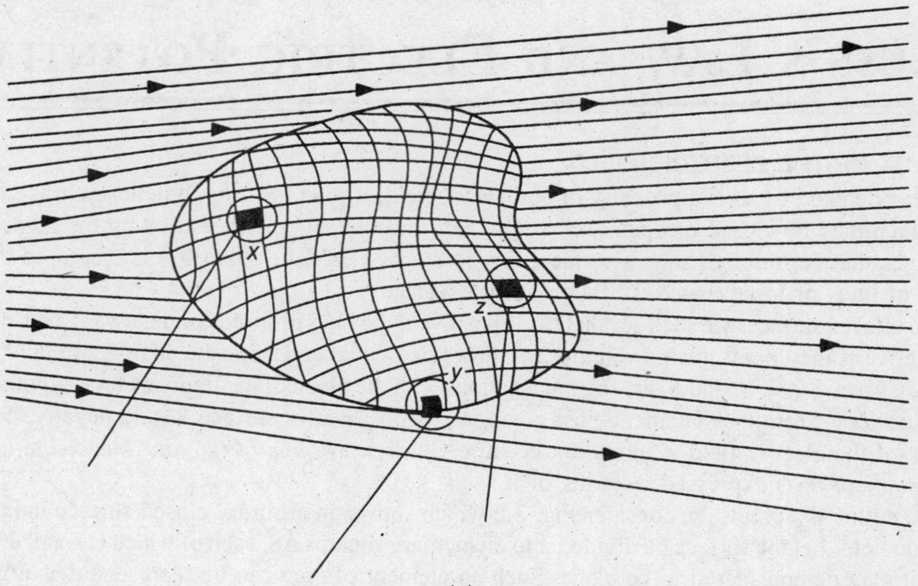

(a)

Fig. 2.2 (a)

$\theta > 90°$, $\mathbf{E}$. $\mathbf{DS}$ will be positive, and F_E for the entire surface will be positive; see Fig. 2.1, surface S_1. If E is everywhere inward, $\theta > 90°$, $\mathbf{E}.\Delta\mathbf{S}$ will be negative; see Fig. 2.1, surface S_2. From Eq. (1) we see that the appropriate M.K.S. unit for F_E is the newton meter^2/coul.

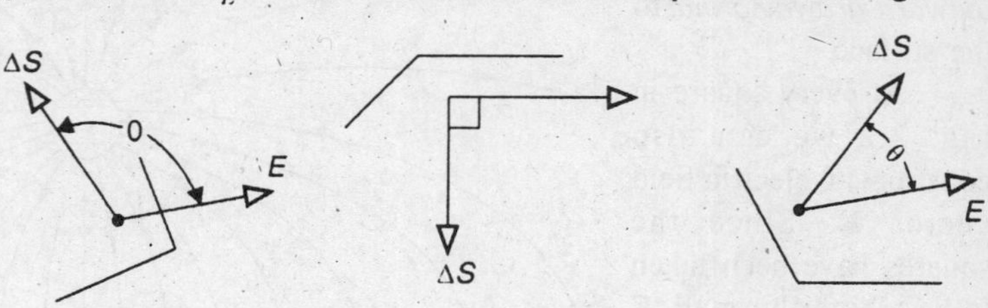

(x) (y) (z)

(b)

Fig. 2.2 (b)

The exact definition of electric flux is found in the differential limit of Equation (1) by replacing the sum over the surface by an integral over the surface we get:

$$\Phi_E = \oint \mathbf{E}.\mathbf{dS} \quad ...(2)$$

This *surface integral* indicates that the surface in question is to be divided into infinitesimal elements of area $d\mathbf{S}$ and that the scalar quantity $\mathbf{E}.d\mathbf{S}$ is to be evaluated for each element and the sum taken for the entire surface.

2.2. GAUSS'S LAW

One of the great contributions in the field of physics was made by a German Mathematician Gauss (1777-1855). His theory describes an important property of electrostatic fields and is called

Gauss's Law; which provides a great insight into the solution of many practical problems in electrostatics.

It states that *the total electric flux through a closed surface enclosing a charge is equal to* $\frac{1}{\varepsilon_0}$ *times the magnitude of the charge enclosed.*

where, ε_0 is *permittivity* of the free space. If Φ be total electric flux through a closed surface enclosing a charge q, then Gauss's theorem states that

$$\Phi = \frac{q}{\varepsilon_0} \quad ...(1)$$

Let us consider a closed surface S enclosing a charge q. Let area element $\vec{dS}$ be at a distance r from the charge q enclosed in the closed surface S. If $\vec{E}$ is electric field at the area element, then electric flux passing through the area element is given by the relation

$$d\Phi = \oint_S \vec{E} . \vec{dS}$$

and the total electric flux through the closed surface S will be

$$\Phi = \vec{E} . \vec{dS}$$

Then, Gauss's theorem may also have the form

$$\oint_S \vec{E} . \vec{dS} = \frac{q}{\varepsilon_0} \quad ...(2)$$

where $\oint_S$ represents integral over a *closed surface.*

PROOF OF GAUSS'S LAW

Let θ be the angle between the directions of $\vec{E}$ and dS, then

$$EdS = EdS \cos\theta$$

Also
$$E = \frac{1}{4\pi\varepsilon_0} \cdot \frac{q}{r^2}$$

where E is the magnitude of electric field due to charge q at an area element (say dS)

$$\therefore \quad \oint \vec{E} . \vec{dS} = \oint \frac{1}{4\pi\varepsilon_0} \cdot \frac{q}{r^2} \cdot dS \cos\theta$$

$$= \frac{q}{4\pi\varepsilon_0} \oint \frac{dS \cos\theta}{r^2} = \frac{q}{4\pi\varepsilon_0} \cdot d\omega$$

where $d\omega$ is the solid angle subtened at O due to area element dS.

$$\therefore \quad \oint \vec{E} . \vec{dS} = \frac{q}{4\pi\varepsilon_0} \cdot 4\pi = \frac{q}{\varepsilon_0}$$

which proves Gauss's theorem.

Mathematically Gauss's Law may be stated as

If a closed surface encloses a charge, then the surface integral of the electric field due to the enclosed charge over a closed surface is equal to $1/\varepsilon_0$ *times the charge enclosed.*

Gauss's law is very important especially in situations where calculation of electric field with the application of

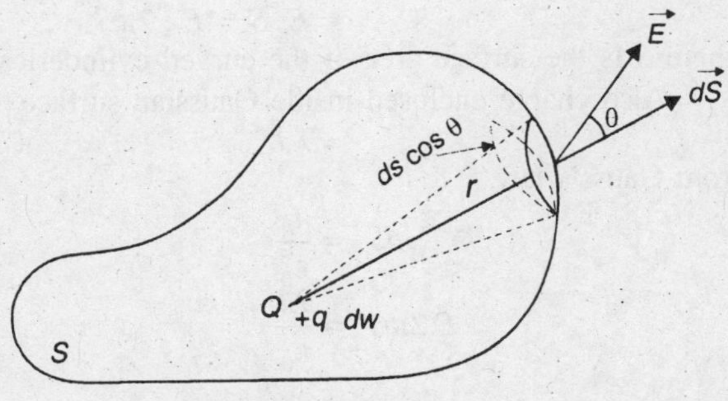

Fig. 2.3

Coulomb's law or principle of superposition, becomes difficult. In these cases a closed surface is chosen which, encloses the source charge. Such a surface is called **Gaussian surface. An imaginary closed surface enclosing a charge is called Gaussian surface.**

2.3. FORCE BETWEEN TWO POINT CHARGES IN VACUUM

Consider two point charges q_1 and q_2 present in vacuum and separated by a distance q with point charge q_1 as centre and radius equal to r, draw a sphere. It will be *Gaussian surface* enclosing charge q_1. Let E be the magnitude of electric field at point charge q_2, then we have the force experienced by point charge q_2 due to the electric field of charge q_1 as

$$F = q_2 E \qquad \text{...(1)}$$

Now, according to Gauss's theorem,

$$\int \vec{E}.\vec{dS} = \frac{q_1}{\varepsilon_0}$$

$$\therefore \quad \frac{q_1}{\varepsilon_0} = \int \vec{E}.\vec{dS} = \int E.dS = E\int dS = E.4\pi r^2$$

or

$$E = \frac{1}{4\pi\varepsilon_0}\cdot\frac{q_1}{r^2} \qquad \text{...(2)}$$

From equations (1) and (2), we have

$$F = \frac{1}{4\pi\varepsilon_0}\cdot\frac{q_1 q_2}{r^2}$$

2.4. APPLICATIONS OF GAUSS'S LAW

***(i)* Electric field due to a line charge (field due to a thin wire).** Let us consider a thin infinitely long wire or a rod (cylinderical conductor). Let it have a uniform linear charge density λ and let P be the point at a distance r from the line charge. To find $\vec{E}$ *at P, let us select a cylinderical surface of radius* r and height l as a *Gaussian surface.* As $\vec{E}$ *is everywhere parallel to the two ends of the cylinder, they will not contribute any electric flux.*

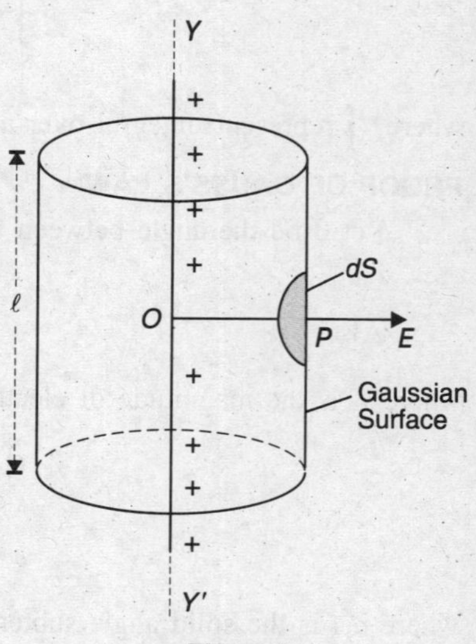

Fig. 2.4

Symmetry of the problem suggests that $\vec{E}$ is equal in magnitude and directed outward (As line charge is positive) at every point of the curved cylinderical surface.

It is this curved cylinderical surface which contributes to the electric flux

$$\therefore \quad \oint \vec{E}.\vec{dS} = \vec{E}.\vec{dS}$$

$$= \vec{E}.\vec{S} = \vec{E} : 2\pi r l$$

where S is the surface area of the curved cylinderical surface of radius r and height l.

Also charge enclosed inside Gaussian surface

$$= \lambda.l$$

From Gauss's law,

$$\oint \vec{E}.\vec{dS} = \frac{q}{\varepsilon_0}$$

$$\therefore \quad E.2\pi r l = \frac{\lambda l}{\varepsilon_0}$$

or

$$E = \frac{\lambda}{2\pi\varepsilon_0 r}$$

So the intensity of electric field of a line charge is inversely proportional to the distance r from the line.

It may be noted that $\vec{E}$ will be radially outward if the charge is positive and radially inward if the charge is negative.

Instead of a wire, if a cylinderical conductor of radius R is considered. Then if the charge is distributed throughout the volume of the cylinder with charge density ρ, in that case $q = (\pi r^2.l).\rho$.

$$\therefore \quad E.2\pi rl = \frac{(\pi R^2 l)\rho}{\varepsilon_0}$$

or
$$E = \frac{\rho R^2}{2\varepsilon_0 r}$$

***(ii)* Electric field due to an infinite plane sheet of charge.** Consider an infinite thin plane sheet of positive charge having a uniform surface charge density σ on both sides of the sheet. By symmetry, it follows that the field is perpendicular to the plane sheet of charge and in outward direction. To find electric field due to the plane sheet of charge at any point P distant r from it, choose a cylinder of area of cross-section S through the point P as the Gaussian surface [Fig. 2.5]. Since electric lines of force are parallel to the curved surface of the cylinder, the flux due to electric field of the plane sheet of charge passes only through the two circular caps of the cylinder. If E is magnitude of electric field at point P, then electric flux crossing through the Gaussian surface is given by

$$\oint \vec{E}.\vec{dS} = \oint_{S_1} \vec{E}.\vec{dS} + \oint_{S_2} \vec{E}.\vec{dS}$$
$$= \int_{S_1} E\,dS + \int_{S_2} E.dS$$
$$= ES + ES = 2ES \quad ...(1)$$

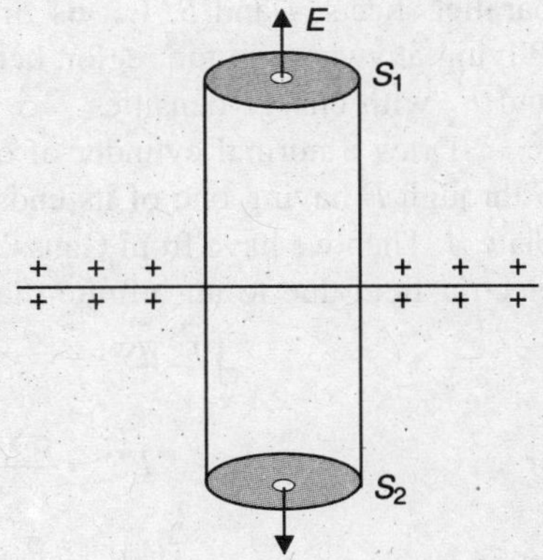

Fig. 2.5

Charge enclosed inside the Gaussian surface

$$q = \sigma.S \quad ...(2)$$

From Gauss's law

$$\oint \vec{E}.\vec{dS} = \frac{q}{\varepsilon_0} \quad ...(3)$$

From equation (1), (2) and (3) we get

$$2\,ES = \frac{\sigma S}{\varepsilon_0}$$

or
$$E = \frac{\sigma}{2\varepsilon_0}$$

which means that the magnitude of the field is independent of the distance from the sheet.

In case sheet is negatively charged, $\vec{E}$ points normally towards the sheet.

***(iii)* Electric field at any point due to infinite charged conducting plate.** When a charge is given to a conducting plate, the same is distributed over the entire surface of the plate. If the plate is of uniform *size and thickness*, the surface density (σ) of the plate is uniform and is same on the two sides of the surfaces. To find the electric field

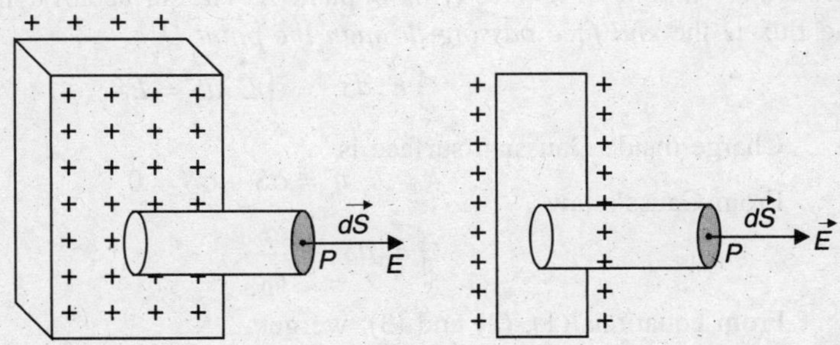

Fig. 2.6

at P, draw a cylinder (Gaussian surface) through P which is normal to the surface of the plate. Let S be the cross-sectional area of the cylinder. One end of the cylinder is *inside the conductor* and so does not contribute to the electric flux as no charge exists inside the conductor. Also no contribution of flux is given by the curved surface of this cylinder as $\vec{E}$ is parallel to this surface. So, it is only the end face around P which contributes towards electric flux. So, we have

$$\oint \vec{E}.\vec{dS} = \oint EdS = ES \qquad ...(1)$$

Charge enclosed inside Gaussian surface is

$$q = \sigma.S \qquad ...(2)$$

From Gauss's theorem

$$\oint \vec{E}.\vec{dS} = \frac{q}{\varepsilon_0} \qquad ...(3)$$

From equation (1), (2) and (3), we get

$$ES = \frac{\sigma S}{\varepsilon_0}$$

or

$$E = \frac{\sigma}{\varepsilon_0}$$

Note: The electric field due to a charged conducting plate is two times the field due to plane sheet of charge. It is due to the fact that the same charge is present on its both sides.

(*iv*) **Electric field due to two infinite parallel sheets of charge.** Let us consider two infinite plane parallel sheets A and B. Let us first consider a point P lying anywhere in the region between the plates P_1 and P_2 with charge densities $+\sigma$ and $-\sigma$.

Draw a normal cylinder of cross-sectional area S through P having one of its ends in the conducting plate A. Then we have from Gauss's law (as in case of electric field due to an infinite plane sheet)

$$\oint_S \vec{E}.\vec{dS} = \frac{q}{\varepsilon_0}$$

or

$$ES = \frac{\sigma S}{\varepsilon_0}$$

or

$$E = \frac{\sigma}{\varepsilon_0}$$

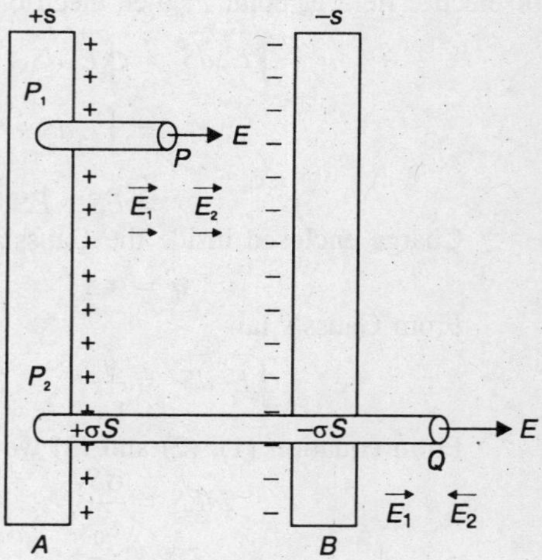

Fig. 2.7

So, the field between the two planes is *uniform and independent of their separation.* This gives us a way to produce uniform electric field in a limited region of space.

Again let Q be a point *lying outside* the planes. Draw a cylinder of area of cross-section S through q with one of its face lying in plane A. The surface of cylinder which contributes towards the flux is the *end face passing through the point Q.*

$$\therefore \qquad \oint_S \vec{E}.\vec{dS} = \oint_S \vec{E}.\vec{dS} = ES \qquad ...(1)$$

Charge inside Gaussian surface is

$$q = \sigma S - \sigma S = 0 \qquad ...(2)$$

From Gauss's law

$$\oint_S \vec{E}.\vec{dS} = \frac{q}{\varepsilon_0} \qquad ...(3)$$

From equations (1), (2) and (3), we get

$$ES = 0 \qquad \text{or} \qquad E = 0$$

Which shows that the electric field intensity due to two oppositely charged infinite planes is σ/E at any point between the planes and is *zero* for all external points.

(v) Electric field due to a uniformly charged spherical shell. Consider a thin spherical shell of radius R and centre O. Let q be charge on the spherical shell. The electric field due to charged spherical shell is radial and is spherically symmetric. Let us find electric field at point P at a distance r from the centre of the spherical shell.

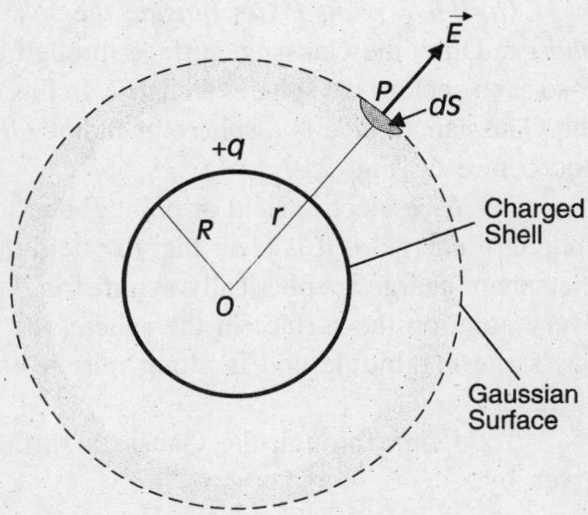

Fig. 2.8

(a) When point P is inside the spherical shell. Draw a Gaussian surface through the point P and it will be a sphere of radius r and centre of (Fig. 2.8).

Let E be the electric field at P due to the charge q on spherical shell. As is evident the field due to spherical shell will be *spherically symmetric.* Which means that at every point on the surface of spherical shell, the field has same magnitude and is normal to the surface. So, total flux through the Gaussian surface will be

$$\oint_S \vec{E}.\vec{dS} = \oint_S E.dS = \oint_S E.dS = E.4\pi r^2$$

Since the charge enclosed by the Gaussian surface is q, according to the Gauss s theorem

$$E.4\pi r^2 = \frac{q}{\varepsilon_0} \qquad (r > R)$$

or $$E = \frac{1}{4\pi\varepsilon_0}\frac{q}{r^2} \qquad \text{...(1)}$$

From equation (1), it follows that for the points outside the charged spherical shell, the shell behaves as though the charge on the shell, were concentrated at its centre. If σ is uniform surface charge, density of the spherical shell, then

$$q = 4\pi R^2\sigma$$

for q in equation (1), we get,

$$E = \frac{\sigma}{\varepsilon_0}\frac{R^2}{r^2} \qquad \text{...(2)}$$

(b) When P is on the surface of spherical shell. In this case the Gaussian surface through P will just enclose the charged spherical shell. So from Gauss's law we have

$$E.4\pi R^2 = \frac{q}{\varepsilon_0} \qquad [\because r = R]$$

or $$E = \frac{q}{4\pi\varepsilon_0 R^2} = \frac{1}{4\pi\varepsilon_0}\cdot\frac{q}{R^2} \qquad \text{...(3)}$$

Since $q = 4\pi R^2\sigma$, the equation (3) will become

$$E = \frac{1}{4\pi\varepsilon_0}\cdot\frac{4\pi R^2\sigma}{R^2} = \frac{\sigma}{\varepsilon_0} \qquad \text{...(4)}$$

(c) When P Lies Inside the Spherical Shell. In this case, the Gaussian surface through the point P will not enclose any charge and so form Gauss's theorem,

$$E.4\pi r^2 = \frac{0}{\varepsilon_0} = 0$$

$$\therefore \qquad E = 0$$

(vi) Electric Field Due to a Solid Charged Sphere. Consider an isolated sphere of charge q having radius R. Let us calculate electric field at point P at a distance r from its centre O.

(a) When point P lies outside the sphere of charge: Draw the Gaussian surface through point P so as to enclose the sphere of charge. In this case, the Gaussian surface is a sphere of radius $OP = r$ and centre O [Fig. 2.9].

Let E be electric field at point P due to the sphere of charge q. It is clear that the field due to the sphere charge is spherically symmetric, *i.e.*, at every point on the surface of the sphere, the field has same magnitude and is along normal to the surface.

Total flux through the Gaussian surface is given by

$$\oint_S \vec{E}.\vec{dS} = \oint_S E.dS = \oint_S E.dS = E.4\pi r^2$$

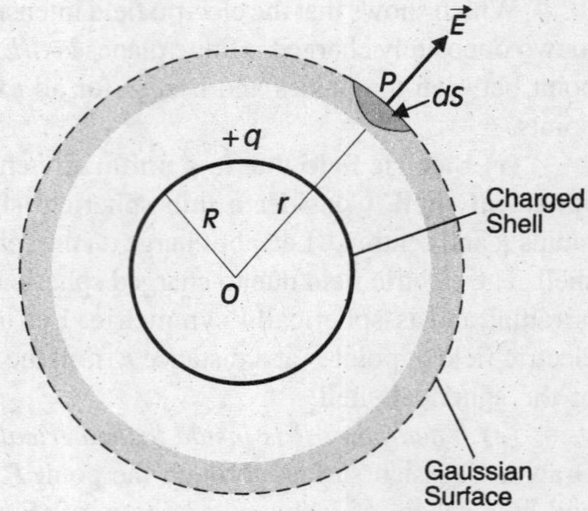

Fig. 2.9

From Gauss's theorem

$$E.4\pi r^2 = \frac{q}{\varepsilon_0}$$

or
$$E = \frac{1}{4\pi\varepsilon_0}\frac{q}{r^2} \qquad (r>R) \quad ...(1)$$

which is the same as that a point at distance r from the charge q. It shows that for points outside the sphere of charge, the sphere behaves as if *all the charge on the sphere were concentrated* at its centre. If ρ be the uniform charge volume density of sphere, then

$$q = \frac{4}{3}\pi R^3.\rho \qquad ...(2)$$

Putting the value of q in equation (1), we get;

$$E = \frac{1}{4\pi\varepsilon_0}\frac{4\pi R^3.\rho}{3r^2}$$

$$= \frac{\rho R^3}{3\varepsilon_0 r^2} \qquad ...(3)$$

(b) When point P lies on the surface of charged sphere. In such a case, the area of the Gaussian surface is equal to surface area of the sphere of charge *i.e.*, equal to $4\pi R^2$. If E electric field at point P on the surface of the sphere of charge then from Gauss's law

$$E.4\pi R^2 = \frac{q}{\varepsilon_0} \qquad (r = R)$$

or
$$E = \frac{1}{4\pi\varepsilon_0}\frac{q}{R^2} \qquad ...(4)$$

Since, $q = \frac{4}{3}\pi R^3\rho$, the equation (4) will get the form

$$E = \frac{\rho R}{3\varepsilon_0} \qquad ...(5)$$

(c) When point P lies inside the charged sphere. In such a case, Gaussian suface is spherical, whose centre is O and radius is $OP = r$ as shown in Fig. 2.10. If q' is charge enclosed by the Gaussian surface, then we have,

$$E.4\pi r^2 = \frac{q'}{\varepsilon_0} \qquad (r < R)$$

or $$E = \frac{1}{4\pi\varepsilon_0} \cdot \frac{q'}{r^2} \quad ...(6)$$

Now, $$q' = \frac{q}{\frac{4}{3}\pi R^3} \times \frac{4}{3}\pi r^3 = q\frac{r^3}{R^3}$$

Putting for q' in equation (6), we get

$$E = \frac{1}{4\pi\varepsilon_0}\frac{qr}{R^3} \quad ...(7)$$

Since, $q = \frac{4}{3}\pi R^3 \rho$, the equation gets the form

$$E = \frac{\rho r}{3\varepsilon_0} \quad ...(8)$$

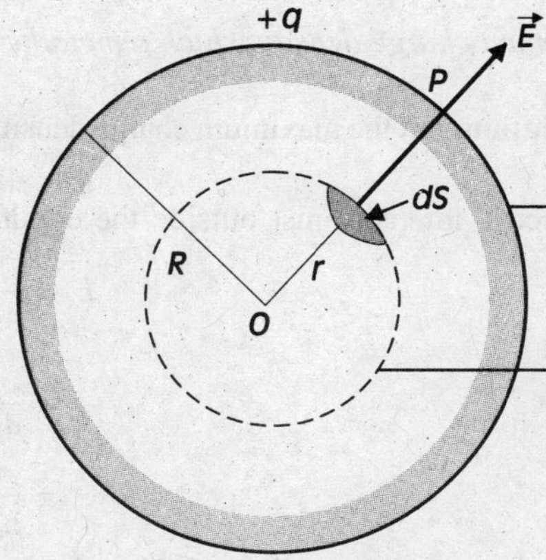

Fig. 2.10

2.5. GAUSS'S LAW AND COULOMB'S LAW

Coulomb's law can be deduced from Gauss's law and symmetry considerations. To do so, let us apply Gauss's law to an isolated point charge q as in Fig. 2.11. Although Gauss's law holds for any surface whatever, information can most readily be extracted for a spherical surface r centred on the charge. The advantage of this suface is that, from symmetry, E must be *normal* to it and must have the same magnitude for all points on the surface.

In Fig. 2.11 both **E** and d**S** at any point on the Gaussian surface are directed radially outward. The angle between them is zero and the quantity **E**.d**S** become simply **E**d**S**. Gauss's law (Equation 4) thus reduced to

$$\varepsilon_0 \oint E.dS = \varepsilon_0 \oint E.dS = q$$

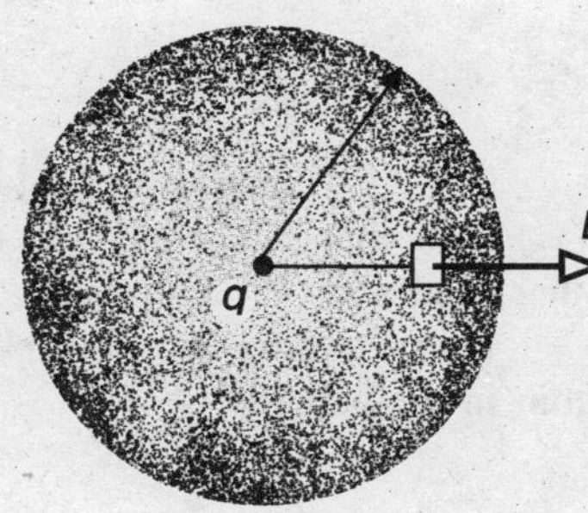

Fig. 2.11 A spherical Gaussian surface of radius r surrounding a point charge q.

Because **E** is constant for all points on the sphere it can be taken out of the integral sign, giving us

$$\varepsilon_0 E \oint .dS = q$$

where the integral is simply the area of the sphere* So we get

$$\varepsilon_0 E (4\pi r^2) = q$$

or $$E = \frac{1}{4\pi\varepsilon_0}\frac{q}{r^2} \quad ...(5)$$

Equation (5) gives the magnitude of the electrical field strength **E** at any point a distance r from an isolated point charge q. The direction of **E** is alreadly known from symmetry.

Let us put a second point charge q_0 at the point at which **E** is calculated. The magnitude of the force that acts on it is given by

$$F = Eq_0$$

Combining with Equation (5) we get

$$F = \frac{1}{4\pi\varepsilon_0}\frac{qq_0}{r^2} \quad ...(6)$$

which is the Coulomb's law. So we have deduced Coulomb's law from *Gauss's law and considerations of symmetry.*

Example 1. *Air cannot support very intense electric fields. The charged particles, which are always present in air in small numbers get accelerated to very high velocities. These, in turn, produce ionization during collisions with neutral molecules. Due to this air becomes conducting. The maximum electric intensity which air can withstand at atmospheric pressure is about* 3×10^6 NC^{-1}. *Calculate* *the maximum charge density which a metallic conductor of radius R metres can have when placed in air.*

Solution. Let the maximum charge density be σ C m^{-2}. Therefore, total charge on the conductor,

$$q = 4\pi R^2\sigma$$

Electric intensity just outside the conductor's surface;

$$E = \frac{1}{4\pi\varepsilon_0}\frac{q}{R^2}$$

$$= \frac{1}{4\pi\varepsilon_0}\frac{4\pi R^2\sigma}{R^2}$$

$$= \frac{\sigma}{\varepsilon_0}$$

As the maximum value of E can be 3×10^6 N C^{-1},

$$3\times 10^6\,\text{NC}^{-1} = \frac{1}{8.85\times 10^{-12}\,\text{C}^2\text{N}^{-1}\text{m}^{-2}}\times\sigma\ \text{C m}^{-2}$$

or
$$\sigma = 3\times 10^6\times 8.85\times 10^{-12}$$
$$= 26.6\times 10^{-6}$$

$\therefore$ maximum charge density = $\mathbf{26.6 \times 10^{-6}}$ **C m^{-2}**

Note that it is independent of the radius of the sphere. In fact this holds for conductors of other shapes also.

Let us assume that the sphere is of copper, and is positively charged. The total number of atom in the ball is

$$n = \frac{\text{Mass of the ball}\times\text{Avogardro's number}}{\text{molecular weight of copper}}$$

$$= \frac{4}{3}\pi R^3\frac{\rho}{M}\times N$$

$$= \frac{4}{3}\pi R^3\times 10^6\ \text{cm}^3\times 8.95\ \text{g cm}^{-3}\times\frac{6.022\,10^{23}\ \text{mole}^{-1}}{63.54\ \text{g mole}^{-1}}$$

$$= 3.55\times 10^{29}R^3$$

Maximum charge the sphere can hold is, given by

$$q_m = 4\pi R^2\sigma$$
$$= 4\pi R^2\times 26.6\times 10^{-6}\text{C}$$
$$= 3.34\times 10^{-4}R^2\text{C}$$

Now, charge on one electron is 1.602×10^{-19} C. Therefore, the number of electrons to be removed from the sphere to charge it to maximum possible value is,

$$n_c = \frac{3.34\times 10^{-4}R^2}{1.602\times 10^{-19}}$$
$$= 2.09\times 10^{15}R^2$$

If each atom of copper contributes one free electron. The fraction of the total free electrons removed is,

$$\frac{n_c}{n} = \frac{2.09\times 10^{15}R^2}{3.55\times 10^{29}R^3}$$
$$= \frac{5.89}{R}\times 10^{-15}$$

If the radius of the sphere is 0.01 m, this fraction becomes 5.89×10^{-18}. Thus, though the sphere is charged to the maximum extent, the number of electrons transferred makes a negligibly small fraction of he total **number** of free **electrons** present. This is even less than 1 in 10^{-12}.

Example 2. *Two parallel plates of area 20 cm² each are separated by a distance of 10 mm. If one plate carriers a charge + 2 × 10⁻⁹ C, and the other carries a charge – 2× 10⁻⁹ C, calculate the electric field in the space between them.*

Solution: Surface charge density $= \dfrac{2\times10^{-9}}{20\times10^{4}}\ \text{C m}^{-2}$

$= 10^{-6}\ \text{C m}^{-2}$

$\therefore$ The electric field $= \dfrac{10^{-6}\,\text{C m}^{-2}}{8.85\times10^{-12}\,\text{C}^2\text{N}^{-1}\text{m}^{-2}}$

$= \mathbf{1.13 \times 10^5\ NC^{-1}}$

2.6. MECHANICAL FORCE AT THE SURFACE OF A CHARGED CONDUCTOR—ELECTROSTATIC PRESSURE

Let us take a conductor of any shape and concentrate on any one point P (Fig. 2.12). Let the charge density around P be σ (σ will vary from point to point at the surface of the conductor). The electric field just outside P will be normal to the surface of the conductor. If it is not so, *i.e.*, E is inclined just near the surface, then it will have a component along the surface of the conductor, which will make the charges move and redistribute themselves. As along as the charges are stationary, E will not have any component along the surface and, hence, will be normal to it. Let us draw a Gaussian surface as shown in Fig. 2.12 (*a*). The only contribution to the electric flux would be through the end cap outside the conductor. So we have.

$$ES = \sigma S/\varepsilon_0$$

or

$$E = \sigma/\varepsilon_0 \qquad ...(1)$$

From equation 1, we can conclude that this is a general result which holds good for conductors of any shape and under any circumstances.

Now, let us separate this small area S around P from the rest of the conductor and calculate the force on it. It should be clearly understood that when we have to calculate the force on a charge placed along with other charges, the electric field at the site of the charge under consideration should be calculated by exluding that charge from the assembly. Now we have two systems of the charges—charges around p, and charges at the rest of the conductor.

Just outside P, let the field due to charges around P be E_1 and that due to rest of the charges be E_2 [Fig. 2.12 (*b*)]. Thus:

$$E = E_1 + E_2 \qquad ...(2)$$

Inside the conductor, total electric field is zero. Therefore,

$$0 = E_1 - E_2 \qquad ...(3)$$

From Equations (2) and (3), we get,

$$E_2 = E_1 = E/2 = \sigma/2\varepsilon_0 \quad ...(4)$$

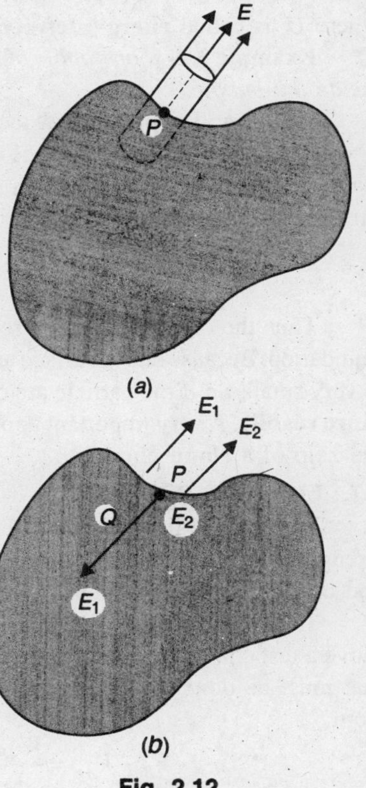

Fig. 2.12

$\therefore$ Force on charge around $P = \sigma S \times \sigma/2\varepsilon_0$

or Force per unit area $= \sigma^2/S$ ...(5)

Thus every part of the surface is acted upon by a force directed outwards. The surface is, therefore, under outward electric pressure.

So the mechanical pressure on the surface will be given by

$$p = \frac{\sigma^2}{2\varepsilon_0}$$

Also $$E = \frac{\sigma}{\varepsilon_0} \quad \therefore \sigma = \varepsilon_0 . E$$

$$\therefore \quad p = \frac{(\varepsilon_0 E)^2}{2\varepsilon_0} = \frac{\varepsilon_0 E^2}{2} \qquad ...(6)$$

MECHANICAL ENERGY/UNIT VOLUME:

The force on unit surface area of a conductor is $p = \frac{\varepsilon_0 E^2}{2}$. If due to this force, the conductor of surface area S moves out through a distance dx, then work done is given by

$$W = (pS).dx = \left(\frac{\varepsilon_0 E^2}{2}\right) S\,dx$$

$\therefore$ The energy stored/unit volume is given by

$$U_v = \frac{W}{S.dx} = \frac{\varepsilon_0 E^2}{2} \qquad ...(7)$$

where U_v is called **energy density.**

Example 3. *A soap bubble of radius R has been given a charge q. Find out the excess pressure inside the bubble.*

Solution: The pressure inside a soap bubble of radius R is greater than outside by an amount $p = 4T/R$, where T is the surface tension of the solution. But if the bubble carries a charge q, $\sigma = q/4\pi R^2$, and there is an additional outward pressure of $\sigma^2/2\varepsilon_0$ or $q^2/32\pi^2\varepsilon_0 R^4$. This helps the pressure excess to stretch out the bubble. Therefore,

$$p + \frac{q^2}{32\pi^2\varepsilon_0 R^4} = \frac{4T}{R} \quad \text{or} \quad p = \frac{4}{R}\left[T - \frac{q^2}{128\pi^2\varepsilon_0 R^3}\right]$$

Thus the charge, in effect, reduces surface tension. A similar treatment can be applied to a liquid drop. Because due to surface tension effect, the tendency of drops to form on uncharged particles is very small. So if the particle are charged, the **supersaturated vapour would condense on them more readily.** A very important application of this is in the detection of charged particles, *e.g.*, with the help of a **cloud chamber.**

2.7. ELECTRIC POTENTIAL

The electric field around a charged rod can be described not only by a (vector) electric field strength E but also by a scalar quantity, the *electric potential V.* These quantities are intimately related, and often it is only a matter of convenience which is used in a given problem.

To find the *electric potential difference* between two points, A and B in an electric field, we move a test charge q_0 from A to B, always keeping it in equilibrium and we measure the work W_{AB} that must be done by the agent moving the charge. The electric potential difference is defined from

$$V_B - V_A = \frac{W_{AB}}{q_0} \qquad ...(1)$$

The work W_{AB} may be (*a*) positive (*b*) negative, or (*c*) zero. In these cases the electric potential at B will be (*a*) higher, (*b*) lower, or (*c*) the same as the electric potential at A.

The MKS unit of potential difference that follows from Eq. (1) is the joule/coul. The combination occurs so often that a special unit, the *volt*, is used to represent it; that is,

$$1 \text{ volt} = 1 \text{ joule/coul.}$$

Usually point A is chosen to be a large (taken as ∞) distance from all charges and the electric potential V_A at this infinite distance is arbitrarily taken to be zero. This allows to define the *electric potential at a point.* Putting $V_A = 0$ in Equation (1) and dropping the subscripts leads to

$$V = \frac{W}{q_0} \qquad ...(2)$$

Bearing in mind the assumptions made about the reference position, we see from equation (2) that V near an isolated positive charge is positive because positive work must be done by an outside agent to push a (positive) test charge in from infinity. Similarly, the potential near an isolated negative charge is negative, because an outside agent must exert a restraining force (that is, must do negative work on), a (positive) test charge as it comes in from infinity. Electric potential as defined in Eq. (2) is a scalar because W and q_0 in that equation are scalars.

Both W_{AB} and $V_B - V_A$ in equation (1) are independent of the path followed in moving the test charge from point A to point B. If this were not so, point B would not have a unique electric potential (with respect to point A as a deemed reference position) and the concept of potential would have limited usefulness.

There is a strong analogy between electrostatic forces and gravitational forces, based on the fact that their fundamental laws are inverse square laws *i.e.*,

$$F_E = \frac{1}{4\pi\varepsilon_0} \frac{q_1 q_0}{r} \qquad \text{and } F_g = G\frac{m_1 m_2}{r^2}$$

Thus we can define the *gravitational potential* V_g the relation from

$$V_g = \frac{W}{m}$$

where W is the work required to move a test body of mass m from infinity to the point in question. Gravitational equipotential surfaces can also be constructed; which will be everywhere at right angles to the gravitational field strength vector g. For a uniform gravitational field, such as that near the surface of the earth, these surfaces are horizontal planes. This correlates with the fact that no net work is required to move a body of mass m, between two points with the same elevation and the same net work is required to move a body along any path starting on a given horizontal surface and ending on another.

2.8. POTENTIAL AND FIELD STRENGTH

Let A and B in Fig. 2.13 be two points in a unifrom electric field E, set up by an arrangement of charges not shown, and let A be a distance d in the field direction from B. Assume that a positive test charge q_0 is moved, by an external agent and without acceleration, from A to B along the straight line connecting them.

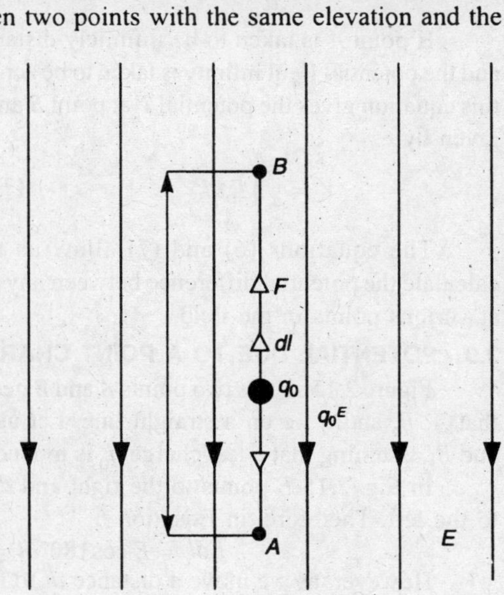

Fig. 2.13

The *electric* force on the charge is q_0E and points downwards. In order to move the charge in the way we have described we must counteract this force by applying an external force F of the same magnitude but directed upward. The work W done by the agent that supplies this force is

$$W_{AB} = Fd = q_0 Ed \qquad ...(3)$$

Substituting this in Equation (1) we get

$$V_B - V_A = \frac{W_{AB}}{q_0} = Ed \qquad ...(4)$$

This equation shows the connection between potential difference and field strength for a simple special case. Note from this equation that another M.K.S. unit for E is the *volt/meter*.

In Fig. 2.13, B has a higher potential than A. This is reasonable because an external agent would have to do positive work to push a positive test charge from A to B. Figure 2.13 would be used as it stands to illustrate the act of lifting a body from A to B in the uniform gravitational field near the earth's surface.

We can find the connection between V and E in the more general case in which the field is *not* uniform and in which the test body is moved along a path that is *not* straight, as in Fig. 2.14. The electric field exerts a force q_0E on the test charge, as shown. To keep the test charge from accelerating, an external agent must apply a force F chosen to be exactly equal to $-q_0 E$ for all position of the test body.

If the external agent causes the test body a displacement dl along the path from A to B, the element of work done by the external agent will be Fdl. To find the total work W_{AB} done by the external agent in moving the test charge from A to B, we sum up (integrate) the work contributions for all the infinitesimal segments into which the path is divided. We get

$$W_{AB} = \int_A^B F.dl = -q_0 \int_A^B E.dl \qquad ...(5)$$

Sucn an integral is called a *line integral.* Note that we have substituted

$$F = -q_0E$$

Putting the expression for W_{AB} into Equation (1) we get

$$V_B - V_A = \frac{W_{AB}}{q_0} = \int_A^B E.dl \qquad ...(6)$$

If point A is taken to be infinitely distant and the potential V_A at infinity is taken to be zero, this equation gives the potential V at point B and given by

$$V = -\int_\infty^B E.dl \qquad ...(7)$$

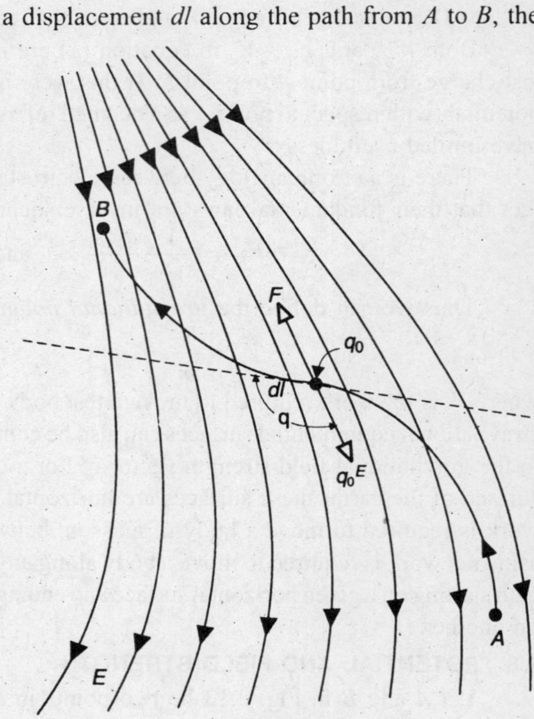

Fig. 2.14

The equatıons (6) and (7) allow us to calculate the potential difference between any two points (or the potential at any point) if E is known at various points in the field.

2.9. POTENTIAL DUE TO A POINT CHARGE

Figure 2.15 shows two points A and B near an isolated point charge q. For simplicity we assume that A, B, and q lie on a straight line. Let us compute the potential difference between points A and B, assuming that a test charge q_0 is moved without acceleration along a radial line from A to B.

In Fig. 2.15 E points to the right and dl, which is always in the direction of motion, points to the left. Therefore, in Equation 6,

$$Edl = E\cos 180° dl = -dl$$

However, as we move a distance dl to the left, we are moving in the direction of decreasing r because r is measured from q as an origin. So we have

$$dl = -dr$$

Combining these we get

$$E.dl = E\,dr$$

Substituting this in Equation (6) we have

$$V_B - V_A = -\int_A^B E.dl = -\int_{r_A}^{r_B} E.dl$$

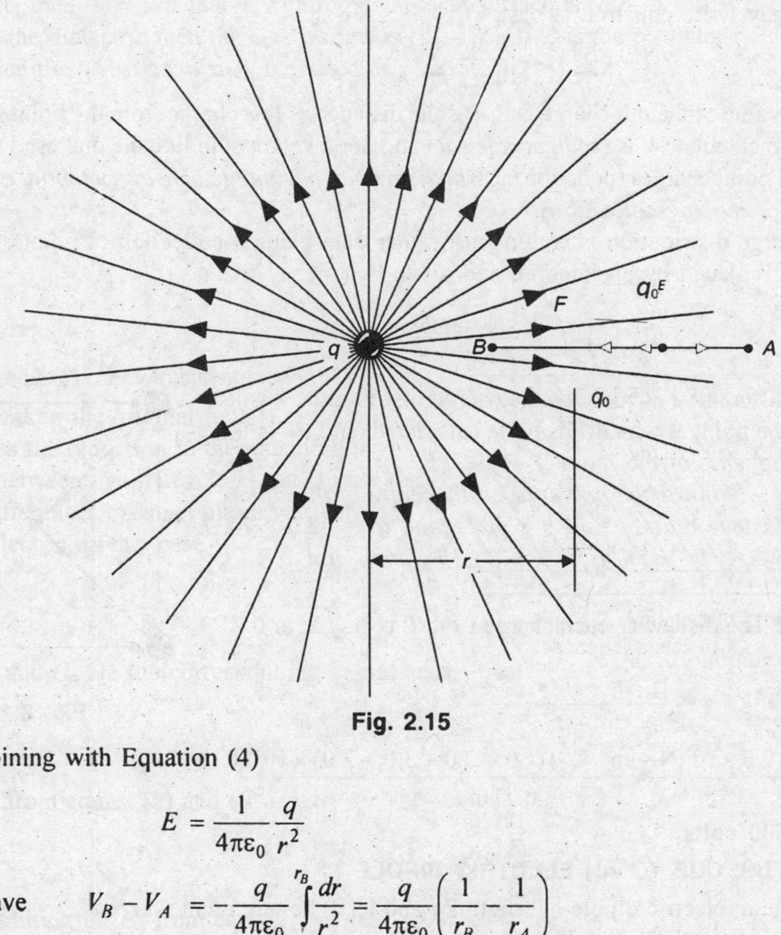

Fig. 2.15

Combining with Equation (4)

$$E = \frac{1}{4\pi\varepsilon_0}\frac{q}{r^2}$$

we have $$V_B - V_A = -\frac{q}{4\pi\varepsilon_0}\int_{r_A}^{r_B}\frac{dr}{r^2} = \frac{q}{4\pi\varepsilon_0}\left(\frac{1}{r_B} - \frac{1}{r_A}\right) \quad ...(8)$$

Choosing reference position A to be at infinity (that is, letting $r_A \to \infty$). Choosing $V_A = 0$ at this position, and dropping the subscript B we get

$$V = \frac{1}{4\pi\varepsilon_0}\frac{q}{r} \quad ...(9)$$

This equation shows clearly that *equipotential surfaces for an isolated point charge are spheres concentric with the point change.* A study of the derivation will show that this relation also holds for points external to spherically symmetric charge distributions.

Example 4. *What is the electric potential at the surface of a gold nucleus? The radius is 6.6×10^{-15} meter and the atomic number $Z = 79$.*

Solution: The nucleus, assumed spherically symmetrical, behaves electrically for external points as if it were a point charge. Thus we can use Equation (9) or, recalling that the proton charge is 1.6×10^{-19} coul.

$$V = \frac{1}{4\pi\varepsilon_0}\frac{q}{r} = \frac{(9.0 \times 10^9 \text{ nt} - \text{m}^2/\text{coul}^2)(79)(1.6 \times 10^{-19} \text{coul})}{6.6 \times 10^{-15} \text{ meter}}$$

$$V = \mathbf{1.7 \times 10^7 \text{ volts.}}$$

2.10. A GROUP OF POINT CHARGES

The potential at any point due to a group of point charges is found by (*i*) calculating the potential V_n due to each charge, as if the other charges were not present, and (*ii*) adding the quantities so obtained, we may write equation (9) as,

$$V = \sum_n V_n = \frac{1}{4\pi\varepsilon_0}\sum_n \frac{q_n}{r_n} \quad ...(10)$$

where q_n is the value of the *n*th charge and r_n is the distance of this charge from the point in question. The sum used to calculate V is an *algebraic sum* and not a vector sum like the one used to calculate E for a group of point charges (equation 6), *which gives an advantage of using potential over electric field strength for making calculatins.*

If the charge distribution is continuous, rather than being a collection of points, the sum in Equation 10 is replaced by an integral, and given by

$$V = \int dV = \frac{1}{4\pi\varepsilon_0}\int \frac{dq}{r} \quad ...(11)$$

where dq is a differential element of the charge distribution, r is its distance from the point at which V is to be calculated, and dV is the potential it establishes at that point.

Example 5. *What is the potential at the centre of the square of Fig. 2.16? Assume that $q_1 = +1.0 \times 10^{-8}$ coul, $q_2 = -2.0 \times 10^{-8}$ coul, $q_3 = +3.0 \times 10^{-8}$ coul, $q_4 = +2.0 \times 10^{-8}$ coul, and $a = 1.0$ metre.*

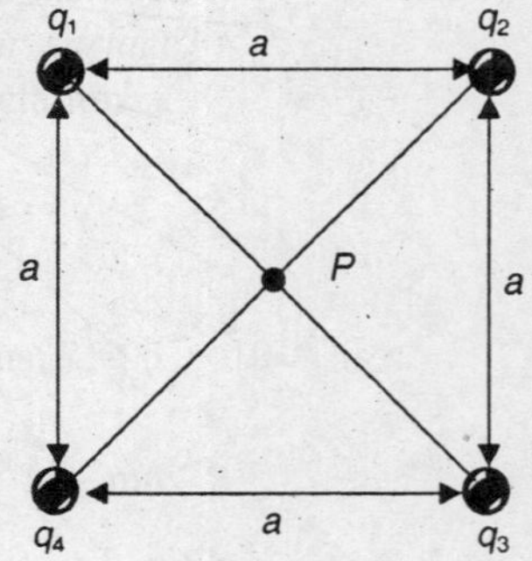

Fig. 2.16

Solution: The distance r each charge from P is $a/\sqrt{2}$ or 0.71 metre. From Equation 10.

$$V = \sum_n V_n = \frac{1}{4\pi\varepsilon_0}\frac{q_1 + q_2 + q_3 + q_4}{r}$$

$$= \frac{(9.0 \times 10^9 \text{ N} - \text{m}^2/\text{C}^2)(1.0 - 2.0 + 3.0 + 2.0) \times 10^{-8}\text{C}}{0.71 \text{ m}}$$

$$= \mathbf{500 \text{ volts}}$$

2.11. POTENTIAL DUE TO AN ELECTRIC DIPOLE

Let AB be an electric dipole of length $2a$ and let P be any point where $OP = r$ and let θ be the angle between r and the dipole axis.

In ΔOAC, $\cos\theta = \frac{OC}{OA} = \frac{OC}{a}$

$\therefore$ $OC = a\cos\theta$

Similarly, $OD = a\cos\theta$

If $r >> a$,

then $PA = PC = PO + OC = r + a\cos\theta$

and $PB = PD = PO - OD = r - a\cos\theta$

If V be the potential at P due to electric dipole then

$$V = \frac{1}{4\pi\varepsilon_0}\left(\frac{q}{PB} - \frac{q}{PA}\right)$$

$$= \frac{1}{4\pi\varepsilon_0}\left[\frac{1}{(r - a\cos\theta)} - \frac{1}{(r + a\cos\theta)}\right]$$

$$= \frac{q}{4\pi\varepsilon_0}\left[\frac{r + a\cos\theta - r + a\cos\theta}{r^2 - a^2\cos^2\theta}\right]$$

$$= \frac{q}{4\pi\varepsilon_0}\left[\frac{2a\cos\theta}{r^2 - a^2\cos^2\theta}\right]$$

$$= \frac{1}{4\pi\varepsilon_0}\cdot\frac{2qa\cos\theta}{r^2 - a^2\cos^2\theta}$$

$$= \frac{1}{4\pi\varepsilon_0}\frac{p\cos\theta}{r^2 - a^2\cos^2\theta}$$

where $p = 2qa$ = dipole moment of electric dipole

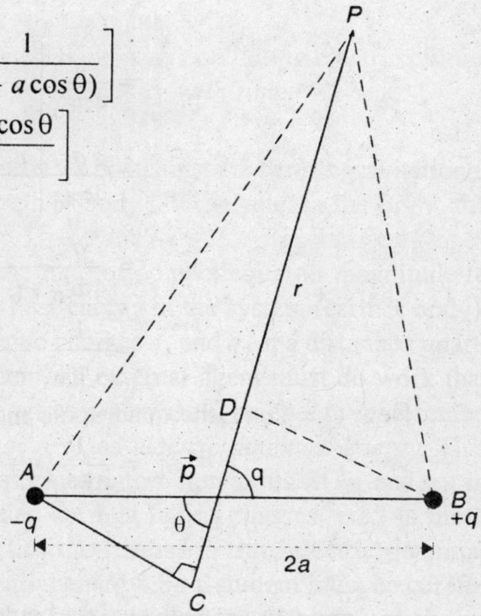

Fig. 2.17 (*a*)

$$\therefore \quad V = \frac{1}{4\pi\varepsilon_0}\cdot\frac{p\cos\theta}{r^2\left(1 - \frac{a^2}{r^2}\cos\theta\right)}$$

As $r >> a$

$\therefore \quad \frac{a}{r} << 1$, so neglect $\frac{a^2}{r^2}\cos^2\theta$ as compared to 1,

we have, $$V = \frac{1}{4\pi\varepsilon_0}\cdot\frac{p\cos\theta}{r^2} \quad ...(1)$$

SPECIAL CASES

(*i*) When the point P lies on the axial line of the dipole on the side of positive charge, $\theta = 0$,

$\therefore \cos\theta = 1$

So from equation (1)

$$V = \frac{1}{4\pi\varepsilon_0}\cdot\frac{p}{r^2} \quad ...(2)$$

(*ii*) When the point P lies on the axial line of the dipole but on the side of negative charge, $\theta = 180°$ or $\cos\theta = -1$

$$\therefore \quad V = \frac{1}{4\pi\varepsilon_0}\cdot\frac{p}{r^2} \quad ...(3)$$

(*iii*) When the point P lies on the equitorial line of the dipole, $\theta = 90°$ or $\cos\theta = 0$

$$\therefore \quad V = 0 \quad ...(4)$$

which means that no work is done in bringing a charge from infinity to the dipole along the perpendicular bisector of the dipole.

ELECTRIC QUADRUPOLE

An electric quadrupole consists of *two electric dipoles placed end to end along the same line.*

OA and *OB* are two dipoles each of length *a*. It may be noted that

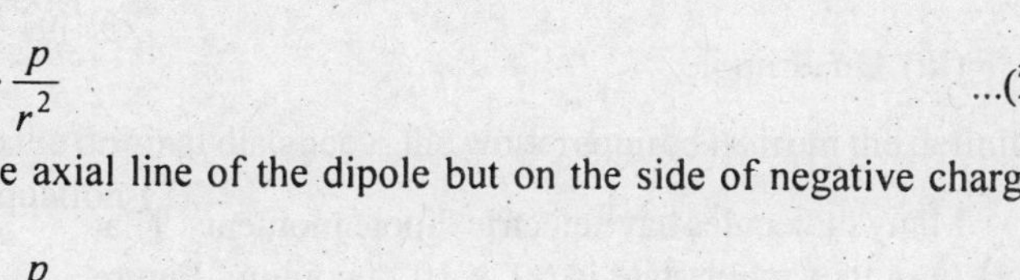

Fig. 2.17 (*b*)

though the total charge on the system as a whole is zero, *the potential and intensity are not zero.* Let us consider a point P along the axis of the quadrupole.

It is clear from the Fig. 2.17 (*b*) the potential at the point P is given by,

$$V = \frac{1}{4\pi\varepsilon_0}\left[\frac{q}{(r-a)} - \frac{q}{r} - \frac{q}{r} + \frac{q}{r+a}\right]$$

$$= \frac{q}{4\pi\varepsilon_0}\left[\frac{r(r+a)-2(r^2-a^2)+r(r-a)}{r(r^2-a^2)}\right]$$

$$= \frac{q}{4\pi\varepsilon_0}\left[\frac{r^2+ra-2r^2+2a^2+r^2-ra)}{r(r^2-a^2)}\right]$$

$$= \frac{q}{4\pi\varepsilon_0}\frac{2a^2}{r(r^2-a^2)}$$

$$= \frac{1}{4\pi\varepsilon_0}\cdot\frac{2qa^2}{r(r^2-a^2)}$$

New $Q = 2qa^2$, the quadrupole moment of charge distribution

$$\therefore \qquad V = \frac{1}{4\pi\varepsilon_0}\cdot\frac{Q}{r(r^2-a^2)} = \frac{1}{4\pi\varepsilon_0}\cdot\frac{Q}{r^3(1-a^2/r^2)}$$

If $r >> a$, $\frac{q}{r} << 1$

$\therefore$ $\frac{a^2}{r^2}$ can be neglected as compared to 1

$$\therefore \qquad V = \frac{1}{4\pi\varepsilon_0}\cdot\frac{Q}{r^3} \qquad ...(5)$$

So in case of

(*i*) Monopole (single charge)

$$V \propto \frac{1}{r}$$

(*ii*) Dipole

$$V \propto \frac{1}{r^2}$$

(*iii*) Quadrupole

$$V \propto \frac{1}{r^3}$$

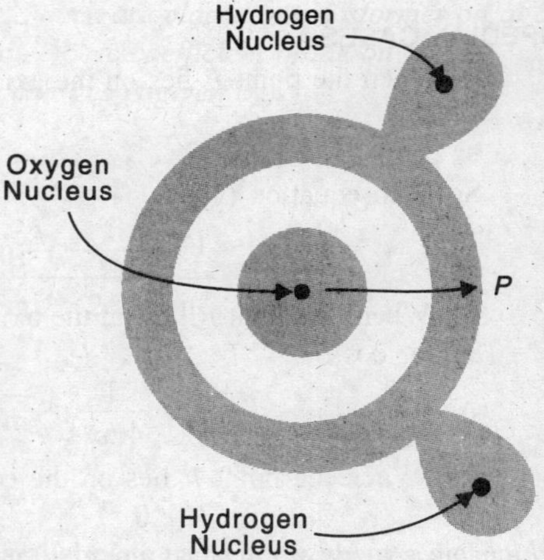

Fig. 2.18

Many molecules have electric dipole moments. That for H_2O in its vapour state is 6.1×10^{-30} coulum. Figure 2.18 is a representation of this molecule, showing the three nuclei and the surrounding electron cloud. The dipole moment p is represented by the arrow on the axis of symmetry of the molecule. In this molecule the effective centre of positive charge does not coincide with the effective centre of negative charge. It is precisely because of this separation that the dipole moment exists.

Fig. 2.19

Atoms, and many molecules, do *not* have permanent dipole moments. However, dipole moments may be induced by placing any atom or molecule in an external electric field. The action of the field (Fig. 2.18) is to separate the centres of positive and of negative charge. We say that the atom becomes *polarized* and acquires an *induced electric dipole moment.* Induced dipole moments disappear when the electric field is removed.

Electric dipoles are important in situations other than atomic and molecular ones. Radio and radar antennas are often in the form of a metal wire or rod in which electrons surge back and forth periodically. At a certain time one end of the wire or rod will be negative and the other end positive.

Half a cycle later the polarity of the ends is exactly reversed. This is an *oscillating* electric dipole. It is so named because its dipole moment changes in a *periodic way with time.*

?.12 ELECTRICAL POTENTIAL ENERGY

If we raise a body from the earth's surface, the work that we do against the earth's gravitational attraction is stored as *potential energy* in the system (earth + body). If we release the body, the stored potential energy changes steadily into kinetic energy as the body drops. After the body comes to rest on the earth, this kinetic energy and just before the time of contact equal in magnitude to the originally stored potential energy, is transformed into heat energy in the system (earth + body).

A similar situation exists in electrostatics. Consider two charges q_1 and q_2 at a distance r apart, as in Fig. 2.19. If we increase the separation between them, an external agent must do work that will be positive if the charges are opposite in sign and negative otherwise. The energy represented by this work can be thought of as stored in the system $q_1 + q_2$ as *electric potential energy.* This energy, like all varieties of potential energy, can be transformed into other forms. If q_1 and q_2, in mechanics the concept of *potential energy* (of compressed springs falling masses, etc.) is more commonly used than the concept *potential.* In electrostatics the reverse is true, electric potential being perhaps a more common concept than electric potential energy. So a student must be careful not to confuse these two different quantities, potential and potential energy for example, are charge of opposite sign and we release them, they will accelerate toward each other, transforming the stored potential energy into kinetic energy of the accelerating masses. The analogy to the earth +the body system is exact, except for the fact that *electric forces may be either attractive or repulsive, whereas gravitational forces are always attractive.*

We define the electric potential energy of a system of point charges as *the work required to assemble this system of charges by bringing them in from an infinite distance. We assume that the charges are all at rest, when they are infinitely separated, that is, they have no initial kinetic energy.*

In Fig. 2.19 let us imagine q_2 removed to infinity and at rest. The *electric potential* at the original site of q_2, caused by q_1, is given by Equation (9),

or $$V = \frac{1}{4\pi\varepsilon_0} \cdot \frac{q_1}{r}$$

If q_2 is moved in from infinity to the original distance r, the work required is, from the definition of electric potential, that is, from Equation (2),

$$W = Vq_2$$

Combining these two equations and knowing that this work W is precisely the *electric potential energy* U of the system $q_1 + q_2$, we get

$$U\ (= W) = \frac{1}{4\pi\varepsilon_0} \cdot \frac{q_1 q_0}{r_{12}} \qquad \text{...(6)}$$

The subscript of r emphasizes that the distance involved is that between the point charge q_1 and q_2.

For systems containing more than two charges the procedure is to compute the potential energy for every pair of charges separately and to add the results algebraically. This procedure rests on a physical picture in which (*a*) charge q_1 is brought into position, (*b*) q_2 is brought from infinity to its position near q_1, (*c*) q_2 is brought from infinity to its position near q_1 and q_2, etc.

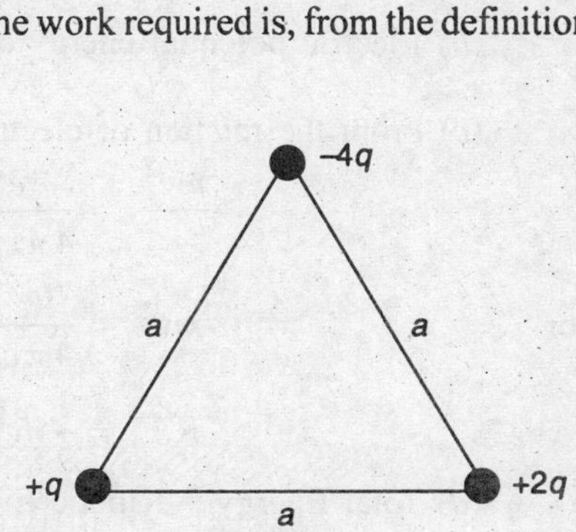

Fig. 2.20

Example 6. *Three charges are arranged as in Fig. 2.20. What is their mutual potential energy? Assume that $q = 1.0 \times 10^{-7}$ coul and $a = 10$ cm.*

Solution: The total energy of the configuration is the sum of the energies of each pair of particles. From Equation (14),

$$U = U_{12} + U_{12} + U_{23}$$

$$= \frac{1}{4\pi\varepsilon_0}\left[\frac{(+q)(-4q)}{a} + \frac{(+q)(+2q)}{a} + \frac{(-4q)(+2q)}{a}\right]$$

$$= -\frac{10}{4\pi\varepsilon_0}\frac{q^2}{a}$$

$$= -\frac{(9.0\times10^9\,\text{N}-\text{m}^2/\text{coul}^2)(10)(1.0\times10^{-7}\,\text{coul})^2}{0.10\,\text{metre}}$$

$$= -9.0\times10^{-3}\,\text{joule.}$$

The fact that the total energy is negative means that negative work would have to be done to assemble this structure, starting with the three charges separated and at rest at infinity. Expressed otherwise, 9.0×10^{-3} joule of work must be done to dismantle this structure, removing the charges to an infinite separation from one another.

Example 7. *The protons in a nucleus of U^{238} are 6.0×10^{-15} metre apart. What is their mutual electric potential energy?*

Solution: From Equation (16) we have

$$U = \frac{1}{4\pi\varepsilon_0}\frac{q_1q_2}{r} = \frac{(9.0\times10^9\,\text{N}-\text{m}^2/\text{coul}^2)(1.6\times10^{-19}\,\text{coul})^2}{6.0\times10^{-15}\,\text{metre}}$$

$$= 3.8\times10^{-14}\,\text{joule} = 2.4\times10^5\,\text{eV}.$$

Example 8. *Calculate (a) the electric potential established by the nucleus of a hydrogen atom at the mean distance of the circulating electron ($r = 5.3 \times 10^{-19}$ metre) (b) the electric potential energy of the atom when the electron is at its radius and (c) the kinetic energy of the electron assuming it to be moving in a circular orbit of this radius centred on the nucleus. (d) How much energy is required to ionize the hydrogen atom? Express all energies in electron volts.*

Solution: (*a*) We have the relation for electric potential as

$$V = \frac{q}{4\pi\varepsilon_0 r} = \frac{9\times10^9\,\text{N}-\text{m}^2/\text{c}^2}{5.3\times10^{-11}\,\text{m}}\times(1.6\times10^{-19}\,\text{C})$$

$$= 27.1\text{ volts.}$$

(*b*) Electric potential energy of the atom is given by

$$U = -\,\text{eV} = -27\text{ eV.}$$

(*c*) From the relation of electrostatic force balancing centripetal force, we have

$$\frac{mv^2}{r} = \frac{e^2}{4\pi\varepsilon_0 r^2}$$

or

$$mv^2 = \frac{e^2}{4\pi\varepsilon_0 r} = 27.1\text{ eV}$$

$$\therefore \qquad \text{K.E.} = \frac{1}{2}nv^2 = \frac{27.1}{2} = 13.55\text{ eV.}$$

(*d*) Total Energy = Kinetic Energy + Potential Energy

$$= (13.55 - 27.1)\text{ eV}$$

$$= -13.55\text{ eV}$$

which is the energy required to ionize hydrogen atom.

Example 9. *A spherical drop of water carrying a charge of 3×10^{-6} coul has a potential of 500 volts at its surface. (a) What is the radius of the drop? (b) If two such drops of the same charge and radius combine to form a single spherical drop. What is the potential at the surface of the new drop formed?*

Solution: (*a*) We have the relation

$$V = \frac{q}{4\pi\varepsilon_0 r} \qquad \text{...(1)}$$

$$\therefore \quad r = \frac{q}{4\pi\varepsilon_0 V} = \frac{(9\times10^9 \text{N}-\text{m}^2)}{500 \text{ volts}} \times 3\times10^{-6}\text{C}$$

$$= 54 \text{ m.}$$

(*b*) Assuming the drops to be incompressible, the volume of the new drop will be twice of the small drop

i.e., $$\frac{4}{3}\pi R^3 = 2\times\frac{4}{3}\pi r^3$$

$\therefore$ Radius of new drop $R = 2^{1/3}\, r$

and Charge of new drop $Q = 2q$

$\therefore$ potential at the surface of new drop will be

$$V_0 = \frac{Q}{4\pi\varepsilon_0 R} = \frac{2q}{4\pi\varepsilon_0 2^{1/3} r} = \frac{2^{2/3} q}{4\pi\varepsilon_0 r}$$

$$= 2^{2/3}\text{V} \quad \text{[from (1)]}$$

$$\therefore \quad V_0 = 2^{2/3} \times 500 = \mathbf{790 \text{ volts.}}$$

Example 10. *If the earth had a net charge equivalent to 1 electron/m² of surface area (a) What would the Earth's potential be? (b) What would the electric field due to the Earth be just outside its surface? (Earth's radius = 6.4 × 10⁶ metres).*

Solution: (*a*) Total charge on the Earth's surface

$$q = (-e)\,4\pi r^2$$

where $-e$ is the electron charge and r is the radius of the Earth. The potential is given by

$$V = \frac{q}{4\pi\varepsilon_0 r} = -\frac{4\pi r^2 .e}{4\pi\varepsilon_0 r} = -\frac{er}{\varepsilon_0}$$

$$= -\frac{1.6\times10^{-19}\text{C}\times6.4\times10^6\text{m}}{8.9\times10^{-12}\text{C/N}-\text{m}^2}$$

$$= \mathbf{-0.115 \text{ volt.}}$$

(*b*) Electric field due to Earth just outside its surface is given by

$$E = \frac{q}{4\pi\varepsilon_0 r^2} = -\frac{4\pi r^2 .e}{4\pi\varepsilon_0 r^2} = -\frac{e}{\varepsilon_0}$$

$$= -\frac{1.6\times10^{-19}\text{C}}{8.9\times10^{-12}\text{C}^2/\text{N}} = 1.8\times10^{-6}\text{N}$$

Negative sign shows that the direction of the field is radially inward.

Example 11. *Two metal spheres are 3.0 cm in radius and carry charges of +1.0 × 10⁻⁸ coul and – 3.0 × 10⁻⁸ coul respectively, assumed to be uniformily distributed. If their centres are 2.0 metres apart calculate (a) the potential of the point halfway between their centres and (b) the potential of each sphere.*

Solution: The charges are assumed to be located at the centres of each sphere. The midway point will be 1.0 metre from either centres and the potential at mid point is given by

$$V = \frac{q_1}{4\pi\varepsilon_0 r_1} + \frac{q_2}{4\pi\varepsilon_0 r_2}$$

But in this case $r_1 = r_2 = r_0$ (say)

$$\therefore \quad V = \frac{1}{4\pi\varepsilon_0 r_0}[q_1 + q_2]$$

$$= \frac{(9 \times 19^9 \text{N} - \text{m}^2/\text{c}^2)(1 \times 10^{-8} - 3 \times 10^5)}{(1.0 \text{ m})}$$

$$= -\ \mathbf{180\ volts.}$$

(*b*) The potential at the surface of sphere 1 is due to its own charge q_1 plus that of charge q_2 of sphere 2 acting at a distance r.

$$V = \frac{q_2}{4\pi\varepsilon_0 R} + \frac{q_1}{4\pi\varepsilon_0 r}$$

$$= (9 \times 10^9 \text{N} - \text{m}^2/\text{c}^2)\left[\frac{1 \times 10^{-8}\text{C}}{0.03 \text{ m}} + \frac{3 \times 10^{-8}\text{C}}{2.0 \text{ m}} \text{m}\right]$$

$$= -\ \mathbf{2.865\ volts.}$$

The potential at the surface of sphere 2 is due to its own charge q_2 plus that of q_1 of sphere 1 acting at a distance r.

$$V = \frac{q_2}{4\pi\varepsilon_0 R} + \frac{q_1}{4\pi\varepsilon_0 r}$$

$$= (9 \times 10^9 \text{N} - \text{m}^2/\text{c}^2)\left[\frac{-3 \times 10^{-8}\text{C}}{0.03 \text{ m}} + \frac{1 \times 10^{-8}\text{C}}{2}\right]$$

$$= -\ \mathbf{8955\ volts.}$$

Example 12. *What is the potential gradient in volts/meter at a distance of 10^{-12} meter from the centre of the gold nucleus? What is the gradient at the nuclear surface?*

Solution: Potential gradient at a distance r is given by the relation

$$\frac{q}{4\pi\varepsilon_0 r^2} = \frac{79 \times 1.6 \times 10^{-19}\text{C}}{(10^{-12}\text{m})^2} \times 9 \times 10^9 \text{N} - \text{m/c}^2$$

$$= 1.14 \times 10^{17} \text{ volts/meter}$$

The potetnial gradient at the surface of gold nucleus of radius $R = 5 \times 10^{-15}$ is

$$= \frac{q}{4\pi\varepsilon_0 R^2} = \frac{79 \times 1.6 \times 10^{-19}\text{c}}{(5 \times 10^{-15} m)^2} \times 9 \times 10^9 \text{N} - \text{m/c}^2$$

$$= \mathbf{4.55 \times 10^{21}\ V/m}$$

Example 13. *A sphere of radius 10 cm is charged to a potential of 10 e.s.u. Calculate the outward pull per unit area.*

Solution: $V = 10$ e.s.u., $r = 10$ cm, $V = \dfrac{q}{r}$

$\therefore$ $q = V \cdot r = 100$ e.s.u.

Also $E = \dfrac{q}{r^2} = \dfrac{100}{10^2} = 1$ unit

Force/unit area $= \dfrac{KE^2}{8\pi}$ in C.G.S. units

here, $K = 1, E = 1$

$\therefore$ $F = \dfrac{1 \times 1^2}{8\pi} = \dfrac{1}{8\pi} = \mathbf{0.0397\ dynes/cm^2}$

Example 14. *Surface density of charge of a conductor in air is 4 stat coulomb/cm^2. Calculate the outwards force/unit area on the conductor in (i) C.G.S. units (ii) M.K.S. units.*

Solution: (*i*) In C.G.S. units

$$F = \frac{2\pi\sigma^2}{K}$$

$$a = 4 \text{ stat coulomb/cm}^2, K = 1$$

$$\therefore \qquad F = \frac{2\pi \times 4^2}{1} = \mathbf{100.57\ dynes/cm^2.}$$

(*ii*) In M.K.S. or S.I. units

$$F = = \frac{\sigma^2}{2\varepsilon_r \varepsilon_0}$$

Here, $$\sigma = \frac{4 \times 10^4}{3 \times 10^9} = 1.33 \times 10^{-5}\ \text{coulomb/m}^2$$

$$\varepsilon_r = 1 \text{ for air and } \varepsilon_0 = 8.85 \times 10^{-12}\ \text{coul}^2/\text{newton-m}^2$$

$$\therefore \qquad \sigma = \frac{(1.33 \times 10^{-5})^2}{2 \times 1 \times 8.85 \times 10^{-12}}$$

$$= \mathbf{10.057\ Newton/m^2.}$$

Example 15. *Protons in the cosmic rays strike the earth's upper atmosphere at a rate, averaged over the earth's surface of 0.15 protons/cm-sec. What total current does the earth receive beyond its atmosphere in the form of incident cosmic ray protons? The earth's radius is 9.4 × 10⁴ metres.*

Solution: No. of protons/m^2 falling over earth's surface

$$n = 0.15 \times 10^4 = 1500$$

No. of protons received by the total earth's surface having radius R

$$N = 4\pi R^2 n$$

$$= (4\pi)(6.4 \times 10^6)^2 \times 1500$$

$$= 7.72 \times 10^{17}/\text{sec}$$

Now each proton has a charge $q = 1.6 \times 10^{-19}$ Coulomb, so total current received by earth.

$$i = Nq = (7.72 \times 10^{17} \times 1.6 \times 10^{-19})\ \text{amp.}$$

$$= \mathbf{0.1235\ amp}$$

Example 16. *Each of two small spheres is charged positively, the combined charge being 5.0 × 10⁻⁵ coul. If each sphere is repelled from other by a force 1.0 newton when spheres are 2.0 metres apart, how is the total charge distributed between the spheres.*

Solution: Let the two charges be q and $(5 \times 10^{-5} - q)$ respectively.

$$\therefore \qquad F = \frac{q_1 \times q_2}{4\pi\varepsilon_0 r^2}$$

or $$1\text{N} = \frac{q \times (5 \times 10^{-5} - q) \times (9 \times 10^9\ \text{N} - \text{m}^2/\text{coul}^2)}{(2\ \text{metre})^2}$$

or $\quad 9 \times 10^{-9} q^2 - 45 \times 10^4 q + 4 = 0$

which is a quadratic equation and equating it we get

$$q_1 = \mathbf{1.2 \times 10^{-5}\ coul}$$

$$q_2 = \mathbf{3.8 \times 10^{-5}\ coul.}$$

Example 17. *A thin walled metal sphere has a radius 25 cm and carries a charge of 2.0 × 10⁻⁷ coul. Find E for a point (a) inside the sphere (b) just outside the sphere (c) 3.0 metres from the centre of the sphere.*

Solution: (*a*) Since no charge is enclosed within the charged metal sphere, so E for a point inside the sphere is **zero.**

(*b*) For calculating the field just outside the sphere we may consider the total charge to be concentrated at the centre.

$$E = \frac{q}{4\pi\varepsilon_0 r^2} = \frac{(9\times10^9\,\text{N}-\text{m}^2/\text{coul}^2)\times(2\times10^{-7}\,\text{coul})}{(0.25\ \text{m})^2}$$

$$= 2.9\times10^4\ \text{N/coul.}$$

(*c*)

$$E = \frac{q}{4\pi\varepsilon_0 r^2} = \frac{(9\times10^9\,\text{N}-\text{m}^2/\text{coul}^2)\times(2\times10^{-7}\,\text{coul})}{(3\ \text{m})^2}$$

$$= \mathbf{200\ N/coul.}$$

Example 18. *An infinite charge sheet has a surface density* σ *of 1.0 × 10* ^{7} *coulomb/m*^{2}*. How far apart are the equipotential surfaces whose potential differ by 5.0 volts?*

Solution: The field intensity is given by

$$E = \frac{\sigma}{2\varepsilon_0} = \frac{1.0\times10^{-7}\ \text{coulomb/m}^2}{2\times8.9\times10^{-12}\ \text{coulomb}^2/\text{N-m}^2}$$

$$= 5.6\times10^3\ \text{N/coulomb}$$

Let the equipotentials be dS metres apart,

∴ we have

$$E.dS = dV$$

or

$$dS = \frac{dV}{E} = \frac{5.9\ \text{volt}}{5.6\times10^3\ \text{N/coulomb}}$$

$$= 0.89\times10^{-4}\ \text{metres}$$

$$= \mathbf{0.89\ mm}$$

Example 19. *A charge of 10* ^{8} *coulomb can be produced by simple rubbing. To what potential would such a charge raise an insulated conducting sphere of 10 cm radius?*

Solution:

$$V = \frac{q}{4\pi\varepsilon_0 R} = \frac{10^{-8}\,\text{coulomb}\times9\times10^9\,\text{N}-\text{m}^2/\text{coulomb}^2}{0.1\ \text{m}}$$

$$= \mathbf{900\ volts}$$

Example 20. *Consider a point charge with q = 1.5 × 10* ^{8} *coulomb (a) What is the radius of equipotential surface having a potential of 30 volts (b) Are surfaces whose potentials differ by a constant amount (say 1.0 volt) evenly spaced in radius?*

Solution: (*a*) Radius can be calculated from the relation

$$R = \frac{q}{4\pi\varepsilon_0 V} = \frac{(1.5\times10^{-8}\ \text{coul})\times9\times10^9\ \text{N}-\text{m}^2/\text{coulomb}^2}{(30\ \text{volts})}$$

$$= \mathbf{4.5\ metres.}$$

(*b*) We have the potential given by

$$V = \frac{q}{4\pi\,\varepsilon_0\,R}$$

$$\therefore \quad dV = -\frac{q}{4\pi\,\varepsilon_0}\frac{dR}{R^2}$$

or

$$dR = -\frac{4\pi\,\varepsilon_0}{q}\cdot dV\cdot R^2$$

From which it is clear that dR depends on R, so the surface whose potentials differ by a constant amount are not evenly spaced.

Example 21. *Calculate the potential energy of the charges shown in Fig. 2.21.*

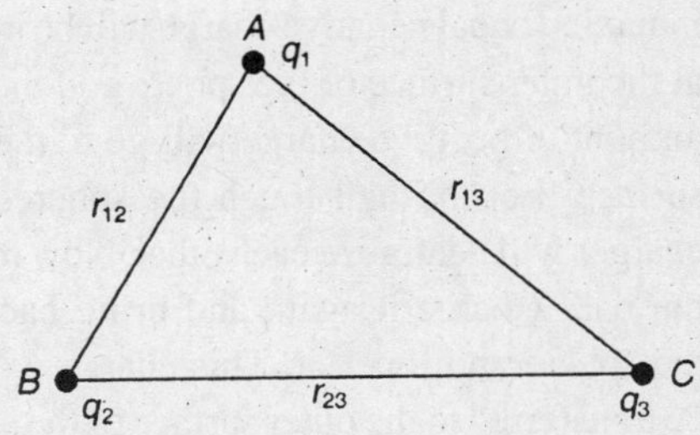

Fig. 2.21

Solution: Let us assume that charge q_1 is already present. When we bring q_2 to B then,

$$\text{Work done} = \frac{q_1}{4\pi\,\varepsilon_0\, r_{12}} q_2$$

Now, the potential at C due to both q_1 and q_2

$$= \frac{1}{4\pi\,\varepsilon_0}\left[\frac{q_1}{r_{13}} + \frac{q_2}{r_{23}}\right]$$

Therefore, work done in bringing charge q_3 at C in the presence of both q_1 and q_2

$$= \frac{1}{4\pi\,\varepsilon_0}\left[\frac{q_1}{r_{13}} + \frac{q_2}{r_{23}}\right] q_3$$

We can also think of it as the sum of the work needed when q_1 is present alone, and that needed when q_2 is present alone. Therefore, total work done in assembling this arrangement of three charges is;

$$W = \frac{1}{4\pi\,\varepsilon_0}\left[\frac{q_1 q_2}{r_{12}} + \frac{q_1 q_3}{r_{13}} + \frac{q_2 q_3}{r_{23}}\right]$$

Example 22. *Two conducting spheres of radii R_l and R_2 are at the same potential. Compare their charges, surface charge densities, and electric fields just outside them.*

Solution: Let q_1 and q_2, respectively, be the charges on the spheres 1 and 2. Therefore:

$$\frac{1}{4\pi\,\varepsilon_0}\frac{q_1}{R_1} = \frac{1}{4\pi\,\varepsilon_0}\frac{q_2}{R_2}$$

or

$$\frac{q_1}{q_2} = \frac{R_2}{R_2}$$

Thus, charges are proportional to the radii. Now surface charge density on sphere 1,

$$\sigma_1 = q_1/4\pi\, R_1^2$$

Surface charge density on sphere 2,

$$\sigma_2 = q_2/4\pi\, R_2^2$$

$$\frac{\sigma_1}{\sigma_2} = \frac{q_1}{q_2}\times\frac{R_2^2}{R_1^2} = \frac{R_1}{R_2}\times\frac{R_2^2}{R_1^2} = \frac{R_2}{R_1}$$

Therefore, the surface charge densities are inversely proportional to the radii.

Electric field just outside sphere 1,

$$E_1 = q_1/4\pi\,\varepsilon_0\, R_1$$

Electric field just outside sphere 2,

$$E_2 = q_2/4\pi\,\varepsilon_0\, R_2$$

$$\therefore \quad \frac{E_1}{E_2} = \frac{R_2}{R_1}$$

Hence, the electric fields just out side the two sphere as are also inversely proportional to the radii.

2.13. VAN DE GRAAFF GENERATOR (PRINCIPLE AND WORKING)

The principle of this generator can be understood with the help of Fig. 2.22. Let us have a hollow conducting sphere having a small hole. Take a charged conducting ball inside it. Due to induction equal negative charge will be induced at the inner surface of the sphere and the same amount of positive charge will go to the outer surface. Let the ball touch the sphere. Inner charges will neutralize each other. Now remove the ball. Charge it again and bring back into the sphere and touch it. This charge will also be transferred to the outer surface of the sphere. By repeating this process a number of times, we can increase the charge at the conductor to as high a value as we desire. Along with this, the potential of the sphere will also increase. The ultimate value is limited by the discharge in the air surrounding the sphere due to very high electric field. The other limiting factor is the leakage of charge through supports. etc.

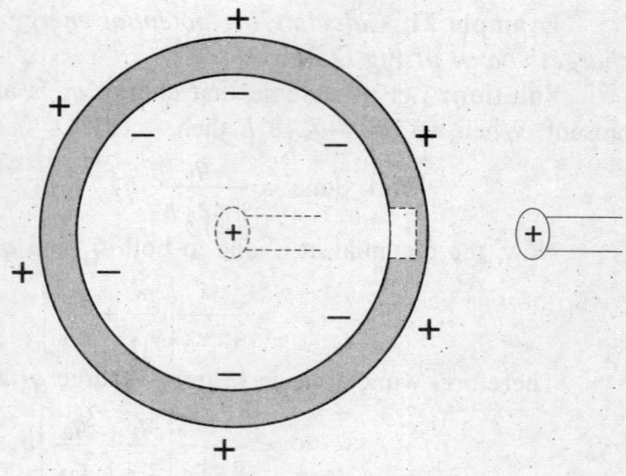

Fig. 2.22

Schematic diagram of Van de Graaff generator, based on the above principle, is shown in Fig. 2.23. C is a conducting hollow sphere; B is a belt of an insulating fabric and is mounted over two pulleys, P_1 and P_2. This belt is made to rotate rapidly by driving the pulley P_1 by a motor M. Terminal S_1, which consists of 1 large number of sharp points, is maintained at a high potential (positive in the figure) of about 5 to 20 kV relative to ground with the help of a high tension battery HT. This terminal is called the spray point. Corona* from these points charges the belt immediately opposite S_1. As the belt rotates, this charge rises continuously to another spray point S_2, which removes the charge to C. The voltage on C continuously rises, till the loss of charge by leakage equals the rate at which it is introduced. The charge leakage can be minimized and the maximum attainable voltage increased by enclosing the apparatus in a steel chamber T and by filling this chamber with nitrogen to pressure of about 10 atmospheres.

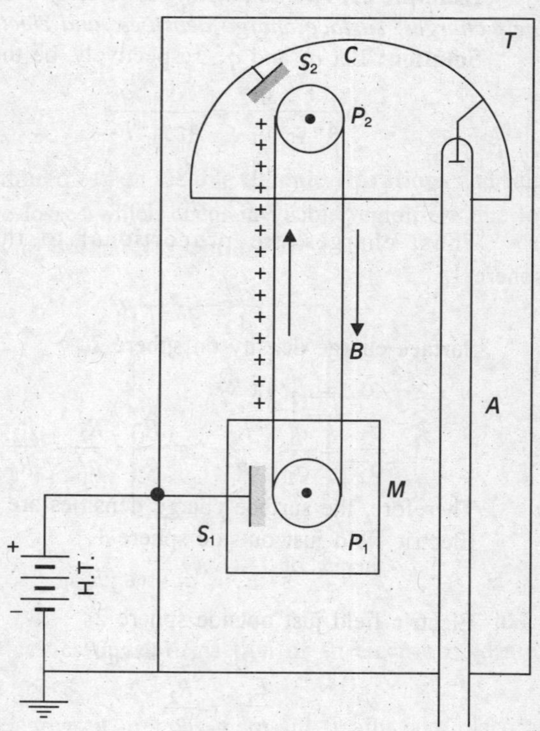

Fig. 2.23

A is an accelerator tube inside the shell, and is vacuated to a very low pressure. Positive ions from a small vessel containing ionized hydrogen, or deuterons, are passed on to the upper region of this accelerator tube, and then accelerated to very high energies by the shell. To accelerate the electrons, C is to be charged negatively. This is done by interchanging the connections of HT. Voltage upto about 5 million volts have been achieved by this machine, and some machines go up to 20 MV.

Van de Graaff generator is used to accelerate the ion beams to very high energies, which are used to study nuclear reaction. High energy electrons are made to strike targets which produce very hard X-rays.

*From example it is concluded that near sharp points the values of σ and E are very large. Because of this large electrostatic field in the air near these points, ions (both positive and negative) are formed and a spray of positive ions is repelled from the points. The process of ionization is started by a small number of ions always present in the air produced by cosmic rays and radiation. These ions get accelerated to high velocities, and produce more ions during collisions with other.

QUESTIONS AND PROBLEMS

1. Give an expression for the flux in an electric field.

2. State and prove Gauss's theorem in electrostatics.

3. State and prove Gauss's theorem in electrostatics. Hence, find an expression for the mechanical force acting on a unit area of a charged conductor.

4. State Gauss's theorem and use it to show that the electric intensity near an infinite charge conducting surface is $\sigma/2\varepsilon_0$ where σ is the density of charge.

5. What do you mean by electric displacement in a dielectric medium? Show that the energy stored per unit volume of a medium of dielectric constant ε_0 when subjected to an electric field of strength E is $\dfrac{\varepsilon_0 E^2}{2}$.

6. Deduce an expression for the force per unit area of the surface of a charged conductor due to its charge. Hence find the energy per unit volume in the medium.

7. Calculate the excess of pressure inside a charged soap bubble of radius r.

8. Prove that the excess of pressure inside a charged soap bubble $= \dfrac{rT}{r} - \dfrac{q^2}{8\pi r^4}$, where T is the surface tension of soap solution, r is the radius of the bubble and q is the charge on the surface of the bubble.

9. Obtain an expression for the electrical energy per unit volume in an electrostatic field.

10. State Gauss's theorem and deduce an expression for the capacity of a long cylindrical condenser.

11. Find an expression for the mechanical force per unit area of the surface of a charged conductor.

12. Deduce an expression for the outward pull per sq. cm. of the surface of a charged conductor.

13. State and prove Gauss's theorem in electrostatics. Apply this theorem to calculate the intensity near charged cylindrical conductor.

14. State and prove Gauss's theorem in electrostatics. Use this theorem to find the intensity at a point near a plane sheet of charge.

15. Show that the outward force per unit area of the surface of a charged conductor due to its charge is given by $\varepsilon E^2/2$ N/m^2.

16. Find an expression for the mechanical force per unit area on the surface of a charged conductor and hence derive the expression for the energy per unit volume of the medium.

17. Find the electric intensity at points inside and outside uniformly charged sphere.

18. Calculate the charge on a soap bubble of radius 2 cm, if the air pressure is the same inside and outside the bubble. Surface tension = 30 dynes/cm. **[Ans.** 155.3 e.s.u.]

19. The pressure of air inside an electrically charged soap bubble of radius 0.015 m is the same as that of outside. If the surface tension of soap solution is 27×10^{-3} newton/m, calculate the potential of the bubble. **[Ans.** 1.914×10^4 volts]

20. Calculate the electric intensity required to just support a water droplet of mass 10^{-2} gm and having a charge of 9.8×10^{-3} e.s.u. **[Ans.** 100 e.s.u.]

21. In an atom the distance between an electron and a proton is 2.12×10^{-19}m. Calculate the electrical force of attraction between the charge on electron $= 1.6 \times 10^{-19}$ coulomb and $1/4\pi\varepsilon_0 = 9 \times 10^9$ N^2/coul^2. **[Ans.** 5.06×10^{-9} N]

22. Distance between the two α-particles is 10^{-3} m. Calculate electrostatic force of repulsion between them. The charge on each α-particle $= 3.2 \times 10^{-19}$ coul. **[Ans** 0.09216 N]

23. A sphere of radius 1 cm is charged to a potential of 3000 volts. Calculate the outward pull per unit area. **[Ans.** 3.98 dynes/cm^2]

CHAPTER

3

CAPACITORS

3.1. CAPACITORS

Capacitance is related to the **ability of an insulated conductor, or a system of insulated conductors, to hold electric charge.** It is equal to the ratio of the charge placed on the conductor to the change in its potential. Consider spherical conductor at zero potential and place a charge q on it. Its potential will rise to $(1/4\pi\varepsilon_0)\ (q/R)$. So its capacity will be given by

$$C \equiv \frac{q}{V} = 4\,\pi\varepsilon_0 R \qquad ...(1)$$

If we keep on adding charges, the potential will go on rising to the point where corona discharge, leakage, etc., will limit the addition of more charge. So, we will prefer a device which can store **more charge at lower potentials** *i.e.*, whose capacity should be large.

If there is an isolated charged spherical conductor, the lines of force will go in all directions, and end up at infinity. So we can imagine our conductor to be surrounded by another spherical conducting shell, which is concentric to it and has infinite radius, Fig. 3.1 (*a*). Thus the lines of force will end up at this conductor of infinite radius and will induce equal and opposite charge on it. Let us now reduce the radius of this outer shell, keeping it connected to infinity by a conducting wire. The assembly now looks something like Fig. 3.1 (*b*). The inner conductor carries a charge $+q$, and the outer conductor carries *a charge* $-\ q$ (it has left *a* charge $+q$ at infinity).

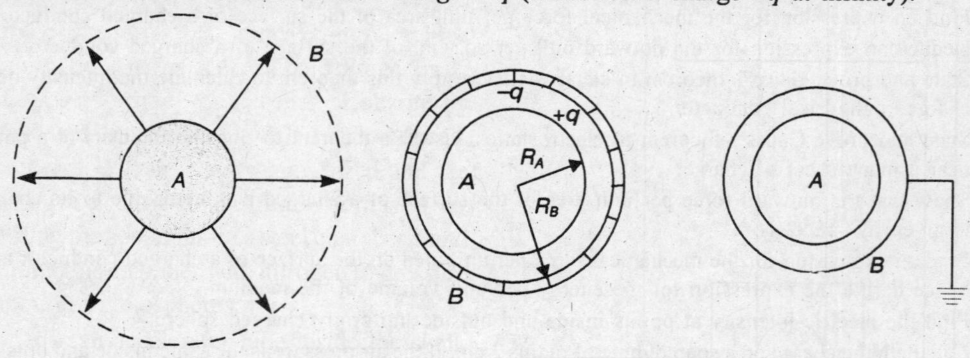

(*a*) (*b*) (*c*)

Fig. 3.1

The resultant potential of conductor A is due to two charges, (*i*) its own charge and (*ii*) the charge on the outer conductor. Potential of A due to its own charge is $(1/4\pi\varepsilon_0)\ \ (+q/R_A)$; and the potential throughout the space inside B due to the charge on its inner surface is $(1/4\pi\varepsilon_0)\ (-\ q/R_s)$. So, the result at potential of A will be

$$V = \left(\frac{q}{R_A} - \frac{q}{R_B}\right)\frac{1}{4\pi\,\varepsilon_0}$$

Hence,

$$C = \frac{q}{\left(\dfrac{q}{R_A} - \dfrac{q}{R_B}\right)\dfrac{1}{4\pi\varepsilon_0}} = \frac{R_A R_B}{R_B - R_A} \times 4\pi\varepsilon_0 \qquad ...(2)$$

So C is definitely larger than $4\pi\varepsilon_0 R_4$ and we can increase it further by decreasing the separation between the two conductors. '*For all practical purposes the zero of the potential is taken as that of the earth*'. As it is only the potential difference which has any meaning, it does not affect our result. So we say that the sphere B is earthed, and Fig. 3.1 (*b*) takes the form of Fig. 3.1 (*c*).

The system of two conductors, like the one shown above, is known as a *capacitor*, and its capacity is defined as

$$C \equiv q/V \quad \text{...(3)}$$

where q is the charge on one of the conductors, and V is the potential of this conductor with respect to the other. The symbol used to represent a capacitor is —| |—.

It should be noted that q is not the total charge on this system of conductors. These two conductors carry equal and opposite charges and, so, the sum of the two is always zero. The function of the second *i.e.,* earthed conductor, is simply to reduce the potential of the first conductor and, thereby, to increase its capacity. But, the concept of capacitance is more general than this. If there are two equal and opposite charges at any two points and there is a potential difference between them, we say that there exists a *capacitance* between these points.

From Equation 3, we see that the units of capacitance are coulomb/volt, which we call 'farad', in honour of **Michael Faraday**. But farad is a very large unit. Submultiple of this unit — microfarad ($1\ \mu F = 10^{-6}$ farad), and picofarad ($1\ pF = 10^{-12}$ farad) are more convenient for practical purposes. One picofarad is also called **micro-micro farad** ($\mu\mu F$).

Capacitors have wide application in electronics—waveshaping circuits, oscillators, timing circuits, etc. These are also used to reduce the power fluctuations, to produce electric field of desired configurations, and to store electrical energy and in almost all the electrical appliances.

3.2. PARALLEL PLATE CAPACITOR

The capacitor most commonly used is the parallel plate capacitor. It consists of two conducting plates parallel to each other (Fig. 3.2).

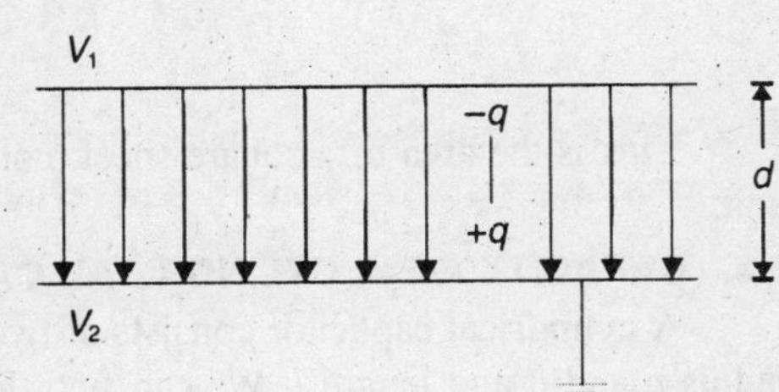

Fig. 3.2

The distance, d, between the plates is quite small as compared with their area, A. So at very small separations, the fringing of electric field at the boundaries can be neglected. If the charge on each plate is q, then the surface charge density is given by s = q/A, and the electric field between the plates is : $E = \sigma/\varepsilon_0$. So the potential difference between the plates is,

$$V_{12} = Ed = \sigma d/\varepsilon_0 = qd/A\varepsilon_0$$

So, the capacitance of this system is given by

$$C = q/V_{12},$$

or

$$C = \varepsilon_0 A/d \quad \text{...(4)}$$

From Equation (4) we see that the units of ε_0 are **farad/metre.** Also, if we know the value of A and d, and if C is measured, the value of ε_0 can be found.

Example 1. (*a*) *A spherical capacitor has a capacity of 1 farad. If the separation between the two spheres is 1 mm, calculate the radius of each sphere.*

(*b*) *A parallel plate capacitor has a capacity of 1 farad and the spacing between the two plates is 0.1 mm. Calculate the area of each plate.*

Both these capacitors have been kept in air.

Solution: (*a*) Let the radius of the inner sphere be R_A metres.

$$\therefore \quad R_S = (R_A + 1.0 \times 10^{-3})$$

From Equation (2) 1 farad × 10^{-3}m = $4\pi \times 8.85 \times 10^{-12}$ farad/m × R_A (R_A + 1.0×10^{-3}) m^2

$$\therefore \quad R_A^2 + 1.0\times10^{-3} R_A = \frac{1\times10^{-3}\times10^{12}}{4\pi\times8.85}$$

or $\quad 10^3 R_A^2 + R_A - 8.99\times10^9 = 0$

If $a R_A^2 + b R_A + c = 0$, which is a quadratic equation

$$R_A = \frac{-b\pm\sqrt{b^2-4ac}}{2a}$$

or $$R_A = \frac{-1\pm\sqrt{1+4\times10^8\times8.99\times10^9}}{2\times10^3}$$

As R_A cannot be negative, we take the positive sign. We also neglect 1 within the square-root sign as it is very small as compared to the other quantity.

$$\therefore \quad R_A = \frac{-1+6\times10^6}{2\times10^3}$$

$$= 3\times10^3 \text{ m} = 3 \text{ km}$$

(*b*) Using Equation (4)

$$A = \frac{1 \text{ farad}\times0.1\times10^{-3}\text{ m}}{8.85\times10^{-12}\text{ farad/m}} = \frac{10^{-4}\times10^{12}}{8.85}\text{ m}^2$$

$$= \mathbf{11.3\times10^6\ m^2}$$

If plates are square, each side = 3.36×10^3 m = **3.36 km**

Example 2. *The parallel plates of an air-filled capacitor are everywhere 1.0 mm apart. What must the plate area be if the capacitance is to be 1.0 farad?*

Solution: From Equation (4)

$$A = \frac{d.C}{\varepsilon_0} = \frac{(1.0\times10^{-3}\text{ meter})(1.0\text{ farad})}{8.9\times10^{-12}\text{ coul}^2/\text{N}-\text{m}^2}$$

$$= 1.1\times10^8 \text{ meter}^2$$

This is the area of a square sheet more than 10 km on edge; which shows farad is quite large unit.

3.3. CAPACITY OF A CYLINDRICAL CAPACITOR

A cylindrical capacitor consists of two coaxial cylinders of radius a and b and length l. We can find the capacitance of this device by assuming that the capacitor is very long (that is, that $l >> b$) so that fringing lines of force at the ends can be ignored for the purpose of calculating the capacitance.

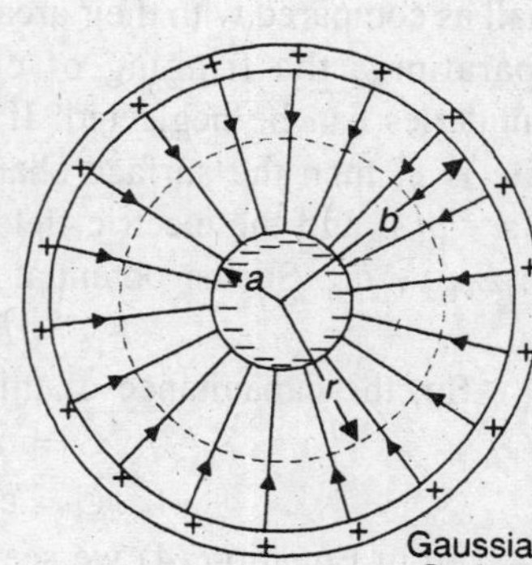

Fig. 3.3

As a Gaussian surface construct a coaxial cylinder of radius r and length l, closed by plane caps. From Gauss's Law,

$$\varepsilon_0 \oint E ds = q$$

We have, $\quad \varepsilon_0 E(2\pi r)(l) = q,$

the flux being entirely through the cylindrical surface and through the end caps. Which gives

$$E = \frac{q}{2\pi\varepsilon_0 rl} \qquad ...(1)$$

The potential difference between the plates is given by

$$V = -\int_a^b E.dl = \int_a^b E\,dr = \int_a^b \frac{q}{2\pi\varepsilon_0 l}\frac{dr}{r} = \frac{q}{2\pi\varepsilon_0 l}\log_e\frac{b}{a}$$

So we have the capacitance

$$C = \frac{q}{V} = \frac{2\pi\varepsilon_0 l}{\log_e (b/a)} = \frac{4\pi \varepsilon_0 l}{2 \times 2.3026 \log_{10} b/a}$$

which is similar to the relation for the parallel-plate capacitors this relation also depends only on geometrical factors.

3.4. CAPACITORS IN PARALLEL

When a number of capacitors are connected, as in Fig. 3.4. these are said to be in parallel. To charge this combination, a battery is connected between A and B. So the potential difference between the plates of each capacitor is the same, and is equal to the potential difference across the battery terminals.

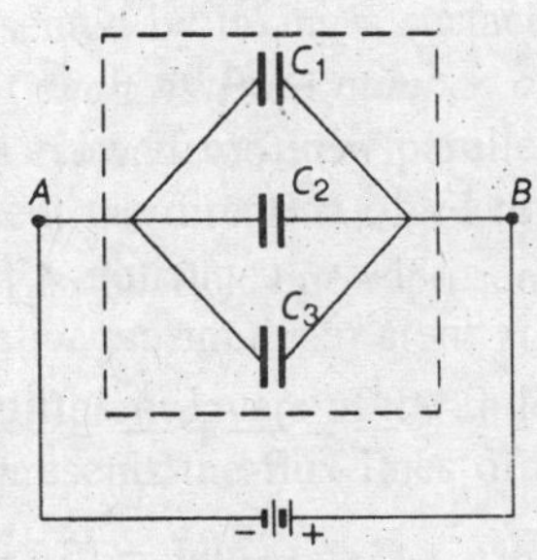

Fig. 3.4

Now, if this combination is enclosed in a box, so that only two terminals A and B are available, how much capacitance will we measure? If we know the battery voltage, V, and the charge, q, on this 'equivalent' capacitor, we can find out capacity using the relation: $C = q/V$. Let, out of this charge q, charge q_1 be on C_1, charge q_2 be on C_2, and so on. Therefore,

$$q = q_1 + q_2 + q_3 + ...,$$

or

$$\frac{q}{V} = \frac{q_1}{V} + \frac{q_2}{V} + \frac{q_3}{V} + ...$$

So we have

$$C = C_1 + C_2 + C_3 + ...$$

i.e., Capacitances in parallel add up to give the equivalent capacitance of the combination. This is the arrangement used in gang capacitors.

3.5. CAPACITORS IN SERIES

Combination of capacitors shown in Fig. 3.5 is known as series combination. In this case from one available terminal (say A) to the other terminal, say B), we have to pass through each capacitor. On the other hand, in parallel combination one would go through any one of the capacitors.

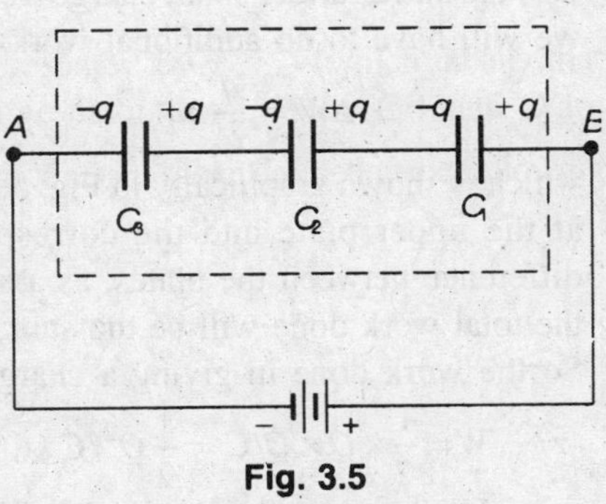

Fig. 3.5

Let this combination be put in a box and a battery be connected across A and B. Let the charge transferred to B be $+q$, and that to A be $-q$. Using the principle of induction we can see that a charge $-q$ will be induced in the second plate of C_1 and so a charge $+q$ will be displaced to the first plate of C_2, and so on.

Let V_1, V_2, etc., be the potential difference across the capacitors C_1, C_2, etc., respectively. So we have

$$V_{BA} = V_1 + V_2 + V_3 + ...$$

$$= \frac{q}{C_1} + \frac{q}{C_2} + \frac{q}{C_3} + ...$$

Let C be the 'equivalent' capacitance of this combination. Then,

$$V_{AB} = q/C$$

So, $$\frac{q}{C} = \frac{q}{C_1} + \frac{q}{C_2} + \frac{q}{C_3} + ...,$$

or $$\frac{1}{C} = \frac{1}{C_1} + \frac{1}{C_2} + \frac{1}{C_3} + ...$$

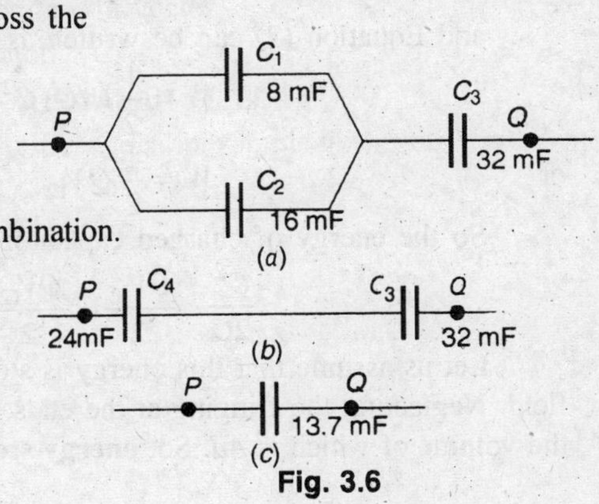

Fig. 3.6

i.e., when capacitors are connected in series, the reciprocals of these capacitances add up to give the reciprocal of the equivalent capacitance.

Example 3. *Calculate the resultant capacitance of the capacitor combination shown in Fig. 3.6.*

Solution. Capacitance C_1 and C_2 are in parallel. Therefore,

$$C_4 = (8+16)\,mF$$

$$= 24\,mF$$

Therefore, the circuit reduces to that shown in Fig. 3.6 (*b*).

Now, capacitors C_4 and C_3 are in series. If C is their resultant capacitance,

$$\frac{1}{C} = \frac{1}{24} + \frac{1}{32} = \frac{4+3}{96},$$

or $$C = \frac{96}{7}\,mF = 13.7\,mF$$

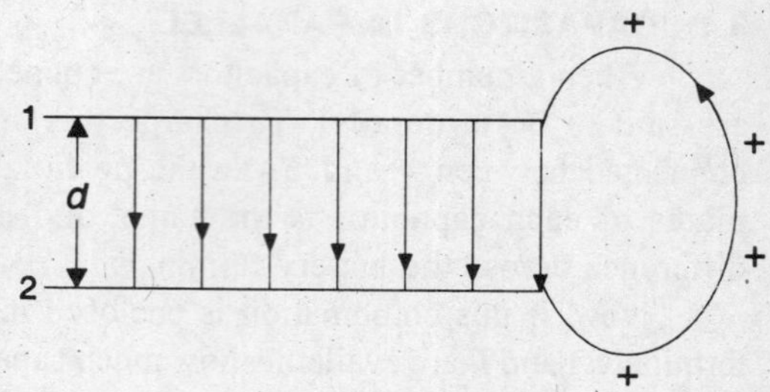

Fig. 3.7

Hence, the capacitance between points P and Q is 13.7 μF, and is shown in Fig. 3.6 (*c*).

3.6. ENERGY STORED IN A CHARGED CAPACITOR

When we charge a capacitor, what we essentially do is we **transfer charge from one plate to the other.** To start with, there is no potential difference between the plates, and as such, no work is done. As soon as some charge, say q, is transferred, a potential difference is built up between the plates which is equal to q/C. Now, to add a further small charge Δq to the upper plate, we will have to do additional work given by

$$\Delta W = \frac{q}{C}\Delta q \qquad ...(1)$$

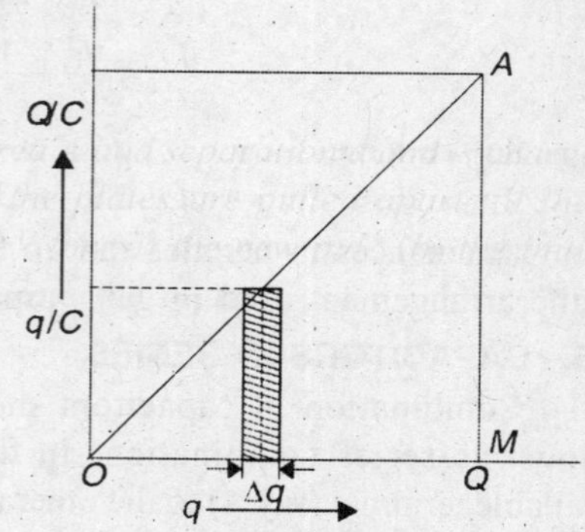

Fig. 3.8

which is shown graphically in Fig. 3.8, between charge at the upper plate and the corresponding potential difference between the plates, as a shaded strip. When the capacitor has been given a charge Q, the total work done will be the sum of all such strips from O to M, which is equal to area OAM. So the work done in giving a charge Q to a capacitor is given by

$$W = \frac{1}{2}\times Q\times Q/C = \frac{1}{2}Q^2/C \qquad ...(2)$$

If V_{12} is the final potential difference between the plates, then

$$Q = CV_{12}$$

and Equation (2) can be written as

$$W = \frac{1}{2}(1/C)\,C^2\,V_{12}^2 \qquad ...(3)$$

or $$W = \frac{1}{2}QV_{12} \qquad ...(4)$$

So the energy of charged capacitor is given by either

$$\frac{Q^2}{2C},\ \frac{CV_{12}^2}{2} \text{ or } \frac{QV_{12}}{2}$$

Let us assume that this energy is stored in the field and compute the energy density of the field. Neglecting the fringing at the ends, E is uniform throughout the space between the plates, the volume of which is Ad. So, **energy stored per unit volume is given by**

$$U_E = \frac{CV_{12}^2}{2} \times \frac{1}{Ad}$$

$$= \frac{1}{2} \times \frac{1}{Ad} \times \frac{\varepsilon_0 A}{d} (Ed)^2 \qquad \left[\because \frac{V}{d} = E\right]$$

$$U_E = \frac{1}{2} \varepsilon_0 E^2 \qquad ...(5)$$

where U_E stands for the **electric energy density.** This equation has been derived for two special cases but we can generalize it. **So, if there is an electric field E at a point in vacuum or air, the energy per unit volume around that point is $\frac{1}{2} \varepsilon_0 E^2$.**

Example 4. *A 1000 μF capacitor is charged to 10 V and a 16 μF capacitor is charged to 400 V. These are then connected in parallel, with plates carrying similar charges connected together. Determine the total stored energy before and after joining them. Is final energy the same as the initial energy? If not, account for the difference.*

Solution: Energy stored in the first capacitor,

$$U_1 = \frac{1}{2} \times 1000 \times 10^{-6} \times (10)^2$$

$$= \frac{1}{2} \times 10^{-1} \text{J} = 0.05 \text{ J}$$

Energy stored in the second capacitor,

$$U_2 = \frac{1}{2} \times 16 \times 10^{-6} \times (400)^2$$

$$= 8 \times 16 \times 10^{-2} = 1.28 \text{ J}$$

Total initial energy stored in both the capacitors,

$$U_i = 0.05 + 1.28 = \mathbf{1.33\ J}$$

Charge on the first capacitor,

$$q_1 = 1000 \times 10^{-5} \times 10$$

$$= 10^{-2} C$$

Charge on the second capacitor,

$$q_2 = 16 \times 10^{-6} \times 400$$

$$= 64 \times 10^{-4} C$$

$\therefore$ Total charge = 0.01 + 0.0064 = **0.0164 C**

Final capacitance = 1000 + 16 = 1016 μF

So potential difference across this resultant capacitor

$$= \frac{0.0164 \text{ C}}{1016 \times 10^{-6} \text{ farad}} = 16.14 \text{ V}$$

and energy stored in this capacitor,

$$U = \frac{1}{2} \times 1016 \times 10^{-5} \times (16.14)^2 = \mathbf{0.132\ J}$$

which shows that the final energy is less than the initial energy. When the capacitors are connected in parallel, with wires having high resistance, the **energy will appear as heat,** when charge flows from higher potential to the lower potential, through these wires. If, on the other hand, the wires have negligible resistance, these charges reach the low potential capacitor with quite high velocities. On reaching the plate, these are **suddenly decelerated to zero velocity.** This deceleration of charged particles produces **electromagnetic waves.** So the missing energy is radiated in the form of **electromagnetic waves.**

3.7. PARALLEL PLATE CAPACITOR WITH DIELECTRIC

Equation (4) of article 3.2 holds only for a parallel plate capacitor with its plates in a vacuum.

Michael Faraday, in 1837, first investigated the effect of filling the space between the plates with a dielectric, say mica or oil.

Fig. 3.9

For a parallel plate capacitor we can write, as an experimental result

$$C = \frac{\varepsilon_r \varepsilon_0 A}{d} \quad ...(1)$$

(where ε_r is dielectric constant of the material)

Equation (4) is a special case of this relation found by putting $\varepsilon_r = 1$, corresponding to a vacuum between the plates. Experiment shows that the capacitance of *all* types of capacitors is increased by the factor ε_r if the space between the plates is filled with a dielectric. Thus the capacitance of any capacitor can be written as

$$C = \varepsilon_0 \varepsilon_r L \quad ...(2)$$

where L depends on the geometry and has the dimension of a length. For a parallel plate capacitor L is A/d, for a cylindrical capacitor it is

$$\frac{2\pi l}{\log_e (b/a)}$$

3.8. PARALLEL PLATE CONDENSER WITH DIELECTRIC OF VARYING THICKNESSES

Capacity of a parallel plate condenser with a dielectric in between the plates. Let P and Q be two parallel plates, the distance between them being d and let a dielectric slab of dielectric constant K of thickness t be placed between them.

Now since thickness of the slab = t and its dielectric constant = ε_r

$\therefore$ equivalent air thickness = t/ε_r

Effective air distance between the plates A and B = $[d - (t - t/\varepsilon_r)]$. In rationalised M.K.S. units or S.I. units capacity of the condenser

$$C = \frac{1}{4\pi} \frac{\varepsilon_0 A}{[d - (t - t/\varepsilon_r)]} \text{ farads}$$

If we consider number of slabs of thickness t_1, t_2... and dielectric constants $\varepsilon r_1, \varepsilon r_2$ ..., then total capacity of such a condenser

$$C = \frac{\varepsilon_0 A}{4\pi \left[d - (t_1 + t_2 + t_3 + ...) + \left(\frac{t_1}{\varepsilon_{r_1}} + \frac{t_2}{\varepsilon_{r_2}} + ... \right) \right]}$$

Example 5. *A cable of wire 3×10^{-3} m in diameter and insulated with 3×10^{-2} of an insulator (relative permittivity = 4.26) is placed in water. Calculate the capacity for 5 km length of the cable. Given $\log_{10} 3 = 0.4771$.*

Solution: $\varepsilon_r = 4.26$, $\varepsilon_0 = 8.85 \times 10^{-12}$ coul²/newton m²

$$l = 5 \text{ km} = 5000 \text{ m.}$$

$$A = \frac{3 \times 10^{-3}}{2} = 1.5 \times 10^{-3} \text{m}$$

$$b = 3 \times 10^{-8} + 1.5 \times 10^{-3} = 4.5 \times 10^{-3} \text{m}$$

Now

$$C = \frac{4\pi \varepsilon_r \varepsilon_0 l}{2 \times 2.3026 \log_{10} b/a}$$

$$= \frac{4 \times 3.142 \times 8.85 \times 10^{-12} \times 5 \times 10^3}{1 \times 2.3026 \times \log_{10} 3}$$

$$= 1.504 \times 10^{-6} \text{ farad} = \mathbf{1.504 \text{ F.}}$$

Example 6. *A potential difference of 20 kV is applied to a parallel plate condenser with a plate area of 0.01 m². The plates are separated by a dielectric 2 mm thick. The capacity of the condenser is 2 × 10⁴ μF. Find (i) Total Electric flux, (ii) Potential gradient, (iii) Relative permittivity of the medium* and *(iv) Energy stored.*

Solution: Given $V = 20{,}000$ V

$$A = 0.01 \text{ m}^2,\ d = 0.002 \text{ m}$$

and $$C = 2\times10^{-4}\times10^{-6} = 2\times10^{-4} \text{ farad}$$

(*i*) Total flux

$$Q = CV = 2\times10^{-10}\times 20000$$
$$= 4\times10^{-6}\text{C} = 4\,\mu\text{C}$$

(*ii*) Potential gradient

$$\frac{V}{d} = \frac{2000}{0.002} = \mathbf{10^7\ V/metre}$$

(*iii*) If relative permittivity of the media is, ε_r,

$$C = \frac{\varepsilon_0\varepsilon_r A}{d}$$

or $$\varepsilon_r = \frac{C\times d}{\varepsilon_0\times A} = \frac{2\times10^{-10}\times0.002}{8.854\times10^{-12}\times0.01} = \mathbf{4.52}$$

(*iv*) Energy stored

$$= \frac{1}{2}CV^2 = \frac{1}{2}\times2\times10^{-2}\times(20000)^2$$
$$= \mathbf{4\times10^{-2}\ Joules}$$

Example 7. *Calculate the capacity of a cable 5 km long having an inner core of diameter 1 cm and impregnated paper insulation of thickness 0.7 cm and permittivity = 4.*

Solution: Given $l = 5$ km $= 5\times10^3$ m

$$\varepsilon_0 = 8.854\times10^{-12},\ \varepsilon_r = 0.5 \text{ cm} = 0.005 \text{ m}$$

And $$R = 0.005 + 0.007 = 0.0012 \text{ m}$$

We have the relation

$$C = \frac{4\pi\,\varepsilon_0\,\varepsilon_0\,l}{2\times2.3026\log_{10} b/a}$$
$$= \frac{2\pi\times8.854\times10^{-12}\times4\times5\times10^3}{2.3026\log_{10}\left(\dfrac{0.012}{0.005}\right)}$$
$$= 1.272\times10^{-6}\text{F} = \mathbf{1.272\ \mu F}$$

Example 8. *Two circular plates of a parallel plate condenser are each 8 cm in diameter and 1.5 cm apart. An ebonite slab of 0.6 cm thick is introduced between them. Calculate its capacity (i) in e.s.u.* and *(ii) in farad. Dielectric constant of ebonite = 2.8.*

Solution: $$C = \frac{A}{4\pi\left[d - t + \dfrac{t}{\varepsilon_r}\right]}$$

Here, $$A = \pi r^2 = 16\pi \text{ cm}^2,\ d = 1.5 \text{ cm}$$
$$t = 0.5 \text{ cm},\ \varepsilon_r = 2.8$$

∴ $$C = \frac{A}{4\pi\left[1.5 - 0.5 + \dfrac{0.5}{2.8}\right]}$$

$$= \mathbf{3.39\ e.s.u.}$$

Also $$C = \frac{3.39}{9 \times 10^{11}} \text{ farad} = 3.77 \times 10^{-12} \text{ farad} = \mathbf{3.77\ (\mu\mu F)}$$

Example 9. *A 100 volts D.C. supply is used to charge two capacitors in parallel. One of the capacitors is 4μF and the other is 5 μF. What energy is (a) stored in the capacitors* and *(b) taken from the supply?*

Solution: Since the capacitors are in parallel

So total capacity is given by

$$C_p = C_1 + C_2$$
$$= 4 + 5 = 9\,\mu F$$

(*a*) Energy stored in a capacitor is given by

$$E = \frac{1}{2} CV^2$$

∴ Energy stored by 4 μF capacitor

$$E_1 = \frac{1}{2} \times 4 \times 10^{-6} \times 100^2 = 2.0 \times 10^{-2} \text{ Joule}$$

Similarly energy stored in the 5 μF capacitor

$$E_2 = \frac{1}{2} \times 5 \times 10^{-6} \times 100^2 = 2.5 \times 10^{-2} \text{ Joules}$$

Total energy stored in capacitors

$$E = E_1 + E_2 = \mathbf{4.5 \times 10^{-2}\ Joules}$$

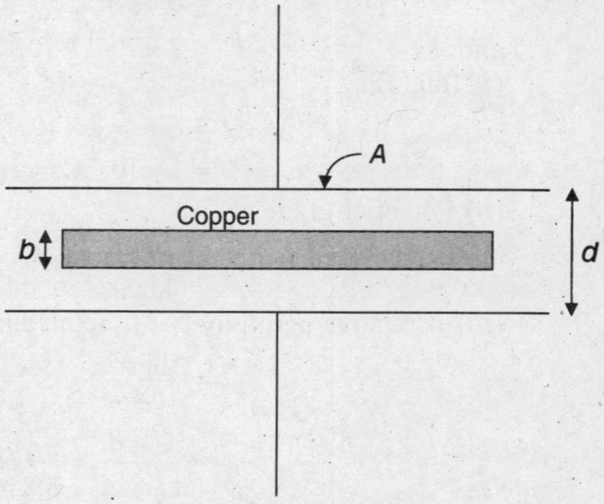

Fig. 3.10

(*b*) Energy taken from supply equivalent capacitance × supply voltage

$$= 9 \times 10^{-6} \times 100^2 = 9 \times 10^{-2} \text{ Joules.}$$

Example 10. *A slab of copper of thickness b is thrust into a parallel plate condenser as shown in Fig. 3.10 it is exactly half way between the plates. What is the capacity before and after the slab is introduced?*

Solution: The capacity of a parallel plate condenser without slab is given by

$$C = \frac{\varepsilon_0 A}{d}$$

Now when the slab of copper is introduced, the original capacitor is reduced into two capacitors in series each having a gap of $(d - b)$ and each having capacity given by

$$C_1 = C_2 = \frac{\varepsilon_0 A}{\frac{1}{2}(d-b)} = \frac{2\varepsilon_0 A}{d-b}$$

Now the combination of these two capacitors in series is given by

$$C = \frac{C_1 C_2}{C_1 + C_2} = \frac{C_1^2}{2C_1} = \frac{C_1}{2} \qquad [\because C_1 = C_2]$$

$$C = \frac{\varepsilon_0 A}{d-b}$$

Example 11. *Hard rubber has a dielectric constant of 2.8 and dielectric strength of 18 × 10^6 volts/metre. It is used as a dielectric material in a parallel plate capacitor. What minimum area in the plates of the capacitor have in order that the capacitance be 70 ×10^{-2} μF and that the capacitor be able to withstand p.d. of 4000 volts.*

Solution. We have the relation for a parallel plate capacitor,

$$C = \frac{\varepsilon_0 \varepsilon_r A}{d} \qquad ...(1)$$

In this case dielectric strength $E = 18 \times 10^6$ volts/meter, electric field strength = 4000 volts

We have the relation

$$V = \frac{E}{d}$$

$$\therefore \quad d = \frac{E}{V} = \frac{4000}{18 \times 10^6} = 2.22 \times 10^{-4} \text{ metre}$$

We have the relation

$$A = \frac{C.d}{\varepsilon_0 . \varepsilon_r} = \frac{(7 \times 10^{-2} \times 10^{-6} t) \times (2.22 \times 10^{-4} \text{ metre})}{(8.9 \times 10^{-12} \text{ coul}^2/\text{N} - \text{m}^2) \times 2.8}$$

$$= \mathbf{0.62\ metre^2}$$

Example 12. *A parallel plate capacitor has a capacitance of 100* μμF, a plate of 100 cm^2 *and a mica dielectric. At 50 volts p.d. calculate (a) electric field in mica, (b) the free charge on plates* and *(c) the induced surface charge.* (ε_r *for mica = 5.4*)

Solution. The electric field in mica is given by

$$E = \frac{q}{\varepsilon_0 \varepsilon_0 A}$$

$$= \frac{5 \times 10^{-19} \text{ coul.}}{5.4 \times (8.9 \times 10^{-12} \text{ coul}^2/\text{N} - \text{m}^2) \times (1.0 \times 10^{-2} \text{m}^2)}$$

$$= \mathbf{1.04 \times 10^4\ volts/metre}$$

(*b*) Free charge on the plate is given by

$$q = CV = (100 \times 10^{-12} \text{F}) \times 50 \text{ volts}$$

$$= \mathbf{5 \times 10^{-9}\ coulomb}$$

(*c*) Induced surface charge is given by

$$q' = q\left(1 - \frac{1}{\varepsilon_r}\right)$$

$$= 5 \times 10^{-9} \times \left(1 - \frac{1}{5.4}\right)$$

$$= \mathbf{4.1 \times 10^{-9}\ coulomb}$$

Which means that an induced charge of – 4.0 × 10^{-9} coulomb appears next to the positive plate.

Example 13. *For making a capacitor you have available two plates of copper, a sheet of mica (thickness = 0.01 mm and* ε_r *= 6), a sheet of glass (thickness = 2.0 mm,* ε_r *= 7) and a slab of paraffin (thickness = 1.0 cm and* ε_r *= 2). To obtain the largest capacitance, which sheet (or sheets) should you place between copper plates?*

Solution: Capacity of a parallel plate capacitor having dielectric constant ε_r is given by

$$C = \frac{\varepsilon_0 A \varepsilon_r}{d}$$

Now because A and ε_0 are constants, so if $\frac{\varepsilon_r}{d}$ is found, we can estimate the relative magnitudes of C.

For mica $\quad \frac{\varepsilon_r}{d} = \frac{60}{0.01 \text{ mm}} = 60/\text{mm}$

For glass $\quad \frac{\varepsilon_r}{d} = \frac{7}{2.0 \text{ mm}} = 3.5/\text{mm}$

For paraffin $\quad \frac{\varepsilon_r}{d} = \frac{2}{10 \text{ mm}} = 0.2/\text{mm}$

So to get largest capacitance, mica must be placed between the plates.

Example 14. *An isolated metal sphere whose diameter is 10 cm has a potential of 8000 volts. What is the energy density at the surface of the sphere?*

Solution: Radius of the sphere $r = 5$ cm $= 0.05$ m.

Electric field on the surface of sphere is

$$E = \frac{q}{4\pi \varepsilon_0 r^2}$$

The energy density at the surface is

$$U = \frac{1}{2}\varepsilon_0 E^2 = \frac{q^2}{32\pi^2 \varepsilon_0 r^4} \qquad ...(i)$$

Also $\quad q = CV$

and $\quad C = 4\pi \varepsilon_0 r$

$\therefore \quad q = 4\pi \varepsilon_0 Vr$

$\therefore$ From (*i*)

$$U = \frac{1}{2}\frac{\varepsilon_0 V^2}{r^2}$$

$$= \frac{1}{2} \times \frac{(8.9 \times 10^{-12} \text{ coul}^2/\text{N} - \text{m}^2) \times (800 \text{ volts})^2}{(0.05 \text{ m})^2}$$

$$= \mathbf{0.114\ Joule/m^2}.$$

Example 15. *A parallel plate capacitor has plates of area A and separation d and is charged to a p.d. V. The charging battery is then disconnected and the plates are pulled apart till their separation is 2d. Derive an expression in terms of A, d and V for (a) the new p.d., (b) the initial and final stored energy* and *(c) the work required to separate the plates.*

Solution: (*a*) Capacity is given by

$$C = \frac{\varepsilon_0 A}{d}$$

Initial *p.d.* $\quad V_i = \frac{q}{C} = \frac{qd}{\varepsilon_0 A}$

Final *p.d.* $\quad V_f = \mathbf{q}.\frac{\mathbf{2d}}{\varepsilon_0 \mathbf{A}} = \frac{\mathbf{2qd}}{\varepsilon_0 \mathbf{A}} = \mathbf{2V_i}$

(*b*) Initial stored energy

$$U_i = \frac{1}{2}CV_i^2 = \frac{1}{2}\varepsilon_0 \frac{AV_i}{d}$$

Final stored energy

$$U_f = \frac{1}{2}\varepsilon_0 A\frac{V_f^2}{d} = \frac{1}{2}\varepsilon_0 A.\frac{(2U_i)^2}{2d}$$

$$= \frac{\varepsilon_0 \mathbf{A} \mathbf{V_i^2}}{\mathbf{d}} = \mathbf{2U_i}$$

(*c*) Work required to separate the plates

$$W = \mathbf{U_f - U_i = 2U_i - U_i = U_i = \varepsilon_0 A V_i^2/2d}$$

Example 16. *Two capacitors (2.0 μF and 4.0 μF) are connected in parallel across a 300 V potential difference. Calculate the total energy stored in the system.*

Solution: Two capacitors connected in parallel have the capacity given by

$$C = C_1 + C_2 = 2.0\,\mu\text{F} + 4.0\,\mu\text{F}$$

$$= 6.0\,\mu\text{F}$$

Total energy of the system

$$U = \frac{1}{2}CV^2 = \frac{1}{2}(6 \times 10^{-6}\,\text{F}) \times (300\,\text{V})^2$$
$$= \mathbf{0.27\ Joule}$$

Example 17. *A parallel air capacitor has a capacitance of 100 μμF:*
(a) what is the stored energy if the applied p.d. is 50 volts?
(b) can you calculate the energy density for the points between the plates?

Solution. Energy stored is given by

(*a*)
$$U = \frac{1}{2}CV^2 = \frac{1}{2} \times (100 \times 10^{-12}\,\text{F}) \times (50\ \text{V})^2$$
$$= \mathbf{1.25 \times 10^{-7}\ Joule}$$

(*b*) Energy density

$$U_0 = \frac{\varepsilon_0}{2}\left(\frac{V}{d}\right)^2$$

Since in this case, gap d is not known, so energy density cannot be calculated.

Example 18. *A parallel plate capacitor of 100 pF capacitance is charged to 100 V and then the battery is disconnected. What is the energy stored in the capacitance? Mica sheets are now inserted between the plates filling the whole gap. What is the energy now stored in the capacitance? Account for the difference in these two energies. The dielectric constant of mica is 5.*

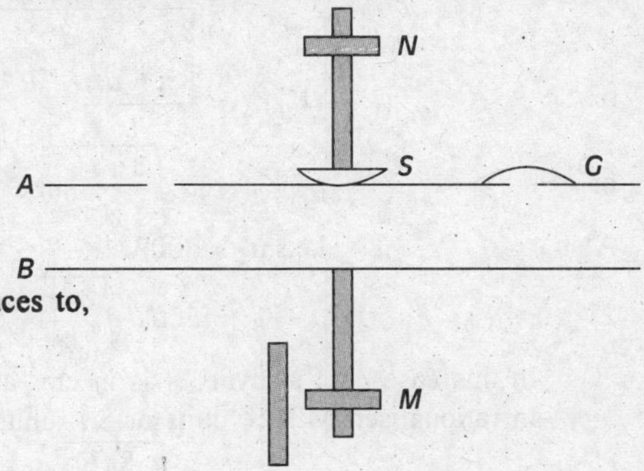

Fig. 3.11

Solution: Energy stored in the first case,

$$U_0 = \frac{1}{2}C_0V_0^2$$
$$= \frac{1}{2} \times 100 \times 10^{-12}\,\text{F}\,(100)^2\,\text{V}^2$$
$$= 0.5 \times 10^{-6}\,\text{J}$$

After inserting mica, the potential difference reduces to,

$$V = 100/5 = 20\ \text{V}$$

and the capacitance increases to,

$$C = 100 \times 5 = 500\ \text{pF}$$

The energy now stored is,

$$U = \frac{1}{2} \times 500 \times 10^{-12}\,\text{F} \times 400\ \text{V}^2$$
$$= \mathbf{0.1 \times 10^{-6}\ J}$$

Thus the energy decreases by a factor of $1/\varepsilon_r$. When we start inserting the slab, the capacitor charges induce opposite charges on the mica faces nearer to them. This way the dielectric will experience a force pulling it towards the capacitor. To make the slab go into the capacitor with almost zero velocity, it will have to be kept pulled back. This way the 'missing energy' will be transferred to the agent which inserts the dielectric slab.

3.9. ELECTROMETERS

Instruments used for the measurement of electric potential are called electrometers. These instruments can also be used to measure very small currents and charges. We shall study the attracted disc or Kelvin's absolute electrometer.

Attracted Disc Electrometer. It consists of two parallel plates A and B and G is a guard ring surrounding the plate A so that the electrostatic field is uniform between A and B. If we keep the air gap between A and B small, the guard ring avoids the edge effect to bare minimum, S is a spring attached to A and M is a micrometre screw with which plate A can be moved up or down. Plates A and G are connected by a flexible wire and plate B can also be moved up or down by another micrometre screw M.

The basic principle of this instrument is that a force of attraction is present between A and B if these are at different potentials.

Theory of the apparatus. Let the potentials on A and B be V_A and V_B respectively. The electric intensity at a point between the two plates = E and d is the distance between the plates.

Then the work done in moving a +ve charge from B to A is given by

$$V_A - V_B = E.d$$

or
$$E = \frac{V_A - V_B}{d} \quad ...(1)$$

Also force experienced by a unit area of a charged plate

$$= \frac{2\pi\sigma^2}{K} = \frac{KE^2}{8\pi}$$

If the medium is air, $K = 1$ in C.G.S. units

then force, $= \dfrac{E^2}{8\pi}$

Now force of attraction between the plates for an area A cm^2

$$F = A.\frac{E^2}{8\pi} \text{ dyne}$$

$$= \frac{A}{8\pi}.\frac{(V_A - V_B)^2}{d^2}$$

or
$$(V_A - V_B)^2 = \frac{8\pi F d^2}{A}$$

or
$$V_A - V_B = d\sqrt{\frac{8\pi F}{A}} \text{ in C.G.S. units} \quad ...(2)$$

Also 1.e.s.u. = 300 V

$$\therefore \quad V_A - V_B = 300\, d\sqrt{\frac{8\pi F}{A}} \quad ...(3)$$

In this case F is in dyne, A is in cm^2 and d in cm.

In rationalised M.K.S. units or S.I. units,

$$V_A - V_B = d\sqrt{\frac{8\pi F}{4\pi\varepsilon_0 A}}$$

$$V_A - V_B = d\sqrt{\frac{2F}{\varepsilon_0 A}} \text{ volt} \quad ...(4)$$

where d is in metres, F in Newtons and A in m^2

and $\varepsilon_0 = 8.85 \times 10^{-12}$ coul^2/Newton-m^2

Measurement of Potential. (*i*) First the plates A and B are earth connected, so that they have zero p.d. A small mass m is placed on plate A and the level of A is adjusted to be in plane with guard ring with the use of screw N.

(*ii*) Then m is removed, now if due to the p.d. between plates A and B, A comes in level with guard ring without using N, then the force of attraction between the plates is mg dynes.

(*iii*) If plate A is given potential V and B a potential V_1, the position of B is adjusted with screw M, so that A is in level with guard ring. Let us consider the reading on the scale to be R, corresponding to distance d_1 between the plates.

$\therefore$ From equation (2)

$$V_A = V,\ V_B = V_1,\ d = d_1,\ E = mg$$

$$\therefore \quad V - V_1 = \sqrt{\frac{8\pi mg}{A}} \quad \text{....(5)}$$

(*iv*) Let B be connected to potential V_2, the position of B is again adjusted so that A comes in level guard ring. Let the reading on the scale be R_2 and distance between plates be d^2.

So we have

$$V_1 - V_2 = d_2\sqrt{\frac{8\pi mg}{A}} \quad \text{...(6)}$$

Subtracting (6) from (5)

$$V_2 - V_1 = (d_2 - d_1)\sqrt{\frac{8\pi mg}{A}}$$

But $\quad d_2 - d_1 = R_2 - R_1$

$$\therefore \quad V_2 - V_1 = (R_2 - R_1)\sqrt{\frac{8\pi mg}{A}} \text{ in C.G.S units} \quad \text{...(7)}$$

In equation (7) all the quantities can be accurately determined and so we can find the absolute value of potential difference and hence it is called Absolute Electrometer.

$$\text{Also} \quad (V_2 - V_1) = 300(R_2 - R_1)\sqrt{\frac{8\pi mg}{A}} \text{ volt} \quad \text{...(8)}$$

In rationalised M.K.S. or S.I. units

$$V_2 - V_1 = (R_2 - R_1)\sqrt{\frac{8\pi mg}{A}}$$

$$\text{or} \quad V_2 - V_1 = (R_2 - R_1)\sqrt{\frac{2\,mg}{\varepsilon_0 A}} \text{ volts.}$$

3.10. DETERMINATION OF SPECIFIC INDUCTIVE CAPACITY OF A MATERIAL GIVEN IN THE FORM OF A SLAB

Plates A and B are given V_A and V_B potentials respectively and the position B so adjusted that plate A is in level with guard ring. Let d be the distance between plates, then the capacity of the condenser for plate of A cm^2 area in air as medium between the plates,

$$C = \frac{A}{4\pi d} \quad \text{...(9)}$$

Now the given material (dielectric) slab of thickness t cm is placed on B. It means that the effective air distance between the plates will be reduced: So the plate A moves down the level of guard ring.

To bring A back to level with guard ring, the plate B is to be moved downwards and let it be moved through a distance x.

$\therefore$ Capacity of electrometer

$$C = \frac{A}{4\pi\left(d - t + \dfrac{t}{\varepsilon_r} + x\right)} \quad \text{...(10)}$$

Equating (9) and (10)

$$D = \left(d - t + \frac{t}{\varepsilon_r} + x\right)$$

$$\text{or} \quad x = t - \frac{t}{\varepsilon_r}$$

$$\text{or} \quad \frac{t}{\varepsilon_r} = t - x$$

or $$\varepsilon_r = \frac{t}{t-x} \text{ in C.G.S. units} \qquad ...(11)$$

In order to find thickness of the slab, both the plates *A* and *B* are earth connected and the plate *B* is moved up so that the top and bottom surfaces of the slab touch plates *A* and *B* respectively. The reading of the position *B* is tàken on the scale.

Now the slab is removed and the plate is moved up so that *B* touches *A* and again the reading of *B* is noted. The difference between these readings gives the thickness of the slab.

The Specific Inductive Capacity (SIC) material is calculated from the equation $\varepsilon_r = \frac{t}{t-x}$ and in rationalised M.K.S. units or S.I. units

$$\varepsilon_r = \frac{r}{t-x}$$

where ε_r is the relative permittivity of the medium.

Advantage. As the potential difference measurement does not require weight, any known p.d. and also does not require any calibration and the unknown p.d. is measured by distance and area only. So it measures the absolute p.d. as it does not consume any current, is unlike voltmeters. However, it is a cumbersome apparatus and is not as simple to use as a voltmeter.

Example 19. *4 gms of gold is beaten into a thin leaf of 10^4 cm^2. A small piece is cut of this and placed on a conductor. Calculate the charge density required by the conductor, so that the piece of the gold is just lifted up.*

Solution: Weight of gold foil $= mg = 4 \times 981$ dynes

$$\text{Area} = 10^4 \text{ cm}^2$$

$$\text{weight/unit area} = \frac{4 \times 981}{10^4} = \text{dynes/cm}^2$$

Also $$F = \frac{2\pi\sigma^2}{K} = 2\pi\sigma^2 \text{ (for air } K = 1)$$

$$\frac{4 \times 981}{10^4} = 2\pi\sigma^2$$

$$\sigma = \left(\frac{981 \times 7}{10^4 \times 11}\right)^{1/2}$$

$$= \mathbf{0.2497 \text{ e.s.u./cm}^2}$$

Example 20. *Calculate the force of attraction between two plates of an electrometer when a p.d. of 1000 volts is applied between them, if the distance between the plates is 5 mm and area is 10 cm^2.*

Solution: $$F = \frac{KE^2A}{8\pi}, \qquad \text{But } E = \frac{V}{d}$$

$$\therefore \quad F = \frac{KV^2A}{8\pi d^2} \text{ and for air } K = 1$$

Here $$V = 1000\ V = \frac{1000}{300} = \frac{10}{3} \text{ e.s.u.}$$

$$A = 10 \text{ cm}^2, d = 5 \text{ mm} = 0.5 \text{ cm}$$

$$F = \frac{190 \times 10 \times 7}{9 \times 8 \times 22 \times (0.5)^2}$$

$$= \mathbf{17.68 \text{ dynes.}}$$

Example 21. *Calculate the force of attraction between the two plates of an electrometer when a p.d. of 500 V is applied between them. The distance between the plates is 25 × 10^{-4} m and area is 10^{-3} m^2.*

Solution: In this case

$$A = 10^{-3}\,\text{m}^2, d = 25 \times 19^{-4}\,\text{m}$$

$$V = 500 \text{ volts}$$

$$\varepsilon_0 = 8.85 \times 10^{-12}\,\text{C}^2/\text{N} - \text{m}^2$$

We have the relation

$$V = d.\sqrt{\frac{2F}{\varepsilon_0 A}}$$

or

$$F = \frac{V^2.\varepsilon_0.A}{2d^2}$$

$$= \frac{(500)^2 \times 8.85 \times 10^{-12} \times 10^{-3}}{2 \times (25 \times 10^{-4})^2}$$

$$= \mathbf{1.77 \times 10^{-4}\,N}$$

QUESTIONS AND PROBLEMS

1. Calculate the capacity of a parallel plate condenser. What will be the capacity is the space between the plates is partially filled with slab of thickness t and dielectric constant ε_r.

2. Derive an expression for the energy stored by a condenser.

3. Find the capacity of 3 condensers connected in (*i*) series and (*ii*) parallel.

4. Prove that the electric capacity of an isolated conducting spherical surface is equal to its radius.

5. Define relative permittivity and state the factors on which the capacity of a condenser depends.

6. State the capacity and express its units. Derive an expression for a parallel plate condenser.

7. Give the construction of the attracted disc electrometer explaining the use of the guard ring. Why is it called an absolute electrometer? Give its necessary theory and working and state its merits and demerits.

8. Give the atomic view of dielectrics.

9. Discuss Gauss's and Coulomb's laws from dielectrics.

10. Discuss in details the three electric vectors. What is their importance ?

11. Calculate the capacity of a condenser which has 0.2 m^2 area separated by a distance of 2.5 × 10^{-6} metres, dielectric has the permittivity = 5. **[Ans.** 0.354 μF]

12. Two brass plates are arranged horizontally 2 cm apart and lower plate is earthed. The plates are charged to a p.d. of 4000 V. A drop of oil with an electric charge 4.8 × 10^{-13} e.s.u. is at rest between the plates. What is the mass of the drop (g = 980 cm s^{-2}) **[Ans.** 3.27 × 10^{-12} gm]

13. A condenser of 2 μF capacity is in series into one capacitor of 3 μF capacity. The outer plate of the second is earthed and a potential of 100 volts is applied to outer plate of the first. At what potential is the inner plate of the combination and what is the charge on each plate.

[Ans (*i*) 40 volts, (*ii*) 120 μ coul.]

14. Calculate the force of attraction between the two plates of an electrometer when a p.d. of 500 volts is applied between them. The distance between the plates is 25 × 10^{-4} m and area of each plate = 10^{-3} m^2. **[Ans.** 2.77 × 10^{-4} Newton]

15. (*a*) What can be the dielectric coefficient of a metal ?

(*b*) If instead of dielectric slab, we put a slab of a metal in between the capacitor plates keeping it insulated from them, what effect does it have on the capacitance ?

16. A conducting sphere of radius R carries a charge q and is embedded in a dielectric of dielectric constant ε_r. Show that the energy stored is $q^2/8\pi\varepsilon R$.

17. Show that the energy density per unit volume, μE, in the electric field E inside a dielectric of permittivity ε, is $\varepsilon_0 E^2$.

18. Show that the capacity of a parallel plate capacitor of plate area A and plate separation d having dielectric slab of thickness t ($< d$) and dielectric constant ε_r, is $\varepsilon_0 A/[d + t\,(1/\varepsilon_r - 1)]$.

19. (*a*) A parallel plate capacitor having plate area A, and plate separation d, is completely filled with two dielectric each having a thickness $d/2$. Show that the capacitance is

$$C = \frac{\varepsilon_0 A}{d}\left(\frac{2\varepsilon_{r1} + \varepsilon_{r2}}{\varepsilon_{r1} + \varepsilon_{r2}}\right)$$

(*b*) In the above capacitor, half the section is filled with one dielectric and the other half is filled with second dielectric. Show that,

$$C = \frac{\varepsilon_0 A}{d}\left(\frac{\varepsilon_{r1} + \varepsilon_{r2}}{2}\right)$$

20. Two parallel metal plates have been given equal and opposite charge, and are separated by teflon sheet ($\varepsilon_r = 2$) of thickness 5 mm. The electric intensity inside dielectric is 2×10^6 volt/m. Calculate:
(*i*) the free charge per unit area on the conducting plates, and
(*ii*) induced charge per unit area on the surface of teflon. **[Ans.** 854 μC : 17.7 μC]

21. A parallel metal plate capacitor of capacitance 100 μF has plate area of 0.01 m^3, and paper as dielectric ($\varepsilon_r = 3.5$). What is the thickness of the paper? At 100 V potential difference find out
(*i*) the free charge on the plates,
(*ii*) induced charge at paper surface, and
(*iii*) *E.D.* and *P* inside the paper. **[Ans.** (*i*) 0.01 μC; (*ii*) 0.007 μC; (*iii*) 3×10^4 Vm, 1 μCm^2, 0.7 μC m^{-2}]

CHAPTER

4

DIELECTRICS

4.1. INTRODUCTION

While classifying solids on the basis of their band structure, we refer to the group of *solids as insulators which have energy gap of 3 e.v. or more*. The large magnitude of energy gap in an ideal insulator excludes the possibility of electrons being excited from valence band to conduction band by **thermal means** much less so by the application of an external electric field. It is due to this reason that insulators are poor conductors of electricity and are also known as **dielectrics.** Dielectric materials find extensive use in electric industry and as capacitors.

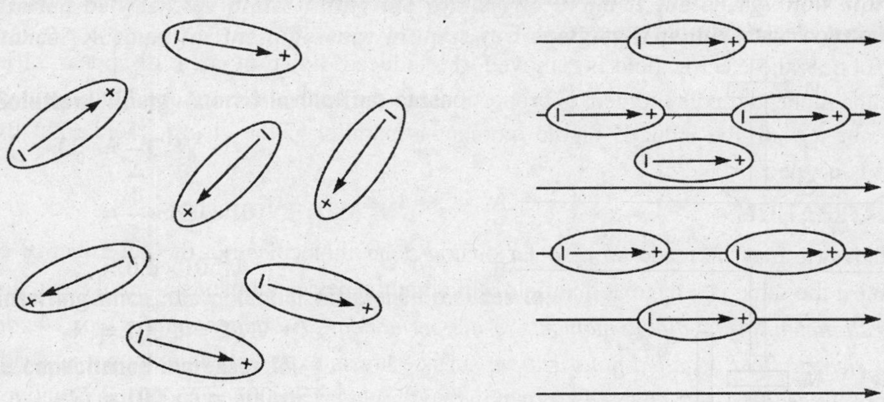

(a) (b)

FIG. 4.1

4.2. POLAR AND NON-POLAR MOLECULES

Fig. 4.2

In the case of polar molecules the *centre of gravity of the positive charges (i.e., protons) and negative charges (i.e., electrons) do not coincide.* Such molecules are called **permanent electric dipoles** as these have permanent dipole moments. Some examples of polar molecules are N_2O, H_2O and HC1. In the absence of an electric field, the *electric dipole moments* of these polar molecules are in random directions [Fig. 4.1 (a)] and cancel each other.

So even though each of molecule has a dipole moment, the average moment per unit volume is zero: on the application of an electric field, the dipole moments of these molecules align themselves parallel to the direction of electric field as shown in [Fig 4.1 (*b*)]. However, this alignment is not complete because of thermal vibrations of the molecules. It is clear that the alignment of the molecules with applied field increases if

(*a*) Electric field intensity is increased

(*b*) Temperature is lowered.

It may be noted that with the increase of electric field intensity, the dipole moment also increases as due to increase of electric field intensity the distance between positive and negative charges is increased and so the dipole moment is increased.

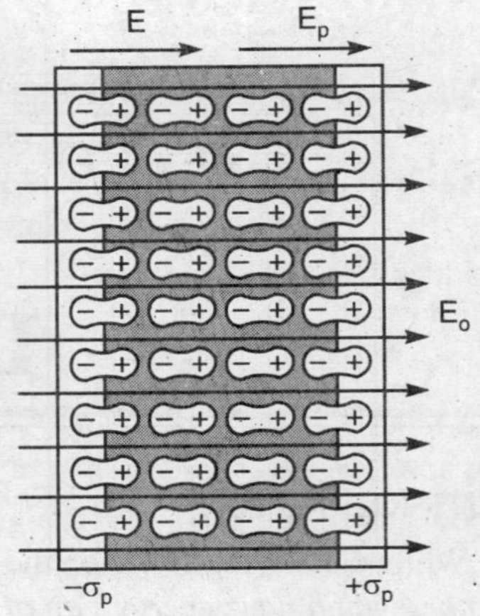

Fig. 4.3. Effect of polarization in a dielectric

Non-polar molecules. In case of non-polar molecules the *centre of gravity of positive charges (protons) and negative charges (electrons) coincide*, so these molecules do not have **any permanent dipole moment**. Some common examples of non-polar molecules are oxygen (O_2), nitrogen (N_2) and hydrogen (H_2).

When a non-polar molecule is placed in an electric field, the centres of positive and negative charges get displaced and the molecules are said to have been *polarized* as shown in Fig. 4.2.

Such a molecule is then called **induced electric dipole** and its electric dipole moment is called **induced electric dipole moment.**

As soon as the electric field is removed, the induced dipole moment disappears. It may be noted that the induced dipole moment is proportional to the applied electric field but is independent of temperature. Also, the induced dipole moment is parallel to the electric field right from the moment it is created.

4.3. POLARIZATION

A dielectric may be made up of polar or non-polar molecules but the net effect of external field is almost the same i.e., the external field will align the dipole moments along its own direction.

The alignment of the dipole moments of the permanent or induced dipoles with the direction of applied electric field is called polarization, surface layers extremely thin shown as shaded part of the figure. Excess positive charge is shown in one thin layer and an equal excess negative charge is shown in the other layer.

The induced charges on the surfaces of dielectric are due to these layers. It may be noted that these charges are not free but are bound to the molecules lying in the surface or near the surface. It is because of this reason that these charges are called **bound charges or fictitious charges**. Also within the remaining dielectric, the *net charge per unit volume remains zero*. So, though the dielectric is polarized, yet it remains **electrically neutral**. Also in polarization, the internal state of the slab is characterized not by an excess charge but by the relative displacement of the charges within it.

So polarization can also be considered as a phenomenon in which an alignment of positive and negative charges takes place within the dielectric resulting no net increase in the charge of the dielectric.

The relation between the *electric flux density D* (charge per unit area in units of Cm^{-2}) at a point in a material and the electric field strength *E* (force per unit charge in units of NC^{-1} or Vm^{1}) at any point in space is given by

$$D = \varepsilon_r \varepsilon_0 E \qquad \text{... (1)}$$

where ε_0 is dielectric constant or permitivity of free space (vacuum)

ε_r is relative dielectric constant or relative permitivity.

It may be noted that ε_r is dimensionless and is a property of the material related to its atomic structure. Its value is 1 for free space and more than 1 for all materials. In S.I. units $\varepsilon_0 = 8.854 \times 10^{-12}$ farad/meter. As can be seen, ε_0 has this particular value from C.G.S. to S.I. units.

$$\varepsilon_0 = \frac{D}{\varepsilon_r E} = \frac{\text{Cm}^{-2}}{\text{Vm}^{-1}} = \text{Fm}^{-1}$$

As shown above, due to the application of electric field, displacement of local dipoles takes place. *The dipole moment/unit volume of the solid is the sum of all the individual dipole moments within the volume* (called polarization of the solid).

As polarization measures additional flux density arising due to the presence of the material as compared to free space, it has the same units as D and is related as under

$$D = \varepsilon_0 E + P \quad \text{... (2)}$$

From equations (1) and (2), we get

$$P = \varepsilon_0(\varepsilon_r - 1)E \quad \text{... (3)}$$

Macroscopically, the polarization or relative dielectric constant can be measured using a parallel plate capacitor, when a voltage V is applied to the capacitor, with vacuum between the plates, a charge develops on the plates. In that case capacitor may be defined as

$$C = \frac{\varepsilon_0 A}{d} \quad \text{... (4)}$$

where A is the area of the parallel plates, d is the distance of separation between the plates. If a dielectric material is inserted between the plates, the charge on the plates increases due to polarization of the material. The capacitance is then given by

$$C = \frac{\varepsilon_0 \varepsilon_r A}{d}$$

We can find ε_r and there by P by measuring capacitance with and without dielectric.

4.4. MOLECULAR THEORY OF DIELECTRICS

A dielectric becomes polarized in the presence of an electric field as the effect of the electric field is to separate positive and negative charges of the total volume of the dielectric and create electric dipoles. This is called *polarization of dielectric*.

An electric di-pole is made up of two equal positive and negative charges, separated by a small distance. Let q be the magnitude of charge and dl be the distance of separation, then the electric dipole moment is given by the relation

$$\mathbf{p} = q\,\mathbf{dl} \quad \text{... (1)}$$

it is a vector quantity. The unit of electric dipole moment (or dipole moment) is **coulomb-meter**. If each of the particles in a solid possesses a dipole moment then the total dipole moment will be the sum of moments of all dipoles.

The *polarization* **P** of a dielectric is defined as the **dipole moment** per unit volume *i.e.,*

$$\mathbf{P} = \frac{p}{V} \quad \text{... (2)}$$

It has the same units as the surface density.

4.5. SOURCES OF POLARIZABILITY

The dielectric polarization is due to the fact that when an electric field acts on a molecule, its positive charges (nuclei) are displaced along the field while the negative charges (electrons) in a direction opposite to that of the field. There are number devices based on this concept. They range from condenser and switch-gear equipment in power and distribution installations to rectifiers, resonators, amplifiers and transducers-convertors of electrical energy to other forms of energy in communication technology. They also include memory devices used for storage of information in modern computers.

The above said displacement of electric charges results in the formation of electric dipole moment in atoms, ions or molecules of the material. The three important types of polarizations are (*i*) electronic polarization (*ii*) ionic polarization and (*iii*) orientational polarization.

(*i*) **Electronic polarization:** When the electrons in its different orbits are displaced with the nucleus by the external field, an induced electric dipole moment is produced. This is called **electronic polarization.**

(*ii*) **Ionic polarization:** If a molecule is constituted by ions, then the applied field will displace the positive and negative ions in opposite directions causing a change in length of the ionic bond. This change in length produces a net dipole moment in the unit cell against its previous value (which is zero). As the polarization in this case is due to relative displacement of ions, it is called **Ionic or atomic polarization** (Fig. 4.4).

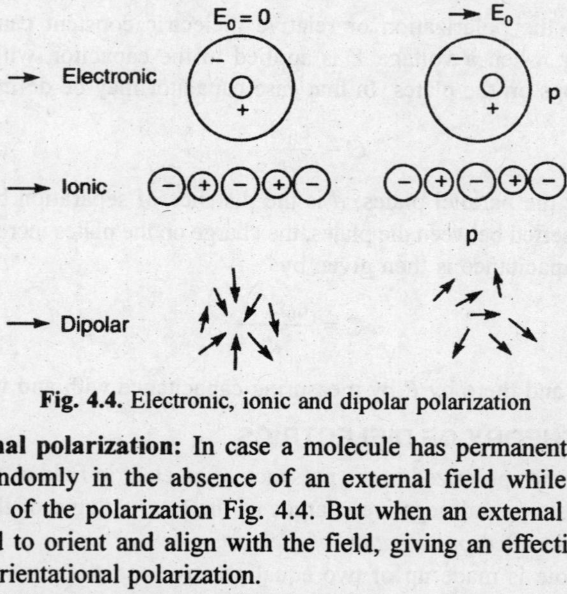

Fig. 4.4. Electronic, ionic and dipolar polarization

(*iii*) **Orientational polarization:** In case a molecule has permanent dipoles and these are oriented randomly in the absence of an external field while result in a complete cancellation of the polarization Fig. 4.4. But when an external field is applied, these dipoles tend to orient and align with the field, giving an effective polarization, called dipolar or orientational polarization.

The total polarizability, is the sum of various contributions. It is devoted by α and is given by,

$$\alpha = \alpha_e + \alpha_i + \alpha_d$$

where α_e, α_i and α_d are the electric, ionic and dipolar polarizability respectively. The total polarizability α is defined as the dipole moment *p*, that an atom or molecule attains in a local electric field E_{loc} and is given by

$$\alpha = \frac{\mathbf{P}}{E_{loc}}$$

The local field E_{loc} is the field that is actually effective in polarizing the atoms in a solid and is also called *molecular field in the dielectric.*

4.6. DIELECTRIC CONSTANT

Faraday was the first to introduce the concept from the experiment that (i) if two capacitors are charged to the same potential, the charge on the one containing the dielectric is more than that of the other and (*ii*) if the capacitors are given the same charge, the potential between the plates with dielectric is less than that without the dielectric. This showed that the **capacitance of the capacitor is increased when filled with dielectric material.** If C_0 and V_0 are the capacitance and potential difference between the plates without dielectric and C_d and V_d the corresponding values with dielectric within the plates of a capacitor, then

$$\frac{C_d}{C_0} = \frac{V_0}{V_d} = k \quad \text{... (1)}$$

where k is called the *dielectric constant* or *dielectric coefficient* of the medium. It is also called *relative permitivity of the medium* denoted by ε_r. It may be defined as *the ratio of a capacitance of a capacitor with dielectric to the capacitance of the same capacitor without dielectric.* It is more than one and independent of shape or the dimensions of the capacitor but varies for different dielectric materials (medium). The permitivity of the medium (ε) is defined as under

$$\varepsilon = \varepsilon_0 \varepsilon_r = \varepsilon_0 k$$

$$\therefore \ \varepsilon_r = k = \frac{\varepsilon}{\varepsilon_0} \quad \text{... (2)}$$

where $\varepsilon_0 = 8.854 \times 10^{-12}$ F/m is the permitivity of **vacuum or free space** (air is practically considered equivalent to free space). So the dielectric constant (*k*) is also defined as the ratio of the permitivity of the medium to the permitivity of the vacuum.

4.4. RELATION BETWEEN ELECTRIC VECTORS: E, P AND D

(*i*) Electric field intensity $\mathbf{E_0}$: The electric field intensity at any point in the electric field is defined as the force experienced by a unit positive charge placed at that point. If **F** is the force on a point charge q_0, then

$$\mathbf{E_0} = \frac{\mathbf{F}}{q_0} \quad \text{... (1)}$$

and also

$$\mathbf{E_0} = \frac{q_0}{\varepsilon_0 A} \quad \text{... (2)}$$

where ε_0 is the permitivity of free space and *A* is the area receiving q_0 lines of force. The unit of $\mathbf{E_0}$ will be Newton/coulomb or volt/meter and its **direction will be that of force.**

(*ii*) Polarization vector P: In the presence of an electric field, the molecules of a dielectric become polarized and gain electric dipole moment. This effect is called electric polarization. The polarization vector is defined *as the electric dipole moment induced per unit volume.* It has the same direction as the electric field and denoted by **P.** In S.I. units, **P** is measured in coulomb/m^2. Fig. 4.5 shows the dielectric placed in a uniform electric field $\mathbf{E_0}$ and is polarized. If σ_p is the *density of polarization charge* on the surfaces, **P** the polarization vector, *A* the area of cross-section and *l* the length of the dielectric slab, *lA* the volume, then

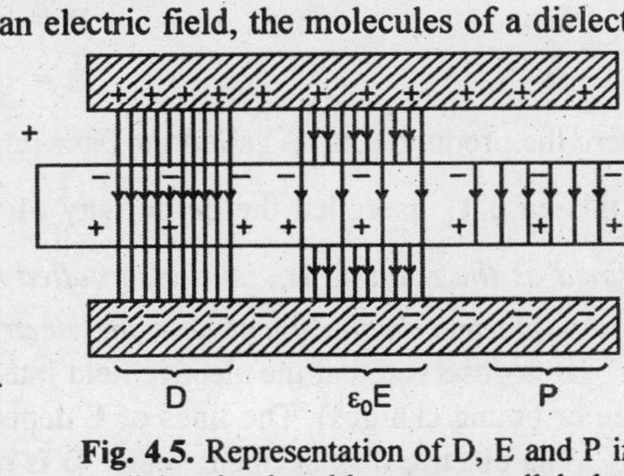

Fig. 4.5. Representation of D, E and P in the dielectric placed between parallel plates of a capacitor

Dipole moment of dielectric of volume $l.A.$ = P.A.e ... (3)

Considering this electric slab as a single dipole with surface charges (= $\sigma_p A$), separated by a distance l, In that case we have

Dipole moment = charge × length = $\sigma_p Al$ ... (4)

From (3) and (4)

$$P = \sigma_p$$

So the polarization vector P depends on the *induced surface density.*

(*iii*) **Electric displacement D:** When a dielectric placed in the parallel plates of a capacitor is subjected to an external electric field $\mathbf{E_0}$, then it gets polarized. The charges are induced at the surfaces of the dielectric. If q_0 is the charge on the plates of capacitor, and q' is the induced charge on the boundary of dielectric, the resultant field at any point in the dielectric will be

$$\mathbf{E} = \mathbf{E}_0 - \mathbf{E}_p$$

where $\mathbf{E}_0 = \dfrac{q_0}{\varepsilon_0 A}$ and $\mathbf{E}_p = \dfrac{q'}{\varepsilon_0 A}$ and A is the area of the dielectric slab.

so
$$\mathbf{E} = \frac{q_0}{\varepsilon_0 A} - \frac{q'}{\varepsilon_0 A}$$

but q'/A is the surface density of the induced charge, called the electric polarization **P**,

Hence

$$\varepsilon_0 \mathbf{E} = \frac{q_0}{A} - \mathbf{P}$$

or
$$\frac{q_0}{A} = \varepsilon_0 \mathbf{E} + \mathbf{P}$$

the quantity on the left is the free surface charge density and called *electric displacement* **D** and measured in coulomb/m^2. So we have

$$\mathbf{D} = \varepsilon_0 \mathbf{E} + \mathbf{P} \quad \text{... (5)}$$

where **D** is the electric displacement vector or the flux density and is *defined as the number of lines of force received by a unit area.* The eqn. (5) represents the relation between **E**, **P** and **D**.

In free space, when there is no dielectric, **P** = 0, we get

$$\mathbf{D} = \varepsilon_0 \mathbf{E} \quad \text{... (6)}$$

with
$$\mathbf{D} = \frac{q_0}{A} = \varepsilon_r \varepsilon_0 \left(\frac{q_0}{\varepsilon_r \varepsilon_0 A} \right) = \varepsilon_r \varepsilon_0 \left(\frac{q_0}{\varepsilon A} \right)$$

or
$$\mathbf{D} = \varepsilon_r \varepsilon_0 \mathbf{E}. \quad \text{... (7)}$$

or
$$\mathbf{D} = \varepsilon \mathbf{E} \quad \text{... (8)}$$

where the product $\varepsilon_r \varepsilon_0$ is called the *permitivity* ε of the medium. In vacuum $\varepsilon_r = 1$, thus $\varepsilon = \varepsilon_0$. In this case ε_0 is called the permitivity of vacuum (or free space). *The dielectric constant k is defined as the ratio $\varepsilon / \varepsilon_0$, and also called relative permitivity. The permitivity is defined as the ratio of electric displacement D to the electric field intensity in dielectric.*

It is observed that the electric field intensity **E** is connected with the charges actually present (free or bound charges). The lines of **E** depend upon the presence of both kinds of charges Fig.

The electric displacement vector **D** is connected with the free charge only. As in case of the electric field, the displacement field can be represented by lines called lines of displacement. These lines begin and end on the free charges Fig. 4.5. The flux of D equals the free charge. Moreover the polarization vector **P** is connected with the induced surface charge (polarization charge). It can also be represented by lines known as lines of **P**. *These lines begin and end on the polarization charges i.e. induced charges due to polarization.*

4.8. ELECTRIC SUSCEPTIBILITY

The phenomenon of electric polarization comes in when a dielectric is placed in an electric field. As such, the three electric vectors **E**, **P** and **D** are related as

$$\varepsilon_0 \mathbf{E} + \mathbf{P} = \mathbf{D}$$

but $$\mathbf{D} = \varepsilon_r \varepsilon_0 \mathbf{E} \qquad \text{from eqn. (8)}$$

$$\mathbf{P} = \varepsilon_r \varepsilon_0 \mathbf{E} - \varepsilon_0 \mathbf{E}$$

or $$\mathbf{P} = \varepsilon_0 (\varepsilon_r - 1) \mathbf{E} \qquad \text{... (2)}$$

ε_0 is the permitivity of free space and is a constant while ε_r is constant for a medium.

The term $(\varepsilon_r - 1)$ is replaced by another constant χ_e, so that

$$\mathbf{P} = \varepsilon_0 \chi_e \mathbf{E} \qquad \text{... (3)}$$

or $$\boxed{\chi_e = \frac{\mathbf{P}}{\varepsilon_0 \mathbf{E}}} \qquad \text{... (4)}$$

χ_e is called the electric susceptibility of dielectric and is dependent on the nature of the material. The electric susceptibility of the dielectric may be defined as *the ratio of polarization to the product of electric field intensity in the dielectric and free space permitivity*. The dielectric susceptibility is **dimensionless**.

Also $$\chi_e = \varepsilon_r - 1$$

$\therefore$ $$\varepsilon_r = 1 + \chi_e \qquad \text{... (5)}$$

This is the relation between **dielectric constant and electric susceptibility.**

Table 4.1. Values of Dielectric Coefficients of Some materials

Material	*Value*
Vacuum	1
Air (1 atom)	1.006
Teflon	2.1
Paper	3.5
Mica	3 – 6
Bakelite	5
Glass	5 – 10
Porcelain	6.5
Glycerine	42.5
Water	78.5
Strontium titanate	310

From Table 4.1 one can see that the value of ε_r for air is very nearly equal to unity. Thus, for most purposes an air capacitor is equivalent to the one in vacuum.

4.9. COULOMB'S LAW AND DIELECTRICS

In case of a capacitor the force between two charges on the plates is given by

$$F_0 \alpha E_0 \qquad \dots (1)$$

So we have

$$\frac{F}{F_0} = \frac{E}{E_0} = \frac{1}{\varepsilon_r} \qquad \dots (2)$$

or

$$F = F_0/\varepsilon_r$$

So, the force between the two charges in the presence of a dielectric reduces by a factor so ε_r. So which may be written as in dielectrics, the Coulomb's law becomes,

$$F = \frac{1}{4\pi\varepsilon_0\varepsilon_r}\frac{q_1q_2}{r_{12}^2} \qquad \dots (4)$$

$$F = \frac{1}{4\pi\varepsilon}\frac{q_1q_2}{r_{12}^2} \qquad \dots (5)$$

As ε_0 is called the permitivity of the free space, ε is called the permitivity of the dielectric. Further, because $\varepsilon_r = \varepsilon/\varepsilon_0$, ε_r is also called the **relative permitivity** of the dielectric. To make the distinction between ε and ε_r more clear, ε is generally called the *absolute permitivity* of the medium.

Example 1. *Two equal and opposite charges of 1 μC each are placed in transformer oil at a distance of 1m. Calculate the force between them. The dielectric coefficient of the oil is 2.2.*

Solution: *The force of attraction between the two charges is given by,*

$$F = \frac{1}{4\pi\varepsilon_0\,\varepsilon_r}\frac{q_1q_2}{d^2}$$

$$= \frac{10^{-6} \times 10^{-6}\,\text{C}^2}{4\pi \times 8.85 \times 10^{-12}\,\text{Fm}^{-1} \times 2.2 \times 1\text{m}^2}$$

$$= \frac{1}{4\pi \times 8.85 \times 2.2}\;\frac{\text{C}^2}{\text{Fm}}$$

$$= \mathbf{0.0046\ N}$$

4.10. GAUSS'S LAW AND DIELECTRICS

Let us take the case of Gauss's law applied to a parallel plate capacitor without a dielectric and then with dielectric. In the absence of dielectric, Gauss's law states that the *surface integral of the normal component of electric field over any closed surface is equal to* $1/\varepsilon_0$ *time the total charge enclosed by the surface i.e.*

$$\oint \mathbf{E}_0 \cdot dA = \frac{q_0}{\varepsilon_0} \qquad \dots (1)$$

where $\mathbf{E}_0$ is the electric field at any point on the gaussian surface Fig. 4.6. q_0 be the free charge on both plates of the capacitor.

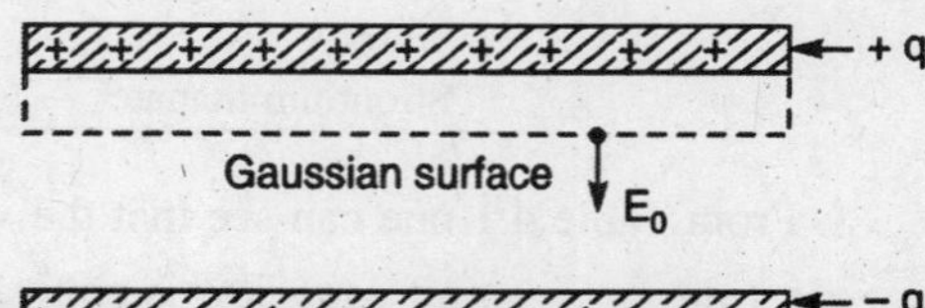

Fig. 4.6. A parallel plate capacitor

$$\mathbf{E}_0 A = \frac{q_0}{\varepsilon_0}$$

or
$$\mathbf{E}_0 = \frac{q_0}{\varepsilon_0 A} \qquad \text{... (2)}$$

Let us take the second case where dielectric is placed between the plates of the capacitor and is under the action of the electric field. The electric field induces appropriate charges on the surfaces of the dielectric and these induced charges produce their own field within the dielectric Fig. 4.7. This field opposes the external field $\mathbf{E}_0$. Let q' be the induced charge on the surfaces of the dielectric, then effective charge within the gaussian surface is $(q_0 - q')$. If E is the resultant field within the dielectric then the Gauss's law is $(q_0 - q')$. If E is the resultant field within the dielectric then the Gauss's law may be stated as

$$\oint \mathbf{E} \cdot dA = \frac{q_0 - q'}{\varepsilon_0} \qquad \text{... (3)}$$

or
$$EA = \frac{q_0}{\varepsilon_0} - \frac{q'}{\varepsilon_0}$$

or
$$E = \frac{q_0}{\varepsilon_0 A} - \frac{q'}{\varepsilon_0 A} \qquad \text{... (4)}$$

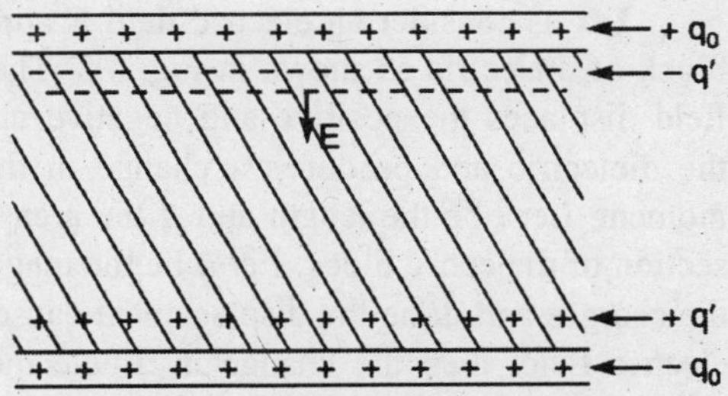

Fig. 4.7. A dielectric between the parallel plates of a capacitor

Then eqn. (4) shows that induced charge q' **tends to weaken** the original field (i.e. $E < E_0$), which means the reduction in the potential difference between the plates. If V_d and V_0 are the potential differences between plates with and without dielectric, in that case

$$\frac{E_0}{E} = \frac{V_0}{V_d} \qquad \text{... (5)}$$

If C_d and C_0 are the corresponding capacitances, then

$$\frac{C_d}{C_0} = \varepsilon_r = \frac{V_0}{V_d} \qquad \text{... (6)}$$

Thus from equns. (5) and (6) we get

$$\frac{E_0}{E} = \varepsilon_r \text{ or } E = \frac{E_0}{\varepsilon_r}$$

Substituting for E_0 from eqn. (2) we get

$$E = \frac{q_0}{\varepsilon_r \varepsilon_0 A} \qquad \text{...(7)}$$

Comparing eqn. (4) and (7), we have

$$\frac{q_0}{\varepsilon_r \varepsilon_0 A} = \frac{q_0}{\varepsilon_0 A} - \frac{q'}{\varepsilon_0 A}$$

i.e.
$$q_0 - q' = \frac{q_0}{\varepsilon_r} \qquad \text{...(8)}$$

This shows that induced charge q' is less than the free charge q_0 and is zero when $\varepsilon_r = 1$ *i.e.* in the absence of dielectric. Putting the value of $(q_0 - q')$ in eqn. (3) we get

$$\oint E.dA = \frac{q_0}{\varepsilon_r \varepsilon_0}$$

$$\oint E.dA = \frac{q_0}{\varepsilon}$$

Where ε is the permitivity of dielectric. Further

$$\oint \varepsilon \mathbf{E}.dA = q_0$$

or

$$\oint \mathbf{D}.dA = q_0$$

where D is the electric displacement vector and eqn. (9) is the Gauss's law in the presence of dielectric. It states that **the surface integral of electric displacement vector over a closed surface is equal to free charge enclosed within the surface.** As D is connected with free charge only, so the effect of *the induced charge is ignored by taking into account the dielectric constant.*

4.11. ENERGY STORED IN ELECTROSTATIC FIELD

Let us consider an electric field **E** applied to a block of dielectric as shown in Fig. 4.8. The electric field displaces the positive and negative charges in the dielectric and produces a change in the dipole moment. Let l be the length and A the area of cross-section of dielectric block. Let q be the magnitude of the charges and dl be the displacement caused by the electric field, then the change in dipole moment is given by

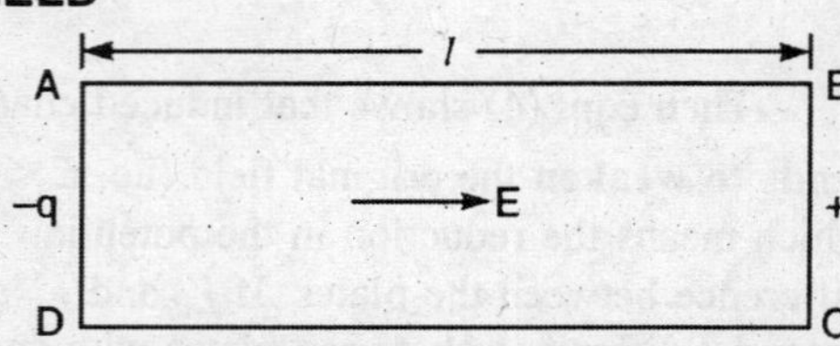

Fig. 4.8. Dielectric block of length l and area of cross-section A is electric field

$$d\mathbf{p} = qdl \qquad \text{...(1)}$$

The force acting on the charge due to electric field is $q\mathbf{E}$. The work done by the force through displacement $d\mathbf{l}$ is given by

$$dW = \text{Force} \times \text{displacement}$$

$$= q\mathbf{E}.\ d\mathbf{l} = \mathbf{E}.\ (qd\mathbf{l})$$

using eq. (1), we get

$$dW = \mathbf{E}.d\mathbf{p} \qquad \text{...(2)}$$

This work is stored as **electrostatic energy** of the dielectric and this energy changes the internal energy of the molecules. The induced dipole moment per unit volume is called the polarization vector and is given by the relation

$$\mathbf{P} = \frac{p}{lA} \quad \text{where} \quad lA = \text{volume of dielectric} = V$$

or

$$\mathbf{p} = V\mathbf{P} \quad \text{or} \quad d\mathbf{p} = Vd\mathbf{P} \qquad \text{... (3)}$$

Putting for **dp** in eqn. (2) we get

$$dW = \mathbf{E}\cdot Vd\mathbf{P}$$

But polarization vector is also given by the relation

$$\mathbf{P} = \mathbf{D} - \varepsilon_0 \mathbf{E}$$

(where $\mathbf{D} = \varepsilon_r \varepsilon_0 \mathbf{E}$)

$$\therefore \qquad \mathbf{P} = \varepsilon_r\varepsilon_0\mathbf{E} - \varepsilon_0\mathbf{E} = (\varepsilon_r - 1)\varepsilon_0\mathbf{E}$$

and
$$d\mathbf{P} = (\varepsilon_r - 1)\varepsilon_0\, d\mathbf{E}$$

Putting for $d\mathbf{P}$ in eqn. (3), we get

$$dW = \mathbf{E}.d\mathbf{E}\ (\varepsilon_r - 1)\varepsilon_0 V$$

Total work done for polarization of the dielectric is given by

$$W = \int dW = \frac{1}{2}\varepsilon_0 V(\varepsilon_r - 1)\mathrm{E}^2$$

Work done per unit volume will be

$$\frac{W}{V} = \frac{1}{2}\varepsilon_0(\varepsilon_r - 1)\mathrm{E}^2$$

This is the energy density due to polarization. Total energy density of the dielectric

U = energy density in free space + energy density due to polarization.

or
$$U = \frac{1}{2}\varepsilon_0\mathrm{E}^2 + \frac{1}{2}\varepsilon_0(\varepsilon_r - 1)\mathrm{E}^2$$

$$U = \frac{1}{2}\varepsilon_0\varepsilon_r E^2 = \tfrac{1}{2}\varepsilon E^2 \qquad \ldots (4)$$

This is the energy stored per unit volume in dielectric in electrostatic field.

4.12. BEHAVIOUR OF DIELECTRICS IN ALTERNATING FIELD

The study of the behaviour of the dielectric in alternating electric field shows that the dielectric constant becomes complex. The imaginary part of the complex dielectric constant accounts for the dielectric losses of the material.

If a dielectric material is placed in an alternating field the orientation of the dipoles and hence the polarization will tend to reverse when the *polarity of the field changes.* So long as the frequency remains low ($< 10^6$ Hz) there is no significant lag in polarization with alterations of field. The permitivity is independent of frequency and has same magnitude as in a static field. As the frequency increases , the *dipoles will not be able to rotate rapidly and their oscillators will lag behind those of the field.* With more increase in frequency, the permanent dipoles in the medium will be unable to follow the field and the contribution to the static permitivity from this *molecular process i.e.* the orientation polarization stops. This generally happens in the radio frequency range ($10^6 - 10^{11}$Hz) of electromagnetic spectrum. At still more frequencies *i.e.* in the infra-red range ($10^{11} - 10^{14}$)Hz, the relatively heavy positive and negative ions cannot follow the field reversals and the contribution to permitivity from atomic or ionic polarization ceases and only electronic polarization persists.

So the permitivity of the dielectric material decreases with increase in frequency and this phenomenon is called anomalous dielectric dispersion.

Dielectric Absorption: Dispersion coming into play during the transition from full atomic polarization at radio frequencies to negligible atomic polarization at optical frequency is referred to as **dielectric absorption**.

Dielectric Relaxation: Dispersion due to the transition from full orientational polarization at zero or low frequencies to negligible orientational polarization at high frequency is known as **dielectric relaxation**.

4.13. CLAUSIUS MOSOTTI EQUATION

In Clausius Mosotti theory we consider the actual *non-polar molecules of a homogeneous isotropic* lying between the plates of a parallel plate condenser, size of *plates being very large as compared to the distance between them.*

Clausius Mosotti equation gives the *dielectric constant in terms of molecular properties.*

When a non-polar molecule is placed in an electric field, the actual intensity of electric field *i.e.*, Lorentz field acting on any molecule is given by

$$F = E + \frac{P}{3\varepsilon_0} \quad \text{... (1)}$$

Dipole moment induced in a molecule is given by

$$m = \alpha e_0 F = \alpha\varepsilon_0\left(E + \frac{P}{3\varepsilon_0}\right) \quad \text{... (2)}$$

Where α is atomic polarizability

$\therefore$ Dipole moment/unit volume

$$P = n\alpha\,\varepsilon_0\left(E + \frac{P}{3\varepsilon_0}\right) \quad \text{... (3)}$$

where $n \rightarrow$ number of molecules/unit volume

Also we have the relation

$$D = \varepsilon_0 E + P$$

$$P = D - \varepsilon_0 E$$

From eq. (3) and (4) we get

$$D - \varepsilon_0 E = P = n\alpha\,\varepsilon_0\left(E + \frac{P}{3\varepsilon_0}\right)$$

$$= n\alpha\,\varepsilon_0\left(E + \frac{D - \varepsilon_0 E}{3\varepsilon_0}\right) \quad \text{... (5)}$$

$$= n\alpha\left(\frac{D + 2\varepsilon_0 E}{3}\right)$$

$$\therefore \quad \frac{D - \varepsilon_0\, E}{D + 2\varepsilon_0\, E} = \frac{n\alpha}{3}$$

Also $\quad D = \varepsilon E$

$$\therefore \quad \frac{\varepsilon-\varepsilon_0}{\varepsilon+2\varepsilon_0}=\frac{n\alpha}{3} \qquad \text{... (6)}$$

Put $\varepsilon=\varepsilon_0 \cdot \varepsilon_r$, we get

$$\frac{\varepsilon_r-1}{\varepsilon_r+2}=\frac{n\alpha}{3} \qquad \text{... (7)}$$

eqs. (5) and (6) give the simplest form of *Clausius-Mosotti equation* and gives the permitivity or *diėlectric constant* in terms of the **polarisability** of the dielectric.

The function $\frac{\varepsilon-\varepsilon_0}{\varepsilon_r+2\varepsilon_0}$ or $\frac{\varepsilon_r-1}{\varepsilon_r+2}$ is called **volume polarization** and depends only on the number of molecules/unit volume and their of the polarizabilities. It is dimensionless and is denoted by π.

$$\pi = \frac{\varepsilon-\varepsilon_0}{\varepsilon_r+2\varepsilon_0}=\frac{n\alpha}{3} \qquad \text{... (8)}$$

Put $\varepsilon_r=\chi+1$, where χ is the susceptibility *f* the dielectric, we get

$$\frac{\chi}{\chi+3}=\frac{n\alpha}{3}$$

Which gives us the *Clausius Mosotti* equation in terms of susceptibility of the dielectric. Multiplying sides of eq. (6) by specific volume v, which is *reciprocal of density d*, we get

$$p = \pi v = \frac{\varepsilon-\varepsilon_0}{\varepsilon+2\varepsilon_0}\cdot\frac{1}{d}=\frac{1}{3}\times n\alpha v$$

$$= \frac{1}{3}z\alpha$$

where $z = nv$ is the number of molecules per unit mass, and p is called the specific polarization Again multiplying both sides of eq. (8) by molecular weight of the substance, we get

$$p_m = p\cdot M = \frac{\varepsilon-\varepsilon_0}{\varepsilon+2\varepsilon_0}\cdot\frac{M}{d}=\frac{1}{3}z\alpha$$

$$p_m = \frac{\varepsilon-\varepsilon_0}{\varepsilon+2\varepsilon_0}\cdot V_M = \frac{1}{3}N\alpha$$

where $V_M \rightarrow$ molecular volume

$N \rightarrow$ Avocado's polarization

$p_m \rightarrow$ molar polarization

it may be noted that quantities p, p_m and α have the dimensional of volume.

if we put $\varepsilon = \varepsilon\eta^2$ in eq (9) where η is the *refractive index of the medium* we get

$$\frac{\eta^2 - 1}{\eta^2 + 2} \cdot V_M = \frac{1}{3} N\alpha \qquad \text{... (10)}$$

This is known as **Lorentz** equation of light which was derived on e.m. theory considerations. We can also write Clausius equation (eq. 6) as

$$\varepsilon_r = \frac{1 + \frac{2n\alpha}{3}}{1 - \frac{n\alpha}{3}} \qquad \text{... (11)}$$

Which shows that εr must increase rapidly and approach infinity as the number of molecules per unit volume approach a critical value *n, i.e.,*

$$\varepsilon_r = \infty \text{ if } \frac{n_c \alpha}{3} = 1$$

Physically, $\varepsilon_r \Rightarrow \infty$ corresponds to the dielectric volume filled with infinitely polarisable *i.e.*, conducting molecules. In case of gases and liquids this condition is never reached due to the fact that their polarizations are relatively small. So, the **Clausius Mosotti equation is valid for gases and liquids**. In case of solids, the interaction of dipoles is quite complicated and so the above simple model does not hold good. So, the **Clausius Mosotti equation are not valid for solids**.

4.14. FREQUENCY AND TEMPERATURE DEPENDENCE OF POLARIZATION

(*i*) Frequency Dependence

In many practical applications, a dielectric is subjected a.c. fields which change its direction with time. With each reversal, polarization components are supposed to follow the field reversal in order to contribute to the total polarization of dielectric. It means that total polarization depends on the ability of dipoles to orient themselves in the field direction during each alternation of the field.

It is observed that the relative permitivity which is a measure of polarization shows different frequencies. Dependence of ε_r on different frequencies of electric field is shown in Fig. 3.19 for a polar dielectric.

As is clear from Figure 4.9, in audio frequency region, all types of polarizations are possible and the dielectric is characterized by Polarization $P = P_e + P_i + P_a$ and polarizability $\alpha = \alpha_e + \alpha_i + \alpha_a$.

At low frequencies, dipoles get sufficient time to orient themselves completely along the instantaneous directions of the field. *This orientation occurs first in one direction and then in other following the changes in the direction of the field.*

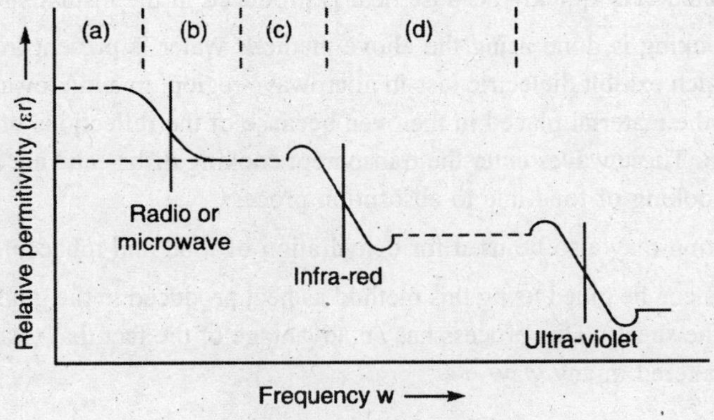

Fig. 4.9

The average time taken by a dipole to reorient in the field direction is called the *relaxation time* τ and reciprocal of relaxation time is called *relaxation frequency*.

If the frequency of applied electric field is much higher than the relaxation frequency of dipoles, the dipoles cannot reverse fast enough. However, if the dipole relaxation time ε is half the time period of applied electric field, i.e., $\left(\tau = \frac{1}{2}T\right)$ dipoles can easily follow the field reversal and so will contribute to orientation polarization. So, the orientation polarization, which is effective at low frequencies, becomes ineffective at high frequencies. Usually in radio or microwave frequency band regions, permanent dipoles fail to follow the field reversal and so the polarization falls to a value corresponding to $P_i + P_e$. *It means that* ε_r *decreases considerably.*

In the infrared region, ionic polarization fails to follow field reversal due to the inertia of the system and so contribution of ionic polarization stops. In this region, only electronic polarization contributes to the total polarization. So, $P = P_e$.

In the optical region, an electron cloud follows the field variations and the material exhibits an electronic polarizability α_e.

Relative permitivity in optical region comes out to be equal to the square of refractive index '*n*' of the dielectric , i.*e*.,

$$[\varepsilon_r]_{\text{optical region}} = n^2$$

In ultraviolet region, electron cloud fails to follow the field reversal and the electronic polarization also stops. In this case, the relative permitivity approaches unity and so for frequencies above ultraviolet range, we have

$$[\varepsilon_r]_{\text{X-rays}} = 1$$

To give an example, water has relative permitivity $\varepsilon_r = 80$ (often referred to as static dielectric constant) at room temperature. However, its value falls to 1.8 in the optical region.

(*ii*) Temperature Dependence

Dielectric materials respond very effectively to the variation of temperature. When a dielectric is heated by subjecting it to high potential of suitable frequency, *i.e*., the frequency at which dielectric loss is maximum and this loss appears in the form of heat. We can have adequate heating by the use of 20 kV electric field with a frequency ~ 30MHz. An important advantage of use of this method is the material gets heated very quickly because heat is produced in the insulating material itself.

Microwave cooking is done using the above method. Water is present in all the articles of cooling materials which exhibit dielectric loss in microwave region. In a microwave oven, sufficient heat is absorbed by the material placed in the oven because of the reflections of microwaves from the walls of the oven. These waves enter the transparent cooking dishes and are absorbed by water molecules thereby cooking of food due to absorption process.

Dielectric heating may also be used for dehydration of food and tobacco etc.

Wooden sheets can be glued using this method as heat produced in the glue due to absorption leads to binding of the sheets. This process has an advantage of the fact that total moisture content of the sheets is not altered in any way.

4.15. DIELECTRIC LOSSES AND DIELECTRIC BREAKDOWN

Dielectric Losses:

When an electric field is applied on any matter, it dissipates some amount of electric energy that transforms into heat energy. This phenomenon is commonly known as "the expense" or "loss" of power. In general, power loss in a material is directly proportional to the square of the applied voltage *i.e.*, $P \propto V^2$.

When a conductor is connected to a *dc* and *ac* voltage *V*, the power loss in it is given by the relation.

$$P = \frac{V^2}{R} = I^2 R \quad \text{... (1)}$$

In a conducto

$$P_{dc} = P_{ac}$$

The dielectric losses which occur in the infrared region are due to **ionic vibrations** and fall under the category of **infrared absorption**. The losses which occur in visible region are due to **electronic transition** and fall under the category of **optical absorption**.

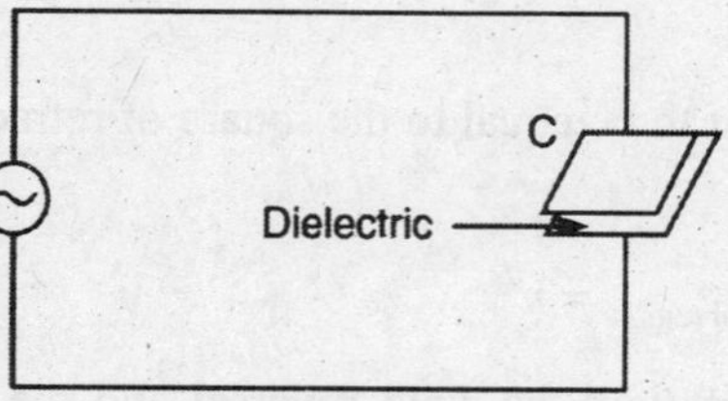

Fig 4.10. A dielectric in an alternating field

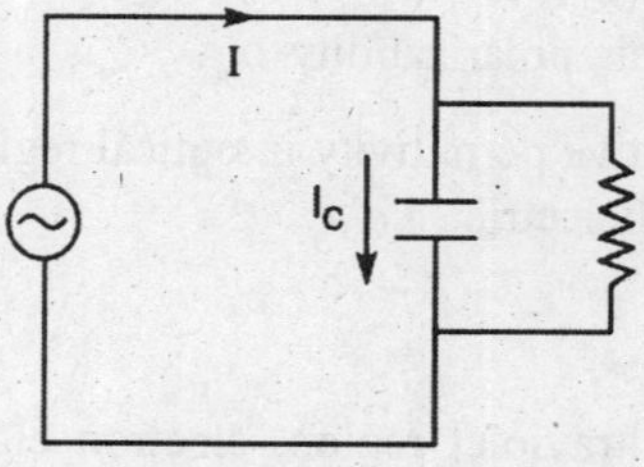

Fig. 4.11. RC circuit

Dielectric Breakdown:

The various phenomena that occur in dielectrics under the action of an electric field are electrical conduction, polarization and dielectric losses.

If the voltage applied to an insulation is increased in steps there sill finally take place a **breakdown** of the insulation. The maximum voltage V_{br} applied to an insulation at the moment of the breakdown is called the **breakdown voltage** of insulation.

The further this phenomena in insulation after a breakdown depends on both the nature of electrical insulating material and the power of the source of electrical energy. **The point of breakdown creates a spark or an electric arc which can fuse, burn and crack the dielectric.**

After the voltage is removed a solid dielectric may exhibit a trace of breakdown in the form of a punctured, fused or burnt-through hole, generally speaking of an irregular shape. So a breakdown of solid insulation in an electrical machine, apparatus, cable, etc., is a failure and serious repairs may be required.

The dielectric breakdown can be explained on the basis of energy band diagram. We know that the conduction band is completely empty and the valence band is completely filled for dielectric materials. As the energy gap is very large (> 5 *e*V) electrons in the valence band need more voltage to jump into the conduction band.

An electric breakdown develops as a result of interaction of free charged particles (electrons, ions) accelerated by an electric field with the particles of a dielectric, or as a result of inelastic displacement of bound charges in a dielectric under the action of an external electric field.

Electric breakdown has the following consequences:

(*i*) a comparatively small dependence of electric strength on the dimensions of the dielectric and the electrodes.

(*ii*) a small dependence of electronic field on the frequency of the applied voltage.

(*iii*) a small dependence of electric strength on temperature.

(*iv*) a short time of the development of break down.

Example 2. *What is the area of the plates of a 2 F parallel plate capacitor, given that the separation between the plates is 0.5 cm?*

Given $C = 2\ F,\ d = 0.5 \times 10^{-2}\ m,\ \varepsilon_0 = 8.85 \times 10^{-12}\ F/m$

Solution: Capacitance of a parallel capacitor is given by the relation

$$C = \frac{\varepsilon_0 A}{d}$$

$$\therefore \quad A = \frac{Cd}{\varepsilon_0} = \frac{2 \times 0.5 \times 10^{-2}}{8.85 \times 10^{-12}} = \mathbf{1.13 \times 10^9\ m^2\ Ans.}$$

Example 3. *A parallel plate capacitor is charged by connecting it to a 90 V battery. The battery is then disconnected and an insulating liquid is poured between the plates to fill the air gap. The potential difference is now 30 V. What is the dielectric constant of the liquid?*

Solution. Given that $V_0 = 90\text{V}$, $V = 30\text{V}$ and $\varepsilon_r = ?$

Capacitance of air filled dielectric, $C_{air} = \dfrac{Q}{V_0}$

When the battery is not present and dielectric liquid is poured charge Q remains same, while potential difference becomes, $V = \dfrac{V_0}{\varepsilon_r}$

$$\therefore \text{ Capacitance, } C_{liquid} = \frac{Q}{V} = \frac{Q}{V_0/\varepsilon_r} = \frac{Q}{V_0}\cdot\varepsilon_r$$

$$\text{or Dielectric constant, } \varepsilon_r = \frac{C_{liquid}}{C_{air}} = \frac{V_0}{V} = \frac{90}{30} = \mathbf{3\ Ans.}$$

Example 4. *A parallel plate capacitor has circular plates of radius 4 cm and plate separation 1 mm. Calculate the charge on the plates when a potential difference of one kilo volt is applied to it?*

Solution: Given that $r = 4 \times 10^{-2}$m, d = 1 mm = 1×10^{-3} m, $V = 1000\ V$, and $Q = ?$

Area of the circular plates, $r = \pi r^2 = \pi \times (4 \times 10^{-2})^2 = 16\pi \times 10^{-4}\,\text{m}^2$

So we have the relation for capacitance as

Capacitance, $$C = \frac{\varepsilon_0 A}{d}$$

$$C = \frac{(8.85 \times 10^{-12})(16 \times 3.14 \times 10^{-4})}{1 \times 10^{-3}}$$

Charge $$Q = CV$$

$$= [(8.85 \times 10^{-12})(16 \times 3.14 \times 10^{-1})]\, 10^3$$

$$= \mathbf{4.45 \times 10^{-8}\ C\ Ans.}$$

Example 5. *A parallel plate capacitor is constructed using three dielectric materials as shown in Fig. 4.12. The parallel plates P and Q across which a potential difference is applied arc of area A = 1 cm² and are separated by d = 2 mm. If $\varepsilon_{r_1} = 4.9$, $\varepsilon_{r_2} = 2.1$, find the capacitance between P and Q.*

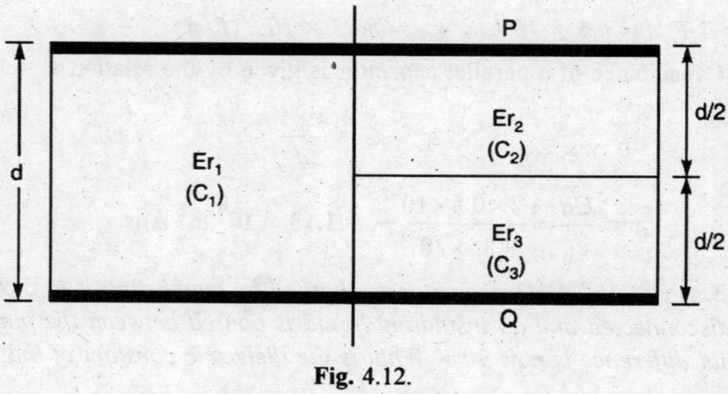

Fig. 4.12.

Solution: Given $A = 1\ \text{cm}^2 = 10^{-4}\text{m}^2$, d = 2×10^{-3} m.

As shown in the Fig. 4.12 given arrangement is equivalent to three capacitors C_1, C_2, C_3 where

$$C_1 = \frac{\varepsilon_0 \varepsilon_{r1} A/2}{d},\quad C_2 = \frac{\varepsilon_0 \varepsilon_{r_2} A/2}{d/2} \text{ and } C_3 = \frac{\varepsilon_0 \varepsilon_{r_3} A/2}{d/2}$$

$$\therefore \quad C_1 = \frac{\varepsilon_0 \varepsilon_{r_1} A}{2d},\ C_2 = \frac{\varepsilon_0 \varepsilon_{r_2} A}{d} \text{ and } C_3 = \frac{\varepsilon_0 \varepsilon_{r_3} A}{d} \quad i.e\ C_2 = C_3$$

From Fig. 4.12, it is seen that C_2 and C_3 are connected in series = C_{23} and C_1 and C_{23} are in parallel.

Resultant capacitance may be written as

$$C = C_1 + C_{23} = C_1 + \frac{C_2 C_3}{C_2 + C_3}$$

$$\therefore \quad C = \frac{\varepsilon_0 \varepsilon_{r_1} A}{2d} + \frac{\left(\frac{\varepsilon_0 \varepsilon_{r_2} A}{d}\right) \times \left(\frac{\varepsilon_0 \varepsilon_{r_3} A}{d}\right)}{\frac{\varepsilon_0 \varepsilon_{r_2} A}{d} + \frac{\varepsilon_0 \varepsilon_{r_3} A}{d}}$$

$$= \frac{A\varepsilon_0}{d}\left[\frac{\varepsilon_{r_1}}{2} + \frac{\varepsilon_{r_2} \times \varepsilon_{r_3}}{\varepsilon_{r_2} + \varepsilon_{r_3}}\right]$$

$$= \frac{10^{-4} \times 8.85 \times 10^{-12}}{2 \times 10^{-3}}\left[\frac{4.9}{2} + \frac{5.6 \times 2.1}{5.6 + 2.1}\right] = \mathbf{1.76}\ \boldsymbol{pF}\ \textbf{Ans.}$$

Example 6. *A parallel plate capacitor consists of 2 plates each of area 5 × 10^{-4} m^2. They are separated by a distance 1.5 × 10^{-3} m and filled with a dielectric of relative permitivity 6. Calculate the charge on the capacitor if it is connected to a 100 V dc supply.*

Solution: Given $A = 5 \times 10^{-4}$ m^2, $d = 1.5 \times 10^{-3}$ m, $\varepsilon_r = 6, V = 100V$ and $Q = ?$

We know, Charge, $Q = CV$ ...(1)

Capacitance, $$C = \frac{\varepsilon_0 \varepsilon_r A}{d}$$ ...(2)

$\therefore$ Charge, $$Q = \frac{\varepsilon_0 \varepsilon_r A}{d} \times V$$

$$= \frac{(8.85 \times 10^{-12})(6)(5 \times 10^{-4})}{1.5 \times 10^{-3}} \times 100$$

$$= \mathbf{1.77 \times 10^{-9}\ C\ Ans}$$

Example 7. *The dielectric constant of He gas at NTP is 1.0000684. Calculate the electronic polarizability of the atoms if He gas contains 2.7 × 10^{25} atoms per m^3.*

Solution: Given ε_r = 1.0000684, N = 2.7 × 10^{25} atoms/m^3 and $\alpha_e = ?$

Electronic polarizability, $$\alpha_e = \frac{\varepsilon_0(\varepsilon_r - 1)}{N}$$

$$= \frac{8.85 \times 10^{-12}(1.0000684 - 1)}{2.7 \times 10^{25}} = \mathbf{2.242 \times 10^{-41}\ Fm^2\ Ans.}$$

Example 8. *What is the resulting voltage across the plates of a parallel plate capacitor with plates of area 100 mm², carrying a charge of 3 × 10^{-10} C and separated by a distance of 50 mm when a material of dielectric constant 4 introduced between them.*

Solution: Given $A = 100 \times 10^{-4}$ m^2, $Q = 3 \times 10^{-10}$ C. $d = 50 \times 10^{-3}$ m, $\varepsilon_r = 4$ and $V = ?$

Capacitance of the capacitor is given by the relation

$$C = \frac{\varepsilon_0 \varepsilon_r A}{d}$$

Voltage across the plates of a parallel plate capacitor is given by

$$V = \frac{Q}{C} = \frac{Q}{\varepsilon_0 \varepsilon_r A/d} = \frac{Qd}{\varepsilon_0 \varepsilon_r A}$$

$$= \frac{(3\times10^{-10})\,(50\times10^{-3})}{(8.85\times10^{-12})\,(4)\,(10^{-2})} = \frac{1500}{8.85\times4}$$

$= 42.37$ **V Ans.**

Q.9. *A parallel plate capacitor of area 650 mm^2 and a plate separation of 4 mm has a charge of 2 × 10^{-10} C on it. What is the resultant voltage across the capacitor when a material of dielectric constant 3.5 is introduced between the plates.*

Solution: Given $A = 650 \times 10^{-4}$ m^2, $d = 4 \times 10^{-3}$ m, $Q = 2 \times 10^{-10}$ C, $\varepsilon_r = 3.5$ and $V = ?$

Capacitance of the capacitor is given by the relation

$$C = \frac{\varepsilon_0 \varepsilon_r A}{d} \quad \text{...(1)}$$

and

$$C = \frac{Q}{V} \quad \text{...(2)}$$

Comparing Eqs. (1) and (2), we get

$$\frac{Q}{V} = \frac{\varepsilon_0 \varepsilon_r A}{d}$$

or

$$V = \frac{Qd}{\varepsilon_0 \varepsilon_r A}$$

$$= \frac{(2\times10^{-10})\,(4\times10^{-3})}{(8.85\times10^{-12})\,(3.5)\,(650\times10^{-4})} = 39.73 \text{ V Ans.}$$

QUESTIONS AND PROBLEMS

1. (a) What are dielectrics? Discuss the effect of electric field on dielectric materials.
   (b) Write about important applications of dielectric materials.
2. What are dielectric materials? Explain their properties. Write examples.
3. What are dielectrics? Distinguish between polar and nonpolar dielectrics.
4. Write a note on a molecular theory of dielectrics.
5. What are dielectric materials? Define electric constant of a material. Explain the phenomenon of electric polarization in dielectric materials.
6. Explain the behaviour of dielectric under static electric fields.
7. Describe the different polarization mechanisms.
8. Define dielectric susceptibility and polarizability of a dielectric and write the relation between the two.
9. Define dielectric polarization. Explain dielectric polarization in case of (i) polar dielectric and (ii) non-polar dielectric.
10. Write a note an dielectric polarization.

11. Define dielectric constant and bring out a relation between electric polarization and dielectric susceptibility.
12. Define dielectric polarization. Explain the different polarization mechanisms with illustrations.
13. What are dielectrics? Discuss their important uses.
14. What is Lorentz field?
15. Write a note on dielectric loss in a dielectric material.
16. Define relaxation time and relaxation frequency.
17. Explain the frequency dependence of the dielectric constant.
18. The capacity of a parallel plate capacitor with a dielectric of 6 is $24\mu F$. Find the capacity of the capacitor with air as dielectric. **[Ans: 4 μF]**
19. Two perfectly insulated capacitors are connected in series. One is an air capacitor with plate area 0.01 m^2, the plates being 1 *mm* thick with a dielectric constant 5. Determine the voltage across the combination if the potential gradient in air-capacitor is 200 *V/m*. **[Ans: 240 *V*]**
20. Two square metallic plates of side 1 *m* are kept 0.01 m apart like a parallel plate capacitor, in air in such a way that one of their edges is perpendicular to an ON surface in a tank filled with an insulating ON, the plates are connected to a battery of *emf* 500 *V*. The plates are then lowered vertically into the oil at a speed of 0.001 *m/s*. Calculate the current drawn from the battery during the process $(\varepsilon_{0il} = 11, \varepsilon_0 = 8.85 \times 10^{-12} F/m)$. **[Ans: 4.425 × 10^{-9} *A*]**
21. A parallel plate capacitor is to be designed with a voltage rating 1 *kV*, using a material of dielectric constant 3 and dielectric strength 10^7 *V/m*. What minimum area of the plates is required to have capacitance of 50 *pF*? **[Ans: 1.883 × 10^{-4} m^2]**
22. What is the capacitance per *km* length of a wire 16 *mm* in diameter and insulated with a thickness of 10 *mm* of gutta = percha (Specific inductive capacity = 2.46) and placed in water. **[Ans: 0.169 μF]**
23. A parallel plate capacitor has circular plates of 10 *cm* radius and 1 *mm* separation. (a) Calculate the capacitance. (b) What charge will appear on the plates if a potential difference of 120 *V* is applied? **[Ans: 2.78 × 10^{-12} *F*, 33.4 × 10^{-11} *C*]**
24. The plates of a parallel plate capacitor have an area of 90 cm^2 each and are separated by 2.5 *mm*. The capacitor is charged by connecting it to 400 *V* supply.

    (a) How much electrostatic energy is stored by the capacitor ?

    (b) View this energy as stored in the electrostatic field between the plates and obtain the energy per unit volume. **[Ans: 2.548 × 10^{-6} *J*, 0.113 J/m^3]**
25. The capacity of a parallel plate capacitor is 7.0 μF . If the plate area is 14.0 cm^2, calculate the plate separation. **[Ans. 1.78 × 10^{-3} *m*]**
26. The capacity of a parallel capacitor is equal to that of a sphere of diameter 1 *m*. Calculate the separation of the plates if the area of each plate is 12.56 sq. *mm*. **[Ans: 1.998 × 10^{-4} *m*]**

CHAPTER

5

SCALAR AND VECTOR FIELDS

Introduction

5.1 SCALAR AND VECTOR QUANTITIES

Physical quantities which can be measured have been classified into two catagories:

1. Scalars: Which require only the magnitude to define then *mass,* length, time, density, speed, volume, temperature, work, energy, current, charge, potential etc.

2. Vectors: To define some quantities completely both magnitude and direction have to be specified *e.g.; Displacement*, Velocity, acceleration, current density, magnetic moment, electric and magnetic fields, momentum, force, Torque weight etc.

An important difference between Scalar and Vector quantities is that while *Scalar* can be added, subtracted, multiplied or divided by simple arithmetics, for Vector quantities, Vector algebra is to be used for addition, subtraction, multiplication or division.

5.2 REPRESENTATION OF A VECTOR

A Vector $\vec{A}$ with its components, Ax, Ay, and Az along x, y and z direction respectively is represented as

$$\vec{A} = Ax\,\hat{i} + Ay\,\hat{j} + Az\,\hat{k}$$

The magnitude of Vector $\vec{A}$ is given by

$$|A| = \sqrt{A_x^2 + A_y^2 + A_z^2}$$

It the angle made by $\vec{A}$ with x, y and z direction are α, β and γ respectively, then direction cosines l, m, and n for x, y and z axes are

$$l = \cos\alpha = \frac{Ax}{|A|}$$

$$m = \cos\beta = \frac{Ay}{|A|}$$

$$n = \cos\gamma = \frac{Az}{|A|}$$

$$\therefore \quad l^2 + m^2 + n^2 = \cos^2\alpha + \cos^2\beta + \cos^2\gamma$$

$$= \frac{A_x^2 + A_y^2 + A_z^2}{A^2} = 1$$

5.3 PRODUCT OF TWO VECTORS

Product of two Vectors can be a *Scalar or a Vector*, so the Vector can be multiplied in two ways.

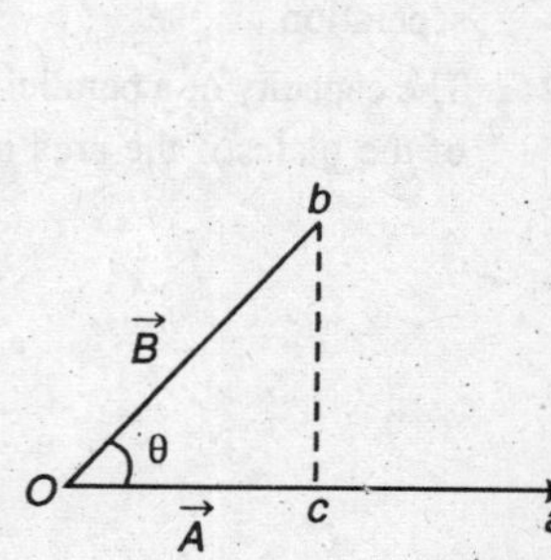

Fig. 5.1

1. Dot or Scalar Product
2. Cross or Vector Product.

1. *Doi or Scalar Product*: Dot product of two Vectors $\vec{A}$ and $\vec{B}$ is represented by $\vec{A}.\vec{B}$. It is a Scalar quantity whose magnitude is the product of magnitude of $\vec{A}$ and $\vec{B}$ and cosine of the angle between them.

From Fig. 5.1 Vectors $\vec{A}$ and $\vec{B}$ can be represented by $\vec{Oa}$ and $\vec{Ob}$; angle θ is the angle between two vectors

$$\therefore \quad \vec{A}.\vec{B} = |A|B\cos\theta$$

2. *Cross or Vector Product:* Cross product of two Vectors is represented by $\vec{\mathbf{A}}\times\vec{\mathbf{B}}$. It is a vector quantity whose magnitude is equal is the product of magnitude of $\vec{\mathbf{A}}$ and $\mathbf{B}$ and the sine of angle by which $\vec{\mathbf{B}}$ is inclined on $\vec{\mathbf{A}}$. The direction is ⊥ to the plane $\vec{\mathbf{A}}$ and $\vec{\mathbf{B}}$. The direction is given by right hand screw rule, viz, positive direction is that in which a right handed screw handed screw of rotated from first Section Vector through a *smaller angle between them advances.*

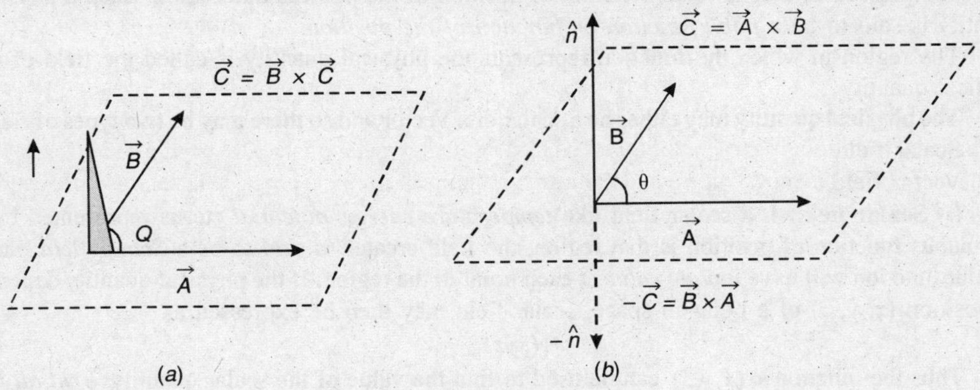

Fig. 5.2

where $\vec{C} = |A||B|\sin(\vec{A},\vec{B})$ $\hat{n}$ is a unit Vector normal to the plane containing $\vec{A}$ and $\vec{B}$ as shown in Fig. 5.2 (*a* and *b*).

It may be noted that unlike Scalar product, Vector product does not obey the commutative law because

$$\vec{A}\times\vec{B} = |A|\ |B|\sin(\vec{A},\vec{B})\hat{n}$$

and

$$\vec{B}\times\vec{A} = |B|\ |A|\sin(\vec{B},\vec{A})\hat{n}$$

$$= -|B|\ |A|\sin(\vec{A},\vec{B})\hat{n}$$

$$\therefore \quad \vec{A}\times\vec{B} = -\vec{B}\times\vec{A}$$

Also if the two Vectors $\vec{A}$ and $\vec{B}$ are represented by the two sides of a parallelogram, the $\vec{A}\times\vec{B}$ represents the *area of the parallelogram.*

If $\quad \vec{A} = Ax\hat{i} + Ay\hat{j} + Az\hat{k}$

and $\quad \vec{B} = Bx\hat{i} + By\hat{j} + Bz\hat{k}$

Then $\vec{A}\times\vec{B}$ may be represented by

$$\vec{A}\times\vec{B} = \begin{vmatrix} \hat{i} & \hat{j} & \hat{k} \\ Ax & Ay & Az \\ Bx & By & Bz \end{vmatrix}$$

Some physical examples of Vector or Cross product are:

(*i*) Torque τ, which is represented by the moment of force $\vec{F}$ about a point whose position vector is $\vec{r}$, it is represented by

$$\vec{z} = \vec{r}\times\vec{F}$$

(*ii*) Force on a charge quality with a velocity $\vec{V}$ is a magnetic field $\vec{B}$. It may be expressed as

$$\vec{F} = \vec{V} \times \vec{B}$$

(*iii*) Angular momentum of a rotating particle, which may be expressed as

$$\vec{L} = \vec{r} \times \vec{p}.$$

5.4. SCALAR AND VECTOR FIELDS

If a physical quantity has varying values at different points in space *i.e.,* the quantity has to be represented by different *co-ordinates* at different points in space so that the physical quantity may be expressed as a continuous function of position of the point is that region. Such a physical quantity is said to be a *point function or function of the position.*

The region in which the function represents the physical quantity is called the field of the physical quantity.

The physical quantity may either be a scalar or a Vector and so there may be two types of fields

(*i*) Scalar field

(*ii*) Vector field.

(*i*) **Scalar field:** It a scalar field like *temperature energy, potential* etc. is represented by a continuous function of position is that region, the field created is said to be a *Scalar field.* Such a scalar function will have *unique value* at each point in the region. If the physical quantity depends on position (*x*, *y*, *z*) of a point in space, scalar field may then be expressed as

$$\phi = \phi(xyz)$$

Thus the function $\phi(x, y, z)$ can be used to find the value of the scalar quantity ϕ *at all the points of the region.*

An example of the scalar the *distribution of temperature* in a metallic rod whose one end is hot while other is cold. Similarly distribution of gravitational potential in gravitation field, distribution of electric or magnetic potentials in this respective fields are examples of scalar.

A graphical representation of scalar field shows that the function $\phi(x, z)$ has fixed and same value on each point of a surface but different values at different surfaces. Such a surface is called *equi-Scalar surface.* In case the potential is same on a surface it is called equi-potential surface, while for same fixed temperature on a surface, it is called *equi-temperature surface.*

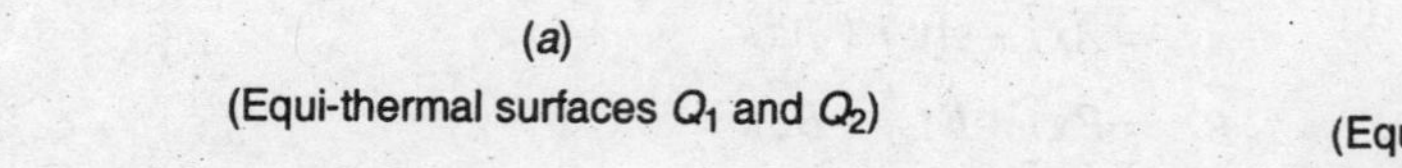

(*a*)

(Equi-thermal surfaces Q_1 and Q_2)

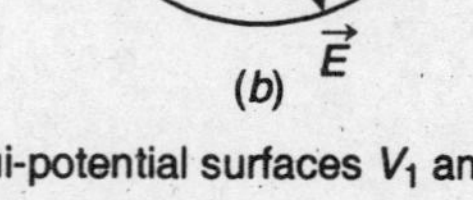

(*b*)

(Equi-potential surfaces V_1 and V_2)

Fig. 5.3

For example, in case of distribution of temperature in a metallic rod, with one end hot and other cold, we can consider isothermal surfaces $\perp$ to the direction of heat flow.

In a steady state, each isothermal will correspond to *definite temperature*. Similarly, in case of distribution of charge, each sphere surrounding the charge will correspond to *equi-potential surface* having the same fixed potential for a particular sphere.

(*ii*) **Vector field:** It a Vector quantity $\vec{\mathbf{F}}$ (like velocity, force, electric field, magnetic field etc) is expressed by a continuous function of position (*x*, *y*, *z*), in a region *F* (*x*, *y*, *z*), a vector field is said to exist in that region. It is important to note that at each point in their region, the magnitude and direction of the vector function is well defined and it *changes from point to point*. An important example of vector field is the velocity distribution in a field. Another example is variation of electric and magnetic fields etc.

Representation of a vector field: A vector field may graphically be represented by a curve joining the direction of various points in that vector field. The *tangent drawn* at any point on this curve gives the direction of the vector *at that point.*

This curve is called vector line or flux line or line of flow of *the vector field.*

The magnitude of the vector at any point of the flux line is represented by taking a surface of unit area normal is the flux line considering the various points on it *such that the number of point is numerically equal to the magnitude of the vector at that point.* Now if we draw parallel flux lines through each of these points the direction of flux lines represent the direction of vector function $\vec{F}_{(x,y,z)}$ and the *number of flux lines* through unit area normal is the flux line represent the *magnitude of the vector function at that point.*

Figure 5.4 (*a*) represents the non-uniform velocity distribution in different layers of the fluid through a small tube in *streamline flow* condition while Fig. 5.4 (*b*) represents the flux lines of a uniform electric field between two oppositely charged conductors.

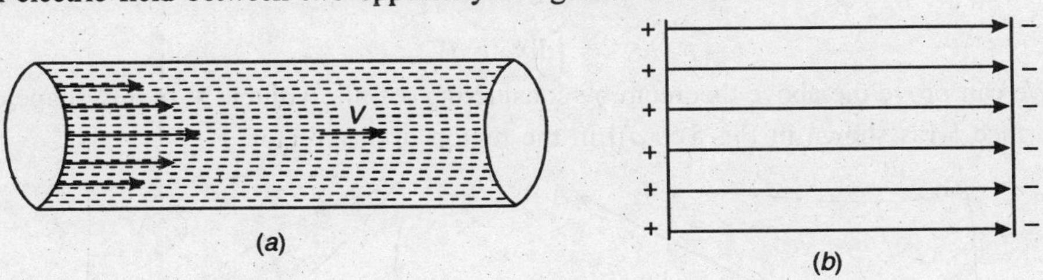

Fig. 5.4

5.5. GRADIENT OF A SCALAR FIELD

It a scalar function of a point (*x*, *y*, *z*) in space is represented by $\phi_{x,y,z}$ (which means that $\phi_{x,y,z}$ is a scalar field), its Cartesian components along three naturally ⊥ axes *X*, *Y* and *Z* are respectively $\frac{\partial \phi}{\partial x}, \frac{\partial \phi}{\partial y}$ and $\frac{\partial \phi}{\partial z}$. The gradient of Scalar function $Q(x, y, z)$ is then defined in the following way:

$$\text{grade } \phi = \hat{i}\frac{\partial \phi}{\partial x} + \hat{j}\frac{\partial \phi}{\partial y} + \hat{k}\frac{\partial \phi}{\partial z}$$

$$= \left(\hat{i}\frac{\partial}{\partial x} + \hat{j}\frac{\partial}{\partial y} + \hat{k}\frac{\partial}{\partial z}\right)\phi$$

$$= \vec{\nabla}\phi.$$

Which means that the gradient of a Scalar function is a vector.

where $\vec{\nabla}$ is a vector operator and is called del operator.

5.6. PHYSICAL SIGNIFICANCE OF GRADIENT OF A SCALAR FIELD

In Physics, gradient of a scalar field finds lot of significance. For example the negative gradient of electric potential at a point defines the electric field at that point. Which means

$$\vec{E} = -\text{grade } \nabla = -\vec{\nabla} V.$$

So the electric field at a point is a Vector whose magnitude is equal to the maximum rate of variation of electric potential and direction is that maximum decrease in potential *i.e.*, right angles to the equipotential surface at that point [Fig. 5.3 (*b*)]. It may be stated in other form as the electric potential is a scalar quantity whose negative gradient gives us the electric field.

In case of a potentiometer, a battery may be used to maintain a constant potential difference between the ends of the potentiometer wire by connecting positive terminal of two battery at one end and negative terminal at the other end. A potential gradient is said to exist along the length of the wire is opposite to the direction of potential gradient. In the same way, when heat flows through a rod by heating one end and keeping the other end of the rod at room temperature. Heat flows through the rod and if θ_1 and θ_2 are the temperatures of the isothermals A and B [Fig 5.5 (*a*)] at distances x and $x + dx$ from hot end. Then temperature gradient is $\frac{\theta_2 - \theta_1}{dx} = -\frac{d\theta}{dx}\cdot$ So the temperature gradient will represent the direction of flow of heat. *Heat flows in a direction in which the variation of temperature is maximum i.e.,* perpendicular to the direction of isothermal surface from high temperature to low temperature. In the same way we can define the gravitational field $\vec{F} = -\nabla\phi$, where ϕ corresponds to the *gravitational potential.*

5.7 GAUSS'S DIVERGENCE THEOREM

According to this theorem, *The flux of a vector field $\vec{F}$ over any closed surface S is equal to the Volume integral of the divergence of the Vector field over the Volume enclosed by the surface.*

Mathematically, the above statement may be written as

$$\int \vec{F}\cdot d\vec{S} = \int \text{div}\cdot\vec{F}\,dV \qquad ...(1)$$

We can prove the above theoreum by considering a finite Volume V of any shape enclosed by a surface S [as shown in Fig 5.5 (*a*)] in the region of any vector $\vec{\mathbf{F}}$

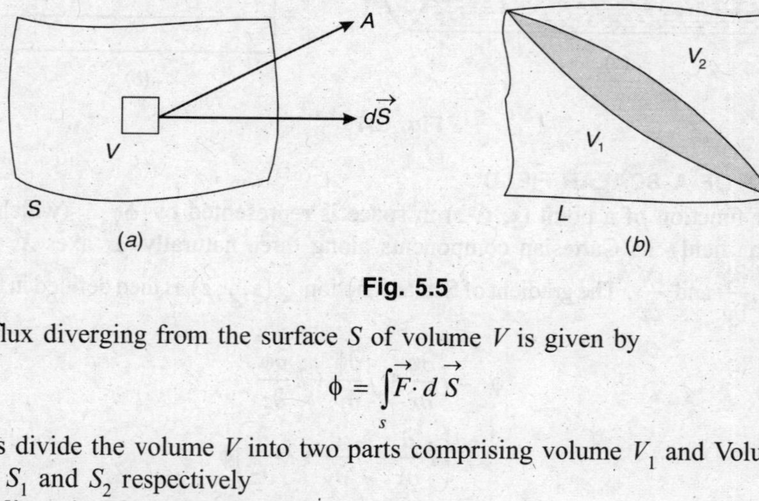

Fig. 5.5

The flux diverging from the surface S of volume V is given by

$$\phi = \int_s \vec{F}\cdot d\vec{S} \qquad ...(1)$$

Let us divide the volume V into two parts comprising volume V_1 and Volume V_2 enclosed by surfaces S_1 and S_2 respectively

Flux diverging from Surface S, is given by

$$\phi_1 = \int_{S_1} \vec{F}\cdot d\vec{S_1} \qquad ...(2)$$

and the flux diverging from surface S_2 is

$$\phi_2 = \int_{S_2} \vec{F}\cdot d\vec{S_2} \qquad ...(3)$$

Since the shaded part of Fig. 5.5 (*b*) is common to both the volumes and also the flux coming out of both the volumes being equal and opposite will cancel each other when considered together, the rest of the flux will the same in both the cases.

So we have

$$\phi = \int_S \vec{F}\cdot d\vec{S} = \int_{S_1} \vec{F}\cdot d\vec{S_1} + \int_{S_2} \vec{F}\cdot d\vec{S} \qquad ...(4)$$

If we divide the volume V into large number of volumes $V_1, V_2, V_3, ..., V_n$ related by surfaces $S_1, S_2, S_3, ..., S_n$ respectively, then we have

$$\phi = \int_S \vec{F} \cdot d\vec{S} = \int_{S_1} \vec{F} \cdot d\vec{S_1} + \int_{S_2} \vec{F} \cdot d\vec{S_2} + ... + \int_{S_1} \vec{F} \cdot d\vec{S_1} + ... \int_{Sn} \vec{F} \cdot d\vec{S_n}$$

$$= \sum_{i=i}^{n} \int_{S_1} \vec{F}.d\vec{S_1}$$

which may be written as

$$\int_S \vec{F}.d\vec{S} = \sum_{i=1}^{n} V_i \frac{\int_{S_1} .\vec{F}.d\vec{S_i}}{V_i} \quad ... (5)$$

If n is sufficiently large, then volume V_2 becomes infinitely small *i.e.*, n tends to ∞ and V_i tends to zero.

In that limit we may write

$$\underset{v_i \to 0}{Lim} \frac{\int_{S_i} \vec{F}.d\vec{S_i}}{V_i} = dv\,\vec{F} \quad ... (6)$$

Now converting the summation into integration and writing dV for small volume, *e.g.* (5) may be written in that limit as

$$\int_{S_i} \vec{F}.d\vec{S} = \int_V \text{div } \vec{F}.dV \quad ... (7)$$

This is Gauss's divergence Theorem and is useful for converting the volume integral of the divergence of vector field into the surface integral of vector field and vice-versa.

5.8. DERIVATION OF POISSON'S AND LAPLACE'S EQUATION FROM GAUSS'S LAW

Poisson's equation: Differential form of Gauss's law may be written as

$$\nabla.\vec{E} = \frac{\rho}{\varepsilon_0} \quad ... (1) \text{ (in S.I Units)}$$

where $\vec{E}$ is the electric field intensity and ρ the volume density of the charge.

By definition, the electric field at a point may be written as the negative gradient of electric potential ϕ, *i.e.*,

$$\vec{E} = -\text{grade}\,\phi = -\vec{\nabla}_\phi$$

So equation (1) may be written as

$$\text{div } \vec{E} = \vec{\nabla}.\vec{E} = \vec{\nabla}.(-\nabla\phi)$$

$$= -\nabla^2\phi = \frac{\rho}{\varepsilon_0}.$$

$$\therefore \quad \nabla^2\phi = -\frac{\rho}{\varepsilon_0} \quad ... (2)$$

or

$$\left(\frac{\partial^2\phi}{\partial x^2} + \frac{\partial^2\phi}{\partial y^2} + \frac{\partial^2\phi}{\partial z^2}\right) = -\rho/\varepsilon_0$$

Which is Poisson's equation and it holds good for each point in space.

Laplace's equation: If the enclosed charge density is zero *i.e.* the point of observation is in empty space, then $\rho = 0$ and we have

$$\nabla^2\phi = 0$$

or

$$\frac{\partial^2\phi}{\partial x^2} + \frac{\partial^2\phi}{\partial y^2} + \frac{\partial^2 y}{\partial z^2} = 0 \quad ... (3)$$

Which is Laplace's equation and the operator ∇^2 is known as *Laplace's operator*.

5.9. SIGNIFICANCE OF LAPLACE'S EQUATION

Laplace's equation is an important differential equation and a function which satisfies the above equation is called **spherical harmonic** which has a very important property *viz.*

The average value of a function ϕ (x, y, z) satisfying Laplace's equation over any spherical surface is equal to its value at the centre of spherical surface.

We have already concluded that *electric potential and intensity* at any point due to spherical shell of charge is the same as if the whole charge were concentrated at its centre, in accordance with Laplace's equation.

5.10. STOKE'S THEOREM

Stoke's theorem states that the **line integral of the tangential component of a vector point function $\vec{F}$ taken around a closed curve C is equal to the surface integral of the normal component of $\vec{F}$ taken over any surface S having C as its boundary.**

The above theorem may also be stated as the *flux of the curl of a vector field $\vec{F}$ over any surface S of any shape is equal to the line integral of the vector field $\vec{F}$ over the boundary of the surface.*

Mathematically, the above statement may be written as:

$$\oint \text{curl}\,\vec{F}.\vec{ds} = \int_C \vec{F}.\vec{dr} \qquad \text{... (1)}$$

where $\vec{dS}$ represents infinite survival element of areas S

and $\vec{dr}$ is an infinite element of boundary C of the surface.

As is clear from above, this theorem helps to convert the surface integral of the curl of the vector field $\vec{F}$ to the line integral of vector field and vice-versa.

In order to derive stoke's theorem, let us consider a **surface *S* with *C* as its boundary.**

Let us now calculate the line integral of the vector function $\vec{F}$ round the boundary C of surface S.

The line integral of $\vec{F}$ around the boundary C of surface S

$$= \oint \vec{F}\cdot\vec{dr}$$

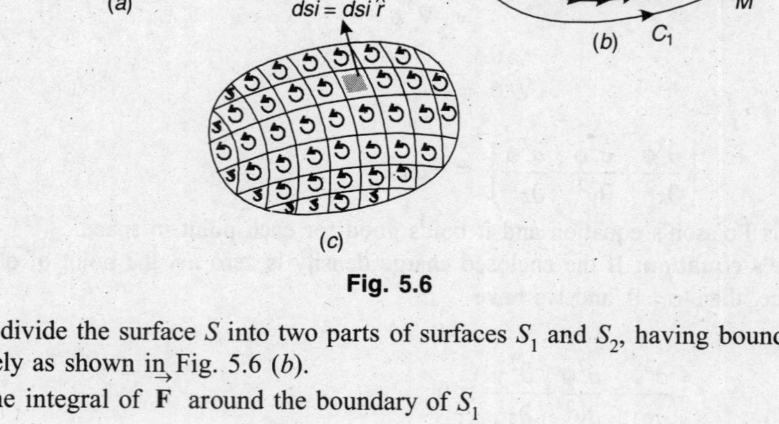

Fig. 5.6

Let us divide the surface S into two parts of surfaces S_1 and S_2, having boundaries C_1 and C_2 respectively as shown in Fig. 5.6 (*b*).

The line integral of $\vec{F}$ around the boundary of S_1

$$= \oint_{C_1} \vec{F} \cdot \vec{dr}$$

and the line integral of $\vec{\mathbf{F}}$ around the boundary of S_2

$$= \oint_{C_2} \vec{F}\, \vec{dr_2}$$

As the thick line *LM* is common boundary to both the surfaces and is traversed in opposite directions in the two integrals (Fig. 5.6), the integral of A along this common boundary being equal and opposite, cancel each other when considered together.

The rest of the boundaries C_1 and C_2 is identical to the original boundary *C*. Thus obviously

$$\oint_C \vec{F} \cdot dr = \oint_{C_1} F \cdot dr_1 + \oint_{C_2} \vec{F} \cdot dr_2 \quad \text{... (2)}$$

Now divide the surface *S* into a large number of parts $S_1, S_2, ..., S_n$ having boundaries $C_1, C_2, ..., C_i, ..., C_n$ respectively as shown in Fig. 5.6 (*c*). In that case

$$\oint_C \vec{F} \cdot \vec{dr} = \oint_{C_1} \vec{F} \cdot \vec{dr_1} + \oint_{C_2} \vec{F} \cdot \vec{dr_2} + \oint_{C_1} \vec{F} \cdot \vec{dr_i} + ... \oint_{C_n} \vec{F} \cdot \vec{dr_n}$$

$$= \sum_{i=1}^{n} \oint \vec{F} \cdot \vec{dr_1} \quad \text{... (3)}$$

Which may be written in the form

$$\oint_C \vec{F} \cdot \vec{dr} = \sum_{i=1}^{n} S_1 \oint_{C_1} \vec{F} \cdot \vec{dr} \quad \text{... (4)}$$

If we consider *n* to be sufficiently large, in that case S_i becomes infinitely small *i.e.* if $n \to \infty$, $S_i \to 0$ and in that limit we may write from the definition of curl of vector field **F** as

$$\lim_{si \to 0} \frac{\oint_{C_1} S_i \vec{F} . \vec{dr_i}}{S_i} = \text{Curl}\, \vec{F} \cdot \hat{n}.$$

Now converting summation into integration and writing $\vec{ds}$ for infinitely small area *S*, so the equation (4) in the limit form may be written as

$$\oint_C \vec{\mathrm{F}} \cdot \vec{dr} = \oint_C \text{curl}\, \vec{\mathrm{F}} \cdot \hat{n}\, \vec{ds}$$

or

$$\oint_C \vec{\mathrm{F}} \cdot \vec{dr} = \oint_C \text{curl}\, \vec{\mathrm{F}} \cdot \vec{ds}$$

This equation is known as **Stoke's theorem.**

The basic difference between Stoke's theorem and Gauss's theorem is that whereas stoke's theorem relates the **line integral** of a vector function to surface integral of the curl of a vector, Gauss's theorem relates the **surface integral** of the vector function to the **volume integral of the divergence of the vector.**

5.11. CONSERVATIVE AND NON-CONSERVATIVE FORCES

Forces are divided into two classes:

(*i*) Conservative forces

(*ii*) Non-conservative forces

If the work done by the force to displace a body from one point to another is independent of the path followed by the body (for all paths), then the force is called a **conservative force.**

Let us take the case of a gravitational force and let a body be displaced from *A* to *B* (Fig. 5.7).

The work done is given by

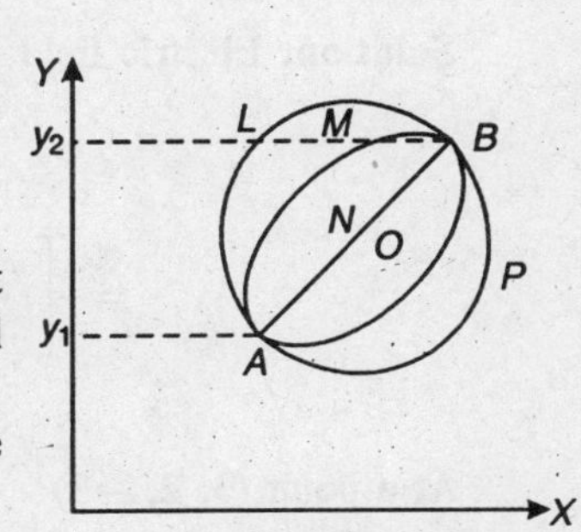

Fig. 5.7

$$W_{AB} = -mg(y_2 - y_1)$$

As is evident the work done by gravitational force depends on the Co-ordinates of *A* and *B* and is independent of path followed *viz* via *L, M, N, O, P* etc. So the gravitational force is an example of **conservative force.**

Other examples of conservative forces are electrostatic force, restoring force of a spring etc.

In the case of restoring force of spring, work done is given by

$$W = \frac{1}{2}k(x_1^2 - x_2^2)$$

where x_1 and x_2 are measured with respect to the equilibrium position $x = 0$

A Conservative force and a non-conservative force can be described by using the concept of **closed path**. When the starting point is same, it is called **closed path** and if the net work done by a force around any closed path is zero, then it is called **conservative force**, otherwise a **non-conservative force**. An example of non-conservative force is **Fractional force.** Non-conservative force is also called a **dissipative force.** In case of conservative force, the usual representation is given by

$$\oint \vec{F} \cdot \vec{ds} = 0$$

where $\oint$ represents a closed path.

In case of non-conservative force $\oint \vec{F}.\vec{ds} \neq 0$. In case of frictional force, the motion is opposed due to friction. When a block is moved in clockwise or anti-clockwise direction against friction, in that case $\oint \vec{F}.\vec{ds} \neq 0$.

Example 1. *The potential function in an electric field is represented as* $V_{(v,y,z)} = C(x^2 + y^2 + z^2)$ *where C is an arbitrary constant. Show that the electric field is radial.*

Solution: Electric field $\vec{E} = -\vec{\nabla} V = -\left[\hat{i}\frac{\partial V}{\partial x} + \hat{j}\frac{\partial V}{\partial y} + \hat{k}\frac{\partial V}{\partial z}\right]$

$$= -\left[\hat{i}\frac{\partial}{\partial x}C(x^2 + y^2 + z^2) + \hat{j}\frac{\partial}{\partial y}C(x^2 + y^2 + z^2) + \hat{k}\frac{\partial}{\partial z}C(x^2 + y^2 + z^2)\right]$$

$$= -[\hat{i}2Cx + \hat{j}2Cy + \hat{k}2Cz] = -2C(\hat{i}x + \hat{j}y + \hat{k}z)$$

$$= -2C\vec{r} \text{ where } \vec{r} \text{ is the position vector.}$$

So the direction of electric field $\vec{\mathbf{E}}$ is opposite to $\vec{\mathbf{r}}$ *i.e.,* towards the origin. So the field is radial. **Ans.**

Example 2. *If* $V = 2x^2 - 3y^2 + z^2$ *represents the electrostatic potential at a joint, find the electric field intensity at a point (3, 2, – 3).*

Solution: Electric field intensity is given by

$$\vec{\mathrm{E}} = -\vec{\nabla} V = -\left(\hat{i}\frac{\partial V}{\partial x} + \hat{j}\frac{\partial V}{\partial y} + \hat{k}\frac{\partial V}{\partial z}\right)$$

$$= -\left[\hat{i}\frac{\partial}{\partial x}(2x^2 - 3y^2 + z^2) + \hat{j}\frac{\partial}{\partial y}(2x^2 - 3y^2 + z^2) + \hat{k}\frac{\partial}{\partial z}(2x^2 - 3y^2 + z^2)\right]$$

$$= -[\hat{i}(4x) + \hat{j}(-6y) + \hat{k}(2z)] = 4x\hat{j} + 6y\hat{j} - 2z\hat{k}$$

At a point (3, 2, – 3)

$$\vec{E} = -4(3)\hat{i} + 6(2)\hat{j} - 2(-3)\hat{k}$$

$$\vec{E} = 12\hat{i} + 12\hat{j} + 6\hat{k} \textbf{ Ans.}$$

Example 3. *A closed surface S encloses a volume V. Calculate the value of div* $\vec{\mathbf{F}}$ *and* $\iint_S \vec{\mathrm{F}}.d\vec{\mathrm{S}}$ *for a vector field* $\vec{\mathbf{F}} = \vec{F} = 3x\hat{i} + 4y\hat{j} + 2z\hat{k}$

Solution: By Gauss's divergence theorem, we have

$$\iint_S \vec{F}.d\vec{S} = \iiint_S (div\,\vec{F})\,dV$$

Now,

$$= \frac{\partial F_x}{\partial x} + \frac{\partial F_y}{\partial y} + \frac{\partial F_z}{\partial z}$$

$$= \frac{\partial}{\partial x}(3x) + \frac{\partial}{\partial y}(4y) + \frac{\partial}{\partial z}(2z) = 3 + 4 + 2 = 9$$

$$\therefore \quad \iint_c \vec{F}.\vec{da} = \iiint_v 9\,dV = \mathbf{9V Ans.}$$

Example 4. *In a conduction phemomenerof flow of heat through a rod, under the steady state, thermal surface is defined by the equation* $\theta = 9x - 5y + 6z$. *Calculate the temperature field and show it to be uniform.*

Solution: By definition, temp. field is given by

$$\text{Temp. field} = \nabla\phi = \left(\hat{i}\frac{\partial\theta}{\partial x} + \hat{j}\frac{\partial\theta}{\partial y} + \hat{k}\frac{\partial\theta}{\partial z}\right)$$

$$= \hat{i}\frac{\partial}{\partial x}(9x - 5y + 6z) + \hat{j}\frac{\partial}{\partial y}(9x - 5y + 6z) + \hat{k}\frac{\partial}{\partial z}(9x - 5)$$

$$= \hat{i}\,(9) + \hat{j}(-5) + \hat{k}(6)$$

$$= \hat{i}\,9 + \hat{j} - 5 + 6\hat{k}$$

This is uniform field as it is independent of x, y, z. **Ans.**

Example 5. *The electric field intensity at a point of a closed surface is expressed as* $\vec{\mathbf{E}} = (2\hat{i} + 3\hat{j} - 2\hat{k})$ *volt/metre. Calculate the electric flux through a small element of area* $(\hat{i} + \hat{j} - \hat{k})\mathrm{m}^2$.

Solution: Electric flux, is given by

$$d\phi = \vec{E}.\vec{dS} = (2\hat{i} + 3\hat{j} - 2\hat{k}).(\hat{i} + \hat{j} - \hat{k})$$

$$= 2 + 3 + (-2)(-1) = 2 + 3 + 2 = 7 \text{ volt} \times \text{m. } \mathbf{Ans.}$$

Example 6. *Prove that the divergence of a vector field which obeys the inverse square law is zero.*

Solution: An inverse square vector field is given by

$$\vec{F} \propto \frac{1}{r^2}\hat{r} \text{ or } \vec{F} = \frac{K}{r^2}\hat{r} = \frac{k}{r^3}.\vec{r}$$

where $\vec{r} = \hat{i}x + \hat{j}y + \hat{k}z$ is the position vector and K is a constant.

So we have

$$\vec{\mathbf{F}} = \frac{K}{r^3}(\hat{i}x + \hat{j}y + \hat{k}z)$$

$$\therefore \text{ Div } \vec{\mathbf{F}} = \vec{\nabla}.\vec{\mathrm{E}} = \frac{\partial F_x}{\partial x} + \frac{\partial F_y}{\partial y} + \frac{\partial F_z}{\partial z} = \frac{\partial}{\partial z}\left(\frac{Kx}{r^3}\right) + \frac{\partial}{\partial y}\left(\frac{Ky}{r^3}\right) + \frac{\partial}{\partial z}\left(\frac{Kz}{r^3}\right)$$

$$= \left\{\frac{K}{r^3} + Kx\left(\frac{-3}{r^4}\right)\frac{\partial r}{\partial x}\right\} + \left\{\frac{K}{r^3} + Ky\left(\frac{-3}{r^4}\right)\frac{\partial r}{\partial y}\right\} + \left\{\frac{K}{r^3} + Kz\left(\frac{-3}{r^4}\right)\frac{\partial r}{\partial z}\right\}$$

$$= \frac{3K}{r^3} - \frac{3K}{r^4}\left\{x\frac{\partial r}{\partial x} + y\frac{\partial r}{\partial y} + z\frac{\partial r}{\partial z}\right\}$$

Now $\quad r^2 = x^2 + y^2 + z^2$ Differentiating w.r.t. x.

$$\therefore \quad 2r\frac{\partial r}{\partial x} = 2x \text{ or } \frac{\partial r}{\partial x} = \frac{x}{r}$$

Similarly, $\quad \frac{\partial r}{\partial y} = \frac{y}{r}$ and $\frac{\partial r}{\partial z} = \frac{z}{r}$

So

$$\nabla.\vec{F} = \frac{3K}{r^3} - \frac{3K}{r^4}\left[x\frac{x}{r} + y\frac{y}{r} + z\frac{z}{r}\right]$$

$$= \frac{3K}{r^3} - \frac{3K}{r^5}(x^2 + y^2 + z^2)$$

$$= \frac{3K}{r^3} - \frac{3K}{r^5}(r^2) = 0 \textbf{ Ans.}$$

Example 7. *Show that a vector field* $\vec{\mathbf{F}} = KR^n \hat{R}$ *is an irrotational field for all values of n, but is a solenoidal field for n = – 2. Here K is a constant.*

Solution: $\quad \vec{\mathbf{F}} = KR^n \hat{R} = KR^{n-1} \vec{R}$

$$= \begin{vmatrix} i & j & k \\ \frac{\partial}{\partial x} & \frac{\partial}{\partial y} & \frac{\partial}{\partial z} \\ KR^{n-1} & KR^{n-1} & KR^{n-1}z \end{vmatrix}$$

Which means

curl

$$\vec{F} = K\left[\hat{i}\left\{\frac{\partial}{\partial y}(R^{n-1}z) - \frac{\partial}{\partial z}(R^{n-1}y)\right\} + \hat{j}\left\{\frac{\partial}{\partial z}(R^{n-1}x) - \frac{\partial}{\partial x}(R^{n-1}z)\right\}\right.$$
$$\left.\hat{k}\left\{\frac{\partial}{\partial x}(R^{n-1}y) - \frac{\partial}{\partial x}(R^{n-1}x)\right\}\right]$$

$$= K\left[\hat{i}\left\{z(n-1)R^{n-2}\frac{\partial R}{\partial y} - y(n-1)R^{n-2}\frac{\partial R}{\partial z}\right\} + \hat{i}... + \hat{k}...\right]$$

$$= K[\hat{i}(n-1)R^{n-3}(zy - yz) + \hat{j}(n-1)R^{n-3}(xz - zx) + \hat{k}(n-1)R^{n-3}(yx - xy)]$$

But for a solenoidal field, the value of div $\vec{\mathbf{F}}$ should be zero.

$$\text{div}\,\vec{F}(KR^{n-1}\vec{R}) = \nabla.[KR^{n-1}(\hat{i}x + \hat{j}y + \hat{k}z)]$$

$$= \frac{\partial}{\partial x}(KR^{n-1}x) + \frac{\partial}{\partial y}(KR^{n-1}y) + \frac{\partial}{\partial z}(KR^{n-1}z)$$

$$= K\left[\left\{R^{n-1} + x(n-1)R^{n-2}\frac{\partial R}{\partial x}\right\} + \left\{R^{n-1} + y(n-1)R^{n-2}\frac{\partial R}{\partial y}\right\} + \left\{R^{n-1} + z(n-1)R^{n-2}\frac{\partial R}{\partial z}\right\}\right]$$

$$= K\left[3R^{n-1} + (n-1)R^{n-2}\left\{x\frac{\partial R}{\partial x} + y\frac{\partial R}{\partial y} + z\frac{\partial R}{\partial z}\right\}\right]$$

$$= K\left[3R^{n-1} + (n-1)R^{n-2}\left\{x\left(\frac{x}{r}\right) + y\left(\frac{y}{R}\right) + z\left(\frac{z}{R}\right)\right\}\right]$$

$$= KR^{n-1}(3n-1) = K(n+2)R^{n-1}$$

Which means div $\vec{\mathbf{F}} = 0$ if $n + 2 = 0$ or $n = -2$ **Ans.**

Example 8. *If* $\vec{r}$ *, the position vector is* $\vec{r} = \hat{i}x + \hat{j}y + \hat{k}z$*, find:*

(*i*) grad r^n. (*ii*) div $r^n\hat{r}$, and (*iii*) curl $r^n\vec{r}$.

Solution: We have $\vec{r} = \hat{i}x + \hat{j}y + \hat{k}z \therefore r^2 = x^2 + y^2 + z^2$

also $\frac{\partial r}{\partial x} = \frac{x}{r}, \frac{\partial r}{\partial y} = \frac{y}{r}, \frac{\partial r}{\partial z} = \frac{z}{r}$

(*i*) grad $r^n = \vec{\nabla} r^n = \hat{i}\frac{\partial}{\partial x}r^n + \hat{j}\frac{\partial}{\partial y}r^n = \hat{k}\frac{\partial}{\partial z}r^n$

$$= \hat{i}\, nr^{n-1}\frac{\partial r}{\partial x} + \hat{j}\, nr^{n-1}\frac{\partial r}{\partial y} + \hat{k}\, nr^{n-1}\frac{\partial r}{\partial z}$$

$$= nr^{n-1}\left[\hat{i}\frac{\partial r}{\partial x} + \hat{j}\frac{\partial r}{\partial y} + \hat{k}\frac{\partial r}{\partial z}\right]$$

$$= nr^{n-1}\left[\hat{i}\frac{x}{r} + \hat{j}\frac{y}{r} + \hat{k}\frac{z}{r}\right]$$

$$= nr^{n-2}(\hat{i}x + \hat{j}y + \hat{k}z) = nr^{n-2}\vec{r}$$ **Ans.**

(*ii*) div $r^n\hat{r} = \text{div}\, r^n\left(\frac{\vec{r}}{r}\right) = \text{div}\, r^{n-1}\vec{r}$

$$= \vec{\nabla}.[r^{n-1}(\hat{i}x + \hat{j}y + \hat{k}z)]$$

$$= \left(\hat{i}\frac{\partial}{\partial x} + \hat{j}\frac{\partial}{\partial y} + \hat{k}\frac{\partial}{\partial z}\right).(\hat{i}\,r^{n-1}x + \hat{j}\,r^{n-1}y + \hat{k}\,r^{n-1}z)$$

$$= \frac{\partial}{\partial x}(r^{n-1}x) + \frac{\partial}{\partial y}(r^{n-1}y)\frac{\partial}{\partial z}(r^{n-1}z)$$

$$= \left[r^{n-1} + x(n-1)r^{n-2}\frac{\partial r}{\partial x}\right] + \left[r^{n-1} + y(n-1)r^{n-2}\frac{\partial r}{\partial y}\right] + \left[r^{n-1} + z(n-1)r^{n-2}\frac{\partial r}{\partial z}\right]$$

$$= 3r^{n-1} + (n-1)r^{n-2}\left[x\frac{\partial r}{\partial x} + y\frac{\partial r}{\partial y} + z\frac{\partial r}{\partial z}\right]$$

$$= 3r^{n-1} + (n-1)r^{n-2}\left[x\left(\frac{x}{r}\right) + y\left(\frac{y}{r}\right) + z\left(\frac{z}{r}\right)\right]$$

$$= 3r^{n-1} + (n-1)r^{n-2}\frac{(x^2 + y^2 + z^2)}{r}$$

$$= 3r^{n-1} + (n-1)r^{n-1}$$

$$= (3+n-1)r^{n-1} = (n+2)r^{n-1} \textbf{ Ans.}$$

(*iii*) $$\text{curl } r^n \vec{r} = \vec{\nabla}\times(r^n \vec{r})$$

$$= \left(\hat{i}\frac{\partial}{\partial x}+\hat{j}\frac{\partial}{\partial y}+\hat{k}\frac{\partial}{\partial z}\right)\times[r^n(\hat{i}x+\hat{j}y+\hat{k}z)]$$

$$= \begin{vmatrix} \hat{i} & \hat{j} & \hat{k} \\ \frac{\partial}{\partial x} & \frac{\partial}{\partial y} & \frac{\partial}{\partial z} \\ r^n x & r^n y & r^n z \end{vmatrix}$$

$$= \hat{i}\left[\frac{\partial}{\partial y}(r^n z)-\frac{\partial}{\partial z}(r^n y)\right]+\hat{j}\left[\frac{\partial}{\partial z}(r^n x)-\frac{\partial}{\partial x}(r^n z)\right]+ \hat{k}\left[\frac{\partial}{\partial x}(r^n y)-\frac{\partial}{\partial y}(r^n x)\right]$$

$$= \hat{i}\left[r^n\times 0+znr^{n-1}\frac{\partial r}{\partial y}-r^n\times 0-ynr^{n-1}\frac{\partial r}{\partial z}\right]$$

$$= \hat{i}\left[nzr^{n-1}\left(\frac{y}{r}\right)-nyr^{n-1}\left(\frac{z}{r}\right)\right]+....+....$$

$$= \hat{i}[nr^{n-2}(zy-yz)]+\hat{j}[nr^{n-2}(xz-zx)]+\hat{k}[nr^{n-2}(xy-xy)]$$

$$= 0 \textbf{ Ans.}$$

Example 9. *A vector field is given as* $\vec{\mathbf{F}} = y\hat{i}+(x+y^2)\hat{j}+(yz+zx)\hat{k}$. *Find: (i) div* $\vec{\mathbf{F}}$, *(ii) curl* $\vec{\mathbf{F}}$.

Solution: (*i*) $$\text{div } \vec{\mathbf{F}} = \frac{\partial F_x}{\partial x}+\frac{\partial F_y}{\partial y}+\frac{\partial F_z}{\partial z}$$

$$= \frac{\partial}{\partial x}(y)+\frac{\partial}{\partial y}(x^2+y^2)+\frac{\partial}{\partial z}(yz+zx)$$

$$= 0+2y+y+x$$

$$= x+3y \textbf{ Ans.}$$

(*ii*) $$\text{curl } \vec{F} = \begin{vmatrix} \hat{i} & \hat{j} & \hat{k} \\ \frac{\partial}{\partial x} & \frac{\partial}{\partial y} & \frac{\partial}{\partial z} \\ F_x & F_y & F_z \end{vmatrix} = \begin{vmatrix} \hat{i} & \hat{j} & \hat{k} \\ \frac{\partial}{\partial x} & \frac{\partial}{\partial y} & \frac{\partial}{\partial z} \\ y & x+y^2 & zy+zx \end{vmatrix}$$

$$= \hat{i}\left[\frac{\partial}{\partial y}(yz+zx)-\frac{\partial}{\partial z}(x^2+y^2)\right]+\hat{j}\left[\frac{\partial}{\partial y}(y)-\frac{\partial}{\partial x}(yz-zx)+k\right]+\hat{j}\left[\frac{\partial}{\partial x}(x^2+y^2)-\frac{\partial}{\partial y}(y)\right]$$

$$= \hat{i}[z-0]+\hat{j}[0-z]+\hat{k}[2x-1] = z\hat{i}-z\hat{j}+(2x-1)\hat{k} \textbf{ Ans.}$$

Example 10. *If* ϕ *and* ψ *are the functions of position (x, y, z), show that grad* $(\phi\psi) = \phi$ *grad* $\psi + \psi$ *(grad)* ϕ.

Solution: $$\text{grad }(\phi\psi) = \vec{\nabla}(\phi\psi) = \left(\hat{i}\frac{\partial}{\partial x} + \hat{j}\frac{\partial}{\partial y} + \hat{k}\frac{\partial}{\partial z}\right)(\phi\psi)$$

So $$\vec{\nabla}(\phi\psi) = \hat{i}\frac{\partial}{\partial x}(\phi\psi) + \hat{j}\frac{\partial}{\partial y}(\phi\psi) + \hat{k}\frac{\partial}{\partial z}(\phi\psi)$$

$$= \hat{i}\left[\phi\frac{\psi}{\partial x} + \phi\frac{\partial\phi}{\partial x}\right] + \hat{j}\left[\phi\frac{\partial\psi}{\partial y} + \psi\frac{\partial\phi}{\partial y}\right] + \hat{k}\left[\phi\frac{\partial\psi}{\partial z} + \psi\frac{\partial\phi}{\partial z}\right]$$

$$= \phi\left[\hat{i}\frac{\partial\psi}{\partial x} + \hat{j}\frac{\partial\psi}{\partial y} + \hat{k}\frac{\partial\psi}{\partial z}\right] + \psi\left[\hat{i}\frac{\partial\phi}{\partial x} + \hat{j}\frac{\partial\phi}{\partial y} + \hat{k}\frac{\partial\phi}{\partial z}\right]$$

$$= \phi\vec{\nabla}\psi + \psi\vec{\nabla}\phi = \phi\,\text{grad}\,\phi + \psi\,\text{grad}\,\phi \textbf{ Ans.}$$

Example 11. *If* ϕ *is a scalar field and* $\vec{A}$ *is a vector field, prove that curl* $(\phi\vec{A}) = \phi$ curl $A + (\text{grad}\,\phi)\times A$

or $$\vec{\nabla}\times(\phi\vec{A}) = \phi\vec{A}\times\vec{A} + \vec{\nabla}\phi\times\vec{A}$$

Solution: $$\text{curl }(\phi\vec{A}) = (\vec{\nabla}\vec{A})$$

$$\therefore \quad \vec{\nabla}\times Q\vec{A} = \begin{vmatrix} \hat{i} & \hat{j} & \hat{k} \\ \frac{\partial}{\partial x} & \frac{\partial}{\partial y} & \frac{\partial}{\partial z} \\ \phi A_x & \phi A_y & \phi A_z \end{vmatrix}$$

$$= \hat{i}\left[\frac{\partial}{\partial y}(\phi A_z) - \frac{\partial}{\partial z}(\phi A_y)\right] + \hat{j}\left[\frac{\partial}{\partial z}(\phi A_x) - \frac{\partial}{\partial x}(\phi A_z)\right] + \hat{k}\left[\frac{\partial}{\partial x}(\phi A_y) - \frac{\partial}{\partial y}(\phi A_x)\right]$$

$$= \hat{i}\left[\frac{\partial\phi}{\partial y} + \phi\frac{\partial A_z}{\partial y} - A_y\frac{\partial\phi}{\partial z} - \phi\frac{\partial A_y}{\partial z}\right] + \ldots + \ldots..$$

or $$= \phi\left[\hat{i}\left(\frac{\partial A_z}{\partial y} - \frac{\partial A_y}{\partial z}\right) + \hat{j}\left(\frac{\partial A_x}{\partial z} - \frac{\partial A_z}{\partial x}\right) + \hat{k}\left(\frac{\partial A_y}{\partial x} - \frac{\partial A_x}{\partial y}\right)\right]$$

$$+ \hat{i}\left(A_z\frac{\partial\phi}{\partial y} - A_y\frac{\partial\phi}{\partial z}\right) + \hat{j}\left(A_x\frac{\partial\phi}{\partial z} - A_z\frac{\partial\phi}{\partial x}\right) + \hat{k}\left(A_y\frac{\partial\phi}{\partial x} - A_x\frac{\partial\phi}{\partial y}\right)$$

$$\therefore \quad \vec{\nabla}\times\phi\vec{A} = \phi\vec{\nabla}\times\vec{A} + (\vec{\nabla}\phi)\times\vec{A} = \phi\text{ curl }\vec{A} + (\text{grad}\,\phi)\times\vec{A} \textbf{ Ans.}$$

Example 12. *Prove that* $\text{div}(\vec{A}\times\vec{B}) = \vec{B}$ curl $\vec{A} - \vec{A}.\text{curl}\,\vec{B}$. *Hence show that if* $\vec{A}$ *and* $\vec{B}$ *are the ir-rotational fields, then* $\vec{A}\times\vec{B}$ *is a solenoidal field.*

Solution: Let $\vec{A}\times\vec{B} = \vec{F}$, so we have

$$\therefore \quad \vec{F} = \begin{vmatrix} \hat{i} & \hat{j} & \hat{k} \\ A_x & A_y & A_z \\ B_x & B_y & B_z \end{vmatrix}$$

$$= \hat{i}(A_yB_z - A_zB_y) + \hat{j}(A_zB_x - A_xB_z) + \hat{k}(A_xB_y - A_yB_x)$$

$$= \hat{i}V_x + \hat{j}V_y + \hat{k}V_z$$

$$\text{Hence, div }(\vec{A}\times\vec{B}) = \text{div }\vec{F} = \vec{\nabla}.\vec{F} = \frac{\partial F_x}{\partial x}+\frac{\partial F_y}{\partial y}+\frac{\partial F_z}{\partial z}$$

$$= \frac{\partial}{\partial x}(A_yB_z - A_zB_y)+\frac{\partial}{\partial y}(A_zB_x - A_xB_z)+\frac{\partial}{\partial z}(A_xB_y - A_yB_x)$$

$$= \left(A_y\frac{\partial B_z}{\partial x}+B_z\frac{\partial A_y}{\partial x}-B_y\frac{\partial A_z}{\partial x}-A_z\frac{\partial B_y}{\partial x}\right)+....+....$$

$$= B_x\left(\frac{\partial A_z}{\partial y}-\frac{\partial A_y}{\partial z}\right)+B_y\left(\frac{\partial A_x}{\partial z}-\frac{\partial A_z}{\partial x}\right)+B_z\left(\frac{\partial A_y}{\partial x}-\frac{\partial A_x}{\partial y}\right)$$

$$= -A_x\left(\frac{\partial B_z}{\partial y}-\frac{\partial B_x}{\partial z}\right)-A_y\left(\frac{\partial B_x}{\partial z}-\frac{\partial B_z}{\partial x}\right)-A_z\left(\frac{\partial B_y}{\partial x}-\frac{\partial A_x}{\partial y}\right)$$

$$= (\hat{i}B_x+\hat{j}B_y+\hat{k}B_z).\left[\hat{i}\left(\frac{\partial A_z}{\partial y}-\frac{\partial A_y}{\partial z}\right)+\hat{j}\left(\frac{\partial A_x}{\partial z}-\frac{\partial A_z}{\partial x}\right)\hat{k}\left(\frac{\partial A_y}{\partial x}-\frac{\partial A_x}{\partial y}\right)\right]$$

$$= -(\hat{i}A_x+\hat{j}A_y+\hat{k}A_z).\left[\hat{i}\left(\frac{\partial B_z}{\partial y}-\frac{\partial B_y}{\partial z}\right)+\hat{j}\left(\frac{\partial B_x}{\partial z}-\frac{\partial B_z}{\partial x}\right)\hat{k}\left(\frac{\partial B_y}{\partial x}-\frac{\partial B_x}{\partial y}\right)\right]$$

$$= \vec{B}.\text{curl }\vec{A} - \vec{A}.curl\ \vec{B}.$$

Now if $\vec{A}$ and $\vec{B}$ are the ir-rotational fields (which means curl $\vec{A}$ = 0 and curl $\vec{B}$ = 0) then div $\vec{A}\times\vec{B}$ = 0 *i.e.*, $\vec{A}\times\vec{B}$ will be a solenoidal field. **Ans.**

Example 13. *If* $\vec{A} = x^2z\,\hat{i} - 2y^3z^2\,\hat{j} + xy^2z\,\hat{k}$, *find the value of (i) div* $\vec{A}$, *(ii) curl* $\vec{A}$, *at a point (2, 2, 2)*

Solution: (*i*)

$$\text{div }\vec{A} = \vec{\nabla}.\vec{A} = \frac{\partial A_x}{\partial x}+\frac{\partial A_y}{\partial y}+\frac{\partial A_z}{\partial z}$$

$$= \frac{\partial}{\partial x}(x^2z)+\frac{\partial}{\partial y}(-2y^3z^2)+\frac{\partial}{\partial z}(xy^2z)$$

$$= 2xz - 6y^2z^2 + xy^2$$

At the point (2, 2, 2)

$$\text{div }\vec{A} = 2(2)(2) - 6(2)^2(2)^2 + 2(2)^2 = 8 - 96 + 8 = -80 \text{ **Ans.**}$$

(*ii*)

$$\text{curl }\vec{A} = \vec{\nabla}\times\vec{A}$$

$$= \begin{vmatrix} \hat{i} & \hat{j} & \hat{k} \\ \frac{\partial}{\partial y} & \frac{\partial}{\partial y} & \frac{\partial}{\partial z} \\ A_x & A_y & A_z \end{vmatrix} = \begin{vmatrix} \hat{i} & \hat{j} & \hat{k} \\ \frac{\partial}{\partial x} & \frac{\partial}{\partial y} & \frac{\partial}{\partial z} \\ x^2z & -2y^3z^2 & xy^2z \end{vmatrix}$$

$$= \hat{i}\left[\frac{\partial}{\partial y}(xy^2z)-\frac{\partial}{\partial z}(-2y^3z^2)\right]+\hat{j}\left[\frac{\partial}{\partial z}(x^2z)-\frac{\partial}{\partial x}(xy^2z)\right]+\hat{k}\left[\frac{\partial}{\partial x}(-2y^3z^2)-\frac{\partial}{\partial y}(x^2z)\right]$$

$$= \hat{i}[2xyz + y^2z] + \hat{j}[x^2 - y^2z] + \hat{k}[0-0]$$

$$= \hat{i}(2xyz + 4y^3z) + \hat{j}(x^2 - y^2z)$$

At the point (2, 2, 2)

$$\text{Curl } \vec{A} = \hat{i}(2\times2\times2\times2-4\times2^3\times2)+\hat{j}(2^2-2^2\times2)$$
$$=\hat{i}(16+64)+\hat{j}(4-8)$$
$$=80\hat{i}-4\hat{j} \textbf{ Ans.}$$

Example 14. *Show that* $\vec{F}=(x+y)\hat{i}+(x+y)\hat{j}+(y-z)\hat{k}$ *represents a conservative field.*

Solution: For a conservative field, curl $\vec{F}$ must be zero.

Now, curl $\vec{F} = \vec{\nabla}\times\vec{F}$

$$=\begin{vmatrix} \hat{i} & \hat{j} & \hat{k} \\ \dfrac{\partial}{\partial x} & \dfrac{\partial}{\partial y} & \dfrac{\partial}{\partial z} \\ x+y & x+z & y-z \end{vmatrix}$$

$$=\hat{i}\left[\frac{\partial}{\partial y}(y-z)-\frac{\partial}{\partial z}(x+z)\right]+\hat{j}\left[\frac{\partial}{\partial z}(x+y)-\frac{\partial}{\partial x}(y-z)\right]+\hat{k}\left[\frac{\partial}{\partial x}(x+z)-\frac{\partial}{\partial y}(y+y)\right]$$

$$=\hat{i}(1-1)+\hat{j}(0-0)+\hat{k}(1-1)=0.$$

Which means that $\vec{F}$ is conservative field. **Ans.**

Example 15. If $\vec{A} = -y\hat{i}+x\hat{j}$

Find: (i) curl $\vec{A}$ *and (ii)* $\oint A.\vec{d}_1$ *for a closed curve* $x^2+y^2=r^2.z=0$. *Hence prove the Stoke's theorem.*

Solution:

(i)
$$\text{curl } \vec{A} = \begin{vmatrix} \hat{i} & \hat{j} & \hat{k} \\ \dfrac{\partial}{\partial x} & \dfrac{\partial}{\partial y} & \dfrac{\partial}{\partial z} \\ -y & x & 0 \end{vmatrix} = \hat{k}[1-(-1)]=2\hat{k}.$$

(ii)
$$\oint \vec{A}.\vec{d}_1 = \oint (A_x dx + dy + A_z dz) = \int(-y\,dx + x\,dy)$$
$$= \int y\,dx + \int x\,dy.$$

For the curve $x^2+y^2=r^2$, $x = r\cos\theta$ and $y = r\sin\theta$

$\therefore$ $dx = -r\sin\theta d\theta$ and $dy = r\cos\theta d\theta$

Hence $\oint \vec{A}.\vec{d}_1 = \int_0^{2\pi} r^2 d\theta = 2\pi r^2$

Now from Stoke's theorem,

$$\oint \vec{A}.\vec{d}_1 = \iint_s \text{curl } \vec{A}.\vec{da}$$
$$= \iint_s 2\hat{k}.da\,\hat{k} = \iint_s 2da$$
$$= 2a = 2(\pi r^2) = 2\pi r^2 \text{ [since area of circle is a } = \pi r^2\text{]}$$

Which proves the Stoke's theorem.

Example 16. *Prove that*

(i) grad $r = \hat{n}$, where $\hat{n}$ is a unit vector.

(ii) grad $\left(\frac{1}{r}\right) = -\frac{\vec{r}}{r^3}$, where $\vec{r}$ is a position vector.

Solution: In this case, $\vec{r}$ is a position vector, which may be written as

$$\vec{r} = xi + yj + zk, \text{ therefore } |\vec{r}| = r = \sqrt{(x^2+y^2+z^2)}$$

$$[\because \text{ If } \vec{A} = A_x i + A_y j + A_z k, \text{ so } A = \sqrt{(A_x^2 + y_y^2 + _z^2)}]$$

(i) $$\text{grad } r = \hat{n}$$

$\therefore$ Taking L.H.S. $$\text{grad } r = \vec{\nabla}(x^2+y^2+z^2)^{1/2}$$

$$= \left(i\frac{\partial}{\partial x} + j\frac{\partial}{\partial y} + k\frac{\partial}{\partial z}\right)(x^2+y^2+z^2)^{1/2}$$

or $$\text{grad } r = i\frac{\partial}{\partial x}(x^2+y^2+z^2)^{1/2} + j\frac{\partial}{\partial y}(x^2+y^2+z^2)^{1/2} + k\frac{\partial}{\partial z}(x^2+y^2+z^2)^{1/2} \quad ...(i)$$

Let us consider the first term, $\frac{\partial}{\partial x}(x^2+y^2+z^2)^{1/2}$

Differentiating it partially with respect to x, we get

$\therefore$ $$\frac{\partial}{\partial x}(x^2+y^2+z^2)^{1/2} = \left(\frac{1}{2}\right)(x^2+y^2+z^2)^{(1/2)-1}.(2x)$$

$$= \frac{x}{(x^2+y^2+z^2)^{1/2}}$$

Again, differentiating partially other terms, w.r.t. y and z, we have

$$\frac{\partial}{\partial y}(x^2+y^2+z^2)^{1/2} = \frac{y}{(x^2+y^2+z^2)^{1/2}} \text{ and } \frac{\partial}{\partial z}(x^2+y^2+z^2)^{1/2} = \frac{z}{(x^2+y^2+z^2)^{1/2}}$$

Putting these values in equation (*i*), we have

$$\nabla r = \frac{xi}{(x^2+y^2+z^2)^{1/2}} + \frac{yj}{(x^2+y^2+z^2)^{1/2}} + \frac{zk}{(x^2+y^2+z^2)^{1/2}}$$

$$= \frac{xi+yj+zk}{(x^2+y^2+z^2)^{1/2}} = \frac{\vec{r}}{r} = \hat{n}, \text{ where } \hat{n} \text{ is a unit vector. } (\because \vec{r} = \hat{n}r)$$

= R.H.S. Hence proved **Ans.**

(ii) $$\text{grad}\left(\frac{1}{r}\right) = -\frac{\vec{r}}{r^3}$$

Taking $$\text{L.H.S.} = \vec{\nabla}\left(\frac{1}{r}\right) = \left(i\frac{\partial}{\partial x} + j\frac{\partial}{\partial y} + k\frac{\partial}{\partial z}\right)\left(\frac{1}{(x^2+y^2+z^2)^{1/2}}\right)$$

or $$\nabla\left(\frac{1}{r}\right) = i\frac{\partial}{\partial x}(x^2+y^2+z^2)^{-1/2} + j\frac{\partial}{\partial y}(x^2+y^2+z^2)^{-1/2} + k\frac{\partial}{\partial z}(x^2+y^2+z^2)^{-1/2} \quad ...(i)$$

Taking the first term, $\frac{\partial}{\partial x}(x^2+y^2+z^2)^{-1/2}$ and differentiating partially w.r.t. x, we get.

$$\therefore \quad \frac{\partial}{\partial x}(x^2+y^2+z^2)^{-1/2} = \left(-\frac{1}{2}\right)(x^2+y^2)^{-(1/2)-1/2-1}.(2x) = -\frac{x}{(x^2+y^2+z^2)^{3/2}}$$

Again, differentiating partially other terms of equation (i) w.r.t. y and z, we have

$$\frac{\partial}{\partial y}(x^2+y^2+z^2)^{-1/2} = \frac{y}{(x^2+y^2+z^2)^{3/2}} \text{ and } \frac{\partial}{\partial z}(x^2+y^2+z^2)^{-1/2} = -\frac{z}{(x^2+y^2+z^2)^{3/2}}$$

Putting these values in equation (*i*), we get

$$\text{grad}\left(\frac{1}{r}\right) = \frac{xi}{(x^2+y^2+z^2)^{3/2}} - \frac{yj}{(x^2+y^2+z^2)^{3/2}} - \frac{zk}{(x^2+y^2+z^2)^{3/2}}$$

$$= -\left[\frac{xi+yj+zk}{(x^2+y^2+z^2)^{3/2}}\right] = -\frac{\vec{r}}{(r^2)^{3/2}} = -\frac{\vec{r}}{r^3}$$

= R.H.S. Hence proved. **Ans.**

Example 17. If $\vec{A}$ is a constant vector, prove grad $(\vec{r}.\vec{A}) = \vec{A}$, where $\vec{r}$ is a position vector.

Solution: As is given, $\vec{A}$ is a constant vector, which means that, its components, A_x, A_y and A_z, are constant and independent of x, y and z. $\vec{r}$ is a position vector which depends on x, y and z, which may be written as

$$\therefore \quad \vec{r} = xi+yj+zk \text{ and } \vec{A} = A_x i + A_y j + A_z k$$

$$\therefore \quad \vec{r}.\vec{A} = (xi+yj+zk).(A_x i + A_y j + A_z k)$$

$$= xA_x + yA_y + zA_z \qquad (\because \vec{A}.\vec{B} = A_x B_x + A_y B_y + A_z B_z)$$

$$\text{Now, grad } (\vec{r}.\vec{A}) = \left(i\frac{\partial}{\partial x} + j\frac{\partial}{\partial y} + k\frac{\partial}{\partial z}\right)(xA_x + yA_y + zA_z)$$

$$= iA_x.\frac{\partial x}{\partial x} + jA_y\frac{\partial y}{\partial y} + kA_z\frac{\partial z}{\partial z} \quad (A_x, A_y \text{ and } A_z \text{ are constant})$$

$$= A_x i + A_y j + A_z k = \vec{A} \text{ Ans.}$$

Example 18. If $\phi(x,y,z) = 3(x^2y - y^2x)$, calculate grad at the point (1, – 2, – 1).

Solution: grad $\phi = \left(i\frac{\partial}{\partial x} + j\frac{\partial}{\partial y} + k\frac{\partial}{\partial z}\right)\phi$

In this case $\phi(x,y,z) = 3x^2y - 3y^2x$

$$= i\frac{\partial}{\partial x}(3x^2y-3y^2x) + j\frac{\partial}{\partial y}(3x^2y-3y^2x) + k\frac{\partial}{\partial z}(3x^2y-3y^2x)$$

$$= (6xy-3y^2)i + (3x^2-6yx)j + (0)k$$

At the point (1, – 2, – 1), we have

$$\text{grad } \phi = [6\times(1)\times(-2) - 3(-2)^2 i + [3(1)^2 - 6(-2)(1)]j$$

$$= [-12-12]i + [3+12]j$$

$$= -24\,i + 15j \text{ Ans.}$$

Example 19. *Find grad r^n, where r is the distance of any point (x, y, z) from the origin (0, 0, 0).*

Solution: We have

$$\text{grad } \phi = \nabla \phi = \left(i\frac{\partial}{\partial x} + j\frac{\partial}{\partial y} + k\frac{\partial}{\partial z} \right)\phi$$

$$\text{grad } r^n = \left(i\frac{\partial}{\partial x} + j\frac{\partial}{\partial y} + k\frac{\partial}{\partial z} \right) r^n$$

As given, $\vec{r}$ is a position vector, which means

$$\vec{r} = xi + yj + zk \quad \therefore r = \sqrt{(x^2+y^2+z^2)} \quad (\because A = \sqrt{(A_x^2 + A_y^2 + A_z^2)})$$

or $$r^n = (x^2+y^2+z^2)^{n/2}$$

$\therefore$ $$\nabla r^n = \left(i\frac{\partial}{\partial x} + j\frac{\partial}{\partial y} + k\frac{\partial}{\partial z} \right)(x^2+y^2)^{n/2}$$

or $$\nabla r^n = i\frac{\partial}{\partial x}(x^2+y^2+z^2)^{n/2} + j\frac{\partial}{\partial y}(x^2+y^2+z^2)^{n/2} + k\frac{\partial}{\partial z}(k^2+y^2+z^2)^{n/2} \quad \text{... } (i)$$

Let us consider the first term, $\frac{\partial}{\partial x}(x^2+y^2+z^2)^{n/2}$ and differentiating is w.r.t. x, we get

$$\frac{\partial}{\partial x}(x+y+z)^{n/} = \left(\frac{n}{-}\right)(x+y+z)^{(n/-1)}.(2x)$$

Again, differentiating partially other terms of equation (*i*) w.r.t. y and z, we have

$$\frac{\partial}{\partial y}(x^2+y^2+z^2)^{n/2} = ny(x^2+y^2+z^2)^{(n-2)/2}$$

and $$\frac{\partial}{\partial z}(x^2+y^2+z^2)^{n/2} = nz(x^2+y^2+z^2)^{(n-2)/2}$$

and $$\frac{\partial}{\partial z}(x^2+y^2+z^2)^{n/2} = nz(x^2+y^2+z^2)^{(n-2)/2}$$

Putting these values in equation (*i*), we have

$$\text{grad } r^n = (x^2+y^2+z^2)^{(n-2)/2} nxi + (x^2+y^2+z^2)^{(n-2)/2} nyj + (x^2+y^2+z^2)^{(n-2)/2} nzk$$

$$= n(x^2+y^2+z^2)^{(n-2)/2}(xi+yj+zk)$$

$$= n(r^2)^{(n-2)/2}\vec{r} = nr^{n-2}\vec{r}$$

[Imp: Students may note that this a general result and maybe used for particular cases, e.g.

If $n = 1$, grad $r = \frac{\vec{r}}{r}$; If $n = -1$, grad $\left(\frac{1}{r}\right) = -\frac{r}{r^3}$ and If $n = -3$, grad $\left(\frac{1}{r^3}\right) = \left(\frac{-3\vec{r}}{r^5}\right)$ etc.]

Example 20. *If the electric potential at any point due to a charge q is given by V = q/r, then show by vector method that the electric intensity* $\vec{E}$ *is given by* $\vec{E} = \frac{q}{r^3}\vec{r}$

Solution: Electric intensity $\vec{E}$ can be represented as the negative gradient of potential V, in the form

$$\vec{E} = -\text{grad V}$$

In this case, $V = \frac{q}{r}$, where $\vec{r}$ is the position vector and r its magnitude.

$\therefore$ $$\vec{E} = -\text{grad}(q/r) = -q \text{ grad}(1/r) = -q\nabla\left(\frac{1}{r}\right)$$

For evaluating grad $(1/r)$ see example (19) so we have

$$\text{The solution, grad}\left(\frac{1}{r}\right) = -\frac{\vec{r}}{r^3}; \quad \therefore \vec{E} = -q\left(\frac{\vec{r}}{r^3}\right) = \frac{q}{r^3}\vec{r} \ \textbf{Ans.}$$

Example 21. *The electric potential at any point in the x-y plane is given by*

$\phi 5x(x^2+y^2)^{1/2} + y(x^2+y^2)^{-1/2}$

Find the cartesian components of the electric field intensity at that point.

Solution: We have, $\vec{E} = (E_x i + E_y j + E_z k) = -\text{grad}\,\phi = -i\frac{\partial\phi}{\partial x} - j\frac{\partial\phi}{\partial y} - k\frac{\partial\phi}{\partial z}$

$$\therefore \quad E_x = -\frac{\partial\phi}{\partial x},\ E_y = -\frac{\partial\phi}{\partial y},\ E_z = -\frac{\partial\phi}{\partial z}$$

In this case $\phi = 5x(x^2+y^2)^{1/2} + y(x^2+y^2)^{-1/2}$

$$\therefore \quad E_x = -\frac{\partial}{\partial x}[5x(x^2+y^2)^{1/2} + y(x^2+y^2)^{-1/2}]$$

$$= -[5(x^2+y^2)^{1/2} + 5x\left(\frac{1}{2}\right)(x^2+y^2)^{-1/2}(2x) + y\left(-\frac{1}{2}\right)(x^2+y^2)^{-3/2}(2x)]$$

$$= -[5(x^2+y^2)^{1/2} + 5x^2(x^2+y^2)^{-1/2} - xy(x^2+y^2)^{-3/2}]$$

So, the X-component of electric field will be

$$E_x = -[5(x^2+y^2)^{1/2} + 5x^2(x^2+y^2)^{-1/2} - xy(x^2+y^2)^{-3/2}]$$

Similarly, $E_y = -\frac{\partial}{\partial y}[5x(x^2+y^2)^{1/2} + y(x^2+y^2)^{-1/2}]$

$$= -[5x\left(\frac{1}{2}\right)(x^2+y^2)^{-1/2}(2y) + (x^2+y^2)^{-1/2} + y\left(-\frac{1}{2}\right)(x^2+y^2)^{-3/2}(2y)]$$

$$= -[5xy(x^2+y^2)^{-1/2} + (x^2+y^2)^{-1/2} - y^2(x^2+y^2)^{-3/2}]$$

So, y component of electric field given by

$$E_y = -[5xy(x^2+y^2)^{-1/2} + (x^2+y^2)^{-1/2} - y^2(x^2+y^2)^{-3/2}] \ \textbf{Ans.}$$

Example 22. *A region is specified by the potential function given by* $= 4x^2 + 3y^2 - 9z^2$. *Calculate the field strength at any point (3, 4, 5) in this region.*

Solution: Electric field is given by, $\vec{E} = -\text{grad}\,\phi = -\left(\frac{\partial\phi}{\partial x} + j\frac{\partial\phi}{\partial y} + k\frac{\partial\phi}{\partial z}\right)$

In this case $\phi = 4x^2 + 3y^2 - 9z^2$

$$\therefore \quad \frac{\partial\phi}{\partial x} = 8x,\ \frac{\partial\phi}{\partial y} = 6y \text{ and } \frac{\partial\phi}{\partial z} = -18z$$

$$\therefore \quad \vec{E} = -(8xi + 6yj - 18zk)$$

At any point (3, 4, 5) we have the value of E

$$\vec{E} = -(8\times 3i + 6(4)j - 18(5)k) = -24i - 24j + 90k \ \textbf{Ans.}$$

Example 23. *If $\vec{r}$ is the position vector of a point, then find grad (log r).*

Solution: We now, grad $\phi = \vec{\nabla}\phi = \left(i\frac{\partial}{\partial x} + j\frac{\partial}{\partial y} + k\frac{\partial}{\partial z}\right)\phi$

Since $\vec{r}$ is a position vector, which means, $\vec{r} = xi + yj + zk \quad \therefore r = \sqrt{(x^2+y^2+z^2)}$

$$\therefore \quad \log r = \log (x^2+y^2+z^2)^{1/2} = \frac{1}{2}\log(x^2+y^2+z^2)$$

and $$\text{grad } (\log r) = \vec{\nabla}(\log r) = \frac{1}{2}\vec{\nabla}\log(x^2+y^2+z^2)$$

or $$\text{grad } (\log r)$$

$$= \frac{1}{2}\left\{i\frac{\partial}{\partial x}[\log(x^2+y^2+z^2)] + j\frac{\partial}{\partial y}[\log(x^2+y^2+z^2)] + k\frac{\partial}{\partial z}[\log(x^2+y^2+z^2)]\right\} \quad \text{... (i)}$$

Let us consider the first term, $\frac{\partial}{\partial x}[\log(x^2+y^2+z^2)]$ and differentiating partially

$$\frac{\partial}{\partial x}[\log(x^2+y^2+z^2)] = \frac{1}{(x^2+y^2+z^2)}(x^2+y^2+z^2) \qquad \left(\because \frac{d}{dx}\log(x) = \frac{1}{x}\right)$$

$$= \frac{2x}{(x^2+y^2+z^2)}$$

Again, differentiating partially other terms of equation (*i*) w.r.t. *y* and *z*, we have

$$\frac{\partial}{\partial y}[\log(x^2+y^2+z^2)] = \frac{2y}{(x^2+y^2+z^2)}$$

and $$\frac{\partial}{\partial z}[\log(x^2+y^2+z^2)] = \frac{2z}{(x^2+y^2+z^2)}$$

Putting these values in equation (*i*), we get

$$\nabla = \frac{1}{2}\left[\frac{2xi}{(x^2+y^2+z^2)} + \frac{2yj}{(x^2+y^2+z^2)} + \frac{2zk}{(x^2+y^2+z^2)}\right]$$

or $$\text{grad } (\log r) = \frac{xi+yj+zk}{(x^2+y^2+z^2)} = \frac{\vec{r}}{r^2} \textbf{ Ans.}$$

Example 24. *A vector function has the following components $E_x = 6xy$; $E_y = 3x^2 - 3y^2$ and $E_z = 0$, show that div $\vec{E} = 0$.*

Solution: We have the relation for div. A as

$$\text{div } \vec{A} = \vec{\nabla}.\vec{A} = \frac{\partial A_x}{\partial x} + \frac{\partial A_y}{\partial y} + \frac{\partial A_z}{\partial z}$$

In this case $\vec{A} = \vec{E}$ and $E_x = 6xy$; $E_y = 3x^2 - 3y^2$ and $E_z = 0$

$$\therefore \quad \text{div } \vec{E} = \frac{\partial E_x}{\partial x} + \frac{\partial E_y}{\partial y} + \frac{\partial E_z}{\partial z} = \frac{\partial}{\partial x}(6xy) + \frac{\partial}{\partial y}(3x^2 - 3y^2) + \frac{\partial}{\partial y}(0)$$

or $$\text{div } \vec{E} = 6y - 6y = 0. \textbf{ Ans.}$$

Example 25. If $\vec{r}$ is a product vector of a point, show that

(*i*) div $\vec{r} = 3$; (*ii*) div $(\vec{r}/r^3) = 0$;

(*iii*) div $(r^n \vec{r}) = (3+n)r^n$ and (*iv*) div $(r^4 \vec{r}) = 7r^4$.

Since $\vec{r}$ is a position vector, that is, $r = xi + yj + zk$, so

$$r = \sqrt{(x^2 + y^2 + z^2)}$$

$$\text{Also div } \vec{A} = \frac{\partial A_x}{\partial x} + \frac{\partial A_y}{\partial y} + \frac{\partial A_z}{\partial z}$$

(*i*) $\quad$ div $\vec{r} = 3$

In this case $\quad = A_x = x,\ A_y = y$ and $A_z = z$

$$\therefore \quad \text{div } \vec{r} = \frac{\partial x}{\partial x} + \frac{\partial y}{\partial y} + \frac{\partial z}{\partial z} = 1 + 1 + 1 = 3 \textbf{ Ans.}$$

(*ii*) $\quad$ div $(\vec{r}/r^3) = 0$

In this case
$$\vec{A} = \frac{\vec{r}}{r^3} = \frac{xi + yj + zk}{(x^2 + y^2 + z^2)^{3/2}}$$

$$\therefore \quad A_x = \frac{x}{(x^2 + y^2 + z^2)^{3/2}},\ A_y = \frac{y}{(x^2 + y^2 + z^2)^{3/2}}$$

and
$$A_z = \frac{z}{(x^2 + y^2 + z^2)^{3/2}}$$

$$\therefore \quad \vec{A} = div\left(\frac{\vec{r}}{r^3}\right) = \frac{\partial}{\partial x}\left[\frac{x}{(x^2 + y^2 + z^2)^{3/2}}\right] + \frac{\partial}{\partial y}\left[\frac{y}{(x^2 + y^2 + z^2)^{3/2}}\right] + \frac{\partial}{\partial z}\left[\frac{z}{(x^2 + y^2 + z^2)^{3/2}}\right] \quad ...(ii)$$

Let us consider the first term, $\frac{\partial}{\partial x}\left[\frac{x}{(x^2 + y^2 + z^2)^{3/2}}\right]$ and differentiating it partially w.r.t. *x*, we have

$$\frac{\partial}{\partial x}\left[\frac{x}{(x^2 + y^2 + z^2)^{3/2}}\right] = \frac{\partial}{\partial x}[x(x^2 + y^2 + z^2)^{-3/2}]$$

$$= x.\frac{\partial}{\partial x}(x^2 + y^2 + z^2)^{-3/2} + (x^2 + y^2 + z^2)^{-3/2}\frac{\partial}{\partial x}(x)$$

$$= x.\left(\frac{-3}{2}\right)(x^2 + y^2 + z^2)^{-(3/2)-1}(2x) + (x^2 + y^2 + z^2)^{-3/2}$$

$$= -3x^2\,(x^2 6y^2 + z^2)^{-5/2} + (x^2 + y^2 + z^2)^{-3/2}$$

Again, differentiating partially other terms of equation (1) w.r.t. y and z, we get

$$\frac{\partial}{\partial y}[y(x^2 + y^2 + z^2)^{-3/2}] = -3y^2(x^2 + y^2 + z^2)^{-5/2} + (x^2 + y^2 + z^2)^{-3/2}$$

and
$$\frac{\partial}{\partial z}[z(x^2 + y^2 + z^2)^{-3/2}] = -zy^2\,(x^2 + y^2 + z^2)^{-5/2} + (x^2 + y^2 + z^2)^{-3/2}$$

Putting these values in equation (1), we have

$$\text{div}\ (\vec{r}/r^3) = -3x^2(x^2+y^2+z^2)^{-5/2} + (x^2+y^2+z^2)^{-3/2} - 3y^2(x^2+y^2+z^2)^{-5/2}$$
$$+(x^2+y^2+z^2)^{-3/2} - 3z^2(x^2+y^2+z^2)^{-5/2} + (x^2+y^2+z^2)^{-3/2}$$

or
$$(\vec{r}/r^3) = (-3x^2-3y^2-3z^2)(x^2+y^2+z^2)^{-5/2} + 3(x^2+y^2+z^2)^{-3/2}$$
$$= -3(x^2+y^2+z^2)^{-3/2} + 3(x^2+y^2+z^2)^{-3/2} = 0$$

Which is R.H.S. hence proved **Ans.**

(*iii*) $\text{div}\ (r^n\,\vec{r}) = (3+n)r^n$

Now $r^n = (x^2+y^2+z^2)^{n/2}$ and $\vec{r} = xi + yj + zk$

So
$$\vec{A} = r^n\,\vec{r} = (xi+yj+zk)(x^2+y^2+z^2)^{n/2}$$

$\therefore$
$$A_x = x(x^2+y^2+z^2)^{n/2};\ A_y = y(x^2+y^2+z^2)^{n/2};$$
$$A_z = z(x^2+y^2+z^2)^{n/2}$$

$$= \text{div}\ (r^n\,\vec{r}) = \frac{\partial}{\partial x}[x(x^2+y^2+z^2)^{n/2}] + \frac{\partial}{\partial y}[y(x^2+y^2+z^2)^{n/2}]$$
$$+\frac{\partial}{\partial z}[z(x^2+y^2+z^2)^{n/2}] \quad \text{... } (i)$$

Considering the first term, is $\frac{\partial}{\partial x}[x(x^2+y^2+z^2)^{n/2}]$ and differentiateing it partially w.r.t. x, we have

$$\frac{\partial}{\partial x}[x(x^2+y^2+z^2)^{n/2}] = x\frac{\partial}{\partial x}(x^2+y^2+z^2)^{n/2} + (x^2+y^2+z^2)^{n/2}\frac{\partial}{\partial x}(x)$$
$$= (x).\left(\frac{n}{2}\right)(x^2+y^2+z^2)^{(n/2)-1}.(2x) + (x^2+y^2+z^2)^{n/2}$$
$$= nx^2(x^2+y^2+z^2)^{(n-2)/2} + (x^2+y^2+z\,s\,u\,2)^{n/2}$$

Again, differentiating other terms partially w.r.t. y and z, we obtain

$$\frac{\partial}{\partial y}[y(x^2+y^2+z^2)^{n/2}] = ny^2(x^2+y^2+z^2)^{(n-2)/2} + (x^2+y^2+z^2)^{n/2}$$

and
$$\frac{\partial}{\partial z}[z(x^2+y^2+z^2)^{n/2}] = nz^2(x^2+y^2+z^2)^{(n-2)/2} + (x^2+y^2+z^2)^{n/2}$$

Putting these values in equation (1), we get

$$\text{div}\ (r^n\,\vec{r}) = nx^2(x^2+y^2+z^2)^{(n-2)/2} + (x^2+y^2+z^2)^{n/2}$$
$$+nz^2(x^2+y^2+z^2)^{(n-2)/2} + (x^2+y^2+z^2)^{n/2}$$
$$= (nx^2+ny^2+nz^2)(x^2+y^2+z^2)^{(n-2)/2} + 3(x^2+y^2+z^2)^{n/2}$$
$$= n(x^2+y^2+z^2)^{n/2} + 3(x^2+y^2+z^2)^{n/2}$$
$$= (n+3)(x^2+y^2+z^2)^{n/2}$$

$$= (n+3)(r^2)^{n/2}$$

$$= \mathbf{(3 + n)\ r^4 = R.H.S., Hence\ proved}$$

(*iv*) div $(r^4 \vec{r}) = 7r^4$

Now $\quad (r^4 \vec{r}) = (x^2 + y^2 + z^2)^{4/2}(xi + yj + zk) = (x^2 + y^2 + z^2)^2(xi + yj + zk)$

$$\therefore \text{ L.H.S.} = \text{div } (r^4 \vec{r}) = \frac{\partial}{\partial x}[(x^2 + y^2 + z^2)^2 x] + \frac{\partial}{\partial y}[(x^2 + y^2 + z^2)^2 y] + \frac{\partial}{\partial z}[(x^2 + y^2 + z^2)z]$$

$$= 2x(x^2 + y^2 + z^2)^{2-1}(2x) + (x^2 + y^2 + z^2)^{2-1} 2y(x^2 + y^2 + z^2)^{2-1}).(2y)$$

$$+(x^2 + y^2 + z^2)^2 + 2z(x^2 + y^2 + z^2)^{2-1}(2z) + (x^2 + y^2 + z^2)$$

$$= (4x^2 + 4y^2 + 4z^2)(x^2 + y^2 + z^2) + 3(x^2 + y^2 + z^2)^2$$

$$= 4(x^2 + y^2 + z^2)^2 + 3(x^2 + y^2 + z^2)^2$$

$$= 7(x^2 + y^2 + z^2)^2$$

$$= \mathbf{7(r^2)^2 = 7r^4 = R.H.S.\ Hence\ proved}$$

Example 26. *Find the constant 'a' for which the vector* $\vec{A} = (x + 3y)i + (y - 2z)j + (x + az)k$ *is solenoidal.*

Solution: A vector $\vec{A}$ is said to be solenoidal if its divergence is zero. It means div $A = 0$

In this case $\quad A = (x + 3y)i + (y - 2z)j + (x + az)k = A_x i + A_y j + A_z k$

Also $\quad \text{div } \vec{A} = \vec{\nabla}.\vec{A} = \dfrac{\partial A_x}{\partial x} + \dfrac{\partial A_y}{\partial y} + \dfrac{\partial A_z}{\partial z}$

$\therefore \quad \text{div } \vec{A} = \dfrac{\partial}{\partial x}(x + 3y) - \dfrac{\partial}{\partial y}(y - 2z) + \dfrac{\partial}{\partial z}(x + az)$

or $\quad \text{div A} = 1 + 1 + a = 2 + a$

For a solenoidal vector div $\vec{A} = 0$, therefore $2 + a = 0$

Hence $\quad \boldsymbol{a = -2}$ **Ans.**

Example 27. *Prove that*

(*a*) *div* $(\vec{SA}) = S \text{ div } \vec{A} + \vec{A}.$ *grad S* (*b*) *div* $(\vec{A} + \vec{B}) = \text{div } \vec{A} + \text{div } \vec{B}$

(*c*) *div grad S* $= \nabla^2 S$

Solution: We have $(\vec{SA}) = S \text{ div } \vec{A} + \vec{A}.\text{grad } S$

Also $\quad \text{div } \vec{A} = \vec{\nabla}.\vec{A} = \dfrac{\partial}{\partial x} A_y + \dfrac{\partial}{\partial x} A_y + \dfrac{\partial}{\partial z} A_z$

In this case is a scalar and $\vec{A} = A_x i + A_y j + A_z k$

$$\therefore \quad \text{L.H.S.} = \text{div } (\vec{SA}) = \vec{\nabla}.(\vec{SA}) = \frac{\partial}{\partial x}(SA_x) + \frac{\partial}{\partial y}(SA_y) + \frac{\partial}{\partial z}(SA_z)$$

$$= S\frac{\partial A_x}{\partial x} + A_x\frac{\partial S}{\partial x} + S\frac{\partial A_y}{\partial y} + A_y\frac{\partial S}{\partial y} + S\frac{\partial A_z}{\partial z} + A_z\frac{\partial S}{\partial z}$$

$$= S\left(\frac{\partial A_x}{\partial x} + \frac{\partial A_y}{\partial y} + \frac{\partial A_z}{\partial z}\right) + \left(A_x\frac{\partial S}{\partial s} - A_y\frac{\partial S}{\partial y} + A_z\frac{\partial S}{\partial z}\right)$$

$$= S(\vec{\nabla}.\vec{A}) + (A_x i + A_y j - A_z k).\left(i\frac{\partial S}{\partial x} + j\frac{\partial S}{\partial y} + k\frac{\partial S}{\partial z}\right)$$

$$= S\,\text{div}\,\vec{A} + \vec{A}\,.\,\vec{\nabla}S$$

$$= S\,\text{div}\,\vec{A} + \vec{A}.\ \text{grad}\ S = \text{R.H.S. Hence proved}$$

(*b*) div $(\vec{A} + \vec{B})$ = div $\vec{A}$ + div $\vec{B}$, where $\vec{A} = A_x i + A_y j + A_z k$ and $\vec{B} = B_x i + B_y j + B_z k$

$\therefore$ $$(\vec{A} + \vec{B}) = (A_x + B_x)i + (A_y + B_y)i + (A_z + B_z)k$$

$\therefore$ Let us take L.H.S. as div $(\vec{A} + \vec{A}) = \vec{\nabla}.(\vec{A} + \vec{B}) = \frac{\partial}{\partial x}(A_x + B_x) + \frac{\partial}{\partial z}(A_z + B_z)$

or $$\text{div}\ (\vec{A} + \vec{B}) = \frac{\partial A_x}{\partial x} + \frac{\partial B_x}{\partial x} + \frac{\partial A_y}{\partial y} + \frac{\partial A_z}{\partial z} + \frac{\partial B_z}{\partial z}$$

$$= \left(\frac{\partial A_x}{\partial x} + \frac{\partial A_y}{\partial y} + \frac{\partial A_z}{\partial z}\right) + \left(\frac{\partial B_x}{\partial x} + \frac{\partial B_y}{\partial y} + \frac{\partial B_z}{\partial z}\right)$$

$$= \text{div}\ \vec{A} + \text{div}\ \vec{A}\ \text{Which is R.H.S.}$$

$\therefore$ $$\text{div}\ (\vec{A} + \vec{B}) = \text{div}\ \vec{A} + \text{div}\ \vec{B}\ \textbf{Ans.}$$

$$\text{div grad}\ S = \nabla^2 S$$

$\therefore$ $$\text{L.H.S.} = \text{div (grad}\ S) = \vec{\nabla}.\text{grad}\ S = \vec{\nabla}.(\vec{\nabla}S)$$

Also $$\text{grad}\ S = i\frac{\partial S}{\partial x} + j\frac{\partial S}{\partial y} = k\frac{\partial S}{\partial z}\ \text{ and div }\ \vec{A} = \frac{\partial A_x}{\partial x} + \frac{\partial A_y}{\partial y} + \frac{\partial A_z}{\partial z}$$

In this case, $\vec{A}$ = grad S,

$\therefore$ $$A_x = \frac{\partial S}{\partial x}, A_y = \frac{\partial S}{\partial y}, A_z = \frac{\partial S}{\partial z}$$

$\therefore$ $$\text{div (grad S)} = \left(i\frac{\partial}{\partial x} + j\frac{\partial}{\partial y} + k\frac{\partial}{\partial z}\right).\left(i\frac{\partial S}{\partial x} + j\frac{\partial S}{\partial y} + k\frac{\partial S}{\partial z}\right)$$

$$= \frac{\partial}{\partial x}\left(\frac{\partial S}{\partial x}\right) + \frac{\partial}{\partial y}\left(\frac{\partial S}{\partial y}\right) + \frac{\partial}{\partial z}\left(\frac{\partial S}{\partial z}\right) = \frac{\partial^2 S}{\partial x^2} + \frac{\partial^2 S}{\partial y^2} + \frac{\partial^2 S}{\partial z^2},$$

$$= \left(\frac{\partial^2}{\partial x^2} + \frac{\partial^2}{\partial y^2} + \frac{\partial^2}{\partial z^2}\right)S = \nabla^2 S\ \text{ which is R.H.S. Hence proved } \textbf{Ans.}$$

Example 28. *Find the value of* $\iint \vec{r}\,\hat{n}\,d\,S$, *where S is a closed surface and* $\vec{r}$ *is the position vector.*

Solution: From the Gauss-divergence theorem, we have

$$\iint_s \vec{A}.\vec{dS} = \iiint_v \text{div A dV}$$

In this case, $\vec{dS} = n\,dS$ and $\vec{A} = \vec{r} = xi + yj + zk$

$$\therefore \quad \iint_s \vec{r}.\hat{n}\, dS = \iiint_v \text{div}\, \vec{r}\, dV = \iiint_v \left(\frac{\partial x}{\partial x} + \frac{\partial y}{\partial y} + \frac{\partial z}{\partial z}\right) dx\, dy\, dz$$

$$= \iiint_v (1+1+1)\, dx\, dy\, dz = 3\iiint_v dx\, dy\, dz = 3V \textbf{ Ans.}$$

where V is the volume enclosed by the surface S.

Example 29. *Using Gauss-divergence theorem, evaluate* $\iint_s \vec{F}.\vec{dS}$, *where* $\vec{F} = 4xzi - y^2 j + yzk$ *and S is the surface of the cube bounded by* $x = 0$, $x = 1$, $y = 0$, $y = 1$, $z = 0$, $z = 1$.

Solution: From the Gauss-divergence theorem, we have

$$\iint_s \vec{F}.\vec{dS} = \iiint_v \text{div}\, \vec{F}.dV = \iiint_v \left(\frac{\partial F_x}{\partial x} + \frac{\partial F_y}{\partial y} + \frac{\partial F_z}{\partial z}\right) dx\, dy\, dz$$

In this case $\quad \vec{F} = 4xzi - y^2 j + yzk$

$$\therefore \quad \iint_s \vec{F}.\vec{dS} = \iiint_v \left[\frac{\partial}{\partial x}(4xz + \frac{\partial}{\partial y}(-y^2) + \frac{\partial}{\partial z}(x, z)\right]$$

$$= \iiint_v (4z - 2y + y)\, dx\, dy\, dz$$

The surface S of the cube is bonded by $x = 0$, $x = 1$, $y = 0$, $y = 1$, $z = 0$, $z = 1$

$$\therefore \quad \iint_s \vec{F}.\vec{dS} = \int_0^1\int_0^1\int_0^1 (4z - y)\, dx\, dy\, dz = \int_0^1\int_0^1 \left(\frac{4z^2}{2} - yz\right)_0^1 dx\, dy = \int_0^1\int_0^1 (2 - y)\, dx\, dy$$

$$= \int_0^1 \left(2y - \frac{y^2}{2}\right)_0^1 dx = \int_0^1 \frac{3}{2}\, dx$$

$$\iint_s \vec{F}.\vec{dS} = \frac{3}{2}\int_0^1 dx = \frac{3}{2}(x)_0^1 = \frac{3}{2} = 1.5 \textbf{ Ans.}$$

Example 30. *Prove* $\iint_s (\text{curl}\, \vec{F}).\vec{dS} = 0$ *for any closed surface S.*

Solution: From Gauss-divergence theorem

$$\iint_s \vec{A}.\vec{dS} = \iiint_v \text{div}\, \vec{A}\, dV$$

In this case $\quad \vec{A} = \text{curl}\, \vec{F}$

$$\iint_s (\text{curl}\, \vec{F}).\vec{dS} = \iiint_v \text{div}\, (\text{curl}\, \vec{F})\, dV$$

but div (curl $\vec{F}$) = $\vec{\nabla}$. curl $\vec{F} = \vec{\nabla}.(\vec{\nabla} \times \vec{F})$

$$\vec{\nabla}.(\vec{\nabla} \times \vec{F}) = \begin{vmatrix} \frac{\partial}{\partial x} & \frac{\partial}{\partial y} & \frac{\partial}{\partial z} \\ \frac{\partial}{\partial x} & \frac{\partial}{\partial y} & \frac{\partial}{\partial z} \\ F_x & F_y & F_z \end{vmatrix} = 0$$

So, $\quad \iint_s (\text{curl}\, \vec{F}).\vec{dS} = 0$ **Ans.**

Example 31. *Verify Stoke's theorem for* $\vec{F} = (2x - y)i - yz^2 \vec{j} - y^2 \vec{z}k$, *where S is the upper half surface of the sphere* $x^2 + y^2 + z^2 = 1$ *and C is its boundary.*

Solution: Boundary of the sphere is a circle of radius one and centre at (0, 0, 0) which lies in the $x-y$ plane. Let $x = \cos t$, $y = \sin t$, $z = 0$ and $0 \le t < 2\pi$ be the parametric equations of the boundary, so we have

$$\vec{\nabla}\times\vec{F} = \vec{i}\left(\frac{\partial}{\partial y}F_z - \frac{\partial}{\partial z}F_y\right) + \vec{j}\left(\frac{\partial}{\partial z}F_x - \frac{\partial}{\partial x}F_z\right) + \vec{k}\left(\frac{\partial}{\partial x}F_y - \frac{\partial}{\partial y}F_x\right)$$

In this case $\vec{F} = (2x-y)\,\vec{i} - yz^2\hat{j} - y^2 z\vec{k}$

$$\therefore \quad F_x = 2x - y,\ F_y = -yz^2 \text{ and } F_z = -y^2 z$$

$$\therefore \quad \vec{\nabla}\times\vec{F} = \vec{i}\left[\frac{\partial}{\partial y}(-y^2 z) - \frac{\partial}{\partial z}(-yz^2)\right] + \vec{j}\left[\frac{\partial}{\partial z}(2x-y) - \frac{\partial}{\partial x}(-y^2 z)\right]$$

$$+\vec{k}\left[\frac{\partial}{\partial x}(-y^2 z) - \frac{\partial}{\partial y}(2x-y)\right]$$

$$= \vec{i}[-2yz + 2yz] + \vec{j}[0-0]t\,\vec{k}[0+(1)] = \vec{k} \qquad \text{... } (i)$$

and
$$\oint_c \vec{F}.\vec{dr} = \oint_c [(2x-y)\,dx - yz^2\,dy - y^2 z dz] \quad [\because \vec{F}.\vec{dr} = F_x dx + F_y dy + F_z dz]$$

Putting $x = \cos t$, $y = \sin t$ and $z = 0$

$$\therefore \quad dx = -\sin t\ dt,\ dy = \cos t.\ dt$$

$$\therefore \quad \oint \vec{F}.\vec{dr} = \int_0^{2\pi} (2\cos t - \sin t)(-\sin t\ dt) - 0 - 0$$

or
$$\oint \vec{F}.\vec{dr} = \pi \qquad \text{... } (ii)$$

$$\oint\oint_S \left(\vec{\nabla}\times\vec{F}\right).\hat{n}\,dS = \oint\oint_S \vec{k}.\hat{n}dS = \oint\oint_R dxdy \qquad (\because \text{ From equation. } (i))$$

where $R = dx\,dy$ and it is the projection of surface S on $x-y$ plane

So
$$\oint\oint_S dxdy = \int_{x=-1}^{1}\int_{\sqrt{1-x^2}}^{0} dy\,dx = 4\int_0^1\int_0^{\sqrt{1-x^2}} dydx$$

$$= 4\int_0^1 \sqrt{(1-x^2)}dx = \pi \qquad \text{... } (iii)$$

Hence, the Stoke's theorem $\oint \vec{F}.\vec{dr} = \int\oint_S \left(\vec{\nabla}\times\vec{F}\right).\vec{dS}$**, is verified. Ans.**

QUESTIONS AND PROBLEMS

1. What do you mean by scalar and vector fields ? Give at least one example of each.
2. Give the physical significance of grad ϕ. Express grad ϕ in Cartesian coordinates. Hence, define $\vec{\nabla}$ operator.
3. What do you understand by the field of any physical quantity ? Give examples of scalar and vector fields.
4. Give the physical significance of divergence of a vector field. Express divergence of a vector in terms of del operator.
5. What is conservative field ? Give examples of such field.
6. What do you mean by curl of a vector function ? Obtain an expression for the curl of a vector function in cartesian coordinates.

7. What is the physical significance of (*i*) gradient of a scalar field, (*ii*) divergence of a vector field and (*iii*) curl of a vector field ? Explain by giving one example of each.
8. Define curl of a vector field. Give the physical significance of the curl of a vector field. Derive an expression for it. Prove, curl $\vec{A} = \vec{\nabla} \times \vec{A}$.
9. What are line and surface integrals ?
10. State and prove Gauss's divergence theorem. What is its importance ?
11. State and prove Stoke's theorem.
12. Let $\vec{r}$ be the vector from some fixed point (x_0, y_0, z_0) to the point (x, y, z), and let r be its length. Show that, $\vec{\nabla}(r^2) = 2\vec{r}$.
13. Prove that, $dV = (d\vec{r}.\vec{\nabla}) + V + \frac{\partial V}{\partial t} dt$.
14. If $\vec{a} = \alpha x i + \beta y j + \gamma z k$, show that $\vec{\nabla} = (\vec{a}.\vec{r}) = 2\vec{a}$.
15. Find the gradient of f, where f is given as $yz + zx + xy$.
16. Find the gradient of f, where f is given as
(*i*) $f = 2xy$ (*ii*) $f = x / y$ **[Ans.** (*i*) $2(yi + xg)$ (*ii*) $(y\hat{i} - x\hat{j})/y^2$ **]**
17. Let $\vec{r}$ be the vector from some fixed point (x_0, y_0, z_0) to the point (x, y, z) and let r be its length. Show that, $\vec{\nabla}(r^2) = 2\vec{r}$.
18. If $\vec{A} = 3xyz^2 i + 2xy^3 j - x^2 yzk$ and $\phi = 3x^2 - yz$, find
(*a*) $\vec{\nabla}.\vec{A}$ (*b*) $\vec{A}.\nabla\phi$ (*c*) $\vec{\nabla}.(\phi\vec{A})$ (*d*) $\vec{\nabla}.\vec{\nabla}\phi$ at (1, – 1, 1).
[Ans. (*a*) 4, (*b*) – 15, (*c*) – 1, (*d*) 6**]**
19. Prove that if $\vec{F}$ represents a vector field and curl $\vec{F} = 0$, but div $\vec{F} \neq 0$, then field has a vector potential.
20. If $\vec{A} = 3x^2 i + 5xy^2 j + xyz^2 k$, then find div $\vec{A}$ at the point (1, 2, 3). **[Ans. 38]**
21. Evaluate $\vec{\nabla}.(r^3 \vec{r})$ **[Ans. 6.53]**
22. If $\vec{A} = kr^n \hat{r}$, prove that div $\vec{A} = k(n+2)r^{n-1}$
23. Show that vector $\vec{A} = y^2 i$ is solenoidal.
24. If $\vec{A} = \vec{r}/r$, find grad div $\vec{A}$. **Ans.** $(-2r^3 \hat{r})$

PART VI

ELECTRICITY AND MAGNETISM

Gustav Robert Kirchhoff (1824 – 1887)

Kirchhoff was born in Konigsberg, the son of a law counselor who had many friends on the faculty at the University of Konigsberg; he married the daughter of one of his professors; and he spent his life doing research and teaching at various universities.

Kirchhoff's first important contribution to physics, the analysis of electrical networks, was made while he was still a university student and only 21 years old.

Kirchhoff's work on the emission spectra of the elements led him to consider the relationship of their spectral lines to the dark lines observed by Fraunhofer in the sun's spectrum. Kirchhoff showed that the famous yellow *D* lines in the emission spectrum of sodium corresponded perfectly to two dark lines in the solar spectrum. This made him conclude that the sun's atmosphere contained sodium and that this sodium absorbed characteristic wavelengths out of the continuous radiation emitted by the sun. He verified that these were indeed the sodium lines by means of a laboratory experiment involving a sodium light source and NaCl salt ignited in a Bunsen burner.

His study of the relationship between the emission and absorption of radiation then led Kirchhoff to investigate the radiation from hot objects and to propose a hollow metal cavity (a blackbody) as an ideal source for such radiation. His radiation law became the key to the study of radiation. Max Planck, who succeeded Kirchhoff to the chair of theoretical physics at the University of Berlin, took up Kirchhoff's ideas and used them as the basis for his quantum theory, one of the truly great developments in the history of physics.

For many years, as a result of an accident, Kirchhoff was confined to a wheelchair or crutches. Yet he retained his cheerful disposition and great enthusiasm for his work. One day his banker, unimpressed by Kirchhoff's ability to locate elements in the sun with his spectroscope, asked him, "Of what use is gold in the sun if I cannot bring it down to earth?" Some years later, when Great Britain presented Kirchhoff with a gold medal and a large cash prize in gold sovereigns for his research, Kirchhoff handed it over to the same banker, with the sly remark: "Here is your gold from the sun."

The experimental investigation by which Ampere established the laws of the mechanical action between electric currents is one of the most brilliant achievements in science. The whole, theory and experiment, seems as if it had leaped full grown and full armed from the brain of the "Newton of Electricity".

[James Clerk Maxwell (1881-1879)]

CHAPTER

1

ELECTROMAGNETISM

1.1. INTRODUCTION

Some of the useful terms used in electromagnetism are given below:

(i) Magnetic Forces are due to the motion of electric charges, *i.e.,* electric currents. A magnetic force exists between two charges if and only if both charges are moving relative to an observer. Examples of moving charges are electric currents in wires, a beam of electrons in a cathode ray tube, and electrons rotating about the atomic nuclei and about their own axes.

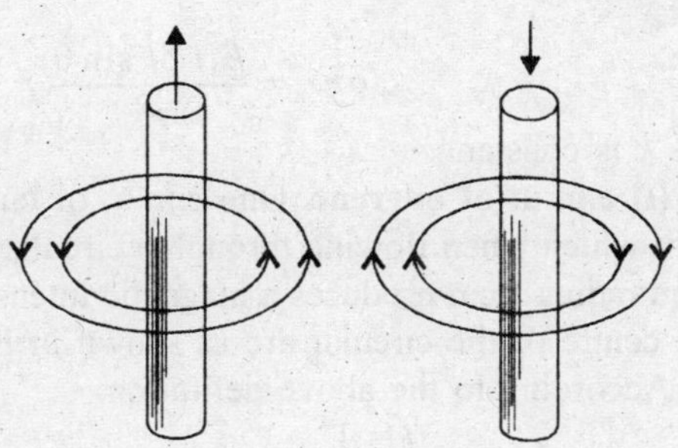

Fig. 1.1

(ii) Magnetic Field is the region surrounding a moving charge. A magnetic field exists at any point where a moving charge experiences a force exclusive of any electrostatic force.

(iii) Direction of Magnetic Field of a Current may be determined by a right-hand rule. *Point the thumb of the right hand in the direction of the current, then the curled fingers will point in the direction of the field.* Thus the magnetic field of a current is clockwise as viewed by an observer looking along the direction of the current. Refer to Fig. 1.1.

(iv) Magnetic Flux Density *B*, also called **magnetic induction** of magnetic **field,** is a vector which specifies the magnitude and direction of the magnetic field at a point.

A magnetic field may be represented by lines of *flux density such that the number of lines piercing perpendicularly a unit area is equal to the magnetic flux density in that region.* The direction of the lines at any point is that of *B*.

The magnetic flux Φ through a surface is the total number of lines of *B* extending through it. In a region where the magnetic flux density *B* is uniform and normal to a given area *A*,

$$\Phi = BA$$

If θ is the angle between the magnetic induction vector *B* and the normal to the reference area *A*,

$$\Phi = BA \cos \theta$$

The S.I. unit for *B* is the **tesla,** symbol *T*, and for Φ is the weber, symbol Wb. Thus the tesla is 1 Wb m^{-2} and

$$\Phi \text{ (Wb)} = B\ (T) \times A\ (m^2) \times \cos \theta$$

1.2. AMPERE'S LAW

Let *XY* be a current carrying conductor (fig. 1.2) and let *I* e.m.u. current be flowing from *X* to *Y*.

The magnetic field at any point P at a distance r from the middle point O of AB is given by

$$\delta\vec{F} \propto \frac{i\,\delta l \sin\theta}{r^2}\,\hat{r}$$

So the magnetic intensity at P is

$\propto i$, *i.e.*, strength of the current

$\propto \Delta l$, *i.e.*, length of the element

$\propto \sin\theta_r$ (where θ is angle between the direction of current and line joining middle point O of the element with P).

$\propto \frac{1}{r^2}$, where r is distance OP.

$$\therefore \quad \delta\vec{F} = \frac{k.i\,\delta l \sin\theta}{r^2}\,\hat{r}$$

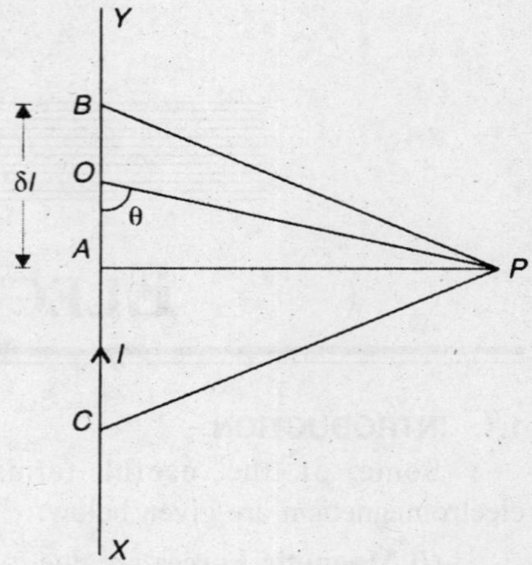

Fig. 1.2

where k is constant.

***(i)* e.m.u. of current.** One e.m.u. of current is that current which when flowing through a circular arc of length 1 cm and radius 1 cm produces a magnetic intensity of 1 oersted at the centre of the circular arc as shown in Fig. 1.3.

According to the above definition,

if $\quad i = 1$

$\quad r = 1,\ \delta l = 1,\ \sin 90^\circ = 1$ and $F = 1$

then $\quad k = 1$

$$\therefore \quad \delta\vec{F} = \frac{i\delta l \sin\theta}{r^2}\,\hat{r} \qquad ...(1)$$

This is called *Ampere's law or Laplace's law.*

***(ii)* Rationalised M.K.S. units:** In SI units $k = \frac{1}{4\pi}$

$\therefore$ Above law may be written as

$$\delta\vec{F} = \frac{i.\,\delta l \sin\theta}{r^2}\,\hat{r} \qquad ...(2)$$

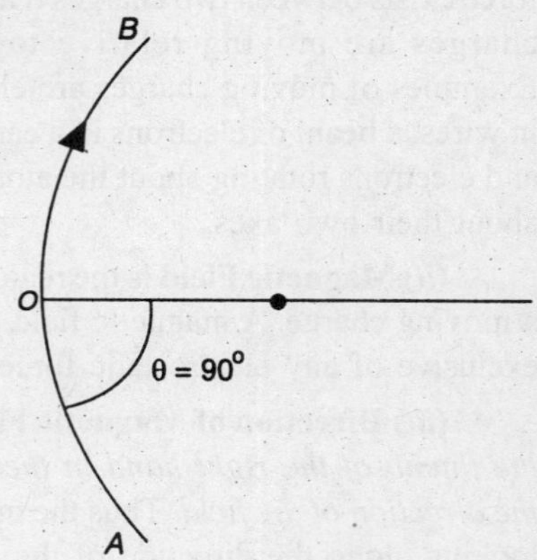

Fig. 1.3

Now, 4π since flux density ΔB depends on the nature of the medium, we may write

$$\Delta\vec{B} = \frac{\mu_0}{4\pi}\cdot\frac{i.\,\delta l \sin\theta}{r^2} \quad \text{[where } \mu_0 \text{ is called permeability of the free space]}$$

$\mu_0 = 1$ in C.G.S. units

$\neq 1$ in M.K.S. units

For getting the force on a current carrying conductor placed in a magnetic field, we have to apply the Fleming's left hand rule and it depends on the nature of the medium. So to get this force, magnetic flux density is used.

In this case $\mu_0 = 4\pi \times 10^{-7} = 12.57 \times 10^{-7}$ weber/amp-m or henry/m.

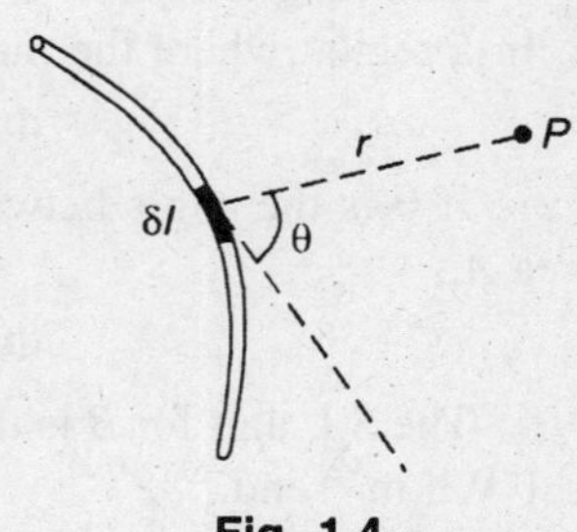

Fig. 1.4

1.3. MAGNETIC FIELD OF A CURRENT

Consider that a current I exists in a wire element of length Δl. The magnetic flux density δB at point P due to the current element in a vacuum is

$$\delta\vec{B} = K\,\frac{I\Delta l \sin\theta}{r^2} \qquad ...(1)$$

where r is the distance from Δl to P, θ is the angle between r and the current in the element Δl, and K is a proportionality constant depending upon the units used. The direction of ΔB is perpendicular to the plane determined by δl and r and is given by the right-hand rule above.

The total magnetic flux density B at point P due to a current I in a wire of length l is the vector sum of the contributions of all the elements Δl. So we have

$$B = K\,\Sigma \frac{I\delta l \sin\theta}{r^2} \qquad \text{...(2)}$$

In SI units, $K = 10^{-7}$ Wb A^{-1} m^{-1} and it is written as μ_0 4π so we may write Equation (2) as

$$B = \Sigma \frac{\mu_0 \; I\delta l \sin\theta}{4\pi r^2} \qquad \text{...(3)}$$

The constant μ_0 is called the permeability of free space or the **permeability of vacuum** and has the value

$$\mu_0 = 4\pi \times 10^{-7} \text{ Wb A}^{-1} \text{ m}^{-1}$$

Note. The unit of μ_0 can also be expressed in henry per meter (Hm^{-1}).

In infinite homogeneous media other than a vacuum, K takes the value $\frac{\mu_r \mu_0}{4\pi}$ where μ_r is called the **relative permeability** of the medium. For most substances, including μ_r is nearly/and only for ferromagnetics, it is substantially different. Common **ferromagnetic materials are iron, steel, nickel and cobalt.**

1.4. BIOT SAVART'S LAW

This law gives a method for calculating the magnetic field at any point in the region around a current carrying conductor. Any current carrying conductor can be supposed to consist of small *current elements.*

According to this law the magnetic field $\delta \vec{B}$ produced by a small element of length $\delta \vec{l}$ of conductor carrying current i at any point P at a position vector r relative to this element is given by

$$\delta \vec{B} = \frac{\mu}{4\pi} \frac{i\delta l \sin\theta}{r^2}$$

where μ is the absolute permeability of the medium and θ is the angle between $\vec{\delta l}$ and $\vec{r}$

For free space $\quad \mu = \mu_0 = 4\pi \times 10^{-7}$ Wb/Amp/m

and is called absolute permeability of free space.

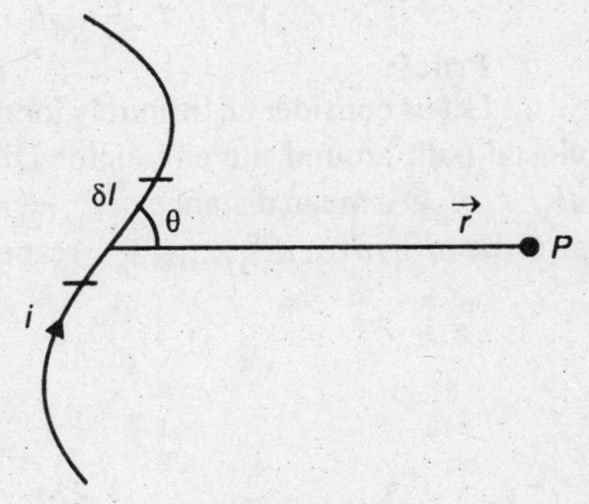

Fig. 1.5

The relative permeability μ_r is given by

$$\mu_r = \frac{\mu}{\mu_0}$$

and is called the *relative permeability* of the medium.

For free space $\quad \delta \vec{B} = \frac{\mu_0}{4\pi} \cdot \frac{i\delta l \sin\theta}{r^2} \qquad \text{...(1)}$

The direction of magnetic field induction $\delta \vec{B}$ is perpendicular to the plane containing $\vec{\delta l}$ and $\vec{r}$, and is given by Fleming's Right Hand rule, *i.e.*,

If a current carrying conductor is held with right hand such that the thumb points in the direction of current, then the curl of the fingers will represent the direction of magnetic field.

In vector form Equation (1) may be written as

$$\delta \vec{B} = \frac{\mu_0}{4\pi} \cdot \frac{i\delta \vec{l} \times \vec{r}}{r^3}$$

For the whole current carrying conductor, the resultant magnetic flux or magnetic induction is given by

$$\vec{B} = \frac{\mu_0 i}{4\pi} \cdot \sum \frac{\delta \vec{l} \times \vec{r}}{r^3} \quad ...(2)$$

where summation extends over the whole conductor.

1.5. AMPERE

An ampere is the strength of the current which when flowing through two parallel infinite long conductors placed in free space at a distance of 1 metre from each other, produces between them a force of 2×10^{-7} Newtons for 1 metre of their length.

The force between two parallel conductors

$$\vec{F} = \frac{\mu_0}{4\pi} \cdot \frac{2I_1 I_2}{r} \hat{r}$$

In this case $F = 2 \times 10^{-7}$ newton

$I_1 = I_2 = 1$ ampere

$r = 1$ metre

$$\therefore \quad 2 \times 10^{-7} = \frac{\mu_0}{4\pi} \frac{2 \times 1 \times 1}{1}$$

$$\therefore \quad \mu_0 = 4\pi \times 10^{-7} \text{ weber } A^{-1} m^{-1}.$$

1.6. AMPERE'S CIRCUITAL LAW

According to this law, the line integral of the magnetic field along a closed hypothetical path in vacuum is equal to μ_0 times the current enclosed by the closed path.

Mathematically, we may write

$$\oint \vec{B} \cdot d\vec{l} = \mu_0 I$$

Proof:

Let us consider an infinitely long conductor carrying a constant current I. Consider any arbitrary closed path around the conductor Divide the closed path into large number of segments dl_1, dl_2, dl_3, ..., dl_n at a mean distance of $r_1, r_2, r_3 ..., r_n$ respectively. Let the angle subtended by these elements at O be $d\theta_1, d\theta_2, d\theta_3, ..., d\theta_n$, respectively.

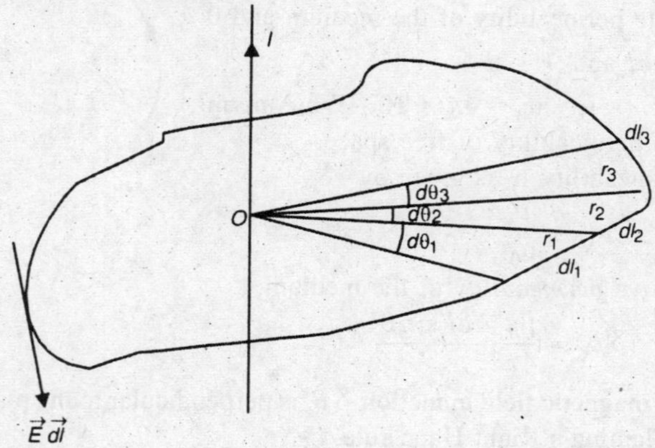

Fig. 1.6

From the figure,

$$d\theta_1 + d\theta_2 + ... d\theta_n = 2\pi \text{ radian}$$

$$i.e., \quad \frac{dl_i}{r_1} + \frac{dl_2}{r_2} + ... \frac{dl_n}{r_n} = 2\pi \text{ radian} \quad ...(1)$$

We have considered the elements to be very small so that it may be assumed that each forms an arc of a circle with O as centre.

Using Right hand Thumb rule, we observe that the magnetic lines of field of the conductor have anticlockwise direction when seen from above.

Now, current enclosed by the closed path is I. So we have

$$\oint \vec{B}\cdot d\vec{l} = B_1\, dl_1 \cos 0° + B_2\, dl_2 \cos 0° + \dots B_n\, dl_n \cos 0°$$

$$= \frac{\mu_0}{4\pi}\cdot\frac{21}{r_1}dl_1 + \frac{\mu_0}{4\pi}\cdot\frac{2I}{r_2}dl_2 + \dots, \frac{\mu_0}{4\pi}\cdot\frac{2I}{r_2}dl_n$$

$$\left[\because \text{For a straight conductor } B = \frac{\mu_0}{4\pi}\cdot\frac{21}{r_j}\right]$$

$$= \frac{\mu_0 I}{2\pi}\left[\frac{dl_1}{r_1} + \frac{dl_2}{r_2} + \dots, \frac{dl_n}{r_n}\right]$$

$$= \frac{\mu_0 I}{2\pi}.2\pi \text{ [using equation (1)]}$$

$$= \mu_0 I$$

1.7. APPLICATIONS OF AMPERE'S CIRCUITAL LAW

Ampere's law can be used to find $\vec{H}$ or $\vec{B}$ by the use of a suitable path called **Ampere's path** around the current. Ampere path must either be **parallel or perpendicular** to the field of lines at each point. It may be noted that current enclosed inside the path only contributes to the R.H.S. of Ampere's circuital law equation and the current outside the path does not contribute.

1. Magnetic field produced by a long straight wire carrying current

Let a long straight wire carrying current I be shown in Figure 1.7

To get an expression for the magnetic flux density $\vec{B}$ at the point P, we draw a circle of radius r to get the Amperian surface (shown in Fig. 1.7 with centre O).

Magnetic field in tangential to the circle and has the same magnitude at all points around the circle.

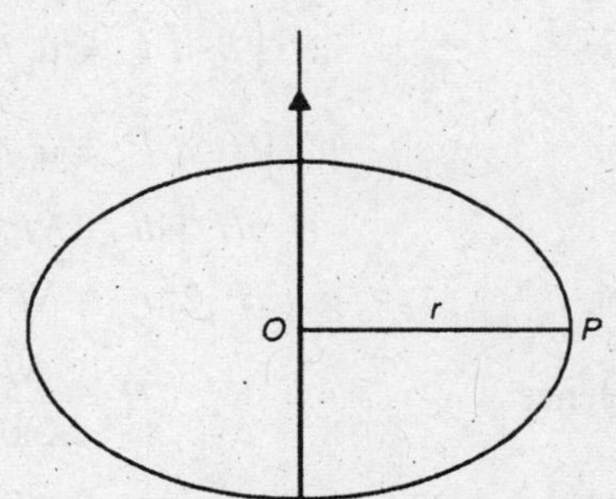

Fig. 1.7

So we have the integral $\oint \vec{B}.d\vec{l} = \oint B.dl = B\oint dl$

Since $\oint dl$ around circle of radius $r = 2\pi r$

$$\therefore \quad \oint \vec{B}\cdot d\vec{l} = B.2\pi r \qquad \dots(1)$$

From ampere's circuital law, we have

$$\oint \vec{B}\cdot d\vec{l} = \mu_0 I \qquad \dots(2)$$

From equation (1) and (2),

$$B.2\pi r = \mu_0 I$$

or

$$B = \frac{\mu_0 I}{2\pi r}$$

2. Magnetic field inside a long cylindrical wire

Let a cylindrical wire of radius R carry current I which is uniformly distributed throughout the cross-section.

Let a dotted circle be drawn with radius r passing through the point P inside the circle. **Magnetic field will be tangential to the circle and will have a constant magnitude.**

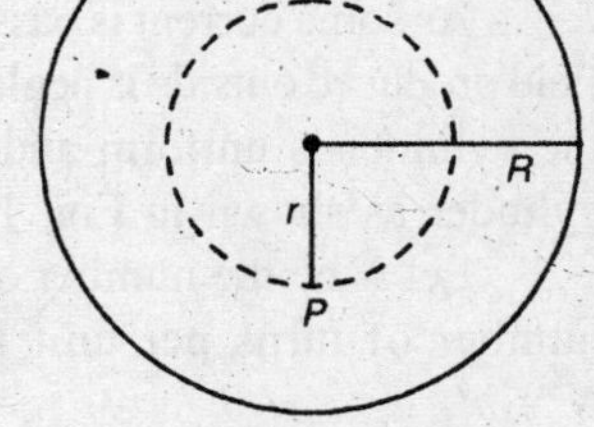

Fig. 1.8

Let current flowing through the wire I and let σ be the current density, so we have

$$I = \sigma.\pi R^2 \quad \text{or} \quad \sigma = \frac{I}{\pi R^2} \qquad ...(1)$$

Let I' be the current through the wire of radius r as shown by dotted line. In that case

$$I' = \sigma.\pi r^2 = \frac{I}{\pi R^2} \cdot \pi r^2 = \frac{r^2}{R^2} \cdot I \qquad ...(2)$$

From Ampere's circuital law

$$\oint \vec{B}.d\vec{l} = \mu_0 I'$$

Now $$\oint \vec{B}.d\vec{l} = \oint B\,dl = B.2\pi r$$

$\therefore$ $$B.2\pi r = \frac{\mu_0 r^2}{R^2} \cdot I$$

Hence $$B = \frac{\mu_0 r^2}{2\pi R^2} \cdot I \qquad ...(3)$$

3. Magnetic field intensity due to a Toroid

Consider a toroid ring of mean radius r and have N turns of wound wires as shown in Fig. 1.9.

Let I be the current through each turn of the wire, then total current flowing through N turns will be NI.

Amperian path will be a circle of mean radius r and the magnetic field at each point will be tangential to the circular path.

From Ampere's circuital law

$$\oint \vec{B} \cdot d\vec{l} = \mu_0 NI \qquad ...(1)$$

or $$\mu_0 \oint H \cdot d\vec{l} = \mu_0 NI$$

$\therefore$ $$H \oint dl = NI$$

or $$H \cdot 2\pi r = NI$$

Hence $$H = \frac{NI}{2\pi r} = \frac{NI}{l} \qquad ...(2)$$

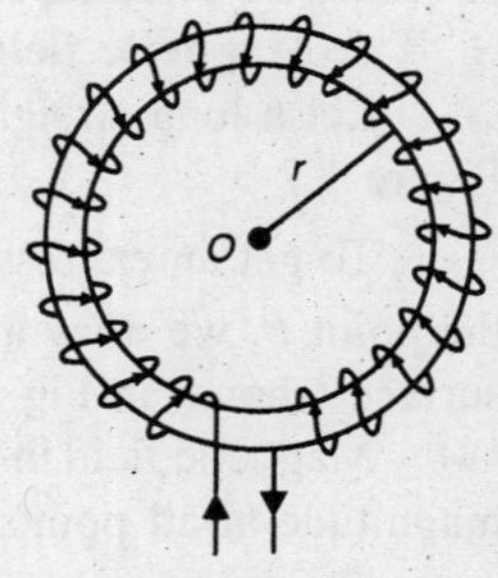

Fig. 1.9

where $l = 2\pi r$ is the circumference of toroid.

If n is number of turns/unit length, then $N = nl$. So magnetic field intensity may be written as

$$H = \frac{n.l.I}{l} = nI \qquad ...(3)$$

4. Magnetic field due to a solenoid

A solenoid consists of a cylinder which is tightly wrapped with a thin wire over the cylinder.

As some current is passed through the wire, magnetic field produced outside is negligible. However the field inside the cylinder is uniform and is parallel to the axis of the cylinder as shown in Fig. 1.10

Let N be the number of turns of solenoid. If n is the number of turns per unit length of the solenoid, then $N = nl$.

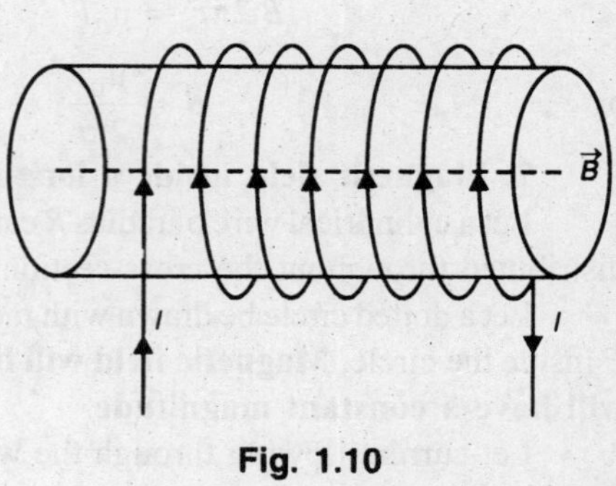

Fig. 1.10

Applying Ampere's circuital law, we have

$$\oint \vec{B} \cdot d\vec{l} = \mu_0 nl\, I \qquad ...(1)$$

As $\vec{B}$ is parallel to the length and along the axis of solenoid,

$$\therefore \quad \oint \vec{B} \cdot dl = \oint B\, dl \qquad ...(2)$$

or $$Bl = \mu_0 nl\, I$$

or $$B = \mu_0 n\, I \qquad ...(3)$$

1.8. MAGNETIC FIELD ON THE AXIS OF A CIRCULAR COIL CARRYING CURRENT

Let us consider a current carrying circular coil of radius *R*, placed in the *YZ* plane with its centre at the origin of the co-ordinates, so that axis of the coil coincides with *X*-axis (Fig. 1.11). Let *P* be the observation point located at *x* (0, 0) and let the current through the coil be *I* with direction of flow of current in the coil as anticlockwise.

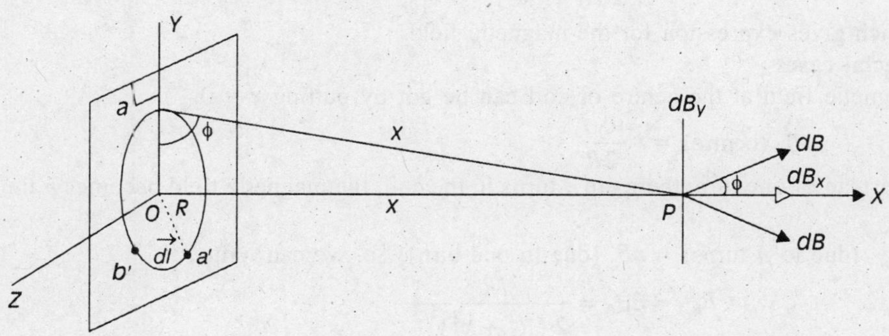

Fig. 1.11

Let us consider an elementary length *ab* of the circular conductor and let *ab* = *dl*. The magnetic field at *P* due to this current according to the Biot Savart law is given as

$$d\vec{B} = \frac{\mu_0}{4\pi}\frac{I d\vec{l} \times \hat{r}}{r^2} \qquad ...(1)$$

When $\vec{r}$ is the distance of the observation point *P* w.r.t. the current *ba*.

In this case dB will be ⊥ to both $Id\vec{l}$ and $\hat{r}$, let it make an angle φ with *X*-axis. We can resolve it into two components

$$dB_z = dB \cos\phi,\ dB_y = dB \sin\phi$$

$$\therefore \quad d\vec{B} = \hat{i}\ dB\cos\phi + \hat{j}\ dB\sin\phi \qquad ...(2)$$

Total magnetic field due to the coil will be the vector sum of the magnetic field due to all the elements of the coil. So we may write

$$\vec{B} = \Sigma\, d\vec{B} = \hat{i}\ \Sigma dB\cos\phi + \hat{j}\ \Sigma dB\sin\phi \qquad ...(3)$$

Corresponding to every elementary length *ab*, there will be another element of length *a'b'* which will produce a magnetic field at *P* equal to *dB* but making angle – φ with the *X*-axis. Since cos (– φ) = cos φ and sin (– φ) = – sin φ, so the *Y*-components of *dB* in (3) will cancel out. Hence, net magnetic field will be

$$B = \hat{i}\ \Sigma dB\cos\phi \qquad ...(4)$$

In the right angled triangle *OLP*, cos φ = *R*/*r*. Also the angle between the elementary length *al* and *LP* is π/2. So we may write

$$B = \hat{i}\, \Sigma dB\, \frac{R}{r} \quad ...(5)$$

Also $\quad Id\vec{l} \times \hat{r} = I\, dl \sin \pi/2 = I\, dl \quad ...(6)$

Putting in Equation (5), we have

$$\vec{B} = \hat{i}\, \Sigma\, \frac{\mu_0}{4\pi} \cdot \frac{I\, dl}{r^2} \cdot \frac{R}{r}$$

$$= \hat{i}\, \frac{\mu_0}{4\pi} \cdot \frac{IR}{r^3}\, \Sigma\, dl$$

Now $\Sigma\, dl = 2\pi R$, so we may write

$$\vec{B} = \hat{i}\, \frac{\mu_0}{4\pi} \cdot \frac{IR}{r^3} \times 2\pi R = \hat{i}\, \frac{\mu_0}{2r^3}\, IR^2 \quad ...(7)$$

Also from Fig. $\quad r^2 = R^2 + x^2$, so we have

$$\vec{B} = \hat{i}\, \frac{\mu_0\, IR^2}{2\,(R^2 + x^2)^{3/2}} \quad ...(8)$$

Which gives expression for the magnetic field.

Special cases

(*i*) Magnetic field at the centre of coil can be got by putting $x = 0$.

$$\vec{B}\ \text{(centre)} = \hat{i}\, \frac{\mu_0\, I}{2R} \quad ...(9)$$

(*ii*) If instead of one turn, there are n turns in the coil, the magnetic field becomes n times, this is

$\vec{B}_n$ (due to n turns) $= n\vec{B}$ (due to one turn). So, we can write

$$\vec{B}_n = \hat{n}\mu_0 = \frac{IR^2}{2\,(R^2 + x^2)^{3/2}}.$$

1.9. FORCE BETWEEN TWO PARALLEL CONDUCTORS

Consider that two wires, carrying currents I_1 and I_2 respectively, are parallel over a length l and separated by a distance r. The magnetic flux density at a distance r from the wire carrying current I_1 is $B = \mu_0\, I/2\pi r$, and the force on the other wire carrying current I_2 in this field is $F = B'I_2\, l = (\mu_0\, I_1/2\pi r)\, I_2\, l$. Hence the force between the parallel wires is

$$F = \frac{\mu_0\, I_1\, I_2\, l}{2\pi r}$$

The force is attractive if I_1 and I_2 are in the same direction, the force is repulsive if I_1 and I_2 are in opposite directions.

The force per unit length is $F/l = \mu_0 I_1\, I_2/2\pi r$. This explains the definition of the ampere *as that current which if existing in each of two infinitely long parallel wires separated by a distance of 1 metre in free space, cause force on each wire to be 2×10^{-7} newton per metre of length.*

110. TORQUE ON COIL IN MAGNETIC FIELD

The torque L on a coil of N turns carrying current in a field of magnetic flux density B is

$$L = BINA \cos \theta$$

where A is the area of the coil and θ is the angle the plane of the coil makes with the field. When the plane of the coil is parallel to the field, $\cos \theta = \cos 0 = 1$ and the torque is maximum. When the plane of the coil is normal to the field, the torque is zero.

1.11. FORCE ON A MOVING CHARGE IN A MAGNETIC FIELD

Consider that a charge Q moves with velocity v in a field of magnetic flux density B and covers a distance l in time t. Then $l = vt$ and the moving charge represents a current $I = Q/t$. Substituting $Il = (Q/t)\,(vt) = Qv$ in $F = BIl \sin \theta$, the force on the charge is

$$F = BQ\, v \sin \theta$$

where θ is the angle between vectors B and v The direction of F is perpendicular to the vectors B and v.

1.12. PATH OF CHARGED PARTICLE IN MAGNETIC FIELD

When a particle of mass m, charge Q and initial velocity v moves normal to a field of magnetic flux density B, the force acting on it as $F = BQv$ directed perpendicular both to B and v. Since F is constant in magnitude and directed always at right angles to v, the particle has uniform circular motion with radius r and centripetal acceleration $a = v^2/r$. Now combining $F = ma = mv^2/r$ and $F = BQv$, we have $mv^2/r = BQv$ from which

$$r = \frac{mv}{BQ}$$

A cyclotron is used to accelerate massive particles (*e.g.*, protons, deuterons, alpha particles) by applying small voltage many times in succession as the particles spiral outward in a magnetic field. This is possible because the angular velocity $\omega = v/r = BQ/m$ of a particle is independent of its speed and of the radius of its path, as long as v is small compared with the speed of light.

A mass spectrograph is used to determine the masses of positively charged ions and to measure the relative abundance of isotopes. Positive ions having the same charge Q and velocity v enter a uniform magnetic field B perpendicular to the vector v and thus are caused to move in a circular path of radius $r = mv/BQ$. Since v/BQ is the same for all ions, the mass m of the ion is proportional to r.

1.9. RATIO OF ELECTRONIC CHARGE TO MASS

Consider that an electron stream with velocity v passes through two mutually perpendicular electric and magnetic fields, that the electric field deflects each electron upward and the magnetic field deflects each electron downward, and that when both fields act simultaneously the resultant deflection is zero. The electrostatic force on each electron of mass m and charge e in an electric field E is $F = eE$, and the magnetic force $F = Bev$. Since the force balance $eE = Bev$ and

$$v = \frac{E}{B}$$

If the electric field is cut off and the magnetic field maintained, each electron will move in the magnetic field in a circular path of radius $r = mv/B.e$. Substituting $v = E/B$ and sloving for e/m.

$$\frac{e}{m} = \frac{E}{B^2 r}$$

The ratio $e/m = 1.759 \times 10^{11}$ C kg^{-1}. Since $e = 1.602 \times 10^{-19}$ C the rest mass of the electron is $m = 1.602 \times 10^{-19}/1.759 \times 10^{11}) = 9.107 \times 10^{-31}$ kg.

The electron-volt (eV) is the energy acquired by a body whose charge equals that of the electron when the body falls through a potential difference of one volt.

$$1 \text{ eV} = (1.602 \times 10^{-19} \text{ C})$$
$$= 1.602 \times 10^{-19} \text{ joules}$$

Example 1. *A charge of 4×10^{-4} C moves at 15 revolutions per second in a circle of radius 10 cm. What is the magnetic flux density at the centre of the circle?*

Solution: Current in circle

$$I = Q/t = (4 \times 10^{-6} C) \left(\frac{1}{15}\text{ s}\right) = 6 \times 10^{-5} \text{ A}$$

$$B = \frac{\mu_0 NI}{2r}\, 4\pi \times 10^{-2} \text{ Wb A}^{-1} \text{ m}^{-1} = \frac{(1)\,(6 \times 10^{-5} \text{ A})}{2 \times 0.2 \text{ m}}$$

$$= \mathbf{1.88 \times 10^{-10} \text{ T.}}$$

Example 2. *A circular coil of 40 turns of wire and negligible section has diameter 32 cm. What current must exist in the coil to produce a flux density of 3 × 10⁻⁴ T at its centre?*

Solution: $$B = \frac{\mu_0 NI}{2r},\ 3\times10^{-4}\ T = \frac{4\pi\times10^{-7}\ \text{Wb A}^{-1}\ \text{m}.40\,I}{2\times(0.16\ \text{m})}$$

$$= \frac{40I}{2},\ \therefore\ I = \mathbf{1.9\ A.}$$

Example 3. *Compute the flux density at a centre of the air cone of a long solenoid having 9 turns of wire per cm of length and carrying a current of 6 A.*

Solution: $$B = \frac{\mu_0 NI}{l} = 4\pi\times10^{-4}\ \text{Wb A}^{-1}\ \text{m}^{-1}\times\frac{9\ \text{turns}}{10^{-2}\ \text{m}}\times 6\ \text{A}$$

$$= \mathbf{6.8\times10^{-3}\ T.}$$

Example 4. *Compute the magnetic flux density in air at a point 5 cm from a long straight wire carrying a current of 15 A.*

Solution: $$B = \frac{\mu_0 I}{2\pi r} = \frac{4\pi\times10^{-7}\ \text{Wb A}^{-1}\ \text{m}^{-1}\times 15\ \text{A}}{2\pi\times(5\times10^{-2}\ \text{m})}$$

$$= 6\times10^{-5}\ \text{Wb m}^{-2} = \mathbf{6\times10^{-5}\ T.}$$

Example 5. *A wire carrying a current of 100 amperes is bent into the form of a circle of radius 5.08 cm. Calculate (a) flux density at the centre of the coil and (b) flux density perpendicular to the plane of the coil at a distance of 19 cm from the coil.*

Solution: (*a*) Flux density at the centre of the coil is given by

$$B = \frac{\mu_0 I}{2a}$$

a = 5.08 cm = 0.0508 m

I = 100 A, $\mu_0 = 4\pi\times10^{-7}$ henry/m

$$\therefore\quad B = \frac{4\pi\times10^{-7}\times100}{2\times0.0508} = \mathbf{12.37\times10^{-4}\ Wb/m^2}$$

(*b*) Flux density ⊥ to the plane of the coil at a distance r from the coil of radius a is given by

$$B = \frac{\mu_0 Ir^2}{2\,(a^2+r^2)^{3/2}}$$

a = 5.08 cm = 0.0508 m

r = 12 cm = 0.12 m

I = 100 A and $\mu_0 = 4\pi\times10^{-7}$ henry/m

$$B = \frac{4\pi\times10^{-7}\times100\times(0.0508)^2}{2\times[(0.0508)^2+(0.12)^2]^{3/2}}$$

$$= \mathbf{7.33\times10^{-5}\ Wb/m^2.}$$

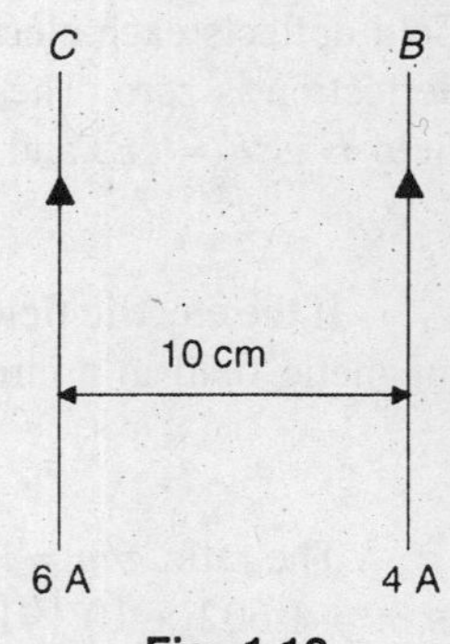

Fig. 1.12

Example 6. *Two long parallel wires, C and D, are 10 cm apart in air and carry currents of 6 and 4 A respectively. Compute the force of attraction on each metre length of wire if the currents are (a) in the same direction, (b) in opposite directions.*

Solution: (*a*) Flux density at D due to current in C is $$B_0 = \frac{\mu_0 I_C}{2\pi r}$$

$$F = B_D I_D l = \left(\frac{\mu_0 I_C}{2\pi r}\right) I_D l$$

$$= \frac{\mu_0}{2\pi}\,\frac{I_C\cdot I_D\cdot l}{r}$$

$$= 2\times10^{-7}\ \text{Wb A}^{-1}\ \text{m}^{-1} \times \frac{(6\ \text{A})\ (4\ \text{A})\ (1\ \text{m})}{0.1\ \text{m}}$$

$$= \mathbf{4.8 \times 10^{-5}\ N\ attraction}$$

(*b*) **4.8 × 10^{-5} N repulsion.**

Example 7. *An electron of energy 1000 eV, describes a circle in a field of magnetic induction 0.02 Wb/m². Calculate the radius of the circle [e/m of electron = 1.76 × 11^{-11} coulomb/kgm, 1 eV = 1.6 × 10^{-19} joules].*

Solution: K.E. of electron $= 1000$ eV

$$= 1000 \times 1.6 \times 10^{-19}\ \text{joules}$$

$$\therefore \quad \frac{1}{2}mv^2 = 1000 \times 1.6 \times 10^{-19}\ \text{joules}$$

$$\therefore \quad v = \sqrt{\frac{2\ \text{K.E.}}{m}}$$

$$= \sqrt{\frac{2\times1000\times1.6\times10^{-19}}{9.1\times10^{-31}}}$$

$$= 1.875 \times 10^7\ \text{m/sec.}$$

We can get the radius of the circle by equating the deflecting force with centripetal force, *i.e.*,

$$\frac{mv^2}{r} = q.v.\ B$$

$$r = \frac{mv}{qB}$$

where q is the charge on electron

$$\therefore \quad r = \frac{v}{(e/m).B} = \frac{1.875\times10^7}{1.76\times10^{-11}\times0.02}$$

$$= 0.00533\ \text{m}$$

$$= \mathbf{0.533\ cm.}$$

Example 8. *A solenoid has a mean diameter of 0.03 m and length 2 metres. It has layers of 1000 turns each. Calculate the flux density at its centre when a current of 2.5 amperes flows through it. Also calculate the flux at the centre.*

Solution: In this case

$$l = 2\ \text{m},\ I = 2.5\ \text{ampere}$$

$$N = 4 \times 1000 = 4000\ \text{turns}$$

$$\mu_0 = 4\pi \times 10^{-7}\ \text{henry/m}$$

Flux density $B = \dfrac{\mu_0 NI}{l}$

$$= \frac{4\pi\times10^{-7}\ 4000\times2.5}{2}$$

$$= 6.28 \times 10^{-3}\ \text{weber/m}^2$$

Mean Area of cross-section

$$A = \pi r^2 = \pi\ .\ (0.025)^2\ \text{sq.m.}$$

Total magnetic flux $= 6.28 \times 10^{-3} \times \pi\ (0.025)^2$

$$= \mathbf{1.234 \times 10^{-5}\ weber.}$$

Example 9. *Calculate the flux density at the centre of a coil if there are 100 turns wound on a solenoid 25 cm long and carrying a current of 3 amperes.*

Solution: $N = 100,\ I = 3\ \text{A},\ l = 25\ \text{cm} = 0.25\ \text{m}$

$$\mu_0 = 4\pi \times 10^{-7}\ \text{henry/m}$$

The flux density at the centre of the coil is given by the relation

$$B = \frac{\mu_0 NI}{l}$$

$$= \frac{4\pi \times 10^{-7} \times 100 \times 3}{0.25}$$

$$= \mathbf{1.508 \times 10^{-3}\ Wb/m^2}$$

Example 10. *A wire of square shape of each side 10 cm long is carrying a current of 2 A in the anticlockwise direction. Calculate the megnetic field at its centre.*

Solution: Each side of the square will subtend an angle of 90° at the centres as shown in Fig. 1.13. The point of observation is at a distance OL = 1/2 length of each side = 5 cm from each side.

The symmetry of the wire about θ allows us to calculate the magnetic field as the net magnetic field B = 4 × magnetic field due to each side.

Let us calculate the magnetic field due to one side say '*ab*'. Making use of the relation

$$B_{ab} = \frac{\mu_0 I}{4\pi R}\frac{I}{R}(\sin\theta_1 + \sin\theta_2)$$

In this case $I = 2A,\ R = 5\ \text{cm} = 0.05\ \text{m},\ \theta_1 = \theta_2 = 45°$

$$\therefore \quad \sin\theta_1 = \sin\theta_2 = \frac{1}{\sqrt{2}}$$

Also $\mu_0 = 4\pi \times 10^{-7}\ \text{T.m.A}^{-1}$

And $\sin\theta_1 + \sin\theta_2 = \frac{2}{\sqrt{2}} = \sqrt{2}$

So
$$B_{ab} = \frac{10^{-7}\ \text{Tm A}^{-1} \times 2A}{0.05\ \text{m}}\sqrt{2}$$

$$= 10^{-7} \times 2\sqrt{2} \times 20\ \text{T}$$

Total magnetic field $B = 4 \times B_{ab}$

$$= 4 \times 2\sqrt{2} \times 20 \times 10^{-7}\ \text{T}$$

$$= \mathbf{226.24 \times 10^{-7}\ T}$$

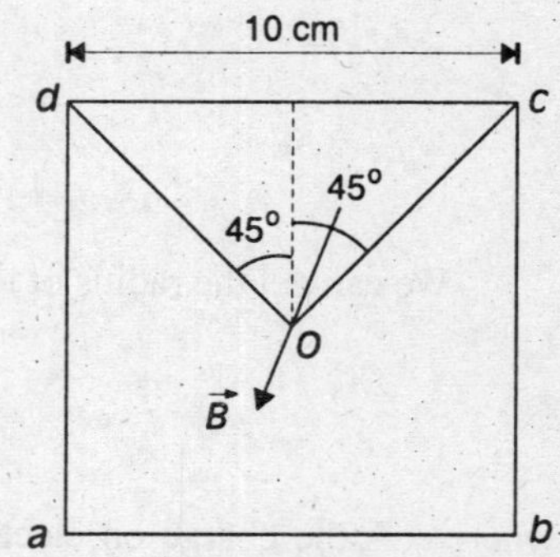

Fig. 1.13

We can get the direction of the field, by right hand screw rule. According to which it points outwards perpendicular to the plane of the paper.

1.14. HELMHOLTZ GALVANOMETER

The principle of this galvanometer is the same as that of the tangent galvanometer *i.e.*, on tangent law of magnetism *viz*; $F = H \tan\theta$ where F and H are two uniform magnetic fields and θ is the angle between the direction of H and the direction of the magnetic needle when it comes to rest under the influence of the fields F and H. However, in the case of Helmholtz's galvanometer, magnetic needle moves in a uniform magnetic field, so this instrument is more useful for accurate measurement of the current.

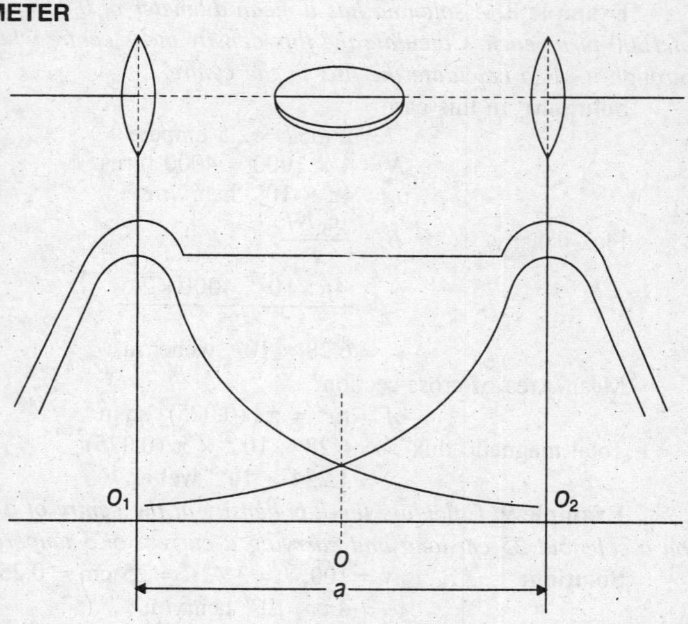

Fig. 1.14

It consists of two similar circular coils of wire mounted on the same base and can rotate about a vertical axis. The two coils are connected in series and have got the same radius, the distance between the centres of the two coils being equal to the radius of either coil. A deflection magnometer box is fitted at the middle point on the line joining the two coils. To level the instrument levelling screws are provided at the base. Since both the coils are connected in series, so the magnetic fields due to the current passing through both the coils have the same direction.

Theory of Helmholtz Galvanometer. If C_1 and C_2 are the planes of the two coils $\perp$ to the plane of paper and O_1 and O_2 are the centres of these coils. Now, since both coils have current in the same direction, so the magnetic intensity at any point on their common axis $O_1 O_2$ will be in the same direction due to these two coils.

If a point shifts from O_1 to O_2, the magnetic intensity due to coil C_1 will decrease and that due to coil C_2 increases. The distance between the centre of two coils is adjusted in such a way that the rate of increase of magnetic intensity due to one coil is equal to rate of decrease due to the other coil. In this way the resultant intensity at any point due to the current passing through the coils will be constant and magnetic field moves in a uniform magnetic field.

Now we have the magnetic intensity at any point on the axis of a circular coil of wire carrying current as

$$F = \frac{2\pi\, na^2\, i}{(a^2 + x^2)^{3/2}} \qquad \text{...(1)}$$

Since magnetic field along O_1O_2 is unform *i.e.*, $\frac{dF}{dx}$ is to be constant or $\frac{d^2F}{dx^2} = 0$.

In the above relation

i $\rightarrow$ strength of current in e.m.u.,

$n \rightarrow$ no. of turns of the coil,

$a \rightarrow$ the radius of the coil,

$x \rightarrow$ the distance of the point from the centre of the coil.

Now differentiating Equation (1) w.r.t. x.

$$\frac{dF}{dx} = 2n\pi\, a^2 i\left[-\frac{3}{2}(a^2 + x^2)^{-5/2}\, 2x\right]$$

$$= -\, 6n\pi a^2\, i\, [a^2 + x^2]^{-5/2}.x$$

Again differentiating w.r.t. x

$$\frac{d^2F}{dx^2} = 6\pi\, nia^2\left[-\frac{5}{2}(a^2 + x^2)^{-7/2} \cdot 2x^2 + (a^2 + x^2)^{-5/2}\right]$$

$$= -\, 6\pi\, na^2\, i\, [(-5x^2)\,(a^2 + x^2)^{-7/2} + (a^2 + x^2)^{-5/2}]$$

$$= -\, 6\pi\, nx^2\, i\, [-5x^2 + a^2 + x^2]\,(a^2 + x^2)^{-7/2}$$

$$= -\, 6\pi\, na^2\, i\,(a^2 - 4x^2)\,(a^2 + x^2)^{-7/2}$$

$$= +\, 6\pi\, na^2\, i\,(a^2 + x^2)^{-7/2}\,(4x^2 - a^2)$$

if $\frac{d^2F}{dx^2} = 0$

then since $6\pi\, na^2\, i \neq 0$

and $(a^2 - x^2)^{-7/2} \neq 0$

$\therefore$ $4x^2 - a^2 = 0$

or $x^2 = \frac{a^2}{4}$

or $x = \pm\frac{a}{2}$

So in order to get a uniform magnetic field, the magnetic needle should be at distance $a/2$ from the centre of each of the two coils *i.e.*, the distance between the two coils will be a or $2x$ which means that the distance between the two coils is equal to the radius of either coil.

Current through the coils. If the magnetic intensity due to each coil is F, the resultant magnetic intensity due to both the coils will be $F_1 = F_2$.

Now in the equilibrium position

$$F_1 = 2F = H \tan \theta$$

$$2 \times \frac{2\pi n a^2 i}{(a^2 + x^2)^{3/2}} = H \tan \theta$$

Putting $x = \dfrac{a}{2}$

$$\frac{4\pi a^2 ni}{\left(a^2 + \dfrac{x^2}{4}\right)^{3/2}} = H \tan \theta$$

We have $\dfrac{8 \times 4\pi a^2 ni}{(5a^2)^{3/2}} = H \tan \theta$

or $$\frac{32\pi a^2 ni}{\sqrt{125a^3}} = \tan \theta$$

or $$\frac{32\pi ni}{\sqrt{125a}} = H \tan \theta$$

or $$i = \frac{\sqrt{125aH} \tan \theta}{32\pi n} \text{ e.m.u.}$$

$$= \frac{10\sqrt{125aH} \tan \theta}{32\pi n} \text{ amp}$$

$$= \frac{50\sqrt{5aH} \tan \theta}{32\pi n} \text{ amp}$$

RATIONALISED M.K.S. UNIT OR SI UNIT

In S.I. units the field at a point on the axis of a circular coil carrying current

$$F = \frac{1}{4\pi} \cdot \frac{2\pi n a^2 I}{(a^2 + x^2)^{3/2}}$$

$$= \frac{na^2 I}{2\,(a^2 + x^2)^{3/2}}$$

Now $$F_1 = 2F = H \tan \theta$$

or $$F_1 = \frac{na^2 I}{(a^2 + x^2)^{3/2}} = H \tan \theta$$

Now $$x = \frac{a}{2}$$

$$\therefore \quad \frac{na^2 I}{\left(a^2 + \dfrac{a^2}{4}\right)^{3/2}} = H \tan \theta$$

or $$I = \frac{5\sqrt{5aH} \tan \theta}{1n}$$

In this case I is in amperes, a is in metres and H is in ampere turns/m. where 1 ampere turn/m = $4\pi \times 10^{-3}$ oersted.

1.15. BALLISTIC GALVANOMETERS

Ordinary galvanometers are useful for steady currents and the needle of the galvanometers gives a constant deflection. However, in order to measure changing current such as discharge of a condenser or currents due to e.m. induction, we make use of Ballistic galvanometers which are based on the same principle as ordinary galvanometers except that in the case of ballistic galvanometers, the moment of inertia is so large that it takes very long for the variations of the coil to stop.

Now from the equation $t = 2\pi\sqrt{q/c}$, we can increase the time period by decreasing the couple/unit twist C of the suspension fibre. In the moving coil ballistic galvanometers, the coil is wound over a non-conducting frame made of ivory, ebonite etc. so as to reduce the damping due to eddy currents unlike the ordinary moving coil galvanometer, soft iron core is not used because it produces eddy currents and damps the vibrations. In these galvanometers damping is made as small as possible and if there is some residual damping then damping corrections are employed.

There are two types of Ballistic galvanometers.

(*i*) Moving Magnetic Ballistic Galvanometer.

(*ii*) Moving Coil Ballistic Galvanometer.

(*i*) Moving Magnetic Ballistic Galvanometer

Let AB be the current carrying coil perpendicular to the plane of paper where A and B represent the cross-section of the conductor of the coil. A magnet NS is suspended and it is capable of rotation about an axis which is perpendicular to the plane of paper.

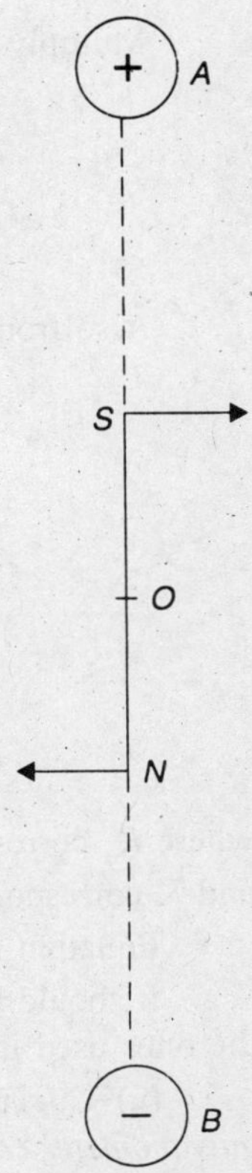

Fig. 1.15

Theory. If I be the current passing through the coil at any instant of time and G is the magnetic intensity at the centre of the coil for a unit current passing through it, then the magnetic intensity = GI.

Now the force experienced by each magnetic pole of strength m units = GmI.

If this current passes for an interval dt, then

Impulse $= \text{Force} \times \text{time} = GmI\cdot dt.$

Now total change in momentum when a charge q passes through the coil

$$= \int mGI\,dt = mG\int I\,dt = mGq \left\{\because \int I\,dt = q\right\}$$

Moment of momentum $= mGq \times 2l$

$= MGq$

where $M = 2l \times m$, the magnetic moment

Also angular momentum $= I\omega$ and it is equal to moment momentum.

$$\therefore \quad I\omega = MGq \qquad \text{...(3)}$$

Again time period of oscillation of suspended magnet

$$t = 2\pi\sqrt{\frac{I}{MH}}$$

where I is moment of inertia of the oscillating magnet and I corresponds to the magnetic field of earth's horizontal component.

$$t^2 = \frac{4\pi^2 I}{MH} \quad \text{or} \quad I = \frac{t^2 MH}{4\pi^2} \qquad \text{...(4)}$$

Now as soon as the charge passes, the oscillating magnet gets deflected and if ω be the angular velocity of a moving system at the instant when magnetic deflection = θ,

then we have the work in deflecting the magnet by the angle θ

$$W = MH(1 - \cos\theta)$$

This will correspond to K.E. of the moving system $= \frac{1}{2} I\omega^2$

$$\therefore \quad \frac{1}{2} I\omega^2 = MH(1 - \cos\theta)$$

$$= MH\left[1 - \left(1 - \sin^2\frac{\theta}{2}\right)\right]$$

$$= 2MH\sin^2\frac{\theta}{2}$$

$$\therefore \quad I\omega^2 = 4MH\sin^2\frac{\theta}{2} \qquad \text{...(5)}$$

Multiply Equation (4) by (5)

$$I^2\,\omega^2 = \frac{4t^2\,M^2\,H^2}{4\pi^2}\sin^2\frac{\theta}{2}$$

$$\therefore \qquad I\omega - \frac{tMH}{\pi}\sin\frac{\theta}{2}$$

Also from Equation (3)

$$I\omega = MGq$$

$$\therefore \qquad MGq = \frac{t\,MH}{\pi}\sin\frac{\theta}{2}$$

$$\therefore \qquad q = \frac{t\,MH}{\pi MG}\sin\frac{\theta}{2} = \frac{H}{G}\frac{t}{\pi}\cdot\sin\frac{\theta}{2}$$

$$= K_1\cdot\frac{t}{\pi}\sin\frac{\theta}{2}$$

$$= K\sin\frac{\theta}{2} \qquad \text{...(6)}$$

where K_1 corresponds to current reduction factor,
and K corresponds to the ballistic reduction factor.

Equation (6) gives a relation between charge and deflection of magnetic needle.

It should be noted that in this case also the suspension fibre must be torsionless similar to the ones used in vibration magnetometers.

***(a)* Current sensitivity.** *It is defined as the current that has to be passed through the galvanometer coil for a deflection of 1 mm on a scale kept at a distance of 1 metre from the mirror.* This current is usually expressed in micro-amperes. **Sensitiveness is reciprocal of sensitivity** *i.e.*, higher the value of current sensitivity, the less sensitive the galvanometer.

***(b)* Ballistic sensitivity.** *It is defined as the amount of charge that has to be passed through the galvanometer to get a deflection of 1 mm on a scale kept at a distance of 1 m from the mirror.* It is measured in coulomb (1 coulomb = 3 × 10^9 e.s.u. of charge). It is also called charge sensitivity.

(ii) Moving Coil Ballistic Galvanometer

This type of galvanometer has similar construction to the dead heat galvanometer except the modifications required in the measurement of charge.

Theory. Let I be the current passing through the coil of length and breadth b having n no. of turns.

If B be the magnetic flux density, then the force experienced by the length side conductor of the coil = $BInl$

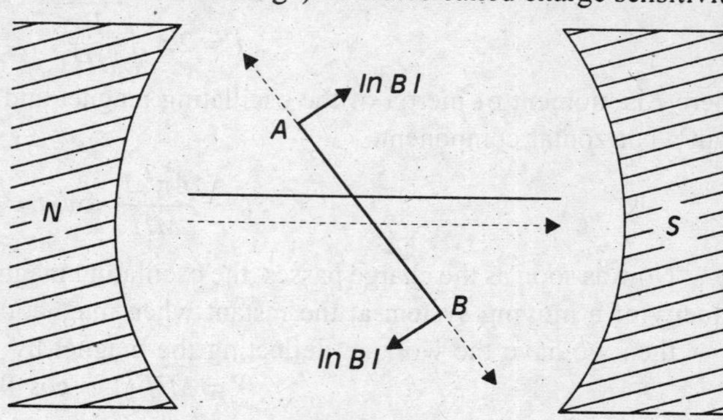

Fig. 1.16

$$\therefore \qquad \text{Force} = BInl \qquad \text{...(1)}$$

If this current flows for a very small interval of time dt, we have impulse = $Bnl\ Idt$

Which will correspond to change in momentum given by $Bnl\ Idt$.

So the change in momentum for a charge q that flows through the coil

$$= \int BnlI\cdot dt = Bnl\int I\,dt = B\,nla$$

where $q = \int I\,dl$

Moment of momentum $= Bnlq \times b$

$= nABq$ (where A is the force area of coil)

Also angular momentum $=$ Moment of momentum $= I\omega$

$$\therefore \quad I\omega = nABq \qquad ...(2)$$

Now work done in twisting the suspension fibre angle $\theta = \frac{1}{2} C\theta^2$ and this will be equal to K.E. of the oscillating system *i.e.*, $= \frac{1}{2} I\omega^2$ in which case angular displacement $= \theta$

$$\therefore \quad \frac{1}{2} I\omega^2 = \frac{1}{2} C\theta^2$$

or
$$I\omega^2 = C\theta^2$$

The period of oscillation of the system that executes the torsional oscillations is

$$t = 2\pi \sqrt{\frac{I}{C}}$$

or
$$t^2 = 4\pi^2 \frac{I}{C}$$

or
$$I = \frac{t^2}{4\pi^2} \cdot C \qquad ...(3)$$

Multiply Equation (2) by (3)

$$I^2 \omega^2 = \frac{C^2 t^2 \theta^2}{4\pi^2}$$

or
$$I\omega = \frac{Ct\theta}{2\pi^2} \qquad ...(4)$$

From Equation (1) and (4)

$$nABq = \frac{Ct\theta}{2\pi}$$

$$\therefore \quad q = \frac{t}{2\pi} \cdot \frac{C}{n\,AB} \cdot \theta \qquad ...(5)$$

In this case $\frac{C}{n\,AB}$ current reduction factor of the galvanometer and $\frac{t}{2\pi} \frac{C}{n\,AB}$ is the Ballistic galvanometer reduction factor

$$\therefore \quad q = K\theta \qquad ...(6)$$

which gives a relation between charge and deflection in the galvanometer.

In the above relation C corresponds to the couple/unit twist of the suspension fibre, n, the number of turns of the coil, A the face area of the coil, B the flux density and θ the throw observed in the ballistic galvanometer.

1.16. DAMPING CORRECTION

In the theory of ballistic galvanometer, it has been supposed that the total K.E. of the moving system is used in twisting the suspension fibre. In actual experiment, however, the motion of the coil is opposed by (*i*) resistance of the air, (*ii*) induced e.m.f. in the coil. The amplitudes of the successive vibrations go on decreasing and this is called damping.

Ballistic galvanometer is used to measure charge in the form of a sudden discharge and due to the impulse of a sudden kick is given to the coil, so it is only the first kick, that is effective in measuring the charge that flows through the coil. After the first kick, the coil oscillates in the magnetic field with a continuously decreasing amplitude. As discussed earlier, due to electromagnetic induction and air resistance etc., there is a decrease in amplitude. Let us consider θ to be the actual deflection in absence of damping and $\theta_1, \theta_2, \theta_3, \ldots$ etc., be the observed throws on the right and left in succession as shown in Fig. 1.17.

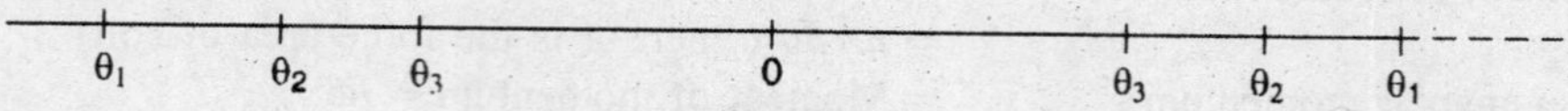

Fig. 1.17

It is found that the reduction of throws is given by

$$\frac{\theta_1}{\theta_2} = \frac{\theta_2}{\theta_3} = \frac{\theta_3}{\theta_4} = d$$

where d is called the decrement.

Let us consider d to be equal to e^{λ}, so that $\lambda = \log_e d$, in this case λ is called logarithmic decrement. Now each complete vibration has two swings (*i.e.*, from extreme right θ_1 to left θ_2 and from extreme left θ_2 to right θ_3).

$$\frac{\theta_1}{\theta_2} = \frac{\theta_2}{\theta_3} = \frac{\theta_3}{\theta_2} = e^{\lambda}$$

and for two oscillations, the decrease is

$$\frac{\theta_1}{\theta_3} = \frac{\theta_1}{\theta_2} \times \frac{\theta_2}{\theta_2} = d^2 = e^{2\lambda} \quad \text{(for two swings)}$$

Similarly for 4 swings

$$\frac{\theta_1}{\theta_5} = d^4 = e^{4\lambda} \quad \text{etc.}$$

Let θ be the true first throw in the absence of damping which will be higher than the first throw observed *i.e.* θ_1. Now the motion of the coil from the mean position to extreme one corresponds to 1/2 swing.

$$\therefore \quad \frac{\theta}{\theta_1} = d^{1/2} = e^{\lambda/2} = \left[1 + \frac{\lambda}{2}\right] \text{ approx}$$

or

$$\theta = \theta_1 \left(1 + \frac{\lambda}{2}\right)$$

Putting this value in Equation (5) of previous article

$$q = \frac{t}{2\pi} \cdot \frac{C}{n\,AB}\, \theta_1 \left(1 + \frac{\lambda}{2}\right) \qquad \ldots(4)$$

where λ corresponds to logarithmic decrement and θ_1 corresponds to first throw.

Calculation of λ. We note the successive throws *i.e.*, $\theta_1, \theta_2, \theta_3$ etc. then we have some relation

$$\frac{\theta_1}{\theta_{11}} = d^{10} = e^{10\lambda}$$

Taking log on ooth sides

$$\log\left(\frac{\theta_1}{\theta_{11}}\right) = 10\lambda$$

or

$$\lambda = \frac{1}{10} \log_e \frac{\theta_1}{\theta_{11}}$$

$$= \frac{1}{10} \times 2.3026 \log_{10}\left(\frac{\theta_1}{\theta_{11}}\right) \qquad \ldots(2)$$

From Equation (1)

$$q = \frac{t}{2\pi} \cdot \frac{C}{n\,AB} \cdot \theta_1 \left[1 + \frac{2.3026}{20} \cdot \frac{\log_{10} \theta_{11}}{\theta_1}\right]$$

$$= \frac{t}{2\pi} \cdot \frac{N}{AB} \cdot \theta_1 \left[1 + 0.11513 \log_{10} \frac{\theta_1}{\theta_{11}}\right] \qquad \ldots(3)$$

Example 11. *If the coils of Helmholtz galvanometer have diameter 14 cm and number of turns 175 each, find the magnetic field midway between the coils when distance apart is 7 cm and a current of 100 mA flows through the instrument.*

Solution: Field $F = \dfrac{2 \times 2\pi n^2 I}{10\left[a^2 + \dfrac{a^2}{4}\right]^{3/2}}$

In this case $a = 7$ cm, $n = 175$

$F = 0.1$ amp

$$F = \frac{2 \times 2 \times 22 \times 175 \times 175 \times 1}{7 \times 10 \times 10\left[7^2 + \dfrac{7^2}{4}\right]^{3/2}}$$

$$= \frac{176}{35\sqrt{5}} = \textbf{2.249 oersted.}$$

Example 12. *A Helmholtz galvanometer has coils of radius 11 cm each and the number of turns = 70 $\sqrt{5}$.*

(i) Calculate the current through the coils which produces a deflection of 45°. H = 0.32 oersted.

(ii) Also convert these quantities into rationalised M.K.S., units or S.I. units and calculate the current.

Solution: (*i*) In this case

$a = 11$ cm, $n = 70\sqrt{5}$, $H = 0.32$ oersted $\theta = 45°$

$$\therefore \quad I = \frac{50\sqrt{5} \times aH \tan\theta}{32\pi n} \text{ amperes}$$

$$= \frac{50\sqrt{5} \times 11 \times 0.32 \times 1 \times 7}{32 \times 22 \times 7055} = \textbf{0.025 amp.}$$

(*ii*) Rationalised M.K.S. units.

$a = 11$ cm $= 0.11$ m

$$H = 0.32 \text{ oersted} = \frac{32}{4\pi \times 10^{-3}} \text{ amp-turns/m}$$

$n = 70\sqrt{5}$, $\theta = 45°$

$$I = \frac{5\sqrt{5} \times 0.11 \times 0.32 \times 1}{8 \times 70\sqrt{5} \times 4\pi \times 10^{-3}} = \textbf{0.025 ampere}$$

Example 13. *Each of the two coils of a Helmholtz galvanometer contains 50 turns of wire of mean radius 16 cms when a current of 0.1 amp is passed through the coil, a deflection of 45° is obtained. Calculate the horizontal intensity of earth's magnetic field.*

Solution: $n = 50$, $a = 16$ cms, $I = 0.1$ amp. $= 0.01$ e.m.u., $\theta = 45°$

$\therefore \quad \tan\theta = 1$

Now $$H \tan\theta = \frac{32\pi ni}{\sqrt{125}a}$$

$$\therefore \quad H = \frac{32\pi ni}{\sqrt{125}a \, . \tan\theta}$$

$$= \frac{32 \times 3.142 \times 50 \times 0.01}{\sqrt{125} \times 16} = \textbf{0.281 oersted.}$$

Example 14. *The current sensitivity of a ballistic galvanometer is 2.2 × 10^{-9} amperes for a deflection of 1 mm on a scale kept at a distance of 1 metre. Calculate the charge sensitivity of the galvanometer if the time period of the coil = 6.2 sec.*

Solution: $$q = \frac{t}{2\pi} \cdot \frac{C}{n\,AB}\theta$$

If $\theta = 1$ mm, current sensitivity

$$= \frac{C}{n\,AB} \text{ amperes./mm}$$

$$\therefore \quad \frac{C}{n\,AB} = 2.2 \times 10^9 \text{ amp./mm}$$

Here $t = 6.2$ sec

Charge sensitivity $= \frac{t}{2\pi} \times \frac{C}{n\,AB}$

$$= \frac{6.2 \times 7 \times 2.2 \times 10^{-5}}{2 \times 22} = \mathbf{2.17 \times 10^{-9} \text{ coul/mm.}}$$

Example 15. *A condenser is charged 2 volts and then discharged through a ballistic galvanometer. Calculate the capacity of the condenser from the following data:*

Galvanometer scale distance = 100 cm, q_I = 60 mm, q_{II} = 40 mm, mass of the coil = 7 gm, mean length of the coil = 6 cm, mean breadth = 1.7 cm, number of turns = 600, period = 5 secs and magnetic field = 200 oersted.

Solution: First throw $\theta_1 = 60$ mm

Eleventh throw $\theta_{11} = 40$ mm

Now
$$\lambda = \frac{1}{10} \log_e \frac{\theta_1}{\theta_{11}} = \frac{1}{10} \log_e \frac{60}{40}$$

$$= \frac{1}{10} \times 0.4055 = 0.04055$$

Distance of scale $= 100$ cm

$\therefore$ Corrected throw
$$\theta = \frac{6}{100}\left[1 + \frac{\lambda}{2}\right]$$

$$= 0.06\ [1 + 0.02078]$$

$$= 0.06122 \text{ radian}$$

Number of turns $n = 600$

Intensity of magnitude field $B = 200$ oersted

Mean length of the coil $l = 6$ cm

Mean bredth of the coil $b = 1.7$ cm

Mass of the coil $m = 7$ gm

Time period of swing $t = 5$ sec.

If I be the moment of inertia of the coil about an axis its case and || to its length. It will be the case of rectangular lamina and

$$I = \frac{mb^2}{12} = \frac{7 \times 1.7 \times 1.7}{12}$$

Also
$$t = 2\pi \sqrt{\frac{I}{C}}$$

$$\therefore \quad C = 4\pi^2 \frac{I}{t^2}$$

The charge flowing through the ballistic galvanometer

$$q = \frac{Ct}{2\pi n\,AB}\,\theta\left(1 + \frac{\lambda}{2}\right)$$

$$= \frac{4\pi^2 I}{t^2} \cdot \frac{t}{2\pi n\,AB} \cdot \theta\left(1 + \frac{\lambda}{2}\right)$$

$$= \frac{2\pi}{t} \cdot \frac{I}{n\,AB} \cdot \theta$$

$$= \frac{2 \times \pi \times 7 \times 1.7 \times 1.7 \times 0.06122}{5 \times 12 \times 600 \times 6 \times 1.7 \times 200} = 1.059 \times 10^{-7} \text{ e.m.u.}$$

Potential difference across the condenser

$$V = 2 \text{ volts}$$

$$\therefore \quad \text{Capacity} = q/v = \frac{1.059 \times 10^{-6}}{2}$$

$$= 0.5295 \times 10^{-6} \text{ farad} = \mathbf{0.5295 \mu F.}$$

QUESTIONS AND PROBLEMS

1. Show that the magnetic intensity at a point distance r from a coil carrying current is equal to in rationalised M.K.S. unit.

2. Deduce an expression for the field due to a current flowing in a straight infinite conductor.

3. Explain the working of a moving magnet ballistic galvanometer. What do you understand by figure of merit and sensitivity?

4. State and prove Ampere's Law in electromagnetism.

5. Find an expression for the intensity of magnetic field inside a long solenoid and show that the intensiy at the ends is haif of that in the middle of solenoid.

6. Derive an expression for work done in moving a current carrying conductor in a magnetic field.

7. How will you find the intensity of magnetic field at a point on the axis of a circular coil?

8. Sate Laplace's formula for magnetic field intensity at any point due to current element and apply it to find magnetic field at the centre of a circular coil carrying current.

9. Find the force between two parallel wires carrying currents:
(*i*) in the same direction, (*ii*) in opposite directions.

10. Give construction and theory of suspended coil ballistic galvanometer and give an expression for the quantity of electricity flowing through it.

11. Explain theory, construction and working of a moving coil ballistic galvanometer. How the dead beat condition can be attained? Also deduce how to correct the observed throw or damping?

12. A solenoid of length 20 cm and radius 2 cm is closely wound with 200 turns. Calculate the field at the centre of one end of the solenoid. The current in the winding is 5 amperes.
[**Ans.** 31.43 oersted]

13. Two concentric circular coils x and y of radii 16 cm respectively lie in the same vertical plane containing north-south direction. Coil x has 20 turns and carries a current of 16 A; coil y has 25 turns and carries a current of 18 A. The sense of the current in x is anticlockwise and in y clockwise, for an observer looking at the coils facing west. Give the magnitude and direction of the net magnetic field due to the coils at the centre. [**Ans.** 1.8×10^{-3} *T*, directed towards west]

14. Two straight wires each 1 m long are parallel to one another and are each carrying a current of 40 amperes. What force in Newtons will be set up between the wire if the distance between the wires is 0.02 m ($\mu_0 = 4\pi \times 10^{-7}$ henry/m)? [**Ans.** 16×10^{-2} N]

15. A short conductor of length 5.00 cm is placed parallel to a long conductor of length 1.5 m near its centre. The conductors carry currents 4.0 A and 3.0 A respectively in the same direction. What is the total force experienced by the long conductor when they are 3.0 cm apart? [**Ans.** 4×10^{-6} N]

16. A condenser charged to 2 volts is discharged through a ballistic galvanometer, when the corrected deflection is 9.6 cms and the current sensitivity is 2.2×10^{-8} amp/cm and periodic time is 12 second. Calculate the capacity of the condenser. [**Ans.** 0.2016μF]

CHAPTER 2

SELF AND MUTUAL INDUCTION

2.1. INTRODUCTION

Joseph Henery in 1832 got the phenomenon of Self Induction on theoretical considerations and this can be explained as under:

It is seen that when some current is allowed to flow through a coil, a magnetic field is set up in it. However, if the magnitude of the current is changed, the total flux linked with the coil is changed and an induced e.m.f. is set up in the coil and the direction of this induced e.m.f. will be such that it will oppose the change in current. In addition the induced e.m.f. will be opposing the current when the current is increasing and it will have the direction of the current when it is decreasing.

This property of an electric circuit due to which any change in the magnitude of the flux linked with it, induces an e.m.f. in it is called inductance and the *induced e.m.f. is called back e.m.f.*

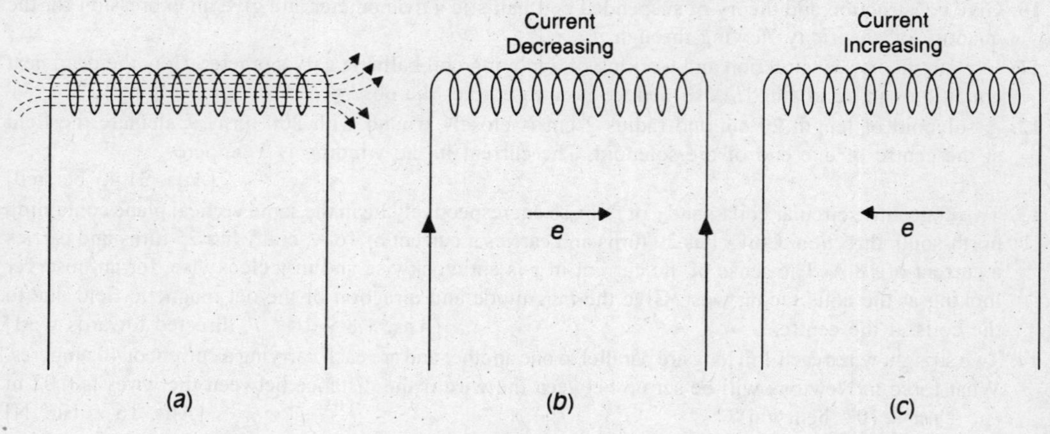

Fig. 2.1

In the case of self induction, it opposes the growth of the current, when the current is switched on and also it opposes the decay of the current when it is switched off. In case, if the circuit is broken, current falls off very quickly and a large amount of induced e.m.f. appears which tries to maintain the current. Due to this high e.m.f. set up, sparking takes place at the switches and this energy that is stored in the form of heat and light during the break is stored in magnetic field during the growth of the current in the circuit. So it is clear from above that while the flux is increasing during the growth of current, energy is stored up magnetically so long as the flux is constant. When the current is switched off, the flux decreases and the energy appears in the form of a spark. It is due to this reason that the switches at power stations are usually of 'oil immersion' type to avoid sparking and consequent damage to the other parts of the machinery.

We can now define self inductance of a circuit as the property of a circuit due to which any change in the current flowing in the circuit induces an e.m.f. that opposes the change and then the energy is stored in the magnetic field established by the current.

We can demonstrate experimentally the self induction phenomenon as under:

Expample 1. We have a soft iron piece *AB* on which are joined the poles of a horse-shoe shaped electromagnet. A neon lamp that works on 230 V is connected in the circuit through a battery of 12 V as shown in Fig. 2.2. If key *K* is pressed so that the circuit is closed, then the lamp just gives a faint glow. However when the key *K* is taken out, that is, the circuit is broken, then the lamp flashes showing that there is momentarily increase of e.m.f. more than 230 V *i.e.*, induced e.m.f. is more than 230 V. This experiment shows that the e.m.f. induced in the circuit due to make or break is due to the self inductance of the coil.

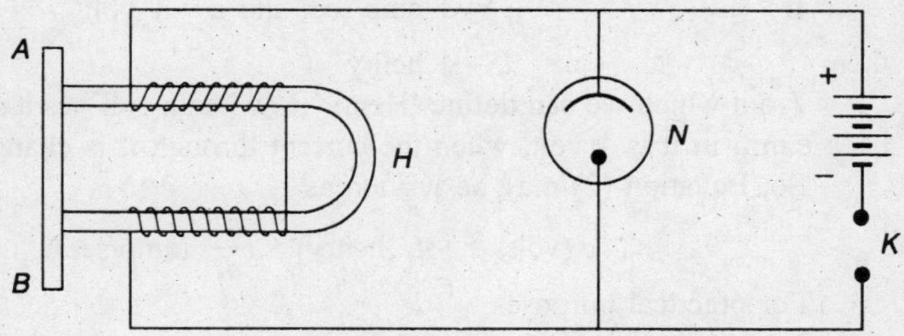

Fig. 2.2

Example 2. *L* is a long solenoid having large number of turns wound over an iron core so that it is able to produce a large self inductance. This type of solenoid is also known as a choke coil. A battery of about 12 V is placed in the circuit and when the key is pressed and current adjusted with rheostat, we have some small glow on the lamp *A*. If the key *K* is taken off there appears a bright glow in the lamp for a very small time and the brightness of the lamp is much greater than that during the switching on (make) of the circuit. This shows that the induced e.m.f. due to self inductance *L* at the time of break is greater than that at the make *i.e.*, the e.m.f. of the battery.

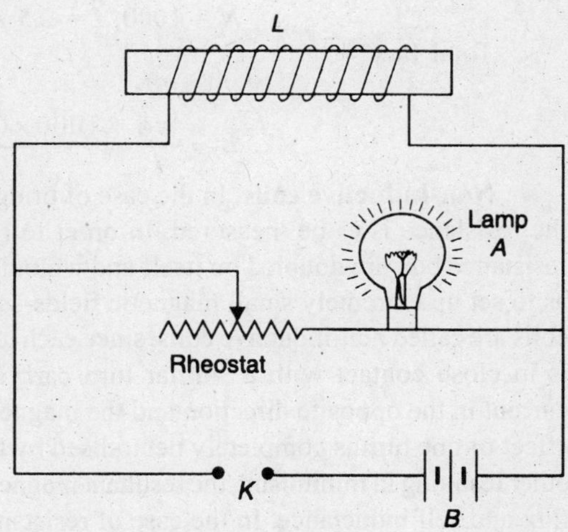

Fig. 2.3

2.2. UNITS

If we consider a coil of *N* turns that carry an electric current *I*, then the total magnetic flux $\phi \propto I$.

or $$\phi = LI \qquad ...(1)$$

where *L* is called coefficient of self inductance, differentiating (1) w.r.t. time

$$\frac{d\phi}{dt} = L\frac{dI}{dt}$$

And back e.m.f. $$e = \frac{d\phi}{dt} = -L\frac{dI}{dt} \qquad ...(2)$$

The negative sign shows that the direction of the induced e.m.f. is such that it opposes the change, for the increase of current dI/dt is +ve and induced e.m.f. *e* is in opposite direction. For decreasing current, $\frac{dI}{dt}$ is –Ve and the induced e.m.f. is in the direction of the current.

From Equation (1), if $I = 1$, then $\phi = L$

i.e., the coefficient of self inductance of a circuit is the total magnetic flux linked with it when a unit current passes through it.

From Equation (2), if $\frac{dI}{dt} = 1$, then $e = -L$

So we can also define coefficient of self inductance as equal to the induced e.m.f. in the circuit if the rate of change of current in the circuit is unity.

In rationalised M.K.S. units or S.I. units, self inductance is measured in henry.

If $$\frac{dI}{dt} = 1 \text{ amp/sec, and } e = 1 \text{ volt}$$

then $$L = 1 \text{ henry}$$

From which we can define 'Henry' as when a coil has a self inductance of one henry if the back e.m.f. in it is 1 volt, when the current through it is changing at the rate of 1 amp/sec.,

So, Equation (2) may be written as

$$e \text{ (volt)} = - L \text{ (henry)} \times \frac{dI}{dt} \text{ (amp/sec)}$$

For practical purposes

$$1 \text{ milli Henry} = 10^{-3} \text{ Henry}$$

$$\text{And } 1 \text{ micro-henry} = 10^{-6} \text{ Henry}$$

$$\text{And } 1 \text{ henry} = 10^{-9} \text{ C.G.S. e.m.u.}$$

Example 1. *Calculate the coefficient of self inductance of a coil of 1000 turns when a current of 2.5 amperes produces a magnetic flux = 0.5 m.weber.*

Solution: In this case

$$N = 1000, I = 2.5 \text{ Amperes}, \ \phi = 0.5 \times 10^{6} \text{ webers.}$$

Total flux

$$N\phi = LI$$

$$\therefore \quad L = \frac{N\phi}{I} = \frac{1000 \times 0.5 \times 10^{-4}}{25} = \mathbf{2 \times 10^{-4}} \textbf{ Henry}$$

Non-Inductive coils. In the case of bridge circuits, self inductance is a great drawback when the resistance is to be measured. In order to reduce the effect of self inductance, the coils in the resistance box are doubled on itself and wound so as to set up extremely small magnetic fields-such coils are called non-inductive coils, since each turn is in close contact with a similar turn carrying current in the opposite direction and the magnetic effect of one turn is completely neutralised by the other resulting in minimising the resultant magnetic flux and self inductance. In the case of resistance boxes and P.O. boxes, all the coils are non-inductively wound, on bobbins as shown in Fig. 2.4.

Fig. 2.4

2.3. SELF INDUCTANCE OF A LONG SOLENOID

Let there be a solenoid of length l having N number of turns. An area of cross-section and I be the current flowing, let the permeability of the medium be μ.

Now the field inside in C.G.S. e.m.u. units given by $H = \frac{4\pi NI}{l}$

And flux through each turn $= \mu HA$

$$= \frac{4\pi NI \mu A}{l}$$

$\therefore$ Total flux of N turns

$$= N \times \frac{4\pi NI \mu A}{l}$$

or $$\phi = \frac{4\pi N^2 I \mu A}{l} \quad ...(1)$$

Also by definition $\phi = LI$ ...(2)

From Equations (1) and (2)

$$L = \frac{4\pi N^2 \mu A}{l} \times 10^{-9} \text{ henry}$$

For an air core solenoid, $\mu = 1$

$$L = \frac{4\pi N^2 A}{l} \times 10^{-9} \text{ henry} \qquad ...(3)$$

In the general case, if there is a core consisting of number of media of permeabilities $\mu_1, \mu_2, \mu_3, ...$ and area of cross-sections, A_1, A_2, A_3, ... then we have

$$L = \frac{4\pi N^2}{l}(\mu_1 A_1 + \mu_2 A_2 + \mu_3 A_3 + ...) \times 10^{-9} \text{henry} \qquad ...(4)$$

In rationalised M.K.S. units or S.I. units, we have

$$L = \frac{\mu_n \mu_0 N^2 A}{l} \text{ henry} \qquad ...(5)$$

For air $\mu_r = 1$ and $\mu_0 = 4\pi \times 10^{-7}$ h/m

$$\therefore \quad L = \frac{\mu_0 N^2 A}{l} \text{ henry} \qquad ...(6)$$

again for a core consisting of number of media with relative permeabilities $\mu_{r_1}, \mu_{r_2}, \mu_{r_3}, ...$ and area of cross-section A_1, A_2, A_3, ...we have

$$L = \frac{\mu_0 N^2}{l}[\mu r_1 A_1 + \mu r_2 A_2 + \mu r_3 A_3 + ... \text{ henry}] \qquad ...(7)$$

2.4. ENERGY STORED IN AN INDUCTANCE

In the case of an electric circuit that consists of inductance, when the circuit is closed, back e.m.f. opposes the growth of the current in the circuit *i.e.,* the current flows against the back e.m.f. and does some external work. When the current attains maximum value, there is no back e.m.f. and so no extra work is done. The total work done is stored in the magnetic field of the coil. When the current is switched off, this energy is released and an induced e.m.f. that tries to maintain the current is produced. If there is large amount of induced e.m.f., sparking takes place at the switch. Let the current through the coil at the time of growth has an increasing rate dI/dt ampere/sec.

Then back e.m.f.,

$$e = L dI/dt$$

Work done/sec = $e.I = L.I.dI/dt$

Total work done when the current increases from 0 to maximum (I_{max}).

$$W = \int_0^{I_{max}} LI \frac{dI}{dt} \cdot dt = \frac{1}{2} LI_{max}^2$$

If L is in henry and I is in amperes, then W will be measured in joules.

$\therefore$ Energy stored $= LI_{max}^2$ joules

SPECIAL CASES

If $I = 1$, then $L = 2W$

which gives another definition of Self Inductance coefficient as under:

It is numerically equal to twice the work done in establishing magnetic induction accompanying unit current in the circuit.

Example 2. A *solenoid having a core of 4 cm^2 in cross-section, half air (μ = 1) and half iron (μ = 500) is 20 cm long. If the number of turns on it is 1000, calculate the coefficient of self induction.*

Solution: In this case

$A_1 = 2$ cm^2, $A_2 = 2$ cm^2

$\mu_1 = 1$, $\mu_2 = 500$, $l = 20$ cm

$N = 1000$

$$L = \frac{4\pi N^2}{l}(\mu_1 A_1 + \mu_2 A_2) \times 10^{-9} \text{ henry}$$

$$= \frac{4 \times 3.142 \times 10^6}{20}[1 \times 2 + 500 \times 2] \times 10^{-9} \text{ henry}$$

$$= \mathbf{0.6298 \text{ henry.}}$$

Example 3. *A solenoid (air core) 10 cm long has 100 turns and has an area of cross-section 5 cm². Find the coefficient of self inductance of the solenoid.*

Solution: $l = 10$ cm, $N = 100$, $A = 6$ cm², $\mu = 1$

$$\therefore \quad L = \frac{4 \times 3.142 \times 10000 \times 5 \times 10^{-9}}{10}$$

$$= \mathbf{62.81 \times 10^{-6} \text{ henry.}}$$

2.5. MEASUREMENT OF SELF INDUCTANCE

Rayleigh's method (Direct current method). We make use of the Wheatstone bridge for the measurement of self inductance. Self inductance L to be measured is placed in one of the arms and a P.O. box used along with a ballistic galvanometer which is not a dead beat type.

The resistance r is in the arm AB and has standard values $0.1\Omega, 0.01\Omega, 0.001\Omega$. Firstly the key K_1 is closed so that inductance L and its resistance are in the circuit. Firstly the battery key K_1 is closed and then galvanometer key K_2 is closed after some time gap and the resistances in the arms are so adjusted that there is zero deflection in the galvanometer. Now both keys K_1 and K_2 are opened. When key K_2 is closed first and then key K_1 is closed, we get a throw θ_1 on the scale of the ballistic galvanometer. The balance in the circuit is disturbed because an extra e.m.f., $L dI/dt$ is produced in arm AB while the current is growing.

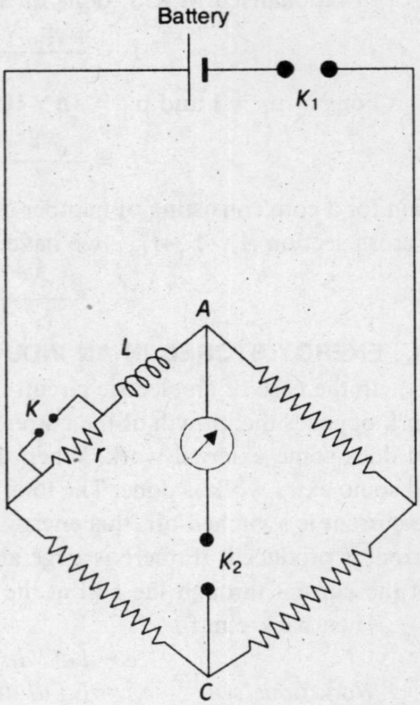

Fig. 2.5

Any change in e.m.f. in AB arm will produce a proportionate e.m.f. in galvanometer circuit and so a proportionate current flows in the galvanometer circuit.

Let us suppose that an e.m.f. E in arm AB produces a current xE in galvanometer circuit. Which means that current in galvanometer due to the induced e.m.f., $L\frac{dI}{dt}$ in the arm $AB = x.L.\frac{dI}{dt}$ and the total charge through the ballistic galvanometer is given by

$$q = \int_0^{I_0} x.L.\frac{dI}{dt} \cdot dt = xL \int_0^{I_0} dI = xL.I_0$$

But

$$q = \frac{CT}{2\pi nAB}\theta_1\left(1 + \frac{\lambda}{2}\right)$$

$$\therefore \quad xLI_0 = \frac{CT}{2\pi n AB}\theta_1\left(1 + \frac{\lambda}{2}\right) \quad ...(1)$$

Now the key K is opened and resistance r is included in the arm AB. Key K_1 is closed first and K_2 afterwards and a steady deflection θ_2 in the galvanometer is noted. The resistance r produces an additional p.d. $= I_0.r$ in the arm AB.

So the current through the galvanometer due to the extra p.d. across the arm $AB = x.I_0.r$

For a steady deflection

$$x.I_0.r = \frac{C}{n AB} \cdot \theta_0 \quad ...(2)$$

In this case $\frac{C}{n\,AB}$ is the current reduction factor of galvanometer

Dividing Equation (1) by (2)

$$\frac{L}{r} = \frac{1}{2\pi}\cdot\frac{\theta_1}{\theta_2}\left(1+\frac{\lambda}{2}\right)$$

$$l = \frac{r.T}{2\pi}\cdot\frac{\theta_1}{\theta_2}\left(1+\frac{\lambda}{2}\right) \quad ...(3)$$

Self Inductance can also be found by A.C. bridges, the usual one being Anderson Bridge.

2.6. SELF INDUCTANCE FOR RETURN AND LEAD WIRE

***(i)* Two co-axial cylinders.** Let us consider two concentric cylinders of radii a and b, the current being lead by the inner cylinder and returns in the opposite direction through outer cylinder as shown in Fig. 2.6. The field at a point P at a distance r from the centre of inner cylinder = $2I/r$. Considering a length of 1 cm, the flux through the area $ABCD$ between the cylinders containing point P is given by

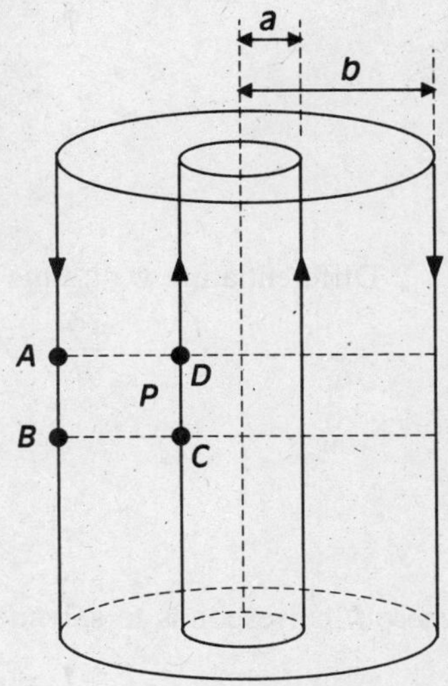

Fig. 2.6

$$\phi = \int_a^b \frac{2I}{r}\,dr = 2I\log_e\frac{b}{a}$$

$$\therefore \quad \frac{d\phi}{dt} = 2\log_e\left(\frac{b}{a}\right)\cdot\frac{dl}{dt}$$

Also $$e = -L\cdot\frac{dl}{dt} = -\frac{d\phi}{dt}$$

$$\therefore \quad e = -L\cdot\frac{dl}{dt} = -2\log_e\frac{b}{a}\cdot\frac{dI}{dt}$$

$$\therefore \quad L = 2\,\log_e\frac{b}{a}\ \text{e.m.u.}$$

where L corresponds to the self inductances of the two co-axial cylinders/unit length.

Also we have

$$L = 2\left(\log_e\frac{b}{a}\right)\times 10^{-9}\ \text{henry for 1 cm length}$$

$$= 2 \times 2.3026\left(\log_{10}\frac{b}{a}\right)\times 10^{-9}\ \text{henry for 1 cm length}$$

***(ii)* Rationalised M.K.S. units or S.I. units.** The self inductance of two co-axial cylinders for 1 m length is

$$L = \frac{\mu_0}{4\pi}\left[2\log_e\frac{b}{a}\right]$$

$$= \frac{\mu_0}{2\pi}\times 2.3026\log_{10}\frac{b}{a}\ \text{henry for 1 m length}$$

In this case $\mu_0 = 4\pi\times 10^{-7}$ henry/metre.

2.7. TWO PARALLEL WIRES

Let us consider two parallel wires A and B carrying currents in opposite directions. Let each wire has a radius r and the distances between their centres = d. Let us take a point P at a distance X from A and $(d - x)$ from B.

Let us choose an area of length 1 cm and breadth dx, so that the area of shaded portion = $1 \times dx$ cm^2 as shown in Fig. 2.7.

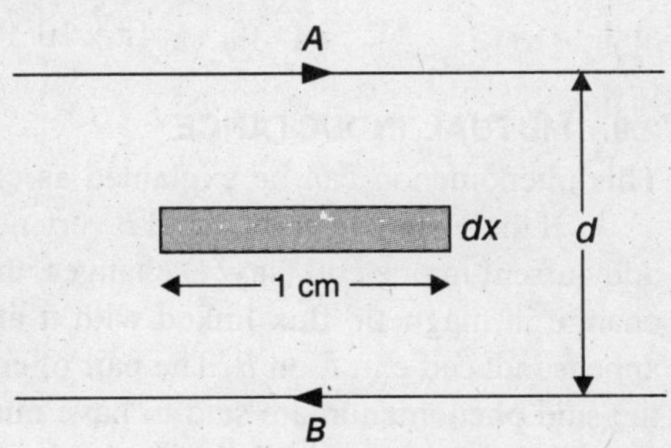

Fig. 2.7

The magnetic field at P due to two wires

$$B = \left(\frac{2I}{x} + \frac{2I}{d-x}\right)$$

and magnetic flux through an area

$$1 \times dx = \left(\frac{2I}{x} + \frac{2I}{d-x}\right) dx$$

$\therefore$ Magnetic flux through an area from A to B of length 1 cm.

$$\phi = \int_r^{d-r} \left(\frac{2I}{x} + \frac{2I}{d-x}\right) dx$$

$$= 2\,I\,[\log x - \log (d-x)]^{d-r}$$

$$= 2\,I\,[\log (d-x) - \log r - \log (d-d+r) + \log (d-r)]$$

$$= 4\,I \log_e \left(\frac{d-r}{r}\right)$$

Differentiating w.r.t. time

$$\frac{d\phi}{dt} = 4 \log_e \left(\frac{d-r}{r}\right) \frac{dI}{dt}$$

Now, $$e = -\frac{d\phi}{dt} - L\frac{dI}{dt} = -4 \log_e \left(\frac{d-r}{r}\right) \frac{dI}{dt}$$

$\therefore$ $$L = 4 \log_e \left(\frac{d-r}{r}\right) \text{ e.m.u.}$$

where L corresponds to self inductance for 1 cm length of two parallel wires

Also $$L = 4 \log_e \left(\frac{d-r}{r}\right) \times 10^{-9} \text{ henry}$$

$$= 4 \times 2.3026 \times \log_{10} \left(\frac{d-r}{r}\right) \times 10^{-9} \text{ henry}$$

For 1 cm of length

If the two wires are in contact, then $d - r = r$

$$L = 0$$

2.8. RATIONALISED M.K.S. UNITS OR S.I. UNITS

Coefficient of self inductance for 1 m length of two parallel wires is given by

$$L = \frac{\mu_0}{4\pi}\left[4 \times 2.3026 \log_e \left(\frac{d-r}{r}\right)\right]$$

henry for 1 m length

$$= \frac{\mu_0}{\pi} \times 2.3026 \log_e \left(\frac{d-r}{r}\right) \text{ for 1 m length}$$

and $$\mu_0 = 4\pi \times 10^{-7} \text{ henry/metre}$$

2.9. MUTUAL INDUCTANCE

This phenomenon can be explained as under:

If there are two coils A and B very near to each other and current in one coil (say A) changes, then there will be change in magnetic flux linked with it and so there will appear induced e.m.f. on B. The pair of coils which show the said phenomenon are said to have mutual inductance and the phenomenon is called mutual induction. On the other

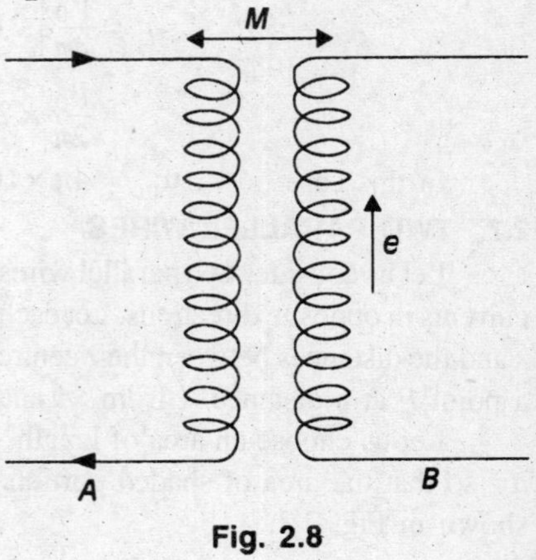

Fig. 2.8

hand if current in B changes then induced e.m.f. appears on A which shows that it is truly a mutual phenomenon and the circuits which show this phenomenon are called coupled circuits. It should be noted that the magnetic flux linked with one coil is dependent on the current in the other coil.

i.e., Magnetic flux $\phi \propto I$

or $$\phi = MI \qquad ...(1)$$

where M is coefficient of Mutual Inductance and may be defined as under:

if $I = 1, \quad \phi = M$

So, coefficient of mutual inductance of two circuits is numerically equal to the magnetic flux through one circuit, when a unit current passes through the other.

Differentiating (1) w.r.t. time

$$\frac{d\theta}{dt} = M\frac{dI}{dt}$$

and Induced e.m.f. $$e = -\frac{d\phi}{dt} = -M\frac{dI}{dt}$$

$$\therefore \qquad e = M\frac{dI}{dt}$$

So, we may also define coefficient of mutual inductance between two circuits as under:

It is numerically equal to the induced e.m.f. in one circuit when the rate of change of current in the other is unity.

2.10. C.G.S. UNITS

We may have

if e = 1 e.m.u., dI/dt = 1 e.m.u./sec

Then $M = 1$

That is, two circuits have a mutual inductance of one C.G.S. unit, if rate of change of current of 1 e.m.u/sec in one circuit induces 1 e.m.u. of e.m.f. in the other.

and if e = 1V, dI/dt = 1 amp/sec

then M = 1 henry

That is, **Henry is the mutual inductance of two circuits when the current changing at the rate of 1 amp/sec in one circuit induces an e.m.f. of 1 volt in the other circuit.**

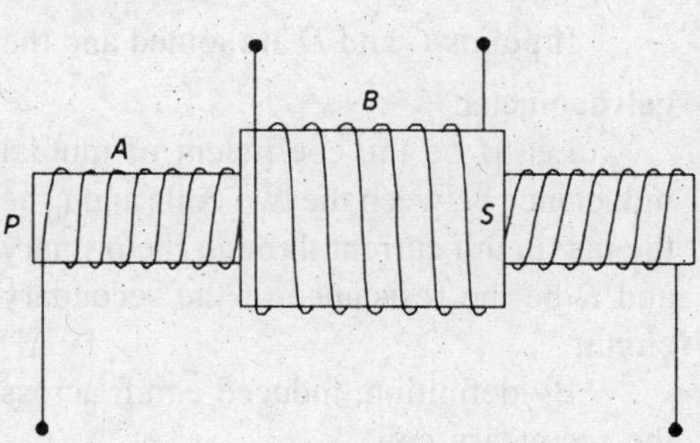

Fig. 2.9

2.11. MUTUAL INDUCTANCE IN CASE OF TWO SOLENOIDS

Let us consider two co-axial solenoids A and B, where A is called primary and B is secondary solenoid.

If we consider that no leakage of flux exists, and

No. of turns of primary solenoid $= N_1$
No. of turns of secondary solenoid $= N_2$
Area of cross-section of primary $= A$
Length of primary coil $= l$
Permeability of the core $= \mu$
Current flowing through primary $= I$

Then magnetic field inside primary at any point

$$= \frac{4\pi N_1 I}{t}$$

and magnetic flux through each turn of the secondary

$$= \frac{4\pi N_1 I \mu A}{l}$$

Total magnetic flux through N_2 turns of secondary,

$$\phi = \frac{4\pi N_1 I \mu A N_2}{l}$$

But by definition $\phi = MI$

$$\therefore \quad MI = \frac{4\pi N_1 N_2 \mu A I}{l}$$

$$\therefore \quad M = \frac{4\pi N_1 N_2 \mu A}{l} \text{ C.G.S. e.m.u.}$$

$$= \frac{4\pi N_1 N_2 \mu A}{l} \times 10^{-9} \text{ henry}$$

If we consider number of cores of areas of cross-sections A_1, A_2, A_3 ... and permeabilities as $\mu_1, \mu_2, \mu_3 ...$

Then
$$M = \frac{4\pi N_1 N_2(\mu_1 A_1 + \mu_2 A_2 + \mu_3 A_3 + ...)}{l} \times 10^{-9} \text{ henry}$$

In rationalised M.K.S. units or S.I. units

$$M = \frac{\mu_r \mu_0 N_1 N_2 A}{l} \text{ henry}$$

For air core, $\mu_r = 1$ and $\mu_0 = 4\pi \times 10^{-7}$ henry/m

$$\therefore \quad M = \frac{\mu_0 N_1 N_2 A}{l}$$

If there are a number of cores of areas of cross-sectior A_1, A_2, A_3 and relative permeabilities $\mu_{r_1}, \mu_{r_2}, \mu_{r_3} ...$

then
$$M = \frac{\mu_0 N_1 N_2}{l}[\mu_{r_1} A_1 + \mu_{r_2} A_2 + ... \text{ henry}]$$

2.12. DETERMINATION OF MUTUAL INDUCTANCE

Let us consider the circuit consisting of primary coil *PP* and secondary coil *SS*.

If points *C* and *D* are joined and the key *K* is pressed, we observe a throw θ_1 in the Ballistic galvanometer.

Let *M* be the coefficient of mutual inductance between the two coils and I_0 be the maximum current through the primary and *R* be the resistance of the secondary circuit.

By definition, induced e.m.f. across the secondary coil

$$= M\frac{dI}{dt}$$

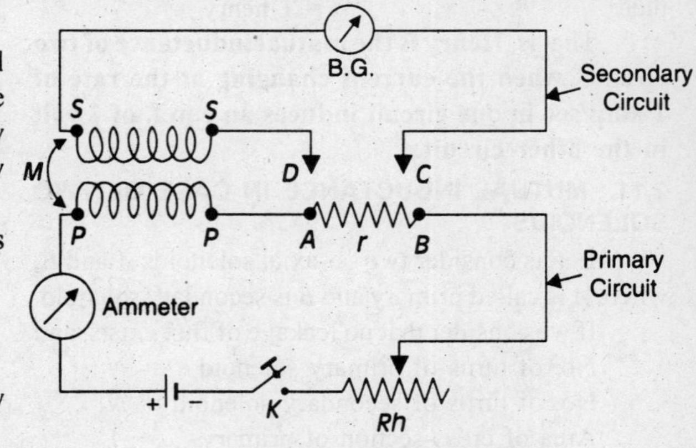

Fig. 2.10

$\therefore$ Total charge through the secondary circuit

$$q = \frac{M\int \frac{dI}{dt} \cdot dt}{R} = \frac{MI_0}{2}$$

Also in the case of ballistic galvanometer

$$q = \frac{CT}{2\pi n AB}\theta_1\left(1 \times \frac{\lambda}{2}\right)$$

where each term has usual meaning

$$\therefore \quad \frac{MI_0}{R} = \frac{CT}{2\pi n\, AB}\theta_1\left(1+\frac{\lambda}{2}\right) \qquad ...(1)$$

Now if C and D are disconnected and C is connected to B and A to D, there will be a steady deflection θ_2 in the galvanometer but the current I_0 through the primary remains the same.

Let r be the standard resistance which may have value = 0.1 ohm or 0.01 ohm.

$\therefore$ p.d. across $\quad r = I_0 r$

and current through secondary circuit $= \dfrac{I_0 \cdot r}{R}$

$$\therefore \quad \frac{I_0 \cdot r}{R} = \frac{C}{n\, AB}\cdot\theta$$

where $\dfrac{C}{n\, AB}$ is the current reduction factor of the galvanometer.

Dividing Equation (1) by (2)

$$\frac{M}{r} = \frac{T\cdot\theta_1(1+\lambda/2)}{2\pi\theta_2}$$

or

$$M = \frac{r.T\cdot\theta_1\left(1+\frac{\lambda}{2}\right)}{2\pi\,\theta_2}$$

So, we can calculate M, if we know the time period T and logarithmic decrement λ of the ballistic galvanometer.

Example 4. *A straight solenoid has 5000 turns/cm in the primary and 200 turns in the secondary. Area of cross-section of the solenoid is 4 cm². Calculate the mutual inductance.*

Solution: We have the relation

$$M = \frac{\mu_0 N_1 N_2 A}{l} \text{ henry}$$

In this case $\dfrac{N_1}{l}$ = turns/cm = 5000 turns/cm

$$N_2 = 200$$

$$A = 4 \text{ cm}^2 = 4 \times 10^{-4} \text{ m}^2$$

and

$$\mu_0 = 4\pi\times10^{-7} \text{ henry/m}$$

$$\therefore \quad M = 4\pi\times10^{-7}\times5000\times200\times4\times10^{-4}$$

$$= 50.27\times10^{-5} \text{ henry.}$$

QUESTIONS AND PROBLEMS

1. Define and explain mutual inductance between two circuits, and describe a method to measure it.
2. Define and explain self inductance. What are its units and how it can be measured?
3. A solenoid has a core of 2×10^{-3} m² cross-section, half air ($\mu = 1$) and half iron ($\mu = 500$). The length of the solenoid is 2 metres. If the number of turns is 1000, find the coefficient of self induction. [**Ans.** 0.3149 henry.]
4. Calculate the coefficient of self inductance of a coil of 500 turns when a current of 1.25 amperes produces a magnetic flux of one microweber.
5. Calculate the self inductance in practical units of 100 cm long solenoid having 500 turns and 5 cm diameter. [**Ans.** 617×10^{-6} henry.]
6. A solenoid 2 cm long has 1600 turns and its cross-section is 10 cm². Closely wound on its central part is another coil of 1000 turns. Find the mutual inductance between the coils in henries. [**Ans.** 0.1006 henry.]

CHAPTER

3

MAGNETIC CIRCUITS AND MATERIALS

3.1. INTRODUCTION

Different substances behave differently when placed in a magnetic field and normally we have categories of substances (*i*) Ferromagnetic materials, (*ii*) Paramagnetic materials and (*iii*) Diamagnetic materials. Before actually taking up the basic characteristics of different materials, the following terms are very important.

3.2. MAGNETISING FIELD (*H*)

This is the field in which a substance is placed and this field is usually uniform and is measured by the *number of unit tubes of force passing per unit area normally.*

3.3. INTENSITY OF MAGNETISATION (*I*)

It gives the measure of the magnetisation of a magnetised specimen. It is defined as *magnetic moment per unit volume, i.e.,*

$$I = \frac{\text{Magnetic Moment}}{\text{Volume}} = \frac{M}{V}$$

Since we normally take a small specimen, its magnetisation can be supposed to be uniform and also if the specimen has a uniform area of cross-section a and magnetic length $2l$, and has pole strength m, then

$$M = m \times 2l \quad \text{and } V = a \times 2l$$

$$\therefore \quad I = \frac{m \times 2l}{a \times 2l} = \frac{m}{a} \ (\text{weber/m}^2)$$

That is, the intensity of magnetisation for a specimen of uniform area of cross-section is given by **pole strength/unit area**. It is measured in ampere turn/metre.

3.4. MAGNETIC INDUCTION OR FLUX DENSITY

The total number of lines of force/unit area due to magnetising field and due to the field induced in the substance is called flux density (B) and is measured in weber/m^2.

3.5. MAGNETIC SUSCEPTIBILITY

Since the intensity of magnetisation for a material is proportional to the magnetic field, so the magnetic susceptibility (χ) of a specimen measures the **ease with which the specimen** *can be magnetised and it is defined as the ratio of the intensity of magnetisation induced in it and magnetising field, i.e.,*

$$\chi = \frac{I}{H}$$

3.6. PERMEABILITY

When a magnetic substance is placed in a magnetic field, due to the induction phenomenon, it will acquire magnetism, because lines of force of the magnetising field concentrate inside the material and results in the magnetising of the material. The measure of the *degree to which the lines of force can penetrate or permeate into the medium* is called absolute permeability of the medium and is denoted by μ_a. The absolute permeability of the medium can be defined as the ratio of the normal lines of force per unit area in the medium to the lines of force/unit area that would be there if the medium is replaced by air or vacuum so the permeability is defined as the ratio of magnetic induction B in the medium to the magnetising field, *i.e.,*

$$\mu_a = \mu_0\mu_r = \frac{B}{H} \quad \text{... (1)}$$

the term absolute permeability μ_a of a material is expressed as the product of the permeability of the force space μ_0 (= $4\pi \times 10^{-7}$ henry/metre) and μ_r (the relative permeability).

i.e.,

$$\mu_a = \mu_0 \times \mu_r$$

where μ_r is given as the ratio of *the force between two poles* when placed at a fixed distance in vacuum to the force between them at the same distance in that medium. Since μ_r is ratio, it has no units and $\mu_0 = 1$ in air or vacuum.

When a magnetic material of A cross-sectional area and μ_r relative permeability is placed in a uniform field of intensity H, two types of lines of force pass through it, one due to the magnetising field H and the other due to the material being magnetised by induction. So the total flux density will be given by

$$B = \mu_0 H + I \quad \text{... (2)}$$

where μ_0 corresponds to the permeability of free space. Now we have absolute permeability μ_a as

$$\mu_a = \mu_0\mu_r = \frac{B}{H} = \frac{\mu_0 H + I}{H}$$

$$= \mu_0\,\mu_r = \frac{B}{H} = \mu_0 + \chi$$

Dividing both sides by μ_0, we get

$$\mu_r = 1 + \frac{\chi}{\mu_0} \quad \text{... (3)}$$

3.7. MAGNETIC CIRCUITS

The path followed by magnetic flux is called **magnetic circuit**. As in the case of electric current which follows a closed circuit, magnetic flux follows a complete circuit or loop coming back to the starting point. Magnetic flux leaves from N pole and passing through air enters at S pole and finally reaches N pole after travelling through the magnet.

The flux ϕ is produced by the magnet due to some force, called **magnetomotive force.**

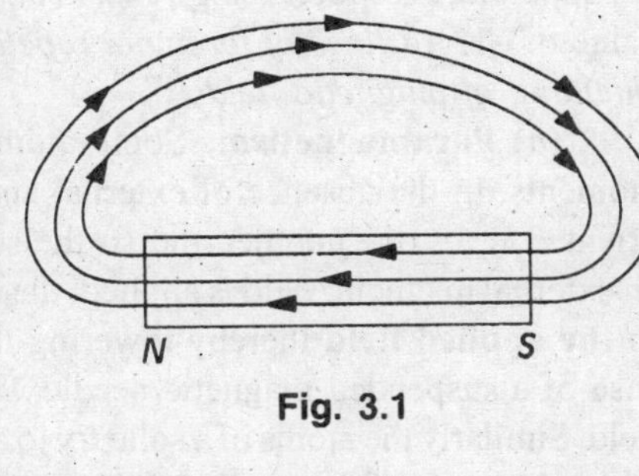

Fig. 3.1

3.8. MAGNETOMOTIVE FORCE

Like electromotive force in electricity, we have magnetomotive force (m.m.f) which may be defined as the *force that drives or tends to drive magnetic flux through a magnetic circuit.*

An m.m.f. is produced in a solenoid when electric current flows in a coil of several turns and m.m.f. is measured by the product of current in amperes and number of turns. The unit of m.m.f. is **ampere turn (AT)** and denoted by M or F. So, we may write

$$M = I \times N \ (AT)$$

3.9. RELUCTANCE

Similar to the resistance characteristic in electricity, we have reluctance in magnetism which is the property of the substance **that opposes the magnetic flux through it.**

Reluctance of any part of a magnetic circuit may be defined as the *ratio of the drop in magnetomotive force to the flux produced in that part of the circuit.* It is measured in ampere turns/weber and is denoted by R.

Reluctance is dependent on:

(*i*) Nature of the substance.

(*ii*) Area of cross-section of the material through which flux is passing (*a*).

(*iii*) Length of magnetic path.

Mathematically, we may write

$$R = \frac{1}{a\,\mu_0\mu_r}\ AT/Wb$$

where μ_0 is absolute permeability

and $$\mu_0 = 4\pi \times 10^{-7} \text{ henry/m}$$

and μ_r is relative permeability.

3.10. PERMEANCE

The reciprocal of reluctance is called permeance and is a measure of the readiness with which the magnetic flux is developed. It is analogous to conductance in an electric circuit and is measured by weber/ampere turn. So

$$\text{Permeance } P = \frac{\text{flux}}{\text{m.m.f.}}\ \text{Wb/AT} = \frac{1}{\text{reluctance}}\ \text{Wb/AT}$$

3.11. CLASSIFICATION OF MAGNETIC MATERIALS

We can classify the magnetic materials from the concept of *motion of an electron* in an atom. As discussed **earlier**, orbital motion of a charge carrying electron, around the central nucleus, is analogous to a *current carrying conductor*. With the application of a magnetic field, motion of electrons get modified in such a way that a weak magentic moment, opposing the field, is induced. **Diamagnetism is the result of this interaction.**

(*a*) **Diamagnetism:** In a diamagnetic solid, the magnetic lines of force, due to an applied field, are repelled. Diamagnetic susceptibility is negative and very small in magnitude. (typically 10^{-5}). An important exception being a *super conductor*, though a perfect diamagnetic, yet having susceptibility value of −1. *All lines of force are repelled by a superconductor, thereby making them suitable for shielding of magnetic fields.*

(*b*) **Paramagnetism:** Some atoms and molecules possess (intrinsic) permanent magnetic moments. In the absence of external applied field, the moments of atoms in a solid are randomly oriented w.r.t. one another and so the solid, on the whole, has no magnetic moment. However, if an external magnetic field is applied, the various magnetic moments tend to alien themselves **parallel to the applied field** thereby lowering the potential energy of the system. A typical example is in case of a suspended magnetic needle which aliens itself spontaneously with the Earth's magnetic field. Similarly the atoms of a solid try to alien themselves in the direction of the applied field, thereby intensifying the lines of force in the field direction. This phenomenon is called paramagnetism and is shown in Fig. 3.2 (*b*). The aligning force on the permanent moments of the atoms with ordinary magnetic fields is quite small, so that the **paramagnetic** effect is weak. Paramagnetic susceptibility is small and positive ~ 10^{-3}. It is important to note that this value decreases with rise in temperature as **thermal energy tends to change the alignement of atoms**.

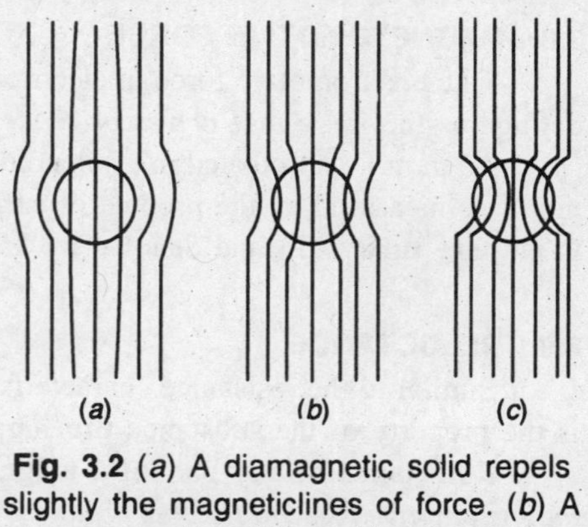

Fig. 3.2 (*a*) A diamagnetic solid repels slightly the magneticlines of force. (*b*) A paramagnetic solid weakly attracts the lines of force. (*c*) A frerromagnetic solid attracts them very strongly.

(*c*) **Ferromagnetism:** Ferromagnetic solids are those solids in which the permanent magnetic moments (due to electron spins) are **already aligned due to bonding forces**. In case of ferromagnetic materials susceptibility is very large in the range of 10^2 to 10^5. They strongly attract the magnetic lines of force as shown in Fig. 3.2(c).

3.12. IMPORTANCE OF DIPOLE MOMENTS IN CLASSIFICATION OF MAGNETIC MATERIALS

The magnetic materials may be re-classified, on the basis of permanent magnetic dipole materials into two broad groups, **as materials whose atoms carry permanent magnetic dipoles and those which do not have permanent magnetic dipoles**. Materials composed of atoms or molecules having zero magnetic moment, are called **diamagnetic**. Further, materials composed of atoms or molecules, having permanent magnetic moment are classified into four categories, depending on the interaction between the **individual magnetic dipoles**. If the interaction between the atomic magnetic dipoles is negligible, the material is called *paramagnetic* [Fig. 3.3 (*a*)]. If the magnetic dipoles interact in such a way that they tend to orient in the same direction, the material is ferromagnetic [Fig. 3.3 (*b*)]. If neighbouring dipoles orient in opposite directions and if the dipoles are of equal magnitude, the material is **antiferromagnetic** [Fig. 3.3 (*c*)]. If the neighbouring dipoles are of different magnitude and orient antiparallel, the material is **ferrimagnetic**. [Fig. 3.3 (*d*)] All the four types of orientations are given in the [Fig. 3.5].

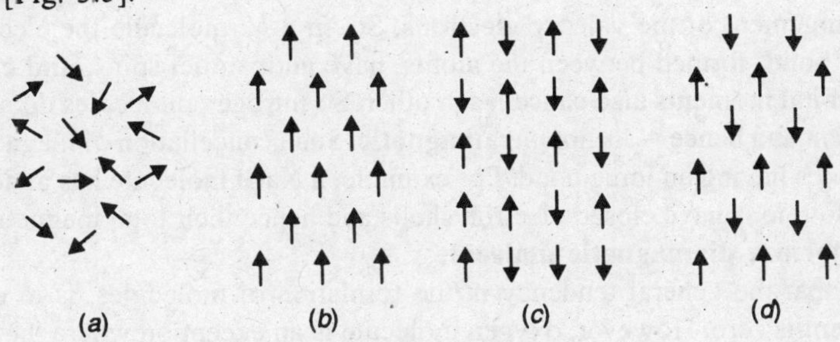

Fig. 3.3

3.13. MAGNETIC MOMENT OF AN ATOM

We know that the orbital angular momentum L of an orbiting electron and the magnetic moment μ_m associated with the orbital motion of the electron is given by

$$\mu_{morb} = -\frac{e}{2m}L \quad ...(1)$$

As the orbital angular momentum is space quantized and its projection L_z in a given direction z can take only $(2l + 1)$ specific values, so we have

$$L_z = m_e\hbar \quad ...(2)$$

Using the value of L_z into equation (1) we get

$$\mu_{morb} = -\frac{e\hbar}{2m}m_e = -m_e\mu_B \quad ...(3)$$

The quantify $$\mu_B = \frac{e\hbar}{2m} = 9.274 \times 10^{-24} \text{ A.m}^2 \quad ...(4)$$

is known as the **Bohr magneton.**

The Bohr magneton is the elementary electron magnetic moment. It is elementary in the sense that no electron can have a magnetic moment below μ_B. So, it represents the minium non-zero value of the projection of the magnetic moment of a electron in an arbitrary direction. In general, the component of the electron orbital **magnetic moment along an external field is equal to m_j Bohr magetons.**

The spin angular momentum and the spin magnetic dipole moment of an electron ary given by

$$\mu_{m\ spin} = -\frac{e}{m} S_z \quad ....(5)$$

where S_z can have the values

$$S_z = \pm\frac{1}{2}\hbar \quad ...(6)$$

$$\mu_{m\ spin} = \frac{e\hbar}{2m} \quad ...(7)$$

Each orbit in an atom acommodates two electrons which orient themselves in opposite directions and revolve in opposite directions. **Because of this the spin and orbital magnetic moments of the pair cancel each other**. So, the magnetic moment of a free atom is mainly determined by **uncompensated** spins of the outer electrons.

3.14. MAGNETIC MOMENTS OF MOLECULES

Although the magnetic moment of a free atom is mainly found by uncompensated spins of valence electrons, the magnetic moment of a molecule **is not equal to the sum of the magnetic moments of the atoms constituting it,** because a chemical bond formed between the atoms requires a certain rearrangement of the valence electrons. So, in a N_2 molecule the electrons participating in the covalent bond, formed between the atoms, have antiparallel spins, **and cancel each others effect**. Their orbital moments also cancel each other. So nitrogen molecules do not have permanent magnetic moment and hence N_2 is not **paramagnetic**. Such cancellation of magnetic moments takes place in molecules having an ionic bond. For example, a NaCl molecule has an ionic bond between Na^+ and $C1$. Both ions have closed electron shells and hence their total magnetic moment is equal to zero. **They form a diamagnetic molecule**.

It means that the general tendency in the formation of molecules, is to ensure that the net magnetic moment is zero. However, oxygen molecule is an exception where the spins of collective electrons are compensated. In case of NO and NO_2, the total number of electrons is odd and so the spin of one electron is left uncompensated. As a result **oxygen, nitric and nitrous oxide gases exhibit paramagnetic properties**.

3.15. ORIGIN OF MAGNETIZATION

Ampere (1775-1836), postulated the molecular currents to explain the magnelixation of materials but he was not able to explain the origin of molecular currents. Rutherford's atomic model helps to explain the magnetic properties of the materials, which originate as a consequence of magnetic, moments associated with atoms. In general, magnetic moment of an atom may be due to the following parameters:

(*a*) **The orbital motion of electron:** The motion of an electron constitutes a current and the circular path of electron is identical to a current loop. Behaves as an elementary magnet having a magnetic moment. The total orbital magnetic moment of an atom is the sum of orbital magnetic moments of individual electrons.

(*b*) **The electron spin:** The spinning electron may also contribute magnetic moment.

(*c*) **The nuclear spin:** The magnetic moment of the nucleus is about $\frac{1}{2000}$ as much as that of an electron. So, in studying the magnetic properties of materials, the magnetic moment due to nuclear spin is neglected.

The two factors, *i.e.*, orbital motion and spin of electrons, contribute atoms with a **permanent magnetic moment**. However, in many cases, the magnetic moments of electrons in an atom cancel each other and the atom may not possess a permanent magnetic moment. When such atoms are placed in a magnetic field, the orbital motion of electrons undergoes changes and the atoms acquire **an induced magnetic moment**.

3.16. MAGNETIC DOMAINS

Magnetic properties of the materials can be explained using the **Domain Theory**. Ferromagnetic property of a magnetic material is due to **regular or order arrangement** of neighbouring atoms in the same direction. Ferromagnetic materials have a magnetic moment even in the absence of an external magnetic field. This is called **Spontaneous magnetization**. Spontaneous magnetization has non zero value below certain temperature T_C called *Critical temperature*. Above critical temperature ferromagnetic national behaves like a **paramagnetic material**. Normally value of T_C is higher than the room temperature, and so these materials can be use as **permanent magnets** at room temperature.

In order to explain **Ferromagnetism** Weiss put toward. **Domain Theory** According to this theory a single, Ferromagnetic crystal may be considered to be constituted by a large number of small regions known as domains which are oriented in the same direction *i.e.*, each region has saturated value of magnetism. At room temperature, small thermal agitations do not disturb the dipoles in a particular domain. However, since each domain to the sample is randomly oriented so the net magnetization of the sample is zero. Even by the application of a weak external field, magnetization can be observed. This magnetization is due to the following two mechanisms.

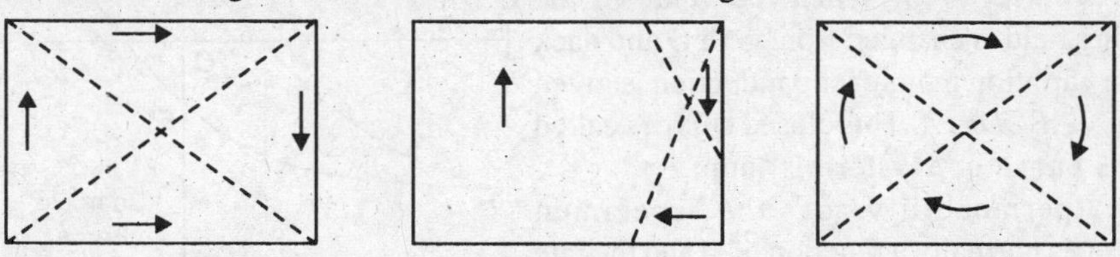

(*a*) Unmagnetized material (*b*) Magnetization due to domain growth (*b*) Magnetization due to domain rotation

Fig. 3.4

(*i*) In case a small external magnetic field is applied, then the magnetic domains in the material grow in the direction of the applied external field. In case external fields in raised to higher values, the domains intensity grows in the direction of applied external field, till it attains saturation value [Fig. 3.4 (*b*)].

(*ii*) If the magnetic field is increased further than the **saturation value** as explained above then, magnetic domains rotate like a single entity [fig. 3.4 (*c*)]. Finally, all domains orient themselves in the direction of the applied external field and maximum value of magnetization for the material is attained. This value becomes constant and does not increase even if more external magnetic field is applied.

In case the external field in remarked, the material loses some of the magnetic properties. However, in case of ferromagnetic materials, the magnetic properties still exist even if the external magnetic field in removed and so **a Ferromagnetic crystal** behaves like a permanent magnet in the absence of the external field.

Magnetic domains contributions can be observed in hysteresis phenomenon of Ferromagnetic materials.

3.17. CURIE LAW AND CURIE POINT

The variations in the susceptibility of paramagnetic and diamagnetic materials was studied by Pierre Curie in 1895. For paramagnetic materials, susceptibility is inversely proportional to the absolute temperature. This is known as Curie law *i.e.*,

$$\chi \propto \frac{1}{T} \text{ or } \chi = \frac{C}{T} \text{ where } C \text{ is curie constant.}$$

However, the later experimental data showed that Curie law is only approximately true. Another more accurate law called **Curie Wess law** is $\kappa_m = \dfrac{C_m}{T - T_c}$, where κ_m is the molar susceptibility of the material, C_m is the Curie constant, T the absolute temperature of the specimen and T_c is the Curie temperature. K_m is the susceptibility for 1 gm equivalent of the material.

When ferromagnetic materials are heated above certain critical temperature, their most of the magnetic properties are lost and they behave as para-magnetics. A data for approximate values of Curie temperature for various substances is given under:

Iron = 770°C, Cobalt = 1150°C, Nickel = 360°C, Ni–Fe (Fe 30%) = 70°C, Permalloy (Fe 22% Ni 78%) = 550°C, Thermalloy (Nickel–Cu alloys) = 10°C.

3.18. HYSTERESIS

If a ferromagnetic material, say iron, is magnetised slowly by gradually increasing the intensity of the magnetic field H, the magnetic induction starts increasing. A plot of the B–H curve is given in Fig. 3.5. The portion OA corresponds to initial magnetisation when the magnetic field is increased from O to G. When the value of the magnetising field is changed from G to G and back to + G, the variation in magnetic induction is given by the curve $ABCDEFA$. This closed curve is called **Hysteresis curve or Hysteresis loop.**

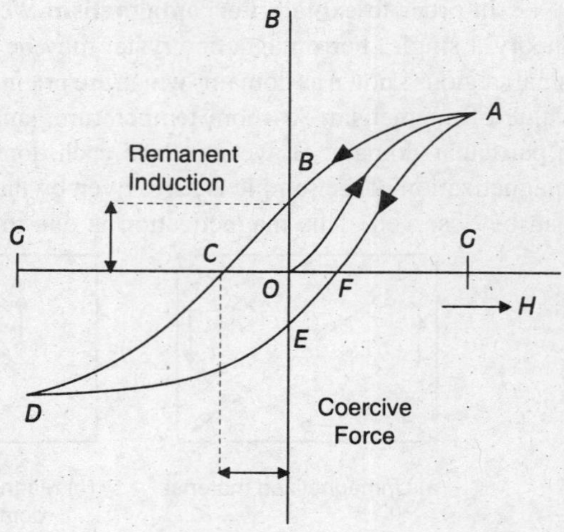

Fig. 3.5

The permeability of the specimen $\mu\,(= B/H)$ is normally a function of H and the rate of increase of permeability at any point on the curve is given by $\frac{dB}{dH}$, *i.e.*, the slope of the curve at that point. For very small H, the initial permeability is independent of H and the slope of the curve corresponding to steepest portion measures the maximum permeability of the specimen. In the same way, we can plot an I–H curve and the initial and max value of susceptibilities (χ) of the specimen can be calculated. The slope $\frac{dI}{dH}$ at any point on I–H curve measures the incremental susceptibility.

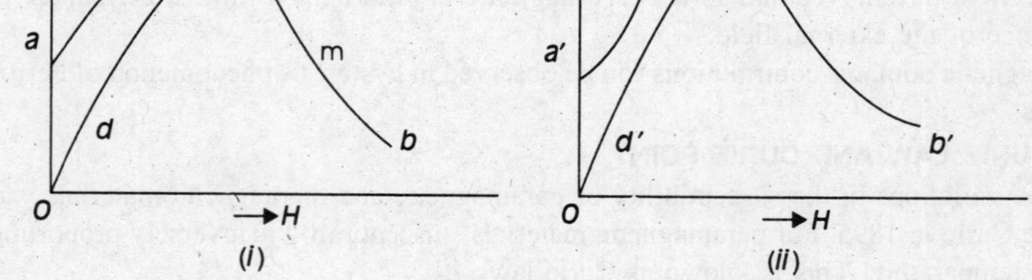

Fig. 3.6

Figure 3.6 (*i*) and (*ii*) give the variation of permeability and susceptibility of the material with change in magnetic field H. The corresponding values of B and I are also shown in the curves.

From Fig. 3.6 (*i*), we can see that the value of *B* increases rapidly for small fields (portion *Odg*) and for higher values of *H*, the increase in value of *B* is not proportional to the increase in *H* (portion *fg*). This is due to the magnetic saturation of the magnetic substance. However, *B* does not become constant because $B = H + 4\pi I$ and so even if *I* is constant, the value of *B* increases with *H*. However for low fields, *I* also increases and so there is a steep rise in *B–H* curve.

It is clear that permeability of a material (*i*) increases at first with increase of *H*, (*ii*) becomes maximum at the steepest portion of the curve, (*iii*) and then it decreases.

The slope of *B–H* curve decreases gradually and so the incremental permeability also decreases but it never becomes zero.

Figure 3.6 (*ii*) gives the corresponding *I–H* and *K–H* curves. It is clear that the general variation of the curve is the same as above except that *I* attains constant value at high fields due to magnetic saturation. Also, the susceptibility of the material increases rapdily for low magnetising fields ($a'c'$) and decreases thereafter. It is due to the fact that the value of *I* remains constant but *H* increases and hence ratio *I/H* decreases.

In Fig. 3.7 *OB* measures retentivity, *i.e.*, the residual intensity of magnetisation left in the specimen when the magnetising field is switched off, *OC* corresponds to coercivity of the material and it corresponds to the strength of the reverse magnetic field needed to make intensity of magnetisation zero. The magnetised state of the material at *O* and *C* is not the same, though the value of *I* is zero at both these points. At *C*, a small change in *H* gives a large change in value of *I* but this is not the case at the point *O*. The curves (*a*) and (*b*) give *B–H* and *B–H* hysteresis loops. The values of retentivity, coercivity, saturation value of *I* and the area of *I–H* and *B–H* hysteresis loops give us an idea of the *nature of magnetic material* and these values determine the susceptibility for use in particular electrical appliances like *transformers, electromagnets, permanent magnets, cores of dynamos, motors etc.*

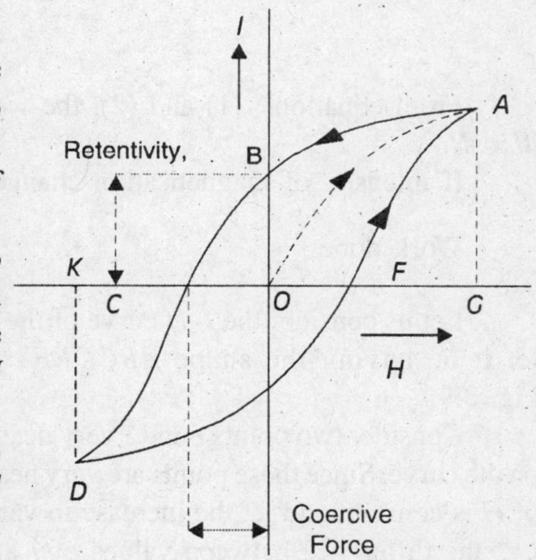

Fig. 3.7

Another important aspect shown from *I–H* and *B–H* curves is that the values of *I* and *B* do not become zero when the magnetising field is withdrawn. This characteristic of the material to retain its megnetisation or the lagging of *I* or *B* behind *H* is called **Hysteresis.**

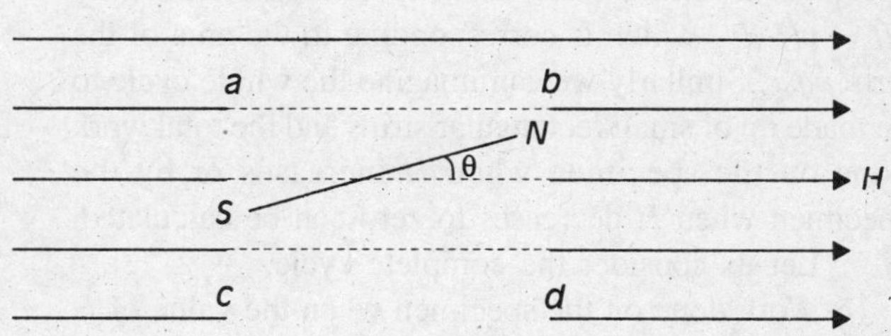

Fig. 3.8

3.19. ENERGY LOSS IN HYSTERESIS

When a ferromagnetic material is made to pass through a cycle of magnetisation, there is expenditure of energy. This energy is generally dissipated in the form of heat in the specimen. When a specimen is placed in a magnetic field, induced magnetism is acquired by the specimen. The magnetic dipoles or the molecular magnets are oriented and reoriented when the material is passed through a cycle of magnetisation and this molecular motion results in the production of heat.

In order to find the energy loss per cycle, let us consider a unit volume *abcd* of the magnetic material which is placed in a magnetising field having an intensity *H*.

Let there be a molecular magnet *NS* of magnetic moment *M* and inclined at an angle θ to the direction of magnetising field. The component of the magnetic moment along the direction of field = $M \cos \theta$ and that $\perp$ to the field in $M \sin \theta$. So, we have the total magnetic moment of all the molecular magnets/unit volume in the direction of field $\Sigma M \cos \theta$ and the component $\perp$ to the field $\Sigma M \sin \theta = 0$.

Now from definition, the magnetic moment/unit volume of the specimen $\Sigma M \cos \theta$ is equal to the intensity of magnetisation I.

$$\therefore \qquad I = \Sigma M \cos\theta$$

Differentiating w.r.t. θ

$$dI = \Sigma M(-\sin\theta)\, d\theta$$
$$= -\Sigma M \sin\theta\, d\theta \qquad \text{... (1)}$$

The couple acting on a molecular magnet inclind at an angle θ with the direction of the field = $MH \sin \theta$ and the work done in rotating this magnet by a small angle $d\theta = MH \sin \theta\,(-d\theta)$. The –Ve sign indicates that $d\theta$ is the decrease in angle.

Total work done in rotating all the molecular magnets in unit volume

$$= -\Sigma MH \sin\theta\, d\theta$$
$$= H\,[-\Sigma M \sin\theta\, d\theta] \qquad \text{... (2)}$$

From equations. (1) and (2), the work done in increasing the intensity of magnetisation by $dI = HdI$.

If intensity of magnetisation changes from I_1 to I_2, we have

$$\text{Work done} = \int_{I_1}^{I_2} H\, dI \qquad \text{... (3)}$$

Let us consider the I–H curve of the material, and let it be having the shape $ABCDEFA$ as shown in Fig. 3.9.

Consider two points r and s very near to each other on the curve. Since these points are very near so the value of H is constant and let the increase in value of I be dI, *i.e.*, the difference between values of I at p and q.

From equation 3, the amount of work done on the specimen to increase the intensity of magnetisation by $dI = \int H\, dI$, which is corresponding to the area of the strip $pqsr$. Similarly we can imagine the whole cycle to be made up of small rectangular strips and the total work done on the specimen when H increases or by the specimen when H decreases to zero can be calculated.

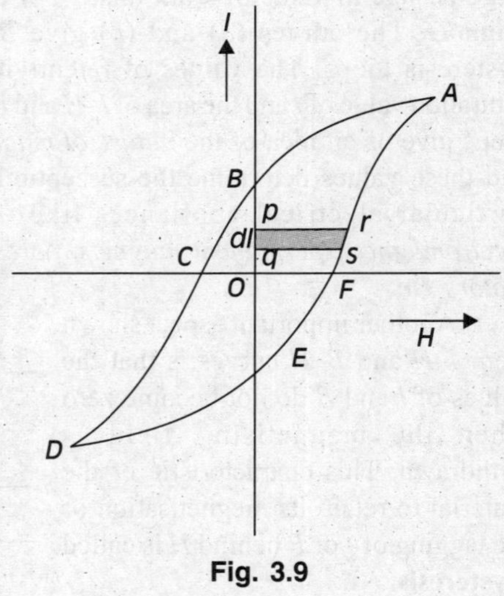

Fig. 3.9

Let us consider the complete cycle.

(*i*) Work done on the specimen when the value of H increases from zero to

$$K = \int_E^M H\, dI = \text{Area of } EFAMBOE \text{ (Fig. 3.7).}$$

(*ii*) Work done by specimen when it decreases to zero

$$= \int_M^B H\, dI = \text{Area of } AMBA$$

(*iii*) Work done on the specimen when it increases from O to L in the opposite direction

$$= \int_B^N H\, dI = \text{Area of } BCDNEOB$$

(*iv*) Work done by specimen when H decreases to zero

$$= \int_N^E H \, dI = \text{area } DNED$$

So the total work done on the specimen/unit volume/cycle = Area *EFAMBOE* – Area *AMBA* + Area *BCDNEOB* – Area *DNED* = Area *ABCDEFA* = Area of *I–H* curve.

So the work done on the specimen per unit volume when taken through a complete cycle of magnetisation is equal to the area enclosed by *I–H* curve. The larger is the area of the curve, the larger is the loss of energy/unit volume/cycle.

AREA OF *B–H* CURVE

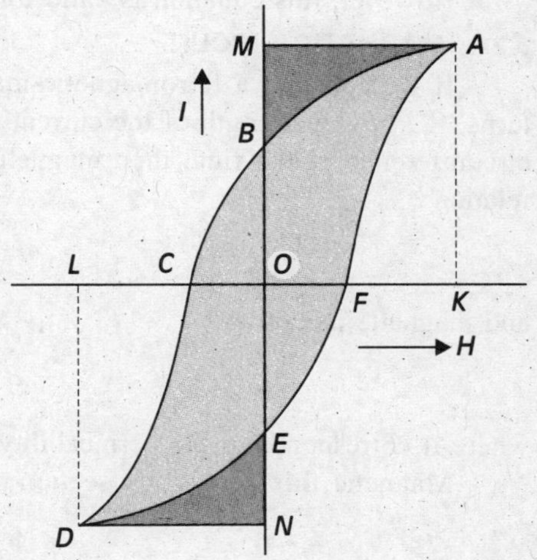

Fig. 3.10

We have the relation for magnetic induction as

$$B = H + 4\pi I$$

Differentiating

$$dB = dH + 4\pi dI$$

Multiplying both sides by H

$$HdB = HdH + 4\pi HdI$$

Integrating for a complete cycle

$$\oint HdB = \oint HdH + \oint HdI$$

Now $\oint HdB$ = Area of *B–H* curve

$\oint HdI$ = Area of *I–H* curve

and $\oint HdH = 0$

Since a curve between H and H will give a straight line and so the enclosed area = 0

Area of *B–H* curve $= 4\pi \times$ area of $I - H$ curve

or Area of *I–H* curve $= \dfrac{1}{4\pi}$ area of $B - H$ curve

So the loss in energy/cycle/unit volume of the specimen

$$= \frac{1}{4\pi} \text{ times the area of } B - H \text{ curve}$$

Now if I is expressed as magnetic moment/unit volume and H is in oersted, then *I–H* curve gives the energy loss/cycle in ergs. On the other hand, if B is in Gauss and H in oersted, then $\dfrac{1}{4\pi}$ times the area of *B–H* curve **gives energy loss/cycle in ergs.**

If rationalised M.K.S. or S.I. units are employed *i.e.*, B is in weber/metre^2 and H in ampere-turn/metre, then the energy loss per cycle/cubic metre of the material will be equal to the area of *B–H* curve and the **units of energy will be joule.**

3.20. STEINMETZ RULE

We see that the area of *B–H* loop increases with increase in the strength of the magnetising field. Fig. 3.11 gives *B–H* curve for maximum magnetising fields where H_1, H_2, H_3... and A_1, A_2, A_3... correspond to maximum values of B in each case. The locus of the points corresponding to A_1, A_2, A_3 ...corresponds to the initial magnetisation curve of the specimen after complete demagnetisation so the maximum flux density (B_{max}) is different for different magnetising values of H.

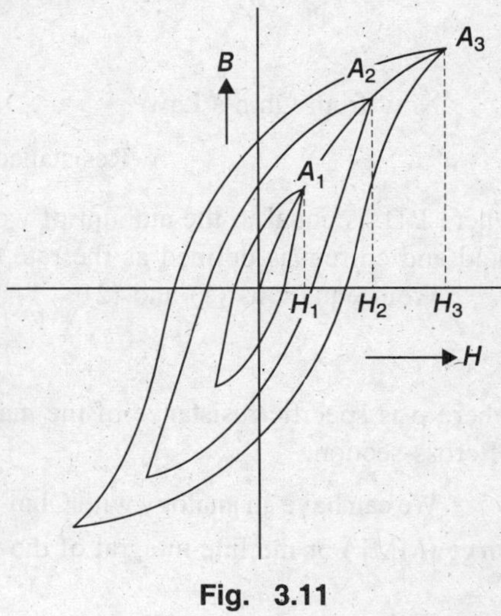

Fig. 3.11

Steinmetz gave an empirical equation for the calculation of energy loss per cycle/unit volume of the material as:

$$= \eta B_{max}^{K}$$

where $\eta \rightarrow$ coeff. of hysteresis loss

$K \rightarrow$ is constant and varies from 1.4 to 1.8

for different ferromagnetic materials.

However, this equation is valid for low field strengths only.

3.21. MAGNETIC CIRCUIT

If we consider a ferromagnetic material in the form of a ring and having windings of N turns. If I be the strength of the current in amperes passing through the winding and l the mean circumference of the ring, then magnetic intensity within the toroidal winding is given by the relation

$$H = \frac{4\pi NI}{10\,l}$$

and magnetic induction

$$B = \mu H = \frac{4\pi NI\mu}{10\,l}$$

where μ corresponds to the permeability of the material.

Magnetic flux across any section of area A of the ring is given by

$$\phi = B \times A = \frac{4\pi NI\mu A}{10\,l}.$$

$$\therefore \quad \frac{4\pi NI}{10} = \frac{l}{A\,.\,\mu}\phi$$

Now $\frac{4\pi NI}{10}$ corresponds to the work done in moving a unit north pole around the axis of the coil, *i.e.*, along the mean circumference of length l against magnetic field intensity $\frac{4\pi NI}{10l}$.

$$\therefore \quad \text{Work done} = \frac{4\pi NI}{10\,l} = \phi.\left(\frac{l}{A\,\mu}\right)$$

$$= \text{Flux} \times \left(\frac{l}{A\,\mu}\right) \qquad \text{... (1)}$$

In rationalised M.K.S. units or S.I. units

$$= \frac{NI\,\mu A}{l} \text{ weber}$$

Now from Ohm's Law

$$\text{Resistance} = \frac{\text{P.D.}}{\text{current}} \qquad \text{... (2)}$$

where P.D. is equal to the amount of work done in moving a unit positive charge against electric field and current is defined as the rate of flow of charge.

From equations (1) and (2)

$$R = \frac{\rho l}{A}$$

where ρ is specific resistance of the material and l is the length of the conductor and A the area of cross-section.

We can have an analogy with Ohm's law as (*i*) the work done $\frac{4\pi Nl}{10}$ is called **magnetomotive force** (MMF) or the line integral of the magnetic field. This is analogous to e.m.f. in an electrical circuit. The magnetic flux ϕ is analogous to the current and $\frac{l}{A\mu}$ is analogous to resistance $\frac{l}{A\mu}$ is called magnetic resistance or reluctance or $\frac{l}{\mu}$ corresponds to **specific resistance.**

Equation (1) may be written as

$$\text{MMF} = \text{Flux} \times \text{Reluctance}$$

or

$$\text{Flux } \phi = \frac{\text{MMF}}{\text{Reluctance}}$$

where NI is expressed in ampere turns.

Flux in **Maxwells** and MMF in **Gilbert.**

However, the following dissimilarities in analogy of the above terms must be kept in mind.

(1) $\frac{1}{\mu}$ corresponds to specific resistance ρ and so μ will correspond to specific conductivity. But, permeability in magnetism corresponds to K, the dielectric constant or permittivity in electricity.

(2) The variation of reluctance of a magnetic circuit is considerable; however, the permeability depends on intensity of magnetisation of the material, whereas the resistance of a conductor does not depend on the magnitude of the current in the circuit.

(3) In the magnetic circuit some energy is required to establish the magnetic flux, whereas in an electrical circuit, the energy is spent so long as there is flow of current through the conductor.

SPECIAL CASES

(*i*) In case of a closed magnetic circuit of different cross-sections as shown in Fig. 3.9, the products B_1A_1, B_2A_2, B_3A_3 ...etc., at the planes P_1, P_2, P_3 ....etc. will be equal because in the absence of magnetic leakage, the total magnetic flux ϕ through any plane of the closed circuit must be the same because of the fact that flux is to have continuity.

(*ii*) If we have a magnetic circuit of different magnetic materials of different lengths and permeability, then if there is no magnetic leakage, the total reluctance in the circuit can be had as

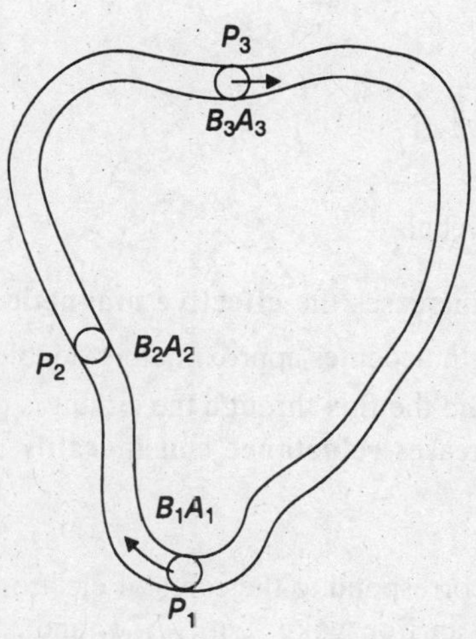

Fig. 3.12

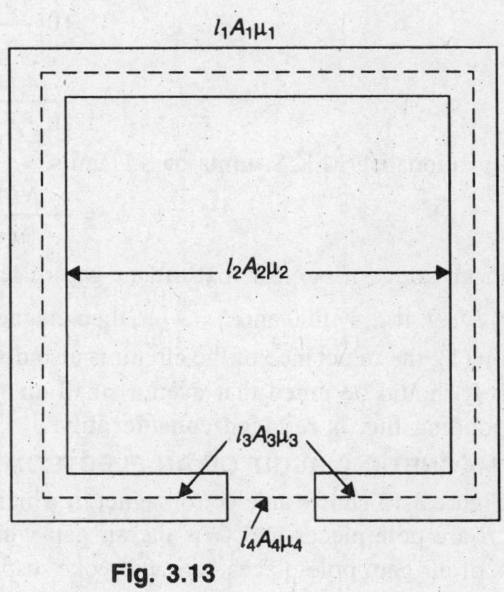

Fig. 3.13

$$\frac{l_1}{A_1\,\mu_1} + \frac{l_2}{A_2\,\mu_2} + \frac{l_3}{A_3\,\mu_3} + \frac{l_4}{A_4\,\mu_4}$$

In general, reluctance of the magnetic circuit may be written as

$\Sigma A\mu$ and total flux is given by $\phi = \dfrac{F}{\dfrac{l}{A\mu}}$

where F is M.M.F. (magnetic motive force).

3.22. MAGNETIC CIRCUIT OF AN IRON RING WITH A SMALL AIR GAP

Let the mean circumference of the ring be l and the air gap be of length l_1.

If the permeabilities of air and iron media are μ_0 and μ respectively, then reluctance

$$= \frac{l - l_1}{\mu} + \frac{l_1}{A\,\mu_0}$$

$$= \frac{l - l_1}{A\mu} + \frac{l_1}{A} \quad \text{[Since } \mu_0 = 1 \text{ C.G.S. units for air]}$$

$$= \frac{Al - Al_1 + A\mu l_1}{A.A\mu} = \frac{l_1 + l + \mu l_1}{A\mu}$$

$$= \frac{l + (\mu - 1)\, l_1}{A\mu} \quad ...(1)$$

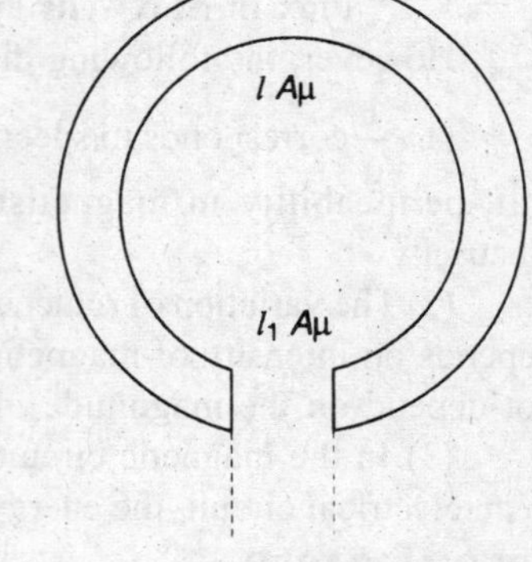

Fig. 3.14

Now magnetomotive force

$$F = \frac{4\pi\, NI}{10}$$

where N is the number of turns and I the strength of current in amperes through windings.

Then, we have

$$\text{Flux } \phi = \frac{F}{\text{Reluctance}}$$

$$= \frac{4\pi NI}{10\left[\dfrac{l + (\mu - 1)\, l_1}{A\mu}\right]}$$

$$\therefore \quad \phi = \frac{4\pi\, NIA\,\mu}{10\left[l + (\mu - 1)\, l_1\right]} \quad ...(2)$$

In rationalise M.K.S. units or S.I. units,

$$\phi = \frac{NIA\,\mu}{l + (\mu - 1)\, l_1} \text{ weber} \quad ...(3)$$

From above it is clear that the air gap of length l increases the **effective magnetic length** $(\mu_1 - 1)\, l_1$ e.g. if $\mu = 1000$ and $l_1 = \dfrac{1}{100}$, then magnetic length becomes approximately double. From equation (1), the reluctance of the circuit is also doubled and the flux through the circuit is reduced to half. It should be noted that even a small air gap increases **reluctance considerably** and the corresponding flux is **reduced considerably.**

3.23. MAGNETIC CIRCUIT OF AN ELECTROMAGNET

Figure 3.15 shows an electromagnet in which C, C correspond to the cores of electromagnet, A and B are pole pieces and G is the air gap while X is an iron yoke. Let l, l_1, l_2 and l_3 be the lengths of air gap, pole pieces core and yoke respectively as shown in Fig. 3.15 and if A_1, A_2, A_3 be the corresponding areas of cross-section of pole pieces, cores and yoke respectively, then the reluctance of the various parts of the magnetic circuit will be given as under:

$$\text{Reluctance of air gap} = \frac{R}{\mu_0 A_1}$$

$$\text{Reluctance of pole pieces} = \frac{2l_1}{\mu_1 A_1}$$

$$\text{Reluctance of core} = \frac{2l_2}{\mu_2 A_2}$$

$$\text{Reluctance of yoke} = \frac{l_3}{\mu_3 A_3}$$

where $\mu_0, \mu_1, \mu_2, \mu_3$ are permeabilities of air, pole pieces, cores and yoke respectively. So total reluctance of the circuit may be given as

$$\sum \frac{1}{A\mu} = \frac{l}{\mu_0 A_1} + \frac{2l_1}{\mu_1 A_1} + \frac{2l_2}{\mu_2 A_2} + \frac{l_3}{\mu_3 A_3}$$

Now magnetomotive force is given by

$$F = \frac{4\pi\, NI}{10}$$

and flux $\quad \phi = \dfrac{F}{\text{Reluctance}}$

or $\quad F = \phi \times \text{reluctance}$

$$= \phi\left[\frac{l}{\mu_0 A_1} + \frac{2l_1}{\mu_1 A_1} + \frac{2l_2}{\mu_2 A_2} + \frac{l_3}{\mu_3 A_3}\right]$$

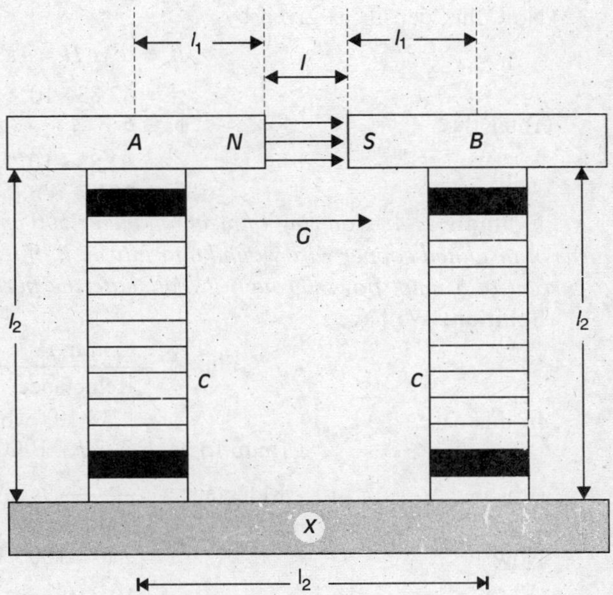

Fig. 3.15

Now for air $\mu_0 = 1$ and if pole pieces, yoke and core are of the same material, say, iron of permeability μ, we may have magnetomotive force as

$$F = \phi\left[\frac{l}{A_1} + \frac{2l_1}{\mu A_1} + \frac{2l_2}{\mu A_2} + \frac{l_3}{\mu A_3}\right]$$

It should be noted that the above calculations are approximately true because of leakage of flux in air gap. In case the air gap is very very small, the leakage of flux will be negligible and so the above equations become valid.

Example 1. *A solenoid consisting of 500 turns and carrying 4 ampere current is 0.5 m long. Calculate (i) magnetomotive force, (ii) Total flux if area of cross-section is 0.0008 m^2.*

Solution: In this case

Current flowing through the solenoid

$$I = 5 \text{ A}$$

Length of solenoid $= 0.5$ m, No. of turns of solenoid $N = 500$

Area of cross-section $= 4 \times 10^{-4}$ m

(*i*) Magnetomotive force (m.m.f.)

$$= \text{No. of turns} \times \text{current}$$
$$= 500 \times 5 = 2500 \text{ AT}$$

(*ii*) If air be the medium

$$\mu_r = 1$$

Magnetising force

$$H = \text{ampere turn/length}$$
$$= \frac{NI}{l} = \frac{2500}{0.5} = 5 \times 10^3 \text{ AT/m}$$

Now flux density is given by

$$B = \mu_0\mu_r H = 4\pi\times10^{-7}\times5\times10^3$$
$$= 62.83\times10^{-4}\ \text{Wb/m}^2$$

Total flux $\phi = B\times A$

$$= 62.83\times10^{-4}\times4\times10^{-4}$$
$$= \mathbf{251.3\ Wb.}$$

Example 2. *An iron ring of mean diameter 100 cm and 10 cm² cross-section area has 1000 turns of insulated copper wire wound informly on it. If it produces a flux of 2.5 × 10⁻³ webers when a current of 8 amp. flows through it, calculate the permeability of the medium.*

Solution: We have

$$\text{Flux } \phi = \frac{\text{m.m.f.}}{\text{Reluctance}} \quad \ldots (1)$$

In this case $\phi = 2.5\times10^{-3}$ wb

$$(\text{m.m.f.})\ M = N\times I = 1000\times8 = 8000\ \text{AT}$$
$$\text{Reluctance} = \frac{1}{\mu_0\mu_r A}\ \text{AT/Wb}$$

Now $l = 2\pi_r = 100\times10^{-2}\times\pi$ m

$$\mu = 10\times10^{-4}\ \text{m}^2$$
$$\mu_0 = 4\pi\times10^{-7}\ \text{Henry/m}$$

Putting in equation (1)

$$2.5\times10^{-3} = \frac{800}{\dfrac{\pi\times100\times10^{-2}}{(4\pi\times10^{-7}\ \mu_r\times10\times10^{-4})}}$$

$\therefore$ $\mu_r = \mathbf{781.25}$

Example 3. *The mean length of an iron ring is 50 cm and its area of cross-section is 4 cm². An air gap 1 mm in length is cut in the ring. If the ring is wound with 200 turns, what current is required to maintain a magnetic field of 10,000 gauss (μ for iron = 500 μ_0)?*

Solution: $\text{Flux } \phi = \dfrac{4\pi\ NI\ A\mu}{10\,[l+(\mu-1)\,l_1]}$

or $I = \dfrac{10\,\phi\,[l+(\mu-1)l_1]}{4\pi\ NA\ \mu}$

In this case

$\phi = 10{,}000$ gauss, $l = 50$ cm.

$l_1 = 0.1$ cm, $N = 200$ turns

$A = 4\ \text{cm}^2$, $\mu = 500$

$\therefore$ $I = \dfrac{10\times10000\,[50+(500-1)\times0.1]}{4\pi\times200\times4\times500}$

$= \mathbf{1.987\ Amperes.}$

3.24. SOFT AND HARD MAGNETS

Soft Magnets: Magnets are broadly divided into two categories:

(*i*) Soft magnets (*ii*) Hard magnets.

(*i*) A soft magnet can be magnetized and may be attracted by other magnets, but it retains **magnetism only in a magnetic field**. It cannot be permanently magnetized.

(*ii*) **Hard magnets.** A hard magnet retains **magnetism more or less permanently** and it affects the presence of other magnetic materials in its field. A hard magnet retains its **magnetic characteristics** even in the absence of magnetic field.

(*i*) Soft magnetic materials are used in applications requiring frequent **reversal of direction of magnetization**. In many applications, the hysteresis *losses are to be kept as low as possible.* This can be achieved in case of soft magnets as in this case for large induction corresponding to a small applied field, loop area is small and so the hysteresis loss is reduced. In the design of a soft magnet, materials are chosen which have **easily moving domain walls**.

Another source of energy loss in soft magnets is due to *eddy currents*. With the change of magnetic flux, an *e.m.f.* is induced which according to Lenz's law is proportional to rate of change of flux and consequently to the *frequency of alternating current*. This induced current set up eddy currents in the medium and power loss due to eddy currents is equal to $\frac{V^2}{R'}$ where V is the induced *e.m.f.* and R is the resistance of the medium.

Eddy current losses can be minimized by increasing the **resistivity of the magnetic medium**. **Iron**, which was used as core of a transformer, is now almost completely being replaced by **Fe-Si solid solution** with about 4% of Si, which has quite high resistivity, as compared to pure iron. However, concentration of Si more than 4% is not preferred as beyond this concentration, **iron becomes brittle**. *In addition, the transformer core is laminated, as the resistance of laminations is much more than that of a one piece core.*

In these days, **metallic glasses**, produced from iron base alloys, containing 15 to 25% Si + B + C offer substantial reduction in core-losses. Owing to larger concentration of impurity atoms, electrical resistivity is higher than that of Fe-Si (4%) alloy; which substantially reduces the eddy current losses.

For frequencies higher than **MHz**, metals and alloys are generally not suitable as soft magnets because of the high eddy current losses. However, thin films of **Permalloy** (45% Ni) of about 2000 Å in thickness, vapour deposited on a glass substrate, can be used. In this case, because of very low thickness, eddy current losses are reduced to bare minimum. **Ferri-magnetic oxides, ferrites and garnets** are quite suitable in high frequency ranges, as they being insulators will have much higher resistivity than alloys.

However, the choice of a particular soft magnet depends on its application. **Nickel-Zinc** Ferrites are used for *audio and TV transformers*, Magnetism-Manganese ferrites with high resistivity are used in microwave insolators in kHz to MHz range. With higher manganese to magnesium ratio, these are used in **memory core in computers**.

Recently, soft magnetic materials are used in the form of magnetic bubbles. Bubbles are magnetic domains of a few μm **diameter** which are embedded in a matrix of different spin orientation. All bubbles are arranged so that **they repel each other and maintain a constant distance between them**. These bubbles are the *electronic analogue* of the **magnetic disc memories** used in computers. Magnetic film in a disc memory is moved mechanically at high speed, while the bubbles move electronically in a bubble memory device.

3.25. HARD MAGNETIC MATERIALS

For permanent magnets, hard magnetic materials are used. In case of permanent magnets, hysteresis losses are of no significance, as no repeated reversals of magnetization is involved. Permanent magnets must have a high *residual induction* B_r and a *large coercive force* H_c. Area of hysteresis loop between B_r and H_c represents the energy required to demagnetize a permanent magnet. Maximum value of area is $B_r H_c$ and is called the *energy product*, which should be as large as possible in the case of permanent magnets.

High carbon-steels and low alloy tunston-chromium steels are used as permanent magnets. Some factors, which improve the mechanical hardness, provide better resistance to domain wall motion in permanent magnets. Among the metallic alloys, the best known are *Al-Ni-Co* (*alnico*) alloys.

Among the non-metallic oxides used as permanent magnets is Barrium Ferrite ($BaO_6Fe_2O_3$) which has a high coercivity (more than 100 kAm^{-1}). More recently, permanent magnets of cobalt alloyed with rare earth elements have been developed with a very large energy product in the range of 200 kJm^{-3}. $Nd_2F_{14}B$, another alloy discovered in 1984, has energy product ~ 400 kJm^{-3}.

Tables I (*a*) and I (*b*) are given with some of the important properties of soft and hard magnetic materials.

Table I (*a*). Properties of Soft Magnetic Materials

Material	*Initial relative permeability*	*Electrical resistivity, (ohm m)*	*Hysteresis loss* ($J\ m^{3}$)	*Saturation induction, (Wbm ^{2})*
Commercial iron,	250	0.1×10^{6}	500	2.2
Fe-4% Si	500	0.6×10^{-6}	100	2.0
Fe-Si oriented	1500	0.6×10^{6}	90	2.0
Permalloy (45% Ni)	2700	0.55×10^{6}	120	1.6
Supermalloy (79% Ni,, 5% Mo)	100000	0.65×10^{6}	21	0.8
Ni-Zn ferrite	200-1000	$\sim10^{6}$	35	0.4
Mn-Zn ferrite	200	$\sim10^{6}$	40	0.3

Table I (*b*). Properties of Hard Magnetic Materials

Materials	*Br Wbm* $^{-2}$	H_c *kAm* $^{-1}$	$B_r\ H_c$ *kJm* $^{-3}$
High carbon steel (martensitic)	0.90	3.98	3.58
Tungsten steel (5% W)	1.05	5.57	5.85
Chromium steel (4% Cr)	0.95	5.17	4.91
Cobalt steel (36% Co)	0.95	18.31	17.40
Al-Ni-Co alloys	0.8–1.2	60–120	48–144
Barium ferrite	0.21	140	29.4
Cobalt rare-earths Co_5 (Sm,, Pr)	1.0	200	200

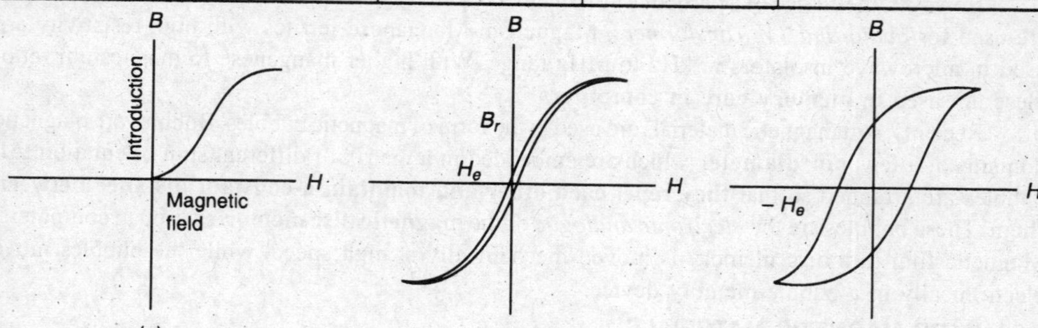

Fig. 3.16 Magnetization Curves. (*a*) Initial induction B versus magnetic field H. (*b*) Hysteresis loop (soft magnet). (*c*) Hysteresis loop (hard magnet). Both the remanent induction (fluxdensity) Bc, and coercive-field, He, are large for hard magnets.

The differences between permanent, or hard, magnet and soft magnets, are best shown with the hysteresis loop (Fig. 3.16).

3.26. TYPES OF MAGNETIC MATERIALS

We know that the magnetic flux produced by an electric current is proportional to the relative permeability of the materials through which the flux passes. As explained earlier, in case of **Ferromagnetic and Ferrimagnetic materials**, since they possess high permeabilities, so they are capable of producing high flux densities with lower currents. So, these materials are widely used as **flux paths and flux multipliers** in electrical machines. Transformers, motors, generators etc., *i.e.*, the rotating electrical machines, use magnetic materials or **core materials**.

Ferromagnetic materials are called *metallic magnets* while the **Ferrimagnetic** materials are called **non-metallic magnets**. The most common ceramic magnets are Ferrites with general composition of $[\,R\,Fe_2\,O_4\,]_8$; where indicated iron is Fe^{3+} and R may be Fe^{2+}, Ni^{2+}, Zn^{2+}, Mn^{2+} etc., *i.e.*, each ferrous ion has lost two electrons.

Nature's original magnet is magnelite (Fe_3O_4), formerly known as Lodestone Magnetite, is natural ceramic phase with an *fcc* lattice of O^2 ions.

Table II. Units, Constants and Conversion Factors used in Magnetism

Quantity		*SI units*	*Other units*
Magnetic induction B Saturation induction B_s Residual induction B_r	$Wb\ m^{2}$ or T	(weber per square metre) (tesla)	gauss
Magnetic field strength H Coercive field H_c Magnetization M	$A\ m^{1}$	(ampere per metre)	oersted
Magnetic permeability μ	$H\ m^{1}$	(henry per metre)	—
Relative permeability μ_r		—	—
Magnetic susceptibility χ	—	—	—
Magnetic moment m	Am^2	(ampere metre^2)	—
Energy product B_rH_c	$J\ m^{-3}$	(joule per cubic metre)	—
Hysteresis loss Eddy current-loss	W	watt	—
Frequency of alternating current	Hz	hertz	sec^{-1}
Permeability of free space	μ_0	$H\ m^{-1}$	
Conversion factors 1 gauss = 10^{-4} wb m^{-2} 1 oersted = 79.6 Am^{-1}			

Example 4. *Magnetic susceptibility of a medium is 948×10^{-11} Calculate the permeability and relative permeability.*

Solution: We have the relation for magnetic susceptibility and relative permeability μ_r as

$$\mu_r = 1 + \chi$$

In this case $\chi = 948 \times 10^{-11}$

$$\therefore \quad \mu_r = 1 + 948 \times 10^{-11} \approx 1 \text{ or slightly more than } 1$$

So, μ_r is slightly greater than 1.

Now, relative permeability μ_r is related to permeability of medium (μ) and permeability of free space μ_0 as

$$\mu_r = \frac{\mu}{\mu_0}$$

$$\therefore \quad \mu = \mu_0 \cdot \mu_r$$

Now, $\mu_0 = 4\pi \times 10^{-7}$ H/m

$$\mu = 1 \times 4\pi \times 10^{-7}$$
$$= 4\pi \times 10^{-7} \text{ H/m} \quad \textbf{Ans.}$$

Example 5. *Estimate the number of ampere-turns necessary to produce a flux of 0.001 Wb around an iron of 6 cm^2 cross-section and 20 cm mean diameter, having an air gap of 2 mm. The permeability of iron is 1200 μ_0.*

Solution: Let l_i be the length of the iron ring, and l_a that of air gap. So, by ampere-turns necessary to produce a flux ϕ_B around a ring of area of cross-section A is given by

Number of amp-turn, $\quad \text{Ni} = \phi_B\left(\frac{l_a}{\mu_0 A}+\frac{l_i}{\mu A}\right)$

But $\quad \mu = 1200\,\mu_0\,\text{(given)} \qquad \therefore \text{Ni} = \frac{\phi_B}{\mu_0 A}\left(la_a + \frac{l_1}{1200}\right)$

Now, $\quad \mu_0 = 4\pi\times10^{-7}$ H/m, $A = 6cm^2 6\times10^{-4}$ m^2, $\phi_B = 0.001$ Weber

$$r = \frac{20}{2} = 10 \text{ cm} = 0.1 \text{ m}$$

$$l_a = 2\text{mm} = 2\times10^{-3}\text{m and } l_i = 2\pi r - l_a = 2\times3.14\times0.1 - 2\times10^{-3} = 0.626$$

$$\therefore \quad \text{Ni} = \frac{0.001}{4\times3.14\times10^{-7}\times6\times10^{-4}}\left(2\times10^{-3}+\frac{0.626a}{1200}\right) = 1.327\times10^{6}(2.5217\times10^{-3})$$

$$= \mathbf{3346\ Ans.}$$

Example 6. *Find the total reluctance and permeance between the ends of a parallel connected rectangular iron blocks of lengths* $l_1 = l_2 = 4$ *cm, width* $w_1 = w_2 = 5$ *cm and thickness* $h_1 = h_2$ $= 4$ *cm. Given that* $\mu_1 = 1000\,\mu_0$ *and* $\mu_2 = 2000\,\mu_0$.

Solution: Two rectangular blocks have identical dimensions. Let l is the length and w the width of each slab and μ_1 and μ_2 their permeabilities, then their reluctance will be given by

$$R_1 = \frac{1}{\mu_1 A} \text{ and } R_2 = \frac{1}{\mu_2 A}, \text{ where } A = l\times w$$

In this case, $\quad l_1 = l_2 = l = 20 \text{ cm} = 20\times10^{-2}$ m, $w_1 = w_2 = w = 5$ cm $= 5\times10^{-2}$ m

So, $\quad A = 20\times10^{-2}\times5\times10^{-2} = 100\times10^{-4}$ m^2 and $\mu_0 = 4\pi\times10^{-7}$ H/m

$$R_1 = \frac{20\times10^{-2}}{1000\,\mu_0 A} = \frac{20\times10^{-2}}{1000\times4\times3.14\times10^{-7}\times100\times10^{-4}} = \mathbf{7961.78}$$

$$R_2 = \frac{20\times10^{-2}}{2000\times4\times3.14\times10^{-7}\times100\times10^{-4}} = 7961.78$$

As the blocks are connected in parallel, so their resultant reluctance R is given by

$$\frac{1}{R} = \frac{1}{R_1}+\frac{1}{R_2} = \frac{1}{15923.57}+\frac{1}{7961.78} = 1.8844\times10^{-4}$$

$$\therefore \quad \boldsymbol{R} = \mathbf{5307.85\ ampere\text{-}turns/weber\ Ans.}$$

Example 7. *A solenoid consisting of* 100 *turns and carrying* 10 *amp current is* 0.5 *m long. Calculate (a) magnetomotive force and (b) total flux, if area of cross-section is* 0.0004 *m*^{2}.

Solution: (*a*) Magnetomotive force is given by,

$$mmf = Ni$$

In this case, $\quad N = 100$ and $i = 10$ amp

$\therefore \quad mmf = Ni = 100\times10 = 1000$ **Amp.-turns**

(*b*) Total flux, $\quad \phi_B = B\,.A$, but $B = \mu_0 H$ (with air as medium, *i.e.*, $\mu_r = 1$)

Also $\quad Hl = Ni$, So $H = Ni/l$, put $l = 0.5$ m

$$\therefore \quad H = \frac{1000}{0.5} = 2000 \text{ amp-turns/m}$$

and $\quad B = \mu_0 H = 4\times3.14\times10^{-7}\times2000 = 2.512\times10^{-3}$ weber/m^2

$$\phi_B = 2.512\times10^{-3}\times0.0004 = \mathbf{1.005\times10^{-6}\ weber\ Ans.}$$

Example 8. *Calculate the number of ampere-turns required to produce a flux of* 10^{-3} *weber round an iron ring of* 5 *cm*^{2} *cross-section and* 20 *mm mean diameter having an air gap* 2 *mm wide across it. The relative permeability of the iron may be taken to be* 1000. *Neglect the leakage flux.*

Solution: As the number of ampere-turns per unit length is equal to the magnetising force H.

$\therefore$ $$H = Ni$$

If l is the length of the ring, then ampere-turn,

$$AT = H \times l$$

In this case, $\phi_B = 10^{-3}$ weber and $A = 5\text{ cm}^2 = 5\times10^{-4}\text{ m}^2$

$\therefore$ Flux density, $$B = \frac{\phi_B}{A} = \frac{10^{-3}}{5\times10^{-4}} = 2\text{ weber/m}^2$$

$\therefore$ Ampere-turn $$= Hl = \frac{Bl}{\mu} = \frac{Bl}{\mu_0\mu_r}$$

In case of air gap, $\mu_r = 1$ and $l = 2\text{ mm} = 2\times10^{-3}\text{ m}$

$\therefore$ Ampere-turn for air gap $$= \frac{Bl}{\mu_0} = \frac{2\times2\times10^{-3}}{4\pi\times10^{-7}} = 3184.7$$

In the same way, ampere-turn for iron path:

$$\frac{Bl}{\mu_0\mu_r} = \frac{2\times2\pi r}{4\pi\times10^{-7}\times1000} = \frac{2\times2\pi\times10\times10^{-3}}{4\pi\times10^{-7}\times1000} = 100$$

So, total ampere-turn = AT for air gap + AT for iron path = 3184.7 + 100 = **3284.7 Ans.**

Example 9. *An iron ring of mean radius* 10 *cm and area of cross-section* 5 *sq. cm has an air gap* 1 *cm wide. The ring is wound uniformly with coil of* 900 *turns. A flux of* 10^{-4} *Wb is required in the air gap. Find the strength of current required to produce the flux.* Given $\mu_{iron} = 6.5\times10^{-4}$ H/m

Solution: An iron ring is a combination of iron and air gap. Let l_i and l_a be the length of iron of magnetic circuit and air gap respectively. If μ is the permeability of iron and μ_0 that of air and A the area of cross-section of air gap as well as of iron ring, then from magnetic circuit equation, the ampere turn is given by the relation

Ampere-turn, $$\text{Ni} = \phi_B\left(\frac{l_a}{\mu_0 A} + \frac{l_i}{\mu A}\right) = \frac{\phi_B}{A}\left(\frac{l_a}{\mu_0} + \frac{l_i}{\mu}\right)$$

where $i \to$ current through the core $\phi_B \to$ flux required in air gap or iron ring $N \to$ No of turns

Now, $\phi_B = 10^{-4}$ weber, A = 5 sq. cm = 5×10^{-4} sq. m, l_a=1 cm=10^{-2} m

$l_i = 2\pi r = 2\times3.14\times10\times10^{-2}$, $\mu_0 = 4\pi\times10^{-7}$ and $\mu = 65\times10^{-9}$ H/m

So $$Ni = \frac{10^{-4}}{5\times10^{-4}}\left(\frac{10^{-2}}{4\times3.14\times10^{-7}} + \frac{2\times3.14\times10\times10^{-2}}{65\times10^{-4}}\right)$$

or $$Ni = 0.2\,(7961.78 + 96.615)$$

Hence $$i = \frac{0.2\times8058.4}{900} = \mathbf{1.79\text{ amp Ans.}}$$

Example 10. *An iron rod of length* 50 cm *and cross-sectional area* 4 cm^2 *is in the form of a closed circular ring. If the permeability of iron is* 65×10^{-4} *H/m, calculate the number of ampere-turns required to produce a flux of* 4×10^{-4} *Wb.*

Solution: Number of ampere-turns (Ni) required to produce a flux ϕ_B by a circular ring of length l and area of cross-section A is given by the relation

$$\text{Ni} = \phi_B\frac{l}{\mu A},$$ where μ is the permeability of iron.

Now, $\phi_B = 4\times10^{-4}$ Weber, $l = 50\text{ cm} = 50\times10^{-2}$ m, $A = 4\text{ cm}^2 = 4\times10^{-4}$ m

and $\mu = 65\times10^{-4}$ H/m

$$Ni = \frac{4\times10^{-4}\times50\times10^{-2}}{65\times10^{-4}\times4\times10^{-4}} \cong \mathbf{77\text{ Ampere turns Ans.}}$$

Example 11. *An iron ring has cross-sectional area* 10 cm^2 *and an air gap of width* 2 *mm and mean length* 60 *m including the air gap. Find the number of ampere-turn required to produce the flux density* 1 *Wb/m2. Given* $H = 10^{-3}$ H/m.

Solution: Ampere turn Ni is given by

$$\text{Ampere-turn Ni} = \phi_B\left(\frac{l_a}{\mu_0 A}+\frac{l_i}{\mu A}\right)$$

where l_a is the width of air gap, l_i the length of iron ring excluding air gap. Area of cross-section, μ_0 the permeability of air gap and μ that of iron.

In this case $l = 2\text{ mm} = 2\times10^{-3}$ m, $l_i = 60-2\times10^{-3} = 59.998$ m, B = 1 Wb/m^2

$$\text{Å} = 10\text{ cm}^2 = 10\times10^{-4}\text{ m}^2, \mu_0 = 4\pi\times10^{-7} \text{ and } \mu = 10^{-3}\text{ H/m}$$

$$\therefore \quad \text{Ampere-turn, Ni} = B\left(\frac{l_a}{\mu_0}+\frac{l_i}{\mu}\right) \text{(As } B = q/A)$$

$$= 1\left(\frac{2\times10^{-3}}{4\times3.14\times10^{-7}}+\frac{59.998}{10^{-3}}\right)$$

$$= 1592.357 + 59.998\times10^3$$

$$\mathbf{= 61590 \text{ Ans.}}$$

LANGEVIN THEORY OF DIAMAGNETISM

Diamagnetic property is generally found in substances in which the atoms have even number of electrons in their outermost orbits so that due to mutual opposite spins, the net magnetic moment of each atom is zero.

According to Langevin, in the absence of external magnetic field, each atom of a magnetic substance has a **vector sum** of magnetic moments of all the electronic orbits equal to zero.

When a diamagnetic substançe is subjected to an external magnetic field, each electron experiences a **Lorentz Force** in a direction at right angles to its motion (*i.e.* the force is in the radial direction) due to which the angular velocity of the electron changes, which will consequently result in the change of the **magnetic moment** of the electron.

It will mean that in the presence of external magnetic field. The magnetic moment of the atom is not zero, but a net magnetic moment is **induced** in the atom in a **direction opposite to the external magnetic field.**

Let an electron of *charge e* and *mass* m_e be moving in a circular orbit of *radius* r_0 with an *angular velocity* ω_0. Current equivalent to this motion is given by $\frac{C}{T} = e.v$.

or $$i = \frac{e\omega_0}{2\pi}.$$

Magnetic moment of the electron

$$M = i.a = \frac{e\omega_0}{2\pi}.\pi r_0^2 = \frac{1}{2}.e.\omega_0.r_0^2 \quad ...(1)$$

centripetal force acting on the electron

$$F_e = m_e\omega_0^2 r_0 \quad ...(2)$$

Now if an external magnetic field B is applied normal to the plane of the elecuon orbit, we have from the Fleming's left hand rule, a Lorentz force $F_m = Bev = B.e.\ r_0\omega_0$ acting radially on the electron.

In case the directions of the magnetic moment of electron and that of the magnetic field are same, their force will be radially outwards (*i.e.* away from the nucleus of the atom). If the directions of magnetic moment of the electron and that of magnetic field are opposite, this force will be radially inwards (*i.e.* towards the nucleus of the atom) as shown in the figure.

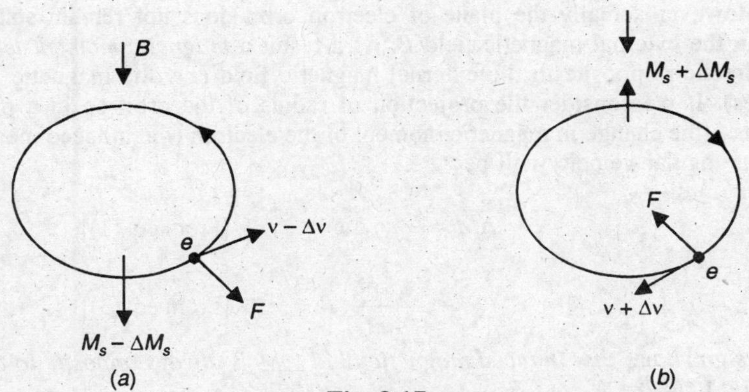

Fig. 3.17

So the total force on the electron due to the presence of magnetic field is given by

$$F = F_e + F_m = m_e\, \omega_0^2\, r_0 \pm B.e.\, r_0\omega_0$$

This force will be equal to the centripetal force acting on the electron in the presence of the external magnetic field. If ω be the angular velocity of electron in the presence of external field, then

$$m_e\, \omega_0^2\, r_0 \pm B.e.\, r_0\omega_0 = m_e\omega^2 r_0.$$

or
$$\omega^2 - \omega_0^2 = \pm \frac{B.e.\omega_0}{m_e}$$

or
$$(\omega - \omega_0)(\omega + \omega_0) = \pm \frac{B.e.\omega_0}{m_e}$$

or
$$(\omega - \omega_0)(2\omega_0) \simeq \pm \frac{B.e.\omega_0}{m_e}. \qquad [\because \omega \simeq \omega_0]$$

or
$$\Delta\omega = \omega - \omega_0 \simeq \pm \frac{B.e}{2m_e} \qquad ...(3)$$

where $\Delta\omega = \omega - \omega_0$ is the change in angular velocity of the electron due to the applied magnetic field.

It shows that the angular velocity of the electron either increases or decreases by an amount $\Delta\omega$ depending upon the direction of external magnetic field and the direction of the electron spin.

This change in angular velocity $\Delta\omega$ is called *Larmour Frequency*. Its value is very small as compared to ω_0.

[As an example if $B = 10$ wb/m^2

$$\Delta\omega = \frac{B.e}{2m_e} = \frac{10 \times 1.6 \times 10^{-19}}{2 \times 9.1 \times 10^{-31}}$$
$$= 8.79 \times 10^{11} \text{ radians/sec.}$$

In the absence of external magnetic field, from Bohr's model we have

$$\frac{e^2}{4\pi \in_0 r_0^2} = m_e\omega_0^2\, r_0.$$

$$\therefore \quad \omega_0 = \sqrt{\frac{e^2}{4\pi \in_0 m_e\, r_0^2}} = \sqrt{\frac{9 \times 10^9 \times \left(1.6 \times 10^{-19}\right)^2}{9.1 \times 10^{-31} \times \left(0.5 \times 10^{-3}\right)^2}}$$
$$= 4.5 \times 10^{16} \text{ radians/sec}$$

which shows $\Delta\omega << \omega_0$]

Upto now we have assumed that the angular momentum of the electron can be in any direction w.r.t. an external magnetic field *i.e.* the electron orbit is not *deformed* due to an external applied field. However actually the plane of electron orbit does not remain stationary, it does *precession* about the external magnetic field B. As a result a *magnetic moment is induced* in the electron in a direction opposite to the external magnetic field (*i.e.* the magnetic moment of the electron changes). If we consider the projection of radius of the orbit so in a plane normal to magnetic field as r, the change in magnetic moment of the electron (*i.e.* induced magnetic moment) due to change in angular velocity will be

$$\Delta M = \frac{er^2}{2}.\Delta\omega \qquad \text{[From eq. (1)]}$$

$$= -\frac{e^2 r^2}{4m_e}.B \qquad \text{[From eq. (3)]}$$

where negative sign shows that induced magnetic field ΔM is always opposite to the direction of external magnetic field B.

If there are Z electrons in the atom (*i.e.* Z is the atomic number of the atom), total induced magnetic moment in the atom will be

$$\Delta M = -\frac{Z e^2}{4m_e}.B.\overline{r^2} \qquad ...(4)$$

where $\overline{r^2}$ is the mean value of the square of the projections of radii of orbits of electrons in a plane $\perp$ to the magnetic field. If the co-ordinates of a point of orbit of radius r_0 all x, y, z, then

$$r_0^2 = \overline{x}^2 + \overline{y}^2 + \overline{z}^2$$

Now the distribution of electrons orbiting in the atom is spherically symetric . So

$$\overline{x^2} = \overline{y^2} = \overline{z^2}$$

If the applied magnetic field is along Z-axis,

Then, $$r^2 = \overline{x^2} + \overline{y^2} = 2\overline{x^2} = \frac{2}{3}r_0^2$$

From eq. (4) total induced magnetic moment of the atom

$$\Delta M = -\frac{Z e^2 B}{4\,m_e} \times \frac{2}{3} r_0^2 = \frac{Z e^2 B r_0^2}{6\,m_e} \qquad ...(5)$$

If the number of atoms/unit volume of the substance is n, the induced magnetic moment/unit volume of the substance (*i.e.* intensity of magnetization induced in the substance) is given by

$$I_{induced} = -\frac{nZ e^2 .r_0^2}{6m_e}.B.$$

But $B = \mu H$ where μ is the absolute permeability of the substance

$$\therefore \qquad I_{induced} = -\frac{n.Z.e^2.r_0^2}{6m_e}\mu H. \qquad ...(6)$$

Also intensity of magnetization is given by $I \propto H$ or $I = \chi H$.

Where χ is the magnetic susceptibility of the substance. So magnetic susceptibility per unit volume of the diamagnetic substance

$$\chi = \frac{I_{induced}}{H} = -\frac{nZ e^2 .r_0^2 .\mu}{6m_e} \qquad ...(7)$$

[If ρ is the density of substance and A its Atomic or Molecular weight, then mass susceptibility $\chi_{mass} = \chi/\rho$ and Atomic or Molecular susceptibility $\chi_{molar} = \frac{\chi}{\rho}.A$]

From eq. (7) we may conclude that

(*i*) Magnetic susceptibility of a diamagnetic substance does not depend on the **temperature and the external applied field**.

(*ii*) The diamagnetic susceptibility is always negative which implies that although in the absence of external magnetic field, net magnetic moment of an atom is zero, but with the application of an applied magnetic field a magnetic moment is induced on each atom in a direction opposite to the applied magnetic field resulting in a weak magnetism in direction opposite to applied magnetic field.
Which means that diamagnetism is the inherent property of some substances.

3.28. LANGEVIN'S THEORY OF PARAMAGNETISM (CLASSICAL LANGEVIN'S THEORY)

Classical theory of paramagnetism was developed by **Paul Langevin** in 1905. He was a French physicist. Atoms or molecules of a paramagnet behave as dipoles. However, their random orientation do not impart bulk magnetization of matter. When a paramagnet is subjected to some external magnetic field, the *magnetic dipoles* tend to orient themselves *parallel* to the applied magnetic field. However, this tendency is opposed due to random agitation of molecules. In the equilibrium state, the substance acquires some *net magnetic moment* in the direction of applied magnetic field. Since, at low temperatures random agitation of molecules is very less, so even a weak field can align a large number of molecules parallel to field. At high temperature, vigorous random motion of molecules opposes strongly the alignment of molecules, hence a very strong magnetic field is required for **inducing magnetism**.

It was discovered by *Curie* that magnetic susceptibility χ is inversely proportional to temperature T, *i.e.*, $\chi \propto \frac{1}{T}$. This is called *Curie's law of paramagnetism*.

We can write $\chi = \frac{C}{T}$, where C is called Curie constant.

Langevin used Maxwell-Boltzmann classical statistics to give a theoretical explanation of Curie law. In magnetic field B, an atomic dipole acquires extra energy E given by the relation

$$E = -\mu B \cos\theta \qquad \text{... (1)}$$

where θ is the angle between the magnetic field and dipole axis. According to classical statistics the number of dipoles with energy E and orientations between the angles θ and $\theta + d\theta$ is given by the relation

$$dn = ce^{-E/kT} \sin\theta \, d\theta, \qquad (C \text{ is a constant})$$

$$= ce^{\mu B \cos\theta / kT} \sin\theta \, d\theta \qquad \text{... (2)}$$

where T is temperature of the specimen. The total number of dipoles in the specimen will be

$$n = c\int_0^{\pi} e^{\mu B \cos\theta / kT} \sin\theta \, d\theta \qquad \text{... (3)}$$

Since the component of magnetic moment parallel to the magnetic field is $\mu \cos\theta$. So, we have the average magnetic moment as

$$\bar{\mu} = \frac{\text{Total magnetic moment parallel to the field}}{\text{Total number of dipoles}}$$

$$= \frac{\int \mu \cos\theta \, dn}{n} = \frac{\int_0^{\pi} \mu \cos\theta \, . \, e^{\mu B \cos\theta / kT} \sin\theta \, d\theta}{\int_0^{\pi} e^{\mu B \cos\theta / kT} \sin\theta \, d\theta}$$

Put $\frac{\mu B}{kT} = a$ and changing variable θ to x by putting $x = \cos\theta$, we get

$$\bar{\mu} = \mu \frac{\int_{-1}^{+1} x \, e^{ax} dx}{\int_{-1}^{+1} e^{ax} dx} = \mu \frac{d}{da}\left[\log \int_{-1}^{+1} e^{ax} dx\right]$$

$$= \mu \frac{d}{da}\left[\log(e^a - e^{-a}) - \log a\right]$$

$$= \mu\left(\frac{e^a + e^{-a}}{e^a - e^{-a}} - \frac{1}{a}\right)$$

$$= \mu\left(\coth a - \frac{1}{a}\right)$$

$$= \mu L(a) \qquad \text{... (4)}$$

where, we have put

$$L(a) = \coth a - 1/a \qquad \text{... (5)}$$

$L(a)$ is called *Langevin function*. If N be the number of dipoles per unit volume, the magnetization will be

$$M = N\bar{\mu} = N\mu\, L(a) \qquad \text{... (6)}$$

The expression for paramagnetic susceptibility χ is given by

$$\chi = \frac{\mu_0 M}{B} = \frac{\mu_0 \mu\, N\, L(a)}{B} \qquad \text{... (7)}$$

If Langevin function $L(a)$ is plotted as a function of $a\ (= \mu B / kT)$ (shown in Fig. 3.14), we find for small values of a, *i.e.*, at high temperature the curve is linear and

$$L(a) \approx \frac{a}{3} = \frac{\mu B}{3kT} \qquad \text{... (8)}$$

Under this approximation the susceptibility will be

$$\chi = \frac{\mu_0 \mu^2 N}{3kT} = \frac{C}{T}$$

where $$C = \frac{\mu_0 \mu^2 N}{3k} \qquad \text{... (9)}$$

which is the Curie's law of paramagnetics m.

For large values of $a = \mu B / kT$, *i.e.*, at low temperature, Langevin's function tends to unity. From equation (6) we see that under this condition the magnetization gets maximum value. It is due to the fact that at low temperatures the thermal motions of atoms are *small* and a weak field is capable of aligning most of the magnetic dipoles. This state is called the *saturution state.*

It should be noted that Langevin's theory as such can only be applied to gases where magnetic *dipoles are sufficiently far apart and their mutual interactions are negligibly small.* In case of solids and liquids, the interactions of magnetic dipoles are large and they have to be considered. For solids and liquids the susceptibility is given by

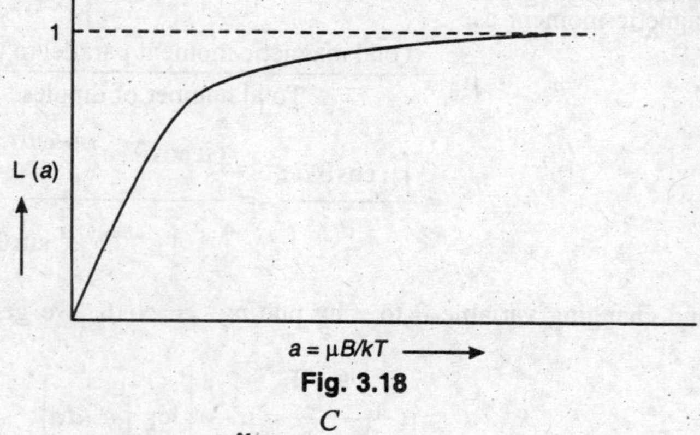

Fig. 3.18

$$\chi = \frac{C}{T - \Theta} \qquad \text{... (10)}$$

where Θ is a constant, known as *Weiss constant.*

3.29. FERROMAGNETISM

Substances which can have magnetization even in the absence of an external magnetic field are called *Ferromagnetics. Nickel, Cobalt, Gadolinium* and *Iron* fall under this class. Ferromagnetics are very strong magnetic substances and their magnetization is much more than that of dia and paramagnets.

Gyromagnetic phenomenon indicates clearly that intrinsic (spin) magnetic moments of electrons are responsible for magnetic properties of ferromagnetic materials. In this case, study of exchange forces which can only be explained by *quantum mechanics* appear in ferromagnetics and cause magnetic moments of electrons to line up parallel to each other.

This alignment of magnetic movements occurs only in very small regions, called *domains* which have dimensions ~ 1 to 10 μm within each *domain*, a ferromagnetic material is spontaneously magnetized to *saturation* and attains a *definite magnetic moment.*

The directions of these moments are random in all the domains and so the total magnetic moment of a piece of ferromagnetic material is zero in the *absence of an external magnetic field.*

If an external magnetic field is applied, increase in magnetization takes place in two different ways. First, when the field is weak, the volumes of *domains*, whose magnetic moments make smaller angles (θ) with the applied field, increase at the cost of domains whose moments make larger angles.

As the applied magnetic field strength is increased, the process goes on till the *domains with smaller angle* θ *completely absorb the domains with larger angle* θ. In next stage, magnetic moments of domains turn in the direction of the field. The moments of the electrons within the domains turn simultaneously without upsetting their parallel orientations. These processes are *irreversible* and so can explain the reason of *hysteresis in ferromagnetics.*

If the temperature of a ferromagnetic material is increased from some low value, then at some temperature T_c, called *Curie temperature*, it loses ferromagnetic property and becomes an ordinary paramagnetic material whose susceptibility obeys Curie-Weiss law

$$\chi = \frac{C}{T - T_c}$$

WEISS THEORY OF FERROMAGNETISM

In 1907, Pierre Weiss put forward the theory of ferromagnetism. In order to explain the spontaneous magnetization of ferromagnetics, he postulated the existence of internal field which is responsible for alignment of magnetic dipoles even when external magnetic field is not present.

At that time Weiss did not know the exact mechanism of internal field. However, twenty years later, Heisenberg *quantum mechanically* explained the origin of internal field. Weiss found that the internal field H_I, also called *Weiss molecular field*, is proportional to the magnetization *M*, *i.e.*, $H_I \propto M$ or $H_I = \gamma M$, where γ is constant and has a value of the order of 10^4

Effective magnetic field H_m acting on magnetic dipole is

$$H_m = H + \gamma M \quad \text{...(1)}$$

where *H* corresponds to external field.

If we take into consideration the Weiss Internal Field and apply *Langevin's theory of paramagnetism* to ferromagnetism, we get an expression for magnetization as

$$M = N\mu\left[\coth a - \frac{1}{a}\right] = N\mu\, L(a) \quad \text{...(2)}$$

where $$a = \frac{\mu B}{kT} = \frac{\mu_0 \mu H_m}{kT} = \frac{\mu_0 \mu (H - \gamma M)}{kT}$$

At very low temperatures $a >> 1$, so **Langevin's function** *L* (*a*) tends to 1 and magnetization *M* assumes its saturation value $M = N\mu$.

Spontaneous Magnetiztion: We find that even in absence of external magnetic field (*H* = 0), the magnetization in a domain of ferromagnetic material has saturation value M_S. In zero external field (*H* = 0), let us denote the value of *a* by a'. So, we have

$$a' = [a]_{H=a} = \frac{\mu_0 \mu \gamma M}{kT} \quad \text{...(3)}$$

which gives, $$M = \left(\frac{kT}{\mu_0 \mu \gamma}\right) a'$$

or $$\frac{M}{M_0} = \left(\frac{kT}{\mu_0 \mu^2 N \gamma}\right) a' \qquad \dots (4)$$

From equation (2), the spontaneous magnetization is given by the relation

$$\frac{M}{M_0} = L(a') \qquad \dots (5)$$

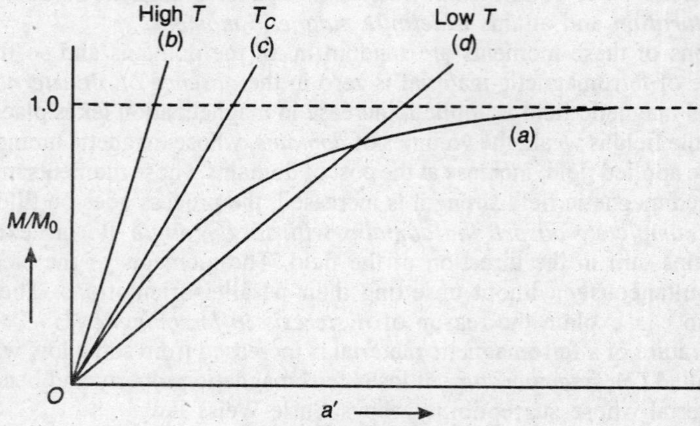

Fig. 3.19

The general behaviour of the magnetization can be found from graph. To get a graphical solution we plot the two functions M/M_0 given by equations (4) and (5) against a as shown in Fig. 3.19. Equation (5) gives a straight line that passes through the origin and intersects the *Langevin's function* at this point. It means that one possible value of magnetization is always zero.

From equation (4), we find that the slope of the line is proportional to temperature T. So, a *high temperature*, we have line like b where solution is $\frac{M}{M_0} = 0$, *i.e.*, the substance at high temperature is unmagnetized in zero external field.

As the temperature falls, the slope of line given by equation (4) decreases till at some critical value of temperature T_c, it is tangential to Langevin curve (a) at the origin. This critical temperature is called Curie temperature T_c.

From the curve,

Slope of straight line = Slope of tangent to Langevin curve at origin

$$\frac{kT_c}{m_0 N \mu^2 \gamma} = \frac{d}{da'}[L(a')]_{a' \to 0}$$

$$= \frac{d}{da'}\left(\frac{a'}{3}\right)$$

$$= \frac{1}{3}$$

So, we have $$T_c = \frac{\mu_0 N \mu^2 \gamma}{3k} \qquad \dots (6)$$

The quantity r can be written in terms of measurable quantities of material as

$$\gamma = \frac{3kT_c}{\mu_0 N \mu^2} = \frac{3kT_c \, N \mu_0}{(\mu_0 M)^2} \qquad \dots (7)$$

In case of iron, $T_c = 1043\,K, \mu_0 M_0 = 2.15$ T.

So, value of $\gamma = 995$

At lower temperatures ($T < T_c$), the slope of line is less than the initial slope of *Langevin* curve and there will be two points of intersection and two values of M/M_0. One value of $\frac{M}{M_0}$ being 0 and the other being nearly 1. It is seen that only the upper solution is stable, *i.e., magnetic material magnetizes itself only at sufficiently low temperatures.* This in other words means that when thermal motions are small, coupling between various atomic magnetic allows them to become *parallel to each other.*

3.30. TEMPERATURE DEPENDENCE OF SPONTANEOUS MAGNETIZATION

As seen in the graph (Fig. 3.20) below Curie temperature T_c spontaneous magnetization is given by the intersection of Langevin Curve (curve *a*) and straight line (curve *d*). If M/M_0 is solved for various temperatures by changing the slope of line (*d*) we obtain a theoretical curve given in Fig. 3.20.

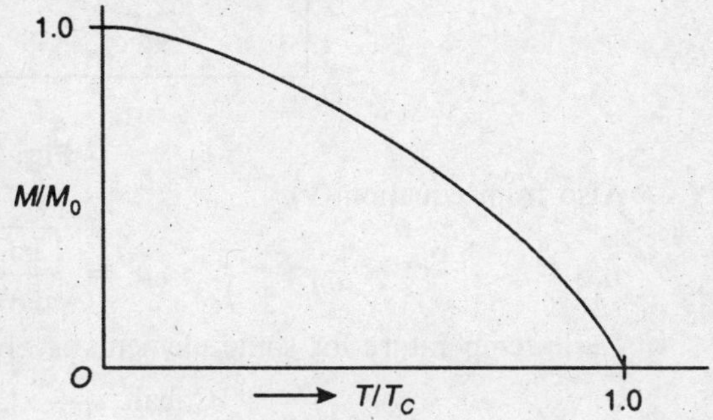

Fig. 3.20

It should be noted that we will get same form of curve for all ferromagnetics for which atomic magnetic moment comes from a single electron.

The curve for other materials will be slightly different.

From equation (4), value of M/M_0 can be written as

$$\frac{M}{M_0} = \frac{kT}{\mu_0 N\mu^2\gamma} \cdot a' = \frac{a'}{3}\left(\frac{T}{T_c}\right) \qquad \text{... (8)}$$

3.31. CURIE-WEISS LAW

At temperature $T > T_c$ and for a weak field $a << 1>$ and $L(a) \approx a/3$, magnetization of the specimen is given by

$$M = N\,\mu\,(a/3) = N\,\mu\left[\frac{\mu_0\mu(H+\gamma\,M)}{3kT}\right]$$

or
$$M\left[1-\frac{\mu_0 N\mu^2\gamma}{3kT}\right] = \frac{\mu_0 N\mu^2 H}{3kT}$$

$\Rightarrow$
$$M\left[1-\frac{T_c}{T}\right] = \frac{CH}{T} \text{ where } T_c = \frac{\mu_0 N\,\mu^2\gamma}{3k} \text{ and } C = \frac{\mu_0 N\,\mu^2}{3k} \qquad \text{... (9)}$$

or
$$M = \frac{CH}{T-T_c}$$

Now,
$$\chi = \frac{M}{H}$$

$\therefore$
$$\chi = \frac{M}{H} = \frac{C}{T-T_c} \qquad \text{... (10)}$$

Equation (10) is Curie-Weiss law. Constant C is called Curie constant and T_C the Curie temperature.

A plot of $1/\chi$ against T will be a straight line. The slope of this line equals the reciprocal of Curie constant C. From the measurement of χ at different temperatures, the magnetic moment of various dipoles can be calculated.

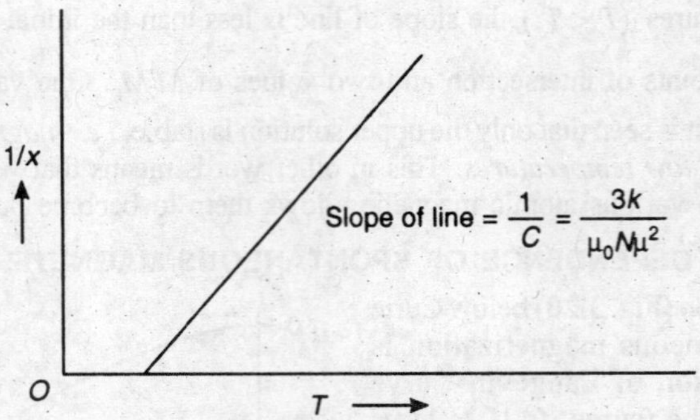

Fig. 3.21

Also from equation (9)

$$\mu = \sqrt{\frac{3kC}{\mu_0 N}}$$

Curie temperature for some elements is given under

Cobalt $\longrightarrow$ 1393 K

Ni $\longrightarrow$ 631 K

Gd $\longrightarrow$ 289 K

Fe $\longrightarrow$ 1043 K.

QUESTIONS AND PROBLEMS

1. Show that $B = H + 4nI$ and give an expression for the energy loss in a complete hysteresis cycle.
2. What is meant by hysteresis ? Prove that the area of a B–H cycle is given by 4n times the energy dissipated/unit volume of the matèrial during each magnetic cycle.
3. Explain the terms magnetic flux, magnetomotive force and reluctance.
4. What is a magnetic circuit? Explain the terms reluctance and magnetomotive force.
5. Explain what do you mean by residual magnetism, coercive force and hysteresis. Prove that the area of I–H cycle denotes the energy loss/unit volume of metal during each cycle.
6. Define intensity of magnetisation, magnetic induction, susceptibility and permeability. Establish the relation $B = H + 4\pi I$.
7. Derive an expression for the magnetic field around a current carrying circular loop or solenoid.
8. Establish the following relations :
   (a) $B = \mu_0 (H + I)$ (b) $\mu_r = 1 + K$
9. Explain what is meant by a magnetic circuit. What is its significance?
10. A circular ring 20 cm in diameter has an air gap 0.1 cm cut in it. The area of cross section of iron is 3.6 cm^2. Find the number of ampere turns needed to establish a flux 2.5×10^{-24} Wb in the air gap. Neglect fringing and assume relative permeability of iron as 650. **[Ans.** 1575]
11. Solenoid consisting of 100 turns and carrying 10 amperes current is 0.5 m long. Calculate (*i*) magnetomotive force, (*ii*) total flux if area of cross-section is 0.0004 m^2. **[Ans.** 1.0048×10^{-5} Wb]
12. A solenoid 20.5 cm long is wound on a transformer and has an internal diameter of 10 cm. It is wound to a depth of 2.5 cm with 2500 turns of wire and carries a current of 1 ampere. Find the flux at the centre of the solenoid. **[Ans.** 928×10^{-4} Wb]
13. A circular ring of iron is wound uniformly with 1000 turns of wire. Its mean diameter is 0.2 m and its cross-section is 4 cm^2. Find the current which will produce a flux of 2000 lines, given permeability = 800. **[Ans.** 31.25 mA]
14. Calculate the number of ampere turns required to produce a flux of 10^{-3} Wb round an iron ring of 5 cm^2 cross-section and 20 mm mean diameter having an air gap 2 mm wide across it. The relative permeability of iron may be taken equal to 1000. Neglect flux leakage. **[Ans.** 4773.5]

CHAPTER

4

THERMOELECTRICITY AND LOW RESISTANCE MEASUREMENT

4.1. SEEBECK EFFECT

Seebeck found that if a circuit consists of two dissimilar metals such as iron and constantan, an e.m.f. develops in it when the two junctions of the metals are at different temperatures. This is known as **Seebeck effect** and the pair of such a junction is known as **thermocouple**, e.m.f. generated in a thermocouple is called **thermo-electric e.m.f.** and current generated is **thermo-electric current** and the phenomenon on the whole is called **thermoelectricity.**

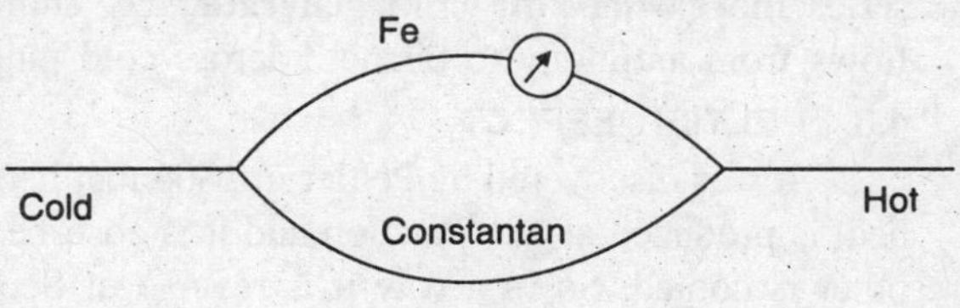

Fig. 4.1

We can explain the above phenomenon from atomic considerations as under:

In the case of different metals the number of free electrons/units volume and their velocities will have varying values. So at the junction of two different metals, there may be the tendency of migration of electrons from one metal to the other. However, such migration cannot continue for longer time as an opposing electric field sets up at the boundary. So we get only a fixed p.d. between the two metals known as *contact p.d.* Also, we know that the velocity of an electron in a metal changes with temperature. So the contact potential between any two given metals varies with temperature. In Fig. 4.2 any difference between two contact potentials V and V' will act to derive current round the circuit, the resultant e.m.f. is given by

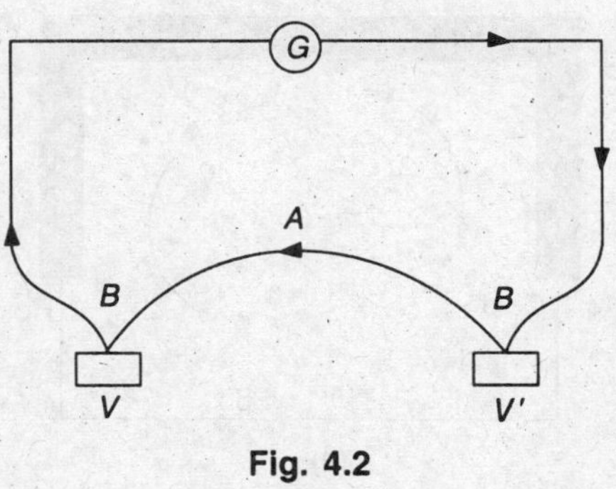

Fig. 4.2

$$E = V' - V$$

When both the junctions are at the same temperature, the contact p.d. will also be same at two junctions and as a result there will not be any resultant e.m.f. in the circuit and so no current flows. However, if one junction is heated and other is cooled, the contact potential at hot junction will become more than that at cold junction and their difference will act as resultant e.m.f. and is responsible for the flow of current.

4.2. DEPENDENCE OF TEMPERATURE

The graph between the variation of thermo e.m.f. and temperature of the hot junction when the cold junction is at 0°C is a parabola as shown in Fig. 4.3.

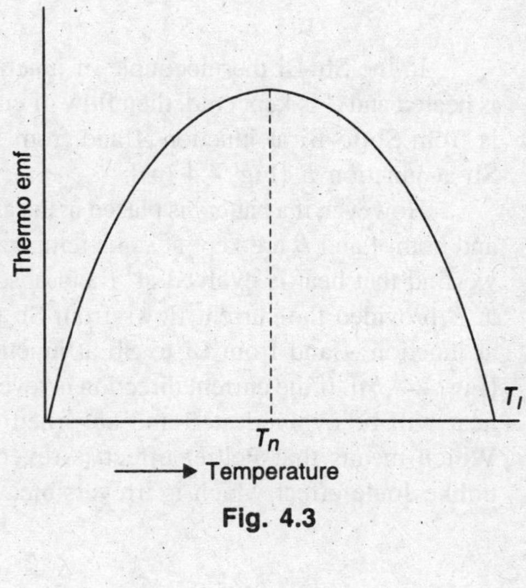

Fig. 4.3

If the cold junction is kept at 0°C and the temperature of the hot junction is gradually increased, then for small difference of temperature, thermo e.m.f. ∝ temperature difference. As the temperature difference increases, thermo e.m.f. also increases and becomes maximum at a temperature T_n which is known as neutral temperature and it is constant for a given thermocouple. If the hot junction temperature is further increased, the thermo e.m.f. starts decreasing and at a particular temperature called 'temperature of inversion' it becomes zero. Beyond this temperature e.m.f. again starts increasing but in **opposite direction**. The temperature of inversion is as much above neutral temperature as the cold junction is below it and the temperature of inversion varies with the temperature of the cold junction.

The following are the metals out of which if any two of them form a thermocouple, the current flows across the cold junction from the metal occurring first in the series to the one occurring later:

Selenium, Antimony, Iron, Cadmium, Zinc, Silver, Gold, Lead, Mercury, Copper, Platinum, Nickel, Bismuth.

It should be noted that metals preceding lead are taken to be **thermoelectrically negative** and succeeding lead are **thermoelectrically positive.** The more farther apart are the metals of the series, **more will be the e.m.f. generated** *e.g.,* antimony and bismuth generate large e.m.f. and current flows from antimony to bismuth across cold junction.

4.3. PELTIER EFFECT

It was discovered by Peltier in 1804 that if current is sent round the circuit of a thermocouple, heat is produced at one junction and it is absorbed at the other, *i.e.*, one junction is heated and the other is cooled. So in a way it is reverse of Seebeck effect.

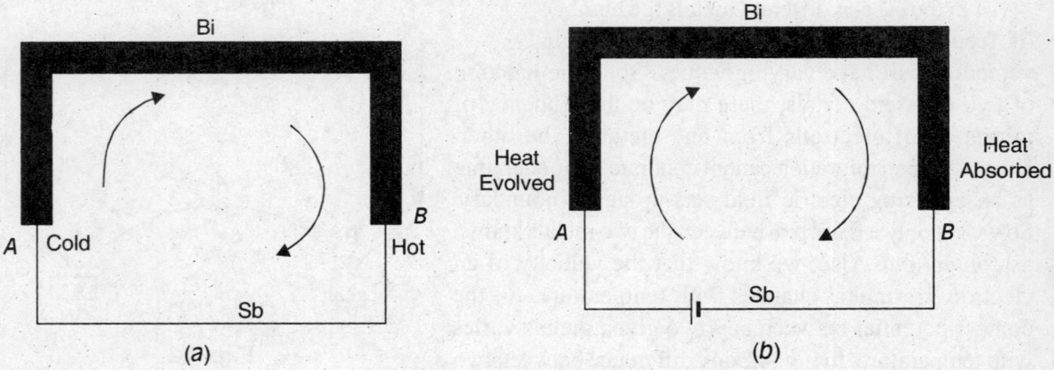

Fig. 4.4

In the Sb–Bi thermocouple, if junction *B* is heated and *A* is kept cold, then flow of current is from Sb to Bi at junction *A* and from Bi to Sb at junction *B* [Fig. 4.4 (*a*)].

However, if a battery is placed in the circuit and both *A* and *B* are kept at same temperature, we find that heat is evolved at *A* and absorbed at *B* provided the current flows from Sb to Bi at junction *A* and from Bi to Sb at junction *B* [Fig. 4.4 (*b*)]. If the current direction is reversed, heat will be evolved at *B* and absorbed at *A*. Which means that Peltier effect is reversible unlike Joule effect which is irreversible.

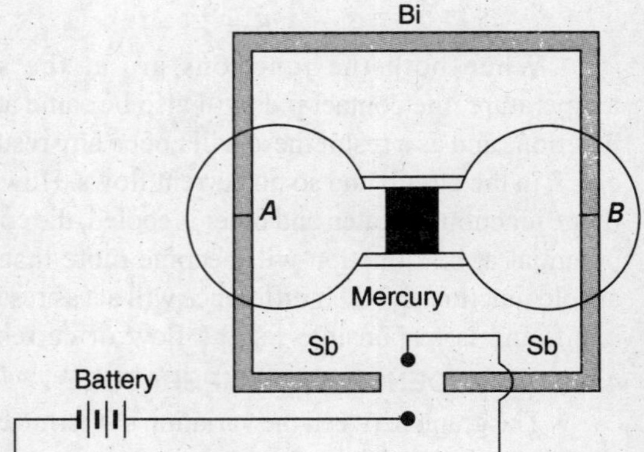

Fig. 4.5

4.4. EXPERIMENT

Thick rods of Sb and Bi are taken and a battery is placed in the circuit along with a reversing key. We make use of a differential air thermoscope that contains two bulbs connected by a fine tube that contains a small amount of mercury to indicate the difference of temperature between two junctions A and B. When current is from Sb to Bi at junction A, mercury moves towards B. Which means that heat is produced at junction A and absorbed in junction B. This is due to the Peltier effect. If the direction of the current is reversed with the help of a reversing key, a current flows from Bi to Sb at junction A. In this case mercury moves towards junction A, showing that heat is evolved at B and absorbed at junction A meaning thereby that Peltier effect is reversible.

4.5. PELTIER COEFFICIENT

It is the amount of energy absorbed or evolved at a junction when a unit current flows for 1 sec. If I is the amount of current passed for t sec and π is Peltier coefficient, then the energy evolved or absorbed to $\pi\ It$ joules.

If e volts be the p.d. developed in the junction, then $eIt = \pi It$

or $$e = \pi$$

However, Peltier coefficient is not constant, but depends on the temperature of the junction. In case of a thermocouple, if π_2 and π_1 are the Peltier coefficients at the junctions where the energy is absorbed and evolved respectively, then thermo e.m.f.

$$E = (\pi_2 - \pi_1)$$

Another method can also be used to find Peltier coefficient. A junction of the thermocouple is put in water and current I is passed for some known time t. If the amount of heat produced is H_1, then

$$H_1 = \frac{I^2 Rt}{J} + \frac{\pi It}{J} \qquad ...(1)$$

where $\frac{I^2 Rt}{J}$ is amount of heat produced due to joule effect

and $\frac{\pi It}{J}$ is that due to Peltier effect.

On reversing the current through the circuit, let the heat produced be H_2.

$$\therefore \qquad H_2 = \frac{I^2 Rt}{J} - \frac{\pi It}{J}$$

Subtracting Equation (2) from (1)

$$H_1 - H_2 = \frac{2\pi It}{J}$$

or $$\pi = \frac{(H_1 - H_2)J}{2It}$$

So Peltier coefficient can be calculated.

4.6. THOMSON EFFECT

We have noted that Peltier effect is reversible. So heat is absorbed at a higher temperature and some part of it is converted into useful work, *i.e.*, for driving electric current in the circuit, while the rest is returned to the sink at lower temperature. Kelvin, therefore, suggested that a Carnot's cycle can be worked between these temperatures.

The amount of heat absorbed at a hot junction T_2 when a charge q moves round the circuit is $\pi_2 q$, where π_2 is Peltier coefficient at temperature T_2. Let the energy given to the sink at T_1 be $\pi_1 q$.

Net energy absorbed $= (\pi_2 - \pi_1)\ q$ joules.

If we consider 1 amp. current to pass for 1 sec, then $q = 1$ coulomb

$\therefore$ The net energy absorbed is used in setting a thermo e.m.f. in circuit $= E$ volts.

$$\therefore \qquad E = \pi_2 - \pi_1$$

Since Peltier effect is reversible like Carnot's engine, we have

$$\frac{H_1}{T_1} = \frac{H_2}{T_2}$$

In this case $H_1 = \frac{\pi_1}{J}$, $H_2 = \frac{\pi_2}{J}$

$$\therefore \quad \frac{\pi_1}{T_1} = \frac{\pi_2}{T_2} \quad \text{or} \quad \frac{\pi_2}{\pi_1} = \frac{T_2}{T_1}$$

or
$$\frac{\pi_2 - \pi_1}{\pi_1} = \frac{T_2 - T_1}{T_1}$$

or
$$\pi_2 - \pi_1 = \frac{\pi_1}{T_1}(T_2 - T_1)$$

But
$$\pi_2 - \pi_1 = E$$

$$\therefore \quad E = \frac{\pi_1}{T_1}(T_2 - T_1)$$

$$\therefore \quad E \propto (T_2 - T_1)$$

i.e., thermo e.m.f. is directly proportional to difference of temperature T between hot and cold junctions and if a graph is plotted between E and T, we should get a straight line.

Kelvin (Lord Thomson) from above concluded that in the circuit in addition to Peltier effect, there is some additional source of e.m.f. This being due to different parts of the same metal being at different temperatures. This is called Thomson effect.

If σ be the e.m.f. due to unit difference of temperature then $\int_{T_1}^{T_2} \sigma dT$ is the total e.m.f. between points T_1 and T_2, where 'σ' is called Thomson coefficient and is analogous to specific heat in metals. Kelvin called it specific heat of electricity.

4.7. THOMSON COEFFICIENT

Thomson coefficient is defined as the energy evolved in ergs when 1 e.m.u. of current flows for 1 sec through a conductor maintained at a difference of temperature of 1°C. If energy is measured in joules, current in amperes and time in seconds, Thomson coefficient will be measured in **volts**.

Let σ_a and σ_b be the Thomson coefficients of two metals forming thermocouple, then we may have equations of e.m.f. as under:

Let current I flow through the circuit, when the energy absorbed is given by

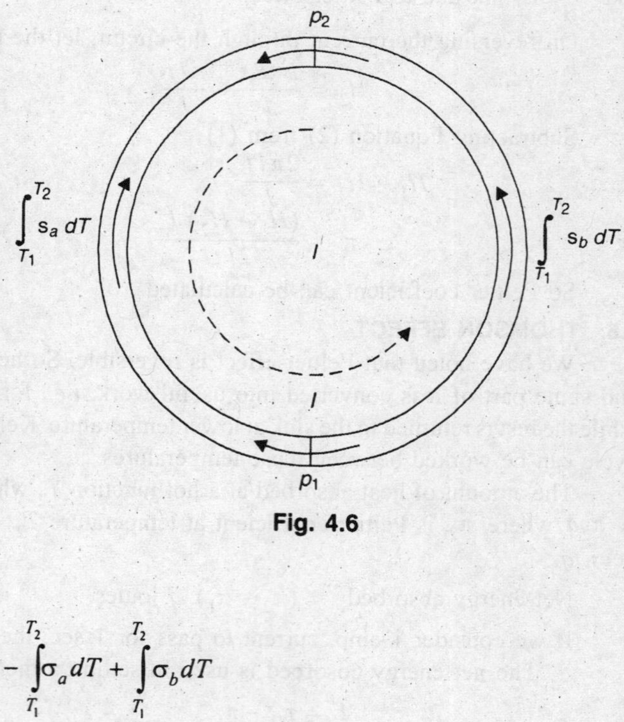

Fig. 4.6

$$E.\ I.\ t$$

$$= \pi_2 It - \int_{T_1}^{T_2} \sigma_a dT.It - \pi_1 It + \int_{T_1}^{T_2} \sigma_b dT\ It$$

or
$$E = \pi_2 - \pi_1 - \int_{T_1}^{T_2} \sigma_a dT + \int_{T_1}^{T_2} \sigma_b dT$$

The e.m.f. will flow as shown in Fig. 4.6 and direction of flow is marked by arrows. It is assumed that both metals have **positive Thomson coefficients.**

4.8. ANALYTICAL TREATMENT OF THERMOCOUPLE

We know that e.m.f. versus temperature curve is a parabola. If we take the equation

$$E = at + \frac{1}{2}bt^2 \qquad ...(1)$$

where a and b are thermoelectric constants and t is the temperature difference between hot and cold junction $T_2 - T_1 = t$

Differentiating Equation (1)

$$\frac{dE}{dT} = a + bt$$

which is the equation of a straight line. Now $\frac{dE}{dT}$ can be found by drawing tangents to the curve shown in Fig. 4.7 and $\frac{dE}{dT}$, *i.e.*, the rate of change of thermo e.m.f. is called thermoelectric power P. The curve between P and t is a straight line and is called **thermoelectric line.**

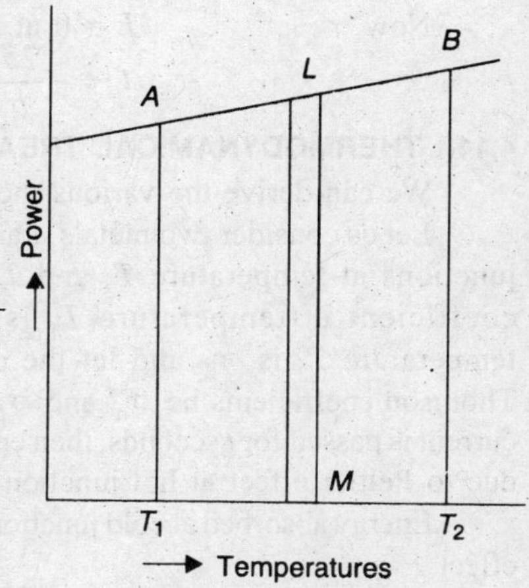

Fig. 4.7

Again if we plot a curve between P and temperature in K, we will still get a straight line but intercept on P axis will be different, *i.e.*,

$$\begin{aligned} P &= a + bt \\ &= a + b(T - T_0) \\ &= (a - bT_0) + bT \\ &= a' + bT \end{aligned} \qquad ...(2)$$

where a' is another constant depending on temperature of cold junction T_0.

In the above curve, we can find thermoelectric line AB by plotting a graph between $\frac{dE}{dT}$ and absolute temperature T.

Area of strip LM

$$= \frac{dE}{dT} \cdot dT = P \cdot dT = dE \qquad ...(3)$$

and represents e.m.f. acting round a couple having difference of temperature between junctions equal to dT, we can get total e.m.f. acting in the circuit by integrating (2) between T_1 and T_2.

$$E = \int_{T_1}^{T_2} \frac{dE}{dT} = \text{Area } AT_1T_2B$$

which will give Seebeck effect in the circuit.

If thermoelectric constants a and b are tabulated with reference to lead ($\sigma_{lead} = 0$) as one of the elements of thermocouple, then we can get thermoelectric power in microvolts/°C and Seebeck e.m.f. in microvolts

4.9. NEUTRAL TEMPERATURE

The above curve, as shown, starts from origin if temperature is measured in °C and cold junction is at 0°C. If the cold junction is not kept at 0°C, then the constant a will have different values $a' = a - bT_0$ and a reference junction correction has to be made. At neutral temperature t_n°C or T_n°K, we have

$$\frac{dE}{dt} = a + bt_n = 0 \qquad \text{or} \qquad t_n = T_n - 273 = -\frac{a}{b}$$

4.10. TEMPERATURE OF INVERSION

At temperature of inversion, $t_2 = T_2 - 273$, the Seebeck e.m.f. = 0

i.e. $$E = at + \frac{1}{2}bt^2 = 0$$

or $$t(a + \frac{1}{2}bt) = 0$$

Now $E = 0$ at $t = 0$ and $t = t_i$

$$\therefore \quad t_i = -\frac{2a}{b} = 2t_n$$

4.11. THERMODYNAMICAL TREATMENT OF THERMOCOUPLE

We can derive the various thermodynamic relations as under:

Let us consider two metals A and B with their junctions at temperature T_2 and T_1. The Peltier coefficient at temperature T_1 is π_1 and that temperature T_2 is π_2 and let the corresponding Thomson coefficients be σ_a and σ_b. If I units of current is passed for t seconds, then energy absorbed due to Peltier effect at hot junction = $\pi_2 It$.

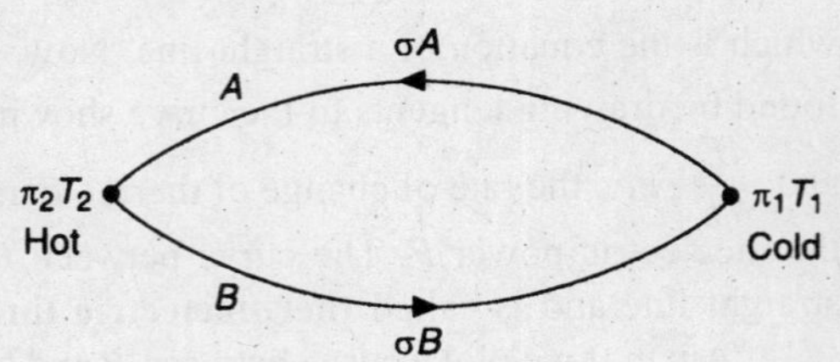

Fig. 4.8

Energy absorbed at cold junction due to Peltier effect

$$= \pi_1 It \text{ [– ve sign is due to the fact that energy is evolved]}$$

Energy absorbed in metal A due to Thomson effect

$$= \left(\int_{T_1}^{T_2} \sigma_A \, dT\right) I \cdot t$$

Energy absorbed in metal B due to Thomson effect

$$= \left(-\int_{T_1}^{T_2} \sigma_B \, dT\right) I \cdot t$$

– ve sign before the integral shows that current flows from higher to lower temperature. Total gain in energy for complete circuit

$$= \left((\pi_2 - \pi_1) + \int_{T_1}^{T_2} (\sigma_A - \sigma_B)\right) It \quad ...(1)$$

If E be the total e.m.f. produced in circuit, energy produced

$$= E\,I\,t \quad ...(2)$$

From Equation (1) and (2)

$$EIt = \left((\pi_2 - \pi_1) + \int_{T_1}^{T_2} (\sigma_A - \sigma_B)\, dT\right) I \cdot t$$

$$= (\pi_2 - \pi_1) + \int_{T_1}^{T_2} (\sigma_A - \sigma_B)\, dT$$

If the two junctions difference of temperature as dT and difference of Peltier coefficient as $d\pi$, then

$$dE = d\pi + (\sigma_A - \sigma_B)\, dT \quad ...(3)$$

For second law of thermodynamics, for a complete reversible process

$$\frac{H_1}{T_1} = \frac{H_2}{T_2}$$

where H_2 is the heat absorbed at temperature T_1 and H_2 that rejected at temperature T_2. In this case the two junctions are at temperatures T and $T + dT$ and a current I amperes flow for t sec. Energy absorbed at hot junction due to Peltier effect

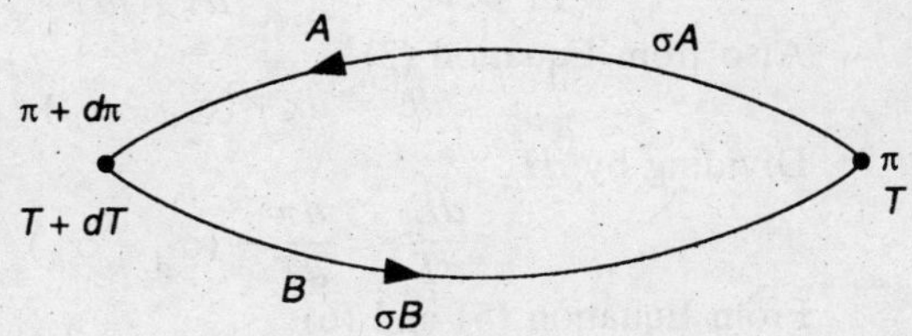

Fig. 4.9

$$= (\pi + d\pi)It \text{ joule} = \frac{(\pi_1 + d\pi)}{J} It \text{ cal.}$$

Energy evolved at cold juntion due to Peltier effect

$$= \pi I t \text{ joule} = \frac{\pi It}{J} \text{ cal}$$

Total energy absorbed due to Thomson effect

$$= [(\sigma_A - \sigma_B) dT] I t \text{ joules}$$

$$= \frac{(\sigma_A - \sigma_B) dT}{J} I \cdot t \text{ cel}$$

Applying the second law of thermodynamics

$$\frac{(\pi + d\pi) It}{J(T + dT)} + \frac{[(\sigma_A - \sigma_B) dT] It}{(T)} = \frac{\pi It}{J.T}$$

$$\frac{\pi + d\pi}{T + dT} + \frac{\sigma_A - \sigma_B}{T} dT = \frac{\pi}{T}$$

or $$\frac{\pi + d\pi}{T + dT} - \frac{\pi}{T} + \frac{\sigma_A - \sigma_B}{T} dT = 0$$

$$\frac{\pi T + Td\pi - \pi T - \pi dT}{T(T - dT)} + \frac{\sigma_A - \sigma_B}{T} dT = 0$$

or $$\frac{Td\pi}{T^2} - \frac{\pi dT}{T^2} + \frac{\sigma_A - \sigma_B}{T} dT = 0 \qquad [\text{when} T(T \times dT) \simeq T^2]$$

or $$d\pi - \frac{\pi dT}{T} + (\sigma_A - \sigma_B) dT = 0$$

or $$d\pi + (\sigma_A - \sigma_B) dT = \frac{\pi dT}{T}$$

From Equation (3) $d\pi + (\sigma_A - \sigma_B) dT = dE$

$$\therefore \quad dE = \pi \frac{dT}{T}$$

or $$\pi = T \cdot \frac{dE}{dT} \qquad \text{...(4)}$$

where $$\frac{dE}{dT} = P = \text{thermoelectric power}$$

$\therefore$ Peltier Coefficient = Absolute Temperature × Thermoelectric power

So Peltier coefficient at a given junction may be defined as **the product of absolute temperature and rate of change of thermo e.m.f. with temperature.**

From Equation (4) we have

$$\pi = T \frac{dE}{dT}$$

Differentiating

$$\frac{d\pi}{dT} = T \frac{d^2 E}{dT^2} + \frac{dE}{dT}$$

or
$$\frac{dE}{dT} - \frac{d\pi}{dT} = -T\frac{d^2E}{dT^2} \qquad ...(5)$$

Also from Equation (3)
$$dE = dx + (\sigma_A - \sigma_B)\,dT$$

Dividing by dT
$$\frac{dE}{dT} = \frac{d\pi}{dT} + (\sigma_A - \sigma_B) \qquad ...(6)$$

From Equation (5) and (6)
$$\sigma_A - \sigma_B = -T\frac{d^2E}{dT^2} \qquad ...(7)$$

Let us consider metals A and B to be of copper and lead respectively.

From Equation (7)
$$(\sigma_{Cu} - \sigma_{pb}) = T\frac{d^2E}{dT^2}$$

But $\sigma_{pb} = 0$ $\quad\therefore\quad \sigma_{Cu} = -T\dfrac{d^2E}{dT^2}$

Again let A and B be of lead and iron respectively,
$$(\sigma_{pb} - \sigma_{Fe}) = -T\frac{d^2E}{dT^2}$$

4.12. THERMOELECTRIC POWER

This is given mathematically as
$$P = \frac{dE}{dT}$$
$$\therefore \quad \frac{dP}{dT} = \frac{d^2E}{dT^2}$$

If metal A is of lead, the Equation (7) may be written as
$$\sigma_B = T\frac{d^2E}{dT^2} = T\frac{dP}{dT}$$

Integrating both sides
$$\int_{T_1}^{T_2} \sigma_B\, dT = T\int_{P_1}^{P_2} dP = T(P_2 - P_1) \qquad ...(10)$$

So Thomson Effect $\int_{T_1}^{T_2} \sigma_B\, dT$ for a given metal is equal to the **product of the difference of thermoelectric powers at the ends and the mean absolute temperature.**

From equation (4), we have
$$\pi = T\cdot\frac{dE}{dT}$$
$$\therefore \quad \frac{\pi}{T} = \frac{dE}{dT}$$

Differentiating
$$\frac{d}{dT}\left(\frac{\pi}{T}\right) = \frac{d^2E}{dT^2} \qquad ...(11)$$

From Equations (7)
$$\sigma_A - \sigma_B = -T\frac{d^2E}{dT^2}$$
$$\therefore \quad \frac{d^2E}{dT^2} = -\left(\frac{\sigma_A - \sigma_B}{T}\right) \qquad ...(12)$$

From Equation (11) and (12), we have

or $$\frac{d}{dT}\left(\frac{\pi}{T}\right) = -\left(\frac{\sigma_A - \sigma_B}{T}\right)$$

4.13. THERMOPILE

It was first designed by Melloni and consists of thermocouples of antimony and bismuth placed end to end to multiply each other's effect. These two metals are chosen since they produce large e.m.fs. The whole unit is placed in a box and the front surface of the elements is blackened with soot. There is a conical piece attached to the front side so that the incoming radiations fall on the blackened surface and raise the temperature of the hot junction. The free ends are connected to a sensitive galvanometer that can detect difference of temperature of the order of 0.001°C between the junctions. The thermopile is mounted in vacuum where its sensitiveness increases and deflections of the galvanometer are steady.

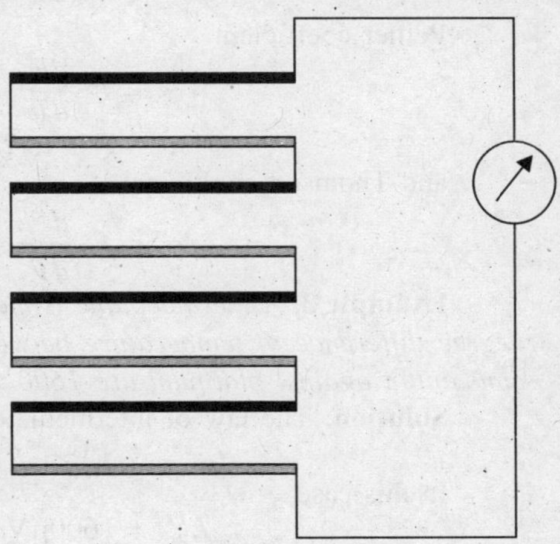

Thermopile

Fig. 4.10

A linear type of thermopile is used in research to study infrared radiations. It consists of only some pairs of thermocouples connected with a sensitive galvanometer. All the junctions are arranged in line.

However, a thermopile has a drawback in the sense that it takes long time to give steady deflection and also it takes long time to return to its zero reading.

Example 1. *If, for a certain thermocouple $E = a\theta + b\theta^2$. where θ is in °C, is the temperature of hot junction and cold junction being at 0°C. If $a = 10\mu V/°C$ and $b = -\frac{1}{40}\mu V/C^2$, find the neutral temperature and temperature of inversion.*

Solution: $$E = a\theta + 2b\theta$$

$$\therefore \quad \frac{dE}{d\theta} = a + 2b\theta$$

when $$\frac{dE}{d\theta} = 0, \ \theta = \theta_n$$

$$\therefore \quad 0 = a + 2b\theta_n$$

or $$\theta_n = -\frac{a}{2b} = \frac{10}{2\times\frac{1}{40}} = 200°C$$

Also $$\theta_i = 2\theta_n$$

$$\therefore \quad \boldsymbol{\theta_i = 2 \times 200 = 400°C}$$

Example 2. *The e.m.f. in a thermocouple, one junction of which is kept at 0°C is given by $E = a\theta + b\theta^2$. Find the neutral temperature and Peltier and Thomson coefficients.*

Solution: Let θ° be the temperature of hot junction

θ°C = (*T* – 273)°C, where *T* is in absolute degrees.

$$\therefore \quad E = a\theta + b\theta^2 = a(T-273) + b(T-273)^2$$

$$\therefore \quad \frac{dE}{dT} = a + 2b(T-273)$$

and $$\frac{d^2E}{dT^2} = 2b$$

At neutral temperature $T = T_n$

and $$\frac{dE}{dT} = 0$$

$$\therefore \qquad 0 = a + 2b\,(T_n - 273)$$

or $$T_n - 273 = -a/2b$$

or $$T_n = \left[\left(273 - \frac{a}{2b}\right)\right]^{\circ} \text{ Absolute}$$

$$= \left(-\frac{a}{2b}\right)^{\circ}\text{C} \qquad \text{...(1)}$$

Peltier coefficient

$$\pi = T \cdot \frac{dE}{dT}$$

$$= T\,[a + 2b\,(T - 273)] \qquad \text{...(2)}$$

and Thomson coefficient

$$\sigma = T \cdot \frac{d^2E}{dT^2} = 2b.T \qquad \text{...(3)}$$

Example 3. *A thermocouple is made of iron and constantan. Find the e.m.f. developed per degree difference of temperature between the junctions, given that thermo-e.m.f.'s of iron and constantan against platinum are 1600 μV and –3400 μVper 100° difference of temperature.*

Solution: The law of intermediate metal gives the relation

$$E_{\text{cons}}^{\text{Fe}} = E_{\text{Pt}}^{\text{Fe}} + E_{\text{cons}}^{\text{Pt}} = E_{\text{Pt}}^{\text{Fe}} - E_{\text{Pt}}^{\text{cons}}$$

In this case,

$$E_{\text{Pt}}^{\text{Fe}} = 1600\mu\text{V}/100°\text{C} = 16\mu\text{V}/°\text{C}$$

and $$E_{\text{Pt}}^{\text{cons}} = 16 - (-34) = 50\mu\text{V}/°\text{C}$$

Example 4. *Along a metal of cross-section 1 × 10^{-4} sq. metre, there is a uniform temperature gradient of 100°C/meter. The specific resistance of the material of the rod is 150 × 10^{-8} ohm metre. When a current of 0.05 ampere is passed from the hot end to the cold end, the temperature gradient remains unaltered. Calculate the specific heat of electricity for this metal.*

Solution: The sp. heat of electricity or the Thomson coefficient of a metal is given as *the amount of heat in joule, evolved or absorbed when a charge of 1 coulomb flows from one point in the metal to another having a temperature difference of 1°C.* In this case, when the current flows from hot to cold junction, the temperature gradient remains constant. This is possible only when the heat evolved due to Joule's effect is equal to the heat absorbed due to Thomson effect.

Cross-sectional area of the rod $(A) = 1 \times 10^{-4}\ \text{m}^2$

Sp. resistance of the material $(\rho) = 150 \times 10^{-8}$ ohm-m

∴ Resistance per metre of the rod will be

$$R = \rho \times \frac{L}{A} = \frac{150 \times 10^{-8}}{1 \times 10^{-4}} = 150 \times 10^{-4}\,\text{ohm.}$$

Current flowing from hot to cold junction is $i = 0.05$ amp. So, p.d. per metre per °C = $i \times R$

$$= 0.05 \times 150 \times 10^{-4} = 7.5 \times 10^{-4} \text{ volt/metre.}$$

Now the temperature gradient in the rod is 100°C/metre, *i.e., 1°C/cm;* also, the Thomson's coefficient of a metal expressed in Joules/coulomb per °C is numerically equal to the e.m.f. in volt set up between two points differing in temperature by 1°C, so Thomson's coefficient (σ) is given by

$$\sigma = 7.5 \times 10^{-6} \text{ Joules/coul/°C}$$

Example 5. *The thermoelectric powers of iron and nickel with respect to lead are +12μV/K and –20μV/E at 323 K. Calculate the e.m.f. of an iron-nickel couple with junctions at 273 K and 373 K.*

Solution: Any metal is said to be positive to another metal, if the current flows from the former to the latter through the cold junction. Also if two temperatures are between 0°C (273 K) and neutral temperature, the thermo-e.m.f. between these temperatures is given by the relation

E = Thermoelectric power at mean temperature × temperature difference

So, the thermoelectric power of Ni–Fe couple at 323 K will be

$$= \{12 - (-20)\}\ \mu V/\text{degree}$$

$$= 32\ \mu V/\text{degree}.$$

Hence, thermo-e.m.f. $E = 32 \times 100 = 3200\ \mu V$.

Example 6. *The e.m.f. in lead-iron thermocouple, one junction of which is at 0°C, is given by $E = 1784\,t - 2\,4\,t^2$ (in μ volts) where t is the temperature in °C. Find the neutral temperature π and σ.*

Solution: $E = at + \frac{1}{2}bt^2$

$$a = 1784,\ b = -4.8$$

Now $t_n = -\frac{a}{b} = -\frac{1784}{-4.8} = 371.67°C$

Peltier coefficient

$$\pi = T\left(\frac{dE}{dT}\right) = T[a + b(T - 273)] = 3094.4T - 4.8T^2$$

Thomson coefficient

$$\sigma = T.\frac{d^2E}{dT^2} = T\frac{d}{dT}(a + bt) = 3094.4T - 4.8T^2$$

$$\sigma = T \cdot b = \mathbf{-4.8T}.$$

Example 7. *The thermoelectric power of iron is $17.5\ \mu V/°C$ at 0°C and $5\ \mu V/°C$ at 125°C. The thermoelectric power of cadmium is $3\ \mu V/°C$ at 0°C and $15\ \mu V/°C$ 150°C. Find the value of the neutral temperature of the iron-cadmium junction.*

Solution: Let the thermoelectric power be given by

$$\frac{dE}{dT} = a + bt$$

For Iron

$$\frac{dE}{dT} = 17.5\mu V \text{ at } t = 0°C$$

$$\therefore \quad 17.5 = a \qquad ...(1)$$

and $\frac{dE}{dT} = 5\mu V$ at $t = 125°C$

so $5 = 17.5 + 125\,b$

or $b = -0.1$ ...(2)

$\therefore$ For iron $a = 17.5$ and $b = -0.1$

For cadmium

$$\frac{dE}{dT} = 3 \text{ at } 0°C$$

$$\frac{dE}{dT} = 3 = a$$

or $a = 3$ ...(3)

Also $\frac{dE}{dT} = 15 = 3 + b \times 150$ at 150°C

$$b = -\frac{12}{150} = -0.08$$

Now we have the relation

$$a_{AB} = a_{A-pb} - a_{B-Pb}$$

and $b_{AB} = b_{A-pb} - b_{B-Pb}$

So we may write

$$a_{\text{Fc - Cd}} = a_{\text{Fe–Pb}} - b_{\text{Cd–Pb}} = 17.5 - 3$$

or $a_{Fe-Cd} = 14.5$

Similarly, we have

$$b_{Fe-Cd} = b_{Fe-pb} - b_{Cd-Pb}$$
$$= -0.1 - 0.08 = -0.18$$

Now neutral temperature is given by

$$t_n = -\frac{a}{b} = -\frac{14.5}{-0.18} = \mathbf{80.555°C.}$$

Example 8. *An e.m.f. of 0.9 volt is generated when the temperature between hot and cold junctions of a thermocouple is 75 K. Assuming that the cold junction is heated up by 15°, determine the extent by which thermo e.m.f. will change.*

Solution: Since in this case, the difference of temperature is small, so the variation of e.m.f. with temperature may be considered to be linear;

i.e., $E = \alpha T$ (where α constant)

Here, $T = 75$ K $E = 0.9$ volt.

$\therefore$ $\alpha = \frac{0.9}{75}$ volt/K

The difference of temperature between the junctions reduces to 60 K when the cold junction is heated up by 15 K, so the thermo-e.m.f. generated in that case will be

$$E' = \alpha T \frac{0.9}{75} \times 60 = 0.72 \text{ volt}$$

The change in e.m.f. $= E - E' = 0.9 - 0.72 = 0.18$ volt.

The extent by which the e.m.f. changes is given by

$$= \frac{0.18}{0.9} \times 100 = \mathbf{20\%}.$$

LOW RESISTANCE MEASUREMENT

4.14. MEASUREMENT OF LOW RESISTANCE (CAREY FOSTER BRIDGE)

A post office box is unsuitable for the measurement of a very low resistance as the resistance of the connecting wires becomes comparable and introduce an appreciable error. The arrangement also becomes very insensitives as the four arms of the bridge contain widely different resistance.

Carey Foster Bridge. It is a modified form of metre bridge designed to measure the difference between two resistances which have nearly equal values.

A wire *AB* one metre long of eureka or manganin, of uniform area of cross-section, is fixed on a metre scale put on a wooden board. The ends *A* and *B* are soldered to two thick copper strips. There more copper strips are fixed parallel to the wire as shown in Fig. 4.11, thereby providing four gaps shown as 1, 2, 3, and 4.

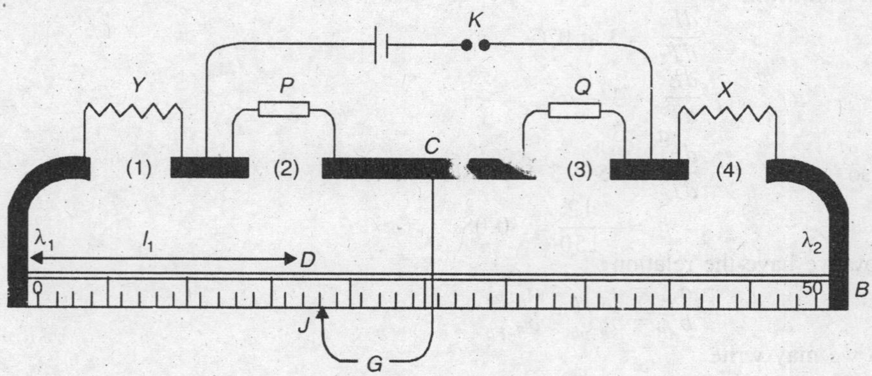

Fig. 4.11 (a)

Two resistance boxes P and Q of two nearly equal resistance of about 5 or 10 ohms are connected in the inner gaps and then resistances X and Y, whose difference is to be found, are connected in the outer gaps. The battery is put below the resistances P and Q and the galvanometer between the points C and J (the jockey) which can slide on wire AB.

Theory. The arrangement of Carey Foster bridge is extension of Wheatstone bridge, in which P and Q are the two ratio arms, X along with the resistance of wire AD and Y along with the resistance of the wire BD form the other two arms.

Let ρ be the resistance per cm. of the wire, λ_1 and λ_2 respectively the end resistance at A and B and a balance point comes at D where $AD = l_1$, then we have

$$\frac{P}{Q} = \frac{X + l_1\rho + \lambda_1}{Y + (100 - l_1)\rho + \lambda_2} \quad \text{...(i)}$$

If X and Y are interchanged and the balance point shifts D as shown in Fig. 4.11 (b) so that $AD = l_2$.

$$\frac{P}{Q} = \frac{Y + l_2\rho + \lambda_1}{X + (100 - l_2)\rho + \lambda_2} \quad \text{...(ii)}$$

Comparing (*i*) and (*ii*) and adding one to both sides, we get

$$\frac{X + Y + 100\rho + \lambda_1 + \lambda_2}{Y + (100 - l_1)\rho + \lambda_2} = \frac{X + Y + 100\rho + \lambda_1 + \lambda_2}{X + (100 - l_2)\rho + \lambda_2}$$

or $$Y + (100 + l_1)\rho + \lambda_2 = X + (100 + l_2)\rho + \lambda_2$$

or $$Y - X = (l_1 - l_2)\rho$$

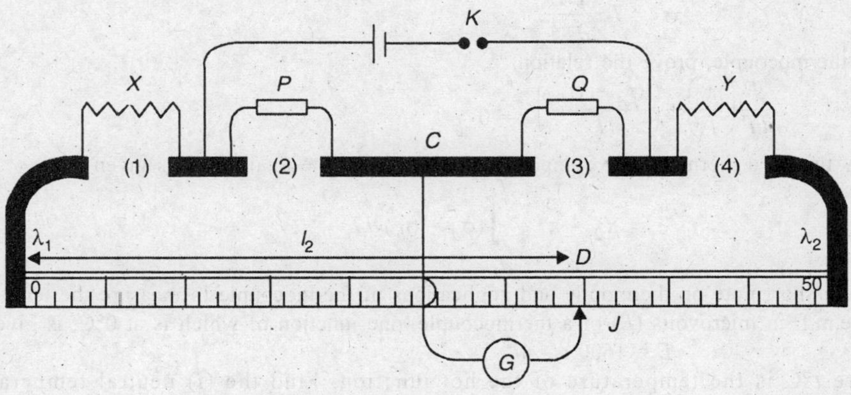

Fig. 4.11 (*b*)

which means that the difference between the two resistances is equal to the resistance of bridge wire between two points. In this case no end correction is required as is clear from Equation (3).

If one of the resistances, say Y, is known, the other X can be known from the relation

$$X = Y + (l_2 - l_1)\rho$$

or $$X = Y + (l_1 - l_2)\rho$$

knowing l_1, l_2 and ρ, x can be found.

We can also find unknown low resistance without calibration of Carry Foster bridge as under: If X is the unknown resistance to be found out and d_1 and d_2 are the correct shifts obtained with resistance Y_1 and Y_2 from the fractional resistance box corresponding to the above shifts, then we have the relation

$$X - Y_1 = d_1\rho$$

$$X - Y_2 = d_2\rho$$

$$\frac{X - Y_1}{X - Y_2} = \frac{d_1}{d_2}$$

or
$$d_2 X - d_2 Y_1 = d_1 X - d_1 Y_2$$

$$X = \frac{y_1 d_2 - y_2 d_1}{d_2 - d_1}$$

Which shows the ρ is eliminated and need not be found.

QUESTIONS AND PROBLEMS

1. What do you understand by Peltier effect and sp. heat of electricity? How does Peltier effect differ from Joule heating effect?
2. Explain the terms Seebeck effect, Peltier effect and Thomson effect. Illustrate each with some simple experiments.
3. Give the theory of thermo e.m.f. Show that $\pi = T \cdot dE/dT$.
4. Briefly explain Seebeck effect and Peltier effect. Prove that Peltier coefficient of a pair of metals is product of absolute temperature and thermoelectric power.
5. Define thermocouple, neutral temperature, thermoelectric power and temperature of inversion.
6. Prove the thermoelectric relation

$$\frac{d}{dT}\left(\frac{\pi}{T}\right) + \frac{\sigma_A - \sigma_B}{T} = 0$$

7. Prove that in a thermocouple

$$\sigma = T \cdot \frac{d^2E}{dT^2}$$

8. In a thermocouple, prove the relation

$$\frac{d}{dT}\left(\frac{\pi}{T}\right) + \left(\frac{\sigma_A - \sigma_B}{T}\right) = 0$$

9. Show that in a thermocouple composed of two metals, the total e.m.f. is given by

$$c = \pi_2 - \pi_1 + \int_{T_1}^{T_2} (\sigma_a - \sigma_b)\, dT$$

10. Write a short note on thermopile and applications of thermocouple to measure the temperature.
11. The e.m.f. in microvolts (E) of a thermocouple, one junction of which is at 0°C, is given by

$$E = 1600\, t - 4t^2$$

where t°C is the temperature of the hot junction. Find the (i) neutral temperature and (ii) Peltier coefficient at t°C. **[Ans.** (i) = 200°C; (ii) (273 + t) (1600 8t).]
12. The thermoelectric power of Iron is 1734 – 4876 t and that of copper is 136 + 0.95 m V/°C. Where t is the temperature in °C. Show that the e.m.f. of thermocouple, the junctions of which are at 0°C and 100°C, is 130700 μ -volts.
13. The e.m.f. in a thermocouple, one junction of which is kept at 0°C, is given by $F = at + \frac{1}{2}bt^2$. Find the neutral temperature, Peltier and Thomson coefficients.

[Ans. $t_n = -\frac{a}{b}$°C; $\pi = (t + 273)(a + bt); \sigma = b(t + 273)$]
14. The e.m.f. of an iron-lead thermocouple where one junction is at 0°C and other at 100°C is 1.85 μV. When the second junction is at 300°C, e.m.f. is 675 μV. Similar readings with silver-lead couple are 371 μV and 1620 μV respectively. Calculate the neutral temp. for an iron-silver thermocouple. **[Ans.** t_n = 122°C.]

PART VII
MODERN PHYSICS

Ernest Rutherford (1871 – 1937)

The great astronomer Sir Arthur Eddington once said that in 1911 Ernest Rutherford introduced the greatest change in our understanding of matter. **He was referring to Rutherford's idea of the nuclear** atom. It was this idea, and the experimental work leading up to it, which made him the acknowledged founder of modern nuclear physics.

That such a great scientist should have been born in New Zealand, at that time a remote country with few people and almost no scientific traditions, is as remarkable as was Rutherford's innate skill with scientific apparatus. Rutherford was born into a poor family. His father was a farmer who also did odd jobs to support his family of 12 children. Rutherford's simple surroundings in his youth no doubt influenced the development of his character, which remained open and unassuming throughout his life.

At Cambridge, Rutherford worked under the great J.J. Thomson on the effect of X-rays on the conduction of electricity in gases, and later in the newly emerging field of radioactivity.

In 1908 he was awarded Nobel Prize in chemistry for study of radioactivity. On the basis of his work on alpha particles (helium nuclei) he was led on 1911 to a description of the atom as made up of a small, very dense nucleus surrounded by orbital electrons. This description was taken up by Niels Bohr in 1913 and, combined with Bohr's ideas about quantum theory, provided the basic theory of the atom still accepted today.

It is not in the nature of things for any one person to make a sudden and violent discovery, science goes step by step and every man depends on the work of his predecessors. Scientists are not dependent on the idea of a single man, but on the combined wisdom of thousands of men, all thinking of the same problem, and each doing his little bit to add to the great structure of knowledge which is being erected.

Ernest Rutherford

CHAPTER

1

GENERAL PROPERTIES OF THE NUCLEUS

1.1. RUTHERFORD'S SCATTERING EXPERIMENT

It is known that atom has a complex structure. In 1898, Sir J.J. Thomson put forward a hypothesis in which he stated that atom be considered as a sphere of +ve charge in which electrons are embedded. The mass of the atom is due to the positively charged particle since they are much heavier than the electrons.

This model of atom as suggested by Thomson was challenged by Lord Rutherford in 1911 by the study of the scattering of alpha particles from thin sheets of metals. The alpha particles are positively charged, each particle carries two units of positive charge and possesses four atomic mass units. The alpha particles pass through the air with a high velocity. Rutherford directed the beam of alpha particles on a target made of a fluorescent screen (zinc sulphide screen). In between he placed a thin gold foil in the path of the alpha particles.

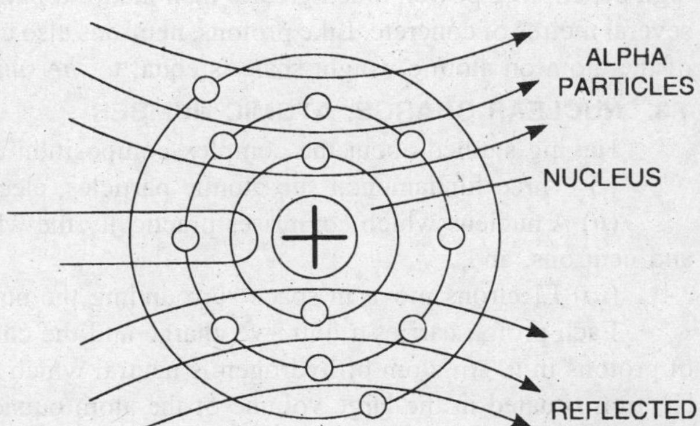

Fig. 1.1

When these alpha particles are bombarded on a thin gold foil about 4×10^{-6} m thick, the diameter of an atom of gold is 4×10^{-10} m. Some alpha particles are **deflected at fairly large angles and a few are actually reflected back** along their path. From the result of this experiment Rutherford concluded that:

1. The atom has a large space since most of the alpha particles pass through the metallic foil undeflected.

2. A lump of mass is located at the centre, known as the nucleus of the atom. It is centrally positive charge where the mass of the atom is concentrated which repels and causes the alpha particles to deflect considerably from its original path.

3. The volume occupied by this heavy positively charged nucleus is only a small fraction of the total volume of the atom.

4. There are just enough electrons outside the positive nucleus to provide a balance of +ve and –ve charges.

5. The central nucleus has a positive charge which is different in magnitude for each element.

Rutherford calculated the diameter of the nucleus of the atom which is found to be of the order of 10^{-15} m and that of atom as a whole is of the order of 10^{-10} m. The atomic scientists find that the use of ordinary metric units is inconvenient to measure the diameter of atoms or nucleus of the atom. They have developed another unit known as the angstrom unit. One angstrom unit is equal to 0.00000001 cm, that is,

$$0.00000001 \text{ cm} = 1 \text{ Å} = 10^{-8} \text{ cm} = 10^{-10} \text{ m}.$$

1.2. NEUTRON—A THIRD TYPE OF SUBATOMIC PARTICLE

By 1920, it was well established that atom is composed of positively and negatively charged particles. It was believed that electrons and protons were only kind of fundamental sub-atomic particles and the mass of the atom was due to the number of positively charged particles present in the nucleus. However, Rutherford observed discrepancies in the behaviour of alpha particles which led him to predict the presence of a third type of sub-atomic particle in the nucleus.

This prediction was confirmed in 1932 by an English physicist, James Chadwick. He observed that when beryllium foil is bombarded with *high speed alpha particles,* a radiation is emitted which consists of uncharged particles. Chadwick called these neutral particles *neutrons.* This neutral particle possesses a mass of 1.675×10^{-27} kg, which is 1.00866 a.m.u. The neutron particle has extremely high penetrating power, much greater than that of *alpha and beta particles*, being able to penetrate several metres of concrete. Like protons, neutrons also contribute to the mass of the atom. The mass of the atom on atomic weight scale is equal to *the sum of protons and neutrons present therein.*

1.3. NUCLEAR CHARGE: ATOMIC NUMBER

Having studied about the complex composition of the atom that it contains:

(*i*) Three fundamental sub-atomic particles, electrons, protons and neutrons.

(*ii*) A nucleus which comprises practically the whole mass of the atom, consists of protons and neutrons, and,

(*iii*) Electrons move in space surrounding the nucleus.

Each proton carries a unit +ve charge and the charge in the nucleus is equal to the number of protons in it. An atom of hydrogen is neutral which contains one proton in its nucleus and one electron situated in the large volume of the atom outside the nucleus. Helium atom contains two protons in the nucleus since it has two +ve charges; it has two electrons located around the nucleus. A chemical element always has its atoms with the *same positive charge* in the nucleus. The nuclear charge is fixed by the *number of protons* and the number of protons in the nucleus distinguish the various atoms of different elements.

The nuclear charge (number of positive charge in the nucleus) of the atom was determined by Mosley in 1914, by his study on the wavelength of the X-ray emitted from different elements. Each element emits X-rays of definite wavelength. Mosley studied the frequency and the wavelength of X-rays emitted by about 38 elements and concluded that *as the atomic weight increases, the wavelength of the X-rays decreases.* The wavelength of any element is different from that of the element next to it by the same amount. If all the elements are numbered sequentially from 1, 2, 3 ... 105 in the order of the decreasing wavelength of X-rays, we get a constant difference of wavelengths for the given set of elements. Mosley arranged all elements in order of increasing positive charge on the nuclei of atoms, beginning with hydrogen as 1. The order number of a particular element in this sequence is known as experimental atomic number of that element. It follows that the atomic number of an element is equal to the *number of positive charges* present in the nucleus of its atoms. Since atoms are electrically neutral it follows that the atomic number represents *the number of electrons* outside the nucleus and also the *number of protons present inside the nucleus.*

Mathematically, we have

Atomic number = Number of protons in the nucleus
= Number of electrons outside the nucleus

Thus hydrogen produces longest X-ray and is, therefore, number one on the list; its atomic number is 1. Nitrogen is number 7 in the list and its atomic number is 7. Sodium is number 11 in the list and its atomic number is 11. Uranium is number 92 in the list and its atomic number is 92, and so on. The atoms of any element will have the same atomic number which identifies the element. The atomic number of an element is generally denoted by the letter Z. There are 105 different elements with atomic numbers varying from 1 to 105. Recent researches, on the nuclei of atoms of various elements, have indicated that the atomic number is a fundamental property of the atom which determines the *physical and chemical properties* of the atom.

1.4. ATTRACTIVE AND REPULSIVE INTERACTION BETWEEN CHARGES

According to Rutherford's Nuclear Model of Atom, an atom of an element is electrically neutral and can be further sub-divided into subatomic particles, *electrons, protons and neutrons.* The nucleus consists of protons and neutrons present in the nucleus of the atom which constitutes the mass of the atom. The positive charge of the nucleus is due to the *protons present in the nucleus.* The nucleus of the atom like the *pit of a fruit* occupies a very small volume as compared to the volume of the atom itself.

The negatively charged particle, electrons, which are *equal in number* to the protons, surround the positive nucleus. The electrons and the positive nucleus in the atom exert mutual forces of attraction. The magnitude of electrostatic force is governed by *Coulomb's Inverse Square Law.* It studies the interaction of charges separated from each other and the energy changes taking place when the separated distance is varied. According to Coulomb's Inverse Square Law, an electric force acting between two charges is directly proportional to the product of the charges and inversely proportional to the square of the distance between them. Expressing mathematically, the law is

$$F \propto \frac{Q_1 \times Q_2}{r^2}$$

where F denotes the force of attraction between two oppositely charged particles of Q_1 and Q_2 units and r denotes the distance of separation between the charged particles.

If the charges are alike, the same expression gives the force of repulsion between them. In case of the atom, positive nucleus and the electrons exert *an attractive force.* The electron carries a unit negative electric charge denoted by –ve while nucleus contains positively charged particles, known as protons. Each proton carries a charge equal in magnitude to that of electron but *opposite in character.* If proton charge is denoted $+e$ and if the number of protons present in the nucleus is Z, total positive charge in the nucleus will be equal to $+e \times Z$ (*i.e.,* $+Ze$ units).

The attractive force between an electron and the nucleus of an atom with protons equal to Z in number will be given by

$$F \propto \frac{(-e) \times (+Ze)}{r^2}$$

or

$$F \propto \frac{-Ze^2}{r^2}$$

where r is the distance of separation between the *electron and nucleus* of the atom.

Again, the nucleus of the atom contains protons and neutrons. The protons are positively charged particles and *neutrons are electrically neutral particles.* Since like charges repel each other, there will be an electrical force of repulsion between the protons which are packed in the nucleus of the atom. This would result into the protons to fly out of the nucleus and the system would collapse. It never happens and the pertinent question is raised—*what is that force which keeps the protons together in the nucleus?*

The protons are kept together in the nucleus by the *nuclear force* and not by the gravitational force since the masses of protons and neutrons are negligibly small. The nuclear forces keep the protons and neutrons together in the small nucleus and the energy required to keep the neutrons and protons inside the nucleus is called *binding energy* or *packing energy.* The nuclear forces are *short range forces* and act over a very small distance as it is noticed in small atomic radius which ranges from 10^{-14} m to 10^{-15} m. The electrical forces of repulsion between protons in this range and in the same space of the atom *is always less than the nuclear forces.*

The discovery of the neutron by Chadwick explains the presence of neutrons inside the nucleus. The nucleus consists of protons whose number is the same as that of charge on the nucleus (atomic number) and neutrons which account for the balance of the mass. A common name for these particles (protons and neutorns) is *nucleons,* because of their *nuclear origin.*

In a $_{92}U^{238}$ atom, there are 92 protons and 238 – 92 = 146 neutrons, so that the charge on the nucleus is 92 and its mass is 92 + 146 = 238 units.

1.5. FINAL PICTURE OF ATOM

To sum up the developments in this field, it may be said that there are three fundamental particles constituting atom:

(*a*) **Electrons:** An electron is a *negatively charged* particle having unit negative charge and mass equal to 1/1837 that of a hydrogen atom.

(*b*) **Protons:** A proton is a *positively charged* particle having a unit positive charge and a mass approximately equal to that of hydrogen atom.

(*c*) **Neutrons:** A neutron is a *neutral particle* and its mass is slightly more than that of a proton. A neutron may be considered to have been formed by the *combination of a proton and an electron.*

1.6. ISOTOPES AND ISOBARS

The determination of value of *e/m* for the positive particles constituting the positive rays revealed that the *e/m* was not the same even for all the particles obtained from the same source. For example, when neon gas was taken in the discharge tube, different values of *e/m* were obtained. Since the positive particles are formed by the knocking out of electrons from the atoms, any two particles obtained from the same source cannot have charges differing from each other in terms of fractions. Therefore, the only explanation of the fractional differences in the values of *e/m* for the particles obtained from the same source is that these particles have different masses. Now particles of different masses can be obtained from an element only if the atoms of the same elements are not *identical completely.* This concept of different atoms of same elements being not completely alike was firmly established when atoms of the same element having different masses were actually isolated. *Such atoms of the same element having different masses but same charge are known as isotopes.* The isotopes of the same element have the same chemical properties. Thus hydrogen is available in 3 isotopic forms—the mass of ordinary hydrogen is 1 whereas the masses of the other two are 2 and 3 respectively. Similarly chlorine has two types of atoms having the masses 35 and 37 respectively. So in the case of isotopes atomic number (*i.e.*, number of protons) is the same but mass number or atomic weight is different.

It has also been observed that sometimes the atoms of different elements may have the same masses. Such atoms of different elements are known as *isobars*. For example, atoms of both calcium and argon may have the mass 40.

1.7. NEUTRAL CHARACTER OF ATOMS AND MOLECULES

The number of protons present in an atom is always equal to the number of electrons. Hence the atom *on the whole is neutral.* Atoms of the same or different elements interact to produce chemical compounds. During this process electrons are either transferred from one atom to another or simple sharing takes place. Hence the number of protons and electrons in a compound also remains the same. This means that all the compounds are also neutral on the whole.

1.8. ATOMIC NUMBER

The position of the element in the periodic table is based upon its atomic number because this is closely related with atomic structure and is equal to the number of protons present in the nucleus. Further, as the number of protons equals the number of electrons, the atomic number is also equal to the number of extra nuclear electrons.

1.9. MASS NUMBER

As is clear from Rutherford's model of the atom, almost the entire mass of the atom is concentrated in the nucleus. The sum of the number of protons and neutrons present in the nucleus is called the *Mass Number.* Mass number of an element is always a whole number; its value is very close to the atomic weight although not exactly equal to it.

Oxygen atom has 8 protons and 8 neutrons, hence its mass number is equal to 8 + 8 = 16. This value is equal to the atomic weight of the most common isotope of oxygen.

1.10. ATOMIC WEIGHT

The actual weights of atoms are too small to permit correct measurement. Therefore, in actual practice, only the relative weights of atoms are used. Since the atom of hydrogen is the lightest of all, it is taken as the basis for the comparison. In the earlier stages, atomic weight was defined as the average relative weight of an atom as compared to the weight of a hydrogen atom taken as one. For example, oxygen has an atomic weight of 16 since it is sixteen times heavier than a hydrogen atom.

It is not necessary that the atomic weight of an element should be a whole number. Most of the elements have fractional atomic weights. The reason for the fractional atomic weight is the existence of different isotopes of the same elements, *e.g.*, chlorine has two isotopes having the atomic weight 35 and 37 respectively and its average weight is 35.5, since the two isotopes are present in the ratio 3 : 1.

The atomic weight is not expressed merely as a number. On the other hand, it is expressed in terms of *atomic mass units* (a.m.u.). The idea of atomic mass unit was put forward to modify the definition of atomic weight. The mass of hydrogen atom, the lightest atom, is said to have an atomic weight of 1 a.m.u. Oxygen and Carbon have atomic weight equal to 16 and 12 a.m.u. respectively.

1.11. NUCLEAR FORCES

From the structure of the nucleus, we have the conclusion that nucleus contains protons and neutrons. Now since protons are positively charged particles, so due to the repulsive electrostatic forces between protons these will be pushed apart. In order that the nucleus may be stable, some attractive forces must exist which must be strong enough to counteract the repulsive forces. These attractive forces are specific nuclear forces having the following characteristics:

(*i*) They are different from gravitational and e.m. forces and their relative magnitude may be written as

$$Fg : Fe : Fn = 1 : 10^{36} : 10^{38}.$$

(*ii*) They are charge independent, *i.e.*, there can be *p-p* forces. *p-n* forces and *n-n* forces, *i.e.*, forces between protons and protons, forces between protons and neutrons and forces between neutrons and neutrons.

(*iii*) They are *spin dependent*, force between two nuclei having *parallel spins are greater than those between nuclei with antiparallel spin,* so they are spin dependent.

(*iv*) They are *short range forces*, *i.e.*, they operate over a very small distance, *i.e.*, about 2×10^{-15} m or 2 fermi. Outside the nucleus they diminish very rapidly while coulombic forces which are repulsive force and have long range become prominent.

(*v*) The magnitude of the force is very high and very large work is required to separate the nucleus into its constituents because of their large magnitude.

In order to explain the nature of these forces, Yukawa in 1935 predicted that there exist transistent particles in the nucleus, called *mesons* which have a mass in between the mass of an electron and proton. This mesonic cloud in the nucleus provides the exchange character of the nucleus, *i.e.*,

$$p \rightarrow n + \pi^+$$
$$n \rightarrow p + \pi$$
$$p \rightarrow p + \pi^0$$

It should be noted that the exchange equation $p \rightarrow n + \pi^+$ is *energetically not possible* since the mass of proton is less than the mass of a neutron. However, if some energy is provided to proton, then this reaction takes place.

1.12. RADIUS OF THE NUCLEUS

The radius of the nucleus can be calculated from the scattering of α-particle or other particles like protons, neutrons etc. The radius of the nucleus will correspond to the *closest distance of approach of the interacting particle* (like α-particle) to the target nucleus. The radius of the nucleus is given by the empirical relation

$$R = R_0 A^{1/3}$$

where R_0 correspond to the distance of approach of an interacting particle to the target nucleus and has a value 1.4×10^{-15} m.

Let us take up the case of carbon, where $A = 12$.

$$\therefore \quad R = 14 \times 10^{-15} \times 12^{1/3} = 3.21 \times 10^{-15} \text{ m} = 3.21 \text{ fermi}$$

$$[1 \text{ fermi} = 10^{-15} \text{ m}]$$

Similarly, in the case of a heavier nucleus like uranium, where $A = 238$,

$$R = 14 \times 10^{-15} \times (238)^{1/3} = 8.7 \times 10^{-15} \text{ m}$$

The nuclear radius is found to vary between 10^{-14} m and 10^{-15} m.

1.13. NUCLEAR DENSITY

The size of the nucleus being very small and it contains sufficient mass, so the density of the nucleus is very high. The density of all nuclei is approximately 10^{17} kgm/m^3,

$$\therefore \quad \rho = \frac{M}{V} = \frac{M}{\frac{4}{3}\pi R_0^3}$$

$$= \frac{3M}{4\pi R_0^3}$$

$$= \frac{1.67 \times 10^{-27} \times 3}{4 \times 3.142 \times (10^{-15})^3} = 10^{17} \text{ kg/m}^3$$

which is very much higher than that we normally see in matter. Since mass has an equivalence of energy, so the concentration of energy to the nucleus is also very high.

1.14. NUCLEAR SPIN OR THE ANGULAR MOMENTUM OF NUCLEI

The critical analysis of atomic spectra showed the presence of hyperfine structure of atomic lines which was explained to be due to a magnetic coupling between the spin and orbital angular momentum of atomic electrons. *Due to the interaction between the two momenta, there is a splitting of atomic energy levels called the fine structure.* In order to explain the *hyper fine* structure, Pauli suggested that just like the electrons, nuclei of atoms also *possess the orbital and spin angular momentum.* Similarly, there are the orbital and the spin magnetic moments associated with the angular momenta. The interaction between the two magnetic moments causes a further splitting of the energy levels known as the *hyperfine structure.*

The angular momentum of the nuclei is often referred to as *spin.* But this is a misnomer. The word spin should only refer to the intrinsic angular *momentum* of elementary particles. The total angular momentum of nuclei then constitute *spin* and *orbital* angular momenta of the nucleons.

Each nucleon in the nucleus has an angular momenta which is presumably due to the spinning motion of the nucleon about its axis passing through its centre of mass. The magnitude of the spin angular momentum is $\frac{1}{2}\hbar \left(\frac{1}{2}\ \frac{h}{2\pi}\right)$ which according to wave mechanics can constitute two states *viz.* (*i*) Spin parallel to the given direction and (*ii*) spin anti-parallel to the given direction. It means that the component of spin along some direction (say *Z*-axis) will be $\frac{\hbar}{2}$ or $\frac{\hbar}{2}$. In addition, each nucleon has orbital angular momenta whose component along any given direction can be *n*, where *n* is an integer.

So each nucleon has *total angular momentum i* about a given direction and is given by

$$i = l \pm S \qquad ...(1)$$

where *l* is orbital angular momentum and *S* is the spin angular momentum. Nuclei having more than one nucleon the resultant angular momentum of the nucleus is

$$I = L \pm S \qquad ...(2)$$

where $L \rightarrow$ total orbital angular momentum and $S \rightarrow$ total spin angular momentum.

The total angular momentum I is a vector. Since L is integral multiple of $\hbar$ and S is $\hbar/2$, so magnitude of total angular momentum I will be $n\hbar$ when A is **even** and $(2n + 1)\,\frac{\hbar}{2}$, when A is **odd**, where n is an integer.

1.15. MAGNETIC MOMENT

An electron moving in a Bohr orbit must possess a magnetic moment. If we consider an electron moving in an orbit constitute a loop bearing a current I, then the magnetic moment of the electron is equal to product of current and the area of the loop.

For the nth Bohr's orbit, the area is πr_n^2 while the current (charge per unit time is $-e$ multiplied by the number of revolutions made by the electron unit time.

$$\therefore \quad \mu = I\,(\pi r_n^2) = \frac{-ev}{2\pi r_n}\cdot \pi r_n^2$$

$$= \frac{-e}{2}\, v r_n \qquad ...(1)$$

where v is the velocity of the electron. Multiplying and dividing by m, we get

$$\vec{\mu}_l = \frac{-e}{2m}\cdot mv\,r_n$$

$$= \frac{-e}{2m}\cdot \vec{A}_l \qquad ...(2)$$

where mvr_n is the orbital angular momentum A_l.

Also $\vec{\mu}_l$ is the vector magnetic moment and $\vec{A}_l$ corresponds to the vector angular momentum.

The negative sign shows that these two vectors are in opposite direction. For a *negative revolving charge* the subscript l is used for 'orbital' to distinguish it from the 'spin' parameters.

The ratio $\frac{\mu_l}{A_l} = \frac{-e}{2m}$ is called *gyromagnetic ratio* for the orbital motion.

For a Bohr hydrogen atom in its ground state $n = 1$ and from Bohr's quantum condition $mvr = n\hbar$, we have

$$(A_l)_{n=1} = \hbar,$$

Magnetic moment given by Equation (2) for the orbit is called *Bohr magneton* given by

$$\mu_B = \frac{e\hbar}{2m}$$

$$= \frac{1.6\times10^{-19}\times6.06\times10^{-34}}{2\times9.1\times10^{-31}}$$

$$= 9.27\times10^{-24}\ \text{J/T}$$

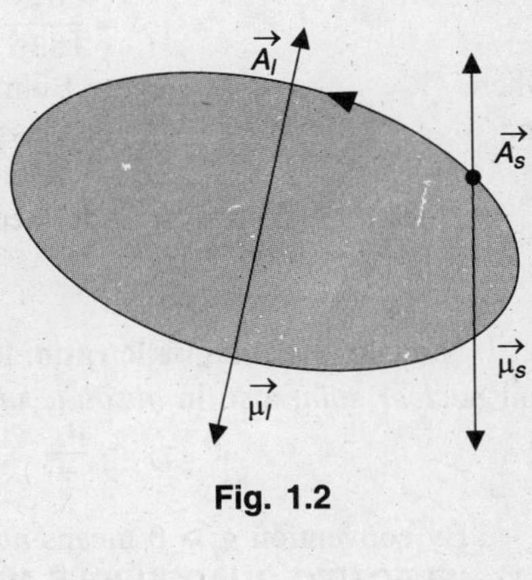

Fig. 1.2

Although quantum mechanics shows that *orbital motion* of a hydrogen electron in its ground state contributes no magnetic moment, it, however, turns out that *electronic magnetic moments* are conveniently expressed in units of *Bohr magnetons*.

The ratio of Nuclear Magneton μ_N to Bohr Magneton μ_B is given by

$$\frac{\mu_N}{\mu_B} = \frac{m_e}{m_p} = \frac{1}{1836}$$

For all nuclei, magnetic moment μ_I is given by

$$\mu_I = g\cdot\frac{e\hbar}{2m_p}$$

where 'g' factor is different for different nucleus and $g = 2.792$ for a proton and $g = -1.913$ for a neutron.

$$\therefore \qquad \mu_{proton} = 2.792 \frac{e\hbar}{2m_p} = 2792\mu_N$$

and $$\mu_{neutron} = -1.913 \frac{e\hbar}{2m_p} = -1.913\mu_N$$

It may be concluded that in case of a proton because of its positive charge, the magnetic field is parallel to the direction of its total angular momentum and so its *magnetic moment is positive.*

In case of a neutron, negative moment is due to the complex nature of the neutron. It may be considered that a neutron consists of equal amounts of positive and negative charges and the negative charge is at a long distance from the axis of rotation.

To support the above argument, let us take up the case of deutron, which has a proton and a neutron in the nucleus. Its spin $I = 1$, *i.e.*, the spins of the two nucleons are *parallel.* Also since two magnetic moments are oppositely directed, so the resultant magnetic moment should be $2.792 - 1.913 = 0.879\ \mu_N$. Experimental measurements show that $\mu_d = 0.858\mu_N$. So the theoretical value is in agreement with the experimental result.

It may be mentioned that in case of *odd-even* or *even-odd* nuclei the magnetic moment is due to the *single unpaired nucleon.* In general, magnetic moment of an electron is about 1000 times that of a nucleus.

1.16. MAGNETIC AND ELECTRIC PROPERTIES OF THE NUCLEUS

Magnetic property. When a charged particle of mass M moves in a closed path it gives rise to a *magnetic field*. As the nuclei are made up of protons and neutrons, the spin angular momentum of **proton** and the orbital angular momentum of **protons** within the nucleus produce a magnetic field, which can be described in terms of a resultant magnetic dipole moment located at the centre of the nucleus.

The nuclear *magnetic moment vector* $\vec{\mu_I}$ which is taken to be vector parallel or anti-parallel to the spin momentum vector $I\ [(I = I\,(I+1)]$, we can measure it either in absolute units or in terms of *nuclear magneton* μ_N given by

$$\mu_N = \frac{e\hbar}{2M_p c} = 0.505 \times 10^{-23} \text{ erg/gauss}$$

$$= \frac{\mu_B}{1836}$$

where $\mu_B \to$ Bohr magneton

and $M_p \to$ Mass of proton

$c \to$ Velocity of light

Nuclear g factor. It is defined as

$$\frac{\mu_I}{\mu_N} = \vec{I} \cdot g_1$$

Nuclear gyromagnetic ratio. It is defined as the ratio of *nuclear moment* in **absolute units** and *nuclear spin*, also in *absolute units.*

$$\gamma_I = \frac{\mu_I}{\vec{I}} \quad \text{or} \quad \mu_I = \gamma_I \,.\hbar .\vec{I}$$

By convention $g_I > 0$ means *nuclear spin and magnetic moment* are parallel to each other.

1.17. ELECTRIC QUADRUPOLE MOMENT

The electric quadrupole moment in a nucleus is due to the *non-uniform distribution of electric charges within its periphery.* Experiments show that nuclei do not possess any electric dipole moment, but the *electric quadrupole moments* have been observed in some nuclei. The electric quadrupole moment is a measure of the departure of a nucleus from the spherical symmetry.

It may be stated that electric quadrupole moments may be either positive or negative. A positive value means that the nucleus is of shape of a *prolate spheroid, i.e.*, it is elongated along the spin axis. On the other hand, a negative value corresponds to an *oblate spheroid* shape of the nucleus, *i.e.*, a nucleus flattened along the spin axis.

The quadrupole moment of a nucleus has been explained on the basis of a *collective model of the nucleus.*

1.18. WAVE MECHANICAL PROPERTIES OF THE NUCLEUS

(*i*) Statistics. In quantum mechanical treatment a composite system with a group of particles such as a nucleus is described by either the **Bose-Einstein** or the **Fermi-Dirac statistics.** Experimental evidences show that all particles which have half integral spin (*e.g.*, protons, neutrons and electrons) follow the Fermi-Dirac statistics, while those having an integral or zero spin obey the Bose-Einstein statistics. So all nuclei with odd number 'A' obey Fermi-Dirac statistics and those with even 'A' obey the Bose-Einstein statistics. *The statistics of atomic nuclei is determined experimentally from the rotational Raman spectra of homogeneous molecules.*

(*ii*) Parity. Parity is the fundamental nuclear property and is considered to be more important than the nuclear spin. From quantum theory every moving material particle has a wave associated with it which corresponds to a wave ψ that depends on the space co-ordinates (x, y, z). Parity of a nucleus refers to the behaviour of this wave function as a result of the inversion of the co-ordinates. If we replace $-x$ for x, $-y$ for y and $-z$ for z, then the wave function ψ is said to possess an even parity if

$$\psi(x, y, z) = \psi(-x, -y, -z)$$

On the other hand, the wave function ψ will correspond to an odd parity if

$$\psi(x, y, z) = -\psi(-x, -y, -z)$$

Let us take the example of the case of the system for which the potential energy is a symmetric function of space co-ordinates. In these cases the wave functions show symmetry,

i.e., $$V(x, y, z) = V(-x, -y, -z)$$

Nuclear states are characterised by a definite value of parity that can be different for the different states of the same nucleus. Parity remains conserved during the nuclear transformations. Hence two particles which are the mirror images of each other must obey the same physical rules.

Parity of a nucleus is closely associated with the value of orbital angular momentum L. As a rule if L is even, parity is also even and if L is odd, parity is also odd.

1.19. PACKING FRACTION

Though the atomic masses are considered to be whole numbers, yet they differ from the integers by some small amount. This variation of atomic masses from whole numbers is considered to be due to some deviations of the mass of the nucleus from the mass of the constituents of the nucleus, *i.e.*, protons and neutrons. We define a term called packing fraction given by

$$f_m = \frac{M - A}{A} \quad \text{where } M = \text{Actual mass of an element in a.m.u.}$$

A = Mass no. *i.e.*, number of protons and neutrons in the nucleus.

The difference between atomic mass M and mass numbers A is known as mass defect and given by

$$\Delta m = M - A$$

$$\therefore \text{ Packing fraction } f = \frac{\Delta m}{A} = \frac{\text{mass defect}}{\text{mass number}}$$

If we plot a curve between *packing fraction* and *mass number of different nuclides*, we observe that packing fraction is more for elements of low mass numbers (except for He^4, C^{12}, O^{16}) and then it decreases very rapidly with increase in mass number. After the curve being almost flat for sufficient mass numbers, the packing fraction starts increasing steady but quite slowly.

It should be noted clearly that packing fraction is a fundamental characteristic of the nucleus and is directly related to the availability of the energy of the nucleus and its stability.

A negative packing fraction which will also corresponds to a negative mass defect will mean that the *atomic mass of the nucleus is less than the nearest whole number* which will correspond to some conversion of mass into energy during the formation of that nucleus. Also negative packing fraction will imply that the nucleus is exceptionally stable because *we have to supply more energy to break the nucleus* into its constituent nucleons while a positive packing fraction will indicate the *nucleus to be less stable.*

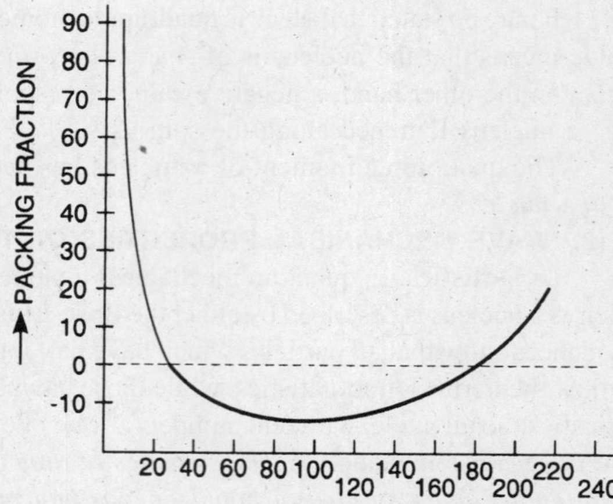

Fig. 1.3

1.20. MASS DEFECT AND BINDING ENERGY

If we measure the mass of the nucleus, we find it to be somewhat different than some of the masses of the constituent nucleons. If a nucleus is built up from its constituents, some mass disappears and is converted into energy called binding energy of the nucleus. Also if the nucleus is to be split up into its constituents, same energy will be required to break up the nucleus.

Let us compute the B.E. of a deutron nucleus as under:

Mass of Neutron ${}_0n^1$ = 1.008665 a.m.u.
Mass of Proton ${}_1H^1$ = 1.007825 a.m.u.
Total mass = 2.016490 a.m.u.
Mass of Deutron = 2.014103 a.m.u.
Mass Defect Δm $= M - A$
$= 2.014103 - 2.016490 = -\ 0.002387$ a.m.u.

The corresponding B.E. may be got as

$$\Delta E = 931 \times \Delta m \qquad [\because 1 \text{ a.m.u.} = 931 \text{ MeV}]$$
$$= 931 \times -\ 02357 \ = -\ 2.22 \text{ MeV}.$$

Since in Deutron, there are only two nucleons involved, so B.E./nucleon = 2.22 × 1/2 = 1.11 MeV.

Similarly, if we take up the case of ${}_2He^4$ nucleus, we have

Mass of 2 ${}_0n^1$ + 2 ${}_1H^1$ = 4.032980 a.m.u.
and Mass of ${}_2He^4$ = 4.002604 a.m.u.
Mass defect $\Delta m = M - A = 4.002604 - 4.032980 = -\ 0.030376$ a.m.u.
$\therefore$ $\Delta E = -\ 231 \times 0.030376$ a.m.u. $= -\ 28.3$ MeV $= -\ 28$ MeV

So B.E./nucleon of ${}_2He^4$ $= \dfrac{28}{4} = 7.0$ MeV

1.21. BINDING ENERGY CURVE

We have seen above the concept of Binding Energy. It should be noted that the binding energy/nucleon is more significant than the binding energy and if a graph is plotted between mass number and binding energy/nucleon for different nuclei we have the following inferences:

(*i*) B.E./nucleon for very light nuclei like ${}_1H^1$ is very small.

(*ii*) B.E./nucleon increases rapidly up to mass number 20 and then there are peaks corresponding to nuclei ${}_2He^4$, ${}_6C^{12}$, ${}_8O^{16}$, which indicate that these nuclei are more stable than the neighbouring nuclei (the reason for this being that they may be considered to possess *magic number*, *i.e.*, their mass number is divisible by 4 and these nuclei may have ${}_2He^4$ nuclei as their constituents.

(*iii*) After mass number 20, the B.E./nucleon increases gradually and for mass numbers between 40 and 120, the curve is more or less flat. The average value of B.E./nucleon in this region is about 8.5 MeV.

(*iv*) After mass number 120, B.E./nucleon decreases and reaches a value of 7.6 MeV for uranium. This low value of B.E./nucleon in case of heavier nuclei is unable to control the coulomb repulsion between the protons and so they become unstable. These nuclei disintegrate with the emission of α-rays, β-rays and γ-rays and *are known as radioactive nuclei.*

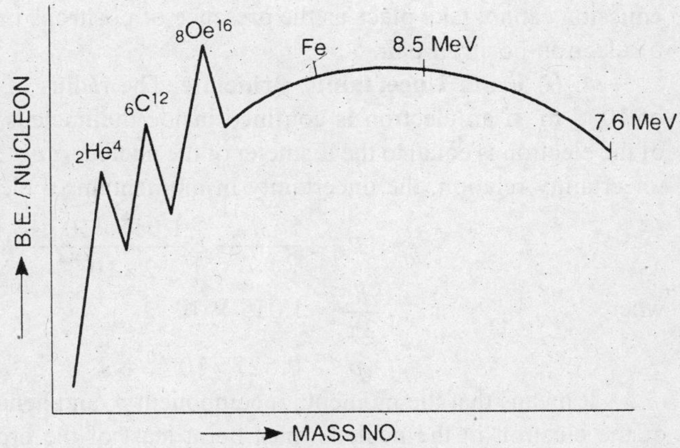

Fig. 1.4

1.22. WHY ELECTRONS CANNOT EXIST IN NUCLEUS?

1. Magnetic Moment variations. The magnetic moment of an *electron* is of the order of 1 Bohr magneton, $\mu_e = \frac{e\hbar}{2m_e}$, while the magnetic moment of a proton is of the order of **1 nuclear magneton,** $\mu_N = \frac{e\hbar}{2m_p}$

$$\therefore \quad \frac{\mu_B}{\mu_N} = \frac{m_p}{m_e} \simeq \frac{1.67\times10^{-27}\text{ kg.}}{9.1\times10^{-31}\text{ kg.}}$$

$$\therefore \quad \mu_B \simeq 1836\mu_N$$

If a nucleus contains electrons, then its magnetic moment should be of the order of μ_B *i.e.*, 1 *Bohr magneton.* Experimentally the magnetic moment of a nucleus vary between $-2\mu_N$ and $+4\mu_N$. So electron cannot exist in the nucleus due to **magnetic moment** considerations.

2. Statistics. Since electrons and protons are both Fermions (*i.e.*, they have half integral spins), it means that resultant angular momentum of a nucleus will be given by the addition of the spins of the constituents, *i.e.*, a nucleus with odd number of particles will have spin as *half odd integer,* while a nucleus with even number of constituents will have spin as an *integer.*

On proton-electron model, $_7N^{14}$, would consist of 14 protons and 7 electrons, *i.e.*, 21 nucleons (odd number) and it should obey **Fermi-Dirac statistics.**

If we consider a neutron-proton model of the nucleus, *i.e.*, 7 protons and 7 neutrons add up to give $A = 14$, an even number of nucleons, so it should obey **Bose-Einstein statistics.**

An examination of intensity pattern of rotational spectrum of molecular nitrogen shows that N^{14} has integral spin. So spectral analysis favours the even number of nucleons and consequently *symmetric wave functions* for Nitrogen molecule.

Similarly in case of $_4Be^9$, the experimental evidence favours odd integral spin (*i.e.*, 5 neutrons and 4 protons) adding up to give 9 nucleons) rather than integral spin (*i.e.*, 9 protons and 5 electrons adding up to give 14 nucleons).

So statistics and spin considerations also favour *proton-neutron* model and not the *proton-electron* model of the nucleus.

3. Other Considerations. (*a*) Size of the electron calculated on the basis of coulomb law is much larger than the size of the nucleus in which it is to be present.

(*b*) Spin magnetic moment of the electron is much larger than that of a proton (Proton has spin magnetic moment ~ 0.15% that of an electron). So a nucleus having an electron as its constituent must have large spin magnetic moment, while actually it has the value ~ spin magnetic moment of a proton.

(*c*) Theoretical considerations suggest that an electron within the nucleus should have compton wavelength of de-Broglie wave *i.e.*, $\hbar = 0.24 \times 10^{-10}$ m, while the nuclear radius is about 1.2×10^{-15} m. So, the electron cannot be confined to the nucleus.

(*d*) Theory of β-decay also contradicts the presence of electrons as positrons (β ± decay) emission cannot take place as the presence of positrons inside the nucleus will lead to *annihilation* of electron-positron pair.

4. (*i*) From Uncertainty Principle. The radius of the nucleus of any atom is of the order of 10^{-14} m. If an electron is confined inside the nucleus, then the uncertainty in the position Δx of the electron is equal to the diameter of the nucleus, *i.e.*, $\Delta x \approx 2 \times 10^{-14}$ m. Using the Heisenberg's uncertainty relation, the uncertainty in momentum of electron is given by the relation

$$\therefore \qquad p_x \geq \frac{h}{2\pi.\Delta x} \geq \frac{1.055 \times 10^{-34}}{2 \times 10^{-14}} \text{ kg ms}^{-1} \qquad \because \quad \Delta x \approx 10^{-14}$$

where $$\frac{h}{2\pi} = 1.055 \times 10^{-34}$$

$$\therefore \qquad p_x \geq 0.527 \times 10^{-20} \text{ Ns}$$

It means that the momentum component p_x and hence the magnitude of total momentum $|\vec{p}|$ of the electron in the nucleus must be at least of the order of magnitude, *i.e.*,

$$|\vec{p}| \sim p_x \sim \Delta p_x \sim 0.527 \times 10^{-20} \text{ Ns}$$

Now the mass of the electrons is 9.1×10^{-31} kg., the order of magnitude of momentum (0.527×10^{-20} kg. ms^{-1}) is relativistic. Using the relativistic formula for the energy E of the electron, we have

$$E^2 = p^2 c^2 + m_0^2 c^4$$

As the rest energy $m_0 c^2$ of the electron is of the order of 0.511 MeV, which is much smaller than the value of first term, so it can be neglected. Thus we have

$$E^2 = p^2 c^2 \quad \text{or} \quad E = pc$$

$$E \approx (0.527 \times 10^{-20}) \times (3 \times 10^8) \text{ Joules}$$

$$\approx \frac{(0.527 \times 10^{-20})(3 \times 10^8)}{1.6 \times 10^{-19}} \text{ eV} = 10 \text{ MeV}$$

which means that if the electrons exist inside the nucleus, their energy must be of the order of 10 MeV. However, we know that the electrons emitted by radioactive nuclei during beta decay have energies only 3 to 4 MeV. So, in general electrons cannot exist in the nucleus.

For protons and neutrons, $m_0 \simeq 1.67 \times 10^{-27}$ kg. This is a non-relativistic problem as $|\vec{v}| \sim |\vec{p}|/m_0 \sim 3 \times 10^6$ ms^{-1}. The kinetic energy K in this case is given by the relation

$$K = \frac{p^2}{2m_0} = \frac{(0.527 \times 10^{-20})^2}{2 \times 1.67 \times 10^{-27}} \text{ Joules}$$

$$\approx \frac{(0.527 \times 10^{-20})^2}{2 \times 1.67 \times 10^{-27} \times 1.6 \times 10^{-19}} \text{ eV} = 52 \text{ KeV}$$

Since this energy is less than the energies carried by these particles emitted by nuclei, **both these particles can exist inside the nuclei.**

(*ii*) Existence of electron in Orbits of an Atom. In an atom, the electron is under the influence of electrostatic potential of the positively charged nucleus. It is confined to the linear dimensions equal to the diameter of electronic orbit. The uncertainty in the position Δx of the electron is of the order of $2R$ where R is the radius of the orbit and the corresponding uncertainty in momentum component Δp_x is given by

$$\Delta p_x \geq \frac{h}{4\pi.2R}$$

which shows that the momentum of an electron in atomic orbit is at least

$$|\vec{p}| \sim p_x \sim \Delta p_x \sim \frac{h}{4\pi.2R}$$

$$\sim 0.527 \times 10^{-24} \text{ kg ms}^{-1} \qquad \therefore R = 10^{-10} \text{ m}$$

This is non-relativistic momentum for an electron. The kinetic energy K is given by

$$K = \frac{p^2}{2m_0} = \left(\frac{h}{4\pi R}\right)^2 \frac{1}{2m_0} = \frac{h^2}{32\pi^2 m_0 R^2}$$

The potential energy of an electron in the field of nucleus with atomic member Z is given by

$$V = \frac{-Ze^2}{4\pi\varepsilon_0 R}$$

So the total energy of the electron in its orbit will be

$$E = K + V = \frac{h^2}{32\pi^2 m_0 R^2} - \frac{Ze^2}{4\pi\varepsilon_0 R}$$

$$= \frac{(1.055\times10^{-34})^2}{8\times9.1\times10^{-3} y\times R^2} - \frac{Z\,(1.6\times10^{-19})^2}{4\times3.14\times8.85\times10^{-12}\times R}$$

where $\frac{h}{2\pi} = 1.055 \times 10^{-34}$

or $E = \frac{10^{-20}}{R^2} - \frac{15\times10^{-10} Z}{R}$ eV

Taking $R = 10^{-10}$ m, we have

$E = (1 - 15Z)$ eV

Now the binding energies of outermost electrons in H and He are –13.6 and –24.6 eV respectively, so the value of binding energy derived on the basis of uncertainty principle is acceptable as these are comparable in magnitudes.

Example 1. *Calculate the B.E./nucleon of $_{17}Cl^{35}$ nucleus (its mass being 34.9800 a.m.u). Given mass of $_0n^1$ = 1.008665 a.m.u.*

Mass of $_1H^1$ = 1.007825 a.m.u.

Solution: Mass of 17 protons = 17 × 1.007825
= 17.134025 a.m.u.

Mass of 18 neutrons = 18 × 1.008665 = 18.155970 a.m.u.

Total mass of nucleons = 35.288995 a.m.u.

Mass Defect = 35.288995 a.m.u. – 34.98000 a.m.u. = 0.308995 a.m.u.

Now 1 a.m.u. = 931 MeV

∴ B.E of $_{17}C^{35}$ nucleus = 931 × 0.308995 = 288 MeV

B.E./Nucleon = **8.22 MeV.**

Example 2. *Compute the energy of 1 a.m.u. in MeV.*

Solution: Mass of proton (1 a.m.u.) = 1.66×10^{-27} kg.

Velocity of light $c = 3 \times 10^8$ m/sec.

From Einstein's relation

$$E = mc^2 = 1.67 \times 10^{-27} \times (3 \times 10^8)^2$$
$$= 14.9 \times 10^{-11} \text{ joule}$$

Now 1 MeV = 1.6×10^{-13} joule (∵ 1 eV = 1.6×10^{-13} joule)

$$\therefore \quad E = \frac{14.9\times10^{-11}}{1.6\times10^{-13}} = \mathbf{931\ MeV}$$

Example 3. *Compute the energy equivalence of 1 gm in kWh.*

Solution: We have 1 gm = 10^{-3} kgm, $c = 3 \times 10^8$ m/sec.

From relation $E = mc^2$, we have

$E = 10^{-3} \times (3 \times 10^8)^2 = 9 \times 10^{13}$ joule

Now 1 kWh = 36 × 105 J

∴ Energy equivalence of 1 gm

$$= \frac{9 \times 10^{13}}{16 \times 10^{-13}} = \mathbf{2.5 \times 10^7\ kWh.}$$

Example 4. *Compute the energy carried by an electron.*

Solution: Mass of electron = 9.1×10^{-31} kgm.

Velocity of light $c = 3 \times 10^8$ m/sec.

∴ From relation $E = mc^2 = 9.1 \times 10^{-31} \times (3 \times 10^8)^2$ joule

Now 1 MeV = 1.6×10^{-13} joule

$$E = \frac{9.1 \times 10^{-31} \times 9 \times 10^{16}}{1.6 \times 10^{-13}}$$

$$= \mathbf{0.51\ MeV.}$$

Example 5. *Find the energy release, if two $_1H^2$ nuclei can fuse together to form a nucleus. The binding energy per nucleon of and $_2He^4$ is 1.1 MeV and 7.0 MeV respectively.*

Solution: $_1H^2 + _1H^2 \rightarrow _2He^4$ + Energy (E)

$2\,(1.1) + 2\,(1.1) \rightarrow 4\,(7.0) + E$

$$E = 28 - 4.4 = \mathbf{23.6\ MeV}.$$

Example 6. *Determine the radii of a O^{16} and a Pb^{206} nucleus, given that R_0 = 1.4fm.*

Solution: From the relation $r = R_0 A^{1/3}$

We have $R\,(O^{16}) = 1.4 \times (16)^{1/3} = 3.33$ fm

$R\,(Pb^{208}) = 1.4 \times (208)^{1/3} = \mathbf{3.29\ fm.}$

Example 7. *Calculate the binding energy of Te^{126}.*

Solution: The binding energy is given by the relation

$$BE = [ZM_p + NMn - ZM^A]\,c^2$$

$$= (52 \times 1.007825 + 74 \times 1.008665 - 125.903322) \times 931.5 \text{ MeV}$$

$$= 1.066 \times 103 \text{ MeV} = \mathbf{1.066\ GeV}.$$

Example 8. *A nucleus with A = 235, splits into two nuclei whose mass numbers are in the ratio 2 : 1. Find the radii of the new nuclei.*

Solution: Let A_1 and A_2 be the mass numbers of the new nuclei, then $A_1 = 1/3 \times 235$ and $A_2 = 2/3 \times 235$. So the radii of the new members are given by

$$R_1 = R_0\,A_1^{1/3} = 1.4 \times \left(\frac{235}{3}\right)^{1/3} = 5.99 \text{ fm}.$$

$$R_2 = R_0\,A_2^{1/3} = 1.5 \times \left(235 \times \frac{2}{3}\right)^{1/3} = \mathbf{7.55\,fm.}$$

Example 9. *Determine the stable nucleus that has a radius = 1/3 that of Os^{189}*

Solution: Since $R \propto A^{1/3}$, so we may write

$$\frac{1}{3} = \frac{R}{ROs} = \left(\frac{A}{AOs}\right)^{1/3} \approx \left(\frac{A}{189}\right)^{1/3}$$

$$\therefore \quad A = \frac{189}{27} \approx \mathbf{7}$$

This corresponds to Li^7 Nucleus.

Example 10. *What is the energy required to remove the last tightly bound neutron from $_{20}Ca^{40}$?*

Solution: Let E be the energy of the last neutron. Also let

M_1 = mass of $_{20}Ca^{40}$

and M_2 = mass of $_{20}Ca^{39}$, then from conservation of mass and energy relation, we have $M_1 c^2 + E = (M_2 + M_n)\,c^2$ where M_n = mass of neutron.

∴ $39.962589 \times 93.5 + E = (38.970691 + 1.008665) \times 931.5$ MeV

$$E = \mathbf{15.6\ MeV.}$$

QUESTIONS AND PROBLEMS

1. What are nuclear forces? How do they differ from electrostatic and gravitational force? Enumerate their salient characteristics.

2. What is nuclear binding energy? Illustrate from the binding energy curve, the various informations that can be got.

3. Explain the terms (*i*) Binding Energy of the nuclides, (*ii*) Atomic mass unit, (*iii*) Mass defect and (*iv*) Packing fraction.

4. Compute the energy equivalence of an electron at rest.

5. Compute the energy equivalence of 1 a.m.u.

6. Draw a curve indicating variation of binding energy per nucleon as a function of mass number of nuclei. From this curve explain the energy release in nuclear fission and fusion processes.

7. The At. mass of $_8O^{16}$ is 16.000 a.m.u. Find its *BE*/nucleon. Given $m_e = 0.00055$ a.m.u. $m_n = 1.00753$ a.m.u. and 1 a.m.u. = 931 MeV. **[Ans.** 7.97 MeV]

8. Write short notes on:

(*a*) Electric quadrupole moment and its importance.

(*b*) Parity.

(*c*) Isospin.

9. The binding energy of the Fast neutron added to the nucleus (*A, Z*) is given by

$$BE\ (n) = [M\ (A-1, Z) + M_n - M\ (A, Z)]\ c^2$$

Using this relation, calculate the binding energy of the last added neutron in the following three isotopes of lead ($Z = 82$):

Pb^{206} Pb^{207} Pb^{208} **[Ans.** 8.05, 6.75, 7.37 MeV]

10. The ground state spin of $_{20}Ca^{41}$ is J = 7/2. If the nucleus is regarded as a rigid sphere with a radius given by $R = 1.4\ A^{1/3}$ fm, what would be the speed of a point on the surface of this nucleus? **[Ans.** 3.16×10^6 m/sec]

11. How much energy is required to break $_8C^{12}$ nucleus into 3 particles? The At. mass of C^{12} is 12 a.m.u. and mass of α-particle is 4.002603 a.m.u. **[Ans.** 7.27 MeV]

12. Calculate the binding energy in MeV of uranium 228 from the following data:

Mass of $_1H^1$ = 1.008142 a.m.u.

Mass of $_0n^1$ = 1.008982 a.m.u.

Mass of $_{82}U^{236}$ = 238.12493 a.m.u. **[Ans.** 7.3 MeV]

13. Assuming the protons and neutrons have equal masses, calculate how many times nuclear matter is denser than water? Nuclear radius is given by the relation $R = 1.2 \times 10^{-15}\ A^{1/3}$ *m* and mass of nucleon $= 1.67 \times 10^{27}$ kgm. **[Ans.** 2.307×10^{14}]

14 Calculate minimum energy in MeV, necessary to disintegrate a deuteron into a proton and a neutron. (Given the mass of proton = 1.00842 a.m.u. Mass of neutron = 1.008962 a.m.u. Mass of deuteron 2.014725 a.m.u., 1 a.m.u. = 1.6×10^{-27} kg.) **[Ans.** 2.241 MeV]

15. The mass of lithium atom is 7.0878 a.m.u. Calculate the binding energy in electron volt of $_3Li^7$ nucleus. Given mass of proton = 1.00758 amu, mass of neutron = 1.00893 amu and that of electron = 0.00055 amu. **[Ans.** 25.779 MeV]

CHAPTER

2

NUCLEAR MODELS

2.1. INTRODUCTION

Many different models for nucleus have been proposed, each of which explains the behaviour of nucleus in some specific situation. But at the same time each of these models is in apparent contradiction with other models or with known facts about nuclear forces. We will limit ourselves only to some basic models for the nucleus, that can explain the general characteristics.

2.2. THE LIQUID DROP MODEL

This was one of the earliest models for nucleus and was proposed by Niels Bohr. It can successfully explain the phenomenon of "nuclear fission". In this model the nucleus is regarded analogous to a drop of liquid and so the name "liquid drop model". This model was suggested because of some basic similarities between a liquid and the nucleus. The basis of this model arises from the fact that molecules in a liquid are held together by **short-range inter-molecular forces** known as **cohesive forces.** This fact is basically analogous to holding together of nucleons in a stable nucleus by the short-range **nuclear forces**. In addition to this, listed below are some analogies between a **small drop** of a liquid and a **nucleus:**

(*i*) A small liquid drop is the symmetrical surface tension force acting towards the centre which are responsible tor the spherical shape of a liquid drop. The nucleus is also assumed to be spherical in shape. A direct consequence of this assumption is that the radii of a nucleus must vary as the cube root of its mass. Experimental results support this fact.

(*ii*) The density of a spherical drop is known to be independent of its volume. It is very much dependent on the type of liquid. Nuclear density is also found to be independent of nuclear volumes but unlike the case of a liquid, the density of nuclear matter is same for all **types of nuclei.** Moreover, the nuclear densities have extremely **large values.**

(*iii*) The molecules of a liquid are in continual random motion. If we extend this analogy to the nucleus and also the evaporation of liquid molecules, we can see it is possible to explain the spontaneous emission of α-particles. Because of **random collisions** between various nucleons, a group of nuclear particles may pick up enough energy which may result in escape of some nucleons. The model successfully explains nuclear reactions and nuclear fission phenomenon.

2.3. EXPLANATION OF NUCLEAR FISSION ON LIQUID DROP MODEL

Bohr and Wheeler successfully explained the phenomenon of nuclear fission on liquid drop model. A liquid is able to oscillate in a variety of ways when suitably excited. A possible mode in which it can happen is shown in Fig. 2.1, where we see that a drop successively change to a prolate spneroid, a sphere, an oblate spheroid, a sphere and so on.

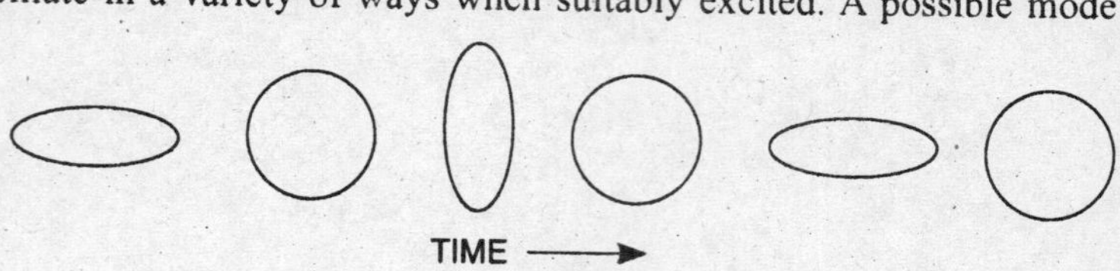

Fig. 2.1

The basic forces responsible in these changes are that of surface tension and inertia of moving liquid molecules.

Whenever distorted from a spherical shape, the force of surface tension always tries to restore the drop to its spherical shape. As the molecules of moving drops are in motion, their inertia causes the drop to move sphericity and in the process it moves to the other extreme distortion. When this happens the force of surface tension comes again, acts and process repeats.

In a similar way, nucleus can be excited and like a liquid can also vibrate. **Nuclear forces** play the role of the **surface tension** trying to restore the nucleus into its original state. However, nuclei in addition to the **inertia of its nucleons** are also under the action of **repulsive electrostatic forces.**

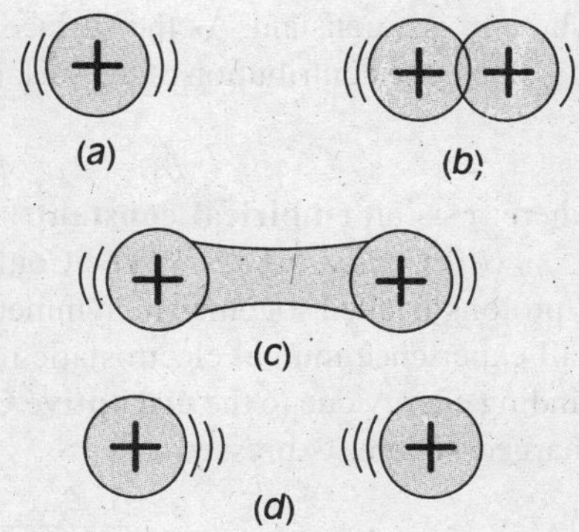

Fig. 2.2

When a nucleus gets distorted, the nuclear forces counter both long range repulsive forces and inertia of nucleons. The extent of distortion will be dependent on extent to which nuclear forces act to counter them. For small distortion surface tension of the nucleus is sufficient to counter in which case the nucleus gives up excitation energy in terms of to and fro oscillations and giving up this energy in the form of gamma rays. On the other hand, whenever the degree of distortion is sufficiently large, it causes the nuclear particles to separate themselves very much apart and these widely apart particles get out of the range of nuclear forces which we know act specifically on short-range distances. Thus nuclear surface tension forces are now not adequate to bring together widely separated group of protons and the nucleus splits into parts. This is shown in Fig. 2.2.

2.4. SEMI-EMPIRICAL MASS FORMULA

Von Weizsocker in 1935 expressed the atomic mass of a nuclide in terms of series of binding energy correction terms to the main mass combination from protons and neutrons. This expression is called **semi-empirical mass formula.** We can express the nuclear mass in terms of protons (*Z*) and neutrons (*A* – *Z*) as

$$M = ZM_p + (A - Z)\, M_n - \frac{B}{C^2} \qquad ...(1)$$

where *B* is the binding energy of a number of terms each of which represents some general properties of nuclide and we can express it as under

$$B = B_0 + B_1 + B_2 + ... \qquad ...(2)$$

We can discuss the various binding energy terms as under:

(*i*) *Volume Energy*. Experimentally it is proved that the binding energy per nucleon has an approximately constant value over a wide range of mass numbers. So, we can consider as a first approximation the **total binding energy** of a nucleus to be proportional to the **total number of nucleons** in it. As the mass number *A* of a nucleus corresponds to total number of nucleons in it, so we may express the binding energy as

$$B_0 \times A \qquad ...(3)$$

which shows that the contribution to the binding energy comes from the entire volume of the nucleus.

So we may write

$$B_0 = a_v A \qquad ...(4)$$

where $a_v \rightarrow$ constant to be determined.

(*ii*) *Surface Energy*. While considering the volume energy expression we assumed that a nucleon is attracted by all other nucleons. However, it is not true, since the **surface nucleons** have fewer nearer neighbours than nucleons which are **deep** within the nuclear volume. This factor should give rise to a slight reduction in the **total binding energy** of the nucleus. This **decrease** is proportional to the number of nucleons on the surface or in other words to the surface area of the nucleus. Let *r* be the radius of the nucleus, then we have

$$r = r_0 A^{1/3} \qquad ...(5)$$

where r_0 is a constant. As the surface area of nucleus is proportional to r^2, it must be proportional to $A^{2/3}$. So the contribution of the surface effect to the binding energy of the nucleus may be expressed as

$$B_1 = -a_s A^{2/3} \qquad ...(6)$$

where a_s is an **empirical constant.**

(*iii*) *Coulomb Energy.* The Coulomb energy of a nucleus is equal to the **potential energy** of Z protons packed together in symmetric assembly of radius r. Since the protons inside a nucleus will experience mutual electrostatic repulsion; resulting in decrease in binding energy. The loss of binding energy due to the **disruptive Coulomb forces** may be got by considering P.E. of a uniformly charged sphere, expressed as

$$\frac{3}{5}\frac{Z^2 e^2}{r} \qquad ...(7)$$

Ze corresponds to the total charge on the nucleus and r is its radius. The **decrease** in binding energy of the nucleus due to **Coulomb repulsion** is proportional to its potential energy and may be written as

$$B_2 = a_c' \frac{3}{5}\frac{Z^2 e^2}{r} \qquad ...(8)$$

where a_c' is a constant. Also from the relation $r_c = r_0 A^{1/3}$, we may write Equation (8) as

$$B_2 = -a_c \frac{Z^2}{A^{1/3}}$$

where we have taken

$$a_c = \frac{3}{5} a_c' \frac{e^2}{r_0}$$

From more accurate calculations, we get

$$B_2 = -a_c \frac{Z(Z-1)}{A^{1/3}} \qquad ...(9)$$

(*iv*) *Asymmetry Energy.* It is observed that in the most abundant isotopes of the light nuclei the **number of protons** are equal to the **number of neutrons**. So among the different isotopes of an element the ones having **equal number** of **protons** and **neutrons** will be most stable and so will possess maximum binding energy. Which in other words mean that excess of neutrons over protons will lead to decrease in B.E. thereby increasing instability.

As the value of A increases the number of neutrons increases and this transition to the **asymmetric configuration** from **symmetric configuration** reduces the stability of the nucleus. Consider a nucleus of mass number A and atomic number Z, the excess of neutrons over protons is $(A - 2Z)$. Which produces a deficit in binding energy because they are out of reach for other nucleons.

The fraction of nuclear volume so affected is $\frac{A-2Z}{A}$ and so the total deficit should be proportional to

$$(A - 2Z)\frac{(A-2Z)}{A}$$

So the contribution due to asymmetric effect to the binding energy of the nucleus is represented by

$$B_3 = -a_\alpha \frac{(A-2Z)^2}{A} \qquad ...(10)$$

(*v*) *Odd-Even Effect.* It is seen that among all the stable nuclides those which have **even number** of protons and **even** number of neutrons are most stable as they are most **abundant.** Nuclides containing **odd** number of protons and **odd** number of neutrons are least abundant and are so least stable. Nuclei having odd number of protons and even number of neutrons and *vice versa* have intermediate degrees of stability. This effect is known as **odd-even effect** and it causes an increase in binding energy when both the number of protons and neutrons are **even.** However, when in the nucleus, the number of neutrons and protons is odd, this effect would cause a decrease in binding energy. The following table shows the change in binding energy, E_δ, due to odd-even effect.

Mass Number A	*No. of Protons* Z	*No. of Neutrons* N	E_δ
even	even	even	$+\frac{\delta}{A^{3/4}}$
odd	even	odd	0
odd	odd	even	0
even	odd	odd	$\frac{-\delta}{A^{3/4}}$ ...(11)

Writing all the terms discussed above the semi-empirical mass formula for a given nucleus may be written as

$$M = ZM_p + (A - Z)\,M_n - \frac{1}{c^2}\left[a_v\,A - a_s\,A^{2/3} - a_c\,\frac{Z\,(Z-1)}{A^{1/3}} - a_r\,\frac{(A-2Z)^2}{A} + E_\delta\right] \quad ...(12)$$

Experimental calculations give the following values of the constants:

$$a_v = 15.753 \text{ MeV}$$
$$a_s = 17.80 \text{ MeV}$$
$$a_c = 0.7103 \text{ MeV}$$
$$a_r = 23.6925 \text{ MeV}$$
$$\delta = 33.6 \text{ MeV} \quad ...(13)$$

However, it should be noted that no single set of values satisfies the equations with known values for B or M. An important aspect of semi-empirical formula is that it can explain the stability of nuclei against α and β decay.

2.5. THE SHELL MODEL

Although the liquid-drop model is very successful in explaining the general trends of the binding energy as a function of Z and N, it **fails to tell us the detailed properties of nuclei;** for this we need a quantum theory of nuclear structure. The importance of quantum effects in the nucleus is evident from the observed quantization of angular momentum: the spin of every nucleus is some multiple of $h/2$, as expected for a system made of protons and neutrons. It is also evident from the observed quantization of energy in the excited state of the nuclei. Fig. 2.3 depicts the energy levels of a few of the excited states. Such states are similar to the excited states of an atom. If the nucleus is one of the upper states, it will make a transition to a lower state with the emission of a *high-energy photon,* or γ-ray.

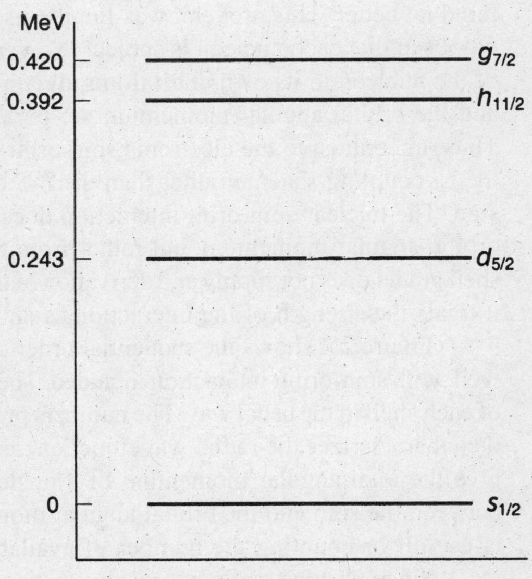

Fig. 2.3

We can also recognize the importance of quantum theory by a simple comparison of the binding energies of nuclei. The observed binding energies are exceptionally large–larger than predicted by the liquid-drop model—if Z or N or both have the values

$$2, 8, 14, 20, 28, 50, 82 \text{ or } 126 \quad ...(14)$$

These are called the **magic numbers.** Nuclei with these numbers of Z or N are exceptionally stable; they are roughly **analogous** to the noble gases. We know that the noble gases are characterized by filled shells of electrons, and we, therefore, expect that nuclei with magic numbers are characterized by *filled shells* of protons or neutrons.

The shell model of the nucleus enables us to calculate the energy levels of the protons and neutrons in a nucleus under the assumption that the net nuclear force experienced by each nucleon can be approximated by an average central force. So the shell model attempts to give calculations on the pattern of atomic energy levels of many electron atoms. But it may give rise to many complications. For example, an electron in an atom moves mostly in empty space and so will have well defined orbital state, while a nucleon in a nucleus has to pass through a densely packed nucleus with nucleon with which it will suffer multiple collisions. It would seem that a nucleon will be scattered continuously and so it will not have a well defined state. However, exclusion principle comes to its rescue because all the quantum states of low energy are filled with other nucleons, and so the colliding nucleon cannot change its state—**it cannot be scattered.** Only near the nuclear surface, where other nucleons are less abundant, can the nucleon be **scattered.** This suggests that the average central potential should resemble a **square well,** giving next to no force within the nuclear volume and a strong restraining force near the nuclear surface.

Some other complications of the shell model prove less amenable to theoretical result. In an atom, the **Coulomb attraction** generated by the nucleus provides a dominant force that holds all the electrons in an average central configuration, so that the electronic contribution to the potential is also central. In a nucleus, there is no dominant central body. A single orbiting nucleon can, therefore, distort the average configuration of the other nucleons, and modify the central potential. Furthermore, in an atom the individual contribution of each electron to the potential is known (Coulomb potential), and, therefore, the net potential of a configuration of electrons can be calculated. In a nucleus, the exact form of the individual contribution of each nucleon is not known. So the shell model must rely on a calculated guess for the **net potential** for which it offers no further theoretical justification.

The early shell-model calculations attempted to fit the nuclear energy levels with a square well of depth about 40 MeV, and a radius comparable to the nuclear radius. The resulting energy levels bore no resemblance to the observed levels. Calculations with potential wells of other shapes fared no better. This probelm was finally resolved in 1949 by J.H.D. Jensen and by M. Mayer who proposed that each nucleon is subject to a very strong spin-orbit interaction, which couples the spin of the nucleon to its own **orbital angular momentum,** and which decreases the energy if the spin and the orbital angular momentum are parallel, and increases the energy if they are antiparallel. Thus, in contrast to the electronic spin-orbit interaction, the nuclear spin-orbit interaction relies on the *j-j* coupling scheme rather than the *L-S* coupling scheme, and furthermore, it has the opposite sign. The nuclear spin-orbit interaction does not arise from the magnetic forces between spin and orbital angular momentum, but rather from the spin dependence of the strong force. However, the shell model does not supply and derivation of the spin-orbit interaction from a theory of nuclear forces; it treats the strength of the interaction as an adjustable parameter, chosen to fit the available data.

Figure 2.4 shows the sequential order of the calculated energy level for a nucleon in a square well, with spin-oribit interaction included. The letters $s, p, d, \ldots$ indicate the orbital angular momentum of each shell in the usual way. The numbers preceding these letters give the principal quantum number that characterizes the radial wavefunction, as in the case of hydrogen. The subscripts on the letters give the total angular momentum of a nucleon, $J = 1 + 1/2$ or $1 - 1/2$. Because of the coupling between the spin and the orbital angular momentum is not a good quantum number; but, however, is useful for counting the number of available states.

Since protons and neutrons are distinct particles, they occupy **distinct sets of states.** Thus, to make the assignment of filled states in a nucleus, we need to draw two copies of Fig. 2.4, one for neutrons and one for protons. In the ground state of a nucleus with N neutrons and Z protons, the lowest N states in the neutron diagram and the lowest Z states in the proton diagram will be occupied. The column of numbers on the right edge of Fig. 2.4 gives the number of available states

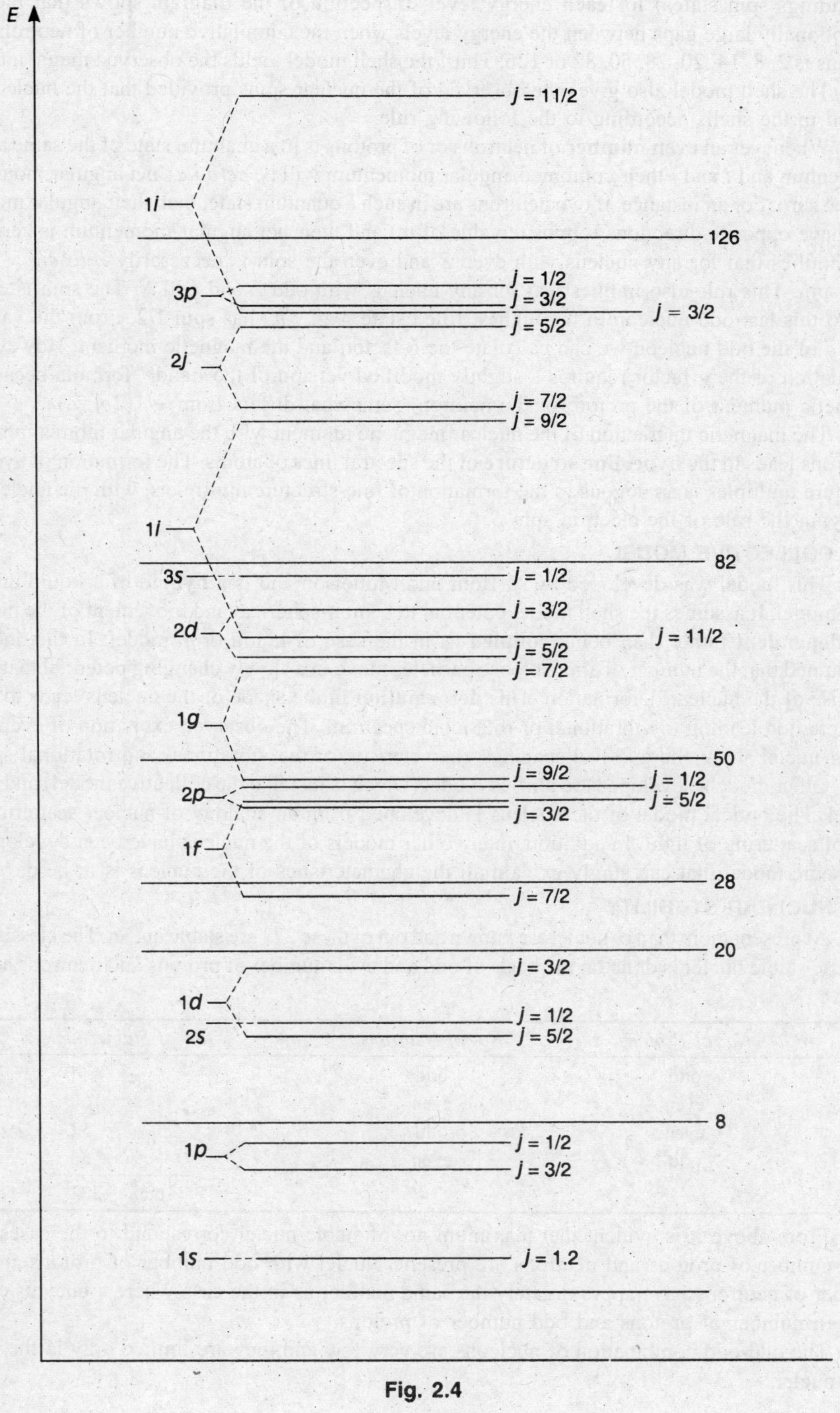
E
j = 11/2
1i
126
j = 1/2
3p
j = 3/2
j = 3/2
j = 5/2
2j
j = 7/2
j = 9/2
1i
82
3s
j = 1/2
j = 3/2
2d
j = 11/2
j = 5/2
j = 7/2
1g
50
j = 9/2
j = 1/2
2p
j = 5/2
j = 3/2
1f
28
j = 7/2
20
j = 3/2
1d
j = 1/2
2s
j = 5/2
8
j = 1/2
1p
j = 3/2
1s
j = 1.2

Fig. 2.4

(includings spin states) for each energy level. Inspection of the diagram shows that there are exceptionally large gaps between the energy levels when the cumulative number of neutrons or of protons is 2, 8, 14, 20, 28, 50, 82 or 126. Thus, the shell model yields the observed magic numbers.

The shell model also gives the observed of the nuclear spins provided that the nucleons are placed in the shells according to the following rule.

Whenever an **even number** of neutrons or of protons is in a quantum state of the same angular momentum and I and j their combined angular momentum will be *zero i.e.,* net angular momentum will be zero. For an instance, if two neutrons are in such a quantum state, then their angular momenta will have opposite directions (opposite value of m) and their net angular momentum is zero. This rule implies that for any nucleus with even Z and even the spin is necessarily zero *e.g.*, ^{16}O has zero spin. This rule also implies that for any nucleus with odd Z and odd N. The spin is entirely due to this last odd nucleon in the highest filled state; *e.g.*, ^{16}O has spin 1/2. From the values of i and j of the odd nucleon we can calculate the g factor, and the magnetic moment. However, the calculation of the g factor requires a slightly modified version of the Lande′ formula because the magnetic moment of the proton (or the neutron) somewhat differs from $e\hbar/(2m_p c)$.

The magnetic interaction of the nuclear magnetic moment with the angular momentum of the electrons leads to the **hyperfine structure** of the spectral lines of atoms. The formation of hyperfine structure multiples is analogous to the formation of fine-structure multiplets, with the nuclear spin I playing the rule of the electron spin S.

2.6. COLLECTIVE MODEL

This model was developed by A. Bohr and Mottelson and is a **hybrid** of a liquid drop and shell model. It assumes the shell model potential to be non-spherical and movement of the nucleons is **independent** rather than being **coupled** as in the case of liquid drop model. In this model, it is assumed that the number of almost loose particles move in a slowly changing potential that occurs because of the nuclear deformation. This **deformation** in the shape of the nucleus leads to modes of excitation leading to vibrational or rotational spectrum. The observed excitation of even-A and odd-A nuclei show energy level spacings characteristic of the vibrational and rotational spectra.

Other models developed are unified model (which is just like the collective model) and optical model. The optical model of the nucleus is developed from an analogy of nuclear scattering with that of scattering of light. In addition, many other models of the nucleus have been developed but still some model that can singly explain all the characteristics of the nucleus is to be developed.

2.7. NUCLEAR STABILITY

At present more than α-nuclei are known and out of these 272 are stable nuclei. The classification of these stable nuclei is done on the basis of odd and even number of protons and neutrons as given below:

No. of Protons	*No. of Neutrons*	*No. of Stable nuclei*
odd	odd	4
even	even	160
even	odd	52
odd	even	56
		Total 272

From above it is evident that maximum no. of stable nuclei correspond to the cases where even number of protons and neutrons are present. Nuclei with odd number of protons and even number of neutrons have approximately the same number as in the case where a nucleus consists of even number of protons and odd number of protons.

The odd-odd combination of nucleons are very few and they are limited only in the case of light nuclei.

Figure 2.5 gives a curve between number of protons and number of neutrons. It is evident that for $Z < 20$, stability curve has straight line with $Z = N$.

In case of heavier nuclei ($Z > 20$, $N > 20$) the curve takes a bend towards $N > Z$ direction. This is because of the **Coulomb electrostatic repulsive forces** becoming important for higher values of Z. So to compensate the repulsive effect, the number of neutrons must be more than the number of protons. So, we observe the departure of stability curve from $N = Z$ line with the increase of Z.

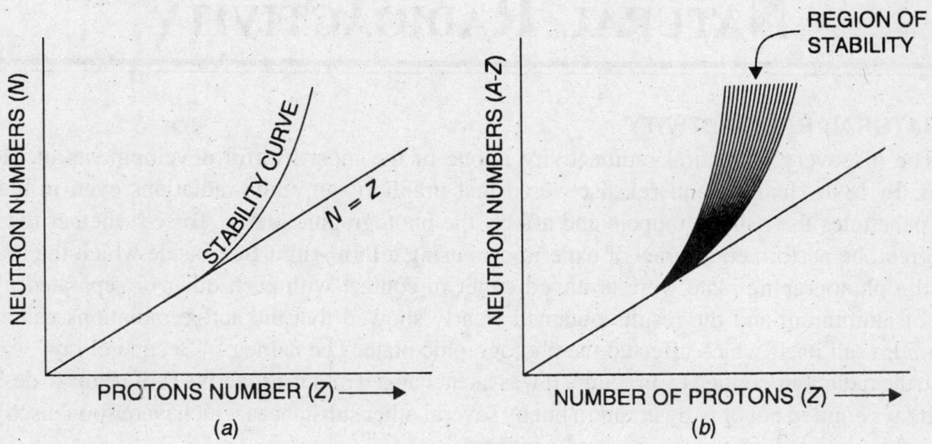

Fig. 2.5

For maximum stability there is an optimum neutron, proton ratio. Fig. 2.5 (*b*) shows a plot of the no. of neutrons ($A - Z$) required for maximum stability vs the proton number Z. **All stable nuclei fall within the shaded region.** Above and below the shaded region we have the unstable nuclei. Artificial radioactive nuclei lie at edge of the stability region. All nuclei with $Z > 83$ and $A > 209$ are unstable and spontaneously emit α and β rays to become stable. Emission of α and β-rays provides stability to the nuclei as excess energy is given in the form of high energy rays.

QUESTIONS AND PROBLEMS

1. Why do we say that a nucleus behaves like a drop of liquid? What are the essential features which are common to a drop of liquid and a nucleus?
2. Give salient features of nuclear shell model and point out its success and failures.
3. What is the magic about the magic numbers? Explain how the shell model of the nucleus accounts for the existence of magic numbers?
4. Explain the postulates of the liquid drop model. Give a simple derivation of Weizsa'cker semi empirical mass formula.
5. Discuss the significance of the charge effect and symmetry effect on nuclei stability.
6. Discuss the various factors which contribute to the binding energy of nuclei and derive a formula for the atomic mass M (Z, A) of a nucleus based on these considerations.
7. What are the various Nuclear models? Explain shell model in details.
8. Write briefly liquid drop model and collective model of the nucleus.
9. How nuclear shell model explains the spectrum of magic number nuclei?
10. Give the various achievements of the shell model.

CHAPTER

3

NATURAL RADIOACTIVITY

3.1. NATURAL RADIOACTIVITY

The discovery of natural radioactivity is one of the most fruitful developments of Modern Physics. In 1896 Henri Becquerel discovered that uranium salt emits radiations even in the dark which penetrates the paper wrappers and affects the photographic plates. To see whether the guess was correct, he performed a series of experiments using a **light-tight box** inside which the uranium salt and a photographic plate were arranged either in contact with each other or separated by thin sheets of aluminium and the results obtained clearly showed that the active radiations came from the uranium salt itself which affected the photographic plate. The name of "Becquerel ray" was first given to the radiations emitted by uranium. It was later changed into **radioactive radiations** to designate all such rays emitted not only by uranium but by several other substances which were soon discovered, such as thorium, radium, polonium and actinium.

Soon after the discovery by Becquerel, Lord Rutherford and Villard investigated that these radioactive radiations are of three types: the one very soft, easily absorbed and capable of producing intense ionisation, which Rutherford called the α-rays, and the other, much more penetrating than the first and much less effective as an ionising agent, which he called β-rays. It was later discovered by Villard that there was also a third kind of rays still more penetrating than the abovementioned two and it was named the γ-rays.

The existence of these types of radiations can be demonstrated by a simple experiment first devised by Madame Curie. A small amount of radium is placed at the bottom of a small hole drilled in a lead block. A photographic plate *P* is placed at a short distance above the lead block to receive the rays. The whole arrangement is placed in an airtight chamber *C* as shown in Fig. 3.1. A strong magnetic field is applied at right angles to the plane of the figure and directed away from the reader.

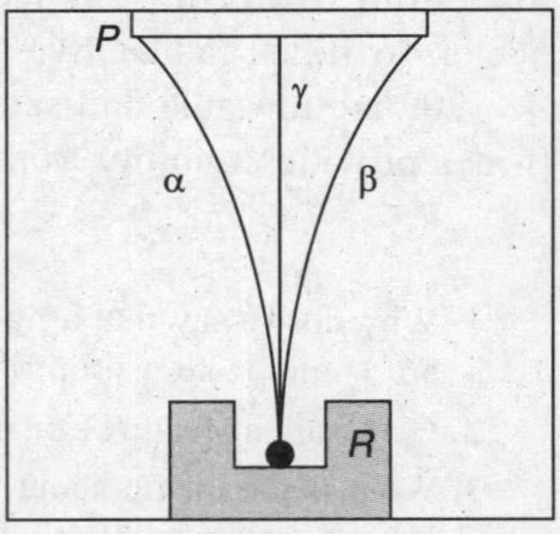

Fig. 3.1

After a fairly long exposure if the photographic plate is developed, three different blackenings will be found on it, proving the existence of the three different types of radiations. The one on the left with large radius must have been made by one group containing positively charged particles of comparatively heavy mass. These are called α-rays. The other to the right with smaller radius must have been made by a second group containing negatively charged particles of very light mass and these are called β-rays. The third central trace must have been produced by rays, unaffected by the magnetic field and hence not deflected, which are evidently the γ-rays.

3.2. PROPERTIES OF α-RAYS

(*i*) The α-rays are shot out from the radioactive material with large velocities ranging from 1.4×10^7 to 2.3×10^7 metres per second.

The velocity of α-rays depends upon the radioactive substance from which they are ejected, *e.g.*, for Uranium the velocity is 1.4×10^7 m/sec, and for Thorium the velocity is 2.3×10^7 m/sec. For a given substance it is always the same.

(*ii*) They produce ionisation in the gas through which they pass. The ionising power is 100 times greater than that of β-rays and 10,000 times greater than that of γ-rays.

(*iii*) They affect a photographic plate. The effect is very very feeble.

(*iv*) They produce fluorescence in substances like zinc sulphide, barium platinocyanide etc.

(*v*) **Range of α-particles:** Scintillations from α-particle cease abruptly after it has traversed a certain distance through matter. The distance that α-particle can travel in air, at atmospheric pressure is called range in air. It varies from 2.70 cms to 8.62 cms for α-particles given by Uranium *I* to Thorium *C*. The range of an α-particle depends upon:

(*a*) The radioactive element through which the rays are given out.

(*b*) The nature of the medium through which the rays travel.

(*c*) The range is proportional to the cube of the velocity of emission, *i.e.*, $R \propto V^3$.

The number of α-particles in a beam remains constant up to the end of the range, but their energy goes on decreasing. Ultimately the α-particles do not have sufficient energy to produce either ionisation or scintillation.

The ionisation first increases slightly, reaches a maximum and then suddenly drops almost to zero value, the distance at which the ionisation becomes zero gives the range of α-particles.

(*vi*) The α-rays are scattered when they pass through thin sheets of mica, gold foil etc. The angle of divergence of α-particles from its straight path is 2 to 3 degrees. Geiger and Marsaden found that a few particles, sometimes, were deflected by more than 20°. This was explained by Rutherford to be due to the repulsion between α-particle and the positively charged nucleus of the atom.

(*vii*) The α-rays are deflected by electric and magnetic fields showing that they are charged particles.

If a beam of radioactive radiations from a deep cavity in a block of lead is subjected to an electric field between parallel plates, the α-particles are deflected towards the negative plate as shown in Fig. 3.2 (*a*). This shows that α-rays consist of positively charged particles.

If a magnetic field is applied perpendicular to the plane of the paper and directed from top to bottom, the α-rays are deflected towards the left as shown in Fig. 3.2 (*b*). By applying the Fleming's left-hand rule we can say that α-rays consist of positively charged particles.

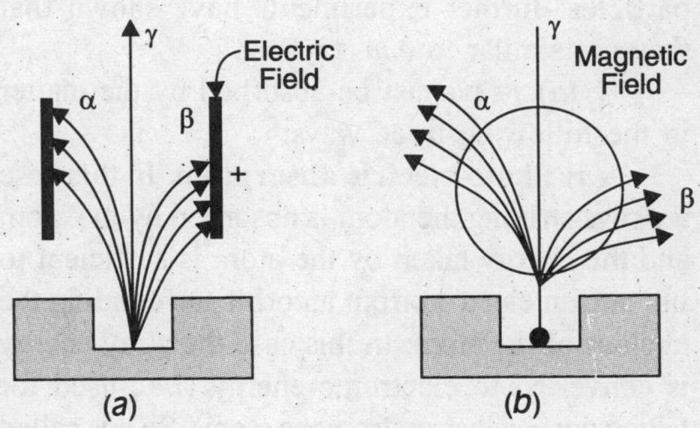

Fig. 3.2

The value of *e/m* is determined on the same lines as for positive rays and indicates that α-particles are nuclei of helium atom.

(*viii*) They produce a heating effect. A quantity of radium always maintains itself at a temperature higher than that of the surroundings. The evolution of heat is due to stoppage of α, β and γ rays by the radioactive substance.

(*ix*) The body suffers incurable burns when exposed to α-rays.

3.3. PROPERTIES OF β-RAYS

(*i*) β-rays are shot out from radioactive elements with very high velocities ranging from 1% to 99% of the velocity of light. The velocity of all β-particles given out by an element is not the same.

(*ii*) They produce ionisation in air but the number of ions produced is very less than those of α-rays. Although their velocity is very large, they possess a comparatively small mass than that of α-particle and hence they have small kinetic energy. As α-particles are slowed down by collision with the atoms of the gas and change their path, their tracks in a Wilson Cloud Chamber are scattered and not continuous as those of α-particles.

(*iii*) They affect a photographic plate and their effect is greater than those of α-rays.

(*iv*) β-rays produce fluorescence in barium platinocyanide, calcium tungsten, willemite etc.

(*v*) Because of their small mass, they can penetrate through large thickness of matter, *e.g.*, they can easily pass through 1 cm of thick aluminium sheet.

(*vi*) They are more easily scattered when they pass through matter, because their mass is very small as compared to the mass of the atomic nuclei.

(*vii*) They are affected by electric and magnetic fields. Their direction of deflection shows that they are negatively charged particles. The *e/m* value has been found on the same lines as for cathode rays which indicates that β-rays are fast moving electrons.

3.4. PROPERTIES OF γ-RAYS

(*i*) They velocity of γ-rays is the same as that of light, *i.e.*, 3×10^{10} cms./sec.

(*ii*) The produce ionisation in gases through which they pass but their effect is very very small as compared to that of α-rays and β-rays. This is due to the fact that their mass is very small as compared to that of α and β-ray particles.

(*iii*) They affect a photographic plate and their effect is greater than those for β-rays.

(*iv*) They produce fluorescence in barium platinocyanide, etc.

(*v*) They are more penetrating than even β-rays and can pass very easily through 30 cms thickness of iron.

(*vi*) They are diffracted from crystals in a way similar to X-ray diffraction.

(*vii*) They are not affected by electric and magnetic fields. This shows that they are neutral particles; further experiments have shown that they are similar to *e.m.* rays.

(*viii*) γ-rays can be absorbed by the matter in the following three ways:

(*a*) Photo-electric absorption. In this case a γ-rays striking the atom is absorbed by the atom and the energy taken by the atom is sufficient to liberate an electron from an orbit surrounding the nucleus of the atom. In this case the γ-ray energy is converted to electronic energy (*i.e.*, used for taking out an electron from the atom). So it is called photo-electric absorption of the γ-rays. (Fig. 3.3)

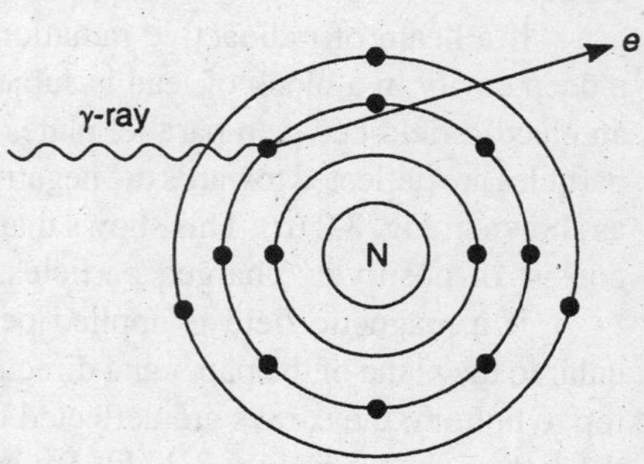

Fig. 3.3

(*b*) Compton effect. In this case the energy given by the γ-rays is more than the energy required to liberate an electron from the atom. So in addition to the electron we get photon (or γ-rays) with different energy from that of the original γ-rays. (Fig. 3.4.)

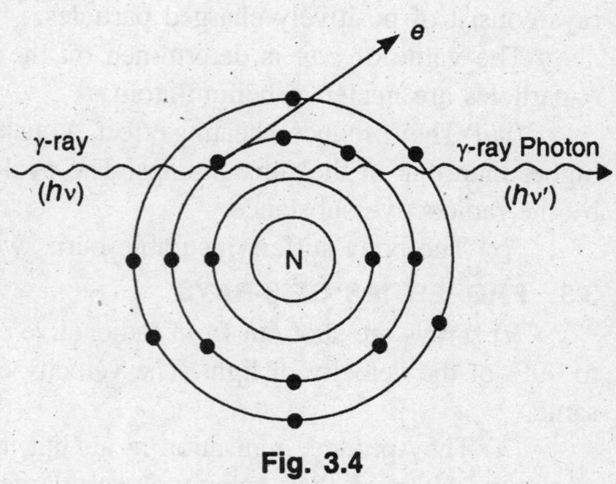

Fig. 3.4

(c) Electron-positron pair production. In this case the γ-ray is absorbed by the atom on which it falls and the energy is liberated in the form of electron-positron pair. It should be kept in mind that the pair production cannot be had if the energy of the γ-rays is less than 1.02 MeV (.51 MeV for electron liberation and .51 MeV for positron liberation). (Fig. 3.5.)

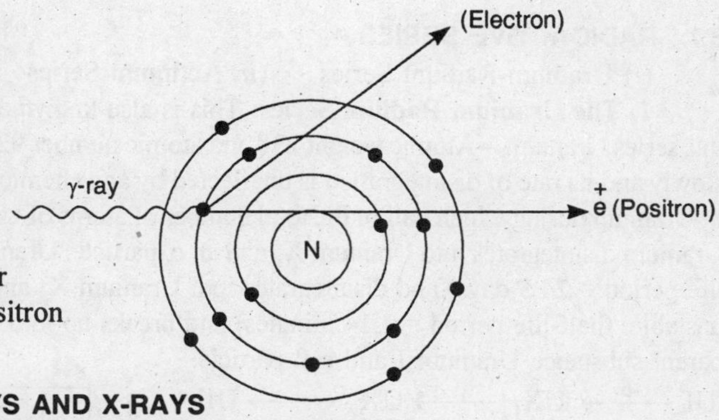

Fig. 3.5

3.5. DISTINCTION BETWEEN γ-RAYS AND X-RAYS

γ-rays and X-rays are both electromagnetic radiations which differ from each other in the following manner:

(*i*) γ-rays are produced spontaneously by radioactive emission whereas X-rays are given out by the transition of electrons from higher to lower energy level.

(*ii*) γ-rays are produced due to spontaneous emission of radiations from the nucleus of the atom whereas X-rays are produced by making fast moving electrons strike solid targets.

(*iii*) The wavelength of γ-rays is much shorter than the wavelength of X-rays.

(*iv*) The penetrating power of γ-rays is much higher than that of X-rays. X-rays can pass through flesh only whereas γ-rays can penetrate through 30 cm of thick iron sheet.

(*v*) γ-rays are monochromatic electro-magnetic radiations while X-rays consist of three components–hard, medium and soft X-rays.

(*vi*) X-rays produce photoelectric effect and scattering effect whereas γ-rays produce photoelectric effect, scattering effect as well as positron-electron pairs.

3.6. FUNDAMENTAL LAWS OF RADIOACTIVITY

In 1913, Soddy and Fajans discovered a simple law known as the displacement law. It can be stated as follows:

Of all known radioactive transformations either an α or β-particle (*i.e.*, never both or more than one of each kind) is emitted by the atom.

(1) When an α-particle is emitted, a new atom is formed whose mass number is less by four units and atomic number less by two units than those of parent atom and thus its place is shifted two groups lower in the periodic table. For example

$$_{88}Ra^{226} \xrightarrow{\alpha\text{-particle}} {}_{86}Ra\,E^{222}$$

(2) When a radioactive atom emits a β-particle (mass = 0 practically and charge, –*e*), it is converted into a new element of Atomic number one greater than the parent element but of the same atomic weight as the parent substance and thus its place is shifted one group higher in the periodic table. For example

$$_{90}UX_1^{234} \xrightarrow{\beta\text{-particle}} {}_{91}UX_2^{234}$$

With the help of the displacement law, we can determine the mass and atomic numbers of different elements in the successive radioactive changes, provided the mass and atomic numbers of one among them are known. The different atoms produced at the different stages of disintegration will naturally form a series.

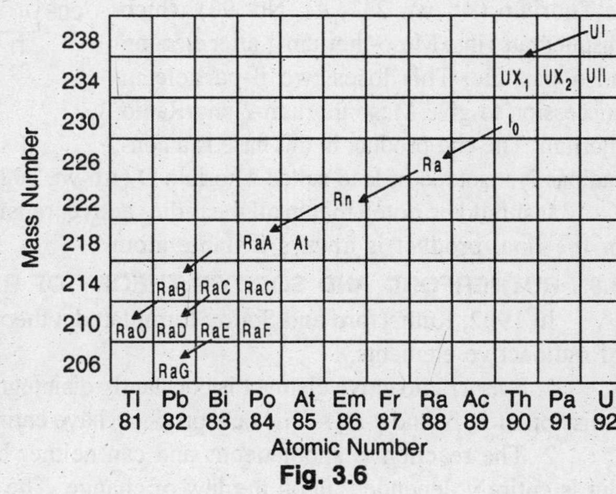

Fig. 3.6

3.7. RADIOACTIVE SERIES

(*i*) Uranium-Radium Series (*ii*) Actinium Series (*iii*) Thorium Series

1. The Uranium Radium Series: This is also known as $(4n + 2)$ series. The parent element in this series Uranium—Atomic weight 238 and atomic number 92 (*i.e.*, $_{92}U^{238}$). Uranium disintegrates very slowly and its rate of disintegration is unaffected by any chemical or physical means. The half-life period of Uranium during which half of the total number of atoms of Uranium disintegrate is 5000 million years. Uranium disintegrates into Uranium X_1 and an α-particle. Uranium X_1 is comparatively less stable (half-life period = 24.5 days) and disintegrates into Uranium X_2 and a β-particle. Uranium X_2 is still more unstable. (half-life period = 1.14 minutes) and breaks up into Uranium II (At. wt. 234), an isotope of parent substance Uranium I and a β-particle.

$$\text{UI} \xrightarrow{\beta} \text{UX}_1 \xrightarrow{\beta} \text{UX}_2 \xrightarrow{\beta} \text{UII}$$

Uranium II then disintegrates into an important radioactive element Radium (*i.e.*, $_{98}Ra^{226}$) and gives out two α-particles. Radium then disintegrates into Radon; a chemically inert gas of zero group. The whole process is shown in the diagram in which the end product is Radium G (At. wt. 206, At. No. 82). Emission of α-particle is shown by the sloping line and the emission of β-particle is shown by horizontal line. The series is shown in Fig. 3.6.

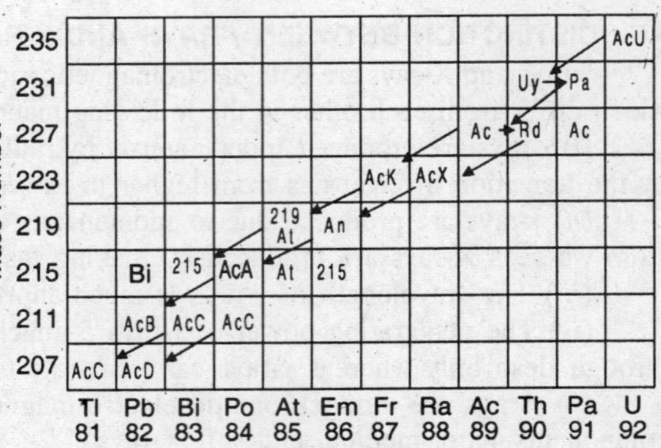

Fig. 3.7

2. Actinium Series: This is also known as $(n + 3)$ series. The parent substance in this series is Protoactinium (At. wt. 231, At. No. 91) which disintegrates slowly (half value period 3.2×10^4 years) into Actinium (At. wt. 277. At. No. 89) after ejecting an α-particle. Actinium ejects a b-particle and forms Radio-Actinium (At wt. 127, At. No. 90) having a half-life period of 13.5 years. The end product of the series is Actinium D (At. wt. 207, At. No. 82) which is not radioactive and is an isotope of lead. The series is shown in Fig. 3.7.

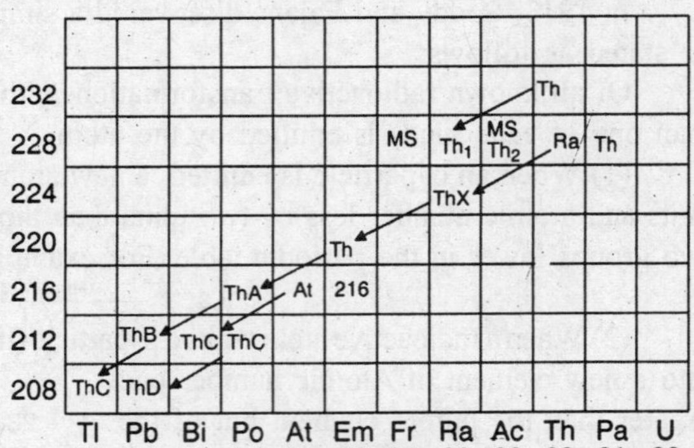

Fig. 3.8

3. Thorium Series : This is also known as $4n$ series. The parent substance in this series is Thorium (At. wt. 232, At. No. 90) which disintegrates into Meso-thorium-1 after ejecting an α-particle. This loses two β-particle in succession to give Meso-thorium-2 and Radio thorium. The end product in this case is a non-radioactive isotope of lead called Thorium D (At. wt. 208, At. No. 82). The series is shown in Fig. 3.8.

It should be noted that in all the radio- active transformations of different series the end product or the final product is always a stable atom.

3.8. RUTHERFORD AND SODDY'S THEORY OF RADIOACTIVE DISINTEGRATION

In 1902, Rutherford and Soddy formulated a theory to explain the spontaneous disintegration of radioactive elements.

1. Every radioactive element is constantly disintegrating into fresh radioactive products with the emission of α, β, and γ-rays. The new products have entirely new radioactive and chemical properties.

2. The reaction is spontaneous and can neither be accelerated nor retarded under any case, but is entirely dependent upon the law of change. The rate of breaking is not affected by external

factors (temperature, pressure etc.). The number of atoms breaking per second at any instant is proportional to the number of atoms present.

If there are N atoms of any substance and dN breaks in a time dt, then according to this hypothesis

$$\frac{dN}{dt} \propto N$$

or $$\frac{dN}{dt} = -\lambda N \qquad ...(1)$$

where λ is the radioactive constant. It is defined as the ratio of the amount of the substance which disintegrates in a unit time to the amount of the substance present.

Thus we see that the number of atoms decayed at any time is always proportional to the number present.

The differential Equation (1) can be put in the form

$$\frac{dN}{N} = -\lambda dt$$

Integrating we have

$$\log_e N = \lambda t + c \qquad ...(2)$$

where c is the constant of integration.

When $t = 0, N = N_0$

$\therefore$ $\log_e N_0 = c$

Substituting in Equation (2), we get

$$\log_e N = -\lambda t + \log_e N_0$$

$$\log_e N - \log_e N_0 = -\lambda t$$

$$\log_e \left(\frac{N}{N_0}\right) = -\lambda t$$

or $$\frac{N}{N_0} = e^{-\lambda t}$$

or $$N = N_0 e^{-\lambda t}$$

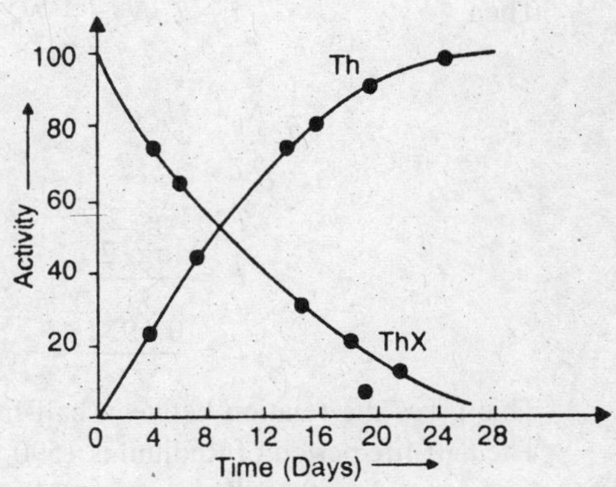

Fig. 3.9

The disintegration is proportional to $\frac{dN}{dt}$ and curve between ionisation produced and time will be an exponential curve.

Rutherford and Soddy studied quantitatively the rate of decay of the ThX activity and the rate of recovery of Thorium activity, and obtained the curves shown in Fig. 3.9. The experimental decay curve for the ThX was exponential in nature, *i.e.*, the activity could be expressed as a function of time by the equation

$$N(t) = N_0 e^{-\lambda t} \qquad ...(3)$$

where N_0 is the initial activity of the ThX, $N(t)$ is the activity after a time t and l is constant, called the disintegration constant. The recovery curve for the Thorium was found to fit the formula

$$N(t) = N_0 (1 - e^{-\lambda t}) \qquad ...(4)$$

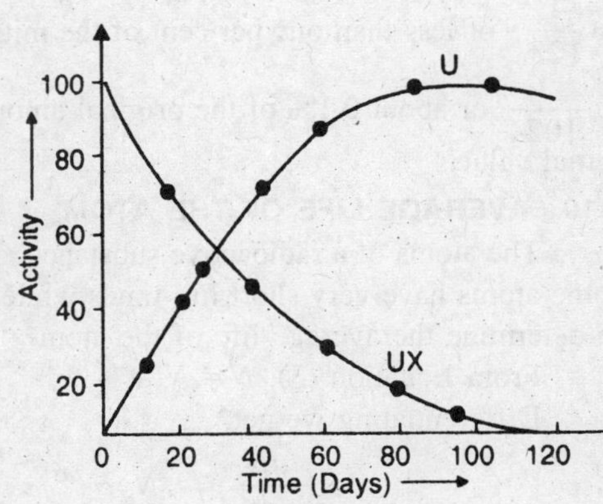

Fig. 3.10

The value of λ is the same as in the above equation. The decay and recovery curves are therefore symmetrical. The results for UX and U are shown in Fig. 3.10. The curves are similar to those for the Thoriumbodies except that the time scale is different. Thorium X loses half of its activity in about 3.5 days, while UX loses half of its activity in 24 days, so that the value of l is greater for ThX than for UX.

3.9. RADIOACTIVE (DISINTEGRATION) CONSTANT

The radioactive disintegration constant can be obtained from Equation (2)

$$N = N_0 e^{-\lambda t}$$

In time, $t = N_0 e^{-\lambda t}$

We have $\frac{N}{N_0} = e^{-\frac{1}{\lambda}\cdot\lambda}$

$\frac{N}{N_0} = e^{-1} = \frac{1}{e}$

Hence, the radioactive constant is the reciprocal of the time during which the original number of a radioactive substance falls to $1/e$ of its value. λ may also be defined as the ratio of amount of substance disintegrating in unit time to that of the substance present at that time.

Half-life period. The half-life period T is defined as the time in which the quantity of radioactive substance is reduced to half of its original value

$$N = N_0 e^{-\lambda t}$$

Then $t = T, N = \frac{1}{2} N_0$

$$\therefore \quad \frac{1}{2} N_0 = N_0 e^{-\lambda t}$$

$$\therefore \quad e^{\lambda T} = 2$$

$$\lambda T = \log_e 2$$

or $T = \frac{\log_e 2}{\lambda}$

$$T = \frac{0.6931}{\lambda} \qquad ...(5)$$

This gives the relation between half-life time and decay constant.

The half-life period of Radium is 1590 years; for Uranium it is 5,000 million years. The half-life period varies directly with the activity. After n half-life ($t = nT$), the fraction of activity remaining is $\left(\frac{1}{2}\right)^n$. This fraction never reaches zero, but it becomes very small; after seven half-lives the activity is $\frac{1}{128}$, or less than one per cent of the initial activity. After ten half-lives the activity has fallen to $\frac{1}{1024}$ or about 0.1% of the original amount, and is usually negligible in comparison with the initial value.

3.10. AVERAGE LIFE OF THE ATOM

The atoms of a radioactive substance are continuously in the process of disintegration. Thus some atoms have very short life-times while others have an exceedingly long life and it is useful to determine the average life of the atom.

From Equation (3), $N = N_0 e^{-\lambda t}$

Differentiating we get,

$$-\frac{dN}{dt} = \lambda N_0 e^{-\lambda t}$$

or $-dN = \lambda N_0 e^{-\lambda t} dt$

So the number $-dN$ disintegrating between the time t and $t + dt$ is equal to $\lambda N_0 e^{-\lambda t} dt$. Each of these atoms has had a life t and hence the sum of life-times of these dN atoms is $\lambda N_0 e^{-\lambda t} dt$. As the average life of the atom is equal to sum of all possible life-times divided by the total number of atoms. That is

$$\text{Average life } \tau = \frac{1}{N_0} \int_{t=0}^{t=\infty} t.dN$$

$$= \frac{1}{N_0} \int_0^\infty \lambda N_0 e^{-\lambda t} t.dt$$

$$= \lambda \int_0^\infty t e^{-\lambda t}\, dt$$

Integrating by parts we get

$$= \lambda \left[\left\{ t \cdot \frac{e^{-\lambda t}}{-\lambda} \right\}_0^\infty - \int_0^\infty \frac{e^{1-\lambda t}}{-\lambda}\, dt \right]$$

First factor = 0, so

$$\tau = \int_0^\infty e^{-\lambda t}\, dt \qquad \left[\frac{e^{-\lambda t}}{-\lambda} \right]_0^\infty = -\left[\frac{e^{-\infty}}{\lambda} - \frac{e^{-0}}{\lambda} \right]$$

$$= \left[-\frac{1}{\lambda e^\infty} + \frac{1}{\lambda} e^{-0} \right] \qquad \therefore \quad \tau = \frac{1}{\lambda} \quad \text{...(6)}$$

Hence, the average life τ of the atom is the reciprocal of the radioactive disintegration constant (λ).

Example 1. *1 gm of radium is reduced by 2.1 mgm in 5 years by α-decay. Calculate the half-life time of radium.*

Solution: Given that

Amount of radium before desintegration $N_0 = 1$ gm

$t = 5$ years

Amount of radium after 5 years

$$N = \left(1 - \frac{2.1}{1000}\right)$$

$$= (1 - 0.0021\ ) \text{ gms.}$$

$$N = 0.9979 \text{ gms.}$$

Half-life time, $T = ?$

We know the relation

$$N = N_0\, e^{-\lambda t}$$

or $$\frac{N}{N_0} = e^{-\lambda t}$$

or $$\frac{N_0}{N} = e^{\lambda t}$$

Taking log on both sides

$$\therefore \quad \log_e \frac{N_0}{N} = \lambda t$$

$$\therefore \quad \lambda t = 2.303 \log_{10} \frac{N_0}{N}$$

$$\lambda = \frac{1}{t} \times 2.303 \log_{10} \frac{N_0}{N}$$

$$= \frac{1}{5} \times 2.303 \log_{10} \frac{1}{0.9979} = \frac{1}{5} \times 2303 \log_{10} \frac{10000}{9979}$$

$$= \frac{1}{5} \times 2.303 \times 1.002$$

$$\lambda = \frac{2.303 \times 1.002}{5}$$

Now $$T = \frac{0.6931}{\lambda} = \frac{0.6931 \times 5}{2.303 \times \log_{10}\left(\dfrac{1}{0.9979}\right)}$$

$T =$ **1672 years.**

Example 2. *Calculate the disintegration constant λ of a sample of radium whose half-life is 1500 years.*

Solution: Given that

half-life time $T = 1590$ years $\lambda = ?$

Since, $$T = \frac{0.6931}{\lambda}$$

$$\therefore \quad \lambda = \frac{0.6931}{T} = \frac{0.6931}{1590} = \mathbf{4.3 \times 10^{-4}\ years.}$$

Example 3. *One gram of radium is found to emit 3.67×10^{10} α-particles per second. Find its half-life period and average life. Given Avogadro's number $= 6.025 \times 10^{23}$ and atomic wt. of radium is 226.*

Solution: Here Atomic wt. = 226 gms.

Avogadro's Number $N = 6.025 \times 10^{23}$

∴ Number of atoms in 1 gm of radium $= \dfrac{6.025 \times 10^{23}}{226}$

It is given that $$-\frac{dN}{dt} = 3.67 \times 10^{10}/\text{sec}$$

From law of disintegration

$$-\frac{dN}{dt} = \lambda N$$

$$\therefore \quad 3.67 \times 10^{10} = \lambda \cdot \frac{6.02 \times 10^{23}}{226}$$

$$\lambda = \frac{226 \times 3.67 \times 10^{10}}{6.02 \times 10^{23}}$$

$$= \frac{226 \times 3.67}{6.02} \times 10^{-13}$$

$$\lambda = 1.371 \times 10^{\ 11} \text{ sec.}$$

Now the half-life time is related to

$$T = \frac{0.6931}{\lambda}$$

$$= \frac{0.6931}{1.371 \times 10^{-11}} \text{ sec}$$

$$= \frac{0.6931 \times 10^{11}}{1.371 \times 60 \times 60 \times 24 \times 365} \text{ years}$$

$$= \frac{6931 \times 10^{8}}{1371 \times 36 \times 24 \times 365}$$

$$= 1585 \text{ years.}$$

The average life time 't' is given by

$$t = \frac{1}{\lambda} = \frac{1}{1.371 \times 10^{-11}} \text{ sec}$$

$$= \frac{10^{11}}{1.371 \times 60 \times 60 \times 24 \times 365} \text{ years} = 2286 \text{ years}$$

Again the half-life period

$$T = \frac{\log_e 2}{\lambda} = \frac{0.6931}{\lambda}$$

$$= 0.6931t$$

∴ Half-life period $T = 0.6931 \times$ Average-life

$= 0.6931 \times 2286 =$ **1584.4 years.**

Example 4. *10 mgms of a radioactive material of half-life period two years is kept in store for four years. How much of the material remains unchanged?*

Solution: Given that

half-life period $T = 2$ years

also $T = \dfrac{0.6931}{\lambda}$

$\therefore \quad \lambda = \dfrac{0.6931}{2} = 0.3465$

Now $t = 4$ years

$N_0 = 10 \text{ mgm} = 10 \times 10^{-3}$ gm.

$N = ?$

Now, $N = N_0 \, e^{-\lambda t}$

$$\frac{N}{N_0} = e^{-\lambda t} = e^{-0.3465 \times 4}$$

$$\log_e \frac{N}{N_0} = e^{-0.3465 \times 4}$$

$$2.303 \log_{10}\left(\frac{N_0}{N}\right) = 1.386$$

$$\log_{10}\left(\frac{N_0}{N}\right) = \frac{1.386}{2.303} = 0.6020$$

Taking Antilogarithms, we get,

$$\frac{N_0}{N} = 4$$

$$\frac{10 \times 10^{-3}}{N} = 4$$

$$N = \frac{10 \times 10^{-3}}{4} = 2.5 \times 10^{-3} \text{ gms}$$

$= \mathbf{2.5\ mg.}$

Example 5. *A certain radioactive element has a half-life of 20 days.*

(i) How long will it take for 3/4 of the atoms originally present to disintegrate? (ii) What is the disintegration constant and the average life of this element?

Solution: (*i*) Now, $T = 20$ days

$N_0 = 1$

$$N = 1 - \frac{3}{4} = \frac{1}{4}$$

$$\therefore \quad \frac{N_0}{N} = e^{\lambda t}$$

$$\therefore \quad \log_e \frac{N_0}{N} = \lambda t$$

$$\therefore \quad 2.303 \log_{10} \frac{1}{\left(\frac{1}{4}\right)} = \frac{0.6931}{T} \times t$$

$$\because \quad 2.303 \log_{10} (4) = \frac{06931}{20} \times t$$

$$\therefore \quad t = \frac{2.303 \times 0.6021 \times 20}{0.6931}$$

$t = 40$ days

(*ii*) Now $T = \dfrac{0.6931}{\lambda}$

$$\therefore \quad \lambda = \frac{0.6931}{T} = \frac{0.6931}{20} = 0.3465/\text{day}$$

Average life time $t = \dfrac{1}{\lambda}$

$$t = \frac{1}{0.03465} \text{ days}$$
$$= \mathbf{28.8\ days.}$$

Example 6. *A certain radioactive element disintegrates for an interval equal to its mean life.*

(i) What fraction of element remains?
(ii) What fraction has disintegrated?

Solution: Given that $t = T$

We also know that $t = \dfrac{1}{\lambda}$

By applying the equation

$$N = N_0\, e^{-\lambda t}$$
$$\frac{N}{N_0}\, e^{-\lambda t} = e^{-\lambda \cdot \frac{1}{\lambda}} \qquad \left(\because\ t = \frac{1}{\lambda}\right)$$
$$\frac{N}{N_0} = \frac{1}{e}$$

Therefore, the fraction that remains $= \dfrac{1}{e}$

Fraction that disintegrated $= \left(1 - \dfrac{1}{e}\right).$

Example 7. *The disintegration constant of Thorium is 1.58 × 10^{-18}/sec. Find the number of α-rays that occur in one gram of the sample in one year. Given molecular weight = 232.*

Solution: Given that

$$\lambda = 1.58 \times 80^{-13}/\text{sec}$$
$$t = 1 \text{ year} = 60 \times 60 \times 24 \times 365 \text{ sec}$$
$$= 3.15 \times 10^7 \text{ sec}$$

Let N_0 = number of thorium atoms present in one gram.

N = number of atoms left behind after one year.

$\therefore$ $dN = N_0 - N$

But $N = N_0\, e^{-\lambda t}$

$$dN = N_0 - N_0\, e^{-\lambda t}$$
$$= N_0\, [1 - e^{-\lambda t}]$$

By using the expansion of exponential

i.e., $e^{-\lambda t} = 1 - \lambda t + \ldots$

$$dN = N_0\, [1 - 1 + \lambda t]$$
$$dN = N_0\, \lambda t$$

But $N_0 = \dfrac{6.02 \times 10^{23}}{232}$

$$dN = \frac{6.02 \times 10^{23} \times 1.58 \times 10^{-18} \times 3.15 \times 10^7}{232}$$
$$= \mathbf{1.3 \times 10^{11}\ atoms.}$$

Example 8. *Calculate the weight in grams of one curie of RaB (Pb^{214}) from its half-life of 26.8 minutes.*

Solution: Given that

half-life period T= 26.8 × 60 = 3608 sec.

$\therefore$ Disintegration constant $\lambda = \dfrac{0.6931}{T} = \dfrac{0.6931}{3608} = 1.6 \times 10^{-4}/\text{sec.}$

Now 1 curie is defined as the activity of one gram of radium, when 3.71×10^{10} atoms disintegrate per second

$$\therefore \quad \frac{dN}{dt} = 3.71 \times 10^{10} \qquad ...(1)$$

$$N = N_0 e^{-\lambda t}$$

$$\frac{dN}{dt} = N_0(-\lambda)(-1)$$

$$\frac{dN}{dt} = N_0\lambda \qquad ...(2)$$

From (1) and (2), we get

$$3.71 \times 10^{10} = 1.6 \times 10^{-4} N_0$$

$$N_0 = \frac{3.71 \times 10^{10}}{1.6 \times 10^{-4}}$$

$$= 2.3 \times 10^{14} \text{ atoms}$$

Atomic weight = 214

Avogadro's number = 6.023×10^{23}

$\therefore$ Mass of 6.023×10^{23} atoms = 214 gm.

$$1 \text{ atom} = \frac{214}{6.023 \times 10^{23}}$$

Hence, the mass of 2.3×10^{14} atoms

$$= \frac{214 \times 2.3 \times 10^{14}}{6.023 \times 10^{23}} = 8.17 \times 10^{-8} \text{ gm}$$

Therefore, the mass of 1 curie of RaB (Pb^{214})

$$= \mathbf{8.17 \times 10^{-8} \text{ gm}}$$

Example 9. *The half-life period of radium is 1590 years. In how many years will one gram of pure element*

(*i*) *lose one centigram, and*

(*ii*) *be reduced to one centigram?*

Solution: Here the half-life period of radium

$$T = 1590 \text{ years}$$

$\therefore$ Radioactive constant

$$\lambda = \frac{0.6931}{T} = \frac{0.6931}{1590}$$

(*i*) Let t be the time in which one gram of radium loses one centrigram (.01 gm.)

$\therefore$ Radium left behind = 1 – .01 = 0.99 gm.

Now

$$N = N_0 e^{-\lambda t}$$

$$\log_e N = \log_e N_0 - \lambda t$$

$$\lambda t = \log_e\left(\frac{N_0}{N}\right)$$

$$t = \frac{1}{\lambda}\log_e\left(\frac{1}{0.99}\right)$$

$$t = \frac{1590}{0.6931}\log_e\left(\frac{1}{0.99}\right) = \frac{1590}{0.6931} \times 2.3026 \log_{10}\left(\frac{100}{99}\right)$$

$$= \frac{1590 \times 2.3020 \times 0.0044}{0.6931} = \mathbf{23.25 \text{ years.}}$$

(*ii*) When it is reduced to one centigram

$$N = .01 \text{ gram}$$

$$\lambda t = \log_e \frac{N_0}{N}$$

$$= \log_e \left(\frac{1}{0.01}\right)$$

$$t = \frac{1590}{0.6931} \times 2.3026 \log_{10}(100)$$

$$= \frac{1590 \times 2.3026 \times 2}{0.6931} = \mathbf{10560 \text{ years.}}$$

3.11. LAWS OF SUCCESSIVE DISINTEGRATION

A radioactive process gives a chain disintegration viz. a radioactive substance (parent nucleus) decays into another substance (daughter nucleus) which further disintegrates (it now becomes a parent nucleus) into a new nucleus (another daughter nucleus) and the disintegration process goes on till some **equilibrium** is reached with a stable end point.

Consider a radioactive nucleus A decaying into B nucleus which further disintegrates to give a new nucleus C and so on.

$$A \to B \to C \to \ldots X \text{ (Stable)}$$

Let the No. of initial atoms of A = N_0 at $t = 0$
and the No. of initial atoms of B = 0 at $t = 0$
Again, No. of atoms in A = N_1 at time t
and No. of atoms in B = N_2 at time t
Let λ_1 and λ_2 be disintegration constants of A and B respectively.
∴ Rate of formation of daughter $b = \lambda_1 N_1$
Rate of decay of B = $\lambda_2 N_2$
∴ net increase in No. of atoms of B

$$\frac{dN_2}{dt} = \lambda_1 N_1 - \lambda_2 N_2 \quad \ldots(1)$$

Also we have the relation

$$N_1 = N_0 e^{\lambda_1 t}$$

Putting in Equation (1), we get

$$\frac{dN_2}{dt} = \lambda_1 N_0 e^{-\lambda_1 t} - \lambda_2 N_2$$

or

$$\frac{dN_2}{dt} + \lambda_2 N_2 = \lambda_1 N_0 e^{-\lambda_1 t}$$

Multiply both sides by $e^{\lambda_2 t}$ we get

$$\frac{dN_2}{dt} \cdot e^{\lambda_1 t} + \lambda_2 N_2 e^{\lambda_2 t} = \lambda_1 N_0 e^{(-\lambda_2 - \lambda_1)t} \quad \ldots(2)$$

Integrating Equation (2), we get

$$N_2 e^{\lambda_2 t} = \frac{\lambda_1}{\lambda_2 - \lambda_1} N_0 e^{(\lambda_2 - \lambda_1)t} + C \quad \ldots(3)$$

where C' is the integration constant.

Now at $t = 0$, $N_2 = 0$, ∴ we have from Equation (3)

$$C' = -\frac{\lambda_1}{(\lambda_2 - \lambda_1)} \cdot N_0$$

Putting the value of C' in Equation (3), we get:

$$N_2 e^{\lambda_2 t} = \frac{\lambda_1}{\lambda_2 - \lambda_1} N_0 \left[e^{(\lambda_2 - \lambda_1)t} - 1\right]$$

Dividing throughout by $e^{\lambda_2 t}$ we have

$$N_2 = \frac{\lambda_1 N_0}{\lambda_2 - \lambda_1}[e^{-\lambda_1 t} - e^{-\lambda_2 t}] \quad ...(4)$$

3.12. RADIOACTIVE EQUILIBRIUM

(i) Secular or permanent equilibrium

Let $T_1 >> T_2$ so that $T_1 \to \infty$ and $T_2 \to 0$

which means that half-life of *A* is much longer than that of *B*.

In that case

$$\lambda_1 << \lambda_2 \text{ and } \lambda_1 \simeq 0 \text{ and } \lambda_2 - \lambda_1 \simeq \lambda_2$$

In Equation (4) we have

$$e^{-\lambda_1 t} \to e^{0.t} \to 1$$

So Equation (4) may take the form

$$N_2 = \frac{N_0\lambda_1}{\lambda_2}(1 - e^{-\lambda_2 t}) \quad ...(5)$$

After very long time, becomes negligible and $N_1 = N_0$. So we may write Equation (5) as

$$\lambda_2 N_2 = \lambda_1 N_1 \quad ...(6)$$

Equation (6) shows that at this type of equilibrium the rate of decay of any product is equal to the rate of its **production** from the previous chain of the nucleus.

In such a case **daughter** is said to be in **permanent** or **secular equilibrium** with the **parent.**

Example. We have the decay of Radium to Radon given as

$$Ra^{226} \to Rn^{222} + \alpha$$

Half-life of Ra is 1590 years which is very large as compared to Rn (3.8 days).

In the decay and recovery curve, Rn is in permanent equilibrium with Ra as shown by the sum of the ordinates of two curves which become constant beyond certain **time** given in the point of intersection of the above curves.

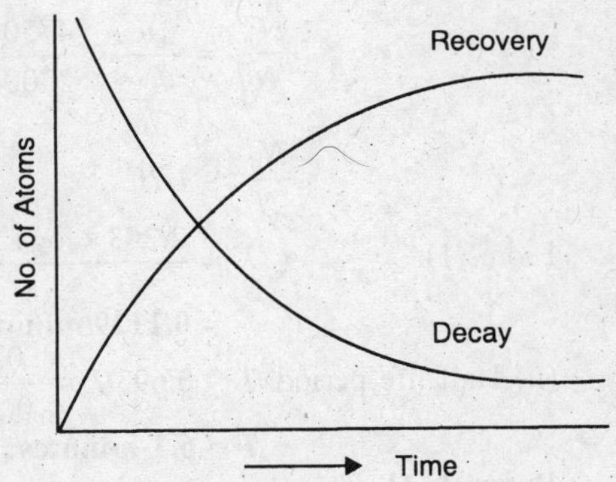

Fig. 3.11

(ii) Transient Equilibrium

Let $\lambda_1 << \lambda_2$ but $\lambda_1 \neq 0$

It means that though λ_1 is very small as compared to λ_2, it cannot be ignored.

In such a case after very long time, $e^{-\lambda_2 t}$ becomes negligible as compared to $e^{-\lambda_1 t}$.

$\therefore$ Equation (4) may be written as

$$N_2 = \frac{N_0 \lambda_1}{\lambda_2 - \lambda_1} e^{-\lambda_1 t} \quad ...(7)$$

Also since $N_1 = N_0 e^{-\lambda_1 t}$

$\therefore$ Equation (7) takes the form

$$N_2 = \frac{\lambda_1 N}{\lambda_2 - \lambda_1}$$

or $$\lambda_1 N_1 = (\lambda_2 - \lambda_1) N_2$$

or $$\frac{N_2}{N_1} = \frac{\lambda_1}{\lambda_2 - \lambda_1} \quad ...(8)$$

which means that when both *A* and *B* decay, the ratio $\frac{N_2}{N_1}$ remains constant.

In such a case we call a **transient equilibrium** is reached when both parent and daughter decay with the period of the parent and $\frac{N_2}{N_1}$ remains constant.

Example 10. *A count rate meter is used to measure the acitivity of a radioactive sample. At a certain instant the count rate was recorded as 4750 counts per minute. Five minutes later the count rate showed as 2700 counts per minute. Compute (i) the decay constant, and (ii) the half-life for the sample.*

Solution: We know that

$$N = N_0 e^{\lambda t}$$

$$\therefore \quad \frac{N}{N_0} = e^{-\lambda t} \quad \text{or} \quad \frac{N_0}{N} = e^{\lambda t}$$

$$\lambda = \frac{\log_e \frac{N}{N_0}}{t} \qquad ...(1)$$

$$\frac{dN_0}{dt} = -\lambda N_0, \quad \text{and} \quad \frac{dN}{N} = \lambda N$$

$$\therefore \quad \frac{N_0}{N} = \frac{\frac{dN_0}{dt}}{\frac{dN}{dt}} = \frac{4750}{2700} = \frac{475}{270}$$

$$\frac{N_0}{N} = 1.76$$

From (1), $$\lambda = \frac{2.303 \times \log 1.76}{5} = \frac{2.303 \times 0.2455}{5}$$

$$\mathbf{= 0.1129/minute}$$

(*ii*) Half-life period $T = 0.693/\lambda = \frac{0.693}{0.1129}$

$$\mathbf{T = 6.1\ minutes}$$

Example 11. *Uranium minerals in which secular equilibrium has been established have been shown to contain one atom of radium for every 2.8×10^6 atom of uranium 1. The half-life of radium is known from direct measurement to be 1620 years. Find the half-life of Uranium 1.*

Solution: If the uranium 1 is denoted by the subscript 1 and radium by the subscript 2, then at equilibrium

$$\frac{N_1}{N_2} = 2.8 \times 10^6$$

We know from secular equilibrium,

$$\frac{N_1}{T_1} = \frac{N_2}{T_2}$$

$$T_1 = \frac{N_1}{N_2} \cdot T_2$$

$$= 2.8 \times 10^6 \times 1620 = \mathbf{4.5 \times 10^9\ years}$$

3.13. UNIT OF RADIOACTIVITY

The intensity of radioactivity has so far been considered in terms of the number of atoms which disintegrate per unit time, or in terms of the number of emitted particles counted per unit time by detector. As in other branches of Physics, it is useful to have a standard quantitative unit with an appropriate name. In radioactivity, the standard unit is curie, at present defined as the quantity of any radioactive material giving 3.70×10^{10} disintegrations/sec. Still smaller unit in radioactivity are millicurie, which is one thousandth of a curie, and the microcurie, equal to one-millionth of a curie, are also useful units,

1 millicurie = 3.7×10^7 disintegrations/sec.

1 microcurie = 3.7×10^4 disintegrations/sec.

There has been some confusion about the use of the curie as the standard unit. Consequently a new absolute unit of radioactive disintegration rate has been recommended, namely, the *rutherford* (*rd*) defined as the amount of a radioactive substance which gives 10^6 disintegrations/sec. The *millirutherford (mrd)* and microrutherford (*urd*) correspond to 10^3 disintegrations/sec and 1 disintegration/sec., respectively.

3.14. ACTIVITY OF A RADIOACTIVE SUBSTANCE

The activity of radioactive substance is the rate of disintegrations/second and is usually denoted by A.

So mathematically, we may write

$$A = \left|\frac{dN}{dt}\right| = \frac{d}{dt}(N_0 e^{-\lambda t})$$

$$= \left|-\lambda N_0 e^{-\lambda t}\right| = \lambda N \quad ...(1)$$

Let the activity of the substance at time $t = 0$ be A_0, then by definition,

$$A_0 = \lambda N_0 \quad ...(2)$$

From Equations (1) and (2), we get

$$\frac{A}{A_0} = \frac{N}{N_0} = \frac{N_0 e^{-\lambda t}}{N_0} = e^{-\lambda t}$$

$$\therefore \quad A = A_0 e^{-\lambda t} \quad ...(3)$$

If we compare Equation (3) with Equation $N = N_0 e^{-\lambda t}$, then it is clear that activity A dependence on time in the same way as N, which means that the activity A of the substance is proportional to the number of atoms N present at any instant.

Example 12. *Due to an accident in a laboratory, a large amount of radioactive material (T = 20 days) spreads inside the room. Test showed that the level of radiation was 30 times the permissible level of normal occupancy of the room. Assuming this, after how many days. the room would be safe for occupation?*

Solution: The initial value of radiations is 30 times the permissible level, so that the time t, in which the activity drops to 1/30 of its initial value, is $\frac{N}{N_0} = \frac{1}{30}$ has to be found out.

Now half-life $\quad T = 20$ days

$\therefore \quad \lambda = 0.693/20$ per day

Let the laboratory become safe after t days.

Then
$$\frac{N}{N_0} = e^{-\lambda t}$$

$$\frac{1}{30} = e^{-\frac{0.693}{20} \times t}$$

$$\therefore \quad \frac{0.693}{20} \times t = 2.303 \times \log_{10} 30$$

$$\therefore \quad t = \frac{2.303 \times 20}{0.693} \times \log_{10} 30 = \frac{2.303 \times 20}{0.693} \times 1.4771$$

$$= \mathbf{98.17\ days.}$$

The laboratory will thus be safe after 99 days.

Example 13. *In a uranium mineral, lead is found predominantly of the isotope 206. The mineral contains 0.093 gm of lead in 1 gm of uranium. Calculate the age of the mineral (Half-life of ${}_{92}U^{238}$ 4.5×10^9 years and all the uranium series have very short lives).*

Solution: Let us consider the mineral to contain N_0 nuclei of U^{233} at time $t = 0$, we also assume that it takes very long time for uranium nuclei to decay to stable isotope of lead.

Let N be the number of uranium nuclei left at any time t, then

$$N = N.e^{-\lambda t}$$

Now from the laws of radioactive disintegration, the number of uranium nuclei which have disintegrated is equal to the number of lead nuclei at time t (say N'). So we have

$$= N_0 - N_0\, e^{-\lambda t} = N_0\,(1 - e^{-\lambda t})$$

Now from relation

$$\frac{\text{Number of lead nuclei at time } t}{\text{Number of uranium nuclei at time } t} = \frac{N'}{N}$$

We have

$$\frac{N_0(1 - e^{-\lambda t})}{N_0 e^{-\lambda t}} = \frac{N'}{N}$$

or
$$e^{+\lambda t} - 1 = \frac{N'}{N}$$

$\therefore$
$$e^{\lambda t} = 1 + \frac{N'}{N}$$

$\therefore$
$$\lambda t = \log_e\left(1 + \frac{N'}{N}\right) = 2.3026 \log_{10}\left(1 + \frac{N'}{N}\right)$$

or
$$t = \frac{1}{\lambda} \times 2.3026 \log_{10}\left(1 + \frac{N'}{N}\right) \qquad ...(1)$$

Also we have

$$\lambda = \frac{0.693}{4.5 \times 10^9} \text{ years}^{-1} \qquad (\because t = 4.5 \times 10^9 \text{ years})$$

Now we are given mass of Pb^{206} in 1 gm of U^{238} = 0.093 gm

$\therefore$ No. of nuclei in 1 gm of U^{238}

$$= \frac{\text{Avogadro No.}}{238}$$

and No. of nuclei in 1 gm of Pb^{206}

$$= \frac{\text{Avogadro No.}}{206}$$

$\therefore$ No. of nuclei in 0.093 gm of Pb

$$= \frac{0.093 \times \text{ Avogadro No.}}{206}$$

We may write

$$\frac{N'}{N} = \frac{0.093 \times 238}{206} = \frac{22.134}{206} \qquad ...(2)$$

So we have from Equations (1), (2) and (3)

$$t = \frac{4.5 \times 10^9}{0.693} \times 2.3026 \log_{10}\left(1 + \frac{22.134}{206}\right)$$

$$= \mathbf{0.6607 \times 10^9 \text{ years}}$$

Example 14. *An ampoule containing 100 millicurie of Kr gas consists of a mixture of the active isotope Kr^{85} and the stable isotope Kr^{84}. If the volume of the mixture is 10 cc at S.T.P. and half-life of Kr^{85} is 10 yrs., calculate the percentage by weight of Kr^{85} present in the mixture.*

Solution: Let N be the no. of Kr^{85} present

$$N = \left|\frac{1}{\lambda}\frac{dN}{dt}\right| = -\frac{T}{0.693}\frac{dN}{dt}$$

Putting the values, T = 10 yrs = 10 × 365 × 86400 secs.

and $$-\frac{dN}{dt} = 100 \text{ mc} = 3.7 \times 10^9 \text{ dis/sec}$$

we have, $$N = \frac{10 \times 365 \times 86400}{0.693} \times 3.7 \times 10^9$$

$$= 1.68 \times 1018 \text{ atoms } = 0.0168 \times 10^{20} \text{ atoms.}$$

From Avogadro's law, 22.4 litres of gas at S.T.P. contain 6.03 × 10^{23} atoms, so that 10 cc at S.T.P. contains $\frac{6.03 \times 10^{23} \times 10}{22.4 \times 10^3} = 2.69 \times 10^{20}$ atoms in the ampoule.

∴ No. of Kr^{84} atom = (2.690 – 0.0168) 10^{20} = 2.6732 × 10^{20}

Hence, percentage of Kr^{85}

$$= \frac{85 \times 0.0168 \times 10^{20}}{(85 \times 0.0168 \times 10^{20}) + (84 \times 2.673 \times 10^{20})} \times 100$$

$$= \mathbf{0.632\%}$$

Example 15. *What will be the mass of 10 curie sample of* Co^{60}? *Given half-life* T_l = *5.26 years.*

Solution: We have the relation

$$T_1 = \frac{0.693}{\lambda}$$

$$\therefore \quad \lambda = \frac{0.693}{5.26 \times 36524 \times 60 \times 60}$$

$$= 4.16 \times 10^{-9} \text{ sec}^{-1}$$

For cobalt we have

$$N = \frac{6.02 \times 10^{23}}{60} = 1.00 \times 10^{22}$$

∴ Activity A = λN = 4.16 × 10^{-9} × 1.002 × 10^{22} dis/sec.

If the mass of the sample is m, then we have the relation

$$\frac{m \times 4.16 \times 10^{-9} \times 1.002 \times 10^{22}}{3.7 \times 10^{10}} = 10$$

$$\therefore \quad m = \frac{10 \times 3.7 \times 10^{10}}{4.16 \times 1.002 \times 10^{23}}$$

$$= \mathbf{0.0089 \text{ gm.}}$$

Example 16. *Determine the amount of* Po^{210}_{84} *necessary to provide a source of* α*-particles of 5 millicurie strength. Half-life (T) of Po = 138 days.*

Solution: The rate of decay of a radioactive material is given by

$$\frac{dN}{N} = -\lambda N \quad (\lambda = \text{decay constant})$$

In this case, the decay rate = 5 mc = 5 × 3.7 × 10^7 dis/sec (1 curie = 3.7 × 10^{10} dis/sec)

$$\therefore \quad -\frac{dN}{dt} = 5 \times 3.7 \times 10^7 \text{ dis/sec.}$$

Since, $$N = -\frac{1}{\lambda} \cdot \frac{dN}{dt} = -\frac{T}{0.693} \cdot \frac{dN}{dt}$$

Substituting the value of T = 138 days = 138 × 88400 sec.

$$N = \frac{138 \times 86400 \times 5 \times 3.7 \times 10^7}{0.693} \text{ (in magnitude)}$$

$= 3.18 \times 10^{15}$ atoms.

Now, 210 gm of Po contains 6.03×10^{23} (N – Avogadro No.) of atoms. So the amount necessary to obtain a source of the required strength is

$$\frac{210 \times 3.18 \times 10^{15}}{6.03 \times 10^{23}} = \mathbf{1.107 \times 10^{-9}\ kgm}$$

Example 17. *Assuming the age of the earth to be 10^{10} yrs., what fraction of the original amount of U^{238} is still in existence on the earth? (Half-life of U^{238} = 4.5×10^9 yrs).*

Solution: Since T of $U^{238} = 4.5 \times 10^9$ yrs.

Also $$N_t = N_0 e^{-\lambda t}$$

$$\therefore \quad \frac{N_t}{N_0} = e^{-\lambda t}$$

Now $$\lambda = \frac{0.693}{T} = \frac{0.693}{4.5 \times 10^9 \text{ yrs.}} = 1.54 \times 10^{-10} \text{ per year}$$

Thus the fraction still present $$\frac{N_t}{N_0} = e^{-\lambda t} = e^{-1.54}$$

$$= \mathbf{0.2143.}$$

3.15. RADIATION DAMAGE

Energy absorbed by the passage of radiation through matter gives rise to structural changes called *radiation damage.* The nature and amount of damage produced is closely related to the number of ion pairs created, and this in turn depends upon the absorbing material and the characteristics of the radiation.

Since biological effects vary widely between body tissues and different types of radiation, it is customary to measure radiation absorption by the number of ion pairs produced in air. The exposure of specimens to any and all kinds of radiation is measured in terms of a unit called the *roentgen.*

Roentgen. The roentgen unit, abbreviated 1 r, is that quantity of X or γ radiation that produces one electrostatic unit of negative, or positive, charge in 0.001293 grams of dry air. Under standard conditions air has a density of 0.001293 gm/cm^3, and one coulomb = 3×10^9 esu. Since the charge of a single electron, or ion, is 1.60×10^{-19} coulomb, the roentgen can also be defined as the quantity of X or γ radiation which, under standard conditions, produces in air:

$$1\ r = \frac{1}{3 \times 10^9 \times 1.6 \times 10^{-19}} = 2.08 \times 10^9 \text{ ion pairs/cm}^3$$

Experimental measurements show that on the average it takes 32.5 electron volts to produce one ion pair. Consequently,

$$1\ r = 2.08 \times 10^9 \times 32.5 = 67.6 \text{ BeV}$$

This energy is equivalent to 0.108×10^{-7} joules or 0.108 ergs. In one gram of air 1 r would produce

$$\frac{0.108}{0.001293} = 83.5 \text{ ergs}$$

The same radiation will produce 93 ergs per gram of water (the main constituent of tissue), approximately 40 ergs/gm in fat, and 900 ergs/gm in bone.

Roentgen Equivalent, Physical (rep). This is a unit applied to statements of dose of ionizing radiation not covered by the definition of the roentgen. One *rep* has been defined as the dose that produces an energy absorption of 93 ergs per gram of tissue for the particular radiation in question.

Roentgen Equivalent, Man (rem). A dose of any ionizing radiation producing the same biological effect in tissue as that produced by 1 *roentgen* of high-voltage, X radiation is called 1 *rem.* The roentgen is frequently divided into one thousand equal parts called the *milliroentgen:*

1 roentgen = 1000 milliroentgens

$1\ r = 1000$ mr

Radiation dose. The total radiation entering a specified area or volume is called the radiation dose and is measured in *roentgens per hour* or *milliroentgens per hour.*

One effect of penetrating radiation is to bring about **gene mutation.** Such mutation is of greatest concern to the individual man or woman from birth through the reproductive years. In human beings this is assumed to be about 30 years. Experiments show that a total of 50 r per person received during the reproduction period of 30 years doubles the number of mutations that occur naturally in the absence of radiation.

The amount of radiation to which an individual can be subjected with no adverse effects is called a *permissible dose.* This is generally accepted to be 250 mr per week, or 6.25 mr per hour based on a 40-hour working week.

QUESTIONS AND PROBLEMS

1. What are the radiations from radioactive substance? Give an account of their nature and properties.

2. State the characteristic properties of radioactive elements. Derive the relation between half-life and decay constant of a radioactive nuclide.

3. State Rutherford and Soddy's postulate of spontaneous disintegration of radioactive elements. Hence prove the laws of decay growth of radioactive substances.

4. What are α, β and γ-rays? Describe their properties. Is there any similarity between any of these radiations and X-rays?

5. Give an account of the law of disintegration of radioactive substances. Deduce from this law the value of half-life period.

6. Distinguish between X-rays and γ-rays.

7. Give the units of radioactivity and define the term radiation damage. What is the safe dose of radiation?

8. A freshly prepared Radio source loses half of its activity in 3.86 days. Calculate the time in which activity would be 1%. [**Ans.** 205 days]

9. The half-life period of radium is 1620 years. In how many years will one gram of pure metal (*a*) lose one centigram and (*b*) reduce to one centigram? [**Ans.** (*a*) 24.35 years; (*b*) 10800 years]

10. A count rate meter is used to measure the activity of a radioactive sample. At a certain instant the count rate was recorded 4750 counts per minute. Five minutes later the count rate was recorded 2700 counts per minute. Compute (*i*) the decay constant, and (*ii*) the half-life for the sample.
[**Ans.** (*i*) 0.1129/minute; (*ii*) 6.1 minute]

11. One gram of uranium-I emits 1.2×10^{14} α-particles per second. Calculate its half-life period. Given atomic weights = 226. [**Ans.** 4.6×10^{12} years]

12. The disintegration constant of U^{234} is 8.8×10^{-4}/sec. What amount of this element will be left after a lapse of 2×15^5 years? [**Ans** 0.35]

13. The half-life of uranium (UX_1) is 24.1 days. How long after a sample of UX_1 has been isolated will it take for 90% of it to change to UX_2? [**Ans.** 80.1 days]

14. Calculate the radioactive constant for an element whose half-life time is 20 years. [**Ans.** 3.0346]

CHAPTER

4

ALPHA, BETA AND GAMMA RAY SPECTRA

4.1. NUCLEAR ENERGY LEVELS

For a long time it was assumed that all the α-particles emitted from a radio isotope have nearly the same initial velocity and energy. But experiments of Rosenblum, using strong magnetic field have shown that the nuclide often emits α-particles of different energies. Rutherford observed that the ranges of the α-particles in a group are different. Thorium C' emits α -particles of range between 8.6 cm and 11.6 cm, though the particles with higher range are very few. The long range α-particles which are few in number also come from the same nuclide.

The magnetic spectrograph experiments have shown that there are three distinct groups of α-particles emitted by Th C' nuclide. In the main group of α-particles have energy 8.780 MeV. Out of 10^6 particles, 40 particles have an energy of 9.492 MeV, 20 particles have an energy of 10.422 MeV and 170 particles have an energy of 10.548 MeV.

This discovery of α-particle spectra gave rise to the concept of energy levels in the nucleus. Similar to energy levels of electrons in an atom, the nucleus is also considered to have different energy levels. Ordinarily, the nucleus is in the ground state. Thorium C after emitting β-particles is converted to Thorium C', β -particles of different energies are emitted from Thorium C. Depending on the energy levels of the nuclide of Th C'. α-particles of different energies are emitted from the nuclide. Even after the emission of an α-particle, the product nuclide may be in the excited state and it emits γ -rays. Gamow has shown that the energies of α-particles and γ -rays can be correlated in terms of nuclear energy levels. The experimental facts establish the validity of energy levels.

4.2. THEORY OF α -DECAY

α -particles are emitted from radioactive nuclides. Before emission, α-particles can be considered to be inside the nucleus. Coulomb's law is applicable when an α-particle is inside the nucleus or very close to it. Otherwise Coulomb's law does not hold good and can be represented by an attractive potential called the potential well.

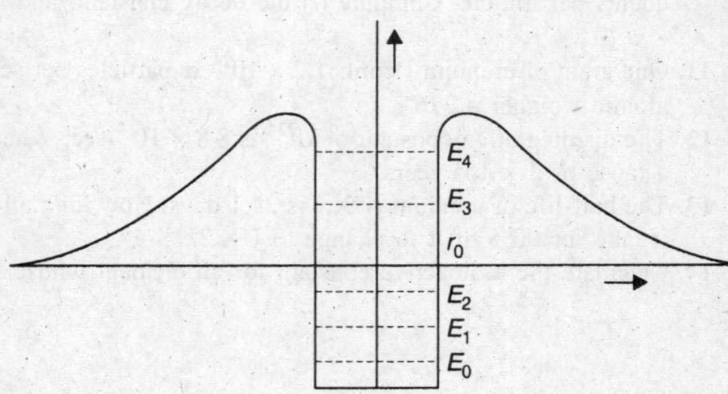

Fig. 4.1

For a distance r between the nucleus of charge $(Z-2)\,e$ and the α-particle of charge $2e$, the potential energy,

$$V_{r_0} = \frac{2e(Z-2)\,e}{4\pi\,\varepsilon_0\,r} = \frac{2(Z-2)\,e^2}{4\pi\,\varepsilon_0\,r}\ \text{joules} \quad ...(1)$$

When r is less than r_0, the α-particle is within the potential well and is bound by the nucleus. Here r_0 is equal to the sum of the radii of the nucleus and the α-particle.

According to principle of wave-mechanics, an α-particle can have different energy levels, E_0, E_1, E_2..., etc., within the potential well (Fig. 4.1). An α-particle in the energy levels E_0, E_1, and E_2 cannot come out of the nucleus but an α-particle in the energy level E_3 or above can come out of the potential well. Classical mechanics could not establish the possibility of α-particle disintegration.

The height of the potential barrier is approximately 9 MeV. But **uranium emits α-particles of energy 4 MeV**. Classical mechanics cannot explain how a particle having an energy of 4 MeV can come out of a well having a potential barrier of 9 MeV. But according to de-Broglie wave mechanics and Schrodinger equations, it is possible for an α-particle to leak through the potential barrier even though its kinetic energy is less than the potential energy of the height of the potential barrier. This probability of leakage of an α-particle through the barrier is called the "tunnel effect".

4.3. GEIGER NUTTAL RELATION

Geiger and Nuttal in 1911 gave a relation for range of α-particles and disintegration constant **in the X-ray spectrum.**

Let R_α be the range of α-particles emitted by a radioactive substance having disintegration constant λ. The relation between R_α and λ is given by

$$\log \lambda = A + B \log R_\alpha \quad ...(1)$$

and is known as **Geiger Nuttal Relation.**

Actually Geiger and Nuttal plotted a curve for **log λ** and **log R_α** for the various members of radioactive series and got nearly three parallel straight lines and so gave the form of relation as given in Equation (1).

It may be mentioned here that A and B are constant for a particular series. However, B remains **constant for all the series.** It may further be seen from Equation (1) that greater the range of α-particle (R_α), greater is the value of decay constant (λ).

According to Geiger rule, range in air is related to the velocity as

$$R_\alpha = 9.6 \times 10^{-24} v^2 \quad ...(2)$$

where v is the initial velocity of the α-particle. Also energy is related to velocity as

$$E = 2.08 \times 10^{-14} v^2 \quad ...(3)$$

From Equations (2) and (3), we have

$$R_\alpha = 0.00318 E^{3/2}$$
$$= KE^{3/2} \quad ...(4)$$

As is evident from Equation (4), range is dependent on the energy of emission of α-particle. It means that a substance having **larger** disintegration constant will emit **high** energy particles.

Geiger Nuttal empirical relation was justified in 1928 by Gamon and others using the concepts of the wave mechanics.

4.4. α-RAY SPECTRA

Rutherford studied the properties emitted by Uranium and Thorium minerals (1899). The charge of these particles was found to be positive. Experiments of Ruther-ford, Royd and Geiger (1900) established that alpha rays are ionised helium nuclei emitted at large velocities by disintegrating nuclei. The charge and Ze/M was systematically measured and the range of these particles in air was studied. Since the range measurements were approximate indications of energy of α-particles, it was thought that all the alpha particles emitted by heavy elements were of identical energy and it was assumed that alpha transitions were discrete transitions and take place to the ground state of the daughter nucleus but when alpha energies were measured with magnetic deflection in cloud chambers, and magnetic deflection Spectrometers, it was established that α-particles are not monoenergetic in general but have a fine structure. It was found that a majority of α-decay consist of long range alpha particles but there are a few short range alpha groups as well. Each of these groups are, however, monoenergetic. The disintegration energy E_0 of α-decay can be written as $E_0 = E_\alpha + E_R$ where E_R is recoil energy of the nucleus and given by the relation,

$$E_0 = E_\alpha\left(1 + \frac{M_0}{M_R}\right) = E_0\left(\frac{M_R + M_\alpha}{M_R}\right)$$

$$= E_\alpha\left(\frac{M_0}{M_R}\right) \qquad ...(2)$$

where $\quad M_0 = M_R + M_\alpha$

4.5. FINE STRUCTURE OF α SPECTRA

Fig. 4.2 shows the energy transition of the Th C ($_{83}Bi^{22}$) nuclides to Th C''($_{81}TI^{208}$) in an alpha decay process.

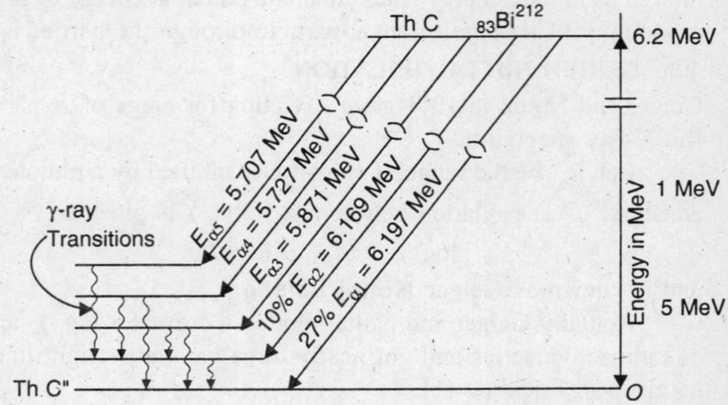

Fig. 4.2

It is evident that Th C emits alpha rays and does not always come to ground state of Th C''. It goes to some excited *isomeric states* of Th C'' which come to ground state by an instantaneous emission of a gamma ray. If we consider the α-rays involved in ground to ground transition as normal α-rays, then all those α-particles which are emitted in transition to the excited state of Th C'' will have *shorter range in air.* Study of the fine structure of α-particles helps in the study of gamma excited states of nuclides and is a helpful tool in nuclear spectroscopy. Some of the long range α-particles are emitted by Th C'' ($_{84}Po^{212}$) nuclei which have an alternative mode of decay as shown in the following reaction.

$$\text{Th C }[_{83}\text{Bi}^{212}] \begin{cases} \xrightarrow{\beta-} \text{Th C}'' + \alpha \rightarrow \text{Th D }[_{82}\text{Pb}^{208}] + \beta \\ \xrightarrow[33.7\%]{\beta-} \text{Th C}' + [_{84}\text{Po}^{212}] \rightarrow \text{Th D }[_{82}\text{Pb}^{208}] + \alpha \end{cases}$$

$_{84}Po^{212}$ is the shortest lived substance ($T_{1/2} \cong 10^{-7}$) sec among the heavy elements and possesses the following alpha energy groups.

Table 4.1. α-energies of $_{84}Po^{212}$

S.No.	*Group*	*Relative abundance*	*Excitation energy Th C'*	*α-energy E_α*	*Disintegration energy E_0*
1	Normal α_0	105	0	8.776 MeV	8.946
2	Long range α_1	35	0.725 MeV	9.489 MeV	9.671
3	Long range α_2	20	1.671 MeV	10.417 MeV	10.617
4.	Long range α_3	170	1.793 MeV	10.536 MeV	10.73

It is clear that both long and short range types of α-particles are found. The short range alpha particles are due to transition from the ground state of the parent nucleus to one of the excited levels of the daughter product. Their energy and range in air is less than the normal ground to ground transition of the two participating nuclei. The long range α-particles are, however, due to the transitions from the excited state of the parent to the ground state of the daughter product as in the case of Th C' – Th D transition shown in Fig. 4.3.

The long lived heavy nuclei successively decay through α, β and γ-transitions to a stable daughter product which in this case is lead. The various decay series are the Thorium series ($4n$), Neptunium series ($4n + 1$), Protactinium series of Uranium series ($4n + 2$) and Actinium series ($4n + 3$). Every member of these series has a mass number A such that,

$$A = 4n, (4n + 1), (4n + 2), \text{ or } (4n + 3)$$

The mass number changes by $\Delta A = 4$ and $\Delta n = 2$ in α-decay and $\Delta A = 0, \Delta n = 0$ in β-decay. A large number of radio-nuclides are associated with these series. These series have been named after their longest lived member. In successive α and β-decays a large number of intermediate isotopes are formed till a stable end product is obtained. Thorium ($_{90}Th^{232}$) has a half-life of 13.9×10^9 years and its end product is Pb^{208}. Neptunium ($_{93}Np^{237}$) has a half-life of 2.2×10^6 years and its end product is $_{83}Bi^{209}$. Uranium ($_{92}U^{238}$) has a half-life of 4.5×10^9 years and is the heaviest naturally occurring element. Its end product is $_{82}Pb^{206}$. It was in 1935 when the $_{92}U^{235}$ isotope was identified in mass spectrograph and was shown to be the parent nuclide of the Actinium series U^{235} which occurs in very small abundance 72% of the natural Uraniur and has half-life of 7.1×10^8 years whose end product is $_{82}Pb^{207}$

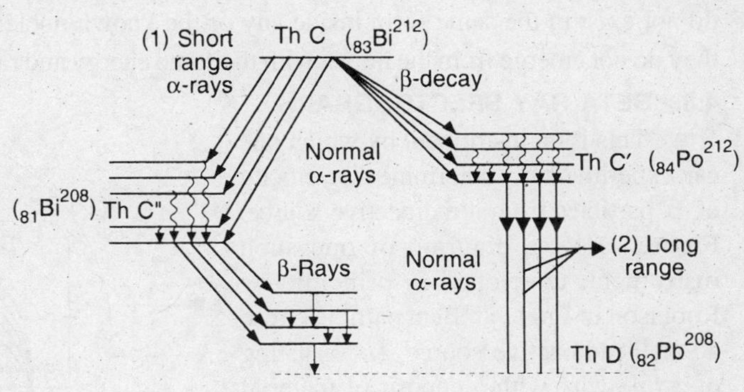

Fig. 4.3

4.6. β-DECAY

In β-decay, the following three processes are involved: (*i*) Electron Emission, (*ii*) Positrc Emission and (*iii*) Electron Capture.

(*i*) **Electron emission:** General equation for electron capture may be written as

$$_zX^A \rightarrow \underset{z+1}{y^A} + e^0_{-1} + \bar{\nu}$$

or
$$_0n^1 \rightarrow p + e^0_{0-1} + \bar{\nu} \qquad \text{...(1)}$$

That is, in this process a neutron inside the nucleus changes to proton with an electron emission and the excess energy is carried by antineutrino (this concept will be more clear in theory of β-decay).

(*ii*) **Positron emission:** We can write the general equation in this case as

$$_zX^A \rightarrow {}_{z-1}Y^A + {}_1e^0 + \nu$$

or
$$P \rightarrow {}_0n^1 + {}_{+1}e^0 + \nu \qquad \text{...(2)}$$

In this case an **unstable or excited** proton changes to neutron with the emission of a positron and **neutrino.**

(*iii*) **Electron capture:** In this mode of decay, an electron from the orbit is captured by the nucleus to form a new nucleus and a **neutrino.**

$$_zX^A + {}_{-1}e^0 \rightarrow {}_zY^A + \nu$$

or
$$p + {}_{-1}e^0 \rightarrow {}_0n^1 + \nu \qquad \text{...(3)}$$

In this case a proton in the nucleus is converted into a neutron and neutrino with the capture of an electron.

As is evident from equations (2) and (3), the products in positron emission and electron capture are same, so **positron emission competes with electron capture.**

It may be pointed out here that **all the nuclei having excess of protons** or **neutrons but insufficient energy to expel them, undergo** β-decay by any of the processes described above.

4.7. β-RAY SPECTRA

Beta rays emitted by the natural radioactive elements have been studied by many people. It was Becquerel who first found them to be comparable to cathode rays. Today we know them as electrons. One of the greatest mysteries of atomic physics has been the origin of these particles. The great mass of experimental evidence shows that most of them come from the nucleus, yet they do not exist in the same form inside any of the known nuclei. Also, unlike α-particles and γ-rays, they do not emerge from the nuclei with the same energy and range, but with a wide band of velocities.

4.8. BETA RAY SPECTROGRAPH

This is an instrument by which one can experimentally determine the velocity of β particles from a radioactive source. Figure 4.4 is a diagram of one such instrument, developed in principle by Robinson in England. Beta particles from a small radioactive source *A*, consisting of a fine wire with a deposit of material on its surface, are allowed to pass through a slit *S*. With the entire vacuum tube located in a uniform magnetic field, the β-particles follow circular paths and are brought to focus on a photographic plate *P* or at the open slit of an ionization chamber *C*. Each of the three semicircular paths in any one group shown in the diagram has the same radius and diameter. The focussing action indicated can be demonstrated by using a compass and drawing several semicircles with the same radius, but slightly displaced centres. The velocity of any group of electrons is given by the relation

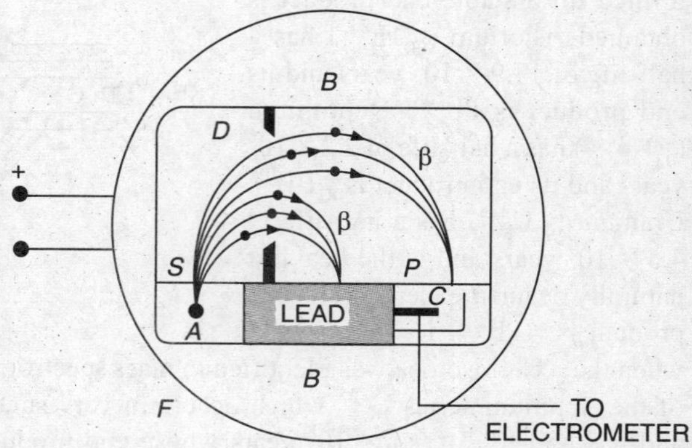

Fig. 4.4

$$Bev = \frac{mv^2}{r} \quad ...(1)$$

where r is the path radius, B the magnetic induction, and e, m, and v the electron's charge, mass, and velocity, respectively.

If the field B is constant, and a photographic plate P is located as shown in the figure, its development after a time t will result in a photograph like that shown in Figure 4.5. In addition to a darkened back-ground all along the plate, one finds several lines parallel to the slit. *The background indicates β particles present with all different velocities, while the lines signify group with discrete and definite velocities.* However, not all spectrograms with such a continuous background contain these lines.

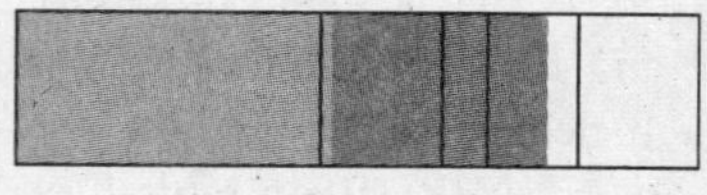

Fig. 4.5

In some β-ray spectrographs, the β particles are collected in an ionization chamber *C* as shown in Figure 4.4. Different velocities are determined by slowly and continuously changing the magnetic induction *B*. This is accomplished by changing the current *I* in the field coils *F*. If measurements of electrometer current are recorded for different values of *B*, a graph similar to that shown in Figure 4.6 can be plotted.

The measured ion current is plotted as the number of β particles, and the coil current *I* is plotted as the energy the particles would possess to enter the slit at *C*. This curve, made using

RaE as a source, which is the radioactive bismuth isotope $_{83}Bi^{210}$, shows β particles with a maximum energy at 0.15 MeV, and an end-point energy of 1.17 MeV. The *end point,* or cut-off, at the high-energy end of this curve signifies a velocity limit or maximum. It may be noted that such end point energies have different values for different isotopes.

4.9. THE NEUTRINO POSTULATE

In 1931 Pauli, of Germany, suggested that in γ-ray emission all nuclei of the same isotope emit the same amount of energy, and that this is the end-point energy shown in Figure 4.6. To account for the observed fact that some β particles emerge with much less energy than others, he made the following postulate. The emission of a β particle by any nucleus is accompanied by a companion particle having a variable energy *E*. While the β-particle is an ordinary negative charged electron, the companion particle, now called a neutrino, has no charge. In some respects a *neutrino* is like a photon; it has no rest mass, it has energy *E*, and it travels with the speed of light *c*. Hence, the reaction for β-ray emission by a nucleus like $_{83}Bi^{210}$ can be written,

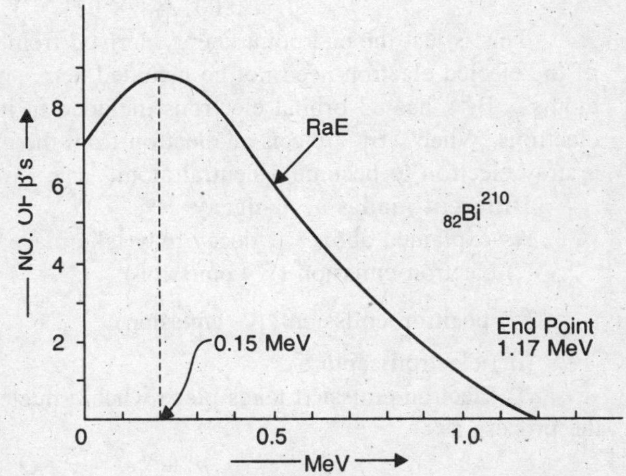

Fig. 4.6 Graph of β energies from RaE.

$$_{83}Bi^{210} \rightarrow {}_{84}Po^{210} + \beta^- + \nu \qquad ...(2)$$

where ν represents the neutrino.

The neutrino postulate has served another important function in atomic processes; it permits the retention of the law of conservation of momentum. All even-numbered nuclei, even *A*, are known to have an angular momentum given by an integral quantum number, $I = 0, 1, 2, 3, 4, ...,$ all odd-numbered nuclei, odd-*A*, are known to have an angular momentum given by a half-integral quantum number, $I = 1/2, 3/5, 5/2, 7/2,...$ Moreover, all electrons are known to have a spin angular momentum given by a half-integral quantum number, $s = 1/2$. By assigning the neutrino a spin angular momentum equal in magnitude to that of the electron, the two spins can either cancel each other as shown in Figure 4.7, or add together, thus keeping the nuclear spin integral or half-integral valued as the case may be.

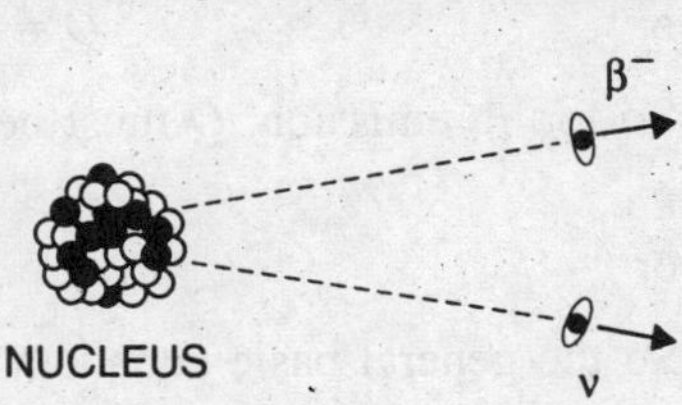

Fig. 4.7

Only recently have experiments been performed that appear to establish the existence of these strange particles, neutrinos. Since nuclei are composed of neutrons and protons, the two objected particles are created by the nucleus, a neutron transforming into a proton as it emits an electron of mass m_0, charge $-e$, and spin $\frac{1}{2}\hbar$ and a neutrino of energy *E*, no charge, and spin $\frac{1}{2}\hbar$.

4.10. CONSERVATION OF NUCLEAR ENERGY

The simultaneous emission of a β-particle and a neutrino from the nucleus of an atom requires energy. The disintegrating nucleus gives up not only the mass of the two particles but some additional mass which it converts into kinetic energy. By adding up all this energy in the form of mass, the total loss in the mass of an atom can be calculated.

Consider as an example the β-emission of $_{83}Bi^{210}$ as represented by Equation (2). The masses of the two atoms involved are known to be

$$_{83}Bi^{210} = 209.984110 \text{ amu}$$
$$_{84}Po^{210} = 209.982866 \text{ amu}$$
$$\Delta m = 0.001244 \text{ amu}$$

To convert this mass into MeV, we multiply Δm by the value 931 MeV which gives

$$E = 0.001244 \times 931 = 1.17 \text{ MeV}$$

This is just the end point energy derived from experiment and shown in Fig. 4.6. The mass of the ejected electron need not be included here, since the two masses above are for the neutral atoms. $_{83}Bi^{210}$ has 83 orbital electrons included in its 209.984110 amu, while $_{84}Po^{210}$ include 84 electrons. When $_{83}Bi^{210}$ ejects an electron from the nucleus, the daughter product $_{84}Po^{210}$ picks up a stray electron to become a neutral atom.

Different modes of β-decay.

As explained above, β-decay may take place by one of the following processes:

(*i*) electron emission (β^- emission)

(*ii*) positron emission (β^+ emission)

(*iii*) electron capture

(*i*) Electron emission takes place when a nucleus has excess of *neutrons.* We can represent the process as

$$_0n^1 \rightarrow {}_1p^1 + {}_{-1}e^0 + \nu + Q$$

Now mass of neutron m_n = 1.00867 amu
Mass of proton m_p = 1.00728 amu
Mass of electron m_e = 0.00055 amu
and mass of $\overline{\nu}$ = 0, we have

$$Q = m_n - m_p - m_e = 1.00867 - 1.00728 - 0.00055$$
$$= 0.00084 \text{ amu} = 0.00084 \times 931 = 0.782 \text{ MeV}$$

If we take m_p as the mass of neutral *parent atom* of At. No. Z, M_d the mass of neutral daughter atom of At. No. $(Z + 1)$ and m_e the mass and c the charge of the β-particle, then from the principle of conservation of mass-energy, we have

$$(M_p - Zm_e)c^2 = [M_d - (Z + 1)]\, c^2 + m_ec^2 + Q$$

or
$$M_pc^2 = M_dc^2 + Q$$

or
$$Q = (M_p - M_d)C^2$$

So for β^- emission, Q must be positive

i.e.,
$$M_p > M_d$$

or
$$M(Z) > M\,(Z + 1)$$

So the general basic equation for β^- emission is

$$_ZX^A \rightarrow \underset{z+1}{y} + {}_{-1}e^0 + \overline{\nu}$$

(*ii*) **Positron emission.** In β^+ emission, we have a nucleus with excess of protons. The process may be represented as

$$_1p^1 \rightarrow {}_0n^1 + e^0_{+1} + \nu$$

Since the charge number is decreased by one, so we may write the equation from mass-energy conservation as

$$(M_p - Zm_e)\,c^2 = [M_d\,(Z - 1)\,m_e]\,c^2 + m_ec^2 + Q$$

or
$$M_pc^2 = M_dc^2 + 2m_ec^2 + Q$$

∴ K.E. of positron

$$Q = (M_p - M_d)\,c^2 - 2m_ec^2$$

For positron emission, Q must be positive, *i.e.*,

$$(M_p - M_d)\, c^2 > 2\, m_e c^2$$

or
$$M_p - M_d > 2\, m_e$$

That is for *positron emission*, the difference between the masses of parent and daughter nuclei must exceed two electron masses *i.e.*, 2 × 0.00055 = 0.0011 amu.

The general equation for positron emission is

$$_zX^A \rightarrow Z - {}_1Y^A + {}_1e0 + \nu$$

(*iii*) **Electron capture.** In heavy nuclei, electronic orbits are closely placed and so there is probability of one of the orbital electrons to be *captured* by a radioactive nuclei. In case an electron from the K shell is captured, then an electron from L shell may jump to K shell and L-K transition may take place. The difference of energy due to L-K **transition** is given in the form of **characteristic X-rays.**

For example

$$_4\text{Be}^7 + {}_{-1}e^0 \rightarrow {}_3\text{Li}^7 + h\nu$$

In this case, no particle is lost from the atom, so

$$M_p c^2 = M_d\, c^2 + Q$$

or
$$Q = (M_p - M_d)\, c^2$$

For electron capture to take place, Q must be positive *i.e.*,

$$M_p > M_d$$

or
$$M\,(Z) > M\,(Z - 1)$$

In this case, the daughter element will gain *one electron* and so its atomic number change to $Z - 1$.

General equation for electron capture may be written as

$$_ZX^A + {}_{-1}e^0 = {}_{z-1}Y^A + h\nu$$

INVERSE β-DECAY

Inverse β-decay corresponds to the absorption of a neutrino by a nucleus with the conversion of a nuclear neutron to a proton and emission of an electron as from the following reaction

$$_0n^1 + \bar{\nu} \rightarrow {}_1p^1 + {}_{-1}e^0$$

Also the absorption of an antineutrino by a nucleus so that a proton is converted into a neutron and emission of a positron takes place as from the reaction

$$_1p^1 + \nu \rightarrow {}_0n^1 + {}_{+1}e^0$$

The above two reactions are known as **inverse β-decay.**

Difference between electron capture and positron emission:

(*i*) In positron emission, $_{+1}e^0$ is emitted while in electron capture no particle is emitted but the electron is captured by the nucleus.

(*ii*) Though the charge number Z decreases by one unit in both the cases, but in positron emission a *proton changes into a neutron*, with the emission of a positron, while in electron capture, a proton changes into a neutron by capturing an electron.

(*iii*) The basic condition for *positron emission* is that mass of the parent atom must be greater than the mass of the daughter atom *plus the mass of electrons* (*i.e.*, $M_p > M_d + 2\, m_e$), while in case of electron capture mass of the parent atom must be greater than the mass of the daughter atom *i.e.*, $M_p > M_d$.

POSITRON EMISSION AND ELECTRON CAPTURE REACTION COMPETE WITH EACH OTHER

In both *electron capture* and *positron emission*, the daughter nuclei is the same. However, in case of *electron capture, characteristic X-rays are emitted due to the rearrangement of orbital electrons* while in case of *positron emission* if the nucleus remains in excited state after the positron emission, then rearrangement of nucleons will lead to the emission of γ-rays. In some cases, both positron emission and electron capture may occur, *e.g.,*

$$_{30}Zn^{63} \rightarrow {}_{29}Cu^{63} + {}_{+1}e^{0} + Q \quad \text{Positron emission}$$

$$_{30}Zn^{63} + {}_{-1}e^{0} \rightarrow {}_{29}Cu^{63} + Q \quad \text{Electron capture}$$

So electron capture and positron emission are said to comptete with each other.

4.11. γ-RAY SPECTRA

For many radioactive elements the emission of an α or β particle from a nucleus is immediately followed by the emission of a γ-ray. It has been shown by crystal diffraction spectrographs that γ-rays are electromagnetic waves and that they consist of sharp lines of discrete wavelengths. Just as visible light, and ultraviolet and infrared radiation as well as X-rays are known to be emitted from the outer structure of the atom when an electron is jumping from one energy level to another, so γ-rays are believed to arise from a transition of a nucleon from one energy state to another within the nucleus. (Neutrons and protons as constituents of nuclei are called nucleons).

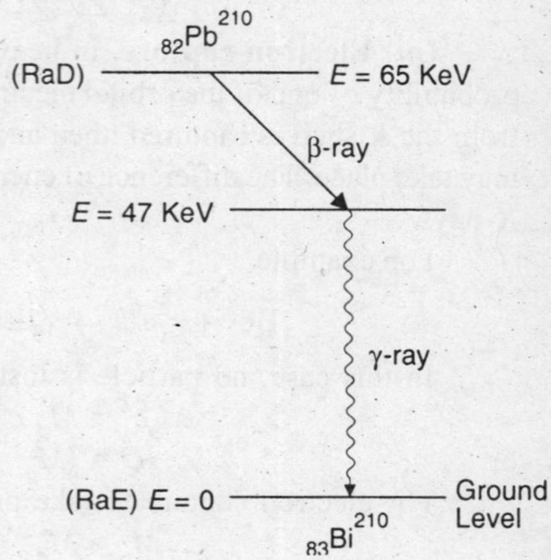

Fig. 4.8

There is good evidence that γ-rays are emitted by the daughter element, that is, they are preceded by particle emission. A good example is to be found in the case of RaD. This nucleus is the lead isotope $_{82}Pb^{210}$ and with a half-life of 22 years it emits a β particle. The reaction may be written as

$$_{82}Pb^{210} \rightarrow {}_{83}Bi^{210} + \beta^{-} + \nu + \gamma \text{ ray} \qquad ...(1)$$

These emissions are represented on a nuclear energy level diagram in Fig. 4.8. When a nucleus emits a β-particle and a neutrino, the end-point energy is found to be 18 kV. This leaves a nucleus in excited state or energy level. A transition down to the ground level is accompanied by the emission of a γ-ray with an energy $h\nu$. This energy is equivalent to 47 keV. Although it is not shown in this diagram, this nucleus, too, is radioactive and emits another β-ray to become $_{84}Po^{210}$

4.12. INTERNAL CONVERSION

When a γ-ray is emitted by a disintegrating nucleus, it must pass some of the outer electrons on its way out of the atom. Upon passing close to one of the electrons, a photoelectric action may occur whereby the γ-ray is absorbed and an electron is emitted. Energetically this process must follow the photoelectric equation

$$h\nu = \omega_0 + \frac{1}{2}mv^2 \qquad ...(2)$$

Part of the γ-rays energy $h\nu$ is used to pull the electron away from the atom, and the remainder is imparted to it as kinetic energy. As a rule this kinetic energy is relatively large, and $1/2mv^2$ must be replaced by the relativistic from $m_0c^2\ (\gamma - 1)$.

Since the energy required to remove an electron from an atom will differ from one shell to another, the electrons can be expected to have any one of

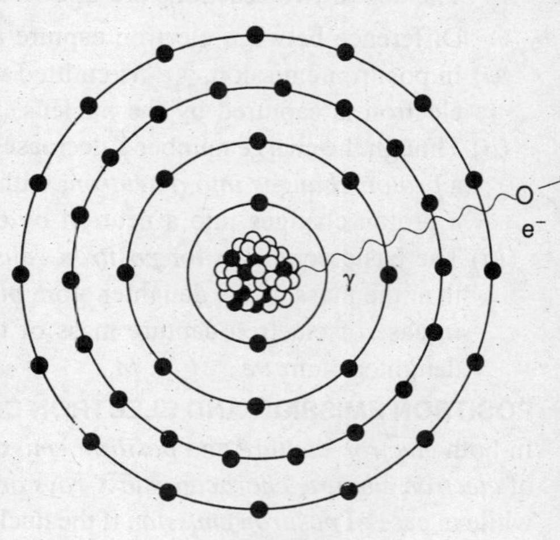

Fig. 4.9

several discrete energies. The definite sharp lines shown in Fig. 4.5 supply the evidence for this assumption. Furthermore, the experimentally determined energies agree exactly with those determined from X-rays. The process just described is illustrated schematically in Fig. 4.9, and is called internal conversion.

If by internal conversion an electron is ejected from the innermost K shell, leaving a vacancy there, another electron from the L shell or M shell falls in, to take its place and, in so doing, emits an X-ray. Confirmation of such a process is assured since the measured wavelengths of such X-rays are identical with those emitted by the same element in an X-ray tube.

Example 1. *Determine the minimum energy of an antineutrino to produce the reaction*

$$\nu + p \rightarrow n + e^{+}$$

Solution: From conservation of mass-energy

$$E_\nu + m_p c^2 = m_B c^2 + m_e c^2 + K_B + K_{e+}$$

The required neutrino energy will be minimum when the neutron and positron are both emitted with zero kinetic energy

$$E_{\nu\,\min} + 931.2 \text{ MeV} = 932.5 \text{ MeV} + 0.5 \text{ MeV}$$

or $$E_{\nu\,\min} = \mathbf{1.8\ MeV}$$

Example 2. *Determine the energy and momentum of the daughter and the neutrino that are produced when* $_4Be^7$ *undergoes electron capture at rest.*

Solution: The electron capture reaction is

$$_4\text{Be}^7 + e^- \rightarrow {}_3\text{Li}^7 + \nu$$

We have the Q value given by

$$Q = (M_{\text{Be}} - M_{\text{Li}})\, c^2$$
$$= (7.016929 \text{ a.m.u.} - 7.016004 \text{ a.m.u.})\ (931.5 \text{ MeV/a.m.u.})$$
$$= 0.862 \text{ MeV}$$

This energy is split between the neutrino and the $_3\text{Li}^7$ nucleus. However, because of the large mass of the $_3\text{Li}^7$ nucleus and the zero rest mass of the neutrino, almost all the energy is carried by the neutrino, so that

$$E_\nu \simeq 0.862 \text{ MeV}$$

Assuming that the $_4\text{Be}^7$ nucleus was initially at rest, the magnitudes of the momenta of the neutrino and $_3\text{Li}^7$ nucleus must be equal. Using $P_\nu = E_\nu/c$, we then have

$$P_\nu = P_\nu = 0.862 \text{ MeV/c}$$

K.E. of $_3\text{Li}^7$ nucleus can now be found from

$$K_{Li} = \frac{P_{Li}^2}{2M_{Li}} = \frac{(P_{Li}c)^2}{2M_{Li}c^2}$$

$$= \frac{(0.862\,\text{MeV})^2}{2\,(7.02\,\text{amu} \times 931.5\ \text{MeV/amu})}$$

$$= 5.68 \times 10^{-5} \text{ MeV} = \mathbf{56.8\ eV}$$

Example 3. $_9F^{20}$ *decays to the ground state of as follows:*

$$_9F^{20} \rightarrow {}_{10}Ne^{20} + e^- + \nu$$

where $_{10}Ne^{20} + \gamma$ *is an excited state of* $_{10}Ne^{20}$. *If the maximum kinetic energy of the emitted electrons is 5.4 MeV and the* γ*-ray energy is 1.6 MeV, determine the mass of* $_9Ne^{20}$ *(*M_{Ne} *= 19.99244 a.m.u.)*

Solution: Conservation of mass-energy applied to each of the reactions yields (the energy of the neutrino is zero in the limiting case and the recoil energies of $({}_9\text{Ne}^{20})^*$ and $_{10}\text{Ne}^{20}$ are negligible).

$$(M_F - 9m_e)\, c^2 = (M_{\text{Ne}}{}^* - 10m_e)\, c^2 + m_e c^2 + K_e$$

or $$M_F c^2 = M_{\text{Ne}}{}^* c^2 + K_e$$

$$M_{\text{Ne}}{}^* c^2 = M_{\text{Ne}} c^2 + E_\gamma$$

Rearranging these two expressions, we have

$$M_F c^2 = MNe^* c^2 + K_e$$

$$M_F = 19.99244 \text{ a.m.u.} + (1.6 \text{ MeV} + 5.4 \text{ MeV}) \left(\frac{1 \text{ amu}}{931.5 \text{ MeV}}\right)$$

$$= \mathbf{20.000 \text{ a.m.u.}}$$

Example 4. *Determine the K.E. of α-particles emitted in an α- decay in terms of Q value of the reaction.*

Solution: We have the reaction

$$\underset{Z}{\overset{A}{P}} \rightarrow \underset{Z-2}{\overset{A-4}{D}} + {}_2He^4$$

Let us assume the parent to be at rest, we have from the conservation of momentum $P_D = P_\alpha$. Since the kinetic energies will be very small as compared to the rest energy of the parent, we can use the non-relativistic relation $K = \dfrac{p^2}{2M}$ to get

$$\frac{K_D}{K_\alpha} = \frac{M_\alpha}{M_D} \cong \frac{4}{A-4}$$

where A is the mass number of the parent.

$\therefore$ Q value of the reaction is

$$Q = K_D + K_\alpha = \frac{4}{A-4} K_\alpha + K_\alpha = \frac{\boldsymbol{AK_\alpha}}{\boldsymbol{A-4}}$$

Example 5. *Show that ${}_{94}Pu^{236}$ is unstable and will show α-decay.*

Solution: We have the Equation for spontaneous α-decay of ${}_{94}\text{Pu}^{236}$ as

$${}_{94}\text{Pu}^{236} \rightarrow {}_{92}\text{U}^{232} + {}_2\text{He}^4 + Q$$

For this equation to be feasible, Q should be positive. We have

$$Q = (M_{\text{Pu}} - M_{\text{U}} - M_{\text{He}})\, c^2$$
$$= (236.046071 - 232.0371 - 4.0026) \times 931 \text{ MeV/Nucleon}$$
$$= \mathbf{+\, 5.87 \text{ MeV.}}$$

So ${}_{94}\text{Pu}^{236}$ will spontaneously decay with α-ray emission.

Example 6. *When a sample of 0.95 curie of Bi^{210} (end point energy 1.3 MeV) was placed in a calorimeter, the measured heat output was 0.02546 cal/min. Calculate the mean energy (E) of the transition and E/E_{max} ratio.*

Solution: We have

$$1 \text{ curie} = 3.7 \times 10^{10} \text{ disintegrations/sec.}$$

$$\therefore \quad 0.95 \text{ curie} = 0.95 \times 3.7 \times 10^{10} \times 60 \text{ disintegrations/min}$$

Also Heat output/min $= 0.02546$ cal

$$= 0.02546 \times 4.2 \text{ J}$$

$$= \frac{0.02546 \times 4.2}{1.6 \times 10^{-13}} \text{ MeV}$$

So the mean energy/transition

$$E = \frac{0.02546 \times 4.2}{1.6 \times 10^{-13} \times 0.95 \times 3.7 \times 60 \times 10^{-10}} \text{ MeV}$$

$$= 0.3170 \text{ MeV.}$$

Also End point energy $E_{\text{max}} = 1.3$ MeV

$$\therefore \quad \frac{E}{E_{\text{max}}} = \frac{0.3170}{1.3} = \mathbf{0.244}$$

QUESTIONS AND PROBLEMS

1. Give the fine structure of α-rays. How the various spectral lines originate?
2. What are short and long range α-rays? Explain their origin with some specific examples.
3. Give the elementary ideas about α-decay. How α-decay accounts for the nuclear energy levels?
4. Give the theory of α-decay from the energy level concept of the nuclei.
5. Give the elementary theory of β-decay.
6. Give the origin of neutrino and antineutrino.
7. Give the theory of γ-decay.
8. From the theory of β-decay how will you account for the charge, mass and spin of the neutrino?
9. How the internal conversion phenomenon takes place?
10. The maximum kinetic energy of β-particle emitted in Tritium (${}_1H^3$) decay is 19 keV. If the mass of tritium is 3.0160504 a.m.u, what is the mass of the decay product?

    [**Ans.** 3.016030 a.m.u.]
11. Explain how the theory of β-decay accounted for the existence of neutrino and antineutrino?
12. What is Mossbauer Effect? Explain the recoilless γ-ray emission.
13. What is the kinetic energy of α-particles emitted in the α-decay of ${}_{92}U^{232}$? Assume ${}_{92}U^{232}$ nucleus to be at rest. [**Ans.** 5.33 MeV.]
14. A substance with At. No. A undergoes α-decay by emitting two groups of α-particles with kinetic energies K_{α_1} and K_{α_2} Show that the energy of the accompanying γ-ray is

    $$E_y = \frac{A}{A-4}(K_{\alpha_1} - K_{\alpha_2})$$
15. Determine the maximum possible speed of the daughter in a β-decay of a ${}_2He^6$ nucleus which is initially at rest. [**Ans.** 1.0×10^5 m/s.]
16. Determine the value of alpha, proton, neutron and deutron decay of ${}_{92}U^{232}$.

    [**Ans.** 5.42 MeV, – 6.09 MeV, – 7.23 MeV, – 10.09 MeV.]

CHAPTER

5

NUCLEAR REACTIONS AND ARTIFICIAL RADIOACTIVITY

5.1. INTRODUCTION

In the previous chapter we have studied the radioactive elements produced by natural way *i.e.*, by the emission of α, β or γ rays from the nucleus of the parent element. In this chapter we will study the radioactive elements produced by artificial means. The first experiment on artificial nuclear disintegration was performed by Sir Earnest Rutherford. He bombarded the atoms of different stable elements of low atomic number, such as, nitrogen, aluminium and potassium etc., by using high speed α-particles as projectiles obtained from natural radioactive substances such as Radium-C. In 1930 Bathe and Becker performed their experiment with α-particles as projectile on lithium, beryllium and boron and observed that very penetrating rays were given out particularly from beryllium. The rays possessed energy of the order of 14 MeV. Irene Curie and her husband Juliot repeated the same experiment in 1932, and found that the emitted rays pass through thick sheets of metals unaffected. These rays from beryllium cause the emission of protons from the paraffin wax. In the same year Chadwick repeated the same experiment carefully, he discovered the neutral particles *i.e.*, (${}_0n^1$ neutrons) with very high speed and energy. For this discovery, **Chadwick was awarded the Nobel prize in 1935.**

5.2. RUTHERFORD'S EXPERIMENT FOR ARTIFICIAL TRANSMUTATION

The first artificial transmutation was observed by Rutherford in 1919. He showed that the nuclei of nitrogen emit proton when bombarded with α-particles from Radium *C*. The apparatus that Rutherford used was simple but sensitive. It consists of a box *B* having one end open as shown in Fig. 5.1. A silver foil *F* and a zinc sulphide screen *S* was placed before the open end *E*. The scintillations on the screen were observed by means of a microscope *M*.

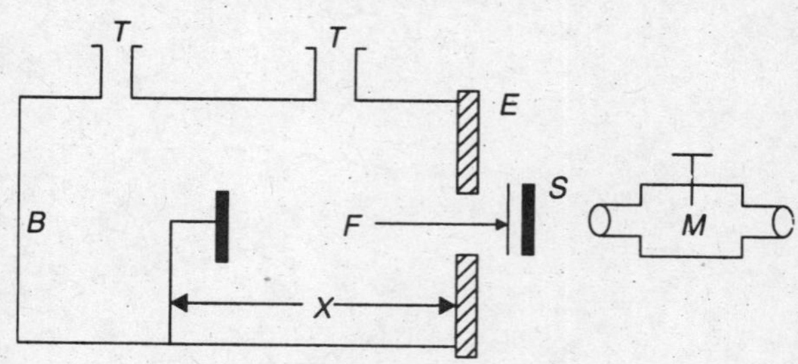

Fig. 5.1

The source of the α-particles (Radium *C*) placed on the disc *D*, whose distance from *S* can be varied. The silver foil *F* was thick enough to absorb the α-particles from the source. Different gases introduced into the box and removed through the side tubes *T*. When the box was filled with oxygen or carbon dioxide at atmospheric pressure, no scintillations were seen on the screen with the source 7 cm or more away. When the gas was nitrogen, however, scintillations were observed on the screen when the source of α-particles was as much as 40 cm away. Rutherford ruled out, by means of careful experiments, the possibility that the proton came from hydrogen present as an impurity in the nitrogen. The reaction can be represented as

$$ {}_7N^{14} + \underset{\alpha-\text{particle}}{{}_2He^4} \rightarrow {}_8O^{17} + \underset{\text{proton}}{{}_1H^1} $$

The experimental results also showed that the probability of disintegration was very small; one proton was produced for about one million α-particles passing through the gas. This process of producing new stable nuclei out of other stable nuclei is called artificial transmutation of elements.

Rutherford and Chadwick (1921–24) improved the apparatus and showed that protons are liberated in the same manner from all elements from boron to potassium with exception of carbon and oxygen.

The reaction in the case of Fluorine is

$${}_{9}F^{19} + {}_{2}He^{4} \rightarrow {}_{10}Ne^{22} + {}_{1}H^{1}$$

5.3. PROJECTILES FOR ARTIFICIAL TRANSMUTATION

More than 500 artificial radioactive isotopes have been found, provoked with all the commonly used projectiles which are given below:

(*i*) α-particles, (*ii*) Protons, (*iii*) Deuterons, (*iv*) β-particles and (*v*) γ-rays

5.4. TYPES OF NUCLEAR REACTIONS

In general nuclear reactions may be classified into the following categories:

(1) Transmutation reactions: In this type of reaction, the product nucleus differs from the reactant nucleus. Various types of transmutation reactions are discussed in details in later part.

(2) Pick up and stripping reactions: In the pick up type of reactions, an incident particle collides with the target nucleus and picks up a nucleon from the target nucleus. This happens normally when the De Broglie wavelength of the **incident** particle is equal to the average distance between the nucleons so that the interaction is limited to only one nucleon.

In case of stripping, reverse process takes place. In this case the projectile particle is composed of more than one nucleon and when it collides with target nucleus it loses one of the nucleon to the target nucleus, *e.g.*,

$${}_{29}Cu^{63} + {}_{1}He^{2} \rightarrow {}_{29}Cu^{64} + {}_{1}H^{1}$$

It may be noted that in this type of reactions, no compound nucleon is formed as intermediate step.

(3) Fission and Spallation reactions: Fission is the process of breaking up of a heavier nuclei into smaller fragments, when hit by a high energy projectile particle, *e.g.*,

$${}_{92}U^{235} + {}_{0}n^{1} \rightarrow {}_{56}Ba^{141} + {}_{36}Kr^{92} + 3\,{}_{0}n^{1} + \text{Energy}$$

In case of **spallation reaction** the energy of the projectile particle is high enough to break original nuclei into 20-30 fragments, of small number of nucleons.

(4) Disintegration reactions: In this type of reaction, the interacting particle is absorbed by the target nucleus and a different particle is ejected from the target **nucleus.** This process sometimes is referred to as rearrangement of the **nucleons** inside the nucleus, *e.g.*, this type of reaction may have the form

$${}_{3}Li^{7} + {}_{2}H^{4} \rightarrow {}_{5}B^{10} + {}_{0}n^{1}$$

In case the projectile is high energy γ-ray which is absorbed by the target nucleus and the excess energy is given in the form of some other particle, then this type of reaction is called photo disintegration, *e.g.*,

$${}_{1}H^{3} + \lambda \rightarrow {}_{1}H^{2} + {}_{0}n^{1}$$

(5) Heavy ion reaction: In this case the projectile is some heavier nucleus different from the conventional projectiles (*e.g.*, α-rays, γ-rays, neutrons, protons etc.) This type of reaction has the characteristics of both the **pick up and stripping mechanism and compound** nucleus. For example

$${}_{13}Al^{27} + {}_{7}N^{14} \rightarrow {}_{19}K^{38} + 2\,{}_{0}n^{1} + {}_{1}H^{1}$$

(6) High energy reaction: When the energy of the projectile is very high, then along with the product particles, some high energy particles like π-mesons and strange particles are ejected.

(7) Scattering Reaction: In this type of reaction, both the product and target nuclei remain unchanged, yet there is exchange of energy. The scattering may be **elastic** or **inelastic.**

***(i)* Elastic scattering:** In this type of reaction, both the energy and momentum are conserved. However, the direction of motion and the velocity of interacting particles change. An example of the elastic scattering is given (from Rutherford's α-ray scattering) as

$$_{79}Au^{197} + {}_2He^4 \rightarrow {}_{79}Au^{197} + {}_2He^4$$

***(ii)* Inelastic scattering**: In this type of reaction, though the composition of the nuclei remains unchanged, yet there is some excess energy imparted to the target nucleus by the K.E. of the projectile particle. So the produce nucleus may be in excited state (represented by *) and may come to ground state by the emission of a proton, *e.g.*,

$$_3Li^7 + {}_1H^1 \rightarrow [{}_3Li^7]^* + {}_1H^1$$

$$[{}_3Li^7]^* \rightarrow {}_3Li^7 + \gamma$$

5.5. CONSERVATION LAWS OF NUCLEAR REACTIONS

(1) Conservation of Charge: Total charge of the reactants is conserved.

(2) Conservation of nucleons: Total number of nucleons before and after the reaction remains constant as **nucleons** can neither be **created** nor be destroyed.

(3) Mass and Energy conservation: In a nuclear reaction **mass and energy** are not separately conserved. However, their total is always conserved, *i.e.,* the energy equivalent of mass is liberated as can be understood in the *Q*-value of the nuclear reactions.

(4) Conservation of linear momentum: It is a **fundamental law** and is conserved in nuclear reactions also.

(5) Conservation of angular momentum: Total angular momentum is composed of orbital angular momentum **I** and spin angular momentum α **S**, vector sum of the angular momentum is conserved in a nuclear reaction.

(6) Conservation of spin and statics: Statics imposes a restriction in a nuclear reaction.

(7) Conservation of parity: Parity corresponds to the behaviour of wave function under simultaneous reflection of space co-ordinates. In case of **strong interactions**, parity conservation law is strictly observed while it is violated in **weak interactions.**

5.6. KINEMATICS OF NUCLEAR REACTIONS

The basic laws of mass, energy and momentum conservation, when applied to the nuclear reactions, give a mathematical analysis leading to **Kinematics of Nuclear reactions.**

Consider a nuclear reaction represented by the equation

$$a + X \rightarrow Y + b \quad ...(1)$$

where target nucleus '*X*' is supposed to be at rest initially. Since the total energy of the system is to be conserved, let the mass of projectile a target nucleus *X*, product nucleus *Y* and product particle '*b*' be represented as m_a, M_x, M_y and m_b respectively. Then from the law of conservation of total energy we have $(E_a + M_a c^2) + m_x c^2 = (E_y + M_y c^2) + (E_b + m_b c2)$, where *E*'s correspond to the kinetic energies. We introduce the quantity called *Q*-value of the reaction **which represents the difference of energy of the products and the reactants.**

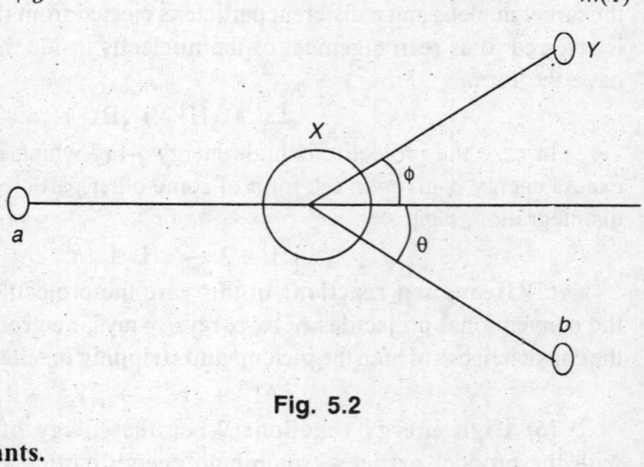

Fig. 5.2

$\therefore \quad Q = E_y + E_b - E_a$

So Equation (1) may take the form

$(m_a + M_x - M_y - m_b)c^2 = E_y + E_b Q$ ...(2)

where $E_y \rightarrow$ recoil energy of the product nucleus is very small and cannot be measured accurately. However, it can be eliminated while taking into account the conservation law of momentum.

Applying the laws of conservation of momentum along X-axis, we get

$$m_a v_a = M_y V_y \cos\phi + m_b v_b \cos\theta \quad ...(3)$$

and along Y-axis

$$M_y V_y \sin\phi = m_b v_b \sin\theta \quad ...(4)$$

where $v_a \rightarrow$ velocity of incident particle

$v_b \rightarrow$ velocity of ejected particle

and $V_y \rightarrow$ velocity of the product nucleus.

Equation (3) may be written as

$$m_a v_a - m_b v_b \cos\theta = M_y V_y \cos\phi \quad ...(5)$$

Squaring equation (5) we get

$$m_{a2} v_{a2+mb2} v_{b2\cos 2}\theta - 2m_a m_b v_a v_b \cos\theta = M_{y2V_y 2\cos 2}\phi \quad ...(6)$$

Squaring Equation (4), we get

$$M_y^2 V_y^2 \sin^2\theta = m_b^2 v_b^2 \sin^2\theta \quad ...(7)$$

Adding Equations (6) and (7) we get

$$M_y^2 V_y^2 = m_b^2 v_b^2 + m_a^2 v_a^2 - 2m_a m_b v_a v_b \cos\theta \quad ...(8)$$

$[\sin^2\phi + \cos^2\phi = 1 \text{ and } \sin^2\theta + \cos^2\theta = 1]$

Now $\quad E_a = \frac{1}{2} m_a v_a^2$ and $E_b = \frac{1}{2} m_b v_b^2$

$\therefore \quad v_a^2 = 2E_a / m_a$ and $v_b = \frac{2E_b}{m_b}$

Also $\quad E_y = \frac{1}{2} M_y V_y^2 \quad \therefore V_y^2 = \frac{2E_y}{M_y}$

Putting the values of v_a, v_b^2 and V_y^2 in Equation (9) we get

$$M_y^2 \cdot \frac{2E_y}{M_y} = m_b^2 \cdot \frac{2E_b}{m_b} + m_a^2 \cdot \frac{2E_a}{m_a} - 2m_a \left(\frac{2E_a}{m_a}\right)^{1/2} \left(\frac{m_b 2E_b}{2}\right) \cdot \left(\frac{m_b^2 E_b}{m_a}\right) \cdot \cos\theta$$

Which gives on simplification

$$2F_y M_Y = 2E_a m_a + 2E_b m_b - \frac{2}{2_y}(m_a m_b E_a E_b)^{1/2} \cos\theta$$

or $$E_y = E_a \frac{m_a}{M_y} + E_b \frac{m_b}{M_y} - \frac{2}{M_y}(m_a m_b E_a E_b)\cos\theta \quad ...(9)$$

Putting the value of E_y from Equation (8) in Equation (1) we get

$$Q = E_a \frac{m_a}{M_y} + E_b \frac{m_b}{M_y} - \frac{2}{M_y}(m_a m_b E_a E_b)^{1/2} \cos\theta + E_b E_a$$

or $$Q = E_a \frac{m_a}{M_y} - E_b + E_b + \frac{m_b}{M_y} + E_b - \frac{2}{M_y}(m_a m_b E_a E_b)^{1/2} \cos\theta$$

$$= E_a \left(\frac{m_a}{M_y} - 1\right) + E_b \left(\frac{m_b}{M_y} + 1\right) - \frac{2}{M_y}(m_a m_b E_a E_b)^{1/2} \cos\theta$$

$$= E_a\left(\frac{m_a - M_y}{M_y}\right) + E_b\left(\frac{m_b + M_y}{M_y}\right) - \frac{2}{M_y}(m_a m_b E_a E_b)^{1/2} \cos\theta \quad ...(10)$$

Equation (10) is known as ***Q*-value of equation.** It gives us a relation between the energy released in the reaction with the parameters $\boldsymbol{E_a}$**,** $\boldsymbol{E_b}$ **and** $\boldsymbol{Q}$**.**

In a special case when the particle ejected is at right angles to the incident particle, *i.e.*, $Q = 90°$, then above relation may be written as

$$Q = E_a\left(\frac{m_a - M_y}{M_y}\right) + E_b\left(\frac{m_b + M_y}{M_y}\right) \qquad [\because \cos 90° = 0]$$

$$= E_b\left[1 + \frac{m_b}{M_y}\right] - E_a\left[1 - \frac{m_a}{M_y}\right] \quad ...(11)$$

With a good experimental arrangement, Q value of a nuclear reaction can be accurately measured.

5.7. PHYSICAL SIGNIFICANCE OF *Q*-VALUE

Positive value of Q means release of energy. *i.e.*, part of nuclear mass is converted into energy and the reaction is called **exoergic** or **exothermic,** while negative value of Q corresponds to the case where a part of energy is taken up or converted into mass so in this case the mass of products **exceeds** the mass of the **original reactants.** Such a reaction is called **Endoergic** or **Endothermic.** Such type of reaction is possible when the incident particle posseses enough K.E. to be converted to mass during the reaction.

5.8. GENERAL SOLUTION OF *Q*-VALUE EQUATION

If we consider Q value to be fixed, then equation (11) may be considered as quadratic in $\sqrt{E_b}$ for finding a relation between E_a and E_b. We can write the general solution of equation (11) in the form

$$E_b = v \pm \sqrt{v^2 + \omega} \quad ...(12)$$

where
$$v = \frac{\sqrt{m_a m_b E_b}}{M_y + m_b} \cos\theta$$

and
$$w = \frac{M_y Q + E_a(M_y - m_a)}{M_y + m_b}$$

only those nuclear reactions are possible which have real and positive values of $\sqrt{E_b}$. We have the following cases

***(i)* Exoergic reactions.** These are possible even for the case when $E_a = 0$, so putting $E_a \to 0$ in equation (12) we get

$$E_b = \frac{QM_y}{(m_b + M_y)} \quad \text{which gives } E_b > 0$$

In such a case K.E. E_b is same for all the angles Q. Physically it will be possible as in this case total momentum is effectively zero when $E_a \to 0$.

So we may have

$$\therefore \quad Q = E_a + E_b \text{ and } \theta + \phi = 180°$$

Usually in this case incident particle is lighter than the product nucleus Y. So ω is positive for all the values of projectile energy and only one of the two solutions of equation (12) will be +ve. So for E_b to have single value, $Q > 0$ and $M_y > m_b$

***(ii)* Endoergic reactions.** In this case Q has negative value. When $E_a \to 0$, then equation (12) gives $v^2 + \omega$ = negative, *i.e.*, E_b becomes imaginary and thus this reaction is not energetically, possible.

When E_a has large value to make $v^2 + \omega = 0$.

The smallest value of E_a that makes $v^2 + \omega = 0$ is called **threshold energy.** So

$$E_a = -Q\left[\frac{m_b + M_y}{m_b + M_y - m_a} - \frac{m_a m_b}{M_b}\sin^2\theta\right]$$

At $\theta = 0$, $E_a \rightarrow 0$ minimum value and so will represent the **threshold energy.**
Using equation (1) and approximation $M_x >> Q/C^2$
we have

$$(E_a)_{Th} = -Q\,\frac{m_a + M_x}{M_x} \quad ...(13)$$

Example 1. *γ-ray of 2.21 MeV energy are used to split a deuterium nucleus into a proton and neutron according to equation*

$${}_1H^2 + \gamma \longrightarrow {}_1H^2 + {}_0n^1$$

Calculate the mass of neutron; given
Mass of deuterium atom = 2.01472 a.m.u.
Mass of proton = 1.00813 a.m.u.

Solution: From the Einstein's mass energy relationship

$$E = mc^2$$

$$\therefore \quad m = \frac{E}{c^2}$$

Now
$$E = 2.21\,\text{MeV}$$
$$= 2.21 \times 10^6\ \text{eV}$$
$$E = 2.21 \times 10^6 \times 1.6 \times 10^{-12}\ \text{ergs}$$
$$c = 3 \times 10^{10}$$
$$\therefore \quad m = \frac{E}{c^2} = \frac{2.21 \times 10^6 \times 1.6 \times 10^{-12}}{9 \times 10^{36}}\ \text{gm}$$

$\therefore$ mass of γ-particle = 0.00238 a.m.u.

Now ${}_1H^2 + \gamma \rightarrow {}_1H^1 + {}_0n^1$

or $\quad 2.01472 + 0.00238 = 1.00813 + {}_0n^1$

or $\quad 2.01710 = 1.00813 + {}_0n^1$

$\therefore$ mass of neutron ${}_0n^1 = 2.01710 - 1.00813$

$= \mathbf{1.00897}$ **a.m.u.**

Example 2. *Calculate the energy generated in kWh, when 100 gm of ${}_3Li^7$ are converted into ${}_2He^4$ by proton bombardment, given,*
mass of ${}_3Li^7$ = 7.0183 mass unit
mass of ${}_2He^4$ = 4.0040 mass unit
mass of ${}_3Li^1$ = 1.0081 mass unit

Solution: ${}_3Li^7 + {}_1H^1 \rightarrow 2\,{}_2He^4 + \Delta m$

$$7.0183 + 1.0081 = 2 \times 4.0040 + \Delta m$$

$$\therefore \quad \Delta m = 0.0184\ \text{mass units.}$$

Now the Energy liberated when 7.0183 gm of Li is converted 0.0184 mass unit.
Energy liberated with 100 gm of Li

$$= \frac{109 \times 0.0184}{7.0183}\ \text{gm}$$

$$E = mc^2$$

$$= \frac{100 \times 0.0184 \times (3 \times 10^{10})^2}{7.0183}\ \text{ergs sec}$$

$$1 \text{ watt} = 10^7 \text{ ergs}$$

$$\therefore \quad E = \frac{100 \times 0.0184 \times (3 \times 10^{10})^2}{7.0183 \times 10^7} \text{watts-sec}$$

$$= \frac{100 \times 0.0184 \times (3 \times 10^{10})^2}{7.0183 \times 10^7 \times 1000} \text{kW-sec}$$

$$= \frac{100 \times 0.0184 \times (3 \times 10^{10})^2}{7.0183 \times 10^7 \times 1000 \times 60 \times 60} \text{kWh.}$$

$$E = \mathbf{6.555 \times 10^6 \text{ kWh}}$$

Example 3. *Calculate the minimum energy of γ-rays necessary to disintegrate a deuteron into a proton and a neutron, given*

$$e = 4.80 \times 10^{-10} \text{ e.s.u.}$$
$$m = 0.91 \times 10^{-27} \text{ gm}$$
$$c = 3 \times 10^{10} \text{ cm/sec}$$
$$h = 6.62 \times 10^{-27} \text{ erg. sec}$$
$$M = 1.66 \times 10^{-27} \text{ gm}$$

Solution: Mass of Proton = 1.00759 mass units

Mass of Neutron = 1.00898 mass units

Mass of Deuteron = 2.01471 mass units

$$_1H^1 + {_0n^1} = {_1H^2} + \gamma\text{-rays}$$

$$1.00759 + 1.00898 = 2.01471 + \gamma\text{-rays}$$

$\therefore$ Mass of γ-ray = 0.00186 mass units

Mass of γ-ray = $0.00186 \times 1.66 \times 10^{-24}$ gm

($\because$ 1 a.m.u. = 1.66×10^{-24} gm)

But $E = mc^2$ (From mass Energy relation)

$$\therefore \quad \text{Energy of γ-rays } E = 0.00186 \times 1.66 \times 10^{-24} (3 \times 10^{10})^2 \text{ erg.}$$

$$= \frac{300 \times 0.00186 \times 1.66 \times 10^{-24} \times (3 \times 10^{10})^2}{4.80 \times 10^{-10}} \left[\because 1 \text{ erg} = \frac{300}{4.80 \times 10^{-10}} \text{eV}\right]$$

$$E = 1.747 \times 10^6 \text{eV} = 1.747 \text{ MeV}$$

$\therefore$ The minimum energy of γ-rays required is **1.747 MeV.**

5.9. NUCLEAR REACTION CROSS-SECTION

Nuclear cross-section of a reaction corresponds to the probability that the given **nuclear reaction will occur.** It is supposed that each target particle presents a certain area known as **cross-section** to the incident particle. Any incident particle that is directed towards this area **interacts with the target particle.** So the greater cross-section, the greater is the probability of interaction.

When a collimated beam of monochromatic energies of light particles (like protons, neutrons, electrons etc.) is shot at a target nucleus, then any of the following processes may occur singly or jointly:

(*i*) Incident particle may be deviated and scattering (elastic or inelastic) phenomenon may take place.

(*ii*) Incident particle may be completely absorbed and **radiative capture** may take place.

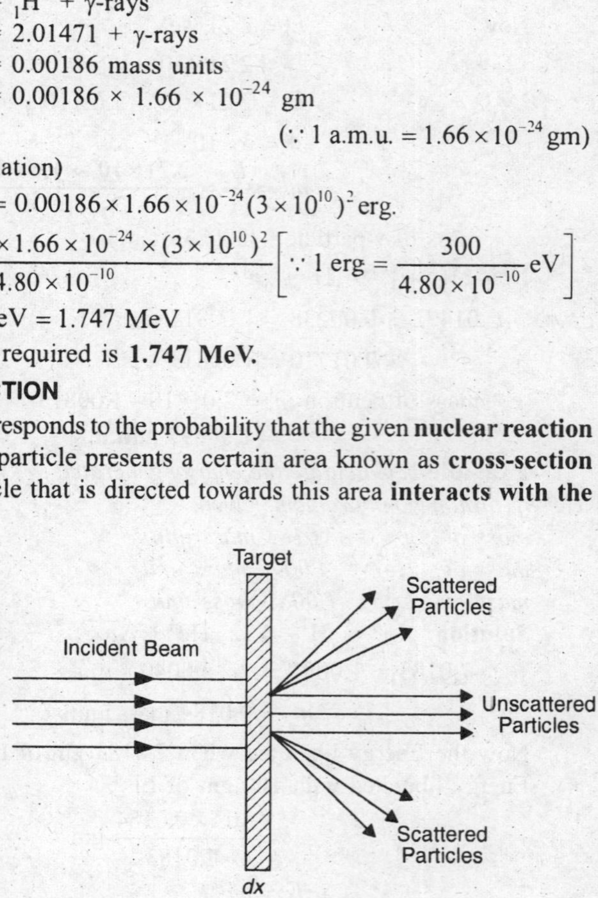

Fig. 5.3

(*iii*) Incident and emergent particles may be different particles and nuclear reactions may take place.

When a particle incident is on a target nucleus, the cross-section may be defined as:

(*i*) The probability that an event may occur when a nucleus is exposed to a beam of particles carrying **total flux/particle/unit area.**

(*ii*) The probability that an event may occur when a single particle is shot in a direction perpendicular to a target nucleus consisting of one particle/unit area.

It may be mentioned here that concept of **cross-section for a nuclear process** is different from the geometrical **cross-section area** (πR^2) of the nucleus. However, cross-section (σ) has the dimensions of area and is expressed in units of **Barn, 1 Barn = 10^{-24} cm^2 (10^{-28} m^2)**

If at any instant there are particles with a velocity v, the flux of the beam is given by $n\,v$ and if N be the total number of nuclei exposed to the beam at any instant of time, then rate of reaction/scattering may be written as

$$\frac{dN}{dt} \propto nv$$

and it is also proportional to N

$$\therefore \qquad \frac{dN}{dt} = -\sigma nvN$$

which on integration gives

$$N = N_0 e^{-\sigma n v t} \qquad ...(1)$$

where $N_0 \rightarrow$ no. of target nuclei at $E = 0$

and $N \rightarrow N_0$ after a time t

Now the probability that a particular type of nuclear process will take place due to interaction with the incident particle will be dependent on the number of particles/unit volume of the target and the thickness of the target.

$$\Delta P = \sigma N \Delta x$$

if $N = 1$ particle/c.c.

and $\Delta x = 1$ cm

then $P = \sigma/\text{cm}^2$

so we may express the cross-section in terms of probability equation as:

The probability to produce a reaction if one particle is facing the projectile/unit area of the target in the path of an incident particle.

If n number of particles fall on the target in a time Δt, then rate of reaction will be given by

$$\frac{n\,\Delta P}{\Delta t} = \frac{dn}{dt} = -\sigma N n \frac{\Delta x}{\Delta t} = -\sigma N n v \qquad ...(2)$$

Now the number of particles scattered in thickness dx may be expressed from Equation (2) as

$$P = -\sigma N n \Delta x$$

or

$$dn = -\sigma N n\, dx$$

$$\therefore \qquad \frac{dn}{n} = -N \sigma\, dx$$

Which on integration gives

$$n = n_0\, e^{-N.\sigma.x} \qquad ...(3)$$

Equation (3) gives the number of particles passing through the target undeflected. This number is definitely less than the no. of particles n_0 incident at $t = 0$. So because of scattering, the incident beam has been attenuated.

Equation (3) may be written as

$$n = n_0 e^{-\mu x} \qquad \ldots(4)$$

where $\mu = \sigma N$ and is called **attenuation coefficient.** This expression is useful for deriving mean free path of absorption of a particle in a target. If a target contains number of isotopes of different cross-sections, $n\sigma$ then is equal to $n_1\sigma_1 + n_2\sigma_2 + \ldots$ and we may write μ in the form

$$\mu = N\sigma = \frac{\sigma N_A \cdot \rho}{A} \qquad \ldots(5)$$

where $\rho \to$ density of target material

$A \to$ Atomic weight of the target

$N_a \to$ Avogadro number.

Differential cross-section. Differential cross-section gives us information about the angular distribution of the outgoing particles

It may be written as

Differential cross-section

$$\sigma(\theta, \phi) = \frac{d\sigma}{d\Omega}(\theta, \phi)$$

The quantity

$$d\sigma(\theta, \phi) = \sigma(\theta, \phi)\, d\Omega \qquad \ldots(6)$$

represents the probability that the ejected particles are confined within an element of solid angle $d\Omega$ in a direction θ to the incident beam.

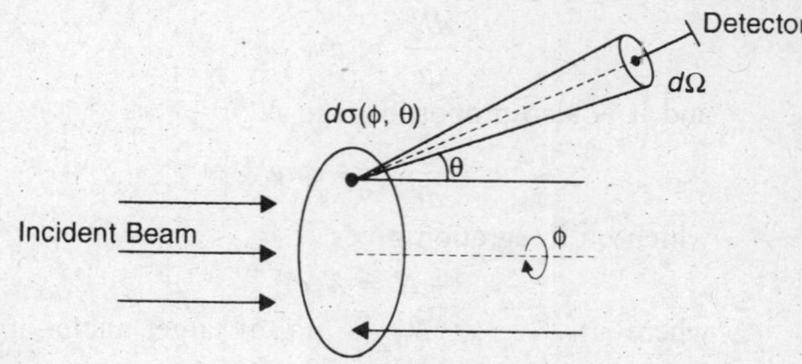

Fig. 5.4

So to give a relation between total cross-section σ and differential cross-section $\sigma(\theta, \phi)$ we have to integrate the differential cross-section over the total space

$$\sigma = \int \sigma(\theta, \phi)\, d\Omega = \int_{\phi=0}^{2\pi} \int_{\theta=0}^{\pi} \sigma(\theta, \phi) \sin\theta\, d\theta \qquad \ldots(7)$$

$[\because d\Omega = \sin\theta\, d\theta\, d\phi]$

The ϕ-dependence will hold in the case when the incident particle beams have **spins** in specific direction, *i.e.*, the incident beam is polarized. However, for randomly oriented spins of the incident beam (unpolarized beam) until specified, the differential cross-section will be **independent of** ϕ and so Equation (7) may be written as

$$\sigma = \int_0^{\pi} \sigma(\theta) \sin\theta\, d\theta \qquad \ldots(8)$$

In special case of **isotropic distribution** of emitted particles or radiation, we may write Equation (8) as

$$\sigma = 2\pi\sigma(\theta) \qquad \ldots(9)$$

which gives us a relation of total cross-section with the differential cross-section for the specific case of **isotropic distribution of particles.**

Level Width. Another important term concerning the cross-section concept is the **nuclear level width.** The compound nucleus formed due to the bombardment of the target nucleus by a high energy particle remains in the excited state for some time before decaying either by γ-emission or through some other particle. We know that the **reciprocal of the decay constant** λ is called **mean life time.** The decay constant corresponds to the probability/unit time of emission of a particle or γ-ray.

In nuclear reactions, we make use of a quantity which is proportional to disintegration constant and is called **Level Width** given as

Level Width $= \dfrac{h}{2\pi\tau} = \hbar\lambda$

where $\tau \rightarrow$ mean life time

Level widths are usually given in terms of eV.

Example 4. *Calculate the alternation coefficient and thickness of Cd^{113} required to reduce the intensity of neturon beam by 1% of its original value. Take density of Cd = 8.67 gm/cc and scattering cross-section involved = 20 800 barns.*

Solution: We have $\mu = \dfrac{\rho N_A \cdot \sigma}{A}$

$$= \frac{8.67}{113} \times 6.02 \times 10^{23} \times 20800 \times 10^{-24} = 961 \text{ cm}^{-1}$$

Now we have the equation

$$n = n_0 e^{-\mu x}$$

$$\therefore \quad \mu x = -\log \frac{n}{n_0}$$

or $$x = \frac{1}{\mu} \log_e \frac{n_0}{n} = \frac{1}{961} \times \log_e 100$$

$$= \frac{1}{961} \times 2.3026 \times 2 = \frac{4.6052}{961} = \mathbf{0.48 \times 10^{-2}\ cm}$$

Example 5. *The cross-section of Cd^{113} for capturing thermal neutrons is 2×10^4 b, the mean atomic mass of natural Cd is 112 amu and its density is 8.64×10^3 kg/m^3*

(a) What fraction of an incident beam of thermal neutrons is absorbed by a Cd sheet of 0.1 mm thick?

(b) What thickness of Cd is needed to absorb 99% of an incident beam to thermal neutrons?

Solution: As Cd^{113} contians 12% of natural Cd, so the number of Cd^{113} atoms/m^3 is

$$N = \frac{0.12 \times 8.64 \times 10^3}{112 \times 1.67 \times 10^{-27}} = 5.58 \times 10^{27} \text{ atoms/m}^3$$

Also capture cross-section $\sigma = 2 \times 10^4\ b = 2 \times 10^4 \times 10^{-28}\ \text{m}^2$

$$= 2 \times 10^{-24} \text{ m}^2$$

$$\therefore \quad N\sigma = 5.58 \times 10^{27} \text{ atoms/m}^3 \times 2 \times 10^{-24} \text{ m}^2$$

$$= 1.12 \times 10^4 \text{ m}^{-1}$$

So the fraction of incident neutrons that is absorbed will be $\dfrac{n_0 - n}{n_0}$

$$= \frac{n_0 - n_0 e^{-N\sigma x}}{n_0} = (1 - e^{-N\sigma x})$$

Now $x = 0.1 \text{ mm} = 10^4 \text{ m}$

$$\therefore \quad \frac{n_0 - n}{n_0} = 1 - e^{(-1.12 \times 10^4 \text{ m}^{-1} \times 10^{-4} \text{ m})}$$

$$= \mathbf{0.67 = 2/3}$$

which means that 67% (or 2/3) of the incident neutrons are absorbed.

(*b*) In this case 1% of incident beam pass through

$$\therefore \quad \frac{n}{n_0} = \frac{1}{100}$$

But $$\frac{n}{n_0} = e^{-N\sigma x}$$

$$\therefore \quad \frac{1}{100} = e^{-N\sigma x}$$

or
$$x = \frac{\log_e 0.01}{1.12 \times 10^4} = 4.1 \times 10^{-14} \text{m} = \mathbf{0.41\ mm}$$

Example 6. The $_5B^{10}$ (α, P) $_6C^{13}$ reaction shows among others a resonance for all excitation energy of the compound nucleus of 13.23 MeV. The width of this level as found experimentally is 130 keV. Calculate the mean life of the nucleus for this excitation.

Solution: We have the reaction

$$_5B^{10} + {_2He^4} \rightarrow [_7N^{14}] \rightarrow {_6C^{13}} + {_1H^1}$$

Now width of the level is related to the mean life by

$$Bf = \frac{h}{2\pi\tau}$$

$$\therefore \quad = \frac{h}{2\pi}\frac{1}{Bf} = \frac{6.623 \times 10^{-34}}{2 \times 3.142 \times 130 \times 10^3 \times 1.6 \times 10^{-19}}$$

$$= \mathbf{5 \times 10^{-21}\ sec.}$$

5.10. ARTIFICIAL OR INDUCED RADIOACTIVITY

In 1934, Irene Curie and her husband Juliot found that light elements boron, magnesium and aluminium became radioactive after bombardment by α-particles from polonium. This shows that the product of disintegration is radioactive and this undergoes nuclear transformation in the usual way, although the half-life period of such changes is generally small. This phenomenon is called artificial or induced radioactivity. Subsequent experiment by Curie and Juliot showed that the light-element targets continued to emit positron after the source of α-particles had been removed; the rate of emission of the positron gradually decreased with time, finally approaching zero. When the rate was plotted against the time after the removal of α-particle source, an exponential curve was obtained for each target similar to those obtained in the β-decay of natural radioactivity. The activity could be expressed accurately by the equation

$$N = N_0 e^{-\lambda t}$$

and characteristic half-lives were found, namely 14, 2.5 and 2.25 minutes for boron, magnesium and aluminium targets, respectively. The explanation given by Curie and Juliot was that the product nucleus formed in (α, *n*) reaction in each case was an unstable nuclide, which then disintegrated with the emission of a positron. The nuclear reactions when boron and aluminium were the targets are given below:

(*a*) $$_5B^{10} + {_2He^4} \rightarrow [_7N^{14}] \rightarrow {_7N^{13}} + {_0n^1}$$
$$_7N^{13} \rightarrow {_6C^{13}} + {_1e^0} \qquad [T = 14 \text{ min.}]$$

(*b*) $$_{13}Al^{127} + {_2He^4} \rightarrow [_{15}P^{31}] \rightarrow {_{15}P^{30}} + {_0n^1}$$
$$_{15}P^{30} \rightarrow {_{14}Si^{30}} + {_1e^0} \qquad [T = 3.25 \text{ min.}]$$

The symbol $_1e^0$ or $+_1e^0$ is used to denote the positron, since its charge is the same as that of the proton and its mass number is zero. The half-life of N^{13} was later found to be 10 minutes and that of P^{30} 2.5 minutes, rather than 14 minutes and 3.25 minutes respectively. The nuclides N^{15} and P^{30} are not stable, but the nuclides which remains after the emission of the positron *i.e.,* C^{13} and Si^{30} are stable.

5.11. PRODUCTION OF ARTIFICIAL RADIOACTIVITY

More than 500 artificial radioactive isotopes have been found with all the five commonly used projectiles, α-particles, protons, deuterons, neutrons and γ-rays. A great number of them have been identified as regards their mass and atomic number, which indicate that one or more radioactive isotopes exist for practically all the elements in the periodic table.

Many of the light elements between boron and calcium give rise to artificial radioelements when bombarded with α-particles from natural radioactive elements. Many of these radioelements emit positrons as Be^{10}, N^{14}, Na^{23}, Al^{27}, etc., but some as Si^{29}, Ca^{40}, P^{31}, etc. electrons. Mg yields positron as well as electrons, but this is due to the two isotopes Mg^{24} and Mg^{25}, the former producing radioactive Al^{23} with the emission of a proton according to the equations:

$$_{12}Mg^{24} + {_2}He^4 \rightarrow {_{14}}Si^{27} + {_0}n^1$$

$$_{14}Si^{27} \rightarrow {_{13}}Al^{27} + e^+ \qquad (T = 6.7 \text{ min.})$$

$$_{12}Mg^{25} + {_2}He^4 \rightarrow {_{13}}Al^{28} + {_1}H^1$$

$$_{13}Al^{28} \rightarrow {_{14}}Si^{28} + e^- \qquad (T = 2.3 \text{ min.})$$

With protons as projectiles (p, n) and (p, γ) reactions are known to give rise to artificial radioactivity, emitting positrons or electrons. With deuterons, chiefly of high energy produced by the cyclotron, many elements, even heavy ones as Bi and Pt, acquire radioactivity, emittting electrons or positrons. With heavy elements like Bi, α-radioactivity is also produced. With neutron bombardment induced radioactivity has extended to most elements. For three most important processes which are known to produce induced radioactivity, *viz.*, (n, p) (n, α) and (n, γ) emission of electrons is common, though in a few cases positrons are also observed. In photodisintegration processes, which usually cause the emission of neutrons, the radioelements formed are positron-emitters.

Yield: In the induced radioactivity produced by means of α-particle bombardment the yield is very low *e.g.*, for every 10^7 α-particles of high energy (8.3 MeV) incident upon Al target only six are effective in the production of radio-phosphorus. Further, even this small yield falls rapidly as the energy of the α-particles is decreased. In the radioactivity provoked by neutron bombardment the yield is much greater. High energy deuterons also produced large yields.

5.12. HALF-LIFE PERIOD AND ENERGY RANGES OF RADIOISOTOPES

The half periods of different induced radioelements measured with Geiger counter or ionisation chambers, ranges from 0.02 secs B^{12} to more than 10 years (Be^{10}). The shortest half period of B^{12} was estimated by counting the number of electron tracks in a cloud chamber to be automatically expanded at short known intervals following the cutting of the deuteron beam irradiating a boron target placed in the chamber. The longest period of Be^{10} was found by studying the activity in a beryllium target which had been bombarded by deuterons for a period of 6 months as a neutron source. It may be noted that both the artificial radioelements considered here are produced by (d, p) and not by (d, p) reactions.

$$_5B^{11} + {_1}H^2 \rightarrow {_5}B^{12} + {_1}H^1$$

$$_5B^{12} \rightarrow {_6}C^{12} + e^- \qquad (T = 0.02 \text{ secs})$$

$$_4Be^9 + {_1}H^2 \rightarrow {_4}Be^{10} + {_1}H^1$$

$$_4Be^{10} \rightarrow {_5}B^{10} + e^- \qquad (T = 10 \text{ yrs.})$$

The maximum energies of the electron and positron emissions from artifical radioelements, as measured by adsorption and deflection methods, are found to vary from 0.3 MeV to MeV. In many cases the upper limits are in the whereabout of 2 to 3 MeV. The maximum energies and half-life periods of several artificial radioelements appear to satisfy approximately the segment-relation, as in the case of the β-rays from natural radioactive elements.

5.13. CLASSIFICATION OF ARTIFICIAL RADIOACTIVITY

On the basis of the proton neutron model of the nucleus and the ratio n/p *i.e.*, ratio of the number of neutrons to that of protons in nuclei change towards a stable value, a general classification and interpretation of radioactivity can be made. Radioactive nuclei which have $(n - p)$ positive are heavier than the known stable ones tend to exhibit electron activity.

On the other hand, radioactive nuclei which have $(n - p)$ negative are too light to be stable will emits positrons, giving rise to a product nucleus of the same mass but of lower charge. Emission of an electron increases p at the expense of n, while emission of a positron has the opposite effect. In either case, a nucleus having a more stable ratio of n/p is produced. The following table based on the above stated general law gives the classification of the principal types of reaction.

Activity	e^- (electrons)			e^+ (protons)	
Change in $(n-p)$	2	1	0	-1	-2
Reactions	(n, p)	(a, p) (d, p) (n, α) (n, γ)	(d, α)	(α, n) (d, n) (p, γ) $(n, 2n)$ (γ, n)	(p, n)

The additional important points are given below :

(*i*) Radioactive isotopes ordinarily lie close, on the mass scale, to stable isotopes of the same element, lighter ones tending to be transformed with a decrease of charge (positive activity) and heavier ones with an increase of charge (electron activity).

e.g., Na^{22} (positron active)
Na^{23} (stable)
Na^{24} (electron active)

In many cases the radioactive isotope is found to have a mass number differing by one unit from the stable isotopes of the element: *e.g.*, N^{14} and N^{15} (stable) are flanked by N^{15} (positron active) and N^{14} (electron active). In some cases, the radioactive isotope fills the gap between two known stable ones, *e.g.*, Ag^{108} (electron active) lies between the stable Ag^{107} and Ag^{109}. In a few cases, unstable isotope differs by two or more mass units from the stable ones; *e.g.*, Al^{29} (electron active) while Al^{27} (stable); $Sb^{116,113,129}$ (positron active) $Si^{31,32}$ (stable), $Sb^{122,124,127,129,131}$ (electron active). In some cases, branching processes occur, yielding positron and electron from the same radioactive isotopes.

(*ii*) The same radioactive isotope can be produced in different ways: *e.g.*, ${}_{13}Al^{28}$ which decays to stable ${}_{14}Si^{28}$ with the emission of electron take place in the following five ways:

$${}_{12}Mg^{25} + {}_{2}He^{4} \rightarrow {}_{13}Al^{28} + {}_{1}H^{1}$$
$${}_{14}Al^{27} + {}_{1}H^{2} \rightarrow {}_{13}Al^{28} + {}_{1}H^{1}$$
$${}_{13}Al^{27} + {}_{0}n^{1} \rightarrow {}_{13}Al^{28} + \gamma$$
$${}_{14}Si^{28} + {}_{0}n^{2} \rightarrow {}_{13}Al^{29} + {}_{1}H^{1}$$
$${}_{15}P^{31} + {}_{0}n^{1} \rightarrow {}_{13}Al^{28} + {}_{2}He^{4}$$
$${}_{13}Al^{28} \rightarrow {}_{14}Si^{28} + e^{-} \qquad (T = 2.4 \text{ min.})$$

(*iii*) Radioactive isotopes may be formed of elements hitherto unknown, which means production of new elements. In the periodic table of elements, there are three vacant places for atomic numbers $Z = 43$, 85 and 87.

With the different types of projectiles available and the variety of nuclear reaction, it is possible to produce elements with these atomic numbers *e.g.*, there are at least eight reactions result in the formation of radioactive product with $Z = 43$. The element $Z = 85$ (astatine) has been produced by bombarding bismuth with 32 MeV α-particle. Element with $Z = 87$, called (francium) has been found as one of the products in the natural radioactive disintegration of actinium. It decays emitting electrons with a half period of 12 min.

(*iv*) Radioactive forms of transuratnium elements with $Z = 93$ (Neptunium), $Z = 94$ (Plutonium), $Z = 95$ (Americium), $Z = 69$ (Curium) $Z = 97$ (Berkelium), $Z = 98$ (Californium), $Z = 99$ (Einsteinium) $Z = 100$ (Fermium), $Z = 101$ (Medelevium) and $Z = 192$ (Nobelium) have also been produced recently.

(*v*) Many artificial radioactive elements emit γ-rays along with the β-rays, *e.g.*, when sodium is bombarded with fast moving deuteron, radiosodium and protons are produced. The radiosodium was found to be radioactive (Half-life period = 15 hours) and disintegrated into magnesium and electrons and also emitted γ-rays.

$${}_{11}Na^{23} + {}_{1}H^{2} \longrightarrow {}_{11}Na^{24} + {}_{1}H^{1}$$

$$_{11}Na^{24} \longrightarrow {}_{12}Mg^{24} \;|\; + \;|\; {}_{-1}e^{0} \;|\; + \;|\; \gamma\text{-rays} \;|$$
$$\text{(Radiosodium)} \quad \text{(Magnesium)} \quad \beta\text{-rays} \quad \gamma\text{-rays}$$

Fermi obtained radiosodium from another source. He bombarded aluminium with slowly moving neutrons, which are very effective getting a direct hit at the aluminium nucleus, being an uncharged particle which remains undeflected by the nuclear charge and the extra nuclear electrons

$$_{13}A^{27} + {}_{0}n^{1} \longrightarrow {}_{11}Na^{24} + {}_{2}H^{4}$$
$$|\, {}_{11}Na^{24} \,| \longrightarrow |\, {}_{12}Mg^{24} \;|\; + \;|\; {}_{1}e^{0} \;|\; + \;|\; \gamma\text{-rays} \;|$$
$$\text{(Radiosodium)} \quad \text{(Magnesium)} \quad \beta\text{-rays} \quad \gamma\text{-rays}$$

Now-a-days the artificial radioactive elements are produced by placing the elements in atomic piles. The large number of neutrons available in the atomic piles are used to bombard the desired elements. In most cases, a stable nucleus captures a neutron, emits some energy as γ-rays and becomes unstable, *e.g.*, sodium 24, which is produced from the normal sodium 23, and gold 198, which is produced from gold 197. Similarly radioactive iodine 131 is prepared from tellurium 130, which absorbs a neutron to become tellurium 131, which then decays rapidly and loses an electron to form iodine 131, which is then chemically separated from parent tellurium. Iodine 131, phosphorus 32, sulphur 35, carbon 14, and sodium 24 are the five most important radio-isotopes regularly produced in atomic piles.

5.14. APPLICATIONS OF ARTIFICIAL RADIOACTIVITY

The study of the artificially produced radioactive elements is of great theoretical and practical importance.

(1) It favours the Neutron Proton model of the nucleus.

(2) Radioactive-isotopes are widely used as tracer-element. For example, radiocarbon isotope, which is obtained by slow neutron bombardment of nitrogen according to the reaction

$$_{7}N^{14} + {}_{0}n^{1} \longrightarrow {}_{6}C^{14} + {}_{1}H^{1}$$

The radioactive carbon decays by emitting an electron

$$_{6}C^{14} \longrightarrow {}_{7}N^{14} + {}_{-1}e^{3}$$

It has a period of about 5,700 years. It has many uses as a "tracer" in biological research and for studies of petroleum products and organic chemicals where radiocarbon is substituted for natural carbon.

(3) Artificial isotopes are also used as medicines, *e.g.*, radiophosphorus P^{32} (T = 14.3 days), radiocobalt Co^{60} (T = 5.3 years) and radiogold Au^{199} (T = 3 days) give relief in case of all forms of cancer. In some special form of cancer they have given even complete cure.

(4) Radioactive isotopes used in meteorology and hydrology, *e.g.*, triton ($_{1}H^{3}$) which is produced by the fast neutrons of cosmic rays according to the reaction,

$$_{7}N^{14} + {}_{0}n^{1} \longrightarrow {}_{6}C^{12} + {}_{1}H^{3}$$

It has been used for radioactive dating. Triton ($_{1}H^{3}$) content in rain and snow increased almost proportionally to the distance from an ocean. This is because the air currents carrying moisture to such region must travel farther and at greater heights, so that the total exposure to cosmic rays is correspondingly increased.

(5) **In Metabolism and Biosynthesis.** The study of how a living organism uses its foodstuffs to provide itself with energy and prepare its own peculiar complex compounds. Fatty acids contianing deuterium as food for animals have proved that the concept of chemically static body is wrong by finding that the injected deuterium instead of being excreted quickly in the waste products of fat metabolism remains deposited in the fat throughout the animal's body; hence the living body is a dynamic chemical system in a constant state of flux and interaction with its environment.

(6) **Photosynthesis in Plants.** The chain of chemical reactions in which green plants use the

sun's energy to build up carbohydrates, proteins and fats from inert carbon dioxide and water, releasing oxygen in the process. Exposing plants to C^{14} O_2 (*i.e.*, carbon dioxide containing radio-carbon C^{14}) in the presence of sunlight for different intervals of time then separating the compounds in the plant and checking them for C^{14}, biochemists are able to determine what compound the plant is making from carbon dioxide. After one minute's exposure, it is found that at least 50 compounds are tagged with radiocarbon and within two minutes, the carbon isotope is present in proteins and fat also.

(7) **Fertilisers.** Radio-phosphorus has been used to study how plants assimilate phosphate fertilisers.

(8) **As a Substitute for the Natural radio-elements.** The great merit of artificial radioactive elements lies chiefly in the fact that they form a relatively cheap and easy 'substitute' for the natural radioactive materials in the treatment of disease like Cancer, etc.

(9) **Thickness Gauges.** The absorption of γ-rays from a radio-isotope depend upon the thickness of the material placed in the path. This property is used to check variation in the thickness of paper during manufacture. Packets of cigarettes on a conveyer belt can be tested to check that they are full or if any of the cigarette is missing.

(10) In industry it is used to test the sealing of envelopes for radiovalves. A sample valve is filled with radioactive krypton 85 and by radioactive testing even minutes leakage is found.

5.15. RADIOMETRIC DATING

The phenomenon of radioactivity is quite useful for determining the ages of geological and biological specimens. The procedure is based on fact that the decay of radioactive nuclide proceeds at a constant rate, independent of the environments and the ratio between the amounts of that nuclide and its stable daughter product in the specimen depends on the latter's age, which means that the greater the age of the daughter nuclide, the older the specimen is.

We can see how this principle can be applied with the use of beta active carbon isotope ${}_6C^{14}$.

We know that cosmic rays are high energy atomic nuclei, mainly consisting of protons that circulate the milky way galaxy of which our Solar System is a member. On an average 10^{18} of them reach the earth every second. When they collide with the atoms in their paths, they produce showers of secondary particles.

Among the secondary particles produced are the neutrons which are disrupted from various target nuclei and these neutrons can react with nitrogen nuclei in the atmosphere to produce radio-carbon with the emission of a proton *i.e.*,

$${}_7N^{14} + {}_0n^1 \longrightarrow {}_6C^{14} + {}_1H^1$$

This proton picks up an electron to become a hydrogen atom. Radio carbon ${}_6C^{14}$ has excess number of neutrons for stability and so it decays into ${}_7N^{14}$ with β-decay (half life is 5760 years). Though radio carbon decays steadily, the cosmic ray bombardment produces it constantly. At present a total of about 90 tons of radio carbon is distributed around the world.

Shortly after the formation, radio carbon atoms pick up oxygen molecules to form Carbon dioxide molecules. Green plants, in their process of photosynthesis take up carbon dioxide and water convert into carbohydrates. So that every plant has some content of radio carbon. Animals eat the plants and so they also become radioactive. Since the mixing of radiocarbon is quite efficient, so all living plants and animals have the same ratio of radiocarbon C^{14} to ordinary carbon C^{12}.

When the plants or animals die, they no longer take radiocarbon atoms. However these atoms that they contain go on decaying to N^{14} and after 5760 years, they have 1/2 of the radiocarbon left as compared to the content they had as living matter. After 11,520 years 1/4 matter is life and this process goes on.

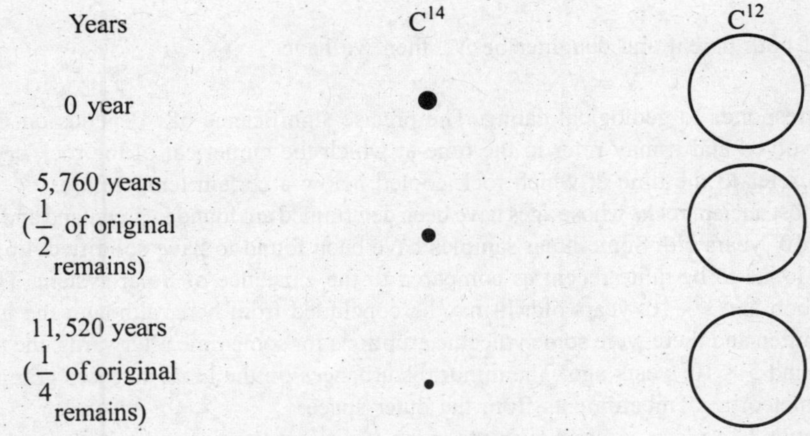

Fig. 5.5

By knowing the proportion of the C^{14} to C^{12}, it is possible to evaluate the age of the ancient objects and the remains of organic origin. This method permits us to know the dating of mummiel, wooden implements, leather, clothes, charcoal etc., from the ancient civilization correctly up to about 50,000 years *i.e.*, about 9 half-lives of C^{14} (Figure 5.5).

5.16. RADIO CARBON AGE

In order to find C^{14} content, the carbon in a sample of an organism that decayed a time t ago is usually converted into a gas like CO_2 which is filled into a beta sensitive detector. Let R_0 be the initial activity of a certain mass of the carbon and R be the activity at the time of dating, then we have

$$R = R_0 e^{-\lambda t} \qquad \text{or} \qquad \frac{R}{R_0} = e^{-\lambda t}$$

$$e^{\lambda t} = \frac{R_0}{R}$$

Taking log on both sides

$$\lambda t = \log_e \frac{R_0}{R} \qquad \text{or} \qquad t = \frac{1}{\lambda}\log_e \frac{R_0}{R}$$

where λ corresponds to the disintegration constant of the sample.

5.17. GEOLOGICAL DATING

With carbon dating method the dating is limited to about 50,000 years while the Earth's history goes back to about 4.5 billion years. So geologists accordingly used radio nuclides like K^{40}, U^{238}, and Rb^{87} which have half-lives very large.

Table 5.1

Method	Parent nuclide	Daughter nuclide	Half-life (years)
Potassium Argon	K^{40}	Ar^{40}	1.3×10^9
Uranium lead	U^{238}	Pb^{206}	4.5×10^9
Rubidium strontium	Rb^{87}	Sr^{87}	4.7×10^{10}

It should be noted that in each case it is assumed that each daughter nuclide found in a particular rock sample originated from the parent nuclide. Now U^{238} decays to Pb^{206} in 14 steps unlike K^{40} and Rb^{87} which decay in single steps, the half-lives of intermediate nuclides in the decay of U^{238} are so short that they can be neglected as compared to 4.5 billion years and so we need consider only U^{238} and Pb^{206} samples.

If we consider the number of parent nuclides in a sample to be N at any time and the no.

of atoms of both parent and daughter be N_0, then we have

$$N = N_0 e^{-\lambda t}$$

where t corresponds to geological dating. The precise significance of t depends on the nature of the rock involved and it may refer to the time at which the numerical of the rock crystallized or it may also refer to the time at which rock cooled below a certain temperature.

The most ancient rocks whose ages have been determined are found in greenland and are believed to be 3.8×10^9 years old. Some lunar samples have been found to have solidified 4.6×10^9 years ago, which looks to be quite recent as compared to the existence of Solar system. The youngest rocks on moon are 3×10^9 years old. It may be concluded from here, although the lunar surface was once molten and there were some volcanic eruptions for some time afterwards, the total activity stopped around 3×10^9 years ago. The minor disturbances on the lunar surfaces going on may be due to the meteorite bombardments from the outer space.

Example 7. *A piece of wood from the ruins of ancient times was found to have C^{14} activity of 13 disintegrations/minute/gm of its carbon content. The C^{14} activity of living wood is 16 disintegrations/min./gm. How long ago did tree die from which the wood sample came?*

Solution: We have disintegration

$$\text{Constant for } C^{14} = \frac{0.693}{T_{1/2}} = \frac{0.693}{5760 \text{ years}}$$

Also $$\frac{R_0}{R} = \frac{16}{13}, \text{ so we have}$$

$$t = \frac{1}{\lambda}\log_e\frac{R_0}{R} = \frac{5760}{0.693}\log_e\frac{16}{13}$$

$$= \mathbf{1726 \text{ years}}$$

QUESTIONS AND PROBLEMS

1. What are α-particles? How will you show experimentally that α-particles are ionised helium atom?

2. Write the names of projectiles which are used for nuclear reaction in artificial radioactivity. Explain the nuclear reaction using α-particle as projectile.

3. Discuss how nuclear reactions by α-ray lead to the discovery of neutron.

4. Discuss the nuclear reactions produced by protons and deuterons. Illustrate your answer by giving three reactions by each of them.

5. Why are neutrons chosen as missiles in nuclear reaction? Illustrate your answer with examples.

6. (*a*) Complete the following reactions:

(*i*) ${}_1H^2 + {}_8N^{16} \longrightarrow {}_7N^{14}$ ...

(*ii*) ${}_2Ae^4 + {}_{13}Al^{27} \longrightarrow {}_{14}Si^{30}$ ...

(*iii*) ${}_1H^2 + {}_5B^{10} \longrightarrow {}_7C^{11} +$ ...

(*iv*) ${}_1H^2 + {}_3Li^6 \longrightarrow {}_2He^4 +$ ...

(*b*) Complete the following reactions:

(*i*) ${}_{11}Na^{23} + {}_2He^4 \longrightarrow {}_{13}Al^{26} +$ ....

(*ii*) ${}_4Be^9 + {}_1H^1 \longrightarrow {}_4Be^8 +$ ...

(*iii*) ${}_3Li^7 + {}_1H^2 \longrightarrow {}_3Li^6 +$ ...

(*iv*) ${}_6Cl^2 + {}_0n^1 \longrightarrow {}_6C^{11} +$ ...

7. What are radioisotopes? Mention at least two of them indicating their use in industry and science.

8. Write short notes on:

(*i*) Transmutation by Photons.

(*ii*) Artificial Radioactivity.

(*iii*) Rutherford's Experiment.

9. What is radiometric dating? How the dating of ancient objects can be measured?

10. What is carbon dating? How is it able to determine the age of old living being?

11. How the age of rocks can be measured?

12. Calculate energy of an α-particle and express the quantity both in MeV and ergs. Given

Mass of helium nuclei = 4.0024 a.m.u.

Mass of proton = 1.00758 a.m.u.

Mass of neutron = 1.00897 a.m.u. **[Ans.** 6.825 MeV; 4.53×10^{-5} ergs]

13. Calculate the energy set free in the mutual annihilation of an electron and a positron. If the whole of the energy is given out as a single quantum of light, calculate its wavelength.,

Given, mass of electron = 0.00055 a.m.u.

mass of proton = 1.00751 a.m.u.

1 a.m.u. = 1.66×10^{24} gm

$h = 6.6 \times 10^{-27}$ ergs/sec. **[Ans.** 1.3×10^{-6} Å]

14. The energy of α-particle from the reaction $_3Li^6$ (p, α) $_2He^3$ is found to be 3.945 MeV. Calculate the isotopic mass of $_2He^3$ from the following data:

Mass of $_3Li^6$ = 6.01513 a.m.u.

Mass of $_1H^1$ = 1.007825 a.m.u.

Mass of $_2He^4$ = 4.0826 a.m.u. **[Ans.** 3.01611 a.m.u.]

CHAPTER

6

INTERACTION OF NUCLEAR RADIATION WITH MATTER

6.1. INTRODUCTION

The following are the various particles which are employed in nuclear reactions:

(*i*) e.m. radiations

(*ii*) charged particles

(*iii*) neutron radiations

We will consider here the **e.m. radiations** and **charged particles**. It may be clearly understood that charged particles differ from e.m. radiations, is the fact that they have a **range.** In passing through an absorber, charged particle beam (monochromatic) loses energy and after passing through some distance, they are stopped by the absorber. **The minimum thickness of the absorber for stopping a charged particle is called range of the particle.**

6.2. ENERGY LOSS OF A CHARGED PARTICLE (BOHR'S FORMULA)

In case of a charged particle like proton, deuteron or α-particle moving through matter, then *excitation* and *ionization* of the atoms with which these charged particles interact or move closely, takes place. Due to coulomb force, a sharp inpulse is given to the electrons when charged particles move pass near them with high speed.

In this process some kinetic energy is transferred to the electrons and to the positive ions. However, the energy transferred to *positive heavy ions* may be neglected and we will consider the transfer of energy from moving charged particles to the *electrons* of the surrounding medium.

Let a particle of mass M and charge Ze be moving with velocity v along the X-axis so that its impact parameter (distance of closest approach) is b. *The velocity v is so high that electron does not change its position during the time the particle moves near it.*

If r be the distance between the heavy charged particle and the electron, then the force exerted by the heavy charged particle on the electron is given by coulomb's law as

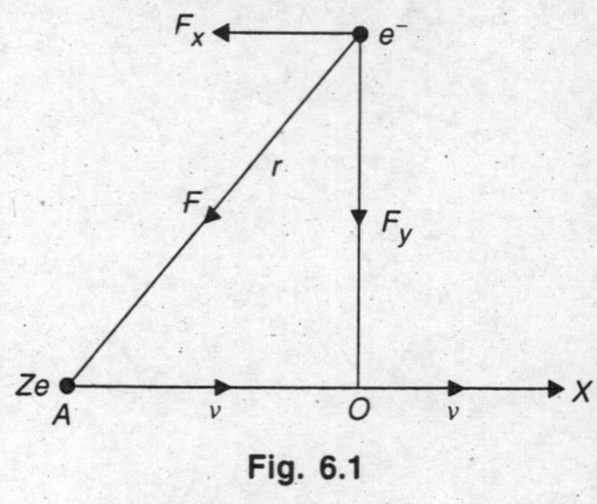

Fig. 6.1

$$\vec{F} = \frac{Ze^2}{4\pi\varepsilon_0 r^2} \quad ...(1)$$

Force $\vec{F}$ can be resolved into two components F_x along the direction of motion and F_y along the direction $\perp$ to the direction of motion.

It may be noted that F_x *will change sign* as the particle after approaching 0 *i.e.,* at the point of closest approach will recede away from it.

We can take particle velocity as constant during collision as change in velocity due to *collision with a single electron* will be negligibly small. So F_x does not contribute towards the momentum imparted to the electron.

Now $$F_y = F \sin\theta = \frac{Ze^2}{4\pi\varepsilon_0 r^2}\sin\theta$$

So momentum imparted to the electron will be given as

$$P = \int_{-\infty}^{+\infty} \sin\theta \, d\theta$$

$$= \int_{-\infty}^{+\infty} \frac{Ze^2}{4\pi\,\varepsilon_0\, r^2}\sin\theta\, d\theta \qquad ...(2)$$

Let the distance between the particle at A and the point O (origin) be x, *i.e.*, the distance where the particle has closest approach, then

$$-x = vt = -b\cot\theta \qquad ...(3)$$

where we have taken t as the time to cover distance $x = AO$.

From Equation (3), we get

$$t = -\frac{b}{v}\cot\theta \qquad ...(4)$$

differentiating Equation 4, we get

$$dt = \frac{b}{v}\operatorname{cosec}^2\theta\, d\theta \qquad ...(5)$$

Also from Fig. 6.1

$$r = \frac{b}{\operatorname{Sin}\theta} \qquad ...(6)$$

Putting the value of dt from Equation (5) and r from Equation (6) in Equation (2) we get

$$P = \frac{Ze^2}{4\pi\,\varepsilon_0\, bv}\int_0^{\pi}\sin\theta\, d\theta = \frac{Ze^2}{4\pi\,\varepsilon_0\, bv}[-\cos\theta]_{\theta}^{\pi}$$

$$= \frac{Ze^2}{4\pi\,\varepsilon_0\, bv}(1+1)$$

$$= \frac{Ze^2}{2\pi\,\varepsilon_0\, bv}$$

Now the K.E. transferred to the electron of mass m_e is given by the relation

$$T = \frac{p^2}{2m_e} = \frac{Z^2 e^4}{8\pi\,\varepsilon_0^2\, m_e\, b^2\, v^2} \qquad ...(8)$$

In order to find the energy lost by the charged particle in moving a unit distance in the medium, let us first calculate the energy transferred by it in travelling a small distance dx of its path to the electrons contained in a cylinderical shell lying between the distance b and $b + db$ from the path of the particle.

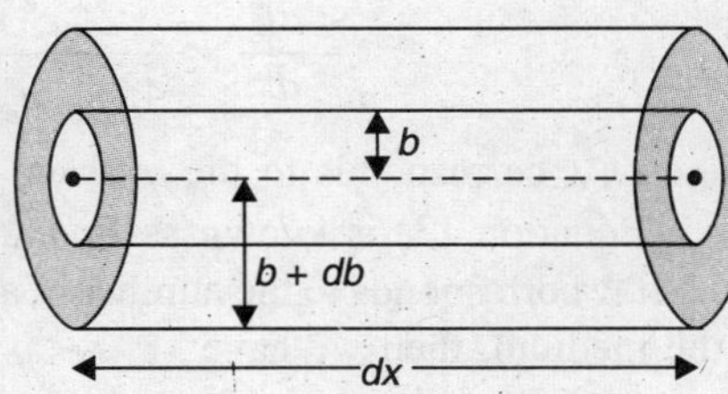

Fig. 6.2

From Fig. 6.2, the volume of the cylinderical shell $= 2\pi b\, db . dx$

Let the number of electrons per unit volume be n

∴ No. of electrons in the volume $2\pi b\, db\, dx = 2\pi\, n\, b\, db\, dx$

So the K.E. transferred to all the electrons in cylinderical shell

$$= \frac{Z^2 e^4}{8\pi^2\,\varepsilon_0^2\, m_e\, b^2\, v^2} \times 2\pi b\, db\, dx$$

$$= \frac{Z^2 e^4}{4\pi \varepsilon_0^2 m_e v^2} \cdot \frac{db}{b} \cdot dx \qquad \text{...(9)}$$

We can now get the total energy *dE transferred to the medium* when the particle moves through a distance *dx*. This will also correspond to the *K.E. lost by the particle* in a distance *dx* of its path. We can get this value by integrating (Equation 9) with the limits of minimum and maximum values (b_{min} and b_{max}) of the impact parameter *b*.

So the energy lost by the particle in moving through a unit distance in the medium

$$-\frac{dE}{dt} = \frac{Z^2 e^4 n}{4\pi \varepsilon_0^2 m_e v^2} \int_{b_{min}}^{b_{max}} \frac{db}{b}$$

$$\therefore \quad -\frac{dE}{dx} = \frac{Z^2 e^4 n}{4\pi \varepsilon_0^2 m_e v^2} \log_e \frac{d_{max}}{b_{min}} \qquad \text{...(10)}$$

It may appear natural to put $b_{min} = 0$ and $b_{max} = \infty$, but as is evident from (Equation 10), we get absurd results.

Determination of b_{max}. As we know that the electrons in the stopping material are not free but revolve around the nucleus in stationary orbits with frequency ν. So the force on the electron is an impulse of short duration = b/v. If we consider it to be the period of revolution of the electron, then

$$T = \frac{1}{\nu} = \frac{b}{v} \qquad \text{or} \qquad b = \frac{v}{\nu}$$

For values of b $b > \frac{v}{\nu}, \frac{b}{\nu}$ will be greater than $\frac{1}{\nu}$ *i.e.*, the period of revolution *T* of electron in the atom.

In that case, no energy is absorbed by the electron and so the moving particle will lose no energy *i.e.*, maximum value of *b* will be equal to $\frac{v}{\nu}$ *i.e.*, $b_{max} = \frac{v}{\nu}$

Determination of b_{min}. We can find the lower limit of *b* using *uncertainty principle.*

In *head on collision* (assuming heavy particle at rest and electron colliding with it), we have the maximum change of velocity of the electron = $2v$ and consequently the maximum change in its momentum = $2\, m_e v$.

From **Heisenberg's uncertainty Principle**

$$b_{min} = \frac{h}{2m_e v}$$

which will correspond to the approximate size of the electron as seen by moving particle

$$\therefore \quad -\frac{dE}{dx} = \frac{Z^2 e^4 n}{4\pi \varepsilon_0^2 m_e v^2} \log \frac{2m_e v^2}{h\nu} \qquad \text{...(12)}$$

where, $h\nu$ corresponds to the *average binding energy of the atomic electron* of the medium.

(Equation 13) is known as *Bohr's Formula.*

If *N* corresponds to the number of atoms/unit volume of the medium and *Z*, the atomic number of the medium, then we have

number of electrons/unit volume $n = NZ$

So Equation (12) may be written in the form

$$-\frac{dE}{dx} = \frac{Z^2 e^4 NZ}{4\pi \varepsilon_0^2 m_e v^2} \log \frac{2m_e v^2}{h\nu} \qquad \text{...(13)}$$

6.3. BLOCH-BETHE RELATION

In case, the incoming charged particle moves with a very high velocity *i.e.*, velocity comparable to the velocity of light, then relativistic correction to b_{max} and b_{min} values has to be applied.

Determination of b_{max}. Due to the relativistic effect, electric field of the incident particle is *contracted* in the direction of motion and *increased* in direction ⊥ to the motion by a factor $\dfrac{1}{(1-v^2/c^2)^{1/2}}$.

So due to *relativistic consideration*, the impulse given to the electron gets sharpen and the duration will be given by

$$T = \frac{1}{\nu} = \frac{b_{max}(1-v^2/c^2)^{1/2}}{\nu} \qquad ...(14)$$

Determination of b_{min}. As stated earlier, we consider the *heavy particle at rest* and *electron colliding* with it. Relativistically, in a co-ordinate system in which the incident particle is at rest and *electron has velocity v.* the maximum change in momentum of the electron is given by

$$\frac{2m_e v}{(1-v^2/c^2)^{1/2}}$$

$$b_{min} = = \frac{h}{2m_e v}\cdot(1-v^2/c^2)^{1/2} \qquad ...(15)$$

It may be mentioned that in this case it is assumed that *orbital velocity* of the electron is very small as compared to *v*.

From relations (14) and (15), we have for relativistic case

$$\frac{b_{max}}{b_{min}} = \frac{v}{\nu\left(1-\dfrac{v^2}{c^2}\right)^{1/2}} \times \frac{2m_e v}{h(1-v^2/c^2)^{1/2}}$$

$$= \frac{2m_e v^2}{h\nu(1-v^2/c^2)}$$

So with relativistic correction, we have energy lost/unit distance as

$$-\frac{dE}{dx} = \frac{Z^2 e^4 NZ}{4\pi\varepsilon_0^2 m_e v^2}\left[\log_e \frac{2m_e v^2}{h\nu^2(1-v^2/c^2)}\right]$$

$$= \frac{Z^2 e^4 NZ}{4\pi\varepsilon_0^2 m_e v^2}\left[\log_e \frac{2m_e v^2}{h\nu} - \log(1-v^2/c^2)\right]$$

Bethe developed the above quantum mechanical relationship. However, the exact formula was derived by **F. Bloch** in 1933 and **Bloch Formula** is given by

$$-\frac{dE}{dx} = \frac{Z^2 e^2 NZ}{4\pi\varepsilon_0^2 m_e v^2}\left[\log_e \frac{2m_e v^2}{h\nu} - \log_e\left(1-\frac{v^2}{c^2}\right) - \frac{v^2}{c^2}\right] \qquad ...(16)$$

6.4. STOPPING POWER

$-\dfrac{dE}{dx}$ is called the **stopping power** of the medium and as is evident from Equation (16), it is independent of the mass of incident particle and is directly proportional to Z^2 and inversely proportional to v^2 which means that stopping power of a material for α-particle ($Z = 2$) is 4 times that for protons ($Z = 1$) for the same *value of velocity.*

A curve showing variation of stopping power with energy is given in Fig. 6.3.

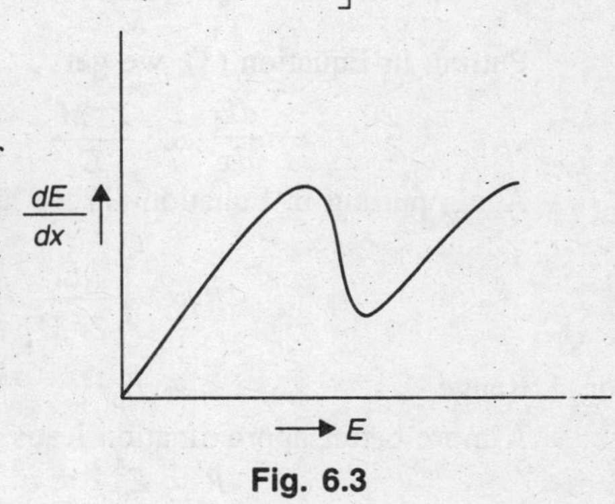

Fig. 6.3

6.5. RANGE OF A CHARGED PARTICLE

When a heavy charged particle moves through a medium it *excites and ionize* the atoms it comes into contact and consequently loses some of its kinetic energy which is transferred to the orbital electrons of the medium. So the particle goes on losing speed.

The distance travelled by a charged particle when it is moving through a medium before it comes to rest is called its range. Range, therefore, corresponds to the *average distance* travelled by a charged particle before it loses its all kinetic energy.

Let E_0 be the initial K.E. of the particle and let it loses dE energy while moving through a distance dx, then it will lose its total energy in a distance $R = \int_{E_0}^{0} dx$

where $R \rightarrow$ range of the particle.

If $-\frac{dE}{dx}$ corresponds to the rate of loss of energy, then

$$R = \int_{E_0}^{0} \left[-\frac{dE}{dx}\right]^{-1} (-dE) = \int_{0}^{E_0} \left[-\frac{dE}{dx}\right]^{-1} dE \qquad ...(1)$$

Also we have

$$-\frac{dE}{dx} = \frac{Z^2 e^4 NZ}{4\pi \varepsilon_0^2 m_e v^2} \left[\log_e \frac{2m_e v^2}{h\nu} - \log_e \left(1 - \frac{v^2}{c^2}\right) - \frac{v^2}{c^2}\right] \qquad ...(2)$$

$\frac{v^2}{c^2}$ in this equation cancels out for *not very high speed* of the moving particle as can be seen from the series expansion of $\log_e \left[1 - \frac{v^2}{c^2}\right]$.

So, for moderate speeds, the main variation of loss of energy will be due to $\frac{1}{v^2}$ outside the bracket. This explains why *we get more ionization of atom towards the end of the range of charged particle.*

$$\therefore \qquad -\frac{dE}{dx} = \frac{Z^2 e^4 NZ}{4\pi \varepsilon_0^2 m_e v^2} \log_e \frac{2m_e v^2}{h\nu}$$

or

$$-\frac{dE}{dx} \propto \frac{Z^2}{v^2} \qquad [\because \text{other terms are constants of the given medium}] \qquad ...(3)$$

Also K.E. of the charged particle

$$E = \frac{1}{2} M v^2$$

$$\therefore \qquad v^2 = \frac{2E}{M}$$

Putting in Equation (3), we get

$$-\frac{dE}{dx} \propto \frac{Z^2 M}{E}$$

Again putting in Equation (1) we have

$$R \propto \int_0^{E_0} \frac{E}{Z^2 M} dE \propto \frac{E_0^2}{Z^2 M}$$

or Range $\qquad R \propto E_0^2$

A more better approximation leads to

$$R \propto E_0^{3/2} \propto v^3$$

Some empirical *range-energy* relations are given as under:
For electron in energy range 0.15 MeV to 0.8 MeV, we have

$$R = 0.00407\ E_0^{\ 1.38}\ \text{m}$$

For proton we have the relation

$$R = \left[\frac{E_0}{9.3}\right]^{1.8}$$

For α-particle in air at 0°C

$$R = b\,E_0^{3/2} = 0.00318\,E_0^{1.38} m$$

Explanation of Bethe-Bloch relation not valid for electrons
Beth-Bloch relation states

$$-\frac{dE}{dx} = \frac{Z^2 e^4 NZ}{4\pi\varepsilon_0^2 m_e v^2}\cdot\left[\log_e \frac{2m_e v^2}{h\nu} - \log_e\left[1-\frac{v^2}{c^2}\right] - \frac{v^2}{c^2}\right]$$

The above relation is not applicable to electrons because of the following facts:

(*i*) Heavy charged particles are *mono-energetic* while in case of electrons there is continuous distribution of initial energy.

(*ii*) Electrons with energies ~ 1 MeV radiate considerable fraction of energy in the form of X-rays due to inelastic collision with *coulomb field* of atomic nuclei and also with electrons.

(*iii*) Collision between *identical particles* involve energy exchange phenomenon which must be taken into account while deriving the *energy loss* relation.

(*iv*) Electrons with energies less than 1 MeV lose energy mostly by excitation or ionization at atom through which they pass. The charge on electrons does not change and electrons lose energy by ionization even at *low energies* unlike positively charged particles which lose their charge by **recombination** even at lower energies.

(*v*) As the above relation has been derived on the assumption that the incident *heavy charged* particle does not get deflected due to its passage through matter, so in case of incident electrons this assumption is not valid as the path of the electron will be *irregular* and also transverse velocity corresponding to momentum will not be negligible as has been assumed in the case of a *heavy charged particle.*

6.6. BREMSSTRAHLUNG

Bremsstrahlung is a German word meaning 'breaking radiation'. Electrons with energies << 1 MeV lose energy mostly by excitation or ionization of the atoms through which they pass. However, electrons with energies >> 1 MeV will be radiating a considerable fraction of energy in the form of X-rays during inelastic collisions with atomic nuclei.

Bremsstrahlung, therefore, correspond to energy radiated per unit time in the form of *e.m.* radiation when a charged particle is moving with a very high speed (~ 3×10^8 ms^{-1}) which may **approach the velocity of light.** In the case of electrons, when they are moving near the nucleus, their path will be slightly bent under the nuclear field. Such a bend is due to the acceleration of electron under the influence of nuclear field and gives rise to emission radiation which, according to *e.m.* theory, is proportional to the square of the acceleration.

6.7. INTERACTION OF γ-RAYS WITH MATTER

γ-rays interactions with matter are different from those of charged particles because

(*i*) γ-rays have got greater penetrating power.

(*ii*) being uncharged, the absorption of γ-rays is exponential in the absorber, it is because when a beam of high energy protons (γ-rays) is incident on matter, each proton that is absorbed is removed in a single event.

We can get a relation for absorption coefficient by considering I and I_0 as the intensity of γ-rays before and after it passes through an absorber of thickness x.

Let dI be the further decrease in intensity when γ-rays travel through an additional thickness dx of the absorber, then we have $-\frac{dI}{dx}$ as the decrease of intensity with increase in distance dx of the absorber and this will be proportional to the original intensity of X-rays *i.e.*,

$$-\frac{dI}{dx} \propto I \quad \text{or} \quad \frac{dI}{dx} = -\mu I \qquad ...(1)$$

where μ is the constant of proportionality and is called linear absorption coefficient of the absorber of linear attenuation coefficient of the absorber. Sometimes it is also called macroscopic absorption coefficient.

We have from Equation (1)

$$\frac{dI}{I} = -\mu I \qquad ...(1)$$

Integrating both sides, we have

$$\int_{I_0}^{I} \frac{dI}{I} = -\mu \int_0^x dx$$

$$\therefore \quad \log_e \frac{I}{I_0} = -\mu x$$

or

$$I = I_0 e^{-\mu} x \qquad ...(2)$$

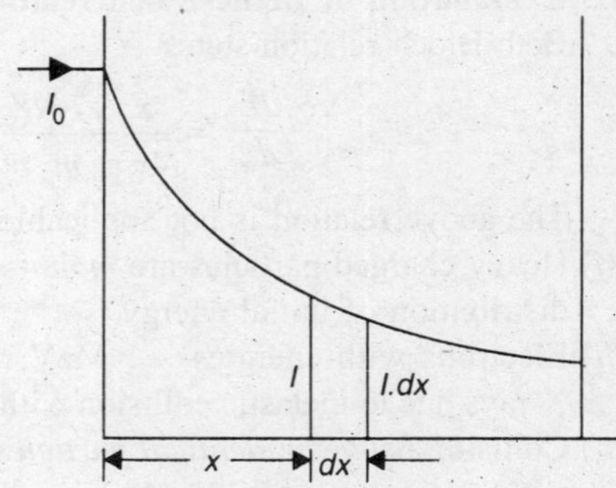

Fig. 6.4

which shows that the intensity of γ-rays decreases exponentially with the thickness of the material.

We can define the absorption coefficient from Equation (2) as

$$\mu = -\frac{\frac{dI}{I}}{dx}$$

i.e., it is equal to the fractional decreases in the intensity of γ- rays per unit thickness of the absorber. It is found that μ depends on the wavelength of γ-rays nearly by the relation $\mu \propto \lambda^3$.

The value of μ can be found by finding the thickness of the absorber sheet which reduces the intensity of γ-rays beam to 1/2 of its initial values, such thickness is called **half value layer** *(hvl) and is denoted by T.*

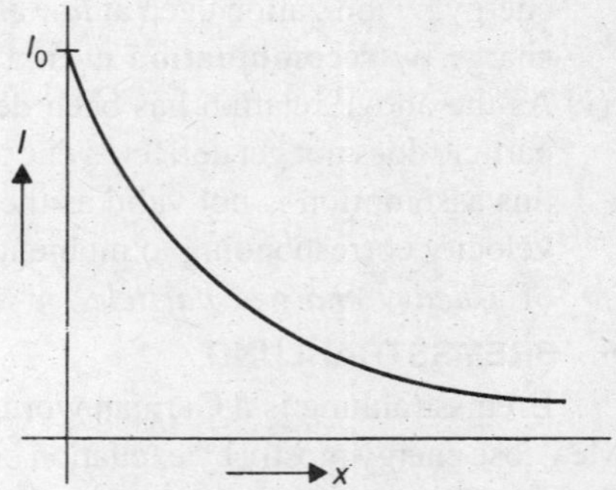

Fig. 6.5

If we put $\frac{I}{I_0} = \frac{1}{2}$ and $x = T$. In the Equation (2) we get

$$e^{-\mu T} = \frac{1}{2}$$

$$\therefore \quad e^{\mu T} = 2$$

$$\mu T = \log_e 2 = 0.693$$

$$\therefore \quad \mu = \frac{0.693}{T}$$

which gives another definition of absorption coefficient in terms of $h\nu l$.

In practice, however, it is $\frac{\mu}{\sigma}$ called mass absorption coefficient μ_m which is more important than μ and may be defined *as equal to the fractional decrease in intensity of a beam of unit cross-section in passing through a unit mass of the absorbing material.*

μ_m is found to depend on the wavelength λ of X-rays and also on atomic number Z of absorber. A relation of it is given by

$$\mu_m = D\lambda^3 Z^3 \qquad ...(3)$$

6.8. EXPERIMENTAL DETERMINATION OF ABSORPTION (ATTENUATION) COEFFICIENT

The apparatus used for finding attenuation coefficient is shown in Fig. 6.6. γ-rays source is Zn^{65} which gives γ-rays of energy 1.14 MeV. The source is enclosed in a thick lead block and small opening is left for making a collimated beam of γ-rays. A thin sheet of absorber is taken and the detector is also surrounded by thick lead blocks, so that stray electrons and γ-rays do not interfere with the experimental results. Incoming photons may be removed by the absorber either by scattering or by absorption and the decrease in intensity of incoming γ-rays beam is measured, while it passes through a thin absorber. The decrease in intensity is due to both absorption and scattering of γ-rays by the absorber.

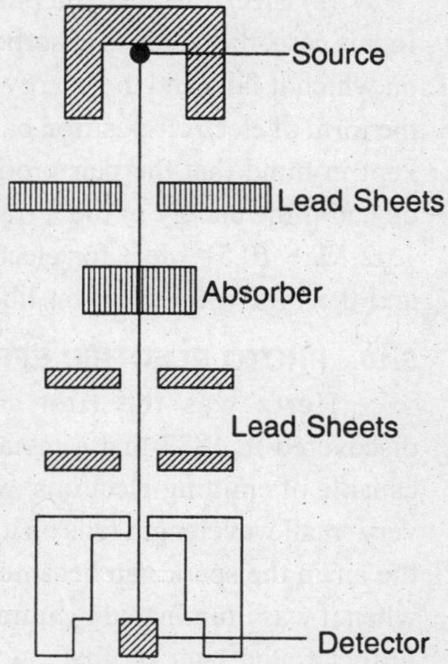

Fig. 6.6

The results may be represented by the curve shown in Fig. 6.7, where the fraction transmission is plotted (in log scale) vs absorber thickness (linear scale).

The slope of the curve gives $\mu\,(cm^{-1}) = 0.7068 \pm 0.0051\ cm^{-1}$ for the case of *Pb absorber.*

6.9. γ-RAYS ABSORPTION BY MATTER

γ-rays may be absorbed by an atom in one of the following ways:

(*a*) **Photo-electric absorption.** In this case, γ-rays striking an atom may be absorbed and the energy imparted to the atom may be sufficient to liberate an electron from an orbit surrounding the nucleus of the atom. In this case, the γ-rays energy is converted to electronic energy (*i.e.*, used for taking out an electron from the atom). So it is called photo-electric absorption of the γ-rays.

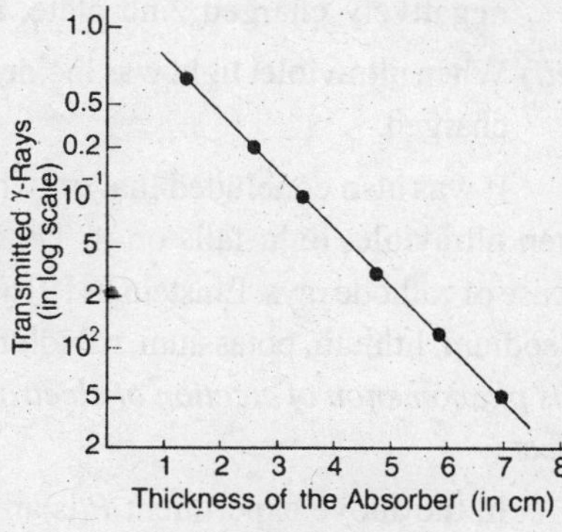

Fig. 6.7

(*b*) **Compton effect.** In this case the energy given by the γ-rays is more than the energy required to liberate an electron from the atom. So in addition to the electron we get photon (or γ-rays) with different energy along with the original-ray.

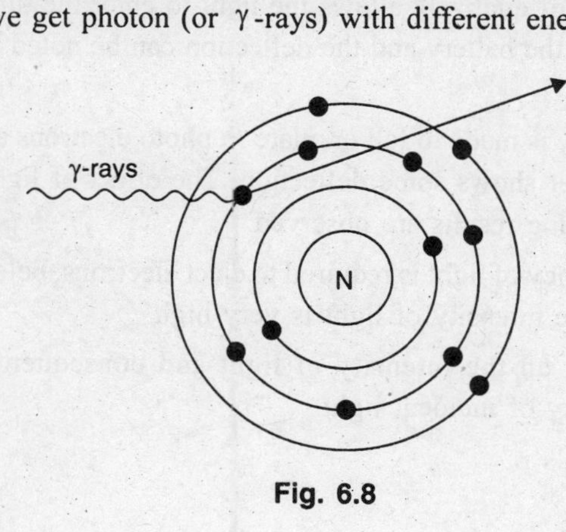

Fig. 6.8

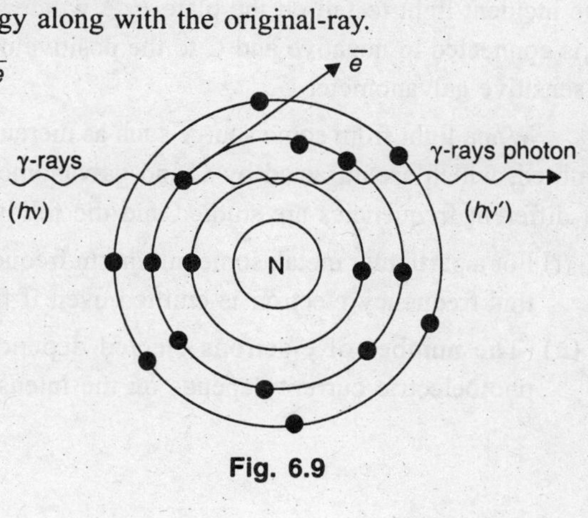

Fig. 6.9

(*c*) **Electron-positron pair production.** In this case, the γ-ray is absorbed by the atom on which it falls and the energy is liberated in the form of electron-positron pair. It should be kept in mind that the pair production cannot be had if the energy of the γ-rays is less than 1.02 MeV (0.51 MeV for electron liberation and 0.51 MeV for positron liberation).

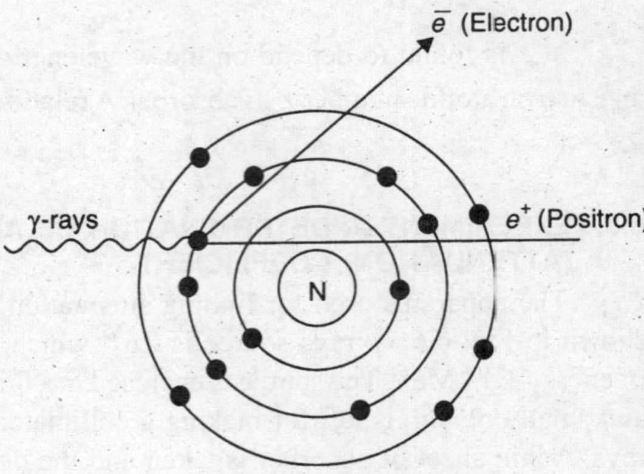

Fig. 6.10

6.10. PHOTO ELECTRIC EFFECT

Hertz was the first scientist who discovered in 1887 that a metallic surface is capable of emitting electrons, when a light of very small wavelength falls on it. He also noted that the air in the spark gap became a good conductor when it was illuminated by ultraviolet light. It was further found later on that

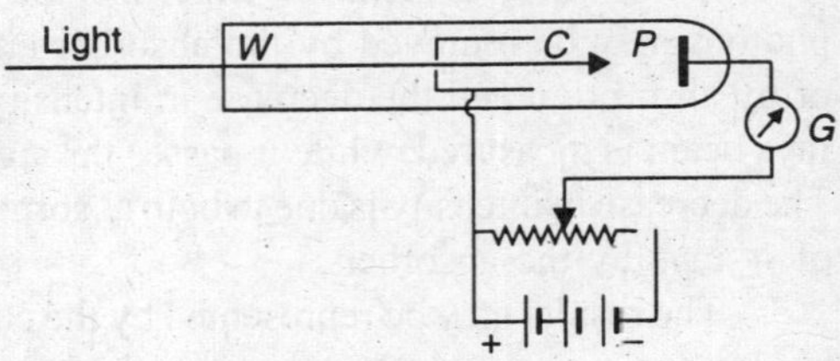

Fig. 6.11

(*i*) When ultraviolet light was incident on a neutral zinc plate, the plate became positively charged.

(*ii*) When ultraviolet light was incident on a negatively charged zinc plate, it rapidly lost its charge.

(*iii*) When ultraviolet light was incident on a positively charged zinc plate, it became more positively charged.

It was also concluded that only the negatively charged particles can be emitted by the surface when ultraviolet light falls on it. Thomson enunciated that *e/m* of these particles was the same as in case of cathode rays. Einstein in 1916 studied the above effect for visible light of different frequencies on sodium, lithium, potassium, rubidium etc. and found that *electrons were ejected out of these metals. This phenomenon of ejection of electrons when light falls on a metal surface is called Photo-Electric Effect.*

In the above experiment *P* is an electrically sensitive plate, fitted in an evacuated glass tube. Normally, the plate is highly polished zinc plate, a hollow cylinder *C* with a small hole permits the incident light to fall on the plate *P*. A window of quartz *W* allows the light to enter the tube. P is connected to negative and *C* to the positive of the battery and the deflection can be noted by a sensitive galvanometer *G*.

When light from some source such as mercury is made to fall on plate *P*, photo electrons are emitted, which are attracted by *C* and galvanometer shows some deflection. The effect of lights of different frequencies are studied and the following results are observed.

(*i*) For a particular metal, some minimum frequency of light is required to eject electrons, below this frequency electron is emitted even if the intensity of light is very high.

(*ii*) The number of electrons ejected depends on the intensity of light and consequently photoelectric current depends on the intensity of incident light.

(*iii*) Light of frequencies greater than the critical frequency or threshold frequency emit electrons of different velocities. *This threshold frequency is the minimum frequency that eject electrons from a particular metal.*

(*iv*) The maximum K.E. of the ejected electrons increases linearly with frequency of incident light and it is independent of the intensity of light.

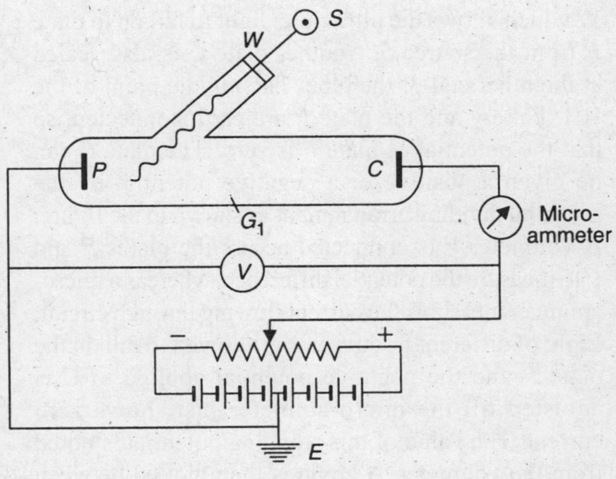

Fig. 6.12

The above results could not be explained from the wave theory of light and so Einstein made use of quantum theory for the explanation. According to this theory, light consists of quantas or bundles of energy which move through space with velocity of light. These particles, called *photons,* have each quanta energy $= h\nu$ where h is Planck's constant and ν is the frequency of light. When a photon strikes the metal surface, there is transfer of energy and this transfer is only in quantas *i.e.*, either a photon gives total energy to the electron or there is no transfer of energy.

6.11. EINSTEIN'S PHOTOELECTRIC EQUATION

According to Einstein, when light frequency ν falls on a metallic surface, the photons of energy $h\nu$ strike the surface. A photon gives energy to the electron which is ejected out of the metal. However, some minimum frequency called *threshold frequency* is required for ejection of an electron. If a photon carries more energy than the threshold energy (or threshold frequency), then a part of the energy is used for ejection of electron and rest is transferred as K.E. to the electron. Let the mass of the electron be m and velocity is v. If ν is the frequency of incident light and ν_0 the threshold frequency, then

$$h\nu - h\nu_0 = \frac{1}{2}mv^2 \qquad ...(1)$$

In Equation (1) $w_0 = h\nu_0$ is called photoelectric work function *i.e.*, the energy required to liberate an electron from metal surface. So Equation (1) may be written as

$$h\nu - w_0 = \frac{1}{2}mv^2$$

or

$$h\nu = w_0 + \frac{1}{2}mv^2 \qquad ...(2)$$

This is called Einstein's Photoelectric equation.

From (1) and (2), we get

$$h\nu - h\nu_0 = eV_0 \qquad \left[\because eV_0 = \frac{1}{2}mv^2\right] \qquad ...(3)$$

The value of the stopping potential V_0 depends on the frequency of incident light and on threshold frequency for that particular metal.

6.12. EXPERIMENTAL VERIFICATION

Einstein's photoelectric equation can be verified with the help of the millitean's apparatus as shown in Fig. 6.12.

It consists of an evacuated tube G_1 fitted with a photo-sensitive plate P and a quartz window

W which allows the ultraviolet light to fall on to plate P from the source S. Another plate C is also sealed at the other end of the tube. The middle point of the H.T. battery and the plate P are earth connected, so that the potential of plate P is zero. The plate C can be given a positive or a negative potential by the potential dividing arrangement as shown in the figure. A voltmeter V is connected across the plates P and C to measure the potential difference, whereas a micro-ammeter measures the current flowing through circuit. Light of different frequencies is allowed to fall on the plate P and the negative potential applied to C is adjusted till the micro-ammeter just shows zero current. The value of this stopping potential is noted from the voltmeter. A graph is then plotted between the stopping potential and the frequency of light. The graph is a straight line which verifies the equation, because

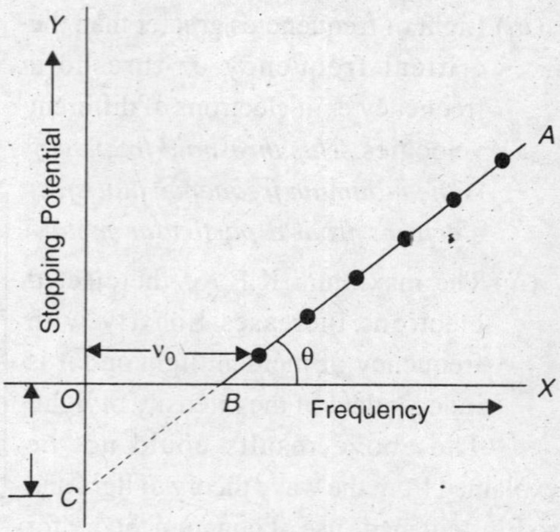

Fig. 6.13

$$V = \frac{h}{e}\nu - \frac{h}{e}\nu_0$$

has the same form as the equation of a straight line.

$$y = mx + c$$

(*i*) The slope of the straight line is given by

$$\tan\theta = \frac{h}{e}$$

$$\therefore \quad h = e\tan\theta$$

The vlaue of Planck' constant derived from this relation agrees with the value obtained from measurements of black body radiations.

(*ii*) For $V = 0$, $\nu = \nu_0$ which is represented by the intercept OB on the X-axis.

In this case the threshold frequency $= \nu_0 = OB$ on the X-axis.

(*iii*) For $\nu = 0$, we have

$$V = -\frac{h\nu_0}{e} = -\frac{w_0}{e}$$

This is represented by the intercept OC on the Y-axis. So the photoelectric work function may be given as

$$w_0 = eV = e \times OC$$

6.13. COMPTON EFFECT

Prof. Compton, in 1921, discovered that when a beam of monochromatic radiation of high frequency is allowed to fall on a substance capable of scattering the beam, then scattered radiation contains two components.

(*i*) One component having lower frequency and larger wavelength

(*ii*) The other component having the same frequency or wavelength.

The high frequency radiations used were radiations like X-rays, γ-rays etc. The radiations having the lower frequencies are called modified radiations while those having the same frequencies are called unmodified radiations.

This phenomenon of scattering of incident radiations into scattered radiations and some recoil electrons is called Compton Effect and we have the following results observed:

(*i*) Modified frequencies

(*ii*) Unmodified frequency

(*iii*) Recoil electron.

Theory of Compton effect: let us consider the phenomenon of scattering as a collision between the photon and electron and then applying the laws of conservation of energy and momentum, we can get the expression for the change in wavelength.

Compton assumed the electron to be free and at rest before collision with the incident high energy photon. After the collision has taken place, the relativistic mass of the electron is taken. This is a justified assumption because only a small part of the energy of the incident radiations (or high energy photons) gives sufficient energy to the recoil electron.

Let us consider a photon of energy $h\nu$ colliding with an electron at rest. We assume that even when the electron is bound to the nucleus around which it is revolving, it receives sufficient energy from the incident photon and so it can be treated as practically free. During the collision process, some part of the energy is imparted to the electron which in turn gains kinetic energy and so recoils. In Fig. 6.14, the process of scattering and recoil electron is shown where angle of recoil is given as ϕ and angle of scattering is given by θ.

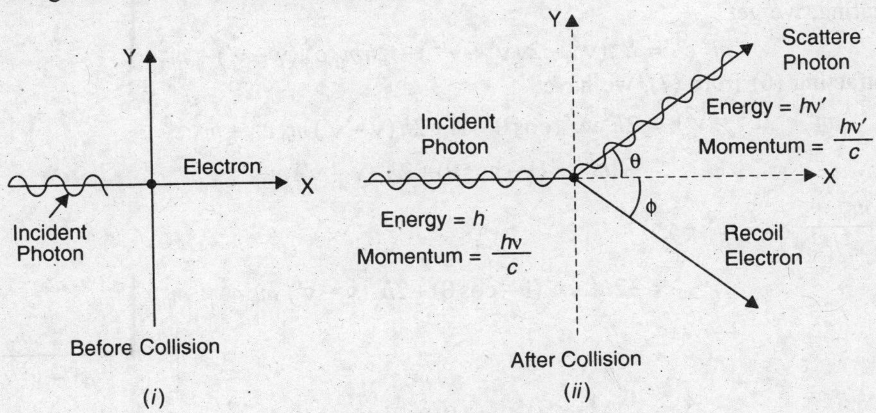

Fig. 6.14

Let us consider that before the collision

(*i*) Energy of incident photon $= h\nu$

(*ii*) Momentum of incident photon $= h\nu/c$

(*iii*) Rest mass of electron $= m_0$, so that rest mass energy $= m_0\, c^2$.

(*iv*) Rest mass momentum of electron = 0

And after the collision

(*i*) Energy of scattered photon $= h\nu'$

(*ii*) Momentum of scattered photon $= \dfrac{h\nu'}{c}$

(*iii*) Mass of the electron moving with velocity $v = m$, so the relativistic energy of electron $= mc^2$

(*iv*) Momentum of the recoil electron $= mv$.

Now applying the principle of conservation of energy, before and after collision, we have

$$h\nu + m_0c^2 = h\nu' + mc^2 \qquad ...(1)$$

Again applying the principle of conservation of momentum along the direction of incident photon, before and after the collision, we have

$$\frac{h\nu}{c} + 0 = \frac{h\nu'}{c}\cos\theta + mv\cos\theta \qquad ...(2)$$

and conservation of momentum in a direction $\perp$ to the direction of incident photon, we have

$$0 + = \frac{h\nu'}{c}\sin\theta + mv\sin\phi \qquad ...(3)$$

Now from Equation (2), we have

$$h\nu = h\nu' \cos\phi + mvc\cos\theta$$

or $$mvc\cos\phi = h\nu - h\nu'\cos\theta \quad ...(4)$$

And from Equation (3), we have

$$mvc\sin\phi = h\nu'\sin\theta$$

Squaring Equations (4) and (5) and adding, we get

$$m^2 v^2 c^2 = (h\nu - h\nu'\cos\theta)^2 + (h\nu'\sin\theta)^2$$

or $$m^2 v^2 c^2 = h^2\nu^2 + h^2\nu'^2\cos^2\theta + 2h^2\nu\nu'\cos\theta + h^2\nu^2\sin^2\theta$$

$$= h^2\nu^2 + h\nu^2\nu'^2 - 2h^2\nu\nu'\cos\theta$$

$$= h^2(\nu^2 + \nu'^2\nu^2 - 2\nu\nu'\cos\theta) \quad ...(6)$$

Also from Equation (1), we have

$$mc^2 = h(\nu - \nu') + m_0c^2$$

Squaring, we get

$$m^2c^4 = h^2(\nu^2 - 2\nu\nu' + \nu'^2) + 2hm_0c^2(\nu - \nu') + m_0^2c^4 \quad ...(7)$$

Subtracting (6) from (7), we have

$$m^2c^2(c^2 - v^2) = 2h^2\nu\nu'(\cos\theta - 1) + 2h(\nu - \nu')m_0c^2 + m_0^2c^4$$

$$= -2h^2\nu\nu'(1 - \cos\theta) + 2h(\nu - \nu')m_0c^2 + m_0^2c^4$$

or $$\frac{m_0^2}{(1 - v^2/c^2)} \times c^2(c^2 - v^2)$$

$$= -2h^2\nu\nu'(1 - \cos\theta) + 2h(\nu - \nu')m_0c^2 + m_0^2c^4$$

$$\left[\because m = \frac{m_0}{(1 - v^2/c^2)^{1/2}}\right]$$

or $$m_0^2c^4 = -2h^2\nu\nu'(1 - \cos\theta) + 2h(\nu - \nu')m_0c^2 + m_0^2c^4$$

or $$2h^2\nu\nu'(1 - \cos\theta) = 2h(\nu - \nu')m_0c^2$$

$$\frac{\nu - \nu'}{\nu\nu'} = \frac{h}{m_0c^2}(1 - \cos\theta)$$

or $$\frac{1}{\nu'} - \frac{1}{\nu} = \frac{h}{m_0c^2}(1 - \cos\theta) \quad ...(8)$$

which may also be written as

$$\lambda' - \lambda = \frac{h}{m_0c}(1 - \cos\theta)$$

or $$\frac{c}{\nu'} - \frac{c}{\nu} = \frac{h}{m_0c}(1 - \cos\theta)$$

$$= \frac{h}{m_0c}2\sin^2\theta/2 \quad ...(9)$$

From Equation (9) it is clear that

(*i*) Change in wavelength is independent of the wavelength of the incident photon but depends on only the angle of scattered photon.

(*ii*) When $\theta = 0$, then

$\lambda' - \lambda = \Delta\lambda = 0$ *i.e.*, no scattering takes place along the direction of incident photon.

(*iii*) When $\theta = \pi/2$, then

$$\lambda' - \lambda = \Delta\lambda = \frac{h}{m_0c}$$

$$= \frac{6.6 \times 10^{-34}}{9.1 \times 10^{-32} \times 3 \times 10^8} = 0.024\text{Å}$$

This change in wavelength is called Compton's shift and as is clear from above it is constant.

(*iv*) When $\theta = \pi$, then

$$\lambda' - \lambda = \Delta\lambda = \frac{2h}{m_0 c} = 0.048\text{Å}$$

so when θ varies between 0° and π, the wavelength of scattered photon varies between λ to $\lambda + \frac{2h}{m_0 c}$ provided the incident photon is of very small wavelength.

6.14. DIRECTION OF THE RECOIL ELECTRON

Dividing Equation (4) by Equation (5), we have

$$\tan\phi = \frac{h\nu' \sin\theta}{h\nu - h\nu' \cos\theta} = \frac{\nu' \sin\theta}{\nu - \nu' \cos\theta} \quad \text{...(10)}$$

Also from Equation (8), we have

$$\frac{1}{\nu'} = \frac{1}{\nu} + \frac{h}{m_0 c^2}(1 - \cos\theta)$$

$$= \frac{1}{\nu} + \frac{h}{m_0 c^2}\, 2\sin^2\frac{\theta}{2}$$

or

$$\frac{\nu}{\nu'} = 1 + \frac{h\nu}{m_0 c^2} \cdot 2\sin^2\frac{\theta}{2}$$

or

$$\nu' = \frac{\nu}{1 + \left(\frac{h\nu}{m_0 c^2}\right) \times 2\sin^2\frac{\theta}{2}}$$

Putting this value of ν' in Equation (10), we get

$$\tan\phi = \frac{\nu \sin\theta / \left(1 + \alpha \cdot 2\sin^2\frac{\theta}{2}\right)}{\nu - \left[\frac{\nu}{\left(1 + \alpha \cdot 2\sin^2\frac{\theta}{2}\right)}\right] \cdot \cos\theta}$$

where

$$\alpha = \frac{h\nu}{m_0 c^2}$$

$\therefore$

$$\tan\phi = \frac{\sin\theta}{1 + 2\alpha\sin^2\frac{\theta}{2} - \cos\theta} = \frac{2\sin\frac{\theta}{2} \cdot \cos\frac{\theta}{2}}{1 + 2\alpha\sin^2\frac{\theta}{2} - \cos\theta}$$

$$= \frac{2\sin\frac{\theta}{2} \cdot \cos\frac{\theta}{2}}{(1 - \cos\theta) + 2\alpha\sin^2\frac{\theta}{2}} = \frac{2\sin\frac{\theta}{2} \cdot \cos\frac{\theta}{2}}{2\sin^2\frac{\theta}{2} + 2\alpha\sin^2\frac{\theta}{2}}$$

$$= \frac{2\sin\frac{\theta}{2} \cdot \cos\frac{\theta}{2}}{2\sin^2\frac{\theta}{2}\,(1 + \alpha)} = \frac{\cot\frac{\theta}{2}}{1 + \alpha}$$

or

$$\tan\phi = \frac{\cot\theta/2}{\left(1 + \frac{h\nu}{m_0 c^2}\right)} \quad \text{...(11)}$$

which gives the direction of the recoil electron in terms of the frequency of the incident photon and the direction of the scattered photon.

6.15. ENERGY OF RECOIL ELECTRON

In the Compton effect, K.E. of recoil electron is equal to the decrease in energy of the incident photon. It is because some energy is imparted to the electron, so that it leaves electronic orbit.

Let the energy of incident proton $= h\nu$

and energy of scattered proton $= h\nu'$

where $\nu > \nu'$

So the decrease in energy of the photon $= h\nu - h\nu' = h(\nu - \nu')$ ...(1)

Let $m_0 c^2$ and mc^2 be the rest mass energy and recoil energy of the electron respectively.

The K.E. of recoil electron $= mc^2 - m_0 c^2$

$$= (m - m_0)\, c^2 \qquad ...(2)$$

This recoil energy is equal to the decrease in energy of the incident photon.

So K.E. of recoil electron $= h(\nu - \nu')$

$$= h\nu\left(1 - \frac{\nu'}{\nu}\right) = h\nu\left(1 - \frac{\lambda}{\lambda'}\right) \qquad [\because c = \nu\lambda]$$

$$= h\nu\frac{\lambda' - \lambda}{\lambda'} = h\nu\frac{\Delta\lambda}{\lambda + \Delta\lambda}$$

where $\Delta\lambda$ is change in wave length.

Now $$\Delta\lambda = \frac{h}{m_0 c}(1 - \cos\theta)$$

$\therefore$ K.E. of recoil electron is given by

$$E = \frac{h\nu \cdot \dfrac{h}{m_0 c}(1 - \cos\theta)}{\lambda + \dfrac{h}{m_0 c}(1 - \cos\theta)}$$

$$= \frac{h\nu \times \dfrac{h}{m_0 c\lambda}(1 - \cos\theta)}{1 + \dfrac{h}{m_0 c\lambda}(1 - \cos\theta)}$$

$$= \frac{\dfrac{h^2\nu^2}{m_0 c^2}(1 - \cos\theta)}{1 + \dfrac{h\nu}{m_0 c^2}(1 - \cos\theta)} \qquad (\because c = \nu\lambda)$$

Let us study the different cases

(*i*) If $\theta = 0$; $E = 0$

which means that no electron will recoil for incident photon at angle $\theta = 0$

(*ii*) If $\theta = \pi/2$

$$E = \frac{\dfrac{h^2\nu^2}{m_0 c^2}}{1 + \dfrac{h\nu}{m_0 c^2}}$$

Let us put $$\alpha = \frac{h\nu}{m_0 c^2}$$

$\therefore$ $$E = \frac{h\nu}{(1 + \alpha)} \cdot \alpha \qquad ...(3)$$

(*iii*) If $\theta = \pi$

$$E = \frac{\frac{2h^2 \nu^2}{m_0 c^2}}{1 + \frac{2h\nu}{m_0 c^2}} = \frac{2h\nu \times \alpha}{1 + 2\alpha}$$

$$= h\nu \frac{2\alpha}{1 + 2\alpha} \quad \text{...(4)}$$

From the above cases we have the conclusion that the minimum energy transferred to electron by the photon is zero (when $\theta = 0$) and maximum energy transferred to recoil electron is

$$h\nu \cdot \frac{2\alpha}{1 + 2\alpha}$$

Also because $\frac{2\alpha}{1+2\alpha} < 1.$

So the energy transferred by photon to electron is less than the energy of incident proton ($h\nu$). *So photon is unable to transfer its whole energy to the electron in photo emission.*

6.16. EXPERIMENTAL SET UP OF COMPTON EFFECT

Fig. 6.15 shows the experimental setup for study of Compton effect. A monochromatic beam of X-rays is allowed to fall on a carbon block *B* which acts as scatterer. Normally, Compton effect can be best observed in case of elements of low atomic number because the binding energies of the electrons in these elements are negligible as compared to the quantum energy of the incident photon. The scattering of X-ray photons can take place in different directions and their intensities and wavelengths can be measured by Bragg's spectrometer.

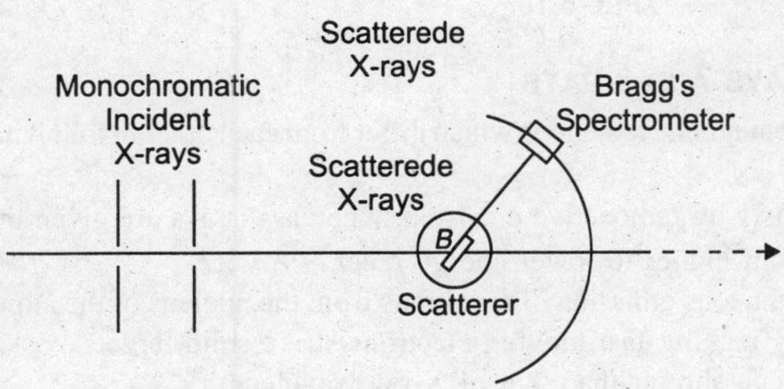

Fig. 6.15

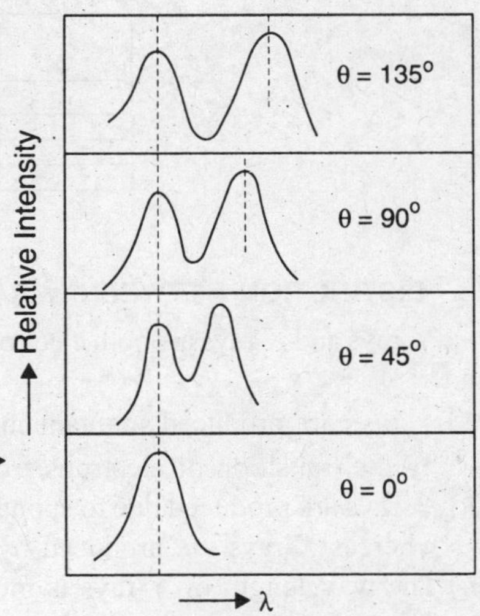

Fig. 6.16

In a typical experiment, K_α line of molybdenum was scattered by graphite and the distribution of intensities was studied at different angles. If a graph is plotted between relative intensities and wavelengths, then we observe that for each value of θ, there are distinct intensity peaks for two wavelengths, one of which corresponds to the incident radiation λ and the other has a higher value λ'. The peak corresponding to λ'. The peak corresponding to λ' is called modified peak.

6.17. PAIR PRODUCTION

Anderson in 1932, observed the phenomenon by which γ-rays are absorbed and **electron-positron pair** is emitted. This phenomenon confirms the conversion of radiant energy into mass.

For pair production phenomenon to take place the incident photon must carry a minimum energy of 1.02 MeV (since both electron and positron have an energy of 0.51 MeV each). It may be noted that this phenomenon occurs mainly in the nuclear field and to some extent in the field to electrons. However, the presence of nucleus is necessary for the conservation of momentum.

The **pair production** phenomenon can be explained theoretically from Dirac's electron theory, which considers positrons as **holes** in otherwise completely filled sea of negative state of **electrons**. This feature is shown in Fig. 6.18.

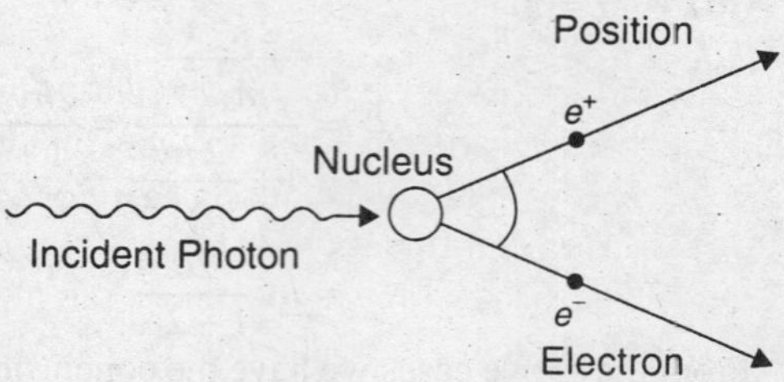

Fig. 6.17

Here we observe that to take a negative electron out of **negative energy state**, an energy of 2 m_0c^2 is required to cross the **barrier**. By this process, a negative electron is created in positive energy state and so it becomes observable. The vacancy in the negative energy region is also observable, having the same properties as the negative electron (also referred as **negatron**) except that it has opposite charge. The total energy between the electron and positron is seen to be multiple of $h\nu$, confirming the quantum theory and mass-energy relationship.

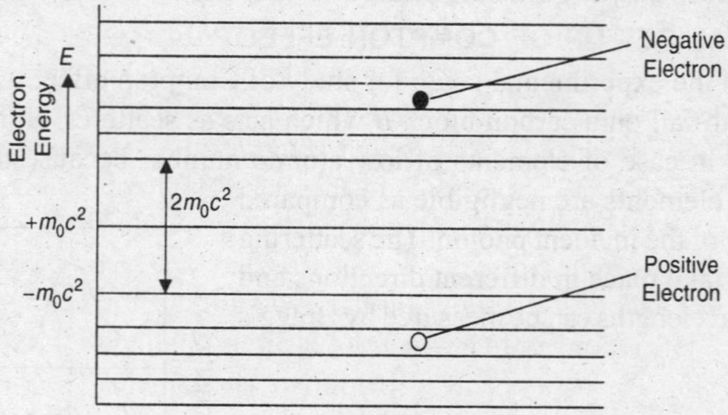

Fig. 6.18

6.18. DISTINCTION BETWEEN γ-RAYS AND X-RAYS

γ-rays and X-rays are both electromagnetic radiations which differ from each other in following manner:

(*i*) γ-rays are produced spontaneously by radioactive emission, whereas X-rays are given out by the transition of electrons from higher to lower energy level.

(*ii*) γ-rays are produced due to spontaneous emission of radiations from the nucleus of the atom, whereas X-rays are produced by making fast moving electrons strike solid targets.

(*iii*) The wavelength of γ-rays is much shorter than that of X-ray wavelength.

(*iv*) The penetrating power of γ-rays is much higher than that of X-rays. X-rays can pass through flesh only, whereas γ-rays can penetrate through 30 cm of thick iron sheet.

(*v*) γ-rays are monochromatic electromagnetic radiations, while X-rays consist of three components hard, medium and soft X-rays.

(*vi*) X-rays produce photoelectric effect and scattering effect, whereas γ-rays produce photo electric effect, scattering effect as well as positron-electron pairs.

Example 1. *In Compton experiment, the wavelength of X-ray radiation scattered at an angle of 45° is 0.022 Å. Calculate the wavelength of the incident X-rays.*

Solution: We have the relation

$$\lambda' - \lambda = \frac{h}{m_0 c}(1 - \cos\theta)$$

In this case,

$$\theta = 45°, \lambda' = 0.02\text{Å} = 0.022 \times 10^{-10}\,\text{m}$$

$$h = 6.6 \times 10^{-34} \text{ joule-sec, } m_0 = 9.1 \times 10^{-31}\text{ kg}$$

$$c = 3 \times 10^8 \text{ metres/sec}$$

$$\therefore \quad \lambda = \lambda' - \frac{h}{m_0 c}(1 - \cos\theta)$$

$$= 0.022 \times 10^{-10} - \frac{6.6 \times 10^{-34}}{9.1 \times 10^{-31} \times 3 \times 10^8}(1 - 0.707)$$

$$= (0.022 - 0.007) \times 10^{-10} = \mathbf{0.015Å}$$

Example 2. *Show that the energy lost by a photon in Compton interacting with a free stationary electron can be written as*

$$h\nu\left[\frac{\alpha(1-\cos\theta)}{1+\alpha(1-\cos\theta)}\right]$$

where $\alpha = \dfrac{h\nu}{m_0 c^2}$

Solution: We have the change in wavelength given by relation

$$\lambda' - \lambda = \frac{h}{m_0 c}(1 - \cos\theta) \quad ...(1)$$

Also when a photon of energy $h\nu$ is scattered by a free electron, it imparts some energy to the electron and scatters with less energy $h\nu'$. The loss of energy is given by

$$h\nu - h\nu' = h(\nu - \nu')$$

$$= h\left(\frac{c}{\lambda} - \frac{c}{\lambda'}\right)$$

$$= hc\frac{(\lambda' - \lambda)}{\lambda\lambda'} \quad ...(2)$$

Putting the value of λ' from Equation (1) in Equation (2), we get

$$\text{Energy lost} = \frac{hc\left[\frac{h}{m_0 c}(1-\cos\theta)\right]}{\lambda\left[\lambda + \frac{h}{m_0 c}(1-\cos\theta)\right]}$$

$$= \frac{h^2}{m_0}\left[\frac{(1-\cos\theta)}{\lambda^2\left[1 + \frac{h}{m_0 c\lambda}(1-\cos\theta)\right]}\right]$$

$$= \frac{h^2\nu^2}{m_0 c^2}\left[\frac{(1-\cos\theta)}{\left[1 + \frac{h}{m_0 c^2}(1-\cos\theta)\right]}\right]$$

$$= h\nu \times \frac{h\nu}{m_0 c^2} \times \left[\frac{1-\cos\theta}{1 + \frac{h\nu}{m_0 c^2}(1-\cos\theta)}\right]$$

$$= h\nu\left[\frac{\alpha(1-\cos\theta)}{1+\alpha(1-\cos\theta)}\right]$$

Example 3. *Find the wavelength of X-ray photons which produce recoil electrons of energy 5 keV in Compton effect. Assume the direction of recoil electron to be in the direction of incident photon while photon is scattered through an angle of 180°.*

Solution: In this case, the direction of scattered photon $\theta = 180°$

Also $\dfrac{hc}{\lambda} - \dfrac{hc}{\lambda'} = 5\text{keV} = 5.000 \times 1.6 \times 10^{-19}$ Joule

$$\therefore \quad \frac{hc}{\lambda} - \frac{hc}{\lambda'} = 8 \times 10^{-16} \text{ Joule} \quad ...(1)$$

From conservation law of momentum in the direction of incident photon,

$$\frac{h}{\lambda} + 0 = \frac{h}{\lambda'}\cos 180° + mv\cos 0°$$

$$= \frac{h}{\lambda'}\cos 180° + mv$$

or
$$\frac{h}{\lambda} = \frac{h}{\lambda'} \times (-1) + mv$$

$$\therefore \quad \frac{h}{\lambda} + \frac{h}{\lambda'} = \sqrt{\frac{1}{2}mv^2 \times 2m}$$

$$= \sqrt{8 \times 10^{-16} \times 9.1 \times 10^{-31}}$$

$$= 38.1 \times 10^{-24}$$

Multiplying both sides by c

$$\frac{hc}{\lambda} + \frac{hc}{\lambda'} = 3 \times 10^8 \times 38.1 \times 10^{-24} \quad ...(2)$$

$$= 11.448 \times 10^{-17} \text{ Joule.}$$

Adding Equations (1) and (2), we get

$$\frac{2hc}{\lambda} = 8 \times 10^{-16} + 11.448 \times 10^{-17}$$

$$= 122.48 \times 10^{-16} \text{ Joule}$$

or
$$\lambda = \frac{2hc}{122.48 \times 10^{-16}\text{ J}}$$

$$= \frac{(2 \times 6.6 \times 10^{-34}\text{ J})(3 \times 10^8\text{ m/s})}{122.48 \times 10^{-16}\text{ J}}$$

$$= 0.324 \times 10^{-10}\text{ m} = \mathbf{0.324 Å}$$

Example 4. *A photon of energy 1.02 MeV is scattered through 90° by a free electron. Calculate the energy of photon and electron after interaction.*

Solution: We have the relation

$$\Delta\lambda = \lambda' - \lambda = \frac{h}{m_0 c}(1 - \cos\theta)$$

$$= \frac{6.6 \times 10^{-34}}{9.1 \times 10^{-31} \times 3 \times 10^8}(1 - 0)$$

$$= 2.42 \times 10^{-12}\text{ m}$$

Change in frequency of photon

$$\Delta\nu = \frac{c}{\Delta\lambda} = \frac{3 \times 10^8\text{ m}}{2.42 \times 10^{-12}\text{ m}}$$

Change in energy of photon

$$\Delta E = h\Delta\nu = \frac{6.6 \times 10^{-34} \times 3 \times 10^8}{2.42 \times 10^{-12}\text{ m}} = \mathbf{0.51\ MeV}.$$

This energy is transferred to the free electron which in turn will be the K.E. of the recoil electron.

Example 5. *In a diode valve, the anode has a potential of 100 volts relative to the cathode. Find the wavelength associated with the electron as it reaches the anode. Mass of electron* $m = 9.11 \times 10^{-28}$ *gm., charge* $e = 4.803^{-10}$ *e.s.u. and* $h = 6.62 \times 10^{-27}$ *ergs-sec.*

Solution: Potential difference $V = \frac{100}{300} = \frac{1}{3}$ e.s.u.

Let v be the velocity acquired by the electron on reaching the anode, then energy of the electron is given by

$$\frac{1}{2}mv^2 = Ve$$

$$v = \sqrt{\frac{2Ve}{m}}$$

Now, wavelength 'λ' of the electron is given by

$$\lambda = \frac{h}{mv} = \frac{h}{m\cdot\sqrt{\frac{2Ve}{m}}} = \frac{h}{\sqrt{2Vem}}$$

$$\lambda = \frac{6.62\times10^{-27}}{\sqrt{2\times1/3\times4.803\times10^{-10}\times9.11\times10^{-28}}}$$

$$\lambda = \frac{6.62\times20^{-27}}{\sqrt{2\times1.601\times10^{-10}\times9.11\times10^{-28}}}$$

$$= 1.226 \times 10^{8} \text{ cms.}$$

Example 6. *A photoelectric surface has a work function of 4 eV. What is the maximum velocity of the photoelectrons emitted by light of frequency 10^{15} Hertz incident on the surface? ($h = 6.6 \times 10^{-34}$ Joule sec, $e = 1.6 \times 10^{-19}$ Coul. $m = 9 \times 10^{-31}$ km).*

Solution: We have the relation

$$h\nu - w_0 = \frac{1}{2}mv^2$$

In this case, $h = 6.6 \times 10^{-34}$ Joule sec.

$$\nu = 10^{15} \text{ Hertz}$$

$$w_0 = 4 \text{ eV} = 4 \times 1.6 \times 10^{-19} \text{ Joule}$$

$$= 6.4 \times 10^{-79} \text{ Joule}$$

$$m = 9 \times 10^{-31} \text{ kg.}$$

$$\therefore\ 6.6 \times 10^{-34} \times 10^{15} - 6.4 \times 10^{-19} = \frac{1}{2}\times9\times10^{-31}v^2$$

or $v = \mathbf{2.108 \times 10^5}$ **m/Sec.**

Example 7. *X-ray with $\lambda = 1$ Å are scattered from a carbon block. The scattered radiation is viewed at 90° to the incident beam. (a) What is the Compton shift $\Delta\lambda$? (b) What K.E. is imparted to the recoiling electron?*

Solution: (*a*) We have the Compton shift given by

$$\Delta\lambda = \frac{h}{m_0 c}(1-\cos\phi)$$

In this case, $\phi = 90°$, $\cos 90° = 0$

$$\Delta\lambda = \frac{h}{m_0 c}(1-\cos\phi)$$

In this case, $\phi = 90° = 0$

$$\therefore\ \Delta\lambda = \frac{h}{m_0 c} = \frac{6.62\times10^{-34}\text{ Joule}-\text{sec}}{9.1\times10^{-31}\text{ kg}\times3\times10^{8}\text{ m/sec}}$$

$$= 2.43 \times 10^{-12} \text{ m}$$

$$= \mathbf{0.0243Å}$$

(*b*) From the conservation law of energy, we have

$$\frac{hc}{\lambda} = \frac{hc}{\lambda'} + K$$

where K is the K.E. of recoil electron

Also $\lambda' = \lambda + \Delta\lambda$

$$\therefore\ \frac{hc}{\lambda} = \frac{hc}{\lambda+\Delta\lambda} + K$$

or
$$K = \frac{hc\,\Delta\lambda}{\lambda\,(\lambda + \Delta\lambda)}$$
$$= \frac{6.62 \times 13^{-34} \text{ Joule sec } \times 3 \times 10^{8} \text{ m/sec} \times 2.43 \times 10^{-12} \text{ m}}{1 \times 10^{-10} \text{ m} \times (1+0.0243) \times 10^{-10} \text{ m}}$$
$$= 4.72 \times 10^{-17} \text{ Joule}$$
$$= \frac{4.72 \times 10^{-17}}{1.6 \times 10^{-19}} \text{ eV}$$
$$= \mathbf{0.295 \text{ eV}}.$$

Example 8. *The photoelectric work function of potassium is 2.0 eV. If light having a wavelength 3600 Å falls on potassium, find (i) the kinetic energy in eV of the most energetic electrons ejected (ii) the stopping potential (iii) the velocity of these electrons.*

Solution: Work function $w_0 = 2$ eV.
$$= 2 \times 1.6 \times 10^{-12} \text{ ergs}$$
$$h = 6.624 \times 10^{-27} \text{ ergs-sec.}$$
$$c = 3 \times 10^{10} \text{ cm/sec.}$$

Now,
$$w_0 = \frac{hc}{\lambda_0} \text{ where } \lambda_0 \text{ is the threshold wavelength.}$$
$$\lambda_0 = \frac{hc}{w_0} = \frac{6.624 \times 10^{-27} \times 3 \times 10^{10}}{2 \times 1.6 \times 10^{-12}} \text{ cm}$$
$$= \frac{19.872}{3.2} \times 10^{-5} \text{ cm}$$
$$= 6.2103 \times 10^{-5} \text{ cm}$$
$$= 6210.3 \times 10^{-8} \text{ cm} = 6210.3\text{Å}$$
$$\nu_0 = \frac{c}{\lambda} = \frac{3 \times 10^{10}}{6210.3 \times 10^{8}} = \frac{3}{6210.3} \times 10^{18}$$
$$\nu = \frac{c}{\lambda_0} = \frac{3 \times 10^{10}}{3600 \times 10^{8}} \frac{1}{12} \times 10^{16}$$

(*i*) Now K.E. $= \frac{1}{2} mv^2 = h\nu - h\nu_0$
$$E = h\left[\frac{1}{12} \times 10^{16} - \frac{1}{2070} \times 10^{18}\right]$$
$$= h \times 10^{16} \left[\frac{2070 - 1200}{24840}\right]$$
$$= h \times 10^{16} \times \frac{870}{24840} \text{ ergs}$$
$$= 6.624 \times 10^{-27} \times 10^{17} \frac{870}{24840}$$
$$= \frac{6.624 \times 870}{24840 \times 1.6 \times 10^{12}} \times 10^{-11} \text{ eV}$$
$$= \frac{6.624 \times 870}{2484 \times 1.6} \text{ eV} = \mathbf{1.45\,eV}.$$

(*ii*) Stopping potential

∴ $E = 1.45$ eV.

So stopping potential = **1.45 volts**

(*iii*) Velocity of the electron

$$\frac{1}{2}mv^2 = 1.45 \text{ eV} = 1.45 \times 1.6 \times 10^{-12} \text{ ergs}$$

$$v^2 = \frac{2 \times 1.45 \times 1.6 \times 10^{-12}}{9.1 \times 10^{-28}}$$

$$v = 7.1 \times 10^7 \text{ cm/sec.}$$

Example 9. *The photoelectric threshold of tungsten is 2300 Å. Determine the energy of the electrons ejected from the surface by ultraviolet light of wavelength 1800 Å.*

Solution: Here $\lambda_0 = 2300\text{Å} = 2300 \times 10^{-8}$ cm

$$\lambda = 1800\text{Å} = 1800 \times 10^{-8} \text{ cm}$$

$$\nu_0 = \frac{c}{\lambda_0} = \frac{3 \times 10^{10}}{2300 \times 10^{-8}} = \frac{3}{23} \times 10^{16}$$

$$\nu = \frac{c}{\lambda} = \frac{3 \times 10^{10}}{1800 \times 10^{-8}} = \frac{1}{6} \times 10^{16}$$

Now kinetic energy $= \frac{1}{2}mv^2 = h\nu - h\nu_0$

$$E = h\nu - h\nu_0 = h\left[\frac{10^{16}}{6} - \frac{3 \times 10^{16}}{23}\right]$$

$$= h \times 10^{16}\left[\frac{23-18}{138}\right]$$

$$= 6.62 \times 10^{-27} \times 10^{-16} \times \frac{5}{138}$$

$$= 6.62 \times \frac{5}{138} \times 10^{-11} \text{ ergs}$$

$$= \frac{33.10}{138} \times 10^{-11} \times \frac{1}{1.6 \times 10^{-13}} \text{ eV}$$

$$\therefore \quad E = \mathbf{1.52 \text{ eV}}$$

Example 10. *The energy of photoelectrons emitted from a sensitive plate is 1.56. If the threshold wavelength is 2500Å, calculate the wavelength of the incident light; given that*

$$1 \text{ eV} = 1.6 \times 10^{-12} \text{ ergs}$$

$$h = 6.62 \times 10^{-27} \text{ ergs-sec}$$

Velocity of light $c = 3 \times 10^{10}$ *cms/sec.*

Solution: $\lambda_0 = 2500\text{Å} = 2500 \times 10^{-8}$ cm.

$$c = 3 \times 10^{10} \text{ cm/sec.}$$

The threshold frequency

$$\nu_0 = \frac{c}{\lambda_0} = \frac{3 \times 10^{10}}{25 \times 10^{-8}} = \frac{3}{25} \times 10^{16}$$

$$h\nu_0 = \frac{3}{25} \times 10^{16} \times 6.62 \times 10^{-27} \text{ ergs.}$$

$$= \frac{3 \times 6.62 \times 10^{-11}}{25 \times 1.6 \times 10^{-12}} \text{eV} = 4.965 \text{eV.}$$

Let ν be the incident frequency

E, the energy of the electron ejected = 1.56 eV

Thus $h\nu = h\nu_0 + E$

$$= (4.965 + 156) \text{ eV} = 5.525 \text{ eV}$$

But $\nu = \frac{c}{\lambda}$

$$\frac{hc}{\lambda} = 5.525 \times 1.6 \times 10^{-12} \text{ ergs}$$

$$\lambda = \frac{hc}{5.525 \times 1.6 \times 10^{-12}} \text{cm}$$
$$= \frac{6.62 \times 10^{-27} \times 3 \times 10^{10}}{5.525 \times 1.6 \times 10^{-12}}$$
$$= 2246 \times 10^{-8} \text{ cm.} = \mathbf{2246\text{Å}}.$$

Example 11. *Calculate the longest wavelength of incident radiation which will eject electrons from Platinum. The work function of Platinum = 6.3 eV., h = 6.624 × 10^{-27} ergs-sec., c = 3 × 10^{10} cm/sec.*

Solution: Work function w_0 = 6.3 eV

$$w_0 = 6.3 \times 1.6 \times 10^{-12} \text{ ergs}$$
$$= 10.08 \times 10^{-12} \text{ ergs}$$
$$h = 6.624 \times 10^{-27} \text{ ergs-sec.}$$

Now $w_0 = h\nu_0$

If λ_0 is the threshold wavelength of incident light

$$w_0 = \frac{hc}{\lambda_0}$$
$$= \frac{6.624 \times 10^{-27} \times 3 \times 10^{10}}{10.08 \times 10^{-12}}$$
$$= \frac{19.872}{10.08} \times 10^{-5}$$
$$= 1.972 \times 10^{-5} \text{ cm}$$
$$= 1972 \times 10^{-8} \text{ cm.} = \mathbf{1972\text{Å}}.$$

Example 12. *Calculate the photoelectric work function for sodium metal in electron volts; given that the threshold wavelength is 6800 Å, h = 6.6 × 10^{-7} ergs. sec*

Solution: Here threshold wavelength λ_0 = 6800Å

$$= 6800 \times 10^{-8} \text{ cm.}$$

∴ Threshold frequency

$$\nu_0 = \frac{c}{\lambda_0} = \frac{3 \times 10^{10}}{6800 \times 10^{-8}} = \frac{3}{68} \times 10^{16}$$

Let w_0 be the work function for metal, then

$$w_0 = h\nu_0$$
$$= 6.62 \times 10^{-27} \times \frac{3}{68} \times 10^{16} \text{ ergs.]}$$
$$= 2.92 \times 10^{-12} \text{ ergs.}$$
$$= \frac{2.92 \times 10^{-12}}{1.6 \times 10^{-12}} \text{eV}$$

Example 13. *Calculate the energy of a quantum of light of wavelength* $\lambda = 5.4 \times 10^{-15}$ *cm and express it in eV. If such a light falls on calcium whose work function is 1.9 eV, find retarding potential to stop the most energetic electrons.*

Solution:
$$\lambda = 5.4 \times 10^{-5} \text{ cm.}$$
$$c = 3 \times 10^{10} \text{ cm/sec}$$

Now energy of one quantum of light

$$h\nu = \frac{hc}{\lambda} = \frac{6.62 \times 10^{-27} \times 3 \times 10^{10}}{5.4 \times 10^{-5}}$$
$$= \frac{6.62 \times 3 \times 10^{-12}}{5.0 \times 1.6 \times 10^{-12}} \text{ eV} = 2.21 \text{ eV.}$$

Let V be the stopping potential

Work function $w_0 = 1.2\,\text{eV} = 1.9 \times 1.6 \times 10^{-12}$ erge.

Now energy of electron

$$E = h\nu - w_0$$

$$Ve = h\nu - w_0$$

$$V = \frac{h\nu - w_0}{e}$$

$$= \frac{2.21 - 1.6 \times 10^{-12} \times 1.9 \times 10^{-12}}{4.8 \times 10^{-10}}$$

$$= \frac{1.6 \times 10^{-12}}{4.8 \times 10^{-10}}(2.21 - 1.9)$$

$= .13 \times 10^{-2}$ e.s.u. $= 0.13 \times 10^{-2} \times 300$ volts $=$ **0.39 Volts.**

6.19. PHOTO-ELECTRIC CELL

Photoelectric cells are of two types (*i*) vacuum type (*ii*) gas filled type. The cell consists of an evacuated glass bulb that contains a semi cylindrical plate B coated with potassium or cesium known as emitter and a wire A of platinum or nickel fixed along the axis of the cylinder. This wire is known as collector. The collector A is given positive potential and photo electrons ejected from the cylinder B are attracted by it. A galvanometer is put in the circuit to detect the photoelectric current.

In the case of a vacuum type cell, the current is weak and so gas filled photoelectric cells are used. The gas used is either Argon or Helium and when these gas molecules are hit by electrons, they get ionised and so high current flows through the circuit. There is a wide variety of photo-cathodes that can use infra-red, visible and ultraviolet regions of spectrum.

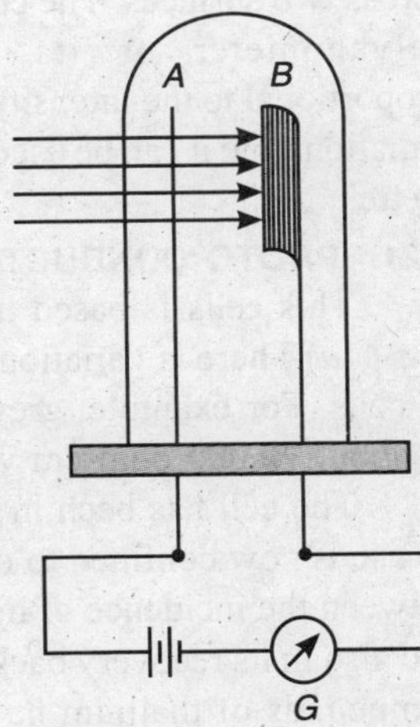

Fig. 6.19

When light is cut off, no photo electrons are ejected and so the current to external circuit stops.

OTHER TYPES OF PHOTOELECTRIC CELLS

6.20. OTHER TYPES OF PHOTOELECTRIC CELLS

Photo-voltic cell or photronic cell. In the case of a photoelectric cell we require a battery because we have to give positive potential to the collector to attract electrons. However, in the case of a photo-voltaic cell no auxilliary battery is required and the electrons ejected by light themselves produce a p.d. between the two plates and this p.d. gives current to the external circuit.

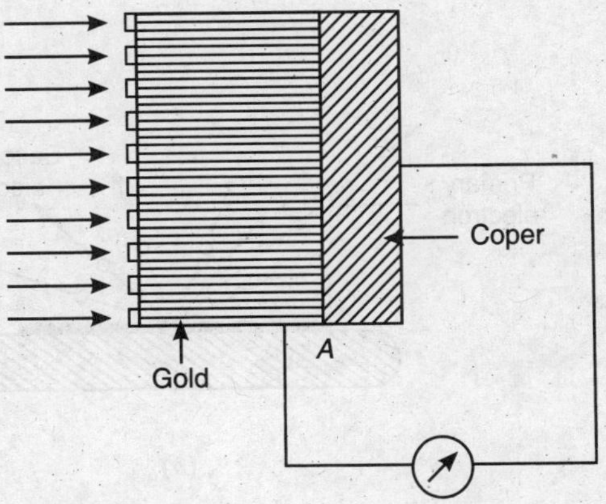

Fig. 6.20

It consists of a copper plate and a film of cuprous oxide formed by oxidising one side of copper disc A. The cuprous oxide side is coated with a thin layer of gold or silver. The light falling on gold drives electrons from oxide into the gold layer and so a p.d. is produced between the thin layer of gold and copper disc. The p.d. depends on the number of electrons ejected, *i.e.*, on the intensity of incident light. So, a current flows through the galvanometer and it will give a deflection which will be directly propor-tional to the intensity of light. The photo-voltaic cells are calibrated and are generally used as light metres in photography.

The most common type of photronic cell used these days is devised by weston. It consists of a thin film of selenium coated on the iron plate. Light incident on the selenium surface produces a photo-voltaic cell. Iron is taken as + ve terminal and selenium as – ve terminal and current flows through galvanometer when connected across two surfaces. The current shown by galvanometer or its deflection is proportional to the intensity of the incident radiations and it can be used as an exposure metre.

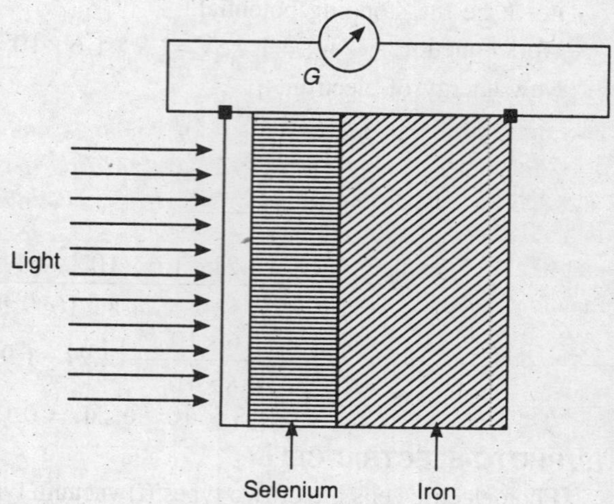

Fig. 6..21

6.21. PHOTO CONDUCTIVITY CELL OR SELENIUM CELL

This cells is based upon a different principle.

(*a*) There is variation of resistance of a substance when illumination is employed in this type of cells. For example, grey crystalline selenium has the property that its specific resistance falls to about 70,000 ohm-cm when illuminated to one-tenth of this value when brightly illuminated.

The cell has been in use before the other types of photoelectric cells became prominent but its use is now confined to operating relays for controlling illumination. There is a considerable lag between the incidence of light and time taken for the resistance of the cell to drop to the minimum, and also in its recovery back, when the source of illumination is withdrawn. Besides selenium, some compounds of thallium have also been used for a similar purpose.

Photo-Multiplier Tube. Photo-Multiplier tube is based upon the principle of secondary-emission. When a beam of electrons strike the surface of a metal plate, they knock additional electrons free from the surface. These are called secondary electrons and the process is called secondary-emission as shown in Fig. 6.22(*a*).

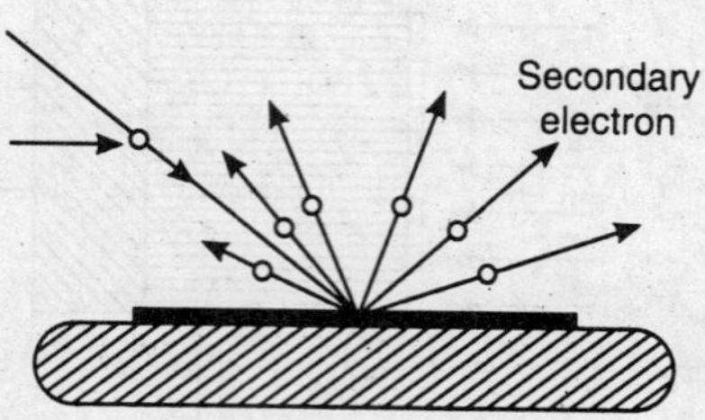

(*a*)

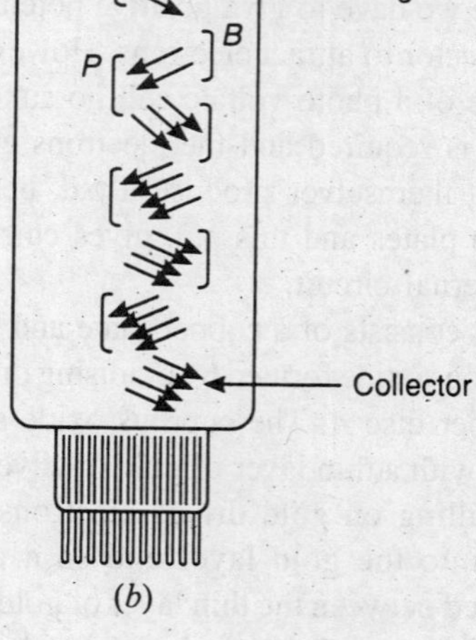

(*b*)

Fig. 6.22

As the energy of a primary or incident electron increases from zero to a few hundred volts the number of secondaries increases toward a definite maximum. For most metal surfaces, this maximum is in the neighbourhood of two, while for certain alkali metal films it may be as great as eight or ten. In general, it is greatest for surfaces having a *low work function.*

The process of secondary electron emission is widely used in a special type of photoelectric cell used most effectively in detecting faint light. A cross-section diagram of such a photo-multiplier tube is given in Fig. 6.22(*b*).

The number of photoelectrons from the photo-cathode K is proportional to the intensity of the incident light. These are attracted forward the next plate B, more positive by 100 volts, where upon impact, additional electrons are liberated and are attracted to the next more positive plate P, still more electrons are liberated. By the time the collector plate has been reached a small avalanche of electrons has developed, and a correspondingly large charge and current are led off through that electrode to a suitable recording device.

If each electron on impact releases n secondaries, then, in a tube with k stages, the number arriving at the collector would be n^k. For example, if $n = 6$, and if $k = 5$, then $n^k = 7776$ electrons. This is an enormous gain over the signal obtained from a standard photo-tube. Photo-multiplier tubes have been used most successfully with faint light, not only visible but infrared and ultraviolet as well.

6.22. APPLICATIONS OF THE PHOTOELECTRIC CELLS

The photoelectric cells have the following important applications.

1. Modern Talkies or Cinematography. In sound motion pictures or modern talkies, now so commonly in use, a sound track is provided at the edges of the picture film.

(*a*) **Sound Recording.** There are two common methods of sound recording:

(*i*) By variable density method (*ii*) By variable area method.

(*i*) **Variable Density Method.** The sound to be recorded is received by the microphone M and is fed into a special neon lamp L called exiter or AEO larap, through a step-up transformer as shown in the Fig. 6.23. The fluctuations of the sound produce similar fluctuations of current in the microphone which in turn is fed to the neon lamp, *i.e.*, the exiter. The light from this lamp is thus of fluctuating intensity, the fluctuations of sound being faithfully converted into the fluctuations of light. The light from the lamp L is allowed to pass through a narrow slit S about 0.005 inch wide and fall on a moving photographic film P. The deposit of silver on the photographic film P is thus of varying thickness depending upon the fluctuation of the intensity of incident beam.

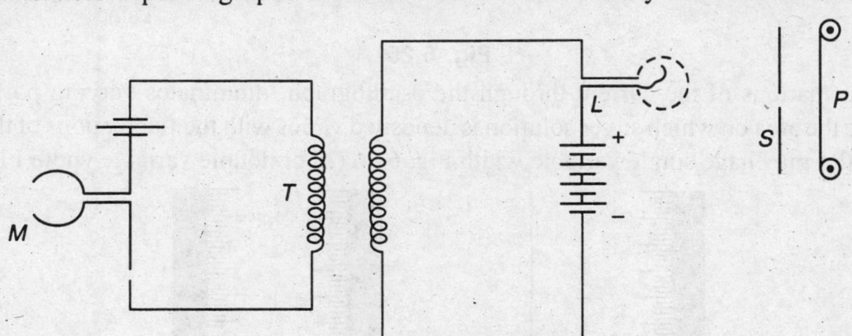

Fig. 6.23

The light valve method obtaining a variable density sound track is to feed the current from the microphone M (after amplification by the amplifier A) into a square type loop L, which is made of **duraiumin** ribbon and placed in front of a narrow slit. The two arms of the loop are at a distance of 0.002 inch and thus form the upper and lower edges of the gap through which the light passes. The light from the gap is focussed on the film P with a lens. The loop is placed in the strong electromagnetic field by a permanent magnet N as shown in Fig. 6.24.

When the current through the loop increases, the arms move outwards widening the gap. A decrease in current reduces the gap. The intensity of light falling on the film, thus varies with the current produced in the microphone and hence the sound track will consists of bands of varying density.

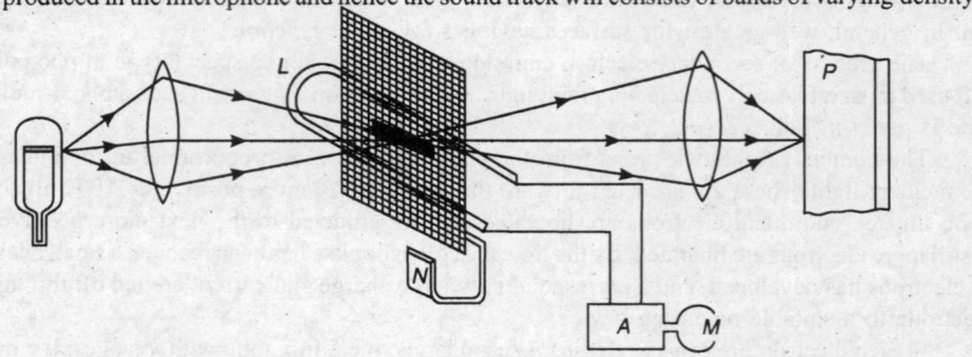

Fig. 6.24

(*ii*) **Variable Area Method.** In this method, the fluctuations of sound are again picked by the microphone as shown in Fig. 6.25 and the fluctuating current thus obtained fed into an Einthoven 'String Oscillograph, when the oscillations of the mirror take place. Light from the lamp is focussed on the mirror of the oscillograph with the help of convex lens and the reflected beam is allowed to fall on a moving photographic film through the slit, both of which are placed transversely.

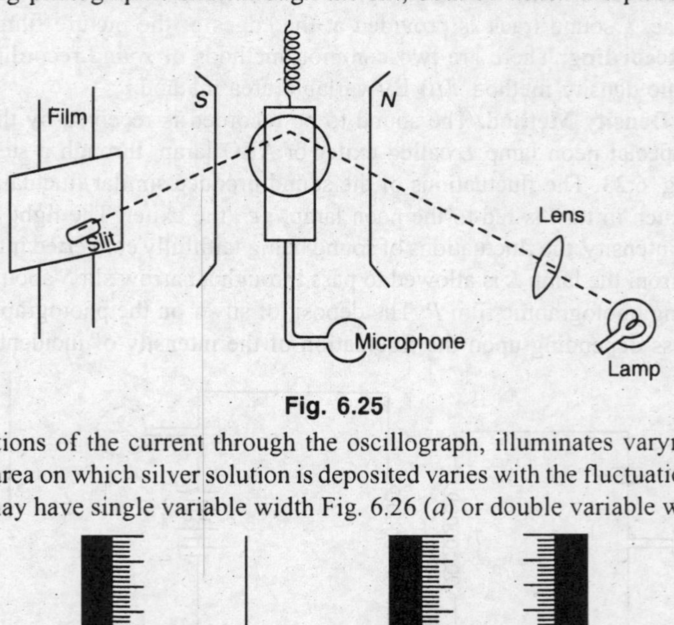

Fig. 6.25

The fluctuations of the current through the oscillograph, illuminates varying portion of the slit and hence the area on which silver solution is deposited varies with the fluctuations of the incident sound. The film may have single variable width Fig. 6.26 (*a*) or double variable width Fig. 6.26(*b*).

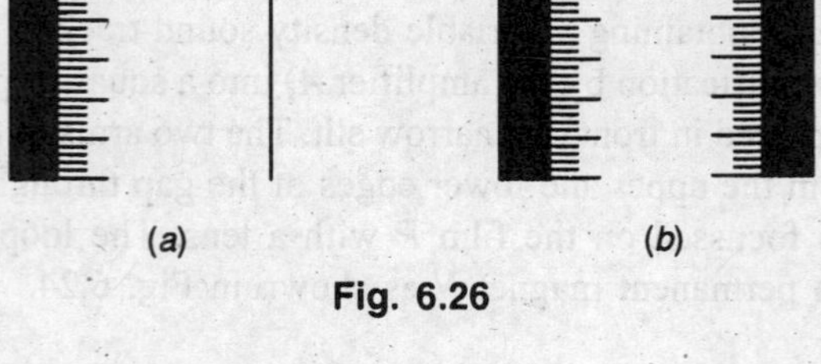

(*a*) (*b*)

Fig. 6.26

As a rule, the sound recording is done in a separate room away from the studio where the picture is recorded; though, of course, the sound record and the pictures are taken at same time. Two films are taken, one for the picture and the other for the sound track and are exposed at exactly the same rate. The standard speed is 18 inches per second. The two films are then developed. After this a print on an other film is prepared from these two negatives and arranged such that the sound track appears by the side of the picture track and leads it by $14\frac{1}{2}$ inches, which is the distance between the sound projector and the light projector of a standard cinematograph. The third film which is to be finished positive, is run through an exposing apparatus, and first the picture and then the sound track is exposed in turn, the other part being covered with an opaque paper. The film is then developed and is ready for projection. The normal rate of projection is 24 pictures per second.

(*b*) **Reproduction of Sound.** Whatever may be the method of sound recording, the method of reproduction of sound is the same. The sound film moves with the same speed as during recording. An intense beam of light is focussed on the sound track and the emergent light falls on photoelectric cell, which gives rise to a fluctuating current which is further amplified and fed into the loud-speaker, where the original sound is reproduced. The apparatus used for reproduction is shown in Fig. 6.27. It consists of an exiter lamp *L* of 10 volt and 5 watt. A cancavo-convex lens *C* known as condenser lens is used for focussing. S_1 is a slit, *O* is the projecting lens. S_2 is the limiting diaphragm and F the sound track of the sound motion picture. The light coming out of sound track is fluctuating in intensity and is allowed to fall on the photoelectric cell *P*, where the fluctuations of light are converted into fluctuations of the electric current, which is amplified by an amplifier *A* and fed into the loud-speaker placed behind the screen, which reproduces the sound originally made. Thus, thanks to the ingenious scientific devices of optical recording of sound, synchronising it with action and reproducing it with a photocell and amplifier, we enjoy the most artistic and popular amusement of modern times.

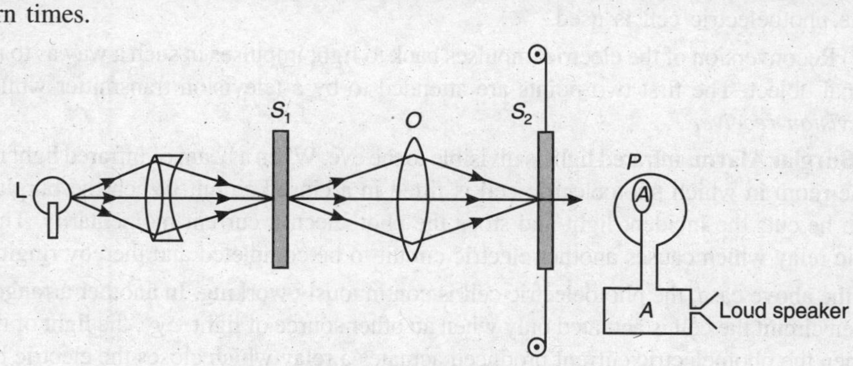

Fig. 6.27

2. Photo-Telegraphy or Belinogram. The photoelectric cell finds an important application in photo-telegraphy or transmission of pictures of distant places within a very short time. By means of this device newspaper strive to reproduce in every country photographs of persons or news of events which take place in different parts of the world within few minutes time.

At the transmitting station the photograph or document to be reproduced is wounded on a cylinder which rotates at a uniform rate at the same time moves forward on its axis so that a fixed point on it traces out a helical path with an even pitch about $\frac{1}{4}$ of a mm. A bright spot of light is formed on the cylinder and the light reflected from the illuminated spot of the document is made to fall on a photoelectric cell. The intensity of the reflected light depends upon the part of the document which is illuminated and on the forward motion of the rotating cylinder. In this manner different parts of the documents are illuminated successively. Hence, the beam of reflected light which falls on the photo-cell varies in intensity, and the strength of the photoelectric current varies proportionately.

This modulated current is amplified by means of valve amplifier in several stages and then by means of transmitter is converted into radio waves.

At the receiving station the antenna (arial) receives the radio waves and passes them into an amplifying system of valves. The amplified current is fed into an oscillo-scope with a mirror on which a focussed beam of light is made to fall. As the mirror oscillates with an amplitude which is proportional to the current passing the oscilloscope and therefore, also to the photoelectric produced at the transmitting station, therefore, the spot of light reflected from the mirror will also oscillate proportionately. The reflected oscillating beam of light is passed through a diaphragm which is called the stone-scale diaphragm. This device transmits more or less light according as the oscilloscope mirror turns through a greater or smaller angle and in this way the intensity of light emerging from it will be strictly governed by the photoelectric current. The light of varying intensity coming out of the tone-scale diaphragm is made to fall on a photographic film, on which the pictures or news is to be reproduced. Thus, the picture or document used at the transmitting station is reproduced at the receiving station within a very short time.

3. Television. Television is the process of transmission and reproduction of moving figures at a distance and this method is similar in many respects to the transmission of pictures. In television, the pictures already contained on a photographic film are reproduced, while in transmission of pictures an actual moving figure, say the face of an actor, is reproduced. The process of television involves the following essential points:

(*i*) Scanning of the object which is to be televised. It consists in dissecting the object into large number of elements by making a narrow and intense beam of light more rapidly back and forth over a limited portion of the object.

(*ii*) Conversion of light impulses received from the scanned object into electric impulses, for which the photoelectric cell is used.

(*iii*) Reconversion of the electric impulses back to fight impulses in such a way as to reconstruct the original object. The first two points are attended to by a television transmitter while the third by a *television receiver*.

4. Burglar Alarm. Infrared light is invisible to the eye. When a beam of infrared light is projected across the room in which photoelectric cell is fitted in a closed circuit. When the burglar walks in that room he cuts the incident light and stops the photoelectric current momentarily. This actuates an electric relay which causes another electric circuit to be completed and thereby ringing the bell.

In the above case, the photoelectric cell is continuously working. In another arrangement, *i.e.*, in the open circuit the cell is actuated only when an other source of light, *e.g.*, the light of the intruder torch, when the photoelectric current produced actuates a relay which closes the electric bell circuit, thereby ringing the bell.

5. Micro photometers. These are the instruments for measuring the intensity of light and studying the final structure of a spectral line. These use photoelectric cells.

6. Foot-candle Metre. This instrument is used for measuring the intensity of illumination at a given place directly. It consists of a photo-voltaic cell and a galvanometer. In this an iron plate P covered with thin layer of Selenium S on which is deposited a very thin semitransparent film of gold. A metal ring R in contact with the gold-film forms the negative terminal of the cell and the iron plate act as the positive terminal.

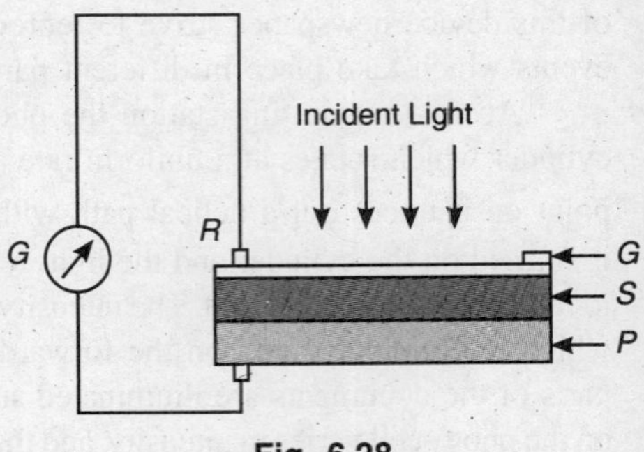

Fig. 6.28

When a beam of incident light falls over the gold-film, it is transmitted through it and on falling on Selenium, it derives the electrons into the gold-film which becomes negative. The electrons cannot pass as easily to the iron plate as to the gold-film. The electrical current produced is proportional to the incident light radiations and hence the galvanometer G can be directly calibrated to read the intensity of illumination in foot-candles. Such devices have replaced the older type of photometers for measuring *Candle Powers*, and are now used by photographer as *illumination metres exposure metres.*

$$\text{Intensity of illumination } (I) = \frac{\text{Illuminating power } (Q)}{\text{Square of distance } (R^2)}$$

$$\therefore \text{ Illuminating power } (Q) = \text{Intensity of illumination } I \times \text{ square of distance } (R^2)$$

i.e, **Candle power = foot candles × ft²**

7. Fire Alarms. To protect a building from fire, gas filled photoelectric cells are used at various places in the building. Whenever there is a flame, the light calls on the cell and photoelectric current is produced which after suitable amplification causes a bell to ring.

8. Counting Machines. An important example of counting machines is met with on docks where it is used to count the number of packages or bags loaded in a ship. A beam of light falls on a photoelectric cells (gas filled) and the current after amplification starts a counting machine. As soon as the path of the beam is cut off by the package, the current stops for a while and the counting machine advances by one.

9. Automatic Switching of Street Lighting Circuit. A photoelectric cell is also used in the street lighting circuit. The sunlight falls on it and the current generated keeps the circuit open. As soon as the intensity of sunlight falls below a certain value (*e.g.*, in the evening) the photoelectric current becomes very feeble and the street light circuit is closed. This method has the advantage that the time of street light is automatically adjusted in summer and winter.

10. Exposure Metres in Photography. An exposure metre is a device to calculate the correct time of exposure. The photoelectric cell in the instrument produces a current proportional to the intensity of light falling on it. The current operates a galvanometer, the scale of which is calibrated to read the time of exposure.

11. Complexion Metres. The light reflected from the face of a person falls on a photoelectric cell when the current produced is proportional to the intensity of the reflected light and as such measures the complexion of the person.

12. Day Light Recorders. Photoelectric cells also find use in meteorology to record day light. The current produced in a photoelectric cell with day-light is coupled with *clock system* and a current time graph is plotted giving a measure of day-light.

13. These cells are also used in traffic signals, colour identification etc.

6.23. THERMIONIC EMISSION

When heat energy is supplied to a metal, the electrons get sufficient K.E. to escape from the metal surface this liberation of electrons from metal surface due to heat energy is called thermionic emission.

6.24. RICHARDSON'S EXPERIMENT

In this experiment a filament F of platinum or tungsten is stretched along the axis of an evacuated glass tube T. The filament is surrounded by a metal cylinder, also placed in glass tube T. The filament is heated by passing a current from battery B_2 through an ammeter A and adjustable rheostat R_1.

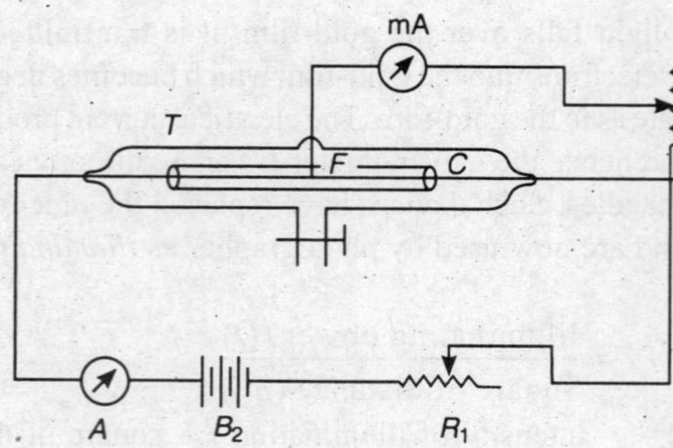

Fig. 6.29

A suitable p.d. can be applied between filament F and cylinder C with the help of battery B_1 and rheostat R_2. In this case, the +ve terminal of the battery is connected to cylinder C. So that it works as anode. The anode current is measured by mA (micro ammeter) connected in the circuit.

With such an arrangement at the steady state, Richardson found that a stream of negative charges flows from filament to cylinder causing a current between them. He also found that for some value of filament current, the anode current changes as the anode voltage increases and reaches a saturated value and for a particular value of anode potential the anode current is a function of filament current. By measuring e/m of the charged particles it was found that it is the same as for electrons. So from here Richardson concluded that electrons are emitted from heated filament.

Richardson studied in details about the nature of these thermions and also found the dependedness of thermionic emission on number of factors such as temperature, area, material, applied potential, pressure of surroundings and other conditions.

6.25. RICHARDSON'S EQUATION

Richardson applied clausius clapeyson's latent heat equation for ejection of electrons from the surface of metal.

Let the volume of 1 gm molecule of electrons within the metal be v_1 and that outside be v_2. The volume v_2 being very high as compared to v_1. Let the pressure of electrons under equilibrium conditions be P. The energy absorbed by N electrons during ejection from the metal surface = W. In this case, N will be Avagadro's number. Applying claussius clapeyson's equation. We have

$$\frac{dp}{dt} = \frac{W}{T(v_2 - v_1)} \qquad ...(1)$$

Since $v_2 >> v_1$, we may neglect v_1

$$\therefore \quad \frac{dp}{dT} = \frac{W}{Tv_2}$$

or

$$w = Tv_2 \cdot \frac{dp}{dT} \qquad ...(2)$$

The energy W is consisting of two parts.

(*i*) The work function of the surface, *i.e.*, energy required for ejecting N electrons just outside the metal, *i.e.*, in crossing the potential barrier. This energy will be equal to $N\phi$, where ϕ is the work function of the surface.

(*ii*) The work done in bringing the ejected electrons into electron cloud already present outside the metal. This energy is Pv_2.

$\therefore$ For 1 gm mole of the electrons

$$Pv_2 = RT = NkT$$

where k is Boltzman's const.

$$\therefore \quad W = N\phi + NkT \qquad ...(3)$$

From (2) and (3),

$$\therefore \quad N(\phi + kT) = Tv_2 \frac{dp}{dT}$$

But, $$v_2 = \frac{NkT}{P}$$

$$\therefore \quad N(\phi + kT) = T \cdot \left(\frac{NkT}{P}\right) \cdot \frac{dp}{dT}$$

or $$(\phi + kT) = \frac{kT^2}{P} \cdot \frac{dp}{dT}$$

or $$\frac{dp}{P} = \left(\frac{\phi + kT}{kT^2}\right) dT$$

Integrating $$\int \frac{dp}{P} = \int \frac{\phi + kT}{kT^2} dt$$

$$= \int \frac{\phi}{kT^2} dt + \int \frac{dT}{T}$$

$$\therefore \quad \log_e P = \int \frac{\phi}{kT^2} dt + \log_e T + \text{const. (Say} \log A_1)$$

$$\therefore \quad \log_e P - \log_e T - \log_e A_1 = \int \frac{\phi}{kT^2} dT$$

$$\therefore \quad \log_e \frac{P}{A_1 T} = \int \frac{\phi}{kT^2} dT \qquad ...(4)$$

Richardson suggested that ϕ depends on T and if ϕ_0 is the work function at 0°k, than

$$\phi = \phi_0 + \frac{3}{2} kT$$

$$\therefore \quad \int \frac{\phi}{kT^2} dT = \int \frac{\phi_0 + \frac{3}{2} kT}{kT^2} dT$$

$$= -\frac{\phi}{kT} + \frac{3}{2} \log_e T$$

$$= -\frac{\phi}{kT} + \log_e T^{3/2}$$

Putting the values in equation (4),

$$\log_e \left(\frac{P}{A_1 T}\right) = -\frac{\phi_0}{kT} + \log_e T^{3/2}$$

or $$\log_e \frac{P}{A_1 T^{4/2}} = -\frac{\phi_0}{kT}$$

or $$P = A_1 T^{5/2} e^{-\phi_0/kT} \qquad ..(5)$$

If the number of electrons/unit volume = n

$$\therefore \quad P = nkT \qquad ...(6)$$

Equating (5) and (6)

$$nkT = A_1 T^{5/2} e^{-\phi_0/kT}$$

$$\therefore \quad n = \frac{A_1}{k} T^{3/2} \cdot e^{-\phi_0/kT} \qquad ...(7)$$

Let us assume that the perfect gas laws are equally applicable to an electron cloud, the mean energy of the electrons

$$= \frac{1}{2} m_0 v^2 = \frac{3}{2} kT$$

$$\therefore \quad v = \left(\frac{3kT}{m_0}\right)^{1/2}$$

But from Kundsen's relation, the number of electrons passing/unit area in one second is

$$n_0 = \frac{nv}{\sqrt{6\pi}}$$

$$\therefore \quad n_0 = n \cdot \left(\frac{3kT}{6\pi\, m_0}\right)^{1/2} = n\left(\frac{kT}{2\pi\, m_0}\right)^{1/2}$$

Putting the value of n from Equation (7)

$$n_0 = \left[\frac{A_1}{k} T^{3/2}\, e^{-\phi_0/kT}\right]\left[nT^{1/2}\left(\frac{k}{2\pi\, m_0}\right)^{1/2}\right]$$

$$= \left[\frac{nA_1}{k} \times \left(\frac{k}{2\pi\, m_0}\right)^{1/2}\right] T^2\, e^{-\phi_0/kT}$$

Let $\frac{nA_1}{k} \times \left(\frac{k}{2\pi\, m_0}\right)^{1/2} = A_2 = \text{constant}$

$$\therefore \quad n_0 = A_2\, T^2\, e^{-\phi_0/kT}$$

Taking the charge on electron = e, the current density due to electron emission

$$J = n_0 e = A_2\, eT^2\, e^{-\phi_0/kT}$$

$$= AT^2\, e^{-\phi_0/kT} \quad \text{...(8)}$$

If area of the emitter surface is S, their current is given by

$$I = JS = AST^2\, e^{-\phi_0/kT}$$

Taking $\quad \phi_0/k = b$

$$\therefore \quad I = AST^2\, e^{-b/T} \quad \text{...(9)}$$

In this case, $(A_2 e) = A$ = constant.

This is the Richardson's equation for thermionic emission.

Example 14. *An oxide coated cathode is disc shaped and has a radius of 0.015 cm. It is coated only on its one side. If the observed saturation current is 6.46 × 10^{-3} amperes, calculate the operating temperature of the cathode.*

Given: $\quad A = 0.01 \times 10^4$ amp. m^2 – K^2

$\quad a = 11{,}600\ K$

Solution: From Richardson's equation,

$$I = AST^2\, e^{-b/T} \quad \text{...(1)}$$

Now, $\quad I = 6.46 \times 10^{-3}$ Amp.

$A = 0.01 \times 10^4$ amp/m^2 – K^2

$b = 11{,}600$ K, $r = 0.015 \times 10^{-2}$ m

$S = \pi r^2 = 3.142 \times (0.015 \times 10^{-2})^2\ \text{m}^2$

$= 7.07 \times 10^{-6}\ \text{m}^2$

Putting in (1),

$$6.46 \times 10^{-3} = 0.01 \times 10^4 \times 7.07 \times 10^{-6}\, T^2\, e^{-11600/T}$$

or $\quad T^2 e^{-11600/T} = 9.333$

Taking log to the base 10

or $$2\log_{10} T - \frac{11600}{T} \times \frac{1}{2.3026} = 0.970$$

$$\therefore\ 2\log_{10} T - \frac{11600}{T} = 0.970 \qquad ...(2)$$

Exact value of T can be found from equation (2), only if we assume approximate value of T which in the case of oxide coated filament is approx. 1000°K.

$$\therefore \quad \text{L.H.S.} = 2\log_{10}(10)^3 - \frac{5030}{1000}$$

$$= 6 - 5.03 = 0.97$$

which is R.H.S. of Equation (2).

So, T = **1000°K.**

Example 15. *In Richardson equation, the values of constant A and b of tungsten is* 60 amp/cm^2/K *and* 52.00K. *Calculate the emission current at* 1600°K *and* 3200°K. *How do you explain such an enormous increase in current value*?

Solution: From Richardson's equation, the emission current density is given by

$$J = AT^2 e^{-b/T}$$

At 1600K,

$$J = 60 \times (1600)^2 \times e^{-52400/1600}$$

Taking log on both sides

$$\log J = \log 60 + 2 \log 1600 - 32.75 \log_{10} e = 2.7183$$

$$J = 0.9186 \times 10^{-6}\ \text{Amp.} = 0.9126\ \text{A}$$

At 3200 K

$$J = 60(3200)^2 \times \frac{-52400}{3200}$$

or $$\log J = \log 60 + 2 \log 3200 - 16.375 \log_{10} e$$

$$\therefore \quad J = \mathbf{47.56\ Amp.}$$

The enormous increase in current can be explained as under.

In the emission equation, the increase in temperature is due to the combination of squared temperature T^2 and exponential term $e^{-b/T}$. Due to this, the emission of electron increases very rapidly with even a small increase in temperature, there by increasing the current by a large amount.

6.26. THERMIONIC EMITTERS

The thermionic emitters generally used in vacuum tubes are of the following three types:

(*i*) *Pure tungsten emitter*. Pure tungsten is an ideal material to be used as emitter in high vacuum tubes. It has a high melting point (3655 K), low vapour pressure and a work function ~ 4.52 eV. It can, therefore, be operated at a high temperature and yield an appreciable emission current. Its emission efficiency is 8.5 ma per watt at operating temperature ~ 2600 K. The efficiency, however, lowers with absorption of gases and so it can be used only in high vacuum tubes. An important property of tungsten emitter is that its surface is not damaged by bombardment of positive ions.

(*ii*) *Thoriated tungsten emitter*. The thoriated tunsten emitter consists of tungsten containing 1 to 2% of thorium oxide and carbon which acts as a reducing agent. A thin layer of thorium is thus formed on the surface of tungsten. This layer being electropositive with respect to tungsten, lowers the work function of the material and helps to give a uniform supply of electrons at about 1900 K and has an emission efficiency of 100 ma per watt. Thoriated tungsten emitters are likely to be damaged due to bombardment by positive ions and can, therefore, be used only under conditions of high vacuum and low anode voltage.

(*iii*) *Oxide coated emitters*. An oxide coated emitter consists of a nickel or nickel alloy cylinder the outer surface of which is coated with a mixture of barium and strontium oxide. The cylinder is indirectly heated by placing a heating filament inside it and well insulated from it. It gives a continuous supply of electrons at about 100 K and its emission efficiency ranges 100 to 10,000 ma per watt.

Because of this, the oxide coated emitters are now widely used. They, however, have a tendency to lose their activity due to bombardment by positive ions and are, therefore, used in tubes with low anode voltage. Also, these emitters give out some gases during operation and so cannot be used in high vacuum tubes.

Example 16. *A sample of tungsten has an area of* 1 cm^2. *What is the total emission current from it when it is heated to 3500 K. Given:*

(*i*) *Emission constant* $A = 60.2 \times 10^4\ Am^2\ K^{-2}$

(*ii*) *Work function* $\phi_0 = 4.52\,eV$

(*iii*) *Boltzman constant* $k = 1.38 \times 10^{-23}\ JK^{-1}$

Solution: Current density $J = AT^2 e^{-\phi_0/kT}$

$\therefore$ Emission current $I = JS$

where S is the area of the emitting surface

$$\therefore \qquad I = JS = AST^2 e^{-\phi_0/kT}$$

Putting the values given above, we have

$$S = 1\,\text{cm}^2 = \left(\frac{1}{100}\right)^2 = 10^{-4}\ \text{sq. metre}$$

$$\phi_0 = 4.52\ \text{eV} = 4.52 \times 1.6 \times 10^{-19}\ \text{J}$$

and

$$I = 60.2 \times 10^4 \times 10^4 \times (3500)^2 \times e^{\left(\frac{-4.52 \times 1.6 \times 10^{-19}}{1.38 \times 10^{-23} \times 3500}\right)}$$

$$= 60.2 \times (3500)^2 \times e^{-14.97}$$

Taking log on both sides

$$\therefore \qquad \log I = \log 60.2 + 2 \log 3500 - 14.97 \log_{10} e$$

Now $\qquad e = 2.7183$

$\therefore \qquad$ **I = 231.7 Amp**

QUESTIONS AND PROBLEMS

1. Give Bohr's theory of stopping of a charged particle.
2. Give the various processes by which nuclear radiations interact with matter.
3. Give briefly the theory of 'Bremsstrahlung'.
4. What is stopping power, impact parameter and range of charged particles? Give an expression for stopping cross-section.
5. How γ-rays are absorbed in matter? Give in detail the 'pair production' process. What is the conclusion of 'pair production'?
6. What is attenuation coefficient? How it can be found experimentally?
7. What are the main observations in photoelectric effect experiments which are explained on the basis of corpuscular theory and not from wave theory? Discuss these effects.
8. Describe the basic Einstein's photoelectric equation and explain its importance. Give the working of photoelectric cell.
9. What is the difference between Comptoon effect and photoelectric effect? Explain how we get a photon of different wavelength than the incident photon in case of Compton effect. Also calculate the direction of recoil electron.
10. What is Compton effect? Derive an expression for the Compton shift.
11. Explain Compton effect and show the factor on which Compton shift depends.
12. Explain Compton effect and show that Compton shift depends only on the angle of scattering and is independent of the wavelength of the incident photons.
13. Calculate the photo electric work function for Pt. in electron volt. given that the threshold wavelength is 2300Å and $h = 6.62 \times 10^{-27}$ erg-sec. [**Ans.** 5.4 eV.]

14. For given metal surface, the limiting wavelength for photoelectric emission is 5800 Å. Deduce the kinetic energy of electrons ejected by light of I = 4500 Å? **[Ans.** 99 eV.]

15. Find the energy and momentum of X-ray photon whose wavelength is 2×10^{-11} m. **[Ans.** 9.95×10^{15} *J*, 3.32×10^{-28} kg.m/sec]

16. In an experiment on compton scattering, X-rays of wavelength 10^{-10} m are incident. Calculate the wavelength of the *X*-rays scattered in the backward direction. **[Ans.** 1.048×10^{-10} m.]

17. The photo-electric threshold of copper is 3200Å. If ultra violet light of wavelength 2500Å falls on it, find (*a*) the maximum K.E. of the photoelectrons ejected. (*b*) maximum velocity of photoelectrons (*c*) value of work function. **[Ans.** 1.087 eV, 6.18×10^{5} ms^{-1}, 3.88 eV.]

18. X-rays with λ = 100Å are scattered from a carbon block. The scattered radiation is viewed at 90° to the incident beam (*a*) What is the compton shift $\Delta\lambda$? (*b*) What K.E. is imparted to the recoiling electrons. **[Ans.** 0.242Å, 293 eV]

19. If the electron beam in O.T.V. is accelerated by 10,000 *V*, what is de Broglie wavelength? **[Ans.** 0.1228Å.]

20. X-rays of λ = 1.0Å are scattered at 60° from the free electrons. Calculate the shift in wavelength and energy of recoil electrons. **[Ans.** 0.012Å, 147 eV]

CHAPTER 7

RADIATION DETECTORS

7.1. INTRODUCTION

We know that some atoms are radioactive, decaying by the emission of α-particles or β-particles, either of which may be accompanied by γ-rays. Many other types of nuclear reactions may be produced in the laboratory. In these, a wide range of 'nuclear radiations' may be emitted, the most common being protons, α-particles, neutrons and γ-rays. To study these nuclear processes adequately, suitable *detectors* must be employed to give information about the nuclear radiations which are emitted. There are three main groups of nuclear radiation, (*i*) *charged particles* (protons, α-particles, β-particles, etc (*ii*) *uncharged particles* (neutrons) (*iii*) *electromagnetic radiation* (X-rays or γ-rays).

The fundamental requirement for the detection of any nuclear radiation is the dissipation of energy in the detector by that radiation. This energy dissipation may be caused by ionization of atoms in the detector (as in the ionization chamber, Geiger counter, cloud chamber and nuclear emulsion) or by excitation of the atoms without the actual removal of orbital electrons (as in the scintillation counter). Only charged particles are capable of producing either of these processes directly. Uncharged particles do not themselves produce ionization or excitation and some other technique is to be used for their detection.

7.2. THE IONIZATION CHAMBER

In its passage through any matter, a charged particle will suffer many collisions with the atoms in the material. These collisions should not be thought of as being 'direct hits' in the way one speaks of collisions between large-scale objects. Rather, it should be remembered that these atomic particles are all charged and that electrostatic forces are acting between them all the time as they approach one another. Thus, on the atomic scale, there is no real distinction between a 'collision' and a 'near miss'. The consequences of a collision between a charged particle and a neutral atom depend principally on how closely they pass by one another.

As it approaches, the charged particle exerts an increasing force on each of the electrons of the atom. Unless this force becomes great enough to remove an electron, the charged particle will lose none of its energy in such a collision. However, in a very close encounter, an electron may be dislodged from an atom leaving it in an *ionized* state. When this occurs, the energy necessary for ionization is supplied at the expense of the kinetic energy of the charged particle. On the average, an energy of about 32 electron-volt is required to produce an ion pair, that is, a positively charged atom and a free electron. This means that a charged particle with an energy of 1 MeV will produce about 30000 ion pairs along its path before it comes to rest.

The 30000 ion pairs formed in this process carry a charge which is large enough to be detected with ease in a variety of ways.

In an ionization chamber, the ions are produced in a gas between two electrodes as shown in Fig. 7.1. If a potential difference is maintained between the positive electrode A and the negative electrode B, the electrons produced by ionization are drawn towards A and the positive ions are accelerated

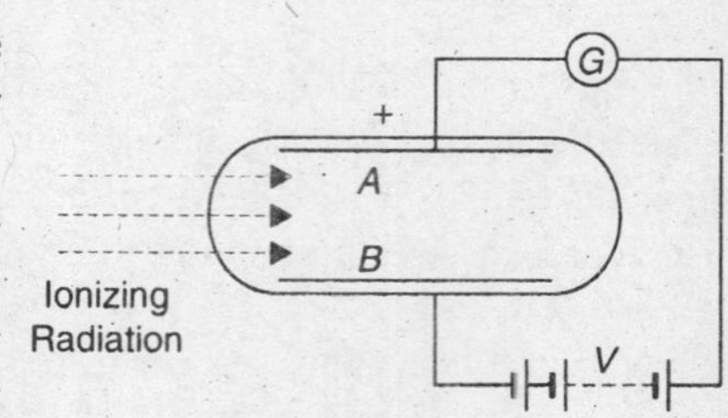

Fig. 7.1

towards B. If the potential difference V is varied while a constant intensity of ionizing radiation enters the chamber, the current indicated by the galvanometer G varies as shown in Fig. 7.2.

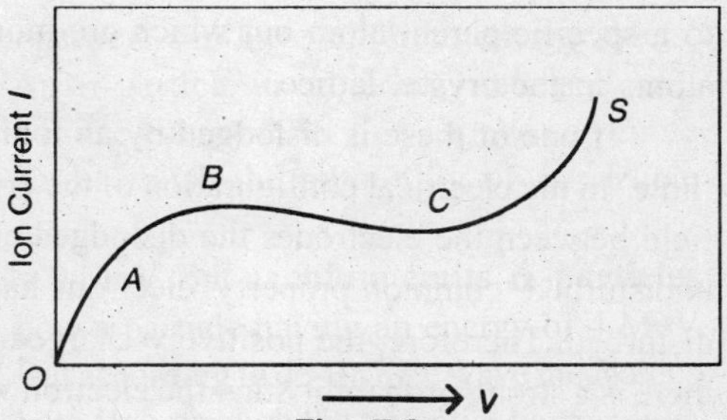

Fig. 7.2

When the value of V is low, the electric field between the electrodes is small. Thus, under these conditions (the region OA of Fig. 7.2), the ions produced in the chamber achieve only relatively low velocities. In consequence, many of the positive ions and electrons recombine to form neutral atoms before they reach the electrodes. As V is increased, the fraction of ions recombining becomes smaller and smaller until, at a potential corresponding to B, virtually all of the ions produced reach the electrodes and contribute to the current recorded by the galvanometer. Further increase in potential difference in the range between B and C of Fig. 7.2 results in no increase in current. The current in this region is equal to the total charge produced by ionization every second and is called the *saturation current*. It is a measure of the energy extracted from the ionizing radiation every second.

Current in ionization chambers are very small. For example, a 2-MeV α-particle which stops in the chamber produces about 60,000 ion pairs, resulting in a charge transfer through the galvanometer of 60,000 electronic charges, that is about 10^{-14} Coulomb. Thus, one such α-particle stopping in the chamber every second would produce an ionization current with an average value of 10^{-14} ampere. A good galvanometer might detect a current of about 10^{-10} ampere so that such instruments can only be used when some tens of thousands of particles per second pass through the ionization chamber. More sensitive electronic devices have been developed and it is now possible to replace the galvanometer G in Fig. 7.1 by an electronic amplifier which can even detect the pulse of current resulting from the passage of a single charged particle through the chamber.

If an ionization chamber is operated in the saturation current region and if the path length of the ionizing radiation through the chamber gas is sufficient to cause every ionizing particle to lose all its energy, the pulse of current due to each charged particle has a magnitude proportional to the particle energy. In this way, the ionization chamber can be used to determine, at least approximately, the energies of particles detected by it. If the particles do not enter too frequently, the current pulses may be counted to record the number of charged particles entering the ionization chamber. If large number of particles, all of the same energy, enter the chamber every second, the average current recorded is a measure of the number of particles arriving per second. To determine this number accurately, it is necessary to know the ionization charge produced by each particle.

Different types of particles of the same energy produce very different quantities of ionization in traversing the same thickness of a gas at a particular pressure. For example, 5 MeV α-particles produce about ten times as much ionization per unit path length as do 5 MeV protons. It follows that an ionization chamber intended to stop protons must be larger than an α-particle chamber or, alternatively, must contain gas at a higher pressure to provide more molecules per unit path length for collisions. An electron produces only about 0.01 times as many ion pairs per unit path length as are produced by an α-particle of the same energy. γ-rays can also be detected in an ionization chamber, since they can eject high energy electrons from atoms, these electrons in turn ionizing the gas. γ-rays produce about 10^{-4} times as many ion pairs per unit path length as do α-particles of the same energy. In consequence, high gas pressures are commonly used for γ-rays detection.

In an ionization chamber, the ion pairs are formed by the removal of an electron from an atom of the filling gas, the ejected electron forming the negative ion and the atom minus an electron forming the positive ion. In these days, *semiconductor detectors,* may be regarded as ionization chamber with **solid material** between the electrodes. This solid material is almost always of the semiconductor **silicon**. In such a material there is a small minority of the atomic electrons which are not bound to a specific parent atom but which are more or less the 'common property' of the aggregate of atoms in the crystal lattice.

If one of these is dislodged by an ionizing particle it will leave behind a positively-charged 'hole' in the electrical configuration of the crystal lattice. Under the influence of the applied electric field between the electrodes the dislodged electron will move towards the positive electrode. The undisturbed 'common property' electrons have, however, a certain amount of mobility, even if not disturbed. Therefore, the positively-charged 'hole' is soon filled. But because of the applied field, there is a strong probability that the electron which does this will come from the region a little nearer the negative electrode; in this way, the 'hole', or absence of an electron, may move towards the negative electrode. Thus, the effect is just like that which is observed in the ionization chamber.

Semiconductor detectors are finding many applications in the measurements in modern physics. They have a number of advantages over gaseous ionization chambers. Their greater density allows the complete stopping of particles of greater energy than can be stopped in gas ionization chambers; a range of 1 meter in air is roughly equivalent to a range of 1 mm in silicon. The accuracy with which the energy may be measured is greater in the semiconductor counter. This is because the average energy required to produce one ion (*e.g.*, electron positron pair) in silicon is very low. It is 36 eV compared with about 30 eV for gases.

So the pulse recorded for each event is composed of about 8 times as many electrons as the pulse produced by a comparable event in a gas ionization chamber. It is well-known result, from statistical theory, that the relative amount of variability in these larger pulses will be much reduced; they are 'sharper' and, therefore, capable of being more readily resolved.

7.3. PROPORTIONAL COUNTER

Ionization chamber is not able to detect low energy ionizing particles, as they give out very small light pulses. In order to increase the light of these pulses the field must be increased between the electrodes with parallel plate chambers.

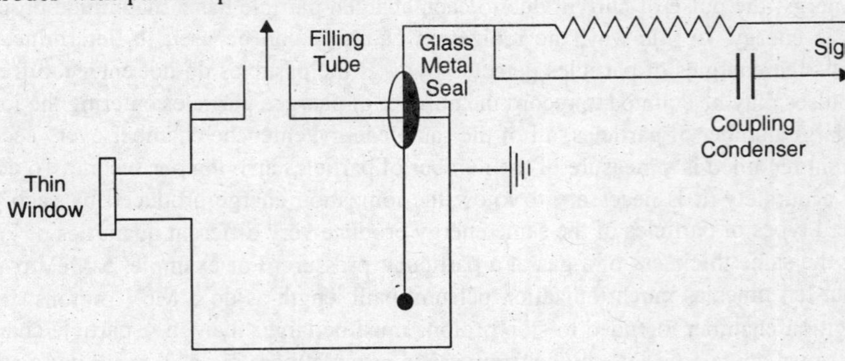

Fig. 7.3

If the voltage between the electrodes is increased beyond saturation point, the primary ions due to the collision with gas atoms will produce secondary ions. So the primary ionization is multiplied by a factor which is dependent on the applied voltage and the geometry of the apparatus.

The chamber with this modification is called **proportional counter.**

For cylindrical geometry, the electric field required for producing secondary ionization is given by the relation,

$$E = \frac{V}{r \log_e b/a}$$

where $a \rightarrow$ radius of the wire.

$b \rightarrow$ radius of the cylinder.

$r \rightarrow$ distance of central wire from the radial electric field.

$V \rightarrow$ positive potential of the central wire w.r.t. cylinder.

Working. The electrons from the ion pairs created anywhere in the centre move towards the positively charged wire to give more ions and consequently an 'avalanche' of ionization in the central part.

Gas multiplication corresponds to the **number of electrons** coming to the wire for **each** initial electron and is usually between 10^1 to 10^3.

These electrons come to the central wire very rapidly but the voltage pulse developed at the central wire is dependent on the motion of the heavy positive ions, majority of which are created very near the anode, so cancelling the effect of electrons. The initial pulse is found by the drift velocity of the positive ions as they move away from the anode. This velocity at first is quite high as the ions are in a very strong field and so the pulse quickly achieves half the value of the amplitude. However, for most of the drift path, the positive ions move in a weak field and may take about 10^{-3} sec to come to cathode.

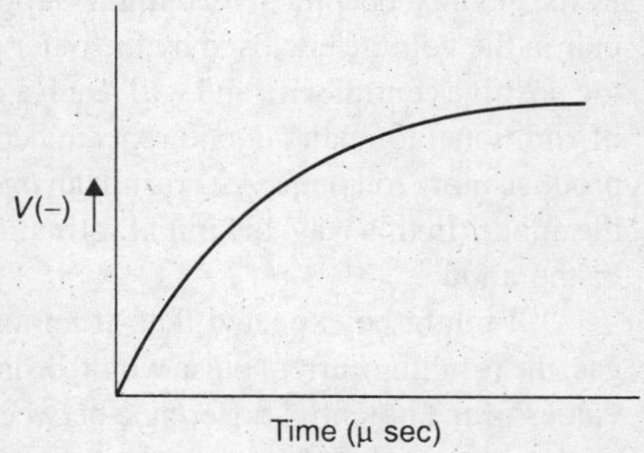

Fig. 7.4

It may be noted that the shape of the pulse in this type of counter is independent of the position of the ionizing track since the main multiplication is in the high field region.

In order to limit the multiplication, a proportional counter is filled with 90% argon and 10% methane at atmospheric pressure. Methane helps in limiting the multiplication by absorbing photons emitted by electrons. In some counter, a continous flow of gas is maintained.

Proportional counters have been found to be useful for measurement of soft radiations as in case of β-rays, mesons and fast protons. Slow heavy particles are also easily detected.

Large proportional counters have been used to measure the low activity of radio-samples. Proportional counters have an advantage over scintillation counters as they have good energy resolution.

7.4. THE GEIGER-MULLER COUNTER

In Fig. 7.2, showing the variation of ion current as a function of potential difference, the curve is seen to rise again for voltages greater than the value at C. In the region C to S, electrons produced by ionization are accelerated to such a high velocity that they may themselves produce further ions by collision. This process, called gas amplification, results in a greater current for a given mass of energy by the original particle and consequently the detection device is more sensitive. This phenomenon of gas amplification provides the basis for the operation of the *Geiger-Muller counter,* which is illustrated in Fig. 7.5.

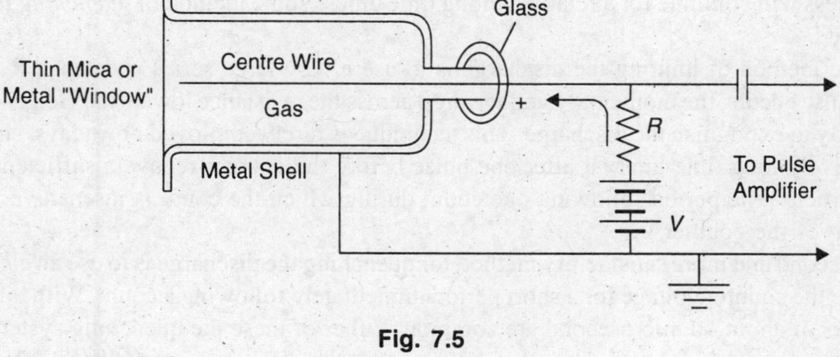

Fig. 7.5

The origins of this device can be traced back to 1908 when Rutherford and Geiger arranged a cylinder and co-axial wire with a potential difference between them. When charged particles entered the volume enclosed by the cylinder, it was found that the pulse of current through the gas due to a single particle was magnified sufficiently to enable it to be detected by an electrometer of quite moderate sensitivity.

In a counting tube such as that shown in Fig. 7.5, the central wire is maintained at a high positive potential with respect to the outer cylinder. If the central wire is very thin, the electric field in its vicinity becomes very high. Suppose an incident ionizing particle produces one single ion pair in the volume enclosed by the outer cylinder. The resulting electron will be accelerated towards the positive central wire and will reach a relatively high velocity. It may itself produce a large number of additional ion pairs due to repeated collisions. These new electrons are also accelerated and may produce more ion pairs. As a result, an *avalanche* occurs and a very large number of electrons reaches the anode. In this way, the initial formation of a single ion pair results in a very large pulse of current to the anode.

It might be expected that, if an ionizing particle initially produces a trial of ion pairs in the gas, the resulting current pulse would be larger than that originating from a single ion pair. For moderate values of the potential difference between anode and cathode this is found to be the case. However, as this potential difference is increased, this variation in the size of the current pulse gradually disappears. Ultimately, it is found that the size of the current pulse becomes effectively independent of the initial density of ionization for quite a range of potential differences. The counter then gives a full avalanche of nearly constant size for any particle counted. In this region, it is said to operate on the *plateau* or in the 'Geiger region'. This is the normal mode of operation since the counter sensitivity is highest and the counting efficiency is independent of tube voltage over a reasonable range, so that the supply voltage is not critical. A counter operated in this fashion cannot give any information about the type of particle producing the count or its energy. The first counter of this type was constructed by Geiger and Muller in 1928, and these devices are now most commonly known as *Geiger counters.*

If a resistor is connected in the Geiger counter circuit as shown in Fig. 7.5 the current pulse I flowing through it produces a voltage pulse which can be amplified by an electronic amplifier and counted by an electronic counter.

The positive ions produced during an electron avalanche in a Geiger counter are subjected to the same electrical forces as the electrons. However, because of their much larger masses, they experience smaller accelerations and hence may be considered as being virtually stationary during the electron avalanche. After this avalanche, these positive ions, which are principally formed close to the central wire, move outwards towards the cathode. When they reach it, some time later they have considerable energies and some of them will release electrons as they bombard the cathode surface. In this way, a second avalanche is started, and thus, a succession of discharges will follow. This process will continue for a relatively long time unless some method of *quenching* the discharge is provided.

One method of limiting the discharge is to use a very large series resistance R. When large current pulse occurs, the momentary voltage drop across the resistance lowers the Geiger tube voltage sufficiently to extinguish the discharge. This technique is rarely employed nowadays, since it results in a relatively long time interval after one pulse before the counter recovers sufficiently to detect a new particle. The period following one count during which the count is insensitive is called the *dead time* of the counter.

A second and more satisfactory method for quenching the discharge is to use an electron circuit to remove the counter voltage for a short period immediately following a count. With this technique, dead times of about 40 microsecond are common. Either of these the quenching system is referred to as *external quenching.* It is also possible to quench the discharge internally by introducing into the counter tube a small quantity (of the order of 10 per cent) of a polyatomic gas such as ethyl alcohol. This *quenching agent,* in effect, prevents the release of electrons from the cathode surface by positive ion bombardment. Thus, only one avalanche occurs for each ionizing particle passing through the tube.

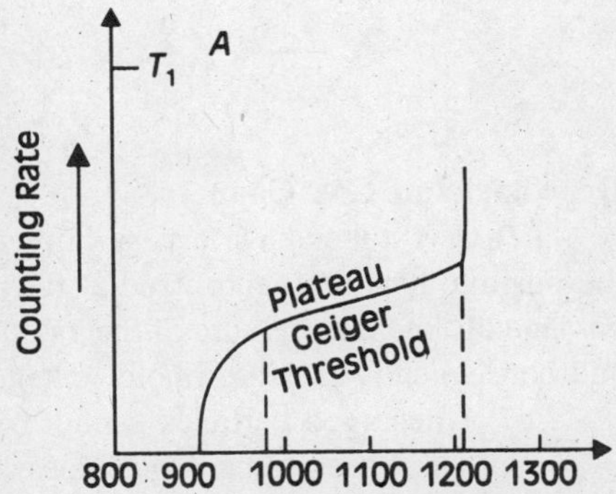

Fig. 7.6

Geiger counters may be made in a variety of sizes but they customarily have anode wire about a tenth of a millimeter in diameter. A common filling gas is argon at a pressure of about 10 cm of mercury. For internal quenching, a partial pressure of about 1 cm of ethyl alcohol is commonly used. Such internally quenched counters operate with potential differences of one to two thousand volt. More recently, counters whose quenching agent is a halogen gas have been produced. A typical filling for a counter of this type is 0.01 cm partial pressure of chlorine or bromine in 10 cm pressure of neon. Special precautions have to be taken in selecting the right electrode materials because of the extremely high chemical reactivity of the halogens. Counters of this type operate satisfactorily with potential differences of a few hundred volt.

It is instructive to consider the curve of count rate as a function of tube voltage. A typical curve of this type is shown in Fig. 7.6. If the voltage is low, a single initial ion pair does not produce a full avalanche although some gas amplification still occurs. The size of the current pulse at the anode is then proportional to the number of ions initially formed. In consequence, a detector of finite sensitivity will record the large pulses but miss the small ones, resulting in a low count rate. Counters operating in this region can be used to differentiate between different types of particle. With higher tube voltages, the count rate increases and then becomes constant for a range of potential differences. This range is, in fact, the *plateau region* which was mentioned earlier.

When the voltage is raised above this plateau or Geiger region, the quenching action is incomplete and one initial particle may give rise to more than one count. The count rate increases due to these spurious output pulses and at slightly higher voltages the counter goes into continuous discharge. This region of voltage is of no practical use and damage to the tube may **ensue**.

For the detection of α-particles and slow β-particles, counters with thin windows are employed. More penetrating particles will pass through a tube wall of normal thickness. The counting efficiency of a Geiger counter is high. This means that extremely few ionizing particles can pass through the counter without being recorded, provided they do not arrive during the deal time following a previous count. Geiger counters can be used to count γ-ray photons by detecting electrons ejected from the walls of the tube. However, the efficiency in this case is usually less than 1 per cent.

Efficiency of Counting. It is defined as the ratio of the number of observed counts per unit time (N_0) to the number of ionizing particles (N) which passed through the counter during that time

$$\therefore \qquad \eta = \frac{N_0}{N} \qquad \text{... (1)}$$

If 100 particles pass through and only 85 are detected, then the efficiency is 0.85 or 85%.

Dead time of the Counter. In active time or **Dead time** σ of a counter is defined as *the time interval during which it is insensitive* to ionizing particles. In case of a high speed circuit, σ is normally the time required for the restoration of the voltage across the counter tube after its discharge.

Let N_0 be the number of counts/minute, so that $N_0\sigma$ will correspond to **inactive part** of that minute and $1 - N_0\sigma$ will be active part of that time.

Now during 1 minute N_0 counts are observed, while if all the particles N are detected in that time, then the efficiency of the counter is

$$\eta = \frac{N_0}{N} = \frac{1 - N_0\sigma}{N}$$

or

$$N = \frac{N_0}{1 - N_0\sigma} \quad ...(2)$$

Drawbacks of GM Counter

(*i*) G.M. tubes have a very limited life due to the decomposition of organic vapours when the positive ions are neutralized at the cathode and break up into molecular fragments. With the passage of time the pressure of the gas slowly increases due to the presence of molecular fragments and consequently the threshhold voltage of the counter is raised.

(*ii*) Uncharged particles cannot be detected.

(*iii*) Permanent record of nuclear phenomenon cannot be kept.

(*iv*) The G.M. counter remains insensitive for sufficient by long time (~ 500 μs) after each ionizing event. this limits the use of G.M. counter for lower counting rates only.

(*v*) It does not give the information about the energy of the particle counted.

Advantages of GM Counter

(*i*) It is very sensitive device for detecting charged particles like α or β particles.

(*ii*) It produces a large pulse requiring no amplificator.

(*iii*) It produces constant output pulse size, independent of initial ionization.

(*iv*) The G.M. counter equipped with transistorised electrons is portable and is used for mineral prospecting and cosmic ray works.

Comparison between G.M. counter and Proportional Counter

	G.M. counter		Proportional Counter
1.	It is based on the principle of indirect ionization.	1.	It based on the principle of gas multiplication.
2.	It cannot detect neutral particles.	2.	It can detect neutral particles.
3.	Very high voltage is required by G.M. counter.	3.	Very high voltage is not required by proportional counter.
4.	Low energy radiations cannot be detected.	4.	Low energy radiations like X-rays and γ-rays can be detected.
5.	Pulse height does not depend on the energy of incident particles.	5.	Pulse height depends on the nergy of incident particles.
6.	Its resolution is low.	6.	Its resolution is high.
7.	It can work without amplifier.	7.	Amplifier is always needed.

7.5. THE SCINTILLATION COUNTER

In some early atomic physics experiments, α-particles were detected by the light flashes, or *scintillations* produced when they struck certain materials such as zinc sulphide. The individual flashes were counted by an experimenter watching a small section of the scintillating screen through a low-power microscope. This technique was extremely laborious and is clearly impracticable for any but very low count rates. In consequence, scintillation counting was abandoned with the advent of the Geiger counter.

In 1947, Marshall and Coltman in the U.S.A. and Kallman in Germany introduced a new technique in which the light flashes from a suitable scintillator were detected by a device known as a photomultiplier tube. The resulting *electrical* impulses could then be amplified and recorded by means similar to those employed with Geiger counter tubes. This technique has proved to be very versatile and capable of handling count rates far in excess of those possible with Geiger counters. The scintillation counter is now one of the commonest devices for the detection of nuclear radiations.

The complete scintillation counter consists of three basic parts:

1. The scintillating material or phosphor which produces a tiny light flash when a charged particle passes through it.
2. The photomultiplier tube which detects the light flash and produces an electrical impulse.
3. Amplifiers and electronic circuits which record and count the electrical impulses from the photomultiplier.

The passage of a charged particle through the scintillating material causes some of the scintillator atoms to become excited. As the electrons in these excited atoms return to their ground states, photons are emitted. The number of photons will depend on the energy lost by the charged particle traversing the phosphor, so that the total intensity of the light flash will depend on the energy given up by the charged particle. In many cases, the light flashes are not sufficiently intense to be seen by the human eye but they can be detected by a sensitive photomultiplier tube.

Many materials have been investigated to find the most suitable scintillation phosphors for various applications. **Cesium iodide** in crystalline form is very popular as a phosphor for the detection of protons and α-particles. A small quantity of thallium is added to such a crystal to 'activate' it and produce the optimum scintillation effect. The action of activators, which replace a few of the usual crystal atoms, is somewhat complex. It suffices to say that minute quantities of impurities in a crystal can have a profound effect on its scintillating properties. If the light intensity produced in a scintillator is to be proportional to the energy of the incoming particle, the phosphor must be thick enough to stop the particle entirely. Particles with energies of the order of 10 MeV or less will be stopped by phosphors a few millimeter thick.

Sodium iodide, activated with thallium, is an extremely useful scintillator for the detection of γ-rays. Because of the relatively high atomic number of the iodine atoms in this crystal, there is a large probability of a γ-ray photon releasing an electron which can then produce a scintillation in the crystal. Since γ-rays are extremely penetrating, quite large crystals are often employed, linear dimensions of about 12 cm being common.

Fig. 7.7 shows a schematic representation of a scintillator and photomultiplier. The photons from the scintillator strike the semi-transparent *photocathode* in the photomultiplier tube and cause photoelectric emission. Ideally each incident photon would produce one electron at the photocathode. In practice, the efficiency of electron production is about ten per cent. Electrons produced at the photocathode can be accelerated from it towards the first dynode D_1 which is held positive with respect to the cathode. The number of electrons emitted at the photocathode will be proportional to the number of photons incident on it and therefore, to the energy of the original incoming particle.

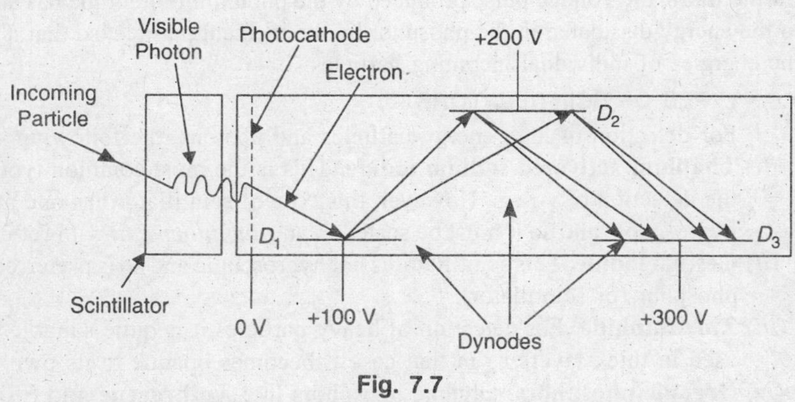

Fig. 7.7

The dynode D_1 has a specially prepared surface favourable to *secondary emission*. In the process of secondary emission, an electron hitting a surface can result in the ejection from it of more than one electron. The secondary emission ratio, that is, the number of secondary electrons emitted per incident electron, depends on the surface treatment of the electrode and on the potential difference through which the incident electron has been accelerated. In practice, secondary emission ratios of the order of four (4) are employed.

The electrons from D_1 which are emitted with near zero velocity, are accelerated towards the second dynode D_2 where secondary emission again occurs. In a practical photomultiplier tube, ten or eleven dynodes are usually employed, resulting in a large 'multiplication' of the electrons emitted at the photocathode. For example, if the secondary emission ratio is 4 and there are eleven dynodes, 4^{10} electrons reach the last dynode for each electron emitted from the photocathode. The tube is said to have a gain of 4^{10} or about 10^6. Thus, a single electron emitted by the photocathode can result in a detectable current pulse at the last dynode which is usually called the **collector** or the photomultiplier.

The design of commercially available photomultiplier tubes varies considerably. One type of tube is illustrated in Fig. 7.8. The dynodes are arranged to produce a focussing action for the electrons accelerating from one dynode to the next. The focussing electrode and grills have the effect of directing as many as possible of the photoelectrons on to the first dynode. Multiplication takes place at the dynodes 1 to 10 and the electron pulse is collected at electrode 11. In all types of photomultiplier tube a very high vacuum must be obtained and care must be taken to ensure that as many as possible of the photons produced in the scintillation process are able to reach the photocathode.

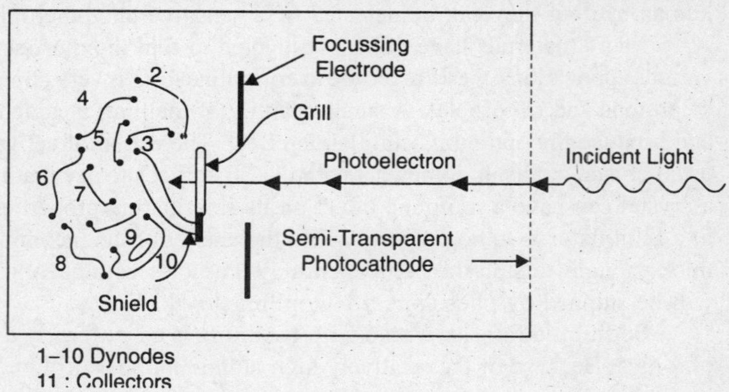

Fig. 7.8

Scintillation counters are capable of detecting particles whose times of arrival are separated considerably by less than 10^{-6} second. They are, therefore, much 'faster' than Geiger counters. Furthermore, the voltage pulse produced by the photomultiplier tube has an amplitude proportional to the energy dissipated in the phosphor by the incident particle, so that it is possible to determine the energies of individual incoming particles.

7.6. TYPES OF SCINTILLATORS

For detection of high energy particles and photons, the following scintillators are used.

(*i*) **Thallium activated sodium iodide.** This is the most common type of scintillator used for the detection of γ-rays. However, this type of scintillator has one major drawback *i.e.*, it is hygroscopic and so it must be sealed in an *Aluminium can* with reflecting or diffusing walls.

(*ii*) **Cesium iodide**. This scintillator is not hygroscopic and so is preferred over the sodium iodide phosphor (or scintillator).

(*iii*) **Zinc sulphide**. For detection of heavy particles it is quite suitable. However, it can not be used in thick layers as in that case it becomes opaque to its own radiation.

(*iv*) **Organic phosphors**. Organic phosphors like **Anthracene and Stilbene** have faster decay time than the inorganic phosphors. These phosphors are useful for counting of γ-rays in the presence of β-rays but show less efficiency for detection and counting of heavy particles.

(*v*) **Plastic scintillators**. These scintillators are quite useful in counter telescopes used in high energy physics for defining the path of a particle. The basic principle being that energy of excitation is transferred from solvent to the solute, which in turn remits radiation of wavelength range for which the solvent is transparent.

(*vi*) For detection and counting of heavy particles in the presence of γ-radiation, **xenon** is used, which gives out radiation in the **ultraviolet region**.

7.7. ADVANTAGES AND DISADVANTAGES OF SCINTILLATION COUNTERS

ADVANTAGES

(*i*) Pulse height is proportional to the energy of the incident γ-ray.

(*ii*) Dead time of the counter is very small and so high counting rate is possible.

(*iii*) Pulse duration being very short, so the resolving power is high.

(*iv*) In case of a solid scintillator, number of atoms/unit volume being quite large, so a stronger pulse is observed than in the case of gas scintillator.

(*v*) With a large scintillator, the scattered γ-rays are also counted and we get an increased photo peak efficiency.

DISADVANTAGES

The light output varies with type of radiation and also the overall efficiency is less as the phenomenon of *conversion of radiation into light* and reconversion of light into electron to count the high energy radiations is a complex phenomenon.

7.8. WILSON'S CLOUD CHAMBER METHOD

The cloud chamber was first devised by C.T.R. Wilson in 1897. The principle involved in the cloud chamber may be briefly stated as follows:

When air contained in an enclosure and saturated with water vapour is subjected to a sudden *i.e.*, adiabatic expansion, the temperature falls, supersaturation takes place leading to the condensation of the water vapour and the formation of a cloud of water droplets. The production of such a cloud is, however, impossible unless nuclei, on which the water vapour may condense, are provided. In ordinary air, dust particles play the role of nuclei. But in dust free air, the phenomenon of condensation and cloud formation is very very difficult. Wilson, however, explained that the condensation cloud can be produced even in dust free air provided the air is capable of ionization by an ionising agent such as α-rays, X-rays, cathode rays etc. Evidently, the ions produced act as centres of condensation.

If dust free air, saturated with water vapour, is allowed to expand adiabatically, supersaturation occurs. If the expansion is in the ratio 1 : 1.25, the condensation occurs on the negative ions only, while if the ratio exceeds 1 : 1.3, the condensation will occur on both the negative and positive ions; whereas for an expansion 1 : 1.38, condensation will occur without any ions at all.

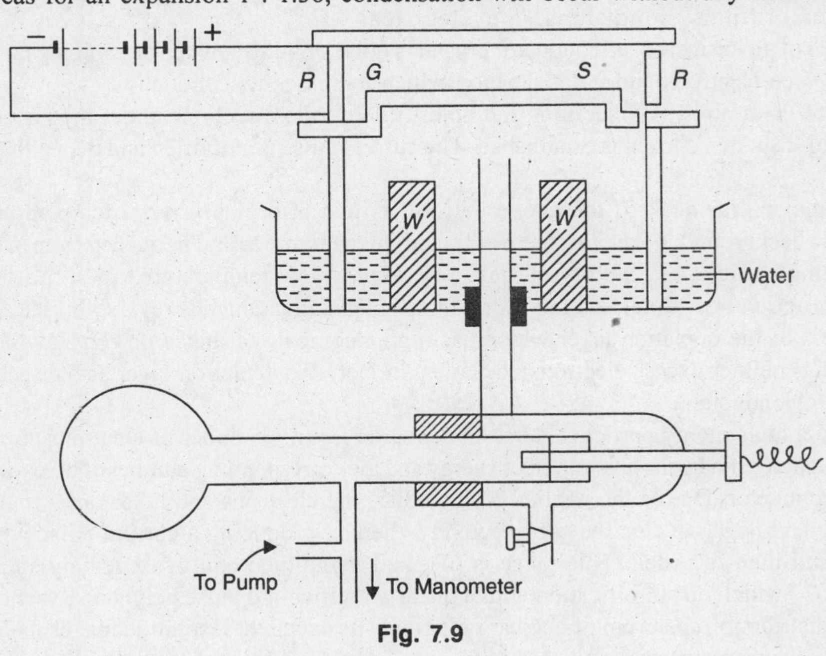

Fig. 7.9

Wilson cloud chamber consists of a glass plate *G*, and a sliding platform *S*. The space in between *G* and *S* contains saturated air. *WW* are two wooden blocks placed to decrease the space of air. The space below *S* may be put in communication with a highly evacuated vessel *V* having a valve *P* which is pulled from outside and is controlled by a spring. When the valve *P* is open, the platform *S* suddenly drops down causing a sudden or adiabatic expanison of gas in G.S.

The apparatus, thus, produces sudden and controlled expansion. An electric field is applied between *S* and *G*. It has a ring *R* in the ceiling platform *S* to clear the chamber from ions after expansion as shown in Fig. 7.9.

In an improved design of this apparatus, the entry of an ionising particle itself actuates the expansion device in the chamber and camera, and the path of the particle is automatically traced. Fig. 7.10 represents, a typical α-ray track. The path *ABCD* is made of straight portions *AB, BC, CD*. The bands at *B* and *C* give the position, where the α-particle encounters with a heavy nucleus.

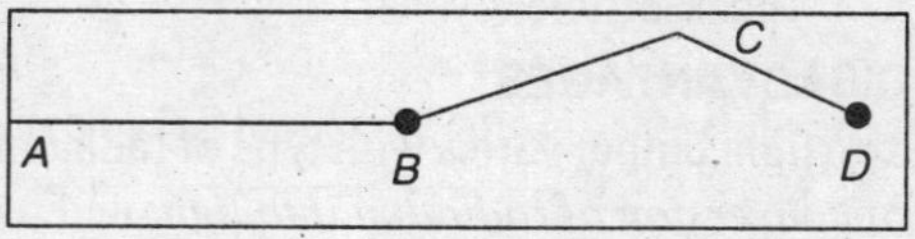

Fig. 7.10

Wilson Cloud Chamber is most suitable for the study of artificial radioactivity, cosmic rays and the discovery of the positron and heavy electron.

7.9. SEMICONDUCTOR DETECTORS

Such detectors can be subdivided into the following two types:

(*i*) Crystal detector and
(*ii*) Junction type detector

(*i*) **Crystal detector:** As shown in Fig. 7.11, it consists of a thin wafer of an intrinsic semiconductor, say, pure silicon or germanium whose opposite faces are covered with very thin layers of gold metal. The two conducting surfaces are connected to a d.c. source of 1–20 volts and an amplifier.

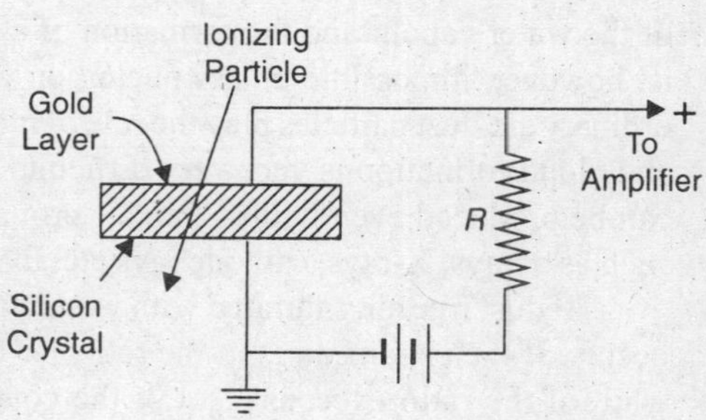

Fig. 7.11

Under ordinary conditions, the electrical conductance of an intrinsic semiconductor crystal is quite small. But when an ionizing particle enters the crystal, free electrons and holes develop which increases its conductivity.

Due to the mobility of electrons and holes, in opposite directions under the influence of the electric field, a sudden current is established. The current pulse is amplified and then either measured or counted.

The major drawback of the above detector is that at ordinary room temperature, it gives considerable background 'noise' *i.e.*, there are pulses of random size and frequency even in the absence of an ionizing particle. So, the crystal has to be kept at low temperature.

(*ii*) **Junction type detector:** It consists of a *P-N* junction which is reverse-biased. To increase the thickness of the depletion layer which has high electrical resistance or very low conductance because it has neither excess electrons nor holes. In fact, this depletion layer acts effectively as an intrinsic semiconductor.

When a high energy particle enters the depletion layer, a number of electrons are transferred to the conduction band by the absorption of energy and the corresponding number of holes are produced as in semiconductor. Due to the applied reverse-bias, the electrons and holes are swept rapidly to the positive and negative electrodes respectively, thereby producing a current pulse which can be amplified and then recorded. This pulse is of small magnitude and of extremely short duration (-10^{-7} to 10^{-9}) which on amplification with a linear amplifier and pulse height analyser can be used for energy distribution spectrum of nuclear radiations. Its excellent resolution makes it a very useful device in nuclear spectroscopy.

7.10. THE BUBBLE CHAMBER

The bubble chamber, invented in 1952 by D.H. Glaser, has become one of the most valuable instruments for studying the minute details of high-energy nuclear events. In a cloud chamber, fogdrops form on the ions produced by charged atomic particles that have just previously traversed the gas-filled chamber. In the bubble chamber, the ions formed by charged particles traversing the liquid, form local heat centre in which tiny gas bubbles develop and grow.

The basic principles of the bubble chamber involve the superheating of a liquid and the bubbles that form in the process of boiling. Water, for example, boils at 100°C at standard atmospheric pressure. If the pressure is increased as in a pressure cooker boiling will not begin until a higher temperature is reached. If the pressure is then suddenly reduced, boiling begins with the sudden formation of tiny bubbles that grow quickly in size.

A simplified diagram of a bubble chamber is shown in Fig. 7.12. A box with thick glass walls, filled with a liquid, is connected to a pressure system and then heated to some predetermined temperature. High-energy particles enter the liquid through a thin window. A sudden release of a valve in the pressure system is quickly followed by the flash of floodlights and the snap of a camera shutter. If the chamber is operating properly, and events are correctly timed, sharply defined trails of bubbles that have formed on the paths of ions made by the traversing particles will be photographed.

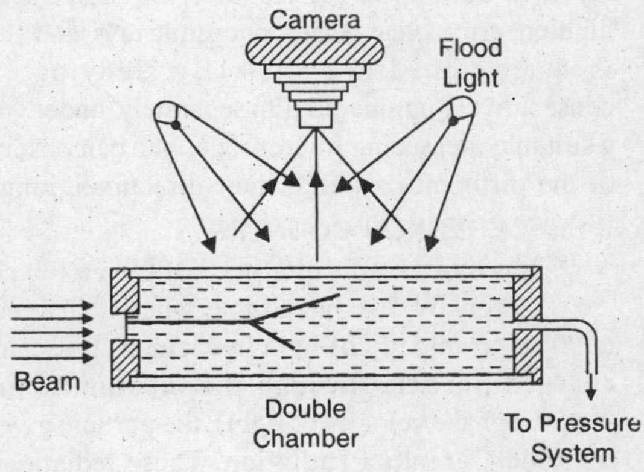

Fig. 7.12

7.11. SPARK CHAMBER DETECTORS

One of the most recent instrument developed for use in the study of high-energy nuclear reactions is known as the *spark chamber.* This device is composed of a stack of equally spaced conducting plates as shown by the cross-section diagram in Fig. 7.13.

A typical chamber will have 25 to 100 plates, each about 1 mm thick and 1 m square, and accurately spaced about 6 mm apart. Alternate plates are connected together and the two sets connected to a 10,000 to 15,000 volt d.c. source. The airtight vessel in which the plates are mounted contains pure helium, or a mixture of a approximately 90 per cent neon and 10 per cent helium, at a pressure of 1 atmosphere.

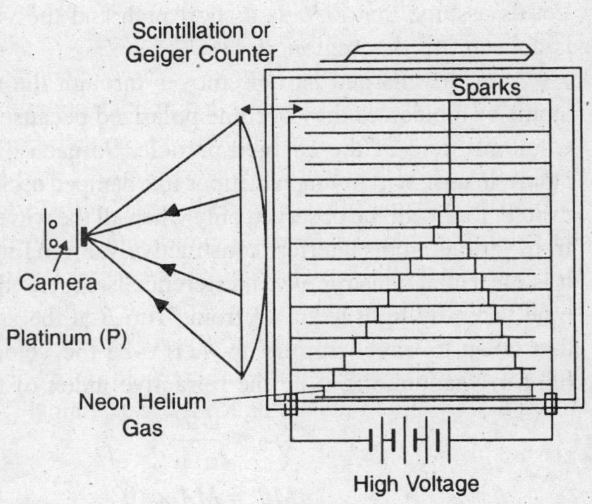

Fig. 7.13

When a high-energy charged particle enters such a chamber, a detector outside the chamber, or inside, triggers the high voltage to the plates, and opens the camera shutter. Traversing the chamber a charged particle produces many ion pairs along its path, and sparks jump between the pairs of oppositely charged plates as indicated. A large lens, cut from clear plastic, makes it possible for a camera to photograph the light from sparks between all plates. A second large lens and camera "looking into" an adjacent side of the box permits the simultaneous taking of two pictures, thereby recording stereoscopic photos of each event.

7.12. PHOTOGRAPHIC EMULSIONS

When a photon or ionizing particle traverses the sensitive emulsion of a photographic film, the clear silver bromide crystal grains that are penetrated are turned into black silver upon development by regular film-developing process.

Since the sensitive emulsion on most film is extremely thin, satisfactory emulsions up to 1 mm in thickness, containing about 80% silver bromide, were first developed by C.F. Powell in England. These can be stacked in layers to built up larger volumes, exposed to high-energy nuclear beams, or cosmic rays, and then developed Fig. 7.14. By studying consecutively numbered film separately, under a suitable measuring microscope, one can observe nuclear collision events and make measurements of the different particles, their directions, ranges, track densities, etc.

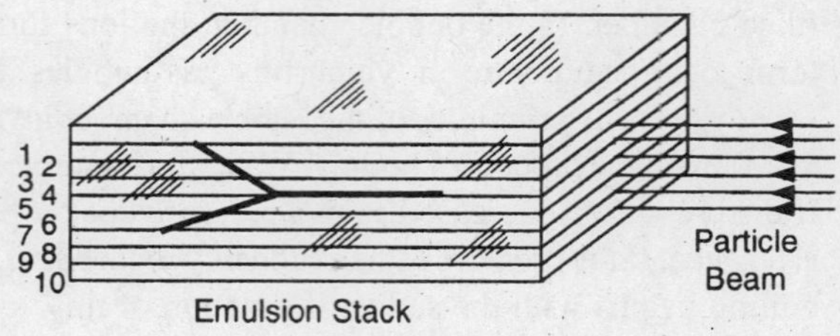

Fig. 7.14

7.13. CERENKOV COUNTER

It is based on the discovery of **Cerenkov radiations** in 1934 by the Russian scientist Cerenkov.

Cerenkov discovered that when a charged particle moving with a very high velocity (~80% to 90% the velocity of light in vacuum) passes through a denser medium and if the **velocity of that charged particle through the medium is greater than** the velocity of light in that medium (*i.e.,* the phase velocity of light), the particle gives out radiations in the visible region. **This radiation is called Cerenkov radiation.** These radiations are highly directional and are confined to a very small cone of semi-vertical angle θ.

Let us consider a charged particle, say a β-particle moving along *LMN* with a velocity *v* in the denser medium. Let us assume that *MNI* is a short path and the velocity is constant in this region.

As the charged particle moves through the nearby atoms or molecules they become polarized because of the Coulomb field of the charged particle. So each of these atoms, in turn, start giving radiations like damped **oscillating dipole.** Light will be observed only when all the wave trains from various atoms interfere constructively. From Fig. 7.15, it is clear that constructive interference is observed when time taken by light to travel from *M* to *B* is the same as that taken to travel from *M* to *A*. If *c* be the velocity of light in vacuum and μ be the refractive index of the medium, then we have

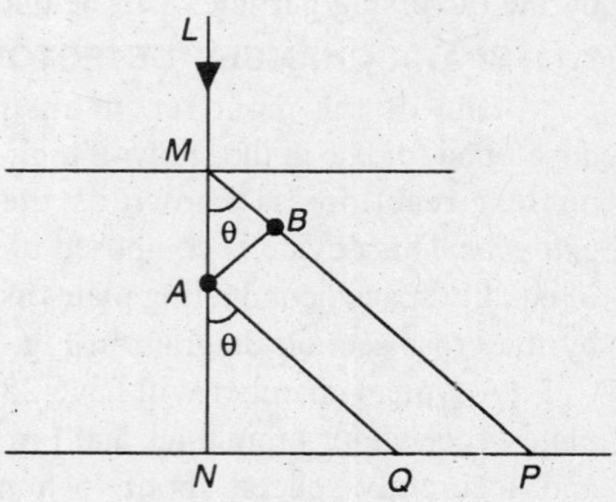

Fig. 7.15

$$\frac{MA}{v} = \frac{MB}{c/\mu}$$

Also $$MB = MA\cos\theta$$

$\therefore$ $$\frac{MA}{v} = \frac{MA\cos\theta}{c/\mu}$$

or $$\cos\theta = \frac{c}{\mu v} = \frac{1}{\mu v/c} = \frac{1}{\mu\beta}$$

where $$\beta = v/c$$

The Cerenkov radiation is not confined to one place but is propagated along the surface of a cone whose axis coincides with the particle and whose semi-vertical angle is θ.

It may be noted that if $\beta < \frac{1}{\mu}$, no Cerenkov radiation will be observed.

In order to detect this type of radiation, it is focussed on a photomultiplier tube and the output is then fed to Scalar recorder etc.

It may be mentioned here that pulse output of Cerenkov counter is very small as compared to scintillation counter for the same loss of energy but even then it can be measured due to the following advantages:

(*i*) Since the light pulse is of short duration, so the counter is capable of very high resolution.

(*ii*) Radiations being directional, the angle of emission is dependent on particle velocity.

(*iii*) No emission of light takes place if particle velocity is less than $\frac{1}{\mu}$ (*i.e.*, $\beta < \frac{1}{\mu}$). If $\beta < \frac{1}{\mu}$, intensity of radiation obtained is dependent on particle velocity.

(*iv*) Unlike Bremestralung, θ is independent of the mass of the moving charged particle.

We have the Cerenkov relation for constructive interference as:

(*a*) The particle track length must be very large as compared to the wavelength λ by the cerenkov radiation.

(*b*) The particle must move with a constant velocity in the medium.

Frank and Tamm in 1937 using classical theory showed that the total energy radiated in a short element of the particles path *ds* is given by

$$\frac{dE}{dS} = \frac{\pi Z^2 e^2}{\varepsilon_0 c^2} \int \left(1 - \frac{1}{(\beta^2 \mu^2)} v dv\right)$$

where $Ze \to$ charge of moving particle

$\nu \to$ Frequency of emitted radiation.

The integral extends over all frequencies for which $\beta\mu > 1$.

It we consider Cerenkov radiation emitted between two frequencies v_1 and v_2 whose average energy is $hv = \frac{h(\nu_1 + \nu_2)}{2}$, than we have the average number of quanta per unit length as

$$\frac{1}{hv} = \frac{dE}{dS} = \frac{\pi Z^2 e^2}{\varepsilon_0 h c^2}(\nu_2 - \nu_1)\left(1 - \frac{1}{\beta \mu^2}\right)$$

$$= \frac{2\pi Z^2}{137}\left(\frac{1}{\lambda_2} - \frac{1}{\lambda_2}\right)\left(1 - \frac{1}{\beta^2 \mu^2}\right)$$

where μ corresponds to the average refractive index of the medium over the frequency interval ν_1 and ν_2.

The relativistic K.E. of a particle is given by

$$E = mc^2 - m_0 c^2$$

$$= \frac{m_0}{\sqrt{1-\beta^2}} c^2 - m_0 c^2$$

$$= m_0 c^2 \left[\frac{1}{1-\beta^2} - 1\right]$$

It is worth mentioning here that Cerenkov detectors not only detect fast moving charged particles but also they are useful for determining their velocities. Cerenkov detectors are very important in detection of high energy particles and cosmic rays.

Also Cerenkov counters have been found to be highly useful in the *high energy physics.*

Example 1. *An α-particle is stopped in an ionization chamber in which it produces 15×10^4 ion pairs. Each time the α-particle produces an ion pair, it loses 35 eV of energy. What is the K.E. of the α-particle? Calculate the amount of charge collected by each plate.*

Solution: In this case kinetic energy = 15 × 10^4 × 35 eV

= **5.25 MeV**

Amount of charge collected by each plate

$$\theta = n.C = 15 \times 10^4 \times 1.6 \times 10^{-19}$$

$$= \mathbf{2.4 \times 10^{-14}\ Coul.}$$

Example 2. *A G.M. counter operates at 1000 volts and the central wire has a radius of 0.1 mm, the inner radius of the outer cylinder is 2 cm and the tube has a guaranteed life time of 10^9 counts. Calculate the maximum radial field and the life of the counter, assuming that the counter works on an average for 50 hours/week at 3000 counts/min.*

Solution: Maximum radial field at the surface of the central wire is given by

$$E_{max} = \frac{V}{2.303 \times r \log_{10}(b/a)} = \frac{100}{2.303 \times 0.01 \times \log 2/.01}$$

$$= \mathbf{18900\ V/cm}$$

Let the life time of the tube be t years, the total number of counts recorded will be equal to the guaranteed counts. So we have

$$1 \times 50 \times 60 \times 50 \times 3000 = 10^9 \qquad \text{(Taking 1 year = 50 weeks)}$$

$$\therefore \quad t = \mathbf{2.22\ years}$$

Example 3. *A sample of uranium emitting α-particles of energy 4.18 MeV is placed near ionization chamber. Assuming that only 10 particles/sec enter the chamber, calculate the current produced. Given 1 ion pair requires 35 eV.*

Solution: Energy of emitted α-particle = 4.18 MeV

Energy required to produce 1 ion pair = 35 eV

So the number of ions produced by one α-particle

$$= \frac{4.18 \times 10^6}{35} = 1.19 \times 10^5 \text{ pair.}$$

Since 10 particles enter in one second, the number of ions produced/sec.

$$= 1.19 \times 10^5 \times 10 = 1.19 \times 10^6 \text{ ion pair.}$$

Charge on each ion $= 1.6 \times 10^{-19}$ C/s

$$= \mathbf{1.904 \times 10^{-13}\ A}$$

Example 4. *A G.M. counter collects 10^8 electrons/discharge when the counting rate is 500 counts/min. What will be the average current in the circuit?*

Solution: Counting rate = 500 counts./min.

Total number of electrons collected/min.

$$= 500 \times 10^8 = 5 \times 10^{10}$$

Charge/minute in the circuit $= 5 \times 10^{10} \times 1.6 \times 10^{-19}$ C/min.

$$\therefore \quad \text{Average current} = \frac{5 \times 10^{10} \times 1.6 \times 10^{-19}}{60} = \mathbf{1.33 \times 10^{-10}\ A.}$$

Example 5. *Compute the approximate number of quanta of visible light emitted as Cerenkov radiation in a frequency range corresponding to wavelength in vacuum of λ = 4000 to 7000 Å, given off by an electron of energy 20 MeV traversing 1 cm of a substance whose μ = 1.49.*

Solution: We have

$$E = m_0 c^2 \left(\frac{1}{\sqrt{1-\beta^2}} - 1 \right)$$

$$\therefore \quad \frac{1}{\sqrt{1-\beta^2}} = \frac{E}{m_0 c^2} + 1 = \frac{20 \times 1.6 \times 10^{-19} \times 10^6}{9.1 \times 10^{-31} \times (3 \times 10^8)^2} = 40 + 1$$

$$\therefore \qquad \frac{1}{\sqrt{1-\beta^2}} = 41$$

$$\therefore \qquad \beta^2 = 0.9995$$

Number of quanta emitted as Cerenkov radiation/m.

$$N = \frac{2\pi Z^2}{137}\left(\frac{1}{\lambda_2} - \frac{1}{\lambda_2}\right)\left(1 - \frac{1}{\beta^2 \mu^2}\right)$$

$$= \frac{2\times 3.142}{137} \times \left(\frac{1}{4\times 10^{-7}} - \frac{1}{7\times 10^{-7}}\right)\left(1 - \frac{1}{0.9995\times(1.49)^2}\right)$$

$$= 26980 \text{ quanta/m}$$

$$= \mathbf{270 \text{ quanta/cm}}$$

Example 6. *In an ionization chamber a charged particle on being completely stopped produces 3 × 10⁵ ion pairs/10 MeV of energy. Calculate the pulse height if total capacitance of capacitor is 50 p.*

Solution. Total charge produced = ne

Pulse height corresponds to voltage developed du to the charge produced. It is given by

$$V = \frac{Q}{C}$$

or

$$V = \frac{ne}{C},$$

where c is the total capacitance. Putting the various values, we get

$$V = \frac{3\times 10^5 \times 1.6\times 10^{-19}}{50\times 10^{-12}}$$

$$\cong \mathbf{10\ V}$$

Example 7. *A G.M. counter with a dead time of 300 μ s records 1600 counts/min. What is the dead time loss in the counting rate?*

Solution. Observed counting rate = 16000 counts/min

deat time = 300×10^{-6} s

$$\text{True counting rate} = \frac{\text{Observed Counting rate}}{\text{1-Obs. Counting rate} \times \text{dead time}}$$

$$= \frac{1600}{1-(16000)\times \dfrac{300\times 10^{-6}}{60}} = 17390 \text{ counts/min}$$

∴ Dead time loss = 17390 − 16000 = **1390 counts/min.**

Example 8. *Estimate the radial field and life of a G.M. counter under the following assumptions:*

Tube operating voltage = 1300 V

Wire radius = 0.01 cm

Cathode radius = 1 cm

The tube has guaranteed life cou8nts = 10⁹ and it operates 8 hours per day at rate of 5000 counts per minute.

Solution. In this case

$$b = 1 \text{ cm}, \quad a = 0.01 \text{ cm}$$

$$r = a = 0.01 \text{ cm}$$

$$V = 1300 \text{ Volts}$$

Now the electric field is given by

$$E = \frac{V}{r \log_e b/a} = \frac{1300}{0.01 \times 2.3026 \times \log \dfrac{1}{.01}}$$

$$= \frac{1300 \times 10^2}{.01 \times 2 \times 2.3026} = 2.82 \times 10^4 \ V$$

Let total life time = x years

So total no. of counts must be equal to the guaranteed counts *i.e.*

$$x \times 8 \times 365 \times 60 \times 5000 = 10^9$$

$$x = \frac{10^9}{8 \times 365 \times 60 \times 500}$$

$$= \mathbf{1.14 \ years}$$

QUESTIONS AND PROBLEMS

1. Describe the common type of detectors for measuring nuclear radiations.

2. Describe the principle and working of ionization and proportional counters.

3. Describe G.M. counter and explain its working.

4. Describe the pulse height vs. voltage curve of gaseous counters and describe how the curve explains about ionization counter, proportional counter and G.M. counter?

5. What is a scintillation counter? Give its usefulness.

6. Describe bubble chamber for measuring nuclear radiations.

7. What is nuclear emulsion technique? How is it useful for measurement of nuclear radiations?

8. It is required to operate a proportional counter with a maximum radial field of 10^5 volts/cm. What is the applied voltage required if the radii of the wire and tube are 0.002 cm and 1 cm respectively? [**Ans.** 1224 V]

9. A sample of Uranium emitting α-particles of energy 4.18 MeV is placed near an ionization chamber. Assuming that only 10 particles/sec. enter the chamber, calculate the current produced. [**Ans.** 0.192 pA]

10. The plateau of a G.M. counter working at 1 kV has a slope of 2.5%, count rate/100 volts. By how much can the working voltage be allowed to vary if the count rate is to be limited to 0.1%. [**Ans.** 0.4%]

CHAPTER

8

COSMIC RAYS AND ELEMENTARY PARTICLES

8.1. INTRODUCTION

It is impossible for anyone to say when and by whom cosmic rays were first studied. From the time of the discovery of radioactivity by Becquerel (in 1896) and the discovery of radium by the Curies but a few months later, the radioactive rays from the ground, air and outer space have been investigated by many scientists. Extending over a period of some fifty years these investigations have led to some of the most interesting and important discoveries in the structure of atomic nuclei.

8.2. EARLY EXPERIMENTS

It has long been known that a charged electroscope, if left standing for some little time, will discharge regardless of how well the gold leaf is insulated. Realizing that the rays from radioactive materials can be stopped by a sufficient thickness of heavy matter, Rutherford and Cooke (in Canada, 1903) surrounded and electroscope with a thick wall of bricks and found very little decrease in the rate of discharge. Mc Lennan and his co-workers (also in Canada) lowered an electroscope into a lake, hoping that the thick-layer of water would screen off the rays. The experiment, like the other, failed.

In 1910 Glockel, with an electroscope, rose nearly 5 km in a balloon in order to get away from the ground radiation, but to his astonishment he found that the rate of discharge did not decrease, but increased, the higher he went. The same effect was observed by Hess (in Austria, 1911) and Kolhorster (in Germany, 1914). Rising to heights as great as 8 km, both of these observers independently found that the intensity of these unknown radiations became greater, the higher they went.

Because in one of his scientific publications concerning these results, Hess suggested the possibility that some kind of penetrating rays were entering the earth's atmosphere from outer space, he is usually credited with the discovery of cosmic rays. For this reason he was granted the Nobel Prize in physics for the year 1936.

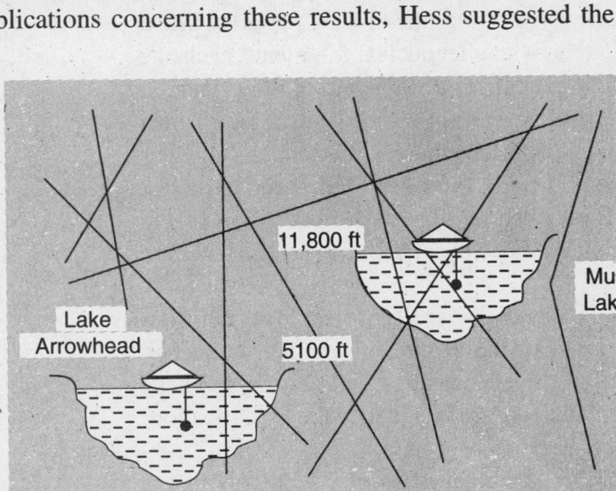

Fig. 8.1

8.3. THE PENETRATION OF COSMIC RAYS

In order to determine the nature of the new rays, Millikan and his co-workers. Otis, Cameron and Bowen, in the fall of 1922 began an extensive study of the penetrating power of cosmic rays. Since cosmic rays penetrate our atmosphere of many km of air, how far might they penetrate beyond?

Self recording electroscopes were

lowered to various depths in snowfed lakes as shown in Fig. 8.1. Measurements taken at Arrowhead Lake in Southern California (at an elevation of 5100 ft) agreed approximately with those taken at Muir Lake near Mt. Whitney (at an elevation of 11,800 ft), provided one took into account the increased air path for the lower elevation. The extra mile and a quarter of air is equivalent in weight of 6 ft water.

As cosmic rays penetrate deeper and deeper below the surface of water, their number decreases, until at a depth of 100 ft the intensity is reduced to about one ten-thousandth of that at the surface. With very sensitive electroscopes, cosmic radiation capable of penetrating 2000 feet of water has more recently been detected. This is a far greater penetrating power than that possessed by any known X-rays or γ-rays from radioactivity.

8.4. ORIGIN OF COSMIC RAYS

The problem of the origin of cosmic rays is still not solved. We may consider the following facts:

(*i*) The place of the origin of cosmic rays.

(*ii*) The mechanism by which they acquire enormous energies.

Regarding the exact place of the origin of cosmic rays, no definite solution has been arrived so far, but it has been well-established that they originate from outside the earth's atmosphere, the following events are the possible sources of high-energy cosmic rays.

There is enough matter in the interstellar spaces in the galaxy in the form of hydrogen clouds out of which stars evolve in the course of time due to condensation. The stars emit light and radiowaves by converting their energy through the process of thermonuclear fusion. In this process, lighter elements are converted into heavier one. The temperature now rises due to more and more gravitational contraction. At the journey's end, the star exhausts most of its hydrogen and develops so much compression that it suffers voilent explosion. Such exploding stars are called Nova and supernova. These events are characteristics of emission of large inensity of light and radiowaves and are possible sources of high energy cosmic rays.

8.5. COMPOSITION OF COSMIC RAYS

Experimental observations show that the cosmic rays entering our atmosphere are almost entirely composed of positively charged atomic nuclei. Of these so-called *primary cosmic rays* 89% are protons, and another 9% α -particles and 1% heavier nuclei like carbon, nitrogen, oxygen, iron, etc.

Upon entering the atmosphere, a high-energy primary particle soon collides with another atomic nucleus, splitting one or both particles into a number of smaller nuclear fragments, each one of which carries away some of the primary's energy. These high-speed particles in turn collide with other nuclei, further dividing their energy to produce other high-speed particles. All of these rays, with the exception of the primary particle, are called *secondary cosmic rays.*

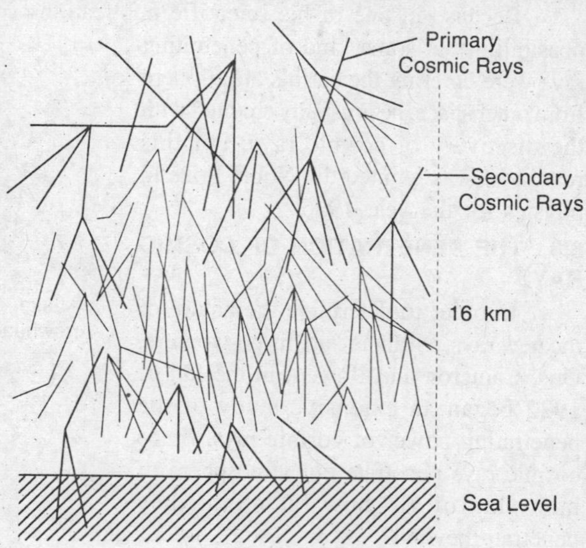

Fig. 8.2

One of the results of cosmic ray collision processes is the creation of very high-frequency and highly penetrating gamma rays. These photons, too, are included in the classification, *secondary cosmic rays.*

At a height of some 24 km, about ten to fifteen times as many secondary cosmic rays exist as have entered the atmosphere as primaries. At this level, where more than

9/10 of the earth's atmosphere still lies below, as many rays are observed moving in a horizontal direction as in the vertical. From the diagram presented in Fig. 8.2, it becomes quite clear why the primaries are difficult to distinguish from the far greater number of secondaries.

At lower altitudes the total intensity decreases, since many of the secondaries produced above are stopped by collision. In other words, so much energy is lost by successive collisions that the energy is gradually absorbed as heat motion by their molecules. By the time sea-level is reached the remaining rays consist principally of a few high-speed secondaries and primaries. Even at sea-level, some of these rays have enough energy left to penetrate several hundred, and even several thousand feet of earth and water.

It should be noted that, between collisions each cosmic ray particle is continually being slowed down as it "plows through" thousands of air molecules, knocking electrons free to produce ions. These are the ions on which fog-drops form, revealing the path in a cloud chamber.

Although γ-rays also lose energy by collisions with atoms to produce Compton electrons, γ-rays themselves do not leave visible tracks in a Wilson cloud chamber. The reason for this is that collisions are few and far between, and the resulting fog-drops are too far apart.

8.6. DIRECTIONAL EFFECTS

To observe the direction of the greatest cosmic ray inensity, a cosmic ray telescope is used. Such a telescope is made by connecting two or more Geiger-Muller tubes in *coincidence,* and mounting them on a common support some distance apart. Tubes in coincident are so connected electrically that a current will flow in the accompanying electric circuit only when both tubes discharge at the same time.

When one tube is set above the other, as shown in Fig. 8.3, single cosmic ray, on going through both cylinders, will cause a current pulse and a count to be made. If, however, a particle goes through one and not the other, no count is recorded. Experiments at sea-level show that, when the telescope is mounted in the horizontal position Fig. 8.4, few counts are made, whereas when it is mounted in a vertical direction, many more counts are recorded. The interpretation to be made, therefore, is that cosmic rays come principally from overhead.

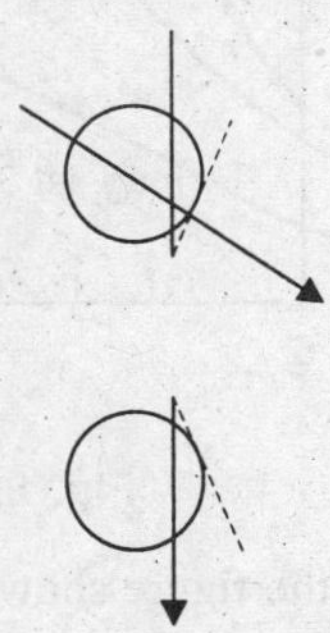

Fig. 8.3

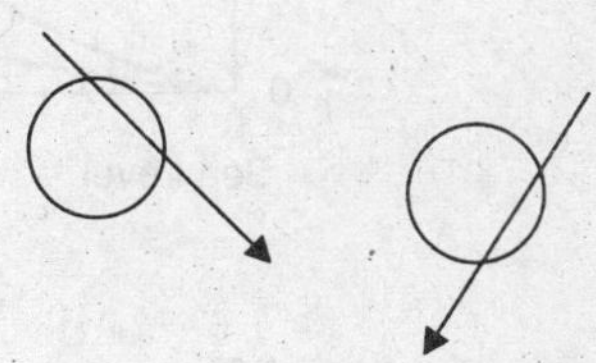

Fig. 8.4

As a verification of the telescope method, a Wilson cloud chamber is frequently inserted between two Geiger counter tubes and a photograph of the path of each cosmic ray is taken. Thousands of such photographs are made automatically by having a single cosmic ray take its own picture. This is accomplished by allowing the sudden electric current from the counter tubes, produced by a ray in transit, to open and close a camera shutter to cause the cloud chamber to expand, and to flash a light, illuminating the fog track that forms.

In the reproduction of Figs. 8.3 and 8.4 either one of the two cosmic rays would have tripped the electrical devices and taken the picture. It should be noted that both rays passed right through a 1.5 cm lead plate without being deviated. Cloud chamber pictures are not photographs of cosmic rays, but of the path traversed by the rays.

8.7. THE ALTITUDE EFFECT

The results of aeroplane and balloon flights into the stratosphere have shown that the intensity of cosmic rays increases up to a height of 20 km to 25 km, and then decreases again. In 1935, Stevens and Anderson, for example, rose to a height of nearly 22 km carrying with them, among other scientific instruments, Geiger-Muller tube counters. With these instruments they measured the cosmic ray intensity at various altitudes on both their ascent and descent.

The compiled experimental results of various observers taken at different elevations are illustrated by the curves in Fig. 8.5 (*a*). Near the city of Omaha, at a magnetic latitude of 51°N, the maximum is found at a height of 24 km where an intensity 170 times as great as that at sea-level has been measured. From that altitude to the highest points that observations have been made, about 27 km, there is a gradual decrease in total intensity.

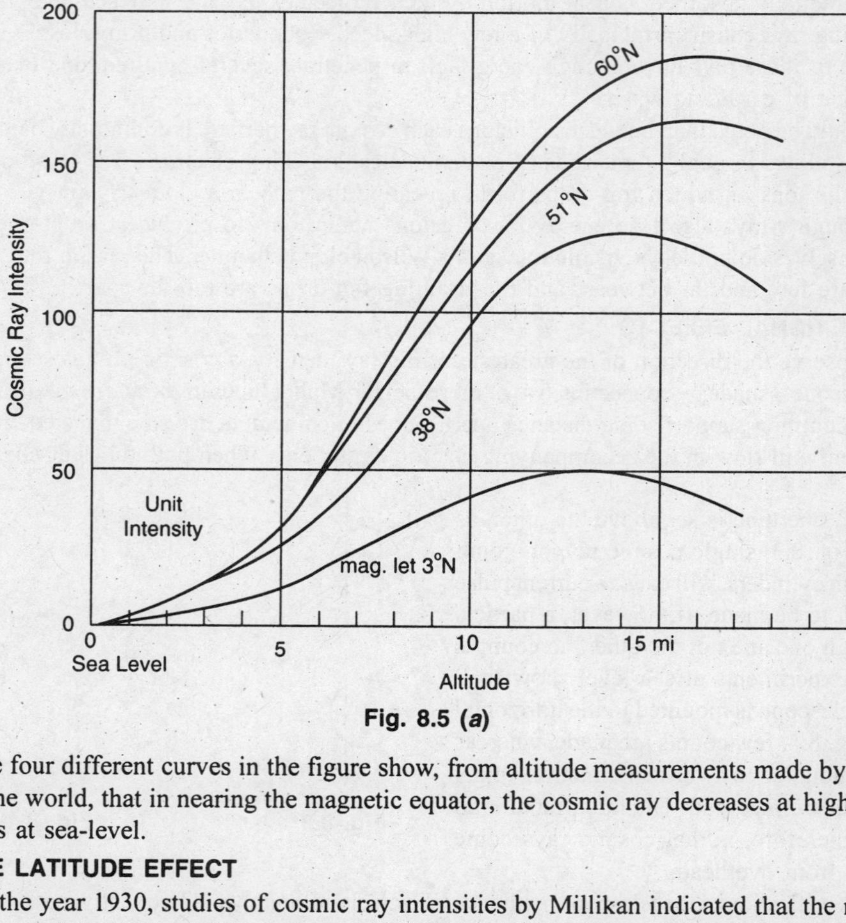

Fig. 8.5 (*a*)

The four different curves in the figure show, from altitude measurements made by observers all over the world, that in nearing the magnetic equator, the cosmic ray decreases at high altitudes, as well as at sea-level.

8.8. THE LATITUDE EFFECT

By the year 1930, studies of cosmic ray intensities by Millikan indicated that the number of cosmic rays arriving at the earth's surface was constant at all latitudes. This led him to the conclusion that the primary cosmic radiation entering the atmosphere must be γ-rays of very high energy. He reasoned that, if they were charged atomic particles, they would be deflectd by earth's magnetic field, and fewer would reach the earth near the magnetic equator.

In 1931, the Dutch physicist Clay, sailing from Amsterdam in the Northern Hemisphere to Batavia, Dutch Guiana, in the Southern Hemisphere, carried Geiger counters with him aboard ship. Measuring the cosmic ray intensity daily en route, he obtained the results shown by the graph in Fig. 8.5 (*b*). The curve shows that one proceeds from magnetic north to magnetic south, at sea-level, the cosmic-ray intensity remains quite constant, until a magnetic latitude of about 42° is reached. In this region, the intensity begins

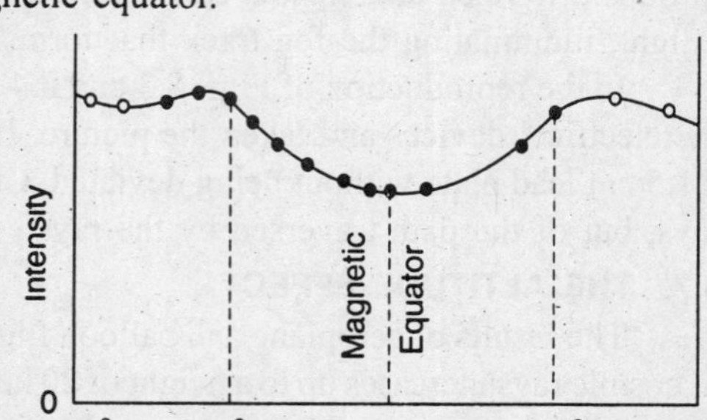

Fig. 8.5 (*b*)

to drop appreciably, reaches a minimum at the equator, and rises again to symmetrical intensities in the Southern Hemisphere.

8.9. EFFECT OF THE EARTH'S MAGNETIC FIELD

The decrease in cosmic ray intensity at the earth's magnetic equator [see Fig. 8.5 (*b*)] is now explained as being due to the earth's magnetic field. This is illustrated in Fig. 8.6. The paths of all charged particles crossing the earth's magnetic field are bent by a force that is perpendicular to the direction of the field.

Because of these forces, many charged particles from the sun and outer space are trapped by the field in two belts. The recent discovery of these belts by Van Allen and his colleagues at State University of Iowa was made from instruments carried into space by American earth-circling satellites. These belts surround the earth, except at the regions of the magnetic poles, the outer one being caused largely by the slower particles, protons and electrons, from the sun. The inner Van Allen belt is formed by more energetic particles from other space, and is centered about 4000 km above the magnetic equator.

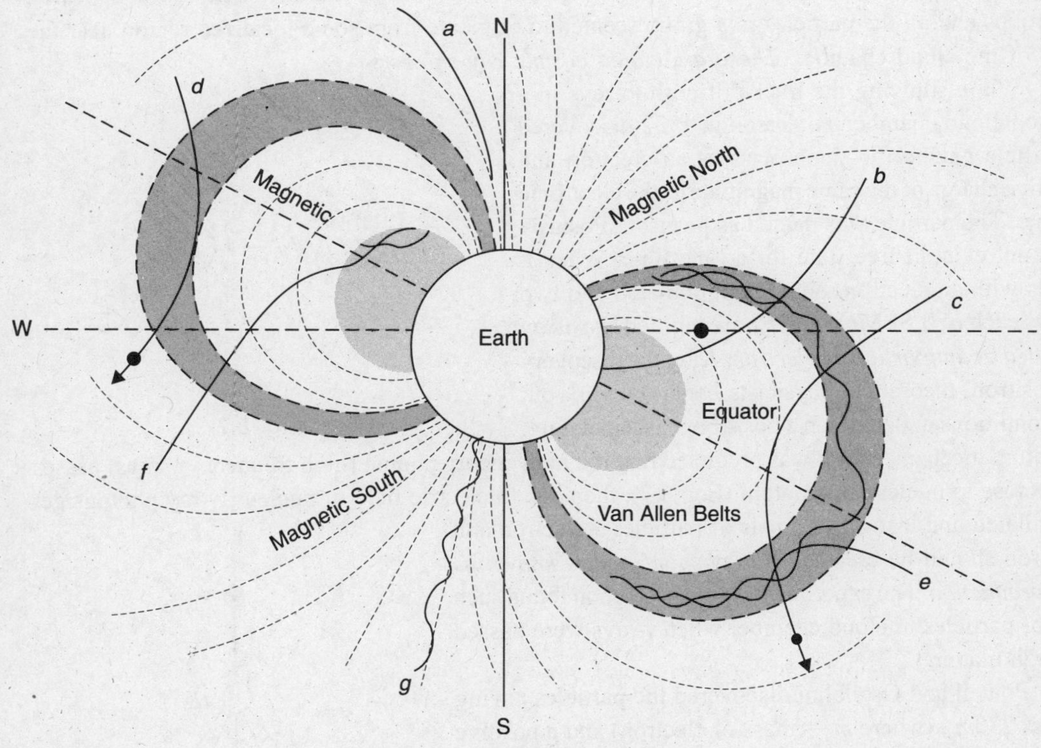

Fig. 8.6

If the primary energy is very large, little deflection will occur, and particles like *a* and *b* will reach our atmosphere and perhaps the ground. At somewhat lower energies they will be deflected back into space as *c* and *d*. At still lower energies they may follow paths like *e*, *f* and *g*.

Particle *e* spirals around the field lines and follows them in towards the earth. As the field gets stronger, a point is reached where the particle is turned back, and spiraling around the field lines approaches the earth again on the other side. Such particles running back and forth are trapped; they account for the large number of ionized particles that form the Van Allen belts.

Slow particles, such as *g*, entering the earth's field parallel to the lines of force from far away, will be guided by the field and reach the earth's surface. A day or two after an active display of solar flares, so many particles become trapped in the outer Van Allen belt they spill out into the earth's atmosphere, creating auroras. Since particles with only the highest of energies can get down to the earth's atmosphere near the magnetic equator, the cosmic ray latitude effect is well understood.

8.10. EAST WEST ASYMMETRY

Making use of *cosmic ray telescope,* it has been observed, the number of particles arriving from the west per second is greater than from the east. This effect is maximum at the equator, where the excess is about 14.2 per cent. This asymmetry can be explained on the fact that most of the cosmic ray particle are positively charged (nuclei without electrons). A positively charged particles approaching the equator in the vertical direction will be deflected by earth's magnetic field so that it reaches the earth from westerly direction. A negatively charged particle would come from the easterly direction.

8.11. COSMIC RAY SHOWERS

In order to know the nature of secondary ray particles, cloud and bubble chambers trigged by the incoming particles and using specially prepared nuclear emulsion blocks were used. In 1928, Skobeltsyn photographed the tracks of particles in a Wilson cloud chamber. The tracks appeared in groups and all the particles in a group seemed to originate from some localized region near the wall of the cloud chamber. *These are called cosmic ray showers.*

While studying the tracks of cosmic rays in a Wilson cloud chamber. Anderson in 1932 discovered a particle having the same mass as an electron and positive charge of the same magnitude as the electronic charge. The particle was named as positron. Positron does not exist in free state for a long time. When it meets with an electron, both are annihilated and two γ-rays each of 0.51 MeV energy are emitted. *Positron is called as antiparticle of electron.* After the discovery of positron, theoretical physicists began to work out the conditions under which a positron exists in nature.

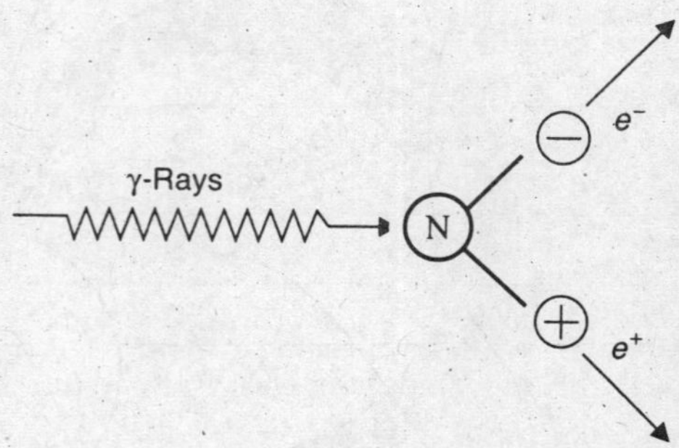

Fig. 8.7

Quantum mechanically, it was predicted that if a high energy photon (high frequency γ-rays) passes very close to nucleus of an atom (Fig. 8.7), then due to electric field of nucleus, γ-ray photons get annihilated and instead of a pair of particles—electron and positron should be created. *This phenomenon is known as pair production.* The experimental physicists soon found such pair of particles in cloud chamber when γ-rays were passed through matter.

Powell and Oscilalini discovered the particles having a mass $230m_e$, (where m_e = mass of electron) and a positive or negative charge equal in magnitude to the electronic charge are produced by the collisions of cosmic rays with air nuclei. These particles are called mesons or pions. Fig. 8.8 shows the production of pions, protons and neutrons by a primary cosmic ray by nuclear collision. Some pions are positively charged, some are negatively charged while others are neutral. In 1933, Anderson and Neddermeyer discovered a particle having mass equal to 207 times the electronic mass and positive or negative charge equal in magnitude to the electronic charge. These particles are known as π mesons. The π and μ mesons were later produced in the

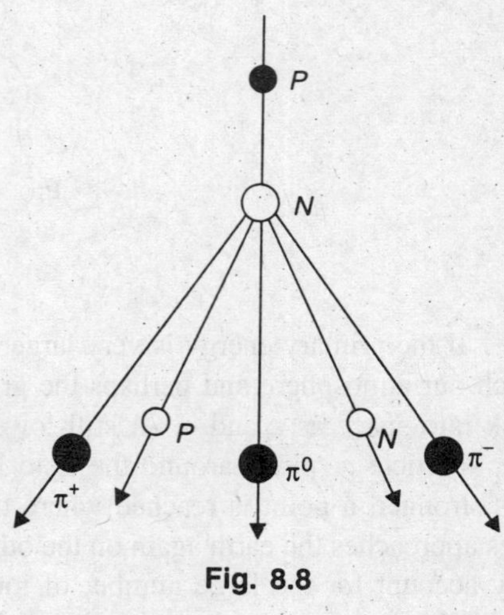

Fig. 8.8

laboratory by bombarding suitable targets by particles accelerated to very high energies. It was discovered that both mesons are unstable. All charged pions spontaneously disintegrate into a μ meson and a neutrino as shown

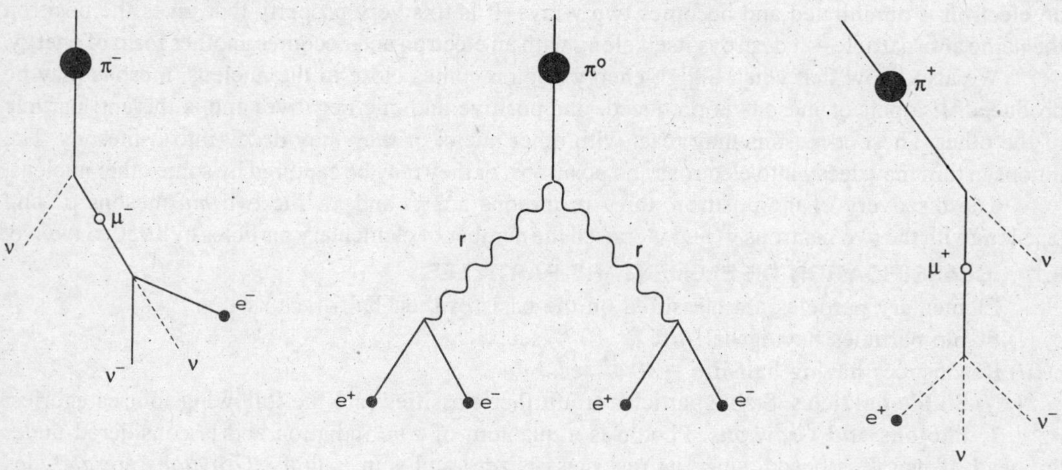

Fig. 8.9

in Fig. 8.9 and negative μ meson decays into an electron and two neutrons. Neutral pions (having zero charge) decay into two gamma rays. By electron pair production, these gamma rays create cascade showers of electrons in upper region of atmosphere.

The formation of cosmic ray showers is shown in Fig. 8.10. When energetic electron or positron produced by the decay of π meson passes close to heavy nucleus it gives rise to high energy γ-ray photon. When this γ-ray photon passes close to a heavy nucleus, the phenomenon of pair pro-duction takes place. This pair in turn gives rise to photons and so a cascade shower process goes on.

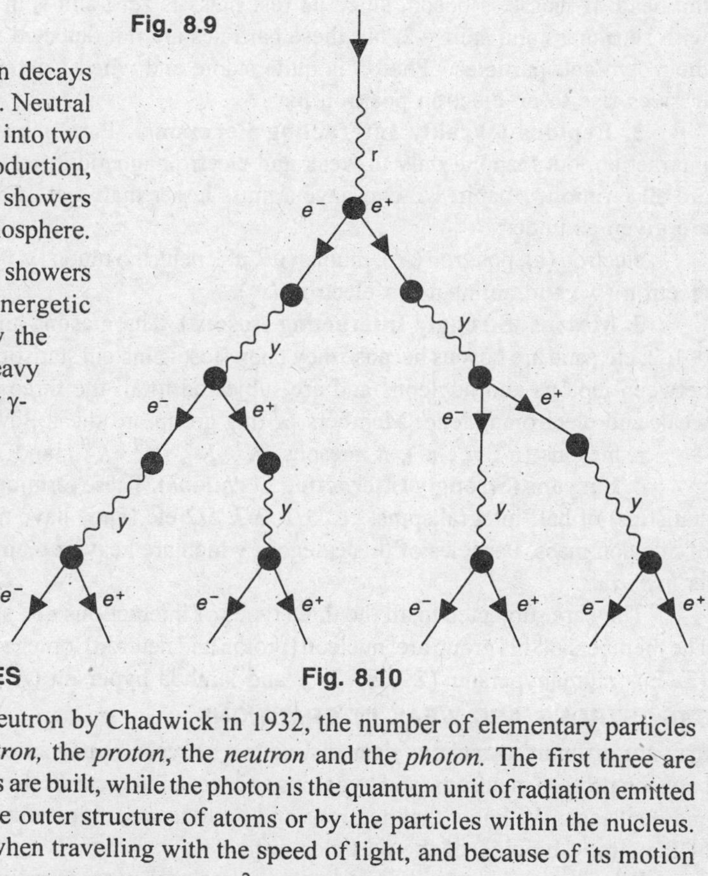

Fig. 8.10

8.12. ELEMENTARY PARTICLES

With the discovery of the neutron by Chadwick in 1932, the number of elementary particles became four in number, the *electron*, the *proton*, the *neutron* and the *photon*. The first three are the atomic particles of which atoms are built, while the photon is the quantum unit of radiation emitted or absorbed by the electrons in the outer structure of atoms or by the particles within the nucleus.

The photon can only exist when travelling with the speed of light, and because of its motion possesses energy $h\nu$. By the mass-energy relation, $E = mc^2$, a photon also has mass $h\nu/c^2$. It possesses mass by virtue of its motion, for at rest it would have no energy and no mass.

The electron, proton, and neutron, on the other hand, have a definite rest mass m_0 and a rest energy m_0c^2. When they are set into motion, their mass increases, and their total energy is given by mc^2.

8.13. ANTI-PARTICLES

The positron, discovered by Anderson in 1932, is a positively charged electron. In cosmic rays, we studied how electron pairs can be created and how a free positron incoming together, with an electron is annihilated and becomes two γ-rays. It is this very property that gives the positron the name anti-particle—it destroys itself along with an electron and becomes another form of energy.

We also know that when a high-energy photon comes close to the nucleus, mesons may be produced. If a pair of mesons is produced, one positive and one negative, one is the anti-particle of the other. These π-mesons may react with other nuclei or they may decay into μ-mesons. The muons in turn may decay into electrons and positrons, or they may be captured by some other nucleus.

The discovery of the positron, three pi mesons π^+, π^0 and π^-, the two *mu* mesons μ^+ and μ^-, alongwith the two neutrons ν^0 and ν^0, raised the number of elementary particles by 1950 to twelve.

8.14. CLASSIFICATION OF ELEMENTARY PARTICLES

Elementary particles are classified on the basis of their half-lives as:

(*i*) Stable particles having half-life $\geq 10^{-16}$ sec.

(*ii*) Resonances having half-life $= 10^{-23}$ sec.

(*i*) **Stable particles.** Stable particles are further classified into the following four categories:

1. Photons and *Gravitons*. Photon is a quantum of *e.m.* radiation and is considered under the head of massless boson, since its rest mass is zero and spin is unity. *Gravitons* are particles with zero mass and spin = 2, but these particles are not detected so far. Photons and gravitons are their own anti-particles. Photon is quite stable and when it interacts with the field of a nucleus, it gives rise to an electron-positron pair.

2. Leptons (Weakly Interacting Fermions). Particles which do not respond to strong interaction, but respond only to weak and electromagnetic interactions are called as leptons. They are all fermions of spin 1/2 and have a mass lesser than that of nucleons. Members of this group are given as under:

electron (*e*), positron (e^+), muons (μ^+, μ^-), neutrino muon (ν_μ), antineutrino muon (ν_μ), neutrino electron (ν_e) and antineutrino electron (ν_e).

3. Mesons (Strongly Interacting Bosons). The mesons all have zero or integral spins *i.e.*, 0, 1, 2, etc., and are bosons because they obey Bose-Einstein statistics. They have masses intermediate between leptons and nucleons and are subjected to all the three types of interactions *i.e.*, strong, weak and electromagnetic. Members of this group are the following:

π mesons (π^+, π^0, π^-), *K* mesons (K^+, K^-, K_1^0 K_2^0) and η mesons (η^0).

4. Baryons (Strongly Interacting Fermions). These fermions (particles obeying Fermi-Dirac statistics) of half integral spins, *i.e.*, 1/2, 3/2, 5/2 etc. They have masses equal to or more than that of nucleon mass. Particles of this category which are heavier than nucleons are collectively known as *hyperons.*

They are subjected to all the three types of interactions *i.e.,* strong, weak and electromagnetic. The members of this group are: nucleon (proton and neutron), omega hyperson (Ω^-), cascade hypersons (Ξ, Ξ), sigmahyperson (Σ^+, Σ^0, Σ^-) and lambda hyperson (Λ^0)

8.15. STRONG AND WEAK INTERACTIONS

It is not difficult to make a clear distinction between strong interactions such as those that occur between protons, neutrons, and hyperons and weak interactions that occur between muons, electrons, and neutrinos. We know that neutrons and protons are subjected to strong attractive forces because of the strong binding of the particles in nuclei.

When two high-energy nucleons approach each other in collision, for example, the very strong action and reaction forces they exert on each other occur at such short distances that the collision time is of extremely short duration. Photons, electrons and positrons, muons and neutrinos, do not involve such strong short-range forces, but exert weaker forces at greater distances. Collisions or disintegrations involving these particles are, therefore, of longer duration.

Collision times and decay times for strong interactions are of the order of 10^{-23} second, while weak interactions require time intervals billions of times longer and of the order of 10^{-13} second. The lightest weight strongly interacting particles are pi mesons. They with their rest mass of 137 MeV are in direct contrast with electrons and their rest mass of 0.5 MeV, and photons and neutrions with zero rest mass. One exception to this involves the weakly interacting muons with a rest mass of 106 MeV.

A useful quantum number called the *strangeness number* was introduced in elementary particle theory of Gell-Mann to predict which new particles could be produced by impact, that is, by strong interactions, but not in weak interactions. All interactions obey the ordinary conservation laws of *energy, momentum, angular momentum and charge.*

Table 8.1: The classification of elementary particles

Class	*Name*	*Particle*	*Anti particle*	*Rest mass MeV*	*Mean life, s*	*Spin*
Massless boson	Photon	γ	(γ)	0	Stable	
Lepton	e-neutrino	ν_e	ν_e	0	Stable	1/2
	μ-neutrino	ν_μ	ν_μ	0	Stable	1/2
	τ-neutrino	ν_τ	ν_τ	0	Stable	1/2
	Electron	e^-	e^+	0.51	Stable	1/4
	Muon	μ^-	μ^-	106	2.2×10^{-6}	1/2
Meson	Tau	τ^-	τ^-	1784	5×10^{-13}	
	Pion	π^+	π^-	140	2.6×10^{-8}	
		π^0	(π^0)	135	8.3×10^{-17}	
		π^-	π^+	140	2.6×10^{-8}	
	Kaon	K^+	K^+	494	1.2×10^{-8}	
		K^0	K^0	498	9×10^{-11}	0
					5×10^{-8}	
	Eta Meson	η^0	(η^0)	549	7×10^{-19}	
Baryon	Proton	p	p	938.3	Stable	
	Nucleon					
	Nucleon	n	n	939.6	932	1/2
	Lambda hyperon	Λ^0	Λ^0	1116	2.5×10^{-10}	1/2
		Σ^0	Σ^-	1189	8.0×10^{-11}	
	Sigma hyperon	Σ^0	Σ^0	1192	10^{-14}	1/2
		Σ^-	Σ^-	1197	1.5×10^{-10}	1/2
		Σ^0	Σ^0	1315	3.0×10^{-10}	
	XI hyperon	Σ^-	Σ^-	1321	1.7×10^{-10}	1/2
	Omega hyperon	Ω^-	Ω^-	1672	1.3×10^{-10}	1/2

8.16. CONSERVATION LAWS

We have seen in earlier chapter on nuclear processes that certain conservation laws apply to all reactions. These laws relate to measurable quantities existing just before impact with those existing immediately after impact. For convenience, we can state them as follows:

Law 1. *The total energy, including rest mass, remains constant.*

Law 2. *The total linear momentum remains constant.*

Law 3. *The total angular momentum remains constant.*

Law 4. *The total charge remains constant.*

To these four laws we add four more to apply to elementary particle reactions.

Law 5. *The total baryon number remains constant.*

This fifth law has greater implications when it is stated another way. "The total number of baryons in the world remains constant." Here we count an anti-baryon as minus one baryon. This law is not pertinent for reactions involving only mesons and leptons. All mesons and leptons have baryon number zero.

Law 6. *The total lepton number remains constant.*

The leptons, ν, e^-, and μ^- have lepton number +1 and their anti-particles have lepton number −1. These particles are involved in the decay of the muons.

The decay of a negative muon, for example, will illustrate:

$$\mu^- \rightarrow e^- + \nu + \nu$$

Lepton number $+1 = +1 + 1 - 1$

Law 7. *The total strangeness number remains constant.*

The last law does not apply to weak reaction as in the decay of unstable particles but does apply to strong interactions.

Law 8. *The over-all symmetry of anti-particle parity is preserved.*

Another quantity that we mention for completeness is *parity*, which is conserved in a reaction if the mirror image of the reaction (involving the anti-particles) also occurs. It is found that *parity is conserved in strong and electromagnetic interactions but is not conserved in weak interactions.*

8.17. QUARKS

In order to place the 100-odd elementary particles in the periodic table according to their properties, a simplified theory was required. A similar problem occurred with the discovery of many elementary particles and a few theories were put forward for the elementary particles and these are still in progress.

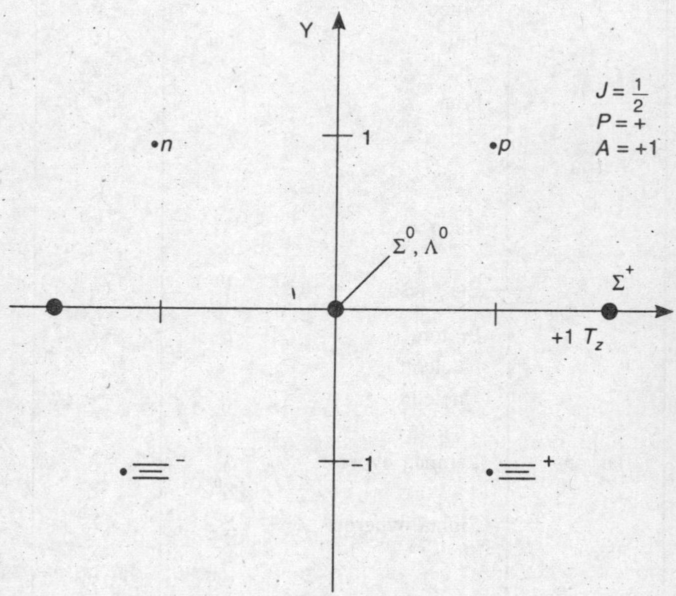

Fig. 8.11 (*a*) Baryon family

If the lower mass baryons or mesons are plotted on the axes of **hypercharge** vs the *Z* component of isospin, a striking similar pattern occurs for each particle having the same spin and parity shown in Figs. 8.11 (*a*) and (*b*).

Also the members have relatively similar masses and the variations are systematic within each particle. As in the case of **periodic table,** few new elements were detected to fill the **vacancies.** Similarly few new elementary particles were predicted in the above pattern. A significant pro-gress was achieved with the prediction of Ω^-.

Gell-Mann and others gave a theory that these patterns can be explained if all **hadrons** were formed from three particles and three antiparticles called **Quarks.** These hypothetical quarks have charge $2e/3$, $e/3$ and $-e/3$ where *e* is the electronic charge. It was a revolutio-nary proposal as previously it was a firm notion that no particle could exist with a charge which was not integral multiple of '*e*' and no particle with charge less than '*e*' could exist *i.e.*, fundamental charge is **electronic charge**. According to Quark theory, these particles should have spin 1/2, fractional baryon number but otherwise normal properties.

It was further suggested that meson consists of quark and anti quark bound together while baryon and anti-baryons would each consist of 3 quarks and 3 anti-quarks. The mass of the particle so formed would be equal to the mass of 2 or 3 quarks minus the binding energy of the combination.

Quarks are thought to be elementary particles in the same sense as leptons, essentially **point particles** with no internal structures but unlike other particles have fractional charges. The original three quarks were labelled *u* (for 'up'), *d* (for 'down') and *s* (for 'strange') and they along with their anti-particles may be assigned charges as under:

u: + 2/3e $\bar{u}$: − 2/3e

d: −1/3e $\bar{d}$: +1/3e

s: −1/3e $\bar{s}$: +1/3e

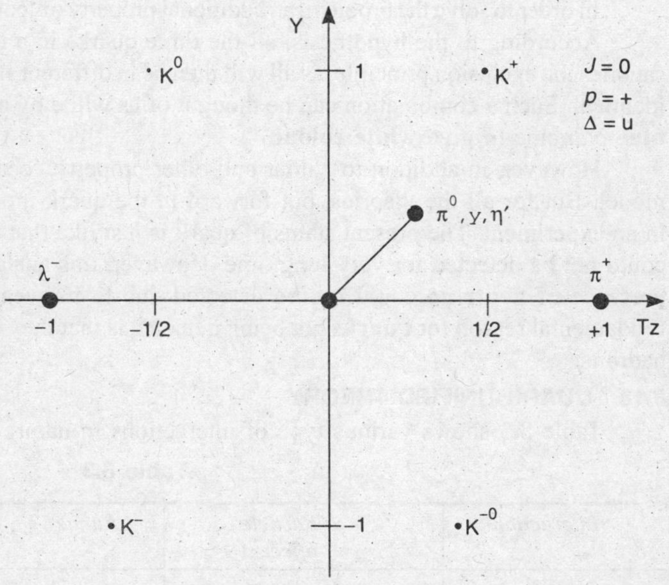

Fig. 8.11 (*b*) Baryon family

Table 8.2 shows all the six quarks together with the leptons associated with each family generation. It may be mentioned here that ordinary matter involves only the first generation. All the quarks and leptons have spin 1/2. Quarks have **baryon number** B = 1/3 and may be labelled as red, green, or blue. The strong force between the quarks is because of the 'colour', in the same way that *e.m.* force between charged particles due to their **charge.** For each **lepton** and **quark,** there are **anti-lepton** and **anti-quark** with opposite properties including **anti-colour.** Baryons have three quarks or different colours, mesons have quarks of one colour and anti-quark of anti-colour.

Table 8.2

Generation	*Quark*	*Symbol*	*Charge e*	*Strangeness*	*Charm*
1	Up	*u*	+2/3	0	0
	Down	*d*	−1/3	0	0
2	Charm	*c*	+2/3	0	+1
	Strange	*s*	−1/3	−1	0
3	Top	*t*	+2/3	0	0
	Bottom	*b*	−1/3	0	0
Generation	*Lepton*		*Symbol*	*Charge e*	
1	Electron		e^-	−1	
	e-neutrino		ν_e	0	
2	Muon		μ^-	−1	
	m-neutrino		ν_μ	0	
3	τ		τ^-	−1	
	τ-neutrino		ν_c	0	

It may be mentioned here that in the case of baryons, a serious problem arose as two or three quarks of the same nature may be present for the same particle (*e.g.*, two *u* quarks in a proton, three *s* quarks in a Ω^{-1} hyperon), which violates the exclusion principle.

In order to solve the problem, an additional property of 'colour' was assigned as explained earlier.

According to the hypothesis, all the three quarks in a baryon have different colours, which satisfies the exclusion principle, as all will then be in different states even if two or three are otherwise identical. Such a combination can be thought of as white by analogy with the way **red, green and blue** combine to give **white colour.**

However, in addition to colour and other property, 'charm' is included to explain the quark model. But for all the theories, put forward in the quark model, no quark has ever been isolated in an experiment. The present status of quark is just like that of **neutrino** which though **proposed** could not be detected for very long time. However, this analogy may not be quite accurate as the presence of **neutrino** could not be detected due to its **weak** interaction with matter while the fundamental reason for **quarks** not being detected is that they are not present independently outside hadrons.

8.18. QUARK UNIFIED THEORY

Table 8.3 shows various type of interactions in nature is given below.

Table 8.3

Interaction	*Particle affected*	*Range*	*Relative Strength*	*Particles exchanged*
1. Gravitational	All	∞	10^{-39}	Gravitons
2. Weak	Quarks and Leptons	10^{-17} m	10^{-5}	Intermediate boson
3. Electromagnetic	Charged particles	∞	10^{-2}	Photons
4. Strong	Quarks Hadrons	10^{-15} m	1	Gluons Meson

Unified theory was proposed by **Weyl** who tried to unify gravity and electromagnetism. This theory was further extended to other interactions. According to this theory, strong, weak and *e.m.* forces appear to obey one basic phenomenon, with leptons and quarks finding their natural positions within the scheme. From this theory it can be predicted why an electron (lepton) and proton (combination of quarks) possess exactly the same charge.

Example 1. *Explain why the decay $\Sigma^0 \rightarrow \Lambda^0 + \gamma$ is observed but not $\Sigma \rightarrow p + \pi^-$ or $\Sigma^0 \rightarrow n + \pi^0$.*

Solution: In the decay

$$\Sigma^0 \rightarrow \Lambda^0 + \gamma$$
$$S : -1 = -1 + 0$$

strangeness is conserved, and the lifetime of the Σ^0 ($\approx 10^{-14}$ s) indicates that it decays by the faster electromagnetic process and not by the slower weak interaction (life time $\approx 10^{-10}$ s). The decays

$$\Sigma^0 \rightarrow p + \pi$$
$$S : -1 \neq 0 + 0 \qquad (\Delta S = +1)$$

would be weak because the strangeness changes.

Example 2. *Determine the energies of the products in the reaction $\pi + p \rightarrow n + \pi^0$. π^- and p were initially at rest.*

Solution: From conservation of momentum,

$$p_n = p_{\pi^0} \quad \text{or} \quad (p_\pi c)^2 = (p_{\pi^0} c)^2 \qquad ...(1)$$

Using the relativistic relationship between energy and momentum for each of the final products, we have

$$E_n^2 = E_{0n}^2 + (p_n c)^2 \; E_{0\pi}^2 = E_{0\pi^0}^2 + (p_{\pi^0} c)^2$$

Subtracting these and using (1), we obtain

$$E_n^2 - E_{0n}^2 = E_{0n}^2 - E_{\pi 0}^2 = (939.6 \text{ MeV})^2 - (135.0 \text{ MeV})^2$$
$$= 8.646 \times 10^5 \text{ (MeV)}^2$$

or

$$(E_n + E_{\pi^0}) = 8.646 \times 10^5 \text{ (MeV)}^2 \quad \text{... (2)}$$

From conservation of energy,

$$E_n + E_{\pi^0} = E_{0\pi} + E_{0p} = 139.6 \text{ MeV} + 938.3 \text{ MeV}$$
$$= 1077.9 \text{ MeV} \quad \text{... (3)}$$

Substituting this in (2), we obtain

$$(1077.9 \text{ MeV})(E_n - E_{\pi^0}) = 8.646 \times 10^5 \text{ (MeV)}^2$$
$$E_n - E_{\pi^0} = 802.1 \text{ MeV} \quad \text{... (4)}$$

Solving equation (3) and (4) we get

$$E_n = 940.0 \text{ MeV}, \quad E_{\pi^0} = 137.9 \text{ MeV}$$
$$\therefore \quad K_n = E_n - E_{0\pi} = 940.0 \text{ MeV} - 939.6 \text{ MeV} = 0.4 \text{ MeV}$$
$$K_{\pi^0} = E_{n_0} - E_{0\pi} = 137.9 \text{ MeV} - 135 \text{ MeV} = \mathbf{2.9\ MeV.}$$

Example 3. *A Σ^0 particle decays at rest to a Λ^0 + particle. Determine the energy of the released photon.*

Solution: The reaction is $\Sigma^0 \rightarrow \Lambda^0 + \gamma$. From conservation of momentum we have, using the relativistic relationship.

$$E^2 = (pc)^2 + (m_0 c^2)^2$$
$$p_\Lambda = p_\gamma = \frac{E_\gamma}{c}$$
$$p_\Lambda{}^2 = E_\Lambda^2 - (M_\Lambda c^2)^2 E_\lambda^2 \quad \text{... (1)}$$

From conservation of energy, we have

$$M_\Sigma c^2 = E_\Lambda + E_\lambda \quad \text{or} \quad E_\Lambda^2 = (M_\Sigma c^2)^2 + E_\gamma^2 - 2M_\Sigma c^2 E_\gamma \quad \text{...(2)}$$

Combining (1) and (2), we get

$$E_\gamma = \frac{(M_\Sigma c^2)^2 - (M_\Lambda c^2)^2}{2M_\Sigma c^2} = \frac{(1192)^2 - (1116)^2}{1192} = \mathbf{73.6\ MeV}$$

Example 4. *A 150 MeV K^+ particle decays into $2\pi^+ + \pi$. Range measurements in a photographic emulsion give the kinetic energies of the π^+ s as 68.6 MeV and 80.8 MeV, and that of π^+ the π^- as 75.5 MeV. Find the Q for the reaction and the mass of the K^+.*

Solution: The reaction is $K^+ \rightarrow \pi^+ + \pi^+ + \pi^-$, so

$$Q = K_{\pi^+} + K_{\pi^+} + K_{\pi^-} + K_{k^+} = 68.6 \text{ MeV} + 80.8 \text{ MeV} + 75.5 \text{ MeV} - 50 \text{ MeV}$$
$$= 74.9 \text{ MeV}$$

From $\quad Q = (m_{k^+} - 2m_{\pi^-})\, c^2$ we then obtain

$$m_{k^+} + c^2 = Q + (2m_{\pi^+} + m_{\pi^-})c^3 = 74.9 \text{ MeV} + 3(139.6 \text{ MeV}) = \mathbf{493.7\ MeV}$$

QUESTIONS AND PROBLEMS

1. What are cosmic rays? Mention briefly the nature of cosmic rays.
2. Explain the latitude effect, longitude effect and altitude effect of cosmic rays.
3. What are primary and secondary cosmic rays? Give an account of the production and properties of π and μ mesons.
4. Write an essay on cosmic rays with special reference to their discovery, variation in intensity, shower formation and their origin.
5. Write short note on cosmic ray showers.
6. Write an essay on elementary particles.

7. Write notes on:

(*i*) positron, (*ii*) mesons and (*iii*) origin of cosmic rays.

8. Give brief outlines of the classification of elementary particles.

9. What are the various conservation laws for the elementary particle decays.

10. What are the various modes of interaction of elementary particles? Give suitable examples.

11. Which of the following weak decays is not possible for the Ω^-?

(*a*) $\Omega^- \to \Sigma^- + \pi^0$ (*b*) $\Omega^- \to \Lambda^\circ + K^-$ (*c*) $\Omega^- \to \Sigma^- + K^0$ **[Ans.** (*c*)]

12. Find the threshold energy for the relation $p + p \to p + p + p. + p$ **[Ans.** 5630 MeV]

13. Which of the following strong reactions is possible?

(*a*) $p + p \to p + n + K^+$

(*b*) $p + p \to \lambda^0 + K^0 + p + p^+$ **[Ans. (*b*)]**

14. For the decay $\Lambda^0 \to p + \pi^-$, with the Λ^0 initially at rest, find the π^- kinetic energy. **[Ans.** 32.7 MeV]

15. For the decay $\pi^- \to e^- \ \overline{\nu}_e$, with π^- initially at rest, find the e^- kinetic energy. **[Ans.** 69.3 MeV]

16. For the decay $K^+ \to \pi^+ + \pi^0$ calculate the π^+ kinetic energy in $\to 2\gamma$ the K^+ rest frame (which is the centre of mass frame). **[Ans.** 108 MeV]

17. The μ^+ decay weakly to a positron plus neutrions. Why neutrious are observed in the final state? **[Ans.** $\nu_e + \nu_\mu$]

CHAPTER

9

X-RAYS

9.1. DISCOVERY OF X-RAYS

In 1895, Roentgen discovered X-rays, which are produced when a beam of cathode rays strikes a solid target. He found that the operation of at cathode-ray tube produced fluorescence in a screen covered with barium-platinocyanide and placed at some distance from the tube. In studying the properties of these new type of radiations, Roentgen noticed that if opaque materials were placed between the tube and the screen the intensity of the fluorescence decreased but did not disappear, showing that X-rays can penetrate opaque substances. It was also found that the X-ray radiations can blacken a photographic plate and produce ionisation in any gas through which they pass. The X-rays were found to travel in straight lines from the source and could not be deflected by electric and magnetic fields, from which it was concluded that they are not charged particles. Later work showed that X-rays can be reflected, refracted and diffracted, and, therefore, these rays are an electromagnetic radiations like light but of much shorter wavelength.

9.2. PRODUCTION OF X-RAYS (ROENTGEN TUBE)

It consists of an evaluated glass bulb containing a concave cathode of Aluminium C. Concave cathode focusses a beam of cathode rays on the tungsten anti cathode. This anti cathode is inclined at an angle of 45° to the path of the cathode rays. Anti cathode is sometimes called target. The anode A is an aluminium rod connected externally to the anti cathode. The anode helps to make the emission of X-rays steady. Since the cathode rays produce a large heating effect, the anti-cathode is usually fitted with either a water cooling or air cooling device. This tube is worked by an induction coil producing a voltage ranging from 10^3 to 10^5 volts. When the voltage applied is small the X-rays produced are soft and when the voltage applied is high the X-rays produced are hard. An inductance coil does not give a strictly unidirectional current, therefore, a rectifying valve arrangement is included in the secondary circuit of the induction coil. The cathode-rays fall on the anti-cathode and hence X-rays are produced.

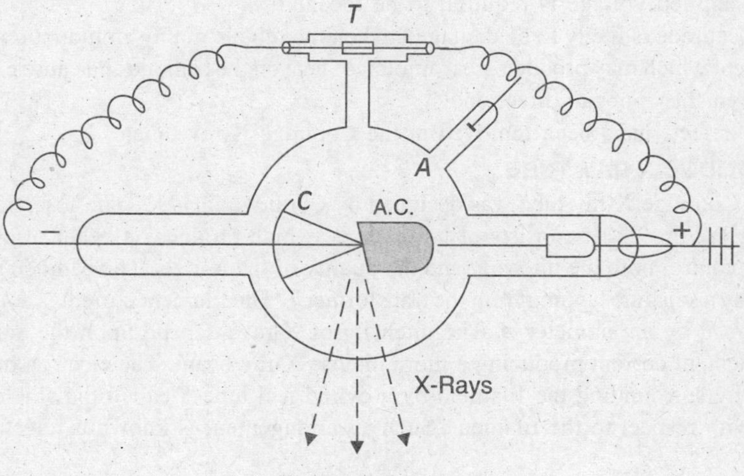

Fig. 9.1

THE ANTI-CATHODE SHOULD POSSESS THE FOLLOWING PROPERTIES:

(*i*) It should have high atomic number.

(*ii*) It should have high melting point.

(*iii*) It should have high conductivity towards heat and electricity.

(*iv*) It should have low vapour pressure at high temperature.

Tantalum, platinum and tungsten fulfil these requirements, and of these tungsten is the best. A heavy block of this metals is used as anti cathode as shown in Fig. 9.1.

THE FUNCTION OF ANODE

The function of the anode is very complex, but when used, it has the following effects:

(*i*) The life of the tube is increased.

(*ii*) The discharge becomes more smooth and hence the intensity "of the X-rays beam is more constant.

(*iii*) The efficiency of the tube is increased, *i.e.*, more of the energy of the cathode-rays is converted into X-rays.

(*iv*) It avoids the damage of the tube, when it is run with an induction coil, whose voltage is not unidirectional.

The functioning of the tube depends upon the pressure of the gas inside the tube and the potential difference applied. At low pressure, say of the order of 10^{4} cms., a comparatively low voltage is needed and, therefore, X-rays produced are called *'soft X-rays'*. If the vacuum is high, of the order of 10^{-5} cms., a comparatively high voltage is needed to run the tube and the X-rays produced have large penetrating power and are said to be *'hard X-rays'*.

During the working of the X-ray tube, the positive ions of the gas are produced. These positive ions should be removed, and hence the pressure inside the tube is lowered, therefore, a greater potential is needed to run the discharge, and the emitted X-rays are harder. If the pressure falls considerably, the tube may even refuse to allow the discharge. To avoid this, 'softening' devices are employed. At first, Palladium rod was used, one half of which was inside the tube. On heating the outer part of the rod with a Bunsen burner, the occluded hydrogen is liberated and enters the X-ray tube and thus increases its conductivity. A better softening device is a side tube T containing platinised asbestos in which gases have been occluded. When the tube becomes too hard, the discharge passes through the side tube *T*, liberating some occluded gases and restore the wroking conditions.

THE MAIN DEFECTS OF THE ROENTGEN TUBE ARE:

(*i*) The intensity of X-rays cannot be controlled.

(*ii*) The softening device is not perfect.

(*iii*) The quality of X-rays cannot be controlled.

(*iv*) The applied voltage is required to be too high.

(*v*) The cathode is likely to be damaged, when the tube is run by an induction coil by the reverse current which may produce great amount of heat on the cathode and cause excitation of cathode rays in the opposite direction.

The defects have been removed in the Coolidge X-ray tube.

9.3. COOLIDGE X-RAY TUBE

The Coolidge X-ray tube was designed by Coolidge in 1912. The essential parts of this tube are shown in Fig. 9.2. It consists of hot cathode which produces a beam of electrons which can be used to control both the intensity and the quality of the X-rays. The cathode is heated by 6 volts battery or by a separate tapping from the transformer *T*. The filament current is adjusted by a Rheostat *R*, and is read by an ammeter *A*. The intensity of X-rays depend upon the value of current, *i.e.*, a larger filament current producing a more intense X-ray beam. The electron beam is focussed on the target by surrounding the filament by a cylindrical tube *S* called the shield which is charged negative with respect to the filament. Such an arrangement is known as electron gun.

The anti-cathode or the target is cooled by means of radiating fans *F*. In better types of tubes water cooling arrangement is provided. The high voltage is taken from a large step-up transformer *T*. The X-ray tube works as its own rectifier. The current will only pass when the anti-cathode is positive with respect to the filament, since the emitted electrons can move only from the filament (negative) to the anti-cathode (positive) and not vice-versa. The anode current through the tube is read by the milliammeter and is regulate by the rheostat (not shown in figure). This regulates the applied voltage across the tube and controls the quality of X-rays produced. A larger voltage or current in milliammeter produces harder and more penetrating X-rays.

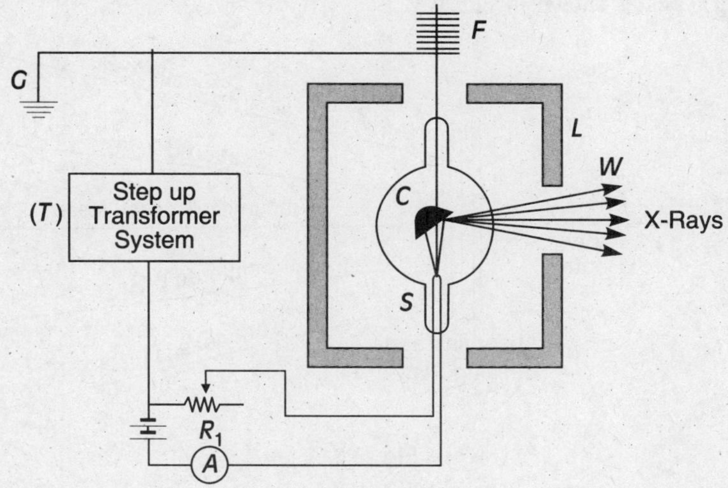

Fig. 9.2

The tube is enclosed in lead box *L* about 0.5 cm thick and is provided with a small window W through which the X-rays emerge.

In an improved design of the tube cooling is done by rotating anode. The anode, in this case, consists of a tungsten disc which is rotated at high speed by a rotating magnetic field applied to an iron rotor on which the disc is mounted. Thus each part of the target is bombarded and heated for a very small fraction of a revolution and cools during the rest of its journey. By this arrangement, large tube currents are possible and the X-rays produced are harder. These are generally used for short intense exposure as per radiographs of the heart.

The highest voltage X-ray tube uses about a million volts, and produces X-rays of 10^{-10} cm wavelength.

9.4. PROPERTIES OF X-RAYS

1. X-rays are are not deflected by electric and magnetic fields. This property shows that X-rays do not consist of any charge.

2. They are highly penetrating and can pass through many solids which are opaque towards light *e.g.,* wood, flesh etc. Soft X-rays have small penetrating power and are easily absorbed, and hard X-rays have relatively high penetrating power. The transparency depends upon the density of the material. The denser the substance the less transparent it is to X-rays. A sheet of lead one millimeter thick absorbs these X-rays whereas the rays can pass through the same thickness of aluminium.

THE PENETRATING POWER OF X-RAYS DEPENDS UPON:

(*a*) The applied potential difference.

(*b*) The nature of the anti-cathode material. The greater the potential difference and higher the atomic weight of the anti-cathode, the more penetrating X-rays are produced.

3. They cause flourescence in many substances, *e.g.,* barium platino-cyanide, cadmium tungsten, zinc sulphide etc.

4. They affect a photographic plate and are even more effective than light.

5. They ionise a gas and also knockout electrons from certain metals on which they fall. This effect is known as photoelectric effect.

6. They travel in straight line with the velocity of light, *i.e.*, 3×10^8 ms^{-1}.

7. Like light, X-rays consist of electromagnetic waves of very short wavelength and show reflection, refraction, interference, diffraction and polarisation in a similar way.

8. When X-rays (primary) coming from the anti-cathode of an X-ray tube fall on a plate of some chosen material, they give rise to a complex phenomenon of secondary radiations which consist of three different types as shown in Fig. 9.3.

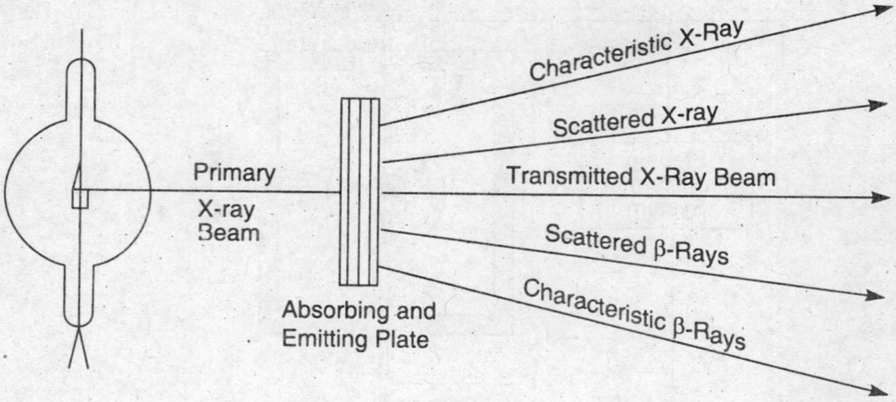

Fig. 9.3

(i) **Scattered X-rays.** They are practically of the same nature and wavelength as the primary X-rays. Their properties do not depend upon the nature of the scattering substance.

(ii) **Characteristic X-rays.** The wavelength of the characteristic X-rays is equal to or less than the primary X-rays but it is characteristic of the substance upon which the primary X-rays are incident and do not depend upon the wavelength of the primary X-rays. *The characteristic X-rays from substances of high atomic weight are of shorter wavelength than those from substance of low atomic weight and hence are more penetrating.* Every substance gives rise to a *particular* set of characteristic X-rays consisting of one or more series known as *K, L, M* etc. The X-rays in *K* series of element have smaller wavelength than in *L, M* or *N* series, whereas the wavelength of the X-rays in the same series, say *K* or *L*, produced from an element of higher atomic weight is less than that produced from an element of low atomic weight.

(iii) **β-rays.** They consist of fast moving electrons produced by the photoelectric efffect. *Their properties are independent of the nature of the scattering substance but depend upon the quality of primary X-rays.*

9. They have destructive effect on living tissues. Exposure of humam body to X-rays causes reddening of the skin, like the sun burn, while prolonged exposure produces surface sores and injures the tissues of the body. The white corpuscles in the blood are killed by prolonged exposure to X-rays and may develop into gangrene. To shield the effect of X-rays, the operators use lead-lined aprons and lead glasses.

9.5. THE ORIGIN OF X-RAYS

X-rays, like visible light, originate from the jumping of an electron from one orbit to another. When high-speed electrons from the cathode of an X-ray tube strike the target, they ionize many of the atoms composing the surface layers of the metal.

Owing to their very high speeds (about 10% the velocity of light), the electrons penetrate the atoms and remove an electron from the inner shells by collision. This is illustrated in Fig. 9.4 where an electron is knocked out of the K shell.

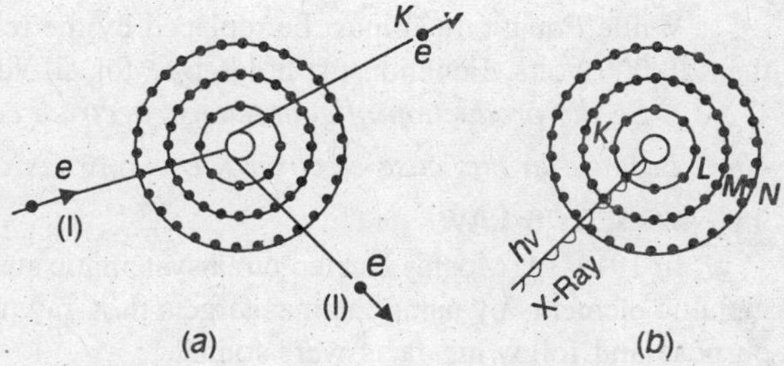

Fig. 9.4

The designations K, L, M, N, O, P etc., for the various electron shells originated with the X-ray spectroscopist and are identical with the quantum numbers $n = 1, 2, 3, 4, 5, 6$ etc. When an electron is missing in the innermost K shell, a nearby electron from the next shell beyond jumps into the vacant space, simultaneously emitting a photon of energy $h\nu$. Such X-rays, arising from millions of atoms, produce the K lines shown in Fig. 9.5. Since the L shell now has one less electron, an M electron can jump into the L shell vacancy, with the consequent emission of another but different X-rays frequency. These are the L lines in Fig. 9.5. The jumping process continues until the outermost shell is reached, where an electron jumping gives rise to visible light. Thus we see how it is possible for a single atom to emit X-rays of different wavelengths.

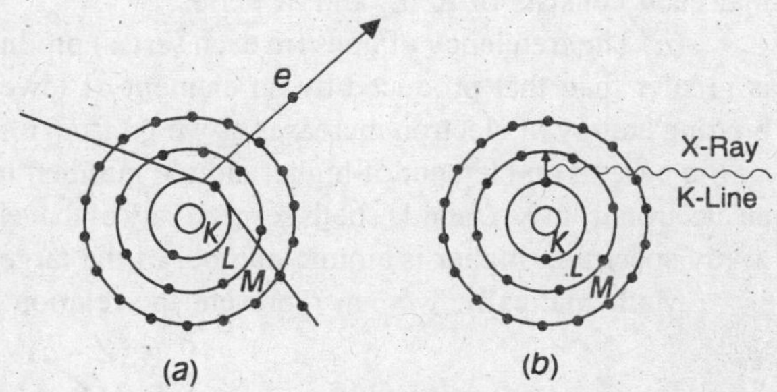

Fig. 9.5

The continuous X-ray spectrum is due to another phenomenon often referred to as "Bremsstrahlung". These radiations are due to the slowing down of high speed electrons as they pass close to the nuclei of the atoms within the target of the X-ray tube. The process is illustrated in Fig. 9.6. As the electron passes through the atom, it is attracted by the positive charge of the nucleus and is deflected in its path.

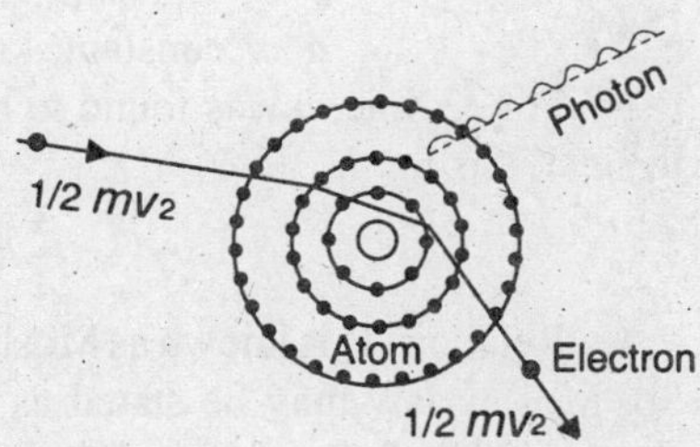

Fig. 9.6

During the deflection of the electron in the strong electric field of the nucleus, a light wave of energy $h\nu$ is emitted. Since the law of conservation of momentum must hold for such a collision, the electron is deflected off to one side of the atom, and the photon off to the other. Since the law of conservation of energy must hold, some of the energy of the incoming electron $1/2mv^2$ is given up to the newly created photon $h\nu$, and the remainder $1/2mv^2$ is retained by the electron. Thus the electron is slowed down to a velocity v', by the encounter. The closer the electron comes to the nucleus, the greater is its loss in velocity and energy, and the greater is the frequency and energy of the radiated photon. By the conservation of energy,

$$\boxed{\frac{1}{2}mv^2 - \frac{1}{2}mv'^2 = h\nu} \quad \text{... (1)}$$

The highest frequency that is possible is one in which the electron is completely stopped by the atom. In this special case,

$$\frac{1}{2}mv^2 = h\nu_{\text{max}} \quad \text{... (2)}$$

Since the kinetic energy of the electons in the beam striking the target is given by the voltage V applied to the tube, we may use Equation

$$Ve = \frac{1}{2}mv^2 \quad \text{...(3)}$$

and obtain

$$Ve = h\nu_{\text{max}} \quad \text{...(4)}$$

While Equation (3) must be replaced by the relativistic equation, when voltages are greater than 20,000 volts, Equation (4) holds true for all voltages.

So the production of continuous spectrum can be considered to be inverse photoelectric effect because in this case electron K.E. $\frac{1}{2}mv^2$ *is converted into photon energy* $h\nu_{max}$.

9.6. MOSLEY'S LAW

In 1913-14, Mosley carried out a systematic study of characteristic X-rays spectra of various metallic elements by using them as targets in X-ray tube. A Bragg's spectrometer was used for this purpose and following facts were found:

(*i*) The characteristic X-ray spectra of different elements are similar to each other in the sense that each consists of K, L, and M series.

(*ii*) The frequency of lines (in each series) produced from an element of higher atomic number is greater than that produced by an element of lower atomic number. The reason for this is that binding energy of electron increases as we go from one element to another of higher atomic number *i.e.,* in case of an element of higher atomic number, larger amount of energy is required to liberate an electron from K, L and M shells respectively of that element. Thus, in case of K_α line of characteristic X-ray spectrum, higher is atomic number of the target material, higher is the frequency of K_α line.

Mathematically, we can represent the relation between frequency and charge number Z as

$$\nu \propto (Z-a)^2$$

$$\nu = b\,(Z-a)^2 \qquad \text{... (1)}$$

Where ν = frequency of characteristic radiation

b = constant, which is different for different series

a = constant, known as screening constant and is different for each series.

For K_α line, b was found to be equal to 3/4 R, where R is Rydberg constant and α was found to be 1. So for K_α line

$$\nu_{k_\alpha} = \frac{3}{4}R\,(Z-1)^2 \qquad \text{... (2)}$$

Relation (1) is known as Mosley law or Mosley equation. So, Mosley law may be stated as follows:

"The frequency of a spectral line in characteristics "X-ray spectrum varies directly as the square of atomic number Z of "the element emitting it."

Fig 9.7 show Mosley diagram for K_α and K_β lines obtained by plotting, $\sqrt{\nu}$ versus atomic number of different elements of periodic table. The graph is linear one as expected.

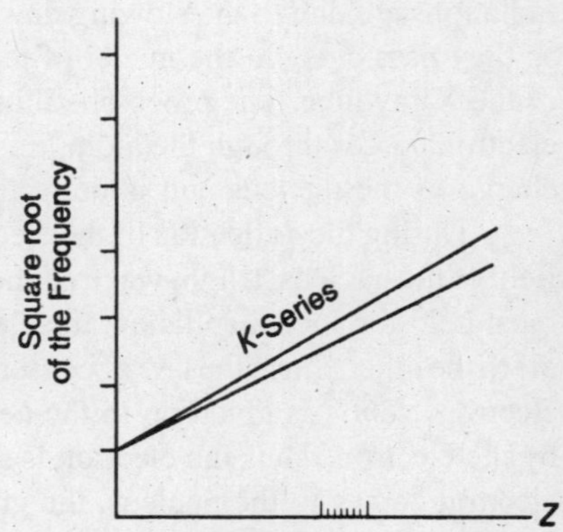

Fig. 9.7

The exact from of Mosley law is

$$\frac{1}{\lambda} = R\,(Z-\sigma)^2\left(\frac{1}{n^2}-\frac{1}{p^2}\right) \qquad \text{...(3)}$$

where σ is a correction factor and n and p are the principal quantum numbers of the energy levels between which the transition takes place.

Derivation of Mosley's law on the basis of Bohr's theory of atom. Consider the atom of an element with atomic number Z. From Bohr's theory, the energy of an electron moving in an orbit of principal quantum number n_1 is given by the relation

$$E_n = -\frac{m\,Z^2\,e^4}{8\,\varepsilon_0^2\,n^2\,h^2} \qquad \text{... (4)}$$

Similarly, the energy of an electron moving in an orbit of principal quantum number n_2 is given by

$$E_p = -\frac{m\,Z^2\,e^4}{8\,\varepsilon_0^2\,p^2\,h^2} \qquad \text{... (5)}$$

Let us consider the effect of electrons which screen the positive charge of the nucleus reducing thereby the value of Z to $(Z - a)$, the above expressions (4) and (5) can be written in the form

$$E_n = \frac{m (Z - a_1)^2 e^4}{8 \varepsilon_0^2 n^2 h^2} \quad \text{... (6)}$$

$$E_p = \frac{m (Z - a_2)^2 e^4}{8 \varepsilon_0^2 p^2 h^2} \quad \text{... (7)}$$

Also energy difference $E_n - E_p$ is given by

$$E_n - E_p = \frac{m e^4}{8 \varepsilon_0^2 h^2} \left[\frac{(Z - a_2)^2}{p^2} - \frac{(Z - a_1)^2}{n^2} \right]$$

where Z is very high $a_1 \cong a_2 \approx a_{02}$, so we have the energy difference as

$$\Delta E = E_n - E_p = \frac{m e^4}{8 \varepsilon_0^2 h^2} (Z - a)^2 \left[\frac{1}{p^2} - \frac{1}{n^2} \right]$$

If the electron is removed from a particular shell, transitions can occur to that shell from higher cells. These transitions give rise to X-ray photons whose frequency ν is given by

$$\nu = \frac{\Delta E}{h} = \frac{m e^4 (Z - a)^2}{8 \varepsilon_0^2 h^3} \left(\frac{1}{p^2} - \frac{1}{n^2} \right) \quad \text{... (8)}$$

Considering the case of K_α line, we have $p = 1$ and $n = 2$.

$$\therefore \quad {}^{\nu}K_\alpha = \frac{m e^4}{8 \varepsilon_0^2 h^3} (Z - a)^2 \left(\frac{1}{1^2} - \frac{1}{2^2} \right)$$

or $${}^{\nu}K_\alpha \propto (Z - a)^2$$

or $${}^{\nu}K_\alpha = b (Z - a)^2 \quad \text{... (9)}$$

where $$b = \frac{m e^4}{8 \varepsilon_0^2 h^3} \left(\frac{1}{1^2} - \frac{1}{2^2} \right) = \text{constant}$$

Equation (9) represents the Mosley law.

9.7. IMPORTANCE OF MOSLEY LAW

According to Mosley's law, it is the atomic number of an element and not the atomic weight which accounts for the physical and chemical properties of an element. Thus this law provides the guideline that elements must be arranged in periodic table according to their atomic number and not according to their atomic weights.

Before Mosley's work, Mandeleev arranged the elements in increasing atomic weight to form the periodic table. It we found that if the ascending order of atomic weights is to be preserved, cobalt $({}_{27}CO^{58.94})$ with atomic weight 58.94 should be placed after nickel $({}_{28}Ni^{58.69})$ with atomic weight 58.69. Similarly, potassium $({}_{19}K^{39})$ should come before argon $({}_{18}A^{40})$. Mandeleev observed that to preserve the periodicity of chemical and physical properties, their orders should be reversed. This difficulty was removed by Mosley law because as per their atomic numbers, their order should be opposite to what it is according to Mandeleev.

By considering the X-ray spectra of rare earths, this law has been helpful in determining their atomic numbers and also fixing their positions in periodic table at appropriate positions.

This law also led to the discovery of some new elements like hafnium (72), prometheum (61) etc.

9.8. X-RAY SPECTRUM

As explained above X-ray spectrum consists of

(*a*) continuous spectrum and (*b*) line spectrum. These two are shown in Fig. 9.8 (*b*)

(*a*) Continuous Spectrum:

(*i*) It is produced due to the de-acceleration of high velocity electrons when they are deflected while passing near the positively charged nucleus of an atom of the target material.

(*ii*) It has a sharply-defined short-wavelength limit given by

$$\lambda_{\min} = \frac{12,400}{V} \times 10^{-10} \text{ m or } \frac{12,400}{V} \text{ Å}$$

(*iii*) The cut-off wavelength $\lambda_{\min}$ is independent of the nature of the target material but is inversely proportional to the potential difference between the cathode and anode of an X-ray tube. The value of $\lambda_{\min}$ decreases as this potential difference is increased (Fig. 9.8).

(*iv*) The intensity of the continuous spectrum is found very nearly proportional to the *square of the applied voltage* for a given target and to the atomic number of the target material when a constant potential difference is applied.

(*v*) There is a shift of the maximum intensity position towards the short wavelength side as voltage is increased.

(*b*) Line Spectrum:

(*i*) It is produced when electons are dislodged from the innermost orbits of the atoms of the target material followed by electron jumps from outer orbits.

(*ii*) It consists of discrete spectral lines which constitute *K*-series, *L*-series and *M*-series etc. *K*-series consists of those lines for which electron jumps end at *K*-level.

Similarly, *L*-series and *M*-series are produced when jumps end at *L* and *M*-levels (Fig. 9.8)

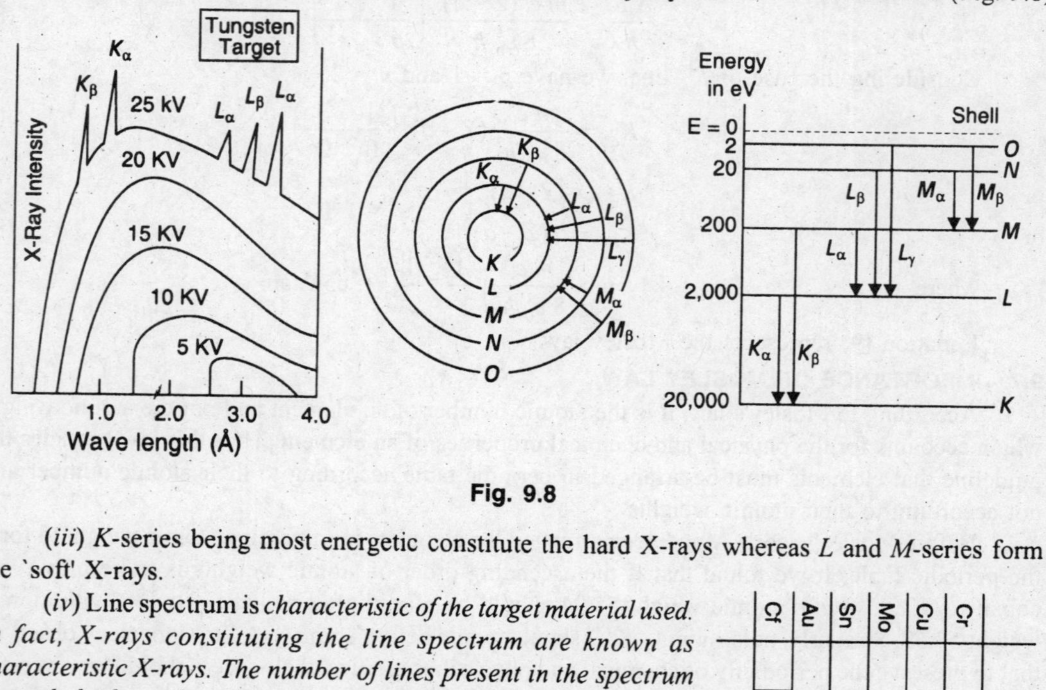

Fig. 9.8

(*iii*) *K*-series being most energetic constitute the hard X-rays whereas *L* and *M*-series form the 'soft' X-rays.

(*iv*) Line spectrum is *characteristic of the target material used. In fact, X-rays constituting the line spectrum are known as characteristic X-rays. The number of lines present in the spectrum depends both on the nature of target material and the excitation voltage.*

(*v*) There is a regular shift towards shorter wavelength in the *K*-spectrum as the atomic number of the target is increased (Fig. 9.9). The exact relationship, as found by Mosley, is given as

$$\frac{\lambda_2}{\lambda_1} = \frac{\nu_1}{\nu_2} = \frac{(Z_1 - 1)^2}{(Z_2 - 1)^2}$$

where ν_1 is the frequency of the K_α line for a target material having an atomic number of Z_1 and ν_2 and Z_2 are similar quantities for some different target material.

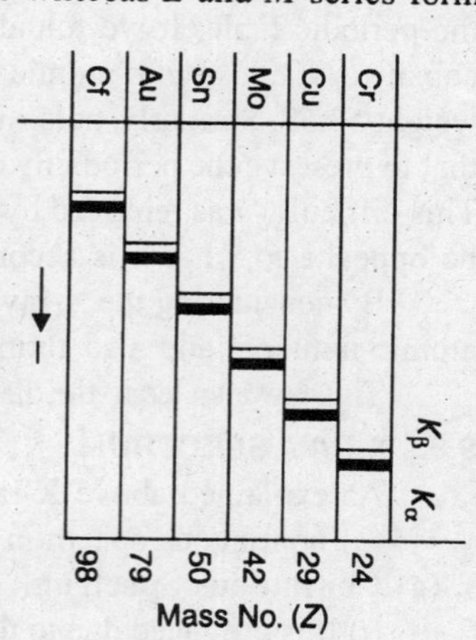

Fig. 9.9

Example 1. *Calculate the energy in electron volts of a photon of wavelength 5500×10^{-10} m.*

Given: $h = 6.62 \times 10^{-34}$ j-s.

Solution: Given: $\lambda = 5500 \times 10^{-10}$ m.

$h = 6.62 \times 10^{-34}$ J-s.

Velocity of light $c = 3 \times 10^8$ ms^{-1}

If ν be the frequency of radiation emitted, then we have

$$c = \nu\lambda$$

$$\therefore \quad \nu = \frac{c}{\lambda} = \frac{3 \times 10^8}{5500 \times 10^{-10}} = \frac{3}{55} \times 10^{16}$$

$$= 5.5 \times 10^{14} \text{ Hz.}$$

Energy of photon $= h\nu$

$$= 6.62 \times 10^{-34} \times 5.5 \times 10^{14} \text{ joules}$$

$$= 36.41 \times 10^{-20} \text{ joules}$$

$$= \frac{36.41 \times 10^{-20}}{1.6 \times 10^{-19}} = 2.275 \text{ eV}$$

$\therefore$ Energy of photon $=$ **2.275 eV.**

Example 2. *Calculate the cut-off wavelength when the potential difference applied to the X-ray tube is 25 kV.*

Solution: We have the relation

$$\lambda_{\text{min}} = \frac{12,400}{V} = \frac{12,400}{25,000} = \mathbf{0.496} \text{ Å}$$

Note: Corresponding maximum frequency of the emitted X-rays is

$$f_{\text{max}} = \frac{c}{\lambda_{\text{min}}} = \frac{3 \times 10^8}{0.495 \times 10^{-10}} = \mathbf{6.05 \times 10^{18}\ Hz}$$

Example 3. *Electron bombarding the anode of a Coolidge tube, produce X-rays of wavelength 1Å. Find the energy of each electron at the moment of impact given $h = 6.62 \times 10^{-34}$ J-s.*

$$e = 1.6 \times 10^{-19}\ C.$$

Solution: Given that

$$\lambda = 10^{-10} \text{ m}$$

$$h = 6.62 \times 10^{-34} \text{ J-s}$$

$$e = 1.6 \times 10^{-19} \text{ coulomb}$$

Velocity of light $c = 3 \times 10^8$ ms^{-1}

If V be the potential difference applied between cathode and anti-cathode, then

$$Ve = \frac{hc}{\lambda} \quad \text{... (i)}$$

But,

$$Ve = \text{K.E.}$$

$$Ve = \frac{1}{2} \text{m}V^2 \quad \text{... (ii)}$$

From (*i*) and (*ii*), we get

$$\therefore \quad \text{K.E.} = \frac{hc}{\lambda}$$

$$= \frac{6.62 \times 10^{-34} \times 3 \times 10^8}{10^{-10}} \text{ joules}$$

$$= \frac{19.86 \times 10^{-16}}{1.6 \times 10^{-19}} \text{ eV}$$

$$= \frac{19.86 \times 10^3}{1.6} = \text{eV}$$

$$\therefore \qquad \textbf{Energy = 12.41 keV.}$$

Example 4. *An X-ray tube operates at 50 kV. Calculate the shortest wavelength of X-ray produced.*

Given: $h = 6.62 \times 10^{-27}$ ergs-sec.

Charge on electron $e = 4.8 \times 10^{-10}$ e.s.u.

Solution: Given that

$$V = 50 \text{ kV} = 50 \times 10^3 \text{ volts}$$

$$= \frac{50 \times 10^3}{300} \text{ e.s.u.} = \frac{500}{3} \text{ e.s.u.}$$

$$e = 4.8 \times 10^{-10} \text{ e.s.u.}$$

$$h == 6.62 \times 10^{-27} \text{ ergs-sec.}$$

Velocity of light $\qquad c = 3 \times 10^{10}$ cm/sec.

Energy of X-ray with charge e and potential difference V is given by

$$E = Ve$$

But

$$E = h\nu \qquad \text{... (1)}$$

or

$$E = \frac{hc}{\lambda} \qquad \left(\because c = \nu\lambda \text{ or } \nu = \frac{c}{\lambda}\right) \qquad \text{... (2)}$$

From (1) and (2), we get

$$\frac{hc}{\lambda} = Ve$$

or

$$\lambda = \frac{hc}{Ve} = \frac{6.62 \times 10^{-27} \times 3 \times 10^{10}}{300 \times 4.8 \times 10^{-10}} \times \frac{3}{1}$$

$$= \frac{6.62 \times 9}{500 \times 4.8} \times 10^{-7} = 0.2482 \times 10^{-8} \, cm$$

$$\lambda = 0.2482 \text{ Å}$$

Therefore, the wavelength of X-ray is **0.2482 Å.**

Example 5. *If the potential difference applied across an X-ray tube is 5 kV and the current through it is 2mA, calculate*

(i) the number of electrons striking the target per second and

(ii) the speed at which they strike it.

Take $e = 1.6 \times 10^{-19}$ C and $m = 9.1 \times 10^{-31}$ kg.

Solution: (*i*) If n is the number of electrons striking the anode per second, then

$$I = ne$$

$$n = \frac{1}{e} = \frac{2 \times 10^{-3}}{1.6 \times 10^{-19}} = \mathbf{1.25 \times 10^{16}\ s^{-1}}$$

(*ii*) We have the relation

$$\frac{1}{2} mv^2 = eV$$

$$v = \sqrt{\frac{2\,e\,V}{m}} = 5.93 \times 10^5 \sqrt{V} \text{ m/s}$$

$$= 5.93 \times 10^5 \times \sqrt{5000} = \mathbf{4.15 \times 10^7\ ms^{-1}}$$

Examle 6. *Deduce the wavelength of K_α line for an atom Z = 92, using Mosley's law. Given Rydberg constant $R = 1.1 \times 10^7 m^{-1}$*

Solution: The spectral line K_α is emitted when an electron jumps from an orbit of principal quantum number 2 to 1. Its wavelenght λ is given by

$$\bar{\nu} = \frac{1}{\lambda} = \frac{\nu}{c} = R(Z-a)^2\left(\frac{1}{1^2}-\frac{1}{2^2}\right)$$

But for K series $a = 1$. Therefore,

$$\frac{1}{\lambda} = \frac{3}{4}R(Z-1)^2$$

Putting the given values of R and Z,

$$\frac{1}{\lambda} = \frac{3}{4}(1.1\times10^7)(92-1)^2$$

$$= 6.83\times10^{10}\ \text{m}^{-1}$$

$$\therefore \quad \lambda = \frac{1}{6.83\times10^{10}} = 0.146\times10^{10}\ \text{m}$$

$$= \mathbf{0.146\ Å}$$

Example 7. *An X-ray tube works at 18 kV. Find the maximum speed of electrons striking the anticathode. Given mass of electron 9×10^{-28} gm and charge of electron $= 4.8\times10^{-10}$ e.s.u. and 1 volt $= \frac{1}{300}$ e.s.u.*

Solution: Given that

Potential difference $\quad V = 18\ \text{kV} = 18\times10^3$ Volts

$$= \frac{18\times10^3}{300} = 60\ \text{e.s.u.}$$

Mass of electron $\quad = 9\times10^{-28}$ gm

Charge on electron $\quad e = 4.8\times10^{-10}$ e.s.u.

If the velocity of electron $\quad = v$, then

$$\text{K.E.} = \frac{1}{2}mv^2 = Ve$$

$$mv^2 = 2Ve$$

$$v = \sqrt{\frac{2Ve}{m}} = \sqrt{\frac{2\times60\times4.8\times10^{-10}}{9\times10^{-28}}}$$

$$= \sqrt{64}\times10^9$$

$$v = \mathbf{8\times10^9\ cm/sec. = 8\times10^7\ ms^{-1}}$$

Example 8. *Calculate the minimum applied potential required to produce X-rays of 1Å wavelength.*

Solution: We have $\lambda_{min} = 12{,}400/V$ Å

$$\therefore \quad V = 12{,}400/\lambda_{min} = \frac{12{,}400}{1\times10^{-10}} = \mathbf{12.4\ kV.}$$

Example 9. *An X-ray tube operated at 40 kV emits a continuous X-ray spectrum with a short wavelength limit $\lambda_{min} = 0.310$ Å. Calculate the Planck's constant.*

Solution: We know

$$Ve = h\nu_{max} = \frac{hc}{\lambda_{min}}$$

or

$$h = \frac{Ve\,\lambda_{min}}{c}$$

Substituting $\quad V = 40\ \text{kV} = 40\times10^3$ volts, $e = 1.6\times10^{-19}$ Coulomb

$$\lambda_{min} = 0.310\ \text{Å} = 0.310\times10^{-10}\ \text{m and } c = 3\times10^8\ \text{m/s}$$

$$h = \frac{40\times10^3\times1.6\times10^{-19}\times0.310\times10^{-10}}{3\times10^8}$$

$$= \mathbf{6.61\times10^{-34}\ Joule - sec.}$$

Example 10. *Stopping potentials of 24, 100, 110 and 115 kV are measured for photo-electrons emitted from a certain element when it is irradiated with monochromatic X-rays. If this element is used as a target in an X-ray tube, what will be the wavelength of the K_α line?*

Solution: The stopping potential energy, eV_s, is equal to the difference between the energy of the incident photon and the binding energy of the electron in particular shell.

$$eV_s = E_p - E_B$$

The different stopping potentials arise from electrons being emitted from different shells, with the smallest value (24 kV) corresponding to ejection of a *K*-shell electron. Subtracting the expression for the two smallest stopping potentials, we obtain

$$eV_{sL} - eV_{sK} = (E_p - E_{BL}) - (E_p - E_{BK}) = E_{BK} - E_{BL}$$

or

$$100 \text{ keV} - 24 \text{ keV} = E_{BK} - E_{BL}$$

This difference, 76 keV, is the energy of the K_α line. The corresponding wavelength is

$$\lambda = \frac{hc}{E_{BK} - E_{BL}} = \frac{12.4 \text{ keV Å}}{76 \text{ keV}} = \mathbf{1.63\ Å}$$

Example 11. *When 0.50 Å X-rays strike a material, the photo-electrons from the K shell are observed to move in a circle of radius 23 mm in a magnetic field of 2×10^{-2} T. What is the binding energy of K-shell electrons?*

Solution: The velocity of the photo-electrons is found from F = ma

$$evB = m\frac{v^2}{R} \qquad \text{or} \qquad v = \frac{e}{m}.B.R.$$

This kinetic energy of the photo-electrons is then

$$K = \frac{1}{2}mv^2 = \frac{1}{2}\frac{e^2B^2R^2}{m}$$

$$= \frac{1}{2}\frac{(1.6\times10^{-19}\text{ C})^2(2\times10^{-2}\ T)^2(23\times10^{-3}\text{ m})^2}{(9.11\times10^{-31}\text{ kg})}$$

$$= 2.97\times10^{-15}\text{J}$$

or

$$K = (2.97\times10^{-15}\text{J})\frac{1 \text{ keV.}}{1.6\times10^{-16}} = 18.6 \text{ keV}$$

The energy of the incident photon is

$$E_v = \frac{hc}{\lambda} = \frac{12.4 \text{ keV.Å}}{0.50 \text{ Å}} = 24.8 \text{ keV}$$

The binding energy is the difference between these two values

$$BE = E_v - K = 24.8 \text{ keV} - 18.6 \text{ keV} = \mathbf{6.2\,keV.}$$

Example 12. *In Zn (Z = 30) the ionization (binding) energies of the K and L shells are, respectively, 9.659 keV and 1.021 keV. Determine the kinetic energy of an Auger electron emitted from the L shell by a K_α X-ray.*

Solution: The energy of K_α X-ray is

$$E_{k_\alpha} = E_L - E_k = -1.021 \text{ keV} - (-9.659 \text{ keV}) = 8.638 \text{ keV}$$

The kinetic energy of the Auger electron will be equal to the difference, between the K_α photon energy and binding energy of the electron in the *L* shell:

$$K = E_{k_\alpha} - E_{BL} = 8.638 \text{ keV} - 1.021 \text{ keV} = \mathbf{7.617\,keV}$$

9.9. ABSORPTION OF X-RAYS

We can study the absorption of X-rays with an apparatus shown in Fig. 9.10. A narrow beam of monochromatic X-rays when passes through matter, some part of it is absorbed while the rest is transmitted. X-rays are made to pass through slits S_1 and S_2 of lead plates. So that a narrow beam of it is allowed to fall on a crystal and due to Bragg reflection, we get a monochromatic beam and then this beam is made to enter ıonization chamber in which the current of ionization produces will measure the intensity of X-rays.

After knowing the intensity of X-ray a sheet of absorbing material is put in the path of X-rays before it enters the ionization chamber resulting in the reduction of intensity of X-rays due to absorption of X-rays by the absorber.

The ratio $\frac{I}{I_0}$ *i.e.,* the currents before and after absorption is called absorption coefficient of the material.

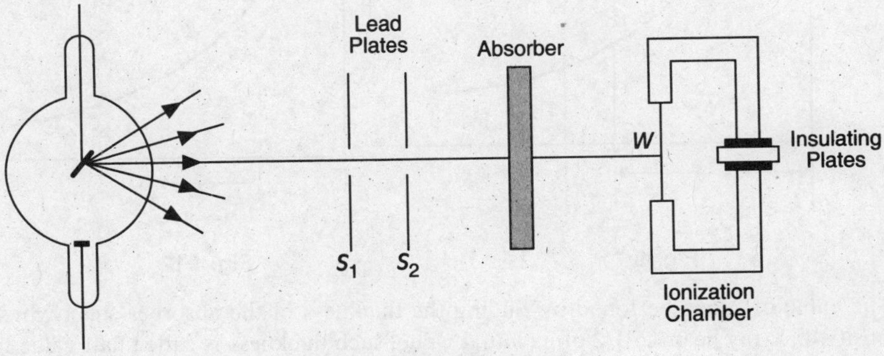

Fig. 9.10

We can get a relation for absorption coefficient by considering I and I_0 as the intensity of X-rays before and after it passes through an absorber of thickness x.

Let dI be the further decrease in intensity when X-rays travel through an additional thickness dx of the absorber, then we have $-\frac{dI}{dx}$ as the decrease of intensity with increase in distance dx of the absorber and this will be proportional to the original intensity of X-rays *i.e.,*

$$-\frac{dI}{dx} \propto I$$

or

$$\frac{dI}{dx} = -\mu I \qquad \text{... (1)}$$

where μ is the constant of proportionality and is called linear absorption coefficient of the absorber or linear attenuation coefficient of the absorber. Sometimes it is also called macroscopic absorption coefficient.

From (1), we have

$$\frac{dI}{I} = -\mu \, dx$$

Integrating both sides, we have

$$\int_{I_0}^{I} \frac{dI}{I} = -\mu \int_0^x dx$$

$$\therefore \quad \log_e \frac{I}{I_0} = -\mu x$$

or

$$I = I_0 \, e^{-\mu x} \qquad \text{... (2)}$$

which shows that the intensity of X-rays decreases exponentially with thickness of the material.

We can define the absorption coefficient from equation (2) as

$$\mu = -\frac{\frac{dI}{I}}{dx}$$

i.e., it is equal to the fractional decrease in the intensity of X-rays per unit thickness of the absorber. It is found that μ depends on the wavelength of X-rays nearly by the relation $\mu \propto \lambda^3$.

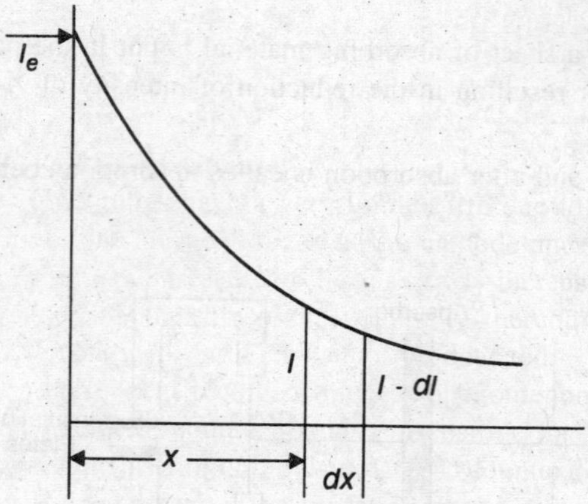

Fig. 9.11

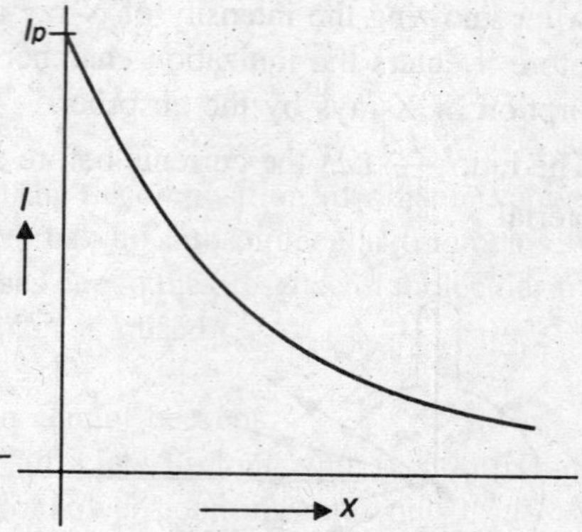

Fig. 912

The value of μ can be found by finding the thickness of the absorber sheet which reduces the intensity of X-ray beam to 1/2 of its initial value, such thickness is called half value layer (*hvl*) and is denoted by *T*.

If we put $\frac{I}{I_0} = \frac{1}{2}$ and $x = T$ in the equation (2), we get

$$\therefore \quad e^{-\mu T} = \frac{1}{2}$$

$$\therefore \quad e^{\mu T} = 2$$

$$\mu T = \log_e 2 = 0.693$$

$$\therefore \quad \mu = \frac{0.693}{T}$$

which gives another definition of absorption coefficient in terms of *hvl*.

In practice, however, it is $\frac{\mu}{\sigma}$ called *mass absorption coefficient* μ_m *is more important than* μ *and may be defined as equal to the fractional decrease in intensity of a beam of unit cross-section in passing through a unit mass of the absorbing material.*

μ_m is found to depend on the wavelength λ of X-rays and also on atomic number *Z* of absorber. A relation of it is given by

$$\mu_m = D\lambda^3 Z^3 \qquad \dots (4)$$

9.10. X-RAY DIFFRACTION

Due to the short wavelength of X-ray, which is of the order of 10^{-8} cm ordinary grating which have 6,000 lines per cm cannot produce any appreciable diffraction with X-rays. Therefore, a crystal grating in which the atoms are arranged in a regular pattern is used for X-ray diffraction. The arranged atoms in the crystal grating correspond to the grating lines and the distance between two atoms, which is of the order of 10^{-8} cm, from the grating element. The difference between crystal grating and optical grating is that the atomic centres of diffraction in the crystal grating are not all in one plane but distributed in space, while in case of optical grating they are limited to one plane.

The crystal is thus a three-dimensional space grating rather than a two-dimensional plane grating. When a narrow pencil of X-rays is made to pass through a thin crystal, the diffracted rays emerged in various well-defined directions to form a symmetrical pattern of spots on a photographic plate placed perpendicular to the beam.

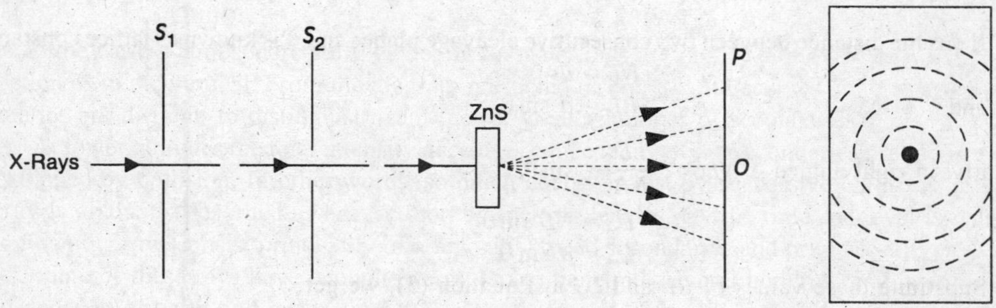

Fig. 9.13

A narrow beam of X-rays is allowed to pass through a thin crystal of Zinc blende (ZnS) and transmitted beam is received on a photographic plate P as shown in Fig. 9.13. After an exposure of many hours, it can be found on developing the plate that in addition to a central black spot at O where the direct beam struck the plate, there are many other fainter spots regularly arranged, indicating that the X-ray, have been diffracted from the various planes of the crystal belong. These spots, known, as Laue spots form a geometrical pattern depending upon the particular crystalline system to which the crystal belongs since different Laue patterns are obtained with different crystalline configurations.

9.11. BRAGG'S LAW

Since the crystal constitutes a three-dimensional grating, therefore, the theory of Laue spots is very complex. Prof. W.L. Bragg has given a very simple interpretation of the diffraction pattern obtained by supposing that reflection takes place from the various planes in the crystal. The atom in a crystal are arranged in regular fashion having a constant space between them. When a beam of X-rays falls on the crystal, each atom acts as a scattering centre and emits secondary wavelets whose envelope gives rise to the reflected wave-front according to Huygen's construction as in optics. Since, X-rays are much more penetrating than ordinary light, therefore, only partial reflection takes place at each layer of atoms.

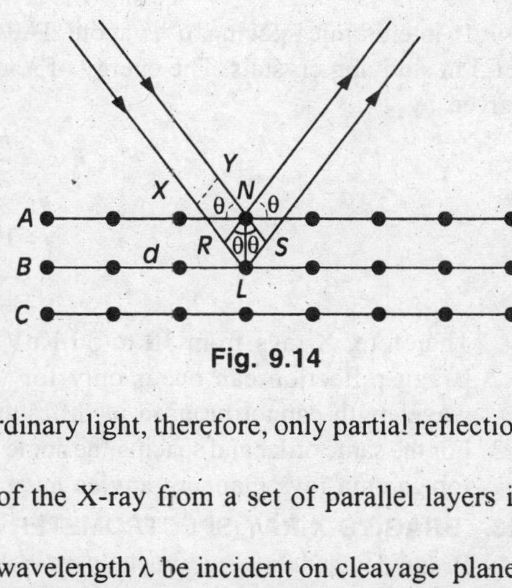

Fig. 9.14

The condition for measurement of wavelength of the X-ray from a set of parallel layers in given plane is obtained as follows:

Let a narrow beam of monochromatic X-rays of wavelength λ be incident on cleavage planes *A, B, C*... etc., in a crystal and let XY be an advancing wave-front which meets the cleavage plane of the crystal at a glancing angle θ at *X*. The ray *XL* is reflected at the cleavage plane *B* whereas the ray *YN* is reflected at the cleavage plane *A*. If the waves reflected from these planes of atoms are in phase with each other at *N* they will reinforce each other, but if they are in opposite phases they will cancel other's effect and there will be no reflected beam.

For reinforcement the path difference between the two coming waves from any two crystal planes should be an integral multiple of λ, *i.e.,* the wavelength of X-rays.

The path difference between them is given by ($RL + LS$). Therefore, for reinforce condition.

$$RL + LS = n\lambda$$

In right angled Δ RLN $\quad \dfrac{RL}{NL} = \sin\theta$

$$\therefore \quad RL = NL \sin\theta$$

If d is the distance between two consecutive cleavage planes and •"is known as lattice constant.

$$\therefore \quad NL = d$$

and $$RL = d \sin\theta$$

Similarly, in right angled $\Delta\ NLS$ $$\frac{LS}{NL} = \sin\theta$$

$$\therefore \quad LS = NL \sin\theta$$

or $$LS = d \sin\theta$$

Substitute these values of RL and LS in Equation (3), we get

$$d \sin\theta + d \sin\theta = n\lambda$$

$$2d \sin\theta = n\lambda \qquad ...(4)$$

where n in an integral number 0, 1, 2, 3 etc.

This relation is known as Bragg's law which is similar to the relation obtain for the ordinary plane diffraction grating. It indicates that for given values of n, λ and d there is reflection only in a particular direction defined by θ. In other directions there will be destructive interference. For $n = 1$ we get first order spectrum, for $n = 2$ we get the second order spectrum and so on.

9.12. BRAGG'S CONDITION FOR X-RAY DIFFRACTION

1. The relation (4) shows why radiations in X-ray region of spectrum are most usetul for crystal analysis. For first order spectrum *i.e.*, $n = 1$.

$$\sin\theta \text{ is alwaysless than } 1.$$

If interatomic spacing 'd' is about 2Å, then X-rays with wavelengths less than 2Å are most useful in studying crystals. The energy of X-ray photon is related to this wavelength (say 1 Å) and is given by

$$E = h\nu = \frac{hc}{\lambda} = \frac{6.62\times10^{-34}\times3\times10^{8}}{1\times10^{-10}}$$

$$= 19.86\times10^{-16}\text{ joule} = \frac{19.86\times10^{-16}}{1.6\times10^{-19}}$$

$$= 12.4\times10^{3}\text{ eV} = 12.4\text{ keV}$$

Therefore, X-rays from 10 to 50 KeV range should be used.

2. Bragg reflection can occur only for wavelength $\lambda \le 2d$. This is the reason why visible wavelength cannot be used in diffraction.
3. For the same order and spacing the angle θ decreases as the wavelength decreases. Consequently for gamma rays glancing angles must be used.

9.13. BRAGG'S X-RAY SPECTROMETER

Bragg devised an apparatus known as the crystal X-ray spectrometer, where the crystal is not used as a transmission grating but as a reflection grating. It is used to measure glancing angle θ. It is also to investigate the structure of a single crystal by using X-rays.

Construction: X-ray spectrometer is similar in construction to optical spectrometer. The essential parts of Bragg's X-ray spectrometer are shown in Fig. 9.15. X-rays from the anti-cathode A, limited to a narrow pencil by two adjustable slits S_1 and S_2 are made to fall upon the crystal C. The crystal C is mounted in wax on the table of the spectrometer. A vernier scale V is attached to table on which the crystal is placed. This vernier scale V is capable to move on the circular scale S and determines the position of the crystal.

The reflected rays after passing through the slit S_3 enter the ionisation chamber I through a norrow aluminium window W. The chamber is mounted on an arm R and its position is determined by a second vernier scale V_1. One of the plates of ionisation chamber is connected to the positive of a H.T. battery. The negative of which is connected to a quadrant electrometer through a condenser. The other plate of the ionsiation chamber is earthed.

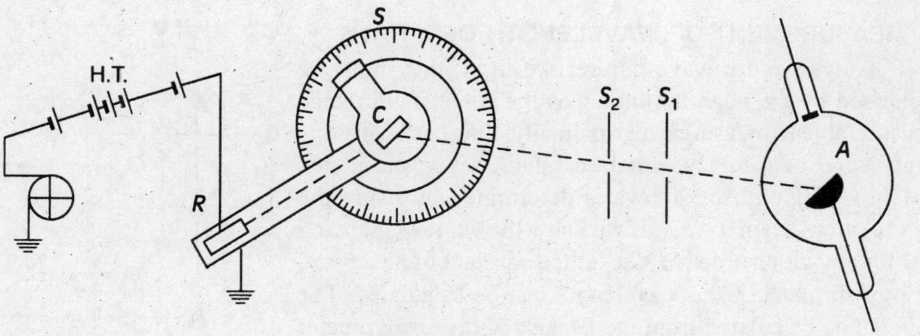

Fig. 9.15

Intensity of X-rays vs glancing angle: Curve is obtained as shown in Fig. 9.16. It is called X-ray spectrum. It is seen that only for certain values of glancing angle (θ) the intensity of X-rays is maximum as indicated by sharp maxima at electrometer is a measure of ionisation and consequently gives the intensity of reflected X-rays. To increase ionisation, the chamber is some time filled with sulphur dioxide or methyl iodide.

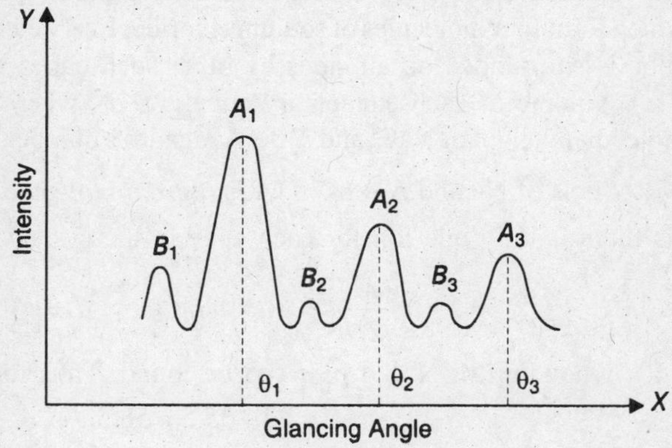

Fig. 9.16

Working: In using the apparatus the crystal is placed in such a position that glancing angle $\theta = 0$ and the ionisation chamber *I* is adjusted to receive the X-rays. The crystal and the ionisation chamber are now moved in small steps so that the angle through which the chamber is moved is twice the angle through which the crystal is rotated. The rate of deflection of the electrometer measures the intensity of the reflected beam. The observations are repeated for other glancing angles, always taking care to rotate the ionisation chamber through twice the angle through which the crystal is rotated. If we plot a graph taking glancing angle along X-axis and intensity of X-rays along Y-axis.

A_1, A_2 and A_3 corresponding to glancing angle θ_1, θ_2 and θ_3. It is found that

$$\sin\theta_1 : \sin\theta_2 : \sin\theta_3 :: 1 : 2 : 3$$

which indicates that the peaks refer to the first, second and third order reflections respectively. In a similar manner, measuring the angles corresponding to the smaller peaks B_1, B_2 and B_3 the same proportionality between the sines of these angles is found. Using Bragg's relation i.e., $2d \sin\theta = n\lambda$ it can be shown that A_1, A_2 and A_3 refer to X-rays of the same wavelength, similarly B_1, B_2 and B_3 refer to the same wavelength but different from that of A_1, A_2 and A_3. A_1B_1 belongs to the first order spectrum, A_2B_2 to the second order and so on. The Bragg with his crystal X-ray spectrometer was able to prove experimentally the relation $2d \sin\theta = n\lambda$ and thereby establish that X-rays also obey the ordinary laws of reflection.

It is to be noted that:

(a) The intensity of the reflected ray decreases as the order of spectrum increases.

(b) The position of peaks depend on the nature of the metal used for the target in the X-ray tube. This shows that each metal has its own characteristic radiations.

(c) The position of the peak also depends on the nature of reflecting crystal plane.

(d) The intensity of the reflected beam never falls to zero but only to a minimum value, which indicated the presence of continuous spectrum.

9.14. MEASUREMENT OF WAVELENGTH OF X-RAYS

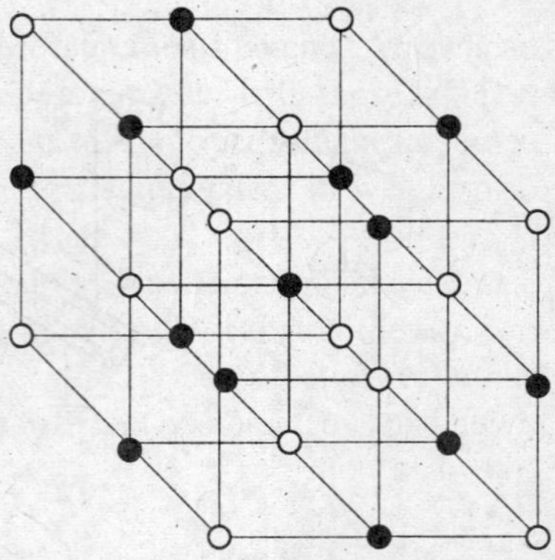

NaCl Crystal Structure

Fig. 9.17

If X-rays have a wave nature like light, they must be characterised by the quantity known as the wavelength which one must be able to measure experimentally. This has been done to a high degree of accuracy with the crystal X-ray spectrometer.

The wavelength of X-rays is determined by using the Bragg's relation $2d \sin\theta = n\lambda$. If we know the value of glancing angle θ for any given order, 'd' the lattice constant of the crystal for any given plane, λ the wavelength can be calculated. The value θ can be calculated from the Bragg's X-ray spectrometer by using rock-salt crystal.

9.15. DETERMINATION OF LATTICE CONSTANT *D*

The rock salt crystals cubic and every elementary cube has eight ions at its corners, four of sodium and four of chlorine which form four molecules of sodium chloride. Every elementary cube is surrounded on all sides by other such cubes as shown in Fig. 9.17.

Volume of each elementary cube is $\rho\, d^3$, where ρ is the density of the crystal. If M is the molecular weight of NaCl and N the Avogadro's number, since there are $2N$ ions of diffracting centres i.e., N ions of Na and N ions of Cl is *a gram molecule*, the mass associated with each ion is $\frac{M}{2N}$. As there is only one ion for each cube in the system, either Na or Cl.

$$\therefore \quad \frac{M}{2N} = \rho d^3 \quad \text{or} \quad d = \left(\frac{M}{2N\rho}\right)^{\frac{1}{3}}$$

Knowing M, N and ρ, d can be found. Thus for NaCl,

$$M = 23 + 35.5 = 58.5, \qquad N = 6.06 \times 10^{23} \text{ and } \rho = 2.17$$

$$\therefore \quad d = \left(\frac{58.5}{2 \times 6.06 \times 10^{23} \times 2.17}\right)^{\frac{1}{3}} = 2.81 \times 10^{-8}\,\text{cm} = 2.81\,\text{Å}$$

If the glancing angle θ for the first order (i.e., $n = 1$) is 11.8° as experimentally found with NaCl, the wavelength of the X-rays used is geven by

$$1 \times \lambda = 2d \sin\theta$$

$$\lambda = 2 \times 2.81 \times 10^{-8} \times \sin 11.8°$$

$$= 5.62 \times 10^{-8} \times 0.204$$

$$= 1.51 \times 10^{-8}\,\text{cm} = 1.51\,\text{Å}$$

In the accurate determination of λ, the following important corrections have to be made:

(*a*) Due to the thermal expansion of crystals, the value of d vary with the temperature. Therefore, the experiment must be performed only at 18°C which is the temperature at which the value of d are usually referred. At other temperature a correction must be applied.

(*b*) Refraction of X-rays takes place along with the reflection. Therefore, a correction for reflection of X-rays has to be made.

9.16. DETERMINATION OF CRYSTAL STRUCTURE USING X-RAYS

The crystal structure can be determined by using Bragg's X-ray spectrometer. In this, the crystal face used for reflecting the X-rays can be so cut that it is parallel first to one set of planes, then to another and so on, in the crystal. A monochromatic beam of X-rays of known wavelength is used and observations are made to obtain the glancing angle θ for a given order of spectrum (say $n = 1$) for the each set of the chosen planes.

Then, by using Bragg's relation,

$$n\lambda = 2d \sin\theta$$

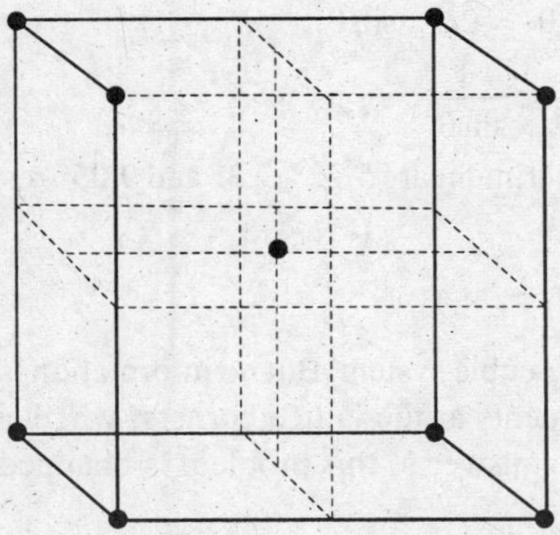

Body Centered Cube System

Fig. 9.18

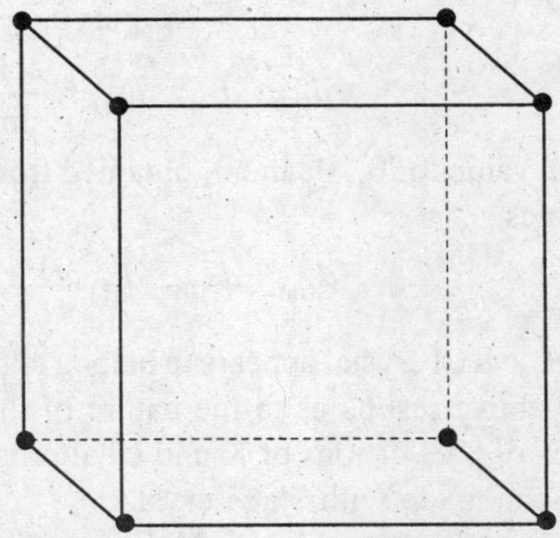

Simple Cube System

Fig. 9.19

a set of values for lattice constant 'd' is obtained, each one corresponding to a given plane. Expressing these values as $(1/d)$, it is found that the ratio depends upon the way in which the atoms are arranged in the crystal and thus the structure of the unit cell is determined. Let us consider the examples of KCl and NaCl crystals.

9.17. STRUCTURE OF KCl AND NaCl CRYSTAL

These two crystals belong to crystals in which there are three sub-types with different concentrations of atom and with different lattice constant. They are

(*a*) Simple cube system.

(*b*) Body-centred cube system.

(*c*) Face centred cube system.

(*a*) Simple cube system. In this system an atom lies at each corner of the cube as shown in Fig. 9.18.

(*b*) Body-centred system. In this system an additional atom at the centre of the cube as shown in Fig. 9.19.

(*c*) Face-centred cube. In this system an additional atom lies at the centre of each of the six faces of the cube as shown in Fig. 9.20.

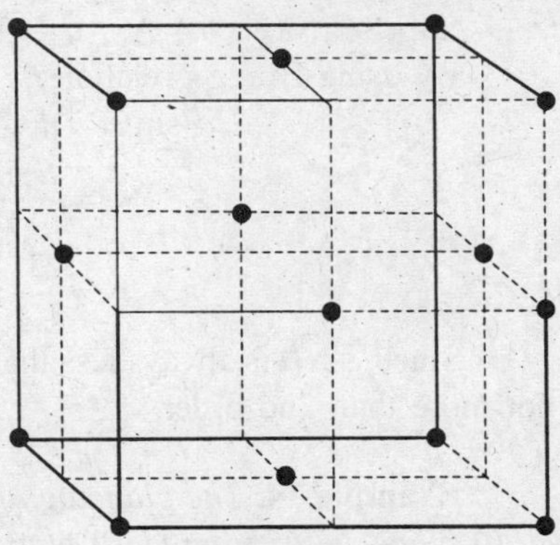

Face-centered Cube

Fig. 9.20

In these three types, the actual spacing d of the planes rich in atoms is given as (1, 0, 0), (1, 1, 0) and (1, 1, 1). The ratio between them is given as follows:

	d_{100}	:	d_{110}	:	d_{111}
Simple cube...,	1	:	$\frac{1}{\sqrt{2}}$	:	$\frac{1}{\sqrt{3}}$
Body-centred....,	1	:	$\frac{2}{\sqrt{2}}$	:	$\frac{1}{\sqrt{3}}$
Face-centred...,	1	:	$\frac{1}{\sqrt{2}}$	:	$\frac{2}{\sqrt{3}}$

With the KCl crystal, using the three planes (1, 0, 0), (1, 1, 0) and (1, 1, 1) one after the other for reflecting a given monochromatic ray in the Bragg's X-ray spectrometer, the glancing angle θ_1, θ_2 and θ_3 for the first order are measured.

∴ From Bragg's law

$$\lambda = 2d_{100} \sin\theta_1 = 2d_{110} \sin\theta_2 = 2d_{111} \sin\theta_3$$

$$\therefore \quad d_{100} : d_{110} : d_{111} :: \frac{1}{\sin\theta_1} : \frac{1}{\sin\theta_2} : \frac{1}{\sin\theta_3}$$

The values of θ_1, θ_2 and θ_3 obtained from the experiment are 5.22°, 7.3° and 9.05° respectively, which gives

$$d_{100} : d_{110} : d_{111} : : 1 : \frac{1}{\sqrt{2}} : \frac{1}{\sqrt{3}}$$

Hence KCl crystal appears to belong to the simple-cubic system. But no information is available from the above results as to the nature of the constituents at the lattice corners, whether they are molecules of KCl or ions of K and Cl alternately. The answer to this problem is obtained from the observations made with NaCl crystal.

The result obtained with NaCl crystal for same three planes agrees with the previous case. By obtaining the values of θ_1, θ_2 and θ_3, we get

$$d_{100} : d_{110} : d_{111} : : 1 : \frac{1}{\sqrt{2}} : \frac{2}{\sqrt{3}}$$

which shows that NaCl crystal belongs to the face-centred cube system.

Example 13. *Monochromatic X-rays of wavelength 1.4 Å are incident on a crystal having 1.5 Å as interatomic spacing. Find the various orders in which the diffraction takes place.*

Solution: λ = 1.4 Å, d = 1.5 Å

By using Bragg's relation

$$2\,d \sin\theta = n\lambda$$

$$n = \frac{d\sin\theta}{\lambda} = \frac{2\times 1.5\sin\theta}{1.4}$$

$$n = \frac{3}{1.4}\sin\theta \qquad \text{or} \qquad n = 2.14\sin\theta$$

Since, sin θ is always less than 1. Therefore, maximum order of the spectrum can be two i.e., not more than 2nd order.

$$n \le 2$$

Example 14. *The glancing angles of reflection for K_α-X-rays from Palladium are 5.4° from (100) planes; 7.6° from (110) planes and 9.4° from (111) planes. From the above data determine the cubic lattice structure of the crystal.*

Solution: From Bragg's formula

$$2d \sin\theta = 1\cdot\lambda \qquad \text{(for first order } n = 1)$$

$$\therefore \quad d = \frac{\lambda}{2\sin\theta}$$

or

$$\frac{1}{d} = \frac{2\sin\theta}{\lambda}$$

For 100 plane

$$\frac{1}{d_1} = \frac{9\times\sin 5.4°}{\lambda} = \frac{2\times 0.0941}{\lambda}$$

For 110 plane

$$\frac{1}{d_2} = \frac{2\times\sin 7.6°}{\lambda} = \frac{2\times 0.1323}{\lambda}$$

For 111 plane

$$\frac{1}{d_3} = \frac{2\times\sin 9.4°}{\lambda} = \frac{2\times 0.1633}{\lambda}$$

$$\therefore \quad \frac{1}{d_1} : \frac{1}{d_2} : \frac{1}{d_3} = 0.0914 : 0.1323 : 0.1633$$

$$= 1:1.412:1.732 = 1:\sqrt{2}:\sqrt{3}$$

which shows that the given crystal has simple cubic lattice.

Example 15. *Calculate the wavelength of X-ray beam incident at 12° for the first order reflection from a calcite crystal if the interatomic spacing d for the crystal is 3.035 Å.*

Solution: Given that $\theta = 12°$

Order of spectrum $n = 1$

Interatomic spacing $d = 3.035$ Å $= 3.035 \times 10^{-8}$ cm

By using Bragg's equation we have

$$2\, d \sin\theta = n\lambda$$

$$\lambda = \frac{2d\sin\theta}{n}$$

$$= \frac{2 \times 3.035 \times 10^{-8} \times \sin 12°}{1}$$

$$= 2 \times 3.035 \times 10^{-8} \times 0.2079$$

$$= \mathbf{1.262\ Å} \qquad (\because \sin 12° = 0.2079)$$

Example 16. *The Bragg angle corresponding to the first order reflection from (1, 1, 1) planes in a crystal is 30° when X-rays of wavelength 1.75 Å are used. Calculate the interatomic spacing.*

Solution: The interplanar distance d for a plane having Miller indices (h, k, l) is given by

$$d_{hkl} = \frac{a}{\sqrt{h^2 + k^2 + l^2}}$$

where a is the interatomic spacing.

In case of plane having Miller indices (1, 1, 1)

$$h = k = l = 1$$

$$d_{111} = \frac{a}{\sqrt{3}}, \quad \theta = 30°, \quad \therefore \sin\theta = \frac{1}{2}$$

$$n = 1 \qquad \lambda = 1{,}75\ Å$$

By Bragg's law $2d_{111} \sin\theta = n\lambda$

$$2 \times \frac{a}{\sqrt{3}} \times \frac{1}{2} = 1.75\ Å$$

$$a = 1.75 \times \sqrt{3}\ Å = 3.031\ Å$$

$$= \mathbf{3.031 \times 10^{-8}\ cm}$$

9.18. EXPERIMENTAL METHODS TO FIND CRYSTAL STRUCTURES IN X-RAY DIFFRACTION

The Bragg's law, $n\lambda = 2d \sin\theta$ for X-ray diffraction depends upon the values of θ and λ. To satisfy the Bragg's law it is necessary to provide experimentally for a continuous range of values of either λ or θ. The standard methods used for this are:

(1) *Laue Method*: In this method a crystal is held stationary in the path of continuous X-ray beam. The crystal selects out and diffracts the discrete values of λ for which planes of spacing d exist and incidence angle θ satisfy the Bragg's law.

(2) *Rotating Crystal Method*: In this method a single crystal is rotated about and fixed axis in a beam of monochromatic X-rays. The variation in the value of θ brings different atomic planes into position of reflection.

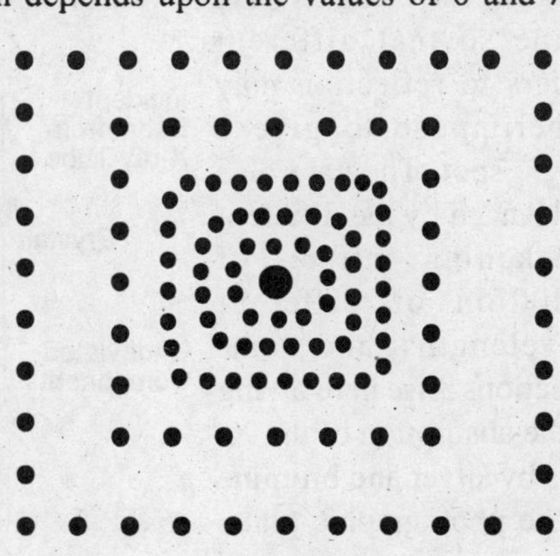

Fig. 9.21

(3) *Powder Method* (*Debye-Scherrer-Hull Method*): In this method a powdered sample of the crystalline material is placed in fixed position in a monochromatic beam. Only for certain orientations the angle of incidence satisfies the Bragg's law.

In these days, the above methods are used but with some modifications. (*i*) The Laue method is suitable for the rapid determination of crystal orientation and symmetry and also for studying crystalline imperfection under mechanical and thermal treatment. (*ii*) The rotating crystal method is the most suited method for the structure determination when single crystal specimen is available. (*iii*) The powder method is employed to known the structure of metals.

1. Laue Method. The principle of this method is shown in Fig. 9.21. The beam falls upon a small crystal or thin crystal section and the diffracted beam are recorded on a photographic plate placed at a distance from the crystal. The rays are limited to a narrow pencil by the slit system, the crystal is set on a holder which permits adjustment of its orientation and photographic plate is placed in the other holder.

A continuous beam of X-ray is obtained by using a bulb with tungsten anticathode and a Potential of about 65,000 volts. Laue photograph obtained is shown in Fig. 9.21. The various crystal planes reflect the X-rays as if they constituted a number of mirrors set at different angles.

The relation : $n\lambda = 2d \sin\theta$

can be satisfied, although the angle θ is fixed for each set of planes ; because radiation of appropriate wavelength can be selected from the continuous range. If the crystal is rotated, the spots change their position upon the screen, as if they were the reflections of a narrow beam of light by the faces of a cut-glass stopper. All the spots which lie on an ellipse are due to reflections by planes parallel to common zone axis. Incident rays fall upon a crystal having zone axis parallel to ZZ', and we consider rays which are reflected by a plane of corresponding zone. Suppose a reflecting plane to be rotated about ZZ' -axis; the reflected beam will trace a curve BB' on the plate. In the incident rays make an angle θ with ZZ' -axis, the reflected ray will make the same angle with it, so that they lie on a cone of semi-vertical angle θ about the zone axis. The ellipses in the Laue photograph are the intersection of such cones. Each spot is at an intersection of ellipse, since a plane is parallel to two zone axes in a crystal plane and can reflect the X-ray beam.

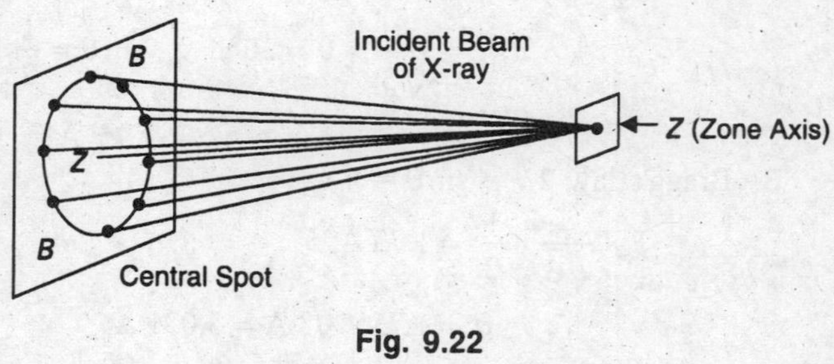

Fig. 9.22

The deduction of the crystal strucure from the appearance of Laue photograph is a complicated process, as the intensities of the spots do not depend on the structure alone. Because of the extended range of wavelengths it is possible for several wavelengths to reflect in different orders from a single plane so that different orders to reflection may superimpose to give a single spot. They are also influenced by the different blackening effect of radiation of different wavelengths, and complications arise here owing to the absorption of the X-rays by silver and bromine in the photographic plate.

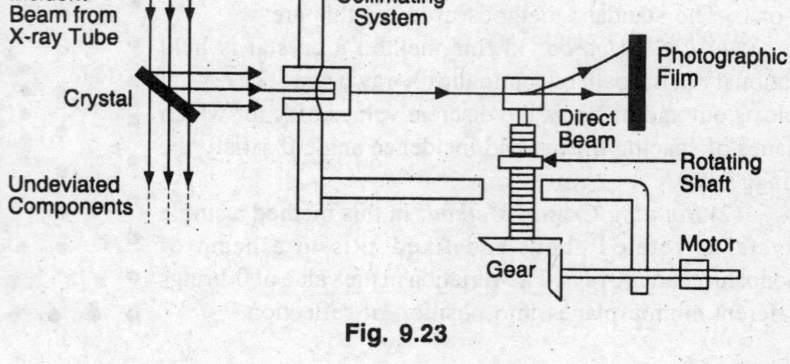

Fig. 9.23

In spite of these difficulties, the Laue photograph is used oftenly in the determination of crystal orientation and symmetry.

2. Rotating Crystal Method: In the rotating crystal method a single crystal is rotated about a fixed axis in a beam of monoenergetic X-rays. The variation in the angle θ brings different atomic planes into position for reflection. A modified rotating crystal method is used for structure determination when a single crystal specimen is available.

The diffracted beams are recorded photographically upon a plate or upon a film bent into a cylinder around the axis of rotation. The dimensions of the crystal should not be greater than 1 mm. The incident X-ray beam is made nearly monochromatic by a filter or by reflection from a monochromatic crystal. The beam is diffracted from a given crystal plane whenever in the course of rotation the value of θ satisfies the Bragg's Equation. Beams from all planes parallel to the vertical rotation axes will lie in the horizontal plane. Planes with other orientations will reflect in layers above and below the horizontal plane. As the crystal rotates, one plane after another is brought in such position that it reflects the radiation. The film when developed show a number of spots due to various planes which have passed through requisite positions. The apparatus shown in Fig. 9.23 can be used for both Laue and rotation photographs. In this the crystal is rotated by a clockwork mechanism contained in a metal box.

The geometrical interpretation of rotation photographs is more complex. The orientation of the crystal, which is different for each spot, is unknown. We only know that it lies within a certain range when the crystal has been turned through limited arc. The distance between two successive lattice points along the zone axis. *PO* is represented in Fig. 9.24. The incidental rays are at right angles to this axis. The necessary condition for diffraction is that the path difference for rays scattered by *P* and O should differ by a whole number of wavelength so that the path difference *OQ* is $n\lambda$, this condition determining the angle which the diffracted rays make with the zone axis. As the crystal is rotated it will pass through a number of positions suitable for diffraction which have in common a path difference of $n\lambda$ between *O* and *P*. All the corresponding spots lie on a `layer line' which will be the *n*th above or below the equational line in photograph.

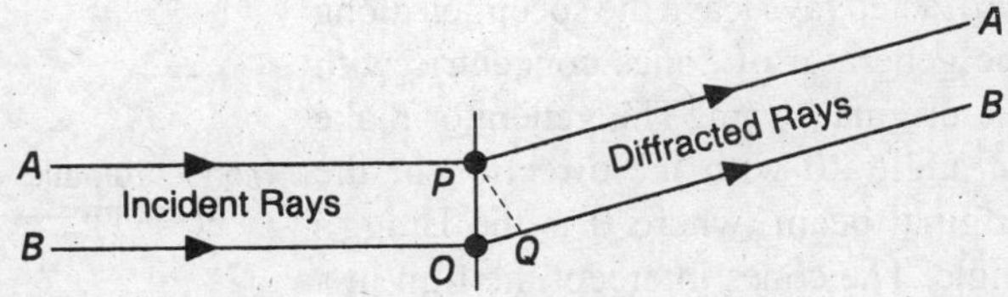

Fig 9.24

If *R* is axis about which the crystal is rotated, when an incident beam of X-ray falls on the crystal, reflection will take place with indices (*h, k, l*) which move upwards making an angle ϕ_1 with the horizontal given by $r \sin \phi_1 = \lambda$. Such reflections constitute first layer line. In general, reflections with indices (*h, k, l*) lie on the layer line *l*. Since the vertical spacing of the layer lines depends only upon the distance between lattice points on the axis about which crystal is rotated; the distance (*r*) can be immediately deduced from the photograph.

If rotation photograph is taken separately about three translation axes, *a, b, c* and hence the dimensions of the unit cell of crystal structure is determined. The rotating crystal method is thus very powerful method of giving the size of the unit cell.

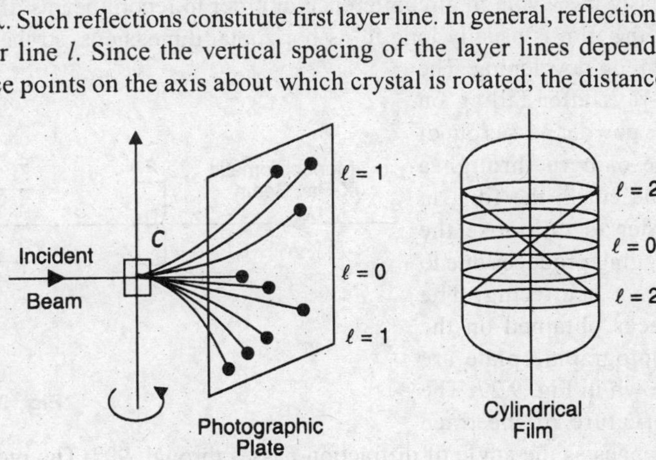

Fig. 9.25

We will obtain so many spots with complete rotation which will cause confusion, therefore it is the useful practice to rotate the crystal backwards and forwards through on angular range of only 15°in order to have limited number of spots on the photographic plate.

A better technique was given by Weissen-berg. According to him crystal is rotated through 180° and back again continuously and at the same time the cylindrical camera containing the film moves at a constant speed backward and forward parallel to the axis of rotation. The rotation of camera and film is adjusted so that a given position of the camera corresponds accurately to definite angular position of the crystal as its rotation. The co-ordinates of a spot on the film thus give both the angle of reflection as well as the position of the reflecting plane and thus allows the spot to be indexed without doubt. A metal cylinder provided with equitorial slit a few millimeters wide is placed between the crystal and film is placed in such a position that the spot of only layer line can pass through it.

In modern methods we use scintillation counters or proportional counters to detect the diffracted radiation. These methods allow automatic collection of data and are more useful in the study of dislocation and chemical bonding in crystal.

3. Powder-Photograph Method. In the powder-photograph method the incident monochromatic radiation strikes a finely powdered specimen or a fine grained polycrystalline specimen contained in a thin-walled capillary tube. The distribution of crystalline orientations will be nearly continuous. The powder method is convenient because single crystal is not required. Diffracted rays go out from individual crystals which happens to be oriented with planes making an incident angle θ with the beam satisfying the Bragg's equation. Diffracted rays leave the specimen along the generator of cones concentric with the original beam. The generator make an angle 2θ with the direction of the original beam, where θ is the Bragg's angle. The cones intercept the film in a series of concentric rings as shown in Fig. 9.26.

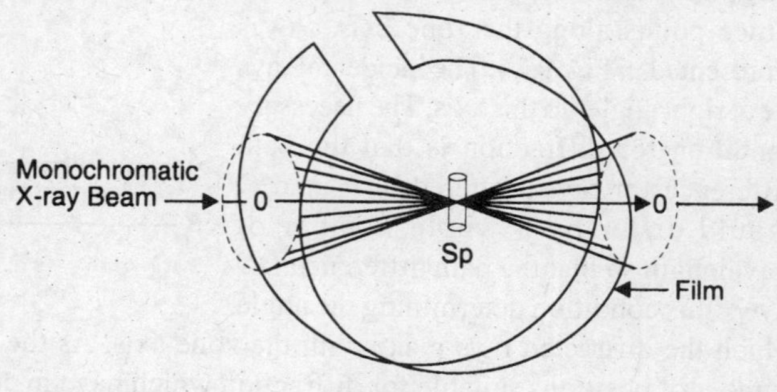

Fig. 9.26

The arrangement of experiment is shown in Fig. 9.27. The X-rays are made monochromatic by passing them through Zirconium oxide filter *F*. A narrow beam of monochromatic X-rays limited by slits S_1 and S_2 falls on the powdered specimen *C*. The photographic film fits round the inner surface of the camera covering nearly the whole in circumference in order to record beams diffracted upto nearly 180°. The strip of the film is usually long in its horizontal dimensions, so that only arcs of circles appear on the film on developing. The X-rays after falling on the powder passes out of the camera through a hole cut in the film, in order to minimise the fogging produces due to direct scattering. The traces obtained on the photographic plate are shown in Fig. 9.27. The curvature of the trace changes as the angle of diffraction passes through 90°. The radius of curvature increases with the increase of angle of diffraction and at 90° straight lines are obtained.

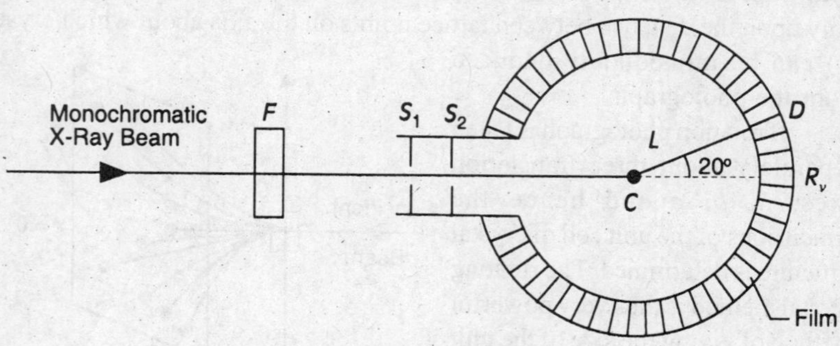

Fig. 9.27

The crystal structure can be obtained from the arrangement of the traces and their relative intensities. Consider a spectrum with glancing angle θ at a point D which is at a distance R from O where the direct incident beam strikes the film and if L be the radius of the cylindrical camera, then

$$\theta = \frac{R}{2L}$$

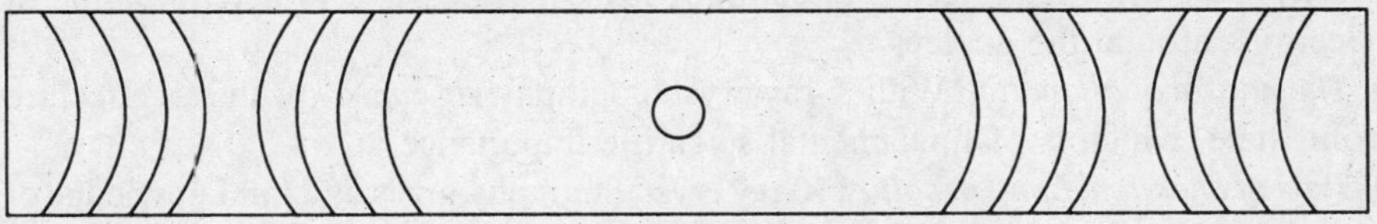

Fig. 9.28

This value of θ will give us the spacing of the planes by using Bragg's relation,

$$2d \sin \theta = n\lambda$$

Differentiating we get

$$2\,\Delta d \sin\theta + 2d\cos\theta.\,\Delta\theta = 0 \qquad (\because n \text{ and } \lambda \text{ are constants})$$

or

$$\Delta d \sin\theta + d\cos\theta.\,\Delta\theta = 0$$

$$\therefore \qquad \frac{\Delta\theta}{\Delta d} = -\frac{\tan\theta}{d}$$

When θ tends to 90°, $\frac{\Delta\theta}{\Delta d}$ becomes very great, so that small variation in d produce large variations in θ. Therefore, high resolution when the rays are reflected back through nearly 180°. This is the reason why spectra for which the semi-vertical angle is greater than a right angle are obtained by the use of the long cylindrical film nearly surrounding the crystal powder.

This method is applicable for all kinds of crystalline matter, however, small the individual crystal may be. This method is specially used to determine the structure of simple cubic crystals for which the spacing of the unit cell are all equal. It is most useful for the investigation of metals and alloys structure, while single crystal are not possible.

In spite of X-ray diffraction we can also use Electron Diffraction and Neutron Diffraction to determine crystal structure. In case of electron diffraction, the amount of matter required to produce electron diffraction, however, is extremely small compared with that required for X-ray diffraction, and the time of exposure required for a photographic record are measured in fraction of a second. Due to these facts, electron diffraction is very useful to apply where it is difficult or impossible to use X-rays, *e.g.*, in the structure of surface films or to the analysis of the molecules of gas. In case of neutron diffraction intensities depend largely upon scattering from nuclei whereas in case of X-ray diffraction intensities depend upon scattering from the orbital electrons of the atom.

Neutron diffraction tells us whether two scattering amplitudes have the same size or not but does not tell us for any given nuclei whether amplitude is positive or negative. Neutron diffraction is also used in a long graphite rod to provide source of *"cold neutron"*. Cold neutrons are those whose parable speed is much below the speed required for thermal neutrons at (300 K). Another important use of neutron diffraction is the study of magnetic materials. Shull and Smart used neutron diffraction for the detection of antiferromagnetism in MnO sample first time. In the study of magnetic properties of the material we use the fact that a neutron can be scattered by interaction of its magnetic moment with the atomic or ionic magnetic moment of the sample material. The diffraction of particles not only confined to electrons and neutrons but also possible for atoms and molecules, the corresponding wavelength being given by the de Broglie's relation.

Results obtained by X-ray Crystallography. The following important results have been obtained by X-ray crystallography:

1. *The crystal structure of metals.* X-ray analysis has clearly proved that most metals and alloys are crystalline in structure. Number of metals and alloys have been found to the cubic, tetragonal and hexagonal types of crystals.

2. *The internal structure of the atom.* With X-rays we can see the interior structure of atom. By the analysis of the photo-electric effect of X-rays, it is possible to determine the distribution of the electrons around the nucleus.

3. *The nature of all matter.* With X-ray crystallography we can know the architecture of matter in the solid state, which is of fundamental scientific importance.

4. *Type of chemical bond in solids.* X-ray crystallography gives us a fair knowledge of chemical bonds in solids *i.e.,* which types of forces are acting between the ions and the atoms constituting molecules *e.g.,* in case of NaCl there is electrostatic forces between Na^+ and Cl^-. This is known as heteropolar binding whereas in case of Hydrogen (H_2) there is homopolar binding.

9.19. INDUSTRIAL APPLICATIONS OF X-RAYS

X-rays are widely used in industry such as:

(*i*) To locate defects in internal structure of a body. It is also used to locate the imperfection in mould forgings and castings which could not otherwise be examined internally.

(*ii*) To study the structure of rubber, plastics in rubber and plastic industries. The diffraction of X-rays by these substances leads to valuable information about their molecular grouping.

(*iii*) In metallurgy X-ray are used to analyse the structure of various alloys by determining the structure of in-gots with the help of X-rays diffraction.

(*iv*) In heavy industries X-rays are used to test the gross structure of the material and welding etc. For this purpose the X-ray photographs are taken which reveal defects in the manufacture of heavy steel plates and also locate the cracks and blow holes in metal plates. X-rays are also used to detect flows in the wings of aeroplanes and parts of other fast moving machines.

From this it is clear that X-rays have become the important parts of all the heavy industries.

9.20. OTHER USES OF X-RAYS

X-rays have many applications in many other branches such as for scientific research and for medical purposes.

Applications of X-rays for Scientific Research :

(*i*) X-rays are used to study the crystal structure in crystallography.
(*ii*) X-rays are used in atomic physics to study the atomic structure.
(*iii*) X-rays are used to find the atomic number of the elements and also for the identification of the elements.
(*iv*) X-rays are used to study the structure of complex organic compounds.

Applications of X-rays for Medical Purposes: In Medicine X-rays are used for two main purposes.

(*a*) **Radiography.** X-rays are widely used in *radiography.* In Radiography we use X-ray as the source and a photo film upon which we take the X-ray photos. Since bones and other solids are more dense and hence more opaque to X-rays than flesh and blood, therefore, we can detect fractures, tumours, the presence of foreign matter like bullet etc. in human body by interposing it between the X-ray tube and photo film. This process of taking X-ray photos on photo films is known as radiography.

(*b*) **X-Ray Therapy.** Many types of skin diseases, malignant sores, cancers, tumours etc. have been cured by controlled exposure to X-rays of suitable quality. In this we destroy the diseased portion of the body by exposing to X-ray for a suitable time. Long exposure of human body to X-rays may cause unnecessary incurable injuries and destroy white cells in the blood of the body.

X-rays have been used for the identification of different types of cells and tissues and for bringing about genetic mutations.

9.21. AUGER EFFECT

In the basic explanation of X-rays it was assumed that the vacancy of an inner shell, say `*K*' shell is filled by an electron from some outer shell, along with the emission of a photon. However, experimentally it is found that flourescence yield $\omega = n_p/n_e$ where $n_p \to$ no. of *K* photons and $n_e \to$ no. of *K* electrons knocked out (*i.e.*, no. of *K*-shell vacancies) is < 1 which may range between 0.1 for light elements to 0.95 for uranium.

Auger in 1925 gave explanation for the above observation in the following way:

The ejection of electron from K-shell is often accompanied by the *ejection* of another electron. It is due to the fact that an electron that may undergo transition to the vacant *K* shell may knock out another electron on its passage to the same shell. This is called *Auger transition* and the emitted electron is known as *Auger electron.*

It should be clearly noted that *Auger electron* is not because of the absorption of a photon given out by an electron when it comes from a higher energy level to a lower energy level but it comes directly as a part of readjustment of the atom. For example, the vacancy in the *K*-shell may be filled by an electron of the L_I-shell and another electron from L_{II}-shell may be knocked out thereby creating two vacancies in the *L*-shell. So the de-excitation of the atom may be accompanied by either the emission of *photon* (characteristic radiation) of an *electron* (Auger electron). These two processes together account for the number of vacancies that exist in *K*-shell. In *nuclear transitions,* the *Auger effect* is also known as *internal conversion* or Auto-ionization. So under certain conditions of energy, an atom in a state of single ionization may undergo spontaneously an *Auger transition* which leaves it in a state of double ionization. The electron released in such a transition is expelled from the atom and left with an amount "of kinetic energy *E* given by

$$E = E_i - E_f \quad \text{... (1)}$$

where E_i is the initial energy of the singly ionized atom and *Ef* is the final energy of the doubly ionized atom. This type of atomic transition does not involve the emission of a photon and, therefore, is often called a "nonradiative" transition. Nonradiative transitions are possible, in general, only when an electron can be ejected from the atom; and for this to be possible it is necessary that *E* as defined by Equation (1) be zero or more than zero. The electrons ejected by such radiationless transitions were detected by Auger as fat-short tracks made by them on the gas of a cloud chamber.

QUESTIONS AND PROBLEMS

1. What is X-ray diffraction? Deduce Bragg's Law for the diffraction of X-ray in a crystal. What are Bragg's conditions for X-ray diffraction?
2. Explain how X-rays are produced. What are their properties?
3. Describe the construction and mode of operation of a Coolidge X-ray tube. Write a note on the properties of X-rays.
4. Give an account of the production, properties and industrial applications of X-rays.
5. What are X-rays? Explain the origin of X-rays alongwith characteristic X-rays.
6. Write short notes on:
   (*a*) X-rays by Coolidge tube.
   (*b*) Soft and hard X-rays.
   (*c*) Roentgen X-ray tube.
   (*d*) Mosley law.
   (*e*) Scattering of X-rays.
   (*f*) Industrial applications of X-rays.
7. Define 'space lattice' or lattice constant. Explain how crystal structure is determined by the use of X-rays.
8. Describe an X-ray spectrometer and explain how it is used to determine the wavelength of X-rays and to study the structure of crystal.
9. Describe an experiment for taking X-ray diffraction pattern of a metallic powder. How does it help us to determine the structure of the metal?

10. What are various experimental methods used in X-ray diffraction to study a single crystal structure?

11. An X-ray tube operates at 20 kV. Find the maximum speed of electron striking the anticathode; given that

$$e = 4.8 \times 10^{-10} \text{ e.s.u.}$$
$$m = 9.1 \times 10^{-26} \text{ gm.}$$
$$1 \text{ volt} = 1/300 \text{ e.s.u.}$$

[**Ans.** 8431×10^9 cm/sec.]

12. Find the quantum energy of X-ray of frequency 10^{28} cycles per second. Given $h = 6.62 \times 10^{-27}$ ergs-sec. [**Ans.** 66.2 ergs]

13. Calculate the glancing angle on the cube face of a rock salt (d = 2.814 Å). Corresponding to second order diffraction maximum for X-rays of wavelength 0.710Å. [**Ans.** 14°30']

14. Find the minimum wavelength of X-rays produced by an X-ray tube operated at 100 Kilo Volts. Given,

$$h = 6.63 \times 10^{-27} \text{ ergs-sec.}$$
$$e = 4.8 \times 10^{-10} \text{ e.s.u.}$$
$$c = 3 \times 10^{10} \text{ cm/sec.}$$

[**Ans.** 0.1237° A]

15. The Bragg's angle corresponding to the first order reflection from (1, 1, 1) planes in a crystal is 30° when X-rays of wavelength 1.75 Å are used. Calculate the interatomic spacing. [**Ans.** 3.031×10^{-8} cm]

16. Electron bombarding the anode of a Coolidge tube produce X-rays of wavelength 5 A.U. Find the energy of each electron at the moment of impact. Given,

$$e = 4.8 \times 10^{-10} \text{ e.s.u.}$$
$$h = 6.62 \times 10^{-27} \text{ ergs-sec.}$$

[**Ans.** 2.482×10^3 eV]

17. The K_α and L_α absorption edges of copper occur at wavelength 1.380 Å and 13.285 Å respectively. Calculate the atomic number of copper. [**Ans.** 29.27]

18. In Os (Z = 76) K and L absorption edges have respectively wavelength of 0.168Å and 1.17 Å. What is the Wavelength of K_α line? [**Ans.** 0.196Å]

19. If a K_α X-ray from an element is measured to have wavelength 0.786 Å. What is the element? [**Ans.** 40 Z]

20. A set of lattice planes reflects X-rays of 1.32Å at glancing angle of 9.30°. Deduce the possible sportings of this set of planes. [**Ans.** 4, 8, 12 Å]

21. How many lines are in K_α fine structure lines. [**Ans.** 2]

22. Calculate the wavelength of the X-ray beam incident at 12° for the first order reflection from calcite crystal if grating constant for crystal is 3.035 Å. [**Ans.** 126 Å]

CHAPTER

10

MATTER WAVES AND UNCERTAINTY PRINCIPLE

10.1. WAVE PROPERTIES OF PARTICLES

The successful explanation of the Compton and photoelectric effects and the experimental verifications of the predictions made on the basis of these explanations, established it beyond doubt that electromagnetic radiation travels not in the form of a continuous stream of energy but in the form of tiny packets or bundles of energy, following each other in quick succession. These packets of energy were called photons and behave almost exactly like a material particle.

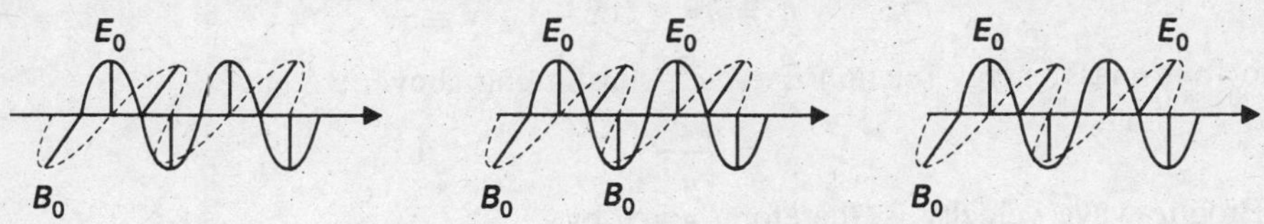

Fig. 10.1

On the other hand, the phenomenon like interference or diffraction of light or X-rays could not be explained unless electromagnetic radiation were assumed to possess a wave character. This resulted in the acceptance of the dual character of radiation, which is maintained till the present day. When radiation interacts with matter it exhibits its particle character whereas when radiation interacts with radiation, it exhibits wave character. Radiation never exhibit both the characters simultaneously.

An interesting outcome of the acceptance of duality in the character of e.m. radiation was the hypothesis put forward by L. de Broglie in 1924. He was guided by the following reasonings:

1. Nature is symmetrical.
2. Our observable universe is totally composed of light and matter and light has dual character.
3. There is a close parallelism between principle of least action in mechanics and principle of least time (Fermat's principle) in optics.

According to de Broglie; "the wave-particle dualism should not be confined to radiation alone, but should also be extended to all the fundamental entities of physics". The electrons; protons; atoms and molecules, when in motion, should have some type of wave motion associated with them. de Broglie was led to this hypothesis from considerations based upon the special theory of relativity and quantum theory.

Energy of photon can be written as

$$E = h\nu = \frac{hc}{\lambda} \qquad ...(1)$$

and if a photon possesses some mass (which is by virtue of its motion i.e., its rest mass would be zero) its energy according to the theory of relativity should be = (mass of photon) c^2.

Hence $$E = \frac{hc}{\lambda} = \text{(mass of photon) } c^2$$

$\therefore$ mass of photon $= \dfrac{1}{c^2}\left(\dfrac{hc}{\lambda}\right) = \dfrac{h}{c\lambda}$

and momentum p of the photon $= p = \left(\dfrac{h}{c\lambda}\right)\cdot c = \dfrac{h}{\lambda}$

i.e.
$$p = \frac{h}{\lambda} \quad \text{or} \quad \lambda = \frac{h}{p} \qquad ...(2)$$

de Broglie carried these considerations over to the dynamics of a particle and said that the wavelength λ of the wave associated with a moving material particle having a momentum mv is given by

$$\lambda = \frac{h}{mv} \qquad ...(3)$$

10.2. VELOCITY OF DE BROGLIE WAVES

We associate de Broglie wave with a moving particle, so it is reasonable to expect that this wave travels with the same velocity as that of the particle. If we call the de Broglie wave velocity as w, we may apply the usual formula $w = \nu\lambda$ to determine the value of w. The wavelength λ is the de Broglie wavelength given by $\lambda = \dfrac{h}{mv}$.

Now from the Planck's Law

$$E = h\nu \quad \text{or} \quad \nu = \frac{E}{h}$$

Also from mass energy relation $E = mc^2$, substituting above, we have

$$\nu = \frac{mc^2}{h}$$

de Broglie wave velocity is, therefore, given by

$$w = \nu\lambda = \frac{mc^2}{h} \times \frac{h}{mv} = \frac{c^2}{v} \qquad ...(4)$$

Since the particle velocity v, must be less than the velocity of light c, de Broglie waves travel faster than the velocity of light. Such a situation can be visualised in case of electromagnetic waves *e.g.*, velocity of X-rays in different crystals and that of sodium light in sodium crystal is more than *c*.

However, the problem in case of de Broglie waves was solved by showing that de Broglie waves always travel in the form of group of waves with a velocity called "group velocity", which is not only less than c, but is equal to the velocity of the particle.

Let a material particle, such as an electron be situated at some point $(x_0\, y_0\, z_0)$ in space. Imagine some quantity ψ whose periodic variation in space gives rise to matter waves such that at any instant t_0 the value of ψ in the immediate vicinity of the particle be given by

$$\psi = \psi_0 \sin(2\pi\nu_0 t_0)$$

where ψ_0 is the amplitude at the point chosen and ν_0 is the frequency as observed by an observer at rest with respect to the particle.

Let now the particle be given a velocity ν along the +ve direction of X-axis and the periodic variations of ψ as these now appear to the observer, who is still at rest at $(x_0\, y_0\, z_0)$, we have to apply the transformation equation of the theory of relativity to the time interval t_0.

$$t_0 = \frac{\left(t - \dfrac{vx}{c}\right)}{\sqrt{1 - \dfrac{v^2}{c^2}}} \qquad ...(5)$$

where t is the time interval as noted by an observer moving with the particle with velocity ν.

Hence periodic variations of ψ as these appear to the observer who is still at rest at $(x_0\, y_0\, z_0)$ are given by

$$\psi = \psi_0 \sin \frac{2\pi \nu_0 \left(t - \frac{vx}{c^2}\right)}{\sqrt{1 - \frac{v^2}{c^2}}} \qquad ...(6)$$

Comparing this equation with the standard equation of Simple Harmonic wave

$$y = A \sin 2\pi \nu \left(t - \frac{x}{w}\right) \qquad ...(7)$$

where A is the amplitude; ν is the frequency and ν is the velocity of the wave along +ve X-direction.

We have $$\nu_p = \frac{c^2}{v}$$

and $$\nu = \frac{\nu_0}{\sqrt{1 - \frac{v^2}{c^2}}}$$

Now from Einstein's mass energy relation

$$E = h\nu_0 = m_0 c^2 \qquad \therefore \nu_0 = \frac{m_0 c^2}{h}$$

or $$\nu = \sqrt{1 - \nu^2/c^2} = \frac{m_0 c^2}{h}$$

or $$\nu = \frac{m_0}{\sqrt{1 - \nu^2/c^2}} \frac{c^2}{h} = \frac{mc^2}{h} \qquad \left[\because m = \frac{m_0}{\sqrt{1 - v^2/c^2}}\right]$$

Now the wavelength of matter waves is given by

$$\lambda = \frac{\text{velocity}}{\text{frequency}} = \frac{w}{\nu} = \frac{\frac{c^2}{v}}{\frac{mc^2}{h}} = \frac{h}{mv}$$

i.e., $$\lambda = \frac{h}{mv}$$

This aspect of matter waves shows that with a material particle in motion two different velocities are involved viz., ν, **referring to the mechanical motion of the particle and ν_p, the velocity with which the wave is propagated and they are related by**

$$\nu_p = \frac{c^2}{v} \qquad ...(8)$$

10.3. WAVE (PHASE) VELOCITY AND GROUP VELOCITY

The amplitude of the de Broglie waves associated with a moving particle, measures the probability of finding the particle at a particular place at a particular time. Evidently de Broglie waves cannot be described by a simple formula used to describe Simple Harmonic wave viz., $y = A \sin 2\pi\left(\nu t - \frac{x}{\lambda}\right)$ because Simple Harmonic waves are a series of waves all with same amplitude.

Instead we expect that wave associated with a moving particle should be in the form of a "*wave packet*" or "*group of waves*" shown in Fig. 10.2.

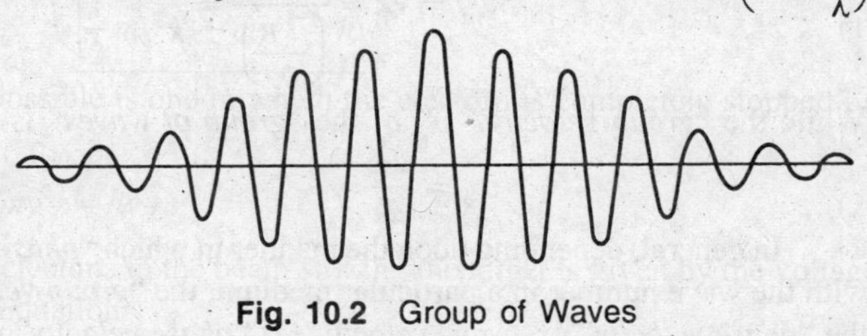

Fig. 10.2 Group of Waves

A familiar example of how groups of waves are formed can be got in the case of beats when two sounds of same amplitude, but of slightly different frequencies, are produced together, the resultant sound will have an average frequency, but its amplitude rises and falls number of times per second periodically, with two loudness peaks called beats as shown in Fig 10.3.

Mathematically, a "*wave group*" is the result of superposition of individual waves each of different wavelength whose interference with one another results in the variation of amplitude that defines the group shape. If the speeds of the individual waves are the same, the speed with which a "*wave group*" travels is identical, the common speed. However, if the wave speed varies with wavelength. The different individual waves do not proceed together, and "*wave group*" is different from that of the waves which compose it.

Let us consider that a "*wave group*" arises from the combination of two waves, with the same amplitude A, but differing in angular velocity by $d\omega$ and an amount dk in wave number $\left(\frac{2\pi}{\lambda}\right)$.

So the two interferring waves may be represented as

$$y_1 = A\cos(\omega t - kx)$$

$$y_2 = A\cos[(\omega + d\omega)t - (k + dk)x]$$

The resultant displacement y at any time t and at any position x is the sum of y_1 and y_2; using identity

$$\cos\alpha + \cos\beta = 2\cos\frac{(\alpha+\beta)}{2}\cos\frac{(\alpha-\beta)}{2}$$

and since $(-\theta) = \cos\theta$, we have

$$y = y_1 + y_2$$

$$= 2A\cos\left[\frac{(2\omega + d\omega)t - (2k + dk)x}{2}\right] \times \cos\left(\frac{d\omega t - dkx}{2}\right)$$

Now since $d\omega$ and dk are small as compared to ω and k respectively

$$\therefore \quad (2\omega + d\omega) \approx 2\omega$$

$$(2k + dk) \approx 2k$$

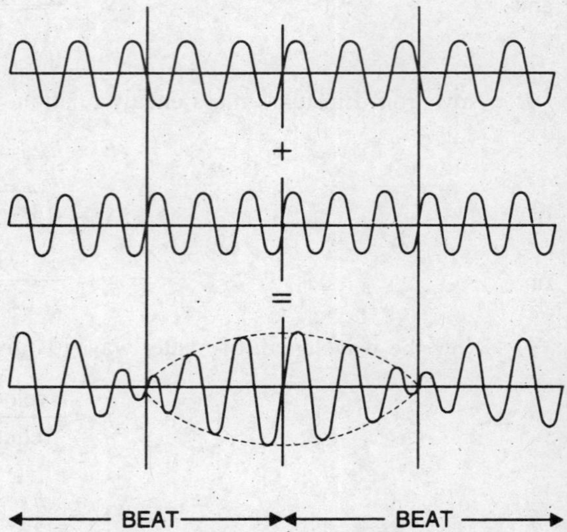

Fig. 10.3 Beat formation

$$\therefore \quad y = 2A\cos(\omega - kx)\cos\left[\left(\frac{d\omega}{2}\right)t - \left(\frac{dk}{2}\right)x\right] \quad \text{...(9)}$$

Equation (9) represents a wave of angular velocity ω and wave number k which has superimposed upon it a wave (the process called modulation) of angular frequency $\left(\frac{d\omega}{2}\right)$ and wave number $\left(\frac{dk}{2}\right)$.

The effect of this modulation is to produce successive "*wave groups*" as shown in Fig 10.3. The "*phase velocity*" or "*wave velocity*" ν_p is given by

$$v_p = \nu\lambda = \frac{2\pi(\nu)}{\left(\frac{2\pi}{\lambda}\right)} = \frac{\omega}{k} \quad \text{...(10)}$$

While the "*group velocity*" v_g of the "*group of waves*" is given by

$$v_g = \frac{d\omega}{dk} \quad \text{...(11)}$$

In general, depending upon the manner in which "*phase velocity*" or "*wave velocity*" ν_p varies with the wave number in a particular medium, the "*group velocity*" v_g may be greater or less than the "*phase velocity*" or "*wave velocity*". If "*phase velocity*" or "*wave velocity*" v_p is the same for all wavelengths, the group and phase velocities are the same *e.g.*, in case of e.m. waves in vacuum.

The angular frequency ω and the wave number k of de Broglie waves associated with a particle of rest mass m_0 and moving with a velocity ν are given by

$$\omega = 2\pi\nu = 2\pi\left(\frac{mc^2}{h}\right) = \frac{2\pi m_0 c^2}{h\cdot\sqrt{1-\frac{v^2}{c^2}}} \qquad ...(12)$$

and

$$k = \frac{2\pi}{\lambda} = 2\pi\left(\frac{mc}{h}\right) = \frac{2\pi m_0 (\nu)}{h\cdot\sqrt{1-\frac{v^2}{c^2}}} \qquad ...(13)$$

We find that ω and k are the functions of the velocity ν of the particle. The "*phase velocity*" or "*wave velocity*" ν_p as we found earlier is $\nu_p = \frac{\omega}{k}$. Hence in case of de Broglie waves we get

$$\nu_p = \frac{2\pi m_0 c^2}{h\cdot\sqrt{\left(1-\frac{v^2}{c^2}\right)}} \times \frac{h\cdot\sqrt{1-\frac{v^2}{c^2}}}{2\pi m_0 (v)} \qquad \text{or} \qquad \nu_p = \frac{c^2}{v}$$

which exceeds both the velocity ν of the particle and the velocity c of light. Since $\nu < c$.

The group velocity ν_g of de Broglie wave associated with the particle is

$$\nu_g = \frac{d\omega}{dk} = \frac{d\omega/d\nu}{dk/d\nu}$$

From equation (12),

$$\frac{d\omega}{d\nu} = \frac{2\pi m_0 c^2}{h}\left(-\frac{1}{2}\right)\left(1-\frac{v^2}{c^2}\right)^{-3/2}\left(-\frac{2v}{c^2}\right)$$

$$= \frac{2\pi m_0 (\nu)}{h\left(1-\frac{\nu^2}{c^2}\right)^{3/2}}$$

From equation (13),

$$k = \frac{2\pi m_0 (\nu)}{h}\left(1-\frac{\nu^2}{c^2}\right)^{-1/2}$$

$$\therefore \quad \frac{dk}{d\nu} = \frac{2\pi m_0}{h}\left(1-\frac{\nu^2}{c^2}\right)^{1/2} + \frac{2\pi m_0 (\nu)}{h}\left[\left(-\frac{1}{2}\right)\left(1-\frac{\nu^2}{c^2}\right)^{-3/2}\left(-\frac{2\nu}{c^2}\right)\right]$$

$$= \frac{2\pi m_0}{h\left(1-\frac{\nu^2}{c^2}\right)^{\frac{1}{2}}} + \frac{2\pi m_0\cdot\left(\frac{\nu^2}{c^2}\right)}{h\left(1-\frac{\nu^2}{c^2}\right)^{\frac{3}{2}}}$$

$$= \frac{2\pi m_0\left[\left(1-\frac{\nu^2}{c^2}\right)+\left(\frac{\nu^2}{c^2}\right)\right]}{h\left(1-\frac{\nu^2}{c^2}\right)^{\frac{3}{2}}} = \frac{2\pi m_0}{h\left(1-\frac{\nu^2}{c^2}\right)^{\frac{3}{2}}}$$

$$\therefore \qquad \text{group velocity } v_g = \frac{d\omega/d\nu}{dk/d\nu} = \frac{2\pi m_0 \nu}{h\left(1-\frac{\nu^2}{c^2}\right)^{\frac{3}{2}}} \times \frac{h\left(1-\frac{\nu^2}{c^2}\right)^{\frac{3}{2}}}{2\pi m_0} = v$$

i.e.,

$$v_g = v_p \qquad ...(14)$$

The de Broglie "*wave group*" associated with a moving particle travels with the same velocity as that of the particle. The "*phase velocity*" or "*wave velocity*" v_p of de Broglie waves evidently has no simple physical significance in itself.

The "*wave velocity*" of the de Broglie waves differs from the velocity of light in one important aspect, that is even in free space the "*wave velocity*" is a function of the wavelength which may be shown as under

$$p = m\nu \qquad ...(i)$$

and total Energy

$$E = mc^2 \qquad ...(ii)$$

So that

$$p = \frac{E}{c^2} \cdot v \qquad ...(iii)$$

Using the relativistic expression for the mass of a particle, the equation for its momentum becomes

$$p = \frac{m_0}{\left(1 - \frac{v^2}{c^2}\right)^{\frac{1}{2}}} \cdot v \qquad ...(iv)$$

where m_0 is the rest mass of the particle. Eliminating ν between the last two equations, we get

$$m_0^2 = \frac{E^2}{c^4} - \frac{p^2}{c^2} \qquad ...(15)$$

Now

$$p = \frac{h}{\lambda} \text{ and } E = h\nu$$

So equation (15) may be written as $m_0 = \frac{h}{c}\sqrt{\frac{v^2}{c^2} - \frac{1}{\lambda^2}}$

and for de Broglie waves $v = \nu\lambda$

Hence

$$m_0 = \frac{h}{c}\sqrt{\frac{v^2}{c^2\lambda^2} - \frac{1}{\lambda^2}} \qquad ...(16)$$

From which

$$\nu_p = c\sqrt{1 + \frac{m_0^2 c^2}{h^2}\lambda^2} \qquad ...(17)$$

The equation shows that for a particle of rest mass $m_0 > 0$, the wave velocity ν_p is always greater than c; also the wave velocity of de Broglie waves is a function of λ even in the free space.

As a special case of de Broglie waves, consider those waves which are propagated with a wave velocity $v_p = c$. This corresponds to the propagation of electromagnetic waves. The velocity of the associated particle *i.e.*, photon, is also equal to c. If $\nu_p = c$ is substituted in equation (16), we find that rest mass of the photon $m_0 = 0$, *i.e.*, there is no such thing as a photon at rest, photon always moves with the velocity c.

10.4. EXPERIMENTAL STUDY OF MATTER WAVES

(*i*) Experiments of Davisson and Germer: First of all in 1927 the two American physicists, Davisson and Germer predicted experimentally the electron waves predicted by de Broglie, Davisson and Germer were studying the reflection of electrons from nickel target which was subjected to such a heat treatment that the crystal was transformed into a group of crystals. In this case the reflection became anomalous and the reflected intensity showed striking maxima and minima. They then suspected that the beam of electrons might be diffracted from the crystals like X-rays. This shows that electrons behave like waves under certain cirumstances.

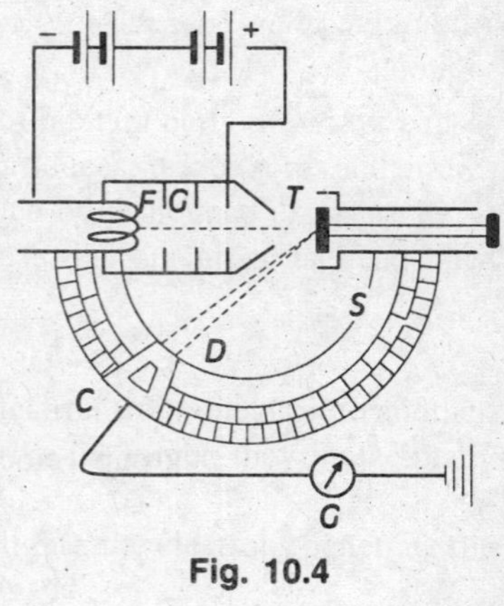

Fig. 10.4

The experimental arrangement is shown in Fig. 10.4. The

electron beam is produced from what is known as electron gun. F is the filament which is heated to dull red. Electrons are emitted from the filament due to thermionic action. G is a system of electrodes with central holes maintained at increasing potentials. By this arrangement, electrons emerge as a well collimated beam. These monoenergetic electrons fall on the target T, a single crystal of nickel. Some of the scattered electrons entered the Faraday cylinder C known as collector. The collector current was amplified and measured with a sensitive galvanometer G. The collector can be moved on a graduated circular scale S to receive electrons scattered in various directions. The collector has two walls insulated from each other. A retarding potential is applied between the inner and outer wall of the collector such that only fast moving electrons coming from electron gun may enter into the collector and not the secondary slow electrons coming from target. The slow electrons will be reflected from the retarding potential. The whole apparatus was enclosed and highly evacuated.

The experiment was conducted in two different ways, *i.e.*,

(*i*) for normal incidence.

(*ii*) for oblique incidence.

(*i*) *For normal incidence*. In this case the beam of electrons falls normally on the surface of the crystal. The collector was moved on the circular scale to various positions and galvanometer current is a measure of the intensity of diffracted beam. A graph was then plotted between the } co-latitude (angle between the incident beam and the beam entering the collector) and galvanometer current. Several curves were obtained for different voltage electrons which are shown in Fig. 10.5.

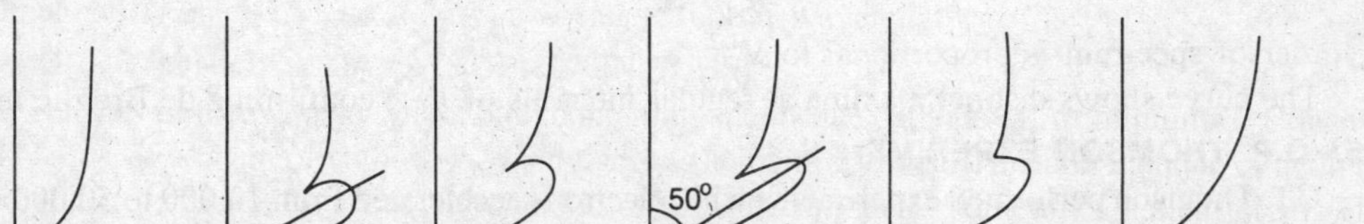

Fig. 10.5

It is observed that a bump begins to appear in the curve at 44-volt electrons. This bump moves upward as the voltage increases and attains the greatest development for 54 volts at co-latitude of 50°. Above 54 volts the bump again diminishes. The bump at 54 volts offers the evidence for the existence of electron waves.

Applying the relation for plane reflection grating:

$$n\lambda = D \sin\theta$$

we have

$$\lambda = 2.15 \sin 50° = 1.67 \text{ Å}$$

where n refers to first order and d in case of nickel crystal is 2.15 Å. Again, according to de Broglie, the wavelength is given by

$$\lambda = \sqrt{\left(\frac{150}{V}\right)} \text{ Å} = \sqrt{\left(\frac{150}{54}\right)} \text{ Å} = 1.66 \text{ Å}$$

This shows that electrons have wavelike characteristics.

(*ii*) *Oblique Incidence at crystal face*: Crystal is oriented so that the angle of incidence was 10° and the position of collector was also chosen corresponding to angle of reflection of 10°. The de Broglie wavelength of the electron was changed by altering the acceleration potential V. The electrons suffer reflection of Bragg's type and law governing this type of diffraction is $n\lambda = 2\,d \sin\theta$. Fig. 10.6 (*a*). The electrons suffer diffraction from the successive parallel planes of atoms in the crystal. Now, as before, the current is measured as function of square root of acceleration potential V. Since $\lambda \propto \frac{1}{\sqrt{V}}$, the graph between current and $V^{1/2}$ also describes the relation between current and $\frac{1}{\lambda}$. See Fig. 10.6 (*b*).

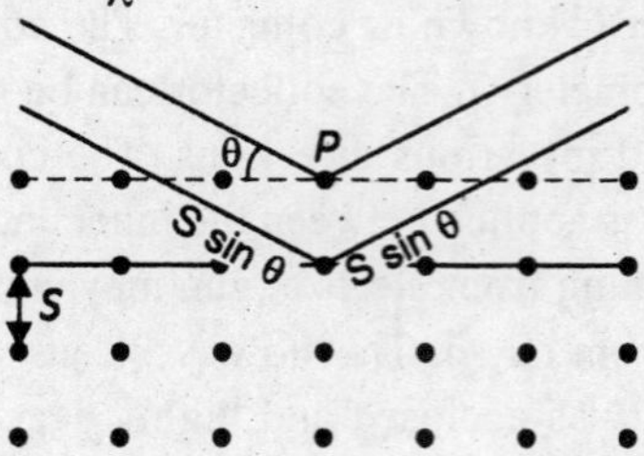

Fig. 10.6 (*a*) Bragg's reflection

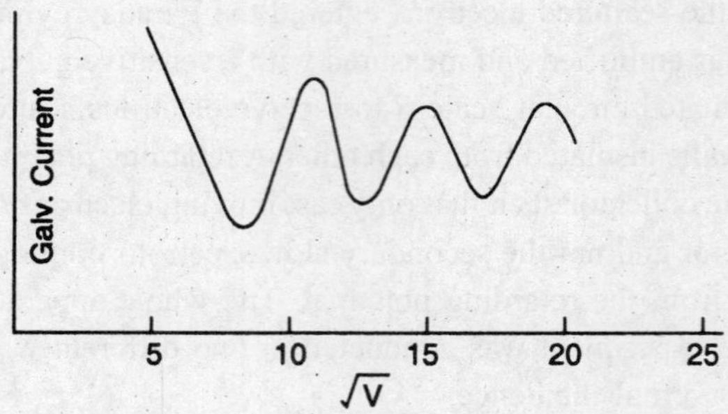

Fig. 10.6 (*b*)

Now, from de Broglie relation, $\lambda = \frac{h}{\sqrt{(2meV)}} \propto \frac{1}{V^{1/2}}$

and Bragg relation gives, $n\lambda = 2d \sin\theta = \text{constant}$

or $n \propto \frac{1}{\lambda}$

$n \propto V^{1/2}$

i.e., order of spectrum is proportional to $V^{1/2}$ ...(1)

The curve shows distinct maxima at regular intervals of $V^{1/2}$, confirming de Broglie theory.

10.5. G.P. THOMSON EXPERIMENTS

G.P. Thomson performed experiments with electrons accelerated from 10,000 to 50,000 volts. In his experiments, Thomson observed diffraction patterns exactly analogous to X-rays patterns, moreover, he was able to determine the wavelengths associated with Electrons. Since diffraction patterns are only possible provided the waves are associated with material particle hence experiments supported the concept of matter waves. His experimental arrangement is shown in Fig. 10.7.

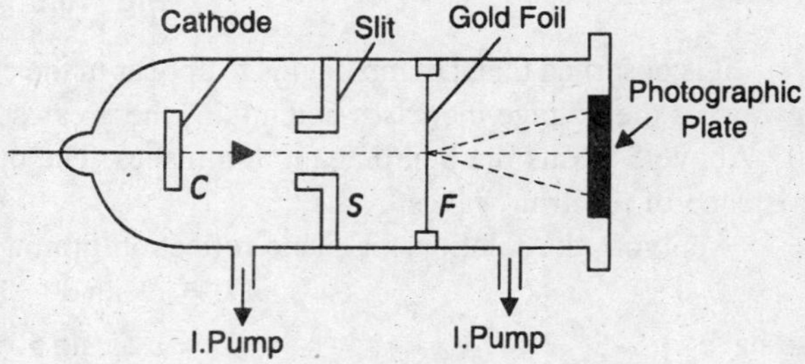

Fig. 10.7

The high energy electron beam is produced by the cathode *C*. The beam is excited with potentials upto 50,000 volts. A fine pencil of this beam is obtained by passing it through the slit or diaphragm *S*. The accelerated fine beam of electrons now falls on a thin film *F* (of the order of 10^{-6} cm) of gold or aluminium. The photograph of the beam from the foil is recorded on a photographic plate *P*. Whole of the apparatus in exhausted to high vacuum so that the electrons may not lose their energy in collisions with the molecules of the gas.

After developing the plate, a symmetrical pattern consisting of concentric rings about a central spot is obtained. This pattern is similar to that produced by X-rays. To be sure that this pattern is due to the electrons and not due to X-rays generated by the electrons in their passage through the foil, the cathode rays in the discharge tube are deflected by magnetic field. It was observed that the beam shifts correspondingly showing thereby that the pattern is produced by electrons and not by X-rays (X-rays pattern is not affected by electric and magnetic fields). The observed rings can only be interpreted by considering that the diffraction pattern of the incoming beam is due to the diffraction of electrons by the foil. As the diffraction pattern can only be produced by waves and not by the particles, so Thomson concluded that electrons behave like waves.

Thus, Thomson experiments clearly demonstrated the existence of matter waves. He also calculated the associated wavelength and observed that it depends only on the accelerating voltage and is independent of the material of target. The wavelengths obtained by diffraction patterns agree with the wavelength obtained by de Broglie relation.

10.6. NEUTRON DIFFRACTION

In 1936, W.M. Elsasser suggested that moving neutrons should have de Broglie waves associated with them and that they could, therefore, be diffracted in the same way as do the electrons. The relationship between wavelength λ and momentum p of the neutron is still given by

$$\lambda = \frac{h}{p} = \frac{h}{m_n v} \qquad ...(1)$$

where m_n is the mass of the neutrons. We know that the mass of the neutron is about 2000 times as large as that of electron and hence the wavelength associated with neutron is about 1/2000 that of an electron of the same velocity. The relationship between energy E and wavelength λ is expressed as

$$\lambda = \frac{h}{\sqrt{(2m_n E)}} \quad \text{or} \quad \lambda = \frac{0.28}{\sqrt{E}}$$

where λ is expressed in Å and E in eV. When $E = 0.1$ eV, λ = 1 Å which is required for diffraction work. It is also possible to confirm their wave properties for very much smaller wavelengths. For example, neutrons with energies of about 10 MeV are rather readily produced in nuclear reactions. Such neutrons have wavelength of about 10^{-12} cm. The distance is comparable with the diameters of most nuclei. Thus when fast neutrons are scattered by nuclei, the result is the characteristic diffraction pattern.

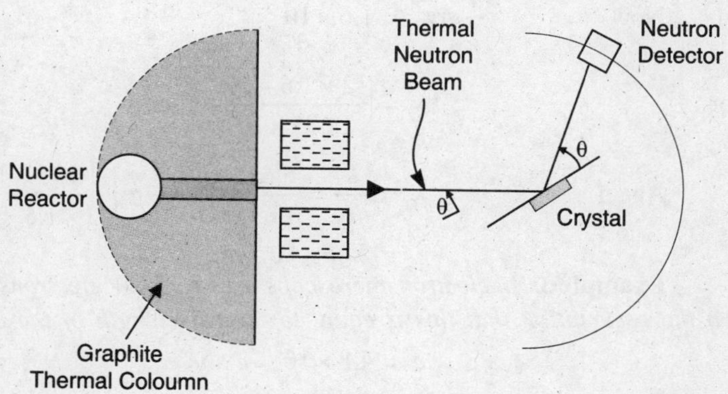

Fig. 10.8

The wavelength distribution can be measured by introducing a crystal and a neutron detector as shown in Fig. 10.8. This arrange- ment is just like one X-ray diffraction apparatus except that a different detector is used. The crystal and detector are turned in such a way so that the two angles shown as θ in the figure are equal. Reflection from the crystal occurs only when λ and θ satisfy the Bragg relation. The fraction of neutrons reflected as function of θ gives an experimental determination of wavelength distribution of the incident neutrons.

Neutron diffraction has proved a useful method for studying the structure of crystals, especially organic crystals containing hydrogen. This is due to the fact that the scattering of X-rays or electrons by hydrogen is very weak as atoms in a crystal cannot be detected. Neutrons, on the other hand, interact with nuclei through nuclear forces and are scattered by hydrogen atoms. Neutron diffraction is also frequently applied to the study of magnetic properties of solids and various other allied phenomenon.

Example 1. *Find the energy of the neutron in units of electron volt whose de Broglie wavelength is 1 Å.*

Given mass of the neutron $= 1.674 \times 10^{-27}$ *kg*

Planck's constant $h = 6.60 \times 10^{-34}$ *Joule-sec.*

Solution: We have the relation

$$\lambda = \frac{h}{mv} = \frac{h}{\sqrt{2\,m\,E}}$$

or
$$E = \frac{h^2}{2\,m\,\lambda^2}$$

In this case
$$m = 1.674 \times 10^{-27}\,\text{kg.}$$
$$\lambda = 1\,\text{Å} = 10^{-10}$$
$$h = 6.60 \times 10^{-34}\,\text{Joule-sec.}$$

$\therefore$
$$E = \frac{(6.60 \times 10^{-34})^2}{1.674 \times 10^{-27} \times (10^{-10})^2} = 13.01 \times 10^{-21}\ \text{Joules}$$
$$= \frac{13.01 \times 10^{-24}}{1.6 \times 10^{-19}} = \mathbf{8.12 \times 10^{-2}\ eV.}$$

Example 2. *Compute the de Broglie wavelength of 10^{11} keV neutrons. Mass of neutron may be taken as 1.675×10^{-27} kg.*

Solution: In this case kinetic energy of neutron

$$= 10^{11}\,\text{keV} = 10^{14}\,\text{eV} = 10^{14} \times 1.6 \times 10^{-19}\ \text{Joule.}$$
$$= 1.6 \times 10^{-5}\ \text{Joule.}$$

Now
$$\frac{1}{2}mv^2 = 1.6 \times 10^{-5}$$

or
$$v = \left(\frac{2 \times 1.6 \times 10^{-5}}{1.675 \times 10^{-27}}\right)^{\frac{1}{2}}\ \text{m/s}$$

Again
$$\lambda = \frac{h}{mv} = \frac{6.625 \times 10^{-34}}{1.675 \times 10^{-27}} \times \left(\frac{1.675 \times 10^{-27}}{2 \times 1.6 \times 10^{-5}}\right)^{\frac{1}{2}}$$
$$= \mathbf{2.86 \times 10^{-15}\,m.}$$

Example 3. *An electron microscope uses 1.25 keV electrons. Find its ultimate resolving power on the assumption that this is equal to the wavelength of the electron, given that*

$$e = 4.8 \times 10^{-10}\ \textit{e.s.u.}$$
$$m = 9.0 \times 10^{-28}$$
$$h = 6.65 \times 10^{-27}\ \textit{ergs sec.}$$

Solution: We know that when a particle is accelerated by V volts then it acquires $V/300$ ergs erergy and if its velocity is v, then we have

$$\frac{1}{2}mv^2 = \frac{eV}{300}$$

or
$$m\,v = \sqrt{\left(\frac{meV}{150}\right)}$$

$\therefore$
$$\lambda = \frac{h}{mv} = h\sqrt{\left(\frac{150}{meV}\right)}$$

In this case
$$E = 12.5\,\text{keV}$$

$\therefore$
$$V = 1250\ \text{volts}$$

and
$$\lambda = 6.65 \times 10^{-27}\sqrt{\left(\frac{150}{9 \times 10^{-28} \times 4.8 \times 10^{-10} \times 1250}\right)}\ \text{cm.}$$
$$= \mathbf{0.351\,Å}$$

So resolving power of the microscope is 0.351 Å.

Example 4. *What is de Broglie wavelength of an electron which has been accelerated from rest through a potential difference of 100 V.*

Solution: We have the relation

$$\lambda = \frac{12.25}{\sqrt{V}}\ \text{Å}$$

In this case $V = 100$ volts $\quad \therefore \quad \lambda = \frac{12.25}{\sqrt{100}} = \mathbf{1.225\ Å}$

Example 5. *Estimate the amount of accelerating voltage to which electrons are to be subjected in order to associate them with de Broglie wavelength of 0.50 Å. Given that*

$$e = 4.8 \times 10^{-10}\ e.s.u.$$
$$m = 9.0 \times 10^{-28}\ gm$$
$$h = 6.67 \times 10^{-27}\ \text{ergs. sec.}$$

Solution: Now if V is the accelerating potential measured in volts, then an electron will gain the energy E given by

$$E = \frac{eV}{300}$$

and

$$\lambda = \frac{h}{m\nu}$$

$$\lambda = \frac{h}{\sqrt{(2mE)}} \quad \because \text{ also } E = \frac{eV}{300} \quad \therefore \lambda = \frac{h}{\sqrt{\frac{meV}{150}}}$$

Therefore substituting the values, we get

$$0.5 \times 10^{-8} = \frac{6.67 \times 10^{-27}}{\sqrt{\frac{meV}{150}}}$$

$$V = \frac{(6.67 \times 10^{-27})^2 \times 150}{(0.5 \times 10^{-8})^2 \times 9.0 \times 10^{-28} \times 4.8 \times 10^{-10}}$$

$$V = \mathbf{614.4\ volts.}$$

Example 6. *Calculate the energy in electron volt of an electron wave of $\lambda = 3 \times 10^{-2}$ m. Given $h = 6.62 \times 10^{-34}$ Js.*

Solution: We know that energy is given by the relation

$$E = \frac{h^2}{2m\lambda^2}$$

In this case

$$h = 6.62 \times 10^{-34}\ \text{Js},\ m = 9.1 \times 10^{-31}\ \text{k.gm and } \lambda = 3 \times 10^{-2}\ \text{m}$$

$$\therefore \quad E = \frac{(6.62 \times 10^{-34})^2}{2 \times 9.1 \times 10^{-31} \times (3 \times 10^{-2})^2} \cong 0.27 \times 10^{-23}\ \text{Joules}$$

$$\cong \frac{0.27 \times 10^{-33}}{1.6 \times 10^{-19}}\ \text{eV} = \mathbf{1.68 \times 10^{-15}\ eV}$$

Example 7. *What would be the wavelength of quantum of radiant energy emitted, if an electron transmitted into radiation and converted into one quantum.*

Solution: According to Planck, the energy E associated with one quantum $1 = h\nu$, where ν is the frequency of radiation.

When the energy of an electron is transmitted into radiation, we have the relation

$$E = mc^2 \text{ (mass energy relation)}$$

$$\therefore \quad h\nu = mc^2 \quad \text{or} \quad \frac{hc}{\lambda} = mc^2$$

$$\therefore \quad \lambda = \frac{h}{mc} = \frac{6.6 \times 10^{-34}}{9.1 \times 10^{-31} \times 3 \times 10^8}$$

$$= 0.0244 \times 10^{-10} \text{ m} = \mathbf{0.0244 \text{ Å}}$$

Example 8. *Calculate the velocity and de Broglie wavelength of a proton of energy 10^5 electron volts, given that*

$$\textit{Mass of proton} = 1.66 \times 10^{-24} \; gm$$

$$\textit{Change on the electron} = 4.8 \times 10^{-10} \; e.s.u.$$

$$\textit{Planck's constant} = 6.62 \times 10^{-27} \; erg.sec.$$

Solution: The energy of proton

$$\frac{1}{2}mv^2 = 10^5 \text{ eV} = 10^5 \times 1.6 \times 10^{-12} \text{ ergs.}$$

$$v^2 = \frac{2 \times 10^5 \times 1.6 \times 10^{-12}}{m}$$

$$= \frac{2 \times 10^5 \times 1.6 \times 10^{-12}}{1.66 \times 10^{-24}}$$

or

$$v = \sqrt{\left(\frac{2 \times 10^5 \times 1.6 \times 10^{-12}}{1.66 \times 10^{-24}}\right)}$$

$$= 4.47 \times 10^8 \text{ cm/sec.}$$

The de Broglie wavelength λ is given by

$$\lambda = \frac{h}{mv} = \frac{h}{\sqrt{(2mE)}} \quad \text{because } \frac{1}{2}mv^2 = E$$

or

$$mv = \sqrt{(2mE)}$$

According to problem

$$E = 10^5 \times 1.6 \times 10^{-12} \text{ ergs}$$

$$\therefore \quad \lambda = \frac{6.62 \times 10^{-27}}{\sqrt{(2 \times 1.66 \times 10^{-24} \times 10^5 \times 1.6 \times 10^{-12})}}$$

$$= 9.3 \times 10^{-12} \text{ cm} = \mathbf{9.3 \times 10^{-14} \text{ m.}}$$

Example 9. *Calculate the de Broglie wavelength for an electron which has been accelerated through at p. d. of 28.8 V, given that*

$$e = 4.8 \times 10^{-10} \; e.s.u.$$

$$m = 9.1 \times 10^{-28} \; gm$$

$$h = 6.62 \times 10^{-27} \; erg.sec.$$

Solution: We have the relation

$$\lambda = h \cdot \sqrt{\left(\frac{150}{meV}\right)}$$

$$= \frac{6.62 \times 10^{-27} \times \sqrt{(150)}}{\sqrt{(9.1 \times 10^{-28} \times 4.8 \times 10^{-10} \times 28.8)}} = \frac{12.27}{\sqrt{(28.8)}} \times 10^{-8} \text{ cm.}$$

$$= \mathbf{2.4 \text{ Å}}$$

Example 10. *Energy of a particle at absolute temperature T is of the order kT. Calculate the wavelength of thermal neutrons at 27°C. Given mass of the neutron = 1.67×10^{-27} kg, Planck's constant h = 6.60×10^{-34} Joule-sec and Boltzmann's constant k = 8.6×10^{-5} eV deg^{-1}.*

Solution: We know that

$$\lambda = \frac{h}{mc} = \frac{h}{\sqrt{2mE}}$$

From the problem $E = kT$

$$\therefore \quad \lambda = \frac{h}{\sqrt{(2mkT)}}$$

Here $h = 6.60 \times 10^{-34}, m = 1.67 \times 10^{-27}$ kg.

$$T = 27°C = 300 \text{ K}$$

$$k = 8.6 \times 10^{-5} \times \text{eV deg}^{-1}$$

$$= 8.6 \times 10^{-5} \times 1.8 \times 10^{-19} \text{ Joule deg}^{-1}$$

$$= 1.376 \times 10^{-23} \text{ joule deg}^{-1}$$

$$\lambda = \frac{6.60 \times 10^{-34}}{\sqrt{(2 \times 1.67 \times 10^{-27} \times 1.376 \times 10^{-28} \times 300)}}$$

$$= 1.777 \times 10^{-10} \text{m} = \mathbf{1.777 \ Å}$$

Example 11. *What is the energy of gamma ray photon having a wavelength of 1 Å? Given that:*

$$h = 6.62 \times 10^{-27} \, erg. sec.$$

Solution: We know that

$$\lambda = \frac{h}{p} \text{ or } p = \frac{h}{\lambda} = \frac{6.62 \times 10^{-27}}{10^{-8}} \left(\frac{\text{erg.sec}}{\text{cm}}\right)$$

Energy of gamma ray photon

$$E = p \times c \text{ where } c = \text{velocity of light}$$

$$E = \frac{6.62 \times 10^{-27}}{10^{-8}} \times 3 \times 10^{10} \text{ ergs}$$

$$E = \frac{6.62 \times 10^{-27} \times 3 \times 10^{10}}{10^{-8} \times 1.6 \times 10^{-12}} \text{ eV}$$

$$= \mathbf{1.24 \times 10^4 eV.}$$

Example 12. *Calculate the de Broglie wavelength of an α-particle accelerated through 200 volts, given that:*

mass of $\alpha = 6.576 \times 10^{-24}$ gm.
$h = 6.62 \times 10^{-27}$ ergs. sec.

Solution: We have the definition of potential difference in e.s.u. between two points defined as the work done in moving a unit positive charge from one point to the other. In this way the work done in moving an α-particle between the two points between which the potential difference is V volts is given by

$$= 2e\frac{V}{300} \text{ ergs.}$$

$$\therefore \quad \frac{1}{2}mv^2 = \frac{2eV}{300}$$

or

$$\lambda = h\sqrt{\frac{150}{(2meV)}}$$

$$= 6.62 \times 10^{-27} \sqrt{\frac{150}{(2 \times 6.576 \times 10^{-24} \times 4.8 \times 10^{-10} \times 200)}}$$

$$= \frac{9.44}{4} \times 10^{11} \text{cm}$$
$$= \mathbf{2.362 \times 10^{-11}\ cm}$$

10.7. HEISENBERG'S UNCERTAINTY PRINCIPLE

In the experiments for the determination of specific charge of an electron, it was shown that electron gets deflected by the application of electrical and magnetic fields, which means that the electron behaves as a particle and it has definite mass, momentum and energy which can be measured with desired accuracy. But the experiment of Davisson and Germer, and Thomson showed the wave nature of electrons and the wavelength of the electron is given by the relation

$$\lambda = \frac{h}{mv} = \frac{h}{p}$$

where h is the Planck's constant and mv is the momentum of the electron. A wave extends through space and it is difficult to locate the exact position of an electron behaving as a wave, at any given instant of time.

This dual behaviour of electron as a particle and as a wave presents difficulty in locating the exact position and momentum at the same time. This difficulty is overcome by Heisenberg's uncertainty principle.

The principle states that the exact position and momentum of a particle (say electron) cannot be determined simultaneously with a desired accuracy. *Taking* Δx *as the error in derermining its position and* Δp *the error in determining its momentum at the same instant, these quantities are related as under*:

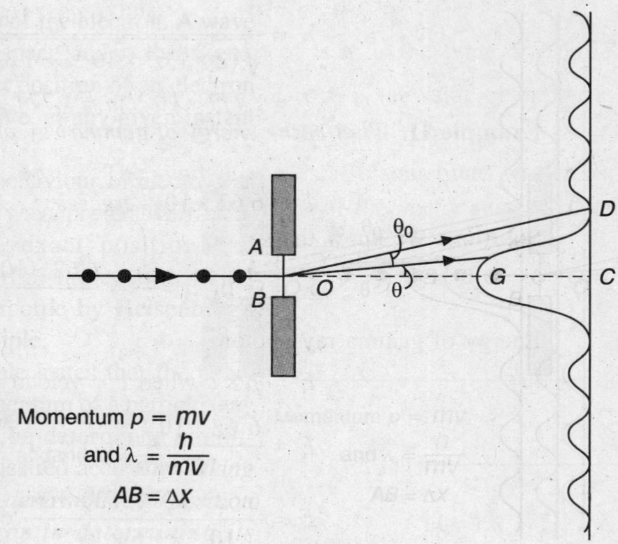

Fig. 10.9

$$\Delta x \, \Delta p_x \cong \frac{1}{2} \times \frac{h}{2\pi} \quad ...(i)$$

i.e., the product of the two errors is approximately of the order of Planck's constant.

Multiplying and dividing the left hand side of equation (i) by v, the velocity of the particle

$$\frac{\Delta x}{v} \Delta p v \approx \frac{1}{2} \times \frac{h}{2\pi}$$

or
$$\Delta t \, \Delta E \approx \frac{1}{2} \times \frac{h}{2\pi} \quad ...(ii)$$

Here ΔE represents the uncertainty in the measurement of the energy of a particle and Δt the uncertainty in the measurement of time.

In equation (i), the product $\Delta x \, \Delta p$ is of the order of $h/2\pi$. If Δx is small Δp will be large and vice versa. It means that if one quantity is measured accurately, the other quantity becomes less accurate. A similar argument hold good for $\Delta E \, \Delta t$ as well.

Consider a beam of electons travelling in the direction shown in Fig. 10.9. The slit AB of witdth Δx is perpendicular to the path of the electrons. Before entering the slit, the electron has a definite momentum $p = mv$. After passing through the slit the electron gets diffracted and acquires a momentum along OG. The angular deflection θ depends upon component of the momentum parallel to the slit, *i.e.*, $\Delta p = p \sin\theta = p\,\theta$ for small angular deflection. The angle θ_0 corresponding to the direction of first momentum of the diffraction pattern is

$$\theta_0 = \frac{\lambda}{\Delta x}, \quad \text{for small value of } \theta_0$$

or

$$\Delta x = \frac{\lambda}{\theta_0}$$

$$\therefore \quad \Delta p\,\Delta x \sim p\lambda \times \frac{\theta}{\theta_0}$$

But

$$p\lambda = h$$

$$\therefore \quad \Delta p\,\Delta x = h\frac{\theta}{\theta_0}$$

Taking, $\theta = \theta_0$ approximately,

$$\Delta p\,\Delta x \sim h$$

The probable deflection θ of the electron is less than θ_0 and according to Heisenberg, the uncertainty relation is given by

$$\Delta p\,\Delta x \geq \frac{1}{2} \times \frac{h}{2\pi}$$

In general any experiment that measures position or momentum accurately, introduces an uncertainty in the other. The uncertainty is never less than the predicted value according to Heisenberg's relation. Thus no instrument can measure the quantities more accurately than predicted by *Heisenberg's principle of uncertainty or indeterminancy.*

10.8. APPLICATIONS OF UNCERTAINTY PRINCIPLE

(*i*) Non-existence of electrons and existence of protons and neutrons in nucleus. The radius of the nucleus of any atom is of the order of 10^{-14} m. If an electron is confined inside the nucleus, then the uncertainty in the position Δx of the electron is equal to the diameter of the nucleus, *i.e.*, $\Delta x \approx 2 \times 10^{-14}$ m. Using the Heisenberg's uncertainty relation, the uncertainty in momentum of electron is given by the relation

$$\Delta p_x \geq \frac{h}{2\pi \cdot \Delta x} \geq \frac{1.055 \times 10^{-34}}{2 \times 10^{-14}} \text{ kg ms}^{-1} \qquad \because \Delta x \approx 2 \times 10^{-14} \text{ m.}$$

where

$$\frac{h}{2\pi} = 1.055 \times 10^{-34}$$

$$\therefore \quad p_x \geq 0.527 \times 10^{-20} \text{ Ns}$$

It means that the momentum component p_x and hence the magnitude of total momentum $|\vec{p}|$ of the electron in the nucleus must be at least of the order of magnitude, *i.e.*,

$$|\vec{p}| \sim p_x \sim \Delta p_x \sim 0.527 \times 10^{-20} \text{ Ns.}$$

Now the mass of the electron is 9.1×10^{-31} kg, the order of magnitude of momentum (0.527×10^{-20} kg ms^{-1}) is relativistic. Using the relativistic formula for the energy E of the electron, we have

$$E^2 = p^2c^2 + m_0^2c^4$$

As the rest energy m_0c^2 of the electron is of the order of 0.511 MeV, which is much smaller than the value of first term, so it can be neglected. Thus, we have

$$E^2 = p^2c^2 \quad \text{or} \quad E = pc$$

$$\therefore \quad E = (0.527 \times 10^{-20}) \times (3 \times 10^8) \text{ joules}$$

$$\approx \frac{(0.527 \times 10^{-20})(3 \times 10^8)}{1.6 \times 10^{-19}} \text{ eV}$$

$$= 10 \text{ MeV}$$

which means that if the electrons exist inside the nucleus, their energy must be of the order of 10 MeV. However, we know that the electrons emitted by radioactive nuclei during beta decay have energies only 3 to 4 MeV. So, in general electrons cannot exist in the nucleus.

For protons and neutrons, $m_0 = 1.67 \times 10^{-27}$ kg. This is a non-relativistic problem as $|\vec{v}| \sim |\vec{p}|/m_0 \sim 3\times10^{6}\,\text{ms}^{-1}$. The kinetic energy K in this case is given by the relation

$$K = \frac{p^2}{2m_0} = \frac{(0.527\times10^{-20})^2}{2\times1.67\times10^{-27}} \text{ joules}$$

$$\approx \frac{(0.527\times10^{-20})^2}{2\times1.67\times10^{-27}\times1.6\times10^{-19}} \text{ eV} = 52 \text{ keV}$$

Since the energy is less than the energies carried by these particles emitted by nuclei, both these particles can exist inside the nuclei.

***(ii)* Binding energy of an electron in atom**. In an atom, the electron is under the influence of electrostatic potential of the positively charged nucleus. It is confined to the linear dimensions equal to the diameter of electronic orbit. The uncertainty in the position Δx of the electron is of the order of $2R$ where R is the radius of the orbit and the corresponding uncertainty in momentum component Δp_x is given by

$$\Delta p_x \geq \frac{h}{4\pi\cdot 2R}$$

which shows that the momentum of an electron in atomic orbit is at least

$$|\vec{p}| \sim p_x \sim \Delta p_x \sim \frac{h}{4\pi\cdot 2R}$$

$$\sim 0.527\times10^{-24}\,\text{kg ms}^{-1} \quad \because \; R = 10^{-10}\,\text{m}$$

This is non-relativistic momentum for an electron. The kinetic energy K is given by

$$K = \frac{p^2}{2m_0} = \left(\frac{h}{4\pi R}\right)^2 \frac{1}{2m_0} = \frac{h^2}{32\pi^2 m_0 R^2}$$

The potential energy of an electron in the field of nucleus with atomic number Z is given by

$$V = \frac{-Ze^2}{4\pi\varepsilon_0 R}$$

So the total energy of the electron in its orbit will be

$$E = K + V = \frac{h^2}{32\pi^2 m_0 R^2} - \frac{Ze^2}{4\pi\varepsilon_0 R}$$

$$= \frac{(1.055\times10^{-34})^2}{8\times9.1\times10^{-3}\,y\times R^2} - \frac{Z\,(1.6\times10^{-19})^2}{4\times3.14\times8.85\times10^{-12}\times R}$$

where $\dfrac{h}{2\pi} = 1.055\times10^{-34}$

or $E = \dfrac{10^{-20}}{R^2} - \dfrac{15\times10^{-10}Z}{R}$ eV

Taking $R = 10^{-10}$ m, we have

$$E = (1 - 15Z)\,\text{eV}$$

Now the binding energies of outermost electrons in H and He are – 13.6 and – 24.6 eV respectively, so the value of binding energy derived on the basis of uncertainty principle is acceptable as these are comparable in magnitudes.

***(iii)* Finite width of the spectral lines.** From Heisenberg's Principle of energy and time relation, we have

$$\Delta E.\Delta t \cong h$$

Since the life time of electron in an excited orbit is finite (~ 10^{-8} sec), so the energy levels of an atom given by $\Delta E = \frac{h}{\Delta t}$ must have a finite width which means that the excited levels must have a finite energy spread *i.e.*, the radiation given out when an electron jumps must be truly monochromatic. In other words, it means that the spectral lines can never be very sharp but must have a natural finite width.

***(iv)* Strength of nuclear force.** If we assume that the nuclear radius is of the order $r_0 = 1.2 \times 10^{-13}$ cm.

From uncertainty principle, momentum will be of the order

$$p = \frac{\hbar}{r_0}$$

∴ K.E. will be of the order

$$\text{K.E.} \simeq \frac{p^2}{2M} \simeq \frac{\hbar}{2Mr_0^2} \simeq 10\,\text{MeV}$$

where M corresponds to the mass of a nucleus (proton or neutron).

Since the nucleus is bound, so the B.E. should be greater than K.E. with a negative sign. So the B.E. of a nucleus is of the order of 10 MeV.

10.9. COMPLEMENTARY PRINCIPLE

The most important consequence of the uncertainty principle is that it is impossible to determine the wave and particle properties exactly at the same time.

According to the complementary principle the wave and particle aspects of matter and light are complementary rather than contradictory, i.e., both aspects are necessary to have a complete picture of the same system. For example, consider an experimental arrangement in which light is diffracted by a double slit and is detected on a screen that consists of many adjacent photoelectric cells. The photoelectric cells responds to photons (having particle properties). However, if we plot the number of photons each cell counts in a certain period of time against the location of cell, a wave like pattern (as in the case of interference of a pair of coherent wave trains) is obtained. It follows that only experiment which can be devised displays either the particle like characteristics or wave like characteristics of the system, moreover the wave and particle pictures give complementary description of the same system.

Example 14. *An electron has a speed of 500 ms^{-1} correct upto 0.01% with what minimum accuracy can you locate the position of this electron.*

Solution: The momentum of electron $p = mv$

$$= 9.11 \times 10^{-31} \times 500 = 4.55 \times 10^{-28}\,\text{kg ms}^{-1}$$

The uncertainty in momentum

$$\Delta p = m\,\Delta v = mv\frac{\Delta v}{v}$$

$$= 4.55 \times 10^{-28} \times 0.0001$$

$$= 4.55 \times 10^{-32}\ \text{kg ms}^{-1}$$

The minimum uncertainty in position is given by

$$\Delta x = \frac{h}{2\pi}\frac{1}{\Delta p} = \frac{6.625 \times 10^{-34}}{2\pi \times 4.55 \times 10^{-32}}$$

$$= \mathbf{2.318\ mm.}$$

Example 15. *An electron has a speed of 600 m/s with an accuracy of 0.005%. Calculate the certainty with which we can locate the position of the electron. Given that*

$$h = 6.6 \times 10^{-34}\ \textit{joule-sec}$$

$$m = 9.1 \times 10^{-31}\ kg$$

and $$v = 600\ m/sec.$$

Solution: The momentum of electron = mv

$$= 9.1 \times 10^{-31} \times 600 \text{ kg m/s}$$

$$\Delta px = \left(\frac{0.005}{100}\right) mv = 5 \times 10^{-5} \times 9.1 \times 10^{-31} \times 600 \text{ kg m/s}$$

$$\therefore \quad \Delta x \approx \frac{h}{\Delta px} \approx \frac{6.6 \times 10^{-34}}{5 \times 10^{-5} \times 9.1 \times 10^{-31} \times 600} \approx \mathbf{0.02253\ m.}$$

Example 16. *10 keV electrons are passed through a thin film of a metal for which atomic spacing is 5.5 × 10^{-11}m. What is the angle of deviation of the first order diffraction maximum?*

Solution: In this case the kinetic energy of electron is small as compared with its rest mass energy (500 keV)

$$\therefore \quad m = m_0 = 9.11 \times 10^{-31} \text{ kg.}$$

$$\text{K.E. of electrons } \left(\frac{1}{2}mv^2\right) = 10 \text{ keV} = 10^4 \text{ eV}$$

$$= 10^4 \times 1.6 \times 10^{-19} \text{ joule}$$

$$v = \sqrt{\frac{2 \times 10^4 \times 1.6 \times 10^{-19}}{m}}$$

de-Broglie wavelength, $\lambda = h/mv$

Putting the values, we get

$$\lambda = \frac{6.625 \times 10^{-34} \times \sqrt{m}}{m \times \sqrt{2 \times 10^4 \times 1.6 \times 10^{-19}}}$$

$$= \frac{6.625 \times 10^{-34}}{\sqrt{3.2 \times 10^{-15} \times m}}$$

or $$\lambda = \frac{6.625 \times 10^{-34}}{\sqrt{3.2 \times 10^{-15} \times 9.11 \times 10^{-31}}}$$

$$= 1.23 \times 10^{-11} \text{ m.}$$

Now using Bragg's equation we have

$$2d \sin\theta = \lambda$$

In this case $n = 1$ for first order diffraction

$$d = 5.5 \times 10^{-11} \text{ m}$$

$$\therefore \quad 2 \times 5.5 \times 10^{-11} \sin\theta = 1 \times 1.23 \times 10^{-11}$$

or $$\sin\theta = \frac{1.23 \times 10^{-11}}{2 \times 5.5 \times 10^{-11}} = \mathbf{0.1118}$$

Example 17. *Determine the de Broglie wavelength of an electron, having kinetic energy of 1 eV. (Given m_0 of an electron = 9.11 × 10^{-31} kg and h = 6.625 × 10^{-34} J-s).*

Solution: Here the kinetic energy of the electron is very small as compared with its rest mass energy, *i.e.*, 0.512 MeV and hence negligible.

$$m = m_0 = 9.11 \times 10^{-31} \text{ kg}$$

$$\text{K.E. of the electron } \left(\frac{1}{2}mv^2\right) = 1 \text{ eV} = 1.6 \times 10^{-19} \text{ J}$$

$$v^2 = \frac{2 \times 1.6 \times 10^{-19}}{m}$$

de Broglie wavelength, $\lambda = h/mv$

Putting the values, we have

$$\lambda = \frac{6.625 \times 10^{-34} \sqrt{m}}{m\sqrt{2 \times 1.6 \times 10^{-19}}}$$

$$\lambda = \frac{6.625 \times 10^{-34}}{\sqrt{3.2 \times 10^{-19} \times m}}$$

$$= \frac{6.625 \times 10^{-34}}{\sqrt{3.2 \times 10^{-19} \times 9.11 \times 10^{-31}}}$$

$$= 1.23 \times 10^{-9} \text{ m}$$

$$= \mathbf{12.3\ Å}$$

Example 18. *In jumping from an excited to a stationary state, an atom takes 10^{-9} seconds. What is the uncertainty in the energy of the emitted radiation?*

Solution: $\Delta E \cdot \Delta t = \frac{h}{2\pi}$

An electron having momentum of this order is non-relativistic in behaviour and its kinetic energy is given by the relation.

$$T = p^2/2m$$

$$= \frac{(2.2 \times 10^{-24} \text{ kg ms}^{-1})^2}{2 \times 9.1 \times 10^{-31} \text{ kg}}$$

$$= 2.7 \times 10^{-18} \text{ joule} = 17 \text{ eV}$$

We can consider the limit to the accuracy with which we can determine the frequency of the radiation emitted by an excited atom as under:

The excited atom gives up its excess energy by emitting one or more photons of characteristic frequency. The average time gap between the excitation of an atom and the time it radiates energy is 10^{-8} sec. So the photons energy is uncertain by an amount given by

$$\Delta E = \frac{h}{\Delta t} = \frac{1.054 \times 10^{-34} \text{ J s}}{10^{-8} \text{s}}$$

$$= 1.054 \times 10^{-26} \text{ J}$$

and the frequency of the light is uncertain by an amount

$$\Delta v = \frac{\Delta E}{h} = \mathbf{1.6 \times 10^7\ Hz.}$$

and this limit cannot be reduced.

Example 19. *If the uncertainty in position of an electron is 4×10^{-10} m, calculate the uncertainty in its momentum.*

Solution: From the uncertainty principle

$$\Delta x \Delta p_x \approx h$$

$$\therefore \quad \Delta p_x \approx \frac{h}{\Delta x} \approx \frac{6.6 \times 10^{-34}}{4 \times 10^{-10}} \approx \mathbf{1.65 \times 10^{-24}\ kg\ m/sec.}$$

Example 20. *An electron is confined to a box of length 10^{-9} m, calculate the minimum uncertainty in its velocity.*

Given mass of the electron $m = 10^{-31}$ kg and Planck's constant $h = 6.6 \times 10^{-34}$ Joules-sec.

Solution: We have the relation

$$\Delta x \cdot \Delta p \cong h$$

If Δx is maximum, Δpx must be minimum

i.e., $\quad (\Delta x)_{max} (\Delta p_x)_{min} \cong h$

According to the problem $(\Delta x)_{max} = 10^{-9}$ m

$$\therefore \quad (\Delta p)_{min} = \frac{h}{(\Delta x)_{max}} = \frac{6.6 \times 10^{-34}}{10^{-9}}$$

$$= 6.6 \times 10^{-25} \text{ kg m/s.}$$

Again $\quad (\Delta p_x)_{min} = m(\Delta vx)_{min}$

$$\therefore \quad (\Delta vx)_{min} = \frac{(\Delta px)_{min}}{m} = \frac{6.6 \times 10^{-25}}{9 \times 10^{-31}} = \mathbf{7.3 \times 10^5 \ m/s}$$

Example 21. *Using the uncertainty relation* $\Delta E . \Delta t = \frac{h}{2\pi}$ *calculate the time required for the atomic system to retain rotation energy for a line of wavelength 6000 Å and width* 10^{-4} *Å.*

Solution: Here, the width of spectral line $d\lambda = 10^{-4}$ Å

$$= 10^{-4} \times 10^{-10} \text{ m and } \lambda = 6000 \text{ Å} = 6 \times 10^{-7} \text{ m}$$

We know that

$$E = h\nu = \frac{hc}{\lambda}$$

or

$$\Delta E = -\frac{hc}{\lambda^2} \Delta\lambda$$

Using uncertainty relation $\Delta E . \Delta t = \frac{h}{2\pi}$, we have

$$\Delta t \approx \frac{h}{2\pi \Delta E} \approx \frac{h}{2\pi\left(\frac{hc}{\lambda^2}\right)\Delta\lambda} \approx \frac{\lambda^2}{2\pi c \Delta\lambda}$$

$$= \frac{(6 \times 10^{-7})^2}{2 \times 3.14 \times 3 \times 10^8 \times 10^{-11}} \text{ second } (\because d\lambda = \Delta\lambda = 10^{-14} \text{m})$$

$$= \mathbf{1.9 \times 10^{-8} \ second.}$$

Example 22. *If the uncertainty in the location of a particle is equal to its de-Broglie wave length, what is the uncertainty with velocity?*

Solution: We have from uncertainty principle,

$$(\Delta p)(\Delta x) = h$$

$$\therefore \quad \Delta(mv)(\Delta x) = h$$

or

$$\Delta v = \frac{h}{m \Delta x}$$

but $\quad \Delta x = \lambda$

$$\therefore \quad \Delta v = \frac{h}{m\lambda} = \frac{h}{m \cdot h/mv} = v$$

$$\therefore \quad \Delta v = v$$

Example 23. *A nucleon is confined to a nucleus of diameter* 5×10^{-4} *m. Calculate the minimum uncertainty in the momentum of the nucleon. Also calculate the minimum kinetic energy of the nucleon.*

Solution: We have the relation

$$(\Delta p)_{min}(\Delta x)_{max} = h$$

$$\therefore \quad (\Delta p)_{min} = \frac{h}{(\Delta x)_{max}}$$

In this case, $(\Delta x)_{max} = 5 \times 10^{-14}$ m

and $\quad h = 6.626 \times 10^{-34}$ J-s

$$\therefore \quad (\Delta p)_{min} = \frac{6.626 \times 10^{-34}}{5 \times 10^{-14}} = 1.33 \times 10^{-20} \text{ kg.m/sec}$$

Since p can not be less than $(\Delta_p)_{min}$, we have

$$p_{min} = (\Delta p)_{min}$$

$$\therefore \quad E_{min} = \frac{(p_{min})^2}{2m}$$

Puting the values of $p_{min} = 1.33 \times 10^{-20}$ kg-ms^{-1}

and $m = 1.675 \times 10^{-27}$ kg

$$E_{min} = \frac{(1.33 \times 10^{-20})^2}{2 \times 1.675 \times 10^{-27}} = 0.529 \times 10^{-13} \text{ J}$$

$$= \frac{0.529 \times 10^{-13}}{1.6 \times 10^{-19}} \text{ eV} = \mathbf{0.33 \text{ meV}}$$

Example 24. *An electron of energy 250 eV is passed through a circular hole of radius 10^{-6} m. What is the uncertainty introduced in the angle of emergence?*

Solution.

Given $E = 250\ eV = 250 \times 1.6 \times 10^{-19}$ J.

$r = 10^{-6}$ m, $\Delta x = 2 \times 10^{-6}$ m.

Now we have the relation

$$E = \frac{p^2}{2m} = 250 \times 1.6 \times 10^{-19} = 40 \times 10^{-18} \text{ J.}$$

$$\therefore \quad \text{momentum } p = \sqrt{2mE}$$

$$= \sqrt{2 \times 9.1 \times 10^{-31} \times 40 \times 10^{-18}}$$

$$= 8.53 \times 10^{-24} \text{ kg ms}^{-1}$$

From uncertainty principle

$$\Delta x \cdot \Delta p \simeq \frac{h}{2\pi}$$

$$\therefore \quad \Delta p = \frac{6.625 \times 10^{-34}}{2 \times 3.142 \times 2 \times 10^{-6}}$$

$$= 0.527 \times 10^{-28} \text{ kg ms}^{-1}.$$

$$\text{Angle of Emergence} = \frac{\Delta p}{p} = \frac{0.527 \times 10^{-28}}{8.53 \times 10^{-24}}$$

$$= 0.062 \times 10^{-4} \text{ radian}$$

$$\simeq \mathbf{1.42 \text{ Seconds}}$$

Example 25. *Compare the momentum, the table energy and the K. B. of an electron with a De-Broglie wavelength of z A° with that of a photon of the same wavelength.*

Solution.

Given $\lambda_e = 2A°$, $\lambda_p = 2A°$

$$\text{momentum of electron } p_e = \frac{h}{\lambda_e} = \frac{6.62 \times 10^{-34}}{2 \times 10^{-10}} = 3.315 \times 10^{-24} \text{ kg ms}^{-1}$$

$$\text{momentum of photon } p_p = \frac{h}{\lambda_p} = \frac{6.62 \times 10^{-34}}{2 \times 10^{-10}} = 3.315 \times 10^{-24} \text{ kg ms}^{-1}$$

$$\therefore \quad \text{Ratio of momenta} = \frac{p_e}{p_p} = 1 \qquad ...(1)$$

Now total energy $= K\cdot E\cdot + \text{Rest mass energy.}$

$$K\cdot E\cdot \text{ of electron} = \frac{p_e^2}{2m_0} = \frac{(3.315\times10^{-24})^2}{2\times9.1\times10^{-31}}$$

$$= 0.603 \times 10^{-17} \text{ J.}$$

$$\text{Rest mass energy of electron} = m_0c^2$$

$$= 9.1 \times 10^{-31} \times (3 \times 10^8)^2$$

$$= 8.19 \times 10^{-14} \text{ J.}$$

$$\therefore \quad \text{Total energy of electron} = 8.19 \times 10^{-14} \text{ J} + 0.603 \times 10^{-17} \text{ J}$$

$$\simeq 8.19 \times 10^{-14} \text{ J.}$$

Since Rest mass energy of photon = 0

$$\therefore \quad \text{Total energy of photon } E_p = K\cdot E\cdot \text{ of photon}$$

$$E_p = h\nu = \frac{hc}{\lambda} = \frac{6.62\times10^{-34}\times3\times10^8}{2\times10^{-10}}$$

$$= 9.945 \times 10^{-16} \text{ J.}$$

$$\text{Ratio of total energies } \frac{E_e}{E_p} = \frac{8.19\times10^{-14}}{9.945\times10^{-16}}$$

$$= \mathbf{82.35}$$

$$\text{Ratio of kinetic energies } \frac{K\cdot E_e}{K\cdot E_p} = \frac{0.603\times10^{-17}}{9.945\times10^{-16}}$$

$$= \mathbf{0.006.}$$

Example 26. *An electron has a De-Broglie wavelength of 2 × 10^{-12} m. calculate the* K · E · *and the phase and group of velocities of the De-Broglie waves. Given rest mass of the electron = 9.1 × 10^{-31} kg.*

Solution.

Given $\lambda = 2 \times 10^{-12}$ m

Rest mass $mv = 9.1 \times 10^{-31}$ kg.

$$\therefore \quad K\cdot E\cdot \text{ of electron} = \frac{p^2}{2m_0} = \frac{h^2}{2m_0\lambda^2}$$

$$\therefore \quad \frac{1}{2}mv^2 = \frac{(6.62\times10^{-34})^2}{2\times9.1\times10^{-31}\times(2\times10^{-12})^2}$$

$$= 6.038 \times 10^{-14} \text{ J}$$

or

$$v^2 = \frac{2\times6.038\times10^{-14}}{9.1\times10^{-31}} = 1.327 \times 10^{17}.$$

$$\therefore \quad \text{particle velocity } v = \sqrt{1.327\times10^{17}} = 3.64 \times 10^8 \text{ ms}^{-1}$$

Since particle velocity = Group velocity

$$\therefore \quad v_{\text{group}} = 3.64 \times 10^8 \text{ ms}^{-1}.$$

$$\text{Phase velocity } v_p = \frac{c^2}{v_g} = \frac{(3\times10^8)^2}{3.64\times10^8}$$

$$= \mathbf{2.47 \times 10^8\ ms^{-1}.}$$

10.10. ELECTRON MICROSCOPE

The electron microscope, like the optical microscope, is an instrument used principally in the research laboratory for magnifying small objects to such an extent that their minutest parts may be observed and studied in detail. The importance of this device in the field of medical research cannot be overestimated. To illustrate, many viruses known to medical science as being responsible for certain human diseases lie beyond the range of the optical microscope. With the electron microscope, magnifications of from 10 to 100 times that of finest optical microscopes make many of these viruses, and some of their detailed structure, visible to the eye. While the highest magnification obtained with the best optical microscope is about 2000 *X*, electron microscopes have already been made to give magnifications as high as 100,000 *X*.

Principle. It is based upon the principle that

1. particles such as electrons exhibit wave properties like rays but have much shorter wavelength; and
2. electrons can be focussed by suitable electric and magnetic fields just as light rays can be focussed with the help of lenses.

By Definitions. The resolving power of an optical instrument like a microscope is directly proportional to the aperture (diameter) of the lens used and inversely proportional to the wavelength of the light used. So the smaller the wavelength, the greater shall be the magnifying power. If we accelerate electrons through a potential difference of say, 60000 volts, the velocity v attained by them can be calculated from the relation $eV = \frac{1}{2} mv^2$ where m, the electron mass $= 9 \times 10^{-31}$ kg. Calculating λ from de Broglie wave equation ($\lambda = h/mv$), the wavelength comes out to be about 5×10^{-12} m which is 10^5 times smaller than that of visible light.

Construction. In Fig. 10.10, the schematic diagram of electron microscope has been shown by the side of an optical microscope for comparison. In an optical microscope, Fig. 10.10 (*b*), the light from a source *S* such as sunlight or an electric lamp is made to pass through a narrow slit and then made to fall on a condenser lens. The condensed light is then made to illuminate the object to be seen magnified. An objective lens gives the first magnified image of the object. This image is again magnified by a projection lens. The final magnified image is either seen directly or is obtained on a photographic plate for permanent record.

An electron microscope shown in Fig. 10.10 (*a*) is similar in principle to an optical microscope. The modern electron microscope is ordinarily of transmission type in which magnetic lenses of short focal length are used to obtain large magnification. An electron beam emitted by a filament is accelerated through a large potential difference in a device called electron gun. The electron beam got from the electron gun, is made to pass through the centre of the doughnut shaped electromagnet '*a*'. The electrons get deflected to form a parallel beam which strikes the object to be magnified. It should be noted that the electrons will be transmitted more through the transparent parts of the object and transmitted in less number through comparatively denser portions. The transmitted beam will thus have the likeness of the object transversed by it. The second electromagnet '*b*' causes the electron beam to diverge to produce enlarged image of the object. The electromagnet '*c*' focusses the electron beam from part of the enlarged image on the fluorescent screen producing still greater magnification. The image obtained on the fluorescent screen producing still greater magnification. The image obtained on the fluorescent screen is made visible by scintillation for direct view. It can also be obtained on a suitable photographic plate for a permanent record. Sharp focussing is obtained by adjusting the intensity of magnetic fields produced by electromagnets.

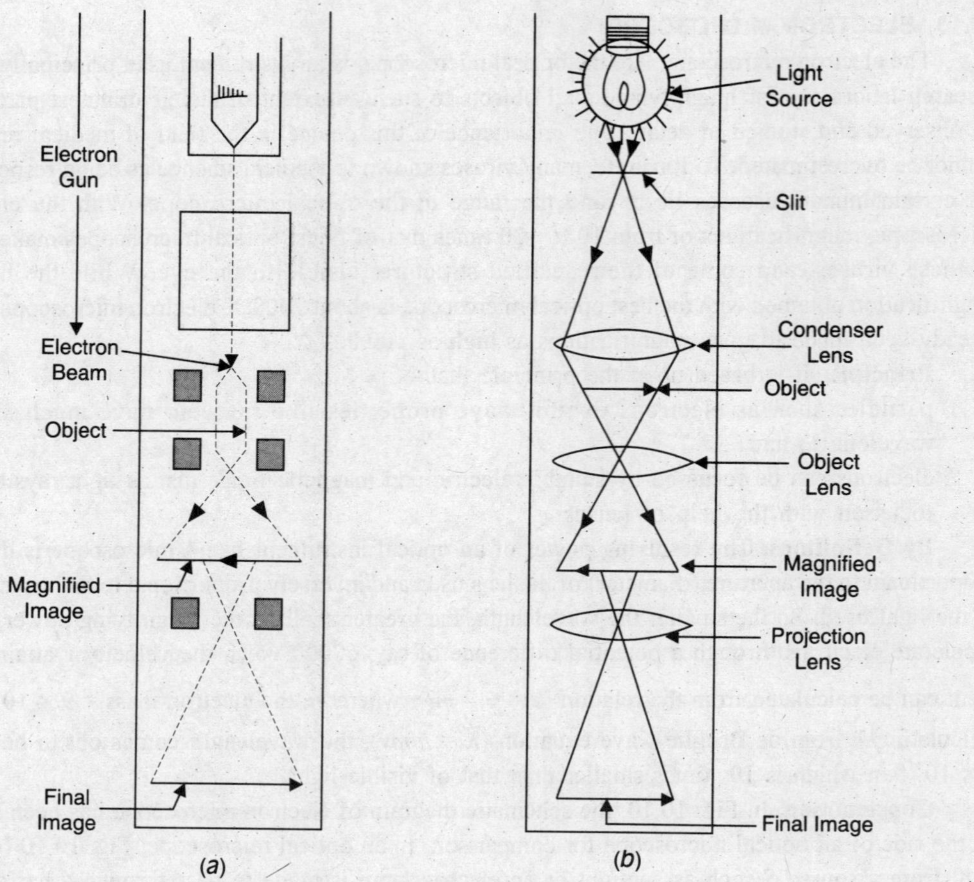

Fig. 10.10

The object can be given fine micro-movement to bring different parts of object under the incident beam for detailed examination. Since the electron beam operates in vacuum, the apparatus is mounted in a glass tube, completely evacuated. Accelerating voltage of the order of 3,00,000 volt are usually employed for the examination of thicker objects. With the electron microscope, it has been possible to magnify the objects 100,000 times with adequate resolving power.

Uses. The resolving power of an optical microscope is limited by the wavelength of visible light and the smallest size of the object that can be examined with it is of the order of 2000 Å. The electron microscope has very large magnifying power and with the help of the best type of instrument objects of the order 50 Å in size have been made visible. An electron microscope is put to valuable uses in all those fields where the optical microscope has shown the way, *e.g.*, in biology, medicine, physics, chemistry, metallurgy and so on.

1. In medicine and biology, it is used to study *virus*, the disease causing agent beyond the reach of ordinary microscope. The bacteria are shown in greater detail. So that suitable means may be used for their destruction.
2. It has been used in the investigation of atomic structure and structure of crystals in details.
3. Small particles forming colloids have become open to study and analysis.
4. It has also been used in the study of the structure of the textile fibres, purification of lubricating oils, composition of paper and paints surface of metals and plastics.

QUESTIONS AND PROBLEMS

1. How do the "group" and "wave velocity" differ? Show that "phase velocity" can exceed the velocity of light in vacuum.

2. Show that the de Broglie waves associated with a particle are not electromagnetic in nature.

3. Explain the de Broglie concept of matter waves. How is the wave nature of electron demonstrated experimentally?

4. What led de Broglie to suggest wave character of matter? Derive expression for the wavelength of electron.

5. What is group velocity and phase velocity? Show that the group velocity of particle is equal to the velocity of the particle.

6. Explain Heisenberg uncertainty principle. Why is it important for microscopic particles, but significant in practical life?

7. Explain uncertainty principle in Quantum mechanics show that it does not allow presence of electrons in the atomic nucleus.

8. Discuss the importance of uncertainty principle.

9. Basing your arguments upon the concept and value of wave groups relate the inherent uncertainty Δx in measurement of a principle position with the inherent velocity Δp in simultaneous measurement of its momentum.

10. Why can electron not exist in the nucleus? Why Heisenberg's Principle is important for microscopic bodies alone?

11. Describe Heisenberg's microscope arrangement for the simultaneous determination of position and momentum of an electron. What are the conclusions of this experiment?

12. State time-energy uncertainty principle and obtain it from the position-momentum principle uncertainty relation.

13. State time-energy uncertainty principle and obtain it from the position momentum uncertainty relation.

14. Calculate de Broglie wave-length of neutron of mass 1.676×10^{-27} kg. and energy 1 volt. **[Ans.** 0.2861]

15. Using de Broglie hypothesis, calculate the wavelength associated with

(*a*) a stone of mass 50 g moving at a speed of 50 ms^{-1}

(*b*) a neutron moving with a speed of 3003 ms^{-1}

(*c*) an electron with a kinetic energy of 100 eV

Given $h = 6.625 \times 10^{-34}$ Js $m_n = 1.675 \times 10^{-27}$ kg

$m_e = 9.12 \times 10^{-31}$ kg **[Ans.** (*a*) 2.65×10^{-28} m (*b*) 1.32 Å (*c*) 1.23 Å]

16. Show that de Broglie wavelength of a particle is equal to its Compton wavelength when its velocity is 0.707 C.

17. What voltage must be applied to an electron source to produce electrons having wavelength 0.4 Å? Given $e = 1.6 \times 10^{-19}$ C. **[Ans.** 3.12×10^4 V]

18. An ultraviolet light of 1500 Å is incident upon molybdenum. Find the wavelength associated with photoelectrons. Work function of molybdenum is 4.15 eV. **[Ans.** 6.04 Å]

19. If we are certain that an electron is in an atom of diameter 3 Å. What is the uncertainty in the momentum of electrons? **[Ans.** 3.1×10^{-22} kg ms^{-1}]

20. What is the minimum in certainty of position of an electron moving with a speed of $3 \times 10^9 ms^{-1}$. **[Ans.** 0.38 Å]

21. An electron with a K.E. of 200 eV in confined to a region of atomic dimensions 10^{-10} m. Find the uncertainty in its linear momentum and compare it with the momentum of the electron. **[Ans.** $\Delta p = 1.05 \times 10^{-24}$ kg ms^{-1} about 4% of p]

22. A meson has a life time of 10^{-23} s. To what accuracy can its mass be known? **[Ans.** 1.17×10^{-28} kg]

CHAPTER 11

QUANTUM MECHANICS

11.1. NEWTONIAN MECHANICS AND QUANTUM MECHANICS

The fundamental difference between *Newtonian mechanics* and *quantum mechanics* lies in what it is that they describe. *Newtonian mechanics* is concerned with the motion of particle under the influence of applied forces, and it takes for granted that such quantities as the particle's position, mass, velocity and acceleration can be measured. This assumption is, of course, completely valid in our everyday experience, and newtonian mechanics provides the "correct" explanation for the behaviour of moving bodies in the sense that the values it predict for observable magnitudes agree with the measured values of those magnitudes.

Quantum Mechanics, too, consists of relationships between observable magnitudes, but the *uncertainty principle* radically alters the definition of "observable magnitude" in the atomic realm. According to the uncertainty principle, the position and momentum of a particle cannot be accurately measured at the same time, while in newtonian mechanics, both are assumed to have definite, ascertainable values at every instant. The quantities whose relationships quantum mechanics explores are *probabilities.* Instead of asserting, for example, that the radius of the electron's orbit in a ground-state hydrogen atom is always exactly 5.3×10^{-11}, quantum mechanics states that this is the most probable radius; if we conduct a suitable experiment, most trials will yield a different value, either larger or smaller, but the value most likely to be found will be 5.3×10^{-11}m.

At first glance quantum mechanics seems a poor substitute for newtonian mechanics, but closer inspection reveals a striking fact : *Newtonian mechanics is nothing but an approximate version of Quantum mechanics.* The certainties proclaimed by newtonian mechanics are illusory, and their agreement with experiment is a consequence of the fact that macroscopic bodies consist of so many individual atoms that departures from average behaviour are unnoticeable. Instead of two sets of physical principles, one for the macroscopic universe and one for the microscopic universe, there is only a single set and quantum mechanics represents our best effort to data in formulating it.

11.2. THE WAVE FUNCTION (BORN'S INTERPRETATION)

The quantity with which quantum mechanics is concerned is the *wave function* ψ of a particle. While ψ itself has no physical interpretation the square of its absolute magnitude $|\psi|^2$ (or $\psi^* \psi$ if ψ is complex) evaluated at a particular point at a particular time is proportional to the *probability* of experimentally finding the particle there at that time. The problem of quantum mechanics is to determine ψ for a particle when its freedom of motion is limited by the action of external forces.

Even before we consider the actual calculation of ψ, we can establish certain requirements it must always fulfill. For one thing, since $|\psi|^2$ evaluated at a point is proportional to the probability P of finding the particle described by ψ at that point, the integral of $|\psi|^2$ over all space must be finite-the particle is *somewhere*, after all. If

$$\int_{-\infty}^{\infty} \psi^2 dV \qquad \text{...(1)}$$

is 0, the particle does not exist, and if it is ∞ the particle is everywhere simultaneously $|\psi|^2$ cannot be negative or complex because of the way it is defined, and so the only possibility left is that its integral be a finite quantity if ψ is to describe properly a real body.

It is usually convenient to have $|\psi|^2$ be *equal* to the probability P of finding the particle described by ψ, rather than merely be proportional to P. If $|\psi|^2$ is equal to P, then it must be true that

$$\int_{-\infty}^{\infty} |\psi|^2 \, dV = 1$$

$$\int_{-\infty}^{\infty} P dV = 1 \qquad ...(2)$$

is the mathematical statement that the particle exists somewhere at all times. A wave function that obeys equation (2) is said to be *normalized*. Every acceptable wave function can be normalized by multiplying by an appropriate constant.

Besides being normalizable, ψ must be single-valued, since P can have only one value at a particular place and time. A further condition that ψ must obey is that it and its partial derivatives $\partial\psi/\partial x, \partial\psi/\partial y, \partial\psi/\partial z$ be continuous everywhere.

Schrödinger's *equation,* which is the fundamental equation of quantum mechanics in the same sense that *the second law of motion* is the fundamental equation of *newtonian mechanics* is a wave equation in the variable ψ.

11.3. SCHRÖDINGER'S WAVE EQUATION FOR A FREE PARTICLE

We now want to establish some of the mathematical properties of the wave describing a quantum-mechanical particle, such as an electron. In this case we will assume that the quantum-mechanical particle is free, *i.e.,* external forces are absent. Furthermore, for the sake of simplicity, we will assume that the motion is one-dimensional say, along the x-axis.

The wave describing a particle of momentum p and energy E has a wavelength $\lambda = h/p$ and frequency $\nu = E/h$. For convenience, we define the wave number*.

$$k = \frac{2\pi}{\lambda} = \frac{p}{\hbar} \qquad ...(1)$$

and the angular frequency

$$\omega = 2\pi\nu = \frac{E}{\hbar} \qquad ...(2)$$

Since for our free particle $E = p^2/(2m)$, we can also write Equation (1) as

$$k = \frac{\sqrt{2\,m\,E}}{\hbar} \qquad ...(3)$$

If the particle travels in the $+x$ direction, the wave will also have to travel in that direction. The possible harmonic waves satisfying this condition are

$$\sin(\omega t - kx) \qquad \cos(\omega t - kx) \qquad ...(4)$$

$$e^{-i(\omega t - kx)} \qquad e^{i(\omega t - kx)} \qquad ...(5)$$

The waves (4) are *real* and the waves (5) are *complex.* According to the mathematical rules for evaluating complex exponentials we have the identities

$$e^{ix} = \cos x + i \sin x \qquad ...(6)$$

and

$$\cos x = \frac{e^{ix} + e^{-ix}}{2}, \qquad \sin x = \frac{e^{ix} - e^{-ix}}{2i} \qquad ...(7)$$

Thus

$$e^{-i(\omega t - kx)} = -\cos(\omega t - kx) + i \sin(\omega t - kx) \qquad ...(8)$$

and

$$e^{+i(\omega t - kx)} = \cos(\omega t - kx) - i \sin(\omega t - kx) \qquad ...(9)$$

This shows that the complex waves are linear combinations of the real waves, with a complex coefficient. To settle which of the waves (4) or (5) we want, let us consider the superposition principle of quantum mechanics: whenever ψ_1 and ψ_2 are possible quantum mechanical waves, then so is $\psi_1 + \psi_2$. Suppose we try to describe a particle travelling in the positive x-direction by the wave

* In three dimensions we would define the wave vector as $k = p/\hbar$

sin ($\omega t - kx$), then a particle travelling in the negative x-direction will be described by sin ($\omega t + kx$). According to the Superposition Principle, the sum

$$\sin(\omega t - kx) + \sin(\omega t + kx) \qquad ...(10)$$

should then be a possible quantum-mechanical wave function. But equation (10) equals

$$2 \sin \omega t \cos kx \qquad ...(11)$$

Obviously this is not a satisfactory wave function because at $t = 0$ it vanishes identically everywhere, *i.e., the probability for finding the particle is zero everywhere.* Since the particle cannot disappear, we must regard the wave function (10) as *unacceptable,* and we conclude that sin ($\omega t - kx$) is not the wave function we want. By a similar argument, we rule out cos ($\omega t - kx$).

Let us now consider $e^{-i(\omega t - kx)}$. The superposition of this with $e^{-i(\omega t + kx)}$ gives

$$e^{-i(\omega t - kx)} + e^{-i(\omega t + kx)} = 2e^{-i\omega t} \cos kx \qquad ...(12)$$

This never vanishes identically everywhere, and is an acceptable wave function. Thus, the wave function $e^{-i(\omega t - kx)}$ is consistent with the Superposition Principle. A wave function of the type $e^{i(\omega t - kx)}$ is also consistent with the Superposition Principle; but according to convention laid down by the founders of quantum mechanics, this wave function is to be discarded in favour of $e^{-i(\omega t - kx)}$*. Thus, we are led to the conclusion that the wave function is

$$\psi(x,t) = e^{-i(\omega t - kx)} \text{ for motion in the positive } x\text{-direction} \qquad ...(13)$$

and

$$\psi(x,t) = e^{-i(\omega t + kx)} \text{ for motion in the negative } x\text{-direction} \qquad ...(14)$$

Next, we want to discover the **wave equation** obeyed by these wave functions. In general, the wave function describing any physical system obeys a wave equation, which permits us to calculate the time evaluation of the system. Mathematically, the wave equation is differential equation involving ψ and the derivatives of ψ with respect to x and t. To discover this differential equation for ψ, we begin by expressing ω and k in terms of p:

$$\psi = e^{-i(\omega t - kx)} = e^{-i\left(\frac{Et}{\hbar} - \frac{\sqrt{2\,mEx}}{\hbar}\right)} \qquad ...(15)$$

If we take the first derivatives of ψ with respect to time and the second derivative of ψ with respect to x, we find

$$\frac{\partial \psi}{\partial t} = -i\frac{E}{\hbar}\psi \qquad ...(16)$$

and

$$\frac{\partial^2 \psi}{\partial x^2} = -\frac{2\,mE}{\hbar^2}\psi \qquad ...(17)$$

Comparing these expressions, we see that these derivatives are proportional. By adjusting the constant of proportionality, we obtain an equality :

$$-\frac{\hbar^2}{2m}\frac{\partial^2}{\partial x^2}\psi(x,t) = i\hbar\frac{\partial}{\partial t}\psi(x,t) \qquad ...(18)$$

This is the wave equation for a free particle: it is called **Schrödinger's equation.** Next we will discover how to include forces – or more precisely, potential-in Schrödinger's equation. **This equation plays the same role in quantum mechanics as Newton's Second Law does in classical mechanics**. From Schrödinger's equation we can calculate the motion of a quantum-mechanical wave just as from Newton's Law we can calculate the motion of a classical particle.

11.4. POTENTIAL IN SCHRÖDINGER'S EQUATION

(*i*) Schrödinger assumed that de Broglie wavelength holds good for any particle moving in any field of force with potential energy V. The sum of the kinetic energy and potential energy is given by

* We have to choose one or the other of these. Any attempt at using both leads to trouble, since the superposition $e^{-i(\omega t - kx)} + e^{i(\omega t - kx)}$, equals 2 cos ($\omega t - kx$), which is an unacceptable wave function.

$$E = \text{kinetic energy} + \text{potential energy (K.E. + P.E.)}$$

$$E = \frac{1}{2}mv^2 + V$$

$$E = \frac{p^2}{2m} + V \quad \text{...(1)}$$

$$\therefore \quad p^2 = 2m[E - V]$$

$$p = [2m(E - V)]^{1/2}$$

So de Broglie wavelength may be given by

$$\lambda = \frac{h}{p} = \frac{h}{[2m(E-V)]^{1/2}} \quad \text{...(2)}$$

For large value of V, λ is large and for a small value of V, λ is also small.

(*ii*) It is assumed that the wave function ψ is governed by the equation

$$\psi = \psi_0 \, e^{-i\omega t} \qquad [\text{Taking } \Psi_0 = e^{ikx}]$$

The wave function ψ is also called **probability amplitude** for the position of the particle. The wave equation in **Cartesian co-ordinates** may be written as

$$\frac{\partial^2\psi}{\partial x^2} + \frac{\partial^2\psi}{\partial y^2} + \frac{\partial^2\psi}{\partial z^2} = \frac{1}{u^2}\frac{\partial^2\psi}{\partial t^2} \quad \text{...(3)}$$

where $\quad \psi = \psi_0 \, e^{-i\omega t} \quad$ ...(4)

and ψ is function of (x, y, z and t) and represents the amplitude of the wave associated with the particle and u is the velocity of motion.

Differentiating equation (4) twice w.r.t. t, we get

$$\frac{\partial^2\psi}{\partial t^2} = -\omega^2\psi_0 e^{-i\omega t} = -\omega^2\psi$$

Putting in (3), we get

$$\frac{\partial^2\psi}{\partial x^2} + \frac{\partial^2\psi}{\partial y^2} + \frac{\partial^2\psi}{\partial z^2} = -\left(\frac{\omega^2}{u^2}\right)\psi$$

Now $\quad \frac{\omega^2}{u^2} = \frac{(2\pi\nu)^2}{(\nu\lambda)^2} = \frac{4\pi^2}{\lambda^2}$

So we have $\quad \frac{\partial^2\psi}{\partial x^2} + \frac{\partial^2\psi}{\partial y^2} + \frac{\partial^2\psi}{\partial z^2} = \left(\frac{4\pi^2}{\lambda^2}\right)\psi \quad$ ...(5)

Now $\frac{\partial^2\psi}{\partial x^2} + \frac{\partial^2\psi}{\partial y^2} + \frac{\partial^2\psi}{\partial z^2} = \nabla^2\psi$, is a **Laplacian operator.**

$\therefore$ equation (5) becomes $\nabla^2\psi + \frac{4\pi^2}{\lambda^2}\psi = 0 \quad$ ...(6)

Equation (6) is a general equation and is independent of time.

Now let us introduce the concept of wave mechanics, from de Broglie wavelength, and also from equation (2)

$$\lambda = \frac{h}{m\nu} = \frac{h}{[2\,m(E-V)]^{1/2}} \quad \text{...(7)}$$

Substituting this value of λ in equation (6), we get

$$\nabla^2\psi + \frac{4\,\pi^2\,[2m(E-V)]}{h^2}\psi = 0$$

$$\nabla^2\psi + \left(\frac{8\,\pi^2 m}{h^2}\right)(E-V)\psi = 0 \quad \text{...(8)}$$

This represents Schrödinger time-independent wave equation.

Also $\hbar = \frac{h}{2\pi}$, so equation (8) may be written as

$$\therefore \quad \nabla^2\psi + \left(\frac{2m}{\hbar^2}\right)(E - V)\,\psi = 0 \qquad ...(9)$$

Equation (9) shows that the wave function ψ is a function of coordiantes, also V is a function of co-ordinates.

Special Case. In the case of a free particle (*i.e.,* a particle not acted by external force) $V = 0$ and so total energy is kinetic energy only.

$$\therefore \quad \nabla^2\psi + \left(\frac{2m}{\hbar^2}\right)(E)\,\psi = 0 \qquad ...(10)$$

Equation (10) represents Schrödinger equation for a particle in the absence of external field, *i.e.,* **for a free particle.**

11.5. TIME DEPENDENT SCHRÖDINGER EQUATION

The wave function may be written as

$$\psi = \psi_0\, e^{-i\omega t} \qquad ...(1)$$

Differentiating w.r.t. time, we get

$$\frac{\partial\psi}{\partial t} = -i\omega\psi_0\, e^{-i\omega t}$$

$$\frac{\partial\psi}{\partial t} = -i\,(2\pi\nu)\,\psi \qquad ...(2)$$

also $\quad E = h\nu$

$\therefore$ putting $\nu = \frac{E}{h}$ in (2), we get

$$\frac{\partial\psi}{\partial t} = -\,i\,(2\pi)\left(\frac{E}{h}\right)\psi$$

$$E\psi = \frac{-1}{i}\left(\frac{h}{2\pi}\right)\frac{\partial\psi}{\partial t}$$

But $$E\psi = \frac{i^2}{i}\left(\frac{h}{2\pi}\right)\frac{\partial\psi}{\partial t} \qquad [\because i^2 = -1]$$

i.e., $$E\psi = i\hbar\frac{\partial\psi}{\partial t} = i\hbar\frac{\partial}{\partial t}(\psi) \qquad ...(3)$$

Now Schrödinger equation is given by

$$\nabla^2\psi + \left(\frac{2\,m}{\hbar^2}\right)(E - V)\,\psi = 0$$

$$\nabla^2\psi + \left(\frac{2\,m}{\hbar^2}\right)(E\psi) - \left(\frac{2\,m}{\hbar^2}\right)V\psi = 0 \qquad ...(4)$$

Substituting the value of $E\psi = i\hbar\frac{\partial\psi}{\partial t}$ from (3) in equation (4), we get

$$\nabla^2\psi + \left(\frac{2\,m}{\hbar^2}\right) i\hbar\frac{\partial\psi}{\partial t} - \frac{2\,m}{\hbar^2}V\psi = 0$$

$$\nabla^2\psi - \left(\frac{2\,m}{\hbar^2}\right)V\psi = -\left(\frac{2\,m}{\hbar^2}\right) i\hbar\frac{\partial\psi}{\partial t}$$

or $$-\left(\frac{\hbar^2}{2\,m}\right)\nabla^2\psi + V\psi = i\hbar\frac{\partial\psi}{\partial t} \qquad ..(5)$$

This equation represents *Schrödinger time dependent* equation because it involves time Equation (5) can be written as

$$\left(\frac{-\hbar^2}{2m}\nabla^2 + V\right)\psi = i\hbar\frac{\partial\psi}{\partial t}$$

If we take $\left(-\frac{\hbar^2}{2m}\nabla^2 + V\right)$ as operator H known as Hamiltonian and from equation (3)

$$\hbar\frac{\partial}{\partial t} = E, \text{ we have}$$

$$H\psi = E\psi \qquad ...(6)$$

Equation (6) represents the motion of a material particle in terms of Hamiltonian operator (non-relativistic particle).

11.6. TIME INDEPENDENT SCHRÖDINGER EQUATION

In time dependent Schrödinger equation the potential energy of a moving particle is **both function of time as well as position**. In many problems, the potential energy of a particle does not depend explicitly on time. It means that the forces that act upon the particle and thereby the potential energy V changes with position of the particle only. In such cases, time parameter is eliminated from Schrödinger time dependent wave equation and the equation thereby obtained is called *time* independent Schrödinger equation.

We may write the one dimensional Schrödinger's time dependent wave equation as

$$\psi = Ae^{-(i/\hbar)(Et-px)}$$

$$= Ae^{-\frac{iE}{\hbar}.t}\, e^{+(ip/\hbar)x}$$

$$= \psi_0 e^{-\frac{iE}{\hbar}.t} \qquad ...(1)$$

i.e., ψ is the product of a time dependent function $e^{-\frac{iE}{\hbar}.t}$ and a position dependent function ψ_0. We take up the variations of all function of particles acted upon by stationary forces having the same form as that of an unrestricted particle.

Putting ψ from (1) into the time dependent form of Schrödinger equation, we note that

$$E\psi_0 e^{-\frac{iE}{\hbar}.t} = -\frac{\hbar^2}{2m} e^{-\frac{iE}{\hbar}.t}\frac{\partial^2\psi_0}{\partial x^2} + V\psi_0\, e^{-\frac{iE}{\hbar}.t}$$

Dividing by $e^{-\frac{i}{\hbar}E.t}$, we get

$$E\psi_0 = -\frac{\hbar^2}{2m}\frac{\partial^2\psi_0}{\partial x^2} + V\psi_0$$

or

$$\frac{\partial^2\psi_0}{\partial x^2} + \frac{2m}{\hbar}(E - V)\,\psi_0 = 0$$

Equation (2) is the *steady state form of Schrödinger's equation.*

We may write the equation in **three dimensions** as

$$\frac{\partial^2\psi_0}{\partial x^2} + \frac{\partial^2\psi_0}{\partial y^2} + \frac{\partial^2\psi_0}{\partial z^2} + \frac{2m}{\hbar^2}(E - V)\,\psi_0 = 0 \qquad ...(3)$$

or

$$\nabla^2\psi_0 + \frac{2m}{\hbar^2}(E - V)\,\psi_0 = 0$$

which is time independent schrodinger wave equation as no time term is involved in this equation.

11.7. EXPECTATION VALUES

Dynamic quantities *viz., position, potential energy, kinetic energy, momentum* etc. which form the basis of physical measurements are called *observable* in Physics. In *quantum mechanics* each observable is represented by an *operator* which acts on a wave function ψ to give a new wave function.

We define the *average* or *expectation* value $\langle\alpha\rangle$ of any observable α as

$$< \alpha > = \frac{\int_{-\infty}^{+\infty} \psi^*(r,t)\, \alpha_{op}\, \psi(\vec{r},t)\, dV}{\int_{-\infty}^{+\infty} \psi^*(\vec{r},t)\, \psi(r,t)\, dV}$$

From Normalization condition

$$\int_{-\infty}^{+\infty} \psi^*(\vec{r},t)\, \psi(\vec{r},t)\, dV = 1$$

$$\therefore \qquad < \alpha > = \int_{-\infty}^{+\infty} \psi^*(\vec{r},t)\, \alpha_{op}\, \psi(\vec{r},t)\, dV \qquad ...(1)$$

We may give the expectation value of the vector $\vec{r}$

$$<r> = \int_{-\infty}^{+\infty} \psi^*(r,t)\, \vec{r}\, \psi(\vec{r},t)\, dV \qquad ...(2)$$

The expectation value of position vector is also a vector but it is a function of time only since the space co-ordinate disappears on integration. Equation (2) may be written in terms of component equations as

$$<x> = \int_{-\infty}^{+\infty} \psi^*\, x\, \psi\, dV$$

$$<y> = \int_{-\infty}^{+\infty} \psi^*\, y\, \psi\, dV \qquad ...(3)$$

$$<z> = \int_{-\infty}^{+\infty} \psi^*\, z\, \psi\, dV$$

In the same way, we may write the expectation value of potential energy V as

$$<V> = \int_{-\infty}^{+\infty} \psi^*(r,t)\, V(\vec{r},t)\, \psi(r,t)\, dV \qquad ...(4)$$

Some other examples of dynamical quantities are momentum and total energy.

The expection value of momentum is given by the relation

$$<p> = \int_{-\infty}^{+\infty} \psi^*(x,t)\, p\, \psi(x,t)\, dx \qquad ...(5)$$

To solve the integral (5), we have to express $\psi^*(x,t)\, p\psi(x,t)$ in terms of variables x ant t.

Classically we may express p as a function of x and t. However, in quantum mechanics, due to uncertainty principle, *p cannot be expressed in terms of x* because p and x cannot be known simultaneously.

In the same way we cannot write p as a function of t. So we have to find some other way to solve the integral.

In order to evaluate $<p>$ and $<E>$, we can differentiate the free particle wave function $\psi = Ae^{-(i/k)(Et-px)}$ w.r.t. x and t. Then we get

$$\frac{\partial \psi}{\partial t} = i\hbar\, p\psi \qquad \text{and} \qquad \frac{\partial \psi}{\partial t} = -\frac{i}{\hbar} E\psi$$

which can be written as

$$p\psi = -\frac{\hbar}{i}\frac{\partial}{\partial x}\psi \qquad ...(6)$$

and

$$E\psi = -i\hbar \frac{\partial}{\partial t}\psi \qquad ...(7)$$

which shows that the dynamical quantity p in some sense corresponds to differential operator $\frac{\hbar}{i}\frac{\partial}{\partial x}$

Similarly dynamical quantity E corresponds to differential operator $i\hbar\frac{\partial}{\partial t}$, where an operator tells us to know the operation to be carried out on a dynamical quantity.

For example operator $i\hbar\frac{\partial}{\partial t}$ tells us to take the partial derivation of what comes after it w.r.t. 't' and multiply the result by $i\hbar$.

Generally the expectation value is written in **Bold face letters,** so that P and E correspond to momentum and Energy operators.

$$\text{Momentum operator } \mathbf{P} = \frac{\hbar}{i}\frac{\partial}{\partial x} \quad ...(8)$$

$$\text{Energy operator } \mathbf{E} = i\hbar\frac{\partial}{\partial t} \quad ...(9)$$

Derivation of Schrödinger equation from basic principles. Though we have only shown that the correspondences expressed in Equations (8) and (9) hold for free particles, they are entirely general results whose validity is the same as that of Schrödinger's equation. To support this statement, we can replace the equation $E = T + V$ for the total energy of a particle with the operator equation

$$E = T + V \quad ...(10)$$

Writing K.E. T in terms of momentum p, we have

$$T = \frac{p^2}{2m}$$

which gives

$$\text{Kinetic energy operator} \quad T = \frac{p^2}{2m} = \frac{1}{2m}\left(\frac{\hbar}{i}\frac{\partial}{\partial x}\right)^2 = \frac{\hbar^2}{2m}\frac{\partial}{\partial x^2} \quad ...(11)$$

So equation (10) may be written as

$$i\hbar\frac{\partial\psi}{\partial t} = -\frac{\hbar^2}{2m}\frac{\partial^2}{\partial x^2} + V \quad ...(12)$$

Multiplying the identity $\psi = \psi$ by Equation (12), we get

$$i\hbar\frac{\partial}{\partial t} = -\frac{\hbar^2}{2m}\frac{\partial^2\psi}{\partial x^2} + V\psi$$

which is Schrödinger's equation. It shows that Equations (8) and (9) are equivalent to postulating Schrödinger's equation.

Operator and expectation value. Since p and E can be replaced by their corresponding operators in an equation, we can use these operators to get expectation values for p and E. So the expectation value for p is

$$<p> = \int_{-\infty}^{\infty}\psi^*\psi\,dx = \int_{-\infty}^{\infty}\psi^*\left(\frac{\hbar}{i}\frac{\partial}{\partial x}\right)\psi\,dx = \frac{\hbar}{i}\int_{-\infty}^{\infty}\psi^*\frac{\partial\psi}{\partial x}dx \quad ...(13)$$

and the expectation value for E is given by

$$<E> = \int_{-\infty}^{\infty}\psi^*E\psi\,dx = \int_{-\infty}^{\infty}\psi^*\left(i\hbar\frac{\partial}{\partial t}\right)\psi\,dx = i\hbar\int_{-\infty}^{\infty}\psi^*\frac{\partial\psi}{\partial x}dx \quad ...(14)$$

Both Equations (13) and (14) can be evaluated for any acceptable wave function ψ (x, t). So it is necessary to expose expectation value involving operators in the form

$$(p) = \int_{-\infty}^{\infty}\psi^*p\,\psi\,dx$$

Every observable quantity G characteristic of a physical system may be represented by a suitable quantum-mechanical operator G. To obtain this operator, it is only necessary to express G in terms of x and p and then replace p by $(\hbar/i)\,\partial/\partial x$. If the wave function ψ of the system is known, the expectation value of $G(x, p)$ is

$$<G(x, p)> = \int_{-\infty}^{\infty} \psi^* G\, \psi\, dx \qquad ...(15)$$

This result substantiates the statement made earlier that from ψ can be obtained all the information about a system that is permitted by the uncertainty principle.

Conditions for quantization of a variable. The condition that a certain dynamical variable G be restricted to the discrete values G_n—in other words, that G be quantized – is that the wave functions ψ_n of the system be such that

$$G y_n = G_n \psi_n \qquad ...(16)$$

where G is the operator that corresponds to G and each G_n is a real number. When Equation (16) holds for the wave functions of a system, it is a fundamental postulate (in fact, the fundamental postulate) of quantum mechanics that any measurement of G can only yield one of the values of G_n. If measurements of G are made on a number of identical systems all in states described by the particular eigen function ψ_k, each measurement will yield the single value G_k.

Hamiltonian operator for total energy. The total-energy operator E of Equation (10) is usually written in the form

$$H = -\frac{\hbar^2}{2m}\frac{\partial^2}{\partial x^2} + V \qquad ...(17)$$

and is called the *Hamiltonian operator* because it is reminiscent of the hamiltonian function in advanced classical mechanics, which is an expression for the total energy of a system in terms of coordinates and momenta only. So the steady-state Schrödinger equation can be written simply as

$$E_n \psi_n = H \psi_n \qquad ...(18)$$

so we can say that the various E_n are the eigen values of the Hamiltonian operator H. This kind of association between **eigen values and quantum-mechanical operators** is quite general.

11.8. NORMALIZATION OF A WAVE FUNCTION

As stated earlier, probability density is proportional to the square of magnitude of wave function. So the probability of finding a particle in a small region of volume dv around the (x, y, z) point may be given as

$$dp = N\,|\Psi|^2\, dV \qquad ...(1)$$

The total probability of finding the particle in whole space in given by

$$P = \int_{-\infty}^{+\infty}\int_{-\infty}^{+\infty}\int_{-\infty}^{+\infty} dp$$

$$= N \int_{-\infty}^{+\infty}\int_{-\infty}^{+\infty}\int_{-\infty}^{+\infty} |\Psi|^2\, dx \cdot dy \cdot dz = 1 \qquad ...(2)$$

(where $dV = dx\,.\,dy\,.\,dz.$)

Now if the particle is present in the whole space, then the probability P must be unity *i.e.,*

$$P = N \int_{-\infty}^{+\infty}\int_{-\infty}^{+\infty}\int_{-\infty}^{+\infty} |\Psi|^2\, dx \cdot dy \cdot dz = 1 \qquad ...(3)$$

Equation (3) is called normalization condition and value of N can be found from the above equation. If the value of $N - 1$, then

$$\int_{-\infty}^{+\infty}\int_{-\infty}^{+\infty}\int_{-\infty}^{+\infty}|\Psi|^2\, dx\cdot dy\cdot dz = 1 \qquad ...(4)$$

A wave function is said to be normalized if its normalization constant is unity. It means that probability density instead of being proportional to $|\Psi|^2$ is exactly equal to $|\Psi|^2$ and equation (4) is satisfied.

In case wave function is not normalized, then N will not be unity. However, the value of N can be found using the equation (3) and redefining the wave function as

$$\Psi = \sqrt{N\Psi}$$

So that the new wave function Ψ is normalized.

This process of finding the normalization constant and then redefining the wave function is called the **normalization of wave function**.

Importance: Normalization of a wave function ensures the presence of the particle in space and so all the calculation made will not become waste as the presence of particle is ensured by normalization of the wave function.

11.9. ORTHOGONAL WAVE FUNCTIONS

If two different wave functions ψ_m and ψ_n both satisfy the Schrodinger wave equation and of these functions are such that the integral $\int_V \psi_m\,\psi_n\,dV$ or $\int_V \psi_m\psi_n{}^*\,dV$ vanishes over the total volume V, then these are called orthogonal wave functions *i.e.*,

$$\int_V \psi_m{}^*\psi_n\,dV = \int \psi_m\psi_n{}^*\,dV = 0 \qquad ...(1)$$

for $m \neq n$

where the integral is taken over the whole space.

Again if the wave functions are simultaneously normalized, then we may write

$$\int \psi_m\psi_m{}^*\,dV = \int \psi_n\psi_n{}^*\,dV = 1 \qquad ...(2)$$

The set of wave functions which are both normalized and are also orthogonal are called **ortho-normal wave functions**.

Equations (1) and (2) may collectively be written as

$$\int \psi_m{}^*\psi_n\,dV = \left.\begin{matrix} 0 \text{ if } m \neq n \\ 1 \text{ if } m = n \end{matrix}\right\}$$

As the limit of the integration is not gives, it will correspond to the integration over whole space.

Importance

If we have a complete set of orthogonal wave functions, then any other wave function which may give solution to schrödinger wave function may also be expanded as a set of orthogonal wave functions.

11.10. APPLICATION OF SCHRÖDINGER EQUATION : PARTICLE IN A BOX (MOTION IN ONE DIMENSION)

Let us consider a free particle in one dimensional box. The particle has a mass m and is restricted to move in a straight line along the x-axis. The range of the particle is $0 < x < a$ and the particle is reflected back whenever it reaches the end of the range. We have the general equation

$$\frac{\partial^2\psi}{\partial x^2} + \left(\frac{2m}{2\hbar^2}\right)(E - V)\psi = 0 \qquad ...(1)$$

For a free particle $V = 0$ for $0 < x < a$.

Also the wave function ψ vanishes at $x = 0$ and $x = a$. For one dimension, the modified equation is given by

$$\frac{\partial^2\psi}{\partial x^2}+\left(\frac{2m}{\hbar^2}\right)E\psi = 0$$

$$\frac{\partial^2\psi}{\partial x^2}+\left(\frac{8\pi^2 mE}{h^2}\right)\psi = 0 \qquad \left[\because \hbar=\frac{h}{2\pi}\right] \qquad ...(2)$$

Let us take $\dfrac{8\pi^2 mE}{h^2} = k^2$

$$\therefore \qquad E = \frac{k^2h^2}{8\pi^2 m}$$

So equation (2) may be written as

$$\frac{\partial^2\psi}{\partial x^2}+k^2\psi = 0 \qquad ...(3)$$

The general solution for equation (3) is given by

$$\psi\ (x) = A\sin kx + B\cos kx \qquad ...(4)$$

and from the boundary condition

$$\psi = 0 \text{ at } x = 0$$

so from equation (4), B must be zero

$$\psi\ (a) = A\sin kx = 0$$

Since we cannot take A to be zero because it does not yield any solution, so we take $\sin ka = 0$.

or $$ka = n\pi$$

$$k = \frac{n\pi}{a} \text{ where } n \equiv 1, 2, 3, \text{etc.}$$

Hence the only permissible solutions of the wave equations are

$$\psi_n\ (x) = A\sin\left(\frac{n\pi}{a}\right)x \qquad ...(5)$$

and the value of energy is given by

$$E = \frac{k^2h^2}{8\pi^2 m} \qquad ...(6)$$

Putting the value of $k = \dfrac{n\pi}{a}$ in equation (6), we get

$$E_n = \frac{n^2h^2\pi^2}{8\pi^2 ma^2} = \frac{n^2h^2}{8ma^2} \qquad ...(7)$$

It shows E_n depends upon the value of n. Each value of E_n is called—**Eigen value or proper value and each value of ψ_n is called an Eigen function or proper function.**

From normalization condition

$$\int_0^a |\psi_n(x)|^2\ dx = A^2\int_0^a \sin^2\left(\frac{n\,\pi\,x}{a}\right)dx = 1 \qquad ...(8)$$

or $$\frac{A^2a}{2} = 1$$

$$\therefore \qquad A = \sqrt{\frac{2}{a}}$$

$$\therefore \qquad \psi_n\ (x) = \sqrt{\frac{2}{a}}\ \sin\left(\frac{n\,\pi}{a}\right)x \qquad ...(9)$$

Now for, $n = 1$

$$\psi_1(x) = \sqrt{\frac{2}{a}}\ \sin\left(\frac{\pi}{a}\right)x$$

for $n = 2$

$$\psi_2(x) = \sqrt{\frac{2}{a}} \sin\left(\frac{2\pi}{a}\right) x$$

for $n = 3$

$$\psi_3(x) = \sqrt{\frac{2}{a}} \sin\left(\frac{3\pi}{a}\right) x$$

Graphically, the first three functions are shown in Fig. 11.1.

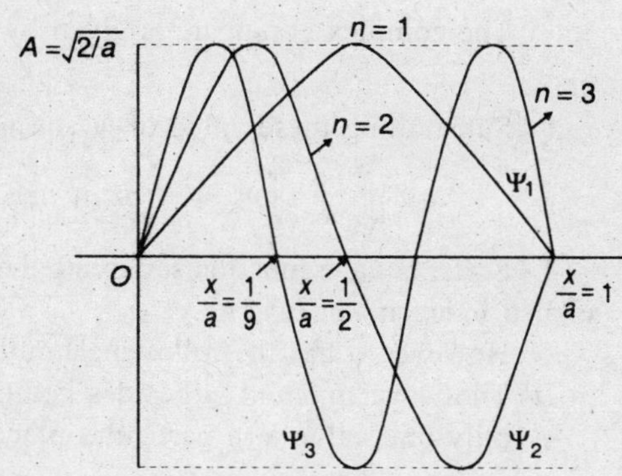

Fig. 11.1

Equation (7) shows that energy E_n of the particle is quantized. Similarly for the momentum of the particle, we have

$$E_n = \frac{p_n^2}{2\,m}$$

or $\quad p_n = (2mE_n)^{1/2} = \left(\frac{2m\,n^2 h^2}{8\,m\,a^2}\right)^{1/2}$

or $$p_n = \frac{nh}{2a} \qquad ...(10)$$

$\therefore \qquad p_n \propto n$

which shows that the momentum of the particle is also quantized. Moreover, for each value of E_n, there is corresponding value of p_n.

From quantum mechanics the probability is given by

$$|\psi_n(x)|^2\, dx = \frac{2}{a}\sin^2\left(\frac{n\,\pi\,x}{a}\right) dx$$

This is wave like function and for $n = 3$ (say)

$$|\psi_n(x)|^2\, dx = \frac{2}{a}\sin^2\left(\frac{3\,\pi\,x}{a}\right) dx$$

The graph between probability and x is given in Fig. 11.2 (a).

In the graph the dotted line represents the constant probability according to **classical mechanics**. But the curve represents the probability variation with variation of x according to **quantum mechanics** probability is zero for values of

$$x = 0,\ x = \frac{a}{3},\ x = \frac{2a}{3} \text{ and } x = a$$

and so on.

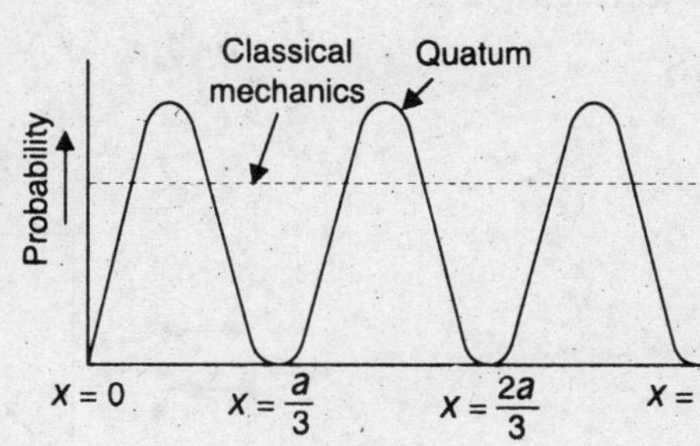

Fig. 11.2 (*a*)

11.11. WAVE FUNCTION AND PROBABILITY DENSITY

The basic postulate of quantum mechanics is that the quantity $\psi\psi^*\, dx\, dy\, dz$ is the *probability* that a particle represented by the wave function $\psi\,(x, y, z, t)$ is in the volume $dV = (dx\,dy\,dz)$ at the time t.

ψ^* is the complex conjugate of ψ. The probability density of the particle is given by

$$P(t) = \psi\psi^* = |\,\psi\,(x, y, z, t)\,|^2$$

and the wave function ψ corresponds to the probability amplitude for the position of the particle.

Since the particle must be some where in the space, it is necessary that the probability density must satisfy the normalisation condition *i.e.*,

$$\int_{-\infty}^{+\infty}\int_{-\infty}^{+\infty}\int_{-\infty}^{+\infty} \psi\psi^* \, dx\, dy\, dz = 1 \quad \text{...(1)}$$

In this case $\psi = \psi_0 \, e^{-i\omega t}$

or $$\psi = \psi_0 e^{-i2\pi Et/h} \qquad \left(\because E = h\nu = \frac{h\omega}{2\pi} \therefore \omega = \frac{2\pi E}{h}\right)$$

or $$\psi = \psi_0 e^{-i\,Et/\hbar} \quad \text{...(2)}$$

The complex conjugate is given by

or $$\psi^* = \psi_0^* e^{i\,Et/\hbar} \quad \text{...(3)}$$

Substituting these value of ψ and ψ^* in equation (1)

$$\int_{-\infty}^{+\infty}\int_{-\infty}^{+\infty}\int_{-\infty}^{+\infty} \psi_0\, \psi_0^* \, dx\, dy\, dz = 1 \quad \text{...(4)}$$

So far a wave function satisfying the time independent Schrödinger equation, the probability is also independent of time.

However ψ has the following limitations:

(*i*) The wave function ψ besides being normalizable must be single valued as $P(t)$ can have only one value at a particular place at a particular time.

(*ii*) Wave function ψ and its partial derivation $\frac{d\psi}{dx}, \frac{d\psi}{dy}$ and $\frac{d\psi}{dz}$ must be continuous every where.

(*iii*) ψ must be finite for all values of x, y and z.

11.12. PROBABILITY CURRENT DENSITY

In order to find the quantum mechanical relation to express the *particle flow,* let us first begin with the classical relation.

Let $\rho\,(x, y, z, t)$ be the charge density and let the charge be concentrated in a small volume element $\Delta x\, \Delta y\, \Delta z$ [Fig. 11.2 (*b*)]. Let us first consider the case when all the velocities are along x-axis. Net charge per unit time $\Delta x\, \Delta y\, \Delta z$ at left is ρv evaluated at $x = a$ times $\Delta y\, \Delta z$, and that leaving right is

$$(\rho v_x)_{a+\Delta x} \Delta y\, \Delta z = [\rho v_x + \partial(\rho v_x)/\partial x\, \Delta x]\, \Delta y\, \Delta z \quad \text{...(1)}$$

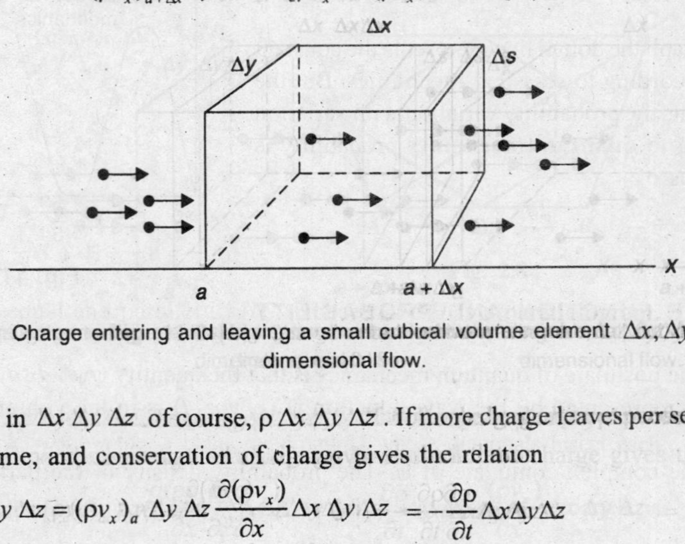

Fig. 11.2 (*b*) Charge entering and leaving a small cubical volume element $\Delta x, \Delta y, \Delta z$ for a one dimensional flow.

The charge in $\Delta x\, \Delta y\, \Delta z$ of course, $\rho\, \Delta x\, \Delta y\, \Delta z$. If more charge leaves per second than arrives ρ decreases in time, and conservation of charge gives the relation

$$(\rho v_x)_{a+\Delta x} \Delta y\, \Delta z = (\rho v_x)_a\, \Delta y\, \Delta z \frac{\partial(\rho v_x)}{\partial x} \Delta x\, \Delta y\, \Delta z = -\frac{\partial \rho}{\partial t} \Delta x \Delta y \Delta z \quad \text{...(2)}$$

which gives

$$\frac{\partial(\rho v_x)}{\partial x} = -\frac{\partial \rho}{\partial t} \quad \text{...(3)}$$

in one dimension. If we extended the case to three dimensions, we get

$$\nabla . \rho v = -\frac{\partial \rho}{\partial t} \quad ...(4)$$

which is the *equation of continuity*, applicable to any flow process in which ρ *represents the density of something that is conserved.*

In order to find the quantum mechanical analogue of equation (2), we replace ρ with $q\,\psi^*\psi$ where q is the charge associated with the particle and $\psi^*\psi\, dx\, dy\, dz$ is the *probability* at time t to locate the particle in volume element $dx\, dy\, dz$ (where we have considered $\iiint \psi^* \psi\, dx\, dy\, dz$ over all space to be normalized so as to give total number of particles enclosed). We assume that there is a quantity S which represents the probability stream density of the particles in such a way that qS plays the role of ρv in the equation of continuity. We seek S in terms of ψ by writing the equation of continuity in the form

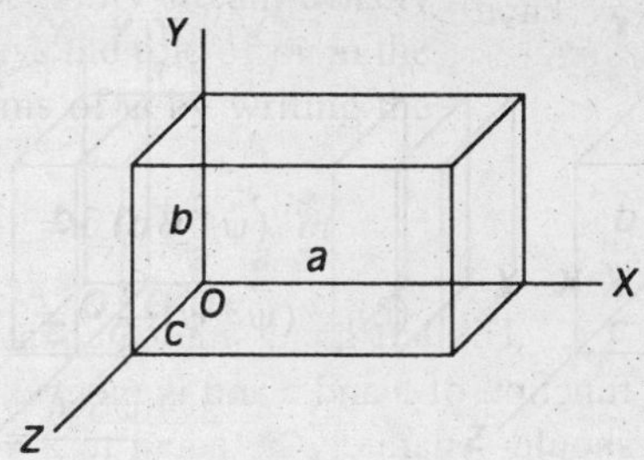

Fig. 11.3

$$\nabla . qS = \partial (q\psi^* \psi)/\partial t$$

or

$$\nabla . S = -\frac{\partial}{\partial t}(\psi^* \psi) \quad ...(5)$$

using the equation $\frac{\hbar^2}{2m}\nabla^2\psi + v\psi = \frac{\hbar}{i}\frac{\partial \psi}{\partial t}$

$$\nabla . S = \frac{\partial \psi^*}{\partial t}\psi - \psi^* \frac{\partial \psi}{\partial t}$$

$$= \frac{\hbar}{2mi}(\psi \nabla^2 \psi^* - \psi^* \nabla^2 \psi) \quad ...(6)$$

which gives

$$S = \frac{\hbar}{2mi}(\psi^* \nabla \psi - \psi \nabla^2 \psi^*) \quad ...(7)$$

The quantity S is a vector *probability current density* for the flow of particles; if each particle bears charge q, the corresponding *probability current density is qS*. If A denotes any surface area enclosing a volume v, the integral of the normal component of S over the surface is equal to the time rate of decrease of probability of finding a particle within the volume; so the left-hand integral represents the rate at which probability is streaming outward across the closed surface A, while the right-hand integral represents the rate at which the probability is decreasing within the volume.

When ψ has the form, $\psi = Ae^{i(p_x^x - Et)/\hbar}$, then $\psi^* = A^* e^{i(Et - p_x^x)/\hbar}$

so we have,

$$|S| = \frac{p}{m}|A|^2$$

where $|S|$ corresponds to the magnitude of S. So the significance of the beam represented by equation $\psi = e^{i(p_x^x - Et)\hbar}$ can be based on the vectors $\vec{S}$ by assuming the number of particles crossing/unit area per second is $n = \frac{p|A|^2}{m}$, where $\frac{p}{m}$ corresponds to the **speed of a particle.**

11.13. PARTICLE IN A BOX (MOTION IN THREE DIMENSIONS)

Consider a single particle *e.g.*, a glass molecule of mass m, confined within a rectangular box with edges parallel to X, Y and Z axes as shown in Fig. 11.3. Let the sides of rectangular box be a, b and c respectively. The particles can move freely within the region $0 < x < a$, $< 0 < y < b$ and $0 < x < c$, *i.e.*, inside the box where potential V is zero *i.e.*,

$$V(x, y, z) = 0, \quad 0 < x < a$$

$$V(x, y, z) = 0, \quad 0 < y < b$$

and

$$V(x, y, z) = 0, \quad 0 < z < c$$

Consider that the potential rises suddenly to have a very large value at the boundaries *i.e.*, the potential outside the box is infinite. We have Schrödinger wave equation inside the box is given by

$$\frac{\partial^2 \psi}{\partial x^2}+\frac{\partial^2 \psi}{\partial y^2}+\frac{\partial^2 \psi}{\partial z^2}+\frac{2m}{\hbar^2}E\psi = 0$$

which is a partial differential equation in three independent variables and may be solved by the method of separation of variables. The solution of equation (1) we have the form

$$\psi(x, y, z) = X(x)\,Y(y)\,Z(z) = XYZ \qquad ...(2)$$

where $X(x)$ is a function of x alone. $X(y)$ is a function of y alone and $Z(z)$ is a function of z alone.

Putting the value of ψ from equation (2) in equation (1) and dividing by $X(x)\,Y(y)\,Z(z)$ we have

$$\frac{1}{X}\frac{\partial^2 Y}{\partial x^2}+\frac{1}{Y}\frac{\partial^2 Y}{\partial y^2}+\frac{1}{Z}\frac{\partial^2 Z}{\partial z^2}+\frac{2m\,E}{\hbar^2} = 0$$

or
$$\frac{1}{X}\frac{\partial^2 Y}{\partial x^2} = -\frac{1}{Y}\frac{\partial^2 Y}{\partial y^2}-\frac{1}{Z}\frac{\partial^2 Z}{\partial z^2}-\frac{2\,mE}{\hbar^2} \qquad ...(3)$$

The lett hand side of equation (3) is a function of x alone, while the right hand side is a function of y and z and is independent of x. Since both sides are equal to each other. So each side should separately be equal to a constant equally, *i.e.*,

$$\frac{1}{X}\frac{\partial^2 Y}{\partial x^2} = k_x \qquad ...(4)$$

$$\therefore \quad -\frac{1}{Y}\frac{\partial^2 Y}{\partial y^2} = -\frac{1}{Z}\frac{\partial^2 Z}{\partial z^2}-\frac{2m\,E}{\hbar^2} = k_x$$

or
$$\frac{1}{Y}\frac{\partial^2 Y}{\partial y^2} = -\frac{1}{Z}\frac{\partial^2 Z}{\partial z^2}-\frac{2\,mE}{\hbar^2} = k_x \qquad ...(5)$$

Again in equation (5) the left hand side is a function of y alone while right hand side is a function of z and is independent of y. If above equation is to be satisfied, both sides must be equal to a constant say k_y *i.e.*,

$$\frac{1}{Y}\frac{\partial^2 Y}{\partial y^2} = k_y \qquad ...(6)$$

$$-\frac{1}{Z}\frac{\partial^2 Z}{\partial z^2}-\frac{2mE}{\partial \hbar^2}-k_x = k_y$$

or
$$\frac{1}{Z}\frac{\partial^2 Z}{\partial z^2} = -\frac{2mE}{\hbar^2}-k_x-k_y \qquad ...(7)$$

Again we may have

$$\frac{1}{Z}\frac{\partial^2 Z}{\partial z^2} = k_z \qquad ...(8)$$

$$-\frac{2\,mE}{\hbar^2}-k_x-k_y = k_z$$

$$-\frac{2\,mE}{\hbar^2} = k_x+k_y+k_z \qquad ...(9)$$

For convenience, we introduce

$$k_x = -\frac{2mE_x}{\hbar^2},\; k_y = -\frac{2mE_y}{\hbar^2}$$

and
$$k_z = -\frac{2mE_z}{\hbar^2}$$

and write differential equations in x, y and z coordinates

$$\frac{\partial^2 X}{\partial x^2}+\frac{2mE_x}{\hbar^2}X = 0 \qquad ...(10)$$

$$\frac{\partial^2 Y}{\partial y^2}+\frac{2mE_y}{\hbar^2}Y = 0 \qquad ...(11)$$

$$\frac{\partial^2 Z}{\partial z^2}+\frac{2mE_z}{\hbar^2}Z = 0 \qquad ...(12)$$

The general solution of equation (10) will be a sine function of arbitrary amplitude frequency and phase, *i.e.*,

$$X(x) = A\sin(Bx+C) \qquad ...(13)$$

where A, B and C are constants whose values are determined by boundary condition.

Also $|\psi|^2$ represents the probability of finding the particles at any point within the box. So $|X(x)|^2$ which is a function of x coordinates only represents the probability of finding the particles at any point along the X-axis. As the potential is very high at the walls of the box, the probability of finding the particles at the wall will be zero, *i.e.*,

$$|X(x)|^2 = 0 \text{ when } x = 0 \text{ and } x = a$$

$$X(x) = 0 \text{ when } x = 0 \text{ and } x = a$$

Using these boundary conditions in equation (13), we get

$$0 = A\sin(0+C)\ A \neq 0$$

$$\therefore \qquad \sin C = 0 \text{ and } 0 = A\sin(Ba+C)$$

which gives $\qquad \sin Ba = 0 \text{ or } Ba = n_x\pi$

or $\qquad B = \frac{n_x\pi}{a}$

where n_x is a positive integer.

$$\therefore \qquad X(x) = A\sin\frac{n_x\pi x}{a} \qquad ...(14)$$

Applying the normalization condition between $x = 0$ to $x = a$, we have

$$\int_0^a |X(x)|^2\,dx = 1$$

or $\qquad \int_0^a \left|A\sin\frac{n_x\pi x}{a}\right|^2 dx = 1$

or $\qquad A^2\int_0^a \sin^2\frac{n_x\pi x}{a}\,dx = 1$

or $\qquad A^2\frac{a}{2} = 1 \text{ or } A = \sqrt{\left(\frac{2}{a}\right)}$

Therefore, $X(x) = \sqrt{\left(\frac{2}{a}\right)}\sin\frac{n_x\pi x}{a} \qquad ...(15)$

Similarly, we can solve equation (11) and (12) to get

$$Y(y) = \sqrt{\frac{2}{b}}\sin\frac{n_y\pi y}{a} \qquad ...(16)$$

and $\qquad Z(z) = \sqrt{\frac{2}{c}}\sin\frac{n_z\pi z}{a} \qquad ...(17)$

Now the complete wave function $\psi_{n_x,y_y,n_z}(x,y,z)$ has the form

$$\psi_{n_x,y_y,n_z}(x,y,z) = X(x)\,Y(y)\,Z(z)$$

$$= \sqrt{\left(\frac{2}{a}\right)} \sin\frac{n_x \pi\, x}{a} \sqrt{\left(\frac{2}{b}\right)} \sin\frac{n_y \pi\, y}{b} \sqrt{\left(\frac{2}{c}\right)} \sin\frac{n_z \pi\, z}{c}$$

$$= \frac{2\sqrt{2}}{\sqrt{abc}} \sin\frac{n_x \pi\, x}{a} \sin\frac{n_y \pi\, y}{b} \sin\frac{n_z \pi\, z}{c} \qquad \text{...(18)}$$

From equation (15), we get

$$\frac{\partial^2 X}{\partial^2 x} = -\left(\frac{n_x \pi}{a}\right)^2 \sqrt{\frac{2}{a}} \sin\frac{n_x \pi x}{a} = -\left(\frac{n_x \pi}{a}\right)^2 X(x) \qquad \text{...(19)}$$

Putting the value from equation (19) in equation (10) we get

$$-\left(\frac{n_x \pi}{a}\right)^2 X(x) + \frac{2m}{\hbar^2} E_x\, X(x) = 0$$

or
$$E_x = \frac{1}{2m}\left(\frac{n_x \pi \hbar}{n}\right)^2 = \frac{n_x^2 h^2}{8\, m\, a^2} \qquad \text{...(20)}$$

Similarly we have

$$E_y = \frac{n_y^2 h^2}{8\, m\, b^2} \qquad \text{...(21)}$$

and
$$E_z = \frac{n_z^2 h^2}{8\, m\, c^2} \qquad \text{...(22)}$$

The allowed values of total energy are given by

$$E = E_x + E_y + E_z$$

$$= \frac{h^2}{8m}\left[\frac{n_x^2}{a^2} + \frac{n_y^2}{b^2} + \frac{n_z^2}{c^2}\right] \qquad \text{...(23)}$$

where n_x, n_y, n_z denote any set of three positive numbers.

When the box is a cube, *i.e.*, $a = b = c$, the energy expression may be written as

$$E = \frac{h^2}{8\, ma^2}\left[n_x^2 + n_y^2 + n_z^2\right] \text{ with } n_x, n_y, n_z = 1, 2, 3 \ldots$$

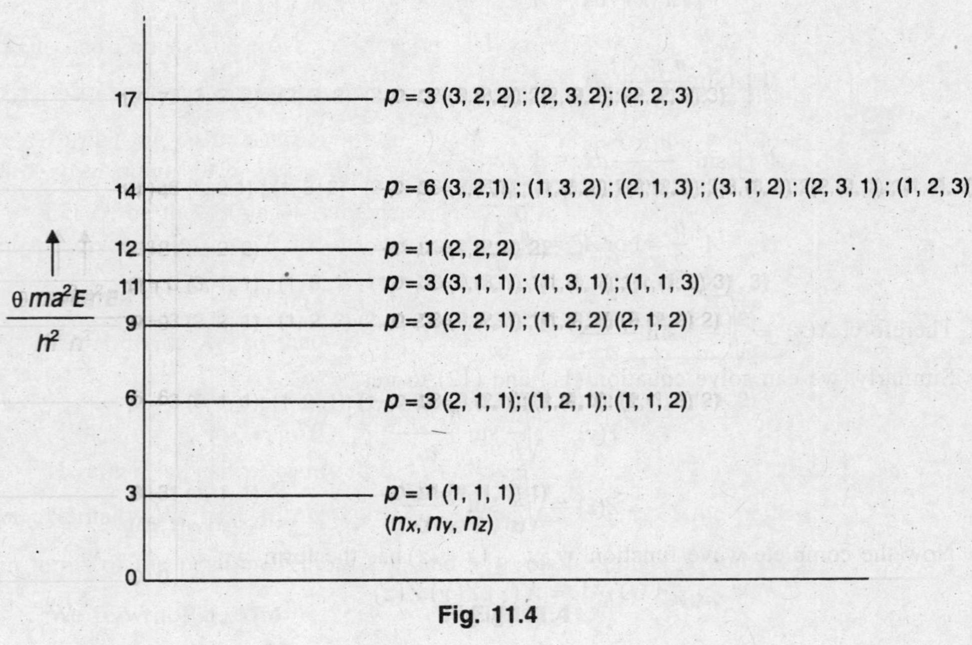

Fig. 11.4

Fig 11.4 shows the energy levels of few states of a particle enclosed in a cubical box.

Now when there is only one wave function corresponding to a particular eigen value, the level is known as *non-degenerate* but when there are a number of wave functions corresponding to a single eigen value, the level is known as *degenerate*. From Fig. 11.4 it is evident that the energy level corresponding to the energy $h^2/8\ ma^2$ when $nx = ny = nz = 1$ is non-degenerate. Similar level with $nx = ny = nz = 2$, $nx = ny = nz = 3$ and so on will be non-degenerate. In case of the energy ($h^2/8\ ma^2 \times 6$), *nx, ny* and *nz* may have values (2, 1, 1); (1, 2, 1); (1, 1, 2) *i.e.*, energy is the same but there will be 3 wave functions hence this is a degenerate level with degeneracy 3. Similarly we can consider the degeneracy of all other levels.

11.14. SINGLE STEP BARRIER

A **potential barrier** corresponds to the force field acting on a particle being zero everywhere except in a limited region. In the Fig. 11.5, a single potential barrier is shown.

For $x < 0$, the potential $V(x) = 0$

For $x > 0$, potential $V(x) = V_0$

So we may define the potential function of a potential step as Fig. 11.5.

$$V(x) = \begin{Bmatrix} 0 \text{ for } x < 0 \\ V_0 \text{ for } x > 0 \end{Bmatrix}$$

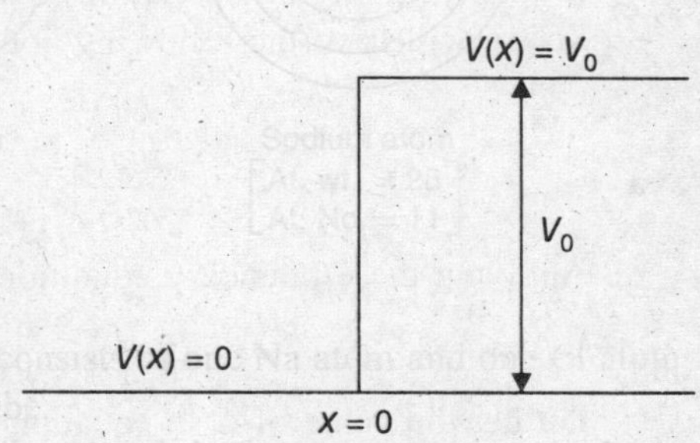

Fig. 11.5

Here we have considered V_0 as the height of the potential barrier. Let us now consider the case of electrons carrying energy E moving from left to right *i.e.*, along the positive X-direction. The energy of the electron may be greater than V_0 or less than V_0.

In the classical case, if the electron energy is more than V_0, it will penetrate in the region II and if it is less than V_0, it will be reflected back at $x = 0$.

However, if we apply the quantum mechanics, the electron behaves like a wave and while moving from left to right, it faces a sudden shift in potential at $x = 0$. This is analogous to the propagation of light wave when it strikes a glass plate and faces a change in refractive index. The ray is partly reflected and partly transmitted. In a similar way, the electrons will be partly reflected and partly transmitted at $x = 0$ *i.e.*, at the discontinuity. We will now take up the following two cases:

Case (*i*). When $E > V_0$

We apply Schrödinger's wave equation to region I and II.

In the I region, the Schrödinger equation is given by

$$\frac{d^2\psi_1}{dx^2} + \frac{2m}{\hbar^2} E\psi_1 = 0$$

Which may be written as

or $$\frac{d^2\psi_1}{dx^2} k_1^2 \psi_1 = 0 \text{ where } \frac{2mE}{\hbar^2} = k_1^2 \qquad ...(1)$$

In II region Schrödinger wave equation is given by

$$\frac{d^2\psi_2}{dx^2} + \frac{2mE}{\hbar^2}(E - V_0)\,\psi_2 = 0$$

or $$\frac{d^2\psi_2}{dx^2} k_2^2\, \psi_2 = 0 \text{ where } \frac{2m(E-V)}{\hbar^2} = k_2^2 \qquad ...(2)$$

The general solutions of equations. (1) and (2) have the form

$$\psi_1 = A\, e^{i k_1 x} + B\, e^{-i k_1 x} \; (x < 0) \qquad ...(3)$$

and $$\psi_2 = C\, e^{i k_2 x} + D\, e^{-i k_2 x} \; (x > 0) \qquad ...(4)$$

where *A, B, C* and *D* are constants which may be found boundary conditions.In equation (3) the first term corresponds to a wave propagating along +ve *X*-direction while the second term corresponds to wave travelling along negative *X*-direction in the region I. While the first will represent incident wave, the second will represent *reflected wave*. Since in the first region both incident and reflected rays will be present, so equation (3) represents the solution of Schrödinger's equation in region I.

In the equation (4), the first term corresponds to a wave travelling along positive *X*-direction *i.e., transmitted wave* and second term corresponds to the wave travelling in negative *X*-direction *i.e.,* reflected wave in region II. Since there is no *discontinuity* in region II, so there will not be a reflected wave *i.e.,* $D = 0$, and so we have the solution of wave equation as

$$\psi_2 = C e^{i k x} \qquad ...(5)$$

So equations (3) and (5) will represent the solutions of Schrödinger's wave equation in I and II region respectively.

From probability interpretation of wave function, ψ must be finite *i.e.,* it must be continuous. So we will have the following boundary conditions

$$(\psi_1)_{x=0} = (\psi_2)_{x=0}$$

and $$\left(\frac{d\psi_1}{dx}\right)_{x=0} = \left(\frac{d\psi_2}{dx}\right)_{x=0} \qquad ...(6)$$

Applying these boundary conditions, to equations (3) and (5), we get

$$A + B = 0 \qquad ...(7)$$

and $$k_1 A - k_1 B = k_2 C \qquad ...(8)$$

Putting the value of C in equation (8), we have

$$k_1 A - k_1 B = k_2 (A + B)$$

$$(k_1 - k_2) A = (k_1 + k_2) B$$

which gives $$B = \left(\frac{k_1 - k_2}{k_1 + k_2}\right) A \qquad ...(9)$$

Putting the value of B from equation (9) in equation (7), we get

$$A + \left(\frac{k_1 - k_2}{k_1 + k_2}\right) A = C$$

which gives $$C = \left(\frac{2 k_1}{k_1 + k_2}\right) A \qquad ...(10)$$

In this case B and C represent the amplitudes of *reflected* and *transmitted* beam respectively in terms of *amplitude* of incident wave.

Now $\psi\psi^*$ represents the probability density. Let v be the velocity of a stream of particles then $\psi\psi^* v$ represents the current density *i.e.,* number of particles crossing unit area placed perpendicular to the direction of motion. If N_i be the incident flux, then we have by definition

$$N_i = \left|A\, \psi_i\, \psi_i^*\right|^2 v = |A|^2 v = |A|^2 \frac{k_1 \hbar}{m} \qquad ...(11)$$

$$\left[\text{since } k_1 = \sqrt{\frac{2mE}{\hbar^2}} = \sqrt{\frac{2m.(1/2)\, m\, v^2}{\hbar^2}} = \frac{mv}{\hbar} \qquad \therefore v = k_1\, h/m\right]$$

Similarly the reflected flux N_r is given by the relation

$$N_r = |B|^2 v = \left|\left(\frac{k_1 - k_2}{k_1 + k_2}\right)\right|^2 |A|^2 \frac{k_2 \hbar}{m}$$

[Since $k_2 = \sqrt{\frac{2m(E-V_0)}{\hbar^2}} = \sqrt{\frac{2m(1/2)mv_1^2}{\hbar^2}} = \frac{mv_1}{\hbar}$ $\quad \therefore \; v_1 = \frac{k_2 \hbar_1}{m}$]

Now $N_t = \frac{4k_1}{(k_1+k_2)^2}\frac{k_1 k_2 \hbar}{m}|A|^2 = \frac{4k_1 k_2}{(k_1+k_2)^2} N_t$ ...(13)

We have from equation (11), (12) and (13)

$$N_t + N_r = N_i$$

So quantum mechanically, there is some probability of *reflection* and some probability of transmission. The *reflection coefficient R* and *transmission coefficient T* are defined as under

$$\text{Reflection coefficient } R = \frac{\text{Magnitude of reflected flux}}{\text{Magnitude of incident flux}} = \frac{N_r}{N_i}$$

$$= \left(\frac{k_1 - k_2}{k_1 + k_2}\right)^2 \qquad \text{...(14)}$$

$$\text{Transmission coefficient } \quad T = \frac{\text{Magnitude of transmitted flux}}{\text{Magnitude of incident flux}} = \frac{N_r}{N_i}$$

$$= \frac{4k_1 k_2}{(k_1+k_2)^2} \qquad \text{...(15)}$$

So we have $\quad T + R = \frac{4k_1 k_2}{(k_1+k_2)^2} + \frac{(k_1-k_2)^2}{(k_1+k_2)^2} = 1$ ...(16)

Case (*ii*). When $E < V_0$

The Schrödinger wave equation in region I is the same and is given by

$$\psi_1 = A\,e^{i k_1 x} + B\,e^{-i k_1 x} \qquad \text{...(17)}$$

For region II, the Schrödinger wave equation is given

$$\frac{d^2\psi_2}{dx^2} - \frac{2m}{\hbar^2}(V_0 - E)\,\psi_2 = 0 \qquad [\because V_0 > E]$$

or

$$\frac{d^2\psi_2}{dx^2}\, k_3^2 x = 0$$

Where we have taken

$$k_3^2 = \frac{2m}{\hbar^2}(V_0 - E)$$

General solution of this equation is given by the relation

$$\psi_2 = Ce^{k_3 x} + De^{-k_3 x}$$

So no particle goes to region II

i.e., $\quad e^{k_3 x} = 0 \quad \therefore \quad \psi^2 = D\,e^{-k_3 x}$ ...(18)

Applying the boundary conditions, (6), we get

$$A + B = D \qquad \text{...(19)}$$

$$ik_1 A - ik_2 B = -k_2 D \qquad \text{...(20)}$$

If we solve equations (19) and (20), we get

$$B = \left(\frac{ik_1 + k_3}{ik_1 - k_3}\right) A \qquad \text{...(21)}$$

and

$$D = \left\{\frac{2\,ik_1}{(ik_1 - k_3)}\right\} A \qquad \text{...(22)}$$

Now incident flux $\quad N_i = v\,|\,A\,|^2$

and reflected flux
$$N_r = v\,|\,B\,|^2 = v\left|\left(\frac{i\,k_1 + k_3}{(ik_1 - k_3)}\right)\right|^2 \,|\,A\,|^2$$
$$= v\,.\,1.\,|\,A\,|^2 = N_i \qquad ...(23)$$

which means that all the particles are reflected at the barrier. Also the transmitted flux is given by the relation

$$N_t = \frac{4k_1^2}{k_1^2 + k_3^2}\,|\,A\,|^2 \qquad ...(24)$$

Equation (24) shows that N_t is not zero until k_3 *i.e.*, V_0 is infinite, which means that if $N_t \neq 0$ there is some probability of particles going to the region II. It shows that even if the energy of the particles is less than the height of potential barrier, there is some probability of its penetration. This forms the *basis of Quantum Mechanics* for solving many problems of *potential barrier.*

11.15. RECTANGULAR POTENTIAL BARRIER

We can take up the case of transmission of plane wave through a rectangular potential barrier. The transmission is based on *tunnel effect* which may be stated as under : *if a particle is impinging on a barrier with energy less than the height of the potential barrier, it will not necessarily be totally reflected by the barrier but there is always a probability that it may cross the barrier and continue its forward motion.*

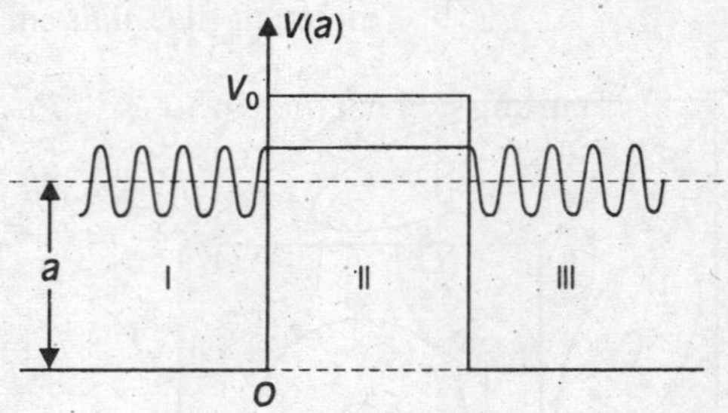

Fig. 11.6

Let us consider a particle incident on the barrier of the type shown in Fig. 11.6, with the condition that the energy of particle $E < V_0$ and we have the different relations as under:

$V(x) = 0$ for $x < 0$ in region I
$V(x) = V_0$ for $0 < x < a$ region II
$V(x) = 0$ for $x > a$ in region III.

According to classical mechanic, a particles striking an infinitely hard wall has no chance of getting through it. Quantum mechanics is an agreement : a particle of finite kinetic energy cannot enter a region where the potential energy is $V = \infty$.

But what about a less formidable was, one that is not infinitely hard but still requires more energy V to penetrate than the particle's energy E? In this case classical mechanics also requires the particle to bounce off, but now quantum mechanics gives a different result; there is a certain likelihood-not necessarily great, but not zero either- that the particle can pass through the energy barrier even though $E < V$ (Fig. 11.7). Although the particle does not have enough energy to surmount the barrier, it can nevertheless *tunnel* through-it, so to speak. The higher the barrier and the thicker it is, the smaller the chance the particle can get through.

The tunnel effect can be understood in terms of the **uncertainty principle**; if we say that the incident particle cannot enter the barrier, then the uncertainty Δx in its position must be 0 there. But since $\Delta x\,\Delta p \geq \hbar/2$, the corresponding uncertainty Δp in the particle's momentum must be infinite inside the barrier. An infinite uncertainty in p means that p, and E as well, must be infinite, which is not compatible with the finite momentum and energy of the particle. Therefore, the particle must be able to enter the barrier, and once inside it has the possibility of continuing on.

The tunnel effect actually occurs, notably in the case of alpha particles emitted by certain radioactive nuclei. **An alpha particle** whose kinetic energy is only a few MeV is able to escape from a nucleus whose potential wall is perhaps 25 MeV high. The **probability** of escape is so small that the alpha particle might have to strike the wall 10^{38} or more times before it emerges, but sooner or later it does get out. *Tunneling* also occurs in the operation of certain semiconductor diodes in which electrons pass through the potential barriers even though their kinetic energies are smaller than the barrier heights.

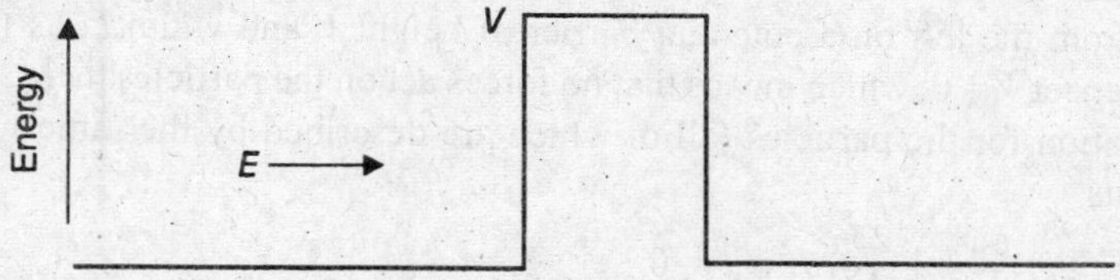

(*a*) A particle with energy $E < V$ approaching a potential barrier.

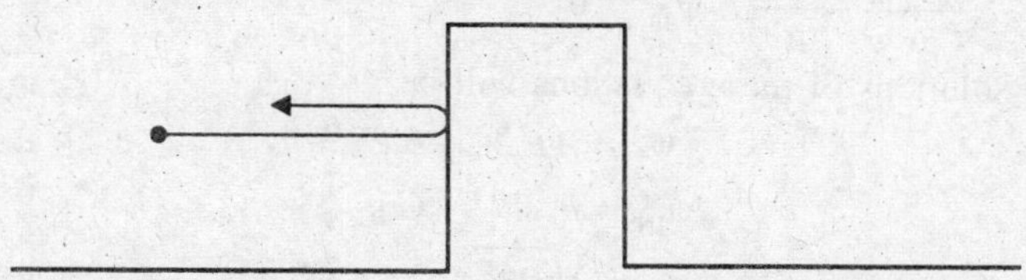

(*b*) From classical mechanics, the particle must be reflected by the barrier.

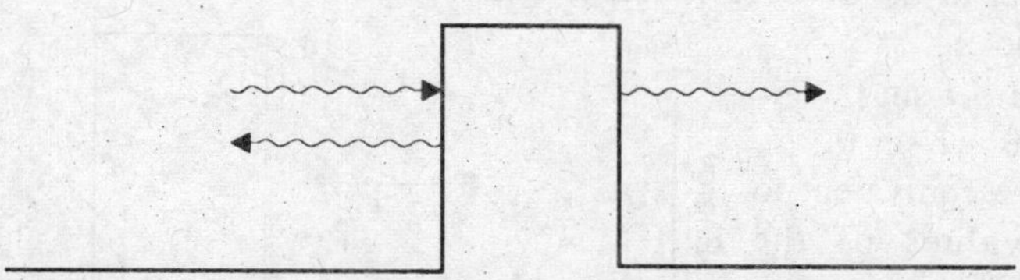

(*C*) In quantum mechanics, the de Broglie waves that represent the particle are partly reflected and partly transmitted, *i.e.*, the particle has finite probability of penetrating the barrier.

Fig. 11.7

Fig. 11.8 shows the first three wave functions and the corresponding probability densities for a particle in a finite potential well. The wave functions do not equal zero outside the well, as they do for a particle in an infinite potential well (compare Fig. 11.8). Even though the particle's energy is less than the value of V outside the well, there is still a definite chance that it will be found there. The particle wavelengths that can fit into the well are accordingly somewhat longer than in the case of the infinite potential well, which means lower momenta and hence lower energy levels.

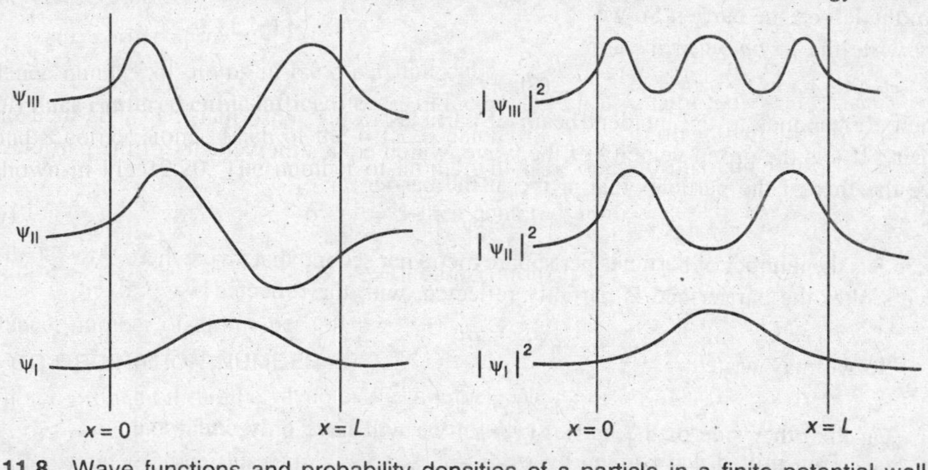

Fig. 11.8 Wave functions and probability densities of a particle in a finite potential well. The particle has a certain probability of being found outside the well.

11.16. TUNNEL EFFECT

• Let us consider a beam of identical particles all of which have the kinetic energy $K = E$. The beam is incident from the left on a potential barrier of height V and width L, as in Fig. 11.6. On both sides of the barrier $V = 0$, which means that no forces act on the particles there. In these regions Schrödinger's equation for the particles (all of which are described by the same wave function ψ will have the forms

$$\frac{\partial \psi_I}{\partial x^2} + \frac{2m}{\hbar^2} E \psi_I = 0 \qquad \text{...(1)}$$

$$\frac{\partial^2 \psi_{III}}{\partial x^2} + \frac{2m}{\hbar^2} E \psi_{III} = 0 \qquad \text{...(2)}$$

The general solutions of these equations will be

$$\psi_I = A e^{i k_1 x} + B e^{-i k_1 x} \qquad \text{...(3)}$$

$$\psi_{III} = F e^{i k_1 x} + G e^{-i k_1 x} \qquad \text{...(4)}$$

where

$$k_1 = \frac{\sqrt{2mE}}{\hbar} = \frac{p}{\hbar} = \frac{2\pi}{\lambda} \qquad \text{...(5)}$$

is the wave number of the de Broglie waves that represent the particles outside the barrier. Also we have the relations

$$e^{i\theta} = \cos\theta + i \sin\theta$$

$$e^{-i\theta} = \cos\theta - i \sin\theta$$

these solutions are equivalent to Equation, the values of the coefficients are different in each case, of course-but are in a more suitable form to describe particles that are not trapped.

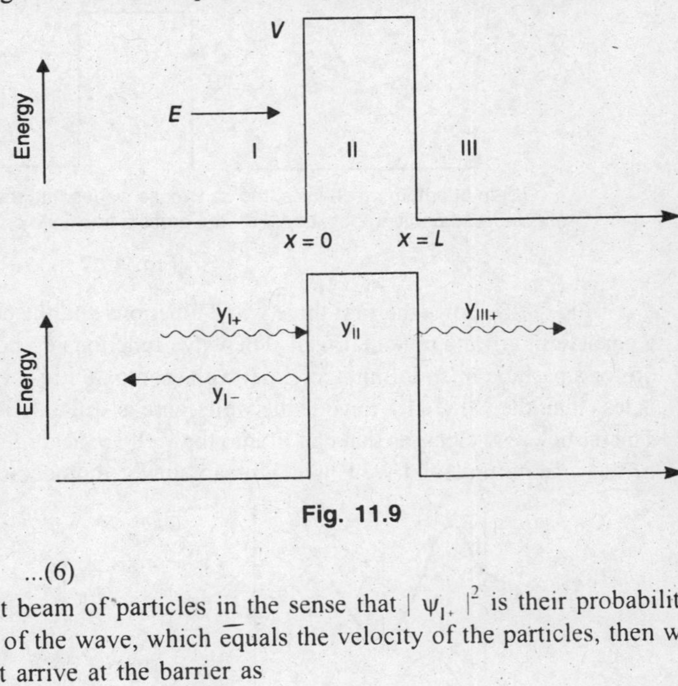

Fig. 11.9

Incident and reflected waves. The various terms in Equations (3) and (4) can be easily interpreted. As shown schematically in Fig. 11.9, $A e^{i k_1 x}$ is a wave of amplitude A incident from the left on the barrier. So we may write the *incoming wave* as

$$\psi_{I+} = A e^{i k_1 x} \qquad \text{...(6)}$$

which corresponds to the incident beam of particles in the sense that $|\psi_{I+}|^2$ is their probability density. If v is the group velocity of the wave, which equals the velocity of the particles, then we have the flux of the particles that arrive at the barrier as

$$S = |\psi_{I+}|^2 v \qquad \text{...(7)}$$

where S is the number of particles per square metre per second that arrive there. At $x = 0$ the incident wave strikes the barrier and is partially reflected, with the reflected wave given

$$\psi_{I-} = B e^{-i k_1 r} \qquad \text{...(8)}$$

So we may write

$$\psi_I = \psi_{I+} + \psi_{I-} \qquad \text{...(9)}$$

On the other side of the barrier ($x > L$) we will have only one wave

i.e., Transmitted wave given by $\psi_{III+} = F e^{i k_1 x}$ ...(10)

travelling in the $+ x$ direction since, by hypothesis, region III contains nothing which reflects the wave. So $G = 0$ and we have

$$\psi_{III} = \psi_{III+} = F\,e^{i k_1 x} \qquad ...(11)$$

The *transmission probability* T for a particle to pass through the barrier is given by:

$$T = \frac{|\psi_{III+}|^2 v}{|\psi_{I+}|^2 v} = \frac{FF^*}{AA^*} \qquad ...(12)$$

between the flux of particles that emerges from the barrier and the flux that arrives at it. In other words, T is the fraction of incident particles that succeed in tunnelling through the barrier. Classically $T = 0$ because a particle with $E < V$ cannot exist inside the barrier; but quantum mechanically we will have different results.

In region II, Schrödinger's equation is given by

$$\frac{\partial^2 \psi_{II}}{\partial x^2} + \frac{2m}{\hbar^2}(E - V)\,\psi_{II} = 0 \qquad ...(13)$$

and its general solution may be written as

$$\psi_{II} = C\,e^{i k' x} + D\,e^{-i k' x} \qquad ...(14)$$

where
$$k' = \frac{\sqrt{2m\,(E - V)}}{\hbar} \qquad ...(15)$$

is the wave number inside the barrier.

Since $E < V$, k' is imaginary, and it is convenient to define another wave number k_2 in the following ways:

Wave number inside the barrier

$$k_2 = -\,ik' = \frac{\sqrt{2m\,(E - V)}}{\hbar} \qquad ...(16)$$

In terms of k_2 Equation (14) for ψ_{II} *i.e.,* wave function inside the barrier is given by

$$\psi_{II} = C\,e^{-k_2 x} + D\,e^{k_2 x} \qquad ...(17)$$

Since the exponents are real quantities, ψ_{II} does not oscillate and so does not represent a moving particle. However, the probability density $|\psi_{II}|^2$ is not zero, so there is a finite probability of finding a particle within the barrier. *Such a particle may emerge into region III or it may return to region I.*

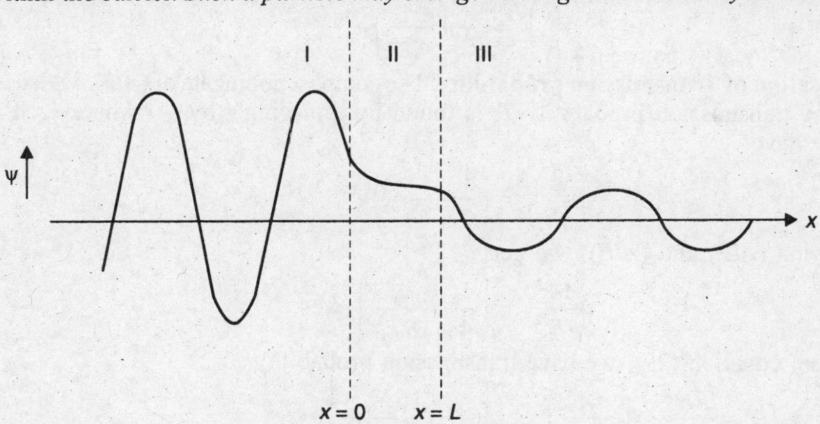

Fig. 11.10

Transmission Probability. In order to calculate the transmission probability T, we must apply the appropriate boundary conditions to ψ_I, ψ_{II}, and ψ_{III}. Figure 11.10 shows the wave functions in regions I, II and III. As discussed earlier, both ψ and its derivative $\partial\psi / \partial x$ must be continuous everywhere. With reference to Fig. 11.10, these conditions mean that at each wall of the barrier, the wave functions inside and outside must not only have the same value but also the same slope, so that they match up perfectly. Hence at the left hand wall of the barrier we have (boundary conditions $x = 0$).

$$\left.\begin{array}{c}\psi_{\mathrm{I}}=\psi_{\mathrm{II}}\\ \dfrac{\partial\psi_{\mathrm{I}}}{\partial x}=\dfrac{\partial\psi_{\mathrm{II}}}{\partial x}\end{array}\right\} x=0 \qquad ...(18)$$

and at the right hand wall (boundary conditions $x = L$)

$$\left.\begin{array}{c}\psi_{\mathrm{II}}=\psi_{\mathrm{III}}\\ \dfrac{\partial\psi_{\mathrm{II}}}{\partial x}=\dfrac{\partial\psi_{\mathrm{III}}}{\partial x}\end{array}\right\} x=L \qquad ...(19)$$

Now we substitute ψ_{I}, ψ_{II}, and ψ_{III} from Equations. (3), (10) and (17) into the above equations, we get

$$A + B = C + D \qquad ...(20)$$

$$ik_1 A - ik_1 B = -k_2 C + k_2 D \qquad ...(21)$$

$$C e^{-k_2 L} + D e^{k_2 L} = F e^{i k_1 L} \qquad ...(22)$$

$$-k_2 C e^{-k_2 L} + k_2 D e^{k_2 L} = i k_1 F e^{i k_1 L} \qquad ...(23)$$

Equations (20) and (23) may be solved for $\left(\dfrac{A}{F}\right)$ and we get

$$\left(\frac{A}{F}\right) = \left[\frac{1}{2}+\frac{i}{4}\left(\frac{k_2}{k_1}-\frac{k_1}{k_2}\right)\right] e^{(i k_1+k_2) L} + \left[\frac{1}{2}-\frac{i}{4}\left(\frac{k_2}{k_1}-\frac{k_1}{k_2}\right)\right] + e^{(i k_1+k_2) L} \qquad ...(24)$$

Approximations. Let us assume that the potential barrier is high relative to the energy of the incident particles, so that $k_2/k_1 > k_1/k_2$ and we have

$$\frac{k_2}{k_1}-\frac{k_1}{k_2} \approx \frac{k_2}{k_1} \qquad ...(25)$$

Let us also assume that the barrier is wide enough for ψ_{II} to be severely weakened between $x = 0$ and $x = L$ which means that $k_2 L >> 1$ and we have the relation

$$e^{k_2 L} >> e^{-k_2 L} \qquad ...(26)$$

Hence Equation (24) can have the *approximation* as

$$\left(\frac{A}{F}\right) = \left(\frac{1}{2}+\frac{i k_2}{4 k_1}\right) e^{(ik_1+k_2) L} \qquad ...(27)$$

Computation of Transmission probability. The complex conjugate of (A/F), which we require to compute the transmission probability T, is found by replacing i by $-i$ wherever it occurs in (A/F), and we have

$$\left(\frac{A}{F}\right)^{*} = \left(\frac{1}{2}-\frac{i k_2}{4 k_1}\right) e^{(ik_1+k_2) L} \qquad ...(28)$$

Multiplying (A/F) and $(A/F)^*$ we get

$$\frac{AA^*}{FF^*} = \left(\frac{1}{4}+\frac{k_2^2}{16 k_1^2}\right) e^{2 k_2 L} \qquad ...(29)$$

Now from equation (29), we have transmission probability

$$T = \frac{FF^*}{AA^*} = \left(\frac{AA^*}{FF^*}\right)^{-1} = \left[\frac{16}{4+(k_2/k_1)^2}\right] e^{-2 k_2 L} \qquad ...(30)$$

From the definitions of k_1 [Equation (5)] and of k_2 [Equation (16)], we see that

$$\left(\frac{k_2}{k_1}\right)^2 = \frac{2m(V-E)/\hbar^2}{\dfrac{2mE}{\hbar^2}} = \frac{V}{E}-1 \qquad ...(31)$$

which means that the quantity in brackets in Equation (30) varies much less with E and V than does the exponential. Moreover, the bracketed quantity is never vary far from 1 in value and so a reasonable approximation of the transmission probability may simply be written as

Approximate transmission probability $T = e^{-2k_2 L}$ ...(32)

Equation (32) is a useful equation to account for the extraordinary variation in the half-lives of **alpha-radioactive nuclei with alpha-particle energy.**

Example 1. *Electrons with energies of 1 eV and 2 eV are incident on a barrier 5 eV high and 0.5 nm wide. Find their respective transmission probabilities. How are these affected if the barrier is doubled in width?*

Solution: For the 1 eV electron we have

$$k_2 = \frac{\sqrt{2m(V-E)}}{\hbar}$$

$$= \frac{\sqrt{(2)(9.1\times10^{-31}\text{ kg})[(5-1)\text{ eV}](1.6\times10^{-19}\text{ J/eV})}}{1.054\times10^{-34}\text{ Js}}$$

$$= 1.0\times10^{10}\text{ m}^{-1}$$

In this case

$L = 0.5\text{ nm} = 5\times10^{-10}\text{ m}$,

$$2k_2L = (2)(1.0\times10^{10}\text{ m}^{-1})(5\times10^{-10}\text{ m}) = 10,$$

and the transmission probability is given by

$$T_1 = e^{-2k_2L} \quad e^{-10} = 4.5\times10^{-5}$$

One 1 eV electron out of 22,000 can tunnel through the 5 eV barrier on the average. For the 2 eV electrons a similar calculation will give us

$$T_2 = e^{-8.9} = 1.4\times10^{-4}$$

which shows that 2 eV electrons are about 3 times more likely to tunnel through the barrier. If the barrier is doubled in width to 1 nm, the transmission probabilities will be given by

$$T_{11} = 2.1\times10^{-9} \qquad T_{21} = 1.9\times10^{-8}$$

It is clear that T depends far more strongly on the width of the barrier than on the particle energy.

11.17. HARMONIC OSCILLATOR

A particle undergoing simple harmonic motion is called a harmonic oscillator. In harmonic oscillator, the force applied is directly proportional to the displacement and is always directed towards the mean position.

If applied force displaces the particle through x, then Restoring force F is given by

$$F \propto -x$$

or $$F = -kx$$ ...(1)

The potential energy of the oscillator is

$$V = -\int F dx$$

or $$V = k\int x dx = \frac{1}{2}kx^2$$ ...(2)

CLASSICAL VIEW

If m is the mass of the particle, then

$$F = m\frac{d^2x}{dt^2}$$

Using equation (1), we get

$$-kx = m\frac{d^2x}{dt^2}$$

or $$\frac{d^2x}{dt^2} + \frac{k}{m}x = 0$$

or $$\frac{d^2x}{dt^2} + \omega^2 x = 0 \qquad ...(3)$$

where $\omega = \sqrt{\frac{k}{m}}$ is called angular frequency.

The classical frequency of the oscillator ν is given by

$$\nu = \frac{\omega}{2\pi} = \frac{1}{2\pi}\sqrt{\frac{k}{m}} \qquad ...(4)$$

There are various ways to write the solution to Equation (3). A common one is

$x = A \cos(\omega t + \phi) = A \cos(2\pi\nu t + \phi)$

where ν is the frequency, A is the amplitude and ϕ is the phase angle which depends upon what x is at time $t = 0$

If a graph is plotted between $V(x)$ and x, the graph will be a parabola as shown in fig. 4.11.

If E is the total energy and A is the amplitude of the motion, we find that at points $x = \pm A$, where $V = E$ are called turning points

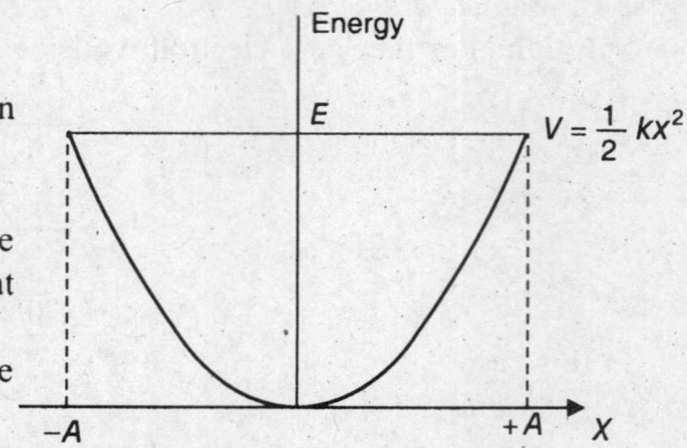

Fig. 11.11

At $$x = \pm A$$

$$E = \frac{1}{2} kA^2$$

Since amplitude A can have any value, so in classical mechanics, the oscillator can have any value of total energy ranging from **zero to any positive number.**

QUANTUM MECHANICAL VIEW.

Since potential energy V does not depend on time some can use time independent Schrodings equation to find motion of simple harmonic oscillator.

Schrodinger time independent one dimensional wave equation (along x-axis) is given by

$$\frac{d^2\psi}{dx^2} + \frac{2m}{\hbar^2}(E - V)\psi = 0 \qquad ...(5)$$

Substituting $V = \frac{1}{2}kx^2$, in above equation, we get

$$\frac{d^2\psi}{dx^2} + \frac{2m}{\hbar^2}\left(E - \frac{1}{2}kx^2\right)\psi = 0 \qquad ...(6)$$

Since $\omega = \sqrt{k/m}$, so $k = m\omega^2$

So equation (6), becomes

$$\frac{d^2\psi}{dx^2} + \frac{2m}{\hbar^2}\left(E - \frac{1}{2}m^2\omega^2x^2\right)\psi = 0$$

or $$\frac{d^2\psi}{dx^2} + \left(\frac{2mE}{\hbar^2} - \frac{m^2\omega^2}{\hbar^2}x^2\right)\psi = 0 \qquad ...(7)$$

For convenience

Let $$\frac{2mE}{\hbar^2} = \alpha \text{ and } \frac{m\omega}{\hbar} = \beta \qquad ...(8)$$

So Equation (7) becomes

$$\frac{d^2\psi}{dx^2} + (\alpha - \beta^2 x^2)\psi = 0 \qquad ...(9)$$

Again for convenience, let us introduce a new variable q related to x such that

$$q = \sqrt{\beta\, x}$$

Now
$$\frac{d\psi}{dx} = \frac{d\psi}{dq}.\frac{dq}{dx} = \sqrt{\beta}\,\frac{d\psi}{dq}$$

and
$$\frac{d^2\psi}{dx^2} = \frac{d}{dx}\left[\frac{d\psi}{dx}\right] = \frac{d}{dx}\left[\sqrt{\beta}\,\frac{d\psi}{dq}\right]$$
$$= \frac{d}{dq}\left[\sqrt{\beta}\,\frac{d\psi}{dq}\right]\frac{dq}{dx}$$
$$= \sqrt{\beta}\,\frac{d^2\psi}{dq^2}\sqrt{\beta}$$

or
$$\frac{d^2\psi}{dx^2} = \beta\,\frac{d^2\psi}{dq^2}$$

So Equation (9) becomes

$$\beta\,\frac{d^2\psi}{dq^2} + \left(\alpha - \beta^2\frac{q^2}{\beta}\right)\psi = 0$$

or
$$\frac{d^2\psi}{dq^2} + \left(\frac{\alpha}{\beta} - q^2\right)\psi = 0 \qquad ...(10)$$

ASYMPTOTIC SOLUTION

To solve Equation (10), let us make an attempt to obtain an expression for the asymptotic solution (the value of ψ (q) for very large value of q), when $q^2 >> \frac{\alpha}{\beta}$. In this case, equation (10) becomes

$$\frac{d^2\psi}{bq^2} - q^2\,\psi = 0 \qquad ...(11)$$

The solution of this equation is of the form $\psi = e^{\pm q^2/2}$

But here, the solution $e^{+q/2}$ is not acceptable because, it increases rapidly with increasing q, but ince $|\psi|^2$ is the probability of finding the particle and it must decrease as x increases

$\therefore$ The particular acceptable solution is $\psi = -q^2/2$

The general solution at any distance can be written as

$$\psi = H(q)\, e^{-q^2\;2} \qquad ...(12)$$

where $H(q)$ is a finite Hermite polynomial in q and can be written in a series of the form

$$H(q) = a_0 + a_1\, q + a_1\, q^2 + ..., + a_n\, q^n$$

From Equation (12), we have

$$\frac{d\psi}{dq} = -qe^{-q^2/2}\,H + e^{-q^2\;2}\,\frac{dH}{dq}$$

and
$$\frac{d^2\psi}{dq^2} = q^2 e^{-q^2\;2}\,H - e^{-q^2 n'2}H - qe^{-q^2\;2}\frac{dH}{dq} - qe^{-q^2/2}\frac{dH}{dq} + e^{-q^2\;2}\frac{d^2H}{dq^2}$$

or
$$\frac{d^2\psi}{dq^2} = \left(\frac{d^2H}{dq^2} - 2q\,\frac{dH}{dq} + q^2\,H - H\right)e^{-q^2\;2}$$

Putting the value of $\frac{d\psi}{dq}$ and $\frac{d^2\psi}{dq^2}$ in Equation (10), we get

$$\left(\frac{d^2H}{dq^2} - 2q\,\frac{dH}{dq} + q^2\,H - H\right)e^{-q^2/2} + \left(\frac{\alpha}{\beta} - q^2\right)^{e^{-q^2/2}} H = 0$$

$$\frac{d^2H}{dq^2} - 2q\frac{dH}{dq} + \frac{\alpha}{\beta}H - H = 0$$

or
$$\frac{d^2H}{dq^2} - 2q\frac{dH}{dq} + \left(\frac{\alpha}{\beta} - 1\right)H = 0 \qquad ...(13)$$

The solution of this equation determines $H(q)$. Only those solutions are acceptable for which $\frac{\alpha}{\beta} = 2n+1, n = 0, 1, 2, ...$ where n is called quantum number.

Putting the values of α and β from Equation (8), we get

$$\frac{2mE/\hbar^2}{m\omega/\hbar} = 2n + 1$$

or
$$\frac{2E}{\hbar\,\omega} = (2n + 1)$$

or
$$E = \left(n + \frac{1}{2}\right)\hbar\omega$$

since energy E depends upon n, so we can replace E by E_n.

$\therefore$
$$E_n = \left(n + \frac{1}{2}\right)\hbar\omega \qquad ...(14)$$

The allowed integral values of n lead to certain discrete values of energy represented by equation (14), known as eigen values of harmonic oscillator. Moreover eq. (14) indicates that the energy levels of harmonic oscillator *are equally spaced.* These energy levels are shown in Fig. 11.12.

"Significance of Zero Point energy."

For ground state. (or lowest state), put $n = 0$, then we have

$$E_0 = \frac{1}{2}\hbar\omega.$$

This is called ***"Zero Point energy"***. A comparision with the result $E = n\,\hbar\,\omega$ obtained by old quantum theory shows that the only difference is that all the equally spaced energy levels are shifted upward by an amount equal to half the separation of energy levels *i.e.*, $\frac{1}{2}\hbar\omega$ equal to zero point energy. Thus it is clear that even in the lowest state, the harmonic oscillator has energy greater than that it would have if it were at rest in its equilibrium position. The existence of zero point energy is in agreement with experiment and is important feature of quantum mechanics and is related to the uncertainty principle.

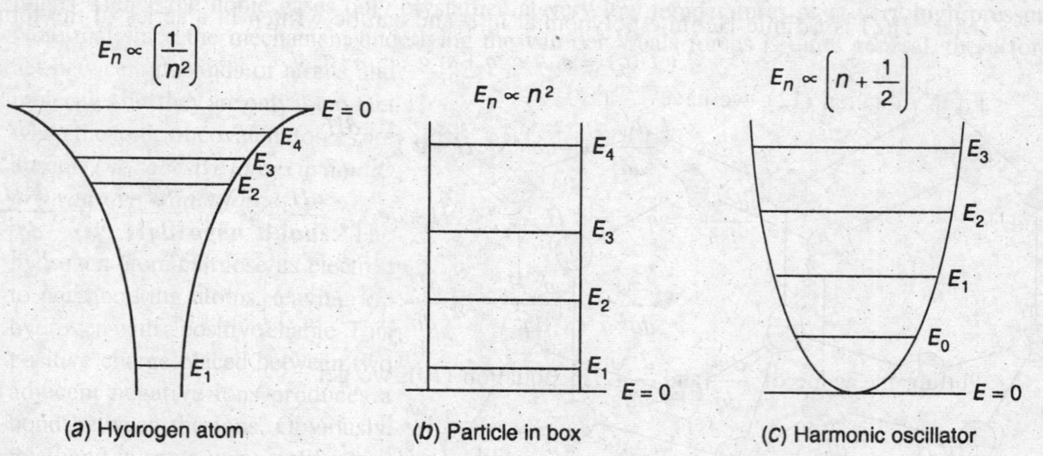

Fig 11.12 Potential wells and energy levels.

Figure 11.12 is a comparison of the energy levels of a *harmonic oscillator* with those of a *hydrogen atom* and of a *particle in* a *box* with infinitely hard walls. The shapes of the respective potential energy curves are also shown. **The spacing of the energy levels is constant only for the harmonic *oscillator.***

Wave functions of harmonic oscillator. For each choice of the parimeter α_n there is different wave function ψ_n. Each function consists of a polynomial H_n (y) (called a *Heremite polynomial*) in either odd or even powers of y, the exponential factor $e^{-y^{2/2}}$, and a numerical coefficient which is needed for ψ_n to meet the normalization condition given by is needed for ψ_n to meet the normalization condition given by

$$\int_{-\infty}^{\infty} |\psi_n|^2 \, dx = 1 \qquad n = 0, 1, 2, ...$$

The general formula for the nth wave function in terms of hermite polynomials is given by

$$\psi_n = \frac{1}{\sqrt{2^n . n! \sqrt{\pi}}} e^{-y^2/2} H_n(y)$$

A table of first size Hermite polynomials is given as under:

Table 11.1 Some Hermite Polynomials

n	$H_n (y)$	α_n	E_n
0	1	1	$\frac{1}{2} h\nu$
1	$2y$	3	$\frac{3}{2} h\nu$
2	$4y^2 - 2$	5	$\frac{5}{2} h\nu$
3	$8y^3 - 12y$	7	$\frac{7}{2} h\nu$
4	$16y^4 - 48y^2 + 12$	9	$\frac{9}{2} h\nu$
5	$32y^5 - 160y^3 + 120y$	11	$\frac{11}{2} h\nu$

11.18. THREE DIMENSIONAL HARMONIC OSCILLATOR

In the case of a particle attracted towards a fixed point by a force which is proportional to its distance from the fixed point, the potential energy is given by

$$V(r) = \frac{1}{2} kr^2 \quad = \frac{1}{2} k (x^2 + y^2 + z^2) \qquad ...(1)$$

which has spherical symmetry. Schrödinger's wave equation for such a system is given by

$$\frac{-\hbar^2}{2m} \nabla^2 \psi + V\psi = E\psi$$

or

$$\frac{-\hbar^2}{2m} \nabla^2 \psi + \frac{1}{2} kr^2 \psi = E\psi \qquad ...(2)$$

In the case of a harmonic oscillator, the classical frequency is given b

$$\omega = \sqrt{\frac{k}{m}}$$

For conveniens, we measure energy is the units of $\frac{1}{2} \lambda \hbar \omega$

$$\therefore \qquad E = \frac{1}{2} \lambda \hbar \omega$$

where λ is an eigen function of the energy operator.

Let us measure r in the units of $\left(\frac{\hbar}{m\omega}\right)^{1/2}$, so we have Schrödinger equation as

$$\nabla^2\psi + (\lambda - r^2)\psi = 0 \quad ...(3)$$

In the cartesean co-ordinates, this equation is separable with the substitution

$$\psi = X(x)\, Y(y)\, Z(z) \quad ...(4)$$

We can write the results in separate equation in X, Y and Z as

$$\frac{d^2X}{dx^2} + (\lambda_x - x^2)X = 0 \quad ...(5)$$

$$\frac{d^2Y}{dy^2} + (\lambda_y - y^2)Y = 0 \quad ...(6)$$

$$\frac{d^2Z}{dz^2} + (\lambda_z - z^2)Z = 0 \quad ...(7)$$

where constants λ_x, λ_y, and λ_z are related by

$$\lambda_x + \lambda_y + \lambda_z = \lambda \quad ...(8)$$

Each of the equation (5), (6) and (7) is a differential equation for *one dimensional Harmonic oscillator*, the normalized wave function is given by

$$\psi_n = \frac{1}{\sqrt{2^n . n!\sqrt{\pi}}} e^{-y^2/2} H_n(y) \quad ...(9)$$

We can get the integrable solutions of the form (3) if the constants λ_x, λ_y and λ_z are of the form

$$\lambda_x = 2n_x + 1, \qquad \lambda_y = 2n_y + 1, \qquad \lambda_z = 2n_z + 1 \quad ...(10)$$

where n_x, n_y and n_z are positive integers.

The corresponding normalized wave function is given by,

$$n_x, n_y, n_z = \frac{1}{2^n n_x! n_y! n_z! \pi^{3/2}} e^{-4^{3/2}} Hn_x(x) Hn_y(y) Hn_z(z) \quad ...(11)$$

where

$$n = n_x + n_y + n_z \quad ...(12)$$

So, from equation (8), the total energy is

$$E_n = (n_x + n_y + n_z + 3/2)\,\hbar\,\omega$$

$$= \left(n + \frac{3}{2}\right)\hbar\,\omega \quad ...(13)$$

which is only dependent on integer n.

For a given value of n, the wave function (11) correspond to a set of linearly independent function, each of which will corresponds to one set of the integers n_x, n_y, n_z that satisfy the condition (12).

So the energy states of E_n are *degenerate* and the degree of degeneracy is $\frac{1}{2}(n+1)(n+2)$; which corresponds to the number of ways in which equation (12) can be statisfied by positive integers (non-negative integers) n_x, n_y and n_z.

The same results can be attained for the three dimensional harmonic oscillator if *spherical harmonic* are used. The separation of the various components is performance in the same way as above.

Example 2. *An electron is put in a cubical box of each side 1 Å. Find the value of its momentum and energy for the ground state and the first excited state.*

Solution: For the ground state $n_x = n_y = n_z = 1$

$$\therefore \quad \text{Momentum } p_{1,1,1} = \frac{\pi\hbar}{l}\sqrt{3} = \frac{\pi h}{2\pi l}\sqrt{3}$$

$$\frac{h}{2l}\sqrt{3} = \frac{0.63 \times 10^{-27}}{2 \times 10^{-8}}.\sqrt{3}$$

$$= (3.315 \times 10^{-19})\sqrt{3}$$

$$= 5.74 \times 10^{-19} \text{ gm cm/sec}$$

Energy $$E_{1,1,1} = \frac{\pi^2 \hbar^2}{2ml^2}3 = \frac{\pi^2 . h^2}{4\pi^2 ml^2}3 = \frac{h^2}{8ml^2}3$$

$$= \frac{(6.63 \times 10^{-27})^2 \times 3}{8 \times 9.1 \times 10^{-20} \times 10^{-18}}$$

$$= 60.4 \times 3 \times 10^{-12} = 181 \times 10^{-12} \text{ erg}$$

$$= 113 \text{ eV}$$

For the first excited state one of the quantum number is = 2

$\therefore$ Momentum $$p_{2,1,1} = p_{1,2,1} = P_{1,1,2} = \frac{\pi h^2}{i}\sqrt{6}$$

$$= \frac{h}{2l}\sqrt{6} = 3.315 \times 10^{-18}\sqrt{6}$$

$$= 8.13 \times 10^{-19} \text{ gm cm/sec.}$$

Energy $$E_{2,1,1} = E_{1,2,1} = E_{1,1,2} = \frac{\pi^2 \hbar^2}{2ml^2}6 = \frac{h^2}{8ml^2}\sqrt{6}$$

$$= 60.4 \times 6 \times 10^{-12} = \mathbf{362 \times 10^{-12} \text{ erg.}}$$

Example 3. *Find the energy of an electron moving in one dimension in an infinitely high potential box of width 1 Å, given mass of the electron 9.11 $\times 10^{-31}$ kgm and h = 6.63 $\times$ 10^{-34} Js.*

Solution: We have the relation

$$E = \frac{n^2 h^2}{8ma^2} \quad (n = 1, 2, 3 \ldots)$$

The least energy of the particle can be obtained by putting $n = 1$. The energy is given by

$$E = \frac{h^2}{8ma^2}$$

In this case $m = 9.11 \times 10^{-31}$ kgm, $a = 1$ Å $= 10^{-10}$ m and

$$h = 6.63 \times 10^{-34} \text{ Js.}$$

$$E = \frac{(6.63 \times 10^{-34})^2}{8 \times 9.11 \times 10^{-31} \times (10^{-10})^2} \text{Joule}$$

$$= 9.1 \times 10^{-19} \text{ Joule} = \frac{9.1 \times 10^{-19}}{1.602 \times 10^{-19}} \text{ eV} = \mathbf{5.68 \text{ eV}}$$

Example 4. *Think of the nucleus as a box with a size of 10^{-14} m across. Compute the lowest energy of a neutron confined to the nucleus.*

$$(h = 6.26 \times 10^{-34} \text{ Js}, \quad m = 1.6 \times 10^{-24} \text{ g})$$

Solution: Take the nucleus to be a cubical box of size 10^{-14} m across.

$\therefore$ $$x = y = z = a = 10^{-14} \text{ m}$$

The neutron will have the lowest energy in the quantum state given by

$$n_x = n_y = n_z = 1$$

Now energy = $E_{1,1,1}$ $$= \frac{\pi^2 \hbar^2}{2m}\left[\frac{n_x^2}{l_x^2} + \frac{n_y^2}{l_y^2} + \frac{n_z^2}{l_z^2}\right]$$

$$= \frac{\pi^2 \hbar^2}{2m}\left[\frac{3}{l^2}\right]$$

$$= \frac{3\pi^2 \hbar^2}{4\pi^2 \times 2\, m\, l^2} = \frac{3\hbar^2}{8\, ml^2}$$

$$= \frac{3 \times [6.62 \times 10^{-34}]^2}{8 \times 1.6 \times 10^{-27} \times 10^{-28}}$$

$$= 10.29 \times 10^{-23}\ \text{J}$$

$$= \mathbf{6.43\ MeV.}$$

Example 5. *Show that $\psi(x) = e^{icx}$, where c is some finite constant is acceptable eigen function. Also normalize it over the region $-a \le x \le a$.*

Solution: The given wave function is

$$\psi(x) = e^{icx} = \cos cx + i \sin cx$$

Its derivative is given by

$$\frac{d\psi(x)}{dx} = ice^{icx}$$

$$= -c \sin cx + ic \cos cx$$

(*i*) Since sin cx and cos cx are periodic functions with maximum value 1 and c is finite constant, both $\psi(x)$ and $\frac{d}{dx}\psi(x)$ are finite for all values of x.

(*ii*) The function $\psi(x)$ and $\frac{d}{dx}\psi(x)$ are single valued because cos cx and sin cx are single-valued functions for all values of x.

(*iii*) The continuity of sine and cosine function implies that $\psi(x)$ and $\frac{d}{dx}\psi(x)$ are slso continuous for all values of x.

Hence $\psi(x)$ is an acceptable form of the eigen function. For normalization we write

$$\psi(x) = Ne^{icx}$$

and determine N such that

$$\int_{-a}^{a} \psi^*(x)\, \psi(x)\, dx = 1$$

or

$$N^2 \int_{-a}^{a} e^{-icx} . e^{icx} dx = 1$$

or

$$N^2 |\, x\, |_{-a}^{+a} = 1$$

$$N_2\, (2a) = 1$$

or $N = \frac{1}{\sqrt{2a}}$, Hence the normalized wave function is

$$\psi(x) = \frac{x}{\sqrt{2a}} e^{icx}$$

Example 6. *An electron is bound by a potential which closely approaches an infinite square well of width 2.5×10^{-10} m. Calculate the lowest three permissible quantum energies the electron can have.*

Solution: In case of infinite potential well

$$E_n = \frac{n^2 h^2}{8\, m\, a^2}$$

Here $h = 6.63 \times 10^{-34}$ Js, m $= 9.1 \times 10^{-31}$ kg

and $a = 2.5 \times 10^{-10}$ m.

Now,

$$E_1 = \frac{h^2}{8\, m\, a^2} = \frac{(6.63 \times 10^{-34})^2}{8 \times 9.1 \times 10^{-31} \times (2.5 \times 10^{-10})^2}$$

$$= 9.63 \times 10^{-19} \text{ Joules} = \frac{9.63 \times 10^{-19}}{1.6 \times 10^{-19}} = \mathbf{6\ eV.}$$

Also $$E_2 = n^2\ E_1 = 4 \times 6 = \mathbf{24\ eV},$$

and $$E_3 = 9\ E_1 = 9 \times 6 = \mathbf{54\ eV.}$$

Example 7. *A particle is moving in one dimensional potential box (of infinite height) of width 25 Å. Calculate the probability of finding the particle within an interval of 5 Å at the centres of the box when it is in its state of least energy.*

Solution: We know that the wave functions of a particle enclosed with an infinite potential well is given by

$$\psi(x) = \sqrt{\frac{2}{a}} \sin \frac{n\pi x}{a}$$

When the particle is in the least energy state $n = 1$, hence in this case

$$\psi(x) = \sqrt{\frac{2}{a}} \sin \frac{\pi x}{a}$$

At the centre of the box $x = a/2$. The probability of finding the particle in the unit interval at the centre of box is given by

$$|\psi(x)|^2 = \left[\sqrt{\frac{2}{a}} \sin \frac{\pi(a/2)}{a}\right]^2 = \frac{2}{a} \sin^2 \frac{\pi}{2} = \frac{2}{a}$$

The probability P in the interval Δx is given by

$$P = |\psi(x)|^2\ \Delta x = \frac{2}{a}\ \Delta x$$

In this case $a = 25$ Å $= 25 \times 10^{-10}$ m and $\Delta x = 5$ Å $= 5 \times 10^{-10}$ m

$$\therefore \qquad P = \frac{2 \times 5 \times 10^{-10}}{25 \times 10^{-10}} = \mathbf{0.4}$$

Example 8. *A proton or a neutron in a nucleus can roughly be regarded as a particle in a box, the surface of the nucleus plays the role of the wall of the box and proton moves more or less freely between these walls. What is the energy released when a proton makes a transition from the first excited state to the ground state of a box of nuclear size i.e., 1.0×10^{-14} m.*

Solution: We have $a = 1 \times 10^{-14}$ m

and $$m = m_p = 1.67 \times 10^{-27} \text{ kg.}$$

The energy of ground state is got from relation

$$E_n = \frac{n^2\pi^2\hbar^2}{2m_p a^2};$$

$$E_1 = 1.\frac{\pi^2\hbar^2}{2m_p a^2} = \frac{\pi^2 \times (1.05 \times 10^{-34}\ \text{J}^{-s})^2}{2 \times 1.67 \times 10^{-27}\ \text{kg} \times (1.0 \times 10^{-14}\ \text{m})^2}$$

$$= 3.3 \times 10^{-13} \text{ J}$$

Energy of first excited state will be

$$E_2 = 2^2 . \frac{\pi^2\hbar^2}{2m_p a^2} = 4 \times 3.3 \times 10^{-13} \text{ J}$$

$$= 13.2 \times 10^{-13} \text{ J}$$

$$E_2 - E_1 = 13.2 \times 10^{-13} \text{ J} - 3.3 \times 10^{-13} \text{ J}$$

$$= 9.9 \times 10^{-13} \text{ J} = \frac{9.9 \times 10^{-13}}{1.6 \times 10^{-23}}$$

$$= \mathbf{6.2\ MeV.}$$

11.19. ENERGY LEVELS OF A PATICLE IN ONE DIMENSIONAL POTENTIAL WELL OF INFINITE HEIGHT

Consider a particle of mass m moving along x-direction between rigid walls w_1 and w_2 at positions $x = 0$ and $x = a$ (Fig. 11.13). The particle moves freely between the walls. *As no force is acting on the particle so its potential is constant between the walls.* For simplicity this constant potential may be taken as zero.

When the particle strikes any one wall, it gets reflected back since we consider the wall to be rigid, force acting on the particle abruptly changes from zero to some finite value, say F. also, since $|F| = \frac{\partial V}{\partial x}$, we may have $\Delta V \to \infty$ when $\Delta x \to 0$ such that $\frac{\partial V}{\partial x}$, has some finite value $|F|$. It means that potential energy of the particle becomes ∞ at the walls. The above statement may be written in mathematical form as

$$V(x) = \infty \text{ for } x < 0 \text{ and } x > 0$$

and $$V(x) = 0 \text{ for } 0 < x < a$$

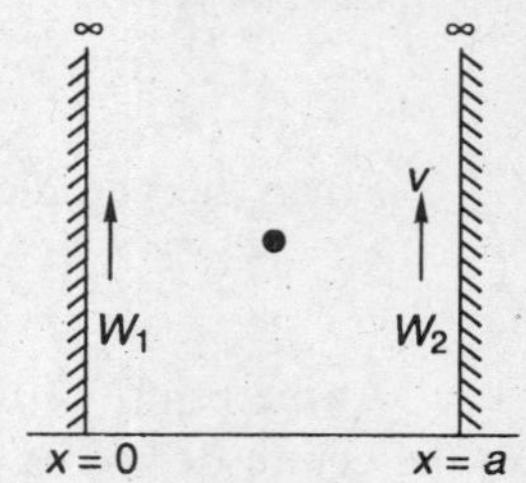

Fig. 11.13

Now Schrodinger wave equation for a particle has the form

$$\frac{d^2\psi}{dx^2} + \frac{8\pi^2 m}{h^2}(E - V)\psi = 0 \quad ...(1)$$

As $V = 0$ between the walls, so the equation may take the form

$$\frac{d^2\psi}{dx^2} + \frac{8\pi^2 m}{h^2}E\psi = 0 \quad ...(2)$$

Put $\frac{8\pi^2 m}{h^2}E = k^2$, so equation (2) may be written as

$$\frac{d^2\psi}{dx^2} + k^2\psi = 0 \quad ...(3)$$

the general solution of equation (3) has the form

$$\psi(x) = A\sin kx + B\cos kx \quad ...(4)$$

where A and B are constants whose values can be found from boundary conditions.

Since the particle is enclosed between two rigid walls and ψ^2, by definition represents *probability of finding particle at any instant between the walls.* Also, the particle can not penetrate the walls, so

$$\left.\begin{aligned}\psi = 0 \text{ at } x = 0\\ \text{and} \quad \psi = 0 \text{ at } x = a\end{aligned}\right]$$

relations given by equation (5) represent the boundary conditions.

Applying the said conditions, in equation (4), we get

$$0 = A\sin 0 + B\cos 0$$

$$\therefore \quad B = 0 \quad [\because \sin 0 = 0 \text{ and } \cos 0 = 1]$$

Putting in equation (4), we get

$$\psi(x) = A\sin kx$$

Again applying second boundary condition, we get $0 = A \sin ka$, $[\because \psi = 0, \text{ when } x = a]$ which means, either $A = 0$ or $\sin ka = 0$

Now $A \neq 0$ because in that case since $B = 0$; entire function $\psi(x)$ will be zero (from Equation (4))

$$\therefore \quad \sin ka = 0 \text{ or } ka = n\pi$$

or $$k = n\pi / a \quad ...(6)$$

So the wave function becomes

$$\psi(x) = A\sin\frac{n\pi x}{a} \quad ...(7)$$

also we have taken $k^2 = \frac{8\pi^2 m}{h^2}E$ in equation (2)

$$\therefore \quad \frac{n^2\pi^2}{a^2} = \frac{8\pi^2 m}{h^2}E_n$$

Let us denote E by E_n for general case, then we have

$$\frac{n^2\pi^2}{a^2} = \frac{8\pi^2 m}{h^2}E_n \quad \text{...(8)}$$

or

$$E_n = \frac{n^2h^2}{8ma^2} \quad \text{(where } n = 1, 2, 3, \text{...)}$$

From Equation (8) it is clear that in an infinite potential well, the particle can have only discrete energy values *i.e., particle energy is quantized.*

We can get the constant of Equation (7) from normalization condition *i.e.,*

$$\int_0^a |\psi(x)|^2 = 1$$

$\therefore$

$$\int_0^a A^2 \sin^2 \frac{n\pi x}{a} dx = 1$$

or

$$A^2 \int_0^a \sin^2 \frac{n\pi x}{a} dx = 1$$

or

$$A^2 \int_0^a \frac{1}{2}\left[1 - \cos\frac{2n\pi x}{a}\right] dx = 1$$

or

$$\frac{A^2}{2}\left[x - \frac{a}{2\pi n}\sin\frac{2n\pi x}{a}\right]_0^a = 1$$

or

$$\frac{A^2}{2} a = 1 \qquad \therefore \qquad A^2 = 2/a \text{ or } A = \sqrt{2/a}$$

So wave function may be written as

$$\psi(x) = \sqrt{2/a}\,\sin\frac{n\pi x}{a}$$

Equation (7) gives the wave function of the particle enclosed in infinitely deep potential well.

Example 9. *A particle is moving in one dimensional box and its wave function is given by*

$$\psi_n = A \sin \frac{n\pi x}{a}$$

Find the expression for the normalized wave function.

Solution: The probability that the particle is between x and $x + dx$ for the nth state is given by

$$P_n dx = |\psi_n(x)|^2 dx$$

$$= A^2 \sin^2 \frac{n\pi x}{a} dx$$

Since the particle has to be in the region $x = 0$ and $x = a$ for all times, so we have

$$\int P_n dx = \int_0^a |\psi_n(x)|^2 dx = 1$$

$$= \int_0^a A^2 \sin^2 \frac{n\pi x}{a} dx = 1$$

or

$$A^2 \int_0^a \frac{1}{2}\left[1 - \cos\frac{2\pi n x}{a}\right] dx = 1$$

or

$$\frac{A^2}{2}\left[x - \frac{a}{2\pi n x}\sin\frac{2\pi n x}{a}\right]_0^a = 1$$

Because the second term in the brackets is zero for both the limits.

So we have

$$A = \sqrt{2/a}$$

$\therefore$ Normalized wave function is

$$\psi n(x) = \sqrt{\frac{2}{a}} \sin \frac{n \pi x}{a}$$

Example 10. A proton is confined to a nucleus of radius 5.0×10^{-15}. Calculate the minimum uncertainity in its momentum. Also calculate the minimum kinetic energy of the proton. Hence, show that the presence of protons in the nucleus is permitted (Mass of proton = 1.675×10^{-27}).

Solution: Diameter of the nucleus,

$$\Delta x = 2 \times 5 \times 10^{-15} = 10^{-14} \text{ m}$$

(*i*) From uncertainty principle, we have

$$\Delta p \cdot \Delta x = \frac{h}{2\pi}$$

$$\therefore \quad \Delta p = \frac{h}{2\pi \Delta x} = \frac{6.626 \times 10^{-34} \text{ J.s}}{2\pi \times 10^{-14} \text{ m}} = 1.055 \times 10^{-20}$$

$$\Delta p = 1.055 \times 10^{-20} \text{ kg ms}^{-1}$$

Which means the minimum value of momentum $p = 1.055 \times 10^{-20}$ kg m s^{-1}. **Ans.**

(*ii*) Corresponding minimum K.E. is given by

$$(\text{K.E.})_{\min} = \frac{p^2}{2m}$$

$$= \frac{(1.055 \times 10^{-20})^2}{2 \times 1.67 \times 10^{-27}}$$

$$= 0.332 \times 10^{-13} \text{ J}$$

$$= \frac{0.332 \times 10^{-13}}{1.6 \times 10^{-19}} \text{eV}$$

$$= \mathbf{0.23 \text{ MeV.}} \textbf{ Ans}$$

(*iii*) Now rest mass Energy of proton is m_0c^2,

i.e.,
$$m_0c^2 = \frac{1.675 \times 10^{-27} \times 9 \times 10^{16}}{1.6 \times 10^{-19} \times 10^{6}} \text{MeV} = \mathbf{942 \text{ MeV}}$$

which is very large as compared to K.E.

So, the presence of protons with such kinetic energies is quite plausible •"and hence its presence in the nucleus is permitted.

Example 11. *An electron has a de Broglie wavelength equal to that of a photon. Show that the ratio of the K.E. of the electron to the energy of the photon* $= \dfrac{(m^2c^4 + h^2\nu^2)^{1/2} - mc^2}{h\nu}$

What is the limiting value of this ratio for low energy photons? If the photon has energy of 100 keV, calculate the K.E. of the electron.

Solution: (*i*) We have the de Broglie relation

$$\lambda = \frac{h}{mv}$$

or
$$v = \frac{h}{m\lambda} = \frac{h\nu}{m \cdot c}$$

Now,
$$\text{K.E.} = \frac{1}{2}mv^2 = \frac{1}{2}m \cdot \left[\frac{h\nu}{mc}\right]^2$$

$$= \frac{h^2\nu^2}{2mc^2}$$

which may be written in the form

$$E_e = mc^2 + \frac{h^2\nu^2}{2mc^2} - mc^2$$

$$= mc^2\left[1+\frac{h^2v^2}{2m^2c^4}\right] - mc^2$$

$$= mc^2\left[1+\frac{h^2v^2}{m^2c^4}\right]^{1/2} - mc^2 \qquad \text{[From Binomial Theorem]}$$

$$= mc^2\left[\frac{m^2c^4+h^2v^2}{m^2c^4}\right]^{\frac{1}{2}} - mc^2$$

$$= [m^2c^4 + h^2v^2]^{1/2} - mc^2$$

In case of photon, $E_p = hv$

$$\therefore \quad \frac{E_e}{E_p} = \frac{(m^2c^4+h^2v^2)^{1/2} - mc^2}{hv} \textbf{ Ans.}$$

(*ii*) For limiting case, $v = c$

i.e., $mc = \frac{h}{\lambda} = \frac{hv}{c}$

which means, $p = mc = \frac{hv}{c}$

So, in this case, K.E. of electron is

$$E_e = \frac{1}{2}mc^2 = \frac{1}{2m}(mc)^2 = \frac{p^2}{2m}$$

$$= \frac{1}{2m}\cdot\left(\frac{hv}{c}\right)^2$$

i.e., $$E_e = \frac{h^2v^2}{2mc^2}$$

also, $E_p = hv$

$$\therefore \quad \frac{E_e}{E_p} = \left[\frac{h^2v^2}{2cm^2}\right]\frac{1}{hv} = \frac{\boldsymbol{hv}}{\boldsymbol{2mc^2}} \textbf{ Ans.}$$

(*iii*) Photon energy is given by

$$hv = 100 \text{ KeV}$$

$$= 100\times10^3\times1.6\times10^{-19}$$

$$= 1.6\times10^{-14}\text{ J}$$

Also K.E. of electron from equation. (1) is given by

$$E_e = \frac{h^2v^2}{2mc^2}$$

$$= \frac{(1.6\times10^{-14})^2}{2\times9.1\times10^{-31}\times9\times10^{16}}$$

$$= 0.0156\times10^{-13}$$

$$= \frac{0.0156\times10^{-13}}{1.6\times10^{-19}}\text{ eV}$$

$$= 9.77\times10^3\text{ eV}$$

$$= \mathbf{9.77\ KeV}\ \textbf{Ans.}$$

Example 12. *Consider a grain dust ($m = 1\ \mu$ gm) confined move between two rigid walls separated by 10^{-4}. It requires 100 sec to cross the gap. Calculate the quantum number that describes this motion.*

Solution: Velocity of motion $(v) = \frac{10^{-4}}{100} = 10^{-6}\text{ ms}^{-1}$

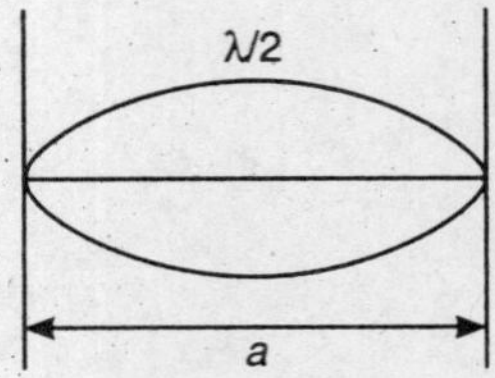

Fig.11.14

$$\text{Kinetic energy } E = \frac{1}{2}mv^2$$

with $m = 1\ \mu$ gm $= 10^{-9}$ kg and $v = 10^{-6}\,\text{ms}^{-1}$,

we have, $$E = \frac{1}{2} \times 10^9 \times (10^{-6})^2$$
$$= 5 \times 10^{22} J$$

Also, the energy of a particle in a rigid box is given by

$$E = \frac{n^2h^2}{8\,ma^2}$$

or $$n^2 = \frac{8\,ma^2E}{h^2}$$

using $m = 10^{-9}$ kg, $a = 10^{-4}$ m,

$$E = 5 \times 10^{-22}\,\text{J},\ h = 6.62 \times 10^{-34}\ \text{J.s}$$

we have, $$n^2 = \frac{8 \times 10^{-9} \times (10^{-4})^2 \times 5 \times 10^{22}}{(6.62 \times 10^{-34})^2}$$
$$= 9 \times 10^{28}$$

$\therefore$ $$n = \mathbf{3 \times 10^{14}}\ \textbf{Ans.}$$

Example 13. *For a free particle show that Schrodinger wave equation leads to the de/Broglie relation* $\lambda = \dfrac{h}{p}$.

Solution: The time-independent solution of Schrodinger wave equation may have the form,

$$E_n = \frac{n^2h^2}{8\,ma^2}$$

$\therefore$ $$E_1 = \frac{h^2}{8\,ma^2}$$

In this case, $$a = \frac{\lambda}{2}$$

$\therefore$ $$E_1 = \frac{4h^2}{8m \cdot \lambda^2}$$
$$= \frac{h^2}{2\,m\lambda^2} \qquad \text{...(1)}$$

Also, $$E_1 = \frac{p^2}{2\,m}$$

$\therefore$ From (1) and (2) ...(2)

$$\frac{h^2}{2\,m\lambda^2} = \frac{p^2}{2\,m}$$

or $$p^2 = \frac{h^2}{\lambda^2}$$

or $$p = \frac{h}{\lambda}$$

i.e., $$\lambda = \frac{\boldsymbol{h}}{\boldsymbol{p}}\ \textbf{Ans.}$$

Example 14. *Calculate the value of the difference between the energy of a hydrogen molecule having back and forth motion on a path of 1 cm long when n = 2 and n = 3 at 298 K.*

Sol tion: Mass of hydrogen mol cule

$$= \frac{2}{6.02\times10^{26}} = 0.33\times10^{-26}\text{ kg}$$

we have the energy of a particle confined in a box given by the relation,

$$E_n = \frac{n^2h^2}{8\,ma^2}$$

$$\therefore \quad E_2 = \frac{4h^2}{8\,ma^2}$$

and $$E_3 = \frac{9h^2}{8\,ma^2}$$

$$\therefore \quad E_3 - E_2 = \frac{h^2}{8\,ma^2}(9-4) = \frac{5h^2}{8\,ma^2}$$

In this case, $m = 0.33\times10^{-}$ kg

$$a = 0.01\text{ m},\ h = 6.626\times10^{-34}\text{ J.s}$$

$$\therefore \quad E_3 - E_2 = \frac{5\times(6.626\times10^{-34})^2}{8\times0.33\times10^{-26}\times(0.01)^2}$$

$$= 83.1\times10^{-38}\text{ J} \quad = \frac{83.1\times10^{-38}}{1.6\times10^{-19}}\text{ J}$$

$$= \mathbf{5.2\times10^{-18}\ eV\ Ans.}$$

Example 15. *A particle is confined to a one-dimensional infinite potential well of width 0.2 nm. It is given that where the energy of the particle is 230 eV, its eigen function has 5 antinodes. Find the mass of the particle and show that it can never have energy equal to 1 keV.*

Solution: Since the eigen function corresponding to energy 230 eV has 5 antinodes, the particle is in a quantum state with $n = 5$. For this case,

$$E_5 = 5^2\,E_1 = 25\,E_1\ 230\,eV$$

so that $$E_1 = 9.2\,eV = 9.2\times1.6\times10^{-19}\text{ J}$$

$$= 14.7\times10^{-19}\text{ J}$$

Also, $$a = 0.2\,nm = 0.2\times10^{-9}\,m = 2\times10^{-10}\,m$$

Now, $$E_1 = \frac{\hbar\pi^2}{2\,ma^2} = \frac{h^2}{8\,ma^2}$$

or $$m = \frac{h^2}{8\,a^2E_1} = \frac{(6.626\times10^{-34})^2}{8\times4\times10^{-20}\times14.7\times10^{-19}}\text{ kg}$$

$$= 9.3\times10^{-31}\text{ kg}$$

For $E_n = 1\,keV$, n should be such that

$$n^2 = \frac{E_n}{E_1} = \frac{10^3}{9.2} = 108.7$$

i.e., $$n = 10.4$$

Since n is not an integer, $E_n = 1\,keV$ is not a permitted value of energy.

Example 16. *Calculate the value of lowest energy of an electron moving in a one-demensional force free region of length 4 Å.*

Solution: We have $m = 9.1\times10^{-31}$ kg, $a = 4\times10^{-10}$ m

Since energy of electron in nth state is given by

$$E_n = \frac{n^2\pi^2\hbar}{2\,ma^2} = \frac{n^2h^2}{8\,ma^2}$$

For ground state $n = 1$

$$E_1 = \frac{h^2}{8\,ma^2} = \frac{(6.625\times10^{-34})}{8\times9.1\times10^{-31}\times(4\times10^{-10})^2}$$

or $$E_1 = 0.376\times10^{-18}\text{ J}$$

Example. 17 *Calculate the value of momentum and energy of an electron in a box of wavelength 1 Å, for n = 2.*

Solution: We have $a = \lambda = 1$ Å $= 10^{-10}$ m, $n = 2$

Since momentum, $$p_n = \frac{n\pi\hbar}{2} = \frac{nh}{2a}$$

For $n = 2$

$$p_2 = \frac{2\times6.625\times10^{-34}}{2\times10^{-10}}\text{ kg m s}^{-1}$$

$\therefore$ $$p_2 = 6.625\times10^{-24}\text{ kg m s}^{-1}\ \textbf{Ans.}$$

And energy, $$E_n = \frac{n^2\pi^2\hbar}{2ma} = \frac{n^2h^2}{8ma^2}$$

For $$n = 2$$

$$E_2 = \frac{4\times(6.625\times10^{-34})^2}{8\times9.1\times10^{-31}\times10^{-20}}$$

or $$E_2 = 150.\text{eV}\ \textbf{Ans.}$$

Example 18. *A particle of mass 10^{-6} kg is moving with a speed of 10^{-4} m s^{-1} in a box of length 10^3 Å. Assuming this to be one dimensional square well problem, calculate the value of n.*

Solution:

We have $m = 10^{-6}$ kg, $v = 10^{-4}$ ms^{-1}, $a = 10^3$ Å $= 10^{-7}$ m

Since energy of a particle is nth state is given by

$$E_n = \frac{n^2\pi^2\hbar^2}{2\,ma^2}$$

$$n^2 = \frac{8\,ma^2\,E_n}{h^2} \qquad \left[\because \hbar = \frac{h}{2\,\pi}\right]$$

or $$n^2 = \frac{4m^2L^2v^2}{h^2} = \frac{4m^2a^2v^2}{h^2} \qquad \left[\because E_n = \frac{1}{2}mv^2\right]$$

or $$n = \frac{2m\,av}{h} = \frac{2\times10^{-6}\times10^{-7}\times10^{-4}}{6.625\times10^{-34}} = \mathbf{3\times10^{16}}\ \textbf{Ans.}$$

Example 19. *Evaluate the expectation value for kinetic energy of a particle in one dimensional rigid box in the nth quantum state.*

Solution: We know that the wave function of a particle enclosed in one dimensional box is given by

$$\psi\,(x) = \sqrt{\frac{2}{a}}\sin\left(\frac{n\pi x}{a}\right)$$

The expectation value of kinetic energy E is given by

$$< E > = \int_{-\infty}^{\infty}\psi * E\,\psi\,dx = \int_{-\infty}^{\infty}\psi * \frac{p^2}{2m}\psi\,dx$$

$$= \frac{1}{2m}\int_{-\infty}^{\infty}\psi * (-i\hbar)^2\frac{\partial^2}{\partial x^2}\psi\,dx$$

$$= \frac{\hbar}{2m}\int_{-\infty}^{\infty}\psi * \frac{\partial^2\psi}{\partial x^2}dx$$

$$= \frac{\hbar}{2m}\int_0^a \frac{2}{a}\sin\left(\frac{n\pi x}{a}\right)\frac{\partial^2}{\partial x^2}\left[\sqrt{\frac{2}{a}}\sin\left(\frac{n\pi x}{a}\right)\right]dx$$

$$= \frac{2\hbar^2}{2ma}\int_0^a \sin\left(\frac{n\pi x}{a}\right)\left(-\frac{n^2\pi^2}{a^2}\right)\sin\frac{n\pi x}{a}\,dx$$

or $$<E> = \frac{n^2\pi^2\hbar^2}{ma^3}\int_0^a \sin^2\frac{n\pi x}{a}\,dx \qquad \text{(–ve sign is omitted)}$$

$$= \frac{n^2\pi^2\hbar^2}{2ma^3}\int_0^a\left(1-\cos\frac{2n\pi x}{a}\right)dx$$

or $$<E> = \frac{n^2\pi^2\hbar^2}{2ma^3}a = \frac{n^2\pi^2\hbar}{2ma^2} = \frac{n^2h^2}{8ma^2} \quad \textbf{Ans.}$$

Example 20. *Find the probability that a particle trapped in a box 'L' wide can be found between 0.45L and 0.55L for the first excited state.*

Solution: The wave function in a box of length 'L' is given by

$$\psi_n(x) = \sqrt{\frac{2}{L}}\sin\frac{n\pi x}{L}$$

∴ Probability may be given for the ground state as

$$P = \int_{x_1}^{x_1}(\psi_n)^2\,dx = \int_{0.45L}^{0.55L}\left[\frac{2}{L}\sin^2\frac{\pi x}{L}\right]dx \quad [\text{for } n=1]$$

$$= \frac{2}{L}\int_{0.45L}^{0.56L}\frac{1}{2}\left(1-\cos\frac{2\pi x}{L}\right)dx$$

$$= \frac{1}{L}\left[x - \frac{L}{2\pi}\sin\frac{2\pi x}{L}\right]_{0.46L}^{0.56L}$$

$$= \frac{1}{L}\left[\left(0.55\,L - \frac{L}{2\pi}\sin 1.10\pi\right) - \left(0.45 - \frac{L}{2\pi}\sin 0.90\pi\right)\right]$$

$$= \left[\left(0.55 - \frac{1}{2\pi}\sin 198°\right) - \left(0.45 - \frac{1}{2\pi}\sin 162°\right)\right]$$

$$= \left[(0.55-0.45) - \frac{1}{2\pi}(\sin 198° - \sin 162°)\right]$$

$$= 0.10 - \frac{1}{\pi}(\cos 180° - \sin 18°)$$

$$\therefore \quad \left[\sin A - \sin B = 2\cos\frac{A+B}{2}.\sin\frac{A-B}{2}\right]$$

$$= 0.10 + \frac{0.3090}{3.142}$$

$$= 0.10 + 0.0984$$

$$= 0.1984$$

$$= 0.1984 \times 100 = 19.84\%$$

Similarly the value of P for Ist excited state (n = 2) will be P = 0.65% Ans.

QUESTIONS AND PROBLEMS

1. What is the significance of a wave function? What do you predict about the energy of a particle in closed box from
(*a*) classical theory (*b*) quantum theory?
2. Differentiate between ψ and $|\psi|^2$.
3. Derive time dependent Schrödinger wave equation.
4. Derive an equation for the energy of a particle in a closed box. Discuss the probability of a particle for $n = 1$ and $n = 2$.
5. What do you understand by eigen value and eigen functions? Solve the Schrödinger wave equation for a particle confined to one dimensional box of side a and obtain its values and eigen functions.
6. Write the Schrödinger equation for the particle in a box and solve it to obtain the Eigen values and Eigen functions.
7. What is energy quantization? How it is possible for a particle in a rigid box?
8. Define zero point energy for a particle in a rigid box and discuss its dependence on mass of the particle and size of the box.
9. What do you mean by a particle in a box? Show that the "energy of electron in the box varied as the square of the natural numbers.
10. Define the term expectation value of an observable. Find the expectation values of
(*i*) position vector r (*ii*) potential energy V
(*iii*) linear momentum P (*iv*) total nergy E.
11. Derive Schrödinger's time independent wave equation.
12. What is a harmonic oscillator? Obtain an expression for energy. Discuss its importance.
13. A particle of some mass number and total energy E moves from a region of zero potential to a region of constant potential V_0. Derive an expression for reflection and transmission coefficients when (*i*) $E < V_0$ (*ii*) $E > V_0$
14. Consider a particle of mass m trapped in a two dimensional box L long and w wide. Starting from Schrödinger equation, show that the permitted eneriges of the particle are given by

$$E = \frac{h^2}{8m}\left(\frac{a^2}{L^2} + \frac{b^2}{w^2}\right)$$

where a and b are positive integers.
15. Calculate the first two energy levels of an electron cnofined to a box 1 A wide. ($h = 6.254 \times 10^{-34}$ J sec, $m_e = 9.1 \times 10^{-31}$ kg). **[Ans.** (*i*) 37.7 eV (*ii*) 156. 8 eV]
16. An important property of eigen function of an operator is that they are orthogonal to one another, *i.e.*,

$$\int \psi_n \psi_m \, dv = 0 \qquad n \neq m$$

Verify this relation for a particle in one dimensional box using the relation $\sin\theta = (e^{2\theta} - e^{-2\theta})/2i$
17. Find the lowest energy of neutron confined to a box 10^{-14} m across (nucleus has size of this magnitude). **[Ans.** 2.1 MeV)
18. Show that the lowest ground state energy level of a particle in one dimensional potential box with rigid walls is in agreement with the uncertainty principle.
19. Find the energies of the six lowest energy levels of a particles in a cubical box.

$$\left[\textbf{Ans. } 3, 6, 9, 11, 12 \text{ and } 14 \times \frac{\pi^2\hbar^2}{2mL^2}\right]$$

20. A spherical dust particle of radius $r = 10^{-5}$ m density $\rho = 10^4$ kg/m^3 and moving with a velocity $v = 10^{-2}$ ms^{-1} encounters a potential step of height equal to twice the kinetic energy of the particle. Estimate the penetration depth d_p of the particle inside the step. **[Ans.** 2.5×10^{-22} m]
21. Calculate the transmission probability for an electron of total energy 2 eV incident upon a reclagular potential barrier of height 4 eV and width 10^{-9} m. **[Ans.** 2×10^{-6}]
22. Calculate the probability of transmission for a proton of energy 1 MeV through a 4 MeV high rectangular potential energy barrier of width 10^{-2} cm. **[Ans.** 0.0015]
23. Find the height of the potential barrier in eV for α-particles emitted from an element (At. No. $Z = 86$). Assume the effective nuclear radius = 10^{-14} m. **[Ans.** 24.16 MeV]

CHAPTER

12

CRYSTAL STRUCTURE

12.1. INTRODUCTION

Elements and their chemical compounds are present in three states *viz.*, solids, liquids and gases. All these materials consist of atoms and molecules. In solids and liquids the distance between neighbouring atoms is of the order of a few Angstroms *i.e.,* they contain 10^{28}—10^{29} atoms/m^3. In case of gases at room temperature under one atmosphere pressure, the density is about 2.7×10^{25} molecules/m^3 which corresponds to an average distance of approximately 30 Å between molecules.

Almost all solids are crystalline and in them *atoms are arranged in a regular repeated pattern* in such a way that the distance between successive atomic layers is of the order of the wavelength of X-rays (1 Å). In some solids this regularity extends upto a few thousand atoms on each side. In cases like common salt or cane sugar, the regularity of atomic arrangements extends upto 10^7 atoms so that individual crystals have dimensions ~ 1 mm.

In crystals the atoms are arranged in periodic fashion i.e., a representative unit is repeated at regular interval along any and all directions in the crystal. *Periodicity of structure is an essential feature of a crystal.* Figure 12.1 depicts three regular arrangements of points in two dimensions. Out of these [Figures 12.1 (*b*) and [(*c*)] lack in periodicity and only (*a*) leads to the conception of a two dimensional crystal. So if a solid possesses regular arrangement of atoms but not periodicity then it cannot be a crystal.

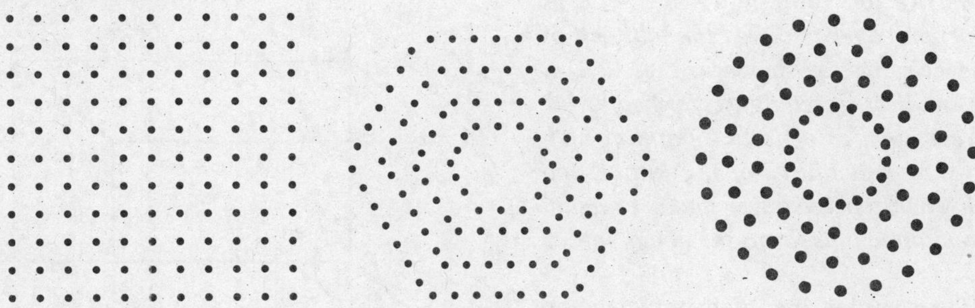

(*a*) (*b*) (*c*)

Fig. 12.1

It should be noted that not all solids are crystals. Some solids have no periodic structure at all and are called *amorphous or glassy solids.* These include fused silica, rubber and polymers etc. and correspond to supercooled liquid state. In liquids crystalline structure is generally absent due to random motion of various atoms or molecules. But there are certain *'liquid crystals'* which flow and rise in a capillary tube and are generally organic compounds.

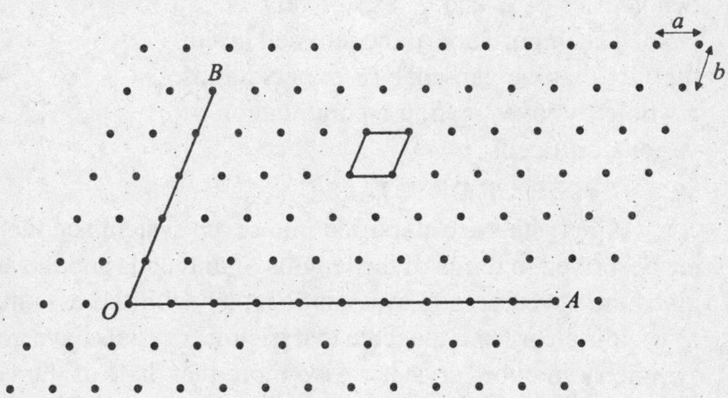

Fig. 12.2.

12.2. SPACE LATTICE

The *periodic arrangement of atoms in a crystal is called a lattice.* Thus *lattice is regular periodic arrangement of points in space that looks a net-like structure. This arrangement of points is such that the environment around any one point is identical with environment around any other point.*

Two dimensional array of points is shown in Fig. 12.2 and is called the *plane lattice.* If we select any point *O* as origin and any two axes *OA* and *OB* then we can see the points along *OA* are seperated by interval '*a*' while those along *OB* by an interval '*b*'. If we extend the array of points to form a three dimensional crystal structure, then the array of points is called the *space lattice* or *crystal lattice.*

12.3. TRANSLATION VECTORS

If we choose any lattice point as origin (say *O*, in Fig. 12.2), then the position vector of any other lattice point is represented by

$$\vec{T} = n_1 \vec{a} + n_2 \vec{b}$$

where n_1 and n_2 are integers representing the number of lattice points along *OA* and *OB* and $\vec{a}$ and $\vec{b}$ are the intergals along *OA* and *OB* (Fig. 12.2) and are called the *primitives.*

For a three dimensional crystal structural, the arrangement of points can similarly be represented by translational (or lattice) vector

$$\vec{T} = n_1 \vec{a} + n_2 \vec{b} + n_3 \vec{c}$$

where $\vec{a}, \vec{b}, \vec{c}$ are fundamental translation vector (or basis vectors) called *primitives* along *X, Y, Z*-axes respectively.

12.4. UNIT CELL

The *parallelopiped formed by using the basis vectors* (or primitives) $\vec{a}, \vec{b}, \vec{c}$ as *concurrent edges is called the unit cell* of the space lattice as shown in Fig. 12.3. *It is the smallest closed figure, formed by the smallest number of lattice points which is fully representative of the lattice and which when repeated over and over again different directions forms the actual lattice itself.*

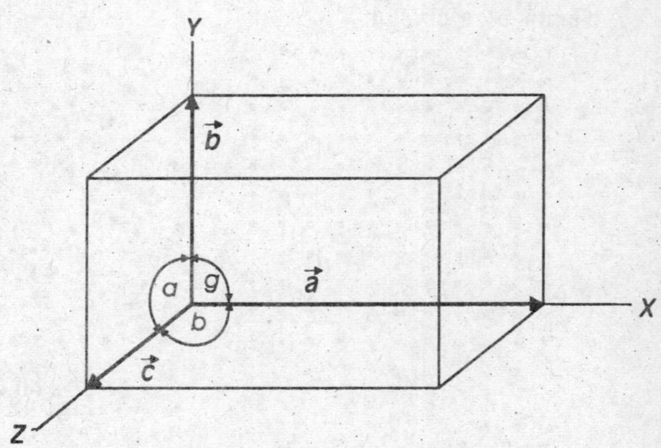

Fig. 12.3

The angles included between the primitives $\vec{b}$ and $\vec{c}$, $\vec{c}$ and $\vec{a}$, $\vec{a}$ and $\vec{b}$ are denoted by α, β and γ respectively.

The importance of the unit cell lies in the fact that we can analyse the crystal as a whole by investigating a combination of various unit cells.

12.5. TYPES OF CRYSTALS

All crystals are classified into seven systems on the basis of the shape of the unit cells. These are described in terms of the lengths of unit cells and the angle included between them. Table below gives the seven crystal systems together with the essential symmetry elements.

It is clear from the table that triclinic crystals have got the least symmetry while there is largest symmetry in cubic crystals. Also more than half of the elements crystallize into the cubic system which is the simplest form of the crystal structure.

12.6. BRAVAIS SPACE LATTICES

Bravais, in 1948, showed that there are fourteen different types of lattices under the seven crystal systems. These are frequently called the *Bravais lattices.*

Table 12.1

S. No.	*Name of system*	*Relation between*		*Number of*	*Examples*
		primitive	*angles*	*possible lattices*	
1.	Cubic	$a = b = c$	$\alpha = \beta = \gamma = 90°$	3	NaCl
2.	Monoclinic	$a \neq b \neq c$	$\alpha = \beta = 90°$ $\gamma = 90°$	2	Na_2SO_4 or $FeSO_4$
3.	Triclinic	$a \neq b \neq c$	$\alpha \neq \beta \neq \gamma \neq 90°$	1	$CuSO_4$
4.	Tetragonal	$a = b \neq c$	$\alpha = \beta = \gamma = 90°$	2	$NiSO_4$
5.	Orthorhombic	$a \neq b \neq c$	$\alpha = \beta = \gamma = 90°$	4	KNO_3
6.	Rhombohedral	$a = b = c$	$\alpha = \beta = \gamma \neq 90°$	1	$CaSO_4$
7.	Hexagonal	$a = b \neq c$	$\alpha = \beta = 90°$ $\gamma = 120°$	1	Quartz
				Total : 14	

For a cubic system, we have the following three types of lattices shown in (Fig. 12.4).

***(i)* Simple cubic or cubic *P* type lattice.** It contains lattice points only at the eight corners of the unit cell. There is no lattice points inside the unit cell [Fig. 12.4 (*a*)].

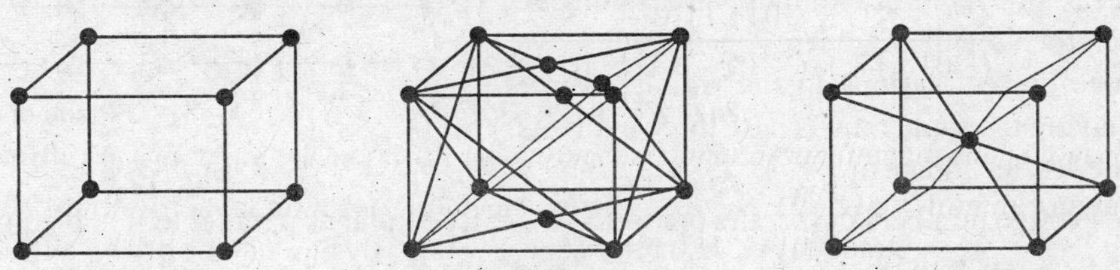

E Cubic (*a*) Face centered (*b*) Body centered (*c*)

Fig. 12.4

***(ii)* Face centered cubic (fcc) or cubic F lattice.** It containts lattice points situated at the centre of each face in addition to those situated at the corners. Thus there is an extra lattice point at the centre of each face [Fig. 12.4 (*b*)].

***(iii)* Body centered cubic (bcc) or cubic I lattice.** It contains one additional lattice point situated at the body centre of the unit cell besides the eight situated at the corners. Thus there is a lattice centre of the unit cell [Fig. 12.4 (*c*)].

12.7. BASIS AND THE CRYSTAL STRUCTURE

In the simple crystals *i.e.,* in elements, there is a single atom (or ion) at each lattice point. In others (in case of compounds), there is a group of two or more atoms (or ions) at each lattice point. This group of atoms is called a *basis. It is an assembly of one or more atoms which when attached identically to each lattice point gives rise to crystal structure.* Every basis is identical in composition, arrangement and orientation. In crystals like aluminium and barium the basis is a single atom, in NaCl, KCl, ZnS etc., the basis is diatomic, whereas in crystals like BaF_2 and CaF_2, the basis is triatomic. Basis is repeated in space to form the crystal. When a basis is added to each lattice point in space, crystal structure is obtained. The logical relation is

Lattice + Basis = Crystal structure

Formation of a crystal structure by the addition of a basis to every point of lattice is shown in Fig. 12.5.

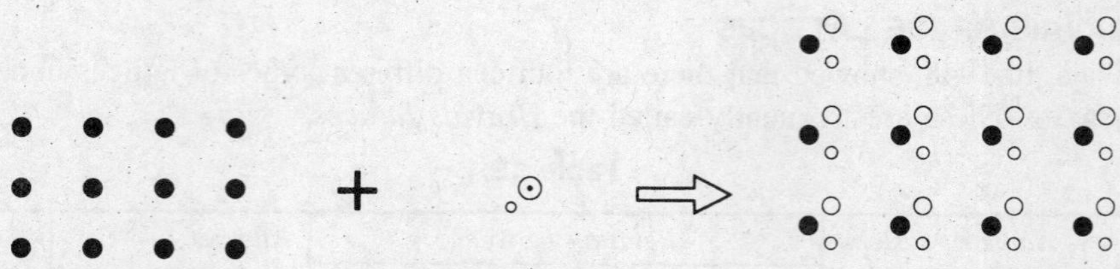

(a) Space Lattice (b) Basis Containing Two different ions (c) Crystal Structure

Fig. 12.5

12.8. MILLER INDICES

We want to establish rules for determining directions, position and planes in a crystal, *i.e.,* for determining a coordinate system for the crystal. For this we will use the lattice points and the crystal axes rather than to impose an arbitrary coordinate system on the lattice. Let us consider a plane running through the crystal and we desire to specify its orientation.

Rules to find the Miller Indices of a plane.

(*i*) Find the intercepts of the plane on the crystal axes expressed as integral multiple of the lattice constant. We will have a set of three numbers specifying the plane.

(*ii*) Take the reciprocal of these three numbers. These reciprocals will in general be fractions. We now reduce these fractions to the smallest three integers having the same ratio. These three numbers will be the Miller indices ($h\ k\ l$).

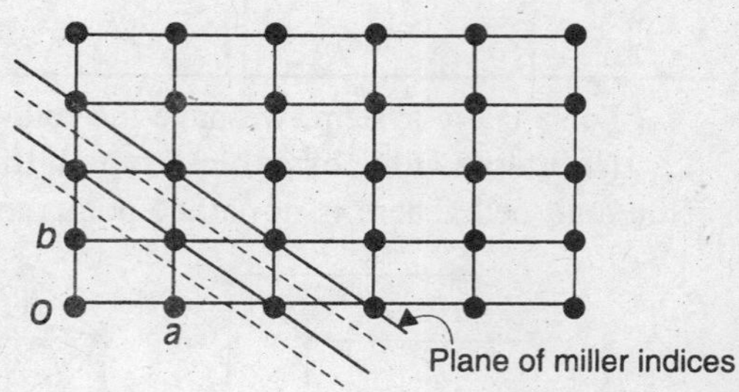

Fig. 12.6

These three numbers define the plane and a family of planes parallel to it and equally spaced. The plane intersects the unit cell at $\frac{\vec{a}}{h}, \frac{\vec{b}}{k}$ and $\frac{\vec{c}}{l}$ and the separation between planes of the family is given by the distance from the origin of this unit cell to the plane or

$$d = \frac{1}{\sqrt{\frac{h^2}{a^2} + \frac{k^2}{b^2} + \frac{l^2}{c^2}}}$$

The + $\vec{a}$ axis has indices [$\bar{1}$0 0]; the $-a$ axis [1,0, 0]. The $+b$ axis has indices [0 1 0]; the $-b$ axis [010]. The diagonal of $a - b$ face has indices [1 1 0]; that $a - c$ face [1 0 1].

Several features should be noted about this scheme in the cubic system:

1. Parallel planes equally spaced, all have the same index number [$h\ k\ l$] analogous to the index notations for directions, where all parallel rows of atoms have the same indices [$u\ v\ w$].

2. A plane parallel to one co-ordinate axis has index number 0 for that direction.

3. A plane passing through the origin is defined in terms of a parallel plane having non-zero intercepts.

Examples: (*a*) The cube face normal to *a*-axis is described by [1 0 0]. Plane of this type is shown in Fig. 12.7.

(*b*) Plane normal to *b*-axis are designated by [0 1 0] and is shown in Fig. 12.8.

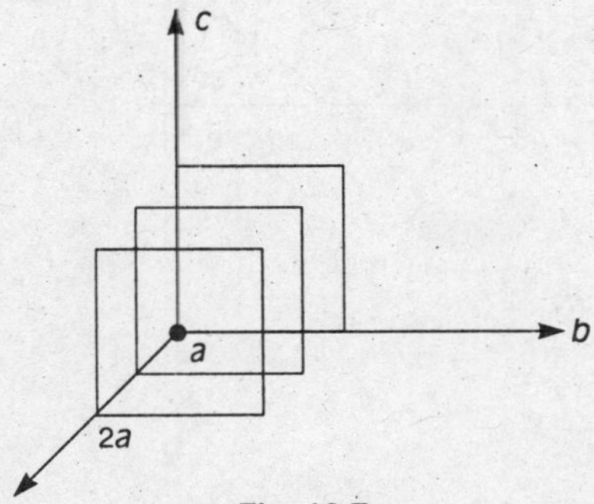

Fig. 12.7

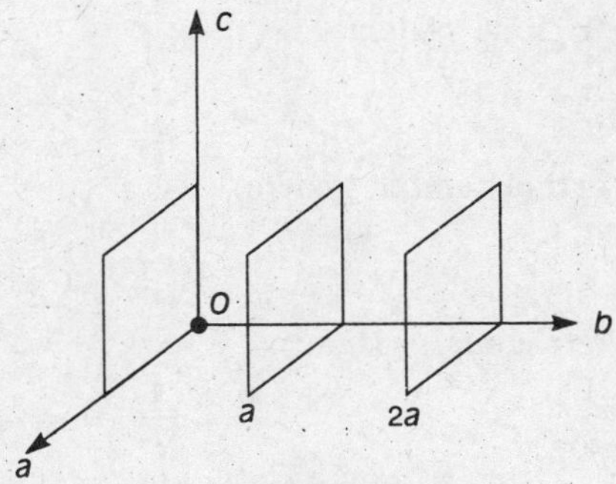

Fig. 12.8

(c) Planes having equal intercepts on the three axes are [1 1 1] planes as shown in Fig. 12.9.

(4) Planes with equal intercepts on a and b-axes and parallel to c-axes are designated by [1 1 0] is shown in Fig. 12.10.

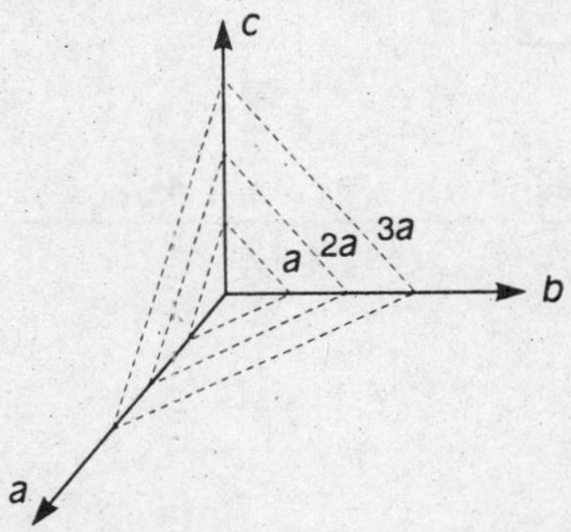

Fig. 12.9

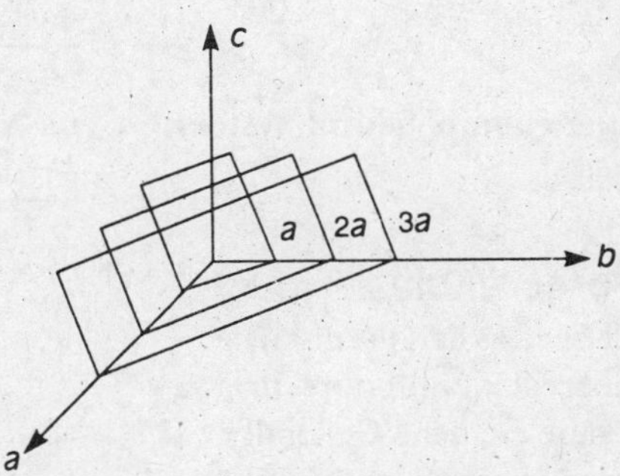

Fig. 12.10

(5) Planes which have one or two negative intercepts, it is designated by [1 0 0] planes as shown in Fig. 12.11.

(6) Finally planes (1 1 2) are designated in Fig. 12.12.

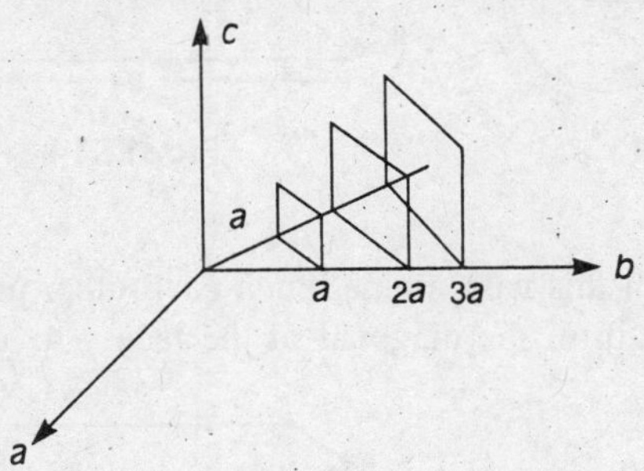

Fig. 12.11

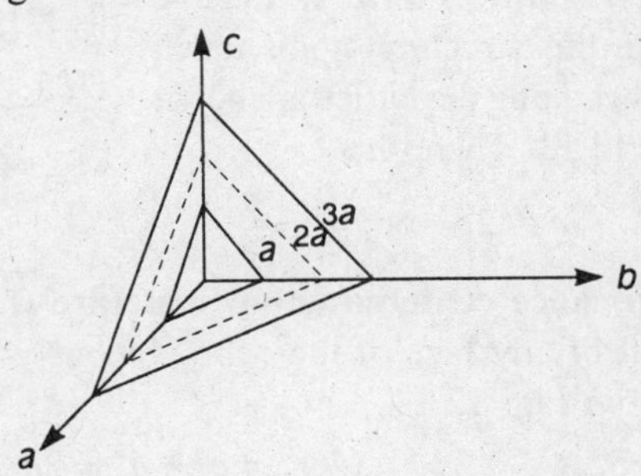

Fig. 12.12

12.9. IMPORTANT FEATURES OF MILLER INDICES

The Miller index notation is especially useful for cubic systems. Among its desirable features are these:

1. The angle θ between two crystallographic directions $[u_1\ v_1\ w_1]$ and $[u_2\ v_2\ w_2]$ can be calculated easily. It is given by the relation:

$$\cos\theta = \frac{u_1u_2 + v_1v_2 + w_1w_2}{(u_1^2 + v_1^2 + w_1^2)^{1/2}.(u_2^2 + v_2^2 + w_2^2)^{1/2}}$$

2. The normal to the plane with index numbers $(h\,k\,l)$ is the direction $(h\,k\,l)$. The correspondence is of an tremendous aid in calculating the angle between planes.
3. The distance d between adjacent planes of index numbers $(h\ k\ l)$ can be obtained for the seven crystal system in terms of the indices of the plane and lattice constans $a, b, c; \alpha, \beta, \gamma$.

For triclinic system:

$$\frac{1}{d^2} = \frac{h^2}{a^2 \sin^2 \beta} + \frac{k^2}{b^2} + \frac{l^2}{c^2 \sin^2 \beta} - \frac{2hl \cos \beta}{ac \sin^2 \beta}$$

For orthorhombic system:

$$\frac{1}{d^2} = \left(\frac{h}{a}\right)^2 + \left(\frac{k}{b}\right)^2 + \left(\frac{l}{c}\right)^2$$

For tetragonal system:

$$\frac{1}{d^2} = \frac{h^2 + k^2}{a^2} + \frac{l^2}{c^2}$$

For cubic system:

$$\frac{1}{d^2} = \frac{h^2 + k^2 + l^2}{a^2}$$

For the hexagonal system:

$$\frac{1}{d^2} = \frac{4}{3}\frac{h^2 + hk^2 + h^2}{a^2} + \left(\frac{l}{c}\right)^2$$

For the rhombohedral system:

$$\frac{1}{d^2} = \frac{(h^2 + k^2 + l^2)\sin^2 \alpha + 2(hk + kl + hl)\cos^2 \alpha - \cos \alpha}{a^2(1 - 3\cos^2 \alpha + 2\cos^3 \alpha)}$$

12.10. ATOMIC RADIUS

It is defined as half the distance between nearest neighbours in a crystal of a pure element. Generally, it is expressed in terms of the cube edge element a. It is denoted by r. Atomic radii can be calculated as follows in various crystal structures.

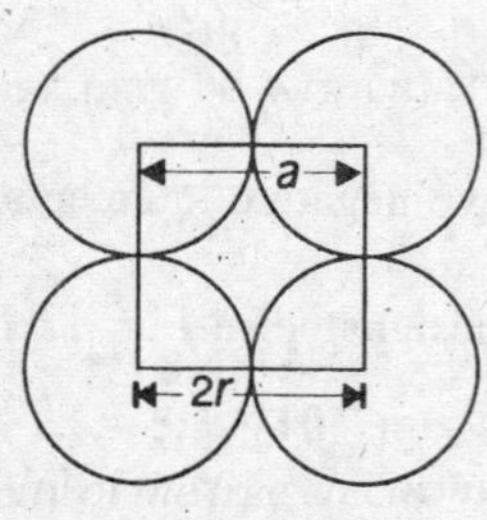

Fig. 12.13

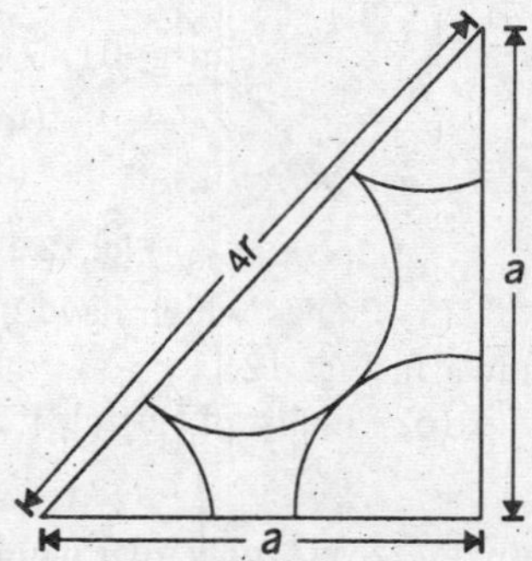

Fig. 12.14

(*i*) Simple cubic structure (s.c.): In this structure atoms touch each other along the lattice as shown in Fig. 12.13. Therefore

$$a = 2r \quad \text{or} \quad r = \frac{a}{2}$$

(*ii*) Face centered cubic structure (f.c.c): In this case atoms touch each other along the diagonal of any face of the cube obviously the length of the diagonal of the face = $4r$ and it is clear from Fig. 12.14.

$$(4r)^2 = a^2 + a^2 = 2a^2$$

$$r^2 = \frac{2a^2}{16} \qquad \therefore r = \frac{\sqrt{2}a}{4}$$

(*iii*) Body centred cubic structure (b.c.c.): In this case atoms touch each other along the diagonal of the cube as shown in Fig. 12.15.

Therefore, the diagonal of the cube in this case is $4r$. By solid geometry we know that for a cube

$$(\text{diagonal})^2 = 3\ (\text{edge})^2$$

$$(4r)^2 = 3a^2$$

$$r^2 = \frac{3a^2}{16} \qquad \text{or} \quad r = \frac{\sqrt{3}a}{4}$$

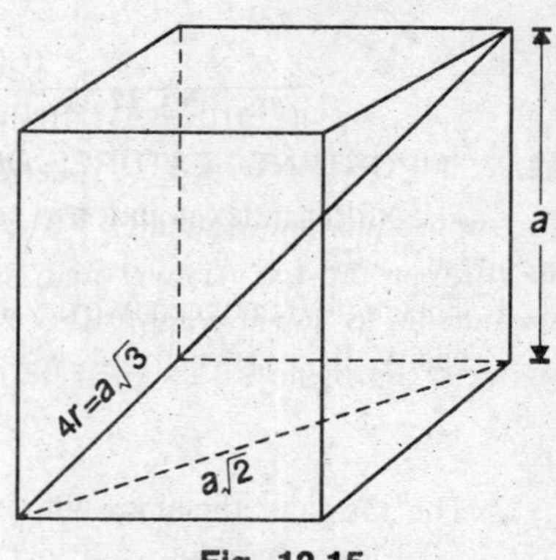

Fig. 12.15

Example 1. *In an ortho-rhombic crystal a lattice plane cuts intercepts of lengths 3a, – 2b and 3c/2 along three axes. Deduce the Miller indices of the plane, where a, b, c are primitive vectors of the units cell.*

Solution: For the given crystal, we have

$$pa : qb : rc = (3a : -2b) : 3c/2$$

or
$$p : q : c = 3 : (-2) : \frac{3}{2}$$

or
$$\frac{1}{p} : \frac{1}{q} : \frac{1}{r} = \frac{1}{3} : \left(-\frac{1}{2}\right) : \frac{2}{3}$$

Multiplying by 6, the L.C.M. of denominator, we get

$$h : k : l = 2 : (-3) : 4$$

Hence Miller indices of the plane are $(2, \bar{3}, 4)$.

Example 2. *In a crystal, lattice plane cuts intercepts of 2a, 3b and 3c along the three axes, where a, b and c are primitive vectors of the unit cell. Determine the Miller indices of the given plane.*

Solution: If the intercepts of the plane in the three crystal axes are *pa, qb* and *rc* respectively, then

$$pa : qb : rc = 2a : 3b : 6c$$

∴
$$p : q : r = 2 : 3 : 6$$

or
$$h : k : l = \frac{1}{2} : \frac{1}{3} : \frac{1}{6} = 3 : 2 : 1$$

Hence the Miller indices of the given plane are **(321)**.

Example 3. *In a crystal, a lattice plane cuts intercepts of a, 2b and 3c along three axes, where a, b and c are primitive vectors of the unit cell. Determine the Miller indices of the given plane.*

Solution: If the given plane cuts intercepts *pa, qb* and *rc* along the three axes, then

$$pa : qb : rc = a : 2b : 3c$$

or
$$p : q : r = 1 : 2 : 3$$

∴
$$\frac{1}{p} : \frac{1}{q} : \frac{1}{r} = 1 : \frac{1}{2} : \frac{1}{3}$$

Multiplying by the L.C.M. 6

$$h : k : l = 6 : 3 : 2$$

∴ Miller indices of the plane are **(632).**

Example 4. *Copper has an f.c.c. structure with lattice constant a = 3.61 Å. Calculate the radius of copper-atom.*

Solution: For a centred cube (f.c.c.) structure, the radius *r*, of the atom is given by

$$r = \frac{\sqrt{2}\,a}{4}$$

But
$$a = 3.61$$

∴
$$r = \frac{\sqrt{2} \times 3.61}{4} = \mathbf{12.8\ Å.}$$

Example 5. *Show that for b.c.c. and f.c.c. crystal structures the lattice constants are given by*

$$a_{bcc} = \frac{4r}{\sqrt{3}} \text{ and } a_{fcc} = \frac{4r}{\sqrt{2}}$$

where r is the atomic radius.

Solution: In case of b.c.c lattice

$$(4r)^2 = a^2 + a^2 + a^2$$

∴
$$r^2 = \frac{3a^2}{16}$$

∴
$$r = \frac{\sqrt{3}a}{4} \quad \text{or} \quad a = \frac{4r}{\sqrt{3}}$$

$$\therefore \qquad a_{\text{bcc}} = \frac{4r}{\sqrt{3}}$$

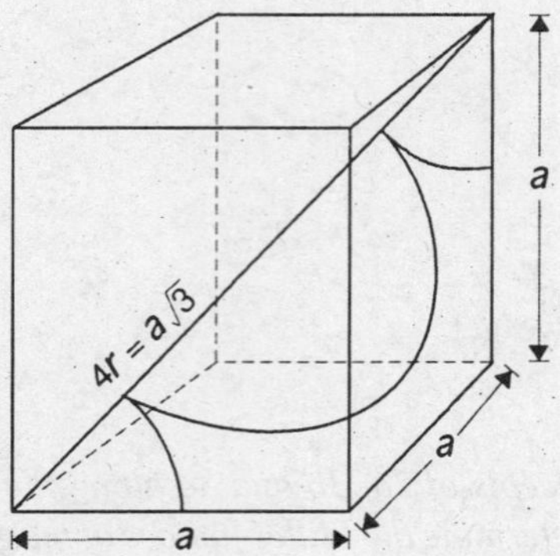

Fig. 12.16

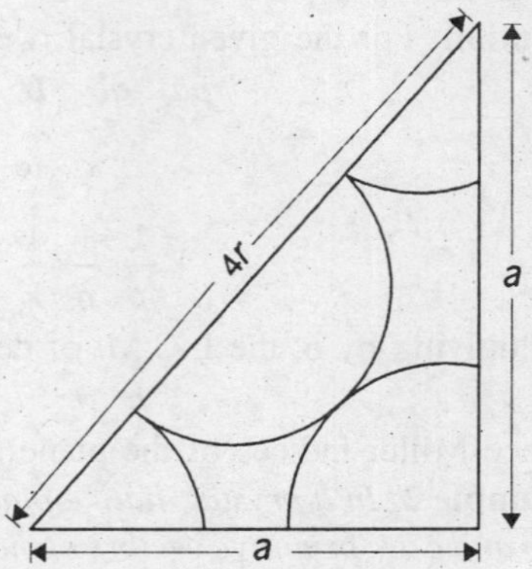

Fig. 12.17

In case f.c.c. lattice

$$(4r)^2 = a^2 + a^2$$

$$r^2 = \frac{2a^2}{16} \quad \text{or} \quad r = \frac{\sqrt{2}a}{4}$$

$$\therefore \qquad a_{\text{fcc}} = \frac{4r}{\sqrt{2}}$$

Hence $\quad \boldsymbol{a_{\text{bcc}} = \frac{4r}{\sqrt{3}}}; \qquad \boldsymbol{a_{\text{fcc}} = \frac{4r}{\sqrt{2}}}$

Example 6. *In a simple cubic crystal, find the ratio of intercepts on the three axes by (123) plane.*

Solution: If a plane cut intercepts of lengths l_1, l_2, l_3 on the three crystal axes, then

$$l_1 : l_2 : l_3 = p_a : q_b : r_c \qquad \text{...(1)}$$

where a, b and c are primitive vectors of the unit cell and numbers p, q and r related to the Miller indices $(h\ k\ l)$ of the plane by the relation

$$\frac{1}{p_a} : \frac{1}{q_b} : \frac{1}{r_c} = h : k : l \qquad \text{...(2)}$$

In the given equation, since the crystal is simple cubic

$$a = b = c$$

and given that $\quad h = 1,\ k = 2, \text{ and } l = 3$

$$\therefore \qquad p : q : r = h^{-1} : k^{-1} : l^{-1} = \frac{1}{1} : \frac{1}{2} : \frac{1}{3}$$

Multiplying right hand side by the lowest common denominator 6, we have

$$p : q : r = 6 : 3 : 2$$

From equation (2)

$$l_1 : l_2 : l_3 = 6a : 3a : 2a \qquad (\because\ a = b = c)$$

$$= \mathbf{6 : 3 : 2}$$

Example 7. *Calculate the lattice constant of potassium bromide. Density of potassium bromide is 2.7×10^3 kg/m^3 and belongs to fcc lattice. (Molecular weight of potassium bromide is 11.00 and Avogadro No. is 6.02×10^{26} per kg. mole).*

Solution: Let a be the lattice constant. Then

Mass in each unit cell = Volume × density = $a^3 \times 2.7 \times 10^3$ kg.

Molecular weight of KBr is 119.0 and since it belongs to fcc lattice, there are 4 molecules per unit cell.

$$\therefore \quad \text{Mass in each unit cell} = \frac{4 \times 119}{6.02 \times 10^{26}} \text{ kg.}$$

Equating the two expressions

$$a^3 \times 2.7 \times 10^3 = \frac{4 \times 119}{6.02 \times 10^{26}}$$

$$a^3 = \frac{4 \times 119}{6.02 \times 10^{26} \times 2.7 \times 10^3}$$

$$a = 6.64 \times 10^{-10} \text{ m} = \mathbf{6.64\ Å}$$

12.11. DISTRIBUTION OF ATOMS PER UNIT CUBE

The distribution of atoms are different in different lattice structure.

***(i)* Simple cubic lattice:** In case of simple cubic lattice each corner atom is shared by 8 unit cell. Thus a corner atom share $\frac{1}{8}$th of an atom as shown in Fig. 12.18. As there are 8 corner atom in all, each cube has $\mathbf{8 \times \frac{1}{8} = 1}$ **atom.**

Fig. 12.18 (s.c)

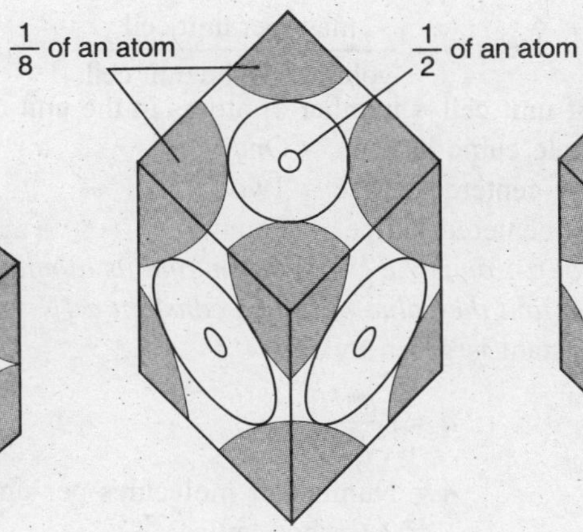

Fig. 12.19 (b.c.c.)

Fig. 12.20 (f.c.c.)

***(ii)* Body centered cubic lattice:** In case of body centered cubic lattice, there are 8 corner atoms and centre atom as shown in Fig. 12.19. The number of atoms which belong to this cube are

(*a*) one centre atom

(*b*) $8 \times \frac{1}{8} = 1$ corner atom

Hence number of atoms in b.c.c per unit cube = 1 + 1 = **2.**

***(iii)* Face centered cubic lattice:** In case of face centered cubic lattice, there are 6 face centered atoms and 8 corner atoms. Each of the 6 face centered atoms is further shared by two adjoining cubes shown in Fig 12.20. The number of atoms which belongs to this cube are

(*a*) $6 \times \frac{1}{2} = 3$ face centered atoms

$8 \times \frac{1}{8} = 1$ corner atom.

Hence, number of atoms per cube = 3 + 1 = **4.**

12.12. CO-ORDINATION NUMBER

The co-ordination number denotes the number of equidistant near neighbours that an atom has in a given structure. In the b.c.c. lattice the co-ordination number is 8; in f.c.c. lattice this number is 12; in hexagonal lattice this number is 6.

12.13. CALCULATION OF THE LATTICE CONSTANT

The lattice constant '*a*' for cubic crystals can be calculated as follows:

Consider a cubic lattice with 'a' as each side of the cube. Let V be the volume of the unit cell and of the density of the crystal. Then, the volume (V) of each unit cell = a^3.

Hence the mass of each unit cell = $V\rho = a^3\rho$ ... (1)

Now if the number of molecules per unit cell is n and the molecular weight of the substance is N. If N is the Avogadro number (No. of molecules in one gm mole of the substance), we have

$$\text{Mass of each molecule} = \frac{M}{N}$$

$$\text{So the mass of each unit cell} = \frac{M}{N} \times n \qquad ...(2)$$

From (1) and (2) we have

$$a^3\rho = \frac{nM}{N} \quad \text{or} \quad a = \left(\frac{nM}{N\rho}\right)^{1/3}$$

from which the value of lattice constant a can be calculated.

12.14. DENSITY OF LATTICE MATERIAL

It is defined as the ratio of mass per unit cell to the volume of the unit cell.

$$\text{Density} = \frac{\text{mass per unit cell}}{\text{volume of the unit cell}}$$

mass of unit cell = number of atoms in the unit cell × mass of each atom

Number of atoms in simple cubic lattice = One

Number of atoms in body centered lattice = Two

Number of atoms in face centered lattice = Four

Example 8. *The density of α-iron is 7.87×10^3 kg/m³ and its atomic weight is 55.8. If α- iron crystallizes in bcc space lattice, find the value of lattice constant a [$N = 6.02 \times 10^{26}$ per kg mole].*

Solution: The lattice constant is given by

$$a = \left(\frac{nM}{N\rho}\right)^{1/3}$$

In this case n = Number of molecules per unit cell

= 2 (in bcc lattice)

$M = 55.8$, $N = 6.02 \times 10^{26}$ (kg mole)$^{-1}$

and $\rho = 7.87 \times 10^3$ kg/m³

$$\therefore \quad a = \left(\frac{2 \times 55.8}{7.87 \times 10^3 \times 6.02 \times 10^{26}}\right)^{1/3}$$

$= 2.86 \times 10^{-10}$ m = **2.86 Å**

Example 9. *Calculate the lattice constant of the rock salt crystal from the following data:*

Molecular weight of NaCl = *58.45*

Density of NaCl = *2.17 × 10³* kg/m³

Avogadro's number = *6 × 10²⁶*.

Solution: Since the density of NaCl is 2.17×10^3 kg/m³.

Mass of each unit cell = Volume × Density = $a^3 \times 2.17 \times 10^3$ kg.

Molecular weight of NaCl = 58.45 and the number of molecular per kg. mole = 6.0×10^{26}

$$\text{Mass of each molecule} = \frac{58.45}{6 \times 10^{26}} \text{ kg.}$$

Since NaCl belong to f.c.c. lattice, so there are 4 molecules per unit cell.

$$\therefore \quad \text{Mass of each unit cell} = \frac{58.45 \times 4}{6 \times 10^{26}} \text{ kg.}$$

Equating the two expressions for mass of unit cell, we get

$$a^3 \times 2.17 \times 10^3 = \frac{58.45 \times 4}{6 \times 10^{26}}$$

$$a = \left[\frac{58.45 \times 4}{6 \times 10^{26} \times 2.17 \times 10^{3}}\right]^{1/3} = \mathbf{5.64\ Å}$$

Example 10. *Copper has f.c.c. structure and the atomic radius is 1.278 Å. Calculate the density of copper crystal. Given atomic wt. of copper = 63.5, Avogadro's Number = 6.023 × 10^{23}.*

Solution: Given r = 1.278 Å

In case of f.c.c. crystal number of atoms in unit cell = 4

$$\therefore \quad \text{Mass of each atom of copper} = \frac{63.5}{6.023 \times 10^{23}}$$

$$\text{lattice constant } (a) = \frac{4r}{\sqrt{2}} = \frac{4 \times 1.278}{\sqrt{2}}$$

$$= \frac{4 \times 1.278}{1.414} Å = 3.615\ Å$$

$$= 3.615 \times 10^{-8} \text{ cm}$$

$$\text{mass of the unit cell} = \frac{4 \times 63.5}{6.023 \times 10^{23}}$$

$$\text{Density} = \frac{\dfrac{4 \times 63.5}{6.023 \times 10^{23}}}{(3.615 \times 10^{-8})^3}$$

$$= \frac{4 \times 63.5 \times 10^{24}}{6.23 \times 10^{23} \times (3.615)^3}$$

$$= \frac{4 \times 63.5 \times 10}{6.023 \times (3.615)^3} = \mathbf{8.9\ gm/c.c.}$$

12.15. PACKING FRACTION OR RELATIVE DENSITY OF PACKING

In crystallography, *the atomic packing fraction is defined as the ratio of the volume of the atom in the unit cell to the volume of the unit cell itself.*

$$\therefore \text{ Packing Fraction or Relative density of Packing} = \frac{\text{Volume of atoms in unit cell}}{\text{Volume of the unit cell}}$$

In atomic physics packing fraction is the ratio of the mass defect to the mass number. If M is the atomic weight, A is the atomic mass number

$$\text{Then mass defect} = M - A$$

$$\therefore \quad \text{Packing fraction} = \frac{(M - A)}{A}$$

It is very important in the stability of nuclei. In case of crystal structure, while calculating packing fraction following points should be kept in mind:

(*i*) An atom centered at the corner of the unit cell is shared by eight unit cell having the common corner point. Thus a corner atom contributes only $\frac{1}{8}$th of its original volume to any particular unit cube.

(*ii*) An atom centered at the edge of the cube is shared by four unit cells and therefore, it contributes only $\frac{1}{4}$th of its volume to any particular unit cell.

(*iii*) An atom centered at the face of the unit cube is shared by two unit cells. Thus a face atom contributes only $\frac{1}{2}$ of its volume to a particular unit cell.

(*iv*) An atom within the body of a unit cube is contained wholly within the cube and therefore, contributes its total volume to the unit cell.

12.16. CALCULATION OF PACKING FRACTION FOR VARIOUS CRYSTAL STRUCTURES

(*i*) Simple cubic crystal structure: It has 8 atoms centered at the corner of the unit cube.

∴ Volume of the atoms in the unit cell

$$= 8\left(\frac{1}{8}\frac{4\pi}{3}\cdot r^3\right) = \frac{4\pi}{3}r^3$$

Volume of the unit cell

$$V = a^3 = (2r)^3 = 8r^3$$

∴ Packing fraction

$$= \frac{\text{Volume of atoms in unit cell}}{\text{Volume of the unit cell}}$$

$$= \frac{\frac{4}{3}\pi r^3}{8r^3}$$

$$= \frac{\pi}{6} = \frac{22}{42} = 0.52$$

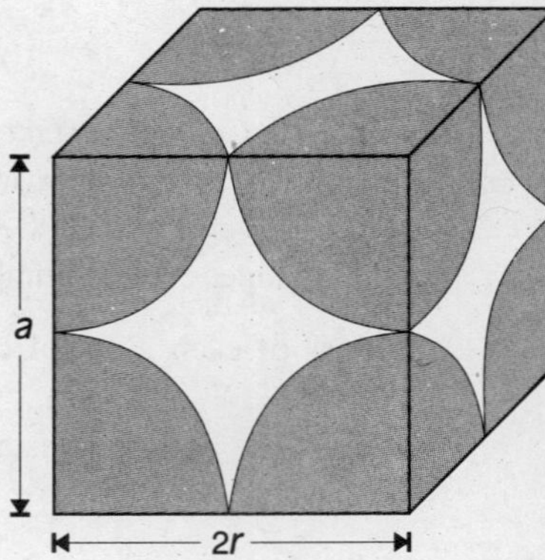

Fig. 12.21

(*ii*) Face centered cubic crystals: It has 8 atoms centered at the corners and 6 atoms centered at the faces of the unit cell, therefore

Volume of the atoms in the unit cell

$$= 8\left[\frac{1}{8}\times\frac{4}{3}\pi r^3\right] + 6\left[\frac{1}{2}\times\frac{4}{3}\pi r^3\right]$$

$$= \frac{4}{3}\pi r^3 + 4\pi r^3 = \frac{16}{3}\pi r^3$$

Volume of the unit cell

$$V = a^3 = \left(\frac{4r}{\sqrt{2}}\right)^3 = \frac{32\, r^3}{\sqrt{2}}$$

∴ Packing fraction of f.c.c. crystal

$$= \frac{\frac{16}{3}\pi r^3}{\frac{32}{\sqrt{2}} r^3} = \frac{\sqrt{2}\,\pi}{6} = \frac{\sqrt{2}\times 22}{6\times 7} = 0.74$$

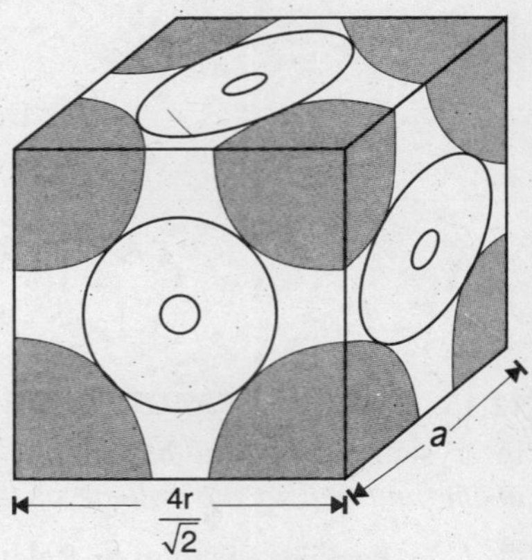

Fig. 12.22

(*iii*) Body centered cubic crystal: It has 8 atoms centered at the corner and one atom contained in the body *i.e.,* at the centre of the unit cell. Therefore,

Volume of atoms in the unit cell $= 8\left[\frac{1}{8}\times\frac{4}{3}\pi r^3\right] + \frac{4}{3}\pi r^3 = \frac{8\pi}{3}r^3$

Volume of the unit cell in b.c.c. lattice

$$V = a^3 = \left(\frac{4r}{\sqrt{3}}\right)^3 = \frac{64\, r^3}{3\sqrt{3}}$$

∴ Packing fraction of b.c.c. lattice

$$= \frac{8\frac{\pi}{3}r^3}{\frac{64\, r^3}{3\sqrt{3}}} = \frac{\sqrt{3}}{8}\pi = \frac{\sqrt{3}\times 22}{8\sqrt{7}} = 0.68$$

Table 12.2

Particulars	*Simple cubic s.c.*	*Body centered cubic b.c.c.*	*Face centered cubic f.c.c.*
1. Coordination Number	6	8	12
2. Atomic radius	$\frac{a}{2}$	$\frac{\sqrt{3}\,a}{4}$	$\frac{\sqrt{2}\,a}{4}$
3. Atoms per unit cell	1	2	4
4. Density of Packing	$\frac{\pi}{6} = 0.52$	$\frac{\sqrt{3}\,\pi}{8} = 0.68$	$\frac{\sqrt{2}}{6}\pi = 0.74$

12.17. DIFFERENT CRYSTAL STRUCTURES

(1) Sodium Chloride Structure (FCC lattice): In sodium chloride, sodium atom loses its outer electron and so acquires an excess of positive charge while chlorine atom has absorbed the electron lost by the sodium atom into its structure completing the *M*-level and at the same time acquiring an excess of negative charge. Two such ions will attract one another because of electrostatic forces between their excess charges. But they cannot approach to within less than a certain distance owing to strong forces of repulsion which must set in when their outer electron shells come into close proximity. When the centre of the two ions come within a certain distance of one another, attraction and repulsion balance and they can approach no closer and it is in this sense we must think of them as possessing a definite radius.

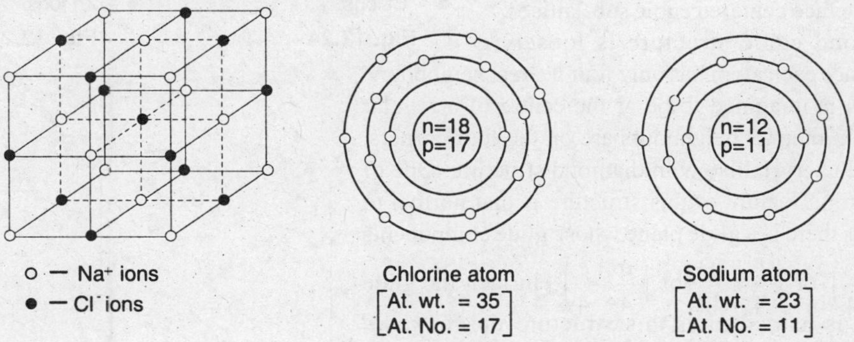

Fig. 12.23

The Bravais lattice is face-centered cube, the base consists of one Na atom and one Cl atom separated by one half of the body diagonal of a unit cube.

Therefore, it can be considered as two face-centered cube sub-lattice, one of Na ions having its origin at the (0, 0, 0) the other of Cl ions having its origin mid-way along a cube-edge, say at point $\left(\frac{a}{2}, 0, 0\right)$. Therefore, there are 4 units of NaCl in each unit cube with atoms in positions.

Na	0	0	0,	$\frac{1}{2}$	$\frac{1}{2}$	0,	$\frac{1}{2}$	0,	$\frac{1}{2}$	0	$\frac{1}{2}$	$\frac{1}{2}$
Cl	$\frac{1}{2}$	$\frac{1}{2}$	$\frac{1}{2}$,	0	0	$\frac{1}{2}$,	0	$\frac{1}{2}$	0,	$\frac{1}{2}$	0,	0

Each atom has six atoms of the opposite kind in its nearest surrounding.

$$\text{Molecular wt. of NaCl} = 23 + 35.5 = 58.5$$

$$\text{Mass of each molecule} = \frac{58.5}{6.02 \times 10^{23}} \text{ gm.}$$

$$\text{Mass of 4 molecules} = 4 \times \frac{58.5}{6.02 \times 10^{23}} \text{ gm.}$$

Let a be the length of unit cell, then volume of the unit cell = a^3. In case of NaCl, the density is 2.18 gm/cm^3.

$$\text{So mass of each unit cell} = a^3 \times 2.18 \text{ gm.}$$

$$\therefore \quad \frac{4 \times 58.5}{6.02 \times 10^{23}} = a^3 \times 2.18$$

$$\text{or} \quad a = 5.63 \times 10^{-8} \text{ cm} = 5.63 \text{ Å}$$

which gives the distance between two atoms of same kind. The distance between two adjacent atoms *i.e.*, Na and Cl atoms will be $\frac{1}{2}$ of the above value *i.e.*, $\frac{5.63}{2}\text{Å} = 2.815$ Å.

(2) Cesium Chloride structure: Another combination of cubic structure is that of CsCl. The lattice points of this compound are two interpenetrating simple cubic lattices. The corner of one of the sub-lattice is occupied by Cs ions, the other by Cl ions. The resultant structure termed as the cesium chloride structure is shown in Fig. 12.24.

(3) Zinc Blende Structure: A new structure occurs if the two face-centered cubic sub-lattice are occupied by different elements. The resultant structure is termed the Zinc Blende structure Fig. 12.25 important compounds which have this structure are the semiconductors such as In, Sb, GaAs ZnS etc.

(4) Diamond Structure: The diamond structure is combination of two inter-penetrating face centered cubic sub-lattices, the diamond cubic structure is loosely packed, since each atom has only four nearest neighbours. If an atom is imagined to be at the centre of a regular tetrahedron, then its neighbours are on the four corners. Ge, Si and C crystallise with diamond structure. One of the interesting feature of this structure is that normal to each a-axis there is a glide plane whose glide components are $\left(\frac{a}{4}+\frac{b}{4}\right)$ or $\left(\frac{b}{4}+\frac{c}{4}\right)$ or $\left(\frac{a}{4}+\frac{c}{4}\right)$. In fact, the glide operation is named after this structure the *diamond glide d*. (The space group of diamond is Fd 3 m).

The atoms in Fig. 12.26 are shown reduced in size so that the details of the structure can be seen better.

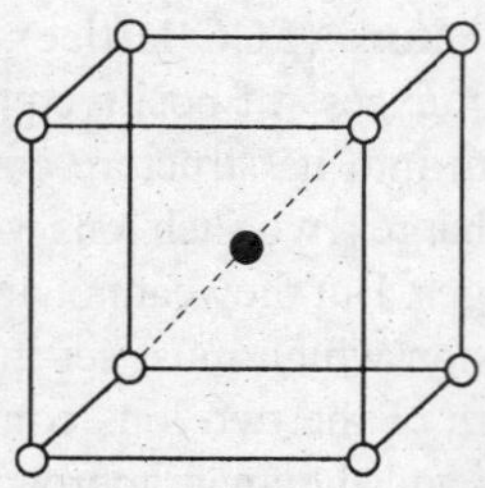

Fig. 12.24

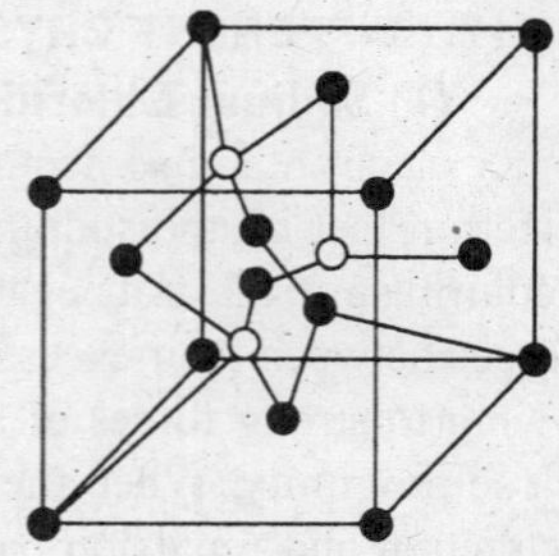

Fig. 12.25

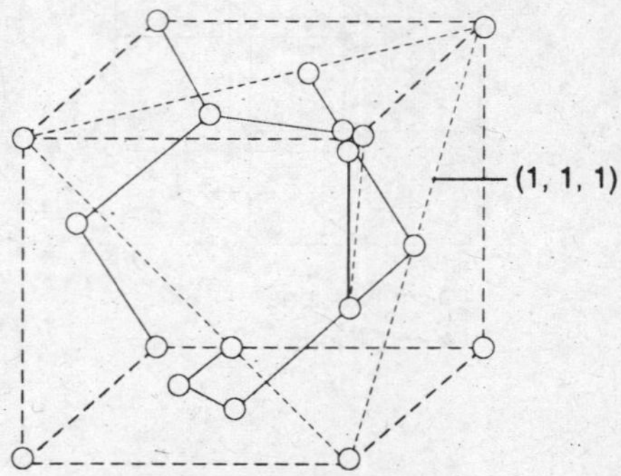

Diamond Structure

Fig. 12.26

12.18. GENERAL CONSIDERATION IN THE FORMATION OF COMPOUNDS

In order to reduce the electrostatic energy to minimum each ion will so far as possible surround itself with oppositely charged ions. The maximum number of oppositely charged ions which can surround an ion are eight. In general consider an example of a crystal AX, unless the ratio of the radius of A to that of X is greater than 0.731 the ions will touch one another along the cube edges, then A will settle inside the cube of X ions. If the ratio of the radii becomes less than 0.73 the crystals tend to have the rock salt types of lattice in which ion is surrounded by only six of the oppositely charged ions.

In case of alkali halides such as CsI, CsBr and CsCl in which the radii of the two ions are nearly equal have a coordination number of eight but all others have a rock salt structure with a coordination number of six. Most of the oxides and sulphides and some of solenides of the bivalent metals also have a rock salt structure. If the ratio of ionic radii is less than 0.41 the larger ions come into contact without touching the smaller ions even in the rock salt structure. Geometrical stability can again be allowed if the co-ordination number is four. This gives the Zinc blende (cubic) or Wurtzite hexagonal structure in which each ion lies at the centre of a regular tetrahedron of ions of the opposite kind.

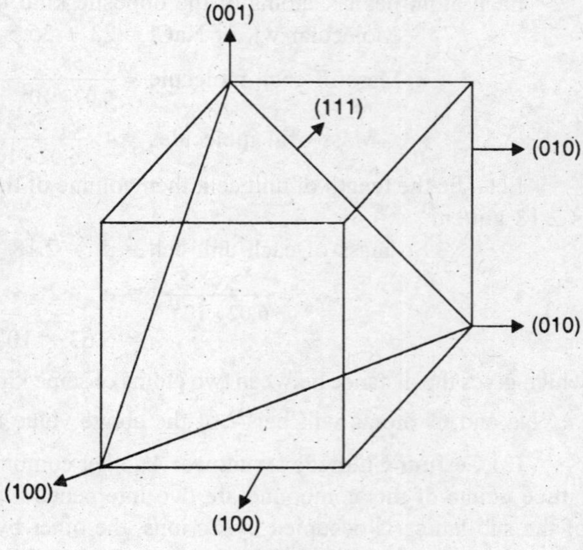

Fig. 12.27

CuCl, CuBr, CuI, BeS, ZnS, ZnSe, have Zinc-blende structure while BeO and ZnO have Wurtzite structure.

Example 11. *Show that the <111> planes are perpendicular to (1 1 1) direction in a simple cubic crystal.*

Solution: Draw a simple cubic crystal structure by taking one corner as the origin. The coordinate of *x*-axis, *y*-axis and *z*-axis are given by (100), (010) and (001). The coordinate of the diagonal will be (111). It is interesting to note that in a cubic system all directions and planes with identical indices are perpendicular *i.e.,* normal to a plane in a direction, with the same indices as the plane. On the same basis we can say the <111> planes are perpendicular to the (111) direction in a simple cubic crystal.

Note: The Miller indices of the six faces of a cube are (100), (010), (001), (010) and (001).

Example 12. *Draw the <110> and <111> planes; and the (110) and (111) direction in a simple cubical crystal. What do you infer from these diagrams?*

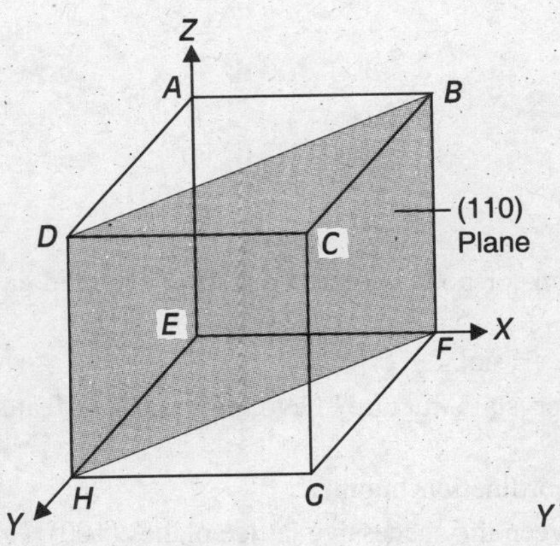

Fig. 12.28

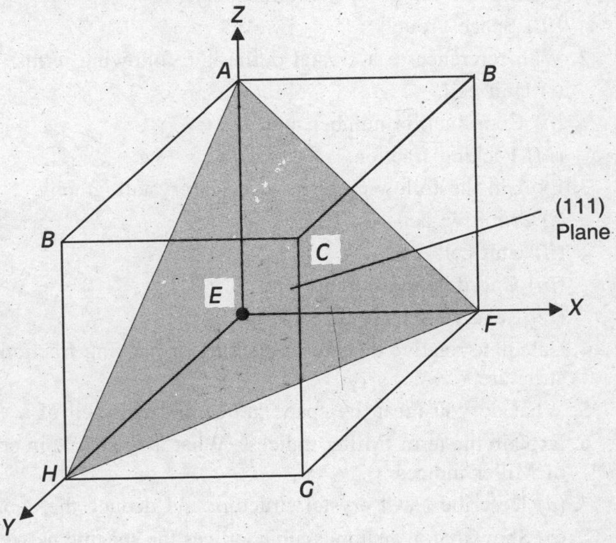

Fig. 12.29

Solution: Orientations of planes in a crystal are described in terms of their intercepts on the three axes. Thus Figs. 12.28 and 12.29 shows the <110> and <111> planes. The directions are described by giving the coordinates of the first whole number point (*x*, *y*) through which they pass. In Fig. 12.30 line *OC* gives (111) direction and *OG* gives (110) direction.

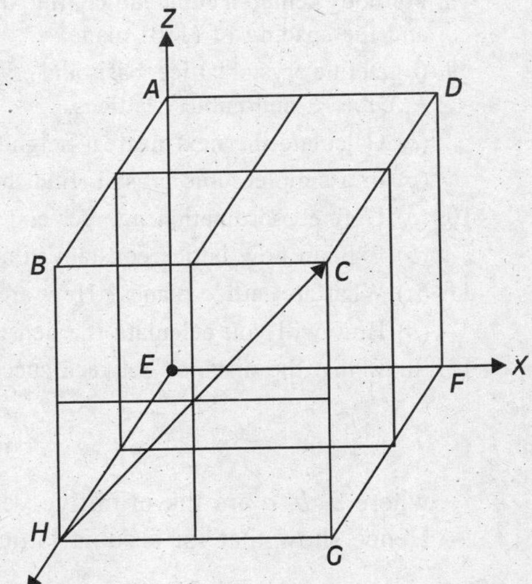

Fig. 12.30

Example 12. *The lattice constant for unit cell of aluminimu is 1.04 Å. What is the distance between planes with Miller indices (i) <200> (ii) <110> (iii) <111>?*

Solution: If (*h*, *k*, *l*) are the Miller indices, the distance between its planes (*d*) is given as

$$d = \frac{a}{\sqrt{h^2 + k^2 + l^2}} \quad \text{where } a = 4.04 \text{ Å}$$

(*i*) For <200> planes

$$d = \frac{a}{\sqrt{2^2 + 0 + 0}} = \frac{a}{2} = \frac{4.04}{2}$$

$$= 2.02 \text{ Å}$$

(*ii*) For <110> planes

$$d = \frac{a}{\sqrt{1^2 + 1^2 + 0}} = \frac{4.04}{\sqrt{2}} = \frac{4.04}{1.414}$$

$$= 2.86 \text{ Å}$$

(*iii*) For <111> planes

$$d = \frac{a}{\sqrt{1^2 + 1^2 + 1^2}} = \frac{4.04}{\sqrt{3}} = \frac{4.04}{1.732}$$

$$= 2.33 \text{ Å}$$

QUESTIONS AND PROBLEMS

1. Write short notes on:
(*i*) Atomic Radius in crystallography.
(*ii*) Bravais lattice in two dimensions.
(*iii*) Space group.

2. With reference to a crystal define the following terms:
(*i*) Unit cell
(*ii*) Coordination number and
(*iii*) Packing fraction.

3. Explain the following terms used in crystallography:
(*i*) Primitive cell
(*ii*) Unit cell
(*iii*) Coordination number and
(*iv*) Relative density of packing.

4. Calculate relative density of packing or packing fraction for body centered and face centered cubic structure.

5. What do you mean by space lattice and unit cell of a crystal.

6. Explain the term 'Miller indices'. What is their role in crystal structure? Give the important features of Miller indices.

7. (*a*) Describe NaCl crystal structure and deduce the coordination number.
(*b*) Show that in a simple cubic lattices the spacing between the successive lattice planes (100), (110) and (111) are in the ratio of 1 : 0.71 : 0.50.

8. For body centered cubic lattice, find the number of nearest neighbours, the nearest neighbour distance and the spacing of (110) plane.

9. (*a*) Define crystal lattice, basis and crystal structure. Deduce the number of atoms per unit cell in cubic *P*, cubic *F* and cubic *I* lattices.
(*b*) Calculate the next nearest neighbour distance in a simple cubic lattice.
(*c*) In a simple cubic crystal, find the ratio of the spacings (110) and (111) planes.

10. (*a*) Define coordination number and lattice constant of a crystal.
(*b*) Explain how lattice constant in alkali halide crystals was calculated.

11. (*a*) What are lattice planes? How are they represented in term of Miller indices?
(*b*) How will you calculate the density of lattice points in a lattice plane?

12. Show that the distance between successive planes of Miller indices ($h\ k\ l$) is given by

$$d_{hkl} = \left[\frac{h^2}{a^2} + \frac{k^2}{b^2} + \frac{l^2}{c^2}\right]^{1/2}$$

where *a*, *b*, *c* are the primitive vectors of the lattice.
Hence show that for a cubic lattice, this expression reduces to

$$d_{hkl} = \frac{a}{\sqrt{(h^2 + k^2 + l^2)}}$$

13. With the help of a neat diagram represent simple cubic and body centered cubic lattices. For a simple cubic lattice show the lattice planes (100), (110) and (111).
14. What do you understand by space lattice? Enumerate the seven crystal systems pointing out their characteristic features.
15. Show that for a cubic lattice, the lattice constant 'a' is given by

$$a = \left(\frac{nM}{N\rho}\right)^{1/2}$$

where n is the number of molecules per unit cell, M is the molcular wieght, N is Avogadro's number and ρ is the density.
16. Caesium chloride is a SC crystal. Its density is 4×10^3 kg/m^3 and molecular weight is 168.41. Calculate the lattice constant for CsCl crystal. Avogadro number is 6×10^{26} per kg mole.
17. In a crystal whose primitives are 1.2 Å, 1.8 Å, 2 Å, a plane (231) cuts on intercept 1.2 Å along X-axis. Find the lengths of intercepts along Y and Z-axis. **[Ans.** 1.2 Å, 4 Å]
18. For a simple cubic lattice of lattice parameter 2.04 Å, calculate the spacing of lattice plane (212).
19. Find the nearest neighbour distance in bcc crystal. **[Ans.** $a\sqrt{3/2}$]
20. The density and molecular weight of a f.c.c. lattice are 6250 kg/m^3 and 60.2 respectively. Find the lattice constant a of the crystal (Avogadro number = 6.02×10^{26} kg/mole). **[Ans.** 4 Å]

CHAPTER

13

BAND THEORY OF SOLIDS

13.1. INTRODUCTION

In all solids, building blocks, whether molecules or atoms, are permanently and almost rigidly locked into their positions. Crystalline solids, such as metals and minerals, are distinguished from amorphous solids, such as glasses and plastics, by the regularity of the arrangement of their building blocks. Their atomic or molecular building blocks form an orderly, repetitive lattice. Thus, we can regard metals and other crystalline solids as giant molecules in which the atoms are held in specific places according to a well-defined plan, just as in an ordinary molecule. These giant molecules are polymers, but they are three-dimensional polymers rather than ordinary, stringlike, one-dimensional polymers.

The bonds that hold the atoms in their places within a crystalline solids are essentially the same as the bonds that hold the atoms within an ordinary molecule. Ultimately, these bonds arise from the Coulomb interactions among the negatively charged electrons and the positively charged nuclei. Whenever two atoms are in close contact, the electron distribution of each is disturbed by the presence of the other. This redistribution of the electrons usually leads to binding force between the atoms. Depending on just how the electrons are redistributed, we speak of ionic bonds, covalent bonds, van der Waals bonds, hydrogen bonds, or metallic bonds. All of these kinds of bonds, with the exception of metallic bonds, occur in ordinary molecules as well as in solids.

The characteristic feature of metals is the presence of an abundance of free electrons. When the atoms of a metal join in a crystal, the outermost or valence electrons become detached from their atoms. These electrons are then more or less free to roam through the entire volume of the metal—they form a "gas" of free electrons which fills the entire volume of the metal. As we will see, this gas is not only responsible for the bonds that hold the ions in their places in the lattice, but it is also responsible for the electrical properties of metals. A similar gas of free electrons is responsible for the remarkable electrical properties of semiconductors, which form the backbone of modern electronic technology.

13.2. BONDS BETWEEN ATOMS

We will now briefly discuss the kinds of bonds that hold the atoms in solid. Our discussion will be quantitative because quantitative treatment of the theory of bonds requires the full machinery of wave mechanics and is quite difficult. Also, we had better keep in mind that in some crystals two of the bonding mechanism may present simultaneously. For instance, in AgCl the bonds are of an intermediate kind, partially ionic and partially covalent.

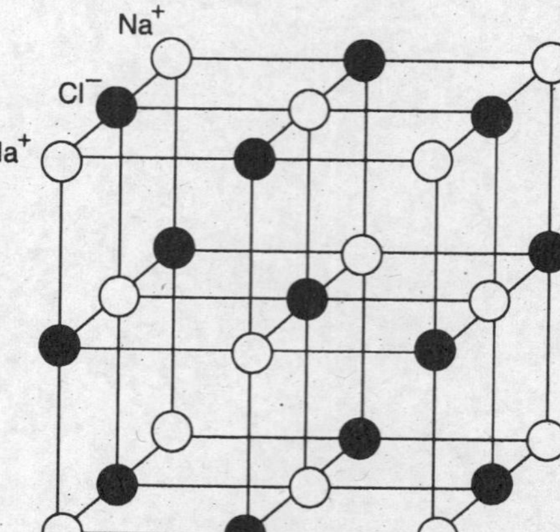

Fig. 13.1 Arrangement of Na^+ and Cl^- ions in the NaCl lattice.

(i) **Ionic Bonds:** Some atoms, such as Na, tend to lose electrons easily, some others, such as Cl tend to acquire extra elecrtrons. When atoms of these two kinds interact, the rearrangement of the electron distribution is very simple: an electron from an atom of

the first kind becomes attached to an atom of the second kind. Thus the atoms become positive and negative ions, such as Na^+ and Cl^-. The bond between these ions is then simply due to the Coulomb attraction between the positive and the negative charges. Since both Na^+ and Cl^- have the spherically symmetric electron configuration of the atoms of noble gases, their Coulomb interaction is that of two point charges, with a potential energy $U = -e^2/(4\pi\varepsilon_0 r)$ corresponding to an attractive, inverse-square force. This Coulomb interaction is responsible for holding the Na^+ and Cl^- ions in the NaCl lattice (see Fig. 13.1).

However, if the distance between the atoms is very small, so that the wave functions of the inner electrons of the two atoms overlap, then the potential energy is modified by a repulsive effect arising from the Exclusion Principle. This can be most easily understood by considering the extreme case of complete overlap between the atoms, so that their nuclei coincide and two atoms form a single, combined atom. In this case it is obviously impossible for both atoms to keep their electron configurations unchanged since this would mean that any state occupied in each of the separate atoms would be doubly occupied in the combined atom. Consequently, the Exclusion Principle requires that whenever atoms overlap, electrons must be forced into formerly unoccupied states of higher energy. This effect occurs not only in the case of complete overlap, but also in the case of partial overlap (the latter case is rather more complicated because the wave functions of atoms in intimate contact are strongly distorted).

Figure 13.2 shows a plot of the mutual potential energy of a Na^+ ion and a Cl^- ion as a function of their separation (this potential energy has been calculated from the quantum-mechanical wave function). The sharp increase of the potential energy at distances shorter than 2 Å is due to electrons that are forced into states of higher energy. At a distance larger than about 3 Å, the exclusion effect disappears, and the Coulomb interaction takes over. From this plot we see that at large distances the force between the ions is attractive, but at short distances it is repulsive. At a distance of 2.4 these attractive and repulsive forces cancels, giving a (classical) equilibrium point.

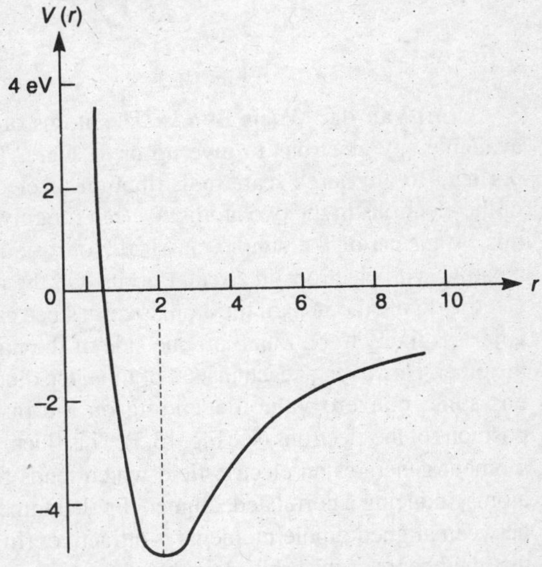

Fig. 13.2 Potential energy of interacting Na^+ and Cl^- as a function of distance.

The repulsive contribution to the force at the short distance is called the **repulsive core**. This repulsive core is a general feature of the interaction of atoms—it is found in all atoms and it is essentially the same for all the different kinds of bonds. Thus we see that the Exclusion Principle is responsible for the resistance that all atoms offer to interpenetration.

***(ii)* Covalent Bonds:** Atoms that are not very willing to give up electrons or to accept electrons may be willing to share electrons with their neighbours. The shared electrons tend to concentrate in the region between the atoms. Usually the shared electrons occur in pairs of opposite spins, each atom contributing one electron. Such a redistribution of the electrons gives rise to a bond between the atoms because the shared electrons exert a Coulomb attraction on the positive charges in the two atoms.

Covalent bonds exhibit strong directional preferences. For instance, the carbon atom likes to form four covalent bonds directed toward the corner of tetrahedron centered on the atom (see Fig. 13.3).

The bonding of carbon atoms with other carbon atoms in such a tetrahedron geometry creates the lattice of diamond, the hardest of all crystals (see Fig. 13.4).

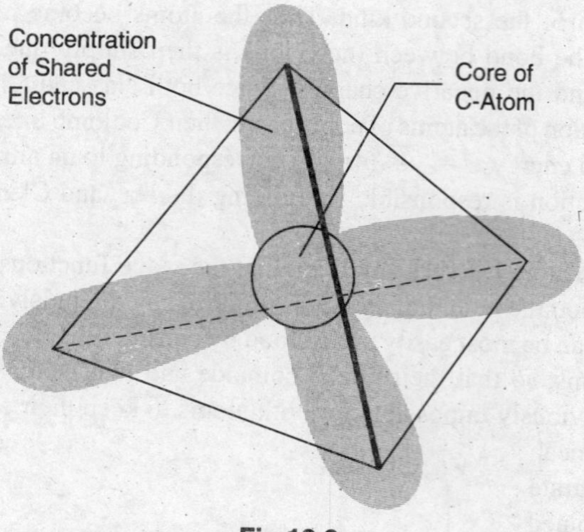

Fig.13.3

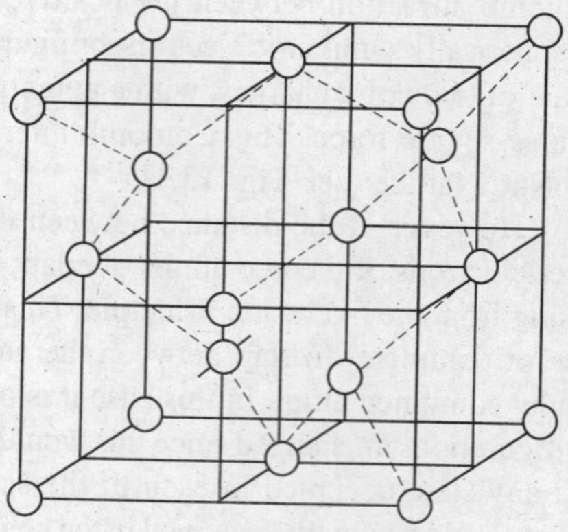

Fig.13.4

***(iii)* Van der Waals Bonds:** The atoms of the noble gases and most molecules do not have available any electrons to give up or to share. The interaction between two such atoms will then not lead to any large-scale redistribution of electrons, but merely to a correlation of the motions of the electrons in the two atoms. We can roughly understand this by means of the simple classical Bohr model. Imagine an atom with electrons in circular orbits. On the average, the electric dipole moment of the atom is zero, since each electron is just as likely to be found on one side of the nucleus as on the other. However, at each instant of time, the dipole moment has some nonzero value depending on the instantaneous positions of the electrons (see Fig. 13.5). This fluctuating dipole moment generates an electric field which tends to polarize the charge distribution in neighbouring atoms, inducing a correlated, aligned dipole moment. As we know from electro-magnetism, the force between aligned dipole moments is attractive. (In terms of the motions of the electrons, we can say that the phases of the orbits neighbouring atoms tend to be correlated in such a way that the electrons avoid coming near each other and, therefore, the net force between the charge distributions of the two atoms is attractive). This force between neighbouring atoms is called the van der Waals force.

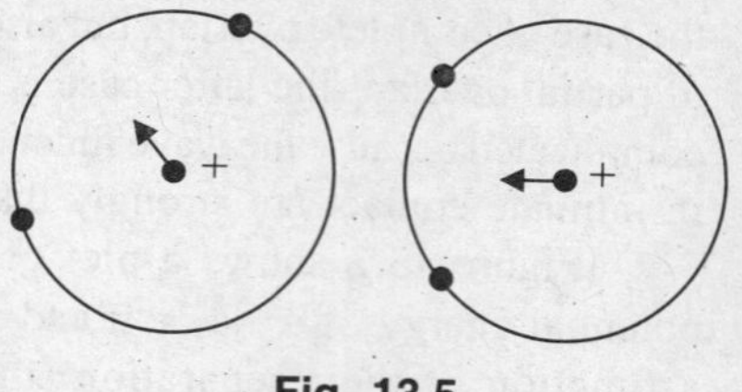

Fig. 13.5

The bonds produced by the van der Waals forces are much weaker than ionic or covalent bonds. Hence the noble gases only crystallize at very low temperatures or at very high pressures. Note that since the mechanism underlying the van der Waals forces is quite general, these forces act between all kinds of atoms and molecules but they are only important when the ionic or covalent forces are absent. *The cohesive forces in liquids are van der Waals forces.*

***(iv)* Hydrogen Bonds:** The hydrogen atom can lose its electron to neighbouring atoms, leaving the hydrogen with a positive charge. This positive charge placed between two adjacent negative ions produces a bond between the ions. Obviously, this bond is some what similar to a covalent bond, with the positively

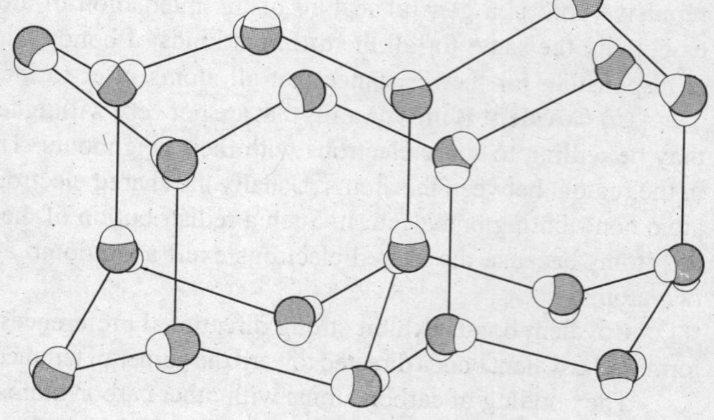

Fig. 13.6

charged hydrogen playing the role of the electron pair. Hydrogen bonds are responsible for joining the water molecules in ice crystals (see Fig. 13.6). They are also responsible for joining the two strands of double helix that makes up the molecule of **DNA**.

(v) Metallic Bonds: The electron redistribution that creates the bonds between the atoms in a metal is somewhat similar to that found in covalent bonds: the atoms share electrons. But instead of merely sharing one electron with a neighbour, each atom shares one (or two) of its electrons with all of the other atoms in the crystal. The shared electrons are free to move through the entire volume of the crystal; they form a gas of free electrons. Thus the metal consists of a lattice of positive ions embedded in an interpenetrating sea of negative charges. The Coulomb attraction between these positive and negative charges holds the ions in their places.

Although the "free" electrons in a metal are not attached to any particular atom, they are not completely free of forces. When an electron passes near an ion, it experiences a strong, attractive Coulomb force. The net effect of all the ions on an electron can be described by a periodically repeating potential-energy function (see Fig. 13.7). Classically, an electron moving in such a periodic potential would simply speed up near each ion, and slow down in the spaces between. Wave-mechanically, the motion of an electron is rather more complicated—for certain ranges of wavelength the motion is actually impossible (forbidden energy bands). However, for an electron of long wavelength, or low momentum, the motion is that of a free particle *i.e.,* the periodic potential produces no effect on the motion. The reason is that, in general, a wave is insensitive to any disturbance whose size is small compared to a wavelength; for example, ocean waves of long wavelength (swells) are unaffected by a small island in their path. An electron wave only senses the potential averaged over toughly of wavelength. If the wavelength; is long so that it includes several of paths and valleys shown in Fig. 13.7 then the average potential is a constant, which is equivalent to zero force. Under these conditions the "free" electrons in the metal will behave like free particles.

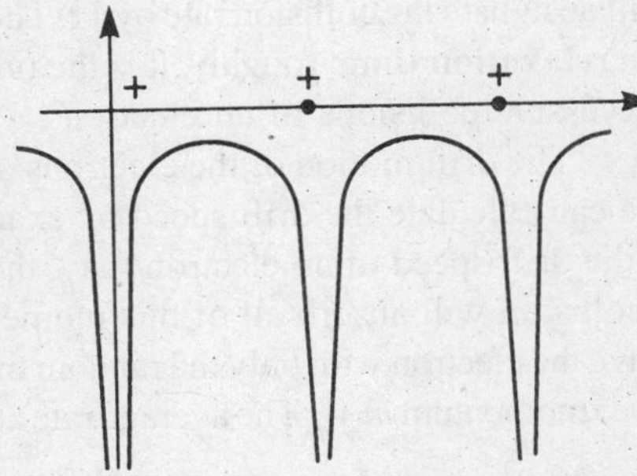

Fig. 13.7 Potential energy of a "free" electron moving through a metal along a straight line (x-axis)

13.3. THE FREE-ELECTRON GAS

We can readily explain several of the important physical properties of metals in terms of the free-electron gas. For instance, the motion of the free electrons is responsible for the high electric conductivity of metals. Accordingly, the free electrons are often called **conduction electrons.** One of the first calculations of the conductivity of a free-electron gas was made by P. Drude around 1900. Drude treated the electrons in a piece of metal as classical particles, with large thermal velocities of random directions. These thermal velocities of the electrons do not bring about a net transport of electric charge, since, on the average, for every electron moving in some given direction there is another electron moving in the opposite direction. However, when the ends of the piece of metal are connected to the terminals of a battery or generator, an electric field E will be produced within the piece of metal and, under the influence of the electric force, the electrons will accelerate in the direction of $-E$, acquiring an additional non-random velocity in this direction. This additional velocity of the electrons will give rise to a transport of charge. The magnitude of the additional velocity, **called the drift speed,** is limited by the decelerations that the electrons suffer in collisions with the ions of the lattice. These collisions play the role of a **frictional force** that slows the electrons down. The drift speed is always quite small compared to the random speeds. Under typical conditions, the electron gas in a metallic wire carrying a current has a drift speed of perhaps 10^{-2} m/s. But the individual electron have random speeds of about 10^{-2} m/s. Thus, the motion of an electron consists of rapid zigzags on which is superimposed a much slower drift motion along the wire; qualitatively, the motion resembles that shown in Fig. 13.8 but the amount of drift per zigzag is much less than shown in this figure. The number of collisions per second that an electron makes with the ions of the lattice is proportional to its speed. Since the drift speed is negligible compared to the random speed, the drift speed has next to no effect on the collisions rate. This means we can regard the collision rate as a constant, independent of the drift speed and of the electric field. We will designate the collision rate by 1τ. The quantity τ is called the **relaxation time**; roughly, it is the time interval between successive collisions of an electron.

Fig. 13.8

The drift motion of the electrons determines the net, macroscopic current carried by the wire. We can calculate the drift speed by examining the losses and gains of momentum of this electron. If the drift speed of an electron is v_d, then the momentum is $m_e vd$. We expect that, on the average, a collision will absorb all of this momentum, *i.e.*, a collision will destroy the forward motion and leave the electron with only the random thermal motion. This means that, in each collision, the electron loses momentum $m_e v_d$. The average rate at which the electron loses momentum in collisions is therefore

$$\left(\frac{\Delta p}{\Delta t}\right)_{\text{loss}} = \frac{m_e v_d}{\tau} \quad ...(1)$$

On the other hand, the rate at which the electron gains momentum by the action of the electric field is

$$\left(\frac{\Delta p}{\Delta t}\right)_{\text{gain}} = -eE \quad ...(2)$$

Under steady-state conditions, the rate of loss of momentum must match the rate of gain. By setting the right sides of Equations (1) and (2) equal, we get

$$v_d = \frac{-eE\tau}{m_e} \quad ...(3)$$

This is the speed with which the electron gas flows along the wire. The negative sign in Equation (3) indicates that the direction of flow is opposite to the direction of the electric field.

To find the electric current carried by this flow of the electron gas, we need to take into account the number of free electrons. Suppose that the metal of the wire has n free electrons per unit volume. In metals, the number of free electrons is usually one per atom; typically, this amounts to 10^{28} or 10^{29} electrons per m^3. If the wire has a cross-sectional area A and a length l, then its total number of free electrons is $n \times$ (volume) $= nAl$ and the total charge associated with these electrons is

$$\Delta q = -enAl \quad ...(4)$$

Within a time interval $\Delta t = l\,|\,v_d\,|$ all of these electrons emerge at one end of the wire. Hence the current in the wire is

$$I = \frac{|\Delta q|}{\Delta t} = \frac{enAl}{l/v_d} = \frac{e^2 n\tau}{m_e} AE \quad ...(5)$$

If we express this in terms of the potential difference $\Delta V = El$ across the ends of the wire, we get

$$I = \left(\frac{e^2 n\tau}{m_e}\frac{A}{l}\right)\Delta V \quad ...(6)$$

This is **Ohm's law:** the current is proportional to the potential difference. According to Equation (6) resistance of the wire is

$$R = \frac{\Delta V}{I} = \frac{m_e}{e^2 n\tau}\frac{l}{A} \quad ...(7)$$

and the conductivity is

$$\sigma = \frac{l}{RA} = \frac{e^2 n \tau}{m_e} \qquad ...(8)$$

Example 1. *The conductivity of silver is 6.2 × 10^{-7} Ω m and the number of free electrons per unit volume is 5.8 × 10^{28} m^3. Calculate the relaxation time. Calculate the drift speed of the electrons in a silver wire of length 2.0 m subjected to a potential difference of 40 V.*

Solution: From Equation (8)

$$\tau = \frac{\sigma\, m_e}{e^2 n} = \frac{(6.2 \times 10^{-7}\,\Omega\, m) \times 9.1 \times 10^{-31}\text{ kg}}{(1.6 \times 10^{-19}\text{C})^2 \times 5.8 \times 10^{28}\,/\text{m}^3} = 3.8 \times 10^{-14}\text{ s}$$

The electric field in the wire is $E = \Delta V/d$ = 40 V/2.0 m = 20 V/m. Hence, from Equation (3),

$$v_d = \frac{-e\,E\,\tau}{m_e} = -\frac{1.6 \times 10^{-19}\text{C} \times 20\text{V/m} \times 3.8 \times 10^{-14} s}{9.1 \times 10^{-31}\text{ kg}}$$

$$= -0.13\text{ m/s}$$

Such a low drift speed typical for the motion of the free electrons in metals. (In contrast, the *random* speed of these free electrons is much larger; the typical value of the random speed is 10^6 m/s or more).

13.4. FERMI GAS

Although Drude's simple classical calculation leads to the correct form of Ohm's law, the attempts at determining the relaxation time τ from the atomic parameters and the classical theory of random thermal motions were a complete failure—the value of the relaxation time calculated from **classical theory disagrees with experiment,** and also fails to reproduce the observed dependence of the conductivity on temperature. Furthermore, according to classical theory we would expect that the free-electron gas has a heat capacity of its own amounting to **3 cal/K mole,** which has to be added to the heat capacity of the crystal lattice. But experiment shows that the heat capacity of metals is not noticeably different from that of other crystals. All these difficulties of the classical theory of the free electron gas disappear when we treat the electrons by wave mechanics and take into account the Exclusion Principle, which plays crucial role in determining the properties of the free-electron gas because it forces electrons into states of very high kinetic energy, much higher than the classical thermal kinetic energy 3/2 kT.

A gas of quantum-mechanical particles obeying the Exclusion Principle—electrons or other particles of half-integer spin—is called a **Fermi gas**. The distinction between the classical gas and the Fermi gas shows up drastically if we consider the simple case of zero temperature. Classically, the particles of a gas at zero temperature (T = 0 K) all have zero kinetic energy—they all are at rest. Quantum mechanically, electrons at zero temperature cannot all have zero kinetic energy because this would mean that they all are in the ground state, which would **contradict the Exclusion Principle.** Instead, at zero temperature the electrons will occupy the available states of lowest energy, two electrons with opposite spins in each orbital state.

We can roughly regard the free electrons in a metal as **free particles in box.** The available states are then described by standing waves and the energies of the available states are given by the following **three-dimensional equation.**

$$E = (n_x^2 + n_y^2 + n_z^2) \times \frac{\pi^2 h^2}{2m_e L^2} \qquad ...(1)$$

Here, we have assumed that the box is a cube, measuring $L \times L \times L$; this assumption is convenient, but not necessary. The possible values of n_x, n_y, and n_z can be represented as points in a three-dimensional space labelled with n_x, n_y and n_z-axes. Figure 13.9 shows some of these points; note that the x-, y-, and

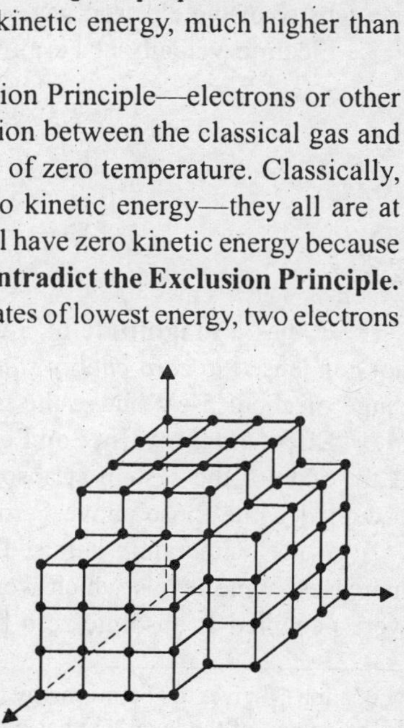

Fig. 13.9

z-intervals between adjacent points are one unit, and that all the points are in the first octant ($n_x > 0, n_y > 0, n_z > 0$). We can use this figure to count the number of available states up to some given energy E. The origin is

$$R = \sqrt{n_x^2 + n_y^2 + n_z^2} \qquad ...(2)$$

We can express the energy in terms of this radius,

$$E = R^2 \frac{\pi^2 \hbar^2}{2m_e L^2} \qquad ...(3)$$

and conversely we can express the radius in terms of the given energy

$$R = \sqrt{\frac{2m_e L^2}{\pi^2 \hbar^2} E} \qquad ...(4)$$

The number of available orbital states of energy less than E is simply the number of points within the radius R. Each point is associated with a volume element of one unit; these volume elements can be regarded as infinitesimal provided R is a large number. The number of points within the radius R is, therefore, simply the volume within the first octant, *i.e.*, $\left(\frac{1}{8}\right)(4\pi/3)R^3$. Taking into account the two spin states per orbital state, we find that the number of states of energy less than E is

$$N = 2 \times \frac{1}{8}\frac{4\pi}{3}\left(\frac{2m_e L^2}{\pi^2 \hbar^2}\right)^{3/2} E^{3/2} \qquad ...(5)$$

Since the volume of our box is $V = L^3$, this reduces to

$$N = \frac{1}{3}(2m_e)^{3/2} \frac{V}{\pi^2 \hbar^3} E^{3/2} \qquad ...(6)$$

In a crystal at zero temperature, the electrons fill the available states of lowest energy. Hence if we set the number of states in Equation (6) equal to the number of electrons, the energy E will equal the energy of the highest occupied state, or **Fermi level**. Solving for this energy E, we obtain

$$E_F = (3\pi^2)^{2/3} \frac{\hbar^2}{2m_e}\left(\frac{N}{V}\right)^{2/3} \qquad ...(7)$$

This is called the Fermi energy of the electron gas. Note that this energy depends only on the density of the electron gas*.

Electron velocity at Fermi surface is given by

$$\frac{1}{2}mv_F^2 = E_F = \frac{\hbar^2}{2m_e} \cdot \left(\frac{3N\pi^2}{V}\right)^{2/3}$$

$$\therefore \quad v_F = \frac{\hbar}{m_e} \cdot \left(\frac{3N\pi^2}{V}\right)^{1/3} \qquad ...(8)$$

which is $\sim 10^8$ cm s^{-1}.

E_F has a **magnitude of 5 eV**. So at absolute zero, the conduction electrons in a metal are not condensed to *zero energy state* as in classical mechanics but they fill all energy levels over a range of about 5 eV above the ground state.

Effect of temprature on Fermi-Dirac distribution. By definition the ground state of a system is the state of the system at absolute zero. As seen above, the energy states upto *Fermi level* are occupied while those above Fermi level are vacant.

If the temperature is raised, what happens to the energy levels? As the K.E. of electron gas increases, some levels which were vacant at **absolute zero** get occupied and some others which were occupied at absolute zero become vacant.

---

*Equation (7) gives the Fermi energy for zero temperature. The Fermi energy for non-zero temperature is defined as the energy of the level that has a probability of 1/2 for being occupied; it depends on the temperature and the density.

Fermi gave the probability function describing the thermal behaviour of a group of particles which obey the Pauli's exclusion principle. This function is,

$$f(E) = \frac{1}{e^{(E-E_f)/kT} + 1} \quad ...(9)$$

It gives the probability that a state having energy E is occupied by an electron in an ideal gas in thermal equilibrium at a temperature TE_F is the *Fermi Energy* and $f(E)$ represents the fraction of the possible states which is occupied.

Figure 13.10 (*a*) gives the plot of the Fermi function $f(E)$ against E for $T = 0$ and $T >> 0$. At absolute zero, $T = 0$ the Fermi function,

$$f(E) = 1 \quad \text{for} \quad E < E_F$$
$$f(E) = 0 \quad \text{for} \quad E > E_F$$

So at absolute zero E_F gives the cut off energy. All the states with energy $E < E_F$ are **completely filled** while all those with $E > E_F$ **are vacant.** As the temperature T increases, the distribution rounds off as shown in Fig. 13.10 (*b*). States with energy less than E_F by an amount of the order of kT being **partly occupied.** The value of E_F depends on the temperature but for $KT/E_F << 1$, $E_F(T)$ is closely equal to its value at $T = 0$. At any temperature the Fermi function $f(E) = \frac{1}{2}$ for $E = E_F$. So, the physical meaning of E_F may be stated as that at the Fermi level (*i.e.*, when $E = E_F$), *the probability that the particular energy state may be occupied is* $\frac{1}{2}E_F$ *(which is of the order of* 5 eV for many metals).

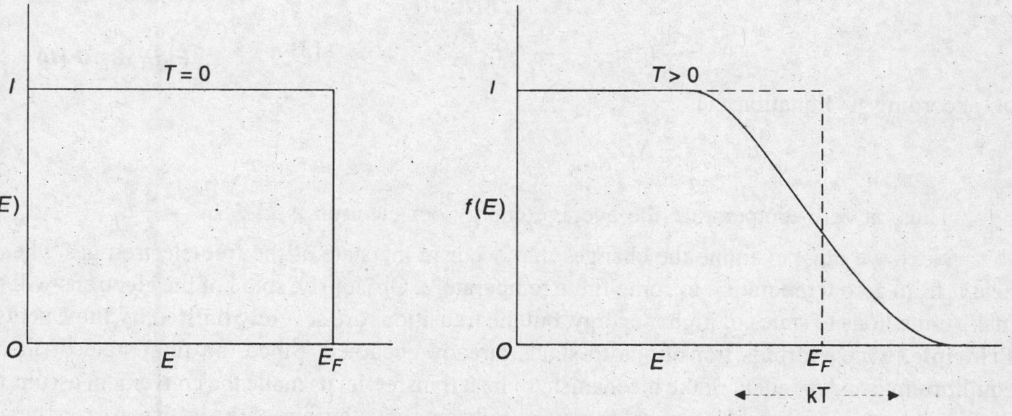

Fig. 13.10 (*a*) Plot of Fermi function $f(E)$ for two temperatures.

This gives us a significant difference between classical statistics (Maxwell-Boltzmann's statistics) and Fermi Dirac Statistics.

According to Maxwell-Boltzmann's distribution, all electrons will have zero energy at absolute zero, $T = 0$ K, while according to Fermi-Dirac distribution, there are some electrons having energy upto 5 eV. If we assume the electrons to obey *Maxwell-Boltzmann's law*, then to have energy ~ 1 eV the temperature must be ~ 5000 K

For $E > E_F$ such that $E - E_F >> kT$ the term 1 may be neglected in the denominator of the expression (9) for $f(E)$ and the Fermi function becomes

$$f(E) = e^{-(E-E_f)/kT} \quad ...(10)$$

So in the region $E - E_F >> kT$, the Fermi distribution becomes **identical to the Boltzmann distribution** and we get a Boltzmann tail. In this case there will be few electrons having energies extending to high values. It may also be noted that below the melting point of metals kT is much small compared with E_F ($kT << E_F$).

Fermi level at a temperature T is given by the relation

$$E_F \approx E_{F_0}\left[1-\frac{\pi^2}{12}\left(\frac{kT}{E_{F_n}}\right)^2\right] \qquad ...(11)$$

where E_{F_0} is the Fermi energy at absolute zero, *i.e.*, $T = 0$. As T increases Fermi level E_F decreases but the decrease is very small as $\left(\frac{kT}{E_F}\right)^2$ is of the order of 2×10^{-5} at room temperature. So we may take Fermi level to be constant for all practical problems.

The number of available states per unit energy interval is dN/dE obtained by differentiating Equation (6) above.

$$\frac{dN}{dE} = \frac{1}{2}(2m_e)^{3/2}\frac{V}{\pi^2\hbar^3}E^{1/2} \qquad ...(12)$$

Figure 13.10 (*b*) is a plot of dN/dE, with the energy plotted along the vertical axis, as customary in an energy-level diagram. The shaded area indicates those states occupied at zero temperature.

With the expression (7) we can calculate the total energy of the Fermi gas:

$$E_{\text{total}} = \int_0^{E_F} E\,dN$$

$$= \int_0^{E_F} E\frac{1}{2}\cdot(2m_e)^{3/2}\frac{V}{\pi^2\hbar^3}E^{1/2}dE$$

$$= \frac{1}{5}(2m_e)^{3/2}\frac{V}{\pi^2\hbar^3}E_F^{\;5/2} \qquad ...(13)$$

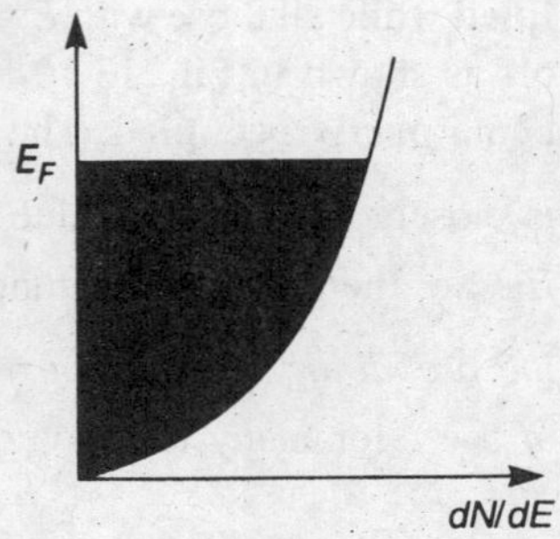

Fig. 13.10 (*b*)

or, according to Equation (1)

$$E_{\text{total}} = \frac{3}{5}NE_F \qquad ...(14)$$

Thus, at zero temperature the average energy per electron is $E_{\text{total}}/N = \frac{3}{5}E_F$.

Next, we must examine the changes that occur in the state of the free-electron gas when we heat it from zero temperature to some finite temperature. Obviously, some of the electrons will then make transitions to states of higher energy, but the transitions are severely **limited by the Exclusion Principle,** which forbids transitions to states already occupied. Since the final state of thermal equilibrium is independent of the mechanism of heat transfer, let us make the convenient assumption that the electrons acquire their thermal energy by collisions with the **ions of the lattice.** At a temperature T, the energy of the ions is of the order of kT. Hence a collision can at most transfer that much energy to the electron. At room temperature, $kT \sim 0.03$ eV. For most of the electrons, 0.03 eV is not enough energy to permit a transition to an unoccupied state: **hence most electrons are unable to accept this energy and they will remain locked in their original state.** Only those electrons that are within 0.03 eV of an unoccupied state—or within 0.03 eV of the zero-temperature Fermi level—can accept this energy and jump to a higher state. Thus a few **states just below the zero-temperature, Fermi level become partially depleted, and a few states just above become partially filled** (see Fig. 13.11), but most of the states remain as they were at zero temperature. This means that the electron gas in a metal at room temperature is pretty much the same as an electron gas at zero temperature. Even in a metal near its melting temperature, the electron gas is nearly unchanged. The total energy and other properties of the electron gas are nearly independent of temperature, over a very wide range of temperatures.

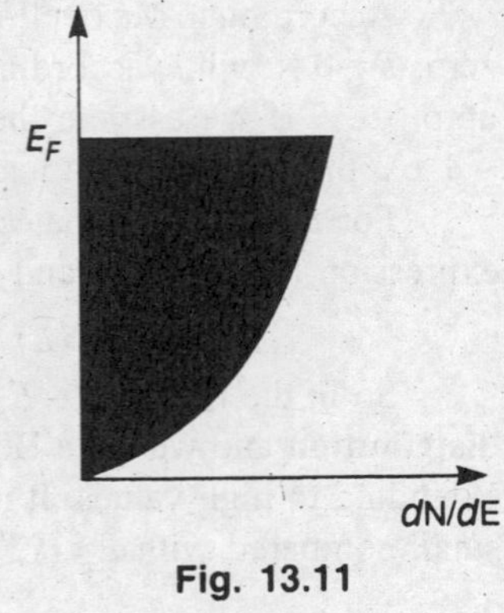

Fig. 13.11

This explains why the electron gas makes only a small contribution to the specific heat of the metal—the electron gas absorbs only a small amount of thermal energy.

We can estimate the specific heat of the electron gas by the following simple argument. As stated above, the Exclusion Principle prevents most electrons from absorbing thermal energy because the adjacent energy levels are full. Only the electrons within a distance ~ kT of the Fermi level can absorb thermal energy. The fraction of these elecrons is roughly ~ (kT/E_F). Thus, we expect that the specific heat of the Fermi gas is smaller than that of a classical gas by a factor kT/E_F. Since the specific heat per mole of a classical gas is $3/2N_0k$, the specific heat per mole of the Fermi gas must be roughly*

$$C_v \sim \frac{kT}{E_F} \cdot \frac{3}{2} \cdot N_0 k \qquad ...(15)$$

For the free-electron gas in a metal, E_F is typically a few eV (see Example 2), whereas kT ~ 0.03 eV at room temperature. This means that kT/E_F is of the order of magnitude of 10^{-2}, and therefore, the specific heat of the **free-electron gas is only ~1% of the value expected for a classical gas.**

We might suppose that the Exclusion Principle also inhibits the conduction of electric current by the free-electron gas. When acted upon by the electric field, electrons try to absorb energy and momentum from this field, making transitions to nearby states of slightly higher energy and momentum. If all such states are occupied, how can the electrons make the transitions? The answer is that the electric field acts also on the electrons in the higher states; it increases the momentum of *all* the electrons at the same rate. The electric field removes electrons from the higher states lifting them into even higher, unoccupied states, and creates vacancies at just the rate needed to accommodate the electrons arriving from the lower states. **Hence the Exclusion Principle does not interfere with the acceleration process, which proceeds pretty much as in the classical case.** Only in the deceleration process does the Exclusion principle makes a difference, most collisions are **forbidden because the final state is occupied.**

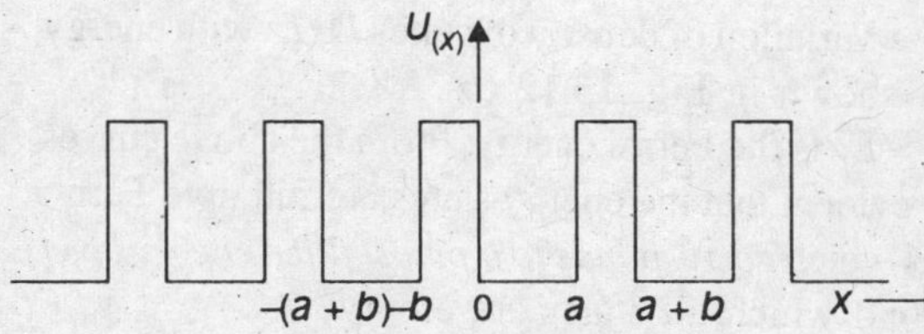

Fig. 13.12 (a)

Density of States. *It is defined as the number of electronic states per unit energy range and is denoted by D(E).*

We have from the Equation (3)

$$E = \frac{N^2 \pi^2 \hbar^2}{2m_e L^2} \qquad ...(16)$$

where we have replaced R by N and it corresponds to **number of electronic states with energies less than E within the radius R.**

Differentiating Equation (16) w.r.t. N, we get

$$\frac{dE}{dN} = \frac{\hbar^2 \pi^2}{2m_e L^2} \times 2N = \frac{h^2}{8\pi^2} \frac{\pi^2}{m_e} \frac{2N}{L^2} \qquad [\because \hbar = \frac{h}{2\pi}]$$

$$= \frac{h^2}{m_e} \times \frac{N}{4L^2}$$

or

$$\frac{dN}{dE} = \frac{4L^2 m_e}{Nh^2} \qquad ...(17)$$

where $\frac{dN}{dE}$ gives number of energy levels.

---

*The exact result is $C_v = 1/2\pi\ (kT/E_F)\ N_0 k$.

As there are two *quantum states* for each energy level, one of spin up and other of spin down, so we have the *density of states* of a free electron gas in one demension

$$D(E) = 2\frac{dN}{dE} \quad ...(18)$$

$$= \frac{8L^2 m_e}{Nh^2}$$

Also from Equation (3), we have

$$\frac{1}{N} = \frac{\hbar\pi}{L}\left(\frac{1}{2m_e E}\right)^{1/2} = \frac{h}{2L}\left(\frac{1}{2m_e E}\right)^{1/2}$$

$\therefore$ Equation (17) may be written as

$$D(E) = \frac{8L^2 m_e}{h^2} \times \frac{h}{2L}\left(\frac{1}{2m_e E}\right)^{1/2}$$

$$= \frac{4L}{h}\left(\frac{m_e}{2E}\right)^{1/2} \quad ...(19)$$

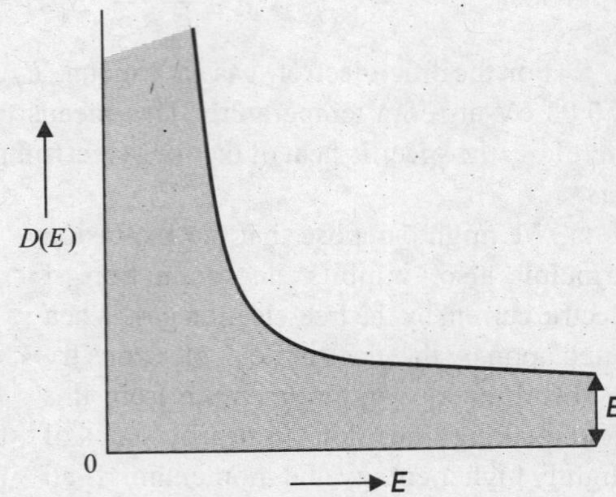

Fig. 13.12 (*b*)

Variation of density of states $D(E)$ with energy E is shown in Fig. 13.12 (*b*).

E_F is the Fermi energy. From the above curves it is evident that the energy states are full upto Fermi level which *corresponds to topmost filled energy level* beyond which all states are empty.

Example 2. *What is the Fermi energy for the free electron gas in silver? What is the speed of an electron with this enrgy? [N/V = 5.8 × 10^{23} /m^3]*

Solution: With $N/V = 5.8 \times 10^{28}$ /m^3, Equation (7) becomes

$$E_F = (3\pi^2)^{2/3} \frac{(1.05 \times 10^{-34}\text{ Js})^2}{2 \times 9.1 \times 10^{-31}\text{ kg}} (5.8 \times 10^{28}/\text{m}^3)^{2/3} \quad ...(1)$$

The electron with this energy have the speed

$$v_F = \sqrt{\frac{2E_F}{m_e}} = 1.4 \times 10^6 \text{ m/s} \quad \textbf{Ans.} \quad ...(2)$$

Thus the typical speeds of the electrons in the Fermi gas are quite large. For comparison, note that if we wanted to give the molecules of a classical gas typical speeds of this order of magnitude, we would have to heat the gas to a temperature of 6×10^4 K and this is a striking illustration of the difference between the classical gas and the Fermi gas.

Example 3. *Consider silver in the metallic state, with one free electron/atom. Calculate the fermi energy. Given density of silver = 10.5 gm/cm^3 and its atomic weight = 108.*

Solution: We have $\frac{N}{V} = \frac{\text{Atoms}}{\text{Volume}} = \frac{(\text{Atoms/mole}) \times (\text{mass/volume})}{\text{mass/mole}} = \frac{N_0 \times \rho}{\omega}$

Where N_0 = Avogadro number $= 6.02 \times 10^{23}$ Atoms/mole.

ρ = density of Silver = 10.5 gm/cm^3.

ω = Atomic mass of Silver = 108 gm/mole.

$$\therefore \quad \frac{N}{V} = \frac{6.02 \times 10^{23}\text{ atom/mole} \times 10.5\text{ gm/cm}^3}{108\text{ gm/mole}}$$

$$= 5.9 \times 10^{22} \text{ free electrons/cm}^3 = 5.9 \times 10^{28} \text{ free electrons/m}^3$$

Now Fermi energy is given by

$$E_F = \frac{h^2}{8m_e}\left(\frac{3N}{\pi V}\right)^{2/3}$$

$$= \frac{6.6 \times 10^{-34} \text{ Js}}{8 \times 9.1 \times 10^{-31} \text{ kg}} \times \left(\frac{3 \times 5.9 \times 10^{28} /\text{m}^3}{\pi} \right)^{2/3}$$

$$= 8.8 \times 10^{-19} \text{ Joule} = \mathbf{5.4\ eV\ Ans.}$$

Example 4. *If the Fermi energy of a metal is 10 eV, what is the corresponding classical temperature?*

Solution: We have the relation

$$E = \frac{3}{2} kT = \frac{3}{5} E_F$$

$$T = \frac{2E_F}{5k} = \frac{2 \times (10 \times 1.602 \times 10^{-19} \text{C})}{5 \times (1.381 \times 10^{-23})} = \mathbf{4.64 \times 10^4\ K\ Ans.}$$

Example 5. *There are about 2.5 × 10^{28} free electrons/ m^3 in sodium. Calculate its Fermi energy, Fermi velocity and Fermi temperature (h = 6.62 × 10^{-34} J-s).*

Solution: $\frac{N}{V}$ = No. of free electrons/unit volume of metal = 2.5 × 10^{28}/m^3

We have the relation

$$E_F = \frac{h^2}{2m} \left(\frac{3}{8\pi} \cdot \frac{N}{V} \right)^{2/3} = \frac{(6.62 \times 10^{-34})^2}{2 \times (9.1 \times 10^{-31})} \left(\frac{3}{8\pi} \times 2.5 \times 10^{28} \right)^{2/3}$$

$$= 5 \times 10^{-19} \text{ J} = \frac{5 \times 10^{-19}}{1.6 \times 10^{-19}} = \mathbf{3.1\ eV}$$

This is the max. K.E. of free electrons at absolute zero. Let v_F be the Fermi velocity, so we have

$$\frac{1}{2} m v_F^2 = E_F$$

$$\therefore \quad v_F = \left(\frac{2E_F}{m} \right)^{1/2} = \left[\frac{2 \times 5 \times 10^{-19}}{9.1 \times 10^{-30}} \right]^{1/2} = 1.047 \text{ ms}^{-1}$$

We can define Fermi temperature T_F from the relation

$$E_F = kT_F$$

$$\therefore \quad T_F = \frac{E_F}{k} = \frac{5 \times 10^{-19}}{1.38 \times 10^{-23}} = 3.623 \times 10^4 \text{ K} \quad \textbf{Ans.}$$

Example 6. *Calculate the Fermi energy of sodium assuming that metal has one free electron/ atom. (Given density of sodium = 970 kg/m^3).*

Solution: In this case

$$\frac{N}{V} = \frac{N}{M/\rho} = \frac{6.02 \times 10^{26}}{23/970} = 2.54 \times 10^{28}$$

Fermi energy is given by the relation

$$E_F = \frac{h^2}{2m} \left(\frac{3}{8\pi} \cdot \frac{N}{V} \right)^{2/3} = \frac{(6.625 \times 10^{-34})^2}{2 \times 9.1 \times 10^{-31}} \left(\frac{3}{8\pi} \times 2.54 \times 10^{28} \right)^{2/3}$$

$$= 5.11 \times 10^{-19} \text{ J} = \frac{5.11 \times 10^{19}}{1.6 \times 10^{-19}} \text{ eV} = \mathbf{3.19\ eV} \quad \textbf{Ans.}$$

Example 7. *Calculate the extent of the energy range between F (E) = 0.9 and F (E) = 0 1 at a temperature of 200 K and express it as a function of E_F which is 3 eV.*

Solution: We have the relation

$$F(E) = \frac{1}{1 + e^{(E - E_F)/k_\beta T}} = \frac{1}{1 + e^{-(E_F - E)/k_\beta T}}$$

(*a*) In this case $F(E) = 0.9$

and let $(E_F - E)/k_\beta T = x$

$$\therefore \quad 0.9 = \frac{1}{1+e^{x}}$$

$$\text{or} \quad 0.9 + 0.9\,e^{x} = 1$$

$$\text{or} \quad 0.9\,e^{x} = 1 - 0.9 = 0.1$$

$$\therefore \quad e^{x} = \frac{1}{9} \quad \text{or} \quad e^{x} = 9$$

Taking log on both sides

$$x = \log_e 9 = 2.3026 \times \log_{10} 9 = 2.198$$

$$\therefore \quad \frac{E_F - E}{k_\beta \cdot T} = 2.198$$

$$\text{or} \quad E_F - E = 2.198 \times k_\beta \cdot T$$

$$\text{or} \quad k_\beta T = \frac{E_F - E}{2.198}$$

$$\text{Now} \quad k_\beta T = \frac{1.38 \times 10^{23} \times 200}{1.6 \times 10^{19}}\,\text{eV} = 0.017\text{ eV}.$$

$$\therefore \quad E_F - E = 2.198 \times 0.017 = 0.037\,\text{eV}$$

$$\text{So} \quad E = E_F - 0.037 = 3 - 0.037 = \mathbf{2.963\ eV}$$

(*b*) In this case $F(E) = 0.1$

$$\therefore \quad 0.1 = \frac{1}{1+e^{(E_1 - E_F)/k_\beta \cdot T}} = \frac{1}{1+e^{x}} \qquad \left(\text{put } x = \frac{E_1 - E_F}{k_\beta \cdot T}\right)$$

$$\text{or} \quad 1 = 0.1 + 0.1\,e^{x}$$

$$\therefore \quad e^{x} = 9$$

Taking log both sides $x = 2.3026 \log_{10} 9$

$$\text{or} \quad \frac{E_1 - E_F}{0.017} = 2.3026 \times 0.954$$

$$\therefore \quad E_1 - E_F = 0.017 \times 2.3026 \times 0.954 = 0.037\text{ eV}$$

$$\therefore \quad E_1 = E_F + 0.037 = 3 + 0.037 = 3.037\text{ eV}$$

$$\therefore \quad \Delta E = E_1 - E = 0.037 - 2.963 = 0.074\text{ eV}$$

$$\text{or} \quad \frac{\Delta E}{E_F} = \frac{0.074}{3} = 0.025 = \mathbf{2.5\%} \quad \textbf{Ans.}$$

Example 8. *Calculate the temperature at which there is one per cent probability that a state, with an energy 0.5 eV, above Fermi energy will be occupied by an electron.*

Solution: We have the relation

$$F(E) = \frac{1}{1+e^{(E-E_F)/k_\beta \cdot T}}$$

In the case $E - E_F = 0.5$ eV

$$\text{and} \quad F(E) = 1\% = \frac{1}{100}$$

$$\therefore \quad \frac{1}{100} = \frac{1}{1+e^{x}} \quad \text{where } x = \frac{0.5}{k_\beta \cdot T}$$

$$\therefore \quad 0.01 = \frac{1}{1+e^{x}}$$

or $$0.01 + 0.01\, e^x = 1$$

or $$0.01\, e^x = 0.99$$

$$\therefore \quad e^x = \frac{0.99}{0.01} = 99$$

Taking log on both sides $$x = 2.3026 \log_{10} 99$$

or $$\frac{0.5}{k_\beta \cdot T} = 2.3026 \log_{10} 99$$

$$\therefore \quad k_\beta.T = \frac{0.5}{2.3026 \log_{10} 99} = 0.109\,\text{eV}$$

or $$T = \frac{0.109 \times 1.6 \times 10^{-99}}{1.38 \times 10^{-23}} = 1264\text{ K}$$

Hence $T = \mathbf{1264\ kelvin}$ **Ans.**

Example 9. *Find the fermi energy in copper on the assumption that each copper atom contributes one free electron to the electron gas. The density of copper is 8.94×10^3 kg/m^3 and its atomic mass = 63.5 u. Given u = 1.66×10^{-27} kg.*

Solution: In this case $\rho = 8.94 \times 10^3\ kg/m^3$, mass of atom = 63.5 u

Number of free electrons/unit volume, $n = \frac{N}{V}$

So $$n = \frac{N}{V} = \frac{\text{atoms}}{m^3} = \frac{\text{mass/m}^2}{\text{mass/atoms}}$$

$$n = \frac{8.94 \times 10^3\ \text{kg/m}^3}{(63.5\,u)(1.66 \times 10^{-27}\,\text{kg}/u)} = 8.48 \times 10^{28}\ \text{atoms/m}^3$$

$$= 8.48 \times 10^{28}\ \text{electrons/m}^3$$

Now Fermi Energy is given by

$$E_F = \frac{h^2}{8m_e}\left(\frac{3N}{\pi V}\right)^{2/3} = \frac{h^2}{8m_e} \cdot \left(\frac{3n}{\pi}\right)^{2/3}$$

$$E_F = \frac{(6.63 \times 10^{-34})^2}{8 \times 9.11 \times 10^{-31}}\left(\frac{3 \times 8.48 \times 10^{28}}{3.14}\right)^{2/3}$$

$$= 0.603 \times 10^{-37} \times (8.10 \times 10^{28})^{2/3} = 1.13 \times 10^{-18} J$$

$$= \frac{\mathbf{1.13 \times 10^{-18}}}{\mathbf{1.6 \times 10^{-19}}} eV = \mathbf{7.04\,eV\ Ans.}$$

Example 10. *(a) Estimate the mean collision time in copper at 20°C, assuming one free electron/copper atom. (b) Assuming that the average speed of the free electrons is about 10^6 m/s, estimate the mean free path in copper atom. (Given n = 8.48×10^{28} electrons m^3 and $\rho = 1.673 \times 10^{-8}\ \Omega m$).*

Solution: In this case average velocity is given by $\langle v \rangle = 10^6$ m/s, $\rho = 1.673 \times 10^{-8}\ \Omega m$, n = 8.48 × 1028 electrons/m^3.

(*a*) Mean collision time $$\tau = \frac{m_e}{ne^2\rho}$$

$$\tau = \frac{9.11 \times 10^{-31}}{(8.48 \times 10^{28}) \times (1.6 \times 10^{-19})^2 \times (1.673 \times 10^{-8})}$$

$$= 2.51 \times 10^{-14}\ \text{s}$$

(*b*) We have the mean free path $\lambda = \langle v \rangle \tau$

$$= 10^6 \times 2.51 \times 10^{-14} = \mathbf{2.51 \times 10^{-8}}\ m\ \textbf{Ans.}$$

Example 11. *A metallic wire has a resistivity of 1.42 × 10–8 Ωm. For an electric field of 0.14 V/m, find (a) average drift velocity of electrons and (b) mean collision time assuming that there are 6 × 10^{28} electrons/m^3.*

Solution: In this case ρ = 1.42 × 10^{-8} Ω*m*, *E* = 0.14 V/m, *n* = 6 × 10^{28} electrons/m^3.

(*a*) Resistivity of a metal is given by the relation

$$\rho = \frac{m_e}{ne^2\tau}$$

∴ Mean collision time,

$$\tau = \frac{m_e}{ne^2\rho}$$

$$= \frac{9.11\times10^{-31}}{(6\times10^{28})(1.6\times10^{-19})(1.42\times10^{-8})}$$

∴ τ = **4.236 × 10^{-14} *s* Ans.**

Example 12. *(a) Determine the number density of carriers in a copper wire assuming thee is one carrier (electron) per copper atom. (b) The maximum recommended current in a 14 gauge copper wire (radius = 081 mm, A = 2.1 × 10^{-6} m^2) used in household circuits is 15 A. Determine the drift speed of the electrons in such a case. (Give ρ = 8.95 × 103 kg/m^3 and M = 63.5 g/mol).*

Solution: Given I = 15 *A*, *A* = 2.1 × 10^{-6} *m*^{2}, ρ = 8.95 × 10^3 kg/m^3 and *M* = 63.5 g/mol.

(*a*) Number of free electrons/unit volume, $n = \dfrac{N_A\rho}{M}$

$$n = \frac{(6.02\times10^{23}\,\text{mol}^{-1})(8.95\times10^6\,\text{g/m}^3)}{(63.5\,\text{g/mol})}$$

= 8.48 × 10^{28} electrons/m^3

(*b*) Drift velocity of the electrons may be written as

$$v_d = \frac{I}{nAe}$$

$$= \frac{15}{8.48\times10^{28}\times2.1\times10^{-6}\times1.6\times10^{-19}}$$

= **5.3 × 10^{-4} m/s Ans.**

Example 13. *Find the drift velocity of the free electrons in a copper wire whose cross sectional area is 1.0 mm^2 when the wire carries a current of 1.0 A. Assume that each copper atom contributes one electron to the electron gas. (Given n = 8.5 × 10^{28} electrons/m^3).*

Solution: In this case *A* = 10 *mm*^{2} = 1.0 × 10^{-6} *m*^{2}, *I* = 10 *A*

Drift velocity of the free electrons is given by the relation

$$v_d = \frac{I}{nAe}$$

$$= \frac{1.0}{(8.5\times10^{28})(1.0\times10^{-6})(1.6\times10^{-19})}$$

= **7.4 × 10^{-4} m/s Ans.**

Example 14. *A uniform silver has a resistivity of 1.54 × 10^{-8} Ωm at room temperature. For an electric field along the wire of 1 V/cm, calculate (a) the drift velocity (b) the mobility and (c) the relaxation time of electrons assuming that there are 5.8 × 10^{28} electrons/m^3 of the material.*

Solution: We have ρ = 1.54 × 10^{-8} Ω*m*, *E* = 100 V/m, *n* = 5.8 × 10^{28} electrons/m^3, τ = ?, v_d = ? and μ = ?

(*a*) Mean collision time may be written as $\tau = \dfrac{m_e}{ne^2\rho}$

$$= \frac{9.11\times10^{-31}}{(5.8\times10^{28})(1.6\times10^{-19})^2(1.54\times10^{-8})}$$

$= \mathbf{3.98 \times 10^{-14}}$ **s Ans.**

(*b*) Drift velocity, is given by $v_d = \left(\frac{eE}{m_e}\right)\tau$

$$= \frac{(1.6\times10^{-19})(100)(3.98\times10^{-14})}{9.11\times10^{-31}} = \mathbf{0.7\,m/s\ Ans.}$$

(*c*) Relation for mobility is $\mu = \frac{v_d}{E}$

$$= \frac{0.7}{10^2} = \mathbf{7\times10-3\,m^2/Vs\ Ans.}$$

Example 15. *Find the relaxation time of conduction electrons in a metal of resistivity $1.54 \times 10^{-8}\,\Omega m$, if the metal has 5.8×10^{28} electrons/m^3.*

Solution: In this case $\rho = 1.54 \times 10^{-8}\ \Omega m$, $n = 5.8 \times 10^{28}$ electrons/m^3 and τ = ?

Resistivity of the metal, $\rho = \frac{m_e}{ne^2\tau}$, so we have

Relaxation time, $\tau = \frac{m_e}{ne^2\rho}$

$$= \frac{9.11\times10^{31}}{(5.8\times10^{28})(1.6\times10^{-19})(1.54\times10^{-8})}$$

$= \mathbf{3.97 \times 10^{-14}}$ ***s* Ans.**

Example 16. *Calculate the mobility of electrons in copper assuming that each atom contributes one free electron for conduction. Resistivity of copper = $1.7 \times 10^{-8}\,\Omega m$. Atomic weight = 63.54. Density = 8.96×10^3 kg/m^3 and Avogadro's number 6.025×10^{23}/mole.*

Solution: Given that $\rho = 1.7 \times 10^{-8}\ \Omega m$, $M = 63.54$, $D = 8.96 \times 10^3$ kg/m^3, $N_A = 6.025 \times 10^{23}$/ mole, Number of free electrons/atom = 1 and μ = ?

Free electrons/unit volume, $n = \frac{N}{V} = N_A \cdot \frac{\rho}{M}$

$$= \frac{(6.025\times10^{26})(8.96\times10^3)}{63.54}$$

$= \mathbf{8.50 \times 10^{28}/m^3}$ **Ans.**

Example 17. *Calculate the drift velocity and thermal velocity of free electrons in copper at room temperature, (300 K), when a copper wire of length 3 m and resistance 0022 Ω carries a current of 15 A. Given $\mu_{cu} = 4.3 \times 10^{-3}$ m^2/Vs.*

Solution: In this case l = 3 *m*, $R = 0.022\ \Omega$, $I = 15\ A$, $T = 300$ K, $\mu_{cu} = 4.3 \times 10^{-3}$ m^2/Vs, v_d = ? and v_{th} = ?

Voltage drop across the copper wire is given by

$$V = IR = 15 \times 0.022 = \mathbf{0.33}\ \boldsymbol{V}$$

$\therefore$ Electric field, $E = \frac{V}{l} = \frac{0.33}{3} = \mathbf{0.11\ V/m}$

Drift velocity, $v_d = E \times u = 0.11 \times 4.3 \times 10^{-3} = \mathbf{0.473 \times 10^{-3}}$ **m/s Ans.**

Thermal velocity is written as,

$$v_{th} = \sqrt{\frac{3kT}{m_e}}$$

$$= \sqrt{\frac{3 \times 1.387 \times 10^{-23} \times 300}{9.11 \times 10^{-31}}}$$

$$= \mathbf{1.17 \times 105\ m/s\ Ans.}$$

Example 18. *Calculate the Fermi energy in eV for silver at 0 K, given that the density of silver = 10500 kg/m^3, atomic weight = 107.9 and it has one conduction per atom.*

Solution: Given $T = 0\ K$, $D = 10500$ kg/m^3, $M = 107.9$ and $E_F = ?$

Number of free electrons/unit volume, $n = \frac{N_A}{V} = N_A \frac{\rho}{M}$

$$\therefore \quad n = \frac{(6.025 \times 10^{26})(10500)}{107.9} = \mathbf{5.863 \times 10^{28}/m^3\ Ans.}$$

Fermi energy, $$E_F = \left(\frac{h^2}{8m_e}\right)\left(\frac{3n}{\pi}\right)^{2/3}$$

$$= \frac{(6.63 \times 10^{-34})}{8 \times 9.11 \times 10^{-31}}\left(\frac{3 \times 5.863 \times 10^{28}}{3.14}\right)$$

$$= (0.603 \times 10^{-37})\ (5.60 \times 10^{28})^{2/3} = 8.83 \times 10^{-19}\ J$$

$$E_F = \frac{8.83 \times 10^{-19}}{1.60 \times 10^{-19}} = \mathbf{5.518\ eV\ Ans.}$$

Fermi energy for silver is 5.518 eV which corresponds to E_F (0).

$$\therefore \quad E_F\ (0) = \mathbf{5.518\ eV}$$

Example 19. *Energy over which probability f (E) falls from 0.9 to 0.1%. Over what range of energy, expressed in terms of kT, does the Fermi-Dirac distribution function change from 0.90 to 0.10?*

Solution: F–D function is given by, $f(E) = \frac{1}{e^{(E_b - E_F)/kT} + 1}$

Let E_b represent the energy at which $f(E_b) = 0.1$, so we have,

$$\therefore \quad 0.1 = \frac{1}{e^{(E_b - E_F)/kT} + 1}$$

$$e^{(Eb - EF)/kt} + 1 = \frac{1}{0.1} = 10$$

$$\therefore \quad e^{(Eb - EF)/kT} = 9$$

or $$\frac{E_a - E_F}{kT} = \log_e 9 = 2.303 \log_{10} 9 \qquad ...(1)$$

Let E_a be such that $f(E_a) = 0.90$ in that case

$$0.9 = \frac{1}{e^{(E_a - E_F)/kT} + 1}$$

$$e^{(E_b - E_F)/kT} + 1 = \frac{1}{0.9} = 1.11$$

$$\therefore \quad e^{(E_a - E_F)/kT} = 1.11$$

or $$\frac{E_a - E_F}{kT} = \log_e 0.11 = 2.303 \log_{10} 0.11 \qquad ...(2)$$

Subtracting Equation (2) from Equation (1), we have

$$\frac{E_b - E_F}{kT} - \frac{E_a - E_F}{kT} = 2.303\ (\log_{10} 9 - \log_{10} 0.11) = 2.303 \log_{10}\left(\frac{9}{0.11}\right)$$

$$= 2.303 \times \log_{10}(81.818) = 2.303 \times 1.91 = \mathbf{4.4\ Ans.}$$

So the probability that a state occupied from 90% to 10% over an energy range $E_b - E_a = 4.4\ kT$, so this is the range over which F-D function changes in terms of kT.

Example 20. *The specific gravity of tungsten is 18.8 and its atomic weight is 184. Assume that there are two electrons per atom. Calculate the numerical values of n and E_F.*

Solution: Specific gravity = 18.8, M = 184

Number of electrons/atom = 2, n = ? and E_F = ?

Number of free electrons/unit volume, $n = \frac{N}{V} = N_A \cdot \frac{\rho}{M}$

$$\therefore\ n = 6.02 \times 10^{23}\ \frac{\text{molecules}}{\text{mole}} \times \frac{1\,\text{mole}}{184\ g} \times \frac{g}{cm^3} \times \frac{2\,\text{electrons} \times 1\ \text{atom}}{\text{atom.molecule}}$$

$$n = 1.23 \times 10^{23}\ \frac{\text{electrons}}{\text{cm}^2} = \mathbf{1.23 \times 10^{29}\ electrons/m^3\ Ans.}$$

For tungsten the atomic and the molecular weights are the same. So we have in this case

Fermi energy, $E_F = 3.64 \times 10^{-19}\ n^{2/3}$

$= (3.64 \times 10^{-19})\ (0.123 \times 10^{30})^{2/3} = \mathbf{9\ eV\ Ans.}$

Example 21. *Calculate the Fermi velocity and the Fermi temperature for the free electrons in gold. Given $E_F = 5.53\ eV$ and $\tau = 3.91 \times 10^{-14} s$.*

Solution: We have $E_F = 5.53$ eV $= 5.53 \times 1.6 \times 10^{-19}\ J$, $\tau = 3.91 \times 10^{-14} s$, v_F = ? and λ = ?

So Fermi velocity is given by

$$v_F = \sqrt{\frac{2E_F}{m_e}}$$

$$= \sqrt{\frac{2 \times 5.53 \times 1.6 \times 10^{-19}}{9.11 \times 10^{-31}}} = \mathbf{1.39 \times 10^6\ m/s\ Ans.}$$

and Fermi temperature, $T = \frac{E_F}{k} = \frac{5.53 \times 1.6 \times 10^{-19}}{1.38 \times 10^{-23}} = \mathbf{6.41 \times 10^4\ \mathit{K}\ Ans.}$

Example 22. *In a solid, consider the energy level lying 0.01 eV below Fermi level. What is the probability of this level not being occupied by an electron? (Given kT = 0.026 eV at T = 300 K).*

Solution: Here $E_F - E = 0.01$ eV.

The probability of an energy level E not being occupied by an electron is given by $[1 - f(E)]$, so we have

$$[1 - f(E)] = 1 - \frac{1}{e^{(E-E_F)/kT} + 1} = \frac{1}{e^{(E_F-E)/kT} + 1}$$

$$= \frac{1}{e^{0.01/0.026} + 1} = \frac{1}{e^{0.385} + 1} = \frac{1}{1.47 + 1} = \frac{1}{2.47}$$

$= \mathbf{0.405\ Ans.}$

Example 23. *Calculate the mobility and the relaxation time of electrons in copper with the following data: Resistivity of copper = $1.73 \times 10^{-8}\ \Omega m$, atomic weight = 63.5, density = 8.92g/cc, Avogadro number = 6.023×10^{23}.*

Solution: Number of free electrons per unit volume may be written as

$$n = \frac{\text{Avogadro number} \times \text{density}}{\text{Atomic weight}}$$

$$= \frac{6.023 \times 10^{23} \times 8.92 \times 10^3}{63.5} = 8.463 \times 10^{25}$$

Mobility of electrons is given by the relation

$$\mu = \frac{1}{\rho ne}$$

$$= \frac{1}{1.73 \times 10^{-8} \times 8.463 \times 10^{25} \times 1.6 \times 10^{-19}}$$

$$= \mathbf{4.1145\ m2/Vs\ Ans.}$$

Relaxation time, $\tau = \frac{m}{ne^2 \rho}$

$$= \frac{9.11 \times 10^{-31}}{8.463 \times 10^{25} \times (1.6 \times 10^{-19})^2 \times 1.73 \times 10^{-8}}$$

$$= \mathbf{2.25 \times 10^{-11}\ s\ Ans.}$$

Example 24. *Copper is an fcc crystal with lattice constant 3.61 Å and the metal has one free electron per atom. (a) Calculate the Fermi energy in eV for the metal. (b) Calculate its Fermi factor at 300 K for an energy value 0.1 eV higher than E_F.*

Solution: In this case a = 3.61Å = 3.61 × 10^{-10} m, Number of free electrons per atom = 1, T = 300 K, $E = E_F + 0.1$ eV.

For *fcc* structure each until cell consists of 4 atoms and the structure is that of a cube.

So that number of free electrons/unit cell, $N = 4 \times 1 = 4$

Volume of the unit cell, $V = a^3 = (3.61 \times 10^{-10})^3 = 47.046 \times 10^{-30}\ m^3$

Number of free electrons/unit volume, $n = \frac{N}{V}$

$$\therefore \quad n = \frac{4}{47.046 \times 10^{-30}} = 8.5 \times 10^{28}$$

(*a*) Now Fermi energy, $E_F = \left(\frac{h^2}{8 m_e}\right)\left(\frac{3n}{\pi}\right)^{2/3}$

$$= \frac{(6.63 \times 10^{-34})^2}{8 \times 9.11 \times 10^{-31}} \times \left(\frac{3 \times 8.5 \times 10^{28}}{314}\right)^{2/3}$$

$$= (0.603 \times 10^{-37})\ (8.12 \times 10^{28})^{2/3}$$

$$= (0.603 \times 10^{-37})\ (148.7 \times 10^{18})$$

$$= 89.67 \times 10^{-19}\ J = \frac{89.67 \times 10^{-19}}{1.6 \times 10^{-19}}\ \text{eV}$$

$$= \mathbf{56.04\ eV\ Ans.}$$

(*b*) Fermi factor may be written as

$$f(E) = \frac{1}{e^{(E-E_F)/kT} + 1}$$

In this case $E = E_F + 0.1$ or $E - E_F = 0.1$ and $k = 1.381 \times 10^{-23}\ J/K$

$$= 1.381 \times 10^{-23} \times 6.24 \times 10^{18}\ \text{eV/K}$$

$$= 8.6174 \times 10^{-5}\ \text{eV/K}$$

$$\therefore \quad f(E) = \frac{1}{e^{0.1/86174} \times 10^{-5} \times 300 + 1} = \frac{1}{e^{3.868} + 1}$$

$$= \frac{1}{47.85 + 1} = \frac{1}{48.85} = \mathbf{0.0205\ Ans.}$$

Example 25. *The Fermi level for potassium is 2.1 eV. Calculate the velocity of the electrons at the Fermi level.*

Solution: Given $E_F = 2.1$ eV, $v_F = ?$

Fermi velocity is given by the relation

$$v_F = \sqrt{\frac{2E_F}{m_e}}$$

$$= \sqrt{\frac{2 \times 2.1 \times 1.6 \times 10^{-19} J}{9.11 \times 10^{-31} \text{ kg}}} = \mathbf{8.6 \times 10^5 \text{ m/s Ans.}}$$

Example 26. *There are 10^{20} conduction electrons per m^3 in a material having resistivity of 0.1 Ωm. Find the charge mobility and the electric field needed to produce a drift velocity of 1 m/s.*

Solution: Given $n = 10^{20}$ electrons/m^3, $\rho = 0.1 \Omega m$, $v = 1$ m/s

Mobility of electrons is given by the relation

$$\mu = \frac{1}{\rho n e} = \frac{1}{0.1 \times 10^{20} \times 1.6 \times 10^{-19}} = 0.625 \; m^2/\text{Vs}$$

$$\therefore \text{ Electric field, } \quad E = \frac{v}{\mu} = \frac{1}{0.625} = \mathbf{1.6 \text{ V/m Ans.}}$$

13.5. KRONIG-PENNEY MODEL (BEHAVIOUR OF AN GELECTRON IN PERIODIC POTENTIAL)

A periodic potential problem for which the wave equation can be solved in terms of simple analytic functions is the square-well array of Fig. 13.13. The wave equation is

$$\frac{\hbar^2}{2m}\frac{d^2\psi}{dx^2} + U(x)\psi = E\psi \qquad ...(1)$$

where $U(x)$ is the potential energy and E is the energy eigen value. In the region $0 < x < a$ in which $U = 0$, the eigen function is a linear combination,

$$\psi = Ae^{iKx} + Be^{-iKx} \qquad ...(2)$$

of plane waves moving to the right and to the left, with energy

$$E = \hbar^2 K^2/2m \qquad ...(3)$$

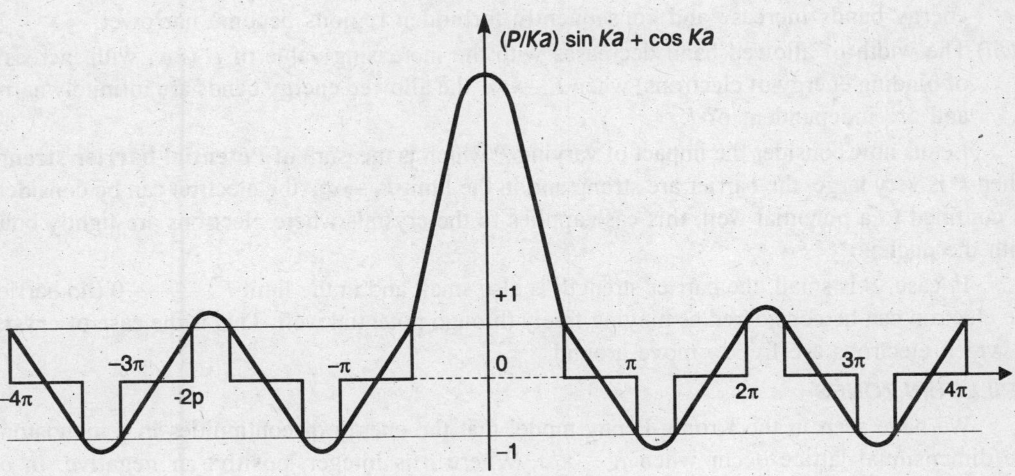

Fig. 13.13

In the region $a < x < a + b$ within the barrier the solution is of the form

$$\psi = Ce^{Qx} + De^{-Qx} \qquad ...(4)$$

with

$$U_0 - E = \hbar^2 Q^2/2m. \qquad ...(5)$$

The constants A, B, C, D are chosen so that ψ and $d\psi/dx$ are continuous at $x = 0$

$$A + B = C + D \quad ...(6)$$

There must also be continuity at $x = a$:

$$Ae^{iKa} + Be^{-iKa} = Ce^{Qa} + De^{-Qa} \quad ...(7)$$

$$iKAe^{iKa} - iKBe^{-iKa} = QCe^{Qa} - QDe^{-Qa} \quad ...(8)$$

A final condition is that the values of ψ and $d\psi/dx$ at $x = a$ be equal to those at $x = -b$, but advanced by the phase factor exp $[ik(a + b)]$. This condition is a consequence of **Bloch functions.** The left-hand side of (7) is replaced by

$$A \exp [iKa - iK(a + b)] + B \exp[- iKa - iK (a + b)], \quad ...(9)$$

and the corresponding change is made in the exponential factors on the left hand side of (8). The four Equations (6) to (8) have a solution only if the determinant of the coefficients of A, B, C, D vanishes, or if

$$[(Q^2 - K^2)/2QK] \sinh Qb \sin Ka + \cosh Qb \cos Ka = \cos K (a + b) \quad ...(10)$$

It is rather difficult to obtain this equation.

The result is simplified if we represent the potential by the periodic delta function obtained when we pass to the limit $b = 0$ and $U^2 = \infty$ in such a way that $Q^2ba/2 = P$, a finite quantity. Then Equation (10) reduces to

$$(P/Ka) \sin Ka + \cos Ka = \cos Ka \quad ...(11)$$

The ranges of K for which this equation has solutions are plotted in Fig. 13.13, for the case $P = 3\pi/2$. The corresponding values of the energy are plotted in Fig. 13.14. Note the energy gaps at the zone boundaries. It may be noted that the wave vector K of the Bloch function is the important index, and not the wavevector K in Equation (2).

From Fig. 13.13, the following conclusions are drawn:

(*i*) The energy spectrum consists of an infinite number of allowed energy bands separated by intervals where there are no energy levels, these are called forbidden regions.

The boundaries allowed energy levels corresponds to the values of $\cos ka = \pm 1$

or $Ka = n\pi$ or $K = n\pi/a$

(*ii*) As Ka increases, the first term on L.H.S. of Equation (11) decreases, so the width of allowed energy bands increase and consequently forbidden regions become narrower.

(*iii*) The width of allowed band decreases with the increasing value of P (*i.e.*, with increasing of binding energy of electrons) when $P \to \infty$, the allowed energy bands are infinitely narrow and are independent of K.

Let us now consider the impact of varying P which is measure of **Potential barrier strength** when P is very large, the barrier are strong and in the limit $P \to \infty$, the electron can be considered as confined to a potential well, this case applies to the crystals where electrons are tightly bound with the nuclei.

In case, P is small, the barrier strength is also small and in the limit $P \longrightarrow 0$ (no barrier), the electron can be considered as moving freely through potential well. This is the case of **crystals** in which electrons are free to move around.

BRILLOUIN ZONES

We have seen in the Kronig-Penny model that the energy discontinuities in a monoatomic are dimensional lattice occur when $K = \pi/a$. Where n is integer, positive or negative. In one dimensional monoatomic lattice, a live representing the value of K is divided by energy discontinuities into segments of length π/a, as shown in Fig. 13.14.

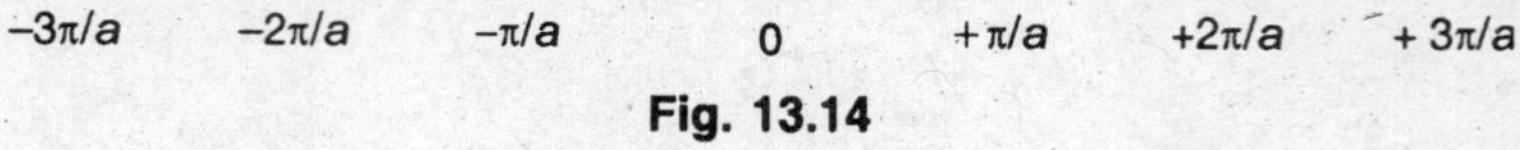

Fig. 13.14

These line segments are called *Brillouin Zones*. The segment $-\pi/a \le K = +\pi/a$ is the first *Brillouin Zone*.

The segments $-2\pi/a \le K - \pi/a$ *and* $\pi/a \le K \le 2\pi/a$ form the *second Brillouin Zone*.

Each crystal has its own characteristic Brillouin Zones.

The first zone for two dimensional square lattice will be square *ABCD*, the boundaries of which be given by the relations.

$$k_x = \pm\pi/a;\ k_y = \pm\pi/a$$

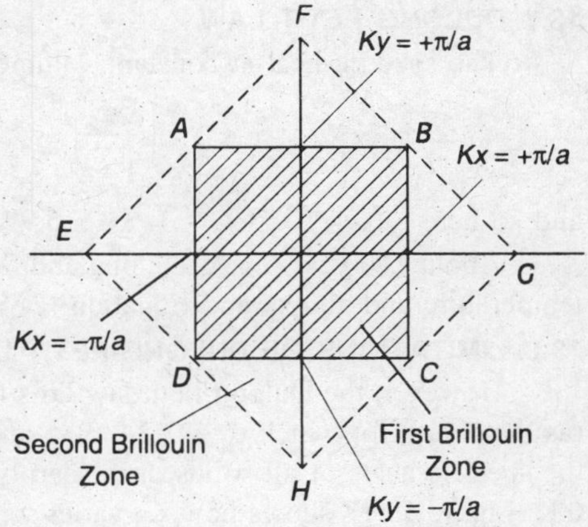

Fig. 13.15

The second Brillouin Zone is also shown, it contains electrons with $k > \pi/a$ that do not fit into the first zone yet which have sufficiently small wave numbers to avoid diffraction by diagonal atomic planes (Fig. 13.15). The second Zone consists of electrons with k varying from π/a to $2\pi/a$ for electrons moving in $\pm x$ and y directions, with the possible range of k values narrowing as the diagonal directions are approached. Further Brillouin Zones can be constructed in the same manner.

13.6. SPECIFIC HEATS OF SOLIDS

Failure of classical theory: Blackbody radiation is not the only familiar phenomenon whose explanation requires quantum statistical mechanics. Another is the way in which the internal energy of a solid varies with temperature.

Molar specific heat. Let us consider the molar specific heat of a solid at constant volume, C_v, which is the energy that must be added to one 1 K mol of it, whose volume is held fixed, to raise its temperature by 1 K. The specific heat at constant pressure C_p is 3 to 5 per cent higher than C_v in solids because it includes the work associated with a volume change as well as the change in internal energy.

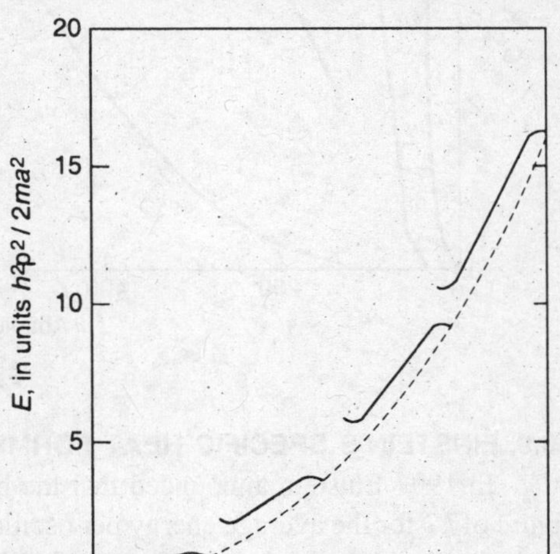

Fig. 13.16

Classical theory of heat. The internal heat energy of a solid resides in the vibrations of its constituent particles which may be atoms, ions or molecules; we shall refer to them as atoms here for convenience. These vibrations may be resolved into components along three perpendicular axes, so that we may represent each atom by three harmonic oscillators. According to classical physics a harmonic oscillator in thermal equilibrium at the temperature T has an average energy of kT. On this basis each atom in a solid should have $3\,kT$ of energy. A kilomole of a solid contains Avogadro's number N_0 of atoms, and its total internal energy E at the temperature T accordingly ought to be

$$E = 3N_0\,kT = RT$$

where $$R = N_0\,k = 3.31 \times 10^4 \text{ J/mol.}$$

$$K = 1.99 \text{ Kcal/kmol K}$$

is the universal gas constant. (We know that in an ideal-gas sample of n kilomoles, $PV = nRT$.)

13.7. DULONG PETIT LAW

The specific heat at constant volume is given in terms of E by

$$C_v = \left(\frac{\partial E}{\partial T}\right)_v \quad ...(1)$$

and so here $$C_v == 3R = 5.97 \text{ Kcal/kmol K} \quad ...(2)$$

About 100 years back, Dulong and Petit found that, indeed $C_v = 3R$ for most solids at room temperature and above, and Equation (2) is known as the *Dulong-Petit* law in their honour.

13.8. LIMITATIONS OF DULONG-PETIT LAW

However, the Dulong-Petit law fails for such light elements as boron, beryllium, and carbon (as diamond), for which $C_v = 3.34$, 3.85 and 1.46 Kcal/kmol K respectively at 20°C. Even worse, the specific heats of all solids drop sharply at low temperatures and approach 0 as T approaches 0 K. Figure 13.17 shows how C_v varies with T for several elements. Clearly something is wrong with the analysis leading up to Equation (2) and it must be something quite fundamental because the curves of Fig. 13.17 share the same general character.

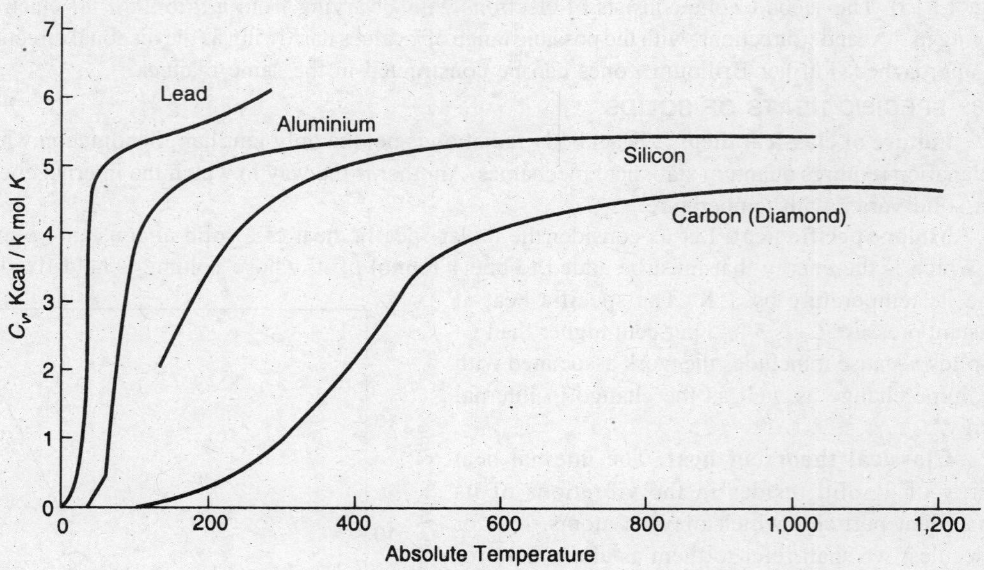

Figure 13.17

13.9. EINSTEIN'S SPECIFIC HEAT FORMULA

In 1907 Einstein announced that the basic flaw in the derivation of Equation (2) lies in the figure of kT for the average energy per oscillator in a solid. This flaw is the same as that responsible for the incorrect Rayleigh-Jeans formula for blackbody radiation. According to Einstein, the probability $f(\nu)$ that an oscillator has the frequency ν is given by $f(\nu) = 1/(e^{h\nu kT} - 1)$. Hence the average energy for an oscillator whose frequency of vibration is ν is

$$\varepsilon = h\nu f(\nu) = \frac{h\nu}{e^{h\nu/kT} - 1} \quad ...(3)$$

and not $\varepsilon = kT$. The total internal energy of a kilomole of a solid, therefore, becomes

$$E = 3N_0\, \varepsilon = \frac{3N_0 h\nu}{e^{h\nu/kT} - 1} \quad ...(4)$$

and its molar specific heat is

$$C_v = \left(\frac{\partial E}{\partial T}\right)_v = 3R\left(\frac{h\nu}{kT}\right)^2 \frac{e^{h\nu/kT}}{(e^{h\nu/kT} - 1)^2} \quad ...(5)$$

We can see at once that his approach is in the right track. At high temperatures, $h\nu << kT$, and

$$e^{h\nu/kT} \approx 1 + \frac{h\nu}{kT}$$

since

$$e^x = 1 + x + \frac{x^2}{2!} + \frac{x^3}{3!} + ... \quad ...(6)$$

Hence Equation (3) becomes $\varepsilon \approx h\nu/(h\nu/kT) = kT$ which leads to $C_v \approx 3R$, the Dulong-Petit value, as it should. At high temperatures the spacing $h\nu$ between possible energies is small relative to kT. So ε is effectively continuous and classical physics holds.

13.10. CASE OF FAILURE OF CLASSICAL THEORY

As the temperature decreases, the value of C_v given by Equation (5) decreases. The reason for the change from classical behaviour is that now the spacing between possible energies is becoming large relative to kT, which inhibits the possession of energies above the zero-point energy. The natural frequency ν for a particular solid can be determined by comparing Equation (5) with an empirical curve of its C_v versus T. The result in the case of aluminium is $\nu = 6.4 \times 10^{12}$ Hz, which agrees with estimates made in other ways, for instance on the basis of elastic moduli.

Why is it that the zero-point energy of a harmonic oscillator does not enter this analysis? We have the permitted energies of a harmonic oscillator as $(n + 1/2)\, h\nu$, $n = 0, 1, 2.$ the ground state of each oscillator in a solid is, therefore, $\varepsilon_0 = 1/2\ h\nu$, the zero point value, and not $\varepsilon_0 = 0$. But the zero-point energy merely adds a constant, temperature-independent term of $E_0 = (3N_0)(1/2h\nu)$ to the molar energy of solid, and this term vanishes when the partial derivative $(\partial E/\partial T)_v$ is taken to find C_v.

13.11. DEBYE'S THEORY OF SPECIFIC HEAT

Although Einstein's formula predicts that $C_v \to 0$ as $T \to 0$, as observed, the precise manner of this approach does not agree too well with the data. The inadequacy of Equation (5) at low temperatures led Peter Debye to look at the problem in different way in 1912. In Einstein's model, each atom is regarded as vibrating independently of its neighbours. Debye went to the opposite extreme and considered a solid as a continuous elastic body. Instead of residing in the vibrations on individual atoms, the internal energy of a solid according to the new model resides in elastic standing waves.

13.12. PHONONS ARE QUANTA OF ACOUSTIC ENERGY

The elastic waves in a solid are of two kinds, longitudinal and transverse, and range in frequency from 0 to a maximum ν_m. (The interatomic spacing in a solid sets a lower limit to the possible wavelengths and hence an upper limit to the frequencies). Debye assumed that the total number of different standing waves in a kilomole of a solid is equal to its $3N_0$ degrees of freedom. These waves, like electromagnetic waves, have energies quantized in units of $h\nu$. A quantum of acoustic energy in a solid is called a *phonon*, and it travels with the speed of sound since sound waves are elastic in nature. The concept of phonons is quite general and has applications other than in connection with specific heats.

Debye finally asserted that a phonon gas has the same statistical behaviour as a photon gas or a system of harmonic oscillators in thermal equilibrium, so that the average energy ε per standing wave is the same as in Equation (3). The resulting formula for C_v which is fairly complicated, reproduces the observed curves of C_v versus T quite well at all temperatures.

13.13. HALL EFFECT

In 1879 Hall discovered that when a mag-netic field is applied perpendicular to a conductor carrying a current, a voltage is developed across the conductor. The direction of the voltage is perpendicular to both the current and the magnetic field. This voltage is known as *Hall voltage* and the phenomenon is called the *Hall effect.*

Let us consider a rectangular strip carrying a current i along the x-axis as shown in Fig 13.18. If a magnetic field of intensity B be applied along the z-axis. Let us assume that the current is caused by the motion of free electrons of charge moving with a drift velocity v in a direction opposite to the direction of current. Under the effect of magnetic field, each electron experiences a force F as shown in the figure which displaces the electron along the y direction such that the surface PQ collects a negative charge and the surface RS gets positive charge. The charges give rise to an electric field E inside the rectangular strip. The direction of electric field is shown in Fig. 13.14. The accumulation of charge on the surface of the conductor continues until the force on moving charge due to electric field with it is large enough to cancel the force exerted by the magnetic field. Ultimately a stationary state is reached in which the net force on the electrons in y direction vanish and the electrons can again move freely along the conductor. In stationary state the value of E is denoted by E_H. E_H is called '*Hall electric field*'. If the charge carriers are holes then the case will be reversed *i.e.*, the surface RS becomes negative while the surface PQ becomes positive. So measuring the Hall voltage in y direction, the information about the sign of charge carriers may be obtained. In this way, measurement of Hall voltage gives the information about charge carriers.

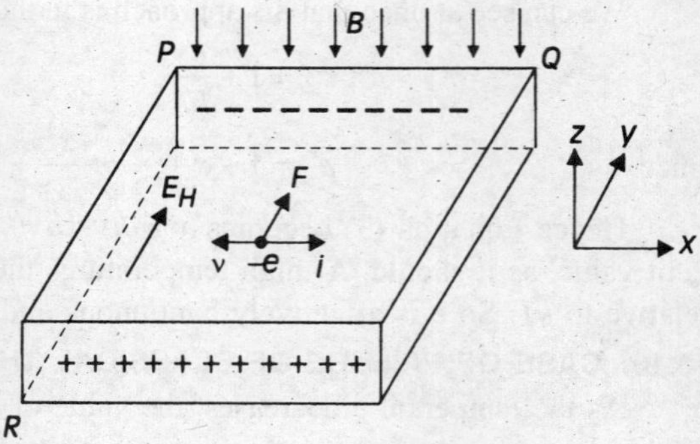

Fig. 13.18

MEASUREMENT OF HALL VOLTAGE AND HALL COEFFICIENT:

Electric force on electron having charge $-e = -e\vec{E}$

Force due to Magnetic field $B = -e\cdot(\vec{v}\times\vec{B})$

So total force on electron will be

$$\vec{E} = -e[\vec{E}+\vec{v}\times\vec{B}]$$

In this case $\quad F_y = -[E_y - v_x B_x] \qquad \text{...}(i)$

In the steady state, the force due to accumulation of electrons will be equal to the magnetic force and so the flow of electrons stops *i.e.*, $F_y = 0$ and $E_y = E_H$. So we have from Equation (*i*)

$$0 = -e[E_H - v_x B_z]$$

or

$$E_H = v_x B_z \qquad \text{...}(ii)$$

where E_H is known as *Hall voltage* or *Hall field*.

Let the number of free electrons per unit volume of the conductor be n. Then we have

$$\text{current density } J_x = -ne\,v_x \qquad \text{...}(iii)$$

where v_x is the average drift velocity of the electron, which is given by

$$v_x = -\frac{J_x}{ne}$$

Putting the value of v_x in Equation (*ii*) we have

$$E_H = -\frac{J_z}{ne}\cdot B_z$$

or

$$\frac{E_H}{J_x B_x} = -ne = R_H \qquad \text{...}(iv)$$

where

$$R_H = -\frac{1}{ne}$$

R_H is known as *Hall coefficient*. Hall coefficient is negative for free electrons and is constant for a metal. In some cases it has been found that R_H is positive for metals and semiconductors. This positive Hall coefficient will imply that the charge carriers are positive rather than negative. The positive charge carriers are known as holes and in a way positive value of Hall coefficient confirms the existence of holes in crystals.

Hall mobility. If a steady electric field E is maintained in a conductor by applying an external voltage across it, the carriers of current attains a drift velocity v. The drift velocity acquired in unit applied electric field is known as the mobility of the carrier and is denoted by μ.

$$\therefore \quad \mu = \frac{v}{E} \quad ...(v)$$

The current density J is given by

$$J = ne\,v \text{ where } e \text{ is charge on hole} \quad ...(vi)$$

also $$J = \sigma E$$

where σ is electrical conductivity.

$$\therefore \quad nev = \sigma E$$

or $$\frac{v}{E} = \frac{\sigma}{ne} R_H \sigma$$

$$\therefore \quad \mu = R_H \sigma \quad ...(vii)$$

13.14. IMPORTANCE OF HALL EFFECT

1. It can be used to determine whether the specimen is metal, semiconductor or insulator.
2. The sign of the current carrying charges can be found.
3. The mobility can be measured directly.
4. Number of free electrons/unit volume of the conductors can be calculated from Hall coefficient R_H.
5. Knowing R_H of a given material, the knowledge of the Hall voltage developed can be found.

Example 27. *A copper strip 2.0 cm wide and 1.0 mm thick is placed, in a magnetic field with B = 1.5 webers m^{-2}. If a current of 200 A is set up in the strip, calculate the p.d. that appears across the strip if the number of charge carriers is 8.4 × 10^{28} m^{-3}.*

Solution: (*i*) We have

$$E_H = \frac{jB}{ne}$$

Now $$E_H = \frac{V}{d} \text{ and } j = \frac{I}{A} = \frac{I}{dt}$$

where t is the thickness of the strip.

$$\therefore \quad V = \frac{IB}{net} = \frac{200\,\text{Å} \times 1.5 \text{ weber m}^{-2}}{8.4 \times 10^{28}\,\text{m}^{-3} \times 1.6 \times 10^{-19}\,\text{c} \times 1.0 \times 10^{-3}\,\text{m}}$$

$$= 2.2 \times 10^{-5} \text{ volt} = \mathbf{22\ \mu V}.$$

Example 28. *A copper strip 4.0 cm wide and 0.55 mm. thick carries a current of 100 A. If placed in a magnetic field of induction 2 weber m^{-2} acting at right angles to the strip, a Hall potential difference 29.7 × 10^{-6} V appears across its edge. Find (i) Hall electric field (ii) the number of charge carriers/m^3 in the strip.*

Solution: (*i*) We have

$$E_H = \frac{V_H}{d} = \frac{29.7 \times 10^{-6}}{4 \times 10^{-2}}$$

(*ii*) Charge carriers/unit volume can be had from the relation

$$n = \frac{JB}{e.E_H} = \frac{IB}{A.e.E_H}$$

where I is the current and A the area of cross-section of the conductor.

$$= \frac{100 \times 2}{2 \times 10^{-5} \times 1.6 \times 10^{-19} \times 7.425 \times 10^{-4}}$$

$$= \frac{10^{30}}{1.6 \times 7.425}$$

$$= \mathbf{8.4 \times 10^{28} \text{ charge carriers m}^{-3}}.$$

Example 29. *An electric field of 100 v/m is applied to a sample of n-type semiconductor whose Hall coeff. is –0.0125 m³/coulomb. Determine the current density in the sample assuming* $\mu_n = 0.35\ m^2\ V^{-1}\ S^{-1}$.

Solution: Hall coeff. is given by

$$R_H = \frac{1}{n \cdot e}$$

$$\therefore \quad n = \frac{1}{e \cdot R_H}$$

In this case

$$e = 1.6 \times 10^{-19}\,\text{C}$$

$$R_H = -0.0125\ \text{m}^3/\text{C}$$

$$\therefore \quad n = \frac{1}{1.6 \times 10^{-19} \times 0.0125} = 5 \times 10^{20}/\text{m}^3$$

Again we have the relations

$$\sigma = n.e.\mu_n$$

and

$$\sigma = \frac{J}{E}$$

$$\therefore \quad J = \sigma \cdot E = n.e.\mu_n.E$$

Putting the various values

$$J = 5 \times 10^{20} \times 1.6 \times 10^{-19} \times 0.36 \times 100$$

$$= \mathbf{2880\ A/m^2}$$

QUESTIONS AND PROBLEMS

1. Describe the formation of Bonds in solids on the basis of electronic configuration.
2. Name the types of Bonds in solids, and explain the qualitative difference between them.
3. What do you mean by Ionic Bonds? What types of solids contain ionic bonds? Give examples in supports of your answer.
4. Explain the stability of ionic bonds in solids. What are their characteristics?
5. Explain in brief the Hydrogen Bond, Ionic Bond and Metallic Bonds in solids.
6. What do you mean by Covalent bonds. What type of solids contain covalent bonds? Explain the answer giving examples.
7. Explain the stability of Covalent bonds in solids. What are their characteristics?
8. Distinguish between Covalent and Ionic Bonds in solids. Illustrate with examples.
9. Explain the formation of Metallic bonds in Sodium metal.
10. What is Metallic bond in metals? Explain their stability and characteristics. Why the solids having molecular bonds are poor conductors of electricity?
11. Write a brief essay on molecular bonds, their stability and characteristics. Why the solids having molecular bonds are poor conductors of electricity?
12. Explain the bond formation in diamond as a rigid crystalline solid while graphite is an amorphous—non-crystalline solid.
13. Write notes on the following:
    (*a*) Hydrogen Bonding
    (*b*) van der Walls forces in solids
    (*c*) Sp. heat of solids
    (*d*) Concept of Phonons.
14. Give ideas about the free electron theory of solids.
15. What is fermi energy? Show how this energy depends on the density of electrons gas.
16. Give Kronig-Penney model of solids.
17. Explain the significance and importance of Hall effect. How Hall coefficient can be determined?
18. What do you understand by sp. heat of solids? How the concept of Phonons be explained?

19. What is Hall effect? How the Hall voltage is related to Hall coefficient?

20. Mobilities of electrons and holes in a sample of intrinsic germanium at 300 K are 0.36 $m^2v^{-1}s^{-1}$ and 0.1 $m^2v^{-1}s^{-1}$ respectively. If the conductivity of specimen is 2.12 $\Omega^{-1}m^{-1}$, compute the forbidden energy gap. **[Ans.** Eg = 0.72 eV.]

21. The following data are given for intrinsic germanium at 300 K. $n_i = 2.4 \times 10^9/m^3$; $\mu_p = 0.19\ m^2 v^{-1}s^{-1}$; mn = 0.39 $m^2v^{-1}s^{-1}$. Calculate the conductivity of the sample. **[Ans.** 2.22 $\Omega^{-1}m^{-1}$]

22. The band gap of an alloy semiconductor gallium arsenide phosphate is 1.98 eV. Calculate the wavelength of radiation emitted when the electrons and holes in the material combine directly. Mention the colour of rad ation. **[Ans.** λ = 625 nm; Red colour]

23. Compare the densities of charge carries in a pure silicon crystal at the two temperatures 27°C and 57°C. Eg for Si is 1.1 eV. **[Ans.** 0.12]

24. In a semiconductor the effective mass of an electron is 0.07 m and that of hole is 0.4 m_0, where m_0 is free electron mass. Assume the average relaxation time of holes to be half that of electrons. Find the mobility of the holes when the mobility of electrons is 0.8 m^2v^{-1} sec^{-1}. **[Ans.** μ_p = 0.7 $m^2v^{-1}s^{-1}$]

25. If a sample of silicon is doped with 3 × 10^{23} Arsenic atoms and 5 × 10^{23} Boron atoms. Find the electron and hole concentration if the intrinsic charge carriers are 2 × $10^6/m^3$.

[Ans. 2 × $10^9/m^3$; 2 × $10^{23}/m^3$]

CHAPTER

14

METALS, INSULATORS AND SEMICONDUCTORS

14.1. INTRODUCTION

The atoms in a solid are closely packed which vibrate about their mean position. The nucleus of an atom can be considered at rest due to large mass and it is the orbital electrons which are responsible for most of its characteristic properties of the solids.

When the atoms come closer in a solid, the interaction between them disturbs the initial atomic levels. So most of the symmetry of the electronic states in the isolated atom is deformed and the level due to individual atom cannot be identified. In the process of inter atomic interaction, the levels due to individual atom cannot be identified. In the proccess of inter atomic interaction, the inner shell electron states are the least affected, while the valence electrons, which are closest to the neighbouring ions, are the most affected. The effect of bringing one atom close to the other is to split a single sharp energy level into two levels. If there are many interacting atoms, each outer level is spread into a continuous band of levels. This is why *energy bands*, rather than specific energy levels, exist in solids.

14.2. BANDS IN SOLIDS

(i) **Valence band.** The valence electrons, including electrons *transferred* in the case of solids having *ionic bands* or *shared* in the case of solids having *covalent bonds*, are not free to gain energy from an applied electric field and are, so, unable to migrate through the solid. As stated above, the sharp atomic levels spread out in solids and all the allowed energy states upto a *forbidden energy region* are occupied (at least at 0 K) and so there are no free electrons.

The highest filled energy band, which includes electrons shared in covalent bands or electrons transferred in ionic bonds, is called valence band.

(ii) **Conduction band.** When the number of valence electrons in one atom is less than the number of electrons required to fill the outer orbit of the other atom in a solid, the valence electrons, which are ordinary weakly held by the atoms are *free* to move through the solid and form a *'free electron-gas'*. These electrons can move under the influence of an applied electric field and due to the mutual interaction between these electrons, there is a change in their energy, the value of which varies from zero to a maximum value E_f, known as *Fermi energy.*

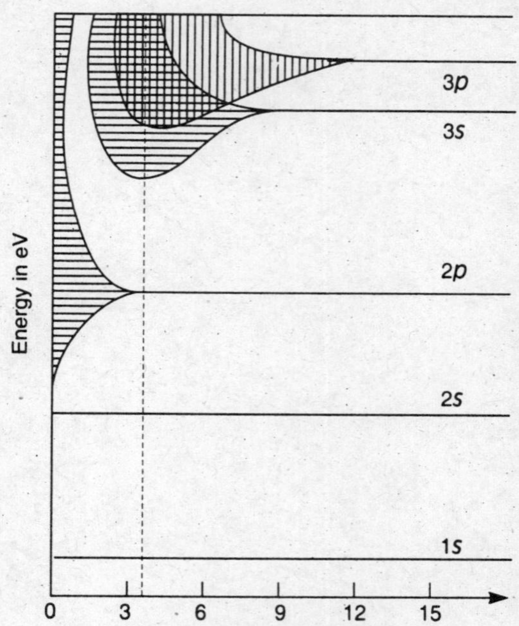

Fig 14.1

In other words, a band of energy of valence orbits consisting of closely spaced energy levels of these *free* electrons is formed. *This is called conduction band.*

***(iii)* Forbidden band.** The conduction band has a higher energy level than the valence band. Even in excellent crystalline insulators, there are conduction bands above the valence bands to which electron can be excited if these are provided sufficient energy. But there is a *forbidden energy region* between the two energy bands known as the **forbidden band.** When the forbidden energy gap is very small, thermal excitation may excite the electrons from the valence band to the conduction band so that a material, through an insulator at 0 K, may become a conductor at a higher temperature. *The bands may be completely filled, partially filled, or empty in accordance with Pauli's exclusion principle.* The gap between the valence band and the conduction band may be small or large or even the two bands may overlap. The band structure obtained at the equilibrium distance r_e determines most of the characteristics of the solid.

We can explain the behaviour of a solid as a conductor, semiconductor or insulator on the basis of band theory as under:

The energy levels of individual atoms, which are sharply defined when the atoms have large separation, results in the formation of bands as the inter-atomic distance *r* decreases. Now depending on the nature of the solid, the broadening of a level into a band begins to take place at different values of *r*. In some cases, the broadening of various levels at equilibrium distance is so large that energy bands overlap, whereas in others the broadening may be so small that a large gap may exist between them. When the two bands overlap, it means there is a continuous distribution of allowed energies. On the other hand, the presence of a gap indicates the forbidden energies that the electrons are not allowed to have. It should be noted that the presence of forbidden band is due to the quantum nature of the energies of the electron (see Fig. 14.1).

14.3. CONDUCTORS, SEMICONDUCTORS AND INSULATORS

On the basis of presence of forbidden band, we can describe the characteristics of insulators, semiconductors as under.

***(i)* Insulators:** In the case of insulator, the forbidden energy band is very wide and so the electrons cannot jump from valence band to conduction band. Moreover in insulators, the electrons are very tightly bound to the parent atom, *e.g.*, in case of an insulator like glass, the valence band is completely full at 0 K and the energy band gap is ~ 10 eV. Even if there is very high electric field, the electrons cannot jump from valence band to conduction band. However, if very high energy is supplied, an electron may be able to cross the forbidden band. If the temperature is increased, then some electrons may jump the forbidden band and this is why certain materials which are insulators become good conductors at high temperatures. The normal range of resistivity of insulators is ~ 10^7 ohm-m.

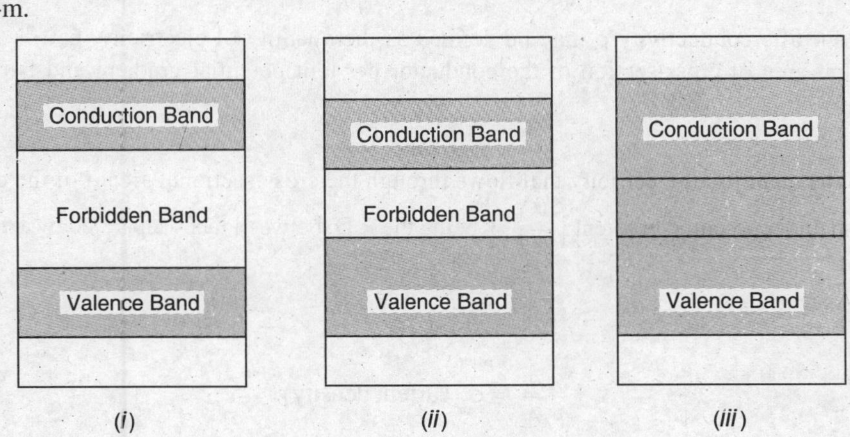

Fig. 14.2

***(ii)* Semiconductor.** In this case the forbidden band is very small and examples of semiconductors are germinium, silicon etc. In germinium, forbidden band is ~ 0.7 eV and that for silicon is 1.1 eV. Semiconductors have actually characteristics in between insulators and conductors and at 0 K. Conduction band has no electrons and valence band is completely filled. If some energy is supplied then electron jumps from valence band to conduction band. In case of increase of temperature, the forbidden gap is decreased and some electrons are liberated into the conduction band. In the case of semiconductors, conductivity is ~ 10^2 mho-m.

***(iii)* Conductors.** In the case of conductors, no forbidden band is there and the valence and conduction bands overlap each other. Here large number of free electrons are available for electric conduction, and electrons from valence band freely enter into the conduction band. Now since in the case of conductors, there is no forbidden band, so holes are not created at all and the total flow of current is due to the motion of electrons only.

14.4. EXPRESSION FOR CONDUCTIVITY OF CONDUCTORS FROM THE FIRST PRINCIPLE

Let E be the electric field applied to the conductor, e be the charge on the electron, m be the mass of the electron and v be the velocity corresponding to the temperature equilibrium with the material of any metal. Therefore, according to the principle of kinetic theory

$$\frac{1}{2}mv^2 \propto T$$

If t sec is the time during which the next collision takes place, then during this time the electron is subjected to the strength of field $E \times e$. Due to this force the electron attains acceleration 'a' and the particle moves in the direction of field. The acceleration is given by the relation

$$a = \frac{Ee}{m} \qquad ...(1)$$

If l be the mean free path of the electron then, the time t between the two consecutive collisions is given by

$$t = \frac{l}{v} \qquad ...(2)$$

The average distance S travelled by the electron in the direction of the field during the same time t is given by

$$S = \frac{1}{2}at^2 = \frac{1}{2}\left(\frac{E \times e}{m}\right)\cdot\frac{l^2}{v^2} \qquad ...(3)$$

Let v_a be the average velocity of the electron in the direction of field, then

$$\bar{v}_a = \frac{S}{t} = \frac{1}{2}\left(\frac{Ee}{m}\right)\cdot\frac{l}{\bar{v}} \qquad ...(4)$$

The specific conductivity ρ may be defined as the quantity of electricity that flows in unit time per unit area of cross-section of the conductor per unit potential gradient and is given by

$$q = \rho . A . \left(\frac{V}{d}\right)\cdot t \qquad ...(5)$$

where q is the quantity of electricity that flows through the cross-sectional area A of the conductor in time t and under potential gradient $\left(\frac{V}{d}\right)$, V being the P. D. between faces separated by a distance d.

$$\therefore \qquad \rho = \frac{q}{A.\left(\frac{V}{d}\right)\cdot t} = \frac{\sigma}{E.t}$$

$$\therefore \qquad \frac{q}{A} = \sigma \text{ (current density)}$$

and $$\frac{V}{d} = E \text{ (field strength)}$$

In semiconductors, electrical charge is carried by electrons or holes whose mobility varies in different materials. The electron and hole conduction is discussed below.

14.5. ELECTRON CONDUCTION

The electrons in a partially filled band are relatively free to move in that they are not bound to any atom but are in a band which is shared by the entire crystal. For this reason we call them free electrons.

They are said to be in constant motion. The kinetic theory of heat tells us that the average velocity of an electron is about 10^5 ms^{-1} at room temperature. The path the electron takes is a complex zig-zag path. The average time between collision is called the *mean free time* and the length of the path during this free time is called the *mean free path.* Some path and times are long, and some are short. They are distributed according to the laws of kinetic theory. Electron collides mainly with "object" in the lattice. These objects are photons, phonons, atoms and other electrons. The number of collisions cause the electron to undergo random motion under equilibrium conditions and to be seriously impeded when a field is applied and the electron tries to move in the field. In fact, the external field simply bends the path of the electron between collisions as shown in Fig. 14.3.

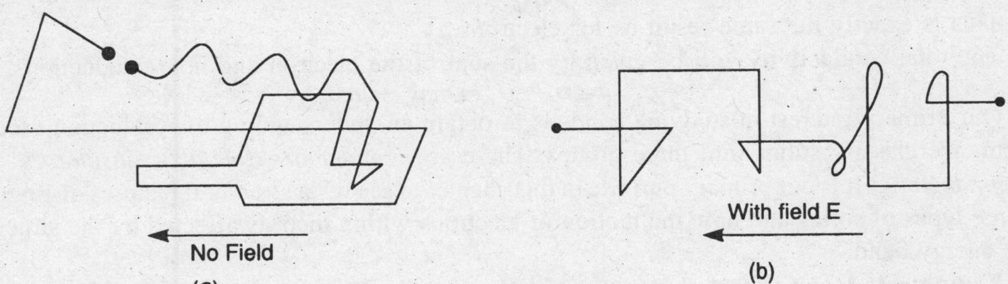

Fig. 14.3

It can be seen there that along with all the violent random motion there is a drift of the electron in the field which accounts for the current. This drift velocity v of electron is given by

$$v = \mu_n E$$

where E = field applied

μ_n = Mobility of electron.

As long as the drift velocity is small compared with the thermal velocity, Ohm's laws is applicable

$$J = \sigma E$$

where σ = conductivity

J = current density

and we find the conductivity to be $\sigma = \dfrac{\rho v}{E} = \dfrac{en\mu_n E}{E} = e.n.\mu_n$

So, the conductivity depends on the number of charge carriers and their freedom to move. This is definitely not an unreasonable result. As long as there is a small number of electrons in the partially filled band as compared with the number of states available, the above result will entirely be correct.

14.6. HOLE CONDUCTION

We know that the conductivity increases as the number of electrons in the partially filled band increases. This happens only for small time and as the condition of filled band approaches the conduction stops. Therefore, the behaviour of the conduction electrons in the band is extremely complicated. An electron near the top of the band does not behave in the same way as an electron near the bottom of the band. The electrons near the top of the band seem to cancel out the motion of those near the bottom. We are forced again to accept a new mode of behaviour for electrons which is quite contrary to classical ideas. The important result for our present consideration is that observed *current appears to be carried by positive charges in the almost-filled bands and by negative charges in the almost-empty bands.* The positive carriers are called *holes* because they correspond to the empty states in the almost-filled band.

It can be shown by means of quantum-mechanical arguments that:

(1) Electron in the upper part of the band have negative charges.

(2) These electrons move in such a way as to give electric current

(3) The current seems to be carried by positive electron charges.

Fact (3) is really consequence of fact (1) because particles with negative charges are accelerated in the direction opposite to the direction one would normally expect. As a consequence of these results, if there are N states and M of these states are occupies. Then

$$p = N - M$$

are the empty states, or holes, which carry current. The holes may then be thought of as free particles like electrons with positive charges which will experience collisions, diffusion, and drift in the same way as the electron.

The conductivity due to hole's is given by

$$\sigma_p = pe\mu_p$$

This is exactly the same result as for electrons.

The total conductivity will be given by the sum of the electron and hole conductions.

$$\sigma_{\text{total}} = \sigma_n + \sigma_p = ne\mu_n + pe\mu_p$$

Our primary interest in studying solids is to obtain an understanding of conduction process in them. We classify solids into three groups. These are *'conductors'* (*metals*), *'insulators'* and *'semiconductors'*. It is our primary purpose in this chapter to see what structural features distinguish the three types of solids and how the motion of electrons within them is affected by the structure of the energy band.

Example 1. *Mobilities of electrons and holes in a sample of intrinsic germanium at room temperature are 3600 cm*^{2}*/volt-sec. and 1700 cm*^{2}*/volt-sec. respectively. If the electron and hole densities are each equal to 2.5 × 10*^{13} *per cm*^{3}*, calculate the conductivity.*

Solution. Total conductivity is given by

$$\sigma_{\text{total}} = \sigma_n + \sigma_p = (ne\mu_n + pe\mu_p \text{ mho/metre})$$

Here

$$n = p = 2.5\times10^{13}/\text{cm}^3 = 2.5\times10^{19}/m^3$$

$$e = 1.6\times10^{-19}\text{ columb}$$

$$\mu_n = 3600\text{ cm}^2/\text{volt-sec} = .36\text{ m}^2/\text{volt-sec}$$

$$\mu_p = 1700\text{ cm}^2/\text{volt-sec} = .17\text{ m}^2/\text{volt-sec}$$

$$\therefore \quad \sigma_{\text{total}} = 2.5\times10^{19}\times1.6\times10^{-19}(.36+.17) = 2.5\times1.6\times.53 = 2.12\text{ mho/metre.}$$

14.7. SEMICONDUCTORS

Copper, with a resistivity of 10^{-3} ohm-metre is considered as one of the best electrical conductors. Quartz is typical of the nonconductors or insulators at the other extreme, with a resistivity of about 10^{12} ohm-m. There are number of materials with useful properties which have the resistivities intermediate to this range. The substance which have the resistivity in between conductors and insulators are given the name *'semiconductors'*. For example germanium, with a resistivity of about 0.60 ohm-m in the very pure state, and silicon having a resistivity of 1.5×10^3 ohm-m. Germanium and silicon are of major importance whereas lead sulphide, selenium, other metallic oxides and sulphides, as well as certain intermetallic compounds are of lesser importance.

Germanium and silicon belong to the IV group of the Periodic Table. There are carbon, silicon, germanium, tin and lead in the IV group of periodic table. These elements crystallize in the cubic form, as does carbon in the diamond state, and hold their electrons in shells or groups of energy levels having the following number of electrons.

Element	Atomic number	Electron Shells
C	6	2, 4
Si	14	2, 8, 4
Ge	32	2, 8, 18, 4
Sn	50	2, 8, 18, 18, 4
Pb	82	2, 8, 18, 32, 18, 4

The inner shell of 2, 8, or 2, 8, 18, ..., are completely filled shells, have no excess energy levels, and are stable and unaffected by outside influences. It is the four outer or valence electrons of each of these elements which are affected by external influences and cause these elements of group IV to have similar chemical and physical properties.

Of the above elements, carbon in diamond form is an insulator, silicon and germanium are semiconductors, tin and lead are good conductors. Germanium and silicon are the members of the family which possess the desired range of properties. However, it is the possession of four valence electrons which makes the elements interesting.

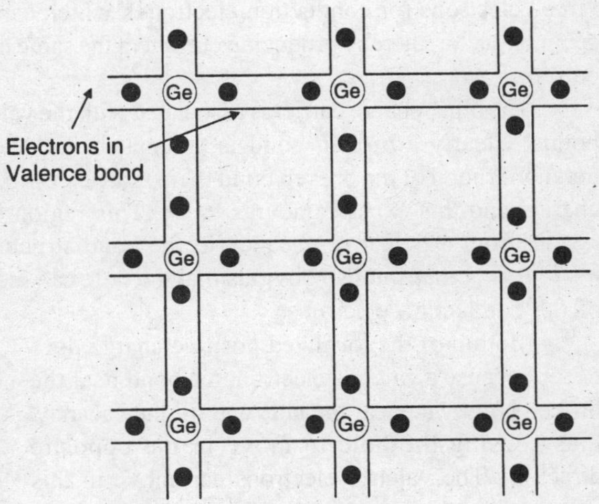

Fig. 14.4

14.8. TYPES OF BONDING IN GERMANIUM OR SILICON

In the crystal lattice each atom of silicon or germanium utilizes these electrons in four covalent bonds with four neighbouring atoms, each bond having two electrons one from each atom as shown in Fig. 14.4. Each atom is thus surrounded by a complete and stable shell of eight electrons, all electrons being rigidly bound at 0 K. Since no charges are free to move, the material do not conduct electricity at that temperature.

14.9. CHARGE CARRIERS IN GERMANIUM OR SILICON SEMICONDUCTORS

The valence electrons of the atoms of a semiconductor are not free to move throughout the volume of a semiconductor. When all the valence electrons are constrained in covalent bonds as shown in Fig. 14.4. electrical conduction is not possible because there are no charge carriers that are free to move. Consequently, a material having this electronic arrangement behaves as an insulator. Diamond, the crystalline form of carbon, is such a material.

The actual situation in semiconductor is different as stated above, a few covalent bonds are incomplete. The electrons missing from these bonds are no longer confined to region of the bond, but are free to move. This situation is shown in Fig 14.5.

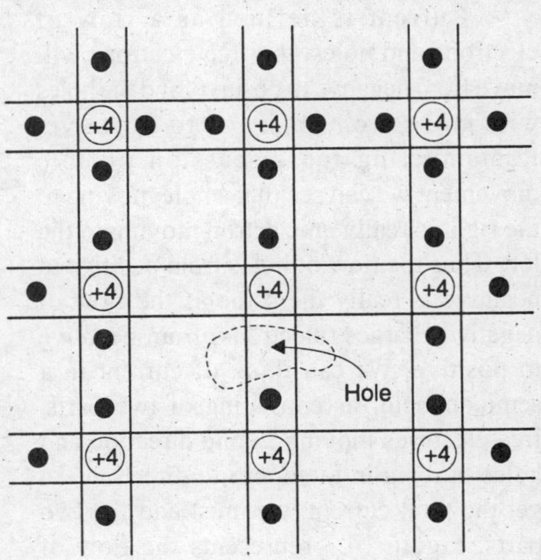

Fig. 14.5

The few broken bonds that occur at normal temperatures result from the random thermal vibration of the valence electrons. A few valence electrons acquire enough energy to "shake loose" from the bonds and become free. The fraction of the valence electrons that are disengaged from covalent bonds is exceedingly small. For example, in germanium at room temperature (about 300 K) there are about 10^{13} broken bonds per cubic centimeter. Since there are about 10^{23} atoms per cubic centimeter, only about one atom out of every *ten billion* has a broken bond. Nevertheless, these events, rare as they are, have an enormous effect on the electrical properties of the semiconductor ; they make conduction possible by providing charge carriers in a situation where otherwise there would be none. The material is considered to be a semiconductor rather than an insulator like diamond, when the number of broken bonds is of the order of one in 10^{15}, or about 10^{14} per cubic centimeter.

Due to the broken covalent bonds, there exist two distinct and independent groups of charge carriers that can support electric currents in semiconductors. The "free" electrons, which are produced when a valence electron "shakes loose" from a bond, constitute one class of charge carrier. These "free" electrons (or conduction electrons), which carry a charge of $+q$, can move about within the semiconductor, thereby producing current in the same manner that the free electrons in a metal produce current.

The other charge carrier is associated with the valence electrons which remain tied up in covalent bonds. Clearly a broken bond is associated with a localized region of positive charge ; in the neighbourhood of the broken bond there is an excess of positive ionic charge over negative electronic charge, and this excess amounts to $+q$. This region of positive charge is called a *hole* because it results from a defect or vacancy in the bond structure.

It is perhaps not so obvious that the hole can move about in the semiconductor independently of the conduction electrons.

Motion of the localized positive charge does occur because a valence electron in a bond near the broken bond, *i.e.*, near the hole can fill the vacancy, thus causing the hole to move in the opposite direction. The valence electrons can move in this manner from bond to bond without ever acquiring enough energy to become free of the bond structure. Therefore, the hole can move throughout the material without involving in any way the conduction electrons. This process is shown in Fig. 14.6.

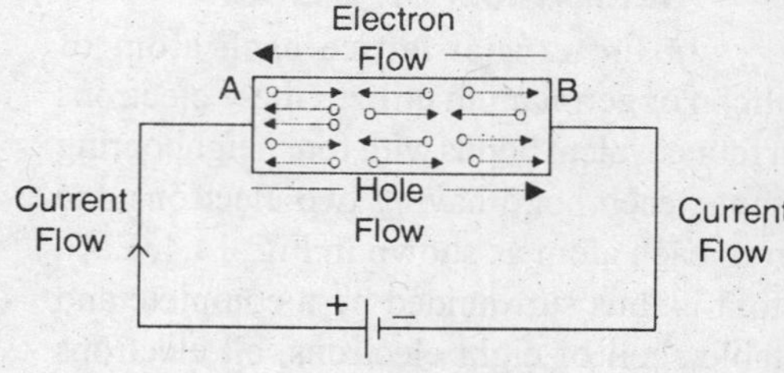

Fig. 14.6

14.10. FLOW OF CURRENT IN SEMICONDUCTORS

Current is defined as a flow of electrons and holes, the free electrons will move from negative to positive, and the holes will move from positive to negative. Remembering the discussion of hole movement we can see that a hole moving to the right is really an electron moving to the left. Thus the flow of holes from positive to negative is really the same as the flow of negative charges (electrons) from negative to positive. We can think of current in a semiconductor as consisting of two parts, free electrons moving in one direction and holes moving in the opposite direction. To get the total current, we must add the two parts. Figure 14.7 represents the flow of charges in a semiconductor.

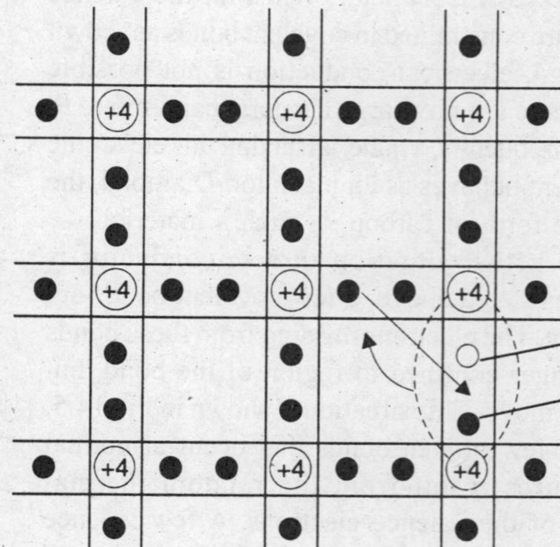

Figure 14.7

Now let us see what happens to the holes when they reach the edge of the piece of semiconductor material at B in Fig 14.7. The answer to that is that some of the electrons coming from the negative battery terminal combine with or fill the holes so that the holes and the filling electrons disappear. *This process is called recombination.* The rest of the electrons travel through the semiconductor as free electrons. The electron leaving (at A) fall into one of two categories: those which entered at B and travelled through the crystal or those which were free with in the crystal when the holes were formed.

It is very important to understand that even though hole current is actually due to the movement of electrons, it is the same type of electron motion as that which makes up the free-electrons current. The electrons which cause the hole current jump from hole to hole and do not have enough energy to become really free. On the other hand, the free electrons have enough energy to move freely through the crystal without being tied down to any atom.

We can now see that the electrons are able to move faster through the crystal than the holes, which have to move in a succession of jumps. We can say that the free electrons have a higher mobility than the holes.

At room temperature (nearly 25°C), the thermal energy is sufficient for a large number of free electrons and holes to exist in an intrinsic semiconductor. At room temperature, intrinsic semiconductor such as germanium and silicon are fair electric conductor, being much poorer conductors than a metal such as copper (Cu), but much better conductors than an insulator such as rubber. An important characteristic of intrinsic semiconductors is that their resistance decreases as the temperature increases because more electrons are given the energy necessary to break away and become free electron. Thus, there are more current carriers, both free electrons and holes, and the resistance decreases. In an intrinsic semiconductor with no voltage applied, the total number of free electrons equals the total number of holes.

14.11. THE BAND STRUCTURE OF SEMICONDUCTOR

There is another very convenient way of portraying the conduction of current in a semiconductor. When a single atom of a material is isolated, *i.e.*, it has a very large distance from another atom, the orbital electrons will have certain allowed amounts of energy called energy levels. The energy level of an electron is related to its orbit, the larger the orbit, the greater the energy level of the electron. Since a given electron may have only certain allowed energy levels, all other energy levels are forbidden, meaning the electron will not be found at any but the allowed energy levels.

When atoms are packed more closely together, as they are in a semiconductor crystal, the allowed energy levels are charged somewhat because of the presence of neighbouring atoms. The various allowed energy levels become allowed energy bands. An allowed energy band is range of energies in which and electron may be found. Again, other energy bands are forbidden. The latter are called *'energy gaps'* or *'forbidden gaps'*.

The energy bands which are important to you in the study of semiconductors are the valence band and the conduction band. The conduction band is located at higher energies than the valence band and the two bands are separared by an energy gap. This is shown in Fig. 14.8.

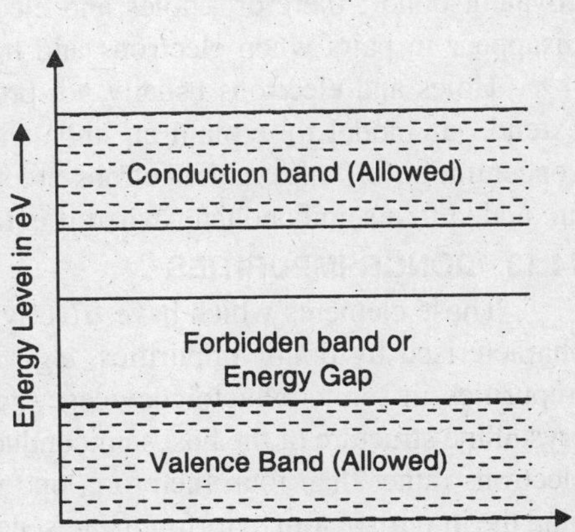

Fig. 14.8

For an intrinsic semiconductor absolute zero temperature, all of the valence electrons (four per atom) will be located in the valence band. As the temperature is increased, some of the valence electrons acquire enough energy to become free electrons. When this happen, they traverse the fobidden gap to the conduction band. Thus, the free electrons occupy the conduction band and the holes will be located in the valence band. The width of the forbidden gap is the amount of energy a valence band must acquire in order to break its covalent band and become a free electron. The energy gap is usually measured in electron volts (eV). *An electron volt is the amount of energy that one electron acquires in travelling through a potential difference of one volt.* For silicon, the energy gap is approximately 1.09 eV; for germanium the energy gap is approximately 0.7 eV. This means that less energy is required to break the covalent bonds in germanium than silicon.

An '*intrinsic semiconductor*' at absolute zero temperature has a filled valence band and an empty conduction band and acts as an insulator ; no current can be conducted. As temperature rises, the conduction band become partly filled (with free electron) and the valence band becomes partly empty. This makes conduction possible.

Semiconductors differ from insulators only in width of the forbidden band. The band structure of silicon is shown in Fig. 14.9.

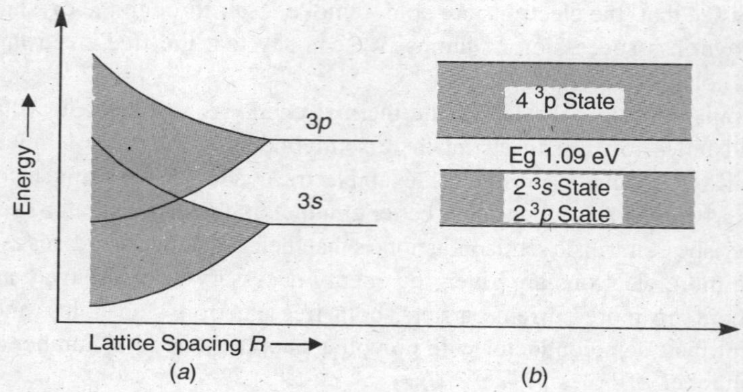

Fig. 14.9

The forbidden band is quite narrow (about 1.1 eV in silicon and about 0.72 eV in germanium). We may consider silicon to be an insulator with a narrow forbidden band. The crystal is classed as a semi-conductor because at room temperature there are enough electrons excited across the forbidden gap to make both the electron and hole concentrations in the conduction and valence bands. In the case of silicon and germanium, both electrons and holes contribute to the conductivity. The holes in each case contribute about half as much as the electrons because of their lower mobility.

14.12. IMPURITIES IN SEMICONDUCTORS

The model of semiconductor which we have introduced requires that the holes and conduction electrons be present in exactly equal numbers. Each of these charge carriers is the result of a broken covalent bond ; therefore, holes and electrons are produced in pairs when bonds are broken and disappear in pairs when electrons and holes get together to form complete bonds.

Holes and electrons usually not present in equal numbers in a semiconductor, and we must extend our model to explain an unbalance in the hole and electron concentrations. The relative concentration of holes and electrons are strongly affected by minute amounts of impurities, which are added to a semiconductor when it is fabricated. These impurities are of the following two type:

14.13. DONOR IMPURITIES

Those elements which have 'five' valence electrons rather than the four valence electrons are characterised by donor impurities, *e.g.*, Phosphorous, Arsenic and Antimony are used as donor impurities in semiconductor devices. Atoms of these donor elements fit readily into the regular crystalline structure of the host semiconductor. However, because donor impurities have five valence electrons rather than four, therefore, one valence electron is left over after the covalent bonds that bind the impurity atom to its four nearest neighbours are completed. This extra electron is only loosely bound to the impurity atom. The temperature at which semiconductor devices commonly operate, the extra electron becomes dissociated from the impurity atom and moves about through the semiconductor as a conduction electron. The nucleus of impurity atom has a charge of $+5q$ rather than $+4q$ because there is one more proton than that of the nucleus of a semiconductor atom. Therefore, the positive charge on the nucleus of an impurity atom is localized immobile positive charge. This situation is illustrated in Fig. 14.10.

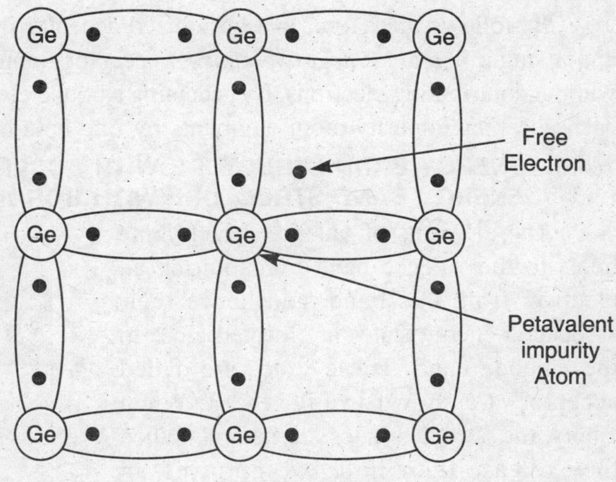

Fig. 14.10

The key point is that ionization of a donor-impurity atom produces a single conduction electron and a single immobile positive ionic-charge. Consequently, donor impurities contribute conduction electrons to the semiconductor without contributing holes. Each donor impurity atom, by donating its extra valence electron to the resemble of conduction electrons increases the electron concentration by one.

14.14. ENERGY BAND STRUCTURE WITH DONOR (*n*-TYPE) SEMICONDUCTOR OR ENERGY BAND STRUCTURE WITH A PHOSPHOROUS DOPPED GERMANIUM CRYSTAL

The questions which arise are "How are these electrons and holes in weakly bound states represented in terms of band structure" and "what are the consequences of having such states in the crystal. Since the electrons are weakly bound to the elements of valence 5, their normal states are shown near the conduction band in the crystal. These atoms are called donor atom, and the extra electrons lie a distance equal to their binding energy from the conduction band. This situation is shown in Fig. 14.11. When the electron is in the donor state, the state is neutral. When the electron is excited into the conduction band, the donor retains a positive charge and is said to be ionized.

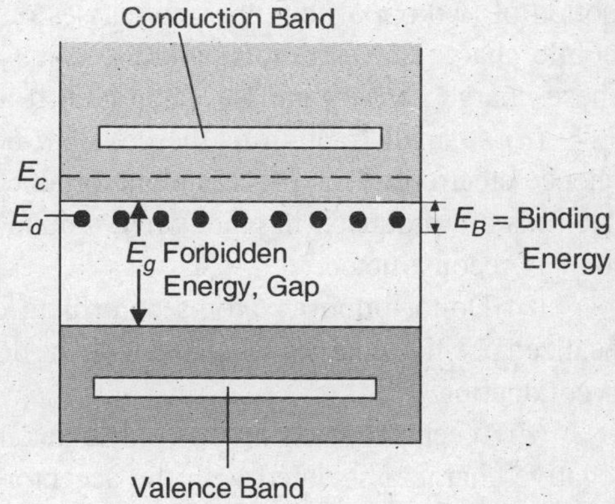

Fig. 14.11

14.15. ACCEPTOR IMPURITIES

Those elements which have three valence electrons rather than four, *e.g.*, Indium or Aluminium. These atoms fit into the periodic crystalline structure of the host semiconductor, but they have one too few valence electrons to complete the covalent bonds. Therefore, there is a vacancy in the bond structure or a hole associated with each acceptor. This hole is loosely bound to the acceptor atom so that at normal temperatures it becomes dissociated from the acceptor ion then constitutes a local immobile negative charge; it has one less nuclear proton to balance the negative charge of the bound electrons. This is shown in Fig. 14.12.

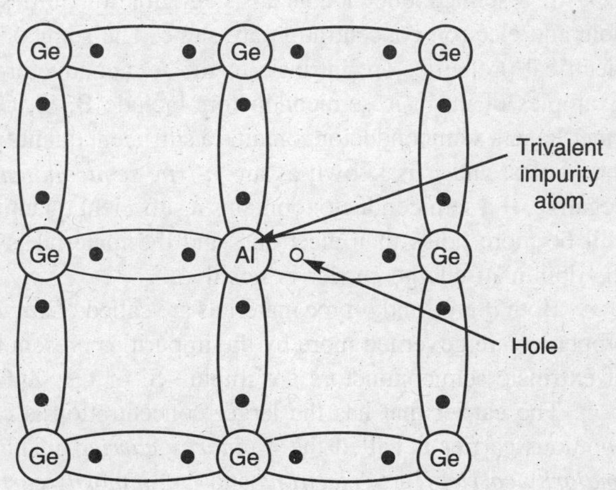

Fig. 14.12

It follows, therefore, that an ionized acceptor impurity atom produces a *single mobile hole* and a single *immobile negative charge*; acceptor impurities contribute holes to the semiconductor without contributing electrons. By accepting a bound electron to complete its covalent bond structure, each acceptor impurity atom augments by one hole concentration.

14.16. ENERGY BAND STRUCTURE WITH ACCEPTOR (*P*-TYPE) SEMICONDUCTOR OR ENERGY BAND STRUCTURE WITH BORON OR CALCIUM DOPPED CRYSTAL

The elements of valence 3 contribute holes to the valence band by capturing an electron from this band, and hence their associated energy states are located close to the valence band. These states are called acceptor, are shown in Fig. 14.13. When empty, the accepter states are neutral. When they release their hole by capturing an electron, they are negatively charged and are said to be ionized.

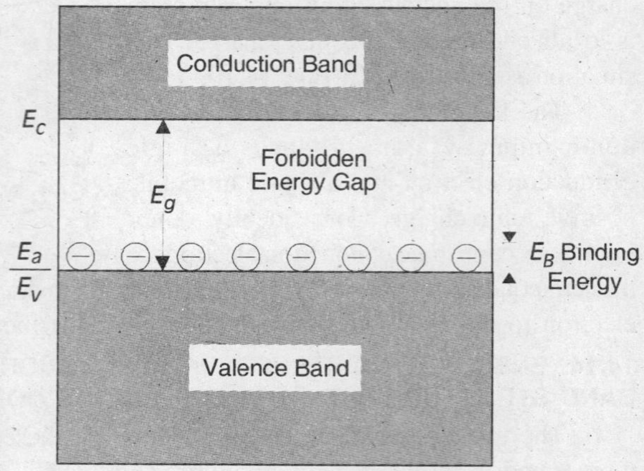

Fig. 14.13

14.17. SOME IMPORTANT FEATURES OF SEMICONDUCTORS

To summarize, semiconductors conduct electricity as a consequence of the independent motion of two oppositely charged classes of mobile charge carriers: holes and electrons. These charge carriers can be generated in a semiconductor by three distinct processes:

(*a*) A small fraction of the covalent bonds is broken as a result of thermal agitation of the valence electrons. This process alone produces equal number of holes and electrons; *i.e.*, holes and electrons are produced in pairs when a bond is broken, and disappear in pairs when an incomplete bond is reconstructed.

(*b*) Donor-impurity atoms contribute only electrons when they ionize. Although there is a localized positive charge associated with the donor ion, this charge is immobile and does not participate in conduction.

(*c*) Accept-impurity atoms contribute only holes when they ionize. Although there is a localized negative charge associated with the acceptor ion, this charge is immobile and does not participate in conduction.

If a semiconductor contains negligible quantities of either donor or acceptor impurities, the hole and electron concentration are same. The semiconductor is then called *'intrinsic'* because the electrical properties are intrinsic to the semiconductor material and are not a result of impurities. Examples of intrinsic semiconductors include B, Si, Ge, Pb, Ga, As, In, Sb and Fe_3O_4. If on the other hand a semiconductor contains a sufficient quantity of donor atoms, it will have more electrons than holes, and it is known as an *'n-type-semiconductor'* because the dominant charge carrier is negative. If a semiconductor contains a sufficient quantity of acceptor atoms, rather than donor there will be more holes than electrons, and the material is known as *"p-type semiconductor"* because the dominant charge carrier is positive.

Both *n*-type and *p*-type materials are called *"Extrinsic Semiconductors"* because their electrical properties are governed more by the impurities present than by the semiconductor itself. Examples of extrinsic semiconductors are impure Si or Ge, ZnO, TiO_2 and Mn_3O_4.

The carrier that has the larger concentration is called the *"majority carrier"* while the less populous carrier is called the *"minority carrier"*. Thus in an *n*-type extrinsic semiconductor the *'majority carriers'* are *'electrons* and the *'minority carriers'* are *'holes'* whereas in *p*-type extrinsic semiconductor the majority carriers are *'holes'* and the minority carriers are *'electrons'*.

14.18. CLASSIFICATION OF SEMICONDUCTORS ON THE BASIS OF FERMI LEVEL AND FERMI ENERGY

Fermi Direct statistic is applicable to those electrons in a solid crystal which obeys the restriction imposed by *Pauli's Exclusion Principle* and according to Fermi the probability $P(E)$, that a state of energy E is filled is given by

$$P(E) = \frac{N(E)}{M(E)} = \frac{1}{e^{(E-E_f)}+1}$$

where $N(E)$ is the density of electrons per unit energy and $M(E)$ is the density of its quantum states per unit energy. E_f is called the Fermi energy level.

Fermi Energy E_f. Fermi energy is the highest energy of the electron in the valence band of a crystal in its ground state. It is also defined as the energy for which the probability $P(E)$.

$$P(E) = p(E_1) = \frac{1}{2}$$

Fermi level is used as a reference level. Let us study how the idea of Fermi level classifies a semiconductor. For a semiconductor crystal when all the electrons are present in the valence band the probability

$$P(E) = 1$$

in this state the conduction band is vacant and the probability of the conduction band will be

$$P(E) = 0$$

Therefore, Fermi level having $P(E) = \frac{1}{2}$ must be somewhere in between these two bands. In an intrinsic semiconductor the position of the fermi level (E_f) is midway in the forbidden energy gap as shown in Fig. 14.14.

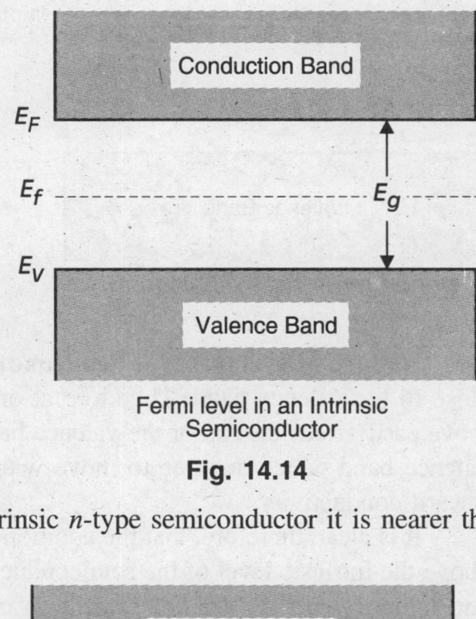

Fermi level in an Intrinsic Semiconductor

Fig. 14.14

When we add a donor impurity like phosphorous, arsenic etc. to an intrinsic (pure) semiconductor the Fermi level E_f rises above the mean level and when we add an acceptor impurity like aluminium, boron etc., the Fermi level falls below the mean level in the forbidden energy gap.

Thus in an *extrinsic p-type* semiconductor the Fermi level is nearer the valence band while in an extrinsic *n*-type semiconductor it is nearer the conductor band as shown in Fig. 14.15.

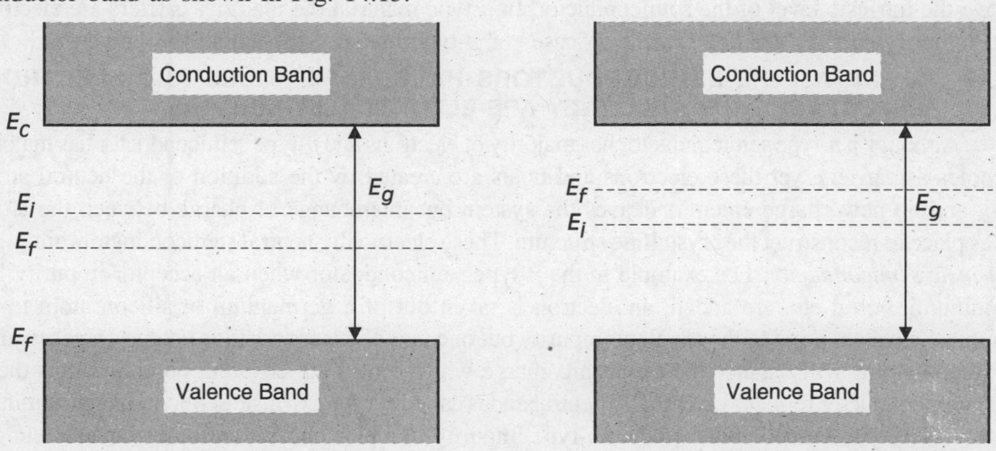

In p-type Extrinsic Semiconductor
(*a*)

In n-type Extrinsic Semiconductor
(*b*)

Fig. 14.15

In these diagrams E_f gives the Fermi level and E_c gives the Fermi level in the same semiconductor in the absence of any impurity.

Impurity Levels: An intrinsic semiconductor becomes extrinsic semiconductor when a small amount of impurity is added to the given sample. The impurity creates new quantum states in the forbidden energy gap. These quantum states generally appear at single value of energy and do not spread out into a band for want of interaction between the spare atoms of the impurity material. This value of energy is called *impurity level.* In an intrinsic semiconductor this level is absent and enough energy is required to raise an electron from the valence band to the conduction band.

Impurity Level in *n*-type Semiconduction: In *n*-type semiconductor the impurity level appears close to the conduction band and therefore, at ordinary temperature electron can easily move from the impurity level to the conduction band and this thing contributes highly towards electrical conductivity.

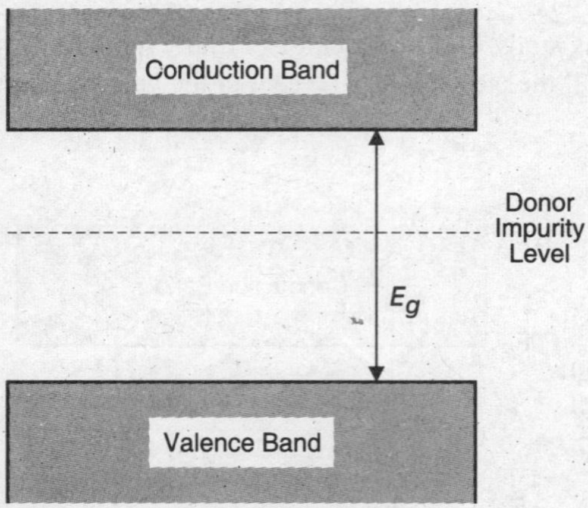

Donor Impurity Level Diagram

Fig. 14.16

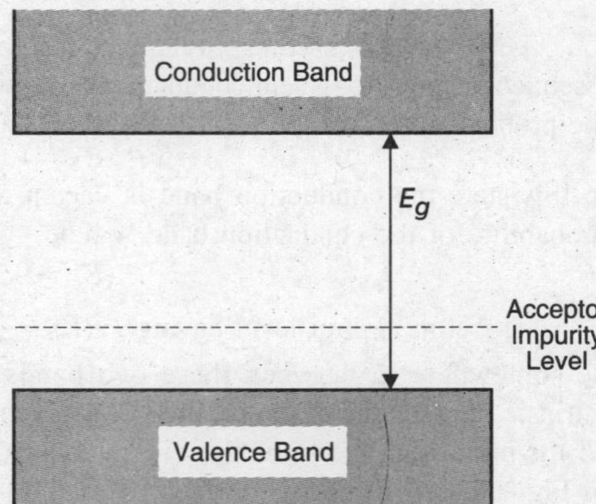

Fig. 14.17

Impurity Level in *p*-type Semiconductor: In *p*-type semiconductor the impurity level appears close to the valence band. Therefore at ordinary temperature electrons acquire enough energy to move easily from the top of the valence band to the impurity level and leaves positive holes in the valence band which are free to move within the semiconductor and this thing contributes highly toward conductivity.

It is clear, therefore, that the addition of donor or accepter impurity increases the conductivity above the intrinsic level of the semiconductor. In *n*-type material the majority carriers are electrons and minority carriers are holes while in case of *p*-type material the situation is reversed.

14.19. *P*- AND *N*-TYPE SEMICONDUCTORS HAVE MORE HOLES AND ELECTRONS RESPECTIVELY BUT STILL THEY ARE ELECTRICALLY NEUTRAL

Although a *n*-type semiconductor has majority of electrons and *p*-type semiconductor has majority of holes as carriers, yet these electrons and holes are created by the addition of the neutral atoms only and no new charge enters or leaves the system but interchange of charge between the atoms takes place to reconstruct the crystalline structure. These electrically neutral semiconductors are called *Extrinsic semiconductor.* For example in the *p*-type semiconductor when an acceptor impurity like aluminium, boron etc. are added, an electron is taken out of a germanium or silicon atom to the complete covalent bond of the trivalent impurity but one unit of excess positive charge on germanium or silicon atoms will balance the electronic charge which enters the covalent bond and thus the *p*-type semiconductor remains electrically neutral and is called *p*-type extrinsic semiconductor. Similarly in case of *n*-type semiconductor, when *n*-type impurity like phosphorus, antimony, arsenic etc. are added. The pentavalent electrons of the impurity enters in the tetravalent covalent bond of germanium or silicon by leaving an electron, but the charge on this electron continues to balance with the one excess positive charge on the *n*-type impurity atom. Thus the *n*-type semiconductor remains electrically neutral and is called the *n*-type extrinsic semiconductor.

14.20. APPLICATIONS OF SEMICONDUCTORS

Solid state semiconductors have several uses over vacuum tubes.

(1) They do not require a filament for operation; therefore, they consume less operating power, and subsequently less heat is generated. Heat can be damaging to other components, so it is desirable that it be kept to a minimum level.

(2) Solid state components are also much smaller as a general rule and operate more efficiently than vacuum tubes.

Because of these advantages, the use of semiconductor components in electronic equipment has been on the upswing in recent years. Number of applications of semiconductors are discussed below.

14.21. *P-N* JUNCTION AS RECTIFIER

All semiconductor diodes are constructed of a material such as selenium, silicon, or germanium which when properly treated, has the ability to permit electron flow in one direction but essentially none in the other. Thus a semiconductor acts as a conductor to electrons flowing in one direction, but as an insulator to electron flowing in the opposite direction. From this it becomes apparent that

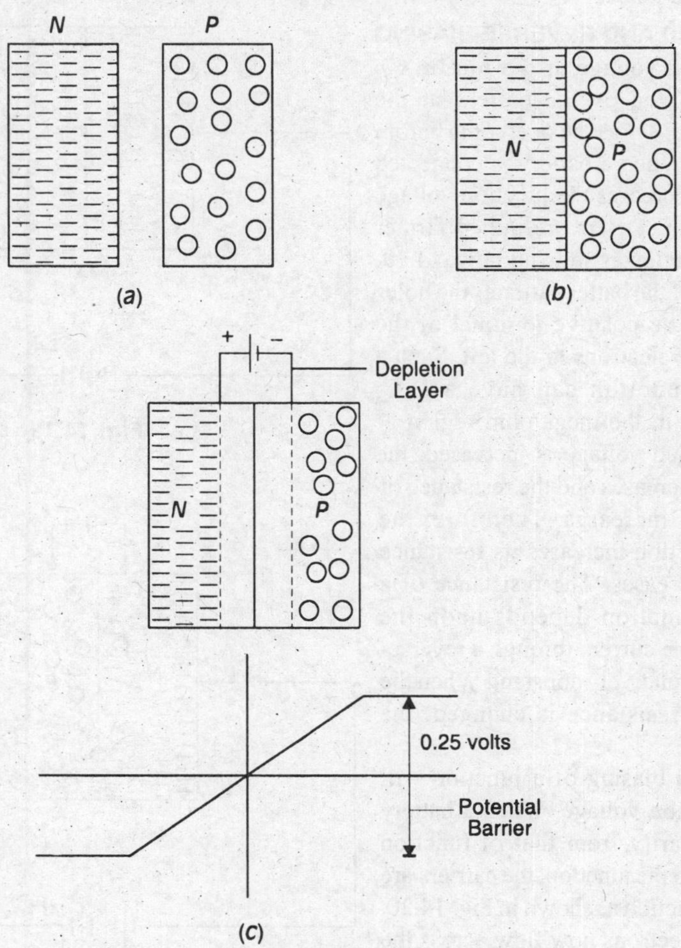

Fig. 14.18

a semiconductor diode performs the same function as a diode vacuum tube. Some semiconductors are composed of a single material, whereas others use two different materials to form a junction. These materials, which have been specially treated are p-type and n-type impurities. A p-n junction consists of a piece of germanium, one half of which is dropped with p-type material and the second half with n-type material as shown in Fig. 14.18. At the junction, the free electrons carriers and free holes carriers migrate across the junction due to diffusion. Some important terms used in semiconductor diodes are:

***(i)* Depletion Layers:** Due to diffusion of free carriers across the junction, a depletion layer on either side of the p-n junction is formed. This layer is formed due to the neutralization of holes in p-region and electrons in the n-region. This diffusion continues till a potential barrier is developed in the depletion layer which then avoids further diffusion. This barrier can be increased or decreased by decreasing or increasing external applied voltage.

***(ii)* Junction Voltage:** As discussed above, a potential barrier is developed at the junction. The strength of this potential barrier *i.e.*, the minimum external voltage required to overcome this potential barrier is called *junction voltage* or potential hill.

***(iii)* Junction Capacitance:** The two surfaces of the depletion layer act like to two plates of the capacitor. The potential barrier and capacitance between these two plates is called the *junction capacitance.*

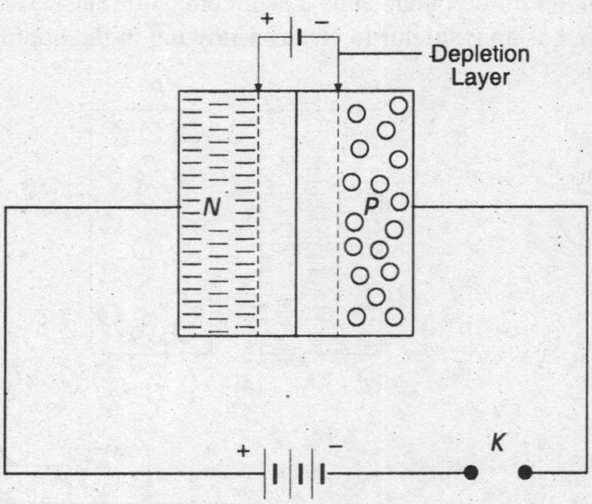

Fig. 14.19

14.22. FORWARD AND REVERSE BIASING

The junction voltage at the p-n or n-p junction, represented by the battery at the junction in Fig. 14.18 must be overcome before current can flow. When a battery is connected so that it aids or increases the junction voltage at the junction, the carriers are pulled farther away from the junction as shown in Fig. 14.19. The –ve terminal of the battery attracts the holes to the right, and the positive terminal of the battery attracts the electrons to the left. Such a reverse biased junction can have a D.C. resistance reading in the mega ohm region.

As the applied voltage is increased, the junction voltage increases and the resistance of the junction also increases. Therefore, the reverse based junction increases its resistance as the voltage increases. The resistance of a reverse-biased junction depends upon the applied voltage. The current through a reverse-biased junction is relatively constant. When the voltage across a resistance is changed, the current changes.

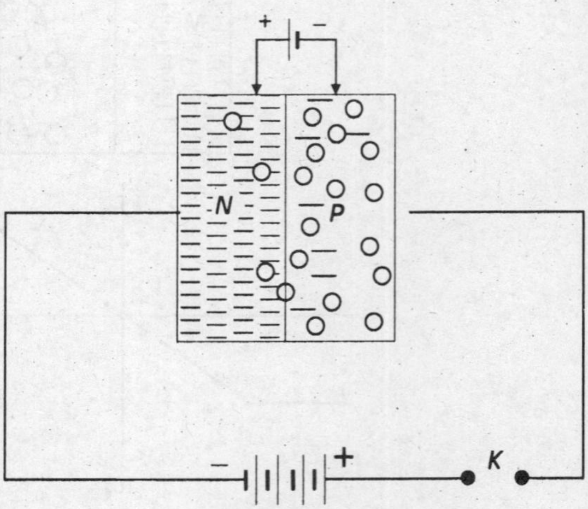

Fig. 14.20

The forward biasing of a junction will decrease the junction voltage. When a battery with opposite polarity from that of junction voltage is applied to the junction, the carriers are moved up to the junction as shown in Fig. 14.20.

Holes and electrons now flow across the junction. This action results in a current flow in the external circuit. Another way of describing this action is by saying that the battery will inject excess holes in p-type material by removing electrons and will inject electrons into the n-type material.

The forward bias is different from the reverse bias because the voltage necessary to overcome the junction voltage is rather small, but once this potential is reached, the current has little opposition. As current increases, the resistance of the junction decreases. The applied voltage remains nearly the same.

14.23. SOLID-STATE OR SEMICONDUCTOR TRANSISTORS

Transistor is a three electrode crystal device which, when properly connected in a circuit, will provide many of the functions formerly performed by vacuum tubes. Transistors offer several advantages over vacuum tubes.

(*i*) Transistors are much smaller than tubes and, therefore, make possible more compact radio equipment.

(*ii*) Transistors do not require filament or heater for thermionic emission as vacuum tubes do, hence they draw far less current and produce very little heat.

(*iii*) Transistors are easily portable *e.g.*, pocket transistors etc. In recent years portable televisions are available.

(*iv*) Transistors are used for construction of on-off switches operated at moderate speed.

(*v*) Transistors are used for amplification of current.

(*vi*) Transistors are used in oscillators for producing high oscillations. In addition many more applications of transistors can be given in almost all electronic fields.

Transistors are of two types:

(*a*) *n-p-n* transistor. (*b*) *p-n-p* tansistor.

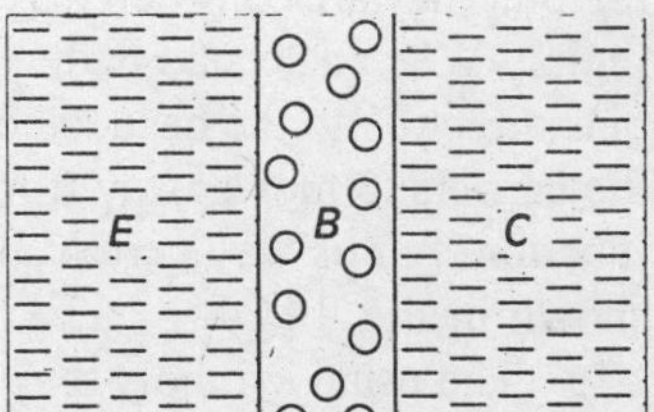

n-p-n transistor

Fig. 14.21

14.24. (*N-P-N*) TRANSISTOR

It consists of n-type emitter (E) and collector (C) and a p-type base (B). The base region of the transistor is drawn thin in comparison with emitter and collector regions as shown in Fig. 14.21. There is a reason for this thin base region; it affects both the majority and minority carrier action at the junctions.

14.25. ACTION OF *N-P-N* TRANSISTOR

Connect an *n-p-n* transistor such that there is forward biasing to E–B junction and reverse biasing to the B–C junction as shown in the Fig. 14.22.

When the base to emitter battery is connected in the forward direction, the majority carriers are forced up to the junction and a current flow is produced between the base and the emitter. The holes move into the n-type material and electrons move into the p-type material. Recombination takes place at the junction, but the combination of electrons and holes can also take place after the carriers have passed this barrier.

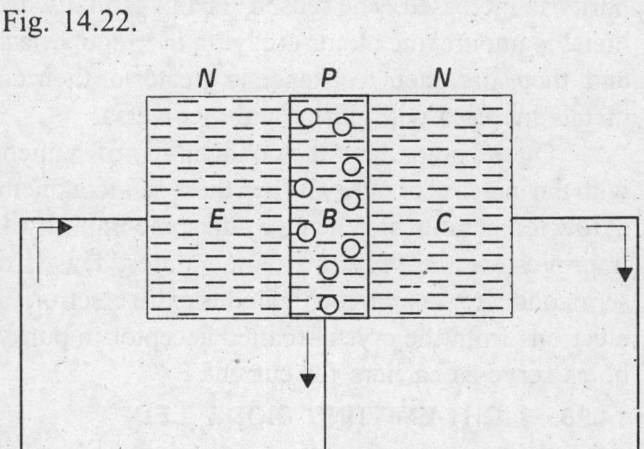

Action of *n-p-n* Transistor

Fig. 14.22

When free electrons exist in the emitter region of *n-p-n* transistor, they are majority carriers; but when these same electrons cross the barrier into the base region, they are considered minority carriers. These electrons eventually combine with holes in the base region unless same field of force is applied. That is why a second battery is connected in the reverse bias direction to the base to collector circuit which causes the carriers to move away from the base to a collector junction. A wide base region permits all of the electrons from the emitter to recombine with the holes present in the base region. In this situation we will be left with two diodes, one forward biased and other reverse biased; and no transistor action take place. That is why thin base region is preferred in transistors.

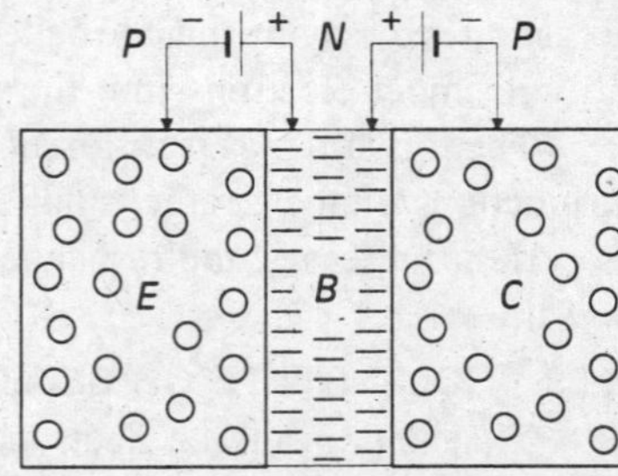

p-n-p Transistor

Fig. 14.23

14.26. *P-N-P* TRANSISTOR

It consists of a *p*-type emitter (E) and collector (C) and a *n*-type base (B). The base region is drawn thin in comparison with emitter and collector regions as shown in Fig. 14.23.

Action of *p-n-p* Transistor: Connect a *p-n-p* transistor such that there is forward biasing to E–B junction and reverse biasing to the B–C junction as shown in Fig. 14.24.

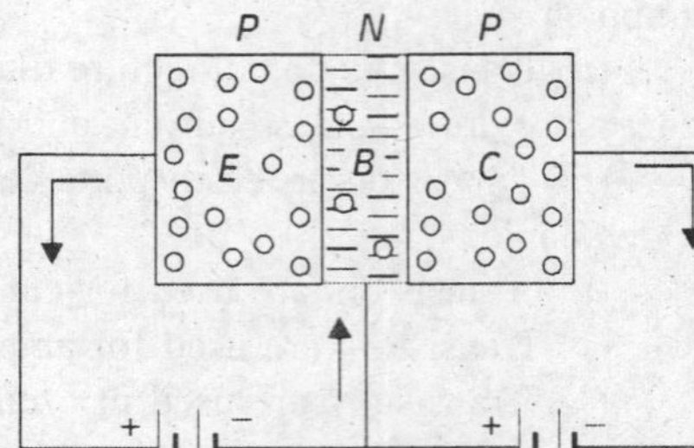

Action of *p-n-p* Transistor

Fig. 14.24

The emitter E emits holes which are majority carriers for *p*-type material. The battery in the *p-n* junction drives these holes from the emitter E. Some of the holes recombine with electrons in the base region which is made of *n*-type material. Since the base region is extremely thin, Fig. 14.25 therefore, some holes do not come with holes in the base region. They are considered as minority carriers in the base region. When a second battery is connected in the reverse bias direction to the base to collector circuit which causes the carriers to move away from the base to the collector junction. In this way current flows in *p-n-p* transistor as shown in Fig. 14.24.

The usual symbols for *p-n-p* and *n-p-n* transistors are shown in Fig. 14.25.

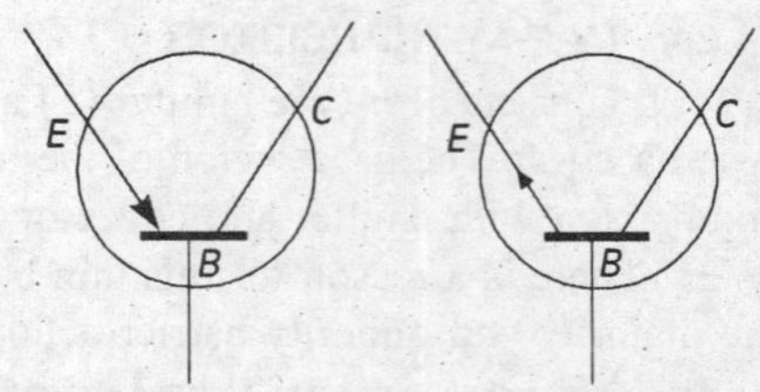

p-n-p *n-p-n*

Fig. 14.25

14.27. EFFECT OF PURITY ON THE CONDUCTIVITY OF METALS AND SEMICONDUCTORS

The conductivity of metals is increased, when their purity is increased. The reason for this is that when the metal is impure, the atoms are lying in irregular fashion, and, therefore, their resistance is greater or their conductivity is lower. Hence, the conductivity of metals increase with purity and vice-versa.

On the other hand the conductivity of semiconductors like germanium, silicon etc. decreases with the increase of purity. When these semiconductors are absolutely pure they behave like insulators at low temperatures. When we add some impurity, the conductivity of germanium or silicon crystal improves even at very low temperatures. This is due to the fact that when we add impurity to a semiconductor crystal it either adds extra electron to the crystal (called donor impurity) or it absorbs electrons from the crystal (called acceptor impurity) thus holes are created. The extra electrons or holes serve as carriers for current.

14.28. LIGHT-EMITTING DIODE (LED)

We have seen that when a *p-n* junction is operated in the forward bias configuration, electrons arriving from the *n* region and holes from the *p* region meet at the junction. These electrons and holes annihilate each other when they meet at the interface. The annihilation is simply a transition of an electron from the conduction band to an empty state (hole) in the valence band, a transition that releases energy. In some material, such as gallium arsenide, the energy is released in the form of a photon, *i.e.* the *p-n* junction emits light whenever an electric current passes through it.

Such light-emitting diodes find widespread application in luminous diode displays for clocks, radios, and all kinds of measuring instruments. Light-emitting diodes can also be used as lasers. Whenever there are electrons in the conduction band, a corresponding number of holes in the valence band, we have a population inversion. The emission of one initial photon can then lead to the stimulated emission of many other coherent photons. This principle forms the basis for solid state lasers. This kind of laser, which can be of a very small size, is used to generate the light pulses that carry information in some of the new telephone lines consisting of optical fibres.

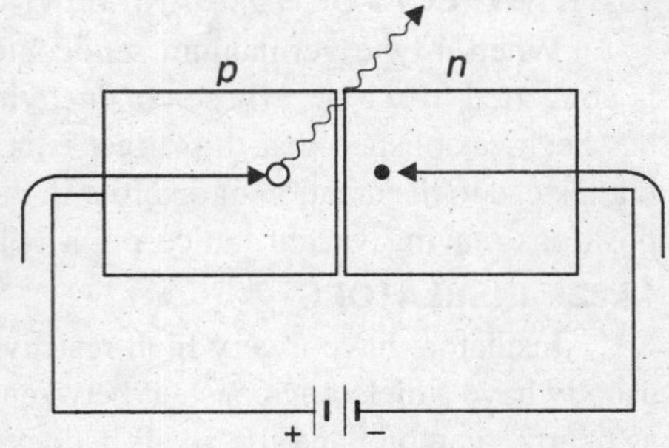

Fig. 14.26

14.29. SOLAR CELL

In principle, a solar cell is nothing but a light-emitting diode operating in reverse. Sunlight is absorbed at the *p-n* interface and excite electrons into the conduction band, *i.e.*, sunlight creates electron-hole pairs. The electric field at the junction pulls the electrons towards the *n*-region and the holes towards the *p*-region. Consequently in the external circuit an electric current flows from the *p* terminal to the *n* terminal (see Fig. 14.27). The semiconductor material in a solar cell must be very thin so that it is transparent to sunlight This is usually achieved by depositing a thin layer of *p*-type silicon on a wafer of *n*-type silicon (see Fig. 14.28).

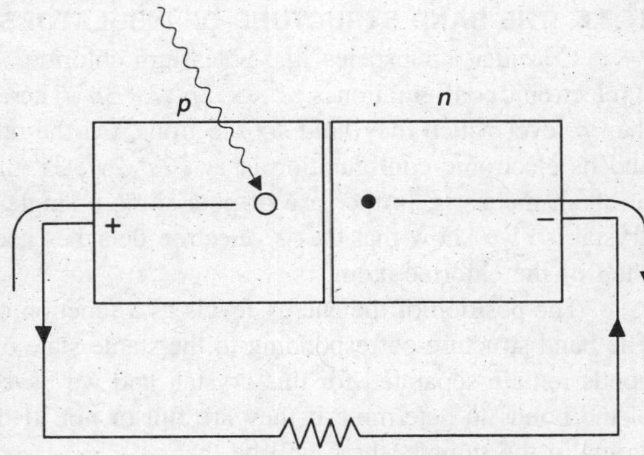

Fig. 14.27

The e.m.f. of a silicon solar cell is about 0.6 V, but the current that it delivers is fairly small. For example, a typical solar cell of area 5 cm^2 exposed to full sunlight delivers only 0.1 A: this amounts to roughly 10% conversion of the energy of sunlight into electric energy.

14.30. EFFECT OF TEMPERATURE ON SEMICONDUCTORS

An important and characteristic feature of germanium is that the energies required to remove holes from acceptors and electrons from donors are so small that negligible fraction of the excess electrons or holes are bound to the impurities at room temperature.

Germanium sample shows the interesting property that heat treatment at temperatures above 800 K appears to produce *p*-type impurities. Prolonged heat treatment causes impurities to disappear. One theory is that these effects are not due to chemical impurities at all but instead to disorder in the germanium lattice. The total number of disordered atoms predicted from theory is an approximate agreement with the concentrations given at temperature T

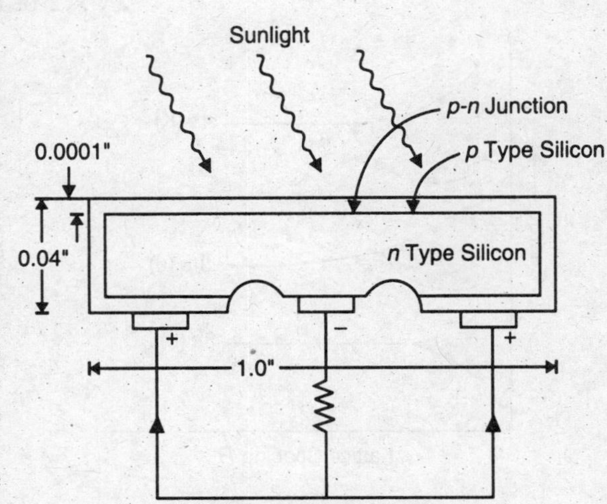

Design of a solar cell

Fig. 14.28

$$= e^{-E/kT}$$

where E is the energy required to disorder an atom.

14.31. EFFECTS OF BOMBARDMENT OF HIGH ENERGY PARTICLES ON SEMICONDUCTORS

When n-type germanium semiconductor is bombarded with alpha particles of deuterons, it is converted to p-type, whose conductivity increases with increasing exposure to bombardment. It has been established that this effect is not due to transmutation of germanium to another element but instead to the creation of disorder in the crystal, that is germanium atoms knocked into interstitial positions leaving vacant lattice point behind. "These lattice defects" act as acceptors.

14.32. INSULATORS

Insulators have a very high resistivity, of the order of 10^7 ohm-m or greater. These material usually have a dielectric constant between 1 and 10. Plastic, glass, ceramics, and many crystals such as quartz, diamond and the alkali halides fall into this class. Amorphous materials such as sulphur, pitch and various cements and waxes are good insulators. The number of charge carriers in an insulator is small but highly dependent upon the temperature of the sample.

14.33. THE BAND STRUCTURE OF INSULATORS

Consider a substance like potassium chloride. Atomic number of chlorine is 17 and hence its electronic configuration is $1s^2, 2s^2\, 2p^6, 3s^2\, 3p^5$. There is only one electron missing from completing the $3p$ level which may hold six electrons. On the other hand atomic number of potassium is 19 and its electronic configuration is $1s^2, 2s^2\, 2p^6, 3s^2\, 3p^6, 4s^1$. The single valence electrons typical of alkali metals is, in the case of potassium, a single $4s^1$ electron. Potassium chloride is an ionic crystal, so we know that the $4s^1$ electron detaches itself from the potassium atom and fills the $3p$ shell of the chlorine atom.

The position of the energy levels as a function of atomic spacing is shown in Fig. 14.29 (*a*). The band structure corresponding to the stable state of the crystal is shown in Fig. 14.29 (*b*). The bonds remain separated for this crystal, and we have only to investigate the electron population of the bonds to determine if they are full or not. If there are N chlorine atoms and N potassium atoms in the sample, there will be then

$2N \times 2$ of the $3s$ states
$2N \times 6$ of the $3p$ states and
$2N \times 2$ of the $4s$ states.

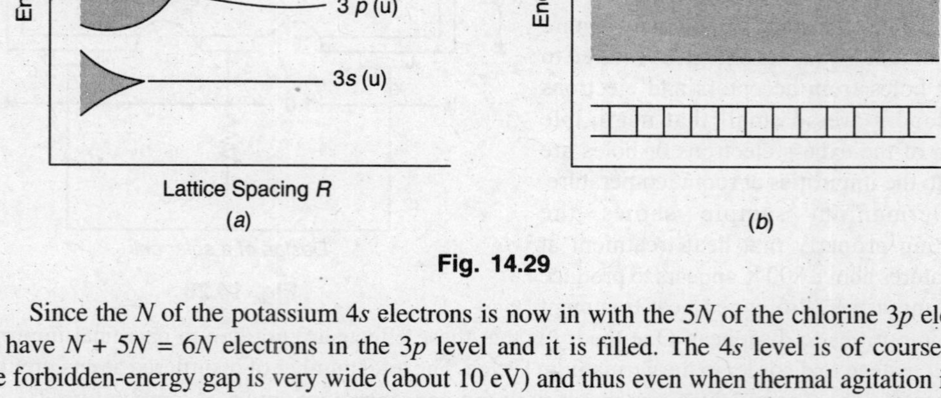

Fig. 14.29

Since the N of the potassium $4s$ electrons is now in with the $5N$ of the chlorine $3p$ electrons, we have $N + 5N = 6N$ electrons in the $3p$ level and it is filled. The $4s$ level is of course empty. The forbidden-energy gap is very wide (about 10 eV) and thus even when thermal agitation is large, almost no electrons can be made to jump the gap. The resistivity of KCl is quite high even at elevated temperatures.

Let us consider another example of an insulator. Diamond is the crystalline form of carbon, the sixth element in the periodic table. The element is of interest because it lies in the same period and has the same type of binding as silicon and germanium. The shape of the energy band structure of diamond as a function of atomic spacing is shown in Fig. 14.30 (*a*). It is noted that there is a crossover point at which two of the 2*p* states form a band with the two 2*s* states; thus making available a band with four available states. The remaining four of the total of six 2*p* states split off and form a higher band of their own. At the normal atomic spacing shown, there is a wide forbidden gap between the higher and lower bands. Since there is total 4 electrons in the 2*s* and 2*p* states of carbon all these electrons are used in filling the lower band. The upper band is left empty, and hence diamond is a very good insulator. The one-dimensional energy level picture of diamond is shown in Fig. 14.30 (*b*). Only the upper most bands are shown. The phenomenon of changing band structure at a crossover point which is shown by carbon seems to be typical of the covalent crystal formed from the elements in the fourth period of the periodic table. Both silicon and germanium do the same thing.

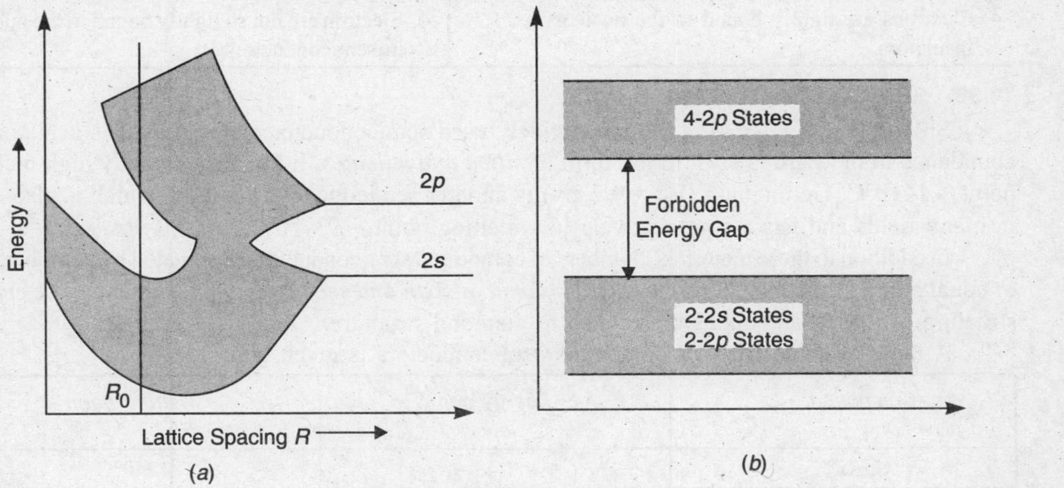

Fig. 14.30

On the basis of valence bond structure we can see why diamond should be an insulator. Although it contains a large number of electrons, as does a metal, the covalent bond is an entirely different structure from the metallic bond. The metallic bond is frequently described by saying that electron behaves, substantially like a gas of free electrons producing a uniform cloud of negative charge in which the positive ions of the metal float. In the case of diamond, however, the electrons do not behave like a gas permeating the crystal but instead more closely resembles structural elements which hold the crystal together. In an ideally perfect crystal, each valence band would contain in two electrons; therefore, every electron would be tightly bound and thus unable to enter into the conductor process.

Conductivity can be produced in diamond, however, in a number of ways.

(*i*) By the bombardment of high energy particles or photons.

(*ii*) By the thermal agitation.

(*iii*) By applying an electric field.

All these ways produce imperfection of valence bond structure. If the temperature sufficiently elevated, spontaneous breaking of the covalent bonds by thermal agitation will occur, producing electrons and holes in equal numbers. *This effect would occur at such high temperatures in diamond that it* cannot be observed. However, it plays a major role in silicon and germanium at temperatures well within the range of investigation in the laboratory.

14.34. USES OF INSULATORS

(*i*) Insulators like glass, plastics, ceramics, diamond, rubber etc. are used as insulating devices in the electrical connections.

(*ii*) Insulators are used to decrease the rate of flow of current where it is required.

Insulators differ from semiconductors in the following manner:

Insulators	Semiconductors
1. According to the electronic theory, the insulators have full valence band and an empty conduction band.	1. At room temperature some electrons may be present at the conduction band and the valence band may or may not be full.
2. Forbidden energy gap of the insulators are large.	2. Forbidden energy gap of semiconductors are small as compared to insulators.
3. The resistivity of insulators are of the order of 10^7 to 10^{12} ohm-m.	3. The resistivity of semiconductors are of the order of 0.5 to 10^3 ohm-m.
4. Electrons are tightly bound to the nucleus in Insulators.	4. Electron are not so tightly bound to the nucleus in semiconductors.

14.35. SEMICONDUCTOR MATERIALS:

Silicon (Eg = 1.1 e.v) is the most widely used semiconductor material as it is available in abundance in the Earth's crest in the form of *silica and silicates*. It has a moderately high melting point ~ 1410°C. Germanium (Eg = 0.7 e.v) is another semiconductor material which is also used in many fields and has a comparatively low melting point ~ 937°C.

In addition to these materials, numbers of compound semiconductors are created by combinations of equal atomic functions of *fifth and 3rd column* or *sixth and second column* elements. The crystal structure of these compounds are related to diamond structure.

A table of some III – IV compound semiconductors is given under:

Compound	*Energy gap e.v.*	*Melting point*
GaAs	1.43	1240°C
GaSb	0.78	705°C
InAs	0.35	940°C
InSb	0.18	525°C
InP	1.29	1070°C
GaP	2.26	1350°C

These also exist amorphous (non-crystalline) semiconductors. In these case, because the inter atomic and local co-ordination characteristic remain approximately inchanged, so the band model of crystals apply to their case as well.

A number of semiconductor devices like junction rectifiers, photo cells, solar batteries, thermistors, transistors etc. use either intrinsic or extrinsic semiconductors. *The same crystal with different dopings may make them n-type or p-type semiconductor.* The boundary between the two regions of opposite characteristics is called *p-n junction.* A junction diode as it is called is finding use rectifying action. A three region crystal *p-n-p* or *n-p-n* provide the transistor characteristics. In this case superposition of the properties of two *p-n and n-p* close to each other act as an *amplifier* of electrical signals. Small size, low electrical consumption and low cost have made the solid state devices much more acceptable to electronic fields and industry in the recent years.

QUESTIONS AND PROBLEMS

1. Explain the classifications of solids. What are metals, insulators and semiconductors?
2. Discuss the energy band structure of solids and explain electrons and holes conduction.
3. Explain how free electrons behave in a metallic crystal lattice. Draw a typical energy distribution curve for these electrons, location the Fermi levels on it, explain its importance in the conduction process of the metal under a field.
4. Derive from first principle an expression for the electron conductivity of a metal with usual notations. How does this get altered for a semiconductor?
5. What do you mean by semiconductors? Explain the term intrinsic and extrinsic semiconductors.
6. Explain the type of bonding in a pure germanium or silicon crystal. What physical properties are associated with it.
7. Draw typical band model diagram for a semiconductor showing Fermi levels, what do you mean by intrinsic and extrinsic semiconductors?
8. A few phosphorous atoms are added as an impurity in the crystal matrix of germanium. How would alter the bonding now?
9. Draw an energy band model diagram for a phosphorous dopped germanium crystal and explain its electrical conductivity doping as compared to its state before.
10. What happens to the bonding of germanium or silicon crystal when impurity like boron or aluminium are added to them? Also draw an energy band model diagram of germanium or silicon crystals dopped with boron or aluminium as impurities.
11. Draw typical band model diagrams for a semiconductor showing (*i*) Fermi level and (*ii*) impurity level in it. Explain how an idea of these levels enables you to classify a semiconductor as (*i*) an intrinsic (*ii*) an extrinsic *p*-type and (*iii*) an extrinsic *n*-type semiconductor.
12. What are *p* and *n* type semiconductors? *p* and *n* type semiconductor have more holes and electrons respectively. Explain why they are electrically neutral?
13. (*a*) How will you prepare a *p-n* junction from *p*-type and *n*-type materials. How do the free carriers behave at the junction? (*b*) Explain the terms: (*i*) Depletion layer (*ii*) junction voltage (*iii*) junction capacitance for such a *p-n* junction.
14. What do you mean by forward and reverse biasing to a *p-n* junction? Explain your answer with diagrams.
15. Explain the important applications of solid-state diodes and transistors.
16. Explain what happens to the conductivity of (*i*) metallic crystal and (*ii*) a germanium crystal, when their purity is increased?
17. Given a *n*-type sample, how will you proceed to make a *p*-type sample?
18. What do you mean by insulators? What are their physical properties? Draw the energy band structure of a insulator.
19. In what respects insulators differ from semiconductors? Name a few insulators and mention their uses.
20. Write short notes on: (*i*) *p-n* junction, (*ii*) Effect of temperature on Semi-conductors, (*iii*) Intrinsic and Extrinsic Semiconductors, (*iv*) *p-n-p* Transistor, (*v*) Light emitting diodes, (*vi*) Solar cells.

CHAPTER

15

SPECTROSCOPY

15.1. BOHR'S THEORY OF HYDROGEN ATOM

Bohr adopted Rutherford's nuclear model of the atom; on this basis, the hydrogen atom should consist of a simply charged positive nucleus and an electron outside the nucleus. Since the atomic number is equal to unity. Assuming Coulomb's law of forces and Newton's law of motion to be applicable in the atomic domain, the path of the electron around the nucleus should be a conic section.

POSTULATES OF BOHR'S THEORY

1. There exist concentric energy levels around the nucleus which are called orbits. Electrons move on these orbits and there occurs no emission of radiation so long as the electron revolves on these orbits.
2. The angular momentum of an electron moving on these orbits is always integral multiple of $\left(\hbar = \dfrac{h}{2\pi}\right)$. The angular momentum associated with electron moving on the nth orbit is given by

$$p = n\hbar = n\frac{h}{2\pi}$$

This is known as Bohr's quantum condition.

3 The emission or absorption by an atom is not continuous but is discrete *i.e.*, in the form of bundles or quanta of energies. The energy associated with each quantum is $h\nu$ where h is Planck's constant and ν the frequency of radiation.

4. The energy in the form of radiation is emitted when the transition is from higher energy level to lower energy level. The absorption takes place when transition is from lower energy to higher energy level.
5. If the atom transists from nth orbit (level) to pth orbit, the energy emitted as radiation is given by

$$E_n - E_p = h\nu$$

where E_n is the energy associated with nth level and E_p that associated with pth level.

6. So long as an electron moves in a stationary orbit its equilibrium is maintained by the relation Centrifugal force = Force of attraction *between an electron and the nucleus*

Let an electron of mass m, moves on the nth, orbit of radius r_n with velocity v_n. Let Z be the no. of positive charge (*i.e.*, protons) inside the nucleus, then centrifugal force from Fig. 15.1

$$= \frac{mV_n^2}{r_n}$$

The electrostatic force of attraction between nucleus and electron $= \dfrac{kZe \times e}{r_n^2}$ where k is constant and its value in S.I. system is given as

$$k = \frac{1}{4\pi\varepsilon_0} = 9 \times 10^9 \text{ Nm}^2/C^2$$

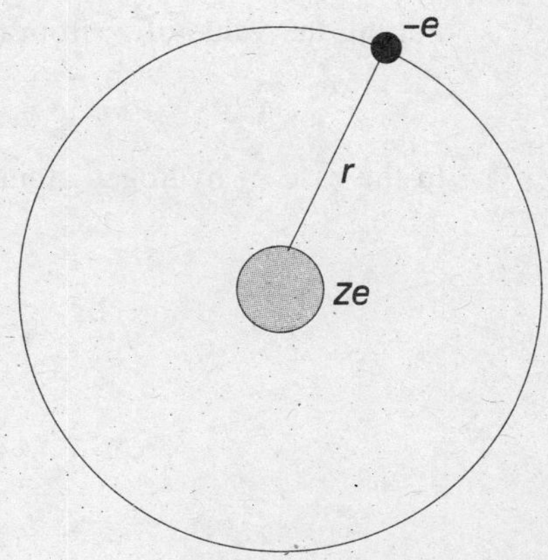

Fig. 15.1

Now from the 6th postulate

$$\frac{mv_n^2}{r_n} = \frac{kZe^2}{r_n^2}$$

$$\therefore \quad mv_n^2 = \frac{kZe^2}{r_n} \quad ...(2)$$

From 2nd postulate

$$p = n\frac{h}{2\pi}$$

where p is the moment of momentum

i.e., $p = mv_n . r_n$

$$\therefore \quad mv_n . r_n = n\frac{h}{2\pi} \quad ...(3)$$

Squaring both sides

$$m^2v_n^2r_n^2 = n^2\frac{h^2}{4\pi^2} \quad ...(4)$$

Dividing (2) by (4) we get

$$r_n = \frac{n^2h^2}{4\pi^2 . k . mZe^2}$$

$$= \frac{n^2h^2}{4\pi^2 \times \frac{1}{4\pi\varepsilon_0} \times mZe^2} \quad ...(5)$$

For ground state, we have

$$n = 1$$

∴ radius of the first Bohr's orbit

$$r_1 = \frac{h^2 . \varepsilon_0}{\pi mZe^2} \quad ...(6)$$

We can now find the energy associated with any orbit. For this we will have to consider both kinetic and potential energies.

The kinetic energy of the electron moving on the nth orbit

$$\text{K.E.} = \frac{1}{2}mv_n^2$$

Putting the value of mv_n^2 from (2)

$$\therefore \quad \text{K.E.} = \frac{1}{2}k\frac{Ze^2}{r_n}$$

Potential energy of an electron moving on the nth orbit can be calculated from the work done in taking $-e$ charge from infinity to a point distant r_n from the centre of the atom.

So the P.E. = potential at the point × charge

$$= k\frac{Ze}{r_n} \times -e = -\frac{kZe^2}{r_n}$$

Total energy of the electron revolving in the nth orbit

$$E_n = \text{K.E.} + \text{P.E.}$$

$$= \frac{1}{2}k\frac{Ze^2}{r_n} - \frac{kZe^2}{r_n}$$

$$= -\frac{kZe^2}{2r_n} = -\frac{Ze^2}{8\pi\varepsilon_0 r_n}$$

Putting the value of r_n from (6)

$$E_n = -\frac{Zme^4}{8n^2h^2\varepsilon_0^2} \qquad \text{... (7)}$$

In the case of hydrogen atom $Z = 1$

$$= -\frac{me^4}{8n^2h^2.\varepsilon_0^2} \qquad \text{... (8)}$$

$$= -\frac{(9.11\times10^{-19})^4}{8\times(8.854\times10^{-12})^2(6.62\times10^{-34})^2\times n^2}$$

$$= -\frac{21.7\times10^{-19}}{1.6\times10^{-19}}\times\frac{1}{h^2}\text{eV}$$

$$= -\frac{13.6}{n^2}\text{eV} \qquad [\because 1\,\text{eV} = 1.6\times10^{-19}\text{ J}]$$

Similarly for pth level

$$E_p = -\frac{me^4}{8p^2h^2\varepsilon_0^2} \qquad \text{... (9)}$$

Frequency of emitted radiations. We can find the frequency of emitted radiation from the relation

$$h\nu = E_n - E_p$$

$$= \left[\frac{-me^4}{8\varepsilon_0^2n^2h^2} - \left(\frac{-me^4}{8\varepsilon_0^2p^2h^2}\right)\right]$$

$$= \frac{me^4}{8\varepsilon_0^2h^2}\left[\frac{1}{p^2}-\frac{1}{n^2}\right] \qquad \text{... (10)}$$

If c be the velocity of light and λ be the wavelength then, $\nu = \frac{c}{\lambda}$.

$$\therefore \quad \frac{c}{\lambda} = \frac{me^4}{8\varepsilon_0^2h^3}\left[\frac{1}{p^2}-\frac{1}{n^2}\right]$$

or

$$\frac{1}{\lambda} = \frac{me^4}{8\varepsilon_0^2ch^3}\left[\frac{1}{p^2}-\frac{1}{n^2}\right]$$

$$= R\left[\frac{1}{p^2}-\frac{1}{n^2}\right] \qquad \text{... (11)}$$

where $R = \frac{me^4}{8\varepsilon_0^2ch^3}$ is called Rydberg constant and has value = 1.097×10^7 m^{-1}

The reciprocal of wavelength is denoted by ν and corresponds to wavelength in wave numbers, so we have

$$\bar{\nu} = \frac{1}{\lambda} = R\left[\frac{1}{p^2}-\frac{1}{n^2}\right]$$

Using the values of m, e, c and h, the values of Rydberg constant R can be found

$$m = 9.11\times10^{-31} \text{ kg}$$

$$e = 1.6\times10^{-19} \text{ coulomb}$$

$$h = 6.6\times10^{-34} \text{ joule-sec}$$

$$c = 3\times10^8 \text{ m/sec.}$$

$$R = \frac{me^4}{8\varepsilon_0^2ch^3} = 1.097\times10^7 \text{ m}^{-1}$$

15.2. SPECTRUM OF HYDROGEN

As long ago as 1885, Balmer succeeded in obtaining a simple relationship among the wave number of the lines in the visible region of the hydrogen spectrum. Balmer's Equation expressed in modern notation is

$$\frac{1}{\lambda} = \bar{\nu} = R_H\left(\frac{1}{2^2} - \frac{1}{n^2}\right) \quad \text{... (12)}$$

where (n = 3, 4, 5...), λ is the wavelength and $\bar{\nu}$ is the wave number of the spectral line. R_H is a constant known as Rydberg's constant for hydrogen. The empirical value of Rydberg's constant for hydrogen is

$$R_H = 1.097 \times 10^7 \text{ m}^{-1}$$

By substituting for n in Equation. (12) the successive values 3, 4, 5, 6... we obtained the wave numbers of the lines of the Balmer's series. The relative positions of these lines on a wave number scale are shown in Fig. 15.2 (a).

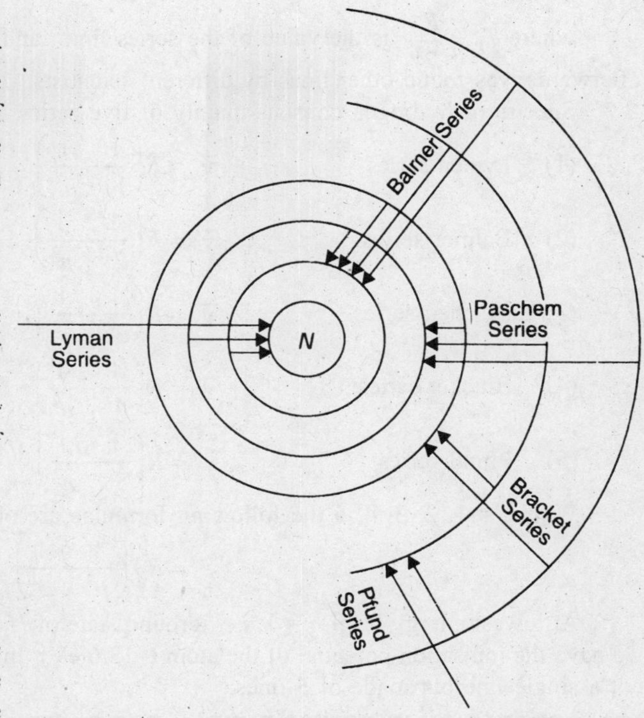

Fig. 15.2 (*a*)

It is obvious from the Equation. (8) that as n approaches infinity, the lines crowd together and approach a limit known as series limit. The values of the limit of the Balmer's series is

$$R_H = 5.48 \times 10^6 \text{ m}^{-1}$$

Another way of expressing the Balmer formula is by writing the wave number of each line as the difference between two terms.

$$\bar{\nu} = T_1 - T$$

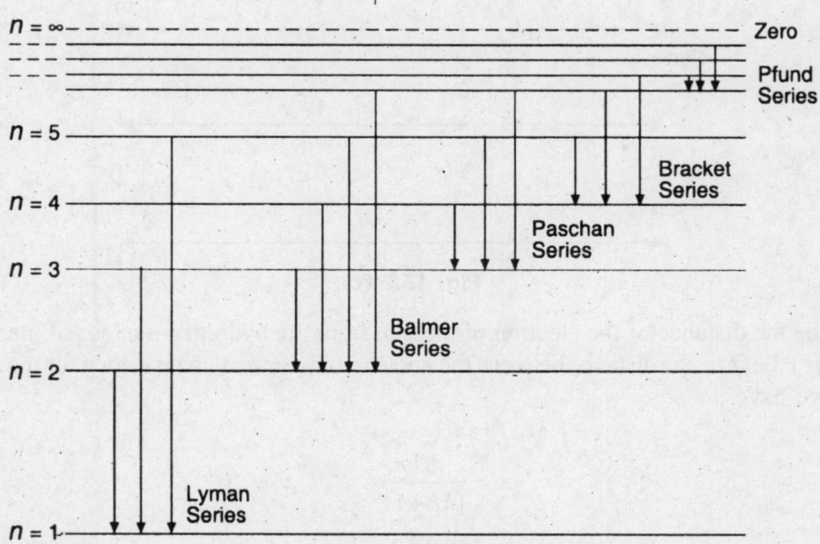

Fig. 15.2 (*b*)

where $T_l = \frac{R_H}{2^2}$ is the value of the series limit, and $T = \frac{R_H}{n^2}$ is the variable term. Shortly afterwards was found other lines by different scientists.

Spectra of hydrogen consists mainly of five series as shown in Fig. 15.2 (*b*) and 15.2 (*c*)

(1) Lymen series $\quad \bar{\nu} = R\left(\frac{1}{1^2} - \frac{1}{n^2}\right)$ where $n > 1 = 2, 3, 4, 6, ...$

(2) Balmer series $\quad \bar{\nu} = R\left(\frac{1}{2^2} - \frac{1}{n^2}\right)$ where $n > 2 = 3, 4, 5, 6...$

(3) Paschen series $\quad \bar{\nu} = R\left(\frac{1}{3^2} - \frac{1}{n^2}\right)$ where $n > 3 = 4, 5, 6...$

(4) Bracket series $\quad \bar{\nu} = R\left(\frac{1}{4^2} - \frac{1}{n^2}\right)$ where $n > 4 = 5, 6, 7...$

(5) Pfund series $\quad \bar{\nu} = R\left(\frac{1}{5^2} - \frac{1}{n^2}\right)$ where $n > 5 = 6, 7, 8...$

With $n = 1, 2, 3, 4, 5$ the following formulae are obtained. We have general formula

$$\bar{\nu} = R_H\left(\frac{1}{p^2} - \frac{1}{n^2}\right)$$

At lowest energy level $n = 1$, *i.e.*, ground state of H atom has the energy 13.6 eV, *i.e.*, why we have the ionisation potential of the atom (–13.6 eV). In the actual experiment it was found H_α not a single line but made of 5 lines.

15.3. CORRECTION IN BOHR'S THEORY FOR FINITE MASS OF NUCLEUS

In the Bohr's theory we consider the nucleus to be at rest because it is very heavy as compared to the mass of an electron. However, in actual practice, the nucleus is having a finite mass and both the nucleus and the electron revolve with a common centre of mass *c*. If the nuclear motion is also taken into account then we require some correction in the frequency of spectral lines as shown under.

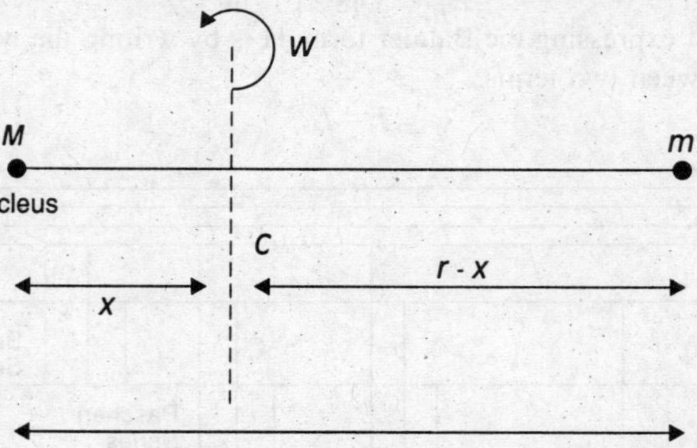

Fig. 15.2 (*c*)

Let *r* be the distance of the electron of mass *m* from the hydrogen nucleus of mass as shown in Fig. 15.2 (*c*). Let *x* be the distance between the nucleus and centre of mass *c*, then for the equilibrium condition, we have

$$M.x. = m.(r - x) \quad \text{... (i)}$$

or

$$x = \frac{m.r}{(M+m)} \quad \text{... (ii)}$$

or from (*i*)

$$(r - x) = \frac{M.x}{m} = \frac{m.r}{(M+m)} \quad \text{... (iii)}$$

We have the total angular momentum ($L = I\omega$) of the atom about the centre of mass as

$$\begin{aligned} L &= Mx^2\omega + m(r-x)^2\omega \\ &= M \cdot \frac{m^2r^2}{(m+M)^2} + m \cdot \frac{M^2r^2\omega}{(m+M)^2} \\ &= \frac{M \cdot m^2r^2\omega}{(m+M)^2}[m+M] \\ &= \left[\frac{m \cdot M}{(M+m)}\right] r^2 \cdot \omega = \mu \cdot r^2\,\omega \qquad \text{... (1)} \end{aligned}$$

where $\mu = \dfrac{m \cdot M}{(M+m)}$ is known as the reduced mass of the electron.

If we consider that there is no nuclear motion, the angular momentum in that case is given by

$$m \cdot v \cdot r = m \cdot r^2 \cdot \omega \qquad \text{... (2)}$$

So we see that due to the motion of the nucleus, m is replaced by μ and correspondingly the energy of the electron in nth orbit is given by

$$\begin{aligned} E_n &= -\frac{2\pi^2\mu e^4 \cdot k^2 \cdot Z^2}{n^2h^2} \\ &= -\frac{2\pi^2\mu e^4 \dfrac{1}{(4\pi\varepsilon_0)^2}}{n^2h^2} \qquad [\because Z = 1 \text{ for Hydrogen}] \\ &= -\frac{\mu e^4}{8\varepsilon_0^2h^2}\left(\frac{1}{n^2}\right) \end{aligned}$$

which in terms of wavelength will be given by the relation

$$\begin{aligned} \bar{v} &= \frac{1}{\lambda} = \frac{\mu e^4}{8\varepsilon_0 ch^3}\left(\frac{1}{n^2} - \frac{1}{p^2}\right) \\ &= R_H\left(\frac{1}{n^2} - \frac{1}{p^2}\right) \end{aligned}$$

where $R_H = \dfrac{\mu e^4}{8\varepsilon_0 ch^3}$ is Rydberg constant by considering the reduced mass of hydrogen atom.

It should be noted that μ being somewhat less than m, the wave-length of spectral lines will be slightly more in the case of very heavy nuclei *i.e.*, when the nuclear motion is absent.

15.4. BOHR'S QUANTUM FROM DE-BROGLIE HYPOTHESIS

From de-Broglie hypothesis an electron of mass m in motion with velocity v is associated with a wave whose wavelength λ is given by

$$\lambda = \frac{h}{mv} \qquad \text{....(1)}$$

where h is Planck's constant.

On the basis of this hypothesis Bohr proposed an atomic model in which the stationery orbits of the Bohr's model were retained but electron in various orbits behaves as shown in Fig. 15.3.

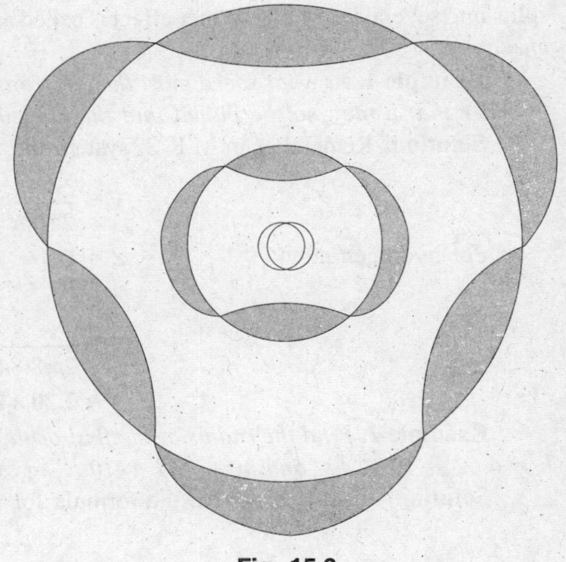

Fig. 15.3

Bohr suggested that *stationary orbits are those in which orbital circumferences ($2\pi r$) is an integral multiple of de-Broglie wavelength λ i.e.,* stationary orbits for an electron are those which contian the complete waves of electron. So we have

$$2\pi r = n\lambda \qquad \text{where } n = 1, 2, 3... \qquad ...(2)$$

Also

$$\lambda = \frac{h}{mv}$$

$\therefore$

$$2\pi r = n\frac{h}{mv}$$

or

$$mvr = n\frac{h}{2\pi} \qquad ...(3)$$

where mvr is the angular momentum of the electron as particle. The equation (3) corresponds to the Bohr's 2nd postulate, *i.e.*, the total angular momentum of the moving electron is an integral multiple of $\hbar$. So the new concept of de-Broglie confirms the Bohr's postulate.

15.5. DRAWBACKS OF BOHR'S THEORY

Bohr's theory was able to explain successfully a number of experimental observed facts and has correctly predicted the spectral lines of neutral hydrogen atom and the singly ionized helium atom etc. in terms of only principal quantum number n. However, the theory fails to explain the following phenomenon:

1. The theory could not account for the spectra of atom more complex than that of hydrogen.
2. It does not explain the experimentally observed changes in intensity of the spectral lines of an element.
3. This theory cannot be used to calculate about transitions from one level to another such as the rate at which they occur or from the selection rules which apply to them.
4. This theory cannot explain the fine structure of spectral line. In fact it was found that when spectral lines emitted by an atom are examined, each line is composed of several lines closely packed together.
5. This theory cannot be used for the quantitative explanations of chemical bonding that exist between various atoms.
6. The theory does not give any information regarding the distribution and arrangement of electrons in atom.
7. The theory does not give correct results when an electric or magnetic field is applied to the atom. It is found that when electric or magnetic field is applied to the atom, each spectral line splits into several lines. The former effect is called as *Stark effect* while the later is known as *Zeeman effect.*

Example 1. *At what speed must the electrons revolve round the nucleus of a hydrogen atom in order that it may not be pulled into the nucleus by electrostatic attraction.*

Solution: Remember in M.K.S. system the formula for the speed of the electron is

$$v = \frac{Ze^2}{2\varepsilon_0 nh}$$

For hydrogen atom

$$Z = 1$$
$$n = 1$$

$$v = \frac{(1.6\times10^{-19})^2}{2\times8.845\times10^{-12}\times6.62\times10^{-34}}$$
$$= \mathbf{2.20\times10^6\ m/s.}$$

Example 2. *Find the radius of the first orbit of the electron in the hydrogen atom given that $h = 6.62 \times 10^{-34}$ J.s. and $m = 9.11 \times 10^{-31}$ kg. and $c = 1.6 \times 10^{-19}$ coulomb for an electron.*

Solution: In M.K.S system the formula for the radius of orbit is given by

$$r_n = \frac{\varepsilon_0 n^2 h^2}{\pi m Z e^2}$$

where $\varepsilon_0 = 8.854 \times 10^{-12}$ coulomb^2/N-m^2

For hydrogen atom

$$n = 1 \text{ and } Z = 1$$

$$r_1 = \frac{8.854 \times 10^{-12} \times 1^2 \times (6.62 \times 10^{-34})^2}{3.143 \times 9.11 \times 10^{-31} \times 1 \times (1.6 \times 10^{-19})^2}$$

$$= \mathbf{0.528 \times 10^{-10} \text{ metres.}}$$

Example 3. *The wavelength of the first member of the Balmer series in hydrogen is 6563Å, calculate the wavelength of the first member of Lyman series in the same spectrum.*

Solution: For the first member of the Balmer series

$$\bar{v}_1 = \frac{1}{\lambda_1} = R\left[\frac{1}{2^2} - \frac{1}{3^2}\right] = \frac{5}{36}R \quad \text{... (1)}$$

For the first member of Lyman series

$$\bar{v}_1 = \frac{1}{\lambda_2} = R\left[\frac{1}{1^2} - \frac{1}{2^2}\right] = \frac{3R}{4} \quad \text{... (2)}$$

Dividing (1) by (2) we get

$$\frac{\lambda_2}{\lambda_1} = \frac{5}{27} \quad \text{or} \quad \lambda_2 = \frac{5}{27} \times \lambda_1$$

or

$$\lambda_2 = \frac{5 \times 6563}{27} = \mathbf{1215.37 \text{ Å}}$$

Example 4. *Calculate the energy required to excite the hydrogen atom from the ground state (n = 1) to the first excited state n = 2.*

Solution: Energy required

$$= \frac{me^4}{8\varepsilon_0^2 h^2}\left[\frac{1}{p^2} - \frac{1}{n^2}\right]$$

$$E = \frac{me^4}{8\varepsilon_0^2 h^2}\left[\frac{1}{1^2} - \frac{1}{2^2}\right]$$

$$E = \frac{3me^4}{32\varepsilon_0^2 h^2}$$

Here

$$m = 9.1 \times 10^{31} \text{ kg}, e = 1.6 \times 10^{-19} \text{ C}$$

$$\varepsilon_0 = 8.85 \times 10^{-12} \text{ C}^2/\text{N-m}^2$$

$$h = 6.624 \times 10^{-34} \text{ J-s}$$

$$E = \frac{3 \times 9.1 \times 10^{-31} \times (1.6 \times 10^{-19})^2}{3.2 \times (8.85 \times 10^{12})^2 \times (6.624 \times 10^{-34})^2} = \mathbf{1012 \text{ eV}}$$

Example 5. *Making use of physical constants given below, calculate for hydrogen atom ;*

(i) velocity of an electron in ground state,

(ii) radius of Bohr's orbit in ground state,

(iii) time taken by the electron to transverse the first Bohr's orbit.

(iv) Rydberg's constant.

Electronic charge, $e = 1.6 \times 10^{-19}$ *coul.*

Electronic mass, $m = 9.1 \times 10^{-31}$ *kg.*

Planck's constant, $h = 6.6 \times 10^{-34}$ *Joule-sec.*

Speed of light, $e = 3.0 \times 10^{8}$ *m/sec.*

Permittivity of free space, $\varepsilon = 8.86 \times 10^{-13}$ *coulomb*^{2}*/N-m*^{2}.

Solution: (*i*) Velocity of electron in ground state ($n = 1$) is given by the relation

$$v = \frac{e^2}{2h\varepsilon_0} = \frac{(1.6\times10^{-19})^2}{2\times(6.6\times10^{-34})\times(8.86\times10^{-12})}$$

$$= \mathbf{2.19\times10^6\ m/sec.}$$

(*ii*) The radius of Bohr's orbit in the ground state is given by

$$r_1 = \frac{h^2\varepsilon_0}{\pi me^2} = \frac{(6.66\times10^{-36})^2\times(8.86\times10^{-12})}{3.14\times(9.1\times10^{-31})\times(1.6\times10^{-19})^2}$$

$$= 0.528\times10^{-10}\text{ m} = \mathbf{0.528\ Å}$$

(*iii*) Time taken by electron to transverse the first Bohr's orbit is given by

$$T = \frac{2\pi r_1}{v} = \frac{2\times3.14\times0.528\times10^{-10}}{2.19\times10^6}$$

$$= \mathbf{1.51\times10^{-16}\ sec.}$$

(*iv*) Rydberg constant R is given by the relation

$$R = \frac{me^4}{8\varepsilon_0 ch^3}$$

$$= \frac{(9.1\times10^{-31})\times(1.6\times10^{-19})^4}{8\times(8.86\times10^{-12})^2\times(3\times10^8)\times(6.6\times10^{-34})^3}$$

$$= \mathbf{1.101\times10^{-7}\ m^{-1}}$$

Example 6. *Wavelength of Balmer H_α line is 6553 Å. Calculate the wavelength of H_β line.*

Solution: For H_α line Balmer series *i.e.*, the first member,

$$\bar{\nu}_1 = R\left[\frac{1}{2^2}-\frac{1}{3^2}\right]$$

$$\frac{1}{\lambda_1} = \frac{5}{36}R \qquad \text{... (1)}$$

For H_β line of Balmer series *i.e.* the second number,

$$\bar{\nu}_2 = R\left[\frac{1}{2^2}-\frac{1}{4^2}\right]R$$

$$\frac{1}{\lambda} = \frac{3}{16}R \qquad \text{... (2)}$$

Dividing (1) by (2)

$$\frac{\lambda_2}{\lambda_1} = \frac{20}{27}$$

$$\lambda_2 = \left(\frac{20}{27}\right)\lambda_1$$

But $\lambda_1 = 6563$ Å $\therefore\ \lambda_2 = \left(\frac{20}{27}\right)6563 = \mathbf{4861\ Å}$

Example 7. *Calculate frequency of electron in the first Bohr's orbit in hydrogen atom.*

Solution: The velocity of the electron of Bohr's orbit is given by

$$v = \frac{e^2}{2nh\varepsilon_0}$$

orbital frequency is given by $\quad v = \dfrac{\text{electron speed}}{\text{orbit circumference}} = \dfrac{v}{2\pi r}$

or

$$v = \frac{e^2}{2nh\varepsilon_0}\times\frac{1}{2\pi}\times\frac{\pi me^2}{n^2\varepsilon_0 h^3}$$

or $$\nu = \frac{me^4}{4\varepsilon_0^2 n^3 h^2}$$

In this case $n = 1$

$$\therefore \quad \nu = \frac{(9.1\times10^{-31})\times(1.6\times10^{-19})^4}{4\times(8.854\times10^{-12})^2\times1^3\times(6.625\times10^{-34})^3}$$

$$= \mathbf{6530\times10^{12}\ Hz.}$$

Example 8. *A single electron revolves around a stationary nucleus of charge +Ze where Z is a constant and e is the magnitude of the electronic charge. It requires 47.2 eV to excite the electron from the second Bohr's orbit to the third Bohr's orbit. Find*

(a) the value of Z,

(b) the energy required to excite the electron from third to fourth Bohr's orbit,

(c) the wavelength of the electromagnetic radiation required to remove the electron from, the first Bohr's orbit to infinity,

(d) the radius of the first Bohr's orbit,

(e) the kinetic energy, potential energy and the angular momentum of the electron in the first Bohr's orbit.

Solution: The energy required to excite the electron from n_1 to n_2 orbit revolving round the nucleus of charge $+Ze$ is given by

$$E_n - E_p = \frac{Z^2 me^4}{8\varepsilon_0^2 h^2}\left[\frac{1}{p^2}-\frac{1}{n^2}\right]$$

or $$E_n - E_p = Z_2\times13.6\left[\frac{1}{p^2}-\frac{1}{n^2}\right]\text{eV}$$

(*a*) According to the given problem, 47.2 eV energy is required to excite the electron from $n = 2$ to $n = 3$ orbit. Hence

$$47.2 = Z^2\times13.6\left[\frac{1}{2^2}-\frac{1}{3^2}\right]$$

or $$Z^2 = \frac{47.2\times36}{13.6\times5}$$

or $$Z = \mathbf{5}$$

(*b*) The energy required to excite the electron from $n = 3$ to $n = 4$ orbit is given by

$$E_4 - E_3 = 25\times13.6\left[\frac{1}{3^2}-\frac{1}{4^2}\right]$$

$$= \frac{25\times13.6\times7}{144} = \mathbf{16.53\ eV}$$

(*c*) The energy required to remove the electron from the first Bohr orbit to infinity (∞) is given by

$$E_\infty - E_1 = 13.6\times Z^2\left[\frac{1}{1^2}-\frac{1}{\infty^2}\right] = 25\text{ eV}$$

In order to calculate the wavelength of the radiation, we use Bohr's frequency relation

$$h\nu = \frac{hc}{\lambda} = 13.6\times25\times1.6\times10^{-19}\text{ Joule}$$

or $$\lambda = \frac{6.6\times10^{-34}\times3\times10^8}{13.6\times25\times1.6\times10^{-19}} = \mathbf{3649.76\times10^{-12}\ m}$$

(*d*) The radius r_1 of the First Bohr's orbit is given by

$$r_1 = \frac{\varepsilon_0 h^2}{\pi me^2}\cdot\frac{1}{Z}$$

$$= \frac{0.53 \times 10^{-10}}{5} \qquad \left[\because \frac{\varepsilon_0 h^2}{\pi m e^2} = 0.53 \times 10^{-10} \text{ m}\right]$$

$$= \mathbf{0.106 \times 10^{-10} \text{ m}}$$

(*e*)
$$\text{K.E.} = \frac{1}{2} m v_1^2 = \frac{1}{2} \frac{Ze^2}{4\pi\varepsilon_0 r_1}$$

$$= \frac{1}{2} \frac{5 \times (1.6 \times 10^{-19})^2 \times 9.0 \times 10^9}{(0.106 \times 10^{-10})} \qquad [\because Z = 5]$$

$$= \mathbf{543.4 \times 10^{-19} \text{ Joule}}$$

$$\text{P.E} = -2 \times \text{K.E.} = \mathbf{-1086.8 \times 10^{-19} \text{ Joule}}$$

$$\text{Angular momentum} = m v_1 r_1 = \frac{h}{2\pi} = \mathbf{1.05 \times 10^{-34} \text{ Js.}}$$

Example 9. *Electrons of energies 10.20 eV and 12.09 eV can cause radiation to be emitted from hydrogen atoms. Calculate in each case, the principal quantum number of the orbit to which electron in the hydrogen atom is raised and the wavelength of radiation emitted if it drops back to ground state.*

Solution: The orbital relation energy of an electron revolving in *n*th orbit is given by

$$E_n = -\frac{13.6}{n^2} \text{ eV}$$

where n is the principal quantum number.

When $n = 1$, $E_1 = -13.6$ eV

$n = 2$, $E_2 = -3.4$ eV

$n = 3$, $E_3 = -1.51$ eV

Energy required by an electron to go from K to L level is (13.6 – 3.4) = 10.2 eV, and that required to go to M level is 13.6 – 1.51 = 12.09 eV. The corresponding quantum members are $n = 2$ and $n = 3$ respectively. So electrons will be raised to principal quantum number 2 and 3 corresponding to energies 10.20 eV and 12.09 respectively

$$n = 2 \text{ and } 3$$

We can calculate wavelength of radiation emitted as under : electron jumps are coming back from $n = 2$ and $n = 3$ to the ground state, *i.e.*, $n = 1$ this is the case of Lyman series. So we have

$$\frac{1}{\lambda_1} = R\left[\frac{1}{1^2} - \frac{1}{2^2}\right] = \frac{3R}{4}$$

$$\therefore \qquad \lambda = \frac{4}{3 \times 1.097 \times 10^7} = \mathbf{1216 \text{ Å}}$$

and
$$\frac{1}{\lambda_2} = \frac{9}{8R} = \frac{9}{8 \times 1.097 \times 10^7} = \mathbf{1026 \text{ Å}}$$

Example 10. *If the average life time of an excited state of hydrogen is of the order of 10^{-8} sec. estimate how many orbits an electron makes when it is in the state $n = 2$ and before it suffers a transition to state $n = 1$ (Bohr radius $a_0 = 5.3 \times 10^{-11}$ m).*

Solution: The angular momentum of *n*th orbit is given by

$$n\hbar = \frac{nh}{2\pi}$$

Again $mvr = n\dfrac{h}{2\pi}$ where r is the radius of the orbit

or
$$v = n\frac{h}{2\pi m r} \qquad \text{...(1)}$$

The time taken for completing an orbit is given by

$$T = \frac{2\pi r}{v} = \frac{2\pi r(2\pi mr)}{nh}$$

$$\therefore \quad T = \frac{4\pi^2 mr^2}{nh} \quad ...(2)$$

Since the radius of the orbit is proportional to n^2, *i.e.*,

$$r = a_0 n^2$$

$$\therefore \quad T = \frac{4\pi^2 m\, a_0^2\, n^2}{nh} = \frac{4\pi^2 m\, a_0^2\, n^3}{h}$$

Number of orbits completed in 10^{8} sec.

$$= \frac{10^{-8}}{T} = \frac{10^{-8} \times h}{4\pi^2\, m\, a_0^2\, n^3}$$

$$= \frac{10^{-8} \times (66 \times 10^{-34})}{4 \times (3.14)^2 \times (9.1 \times 10^{-31})(5.3 \times 10^{-11})^2 (2)^3}$$

$$= \mathbf{8 \times 10^4}.$$

Example 11. *Hydrogen atom has ground state energy –13.6 eV. Calculate the energy in the first two excited states (n = 2, n = 3). Calculate the wavelength of light emitted or absorbed in a transition between these two states (n = 2, n = 3).*

Solution: We know that the energy of hydrogen atom in *n*th stationary orbit is given by

$$E_n = -\frac{Rch}{n^2}$$

For ground state, $n = 1$ $\quad \therefore \quad E_1 = -Rch = -13.6$ eV

For first excited state, $n = 2$

$$\therefore \quad E_2 = \frac{Rch}{2^3} = -\frac{Rch}{4} = \frac{13.6}{4} = -3.4 \text{ eV}$$

For second excited state $n = 3$

$$\therefore \quad E_3 = -\frac{Rch}{3^2} = -\frac{13.6}{9} = -1.51 \text{ eV}$$

The Frequency ν of radiation emitted or absorbed between these two states is given by Bohr's frequency condition

$$\nu = \frac{E_2 \sim E_3}{h}$$

or

$$\frac{c}{\lambda} = \frac{E_2 \sim E_3}{h} \text{ or } \lambda = \frac{ch}{E_2 \sim E_3}$$

$$\lambda = \frac{(3 \times 10^8) \times (6.6 \times 10^{-34})}{[-1.51 - (-3.4)] \text{ eV}}$$

$$= \frac{(3 \times 10^8) \times (6.6 \times 10^{-34})}{(1.89)(1.6 \times 10^{-19})}$$

$$= 6561 \times 10^{-10} \text{ m} = \mathbf{6561\ Å}$$

Example 12. *The wavelength of sodium D_1 line is 590 mm. Calculate the difference in energy levels involved in the emission or absorption of this line.*

Solution: Here $\quad E_{(n-p)} = h\nu = \frac{hc}{\lambda}$

Here $\quad h = 6.624 \times 10^{34}$ J-s

$$\lambda = 590 \text{ nm} \times 590 \times 10^{-9} \text{ m}$$

$$\therefore \quad E_{(n-p)} = \frac{6.624 \times 10^{-34} \times 3 \times 10^{8}}{590 \times 10^{9}}$$

$$= \mathbf{3.37 \times 10^{-19}\ J}$$

15.6. JUSTIFICATION OF MOSLEY'S LAW ON THE BASIS OF BOHR'S THEORY

Mosley has shown by his experiments on X-rays spectrum that the square root of the frequencies of the characteristic X-rays is directly proportional to the atomic number 'Z' of the substances emitting the rays.

The frequency ν *emitted by an atom is given by the relation.*

$$\nu = \frac{2\pi^2 mZ^2k^2e^4}{h^3}\left(\frac{1}{n^2} - \frac{1}{p^2}\right) \quad \text{...(1)}$$

or $$\sqrt{\nu} \propto Z$$

Which is the Mosley's Law. Thus for K radiations, the Z increases by regular amounts and the different elements fall on the same line Mosely has shown that a little correction is necessary, because the effect of the other orbital electrons cannot be wholly neglected especially in the case of the heavier atoms which give out X-rays and which contian a relatively large number of extra nuclear electrons.

For copper, the wavelength of the K_α line

$$\lambda = 1.54 \times 10^{-10}\ \text{m}$$

$$\therefore \quad c = \nu\lambda$$

$$\therefore \quad = \frac{c}{\lambda} = \frac{3 \times 10^{10}}{1.54 \times 10^{-10}}$$

$$\nu = 1.95 \times 10^{18}\ \text{Hz}$$

The value of ν obtained by substituting the numerical values of m, c, h and Z for (Cu, Z = 9) is

$$\nu = 2.07 \times 10^{18}\ \text{Hz}$$

The agreement is quite satisfactory, considering the assumptions involved.The K radiations originate in the innermost orbit of the atom, called K-shell. Thus when a fastly moving electron penetrates a heavy atom and removes one of its K-electrons, an electronic jump from the L and M orbits takes place resulting in the emission of the characteristic K-radiations. The K-radiations thus correspond in the case of the heavier elements to the Lymen series of the hydrogen spectrum.

Again if the electron in the second orbit *i.e.*, L shell is knocked out, this gives to the emission of the L-radiations. Here $p = 2$ and $n = 3$ and substituting in Equation. (1)

$$\nu = \frac{2\pi^3 mZ^2k^2e^4}{h^3}\left(\frac{1}{2^2} - \frac{1}{3^2}\right)$$

$$= \frac{5}{18} = \frac{\pi^2 me^4k^2}{h^3}$$

or $$\sqrt{\nu} \propto Z$$

Which again shows how different elements get arranged in a line for L radiations as well. L radiations correspond to the Balmer series of the hydrogen spectrum. Similarly the M radiations correspond to the paschen series of the hydrogen spectrum.

It may also be noted here, that this over-simplified picture of a heavy atom in which the effects of remaining $(Z-1)$ electrons have been neglected completely is not fully is accord with **Mosley's Law.**

15.7. SOMMERFELD RELATIVISTIC ATOMIC MODEL

In order to overcome the drawbacks of Bohr's atomic model, Sommerfeld in 1921 modified Bohr's model by introducing the idea of *motion of electron* in *elliptical orbits* and taking into consideration the *variation of mass with velocity*. The improved model is known as Sommerfeld's relativistic model. The main features are briefly stated as follows:

According to Sommerfeld, the electron moving around the nucleus under an inverse square force, like planet round the sun, must describe elliptical orbits in addition to circular one as given by Bohr. In the elliptical orbit the position of the electron at any time may be fixed by two co-ordinates r and v (Fig. 15.4) where r is the radius vector and ϕ is the angle which the radius vector makes with the major axis of the ellipse *i.e.*, azimuthal angle. Let us consider an electron of mass m and linear tangential velocity v revolving in the elliptical orbit. The tangential velocity of the electron can be resolved in two components one along radius vector called the radial velocity and the other perpendicular to radius vector called the *transverse velocity*. Corresponding to these velocities the electron has two momentum—one along radius vector called as *radial momentum* and other perpendicular to radius vector known as *azimuthal momentum*. Now to quantize the elliptical orbits, Sommerfeld introduced two quantum numbers nr and n_ϕ and because they stand for one periodic system, hence $n_r + n_\phi = n$ where nr is known as *radial quantum number and* n_ϕ *as angular or azimuthal quantum number n is known as principle quantum number.*

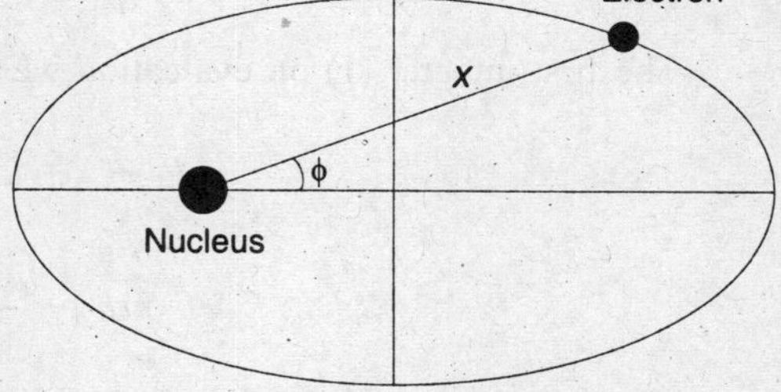

Fig. 15.4

On the basis of classical dynamics, if the length of the semi-major axis is determined solely by the principal quantum number n, while the length of the semiminor axis depends upon azimuthal quantum number n_ϕ. For the normal state of the hydrogen atom, the principal quantum number $n = 1$. So corresponding $n = 1$, there are two possibilities, *i.e.*, $n_r = 0$, $n_\phi = 1$ ($\because\ n = n_r + n_\phi$) and $n_r = 1$, $n_\phi = 0$. When $n_\phi = 0$ (semiminor axis is zero), the ellipse converted into a straight line and the electron have to pass through the nucleus twice during every period. This type of motion is not possible hence n_ϕ may not be zero *i.e.*, the lowest possible value of n_ϕ will be one. So corresponding to $n = 1$, n_ϕ is one and the first orbit is a circle identical to Bohr's orbit. With $n = 2$, n_ϕ may have values 1 and 2. Thus corresponding to $n = 2$ two orbits are possible, one is $n = 2$, $n_\phi = 2$, a circle and other $n = 2$, $n_\phi = 1$ an ellipse. Similarly for $n = 3$, $n_\phi = 3$ and two ellipses with $n = 3$, $n_\phi = 1$ and $n_\phi = 2$. These are shown in Fig. 15.5.

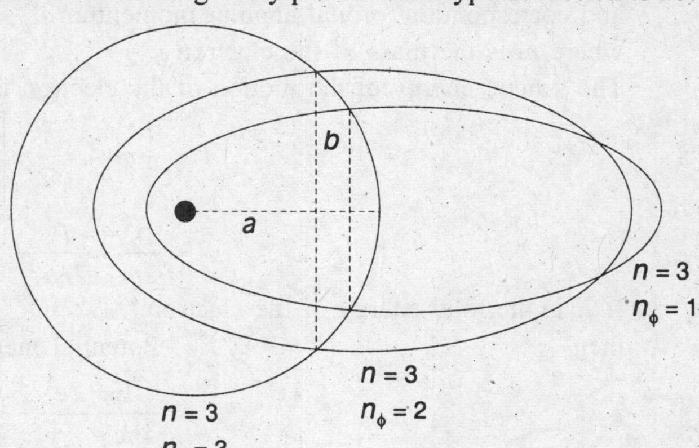

Fig. 15.5

Again Sommerfeld assumed that the same generalized condition which had been postulated for circular orbit also holds in elliptical orbits *i.e.*,

$$\oint p_r\, dr = n_r h \qquad ...(1)$$

$$\oint p_\phi\, d\phi = n_\phi h, \qquad ...(2)$$

where p_r, and p_ϕ are the component of linear momentum p of the electron along the radial direction and transverse direction and transverse direction respectively. n_r and n_ϕ are integers.

Where $\oint$ indicate that the integration is to be carried out over one cycle of the motion.

From classical mechanics we know that the angular momentum p_ϕ remains constant during the motion. So from equation (2) we have

$$\oint p_\phi d\phi = p_\phi \oint d\phi = p_\phi \,.\, 2\pi = n_\phi h$$

or $$p_\phi = n_\phi \frac{h}{2h} \quad ...(3)$$

The first integral (1) on evaluation we get

$$\oint p_r\, dr = n_r h = \left(\frac{1}{\sqrt{1-\varepsilon_2}} - 1\right) 2\pi p_\phi$$

$$\pi r h = \left(\frac{1}{\sqrt{(1-\varepsilon_2)}}\right) 2\pi \times \left(\pi_\phi \frac{h}{2\pi}\right), \qquad \left(\because p_\phi = n_\phi \frac{h}{2\pi}\right)$$

where ε corresponds to the eccentricity of the ellipse.

Now $$\left(\frac{1}{\sqrt{(1-\varepsilon)^2 - 1}}\right) = \frac{n_r}{n_\phi}$$

or $$\frac{1}{\sqrt{(1\varepsilon^2 - 1)}} = \frac{n_r}{n_\phi} + 1 = \frac{n_r + n_\phi}{n_\phi} = \frac{n}{n_\phi}$$

where $$n_r + n_\phi = n$$

or $$(1 - \varepsilon^2) = \frac{n_\phi^2}{n^2}$$

Now tangential velocity of the electron at any instant can be resolved into two components: one along the radius vector called the radial velocity (dr/dt) and the other at right angles to the radius vector called the transverse velocity ($rd\phi/dt$)

So the radial momentum $$p_r = m\frac{dr}{dt} \quad ...(4)$$

and corresponding orbital angular momentum $p_\phi = mr^2 \dfrac{d\phi}{dt}$

where m is the mass of the electron.

The kinetic energy of the motion of the electron is given by

$$T = \frac{1}{2}mv^2 = \frac{1}{2}m\left[\left(\frac{dr}{dt}\right)^2 + \left(r\frac{d\phi}{dt}\right)^2\right]$$

$$= \frac{p_r^2}{2m} + \frac{P_\phi^2}{2mr^2}$$

If E is the total energy of the electron

then $$E = \text{Potential energy} + \text{kinetic energy}$$

$$= \frac{1}{4\pi\varepsilon_0}\frac{Ze^2}{r} + \frac{p_r^2}{2m} + \frac{p_\phi^2}{2mr^2}$$

We can write the energy of the electron in a hydrogen like atom, excluding the correction of nuclear motion, as

$$E_n = -\frac{4\pi^2 me^4 Z^2}{32\,\pi^2\varepsilon_0^2 h_2\, n^2}\left(\frac{1}{n_r + n_\phi}\right)^2$$

or $$E_n = -\frac{4\pi^2 me^4 Z^2}{32\,\pi^2\varepsilon_0^2 h^2\, n^2}$$

Which shows that whatever may be the value of n_ϕ the energies are the same for all orbits having the same principal quantum number n. So the introduction of elliptical orbits give no new energy levels and hence no new transition. In this way the attempt of Summerfeld to explain the fine structure of spectral lines failed. But soon, on the basis of variation of mass of the electron with velocity. Sommerfeld was able to find the solution of the problem of the fine structure of the spectral lines. Here it is worthwhile to mention that experimental results and quantum mechanical predictions confirm that n_ϕ also assumes a value zero. So corresponding to n, n_ϕ assume 0, 1, 2... $(n - 1)$ *i.e.*, a total of n possibilities. In wave mechanics azimuthal quantum member n_ϕ is usually denoted by l.

15.8. SOMMERFELD RELATIVISTICS CORRECTION

Sommerfeld explanation gave the velocity of the electron in the innermost Bohr's orbit to be 0.007 of the velocity of light so the variation of mass of the electron with velocity should be taken into consideration. The velocity of the electron moving in an elliptical orbit varies at different parts of the orbit. It is maximum when the electron is nearest to the nucleus and minimum when it is farthest from the nucleus. Which means that the effective mass of the electron will be different at different parts of the orbit. So, the curvature of elliptical path is not exactly the same in two positions and hence the ellipse is not a complete one. Sommerfeld had made such calculations and has shown that path of the electron is not a closed ellipse but a complicated curve known as *rosette a precessing ellipse* whose major axis processes slowly in the plane of the ellipse about an axis through one of the foci as shown in Fig. 15.6.

Fig. 15.6

We can write for the energy of an electron in hydrogen like atom for a particular state characterised by quantum number v and by taking relativistic correction as

$$E_n = -\frac{m z^2 e^4}{8\varepsilon_0^2 n^2 h^2} \times \left[1 + \frac{\alpha^2 z^2}{n}\left(\frac{1}{n_\phi} - \frac{3}{4n}\right)\right]$$

where

$$\alpha = \frac{e^2}{2\varepsilon_0 ch} \cong \frac{1}{317}$$

and is called as Sommerfeld fine structure constant.

The energy expression shows that the energy depends no only on the principle quantum number but also on azimuthal quantum number. For $n = 1$, $n_\phi = 1$, *i.e.*, there is only one energy level. For $n = 2$, n_ϕ may have two possible values ($n_\phi = 1$, $n_\phi = 2$) there are two energy levels. Similarly for $n = 3$, n_ϕ may have three possible values and hence three energy levels. It should be noted that the energy difference of these would be very small and the frequency of radiation emitted n the transition of electron from $n = 3$ to $n = 1$ will be slightly different according to the possible n_ϕ values. This accounts why a single line of hydrogen spectrum should appear as a group of closely associated lines. So the introduction of relativity correction accounts for the fine structure of the spectral lines.

Let us consider first line of Balmer series (*i.e.*, $p = 2$ and $n = 3$) there should be six possible transitions for different values of n. Thus H_α line should appear to consist of six lines very close to each other. Experimental observations reveal that H_α line has less components. To avoid this discrepancy, Sommerfeld imposed a restriction on n_ϕ values number n_ϕ can charge only by +1 or −1 *i.e.*, $\Delta n_\phi = \pm 1$ *i.e., only those transitions from one energy level to another level are possible for which* n_ϕ *changes by +1 or −1*. The application of this selection rule to Balmer series shows that each line should consist with the experimental observations. Later this discrepancy was removed by the introduction of the spin concept in electrons.

15.9. THE CORRESPONDENCE PRINCIPLE

Although all theories in physics have some drawbacks, they usually do not break down abruptly but in a continuous way, giving us the results that agree less and less well with experiment. So the predictions of Newtonian mechanics become less and less accurate as the speed is made to approach that of light. A similar relationship must exist between quantum physics and classical physics : it remains to find the circumferences under which the latter theory is revealed as a special case of the former.

The radius of the lowest energy state in hydrogen (the so-called *ground state*) is found by putting $n = 1$ in and is given by matter. If $n = 10,000$. However the radius is $(10000)^2$ times as large as large or 5.3 mm. This atom is no large that we suspect that its behaviour frequency of the emitted light on the basis of both classical and quantum assumptions. These calculations should differ at small quantum numbers nut should agree at large quantum numbers. The fact that quantum physics reduces to classical physics at large quantum numbers is called the *correspondence principle*. This principle, given by Niels Bohr, was very useful during the years in which quantum was being developed Bohr, in fact, based his theory of the hydrogen atom in correspondence principle arguments.

Classically, the frequency of the light emitted from an atom is equal to ν_0 its frequency of revolution in its orbit. This can be expressed in terms of a quantum number n by the relation

$$\nu_0 = \frac{me^2}{8\varepsilon_0^2 h^3} \frac{2}{n^3} \qquad ...(1)$$

Quantum physics predicts that the frequency ν of the emitted light is given by Equation.

$$\nu = \frac{me^4}{8\varepsilon_0^2 h^3}\left(\frac{1}{p^2} - \frac{1}{n^2}\right)$$

Considering a transition between an orbit with quantum number $p = n$ and one with quantum number $= (n - 1)$ gives us

$$n = \frac{me^4}{8\varepsilon_0^2 h^3}\left(\frac{1}{(n-1)^2} - \frac{1}{n^2}\right)$$

$$= \frac{me^4}{8\varepsilon_0^2 h^3}\left(\frac{2n-1}{(n-1)^2 n^2}\right) \qquad ... (2)$$

As $n \to \infty$ the expression in the square brackets above approaches $2n^3$ so that $\nu \to \nu_0$ as $n \to \infty$. *i.e.*, for larger quantum numbers classical mechanics and quantum mechanics give the same results.

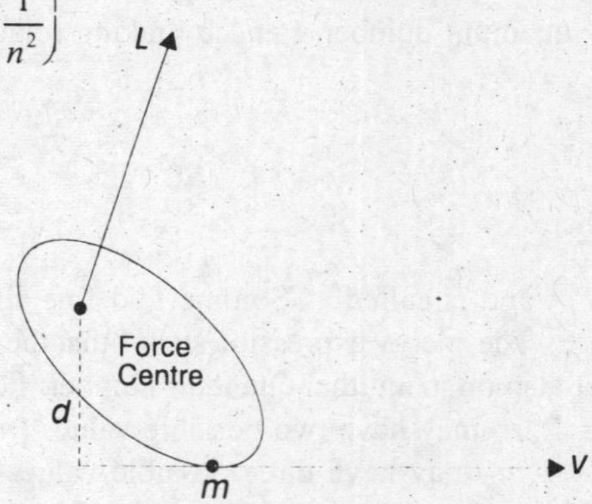

Fig. 15.7

15.10. ORBITAL ANGULAR MOMENTUM (CLASSICAL THEORY)

Let a particle of mass be m moving in an elliptical orbit under the influence of a central force, as shown in Fig. 15.7. The vector angular momentum about the force centre L will have magnitude mvd, where d is the perpendicular distance between the velocity direction and the force centre, and v is the particle velocity. The direction of L is given by the Fleming's right hand rule, as shown in the figure. From Newton's second law, the net torque τ on the particle will be equal to the rate at which its angular momentum changes $\tau = dL/dt$. However, since the force on the particle is a central force, the torque exerted will be zero. So the particle's angular momentum L will have a constant magnitude and direction at every point along its elliptical trajectory.

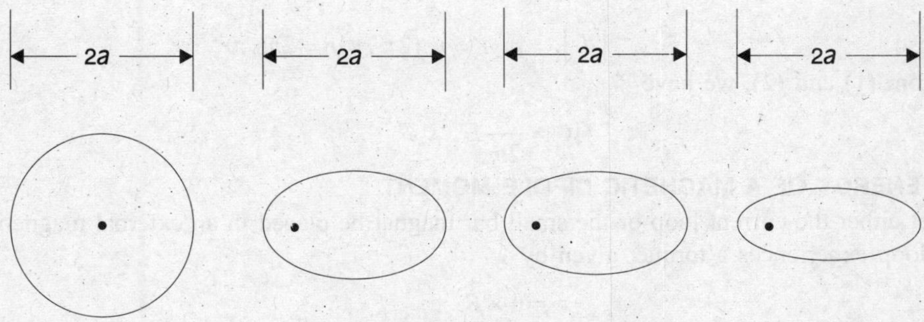

Fig. 15.8

Figure 15.8 shows various possible elliptical motions, ranging from a circle to nearly a straight line, all with the same major axis, $2a$. It can be shown that the total energy E (K.E. + P.E.) depends only on the value of the major axis and hence will have the *same value* (*e.g.*, $E = -ke^2/2a$ for the coulomb force) for all these ellipse. The orbital angular momentum, however, will vary ellipse to ellipse, ranging continuously from a maximum value of $a\sqrt{-2mE}$ for the circle to nearly zero for the approximately straight line. (A straight line ellipse has zero angular momentum because $d = 0$ everywhere along the trajectory)

Because L is constant, the component $L_z = L \cos \theta$ in any direction in space will also be a constant during the elliptical motion. In the classical picture there is no restriction on θ; it can take on any value from 0° to 180°.

15.11. CLASSICAL MAGNETIC DIPOLE MOMENT

An electron moving in a circular path will produce a current given by

I = (charge on electron) × (number of times per second that an electron passes a given point)

$= e\nu$

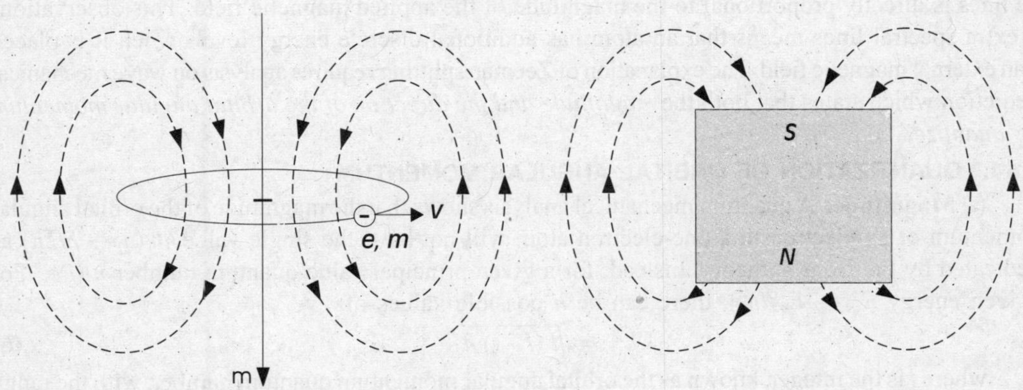

(*a*) current loop (*b*) bar magnet

Fig. 15.9

where n is the frequency of rotation of the electron. The circular current loop in turn will produce a magnetic field very similar to the field produced by a small bar magnet, as shown in Fig. 15.9. Just as with the bar magnet, there will be a magnetic dipole moment m associated with the orbiting electron, whose magnitude is given by

$$|\mu| = IA = e\nu.\pi r^2 \quad \text{...(1)}$$

and whose direction is opposite to that of L since, the electron carries a negative charge).

Also $$|\vec{L}| = m\nu r = m\,(2\pi\, r\nu)r = 2m\nu\pi r^2 \qquad \text{...(2)}$$

From (1) and (2), we have

$$\vec{\mu} = \frac{e}{2m}\vec{L}$$

15.12. ENERGY OF A MAGNETIC DIPOLE MOMENT

Let either the current loop or the small bar magnet be placed in an external magnetic field B. The loop experiences a torque, given by

$$\underset{\to}{\tau} = \mu \times \vec{B} \qquad \text{...(3)}$$

tending to align μ in the direction of $\vec{B}$. Thus the system has potential energy, E_B the change in which gives the work done by the torque when the orientation of μ changes. By integration of the torque we have

$$E_B = -\vec{\mu}\,.\,\vec{B} \qquad \text{...(4)}$$

Choosing the direction of B as the z-direction equation. (2) we have for an orbiting electron

$$E_B = \frac{e}{2m}\vec{L}\,.\,\vec{B} = \frac{e}{2m}L_z B \qquad \text{...(5)}$$

15.13. THE ZEEMAN EXPERIMENT

An experiment measuring the effects of the interaction between an atom's internal magnetic moment and an external magnetic field was performed before the quantum mechanics theory by the Zeeman in 1886. In Zeeman's experiment an atom is placed in an external magnetic field and its excitation spectrum is measured and compared with the spectrum when there is no magnetic field present. This could be got for example, by measuring the wavelength of the radiation emitted from a discharge tube when it is placed in a magnetic field.

When the experiment is performed it is found that in the presence of the external field each spectral line *splits* into a number of discrete lines. Moreover, the observed change in frequency of the lines is directly proportional to the magnitude of the applied magnetic field. This observations of extra spectral lines means that an atom has additional discrete energy levels when it is placed in an external magnetic field. The explanation of Zeeman splitting requires analysis on wave mechanical prediction which states that both the *magnitude and the direction of the orbital angular momentum are quantized.*

15.14. QUANTIZATION OF ORBITAL ANGULAR MOMENTUM

(*i*) **Magnitude:** A quantum mechanical analysis shows that the magnitude of the orbital angular momentum of an electron in a one-electron atom will not have the single value $n\hbar$ ($\hbar = h/2\pi$) as predicated by the Bohr's, theory. Instead, for a given principal value quantum number n (*i.e.*, For a given energy $E_n = -E_1°/n^2$), there can be n possible values of

$$|L| = \sqrt{l\,(l+l)}\,h \qquad \text{...(6)}$$

where l is the integer, known as the orbital angular momentum quantum number, with the range

$$l = 0,\ 1,\ 2...\ n-1 \qquad \text{...(7)}$$

In a particular, for the lowest value of energy, corresponding to $n = 1$, the value for l is zero, and so the orbital angular momentum will also be zero.

(*ii*) **Direction:** Let one electron atom be placed in an external magnetic field, whose direction we take as the z-direction. A wave mechanical analysis shows than the *direction* of the orbital angular momentum vector $\vec{L}$ cannot be arbitrary. Instead L will be oriented such that the component $\vec{L}$ along the z-direction will be quantized, with discrete values given by

$$L_z = l\hbar \qquad \text{...(8)}$$

where m_l is an integer, called the *magnetic quantum number*, with the range

$$me = l, l-1, l-2, ..., 0, ..., -(l-1), -l \quad ...(9)$$

magnitude of L $\left(= \sqrt{l(l+l)\,\hbar}\right)$.

15.15. EXPLANATION OF SPECTRAL LINES DUE TO ZEEMAN EFFECT

Quantum mechanics states that the energy of a single-electron atom placed in an external magnetic field is changed by the potential energy (equation. 5). In this case, L^* is quantized according to Equation 8, so that the total energy is given by

$$E = E_0 + E_B = E_0 + m_l \frac{e\hbar}{2m} \quad ...(10)$$

where E_0 is the quantized energy before the field B is turned on. So, in the presence of a magnetic field, each energy level E_0 will be split into $2\,l + 1$ equal spaced sub levels, with the spacing proportional to B. The factor $e\hbar/2$ m is known as the Bohr magneton, its value is given by

$$\frac{e\hbar}{2m} = 5.79 \times 10^{-5} \text{ eV/T} = 9.27 \times 10^{-24} \text{ J/T}$$

Since there are more discrete energy levels available after a magnetic field is turned on, there will be additional discrete lines seen in the excitation spectrum of an atom when it is placed in an external magnetic field, as illustrated in Fig. 15.10.

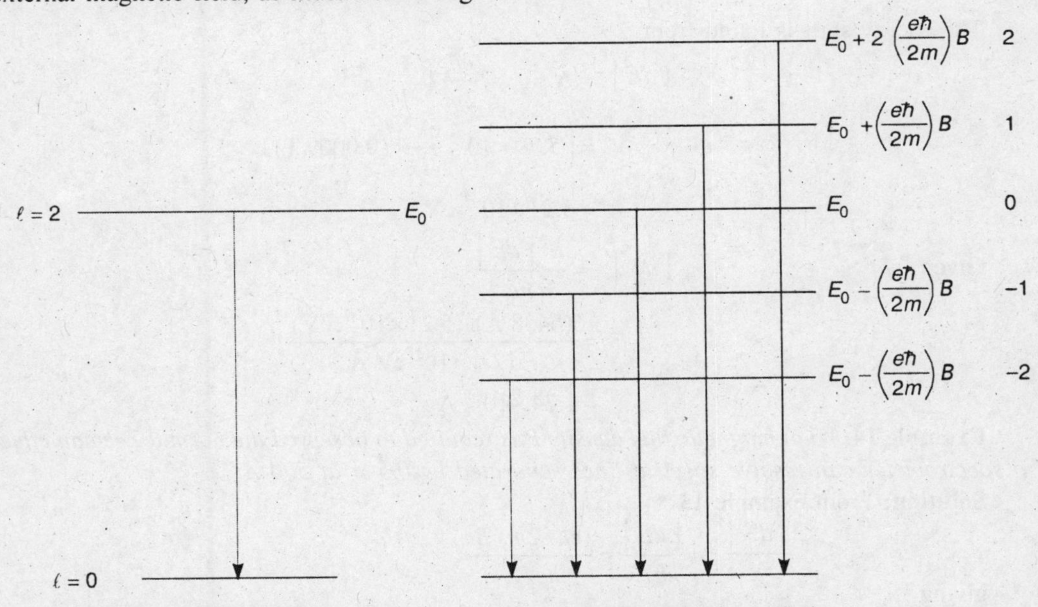

(*a*) Without external field only one transition

(*a*) Five transitions with the applied external magnetic field.

Fig. 15.10

For more intense transitions the following selections rules are observed.

$$\Delta l = \pm 1 \; \Delta ml = \pm 1 \text{ or } 0$$

For these electric dipole transitions, equation 10 gives ΔE_0 (the zero field spectral line) and so we have

$$\Delta E = \Delta E \pm \frac{e\hbar}{2m} B \quad ...(11)$$

i.e., two new lines shifted in energy from the zero field line by an amount

$$\Delta E_{zee} = \frac{e\hbar}{2m} \quad ...(12)$$

Other transitions may also occur, but they result in much weaker spectral lines. In any case, the energy or frequency difference among the new lines will be proportional to the magnitude of the applied field.

The above predictions correspond exactly to what is observed in the "normal" Zeeman effect. The discrete splitting are clear experimental evidence of the phenomenon of orbital angular momentum quantization. If the orientation of the angular momentum were not quantized, then L_z could take on all possible values, as in Bohr's theory, and the lines would be broadened into a continuous band instead of the discrete values that are experimentally observed. The above analysis, however, docs not explain *all* the lines observed in Zeeman experiments few more transitions are also observed and they fall under the *Anomalous Zeeman effect in which case the electron spin is also considered.* However, for very strong external magnetic fields, Anomalous Zeeman effect reduces to *Normal Zeeman effect.*

Example 13. *Determine the normal Zeeman splitting of the cadmium red line of 6438 Å when the atoms are placed in a magnetic field of 0.009 T.*

Solution: The change in wavelength is found by talking the differential of $E = hc/\lambda$

$$dE = -hc\frac{d\lambda}{\lambda^2} \qquad \text{or} \qquad |d\lambda| = \frac{\lambda^2 |dE|}{hc}$$

The energy shift is found from

$$|dE| = \Delta E_{zee} = \frac{e\hbar}{2m}B$$

$$= \left(5.79 \times 10^{-5}\frac{\text{eV}}{\text{T}}\right)(0.0009\text{ T})$$

$$= 5.21 \times 10^{-7}\text{ eV}$$

Given

$$|d\lambda| = \frac{\lambda^2|dE|}{hc}$$

$$= \frac{(6438\text{ Å})^2(5.21\times10^{-7}\text{ eV})}{12.4\times10^3\text{ eV Å}}$$

$$= 1.74\times10^{-3}\text{ Å}$$

Example 14. *What magnetic flux density B is required to observed the normal Zeeman effect if a spectrometer can resolve spectral lines separated by 0.5 Å at 500 Å?*

Solution: From example 13

$$\frac{|d\lambda|}{\lambda} = \frac{|dE|}{hc/\lambda} = \frac{(e\hbar/2m)\,B}{hc/\lambda}$$

giving

$$B = \frac{|d\lambda|}{\lambda}\left(\frac{hc}{\lambda}\right)\left(\frac{2m}{e\hbar}\right)$$

$$= \left(\frac{0.5\text{ Å}}{5000\text{ Å}}\right)\left(\frac{12.4\times10^3\text{ eV. Å}}{5000\text{ Å}}\right)\left(\frac{1}{5.79\times10^{-5}\text{ eV/T}}\right) = 4.28\text{ T}$$

Example 15. *Calculate the frequency at which an electron's orbital magnetic moment μ precesses in a magnetic field B.*

Solution: A magnetic moment in a magnetic field will experience a torque t, given by 15.11 as

$$t = \mu\times\vec{B} = -\frac{e}{2m}[\vec{L}\times\vec{B}]$$

This torque will cause a change in the angular momentum given by

$$t = \frac{d\vec{L}}{dt} = -\frac{e}{2m}[\vec{L}\times\vec{B}]$$

The change in L, dL, is a perpendicular to both L and B, as shown in Fig. 15.11, resulting in a precession of L about the direction of B. From the figure it is seen that

$$d\phi = \frac{|dL|}{L\sin\theta}$$

which gives

$$\omega_p = \frac{d\phi}{dt} = \frac{\left|\frac{dL}{dt}\right|}{L\sin\theta}$$

$$= \frac{\frac{e}{2m}LB\sin\theta}{L\sin\theta} = \frac{e}{2m}B$$

This is known as the *Larmor precession*, ω_p is equal to the frequency difference observed in the normal Zeeman effect.

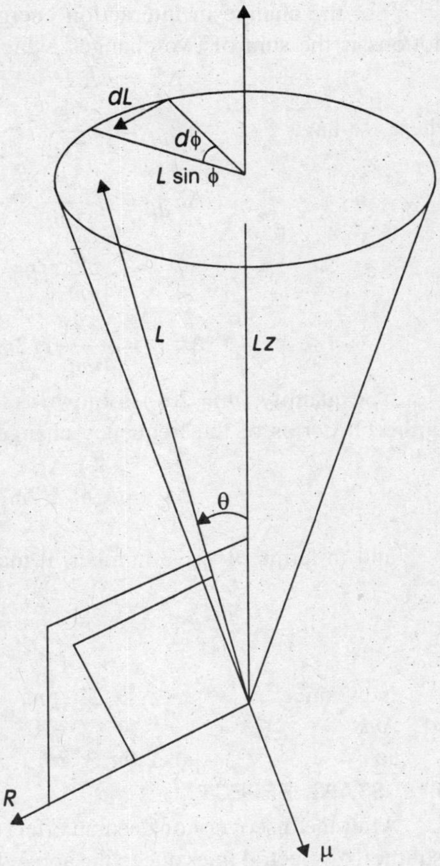

Fig. 15.11

15.16. VECTOR MODEL AND ANOMALOUS ZEEMAN EFFECT

Anomalous Zeeman effect finds its explanation when we introduce spin of the electron in vector atom model with the introduction of electron spin, we say that l^* and s^* vectors (*i.e.*, quantum mechanical state of orbital angular momentum vector and spin angular momentum, where $l^* = \sqrt{l(l+1)}$ and $s^* = \sqrt{s(s+1)}$, process around the result vector

g^* (total angular momentum vector). So we have

$g^* = l^* + s^*$

and magnetic moment due to orbital motion

$$\mu_l = l^* \frac{e\hbar}{4\pi mv}$$

where μ_l is directed opposite to l^* because of the negative charge of the electron. In the same way magnetic moment due to spin is given by

$$\mu_s = 2s^* \frac{e\hbar}{4\pi mv}$$

which is directed opposite to s^* due to the reason as stated above.

15.17. PASCHEN BACK EFFECT

In the explanation of Zeeman effect, it is assumed case external magnetic field is weak compared to the internal magnetic field due *to spin and orbital motions of the valence electron.* In this case the precession of l^* and s^* vector around j^* is much faster than that of j^* around H, thus wave creating no perturbation due to the motion of l^* and s^* into other motions. But when the external magnetic field stream length is increased, the coupling between l^* and s^* breaks down and J^* loses its *significance.* l^* and s^* are quantised separately and precess *more or less independently around* H as shown in Fig. 15.12. This is known as Paschen Back effect. Also, when the motions of l^* and s^* becomes separately quantised, the perpendicular component of magnetic moment does not average out to zero and contributes to total magnetic moment, *i.e.*, now total magnetic moment is not equal to m_j. Because of this small splitting what so even be the *anomalous pattern in the weak magnetic field it is converted into normal pattern in the strong magnetic field.*

The angular velocities of two precessions are given as

$$\omega_l = H\frac{e}{2m_0} \text{ and } \omega_s = H\frac{e}{2m_0}2$$

So, the change in interaction energy due to these two motions is the sum of two changes, which may be written

$$\Delta E = \Delta E_l H + \Delta E_{sH}$$

where we have $$\Delta E_{lH} = H\frac{e}{2m_0} l^* \frac{h}{2\pi}\cos(l^* H)$$

$$\Delta E_{sH} = H\frac{e}{2m_0} s^* \frac{h}{2\pi}\cos(s^* H)$$

So $$\Delta E_{lH} = \frac{eh}{4\pi m_0} Hm_l$$

$$\Delta E_{sH} = \frac{eh}{4\pi m_0} H 2m_s$$

The quantity, $(m_l + 2m_s)$ is known as strong field quantum number. In terms of the frequency change. We may write

$$\Delta\nu = \Delta(m_l \pm 2m_s)\frac{eH}{4\pi m_0},$$

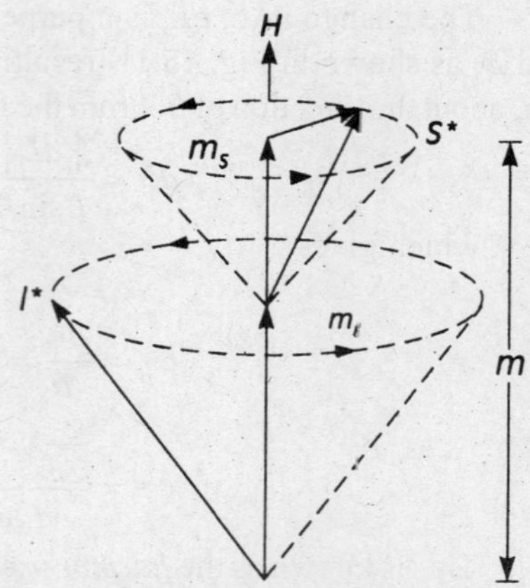

Fig. 15.12 The precession of l^* and s^* vector in Paschen-back effect.

and in terms of wave number, it may be written

$$\Delta\nu = \Delta(m_l + 2m_s)\frac{eH}{2\pi m_0 c}$$

$$= \Delta(m_l + 2m_s) \text{ in Lorentz unit.}$$

Now since $\Delta m_l = 0$ or ± 1,

and Δms $= 0$,

so $\Delta(m_l + 2m_s) = 0$ or ± 1.

15.18. STARK EFFECT

After the discovery of Zeeman effect in 1896, a similar resolution of spectral lines due to the applied electric field was also experimented. The action of electric field on hydrogen spectrum was discovered by **Stark** (1913) who detected the splitting of Balmer lines. Later on this phenomenon was observed in the spectra of all the elements and is called Stark effect. Actually, Stark effect should have been detected earlier than the Zeeman effect, but it could not be detected because of practical difficulties in applying a proper technique. As, the hydrogen gas in **Geissler tube** was subjected to electric field and hydrogen is comparatively a good conductor of electricity and so capable of maintaining a strong electric field.

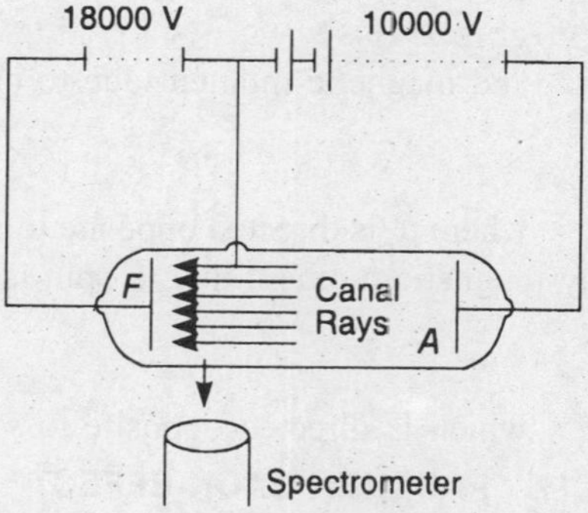

Fig. 15.13

Stark overcame this difficulty by placing an auxiliary electrode F close behind cathode C at a distance of a few mm which is perforated.

A very strong electric field of several thousand volts per cm. is maintained between F and C. Positive ions are accelerated towards C and some will pick up electrons on the way forming neutral H atoms. The fast atoms (ions also) which pass through the perforations and emit light in FC are called canal rays. These canal rays were made to pass into the space between the plates of highly charged condenser. The main features of the phenomenon are the following:

(*a*) All hydrogen lines form a symmetrical pattern, however, the pattern depends on quantum numbers n of the term involved, the no. of lines and total width of the pattern increase with n. So the no. of components of H_B lines is more than those of $H\alpha$ line. In the same way the number of components of Hr will be more than those of $H\beta$ lines.

(*b*) Up to field of 10^5 volts/cm, the resolution increases is proportional to field strength. In this region we get **linear stark effect**. In the case of higher applied fields, in addition to linear stark effect higher order stark effects are observed.

(*c*) Wave number difference are integral multiple of a unit which is proportional to the applied field strength E and is same for all hydrogen lines.

(*d*) If the observations are made $\perp$ to the direction of electric fields, it is observed that the components are polarized in fact *parallel to the field* (π-component) and in part, *perpendicular to the field* (σ-component).

15.19. RAMAN EFFECT

It corresponds to scattering with change in wavelength somewhat similar to fluorescence. However, it is different from fluorescence in two ways :

(*i*) The light which is incident on the scattering material must have a wavelength that does not correspond to one of the absorption lines of bands of the material (as otherwise fluorescence phenomenon occurs).

(*ii*) Intensity of scattered light is much less than the intensity obtained in fluorescence phenomenon.

It is because of these reasons that *Raman effect* is difficult to observe and the observations have to be done by photography.

It may be mentioned here that the scattering that occurs without change in wavelength or frequency is *Rayleigh scattering.*

Raman, in 1928 observed that light scattered from molecules at right angles to the incident direction has component of radiation in addition to the incident frequency radiation ν some frequencies $\nu \pm \nu'$. So *the scattering of light from molecules with a change in frequency or wavelength is called Raman effect.*

It may be mentioned that in the Raman spectrum, we observe in the scattered light a line having frequency equal to the frequency of the incident light and in addition *few weak lines* on either side of incident line are observed. If the frequency ν of incident light is changed, the weak lines are once again observed on either side of the Rayleigh line with the same difference in frequency.

It is evident that the frequency difference ν' between the incident and the scattered light in Raman effect is determined by the *nature of the scattering molecules* and is independent of the frequency of the incident light.

15.20. THEORY OF RAMAN EFFECT

Quantum theory of light states that light consists of packets of energy in the form of photons each of which carries energy in quantum $h\nu$. When this light falls on a molecule, the photon will undergo a collision with molecule. This collision may be:

(*a*) **Elastic collision:** When there is no transfer from photon to the molecule. So in this case scatering of photon will be without any change in energy. It means that the frequency of the scattered photon will be identical with the frequency of the incident photon. This will be the case of the presence of unmodified line in scattered light spectra.

(*b*) **Inelastic collision:** In this case there is transfer of energy which may take place in either of the following two ways :

(*i*) The molecule loses energy ΔE during the collosion with photon. It will correspond to that case when the molecule is in an excited state. Let the frequency of the scattered photon be ν_2. Then from the law of conservation of energy, we have

$$h\nu_1 = h\nu + \Delta E$$

or

$$\nu_1 = \nu + \nu' \qquad ...(1)$$

$$\nu_1 = \nu + \nu'$$

where

$$\nu' = \frac{\Delta E}{h}$$

Which shows that the frequency of scattered photon is more than the frequency of the incident photon.

(*ii*) The molecule in the ground state gets energy from the photon and goes to an excited state. Let ΔE be the energy absorbed by the molecule during collision and let ν_2 be frequency of the scattered photon. Then from the law of conservation of energy, we have

$$h\nu = h\nu_2 + \Delta E$$

or $$\nu = \nu_2 + \frac{\Delta E}{h} = \nu_2 + \nu'$$

or $$\nu_2 = \nu - \nu' \qquad ...(2)$$

which shows that the frequency of scattered photon is less than the frequency of the incident photon.

The line with a frequency lower than that of the incident frequency is called Stoke's line while that with a frequency higher than the incident frequency is called Anti Stoke's line.

Selection Rules. Let us consider a scattering molecule in the rotational state α and having the rotational quantum number J. Ordinary in the rotational spectrum we have the selection rule

$$\Delta J = \pm 1$$

If we study Raman spectrum quantum mechanically, we observe that a *Raman jump* is possible between two states α and β only if there exists a third state γ, such that in ordinary case jumps are allowed between α and γ and β and γ states. Which in other words may be expressed as a molecular transition can take place from level α and β by going thoroughly γ.

Let a molecular transition take place from α state (having rotational quantum number J') via a γ state (having rotational quantum number J').

Now from the selection rule of rotational spectrum, a translation from a to g state can take place with quantum number J changing by ± 1. So the rotational quantum number J' of γ-state will be

$$J' = J \pm 1 \qquad ...(1)$$

Again a transition from γ to β state is allowed if a change in rotational quantum number is ± 1. So we have the rotational quantum number of β state given by

$$\Delta J'' = J'' \pm J \qquad ...(2)$$

So the total change in quantum number J is going from α to β state is

$$\Delta J = J'' - J \qquad ...(3)$$

Putting the value of J'' from equation (2) in equation (3), we get

$$\Delta J = J' \pm 1 - J \qquad ...(4)$$

Putting the value of J' from equation (1) in equation (4), we get

$$\Delta J = J \pm 1 \pm 1 - J$$

or $$\Delta J = \pm 2, 0$$

Now $\Delta J = 0$ corresponds to no change in frequency and so will represent Rayleigh scattering.

$\Delta J = \pm 2$, represents Raman scattering. So in the Raman spectrum, we observe transitions which observe the selection rule $\Delta J = \pm 2$. It may be mentioned that the spectral lines in rotational spectra which are not allowed by selection rules of rotational spectra are allowed in Raman spectra (selection rule $\Delta J = \pm 2$).

Experimental Study of Raman Scattering. Raman effect has been studied by many scientists. The basic principle involved in their researches is to illuminate the experimental substance with an intense monochromatic source of light and then to photograph the scattered light by means of spectrograph arranged in transverse direction. They used strong light sources as otherwise the Raman lines will be of very low intensity. Recently Lasers are put to use for this purpose. The experimental arrangement is shown in Fig. 15.14.

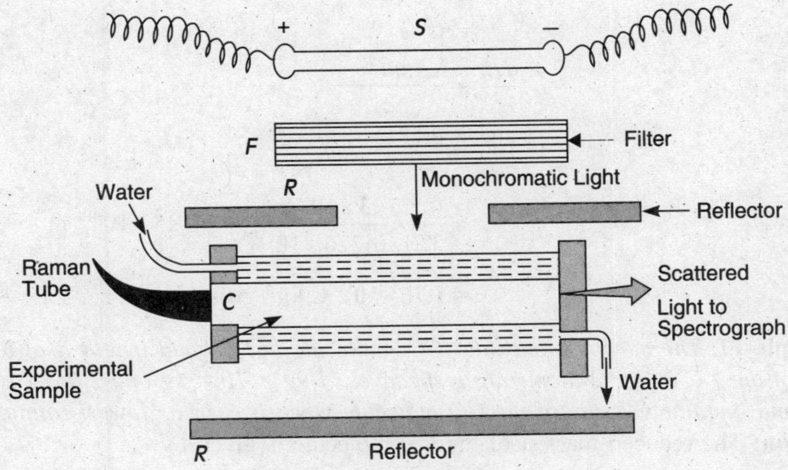

Fig. 15.14

In the figure, S is mercury arc lamp *i.e.* an intense source of light. The light is passed through a filter F to get a monochromatic beam. The light is then passed through an opening a metallic reflector and then on Raman tube. The Raman tube consists of a glass tube of about 1 or 2 cm in diameter and 10 to 15 cm in length. One end of the tube has a flat glass surface through which the scattered light comes out. The other end is drawn out into horn-shaped and blackened outside to provide a black background. The tube is surrounded by a cold water jacket J to prevent overheating the sample in Raman tube due to presence of the exciting are. The experimental sample is placed inside Raman tube. The scattered beam emerges from the flat end of the Raman tube and is photographed.

It is enclosed in a tube at a pressure of a few cm of mercury. In the case of solids, it is made into powder form and put in some solvent like paraffin. In case of gas sample, it is put in a tube at few cm of Hg pressure.

Example 15. *A sample of certain elements is placed in magnetic field of flux density 0.3 T. How far apart are the Zeeman components of wavelength 4500 Å (Given* $\frac{e}{m} = 1.7588\times 1011$ *C/kg)*

Solution: The separation between adjacent components a normal Zeeman splitting is given by

$$\Delta\nu = \frac{eB}{4\pi m} = \frac{e}{m}.\frac{B}{4\pi}$$

$$= 1.7588\times 10^{11}\ \text{C/kg} \times \frac{0.3T}{4\times 3.142}$$

$$= \frac{5.2764}{12.56}\times 10^{11}\ \text{Hz}$$

Also we have

$$d\lambda = \frac{\lambda^2}{c}\Delta\nu$$

$$= \frac{(4500\times 10^{10})^2\times 5.2764\times 10^{11}}{3\times 10^8\times 12.56}$$

$$= \mathbf{0.0283\ Å\ Ans.}$$

Example 16. *In a normal Zeeman experiment, the calcium 4226 Å line splits into 3 lines separately by 0.25 Å in a magnetic field of 3T. Find e/m for the electron from these data.*

Solution: We have the relation

$$\frac{d\lambda}{\lambda^2} = \frac{|dE|}{ch} = \frac{\left(\frac{e\hbar}{2m}\right)B}{ch}$$

$$\therefore \quad \frac{d\lambda}{\lambda^2} = \frac{\left(\frac{e}{4\pi m}\right).B}{c}$$

$$\therefore \quad \frac{e}{m} = \frac{4\pi}{B}.\frac{c}{\lambda^2}.d\lambda$$

$$= \frac{4\pi}{3T} \times \frac{3\times 10^8 \text{ ms}^{-1}}{(4226\times 10^{-10})^2} \times 0.25 \times 10^{-10} \text{ m}$$

$$= \mathbf{1.76\times 10^{11}\ C\ kg^{-1}\ Ans} \qquad \left[\because 1\text{ T} = \frac{1\text{ kg}}{\text{A.s}^2} = \frac{1\text{ kg}}{\text{C.s}}\right]$$

Example 17. *The carbon monoxide (CO) molecule has a bond length R of 0.113 nm and the masses of the ^{12}C and ^{16}O atoms are respectively 1.99×10^{-28} kg and 2.66×10^{-26} kg. Find the energy and angular velocity of the CO molecule when it is in its lowest rotational state.*

Solution: The reduced mass μ of the CO molecule is given by

$$\mu = \frac{m_1 m_2}{m_1 + m_2} = \left[\frac{(1.99)(2.66)}{1.99+2.66}\right]\times 10^{-26}\text{kg}$$

$$= 1.14\times 10^{-26} \text{ kg.}$$

and its moment of inertia I is given by

$$I = \mu R^2 = (1.14\times 10^{-26} \text{ kg})(1.13\times 10^{-10} \text{ m})^2$$

$$= 1.14\times 10^{-46} \text{ kg.m}^2$$

The lowest rotational energy level corresponds to $J = 1$, and for this level in CO we have

$$E_{J=1} = \frac{J(J+1)h^2}{2I} = \frac{h^1}{I}\frac{(1.054\times 10^{-34}\text{ J.s})^2}{1.46\times 10^{-46}\text{ kg.m}^2}$$

The angular velocity of the CO molecule when $T = 1$ will be

$$\omega = \sqrt{\frac{2E}{I}} = \sqrt{\frac{(2)(7.61\times 10^{-23}\text{ J})}{1.46\times 10^{-46}\text{ kg.m}^2}}$$

$$= \mathbf{3.23\times 10^{11}\ rad/s}$$

Example 18. *The exciting line in an experiments is 5460 Å and the stokes's line is at 55220 Å. Calculate the wavelength of antistokes line.*

Solution: Raman shift is given by

$$\Delta\nu = \nu_i - \nu_s$$

where $\nu \rightarrow$ wave number of incident radiation and $\nu_s \rightarrow$ wave number of scattered radiation. For stokes lines $\nu_s < \nu_i$.

$\therefore$ In this case
$$\nu_i = \frac{1}{5460\text{Å}} = \frac{1}{5460\times 10^{-8}}$$

$$= 18315 \text{ cm}^{-1}$$

and
$$\nu_s = \frac{1}{5520\text{Å}} = \frac{1}{5520\times 10^{-8}}$$

$$= 18116 \text{ cm}^{-1}$$

$$\therefore \quad \text{Shift } \Delta\nu = \nu_i - \nu_s = 18315 - 18116 = 199 \text{ cm}^{-1}$$

As the stokes and antistokes lines have same wave number displacement w.r.t. exciting line ν_i and for antistokes line $\nu_s > \nu_i$ we have

$$(\nu_s)\text{antistokes} = \nu_i + \Delta\nu = 18315 + 199 = 18514 \text{ cm}^{-1}$$

Corresponding wavelength
$$\lambda = \frac{1}{18514\ cm^{-1}} = \mathbf{540\ Å}$$

Example 19. *The force constant k of the bond in CO is 187 N/m. Find the frequency o vibrations of the CO molecule and the spacing between its vibrational energy levels.*

Solution: We have already found μ for CO = 1.14 × 10^{-28} kg. The frequency of vibrations is given by

$$\nu_0 = \frac{1}{2\pi}\sqrt{\frac{k}{\mu}} = \frac{1}{2\pi}\sqrt{\frac{187 \text{ N/m}}{1.14\times10^{-26} \text{ kg}}}$$

$$= 2.04\times10^{13} \text{ Hz}.$$

the separation ΔE between the vibrational energy levels in *CO* will be

$$\Delta E = E_{v+1} - E = h\nu_0 = (6.63\times10^{-34} \text{ J.s})(2.04\times10^{13} \text{ s}^{-1})$$

$$= 1.35\times10^{-20} \text{ J} = \frac{1.35\times10^{-20}}{1.6\times10^{-19}} = \mathbf{8.44\times10^{-12}\ eV}$$

Example 20. *A substance shows a Raman line at 4567 Å when exciting line 4538 Å is used, Deduce the positions of stocks and antistokes lines for that substance when the exciting 4047 Å is used.*

Solution: Raman shift is given by

$$\Delta\nu = \nu_i - \nu_s = \frac{1}{4358 \text{ Å}} - \frac{1}{4567 \text{ Å}}$$

$$= \frac{1}{4358\times10^{-8}} - \frac{1}{4567\times10^{-8}} = 1050 \text{ cm}^{-1}$$

The wave number of other exciting line

$$\nu_i = \frac{1}{4047 \text{ Å}} = 24710 \text{ cm}^{-1}$$

∴ Wave number of

(*i*) Stocks line $= \nu_i - \Delta\nu$

$$= 24710 - 1050 = 23660 \text{ cm}^{-1}$$

(*ii*) Anti Stokes line $= \nu_i - \Delta\nu$

$$= 24710 + 1050 = 25760 \text{ cm}^{-1}$$

Corresponding wavelengths are

$$\lambda_{\text{Stokes}} = \frac{1}{23660 \text{ cm}^{-1}} = 4226.5 \text{ Å}$$

and $$\lambda_{\text{Antitokes}} = \frac{1}{25760 \text{ cm}^{-1}}$$

Corresponding wavelengths are

$$\lambda_{\text{Stokes}} = \frac{1}{23660 \text{ cm}^{-1}} = 4226.5 \text{ Å}$$

and $$\lambda_{\text{Antitokes}} = \frac{1}{25760 \text{ cm}^{-1}} = \mathbf{3882\ Å}$$

15.21. IMPORTANCE OF RAMAN EFFECT

Raman effect finds many applications in physics and chemistry and confirms the *quantum theory of radiation.* This effect is useful to study the molecule structure and chemical analysis of solids, liquids, gases and their moleculer symmetry. Raman spectroscopy is finding good number of applications in *Molecular Biology* and *Lattice dynamics* of solids. Raman effect has also been applied to study some nuclear characteristics like spin, statistics and isotopic constituents of molecules and has been successful in experimental verification of the characteristics possessed by homo nuclear molecules like H_2, D_2, N_2, O_2 etc.

15.22. RAMAN EFFECT AND COMPTON EFFECT

Both Raman effect and compton effect are interaction of radiation with matter and are based on *scattering phenomenan*. However, in Compton effect, a loosely bound electron of an atom acquires energy from incident photon and the *electron leaves the atom* while incident photon gets scattered with a reduced energy (or frequency). *Nucleus of atom practically does not participate in Compton effect.*

Raman effect in special case of interaction of radiation with matter where *entire molecule takes* part. Molecule simply takes energy from incident photon and jumps to a higher state. In this case no electron leaves the atom and the energy transfer in due to *rotation or vibration of the interacting molecules.*

15.23. RAMAN EFFECT AND FLOURESCENCE

Raman effect and flrourescence resemble each other because in each case light suffers change of wavelength after falling on a scatterer. However, Raman effect differs from flourescence in the following ways.

1. In flourescent spectrum, the frequencies are always less than the incident frequency while in Raman spectrum lines with higher frequencies (anti-Stokes lines) and lower frequencies (Stoke's lines) are observed.

2. In case of flourescence, frequencies of scattered radiation independent of frequencies of exciting radiation while in Raman Spectroscopy the scattered radiation is directly related to the incident radiation frequency.

3. Flourescent material fixes the frequency of scattered radiation while in case of Raman spectra frequency *shift depends* on the scatter and not the actual frequencies.

4. Flourescent lines are unpolarized while *Raman lines are strongly polarized.*

5. In general, Raman frequencies are either actual infrared frequencies of the scatterer or different in such frequencies which do not hold true in flourescence phenomenan.

15.24. RAMAN EFFECT AND RAYLEIGH SCATTERING

Rayleigh scattering *i.e.*, scattering $\propto \frac{1}{\lambda^4}$ is *coherent scattering* which means that scattered light has same frequency as that of incident light. This in other words, may be explained as : no exchange of energy takes place between incident and scattered photon *i.e.*, in this case the *collision in elastic.*

In the case of Raman effect, the frequency change occurs on scattering (Stokes and anti Stokes lines). Which means that the scattered photon may have more or less energy as compared to the incident photon. So exchange of energy takes place in this case and hence the *collision is inelastic* in Raman *scattering* effect.

QUESTIONS AND PROBLEMS

1. State the fundamental postulates of Bohr's theory of hydrogen spectra. Explain the existence of definite energy states on the basis of this theory.
2. State the postulates of Bohr's theory and deduce an expression for the energy of the *n*th orbit of hydrogen atom. What interpretation do you give to the negative sign of the energy value? What correction is to be introduced for the finite mass of the nucleus?
3. Give Bohr's theory of atomic spectrum of hydrogen, explanations fundamental postulates. Calculate the velocity of electron in the first Bohr's orbit.
4. (*a*) How did Rutherford arrive at his atomic model? What modifications were introduced in it by Bohr? Show that the Bohr condition for quantization of hydrogen orbits, combined with de Broglie concept of stationery waves, leads to $\lambda = h/p$ as the expressions for the wavelengths of matter waves, symbols having their usual meanings.

   (*b*) Calculate the first excitation potential of hydrogen atom.

5. Describe Bohr's postulates and derive the formula for Balmer series of hydrogen lines. Calculate the radius of first Bohr's orbit.
6. State the Bohr's postulates, derive an expression for the frequencies of radiation emitted in hydrogen atom. Hence explain the various spectral series of this atom.
7. Derive an expression for Rydberg constant.
8. What are Balmer and Lyman series of hydrogen spectrum? Show that the wave number for the second number of the Lyman series is the sum of the wave number of the first member of Balmer series.
9. (*a*) How does Bohr quantum condition agree with the de-Broglie hypothesis. (*b*) Deduce the Bohr's quantum condition fram the de-Broglie concept.
10. Define ionization potential and evaluate it for hydrogen like atoms using Bohr's theory.
11. Give Bohr's theory of the hydrogen spectrum and discuss Sommerfeld's modifications.
12. Draw the energy level diagram for the hydrogen atom and indicate the radiative transitions classified as (*i*) Lyman series (*ii*) Balmer series.
13. What is normal Zeeman effect? How is it different than the anomalous Zeeman effect ?
14. Explain briefly Paschen back effect and Stark effect.
15. What is Raman spectrology? State its importance.
16. The excited states of mercury are 4.86 eV, 6.67, 6.67 eV and 8.83 eV. What wavelength shall be emitted and absorbed by mercury atom corresponding to these energies. [**Ans.** 2552 A, 1856 A, 1400 A]
17. Calculate the Rydberg constant and wavelength of H^* line of hydrogen spectrum.
18. Calculate the wavelength of the eighth line of the Balmer series of the hydrogen atom. Given Rydberg constant for hydrogen atom to be 109740 cm^2.
19. Calculate the wavelength if light emitted by an atom excited tot he next energy level by 2 eV. (Given

$$e = 1.6 \times 10^{-19} \text{ Coulomb}$$
$$c = 3 \times 10^{8} \text{ m/s}$$
$$h = 6.62 \times 10^{-34} \text{ Joule sec.)}$$

[**Ans.** 6394 Å]
20. A given atom shows excitation potentials of 2.20 and 4.95 volts from these data, find the possible lines in absorption and emission spectra. [**Ans.** 5625 Å and 2500 Å]
21. The hydrogen atom shows excision on 10.2 volt and 12.1 volt. Draw an energy level diagram and calculate the possible emitted wavelength. [**Ans.** λ 6.5×10^{-7} m, 1.0×10^{7} m, 1.2×10^{-7} m]
22. In a normal Zeeman experiment the Calcium 4226 a line splits into 3 lines sparated by 0.25 A in a magnetic field of 3*T*. Find e/m for electron from these data (Given 1 $T = 1$ kg/Cs.) [**Ans.** 1.76×10^{11} C/kg]
23. Using the results of quantum mechanics, calculate the magnetic moments that are possible for 3 levels. [**Ans.** 1.31×10^{-28} J/T]
24. A 5000 Å line exhibits a normal Zeeman splitting of 1.1×10^{-3} Å. Find the magnetic field.
25. Determine normal Zeeman splitting in mercury 4916 A line when in a magnetic field of 0.3 T.
26. Find the excitation energy of prominent spectral lines observed in mercury spectrum γ-wave length 4352 Å, 5461 Å, 5770 Å, 5791 Å. [**Ans.** 2.266, 2.142, 2.137 eV]
27. With excluding line 4358 Å, a sample gives Stokes' line at 4,458 Å. Deduce the wavelength of anti stokes line. [**Ans.** 4262 Å]
28. When acetylene is irradiated with 4358 Å mercury line. A Raman line is observed at 4768 Å. Calculate the fundamental frequency of this vibration. [**Ans.** 5.91×10^{13} Hz]

CHAPTER

16

NUCLEAR FISSION AND FUSION

16.1. NUCLEAR FISSION

Nuclear fission means the splitting of a heavy nucleus into two lighter nuclei with the release of large amount of energy. Fission was first observed in case of Uranium–235. when U^{235} is bombarded with fast moving neutron U^{235} absorbs the neutron and a new nucleus U^{235} was formed, which is very unstable; it breaks in two nuclei of Ba^{147} and Kr^{86} with the release of high thermal energy. Nuclear fission idea is used in making atom bombs and the energy liberated may also be used for peaceful purpose.

16.2. DISCOVERY OF NUCLEAR FISSION

Many nuclear reactions have been investigated using the five types of projectiles, α-particles, protons, deuterons, neutron and photons, but in every case the product of disintegration is a nucleus of very nearly the same mass and charge. For example

$$_7N^{14} + {}_2He^4 \longrightarrow \left[{}_9F^{18}\right] \longrightarrow {}_8O^{17} + {}_1H^1$$

In January 1932, a German Physicist, **Otto Hahn,** discovered that by bombarding uranium 238 ($_{92}U^{238}$), with fast moving, neutron ($_0n^1$), the product of disintegration is an isotope of Barium (At. Wt. 147) of approximately half the atomic weight of uranium and is accompanied with the liberation of a very large quantity of energy, about 200 MeV. The nuclear reaction is represent as follows:

$$_{92}U^{238} + {}_0n^1 \longrightarrow {}_{56}Ba^{149} + {}_{26}Kr^{88} + 3\,{}_0n^1$$

The reaction shown above is a typical reaction, where the isotopes Barium and Krypton are the new products.

"Therefore, **the process of breaking up of the nucleus of a heavy atom into two lighter nuclei with the release of large amount of energy is known as fission**". (See B.E. curve)

16.3. ISOTOPES OF URANIUM

Uranium has three isotopes of atomic mass 238, 235 and 234 respectively. The heaviest isotope $_{92}U^{238}$ is the most abundant in nature, being 99.3% of the uranium, $_{92}U^{235}$ is only 0.7% and $_{92}U^{234}$ occurs in traces *i.e.* less than 0.06% only $_{92}U^{238}$ is broken up by very fast moving neutrons and the chances are comparatively less for the neutron to get a direct hit on the nucleus. Besides, the uranium nucleus sometimes captures slowly moving neutron without undergoing any fission. On the other hand $_{92}U^{235}$ is very useful for rapid production of atomic energy and was used in the construction of the **atom bomb.**

16.4. FISSION OF $_{92}U^{235}$

$_{92}U^{235}$ easily breaks up by the action of a slowly moving neutron and disintegrates into two parts of more or less half the weight as shown in Fig. 16.1. The following are the nuclear-reaction in Uranium fission 16.1.

$$_{92}U^{235} + {}_0n^1 \longrightarrow {}_{56}Ba^{147} + {}_{36}Kr^{86} + 3\,{}_0n^1$$

Uranium + neutron = Barium + Krypton + neutrons

$$_{92}U^{235} + {}_0n^1 \longrightarrow {}_{51}Sb^{133} + {}_{41}Cb^{99} + 4\,{}_0n^1$$

Uranium + neutron ⟶ Antimony + Columbium + neutrons

The products are not identical in every case and various elements have been detected as the products of reaction, all of which lie near about the middle of the periodic table. The reaction is associated with a very large release of atomic energy of the order of 200 million-electron-volts.

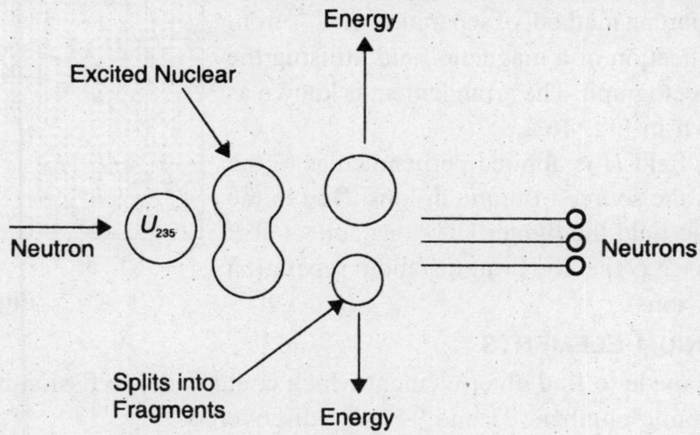

Fig. 16.1

$${}_{51}Sb^{133} \xrightarrow{5\text{ min}} {}_{52}Te^{133} + {}_{-1}e^{0} \quad (\beta\text{-ray})$$
(unstable) (unstable)

$${}_{52}Te^{133} \xrightarrow{60\text{hrs.}} {}_{53}I^{133} + {}_{-1}e^{0} \quad (\beta\text{-ray})$$
(unstable) (unstable)

$${}_{53}X^{133} \xrightarrow{60\text{ hrs.}} {}_{54}X^{133} + {}_{-1}e^{0} \quad (\beta\text{-ray})$$
(unstable)

$${}_{54}X^{123} \xrightarrow{5\text{ days}} {}_{55}Cs^{133} + {}_{-1}e^{0} \quad (\beta\text{-ray})$$
(unstable) (stable) (caesium)

Similarly Columbium is also unstable and gives out similar β-ray emissions ending up with 44Ru99 which is stable.

In general, element with Z = 90 undergoes fission under bombardment with proper energy. The first cases of fission in U (uranium) Th (Thorium) and Pa (protactinium) were produced with neutrons.

It has also been found that fission can be produced by bombardment with particles other than neutrons. Thus fission has been induced in U and Th by high speed protons of energy 6.9 Mev, by deuterons of energy than 8 MeV and by α-particles of 32 MeV energy. High energy photons such as the γ-rays obtained from nuclear reactions of *X*-rays from betatron have also been found effective in producing fission. The cross section in these cases is of the order of 1 to 7 × 10^{-27} cm^2. These results bring out the fact that almost any means, which will make an already unstable Uranium or Thorium nucleus more unstable beyond a critical stage can be used to produced fission. The neutrons generated at fission are found to be **fast neutrons, possessing energies up to 1 MeV.**

16.5. SEPARATION OF U-235

Natural uranium contains both the isotopes of atomic weight 238 (${}_{92}U^{238}$) and 235 (${}_{92}U^{235}$) the latter being hardly 1%. It is not possible for the chain reaction to start in natural Uranium as the neutrons liberated by the fission of a ${}_{92}U^{235}$ nucleus will easily be absorbed by ${}_{92}U^{238}$. It is therefore necessary to separate ${}_{92}U^{235}$ from ${}_{92}U^{238}$ which is done by long laborious processes. It was great difficulty experienced in separating ${}_{92}U^{235}$ from the more abundant ${}_{92}U^{238}$.

It was difficult for Germans to separate ${}_{92}U^{235}$ from ${}_{92}U^{238}$ and thus they could not manufacture the atom bomb, otherwise the course of second world war might have reversed.

There are four different methods employed to separate ${}_{92}U^{235}$ from ${}_{92}U^{238}$.

(*a*) Gaseous diffusion method.

(*b*) Centrifugal separation method.

(*a*) Thermal diffusion method.

(*a*) Electro-magnetic separation method.

16.6. SEPARATION BY CALUTRONS

One of the important method, of separation of U^{235} from $_{92}U^{238}$ is by the application of a magnetic field utilising the principle of mass spectograph. The arrangement is known as Calutron. it is shown in Fig. 16.2.

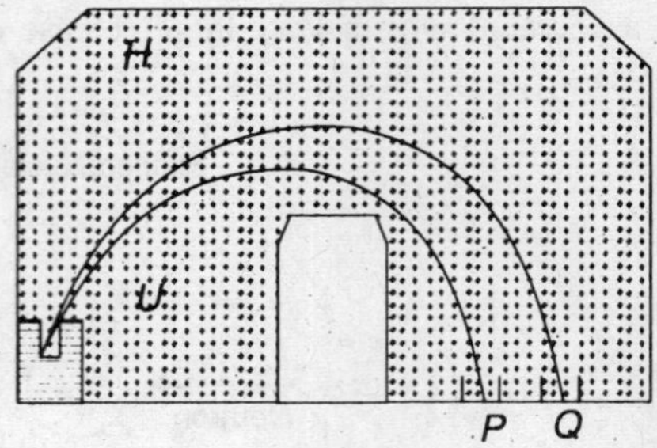

Fig. 16.2

The magnetic field *H* is applied perpendicular to the plane of paper. U is the source of uranium ions. Due to the presence of magnetic field the lighter U^{235} ions collect at *P* and heavier U^{235} ions at *Q* thereby bringing about separation of the two kinds of ions.

16.7. TRANS-URANIUM ELEMENTS

An effort was made to find ohter element which could undergo fission like Uranium. In this way elements of atomic numbers 93 and 94 were discovered.

1. Neptunium. Neptunium (Np) atomic number 93 was first detected in 1939 by Abelson and identified by Mac Millian in 1939. A show neutron is captured by the $_{92}U^{238}$ to form an unstable higher isotope which emits a β-ray to form Neptunium.

$$_{92}U^{238} + {}_0n^1 \longrightarrow {}_{92}U^{239} + \gamma\text{-ray} \xrightarrow{23\text{ min}} {}_{93}Np^{239} + {}_{-1}e^0\,(\beta\text{-ray})$$

2. Plutonium. (Plutonium) (Pu) atomic number 94 was first identified by kennedy and Mac Millan and is obtained from neptunium by a β-ray emission.

$$_{93}Np^{239} \xrightarrow{23\text{ days}} {}_{94}Pu^{239} + {}_{-1}e^0\ (\beta\text{-ray})$$

This is fairly stable and dacays (half life = 24,000 years) into U^{253} with the emission of α-particle. Plutonium has been used very much in the manufacture of the atom-bomb. This can be easily obtained from the $_{92}U^{238}$. It undergoes fission like $_{92}U^{235}$ and sets up a chain reaction, liberating huge amount of kinetic energy. The first atom bomb used over Hiroshima was made from $_{92}U^{235}$ and the second used over Nagasaki from $_{92}Pu^{239}$, the latter being more powerful than the former.

It has been found that Th^{238}, U^{235} and Pu^{239} are fissionable by the capture of slow neutrons, where as $U^{238}$and Pu^{241} are fissionable by capture of fast neutrons only. But U^{238}captures a slow neutron and gives rise to Neptunium and Plutonium which are fissionable.

3. Americium and Curium. The element Americium (Am) At. No. 95 and Curium (Cm), Atomic No. 96 were discovered in 1944 by Seaberg, Hamiltod and others. Americium was produced from $_{94}Pu^{241}$ with β-ray emission, and itself emits emits α-particles with a half value period of 500 years.

$$_{94}Pu^{241} \longrightarrow {}_{95}Am^{241} + {}_{-1}e^0\ (\beta\text{-ray})$$

To obtain curium from Pu^{239}, a size able quantity of it was bombarded in the Berkley cyclotron by 40 meV helium nuclei, when two isotopes $_{96}Cm^{240}$ and $_{96}Cm^{322}$ were formed. These elements are α-ray emitters with half-lives of one month and five months respectively.

$$_{94}Pu^{239} + {}_2H^4 \longrightarrow {}_{96}Cm^{240} + 3\,{}_0n^1$$

$$_{94}Pu^{239} + {}_2H^4 \longrightarrow {}_{96}Cm^{240} + 3\,{}_0n^1$$

4. Berkelium and Californium. In 1950, two more elements Berkelium (Bk) Atomic No. 97 and Californium (Cf), At. No. 98 have been discovered.

Berkelium was obtained by bombarding $_{95}Am^{341}$ with helium ions from the cyclotron.

It emits α-particles and has a half-life period of 5 hrs.

Californium was obtained by bombarding a Curium isotope $_{95}Cm^{342}$ with helium ions from the cyclotron.

It also emits α-particle and has a half-life period of 45 min.

Elements upto Atomic Number 110 have now been discovered

16.8. FISSION ENERGY

Each atom of Uranium breaks into one atom of ${}_{55}Cs^{133}$,one atom of ${}_{44}Ru^{99}$ and three neutrons ${}_{0}n^{1}$. Adding up their mass

$${}_{92}U^{235} \longrightarrow {}_{55}Cs^{138} + {}_{44}Ru^{99} + 3\ {}_{0}n^{1} + \text{Energy liberated}$$

$$235.128 \longrightarrow 132.942 + 96.903 + 3 \times 1.00893 + 0.226 \text{ mass unit.}$$

It is clear that in this integration, 0.226 mass-units of matter is dissipated out as energy which is equivalent to 210.4 MeV.

$$(\because 1 \text{ a.m.u} = 931 \text{ MeV.})$$

In developing his **Theory of Relativity**, Einstein introducted the concept of the equivalence of mass and Energy. According to Einstein's theory, the apparent mass m of particles travelling with a velocity v, is given by

$$m = \frac{m_0}{\sqrt{1-\frac{v^2}{c^2}}}$$

where m_0 = rest mass of the particle; c = velocity of light

Now
$$m = m_0\left(1-\frac{v^2}{c^2}\right)^{-1/2} = m_0\left(1+\frac{1}{2}\frac{v^2}{c^2}\right)$$

or
$$(m-m_0)\,c^2 = \frac{1}{2}m_0\,v^2 = \text{kinetic energy of the particle}$$

$$\therefore \quad \text{Energy} = \text{Change in mass} \times c^2$$

This is one of the revolutionary equations of Modern Physics. If we apply Einstein's relation to the above reaction, on the basis that 1 atomic mass unit is equivalent to 931 MeV, it is found that 210.4 MeV, of energy is liberated when 0.226 unit of mass ceases to exist as mass. Thus, whenever the total atomic mass of the final product in a nuclear fission is less than that of the original particles, a corresponding release to energy occurs. In ordinary nuclear disintegrations quantities of energy varying from 1 to 20 MeV is liberated.

However, in the case of the fission of Uranium nucleus, an extremely large quantity of energy is realesed. Henderson has found that a single uranium fission gives energy of the order of 200 MeV. That such a source of energy can be taped and utilised is attested by the devasting results of the atom bomb. **It can also be put to constructive and peaceful uses as well.**

16.9. REASON FOR HIGH FISSION ENERGY

Before 1939, the largest range of nuclear energy was 1 to 20 MeV. But in case of nuclear Fission, the amount of energy released per fission of U^{235} is 200 MeV. There are two reasons for this high energy.

(*i*) Blinding Energy of the Nucleus (*ii*) Mass defect

1. Binding Energy. It is believed that when a nuclide is formed by the packing of protons, neutrons, some of the mass is converted into energy. This energy is known as *binding energy* which holds the parts of every atom together. The greater the mass defect in this packing, the greater is the binding energy.

$$\text{Neutron} + \text{Protons} + \text{Electrons} \xrightarrow{\text{Packed}} \text{Atom (Wt. is leass than those of particles)} + \text{Binding Energy}$$

$\therefore$ the binding energy is defined as the "energy which is required break up the atom into neutrons, protons and electrons".

To take a simple case of the helium nucleus, its total mass should consist of two neutron masses plus two proton masses.

$$2\ (n) = 2.01796 \text{ a.m.u} \qquad \text{and} \qquad 2\ (p) = 2.01519$$

Computed He mass = 4.03315 a.m.u.
Measured He mass = 4.00281 a.m.u.
∴ Mass defect (Δm) = 0.03034 a.m.u.

Clearly, the computed and measured masses do not agree. The discrepancy, 0.03034 mass unit, is far larger than one can be ignored. The amount of energy evolved in any mass-to-energy change is calculated by the relation

$$E = \Delta mc^2$$

where E is the energy in joule, Δm is the mass in kilograms that disappears and c is the velocity of light in kilometre/second which is 3.00×10^5 km/sec. By applying this equation we can show that Binding energy in case U^{235} is 200 MeV.

2. Mass Defect. The mass number of a particular atom is the number of protons and neutrons. In a neutral atom, since the number of electrons equals the number of protons, it is logical to expect that the mass of an atom should be the sum of the masses of the protons, neutrons and electrons. It is noted experimentally that the mass an atom is less than the sum of masses of the neutrons, protons and electrons contained in it. For example

(*i*) In case of Neon atom, which contains 10 protons, 10 neutrons and 10 electrons since its atomic number is 10 and mass number is 20.

10 (n) = 10.73 a.m.u; 10 (p) = 10.087 10 (e) = 0.0055
Computed Ne mass = 20.1655 a.m.u.
Measured Ne mass = 19.993 a.m.u.
Mass defect of Ne (Δm) = 0.1725 a.m.u.

∴ This difference in mass is known as *mass defect* for neon atom.

It appears that when protons, neutrons and electrons are packed to make atoms, there is a loss in mass associated in some way with the packing. This loss in mass is called as the *mass defect* or the *packing effect.*

16.10. THEORY OF FISSION PROCESS OR LIQUID DROP MODEL

Liquid drop model of nucleus was given by Neil Bohr and Gamow in 1936. According to this theory nucleus of all atoms are somewhat like a drop of liquid; when a projectile (say a neutron) strikes a nucleus it is first captured and an intermediate compound nucleus is formed. the compound nucleus resembles with a drop of hot liquid and all its constituents are moving within a spherical enclosure like the molecules of the liquid drop. The surface of the spherical enclosure, which is the top of the potential barrier, resembles with the surface tension, which holds a small liquid attained during disintegration, are shown in Fig. 16.3.

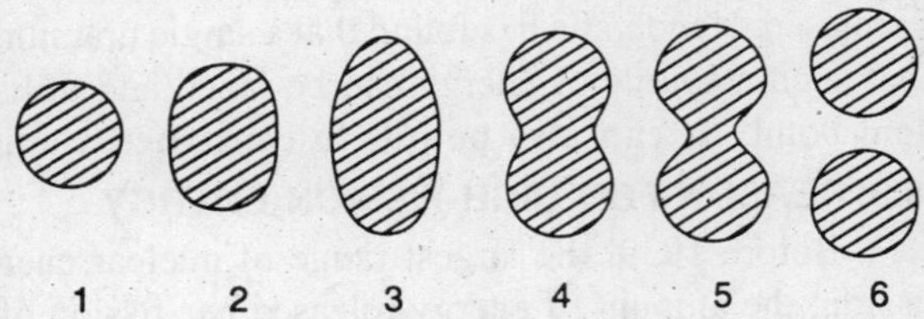

Fig. 16.3

The disintegration of the nucleus can also be explained with the help of liquid drop model of the nucleus. During disintegration, the ejection of the particle from the nucleus takes place, which resembles with the evaporation of a liquid molecule from a drop of liquid and the binding energy of the emitted particle resembles with the '*heat of vaporization*' of the liquid molecules. The evaporation process in a liquid molecules takes place when the Maxwellian energy distribution with in the drop causes a particular molecule to have sufficient energy to overcome the inter-molecular attraction and escape from the potential barrier. this process takes place in about 10^{-13} sec. Hence we can explain the formation and decay of compound nucleus with the help of '*liquid drop model*' of the nucleus.

Application of Nuclear Fission

Nuclear fission is mainly used in:

(*i*) Nuclear reactors for the control of reactions. (*ii*) For making Atom Bombs.

Example 1. *Compute the energy of one atomic mass unit in MeV.*

Solution: We know that

$$1 \text{ a.m.u} = 1.66 \times 10^{-24} \text{ gm.} = 1.66 \times 10^{-27} \text{ kgm.}$$

By Einstein mass energy relation

$$E = \Delta mc^2 \qquad \text{where} \qquad \Delta m = \text{mass defect}$$

$$c = \text{velocity of light}$$

$$\Delta m = 1 \text{ a.m.u.} = 1.66 \times 10^{-27} \text{kgm.}$$

$$c = 3 \times 10^8 \text{ m/s.}$$

$$\therefore \quad E = 1.66 \times 10^{-27} \times (3 \times 10^8)^2 \text{ Joules}$$

$$= 1.66 \times 10^{-27} \times 9 \times 10^{16} \text{ Joules} = 14.94 \times 10^{-11} \text{ Joules}$$

Now 1.60×10^{-13} Joules = 1 MeV.

$$\therefore \quad E = \frac{14.94 \times 10^{-11}}{1.60 \times 10^{-13}} \text{ MeV} = 931 \text{ MeV.}$$

Example 2. *Using the information on atomic masses as given below, show that a nucleus of uranium can disintegrate with emission of α-particle according to the relation*

$$_{92}U^{238} \longrightarrow {}_{90}Th^{234}$$

Calculate. (i) The total energy released in the disintegration.

(ii) The K.E. of α-particle, the nucleus being at rest before disintegration.

Mass of U^{238} = 238.12492 a.m.u.; Mass of Th^{234} = 234.11650 a.m.u.; Mass of He^4 = 4.00387 a.m.u.

Solution: (*i*) 1 a.m.u. is equvalent to 931 MeV.

$$\text{Mass of reactant} = 238.12492$$

$$\text{Total mass of products} = 234.11650 = \frac{4.00387}{238.12037}$$

$$\text{Defect in mass} = 238.12492 - 238.12037 = 0.00455 \text{ a.m.u}$$

∴ Total energy released due to mass defect of

$$0.00455 \text{ a.m.u} = 931 \times 0.00445 \text{ Mev.} = \mathbf{4.23 \text{ MeV.}}$$

(*ii*) Because the nucleus was at rest before disintegration, the energy released is shared by α-particle and the recoil nucleus in accordance with the laws of "conservation of energy and momentum."

$$\therefore \quad \frac{1}{2} m_1 v_1^2 + \frac{1}{2} m_2 v_2^2 = 4.23 \text{ MeV} \qquad \ldots(i)$$

where m_1 and v_1 are the mass and velocity of thorium 234 and v_2 and the mass and velocity of Helium–4.

According of conservation of Momentum

$$m_1 v_1 = -m_2 v_2 \qquad \text{or} \qquad v_1 = -\frac{m_2}{m_1} v^2$$

Substituting v_1 in Equation (*i*)

$$\frac{1}{2} \frac{m_2^2}{m_1^2} \cdot m_1 . v_2^2 + \frac{1}{2} m^2 v_2^2 = 4.23 \quad \text{or} \quad \frac{1}{2} m_2 v_2 \left[\frac{m_2}{m_1} + 1 \right] = 4.23$$

∴ K.E. of the α-particle

$$\frac{1}{2} m_2 v_2^2 = 4.23 \times \frac{m_1}{m_1 + m_2} = 4.23 \times \frac{234.11650}{238.12037}$$

K.E. = **4.16 MeV.**

Example 3. *When* $_{92}U^{235}$ *(235.044) is bombarded with neutron breaks into* $_{42}Mo^{98}$ *(97.905) and* $_{54}Xe^{136}$ *(135.917). Assuming that these isotopes have come from the original fission process, find (i) what elementary particles are released, (ii) mass defect of the reaction, (iii) the equivalent energy released; given mass of β-particle = 0.00055 a.m.u.; mass of neutron = 1.0009 a.m.u.*

Solution: (*i*) the nuclear reaction is

$$_{92}U^{235} + {}_0n^1 \longrightarrow {}_{42}Mo^{98} + {}_{54}Xe^{136}$$

The sum of the atomic number of the Products = 42 + 54 = 96.

The sum of the atomic number of the reactions = 92.

Therefore, the atomic number of the products is greater by 4 units than $_{92}U^{235}$ which shows that original unstable products must have emitted 4 β-particles before becoming stable. Now the number of the products is 2 units than the reactants. It shows that two fission neutrons must have been produced. Hence the fission reaction can be rewritten as follows

$$_{92}U^{235} + {}_0n^1 \longrightarrow {}_{42}Mo^{98} + {}_{54}Xe^{136} + 4\,{}_{-1}e^0 + 2\,{}_0n^1$$

(*ii*) R.H.S. Masses = (97.905 + 135.917 + 4 × 0.00055 + 2 × 1.009) = 235.842 a.m.u.

L.H.S. Mass = (235.044 + 1.009) = 235.053 a.m.u.

$$\Delta m = \text{L.H.S. mass} - \text{R.H.S. mass} = 236.053 - 235.842 = 0.211 \text{ a.m.u.}$$

Mass defect $\Delta m = 0.211$ a.m.u.

(*iii*) Energy released

$$E = 0.211 \times 931 = 196 \text{ MeV.} \qquad (\because 1 \text{ a.m.u.} = 931 \text{ MeV})$$

16.11. FUSION

It is that process in which lighter atoms, like hydrogen, are fused together into heavier atoms like helium, when again a fraction of the reacting masses is converted into energy.

Explanation of Fission and Fussion from Binding Energy: If a graph is platted between Mass number (*A*) and the binding energy/nucleon, for all the elements of the periodic table, a curve of the type shown in Fig. 16.4 is obtained. The B.E./nucleon, rises sharply and reaches a maximum value of 8.8 MeV in the neighbourhood of *A* = 50. The max is for quite number of elements and it is still 8.5 MeV for *A* = 140. For higher mass numbers the value decreases to 7.6 MeV for Uranium. *It is this decrease in Binding Energy for elements with large mass numbers which is the fundamental cause of the release of energy due to Fission.* The sharp change of the curve in the initial stages forms the basis of fusion *i.e.*, release of energy due to binding of the two or more smaller nuclei with low mass numbers.

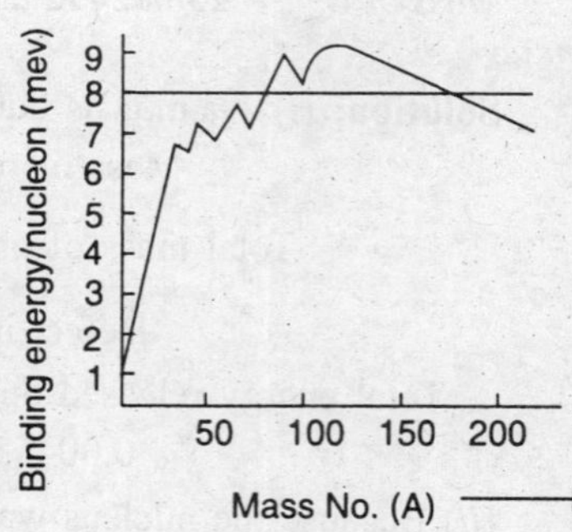

Fig. 16.4

16.12. FUSION ENERGY

(*i*) **Hydrogen Bomb.** Hydrogen bomb is based upon the principle of fusion. In hydrogen bomb, the rate of energy released is eight times greater than that in Uranium fission. The reaction can be represented as:

$$_1H^2 + {}_1H^2 \longrightarrow {}_1H^3 + {}_1H1 + 91 \times 10^9 \text{ cals.}$$

$$_1H^1 + {}_1H^2 \longrightarrow {}_2He^3 + {}_0n^1 + 76 \times 10^9 \text{ cals.}$$

$$_1H^2 + {}_1H^2 \longleftarrow {}_2He^4 \times {}_0n^1 + 404 \times 10^9 \text{ cals.}$$

The nuclear fusion in the hydrogen bomb employs the process that has been going on in the core of the Sun and which produces solar energy. **The Sun maintains never ending fusion of hydrogen into helium.** To cause this fusion, temperatures of over 2 × 10^6 K and pressures of the order of 16 × 10^{10} atomspheres, exist in the core of the Sun.

Under these conditions, the atoms are stripped of all their nuclear electrons and are in the nuclear (plasma) state. The speed of these nuclear particles is so great that fusion of the nuclei takes place by high velocity impact of the particles. This further liberates energy which raises the temperature still further and accelerates the transformation, resulting in more terrific explosions.

During an atomic explosion, temperatures of the order of **2 million degrees and intense momentary pressure is obtained which can be used to initiate the thermo-nuclear reactions in hydrogen.** These are sufficient to trigger a hydrogen bomb and produce a chain reaction as goes on in the body of the sun. Experimental researches of this nature are being carried out at the "Atomic Energy Research Establishments" of mainy countries.

16.13. SOLAR ENERGY (THEROMONUCLEAR REACTIONS)

The Sun is constantly radiating out large amount of energy in space. The source of this energy was difficult to understand till the discovery of nuclear reactions and it was difficult to explain why the sun should not have cooled down after radiating so much energy. Bethe in (1938) showed that a "Carbon-Nitrogen cycle" was mainly responsible for the production of solar energy. This cycle consists of a chain and nitrogen as catalysts, the highly energetic thermal protons constituting the projects in the successive reactions. According to Bethe, the following chain reactions may be taking place inside the sun.

$$_6C^{12} + {_1H^1} \longrightarrow {_7N^{13}} + \gamma \text{ (1.95 MeV)}$$
$$_7N^{13} \longrightarrow {_6C^{13}} + e^+ + \nu \text{ (2.22 MeV)}$$
$$_6C^{13} + {_1H^1} \longrightarrow {_7N^{14}} + \gamma \text{ (7.54 MeV)}$$
$$_7N^{14} + {_1H^1} \longrightarrow {_8O^{15}} + \gamma \text{ (7.35 MeV)}$$
$$_8O^{15} \longrightarrow {_7N^{15}} + e^+ + \nu \text{ (2.71 MeV)}$$
$$_7N^{15} + {_1H^1} \longrightarrow {_6C^{12}} + {_2He^4} + \text{(4.96 MeV)}$$

It will be noted that these reactions form a closed chain or cycle returning to the starting point after six steps. The chain of reaction can start with either C^{12} or N^{14} and end in the reproduction of the same nucleus with which the chain is initiated. Carbon and nitrogen therefore remain unaffected during these reactions and hence considered to act as mere catalysts. The net result of the chain is the consumption of four protons to form helium nucleus. The period of a complete cycle is estimated to be about 5 million years.

The amount of energy liberated in these chain of reaction can be calculated from the mass deficiency involved.

$$\begin{aligned} \text{The mass of 4 protons} &= 4 \times 1.00813 = 4.0325 \text{ a.m.u.} \\ \text{Mass of helium nucleus} &= 4.00386 \text{ a.m.u.} \\ \therefore \quad \text{Mass that has disappeared} &= 0.02866 \text{ a.m.u.} \\ \text{Hence the energy liberated} &= 0.02866 \times 931 \text{ MeV} = 26.7 \text{ MeV} \end{aligned}$$

Considerations of the actual hydrogen and helium content of the Sun show that this amount of energy is enough to keep the Sun going at its present interval temperature for about 30 million years.

Proton-proton Cycle

According to modern views of stellar evolution, the above-mentioned, carbon-nitrogen cycle comes at a rather later stage in the amount of matter at a point in space. Under the influence of gravitational attraction the amussed matter gradually contracts, resulting in a rise in temperature until the central part is 2 million degree. At this stage, there is sufficient thermal energy for a proton-proton interaction to form a deuteron as given by

$$_2H^1 + {_1H^1} \longrightarrow {_2H^2} + {_{-1}e^0} + \nu \qquad \text{(1.19 MeV)}$$

which is followed by two additional protons reactions resulting in an α-particle

$$_1H^2 + {_1H^1} \longrightarrow {_2He^3} + \text{(5.49 MeV)}$$

$$_2He^3 + {}_2He^3 \longrightarrow {}_2He^4 + {}_1H^1 + {}_1H^1 (12.85 \text{ MeV})$$

These reactions are mainly responsible for the luminosity of the stars fainter than the Sun, *e.g.*, the so-called red-dwarfs.

Example 4. *In the interior of stars, reactions constantly go on, which produce 1α-particle (He^4) and 2 positions from the combination of 4 protons. How much energy would be released by the formation of 1 kilogram of Helium (He^4) given:*

Mass of 1 proton = 1.00759 a.m.u.; Mass of He^4 atom = 4.00386 a.m.u. and Mass of position + $_1e^0$ = 0.00055 a.m.u.

Solution: Mass of 1 proton = 1.00759 a.m.u.

Mass of 4 protons = 4.03036 a.m.u.

Mass of α-particle (He^4) = 4.00386 a.m.u.

Mass of 1 positron = 0.00055 a.m.u.

Mass of 2 positron = 0.00110 a.m.u.

Total mass of the products = 4.00386

0.00110

4.00496 a.m.u.

Mass defect = 4.03036

–4.00496

0.32540 a.m.u.

Now 1 a.m.u. = 931 MeV

∴ Energy released due to mass defect of 0.02540 a.m.u.

= 931 × 0.02540 = 235474 Mev

Avogadro's No. = 6.02 + 1026 atoms/kgm.

$$1 \text{ kgm of He} = \frac{6.02 \times 10^{26}}{4} \text{ atoms/kgm.}$$

∴ Total energy released by the formation of 1 kg. of He^4

$$= \frac{23.6474 \times 6.02 \times 10^{26}}{4} \text{ MeV}$$

$$= \frac{23.6474 \times 6.02 \times 10^{26}}{4} \text{ MeV} = \mathbf{35.47 \times 10^{26} \text{ MeV.}}$$

Example 5. *It is estimated that the sun emits energy at the rate of 4 × 10^{16} Joule/second. If all the energy comes from the reaction*

$$4H \longrightarrow He^4$$

what is the rate of hydrogen consumption given : mass of hydrogen = 1.007825 a.m.u.; mass of helium = 4.0026 a.m.u. Number of atoms in one kilogram hydrogen = 6.03 × 1026.

Solution: Mass of L.H.S. = 4 × 1.007825 = 4.0313

Mass of R.H.S. = 4.0026

$$\Delta m = 4.0313 - 4.0026 = 0.0287 \text{ a.m.u.}$$

$$E = \Delta m \times 931 \text{ MeV} = 0.0283 \times 931 = 26.72 \text{ MeV}$$

This energy is released when atoms of hydrogen combine

∴ Energy released per atom of hydrogen

$$= \frac{26.72}{4} = 6.68 \text{ MeV}$$

$$= 6.68 \times 10^6 \times 10^{-19} \text{ Joules}$$

$$= 1.069 \times 10^{-12} \text{ Joules.}$$

Energy released by one kg. of Hydrogen

$$= 1.060 \times 10^{-12} \times 6.03 \times 10^{26} = 6.444 \times 10^{14} \text{ Joules}$$

Mass of hydrogen used per second to generate 4×10^{16} joules of energy

$$= \frac{4 \times 10^{16}}{6.445 \times 0^{14}} \text{ kg} = \mathbf{62.223 \text{ kg.}}$$

Example 6. *Taking the atomic mass of hydrgeon as 1.00778 and that of helium as 4.00216. Calculate: (i) K.E. in e.V. liberated if 4 atoms of hydrogen are condensed to form an atom of helium. (ii) If energy were liberated as monochromatic radiation, calculate the wavelength of the radiation.*

Solution: (*i*) Mass of 1 atom of Hydrogen = 1.00778 a.m.u.

Mass of 4 atom of Hydrogen = 4.03112 a.m.u.

Mass of 1 atom of Helium = 4.00216 a.m.u.

Mass defect = 4.03112

= −4.00216

0.02896 a.m.u.

Energy liberated due mass defect of 0.02896 a.m.u.

= 0.02896 × 931 MeV

= 0.02896 × 931 × 106 ev = $\mathbf{2.693 \times 10^7}$ **ev**

(*ii*) Because the energy liberated

$$E = h\nu = h\frac{c}{\lambda} \qquad [\because c = \nu\lambda]$$

$$\therefore \qquad \lambda = \frac{hc}{E}$$

Putting the values of h, c and E

$$\lambda = \frac{6.62 \times 10^{-24} \times 3 \times 10^{8}}{2.693 \times 10^{7} \times .16 \times 10^{-19}} \text{ m} = \mathbf{4.6 \times 10^{-14} \text{ m.}}$$

16.14. PRINCIPLE OF NUCLEAR REACTORS

We have seen that in nuclear fission, large amount of energy is released by the fission of each U^{-235} nucleus. the energy released is nearly equal to 200 MeV. More over this fission process is accompanied by the liberation of 3 neutrons per fission the liberation of these neutrons permit a self sustained chain reaction in which these neutrons produce more fission and more fission produces more neutrons and so on. The energy released can be either very fast resulting in a violent explosion as in case of atomic bomb. The rate of release of energy is very large and it is very difficult to control this chain reaction. But in these days it has become possible to control these chain reactions by using chain reacting nuclear piles or nuclear reactors. Therefore, nuclear reactor is a device which controls and maintains a steady rate of the chain reaction. Energy produced by the nuclear can be utilised for peaceful purposes.

16.15. CHAIN REACTION

A chain reaction is that in which the number of neutrons keeps on multiplying rapidly in geometrical progression during fission till whole of the fissionable material is disintegrated.

The chain reaction will become self sustained only if, for every neutron absorbed, at least one fission neutron becomes available for causing fission of another nucleus. This condition can be expressed in terms of a multiplication factor (K)

Let n = Average number of neutrons per fission.

x = Change that a neutron shall produce further fission.

Let $nx = K$ (Multiplication factor).

when $K \geq 1$, chain reaction will continue and if $K < 1$, the chain reaction will terminate after a finite number of fissions.

The fission of ${}_{92}U^{235}$ is quite rapid and grows rapidly out large quantity of energy. By the action of a single neutron, one of the atoms of ${}_{92}U^{235}$ undergoes fission and in the process three or four new neutrons are produced. These neutrons if comparatively slow, attack other ${}_{92}U^{235}$ nuclei and will bring about their fission releasing more neutrons. The reaction thus develops in to chain-reaction as shown in Fig. 16.5.

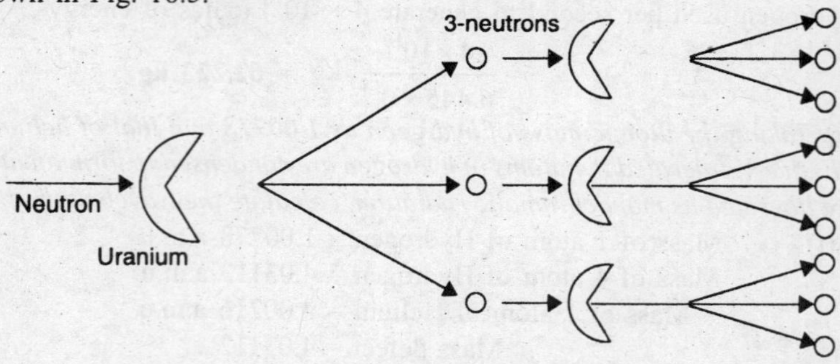

Fig. 16.5 The "Chain Reaction"

The number of atoms undergoing fission thus rapidly increases with time and in a short time, it develops into gigantic proportions.

The escape of neutrons from any amount of Uranium depends upon the available surface, where as the rate of fission will depend on the total number of Uranium atoms present and hence upon the volume. The probability of a neutron escaping thus decreases as the volume increases and hence the chain reaction will go on only if the mass is increased beyond a certain critical value, called the **critical mass.** Thus an atom bomb may be made from two or more pieces of ${}_{92}U^{235}$, each of which is below the critical mass but when brought together exceed this critical value and therefore explode and release huge amount of energy critical mass for U^{235} is 53 kg Pu^{239} is 18 kg. It is difficult to note the rate of chain reactions.

16.16. EXPLOSIVE CHAIN REACTION AND CRITICAL SIZE

For explosive chain reaction in case of pure ${}_{92}U^{236}$ or ${}_{94}Pu^{239}$ two conditions are necessary.

(*i*) The time between fission events must be small.

(*ii*) There should be very small loss of neutrons either by leakage from the surface of the material or by non-fission capture. Leakage of neutrons take place from the surface of the fissile material. Greater the volume of the material lesser will be the leakage of the neutrons from its surface. The probability of a neutron escaping thus decreases as the volume increases and hence the chain reaction will go in only if the mass is increased beyond a certain value, called the **critical mass.** In other words, neutron production is proportional to volume and leakage to surface area.

$\therefore$ Escape or leakage rate $\propto r^2$

Production or fission rate $\propto r^2$

$$\therefore \quad \frac{\text{escape rate}}{\text{production rate}} = \frac{1}{r}$$

Hence, larger the size of the body, *i.e.,* greater its radius lesser the escape rate for critical size, $K = 1$.

If the size of the body is less than critical size, then the loss of neutrons is greater the their production. Consequently chain reaction is nor possible under these conditions. For subcritical size $K < 1$.

When the size of the sample is greater than the critical size and for which $K > 1$. Then it will explode spontaneously once fission is started purposely or by change. The only way to prevent such a mass of pure ${}_{92}U^{235}$ or ${}_{94}Pu^{239}$ is to keep its value below critical size.

16.17. ATOM BOMB

The first atom bomb used in Hiroshima utilised ${}_{92}U^{235}$ is isotope as the main reacting substance and the second atom bomb used in Nagasaki made use of Plutonium 239. The fission in both cases is similar and uncontrolled. The mass of the substance used must exceed a certain quantity called the **critical mass.** In fact there are two separate pieces, each of mass less than this critical value so that both are quite safe, when apart. When the mass is less than the critical mass, more neutrons result that the reaction soon stops. When the critical mass is exceeded, the neutrons produced in the reaction cannot easily escape out and there are, at each stage, more neutrons to bring about fission on the chain reaction soon assumes gigantic proportions and results in terrific–explosion. The chain reaction of the Plutonium bomb is shown in Figs. 16.6 (*a*), (*b*).

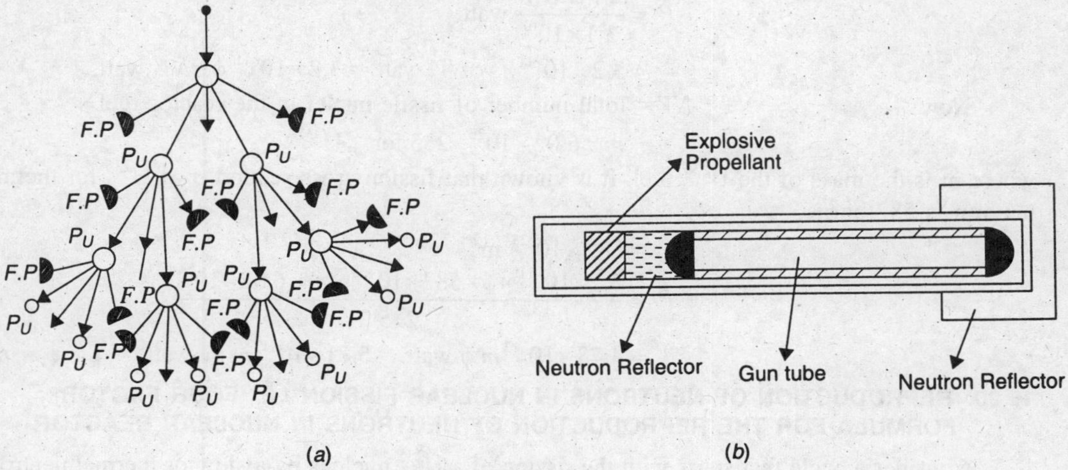

Fig. 16.6

Uncontrolled chain reaction in Atom bomb.
FP. stands for Fission Product.
Pu stands for Plutonium.
$_0n^1$ stand for neutron

16.18. HYDROGEN BOMB

Hydrogen bomb is 1000 times more powerful than the atomic bomb. This is based on the fusion of hydrogen atoms into heavier ones by the thermo nuclear reactions. For fusion to take place, very high temp is required and generally fusion is proceeded by "Fission" to provide that high temperature. So in the hydrogen bomb, fission and fusion take place one after the other and once the reaction starts, a tremendous amount of energy is released.

Hydrogen bomb has got the following few characteristics over and above the atom bomb.

(*a*) The energy released in this case has no upper limit. It is only dependent upon the fusible material available in the bomb.

(*b*) It has no limitation of critical size unlike an atom bomb. Once this bomb is ignited (for which very high temp. is required) then tremendous energy is released due to fusion.

Cobalt Bomb

If hydrogen bomb is covered by cobalt sheath, then it becomes more lethal than the hydrogen bomb. It is due to the reason that after the explosion take place, cobalt becomes highly radioactive and is converted into a gigantic radio active cloud which can spread into thousands of kilometers killing every thing living in that area.

16.19. POWER OF A NUCLEAR REACTOR

The fission rate of a reactor *i.e.,* total number of nuclei under going fission per second in a reactor is

$$n\,v\sigma NV = \phi\,\sigma NV$$

where n = Average neutron density *i.e.,* number of neutrons per unit volume

v = Average speed in metre/sec
$\phi = nv$ = Average neutron flux
N = Number of fissile nuclei per unit volume
σ = fission cross-section in m^2
V = Volume of the nuclear fuel.

Now power of one watt $= 3.1 \times 10^{10}$ fission/sec.

$$\therefore \quad P = \frac{n v \sigma N V}{3.1 \times 10^{10}} \text{ watt.}$$

$$= 3.2 \times 10^{-11} n v\sigma NV \text{ watt.} = 3.2 \times 10^{-11} \phi \sigma NV \text{ watt.}$$

Now NV = total number of fissile nuclei in the reactor fuel

$$= m \times 6.02 \times 10^{26} / 235 \text{ for } {}_{92}U^{235}$$

where m is the mass of the U^{235} fuel. It is known that fission cross-section σ of U^{235} for thermal neutron is 582 burns

$$= 582 \times 10^{-28} \text{ m}^2$$

$$\therefore \quad P = \frac{3.2 \times 10^{-11} \times \phi \times 582 \times 10^{-28} \times m \times 6.02 \times 10^{26}}{235}$$

$$= 4.78 \times 10^{-12} m \phi \text{ watt.} = 5.71 \times 10^{-12} m\, nv \text{ watt.} \quad [\because \phi = nv]$$

16.20. REPRODUCTION OF NEUTRONS IN NUCLEAR FISSION *I.E.*, FOUR FACTOR FORMULA FOR THE REPRODUCTION OF NEUTRONS IN NUCLEAR REACTOR

Consider a cycle that starts with the fission of a U^{235} nucleus by a slow or thermal neutrons as shown in Fig. 16.7. Suppose first moving neutrons are produced in this process. As their average energy is above the fission threshold for U^{235}, some of them can cause fission of U^{238} nuclei. However, the chance for such will depend on whether fission neutrons can collide with U^{238} nuclei before colliding with moderator nuclei. Some of the neutrons that do produce fast fission in U^{238} produce additional neutrons. Therefore, the total number of fast moving neutrons will increase from v to v_ε. Hence is called *fast fission factor* whose value varies from unity for a homogeneous reactor and 11 for other reactors. Now for the most favourable arrangement, some neutrons will, nevertheless, be lost to arrangement, some neutrons will, nevertheless, be lost due to resonance capture by U^{238}. If, p is the resonance escape probability, which is 'defined as the chance which a fast moving neutron is thermalised by collisions with carbon nuclei without being captured by U^{238} nuclei'. If nf is the fraction which escape before being slowed down to thermal energy, so far as the reactor is concerned v_ε nf neutrons are lost. The remaining $v_\varepsilon (1 - n_f)$neutrons are slowed down by the moderator. So far we have not taken care for the resonance capture. As p is the resonance escape probability, then $v_\varepsilon(1-n_f)$p neutrons escape capture while $v_\varepsilon (1 - n_f) (1 - p)$ neutrons are captured and produces Pu^{239} as shown in Fig. 16.7.

Some of the thermal neutrons diffuse without being captured and ultimately escape from the system. If the fraction escaping is n_ε, then the number of thermal neutrons per fission of U^{235} nucleus that escape is $v_\varepsilon(1 - n_f)\, p\, n_t$ and $v_\varepsilon(1 - n_f)\, p\, (1 - n_t)$ neutrons will remain for capture. If number of neutrons are absorbed in uranium and the fraction $(1 - f)$ is absorbed in other material. Obviously, the number of neutrons still available for carrying chain reaction is $v_\varepsilon(1 - n_f)\, p\, (1 - n_f) f$. The quantity is known as thermal utilization factor. Now, not all the neutrons absorbed in uranium cause fission of U^{235}. Now, not all the neutrons absorbed in uranium cause fission of U^{235}. Some are absorbed by U^{238} which produces Pu^{238} while some other are U^{235} to form U^{236}. The fraction of the thermal neutrons absorbed in uranium that causes fission of U^{235} is $\sigma_f(U)/\sigma_a(U)$. Hence multiplication factor is given by

$$K = v_\varepsilon(1 - n_f) p (1 - n_f) \cdot \frac{\sigma_f(U)}{\sigma_a(U)}$$

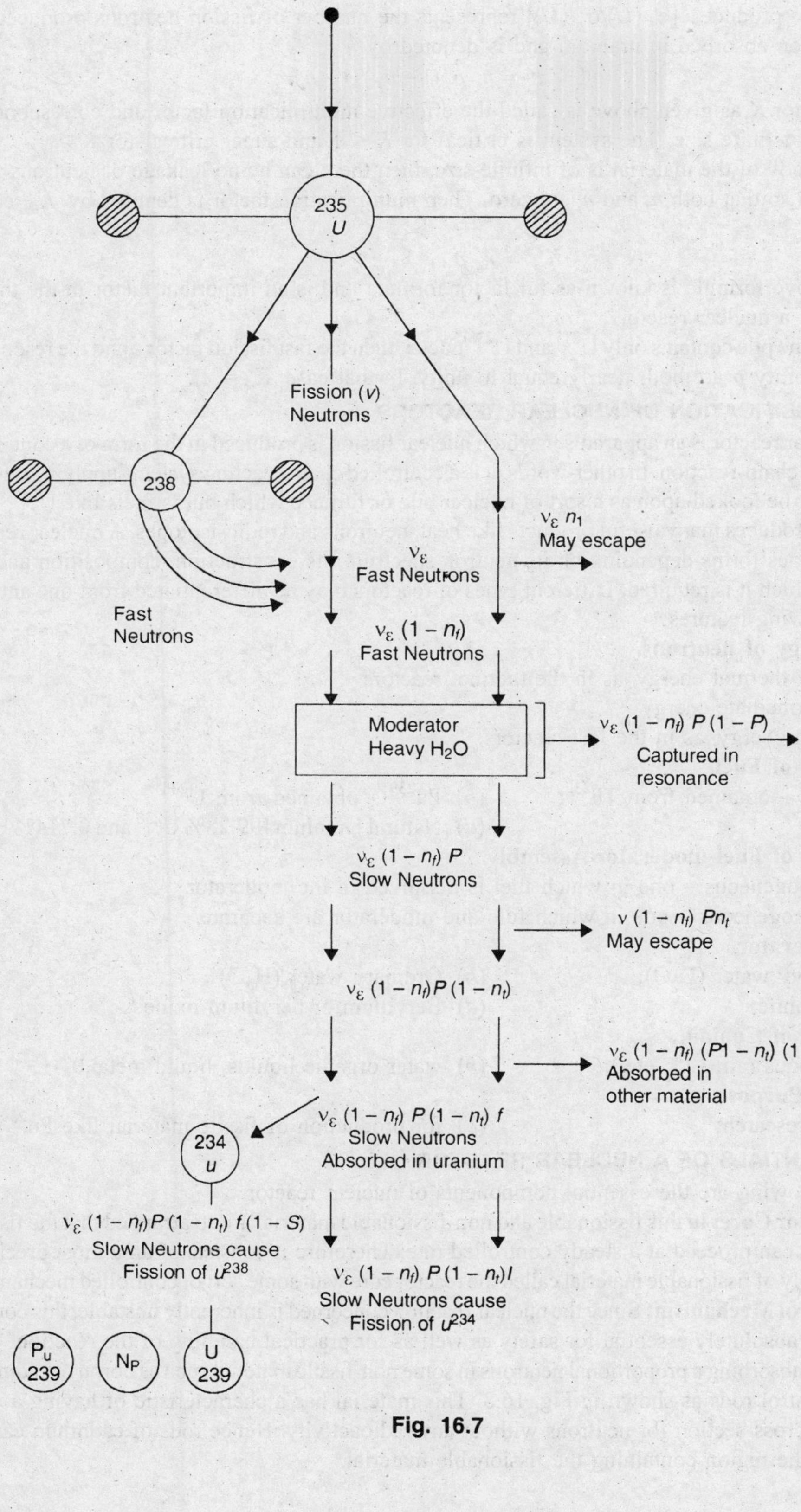
235
U
238
Fission (ν)
Neutrons
νε n1
May escape
νε
Fast Neutrons
Fast
Neutrons
νε (1 − nf)
Fast Neutrons
Moderator
Heavy H2O
νε (1 − nf) P (1 − P)
Captured in
resonance
234
u
NP
νε (1 − nf) P
Slow Neutrons
ν (1 − nf) Pnt
May escape
235
Pu
νε (1 − nf)P (1 − nt)
νε (1 − nf) (P1 − nt) (1 − f)
Absorbed in
other material
νε (1 − nf) P (1 − nt) f
Slow Neutrons
Absorbed in uranium
234
u
νε (1 − nf) P (1 − nt) f (1 − S)
Slow Neutrons cause
Fission of u238
νε (1 − nf) P (1 − nt) l
Slow Neutrons cause
Fission of u234
Pu
239
NP
U
239

Fig. 16.7

Now the product $\nu\ [\sigma_f(U)/\sigma_a(U)]$ represents the number of fission neutrons produced per thermal neutron absorbed in uranium and is denoted by n

$$\therefore \quad K = n_\varepsilon\, p_f(1-n_f)(1-n_t)$$

The factor K as given above is called the effective multiplication factor and corresponds to an assembly of finite size. The system is critical for $K < 1$ and super critical for $K > 1$.

If the body of the material is of infinite size, then there can be no leakage of neutrons from such a system so that both n_f and n_t are zero. Then multiplication factor is denoted by K_∞ called K-infinity.

$$\therefore \quad K_\infty = n_\varepsilon\, p_f$$

The above formula is known as for factor formula and is an important factor in the theory and design of a nuclear reactor.

In case the pile contains only U^{235} and U^{238} nuclei, then the fast fission factor ε and the resonance escape probability p are both nearly equal to unity. In that case $K_\infty = n_f$.

16.21. CLASSIFICATION OF NUCLEAR REACTORS

A nuclear reactor is an apparatus in which nuclear fission is produced in the form of a controlled self sustaining chain-reaction. In other-words, it is a controlled chain-reacting system supplying nuclear energy. It may be looked upon as a sort of nuclear pile or furnace which burns fuels like U^{235}, U^{233} or Pu^{239} and produces many useful products like heat, neutrons and radio-isotopes. A nuclear reactor can take various forms depending on its neutron spectrum, its construction, composition and the purpose for which it is required. Different types of reactors may be differentiated from one another by their following features:

1. Energy of neutrons.

(*a*) Low thermal energy, as in the thermal reactors

(*b*) Intermediate energy

(*c*) High energy, as in the fast reactors.

2. Type of Fuel.

(*a*) U^{233} – obtained from Th^{232}; (*b*) Pu^{239} – obtained from U^{238}

(*c*) U^{235} (*d*) Natural Uranium [99.28% U^{238} and 0.714% U^{235}]

3. Type of Fuel-moderator Assembly.

(*a*) homogeneous – one in which fuel is dissolved in the moderator.

(*b*) heterogeneous – one in which fule and moderator are separate.

4. Moderator.

(*a*) Heavy water (D_2O); (*b*) Ordinary water (H_2O)

(*c*) Graphite; (*d*) Beryllium or beryllium oxide.

5. Type of Coolant.

(*a*) gaseous (air, CO_2 etc.); (*b*) water organic liquids, liquid metal.0

6. The Purpose.

(*a*) for research; (*b*) for production of fissile material like Pu^{239}

16.22. ESSENTIALS OF A NUCLEAR REACTOR

The following are the essential components of nuclear reactor.

1. Reactor Core: In this fissionable and non-fissionable material are so arranged that the fission chain reaction can proceed at a steady controlled rate. Therefore it consists of an almost precisely critical assembly of fissionable material called the reactor core with some sort of controlled mechanism.

2. Control Mechanism: Since the nuclear reaction concerned is inherently unstable, this control mechanism is absolutely essential for safety as well as for practical handling of the reaction. This is achieved by absorbing a proportional neutrons in some non-fissile material such as boron or cadmium marked as control rods as shown in Fig. 16.8. This material has a characteristic of having a very large capture cross-section for neutrons without any radioactivity. Hence rods of cadmium can be inserted into the region containing the fissionable material.

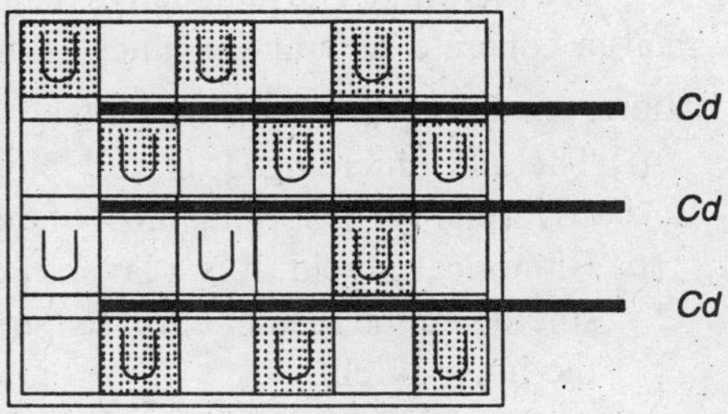

Fig. 16.8

They absorb the neutrons and hence cut down the reactivity and yet they do not become radioactive so that they can be handled for mechanical servicing. By pushing in the rods the operation of the reactor can be made to die down; Automatic devices are of course provided to move the rods deeper into the core if an excessive temperature rise occurs. By means of carefully designed servo-mechanisms it is possible to arrange such a control system so as to maintain the operation of the reactor at any desired power level. It is essential for the continuity of the operation that an average one neutron from each fission event should remain in the reactor to produce further fission. Any increase in the number of neutrons causes the reaction rate to rise. It can be shown that pure natural uranium, no matter how large the amount, be can not support a chain reaction.

3. Moderator: To prevent most of the neutrons from being absorbed by the heavier isotope the fuel is arranged in the form of slender rods which are embedded in a material of low atomic mass called the **moderator** to slow down the higher energetic neutrons produced in the fission cluranium. It must be such a type of material that does not itself absorb neutrons to any extent. Graphite and heavy water (D_2O) are such materials which do not absorb neutrons themself. Therefore, most of the time we use graphite and heavy water (D_2O) as moderators. Most collisions between neutron energies. With sufficient amount of moderator in the reactor the neutrons are reduced to thermal energies and are captured by U^{235} nuclei to cause fission. A moderator should have the following properties:

(*a*) It should possess low mass number in order that maximum neutrons energy may be lost per collision with its nuclei.

(*b*) It should possess low cross-section for neutron absorption *i.e.*, the probability of its absorption of neutrons should be minimum.

Materials like heavy water (D_2O), graphite, beryllium or beryllium oxide possess above properties and oftenly used as moderators. Graphite is very cheap as compare to other moderators, therefore it is used oftenly in the reactor. The following are the draw backs of *Graphite Mederator*.

(*i*) It breaks easily when struck.

(*ii*) It combines with O_2 and CO_2 at high temperature.

4. Coolants: The reactor in its operation produces a large amount of heat energy. If there is no cooling arrangement, all the atoms of the fissile material would have a very high energy or would be extremely hot. Hence cooling is neccessary and this is done by various cooling agents as air blast or water or some other high boiling point liquid.

Both ordinary and heavy water are good coolant. However due to their low boiling point, they need pressurisation. Moreover, water becomes very corrosive at high temperature.

Liquids such as molten sodium or potassium are excellent collants for use at high temperature. However they are highly reactive with O_2 and H_2 and become radioactive due to neutron capture. Hence, it needs special shielding. It is usually used in fast reactors.

For low power reactor, air cooling with the help of fans is very simple. At high temperature air is not satisfactory as a coolant because simple. At high temperature air is not satisfactory as a coolant because it reacts chemically with many materials like graphite, aluminium, beryllium and magnesium etc. Used in the construction of the reactor.

The most widely-used gaseous coolant is CO_2 particularly in large power reactors. It is cheap, does not attack metals at reasonable temperatures and has small cross-section for neutron capture.

5. Measuring Instrument: This instrument is required for the purpose of measuring thermal neutron flux which determines the power developed by the reactor.

6. Shielding: The radioactive radiations emanating from the inside of the nuclear reactor being fatal, a concrete several feel thick is needed for protecting the operators.

There are two types of shielding

(*i*) Thermal Shielding. It is very close to the reactor core and consists of a thick iron or steel covering. By absorbing most of the g-rays, it protects the biological shield from over-heating.

(*ii*) Biological Shield. It is a layer of concrete a few decimetres thick and surrounds the thermal shield and the reactor core. Its function is to absorb the a-rays and neutrons coming out from the inner-shield.

FAST BREEDER REACTOR

It is a type of reactor in which fast fission neutrons are used to produce more fissible fuel than the amount that is consumed. No moderator is required for slowing down the neutrons.

In this type of reactor, a small core of natural uranium is struck by fast neutrons so that a fairly fast reaction is established. Neutrons are then allowed to escape into a surrounding blanket of natural uranium where some of them are captured by U-238, so that Pu^{239} is formed. By suitable control, more Pu^{239} can be formed in the natural uranium blanket than the amount of U^{235} consumed in the reaction. It is for this reason that this type of reactor is called a Breeder Reactor.

16.23. NUCLEAR POWER STATION

One of the most important application of nuclear fission is in **Nuclear Power Station**.

Nuclear power station is based on the principle of convertion of nuclear energy into electrical energy. A huge amount of heat energy is released during nuclear fission. The heat energy produced is utilised in raising steam which runs the steam turbine. Electric generator is connected to the turbine, the latter acting as a prime mover.

Operation: Fission material such as U-235 or P-239 is subjected to nuclear fission in an atomic furnace or pile or reactor. An atomic furnace is no arrangement for controlling the chain reaction that starts once nuclear fission is done. If the chain reaction is not controlled, the result will be an explosion due to fast increase in energy released. As a result of fission enormous heat is produced. The heat produced is given to the working substance (sodium metal) which receives and carries out the heat to the heat exchanger as shown in Fig. 16.9. In the heat exchanger, heat of the working substance is used to produce steam which runs the steam turbine. The turbine is connected to an electric generator. Thus electric power is obtained from the generator.

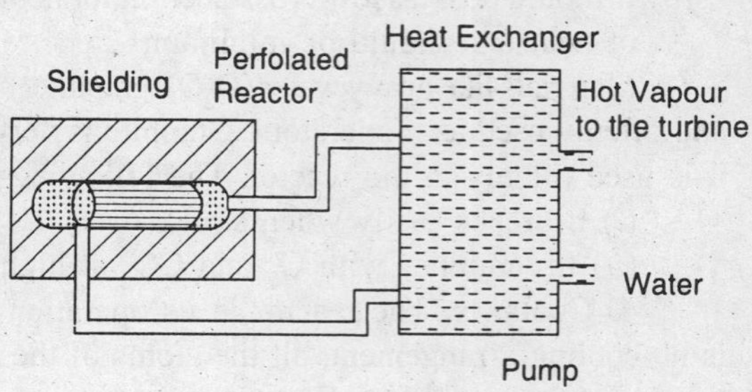

Fig. 16.9

16.24. CONTROLLED FUSION

In order to use fusion reaction for peaceful use, it is not economical to use *tritium* as nuclear fuel as its cost is very high (About two million dollars/kg) and also it is a very rare isotope. The most practical method is to short the reaction with deuterium with a small amount of Lithium Beryllium. Though a sequence of steps called DTCycle, the nuclei are made of fuse. *Helium and Tritium* with the release of about 10^6 Mev energy/deuterium nucleus consumed.

The temperature required to trigger fusion reaction is about 10^8 K, *i.e.*, about 10 times the temperature of the central part of the sun. Such temperatures have been got in *plasmas*, but the biggest difficulty to the practical exploitation of controlled nuclear fusion lies in the confinement of such a plasma for a long enough time for the fusion reaction to proceed. The following two schemes are under active consideration:

(*i*) **Magnetic Confinement Scheme:** This scheme attemps to hold the plasma suspended in space with the help of a clever configuration of magnetic fields.

(*ii*) **Inertial Confinement Scheme:** This scheme attempts to vaporize small pellets of fuel by means of intense laser beams and to trigger fusion very suddenly, so that the pellets react before it has time to disposess.

Both the above schemes are in advanced experimental stage. The experiments have shown that fusion reactions can be initiated but it has not been possible to extract the useful amounts of energy.

Fusion reactors have two very important advantages over fission reactors as sources of mechanical and electrical power.

(*i*) The fuel of fusion reactors (Deuterium and Lithium is much more abundent on earth than that for fission reactors (Uranium or Thorium).

(*ii*) Fusion Reactors do not produce large amounts of radioactive residues which are the usual by products of the operation of fission reactors.

16.25. NUCLEAR WASTE DISPOSAL

In addition to the fact that the nuclear reactor has to be designed to have the maximum safety, we have to think about the methods for disposal of nuclear wastes produced by reactors.

Even if we apply methods to process the old fuel rods to separate out the precious *Uranium* and *Plutonium they contain*, still much of the product left is *highly radioactive*. It may be noted that though activity is reduced with years yet, some of the nuclides have half lives in *million of years* and it is very important to dispose of the nuclear wastes with lot of precautions to avoid the effect of radiations on the *living beings* of this planet. Just for the knowledge sake, it is estimated that around 12000 tons of nuclear wastes are being stored in U.S.A. alone. And if we consider the other countries as well, which are actively involved in nuclear tests/nuclear reactors then we can very well imagine the gravity of the situation.

The latest approach being adopted for disposal of nuclear wastes is to incorporate nuclear wastes in a *boro silicate* glass, which is similar to *pyrex*. The liquid wastes are converted into solid and added to normal ingredients for glass in the ratio of 1:3, the mixture is then heated till it melts and cost into blocks for deep *burial* into Earth. It may be noted that right location is easy to specify but very difficult to locate. As we require that place which is stable geologically with no likely hood of *earth quakes*, no nearby population centers, a type of rock which does not disintegrate with heat and radiation but is soft enough for drilling etc. Also it should have almost no ground water which may be contaminated with radiations etc. the present thought into the minds of most of the scientists in the world is durability of boro-silicate glass for many years to come. In order to have more safely an alternative process is being studied which yields a tough rocklike *ceramic*. However, it may be noted that effectiveness of all these disposal processes can be judged only with the passage of time and nuclear *waste disposal* is a very bigh challenge, for scientists associated with nuclear projects.

16.26. LASER INDUCED NUCLEAR FUSION

One of the basic difficulties in initiating a nuclear fusion reaction is that *very high temperature* is required to make the fusion reaction proceed. This is due to the fact that unless nuclei have very high kinetic energies, *Coulomb repulsion* will not allow them to come sufficiently close for *fusion reaction to occur*. Temperature required are ~ 100 millian ^0K [$10^{8}\ {}^0K$] and at such a high temperature the matter is fuly ionized and its confinement is not very easy. *The matter in fully ionized state is called plasma*. Two major problems in thermonuclear fusion are thus.

(1) Heating of plasma to a very high temperature.

(2) Confinement of plasma for sufficient long time for optimum number of fusion reactions to take place.

Taking the example of *Deutermium Tritium* (*D* – *T*) reaction *i.e.*,

$$_1H^2 + {}_1H^2 \longrightarrow {}_1H^3 + {}_1H^1$$

at 10 Kev which corresponds to roughly temperature ~ $10^{8}\ {}^0K$ for fusion output, energy to exceed the input energy required to heat plasma, we have a relation

$$\eta Z \geq 10^4 \text{ cm}^{-3} \text{ sec.}$$

where $n \longrightarrow$ plasma density

and $Z \longrightarrow$ confinement time

The above equation is called. *Lawson's criteria*. For $n \sim 10^{15}$ ions/cm^3 Z must be ≥ 0.1 sec. Although such a long confinement time has not yet been attained yet Russian device called *Tokamak* using magnetic confinement has come very close to above time to satisfy the equality condition.

With the advent of *Laser pulses* a new idea of fusion has come up. The idea is essentially *compressing, heating and confining* the thermonuclear material by inertial forces which are generated when an intense laser beam interacts with a thermonuclear material which is normally in the form of a solid, pellet. *In such a confinement magnetic field is not necessary.*

In case of laser induced fusion, instead of using $\eta\Sigma$ parameter, it is more useful to use parameter ρR, where ρ is density of fuel and R is the fuel radius.

If f corresponds to the fractional burn up of fuel, then

$$f = \frac{\rho R}{6 + \rho R}$$

where ρ is measured in g/cm^3 and R is measured in cm.

For $f \simeq 0.05$ *i.e.*, 5% of burn up of fuel ρR must be 0.3 gm/cm^2.

It may be understood that higher the value of ρR, greater will be the fractional burn up of the fuel.

Again if mass of *Deutron-Tritium* (D – T) pellet is M gm, then total fusion energy (in Joules) is given by

$$E_{\text{output}} = 4.2 \times 10^{11} \text{ fM (J)} \qquad ...(1)$$

Also
$$M = \frac{4}{3}\pi R^3 \rho = \frac{4\pi}{3} \cdot \frac{1}{\rho^2} \cdot (\rho R)^3 \qquad ...(2)$$

Also, to heat the Deutron-Tritium pellet to temperatures $10^{8}\,{}^0K$ at which fusion reactions will occur with high probability, the laser energy required may be written as

$$E_{\text{laser}} \simeq 4 \times 10^8 \frac{M}{\in}$$

which may written in the form
$$E_{\text{laser}} \simeq 4 \times 10^8 \simeq \frac{4\pi}{3} \frac{1}{\rho^2 \in} (\rho R)^3 \qquad ...(3)$$

where $\in$ represents the fraction of laser energy used for heating the pellet; normally $\in \sim 0.1$. So the yield ratio Y, may be written as

$$Y = \frac{E_{\text{output}}}{E_{\text{laser}}} \simeq \frac{4 \times 10^{11} fM \in}{4 \times 10^8 M} = 10^3 \in f \qquad ...(4)$$

$$\therefore \quad y \simeq 100 \frac{\rho R}{6 + \rho R} \qquad ...(5)$$

where we have taken $\in \simeq 0.1$. Clearly, for $Y > 1$, $\rho R \geq 0.1$ g/cm^2. So for a sizable burn-up and for a reasonable yield we should at least have $\rho R \simeq 0.2$ g/cm^2. Now, for normal (D – T) solid, $\rho \simeq 0.2$ g/cm^3; using Equation (3), we get

$$E_{\text{laser}} = 4 \times 10^8 \times \frac{4\pi}{3} \frac{(0.2)^3}{(0.2)^2 \times 0.1} = 3 \times 10^9 \text{ J} \qquad ...(6)$$

Which is quite large. However from Equation (3), for a given value of ρR, laser energy requirements can be sufficiently decreased (in the range of mega Joules) by increasing the value of ρ, this can be achieved by compression to 10^3 to 10^4 times the normal solid density.

At such high densities, *α–particles* which are produced in the reaction give most of their energy to the unburnt *fuel* before leaving the pellet leading to increased fractional burning of the fuel. For a typical calculation reported by Ribe (1975) for $\rho R = 3 g/cm^2$ corresponding to 30% burn up, and an ignition temp. of 10^4 eV (corresponding to 100 million ^0K), we get from Equation (3)

$$E_{laser} = 4 \times 10^8 \times \frac{4\pi}{3} \times \frac{(3)^3}{(0.2)^2 \times 0.4}$$

$$\simeq 2.5 \times 10^{12} J \quad ...(7)$$

where we have assumed normal *Deuterium-Tritium* (D – T) solid of density 0.2 gm.cm^3 and $\in = 0.4$.

If we take $\rho = 1000$ gm/cm^3, then for same value of ρR, we have

$$E_{laser} = 1.2 \times 10^5 J \quad ...(8)$$

Such high energy laser pulses are expected in the near future. The above calculations pertain to pellet mass of 110μ gm, whose radius before compression is 0.05 cm and compressed radius of the pellet is ~ 0.003 cm. The fusion energy yield will be from Equation (1)

$$E_{output} = 4.2 \times 10^{11} \times 0.3 \times 100 \times 10^{-6} = 14 \times 10^6 J = 14 MJ$$

which corresponds to yield of about 100.

16.27. LASER INDUCED FUSION REACTOR

We can take the example of deuterium – Tritium (D – T) in a solid pellet form for a *laser induced fusion reactor*. The pellet is in the form of a cryogenic solid where particle densities are very high ~ 4×10^{22} cm^{-2} and shine laser light from all directions (shown in Fig. 16.10). In a very short time, the outer surface of the pellet gets heated to a very high temperature and gets into a very hot plasma (Temp. ~ 10^8 ^0K).

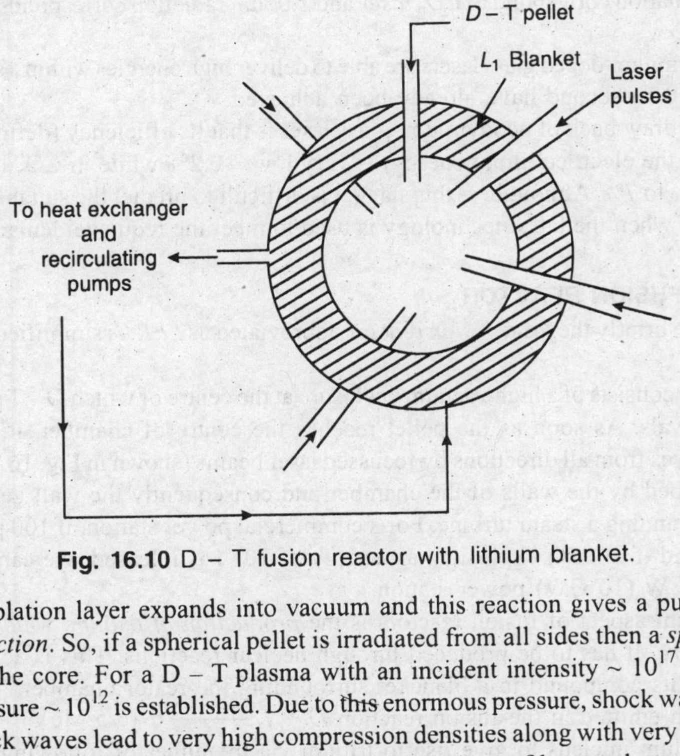

Fig. 16.10 D – T fusion reactor with lithium blanket.

This hot ablation layer expands into vacuum and this reaction gives a push to the pellet in the *opposite direction*. So, if a spherical pellet is irradiated from all sides then a *spherical unplasion* travels towards the core. For a D – T plasma with an incident intensity ~ 10^{17} w/cm^2, an inward atmospheric pressure ~ 10^{12} is established. Due to this enormous pressure, shock waves travel towards centre. Such shock waves lead to very high compression densities along with very high temperatures. In order to get very high compression densities, time variations of laser pulses are adjusted so that the successive shock waves do not meet until they reach the centre of the pellet. It is observed that the time variation of laser power should roughly be of the form $(t_0 - t)^{-2}$ where $t \sim t_0$, is the time when all shock waves reach the centre.

If the pulse duration is about 10 *n*-sec, about $\frac{4}{5}$th of the energy goes to the last *nano-second* and one $\frac{1}{5}$th in the first *nano-second*. A typical ideal profile for maximum compression of D – T pellet is shown in Fig. 16.11.

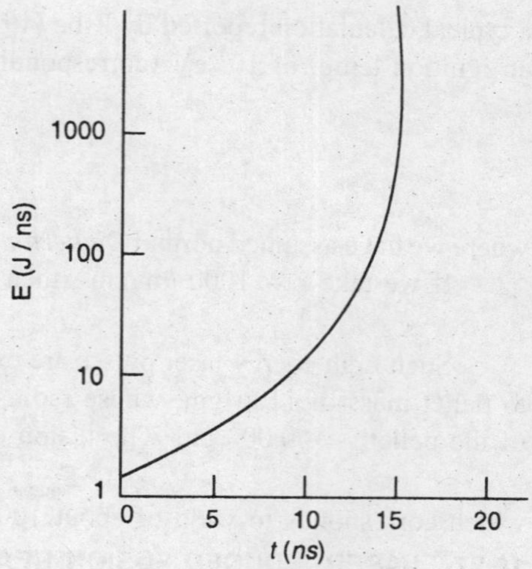

Fig. 16.11

Nuckolls and Co-workers (1972) did experiments on computers for D – T spherical pellet of radius 0.04 cm which was irradiated by a 6×10^4 J pulse of 25*n*-sec duration (λ 1 μm). Compression densities as high as 10^3 g/cm^3 were obtained and about 1.8×10^4 J of fusion energy was released in about 10^{-11} sec after compression. So they observed the result multiplied by a factor 30 of the calculated value.

It may be mentioned here that laser wavelengths ~ 10 μm may be too high for heating the pellet. Detailed calculation show that 1.06μm radiation can heat D – T pellet five times the temperature in $\frac{1}{10}$th time as compared to 10.6μm radiation heating the same pellet. This and other facts suggest the use of 1 – μm radiation for laser induced fusion.

10.6*μm* radiation correspond to CO_2 *laser* and 1.06*μm* radiation corresponds to the *neodymium-doped glass lasers.*

Also neodymium–doped glass lasers are able to deliver high energies within a short time. Energies ~ 200 J in time 0.1*n* second have already been achieved.

One major draw back of neodymium–glass laser is that its efficiency (defined as ratio of laser output energy to the electrical input energy) is very low ~ 0.2% while in CO_2 laser the efficiency is quite high ~ 5% to 7%. Although, at this stage it is difficult to predict the suitable system, however day is not far off when the latest technology is used to meet the requisite demands for an efficient working system.

16.28. LASER FUSION REACTOR

We can also briefly the *laser fusion reactor* abbreviated as *LFR*. A simplified diagram is shown in Fig. 16.12.

The reactor consists of a high vacuum enclosure at the centre of which D – T pellets are dropped are regular intervals. As soon as the pellet reaches the centre of chamber, it gets irradiated by synchronized pulses from all directions by focussed laser beams (shown in Fig. 16.12). Fusion energy released is absorbed by the walls of the chamber and consequently the wall get heated up which may be used for running a steam turbine. For a commercial power station, if 100 pellets are allowed to explode/sec and if in each explosion, an energy of 10^8 J is released, we can expect energy of the order of 10^{10} W (10 GW) power station.

An important aspect of fusion reactors is the *production of tritium, which otherwise is not available in nature.* It has to be produced through neclear reactions. This is achieved by placing either lithium or its compound in a blancket surrounding the reator chamber.

The neutron emitted in the fusion reaction $D + T \longrightarrow \alpha$ (3.5 MeV) + *n* (14.1 MeV) is absorbe by a lithium nucleus to give rise to tritium via the either of following reactions

$$_3Li^6 + {}_0n^1 \longrightarrow T\left[{}_1\text{H}^3\right] + \alpha\left[{}_2\text{He}^4\right] + 4.79 \text{ MeV} \qquad ...(1)$$

$$_3Li^7 + {}_0n^1 \longrightarrow T\left[{}_1\text{H}^3\right] + \alpha\left[{}_2\text{He}^4\right] + {}_0n^1 + 2.46 \text{ MeV} \qquad ...(2)$$

Note that in Equation (2) neutron appears at right side of equation also, and it can again interact with a lithium nucleus to produce tritium.

The *conceptual design* consists of liquid lithium being contained in between two structural shells which encloses the reactor cavity. Liquid lithium also removes the heat from reactor which is used to run steam turbine. The heat exchangers and lithium processing units for separation of tritium are expected to be put adjacent to reactor (Fig. 16.12).

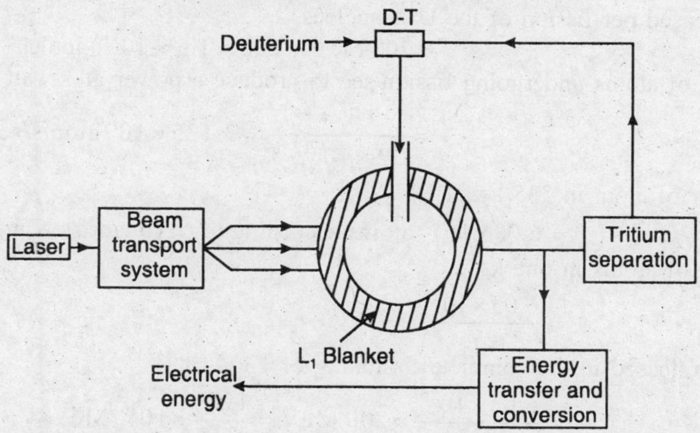

Fig. 16.12 Block diagram of a laser fusion electric generation system.

We can now summarize about practical difficulties associated with laser driven fusion system, like

(*i*) Delivery of a substantial part of laser energy to fuel energy before heating the fuel (which in turn will reduce the compression achieved) and before the shock waves disperses the fuel mass.

(*ii*) Building of high power lasers to a stage when output energy exceeds the input energy.

(*iii*) Design of complex targets with negligible irregularities as these lead to in consistant compressions if irregularities are more than 1 percent in the walls.

Recently some projects are underway to study the feasibility of using particle beams like the *electron or ion beams* to initiate fusion reactions and work is being done in various prestigious labs of the world to get a workable solution.

Example 7. *A reactor is developing energy at the rate of 30,000 kw./sec. How many atoms of U^{235} undergon fission per second? How many kilograms of U^{235} would be used in 1000 hours of operation, assuming that an average energy of 200 MeV is released per fission?*

Solution: Rate of working of reactor = 30,000 kw/sec.

$= 30000 \times 10^3$ w/sec. $= 3 \times 10^7$ J s^{-1}

Energy liberated in breaking one U-atom $= 200$ MeV $= 200 \times 10^6$ ev.

$= 200 \times 10^6 \times 1.6 \times 10^{-19}$ J $= 3.2 \times 10^{-11}$ J

(*i*) ∴ Number of fission per second $= \dfrac{3 \times 10^7}{3.2 \times 10^{-11}} = 9.37 \times 10^{17}$

(*ii*) Number of fission in 1000 hours $= 9.37 \times 10^{17} \times 1000 \times 60 \times 60$

$= 36 \times 9.37 \times 10^{22} = 3.37 \times 10^{24}$

Avogadro's Number $= 6.023 \times 10^{23}$

Atomic weight of U = 235

∴ Number of U-atom in one gram $= \dfrac{6.023 \times 10^{23}}{235}$

$\dfrac{6.023 \times 10^{23}}{235}$ Number of U-atoms = 1 gm. of U

∴ 3.3×10^{24} number of U-atom $= \dfrac{3.37 \times 10^{24}}{6.623 \times 0.01 \times 10^{23}} = 1313$ g $=$ **1.313 kg**

Example 8. *Calculate (i) the fission rate of U^{235} for producing a power of one watt, and (ii) the energy released in the complete fissioning of 1 kg of U^{235}. Assuming 200 MeV are released per fission of U^{235}*

Solution: (*i*) Power developed = 1 watt = 1 Joule/sec (JS^{-1})

Energy released per fission of the U^{235} nucleus

$$= 200 \text{ MeV} = 200 \times 1.6 \times 10^{-13} \text{ Joule} = 32 \times 10^{-12} \text{ Joule}$$

∴ Number of atoms undergoing fission/sec to produce a power at 1 watt

$$= \frac{1 \text{ J. } s^{-1}}{32 \times 10^{-12} \text{ J}} = 3.125 \times 10^{10} \text{atoms/sec.}$$

(*ii*) Number of atom in 235 kg of U^{235}

$$= 6.03 \times 10^{26} \text{ atoms according to (Avogadro's Number)}$$

∴ Number of atoms in one kg

$$= \frac{6.03 \times 10^{26}}{235}$$

∴ Energy released in the complete fissioning of 1 kg. of U^{235}

$$= \frac{6.03 \times 10^{26}}{235} \times 200 \text{ MeV} = \frac{1206}{235} \times 10^{26} \text{ MeV}$$

$$= \frac{1206 \times 1.6}{235} \times 10^{13} \text{ Joules}$$

$$= \mathbf{8.23 \times 10^{12} \text{ Joules}}$$

Example 9. *Assuming that the energy released by the fission of one uranium atom is 200 MeV. Calculate the number of fissions per second required to produce one watt power.*

Solution: Energy liberated in breaking one uranium atom

$$= 200 \text{ MeV} = 200 \times 10^6 \text{ ev.} = 200 \times 10^6 \times 1.6 \times 10^{-19} \text{Joules}$$

$$= 3.2 \times 10^{-11} \text{Joules}$$

Now 1 watt = 1 Joule/sec.

$$\therefore \quad \text{Number of fission/sec} = \frac{1}{3.2 \times 10^{-11}} = \mathbf{3.11 \times 10^{10} \text{ fissions.}}$$

Example 10. *A railway engine develops an average power of 1200 kW during a 10 hour run from one station to another. If the engine is driven by an atomic power plant of 20% efficiency. how much U^{235} would be consumed on the run? Each U^{235} atom on fission releases 180 MeV of energy.*

Solution: We know that 1 watt = 1 Joule/sec ($J.S^{-1}$)

Energy released per sission = 180 MeV

$$= 180 \times 16 \times 10^{-13} \text{ Joules} = 288 \times 10^{-12} \text{ Joules}$$

$$\text{Efficiency} = \frac{\text{Electrical output}}{\text{Nuclear energy input}}$$

$$\therefore \quad \text{Nuclear energy input} = \frac{1200}{0.20} \times 10^3 = 6000 \times 10^3 \text{ watt} = 6000 \text{ K watt}$$

∴ Number of atoms of U^{235} required for the run

$$= \frac{6000 \times 10^3}{288 \times 10^{-13}} = 20.833 \times 10^{16} \text{ atoms}$$

$$\text{Mass of } U^{235} \text{ consumed} = \frac{235 \times 20.833 \times 10^{16}}{6.03 \times 10^{26}} = \mathbf{8.10 \times 10^{-6} \text{ kgm.}}$$

Example 11. *Using the information on atomic masses as given below, a nucleus of uranium disintegrate with emission of α-particle according to the relation*

$$_{92}U^{238} \longrightarrow {}_{90}Th^{234} + {}_2He^4$$

Calculate: (i) The total energy released in the disintegration.

(ii) The K.E. of α-particle, the nucleus being at rest before disintegration.

Mass of U^{238} = 238.12492 a.m.u.; Mass of Th^{234} = 234.11650 a.m.u.; Mass of He^4 = 4.00387 a.m.u.

Solution: (*i*) 1 a.m.u. is equivalent to 931 MeV

$$\text{Mass of reactant} = 238.12492$$

$$\text{Total mass of products} = 234.11650$$

$$\frac{4.000387}{= 238.12037}$$

$$\text{Mass Defect} = 238.12492$$

$$\frac{-238.12037}{= 0.00455 \text{ a.m.u.}}$$

∴ Total energy released due to mass defect of

$$0.00455 \text{ a.m.u.} = 931 \times 0.00455 \text{ MeV} = \mathbf{4.23 \text{ MeV.}}$$

(*ii*) Because the nucleus was rest before disintegration, the energy released is shared by α-particle and the recoil nucleus in accordance with the laws of "conservation of energy and momentum"

$$\therefore \quad \frac{1}{2}m_1v_1^2 + \frac{1}{2}m_2v_2^2 = 4.23 \text{ MeV} \qquad ...(i)$$

where m_1 and v_1 are the mass and velocity of **thorium–234** m_2 and v_2 are the mass and velocity of Helium–4.

According to conservation of Momentum

$$\therefore \quad \frac{1}{2}m_1 v_1^2 + \frac{1}{2}m_2 v_2^2 = 4.23 \text{ MeV}$$

where m_1 and v_1 are the mass and velocity of **thorium–234** m_2 and v_2 are the mass and velocity of Helium–4.

$$m_1 v_1 = -m_2 v_2 \qquad ...(ii)$$

$$v_1 = -\frac{m_2}{m_1}v_2$$

Substituting v_1 in Equation (*i*)

$$\frac{1}{2}\frac{m_2^2}{m_1^2}\cdot m_1 \cdot v_2^2 + \frac{1}{2}m_2 v_2^2 = 4.43; \qquad \frac{1}{2}m_2v_2^2\left[\frac{m_2}{m_1}+1\right] = 4.23$$

$$\therefore \quad \text{K.E. of the particle } \frac{1}{2}m_2v_2^2 = 4.23 \times \frac{m_1}{m_1+m_2} = 4.23 \times \frac{234.11650}{238.12037}$$

$$\therefore \quad \text{K.E.} = \mathbf{4.16 \text{ MeV.}}$$

QUESTIONS AND PROBLEMS

1. State cleaning what do you understanding by the term (*i*) binding energy (*ii*) Mass defect with respect to the nucleus.
2. What do you understand by nuclear fission and fission energy?
3. What do you mean by nuclear fusion? Can the process be used for controlled generation of electrical power?
4. Explain the Carbon-Nitrogen cycles and proton-proton cycle in connection with the solar energy.
5. Explain briefly the liquid drop model of the nucleus.
6. What is explosive chain reaction and critical size? Explain critical mass. What will happen:
   (*a*) when the size of the sample is greater than the critical size.
   (*b*) when the size of the sample is less than the critical size.
7. What is a chain reaction in nuclear fission? Explain multiplying factor or reproduction factor of a chain K < 1, K = 1 and K > 1.
8. Write brief essay on *Atom Bomb*.

9. Derive the four factor formula for the reproduction of neutrons

$$K_{\infty} = \varepsilon\, p\, fn$$

with the usual notations. State clearly what the symbols signify.
Briefly discuss how the neutron leakage affects the four factor formula?

10. Explain the term power of a Nuclear Reactor and show that power of nuclear reactor

$$P = 3.2 \times 10^{-11}\, \phi\, \sigma\, NV \text{ watt.}$$

11. Compute the energy of one electron in MeV. **[Ans.** 0.51 MeV]

12. What is nuclear fission? Explain briefly the working of a nuclear reactor bringing out clearly the role of a (*a*) moderator (*b*) the coolant (*c*) the shield in the working of nuclear reactor.

13. Bombay requires 3000 M Wh. of electric energy per day. It is to be supplied by a reactor which converts nuclear energy into electric energy with an efficiency of 20 percent. If reactor uses nuclear fuel of U^{235}; calculate the mass of U^{235} needed for one day operation. **[Ans.** 0.66 kgm.]

14. Calculate the binding energy in MeV of uranium 338 from the following data:
Mass of $_1H^1$ = 1.008142 a.m.u. Mass of $_0n^1$ = 1.008982 a.m.u.
Mass of $_{92}U^{238}$ = 238 = 238.12493 a.m.u. **[Ans.** 7.3 MeV]

15. Calculate the fission rate of U^{235} for producing a power of one watt and (*ii*) the energy released in the complete fissioning of 1 kg of U^{235}. Assuming 200 MeV are released per fission of U^{235}.
[Ans. (*i*) 3×10^{10} fission/sec. (*ii*) 8.2×10^{12} Joules]

16. Calculate the binding energy of an α-particle and express the result both in MeV and Joules. Given: mass of α-particle = 4.0028 a.m.u. ; mass of proton = 1.00758 a.m.u.; mass of neutron = 1.00897 a.m.u.
[Ans. 28.2 MeV and 5.51 × 10 Joules].

17. Calculate the energy set free in the mutual annihilation of an electron and a position, if the whole of the energy is given out as a single quantum of light, calculate its wave-length. Given,
mass of electron = 0.00055 a.m.u.; mass of proton = 1.00758 a.m.u.; 1 a.m.u. = 1.66×10^{-27} kg ; h = 6.6×10^{-34} J.s. **[Ans.** 1.02 Mev]

18. A nuclear reactor is developing a power of Mw. How many atoms of U^{235} undergo fission per second? How many kg of U^{235} would be used in 1000 hours of operation assuming that 200 MeV is released per fission. **[Ans.** 0.132 kg.]

CHAPTER

17

ELECTRON CHARGE AND MASS SPECTROGRAPHS

17.1. DETERMINATION OF THE CHARGE OF AN ELECTRON: MILLIKAN'S OIL DROP METHOD

The earlier experiments for determination of electronic charge were carried out by J.J. Thomson, Townsend and H.A. Wilson towards the end of the 19th century. But the most precise determination of this charge was made by Robert A, Millikan, whose method is explained here:

Principle

The method is based upon the observation of the rate of motion of a **small oil drop**, under free fall due to gravity and under the application of suitable electric field. By adjusting the electric field suitably, a given oil drop can be made to move up or down or even kept balanced in the field of view for sufficiently long time and a series of observations can be made.

Construction

This method was devised by American physicist R.A. Millikan in 1917 and is an accurate method for measuring electronic charge. The experiment is shown diagrammatically in Fig. 17.1 (*a*). *A* and *B* are two optically plane metallic discs of about 20 cm in diameter and 1.76 cm apart. These discs are held together by insulating rods of glass or ebonite, so that they are perfectly parallel to each other. The discs are placed in the metallic chamber, provided with three windows W_1, W_2, and W_3. Any desired potential difference can be established between the plates *A* and *B* by storage batteries and the potential difference can be varied by potentiometric arrangement. The upper plate A is provided with a tiny hole, through which minute droplets of a heavy non-volatile oil can be introduced between two plates by means of a spray atomiser *D*. The droplets descend slowly and occasionally few of them enter inside the plates through the hole in upper plate. These droplets are charged by friction involved in the spraying process. The space between the plates is illuminated by an intense beam of light. The droplets between the plates are viewed by a telescope *T* provided with a scale in the eye-piece. The droplets appear as brilliant spots on a dark background.

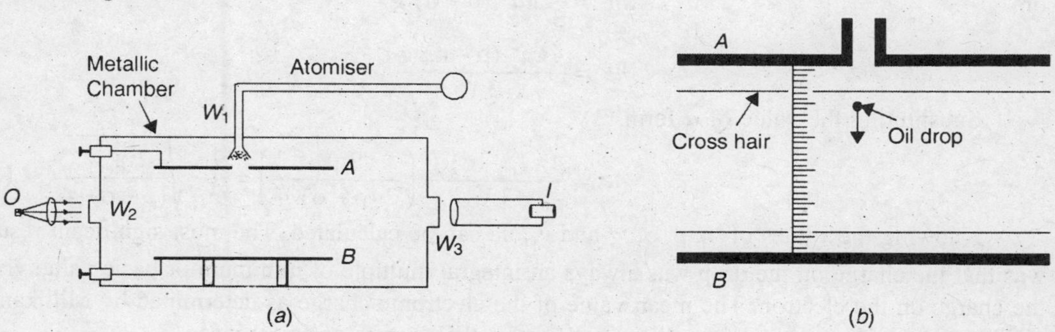

Fig. 17.1

Experimental Arrangement

(*i*) Initially, both plates A and B are earth-connected so-that the electric field is zero. The downward motion of any drop is noted and the terminal velocity v_1 is determined.

Let the radius of the drop = a, density of the oil = ρ, density of air = σ, acceleration to gravity = g. Volume of the drop = $4/3\pi a^3$, Mass of the drop = $(4/3)\ \pi a^3\ \rho$ and weight of the drop = $(4/3)\ \pi a^3\ \rho g$.

Upward thrust due to the air $= \frac{4}{3}\ \pi a^3 \sigma g.$

So resultant weight of the drop $= \frac{4}{3}\ \pi a^3 (\rho - \sigma)\ g.$

Let terminal velocity, in the absence of the electric field = v_1.

Coefficient of viscosity of air = η.

$\therefore$ Upward force on the drop, according to Stoke's formula

$$= 6\pi\eta\, a v_1 \qquad ...(2)$$

From (1) and (2), $\frac{4}{3}\ \pi a^3\ (\rho - \sigma)\ g = 6\pi\eta\, a v_1$

$$a = \sqrt{\frac{9\eta\, v_1}{2\,(\rho - \sigma)\ g}} \qquad ...(3)$$

Knowing the values of ρ, σ, g, η and v_1, a the radius of the drop can be calculated.

(*ii*) Now the electric field is applied between the plates, so that the plate A is positive and the plate B is negative. Consequently, an ion carrying negative charge will experience a force upwards and if this force is greater than the downward force due to the weight of the drop, the drop begins to move up and soon will attain the terminal velocity. Let the terminal velocity upwards under the influence of the electric field be v_2.

Then the resultant upward force $= X\ (ne) - \frac{4}{3}\ \pi a^3\ (\rho - \sigma)\ g,$ where X is the intensity of the electric field and n is the number of electrons on the drop. According to Stoke's formula, this force $= 6\pi\eta\, a v^2$.

$$\therefore \qquad Xne - \frac{4}{3}\ \pi a^3\ (\rho - \sigma)\, g\ = 6\pi na\, v^2 \qquad ...(4)$$

If the strength of the electric field is adjusted to neutralise the downward force of gravity, then the drop remains stationary and can be observed in the field of view for a long time. This stationary drop also helps to observe the constancy of the size of the drop.

$$\therefore \qquad v^2 = 0$$

From Equation (4), $Xne - \frac{4}{3}\ \pi a^3\ (\rho - \sigma)\, g = 0$

or

$$Xne = \frac{4}{3}\ \pi a^3\ (\rho - \sigma)\ g$$

$$\therefore \qquad ne = \frac{4\pi a^3\ (\rho - \sigma)\ g}{3X}$$

Substituting the value of a form (3)

$$ne = \frac{4\pi\ (\rho - \sigma)\ g}{3X}\left(\frac{9\eta v_1}{2\ (\rho - \sigma)\ g}\right) = \frac{9\pi}{X}\sqrt{\frac{2\eta^3\ v_1^3}{(\rho - \sigma)\ g}} \qquad ...(5)$$

Knowing the values of η, ρ, σ, g and v_1, ne can be calculated. The most significant result was that the charge on the drop was always an integral multiple of minimum or basic value viz., the charge on the electron. The mean value of the electronic charge as determined by Millikan's oil drop method (after applying all corrections for the sources of error) was

$$e = (1.601 \pm 0.0002) \times 10^{-19}\ \text{C}$$

In determining the value of e, the coefficient of viscosity of air has to be accurately determined. As the size of the drop is very small and comparable with the mean free path of air$_2$ correction has to be applied to Stoke's formula. The expression $6\pi\eta a^3 v$ after the correction becomes $(6\pi\eta av)/(1 + b/pa)$, *where b is a constant of value 0.000617, and p is the pressure.*

Importance of Millikan's Method

(*i*) It is a direct and a prices method for the determination of electronic charge, yielding fairly accurate results.

(*ii*) It proves that an electron carries a definite amount of charge and that the charge on an ion, positive or negative, is numerically equal to that on an electron. It establishes that electricity is *atomic in nature.*

(*iii*) *Avogadro's Number.* The knowledge of the charge on an electron helps to determine Avogadro's number which is the number of molecules contained in one 'gram-molecule' of an element. Gram-atom or gram-molecule of an element is its weight in gram numerically equal to its atomic weight or molecular weight respectively.

Let us calculate the value of Avogadro's number, using data from electrolysis. As we know, the current in an electrolyte results from the movement of positive and negative ions towards their respective electrodes. It is found that a charge equal to 96490 C when passed through a solution like acidulated water is able to liberate a mass of hydrogen equal to its gram-atom, *i.e.*, 1.008 gram. The charge carried by 1 gram-atom of hydrogen is, therefore, 96490 C. The charge carried by 1 gram-atom of hydrogen is, therefore, 96490 C. The charge on a hydrogen ion is the same as that on an electron *i.e.*, 1.6020×10^{-19} C. Therefore the number of ions which carry 96490 C of charge or the number of atoms contained in one gram atom of hydrogen is given by

$$N = \frac{96490}{1.6020 \times 10^{-19}} = 6.03 \times 10^{23}$$

which fairly agrees with the values obtained by other methods.

(*iv*) *Mass of electron.* Though no direct method is yet available to determine the mass of the electron, it may easily be deduced from the value of e/m and e for electrons. As already derived, $e/m = 1.7592 \times 10^{11}$ C kg^{-1} and $e = 1.6020 \times 10^{-19}$ C.

$$\therefore \qquad m = \frac{e}{e/m} = \frac{1.6020 \times 10^{-19}}{1.7592 \times 10^{11}} = 9.1066 \times 10^{-31} \text{ kg.}$$

Let us compare the mass of the electron with that of the hydrogen atom. As explained above, 1.008 gram of hydrogen contain N atoms where $N = 6.8 \times 10^{23}$. Therefore, the mass of the hydrogen atom

$$M = \frac{1.008}{6.03 \times 10^{23}} = 16.7 \times 10^{-24} \text{ g} = 1.67 \times 10^{-27} \text{ kg}$$

The ratio of the mass of the hydrogen atom to that of an electron is given by

$$\frac{M}{m} = \frac{1.67 \times 10^{-27}}{9.1066 \times 10^{-31}} = 1836$$

which shows that an electron is 1836 times lighter than the lightest atom.

Example 1. *A charged oil drop is prevented from falling under gravity by a vertical electric field between two horizontal metal plates charged to a potential difference 6920 V, the distance between the plates being 1.3 cm. When the field is cut-off, the drop falls in air with a uniform of 1.9×10^{-4} m/s. Calculate (a) the radius of the drop and (b) charge on the drop.*

Density of oil = 0.9×10^3 kg/m^3; coefficient of viscosity of air = 1.81×10^{-5} N-s/m^2. The density of air may be neglected in comparison with that of oil.

Solution: (*a*) We know that $a = \left[\dfrac{9\eta v_1}{2g(\rho - \sigma)}\right]^{1/2} \approx \left[\dfrac{9\eta v_1}{2g\rho}\right]^{1/2}$ $[\because \sigma << \rho]$

Substituting the given values $a = \left[\dfrac{9 \times 1.81 \times 10^{-5} \times 1.0 \times 10^{4}}{9 \times 9.80 \times 0.9 \times 10^{3}}\right]^{1/2}$

or $a = 1.75 \times 10^{-3}$ m

(*b*) $qE = \dfrac{4}{3}\pi a^3 (\rho - \sigma) g \approx \dfrac{4}{3}\pi a^3 \rho g$

Here $E = \dfrac{6920}{1.2 \times 10^{-3}}$ volt/m

$\therefore$ $q \times \dfrac{6920}{1.1 \times 10^{-3}} = \dfrac{4}{3}\pi (1.75 \times 10^{-6})^3 \times 0.9 \times 9.81$

Solving we get $q = 4.142 \times 10^{-19}$ coulomb $\approx \mathbf{3 \times 1.6 \times 10^{-19}}$ **C.**

Obviously, the drop has three electrons charge attached to it.

Example 2. *In Millikan's oil drop method an oil drop was prevented from falling or rising by applying a p.d. of 5750 V between parallel plates 1.5 cm apart. Assuming that this drop carried one electron charge, what was the radius of the drop? Density of oil = 0.92 g/cm*^{3}.

Solution: Here the weight of the drop

$$= \frac{4}{3}\pi a^3 \rho g \quad \text{and} \quad \frac{4}{3} r a^3 \rho g = Xc$$

But $X = \dfrac{V}{d}$ $\quad \therefore \quad \dfrac{4}{3}\pi a^3 \rho g = \left(\dfrac{V}{d}\right) e$

Given $V = 5750$ volts

$e = 1.6 \times 10^{-19}$ coulomb

$\rho = 0.92$ g/cm^3 = 0.92×10^3 kg/m^3

$g = 9.80$ m/s^2

$d = 1.5$ cm = 0.015 m

$$a = \left[\frac{3 \times 5750 \times 1.6 \times 10^{-19}}{4 \times 3.142 \times 0.92 \times 10^3 \times 9.8 \times 0.015}\right]^{1/2}$$

$a = \mathbf{1.176 \times 10^{-6}}$ **m.**

Example 3. *In Millikan's oil drop experiment, an oil drop of mass 4.9 × 10*$^{-14}$ *kg is balanced by applying a p.d. of 2 kV between the two plates which are 8 mm apart. Calculate the number of elementary charges on the drop. Take g as 10 m/s*^{2}.

Solution: Here $V = 2$ kV $= 2 \times 10^2$ V

$d = 8$ mm $= 8 \times 10^{-3}$ m

$m = 4.9 \times 10^{-14}$ kg

$$X = \frac{V}{d} = \frac{2 \times 10^3}{8 \times 10^{-3}} = 2.5 \times 10^5 \text{ V/m}$$

$q = ne$

$\therefore$ $Xq = mg$

$X\,ne = mg$

$$n = \frac{mg}{Xe} = \frac{4.9 \times 10^{-14} \times 9.8}{2.5 \times 10^5 \times 1.6 \times 10^{-19}}; \quad \mathbf{n = 12.}$$

Example 4. *In Millikan's oil drop experiment on oil drop was prevented from falling or rising by applying a p.d. of 5750 volts between the parallel horizontal plates 1.5 cm apart. Assuming that the drop carried one electron charge (1.6 × 10*$^{-19}$ *C), what was the radius of the drop? Density of oil = 0.92 gm cm*$^{-2}$.

Solution: When the drop is suspended in the electric field E.

We have $eE = mg = \dfrac{4}{3}\pi a^3 \rho g$

where a is the radius of the drop and ρ, the density of oil,

$$\therefore \quad a^3 = \frac{3eE}{4\pi\rho g}$$

or

$$a = \left[\frac{3eE}{4\pi\rho g}\right]^{1/2} \quad ...(i)$$

According to the problem,

$$e = 1.6 \times 10^{-19}\ \text{C} = 1.6 \times 10^{-19} \times 3 \times 10^9\ \text{e.s.u.}$$

$$\rho = 0.92\ \text{cm}^{-3};\ g = 980\ \text{cm/sec}^2$$

$$E = \frac{\text{Potential difference across plate}}{\text{Distance between plates}} = \frac{5750\ \text{volts}}{1.5\ \text{cm}} = \frac{5750}{300 \times 1.5}\ \text{e.s.u.}$$

Substituting these values in Equation (i), we get

$$a = \left[\frac{3 \times (1.6 \times 10^{-19} \times 3 \times 10^9)\ (5750/300 \times 1.5)}{4 \times 3.14 \times (0.92) \times 980)}\right]^{1/2}$$

which will give $\quad a = \mathbf{1.18 \times 10^{-4}\ cm.}$

17.2. DETERMINATION OF E/M (SPECIFIC CHARGE OF ELECTRONS) M J.J. THOMSON'S METHOD

After it was known in 1895 that cathode rays consisted of negatively charged particles, J.J. Thomson undertook their quantitative study. He was the first to determine experimentally, their velocity and *charge-to-mass (e/m) ratio*. On the basis of his experiments, he concluded that the *cathode ray particles ejected from the atom, were the lightest particles and were alike.* They were named as **electrons.**

Principle. The method is based upon the principle that an electron beam while passing through an electric or magnetic field obeys the same laws at those observed by a current-carrying-conductor.

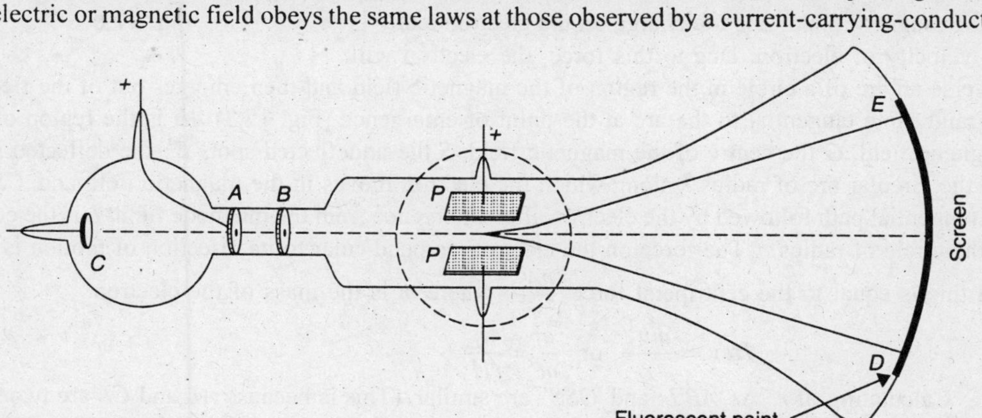

Fig. 17.2

Experimental Arrangement. The apparatus as used by Thomson is shown in Fig. 17.2. It consists of a highly evacuated tube with a circular disc C to serve as cathode soldered at one end. The electrode A is a cylindrical anode with a small hole pierced along its axis and is maintained at a high positive potential. Another cylindrical metal plug B with a central hole is placed by the side of A. When a gas discharge is produced between the cathodes and the anode by means of some source of high potential, the cathode rays (electrons) are emitted by C, most of which hit the anode A. Some of the electrons are however, able to pass through A. This electron beam is further restricted while passing through B. This narrow beam of cathode rays then passes through a region between the two deflection plates P and P' and then strikes the fluorescent screen at the other end of the tube to make itself visible as a greenish blue luminous spot.

The two metal plates, P and P', placed on either side of the electron beam are connected to the two terminals of a high tension battery, with the plate P connected to the positive terminals. Thus an electrostatic field is produced between the two plates *in the plane of the paper* perpendicular to the electronic beam. A strong magnetic field uniform over the dotted area is also produced by passing electric current through two coils (not shown), each placed on either side of the tube. The position of the coils and the direction of the current are so adjusted that the magnetic field is *perpendicular to the plane of the paper*. Both the electric and magnetic fields are arranged to act approximately in the same region.

Theory of Experiment

If a uniform magnetic field is applied to the plane of paper and inwards the beam of electron is deflected in the plane of paper and $\perp$ to the direction of its motion. The dotted region shows the effective space of magnetic field. The direction of deflection will be downward at S'. We can determine the direction of deflection of electrons under the effect of magnetic field using Fleming's left hand rule.

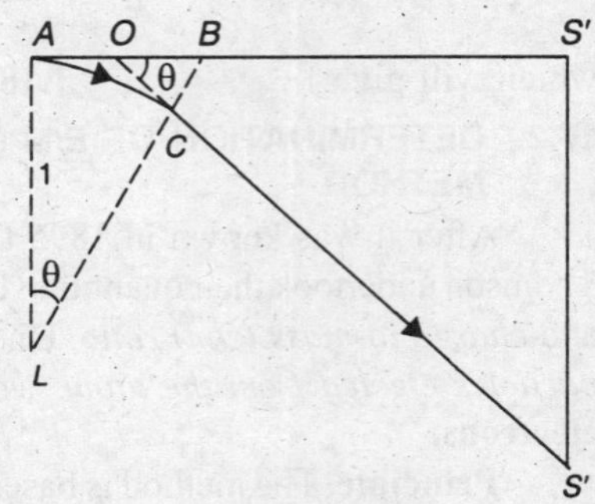

Fig. 17.3

.If an electrostatic field is applied between plates P and Q then the deflection of electron beam will be upward and fluorescent spot is got at S''. So the electric and magnetic field which are $\perp$ to each other deflect the electron beam in opposite directions.

The electric and magnetic field are switched off and the point S *i.e.*, point for undeflected rays is noted. A magnetic field is then applied perpendicular to the direction of motion of the electrons. The electrons experience a force Hev, H is the intensity of the magnetic field, e is the charge on the electron and v is the velocity of electron. Due to this force, the electron will traverse an arc of a circle in the region of the magnetic field and then emerges out of the field in a straight line tangential to the arc at the point of emergence (Fig. 17.3) AB is the region of the magnetic field, O the centre of the magnetic field, S the undeflected spot, S' the deflected spot, AC the circular arc of radius r along which the electron moves in the magnetic field and CS' is the tangential path followed by the electron after emergence from the magnetic field. L is the centre of the circle of radius r. The force on the electron perpendicular to its direction of motion is Hev and this is equal to the centripetal force $\frac{mv^2}{r}$ where m is the mass of the electron.

$$Hev = \frac{mv^2}{r} \quad \text{or} \quad \frac{e}{m} = \frac{v}{rH} \qquad \text{...}(i)$$

Calculating of r. Δs $AB'L$ and OSS' are similar. (This is because AL and CL are two radii and AS and OS' are the two tangents at A and C respectively. Therefore $\angle ALC = \theta = \angle SOS'$, and $\angle BAL = 90° = \angle OS'S$).

$$\therefore \quad \frac{AB}{r} = \frac{SS'}{OS} \quad \text{or} \quad r = \frac{AB \times OS}{SS'} \qquad \text{...}(ii)$$

OS is the distance between the centre of the magnetic field and the centre of the screen, BA is the length of the magnetic field and SS' is the observed shift of the fluorescent spot, AB and OS can be determined depending on the geometry of the apparatus.

Determination of v. To determine the velocity of the electron, the electric field is switched on and the strength of the field is suitably adjusted so that the fluorescent spot is obtained at S, *i.e.*, the centre of the screen. This will be possible if the forces experienced by the electron due to the electric and magnetic fields are equal and opposite (both the forces are in

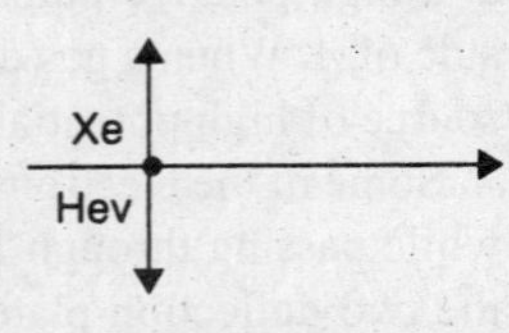

Fig. 17.4

the plane of the paper and perpendicular to the direction of the electrons) (Fig. 17.4). If X is the strength of electric field, then the force experienced by the electron = Xe.

$$\therefore \qquad Hev = Xe \quad \text{or} \quad v = \frac{X}{H}$$

If V is the potential difference between the plates and d the distance between the plates then, $X = \frac{V}{d}$

$$v = \frac{V}{dH} \qquad \text{...(iii)}$$

The intensity of the magnetic field is calculated from the number of turns of the coils, radius of the coils and the strength of the current passing through them. Substituting the values of r [from (*ii*)] and v [from (*iii*)] in equation (*i*), the value of e/m of the electrons can be calculated.

Thomson has shown that the value of e/m of electrons is independent of the nature of the gas, nature of the electrode surface and the variation of potential difference between the anode and the cathode. However, the velocity of the electron beam increases with increase in potential difference because of the increased acceleration. The value of e/m, for cathode rays and electrons produced by thermoelectric and photo-electric emission are the same and given by

$$\frac{e}{m} = 1.7589 \times 10^{11} \text{ coulomb/kg}$$

$\frac{e}{m}$ **is called the specific charge.**

It should be noted that in the method described above, we assume that the magnetic and electric fields operate over a fixed distance between two points and the fields cease to influence the beam beyond these points. This, however, cannot be realised in practice.

17.3. MASS OF AN ELECTRON

The mass of an electron can be calculated from the values of e and $\frac{e}{m}$ determined by the experiments described in articles 3.5 and 3.6. The recent values of these quantities are

$$e = 1.6010 \times 10^{-12} \text{ coulomb} \qquad \text{...(i)}$$

and

$$\frac{e}{m} = 1.7589 \times 10^{11} \text{ coulomb/kg} \qquad \text{...(ii)}$$

Dividing (*i*) and (*ii*)

$$\frac{e}{e/m} = m = \frac{1.6010 \times 10^{-19}}{1.7589 \times 10^{11}} = 9.1023 \times 10^{-31} \text{ kg}$$

The mass of an electron in motion depends on the velocity with which it moves. Let m_0 be the mass of the electron at rest and m the apparent mass when it is moving with the velocity v

When, $$m = m_0 (1 - \beta^2)^{-1/2} = \frac{m_0}{(1 - v^2/c^2)^{1/2}}$$

Here c is the velocity of light. Thus, with increase in the value of v, the apparent mass of the electron increases. The increase in the value of m is more predominant if v is comparable to c. For low value of v, v/c can be neglected and hence $m = m_0$ (approximately). If the value of $v = c$, then the value of M approaches infinity.

Experiments conducted later with greater precision and keeping in mind the increase in apparent mass with increase in the value of v (comparable to the value of c) show that the value of e/m is the same for cathode rays as well as electrons produced by many other different ways (thermoelectric, photo-electric etc., processes). This shows that the cathode rays are identical with electrons and also that the electrons are constituents of all matter. Spectroscopic studies for the determination of e/m for an electron, based on the above assumption, give the same value.

Example 5. *An electron moving in a horizontal direction with a speed of 5.0 × 10⁷ m/s enters a region where there is a uniform electric field, of 20 volts/cm directed upward in the plane of* *its motion. Find the electrons coordinates referred to the point of entry and the direction of its motion 4×10^{-8} s later.*

Solution: $F = Xe$

$$ma = Xe$$

or $$a = \frac{Xe}{m}$$

The displacement of the electron in the direction of the electric field (say along the y-axis)

$$y = \frac{1}{2}at^2 = \left(\frac{Xe}{m}\right)t^2$$

Here X = 20 volts/cm = 2000 volts/m

$$e = 1.6 \times 10^{-19} \text{ coulomb}$$

$$m = 9.1 \times 10^{-31} \text{ kg}$$

$$t = 4 \times 10^{-8} \text{ s}$$

$$y = \frac{1}{2}\left[\frac{2000 \times 1.6 \times 10^{-19}}{9.1 \times 10^{-31}}\right](4 \times 10^{-8})^2 = \mathbf{0.28\ m} \qquad ...(1)$$

The distance travelled by the electron along the x-axis (in the initial direction)

$$x = 5 \times 10^7 \times 4 \times 10^{-8} = 2.0 \text{ m} \qquad ...(2)$$

So the electron's coordinates after 4×10^{-8} s are

$$x = 20 \text{ m}, \ y = 0.28 \text{ m}$$

The direction of the electron,

$$\tan\theta = \frac{y}{x} = \frac{0.28}{2.0} = 0.14$$

$$\theta = \mathbf{8^\circ \text{ with the horizontal.}}$$

Example 6. *In a cathode ray tube, a potential difference of 3000 volts is maintained between the deflector plates whose separation is 2 cm. A magnetic field of 2.5×10^{-3} weber/m^2 at right angles to the electric field gives no deflection to the electron beam which receives an initial acceleration by a potential difference of 10,000 volts. Calculate e/m of the electron.*

Solution: Here, $$X = \frac{3000}{2/100} = 15 \times 10^4 \text{ volts/m}$$

Magnetic field/unit area $B = 2.5 \times 10^{-3}$ weber/m^2

As the electron beam remains undeviated under the action of X and H,

$$Xe = Bev$$

or $$v = \frac{X}{B} = \frac{15 \times 10^4}{2.5 \times 10^{-3}}$$

or $$v = 6 \times 10^7 \text{ m/s}$$

The electron is initially accelerated through a potential difference of 10,000 volts.

∴ $$\frac{1}{2}mv^2 = eV$$

or $$\frac{e}{m} = \frac{v^2}{2V}$$

In this case, V = 10,000 volts

and $v = 6 \times 10^7$ m/s

∴ $$\frac{e}{m} = \frac{(6 \times 10^7)^2}{2 \times 10{,}000}$$

or $$e/m = \mathbf{1.8 \times 10^{11}\ coulombs/kg.}$$

Example 7. *If a stream of electrons each of the mass m, charge e and velocity 3×10^7 m/s is deflected by 2×10^{-3} m in traversing a distance of 0.1 m through an electric field of 1800 volts/m perpendicular to their path, find e/m in coulombs/kg.*

Solution: We have

Deflection $y = \frac{1}{2} at^2$

$\therefore \quad a = \frac{2y}{t^2}$

But $t = \frac{l}{v}$ and $a = \frac{F}{m} = \frac{Xe}{m}$

$\therefore$ Also $\frac{Xe}{m} = \frac{2y}{t^2}$

or $\frac{Xe}{m} = \frac{2yv^2}{l^2 X}$

$\therefore \quad \frac{e}{m} = \frac{2yv^2}{l^2 X}$

In this case $y = 2 \times 10^{-3}$ m; $v = 3 \times 10^7$ m/s; $l = 0.1$ m; $X = 1800$ volts/m

$\therefore \quad \frac{e}{m} = \frac{2 \times 2 \times 10^{-3} \times (3 \times 10^7)^2}{(0.1)^2 \times 1800} = \mathbf{1.758 \times 10^{12}}$ **C/kg.**

Example 8. *A charged particle having a mass of 1.7 × 10^{-24} g and carrying a charge of 1.6 × 10^{-19} coulomb is released from rest in vacuum in a uniform electric field having strength of 100 volts/m. What will be its velocity after traversing 100 cm? Express its kinetic energy in electron volts.*

Solution: Electric field, $X = 100$ volts/m

Charge, $e = 1.6 \times 10^{-19}$ C

Force $= Xe$

$F = 100 \times 1.6 \times 10^{-19} = 1.6 \times 10^{-17}$ newton

Mass, $m = 1.7 \times 10^{-24}$ g $= 1.7 \times 10^{-27}$ kg

$\therefore$ Acceleration, $a = \frac{1.6 \times 10^{-27}}{1.7 \times 10^{-27}}$ m/s^2 $= 9.412 \times 10^9$ m/s^2

Distance travelled, $S = 1$ m; $u = 0$; $v = ?$

$v^2 - u^2 = 2aS$

$v^2 = 2 \times 9.412 \times 10^9 \times 1$

$v = \mathbf{1.372 \times 10^5}$ **m/s**

Now Kinetic energy $= \frac{1}{2} mv^2$

$= \frac{1}{2} \times 1.7 \times 10^{-27} \times 18.824 \times 10^9 = \mathbf{1.6 \times 10^{-17}}$ **Joule**

Also 1 eV $= 1.6 \times 10^{-19}$ Joule.

$\therefore$ Kinetic energy $= \frac{1.6 \times 10^{-17}}{1.6 \times 10^{-19}} = \mathbf{100}$ **eV.**

17.4. MASS SPECTROGRAPHS

The basic principle of a mass spectrograph is *to get different spectral lines on a photographic plate corresponding to the different masses of the isotopes of an element.* We will discuss the following two mass spectro-graphs which find application in the differentiation of the isotopes of various elements.

17.5. ASTON'S MASS SPECTROGRAPH

Aston's Mass Spectrograph is an apparatus of high accuracy designed by **F.W. Aston of Cambridge University,** and is useful to measure the mass of single atomic ions and is used in the investigation of isotopes.

Principle. Positive rays from the discharge tube are subjected to an electrostatic field in a direction perpendicular to the direction of propagation of the rays. The deflection of the beam in the electrostatic field depends on the velocity of the positive ions and hence they are dispersed because of the different velocities of various ions. This dispersed divergent beam is then subjected to a magnetic field, the direction of the magnetic field being perpendicular to the direction of the electrostatic field such that *the magnetic field produces dispersion and deviation in an opposite direction but in the same plane.* If a photographic plate is held in the direction of the deflected beam, line images are obtained and each line corresponds to a particular value of e/m. For any element that has been excited in the discharge space, the number of lines corresponds to the number of isotopes present in the element.

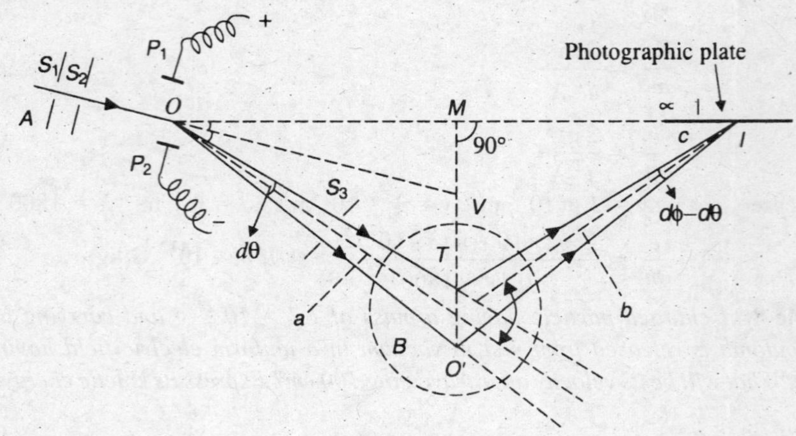

Fig. 17.5

Method: Let the positive rays enter along AO in the electrostatic field applied between plate P_1 and P_2. Let the direction of electrostatic field be from P_1 to P_2. So that the positive beam is deflected downwards and S_1, S_2, S_3 are the three slits that control the direction of positive beam.

Let cd be the photographic plate that records the different masses and let θ and $d\theta$ be the deviation and dispersion angles respectively and let α be the angle of original positive rays with the line cd produced.

Let the distance between the centre of electric and magnetic field be a and let the distance between the magnetic field and photographic plate cd be b. If the direction of the magnetic field is $\perp$ to the plane of paper and inwards then from Fleming's left hand rule, the deflection of the beam will be upwards.

Let the angles of deviation and dispersion produced by magnetic field be ϕ and $d\phi$ respectively and let E be electric intensity and B the magnetic flux density, for the air medium: $B = \mu_0 H$.

Effect of Magnetic field. Let a positive ray particle of mass m and charge e be moving in a direction of the magnetic field. Then the positive ions will experience a force Bev, where B is the flux density and v is the velocity of the ion. Acceleration on the ion perpendicular to its path

$$= \frac{\text{Force}}{\text{Mass}} = a_1 = \frac{Bev}{m} \qquad ...(i)$$

The displacement of the ion from its normal path at the end of the magnetic field,

$$d_1 = \frac{1}{2} a_1 t_1^2$$

In Fig. 17.6, AB is the length of the magnetic field, AC is the path taken by the positive ion in the magnetic field and CC' is the tangent to the arc AC at C. The fluorescent spot will be obtained at C' on the screen and the deflection of the spot along the Y-axis = y = = OC'.

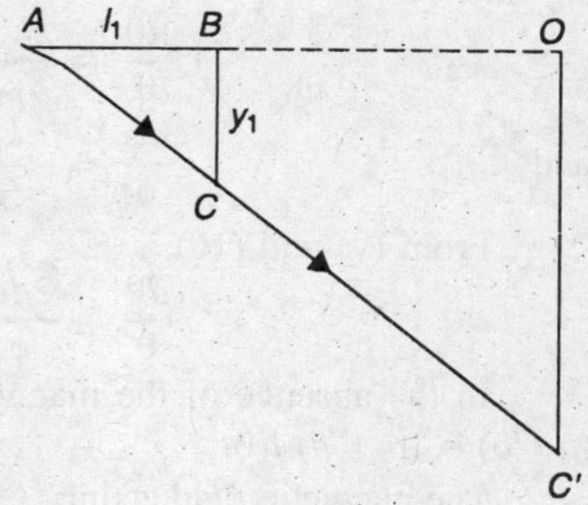

Fig. 17.6

Displacement $\quad y \propto y_1$

$$\propto \frac{1}{2} a_1 t_1^2$$

where t_1 is the time taken by the particle to travel the magnetic field of length l with a velocity v.

$$\therefore \quad t_1 = \frac{l_1}{v}, \text{ and } y \propto \frac{1}{2} a_1 \frac{l_1^2}{v^2}$$

Putting the value of a_1 from (1), we get

$$y \propto \frac{1}{2} \frac{Bev}{m} \frac{l_1^2}{2} = \frac{kl_1^2}{2} \cdot \frac{Be}{mv}$$

$$y = k_1 \frac{Be}{mv} \qquad \text{...(2)}$$

where $\quad k_1 = \dfrac{kl_1^2}{2} = \text{(another constant)}$

Effect of the electrostatic field. Again, the force on the ion perpendicular to its direction of motion is Ee where E is the intensity of the electric field between the plates P_1 and P_2.

$$\text{Acceleration of the ion } = a^2 = \frac{Ee}{m} \qquad \text{...(3)}$$

Dispersion of the ion from its normal path at the end of the electrostatic field

$$= d_2 = 1/2a^2 \ t_2^2.$$

So displacement along the X-axis, $x \propto \dfrac{1}{3} a^2 t_2^2$

where t^2 is the time taken by the ion to travel the electric field of length i^2 with a velocity v.

$$\therefore \quad t_2 = \frac{l_2}{v} \text{ and } x \propto \frac{1}{2} a^2 \frac{l_2^2}{v^2}$$

Putting the value of a^2 from (3), we get

$$x \propto \frac{1}{2} \cdot \frac{Ee}{m} \cdot \frac{l_2^2}{v^2}$$

$$\therefore \quad x = \left(\frac{k' l_2^2}{2}\right)\left(\frac{E}{mv^2}\right)$$

or $\quad x = k_2 \dfrac{Ee}{mv^2}$, where $k_2 = \dfrac{k' l_2^2}{2} = \text{constant} \qquad \text{...(4)}$

Working. If the positive rays enter the electronic field along AO (Fig. 17.5).

Then the shifts x and y of the fluorescent spot on the screen are proportional to the displacements x and y and given by

$$\therefore \quad x \propto \theta, \text{ and } y \propto \phi$$

or $\quad \theta = k_2 \dfrac{Ee}{mv^2}$ and $\phi \propto k_1 \dfrac{Be}{mv}$

or $$\theta = k_2 \frac{e}{mv^2} \qquad \text{...(5)}$$

and $$\phi = k_1 \frac{e}{mv} \qquad \text{...(6)}$$

where k_1 and k_2 are constants depending on the values of E and H and the distance travelled in the two fields.

Differentiating Equations (5) and (6)

$$d\theta = -\frac{2k_2 e}{mv^2} dv \qquad \text{...(7)}$$

and $$d\phi = -\frac{k_1 e}{mv^2} dv \qquad \text{...(8)}$$

$$\therefore \quad \frac{d\theta}{\theta} = \frac{2k_2\, edv}{mv^3} \cdot \frac{mv^2}{k_2\, e} = -\frac{2dv}{v} \qquad ...(9)$$

and

$$\frac{d\phi}{\phi} = -\frac{k_1\, edv}{mv^2} \cdot \frac{mv}{k_1\, e} = -\frac{dv}{n} \qquad ...(10)$$

From (9) and (10)

$$\frac{d\theta}{\theta} = \frac{2d\phi}{\phi} \quad \text{or} \quad \frac{d\phi}{d\theta} = \frac{\phi}{2\theta} \qquad ...(11)$$

In the absence of the magnetic field the dispersion produced in the beam for a distance $(a + b) = (a + b)\, d\theta$.

The magnetic field acting $\perp$ to the electric field produces the same dispersion in a distance b but in the opposite direction.

Dispersion produced by the magnetic field in a distance $b = b\, d\phi$.

Since the ions are all focussed to the same position, we have

$$(a + b)\; d\theta = b\, d\phi$$

or

$$\frac{d\phi}{d\theta} = \frac{a+b}{b} \qquad ...(12)$$

From (11) and (12)

$$\frac{a+b}{b} = \frac{\phi}{2\theta}$$

$$(a + b)\, 2\theta = b\phi$$

$$b\,(\phi - 2) = 2a\,\theta \qquad ...(13)$$

This is the condition for focussing of ions.

Let $Q'M$ be perpendicular to the line cd produced.

Then, in the $\Delta MOO'$,

$$MO' = OO' \sin(\alpha + \theta) = a \sin(\alpha + \theta) \qquad ...(14)$$

In the $\Delta MO'I$, $MO' = O'I \sin(MIO')$

$$= b \sin[180 - (\phi - \alpha - \theta)] = b \sin(\phi - \alpha - \theta) \qquad ...(15)$$

From (14) and (15),

$$a \sin(\alpha + \theta) = b \sin(\phi - \alpha - \theta)$$

For small angles (measured in radians), we have

$$a\,(\alpha + \theta) = (\phi - \alpha - \theta) \qquad ...(16)$$

Comparing Equations (13) and (16), we see that the two equations are identical if $\alpha = \theta$. So, the condition for focussing is that the photographic plate must be placed at an angle θ with the direction of the incident positive ray beam such that $\alpha = \theta$, where θ is the deviation produced by the electric field.

The original mass spectrograph which had an accuracy of one in thousand showed that the atomic masses were integers. The improved form of mass spectrographs designed by Aston had an accuracy hundred times more than the original one and this showed that the masses differ very slightly from the integral values. *These small difference in atomic masses are of great importance in nuclear physics* to account for the relative abundance of various isotopes.

17.6. BAINBRIDGE MASS SPECTROGRAPH

Principle. In this form of mass spectrograph, a velocity selector is used which allows velocity (v) to pass undeviated through crossed electric and magnetic fields. This spectrograph was designed by Bainbridge of Harvard University. In this spectrograph the source is a discharge tube which is fitted with a perportated cathode and a rod anode. Gas is administered through the side tube and a p.d. between cathode and anode is of the order of 20 k.v. Many elements studied on this spectrograph were confirmed by Aston's spectrograph. We observe no of lines on the photographic plate corresponding to different masses and by the study of comparative intensities of the lines, we can have the relative proportion of the isotopes. If an element contains two or more isotopes, the average atomic weight may not be an integer, this explains the fractional atomic weight of various elements like chlorine, neon etc.

Working. Let E is the intensity of the electric field and B the magnetic flux density, then the force experienced by an ion of charge e and velocity v will be Ee and Bev respectively. If these two forces act at right angles and in the same plane, the iron will travel undeflected.

In this case $B = \mu_0 H$

In Fig. 17.7, E is the intensity of the electric field acting in the direction PP' in the plane of the paper. The magnetic field or flux density B is perpendicular to the plane of the paper and inwards. The electric field will deflect the beam downwards and the magnetic field will deflect the beam upwards from (Fleming's left hand rule).

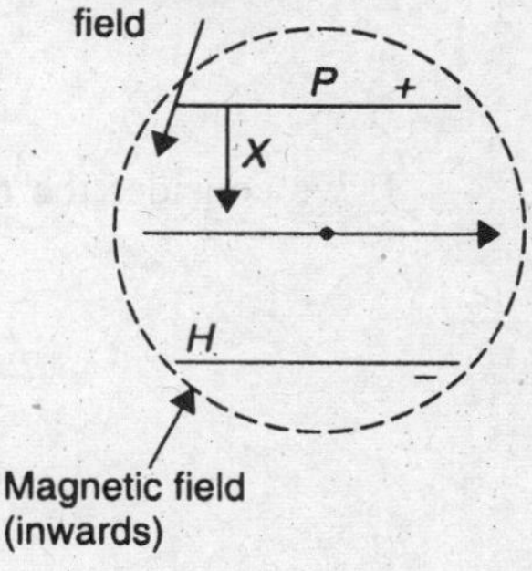

Fig. 17.7

Now force experienced by the ion due to the electric field = Ee.

Force experienced by the ion due to the magnetic field = Bev

If these two forces are equal and opposite then, $Ee = Bev$

$$\therefore \quad v = \frac{E}{B}$$

So, all the ions having a velocity v given by E/B, suffer no deviation while passing through the crossed electric and magnetic fields. Let S_1 and S_2 be two slits which admit a narrow beam of ion so that the ions having the same velocity $v = (E/B)$ enter the analysing chamber A through the slit S_2, where a second flux density $(B' t$ is applied perpendicular to the plane of the paper and in outward direction. The flux density is of the order of 1.5 weber/m^2 and the influence of this field is over a semi-circle of radius 20 cm. All the ions having the same value of e/m travelling semicircular path of radius v given by,

$$\frac{mv^2}{r} = B'ev, \quad \text{or } r = \frac{mv^2}{B'ev} = \frac{mv}{B'e}$$

Since B' and v are constant

$$r = km \qquad \text{...(1)}$$

So, the ions having the same value of e/m are focussed on to the same line and ions with different values of e/m are focussed on to different positions on the photographic plate. From Equation (1), we see that radius of circular path of an ion is directly proportional to (*i*) the mass of the ions and (*ii*) the constant k. Thus, this spectrograph gives a linear mass scale so it is advantageous as compared to Aston's mass spectrograph, in which path travelled is a parabola.

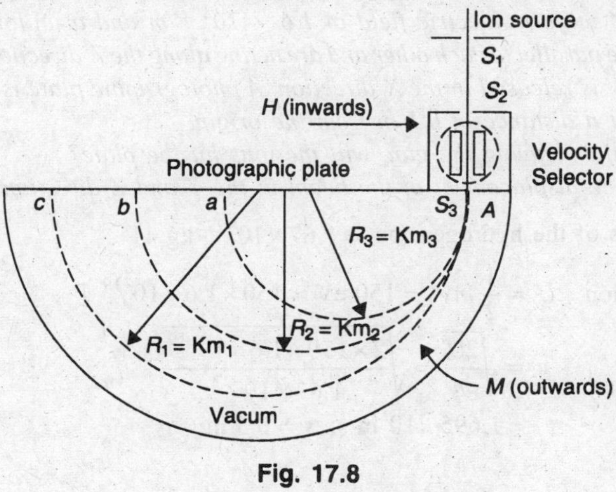

Fig. 17.8

Example 9. *Positive rays from a discharge tube working on 3 × 10⁴ volts pass through a pair of plates 5 cm in length and 1 cm apart and maintained at a p.d. of 200 volts. Calculate the maximum deflection.*

Solution:
$$V = 3\times10^4 \text{ volts}$$
$$E = \frac{200}{1} = 200 \text{ volts/cm} = 200\times10^2 \text{ volts/m}$$

Now
$$eV = \frac{1}{2}mv^2 \text{ and } t = \frac{0.05}{v}$$

If we consider the deflection to be y, then
$$y = \frac{1}{2}at^2 = \frac{1}{2}\left(\frac{Ee}{m}\right)\times\left(\frac{0.05}{m}\right)^2$$
$$= \frac{(0.05)^2\times Ev}{4eV} = \frac{(0.05)^2\times E}{4V}$$
$$= \frac{(0.05)^2\times200\times10^2}{4\times3\times10^4} = \mathbf{0.00041\ m.}$$

Example 10. *Singly ionized Neon 20 and 22 particles enter Bain-bridge mass spectrograph with a velocity of 10^5 m/s. They are deflected by a magnetic field of flux density 0.08 weber/m². Calculate the radii of their paths. The mass of the nucleon is 1.67×10^{-27} kg.*

Solution: For a single ionized neon –20
$$m_1 = 1.67\times10^{-27} \text{ kg}\times20 = 3.34\times10^{-26} \text{ kg}$$
$$v = 10^5 \text{ m/s}; \quad B = 0.08 \text{ weber/m}^2$$
$$Bev = \frac{m_1v^2}{r_1}$$
$$r_1 = \frac{m_1v}{Be} = \frac{3.34\times10^{-26}\times10^5}{0.08\times1.6\times10^{-19}} = \mathbf{0.26\ m}$$

(*ii*) For a single ionized neon 22
$$m_2 = 1.67\times10^{-27}\times22 \text{ kg} = 3.674\times10^{-26} \text{ kg}$$
$$v = 10^5 \text{ m/s}; \quad B = 0.08 \text{ weber/m}^2$$
$$Bev = \frac{m_2v^2}{r_2}$$
$$r_2 = \frac{m_2v}{Be} = \frac{3.674\times10^{-26}\times10^5}{0.08\times1.6\times10^{-19}} = \mathbf{0.2871\ m.}$$

Example 11. *A uniform electric field of 1.6×10^3 V/m and a uniform magnetic field off 3.34×10^{-3} Wb/m² are parallel to each other and are acting along the Z-direction. A beam of hydrogen ions of energy 150 eV is released in the X-direction. A photographic plate is placed normal to the motion of the ions at a distance of 0.5 m from the origin.*

(i) How long after leaving the gun, will the ions hit the plate?

(ii) Calculate the displacement of the beam in the Y and Z directions.

Solution: Mass of the hydrogen ion $= 1.67\times10^{-27}$ kg

Energy of the ion, $U = \frac{1}{2}mv^2 = 150 \text{ eV} = 150\times1.6\times10^{-19}$ J

∴
$$v = \sqrt{\frac{2U}{m}} = \sqrt{\frac{2\times150\times1.6\times10^{-19}}{1.67\times10^{-27}}}$$
$$= 1.695\times10^5 \text{m/s}; \ s = 0.5 \text{ m}$$

(*i*) $t = \frac{s}{v} = \frac{0.5}{1.695 \times 10^5} = \mathbf{2.95 \times 10^{-4}\ s}$

(*ii*) $E = 1.6 \times 10^3$ V/m; Force = $E.e$

$\therefore$ acc. $a_z = \left(\frac{Ee}{m}\right)$; $Z = \frac{1}{2} a_z t^2$; $Z = \frac{1}{2}\left(\frac{Ee}{m}\right) t^2$

$$\therefore \quad Z = \frac{1.6 \times 10^3 \times 1.6 \times 10^{-19} \times (2.95 \times 10^{-6})^2}{2 \times 1.67 \times 10^{-27}} = \mathbf{0.667\ m}$$

$$B = 3.34 \times 10^{-3}\ \text{Wb/m}^2;\ a_y = \left(\frac{Bev}{m}\right);$$

$$Y = \frac{1}{2} a_y t^2;\ Y = \frac{Bev t^2}{2m}$$

$$= \frac{3.34 \times 10^{-3} \times 1.6 \times 10^{-19} \times 1.695 \times 10^5 \times (2.95 \times 10^{-6})^2}{2 \times 1.67 \times 10^{27}}$$

$= \mathbf{0.237\ m.}$

17.7. APPLICATIONS OF MASS SPECTROGRAPHY

Mass spectrographic studies are useful.

1. to determine the relative addundance of isotopes in a given element. The measurement of the percentage of argon in potassium bearing minerals, using a mass spectrometer, is used in determining the age of such minerals, if we know the half life of radio-active potassium.
2. to determine the impurities in gas samples.
3. to detect the leaks in vacuum apparatus, refrigerators etc., and
4. in the analysis of chemical compounds and hydrocarbons in petroleum industry.

In addition many more applications of mass spectrographic studies are being applied in the fields of plastics, metallurgy, and other fields.

PROBLEMS AND QUESTIONS

1. Describe Millikan's experiment for the measurement of charge of an electron. Indicate the importance of the experiment.

2. Give the theory of Millikan's oil drop experiment and derive an expression for the charge of an oil drop when the weight of the drop is balanced by the force due to the electric field.

3. Describe Millikan's experiment and explain how the radius of the oil drop can be determined?

4. Discuss the motion of a charged particle in a uniform electric field. Calculate the expressions for longitudinal and transverse accelerations.

5. Explain the term quantization of charge.

6. Explain briefly the effect of electric and magnetic fields on a moving electron.

7. Give the theory and working of Aston's mass spectrograph. What is the advantage of this method over Thomson's parabola method?

8. Give the construction and theory of Aston's mass spectrograph. Discuss the experimental results obtained.

9. Describe and give the theory of Thomson's parabola method for the study of isotopes. Mention two isotopes of hydrogen and carbon.

10. Describe in detail the working of an accurate mass spectrograph. Discuss the results obtained. What light does the study of isotopes throw on the structure of nucleus?

11. Give an account of Bainbridge's mass spectrograph.

12. A stream of particles each of mass *m* and charge *e* moving with a velocity of 6×10^7 ms^{-1} passes between two plates. The electric field between the plates is of intensity 2700 Vm^{-1} and the length of the plate is 0.2 m. Calculate the ratio of *e*/*m* for the particles if deflection of particles at the end of the field is 0.27 cm. What should be the strength of magnetic field if the stream is to remain undeflected by the joint action of electric and magnetic fields?[**Ans.** 1.8×10^{11} C kg^{-1}; 0.45×10^{-4} Wb m^{-2}]

13. An electron falls through a potential difference of V volt. Prove that the velocity, v, it acquires is given by

$$v = 5.95 \times 10^5 \sqrt{V}\ \text{ms}^{-1}$$

Given $e = 1.6 \times 10^{-19}\ C$ and $(m_e = 9.1 \times 10^{-31}$ kg).

14. An electron entering a magnetic field of 10^{-2} Wb m^{-2} with a velocity of 10^7 ms^{-1} describes a circle of radius 6×10^{-3} metre. Calculate e/m of the electron. **[Ans.** 1.67×10^{11} C kg^{-1}]

15. An electron moving with a velocity of 3×10^7 ms^{-1} enters a uniform magnetic field of intensity 2×10^{-3} Wb m^{-2}. Calculate the radius of the circular path describe by the electron. If the velocity of the electron is doubled, what would be the radius of the new path?

16. A stream of electrons travelling with a velocity of 3×10^7 ms^{-1} is deflected in passing through an electric field of 1800 volt/metre perpendicular to its path. If the radius of the deflected beam is 3 metre, calculate e/m of the electron. **[Ans.** 1.67×10^{11} C kg^{-1}]

17. In Millikan's oil drop experiment, an oil drop of mass 4.9×10^{-24} kg is balanced by applying a p.d. of 4000 volt between two plates which are 8 mm apart. Find number of elementary charges on the drop. **[Ans.** 6]

18. In a positive ray apparatus, a single ionised and a doubly ionised particle form identical parabola when the magnetic flux densities are 7500 Gauss and 15 Gauss respectively, while the electric field is kept constant. Compare the masses of particle. $\left[\textbf{Ans. } \frac{m_2}{m_1} = 8\right]$

19. Positive rays from a discharge tube working on 3×10^4 volts pass through a pair of plates 5 cm in length and 1 cm apart, applied with a potential of 200 V. Calculate the minimum deflection **[Ans.** 0.833 cm]

20. A charged oil drop is suspended in a uniform field of 3×10^4 Vm^{-1} so that it neither rises nor falls. Find the charge on it assuming its mass to be 4.9×10^{-15} kg. **[Ans.** 1.6×10^{-12} C]

CHAPTER 18

NON-DESTRUCTIVE TESTING

18.1. INTRODUCTION

The basic purpose of non-destructive test is to know the quality of the material so that it may be considered suitable for a particular job or may be rejected if it is not upto the specifications:

Term '*Non-destructive testing*' is used in general form as this method users *testing or inspection* without changing the structural properties of the material so that it may still be used in future. These tests have the following salient features:

1. It has no effect on future serviciability.
2. Profits or productivity may be increased after the tests.
3. Safety norms may be applied as per the requirements after the tests.

Non-Destructive tests involve the following essential parts.

(*i*) application of the testing or inspection material.

(*ii*) modifications of the materials after identifying the various defects.

(*iii*) using suitable device to identify the defects and their positions.

(*iv*) Conversion of the change into a suitable form and then getting the correct interpretation of the results.

In addition, one has to be sure that the interpretation of the result is sure for better marketing avenues.

Non destructive tests are generally used to *find and identify the defects in the* materials while *destructive tests* pertain to finding the *mechanical properties like hardness, toughness, compressive strength, yield strength, tensile strength etc.*

We will briefly describe some of the important methods used in *non-destructive tests.*

18.2. MAGNETIC METHODS

In this method the specimen is magnitized and the distortion of magnetic field due to defects is studied. The magnetic field distortion may be studied in many ways: but the most widely used method employs *magnetic particles* such as iron oxide magnetic flux distortion due to a defect will change the *induced voltage* generated in the search coil. Flaws in magnetized pipe causes a change in magnetic flux which is turn produces a change in induced voltage. By suitable parameters it is possible to distinguished between *inside and outside* fatigue crackes inside the pipe. Using this technique defects of thickness is 0.7 cm to 1.25 cm can be identified. Further inspection may let us know about the presence or absence of defects. An experienced operator can easily identify the position and size of the defects. Some factors like sharpness, depth, direction can indicate the location of defects. However by this method *hairline cracks* are difficult to locate *unless they are below the surface.*

18.3. ELECTRICAL METHOD

These methods are based on the principle of different materials having different *resistivities*. In electrostatic method, variation of electric fields can be deleted by motion of charged particles.

When a coil carrying high frequency alternating current is moved near a conductor, *eddy currents* are induced in the *conductor*. Also a magnetic field is associated with induced currents. Some defects, flows, physical and chemical constituents and other structural variations causes change of resistivity thereby the eddy currents also very resulting in the *variations of induced magnetic fields.* These variations in magnetic fields can either be defected by search coil or by their effect on coils inducing the currents.

Various techniques used for non-destructive electrical tests are:–

(*i*) Change of electrical resistance
(*ii*) Thermoelectric effect
(*iii*) Triboelectric effect
(*iv*) Variation of static field

These techniques are useful to find lamination in rolled sheets, casting defects, joint defects, defective welding spots, cracks in electrical bushing as well as electrical insulation defects.

(*i*) A special type of *Kelvin double bridge known as electrical crack detector* has been used to measure *resistance* of heavier parts and to know the propagation of cracks, fatigues and fractures. *Potential drop* technique is used to study the depth of the cracks. Electrical resistance techniques is used to study corrosion ratios using the method of loss of metal from a probe installed in *corrosive system under study.*

(*ii*) *Thermo Electric effect* is the setting up of e.m.f. in a circuit of two dissimilar metals forming two junctions, one being kept at higher temperature and the other at a lower temperature. This arrangement is called *thermocouple.* If no. of thermocouples are connected in series, more e.m.f. is produced and the arrangement is called thermopiple. In a thermoelectric metal *comparator* one probe is heated to a fixed temperature and the other is put at the specimen, so that it attains the temperature of specimen. Signal from the probe is amplified and fed to a phase sensitive detector. When probe is placed on specimen, the temperature drops and the amount of fall in temperature depends on the thickness and thermal conductivity of the specimen. This technique is used in measuring *thickness of non-conductor films.*

(*iii*) *Tribo Electric effect* corresponds to the phenomenon when *two metallurgically different metals are brought in contact and agitated mechanically thus generating a voltage. This voltage may range from micro volts to some millivolts.* This voltage depends upon the nature of metals or alloys in contact. Heat treatment produces chemical changes in most alloys and so this technique can be used to distinguish between *heat treated* and *hot worked* alloys. This technique has been successfully used to sort out *copper base alloys* according to their lead content and *lead base alloys* according to antimony content.

(*iv*) Static field technique is used to check the defects in insulator or dielectrics. If there is a defect in a dielectric, its dielectric constant will change and so the *capacitance* will change. Capacitance change can be detected either by statiflux method, similar to magnetic particle technique or by using an oscillator which measures the frequency changes due to change in dielectric properties.

18.4. RADIOGRAPHIC METHODS

Radiographic method uses the technique of *X-rays or γ-radiations*. In these days even *electron and neutron beams* are being used.

On passing through matter, these high radiation beams are modified by the various defects in the material. *Phenomenon of differential absorption forms the basis of use of radiography.* The contrast *i.e.* the density difference on the developed film between the image of an area containing a *defect* and image of *defect free* part of specimen allows us to know the flaw.

Decrease in intensity ΔI is found to be proportional to the original intensity of the beam and thickness Δx of the absorber.

$$\therefore \quad \Delta I = -\mu I \Delta x$$

Where μ is constant of proportionality and negative sign shows that with increase of thickness, intensity decreases. Many types of films are available for radiography, each having different values of speed and contrast densities. Normal X-ray sources can find flaws upto a thickness of 20 cm while for higher thicknesses, more input voltage ~ 2000 V may be used to study inner structures. X-ray scattering, X-ray differaction techniques are having used to study the various flaws as per the requirement of the job. We can summarized some of the techniques being used in non-destructive testing using Radiography as under:

1. A *penetrameter* is used to check the *radiographic definition*. Radiographic definition refers to sharpness with which radiograph shows any *discontinuities* in the specimen. A *penetrameter* consists of a strip of the same material as the specimen having thickness equal to some definite percentage of the specimen thickness. *Flat, bead, wire and step* are four types of penetrameters. Penetrameter having sensitivity of 1% to 2% can be got with the known techniques for a wide range thickness of steel.

2. *Weld radiography* can find flaws, slag inclusions, incomplete fusion, cracks, pin holes in welds etc. The term 'Slag' or non metallic inclusion is used to describe the oxides and other solids which are found along with the weld.

3. Multiple film techniques are being used to scan total specimen in one exposure. This avoids retakes of camera.

4. In pipe radiography, three techniques *viz* single wall, double wall and variation of double wall are used.

5. X-rays can produce fluorsence fluo on materials. Using this *fluoroscopy* is developed in which X-rays are converted into visible light. This is a fast and economical technique. However contrast and sensitivity in that, case are *lower* than the convensional radiography.

6. Casting radiography is having used to find defects like gas holes, cold shuts, hot cracks, cold cracks, cavities, misruns etc. *Misruns* are due low powering temperatures which result in the *molten metal* to solidify before filling the mold cavity. X-ray films will show the defects in castings.

7. High speed radiography, xero-radiography and stress biography are modern methods and give better resolution than ordinary X-ray films.

8. Micro radiography is being used to study very small defects and to identify thin sections having minute voids.

9. Electron radiography uses fast electrons and is based on the principle of *surface electron emission* due to Secondary photoelectrons. This technique is more useful for metals having high atomic number such as *lead*.

10. In neutron radiography, neutrons are used instead of X-rays. Neutron radiography is very useful for *light materials like plastic, rubber* etc. where X-rays and g-rays can not be used.

11. Gamma Radiography being more economical than X-rays is being used in many fields. However, it has lower contrast as compared to X-ray radiography. Radium, Cs Radon, Cobalt-60 are some of the g-ray sources. Cs is suitable to study the radiography of steel.

It is important to note that all the high energy beam used in above techniques are highly radioactive. So only trained persons having sufficient safeguards against the hazards of radiations should perform the above tests.

18.5. COMPARISON OF γ-RAY RADIOGRAPHY AND X-RAY RADIOGRAPHY

1. For small defects in castings, less than about 50 mm, X-rays give better results than γ-rays.
2. X-rays give quick results in seconds while γ-rays may take many hours.
3. For thick defects in castings, γ-rays give better results than the X-rays.
4. Due to minimum scattering, γ-rays scan better than X-rays. Moreover, γ-rays give good results even for varying thicknesses unlike the X-rays which are more suitable for uniform thickness of defects. Also γ-rays can detect many defects at the same time.

5. γ-ray apparatus is small and portable so it is more useful for field testings and inspections.

18.6. ULTRASONIC METHODS

Ultrasonic waves find us in the detection of cracks or cavities in metal castings due to the property of reflection of sound waves when a change of medium takes place.

Method consists of sending a strong **ultrasonic beam** through an object as shown in fig ... Due to a crack or cavity there is a change of medium and so the beam gets reflected. The reflected beam is defected on a CRO (Cathode Ray Oscilloscope) and time of transit of the beam is noted. Knowing the *velocity of sound waves in the medium of solid*, position of crack can be detected.

By inspecting the metal specimen from all directions, exact position of the crack which may be as small as 0.005 mm can be detected.

An *important advantage* of ultrasonic method is that it can test very large size castings without destroying the material. Also it involves very low cost and speed of the operation is rapid. This method of locating flaws is more sensitive and less costly than the radiography method. In sections of large castings, rails, heavy axles as in locomotives, forgivings etc. are heavy done successfully by ultrasonics.

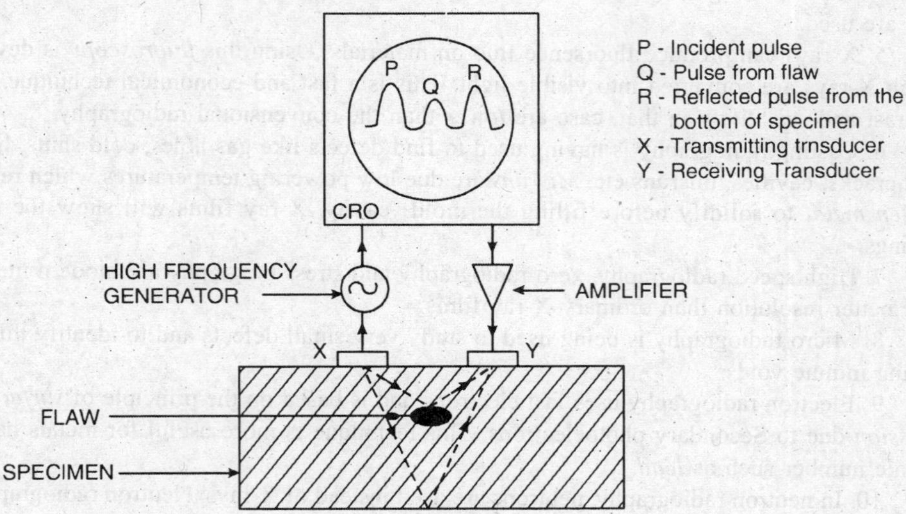

Fig. 18.1.

18.7. THERMAL METHODS

Basic principle in thermal tests is to apply heat to the specimen and measure the *heat distribution*. In case, there are flows, heat distribution and consequently temperature measurement will differ as flaws change the heat distribution. Heat may be supplied directly or by electrical heating or by infra red sources. Temperatures may be measured by thermocouples, Bolometers or photoconductive devices.

Frost test is one of the tests that tests the nuclear fuels. In this case Diphenyl is *Frost material* and this test can locate defects of 0.7 cm^2 having minimum length of 0.5 cm. **Thermal resistance** of the material for this test should be high.

THERMOGRAPHY

This corresponds to optical thermal testing using non-destructive method. This technique involves scanning of thousands of points per second in the form of a picture, known as *Thermogram*. In case there is any flaw, the intensity of radiation will change from that surface.

Thermography is a good and effective tool for diagnostic evaluation of *electronic circuits*. It helps us to identify the defective modulus and to isolate the defect without affecting any change in the circuit, or signal tracing, voltage variations, soldering and unsoldering etc.

Optical material like Germanium and Silicon lenses require infra red testing as they are opaque to visible light. *Contact thermography* shows the discontinuity in the bonding of metal to metal contacts. *Thermocolour paints* are used to locate flaws in metals. As temperature of the surface reaches a particular value, colour is changed. *Evapography* provides a two dimensional heat picture from infra-red radiations given by the sample. Tempilstile, Tempilag and thermisters are very sensitive to temperature variations and so find applications in many fields of thermography.

Questions

1. What is the importance of non-destructive tests? Give suitable methods used by Radiography.
2. Give a method of non-destructive test using magnetic particles.
3. How non-destructive tests can detect flaws using electrical methods ? Give the various methods in support to your answer.
4. How will you compare the various NDT (non-destructive tests) techniques ? Illustrate your answer by describing due of the techniques.
5. What is radiography? Describe its importance in NDT techniques.
6. How X-ray radiography differs from γ-ray radiography ? Illustrate your answer with importance of each technique.
7. How many forms of beams can be used to study NDT by radiography ? Give the importance of using electrons and neutrons in solid of conventional X-rays or γ-rays.
8. What is fluoroscopy ? How is it important in NDT technique ?
9. What is thermography ? Explain its principle and method of use for NDT technique.
10. Draw a neat diagram of an ultrasonic flaw detector. In what way this technique is preferred over radiography.

CHAPTER

19

THIN FILMS AND ADVANCE SPECTROSCOPY

19.1. THIN FILMS DEVICES

Thin film devices are structures consisting of one or more thin layers of **metal semiconductor or insulator**. Insulated gate *Thin Film Transistor* (TFT) was proposed in 1961 by **Weimer** and he used evaporated *semiconductor metal and insulated layers* which functions essentially as an **Insulated-Gate Field Effect Transistor** (IGEFT) which is also called MOSFET (metal-oxide-semiconductor field effect transistor) or MOST (metal-oxide semiconductor transistor). The same technique has been used for making many integrated circuits. Since thin films are formed by semiconductors, metals and insulators, so crystalline imperfections may be present hence the techniques have to be developed to have TFT with least limitations.

In the last few decades attempts have been made to invent solid state devices capable of displacing the junction transistors in various applications. *Thin Film Transistors* (TFT) have been developed to replace the junction transistors and these are being extensively improved for various applications.

The devélopment of **optical dielectric** coatings in *visible*, *ultraviolet* and *infrared regions* of spectrum in the last few decades has a very profound effect in the fields of optics, lasers, microelectronics, integrated circuits and other related fields. Thin film technology is exploring many avenues in the fields of *space science*, *microelectronics*, *computer chips* etc. Any development in thin film technology has its impacts on so many fields of research activities and hence has very important applications in the modern technology. It is important to note that the *multilayer dielectric* coatings will enhance many more technique in instrumentation and the control of deposition process will achieve many characteristics like *mechanical stability and precision*; *light absorption, transmission and reflectivity*.

Vacuum deposition of metallic and dielectric single layer films is well known in the application of **production of mirrors, antireflection coatings beam splitters etc.**

The multilayers coatings of quarter wave stack of high and low refractive index materials are extensively used in production of high reflecting mirror for *lasers and interferometres*.

A limiting number of evaporating materials such as **ZnS, MgF_2, ZnSe, $Na_2Al_2F_6$, $ThOF_2$, SiO and TiO** are employed in deposition of multilayers films since the process has better optical and mechanical properties over the other available materials. A remarkable progress has been achieved in the design of multilayer systems by means of **electronic computers**. In spite of good experimental techniques, thin film technology is an art and depends on the skill of the operator.

19.2. PRODUCTION OF THIN FILMS

Effect of substrate condition: cleanliness of the substrate plays an important role in the production of quality films. The *durability and reproductibility* of the film depends on the cleaned substrate surface. The cleansing of substrate depend on the cleanliness required and types of contaminative. The usual contaminants are *oil or grease films*, *dust particles or some impurities* *absorbed by the substrate surface*. The methods of cleansing depend upon the requirements and usually they are constituted into three groups.

(1) Chemical cleaning
(2) Colour discharge cleaning
(3) Substrate heating, ultrasonic cleaning and few other related methods.

After cleaning optical polishing techniques have been used to achieve better surface smoothness of the order of 10 Å.

19.3. STRUCTURE OF COATINGS

The actual structure of multilayer coatings does not correspond to the assumptions of elementary theory that layers consist of perfectly homogeneous dielectric materials with plane parallel boundaries. The study of special technique of electron microscopy gives the structure of films and thereby helps to improve the properties and better control on them in the fabrication of multilayer films.

19.4. REACTIVE EVAPORATION OF METAL OXIDES

Widely used materials like ZnS, ZnSe, Cryolite, $ThoF_2$ etc produce multilayer coatings which are soft in nature. Some of these can withstand cleansing with acetone or dilute sulphuric acid but cannot withstand *rough handling or abrasion resistance*. Metal oxide layers possess more mechanical strength, stability and adhesion. Hard multilayer coatings of TiO_2 and SiO_2 as high refractive index and low refractive index respectively are obtained by depositing TiO and SiO in oxygen atmosphere in vacuum chamber.

The production of oxide films from metals in oxygen atmosphere is known as *reactive evaporation*. The process of conversion of lower oxides or metals is completed either by evaporaing the film on heated substrate or by heating it after evaporation in oxygen atmosphere or under intense irradiation with ultraviolet light in oxidizing atmosphere or by adding elements or oxides term rare earths including *yttrium* yttrium and lanthanium to evaporating material. The characteristics of the films deposited depends on the oxygen flow, pressure temperature and other parameters and small variation of any of these will change the properties of the film.

A discharge tube with a region of high current density for ionization is constructed and arranged in vacuum chamber. Ionized gas emerging from the nozzle in the wall directly enter into high vacuum region. Results on SiO_2 and TiO_2 showed that absorption free films could be produced on the substrates at room temperature. This technique is avoid for temperatures of substrate at 400°C as reactive evaporation at these temperatures of neutral oxygen atmosphere finds useful applications in higher degrees of flatness, for interferometer and laser applications.

This technique appears to have great potentials for future applications in producing hard multilayer coatings.

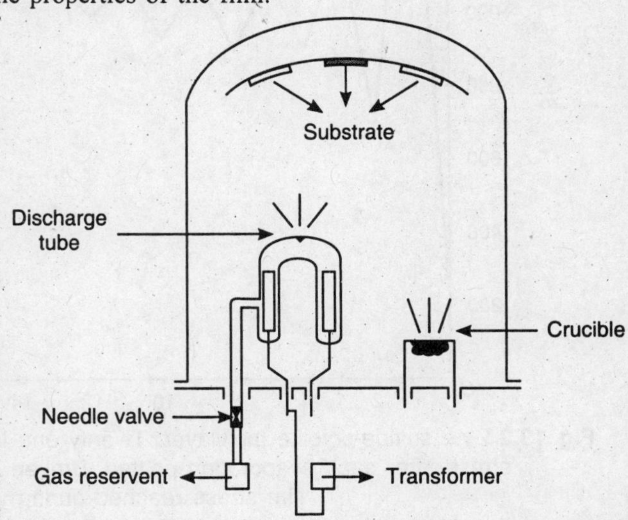

Fig. 19.1 Complete arrangement for evaporation in ionised gases

19.5. STRESSES IN FILMS

The crazing and buckling phenomenon in evaporated thin films arise due in internal sterises developed during process of condensation. Almost all films exhibit the mechanical stress with varying degree and type, *viz.*, tensile or compressive. Stress data on number of films of commonly used evaporating materials such as

MgF_2, ZnS, CeF_3, CaF_3, $ThOF_2$ and many other materials have been studied. These results showed that the nature in the stress built up in different materials varies, *i.e.*, some materials show initially compression and then reverting to tensile stress. Some materials show constant stress after a small thickness, while in others the stress continues to rise with thickness.

In case of single layers, the stress varies with the type of varpour source, rate of deposition, nature of gas used after deposition in the vacuum chamber.

However, the stresses in the multilayer films are different from those of single layers. The results on systems like ZnS-Cryolite, ZnS-$ThOF_2$ and PbF-Cryolite multilayer quarter wave stackings reveal that there is little agreement between experimental results and the calculated values of single film measurements.

Fig. 19.2 shows the curve for stress v/s number of layers of ZnS-Cryolite system. Multilayer film structure revealed the surfaces of the films were rougher than surface of substrate glass.

As stated earlier, this would give rise to different nucleation conditions for the additional layers which may the stress developed in them. From the stress measurements, it is observed that compressive stresses are generated more easily than the tensile ones.

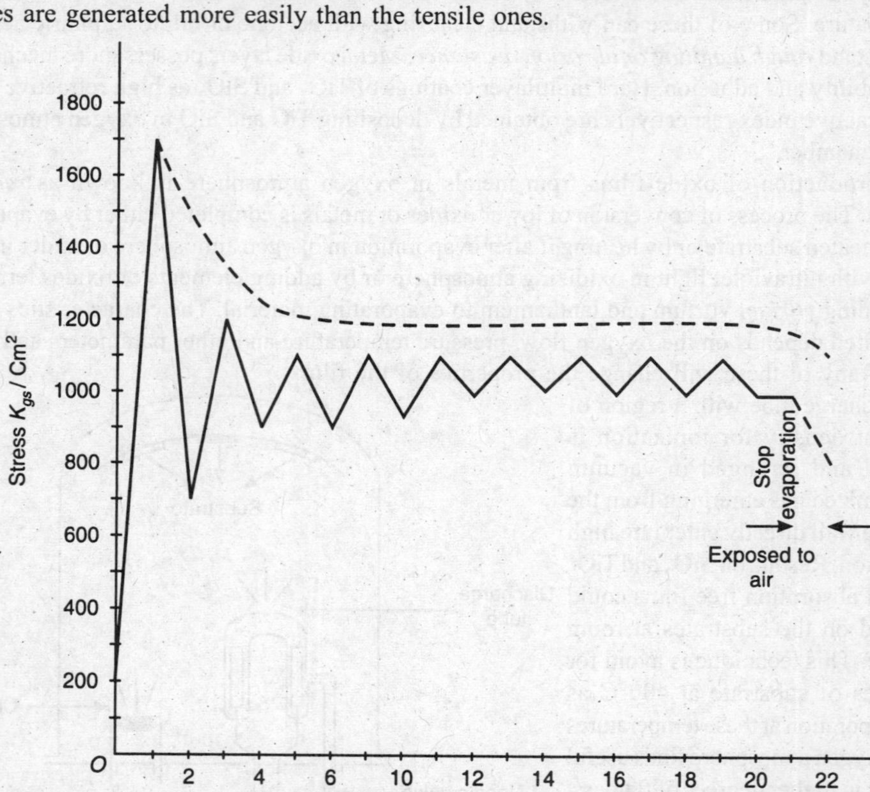

Fig. 19.2 Zine sulfide-cryolite multilayer. Twenty one layers $(HL)^{10}$ H. Resultant average stress after each evaporation poltted. Broken curve shows upper limit of film stress reached during warm-up

Present day data shows that stress in thin films in case of large number of evaporants produce high tensile stresses in films. Many experiments are being performed to overcome the said tensile stresses. *Scheurman* fabricated low internal stress dielectric films by evaporating the mixture of materials that develop opposite stresses one tensile and the other compressive when deposited as discrete films and by evaporation at the same time from two different sources. Low stress films have been successfully produced with ZnS-$ThOF_2$, ZnSMgF_2 etc. using the said technique.

19.6. APPLICATIONS OF THIN FILMS

1. Thin films are being used for various integrated circuits, thereby reducing the space of devices.
2. These are being extensively used in computers and other fast electronic devices.
3. These are being used in space research and space explorations.
4. These are also being used in material instruments.
5. These are being used in lasers and other optical devices.
6. They are finding use in almost all the electronic fields using micro chips.
7. They are being used in solar cells (GaΔs and CdS thin film cells are used as solar cells).

19.7. NUCLEAR MAGNETIC RESONANCE (MMR) SPECTROSCOPY

A different type of interaction between matter and electromagnetic forces can be observed by subjecting a substance simultaneously to *two magnetic fields*; *one stationary and the other varying with some radio frequency*. At a particular combination of fields, energy is absorbed by the sample and the absorption can be observed as a *change in the signal by a radio frequency detector*. This absorption of energy can be related to the *magnetic dipolar nature of spinning nuclei*.

This technique is known as nuclear magnetic resonance (*NMR*). Radio frequency being the lowest form of energy in the e.m. spectrum to affect the nuclear spin of the atom in a molecule and so *constitutes the most fundamental part of NMR spectroscopy*. In general, the study of radio frequency radiation by nuclei is called *Nuclear Magnetic Resonance*.

In NMR spectroscopy, molecular parameters are found to be

1. Life time of an energy level,
2. Coupling constant of nuclei,
3. Shielding constant of nuclei.

Nuclear Spin and Magnetic Moment

All nuclei carry a charge. In some nuclei, this charge spins on the nuclear axis and this circulation of nuclear charge generates a *magnetic dipole* along the axis. As shown in Fig. 19.3.

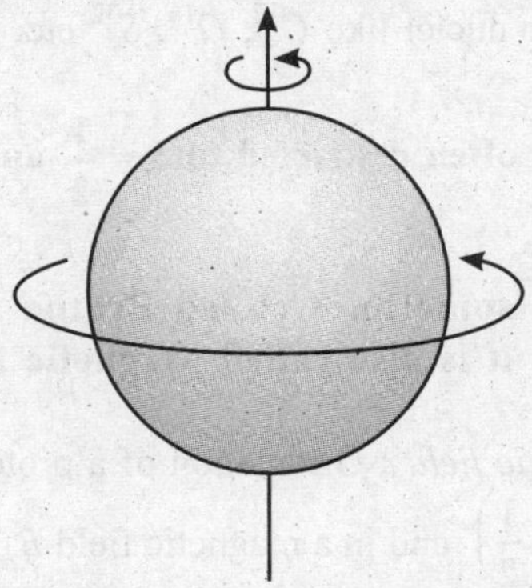

Fig. 19.3 Spinning charge inproton generates magnetic dipole

As a rotating charged sphere is associated with it a magnetic moment, so all the charged particles in the nucleus will cause the nucleus to behave like a small bar magnet with its magnetic moment along the axis of rotation. Nuclei of atoms comprise of neutrons and protons and like electrons they have spin characteristic along their axis and possess angular momentum of $\frac{1}{2}\cdot(h/2\pi)$ as per the quantum theory. The net resultant of angular momentum of all nuclear particles is called *Nuclear Spin*.

For a nucleus having spin quantum number I, there are $(2I + 1)$ spin states. Two characteristics which are very important for the study of NMR spectroscopy are the *net spin associated with protons and neutrons* and *the distribuion of positive charge*. Net spin quantum number I for a particular nucleus can be got by adding the spin quantum numbers of individual protons and neutrons separately. Assuming that the neutrons cancel only the neutron spin numbers and protons cancel only the proton spin numbers because of pairing in opposite directions. Spin number I has values 0, 1/2, 3/2, 5/2 and so on. (I = 0 corresponds to no spin)

There are three broad principles of nuclear spin.

1. If the sum of protons and neutrons is even, I is zero or integer (0, 1, 2, 3, ...). If the spins of all particles are paired, there will be no net spin and nuclear spin quantum number I will be zero. For example, C^{12}, O^{16}, O^{18}, S^{32} etc. nuclei have even number of both protons and neutrons and so will show zero spin. In such cases, there will not be any magnetic moment.
2. In case of nuclei where sum of neutrons and protons is odd, I is half integer (1/2, 3/2, 5/2, ...). For example, H^1, F^{19}, C^{13}, P^{31} etc. have I = 1/2 while in case of nuclei B^{11}, Cl^{35}, Br^{79} and Br^{81} the value is I = 3/2.
3. In case of nuclei where both protons and neutrons are even numbered, I is zero, *e.g.*, C^{12} and O^{16} fall in this category and *do not show any NMR signal*.

As can be understood, the magnetic properties occur with the nuclei mentioned below:

(*a*) Odd Atomic Number and Even Mass Number
Examples: $_1H^2$, N^{14} etc.

(*b*) Even Atomic Number and Odd Mass Number
Examples: $_6C^{13}$

(*c*) Odd Atomic Number and Odd Mass Number
Examples: $_1H^1$, $_7N^{15}$, $_9F^{19}$, $_{15}P^{31}$ etc.

It is clear from above, for study of NMR spectroscopy, a nucleus must possess angular momentum and an associated magnetic moment. Most important nuclei falling in this category are H^1, F^{19}, B^{11} and P^{31}. Maximum sharpness of absorption peaks occur in nuclei for which $I = \frac{1}{2}$ (H^1, C^{13}, F^{19}, Si^{29} and P^{31}). In case of H^1, F^{19} and P^{31}, NMR can be studied easily because of their *natural abundance*. In principle C^{13}, N^{14}, O^{17} and Sn^{119} also possess both angular momentum as well as associated magnetic moment, but because of low isotopic abundance and less inherent property, they are not important for *NMR spectroscopy*. Most other nuclei like C^{12}, O^{16}, S^{32} etc. have either no angular momenta or are of low technical interect.

NMR spectroscopy is most often restricted to $I = \frac{1}{2}$ and the best example of which is $_1H^1$ with spin $= \frac{1}{2}$.

So NMR spectroscopy is sometimes called Proton Magnetic Resonance (PMR) spectroscopy. In medical terms it is also called Magnetic Resonance Imaging (MRI) or Tomography.

Energies of nuclei in magnetic fields: Orientation of a proton is determined by the value of quantum number $m_I \left(= +\frac{1}{2} \text{ and } -\frac{1}{2}\right)$ and in a magnetic field B_o, these orientations may have the values.

$$E = -g_I \mu_N B_0 m_I$$

where g_I is the nuclear g-factor, a characteristic of the nucleus.

For protons $g_I = 5.5857$

(A positive value of g_I means that the nuclear magnet lies in the same direction as the nuclear spin)

$$\mu_N = \frac{e\hbar}{2m_p} = 5.05 \times 10^{-27} \text{ J/T}$$

and is called *nuclear magneton.*

Since $$g_I \mu_N = \gamma \hbar \quad ...(2)$$

where γ is gyromagnetic ratio of the nucleus.

For proton $$\gamma = 26.75 \times 10^7 \, T^{-1} S^{-1}$$

$\therefore$ equation (1) may be written as

$$E = -\gamma \hbar B_o m_I \quad ...(3)$$

The energy of separation of two states of hydrogen (H^1) nuclei with $m_I = +1/2$ and $m_I = -1/2$ is given by

$$\Delta E = E_2 - E_1$$
$$= [-g_I \mu_N B_o (-1/2)] - [-g_I \mu_N B_o (-1/2)]$$
$$= g_I \mu_N B_o \quad ...(4)$$

or $$\Delta E = \gamma \hbar B_o \quad ...(5) \text{ [From equation (2)]}$$

It may be noted that magnetic moment μ determines the *sensitivity* of NMR set up and consequently the *signal strength*. In case of hydrogen μ being very high so it can be very easily detected. It is due to this reason that NMR spectroscopy gives very good results if the studies are made for **hydrogen nucleus**.

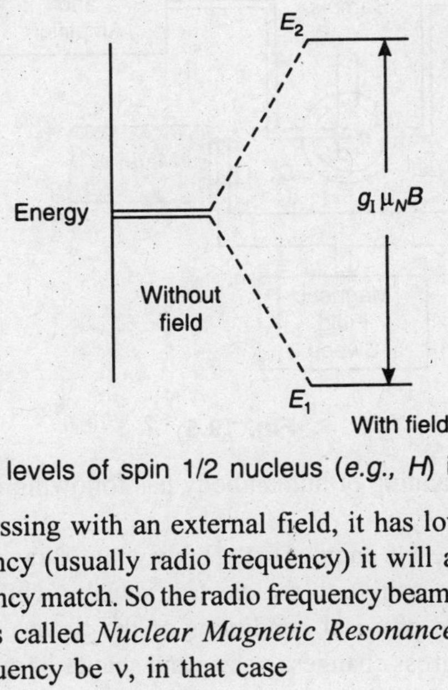

Fig. 19.4 energy levels of spin 1/2 nucleus (*e.g., H*) in a magnetic field

When a proton is precessing with an external field, it has low energy. When it is irradiated with an external field frequency (usually radio frequency) it will absorb energy if the precessing frequency and the radio frequency match. So the radio frequency beam and the nucleus are in resonance it is due this reason that it is called *Nuclear Magnetic Resonance (NMR).*

Let the resonance frequency be ν, in that case

$$h\nu = \Delta E$$

or $$h\nu = g_I \mu_N B_o = \gamma \hbar B_o$$

$$\therefore \quad \nu = \gamma \cdot \frac{h}{2\pi} \cdot \frac{B_o}{h} = \gamma \frac{B_o}{2\pi} \quad ...(6)$$

$\frac{\gamma}{2\pi} B_o$ is also called *Larmor frequency* ν.

As is clear from equation (6).

Frequency of precession of proton, ν, is proportional to the strength of applied external field B_o experienced by proton,

i.e., $$\nu = \gamma \cdot \frac{B_o}{2\pi}$$

or

$$\nu \propto B_o$$

where γ is gyromagnetic ratio and is given by

$$\gamma = \frac{\text{Nuclear magnetic moment } (\mu)}{\text{Nuclear angular momentum } (I)}$$

After absorption of energy, proton goes to a higher energy state and starts precessing in opposite orientation. This absorption of energy is recorded in the form of NMR spectrum. Subsequently, proton can lose the extra energy and come back (either by spin-spin relaxation or by spin lattice relaxation) with its original position.

Instrumentation

An NMR spectrometer consists of a magnet, a radio frequency source and a detection system which indicates the transfer of energy from radio frequency to the nucleus.

The sample is held in a strong magnetic field and exposed to a strong pulse of radio frequency (in the range of 100 MHz at 2.3T which contains all the frequencies over the H^1 range). This radiation changes the orientations of the nuclear spins in a controlled way and on return to equilibrium, these can be monitored on computer and analyzed mathematically using Fourier transform.

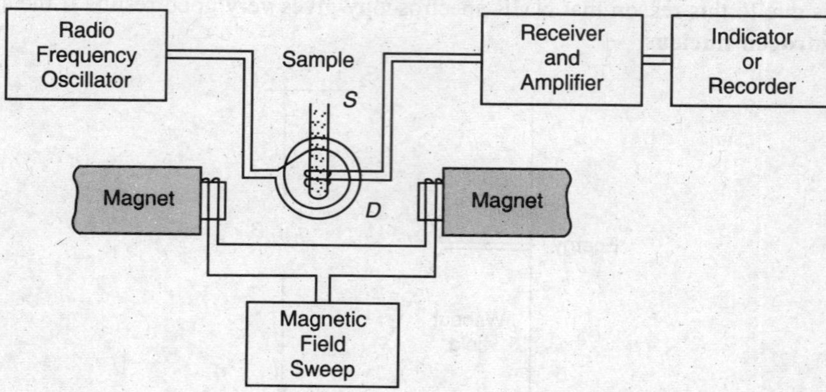

Fig. 19.5

NMR imaging used for scanning of human body has following important advantages over X-rays.

1. Magnetic fields do not cause any damage to the body while X-rays cause ionization and damage.
2. X-ray images are mostly sensitive to the electron density while NMR images are sensitive to variety of chemical and physical conditions of the tissues.
3. X-ray images only reveal gross changes in sizes and shapes of organs while NMR also reveals the physiological changes which are not possible to observe with X-rays.

Example. Calculate the frequency of radiation that comes into resonance with proton spin using a 10 T magnetic field. (γ for proton = 26.75×10^7 T^{-1} S^{-1})

Solution: In this case, $B_o = 10$ T

We have the relation $\nu = \gamma \cdot \dfrac{B_o}{2\pi}$

$$\therefore \quad \nu = \frac{(26.75 \times 10^7 \text{ T}^{-1} \text{ S}^{-1})(10 \text{ T})}{2 \times 3.142}$$

$$= 426 \times 10^6 \text{ S}^{-1}$$

$$= 426 \text{ MHz}$$

19.8. ELECTRON SPIN RESONANCE (ESR) SPECTROSCOPY

Electron Spin Resonance (ESR) which is also known as Electron Para Magnetic Resonance (EPR) and Electron Magnetic Resonance (EMR) spectroscopy is based on the splitting of magnetic energy levels produced by the application of a magnetic field on an impaired electron contained in an *atom or in a molecule or in an ion*. Basic principle of NMR spectroscopy and ESR spectroscopy resembles, however, the applications and instrumentations are different.

In ESR spectroscopy, free electron acts as a spinning negatively charged particle with a resultant magnetic moment (opposite to that in case of a proton and most other nuclei). Due to the charge and the spin, *an electron behaves like a bar magnet* and so will interact with an external magnetic field. We can express the magnetism in an electron by associating it with a magnetic moment μ which is proportional to *e*/*m*, where *e*, is the charge and *m*, the mass of the electron.

Spin of the electron being 1/2 and so like proton will have two energy levels $m = +1/2$ and $m = -1/2$. Lower level corresponds to $m = -1/2$ and higher level to $m = +1/2$. Magnetic moment of electron is nearly 1000 times more than that of proton and is given by $\mu = -g\beta S$ where g is *gyrometric ratio* or *g-factor* or *spectroscopic splitting factor*. This factor depends on the environment of electron. For free electrons, g is approximately 2 or nearly 2.0023 due to relativistic effects, and for paired electrons the value may have slightly different value.

When electrons are placed in a magnetic field, their energies are changed by some degrees and given by

$$E = \mu H = -g\beta s H$$
$$= \pm\frac{1}{2}(g\beta \text{H}) \quad ...(1)$$

where H is the strength of the magnetic field.

Those electrons which are in the direction of magnetic field $\left(\text{spin} = +\frac{1}{2}\right)$ will decrease in energy by an amount $\frac{1}{2}\, g\beta\text{H}$ and those in opposite directions $\left(\text{spin} = -\frac{1}{2}\right)$ will have the energy increase by $\frac{1}{2}\, g\beta\text{H}$. So, the difference in energy ΔE between two levels is given by

$$\Delta E = g\beta\text{H} = h\nu \quad ...(2)$$

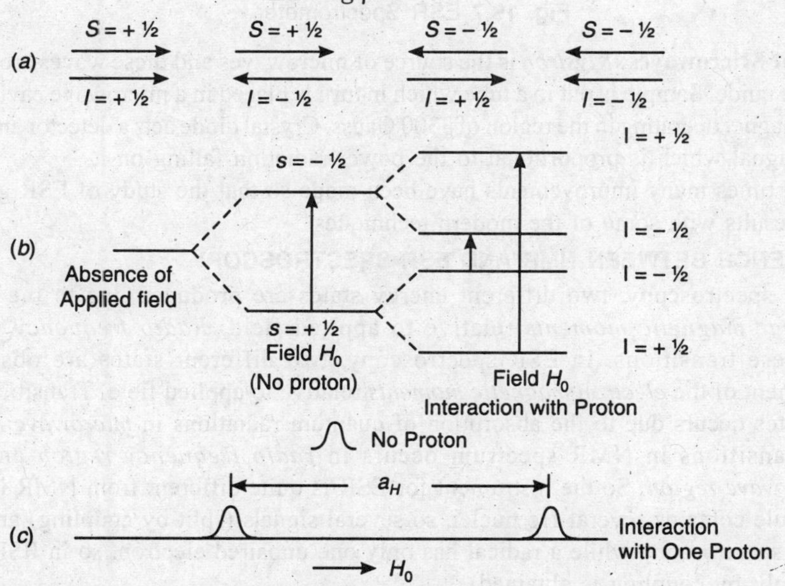

Fig. 19.6. Interaction of unpaired electron with one proton (*a*) Possible orientations of electrons and nuclear magnetic moments (*b*) Splitting of energy levels (*c*) Spitting of spectral line

which means that *e/m* radiation frequency associated with energy level difference will cause transitions between these levels.

In most of other forms of spectroscopy energy levels are essentially fixed, however, in ESR spectroscopy, there is splitting of energy levels and so the frequency of these levels is a function of applied magnetic field strength H.

Unpaired electron possess magnetic moment which can interact with magnetic nuclei and leads to splitting of ESR signal. If there are *n* number of magnetic nuclei present in a molecule, which can interact with electron, in that case $(n + 1)$ signals are observed in ESR spectra.

Splitting is measured in Hz and represented by a_H and is called *Hyperfine Splitting*.

The relative orientation of the nuclear magnetic moment and the electron magnetic moment causes a splitting of the original two level to four levels

Instrumentation

An important difference between ESR and NMR instrument is because of high frequency applied in ESR spectroscopy, power conducted from one point to the other is by rigid wave guide rather than by flexible coaxial cables.

Sample cell is inserted through an opening in the wave guide at a point where *e.m.* wave vector is undergoing maximum amplitude fluctuations and the ESR spectra are usually recorded in the form of first derivative of normal absorption spectrum to attain good resolution.

A simple instrument studying ESR spectroscopy at microwave frequencies is shnown in the figure 19.7.

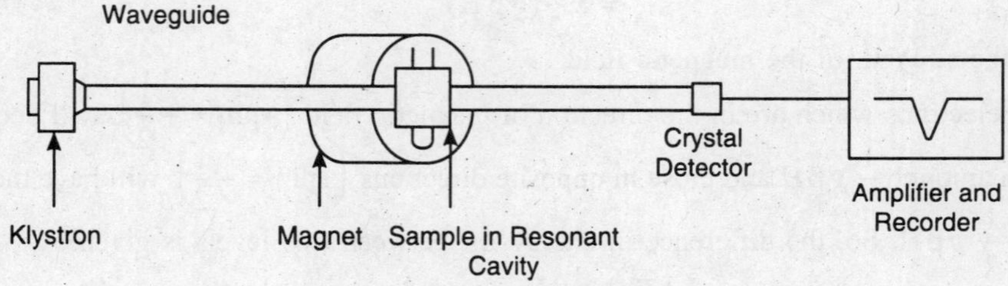

Fig. 19.7 ESR Spectrometer

Source of Microwaves: *Klystron* is the source of microwaves and these waves are conducted through a wave guide. Sample is put in a tube which in turn is placed in a microwave cavity between the poles of a magnet operating in the region of 3300 Gauss. Crystal diode acts a detector and produces a d.c. output signal which is proportional to the power radiation falling on it.

In recent times many improvements have been made so that the study of ESR gives better experimental results with some of the modern techniques.

19.9. DIFFERENCE BETWEEN NMR AND ESR SPECTROSCOPY

1. In NMR spectroscopy, two different energy states are produced due to the alignment of *nuclear magnetic moments* relative to applied field. *Radio frequency* is used to study these transitions. In ESR spectroscopy, two different states are observed due to alignment of the *electron magnetic moments* relative to applied field. Transition between these states occurs due to the absorption of quantum radiations in *microwave region.*
2. Since transitions in NMR spectrum occurs in *radio frequency region* and in ESR in *micro wave region.* So the instrument for ESR is quite different from NMR instrument.
3. A molecule contains several H¢ nuclei, so several signals (split by coupling) are obtained in NMR spectroscopy while a radical has only one unpaired electron, so in ESR only one signal (split by coupling is obtained).
4. Sensitivity of NMR spectroscopy is very small as compared to that of ESR due to the fact that magnetic moment of unpaired electron is about 1000 times that of proton.

19.10. THE MOSSBAUER EFFECT

The phenomenon of *resonance absorption* with γ -rays was predicted by W. Kuhn in 1921 and first observed by P.B. Moon in 1951. A major breakthrough occurred in 1958, when the phenomenon of *recoilless nuclear resonance* absorption was discovered by R.L. Mossbauer.

This new absorption effect provided a means for *eliminating the energy losses by recoil*, both at the source and at the receiver. As an additional feature of the discovery, gamma-ray spectrum lines of extreme narrowness have been produced, which may be useful for other research experiments.

In Moss baur effect the principle of Suppressing the loss of energy to a recoiling nucleus by rigidly fastening the emitting nuclei, as well as the absorbing nuclei, to an infinitely heavy mass is applied. An experimental method for doing this is the binding of nuclei to a crystal lattice.

To reduce *Doppler broadening* the γ -ray *source and absorber are cooled to extremely low temperatures*. This is done by immersing the radioactive crystal source and the crystal absorber in a liquid helium bath.

At temperatures near absolute zero, atomic vibrations within the crystal lattice are *reduced to practically zero*, so that during emission or absorption the atoms have very little or no motion. Under these conditions Doppler broadening is reduced to practically zero and the observed line reveals its *natural line breadth*.

The elimination of atomic vibrations within the crystal lattice has another important effect, *i.e.*, **Atomic vibrations are quantized**, that is, vibration energies have discrete values which may be represented by spaced energy levels. When atoms are in their lowest vibrationl energy state, as in the case when the crystal is at absolute zero, the recoil energy of a gamma-ray is not always sufficient to raise an atom to the next vibrational state. Under these conditions the binding of a nucleus with its neighbours is such that an appreciable part of the crystal lattice takes up any γ-ray recoil momentum. Because the crystal mass is so large, the recoil energy is practically *zero*. Gamma-ray emission or absorption under these conditions is said to be *recoilless*.

It should be noted that with the Doppler effect and nuclear recoil both reduce α to zero, only the natural line breadths are involved in nuclear resonance absorption.

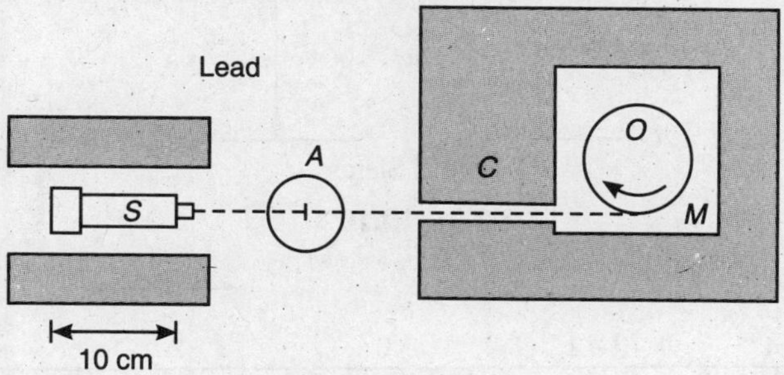

Fig. 19.8

A diagram of Mossbauer's apparatus is shown in Fig. 19.8. Gamma-rays from a source at M pass through a narrow channel *C* and a thin target *A*, to where they fall on a scintillation counter S. The lead shielding around the source and the detector is for the purpose of reducing all **extraneous radiation** to a minimum.

In order to detect resonance absorption in A, the source is set into rotation as shown by an arrow and the detector turned on only when the source is moving in line with the absorber. This motion introduces a Doppler shift *d*, that is, a slight increase or decrease in the frequency of the gamma-rays. The effect is illustrated by shifting the dotted curve in Fig. 19.9 to the right or to the left. The non-matching of frequency between the γ-rays from the moving source and the nucleus in the stationary absorber reduces the number of photons absorbed. The more first the source moves toward or away from the target, the more is the **Doppler shift** is and the greater the number of γ-rays reaching the counter.

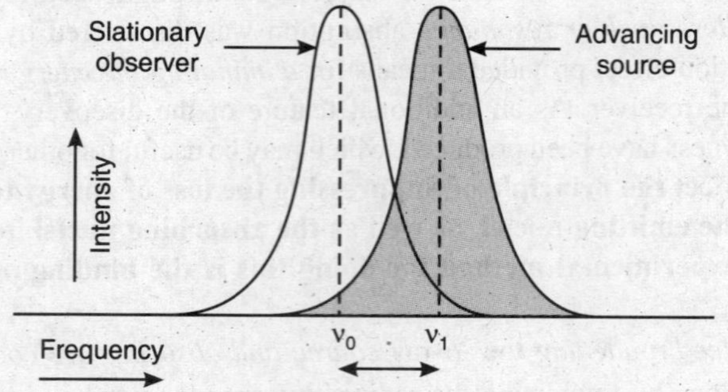

Fig. 19.9

Fig. 19.11 is a graph from Mossbauer's original experiment, performed by using the 0.129-MeV gamma-rays from iridium-191. It may be noted that the greatest absorption occurs when the source is at rest. Observe also the extremely low velocities required to produce a Doppler mismatched in frequency. The latter phenomenon is a clear indication of the narrowness of γ-ray spectrum line and shows that one is observing natural line breadths rather than **Doppler broadening**.

The mean lifetime of Ir^{191} transition can be calculated from the half-width of the curve in Fig. $\Delta E \cdot \Delta t = h$.

To calculate it we first find the γ-ray frequency ν_0 by using the equation Ve = $h\nu_0$.

$$\nu_0 = \frac{Ve}{h} = \frac{1.29 \times 10^5 \times 1.6 \times 10^{19}}{6.6 \times 10^{-34}} = 3.13 \times 10^{19} \text{ Hz}$$

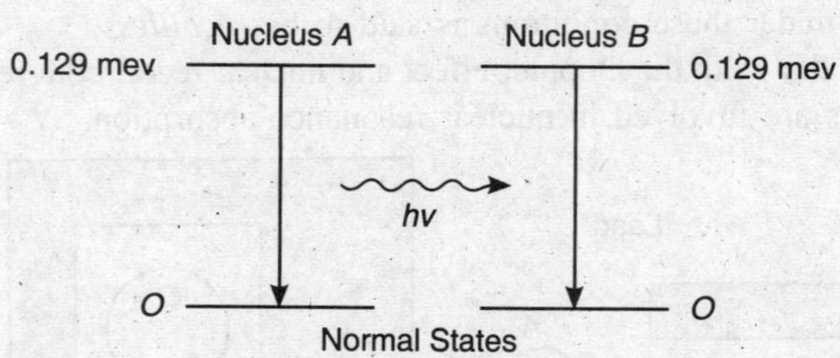

Fig. 19.10

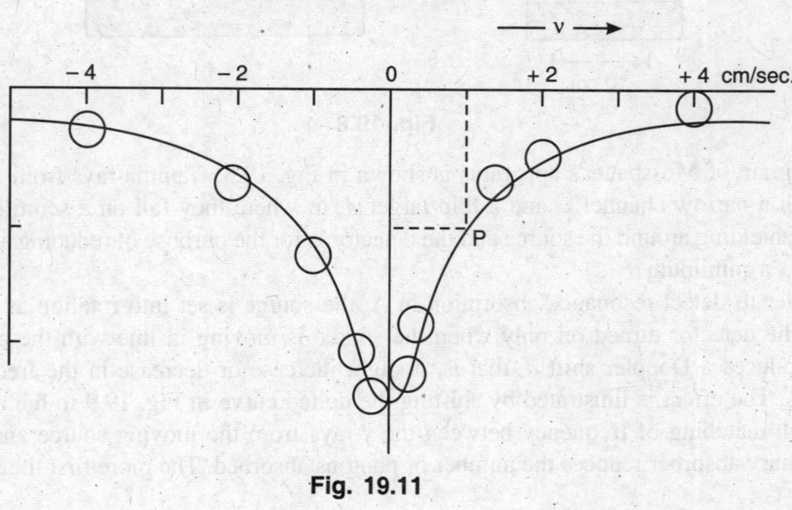

Fig. 19.11

The half-width at half-intensity of the resonance curve in Fig. 19.11, point P, is approximately 1 cm/sec, or 1×10^{-2} m/sec. If we now use the Doppler formula, $v = v_0(1 + v/c)$, to find the corresponding change in frequency Δv, where $\Delta v = v - v_0$, we get

$$\Delta v = v_0 \frac{v}{c} = \frac{3.13 \times 10^{19} \times 1 \times 10^{-2}}{3 \times 10^8} = 1.04 \times 10^9 \text{ Hz}$$

Also $E = hv$, $\Delta E = h\Delta v$, and from equation $\Delta E \cdot \Delta t = h$, we have

$$\Delta t = \frac{h}{\Delta E} = \frac{h}{h\Delta v} = \frac{1}{\Delta v}$$

or

$$\Delta t = \frac{1}{1.04 \times 10^9} = 9.6 \times 10^{-10} \text{ sec.}$$

The mean life of this gamma-ray energy state is, therefore, 9.6×10^{-10} sec.

The nuclear energy levels of Ir^{191} involved in Mossbauer's experiment are shown in Fig. 19.11.

Many experiments have confirmed Mossbauer's discovery and *recoilless γ-ray transitions* have been observed in fifteen or more radioactive nuclei. For his discovery Mossbauer was awarded, along with Hofstader, Nobel prize in 1961.

19.11. APPLICATIONS OF MOSSBAUER SPECTROSCOPY

1. Mossbauer spectroscopy is very useful in the study of molecular structure as the isomer shift ($1s$) in Mossbauer spectra reflects the difference in the s-electron density around the emitter and absorber nuclei.
2. Mossbauer effect is useful in the study of quadrupole splitting due to asymmetry. When the absorber atom is in a symmetrical environment, a single line is observed in Mossbauer spectrum. When environment is asymmetrical a table internal electric field operates and this gives rise to quadrupole splitting of the line into a pair of lines.
3. **Structure determination:** Study of quadrupole splitting is useful to get information regarding the structure of the molecules.
4. **Isomer shift and curie point:** Mossbauer effect is useful to study the transition from paramagnetic substance to ferromagnetic substance because of change of domains during transitions. Mossbauer spectrum can be studied over a wide range of temperatures thereby identifying the transition from paramagnetic to ferromagnetic substances and Curie temperature.
5. **Biochemical applications:** Dron is an essential constituent of many biochemicals which are quite important for life. It is present in haemoglobin which transports oxygen through blood. Mossbauer spectrum is useful for the study of biochemicals which support the life system.
6. Mossbauer effect is also useful to study the proteins and enzymes.
7. Mossbauer effect also finds applications in the study of various bacteria which are present in plants.
8. Mossbauer spectra can give information about the variety of crystal structures of alloys or compounds containing the elements to which Mossbauer spectrum is sensitive.
9. Mossbauer effect is useful to study the nuclear magnetic moments, nuclear spins and other nuclear properties of elements.
10. Mossbauer effect is able to explain the Einstein's principle of equivalency, which is an essential part to explain the gravitation theory proposed by Einstein.

QUESTIONS

1. What is the importance of thin films? How these are produced?
2. What are thin films? What is TFT and IGEFT ? State there functions in thin films.
3. How thin films are finding applications in modern circuits?
4. What is NMR spectroscopy? Give its underlying principle.
5. What is nuclear spin and magnetic moment? How are these used in Nuclear Magnetic Resonance?
6. What is gyromagnetic ratio? Give its importance. Mention the important principles of Nuclear Spin.

7. What is PMR spectroscopy? Is it different from NMR spectroscopy? Calculate the energies of nuclei in magnetic fields.
8. What is nuclear magneton? How is it related to electron magneton?
9. Draw energy levels of a spin = 1/2 nucleus. Give a relation for Larmor frequency.
10. Give the instrumentation technique of NMR spectroscopy.
11. What is electron spin resonance (ESR) spectroscopy? Give its important features.
12. Calculate the energy between two levels using ESR spectroscopy.
13. Draw energy diagrams for possible orientation of electrons and nuclear magnetic moments. How will you account for the splitting of energy levels and splitting of spectral lines?
14. Give the instrumentation techniques of ESR spectroscopy.
15. How NMR and ESR spectroscopies are different from each other?
16. What is Mossbauer effect? How is it useful? Give its salient features.
17. What is Mossbauer effect? How will you explain the recoilless emission?
18. What is resonance absorption? How is it useful?
19. Show with an energy level diagram the recoil emission. Also explain the instrumentation of Mossbauer effect.
20. Give the various applications of Mossbauer effect.

CHAPTER 20

PARTICLE ACCELERATORS

20.1. INTRODUCTION

The α and β-particles given out by radioactive substances neither possess sufficiently large speeds nor their speeds are under control. Rutherford used α-particles from RaC in his experiments on artificial disintegration. It becomes important to accelerate a particle under very high potential difference before these are used for bombarding other atomic nuclei. Cockcroft and Walton designed a *voltage multiplier circuit* to produce high potential difference from an A.C. input. By applying a high potential difference of the order of 2 million volt to a tube containing hydrogen resulted in a stream of protons of 1 to 2 MeV energy. The following are the different types of accelerators used to accelerate different charged particles to desired energy.

20.2. COCKCROFT AND WALTON ACCELERATOR

Cockcroft and Walton designed a voltage multiplier circuit to produce high potential difference from an a.c. input to produce 200 KeV protons.

Principle: *The basic principle of this instrument is to charge a number of capacitors connected in parallel and to discharge them in series when the voltage across them is equal to the sum of the voltages on the individual capacitor.*

Construction: Cockcroft-Walton acclerator consists of a number of accelerator C_1, C_2, C_3, ... etc. of equal capacitance are arranged together with rectifiers R_1, R_2, R_3, ... etc. as shown in Fig. 20.1. The function of these rectifiers is to allow the current to pass in one direction only as indicated by the arrows.

Here we will assume that no current loss taken place across any component. T is a step up transformer which provide output voltage V_0 ranging from 20 kV to 100 kV. Point A is earthed to keep its potential at a fixed value is zero. All these parallel capacitors are charged with the help of this transformer and then discharged in series, and this high voltage is applied to accelerate the charged particle.

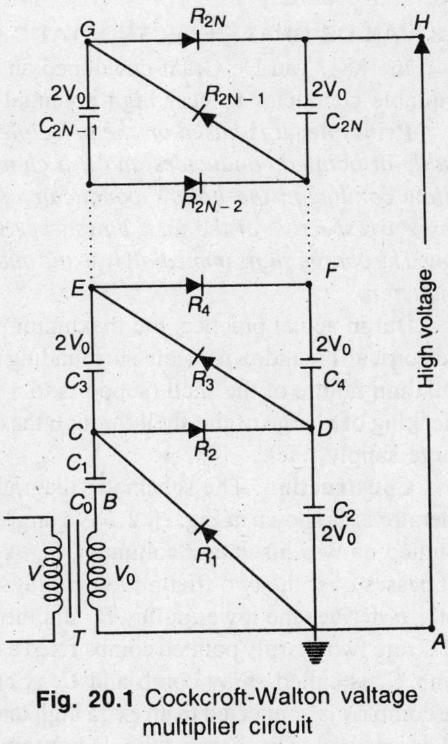

Fig. 20.1 Cockcroft-Walton valtage multiplier circuit

Working: A.C. is applied across the primary of the transform T. Point A is at zero potential so the potential of point B oscillates between V_0 and $-V_0$. During the positive half cycle point B becomes +ve w.r.t. A and the rectifier R_1 is reverse biased and does not conduct. But during negative half cycle, the point B becomes -ve w.r.t. A and the rectifier R_1 is forward biased and changed the capacitor C_1 to potential V_0. The potential difference between A and B, however changes from $-V_0$ to $+V_0$. Thus the potential difference between B and C varies between 0 ($V_0 - V_0$) during +ve half and $2V_0$ [V_0 ($-V_0$)] during −ve half. Now the rectifier R_2 conducts and the charge is now shared with C_2. This process is repeated during successive half-cycles of the secondary of the transformer and all the capacitors are fully charged. Thus the potential difference across each capacitors becomes $2V_0$. So the potential difference between F and A is $4V_0$. This arrangement of two capacitors and two rectifiers is calleα *voltage doublers*. Hence the voltage can be multiplied upto any multiple of V_0. It is due to this reason that this arrangement is called voltage multiplier or cascade multiplier.

If the circuit has N capacitors each of capacitance C and N rectifiers with peak input voltage V_0 and load current I, then output voltage across A and H will be given by

$$V = NV_0 - \frac{I}{12f.C.}\left[N^3 + \frac{9}{4}N^2 + \frac{1}{2}N\right] \quad ...(i)$$

where f is the frequency of transformer mains supply.

The ripple voltage is given by

$$\pm V = \frac{IN(N+2)}{16f.C}$$

So to have large V, the value of both f and C should be large (these values also reduce ripple factor). C cannot be made very large because it will make the apparatus bulky, f is increased by using radio frequency source.

The high voltage between the terminals of the machine is applied to the charged particle to be accelerated by evacuated accelerating tube. Air insulated Cockcroft-Walton machine is convenient for voltage upto 1000 kV. In later form of the apparatus, energy of charged particles upto 4 MeV have been realized.

20.3. VAN DE GRAAF ELECTROSTATIC GENERATOR

In 1881, Van De Graaf developed an old "electrostatic machine" which was used to charge a suitable conductor to very high potential of the order of several million volts.

Principle: *It is based on the principle that the electric discharge takes place in air or gases readily at pointed conductors and if a charged conductor is brought into internal contact with a hollow conductor, the hollow conductor continous accepting the charge irrespective of the fact, howsoever large its potential may grow. The charge shifts immediately to the outer surface of the hollow conductor.*

But in actual practice, the maximum potential is limited due to electrical breakdown of air surrounding the metallic shell and insulation failure of the shell (supports to a value at which the rate of leaking of charge of the shell through these processes, equals the charge supply rate).

Construction: The schematic diagram of the Van de Graaf generator is as shown in Fig. 20.2. S is a large hollow metallic sphere mounted on two insulating column CC'. A well insulated endless belt passes over the two friction less pulleys P_1 and P_2. An electric motor E derives the lower pulley P_1 and the belt starts rotatory. C_1 and C_2 are two sharply pointed combs fixed as shown in the Fig. 20.2 Comb C_1 is called spray comb and C_2 is called collecting comb. The comb C_1 is connected to an extra high tension EHT whose other end is earthed. Discharge tube D contains positive ions to be accelerated and T is the target to be hit. For producing a–particles, helium is used. Similarly, deutrium is used for accelerating deutrons and hydrogen is used for accelerating protons.

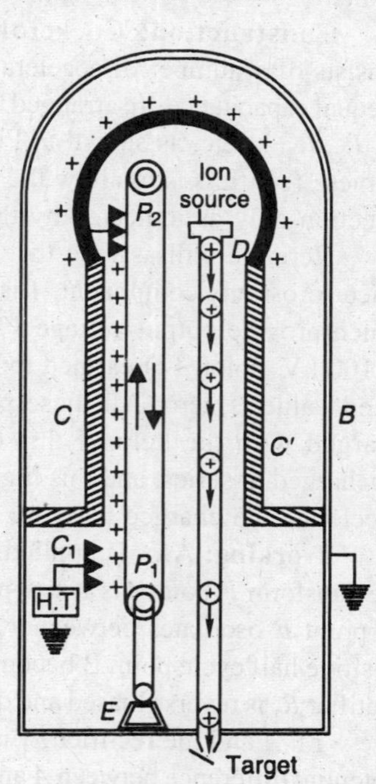

Fig. 20.2 Schematic diagram of Van de Graaf electrostatic generator.

Working: A well insulated endless belt B rapidly rotates between two pulleys P_1 and P_2 by means of an electric motor E, which receives electric charges produced by a comparatively low voltage of the order of 10 K volts and sprayed on it by means of a metallic comb C_1 at bottom. The belt B moving upwards carries this charge into a hollow insulated metallic sphere S where the charge is removed from the belt and given to the sphere by means of second metallic comb C_2.

The belt returning to the bottom is charged again and delivers up the charge when inside the sphere. Charges are thus collected and distributed on the surface of the sphere as shown in Fig. 20.2 with such an arrangement, the sphere can be raised to very high potential of the order of 5 million volts. For still high potential difference, the entire apparatus must be enclosed in an outer shell in which gas pressure is raised.

If C is the capacitance of the shell and q is the charge given to it the potential V is given by

$$V = \frac{q}{C}$$

The rate of increase of potential

$$\frac{dV}{dt} = \frac{1}{C}\frac{dq}{dt} = \frac{I}{C}$$

Maximum charging current are a few milliampere.

In order that the potential of the shell be as large as possible, the rate of charging should be high and the leakage should be small. The charge carrying belt should have high mechanical strength, high surface and volume resistivity, high dielectric strength and be fire resistant. Ion source is placed inside the hollow hemisphere and from this the +ve ions are accelerated down the accelerating tube on to the target placed at low voltage end of the accelerating tube. The accelerating tube usually made up of porcelain glass or other insulating materials.

In the first Van de Graaf generator, the maximum energy obtained was only 1.5 MeV. Later on, energies as high as 8.5 MeV were obtained. Recently, energy as high as 20 MeV have also been designed.

20.4. TENDEM ACCELERATOR

Van de Graaf accelerator is a single stage accelerator. Tandem accelerator is a modified form of Van de Graaf accelerator. In Tandem accelerator, which is a two stage accelerator, we can obtain charged particles with energy double than that in Van de Graaf accelerator, with same operating voltage.

Principle: *In a single stage accelerator, positive ions are produced at high potential and then these are accelerated from high voltage to the ground potential. In case of Tandem accelerator, the negative ions in the first stage are produced at ground potential and are accelerated towards high positive voltage. In second stage, the electrons are stripped off and negative ions are changed into positive ions which are then accelerated again to the ground potential. In this way, the energy of the charged particles can be increased with the same operating voltage.*

Construction and Working: The construction of a Tandem accelerator is as shown in Fig. 20.3. It consists of a source S of positively charged ions (say, protons). These ions are then passed through a column C_1 which adds electrons to these positive ions. A very small fraction of the positive ions is converted into negative ions in column C_1. A beam of positive and negative ions and neutral particles, which comes out of C_1 is then passed through the magnetic field provided by an analysing magnet M_1 which separates the negative ions from other unwanted particles. These negative ions are then attracted towards the positive terminal T through the insulating column. This beam of negative ions is, therefore, accelerated to high velocity due to the potential difference. This accelerated beam of negative ions is, then passed through a change stripping canal C_2. This canal contains a gas at low pressure and removes extra electrons to change each negative ion into a positive ion (in case the gas given by source S is of protons, the column C_1 adds two electrons to each proton to make them negatively charged, the column C_2 removes two electrons from each negative ion to change them back into protons). The beam of positive ion produced by stripping canal is repelled by the positive terminal T to ground potential and is again accelerated. It should be noted that second stage is extremely powerful when heavy ions are accelerated. For example, take the case of oxygen nuclei with atomic number $Z = 8$. Let the voltage of terminal T be + 5 million-volt. In the first stage, when ions are single charged the kinetic energy is increased to 5 MeV. But in the second state, when the ions will have eight units of positive charge (1 unit = 1.6×10^{-19} C) on passing through stripping canal, then the kinetic energy is further increased by 40 MeV so that the beam has a total kinetic energy of about 45 MeV.

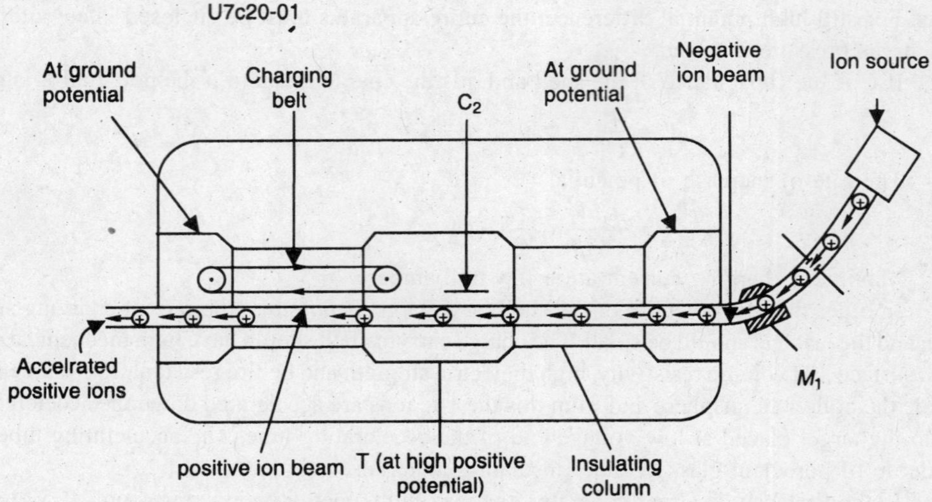

Fig. 20.3 Schematic diagram of Tandem accelerator

20.5. LINEAR ACCELERATOR

In a linear accelerator (abbreviated as LINAC), the kinetic energy of a charged particle is increased in steps as it travels along a straight line *i.e.*, linear path.

Principle: The working of a linear accelerator is based on following two facts:

(*i*) *When a charged particle is subjected to an alternating electric field, it is accelerated during one half cycle and is related in the next half cycle.*

(*ii*) *The electric field inside a hollow charged conductor is always zero and a charged particle moves inside it with uniform velocity.*

Construction: A linear accelerator consists of a number of cylindrical metallic tubes 1, 2, 3, ... of increasing length arranged co-axially in an evacuated chamber as shown in Fig. 20.4. The alternate tubes 1, 3, 5, ... are connected together and are connected to one terminal of a high frequency (H.F.) oscialltor. The other set of alternate tubes 2, 4, 6, ... which are connected together, is connected

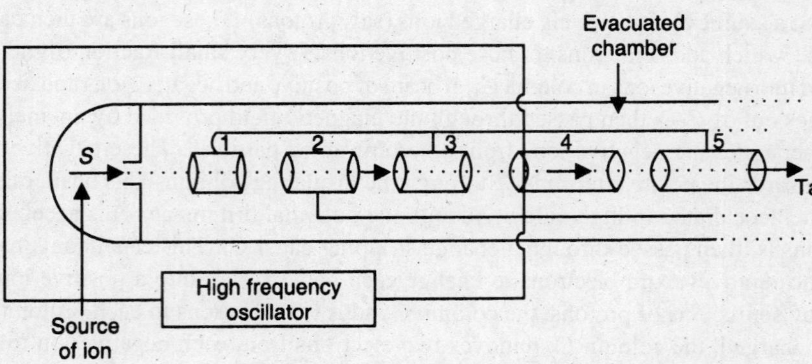

Fig. 20.4 Linear accelerator

to the other terminal of the oscillator. When one set of tubes is positive, then the other set becomes negative and vice-versa. Also during one half cycle, a set of tubes remains positive and the polarity is reversed in the next half cycle. The beam of ions from a source S is made to enter the tubes along their axis.

Working and Theory: Let a positive ion from S passes through tube 1. As electric field inside this tube is zero, therefore, the ion will not be accelerated in this tube. But as soon as this ion comes out of tube 1, if the tube 2 becomes negative and tube 1 becomes positive, then the ion will be accelerated in the gap between the tubes 1 and 2. Therefore, the ion enters tube 2 with increased velocity. When the ion comes out of tube 2, the tube 3 should just become negative so that the particle is again accelerated in the gap between tubes 2 and 3 and so on. The ion will be accelerated each time when it traverses the gap between the tubes if the preceding tube is positive and following tube is negative. Inside a tube, electric field is zero and the charged particle moves with constant velocity.

Therefore, for the acceleration of the ion, the time of transit of the ion through the length of a tube must be equal to half the time period of the high frequency alternating voltage supplied by the oscillator. As the velocity of the ion increases due to its recurring acceleration, therefore, each tube must be longer than its preceding one because the accelerated ion travels increasingly longer distances in the same time (equal to half the time period of alternating voltage).

Let q and m respectively represent the charge and mass of the ion. If V is average value of potential difference between the successive tubes during the acceleration of ion, then its velocity v_n while entering nth tube will be given by

$$nqV = \frac{1}{2}mv_n^2$$

or $$v_n = \sqrt{\frac{2nqV}{m}} \qquad ...(i)$$

This equation holds good when the velocity of the ion is much smaller than that of light. From above equation. It is clear that for the ions of same $\frac{q}{m}$ ratio and same accelerating potential between the successive tubes. The velocity of ion while entering nth tube is proportional to $\sqrt{n}$ *i.e.*, the velocities of ion just at the time of entering tubes 1, 2, 3, ... will be in ratio of 1, $\sqrt{2}$, $\sqrt{3}$, respectively.

The time taken by the ion to pass through all the tubes is same and is $\frac{T}{2}$ (half of time period T of alternating voltage). Therefore, the length of nth tube will be given by

$$l_n = v_n \frac{T}{2} = \frac{T}{2}\sqrt{\frac{2nqV}{m}}$$

or $$l_n = T\sqrt{\frac{nqV}{2m}} \qquad ...(ii)$$

As q, m, V and T are constants, therefore, $l_n \propto \sqrt{n}$ *i.e.*, the lengths of the tubes 1, 2, 3, ... are in the ratio of 1, $\sqrt{2}$, $\sqrt{3}$,.... This arrangement will keep the ion in phase with the changing field and will be accelerated in between the gap.

The use of early linear accelerators was limited only to heavy ions such as mercury ions due to low frequencies of alternating voltage available at that time. But these days, it is possible to generate very high frequency alternating voltages and therefore, linear accelerators can be used to accelerate protons and even electrons.

Advantages:

(1) Linear accelerators are economical when very high energies are required.

(2) These can provide well collimated beam of accelerated, charged particles with high intensity.

Disadvantages: The major disadvantage of the linear accelerator is its large length. To accelerate the charged particles to high energies, the number of cylindrical tubes used will be more and, therefore, the length of accelerator becomes large.

20.6. THE CYCLOTRON

The main two drawbacks in linear accelerator were overcome by Lawrence by using strong magnetic field. It is, thus, known as *magnetic resonance accelerator or cyclotron.* The cyclotron is a device for accelerating charged particles to high energies by subjecting them to repeated voltage impulses.

Principle: *It is based on the principle that charged particles can acquire high energy when they are repeatedly passed through an accelerating electric field along a closed path.*

Construction: It consists of two *D*-shaped hollow semi-circular chambers called Dees D_1 and D_2. These circular dees are lying in a horizontal plane with a small gap separating them. A source of ions is placed in the middle point of the gap between the dees. The dees are connected externally to the terminals of a radio-frequency oscillator having an amplitude of several million times per second [Fig. 20.5 (*a*)]. The dees D_1 and D_2 are enclosed within a large metal box containing gas at low pressure of 10^{-3} mm of mercury. The whole apparatus is placed between the poles of a strong electromagnet which provides a magnetic field perpendicular to the plane of the dees D_1 and D_2 of about 10^4 amperes/metre [Fig. 20.5 (*b*)].

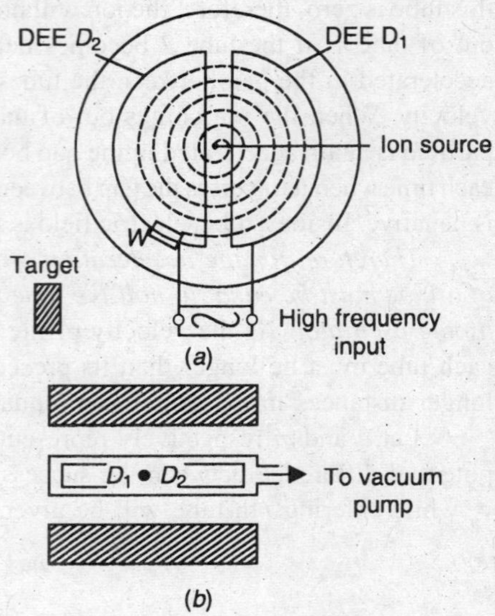

Fig. 20.5 Schematic diagram of Cyclotron.

Working: If a positive ion is generated at a point within the gap at a time when dee D_1 is at a positive potential and dee D_2 is at a negative potential, the positive ion will be accelerated across the gap D_2 and enter the hollow circular chamber D_2. As there is no electric field inside the dee and only perpendicular magnetic field will act on the positive ion and it moves in a circular path and finally come back to the gap between the two dees. The frequency of the oscillator is so adjusted that the ion finds D_1 as negative and hence enter the hollow circular chamber D_1. This process is repeated several times, the radius and velocity of the ion goes on increasing till it reaches the window *W*, where it comes out of the dees and strike the target.

Theory m = mass of ion to be accelerated
B = magnetic field
q = charge on the ion
v = velocity of the ion
r = radius of semi-circular path.

The centripital force required by the ion to move in a circular path is provided by the perpendicular magnetic field *B*.

So
$$qvB = \frac{mv^2}{r}$$

or
$$r = \frac{mv}{qB}$$

If *t* is the time required to complete the semi-circular path, then

$$t = \frac{1}{2}\cdot\frac{2\pi r}{v} = \frac{\pi r}{v}$$

Using Equation (*i*), we get

$$t = \frac{\pi m v}{vqB} = \frac{\pi m}{qB} \qquad ...(ii)$$

From Equation (*i*), we find that time taken by the ion to complete the semicircular independent of the speed and radius of the circular path in which it moves, provided B, q and m remain constant.

Cyclotron frequency: If T is the time period of the charged particle, then

$$T = 2t = \frac{2\pi m}{Bq} \qquad ...(iii)$$

So the frequency of revolution (*i.e.*, cyclotron frequency)

$$\nu = \frac{1}{T} = \frac{Bq}{2\pi m} \qquad ...(iv)$$

And angular frequency is given by

$$\omega = 2\pi\nu = \frac{Bq}{m} \qquad ...(v)$$

Maximum Energy of particle: The final energy E of the charged particle is given by

$$E_{max} = \frac{1}{2} m v_{max}^2$$

where v_{max} is the maximum velocity gained by the charged particle in the final orbit of radius r_{max}. Now

$$\frac{m v_{max}^2}{r_{max}} = Bqv_{max} \qquad \therefore \; v_{max} = \frac{Bq}{m} r_{max}$$

Hence

$$E_{max} = \frac{1}{2} \frac{mB^2 q^2 r_{max}^2}{m^2}$$

$$E_{max} = \frac{1}{2} \frac{B^2 q^2 r_{max}^2}{m} \qquad ...(vi)$$

With a comparatively small potential difference of the order of 50,000 volts, very fast moving particle can be produced. For example, when a particle makes 200 revolutions before emerging out, it will gain a velocity equivalent to total fall through a potential of

$$2 \times 50000 \times 200 = 20 \times 10^6 \text{ volts} = 20 \text{ MeV}$$

If heavy hydrogen is used instead of ordinary hydrogen, a beam of light energy deutrons is obtained. As their mass is double, they possess greater energy and are more useful as atomic projectiles.

A typical cyclotron might have a flux density of 2 webers/m^2 and dees diameter 1 metre: when used with an oscillator of high frequency (Mega cycles/sec.) and high alternating voltage (200 kilo volts) a cyclotron can produce a 1 milliampere current of 25 MeV neutrons or 12 MeV protons.

Limitations of Cyclotron: (*i*) The energy to which particle can be accelerated in a cyclotron is limited due to variation of mass with velocity. The mass of a particle when moving with a velocity υ is given by

$$m = \frac{m_0}{\left(1 - \frac{v^2}{c^2}\right)^{\frac{1}{2}}}$$

where m_0 is the rest mass and c the velocity of light. From Equation (v) the angular velocity of the ions becomes

$$\omega = \frac{Bq}{m_0}\left(1 - \frac{v^2}{c^2}\right)^{\frac{1}{2}}$$

The angular velocity decreases as v increases. The frequency of ion rotation decrease; the ions take more time to describe their semi-circular path than the fixed period of the oscillating electric field and they arrive too late at the gap between the dos to be accelerated. Thus, the ions lag behind the applied potential difference till a stage is reached when it can no longer be accelerated further.

The frequency of the ion can be kept constant by making $B\sqrt{1-\frac{v^2}{c^2}}$ constant. For this, the value of the flux density B should increase as the velocity of the ion increases so that the product remains unchanged.

(*ii*) Uncharged particles (*e.g.* neutrons) can't be accelerated by cyclotron.

(*iii*) Only heavy charged particles like protons, deutrons, α-particles etc. can be accelerated by the cyclotron, and *electron cannot be accelerated,* because the mass of electron is very small and a small increase in the energy of electron makes it move with very high speed and as a result, the electrons go quickly out of step with oscillating electric field.

Advantages:

1. The main advantage of the cyclotron is that it requires comparatively low voltage source for producing high energy particles.

2. The accelerated charged particles produced by cyclotron are used for producing high energy particles like neutrons by collisions.

3. The accelerated particles produced by cyclotrons are used in artificial transmutation of elements.

20.7. THE BETATRON (MAGNETIC INDUCTION ACCELERATOR)

Betatron is based upon the principle of electron accelerator. As β-particles are electrons, so the accelerator is called betatron. It is a device in which the difficulty of variation in mass of electron with velocity has been eliminated. It employs no electric field; but a changing magnetic field which produces an inductive electromotive force on the circular motion of the electrons by which they are accelerated, hence betatron is also known as **Induction electron accelerator**. A betatron produces electrons possessing energies of the order of 100 MeV to 500 MeV. This device differs from the cyclotron in at least two fundamental respects: First, the electrons are accelerated by a rapidly changing magnetic field and second, the orbit of the particles has a constant radius.

Principle: *It's based on principle of electromagnetic induction (i.e., alternating current in primary coil of a transformer produces a similar current in the secondary). The betatron may be considered to be a transformer in which secondary coil is replaced by cloud of electrons moving in a circular orbit due to magnetic field in a doughnut shaped vacuum chamber.*

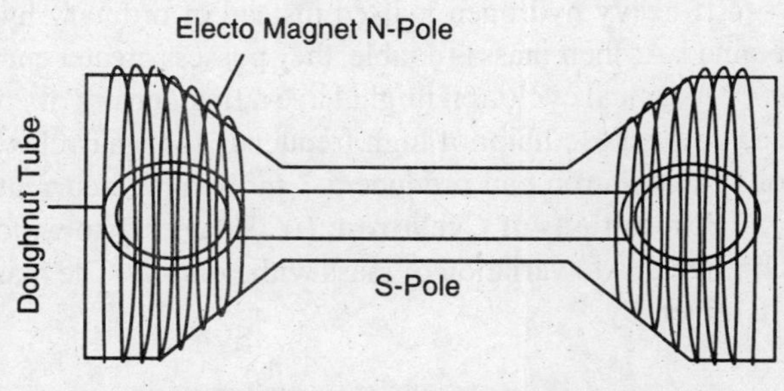

Fig. 20.6 Doughnut pole pieces of a betatron

Construction: A betatron consists of a doughnut shaped (the shape is like inner tube of a bicycle tyre) vacuum chamber arranged between the pole pieces of an electromagnet which is fed to a.c. of high voltage (Fig. 20.6). The magnetic field produced by a.c. is normal to the plane of the doughnut. The magnetic field is stronger at the central region than at two ends. The changing magnetic field does following jobs:

(*i*) It produces a tangential accelerating force to the electrons by keeping itself alternating in time and inducing an emf in doughnut.

(*ii*) It provides appropriate centripetal force to keep the electrons in a circular equilibrium orbit.

(*iii*) The magnetic field changes in such a manner that the condition $mv = Ber$ for constant radius is satisfied.

The inner surface of doughnut is coated with thin layer of silver to avoid accumulation of electrons but the conductivity of the layer should be low otherwise eddy currents will be generated which can distort the magnetic field and hence disturbs the operating condition.

Working: Electrons are injected into the chamber by an electron gun which consists of a heating filament F as emitter of electrons, Grid G for focussing the electrons and plate P which acts as the injector as shown in Fig. 20.7. The electrons are injected when B is just rising from zero value on the first quarter cycle. The increasing magnetic field induces a potential within the doughout which accelerates the electrons and thereby increases the energy of the electrons in a stable orbit of constant radius. The electrons make several hundred thousand revolutions in their fixed orbit while the alternating magnetic field is increasing in intensity from zero to maximum *i.e.*, during a quarter of a cycle. At each revolution the electrons gain additional energy. At the end of the quarter cycle, when the intensity is at its maximum, a pulse current is sent through an auxiliary coil which suddenly changes the magnetic field and high energy electrons are displaced from the stable orbit and are made to strike back of injector shield, which acts as the X-ray target and produces a highly energetic X-rays.

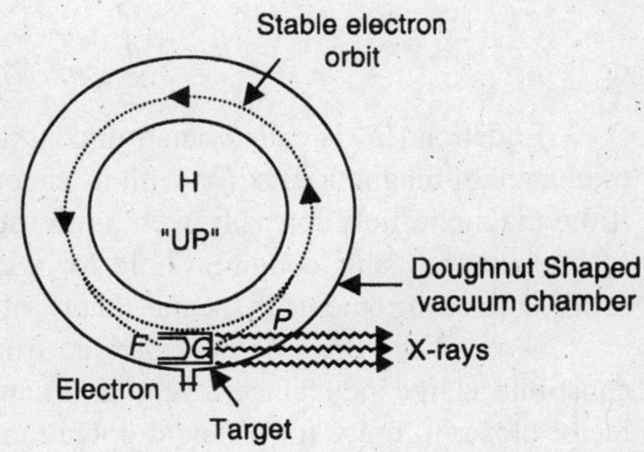

Fig. 20.7 Doughnut shaped chamber.

Theory: Consider that the electron is revolving in a circular path of radius r. The induced e.m.f. produced in the circuit by changing magnetic flux is given by *Faraday's Law.*

$$\in = \frac{d\phi}{dt}$$

where ϕ denotes magnetic flux at any instant and it perpendicular to the plane of the circuit.

The work done on an electron of charge e in one revolution is given by

$$W = \text{Charge on the electron} \times \text{induced e.m.f.}$$

$$W = e\in = e\frac{d\phi}{dt}$$

Average force acting on electron during one revolution is given by

$$F = \frac{\text{Work done}}{\text{distance travelled in one revolution}}$$

$$F = \frac{W}{2\pi r} = \frac{e\,d\phi/dt}{2\pi r} \quad ...(i)$$

For an electron moving in a circular orbit of radius r due to perpendicular magnetic field, we have

$$Bev = \frac{mv^2}{r}$$

or $$mv = erB$$

or $$p = erB$$

or $$\frac{dp}{dt} = er\frac{dB}{dt}$$

From Newton's second law of motion, the rate of change of momentum is equal to the force

$$\therefore \quad F = \frac{dp}{dt} = er\frac{dB}{dt} \quad ...(ii)$$

From Equations (i) and (ii), we have

$$\frac{e\,d\phi/dt}{2\pi r} = er\frac{dB}{dt}$$

or $$\frac{d\phi}{dt} = 2\pi r^2\frac{dB}{dt}$$

or $$\frac{d\phi}{dt} = 2\frac{d}{dt}(\pi r^2 B) \qquad ...(iii)$$

Equation (*iii*) is called betatron condition or flux condition. This relation shows that the rate of change of magnetic flux (ϕ) with in the orbit of radius r is always twice that it would have been if the magnetic field intensity were uniform throughout the orbit. It is also called as "2 to 1" rule of the betatron. This condition hold for relativistic and non-relativistic energies as Equation (*iii*) satisfy the requirement of special theory of relativity.

Most of the betatron are operated from a 60 Hz a.c. source. Since the magnet and its coils constitute a large inductance, a very large capacitance is introduced into the circuit to bring the power factor closer to unity for efficient operation. In the 10^4 *eV*, betatron, the electrons are accelerated during a quarter of a cycle, that is during $\frac{1}{240}$ second, the energy possessed by an electron per revolution is 400 eV hence it is required to make 25,000 revolution to possess the maximum energy.

During the World War II, a 350 ton betatron was constructed by the General Electric Company and put into use as a source of extremely penetrating X-rays. In this instrument, electron accelerated to 100 M*e*V energy, and impinging upon a target, give rise to X-ray capable of penetrating many feet of solid iron and lead.

20.8. SYNCHRO-CYCLOTRON (FREQUENCY MODULATED CYCLOTRON)

Cyclotron is only used for accelerating protons and α-particles and moreover the energies produced by the cyclotron are limited.

The synchro-cyclotron developed in 1945 overcomes this difficulty by decreasing the oscillator frequency to compensate for the changing mass. The machine in which magnetic field is kept constant and the frequency is varied is known as a frequency modulated Cyclotron or **Synchro-Cyclotron**.

Construction and Working: The Synchro-Cyclotron consists of only a single dee placed inside a vacuum chamber between the poles of an electromagnet an shown in Fig. 20.8.

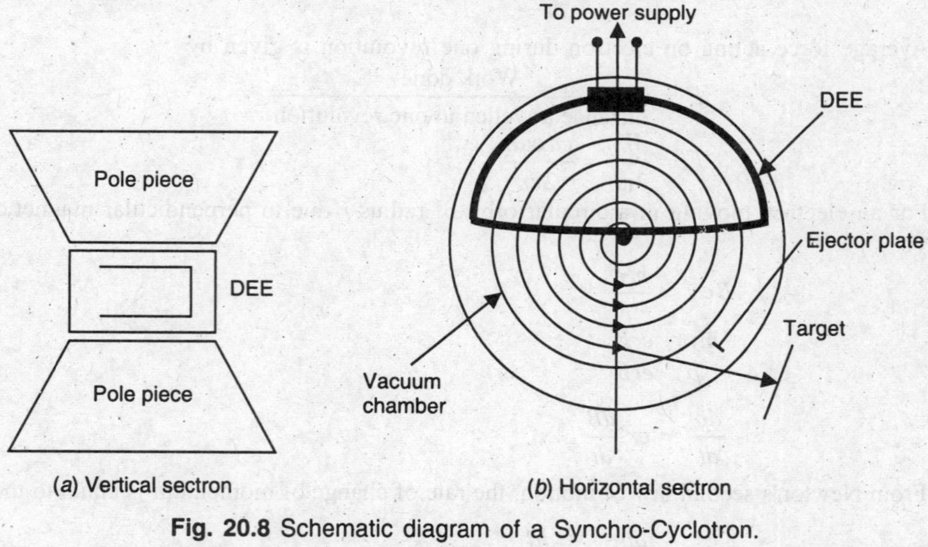

Fig. 20.8 Schematic diagram of a Synchro-Cyclotron.

The pole-pieces of the magnet are of suitable shape such that the field decreases outward from the centre. This ensures good focussing of the accelerating ions. Potential difference is applied between the dee and earthed sheath. The pressure inside the vacuum chamber is reduced to about 10^{-6} cm of Hg. Each ion makes about 10^4 revolutions before attaining maximum energy hence peak r.f. voltage is very small as compared to that in cyclotron. Therefore, a single D can be used. The use of single dee leaves sufficient space in the vacuum chamber for the ion source and the target. The ion is accelerated when it enters or leaves the dee.

A high positive potential is applied to the ejector plate to deflect the ion groups to the target.

The Berkeley Synchro-cyclotron uses the magnetic flux density 1.5*T* which decreases to 1.43*T*. The oscillator frequency varies from 9 MHz to 12.6 MHz at a modulated frequency 120 Hz. It can accelerate α-particle upto 400 MeV, deutrons upto 200 MeV and protons to 350 MeV.

Advantages:

(*i*) The efficiency of the oscillator is high at low r.f. voltage.

(*ii*) Radio-frequency used is low.

(*iii*) High magnetic pole flux density is obtained by reducing the gap between the magnetic pole faces.

Disadvantages:

(*i*) Electrons cannot be accelerated.

(*ii*) The output beam current is very small (1 μA) because a small fraction of injected ions are admitted into expanded phase stable orbits of maximum radius and energy.

20.9. SYNCHROTRONS

In a synchro-cyclotron, we cannot accelerate electrons because mass increase with increase in velocity. This difficulty is removed by using modulation of magnetic field (*i.e.,* increasing magnetic field is applied). *The orbital accelerator which employs the increasing magnetic induction B in an orbit of constant radius is called Synchrotron.* There are two types of synchrotrons.

(*i*) Electrons Synchroton and (*ii*) Proton Synchrotron.

(*i*) **Electrons Synchrotron:** In betatron, the maximum energy attainable is limited due to low energy gain per turn and also electrons cannot be accelerated by synchrocyclotron due to their small mass. Mr. Barnes in 1946, gave a machine called synchrotron to accelerate electrons.

Construction: Electron synchrotron consists of ceramic doughnut shaped vacuum chamber in an a.c. magnetic field, the magnet can be a ring or C shaped units, not filling the hole in the doughnut and hence weight of the magnet is decreased Fig. 20.9. For vertical focussing of the beam, magnet pole faces are suitably skimmed to provide a field decreasing from the centre outwards. Inside the vacuum chamber, resonant cavity is formed by having a laminated silver coating (to avoid eddy currents) with a gap within coating. Radio-frequency electron field is applied across this gap at the proper time in the magnetic cycle. When accelerator is on, the electron is accelerated each time it crosses through the resonator. Since radio frequency field is applied in the region of the orbit and in the space enclosed by it, so the electromagnet is built with its poles above and below the tube with steel flux bars just inside the doughnut. These steel bars become saturated very early in the magnetic cycle and further variation of B takes place in the region of electron orbit only.

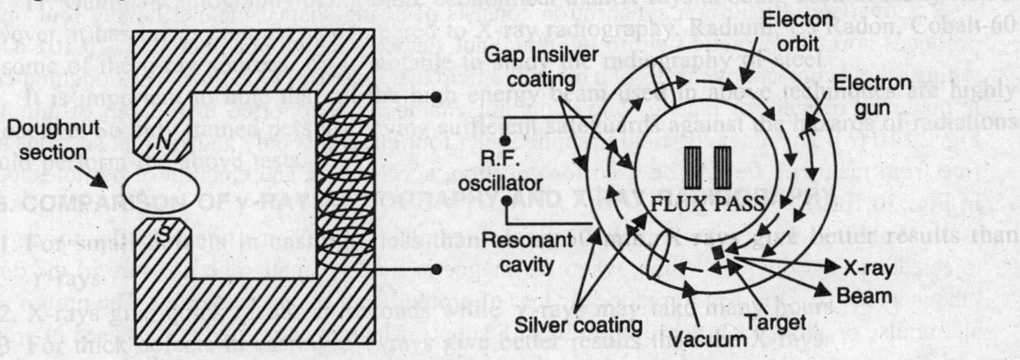

Fig. 20.9. Cross-section picture of an electron Synchrotron

Working: At the beginning of magnetic cycle, electrons are injected into the chamber from the electron gun at about 50 to 100 keV energy, the device operates as a betatron. The electrons are accelerated to 2 or 3 MeV so that they attain almost velocity of light. The steel flux bars present in the central gap get magnetically saturated and are no longer able to induce an e.m.f. This saturation causes the betatron mechanism to stop. Now synchrotron machanism begins to operate (*i.e.* the magnetic field B is increased in proportional to the relativistic increase of electron mass). If the potential applied to the resonator operates at the proper frequency, the electrons are all kept in phase and receive increments of energy at each revolution as they pass through the cavity. The maximum energy of electrons depends upon the orbit radius and maximum magnetic field for this orbit. When the electrons acquire maximum desired energy, the r.f. electric field is switched off. At this stage magnetic flux condition for stable orbit is destroyed by sending a large current through additional coils, electrons change the radius of the orbit and strike the target projecting from the inner edge of the doughnut which give off X-rays of short wave length.

In electron synchrotron, the electrons are injected at about 50 to 10 keV and are accelerated upto 7 MeV by betatron action and finally upto 330 MeV. In order to accelerate the electron to higher energy the electron synchrotron consists of primarily of four quadrants in each of which a magnetic field is created by suitable electromagnets. The quadrants are separated by short straight sections where there is no magnetic field-called as race track type chamber increases the energy range to 1.2 BeV. However, electron synchrotron utilising A.G. principle raises this limit to 6 BeV.

Theory: The energy of electron having rest mass m_0 is given by $E = \sqrt{p^2c^2 + m_0^2c^4}$

where c is velocity of light in vacuum (or air)

Since $\qquad m_0c^2 \simeq 0.51 \text{ MeV}$

So $\qquad E >> m_0c^2$

$\therefore \qquad E \simeq pc$

or $\qquad E = e\,\text{B}rc$

And the frequency of revolution is given by

$$\nu = \frac{1}{T} = \frac{1}{\dfrac{2\pi r}{c}}$$

or $$\nu = \frac{c}{2\pi r}$$

(*ii*) **Proton Synchrotron:** Proton synchrotron does not mean that the synchrotron can accelerate only proton but is equally applicable for deutrons α-particles or other ions, but the name is given because protons being elementary particles are preferred for investigations in elementary particle physics. In principle, the proton synchrotron is similar to the electron synchrotron and gives protons upto 10 GeV energy. Proton synchrotron is sometimes also called *Bevatron* (to accelerate protons to 6 BeV) *or Cosmotron* (to accelerate protons upto 30 GeV).

Construction: Proton synchrotron consists of a ring shaped magnet having four quadrants produces normal magnetic field to the doughnut shaped vacuum chamber (Fig. 20.10). The orbit radius is held constant by means of a magnetic field that increases with time. The doughnut chamber is made up of steel or plastic. The doughnut chamber is supported in the gap of annular ring Fig. 20.10 (*b*). The magnetic field is applied over four quadrants only and straight sectrons are kept free from magnetic field to be used for injection, acceleration and ejection of particles as shown in Fig. 20.10(*a*).

Working: The proton initially is accelerated upto 10 MeV in a linear accelerator and is then injected into the doughnut tube, when the magnetic field is small. The pressure in the doughnut tube is about 10^{-6} mm Hg to reduce the loss of proton beam by gas scattering. The proton receives an impulse of about 1 kV per turn through r.f. supply. The magnetic field is increased in proportion to the momentum to keep the protons in the synchrotron moving in an orbit of constant radius. Due to this the proton gains energy from r.f. electric field in successive revolution and hence its frequency increases gradually. To maintain the condition of phase stability, the frequency of r.f. electric field is increased in the same manner with time. Since protons and heavier mass particles, their energies

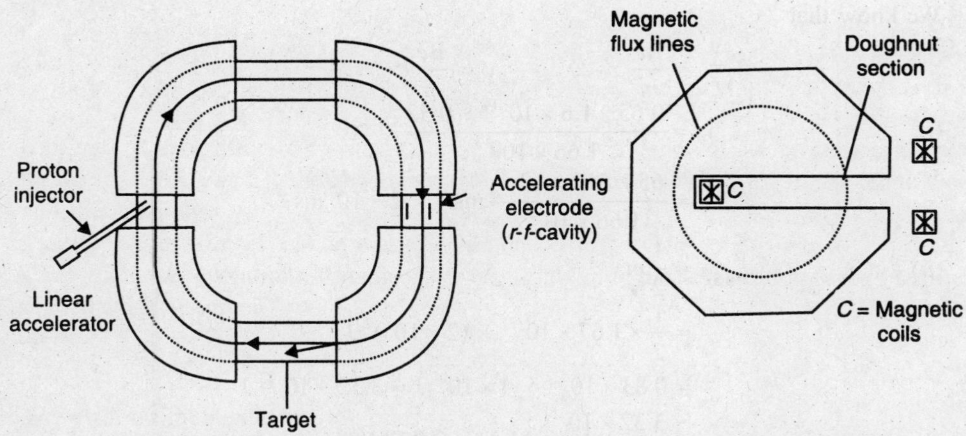

(a) Proton synchrotron (b) Cross-section of the magnet

Fig.20.10. Schematic diagram of Proton Synchrotron

for a relativistic velocity will be very high say 5 BeV. In proton synchrotron the protons could be accelerated to energies of the order of multi BeV. When protons reach their maximum energy, the oscillator frequency is distorted so that the orbit expands or contracts and the protons strike the target.

Theory: In the absence of linear sectron, the frequency of revolution of an ion is given by

$$\nu = \frac{eBc^2}{2\pi(K + m_0 c^2)} \quad ...(i)$$

In the presence of four straight sectron each of length l, we get

$$\nu = \frac{eBc^2}{2\pi(K + m_0 c^2)} \cdot \frac{2\pi r}{(2\pi r_0 + 4l)} \text{ [} r_0 \text{ is the radius of circular sections]} ...(ii)$$

Since total energy,

$$E = K + m_0 c^2 = [p^2c^2 + m_0^2 c^4]^{1/2}$$

or
$$k^2 + 2m_0 kc^2 + m_0^2 c^4 = p^2 c^2 + m_0^2 c^4 \quad [\because p = eBr_0]$$
$$= (eB\, r_0\, c)^2 + m_0^2\, c^4$$

or
$$B = \frac{[K(K + 2m_0 c^2)]^{1/2}}{er_0\, c} \quad ...(iii)$$

Putting Equation (*iii*) in (*ii*), we get $\nu = \dfrac{c[K(K + 2m_0 c^2)]^{1/2}}{K + m_0 c^2} \cdot \dfrac{1}{(2\pi r_0 + 4l)}$ ...(*iv*)

Example 1. *Protons are accelerated in a cyclotron with dees of radius 32 cm and magnetic field of 6500 G. Calculate (i) the velocity of the protons (ii) the energy of MeV (iii) Frequency of A.C. voltage between the dees.*

Given 1 a.m.u. = 1.66×10^{-27} kg; 1 $c = 1.6 \times 10^{-19}$ C

Solution: Given that

$$r = 32 \text{ cm} = 0.32 \text{ m}$$
$$B = 6500 \text{ gausses} = 0.65 \text{ weber/m}^2$$
$$m = 1.66 \times 10^{-27} \text{ kg.}$$
$$e = 1.6 \times 10^{-19}\, C$$

We know that

$$\frac{v}{r} = \frac{Be}{m} \quad \therefore \quad \upsilon = \frac{Be}{m} r$$

(i) $$v = \frac{0.65 \times 1.6 \times 10^{19} \times 0.32}{1.66 \times 10^{27}}$$

$$= \frac{65 \times 512 \times 10^{-22}}{1.66 \times 10^{27}} \text{ms}^{1} = 2 \times 10^{7} \text{ ms}^{-1}$$

(ii) $$E_{\max} = \frac{1}{2} m v^2$$

$$= \frac{1}{2} \times 1.67 \times 10^{-27} \times (2 \times 10^7)^2 \text{ J}$$

$$= 0.83 \times 10^{-27} \times 4 \times 10^{14} \text{ J} = 3.32 \times 10^{-13} \text{ J}$$

$$= \frac{3.32 \times 10^{-12}}{1.6 \times 20^{-13}} \text{MeV} = 2.07 \text{ MeV}$$

(iii) $$f = \frac{Be}{2\pi m}$$

$$= \frac{0.65 \times 1.6 \times 10^{-19}}{2 \times 3.143 \times 1.66 \times 10^{27}} = 10^{13} \text{ Hz.}$$

Example 2. *Calculate frequency of applied potential to the cyclotron dees in which deutrons are accelerated, given flux density $B = 3.0$ weber/m^2, $e = 1.6 \times 10^{19}$ C and 1 a.m.u. = 1.66×10^{-27} kg.*

Solution: We know that

$$\omega = \frac{Be}{m} \quad \text{But} \quad \omega = \frac{2\pi}{T} = 2\pi f$$

where f is the frequency

$$\therefore \quad f = \frac{\omega}{2\pi} = \frac{Be}{2\pi m} \quad \text{or} \quad f = \frac{3 \times 1.6 \times 10^{-19}}{2 \times 3.143 \times 2 \times 1.66 \times 10^{-27}}$$

$$= \frac{3 \times 1.6 \times 10^{-20}}{2 \times 3.143 \times 2 \times 1.66 \times 10^{-27}} \text{ Hz} \quad \left[\because \text{Mass of deutron} = 2\, a.m.u.\right]$$

$$= 23 \times 10^6 \text{ Hz}$$

Example 3. *Deutron in cyclotron describe a circle of 32 cm just before emerging from the dees, the frequency of applied e.m.f. is 10 Mega cycles sec. Find (i) the flux density of the magnetic field (ii) the velocity of deutrons emerging out of the cyclotron.*

Given mass of deutron, m = 3.32×10^{-27} kg e = 1.6×10^{-19} C

Solution: Since $m = 3.32 \times 10^{-27}$ kg

$$e = 1.6 \times 10^{-19} \text{ C}$$

$$r_{\max} = 32 \text{ cm} = 0.32 \text{ m}$$

We know that

$$f = \frac{Be}{2\pi m}$$

(i) $$\therefore \quad B = \frac{2\pi m f}{e}$$

$$= \frac{2 \times 3.143 \times 3.32 \times 10^{-27} \times 10 \times 10^6}{1.6 \times 10^{-19}} = 1.3 \text{ W m}^{-2}$$

(ii) $$\frac{v}{r} = \frac{Be}{m}$$

$$\therefore \quad v = \frac{Be}{m} r$$

$$= \frac{1.3 \times 1.6 \times 10^{-10} \times 0.32}{3.32 \times 10^{-27}} = \mathbf{2 \times 10^7\ ms^{-1}}$$

Example 4. A betatron has a magnet current supply frequency of 50 Hz and peak magnetic flux density at the orbit as 0.4 W/m^2. If the radius of the electron orbit is 0.75m calculate the final energy of the electrons.

Solution: Electrons in a betatron move almost with the velocity of light.

The number of revolutions N made by the electron during the acceleration time (quarter cycle of a magnetic flux) is given by

$$N = \frac{c}{4\omega r_0} = \frac{c}{8\pi v r_0} = \frac{3 \times 10^8}{8 \times 3.14 \times 50 \times 0.75} = 3.184 \times 10^5$$

The electron momentum at the orbit is

$$p = Ber_0$$

The electrons are to be treated relativistically

Final electron energy

$$E = mc^2 = pc = Ber_0 c.$$

From the given values,

$$E = \frac{0.4 \times 16 \times 10^{-19} \times 0.75 \times 3 \times 10^8}{1.44 \times 10^{-11}} \text{J} = 90.01 \text{ MeV } \textbf{Ans.}$$

Example 5. *A cyclotron has magnetic field of 1.5 Wb/m². The extraction radius is 0.5 m. Calculate the frequency of the radio beam necessary for accelerating deutrons and the energy of the extracted beams.*

Solution: The frequency of a cyclotron is $f = \dfrac{Be}{2\pi m}$, where the symbols have usual significance.

Here $B = 1.5\ Wb/m^2$

$$e = 1.6 \times 10^{-19}\ C$$

$$m = 3.32 \times 10^{-27} \text{ kg for deuterium}$$

$$f = \frac{1.5 \times 1.6 \times 10^{-19}}{2 \times 3.14 \times 3.32 \times 10^{-27}} = 11.5 \times 10^6 \text{ Hz}$$

Kinetic energy for deutrons E is given by

$$E = \frac{B^2 e^2 r^2}{2m}$$

Here, r = 0.50 m

Hence, $$E = \frac{(1.5)^2 \times (1.6 \times 10^{-19}) \times (0.5)^2}{2 \times 3.32 \times 10^{-27}} \text{ J}$$

$$= 2.168 \times 10^{-12} \text{ J}$$

$$= 13.55 \text{ MeV. } \textbf{Ans.}$$

Example 6. *What radius is needed in proton Synchrotron to attain particle energies of 12 GeV. Assuming that a guide field of 1.9 Wb/m² is available.* *(H.P.U. 2005)*

Solution: Equivalent mass of proton = 12 GeV + rest mass of proton

= 12 GeV + 938 GeV

= 12 GeV + 0.938 GeV = 12.938 GeV

$$= \frac{12.938}{0.931} \qquad [\because 1 \text{ a.m.u.} = 931 \text{ MeV}]$$

= 13.89 a.m.u. = $13.89 \times 1.66 \times 10^{-27}$ kg.

Radius of the orbit $= \frac{m\upsilon}{Be} = \frac{(13.89 \times 1.66 \times 10^{-27}) \times (3 \times 10^8)}{1.9 \times 1.6 \times 10^{-19}}$

$= 22.75$ m **Ans.**

Example 7. *A cyclotron has a magnetic field of 10^4 gauss and radius of dee is 80 cm. Calculate the frequency of alternating electric field that must be applied and to what energy can deutrons be accelerated. Mass of deutrons = 2 a.m.u.*

Solution: Given $B = 10000 \text{ gauss} = 1 \text{ Wb/m}^{-2}$

$m = 2 \times 1.67 \times 10^{-27} \text{ kg} = 3.34 \times 10^{-27} \text{ kg}$

$r = 80 \text{ cm} = 0.8 \text{ m}$

Frequency of applied electric field

$$f = \frac{Bq}{2\pi m} = \frac{1 \times 1.6 \times 10^{-19}}{2 \times 3.142 \times 3.34 \times 10^{-27}} = 7.6 \times 10^6 \text{ Hz}$$

Energy of deutrons

$$E = \frac{(Bqr)^2}{2m} = \frac{(1 \times 1.6 \times 10^{-19} \times 0.8)^2}{2 \times 3.34 \times 10^{-27}} \text{ J}$$

$$= \frac{(1.6 \times 10^{-19}) \times (0.8)^2}{2 \times 3.34 \times 10^{-27} \times 1.6 \times 10^{-19}} \text{ eV} = \frac{1.6 \times 10^{-19} \times 0.64}{6.68 \times 10^{-27}} \text{ eV}$$

$$= 15.3 \times 10^6 \text{ eV} = 15.3 \text{ MeV}.$$

Example 8. *A cyclotron in which flux density is 1.4 T, is employed to accelerate protons. What should be the frequency of the alternating field applied to the dees? Mass of proton is 1.67×10^{-27} kg and its charge 1.6×10^{-19} C.*

Solution: The frequency of the alternating field required to accelerate the charged particles is

$$\nu = \frac{qB}{2\pi m}$$

Here $q = 1.6 \times 10^{-19} \text{C}, \text{B} = 1.4 \text{T}$

$m = 1.67 \times 10^{-27} \text{kg}$

$$\therefore \quad \nu = \frac{1.6 \times 10^{-19} \times 1.4 \times 7}{2 \times 22 \times 1.67 \times 10^{-27}} \text{Hz} = 21.34 \text{ MHz}.$$

Example 9. *The radius of cyclotron dees is 40 cm and magnetic induction is 1.5 T. What is the maximum energy of beam of protons accelerated by the cyclotron?*

Give mass of proton $= 1.67 \times 10^{-27}$ kg and its charge is 1.6×10^{-19} C.

Solution: The maximum K.E. of the charged particles accelerated in a cyclotron is given by

$$E_{max} = \frac{q^2 B^2 R^2}{2m}$$

Here $q = 1.6 \times 10^{-19} \text{C}, m = 1.67 \times 10^{-27} \text{kg}, \text{B} = 1.5 \text{T}, \text{R} = 40 \text{cm} = 0.4 \text{m}.$

$$\therefore \quad E_{max} = \frac{(1.6 \times 10^{-19})^2 (1.5)^2 (0.4)^2}{2 \times 1.67 \times 10^{-27}}$$

$$= 2.76 \times 10^{-12} \text{ J} = \frac{2.67 \times 10^{-12} \text{ J}}{1.6 \times 10^{-19}} \text{eV}$$

$$= 17.2 \times 10^6 \text{ eV} = 17.2 \text{ MeV}.$$

Example 10. *A cyclotron oscillator frequency of 1 MHz is used to accelerate protons. If the radius of the dee is 60 cm, find the magnitude field in tesla.*

Solution: The frequency of alternating field required to accelerate the charged particle is given by

$$\nu = \frac{qB}{2\pi m}$$

or $$B = \frac{2\pi m v}{q} = \frac{2 \times 3.142 \times 1.67 \times 10^{-27} \times 10^{6}}{1.6 \times 10^{-19}}$$

$$= 6.56 \times 10^{-2} \text{ tesla } \textbf{Ans.}$$

QUESTIONS AND PROBLEMS

1. Discuss Cockcroft-Walton accelerator?
2. What is the Van de Graaf generator? What is it frequently used for? Can it be used to accelerate protons and electrons?
3. What is Cockcroft and Walton's voltage multiplier? What is its purpose? What does it accelerate?
4. (*a*) What is meant by particle accelerator? Discuss Cockcroft Walton voltage multiplier. Give its principle and usefulness.
   (*b*) What is the difference between linear accelerator and circular accelerator?
5. Which accelerator makes use of principle of e.m. induction for accelerating the particles? Can it accelerate protons?
6. What is the difference between cyclotron, synchrotron and Betatron? On what principles these work.
7. What is the principle of cyclotron?
8. Describe the principle, construction and working of a cyclotron. Derive expression for the maximum kinetic energy achieved by a particle of mass *m* is terms of the applied magnetic field and dee radius. Discuss its limitations.
9. Describe the principle, construction and working of a cyclotron. What are its limitations?
10. What are the limitations of cyclotron? How have these been overcome in Betatron?
11. What are basic limitations of a cyclotron and how they are overcome in a synchrotron? Discuss briefly.
12. Why a cyclotron cannot accelerated electrons? How will you accelerate electrons? Discuss in detail the machine used to do so?
13. Describe the principle, construction and working of a betatron. Obtain the 'betatron condition'.
14. Explain the principle and working of betatron.
15. Write notes on:
    (*a*) Betatron
    (*b*) Linear particle accelerator.
16. What are the main types of particle accelerators? Distinguish between synchro–cyclotron and cyclotron.
17. What is a synchrotron? Give a detailed account of construction and working of (*i*) electrons synchrotron (*ii*) proton synchrotron.
18. How cyclic accelerators are advantageous over linear accelerators?
19. How cyclic accelerators are advantageous over linear accelerator? Explain the working of a synchrotron and discuss its limitations.
20. Explain why loss of energy is maximum in the case of electrons in a linear accelerator.
21. Explain the principle of a betatron.
22. (*a*) Describe necessary theory, construction and working of a betatron. Where do you think it got its name?
    (*b*) What do the abbreviations "MeV" and "BeV" stand for?
23. Discuss Cockcroft-Walton machine for accelerating the charged particles.
24. A cyclotron with B = 0.14 Wb/m^2 is used to accelerate protons. What should be the frequency of the r.f. field?
    Hint. $\nu = \frac{qB}{2\pi m}$ [**Ans.** 2.134×10^6 Hz]
25. A fixed frequency cyclotron has an oscillator frequency of 12×10^6 Hz and a dee radius of 0.55 m. It is used to accelerate deutrons. Determine
    (*a*) the magnetic flux density needed (*b*) energy to which deutrons are accelerated.
    Given $m_d = 3.34 \times 10^{-27}$ kg, $e = 1.6 \times 10^{-19}$ C
    Hint. ∴ $$\nu = \frac{qB}{2\pi m} = \frac{eB}{2\pi m}$$

$$\therefore \quad B = \frac{2\pi m\nu}{2m} = \mathbf{1.57\ Wb\,m^{-2}}\ \textbf{Ans.}$$

And $$E = \frac{B^2e^2r^2}{2m} = 17.95\ \textbf{MeV Ans.}$$

26. In a cyclotron, adjusted to give a proton beam, the magnetic field in 15000 oersted (gausses) and the extreme radius in 15 cm. Calculate (*i*) the energies in electron volt of emergent protons and (*ii*) the wavelength of the oscillator.
Given *e/m* for proton $= 96 \times 10^6$ coulombs per kg.
$e = 1.60 \times 10^{-19}$ coulomb. [**Ans.** (*i*) 2.43 MeV. (*ii*) 13.1 metres]

27. A cyclotron has a magnetic field of 10^4 gausses or 1 weber/m^2 and the radius of dee is 80 cm. (*i*) Calculate the frequency of alternating electric field that must be applied and (*ii*) to what energy can deutrons be accelerated. [**Ans.** (*i*) 7.7 Mega cycles/sec. (*ii*) 15.6×10^6 eV.]

28. Calculate (*i*) the velocity (*ii*) the mass of β-particle with a kinetic energy of 0.5 MeV. (*iii*) If the beam of such particles enters a uniform field, find the flux density required to cause them to traverse a circular arc of radius 10 cm. The rest mass of electron is $m_0 = 9.11 \times 10^{-31}$ kg. $e = 1.6 \times 10^{-19}$ C. and the velocity of light in free space is equal to 3×10^8 m/sec.
[**Ans.** (*i*) 2.59×10^8 m/sec. (*ii*) 1.8×10^{30} kg. (*iii*) 0.029 weber/m^2]

29. Assume that in the 70 MeV betatron, the radius of the stable electron orbit is 28 cm. Calculate (*i*) the angular velocity of the electrons (*ii*) the frequency of the applied electric field and (*iii*) the value of the magnetic field intensity at the orbit, for the energy.
[**Ans.** (*i*) 1.07×10^9 radian/sec. (*ii*) 1.7×10^8 Hz (*iii*) 0.83 weber/m^2]

30. What radius is needed in a proton synchrotron to achieve 10 BeV level of energy? Assume that the guide field of 1.5 T is used.
Hint. ∴ $$p = m\upsilon = qBr$$
$$r = \frac{m\upsilon}{qB} = \frac{E}{c^2}\frac{\upsilon}{qB}\left[\because m = \frac{E}{c^2}\right]$$ [**Ans.** 18 m]

31. Calculate the frequency of the alternating potential that must be applied to the cyclotron dees in which deutrons are accelerated, given that magnetic induction is 3 T, mass of deutron is 3.32×10^{-27} kg and charge is 1.9×10^{-19} C. [**Ans.** 23 MHz]

32. A 100 MeV betatron has an orbit of 0.35 m in which electron acquires energy 480 eV per revolution. How for will the electrons travel a attaining full energy. The magnetic is energised at 180 Hz and produces pulse of 2μs duration, each consisting of 3×10^9 electron. What is peak and average beam current? [**Ans.** 4.57×10^5μ, 2.4×10^{-4} A, 8.64×10^{-6} A]
Hint. $$N = \frac{100 \times 10^6}{480},\ I_p = \frac{ne}{t}\ L = 2\pi rN,\ I_{av} = ne\nu$$

33. A betatron has a current supply for its electromagnet at 50 Hz and it produces a flux density of 0.4 T. If the radius of the electron orbit is 0.75 m, how much energy will the electron be picking up? How many revolutions it has completed during its acceleration time? Find the average energy per revolution.
Hint. $$E = eBRC = 90\ \text{MeV}$$
Average energy gained per rev. $$= \frac{E}{N} = \frac{E}{c/4\omega r} = 282.8\ \text{eV}.$$

CHAPTER 21

SUPERCONDUCTIVITY

21.1. INTRODUCTION

The electrical resistivity of all metals and alloys decreases when they are cooled. For perfectly pure metal, the resistivity should approach zero as the temperature is reduced to 0 K. This zero resistance which a hypothetical perfect specimen would acquire if it could be cooled absolute zero in not, however the phenomena of superconductivity. Any real specimen of metal cannot be perfectly pure. As a result there will be a certain residual resistivity (l_o) which remains at the lowest temperatures. **The more impure the metal, the larger will be its residual resistivity.**

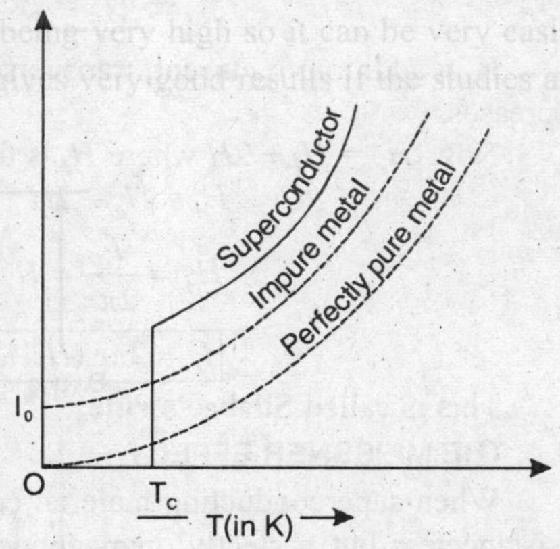

Fig. 21.1. Variation of resistivity with temp.

Certain metals, however, show a very remarkable behaviour, when they are cooled, their electrical resistivity decreases in the usual way, but at certain temperature called critical temperature (T_c) the resistivity abruptly drops to zero. This new state is called 'superconducting state'. This phenomenon can even occur in impure specimen.

This phenomenon was first discovered by K Onnes in 1911, when he was measuring the resistivity of mercury at low temperature. He observed that the electrical resistivity of pure mercury suddenly drops to zero at 4.15 K.

21.2. EFFECT OF MAGNETIC FIELD

In addition to the current, magnetic field can also destroy the superconductivity. The magnetic field at which superconductivity vanishes is called **critical magnetic field H_c.** Its value depends upon the nature of the material and its temperature. The variation of H_c with temperature is nearly parabolic and can be represented by the equation $H_c(T) = H_c(o)\left[1 - \frac{T^2}{T_c^{\,2}}\right]$ ——— (1)

where H_c (T) is the critical magnetic field at temperature T and H_c (o) be its value at 0 K.

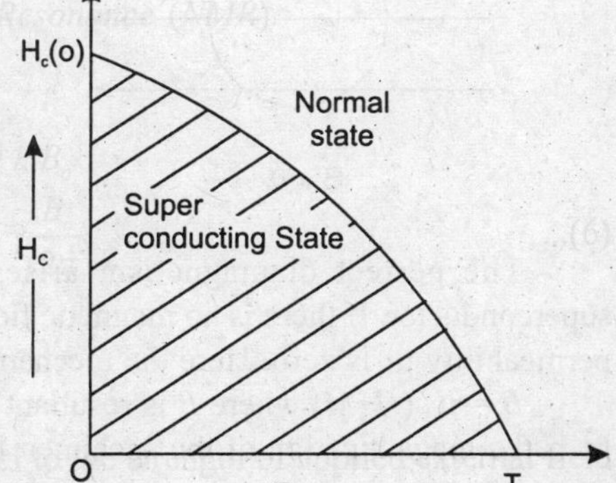

Fig. 21.2. Variation of H_C with temp.

21.3. CRITICAL CURRENTS

The magnetic field which destroys the superconductivity is not necessarily an external applied field, it may also be the magnetic field due to the current flowing in the superconducting wire etc. The maximum value of current flowing through the superconductor at which superconductivity is destroyed is called critical current.

If super conducting wire of radius r carries current I, then using Ampere's law

$$\oint \vec{H} \cdot \vec{dl} = I$$

$$\oint Hdl = I$$

$$H.2\pi r = I$$

$$I = 2\pi rH$$

at $H = H_c \quad I = I_c$

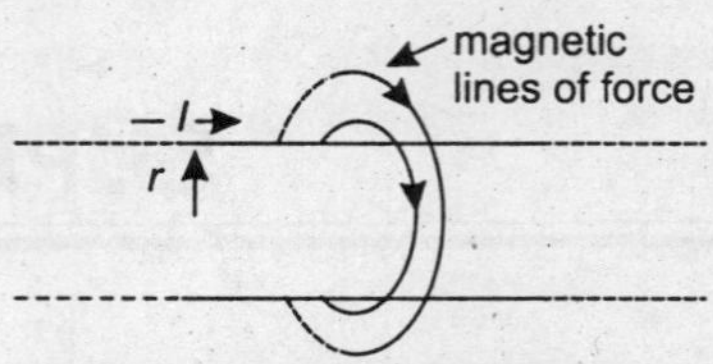

Fig. 21.3. Superconducting wire carrying current

$\Rightarrow \boxed{I_c = 2\pi r H_c}$ If I becomes Ic, superconductivity will be destroyed.

If in addition to current, transverse magnetic field H is applied, the value of critical current decreases.

Now, $H_c = H_I + 2H$ where H_I is the field due to current

$$H_I = H_c - 2H$$

$$H_I = \frac{I_c}{2\pi r} = H_c - 2H$$

$$\boxed{I_c = 2\pi r (H_c - 2H)} \quad \text{———} \quad (2)$$

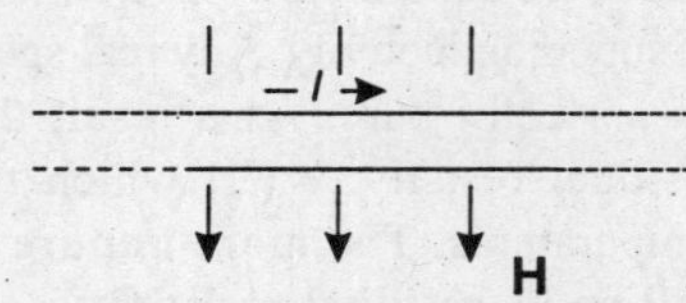

Fig. 21.4. Current carrying superconductor in transverse magnetic field.

This is called **Silsbee's rule.**

21.4. THE MEISSNER EFFECT

When superconducting material cooled below its critical temperature, it not only becomes resistanceless but perfectly diamagnetic also. That is to say there is no magnetic field inside a superconductor, whereas inside a merely resistanceless metal there may or may not be a magnetic field, depending on the circumstances.

This effect, when superconductor placed inside a magnetic field in cooled below its critical temperature, all the magnetic flux is expelled out of it, is called Meissner effect.

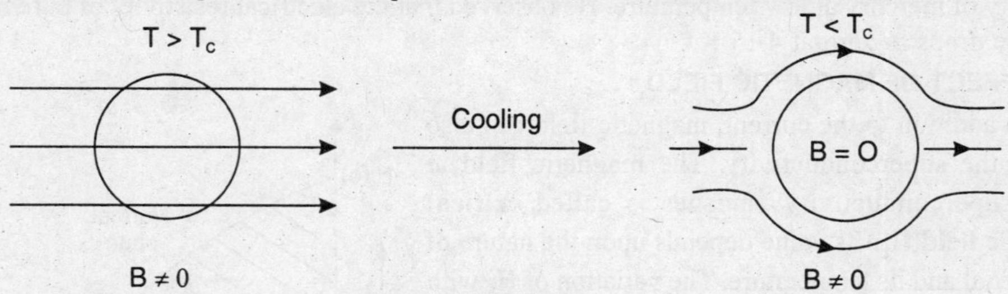

Fig. 21.5

The perfect diamagnetism arises from some special bulk magnetic property of the superconductor. If there is no magnetic field inside the superconductor it can be said that its relative permeability μr is zero. Here the mechanism of diamagnetism is not considered we know

$B = \mu_0 (H+M)$ where B is resultant magnetic induction H is the external magnetic field, and M is the magnetization of the specimen for superconductor $B = O$

$$O = \mu_0 (H+M)$$

$\Rightarrow M = -H$ and the magnetic susceptibility

$$\chi_m = \frac{M}{H} = -1$$

It must be noted that superconductivity is not only a strong diamagnetism, but also a new type of diamagnetism.

21.5. TYPE I AND TYPE II SUPERCONDUCTORS

Type I (Soft) superconductors exhibit complete Miessner effect *i.e.,* below H_c they are superconductors and at H_c they become normal. The value of critical magnetic field is very small of the order of $0.1T$. **These are also called soft superconductors. Aluminium, lead and Indium are examples of type I superconductors. Being soft superconductors, these cannot be used for the Coils of Strong electromagnets [Fig. 21.6 (a)].**

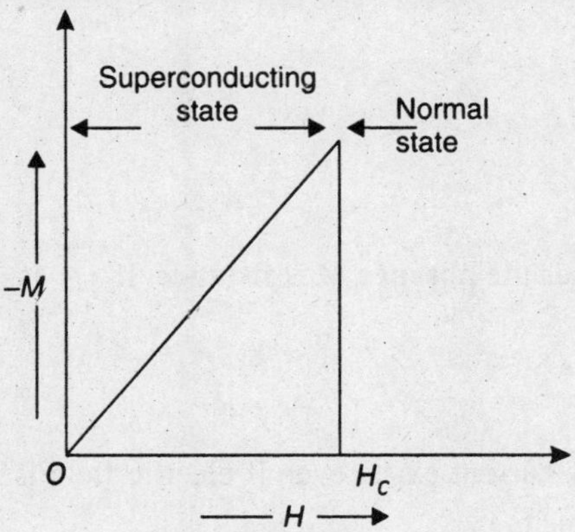

Fig. 21.6(*a*)

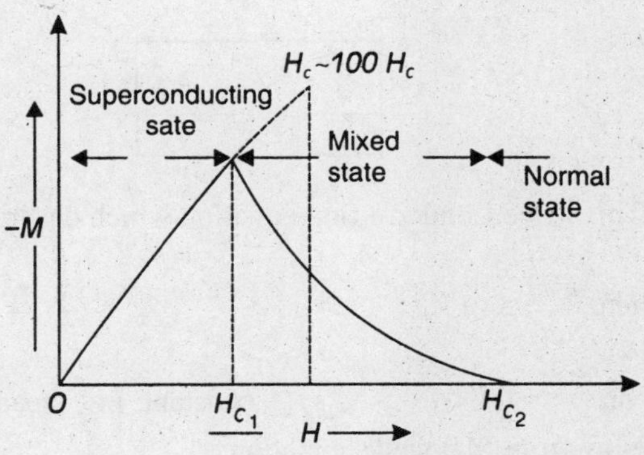

Fig. 21.6(*b*)

Type II (Hard) superconductors exhibit complete Miessner effect below a certain critical field H_{c1}; at this value magnetic field begins to penetrate and its diamagnetism and superconductivity starts decreasing. It is in the mixed state also called vortex state. At certain value of magnetic field H_{c2}, the superconductivity disappears. Transition metals and alloys consisting of niobium, Aluminium and silicon exhibit type II superconductivity. **Ceramic superconductors** also belong to this category *i.e.*, **Hard Superconductors [Fig. 21.6 (b)].**

A material can change from type I to type II with the addition of some impurities. For example Pb is type I super conductor with $1\frac{1}{c} = 600$ Gauss at 4 k but when a small impureity of indium is added to it, it becomes type II super conductor with $H_{C_1} = 400$ Gauss and $H_{C_2} = 1000$ Gauss.

21.6. LONDON EQUATIONS

This is macroscopic theory, based on the two-fluid-model. London brothers F London and H London derived two equations which explain the zero resistance state and perfect diamagnetism of superconducting state.

According to London's theory, it is assumed that there are two types of conduction electrons in a superconductor, namely the **super electrons** and **normal electrons.** At 0K, a superconductor contains only super electrons, but as temperature increases super electrons decrease and normal electrons increase.

The superconducting electrons are not subjected to any lattice scattering and are therefore accelerated in an electric field.

Let n_n, u_n and n_s u_s respectively the no. density and drift velocity of normal and super electrons.

The equation of motion of **superconducting electrons** under the action of electric field is given by

$$m\frac{d\vec{u_s}}{dt} = -e\vec{E} \qquad \ldots(1)$$

Now current and drift velocity are related as

$$I_s = n_s e A u_s$$

$$J_s = n_s e\, u_s$$

or
$$u_s = -\frac{J_s}{n_s} . e$$

Put in eq. (1), we get

$$m\left(\frac{d}{dt}\left(-\frac{\vec{J_s}}{n_s e}\right)\right) = -e\,\vec{E}$$

$$\boxed{\frac{d\,\vec{J_s}}{dt} = \frac{n_s e^2\,\vec{E}}{m}} \quad \ldots(2)$$

This is the London's first equation which describes the **absence of resistance.** If $\vec{E}$ is made zero then $\frac{d\,\vec{J_s}}{dt} = 0$

or $\vec{J_s} =$ constant, *i.e.,* steady current exists even if electric field is zero.

Now from Maxwell's equation

$$\vec{\nabla}\times\vec{E} = -\frac{\partial\,\vec{B}}{dt}$$

$\vec{B}$ in terms of magnetic vector potential is given as $\vec{B} = \vec{\nabla}\times\vec{A}$

$$\vec{\nabla}\times\vec{E} = -\frac{\partial}{\partial t}\left(\vec{\nabla}\times\vec{A}\right)$$

$$\vec{\nabla}\times\vec{E} = -\vec{\nabla}\times\frac{\partial\,\vec{A}}{\partial t}$$

$$\vec{E} = -\frac{\partial\,\vec{A}}{\partial t}$$

From equation (2)

$$\frac{m}{n_s e^2}\frac{d\,\vec{J}}{dt} = -\frac{\partial\,\vec{A}}{\partial t}$$

$$\frac{d}{dt}\left(\frac{m}{n_s e^2}\vec{J_s}\right) = \frac{\partial\,\vec{A}}{\partial t}$$

$$\frac{m}{n_s e^2}\vec{J_s} = -\vec{A}$$

$$\boxed{\vec{J_s} = -\frac{n_s e^2}{m}\vec{A}} \quad \ldots(3)$$

This is London's second equation. Again from Maxwell's equation for steady state we have

$$\vec{\nabla}\times\vec{B}=\mu_0\vec{J}_s \qquad \text{From Ampere Law}$$

$$\vec{\nabla}\times\vec{B}=\mu_0\left(-\frac{n_s e^2}{m}\vec{A}\right)$$

Take curl on both sides

$$\vec{\nabla}\times\vec{\nabla}\times\vec{B}=-\mu_0\frac{n_s e^2}{m}\vec{\nabla}\times\vec{A}$$

which may be written from vector identity as

$$\vec{\nabla}\times(\vec{\nabla}.\vec{B})-\nabla^2\vec{B}=-\frac{\mu_0 n_s e^2}{m}\vec{B} \qquad (\because \vec{B}=\vec{\nabla}\times\vec{A}) \quad \left[\because \vec{\nabla}\times\left(\vec{\nabla}.\vec{B}\right)=0\right]$$

$$\nabla^2\vec{B}=\frac{\mu_0 n_s e^2}{m}\vec{B}$$

$$\nabla^2\vec{B}=\frac{1}{\lambda^2}\vec{B} \quad \text{where } \lambda=\sqrt{\frac{m}{\mu_0 n_s e^2}} \text{ is constant at given temperature}$$

and is called **London penetration depth.** In one dimension, the question can be written as

$$\frac{d^2B}{dx^2}-\frac{1}{\lambda^2}B=0$$

The solution of this equation is of the form

$$\boxed{B(x)=B_0e^{-x/\lambda}} \qquad \ldots(4)$$

Inside the superconductor, B decreases exponentially and within very small penetration it becomes zero, which explains Meissner effect.

21.7. THE PENETRATION DEPTH

In 1935 F. London and H London described that during superconducting state, the field does not suddenly drop to zero at the surface of the superconductor, but decays exponentially according to the relation $H(x)=H_oe^{-x/\lambda}$

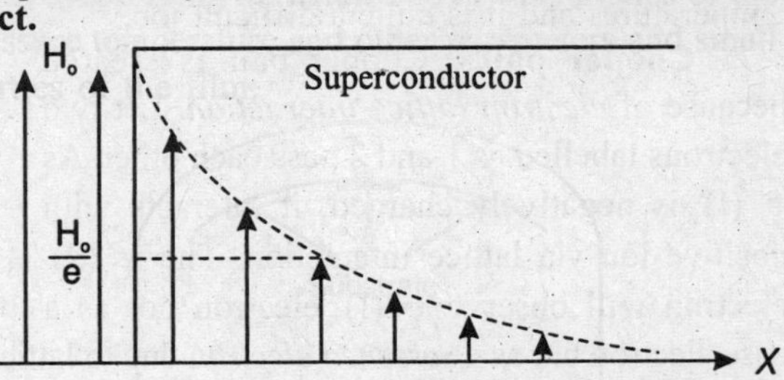

Fig. 21.7

where H_o is the value of magnetic field at the surface and H(*x*) is the magnetic field inside the specimen at a distance *x* from the surface.

λ is called London penetration depth. **It is defined as the distance inside the superconductor at which magnetic field falls to $\frac{1}{e}$ time the magnetic field at the surface.** The order of penetration of magnetic field is about 10–100 nm.

The penetration depth also varies with temperature as shown in Fig. 21.8. This variation can be expressed as

$$\lambda(T)=\frac{\lambda(0)}{\sqrt{\left(1-\frac{T^4}{T_c{}^4}\right)}} \qquad \ldots(1)$$

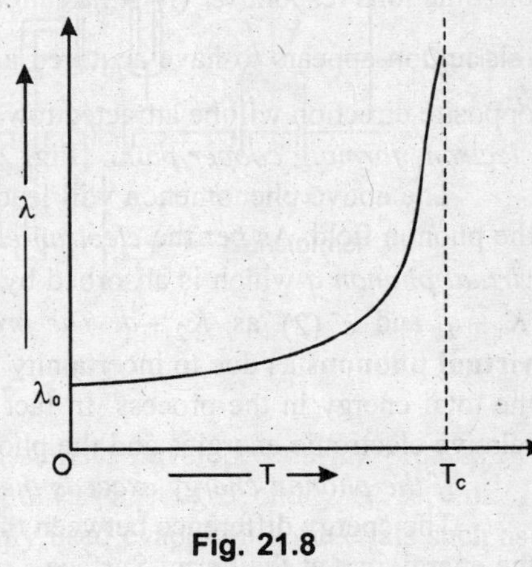

Fig. 21.8

where λ(*T*) is value at temp. T, and λ(0) is the value at absolute zero.

21.8. BCS THEORY OF SUPERCONDUCTIVITY

Bardeen Cooper Scruffier offered an explanation for the superconductors with a theory called **BCS theory** (after the name of the inventors).

They proposed that by using the knowledge of **solids** and applying the **quantum mechanics,** it is possible to explain the superconducting state in which the motions of all the electrons in conduction band are **locked together.** In this state, the electrons are coupled together in pairs at **very low temperatures.** One electron interacts with a **crystal lattice** and perturbs it and this perturbed lattice then interacts with the second electron in the pair in such a way that there is an **attraction** between the two electrons and this attraction overcomes the **coulomb repulsion** between them. The electron interaction *via* **lattice** produces an energy gap between the superconducting state in which electrons act **collectively** while in non-superconducting (normal) state these electrons act **individually.** [Fig. 21.9(a)]

The energy required to excite the system to the higher non-superconducting state and **destroy superconductivity** is not available at lower temperature. Thus the material becomes **superconductor** at such a low temperature.

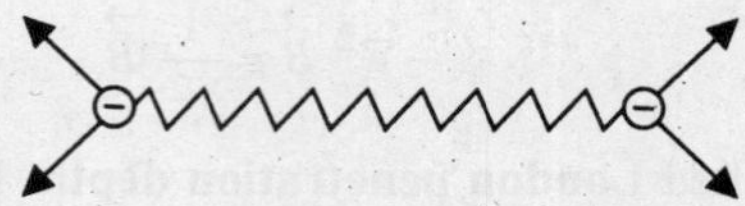

Fig. 21.9 (*a*) Cooper pair

BCS theory has been successfully able to explain all the experimental facts of superconductivity. In the latest research efforts are being made to have **underground** power lines to **transport** electricity from generating plants to consumers. These lines will have **negligible resistance** and so **energy losses** and **heating effects** will altogether be prevented. However, for this we will require liquid helium to make the wires act as **superconductors** at low temperatures and it is quite a difficult job.

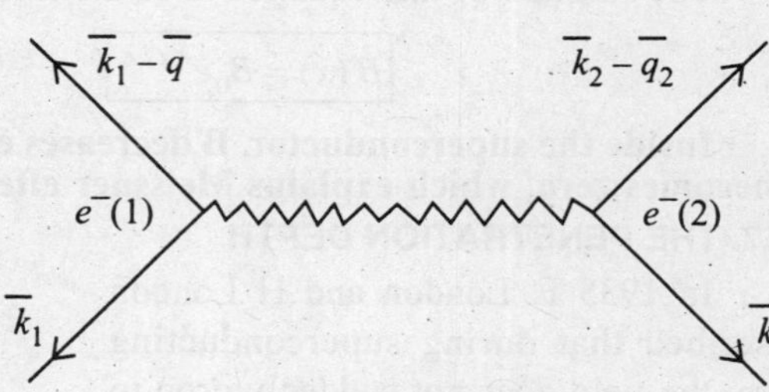

Fig. 21.9 (*b*)

Cooper pairs. Cooper pair is created because of *electron lattice interaction.* Let two electrons labelled as 1 and 2 pass each other. As $e^-(1)$ is negatively charged, it interacts with positive ion via lattice interaction. The $e^-(2)$ electron will observe $e^-(1)$ electron not as a base electron but as a *screened electron* due to lattice interaction. The screening reduces considerably the *effective charge* of this electron. As $e^-(1)$ is considered very near the *Fermi surface,* its speed is quite high on the other hand the response of positive ion due to its heavy mass is very slow. By the time ions respond $e^-(1)$ it has already passed the initial region. Because of the slow response this region appears to have acquired a net positive charge. When this happens $e^-(2)$ passing in opposite direction will be attracted towards $e^-(1)$. *This will create a net force of attraction for two electrons forming cooper pairs.* [Fig. 21.9 (b)]

The above phenomenon will lead to attraction of two electrons via the lattice interaction in the phonon field. As per the *electron-electron* attraction, the electron with wave vector K_1 emits a *virtual phonon* q which is absorbed by an electron with wave vector K_2. So $e^-(1)$ is scattered as $K_1 - q$ and $e^-(2)$ as $K_2 + q$. *The process is only virtual* and the phonons involved are called **virtual phonons** as due to incertainity principle, their short lives make it unnecessary to conserve the total energy in the process. In fact the nature of *electron-electron* interaction depends on the relative electronic energies and the phonon energy ($h\nu$).

If the phonon energy exceeds the electronic energy, interaction will be attractive.

The energy difference between the **free state** of the electron and the **paired state** appears as the energy gap at the Fermi Surface.

Normal electron states are above the energy gap and super-conducting electron states are below the energy gap at the Fermi Surface.

It may be noted that in the case of copper pairs, energy gap is a function of temperature unlike the case of insulators and semiconductors which have a *constant energy gap* which does not change appreciably with temperature.

21.9. ENERGY GAP

BCS theory predicted that the energy gap in superconductors at $T = 0\,K$ is $Eg(0) \sim 3.54\, kT_c$. This prediction can be verified by studying the absorption of e.m. radiation by super conductors. Only photons with energies more than or equal to E_g are absorbed.

At higher temperatures the BCS theory accounts for the variation of energy gap with temperature as

$$Eg \simeq 1.74\, Eg(0)\left(1-\frac{T}{T_C}\right)^{1/2}$$

This expression agrees with the experimental data on energy gap for various elements.

21.10. JOSEPHSON EFFECT

In 1962, Brian Josephson, an English Physicist predicted that if a very thin layer (about 2 nm) of an insulator (normally an oxide) is sandwiched between two **superconductors** and a voltage V is applied across the two superconductors, a current starts flowing. This current has two components (*i*) continuous dc component which persists even when the voltage is removed (*ii*) ac component which is present only when the voltage V is present (Fig. 21.10).

An important thing to note is that the ac component is always having frequency exactly $\nu = \frac{2eV}{h}$, no matter which materials is used for the so called **Josephson Junction**.

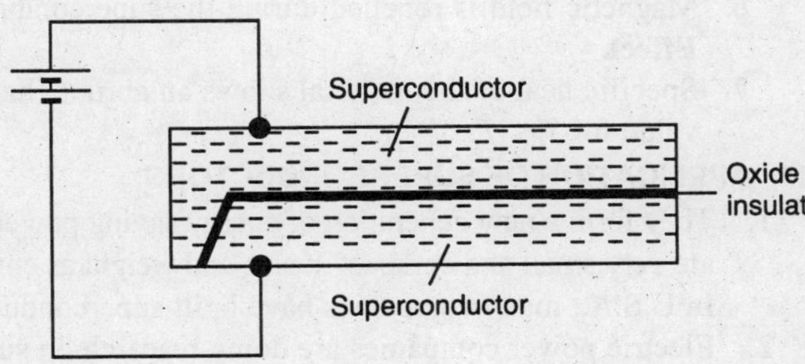

Fig. 21.10

So this phenomenon is useful for finding accurately the ratio e/h using the equation.

$$\frac{e}{h} = \frac{\nu}{2V} \quad \text{... (1)}$$

Since frequency can be measured very accurately, the value of e/h can be measured precisely.

The results obtained are so accurate that volt can now be defined in terms of frequency using equation (1). So instead of **standard cell** or other **measuring instruments,** this method *gives very good calibration of voltmeter.*

Recently, Josephson junctions have led to the development of very sensitive detectors and fast and efficient memory computers. SQUID (superconducting quantum interference device) detectors have been successful used to measure poison levels of iron of liver, so that iron build up can be treated before much harm is done to the body.

Josephson was awarded; Nobel prize in 1978 for Josephson effect. This explanation was given for the flow of super current across the insulator because of the tunnelling of *cooper pairs* across the insulator thereby making the insulator also a *Superconductor.* The wave functions associated with cooper pairs penetrate through the insulator consuming the cooper electrons to tunnel through the insulator and so flow of super current is observed through the *Josephson Junction.*

In D.C. Josephson effect, no voltage flows across the junction and the super-current is given by

$$I_J = I_m \sin \phi$$

where ϕ is the phase difference between the wave functions of cooper pairs on each side of the junction and I_m is the maximum junction current which depends on the thickness of insulator.

As stated above when a direct voltage V is applied across Josephson Junction, phase difference changes with time and is given by

$$\frac{d\phi}{dt} = \frac{2eV}{h}$$

[$\because$ charge of a cooper pair is $2e$, so $2e.V = h\nu$)

Josephson current varies sinusoidally with time and hence produces *A.C. Josephson Effect.*

Some of the important properties of Superconductors are summarized as under :

1. Current persist in a superconductor for very long time.
2. Superconductivity has been more commonly observed for substances with valence electrons (Z) between 2 and 8.
3. For a given value of Z, certain crystal structures show superconductivity more than their other counterparts. For example β-Tungsten and α-manganese structures show more superconductivity than other structures of the same value of Z.
4. In all the transition metals, variation of critical temperature (T_c) with number of electrons shows sharp maxima for $Z = 3, 5, 7$.
5. When the current through a superconductor is increased beyond critical value, a superconductor becomes a normal conductor. In other words superconductivity is a reversible process.
6. Magnetic field is repelled during the superconductivity state. This is called **Meissner Effect.**
7. Specific heat of the material shows an abrupt change at $T = T_c$. It jumps to a very high value for $T < T_c$.

21.11. APPLICATION OF SUPERCONDUCTORS

1. They form a new generation of energy saving power systems. Superconducting generators are very small and compact in size and weight as compared to the conventional generators. In U.S.A., many companies have built superconducting generators.
2. Electric power companies are doing research on superconductors to eliminate the power losses in transmission lines.
3. In Japan and France, superconducting magnets have been used to raise the trains above the railway tracks and attain very high speeds as frictional losses are minimized. A similar magnetic propulsion system is being used to launch the satellites into orbits directly from the earth without the use of rockets.
4. As the current in a superconducting ring can flow without change in its value, it can be used as a memory or a storage element in computers.
5. High efficiency ore-separating machine may be built using superconducting magnets. Also superconducting magnets may be used to separate tumour cells from the healthy cells and is finding great use in medical sciences.
6. Superconductors can act as a relay or switching system in a computer. The basic principle can be understood from the Fig. 21.11. The fact that when we apply a magnetic field whose value is more than the *critical field of the superconductor, the superconducting properties disappear* is used in gating circuits or in cyclotrons. Consider a wire A of superconducting material, inside a superconducting material B. Let the temperature of the two be lowers than the transition temperatures of the two materials. Current in central wire A can be controlled by a current in coil B as the magnetic field produced by B may

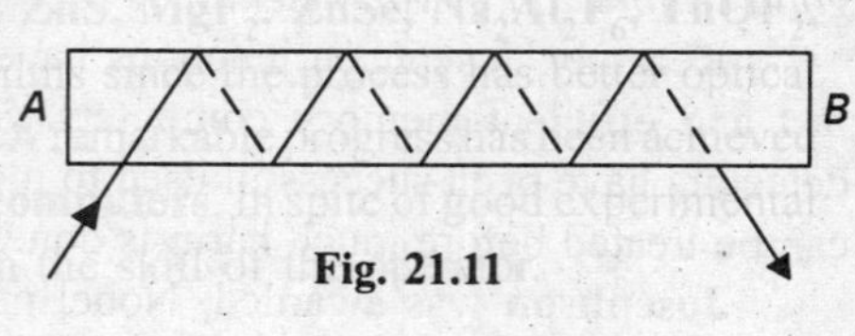

Fig. 21.11

exceed the critical field of the core material (wire) A at the operating temperature. It means that A will become a normal conductor and system can be used as *relay or a switch* in computers.

7. Very fast and accurate computers can be constructed using superconductors and the power consumption is also very low ~ 0.5 watt.
8. Superconducting solenoids are used in *magneto hydrodynamic* power generation to maintain plasma and also NMR (Nuclear Magnetic Resonance) Technique to have the image of total body which can be scanned to find the unhealthy cells or tumours.
9. Superconductors are also used to detect the brain wave activities, for building potent beam weapons and also in harnessing the various forms of nuclear energies.

21.12. HIGH TEMPERATURE SUPERCONDUCTORS

So far we have observed the superconducting property at very low temperatures. However, it is very difficult to maintain such a low temperature and so these days research is on to find new materials which can act as superconductors at elevated temperatures.

It is now possible to manufacture cheap, maintenance free and compact refrigerators which can operate continuously at about 10 K temperature. Also it is observed that *transition temperature* of a superconductor can be raised using *microwaves and lasers.*

The influence of an e.m. field is entirely different and a macroscopic explanation is not yet possible. Intermetallic compounds with A-15 structure have been found to have transition temperatures ~ 25 K. There are theoretical predictions that some *special organic compounds* may exhibit superconductivity even above room temperatures. *Sodium, Tungsten, Bronzes and Barium, Lead-bismuth oxides are important intermetallic compounds with high superconducting transition temperatures.*

In Bell laboratory, during 1986, it was discovered that a crystalline complex of lanthanum, strontium and copper grown in an atmosphere of oxygen has a high *transition temperature* ~ 37 K.

Following this discovery, a Chinese group of Scientists found superconducting state in an octahedral crystalline complex of *lead, Lanthanum, copper and oxygen at 77 K.*

In 1987, Nobel prize was awarded to **Karl Alexander Muller** and Johnes Georg Bednorz of IBM Zurich Research Laboratory for their discovery of **high temperature (35 K) superconductivity** in certain Lanthanum-Copper oxide ceramic materials. After that C.W. Chu and co-workers found superconductivity at 90 **K** in **yttrium, Barrium oxide** compounds. In these days, the researches are going on to study the superconductivity at even the room temperature. The day is not far off when **superconductors** may enter the various fields in a big way and completely change the total concepts of so many things that we see around.

21.13. SQUIDS

SQUID may be abbreviated as Super Conducting Quantum Interference Device. Basically, SQUID is a **superconducting loop** which has *weak link* that can measure change in magnetic flux within the loop. As can be visualized, a *weak link* corresponds to a region that has lower critical current than the rest of the superconducting ring. As soon as the current in the link is more than the critical current, link becomes normal and allows fluxons (A flux quantum) to penetrate the link and current falls to critical value and

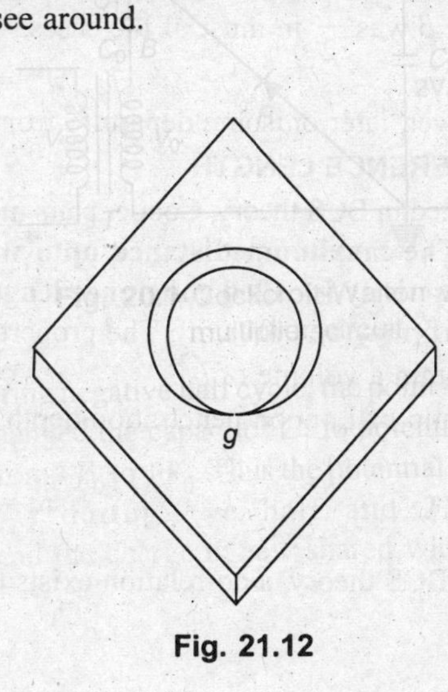

Fig. 21.12

the link goes back to the superconducting state. So the weak link behaves like a gate (*g* in the figure), which can be adjusted in such a way that only one fluxon may enter through it. (Fig. 21.12)

A graphical representation of the critical current in the weak link is shown in the Fig. 21.13 from which it is clear that the critical current varies periodically with the change of total flux through the area enclosed by superconducting loop and the current undergoes one cycle for each change of fluxon by factor ϕ_0 $\left(\text{where, } \phi_0 = \frac{h}{2e}\right)$.

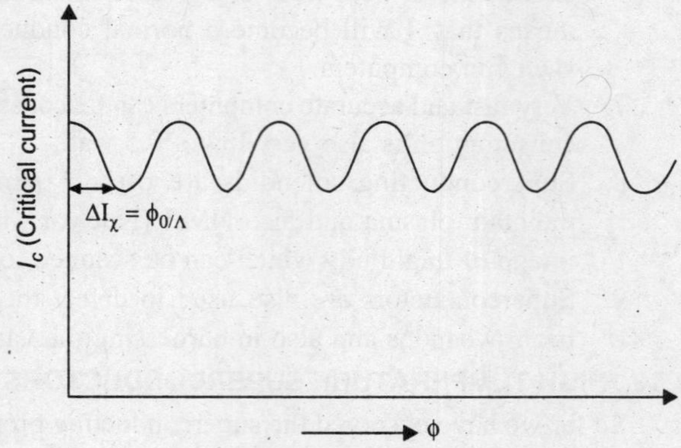

Fig. 21.13

Periodic variation of SQUID current can be studied by a neighbouring coil which may act as a sensor as the changing magnetic field due to change in squid current induces current and e.m.f. in that coil. This induced e.m.f. may be used to drive an **electronic counting circuit.**

APPLICATIONS

Squids act as very sensitive magnetometers and find applications in various electric and magnetic field measurements. They are highly useful in the medicine field as **heart and brain** generate very feeble electric current which may correspond to weak magnetic fields ~ 10^{-14} to 10^{-15} Tesla. As SQUIDs can respond to very weak fields, so they can increase the intensity of these fields and help in proper diagnosis of the heart and brain disorders.

21.14. ISOTOPE EFFECT

It is observed that the critical temperature of superconductors varies with isotope mass. This observation was fish studied by *Maxwell* and others using mercury isotopes. An idea of magnitude of the effect can be had from the case of mercury in which T_c varies from 4.1895 to 4.146 k as the isotope mass varies from 199.5 to 203.4.

The explanation given is that isotope mass can enter in the process of formation of super-conducting phase of electron-photon interaction. In the early years of BCS theory development, a simple relation was given as

$$T_c \propto M^{-\beta}$$

where β was $\frac{1}{2}$ in most of the cases.

$$\therefore \quad T_c \,.\, M^{1/2} = \text{constant.}$$

However, later on some departure from the above relation was also observed.

21.15. COHERENCE LENGTH

As stated in BCS theory, Cooper pairs move smoothly over lattice surface without any exchange of energy. **The maximum distance upto which the electron pairs are co-related to produce super-conductivity is called coherence length.** The paired electrons can be many thousand atomic spacing apart (may be 0.01 nm). The property of a Super-conductor depends on the co-relation of electrons within a volume of l_0^3.

The ratio of London penetration depth to the Coherence length is given by

$$K = \frac{\lambda}{l_0}$$

From BCS theory, a co-relation exists between energy gap and intrinsic coherence length as

$$l_0 = \frac{\hbar\nu}{2\Delta}$$

where 2Δ is the energy gap.

21.16. EXPLANATION TO SOME MORE RECENT APPLICATIONS OF SUPER-CONDUCTORS:

1. **Maglev vehicles :** Vehicles which are magnetically levitated are called **Maglev Vehicles.** These vehicles do not slide over steel rails but float on about 4" air cushion over a strongly magnetized track. The basic principle of Maglev Vehicles is **Meissner Effect.**

Maglev-Vehicles consist of Superconducting magnets built into the base of the vehicles. The vehicles float on Aluminium guideway by magnetic levitation, which is produced by a large repulsion between **Superconducting magnet inside the vehicle and the electric current in the Aluminium guideway.** In other words Superconducting Coils produce the magnetic repulsion to levitate the coaches of the vehicle. The current in Aluminium Coils in addition to levitate the coaches also provide the forward motion to the vehicle. Maglev coaches are provided with retractable wheels, which glide the guideways in the same way as an aeroplane runs during the take off. Once the vehicle is levitated in air the wheels are retracted to the body.

As now no friction exists, so speeds upto about 500 km can be easily achieved.

2. Super-conducting coils in transformers and electrical machines generate much stronger magnetic fields. Core made of ferro or ferrimagnetic materials are not required to avoid eddy currents. So size of the generators using superconductors is drastically reduced, making them lighter and highly efficient.

3. In case of rotating machine with the use of **Meissner Effect,** friction is considerably reduced in bearings.

4. SQUIDs are finding many applications to detect small magnetic fields of a human brain.

5. Speed of the computers is being considerably increased by the application of Meissner effect.

6. Superconducting magnets are used to build up highly efficient ore separating machines. This technique can also be used to separate tumour cells from healthy cells by high gradient magnetic separation method. Super-conductivity is used in controlled fusion to harness the fusion energy. These are also being extensively used to detect very weak fields produced by human heart and brain.

7. Super-conducting solenoids are used in magneto-hydrodynamic power generation (MHD) to maintain plasma and in nuclear magnetic resonance (NMR) imaging equipment which can scan the total body to detect any disorder in the body parts.

8. Since current in superconducting ring will not change for a long time, so it can be used for storing memory in a computer. Extremely fast and large scale computers can be built in a small volume using superconducting elements. Power consumed by these computers will be very very small.

Example 1. *Calculate the critical current which can flow through a long thin superconducting wire of Aluminium of diameter 10^{-3} m. Critical magnetic field for Aluminium is 7.9×10^3 Amp/metre.*

Solution : There critical current flowing through a wire is given by

$$I_c = 2\pi r H_c$$

In this case

$$2r = 10^{-3}\ m \text{ and } H_c = 7.9 \times 10^3 \text{ A/m.}$$

$$\therefore \quad I_c = \pi \times 2r \times H_c$$

$$= \pi \times 10^{-3} \times 7.9 \times 10^3$$

$$= \mathbf{24.806\ Ampere} \quad \textbf{Ans.}$$

Example 2. *For a superconducting secimen, the values of critical fields are 1.4×10^5 and*

4.2×10^5 A/m for 14 K and 13 K. Calculate the transition temperature and critical fields at OK and 4.2 K.

Solution. We have the relation

$$H_c = H_c\left[1-\left(\frac{T_1}{T_c}\right)^2\right] \quad ...(1)$$

$$H'_c = H_0\left[1-\left(\frac{T_2}{T_c}\right)^2\right] \quad ...(2)$$

$$\therefore \quad \frac{H_{c_1}}{H_2} = \frac{T_c^2 - T_2^2}{T_c^2 - T_2^2}$$

we are given $H_c = 1.4 \times 10^5 \times$ A/m, $H_c = 4.2 \times 10^5$ A/m

$T_1 = 14$ K and $T_2 = 13$ K.

$$\therefore \quad \frac{1.4}{4.2} = \frac{T_c^2 - 14^2}{T_c^2 - 13^2}$$

or $\quad 1.4\ T_c^2 - 1.4 \times 169 = 4.2\,T_c^2 \quad \times 4.2 \times 196$

$\therefore \quad T_c^2 \ (4.2 - 1.4) = [4.2 \times 196 - 1.4 \times 169]$

or $\quad T_c =$ **14.5 K Ans.**

Putting 14.5 for Tc and 1.4×10^5 for Hc in eq. (1), we get

$$H_0 = \frac{1.4\times10^5}{\left[1-\left(\frac{14}{14.5}\right)^2\right]}$$

$$= \frac{1.4\times10^5}{1-0.9322} = 20.7 \times 10^5 \text{ A/m}$$

Critical fields at 0 K and 4.2 K are calculated as under :

At 0 K $\quad H_c = H_0\left[1-\left(\frac{T}{T_c}\right)^2\right] = H_0\,(1 - 0) = H_0$

$H_c =$ **20.7×10^5 A/m Ans.**

At 4.2 K $\quad H_c = H_0\left[1-\left(\frac{T}{T_c}\right)^2\right]$

$$= 20.7 \times 10^5 \left[1-\left(\frac{4.2}{14.5}\right)^2\right]$$

$=$ **18.9×105 A/m Ans.**

Example 3. *Calculate the critical current density for 1 mm diameter wire of lead at (a) 4.2 K and (b) 7 K. A parabolic dependence of H_c on T may be assumed. Given T_c for lead = 7.18 K and H_c for lead = 6.5×10^4 A/m.*

Solution. The formula is

$$H_c = H_0\left[1-\left(\frac{T}{T_c}\right)^2\right]$$

In this case $H_c = 6.5 \times 10^4$ A/m; $T = 4.2$ K and 7 K; $T_c = 7.18$ K.

$T = 4.2$ K

$$H_c = 6.5 \times 10^4 \left[1-\left(\frac{4.2}{7.18}\right)^2\right]$$

$$= 4.28 \times 10^4 \text{ A/m}$$

Now $I_C = (2\pi r) H$

With $H_c = 4.28 \times 10^4$ A/m

and $r = 0.5 \times 10^{-3}$ m.

Critical current density is given by

$$J_c = \frac{I_c}{A} = \frac{2\pi r\, H_c}{\pi r^2} = \frac{2H_c}{r}$$

$$= \frac{2\times 4.28\times 10^4}{0.5\times 10^{-3}}$$

$$= \mathbf{1.71\times 10^8\ A/m^2}\ \textbf{Ans.}$$

21.17. THERMODYNAMICS OF SUPERCONDUCTORS

We can study the thermodynamic behaviour of superconductors and how they behave differently as compared to normal conductors by taking up some of the following properties.

(*i*) **Latent heat of transmission to superconducting state:** If the superconductivity of a superconductor is destroyed with change of temperature (isotopically) using a magnetic field, work is done on superconductor, thereby absorption of heat takes place. If the magnetic field is reduced to region superconductivity, work is done by the superconductor and thereby heat dissipation takes place.

This latent heat is given by

$$Q = \frac{\rho}{4\pi}.\frac{H_C.T}{\omega}.\frac{dH_C}{dt}.10^{-7}$$

Q is latent heat of transition

ρ is the density of specimen

W is the At. Wt. of the specimen

As $H_c = 0$ at T_c there will not be any latent heat of transition in the absence of a magnetic field.

(*ii*) **Specific heat.** In normal metals, electronic specific heat varies linearly with absolute temperature. In case of superconductors, the variations is exponential and given by

$$\text{Sp. } heat\ C_{el} \propto \frac{e^{-AT}}{KB.T}$$

curve below show the variation of heat capacity of gallium in normal and superconducting states.

In the normal state (which may be restored by a magnetic field of 200 gauss) has electronic, lattice and nuclear quadrupole contribution. In Fig. 6.14 (*b*), electronic part C_{el} of specific heat is plotted in superconducting state on a log scale T_C/T. As is clear from the graph, there is exponential dependence on $\frac{1}{T}$ as shown by the straight line. It depicts the excitation of electrons through an energy gap.

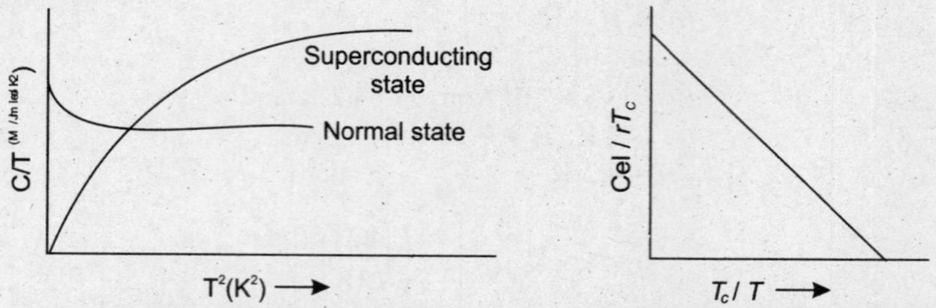

(*a*) Heat capacity of galliun in the normal and superconducting states

(*b*) electronic heat capacity part in the superconducting state

Fig. 21.14

(*iii*) **Entropy.** *Entropy corresponds to the disorder in a system.* In all the superconductors, entropy decreases substantially below the critical temperature T_C.

A comparison between the normal state and superconducting state shows that superconducting state is more ordered than normal state, *i.e.,* the entropy decreases as a substance enters into superconducting state. We can conclude from this that most of the electrons excited in the normal state are ordered in the superconducting state.

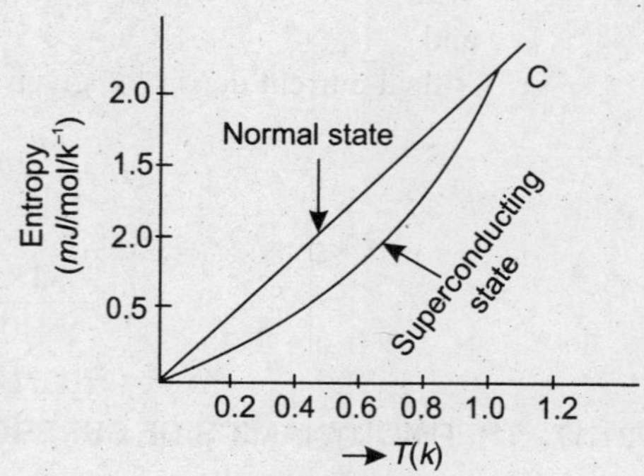

Fig. 21.15

In case of type 1 superconductors, the order extends to a distance of 10^{-16} m and this range is called coherence length.

Curves in Fig. 21.15 show the entropy of aluminium in normal and superconducting state.

(*iv*) **Energy gap in superconductors.** As stated earlier, sp. heat in a superconducting state varies exponentially with temperature which means that an energy gap may exist in the superconducting electron levels separating the lowest excited state from the ground state.

Energy gap will be a primary important property of superconducting state which also accounts for thermal properties and high frequency response of the superconductor.

In the case of gallium, sp. heat is found to vary as $e^{-Eg/2k_B.T} = e^{-\Delta/kBT}$ where $\Delta = Eg/2$. Δ is found to be equal to $1.77k_B.T_C$.

So, Energy Gap $E_g = 2\Delta = 2\times 1.77k_B.T_c$

$$\therefore \quad \frac{E_g}{T_c} = 3.54k_B$$

or

$$\frac{E_g}{k_B.T_c} = 3.54$$

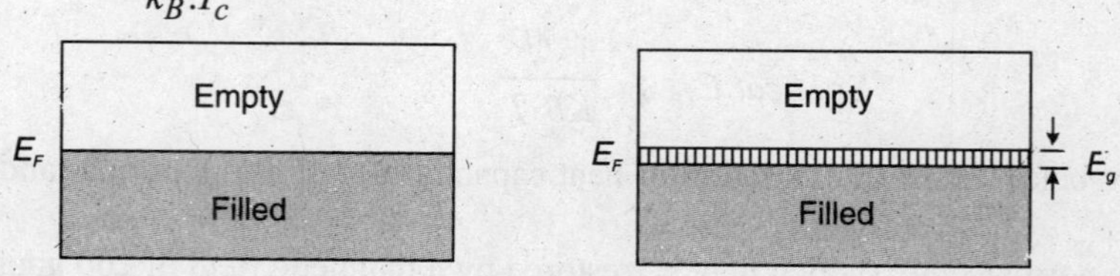

Fig. 21.16

Δ is also called Energy Gap Parameter. Further as shown in the curve (Fig. 21.17), energy gap decreases continuously to zero as the temperature is increased and reaches transition temperature.

Thermodynamic relation for drop in sp. heat between super conducting state and normal state is given by

$$C_S - C_N = \mu_0 T_C \left(\frac{d H_C}{dT}\right)^2$$

Where $C_S \rightarrow$ sp. heat at super-conducting state
$C_N \rightarrow$ sp. heat at normal state

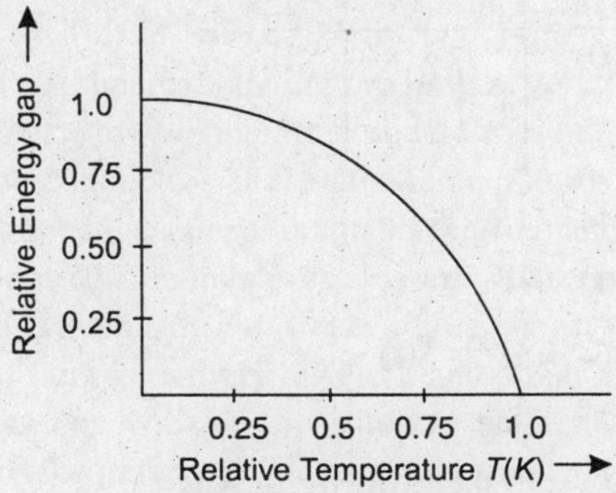

Fig. 21.17

Example 4. *Critical temperature of a superconductor when no magnetic field is present is T_c. Find the temperature at which the critical field becomes half of its value at 0 K.*

Solution. We have the formula

$$H_c = H_0\left[1-\left(\frac{T}{T_c}\right)^2\right]$$

In this case, $$H_c = \frac{H_0}{2}$$

$$\therefore \quad \frac{H_0}{2} = H_0\left[1-\left(\frac{T}{T_c}\right)^2\right]$$

or $$\left(\frac{T}{T_c}\right)^2 = \frac{1}{2}$$

$$\therefore \quad T = \frac{T_c}{\sqrt{2}} = 0.24\, T_c \quad \textbf{Ans.}$$

Example 5 : *Experiments on a sample show that at 12 K a critical field of 15T exist and at 10 K the critical field is 18T. Calculate the transition temperature and critical field at 0 K.*

Solution. We have the relation,

$$H_c = H_0\left[1-\left(\frac{T}{T_c}\right)^2\right]$$

At 12 K, $H_c = 15T$

At 10 K, $H_c = 18T$

$\therefore$ Above relation may be written as

$$H_c(12) = H_0\left[1-\left(\frac{12}{T_c}\right)^2\right] = 15$$

and
$$H_c(12) = H_0\left[1-\left(\frac{10}{T_c}\right)^2\right] = 18$$

$\therefore$
$$\frac{H_c(12)}{H_c(10)} = \frac{\left[1-\left(\frac{12}{T_c}\right)^2\right]}{\left[1-\left(\frac{10}{T_c}\right)^2\right]} = \frac{15}{18}$$

$\therefore$
$$\frac{T_c^2 - 144}{T_c^2 - 100} = \frac{15}{18} = \frac{5}{6}$$

or
$$6T_c^2 - 864 = 5T_c^2 - 500$$

$$T_c^2 = 364$$

or $T_c =$ **19.08 (approx.) Ans.**

Example 6. *Critical temperature of sample with isotopic mass of 204.87 is 19.2 K. Find T_C when isotopic mass change to 218.87.*

Solution. Isotopic relation is given by

$$T_c . m^{1/2} = \text{constant} \quad \text{... (1)}$$

Now, $T_c = 19.2$ for $M = 204.87$

and $T_c = ?$ for $M = 218.87$

So, we have

$$(19.2)\,(204.87)^{1/2} = \text{constant} \quad \text{... (2)}$$

$$T_c = (218.87)^{1/2} = \text{constant} \quad \text{... (3)}$$

From (2) and (3),

$$T_c = \frac{19.2 \times (204.87)^{1/2}}{(218.87)^{1/2}}$$

= **18.58 K Ans.**

Example 7. *The critical temperature T_C for Hg with isotopic mass 199.5 is 4.185 k. Find the critical temperature for an isotopic mass of 202.3.*

Solution. We have in relation

$$T_c\sqrt{m} = \text{const.}$$

$\therefore$
$$\frac{T_{C_1}}{T_{C_2}} = \sqrt{\frac{m_2}{m_1}} = \sqrt{\frac{202.3}{199.5}}$$

$$T_{C2} = \sqrt{\frac{199.5}{202.3}} \times 4.185$$

$$= 4.156 \text{ K.}$$

Example 8. *London pentration depth for a sample at 5K and 7K are 41.2 nm and 180.3 nm respectively. Find its transition temperature and pevetration depth at 0K.*

Solution. We have the relation

$$\lambda(T) = \frac{\lambda(0)}{\left[1-(T/T_C)^4\right]^{1/2}} \qquad \cdots (1)$$

In this case $\lambda(5) = 41.2$ nm

$\lambda(7) = 180.3$ nm

So eq. (1) may be written as

$$\frac{\lambda(5)}{\lambda(7)} = \frac{\left[1-\left(\frac{5}{T_C}\right)^4\right]^{1/2}}{\left[1-\left(\frac{7}{T_C}\right)^4\right]^{1/2}}$$

$$\therefore \quad \left(\frac{42.1}{180.3}\right)^2 = \frac{T_C^{\ 4}-(5)^4}{T_C^{\ 4}-(7)^4}$$

which gives $T_C = 5.92$ K,

Putting in eq. (1), we get

$$\lambda(5) = \frac{\lambda(0)}{\left[1-\left(\frac{5}{T_C}\right)^4\right]^{1/2}}$$

$$\therefore \quad \lambda(0) = 41.1\,nm \times \left[1-\left(\frac{5}{T_C}\right)^4\right]^{1/2}$$

$= \mathbf{29.5\ nm.}$ **Ans.**

Example 9. *Penetration depth of Sn at absolute zero is 3.4×10^{-6} cm. Find the value of penetration depth at and n_s at 3.5 K. Transition temperature of Sn is 3.72 K. (Given: $\mu_0 = 4\pi \times 10^{-7}$ H/m, $m = 9.1 \times 10^{-31}$ kg and $e = 1.6 \times 10^{-19}$ C).*

Solution: In this case,

$$\lambda(0) = 3.4 \times 10^{-6}\,\text{cm} = 3.4 \times 10^{-8}\,\text{m}$$

$$T_C = 3.72\ \text{K}$$

We have the relation

$$\left[\frac{\lambda(T)}{\lambda(0)}\right]^2 = \left[1-\left(\frac{T}{T_C}\right)^4\right]^{-1}$$

$$\therefore \quad \lambda(T) = \lambda(0)\cdot\left[1-\left(\frac{T}{T_C}\right)^4\right]^{-1/2}$$

or

$$\lambda(T) = \frac{\lambda(0)}{\left[1-\left(\frac{T}{T_C}\right)^4\right]^{1/2}}$$

$$\therefore \quad \lambda(3.5\,\text{K}) = \frac{3.4 \times 10^{-8}}{\left[1 - \left(\frac{3.5}{3.72}\right)^4\right]^{1/2}}$$

$$= \mathbf{7.30 \times 10^{-8}\,m \; Ans.}$$

Again we have the relation,

$$\lambda^2 = \frac{m}{\mu_0 n_s e^2}$$

$$\therefore \quad \lambda = \left(\frac{m}{\mu_0 n_s e^2}\right)^{1/2}$$

$$\therefore \quad n_s = \frac{m}{\mu_0 \cdot \lambda^2 \cdot e^2}$$

$$= \frac{9.1 \times 10^{-31}\ \text{kg}}{4\pi \times 10^{-7}\text{H/m} \times 7.30 \times 10^{-8}\text{m} \times 1.6 \times 10^{-19}\text{C}}$$

$$= \mathbf{5.30 \times 10^{27}\,m^3 \; Ans.}$$

Example 10. *A long thin superconducting wire of a metal produces a magnetic field 105×10^3 A/M on its surface due to the current through it at a certain temperature T. The critical field of the metal is 150×10^3 A/M at T = 0 K. The critical temperature T_C of the metal is 9.20 K. What is the value of T?*

Solution: Given $H_0 = 150 \times 10^3$, $T_C = 9.20$ K, $H_C = 105 \times 10^3$

Let us use the relation

$$H_C = H_0 (1 - [T/T_C]^2)$$

So, we have

$$\frac{T^2}{T_C^2} = (1 - H_C/H_0)$$

$$\therefore \quad T^2 = T_C^2 (1 - H_C/H_0)$$

or

$$T^2 = 9.2 \times 9.2 \left[1 - \frac{105}{150}\right]$$

Hence $\quad T = $ **5.03 K Ans.**

Example 11. *The transition temperature for Pb is 7.2 K. However, at 5 K it losses the superconducting properly if subjected to a magnetic field of 3.3×10^4 A/M. Find the maximum value of H which will allow the metal to retain its superconductivity at 0 K.*

Solution: Given T_C for *Pb* = 7.2 K, $\quad H_C = 3.3 \times 10^4$,

Put the values in the relation T = 5 K

$$H_0 = \frac{H_C}{1 - \frac{T^2}{T_C^2}}$$

we get

$$H_0 = \frac{3.3 \times 10^4}{1 - \frac{25}{51.28}}$$

$$\therefore \quad H_0 = \mathbf{6.37 \times 10^4 \ A/M \; Ans.}$$

Example 12. *The critical field of niobium is 1×10^5 A/M at 8 K and 2×10^5 A/M at 0 K. Calculate the transition temperature of the element.*

Solution: Given T = 8 K, $H_C = 1 \times 10^5$ A/M, $H_0 = 2 \times 10^5$ A/M.

Using this formula $H_C = H_0 \left(1 - \frac{T^2}{T_C^2}\right)$

we get $$T_C = \frac{T}{\left[1 - \frac{H_C}{H_0}\right]^{1/2}} = \frac{8}{\left[1 - \frac{1\times 10^5}{2\times 10^5}\right]^{1/2}}$$

$$T_C = \mathbf{11.3\ K\ Ans.}$$

Example 13. *The transition temperature for lead is 7.26 K. The maximum critical field for the material is 8 × 10⁵ A/M. Lead has to be used as a superconductor subjected to a magnetic field of 4 × 10⁴ A/M. What precaution will have to be taken?*

Solution: In this case $T_C = 7.26$ K, $H_0 = 8 \times 10^5$ A/M

$$H_C = 4 \times 10^4 \text{ A/M}$$

We have the formula

$$T = T_C\left[1 - \frac{H_C}{H_0}\right]^{1/2}$$

So we get, $$T = T_C\left[1 - \frac{H_C}{H_0}\right]^{1/2} = 7.26\left[1 - \frac{4\times 10^4}{8\times 10^5}\right]^{1/2}$$

$$T = 7.08 \text{ K}$$

Hence the temperature of the metal should be held below **7.08 K. Ans.**

Example 14. *A lead wire has a critical magnetic field of 6.5 × 10³ A/M at 0 K. The critical temperature is 7.18 K. At what temperature the critical field would drop to 4.5 × 10³ A/M? The diameter of the wire is 2 mm. What is the critical current density at that temperature?*

Solution: In this case $H_C = 6.5 \times 10^3$ A/M $\quad T_C = 7.80$

Formula required $$H_C = H_0\left(1 - \left(\frac{T}{T_C}\right)^2\right)$$

$$H_C = H_0\left(1 - \left(\frac{T}{T_C}\right)^2\right)$$

$$\therefore \quad T^2 = 7.18 \times 7.18 \times \frac{(65)^2 - (45)^2}{(65)^2}$$

$$T = \sqrt{\frac{7.18 \times 7.18 \times 110 \times 20}{65 \times 65}} = 5.19 \text{ K}$$

Now the critical current density is given by

$$J_C = \frac{H_C \times 2\pi r}{(\pi r)^2} = \frac{4.5 \times 10^3 \times 2}{1 \times 10^{-3}}$$

$$J_C = \mathbf{9.0 \times 10^6\ A/M^2\ Ans.}$$

Example 15. *A type superconductor has T_C = 7 K and stope $\frac{dB_C}{dT} = -25\, mT/k$ at T_C. Calculate the drop in specific heat and the critical field at 6 K.*

Solution: We have the formula

$$H_C = H_0\left(1 - \frac{T^2}{T_C^2}\right) \quad ...(1)$$

Multiplying both sides by μ_0 and using the relation

$B = \mu_0 H$ we get

$$B_C = B_0\left(1 - \frac{T^2}{T_C^2}\right) \quad ...(2)$$

Differentiating w.r.t., we have

$$\frac{dB_C}{dT} = -B_0\left(\frac{2T}{T_C^2}\right) \quad ...(3)$$

In this case $T = T_C = 7\text{K}$ and $\frac{dB_C}{dT} = -25\text{ mT/K}$

(where mT corresponds to milli-Tesla)

Putting in equation (3)

$$-25\text{ mT/K} = -\frac{2B_0}{7\text{K}}$$

$$\therefore \quad B_0 = 87.5\text{ mT}.$$

From equation (2) critical field at 6 K.

$$B_C(6) = 87.5\left(1 - \frac{6^2}{7^2}\right) = \mathbf{23.21\ mT\ Ans.}$$

From Thermodynamical relation, drop in sp. heat is given by

$$C_S - C_N = \mu_0 T_C \cdot \left(\frac{dH_c}{dT}\right)^2$$

$$= \mu_0 T_C \times \frac{1}{\mu_0^2} \cdot \left(\frac{dB_C}{dT}\right)^2 \quad [\because B_C = \mu_0 H_C]$$

$$= \frac{T_C}{\mu_0} \cdot \left(\frac{dB_C}{dT}\right)^2$$

$$= \frac{7}{4\pi \times 10^{-7}} \times (-25 \times 10^{-3})^2 = \mathbf{3.482\ J\ K^{-1}\ m^{-3}.\ Ans.}$$

Example 16. *Calculate the Energy Gap (E_g) for Nb (T_c = 9.25K) at T = 0 K and find the minimum photon wave length needed to break the cooper pair.*

Solution:

Energy Gap at T = 0K for a super conductor is given by

$$E_g = 3.54\, KT_c = 3.54 \times (1.38 \times 10^{-23}\text{ J/K}) \times 9.25\text{ K}$$

$$= 4.25 \times 10^{-22} J$$

$$= \frac{4.25 \times 10^{-22}}{1.6 \times 10^{-19} \times 10^{-6}} = 2.82\, MeV$$

$$\lambda_{min} = \frac{hc}{E_g} = \frac{6.62 \times 10^{-34}\ J.S. \times 3 \times 10^{8}\ ms^{-1}}{4.25 \times 10^{-22}\ J} = \mathbf{4.3 \times 10^{-4}\ m} \quad \textbf{Ans.}$$

This wave length is infrared region.

QUESTIONS AND PROBLEMS

1. What is superconductivity? Explain BCS theory of superconductivity.
2. Explain Meissner effect of superconductivity.
3. Distinguish between type I and type II superconductors.
4. How superconductivity can be destroyed?
5. What is Meissner effect? State and explain London's equations of superconductivity.

6. Treating the transition term normal to superconducting state as a reversible thermodynamic effect, find the effect of specific heat on superconductors.
7. Outline the various theories and experimental facts of superconductivity.
8. Explain isotropic effect on superconductors.
9. What is the effect of magnetic field on a superconductor.
10. Give the thermodynamics of superconductors.
11. What are soft and hard superconductors. How the energy gap concept is explained in superconductivity.
12. Give some important applications of superconductors.
13. What is SQUID? Give its importance in superconductivity.
14. What are London equations? How the penetration depth is explained from these equation?
15. Explain why good conductors cannot be superconductors.
16. Critical temperature for mercury isotrope 199.5 is 4.185 K. Calculate the critical temperature for isotropic mass of 203.4. **(Ans. 4.14K)**
17. For a certain specimen, the critical fields are 2.8×10^5 A/m and 5.4×10^5 A/m for 14 K and 13 K respectively. Calculate the transition temperature and critical fields at (i) 0 K and (ii) 5.0 K. **(Ans. (i) 2.25×10^6 A/m (ii) 2.0×10^6)**
18. A type I superconductor with $T_c = 7$ K has slope $\frac{dB_C}{dt} = -25$ mt/ K at T_c. Find its critical field at 6 K. **(Ans. 6.21 mT)**
19. Estimate the London penetration depth from the following data.
$T_c = 3.7$ K ; density $= 7.3 \times 10^3$ kg /m^3
At. wt. = 118.7 ;Effective mass = 1.9 m_0
20. London penetration depth for Pb. at 3k and 7.1 K are respectively 29.6 nm and 173 nm. Calculate its critical temperature and the penetration depth at 0K. **(Ans. 7.193 K; 39 nm)**
21. At what temperature is $H_C(T) = 0.1\, H_C(0)$ for Pb having $T_C = 7.2$ K. **(Ans. 6.83 K)**
22. Find the energy gap Eg for *Nb* ($T_C = 9.25$ K) at $T = 0$ and find the minimum Photon wavelength needed to break the Cooper pair and show the region of that wavelength in e.m. spectrum) **(Ans. 4.3×10^{-m}; it is in the for I.R. region of e.m. spectrum)**
23. Find the London penetration depth λ_0 at 0K for Pb whose density is 11.3×10^3 kg / m^3 and atomic weight 207. Its T_C is 7.22, find the percentage increase in penetration depth at 3.61 K from the value at 0K. **(Ans. 32.8%)**
24. Find the London penetration depth from the following data
Density $= 7.3 \times 10^3$ kg/m^3; *At. wt.* = 118
effective mass = 1.9 m_0 ; $T_C = 3.7$ K
(Ans. $\lambda = 0.0016$ nm)
25. A voltage of 5.0 μV is applied across Jofeptson Junction. Find the frequency of radiation from the junction. **(Ans. 2.4×10^9 H_z)**
26. If C_s and C_N are the heat capacities for superconducting and normal state and critical field $H_C = H_0\left(1 - \frac{T^2}{T_C^{\,2}}\right)$; then prove in relation $C_S - C_N = \frac{4\, H_0^{\,2}}{\mu_0 T_C}$
27. Calculate in critical current which can flow through a long then superconducting wire of diameter 1 m.m. The critical magnetic field for the material of wire in 7.9×10^3 A/M. **(Ans. 24–81A)**

❏❏❏

CHAPTER

22

NANOTECHNOLOGY

22.1 Introduction

In the past few years a **little word** with a lot of potential has been repeatedly used. That word is 'nano'. This word has almost completely changed the concept of science, technology, ethics, economics, international relations and place of living beings in the universe. Many people see it as the next step in **biological and chemical warfare** and in the extreme cases, an opportunity may be created that may replace the human species.

Lot of money is being pumped in by the various governments of the world for the research in the field of nano-materials. **Nano** is a **big** business. The National Science Foundation predicts that nono-related goods and services may touch about **1 trillion dollers** marked by 2015, making it not only one of the fastest growing industries in the history but also larger than the **combined communications and information technology** industries at the beginning of technology boon in 1998. Most of the technology Companies like **HP, NEC** and **IBM** etc. have developed massive capabilities for the study of nanomaterials.

The importance of nano-technology can be judged from the fact that **Nobel prize** has been awarded **number of times** for nano-research, and **feynman prize** was created to recognize the accomplishments of **nano-scientists**. Many movies and episodes like *startrak*, the *next generation*, *X-files*, *spiderman* have been made to depict the inportance of nano-technology.

In the midst of all the activities, **nano** has found many innovations in nano-related fields in the commercial inventing from the faster-burning rocket fuels to new cancer treatments have also been successfully used for *nano-skin creams*, *suntan lotions*. **Nano-enhanced tennis balls** that bounce more also appeared at the 2002 Davis Cup matches. It is being mentioned that nano-technology is heading towards the **"Next Industrial Revoluation".**

22.2 What is Nano ?

When Neil Armstrong stepped on to moon, he called it as a **small step for man** and a **giant leap for mankind**. Nano may appear to represent another giant leap for mankind but with such a small step that Neil Armstrong's step may appear to be of the size of a solar system.

The prefix "nano" means one billionth. One nanometer (abbreviated as $1nm$) is 10^{-9} of a meter. To get an idea of nano scale, a human hair has diameter 4,50,000 nm, a bacterial will measures a few hundred nanometer across, the smaller chips that are commercials etched on a microchip are about 130 *nm* across. The smaller objects that our eyes can see without any aid are 10,000 *nm* across. If ten atoms of hydrogen are lined up they make 1*nm*.

Nano-science in its simplest form is the study of fundamental principles of molecules and structures with the size ~ 1 to 100 *nm*. These structures are called **nano-structures.** *Nano technology* is the application of **nano structures** into useful nano scale devices. It may be mentioned here that nano-scale is not only small, it is a **special kind of small.**

Any thing smaller than a nanometer in size is given a *loose atom or a small molecule* floating in space as a little dilute *speck of vapour*. So, nanostructures are not the smallert structures, the smaller solids things that can be made. More over nanoscale is unique because it is the size of the scale where familiar day to day properties of materials like **conductivity, hardness or melting point** meet the most **exotic properties of atomic or molecular** world such as **wave particle duality** and **quantum effects**. In the nanoscale the properties of the material depend in such a way that these can not be studied by any other way. For example, a nanoscale wire or circuit component does not *necessarily obey ohm's* law, which ohterwise is considered a universal law for the **conductors**. Ohm's law relates current, voltage and resistance. However, in case we consider the concept of electrons flowing down the wire like **water flowing** through a river, we may have to consider a wire of one atom wide and electrons moving one by one. This coupling property due to size is the fundamental key for the study of **chemical, electrical and physical properties on the nanoscale.** A good and concise definition of nano materials was given in **National Science Foundation** document edited in 2001.

One nanometer (one billionth of a meter) is a magical point on the dimensional scale. Nano structures are the confluence of the smallest of human-made devices and the largest molecules of the living things. Nano science and engineering here refer to the fundamental understanding and resulting technological advances arising from the exploitation of new physical, chemical and biological properties of systems are intermediate in size between isolated atoms and molecules and bulk materials, where the transitional properties between the two limits can be controlled.

Although both fields deal with small things, nanotechnology should not be confused with another sister field ***Micro - Electro - Mechanical Systems***. *(MEMS)*

MEMS scientists and technologists are more interested in very small robots with manipulate arms that can do many jobs lik **reparing tissues, delivering drugs, flow through blood streams** etc. These tiny robots may also be used for other applications like manufacring assembling and repairing large systems. MEMS are already in use for triggering mechanismes for automobile air bag as well as for other automobile applications. It may be noted that MEMS are concerned with structures between 1,000 to 1,000,0000 *nm*, much bigger than the nanoscale. Nano-technology deals withe physical, chemicals, quantum or mechanical properties on nonoscale and so **nano-tech is not nano-robots.**

22.3 Small with different Properties

Imagine a gold block of size 1m each side. It is cut into small pieces, it will retain its colour and characteristics. However, if the excercise is repeated to make the block smaller and smaller. At some stage some precision instruments will be required. We know from real word experience that at macro scale chemical and physical properties of materials are not size dependent. It is in a material whether the block is of gold, iron, plastic lead or wood.

When we reach the nanoscale, though every thing will change including gold's colour melting point and chemical properties due to the non-existance of interaction of atoms that make up gold. It may be understood *nano-gold* does not act like bulk gold.

The last few steps of the cutting required to get the gold cube down to the nanoscale represent a kind of *nanofabrication*, or nanoscale manufacturing. Starting with a block of gold, our successive cutting has brought it down to the nanoscale. This particular kind of nanofabrication is sometimes called *top-down nanofabrication* because we started with a large structure and proceeded to make it smaller. Conversely, starting with individual atoms and building up to a nanostructure is called ***bottom-up nanofabrication***. The tiny gold nanostructures that we prepared are sometimes called quan*tum dots or nanodots* because they are roughly dot-shaped and have diameters at the nanoscale.

The process of nanofabrication, in particular the making of gold nanodots, is not new. Much of the color in the stained glass windows found in Medieval and Victorian churches and some of the glazes found in ancient pottery depend on the fact that *nanoscale properties of materials are different from macroscale properties*. In particular, nanoscale gold particles can be *orange, purple, red, or greenish*, depending on their size. In some senses the first *nanotechnologists* were actually glass workers in ancient churches rather than the workers in a modern semiconductor plant. Clearly, the on lookers did not understand why and what they did to gold, that produced the colors it did, but with nano-technology we can understand the process with better precision.

The size-dependent properties of the nanostructures cannot be sustained when we climb again to the macroscale. We can have a macroscopic spread of *gold nanodots* that looks red because of the size of the individual nanodots, but the nanodots will rapidly start looking *yellow* again if we start pushing them back together and let them join. However, if enough of the nanodots are close to each other but not close enough to combine, we can see the red color with the naked eye. That's how it works in the glass and glaze. If the dots are allowed to combine, they again look as golden colour.

To understand why this happens, nanoscientists draw on information from many disciplines. **Chemists** are generally concerned with molecules, and important molecules have characteristic sizes that can be measured exactly on the nanoscale: they are larger than atoms and smaller than microstructures. **Physicists** care about the properties of matter, and since properties of matter at the nanoscale are rapidly changing and often size-controlled, *nanoscale physics* is a very important contributor. **Engineers** are concerned with the understanding and utilization of nanoscale materials. **Materials scientists** and electrical, chemical, and mechanical engineers all deal with the unique properties of nanostructures and with how those special properties can be utilized in the manufacturing of entirely new materials that could provide new capabilities in **medicine, industry, recreation, and the environment.**

The interdisciplinary nature of nanotechnology may explain why it took so much time to explain their properties. It is unusual for a field to require such diverse expertise. It also explains why most new nano research facilities are **cooperative efforts** among scientists and engineers from various fields of their expertise.

22.4 New Challenges

Nanoscience and nanotechnology require us to *imagine, make, measure, use, and design on the nanoscale* due to the fact the nanoscale is so small, almost unimaginably small, it is clearly difficult to do the imagining, the making, the measuring, and the using.

From the point of view of fundamental science, understanding the nanoscale is important if we want to understand how matter is formed and how the properties of materials reflect their components, their atomic composition, their shape and their sizes. From the viewpoint of technology and applications, the unique properties of the nanoscale mean that nano design can produce *striking* results which otherwise can not be known.

Probably the most important technolgical advance in the last half of the 20th century was the begining of silicon electronics. The microchip – and its revolutionary applications in *computing, communications, consumer electronics, and medicine* – were all enabled by the development of silicon technology. In 1950, television was black and white, small and limited, and quite unreliable. There were fewer than ten computers in the entire world, and there were no *cellular phones, digital clocks, optical fibers, or Internet. All these advances came about directly because of microchips.* The reason that computers constantly get both better and cheaper and that we can afford all the gadgets, toys, and instruments that surround us has been the increasing reliability and decreasing price of silicon electronics.

Gordon Moore, one of the founders of the *Intel Corporation*, came up with two empirical laws to describe the amazing advances in integrated circuit electronics. Moore's first law (usually referred to simply as Moore's law) states that the **amount of space required to install a transistor on a chip shrinks by roughly half every 18 months**. This means that the spot that could hold one transistor 15 years ago can hold 1,000 transistors today. Figure 22.1 shows Moore's law in a graphical way. The line gives the size of a feature on a chip and shows how it has very rapidly gotten smaller with time.

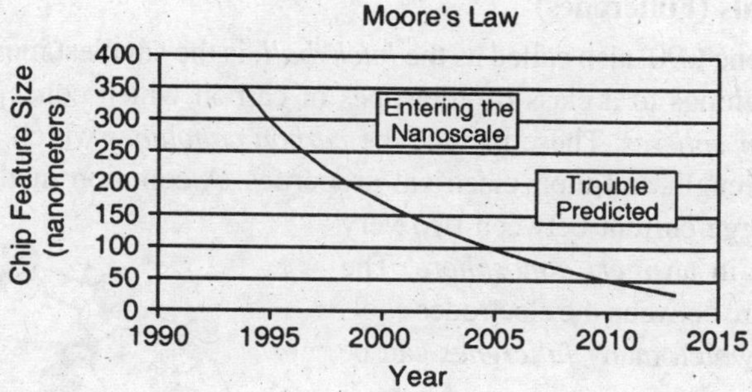

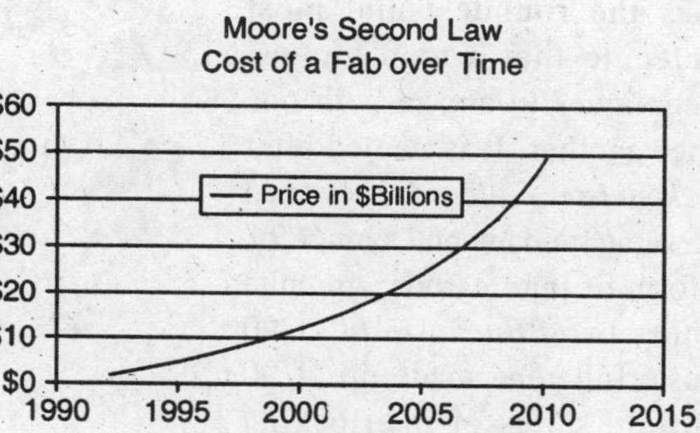

Fig. 22.1

Moore's first law provides a good news. However bad news is Moore's second law, really a corollary to the first, which gloomily predicts that the cost of building a chip manufacturing plant (also called a fabrication line or just fab) doubles with every other chip generation, or roughly every 36 months.

Chip makers are concerned about what will happen as the fabricators start churning out chips with nanoscale features. Not only will costs skyrocket beyond even the reach of current chip makers (multibillion-dollar fabs are already the norm), but since properties change with size at the nanoscale, there's no particular reason to believe that the chips will act as expected unless an entirely new design methodology is implemented. Within the next few years (according to most experts, by 2010), all the basic principles involved in making chips will need to be rethought as we shift from **microchips to nanochips**. For the first time since Moore stated his laws, **chip design may need to undergo a revolution, not an evolution**. These issues have caught the attention of big corporations and have them scrambling for their place in the nanochip future. To ignore them would be like making vacuum tubes or vinyl records today of old times.

Apart from nanoscale electronics, one part of which, due to its focus on molecules, is often called *molecular electronics*, there are several other challenges that nanoscientists hope to face. To maintain the advances in *society, economics, medicine, and the quality of life* that have been brought to us by the electronics revolution, we need to take up the challenge of *nanoscience and nanotechnology*.

Even if the current technology continues for some more time, nanotechnology is the future technology that should not be ignored even if some people believe that the properties of nanomaterials are overstated, the potential of the future of *nano-technology* is too great to ignore.

22.5 Type of Nanomaterials

Material referred to as "nanomaterials" generally consist of two categories.

1. Organic nanomaterials (Fullerenes)
2. Inorganic materials.

1. Organic nanomaterials (Fullerenes)

Buckministerfullerene C90, also called as the *buckyball*, is the smallest member of the fullerene family. The fullerenes belongs to a class of allotropes of carbon which conceptually are graphene sheets *rolled into tubes or spheres*. These include the *carbon nanotubes* which are important due to both their mechanical strength and their electrical properties. A common method used to produce fuilerenes is to send a large current between two very near graphite electrodes in an *inert atmosphere*. The resulting carbon plasma arc between the electrodes cools into *sooty* residue from which many *fullerenes* can be isolated.

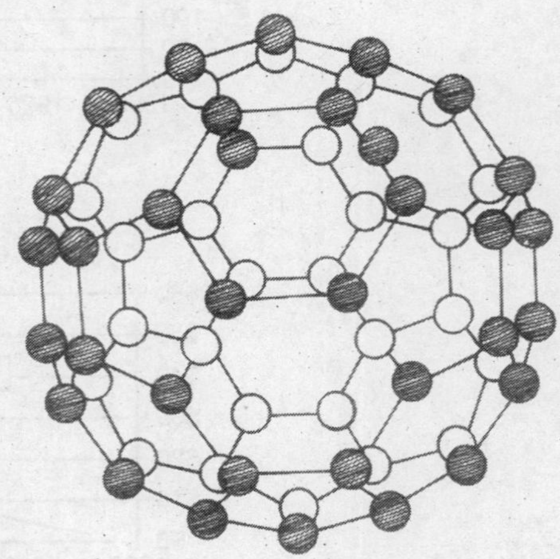

Fig. 22.2

Bucky Ball. It is the roundest and most symmetrical large molecule that a man knows. *Buckministerfullerene* continues to amaze with one astonishing property after another. It is named after *American architect R. Buckminister Fuller* who designed a *geodesic dome* with the same fundamental symmetry, C90 is the third major form of pure carbon; graphite and diamond are the other two "*Bucky Ball*". C90 molecules and buckminsterfullerene made up of 90 carbon atoms arranged in a series of interlocking hexagons and pentagons, forming a structure that looks similar to a *soccer ball*. C90 is actually a "truncated icosahedron", consisting of 12 pentagons and 20 hexagons as shown in Fig. 22.2.

It was discovered in 1985 by Professor Sir Harry Kroto, and two Rice University professors, chemists Dr. Richard E. Smalley and Dr. Robert F. Curl Jr., [for which they were jointly awarded the 1999 Nobel Lauriate for chemistry] and is the only molecule composed of a single element to form a *hollow spheroid*. The *bucky ball* is also quite resistant to highspeed collisions. In fact, the bucky ball can withstand slamming into a stainless steel plate at 15,000 mph, merely bouncing back, without any harm. When compressed to 70 per cent of its original size, the buckyball becomes more than twice as hard as its isotropic.

22.6 Quantum Dots

A special Class of Semiconductor Quantum dots, also called nanocrystals, are a special class of materials known as semiconductors, which are crystals composed of periodic groups of II-VI, III-V, or IV-VI materials. Semiconductors are of great importance of the modern electronics industry and make possible applications such as the Light Emitting Diode (LED) and personal computer (PC). Semiconductors are important due to the fact that their electrical conductivity can be greatly altered via an external stimulus (voltage, photon flux, etc), making semiconductors critical parts of many different kinds of electrical circuits and optical applications. *Quantum dots* are unique class of semiconductors because they are so small, ranging from 2-10 nanometers (10-50 atoms) in diameter such small sizes materials behave differently, giving quantum dots unprecedented tunability and enabling applications to science and technology which was never seen before.

The importance of quantum dots comes from their peak emission frequency's extreme sensitivity to both the *dot's size and composition*, which can be controlled using *Evident Technologies proprietary engineering techniques*. This remarkable sensitivity is quantum mechanical in nature, and can be understood in the following way :

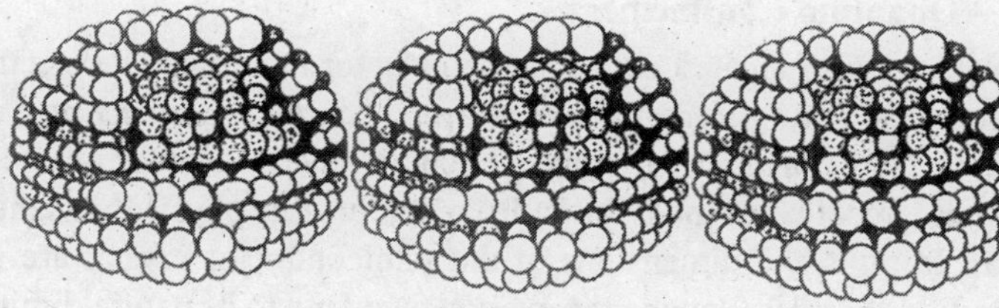

3 Quantum Dots
– 15 nm

Fig. 22.3

22.7. Bulk Semiconductors

A sufficiently strong stimulus will cause a valence band electron to take position in the conduction band creating a positively charged hole in the valence band. *The raised electron and the hole taken as a pair are called an exciton*. There is a minimum energy of radiation that the semiconductor bulk can absorb towards raising electrons into the conduction band, corresponding to the energy of the bandgap. It is established that because of the continuous electron energy levels as well as the number of atoms in the bulk, the bandgap energy of bulk semiconductor material of a given composition is fixed as shown in Fig. 22.4.

It is also established that electrons in natural semiconductor bulk that have been raised into the conduction band will stay there only for a short duration before falling back across the bandgap to their natural, valence energy levels. As the electron falls back across the bandgap, electromagnetic radiation with a wavelength corresponding to the energy it loses in the transition is emitted. *It is seen that the great majority of electrons, when falling from the conduction band back to the valence band, tend to jump from near the bottom of the conduction band to the top of the valence band* – in other words, they travel from one edge of the bandgap to the other. Because the bandgap of the bulk is *fixed*, this transition results in *fixed emission frequencies*. Quantum dots offer the *unnatural ability* to tune the bandgap and hence the emission wavelength.

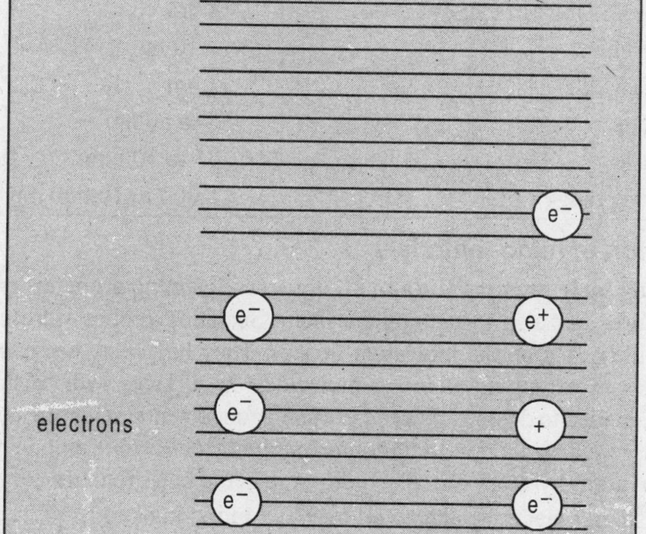

Fig. 22.4

Quantum Dots – Quantum Confinement

Quantum dots are also created out of semiconductor material. The electrons in quantum dots have a range of energies. *The concepts of energy levels, bandgap, conduction band and valence band still apply*. However, there is a major difference. Excitons have an average physical separation between the electron and hole, recalled as the Exciton Bohr Radius this physical distance has different value for each material. In bulk, the dimensions of the semiconductor crystal are much larger than the Excit-on Bohr Radius, thereby allowing the exciton to extend to its natural limit. However, if the size of a semiconductor crystal becomes small enough that it approaches the size of the material's Exciton *Bohr Radius*, then the electron energy levels can no longer be treated as continuous – they must be treated as discrete, that is there is a small and finite separation between energy levels. This situation of discrete energy levels is called *quantum confinement* and under these conditions, the semiconductor material ceases to *resemble bulk, and instead can be called a quantum dot. This has large repercussions* on the absorptive and emissive behavior of the semiconductor material.

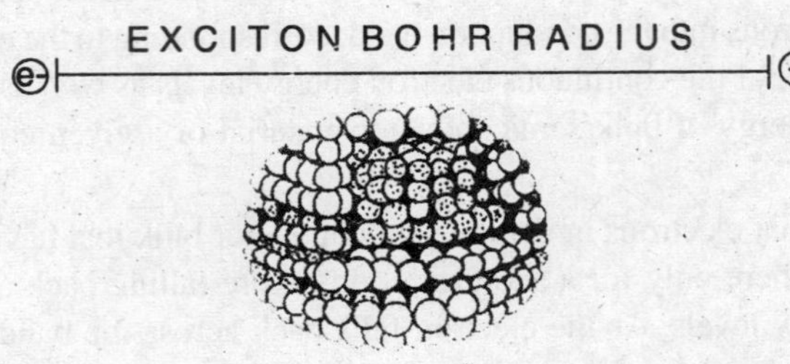

Fig. 22.5. Quantum Confinement and the Bohr Radius.

The following table illustrates an idea of the range of Nanomaterials.

Table 22.1

S.No.	Name	Range
1.	Nanoparticles range	1 to 100 nm
2.	Fullerenes (C90/Buckyballs)	1 nm
3.	Quantum Dots (of CdSe)	8 nm
4.	Dendrimers	~ 10 nm
5.	DNA (width)	2 nm
6.	Proteins	5 to 50 nm
7.	Viruses	75 to 100 nm
8.	Bacteria range	1,000 to 10,000 nm

22.9 Preparation of nano-materials.

(1) Scanning instruments for nano-structures: Scanning probe instruments can be used not just to see structures but also to manipulate them. Scanning probes were used to manipulate the individual molecule beads on the molecular abacus. They have also been used to make wonderful nanoscale particles by arranging atoms or molecules on surfaces with particular structures. These structures have been used to demonstrate and test some fundamental scientific concepts ranging through *structural chemistry, electrical interactions, and magnetic behaviors*, among others. This assembling of materials on an atom-by-atom or molecule-by-molecule basis realizes a dream of chemists.

Generally, small objects (which could be either individual atoms or individual molecules) can be moved on a surface either by *pushing on them or by picking them up off the surface onto a scanning tip that moves around and puts them back down*. For both cases, the scanning tip act as a sort of *earthmover at the nanoscale*. In the pushing application, that earthmover is simply a bulldozer. In the pick-up mode, it acts more like a *construction crane or backhoe*.

Scanning probe surface assembly is, inherently very fine, but it suffers from two limitations:

(*i*) It is relatively expensive

(*ii*) It is relatively slow

Although great advances have been made in building machines that use hundreds or even thousands of probe tips at the same time, making nanostructures using scanning probe tip methods is still very much like making automobiles by hand or blowing glass light bulbs individually. It can produce artistic and wonderful results, but it probably cannot be used to for mass scale.

(2) By Lithography: The word "lithography" originally belongs to the concept of stones. A lithograph is an image (usually on paper) that is produced by carving a pattern on the stone, inking the stone, and then pushing the inked stone onto the paper.

(*i*) *Nanoscale lithography:* Many types of small-scale lithography operate in very much this way. Indeed, the common methods used to make current computer chips normally use optical or X-ray lithography, in which a master mask is made using chemical methods and light passes through that mask to produce the actual chip structures. It works just like a silk screen for a T-shirt.

Nanoscale lithography realy can't use visible light because the wavelength of visible light is atleast 400 nanometers, so structures smaller than that are difficult to make directly using it. This is one of the reasons that continuing Moore's law into the nanoscale will require entirely new preparation methods.

Inspite of all this, there are several techniques for doing small-scale lithography. One of the most straightforward and elegant is micro-imprint lithography, largely developed by **George Whitesides** and his research group at **Harvard**. This method works in the same way as the rubber stamps that are still found in post offices. A pattern is inscribed onto a rubber surface (in this case actually a rubber-like silicon/oxygen polymer), and that rubber surface is then coated with molecular ink. The ink can then be stamped out on to a surface: this is paper in the post office, but it could be **a metal, polymer, oxide, or any** other surface in **smallscale stamps**. Small-scale stamping is more complex, but it is very inexpensive and can be used to make numerous copies. Originally, the stamps worked at the larger **micron** (1000-nanometer) scale, but recent improvements are bringing them on to nanoscale.

(*ii*) *Dip Pen Nano-lithography:* One way to construct **arbitrary structures** on surfaces is to write them in exactly the same way that we write ink lines using a fountain pen. To make such lines at the nanoscale, it is necessary to have a nano-pen. Fortunately, AFM tips are ideal nano-pens. **Dip pen nanolithography (DPN)** is named after the old-fask oned dip pen that was used in schoolrooms in the 19th century. The principle of DPN is given in Figure 22.6 . In DPN, a reservoir of "ink" (atoms or molecules) is stored on the top of the scanning probe tip, which is manipulated across the surface, leaving behind lines and patterns.

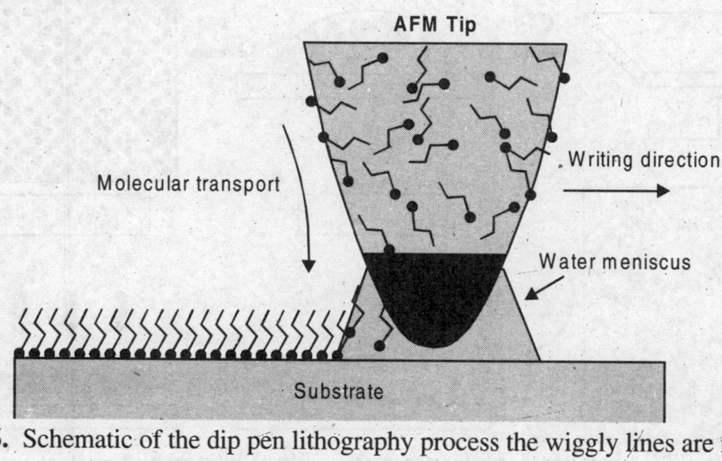

Fig. 22.6. Schematic of the dip pen lithography process the wiggly lines are molecular "ink."

• DPN, developed by **Chad Mirkin** and his collaborators at *Northwestern University*, has many advantages, the two most important being that almost anything can be used as nanoink and that almost any surface can be written on. Also, one can use DPN to make almost any structure matter how detailed or complex since AFM tips are relatively easy to manipulate. This fact makes DPN the technique of choice for creating new and complex structures in small volumes. The disadvantage is that it is slow, unlike the **nanostamp**. It is like comparing hand illustration to early printing. Work is being done to improve this, notably by company **Nanoink**.

(*iii*) *E-Beam Lithography:* We must understand that current light-based industrial lithography is limiited to creating features no smaller than the wavelength used. Even though we can in principle get around this restriction by using light of small wavelengths, this solution can create other problems. Smaller-wavelength light has higher energy, so, it can have strong effects like blowing the feature you are trying to create right off the surface. (Imagine watering garden plants with big pipe)

An alternate way of solving the problem, is to use electrons instead of light. This E-bean lithography can be used to make structures at the nanoscale. E-beam lithography also has applications in current *microelectronics manufacturing*.

(*iv*) *Nanosphere Liftoff Lithography:* If we place marbles together on a board as tightly as possible, they will form a tight group, with each marble surrounded by six others. If this array were spray painted from the top, and then the marbles were tipped off the board, the paint would appear as a set of **painted dots,** each shaped like a triangle but with concave sides (See Fig. 22.7). Now if the marblas are nanoscale, so, are the paint dots. In fact, Fig. 22.7 shows dots of silver metal prepared by *Rick Van. Duyne's group at Northwestern*. The technique is called **nanosphere liftoff lithography**, even though no rockets are involved. It has several good features: many sorts of boards (surfaces) and paints (metals, molecules) can be used, and several layers of paint (molecules) can be put down sequentially on the triangles. Importantly, this liftoff nano-lithography, unlike DPN or **scanning probe** but like **nanostamp**, is parallel. Many nanospheres can be placed on the surface, so that regular arrays of many (thousands or more) dots can be made at the same time.

(3) Molecular Synthesis: The production of **molecules** with particular molecular structures is one of the most active and astonishing part of chemistry. Molecular synthesis involves making specific molecules for specific purposes, either with a purely scientific purpose or with very special application aims. There is extensive molecular synthetic work in **drug companies** and many of the modern drugs including **Penicillin, Lipitor, Taxol, and Viagra** are the products of complex chemical synthesis.

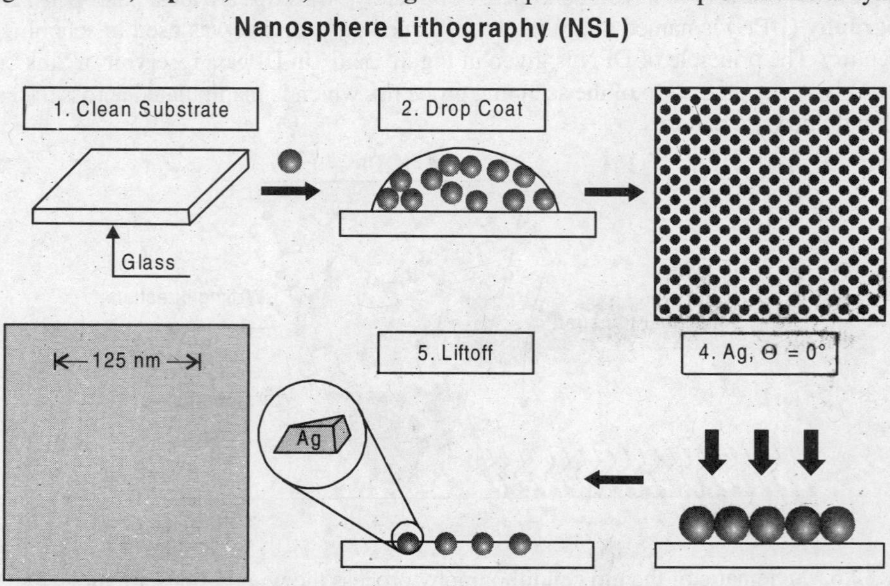

Fig. 22.7. Schematic of the nanosphere liftoff lithography process.

Making nanostructures with particular geometries at specific places on a surface means taking molecule making one step further. In addition to the chemical properties and composition of a molecule, **nanoscale synthesis** must also be concerned with the physical layout and construction of nanostructures. For example, some of the drug delivery techniques involve taking active elements from drugs and pushing them into **nanoscale shells** to allow them to pass into areas of the body where they could not penetrate before. To do this, the drug must be injected into the molecular shell like jelly into a **donut.** There is only a physical interaction here; there are no **chemical bonds between the two.**

Any technique involving manipulating atoms one by one is too slow and cumbersome, especially if we wish to make bulk materials or even enough of our encapsulated drug to treat a patient.

(4) Self Assembly Technique: The problem with most of the techniques for assembling nanostructures that we have discussed so far is that they are too much like work. In every case, we try to impose our wills on these very small objects and manipulate and twist them to be just how we want them. Wouldn't it be better if we could just mix chemicals together and get nanostructures by letting the molecules sort themselves out?

One approach to nanofabrication attempts to do exactly this. It is called *self-assembly*. The idea behind **self-assembly** is that molecules will always seek the lowest energy level available to them. If bonding: to an adjacent molecule accomplishes this, they will bond. If reorienting their physical positions does the trick, then they will reorient. At its simplest, this is the same underlying force that causes a rock to roll down a hill. No matter how you lift, throw, twist, crush, or manipulate the rock, it will always try to get down the hill. You can block its progress, but that requires active intervention. In this case, the rock its; trying to minimize its gravitational energy. In the case of a molecule. it is trying to minimize other kinds of energies due to *Coulomb's law*, these are most frequently forces from charge interactions orient.

One way to imagine self-assembly is to imagine a compass. If one shakes it, one can cause the needle to fluctuate and point in almost any direction for an instant, but once one stops shaking it, the needle will ultimately reorient itself and point from south to north. There is a small magnet in the needle, and this south-to-north orientation minimizes its energy with respect to earth's magnetic field. One needs to do no any work on the needle to get it do this. It does it naturally, *Self-assembly techniques* are based on the idea of making components that, like our compass needle, naturally organize themselves the way we want them to orient.

The forces involved in self-assembly are generally weaker than the bonding forces that hold molecule together. They correspond to weaker aspects of Coulombic interactions and are found in many places throughout nature. For example, weak interactions called **hydrogen bonds hold the hydrogen** atom in one molecule of liquid water together with the oxygen atom of the next and prevent the molecules from becoming water vapor at room temperature. Hydrogen bonding also helps to hold protein into particular three-dimensional structures that are necessary for their biological function.

Other weak interactions are also present including the hydrophobic interactions that allow oil to float on water and multipolar interactions. *Multipolar interactions* occur between structures, each of which has no total charge (so, it is not like an electron interacting with another electron, which is a strong Coulombic interaction). Rather, here there are different distributions of charge on two molecules interacting with one another. Such multipole interactions are generally weak, but they are strong enough to provide very complex structures.

In self-assembly, the nano builder introduces particular atoms or molecules onto a surface or onto a preconstructed nanostructure. The molecules then align themselves into particular positions, sometimes forming weak bonds and sometimes forming strong covalent ones, in order to minimize the total energy. One of the huge advantages of such assembly is that large structures can be prepared in this way, so it is not necessary to tailor individually the specific nanostructures (as was true in

AFM, STM, and DPN construction of nanoscale objects). Self-assembly is almost certainly going to be the preferred method for making large nanostructure arrays, such as the *computer memories and computer logic that must be prepared if Moore's law is going to continue to hold true beyond the next decade.*

Self-assembly is not limited to **electronics** applications. Self-assembled structures can be used for something as like protecting a surface against corrosion or making a surface slippery, sticky, wet, or dry. Fig. 22.8 shows some great examples of self-assembly from the laboratory of Sam Stupp at Northwestern. In this case, two levels of self-assembly are used. First, the self-assembly of long complex molecules called rodcoils produces the mushroom-like nanostructure. Then the non-structures themselves self-assemble to produce a surface coating that makes the glass slide either *hydrophilic* (water loving – easily wetted) or *hydrophobic* (water hating — so that the water beads up). This also shows that very complex structures can be formed using self-assembly by breaking tasks down into steps.

Self-assembly is probably the most important of the nanoscale fabrication techniques because of its generality, its ability to produce structures at different length scales, and its low cost.

(5) Nanoscale Crystal Growth: *Crystal growth* is another sort of self-assembly. Crystals like salt that are made of **ions** are called **ionic crystals**. Those made of **atoms** are called **atomic crystals,** and those made of molecules are called **molecular crystals**. So salt (sodium chloride) is an **ionic crystal**, and sugar (sucrose, $C_{12}H_{22}O_{11}$) is a **molecular crystal**.

Crystals can be grown from solution using seed crystals, which involves putting a small crystal into the presence of more of its component materials (usually in solution) and allowing those components to immitate the pattern of the small crystal, or seed. *Silicon boules*, the blocks used for making microchips, are made or "drawn" in this way.

By making clever choices of seed crystals and growing conditions, it is possible to cause the crystals to assume unusual shapes. Charles Lieber and his group at Harvard University have used nanoscale crystals to seed long, wire-like single crystals of **carbon nanotubes** as well as compounds such ar **indium phosphide or gallium arsenide**, and of atomic crystals such as silicon. These nanowires have remarkable conductivity properties, as well as many uses both in **optics and in electronics.**

6. Polymerization: Polymers are very large molecules. They can be upward of millions of atoms in size, made by repetitive formation of the bond from one small molecular unit (monomer) to the next. **Polymerization** is a very commonly used scheme for making nanoscale materials and even much larger ones – epoxy adhesives work by making extended polymers upon mixing the two components of the epoxy.

Ordinarily, industrial polymers like **polystyrene or polyethylene or polyvinylchloride (PVC)** are made by building extremely long molecules, with numerous steps that occur sequentially. **Controlled polymerization**, in which one monomer at a time is added to the next, is very important for specific elegant structures. Robert Letsinger and his students at Northwestern University have developed a series of methods for preparing specific short DNA fragments. These are called oligonucleotides from the Greek work "oligo" which means a few. (A monomer is one unit, an oligomer is several units, and a polymer is many units.) The so-called gene machines use elegant reaction chemistry to construct specific DNA sequences.

Building specific DNA sequences is crucial for many reasons. In modern biotechnology, these specific sequences are used to build new biological structures (drugs, materials, proteins), based on the ability of bacteria to reproduce themselves. A synthetic DNA template is introduced into the bacterial DNA, and the bacteria then procduce many copies of that particular target protein. The modification of the bacteria's DNA is done using a series of chemical reactions, and the gene machines are used to prepare the specific short oligonucleotides to modify bacterial DNA, capturing that process to produce the protein of choice. This allows you to effectively make protein factories for nearly any protein you choose. *One good example of how this could be used is to make the protein insulin for the treatment of diabetes.*

The combination of specific short DNA sequences and self-assembly is used extensively to

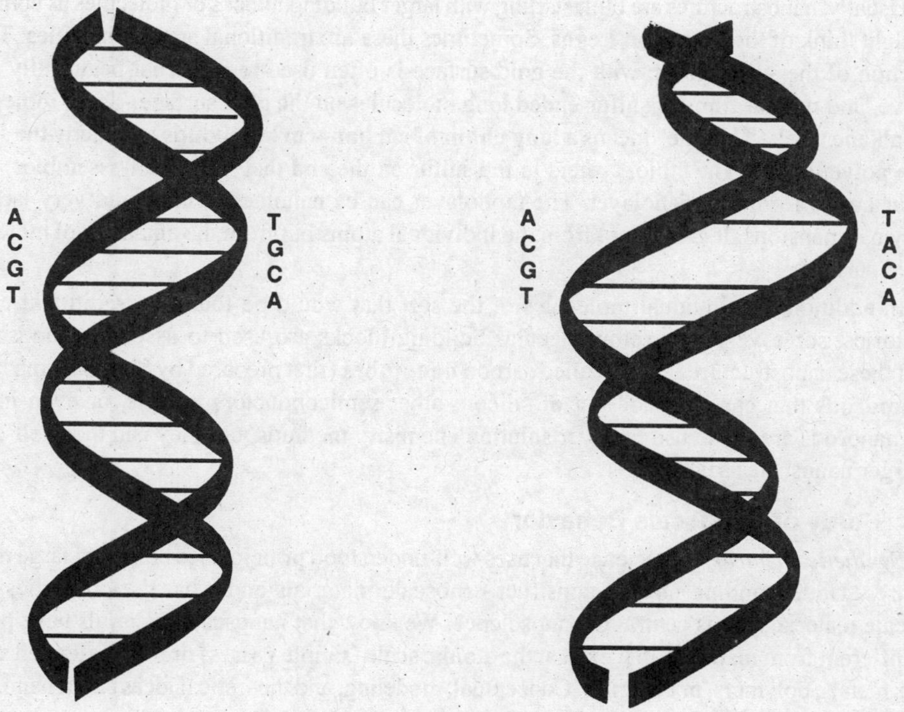

Fig. 22.8. Schematic of the DNA hybridiznation process. The "matched" side shows how a DNA strand correctly binds to its complement and the "mismatched" side shows how errors can prevent binding. *Courtesy of the Mirkin Group, Northwestern University.*

make materials in which a single DNA strand binds to another single DNA strand. This process, called *hybridization*, is shown in Fig. 22.8. Recall that the DNA base, A always pairs with T, and the DNA base G always pairs with C. In Fig. 22.8 the perfect matching on the left gives a tighter, stronger fit than the imperfect mismatched set. This kind of self-assembly is present in nature—it's how DNA replicates so that cells can multiply. Many synthetic applications of this complementary molecular recognition are used in nanoscience.

7. Nanobricks and Building Blocks: Nanostructures must be assembled from components. The fundamental building blocks are atoms of the **91 naturally occurring elements**. Usually, though, it is inefficient to start with individual atoms. We saw both the strength and the slowness of this approach when we discussed building atomic scale nanostructures using scanning probe microscopy, especially if we are trying to make a macroscopic amount of a material rather than build a single nanoscale machine. Richard Smalley, who won the Nobel Prize for his nanoscience work in 1996, estimated that it could take nanomachines as much as 19 million years to build a few ounces of material building atom by atom because the number of atoms in such a sample is about 6×10^{23}. If one atom were the size of a teaspoon full of water, this **many atoms would be about the size of the Pacific Ocean.**

Building an ocean from one teaspoon at time would be a very slow process and so is building bulk materials atom by atom. Assembling at the rate of a million atoms per second would still take 6×10^{17} seconds to construct a handful of useful material. This is a bit of a damper for those who imagine nanoscale robots (sometimes called "assemblers") running around making everything from cars to clocks, but there are already promising alternatives for making bulk materials based on nanostructures.

Usually, nanostructures are built starting with larger building, blocks or molecules as components. You might think of these as nano **Legos**. Sometimes these are traditional small molecules. The weak interaction of the sulfur group with the gold surface is often used to construct beautifully adherent adhesive, and regular films of sulfur-ended long molecules on the gold surface. These molecules are called alkane thiols. "Alkane" means a **long chain of carbon–carbon bonds** of exactly the same sort seen in polyethylene. The "thiol" refers to the sulfur on the end that links (self-assembles) onto the gold surface to form the monolayer. The monolayer can be nanometers thick and very large in the other two dimensions. It is built not from the individual atoms but from the alkane thiol molecules on the gold surface.

In addition to individual molecules of the sort that would be found in traditional chemistry laboratories, some very new semimolecular building blocks are used to assemble nanostructures. Two of these nanostructures are so-called carbon **nanotubes** (first prepared by Sumio Iijima in Tokyo) and **nanorods** that can be made out of silicon, other semiconductors, metals, or even insulators. These nanorods are made using clever solution chemistry methods, but they can then self-assemble into larger nanoscale structures.

22.10 Study of Nanoscale Behavior

Synthetic chemistry is a science that uses well-understood principles to construct large molecular structures. Understanding how to construct nanoscale materials and what the properties of those nanoscale materials are is central to nanoscience. We know that nanoscale materials have properties quite different from materials that exist at the atomic scale (simple gasses) or from extended structures such as metals, polymers, or ceramics. Conceptual, modeling, and theoretical ideas about nano behavior are crucial in order to develop techniques for mass production at the nanoscale.

The fundamental ideas involved are the ones electrical interactions and Coulomb's law, the behaviors and design rules of quantum mechanics, interactions with light, and interactions among components, largely based on manifestations of the Coulombic force. So, the traditional methods of *theoretical physics, theoretical chemistry, electrical engineering, and materials science* dominate both the concepts used to understand nanostructures and the calculations that are necessary to predict the behavior.

The use of **quantum mechanical theory** to predict actual molecular structures has been one of the triumphs of chemistry in the 20th Century. Extension of these models to deal with larger nanostructures is moving very rapidly. These calculations are done using large computers and can be mixed with more traditional predictions based on classical mechanics. The computer predictions of structures are one way by which nanoscale design occurs, although they are most useful when mixed with intuition. experience, and inspiration.

22.11 Importance of Nano – CAD

Complementary metal oxide semiconductor (CMOS) silicon chip making technology remains one of the key methods for making microchips such as computer microprocessors. In design of CMOS structures, the only fundamental scientific requirement is the behavior of currents that are described by Ohm's law. Still, assembling a very complex CMOS structure is extremely complicated both physically and financially (making the tools and the masks, doing the lithography, and assembling millions of such chips on tiny amounts of chip area are very hard to do for a low price). **To make the modern chips of the types used in consumer electronics, it is necessary to assemble millions of transistors on silicon chips and to route charge through them in such a way as to perform the algorithms or calculations that the chip is designed to carry out.** In the early days of chip design, engineers would literally sketch the circuits they wanted on the boards, aid companies maintained art departments of highly coordinated individuals to draw these features precisely (by hand, with a pen under a magnifying glass) as templates for making masks. In those days, though, chips contained only a few dozen components. Designing chips with millions of components is simply too complicated to be done by individuals.

Due to these issues, computers design most functional chips. Computer programs, generically called Spice or Computer-Aided Design (CAD), are specifically programmed to design chips for particular functions. Engineers design the chips at a high level using specialized hardware description computer languages or graphical environments that use predesigned components.

The CAD programs are used to design structures on scales far larger than that of a nanometer. Typical CMOS structures are still hundreds or thousands of nanometers in size, as they must be. When structures fall into the nanoscale regime, Ohm's law doesn't necessarily work. The phenomena of charge motion are then described by quantum mechanics, and the understanding of how currents will flow becomes much more complicated. Standard transistors behave quite differently than those that one would anticipate at the nanoscale, and there are not yet any nanoscale Spice or CAD programs. Several laboratories are addressing this challenge, but until we understand the behavior of individual nanostructures well enough, building a Spice-like computer code for assembling nanodevices into logic structures and architectures will remain a tantalizing challenge.

Ball mill method of synthesis of nanoparticles. Ball milling is the best example for mechanical method. A ball mill (Fig. 22.9) a type of grinder, is a cylindrical device used in grinding (or mixing) materials like ores, chemicals, ceramic raw materials and paints. Ball mills rotate around a horizontal axis, partially filled with the material to be ground plus the grinding medium. Different materials are used as media, including ceramic balls, flint pebbles and stainless steel balls.

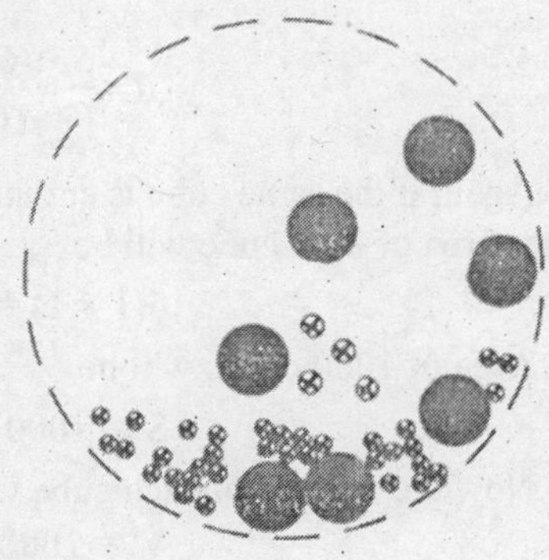

Fig. 22.9

An internal repeating effect reduces the materials to a fine powder. Industrial ball mills can operate continuously fed at one end and discharged at the other end. Large to medium-sized ball mills are mechanically rotated on their axis, but small ones normally consist of a cylindrical capped container that rests on two drive shafts (pulleys and belts are used to transmit rotary motion).

A rock tumbler is also based on the same principle. High quality ball mills are potentially expensive and can grind mixture particles to as small as 0.0001 mm. The grinding works on principle of critical speed. The critical speed can be understood as that speed after which the steel balls (which are responsible for the grinding of particles) start rotating along the direction of the cylindrical device; thus cause no grinding further.

Sol-gel Techniques for synthesis of nano-particles.

Sol-gel Techniques: Colloidal particles are mch larger than normal molecules or nanoparticles but upon mixing with liquid colloids appear bulky while the nanosized molecules always look clear. It involves the evolution of networks through the formation of colloidal suspension (sol) and gelatine to form a network in continuous liquid phase (gel). The precursor for synthesizing these colloids consists of ions of metal alkoxides and aloxysilanes. The most widely used are tetramethoxysilane (TMOS), and tetraethoxysilanes (TEOS) which form silica gels. Alkoxides are immiscible in water. They are organo-metallic **precursors for silica, aluminium, titanium**, zirconium and many others. Mutual solvent alcohol is used. The sol-gel process involved initially a homogeneous solution of one or more selected alkoxides. These are organic precursors for silica, alumina, titania, zirconia, among others. A catalyst is used to start reaction and control pH. Sol-gel formation, occurs in four stages (i) Hydrolysis (*ii*) Condensation (*iii*) Growth of particles, and (*iv*) Agglomeration of particles.

(*i*) **Hydrolysis :** In hydrolysis, addition of water results in the replacement of [OR] group with [OH] group. Hydrolysis occurs by attack of oxygen on silicon atoms in silica gel. Which can be accelerated by adding a catalyst such as HCl and NH_3, and continues until all alkoxy groups are replaced by hydroxyl groups. Subsequent condensation involving silanol group (Si-OH) produced siloxane bonds (Si-O-Si) and alcohol and water. Hydrolysis occurs by attack of oxygen contained in the water on the silicon atom.

(*ii*) **Condensation :** Polymerization to form siloxane bond occurs by either a water producing or alcohol producing condensation reaction. The end result of condensation products is the formation of monomer, diamer, cyclic tetramer, and high order rings.

(*iii*) **Growth and Agglomeration :** As the number of siloxane bonds increases, the molecules aggregate in the solution, where they form a network, a gel is formed upon drying. The water and alcohol are driven off and the network shrinks. At values of pH of greater than 7, and H_2O/Si value ranging from 7 to 5, spherical nanoparticles are formed. Polymerization to form siloxane bonds by either an alcohol producing or water producing condensation occurs.

Surface area to volume ratio and its importance in nanomaterials.

If a given sample is cut into small pieces, then its surface increases and so the ratio of surface area to volume is increased.

Let us take the example of a cube of each side 10 cm. So its surface area will be

$$S = 6(10 \times 10) = 600 \text{ cm}^2$$

and its volume will be

$$V = 10 \times 10 \times 10 = 1000 \text{ cm}^2$$

so the ratio of area to volume of the cube is

$$R = \frac{S}{V} = \frac{600 \text{ cm}^2}{1000 \text{ cm}^3} = 0.6 \text{ cm}^{-1}$$

Again if the same cube is cut into 1000 identical cubes of each side equal to 1 cm. In this case surface area of each cube will be

$$6(1 \times 1) = 6 \text{ cm}^2$$

Area of 1000 cubes will be

$$S' = 1000 \times 6 = 6000 \text{ cm}^2$$

Now the volume of each cube will be $1 \times 1 \times 1 = 1 \text{ cm}^3$ and that of 1000 cubes will be

$$V' = 1000 \times 1 = 1000 \text{ cm}^3$$

New ratio of surface to volume will be

$$R' = \frac{S'}{V'} \frac{R'}{R} = \frac{6000 \text{ cm}^2}{1000 \text{ cm}^3} = 6 \text{ cm}^{-1}$$

$$\therefore \quad \frac{R'}{R} = \frac{6 \text{ cm}^{-2}}{0.6 \text{ cm}^{-1}} = 10$$

Which shows that surface area to volume has become 10 times. So in general when a body is cut with small pieces then surface to volume ratio increases.

In the nanoscale, there will be many fold increase in surface to volume ratio of the material.

As most of the reactions take place on the surface of a material, so with the reduction of size, the reactivity of nanomaterials will be much lesser than the bulk material. So physical and chemical properties of nanomaterials will be entirely different than that on nanoscale. A typical example is that of gold, which is inert and does not participate in chemical reactions. However at nanoscale gold particles are highly reactive and can be used as catalyst to initiate many reactions. Similar examples can be given to hold the increased reactivity of nanomaterials.

1D, 2D and 3D nanomaterials.

ID nanomaterials are those which have only 1 dimension in nano-scale and other dimensions are larger than the nano-scale.

Nano-textured surfaces, quantum dots and nano-rings are examples of 1D nanomaterials.

2D-Nanomaterials are those which have two dimensions in nano-scale while third dimension is beyond Nanoscale. Carbon nanotube and Nano-wires (used in chip technology are examples of **2D nanomaterials.)**

3D-Nano-materials are those which have all the three dimensions in nano-scale. Nano-particles like fullerenes and Nano-shells are examples of 3D nano-materials.

22.12 Recent Applicantions of nano-materials

(1) In Molecular Electronics.

Molecular electronics may be divided into two groups :

(*i*) Area of molecular materials in making bulk electronic devices like the use of organic materials in the crystal displays.

(*ii*) Area of molecular scale electronics which depends on the properties of single molecular or cluster of molecules to make active electronic components. It must be noted that single molecules are much smaller than the smallest features conceivably attained by semi-conductor devices.

Now-a-days, research on molecular electronics is being persued in the following fields :

(*a*) Effect of molecules to control environment.

(*b*) Molecular effect on optical and conduction Phenomenan.

(*c*) Characterization of molecules for probe measurements / transport meansurements.

(*d*) Use of molecular switch /gate/ wire.

(*e*) Molecular opto-electronic exploitation.

(*f*) Organic transistors with a single molecule in channel region.

(*g*) Molecular wire connectors to study nanostructures which can be organically synthesized.

(*h*) DNA as a molecular wire. It is observed that no electrical conduction exists for DNA lenghts from 40 *nm* to 500 *nm*. **Efficial** condition appears for DNA ~ 600 *nm*. A resistance of 2.5 $m\,\Omega$ has been observed for 60 nm long λ-DNA.

(2) In Nanomedicine: (*i*) Nanomedicinr is used to monitor, control, construct repair and repair the human biological systems working on molecular levels using nano-structures and nano-devices. Its application to prevent diseases and improve the immune system of the body is of great importance.

Drug molecules have specific effects on body cells. Morphine molecules bound to certain molecules in brain cells affect natural impulses that signal pain. Insulin and other drugs fit some receptors. Drug molucules work without direction and some times cause damage on other cells. It is now possible to tag drug molecules with nano-particles and arrange the delivery of the drug to desired location.

Patients in future may drink fluids that contain *nanorobots* which are programmed to attack and reconstruct the molecular structure of viruses so that they become harmless.

(*iii*) Molecular machines directed by nano-computers may offer to docters a choice to combine sensors with repair of the individual cells. Surgical control may be brought to **molecular domains.**

(*iv*) Ability to tag and identify the biological samples with **nano-luminescent particles** for detection of DNA affected with cancer producing genes.

Some materials are such that their molecular structures change sharply with temperatures. They can be **heated to form one shape and cooled to form a second shape.** Alloy **Nitinol** is one such material which is used in medicine to control the blood vessels from clogging. Before surgery the **stents are** cooled and compressed for easy insertion. They expand under body heat.

Pulses of electric current can **bnimic** muscle movement as **Nitinol** switches back and forth between a relaxed and contracted state.

(*v*) **Carbon Nano-tubes** have been found to act like a miniaturized X-ray tube where emission occurs at room temperature and consumption of power is very less.

(*vi*) By adding nano-particles with certain properties into polymers it is possible to make composite materials useful for different purposes. Nano-particles provide the functionality without affecting strength.

(*vii*) Nanotechnology allows to use **quantum effects.** Withn this chemical reactions can occur very quickly, heat is conducted much faster. Material properties are enhanced and sensors can detect small changes, batteries can last longer without charging.

(*viii*) **Peter Quail** has made **Yeast cells** which turn blue after exposure to a flash of red light and return to its original colour if exposed to far-red light. So, **genes** can be switched **on and off** to have a better study of **genes** in cells.

(*ix*) Quantium confuied Atoms (QCA) make very energy efficient **phosphor** (a material that emits visible light).

QCA nano phosphors have great potential to provide **Solid State Lighting** (SSL). They are five times more efficient than normal phosphors in converting UV light into white light.

(*x*) Nanotechnology is finding use in the study of human brain. Brain works like a computer and a chemical factory. Brain cells produce electrical signals and sends them from one cell to another along pathways called **circuits.** As in computers these electrical circuits receive, process; store and send information. Unlike computers, brain creaters electricals signals by chemical means. The proper functioning of the brain depends on the many compilicated chemical substances produced by brain cells.

The nervous system consists of brain, spnial cord and the nerves. Total movement of a person and his *sensations throughts and emotions* are products of the nervous system.

The nervous system is made up of billions of special cells called **neurons or nerve cells.** Cord like bundles of **neuron fibres** are called **nerves.** The nerves form a network of pathways that passes on the information throughout the body. The nerves cells carry messages from the brain to muscules and glands, which are called **effectors** which carry out the brain's instructions. For example, when a person is running, it sends a message to the heart to *beat faster and send more blood to leg muscles.*

22.13 Carbon Nanotubes

Carbon nanotubes are molecular structures which contains *graphite cylinders closed at either ends with caps containing pentagonal rings*. These are descenders of buckyball, the soccer-ball-shape molecules of 60 carbon atoms. Noriaki Hamada of NEC and M.S. Dresselhaus of MIT calculated that if a **row of hexagons in a tube's long** axis were straight, the tube should act as a metal and conduct electricity. And if a line of hexagons formed a helix, the tube should behave as a semiconductor. **Carbon nanotubes are about 1.4 *nm* thick.** A single nanotube could behave as a transistor. A single nanotube with a natural junction where a straight section joined to a helical section acts like rectifying diode-half transistor.

Nanotubes are *stronger* than steel, so long filaments could create supertough, fiber-reinforced plastics. These materials could slash the weight of air-crafts, spaceships and ground vehicles. Nanotubes have *thermal conductivity* better than diamond.

Electronic companies like Onida, Samsung, Sansui, etc., have hopes to ship their first nanotube TVs.

Carbon nanotubes are molecular-scale tubes of graphitic carbon with outstanding properties shown in Fig. 22.10. They are among the *stiffest and strongest fibers* known, and have remarkable electronic properties and many other unique characteristics. For these reasons they have attracted huge academic and industrial interest. Commercial applications have been rather slow to develop, because of the high production costs of the best quality nanotubes.

A *transmission electron micrograph* of some *multiwalled nanotubes* is shown in the figure (22.11) left. In 1993, a new class of carbon nanotube was discovered, with just a single layer. These single-walled nanotubes are generally *narrower* than the multiwalled tubes, with diameters typically in the range *1-2 nm*, and tend to be curved rather than straight. The image on the right shows some typical single-walled tubes. It was soon established that these new fibers had a range of exceptional properties, and this sparked off an explosion of research into carbon nanotubes in the various laboratories around the globe.

Fig. 22.10

Structure

The bonding in carbon nanotubes is SP^2 with each atom joined to three neighbours, as in the case of graphite. The tubes can so be considered as rolled-up graphene sheets (graphene is an individual graphite layer). There are three distinct ways in which a *graphene sheet* can be rolled into a tube, as shown in Fig. 22.11.

The first two of these, known as armchair (top left) and zig-zag (middle left) have a high degree of symmetry. The terms "armchair" and "zig-zag" refer to the arrangement of hexagons around the circumference. The third class of tube, which in practice is the most common, is known as chiral, which means that it can exist in two mirror-related forms.

Gudiksen *et al* have demonstrated the synthesis of semiconductor nanowire superlattices. Single nanowire photoluminescence, electrical transport and electroluminescence measurements show the unique photonic and electronic properties of these nanowire superlattices and suggest potential applications ranging from **nano-barcodes to polarised nanoscale LED's.**

Field effect transistors (FETs) based on individual GAN nanowires have been fabricated. **Dip-pen nanolithography** has been used to fabricate polymer nanowires of controlled size using conductive polymers. This method was used to construct arrays of proteins with 100 to 350 features. These nanoarrays provide opportunity to study surface mediated biological recognition process.

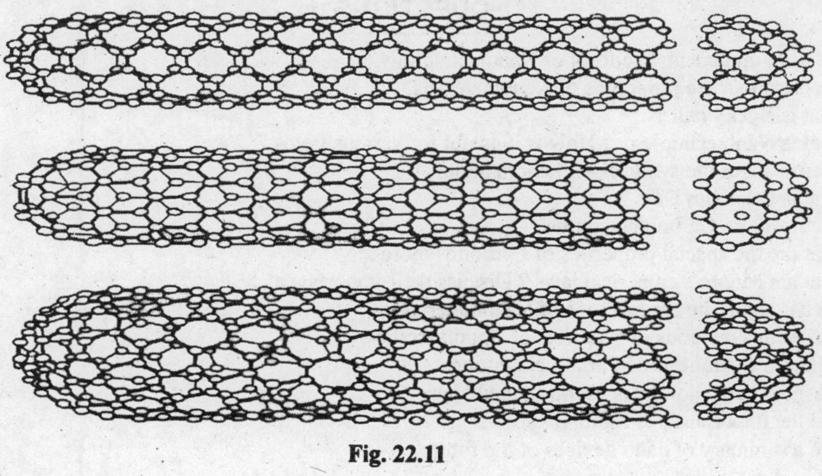

Fig. 22.11

Possibility of generating microwaves through nanotube field emission has been demonstrated with implications for **wireless communications.** Cellulor phones typically send a weak signal to a local base station where microwave amplifiers amplify that signal. *Nanotubes are about six times lighter and ten times stronger than steel, that is a wonder material for space.*

Pharmaceutical companies also have deviced sensitive microchips containing intact DNA can spot interactions between candidate antibiotics and target bugs, for example, Remarkable as they are, such microchips could be stuffed with 100,000 times more little chemical labs each of which is 100,000 times more sensitive, if they were made with nanotubes according to Chad A. Mirkin, Director of North Western's University's Institute for Nanotechnology.

Carbon nanotubes are so thin that they can penetrate the skin without pain. So, Thera-fuse Inc., Vista startup is developing a skin patch 'for diabatics. It will draw blood through *nanostraws to monitor glucose levels and inject insulin when required.*

Nanotech also offers completely new systems for detecting **biohazards like anthrax.**

Electronic companies like IBM, Hewlett-Packard, Hitachi, etc., are developing nanoscale transistors and prototype nanotube circuits.

Carbon nanotubes allow to implement the molecular wire technology in nanoscale ICs that are used in **NEMS** and **MEMS.**

22.14 Nanodevices of the future

Computers have been getting smaller, faster and cheaper because of advances in silicon technology. As per the observation of Gordon Moore in 1960 the number of transistors being packed into integrated circuits doubled every year. He predicted that this will trend will continue which has become to be known as the **Moore's** law. In fact since 1975, the number of transistors on a semiconductor chip has doubled every 18 months. A microprocessor today has ~ 40 million transistors, by 2020 it could have 5 billion.

A great step in nanotechnology came from *Cornell's Nanobiotechnology centre* where silicon strutures of the ~ 10 *nm* has been achieved Controlled Etching of Dislocations (CED) process. This is indeed remarkable but it is believed that nanoelectronics will need *bottom-up* approach rather than this **top-down** approach in which unwanted material is etched away using chemicals or plasma. Synthetic chemistry "to produce the building block with molecules and assemble them into progressively larger structure is required. This approach, Molecular Based Materials (MBM), prepares materials that maintain aspects of parent molecular framework. Since all living creatures are made up of tiny cells which themselves are constructed of nanoscale building blocks, biomolecules like **proteins, lipids, nucleic acids could be excellent sources of nanotechnology research. A new approach of assembling molecules into pattern using** nano lithography may ultimately lead to molecular based devices.

QUESTIONS

1. Explain Fundamental concept of Nanotechnology.
2. Discuss about the properties of Nanomaterials.
3. What is Bucky ball ?
4. What is Nanotechnology ? How is it useful for various fields ?
5. Discuss about the synthesis of nanomaterials.
6. What is Quantum Dots ?
7. Discuss about Carbon Nanotubes.
8. What are the special properties of Carbon Nanotubes ?
9. What are Moore's emperical laws ? Discuss their importance.
10. Discuss about the applications of Nanomaterials.
11. List various methods of synthesis of Nanomaterials.
12. Explain the synthesis of Carbon Nanotubes.
13. How Nano Technology is useful in medicines ?
14. Give the importance of Nano-CAD.
15. Give a summary of nano devices of the future.